Complete Solutions Manual

for

AF700660

# Single Variable Calculus: Concepts & Contexts

Third Edition

**Jeffery A. Cole**
*Anoka-Ramsey Community College*

THOMSON
BROOKS/COLE

Australia • Canada • Mexico • Singapore • Spain • United Kingdom • United States

Printed in the United States of America
1 2 3 4 5 6 7 08 07 06 05 04

Printer: West Group

ISBN: 0-534-41025-1

For more information about our products, contact us at:
**Thomson Learning Academic Resource Center**
**1-800-423-0563**

For permission to use material from this text or product, submit a request online at **http://www.thomsonrights.com.**
Any additional questions about permissions can be submitted by email to **thomsonrights@thomson.com.**

**Thomson Higher Education**
**10 Davis Drive**
**Belmont, CA 94002-3098**
**USA**

**Asia (including India)**
Thomson Learning
5 Shenton Way
#01-01 UIC Building
Singapore 068808

**Australia/New Zealand**
Thomson Learning Australia
102 Dodds Street
Southbank, Victoria 3006
Australia

**Canada**
Thomson Nelson
1120 Birchmount Road
Toronto, Ontario M1K 5G4
Canada

**UK/Europe/Middle East/Africa**
Thomson Learning
High Holborn House
50–51 Bedford Road
London WC1R 4LR
United Kingdom

**Latin America**
Thomson Learning
Seneca, 53
Colonia Polanco
11560 Mexico
D.F. Mexico

**Spain (including Portugal)**
Thomson Paraninfo
Calle Magallanes, 25
28015 Madrid, Spain

# PREFACE

This *Complete Solutions Manual* contains solutions to all exercises in the texts *Calculus: Concepts and Contexts, Single Variable,* Third Edition, and Chapters 1–8 of *Calculus: Concepts and Contexts*, Third Edition, by James Stewart. A student version of this manual is also available; it contains solutions to the odd-numbered exercises in each section, the review sections, the True-False Quizzes, and the Problem Solving sections, as well as solutions to all the exercises in the Concept Checks. No solutions to the Projects appear in the student version. It is my hope that by browsing through the solutions, professors will save time in determining appropriate assignments for their particular class.

Some nonstandard use of notation is used in order to save space. If you see a symbol which you don't recognize, refer to the Table of Abbreviations and Symbols on page v.

I appreciate feedback concerning errors, solution correctness or style, and manual style. Any comments may be sent directly to me at the address below, at jeff.cole@anokaramsey.edu, or in care of the publisher: Thomson Brooks/Cole, 10 Davis Drive, Belmont, CA 94002.

I would like to thank Jim Stewart, for his guidance; Craig Chamberlain, of Palomar College, for his careful assistance with most of the new solutions; Brian Betsill, Kathi Townes, and Jenny Turney, of TECH-arts, for their production services; and Bob Pirtle and Stacy Green, of Thomson Brooks/Cole, for entrusting me with this project as well as for their patience and support.

I dedicate this book to my wife, Joan.

Jeffery A. Cole
Anoka-Ramsey Community College
11200 Mississippi Blvd. NW
Coon Rapids, MN 55433

# ABBREVIATIONS AND SYMBOLS

| | |
|---|---|
| CD | concave downward |
| CU | concave upward |
| D | the domain of $f$ |
| FDT | First Derivative Test |
| HA | horizontal asymptote(s) |
| I | interval of convergence |
| IP | inflection point(s) |
| R | radius of convergence |
| VA | vertical asymptote(s) |
| $\stackrel{\text{CAS}}{=}$ | indicates the use of a computer algebra system. |
| $\stackrel{\text{H}}{=}$ | indicates the use of l'Hospital's Rule. |
| $\stackrel{j}{=}$ | indicates the use of Formula $j$ in the Table of Integrals in the back endpapers. |
| $\stackrel{\text{s}}{=}$ | indicates the use of the substitution $\{u = \sin x, du = \cos x\, dx\}$. |
| $\stackrel{\text{c}}{=}$ | indicates the use of the substitution $\{u = \cos x, du = -\sin x\, dx\}$. |

# CONTENTS

## 8 Infinite Sequences and Series 537



## Appendixes 625

# 1 □ FUNCTIONS AND MODELS

## 1.1 Four Ways to Represent a Function

In exercises requiring estimations or approximations, your answers may vary slightly from the answers given here.

**1.** (a) The point $(-1, -2)$ is on the graph of $f$, so $f(-1) = -2$.

(b) When $x = 2$, $y$ is about 2.8, so $f(2) \approx 2.8$.

(c) $f(x) = 2$ is equivalent to $y = 2$. When $y = 2$, we have $x = -3$ and $x = 1$.

(d) Reasonable estimates for $x$ when $y = 0$ are $x = -2.5$ and $x = 0.3$.

(e) The domain of $f$ consists of all $x$-values on the graph of $f$. For this function, the domain is $-3 \leq x \leq 3$, or $[-3, 3]$. The range of $f$ consists of all $y$-values on the graph of $f$. For this function, the range is $-2 \leq y \leq 3$, or $[-2, 3]$.

(f) As $x$ increases from $-1$ to 3, $y$ increases from $-2$ to 3. Thus, $f$ is increasing on the interval $[-1, 3]$.

**2.** (a) The point $(-4, -2)$ is on the graph of $f$, so $f(-4) = -2$. The point $(3, 4)$ is on the graph of $g$, so $g(3) = 4$.

(b) We are looking for the values of $x$ for which the $y$-values are equal. The $y$-values for $f$ and $g$ are equal at the points $(-2, 1)$ and $(2, 2)$, so the desired values of $x$ are $-2$ and 2.

(c) $f(x) = -1$ is equivalent to $y = -1$. When $y = -1$, we have $x = -3$ and $x = 4$.

(d) As $x$ increases from 0 to 4, $y$ decreases from 3 to $-1$. Thus, $f$ is decreasing on the interval $[0, 4]$.

(e) The domain of $f$ consists of all $x$-values on the graph of $f$. For this function, the domain is $-4 \leq x \leq 4$, or $[-4, 4]$. The range of $f$ consists of all $y$-values on the graph of $f$. For this function, the range is $-2 \leq y \leq 3$, or $[-2, 3]$.

(f) The domain of $g$ is $[-4, 3]$ and the range is $[0.5, 4]$.

**3.** From Figure 1 in the text, the lowest point occurs at about $(t, a) = (12, -85)$. The highest point occurs at about $(17, 115)$. Thus, the range of the vertical ground acceleration is $-85 \leq a \leq 115$. In Figure 11, the range of the north-south acceleration is approximately $-325 \leq a \leq 485$. In Figure 12, the range of the east-west acceleration is approximately $-210 \leq a \leq 200$.

**4.** *Example 1:* A car is driven at 60 mi/h for 2 hours. The distance $d$ traveled by the car is a function of the time $t$. The domain of the function is $\{t \mid 0 \leq t \leq 2\}$, where $t$ is measured in hours. The range of the function is $\{d \mid 0 \leq d \leq 120\}$, where $d$ is measured in miles.

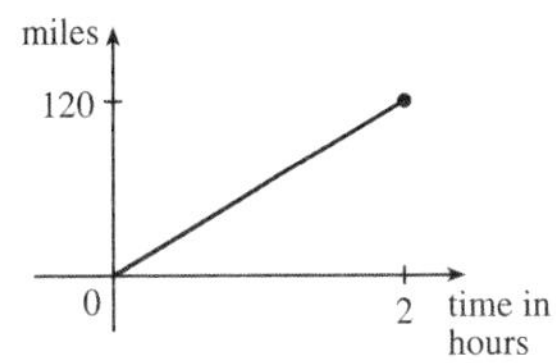


*Example 2:* At a certain university, the number of students $N$ on campus at any time on a particular day is a function of the time $t$ after midnight. The domain of the function is $\{t \mid 0 \leq t \leq 24\}$, where $t$ is measured in hours. The range of the function is $\{N \mid 0 \leq N \leq k\}$, where $N$ is an integer and $k$ is the largest number of students on campus at once.

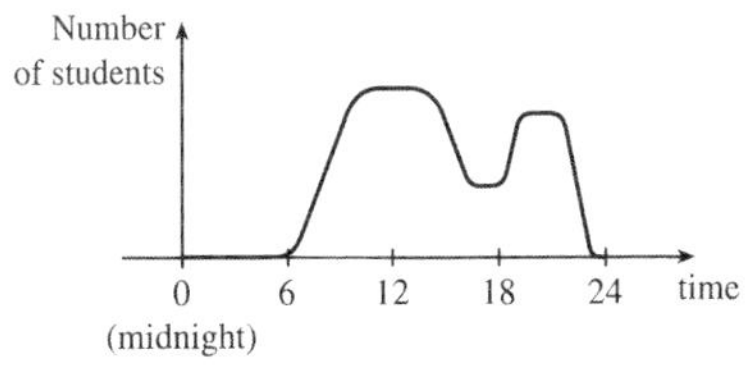


*Example 3:* A certain employee is paid \$8.00 per hour and works a maximum of 30 hours per week. The number of hours worked is rounded down to the nearest quarter of an hour. This employee's gross weekly pay $P$ is a function of the number of hours worked $h$. The domain of the function is $[0, 30]$ and the range of the function is $\{0, 2.00, 4.00, \ldots, 238.00, 240.00\}$.

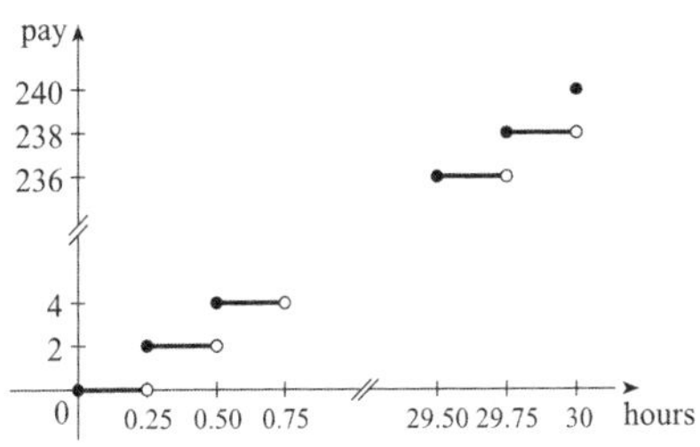

**5.** No, the curve is not the graph of a function because a vertical line intersects the curve more than once. Hence, the curve fails the Vertical Line Test.

**6.** Yes, the curve is the graph of a function because it passes the Vertical Line Test. The domain is $[-2, 2]$ and the range is $[-1, 2]$.

**7.** Yes, the curve is the graph of a function because it passes the Vertical Line Test. The domain is $[-3, 2]$ and the range is $[-3, -2) \cup [-1, 3]$.

**8.** No, the curve is not the graph of a function since for $x = 0, \pm 1$, and $\pm 2$, there are infinitely many points on the curve.

**9.** The person's weight increased to about 160 pounds at age 20 and stayed fairly steady for 10 years. The person's weight dropped to about 120 pounds for the next 5 years, then increased rapidly to about 170 pounds. The next 30 years saw a gradual increase to 190 pounds. Possible reasons for the drop in weight at 30 years of age: diet, exercise, health problems.

**10.** The salesman travels away from home from 8 to 9 A.M. and is then stationary until 10:00. The salesman travels farther away from 10 until noon. There is no change in his distance from home until 1:00, at which time the distance from home decreases until 3:00. Then the distance starts increasing again, reaching the maximum distance away from home at 5:00. There is no change from 5 until 6, and then the distance decreases rapidly until 7:00 P.M., at which time the salesman reaches home.

**11.** The water will cool down almost to freezing as the ice melts. Then, when the ice has melted, the water will slowly warm up to room temperature.

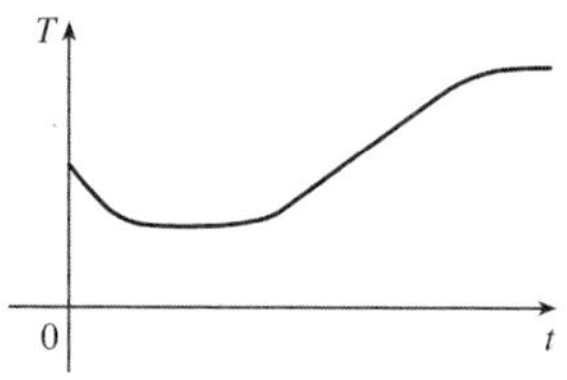


**12.** The summer solstice (the longest day of the year) is around June 21, and the winter solstice (the shortest day) is around December 22.

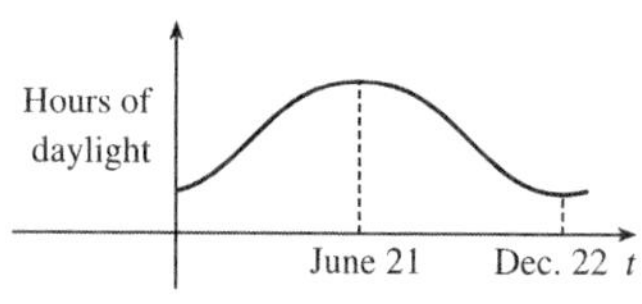


**13.** Of course, this graph depends strongly on the geographical location!

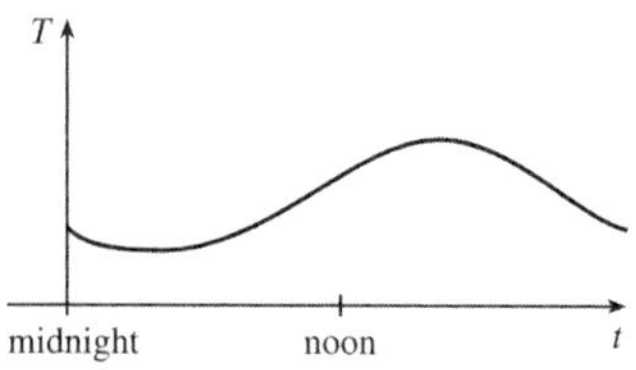


**14.** The value of the car decreases fairly rapidly initially, then somewhat less rapidly.

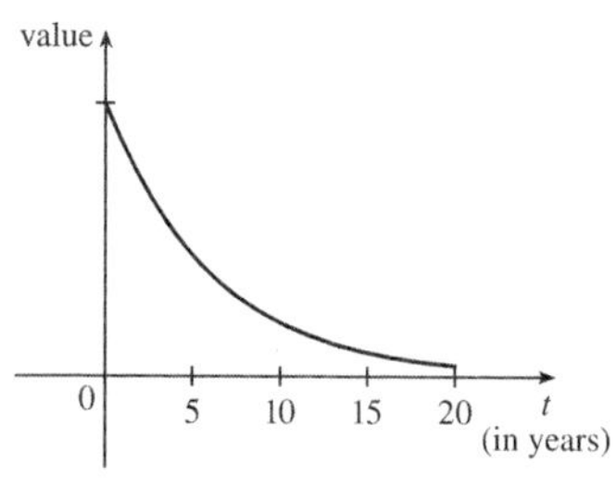


**15.** As the price increases, the amount sold decreases.

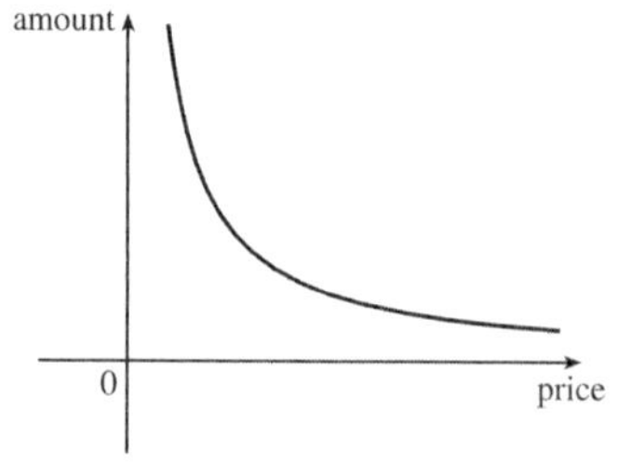

**16.** The temperature of the pie would increase rapidly, level off to oven temperature, decrease rapidly, and then level off to room temperature.

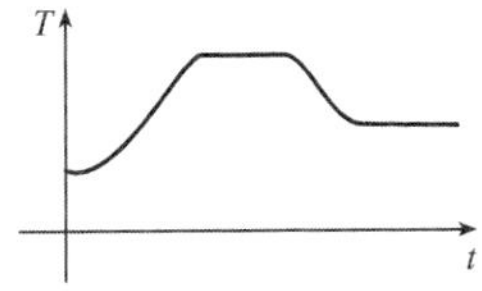


**17.**

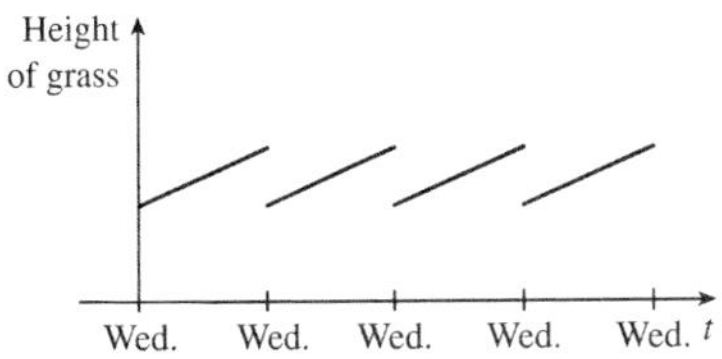


**18.** (a)

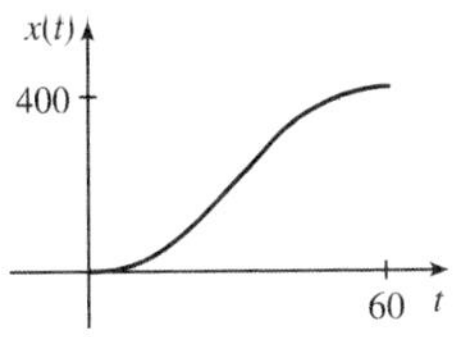


(b)

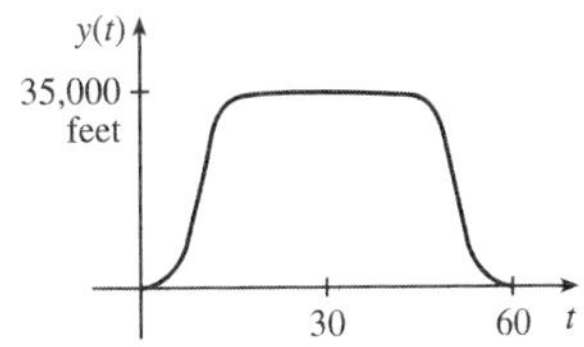


(c)

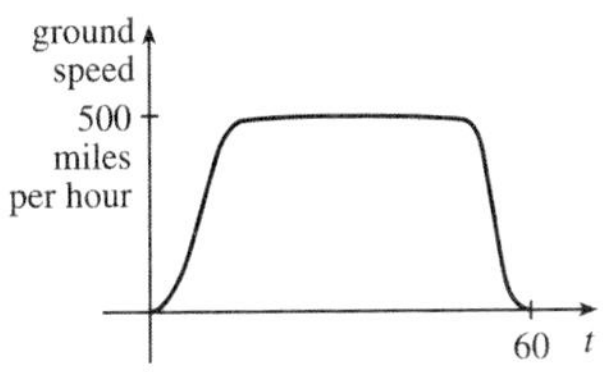


(d)

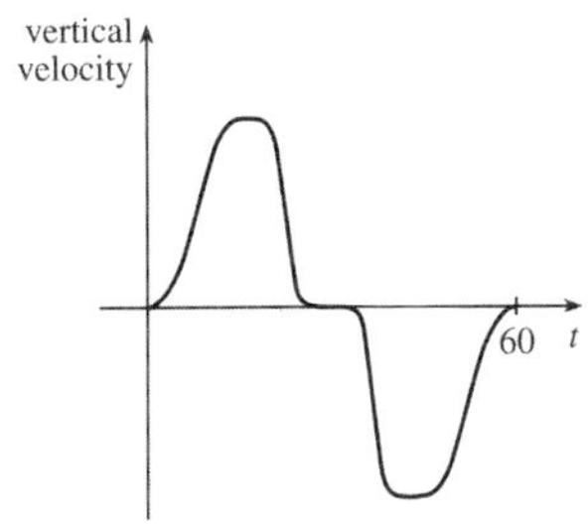


**19.** (a)

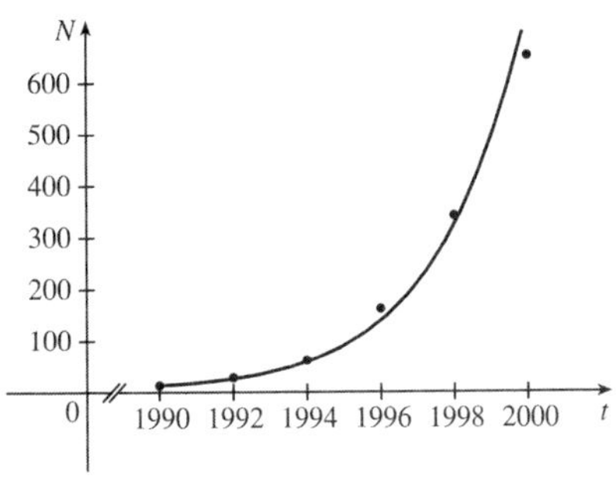


(b) From the graph, we estimate the number of cell-phone subscribers worldwide to be about 92 million in 1995 and 485 million in 1999.

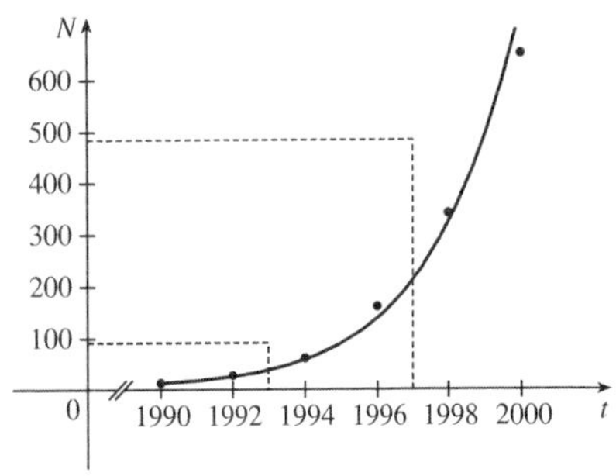


**20.** (a)

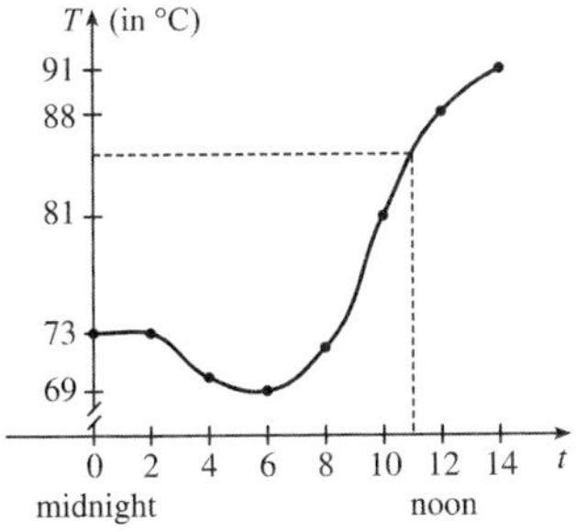


(b) From the graph in part (a), we estimate the temperature at 11:00 A.M. to be about 84.5°C.

**21.** $f(x) = 3x^2 - x + 2.$

$f(2) = 3(2)^2 - 2 + 2 = 12 - 2 + 2 = 12.$

$f(-2) = 3(-2)^2 - (-2) + 2 = 12 + 2 + 2 = 16.$

$f(a) = 3a^2 - a + 2.$

$f(-a) = 3(-a)^2 - (-a) + 2 = 3a^2 + a + 2.$

$f(a+1) = 3(a+1)^2 - (a+1) + 2 = 3(a^2 + 2a + 1) - a - 1 + 2 = 3a^2 + 6a + 3 - a + 1 = 3a^2 + 5a + 4.$

$2f(a) = 2 \cdot f(a) = 2(3a^2 - a + 2) = 6a^2 - 2a + 4.$

$f(2a) = 3(2a)^2 - (2a) + 2 = 3(4a^2) - 2a + 2 = 12a^2 - 2a + 2.$

$f(a^2) = 3(a^2)^2 - (a^2) + 2 = 3(a^4) - a^2 + 2 = 3a^4 - a^2 + 2.$

$$\begin{aligned}[f(a)]^2 &= \left[3a^2 - a + 2\right]^2 = \left(3a^2 - a + 2\right)\left(3a^2 - a + 2\right)\\ &= 9a^4 - 3a^3 + 6a^2 - 3a^3 + a^2 - 2a + 6a^2 - 2a + 4 = 9a^4 - 6a^3 + 13a^2 - 4a + 4.\end{aligned}$$

$f(a+h) = 3(a+h)^2 - (a+h) + 2 = 3(a^2 + 2ah + h^2) - a - h + 2 = 3a^2 + 6ah + 3h^2 - a - h + 2.$

**22.** A spherical balloon with radius $r+1$ has volume $V(r+1) = \frac{4}{3}\pi(r+1)^3 = \frac{4}{3}\pi(r^3 + 3r^2 + 3r + 1)$. We wish to find the amount of air needed to inflate the balloon from a radius of $r$ to $r+1$. Hence, we need to find the difference $V(r+1) - V(r) = \frac{4}{3}\pi(r^3 + 3r^2 + 3r + 1) - \frac{4}{3}\pi r^3 = \frac{4}{3}\pi(3r^2 + 3r + 1).$

**23.** $f(x) = 4 + 3x - x^2$, so $f(3+h) = 4 + 3(3+h) - (3+h)^2 = 4 + 9 + 3h - (9 + 6h + h^2) = 4 - 3h - h^2$,

and $\dfrac{f(3+h) - f(3)}{h} = \dfrac{(4 - 3h - h^2) - 4}{h} = \dfrac{h(-3-h)}{h} = -3 - h.$

**24.** $f(x) = x^3$, so $f(a+h) = (a+h)^3 = a^3 + 3a^2h + 3ah^2 + h^3$,

and $\dfrac{f(a+h) - f(a)}{h} = \dfrac{(a^3 + 3a^2h + 3ah^2 + h^3) - a^3}{h} = \dfrac{h(3a^2 + 3ah + h^2)}{h} = 3a^2 + 3ah + h^2.$

**25.** $$\frac{f(x) - f(a)}{x - a} = \frac{\dfrac{1}{x} - \dfrac{1}{a}}{x - a} = \frac{\dfrac{a - x}{xa}}{x - a} = \frac{a - x}{xa(x - a)} = \frac{-1(x - a)}{xa(x - a)} = -\frac{1}{ax}$$

**26.** $$\begin{aligned}\frac{f(x) - f(1)}{x - 1} &= \frac{\dfrac{x+3}{x+1} - 2}{x - 1} = \frac{\dfrac{x + 3 - 2(x+1)}{x+1}}{x - 1} = \frac{x + 3 - 2x - 2}{(x+1)(x-1)}\\ &= \frac{-x + 1}{(x+1)(x-1)} = \frac{-(x-1)}{(x+1)(x-1)} = -\frac{1}{x+1}\end{aligned}$$

**27.** $f(x) = x/(3x - 1)$ is defined for all $x$ except when $0 = 3x - 1 \Leftrightarrow x = \frac{1}{3}$, so the domain is $\left\{x \in \mathbb{R} \mid x \neq \frac{1}{3}\right\} = \left(-\infty, \frac{1}{3}\right) \cup \left(\frac{1}{3}, \infty\right).$

**28.** $f(x) = (5x + 4)/(x^2 + 3x + 2)$ is defined for all $x$ except when $0 = x^2 + 3x + 2 \Leftrightarrow 0 = (x+2)(x+1) \Leftrightarrow x = -2$ or $-1$, so the domain is $\{x \in \mathbb{R} \mid x \neq -2, -1\} = (-\infty, -2) \cup (-2, -1) \cup (-1, \infty).$

**29.** $f(t) = \sqrt{t} + \sqrt[3]{t}$ is defined when $t \geq 0$. These values of $t$ give real number results for $\sqrt{t}$, whereas any value of $t$ gives a real number result for $\sqrt[3]{t}$. The domain is $[0, \infty)$.

**30.** $g(u) = \sqrt{u} + \sqrt{4-u}$ is defined when $u \geq 0$ and $4 - u \geq 0 \quad \Leftrightarrow \quad u \leq 4$. Thus, the domain is $0 \leq u \leq 4 = [0, 4]$.

**31.** $h(x) = 1 \big/ \sqrt[4]{x^2 - 5x}$ is defined when $x^2 - 5x > 0 \quad \Leftrightarrow \quad x(x-5) > 0$. Note that $x^2 - 5x \neq 0$ since that would result in division by zero. The expression $x(x-5)$ is positive if $x < 0$ or $x > 5$. (See Appendix A for methods for solving inequalities.) Thus, the domain is $(-\infty, 0) \cup (5, \infty)$.

**32.** $h(x) = \sqrt{4 - x^2}$. Now $y = \sqrt{4 - x^2} \quad \Rightarrow \quad y^2 = 4 - x^2 \quad \Leftrightarrow \quad x^2 + y^2 = 4$, so the graph is the top half of a circle of radius 2 with center at the origin. The domain is $\{x \mid 4 - x^2 \geq 0\} = \{x \mid 4 \geq x^2\} = \{x \mid 2 \geq |x|\} = [-2, 2]$. From the graph, the range is $0 \leq y \leq 2$, or $[0, 2]$.

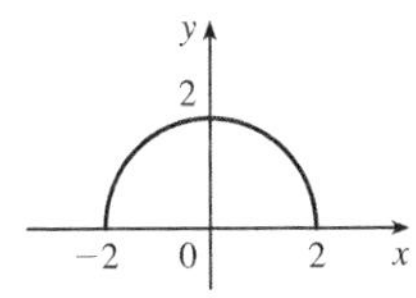


**33.** $f(x) = 5$ is defined for all real numbers, so the domain is $\mathbb{R}$, or $(-\infty, \infty)$. The graph of $f$ is a horizontal line with $y$-intercept 5.

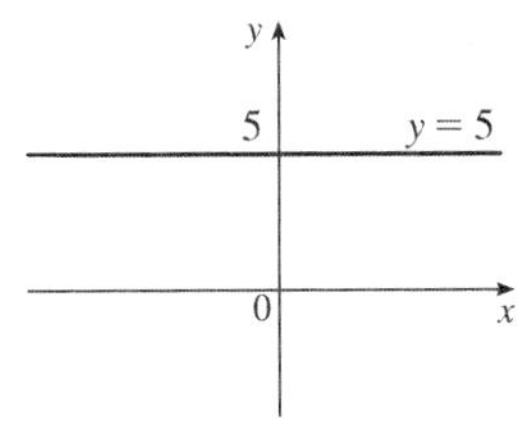


**34.** $F(x) = \frac{1}{2}(x + 3)$ is defined for all real numbers, so the domain is $\mathbb{R}$, or $(-\infty, \infty)$. The graph of $F$ is a line with $x$-intercept $-3$ and $y$-intercept $\frac{3}{2}$.

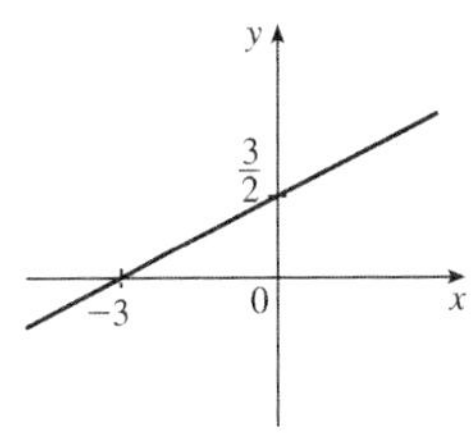


**35.** $f(t) = t^2 - 6t$ is defined for all real numbers, so the domain is $\mathbb{R}$, or $(-\infty, \infty)$. The graph of $f$ is a parabola opening upward since the coefficient of $t^2$ is positive. To find the $t$-intercepts, let $y = 0$ and solve for $t$. $0 = t^2 - 6t = t(t-6) \quad \Rightarrow \quad t = 0$ and $t = 6$. The $t$-coordinate of the vertex is halfway between the $t$-intercepts, that is, at $t = 3$. Since $f(3) = 3^2 - 6 \cdot 3 = -9$, the vertex is $(3, -9)$.

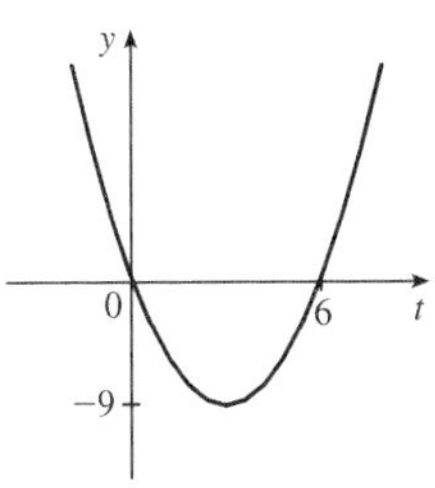


**36.** $H(t) = \dfrac{4 - t^2}{2 - t} = \dfrac{(2+t)(2-t)}{2-t}$, so for $t \neq 2$, $H(t) = 2 + t$. The domain is $\{t \mid t \neq 2\}$. So the graph of $H$ is the same as the graph of the function $f(t) = t + 2$ (a line) except for the hole at $(2, 4)$.

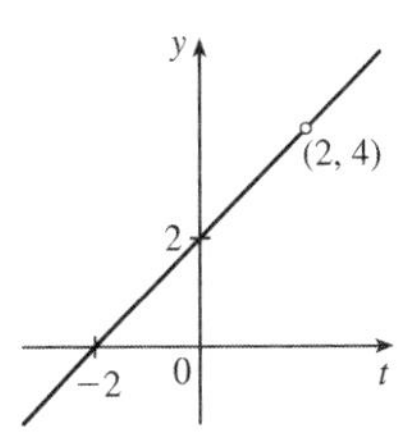


**37.** $g(x) = \sqrt{x - 5}$ is defined when $x - 5 \geq 0$ or $x \geq 5$, so the domain is $[5, \infty)$. Since $y = \sqrt{x-5} \quad \Rightarrow \quad y^2 = x - 5 \quad \Rightarrow \quad x = y^2 + 5$, we see that $g$ is the top half of a parabola.

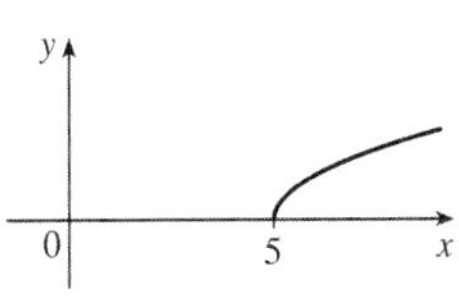

**38.** $F(x) = |2x+1| = \begin{cases} 2x+1 & \text{if } 2x+1 \geq 0 \\ -(2x+1) & \text{if } 2x+1 < 1 \end{cases}$

$= \begin{cases} 2x+1 & \text{if } x \geq -\frac{1}{2} \\ -2x-1 & \text{if } x < -\frac{1}{2} \end{cases}$

The domain is $\mathbb{R}$, or $(-\infty, \infty)$.

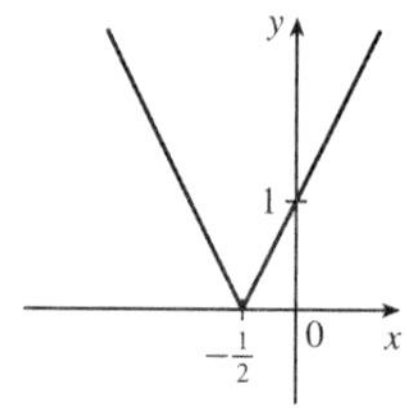


**39.** $G(x) = \dfrac{3x + |x|}{x}$. Since $|x| = \begin{cases} x & \text{if } x \geq 0 \\ -x & \text{if } x < 0 \end{cases}$, we have

$$G(x) = \begin{cases} \dfrac{3x+x}{x} & \text{if } x > 0 \\ \dfrac{3x-x}{x} & \text{if } x < 0 \end{cases} = \begin{cases} \dfrac{4x}{x} & \text{if } x > 0 \\ \dfrac{2x}{x} & \text{if } x < 0 \end{cases} = \begin{cases} 4 & \text{if } x > 0 \\ 2 & \text{if } x < 0 \end{cases}$$

Note that $G$ is not defined for $x = 0$. The domain is $(-\infty, 0) \cup (0, \infty)$.

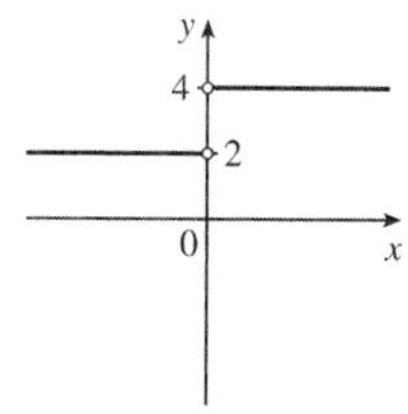


**40.** $g(x) = \dfrac{|x|}{x^2}$. Since $|x| = \begin{cases} x & \text{if } x \geq 0 \\ -x & \text{if } x < 0 \end{cases}$, we have

$$g(x) = \begin{cases} \dfrac{x}{x^2} & \text{if } x > 0 \\ \dfrac{-x}{x^2} & \text{if } x < 0 \end{cases} = \begin{cases} \dfrac{1}{x} & \text{if } x > 0 \\ \dfrac{1}{-x} & \text{if } x < 0 \end{cases}$$

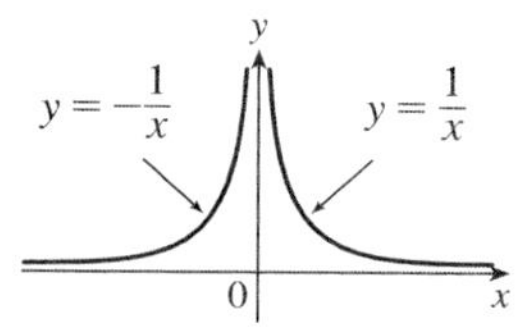


Note that $g$ is not defined for $x = 0$. The domain is $(-\infty, 0) \cup (0, \infty)$.

**41.** $f(x) = \begin{cases} x+2 & \text{if } x < 0 \\ 1-x & \text{if } x \geq 0 \end{cases}$

The domain is $\mathbb{R}$.

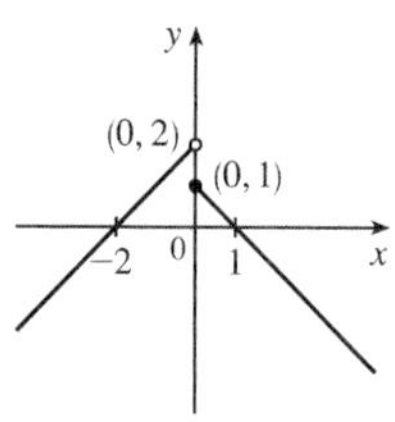


**42.** $f(x) = \begin{cases} 3 - \frac{1}{2}x & \text{if } x \leq 2 \\ 2x-5 & \text{if } x > 2 \end{cases}$

The domain is $\mathbb{R}$.

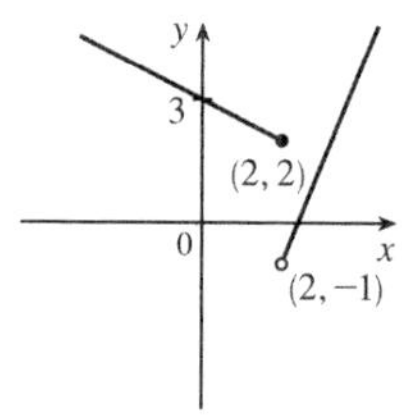


**43.** $f(x) = \begin{cases} x+2 & \text{if } x \leq -1 \\ x^2 & \text{if } x > -1 \end{cases}$

Note that for $x = -1$, both $x+2$ and $x^2$ are equal to 1.

The domain is $\mathbb{R}$.

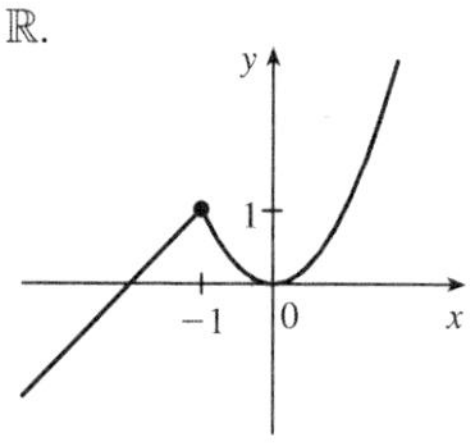


**44.** $f(x) = \begin{cases} -1 & \text{if } x \leq -1 \\ 3x+2 & \text{if } -1 < x < 1 \\ 7-2x & \text{if } x \geq 1 \end{cases}$

The domain is $\mathbb{R}$.

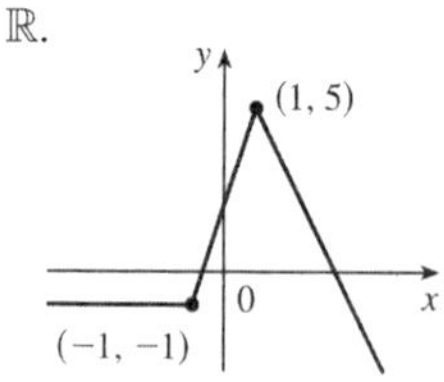

**45.** Recall that the slope $m$ of a line between the two points $(x_1, y_1)$ and $(x_2, y_2)$ is $m = \dfrac{y_2 - y_1}{x_2 - x_1}$ and an equation of the line connecting those two points is $y - y_1 = m(x - x_1)$. The slope of this line segment is $\dfrac{-6-1}{4-(-2)} = -\dfrac{7}{6}$, so an equation is $y - 1 = -\frac{7}{6}(x+2)$. The function is $f(x) = -\frac{7}{6}x - \frac{4}{3}$, $-2 \le x \le 4$.

**46.** The slope of this line segment is $\dfrac{3-(-2)}{6-(-3)} = \dfrac{5}{9}$, so an equation is $y + 2 = \frac{5}{9}(x+3)$.
The function is $f(x) = \frac{5}{9}x - \frac{1}{3}$, $-3 \le x \le 6$.

**47.** We need to solve the given equation for $y$. $x + (y-1)^2 = 0 \Leftrightarrow (y-1)^2 = -x \Leftrightarrow y - 1 = \pm\sqrt{-x} \Leftrightarrow$ $y = 1 \pm \sqrt{-x}$. The expression with the positive radical represents the top half of the parabola, and the one with the negative radical represents the bottom half. Hence, we want $f(x) = 1 - \sqrt{-x}$. Note that the domain is $x \le 0$.

**48.** $(x-1)^2 + y^2 = 1 \Leftrightarrow y = \pm\sqrt{1-(x-1)^2} = \pm\sqrt{2x - x^2}$.
The top half is given by the function $f(x) = \sqrt{2x - x^2}$, $0 \le x \le 2$.

**49.** For $-1 \le x \le 2$, the graph is the line with slope 1 and $y$-intercept 1, that is, the line $y = x + 1$. For $2 < x \le 4$, the graph is the line with slope $-\frac{3}{2}$ and $x$-intercept 4 [which corresponds to the point $(4, 0)$], so $y - 0 = -\frac{3}{2}(x-4) = -\frac{3}{2}x + 6$. So the function is $f(x) = \begin{cases} x+1 & \text{if } -1 \le x \le 2 \\ -\frac{3}{2}x + 6 & \text{if } 2 < x \le 4 \end{cases}$

**50.** For $x \le 0$, the graph is the line $y = 2$. For $0 < x \le 1$, the graph is the line with slope $-2$ and $y$-intercept 2, that is, the line $y = -2x + 2$. For $x > 1$, the graph is the line with slope 1 and $x$-intercept 1, that is, the line $y = 1\,(x-1) = x - 1$. So the function is $f(x) = \begin{cases} 2 & \text{if } x \le 0 \\ -2x+2 & \text{if } 0 < x \le 1 \\ x - 1 & \text{if } 1 < x \end{cases}$

**51.** Let the length and width of the rectangle be $L$ and $W$. Then the perimeter is $2L + 2W = 20$ and the area is $A = LW$. Solving the first equation for $W$ in terms of $L$ gives $W = \dfrac{20 - 2L}{2} = 10 - L$. Thus, $A(L) = L(10 - L) = 10L - L^2$. Since lengths are positive, the domain of $A$ is $0 < L < 10$. If we further restrict $L$ to be larger than $W$, then $5 < L < 10$ would be the domain.

**52.** Let the length and width of the rectangle be $L$ and $W$. Then the area is $LW = 16$, so that $W = 16/L$. The perimeter is $P = 2L + 2W$, so $P(L) = 2L + 2(16/L) = 2L + 32/L$, and the domain of $P$ is $L > 0$, since lengths must be positive quantities. If we further restrict $L$ to be larger than $W$, then $L > 4$ would be the domain.

**53.** Let the length of a side of the equilateral triangle be $x$. Then by the Pythagorean Theorem, the height $y$ of the triangle satisfies $y^2 + \left(\frac{1}{2}x\right)^2 = x^2$, so that $y^2 = x^2 - \frac{1}{4}x^2 = \frac{3}{4}x^2$ and $y = \frac{\sqrt{3}}{2}x$. Using the formula for the area $A$ of a triangle, $A = \frac{1}{2}(\text{base})(\text{height})$, we obtain $A(x) = \frac{1}{2}(x)\left(\frac{\sqrt{3}}{2}x\right) = \frac{\sqrt{3}}{4}x^2$, with domain $x > 0$.

**54.** Let the volume of the cube be $V$ and the length of an edge be $L$. Then $V = L^3$ so $L = \sqrt[3]{V}$, and the surface area is $S(V) = 6\left(\sqrt[3]{V}\right)^2 = 6V^{2/3}$, with domain $V > 0$.

**55.** Let each side of the base of the box have length $x$, and let the height of the box be $h$. Since the volume is 2, we know that $2 = hx^2$, so that $h = 2/x^2$, and the surface area is $S = x^2 + 4xh$. Thus, $S(x) = x^2 + 4x(2/x^2) = x^2 + (8/x)$, with domain $x > 0$.

**56.** The area of the window is $A = xh + \frac{1}{2}\pi\left(\frac{1}{2}x\right)^2 = xh + \dfrac{\pi x^2}{8}$, where $h$ is the height of the rectangular portion of the window.

The perimeter is $P = 2h + x + \frac{1}{2}\pi x = 30 \quad \Leftrightarrow \quad 2h = 30 - x - \frac{1}{2}\pi x \quad \Leftrightarrow \quad h = \frac{1}{4}(60 - 2x - \pi x)$. Thus,

$$A(x) = x\,\frac{60 - 2x - \pi x}{4} + \frac{\pi x^2}{8} = 15x - \tfrac{1}{2}x^2 - \tfrac{\pi}{4}x^2 + \tfrac{\pi}{8}x^2 = 15x - \tfrac{4}{8}x^2 - \tfrac{\pi}{8}x^2 = 15x - x^2\left(\frac{\pi + 4}{8}\right).$$

Since the lengths $x$ and $h$ must be positive quantities, we have $x > 0$ and $h > 0$. For $h > 0$, we have $2h > 0 \quad \Leftrightarrow$

$30 - x - \frac{1}{2}\pi x > 0 \quad \Leftrightarrow \quad 60 > 2x + \pi x \quad \Leftrightarrow \quad x < \dfrac{60}{2 + \pi}$. Hence, the domain of $A$ is $0 < x < \dfrac{60}{2 + \pi}$.

**57.** The height of the box is $x$ and the length and width are $L = 20 - 2x$, $W = 12 - 2x$. Then $V = LWx$ and so

$V(x) = (20 - 2x)(12 - 2x)\,(x) = 4(10 - x)(6 - x)(x) = 4x\left(60 - 16x + x^2\right) = 4x^3 - 64x^2 + 240x$.

The sides $L$, $W$, and $x$ must be positive. Thus, $L > 0 \quad \Leftrightarrow \quad 20 - 2x > 0 \quad \Leftrightarrow \quad x < 10$;

$W > 0 \quad \Leftrightarrow \quad 12 - 2x > 0 \quad \Leftrightarrow \quad x < 6$; and $x > 0$. Combining these restrictions gives us the domain $0 < x < 6$.

**58.**

$$C(x) = \begin{cases} \$2.00 & \text{if } 0.0 < x \le 1.0 \\ 2.20 & \text{if } 1.0 < x \le 1.1 \\ 2.40 & \text{if } 1.1 < x \le 1.2 \\ 2.60 & \text{if } 1.2 < x \le 1.3 \\ 2.80 & \text{if } 1.3 < x \le 1.4 \\ 3.00 & \text{if } 1.4 < x \le 1.5 \\ 3.20 & \text{if } 1.5 < x \le 1.6 \\ 3.40 & \text{if } 1.6 < x \le 1.7 \\ 3.60 & \text{if } 1.7 < x \le 1.8 \\ 3.80 & \text{if } 1.8 < x \le 1.9 \\ 4.00 & \text{if } 1.9 < x < 2.0 \end{cases}$$

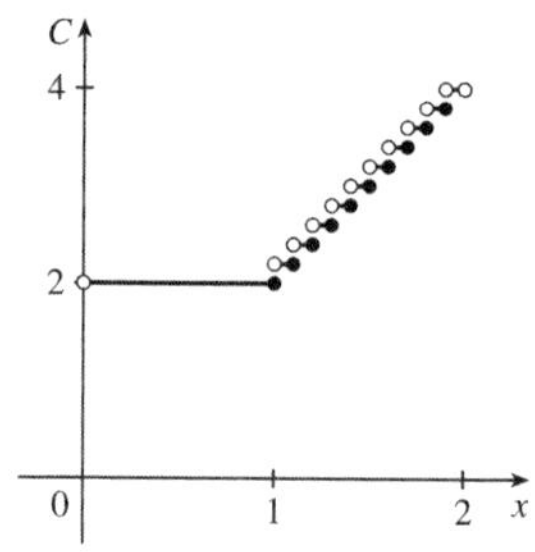


**59.** (a)

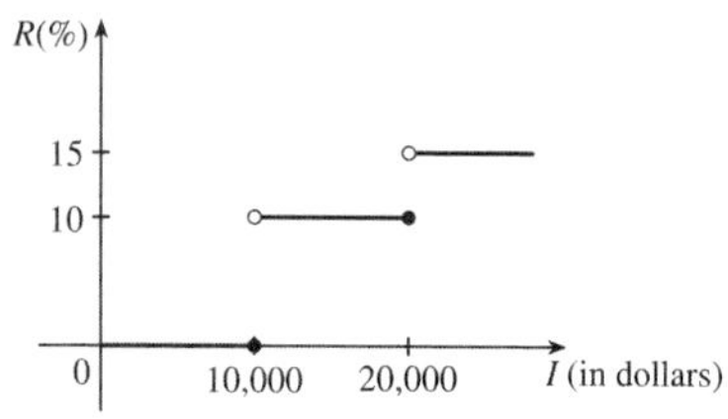


(b) On \$14,000, tax is assessed on \$4000, and 10%(\$4000) = \$400.

On \$26,000, tax is assessed on \$16,000, and

10%(\$10,000) + 15%(\$6000) = \$1000 + \$900 = \$1900.

(c) As in part (b), there is \$1000 tax assessed on \$20,000 of income, so the graph of $T$ is a line segment from $(10{,}000, 0)$ to $(20{,}000, 1000)$. The tax on \$30,000 is \$2500, so the graph of $T$ for $x > 20{,}000$ is the ray with initial point $(20{,}000, 1000)$ that passes through $(30{,}000, 2500)$.

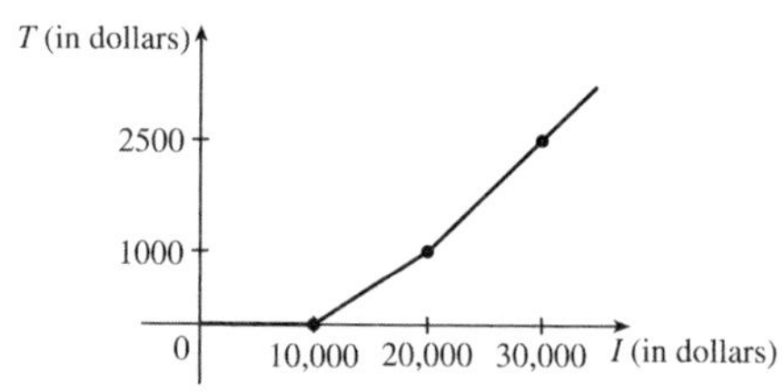


**60.** One example is the amount paid for cable or telephone system repair in the home, usually measured to the nearest quarter hour. Another example is the amount paid by a student in tuition fees, if the fees vary according to the number of credits for which the student has registered.

**61.** $f$ is an odd function because its graph is symmetric about the origin. $g$ is an even function because its graph is symmetric with respect to the $y$-axis.

**62.** $f$ is not an even function since it is not symmetric with respect to the $y$-axis. $f$ is not an odd function since it is not symmetric about the origin. Hence, $f$ is *neither* even nor odd. $g$ is an even function because its graph is symmetric with respect to the $y$-axis.

**63.** (a) Because an even function is symmetric with respect to the $y$-axis, and the point $(5, 3)$ is on the graph of this even function, the point $(-5, 3)$ must also be on its graph.

(b) Because an odd function is symmetric with respect to the origin, and the point $(5, 3)$ is on the graph of this odd function, the point $(-5, -3)$ must also be on its graph.

**64.** (a) If $f$ is even, we get the rest of the graph by reflecting about the $y$-axis.

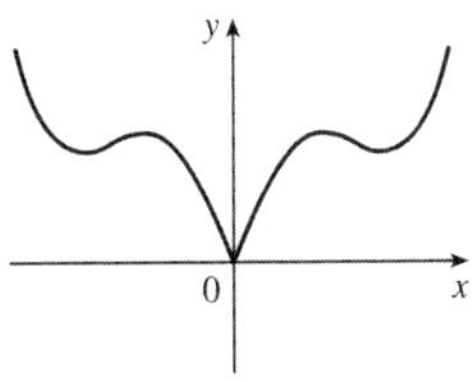


(b) If $f$ is odd, we get the rest of the graph by rotating $180°$ about the origin.

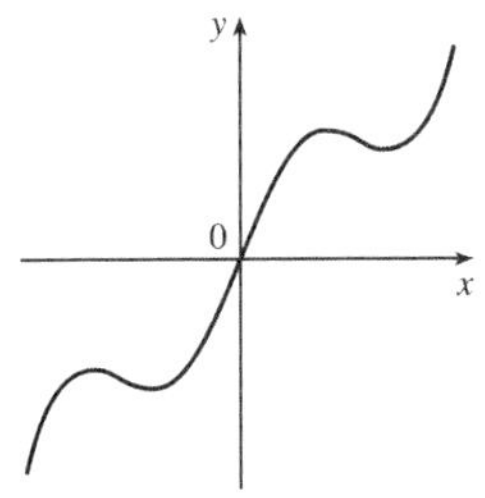


**65.** $f(x) = \dfrac{x}{x^2+1}$.

$$f(-x) = \frac{-x}{(-x)^2+1} = \frac{-x}{x^2+1} = -\frac{x}{x^2+1} = -f(x).$$

So $f$ is an odd function.

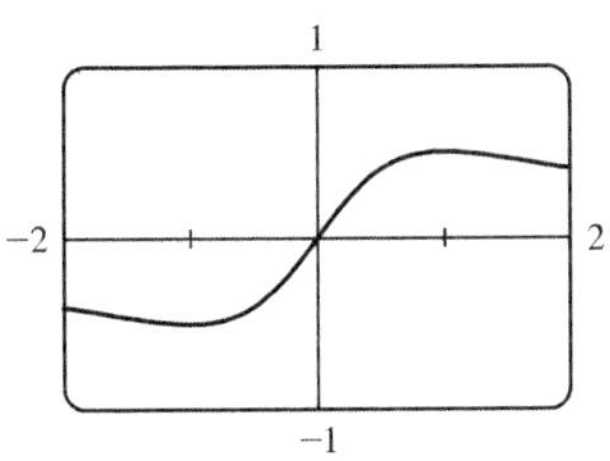


**66.** $f(x) = \dfrac{x^2}{x^4+1}$.

$$f(-x) = \frac{(-x)^2}{(-x)^4+1} = \frac{x^2}{x^4+1} = f(x).$$

So $f$ is an even function.

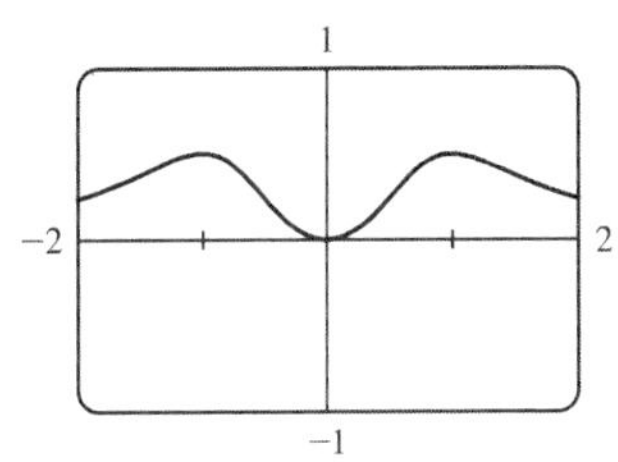


**67.** $f(x) = \dfrac{x}{x+1}$, so $f(-x) = \dfrac{-x}{-x+1} = \dfrac{x}{x-1}$.

Since this is neither $f(x)$ nor $-f(x)$, the function $f$ is neither even nor odd.

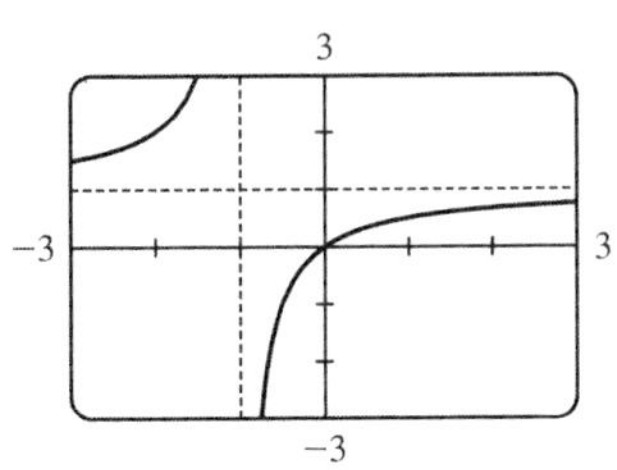


**68.** $f(x) = x\,|x|$.

$$\begin{aligned} f(-x) &= (-x)\,|-x| = (-x)\,|x| = -(x\,|x|) \\ &= -f(x) \end{aligned}$$

So $f$ is an odd function.

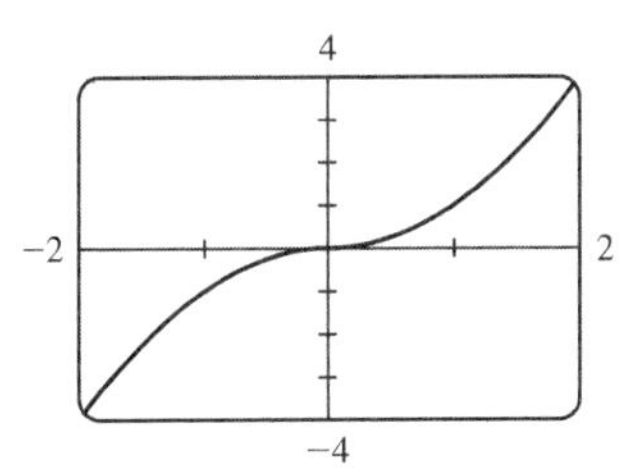

**69.** $f(x) = 1 + 3x^2 - x^4$.

$f(-x) = 1 + 3(-x)^2 - (-x)^4 = 1 + 3x^2 - x^4 = f(x)$.

So $f$ is an even function.

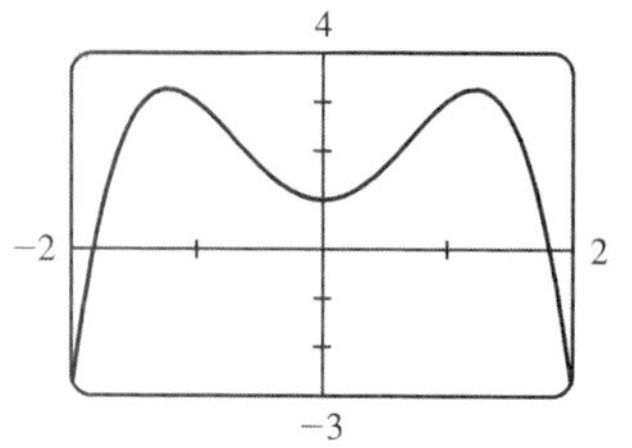


**70.** $f(x) = 1 + 3x^3 - x^5$, so

$$f(-x) = 1 + 3(-x)^3 - (-x)^5 = 1 + 3(-x^3) - (-x^5)$$
$$= 1 - 3x^3 + x^5$$

Since this is neither $f(x)$ nor $-f(x)$, the function $f$ is neither even nor odd.

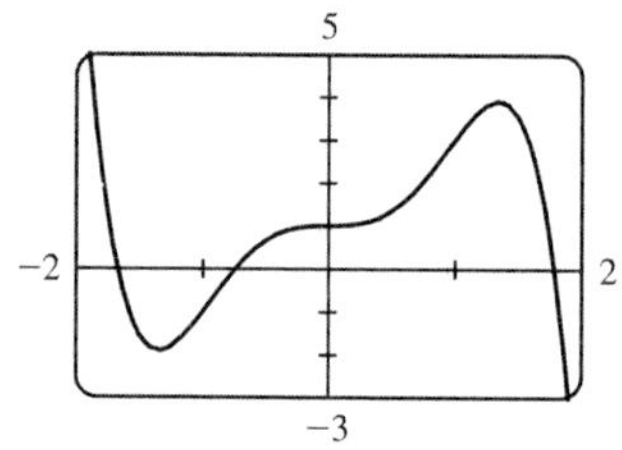


## 1.2 Mathematical Models: A Catalog of Essential Functions

**1.** (a) $f(x) = \sqrt[5]{x}$ is a root function with $n = 5$.

(b) $g(x) = \sqrt{1 - x^2}$ is an algebraic function because it is a root of a polynomial.

(c) $h(x) = x^9 + x^4$ is a polynomial of degree 9.

(d) $r(x) = \dfrac{x^2 + 1}{x^3 + x}$ is a rational function because it is a ratio of polynomials.

(e) $s(x) = \tan 2x$ is a trigonometric function.

(f) $t(x) = \log_{10} x$ is a logarithmic function.

**2.** (a) $y = (x - 6)/(x + 6)$ is a rational function because it is a ratio of polynomials.

(b) $y = x + x^2/\sqrt{x - 1}$ is an algebraic function because it involves polynomials and roots of polynomials.

(c) $y = 10^x$ is an exponential function (notice that $x$ is the *exponent*).

(d) $y = x^{10}$ is a power function (notice that $x$ is the *base*).

(e) $y = 2t^6 + t^4 - \pi$ is a polynomial of degree 6.

(f) $y = \cos\theta + \sin\theta$ is a trigonometric function.

**3.** We notice from the figure that $g$ and $h$ are even functions (symmetric with respect to the $y$-axis) and that $f$ is an odd function (symmetric with respect to the origin). So (b) $[y = x^5]$ must be $f$. Since $g$ is flatter than $h$ near the origin, we must have (c) $[y = x^8]$ matched with $g$ and (a) $[y = x^2]$ matched with $h$.

**4.** (a) The graph of $y = 3x$ is a line (choice $G$).

(b) $y = 3^x$ is an exponential function (choice $f$).

(c) $y = x^3$ is an odd polynomial function or power function (choice $F$).

(d) $y = \sqrt[3]{x} = x^{1/3}$ is a root function (choice $g$).

**5.** (a) An equation for the family of linear functions with slope 2 is $y = f(x) = 2x + b$, where $b$ is the $y$-intercept.

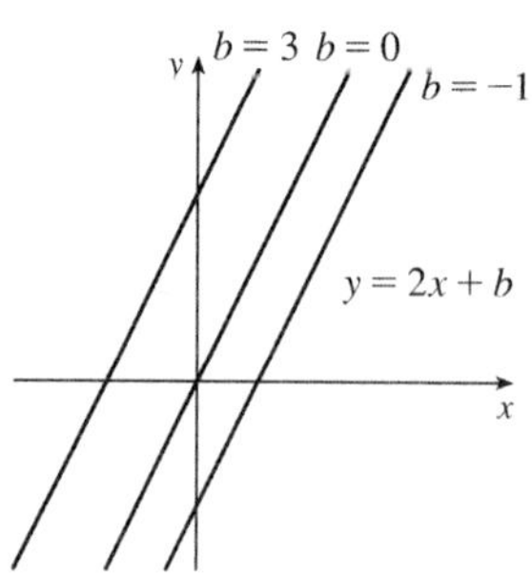

(b) $f(2) = 1$ means that the point $(2, 1)$ is on the graph of $f$. We can use the point-slope form of a line to obtain an equation for the family of linear functions through the point $(2, 1)$. $y - 1 = m(x - 2)$, which is equivalent to $y = mx + (1 - 2m)$ in slope-intercept form.

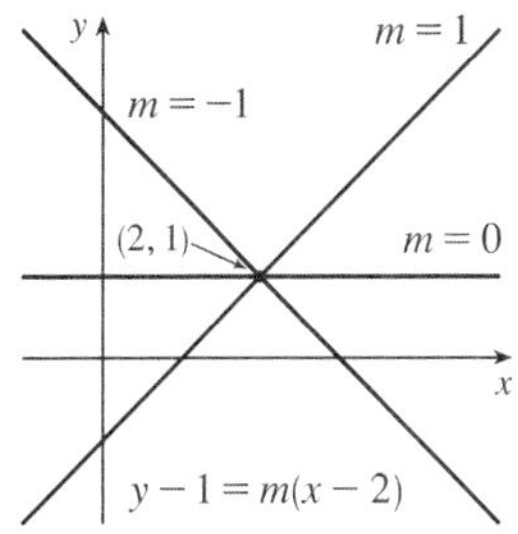


(c) To belong to both families, an equation must have slope $m = 2$, so the equation in part (b), $y = mx + (1 - 2m)$, becomes $y = 2x - 3$. It is the *only* function that belongs to both families.

**6.** All members of the family of linear functions $f(x) = 1 + m(x + 3)$ have graphs that are lines passing through the point $(-3, 1)$.

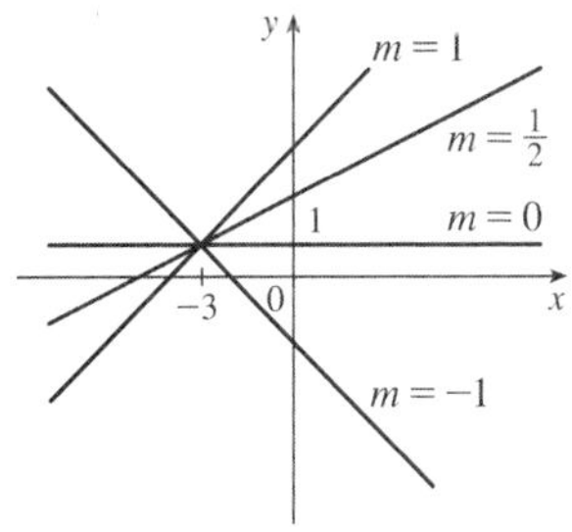


**7.** All members of the family of linear functions $f(x) = c - x$ have graphs that are lines with slope $-1$. The $y$-intercept is $c$.

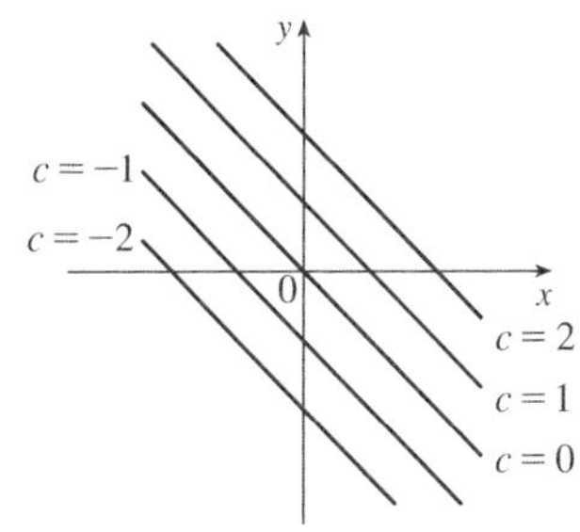


**8.** The vertex of the parabola on the left is $(3, 0)$, so an equation is $y = a(x - 3)^2 + 0$. Since the point $(4, 2)$ is on the parabola, we'll substitute 4 for $x$ and 2 for $y$ to find $a$. $2 = a(4 - 3)^2 \Rightarrow a = 2$, so an equation is $f(x) = 2(x - 3)^2$.

The $y$-intercept of the parabola on the right is $(0, 1)$, so an equation is $y = ax^2 + bx + 1$. Since the points $(-2, 2)$ and $(1, -2.5)$ are on the parabola, we'll substitute $-2$ for $x$ and 2 for $y$ as well as 1 for $x$ and $-2.5$ for $y$ to obtain two equations with the unknowns $a$ and $b$.

$$(-2, 2): \quad 2 = 4a - 2b + 1 \Rightarrow 4a - 2b = 1 \quad \textbf{(1)}$$

$$(1, -2.5): \quad -2.5 = a + b + 1 \Rightarrow a + b = -3.5 \quad \textbf{(2)}$$

$2 \cdot$ **(2)** + **(1)** gives us $6a = -6 \Rightarrow a = -1$. From **(2)**, $-1 + b = -3.5 \Rightarrow b = -2.5$, so an equation is $g(x) = -x^2 - 2.5x + 1$.

**9.** Since $f(-1) = f(0) = f(2) = 0$, $f$ has zeros of $-1$, 0, and 2, so an equation for $f$ is $f(x) = a[x - (-1)](x - 0)(x - 2)$, or $f(x) = ax(x + 1)(x - 2)$. Because $f(1) = 6$, we'll substitute 1 for $x$ and 6 for $f(x)$.
$6 = a(1)(2)(-1) \Rightarrow -2a = 6 \Rightarrow a = -3$, so an equation for $f$ is $f(x) = -3x(x + 1)(x - 2)$.

**10.** (a) For $T = 0.02t + 8.50$, the slope is 0.02, which means that the average surface temperature of the world is increasing at a rate of 0.02 °C per year. The $T$-intercept is 8.50, which represents the average surface temperature in °C in the year 1900.

(b) $t = 2100 - 1900 = 200 \Rightarrow T = 0.02(200) + 8.50 = 12.50$ °C

**11.** (a) $D = 200$, so $c = 0.0417D(a+1) = 0.0417(200)(a+1) = 8.34a + 8.34$. The slope is 8.34, which represents the change in mg of the dosage for a child for each change of 1 year in age.

(b) For a newborn, $a = 0$, so $c = 8.34$ mg.

**12.** (a)

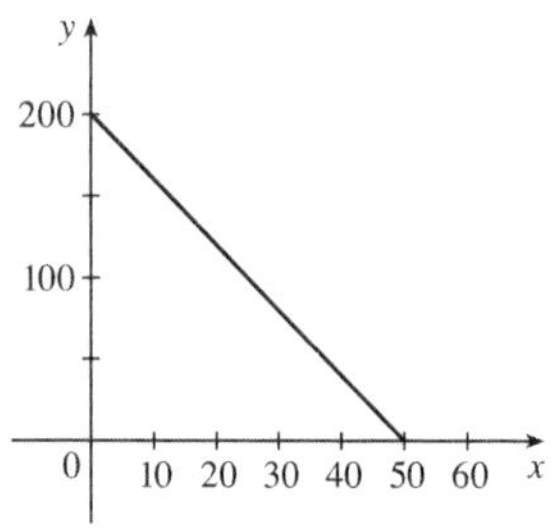


(b) The slope of $-4$ means that for each increase of 1 dollar for a rental space, the number of spaces rented *decreases* by 4. The $y$-intercept of 200 is the number of spaces that would be occupied if there were no charge for each space. The $x$-intercept of 50 is the smallest rental fee that results in no spaces rented.

**13.** (a)

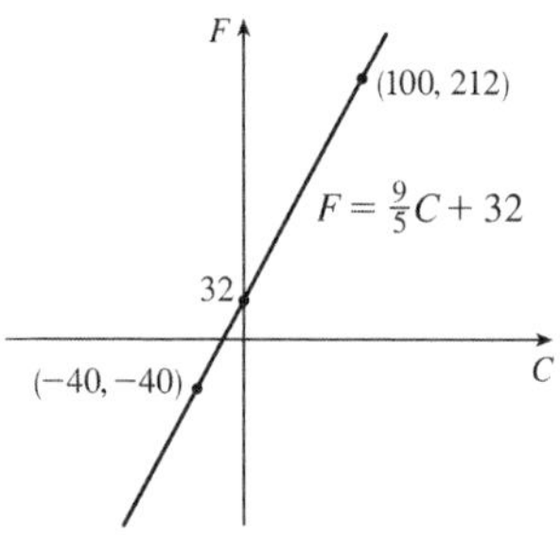


(b) The slope of $\frac{9}{5}$ means that $F$ increases $\frac{9}{5}$ degrees for each increase of $1°$C. (Equivalently, $F$ increases by 9 when $C$ increases by 5 and $F$ decreases by 9 when $C$ decreases by 5.) The $F$-intercept of 32 is the Fahrenheit temperature corresponding to a Celsius temperature of 0.

**14.** (a) Let $d =$ distance traveled (in miles) and $t =$ time elapsed (in hours). At $t = 0$, $d = 0$ and at $t = 50$ minutes $= 50 \cdot \frac{1}{60} = \frac{5}{6}$ h, $d = 40$. Thus we have two points: $(0, 0)$ and $\left(\frac{5}{6}, 40\right)$, so $m = \dfrac{40 - 0}{\frac{5}{6} - 0} = 48$ and so $d = 48t$.

(b)

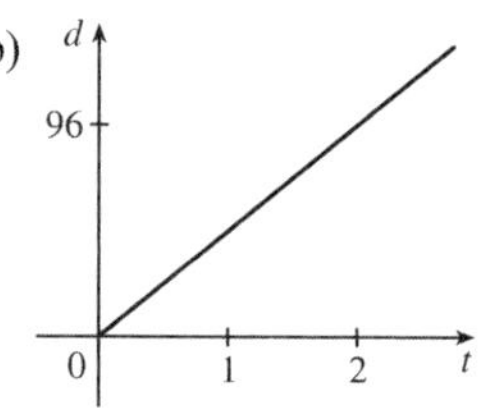


(c) The slope is 48 and represents the car's speed in mi/h.

**15.** (a) Using $N$ in place of $x$ and $T$ in place of $y$, we find the slope to be $\dfrac{T_2 - T_1}{N_2 - N_1} = \dfrac{80 - 70}{173 - 113} = \dfrac{10}{60} = \dfrac{1}{6}$. So a linear equation is $T - 80 = \frac{1}{6}(N - 173) \Leftrightarrow T - 80 = \frac{1}{6}N - \frac{173}{6} \Leftrightarrow T = \frac{1}{6}N + \frac{307}{6}$ $\left[\frac{307}{6} = 51.1\overline{6}\right]$.

(b) The slope of $\frac{1}{6}$ means that the temperature in Fahrenheit degrees increases one-sixth as rapidly as the number of cricket chirps per minute. Said differently, each increase of 6 cricket chirps per minute corresponds to an increase of $1°$F.

(c) When $N = 150$, the temperature is given approximately by $T = \frac{1}{6}(150) + \frac{307}{6} = 76.1\overline{6}\,°\text{F} \approx 76\,°\text{F}$.

**16.** (a) Let $x$ denote the number of chairs produced in one day and $y$ the associated cost. Using the points $(100, 2200)$ and $(300, 4800)$, we get the slope $\frac{4800-2200}{300-100} = \frac{2600}{200} = 13$. So $y - 2200 = 13(x - 100) \Leftrightarrow$ $y = 13x + 900$.

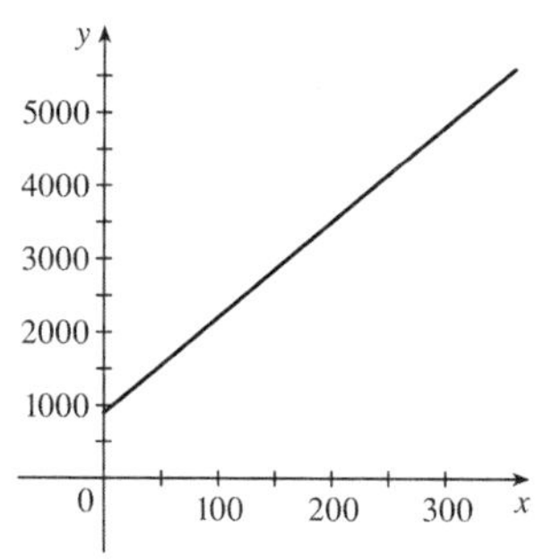


(b) The slope of the line in part (a) is 13 and it represents the cost (in dollars) of producing each additional chair.

(c) The $y$-intercept is 900 and it represents the fixed daily costs of operating the factory.

**17.** (a) We are given $\dfrac{\text{change in pressure}}{\text{10 feet change in depth}} = \dfrac{4.34}{10} = 0.434$. Using $P$ for pressure and $d$ for depth with the point $(d, P) = (0, 15)$, we have the slope-intercept form of the line, $P = 0.434d + 15$.

(b) When $P = 100$, then $100 = 0.434d + 15 \quad \Leftrightarrow \quad 0.434d = 85 \quad \Leftrightarrow \quad d = \frac{85}{0.434} \approx 195.85$ feet. Thus, the pressure is $100\ \text{lb/in}^2$ at a depth of approximately 196 feet.

**18.** (a) Using $d$ in place of $x$ and $C$ in place of $y$, we find the slope to be $\dfrac{C_2 - C_1}{d_2 - d_1} = \dfrac{460 - 380}{800 - 480} = \dfrac{80}{320} = \dfrac{1}{4}$.

So a linear equation is $C - 460 = \frac{1}{4}(d - 800) \quad \Leftrightarrow \quad C - 460 = \frac{1}{4}d - 200 \quad \Leftrightarrow \quad C = \frac{1}{4}d + 260$.

(b) Letting $d = 1500$ we get $C = \frac{1}{4}(1500) + 260 = 635$.

The cost of driving 1500 miles is \$635.

(c)

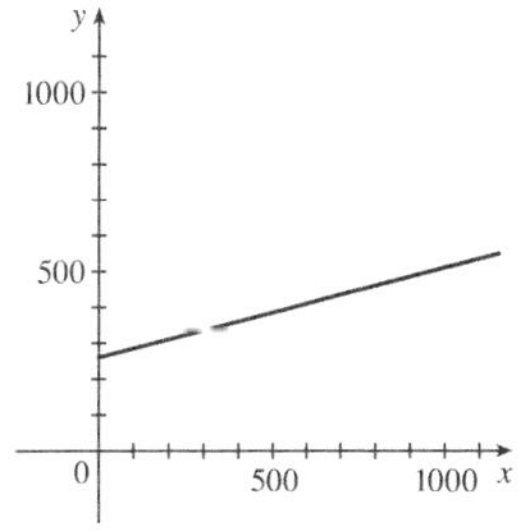


The slope of the line represents the cost per mile, \$0.25.

(d) A linear function gives a suitable model in this situation because you have fixed monthly costs such as insurance and car payments, as well as costs that increase as you drive, such as gasoline, oil, and tires, and the cost of these for each additional mile driven is a constant.

(e) Letting $d = 1500$ we get $C = \frac{1}{4}(1500) + 260 = 635$.

The cost of driving 1500 miles is \$635.

**19.** (a) The data appear to be periodic and a sine or cosine function would make the best model. A model of the form $f(x) = a\cos(bx) + c$ seems appropriate.

(b) The data appear to be decreasing in a linear fashion. A model of the form $f(x) = mx + b$ seems appropriate.

**20.** (a) The data appear to be increasing exponentially. A model of the form $f(x) = a \cdot b^x$ or $f(x) = a \cdot b^x + c$ seems appropriate.

(b) The data appear to be decreasing similarly to the values of the reciprocal function. A model of the form $f(x) = a/x$ seems appropriate.

Some values are given to many decimal places. These are the results given by several computer algebra systems — rounding is left to the reader.

**21.** (a)

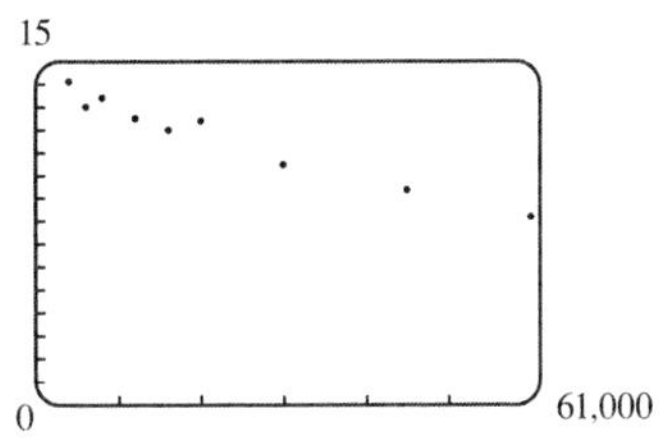


A linear model does seem appropriate.

(b) Using the points (4000, 14.1) and (60,000, 8.2), we obtain

$$y - 14.1 = \frac{8.2 - 14.1}{60{,}000 - 4000}(x - 4000) \text{ or, equivalently,}$$

$$y \approx -0.000105357x + 14.521429.$$

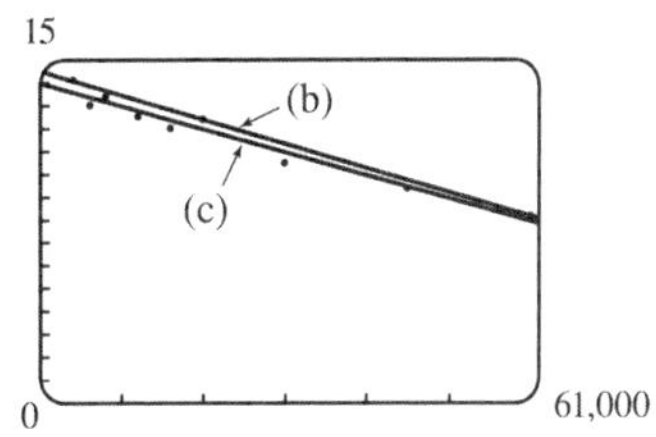


(c) Using a computing device, we obtain the least squares regression line $y = -0.0000997855x + 13.950764$.

The following commands and screens illustrate how to find the least squares regression line on a TI-83 Plus.

Enter the data into list one (L1) and list two (L2). Press STAT 1 to enter the editor.

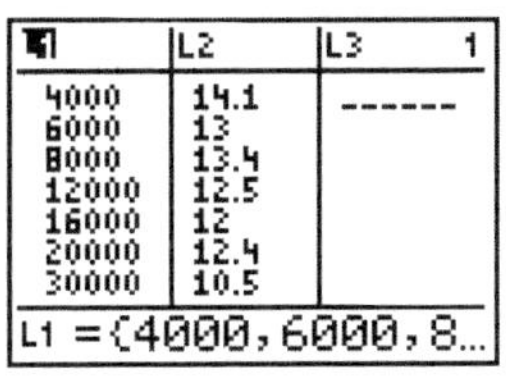


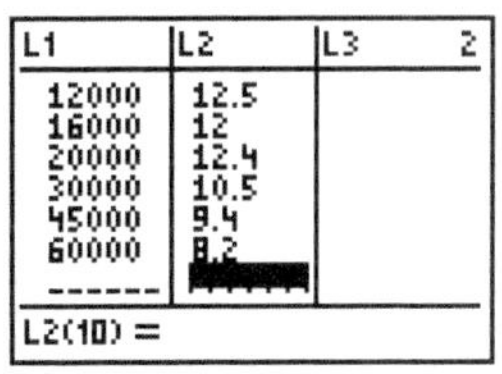

Find the regession line and store it in $Y_1$. Press [2nd] [QUIT] [STAT] [▸] [4] [VARS] [▸] [1] [1] [ENTER].

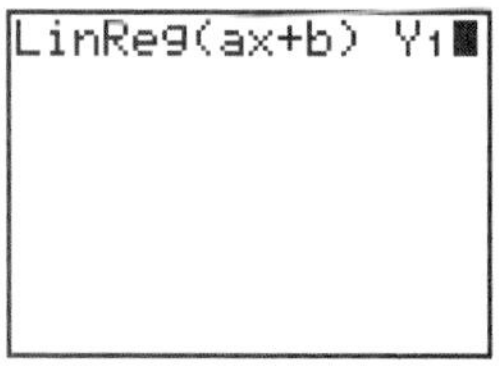

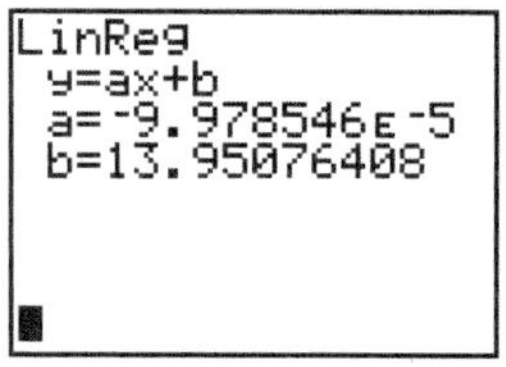

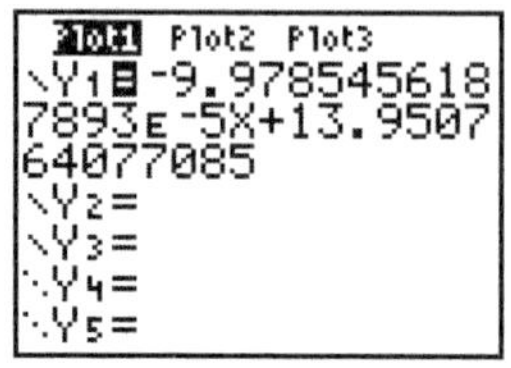


Note from the last figure that the regression line has been stored in $Y_1$ and that Plot1 has been turned on (Plot1 is highlighted). You can turn on Plot1 from the Y= menu by placing the cursor on Plot1 and pressing [ENTER] or by pressing [2nd] [STAT PLOT] [1] [ENTER].

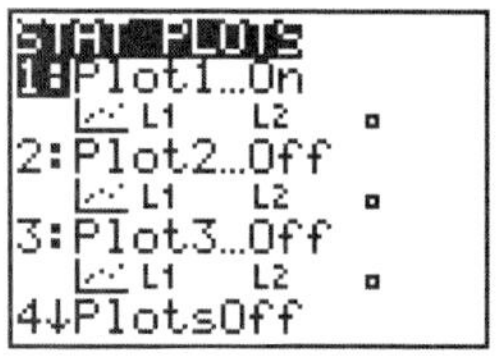

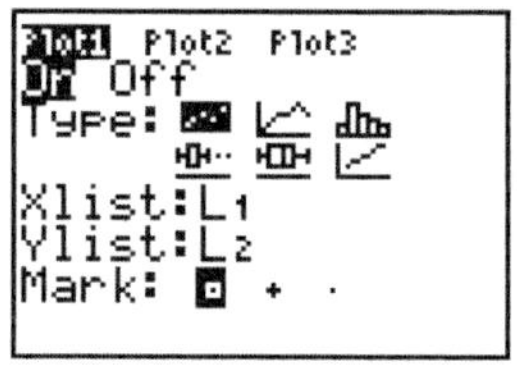


Now press [ZOOM] [9] to produce a graph of the data and the regression line. Note that choice 9 of the ZOOM menu automatically selects a window that displays all of the data.

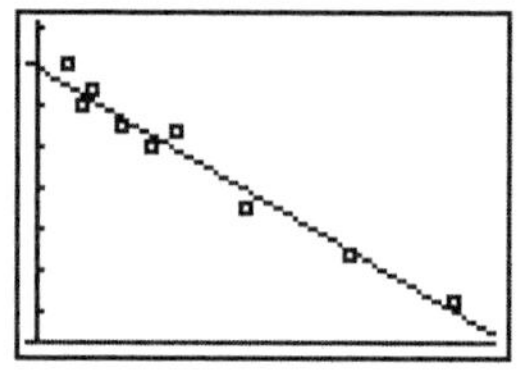

(d) When $x = 25{,}000$, $y \approx 11.456$; or about 11.5 per 100 population.

(e) When $x = 80{,}000$, $y \approx 5.968$; or about a 6% chance.

(f) When $x = 200{,}000$, $y$ is negative, so the model does not apply.

**22.** (a)

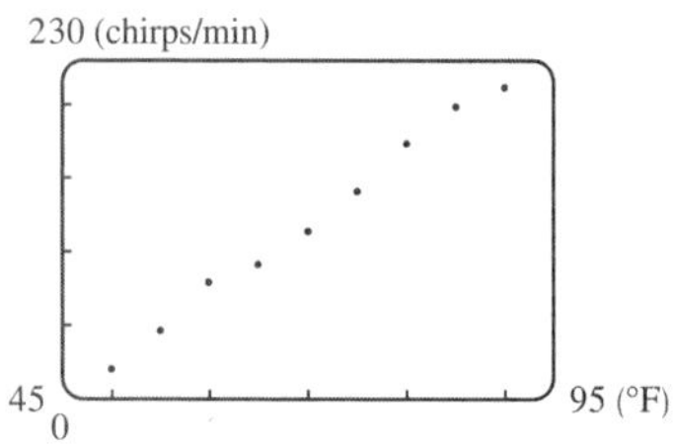


(b)

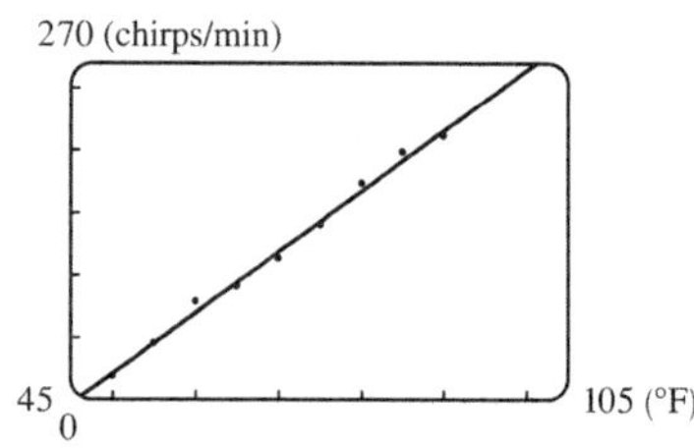


Using a computing device, we obtain the least squares regression line $y = 4.85\overline{6}x - 220.9\overline{6}$.

(c) When $x = 100°$ F, $y = 264.7 \approx 265$ chirps/min.

**23.** (a)

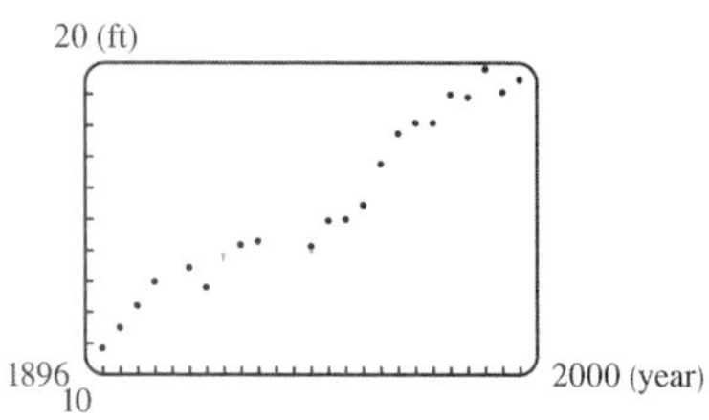


A linear model does seem appropriate.

(b)

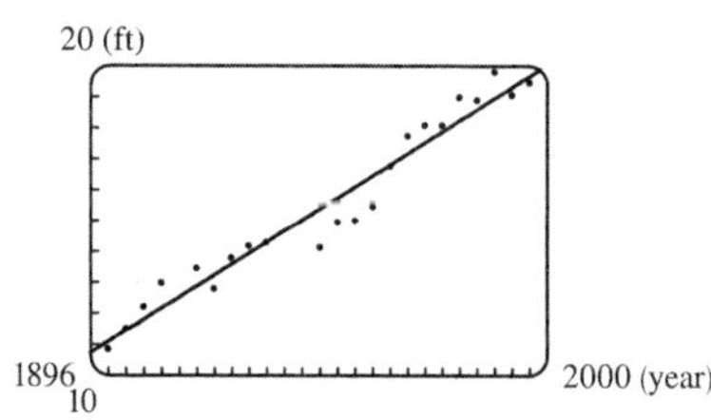


Using a computing device, we obtain the least squares regression line $y = 0.089119747x - 158.2403249$, where $x$ is the year and $y$ is the height in feet.

(c) When $x = 2000$, the model gives $y \approx 20.00$ ft. Note that the actual winning height for the 2000 Olympics is *less than* the winning height for 1996—so much for that prediction.

(d) When $x = 2100$, $y \approx 28.91$ ft. This would be an increase of 9.49 ft from 1996 to 2100. Even though there was an increase of 8.59 ft from 1900 to 1996, it is unlikely that a similar increase will occur over the next 100 years.

**24.** By looking at the scatter plot of the data, we rule out the linear and logarithmic models.

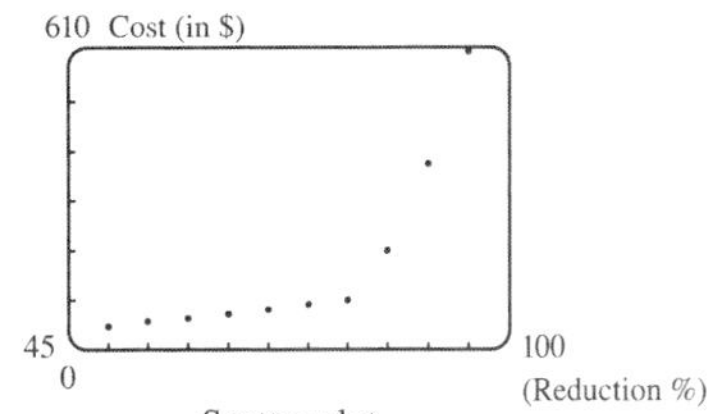

Scatter plot

We try various models.

Quadratic: $y = 0.49\overline{6}x^2 - 62.28\overline{93}x + 1970.6\overline{39}$

Cubic: $y = 0.0201243201x^3 - 3.88037296x^2 + 247.6754468x - 5163.935198$

Quartic: $y = 0.0002951049x^4 - 0.0654560995x^3 + 5.27525641x^2 - 180.2266511x + 2203.210956$

Exponential: $y = 2.41422994\,(1.054516914)^x$

Power: $y = 0.000022854971x^{3.616078251}$

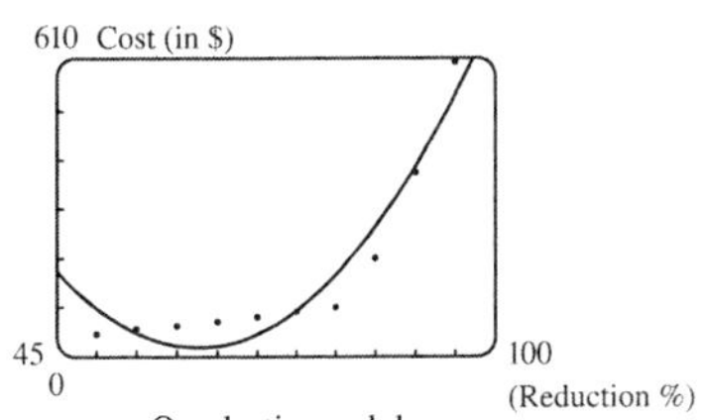

Quadratic model

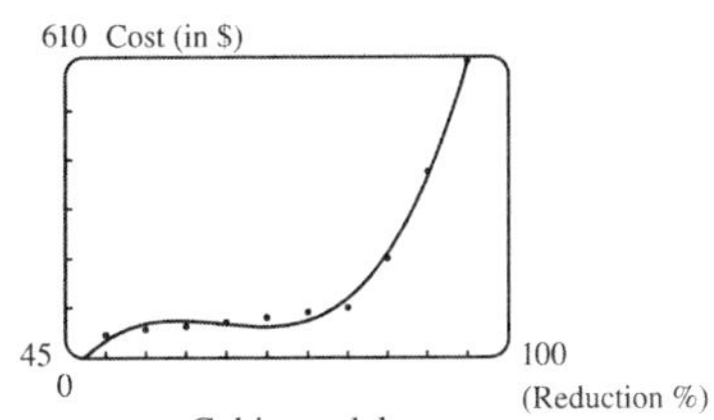

Cubic model

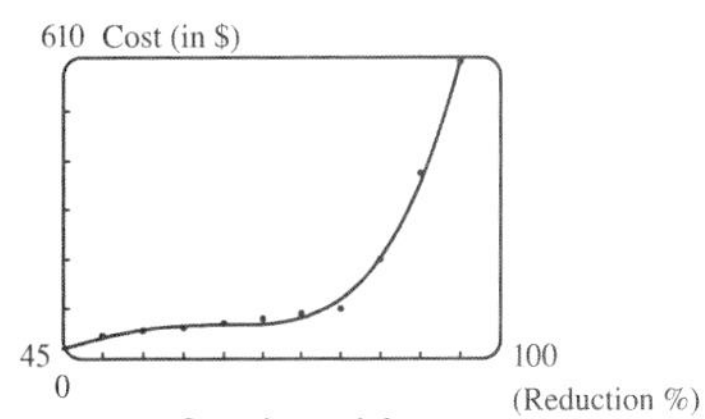

Quartic model

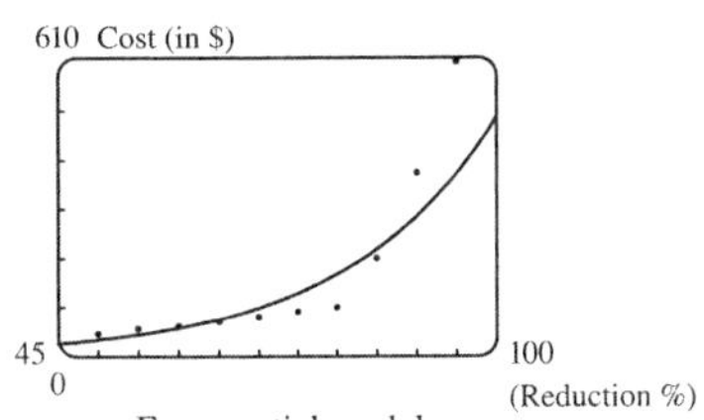

Exponential model

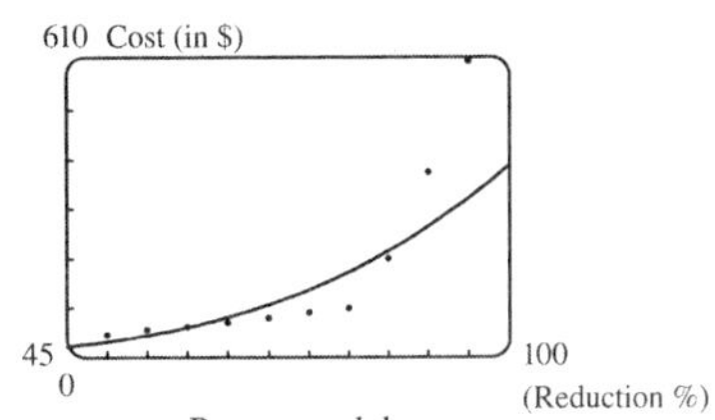

Power model

After examining the graphs of these models, we see that the cubic and quartic models are clearly the best.

**25.**

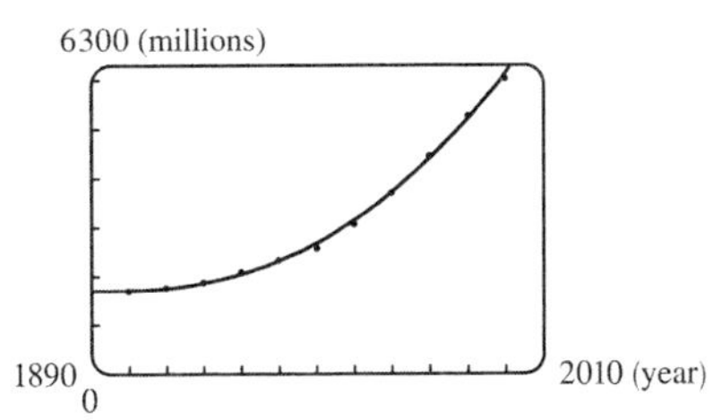


Using a computing device, we obtain the cubic function $y = ax^3 + bx^2 + cx + d$ with $a = 0.0012937$, $b = -7.06142$, $c = 12{,}823$, and $d = -7{,}743{,}770$. When $x = 1925$, $y \approx 1914$ (million).

**26.** (a) $T = 1.000396048d^{1.499661718}$

(b) The power model in part (a) is approximately $T = d^{1.5}$. Squaring both sides gives us $T^2 = d^3$, so the model matches Kepler's Third Law, $T^2 = kd^3$.

## 1.3 New Functions from Old Functions

**1.** (a) If the graph of $f$ is shifted 3 units upward, its equation becomes $y = f(x) + 3$.

(b) If the graph of $f$ is shifted 3 units downward, its equation becomes $y = f(x) - 3$.

(c) If the graph of $f$ is shifted 3 units to the right, its equation becomes $y = f(x - 3)$.

(d) If the graph of $f$ is shifted 3 units to the left, its equation becomes $y = f(x + 3)$.

(e) If the graph of $f$ is reflected about the $x$-axis, its equation becomes $y = -f(x)$.

(f) If the graph of $f$ is reflected about the $y$-axis, its equation becomes $y = f(-x)$.

(g) If the graph of $f$ is stretched vertically by a factor of 3, its equation becomes $y = 3f(x)$.

(h) If the graph of $f$ is shrunk vertically by a factor of 3, its equation becomes $y = \frac{1}{3}f(x)$.

**2.** (a) To obtain the graph of $y = 5f(x)$ from the graph of $y = f(x)$, stretch the graph vertically by a factor of 5.

(b) To obtain the graph of $y = f(x - 5)$ from the graph of $y = f(x)$, shift the graph 5 units to the right.

(c) To obtain the graph of $y = -f(x)$ from the graph of $y = f(x)$, reflect the graph about the $x$-axis.

(d) To obtain the graph of $y = -5f(x)$ from the graph of $y = f(x)$, stretch the graph vertically by a factor of 5 and reflect it about the $x$-axis.

(e) To obtain the graph of $y = f(5x)$ from the graph of $y = f(x)$, shrink the graph horizontally by a factor of 5.

(f) To obtain the graph of $y = 5f(x) - 3$ from the graph of $y = f(x)$, stretch the graph vertically by a factor of 5 and shift it 3 units downward.

**3.** (a) (graph 3) The graph of $f$ is shifted 4 units to the right and has equation $y = f(x - 4)$.

(b) (graph 1) The graph of $f$ is shifted 3 units upward and has equation $y = f(x) + 3$.

(c) (graph 4) The graph of $f$ is shrunk vertically by a factor of 3 and has equation $y = \frac{1}{3}f(x)$.

(d) (graph 5) The graph of $f$ is shifted 4 units to the left and reflected about the $x$-axis. Its equation is $y = -f(x + 4)$.

(e) (graph 2) The graph of $f$ is shifted 6 units to the left and stretched vertically by a factor of 2. Its equation is $y = 2f(x + 6)$.

**4.** (a) To graph $y = f(x + 4)$ we shift the graph of $f$, 4 units to the left.

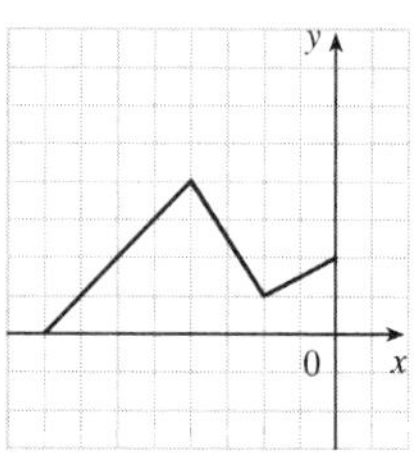


The point $(2, 1)$ on the graph of $f$ corresponds to the point $(2 - 4, 1) = (-2, 1)$.

(b) To graph $y = f(x) + 4$ we shift the graph of $f$, 4 units upward.

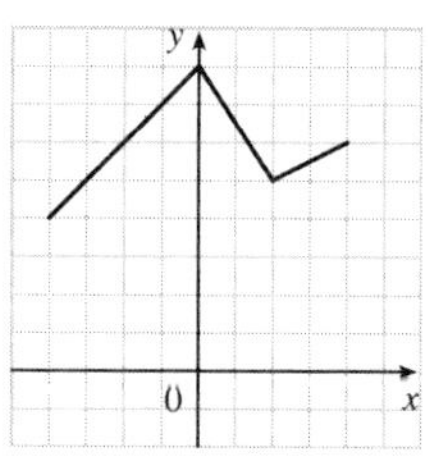


The point $(2, 1)$ on the graph of $f$ corresponds to the point $(2, 1 + 4) = (2, 5)$.

(c) To graph $y = 2f(x)$ we stretch the graph of $f$ vertically by a factor of 2.

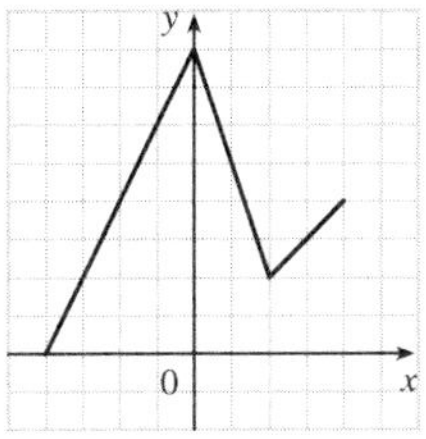


The point $(2, 1)$ on the graph of $f$ corresponds to the point $(2, 2 \cdot 1) = (2, 2)$.

(d) To graph $y = -\frac{1}{2}f(x) + 3$, we shrink the graph of $f$ vertically by a factor of 2, then reflect the resulting graph about the $x$-axis, then shift the resulting graph 3 units upward.

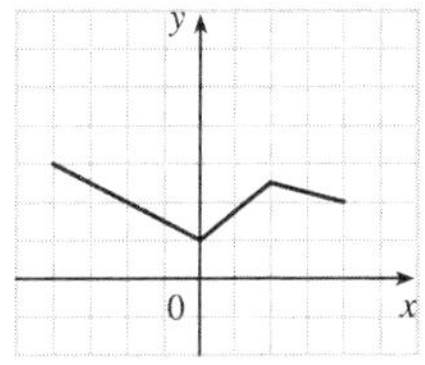


The point $(2, 1)$ on the graph of $f$ corresponds to the point $\left(2, -\frac{1}{2} \cdot 1 + 3\right) = (2, 2.5)$.

**5.** (a) To graph $y = f(2x)$ we shrink the graph of $f$ horizontally by a factor of 2.

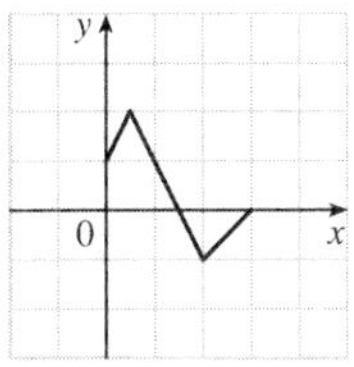


The point $(4, -1)$ on the graph of $f$ corresponds to the point $\left(\frac{1}{2} \cdot 4, -1\right) = (2, -1)$.

(b) To graph $y = f\left(\frac{1}{2}x\right)$ we stretch the graph of $f$ horizontally by a factor of 2.

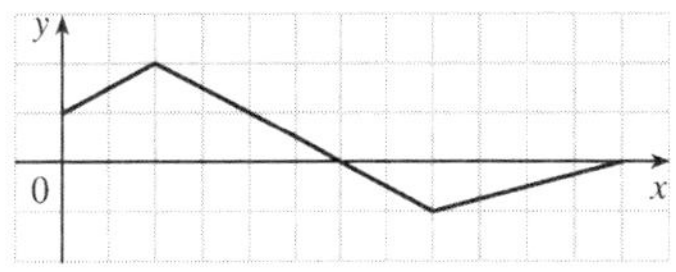


The point $(4, -1)$ on the graph of $f$ corresponds to the point $(2 \cdot 4, -1) = (8, -1)$.

(c) To graph $y = f(-x)$ we reflect the graph of $f$ about the $y$-axis.

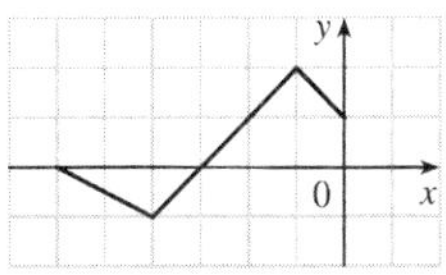


The point $(4, -1)$ on the graph of $f$ corresponds to the point $(-1 \cdot 4, -1) = (-4, -1)$.

(d) To graph $y = -f(-x)$ we reflect the graph of $f$ about the $y$-axis, then about the $x$-axis.

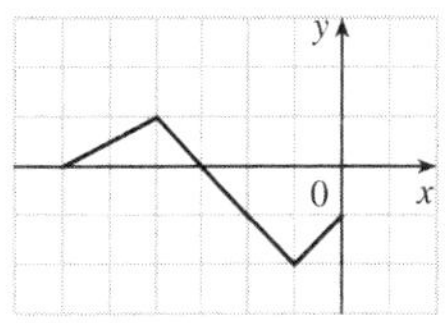


The point $(4, -1)$ on the graph of $f$ corresponds to the point $(-1 \cdot 4, -1 \cdot -1) = (-4, 1)$.

**6.** The graph of $y = f(x) = \sqrt{3x - x^2}$ has been shifted 2 units to the right and stretched vertically by a factor of 2. Thus, a function describing the graph is

$$y = 2f(x-2) = 2\sqrt{3(x-2) - (x-2)^2} = 2\sqrt{3x - 6 - (x^2 - 4x + 4)} = 2\sqrt{-x^2 + 7x - 10}$$

**7.** The graph of $y = f(x) = \sqrt{3x - x^2}$ has been shifted 4 units to the left, reflected about the $x$-axis, and shifted downward 1 unit. Thus, a function describing the graph is

$$y = \underbrace{-1 \cdot}_{\substack{\text{reflect} \\ \text{about } x\text{-axis}}} \quad f\underbrace{(x+4)}_{\substack{\text{shift} \\ \text{4 units left}}} \quad \underbrace{-\ 1}_{\substack{\text{shift} \\ \text{1 unit left}}}$$

This function can be written as

$$y = -f(x+4) - 1 = -\sqrt{3(x+4) - (x+4)^2} - 1 = -\sqrt{3x + 12 - (x^2 + 8x + 16)} - 1 = -\sqrt{-x^2 - 5x - 4} - 1$$

**8.** (a) The graph of $y = 2\sin x$ can be obtained from the graph of $y = \sin x$ by stretching it vertically by a factor of 2.

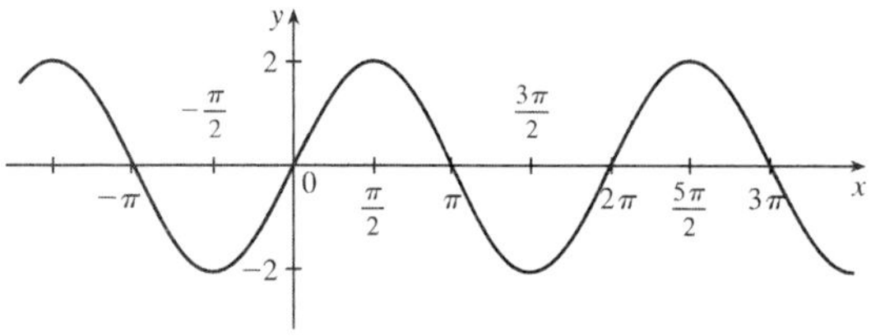


(b) The graph of $y = 1 + \sqrt{x}$ can be obtained from the graph of $y = \sqrt{x}$ by shifting it upward 1 unit.

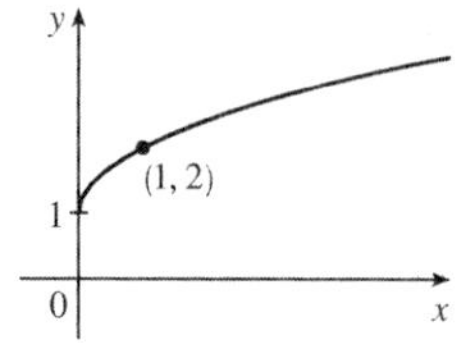


**9.** $y = -x^3$: Start with the graph of $y = x^3$ and reflect about the $x$-axis. Note: Reflecting about the $y$-axis gives the same result since substituting $-x$ for $x$ gives us $y = (-x)^3 = -x^3$.

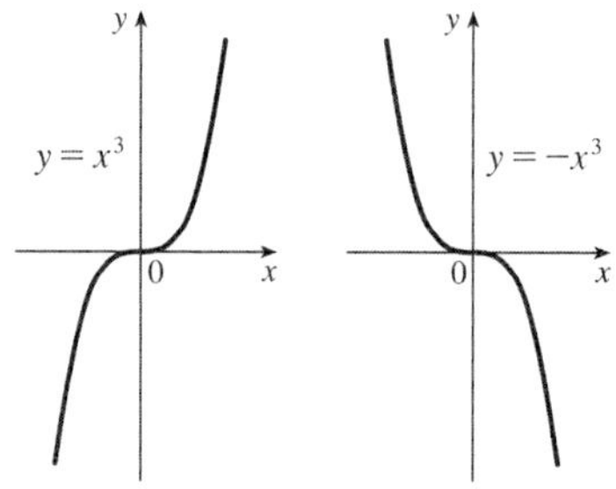


**10.** $y = 1 - x^2 = -x^2 + 1$: Start with the graph of $y = x^2$, reflect about the $x$-axis, and then shift 1 unit upward.

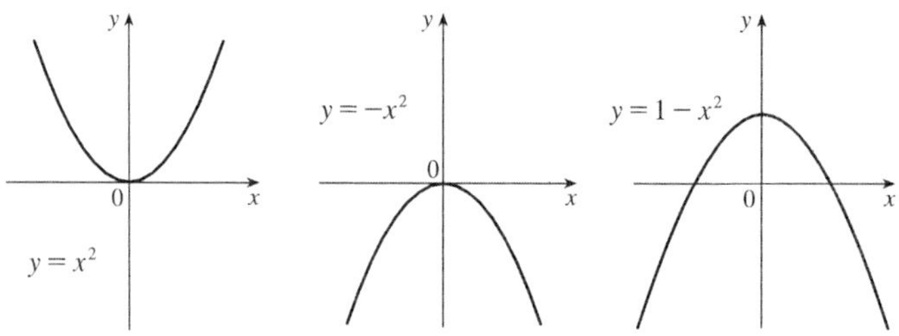


**11.** $y = (x + 1)^2$: Start with the graph of $y = x^2$ and shift 1 unit to the left.

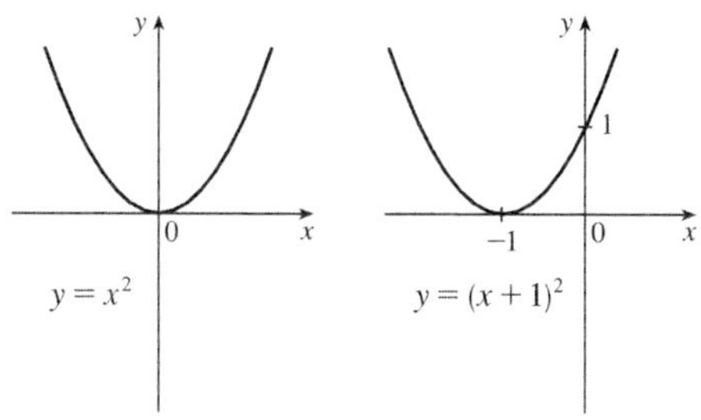


**12.** $y = x^2 - 4x + 3 = (x^2 - 4x + 4) - 1 = (x - 2)^2 - 1$: Start with the graph of $y = x^2$, shift 2 units to the right, and then shift 1 unit downward.

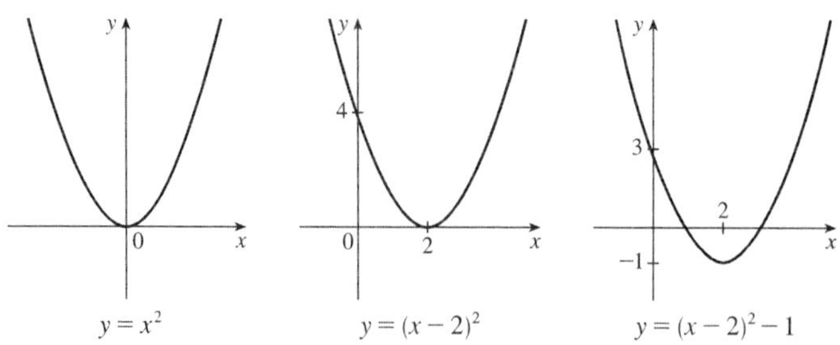

**13.** $y = 1 + 2\cos x$: Start with the graph of $y = \cos x$, stretch vertically by a factor of 2, and then shift 1 unit upward.

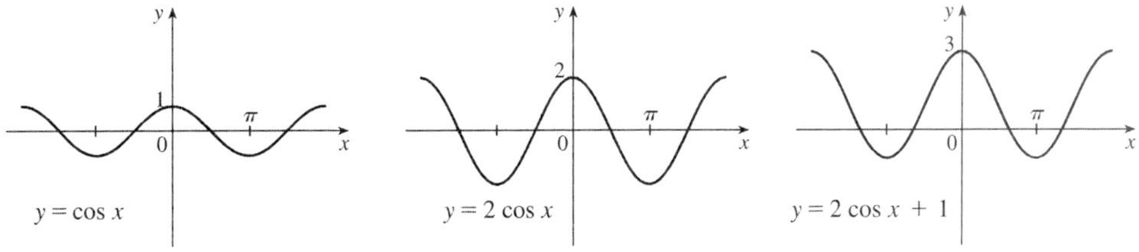


**14.** $y = 4\sin 3x$: Start with the graph of $y = \sin x$, compress horizontally by a factor of 3, and then stretch vertically by a factor of 4.

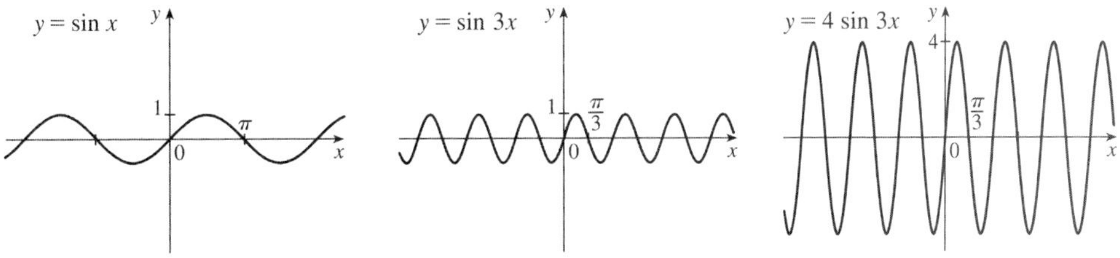


**15.** $y = \sin(x/2)$: Start with the graph of $y = \sin x$ and stretch horizontally by a factor of 2.

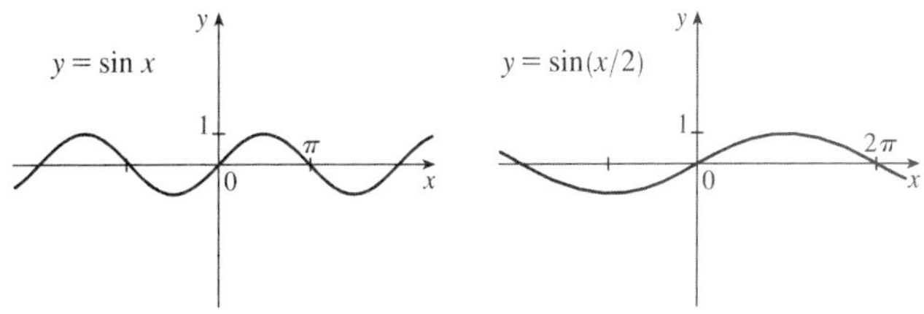


**16.** $y = 1/(x - 4)$: Start with the graph of $y = 1/x$ and shift 4 units to the right.

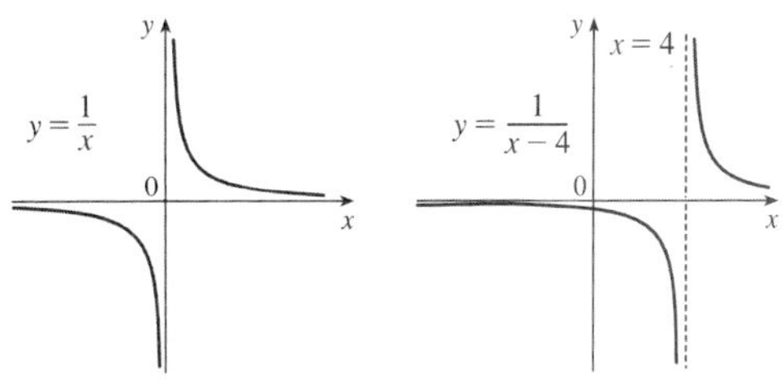


**17.** $y = \sqrt{x + 3}$: Start with the graph of $y = \sqrt{x}$ and shift 3 units to the left.

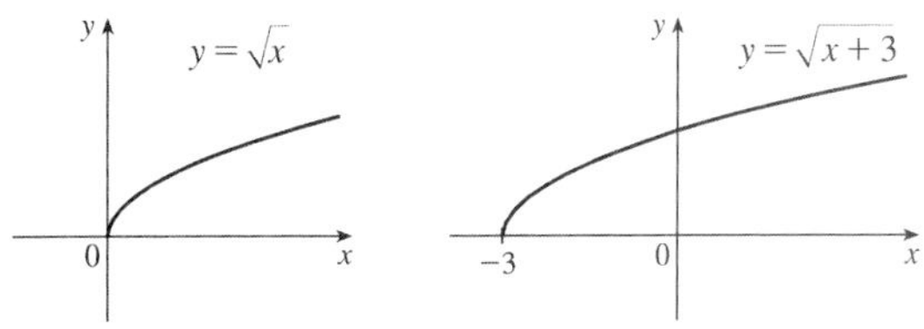


**18.** $y = (x + 2)^4 + 3$: Start with the graph of $y = x^4$, shift 2 units to the left, and then shift 3 units upward.

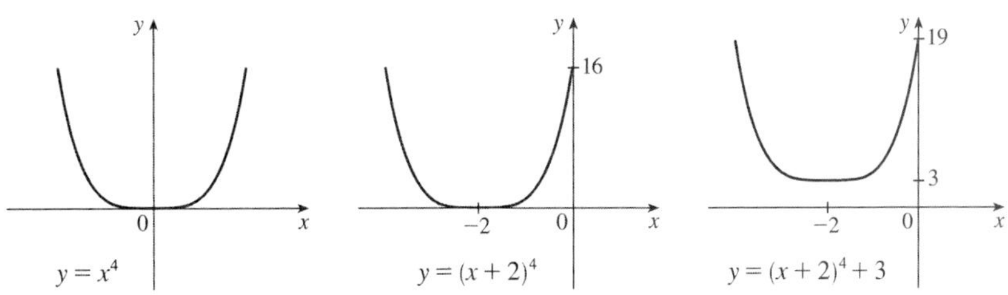

**19.** $y = \frac{1}{2}(x^2 + 8x) = \frac{1}{2}(x^2 + 8x + 16) - 8 = \frac{1}{2}(x + 4)^2 - 8$: Start with the graph of $y = x^2$, compress vertically by a factor of 2, shift 4 units to the left, and then shift 8 units downward.

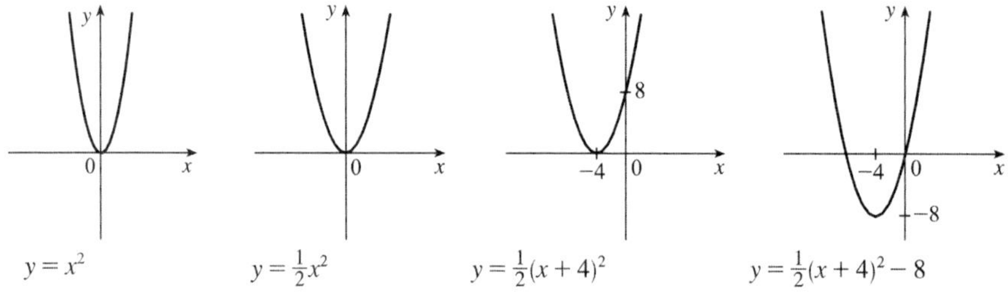


**20.** $y = 1 + \sqrt[3]{x - 1}$: Start with the graph of $y = \sqrt[3]{x}$, shift 1 unit to the right, and then shift 1 unit upward.

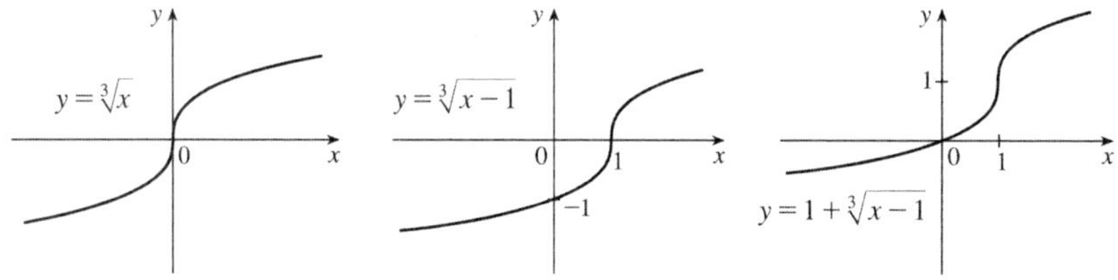


**21.** $y = 2/(x + 1)$: Start with the graph of $y = 1/x$, shift 1 unit to the left, and then stretch vertically by a factor of 2.

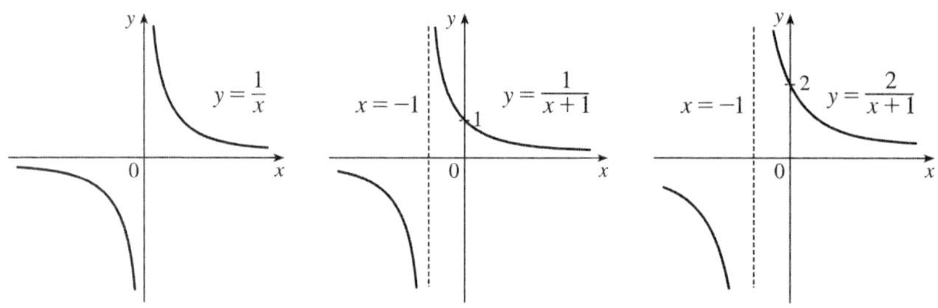


**22.** $y = \frac{1}{4}\tan(x - \frac{\pi}{4})$: Start with the graph of $y = \tan x$, shift $\frac{\pi}{4}$ units to the right, and then compress vertically by a factor of 4.

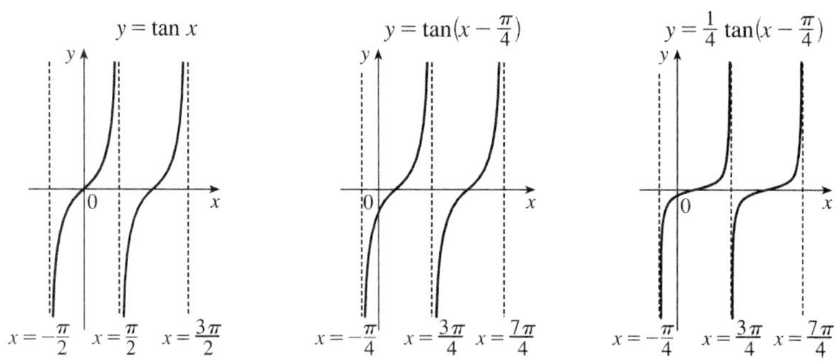


**23.** $y = |\sin x|$: Start with the graph of $y = \sin x$ and reflect all the parts of the graph below the $x$-axis about the $x$-axis.

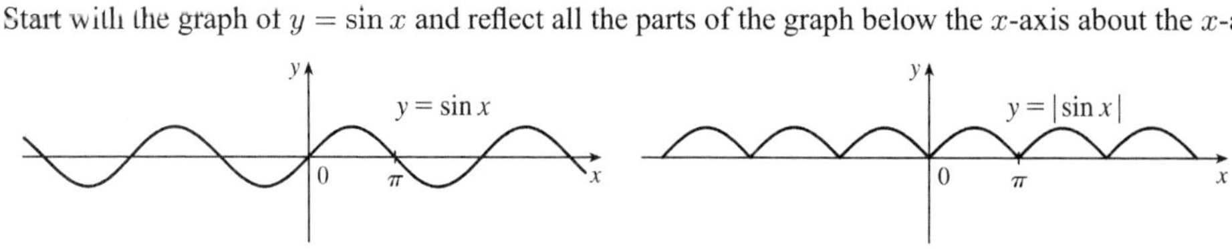

**24.** $y = |x^2 - 2x| = |x^2 - 2x + 1 - 1| = |(x-1)^2 - 1|$: Start with the graph of $y = x^2$, shift 1 unit right, shift 1 unit downward, and reflect the portion of the graph below the $x$-axis about the $x$-axis.

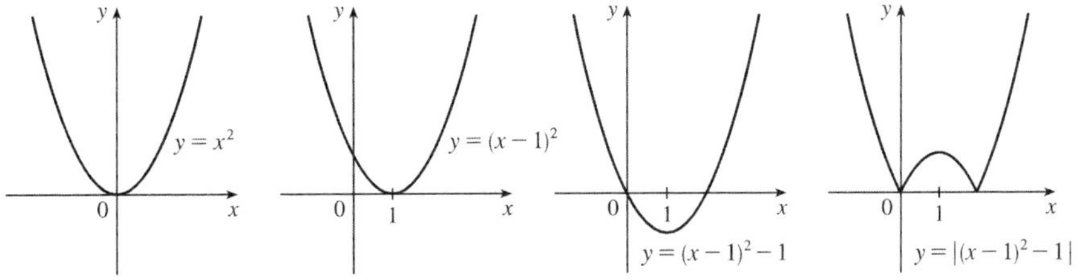


**25.** This is just like the solution to Example 4 except the amplitude of the curve (the 30°N curve in Figure 9 on June 21) is $14 - 12 = 2$. So the function is $L(t) = 12 + 2\sin\left[\frac{2\pi}{365}(t - 80)\right]$. March 31 is the 90th day of the year, so the model gives $L(90) \approx 12.34$ h. The daylight time (5:51 A.M. to 6:18 P.M.) is 12 hours and 27 minutes, or 12.45 h. The model value differs from the actual value by $\frac{12.45 - 12.34}{12.45} \approx 0.009$, less than 1%.

**26.** Using a sine function to model the brightness of Delta Cephei as a function of time, we take its period to be 5.4 days, its amplitude to be 0.35 (on the scale of magnitude), and its average magnitude to be 4.0. If we take $t = 0$ at a time of average brightness, then the magnitude (brightness) as a function of time $t$ in days can be modeled by the formula $M(t) = 4.0 + 0.35\sin\left(\frac{2\pi}{5.4}t\right)$.

**27.** (a) To obtain $y = f(|x|)$, the portion of the graph of $y = f(x)$ to the right of the $y$-axis is reflected about the $y$-axis.

(b) $y = \sin|x|$

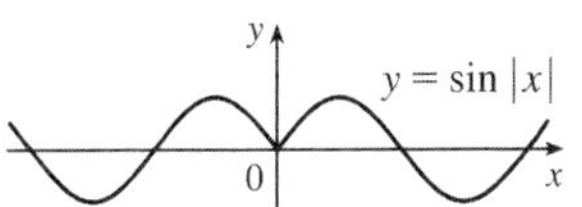


(c) $y = \sqrt{|x|}$

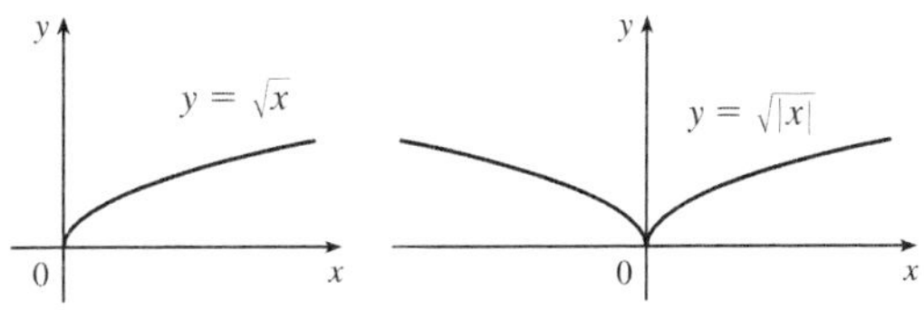


**28.** The most important features of the given graph are the $x$-intercepts and the maximum and minimum points. The graph of $y = 1/f(x)$ has vertical asymptotes at the $x$-values where there are $x$-intercepts on the graph of $y = f(x)$. The maximum of 1 on the graph of $y = f(x)$ corresponds to a minimum of $1/1 = 1$ on $y = 1/f(x)$. Similarly, the minimum on the graph of $y = f(x)$ corresponds to a maximum on the graph of $y = 1/f(x)$. As the values of $y$ get large (positively or negatively) on the graph of $y = f(x)$, the values of $y$ get close to zero on the graph of $y = 1/f(x)$.

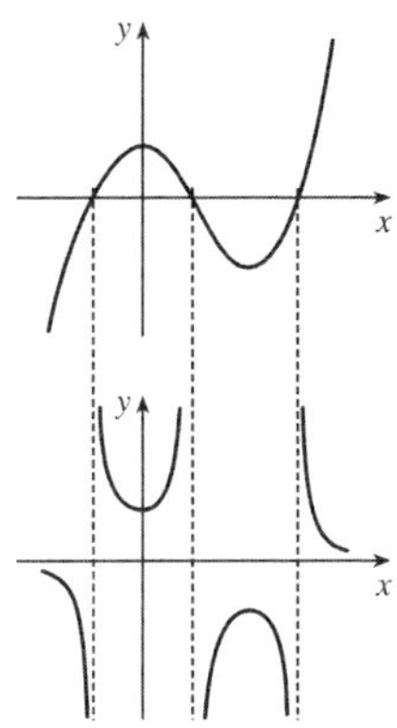

**29.** Assuming that successive horizontal and vertical gridlines are a unit apart, we can make a table of approximate values:

| $x$ | 0 | 1 | 2 | 3 | 4 | 5 | 6 |
|---|---|---|---|---|---|---|---|
| $f(x)$ | 2 | 1.7 | 1.3 | 1.0 | 0.7 | 0.3 | 0 |
| $g(x)$ | 2 | 2.7 | 3 | 2.8 | 2.4 | 1.7 | 0 |
| $f(x) + g(x)$ | 4 | 4.4 | 4.3 | 3.8 | 3.1 | 2.0 | 0 |

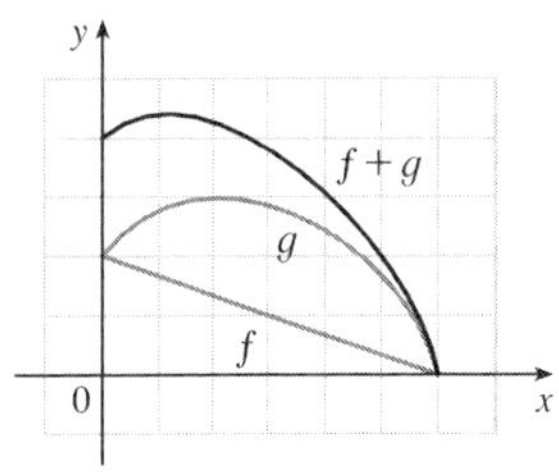


Connecting the points $(x, f(x) + g(x))$ with a smooth curve gives an approximation to the graph of $f + g$. Extra points can be plotted between those listed above if necessary.

**30.** First note that the domain of $f + g$ is the intersection of the domains of $f$ and $g$; that is, $f + g$ is only defined where both $f$ and $g$ are defined. Taking the horizontal and vertical units of length to be the distances between successive vertical and horizontal gridlines, we can make a table of approximate values as follows:

| $x$ | $-2$ | $-1$ | $0$ | $1$ | $2$ | $2.5$ | $3$ |
|---|---|---|---|---|---|---|---|
| $f(x)$ | $-1$ | $2.2$ | $2.0$ | $2.4$ | $2.7$ | $2.7$ | $2.3$ |
| $g(x)$ | $1$ | $-1.3$ | $-1.2$ | $-0.6$ | $0.3$ | $0.5$ | $0.7$ |
| $f(x)+g(x)$ | $0$ | $0.9$ | $0.8$ | $1.8$ | $3.0$ | $3.2$ | $3.0$ |

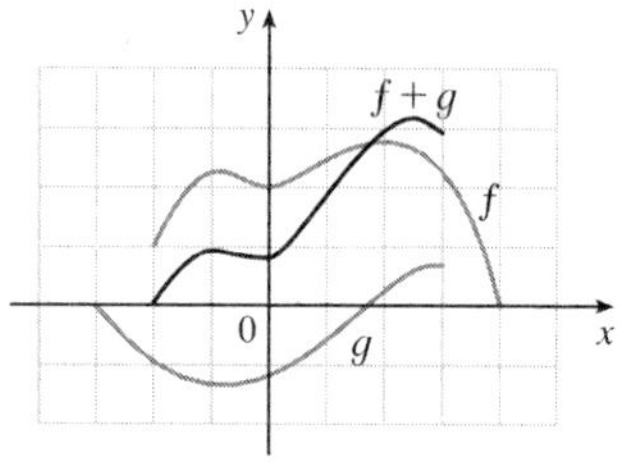


Extra values of $x$ (like the value 2.5 in the table above) can be added as needed.

**31.** $f(x) = x^3 + 2x^2$; $g(x) = 3x^2 - 1$. $D = \mathbb{R}$ for both $f$ and $g$.

$(f+g)(x) = (x^3 + 2x^2) + (3x^2 - 1) = x^3 + 5x^2 - 1, \quad D = \mathbb{R}.$

$(f-g)(x) = (x^3 + 2x^2) - (3x^2 - 1) = x^3 - x^2 + 1, \quad D = \mathbb{R}.$

$(fg)(x) = (x^3 + 2x^2)(3x^2 - 1) = 3x^5 + 6x^4 - x^3 - 2x^2, \quad D = \mathbb{R}.$

$\left(\dfrac{f}{g}\right)(x) = \dfrac{x^3 + 2x^2}{3x^2 - 1}, \quad D = \left\{x \mid x \neq \pm\dfrac{1}{\sqrt{3}}\right\}$ since $3x^2 - 1 \neq 0$.

**32.** $f(x) = \sqrt{1+x}, \quad D = [-1, \infty); \quad g(x) = \sqrt{1-x}, \quad D = (-\infty, 1].$

$(f+g)(x) = \sqrt{1+x} + \sqrt{1-x}, \quad D = (-\infty, 1] \cap [-1, \infty) = [-1, 1].$

$(f-g)(x) = \sqrt{1+x} - \sqrt{1-x}, \quad D = [-1, 1].$

$(fg)(x) = \sqrt{1+x} \cdot \sqrt{1-x} = \sqrt{1-x^2}, \quad D = [-1, 1].$

$\left(\dfrac{f}{g}\right)(x) = \dfrac{\sqrt{1+x}}{\sqrt{1-x}}, \quad D = [-1, 1)$. We must exclude $x = 1$ since it would make $\dfrac{f}{g}$ undefined.

**33.** $f(x) = x, \quad g(x) = 1/x$

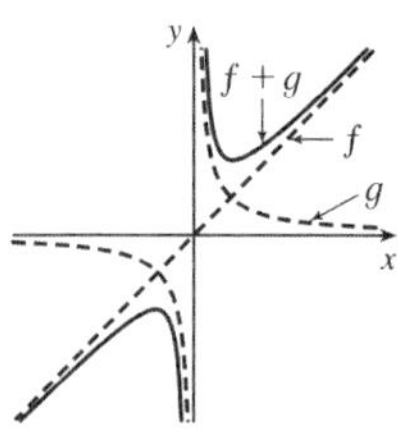


**34.** $f(x) = x^3, \quad g(x) = -x^2$

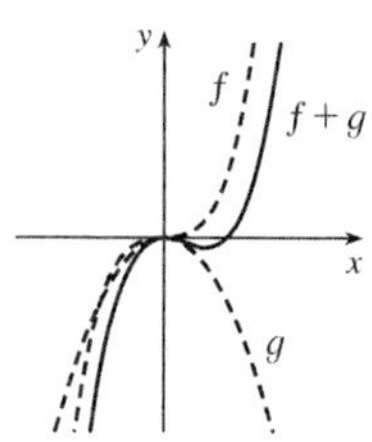


**35.** $f(x) = x^2 - 1$, $D = \mathbb{R}$; $g(x) = 2x + 1$, $D = \mathbb{R}$.

$(f \circ g)(x) = f(g(x)) = f(2x+1) = (2x+1)^2 - 1 = (4x^2 + 4x + 1) - 1 = 4x^2 + 4x$, $D = \mathbb{R}$.

$(g \circ f)(x) = g(f(x)) = g(x^2 - 1) = 2(x^2 - 1) + 1 = (2x^2 - 2) + 1 = 2x^2 - 1$, $D = \mathbb{R}$.

$(f \circ f)(x) = f(f(x)) = f(x^2 - 1) = (x^2 - 1)^2 - 1 = (x^4 - 2x^2 + 1) - 1 = x^4 - 2x^2$, $D = \mathbb{R}$.

$(g \circ g)(x) = g(g(x)) = g(2x+1) = 2(2x+1) + 1 = (4x+2) + 1 = 4x + 3$, $D = \mathbb{R}$.

**36.** $f(x) = 1 - x^3$, $D = \mathbb{R}$; $g(x) = 1/x$, $D = \{x \mid x \neq 0\}$.

$(f \circ g)(x) = f(g(x)) = f(1/x) = 1 - (1/x)^3 = 1 - 1/x^3$, $D = \{x \mid x \neq 0\}$.

$(g \circ f)(x) = g(f(x)) = g(1 - x^3) = 1/(1 - x^3)$, $D = \{x \mid 1 - x^3 \neq 0\} = \{x \mid x \neq 1\}$.

$(f \circ f)(x) = f(f(x)) = f(1 - x^3) = 1 - (1 - x^3)^3 \quad [= x^9 - 3x^6 + 3x^3]$, $D = \mathbb{R}$.

$(g \circ g)(x) = g(g(x)) = g(1/x) = 1/(1/x) = x$, $D = \{x \mid x \neq 0\}$ because 0 is not in the domain of $g$.

**37.** $f(x) = \sin x, \quad D = \mathbb{R}; \quad g(x) = 1 - \sqrt{x}, \quad D = [0, \infty)$.

$(f \circ g)(x) = f(g(x)) = f(1 - \sqrt{x}) = \sin(1 - \sqrt{x})$, $D = [0, \infty)$.

$(g \circ f)(x) = g(f(x)) = g(\sin x) = 1 - \sqrt{\sin x}$.

For $\sqrt{\sin x}$ to be defined, we must have

$$\sin x \geq 0 \quad \Leftrightarrow \quad x \in [0, \pi] \cup [2\pi, 3\pi] \cup [-2\pi, -\pi] \cup [4\pi, 5\pi] \cup [-4\pi, -3\pi] \cup \ \ldots,$$

so $D = \{x \mid x \in [2n\pi, \pi + 2n\pi]$, where $n$ is an integer$\}$.

$(f \circ f)(x) = f(f(x)) = f(\sin x) = \sin(\sin x)$, $D = \mathbb{R}$.

$(g \circ g)(x) = g(g(x)) = g(1 - \sqrt{x}) = 1 - \sqrt{1 - \sqrt{x}}$,

$$D = \{x \geq 0 \mid 1 - \sqrt{x} \geq 0\} = \{x \geq 0 \mid 1 \geq \sqrt{x}\} = \{x \geq 0 \mid \sqrt{x} \leq 1\} = [0, 1].$$

**38.** $f(x) = 1 - 3x, \quad D = \mathbb{R}; \quad g(x) = 5x^2 + 3x + 2, \quad D = \mathbb{R}$.

$$\begin{aligned}(f \circ g)(x) &= f(g(x)) = f(5x^2 + 3x + 2) = 1 - 3(5x^2 + 3x + 2) \\ &= 1 - 15x^2 - 9x - 6 = -15x^2 - 9x - 5, \quad D = \mathbb{R}.\end{aligned}$$

$$\begin{aligned}(g \circ f)(x) &= g(f(x)) = g(1 - 3x) = 5(1 - 3x)^2 + 3(1 - 3x) + 2 = 5(1 - 6x + 9x^2) + 3 - 9x + 2 \\ &= 5 - 30x + 45x^2 - 9x + 5 = 45x^2 - 39x + 10, \quad D = \mathbb{R}.\end{aligned}$$

$(f \circ f)(x) = f(f(x)) = f(1 - 3x) = 1 - 3(1 - 3x) = 1 - 3 + 9x = 9x - 2, \quad D = \mathbb{R}$.

$$\begin{aligned}(g \circ g)(x) &= g(g(x)) = g(5x^2 + 3x + 2) = 5(5x^2 + 3x + 2)^2 + 3(5x^2 + 3x + 2) + 2 \\ &= 5(25x^4 + 30x^3 + 29x^2 + 12x + 4) + 15x^2 + 9x + 6 + 2 \\ &= 125x^4 + 150x^3 + 145x^2 + 60x + 20 + 15x^2 + 9x + 8 \\ &= 125x^4 + 150x^3 + 160x^2 + 69x + 28, \quad D = \mathbb{R}.\end{aligned}$$

**39.** $f(x) = x + \dfrac{1}{x}, \quad D = \{x \mid x \neq 0\}; \quad g(x) = \dfrac{x + 1}{x + 2}, \quad D = \{x \mid x \neq -2\}$.

$$\begin{aligned}(f \circ g)(x) &= f(g(x)) = f\left(\frac{x+1}{x+2}\right) = \frac{x+1}{x+2} + \frac{1}{\dfrac{x+1}{x+2}} = \frac{x+1}{x+2} + \frac{x+2}{x+1} \\ &= \frac{(x+1)(x+1) + (x+2)(x+2)}{(x+2)(x+1)} = \frac{(x^2 + 2x + 1) + (x^2 + 4x + 4)}{(x+2)(x+1)} = \frac{2x^2 + 6x + 5}{(x+2)(x+1)}\end{aligned}$$

Since $g(x)$ is not defined for $x = -2$ and $f(g(x))$ is not defined for $x = -2$ and $x = -1$, the domain of $(f \circ g)(x)$ is $D = \{x \mid x \neq -2, -1\}$.

$$(g \circ f)(x) = g(f(x)) = g\left(x + \frac{1}{x}\right) = \frac{\left(x + \dfrac{1}{x}\right) + 1}{\left(x + \dfrac{1}{x}\right) + 2} = \frac{\dfrac{x^2 + 1 + x}{x}}{\dfrac{x^2 + 1 + 2x}{x}} = \frac{x^2 + x + 1}{x^2 + 2x + 1} = \frac{x^2 + x + 1}{(x+1)^2}$$

Since $f(x)$ is not defined for $x = 0$ and $g(f(x))$ is not defined for $x = -1$, the domain of $(g \circ f)(x)$ is $D = \{x \mid x \neq -1, 0\}$.

$$(f \circ f)(x) = f(f(x)) = f\left(x + \frac{1}{x}\right) = \left(x + \frac{1}{x}\right) + \frac{1}{x + \frac{1}{x}} = x + \frac{1}{x} + \frac{1}{\frac{x^2+1}{x}} = x + \frac{1}{x} + \frac{x}{x^2+1}$$

$$= \frac{x(x)(x^2+1) + 1\,(x^2+1) + x(x)}{x(x^2+1)} = \frac{x^4 + x^2 + x^2 + 1 + x^2}{x(x^2+1)}$$

$$= \frac{x^4 + 3x^2 + 1}{x(x^2+1)}, \quad D = \{x \mid x \neq 0\}.$$

$$(g \circ g)(x) = g(g(x)) = g\left(\frac{x+1}{x+2}\right) = \frac{\frac{x+1}{x+2} + 1}{\frac{x+1}{x+2} + 2} = \frac{\frac{x+1+1(x+2)}{x+2}}{\frac{x+1+2(x+2)}{x+2}} = \frac{x+1+x+2}{x+1+2x+4} = \frac{2x+3}{3x+5}$$

Since $g(x)$ is not defined for $x = -2$ and $g(g(x))$ is not defined for $x = -\frac{5}{3}$, the domain of $(g \circ g)(x)$ is $D = \left\{x \mid x \neq -2, -\frac{5}{3}\right\}$.

**40.** $f(x) = \sqrt{2x+3}$, $D = \left\{x \mid x \geq -\frac{3}{2}\right\}$; $g(x) = x^2 + 1$, $D = \mathbb{R}$.

$(f \circ g)(x) = f(x^2+1) = \sqrt{2(x^2+1)+3} = \sqrt{2x^2+5}$, $D = \mathbb{R}$.

$(g \circ f)(x) = g\left(\sqrt{2x+3}\right) = \left(\sqrt{2x+3}\right)^2 + 1 = (2x+3) + 1 = 2x + 4$, $D = \left\{x \mid x \geq -\frac{3}{2}\right\}$.

$(f \circ f)(x) = f\left(\sqrt{2x+3}\right) = \sqrt{2\left(\sqrt{2x+3}\right)+3} = \sqrt{2\sqrt{2x+3}+3}$, $D = \left\{x \mid x \geq -\frac{3}{2}\right\}$.

$(g \circ g)(x) = g(x^2+1) = (x^2+1)^2 + 1 = (x^4 + 2x^2 + 1) + 1 = x^4 + 2x^2 + 2$, $D = \mathbb{R}$.

**41.** $(f \circ g \circ h)(x) = f(g(h(x))) = f(g(x+3)) = f((x+3)^2 + 2)$

$= f(x^2 + 6x + 11) = \sqrt{(x^2+6x+11) - 1} = \sqrt{x^2+6x+10}$

**42.** $(f \circ g \circ h)(x) = f(g(h(x))) = f\left(g\left(\sqrt{x+3}\right)\right) = f\left(\cos\sqrt{x+3}\right) = \dfrac{2}{\cos\sqrt{x+3}+1}$

**43.** Let $g(x) = x^2 + 1$ and $f(x) = x^{10}$. Then $(f \circ g)(x) = f(g(x)) = f(x^2+1) = (x^2+1)^{10} = F(x)$.

**44.** Let $g(x) = \sqrt{x}$ and $f(x) = \sin x$. Then $(f \circ g)(x) = f(g(x)) = f\left(\sqrt{x}\right) = \sin\left(\sqrt{x}\right) = F(x)$.

**45.** Let $g(t) = \cos t$ and $f(t) = \sqrt{t}$. Then $(f \circ g)(t) = f(g(t)) = f(\cos t) = \sqrt{\cos t} = u(t)$.

**46.** Let $g(t) = \tan t$ and $f(t) = \dfrac{t}{1+t}$. Then $(f \circ g)(t) = f(g(t)) = f(\tan t) = \dfrac{\tan t}{1 + \tan t} = u(t)$.

**47.** Let $h(x) = x^2$, $g(x) = 3^x$, and $f(x) = 1 - x$. Then

$(f \circ g \circ h)(x) = f(g(h(x))) = f(g(x^2)) = f\left(3^{x^2}\right) = 1 - 3^{x^2} = H(x)$.

**48.** Let $h(x) = |x|$, $g(x) = 2 + x$, and $f(x) = \sqrt[8]{x}$. Then

$(f \circ g \circ h)(x) = f(g(h(x))) = f(g(|x|)) = f\,(2 + |x|) = \sqrt[8]{2+|x|} = H(x)$.

**49.** Let $h(x) = \sqrt{x}$, $g(x) = \sec x$, and $f(x) = x^4$. Then

$(f \circ g \circ h)(x) = f(g(h(x))) = f(g(\sqrt{x})) = f(\sec\sqrt{x}) = (\sec\sqrt{x})^4 = \sec^4\left(\sqrt{x}\right) = H(x)$.

**50.** (a) $f(g(1)) = f(6) = 5$

(b) $g(f(1)) = g(3) = 2$

(c) $f(f(1)) = f(3) = 4$

(d) $g(g(1)) = g(6) = 3$

(e) $(g \circ f)(3) = g(f(3)) = g(4) = 1$

(f) $(f \circ g)(6) = f(g(6)) = f(3) = 4$

**51.** (a) $g(2) = 5$, because the point $(2, 5)$ is on the graph of $g$. Thus, $f(g(2)) = f(5) = 4$, because the point $(5, 4)$ is on the graph of $f$.

(b) $g(f(0)) = g(0) = 3$

(c) $(f \circ g)(0) = f(g(0)) = f(3) = 0$

(d) $(g \circ f)(6) = g(f(6)) = g(6)$. This value is not defined, because there is no point on the graph of $g$ that has $x$-coordinate 6.

(e) $(g \circ g)(-2) = g(g(-2)) = g(1) = 4$

(f) $(f \circ f)(4) = f(f(4)) = f(2) = -2$

**52.** To find a particular value of $f(g(x))$, say for $x = 0$, we note from the graph that $g(0) \approx 2.8$ and $f(2.8) \approx -0.5$. Thus, $f(g(0)) \approx f(2.8) \approx -0.5$. The other values listed in the table were obtained in a similar fashion.

| $x$ | $g(x)$ | $f(g(x))$ |
|---|---|---|
| $-5$ | $-0.2$ | $-4$ |
| $-4$ | $1.2$ | $-3.3$ |
| $-3$ | $2.2$ | $-1.7$ |
| $-2$ | $2.8$ | $-0.5$ |
| $-1$ | $3$ | $-0.2$ |

| $x$ | $g(x)$ | $f(g(x))$ |
|---|---|---|
| 0 | $2.8$ | $-0.5$ |
| 1 | $2.2$ | $-1.7$ |
| 2 | $1.2$ | $-3.3$ |
| 3 | $-0.2$ | $-4$ |
| 4 | $-1.9$ | $-2.2$ |
| 5 | $-4.1$ | $1.9$ |

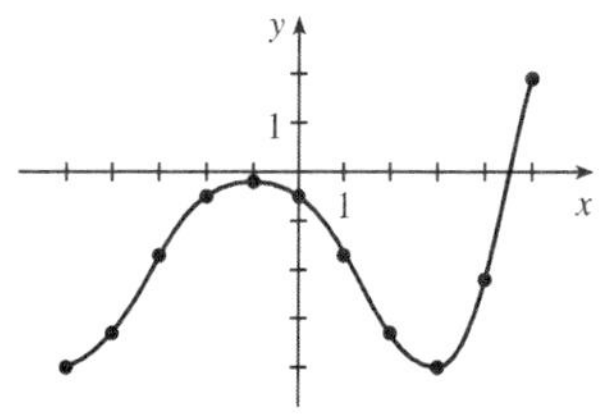


**53.** (a) Using the relationship *distance* = *rate* · *time* with the radius $r$ as the distance, we have $r(t) = 60t$.

(b) $A = \pi r^2 \;\Rightarrow\; (A \circ r)(t) = A(r(t)) = \pi(60t)^2 = 3600\pi t^2$. This formula gives us the extent of the rippled area (in $\text{cm}^2$) at any time $t$.

**54.** (a) The radius $r$ of the balloon is increasing at a rate of 2 cm/s, so $r(t) = (2 \text{ cm/s})(t \text{ s}) = 2t$ (in cm).

(b) Using $V = \frac{4}{3}\pi r^3$, we get $(V \circ r)(t) = V(r(t)) = V(2t) = \frac{4}{3}\pi(2t)^3 = \frac{32}{3}\pi t^3$.
The result, $V = \frac{32}{3}\pi t^3$, gives the volume of the balloon (in $\text{cm}^3$) as a function of time (in s).

**55.** (a) From the figure, we have a right triangle with legs 6 and $d$, and hypotenuse $s$.
By the Pythagorean Theorem, $d^2 + 6^2 = s^2 \;\Rightarrow\; s = f(d) = \sqrt{d^2 + 36}$.

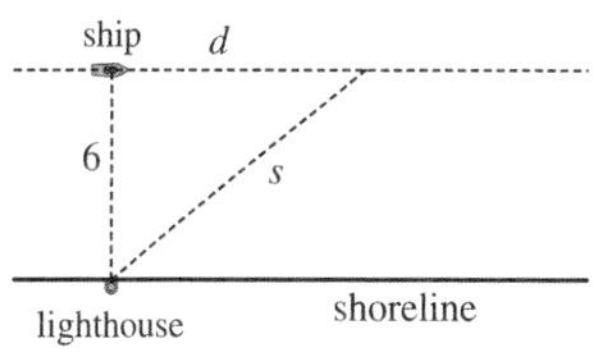


(b) Using $d = rt$, we get $d = (30 \text{ km/hr})(t \text{ hr}) = 30t$ (in km). Thus, $d = g(t) = 30t$.

(c) $(f \circ g)(t) = f(g(t)) = f(30t) = \sqrt{(30t)^2 + 36} = \sqrt{900t^2 + 36}$. This function represents the distance between the lighthouse and the ship as a function of the time elapsed since noon.

**56.** (a) $d = rt \;\Rightarrow\; d(t) = 350t$

(b) There is a Pythagorean relationship involving the legs with lengths $d$ and 1 and the hypotenuse with length $s$: $d^2 + 1^2 = s^2$. Thus, $s(d) = \sqrt{d^2 + 1}$.

(c) $(s \circ d)(t) = s(d(t)) = s(350t) = \sqrt{(350t)^2 + 1}$

**57.** (a)

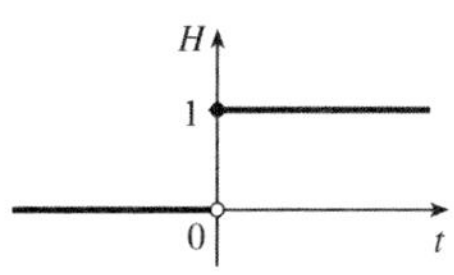


$$H(t) = \begin{cases} 0 & \text{if } t < 0 \\ 1 & \text{if } t \geq 0 \end{cases}$$

(b)

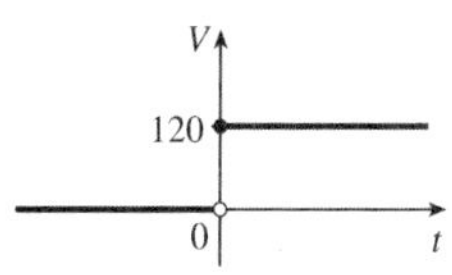


$$V(t) = \begin{cases} 0 & \text{if } t < 0 \\ 120 & \text{if } t \geq 0 \end{cases}$$ so $V(t) = 120H(t)$.

(c) 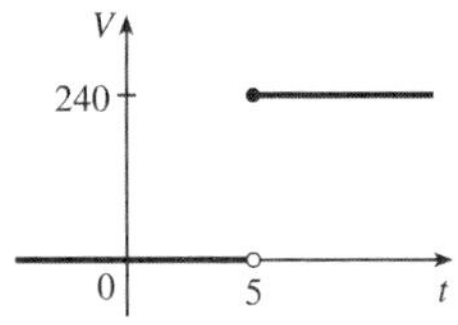


Starting with the formula in part (b), we replace 120 with 240 to reflect the different voltage. Also, because we are starting 5 units to the right of $t = 0$, we replace $t$ with $t - 5$. Thus, the formula is $V(t) = 240H(t - 5)$.

**58.** (a) $R(t) = tH(t)$
$$= \begin{cases} 0 & \text{if } t < 0 \\ t & \text{if } t \geq 0 \end{cases}$$

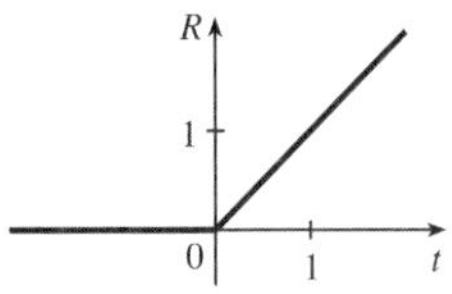


(b) $V(t) = \begin{cases} 0 & \text{if } t < 0 \\ 2t & \text{if } 0 \leq t \leq 60 \end{cases}$

so $V(t) = 2tH(t)$, $t \leq 60$.

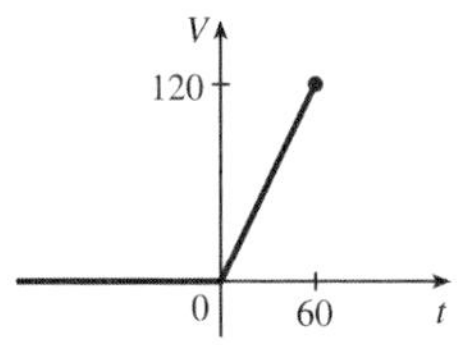


(c) $V(t) = \begin{cases} 0 & \text{if } t < 7 \\ 4\,(t-7) & \text{if } 7 \leq t \leq 32 \end{cases}$

so $V(t) = 4(t - 7)H(t - 7)$, $t \leq 32$.

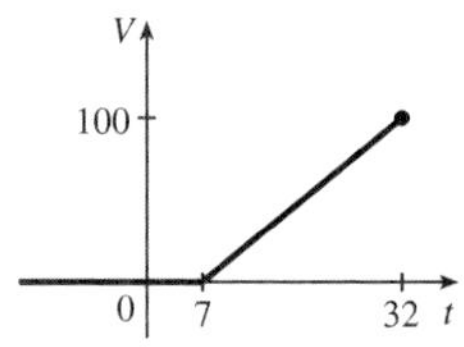


**59.** If $f(x) = m_1x + b_1$ and $g(x) = m_2x + b_2$, then

$(f \circ g)(x) = f(g(x)) = f(m_2x + b_2) = m_1(m_2x + b_2) + b_1 = m_1m_2x + m_1b_2 + b_1$.

So $f \circ g$ is a linear function with slope $m_1m_2$.

**60.** If $A(x) = 1.04x$, then

$(A \circ A)(x) = A(A(x)) = A(1.04x) = 1.04(1.04x) = (1.04)^2x$,

$(A \circ A \circ A)(x) = A((A \circ A)(x)) = A((1.04)^2x) = 1.04(1.04)^2x = (1.04)^3x$, and

$(A \circ A \circ A \circ A)(x) = A((A \circ A \circ A)(x)) = A((1.04)^3x) = 1.04(1.04)^3x, = (1.04)^4x$.

These compositions represent the amount of the investment after 2, 3, and 4 years.

Based on this pattern, when we compose $n$ copies of $A$, we get the formula $\underbrace{(A \circ A \circ \cdots \circ A)}_{n\ A\text{'s}}(x) = (1.04)^n x$.

**61.** (a) By examining the variable terms in $g$ and $h$, we deduce that we must square $g$ to get the terms $4x^2$ and $4x$ in $h$. If we let $f(x) = x^2 + c$, then $(f \circ g)(x) = f(g(x)) = f(2x + 1) = (2x + 1)^2 + c = 4x^2 + 4x + (1 + c)$. Since $h(x) = 4x^2 + 4x + 7$, we must have $1 + c = 7$. So $c = 6$ and $f(x) = x^2 + 6$.

(b) We need a function $g$ so that $f(g(x)) = 3(g(x)) + 5 = h(x)$. But $h(x) = 3x^2 + 3x + 2 = 3(x^2 + x) + 2 = 3(x^2 + x - 1) + 5$, so we see that $g(x) = x^2 + x - 1$.

**62.** We need a function $g$ so that $g(f(x)) = g(x + 4) = h(x) = 4x - 1 = 4(x + 4) - 17$. So we see that the function $g$ must be $g(x) = 4x - 17$.

**63.** (a) If $f$ and $g$ are even functions, then $f(-x) = f(x)$ and $g(-x) = g(x)$.

(i) $(f + g)(-x) = f(-x) + g(-x) = f(x) + g(x) = (f + g)(x)$, so $f + g$ is an *even* function.

(ii) $(fg)(-x) = f(-x) \cdot g(-x) = f(x) \cdot g(x) = (fg)(x)$, so $fg$ is an *even* function.

(b) If $f$ and $g$ are odd functions, then $f(-x) = -f(x)$ and $g(-x) = -g(x)$.

(i) $(f + g)(-x) = f(-x) + g(-x) = -f(x) + [-g(x)] = -[f(x) + g(x)] = -(f + g)(x)$, so $f + g$ is an *odd* function.

(ii) $(fg)(-x) = f(-x) \cdot g(-x) = -f(x) \cdot [-g(x)] = f(x) \cdot g(x) = (fg)(x)$, so $fg$ is an *even* function.

**64.** If $f$ is even and $g$ is odd, then $f(-x) = f(x)$ and $g(-x) = -g(x)$. Now

$(fg)(-x) = f(-x) \cdot g(-x) = f(x) \cdot [-g(x)] = -[f(x) \cdot g(x)] = -(fg)(x)$, so $fg$ is an *odd* function.

**65.** We need to examine $h(-x)$.

$$h(-x) = (f \circ g)(-x) = f(g(-x)) = f(g(x)) \quad \text{[because } g \text{ is even]} \quad = h(x)$$

Because $h(-x) = h(x)$, $h$ is an even function.

**66.** $h(-x) = f(g(-x)) = f(-g(x))$. At this point, we can't simplify the expression, so we might try to find a counterexample to show that $h$ is not an odd function. Let $g(x) = x$, an odd function, and $f(x) = x^2 + x$. Then $h(x) = x^2 + x$, which is neither even nor odd.

Now suppose $f$ is an odd function. Then $f(-g(x)) = -f(g(x)) = -h(x)$. Hence, $h(-x) = -h(x)$, and so $h$ is odd if both $f$ and $g$ are odd.

Now suppose $f$ is an even function. Then $f(-g(x)) = f(g(x)) = h(x)$. Hence, $h(-x) = h(x)$, and so $h$ is even if $g$ is odd and $f$ is even.

## 1.4 Graphing Calculators and Computers

**1.** $f(x) = \sqrt{x^3 - 5x^2}$

(a) $[-5, 5]$ by $[-5, 5]$

(There is no graph shown.)

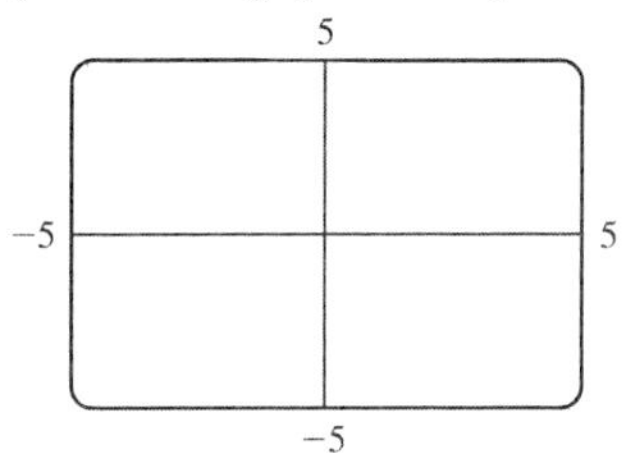


(b) $[0, 10]$ by $[0, 2]$

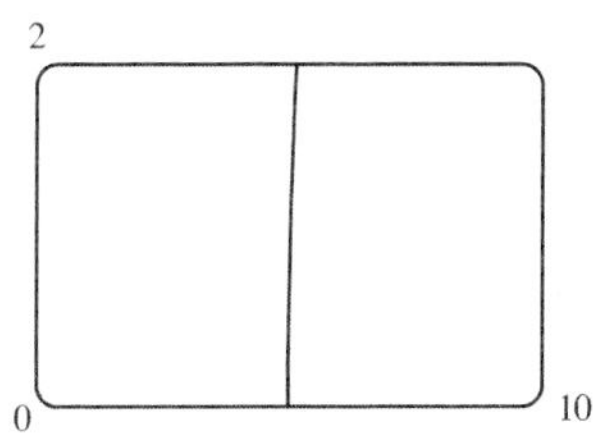


(c) $[0, 10]$ by $[0, 10]$

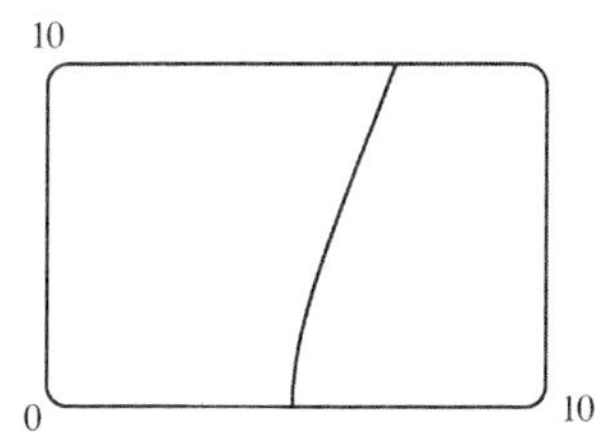


The most appropriate graph is produced in viewing rectangle (c).

**2.** $f(x) = x^4 - 16x^2 + 20$

(a) $[-3, 3]$ by $[-3, 3]$

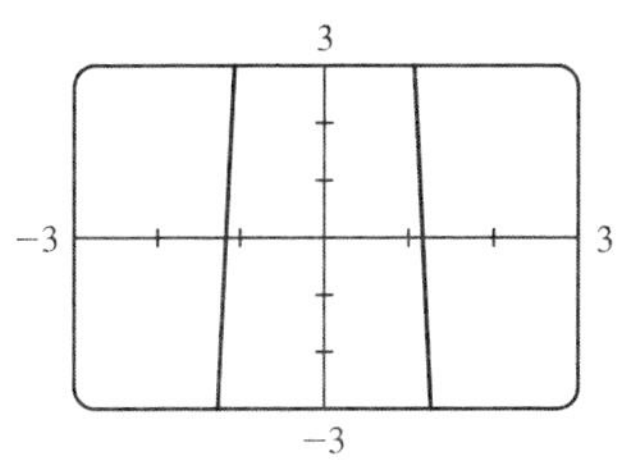


(b) $[-10, 10]$ by $[-10, 10]$

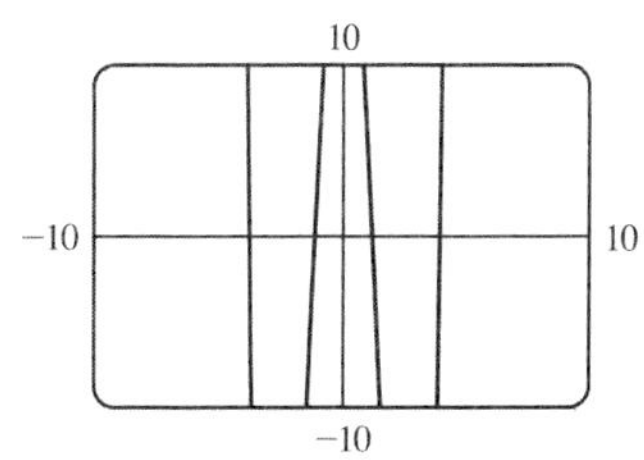


(c) $[-50, 50]$ by $[-50, 50]$

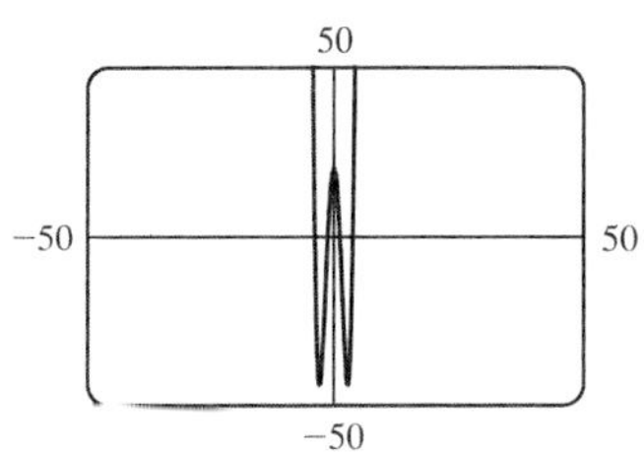


(d) $[-5, 5]$ by $[-50, 50]$

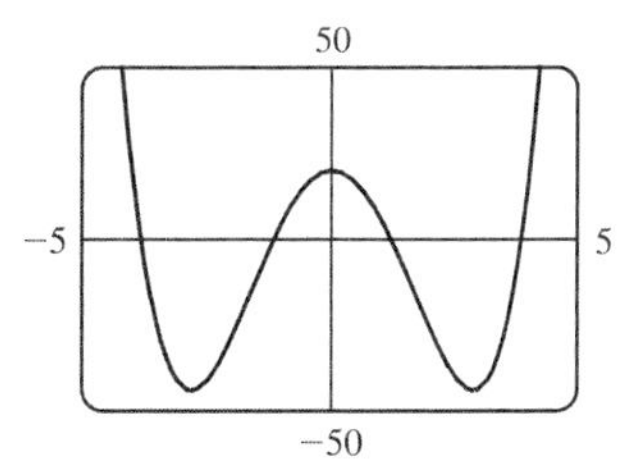


The most appropriate graph is produced in viewing rectangle (d).

3. Since the graph of $f(x) = 5 + 20x - x^2$ is a parabola opening downward, an appropriate viewing rectangle should include the maximum point.

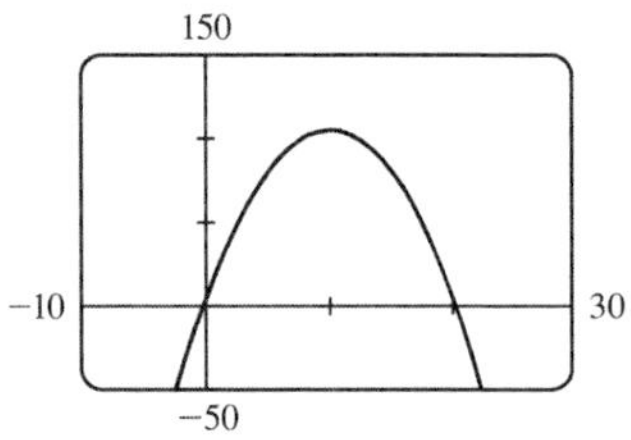


4. An appropriate viewing rectangle for $f(x) = x^3 + 30x^2 + 200x$ should include the high and low points.

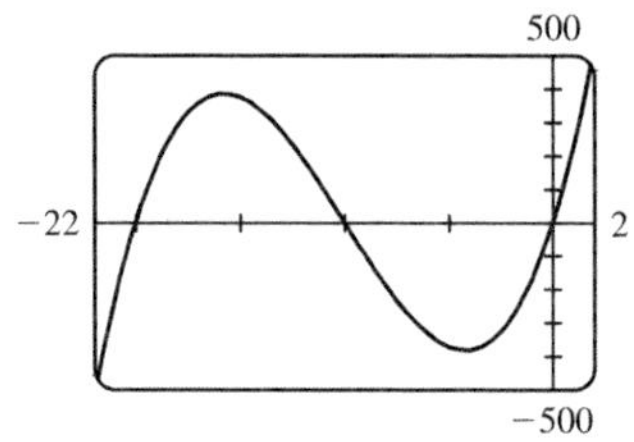


5. $f(x) = \sqrt[4]{81 - x^4}$ is defined when $81 - x^4 \geq 0 \quad \Leftrightarrow \quad x^4 \leq 81 \quad \Leftrightarrow \quad |x| \leq 3$, so the domain of $f$ is $[-3, 3]$. Also $0 \leq \sqrt[4]{81 - x^4} \leq \sqrt[4]{81} = 3$, so the range is $[0, 3]$.

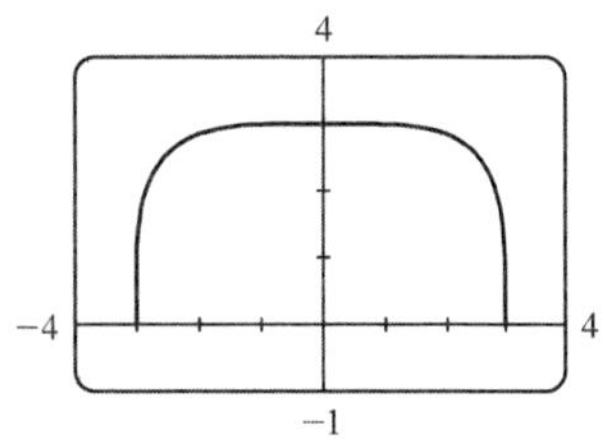


6. $f(x) = \sqrt{0.1x + 20}$ is defined when $0.1x + 20 \geq 0$ $\Leftrightarrow \quad x \geq -200$, so the domain of $f$ is $[-200, \infty)$.

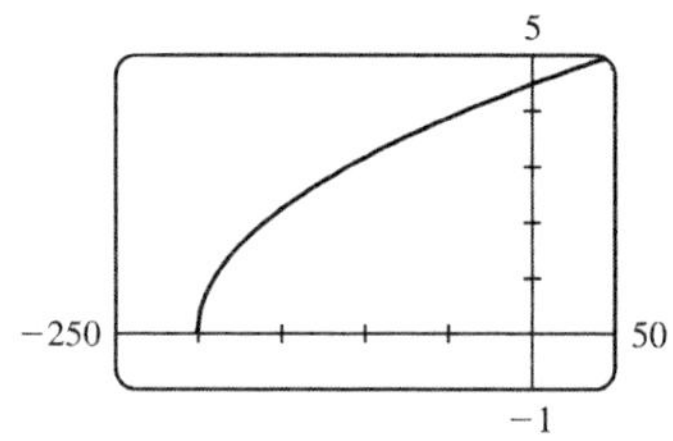


7. The graph of $f(x) = x^2 + (100/x)$ has a vertical asymptote of $x = 0$. As you zoom out, the graph of $f$ looks more and more like that of $y = x^2$.

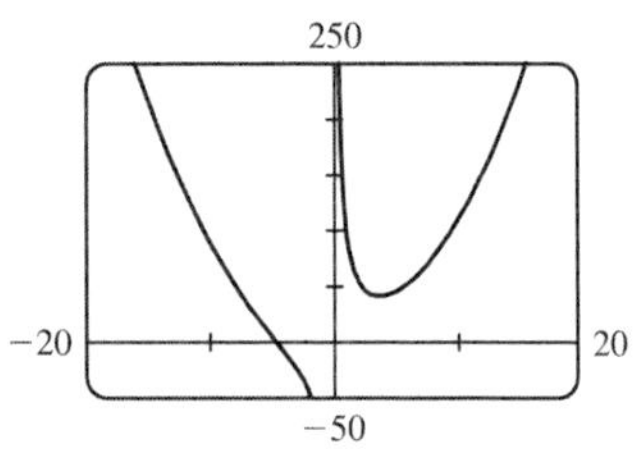


8. The graph of $f(x) = x/(x^2 + 100)$ is symmetric with respect to the origin.

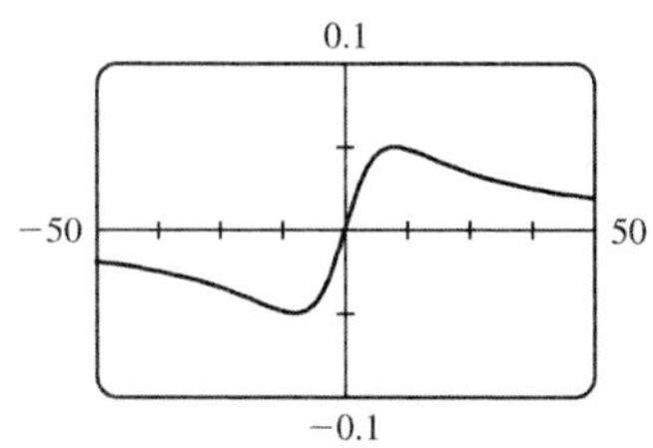


9. The period of $g(x) = \sin(1000x)$ is $\frac{2\pi}{1000} \approx 0.0063$ and its range is $[-1, 1]$. Since $f(x) = \sin^2(1000x)$ is the square of $g$, its range is $[0, 1]$ and a viewing rectangle of $[-0.01, 0.01]$ by $[0, 1.5]$ seems appropriate.

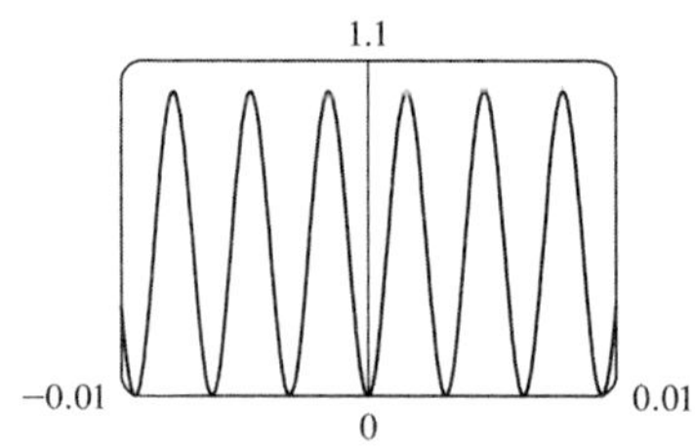

**10.** The period of $f(x) = \cos(0.001x)$ is $\frac{2\pi}{0.001} \approx 6300$ and its range is $[-1, 1]$, so a viewing rectangle of $[-10{,}000, 10{,}000]$ by $[-1.5, 1.5]$ seems appropriate.

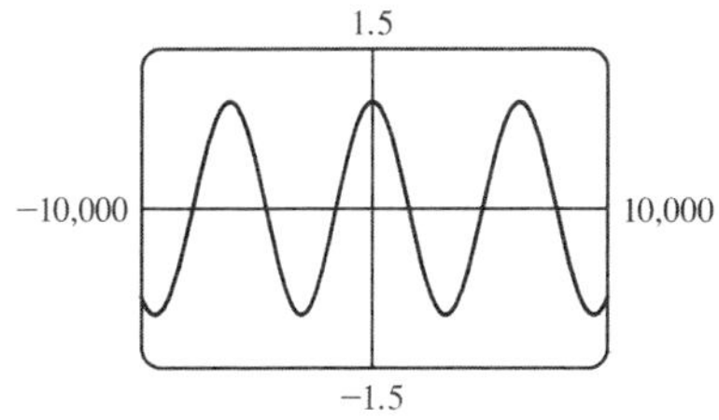


**11.** The domain of $y = \sqrt{x}$ is $x \geq 0$, so the domain of $f(x) = \sin\sqrt{x}$ is $[0, \infty)$ and the range is $[-1, 1]$. With a little trial-and-error experimentation, we find that an Xmax of 100 illustrates the general shape of $f$, so an appropriate viewing rectangle is $[0, 100]$ by $[-1.5, 1.5]$.

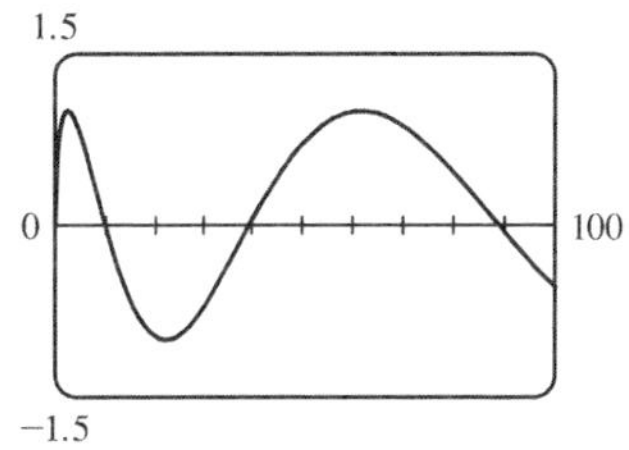


**12.** One period of $y = \sec x$ occurs on the interval $\left(-\frac{\pi}{2}, \frac{\pi}{2}\right) \cup \left(\frac{\pi}{2}, \frac{3\pi}{2}\right)$.
$-\frac{\pi}{2} < 20\pi x < \frac{3\pi}{2} \quad \Rightarrow \quad -\frac{1}{40} < x < \frac{3}{40}$, or equivalently, $-0.025 < x < 0.075$.

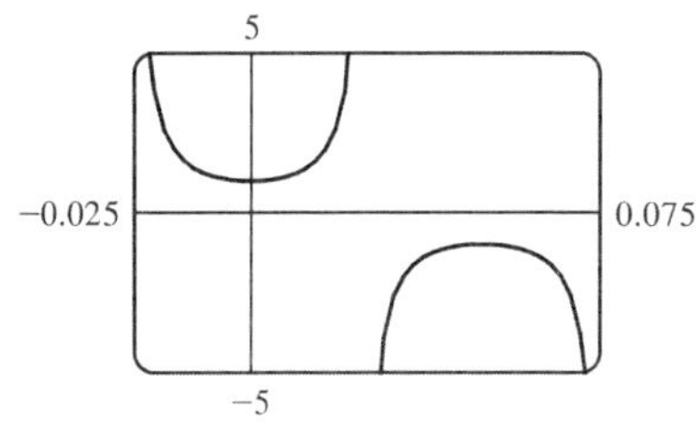


**13.** The first term, $10\sin x$, has period $2\pi$ and range $[-10, 10]$. It will be the dominant term in any "large" graph of $y = 10\sin x + \sin 100x$, as shown in the first figure. The second term, $\sin 100x$, has period $\frac{2\pi}{100} = \frac{\pi}{50}$ and range $[-1, 1]$. It causes the bumps in the first figure and will be the dominant term in any "small" graph, as shown in the view near the origin in the second figure.

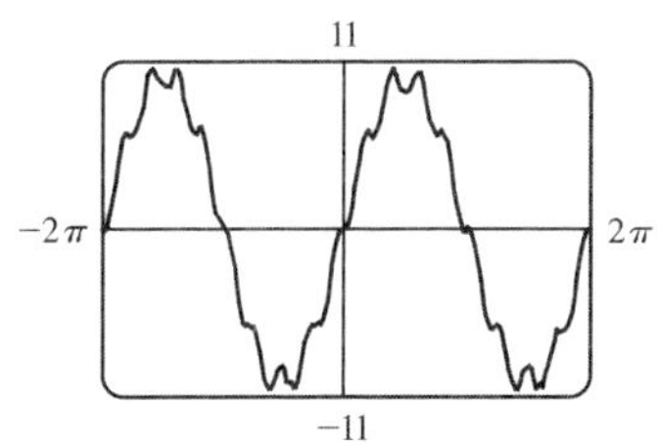


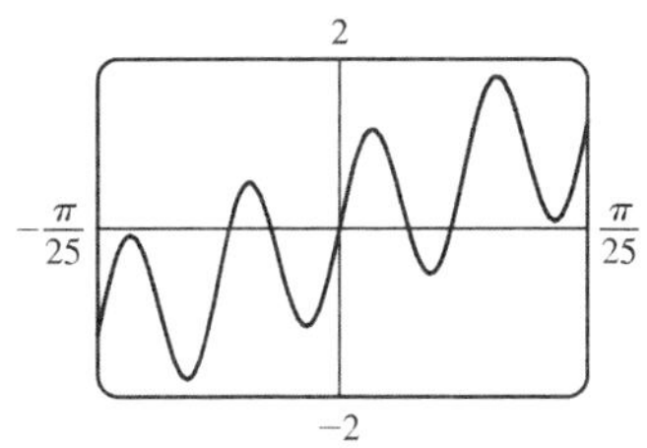


**14.** $y = x^2 + 0.02\sin(50x)$

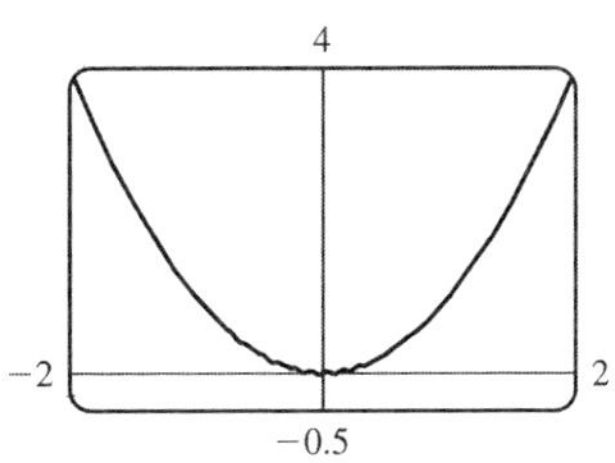


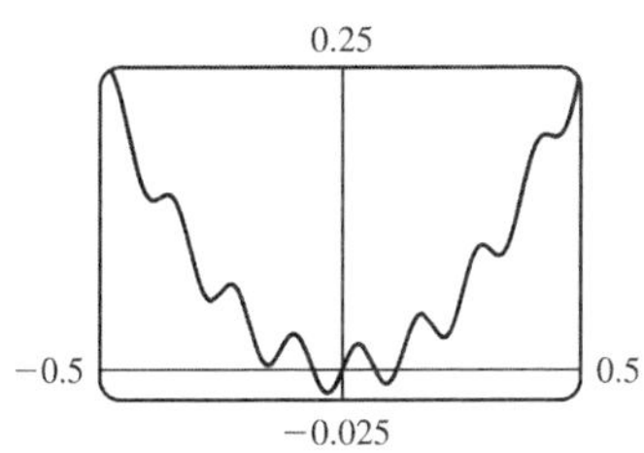

**15.** We must solve the given equation for $y$ to obtain equations for the upper and lower halves of the ellipse.

$$4x^2 + 2y^2 = 1 \quad \Leftrightarrow \quad 2y^2 = 1 - 4x^2 \quad \Leftrightarrow \quad y^2 = \frac{1 - 4x^2}{2} \quad \Leftrightarrow$$

$$y = \pm\sqrt{\frac{1 - 4x^2}{2}}$$

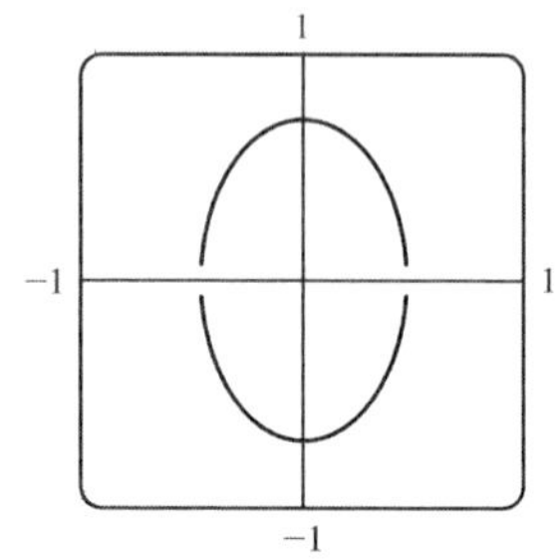


**16.** $y^2 - 9x^2 = 1 \quad \Leftrightarrow \quad y^2 = 1 + 9x^2 \quad \Leftrightarrow \quad y = \pm\sqrt{1 + 9x^2}$

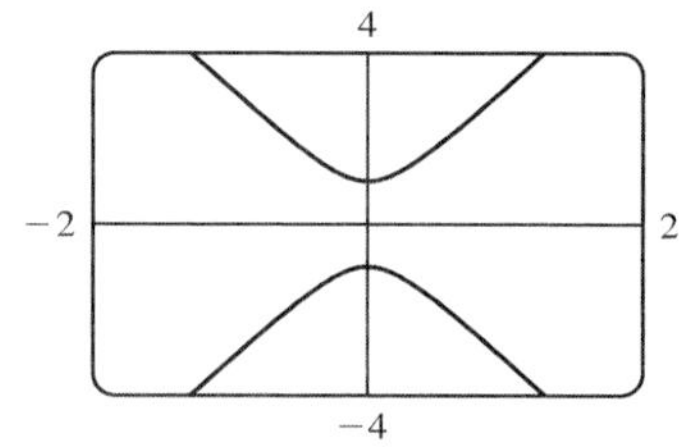


**17.** From the graph of $f(x) = x^3 - 9x^2 - 4$, we see that there is one solution of the equation $f(x) = 0$ and it is slightly larger than 9. By zooming in or using a root or zero feature, we obtain $x \approx 9.05$.

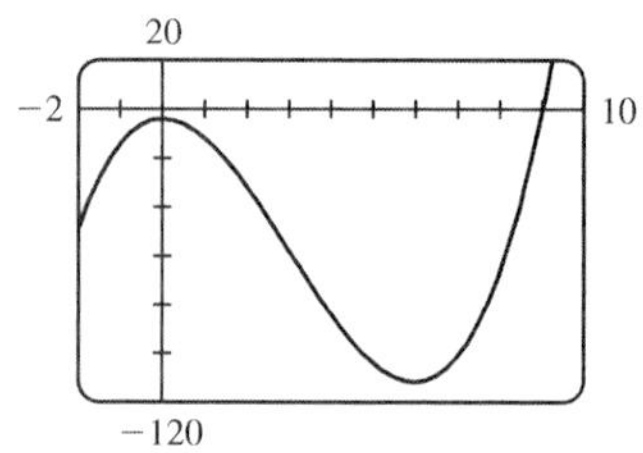


**18.** We see that the graphs of $f(x) = x^3$ and $g(x) = 4x - 1$ intersect three times. The $x$-coordinates of these points (which are the solutions of the equation) are approximately $-2.11, 0.25$, and $1.86$. Alternatively, we could find these values by finding the zeros of $h(x) = x^3 - 4x + 1$.

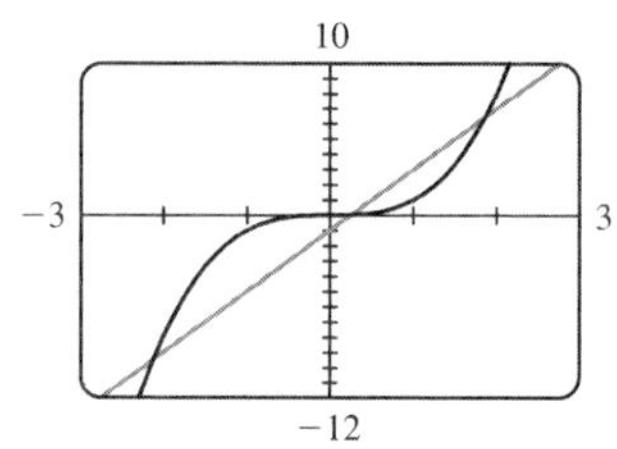


**19.** We see that the graphs of $f(x) = x^2$ and $g(x) = \sin x$ intersect twice. One solution is $x = 0$. The other solution of $f = g$ is the $x$-coordinate of the point of intersection in the first quadrant. Using an intersect feature or zooming in, we find this value to be approximately 0.88. Alternatively, we could find that value by finding the positive zero of $h(x) = x^2 - \sin x$.

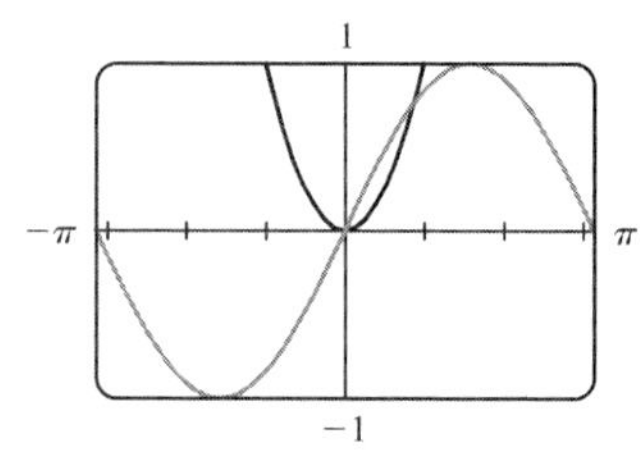


*Note*: After producing the graph on a TI-83 Plus, we can find the approximate value 0.88 by using the following keystrokes: [2nd] [CALC] [5] [ENTER] [ENTER] 1 [ENTER]. The "1" is just a guess for 0.88.

**20.** (a)

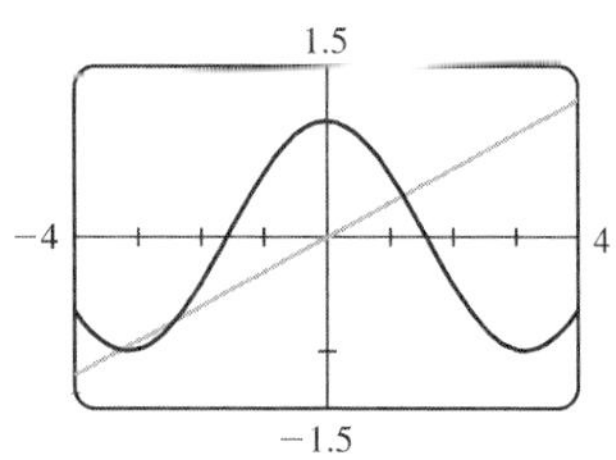


The $x$-coordinates of the three points of intersection are $x \approx -3.29, -2.36$ and $1.20$.

(b) Using trial and error, we find that $m \approx 0.3365$. Note that $m$ could also be negative.

**21.** $g(x) = x^3/10$ is larger than $f(x) = 10x^2$ whenever $x > 100$.

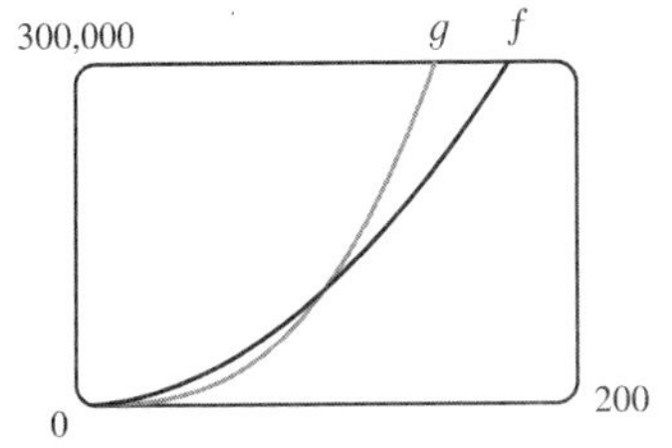


**22.** $f(x) = x^4 - 100x^3$ is larger than $g(x) = x^3$ whenever $x > 101$.

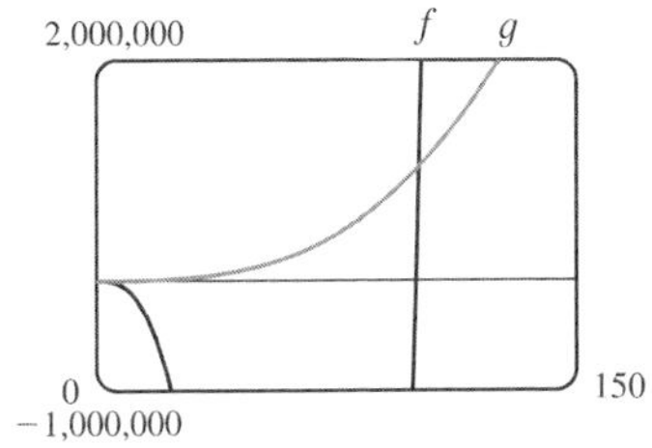


**23.**

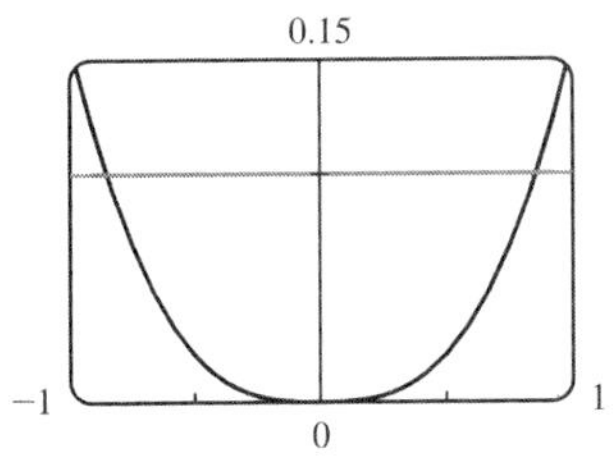


We see from the graphs of $y = |\sin x - x|$ and $y = 0.1$ that there are two solutions to the equation $|\sin x - x| = 0.1$: $x \approx -0.85$ and $x \approx 0.85$. The condition $|\sin x - x| < 0.1$ holds for any $x$ lying between these two values.

**24.** $P(x) = 3x^5 - 5x^3 + 2x$, $Q(x) = 3x^5$. These graphs are significantly different only in the region close to the origin. The larger a viewing rectangle one chooses, the more similar the two graphs look.

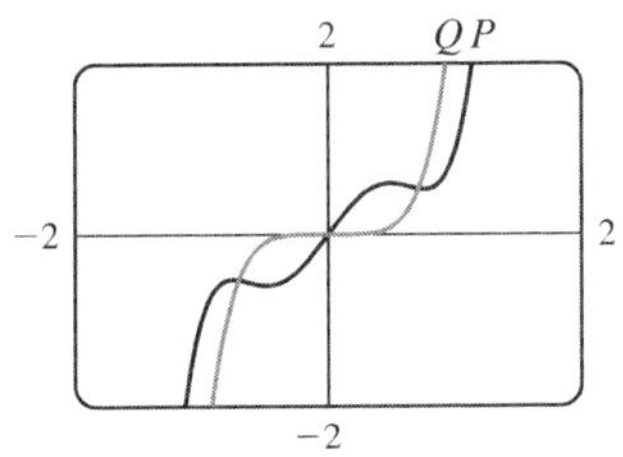


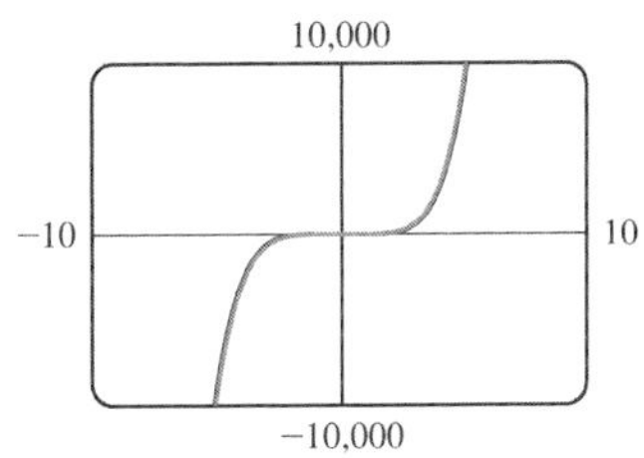


**25.** (a) The root functions $y = \sqrt{x}$, $y = \sqrt[4]{x}$ and $y = \sqrt[6]{x}$

(b) The root functions $y = x$, $y = \sqrt[3]{x}$ and $y = \sqrt[5]{x}$

(c) The root functions $y = \sqrt{x}$, $y = \sqrt[3]{x}$, $y = \sqrt[4]{x}$ and $y = \sqrt[5]{x}$

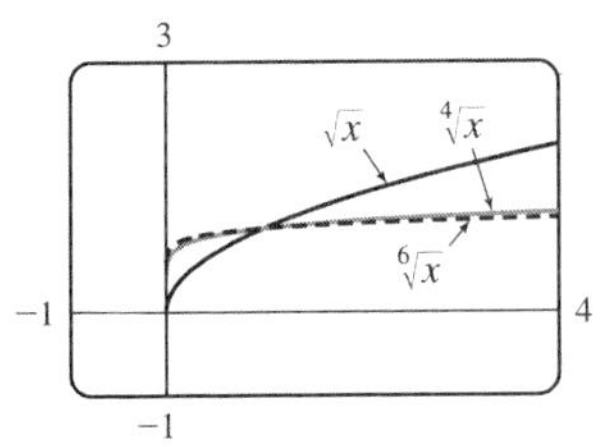


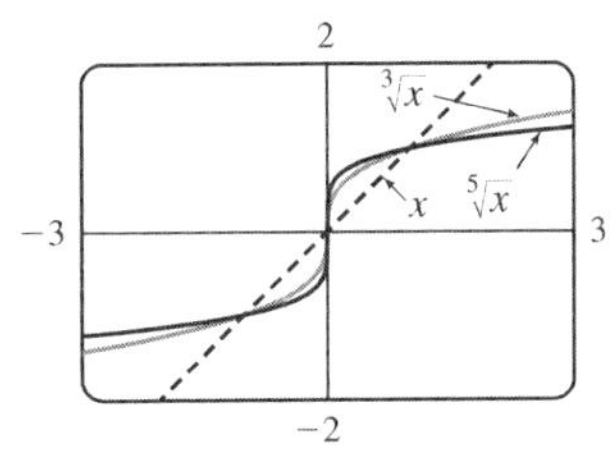


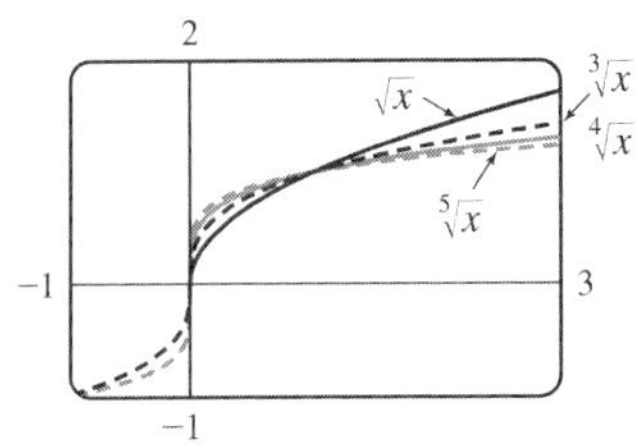


(d) • For any $n$, the $n$th root of 0 is 0 and the $n$th root of 1 is 1; that is, all $n$th root functions pass through the points $(0, 0)$ and $(1, 1)$.

- For odd $n$, the domain of the $n$th root function is $\mathbb{R}$, while for even $n$, it is $\{x \in \mathbb{R} \mid x \geq 0\}$.
- Graphs of even root functions look similar to that of $\sqrt{x}$, while those of odd root functions resemble that of $\sqrt[3]{x}$.
- As $n$ increases, the graph of $\sqrt[n]{x}$ becomes steeper near 0 and flatter for $x > 1$.

**26.** (a) The functions $y = 1/x$ and $y = 1/x^3$

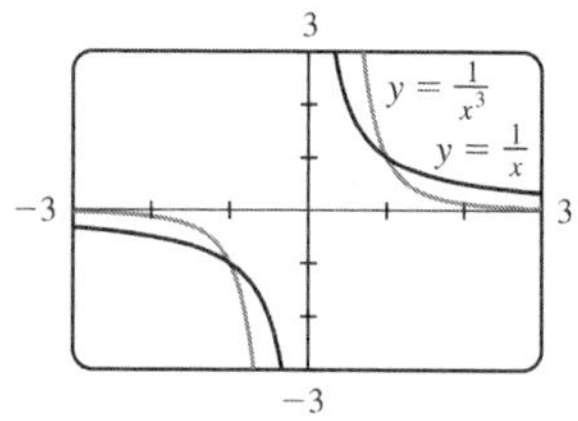


(b) The functions $y = 1/x^2$ and $y = 1/x^4$

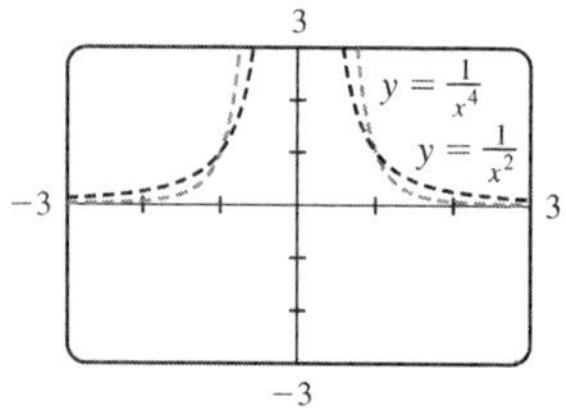


(c) The functions $y = 1/x$, $y = 1/x^2$, $y = 1/x^3$ and $y = 1/x^4$

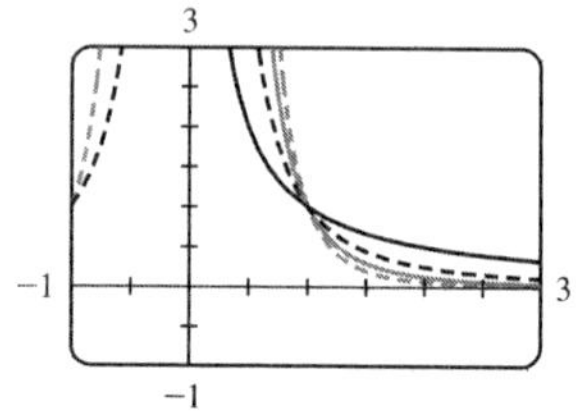


(d) • The graphs of all functions of the form $y = 1/x^n$ pass through the point $(1, 1)$.

• If $n$ is even, the graph of the function is entirely above the $x$-axis. The graphs of $1/x^n$ for $n$ even are similar to one another.

• If $n$ is odd, the function is positive for positive $x$ and negative for negative $x$. The graphs of $1/x^n$ for $n$ odd are similar to one another.

• As $n$ increases, the graphs of $1/x^n$ approach 0 faster as $x \to \infty$.

**27.** $f(x) = x^4 + cx^2 + x$. If $c < -1.5$, there are three humps: two minimum points and a maximum point. These humps get flatter as $c$ increases, until at $c = -1.5$ two of the humps disappear and there is only one minimum point. This single hump then moves to the right and approaches the origin as $c$ increases.

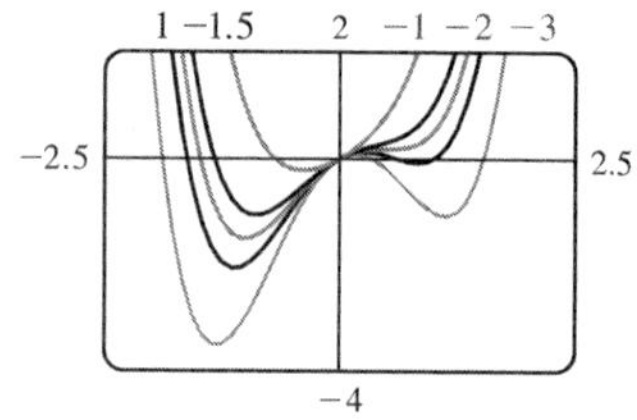


**28.** $f(x) = \sqrt{1 + cx^2}$. If $c < 0$, the function is only defined on $\left[-1/\sqrt{-c}\,, 1/\sqrt{-c}\,\right]$, and its graph is the top half of an ellipse. If $c = 0$, the graph is the line $y = 1$. If $c > 0$, the graph is the top half of a hyperbola. As $c$ approaches 0, these curves become flatter and approach the line $y = 1$.

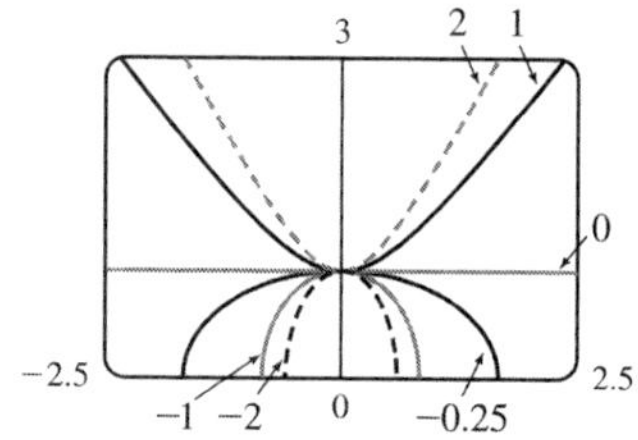


**29.** $y = x^n 2^{-x}$. As $n$ increases, the maximum of the function moves further from the origin, and gets larger. Note, however, that regardless of $n$, the function approaches 0 as $x \to \infty$.

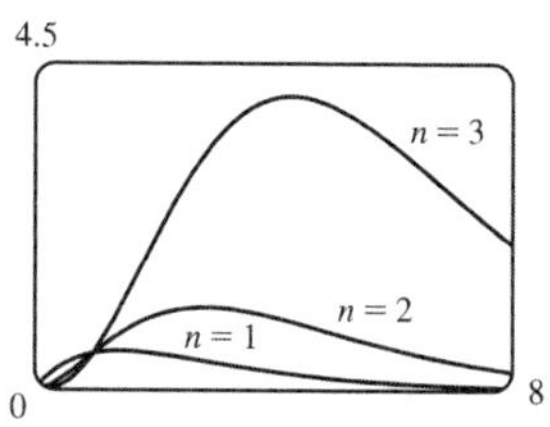


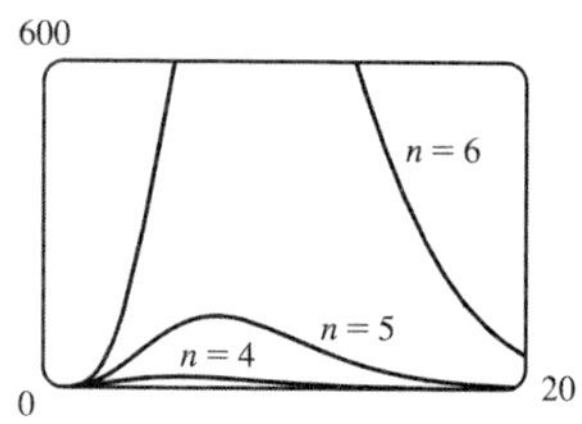


**30.** $y = \dfrac{|x|}{\sqrt{c - x^2}}$. The "bullet" becomes broader as $c$ increases.

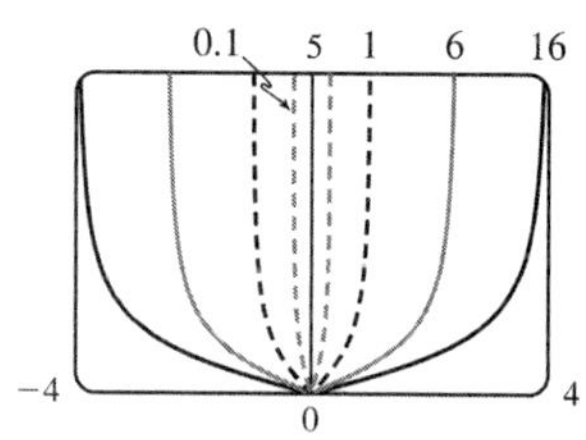


**31.** $y^2 = cx^3 + x^2$

If $c < 0$, the loop is to the right of the origin, and if $c$ is positive, it is to the left. In both cases, the closer $c$ is to 0, the larger the loop is. (In the limiting case, $c = 0$, the loop is "infinite", that is, it doesn't close.) Also, the larger $|c|$ is, the steeper the slope is on the loopless side of the origin.

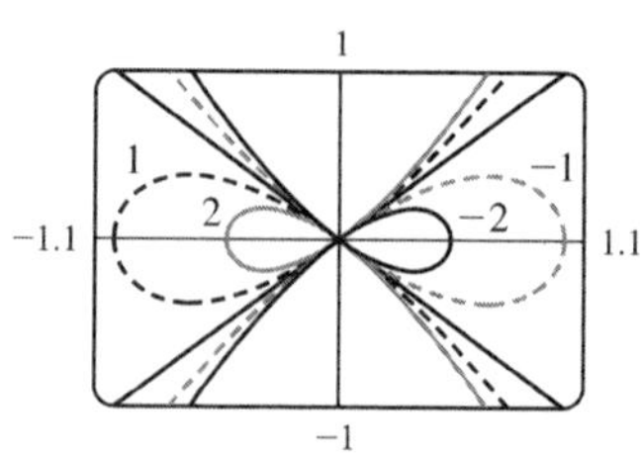

**32.** (a) $y = \sin(\sqrt{x}\,)$

This function is not periodic; it oscillates less frequently as $x$ increases.

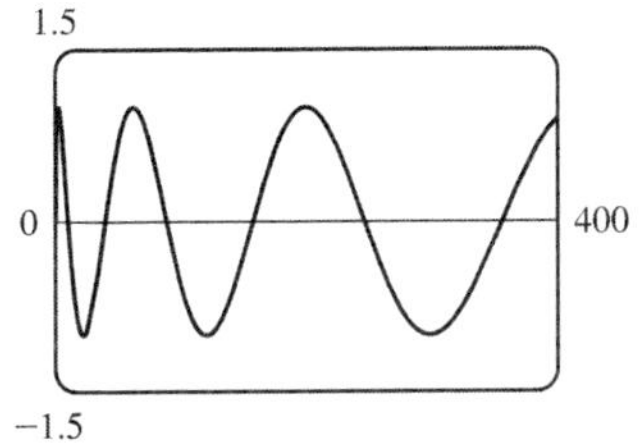


(b) $y = \sin(x^2)$

This function oscillates more frequently as $|x|$ increases. Note also that this function is even, whereas $\sin x$ is odd.

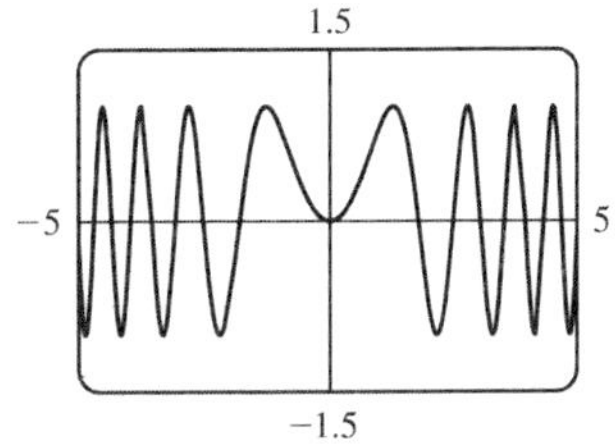


**33.** The graphing window is 95 pixels wide and we want to start with $x = 0$ and end with $x = 2\pi$. Since there are 94 "gaps" between pixels, the distance between pixels is $\frac{2\pi - 0}{94}$. Thus, the $x$-values that the calculator actually plots are $x = 0 + \frac{2\pi}{94} \cdot n$, where $n = 0, 1, 2, \ldots, 93, 94$. For $y = \sin 2x$, the actual points plotted by the calculator are $\left(\frac{2\pi}{94} \cdot n, \sin\left(2 \cdot \frac{2\pi}{94} \cdot n\right)\right)$ for $n = 0, 1, \ldots, 94$. For $y = \sin 96x$, the points plotted are $\left(\frac{2\pi}{94} \cdot n, \sin\left(96 \cdot \frac{2\pi}{94} \cdot n\right)\right)$ for $n = 0, 1, \ldots, 94$. But

$$\begin{aligned}\sin\left(96 \cdot \tfrac{2\pi}{94} \cdot n\right) &= \sin\left(94 \cdot \tfrac{2\pi}{94} \cdot n + 2 \cdot \tfrac{2\pi}{94} \cdot n\right) = \sin\left(2\pi n + 2 \cdot \tfrac{2\pi}{94} \cdot n\right) \\ &= \sin\left(2 \cdot \tfrac{2\pi}{94} \cdot n\right) \quad \text{[by periodicity of sine]}, \quad n = 0, 1, \ldots, 94\end{aligned}$$

So the $y$-values, and hence the points, plotted for $y = \sin 96x$ are identical to those plotted for $y = \sin 2x$.

*Note:* Try graphing $y = \sin 94x$. Can you see why all the $y$-values are zero?

**34.** As in Exercise 33, we know that the points being plotted for $y = \sin 45x$ are $\left(\frac{2\pi}{94} \cdot n, \sin\left(45 \cdot \frac{2\pi}{94} \cdot n\right)\right)$ for $n = 0, 1, \ldots, 94$. But

$$\begin{aligned}\sin\left(45 \cdot \tfrac{2\pi}{94} \cdot n\right) &= \sin\left(47 \cdot \tfrac{2\pi}{94} \cdot n - 2 \cdot \tfrac{2\pi}{94} \cdot n\right) = \sin\left(n\pi - 2 \cdot \tfrac{2\pi}{94} \cdot n\right) \\ &= \sin(n\pi)\cos\left(2 \cdot \tfrac{2\pi}{94} \cdot n\right) - \cos(n\pi)\sin\left(2 \cdot \tfrac{2\pi}{94} \cdot n\right) \quad \text{[Subtraction formula for the sine]} \\ &= 0 \cdot \cos\left(2 \cdot \tfrac{2\pi}{94} \cdot n\right) - (\pm 1)\sin\left(2 \cdot \tfrac{2\pi}{94} \cdot n\right) = \pm\sin\left(2 \cdot \tfrac{2\pi}{94} \cdot n\right), \quad n = 0, 1, \ldots, 94\end{aligned}$$

So the $y$-values, and hence the points, plotted for $y = \sin 45x$ lie on either $y = \sin 2x$ or $y = -\sin 2x$.

## 1.5 Exponential Functions

**1.** (a) $f(x) = a^x,\ a > 0$ (b) $\mathbb{R}$ (c) $(0, \infty)$ (d) See Figures 4(c), 4(b), and 4(a), respectively.

**2.** (a) The number $e$ is the value of $a$ such that the slope of the tangent line at $x = 0$ on the graph of $y = a^x$ is exactly 1.

(b) $e \approx 2.71828$ (c) $f(x) = e^x$

**3.** All of these graphs approach 0 as $x \to -\infty$, all of them pass through the point $(0, 1)$, and all of them are increasing and approach $\infty$ as $x \to \infty$. The larger the base, the faster the function increases for $x > 0$, and the faster it approaches 0 as $x \to -\infty$.

*Note:* The notation "$x \to \infty$" can be thought of as "$x$ becomes large" at this point. More details on this notation are given in Chapter 2.

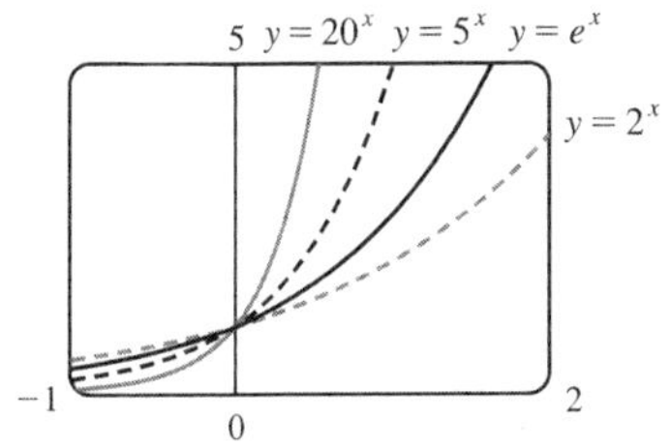

**4.** The graph of $e^{-x}$ is the reflection of the graph of $e^x$ about the $y$-axis, and the graph of $8^{-x}$ is the reflection of that of $8^x$ about the $y$-axis. The graph of $8^x$ increases more quickly than that of $e^x$ for $x > 0$, and approaches 0 faster as $x \to -\infty$.

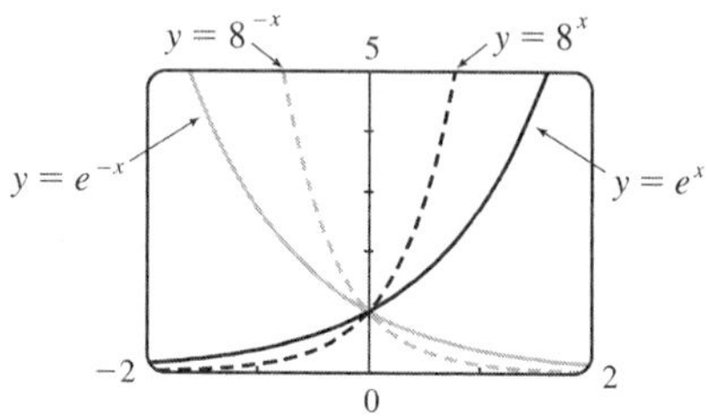


**5.** The functions with bases greater than 1 ($3^x$ and $10^x$) are increasing, while those with bases less than 1 $\left[\left(\frac{1}{3}\right)^x \text{ and } \left(\frac{1}{10}\right)^x\right]$ are decreasing. The graph of $\left(\frac{1}{3}\right)^x$ is the reflection of that of $3^x$ about the $y$-axis, and the graph of $\left(\frac{1}{10}\right)^x$ is the reflection of that of $10^x$ about the $y$-axis. The graph of $10^x$ increases more quickly than that of $3^x$ for $x > 0$, and approaches 0 faster as $x \to -\infty$.

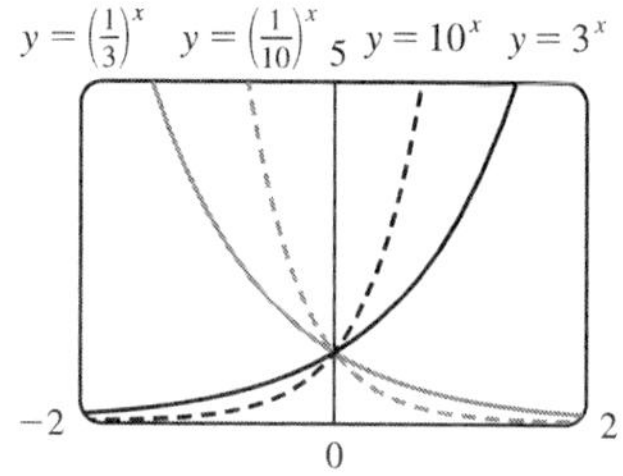


**6.** Each of the graphs approaches $\infty$ as $x \to -\infty$, and each approaches 0 as $x \to \infty$. The smaller the base, the faster the function grows as $x \to -\infty$, and the faster it approaches 0 as $x \to \infty$.

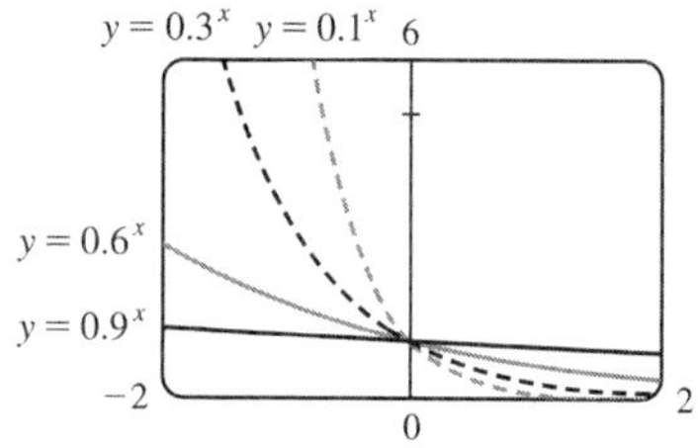


**7.** We start with the graph of $y = 4^x$ (Figure 3) and then shift 3 units downward. This shift doesn't affect the domain, but the range of $y = 4^x - 3$ is $(-3, \infty)$. There is a horizontal asymptote of $y = -3$.

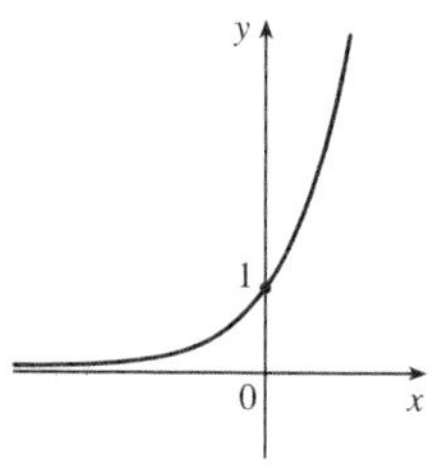


$y = 4^x$

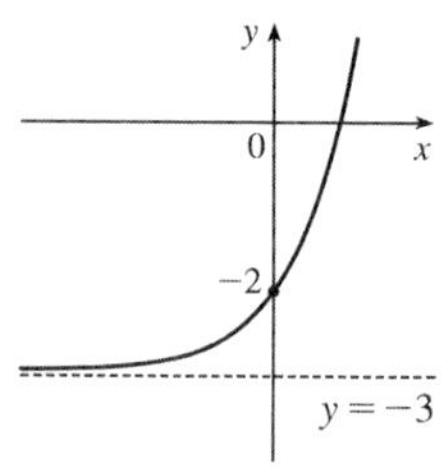


$y = 4^x - 3$

**8.** We start with the graph of $y = 4^x$ (Figure 3) and then shift 3 units to the right. There is a horizontal asymptote of $y = 0$.

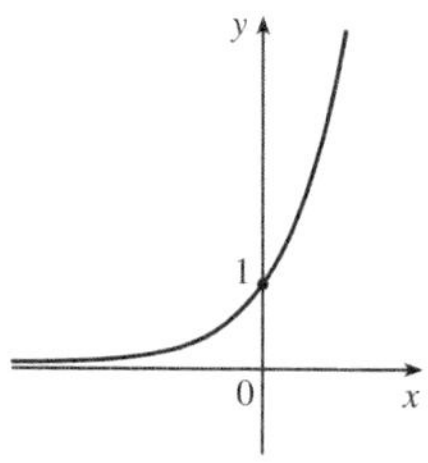


$y = 4^x$

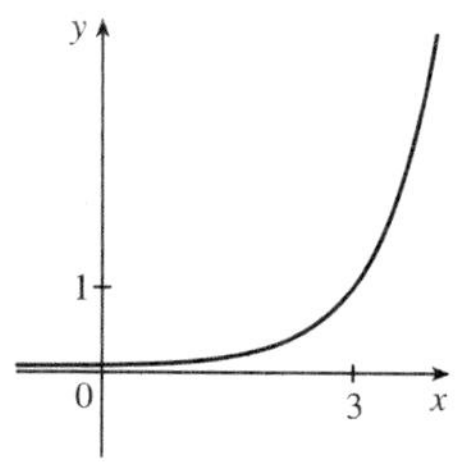


$y = 4^{x-3}$

**9.** We start with the graph of $y = 2^x$ (Figure 3), reflect it about the $y$-axis, and then about the $x$-axis (or just rotate 180° to handle both reflections) to obtain the graph of $y = -2^{-x}$. In each graph, $y = 0$ is the horizontal asymptote.

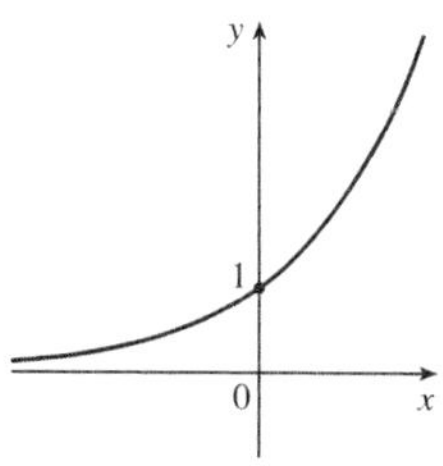


$y = 2^x$

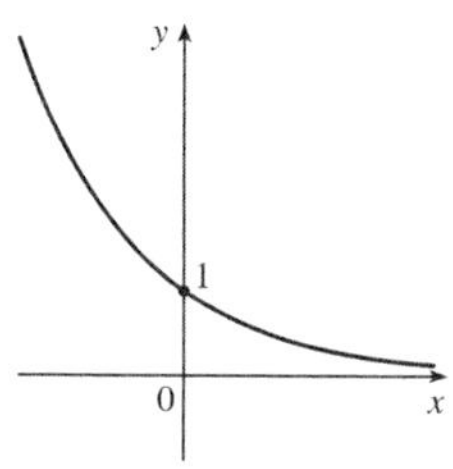


$y = 2^{-x}$

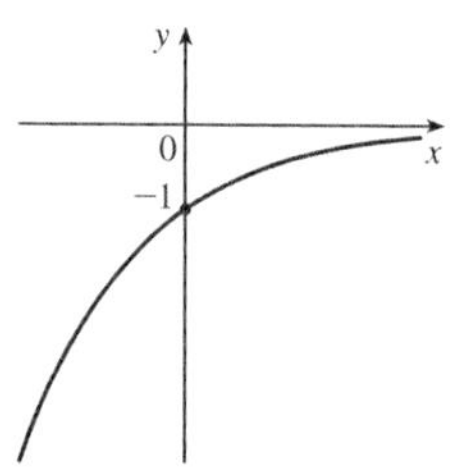


$y = -2^{-x}$

**10.** We start with the graph of $y = e^x$ (Figure 13), vertically stretch by a factor of 2, and then shift 1 unit upward. There is a horizontal asymptote of $y = 1$.

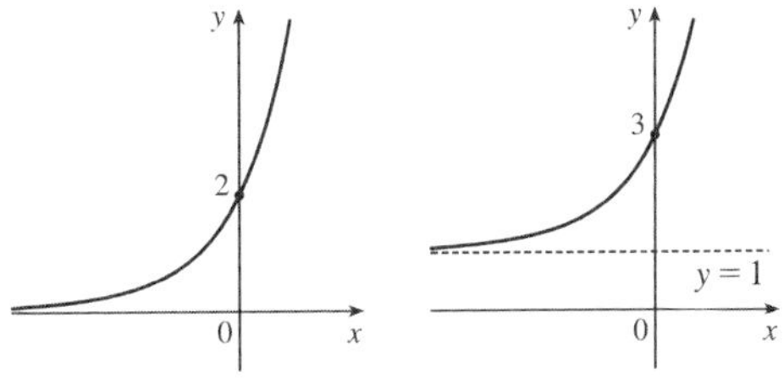


$y = 2e^x$ $\qquad$ $y = 1 + 2e^x$

**11.** We start with the graph of $y = e^x$ (Figure 13) and reflect about the $y$-axis to get the graph of $y = e^{-x}$. Then we compress the graph vertically by a factor of 2 to obtain the graph of $y = \frac{1}{2}e^{-x}$ and then reflect about the $x$-axis to get the graph of $y = -\frac{1}{2}e^{-x}$. Finally, we shift the graph upward one unit to get the graph of $y = 1 - \frac{1}{2}e^{-x}$.

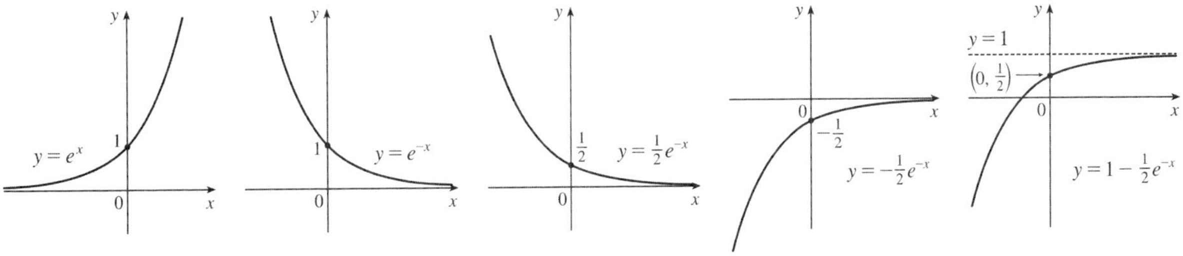


**12.** We start with the graph of $y = e^x$ (Figure 13) and reflect about the $x$-axis to get the graph of $y = -e^x$. Then shift the graph upward one unit to get the graph of $y = 1 - e^x$. Finally, we stretch the graph vertically by a factor of 2 to obtain the graph of $y = 2(1 - e^x)$.

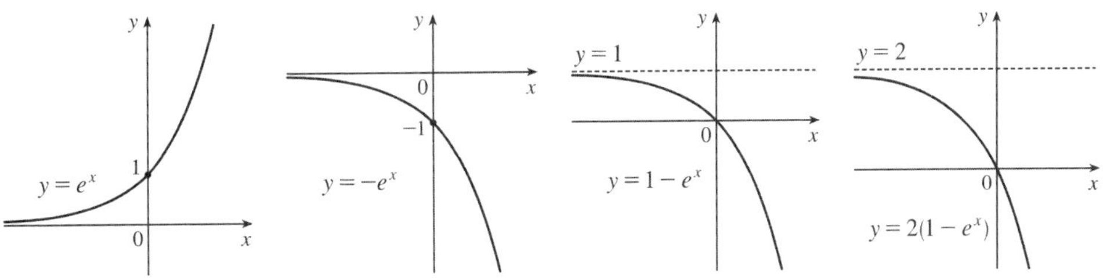


**13.** (a) To find the equation of the graph that results from shifting the graph of $y = e^x$ 2 units downward, we subtract 2 from the original function to get $y = e^x - 2$.

(b) To find the equation of the graph that results from shifting the graph of $y = e^x$ 2 units to the right, we replace $x$ with $x - 2$ in the original function to get $y = e^{(x-2)}$.

(c) To find the equation of the graph that results from reflecting the graph of $y = e^x$ about the $x$-axis, we multiply the original function by $-1$ to get $y = -e^x$.

(d) To find the equation of the graph that results from reflecting the graph of $y = e^x$ about the $y$-axis, we replace $x$ with $-x$ in the original function to get $y = e^{-x}$.

(e) To find the equation of the graph that results from reflecting the graph of $y = e^x$ about the $x$-axis and then about the $y$-axis, we first multiply the original function by $-1$ (to get $y = -e^x$) and then replace $x$ with $-x$ in this equation to get $y = -e^{-x}$.

**14.** (a) This reflection consists of first reflecting the graph about the $x$-axis (giving the graph with equation $y = -e^x$) and then shifting this graph $2 \cdot 4 = 8$ units upward. So the equation is $y = -e^x + 8$.

(b) This reflection consists of first reflecting the graph about the $y$-axis (giving the graph with equation $y = e^{-x}$) and then shifting this graph $2 \cdot 2 = 4$ units to the right. So the equation is $y = e^{-(x-4)}$.

**15.** (a) The denominator $1 + e^x$ is never equal to zero because $e^x > 0$, so the domain of $f(x) = 1/(1 + e^x)$ is $\mathbb{R}$.

(b) $1 - e^x = 0 \quad \Leftrightarrow \quad e^x = 1 \quad \Leftrightarrow \quad x = 0$, so the domain of $f(x) = 1/(1 - e^x)$ is $(-\infty, 0) \cup (0, \infty)$.

**16.** (a) The sine and exponential functions have domain $\mathbb{R}$, so $g(t) = \sin(e^{-t})$ also has domain $\mathbb{R}$.

(b) The function $g(t) = \sqrt{1 - 2^t}$ has domain $\{t \mid 1 - 2^t \geq 0\} = \{t \mid 2^t \leq 1\} = \{t \mid t \leq 0\} = (-\infty, 0]$.

**17.** Use $y = Ca^x$ with the points $(1, 6)$ and $(3, 24)$. $6 = Ca^1 \quad \left[C = \frac{6}{a}\right]$ and $24 = Ca^3 \quad \Rightarrow \quad 24 = \left(\frac{6}{a}\right)a^3 \quad \Rightarrow$

$4 = a^2 \quad \Rightarrow \quad a = 2 \quad$ [since $a > 0$] and $C = \frac{6}{2} = 3$. The function is $f(x) = 3 \cdot 2^x$.

**18.** Given the $y$-intercept $(0, 2)$, we have $y = Ca^x = 2a^x$. Using the point $\left(2, \frac{2}{9}\right)$ gives us $\frac{2}{9} = 2a^2 \quad \Rightarrow \quad \frac{1}{9} = a^2 \quad \Rightarrow \quad a = \frac{1}{3}$ [since $a > 0$]. The function is $f(x) = 2\left(\frac{1}{3}\right)^x$ or $f(x) = 2(3)^{-x}$.

**19.** If $f(x) = 5^x$, then $\dfrac{f(x+h) - f(x)}{h} = \dfrac{5^{x+h} - 5^x}{h} = \dfrac{5^x 5^h - 5^x}{h} = \dfrac{5^x\left(5^h - 1\right)}{h} = 5^x\left(\dfrac{5^h - 1}{h}\right)$.

**20.** Suppose the month is February. Your payment on the 28th day would be $2^{28-1} = 2^{27} = 134{,}217{,}728$ cents, or \$1,342,177.28. Clearly, the second method of payment results in a larger amount for any month.

**21.** $2$ ft $= 24$ in, $f(24) = 24^2$ in $= 576$ in $= 48$ ft. $g(24) = 2^{24}$ in $= 2^{24}/(12 \cdot 5280)$ mi $\approx 265$ mi

**22.** We see from the graphs that for $x$ less than about 1.8, $g(x) = 5^x > f(x) = x^5$, and then near the point $(1.8, 17.1)$ the curves intersect. Then $f(x) > g(x)$ from $x \approx 1.8$ until $x = 5$. At $(5, 3125)$ there is another point of intersection, and for $x > 5$ we see that $g(x) > f(x)$. In fact, $g$ increases much more rapidly than $f$ beyond that point.

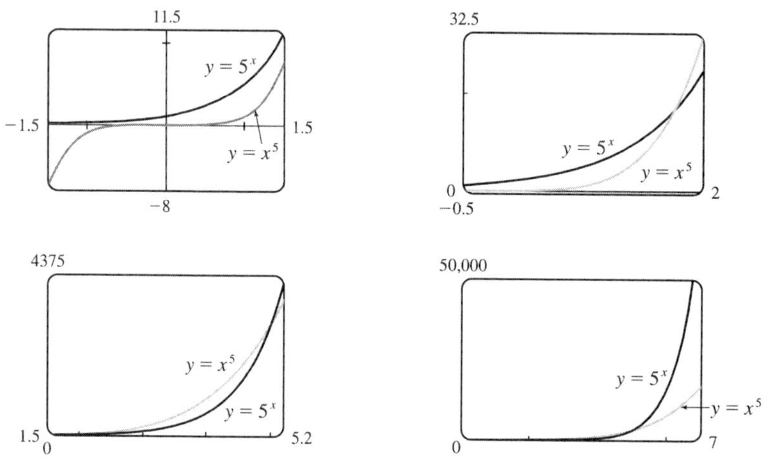


**23.** The graph of $g$ finally surpasses that of $f$ at $x \approx 35.8$.

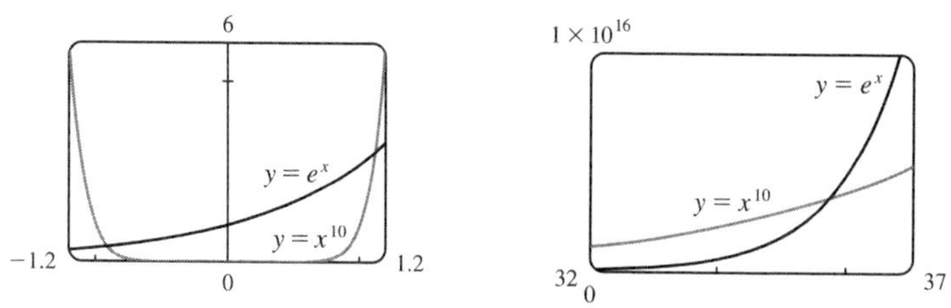

**24.** We graph $y = e^x$ and $y = 1{,}000{,}000{,}000$ and determine where $e^x = 1 \times 10^9$. This seems to be true at $x \approx 20.723$, so $e^x > 1 \times 10^9$ for $x > 20.723$.

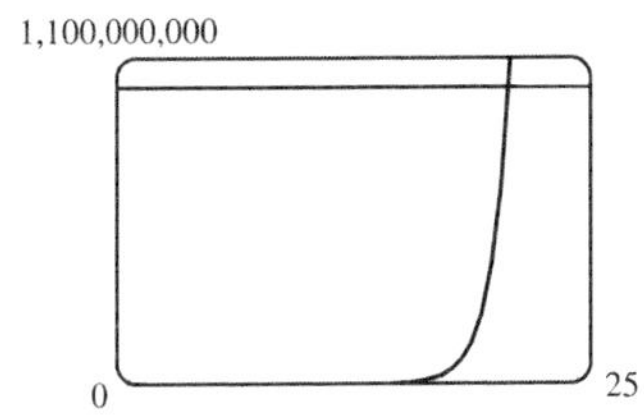


**25.** (a) Fifteen hours represents 5 doubling periods (one doubling period is three hours). $100 \cdot 2^5 = 3200$

(b) In $t$ hours, there will be $t/3$ doubling periods. The initial population is 100, so the population $y$ at time $t$ is $y = 100 \cdot 2^{t/3}$.

(c) $t = 20 \quad \Rightarrow \quad y = 100 \cdot 2^{20/3} \approx 10{,}159$

(d) We graph $y_1 = 100 \cdot 2^{x/3}$ and $y_2 = 50{,}000$. The two curves intersect at $x \approx 26.9$, so the population reaches 50,000 in about 26.9 hours.

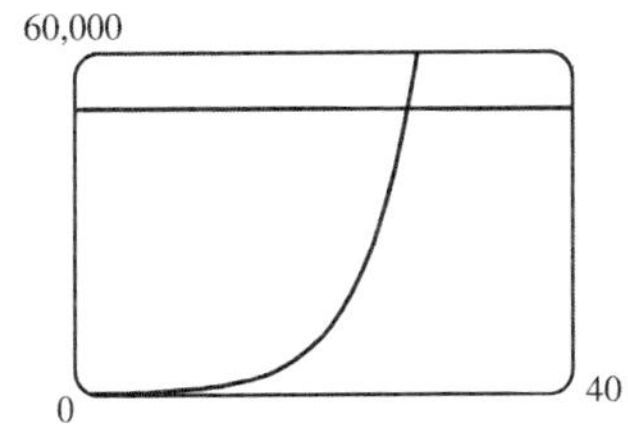


**26.** (a) Three hours represents 6 doubling periods (one doubling period is 30 minutes). $500 \cdot 2^6 = 32{,}000$

(b) In $t$ hours, there will be $2t$ doubling periods. The initial population is 500, so the population $y$ at time $t$ is $y = 500 \cdot 2^{2t}$.

(c) $t = \frac{40}{60} = \frac{2}{3} \quad \Rightarrow \quad y = 500 \cdot 2^{2(2/3)} \approx 1260$

(d) We graph $y_1 = 500 \cdot 2^{2t}$ and $y_2 = 100{,}000$. The two curves intersect at $t \approx 3.82$, so the population reaches 100,000 in about 3.82 hours.

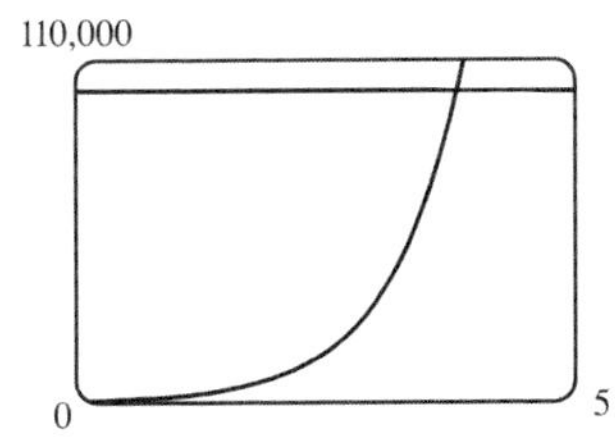


**27.** (a) Fifteen days represents 3 half-life periods (one half-life period is 5 days). $200\left(\frac{1}{2}\right)^3 = 25$ mg

(b) In $t$ hours, there will be $t/5$ half-life periods. The initial amount is 200 mg, so the amount remaining after $t$ days is $y = 200\left(\frac{1}{2}\right)^{t/5}$, or equivalently, $y = 200 \cdot 2^{-t/5}$.

(c) $t = 3$ weeks $= 21$ days $\quad \Rightarrow \quad y = 200 \cdot 2^{-21/5} \approx 10.9$ mg

(d) We graph $y_1 = 200 \cdot 2^{-t/5}$ and $y_2 = 1$. The two curves intersect at $t \approx 38.2$, so the mass will be reduced to 1 mg in about 38.2 days.

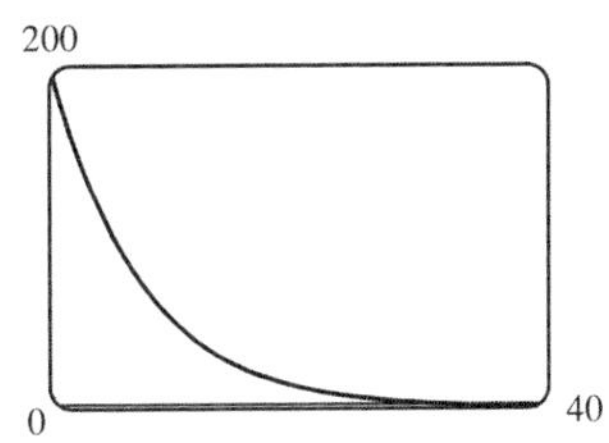


**28.** (a) Sixty hours represents 4 half-life periods. $2 \cdot \left(\frac{1}{2}\right)^4 = \frac{1}{8}$ g

(b) In $t$ hours, there will be $t/15$ half-life periods. The initial mass is 2 g, so the mass $y$ at time $t$ is $y = 2 \cdot \left(\frac{1}{2}\right)^{t/15}$.

(c) 4 days $= 4 \cdot 24 = 96$ hours. $t = 96 \quad \Rightarrow \quad y = 2 \cdot \left(\frac{1}{2}\right)^{96/15} \approx 0.024$ g

(d) $y = 0.01 \quad \Rightarrow \quad t \approx 114.7$ hours

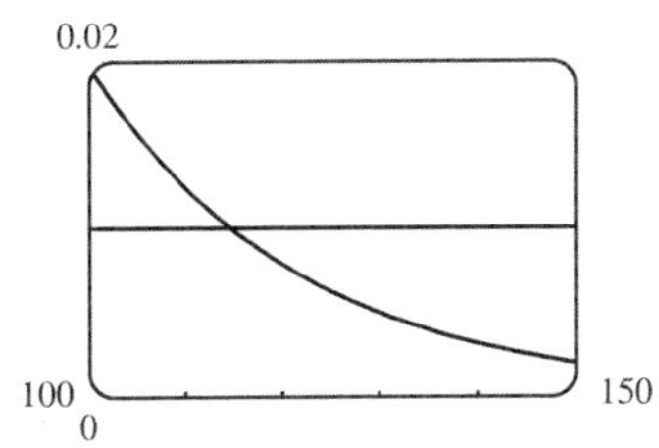


**29.** An exponential model is $y = ab^t$, where $a = 3.154832569 \times 10^{-12}$ and $b = 1.017764706$. This model gives $y(1993) \approx 5498$ million and $y(2010) \approx 7417$ million.

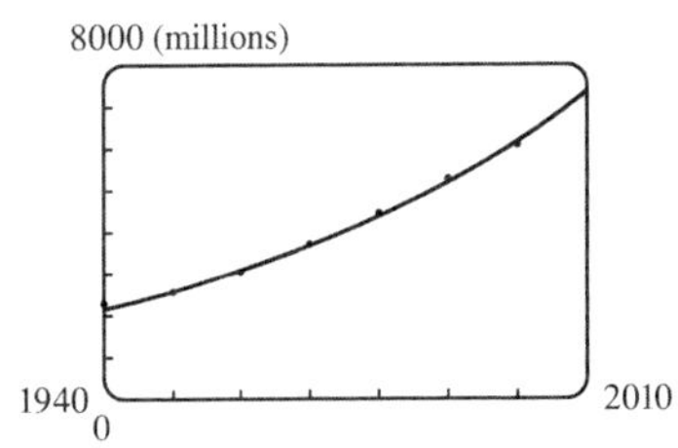

**30.** An exponential model is $y = ab^t$, where $a = 1.9976760197589 \times 10^{-9}$ and $b = 1.0129334321697$. This model gives $y(1925) \approx 111$ million, $y(2010) \approx 330$ million, and $y(2020) \approx 375$ million.

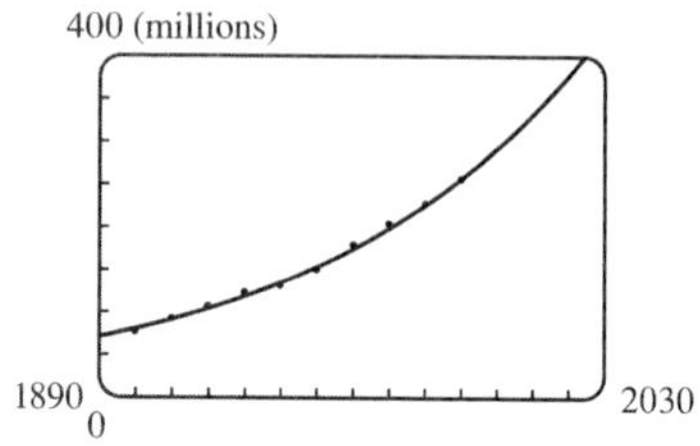


**31.**

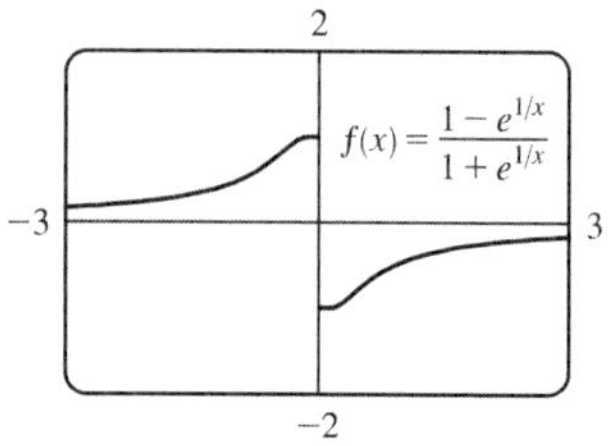


From the graph, it appears that $f$ is an odd function ($f$ is undefined for $x = 0$). To prove this, we must show that $f(-x) = -f(x)$.

$$f(-x) = \frac{1 - e^{1/(-x)}}{1 + e^{1/(-x)}} = \frac{1 - e^{(-1/x)}}{1 + e^{(-1/x)}} = \frac{1 - \dfrac{1}{e^{1/x}}}{1 + \dfrac{1}{e^{1/x}}} \cdot \frac{e^{1/x}}{e^{1/x}} = \frac{e^{1/x} - 1}{e^{1/x} + 1}$$

$$= -\frac{1 - e^{1/x}}{1 + e^{1/x}} = -f(x)$$

so $f$ is an odd function.

**32.** We'll start with $b = -1$ and graph $f(x) = \dfrac{1}{1 + ae^{bx}}$ for $a = 0.1$, 1, and 5.

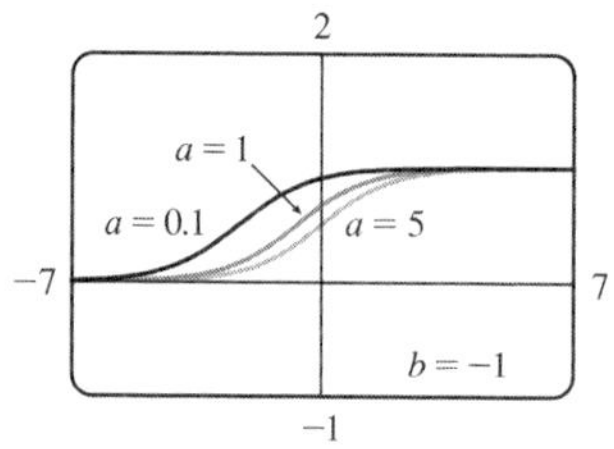


From the graph, we see that there is a horizontal asymptote $y = 0$ as $x \to -\infty$ and a horizontal asymptote $y = 1$ as $x \to \infty$. If $a = 1$, the $y$-intercept is $\left(0, \frac{1}{2}\right)$. As $a$ gets smaller (close to 0), the graph of $f$ moves left. As $a$ gets larger, the graph of $f$ moves right.

As $b$ changes from $-1$ to 0, the graph of $f$ is streched horizontally. As $b$ changes through large negative values, the graph of $f$ is compressed horizontally. (This takes care of negatives values of $b$.)

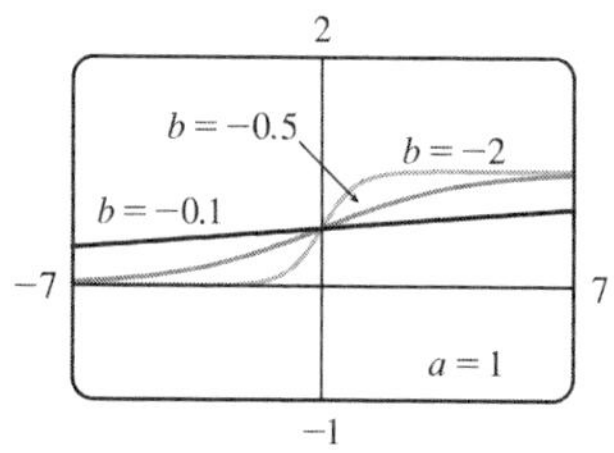


If $b$ is positive, the graph of $f$ is reflected through the $y$-axis.

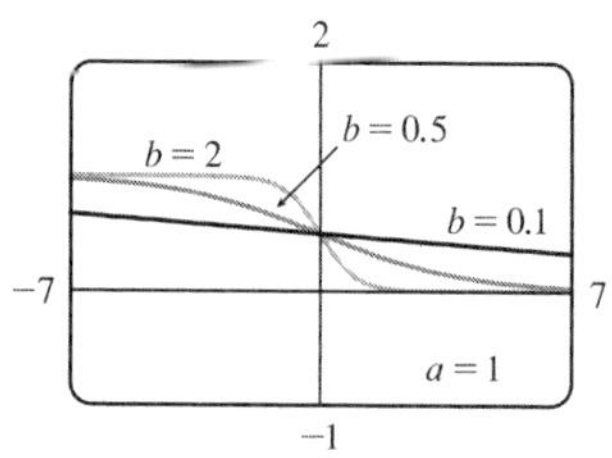


Last, if $b = 0$, the graph of $f$ is the horizontal line $y = 1/(1 + a)$.

## 1.6 Inverse Functions and Logarithms

**1.** (a) See Definition 1.

(b) It must pass the Horizontal Line Test.

**2.** (a) $f^{-1}(y) = x \quad \Leftrightarrow \quad f(x) = y$ for any $y$ in $B$. The domain of $f^{-1}$ is $B$ and the range of $f^{-1}$ is $A$.

(b) See the steps in (5).

(c) Reflect the graph of $f$ about the line $y = x$.

**3.** $f$ is not one-to-one because $2 \neq 6$, but $f(2) = 2.0 = f(6)$.

**4.** $f$ is one-to-one since for any two different domain values, there are different range values.

**5.** No horizontal line intersects the graph of $f$ more than once. Thus, by the Horizontal Line Test, $f$ is one-to-one.

**6.** The horizontal line $y = 0$ (the $x$-axis) intersects the graph of $f$ in more than one point. Thus, by the Horizontal Line Test, $f$ is not one-to-one.

**7.** The horizontal line $y = 0$ (the $x$-axis) intersects the graph of $f$ in more than one point. Thus, by the Horizontal Line Test, $f$ is not one-to-one.

**8.** No horizontal line intersects the graph of $f$ more than once. Thus, by the Horizontal Line Test, $f$ is one-to-one.

**9.** The graph of $f(x) = x^2 - 2x$ is a parabola with axis of symmetry $x = -\dfrac{b}{2a} = -\dfrac{-2}{2(1)} = 1$. Pick any $x$-values equidistant from 1 to find two equal function values. For example, $f(0) = 0$ and $f(2) = 0$, so $f$ is not one-to-one.

**10.** The graph of $f(x) = 10 - 3x$ is a line with slope $-3$. It passes the Horizontal Line Test, so $f$ is one-to-one.
*Algebraic solution*: If $x_1 \neq x_2$, then $-3x_1 \neq -3x_2 \quad \Rightarrow \quad 10 - 3x_1 \neq 10 - 3x_2 \quad \Rightarrow \quad f(x_1) \neq f(x_2)$, so $f$ is one-to-one.

**11.** $g(x) = 1/x$. $\quad x_1 \neq x_2 \quad \Rightarrow \quad 1/x_1 \neq 1/x_2 \quad \Rightarrow \quad g(x_1) \neq g(x_2)$, so $g$ is one-to-one.
*Geometric solution*: The graph of $g$ is the hyperbola shown in Figure 14 in Section 1.2. It passes the Horizontal Line Test, so $g$ is one-to-one.

**12.** $g(x) = \cos x$. $\quad g(0) = 1 = g(2\pi)$, so $g$ is not one-to-one.

**13.** A football will attain every height $h$ up to its maximum height twice: once on the way up, and again on the way down. Thus, even if $t_1$ does not equal $t_2$, $f(t_1)$ may equal $f(t_2)$, so $f$ is not 1-1.

**14.** $f$ is not 1-1 because eventually we all stop growing and therefore, there are two times at which we have the same height.

**15.** Since $f(2) = 9$ and $f$ is 1-1, we know that $f^{-1}(9) = 2$. Remember, if the point $(2, 9)$ is on the graph of $f$, then the point $(9, 2)$ is on the graph of $f^{-1}$.

**16.** (a) First, we must determine $x$ such that $f(x) = 3$. By inspection, we see that if $x = 0$, then $f(x) = 3$. Since $f$ is 1-1 ($f$ is an increasing function), it has an inverse, and $f^{-1}(3) = 0$.

(b) By the second cancellation equation in (4), we have $f\left(f^{-1}(5)\right) = 5$.

**17.** First, we must determine $x$ such that $g(x) = 4$. By inspection, we see that if $x = 0$, then $g(x) = 4$. Since $g$ is 1-1 ($g$ is an increasing function), it has an inverse, and $g^{-1}(4) = 0$.

**18.** (a) $f$ is 1-1 because it passes the Horizontal Line Test.

(b) Domain of $f = [-3, 3] =$ Range of $f^{-1}$.
Range of $f = [-1, 3] =$ Domain of $f^{-1}$.

(c) Since $f(0) = 2$, $f^{-1}(2) = 0$.

(d) Since $f(-1.7) \approx 0$, $f^{-1}(0) = -1.7$.

**19.** We solve $C = \frac{5}{9}(F - 32)$ for $F$: $\frac{9}{5}C = F - 32 \quad \Rightarrow \quad F = \frac{9}{5}C + 32$. This gives us a formula for the inverse function, that is, the Fahrenheit temperature $F$ as a function of the Celsius temperature $C$. $F \geq -459.67 \quad \Rightarrow \quad \frac{9}{5}C + 32 \geq -459.67 \quad \Rightarrow$ $\frac{9}{5}C \geq -491.67 \quad \Rightarrow \quad C \geq -273.15$, the domain of the inverse function.

**20.** $m = \dfrac{m_0}{\sqrt{1 - v^2/c^2}} \Rightarrow 1 - \dfrac{v^2}{c^2} = \dfrac{m_0^2}{m^2} \Rightarrow \dfrac{v^2}{c^2} = 1 - \dfrac{m_0^2}{m^2} \Rightarrow v^2 = c^2\left(1 - \dfrac{m_0^2}{m^2}\right) \Rightarrow v = c\sqrt{1 - \dfrac{m_0^2}{m^2}}$.

This formula gives us the speed $v$ of the particle in terms of its mass $m$, that is, $v = f^{-1}(m)$.

**21.** $f(x) = \sqrt{10 - 3x} \Rightarrow y = \sqrt{10 - 3x}\ \ (y \geq 0) \Rightarrow y^2 = 10 - 3x \Rightarrow 3x = 10 - y^2 \Rightarrow x = -\frac{1}{3}y^2 + \frac{10}{3}$.
Interchange $x$ and $y$: $y = -\frac{1}{3}x^2 + \frac{10}{3}$. So $f^{-1}(x) = -\frac{1}{3}x^2 + \frac{10}{3}$. Note that the domain of $f^{-1}$ is $x \geq 0$.

**22.** $y = f(x) = \dfrac{4x - 1}{2x + 3} \Rightarrow y(2x + 3) = 4x - 1 \Rightarrow 2xy + 3y = 4x - 1 \Rightarrow 3y + 1 = 4x - 2xy \Rightarrow$
$3y + 1 = (4 - 2y)x \Rightarrow x = \dfrac{3y + 1}{4 - 2y}$. Interchange $x$ and $y$: $y = \dfrac{3x + 1}{4 - 2x}$. So $f^{-1}(x) = \dfrac{3x + 1}{4 - 2x}$.

**23.** $y = f(x) = e^{x^3} \Rightarrow \ln y = x^3 \Rightarrow x = \sqrt[3]{\ln y}$. Interchange $x$ and $y$: $y = \sqrt[3]{\ln x}$. So $f^{-1}(x) = \sqrt[3]{\ln x}$.

**24.** $y = f(x) = 2x^3 + 3 \Rightarrow y - 3 = 2x^3 \Rightarrow \dfrac{y - 3}{2} = x^3 \Rightarrow x = \sqrt[3]{\dfrac{y - 3}{2}}$. Interchange $x$ and $y$: $y = \sqrt[3]{\dfrac{x - 3}{2}}$.
So $f^{-1}(x) = \sqrt[3]{\dfrac{x - 3}{2}}$.

**25.** $y = f(x) = \ln(x + 3) \Rightarrow x + 3 = e^y \Rightarrow x = e^y - 3$. Interchange $x$ and $y$: $y = e^x - 3$. So $f^{-1}(x) = e^x - 3$.

**26.** $y = f(x) = \dfrac{1 + e^x}{1 - e^x} \Rightarrow y - ye^x = 1 + e^x \Rightarrow y - 1 = ye^x + e^x \Rightarrow y - 1 = e^x(y + 1) \Rightarrow$
$e^x = \dfrac{y - 1}{y + 1} \Rightarrow x = \ln\left(\dfrac{y - 1}{y + 1}\right)$. Interchange $x$ and $y$: $y = \ln\left(\dfrac{x - 1}{x + 1}\right)$. So $f^{-1}(x) = \ln\left(\dfrac{x - 1}{x + 1}\right)$.
Note that the domain of $f^{-1}$ is $|x| > 1$.

**27.** $y = f(x) = x^4 + 1 \Rightarrow y - 1 = x^4 \Rightarrow x = \sqrt[4]{y - 1}$ (not $\pm$ since $x \geq 0$). Interchange $x$ and $y$: $y = \sqrt[4]{x - 1}$. So $f^{-1}(x) = \sqrt[4]{x - 1}$. The graph of $y = \sqrt[4]{x - 1}$ is just the graph of $y = \sqrt[4]{x}$ shifted right one unit. From the graph, we see that $f$ and $f^{-1}$ are reflections about the line $y = x$.

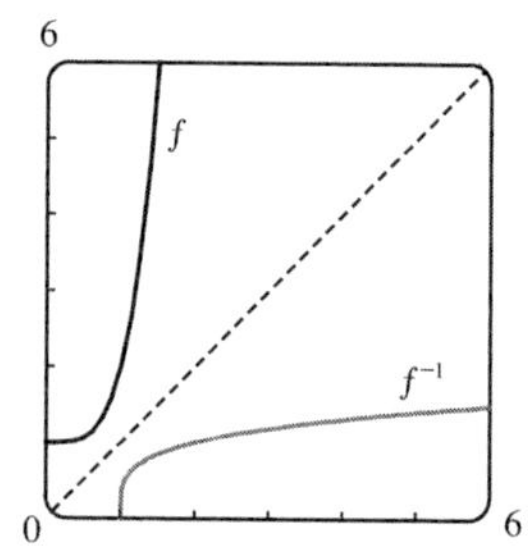


**28.** $y = f(x) = 2 - e^x \Rightarrow e^x = 2 - y \Rightarrow x = \ln(2 - y)$. Interchange $x$ and $y$: $y = \ln(2 - x)$. So $f^{-1}(x) = \ln(2 - x)$. From the graph, we see that $f$ and $f^{-1}$ are reflections about the line $y = x$.

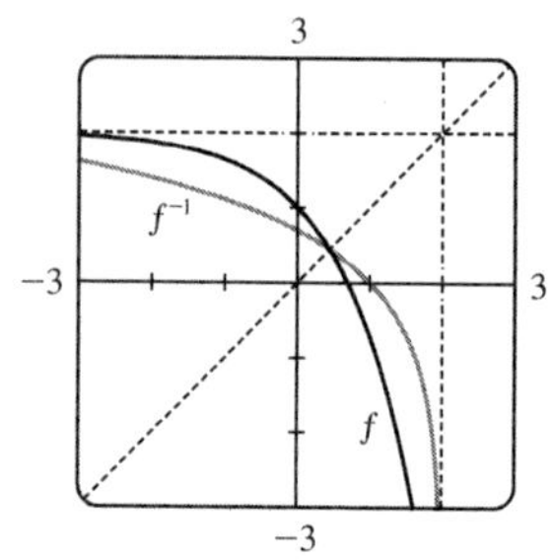


**29.** Reflect the graph of $f$ about the line $y = x$. The points $(-1, -2)$, $(1, -1)$, $(2, 2)$, and $(3, 3)$ on $f$ are reflected to $(-2, -1)$, $(-1, 1)$, $(2, 2)$, and $(3, 3)$ on $f^{-1}$.

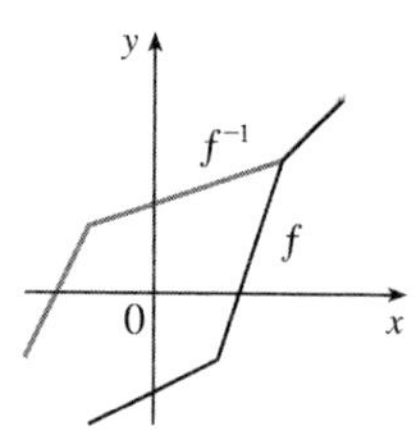

30. Reflect the graph of $f$ about the line $y = x$.

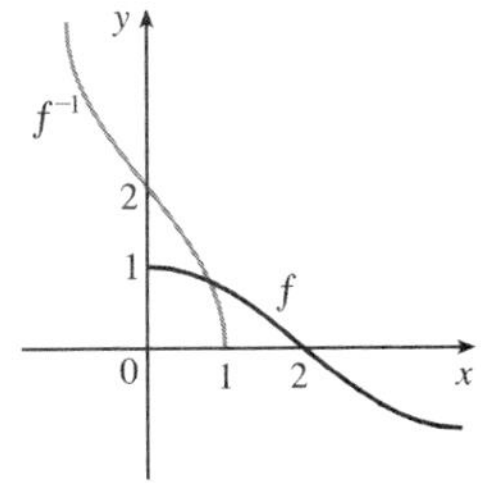


31. (a) It is defined as the inverse of the exponential function with base $a$, that is, $\log_a x = y \Leftrightarrow a^y = x$.
(b) $(0, \infty)$ (c) $\mathbb{R}$ (d) See Figure 11.

32. (a) The natural logarithm is the logarithm with base $e$, denoted $\ln x$.
(b) The common logarithm is the logarithm with base 10, denoted $\log x$.
(c) See Figure 13.

33. (a) $\log_2 64 = 6$ since $2^6 = 64$. (b) $\log_6 \frac{1}{36} = -2$ since $6^{-2} = \frac{1}{36}$.

34. (a) $\log_8 2 = \frac{1}{3}$ since $8^{1/3} = 2$. (b) $\ln e^{\sqrt{2}} = \sqrt{2}$

35. (a) $\log_{10} 1.25 + \log_{10} 80 = \log_{10}(1.25 \cdot 80) = \log_{10} 100 = \log_{10} 10^2 = 2$
(b) $\log_5 10 + \log_5 20 - 3\log_5 2 = \log_5(10 \cdot 20) - \log_5 2^3 = \log_5 \frac{200}{8} = \log_5 25 = \log_5 5^2 = 2$

36. (a) $2^{(\log_2 3 + \log_2 5)} = 2^{\log_2 15} = 15$ [*Or:* $2^{(\log_2 3 + \log_2 5)} = 2^{\log_2 3} \cdot 2^{\log_2 5} = 3 \cdot 5 = 15$]
(b) $e^{3\ln 2} = e^{\ln(2^3)} = e^{\ln 8} = 8$ [*Or:* $e^{3\ln 2} = \left(e^{\ln 2}\right)^3 = 2^3 = 8$]

37. $2\ln 4 - \ln 2 = \ln 4^2 - \ln 2 = \ln 16 - \ln 2 = \ln \frac{16}{2} = \ln 8$

38. $\ln x + a\ln y - b\ln z = \ln x + \ln y^a - \ln z^b = \ln(x \cdot y^a) - \ln z^b = \ln(xy^a/z^b)$

39. $\ln(1+x^2) + \frac{1}{2}\ln x - \ln\sin x = \ln(1+x^2) + \ln x^{1/2} - \ln\sin x = \ln[(1+x^2)\sqrt{x}] - \ln\sin x = \ln\dfrac{(1+x^2)\sqrt{x}}{\sin x}$

40. (a) $\log_{12} 10 = \dfrac{\ln 10}{\ln 12} \approx 0.926628$ (b) $\log_2 8.4 = \dfrac{\ln 8.4}{\ln 2} \approx 3.070389$

41. To graph these functions, we use $\log_{1.5} x = \dfrac{\ln x}{\ln 1.5}$ and $\log_{50} x = \dfrac{\ln x}{\ln 50}$.
These graphs all approach $-\infty$ as $x \to 0^+$, and they all pass through the point $(1, 0)$. Also, they are all increasing, and all approach $\infty$ as $x \to \infty$. The functions with larger bases increase extremely slowly, and the ones with smaller bases do so somewhat more quickly. The functions with large bases approach the $y$-axis more closely as $x \to 0^+$.

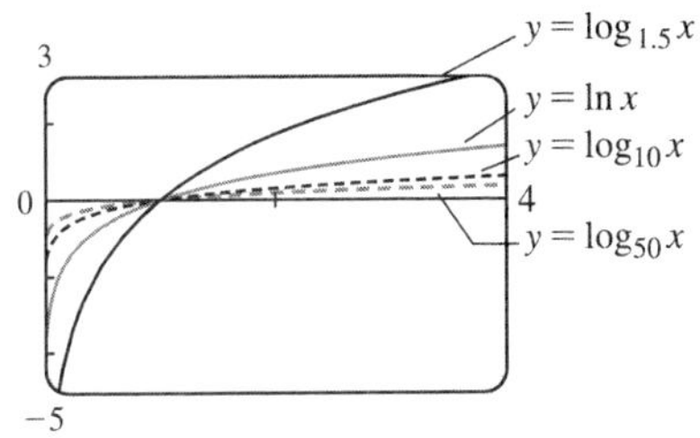


42. We see that the graph of $\ln x$ is the reflection of the graph of $e^x$ about the line $y = x$, and that the graph of $\log_{10} x$ is the reflection of the graph of $10^x$ about the same line. The graph of $10^x$ increases more quickly than that of $e^x$. Also note that $\log_{10} x \to \infty$ as $x \to \infty$ more slowly than $\ln x$.

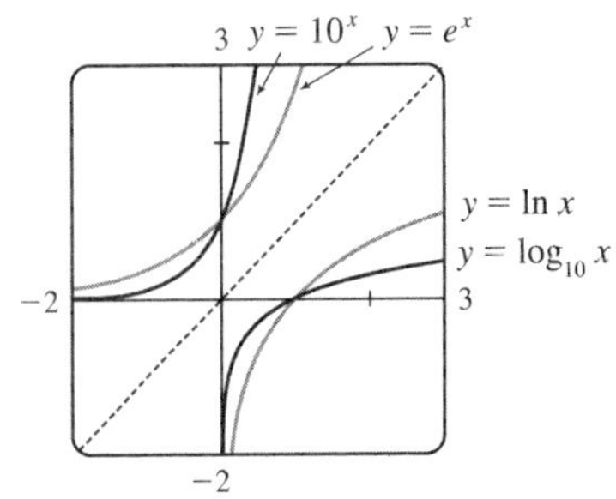


43. 3 ft = 36 in, so we need $x$ such that $\log_2 x = 36 \Leftrightarrow x = 2^{36} = 68{,}719{,}476{,}736$. In miles, this is
$68{,}719{,}476{,}736 \text{ in} \cdot \dfrac{1 \text{ ft}}{12 \text{ in}} \cdot \dfrac{1 \text{ mi}}{5280 \text{ ft}} \approx 1{,}084{,}587.7$ mi.

**44.**

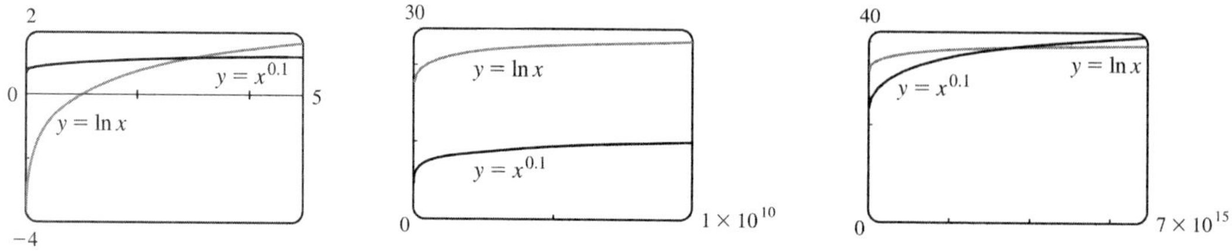


From the graphs, we see that $f(x) = x^{0.1} > g(x) = \ln x$ for approximately $0 < x < 3.06$, and then $g(x) > f(x)$ for $3.06 < x < 3.43 \times 10^{15}$ (approximately). At that point, the graph of $f$ finally surpasses the graph of $g$ for good.

**45.** (a) Shift the graph of $y = \log_{10} x$ five units to the left to obtain the graph of $y = \log_{10}(x+5)$. Note the vertical asymptote of $x = -5$.

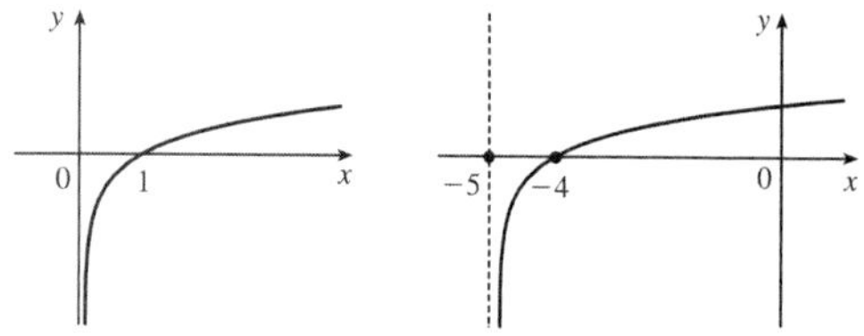

$y = \log_{10} x$ $\qquad$ $y = \log_{10}(x+5)$

(b) Reflect the graph of $y = \ln x$ about the $x$-axis to obtain the graph of $y = -\ln x$.

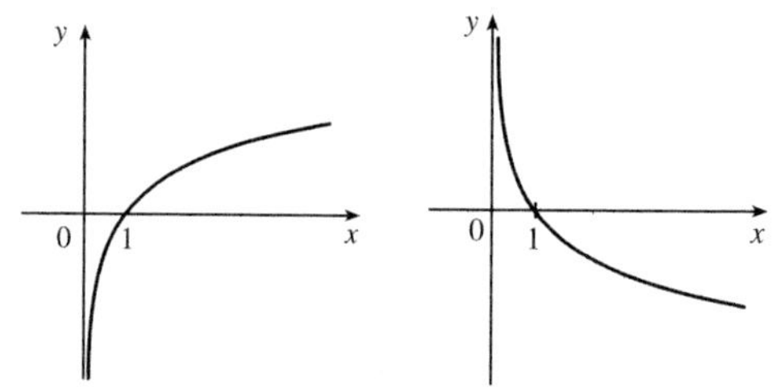

$y = \ln x$ $\qquad$ $y = -\ln x$

**46.** (a) Reflect the graph of $y = \ln x$ about the $y$-axis to obtain the graph of $y = \ln(-x)$.

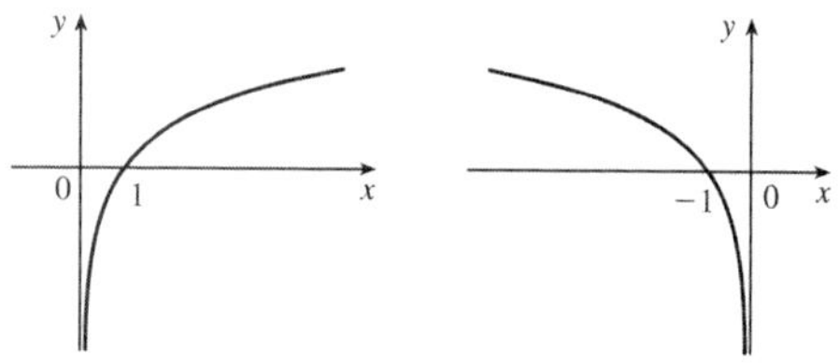

$y = \ln x$ $\qquad$ $y = \ln(-x)$

(b) Reflect the portion of the graph of $y = \ln x$ to the right of the $y$-axis about the $y$-axis. The graph of $y = \ln|x|$ is that reflection in addition to the original portion.

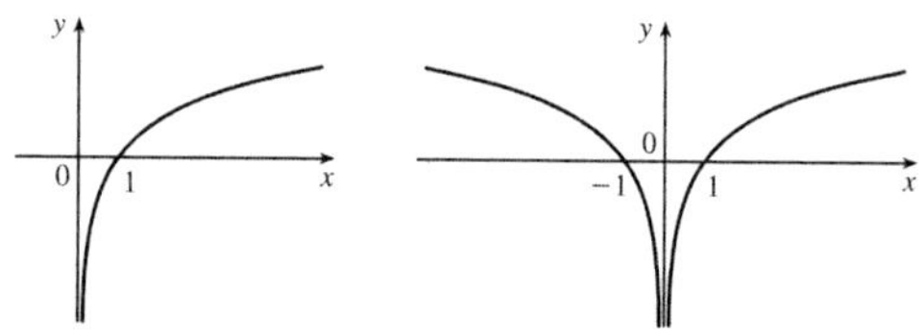

$y = \ln x$ $\qquad$ $y = \ln|x|$

**47.** (a) $2\ln x = 1 \;\Rightarrow\; \ln x = \frac{1}{2} \;\Rightarrow\; x = e^{1/2} = \sqrt{e}$

(b) $e^{-x} = 5 \;\Rightarrow\; -x = \ln 5 \;\Rightarrow\; x = -\ln 5$

**48.** (a) $e^{2x+3} - 7 = 0 \;\Rightarrow\; e^{2x+3} = 7 \;\Rightarrow\; 2x + 3 = \ln 7 \;\Rightarrow\; 2x = \ln 7 - 3 \;\Rightarrow\; x = \frac{1}{2}(\ln 7 - 3)$

(b) $\ln(5 - 2x) = -3 \;\Rightarrow\; 5 - 2x = e^{-3} \;\Rightarrow\; 2x = 5 - e^{-3} \;\Rightarrow\; x = \frac{1}{2}(5 - e^{-3})$

**49.** (a) $2^{x-5} = 3 \;\Leftrightarrow\; \log_2 3 = x - 5 \;\Leftrightarrow\; x = 5 + \log_2 3.$

*Or:* $2^{x-5} = 3 \;\Leftrightarrow\; \ln(2^{x-5}) = \ln 3 \;\Leftrightarrow\; (x-5)\ln 2 = \ln 3 \;\Leftrightarrow\; x - 5 = \dfrac{\ln 3}{\ln 2} \;\Leftrightarrow\; x = 5 + \dfrac{\ln 3}{\ln 2}$

(b) $\ln x + \ln(x-1) = \ln(x(x-1)) = 1 \;\Leftrightarrow\; x(x-1) = e^1 \;\Leftrightarrow\; x^2 - x - e = 0$. The quadratic formula (with $a = 1$, $b = -1$, and $c = -e$) gives $x = \frac{1}{2}\left(1 \pm \sqrt{1 + 4e}\right)$, but we reject the negative root since the natural logarithm is not defined for $x < 0$. So $x = \frac{1}{2}\left(1 + \sqrt{1 + 4e}\right)$.

**50.** (a) $\ln(\ln x) = 1 \;\Leftrightarrow\; e^{\ln(\ln x)} = e^1 \;\Leftrightarrow\; \ln x = e^1 = e \;\Leftrightarrow\; e^{\ln x} = e^e \;\Leftrightarrow\; x = e^e$

(b) $e^{ax} = Ce^{bx} \;\Leftrightarrow\; \ln e^{ax} = \ln[C(e^{bx})] \;\Leftrightarrow\; ax = \ln C + bx + \ln e^{bx} \;\Leftrightarrow\; ax = \ln C + bx \;\Leftrightarrow$

$ax - bx = \ln C \;\Leftrightarrow\; (a - b)x = \ln C \;\Leftrightarrow\; x = \dfrac{\ln C}{a - b}$

**51.** (a) $e^x < 10 \;\Rightarrow\; \ln e^x < \ln 10 \;\Rightarrow\; x < \ln 10 \;\Rightarrow\; x \in (-\infty, \ln 10)$

(b) $\ln x > -1 \;\Rightarrow\; e^{\ln x} > e^{-1} \;\Rightarrow\; x > e^{-1} \;\Rightarrow\; x \in (1/e, \infty)$

**52.** (a) $2 < \ln x < 9 \;\Rightarrow\; e^2 < e^{\ln x} < e^9 \;\Rightarrow\; e^2 < x < e^9 \;\Rightarrow\; x \in (e^2, e^9)$

(b) $e^{2-3x} > 4 \;\Rightarrow\; \ln e^{2-3x} > \ln 4 \;\Rightarrow\; 2 - 3x > \ln 4 \;\Rightarrow\; -3x > \ln 4 - 2 \;\Rightarrow\; x < -\frac{1}{3}(\ln 4 - 2) \;\Rightarrow$

$x \in \left(-\infty, \frac{1}{3}(2 - \ln 4)\right)$

**53.** (a) For $f(x) = \sqrt{3 - e^{2x}}$, we must have $3 - e^{2x} \geq 0 \;\Rightarrow\; e^{2x} \leq 3 \;\Rightarrow\; 2x \leq \ln 3 \;\Rightarrow\; x \leq \frac{1}{2}\ln 3$. Thus, the domain of $f$ is $(-\infty, \frac{1}{2}\ln 3]$.

(b) $y = f(x) = \sqrt{3 - e^{2x}}$ [note that $y \geq 0$] $\Rightarrow\; y^2 = 3 - e^{2x} \;\Rightarrow\; e^{2x} = 3 - y^2 \;\Rightarrow\; 2x = \ln(3 - y^2) \;\Rightarrow$ $x = \frac{1}{2}\ln(3 - y^2)$. Interchange $x$ and $y$: $y = \frac{1}{2}\ln(3 - x^2)$. So $f^{-1}(x) = \frac{1}{2}\ln(3 - x^2)$. For the domain of $f^{-1}$, we must have $3 - x^2 > 0 \;\Rightarrow\; x^2 < 3 \;\Rightarrow\; |x| < \sqrt{3} \;\Rightarrow\; -\sqrt{3} < x < \sqrt{3} \;\Rightarrow\; 0 \leq x < \sqrt{3}$ since $x \geq 0$. Note that the domain of $f^{-1}$, $[0, \sqrt{3})$, equals the range of $f$.

**54.** (a) For $f(x) = \ln(2 + \ln x)$, we must have $2 + \ln x > 0 \;\Rightarrow\; \ln x > -2 \;\Rightarrow\; x > e^{-2}$. Thus, the domain of $f$ is $(e^{-2}, \infty)$.

(b) $y = f(x) = \ln(2 + \ln x) \;\Rightarrow\; e^y = 2 + \ln x \;\Rightarrow\; \ln x = e^y - 2 \;\Rightarrow\; x = e^{e^y - 2}$. Interchange $x$ and $y$: $y = e^{e^x - 2}$. So $f^{-1}(x) = e^{e^x - 2}$. The domain of $f^{-1}$, as well as the range of $f$, is $\mathbb{R}$.

**55.** We see that the graph of $y = f(x) = \sqrt{x^3 + x^2 + x + 1}$ is increasing, so $f$ is 1-1.

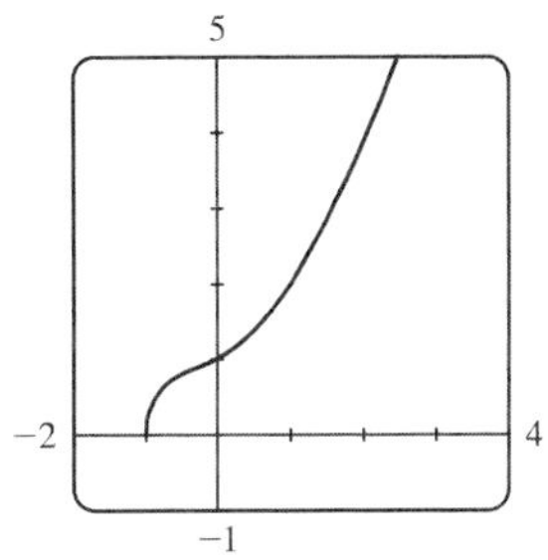


Enter $x = \sqrt{y^3 + y^2 + y + 1}$ and use your CAS to solve the equation for $y$.

Using Derive, we get two (irrelevant) solutions involving imaginary expressions, as well as one which can be simplified to the following:

$y = f^{-1}(x) = -\frac{\sqrt[3]{4}}{6}\left(\sqrt[3]{D - 27x^2 + 20} - \sqrt[3]{D + 27x^2 - 20} + \sqrt[3]{2}\right)$

where $D = 3\sqrt{3}\sqrt{27x^4 - 40x^2 + 16}$.

Maple and Mathematica each give two complex expressions and one real expression, and the real expression is equivalent to that given by Derive. For example, Maple's expression simplifies to $\dfrac{1}{6}\dfrac{M^{2/3} - 8 - 2M^{1/3}}{2M^{1/3}}$, where $M = 108x^2 + 12\sqrt{48 - 120x^2 + 81x^4} - 80$.

**56.** (a) If we use Derive, then solving $x = y^6 + y^4$ for $y$ gives us six solutions of the form $y = \pm\frac{\sqrt{3}}{3}\sqrt{B - 1}$, where $B \in \left\{-2\sin\dfrac{A}{3}, 2\sin\left(\dfrac{A}{3} + \dfrac{\pi}{3}\right), -2\cos\left(\dfrac{A}{3} + \dfrac{\pi}{6}\right)\right\}$ and $A = \sin^{-1}\left(\dfrac{27x - 2}{2}\right)$. The inverse for $y = x^6 + x^4$ $(x \geq 0)$ is $y = \frac{\sqrt{3}}{3}\sqrt{B - 1}$ with $B = 2\sin\left(\dfrac{A}{3} + \dfrac{\pi}{3}\right)$, but because the domain of $A$ is $\left[0, \frac{4}{27}\right]$, this expression is only valid for $x \in \left[0, \frac{4}{27}\right]$.

Happily, Maple gives us the rest of the solution! We solve $x = y^6 + y^4$ for $y$ to get the two real solutions $\pm\dfrac{\sqrt{6}}{6}\dfrac{\sqrt{C^{1/3}\left(C^{2/3} - 2C^{1/3} + 4\right)}}{C^{1/3}}$, where $C = 108x + 12\sqrt{3}\sqrt{x\,(27x - 4)}$, and the inverse for $y = x^6 + x^4$ $(x \geq 0)$ is the positive solution, whose domain is $\left[\frac{4}{27}, \infty\right)$.

Mathematica also gives two real solutions, equivalent to those of Maple.

The positive one is $\dfrac{\sqrt{6}}{6}\left(\sqrt[3]{4}D^{1/3} + 2\sqrt[3]{2}D^{-1/3} - 2\right)$, where $D = -2 + 27x + 3\sqrt{3}\sqrt{x}\sqrt{27x - 4}$. Although this expression also has domain $\left[\frac{4}{27}, \infty\right)$, Mathematica is mysteriously able to plot the solution for all $x \geq 0$.

(b)

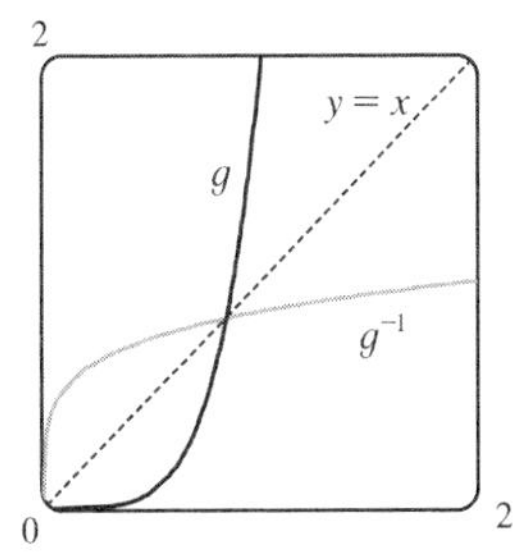

**57.** (a) $n = 100 \cdot 2^{t/3} \quad \Rightarrow \quad \dfrac{n}{100} = 2^{t/3} \quad \Rightarrow \quad \log_2\left(\dfrac{n}{100}\right) = \dfrac{t}{3} \quad \Rightarrow \quad t = 3\log_2\left(\dfrac{n}{100}\right)$. Using formula (10 ), we can write this as $t = f^{-1}(n) = 3 \cdot \dfrac{\ln(n/100)}{\ln 2}$. This function tells us how long it will take to obtain $n$ bacteria (given the number $n$).

(b) $n = 50{,}000 \quad \Rightarrow \quad t = f^{-1}(50{,}000) = 3 \cdot \dfrac{\ln\left(\frac{50{,}000}{100}\right)}{\ln 2} = 3\left(\dfrac{\ln 500}{\ln 2}\right) \approx 26.9$ hours

**58.** (a) $Q = Q_0\left(1 - e^{-t/a}\right) \quad \Rightarrow \quad \dfrac{Q}{Q_0} = 1 - e^{-t/a} \quad \Rightarrow \quad e^{-t/a} = 1 - \dfrac{Q}{Q_0} \quad \Rightarrow \quad -\dfrac{t}{a} = \ln\left(1 - \dfrac{Q}{Q_0}\right) \quad \Rightarrow$ $t = -a\ln(1 - Q/Q_0)$. This gives us the time $t$ necessary to obtain a given charge $Q$.

(b) $Q = 0.9Q_0$ and $a = 2 \quad \Rightarrow \quad t = -2\ln\left(1 - 0.9(Q_0/Q_0)\right) = -2\ln 0.1 \approx 4.6$ seconds.

**59.** (a) To find the equation of the graph that results from shifting the graph of $y = \ln x$ 3 units upward, we add 3 to the original function to get $y = \ln x + 3$.

(b) To find the equation of the graph that results from shifting the graph of $y = \ln x$ 3 units to the left, we replace $x$ with $x + 3$ in the original function to get $y = \ln(x + 3)$.

(c) To find the equation of the graph that results from reflecting the graph of $y = \ln x$ about the $x$-axis, we multiply the original equation by $-1$ to get $y = -\ln x$.

(d) To find the equation of the graph that results from reflecting the graph of $y = \ln x$ about the $y$-axis, we replace $x$ with $-x$ in the original equation to get $y = \ln(-x)$.

(e) To find the equation of the graph that results from reflecting the graph of $y = \ln x$ about the line $y = x$, we interchange $x$ and $y$ in the original equation to get $x = \ln y \quad \Leftrightarrow \quad y = e^x$.

(f) To find the equation of the graph that results from reflecting the graph of $y = \ln x$ about the $x$-axis and then about the line $y = x$, we first multiply the original equation by $-1$ [to get $y = -\ln x$] and then interchange $x$ and $y$ in this equation to get $x = -\ln y \quad \Leftrightarrow \quad \ln y = -x \quad \Leftrightarrow \quad y = e^{-x}$.

(g) To find the equation of the graph that results from reflecting the graph of $y = \ln x$ about the $y$-axis and then about the line $y = x$, we first replace $x$ with $-x$ in the original equation [to get $y = \ln(-x)$] and then interchange $x$ and $y$ to get $x = \ln(-y) \quad \Leftrightarrow \quad -y = e^x \quad \Leftrightarrow \quad y = -e^x$.

(h) To find the equation of the graph that results from shifting the graph of $y = \ln x$ 3 units to the left and then reflecting it about the line $y = x$, we first replace $x$ with $x + 3$ in the original equation [to get $y = \ln(x + 3)$] and then interchange $x$ and $y$ in this equation to get $x = \ln(y + 3) \quad \Leftrightarrow \quad y + 3 = e^x \quad \Leftrightarrow \quad y = e^x - 3$.

**60.** (a) If the point $(x, y)$ is on the graph of $y = f(x)$, then the point $(x - c, y)$ is that point shifted $c$ units to the left. Since $f$ is 1-1, the point $(y, x)$ is on the graph of $y = f^{-1}(x)$ and the point corresponding to $(x - c, y)$ on the graph of $f$ is $(y, x - c)$ on the graph of $f^{-1}$. Thus, the curve's reflection is shifted *down* the same number of units as the curve itself is shifted to the left. So an expression for the inverse function is $g^{-1}(x) = f^{-1}(x) - c$.

(b) If we compress (or stretch) a curve horizontally, the curve's reflection in the line $y = x$ is compressed (or stretched) *vertically* by the same factor. Using this geometric principle, we see that the inverse of $h(x) = f(cx)$ can be expressed as $h^{-1}(x) = (1/c)\, f^{-1}(x)$.

## 1.7 Parametric Curves

**1.** $x = 1 + \sqrt{t}, \quad y = t^2 - 4t, \quad 0 \le t \le 5$

| $t$ | 0 | 1 | 2 | 3 | 4 | 5 |
|---|---|---|---|---|---|---|
| $x$ | 1 | 2 | $1+\sqrt{2}$<br>2.41 | $1+\sqrt{3}$<br>2.73 | 3 | $1+\sqrt{5}$<br>3.24 |
| $y$ | 0 | $-3$ | $-4$ | $-3$ | 0 | 5 |

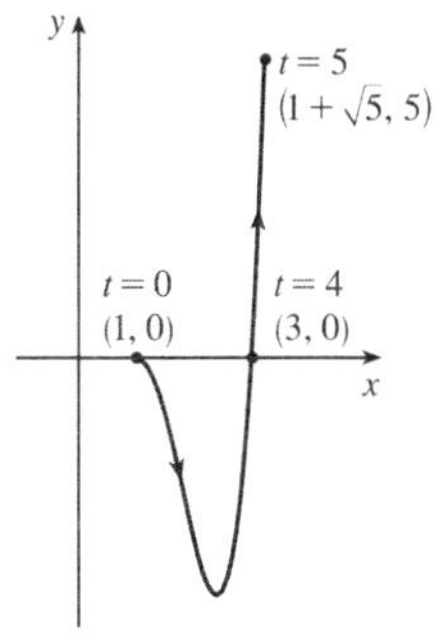


**2.** $x = 2\cos t, \quad y = t - \cos t, \quad 0 \le t \le 2\pi$

| $t$ | 0 | $\pi/2$ | $\pi$ | $3\pi/2$ | $2\pi$ |
|---|---|---|---|---|---|
| $x$ | 2 | 0 | $-2$ | 0 | 2 |
| $y$ | $-1$ | $\pi/2$<br>1.57 | $\pi+1$<br>4.14 | $3\pi/2$<br>4.71 | $2\pi-1$<br>5.28 |

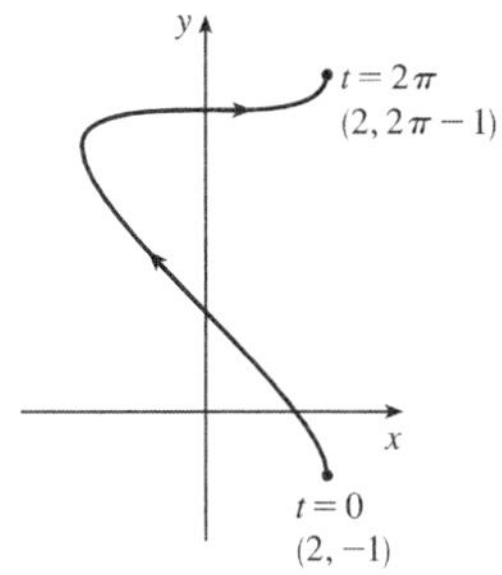


**3.** $x = 5\sin t, \quad y = t^2, \quad -\pi \le t \le \pi$

| $t$ | $-\pi$ | $-\pi/2$ | 0 | $\pi/2$ | $\pi$ |
|---|---|---|---|---|---|
| $x$ | 0 | $-5$ | 0 | 5 | 0 |
| $y$ | $\pi^2$<br>9.87 | $\pi^2/4$<br>2.47 | 0 | $\pi^2/4$<br>2.47 | $\pi^2$<br>9.87 |

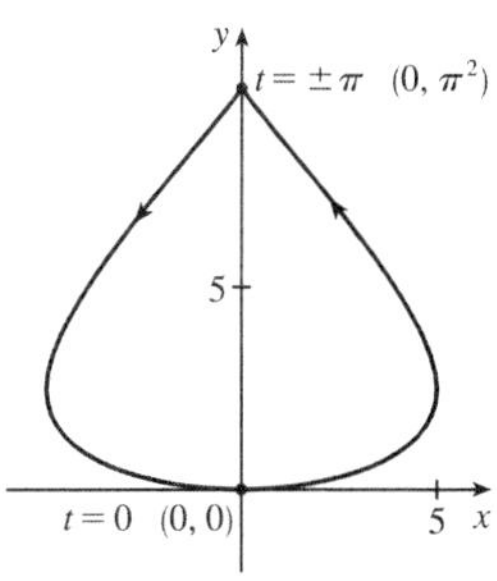


**4.** $x = e^{-t} + t, \quad y = e^t - t, \quad -2 \le t \le 2$

| $t$ | $-2$ | $-1$ | 0 | 1 | 2 |
|---|---|---|---|---|---|
| $x$ | $e^2-2$<br>5.39 | $e-1$<br>1.72 | 1 | $e^{-1}+1$<br>1.37 | $e^{-2}+2$<br>2.14 |
| $y$ | $e^{-2}+2$<br>2.14 | $e^{-1}+1$<br>1.37 | 1 | $e-1$<br>1.72 | $e^2-2$<br>5.39 |

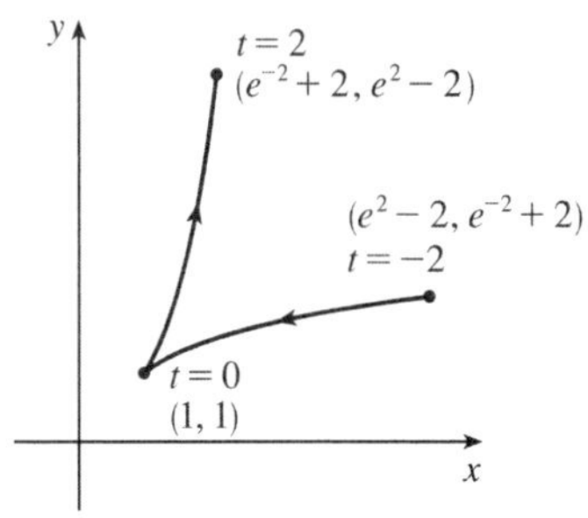


**5.** $x = 3t - 5, \quad y = 2t + 1$

(a)

| $t$ | $-2$ | $-1$ | 0 | 1 | 2 | 3 | 4 |
|---|---|---|---|---|---|---|---|
| $x$ | $-11$ | $-8$ | $-5$ | $-2$ | 1 | 4 | 7 |
| $y$ | $-3$ | $-1$ | 1 | 3 | 5 | 7 | 9 |

(b) $x = 3t - 5 \;\Rightarrow\; 3t = x + 5 \;\Rightarrow\; t = \frac{1}{3}(x+5) \;\Rightarrow$

$y = 2 \cdot \frac{1}{3}(x+5) + 1$, so $y = \frac{2}{3}x + \frac{13}{3}$.

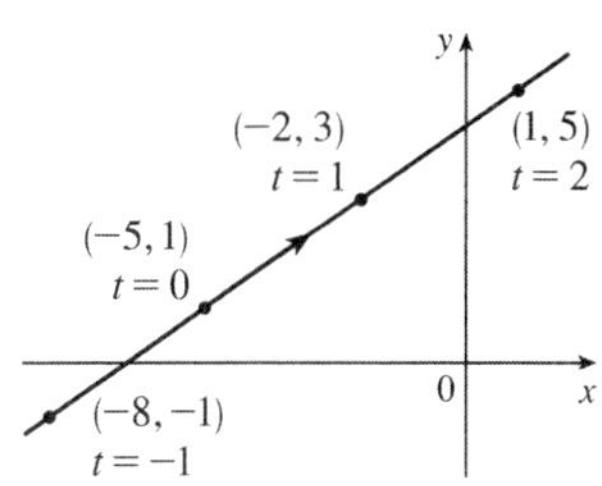

**6.** $x = 1 + 3t, \quad y = 2 - t^2$

(a)

| $t$ | $-3$ | $-2$ | $-1$ | $0$ | $1$ | $2$ | $3$ |
|---|---|---|---|---|---|---|---|
| $x$ | $-8$ | $-5$ | $-2$ | $1$ | $4$ | $7$ | $10$ |
| $y$ | $-7$ | $-2$ | $1$ | $2$ | $1$ | $-2$ | $-7$ |

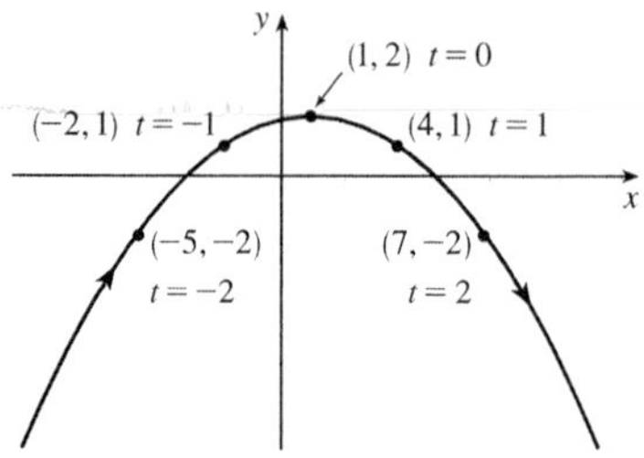


(b) $x = 1 + 3t \;\Rightarrow\; t = \frac{1}{3}(x-1) \;\Rightarrow\; y = 2 - \left[\frac{1}{3}(x-1)\right]^2$,

so $y = -\frac{1}{9}(x-1)^2 + 2$.

**7.** $x = \sqrt{t}, y = 1 - t$

(a)

| $t$ | $0$ | $1$ | $2$ | $3$ | $4$ |
|---|---|---|---|---|---|
| $x$ | $0$ | $1$ | $1.414$ | $1.732$ | $2$ |
| $y$ | $1$ | $0$ | $-1$ | $-2$ | $-3$ |

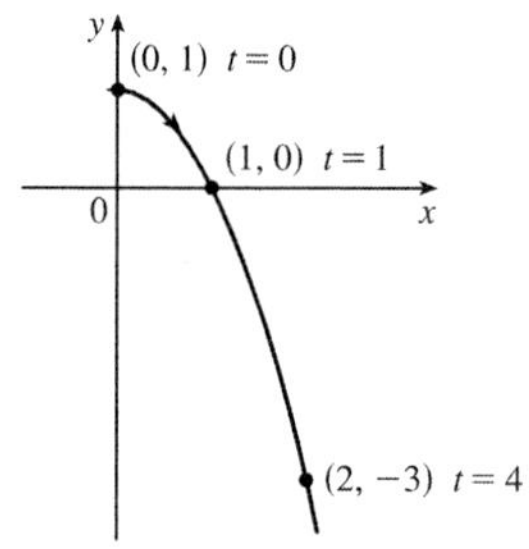


(b) $x = \sqrt{t} \;\Rightarrow\; t = x^2 \;\Rightarrow\; y = 1 - t = 1 - x^2$.

Since $t \geq 0$, $x \geq 0$.

**8.** $x = t^2, y = t^3$

(a)

| $t$ | $-2$ | $-1$ | $0$ | $1$ | $2$ |
|---|---|---|---|---|---|
| $x$ | $4$ | $1$ | $0$ | $1$ | $4$ |
| $y$ | $-8$ | $-1$ | $0$ | $1$ | $8$ |

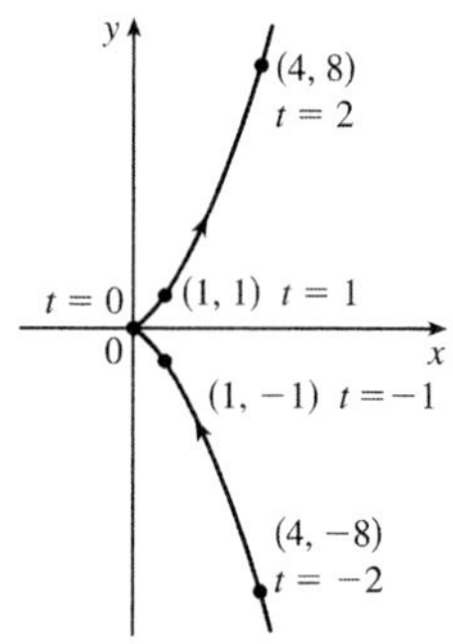


(b) $y = t^3 \;\Rightarrow\; t = \sqrt[3]{y} \;\Rightarrow\; x = t^2 = \left(\sqrt[3]{y}\right)^2 = y^{2/3}$.

$t \in \mathbb{R}, y \in \mathbb{R}, x \geq 0$.

**9.** (a) $x = \sin\theta, y = \cos\theta, 0 \leq \theta \leq \pi$.

$x^2 + y^2 = \sin^2\theta + \cos^2\theta = 1$. Since $0 \leq \theta \leq \pi$, we have $\sin\theta \geq 0$, so $x \geq 0$. Thus, the curve is the right half of the circle $x^2 + y^2 = 1$.

(b)

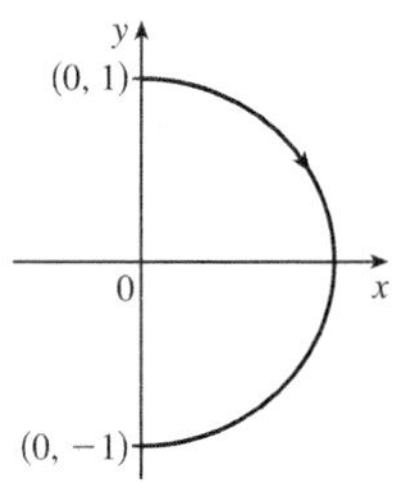


**10.** (a) $x = 4\cos\theta, y = 5\sin\theta, -\pi/2 \leq \theta \leq \pi/2$.

$\left(\frac{x}{4}\right)^2 + \left(\frac{y}{5}\right)^2 = \cos^2\theta + \sin^2\theta = 1$, which is an ellipse with $x$-intercepts $(\pm 4, 0)$ and $y$-intercepts $(0, \pm 5)$. We obtain the portion of the ellipse with $x \geq 0$ since $4\cos\theta \geq 0$ for $-\pi/2 \leq \theta \leq \pi/2$.

(b)

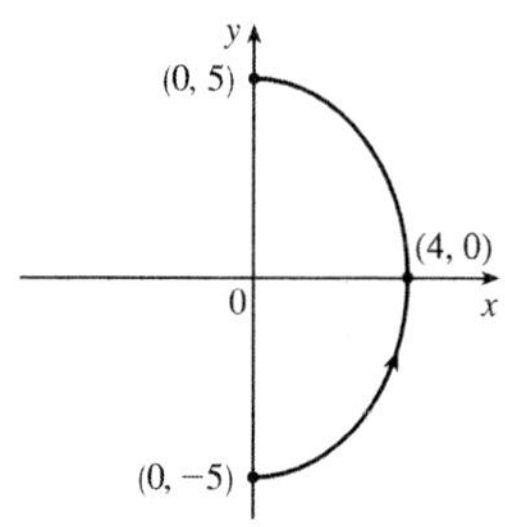


**11.** (a) $x = \sin t, y = \csc t, 0 < t < \frac{\pi}{2}$. $y = \csc t = \dfrac{1}{\sin t} = \dfrac{1}{x}$. For $0 < t < \frac{\pi}{2}$, we have $0 < x < 1$ and $y > 1$. Thus, the curve is the portion of the hyperbola $y = 1/x$ with $y > 1$.

(b)

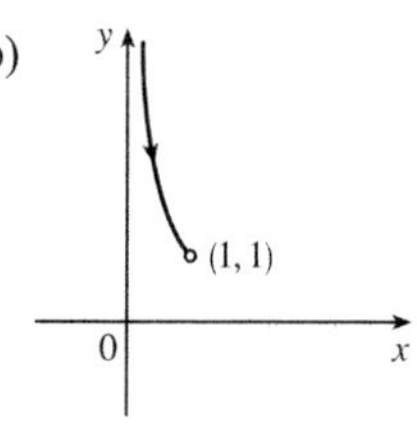

**12.** (a) $x = \sec\theta$, $y = \tan\theta$, $-\frac{\pi}{2} < \theta < \frac{\pi}{2}$. $x^2 - y^2 = \sec^2\theta - \tan^2\theta = 1$, $x \geq 1$, or $x = \sqrt{y^2 + 1}$. Thus, the curve is the right branch of the hyperbola $x^2 - y^2 = 1$.

(b)

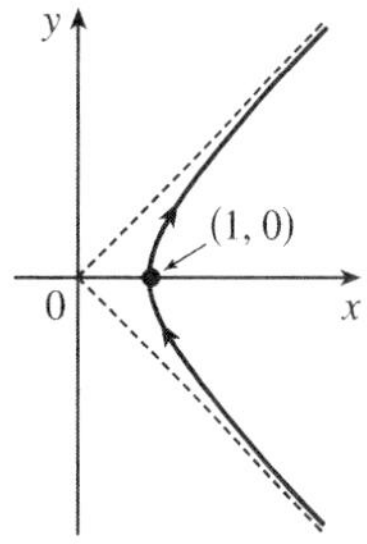


**13.** (a) $x = e^{2t} \;\Rightarrow\; 2t = \ln x \;\Rightarrow\; t = \frac{1}{2}\ln x$.

$y = t + 1 = \frac{1}{2}\ln x + 1$.

(b)

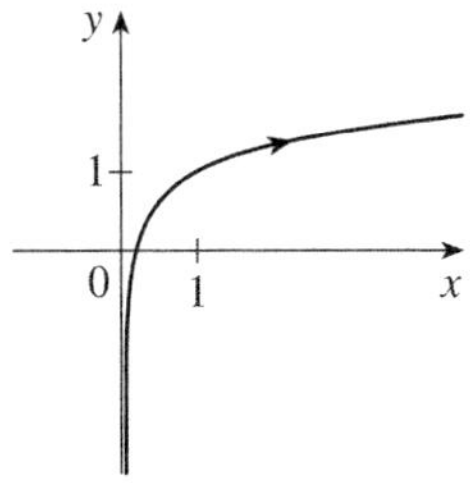


**14.** (a) $x = 1 + \cos\theta \;\Rightarrow\; \cos\theta = x - 1$.

$y = 2\cos\theta - 1 = 2(x - 1) - 1 = 2x - 3$,

$0 \leq x \leq 2$.

(b)

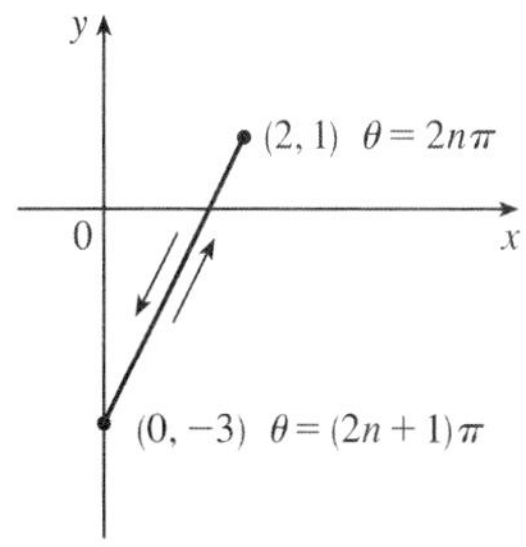


**15.** By Example 4 with $r = 2$, $h = 3$, and $k = 1$, the motion of the particle takes place on a circle centered at $(3, 1)$ with a radius of 2. As $t$ goes from $\frac{\pi}{2}$ to $\frac{3\pi}{2}$, the particle starts at the point $(3, 3)$ and moves counterclockwise to $(3, -1)$ [one-half of a circle].

**16.** $x = 2\sin t$, $y = 4 + \cos t \;\Rightarrow\; \sin t = \dfrac{x}{2}$, $\cos t = y - 4$. $\sin^2 t + \cos^2 t = 1 \;\Rightarrow\; \left(\dfrac{x}{2}\right)^2 + (y - 4)^2 = 1$. The motion of the particle takes place on an ellipse centered at $(0, 4)$. As $t$ goes from 0 to $\frac{3\pi}{2}$, the particle starts at the point $(0, 5)$ and moves clockwise to $(-2, 4)$ [three-quarters of an ellipse].

**17.** $x = 5\sin t$, $y = 2\cos t \;\Rightarrow\; \sin t = \dfrac{x}{5}$, $\cos t = \dfrac{y}{2}$. $\sin^2 t + \cos^2 t = 1 \;\Rightarrow\; \left(\dfrac{x}{5}\right)^2 + \left(\dfrac{y}{2}\right)^2 = 1$. The motion of the particle takes place on an ellipse centered at $(0, 0)$. As $t$ goes from $-\pi$ to $5\pi$, the particle starts at the point $(0, -2)$ and moves clockwise around the ellipse 3 times.

**18.** $y = \cos^2 t = 1 - \sin^2 t = 1 - x^2$. The motion of the particle takes place on the parabola $y = 1 - x^2$. As $t$ goes from $-2\pi$ to $-\pi$, the particle starts at the point $(0, 1)$, moves to $(1, 0)$, and goes back to $(0, 1)$. As $t$ goes from $-\pi$ to 0, the particle moves to $(-1, 0)$ and goes back to $(0, 1)$. The particle repeats this motion as $t$ goes from 0 to $2\pi$.

**19.** We must have $1 \leq x \leq 4$ and $2 \leq y \leq 3$. So the graph of the curve must be contained in the rectangle $[1, 4]$ by $[2, 3]$.

**20.** (a) From the first graph, we have $1 \leq x \leq 2$. From the second graph, we have $-1 \leq y \leq 1$. The only choice that satisfies either of those conditions is III.

(b) From the first graph, the values of $x$ cycle through the values from $-2$ to 2 four times. From the second graph, the values of $y$ cycle through the values from $-2$ to 2 six times. Choice I satisfies these conditions.

(c) From the first graph, the values of $x$ cycle through the values from $-2$ to 2 three times. From the second graph, we have $0 \leq y \leq 2$. Choice IV satisfies these conditions.

(d) From the first graph, the values of $x$ cycle through the values from $-2$ to 2 two times. From the second graph, the values of $y$ do the same thing. Choice II satisfies these conditions.

**21.** When $t = -1$, $(x, y) = (0, -1)$. As $t$ increases to 0, $x$ decreases to $-1$ and $y$ increases to 0. As $t$ increases from 0 to 1, $x$ increases to 0 and $y$ increases to 1. As $t$ increases beyond 1, both $x$ and $y$ increase. For $t < -1$, $x$ is positive and decreasing and $y$ is negative and increasing. We could achieve greater accuracy by estimating $x$- and $y$-values for selected values of $t$ from the given graphs and plotting the corresponding points.

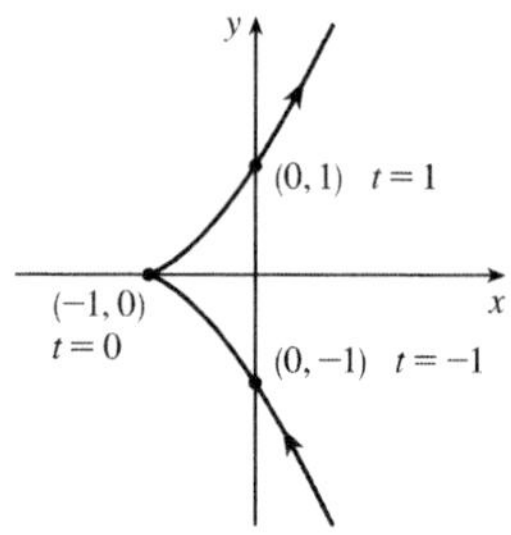


**22.** For $t < -1$, $x$ is positive and decreasing, while $y$ is negative and increasing (these points are in Quadrant IV). When $t = -1$, $(x, y) = (0, 0)$ and, as $t$ increases from $-1$ to 0, $x$ becomes negative and $y$ increases from 0 to 1. At $t = 0$, $(x, y) = (0, 1)$ and, as $t$ increases from 0 to 1, $y$ decreases from 1 to 0 and $x$ is positive. At $t = 1$, $(x, y) = (0, 0)$ again, so the loop is completed. For $t > 1$, $x$ and $y$ both become large negative. This enables us to draw a rough sketch. We could achieve greater accuracy by estimating $x$- and $y$-values for selected values of $t$ from the given graphs and plotting the corresponding points.

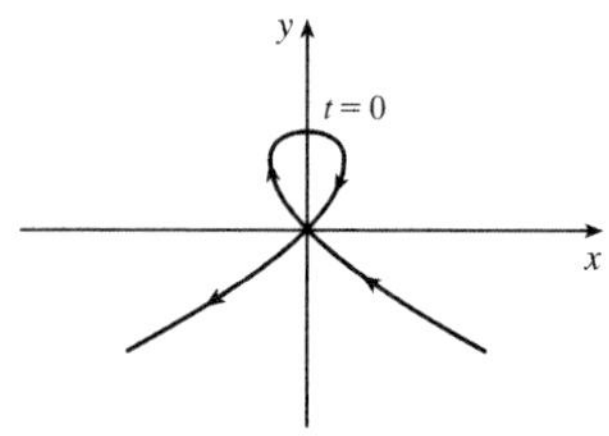


**23.** When $t = 0$ we see that $x = 0$ and $y = 0$, so the curve starts at the origin. As $t$ increases from 0 to $\frac{1}{2}$, the graphs show that $y$ increases from 0 to 1 while $x$ increases from 0 to 1, decreases to 0 and to $-1$, then increases back to 0, so we arrive at the point $(0, 1)$. Similarly, as $t$ increases from $\frac{1}{2}$ to 1, $y$ decreases from 1 to 0 while $x$ repeats its pattern, and we arrive back at the origin. We could achieve greater accuracy by estimating $x$- and $y$-values for selected values of $t$ from the given graphs and plotting the corresponding points.

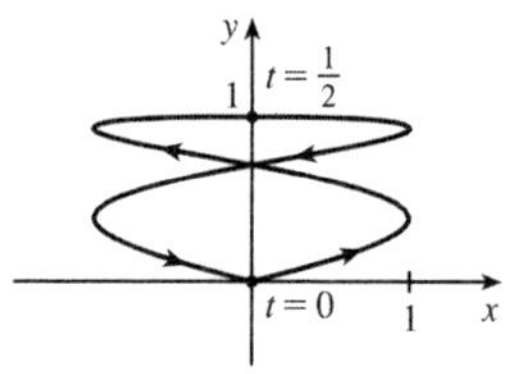


**24.** (a) Note that as $t \to -\infty$, we have $x \to -\infty$ and $y \to \infty$, whereas when $t \to \infty$, both $x$ and $y \to \infty$. This description fits only IV. [But also note that $x(t)$ increases, then decreases, then increases again.]

(b) Note that as $t \to \pm\infty$, $y \to -\infty$. This is only the case with VI.

(c) If $t = 0$, then $(x, y) = (\sin 0, \sin 0) = (0, 0)$. Also, $|x| = |\sin 3t| \leq 1$ for all $t$, and $|y| = |\sin 4t| \leq 1$ for all $t$. The only graph which includes the point $(0, 0)$ and which has $|x| \leq 1$ and $|y| \leq 1$, is V.

(d) Note that as $t \to -\infty$, both $x$ and $y \to -\infty$, and as $t \to \infty$, both $x$ and $y \to \infty$. This description fits only III. (Also note that, since $\sin 2t$ and $\sin 3t$ lie between $-1$ and 1, the curve never strays very far from the line $y = x$.)

(e) Note that both $x(t)$ and $y(t)$ are periodic with period $2\pi$ and satisfy $|x| \leq 1$ and $|y| \leq 1$. Now the only $y$-intercepts occur when $x = \sin(t + \sin t) = 0 \quad \Leftrightarrow \quad t = 0$ or $\pi$. So there should be two $y$-intercepts: $y(0) = \cos 1 \approx 0.54$ and $y(\pi) = \cos(\pi - 1) \approx -0.54$. Similarly, there should be two $x$-intercepts: $x\left(\frac{\pi}{2}\right) = \sin\left(\frac{\pi}{2} + 1\right) \approx 0.54$ and $x\left(\frac{3\pi}{2}\right) = \sin\left(\frac{3\pi}{2} - 1\right) \approx -0.54$. The only curve with these $x$- and $y$-intercepts is I.

(f) Note that $x(t)$ is periodic with period $2\pi$, so the only $y$-intercepts occur when $x = \cos t = 0$ $\Leftrightarrow \quad t = \frac{\pi}{2}$ or $\frac{3\pi}{2}$. Also, the graph is symmetric about the $x$-axis, since $y(-t) = \sin(-t + \sin 5(-t)) = \sin(-t - \sin 5t) = -\sin(t + \sin 5t) = -y(t)$, and $x(-t) = \cos(-t) = \cos t = x(t)$. The only graph which has only two $y$-intercepts, and is symmetric about the $x$-axis, is II.

**25.** As in Example 6, we let $y = t$ and $x = t - 3t^3 + t^5$ and use a $t$-interval of $[-2\pi, 2\pi]$.

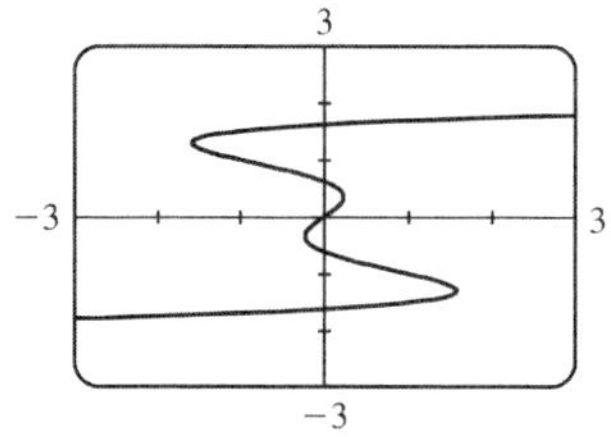


**26.** We use $x_1 = t$, $y_1 = t^5$ and $x_2 = t\,(t-1)^2$, $y_2 = t$ with $-2\pi \leq t \leq 2\pi$. There are 3 points of intersection; $(0,0)$ is fairly obvious. The point in quadrant III is approximately $(-0.8, -0.4)$ and the point in quadrant I is approximately $(1.1, 1.8)$.

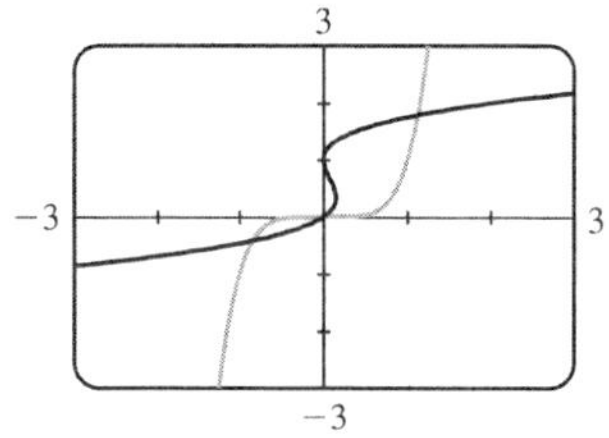


**27.** (a) $x = x_1 + (x_2 - x_1)t$, $y = y_1 + (y_2 - y_1)t$, $0 \leq t \leq 1$. Clearly the curve passes through $P_1(x_1, y_1)$ when $t = 0$ and through $P_2(x_2, y_2)$ when $t = 1$. For $0 < t < 1$, $x$ is strictly between $x_1$ and $x_2$ and $y$ is strictly between $y_1$ and $y_2$. For every value of $t$, $x$ and $y$ satisfy the relation $y - y_1 = \dfrac{y_2 - y_1}{x_2 - x_1}(x - x_1)$, which is the equation of the line through $P_1(x_1, y_1)$ and $P_2(x_2, y_2)$.

Finally, any point $(x, y)$ on that line satisfies $\dfrac{y - y_1}{y_2 - y_1} = \dfrac{x - x_1}{x_2 - x_1}$; if we call that common value $t$, then the given parametric equations yield the point $(x, y)$; and any $(x, y)$ on the line between $P_1(x_1, y_1)$ and $P_2(x_2, y_2)$ yields a value of $t$ in $[0, 1]$. So the given parametric equations exactly specify the line segment from $P_1(x_1, y_1)$ to $P_2(x_2, y_2)$.

(b) $x = -2 + [3 - (-2)]t = -2 + 5t$ and $y = 7 + (-1 - 7)t = 7 - 8t$ for $0 \leq t \leq 1$.

**28.** For the side of the triangle from $A$ to $B$, use $(x_1, y_1) = (1, 1)$ and $(x_2, y_2) = (4, 2)$. Hence, the equations are

$$x = x_1 + (x_2 - x_1)\,t = 1 + (4 - 1)\,t = 1 + 3t,$$

$$y = y_1 + (y_2 - y_1)\,t = 1 + (2 - 1)\,t = 1 + t.$$

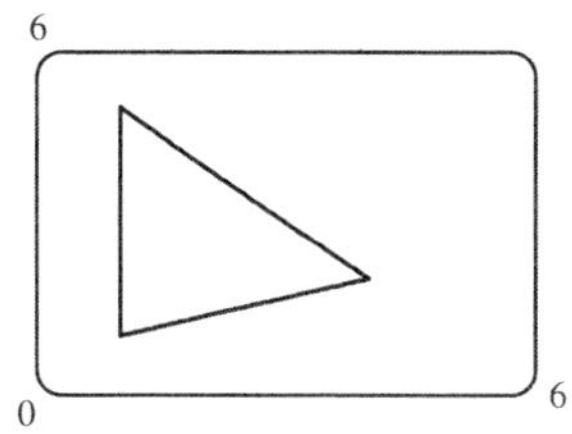


Graphing $x = 1 + 3t$ and $y = 1 + t$ with $0 \leq t \leq 1$ gives us the side of the triangle from $A$ to $B$. Similarly, for the side $BC$ we use $x = 4 - 3t$ and $y = 2 + 3t$, and for the side $AC$ we use $x = 1$ and $y = 1 + 4t$.

**29.** The circle $x^2 + y^2 = 4$ can be represented parametrically by $x = 2\cos t$, $y = 2\sin t$; $0 \leq t \leq 2\pi$. The circle $x^2 + (y-1)^2 = 4$ can be represented by $x = 2\cos t$, $y = 1 + 2\sin t$; $0 \leq t \leq 2\pi$. This representation gives us the circle with a counterclockwise orientation starting at $(2, 1)$.

(a) To get a clockwise orientation, we could change the equations to $x = 2\cos t$, $y = 1 - 2\sin t$, $0 \leq t \leq 2\pi$.

(b) To get three times around in the counterclockwise direction, we use the original equations $x = 2\cos t$, $y = 1 + 2\sin t$ with the domain expanded to $0 \leq t \leq 6\pi$.

(c) To start at $(0, 3)$ using the original equations, we must have $x_1 = 0$; that is, $2\cos t = 0$. Hence, $t = \frac{\pi}{2}$. So we use $x = 2\cos t$, $y = 1 + 2\sin t$; $\frac{\pi}{2} \leq t \leq \frac{3\pi}{2}$.

Alternatively, if we want $t$ to start at 0, we could change the equations of the curve. For example, we could use $x = -2\sin t$, $y = 1 + 2\cos t$, $0 \leq t \leq \pi$.

**30.** (a) Let $x^2/a^2 = \sin^2 t$ and $y^2/b^2 = \cos^2 t$ to obtain $x = a\sin t$ and $y = b\cos t$ with $0 \le t \le 2\pi$ as possible parametric equations for the ellipse $x^2/a^2 + y^2/b^2 = 1$.

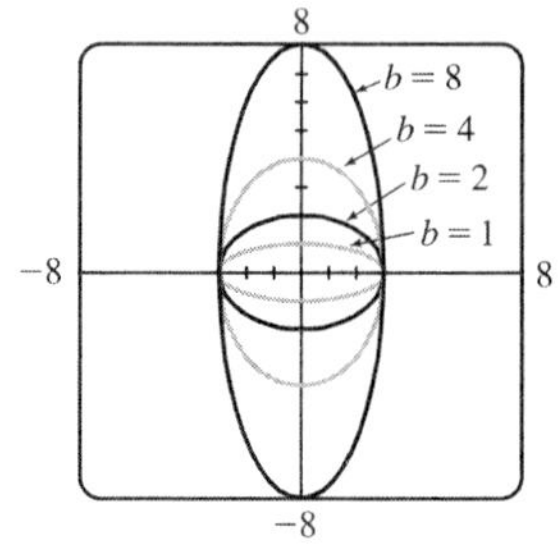


(b) The equations are $x = 3\sin t$ and $y = b\cos t$ for $b \in \{1, 2, 4, 8\}$.

(c) As $b$ increases, the ellipse stretches vertically.

**31.** *Big circle:* It's centered at $(2, 2)$ with a radius of 2, so by Example 4, parametric equations are

$$x = 2 + 2\cos t, \qquad y = 2 + 2\sin t, \qquad 0 \le t \le 2\pi$$

*Small circles:* They are centered at $(1, 3)$ and $(3, 3)$ with a radius of $0.1$. By Example 4, parametric equations are

$$\textit{(left)} \qquad x = 1 + 0.1\cos t, \qquad y = 3 + 0.1\sin t, \qquad 0 \le t \le 2\pi$$

and

$$\textit{(right)} \qquad x = 3 + 0.1\cos t, \qquad y = 3 + 0.1\sin t, \qquad 0 \le t \le 2\pi$$

*Semicircle:* It's the lower half of a circle centered at $(2, 2)$ with radius 1. By Example 4, parametric equations are

$$x = 2 + 1\cos t, \qquad y = 2 + 1\sin t, \qquad \pi \le t \le 2\pi$$

To get all four graphs on the same screen with a typical graphing calculator, we need to change the last $t$-interval to $[0, 2\pi]$ in order to match the others. We can do this by changing $t$ to $0.5t$. This change gives us the upper half. There are several ways to get the lower half—one is to change the "+" to a "−" in the $y$-assignment, giving us

$$x = 2 + 1\cos(0.5t), \qquad y = 2 - 1\sin(0.5t), \qquad 0 \le t \le 2\pi$$

**32.** If you are using a calculator or computer that can overlay graphs (using multiple $t$-intervals), the following is appropriate.

*Left side:* $x = 1$ and $y$ goes from 1.5 to 4, so use

$$x = 1, \qquad y = t, \qquad 1.5 \le t \le 4$$

*Right side:* $x = 10$ and $y$ goes from 1.5 to 4, so use

$$x = 10, \qquad y = t, \qquad 1.5 \le t \le 4$$

*Bottom:* $x$ goes from 1 to 10 and $y = 1.5$, so use

$$x = t, \qquad y = 1.5, \qquad 1 \le t \le 10$$

*Handle:* It starts at $(10, 4)$ and ends at $(13, 7)$, so use

$$x = 10 + t, \qquad y = 4 + t, \qquad 0 \le t \le 3$$

*Left wheel:* It's centered at $(3, 1)$, has a radius of 1, and appears to go about $30^\circ$ above the horizontal, so use

$$x = 3 + 1\cos t, \qquad y = 1 + 1\sin t, \qquad \tfrac{5\pi}{6} \le t \le \tfrac{13\pi}{6}$$

*Right wheel:* Similar to the left wheel with center $(8, 1)$, so use

$$x = 8 + 1\cos t, \qquad y = 1 + 1\sin t, \qquad \tfrac{5\pi}{6} \le t \le \tfrac{13\pi}{6}$$

If you are using a calculator or computer that cannot overlay graphs (using one $t$-interval), the following is appropriate. We'll start by picking the $t$-interval $[0, 2.5]$ since it easily matches the $t$-values for the two sides. We now need to find parametric equations for all graphs with $0 \le t \le 2.5$.

*Left side:* $x = 1$ and $y$ goes from 1.5 to 4, so use

$$x = 1, \qquad y = 1.5 + t, \qquad 0 \le t \le 2.5$$

*Right side:* $x = 10$ and $y$ goes from 1.5 to 4, so use

$$x = 10, \qquad y = 1.5 + t, \qquad 0 \le t \le 2.5$$

*Bottom:* $x$ goes from 1 to 10 and $y = 1.5$, so use

$$x = 1 + 3.6t, \qquad y = 1.5, \qquad 0 \le t \le 2.5$$

To get the $x$-assignment, think of creating a linear function such that when $t = 0$, $x = 1$ and when $t = 2.5$, $x = 10$. We can use the point-slope form of a line with $(t_1, x_1) = (0, 1)$ and $(t_2, x_2) = (2.5, 10)$.

$x - 1 = \dfrac{10 - 1}{2.5 - 0}(t - 0) \;\Rightarrow\; x = 1 + 3.6t.$

*Handle:* It starts at $(10, 4)$ and ends at $(13, 7)$, so use

$$x = 10 + 1.2t, \qquad y = 4 + 1.2t, \qquad 0 \le t \le 2.5$$

$(t_1, x_1) = (0, 10)$ and $(t_2, x_2) = (2.5, 13)$ gives us $x - 10 = \dfrac{13 - 10}{2.5 - 0}(t - 0) \;\Rightarrow\; x = 10 + 1.2t.$

$(t_1, y_1) = (0, 4)$ and $(t_2, y_2) = (2.5, 7)$ gives us $y - 4 = \dfrac{7 - 4}{2.5 - 0}(t - 0) \;\Rightarrow\; y = 4 + 1.2t.$

*Left wheel:* It's centered at $(3, 1)$, has a radius of 1, and appears to go about $30^\circ$ above the horizontal, so use

$$x = 3 + 1\cos\left(\tfrac{8\pi}{15}t + \tfrac{5\pi}{6}\right), \qquad y = 1 + 1\sin\left(\tfrac{8\pi}{15}t + \tfrac{5\pi}{6}\right), \qquad 0 \le t \le 2.5$$

$(t_1, \theta_1) = \left(0, \frac{5\pi}{6}\right)$ and $(t_2, \theta_2) = \left(\frac{5}{2}, \frac{13\pi}{6}\right)$ gives us $\theta - \frac{5\pi}{6} = \dfrac{\frac{13\pi}{6} - \frac{5\pi}{6}}{\frac{5}{2} - 0}(t - 0) \;\Rightarrow\; \theta = \frac{5\pi}{6} + \frac{8\pi}{15}t.$

*Right wheel:* Similar to the left wheel with center $(8, 1)$, so use

$$x = 8 + 1\cos\left(\tfrac{8\pi}{15}t + \tfrac{5\pi}{6}\right), \qquad y = 1 + 1\sin\left(\tfrac{8\pi}{15}t + \tfrac{5\pi}{6}\right), \qquad 0 \le t \le 2.5$$

**33.** (a) $x = t^3 \;\Rightarrow\; t = x^{1/3}$, so $y = t^2 = x^{2/3}$. We get the entire curve $y = x^{2/3}$ traversed in a left to right direction.

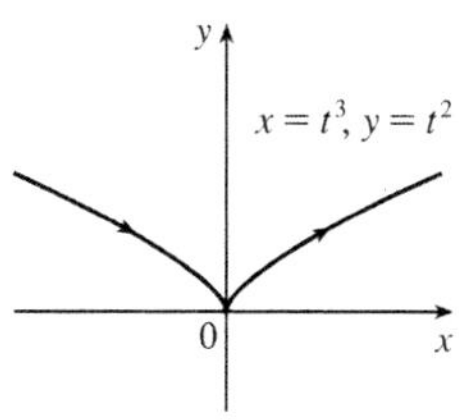


(b) $x = t^6 \;\Rightarrow\; t = x^{1/6}$, so $y = t^4 = x^{4/6} = x^{2/3}$. Since $x = t^6 \ge 0$, we only get the right half of the curve $y = x^{2/3}$.

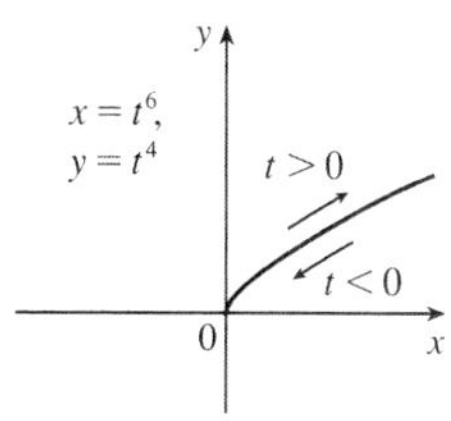

(c) $x = e^{-3t} = (e^{-t})^3$ [so $e^{-t} = x^{1/3}$],

$y = e^{-2t} = (e^{-t})^2 = (x^{1/3})^2 = x^{2/3}$.

If $t < 0$, then $x$ and $y$ are both larger than 1. If $t > 0$, then $x$ and $y$ are between 0 and 1. Since $x > 0$ and $y > 0$, the curve never quite reaches the origin.

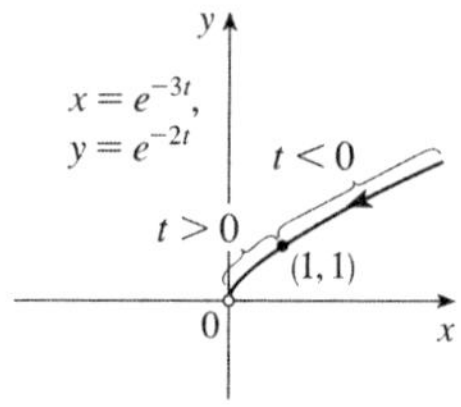


**34.** (a) $x = t$, so $y = t^{-2} = x^{-2}$. We get the entire curve $y = 1/x^2$ traversed in a left-to-right direction.

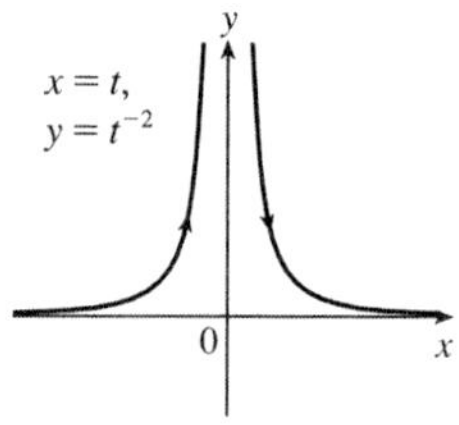


(b) $x = \cos t$, $y = \sec^2 t = \dfrac{1}{\cos^2 t} = \dfrac{1}{x^2}$. Since $\sec t \geq 1$, we only get the parts of the curve $y = 1/x^2$ with $y \geq 1$. We get the first quadrant portion of the curve when $x > 0$, that is, $\cos t > 0$, and we get the second quadrant portion of the curve when $x < 0$, that is, $\cos t < 0$.

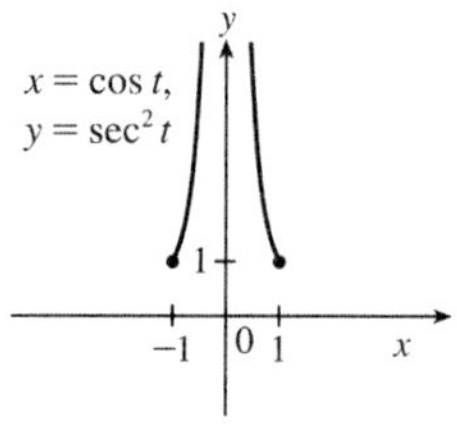


(c) $x = e^t$, $y = e^{-2t} = (e^t)^{-2} = x^{-2}$. Since $e^t$ and $e^{-2t}$ are both positive, we only get the first quadrant portion of the curve $y = 1/x^2$.

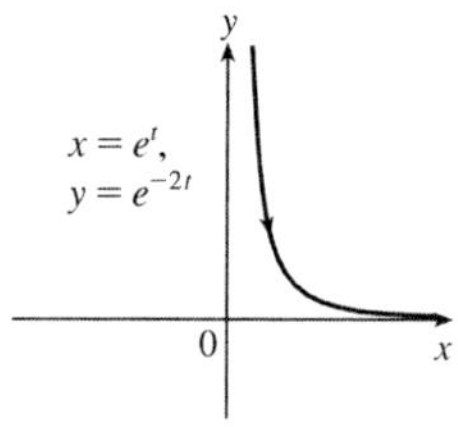


**35.** The case $\frac{\pi}{2} < \theta < \pi$ is illustrated. $C$ has coordinates $(r\theta, r)$ as in Example 8, and $Q$ has coordinates $(r\theta, r + r\cos(\pi - \theta)) = (r\theta, r(1 - \cos\theta))$ [since $\cos(\pi - \alpha) = \cos\pi\cos\alpha + \sin\pi\sin\alpha = -\cos\alpha$], so $P$ has coordinates $(r\theta - r\sin(\pi - \theta), r(1 - \cos\theta)) = (r(\theta - \sin\theta), r(1 - \cos\theta))$ [since $\sin(\pi - \alpha) = \sin\pi\cos\alpha - \cos\pi\sin\alpha = \sin\alpha$]. Again we have the parametric equations $x = r(\theta - \sin\theta)$, $y = r(1 - \cos\theta)$.

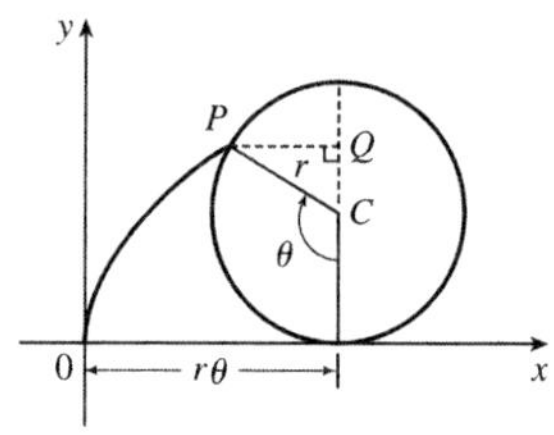


**36.** The first two diagrams depict the case $\pi < \theta < \frac{3\pi}{2}$, $d < r$. As in Example 8, $C$ has coordinates $(r\theta, r)$. Now $Q$ (in the second diagram) has coordinates $(r\theta, r + d\cos(\theta - \pi)) = (r\theta, r - d\cos\theta)$, so a typical point $P$ of the trochoid has coordinates $(r\theta + d\sin(\theta - \pi), r - d\cos\theta)$. That is, $P$ has coordinates $(x, y)$, where $x = r\theta - d\sin\theta$ and $y = r - d\cos\theta$. When $d = r$, these equations agree with those of the cycloid.

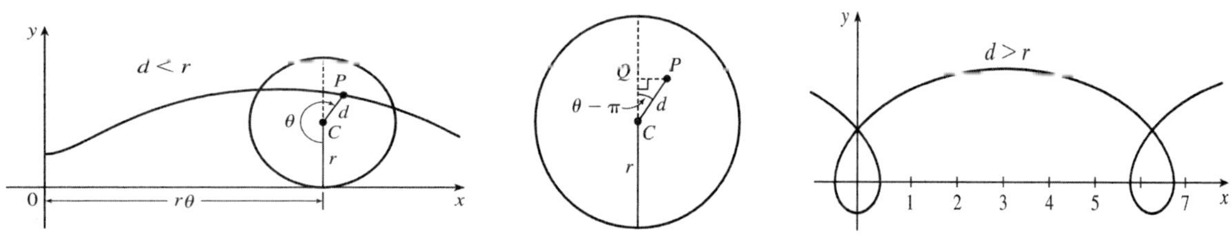

**37.** It is apparent that $x = |OQ|$ and $y = |QP| = |ST|$. From the diagram, $x = |OQ| = a\cos\theta$ and $y = |ST| = b\sin\theta$. Thus, the parametric equations are $x = a\cos\theta$ and $y = b\sin\theta$. To eliminate $\theta$ we rearrange: $\sin\theta = y/b \Rightarrow \sin^2\theta = (y/b)^2$ and $\cos\theta = x/a \Rightarrow \cos^2\theta = (x/a)^2$. Adding the two equations: $\sin^2\theta + \cos^2\theta = 1 = x^2/a^2 + y^2/b^2$. Thus, we have an ellipse.

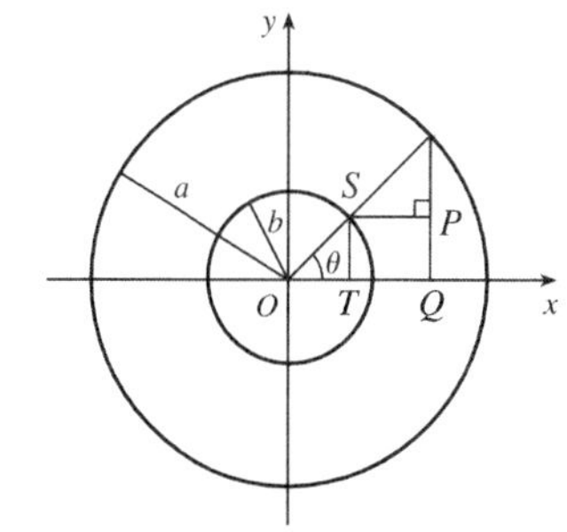


**38.** $C = (2a\cot\theta, 2a)$, so the $x$-coordinate of $P$ is $x = 2a\cot\theta$. Let $B = (0, 2a)$. Then $\angle OAB$ is a right angle and $\angle OBA = \theta$, so $|OA| = 2a\sin\theta$ and $A = ((2a\sin\theta)\cos\theta, (2a\sin\theta)\sin\theta)$. Thus, the $y$-coordinate of $P$ is $y = 2a\sin^2\theta$.

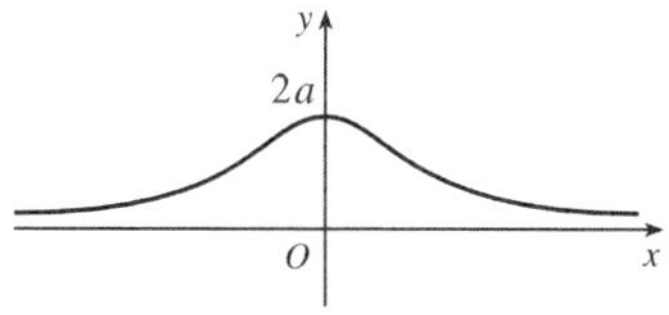


**39.** (a)

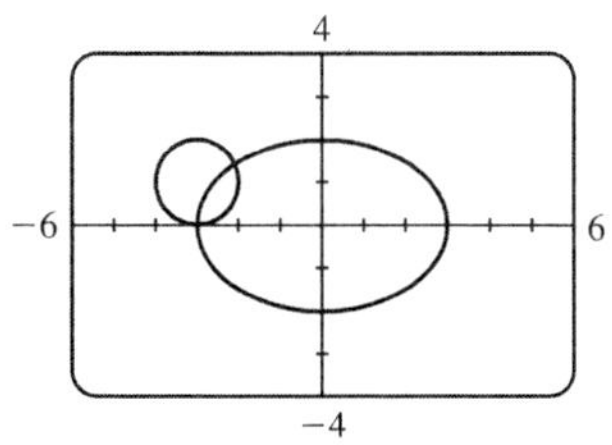


There are 2 points of intersection: $(-3, 0)$ and approximately $(-2.1, 1.4)$.

(b) A collision point occurs when $x_1 = x_2$ and $y_1 = y_2$ for the same $t$. So solve the equations:

$$3\sin t = -3 + \cos t \quad \textbf{(1)}$$
$$2\cos t = 1 + \sin t \quad \textbf{(2)}$$

From **(2)**, $\sin t = 2\cos t - 1$. Substituting into **(1)**, we get $3(2\cos t - 1) = -3 + \cos t \Rightarrow 5\cos t = 0 \;(*) \Rightarrow \cos t = 0 \Rightarrow t = \frac{\pi}{2}$ or $\frac{3\pi}{2}$. We check that $t = \frac{3\pi}{2}$ satisfies **(1)** and **(2)** but $t = \frac{\pi}{2}$ does not. So the only collision point occurs when $t = \frac{3\pi}{2}$, and this gives the point $(-3, 0)$. [We could check our work by graphing $x_1$ and $x_2$ together as functions of $t$ and, on another plot, $y_1$ and $y_2$ as functions of $t$. If we do so, we see that the only value of $t$ for which *both* pairs of graphs intersect is $t = \frac{3\pi}{2}$.]

(c) The circle is centered at $(3, 1)$ instead of $(-3, 1)$. There are still 2 intersection points: $(3, 0)$ and $(2.1, 1.4)$, but there are no collision points, since $(*)$ in part (b) becomes $5\cos t = 6 \Rightarrow \cos t = \frac{6}{5} > 1$.

**40.** (a) If $\alpha = 30°$ and $v_0 = 500$ m/s, then the equations become $x = (500\cos 30°)t = 250\sqrt{3}t$ and $y = (500\sin 30°)t - \frac{1}{2}(9.8)t^2 = 250t - 4.9t^2$. $y = 0$ when $t = 0$ (when the gun is fired) and again when $t = \frac{250}{4.9} \approx 51$ s. Then $x = \left(250\sqrt{3}\right)\left(\frac{250}{4.9}\right) \approx 22{,}092$ m, so the bullet hits the ground about 22 km from the gun. The formula for $y$ is quadratic in $t$. To find the maximum $y$-value, we will complete the square:

$$y = -4.9\left(t^2 - \tfrac{250}{4.9}t\right) = -4.9\left[t^2 - \tfrac{250}{4.9}t + \left(\tfrac{125}{4.9}\right)^2\right] + \tfrac{125^2}{4.9} = -4.9\left(t - \tfrac{125}{4.9}\right)^2 + \tfrac{125^2}{4.9} \le \tfrac{125^2}{4.9}$$

with equality when $t = \frac{125}{4.9}$ s, so the maximum height attained is $\frac{125^2}{4.9} \approx 3189$ m.

(b)

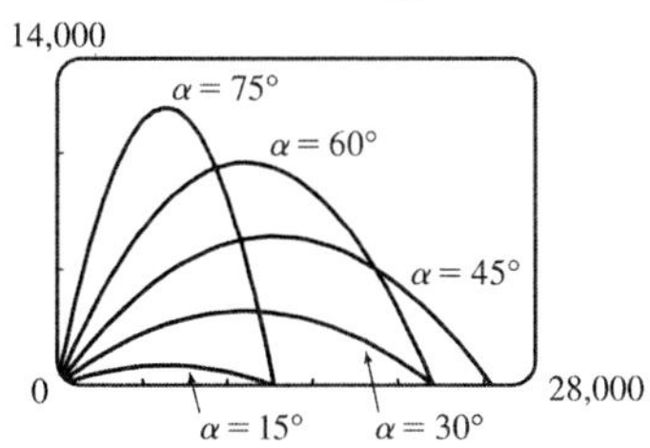


As $\alpha$ $(0° < \alpha < 90°)$ increases up to 45°, the projectile attains a greater height and a greater range. As $\alpha$ increases past 45°, the projectile attains a greater height, but its range decreases.

(c) $x = (v_0 \cos\alpha)t \Rightarrow t = \dfrac{x}{v_0 \cos\alpha}$.

$$y = (v_0 \sin\alpha)t - \tfrac{1}{2}gt^2 \Rightarrow y = (v_0 \sin\alpha)\frac{x}{v_0 \cos\alpha} - \frac{g}{2}\left(\frac{x}{v_0 \cos\alpha}\right)^2 = (\tan\alpha)x - \left(\frac{g}{2v_0^2 \cos^2\alpha}\right)x^2,$$ which is the equation of a parabola (quadratic in $x$).

**41.** $x = t^2, y = t^3 - ct$. We use a graphing device to produce the graphs for various values of $c$ with $-\pi \le t \le \pi$. Note that all the members of the family are symmetric about the $x$-axis. For $c < 0$, the graph does not cross itself, but for $c = 0$ it has a cusp at $(0, 0)$ and for $c > 0$ the graph crosses itself at $x = c$, so the loop grows larger as $c$ increases.

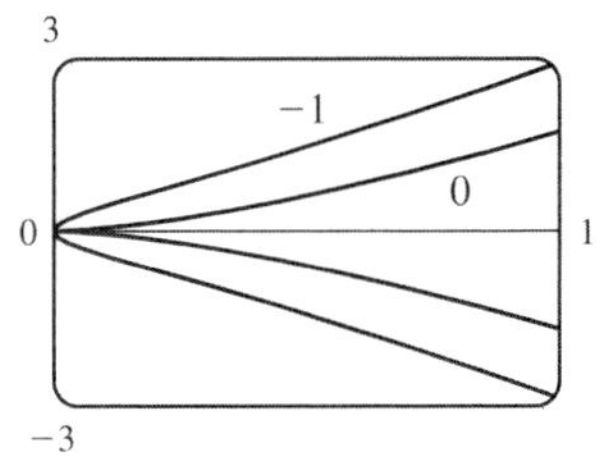


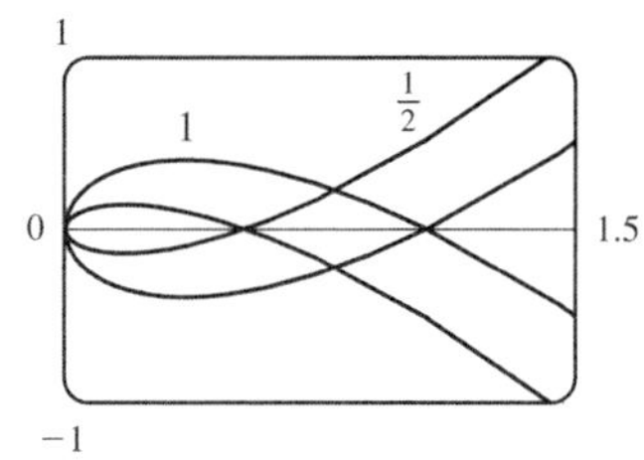


**42.** $x = 2ct - 4t^3, y = -ct^2 + 3t^4$. We use a graphing device to produce the graphs for various values of $c$ with $-\pi \le t \le \pi$. Note that all the members of the family are symmetric about the $y$-axis. When $c < 0$, the graph resembles that of a polynomial of even degree, but when $c = 0$ there is a corner at the origin, and when $c > 0$, the graph crosses itself at the origin, and has two cusps below the $x$-axis. The size of the "swallowtail" increases as $c$ increases.

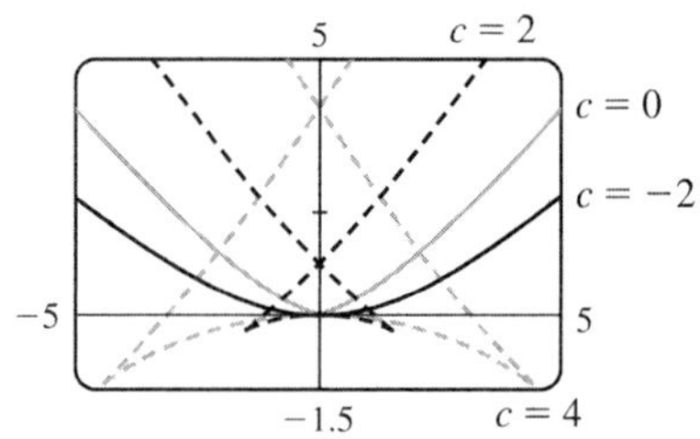


**43.** Note that all the Lissajous figures are symmetric about the $x$-axis. The parameters $a$ and $b$ simply stretch the graph in the $x$- and $y$-directions respectively. For $a = b = n = 1$ the graph is simply a circle with radius 1. For $n = 2$ the graph crosses itself at the origin and there are loops above and below the $x$-axis. In general, the figures have $n - 1$ points of intersection, all of which are on the $y$-axis, and a total of $n$ closed loops.

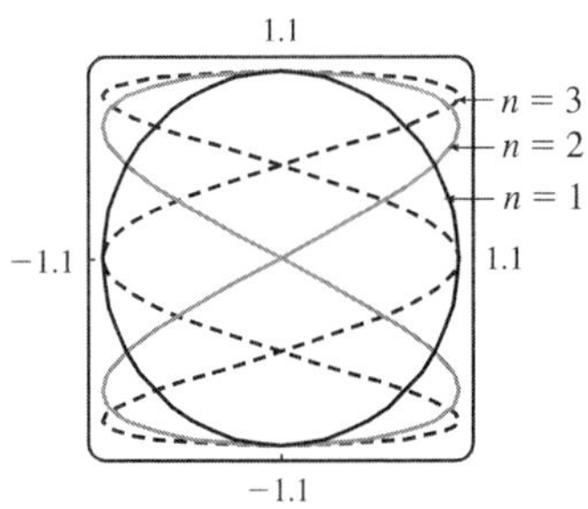


$a = b = 1$

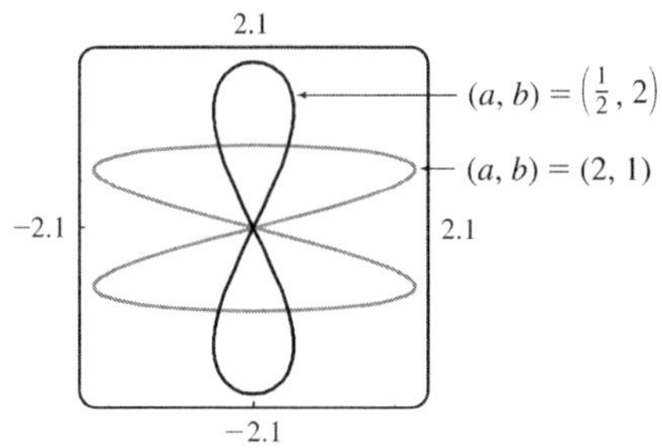


$n = 2$

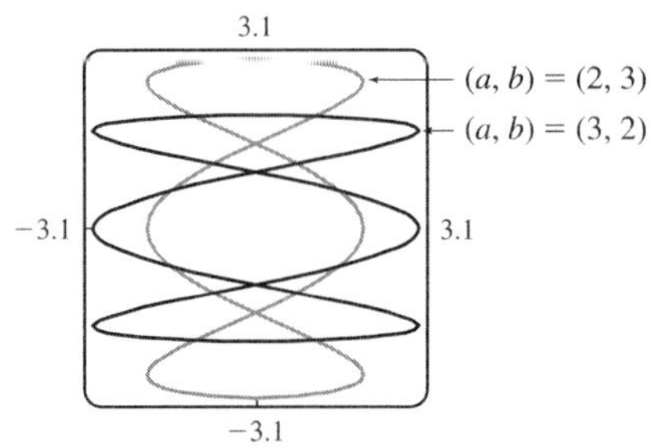


$n = 3$

**44.** We use $-\pi \le t \le \pi$ in the viewing rectangle $[-4, 2] \times [-3, 3]$. We first observe that for $c = 0$, we obtain a circle with center $\left(-\frac{1}{2}, 0\right)$ and radius $\frac{1}{2}$. As the value of $c$ increases, there is a larger outer loop and a smaller inner loop until $c = 1$, when we obtain a curve with a dent (called a **cardioid**). As $c$ increases, we get a curve with a dimple (called a **limaçon**) until $c = 2$. For $c > 2$, we have convex limaçons. For negative values of $c$, we obtain the same graphs as for positive $c$, but with different values of $t$ corresponding to the points on the curve.

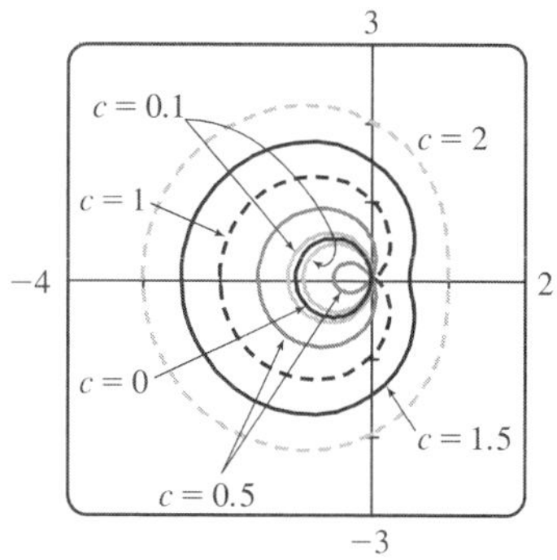


## LABORATORY PROJECT Running Circles Around Circles

**1.** The center $Q$ of the smaller circle has coordinates $((a-b)\cos\theta, (a-b)\sin\theta)$. Arc $PS$ on circle $C$ has length $a\theta$ since it is equal in length to arc $AS$ (the smaller circle rolls without slipping against the larger.) Thus, $\angle PQS = \frac{a}{b}\theta$ and $\angle PQT = \frac{a}{b}\theta - \theta$, so $P$ has coordinates

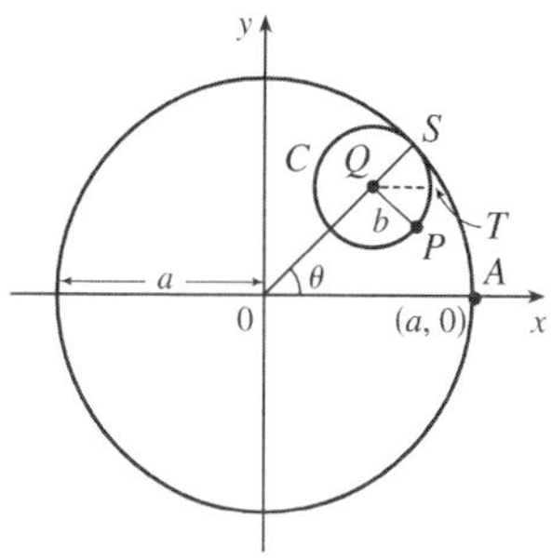


$$x = (a-b)\cos\theta + b\cos(\angle PQT) = (a-b)\cos\theta + b\cos\left(\frac{a-b}{b}\theta\right)$$

and

$$y = (a-b)\sin\theta - b\sin(\angle PQT) = (a-b)\sin\theta - b\sin\left(\frac{a-b}{b}\theta\right)$$

**2.** With $b = 1$ and $a$ a positive integer greater than 2, we obtain a hypocycloid of $a$ cusps. Shown in the figure is the graph for $a = 4$. Let $a = 4$ and $b = 1$. Using the sum identities to expand $\cos 3\theta$ and $\sin 3\theta$, we obtain

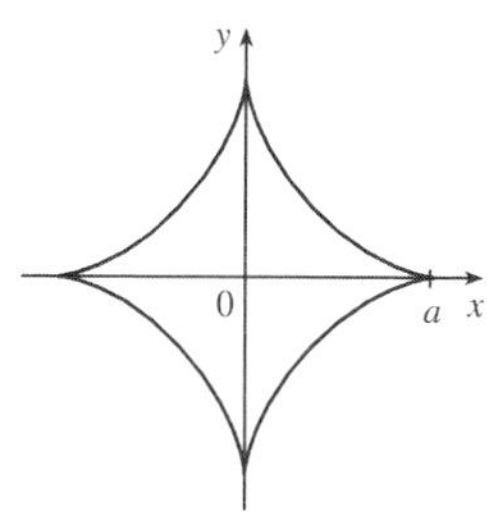


$$x = 3\cos\theta + \cos 3\theta = 3\cos\theta + \left(4\cos^3\theta - 3\cos\theta\right) = 4\cos^3\theta$$

and

$$y = 3\sin\theta - \sin 3\theta = 3\sin\theta - \left(3\sin\theta - 4\sin^3\theta\right) = 4\sin^3\theta$$

**3.** The following graphs are obtained with $b = 1$ and $a = \frac{1}{2}, \frac{1}{3}, \frac{1}{4}$, and $\frac{1}{10}$ with $-2\pi \le \theta \le 2\pi$. We conclude that as the denominator $d$ increases, the graph gets smaller, but maintains the basic shape shown.

Letting $d = 2$ and $n = 3$, 5, and 7 with $-2\pi \le \theta \le 2\pi$ gives us the following:

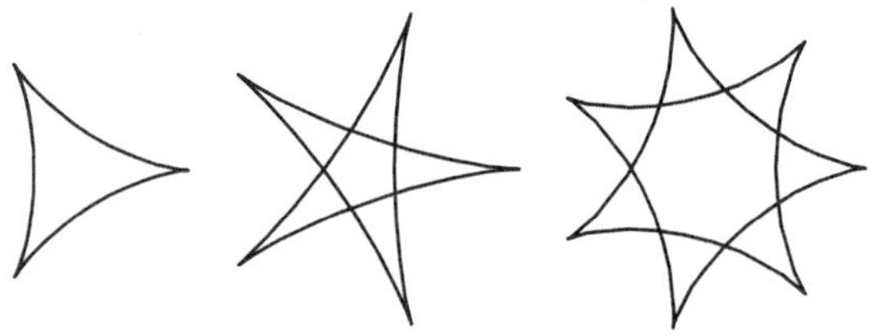

So if $d$ is held constant and $n$ varies, we get a graph with $n$ cusps (assuming $n/d$ is in lowest form). When $n = d + 1$, we obtain a hypocycloid of $n$ cusps. As $n$ increases, we must expand the range of $\theta$ in order to get a closed curve. The following graphs have $a = \frac{3}{2}, \frac{5}{4}$, and $\frac{11}{10}$.

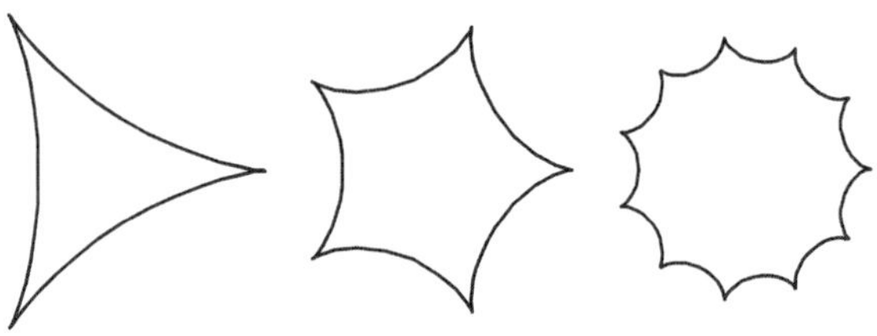

**4.** If $b = 1$, the equations for the hypocycloid are

$$x = (a-1)\cos\theta + \cos((a-1)\theta) \qquad y = (a-1)\sin\theta - \sin((a-1)\theta)$$

which is a hypocycloid of $a$ cusps (from Problem 2). In general, if $a > 1$, we get a figure with cusps on the "outside ring" and if $a < 1$, the cusps are on the "inside ring". In any case, as the values of $\theta$ get larger, we get a figure that looks more and more like a washer. If we were to graph the hypocycloid for all values of $\theta$, every point on the washer would eventually be arbitrarily close to a point on the curve.

$a = \sqrt{2}$

$-10\pi \le \theta \le 10\pi$

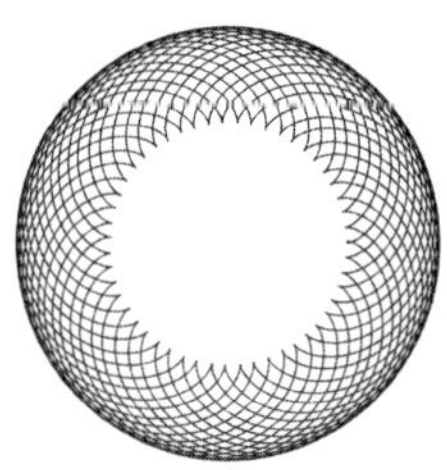

$a = e - 2$

$0 \le \theta \le 446$

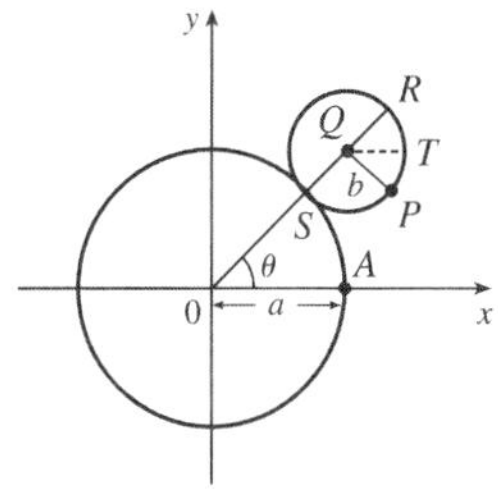


**5.** The center $Q$ of the smaller circle has coordinates $((a+b)\cos\theta, (a+b)\sin\theta)$. Arc $PS$ has length $a\theta$ (as in Problem 1), so that $\angle PQS = \dfrac{a\theta}{b}$, $\angle PQR = \pi - \dfrac{a\theta}{b}$, and $\angle PQT = \pi - \dfrac{a\theta}{b} - \theta = \pi - \left(\dfrac{a+b}{b}\right)\theta$ since $\angle RQT = \theta$. Thus, the coordinates of $P$ are

$$x = (a+b)\cos\theta + b\cos\left(\pi - \frac{a+b}{b}\theta\right) = (a+b)\cos\theta - b\cos\left(\frac{a+b}{b}\theta\right)$$

and

$$y = (a+b)\sin\theta - b\sin\left(\pi - \frac{a+b}{b}\theta\right) = (a+b)\sin\theta - b\sin\left(\frac{a+b}{b}\theta\right).$$

**6.** Let $b = 1$ and the equations become

$$x = (a+1)\cos\theta - \cos((a+1)\theta) \qquad y = (a+1)\sin\theta - \sin((a+1)\theta)$$

If $a = 1$, we have a cardioid. If $a$ is a positive integer greater than 1, we get the graph of an "$a$-leafed clover", with cusps that are $a$ units from the origin. (Some of the pairs of figures are not to scale.)

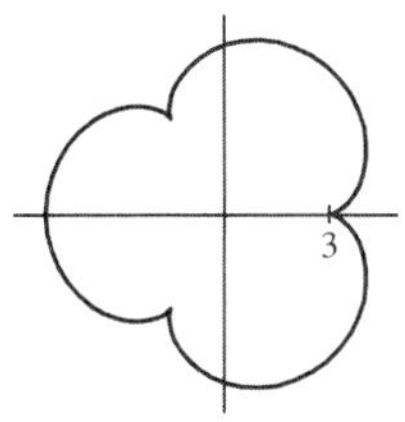

$a = 3$, $-2\pi \le \theta \le 2\pi$

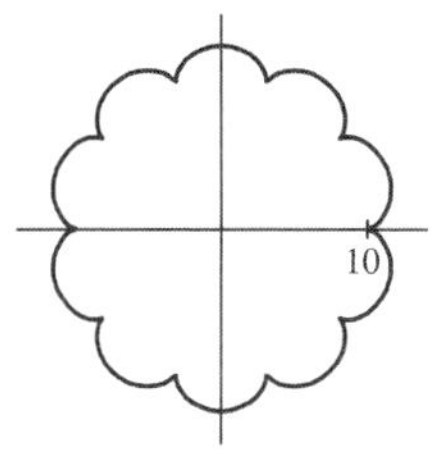

$a = 10$, $-2\pi \le \theta \le 2\pi$

If $a = n/d$ with $n = 1$, we obtain a figure that does not increase in size and requires $-d\pi \le \theta \le d\pi$ to be a closed curve traced exactly once.

$a = \frac{1}{4}$, $-4\pi \le \theta \le 4\pi$

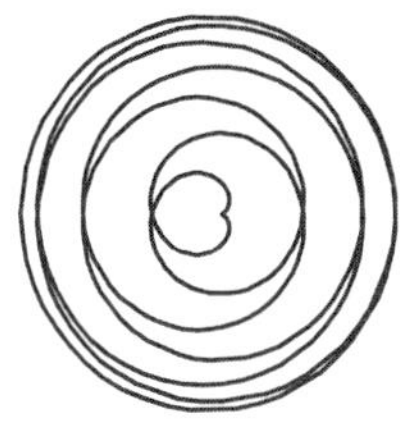
$a = \frac{1}{7}$, $-7\pi \le \theta \le 7\pi$

Next, we keep $d$ constant and let $n$ vary. As $n$ increases, so does the size of the figure. There is an $n$-pointed star in the middle.

$a = \frac{2}{5}$, $-5\pi \le \theta \le 5\pi$

$a = \frac{7}{5}$, $-5\pi \le \theta \le 5\pi$

Now if $n = d + 1$ we obtain figures similar to the previous ones, but the size of the figure does not increase.

$a = \frac{4}{3}$, $-3\pi \le \theta \le 3\pi$

$a = \frac{7}{6}$, $-6\pi \le \theta \le 6\pi$

If $a$ is irrational, we get washers that increase in size as $a$ increases.

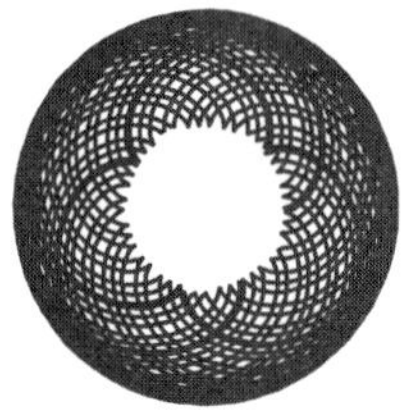

$a = \sqrt{2}, 0 \leq \theta \leq 200$

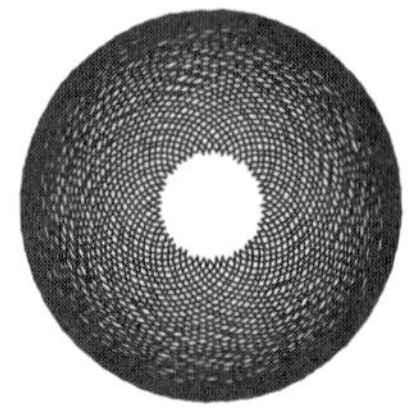

$a = e - 2, 0 \leq \theta \leq 446$

## 1 Review

### CONCEPT CHECK

1. (a) A **function** $f$ is a rule that assigns to each element $x$ in a set $A$ exactly one element, called $f(x)$, in a set $B$. The set $A$ is called the **domain** of the function. The **range** of $f$ is the set of all possible values of $f(x)$ as $x$ varies throughout the domain.

   (b) If $f$ is a function with domain $A$, then its **graph** is the set of ordered pairs $\{(x, f(x)) \mid x \in A\}$.

   (c) Use the Vertical Line Test on page 17.

2. The four ways to represent a function are: verbally, numerically, visually, and algebraically. An example of each is given below.

   **Verbally:** An assignment of students to chairs in a classroom (a description in words)
   **Numerically:** A tax table that assigns an amount of tax to an income (a table of values)
   **Visually:** A graphical history of the Dow Jones average (a graph)
   **Algebraically:** A relationship between distance, rate, and time: $d = rt$ (an explicit formula)

3. (a) An **even function** $f$ satisfies $f(-x) = f(x)$ for every number $x$ in its domain. It is symmetric with respect to the $y$-axis.

   (b) An **odd function** $g$ satisfies $g(-x) = -g(x)$ for every number $x$ in its domain. It is symmetric with respect to the origin.

4. A function $f$ is called **increasing** on an interval $I$ if $f(x_1) < f(x_2)$ whenever $x_1 < x_2$ in $I$.

5. A **mathematical model** is a mathematical description (often by means of a function or an equation) of a real-world phenomenon.

6. (a) Linear function: $f(x) = 2x + 1$, $f(x) = ax + b$

   (b) Power function: $f(x) = x^2$, $f(x) = x^a$

   (c) Exponential function: $f(x) = 2^x$, $f(x) = a^x$

   (d) Quadratic function: $f(x) = x^2 + x + 1$, $f(x) = ax^2 + bx + c$

   (e) Polynomial of degree 5: $f(x) = x^5 + 2$

   (f) Rational function: $f(x) = \dfrac{x}{x+2}$, $f(x) = \dfrac{P(x)}{Q(x)}$ where $P(x)$ and $Q(x)$ are polynomials

7. 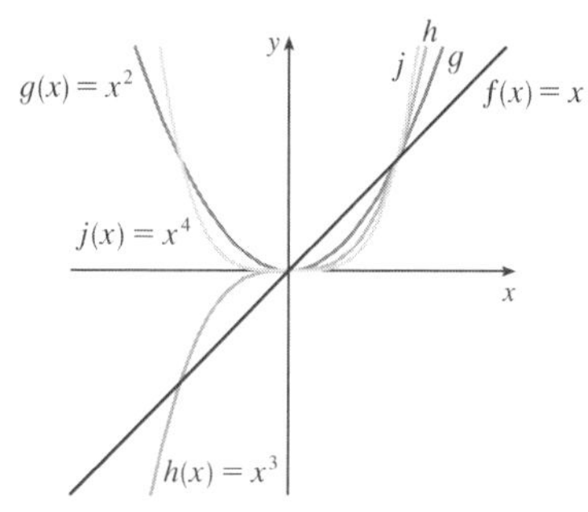


8. (a)

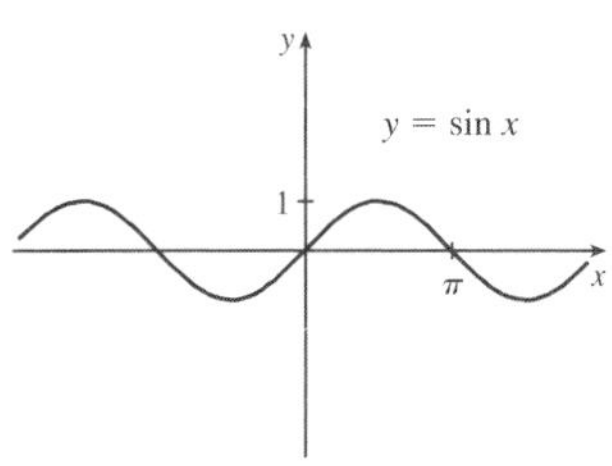


(b)

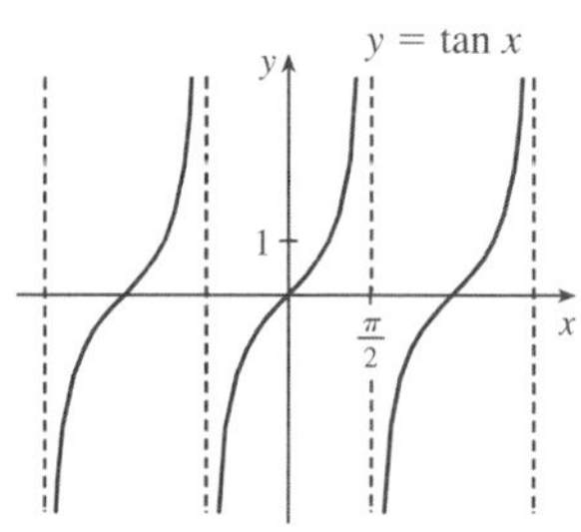


(c)

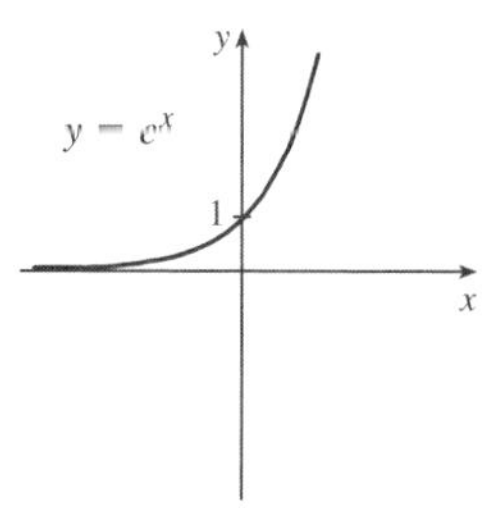

(d)

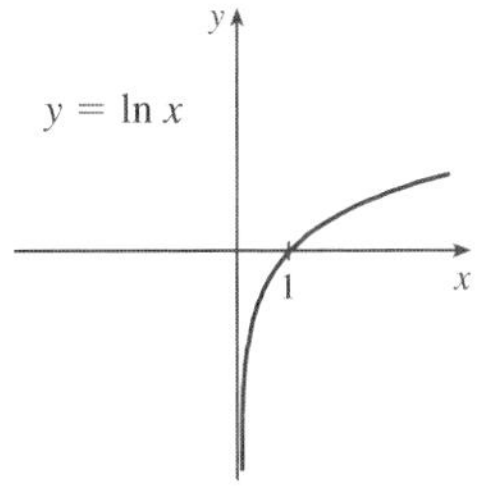


(e)

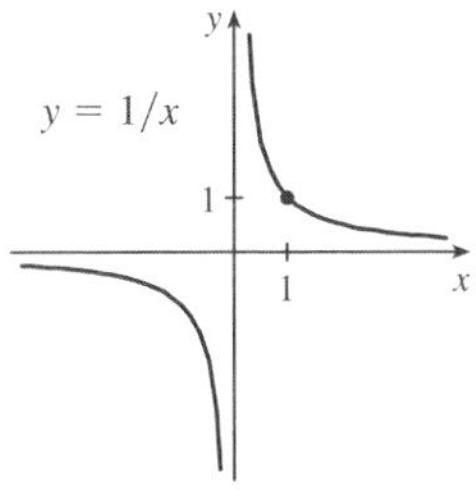


(f)

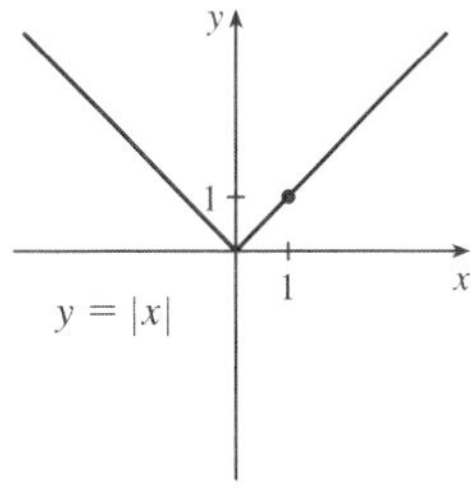


(g)

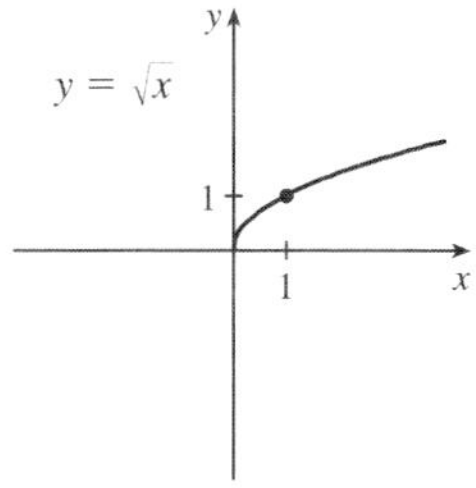


**9.** (a) The domain of $f + g$ is the intersection of the domain of $f$ and the domain of $g$; that is, $A \cap B$.

(b) The domain of $fg$ is also $A \cap B$.

(c) The domain of $f/g$ must exclude values of $x$ that make $g$ equal to 0; that is, $\{x \in A \cap B \mid g(x) \neq 0\}$.

**10.** Given two functions $f$ and $g$, the **composite** function $f \circ g$ is defined by $(f \circ g)(x) = f(g(x))$. The domain of $f \circ g$ is the set of all $x$ in the domain of $g$ such that $g(x)$ is in the domain of $f$.

**11.** (a) If the graph of $f$ is shifted 2 units upward, its equation becomes $y = f(x) + 2$.

(b) If the graph of $f$ is shifted 2 units downward, its equation becomes $y = f(x) - 2$.

(c) If the graph of $f$ is shifted 2 units to the right, its equation becomes $y = f(x - 2)$.

(d) If the graph of $f$ is shifted 2 units to the left, its equation becomes $y = f(x + 2)$.

(e) If the graph of $f$ is reflected about the $x$-axis, its equation becomes $y = -f(x)$.

(f) If the graph of $f$ is reflected about the $y$-axis, its equation becomes $y = f(-x)$.

(g) If the graph of $f$ is stretched vertically by a factor of 2, its equation becomes $y = 2f(x)$.

(h) If the graph of $f$ is shrunk vertically by a factor of 2, its equation becomes $y = \frac{1}{2}f(x)$.

(i) If the graph of $f$ is stretched horizontally by a factor of 2, its equation becomes $y = f(\frac{1}{2})x$.

(j) If the graph of $f$ is shrunk horizontally by a factor of 2, its equation becomes $y = f(2x)$.

**12.** (a) A function $f$ is called a *one-to-one function* if it never takes on the same value twice; that is, if $f(x_1) \neq f(x_2)$ whenever $x_1 \neq x_2$. (Or, $f$ is 1-1 if each output corresponds to only one input.)

Use the Horizontal Line Test: A function is one-to-one if and only if no horizontal line intersects its graph more than once.

(b) If $f$ is a one-to-one function with domain $A$ and range $B$, then its *inverse function* $f^{-1}$ has domain $B$ and range $A$ and is defined by

$$f^{-1}(y) = x \quad \Leftrightarrow \quad f(x) = y$$

for any $y$ in $B$. The graph of $f^{-1}$ is obtained by reflecting the graph of $f$ about the line $y = x$.

**13.** (a) A parametric curve is a set of points of the form $(x, y) = (f(t), g(t))$, where $f$ and $g$ are continuous functions of a variable $t$.

(b) Sketching a parametric curve, like sketching the graph of a function, is difficult to do in general. We can plot points on the curve by finding $f(t)$ and $g(t)$ for various values of $t$, either by hand or with a calculator or computer. Sometimes, when $f$ and $g$ are given by formulas, we can eliminate $t$ from the equations $x = f(t)$ and $y = g(t)$ to get a Cartesian equation relating $x$ and $y$. It may be easier to graph that equation than to work with the original formulas for $x$ and $y$ in terms of $t$.

(c) See the margin note on page 74.

## TRUE-FALSE QUIZ

**1.** False. Let $f(x) = x^2$, $s = -1$, and $t = 1$. Then $f(s + t) = (-1 + 1)^2 = 0^2 = 0$, but $f(s) + f(t) = (-1)^2 + 1^2 = 2 \neq 0 = f(s + t)$.

**2.** False. Let $f(x) = x^2$. Then $f(-2) = 4 = f(2)$, but $-2 \neq 2$.

**3.** False. Let $f(x) = x^2$. Then $f(3x) = (3x)^2 = 9x^2$ and $3f(x) = 3x^2$. So $f(3x) \neq 3f(x)$.

**4.** True. If $x_1 < x_2$ and $f$ is a decreasing function, then the $y$-values get smaller as we move from left to right. Thus, $f(x_1) > f(x_2)$.

**5.** True. See the Vertical Line Test.

**6.** False. Let $f(x) = x^2$ and $g(x) = 2x$. Then $(f \circ g)(x) = f(g(x)) = f(2x) = (2x)^2 = 4x^2$ and $(g \circ f)(x) = g(f(x)) = g(x^2) = 2x^2$. So $f \circ g \neq g \circ f$.

**7.** False. Let $f(x) = x^3$. Then $f$ is one-to-one and $f^{-1}(x) = \sqrt[3]{x}$. But $1/f(x) = 1/x^3$, which is not equal to $f^{-1}(x)$.

**8.** True. We can divide by $e^x$ since $e^x \neq 0$ for every $x$.

**9.** True. The function $\ln x$ is an increasing function on $(0, \infty)$.

**10.** False. Let $x = e$. Then $(\ln x)^6 = (\ln e)^6 = 1^6 = 1$, but $6 \ln x = 6 \ln e = 6 \cdot 1 = 6 \neq 1 = (\ln x)^6$.

**11.** False. Let $x = e^2$ and $a = e$. Then $\dfrac{\ln x}{\ln a} = \dfrac{\ln e^2}{\ln e} = \dfrac{2 \ln e}{\ln e} = 2$ and $\ln \dfrac{x}{a} = \ln \dfrac{e^2}{e} = \ln e = 1$, so in general the statement is false. What *is* true, however, is that $\ln \dfrac{x}{a} = \ln x - \ln a$.

**12.** False. The first pair of equations yields the portion of the parabola $y = x^2$ with $x \geq 0$, whereas the second pair of equations traces out the whole parabola $y = x^2$.

## EXERCISES

**1.** (a) When $x = 2$, $y \approx 2.7$. Thus, $f(2) \approx 2.7$.

(b) $f(x) = 3 \quad \Rightarrow \quad x \approx 2.3, 5.6$

(c) The domain of $f$ is $-6 \leq x \leq 6$, or $[-6, 6]$.

(d) The range of $f$ is $-4 \leq y \leq 4$, or $[-4, 4]$.

(e) $f$ is increasing on $[-4, 4]$, that is, on $-4 \leq x \leq 4$.

(f) $f$ is not one-to-one since it fails the Horizontal Line Test.

(g) $f$ is odd since its graph is symmetric about the origin.

**2.** (a) When $x = 2$, $y = 3$. Thus, $g(2) = 3$.

(b) $g$ is one-to-one because it passes the Horizontal Line Test.

(c) When $y = 2$, $x \approx 0.2$. So $g^{-1}(2) \approx 0.2$.

(d) The range of $g$ is $[-1, 3.5]$, which is the same as the domain of $g^{-1}$.

(e) We reflect the graph of $g$ through the line $y = x$ to obtain the graph of $g^{-1}$.

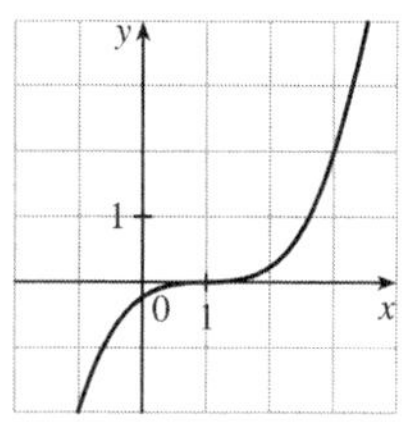


**3.** (a)

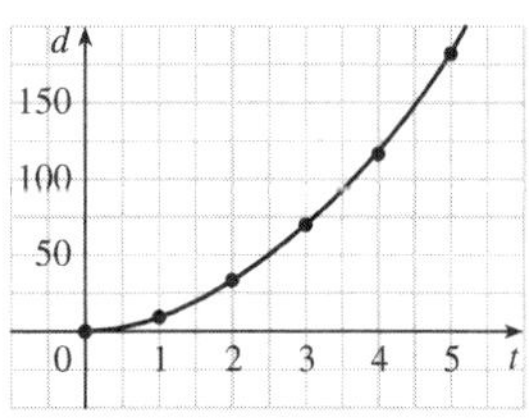


(b) From the graph, we see that the distance traveled after 4.5 seconds is slightly less than 150 feet.

**4.** There will be some yield with no fertilizer, increasing yields with increasing fertilizer use, a leveling-off of yields at some point, and disaster with too much fertilizer use.

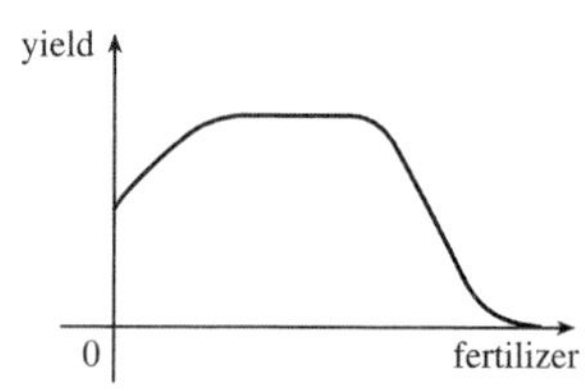

**5.** $f(x) = \sqrt{4 - 3x^2}$. Domain: $4 - 3x^2 \geq 0 \Rightarrow 3x^2 \leq 4 \Rightarrow x^2 \leq \frac{4}{3} \Rightarrow |x| \leq \frac{2}{\sqrt{3}}$.

Range: $y \geq 0$ and $y \leq \sqrt{4} \Rightarrow 0 \leq y \leq 2$.

**6.** $g(x) = \dfrac{1}{x+1}$. Domain: $x + 1 \neq 0 \Rightarrow x \neq -1$. Range: all reals except 0 ($y = 0$ is the horizontal asymptote for $g$.)

**7.** $y = 1 + \sin x$. Domain: $\mathbb{R}$. Range: $-1 \leq \sin x \leq 1 \Rightarrow 0 \leq 1 + \sin x \leq 2 \Rightarrow 0 \leq y \leq 2$.

**8.** $y = \ln \ln x$. Domain: We must have $\ln x > 0 \Rightarrow x > e^0 \Rightarrow x > 1$. Range: $\ln x > 0$, so $\ln(\ln x)$ takes on all real numbers and, hence, the range is $\mathbb{R}$.

**9.** (a) To obtain the graph of $y = f(x) + 8$, we shift the graph of $y = f(x)$ up 8 units.

(b) To obtain the graph of $y = f(x + 8)$, we shift the graph of $y = f(x)$ left 8 units.

(c) To obtain the graph of $y = 1 + 2f(x)$, we stretch the graph of $y = f(x)$ vertically by a factor of 2, and then shift the resulting graph 1 unit upward.

(d) To obtain the graph of $y = f(x - 2) - 2$, we shift the graph of $y = f(x)$ right 2 units (for the "$-2$" inside the parentheses), and then shift the resulting graph 2 units downward.

(e) To obtain the graph of $y = -f(x)$, we reflect the graph of $y = f(x)$ about the $x$-axis.

(f) To obtain the graph of $y = f^{-1}(x)$, we reflect the graph of $y = f(x)$ about the line $y = x$ (assuming $f$ is one–to-one).

**10.** (a) To obtain the graph of $y = f(x - 8)$, we shift the graph of $y = f(x)$ right 8 units.

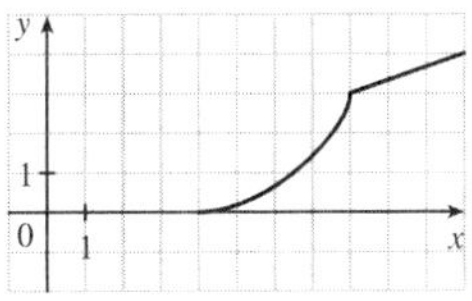


(b) To obtain the graph of $y = -f(x)$, we reflect the graph of $y = f(x)$ about the $x$-axis.

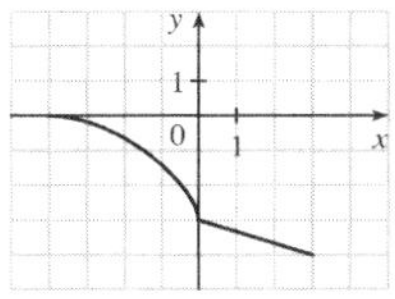


(c) To obtain the graph of $y = 2 - f(x)$, we reflect the graph of $y = f(x)$ about the $x$-axis, and then shift the resulting graph 2 units upward.

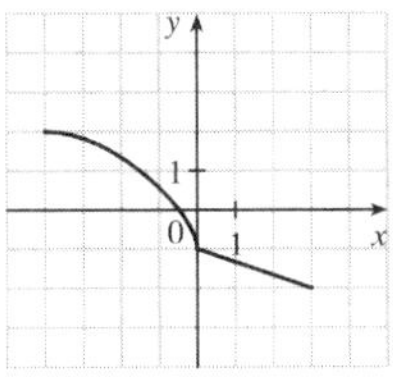


(d) To obtain the graph of $y = \frac{1}{2}f(x) - 1$, we shrink the graph of $y = f(x)$ by a factor of 2, and then shift the resulting graph 1 unit downward.

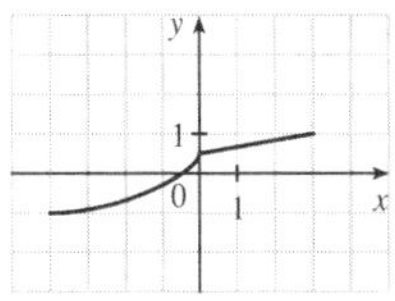


(e) To obtain the graph of $y = f^{-1}(x)$, we reflect the graph of $y = f(x)$ about the line $y = x$.

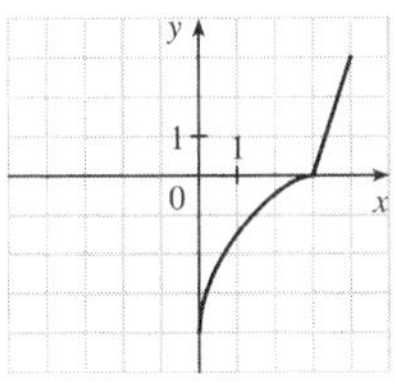


(f) To obtain the graph of $y = f^{-1}(x + 3)$, we reflect the graph of $y = f(x)$ about the line $y = x$ [see part (e)], and then shift the resulting graph left 3 units.

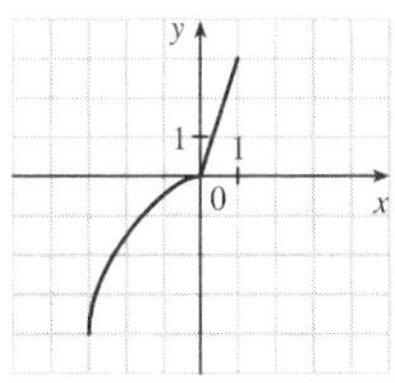

**11.** $y = -\sin 2x$: Start with the graph of $y = \sin x$, compress horizontally by a factor of 2, and reflect about the $x$-axis.

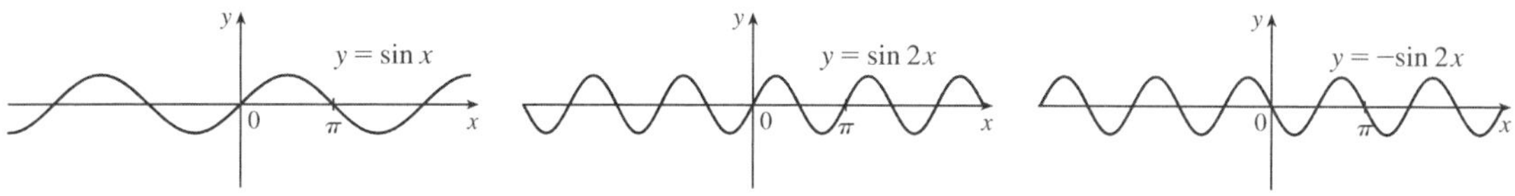


**12.** $y = 3\ln(x-2)$: Start with the graph of $y = \ln x$, shift 2 units to the right, and stretch vertically by a factor of 3.

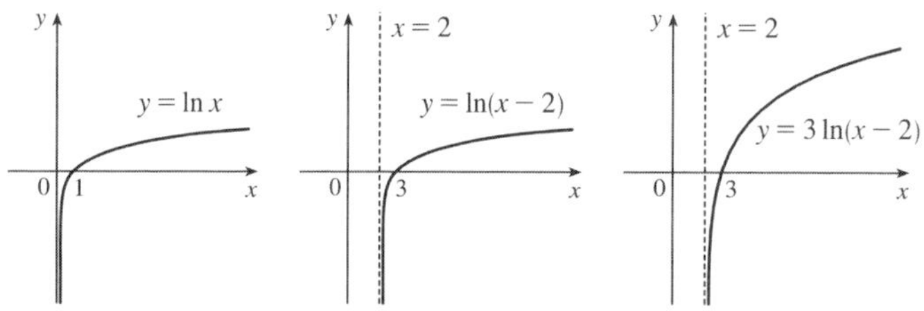


**13.** $y = (1 + e^x)/2$: Start with the graph of $y = e^x$, shift 1 unit upward, and compress vertically by a factor of 2.

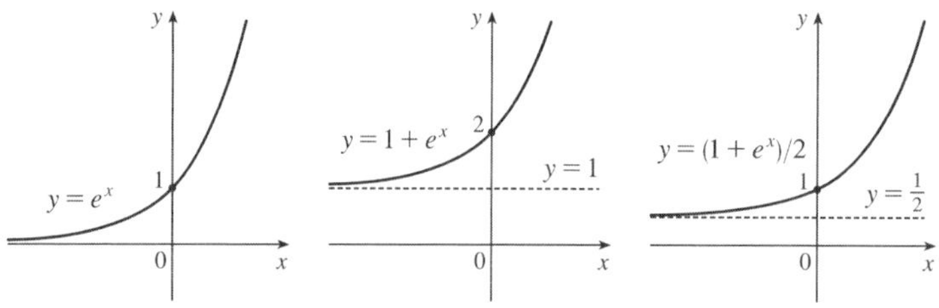


**14.** $y = 2 - \sqrt{x}$: Start with the graph of $y = \sqrt{x}$, reflect about the $x$-axis, and shift 2 units upward.

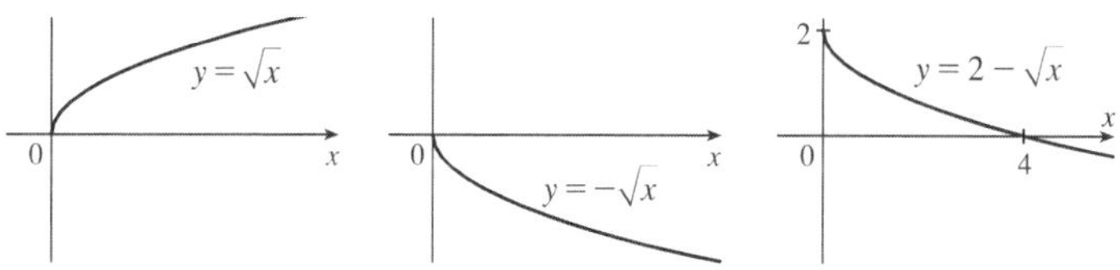


**15.** $f(x) = \dfrac{1}{x+2}$:

Start with the graph of $f(x) = 1/x$ and shift 2 units to the left.

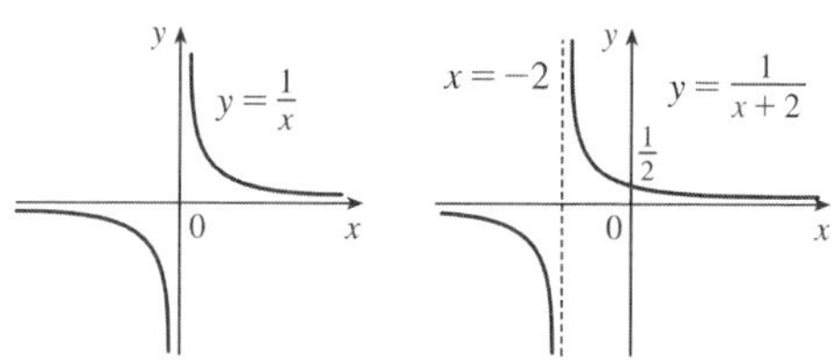


**16.** $f(x) = \begin{cases} -x & \text{if } x < 0 \\ e^x - 1 & \text{if } x \geq 0 \end{cases}$

On $(-\infty, 0)$, graph $y = -x$ (the line with slope $-1$ and $y$-intercept 0) with open endpoint $(0, 0)$.

On $[0, \infty)$, graph $y = e^x - 1$ (the graph of $y = e^x$ shifted 1 unit downward) with closed endpoint $(0, 0)$.

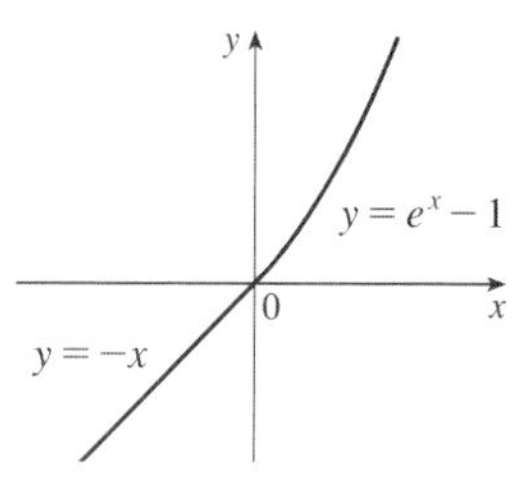

**17.** (a) The terms of $f$ are a mixture of odd and even powers of $x$, so $f$ is neither even nor odd.

(b) The terms of $f$ are all odd powers of $x$, so $f$ is odd.

(c) $f(-x) = e^{-(-x)^2} = e^{-x^2} = f(x)$, so $f$ is even.

(d) $f(-x) = 1 + \sin(-x) = 1 - \sin x$. Now $f(-x) \neq f(x)$ and $f(-x) \neq -f(x)$, so $f$ is neither even nor odd.

**18.** For the line segment from $(-2, 2)$ to $(-1, 0)$, the slope is $\dfrac{0-2}{-1+2} = -2$, and an equation is $y - 0 = -2\,(x+1)$ or, equivalently, $y = -2x - 2$. The circle has equation $x^2 + y^2 = 1$; the top half has equation $y = \sqrt{1-x^2}$ (we have solved for positive $y$.) Thus, $f(x) = \begin{cases} -2x - 2 & \text{if } -2 \leq x \leq -1 \\ \sqrt{1-x^2} & \text{if } -1 < x \leq 1 \end{cases}$

**19.** $f(x) = \ln x, \quad D = (0, \infty); \quad g(x) = x^2 - 9, \quad D = \mathbb{R}$.

$(f \circ g)(x) = f(g(x)) = f(x^2 - 9) = \ln(x^2 - 9)$.

Domain: $x^2 - 9 > 0 \;\Rightarrow\; x^2 > 9 \;\Rightarrow\; |x| > 3 \;\Rightarrow\; x \in (-\infty, -3) \cup (3, \infty)$

$(g \circ f)(x) = g(f(x)) = g(\ln x) = (\ln x)^2 - 9$. Domain: $x > 0$, or $(0, \infty)$

$(f \circ f)(x) = f(f(x)) = f(\ln x) = \ln(\ln x)$. Domain: $\ln x > 0 \;\Rightarrow\; x > e^0 = 1$, or $(1, \infty)$

$(g \circ g)(x) = g(g(x)) = g(x^2 - 9) = (x^2 - 9)^2 - 9$. Domain: $x \in \mathbb{R}$, or $(-\infty, \infty)$

**20.** Let $h(x) = x + \sqrt{x}$, $g(x) = \sqrt{x}$, and $f(x) = 1/x$. Then $(f \circ g \circ h)(x) = \dfrac{1}{\sqrt{x + \sqrt{x}}} = F(x)$.

**21.**

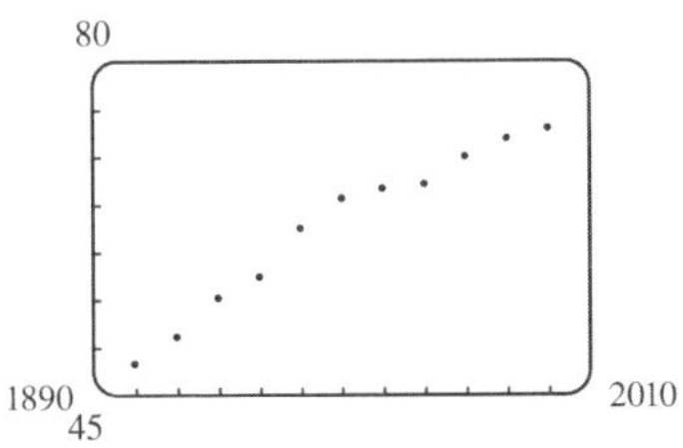


Many models appear to be plausible. Your choice depends on whether you think medical advances will keep increasing life expectancy, or if there is bound to be a natural leveling-off of life expectancy. A linear model, $y = 0.2493x - 423.4818$, gives us an estimate of 77.6 years for the year 2010.

**22.** (a) Let $x$ denote the number of toaster ovens produced in one week and $y$ the associated cost. Using the points $(1000, 9000)$ and $(1500, 12{,}000)$, we get an equation of a line:

$$y - 9000 = \frac{12{,}000 - 9000}{1500 - 1000}\,(x - 1000) \;\Rightarrow$$

$$y = 6\,(x - 1000) + 9000 \;\Rightarrow\; y = 6x + 3000.$$

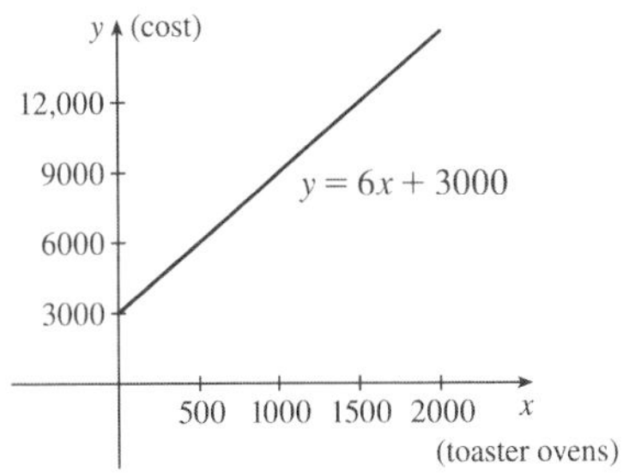


(b) The slope of 6 means that each additional toaster oven produced adds \$6 to the weekly production cost.

(c) The $y$-intercept of 3000 represents the overhead cost—the cost incurred without producing anything.

**23.** We need to know the value of $x$ such that $f(x) = 2x + \ln x = 2$. Since $x = 1$ gives us $y = 2$, $f^{-1}(2) = 1$.

**24.** $y = \dfrac{x+1}{2x+1}$. Interchanging $x$ and $y$ gives us $x = \dfrac{y+1}{2y+1} \;\Rightarrow\; 2xy + x = y + 1 \;\Rightarrow\; 2xy - y = 1 - x \;\Rightarrow$

$y(2x - 1) = 1 - x \;\Rightarrow\; y = \dfrac{1-x}{2x-1} = f^{-1}(x)$.

**25.** (a) $e^{2\ln 3} = \left(e^{\ln 3}\right)^2 = 3^2 = 9$

(b) $\log_{10} 25 + \log_{10} 4 = \log_{10}(25 \cdot 4) = \log_{10} 100 = \log_{10} 10^2 = 2$

**26.** (a) $e^x = 5 \;\;\Rightarrow\;\; x = \ln 5$

(b) $\ln x = 2 \;\;\Rightarrow\;\; x = e^2$

(c) $e^{e^x} = 2 \;\;\Rightarrow\;\; e^x = \ln 2 \;\;\Rightarrow\;\; x = \ln(\ln 2)$

**27.** (a) After 4 days, $\frac{1}{2}$ gram remains; after 8 days, $\frac{1}{4}$ g; after 12 days, $\frac{1}{8}$ g; after 16 days, $\frac{1}{16}$ g.

(b) $m(4) = \dfrac{1}{2}, m(8) = \dfrac{1}{2^2}, m(12) = \dfrac{1}{2^3}, m(16) = \dfrac{1}{2^4}$. From the pattern, we see that $m(t) = \dfrac{1}{2^{t/4}}$, or $2^{-t/4}$.

(c) $m = 2^{-t/4} \;\;\Rightarrow\;\; \log_2 m = -t/4 \;\;\Rightarrow\;\; t = -4\log_2 m$; this is the time elapsed when there are $m$ grams of $^{100}$Pd.

(d) $m = 0.01 \;\;\Rightarrow\;\; t = -4\log_2 0.01 = -4\left(\dfrac{\ln 0.01}{\ln 2}\right) \approx 26.6$ days

**28.** (a)

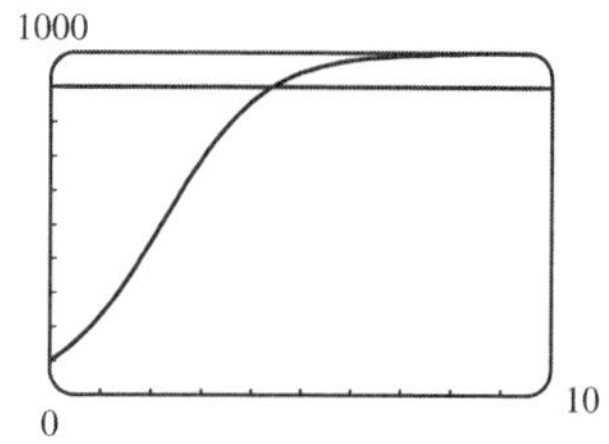


The population would reach 900 in about 4.4 years.

(b) $P = \dfrac{100{,}000}{100 + 900e^{-t}} \;\Rightarrow\; 100P + 900Pe^{-t} = 100{,}000 \Rightarrow\; 900Pe^{-t} = 100{,}000 - 100P \;\Rightarrow$

$e^{-t} = \dfrac{100{,}000 - 100P}{900P} \;\Rightarrow\; -t = \ln\left(\dfrac{1000 - P}{9P}\right) \;\Rightarrow\; t = -\ln\left(\dfrac{1000 - P}{9P}\right)$, or $\ln\left(\dfrac{9P}{1000 - P}\right)$; this is the time required for the population to reach a given number $P$.

(c) $P = 900 \;\;\Rightarrow\;\; t = \ln\left(\dfrac{9 \cdot 900}{1000 - 900}\right) = \ln 81 \approx 4.4$ years, as in part (a).

**29.** $f(x) = \ln(x^2 - c)$. If $c < 0$, the domain of $f$ is $\mathbb{R}$. If $c = 0$, the domain of $f$ is $(-\infty, 0) \cup (0, \infty)$. If $c > 0$, the domain of $f$ is $(-\infty, -\sqrt{c}) \cup (\sqrt{c}, \infty)$. As $c$ increases, the dip at $x = 0$ becomes deeper. For $c \geq 0$, the graph has asymptotes at $x = \pm\sqrt{c}$.

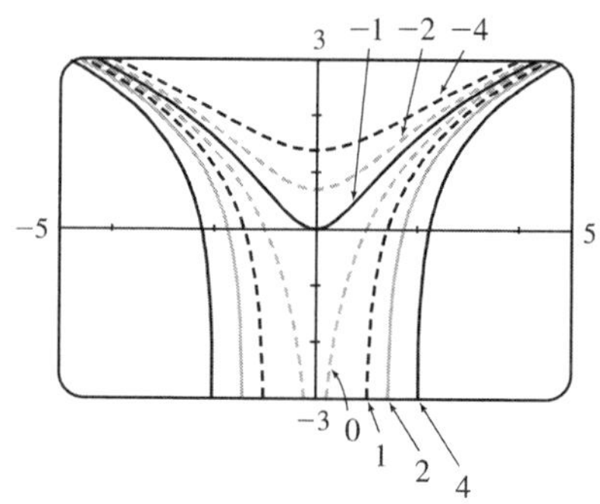


**30.**

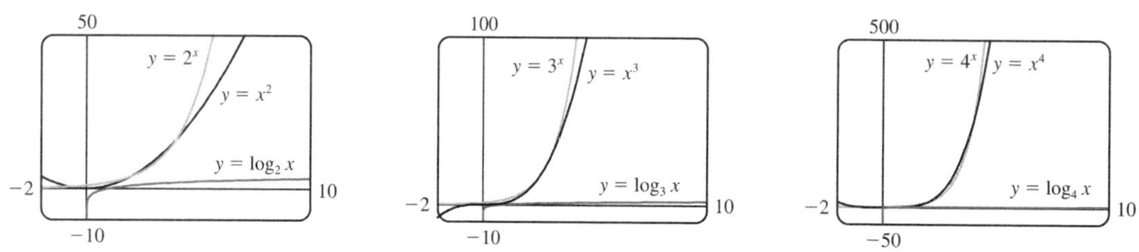


For large values of $x$, $y = a^x$ has the largest $y$-values and $y = \log_a x$ has the smallest $y$-values. This makes sense because they are inverses of each other.

**31.** (a)

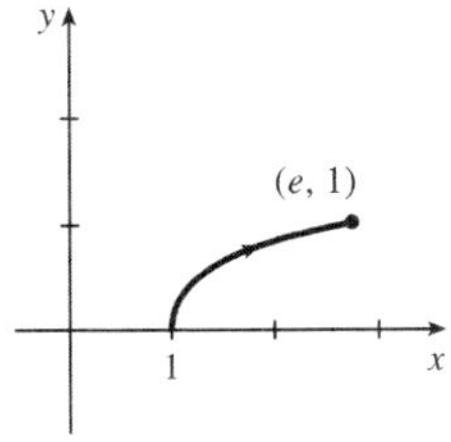


$0 \le t \le 1 \quad \Rightarrow \quad 0 \le y \le 1$ and $1 \le x \le e$.

(b) $x = e^t \quad \Rightarrow \quad t = \ln x$; $y = \sqrt{t}$ so $y = \sqrt{\ln x}$.

**32.** (a) $(x-2)^2 + y^2 = 4 \Rightarrow \dfrac{(x-2)^2}{4} + \dfrac{y^2}{4} = 1$. Let $\dfrac{(x-2)^2}{4} = \sin^2 t$ and $\dfrac{y^2}{4} = \cos^2 t$ (since $\sin^2 t + \cos^2 t = 1$). Solving for $x$ and $y$ gives $x = 2 \pm 2\sin t$ and $y = \pm 2\cos t$. We want to move from $(2, 2)$ to $(2, -2)$ and pass through $(0, 0)$. When $t = 0$, we want $y = 2$, so choose $y = 2\cos t$. When $t = \frac{\pi}{2}$, we want $x = 0$, so choose $x = 2 - 2\sin t$. Thus, parametric equations are $x = 2 - 2\sin t$, $y = 2\cos t$, $0 \le t \le \pi$. Another possibility is $x = 2 + 2\cos t$, $y = 2\sin t$, $\frac{\pi}{2} \le t \le \frac{3\pi}{2}$.

(b)

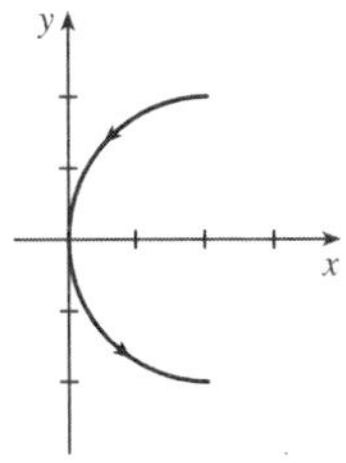


**33.** We sketch $x = t$, $y = 2t + \ln t$ (the function) and$x = 2t + \ln t$, $y = t$ (its inverse) for $t > 0$.

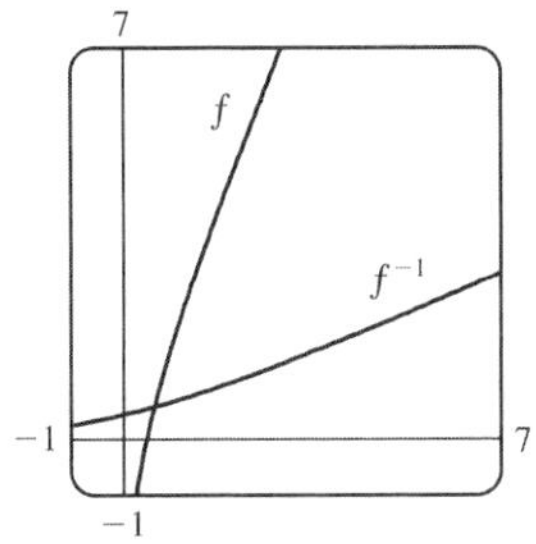


**34.** (a) Let $\theta$ be the angle of inclination of segment $OP$. Then $|OB| = \dfrac{2a}{\cos\theta}$.

Let $C = (2a, 0)$. Then by use of right triangle $OAC$ we see that $|OA| = 2a\cos\theta$.

Now

$$|OP| = |AB| = |OB| - |OA| = 2a\left(\frac{1}{\cos\theta} - \cos\theta\right)$$

$$= 2a\frac{1-\cos^2\theta}{\cos\theta} = 2a\frac{\sin^2\theta}{\cos\theta} = 2a\sin\theta\tan\theta$$

So $P$ has coordinates $x = 2a\sin\theta\tan\theta \cdot \cos\theta = 2a\sin^2\theta$ and $y = 2a\sin\theta\tan\theta \cdot \sin\theta = 2a\sin^2\theta\tan\theta$.

(b)

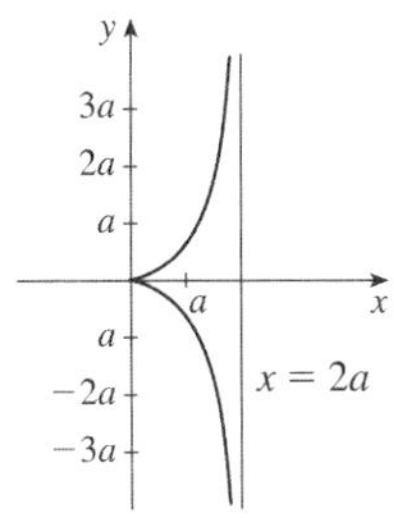

## PRINCIPLES OF PROBLEM SOLVING

1. 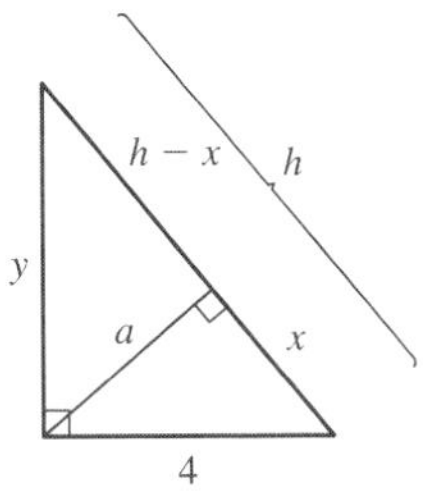


By using the area formula for a triangle, $\frac{1}{2}$ (base) (height), in two ways, we see that $\frac{1}{2}(4)(y) = \frac{1}{2}(h)(a)$, so $a = \dfrac{4y}{h}$. Since $4^2 + y^2 = h^2$, $y = \sqrt{h^2 - 16}$, and $a = \dfrac{4\sqrt{h^2 - 16}}{h}$.

2. 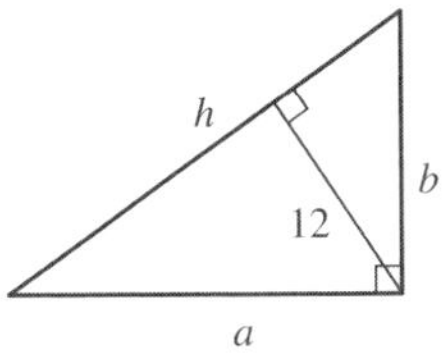


Refer to Example 1, where we obtained $h = \dfrac{P^2 - 100}{2P}$. The 100 came from 4 times the area of the triangle. In this case, the area of the triangle is $\frac{1}{2}(h)(12) = 6h$. Thus, $h = \dfrac{P^2 - 4(6h)}{2P} \;\Rightarrow\; 2Ph = P^2 - 24h \;\Rightarrow$ $2Ph + 24h = P^2 \;\Rightarrow\; h(2P + 24) = P^2 \;\Rightarrow\; h = \dfrac{P^2}{2P + 24}$.

3. $|2x - 1| = \begin{cases} 2x - 1 & \text{if } x \geq \frac{1}{2} \\ 1 - 2x & \text{if } x < \frac{1}{2} \end{cases}$ and $|x + 5| = \begin{cases} x + 5 & \text{if } x \geq -5 \\ -x - 5 & \text{if } x < -5 \end{cases}$

Therefore, we consider the three cases $x < -5$, $-5 \leq x < \frac{1}{2}$, and $x \geq \frac{1}{2}$.

If $x < -5$, we must have $1 - 2x - (-x - 5) = 3 \;\Leftrightarrow\; x = 3$, which is false, since we are considering $x < -5$.

If $-5 \leq x < \frac{1}{2}$, we must have $1 - 2x - (x + 5) = 3 \;\Leftrightarrow\; x = -\frac{7}{3}$.

If $x \geq \frac{1}{2}$, we must have $2x - 1 - (x + 5) = 3 \;\Leftrightarrow\; x = 9$.

So the two solutions of the equation are $x = -\frac{7}{3}$ and $x = 9$.

4. $|x - 1| = \begin{cases} x - 1 & \text{if } x \geq 1 \\ 1 - x & \text{if } x < 1 \end{cases}$ and $|x - 3| = \begin{cases} x - 3 & \text{if } x \geq 3 \\ 3 - x & \text{if } x < 3 \end{cases}$

Therefore, we consider the three cases $x < 1$, $1 \leq x < 3$, and $x \geq 3$.

If $x < 1$, we must have $1 - x - (3 - x) \geq 5 \;\Leftrightarrow\; 0 \geq 7$, which is false.

If $1 \leq x < 3$, we must have $x - 1 - (3 - x) \geq 5 \;\Leftrightarrow\; x \geq \frac{9}{2}$, which is false because $x < 3$.

If $x \geq 3$, we must have $x - 1 - (x - 3) \geq 5 \;\Leftrightarrow\; 2 \geq 5$, which is false.

All three cases lead to falsehoods, so the inequality has no solution.

5. $f(x) = \left|x^2 - 4|x| + 3\right|$. If $x \geq 0$, then $f(x) = \left|x^2 - 4x + 3\right| = |(x - 1)(x - 3)|$.

*Case (i):* If $0 < x \leq 1$, then $f(x) = x^2 - 4x + 3$.

*Case (ii):* If $1 < x \leq 3$, then $f(x) = -(x^2 - 4x + 3) = -x^2 + 4x - 3$.

*Case (iii):* If $x > 3$, then $f(x) = x^2 - 4x + 3$.

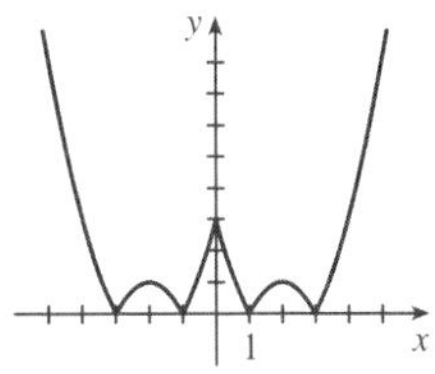


This enables us to sketch the graph for $x \geq 0$. Then we use the fact that $f$ is an even function to reflect this part of the graph about the $y$-axis to obtain the entire graph. Or, we could consider also the cases $x < -3$, $-3 \leq x < -1$, and $-1 \leq x < 0$.

**6.** $g(x) = |x^2 - 1| - |x^2 - 4|$.

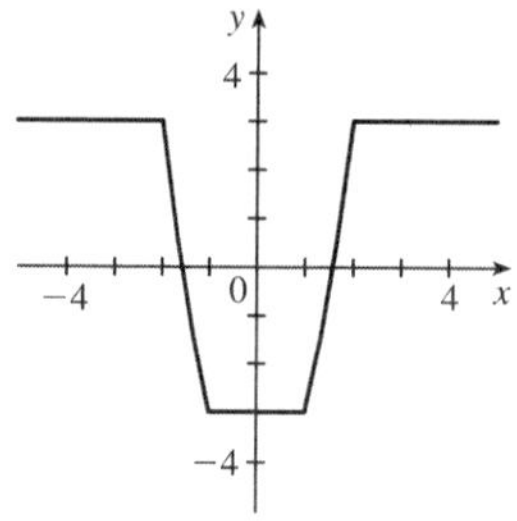


$$|x^2 - 1| = \begin{cases} x^2 - 1 & \text{if } |x| \geq 1 \\ 1 - x^2 & \text{if } |x| < 1 \end{cases} \quad \text{and} \quad |x^2 - 4| = \begin{cases} x^2 - 4 & \text{if } |x| \geq 2 \\ 4 - x^2 & \text{if } |x| < 2 \end{cases}$$

So for $0 \leq |x| < 1, g(x) = 1 - x^2 - (4 - x^2) = -3$, for $1 \leq |x| < 2, g(x) = x^2 - 1 - (4 - x^2) = 2x^2 - 5$, and for $|x| \geq 2, g(x) = x^2 - 1 - (x^2 - 4) = 3$.

**7.** Remember that $|a| = a$ if $a \geq 0$ and that $|a| = -a$ if $a < 0$. Thus,

$$x + |x| = \begin{cases} 2x & \text{if } x \geq 0 \\ 0 & \text{if } x < 0 \end{cases} \quad \text{and} \quad y + |y| = \begin{cases} 2y & \text{if } y \geq 0 \\ 0 & \text{if } y < 0 \end{cases}$$

We will consider the equation $x + |x| = y + |y|$ in four cases.

| (1) $x \geq 0, y \geq 0$ | (2) $x \geq 0, y < 0$ | (3) $x < 0, y \geq 0$ | (4) $x < 0, y < 0$ |
|---|---|---|---|
| $2x = 2y$ | $2x = 0$ | $0 = 2y$ | $0 = 0$ |
| $x = y$ | $x = 0$ | $0 = y$ | |

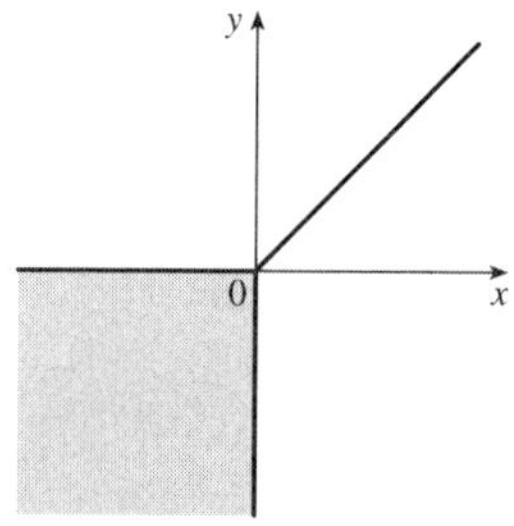


Case 1 gives us the line $y = x$ with nonnegative $x$ and $y$.

Case 2 gives us the portion of the $y$-axis with $y$ negative.

Case 3 gives us the portion of the $x$-axis with $x$ negative.

Case 4 gives us the entire third quadrant.

**8.** $x^4 - 4x^2 - x^2y^2 + 4y^2 = 0 \iff x^2(x^2 - 4) - y^2(x^2 - 4) = 0 \iff (x^2 - y^2)(x^2 - 4) = 0 \iff (x + y)(x - y)(x + 2)(x - 2) = 0$.

So the graph of the equation consists of the graphs of the four lines $y = -x$, $y = x$, $x = -2$, and $x = 2$.

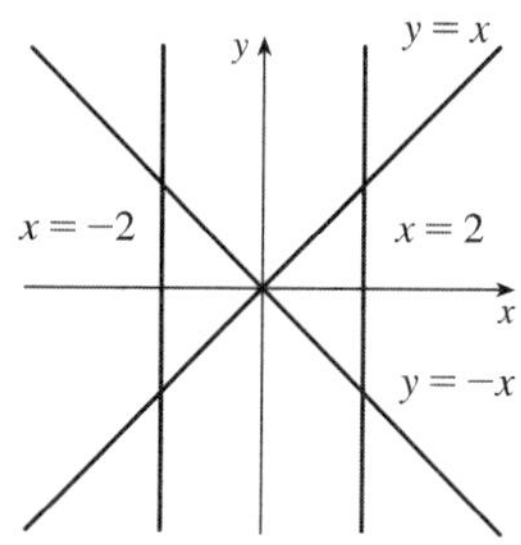


**9.** $|x| + |y| \leq 1$. The boundary of the region has equation $|x| + |y| = 1$. In quadrants I, II, III, and IV, this becomes the lines $x + y = 1, -x + y = 1, -x - y = 1$, and $x - y = 1$ respectively.

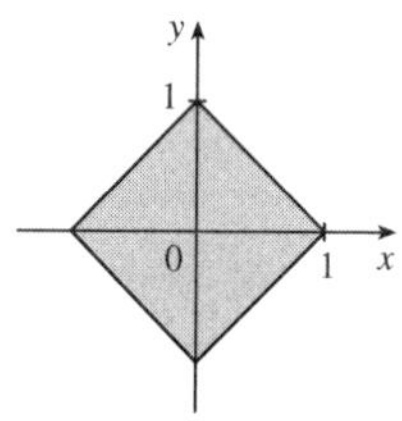


**10.** $|x - y| + |x| - |y| \leq 2$

| | | | | | | | |
|---|---|---|---|---|---|---|---|
| *Case (i):* | $x > y > 0$ | $\Leftrightarrow$ | $x - y + x - y \leq 2$ | $\Leftrightarrow$ | $x - y \leq 1$ | $\Leftrightarrow$ | $y \geq x - 1$ |
| *Case (ii):* | $y > x > 0$ | $\Leftrightarrow$ | $y - x + x - y \leq 2$ | $\Leftrightarrow$ | $0 \leq 2$ (true) | | |
| *Case (iii):* | $x > 0$ and $y < 0$ | $\Leftrightarrow$ | $x - y + x + y \leq 2$ | $\Leftrightarrow$ | $2x \leq 2$ | $\Leftrightarrow$ | $x \leq 1$ |
| *Case (iv):* | $x < 0$ and $y > 0$ | $\Leftrightarrow$ | $y - x - x - y \leq 2$ | $\Leftrightarrow$ | $-2x \leq 2$ | $\Leftrightarrow$ | $x \geq -1$ |
| *Case (v):* | $y < x < 0$ | $\Leftrightarrow$ | $x - y - x + y \leq 2$ | $\Leftrightarrow$ | $0 \leq 2$ (true) | | |
| *Case (vi):* | $x < y < 0$ | $\Leftrightarrow$ | $y - x - x + y \leq 2$ | $\Leftrightarrow$ | $y - x \leq 1$ | $\Leftrightarrow$ | $y \leq x + 1$ |

*Note:* Instead of considering cases (iv), (v), and (vi), we could have noted that the region is unchanged if $x$ and $y$ are replaced by $-x$ and $-y$, so the region is symmetric about the origin. Therefore, we need only draw cases (i), (ii), and (iii), and rotate through $180^\circ$ about the origin.

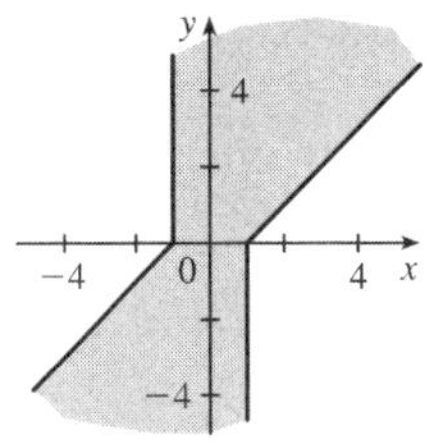


**11.** $(\log_2 3)(\log_3 4)(\log_4 5)\cdots(\log_{31} 32) = \left(\dfrac{\ln 3}{\ln 2}\right)\left(\dfrac{\ln 4}{\ln 3}\right)\left(\dfrac{\ln 5}{\ln 4}\right)\cdots\left(\dfrac{\ln 32}{\ln 31}\right) = \dfrac{\ln 32}{\ln 2} = \dfrac{\ln 2^5}{\ln 2} = \dfrac{5\ln 2}{\ln 2} = 5$

**12.** (a) $f(-x) = \ln\left(-x + \sqrt{(-x)^2+1}\right) = \ln\left(-x+\sqrt{x^2+1}\cdot\dfrac{-x-\sqrt{x^2+1}}{-x-\sqrt{x^2+1}}\right)$

$$= \ln\left(\frac{x^2-(x^2+1)}{-x-\sqrt{x^2+1}}\right) = \ln\left(\frac{-1}{-x-\sqrt{x^2+1}}\right) = \ln\left(\frac{1}{x+\sqrt{x^2+1}}\right)$$

$$= \ln 1 - \ln\left(x+\sqrt{x^2+1}\right) = -\ln\left(x+\sqrt{x^2-1}\right) = -f(x)$$

(b) $y = \ln\left(x+\sqrt{x^2+1}\right)$. Interchanging $x$ and $y$, we get $x = \ln\left(y+\sqrt{y^2+1}\right) \;\Rightarrow\; e^x = y+\sqrt{y^2+1} \;\Rightarrow$

$e^x - y = \sqrt{y^2+1} \;\Rightarrow\; e^{2x} - 2ye^x + y^2 = y^2+1 \;\Rightarrow\; e^{2x}-1 = 2ye^x \;\Rightarrow\; y = \dfrac{e^{2x}-1}{2e^x} = f^{-1}(-x)$

**13.** $\ln\left(x^2-2x-2\right) \le 0 \;\Rightarrow\; x^2-2x-2 \le e^0 = 1 \;\Rightarrow\; x^2-2x-3 \le 0 \;\Rightarrow\; (x-3)(x+1)\le 0 \;\Rightarrow\; x\in[-1,3]$.

Since the argument must be positive, $x^2-2x-2>0 \;\Rightarrow\; \left[x-\left(1-\sqrt{3}\right)\right]\left[x-\left(1+\sqrt{3}\right)\right] > 0 \;\Rightarrow$

$x \in \left(-\infty, 1-\sqrt{3}\right)\cup\left(1+\sqrt{3},\infty\right)$. The intersection of these intervals is $\left[-1, 1-\sqrt{3}\right)\cup\left(1+\sqrt{3},3\right]$.

**14.** Assume that $\log_2 5$ is rational. Then $\log_2 5 = m/n$ for natural numbers $m$ and $n$. Changing to exponential form gives us $2^{m/n} = 5$ and then raising both sides to the $n$th power gives $2^m = 5^n$. But $2^m$ is even and $5^n$ is odd. We have arrived at a contradiction, so we conclude that our hypothesis, that $\log_2 5$ is rational, is false. Thus, $\log_2 5$ is irrational.

**15.** Let $d$ be the distance traveled on each half of the trip. Let $t_1$ and $t_2$ be the times taken for the first and second halves of the trip. For the first half of the trip we have $t_1 = d/30$ and for the second half we have $t_2 = d/60$. Thus, the average speed for the entire trip is $\dfrac{\text{total distance}}{\text{total time}} = \dfrac{2d}{t_1+t_2} = \dfrac{2d}{\dfrac{d}{30}+\dfrac{d}{60}}\cdot\dfrac{60}{60} = \dfrac{120d}{2d+d} = \dfrac{120d}{3d} = 40$. The average speed for the entire trip is 40 mi/h.

**16.** Let $f = \sin$, $g = x$, and $h = x$. Then the left-hand side of the equation is $f\circ(g+h) = \sin(x+x) = \sin 2x = 2\sin x\cos x$; and the right-hand side is $f\circ g + f\circ h = \sin x + \sin x = 2\sin x$. The two sides are not equal, so the given statement is false.

**17.** Let $S_n$ be the statement that $7^n - 1$ is divisible by 6.

- $S_1$ is true because $7^1 - 1 = 6$ is divisible by 6.
- Assume $S_k$ is true, that is, $7^k - 1$ is divisible by 6. In other words, $7^k - 1 = 6m$ for some positive integer $m$. Then $7^{k+1} - 1 = 7^k\cdot 7 - 1 = (6m+1)\cdot 7 - 1 = 42m + 6 = 6(7m+1)$, which is divisible by 6, so $S_{k+1}$ is true.
- Therefore, by mathematical induction, $7^n - 1$ is divisible by 6 for every positive integer $n$.

**18.** Let $S_n$ be the statement that $1 + 3 + 5 + \cdots + (2n - 1) = n^2$.

- $S_1$ is true because $[2(1) - 1] = 1 = 1^2$.
- Assume $S_k$ is true, that is, $1 + 3 + 5 + \cdots + (2k - 1) = k^2$. Then

$$\begin{aligned} 1 + 3 + 5 + \cdots + (2k - 1) + [(2k + 1) - 1] &= 1 + 3 + 5 + \cdots + (2k - 1) + (2k + 1) \\ &= k^2 + (2k + 1) = (k + 1)^2 \end{aligned}$$

  which shows that $S_{k+1}$ is true.
- Therefore, by mathematical induction, $1 + 3 + 5 + \cdots + (2n - 1) = n^2$ for every positive integer $n$.

**19.** $f_0(x) = x^2$ and $f_{n+1}(x) = f_0(f_n(x))$ for $n = 0, 1, 2, \ldots$.

$f_1(x) = f_0(f_0(x)) = f_0(x^2) = (x^2)^2 = x^4$, $f_2(x) = f_0(f_1(x)) = f_0(x^4) = (x^4)^2 = x^8$,

$f_3(x) = f_0(f_2(x)) = f_0(x^8) = (x^8)^2 = x^{16}, \ldots$. Thus, a general formula is $f_n(x) = x^{2^{n+1}}$.

**20.** (a) $f_0(x) = 1/(2 - x)$ and $f_{n+1} = f_0 \circ f_n$ for $n = 0, 1, 2, \ldots$.

$$f_1(x) = f_0\left(\frac{1}{2 - x}\right) = \frac{1}{2 - \dfrac{1}{2 - x}} = \frac{2 - x}{2(2 - x) - 1} = \frac{2 - x}{3 - 2x},$$

$$f_2(x) = f_0\left(\frac{2 - x}{3 - 2x}\right) = \frac{1}{2 - \dfrac{2 - x}{3 - 2x}} = \frac{3 - 2x}{2(3 - 2x) - (2 - x)} = \frac{3 - 2x}{4 - 3x},$$

$$f_3(x) = f_0\left(\frac{3 - 2x}{4 - 3x}\right) = \frac{1}{2 - \dfrac{3 - 2x}{4 - 3x}} = \frac{4 - 3x}{2(4 - 3x) - (3 - 2x)} = \frac{4 - 3x}{5 - 4x}, \ldots$$

Thus, we conjecture that the general formula is $f_n(x) = \dfrac{n + 1 - nx}{n + 2 - (n + 1)x}$.

To prove this, we use the Principle of Mathematical Induction. We have already verified that $f_n$ is true for $n = 1$. Assume that the formula is true for $n = k$; that is, $f_k(x) = \dfrac{k + 1 - kx}{k + 2 - (k + 1)x}$. Then

$$\begin{aligned} f_{k+1}(x) = (f_0 \circ f_k)(x) = f_0(f_k(x)) = f_0\left(\frac{k + 1 - kx}{k + 2 - (k + 1)x}\right) &= \frac{1}{2 - \dfrac{k + 1 - kx}{k + 2 - (k + 1)x}} \\ = \frac{k + 2 - (k + 1)x}{2[k + 2 - (k + 1)x] - (k + 1 - kx)} &= \frac{k + 2 - (k + 1)x}{k + 3 - (k + 2)x} \end{aligned}$$

This shows that the formula for $f_n$ is true for $n = k + 1$. Therefore, by mathematical induction, the formula is true for all positive integers $n$.

(b) From the graph, we can make several observations:

- The values at $x = a$ keep increasing as $k$ increases.
- The vertical asymptote gets closer to $x = 1$ as $k$ increases.
- Thc horizontal asymptote gets closer to $y - 1$ as $k$ increases.
- The $x$-intercept for $f_{k+1}$ is the value of the vertical asymptote for $f_k$.
- The $y$-intercept for $f_k$ is the value of the horizontal asymptote for $f_{k+1}$.

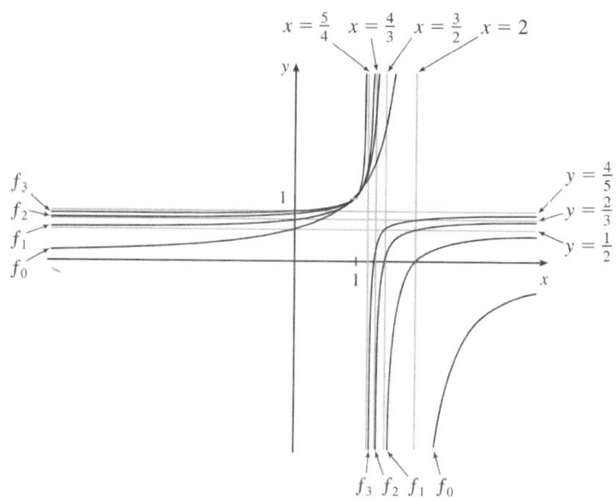

# 2 □ LIMITS AND DERIVATIVES

## 2.1 The Tangent and Velocity Problems

**1.** (a) Using $P(15, 250)$, we construct the following table:

| $t$ | $Q$ | slope $= m_{PQ}$ |
|---|---|---|
| 5 | $(5, 694)$ | $\frac{694-250}{5-15} = -\frac{444}{10} = -44.4$ |
| 10 | $(10, 444)$ | $\frac{444-250}{10-15} = -\frac{194}{5} = -38.8$ |
| 20 | $(20, 111)$ | $\frac{111-250}{20-15} = -\frac{139}{5} = -27.8$ |
| 25 | $(25, 28)$ | $\frac{28-250}{25-15} = -\frac{222}{10} = -22.2$ |
| 30 | $(30, 0)$ | $\frac{0-250}{30-15} = -\frac{250}{15} = -16.\overline{6}$ |

(b) Using the values of $t$ that correspond to the points closest to $P$ ($t = 10$ and $t = 20$), we have

$$\frac{-38.8 + (-27.8)}{2} = -33.3$$

(c) From the graph, we can estimate the slope of the tangent line at $P$ to be $\frac{-300}{9} = -33.\overline{3}$.

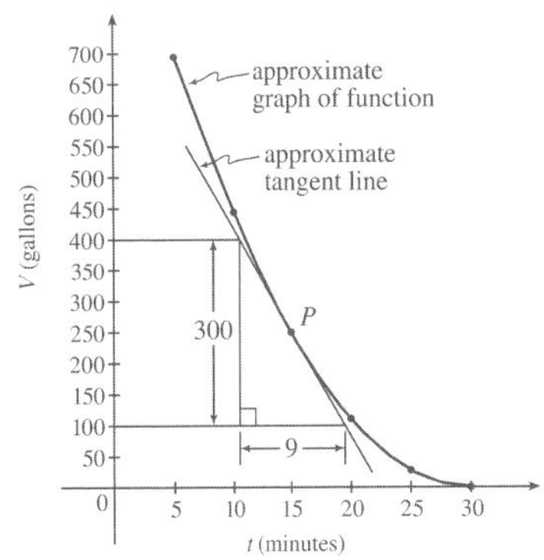


**2.** (a) Slope $= \frac{2948 - 2530}{42 - 36} = \frac{418}{6} \approx 69.67$

(b) Slope $= \frac{2948 - 2661}{42 - 38} = \frac{287}{4} = 71.75$

(c) Slope $= \frac{2948 - 2806}{42 - 40} = \frac{142}{2} = 71$

(d) Slope $= \frac{3080 - 2948}{44 - 42} = \frac{132}{2} = 66$

From the data, we see that the patient's heart rate is decreasing from 71 to 66 heartbeats/minute after 42 minutes. After being stable for a while, the patient's heart rate is dropping.

**3.** (a)

| | $x$ | $Q$ | $m_{PQ}$ |
|---|---|---|---|
| (i) | 0.5 | $(0.5, 0.333333)$ | 0.333333 |
| (ii) | 0.9 | $(0.9, 0.473684)$ | 0.263158 |
| (iii) | 0.99 | $(0.99, 0.497487)$ | 0.251256 |
| (iv) | 0.999 | $(0.999, 0.499750)$ | 0.250125 |
| (v) | 1.5 | $(1.5, 0.6)$ | 0.2 |
| (vi) | 1.1 | $(1.1, 0.523810)$ | 0.238095 |
| (vii) | 1.01 | $(1.01, 0.502488)$ | 0.248756 |
| (viii) | 1.001 | $(1.001, 0.500250)$ | 0.249875 |

(b) The slope appears to be $\frac{1}{4}$.

(c) $y - \frac{1}{2} = \frac{1}{4}(x - 1)$ or $y = \frac{1}{4}x + \frac{1}{4}$.

**4.** (a)

| | $x$ | $Q$ | $m_{PQ}$ |
|---|---|---|---|
| (i) | 2.5 | $(2.5, 0.707107)$ | 0.585786 |
| (ii) | 2.9 | $(2.9, 0.948683)$ | 0.513167 |
| (iii) | 2.99 | $(2.99, 0.994987)$ | 0.501256 |
| (iv) | 2.999 | $(2.999, 0.999500)$ | 0.500125 |
| (v) | 3.5 | $(3.5, 1.224745)$ | 0.449490 |
| (vi) | 3.1 | $(3.1, 1.048809)$ | 0.488088 |
| (vii) | 3.01 | $(3.01, 1.004988)$ | 0.498756 |
| (viii) | 3.001 | $(3.001, 1.000500)$ | 0.499875 |

(b) The slope appears to be $\frac{1}{2}$.

(c) $y - 1 = \frac{1}{2}(x - 3)$ or $y = \frac{1}{2}x - \frac{1}{2}$.

(d)

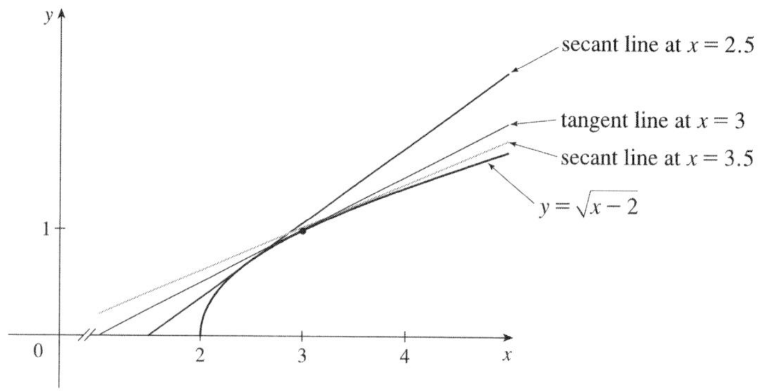


**5.** (a) $y = y(t) = 40t - 16t^2$. At $t = 2$, $y = 40(2) - 16(2)^2 = 16$. The average velocity between times 2 and $2 + h$ is

$$v_{\text{ave}} = \frac{y(2+h) - y(2)}{(2+h) - 2} = \frac{\left[40(2+h) - 16(2+h)^2\right] - 16}{h} = \frac{-24h - 16h^2}{h} = -24 - 16h, \text{ if } h \neq 0.$$

(i) $[2, 2.5]$: $h = 0.5$, $v_{\text{ave}} = -32$ ft/s

(ii) $[2, 2.1]$: $h = 0.1$, $v_{\text{ave}} = -25.6$ ft/s

(iii) $[2, 2.05]$: $h = 0.05$, $v_{\text{ave}} = -24.8$ ft/s

(iv) $[2, 2.01]$: $h = 0.01$, $v_{\text{ave}} = -24.16$ ft/s

(b) The instantaneous velocity when $t = 2$ ($h$ approaches 0) is $-24$ ft/s.

**6.** The average velocity between $t$ and $t + h$ seconds is

$$\frac{58(t+h) - 0.83(t+h)^2 - \left(58t - 0.83t^2\right)}{h} = \frac{58h - 1.66th - 0.83h^2}{h} = 58 - 1.66t - 0.83h \text{ if } h \neq 0.$$

(a) Here $t = 1$, so the average velocity is $58 - 1.66 - 0.83h = 56.34 - 0.83h$.

(i) $[1, 2]$: $h = 1$, 55.51 m/s

(ii) $[1, 1.5]$: $h = 0.5$, 55.925 m/s

(iii) $[1, 1.1]$: $h = 0.1$, 56.257 m/s

(iv) $[1, 1.01]$: $h = 0.01$, 56.3317 m/s

(v) $[1, 1.001]$: $h = 0.001$, 56.33917 m/s

(b) The instantaneous velocity after 1 second is 56.34 m/s.

**7.** (a) (i) On the interval $[1, 3]$, $v_{\text{ave}} = \dfrac{s(3) - s(1)}{3 - 1} = \dfrac{10.7 - 1.4}{2} = \dfrac{9.3}{2} = 4.65$ m/s.

(ii) On the interval $[2, 3]$, $v_{\text{ave}} = \dfrac{s(3) - s(2)}{3 - 2} = \dfrac{10.7 - 5.1}{1} = 5.6$ m/s.

(iii) On the interval $[3, 5]$, $v_{\text{ave}} = \dfrac{s(5) - s(3)}{5 - 3} = \dfrac{25.8 - 10.7}{2} = \dfrac{15.1}{2} = 7.55$ m/s.

(iv) On the interval $[3, 4]$, $v_{\text{ave}} = \dfrac{s(4) - s(3)}{4 - 3} = \dfrac{17.7 - 10.7}{1} = 7$ m/s.

(b) 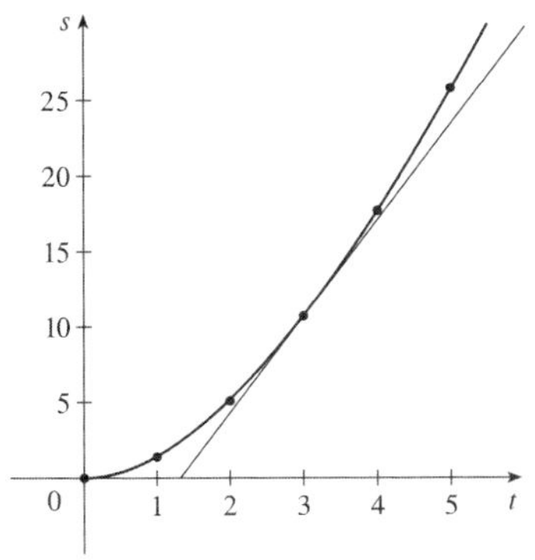


Using the points $(2, 4)$ and $(5, 23)$ from the approximate tangent line, the instantaneous velocity at $t = 3$ is about $\dfrac{23 - 4}{5 - 2} \approx 6.3$ m/s.

**8.** (a) (i) $s = s(t) = 2\sin \pi t + 3\cos \pi t$. On the interval $[1, 2]$, $v_{\text{ave}} = \dfrac{s(2) - s(1)}{2 - 1} = \dfrac{3 - (-3)}{1} = 6$ cm/s.

(ii) On the interval $[1, 1.1]$, $v_{\text{ave}} = \dfrac{s(1.1) - s(1)}{1.1 - 1} \approx \dfrac{-3.471 - (-3)}{0.1} = -4.71$ cm/s.

(iii) On the interval $[1, 1.01]$, $v_{\text{ave}} = \dfrac{s(1.01) - s(1)}{1.01 - 1} \approx \dfrac{-3.0613 - (-3)}{0.01} = -6.13$ cm/s.

(iv) On the interval $[1, 1.001]$, $v_{\text{ave}} = \dfrac{s(1.001) - s(1)}{1.001 - 1} \approx \dfrac{-3.00627 - (-3)}{1.001 - 1} = -6.27$ cm/s.

(b) The instantaneous velocity of the particle when $t = 1$ appears to be about $-6.3$ cm/s.

**9.** (a) For the curve $y = \sin(10\pi/x)$ and the point $P(1, 0)$:

| $x$ | $Q$ | $m_{PQ}$ |
|---|---|---|
| 2 | $(2, 0)$ | 0 |
| 1.5 | $(1.5, 0.8660)$ | 1.7321 |
| 1.4 | $(1.4, -0.4339)$ | $-1.0847$ |
| 1.3 | $(1.3, -0.8230)$ | $-2.7433$ |
| 1.2 | $(1.2, 0.8660)$ | 4.3301 |
| 1.1 | $(1.1, -0.2817)$ | $-2.8173$ |

| $x$ | $Q$ | $m_{PQ}$ |
|---|---|---|
| 0.5 | $(0.5, 0)$ | 0 |
| 0.6 | $(0.6, 0.8660)$ | $-2.1651$ |
| 0.7 | $(0.7, 0.7818)$ | $-2.6061$ |
| 0.8 | $(0.8, 1)$ | $-5$ |
| 0.9 | $(0.9, -0.3420)$ | 3.4202 |

As $x$ approaches 1, the slopes do not appear to be approaching any particular value.

(b) 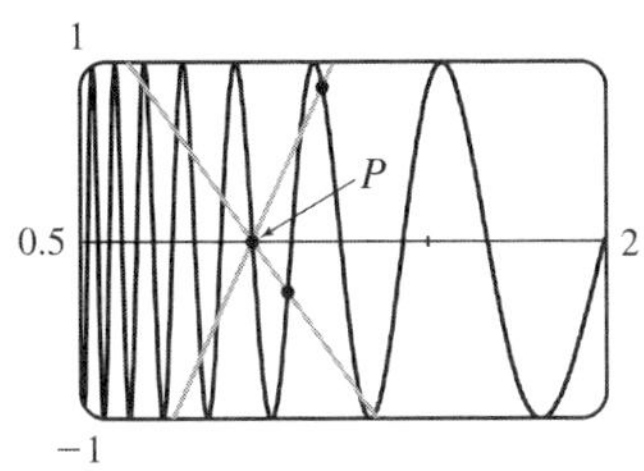


We see that problems with estimation are caused by the frequent oscillations of the graph. The tangent is so steep at $P$ that we need to take $x$-values much closer to 1 in order to get accurate estimates of its slope.

(c) If we choose $x = 1.001$, then the point $Q$ is $(1.001, -0.0314)$ and $m_{PQ} \approx -31.3794$. If $x = 0.999$, then $Q$ is $(0.999, 0.0314)$ and $m_{PQ} = -31.4422$. The average of these slopes is $-31.4108$. So we estimate that the slope of the tangent line at $P$ is about $-31.4$.

## 2.2 The Limit of a Function

1. As $x$ approaches 2, $f(x)$ approaches 5. [Or, the values of $f(x)$ can be made as close to 5 as we like by taking $x$ sufficiently close to 2 (but $x \neq 2$).] Yes, the graph could have a hole at $(2, 5)$ and be defined such that $f(2) = 3$.

2. As $x$ approaches 1 from the left, $f(x)$ approaches 3; and as $x$ approaches 1 from the right, $f(x)$ approaches 7. No, the limit does not exist because the left- and right-hand limits are different.

3. (a) $f(x)$ approaches 2 as $x$ approaches 1 from the left, so $\lim\limits_{x \to 1^-} f(x) = 2$.

   (b) $f(x)$ approaches 3 as $x$ approaches 1 from the right, so $\lim\limits_{x \to 1^+} f(x) = 3$.

   (c) $\lim\limits_{x \to 1} f(x)$ does not exist because the limits in part (a) and part (b) are not equal.

   (d) $f(x)$ approaches 4 as $x$ approaches 5 from the left and from the right, so $\lim\limits_{x \to 5} f(x) = 4$.

   (e) $f(5)$ is not defined, so it doesn't exist.

4. (a) $\lim\limits_{x \to 0} f(x) = 3$ (b) $\lim\limits_{x \to 3^-} f(x) = 4$ (c) $\lim\limits_{x \to 3^+} f(x) = 2$

   (d) $\lim\limits_{x \to 3} f(x)$ does not exist because the limits in part (b) and part (c) are not equal.

   (e) $f(3) = 3$

5. (a) $\lim\limits_{t \to 0^-} g(t) = -1$ (b) $\lim\limits_{t \to 0^+} g(t) = -2$

   (c) $\lim\limits_{t \to 0} g(t)$ does not exist because the limits in part (a) and part (b) are not equal.

   (d) $\lim\limits_{t \to 2^-} g(t) = 2$ (e) $\lim\limits_{t \to 2^+} g(t) = 0$

   (f) $\lim\limits_{t \to 2} g(t)$ does not exist because the limits in part (d) and part (e) are not equal.

   (g) $g(2) = 1$ (h) $\lim\limits_{t \to 4} g(t) = 3$

6. $\lim\limits_{x \to a} f(x)$ exists for all $a$ except $a = \pm 1$.

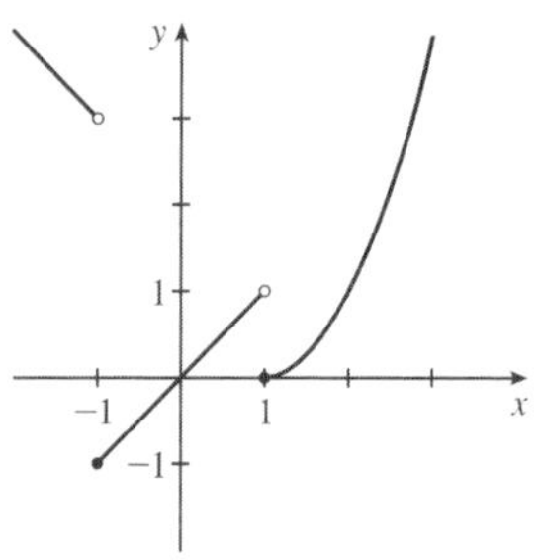


7. 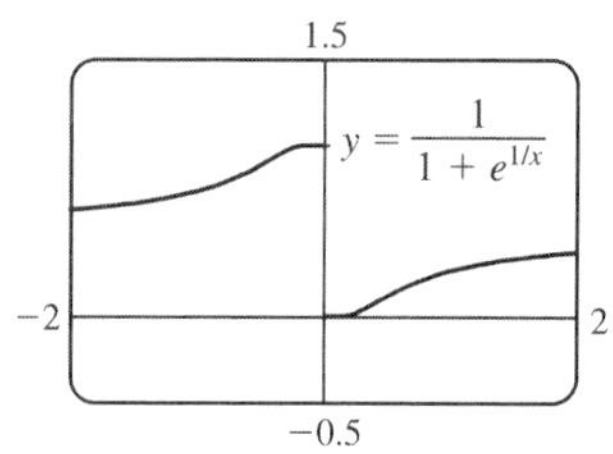


   (a) $\lim\limits_{x \to 0^-} f(x) = 1$

   (b) $\lim\limits_{x \to 0^+} f(x) = 0$

   (c) $\lim\limits_{x \to 0} f(x)$ does not exist because the limits in part (a) and part (b) are not equal.

8. $\lim\limits_{t \to 12^-} f(t) = 150$ mg and $\lim\limits_{t \to 12^+} f(t) = 300$ mg. These limits show that there is an abrupt change in the amount of drug in the patient's bloodstream at $t = 12$ h. The left-hand limit represents the amount of the drug just before the fourth injection. The right-hand limit represents the amount of the drug just after the fourth injection.

9. $\lim\limits_{x \to 1^-} f(x) = 2$, $\lim\limits_{x \to 1^+} f(x) = -2$, $f(1) = 2$

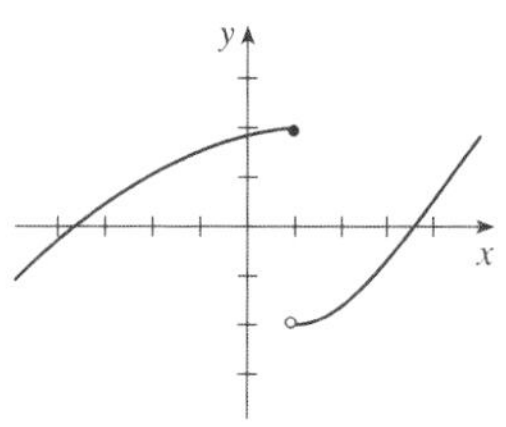


10. $\lim\limits_{x \to 0^-} f(x) = 1$, $\lim\limits_{x \to 0^+} f(x) = -1$, $\lim\limits_{x \to 2^-} f(x) = 0$, $\lim\limits_{x \to 2^+} f(x) = 1$, $f(2) = 1$, $f(0)$ is undefined

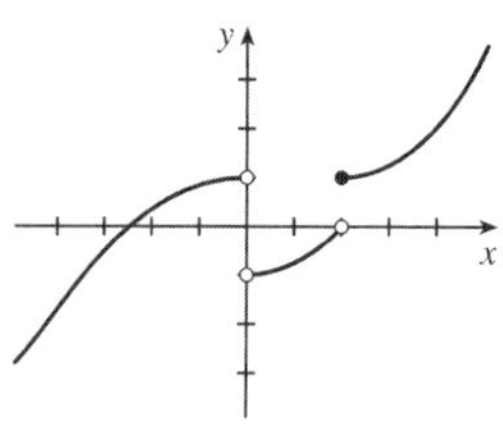

**11.** $\lim_{x \to 3^+} f(x) = 4$, $\lim_{x \to 3^-} f(x) = 2$, $\lim_{x \to -2} f(x) = 2$,

$f(3) = 3$, $f(-2) = 1$

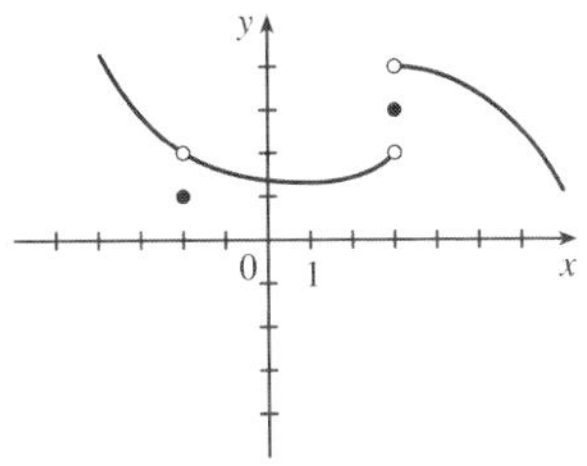


**12.** $\lim_{x \to 1} f(x) = 3$, $\lim_{x \to 4^-} f(x) = 3$, $\lim_{x \to 4^+} f(x) = -3$,

$f(1) = 1$, $f(4) = -1$

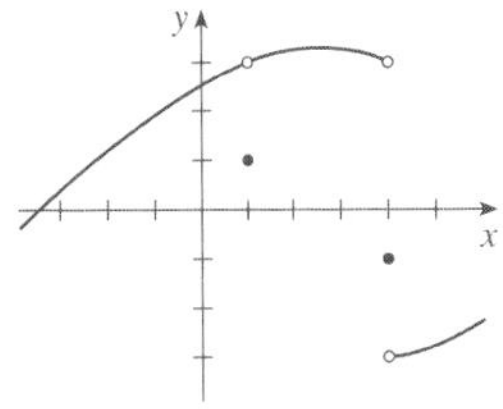


**13.** For $f(x) = \dfrac{x^2 - 2x}{x^2 - x - 2}$:

| $x$ | $f(x)$ |
|---|---|
| 2.5 | 0.714286 |
| 2.1 | 0.677419 |
| 2.05 | 0.672131 |
| 2.01 | 0.667774 |
| 2.005 | 0.667221 |
| 2.001 | 0.666778 |

| $x$ | $f(x)$ |
|---|---|
| 1.9 | 0.655172 |
| 1.95 | 0.661017 |
| 1.99 | 0.665552 |
| 1.995 | 0.666110 |
| 1.999 | 0.666556 |

It appears that $\lim_{x \to 2} \dfrac{x^2 - 2x}{x^2 - x - 2} = 0.\bar{6} = \frac{2}{3}$.

**14.** For $f(x) = \dfrac{x^2 - 2x}{x^2 - x - 2}$:

| $x$ | $f(x)$ |
|---|---|
| 0 | 0 |
| −0.5 | −1 |
| −0.9 | −9 |
| −0.95 | −19 |
| −0.99 | −99 |
| −0.999 | −999 |

| $x$ | $f(x)$ |
|---|---|
| −2 | 2 |
| −1.5 | 3 |
| −1.1 | 11 |
| −1.01 | 101 |
| −1.001 | 1001 |

It appears that $\lim_{x \to -1} \dfrac{x^2 - 2x}{x^2 - x - 2}$ *does not exist* since

$f(x) \to -\infty$ as $x \to -1^-$ and $f(x) \to \infty$ as $x \to -1^+$.

**15.** For $f(x) = \dfrac{e^x - 1 - x}{x^2}$:

| $x$ | $f(x)$ |
|---|---|
| 1 | 0.718282 |
| 0.5 | 0.594885 |
| 0.1 | 0.517092 |
| 0.05 | 0.508439 |
| 0.01 | 0.501671 |

| $x$ | $f(x)$ |
|---|---|
| −1 | 0.367879 |
| −0.5 | 0.426123 |
| −0.1 | 0.483742 |
| −0.05 | 0.491770 |
| −0.01 | 0.498337 |

It appears that $\lim_{x \to 0} \dfrac{e^x - 1 - x}{x^2} = 0.5 = \frac{1}{2}$.

**16.** For $f(x) = x \ln(x + x^2)$:

| $x$ | $f(x)$ |
|---|---|
| 1 | 0.693147 |
| 0.5 | −0.143841 |
| 0.1 | −0.220727 |
| 0.05 | −0.147347 |
| 0.01 | −0.045952 |
| 0.005 | −0.026467 |
| 0.001 | −0.006907 |

It appears that $\lim_{x \to 0^+} x \ln(x + x^2) = 0$.

**17.** For $f(x) = \dfrac{\sqrt{x+4}-2}{x}$:

| $x$ | $f(x)$ |
|---|---|
| 1 | 0.236068 |
| 0.5 | 0.242641 |
| 0.1 | 0.248457 |
| 0.05 | 0.249224 |
| 0.01 | 0.249844 |

| $x$ | $f(x)$ |
|---|---|
| −1 | 0.267949 |
| −0.5 | 0.258343 |
| −0.1 | 0.251582 |
| −0.05 | 0.250786 |
| −0.01 | 0.250156 |

It appears that $\lim\limits_{x\to 0} \dfrac{\sqrt{x+4}-2}{x} = 0.25 = \frac{1}{4}$.

**18.** For $f(x) = \dfrac{\tan 3x}{\tan 5x}$:

| $x$ | $f(x)$ |
|---|---|
| ±0.2 | 0.439279 |
| ±0.1 | 0.566236 |
| ±0.05 | 0.591893 |
| ±0.01 | 0.599680 |
| ±0.001 | 0.599997 |

It appears that $\lim\limits_{x\to 0} \dfrac{\tan 3x}{\tan 5x} = 0.6 = \frac{3}{5}$.

**19.** For $f(x) = \dfrac{x^6-1}{x^{10}-1}$:

| $x$ | $f(x)$ |
|---|---|
| 0.5 | 0.985337 |
| 0.9 | 0.719397 |
| 0.95 | 0.660186 |
| 0.99 | 0.612018 |
| 0.999 | 0.601200 |

| $x$ | $f(x)$ |
|---|---|
| 1.5 | 0.183369 |
| 1.1 | 0.484119 |
| 1.05 | 0.540783 |
| 1.01 | 0.588022 |
| 1.001 | 0.598800 |

It appears that $\lim\limits_{x\to 1} \dfrac{x^6-1}{x^{10}-1} = 0.6 = \frac{3}{5}$.

**20.** For $f(x) = \dfrac{9^x - 5^x}{x}$:

| $x$ | $f(x)$ |
|---|---|
| 0.5 | 1.527864 |
| 0.1 | 0.711120 |
| 0.05 | 0.646496 |
| 0.01 | 0.599082 |
| 0.001 | 0.588906 |

| $x$ | $f(x)$ |
|---|---|
| −0.5 | 0.227761 |
| −0.1 | 0.485984 |
| −0.05 | 0.534447 |
| −0.01 | 0.576706 |
| −0.001 | 0.586669 |

It appears that $\lim\limits_{x\to 0} \dfrac{9^x - 5^x}{x} = 0.59$. Later we will be able to show that the exact value is $\ln(9/5)$.

**21.** (a) From the graphs, it seems that $\lim\limits_{x\to 0} \dfrac{\tan 4x}{x} = 4$.

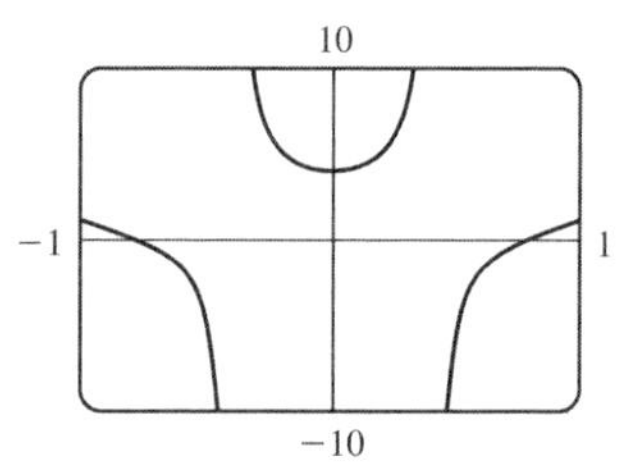


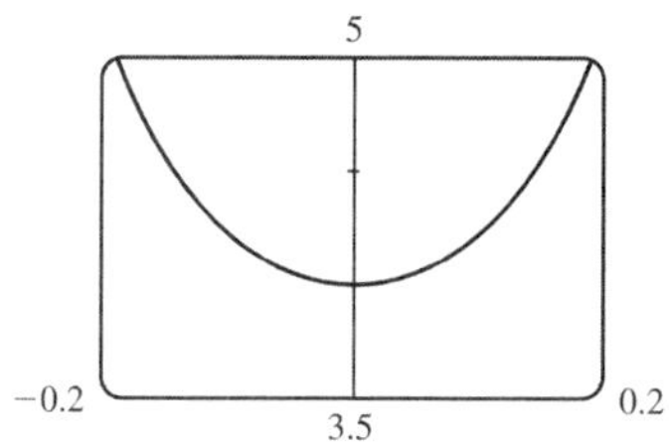


(b)

| $x$ | $f(x)$ |
|---|---|
| ±0.1 | 4.227932 |
| ±0.01 | 4.002135 |
| ±0.001 | 4.000021 |
| ±0.0001 | 4.000000 |

**22.** (a) From the following graphs, it seems that $\lim\limits_{x\to 0} \dfrac{6^x - 2^x}{x} \approx 1.10$.

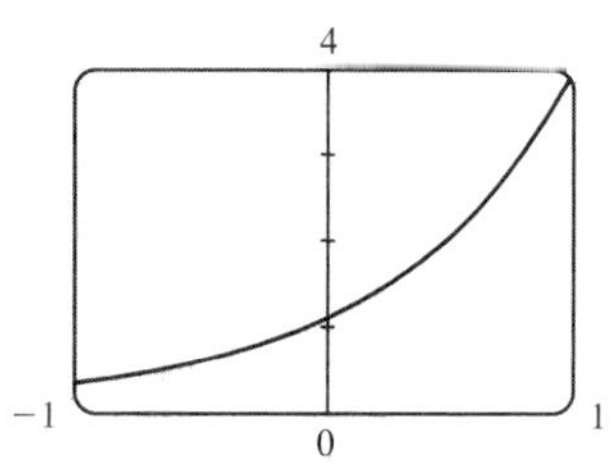


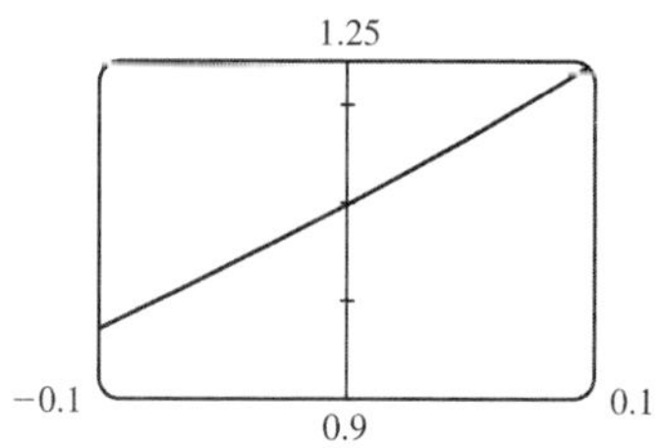


(b)

| $x$ | $f(x)$ |
|---|---|
| −0.01 | 1.085052 |
| −0.001 | 1.097248 |
| −0.0001 | 1.098476 |
| 0.0001 | 1.098749 |
| 0.001 | 1.099978 |
| 0.01 | 1.112353 |

**23.** (a) Let $h(x) = (1+x)^{1/x}$.

| $x$ | $h(x)$ |
|---|---|
| $-0.001$ | 2.71964 |
| $-0.0001$ | 2.71842 |
| $-0.00001$ | 2.71830 |
| $-0.000001$ | 2.71828 |
| 0.000001 | 2.71828 |
| 0.00001 | 2.71827 |
| 0.0001 | 2.71815 |
| 0.001 | 2.71692 |

It appears that $\lim_{x \to 0} (1+x)^{1/x} \approx 2.71828$, which is approximately $e$.

In Section 3.7 we will see that the value of the limit is exactly $e$.

(b)

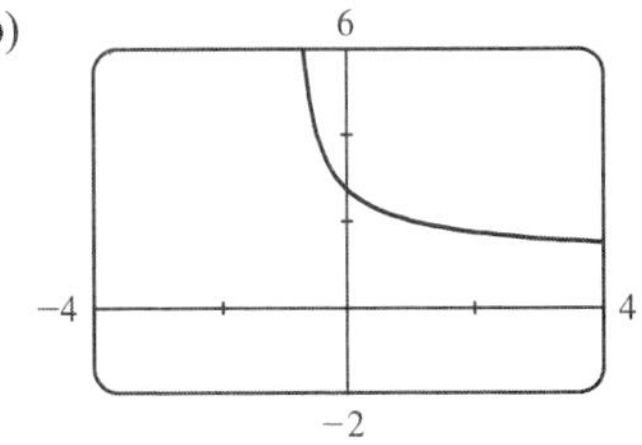


**24.** For the curve $y = 2^x$ and the points $P(0,1)$ and $Q(x, 2^x)$:

| $x$ | $Q$ | $m_{PQ}$ |
|---|---|---|
| 0.1 | $(0.1, 1.0717735)$ | 0.71773 |
| 0.01 | $(0.01, 1.0069556)$ | 0.69556 |
| 0.001 | $(0.001, 1.0006934)$ | 0.69339 |
| 0.0001 | $(0.0001, 1.0000693)$ | 0.69317 |

The slope appears to be about 0.693.

**25.** For $f(x) = x^2 - (2^x/1000)$:

(a)

| $x$ | $f(x)$ |
|---|---|
| 1 | 0.998000 |
| 0.8 | 0.638259 |
| 0.6 | 0.358484 |
| 0.4 | 0.158680 |
| 0.2 | 0.038851 |
| 0.1 | 0.008928 |
| 0.05 | 0.001465 |

It appears that $\lim_{x \to 0} f(x) = 0$.

(b)

| $x$ | $f(x)$ |
|---|---|
| 0.04 | 0.000572 |
| 0.02 | $-0.000614$ |
| 0.01 | $-0.000907$ |
| 0.005 | $-0.000978$ |
| 0.003 | $-0.000993$ |
| 0.001 | $-0.001000$ |

It appears that $\lim_{x \to 0} f(x) = -0.001$.

**26.** For $h(x) = \dfrac{\tan x - x}{x^3}$:

(a)

| $x$ | $h(x)$ |
|---|---|
| 1.0 | 0.55740773 |
| 0.5 | 0.37041992 |
| 0.1 | 0.33467209 |
| 0.05 | 0.33366700 |
| 0.01 | 0.33334667 |
| 0.005 | 0.33333667 |

(b) It seems that $\lim_{x \to 0} h(x) = \frac{1}{3}$.

(c)

| $x$ | $h(x)$ |
|---|---|
| 0.001 | 0.33333350 |
| 0.0005 | 0.33333344 |
| 0.0001 | 0.33333000 |
| 0.00005 | 0.33333600 |
| 0.00001 | 0.33300000 |
| 0.000001 | 0.00000000 |

Here the values will vary from one calculator to another. Every calculator will eventually give *false values*.

(d) As in part (c), when we take a small enough viewing rectangle we get incorrect output.

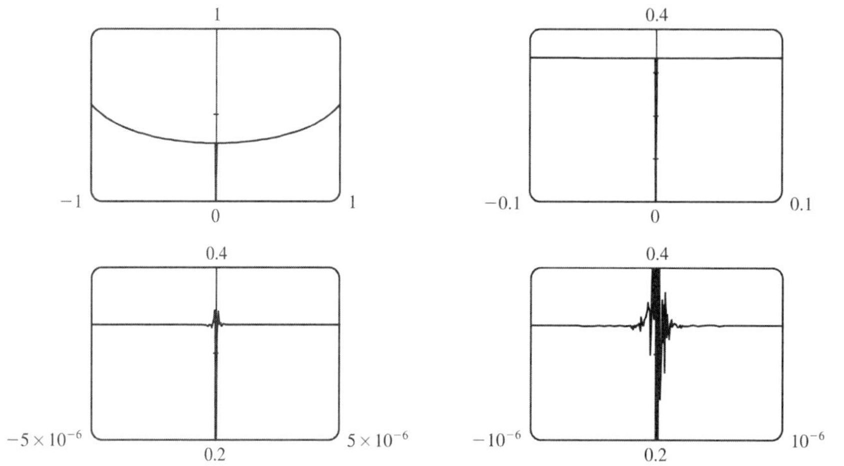

**27.** 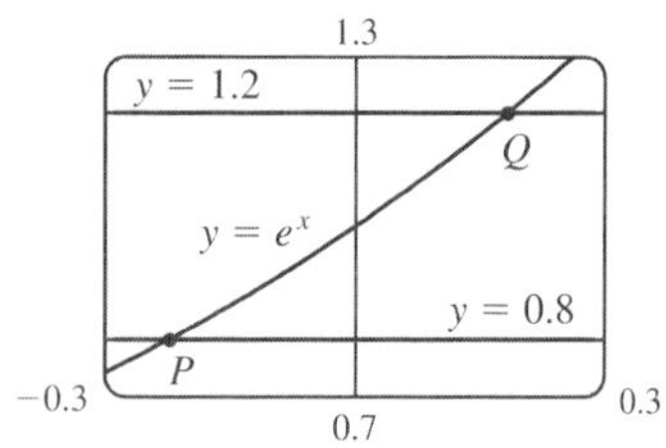


We need to have $0.8 < e^x < 1.2$. From the graph we obtain the approximate points of intersection $P(-0.2231436, 0.8)$ and $Q(0.18232156, 1.2)$. So if $x$ is within 0.182 of 0, then $y$ will be within 0.2 of 1. If we must have $e^x$ within 0.1 of 1, we get $P(-0.1053605, 0.9)$ and $Q(0.09531018, 1.1)$. We would then need $x$ to be within 0.095 of 0.

**28.** (a) Let $y = (x^3 - 1)/(\sqrt{x} - 1)$.

From the table and the graph, we guess that the limit of $y$ as $x$ approaches 1 is 6.

| $x$ | $y$ |
|---|---|
| 0.99 | 5.92531 |
| 0.999 | 5.99250 |
| 0.9999 | 5.99925 |
| 1.01 | 6.07531 |
| 1.001 | 6.00750 |
| 1.0001 | 6.00075 |

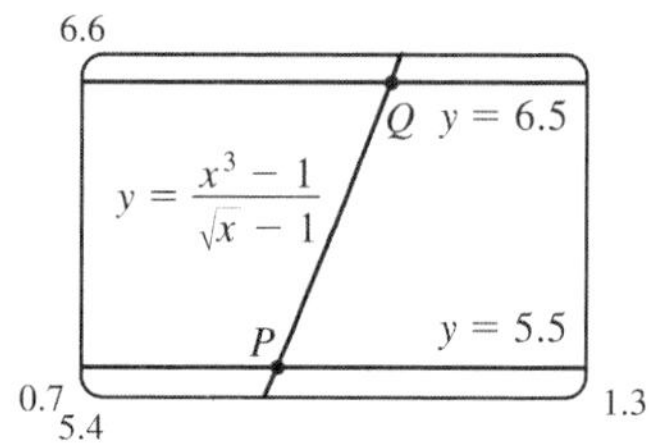


(b) We need to have $5.5 < \dfrac{x^3 - 1}{\sqrt{x} - 1} < 6.5$. From the graph we obtain the approximate points of intersection $P(0.9313853, 5.5)$ and $Q(1.0649004, 6.5)$. Now $1 - 0.9313853 \approx 0.0686$ and $1.0649004 - 1 \approx 0.0649$, so by requiring that $x$ be within 0.0649 of 1, we ensure that $y$ is within 0.5 of 6.

## 2.3 Calculating Limits Using the Limit Laws

**1.** (a) $\lim\limits_{x\to a} [f(x) + h(x)] = \lim\limits_{x\to a} f(x) + \lim\limits_{x\to a} h(x) = -3 + 8 = 5$

(b) $\lim\limits_{x\to a} [f(x)]^2 = \left[\lim\limits_{x\to a} f(x)\right]^2 = (-3)^2 = 9$

(c) $\lim\limits_{x\to a} \sqrt[3]{h(x)} = \sqrt[3]{\lim\limits_{x\to a} h(x)} = \sqrt[3]{8} = 2$

(d) $\lim\limits_{x\to a} \dfrac{1}{f(x)} = \dfrac{1}{\lim\limits_{x\to a} f(x)} = \dfrac{1}{-3} = -\dfrac{1}{3}$

(e) $\lim\limits_{x\to a} \dfrac{f(x)}{h(x)} = \dfrac{\lim\limits_{x\to a} f(x)}{\lim\limits_{x\to a} h(x)} = \dfrac{-3}{8} = -\dfrac{3}{8}$

(f) $\lim\limits_{x\to a} \dfrac{g(x)}{f(x)} = \dfrac{\lim\limits_{x\to a} g(x)}{\lim\limits_{x\to a} f(x)} = \dfrac{0}{-3} = 0$

(g) The limit does not exist, since $\lim\limits_{x\to a} g(x) = 0$ but $\lim\limits_{x\to a} f(x) \neq 0$.

(h) $\lim\limits_{x\to a} \dfrac{2f(x)}{h(x) - f(x)} = \dfrac{2 \lim\limits_{x\to a} f(x)}{\lim\limits_{x\to a} h(x) - \lim\limits_{x\to a} f(x)} = \dfrac{2(-3)}{8 - (-3)} = -\dfrac{6}{11}$

**2.** (a) $\lim_{x\to 2}[f(x)+g(x)] = \lim_{x\to 2} f(x) + \lim_{x\to 2} g(x) = 2 + 0 = 2$

(b) $\lim_{x\to 1} g(x)$ does not exist since its left- and right-hand limits are not equal, so the given limit does not exist.

(c) $\lim_{x\to 0}[f(x)g(x)] = \lim_{x\to 0} f(x) \cdot \lim_{x\to 0} g(x) = 0 \cdot 1.3 = 0$

(d) Since $\lim_{x\to -1} g(x) = 0$ and $g$ is in the denominator, but $\lim_{x\to -1} f(x) = -1 \neq 0$, the given limit does not exist.

(e) $\lim_{x\to 2} x^3 f(x) = \left[\lim_{x\to 2} x^3\right]\left[\lim_{x\to 2} f(x)\right] = 2^3 \cdot 2 = 16$

(f) $\lim_{x\to 1}\sqrt{3+f(x)} = \sqrt{3 + \lim_{x\to 1} f(x)} = \sqrt{3+1} = 2$

**3.**
$$\begin{aligned}\lim_{x\to -2}(3x^4+2x^2-x+1) &= \lim_{x\to -2} 3x^4 + \lim_{x\to -2} 2x^2 - \lim_{x\to -2} x + \lim_{x\to -2} 1 && \text{[Limit Laws 1 and 2]}\\ &= 3\lim_{x\to -2} x^4 + 2\lim_{x\to -2} x^2 - \lim_{x\to -2} x + \lim_{x\to -2} 1 && [3]\\ &= 3(-2)^4 + 2(-2)^2 - (-2) + (1) && \text{[9, 8, and 7]}\\ &= 48+8+2+1 = 59\end{aligned}$$

**4.**
$$\begin{aligned}\lim_{t\to -1}(t^2+1)^3(t+3)^5 &= \lim_{t\to -1}(t^2+1)^3 \cdot \lim_{t\to -1}(t+3)^5 && \text{[Limit Law 4]}\\ &= \left[\lim_{t\to -1}(t^2+1)\right]^3 \cdot \left[\lim_{t\to -1}(t+3)\right]^5 && [6]\\ &= \left[\lim_{t\to -1} t^2 + \lim_{t\to -1} 1\right]^3 \cdot \left[\lim_{t\to -1} t + \lim_{t\to -1} 3\right]^5 && [1]\\ &= \left[(-1)^2+1\right]^3 \cdot \left[-1+3\right]^5 = 8 \cdot 32 = 256 && \text{[9, 7, and 8]}\end{aligned}$$

**5.**
$$\begin{aligned}\lim_{x\to 8}(1+\sqrt[3]{x})(2-6x^2+x^3) &= \lim_{x\to 8}(1+\sqrt[3]{x}) \cdot \lim_{x\to 8}(2-6x^2+x^3) && \text{[Limit Law 4]}\\ &= \left(\lim_{x\to 8} 1 + \lim_{x\to 8}\sqrt[3]{x}\right) \cdot \left(\lim_{x\to 8} 2 - 6\lim_{x\to 8} x^2 + \lim_{x\to 8} x^3\right) && \text{[1, 2, and 3]}\\ &= (1+\sqrt[3]{8}) \cdot (2 - 6 \cdot 8^2 + 8^3) && \text{[7, 10, 9]}\\ &= (3)(130) = 390\end{aligned}$$

**6.**
$$\begin{aligned}\lim_{u\to -2}\sqrt{u^4+3u+6} &= \sqrt{\lim_{u\to -2}(u^4+3u+6)} && [11]\\ &= \sqrt{\lim_{u\to -2} u^4 + 3\lim_{u\to -2} u + \lim_{u\to -2} 6} && \text{[1, 2, and 3]}\\ &= \sqrt{(-2)^4 + 3(-2) + 6} && \text{[9, 8, and 7]}\\ &= \sqrt{16-6+6} = \sqrt{16} = 4\end{aligned}$$

**7.**
$$\begin{aligned}\lim_{x\to 1}\left(\frac{1+3x}{1+4x^2+3x^4}\right)^3 &= \left(\lim_{x\to 1}\frac{1+3x}{1+4x^2+3x^4}\right)^3 && [6]\\ &= \left[\frac{\lim_{x\to 1}(1+3x)}{\lim_{x\to 1}(1+4x^2+3x^4)}\right]^3 && [5]\\ &= \left[\frac{\lim_{x\to 1} 1 + 3\lim_{x\to 1} x}{\lim_{x\to 1} 1 + 4\lim_{x\to 1} x^2 + 3\lim_{x\to 1} x^4}\right]^3 && \text{[2, 1, and 3]}\\ &= \left[\frac{1+3(1)}{1+4(1)^2+3(1)^4}\right]^3 = \left[\frac{4}{8}\right]^3 = \left(\frac{1}{2}\right)^3 = \frac{1}{8} && \text{[7, 8, and 9]}\end{aligned}$$

**8.** (a) The left-hand side of the equation is not defined for $x = 2$, but the right-hand side is.

(b) Since the equation holds for all $x \neq 2$, it follows that both sides of the equation approach the same limit as $x \to 2$, just as in Example 3. Remember that in finding $\lim_{x\to a} f(x)$, we never consider $x = a$.

**9.** $\lim_{x\to 2}\dfrac{x^2+x-6}{x-2}=\lim_{x\to 2}\dfrac{(x+3)(x-2)}{x-2}=\lim_{x\to 2}(x+3)=2+3=5$

**10.** $\lim_{x\to -4}\dfrac{x^2+5x+4}{x^2+3x-4}=\lim_{x\to -4}\dfrac{(x+4)(x+1)}{(x+4)(x-1)}=\lim_{x\to -4}\dfrac{x+1}{x-1}=\dfrac{-4+1}{-4-1}=\dfrac{-3}{-5}=\dfrac{3}{5}$

**11.** $\lim_{x\to 2}\dfrac{x^2-x+6}{x-2}$ does not exist since $x-2\to 0$ but $x^2-x+6\to 8$ as $x\to 2$.

**12.** $\lim_{x\to 4}\dfrac{x^2-4x}{x^2-3x-4}=\lim_{x\to 4}\dfrac{x(x-4)}{(x-4)(x+1)}=\lim_{x\to 4}\dfrac{x}{x+1}=\dfrac{4}{4+1}=\dfrac{4}{5}$

**13.** $\lim_{t\to -3}\dfrac{t^2-9}{2t^2+7t+3}=\lim_{t\to -3}\dfrac{(t+3)(t-3)}{(2t+1)(t+3)}=\lim_{t\to -3}\dfrac{t-3}{2t+1}=\dfrac{-3-3}{2(-3)+1}=\dfrac{-6}{-5}=\dfrac{6}{5}$

**14.** $\lim_{x\to -1}\dfrac{x^2-4x}{x^2-3x-4}$ does not exist since $x^2-3x-4\to 0$ but $x^2-4x\to 5$ as $x\to -1$.

**15.** $\lim_{h\to 0}\dfrac{(4+h)^2-16}{h}=\lim_{h\to 0}\dfrac{(16+8h+h^2)-16}{h}=\lim_{h\to 0}\dfrac{8h+h^2}{h}=\lim_{h\to 0}\dfrac{h(8+h)}{h}=\lim_{h\to 0}(8+h)=8+0=8$

**16.** $\lim_{h\to 0}\dfrac{\sqrt{1+h}-1}{h}=\lim_{h\to 0}\dfrac{\sqrt{1+h}-1}{h}\cdot\dfrac{\sqrt{1+h}+1}{\sqrt{1+h}+1}=\lim_{h\to 0}\dfrac{(1+h)-1}{h\left(\sqrt{1+h}+1\right)}=\lim_{h\to 0}\dfrac{h}{h\left(\sqrt{1+h}+1\right)}$

$=\lim_{h\to 0}\dfrac{1}{\sqrt{1+h}+1}=\dfrac{1}{\sqrt{1}+1}=\dfrac{1}{2}$

**17.** By the formula for the sum of cubes, we have

$$\lim_{x\to -2}\frac{x+2}{x^3+8}=\lim_{x\to -2}\frac{x+2}{(x+2)(x^2-2x+4)}=\lim_{x\to -2}\frac{1}{x^2-2x+4}=\frac{1}{4+4+4}=\frac{1}{12}.$$

**18.** $\lim_{x\to -1}\dfrac{x^2+2x+1}{x^4-1}=\lim_{x\to -1}\dfrac{(x+1)^2}{(x^2+1)(x^2-1)}=\lim_{x\to -1}\dfrac{(x+1)^2}{(x^2+1)(x+1)(x-1)}=\lim_{x\to -1}\dfrac{x+1}{(x^2+1)(x-1)}=\dfrac{0}{2(-2)}=0$

**19.** $\lim_{x\to 7}\dfrac{\sqrt{x+2}-3}{x-7}=\lim_{x\to 7}\dfrac{\sqrt{x+2}-3}{x-7}\cdot\dfrac{\sqrt{x+2}+3}{\sqrt{x+2}+3}=\lim_{x\to 7}\dfrac{(x+2)-9}{(x-7)\left(\sqrt{x+2}+3\right)}$

$=\lim_{x\to 7}\dfrac{x-7}{(x-7)\left(\sqrt{x+2}+3\right)}=\lim_{x\to 7}\dfrac{1}{\sqrt{x+2}+3}=\dfrac{1}{\sqrt{9}+3}=\dfrac{1}{6}$

**20.** $\lim_{h\to 0}\dfrac{(3+h)^{-1}-3^{-1}}{h}=\lim_{h\to 0}\dfrac{\dfrac{1}{3+h}-\dfrac{1}{3}}{h}=\lim_{h\to 0}\dfrac{3-(3+h)}{h(3+h)3}=\lim_{h\to 0}\dfrac{-h}{h(3+h)3}$

$=\lim_{h\to 0}\left[-\dfrac{1}{3(3+h)}\right]=-\dfrac{1}{\lim_{h\to 0}[3(3+h)]}=-\dfrac{1}{3(3+0)}=-\dfrac{1}{9}$

**21.** $\lim_{x\to -4}\dfrac{\dfrac{1}{4}+\dfrac{1}{x}}{4+x}=\lim_{x\to -4}\dfrac{\dfrac{x+4}{4x}}{4+x}=\lim_{x\to -4}\dfrac{x+4}{4x(4+x)}=\lim_{x\to -4}\dfrac{1}{4x}=\dfrac{1}{4(-4)}=-\dfrac{1}{16}$

**22.** $\lim_{t\to 0}\left(\dfrac{1}{t}-\dfrac{1}{t^2+t}\right)=\lim_{t\to 0}\dfrac{(t^2+t)-t}{t(t^2+t)}=\lim_{t\to 0}\dfrac{t^2}{t\cdot t(t+1)}=\lim_{t\to 0}\dfrac{1}{t+1}=\dfrac{1}{0+1}=1$

**23.** (a)

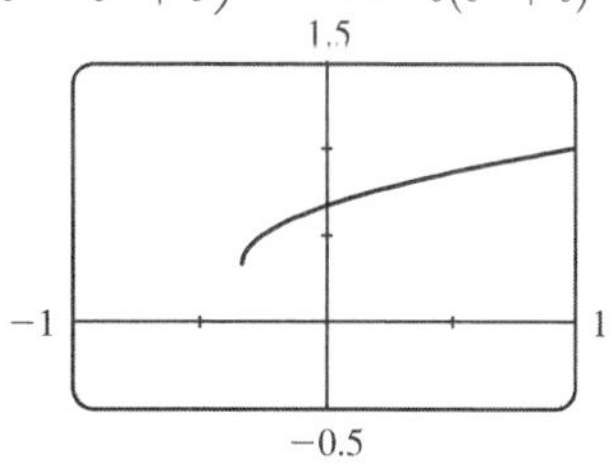


$\lim_{x\to 0}\dfrac{x}{\sqrt{1+3x}-1}\approx\dfrac{2}{3}$

(b)

| $x$ | $f(x)$ |
|---|---|
| −0.001 | 0.6661663 |
| −0.0001 | 0.6666167 |
| −0.00001 | 0.6666617 |
| −0.000001 | 0.6666662 |
| 0.000001 | 0.6666672 |
| 0.00001 | 0.6666717 |
| 0.0001 | 0.6667167 |
| 0.001 | 0.6671663 |

The limit appears to be $\dfrac{2}{3}$.

(c) $\displaystyle\lim_{x\to 0}\left(\frac{x}{\sqrt{1+3x}-1}\cdot\frac{\sqrt{1+3x}+1}{\sqrt{1+3x}+1}\right)=\lim_{x\to 0}\frac{x\left(\sqrt{1+3x}+1\right)}{(1+3x)-1}=\lim_{x\to 0}\frac{x\left(\sqrt{1+3x}+1\right)}{3x}$

$$=\frac{1}{3}\lim_{x\to 0}\left(\sqrt{1+3x}+1\right)\qquad\text{[Limit Law 3]}$$

$$=\frac{1}{3}\left[\sqrt{\lim_{x\to 0}(1+3x)}+\lim_{x\to 0}1\right]\qquad\text{[1 and 11]}$$

$$=\frac{1}{3}\left(\sqrt{\lim_{x\to 0}1+3\lim_{x\to 0}x}+1\right)\qquad\text{[1, 3, and 7]}$$

$$=\frac{1}{3}\left(\sqrt{1+3\cdot 0}+1\right)\qquad\text{[7 and 8]}$$

$$=\frac{1}{3}(1+1)=\frac{2}{3}$$

**24.** (a)

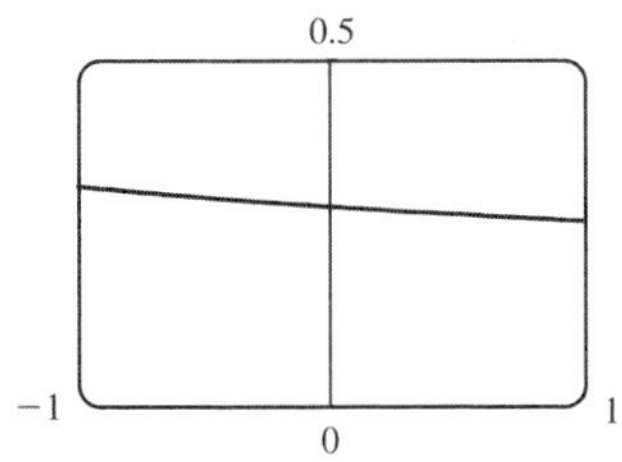


$\displaystyle\lim_{x\to 0}\frac{\sqrt{3+x}-\sqrt{3}}{x}\approx 0.29$

(b)

| $x$ | $f(x)$ |
|---|---|
| $-0.001$ | $0.2886992$ |
| $-0.0001$ | $0.2886775$ |
| $-0.00001$ | $0.2886754$ |
| $-0.000001$ | $0.2886752$ |
| $0.000001$ | $0.2886751$ |
| $0.00001$ | $0.2886749$ |
| $0.0001$ | $0.2886727$ |
| $0.001$ | $0.2886511$ |

The limit appears to be approximately $0.2887$.

(c) $\displaystyle\lim_{x\to 0}\left(\frac{\sqrt{3+x}-\sqrt{3}}{x}\cdot\frac{\sqrt{3+x}+\sqrt{3}}{\sqrt{3+x}+\sqrt{3}}\right)=\lim_{x\to 0}\frac{(3+x)-3}{x\left(\sqrt{3+x}+\sqrt{3}\right)}=\lim_{x\to 0}\frac{1}{\sqrt{3+x}+\sqrt{3}}$

$$=\frac{\lim_{x\to 0}1}{\lim_{x\to 0}\sqrt{3+x}+\lim_{x\to 0}\sqrt{3}}\qquad\text{[Limit Laws 5 and 1]}$$

$$=\frac{1}{\sqrt{\lim_{x\to 0}(3+x)}+\sqrt{3}}\qquad\text{[7 and 11]}$$

$$=\frac{1}{\sqrt{3+0}+\sqrt{3}}\qquad\text{[1, 7, and 8]}$$

$$=\frac{1}{2\sqrt{3}}$$

**25.** Let $f(x)=-x^2$, $g(x)=x^2\cos 20\pi x$ and $h(x)=x^2$.

Then $-1\le\cos 20\pi x\le 1\;\Rightarrow\;-x^2\le x^2\cos 20\pi x\le x^2\;\Rightarrow$

$f(x)\le g(x)\le h(x)$. So since $\lim_{x\to 0}f(x)=\lim_{x\to 0}h(x)=0$, by the Squeeze

Theorem we have $\lim_{x\to 0}g(x)=0$.

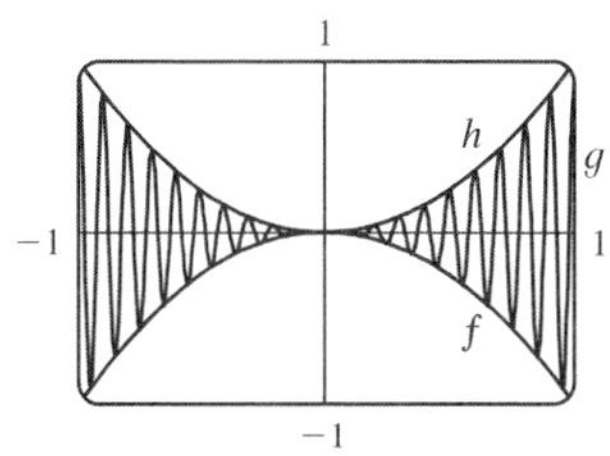


**26.** Let $f(x)=-\sqrt{x^3+x^2}$, $g(x)=\sqrt{x^3+x^2}\,\sin(\pi/x)$, and $h(x)=\sqrt{x^3+x^2}$.

Then $-1\le\sin(\pi/x)\le 1\;\Rightarrow$

$-\sqrt{x^3+x^2}\le\sqrt{x^3+x^2}\,\sin(\pi/x)\le\sqrt{x^3+x^2}\;\Rightarrow\;f(x)\le g(x)\le h(x)$. So

since $\lim_{x\to 0}f(x)=\lim_{x\to 0}h(x)=0$, by the Squeeze Theorem we have $\lim_{x\to 0}g(x)=0$.

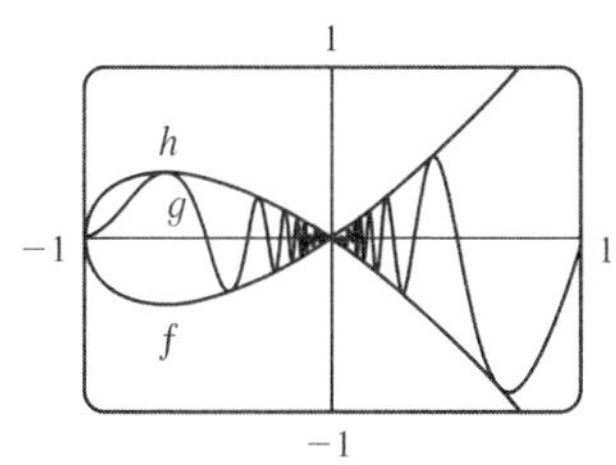

**27.** We have $\lim\limits_{x\to 4}(4x-9) = 4(4) - 9 = 7$ and $\lim\limits_{x\to 4}\left(x^2 - 4x + 7\right) = 4^2 - 4(4) + 7 = 7$. Since $4x - 9 \le f(x) \le x^2 - 4x + 7$ for $x \ge 0$, $\lim\limits_{x\to 4} f(x) = 7$ by the Squeeze Theorem.

**28.** We have $\lim\limits_{x\to 1}(2x) = 2(1) = 2$ and $\lim\limits_{x\to 1}(x^4 - x^2 + 2) = 1^4 - 1^2 + 2 = 2$. Since $2x \le g(x) \le x^4 - x^2 + 2$ for all $x$, $\lim\limits_{x\to 1} g(x) = 2$ by the Squeeze Theorem.

**29.** $-1 \le \cos(2/x) \le 1 \quad\Rightarrow\quad -x^4 \le x^4\cos(2/x) \le x^4$. Since $\lim\limits_{x\to 0}\left(-x^4\right) = 0$ and $\lim\limits_{x\to 0} x^4 = 0$, we have $\lim\limits_{x\to 0}\left[x^4\cos(2/x)\right] = 0$ by the Squeeze Theorem.

**30.** $-1 \le \sin(\pi/x) \le 1 \quad\Rightarrow\quad e^{-1} \le e^{\sin(\pi/x)} \le e^1 \quad\Rightarrow\quad \sqrt{x}/e \le \sqrt{x}\,e^{\sin(\pi/x)} \le \sqrt{x}\,e$. Since $\lim\limits_{x\to 0^+}\left(\sqrt{x}/e\right) = 0$ and $\lim\limits_{x\to 0^+}\left(\sqrt{x}\,e\right) = 0$, we have $\lim\limits_{x\to 0^+}\left[\sqrt{x}\,e^{\sin(\pi/x)}\right] = 0$ by the Squeeze Theorem.

**31.** $|x-3| = \begin{cases} x-3 & \text{if } x-3 \ge 0 \\ -(x-3) & \text{if } x-3 < 0 \end{cases} = \begin{cases} x-3 & \text{if } x \ge 3 \\ 3-x & \text{if } x < 3 \end{cases}$

Thus, $\lim\limits_{x\to 3^+}(2x + |x-3|) = \lim\limits_{x\to 3^+}(2x + x - 3) = \lim\limits_{x\to 3^+}(3x - 3) = 3(3) - 3 = 6$ and $\lim\limits_{x\to 3^-}(2x + |x-3|) = \lim\limits_{x\to 3^-}(2x + 3 - x) = \lim\limits_{x\to 3^-}(x + 3) = 3 + 3 = 6$. Since the left and right limits are equal, $\lim\limits_{x\to 3}(2x + |x-3|) = 6$.

**32.** $|x+6| = \begin{cases} x+6 & \text{if } x+6 \ge 0 \\ -(x+6) & \text{if } x+6 < 0 \end{cases} = \begin{cases} x+6 & \text{if } x \ge -6 \\ -(x+6) & \text{if } x < -6 \end{cases}$

We'll look at the one-sided limits.

$$\lim_{x\to -6^+}\frac{2x+12}{|x+6|} = \lim_{x\to -6^+}\frac{2(x+6)}{x+6} = 2 \quad\text{and}\quad \lim_{x\to -6^-}\frac{2x+12}{|x+6|} = \lim_{x\to -6^-}\frac{2(x+6)}{-(x+6)} = -2$$

The left and right limits are different, so $\lim\limits_{x\to -6}\dfrac{2x+12}{|x+6|}$ does not exist.

**33.** Since $|x| = -x$ for $x < 0$, we have $\lim\limits_{x\to 0^-}\left(\dfrac{1}{x} - \dfrac{1}{|x|}\right) = \lim\limits_{x\to 0^-}\left(\dfrac{1}{x} - \dfrac{1}{-x}\right) = \lim\limits_{x\to 0^-}\dfrac{2}{x}$, which does not exist since the denominator approaches 0 and the numerator does not.

**34.** Since $|x| = x$ for $x > 0$, we have $\lim\limits_{x\to 0^+}\left(\dfrac{1}{x} - \dfrac{1}{|x|}\right) = \lim\limits_{x\to 0^+}\left(\dfrac{1}{x} - \dfrac{1}{x}\right) = \lim\limits_{x\to 0^+} 0 = 0$.

**35.** (a) (i) If $x \to 1^+$, then $x > 1$ and $g(x) = x - 1$. Thus, $\lim\limits_{x\to 1^+} g(x) = \lim\limits_{x\to 1^+}(x - 1) = 1 - 1 = 0$.

(ii) If $x \to 1^-$, then $x < 1$ and $g(x) = 1 - x^2$. Thus, $\lim\limits_{x\to 1^-} g(x) = \lim\limits_{x\to 1^-}(1 - x^2) = 1 - 1^2 = 0$.

Since the left- and right-hand limits of $g$ at 1 are equal, $\lim\limits_{x\to 1} g(x) = 0$.

(iii) If $x \to 0$, then $-1 < x < 1$ and $g(x) = 1 - x^2$. Thus, $\lim\limits_{x\to 0} g(x) = \lim\limits_{x\to 0}(1 - x^2) = 1 - 0^2 = 1$.

(iv) If $x \to -1^-$, then $x < -1$ and $g(x) = -x$. Thus, $\lim\limits_{x\to -1^-} g(x) = \lim\limits_{x\to -1^-}(-x) = -(-1) = 1$.

(v) If $x \to -1^+$, then $-1 < x < 1$ and $g(x) = 1 - x^2$. Thus,

$$\lim_{x\to -1^+} g(x) = \lim_{x\to -1^+}(1 - x^2) = 1 - (-1)^2 = 1 - 1 = 0$$

(vi) $\lim\limits_{x\to -1} g(x)$ does not exist because the limits in part (iv) and part (v) are not equal.

(b)

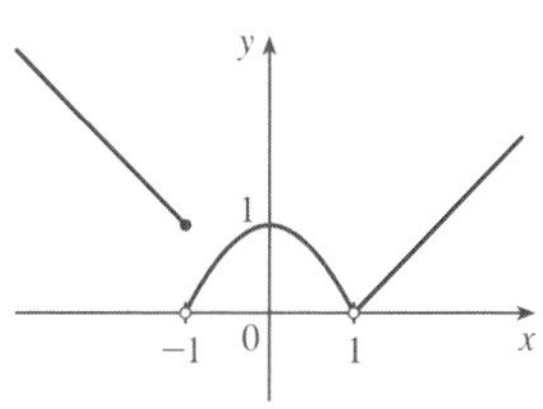

**36.** (a) (i) $\lim\limits_{x\to1^+}\dfrac{x^2-1}{|x-1|}=\lim\limits_{x\to1^+}\dfrac{x^2-1}{x-1}=\lim\limits_{x\to1^+}(x+1)=2$

(ii) $\lim\limits_{x\to1^-}\dfrac{x^2-1}{|x-1|}=\lim\limits_{x\to1^-}\dfrac{x^2-1}{-(x-1)}=\lim\limits_{x\to1^-}-(x+1)=-2$

(b) No, $\lim\limits_{x\to1}F(x)$ does not exist since $\lim\limits_{x\to1^+}F(x)\neq\lim\limits_{x\to1^-}F(x)$.

(c)

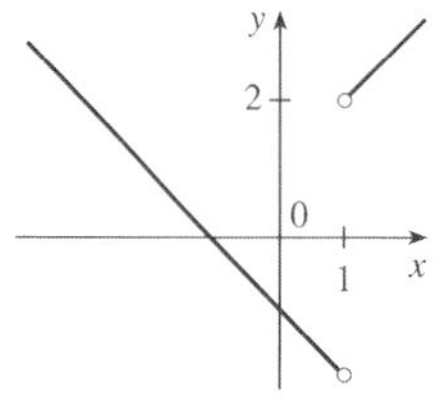


**37.** (a) (i) $[\![x]\!]=-2$ for $-2\le x<-1$, so $\lim\limits_{x\to-2^+}[\![x]\!]=\lim\limits_{x\to-2^+}(-2)=-2$

(ii) $[\![x]\!]=-3$ for $-3\le x<-2$, so $\lim\limits_{x\to-2^-}[\![x]\!]=\lim\limits_{x\to-2^-}(-3)=-3$.

The right and left limits are different, so $\lim\limits_{x\to-2}[\![x]\!]$ does not exist.

(iii) $[\![x]\!]=-3$ for $-3\le x<-2$, so $\lim\limits_{x\to-2.4}[\![x]\!]=\lim\limits_{x\to-2.4}(-3)=-3$.

(b) (i) $[\![x]\!]=n-1$ for $n-1\le x<n$, so $\lim\limits_{x\to n^-}[\![x]\!]=\lim\limits_{x\to n^-}(n-1)=n-1$.

(ii) $[\![x]\!]=n$ for $n\le x<n+1$, so $\lim\limits_{x\to n^+}[\![x]\!]=\lim\limits_{x\to n^+}n=n$.

(c) $\lim\limits_{x\to a}[\![x]\!]$ exists $\Leftrightarrow$ $a$ is not an integer.

**38.** (a)

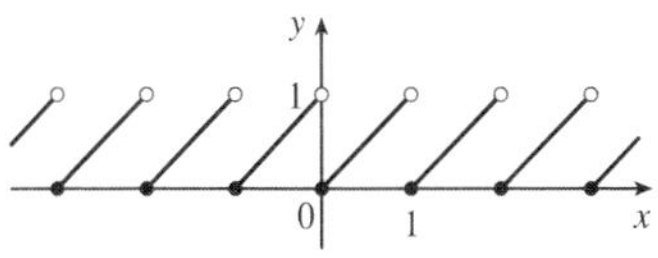


(b) (i) $\lim\limits_{x\to n^-}f(x)=\lim\limits_{x\to n^-}(x-[\![x]\!])=\lim\limits_{x\to n^-}[x-(n-1)]$

$=n-(n-1)=1$

(ii) $\lim\limits_{x\to n^+}f(x)=\lim\limits_{x\to n^+}(x-[\![x]\!])=\lim\limits_{x\to n^+}(x-n)=n-n=0$

(c) $\lim\limits_{x\to a}f(x)$ exists $\Leftrightarrow$ $a$ is not an integer.

**39.** The graph of $f(x)=[\![x]\!]+[\![-x]\!]$ is the same as the graph of $g(x)=-1$ with holes at each integer, since $f(a)=0$ for any integer $a$. Thus, $\lim\limits_{x\to2^-}f(x)=-1$ and $\lim\limits_{x\to2^+}f(x)=-1$, so $\lim\limits_{x\to2}f(x)=-1$. However, $f(2)=[\![2]\!]+[\![-2]\!]=2+(-2)=0$, so $\lim\limits_{x\to2}f(x)\neq f(2)$.

**40.** $\lim\limits_{v\to c^-}\left(L_0\sqrt{1-\dfrac{v^2}{c^2}}\right)=L_0\sqrt{1-1}=0$. As the velocity approaches the speed of light, the length approaches 0.

A left-hand limit is necessary since $L$ is not defined for $v>c$.

**41.** Since $p(x)$ is a polynomial, $p(x)=a_0+a_1x+a_2x^2+\cdots+a_nx^n$. Thus, by the Limit Laws,

$$\lim_{x\to a}p(x)=\lim_{x\to a}\left(a_0+a_1x+a_2x^2+\cdots+a_nx^n\right)=a_0+a_1\lim_{x\to a}x+a_2\lim_{x\to a}x^2+\cdots+a_n\lim_{x\to a}x^n$$

$$=a_0+a_1a+a_2a^2+\cdots+a_na^n=p(a)$$

Thus, for any polynomial $p$, $\lim\limits_{x\to a}p(x)=p(a)$.

**42.** Let $r(x)=\dfrac{p(x)}{q(x)}$ where $p(x)$ and $q(x)$ are any polynomials, and suppose that $q(a)\neq0$. Thus,

$$\lim_{x\to a}r(x)=\lim_{x\to a}\frac{p(x)}{q(x)}=\frac{\lim\limits_{x\to a}p(x)}{\lim\limits_{x\to a}q(x)}\quad\text{[Limit Law 5]}\quad=\frac{p(a)}{q(a)}\quad\text{[Exercise 41]}\quad=r(a).$$

**43.** Let $f(x)=[\![x]\!]$ and $g(x)=-[\![x]\!]$. Then $\lim\limits_{x\to3}f(x)$ and $\lim\limits_{x\to3}g(x)$ do not exist [Example 9]

but $\lim\limits_{x\to3}[f(x)+g(x)]=\lim\limits_{x\to3}([\![x]\!]-[\![x]\!])=\lim\limits_{x\to3}0=0$.

**44.** Let $f(x)=H(x)$ and $g(x)=1-H(x)$, where $H$ is the Heaviside function defined in Exercise 1.3.57.

Thus, either $f$ or $g$ is 0 for any value of $x$. Then $\lim\limits_{x\to0}f(x)$ and $\lim\limits_{x\to0}g(x)$ do not exist, but $\lim\limits_{x\to0}[f(x)g(x)]=\lim\limits_{x\to0}0=0$.

**45.** Since the denominator approaches 0 as $x \to -2$, the limit will exist only if the numerator also approaches 0 as $x \to -2$. In order for this to happen, we need $\lim_{x \to -2} (3x^2 + ax + a + 3) = 0 \Leftrightarrow$

$3(-2)^2 + a(-2) + a + 3 = 0 \Leftrightarrow 12 - 2a + a + 3 = 0 \Leftrightarrow a = 15$. With $a = 15$, the limit becomes

$$\lim_{x \to -2} \frac{3x^2 + 15x + 18}{x^2 + x - 2} = \lim_{x \to -2} \frac{3(x+2)(x+3)}{(x-1)(x+2)} = \lim_{x \to -2} \frac{3(x+3)}{x-1} = \frac{3(-2+3)}{-2-1} = \frac{3}{-3} = -1.$$

**46.** *Solution 1:* First, we find the coordinates of $P$ and $Q$ as functions of $r$. Then we can find the equation of the line determined by these two points, and thus find the $x$-intercept (the point $R$), and take the limit as $r \to 0$. The coordinates of $P$ are $(0, r)$. The point $Q$ is the point of intersection of the two circles $x^2 + y^2 = r^2$ and $(x-1)^2 + y^2 = 1$. Eliminating $y$ from these equations, we get $r^2 - x^2 = 1 - (x-1)^2 \Leftrightarrow r^2 = 1 + 2x - 1 \Leftrightarrow x = \frac{1}{2}r^2$. Substituting back into the equation of the shrinking circle to find the $y$-coordinate, we get $\left(\frac{1}{2}r^2\right)^2 + y^2 = r^2 \Leftrightarrow y^2 = r^2\left(1 - \frac{1}{4}r^2\right) \Leftrightarrow y = r\sqrt{1 - \frac{1}{4}r^2}$ (the positive $y$-value). So the coordinates of $Q$ are $\left(\frac{1}{2}r^2, r\sqrt{1 - \frac{1}{4}r^2}\right)$. The equation of the line joining $P$ and $Q$ is thus

$y - r = \dfrac{r\sqrt{1 - \frac{1}{4}r^2} - r}{\frac{1}{2}r^2 - 0}(x - 0)$. We set $y = 0$ in order to find the $x$-intercept, and get

$$x = -r\frac{\frac{1}{2}r^2}{r\left(\sqrt{1 - \frac{1}{4}r^2} - 1\right)} = \frac{-\frac{1}{2}r^2\left(\sqrt{1 - \frac{1}{4}r^2} + 1\right)}{1 - \frac{1}{4}r^2 - 1} = 2\left(\sqrt{1 - \frac{1}{4}r^2} + 1\right).$$

Now we take the limit as $r \to 0^+$: $\lim_{r \to 0^+} x = \lim_{r \to 0^+} 2\left(\sqrt{1 - \frac{1}{4}r^2} + 1\right) = \lim_{r \to 0^+} 2\left(\sqrt{1} + 1\right) = 4$. So the limiting position of $R$ is the point $(4, 0)$.

*Solution 2:* We add a few lines to the diagram, as shown. Note that $\angle PQS = 90°$ (subtended by diameter $PS$). So $\angle SQR = 90° = \angle OQT$ (subtended by diameter $OT$). It follows that $\angle OQS = \angle TQR$. Also $\angle PSQ = 90° - \angle SPQ = \angle ORP$. Since $\triangle QOS$ is isosceles, so is $\triangle QTR$, implying that $QT = TR$. As the circle $C_2$ shrinks, the point $Q$ plainly approaches the origin, so the point $R$ must approach a point twice as far from the origin as $T$, that is, the point $(4, 0)$, as above.

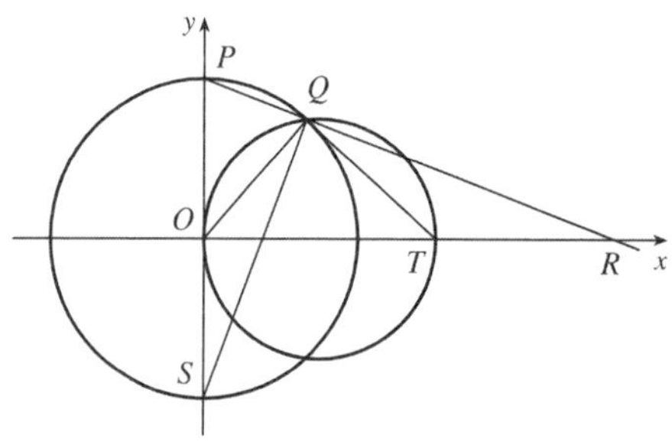


## 2.4 Continuity

1. From Definition 1, $\lim_{x \to 4} f(x) = f(4)$.
2. The graph of $f$ has no hole, jump, or vertical asymptote.
3. (a) The following are the numbers at which $f$ is discontinuous and the type of discontinuity at that number: $-4$ (removable), $-2$ (jump), 2 (jump), 4 (infinite).

   (b) $f$ is continuous from the left at $-2$ since $\lim_{x \to -2^-} f(x) = f(-2)$. $f$ is continuous from the right at 2 and 4 since $\lim_{x \to 2^+} f(x) = f(2)$ and $\lim_{x \to 4^+} f(x) = f(4)$. It is continuous from neither side at $-4$ since $f(-4)$ is undefined.

**4.** $g$ is continuous on $[-4,-2)$, $(-2,2)$, $[2,4)$, $(4,6)$, and $(6,8)$.

**5.** The graph of $y=f(x)$ must have a discontinuity at $x=3$ and must show that $\lim\limits_{x\to 3^-} f(x)=f(3)$.

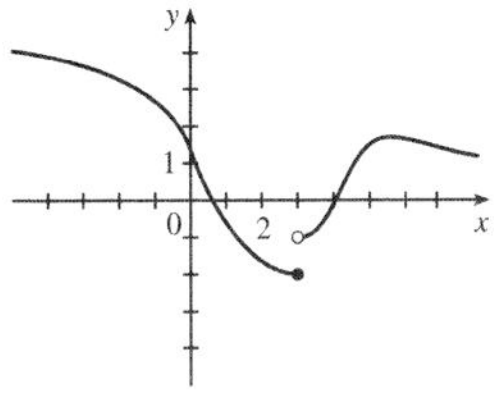


**6.**

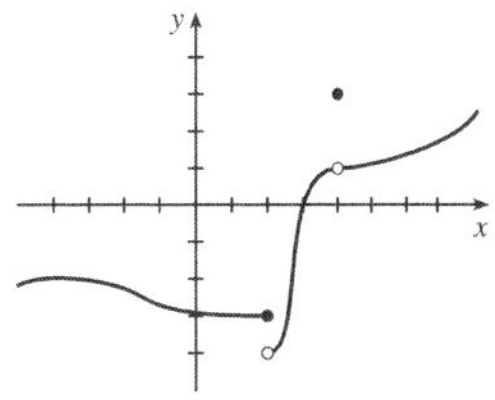


**7.** (a)

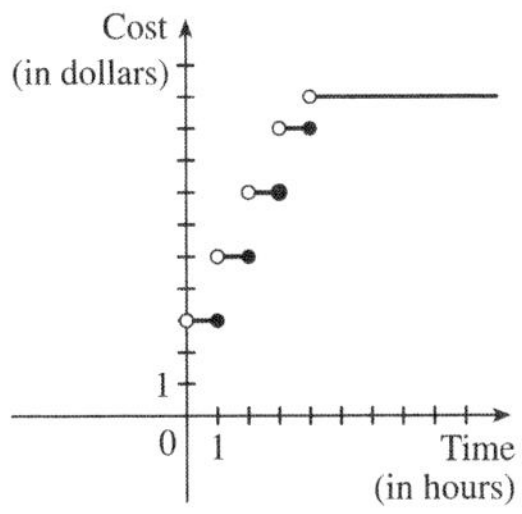


(b) There are discontinuities at times $t=1$, 2, 3, and 4. A person parking in the lot would want to keep in mind that the charge will jump at the beginning of each hour.

**8.** (a) Continuous; at the location in question, the temperature changes smoothly as time passes, without any instantaneous jumps from one temperature to another.

(b) Continuous; the temperature at a specific time changes smoothly as the distance due west from New York City increases, without any instantaneous jumps.

(c) Discontinuous; as the distance due west from New York City increases, the altitude above sea level may jump from one height to another without going through all of the intermediate values—at a cliff, for example.

(d) Discontinuous; as the distance traveled increases, the cost of the ride jumps in small increments.

(e) Discontinuous; when the lights are switched on (or off), the current suddenly changes between 0 and some nonzero value, without passing through all of the intermediate values. This is debatable, though, depending on your definition of current.

**9.** Since $f$ and $g$ are continuous functions,

$$\begin{aligned}\lim_{x\to 3}[2f(x)-g(x)] &= 2\lim_{x\to 3}f(x)-\lim_{x\to 3}g(x) && \text{[by Limit Laws 2 and 3]}\\ &= 2f(3)-g(3) && \text{[by continuity of } f \text{ and } g \text{ at } x=3]\\ &= 2\cdot 5-g(3)=10-g(3)\end{aligned}$$

Since it is given that $\lim\limits_{x\to 3}[2f(x)-g(x)]=4$, we have $10-g(3)=4$, so $g(3)=6$.

**10.** $\lim\limits_{x\to 4} f(x)=\lim\limits_{x\to 4}\left(x^2+\sqrt{7-x}\right)=\lim\limits_{x\to 4}x^2+\sqrt{\lim\limits_{x\to 4}7-\lim\limits_{x\to 4}x}=4^2+\sqrt{7-4}=16+\sqrt{3}=f(4)$.

By the definition of continuity, $f$ is continuous at $a=4$.

**11.** $\lim\limits_{x\to -1} f(x)=\lim\limits_{x\to -1}\left(x+2x^3\right)^4=\left(\lim\limits_{x\to -1}x+2\lim\limits_{x\to -1}x^3\right)^4=\left[-1+2(-1)^3\right]^4=(-3)^4=81=f(-1)$.

By the definition of continuity, $f$ is continuous at $a=-1$.

**12.** For $-4<a<4$ we have $\lim\limits_{x\to a} f(x)=\lim\limits_{x\to a}x\sqrt{16-x^2}=\lim\limits_{x\to a}x\sqrt{\lim\limits_{x\to a}16-\lim\limits_{x\to a}x^2}=a\sqrt{16-a^2}=f(a)$,

so $f$ is continuous on $(-4,4)$. Similarly, we get $\lim\limits_{x\to 4^-}f(x)=0=f(4)$ and $\lim\limits_{x\to -4^+}f(x)=0=f(-4)$,

so $f$ is continuous from the left at 4 and from the right at $-4$. Thus, $f$ is continuous on $[-4,4]$.

**13.** $f(x) = \ln|x - 2|$ is discontinuous at 2 since $f(2) = \ln 0$ is not defined.

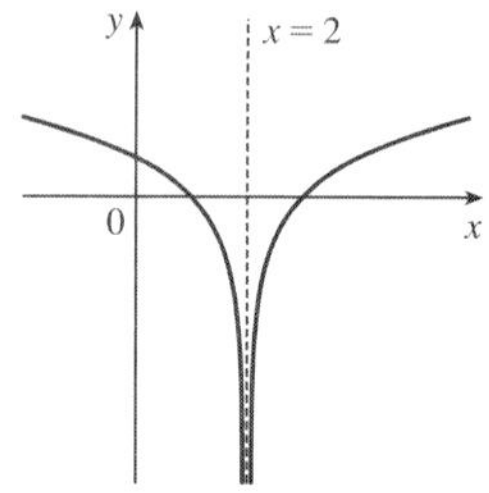


**14.** $f(x) = \begin{cases} 1/(x-1) & \text{if } x \neq 1 \\ 2 & \text{if } x = 1 \end{cases}$ is discontinuous at 1 because $\lim_{x \to 1} f(x)$ does not exist.

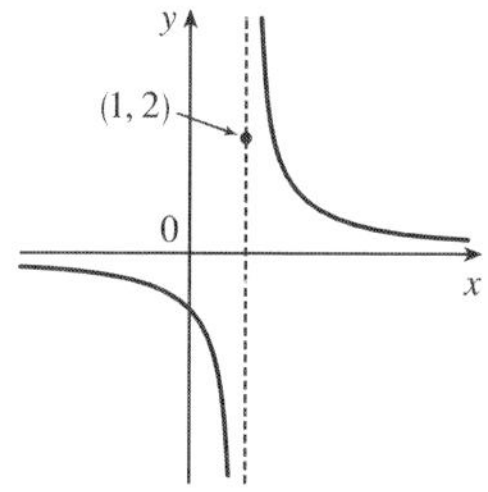


**15.** $f(x) = \begin{cases} e^x & \text{if } x < 0 \\ x^2 & \text{if } x \geq 0 \end{cases}$

The left-hand limit of $f$ at $a = 0$ is $\lim_{x \to 0^-} f(x) = \lim_{x \to 0^-} e^x = 1$. The right-hand limit of $f$ at $a = 0$ is $\lim_{x \to 0^+} f(x) = \lim_{x \to 0^+} x^2 = 0$. Since these limits are not equal, $\lim_{x \to 0} f(x)$ does not exist and $f$ is discontinuous at 0.

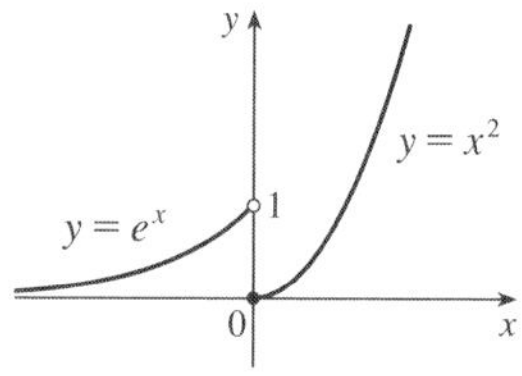


**16.** $f(x) = \begin{cases} \dfrac{x^2 - x}{x^2 - 1} & \text{if } x \neq 1 \\ 1 & \text{if } x = 1 \end{cases}$

$\lim_{x \to 1} f(x) = \lim_{x \to 1} \dfrac{x^2 - x}{x^2 - 1} = \lim_{x \to 1} \dfrac{x(x-1)}{(x+1)(x-1)} = \lim_{x \to 1} \dfrac{x}{x+1} = \dfrac{1}{2}$,

but $f(1) = 1$, so $f$ is discontinous at 1.

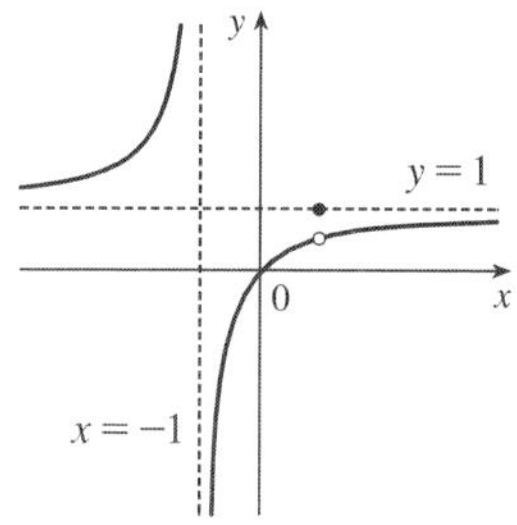


**17.** $F(x) = \dfrac{x}{x^2 + 5x + 6}$ is a rational function. So by Theorem 5 (or Theorem 7), $F$ is continuous at every number in its domain, $\{x \mid x^2 + 5x + 6 \neq 0\} = \{x \mid (x+3)(x+2) \neq 0\} = \{x \mid x \neq -3, -2\}$ or $(-\infty, -3) \cup (-3, -2) \cup (-2, \infty)$.

**18.** By Theorem 7, the trigonometric function $\sin x$ and the polynomial function $x + 1$ are continuous on $\mathbb{R}$. By part 5 of Theorem 4, $h(x) = \dfrac{\sin x}{x + 1}$ is continuous on its domain, $\{x \mid x \neq -1\}$.

**19.** By Theorem 5, the polynomials $x^2$ and $2x - 1$ are continuous on $(-\infty, \infty)$. By Theorem 7, the root function $\sqrt{x}$ is continuous on $[0, \infty)$. By Theorem 9, the composite function $\sqrt{2x - 1}$ is continuous on its domain, $[\frac{1}{2}, \infty)$. By part 1 of Theorem 4, the sum $R(x) = x^2 + \sqrt{2x - 1}$ is continuous on $[\frac{1}{2}, \infty)$.

**20.** By Theorem 5, the polynomial $x^2 - 1$ is continuous on $(-\infty, \infty)$. By Theorem 7, $\sin^{-1}$ is continuous on its domain, $[-1, 1]$. By Theorem 9, $\sin^{-1}(x^2 - 1)$ is continuous on its domain, which is $\{x \mid -1 \leq x^2 - 1 \leq 1\} = \{x \mid 0 \leq x^2 \leq 2\} = \{x \mid |x| \leq \sqrt{2}\} = [-\sqrt{2}, \sqrt{2}]$.

**21.** By Theorem 5, the polynomial $t^4 - 1$ is continuous on $(-\infty, \infty)$. By Theorem 7, $\ln x$ is continuous on its domain, $(0, \infty)$. By Theorem 9, $\ln(t^4 - 1)$ is continuous on its domain, which is $\{t \mid t^4 - 1 > 0\} = \{t \mid t^4 > 1\} = \{t \mid |t| > 1\} = (-\infty, -1) \cup (1, \infty)$.

**22.** By Theorem 7, $\sqrt{x}$ is continuous on $[0, \infty)$. By Theorems 7 and 9, $e^{\sqrt{x}}$ is continuous on $[0, \infty)$. Also by Theorems 7 and 9, $\cos\left(e^{\sqrt{x}}\right)$ is continuous on $[0, \infty)$.

**23.** The function $y = \dfrac{1}{1 + e^{1/x}}$ is discontinuous at $x = 0$ because the left- and right-hand limits at $x = 0$ are different.

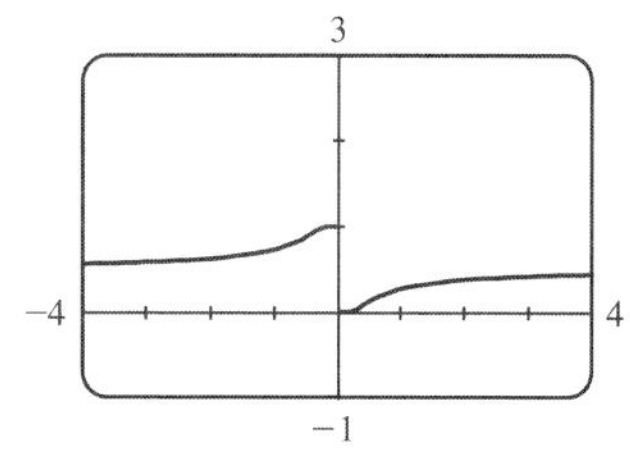


**24.** The function $y = \tan^2 x$ is discontinuous at $x = \frac{\pi}{2} + \pi k$, where $k$ is any integer. The function $y = \ln(\tan^2 x)$ is also discontinuous where $\tan^2 x$ is 0, that is, at $x = \pi k$. So $y = \ln(\tan^2 x)$ is discontinuous at $x = \frac{\pi}{2} n$, $n$ any integer.

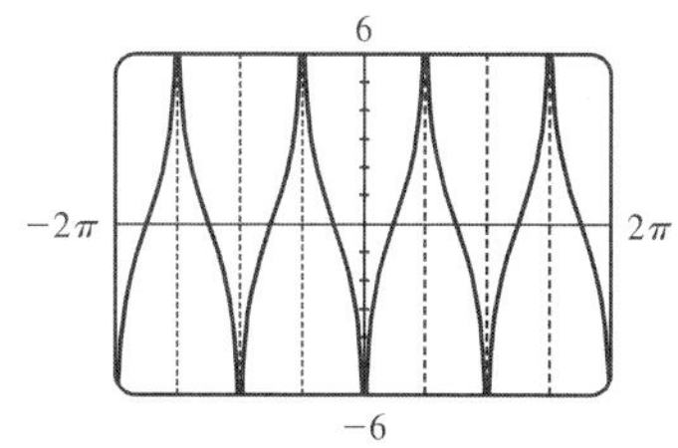


**25.** Because we are dealing with root functions, $5 + \sqrt{x}$ is continuous on $[0, \infty)$, $\sqrt{x + 5}$ is continuous on $[-5, \infty)$, so the quotient $f(x) = \dfrac{5 + \sqrt{x}}{\sqrt{5 + x}}$ is continuous on $[0, \infty)$. Since $f$ is continuous at $x = 4$, $\lim\limits_{x \to 4} f(x) = f(4) = \frac{7}{3}$.

**26.** Because $x$ is continuous on $\mathbb{R}$, $\sin x$ is continuous on $\mathbb{R}$, and $x + \sin x$ is continuous on $\mathbb{R}$, the composite function $f(x) = \sin(x + \sin x)$ is continuous on $\mathbb{R}$, so $\lim\limits_{x \to \pi} f(x) = f(\pi) = \sin(\pi + \sin \pi) = \sin \pi = 0$.

**27.** Because $x^2 - x$ is continuous on $\mathbb{R}$, the composite function $f(x) = e^{x^2 - x}$ is continuous on $\mathbb{R}$, so $\lim\limits_{x \to 1} f(x) = f(1) = e^{1 - 1} = e^0 = 1$.

**28.** Because arctan is a continuous function, we can apply Theorem 8.

$$\lim_{x \to 2} \arctan\left(\frac{x^2 - 4}{3x^2 - 6x}\right) = \arctan\left(\lim_{x \to 2} \frac{(x + 2)(x - 2)}{3x(x - 2)}\right) = \arctan\left(\lim_{x \to 2} \frac{x + 2}{3x}\right) = \arctan \tfrac{2}{3} \approx 0.588$$

**29.** $f(x) = \begin{cases} x^2 & \text{if } x < 1 \\ \sqrt{x} & \text{if } x \geq 1 \end{cases}$

By Theorem 5, since $f(x)$ equals the polynomial $x^2$ on $(-\infty, 1)$, $f$ is continuous on $(-\infty, 1)$. By Theorem 7, since $f(x)$ equals the root function $\sqrt{x}$ on $(1, \infty)$, $f$ is continuous on $(1, \infty)$. At $x = 1$, $\lim\limits_{x \to 1^-} f(x) = \lim\limits_{x \to 1^-} x^2 = 1$ and $\lim\limits_{x \to 1^+} f(x) = \lim\limits_{x \to 1^+} \sqrt{x} = 1$. Thus, $\lim\limits_{x \to 1} f(x)$ exists and equals 1. Also, $f(1) = \sqrt{1} = 1$. Thus, $f$ is continuous at $x = 1$. We conclude that $f$ is continuous on $(-\infty, \infty)$.

**30.** $f(x) = \begin{cases} \sin x & \text{if } x < \pi/4 \\ \cos x & \text{if } x \geq \pi/4 \end{cases}$

By Theorem 7, the trigonometric functions are continuous. Since $f(x) = \sin x$ on $(-\infty, \pi/4)$ and $f(x) = \cos x$ on $(\pi/4, \infty)$, $f$ is continuous on $(-\infty, \pi/4) \cup (\pi/4, \infty)$. $\lim\limits_{x \to (\pi/4)^-} f(x) = \lim\limits_{x \to (\pi/4)^-} \sin x = \sin \frac{\pi}{4} = 1/\sqrt{2}$ since the sine function is continuous at $\pi/4$. Similarly, $\lim\limits_{x\to(\pi/4)^+} f(x) = \lim\limits_{x\to(\pi/4)^+} \cos x = 1/\sqrt{2}$ by continuity of the cosine function at $\pi/4$. Thus, $\lim\limits_{x\to(\pi/4)} f(x)$ exists and equals $1/\sqrt{2}$, which agrees with the value $f(\pi/4)$. Therefore, $f$ is continuous at $\pi/4$, so $f$ is continuous on $(-\infty, \infty)$.

**31.** $f(x) = \begin{cases} x+2 & \text{if } x < 0 \\ e^x & \text{if } 0 \le x \le 1 \\ 2-x & \text{if } x > 1 \end{cases}$

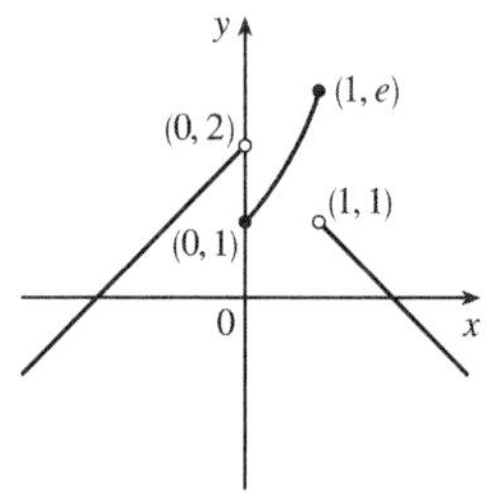


$f$ is continuous on $(-\infty, 0)$ and $(1, \infty)$ since on each of these intervals it is a polynomial; it is continuous on $(0, 1)$ since it is an exponential.

Now $\lim\limits_{x\to0^-} f(x) = \lim\limits_{x\to0^-} (x+2) = 2$ and $\lim\limits_{x\to0^+} f(x) = \lim\limits_{x\to0^+} e^x = 1$, so $f$ is discontinuous at 0. Since $f(0) = 1$, $f$ is continuous from the right at 0. Also $\lim\limits_{x\to1^-} f(x) = \lim\limits_{x\to1^-} e^x = e$ and $\lim\limits_{x\to1^+} f(x) = \lim\limits_{x\to1^+} (2-x) = 1$, so $f$ is discontinuous at 1. Since $f(1) = e$, $f$ is continuous from the left at 1.

**32.** By Theorem 5, each piece of $F$ is continuous on its domain. We need to check for continuity at $r = R$.

$\lim\limits_{r\to R^-} F(r) = \lim\limits_{r\to R^-} \dfrac{GMr}{R^3} = \dfrac{GM}{R^2}$ and $\lim\limits_{r\to R^+} F(r) = \lim\limits_{r\to R^+} \dfrac{GM}{r^2} = \dfrac{GM}{R^2}$, so $\lim\limits_{r\to R} F(r) = \dfrac{GM}{R^2}$. Since $F(R) = \dfrac{GM}{R^2}$, $F$ is continuous at $R$. Therefore, $F$ is a continuous function of $r$.

**33.** $f(x) = \begin{cases} cx^2 + 2x & \text{if } x < 2 \\ x^3 - cx & \text{if } x \ge 2 \end{cases}$

$f$ is continuous on $(-\infty, 2)$ and $(2, \infty)$. Now $\lim\limits_{x\to2^-} f(x) = \lim\limits_{x\to2^-} \left(cx^2 + 2x\right) = 4c + 4$ and $\lim\limits_{x\to2^+} f(x) = \lim\limits_{x\to2^+} \left(x^3 - cx\right) = 8 - 2c$. So $f$ is continuous $\Leftrightarrow$ $4c + 4 = 8 - 2c$ $\Leftrightarrow$ $6c = 4$ $\Leftrightarrow$ $c = \frac{2}{3}$. Thus, for $f$ to be continuous on $(-\infty, \infty)$, $c = \frac{2}{3}$.

**34.**

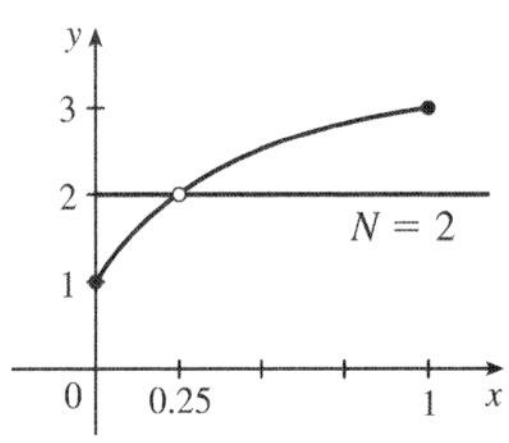


$f$ does not satisfy the conclusion of the Intermediate Value Theorem.

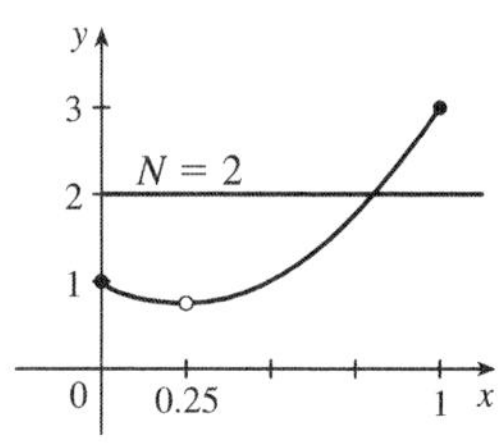


$f$ does satisfy the conclusion of the Intermediate Value Theorem.

**35.** $f(x) = x^2 + 10\sin x$ is continuous on the interval $[31, 32]$, $f(31) \approx 957$, and $f(32) \approx 1030$. Since $957 < 1000 < 1030$, there is a number $c$ in $(31, 32)$ such that $f(c) = 1000$ by the Intermediate Value Theorem. *Note:* There is also a number $c$ in $(-32, -31)$ such that $f(c) = 1000$.

**36.** $f(x) = x^2$ is continuous on the interval $[1, 2]$, $f(1) = 1$, and $f(2) = 4$. Since $1 < 2 < 4$, there is a number $c$ in $(1, 2)$ such that $f(c) = c^2 = 2$ by the Intermediate Value Theorem.

**37.** $f(x) = x^4 + x - 3$ is continuous on the interval $[1, 2]$, $f(1) = -1$, and $f(2) = 15$. Since $-1 < 0 < 15$, there is a number $c$ in $(1, 2)$ such that $f(c) = 0$ by the Intermediate Value Theorem. Thus, there is a root of the equation $x^4 + x - 3 = 0$ in the interval $(1, 2)$.

**38.** $f(x) = \sqrt[3]{x} + x - 1$ is continuous on the interval $[0, 1]$, $f(0) = -1$, and $f(1) = 1$. Since $-1 < 0 < 1$, there is a number $c$ in $(0, 1)$ such that $f(c) = 0$ by the Intermediate Value Theorem. Thus, there is a root of the equation $\sqrt[3]{x} + x - 1 = 0$, or $\sqrt[3]{x} = 1 - x$, in the interval $(0, 1)$.

**39.** $f(x) = \cos x - x$ is continuous on the interval $[0, 1]$, $f(0) = 1$, and $f(1) = \cos 1 - 1 \approx -0.46$. Since $-0.46 < 0 < 1$, there is a number $c$ in $(0, 1)$ such that $f(c) = 0$ by the Intermediate Value Theorem. Thus, there is a root of the equation $\cos x - x = 0$, or $\cos x = x$, in the interval $(0, 1)$.

**40.** $f(x) = \ln x - e^{-x}$ is continuous on the interval $[1, 2]$, $f(1) = -e^{-1} \approx -0.37$, and $f(2) = \ln 2 - e^{-2} \approx 0.56$. Since $-0.37 < 0 < 0.56$, there is a number $c$ in $(1, 2)$ such that $f(c) = 0$ by the Intermediate Value Theorem. Thus, there is a root of the equation $\ln x - e^{-x} = 0$, or $\ln x = e^{-x}$, in the interval $(1, 2)$.

**41.** (a) $f(x) = \cos x - x^3$ is continuous on the interval $[0, 1]$, $f(0) = 1 > 0$, and $f(1) = \cos 1 - 1 \approx -0.46 < 0$. Since $1 > 0 > -0.46$, there is a number $c$ in $(0, 1)$ such that $f(c) = 0$ by the Intermediate Value Theorem. Thus, there is a root of the equation $\cos x - x^3 = 0$, or $\cos x = x^3$, in the interval $(0, 1)$.

(b) $f(0.86) \approx 0.016 > 0$ and $f(0.87) \approx -0.014 < 0$, so there is a root between 0.86 and 0.87, that is, in the interval $(0.86, 0.87)$.

**42.** (a) $f(x) = \ln x - 3 + 2x$ is continuous on the interval $[1, 2]$, $f(1) = -1 < 0$, and $f(2) = \ln 2 + 1 \approx 1.7 > 0$. Since $-1 < 0 < 1.7$, there is a number $c$ in $(1, 2)$ such that $f(c) = 0$ by the Intermediate Value Theorem. Thus, there is a root of the equation $\ln x - 3 + 2x = 0$, or $\ln x = 3 - 2x$, in the interval $(1, 2)$.

(b) $f(1.34) \approx -0.03 < 0$ and $f(1.35) \approx 0.0001 > 0$, so there is a root between 1.34 and 1.35, that is, in the interval $(1.34, 1.35)$.

**43.** (a) Let $f(x) = 100e^{-x/100} - 0.01x^2$. Then $f(0) = 100 > 0$ and $f(100) = 100e^{-1} - 100 \approx -63.2 < 0$. So by the Intermediate Value Theorem, there is a number $c$ in $(0, 100)$ such that $f(c) = 0$. This implies that $100e^{-c/100} = 0.01c^2$.

(b) Using the intersect feature of the graphing device, we find that the root of the equation is $x = 70.347$, correct to three decimal places.

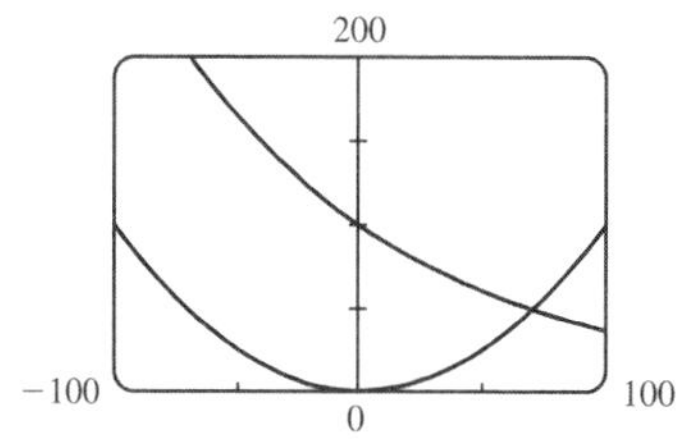


**44.** (a) Let $f(x) = \arctan x + x - 1$. Then $f(0) = -1 < 0$ and $f(1) = \frac{\pi}{4} > 0$. So by the Intermediate Value Theorem, there is a number $c$ in $(0, 1)$ such that $f(c) = 0$. This implies that $\arctan c = 1 - c$.

(b) Using the intersect feature of the graphing device, we find that the root of the equation is $x = 0.520$, correct to three decimal places.

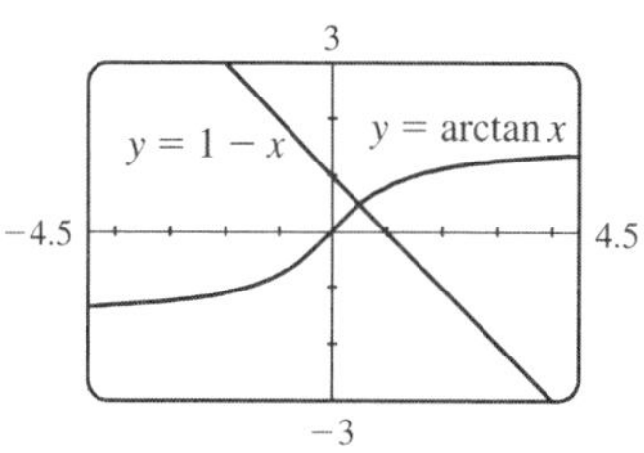


**45.** $$\lim_{h\to 0} \sin(a + h) = \lim_{h\to 0} (\sin a \cos h + \cos a \sin h) = \lim_{h\to 0} (\sin a \cos h) + \lim_{h\to 0} (\cos a \sin h)$$
$$= \left(\lim_{h\to 0} \sin a\right)\left(\lim_{h\to 0} \cos h\right) + \left(\lim_{h\to 0} \cos a\right)\left(\lim_{h\to 0} \sin h\right) = (\sin a)(1) + (\cos a)(0) = \sin a$$

**46.** As in the previous exercise, we must show that $\lim_{h\to 0} \cos(a+h) = \cos a$ to prove that the cosine function is continuous.

$$\begin{aligned}\lim_{h\to 0} \cos(a+h) &= \lim_{h\to 0} (\cos a \cos h - \sin a \sin h) = \lim_{h\to 0} (\cos a \cos h) - \lim_{h\to 0} (\sin a \sin h) \\ &= \left(\lim_{h\to 0} \cos a\right)\left(\lim_{h\to 0} \cos h\right) - \left(\lim_{h\to 0} \sin a\right)\left(\lim_{h\to 0} \sin h\right) \\ &= (\cos a)(1) - (\sin a)(0) = \cos a\end{aligned}$$

**47.** If there is such a number, it satisfies the equation $x^3 + 1 = x \Leftrightarrow x^3 - x + 1 = 0$. Let the left-hand side of this equation be called $f(x)$. Now $f(-2) = -5 < 0$, and $f(-1) = 1 > 0$. Note also that $f(x)$ is a polynomial, and thus continuous. So by the Intermediate Value Theorem, there is a number $c$ between $-2$ and $-1$ such that $f(c) = 0$, so that $c = c^3 + 1$.

**48.** (a) $\lim_{x\to 0^+} F(x) = 0$ and $\lim_{x\to 0^-} F(x) = 0$, so $\lim_{x\to 0} F(x) = 0$, which is $F(0)$, and hence $F$ is continuous at $x = a$ if $a = 0$. For $a > 0$, $\lim_{x\to a} F(x) = \lim_{x\to a} x = a = F(a)$. For $a < 0$, $\lim_{x\to a} F(x) = \lim_{x\to a} (-x) = -a = F(a)$. Thus, $F$ is continuous at $x = a$; that is, continuous everywhere.

(b) Assume that $f$ is continuous on the interval $I$. Then for $a \in I$, $\lim_{x\to a} |f(x)| = \left|\lim_{x\to a} f(x)\right| = |f(a)|$ by Theorem 8. (If $a$ is an endpoint of $I$, use the appropriate one-sided limit.) So $|f|$ is continuous on $I$.

(c) No, the converse is false. For example, the function $f(x) = \begin{cases} 1 & \text{if } x \geq 0 \\ -1 & \text{if } x < 0 \end{cases}$ is not continuous at $x = 0$, but $|f(x)| = 1$ is continuous on $\mathbb{R}$.

**49.** Define $u(t)$ to be the monk's distance from the monastery, as a function of time, on the first day, and define $d(t)$ to be his distance from the monastery, as a function of time, on the second day. Let $D$ be the distance from the monastery to the top of the mountain. From the given information we know that $u(0) = 0$, $u(12) = D$, $d(0) = D$ and $d(12) = 0$. Now consider the function $u - d$, which is clearly continuous. We calculate that $(u - d)(0) = -D$ and $(u - d)(12) = D$. So by the Intermediate Value Theorem, there must be some time $t_0$ between 0 and 12 such that $(u - d)(t_0) = 0 \Leftrightarrow u(t_0) = d(t_0)$. So at time $t_0$ after 7:00 A.M., the monk will be at the same place on both days.

## 2.5 Limits Involving Infinity

**1.** (a) As $x$ approaches 2 (from the right or the left), the values of $f(x)$ become large.

(b) As $x$ approaches 1 from the right, the values of $f(x)$ become large negative.

(c) As $x$ becomes large, the values of $f(x)$ approach 5.

(d) As $x$ becomes large negative, the values of $f(x)$ approach 3.

**2.** (a) The graph of a function can intersect a vertical asymptote in the sense that it can meet but not cross it.

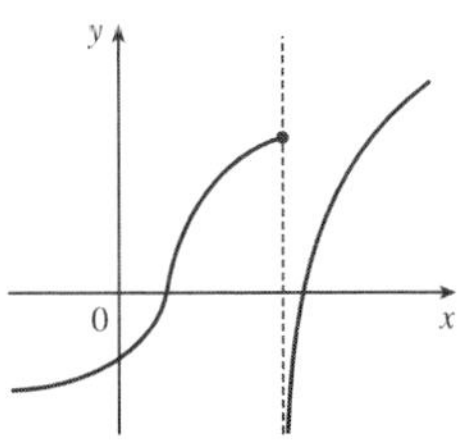


The graph of a function can intersect a horizontal asymptote. It can even intersect its horizontal asymptote an infinite number of times.

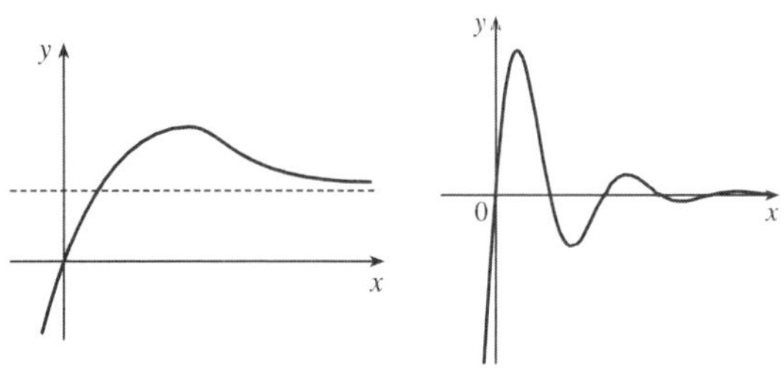

(b) The graph of a function can have 0, 1, or 2 horizontal asymptotes. Representative examples are shown.

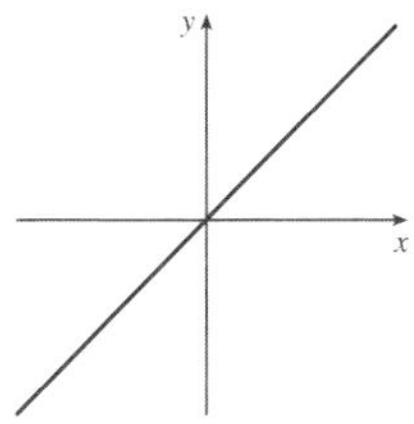
No horizontal asymptote

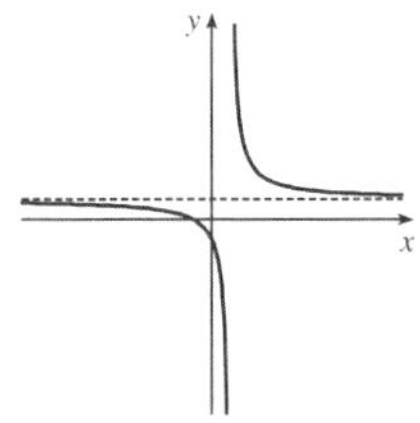
One horizontal asymptote

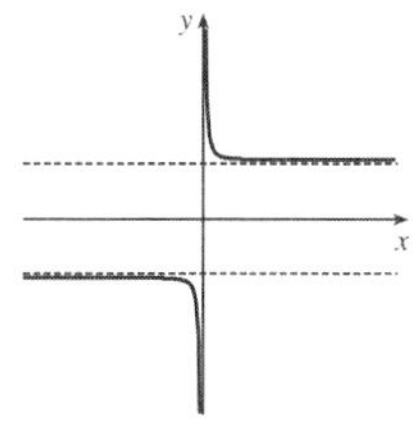
Two horizontal asymptotes

**3.** (a) $\lim_{x\to 2} f(x) = \infty$ (b) $\lim_{x\to -1^-} f(x) = \infty$ (c) $\lim_{x\to -1^+} f(x) = -\infty$

(d) $\lim_{x\to \infty} f(x) = 1$ (e) $\lim_{x\to -\infty} f(x) = 2$ (f) Vertical: $x = -1$, $x = 2$; Horizontal: $y = 1$, $y = 2$

**4.** (a) $\lim_{x\to \infty} g(x) = 2$ (b) $\lim_{x\to -\infty} g(x) = -2$ (c) $\lim_{x\to 3} g(x) = \infty$

(d) $\lim_{x\to 0} g(x) = -\infty$ (e) $\lim_{x\to -2^+} g(x) = -\infty$ (f) Vertical: $x = -2$, $x = 0$, $x = 3$; Horizontal: $y = -2$, $y = 2$

**5.** $f(0) = 0,\quad f(1) = 1,$

$\lim_{x\to\infty} f(x) = 0,$

$f$ is odd

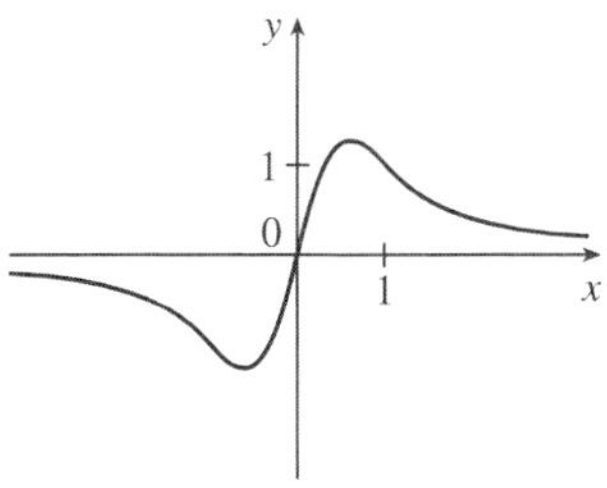


**6.** $\lim_{x\to 0^+} f(x) = \infty,\quad \lim_{x\to 0^-} f(x) = -\infty,$

$\lim_{x\to\infty} f(x) = 1,\quad \lim_{x\to -\infty} f(x) = 1$

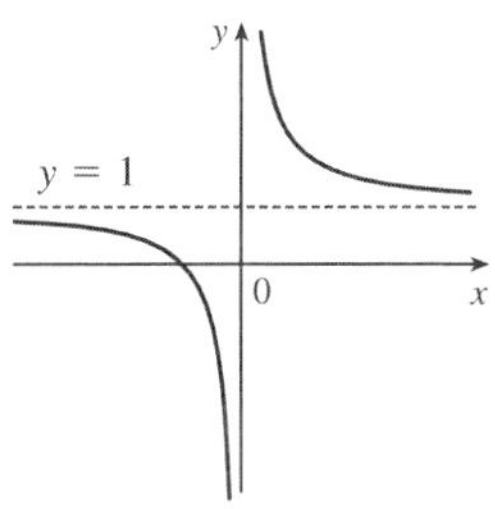


**7.** $\lim_{x\to 2} f(x) = -\infty,\quad \lim_{x\to\infty} f(x) = \infty,$

$\lim_{x\to -\infty} f(x) = 0,\quad \lim_{x\to 0^+} f(x) = \infty,$

$\lim_{x\to 0^-} f(x) = -\infty$

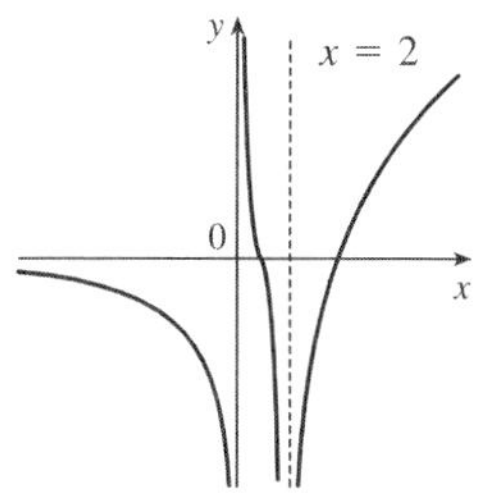


**8.** $\lim_{x\to -2} f(x) = \infty,$

$\lim_{x\to -\infty} f(x) = 3,$

$\lim_{x\to\infty} f(x) = -3$

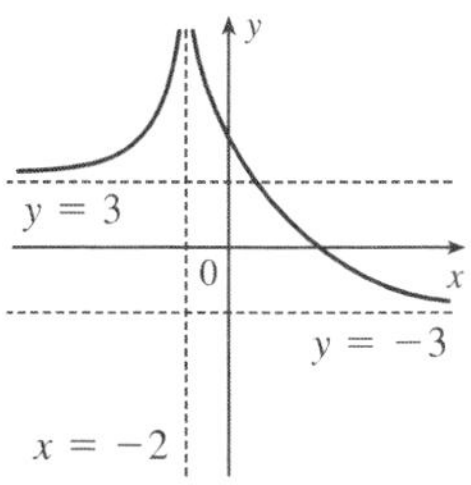


**9.** $f(0) = 3,\quad \lim_{x\to 0^-} f(x) = 4,$

$\lim_{x\to 0^+} f(x) = 2,$

$\lim_{x\to -\infty} f(x) = -\infty,\quad \lim_{x\to 4^-} f(x) = -\infty,$

$\lim_{x\to 4^+} f(x) = \infty,\quad \lim_{x\to\infty} f(x) = 3$

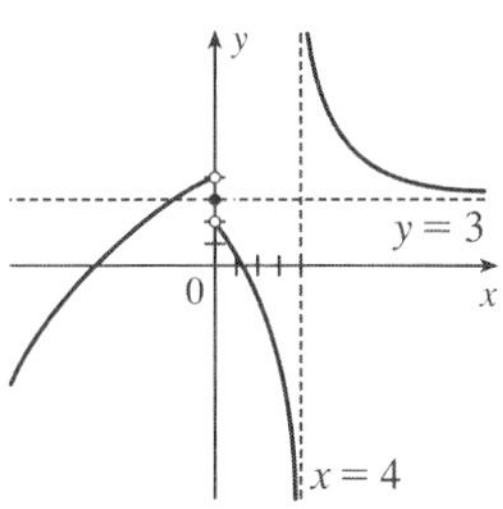


**10.** $\lim_{x\to 3} f(x) = -\infty,\quad \lim_{x\to\infty} f(x) = 2,$

$f(0) = 0,\quad f$ is even

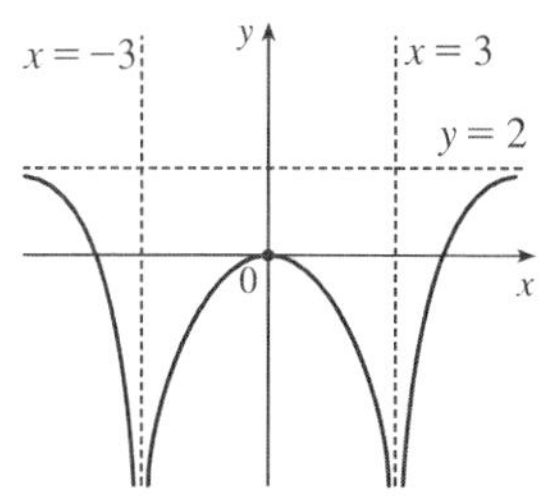

**11.** If $f(x) = x^2/2^x$, then a calculator gives $f(0) = 0$, $f(1) = 0.5$, $f(2) = 1$, $f(3) = 1.125$, $f(4) = 1$, $f(5) = 0.78125$, $f(6) = 0.5625$, $f(7) = 0.3828125$, $f(8) = 0.25$, $f(9) = 0.158203125$, $f(10) = 0.09765625$, $f(20) \approx 0.00038147$, $f(50) \approx 2.2204 \times 10^{-12}$, $f(100) \approx 7.8886 \times 10^{-27}$.

It appears that $\lim_{x\to\infty} (x^2/2^x) = 0$.

**12.** (a) $f(x) = 1/(x^3 - 1)$

| $x$ | $f(x)$ |
|---|---|
| 0.5 | −1.14 |
| 0.9 | −3.69 |
| 0.99 | −33.7 |
| 0.999 | −333.7 |
| 0.9999 | −3333.7 |
| 0.99999 | −33,333.7 |

| $x$ | $f(x)$ |
|---|---|
| 1.5 | 0.42 |
| 1.1 | 3.02 |
| 1.01 | 33.0 |
| 1.001 | 333.0 |
| 1.0001 | 3333.0 |
| 1.00001 | 33,333.3 |

From these calculations, it seems that $\lim_{x\to 1^-} f(x) = -\infty$ and $\lim_{x\to 1^+} f(x) = \infty$.

(b) If $x$ is slightly smaller than 1, then $x^3 - 1$ will be a negative number close to 0, and the reciprocal of $x^3 - 1$, that is, $f(x)$, will be a negative number with large absolute value. So $\lim_{x\to 1^-} f(x) = -\infty$.

If $x$ is slightly larger than 1, then $x^3 - 1$ will be a small positive number, and its reciprocal, $f(x)$, will be a large positive number. So $\lim_{x\to 1^+} f(x) = \infty$.

(c) It appears from the graph of $f$ that $\lim_{x\to 1^-} f(x) = -\infty$ and $\lim_{x\to 1^+} f(x) = \infty$.

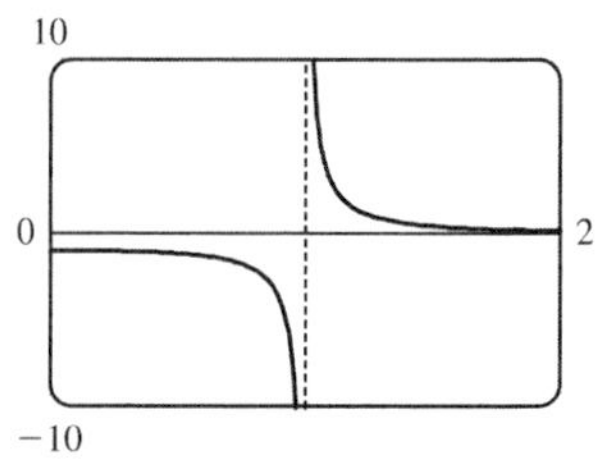


**13.** Vertical: $x \approx -1.62$, $x \approx 0.62$, $x = 1$;

Horizontal: $y = 1$

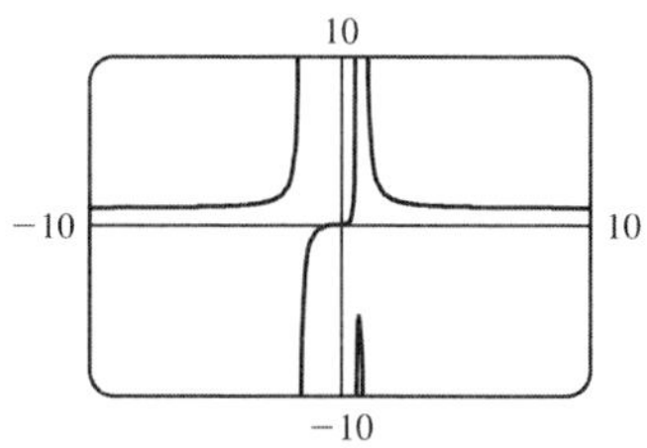


**14.** (a) From a graph of $f(x) = (1 - 2/x)^x$ in a window of $[0, 10{,}000]$ by $[0, 0.2]$, we estimate that $\lim_{x\to\infty} f(x) = 0.14$ (to two decimal places.)

(b)

| $x$ | $f(x)$ |
|---|---|
| 10,000 | 0.135308 |
| 100,000 | 0.135333 |
| 1,000,000 | 0.135335 |

From the table, we estimate that $\lim_{x\to\infty} f(x) = 0.1353$ (to four decimal places.)

**15.** $\displaystyle\lim_{x\to -3^+} \frac{x+2}{x+3} = -\infty$ since the numerator is negative and the denominator approaches 0 from the positive side as $x \to -3^+$.

**16.** $\displaystyle\lim_{x\to 5^-} \frac{e^x}{(x-5)^3} = -\infty$ since the numerator is positive and the denominator approaches 0 from the negative side as $x \to 5^-$.

**17.** $\lim_{x\to 1}\dfrac{2-x}{(x-1)^2}=\infty$ since the numerator is positive and the denominator approaches 0 through positive values as $x\to 1$.

**18.** $\lim_{x\to\pi^-}\cot x=\lim_{x\to\pi^-}\dfrac{\cos x}{\sin x}=-\infty$ since the numerator is negative and the denominator approaches 0 through positive values as $x\to\pi^-$.

**19.** Let $t=x^2-9$. Then as $x\to 3^+$, $t\to 0^+$, and $\lim_{x\to 3^+}\ln(x^2-9)=\lim_{t\to 0^+}\ln t=-\infty$ by (3).

**20.** $$\lim_{x\to\infty}\frac{3x+5}{x-4}=\lim_{x\to\infty}\frac{(3x+5)/x}{(x-4)/x}=\lim_{x\to\infty}\frac{3+5/x}{1-4/x}=\frac{\lim_{x\to\infty}3+5\lim_{x\to\infty}\frac{1}{x}}{\lim_{x\to\infty}1-4\lim_{x\to\infty}\frac{1}{x}}=\frac{3+5(0)}{1-4(0)}=3$$

**21.** Divide both the numerator and denominator by $x^3$ (the highest power of $x$ that occurs in the denominator).

$$\lim_{x\to\infty}\frac{x^3+5x}{2x^3-x^2+4}=\lim_{x\to\infty}\frac{\dfrac{x^3+5x}{x^3}}{\dfrac{2x^3-x^2+4}{x^3}}=\lim_{x\to\infty}\frac{1+\dfrac{5}{x^2}}{2-\dfrac{1}{x}+\dfrac{4}{x^3}}=\frac{\lim_{x\to\infty}\left(1+\dfrac{5}{x^2}\right)}{\lim_{x\to\infty}\left(2-\dfrac{1}{x}+\dfrac{4}{x^3}\right)}$$

$$=\frac{\lim_{x\to\infty}1+5\lim_{x\to\infty}\dfrac{1}{x^2}}{\lim_{x\to\infty}2-\lim_{x\to\infty}\dfrac{1}{x}+4\lim_{x\to\infty}\dfrac{1}{x^3}}=\frac{1+5(0)}{2-0+4(0)}=\frac{1}{2}$$

**22.** $$\lim_{t\to-\infty}\frac{t^2+2}{t^3+t^2-1}=\lim_{t\to-\infty}\frac{(t^2+2)/t^3}{(t^3+t^2-1)/t^3}=\lim_{t\to-\infty}\frac{1/t+2/t^3}{1+1/t-1/t^3}=\frac{0+0}{1+0-0}=0$$

**23.** First, multiply the factors in the denominator. Then divide both the numerator and denominator by $u^4$.

$$\lim_{u\to\infty}\frac{4u^4+5}{(u^2-2)(2u^2-1)}=\lim_{u\to\infty}\frac{4u^4+5}{2u^4-5u^2+2}=\lim_{u\to\infty}\frac{\dfrac{4u^4+5}{u^4}}{\dfrac{2u^4-5u^2+2}{u^4}}=\lim_{u\to\infty}\frac{4+\dfrac{5}{u^4}}{2-\dfrac{5}{u^2}+\dfrac{2}{u^4}}$$

$$=\frac{\lim_{u\to\infty}\left(4+\dfrac{5}{u^4}\right)}{\lim_{u\to\infty}\left(2-\dfrac{5}{u^2}+\dfrac{2}{u^4}\right)}=\frac{\lim_{u\to\infty}4+5\lim_{u\to\infty}\dfrac{1}{u^4}}{\lim_{u\to\infty}2-5\lim_{u\to\infty}\dfrac{1}{u^2}+2\lim_{u\to\infty}\dfrac{1}{u^4}}=\frac{4+5(0)}{2-5(0)+2(0)}=\frac{4}{2}=2$$

**24.** $$\lim_{x\to\infty}\frac{x+2}{\sqrt{9x^2+1}}=\lim_{x\to\infty}\frac{(x+2)/x}{\sqrt{9x^2+1}/\sqrt{x^2}}=\lim_{x\to\infty}\frac{1+2/x}{\sqrt{9+1/x^2}}=\frac{1+0}{\sqrt{9+0}}=\frac{1}{3}$$

**25.** $$\lim_{x\to\infty}\left(\sqrt{9x^2+x}-3x\right)=\lim_{x\to\infty}\frac{\left(\sqrt{9x^2+x}-3x\right)\left(\sqrt{9x^2+x}+3x\right)}{\sqrt{9x^2+x}+3x}=\lim_{x\to\infty}\frac{\left(\sqrt{9x^2+x}\right)^2-(3x)^2}{\sqrt{9x^2+x}+3x}$$

$$=\lim_{x\to\infty}\frac{(9x^2+x)-9x^2}{\sqrt{9x^2+x}+3x}=\lim_{x\to\infty}\frac{x}{\sqrt{9x^2+x}+3x}\cdot\frac{1/x}{1/x}$$

$$=\lim_{x\to\infty}\frac{x/x}{\sqrt{9x^2/x^2+x/x^2}+3x/x}=\lim_{x\to\infty}\frac{1}{\sqrt{9+1/x}+3}=\frac{1}{\sqrt{9}+3}=\frac{1}{3+3}=\frac{1}{6}$$

**26.** $$\lim_{x\to\infty}\left(\sqrt{x^2+ax}-\sqrt{x^2+bx}\right)=\lim_{x\to\infty}\frac{\left(\sqrt{x^2+ax}-\sqrt{x^2+bx}\right)\left(\sqrt{x^2+ax}+\sqrt{x^2+bx}\right)}{\sqrt{x^2+ax}+\sqrt{x^2+bx}}$$

$$=\lim_{x\to\infty}\frac{(x^2+ax)-(x^2+bx)}{\sqrt{x^2+ax}+\sqrt{x^2+bx}}=\lim_{x\to\infty}\frac{[(a-b)x]/x}{\left(\sqrt{x^2+ax}+\sqrt{x^2+bx}\right)/\sqrt{x^2}}$$

$$=\lim_{x\to\infty}\frac{a-b}{\sqrt{1+a/x}+\sqrt{1+b/x}}=\frac{a-b}{\sqrt{1+0}+\sqrt{1+0}}=\frac{a-b}{2}$$

**27.** $\lim_{x\to\infty} \cos x$ does not exist because as $x$ increases $\cos x$ does not approach any one value, but oscillates between 1 and $-1$.

**28.** Since $0 \le \sin^2 x \le 1$, we have $0 \le \frac{\sin^2 x}{x^2} \le \frac{1}{x^2}$. Now $\lim_{x\to\infty} 0 = 0$ and $\lim_{x\to\infty} \frac{1}{x^2} = 0$, so by the Squeeze Theorem, $\lim_{x\to\infty} \frac{\sin^2 x}{x^2} = 0$.

**29.** Since $-1 \le \cos x \le 1$ and $e^{-2x} > 0$, we have $-e^{-2x} \le e^{-2x}\cos x \le e^{-2x}$. We know that $\lim_{x\to\infty}(-e^{-2x}) = 0$ and $\lim_{x\to\infty}(e^{-2x}) = 0$, so by the Squeeze Theorem, $\lim_{x\to\infty}(e^{-2x}\cos x) = 0$.

**30.** $\lim_{x\to\infty} \tan^{-1}(x^2 - x^4) = \lim_{x\to\infty} \tan^{-1}(x^2(1-x^2))$. If we let $t = x^2(1-x^2)$, we know that $t \to -\infty$ as $x \to \infty$, since $x^2 \to \infty$ and $1 - x^2 \to -\infty$. So $\lim_{x\to\infty} \tan^{-1}(x^2(1-x^2)) = \lim_{t\to-\infty} \tan^{-1} t = -\frac{\pi}{2}$.

**31.** $\lim_{x\to-\infty}(x^4 + x^5) = \lim_{x\to-\infty} x^5(\frac{1}{x} + 1)$ [factor out the largest power of $x$] $= -\infty$ because $x^5 \to -\infty$ and $1/x + 1 \to 1$ as $x \to -\infty$.

**32.** If we let $t = \tan x$, then as $x \to (\pi/2)^+$, $t \to -\infty$. Thus, $\lim_{x\to(\pi/2)^+} e^{\tan x} = \lim_{t\to-\infty} e^t = 0$.

**33.** 
$$\lim_{x\to\infty} \frac{x + x^3 + x^5}{1 - x^2 + x^4} = \lim_{x\to\infty} \frac{(x + x^3 + x^5)/x^4}{(1 - x^2 + x^4)/x^4} \quad \text{[divide by the highest power of } x \text{ in the denominator]}$$
$$= \lim_{x\to\infty} \frac{1/x^3 + 1/x + x}{1/x^4 - 1/x^2 + 1} = \infty$$

because $(1/x^3 + 1/x + x) \to \infty$ and $(1/x^4 - 1/x^2 + 1) \to 1$ as $x \to \infty$.

**34.** (a)

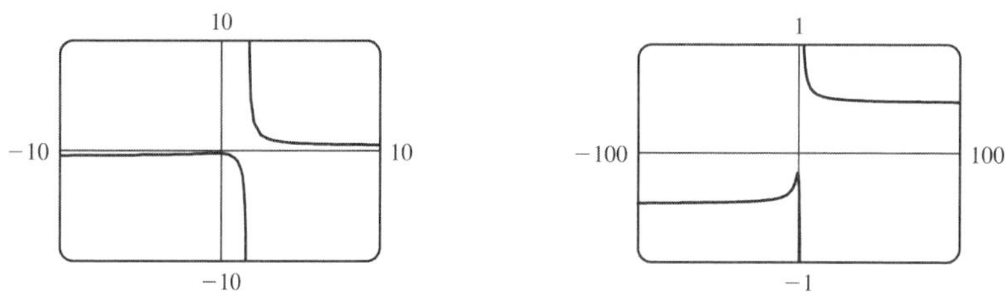


From the graph, it appears at first that there is only one horizontal asymptote, at $y \approx 0$, and a vertical asymptote at $x \approx 1.7$. However, if we graph the function with a wider viewing rectangle, we see that in fact there seem to be two horizontal asymptotes: one at $y \approx 0.5$ and one at $y \approx -0.5$. So we estimate that

$$\lim_{x\to\infty} \frac{\sqrt{2x^2+1}}{3x-5} \approx 0.5 \qquad \text{and} \qquad \lim_{x\to-\infty} \frac{\sqrt{2x^2+1}}{3x-5} \approx -0.5$$

(b) $f(1000) \approx 0.4722$ and $f(10{,}000) \approx 0.4715$, so we estimate that $\lim_{x\to\infty} \frac{\sqrt{2x^2+1}}{3x-5} \approx 0.47$.

$f(-1000) \approx -0.4706$ and $f(-10{,}000) \approx -0.4713$, so we estimate that $\lim_{x\to-\infty} \frac{\sqrt{2x^2+1}}{3x-5} \approx -0.47$.

(c) $\lim_{x\to\infty} \frac{\sqrt{2x^2+1}}{3x-5} = \lim_{x\to\infty} \frac{\sqrt{2+1/x^2}}{3-5/x}$ [since $\sqrt{x^2} = x$ for $x > 0$] $= \frac{\sqrt{2}}{3} \approx 0.471404$.

For $x < 0$, we have $\sqrt{x^2} = |x| = -x$, so when we divide the numerator by $x$, with $x < 0$, we get $\frac{1}{x}\sqrt{2x^2+1} = -\frac{1}{\sqrt{x^2}}\sqrt{2x^2+1} = -\sqrt{2+1/x^2}$. Therefore,

$$\lim_{x\to-\infty} \frac{\sqrt{2x^2+1}}{3x-5} = \lim_{x\to-\infty} \frac{-\sqrt{2+1/x^2}}{3-5/x} = -\frac{\sqrt{2}}{3} \approx -0.471404.$$

**35.** $\lim\limits_{x\to\infty} \dfrac{2x^2+x-1}{x^2+x-2} = \lim\limits_{x\to\infty} \dfrac{\dfrac{2x^2+x-1}{x^2}}{\dfrac{x^2+x-2}{x^2}} = \lim\limits_{x\to\infty} \dfrac{2+\dfrac{1}{x}-\dfrac{1}{x^2}}{1+\dfrac{1}{x}-\dfrac{2}{x^2}} = \dfrac{\lim\limits_{x\to\infty}\left(2+\dfrac{1}{x}-\dfrac{1}{x^2}\right)}{\lim\limits_{x\to\infty}\left(1+\dfrac{1}{x}-\dfrac{2}{x^2}\right)}$

$$= \frac{\lim\limits_{x\to\infty} 2 + \lim\limits_{x\to\infty}\dfrac{1}{x} - \lim\limits_{x\to\infty}\dfrac{1}{x^2}}{\lim\limits_{x\to\infty} 1 + \lim\limits_{x\to\infty}\dfrac{1}{x} - 2\lim\limits_{x\to\infty}\dfrac{1}{x^2}} = \frac{2+0-0}{1+0-2(0)} = 2,$$ so $y=2$ is a horizontal asymptote.

$y = f(x) = \dfrac{2x^2+x-1}{x^2+x-2} = \dfrac{(2x-1)(x+1)}{(x+2)(x-1)}$, so

$\lim\limits_{x\to-2^-} f(x) = \infty$, $\lim\limits_{x\to-2^+} f(x) = -\infty$, $\lim\limits_{x\to1^-} f(x) = -\infty$, and

$\lim\limits_{x\to1^+} f(x) = \infty$. Thus, $x=-2$ and $x=1$ are vertical asymptotes.

The graph confirms our work.

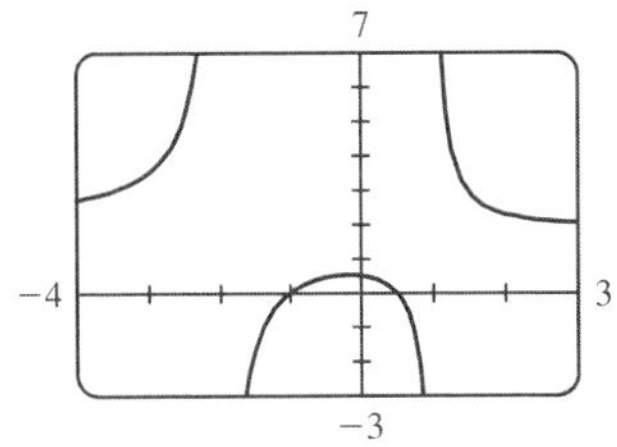


**36.** $\lim\limits_{x\to\infty} \dfrac{2e^x}{e^x-5} = \lim\limits_{x\to\infty} \dfrac{2e^x}{e^x-5}\cdot\dfrac{1/e^x}{1/e^x} = \lim\limits_{x\to\infty}\dfrac{2}{1-(5/e^x)} = \dfrac{2}{1-0} = 2$, so $y=2$ is a horizontal asymptote.

$\lim\limits_{x\to-\infty} \dfrac{2e^x}{e^x-5} = \dfrac{2(0)}{0-5} = 0$, so $y=0$ is a horizontal asymptote. The denominator is zero (and the numerator isn't) when

$e^x-5=0 \quad\Rightarrow\quad e^x=5 \quad\Rightarrow\quad x=\ln 5$. $\lim\limits_{x\to(\ln 5)^+} \dfrac{2e^x}{e^x-5} = \infty$ since the numerator approaches 10 and the denominator

approaches 0 through positive values as $x\to(\ln 5)^+$. Similarly,

$\lim\limits_{x\to(\ln 5)^-} \dfrac{2e^x}{e^x-5} = -\infty$. Thus, $x=\ln 5$ is a vertical asymptote.

The graph confirms our work.

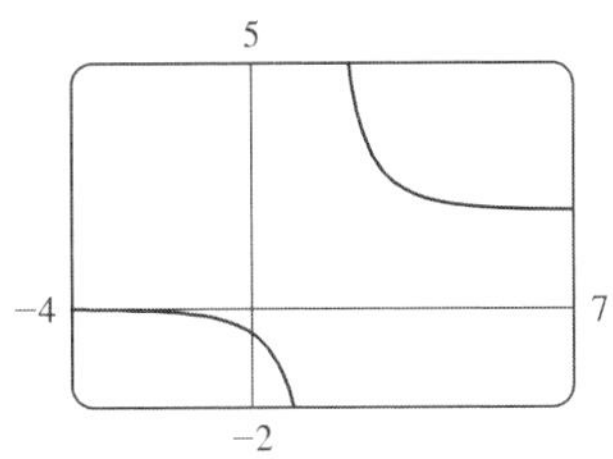


**37.** (a)

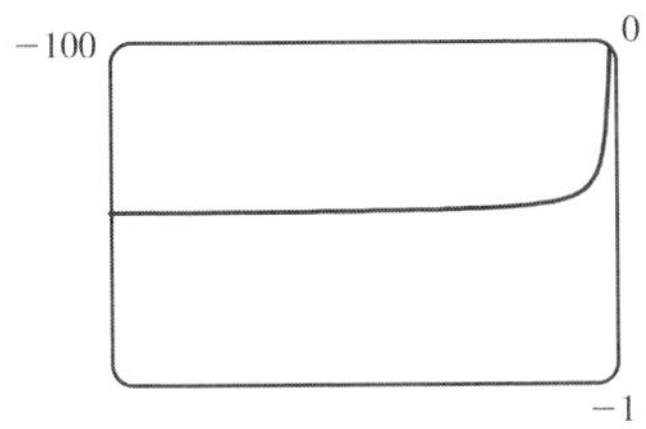


From the graph of $f(x)=\sqrt{x^2+x+1}+x$, we estimate the value of $\lim\limits_{x\to-\infty} f(x)$ to be $-0.5$.

(b)

| $x$ | $f(x)$ |
|---|---|
| −10,000 | −0.4999625 |
| −100,000 | −0.4999962 |
| −1,000,000 | −0.4999996 |

From the table, we estimate the limit to be $-0.5$.

(c) $\lim\limits_{x\to-\infty}\left(\sqrt{x^2+x+1}+x\right) = \lim\limits_{x\to-\infty}\left(\sqrt{x^2+x+1}+x\right)\left[\dfrac{\sqrt{x^2+x+1}-x}{\sqrt{x^2+x+1}-x}\right] = \lim\limits_{x\to-\infty}\dfrac{(x^2+x+1)-x^2}{\sqrt{x^2+x+1}-x}$

$$= \lim_{x\to-\infty}\frac{(x+1)(1/x)}{\left(\sqrt{x^2+x+1}-x\right)(1/x)} = \lim_{x\to-\infty}\frac{1+(1/x)}{-\sqrt{1+(1/x)+(1/x^2)}-1}$$

$$= \frac{1+0}{-\sqrt{1+0+0}-1} = -\frac{1}{2}$$

Note that for $x<0$, we have $\sqrt{x^2}=|x|=-x$, so when we divide the radical by $x$, with $x<0$, we get

$\dfrac{1}{x}\sqrt{x^2+x+1} = -\dfrac{1}{\sqrt{x^2}}\sqrt{x^2+x+1} = -\sqrt{1+(1/x)+(1/x^2)}$.

**38.** (a)

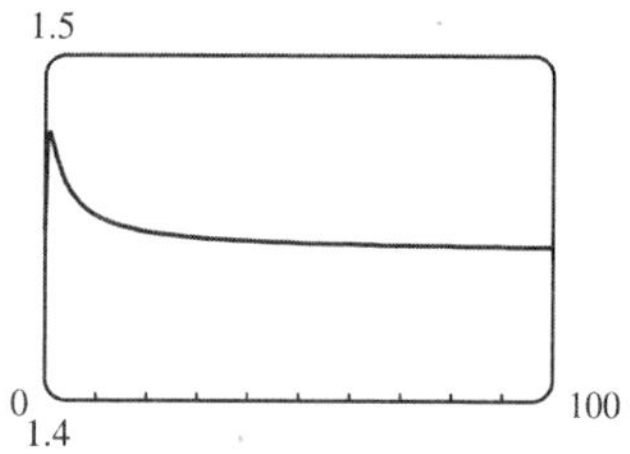


From the graph of $f(x) = \sqrt{3x^2 + 8x + 6} - \sqrt{3x^2 + 3x + 1}$, we estimate (to one decimal place) the value of $\lim\limits_{x \to \infty} f(x)$ to be 1.4.

(b)

| $x$ | $f(x)$ |
|---|---|
| 10,000 | 1.44339 |
| 100,000 | 1.44338 |
| 1,000,000 | 1.44338 |

From the table, we estimate (to four decimal places) the limit to be 1.4434.

(c)
$$\lim_{x\to\infty} f(x) = \lim_{x\to\infty} \frac{\left(\sqrt{3x^2+8x+6} - \sqrt{3x^2+3x+1}\right)\left(\sqrt{3x^2+8x+6} + \sqrt{3x^2+3x+1}\right)}{\sqrt{3x^2+8x+6} + \sqrt{3x^2+3x+1}}$$
$$= \lim_{x\to\infty} \frac{(3x^2+8x+6) - (3x^2+3x+1)}{\sqrt{3x^2+8x+6} + \sqrt{3x^2+3x+1}} = \lim_{x\to\infty} \frac{(5x+5)(1/x)}{\left(\sqrt{3x^2+8x+6} + \sqrt{3x^2+3x+1}\right)(1/x)}$$
$$= \lim_{x\to\infty} \frac{5+5/x}{\sqrt{3+8/x+6/x^2} + \sqrt{3+3/x+1/x^2}} = \frac{5}{\sqrt{3}+\sqrt{3}} = \frac{5}{2\sqrt{3}} = \frac{5\sqrt{3}}{6} \approx 1.443376$$

**39.** From the graph, it appears $y = 1$ is a horizontal asymptote.

$$\lim_{x\to\infty} \frac{3x^3+500x^2}{x^3+500x^2+100x+2000} = \lim_{x\to\infty} \frac{\dfrac{3x^3+500x^2}{x^3}}{\dfrac{x^3+500x^2+100x+2000}{x^3}}$$
$$= \lim_{x\to\infty} \frac{3+(500/x)}{1+(500/x)+(100/x^2)+(2000/x^3)} = \frac{3+0}{1+0+0+0} = 3,$$

so $y = 3$ is a horizontal asymptote. The discrepancy can be explained by the choice in the viewing window. Try $[-100{,}000, 100{,}000]$ by $[-1, 4]$ to get a graph that lends credibility to our calculation that $y = 3$ is a horizontal asymptote.

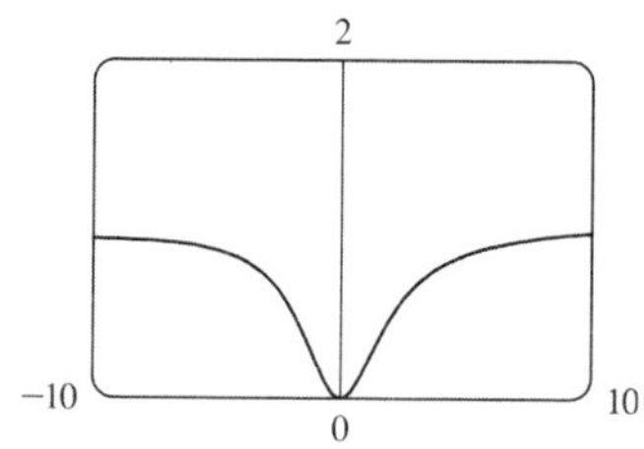


**40.** Since the function has vertical asymptotes $x = 1$ and $x = 3$, the denominator of the rational function we are looking for must have factors $(x-1)$ and $(x-3)$. Because the horizontal asymptote is $y = 1$, the degree of the numerator must equal the degree of the denominator, and the ratio of the leading coefficients must be 1. One possibility is $f(x) = \dfrac{x^2}{(x-1)(x-3)}$.

**41.** Let's look for a rational function.

(1) $\lim\limits_{x\to\pm\infty} f(x) = 0 \;\Rightarrow\;$ degree of numerator < degree of denominator

(2) $\lim\limits_{x\to 0} f(x) = -\infty \;\Rightarrow\;$ there is a factor of $x^2$ in the denominator (not just $x$, since that would produce a sign change at $x = 0$), and the function is negative near $x = 0$.

(3) $\lim\limits_{x\to 3^-} f(x) = \infty$ and $\lim\limits_{x\to 3^+} f(x) = -\infty \;\Rightarrow\;$ vertical asymptote at $x = 3$; there is a factor of $(x-3)$ in the denominator.

(4) $f(2) = 0 \;\Rightarrow\;$ 2 is an $x$-intercept; there is at least one factor of $(x-2)$ in the numerator.

Combining all of this information and putting in a negative sign to give us the desired left- and right-hand limits gives us $f(x) = \dfrac{2-x}{x^2(x-3)}$ as one possibility.

**42.** (a) In both viewing rectangles,

$$\lim_{x\to\infty} P(x) = \lim_{x\to\infty} Q(x) = \infty \text{ and}$$

$$\lim_{x\to-\infty} P(x) = \lim_{x\to-\infty} Q(x) = -\infty.$$

In the larger viewing rectangle, $P$ and $Q$ become less distinguishable.

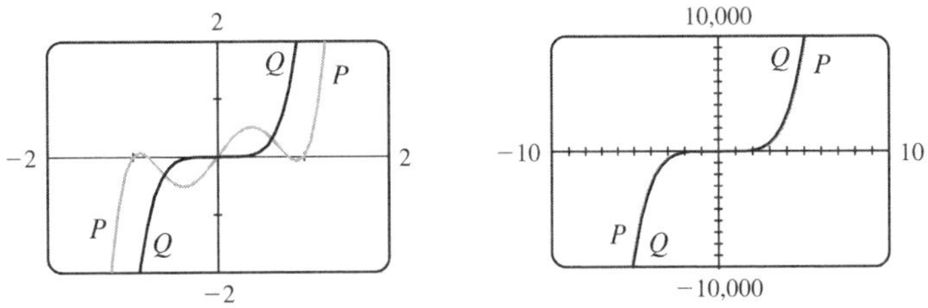


(b) $\displaystyle\lim_{x\to\infty}\frac{P(x)}{Q(x)} = \lim_{x\to\infty}\frac{3x^5-5x^3+2x}{3x^5} = \lim_{x\to\infty}\left(1-\frac{5}{3}\cdot\frac{1}{x^2}+\frac{2}{3}\cdot\frac{1}{x^4}\right) = 1-\tfrac{5}{3}(0)+\tfrac{2}{3}(0) = 1 \Rightarrow$

$P$ and $Q$ have the same end behavior.

**43.** Divide the numerator and the denominator by the highest power of $x$ in $Q(x)$.

(a) If $\deg P < \deg Q$, then the numerator $\to 0$ but the denominator doesn't. So $\displaystyle\lim_{x\to\infty}[P(x)/Q(x)] = 0$.

(b) If $\deg P > \deg Q$, then the numerator $\to \pm\infty$ but the denominator doesn't, so $\displaystyle\lim_{x\to\infty}[P(x)/Q(x)] = \pm\infty$ (depending on the ratio of the leading coefficients of $P$ and $Q$).

**44.**

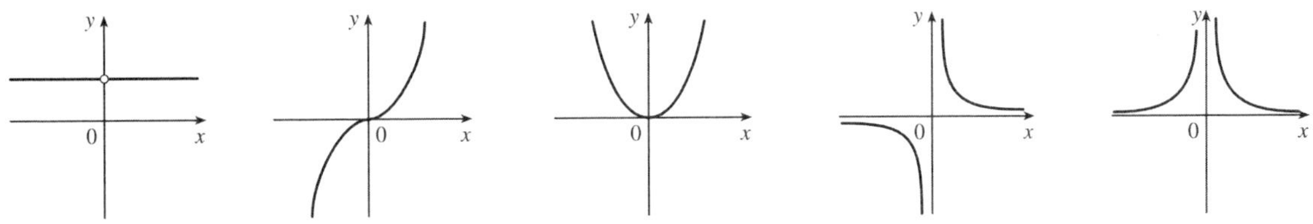


(i) $n = 0$ (ii) $n > 0$ ($n$ odd) (iii) $n > 0$ ($n$ even) (iv) $n < 0$ ($n$ odd) (v) $n < 0$ ($n$ even)

From these sketches we see that

(a) $\displaystyle\lim_{x\to 0^+} x^n = \begin{cases} 1 & \text{if } n = 0 \\ 0 & \text{if } n > 0 \\ \infty & \text{if } n < 0 \end{cases}$

(b) $\displaystyle\lim_{x\to 0^-} x^n = \begin{cases} 1 & \text{if } n = 0 \\ 0 & \text{if } n > 0 \\ -\infty & \text{if } n < 0,\ n \text{ odd} \\ \infty & \text{if } n < 0,\ n \text{ even} \end{cases}$

(c) $\displaystyle\lim_{x\to\infty} x^n = \begin{cases} 1 & \text{if } n = 0 \\ \infty & \text{if } n > 0 \\ 0 & \text{if } n < 0 \end{cases}$

(d) $\displaystyle\lim_{x\to-\infty} x^n = \begin{cases} 1 & \text{if } n = 0 \\ -\infty & \text{if } n > 0,\ n \text{ odd} \\ \infty & \text{if } n > 0,\ n \text{ even} \\ 0 & \text{if } n < 0 \end{cases}$

**45.** $\displaystyle\lim_{x\to\infty}\frac{5\sqrt{x}}{\sqrt{x-1}}\cdot\frac{1/\sqrt{x}}{1/\sqrt{x}} = \lim_{x\to\infty}\frac{5}{\sqrt{1-(1/x)}} = \frac{5}{\sqrt{1-0}} = 5$ and

$\displaystyle\lim_{x\to\infty}\frac{10e^x-21}{2e^x}\cdot\frac{1/e^x}{1/e^x} = \lim_{x\to\infty}\frac{10-(21/e^x)}{2} = \frac{10-0}{2} = 5$. Since $\dfrac{5\sqrt{x}}{\sqrt{x-1}} < f(x) < \dfrac{10e^x-21}{2e^x}$, we have $\displaystyle\lim_{x\to\infty} f(x) = 5$ by the Squeeze Theorem.

**46.** $\displaystyle\lim_{v\to c^-} m = \lim_{v\to c^-}\frac{m_0}{\sqrt{1-v^2/c^2}}$. As $v\to c^-$, $\sqrt{1-v^2/c^2}\to 0^+$, and $m\to\infty$.

**47.** (a) After $t$ minutes, $25t$ liters of brine with 30 g of salt per liter has been pumped into the tank, so it contains $(5000 + 25t)$ liters of water and $25t \cdot 30 = 750t$ grams of salt. Therefore, the salt concentration at time $t$ will be $C(t) = \dfrac{750t}{5000 + 25t} = \dfrac{30t}{200 + t} \dfrac{\text{g}}{\text{L}}$.

(b) $\lim\limits_{t\to\infty} C(t) = \lim\limits_{t\to\infty} \dfrac{30t}{200 + t} = \lim\limits_{t\to\infty} \dfrac{30t/t}{200/t + t/t} = \dfrac{30}{0 + 1} = 30$. So the salt concentration approaches that of the brine being pumped into the tank.

**48.** (a) $\lim\limits_{t\to\infty} v(t) = \lim\limits_{t\to\infty} v^*\left(1 - e^{-gt/v^*}\right) = v^*(1 - 0) = v^*$

(b) We graph $v(t) = 1 - e^{-9.8t}$ and $v(t) = 0.99v^*$, or in this case, $v(t) = 0.99$. Using an intersect feature or zooming in on the point of intersection, we find that $t \approx 0.47$ s.

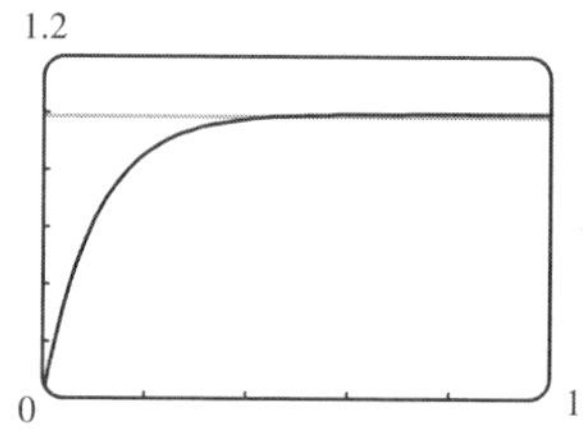


**49.** (a) If $t = -x/10$, then $x = -10t$ and as $x \to \infty$, $t \to -\infty$.

Thus, $\lim\limits_{x\to\infty} e^{-x/10} = \lim\limits_{t\to-\infty} e^t = 0$ by Equation 8.

(b) $y = e^{-x/10}$ and $y = 0.1$ intersect at $x_1 \approx 23.03$.

If $x > x_1$, then $e^{-x/10} < 0.1$.

(c) $e^{-x/10} < 0.1 \quad\Rightarrow\quad -x/10 < \ln 0.1 \quad\Rightarrow$

$x > -10 \ln \frac{1}{10} = -10 \ln 10^{-1} = 10 \ln 10$

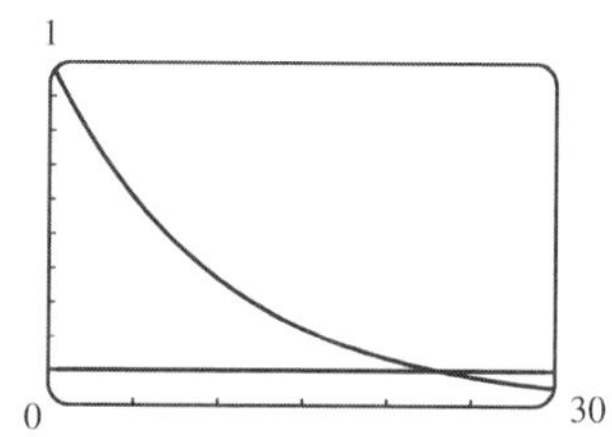


**50.** (a) $\lim\limits_{x\to\infty} f(x) = \lim\limits_{x\to\infty} \dfrac{4x^2 - 5x}{2x^2 + 1} = \lim\limits_{x\to\infty} \dfrac{4 - 5/x}{2 + 1/x^2} = \dfrac{4}{2} = 2$

(b) $f(x) = 1.9 \Longrightarrow x \approx 25.3744$, so $f(x) > 1.9$ when $x > N = 25.4$.

$f(x) = 1.99 \Longrightarrow x \approx 250.3974$, so $f(x) > 1.99$ when $x > N = 250.4$.

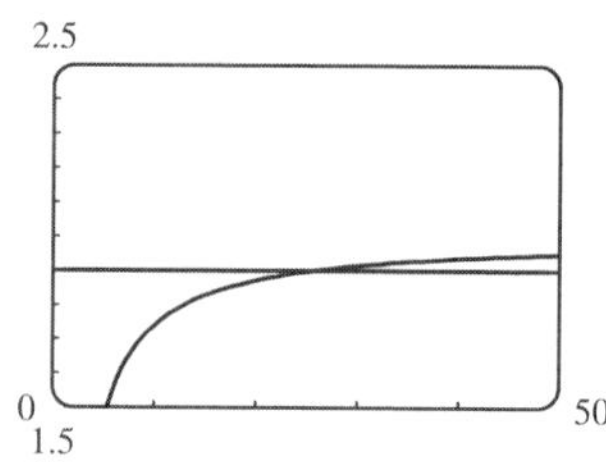


## 2.6 Tangents, Velocities, and Other Rates of Change

**1.** (a) This is just the slope of the line through two points: $m_{PQ} = \dfrac{\Delta y}{\Delta x} = \dfrac{f(x) - f(3)}{x - 3}$.

(b) This is the limit of the slope of the secant line $PQ$ as $Q$ approaches $P$: $m = \lim\limits_{x\to 3} \dfrac{f(x) - f(3)}{x - 3}$.

**2.** (a) Average velocity $= \dfrac{\Delta s}{\Delta t} = \dfrac{f(a + h) - f(a)}{(a + h) - a} = \dfrac{f(a + h) - f(a)}{h}$

(b) Instantaneous velocity $= \lim\limits_{h\to 0} \dfrac{f(a + h) - f(a)}{h}$

**3.** The slope at $D$ is the largest positive slope, followed by the positive slope at $E$. The slope at $C$ is zero. The slope at $B$ is steeper than at $A$ (both are negative). In decreasing order, we have the slopes at: $D$, $E$, $C$, $A$, and $B$.

**4.** The curve looks more like a line as the viewing rectangle gets smaller.

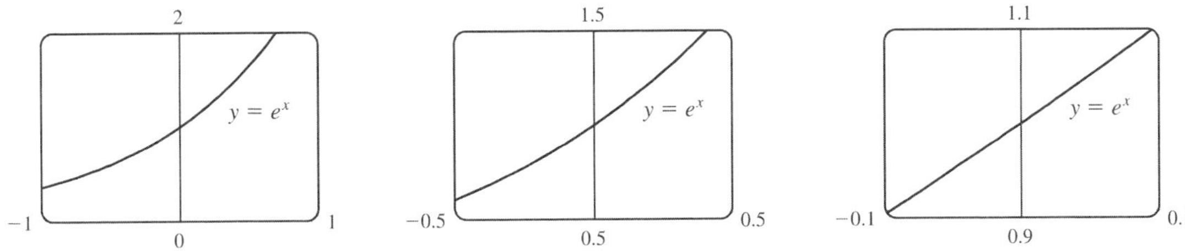


**5.** (a) (i) Using Definition 1,

$$m = \lim_{x\to a}\frac{f(x)-f(a)}{x-a}\ \lim_{x\to -3}\frac{f(x)-f(-3)}{x-(-3)} = \lim_{x\to -3}\frac{(x^2+2x)-(3)}{x-(-3)} = \lim_{x\to -3}\frac{(x+3)(x-1)}{x+3}$$

$$= \lim_{x\to -3}(x-1) = -4$$

(ii) Using Equation 2,

$$m = \lim_{h\to 0}\frac{f(a+h)-f(a)}{h} = \lim_{h\to 0}\frac{f(-3+h)-f(-3)}{h} = \lim_{h\to 0}\frac{[(-3+h)^2+2(-3+h)]-(3)}{h}$$

$$= \lim_{h\to 0}\frac{9-6h+h^2-6+2h-3}{h} = \lim_{h\to 0}\frac{h(h-4)}{h} = \lim_{h\to 0}(h-4) = -4$$

(b) Using the point-slope form of the equation of a line, an equation of the tangent line is $y-3=-4(x+3)$. Solving for $y$ gives us $y=-4x-9$, which is the slope-intercept form of the equation of the tangent line.

(c)

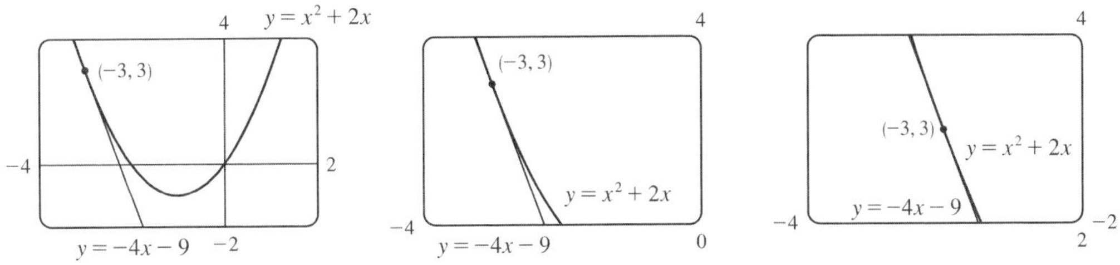


**6.** (a) (i) $m = \lim_{x\to -1}\frac{f(x)-f(-1)}{x-(-1)} = \lim_{x\to -1}\frac{x^3-(-1)}{x+1} = \lim_{x\to -1}\frac{(x+1)(x^2-x+1)}{x+1} = \lim_{x\to -1}(x^2-x+1) = 3$

(ii) $m = \lim_{h\to 0}\frac{f(-1+h)-f(-1)}{h} = \lim_{h\to 0}\frac{(-1+h)^3-(-1)}{h} = \lim_{h\to 0}\frac{h^3-3h^2+3h-1+1}{h}$

$$= \lim_{h\to 0}(h^2-3h+3) = 3$$

(b) $y-(-1)=3\,[x-(-1)] \iff y+1=3x+3 \iff y=3x+2$

(c)

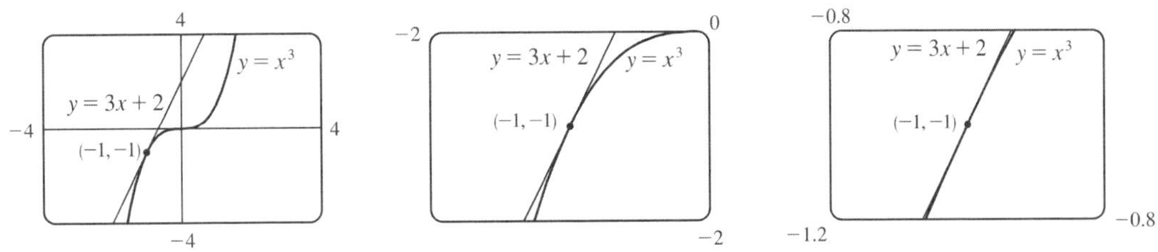


**7.** Using (1) with $f(x)=\dfrac{x-1}{x-2}$ and $P(3,2)$,

$$m = \lim_{x\to a}\frac{f(x)-f(a)}{x-a} = \lim_{x\to 3}\frac{\dfrac{x-1}{x-2}-2}{x-3} = \lim_{x\to 3}\frac{\dfrac{x-1-2(x-2)}{x-2}}{x-3} = \lim_{x\to 3}\frac{3-x}{(x-2)(x-3)} = \lim_{x\to 3}\frac{-1}{x-2} = \frac{-1}{1} = -1.$$

Tangent line: $y-2=-1(x-3) \iff y-2=-x+3 \iff y=-x+5$

**8.** Using (1),

$$m = \lim_{x \to -1} \frac{(2x^3 - 5x) - 3}{x - (-1)} = \lim_{x \to -1} \frac{2x^3 - 5x - 3}{x + 1} = \lim_{x \to -1} \frac{(2x^2 - 2x - 3)(x + 1)}{x + 1} = \lim_{x \to -1} (2x^2 - 2x - 3) = 1.$$

Tangent line: $y - 3 = 1\,[x - (-1)] \;\Leftrightarrow\; y = x + 4$

**9.** Using (1), $m = \lim\limits_{x \to 1} \dfrac{\sqrt{x} - \sqrt{1}}{x - 1} = \lim\limits_{x \to 1} \dfrac{(\sqrt{x} - 1)(\sqrt{x} + 1)}{(x - 1)(\sqrt{x} + 1)} = \lim\limits_{x \to 1} \dfrac{x - 1}{(x - 1)(\sqrt{x} + 1)} = \lim\limits_{x \to 1} \dfrac{1}{\sqrt{x} + 1} = \dfrac{1}{2}.$

Tangent line: $y - 1 = \frac{1}{2}(x - 1) \;\Leftrightarrow\; y = \frac{1}{2}x + \frac{1}{2}.$

**10.** Using (1), $m = \lim\limits_{x \to 0} \dfrac{\dfrac{2x}{(x+1)^2} - 0}{x - 0} = \lim\limits_{x \to 0} \dfrac{2x}{x(x+1)^2} = \lim\limits_{x \to 0} \dfrac{2}{(x+1)^2} = \dfrac{2}{1^2} = 2.$

Tangent line: $y - 0 = 2(x - 0) \;\Leftrightarrow\; y = 2x$

**11.** (a) Using (2) with $y = f(x) = 3 + 4x^2 - 2x^3$,

$$\begin{aligned}
m &= \lim_{h \to 0} \frac{f(a+h) - f(a)}{h} = \lim_{h \to 0} \frac{3 + 4(a+h)^2 - 2(a+h)^3 - (3 + 4a^2 - 2a^3)}{h} \\
&= \lim_{h \to 0} \frac{3 + 4(a^2 + 2ah + h^2) - 2(a^3 + 3a^2h + 3ah^2 + h^3) - 3 - 4a^2 + 2a^3}{h} \\
&= \lim_{h \to 0} \frac{3 + 4a^2 + 8ah + 4h^2 - 2a^3 - 6a^2h - 6ah^2 - 2h^3 - 3 - 4a^2 + 2a^3}{h} \\
&= \lim_{h \to 0} \frac{8ah + 4h^2 - 6a^2h - 6ah^2 - 2h^3}{h} = \lim_{h \to 0} \frac{h(8a + 4h - 6a^2 - 6ah - 2h^2)}{h} \\
&= \lim_{h \to 0} (8a + 4h - 6a^2 - 6ah - 2h^2) = 8a - 6a^2
\end{aligned}$$

(b) At $(1, 5)$: $m = 8(1) - 6(1)^2 = 2$, so an equation of the tangent line is $y - 5 = 2(x - 1) \;\Leftrightarrow$ $y = 2x + 3$.

At $(2, 3)$: $m = 8(2) - 6(2)^2 = -8$, so an equation of the tangent line is $y - 3 = -8(x - 2) \;\Leftrightarrow$ $y = -8x + 19$.

(c)

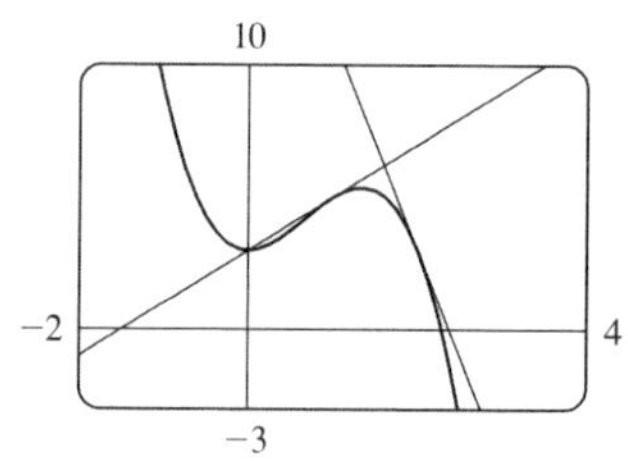


**12.** (a) Using (1),

$$\begin{aligned}
m &= \lim_{x \to a} \frac{\dfrac{1}{\sqrt{x}} - \dfrac{1}{\sqrt{a}}}{x - a} = \lim_{x \to a} \frac{\dfrac{\sqrt{a} - \sqrt{x}}{\sqrt{ax}}}{x - a} = \lim_{x \to a} \frac{(\sqrt{a} - \sqrt{x})(\sqrt{a} + \sqrt{x})}{\sqrt{ax}\,(x - a)(\sqrt{a} + \sqrt{x})} \\
&= \lim_{x \to a} \frac{a - x}{\sqrt{ax}\,(x - a)(\sqrt{a} + \sqrt{x})} = \lim_{x \to a} \frac{-1}{\sqrt{ax}\,(\sqrt{a} + \sqrt{x})} = \frac{-1}{\sqrt{a^2}\,(2\sqrt{a})} = -\frac{1}{2a^{3/2}} \text{ or } -\frac{1}{2}a^{-3/2}
\end{aligned}$$

(b) At $(1, 1)$: $m = -\frac{1}{2}$, so an equation of the tangent line is $y - 1 = -\frac{1}{2}(x - 1) \;\Leftrightarrow\; y = -\frac{1}{2}x + \frac{3}{2}$.

At $(4, \frac{1}{2})$: $m = -\frac{1}{16}$, so an equation of the tangent line is $y - \frac{1}{2} = -\frac{1}{16}(x - 4) \;\Leftrightarrow\; y = -\frac{1}{16}x + \frac{3}{4}$.

(c)

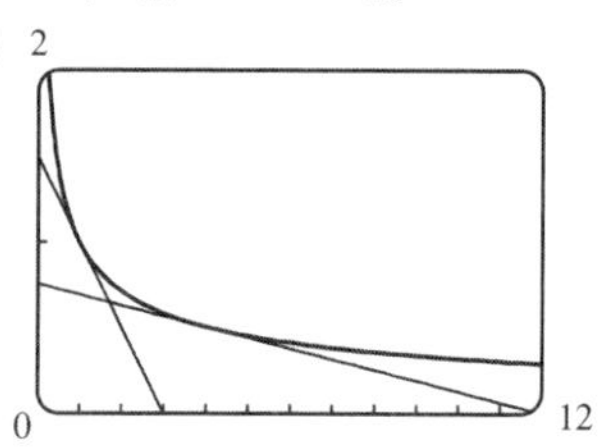

**13.** (a) The particle is moving to the right when $s$ is increasing; that is, on the intervals $(0, 1)$ and $(4, 6)$. The particle is moving to the left when $s$ is decreasing; that is, on the interval $(2, 3)$. The particle is standing still when $s$ is constant; that is, on the intervals $(1, 2)$ and $(3, 4)$.

(b) The velocity of the particle is equal to the slope of the tangent line of the graph. Note that there is no slope at the corner points on the graph. On the interval $(0, 1)$, the slope is $\frac{3-0}{1-0} = 3$. On the interval $(2, 3)$, the slope is $\frac{1-3}{3-2} = -2$. On the interval $(4, 6)$, the slope is $\frac{3-1}{6-4} = 1$.

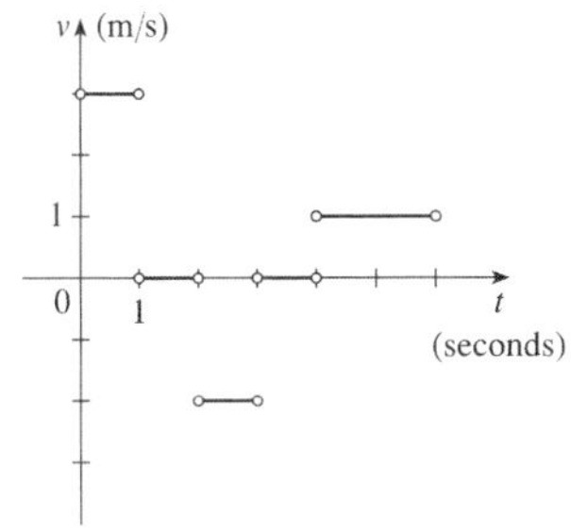


**14.** Let $a$ denote the distance traveled from 1:00 to 1:02, $b$ from 1:28 to 1:30, and $c$ from 3:30 to 3:33, where all the times are relative to $t = 0$ at the beginning of the trip.

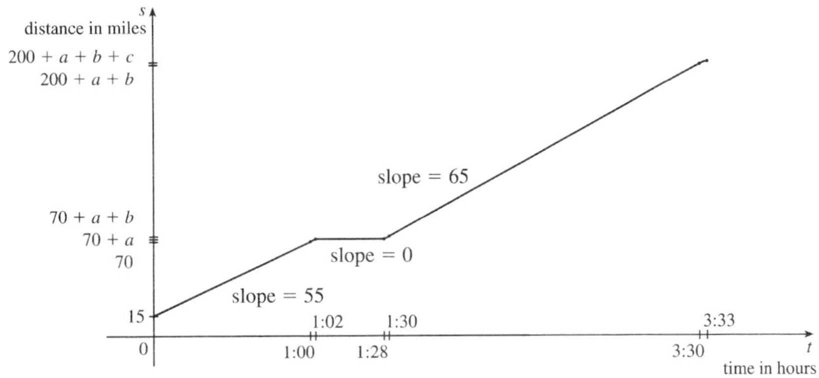


**15.** (a) Since the slope of the tangent at $t = 0$ is 0, the car's initial velocity was 0.

(b) The slope of the tangent is greater at $C$ than at $B$, so the car was going faster at $C$.

(c) Near $A$, the tangent lines are becoming steeper as $x$ increases, so the velocity was increasing, so the car was speeding up. Near $B$, the tangent lines are becoming less steep, so the car was slowing down. The steepest tangent near $C$ is the one at $C$, so at $C$ the car had just finished speeding up, and was about to start slowing down.

(d) Between $D$ and $E$, the slope of the tangent is 0, so the car did not move during that time.

**16.** (a) **Runner A** runs the entire 100-meter race at the same velocity since the slope of the position function is constant. **Runner B** starts the race at a slower velocity than runner A, but finishes the race at a faster velocity.

(b) The distance between the runners is the greatest at the time when the largest vertical line segment fits between the two graphs—this appears to be somewhere between 9 and 10 seconds.

(c) The runners had the same velocity when the slopes of their respective position functions are equal—this also appears to be at about 9.5 s. Note that the answers for parts (b) and (c) must be the same for these graphs because as soon as the velocity for runner B overtakes the velocity for runner A, the distance between the runners starts to decrease.

**17.** Let $s(t) = 40t - 16t^2$.

$$v(2) = \lim_{t\to 2} \frac{s(t) - s(2)}{t - 2} = \lim_{t\to 2} \frac{(40t - 16t^2) - 16}{t - 2} = \lim_{t\to 2} \frac{-16t^2 + 40t - 16}{t - 2} = \lim_{t\to 2} \frac{-8(2t^2 - 5t + 2)}{t - 2}$$

$$= \lim_{t\to 2} \frac{-8(t-2)(2t-1)}{t-2} = -8 \lim_{t\to 2}(2t - 1) = -8(3) = -24$$

Thus, the instantaneous velocity when $t = 2$ is $-24$ ft/s.

**18.** (a) $v(1) = \lim\limits_{h \to 0} \dfrac{H(1+h) - H(1)}{h} = \lim\limits_{h \to 0} \dfrac{(58 + 58h - 0.83 - 1.66h - 0.83h^2) - 57.17}{h}$

$= \lim\limits_{h \to 0} (56.34 - 0.83h) = 56.34 \text{ m/s}$

(b) $v(a) = \lim\limits_{h \to 0} \dfrac{H(a+h) - H(a)}{h} = \lim\limits_{h \to 0} \dfrac{(58a + 58h - 0.83a^2 - 1.66ah - 0.83h^2) - (58a - 0.83a^2)}{h}$

$= \lim\limits_{h \to 0} (58 - 1.66a - 0.83h) = 58 - 1.66a \text{ m/s}$

(c) The arrow strikes the moon when the height is 0, that is, $58t - 0.83t^2 = 0 \Leftrightarrow t(58 - 0.83t) = 0 \Leftrightarrow$ $t = \dfrac{58}{0.83} \approx 69.9$ s (since $t$ can't be 0).

(d) Using the time from part (c), $v\left(\dfrac{58}{0.83}\right) = 58 - 1.66\left(\dfrac{58}{0.83}\right) = -58 \text{ m/s}$. Thus, the arrow will have a velocity of $-58$ m/s.

**19.** $v(a) = \lim\limits_{h \to 0} \dfrac{s(a+h) - s(a)}{h} = \lim\limits_{h \to 0} \dfrac{\dfrac{1}{(a+h)^2} - \dfrac{1}{a^2}}{h} = \lim\limits_{h \to 0} \dfrac{\dfrac{a^2 - (a+h)^2}{a^2(a+h)^2}}{h} = \lim\limits_{h \to 0} \dfrac{a^2 - (a^2 + 2ah + h^2)}{ha^2(a+h)^2}$

$= \lim\limits_{h \to 0} \dfrac{-(2ah + h^2)}{ha^2(a+h)^2} = \lim\limits_{h \to 0} \dfrac{-h(2a+h)}{ha^2(a+h)^2} = \lim\limits_{h \to 0} \dfrac{-(2a+h)}{a^2(a+h)^2} = \dfrac{-2a}{a^2 \cdot a^2} = \dfrac{-2}{a^3} \text{ m/s}$

So $v(1) = \dfrac{-2}{1^3} = -2 \text{ m/s}$, $v(2) = \dfrac{-2}{2^3} = -\dfrac{1}{4} \text{ m/s}$, and $v(3) = \dfrac{-2}{3^3} = -\dfrac{2}{27} \text{ m/s}$.

**20.** (a) The average velocity between times $t$ and $t + h$ is

$$\frac{s(t+h) - s(t)}{(t+h) - t} = \frac{(t+h)^2 - 8(t+h) + 18 - (t^2 - 8t + 18)}{h} = \frac{t^2 + 2th + h^2 - 8t - 8h + 18 - t^2 + 8t - 18}{h}$$

$$= \frac{2th + h^2 - 8h}{h} = (2t + h - 8) \text{ m/s}.$$

(i) $[3, 4]$: $t = 3$, $h = 4 - 3 = 1$, so the average velocity is $2(3) + 1 - 8 = -1$ m/s.

(ii) $[3.5, 4]$: $t = 3.5$, $h = 0.5$, so the average velocity is $2(3.5) + 0.5 - 8 = -0.5$ m/s.

(iii) $[4, 5]$: $t = 4$, $h = 1$, so the average velocity is $2(4) + 1 - 8 = 1$ m/s.

(iv) $[4, 4.5]$: $t = 4$, $h = 0.5$, so the average velocity is $2(4) + 0.5 - 8 = 0.5$ m/s.

(b) $v(t) = \lim\limits_{h \to 0} \dfrac{s(t+h) - s(t)}{h} = \lim\limits_{h \to 0} (2t + h - 8) = 2t - 8$, so $v(4) = 0$.

(c)

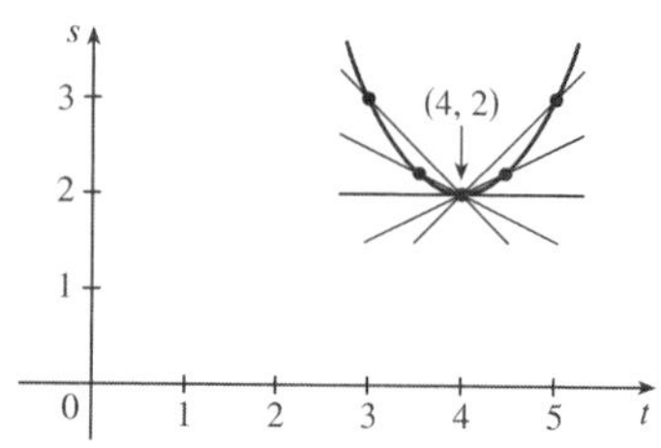


**21.** The sketch shows the graph for a room temperature of 72° and a refrigerator temperature of 38°. The initial rate of change is greater in magnitude than the rate of change after an hour.

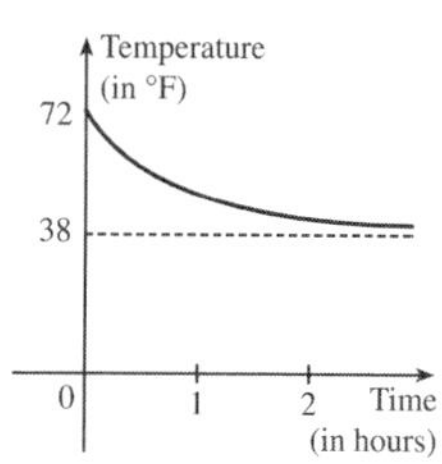


**22.** The slope of the tangent (that is, the rate of change of temperature with respect to time) at $t = 1$ h seems to be about $\dfrac{75 - 168}{132 - 0} \approx -0.7\,^\circ\text{F/min}$.

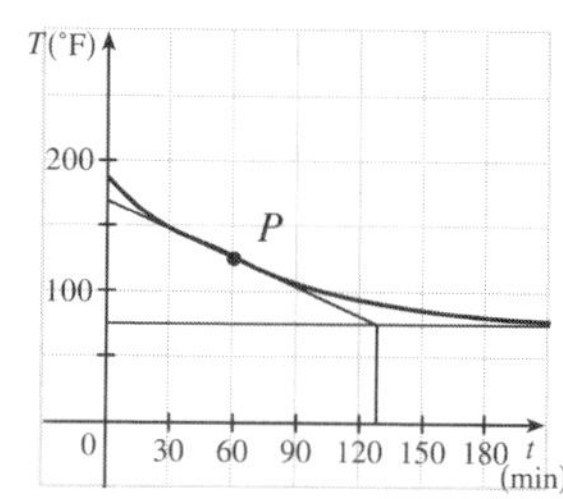

**23.** (a) (i) $[20, 23]$: $\dfrac{7.9 - 11.5}{23 - 20} = -1.2\,^\circ\text{C/h}$

(ii) $[20, 22]$: $\dfrac{9.0 - 11.5}{22 - 20} = -1.25\,^\circ\text{C/h}$

(iii) $[20, 21]$: $\dfrac{10.2 - 11.5}{21 - 20} = -1.3\,^\circ\text{C/h}$

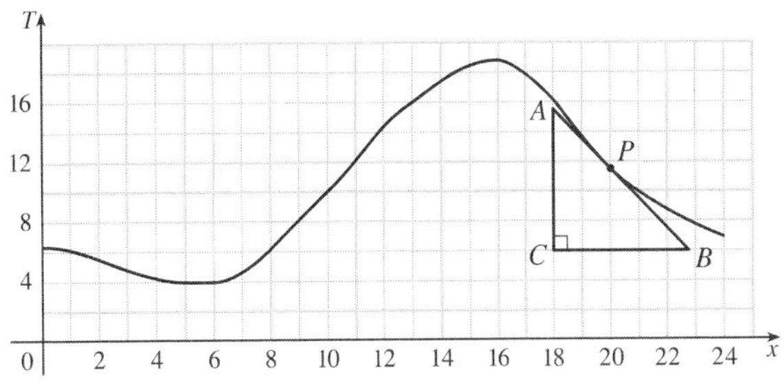


(b) In the figure, we estimate $A$ to be $(18, 15.5)$ and $B$ as $(23, 6)$. So the slope is $\dfrac{6 - 15.5}{23 - 18} = -1.9\,^\circ\text{C/h}$ at 8:00 P.M.

**24.** (a) (i) $[1996, 2000]$: $\dfrac{P(2000) - P(1996)}{2000 - 1996} = \dfrac{30{,}791 - 29{,}672}{4} = \dfrac{1119}{4} = 279.75$ thousand/year

(ii) $[1998, 2000]$: $\dfrac{P(2000) - P(1998)}{2000 - 1998} = \dfrac{30{,}791 - 30{,}248}{2} = \dfrac{543}{2} = 271.5$ thousand/year

(iii) $[2000, 2002]$: $\dfrac{P(2002) - P(2000)}{2002 - 2000} = \dfrac{31{,}414 - 30{,}791}{2} = \dfrac{623}{2} = 311.5$ thousand/year

(b) Using the values from (ii) and (iii), we have $\dfrac{271.5 + 311.5}{2} = 291.5$ thousand/year.

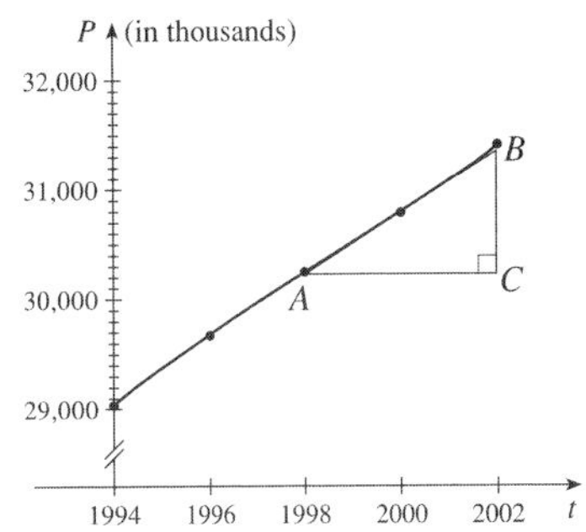


(c) Estimating $A$ as $(1998, 30{,}230)$ and $B$ as $(2002, 31{,}350)$, the slope at 2000 is $\dfrac{31{,}350 - 30{,}230}{2002 - 1998} = \dfrac{1120}{4} = 280$ thousand/year.

**25.** (a) (i) $[2000, 2002]$: $\dfrac{P(2002) - P(2000)}{2002 - 2000} = \dfrac{77 - 55}{2} = \dfrac{22}{2} = 11$ percent/year

(ii) $[2000, 2001]$: $\dfrac{P(2001) - P(2000)}{2001 - 2000} = \dfrac{68 - 55}{1} = 13$ percent/year

(iii) $[1999, 2000]$: $\dfrac{P(2000) - P(1999)}{2000 - 1999} = \dfrac{55 - 39}{1} = 16$ percent/year

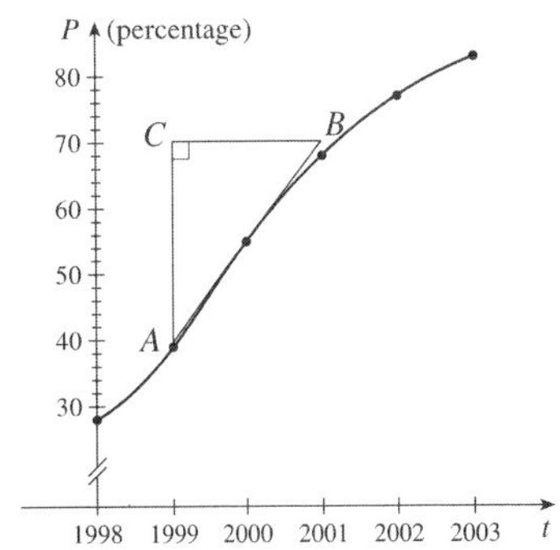


(b) Using the values from (ii) and (iii), we have $\dfrac{13 + 16}{2} = 14.5$ percent/year.

(c) Estimating $A$ as $(1999, 40)$ and $B$ as $(2001, 70)$, the slope at 2000 is

$$\frac{70 - 40}{2001 - 1999} = \frac{30}{2} = 15 \text{ percent/year.}$$

**26.** (a) (i) $[2000, 2002]$: $\dfrac{P(2002) - P(2000)}{2002 - 2000} = \dfrac{5886 - 3501}{2} = \dfrac{2385}{2} = 1192.5$ locations/year

(ii) $[2000, 2001]$: $\dfrac{P(2001) - P(2000)}{2001 - 2000} = \dfrac{4709 - 3501}{1} = 1208$ locations/year

(iii) $[1999, 2000]$: $\dfrac{P(2000) - P(1999)}{2000 - 1999} = \dfrac{3501 - 2135}{1} = 1366$ locations/year

(b) Using the values from (ii) and (iii), we have $\dfrac{1208 + 1366}{2} = 1287$ locations/year.

(c) Estimating $A$ as $(1999, 2035)$ and $B$ as $(2001, 3720)$, the slope at 2000 is

$$\frac{3720 - 2035}{2001 - 1999} = \frac{1685}{2} = 842.5 \text{ locations/year.}$$

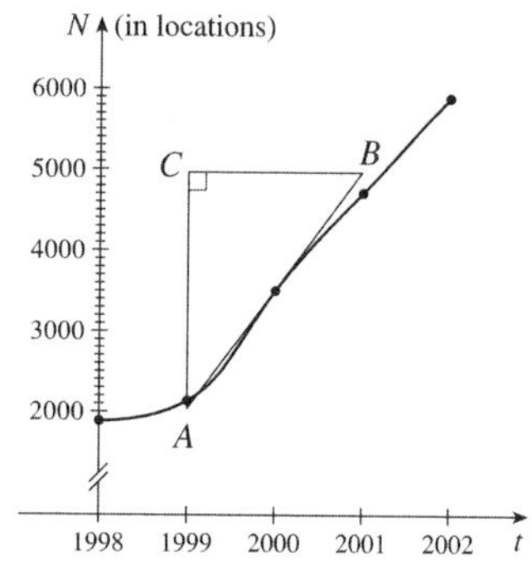


**27.** (a) (i) $\dfrac{\Delta C}{\Delta x} = \dfrac{C(105) - C(100)}{105 - 100} = \dfrac{6601.25 - 6500}{5} = \$20.25/\text{unit}.$

(ii) $\dfrac{\Delta C}{\Delta x} = \dfrac{C(101) - C(100)}{101 - 100} = \dfrac{6520.05 - 6500}{1} = \$20.05/\text{unit}.$

(b) $\dfrac{C(100 + h) - C(100)}{h} = \dfrac{\left[5000 + 10(100 + h) + 0.05(100 + h)^2\right] - 6500}{h} = \dfrac{20h + 0.05h^2}{h}$

$= 20 + 0.05h,\ h \neq 0$

So the instantaneous rate of change is $\displaystyle\lim_{h \to 0} \frac{C(100 + h) - C(100)}{h} = \lim_{h \to 0} (20 + 0.05h) = \$20/\text{unit}.$

**28.** $\Delta V = V(t + h) - V(t) = 100{,}000\left(1 - \dfrac{t + h}{60}\right)^2 - 100{,}000\left(1 - \dfrac{t}{60}\right)^2$

$= 100{,}000\left[\left(1 - \dfrac{t + h}{30} + \dfrac{(t + h)^2}{3600}\right) - \left(1 - \dfrac{t}{30} + \dfrac{t^2}{3600}\right)\right] = 100{,}000\left(-\dfrac{h}{30} + \dfrac{2th}{3600} + \dfrac{h^2}{3600}\right)$

$= \dfrac{100{,}000}{3600} h\,(-120 + 2t + h) = \dfrac{250}{9} h\,(-120 + 2t + h)$

Dividing $\Delta V$ by $h$ and then letting $h \to 0$, we see that the instantaneous rate of change is $\frac{500}{9}(t - 60)$ gal/min.

| $t$ | Flow rate (gal/min) | Water remaining $V(t)$ (gal) |
|---|---|---|
| 0 | $-3333.\overline{3}$ | $100,000$ |
| 10 | $-2777.\overline{7}$ | $69,444.\overline{4}$ |
| 20 | $-2222.\overline{2}$ | $44,444.\overline{4}$ |
| 30 | $-1666.\overline{6}$ | $25,000$ |
| 40 | $-1111.\overline{1}$ | $11,111.\overline{1}$ |
| 50 | $-555.\overline{5}$ | $2,777.\overline{7}$ |
| 60 | 0 | 0 |

The magnitude of the flow rate is greatest at the beginning and gradually decreases to 0.

## 2.7 Derivatives

**1.** 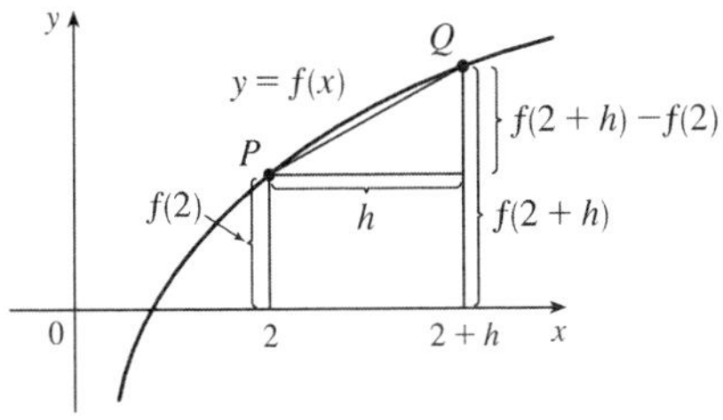


The line from $P(2, f(2))$ to $Q(2+h, f(2+h))$ is the line that has slope $\dfrac{f(2+h)-f(2)}{h}$.

**2.** As $h$ decreases, the line $PQ$ becomes steeper, so its slope increases. So

$0 < \dfrac{f(4)-f(2)}{4-2} < \dfrac{f(3)-f(2)}{3-2} < \lim\limits_{x\to 2}\dfrac{f(x)-f(2)}{x-2}$. Thus, $0 < \frac{1}{2}\left[f(4)-f(2)\right] < f(3)-f(2) < f'(2)$.

**3.** $g'(0)$ is the only negative value. The slope at $x=4$ is smaller than the slope at $x=2$ and both are smaller than the slope at $x=-2$. Thus, $g'(0) < 0 < g'(4) < g'(2) < g'(-2)$.

**4.** (a) Since $g(5)=-3$, the point $(5,-3)$ is on the graph of $g$. Since $g'(5)=4$, the slope of the tangent line at $x=5$ is 4. Using the point-slope form of a line gives us $y-(-3)=4(x-5)$, or $y=4x-23$.

(b) Since $(4,3)$ is on $y=f(x)$, $f(4)=3$. The slope of the tangent line between $(0,2)$ and $(4,3)$ is $\frac{1}{4}$, so $f'(4)=\frac{1}{4}$.

**5.** We begin by drawing a curve through the origin with a slope of 3 to satisfy $f(0)=0$ and $f'(0)=3$. Since $f'(1)=0$, we will round off our figure so that there is a horizontal tangent directly over $x=1$. Lastly, we make sure that the curve has a slope of $-1$ as we pass over $x=2$. Two of the many possibilities are shown.

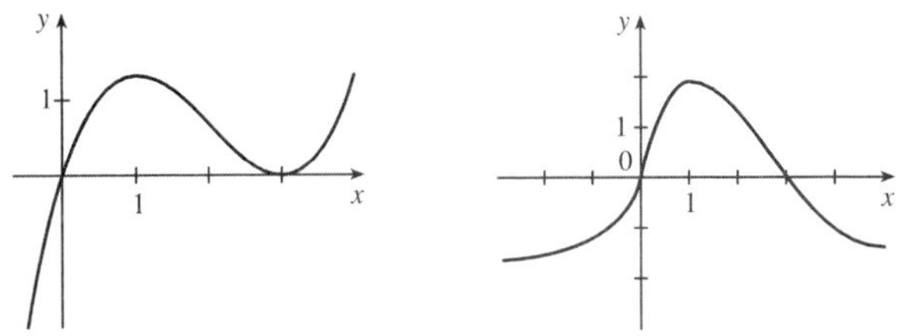


**6.** We begin by drawing a curve through the origin with a slope of 0 to satisfy $g(0)=0$ and $g'(0)=0$. The curve should have a slope of $-1$, 3, and 1 as we pass over $x=-1$, 1, and 2, respectively.

*Note:* In the figure, $y'=0$ when $x\approx -1.27$ or 2.13.

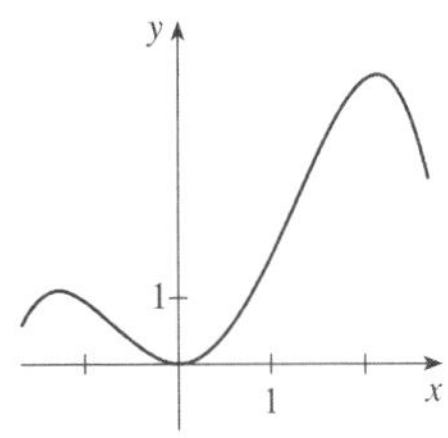


**7.** Using Definition 2 with $f(x)=3x^2-5x$ and the point $(2,2)$, we have

$$\begin{aligned} f'(2) &= \lim_{h\to 0}\frac{f(2+h)-f(2)}{h} = \lim_{h\to 0}\frac{\left[3(2+h)^2-5(2+h)\right]-2}{h} \\ &= \lim_{h\to 0}\frac{(12+12h+3h^2-10-5h)-2}{h} = \lim_{h\to 0}\frac{3h^2+7h}{h} = \lim_{h\to 0}(3h+7) = 7 \end{aligned}$$

So an equation of the tangent line at $(2,2)$ is $y-2=7(x-2)$ or $y=7x-12$.

**8.** Using Definition 2 with $g(x)=1-x^3$ and the point $(0,1)$, we have

$$g'(0) = \lim_{h\to 0}\frac{g(0+h)-g(0)}{h} = \lim_{h\to 0}\frac{\left[1-(0+h)^3\right]-1}{h} = \lim_{h\to 0}\frac{(1-h^3)-1}{h} = \lim_{h\to 0}(-h^2) = 0$$

So an equation of the tangent line is $y-1=0(x-0)$ or $y=1$.

**9.** (a) Using Definition 2 with $F(x) = 5x/(1+x^2)$ and the point $(2, 2)$, we have

$$\begin{aligned} F'(2) &= \lim_{h\to 0} \frac{F(2+h) - F(2)}{h} = \lim_{h\to 0} \frac{\dfrac{5(2+h)}{1+(2+h)^2} - 2}{h} \\ &= \lim_{h\to 0} \frac{\dfrac{5h+10}{h^2+4h+5} - 2}{h} = \lim_{h\to 0} \frac{\dfrac{5h+10-2(h^2+4h+5)}{h^2+4h+5}}{h} \\ &= \lim_{h\to 0} \frac{-2h^2-3h}{h(h^2+4h+5)} = \lim_{h\to 0} \frac{h(-2h-3)}{h(h^2+4h+5)} = \lim_{h\to 0} \frac{-2h-3}{h^2+4h+5} = \frac{-3}{5} \end{aligned}$$

So an equation of the tangent line at $(2, 2)$ is $y - 2 = -\frac{3}{5}(x-2)$ or $y = -\frac{3}{5}x + \frac{16}{5}$.

(b) 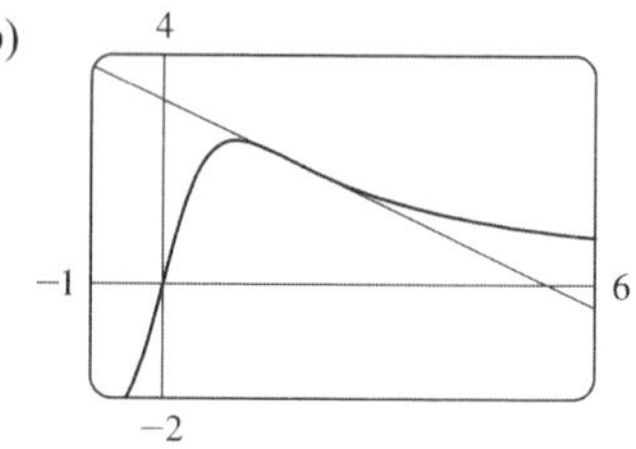


**10.** (a) Using Definition 2 with $G(x) = 4x^2 - x^3$, we have

$$\begin{aligned} G'(a) &= \lim_{h\to 0} \frac{G(a+h) - G(a)}{h} = \lim_{h\to 0} \frac{[4(a+h)^2 - (a+h)^3] - (4a^2 - a^3)}{h} \\ &= \lim_{h\to 0} \frac{4a^2 + 8ah + 4h^2 - (a^3 + 3a^2h + 3ah^2 + h^3) - 4a^2 + a^3}{h} = \lim_{h\to 0} \frac{8ah + 4h^2 - 3a^2h - 3ah^2 - h^3}{h} \\ &= \lim_{h\to 0} \frac{h(8a + 4h - 3a^2 - 3ah - h^{2})}{h} = \lim_{h\to 0}(8a + 4h - 3a^2 - 3ah - h^2) = 8a - 3a^2 \end{aligned}$$

(b) At the point $(2, 8)$, $G'(2) = 16 - 12 = 4$, and an equation of the tangent line is $y - 8 = 4(x-2)$, or $y = 4x$. At the point $(3, 9)$, $G'(3) = 24 - 27 = -3$, and an equation of the tangent line is $y - 9 = -3(x-3)$, or $y = -3x + 18$.

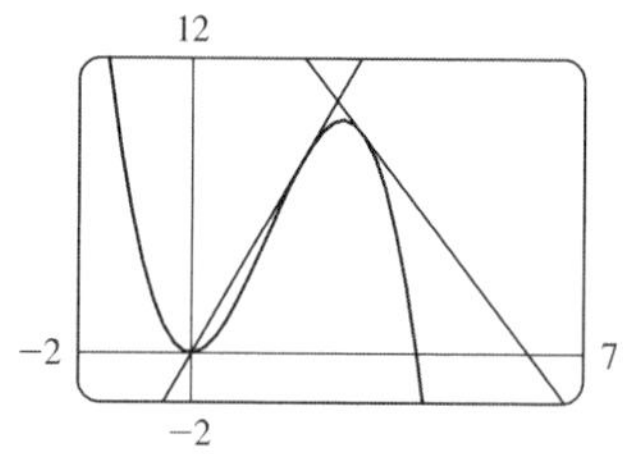


**11.** (a) $f'(1) = \lim\limits_{h\to 0} \dfrac{f(1+h) - f(1)}{h} = \lim\limits_{h\to 0} \dfrac{3^{1+h} - 3^1}{h}$.

So let $F(h) = \dfrac{3^{1+h} - 3}{h}$. We calculate:

| $h$ | $F(h)$ | $h$ | $F(h)$ |
|---|---|---|---|
| 0.1 | 3.484 | −0.1 | 3.121 |
| 0.01 | 3.314 | −0.01 | 3.278 |
| 0.001 | 3.298 | −0.001 | 3.294 |
| 0.0001 | 3.296 | −0.0001 | 3.296 |

We estimate that $f'(1) \approx 3.296$.

(b) 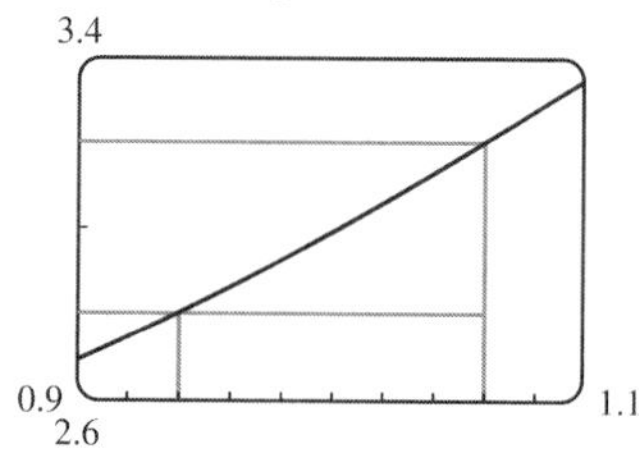


From the graph, we estimate that the slope of the tangent is about $\dfrac{3.2 - 2.8}{1.06 - 0.94} = \dfrac{0.4}{0.12} \approx 3.3$.

**12.** (a) $g'\left(\frac{\pi}{4}\right) = \lim\limits_{h\to 0} \dfrac{g\left(\frac{\pi}{4} + h\right) - g\left(\frac{\pi}{4}\right)}{h} = \lim\limits_{h\to 0} \dfrac{\tan\left(\frac{\pi}{4} + h\right) - \tan\left(\frac{\pi}{4}\right)}{h}$

So let $G(h) = \dfrac{\tan\left(\frac{\pi}{4} + h\right) - 1}{h}$. We calculate:

| $h$ | $G(h)$ | $h$ | $G(h)$ |
|---|---|---|---|
| 0.1 | 2.2305 | −0.1 | 1.8237 |
| 0.01 | 2.0203 | −0.01 | 1.9803 |
| 0.001 | 2.0020 | −0.001 | 1.9980 |
| 0.0001 | 2.0002 | −0.0001 | 1.9998 |

We estimate that $g'\left(\frac{\pi}{4}\right) = 2$.

(b) 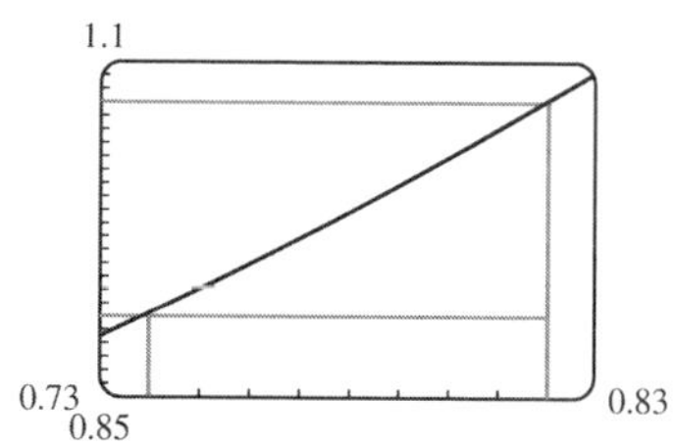


From the graph, we estimate that the slope of the tangent is about $\dfrac{1.07 - 0.91}{0.82 - 0.74} = \dfrac{0.16}{0.08} = 2$.

**13.** Use Definition 2 with $f(x) = 3 - 2x + 4x^2$.

$$f'(a) = \lim_{h\to 0} \frac{f(a+h) - f(a)}{h} = \lim_{h\to 0} \frac{[3 - 2(a+h) + 4(a+h)^2] - (3 - 2a + 4a^2)}{h}$$

$$= \lim_{h\to 0} \frac{(3 - 2a - 2h + 4a^2 + 8ah + 4h^2) - (3 - 2a + 4a^2)}{h}$$

$$= \lim_{h\to 0} \frac{-2h + 8ah + 4h^2}{h} = \lim_{h\to 0} \frac{h(-2 + 8a + 4h)}{h} = \lim_{h\to 0}(-2 + 8a + 4h) = -2 + 8a$$

**14.** $$f'(a) = \lim_{h\to 0} \frac{f(a+h) - f(a)}{h} = \lim_{h\to 0} \frac{[(a+h)^4 - 5(a+h)] - (a^4 - 5a)}{h}$$

$$= \lim_{h\to 0} \frac{(a^4 + 4a^3h + 6a^2h^2 + 4ah^3 + h^4 - 5a - 5h) - (a^4 - 5a)}{h} = \lim_{h\to 0} \frac{4a^3h + 6a^2h^2 + 4ah^3 + h^4 - 5h}{h}$$

$$= \lim_{h\to 0} \frac{h(4a^3 + 6a^2h + 4ah^2 + h^3 - 5)}{h} = \lim_{h\to 0}(4a^3 + 6a^2h + 4ah^2 + h^3 - 5) = 4a^3 - 5$$

**15.** Use Definition 2 with $f(t) = (2t + 1)/(t + 3)$.

$$f'(a) = \lim_{h\to 0} \frac{f(a+h) - f(a)}{h} = \lim_{h\to 0} \frac{\dfrac{2(a+h)+1}{(a+h)+3} - \dfrac{2a+1}{a+3}}{h} = \lim_{h\to 0} \frac{(2a + 2h + 1)(a+3) - (2a+1)(a+h+3)}{h(a+h+3)(a+3)}$$

$$= \lim_{h\to 0} \frac{(2a^2 + 6a + 2ah + 6h + a + 3) - (2a^2 + 2ah + 6a + a + h + 3)}{h(a+h+3)(a+3)}$$

$$= \lim_{h\to 0} \frac{5h}{h(a+h+3)(a+3)} = \lim_{h\to 0} \frac{5}{(a+h+3)(a+3)} = \frac{5}{(a+3)^2}$$

**16.** $$f'(a) = \lim_{h\to 0} \frac{f(a+h) - f(a)}{h} = \lim_{h\to 0} \frac{\dfrac{(a+h)^2+1}{(a+h)-2} - \dfrac{a^2+1}{a-2}}{h}$$

$$= \lim_{h\to 0} \frac{(a^2 + 2ah + h^2 + 1)(a-2) - (a^2+1)(a+h-2)}{h(a+h-2)(a-2)}$$

$$= \lim_{h\to 0} \frac{(a^3 - 2a^2 + 2a^2h - 4ah + ah^2 - 2h^2 + a - 2) - (a^3 + a^2h - 2a^2 + a + h - 2)}{h(a+h-2)(a-2)}$$

$$= \lim_{h\to 0} \frac{a^2h - 4ah + ah^2 - 2h^2 - h}{h(a+h-2)(a-2)} = \lim_{h\to 0} \frac{h(a^2 - 4a + ah - 2h - 1)}{h(a+h-2)(a-2)} = \lim_{h\to 0} \frac{a^2 - 4a + ah - 2h - 1}{(a+h-2)(a-2)}$$

$$= \frac{a^2 - 4a - 1}{(a-2)^2}$$

**17.** Use Definition 2 with $f(x) = 1/\sqrt{x+2}$.

$$f'(a) = \lim_{h\to 0} \frac{f(a+h) - f(a)}{h} = \lim_{h\to 0} \frac{\dfrac{1}{\sqrt{(a+h)+2}} - \dfrac{1}{\sqrt{a+2}}}{h} = \lim_{h\to 0} \frac{\dfrac{\sqrt{a+2} - \sqrt{a+h+2}}{\sqrt{a+h+2}\,\sqrt{a+2}}}{h}$$

$$= \lim_{h\to 0} \left[\frac{\sqrt{a+2} - \sqrt{a+h+2}}{h\sqrt{a+h+2}\,\sqrt{a+2}} \cdot \frac{\sqrt{a+2} + \sqrt{a+h+2}}{\sqrt{a+2} + \sqrt{a+h+2}}\right] = \lim_{h\to 0} \frac{(a+2) - (a+h+2)}{h\sqrt{a+h+2}\,\sqrt{a+2}\left(\sqrt{a+2} + \sqrt{a+h+2}\right)}$$

$$= \lim_{h\to 0} \frac{-h}{h\sqrt{a+h+2}\,\sqrt{a+2}\left(\sqrt{a+2} + \sqrt{a+h+2}\right)} = \lim_{h\to 0} \frac{-1}{\sqrt{a+h+2}\,\sqrt{a+2}\left(\sqrt{a+2} + \sqrt{a+h+2}\right)}$$

$$= \frac{-1}{\left(\sqrt{a+2}\right)^2\left(2\sqrt{a+2}\right)} = -\frac{1}{2(a+2)^{3/2}}$$

**18.** $$f'(a) = \lim_{h\to 0} \frac{f(a+h) - f(a)}{h} = \lim_{h\to 0} \frac{\sqrt{3(a+h)+1} - \sqrt{3a+1}}{h}$$

$$= \lim_{h\to 0} \frac{\left(\sqrt{3a+3h+1} - \sqrt{3a+1}\right)\left(\sqrt{3a+3h+1} + \sqrt{3a+1}\right)}{h\left(\sqrt{3a+3h+1} + \sqrt{3a+1}\right)} = \lim_{h\to 0} \frac{(3a+3h+1) - (3a+1)}{h\left(\sqrt{3a+3h+1} + \sqrt{3a+1}\right)}$$

$$= \lim_{h\to 0} \frac{3h}{h\left(\sqrt{3a+3h+1} + \sqrt{3a+1}\right)} = \lim_{h\to 0} \frac{3}{\sqrt{3a+3h+1} + \sqrt{3a+1}} = \frac{3}{2\sqrt{3a+1}}$$

Note that the answers to Exercises 19 – 24 are not unique.

**19.** By Definition 2, $\lim_{h\to 0}\frac{(1+h)^{10}-1}{h}=f'(1)$, where $f(x)=x^{10}$ and $a=1$.

*Or:* By Definition 2, $\lim_{h\to 0}\frac{(1+h)^{10}-1}{h}=f'(0)$, where $f(x)=(1+x)^{10}$ and $a=0$.

**20.** By Definition 2, $\lim_{h\to 0}\frac{\sqrt[4]{16+h}-2}{h}=f'(16)$, where $f(x)=\sqrt[4]{x}$ and $a=16$.

*Or:* By Definition 2, $\lim_{h\to 0}\frac{\sqrt[4]{16+h}-2}{h}=f'(0)$, where $f(x)=\sqrt[4]{16+x}$ and $a=0$.

**21.** By Equation 3, $\lim_{x\to 5}\frac{2^x-32}{x-5}=f'(5)$, where $f(x)=2^x$ and $a=5$.

**22.** By Equation 3, $\lim_{x\to\pi/4}\frac{\tan x-1}{x-\pi/4}=f'(\pi/4)$, where $f(x)=\tan x$ and $a=\pi/4$.

**23.** By Definition 2, $\lim_{h\to 0}\frac{\cos(\pi+h)+1}{h}=f'(\pi)$, where $f(x)=\cos x$ and $a=\pi$.

*Or:* By Definition 2, $\lim_{h\to 0}\frac{\cos(\pi+h)+1}{h}=f'(0)$, where $f(x)=\cos(\pi+x)$ and $a=0$.

**24.** By Equation 3, $\lim_{t\to 1}\frac{t^4+t-2}{t-1}=f'(1)$, where $f(t)=t^4+t$ and $a=1$.

**25.** $v(5)=f'(5)=\lim_{h\to 0}\frac{f(5+h)-f(5)}{h}=\lim_{h\to 0}\frac{[100+50(5+h)-4.9(5+h)^2]-[100+50(5)-4.9(5)^2]}{h}$

$=\lim_{h\to 0}\frac{(100+250+50h-4.9h^2-49h-122.5)-(100+250-122.5)}{h}=\lim_{h\to 0}\frac{-4.9h^2+h}{h}$

$=\lim_{h\to 0}\frac{h(-4.9h+1)}{h}=\lim_{h\to 0}(-4.9h+1)=1\text{ m/s}$

The speed when $t=5$ is $|1|=1$ m/s.

**26.** $v(5)=f'(5)=\lim_{h\to 0}\frac{f(5+h)-f(5)}{h}=\lim_{h\to 0}\frac{[(5+h)^{-1}-(5+h)]-(5^{-1}-5)}{h}$

$=\lim_{h\to 0}\frac{\frac{1}{5+h}-5-h-\frac{1}{5}+5}{h}=\lim_{h\to 0}\frac{\frac{1}{5+h}-h-\frac{1}{5}}{h}=\lim_{h\to 0}\frac{\frac{5-5h(5+h)-(5+h)}{5(5+h)}}{h}$

$=\lim_{h\to 0}\frac{5-25h-5h^2-5-h}{5h(5+h)}=\lim_{h\to 0}\frac{-5h^2-26h}{5h(5+h)}=\lim_{h\to 0}\frac{h(-5h-26)}{5h(5+h)}=\lim_{h\to 0}\frac{-5h-26}{5(5+h)}=\frac{-26}{25}\text{ m/s}$

The speed when $t=5$ is $\left|-\frac{26}{25}\right|=\frac{26}{25}=1.04$ m/s.

**27.** (a) $f'(x)$ is the rate of change of the production cost with respect to the number of ounces of gold produced. Its units are dollars per ounce.

(b) After 800 ounces of gold have been produced, the rate at which the production cost is increasing is \$17/ounce. So the cost of producing the 800th (or 801st) ounce is about \$17.

(c) In the short term, the values of $f'(x)$ will decrease because more efficient use is made of start-up costs as $x$ increases. But eventually $f'(x)$ might increase due to large-scale operations.

**28.** (a) $f'(5)$ is the rate of growth of the bacteria population when $t=5$ hours. Its units are bacteria per hour.

(b) With unlimited space and nutrients, $f'$ should increase as $t$ increases; so $f'(5)<f'(10)$. If the supply of nutrients is limited, the growth rate slows down at some point in time, and the opposite may be true.

**29.** (a) $f'(v)$ is the rate at which the fuel consumption is changing with respect to the speed. Its units are $(\text{gal/h})/(\text{mi/h})$.

(b) The fuel consumption is decreasing by $0.05\ (\text{gal/h})/(\text{mi/h})$ as the car's speed reaches $20\ \text{mi/h}$. So if you increase your speed to $21\ \text{mi/h}$, you could expect to decrease your fuel consumption by about $0.05\ (\text{gal/h})/(\text{mi/h})$.

**30.** (a) $f'(8)$ is the rate of change of the quantity of coffee sold with respect to the price per pound when the price is \$8 per pound. The units for $f'(8)$ are pounds/(dollars/pound).

(b) $f'(8)$ is negative since the quantity of coffee sold will decrease as the price charged for it increases. People are generally less willing to buy a product when its price increases.

**31.** $T'(10)$ is the rate at which the temperature is changing at 10:00 A.M. To estimate the value of $T'(10)$, we will average the difference quotients obtained using the times $t = 8$ and $t = 12$. Let $A = \dfrac{T(8) - T(10)}{8 - 10} = \dfrac{72 - 81}{-2} = 4.5$ and

$B = \dfrac{T(12) - T(10)}{12 - 10} = \dfrac{88 - 81}{2} = 3.5$. Then $T'(10) = \lim\limits_{t \to 10} \dfrac{T(t) - T(10)}{t - 10} \approx \dfrac{A + B}{2} = \dfrac{4.5 + 3.5}{2} = 4\ {}^\circ\text{F/h}$.

**32.** **For 1910:** We will average the difference quotients obtained using the years 1900 and 1920.

Let $A = \dfrac{E(1900) - E(1910)}{1900 - 1910} = \dfrac{48.3 - 51.1}{-10} = 0.28$ and

$B = \dfrac{E(1920) - E(1910)}{1920 - 1910} = \dfrac{55.2 - 51.1}{10} = 0.41$.

Then $E'(1910) = \lim\limits_{t \to 1910} \dfrac{E(t) - E(1910)}{t - 1910} \approx \dfrac{A + B}{2} = 0.345$. This means that life expectancy at birth was increasing at about 0.345 year/year in 1910.

**For 1950:** Using data for 1940 and 1960 in a similar fashion, we obtain $E'(1950) \approx [0.31 + 0.10]/2 = 0.205$. So life expectancy at birth was increasing at about 0.205 year/year in 1950.

**33.** (a) $S'(T)$ is the rate at which the oxygen solubility changes with respect to the water temperature. Its units are $(\text{mg/L})/{}^\circ\text{C}$.

(b) For $T = 16^\circ\text{C}$, it appears that the tangent line to the curve goes through the points $(0, 14)$ and $(32, 6)$. So $S'(16) \approx \dfrac{6 - 14}{32 - 0} = -\dfrac{8}{32} = -0.25\ (\text{mg/L})/{}^\circ\text{C}$. This means that as the temperature increases past $16^\circ\text{C}$, the oxygen solubility is decreasing at a rate of $0.25\ (\text{mg/L})/{}^\circ\text{C}$.

**34.** (a) $S'(T)$ is the rate of change of the maximum sustainable speed of Coho salmon with respect to the temperature. Its units are $(\text{cm/s})/{}^\circ\text{C}$.

(b) For $T = 15^\circ\text{C}$, it appears the tangent line to the curve goes through the points $(10, 25)$ and $(20, 32)$. So $S'(15) \approx \dfrac{32 - 25}{20 - 10} = 0.7\ (\text{cm/s})/{}^\circ\text{C}$. This tells us that at $T = 15^\circ\text{C}$, the maximum sustainable speed of Coho salmon is changing at a rate of $0.7\ (\text{cm/s})/{}^\circ\text{C}$. In a similar fashion for $T = 25^\circ\text{C}$, we can use the points $(20, 35)$ and $(25, 25)$ to obtain $S'(25) \approx \dfrac{25 - 35}{25 - 20} = -2\ (\text{cm/s})/{}^\circ\text{C}$. As it gets warmer than $20\ {}^\circ\text{C}$, the maximum sustainable speed decreases rapidly.

**35.** Since $f(x) = x\sin(1/x)$ when $x \neq 0$ and $f(0) = 0$, we have

$f'(0) = \lim\limits_{h \to 0} \dfrac{f(0 + h) - f(0)}{h} = \lim\limits_{h \to 0} \dfrac{h\sin(1/h) - 0}{h} = \lim\limits_{h \to 0} \sin(1/h)$. This limit does not exist since $\sin(1/h)$ takes the values $-1$ and $1$ on any interval containing 0. (Compare with Example 4 in Section 2.2.)

**36.** Since $f(x) = x^2\sin(1/x)$ when $x \neq 0$ and $f(0) = 0$, we have

$f'(0) = \lim\limits_{h \to 0} \dfrac{f(0 + h) - f(0)}{h} = \lim\limits_{h \to 0} \dfrac{h^2\sin(1/h) - 0}{h} = \lim\limits_{h \to 0} h\sin(1/h)$. Since $-1 \le \sin\dfrac{1}{h} \le 1$, we have

$-|h| \le |h|\sin\dfrac{1}{h} \le |h| \quad\Rightarrow\quad -|h| \le h\sin\dfrac{1}{h} \le |h|$. Because $\lim\limits_{h \to 0}(-|h|) = 0$ and $\lim\limits_{h \to 0}|h| = 0$, we know that

$\lim\limits_{h \to 0}\left(h\sin\dfrac{1}{h}\right) = 0$ by the Squeeze Theorem. Thus, $f'(0) = 0$.

## 2.8 The Derivative as a Function

**1.** It appears that $f$ is an odd function, so $f'$ will be an even function—that is, $f'(-a) = f'(a)$.

(a) $f'(-3) \approx 1.5$ (b) $f'(-2) \approx 1$

(c) $f'(-1) \approx 0$ (d) $f'(0) \approx -4$

(e) $f'(1) \approx 0$ (f) $f'(2) \approx 1$

(g) $f'(3) \approx 1.5$

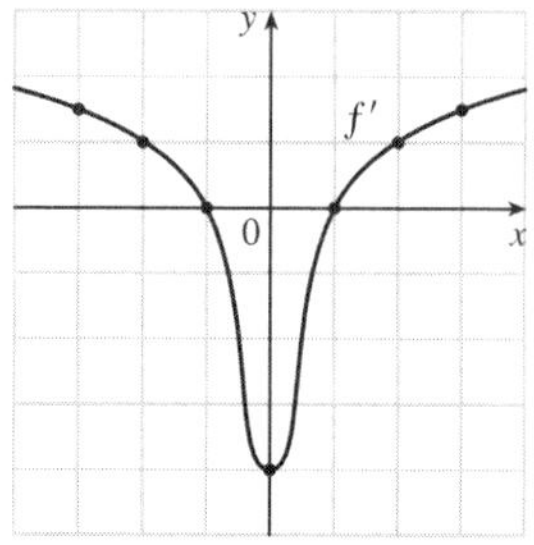


**2.** *Note:* Your answers may vary depending on your estimates. By estimating the slopes of tangent lines on the graph of $f$, it appears that

(a) $f'(0) \approx -3$ (b) $f'(1) \approx 0$

(c) $f'(2) \approx 1.5$ (d) $f'(3) \approx 2$

(e) $f'(4) \approx 0$ (f) $f'(5) \approx -1.2$

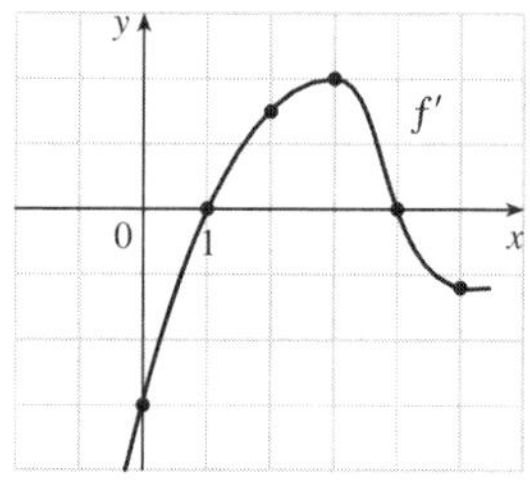


**3.** (a)$'$ = II, since from left to right, the slopes of the tangents to graph (a) start out negative, become 0, then positive, then 0, then negative again. The actual function values in graph II follow the same pattern.

(b)$'$ = IV, since from left to right, the slopes of the tangents to graph (b) start out at a fixed positive quantity, then suddenly become negative, then positive again. The discontinuities in graph IV indicate sudden changes in the slopes of the tangents.

(c)$'$ = I, since the slopes of the tangents to graph (c) are negative for $x < 0$ and positive for $x > 0$, as are the function values of graph I.

(d)$'$ = III, since from left to right, the slopes of the tangents to graph (d) are positive, then 0, then negative, then 0, then positive, then 0, then negative again, and the function values in graph III follow the same pattern.

Hints for Exercises 4–11: First plot $x$-intercepts on the graph of $f'$ for any horizontal tangents on the graph of $f$. Look for any corners on the graph of $f$—there will be a discontinuity on the graph of $f'$. On any interval where $f$ has a tangent with positive (or negative) slope, the graph of $f'$ will be positive (or negative). If the graph of the function is linear, the graph of $f'$ will be a horizontal line.

**4.**

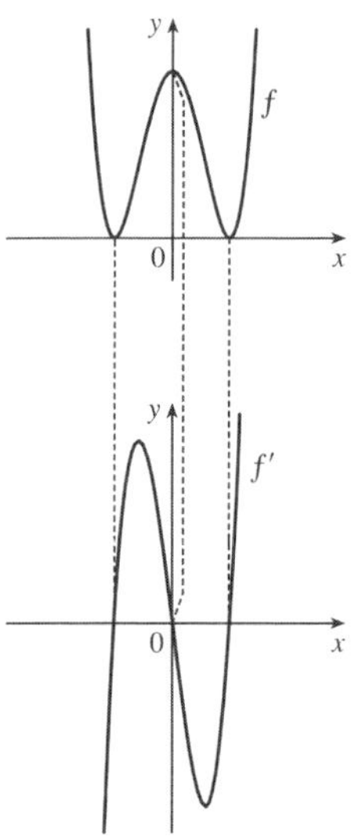


**5.**

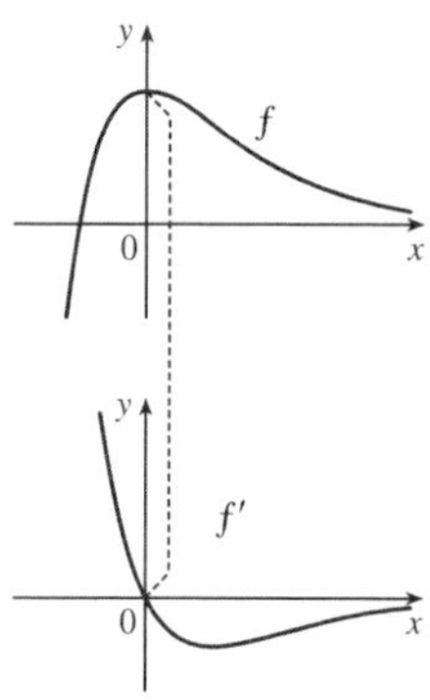


**6.**

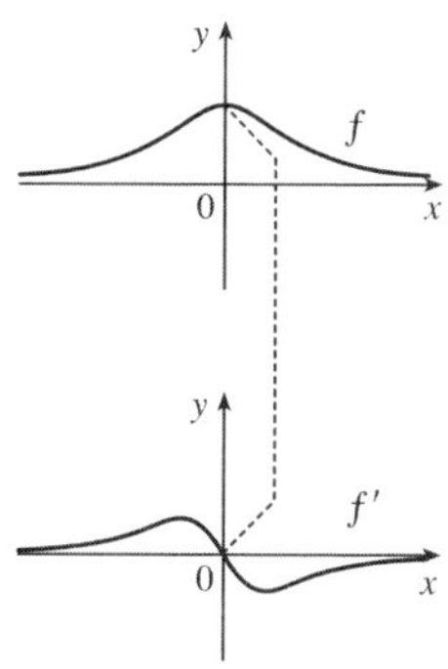

**7.**

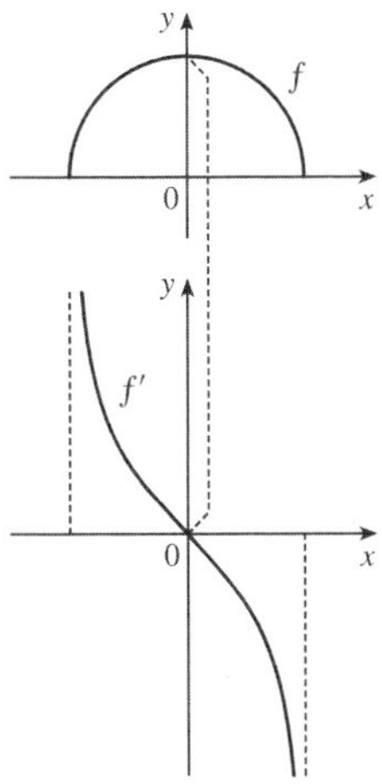


**8.**

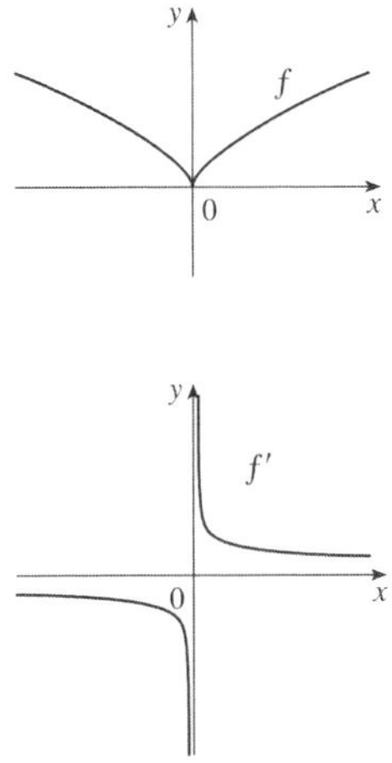


**9.**

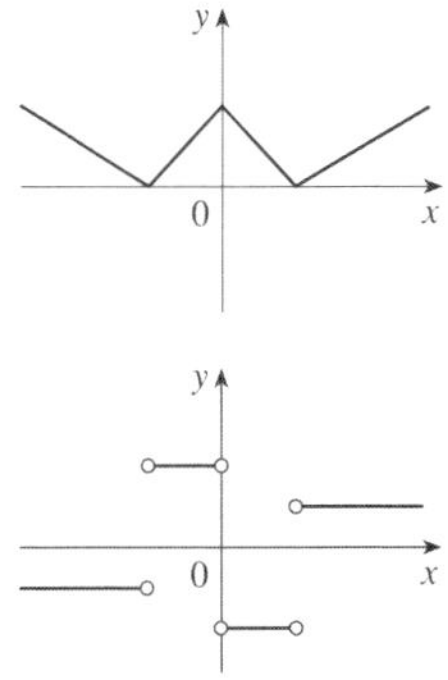


**10.**

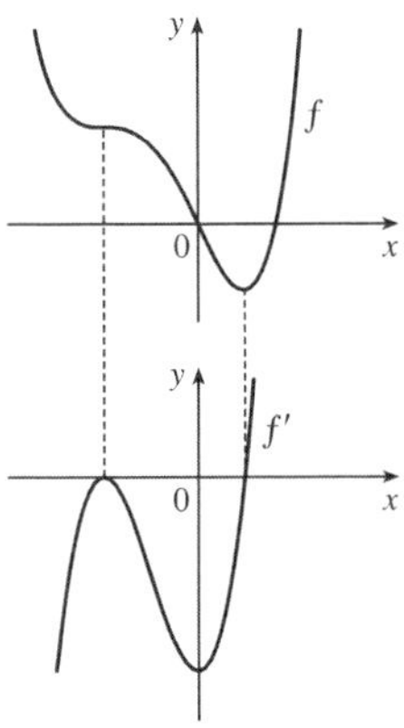


**11.**

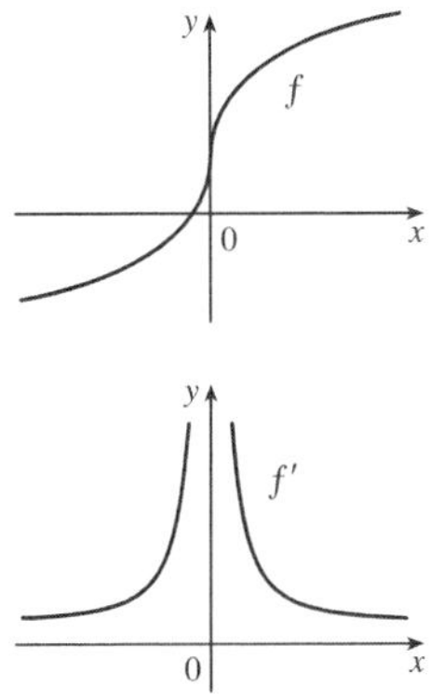


**12.** The slopes of the tangent lines on the graph of $y = P(t)$ are always positive, so the $y$-values of $y = P'(t)$ are always positive. These values start out relatively small and keep increasing, reaching a maximum at about $t = 6$. Then the $y$-values of $y = P'(t)$ decrease and get close to zero. The graph of $P'$ tells us that the yeast culture grows most rapidly after 6 hours and then the growth rate declines.

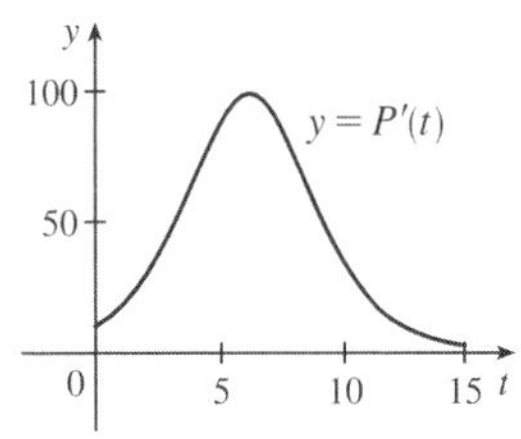


**13.** It appears that there are horizontal tangents on the graph of $M$ for $t = 1963$ and $t = 1971$. Thus, there are zeros for those values of $t$ on the graph of $M'$. The derivative is negative for the years 1963 to 1971.

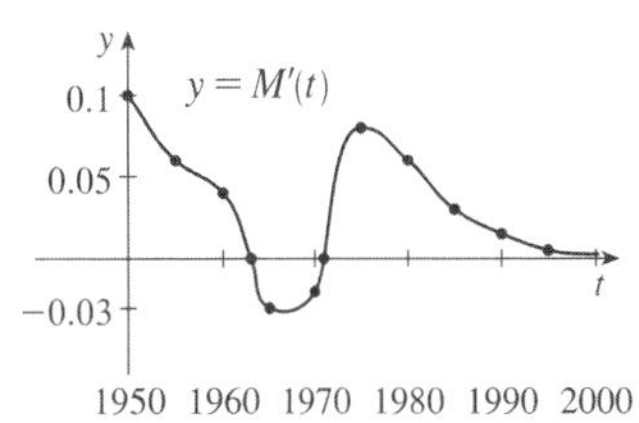


**14.** See Figure 1 in Section 3.4.

**15.**

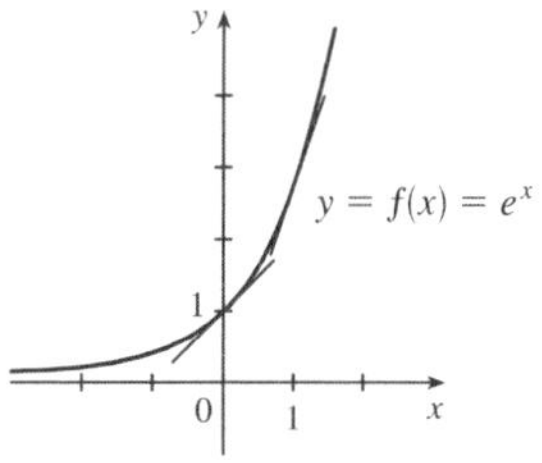


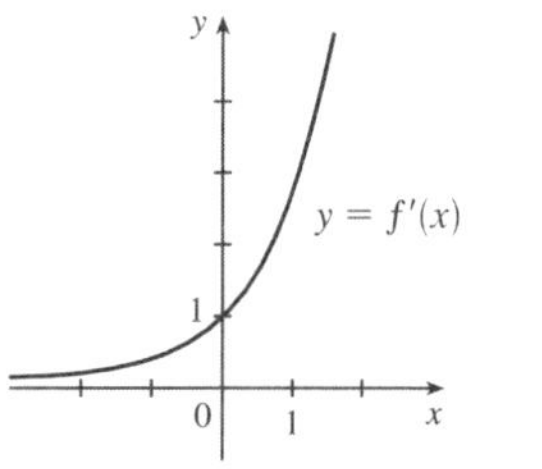


The slope at 0 appears to be 1 and the slope at 1 appears to be 2.7. As $x$ decreases, the slope gets closer to 0. Since the graphs are so similar, we might guess that $f'(x) = e^x$.

**16.**

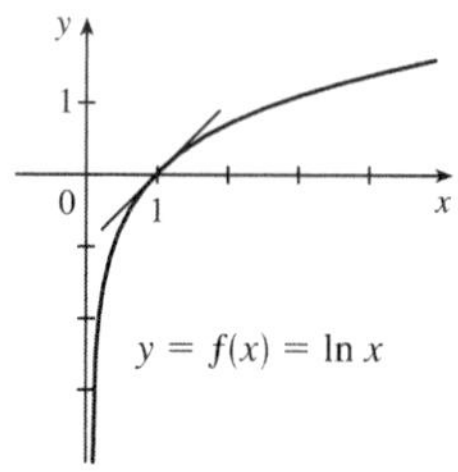


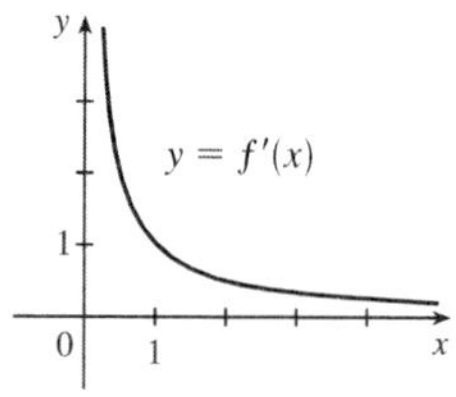


As $x$ increases toward 1, $f'(x)$ decreases from very large numbers to 1. As $x$ becomes large, $f'(x)$ gets closer to 0. As a guess, $f'(x) = 1/x^2$ or $f'(x) = 1/x$ make sense.

**17.** (a) By zooming in, we estimate that $f'(0) = 0$, $f'\left(\frac{1}{2}\right) = 1$, $f'(1) = 2$, and $f'(2) = 4$.

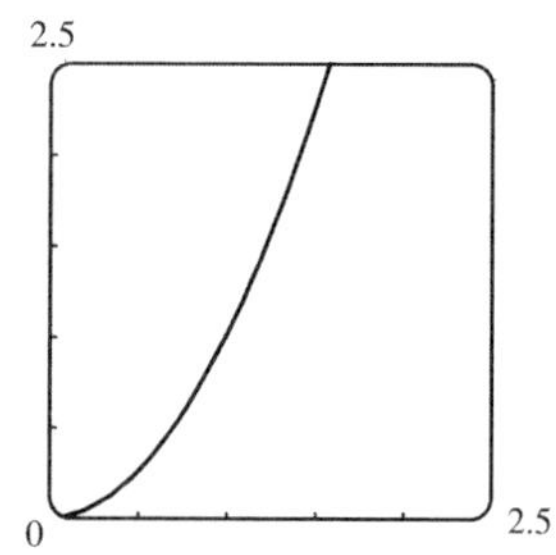


(b) By symmetry, $f'(-x) = -f'(x)$. So $f'\left(-\frac{1}{2}\right) = -1$, $f'(-1) = -2$, and $f'(-2) = -4$.

(c) It appears that $f'(x)$ is twice the value of $x$, so we guess that $f'(x) = 2x$.

(d) $$f'(x) = \lim_{h\to 0} \frac{f(x+h) - f(x)}{h} = \lim_{h\to 0} \frac{(x+h)^2 - x^2}{h}$$
$$= \lim_{h\to 0} \frac{(x^2 + 2hx + h^2) - x^2}{h} = \lim_{h\to 0} \frac{2hx + h^2}{h} = \lim_{h\to 0} \frac{h(2x+h)}{h} = \lim_{h\to 0} (2x+h) = 2x$$

**18.** (a) By zooming in, we estimate that $f'(0) = 0$, $f'\left(\frac{1}{2}\right) \approx 0.75$, $f'(1) \approx 3$, $f'(2) \approx 12$, and $f'(3) \approx 27$.

(b) By symmetry, $f'(-x) = f'(x)$. So $f'\left(-\frac{1}{2}\right) \approx 0.75$, $f'(-1) \approx 3$, $f'(-2) \approx 12$, and $f'(-3) \approx 27$.

(c)

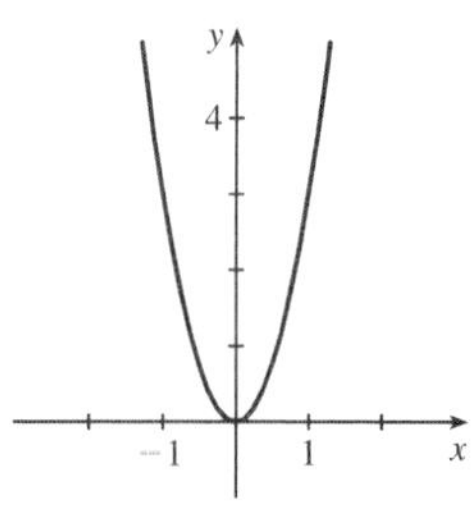


(d) Since $f'(0) = 0$, it appears that $f'$ may have the form $f'(x) = ax^2$. Using $f'(1) = 3$, we have $a = 3$, so $f'(x) = 3x^2$.

(e) $$f'(x) = \lim_{h\to 0} \frac{f(x+h) - f(x)}{h} = \lim_{h\to 0} \frac{(x+h)^3 - x^3}{h} = \lim_{h\to 0} \frac{(x^3 + 3x^2h + 3xh^2 + h^3) - x^3}{h}$$
$$= \lim_{h\to 0} \frac{3x^2h + 3xh^2 + h^3}{h} = \lim_{h\to 0} \frac{h(3x^2 + 3xh + h^2)}{h} = \lim_{h\to 0} (3x^2 + 3xh + h^2) = 3x^2$$

**19.** $f'(x) = \lim_{h\to 0} \dfrac{f(x+h)-f(x)}{h} = \lim_{h\to 0} \dfrac{\left[\frac{1}{2}(x+h)-\frac{1}{3}\right]-\left(\frac{1}{2}x-\frac{1}{3}\right)}{h} = \lim_{h\to 0} \dfrac{\frac{1}{2}x+\frac{1}{2}h-\frac{1}{3}-\frac{1}{2}x+\frac{1}{3}}{h}$

$= \lim_{h\to 0} \dfrac{\frac{1}{2}h}{h} = \lim_{h\to 0} \frac{1}{2} = \frac{1}{2}$

Domain of $f$ = domain of $f' = \mathbb{R}$.

**20.** $f'(x) = \lim_{h\to 0} \dfrac{f(x+h)-f(x)}{h} = \lim_{h\to 0} \dfrac{\left[1.5(x+h)^2-(x+h)+3.7\right]-\left(1.5x^2-x+3.7\right)}{h}$

$= \lim_{h\to 0} \dfrac{1.5x^2+3xh+1.5h^2-x-h+3.7-1.5x^2+x-3.7}{h} = \lim_{h\to 0} \dfrac{3xh+1.5h^2-h}{h}$

$= \lim_{h\to 0}(3x+1.5h-1) = 3x-1$

Domain of $f$ = domain of $f' = \mathbb{R}$.

**21.** $f'(x) = \lim_{h\to 0} \dfrac{f(x+h)-f(x)}{h} = \lim_{h\to 0} \dfrac{\left[(x+h)^3-3(x+h)+5\right]-\left(x^3-3x+5\right)}{h}$

$= \lim_{h\to 0} \dfrac{\left(x^3+3x^2h+3xh^2+h^3-3x-3h+5\right)-\left(x^3-3x+5\right)}{h} = \lim_{h\to 0} \dfrac{3x^2h+3xh^2+h^3-3h}{h}$

$= \lim_{h\to 0} \dfrac{h\left(3x^2+3xh+h^2-3\right)}{h} = \lim_{h\to 0}\left(3x^2+3xh+h^2-3\right) = 3x^2-3$

Domain of $f$ = domain of $f' = \mathbb{R}$.

**22.** $f'(x) = \lim_{h\to 0} \dfrac{f(x+h)-f(x)}{h} = \lim_{h\to 0} \dfrac{\left(x+h+\sqrt{x+h}\right)-\left(x+\sqrt{x}\right)}{h}$

$= \lim_{h\to 0}\left(\dfrac{h}{h}+\dfrac{\sqrt{x+h}-\sqrt{x}}{h}\cdot\dfrac{\sqrt{x+h}+\sqrt{x}}{\sqrt{x+h}+\sqrt{x}}\right) = \lim_{h\to 0}\left[1+\dfrac{(x+h)-x}{h\left(\sqrt{x+h}+\sqrt{x}\right)}\right]$

$= \lim_{h\to 0}\left(1+\dfrac{1}{\sqrt{x+h}+\sqrt{x}}\right) = 1+\dfrac{1}{\sqrt{x}+\sqrt{x}} = 1+\dfrac{1}{2\sqrt{x}}$

Domain of $f = [0,\infty)$, domain of $f' = (0,\infty)$.

**23.** $g'(x) = \lim_{h\to 0} \dfrac{g(x+h)-g(x)}{h} = \lim_{h\to 0} \dfrac{\sqrt{1+2(x+h)}-\sqrt{1+2x}}{h}\left[\dfrac{\sqrt{1+2(x+h)}+\sqrt{1+2x}}{\sqrt{1+2(x+h)}+\sqrt{1+2x}}\right]$

$= \lim_{h\to 0} \dfrac{(1+2x+2h)-(1+2x)}{h\left[\sqrt{1+2(x+h)}+\sqrt{1+2x}\right]} = \lim_{h\to 0} \dfrac{2}{\sqrt{1+2x+2h}+\sqrt{1+2x}} = \dfrac{2}{2\sqrt{1+2x}} = \dfrac{1}{\sqrt{1+2x}}$

Domain of $g = \left[-\frac{1}{2},\infty\right)$, domain of $g' = \left(-\frac{1}{2},\infty\right)$.

**24.** $f'(x) = \lim_{h\to 0} \dfrac{f(x+h)-f(x)}{h} = \lim_{h\to 0} \dfrac{\dfrac{3+(x+h)}{1-3(x+h)}-\dfrac{3+x}{1-3x}}{h} = \lim_{h\to 0} \dfrac{(3+x+h)(1-3x)-(3+x)(1-3x-3h)}{h(1-3x-3h)(1-3x)}$

$= \lim_{h\to 0} \dfrac{\left(3-9x+x-3x^2+h-3hx\right)-\left(3-9x-9h+x-3x^2-3hx\right)}{h(1-3x-3h)(1-3x)}$

$= \lim_{h\to 0} \dfrac{10h}{h(1-3x-3h)(1-3x)} = \lim_{h\to 0} \dfrac{10}{(1-3x-3h)(1-3x)} = \dfrac{10}{(1-3x)^2}$

Domain of $f$ = domain of $f' = \left(-\infty,\frac{1}{3}\right)\cup\left(\frac{1}{3},\infty\right)$.

**25.** $G'(t) = \lim\limits_{h \to 0} \dfrac{G(t+h) - G(t)}{h} = \lim\limits_{h \to 0} \dfrac{\dfrac{4(t+h)}{(t+h)+1} - \dfrac{4t}{t+1}}{h} = \lim\limits_{h \to 0} \dfrac{\dfrac{4(t+h)(t+1) - 4t(t+h+1)}{(t+h+1)(t+1)}}{h}$

$= \lim\limits_{h \to 0} \dfrac{(4t^2 + 4ht + 4t + 4h) - (4t^2 + 4ht + 4t)}{h(t+h+1)(t+1)} = \lim\limits_{h \to 0} \dfrac{4h}{h(t+h+1)(t+1)}$

$= \lim\limits_{h \to 0} \dfrac{4}{(t+h+1)(t+1)} = \dfrac{4}{(t+1)^2}$

Domain of $G$ = domain of $G' = (-\infty, -1) \cup (-1, \infty)$.

**26.** (a)

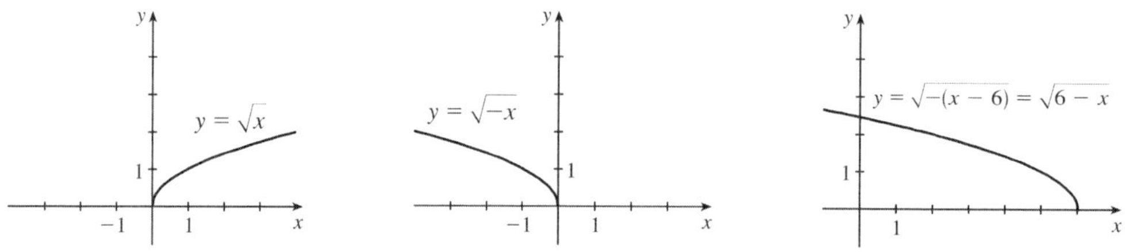


(b) Note that the third graph in part (a) has small negative values for its slope, $f'$; but as $x \to 6^-$, $f' \to -\infty$. See the graph in part (d).

(c) $f'(x) = \lim\limits_{h \to 0} \dfrac{f(x+h) - f(x)}{h}$

$= \lim\limits_{h \to 0} \dfrac{\sqrt{6-(x+h)} - \sqrt{6-x}}{h} \left[ \dfrac{\sqrt{6-(x+h)} + \sqrt{6-x}}{\sqrt{6-(x+h)} + \sqrt{6-x}} \right]$

$= \lim\limits_{h \to 0} \dfrac{[6-(x+h)] - (6-x)}{h\left[\sqrt{6-(x+h)} + \sqrt{6-x}\right]} = \lim\limits_{h \to 0} \dfrac{-h}{h\left(\sqrt{6-x-h} + \sqrt{6-x}\right)}$

$= \lim\limits_{h \to 0} \dfrac{-1}{\sqrt{6-x-h} + \sqrt{6-x}} = \dfrac{-1}{2\sqrt{6-x}}$

Domain of $f = (-\infty, 6]$, domain of $f' = (-\infty, 6)$.

(d)

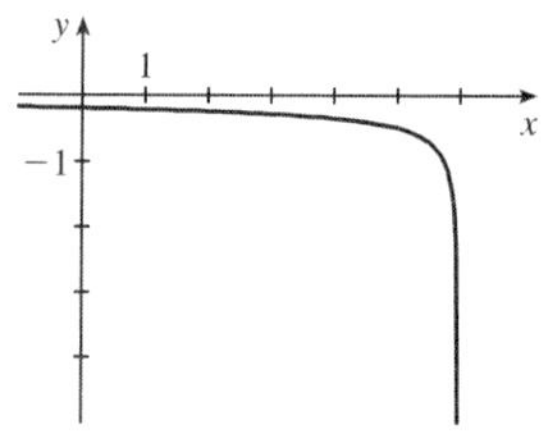

**27.** (a) $f'(x) = \lim\limits_{h \to 0} \dfrac{f(x+h) - f(x)}{h} = \lim\limits_{h \to 0} \dfrac{[(x+h)^4 + 2(x+h)] - (x^4 + 2x)}{h}$

$= \lim\limits_{h \to 0} \dfrac{x^4 + 4x^3h + 6x^2h^2 + 4xh^3 + h^4 + 2x + 2h - x^4 - 2x}{h}$

$= \lim\limits_{h \to 0} \dfrac{4x^3h + 6x^2h^2 + 4xh^3 + h^4 + 2h}{h} = \lim\limits_{h \to 0} \dfrac{h(4x^3 + 6x^2h + 4xh^2 + h^3 + 2)}{h}$

$= \lim\limits_{h \to 0} (4x^3 + 6x^2h + 4xh^2 + h^3 + 2) = 4x^3 + 2$

(b) Notice that $f'(x) = 0$ when $f$ has a horizontal tangent, $f'(x)$ is positive when the tangents have positive slope, and $f'(x)$ is negative when the tangents have negative slope

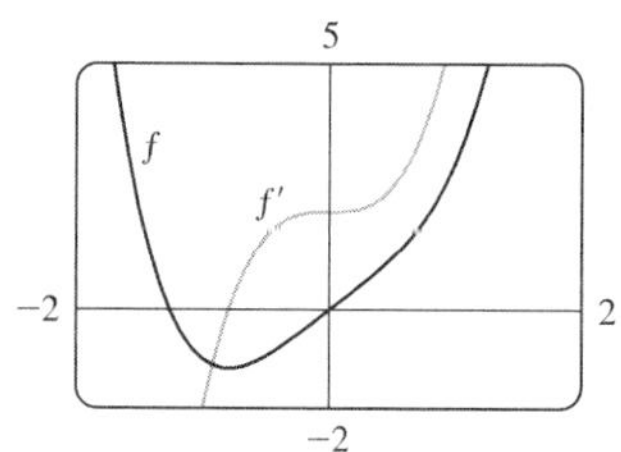

**28.** (a) $f'(t) = \lim\limits_{h\to 0} \dfrac{f(t+h)-f(t)}{h} = \lim\limits_{h\to 0} \dfrac{\left[(t+h)^2 - \sqrt{t+h}\right] - (t^2 - \sqrt{t})}{h}$

$$= \lim_{h\to 0} \frac{t^2 + 2ht + h^2 - \sqrt{t+h} - t^2 + \sqrt{t}}{h} = \lim_{h\to 0} \left(\frac{2ht+h^2}{h} + \frac{\sqrt{t}-\sqrt{t+h}}{h}\right)$$

$$= \lim_{h\to 0} \left(\frac{h(2t+h)}{h} + \frac{\sqrt{t}-\sqrt{t+h}}{h} \cdot \frac{\sqrt{t}+\sqrt{t+h}}{\sqrt{t}+\sqrt{t+h}}\right)$$

$$= \lim_{h\to 0} \left(2t + h + \frac{t-(t+h)}{h\left(\sqrt{t}+\sqrt{t+h}\right)}\right) = \lim_{h\to 0} \left(2t + h + \frac{-h}{h\left(\sqrt{t}+\sqrt{t+h}\right)}\right)$$

$$= \lim_{h\to 0} \left(2t + h + \frac{-1}{\sqrt{t}+\sqrt{t+h}}\right) = 2t - \frac{1}{2\sqrt{t}}$$

(b) Notice that $f'(t) = 0$ when $f$ has a horizontal tangent, $f'(t)$ is positive when the tangents have positive slope, and $f'(t)$ is negative when the tangents have negative slope.

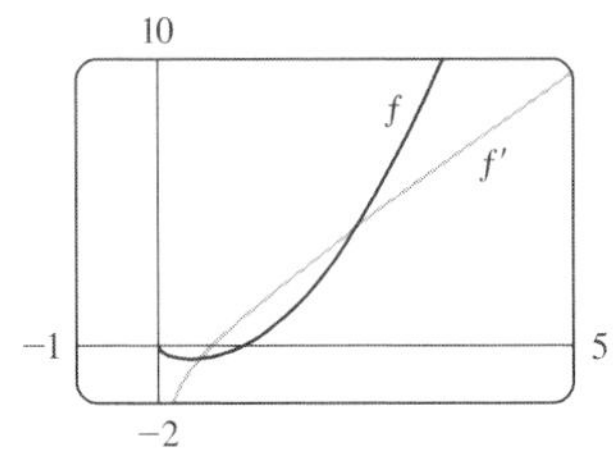


**29.** (a) $U'(t)$ is the rate at which the unemployment rate is changing with respect to time. Its units are percent per year.

(b) To find $U'(t)$, we use $\lim\limits_{h\to 0} \dfrac{U(t+h)-U(t)}{h} \approx \dfrac{U(t+h)-U(t)}{h}$ for small values of $h$.

**For 1993:** $U'(1993) \approx \dfrac{U(1994)-U(1993)}{1994-1993} = \dfrac{6.1-6.9}{1} = -0.80$

**For 1994:** We estimate $U'(1994)$ by using $h=-1$ and $h=1$, and then average the two results to obtain a final estimate.

$h=-1 \;\Rightarrow\; U'(1994) \approx \dfrac{U(1993)-U(1994)}{1993-1994} = \dfrac{6.9-6.1}{-1} = -0.80;$

$h=1 \;\Rightarrow\; U'(1994) \approx \dfrac{U(1995)-U(1994)}{1995-1994} = \dfrac{5.6-6.1}{1} = -0.50.$

So we estimate that $U'(1994) \approx \frac{1}{2}[(-0.80)+(-0.50)] = -0.65$.

| $t$ | 1993 | 1994 | 1995 | 1996 | 1997 | 1998 | 1999 | 2000 | 2001 | 2002 |
|---|---|---|---|---|---|---|---|---|---|---|
| $U'(t)$ | −0.80 | −0.65 | −0.35 | −0.35 | −0.45 | −0.35 | −0.25 | 0.25 | 0.90 | 1.10 |

**30.** (a) $P'(t)$ is the rate at which the percentage of Americans under the age of 18 is changing with respect to time. Its units are percent per year (%/yr).

(b) To find $P'(t)$, we use $\lim\limits_{h\to 0} \dfrac{P(t+h)-P(t)}{h} \approx \dfrac{P(t+h)-P(t)}{h}$ for small values of $h$.

**For 1950:** $P'(1950) \approx \dfrac{P(1960)-P(1950)}{1960-1950} = \dfrac{35.7-31.1}{10} = 0.46$

**For 1960:** We estimate $P'(1960)$ by using $h=-10$ and $h=10$, and then average the two results to obtain a final estimate.

$h=-10 \;\Rightarrow\; P'(1960) \approx \dfrac{P(1950)-P(1960)}{1950-1960} = \dfrac{31.1-35.7}{-10} = 0.46$

$h=10 \;\Rightarrow\; P'(1960) \approx \dfrac{P(1970)-P(1960)}{1970-1960} = \dfrac{34.0-35.7}{10} = -0.17$

So we estimate that $P'(1960) \approx \frac{1}{2}[0.46+(-0.17)] = 0.145$.

| $t$ | 1950 | 1960 | 1970 | 1980 | 1990 | 2000 |
|---|---|---|---|---|---|---|
| $P'(t)$ | 0.460 | 0.145 | −0.385 | −0.415 | −0.115 | 0.000 |

(c)

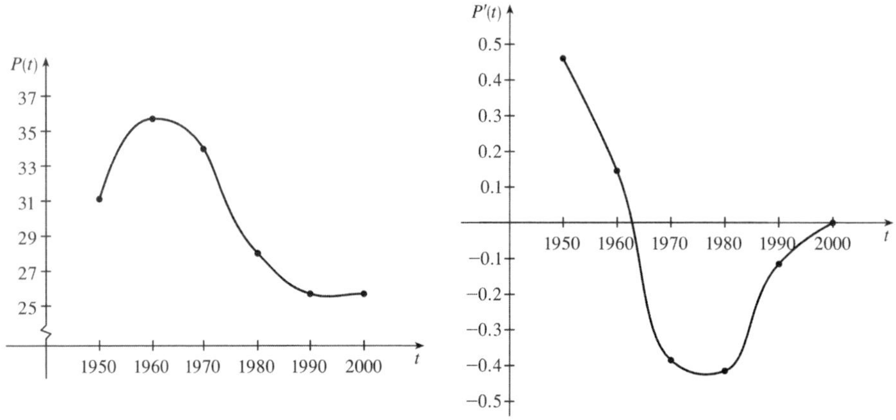


(d) We could get more accurate values for $P'(t)$ by obtaining data for the mid-decade years 1955, 1965, 1975, 1985, and 1995.

**31.** $f$ is not differentiable at $x = -4$, because the graph has a corner there, and at $x = 0$, because there is a discontinuity there.

**32.** $f$ is not differentiable at $x = 0$, because there is a discontinuity there, and at $x = 3$, because the graph has a vertical tangent there.

**33.** $f$ is not differentiable at $x = -1$, because the graph has a vertical tangent there, and at $x = 4$, because the graph has a corner there.

**34.** $f$ is not differentiable at $x = -1$, because there is a discontinuity there, and at $x = 2$, because the graph has a corner there.

**35.** As we zoom in toward $(-1, 0)$, the curve appears more and more like a straight line, so $f(x) = x + \sqrt{|x|}$ is differentiable at $x = -1$. But no matter how much we zoom in toward the origin, the curve doesn't straighten out—we can't eliminate the sharp point (a cusp). So $f$ is not differentiable at $x = 0$.

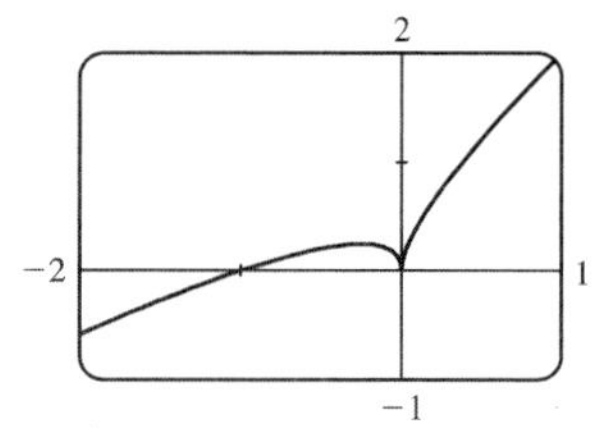


**36.** As we zoom in toward $(0, 1)$, the curve appears more and more like a straight line, so $f$ is differentiable at $x = 0$. But no matter how much we zoom in toward $(1, 0)$ or $(-1, 0)$, the curve doesn't straighten out—we can't eliminate the sharp point (a cusp). So $f$ is not differentiable at $x = \pm 1$.

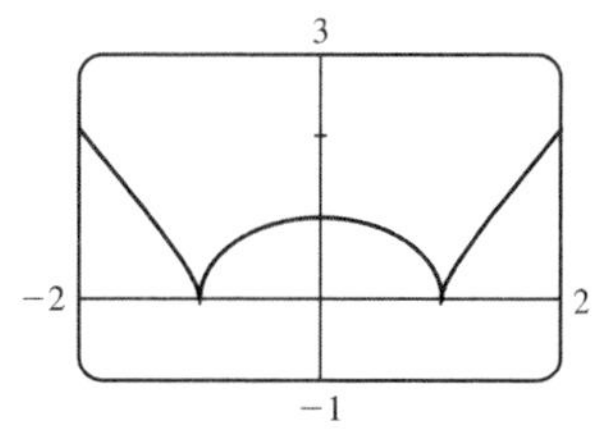


**37.** $a = f$, $b = f'$, $c = f''$. We can see this because where $a$ has a horizontal tangent, $b = 0$, and where $b$ has a horizontal tangent, $c = 0$. We can immediately see that $c$ can be neither $f$ nor $f'$, since at the points where $c$ has a horizontal tangent, neither $a$ nor $b$ is equal to 0.

**38.** Where $d$ has horizontal tangents, only $c$ is 0, so $d' = c$. $c$ has negative tangents for $x < 0$ and $b$ is the only graph that is negative for $x < 0$, so $c' = b$. $b$ has positive tangents on $\mathbb{R}$ (except at $x = 0$), and the only graph that is positive on the same domain is $a$, so $b' = a$. We conclude that $d = f$, $c = f'$, $b = f''$, and $a = f'''$.

**39.** We can immediately see that $a$ is the graph of the acceleration function, since at the points where $a$ has a horizontal tangent, neither $c$ nor $b$ is equal to 0. Next, we note that $a = 0$ at the point where $b$ has a horizontal tangent, so $b$ must be the graph of the velocity function, and hence, $b' = a$. We conclude that $c$ is the graph of the position function.

**40.** $a$ must be the jerk since none of the graphs are 0 at its high and low points. $a$ is 0 where $b$ has a maximum, so $b' = a$. $b$ is 0 where $c$ has a maximum, so $c' = b$. We conclude that $d$ is the position function, $c$ is the velocity, $b$ is the acceleration, and $a$ is the jerk.

**41.** $f'(x) = \lim_{h\to 0} \dfrac{f(x+h)-f(x)}{h} = \lim_{h\to 0} \dfrac{\left[1+4(x+h)-(x+h)^2\right]-(1+4x-x^2)}{h}$

$= \lim_{h\to 0} \dfrac{(1+4x+4h-x^2-2xh-h^2)-(1+4x-x^2)}{h} = \lim_{h\to 0} \dfrac{4h-2xh-h^2}{h} = \lim_{h\to 0}(4-2x-h) = 4-2x$

$f''(x) = \lim_{h\to 0} \dfrac{f'(x+h)-f'(x)}{h} = \lim_{h\to 0} \dfrac{[4-2(x+h)]-(4-2x)}{h} = \lim_{h\to 0} \dfrac{-2h}{h} = \lim_{h\to 0}(-2) = -2$

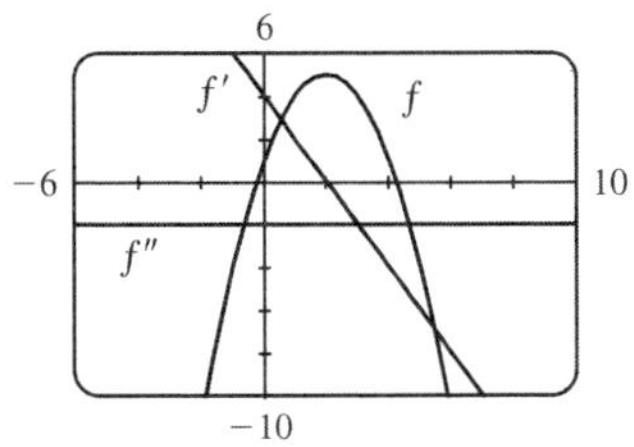


We see from the graph that our answers are reasonable because the graph of $f'$ is that of a linear function and the graph of $f''$ is that of a constant function.

**42.** $f'(x) = \lim_{h\to 0} \dfrac{f(x+h)-f(x)}{h} = \lim_{h\to 0} \dfrac{\dfrac{1}{x+h}-\dfrac{1}{x}}{h} = \lim_{h\to 0} \dfrac{x-(x+h)}{hx(x+h)} = \lim_{h\to 0} \dfrac{-h}{hx(x+h)} = \lim_{h\to 0} \dfrac{-1}{x(x+h)} = -\dfrac{1}{x^2}$

$f''(x) = \lim_{h\to 0} \dfrac{f'(x+h)-f'(x)}{h} = \lim_{h\to 0} \dfrac{-\dfrac{1}{(x+h)^2}-\left(-\dfrac{1}{x^2}\right)}{h} = \lim_{h\to 0} \dfrac{-x^2+(x+h)^2}{hx^2(x+h)^2} = \lim_{h\to 0} \dfrac{2hx+h^2}{hx^2(x+h)^2}$

$= \lim_{h\to 0} \dfrac{2x+h}{x^2(x+h)^2} = \dfrac{2x}{x^4} = \dfrac{2}{x^3}$

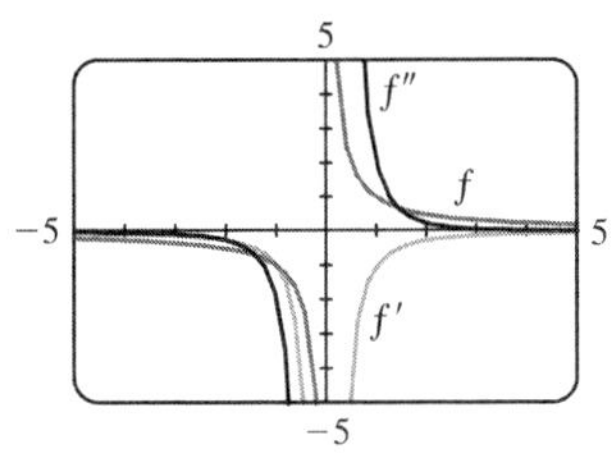


We see from the graph that our answers are reasonable because the graph of $f'$ is that of an even function and is negative for all $x \neq 0$, and the graph of $f''$ is that of an odd function (negative for $x < 0$ and positive for $x > 0$).

**43.** $f'(x) = \lim_{h\to 0} \dfrac{f(x+h)-f(x)}{h} = \lim_{h\to 0} \dfrac{\left[2(x+h)^2-(x+h)^3\right]-(2x^2-x^3)}{h}$

$= \lim_{h\to 0} \dfrac{h(4x+2h-3x^2-3xh-h^2)}{h} = \lim_{h\to 0}(4x+2h-3x^2-3xh-h^2) = 4x-3x^2$

$f''(x) = \lim_{h\to 0} \dfrac{f'(x+h)-f'(x)}{h} = \lim_{h\to 0} \dfrac{\left[4(x+h)-3(x+h)^2\right]-(4x-3x^2)}{h} = \lim_{h\to 0} \dfrac{h(4-6x-3h)}{h}$

$= \lim_{h\to 0}(4-6x-3h) = 4-6x$

$f'''(x) = \lim_{h\to 0} \dfrac{f''(x+h)-f''(x)}{h} = \lim_{h\to 0} \dfrac{[4-6(x+h)]-(4-6x)}{h} = \lim_{h\to 0} \dfrac{-6h}{h} = \lim_{h\to 0}(-6) = -6$

$f^{(4)}(x) = \lim_{h\to 0} \dfrac{f'''(x+h)-f'''(x)}{h} = \lim_{h\to 0} \dfrac{-6-(-6)}{h} = \lim_{h\to 0} \dfrac{0}{h} = \lim_{h\to 0}(0) = 0$

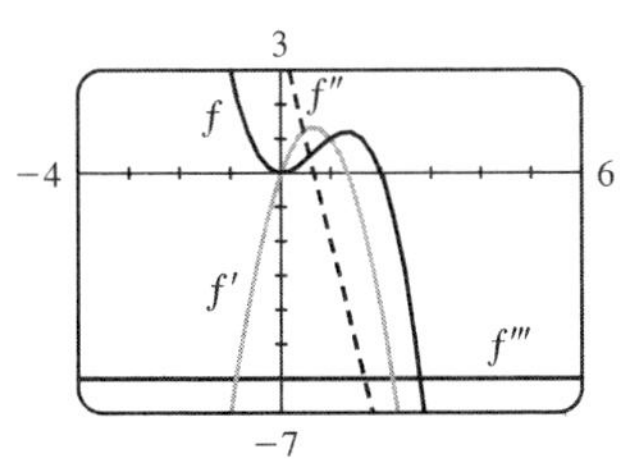


The graphs are consistent with the geometric interpretations of the derivatives because $f'$ has zeros where $f$ has a local minimum and a local maximum, $f''$ has a zero where $f'$ has a local maximum, and $f'''$ is a constant function equal to the slope of $f''$.

**44.** (a) Since we estimate the velocity to be a maximum at $t = 10$, the acceleration is 0 at $t = 10$.

(b) Drawing a tangent line at $t = 10$ on the graph of $a$, $a$ appears to decrease by $10\ \text{ft/s}^2$ over a period of 20 s. So at $t = 10$ s, the jerk is approximately $-10/20 = -0.5\ (\text{ft/s}^2)/\text{s}$ or $\text{ft/s}^3$.

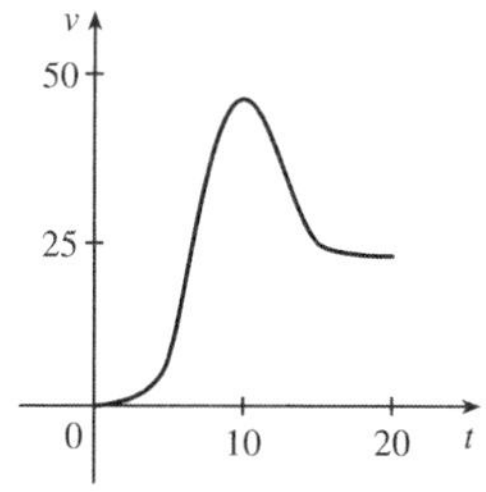


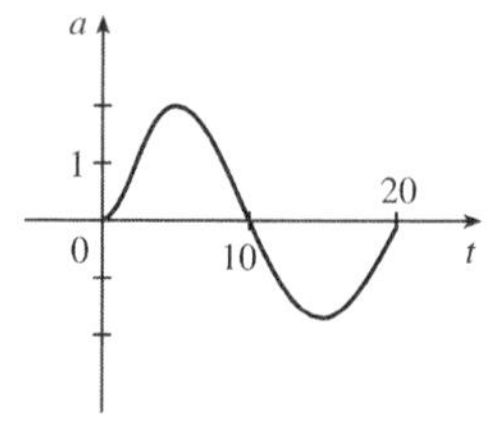


**45.** (a) Note that we have factored $x - a$ as the difference of two cubes in the third step.

$$f'(a) = \lim_{x \to a} \frac{f(x) - f(a)}{x - a} = \lim_{x \to a} \frac{x^{1/3} - a^{1/3}}{x - a} = \lim_{x \to a} \frac{x^{1/3} - a^{1/3}}{(x^{1/3} - a^{1/3})(x^{2/3} + x^{1/3}a^{1/3} + a^{2/3})}$$

$$= \lim_{x \to a} \frac{1}{x^{2/3} + x^{1/3}a^{1/3} + a^{2/3}} = \frac{1}{3a^{2/3}} \text{ or } \tfrac{1}{3}a^{-2/3}$$

(b) $f'(0) = \lim\limits_{h \to 0} \dfrac{f(0 + h) - f(0)}{h} = \lim\limits_{h \to 0} \dfrac{\sqrt[3]{h} - 0}{h} = \lim\limits_{h \to 0} \dfrac{1}{h^{2/3}}$. This function increases without bound, so the limit does not exist, and therefore $f'(0)$ does not exist.

(c) $\lim\limits_{x \to 0} |f'(x)| = \lim\limits_{x \to 0} \dfrac{1}{3x^{2/3}} = \infty$ and $f$ is continuous at $x = 0$ (root function), so $f$ has a vertical tangent at $x = 0$.

**46.** (a) $g'(0) = \lim\limits_{x \to 0} \dfrac{g(x) - g(0)}{x - 0} = \lim\limits_{x \to 0} \dfrac{x^{2/3} - 0}{x} = \lim\limits_{x \to 0} \dfrac{1}{x^{1/3}}$, which does not exist.

(b) $$g'(a) = \lim_{x \to a} \frac{g(x) - g(a)}{x - a} = \lim_{x \to a} \frac{x^{2/3} - a^{2/3}}{x - a} = \lim_{x \to a} \frac{(x^{1/3} - a^{1/3})(x^{1/3} + a^{1/3})}{(x^{1/3} - a^{1/3})(x^{2/3} + x^{1/3}a^{1/3} + a^{2/3})}$$

$$= \lim_{x \to a} \frac{x^{1/3} + a^{1/3}}{x^{2/3} + x^{1/3}a^{1/3} + a^{2/3}} = \frac{2a^{1/3}}{3a^{2/3}} = \frac{2}{3a^{1/3}} \text{ or } \tfrac{2}{3}a^{-1/3}$$

(c) $g(x) = x^{2/3}$ is continuous at $x = 0$ and $\lim\limits_{x \to 0} |g'(x)| = \lim\limits_{x \to 0} \dfrac{2}{3\,|x|^{1/3}} = \infty$. This shows that $g$ has a vertical tangent line at $x = 0$.

(d)

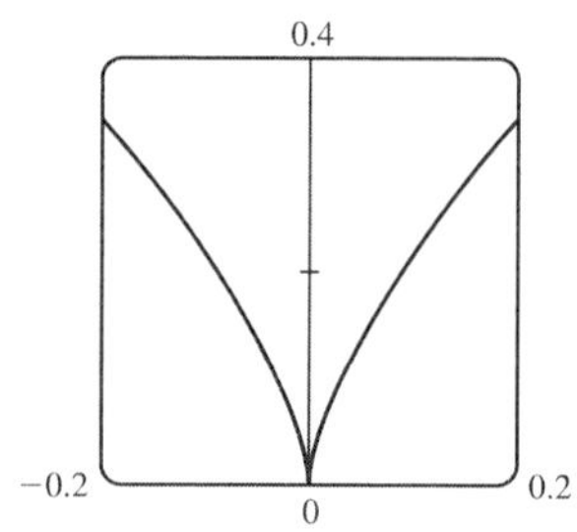


**47.** $f(x) = |x - 6| = \begin{cases} x - 6 & \text{if } x - 6 \geq 6 \\ -(x - 6) & \text{if } x - 6 < 0 \end{cases} = \begin{cases} x - 6 & \text{if } x \geq 6 \\ 6 - x & \text{if } x < 6 \end{cases}$

So the right-hand limit is $\lim\limits_{x \to 6^+} \dfrac{f(x) - f(6)}{x - 6} = \lim\limits_{x \to 6^+} \dfrac{|x - 6| - 0}{x - 6} = \lim\limits_{x \to 6^+} \dfrac{x - 6}{x - 6} = \lim\limits_{x \to 6^+} 1 = 1$, and the left-hand limit is $\lim\limits_{x \to 6^-} \dfrac{f(x) - f(6)}{x - 6} = \lim\limits_{x \to 6^-} \dfrac{|x - 6| - 0}{x - 6} = \lim\limits_{x \to 6^-} \dfrac{6 - x}{x - 6} = \lim\limits_{x \to 6^-} (-1) = -1$. Since these limits are not equal, $f'(6) = \lim\limits_{x \to 6} \dfrac{f(x) - f(6)}{x - 6}$ does not exist and $f$ is not differentiable at 6.

However, a formula for $f'$ is $f'(x) = \begin{cases} 1 & \text{if } x > 6 \\ -1 & \text{if } x < 6 \end{cases}$

Another way of writing the formula is $f'(x) = \dfrac{x - 6}{|x - 6|}$.

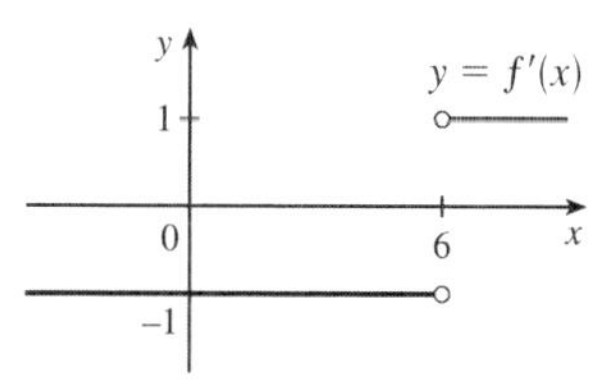

**48.** $f(x) = [\![x]\!]$ is not continuous at any integer $n$, so $f$ is not differentiable at $n$ by the contrapositive of Theorem 4. If $a$ is not an integer, then $f$ is constant on an open interval containing $a$, so $f'(a) = 0$. Thus, $f'(x) = 0$, $x$ not an integer.

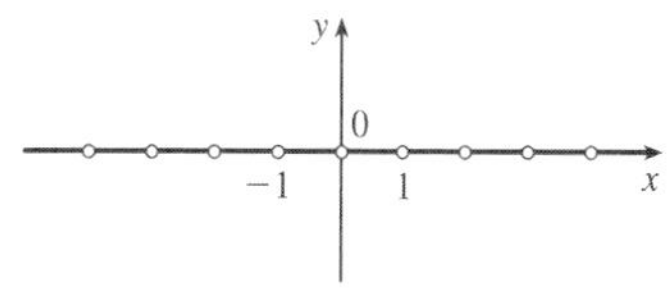


**49.** (a) If $f$ is even, then

$$\begin{aligned} f'(-x) &= \lim_{h\to 0} \frac{f(-x+h)-f(-x)}{h} = \lim_{h\to 0} \frac{f[-(x-h)]-f(-x)}{h} \\ &= \lim_{h\to 0} \frac{f(x-h)-f(x)}{h} = -\lim_{h\to 0} \frac{f(x-h)-f(x)}{-h} \quad [\text{let } \Delta x = -h] \\ &= -\lim_{\Delta x\to 0} \frac{f(x+\Delta x)-f(x)}{\Delta x} = -f'(x) \end{aligned}$$

Therefore, $f'$ is odd.

(b) If $f$ is odd, then

$$\begin{aligned} f'(-x) &= \lim_{h\to 0} \frac{f(-x+h)-f(-x)}{h} = \lim_{h\to 0} \frac{f[-(x-h)]-f(-x)}{h} \\ &= \lim_{h\to 0} \frac{-f(x-h)+f(x)}{h} = \lim_{h\to 0} \frac{f(x-h)-f(x)}{-h} \quad [\text{let } \Delta x = -h] \\ &= \lim_{\Delta x\to 0} \frac{f(x+\Delta x)-f(x)}{\Delta x} = f'(x) \end{aligned}$$

Therefore, $f'$ is even.

**50.** (a)

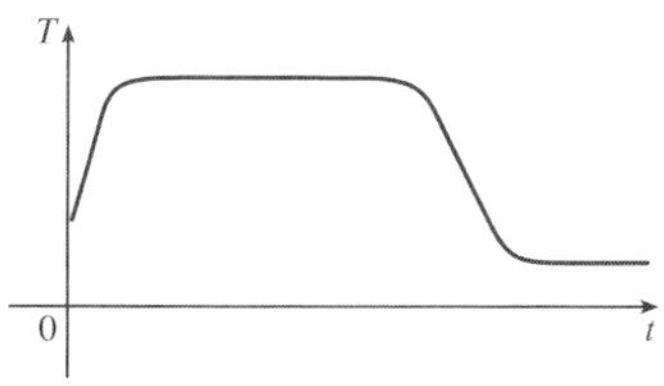


(b) The initial temperature of the water is close to room temperature because of the water that was in the pipes. When the water from the hot water tank starts coming out, $dT/dt$ is large and positive as $T$ increases to the temperature of the water in the tank. In the next phase, $dT/dt = 0$ as the water comes out at a constant, high temperature. After some time, $dT/dt$ becomes small and negative as the contents of the hot water tank are exhausted. Finally, when the hot water has run out, $dT/dt$ is once again 0 as the water maintains its (cold) temperature.

(c)

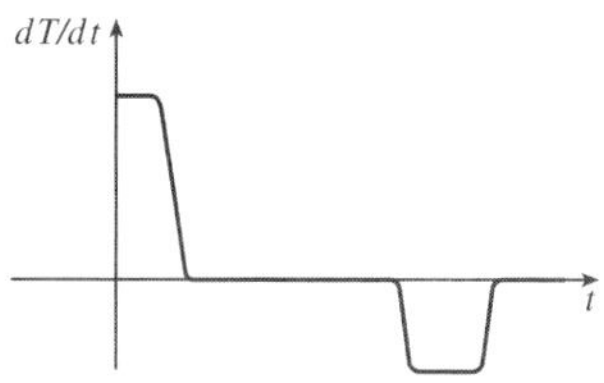


**51.**

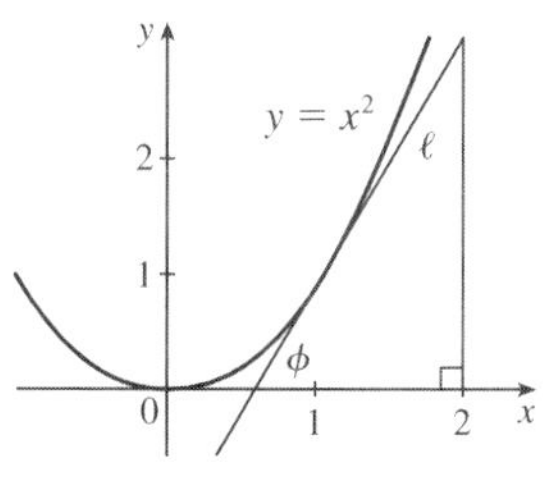


In the right triangle in the diagram, let $\Delta y$ be the side opposite angle $\phi$ and $\Delta x$ the side adjacent angle $\phi$. Then the slope of the tangent line $\ell$ is $m = \Delta y/\Delta x = \tan\phi$. Note that $0 < \phi < \frac{\pi}{2}$. We know (see Exercise 17) that the derivative of $f(x) = x^2$ is $f'(x) = 2x$. So the slope of the tangent to the curve at the point $(1, 1)$ is 2. Thus, $\phi$ is the angle between 0 and $\frac{\pi}{2}$ whose tangent is 2; that is, $\phi = \tan^{-1} 2 \approx 63°$.

## 2.9 What Does $f'$ Say about $f$?

1. (a) Since $f'(x) > 0$ on $(1, 5)$, $f$ is increasing on this interval. Since $f'(x) < 0$ on $(0, 1)$ and $(5, 6)$, $f$ is decreasing on these intervals.

   (b) Since $f'(x) = 0$ at $x = 1$ and $f'$ changes from negative to positive there, $f$ changes from decreasing to increasing and has a local minimum at $x = 1$. Since $f'(x) = 0$ at $x = 5$ and $f'$ changes from positive to negative there, $f$ changes from increasing to decreasing and has a local maximum at $x = 5$.

   (c) Since $f(0) = 0$, start at the origin. Draw a decreasing function on $(0, 1)$ with a local minimum at $x = 1$. Now draw an increasing function on $(1, 5)$ and the steepest slope should occur at $x = 3$ since that's where the largest value of $f'$ occurs. Lastly, draw a decreasing function on $(5, 6)$ making sure you have a local maximum at $x = 5$.

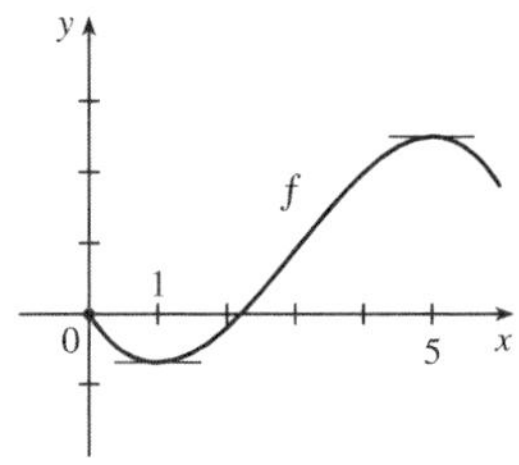


2. (a) $f'(x) > 0$ and $f$ is increasing on $(0, 1)$ and $(3, 5)$. $f'(x) < 0$ and $f$ is decreasing on $(1, 3)$ and $(5, 6)$.

   (b) Since $f'(x) = 0$ at $x = 1$ and $x = 5$ and $f'$ changes from positive to negative at both values, $f$ changes from increasing to decreasing and has local maxima at $x = 1$ and $x = 5$. Since $f'(x) = 0$ at $x = 3$ and $f'$ changes from negative to positive there, $f$ changes from decreasing to increasing and has a local minimum at $x = 3$.

   (c)

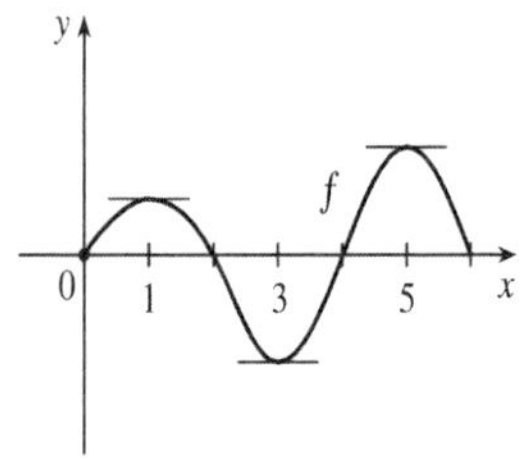


3. The derivative $f'$ is increasing when the slopes of the tangent lines of $f$ are becoming larger as $x$ increases. This seems to be the case on the interval $(2, 5)$. The derivative is decreasing when the slopes of the tangent lines of $f$ are becoming smaller as $x$ increases, and this seems to be the case on $(-\infty, 2)$ and $(5, \infty)$. So $f'$ is increasing on $(2, 5)$ and decreasing on $(-\infty, 2)$ and $(5, \infty)$.

4. (a)

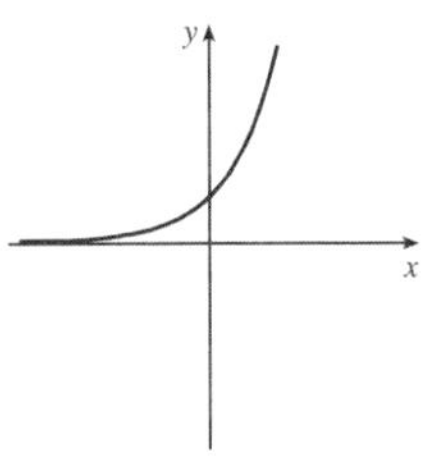


   (b)

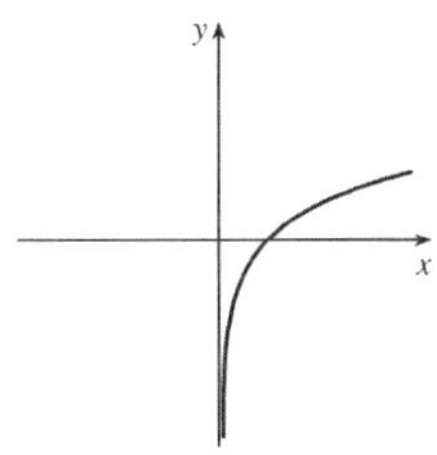


   (c) In part (a), the graph of $y = e^x$ is a curve whose slope is always positive and increasing. In part (b), the graph of $y = \ln x$ is a curve whose slope is always positive and decreasing.

5. If $D(t)$ is the size of the deficit as a function of time, then at the time of the speech $D'(t) > 0$, but $D''(t) < 0$ because $D''(t) = (D')'(t)$ is the rate of change of $D'(t)$.

6. (a) The rate of increase of the population is initially very small, then gets larger until it reaches a maximum at about $t = 8$ hours, and decreases toward 0 as the population begins to level off.

   (b) The rate of increase has its maximum value at $t = 8$ hours.

   (c) The population function is concave upward on $(0, 8)$ and concave downward on $(8, 18)$.

   (d) At $t = 8$, the population is about 350, so the inflection point is about $(8, 350)$.

**7.** (a) The rate of increase of the population is initially very small, then increases rapidly until about 1932 when it starts decreasing. The rate becomes negative by 1936, peaks in magnitude in 1937, and approaches 0 in 1940.

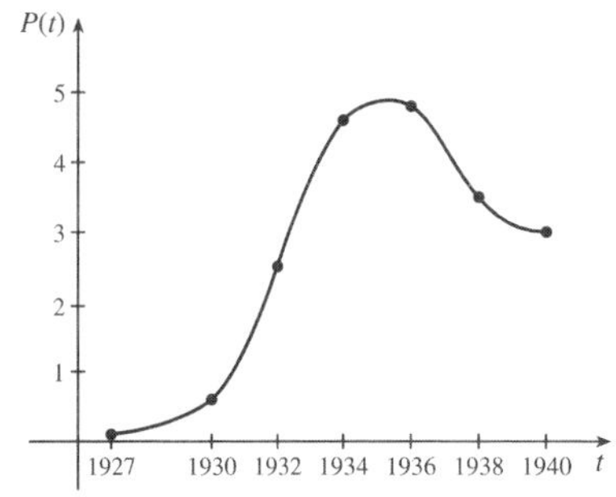


(b) Inflection points (IP) appear to be at $(1932, 2.5)$ and $(1937, 4.3)$. The rate of change of population density starts to decrease in 1932 and starts to increase in 1937. The rates of population increase and decrease have their maximum values at those points.

**8.** (a) If the position function is increasing, then the particle is moving toward the right. This occurs on $t$-intervals $(0, 2)$ and $(4, 6)$. If the function is decreasing, then the particle is moving toward the left—that is, on $(2, 4)$.

(b) The acceleration is the second derivative and is positive where the curve is concave upward. This occurs on $(3, 6)$. The acceleration is negative where the curve is concave downward—that is, on $(0, 3)$.

**9.** Most students learn more in the third hour of studying than in the eighth hour, so $K(3) - K(2)$ is larger than $K(8) - K(7)$. In other words, as you begin studying for a test, the rate of knowledge gain is large and then starts to taper off, so $K'(t)$ decreases and the graph of $K$ is concave downward.

**10.** At first the depth increases slowly because the base of the mug is wide. But as the mug narrows, the coffee rises more quickly. Thus, the depth $d$ increases at an increasing rate and its graph is concave upward. The rate of increase of $d$ has a maximum where the mug is narrowest; that is, when the mug is half full. It is there that the inflection point (IP) occurs. Then the rate of increase of $d$ starts to decrease as the mug widens and the graph becomes concave down.

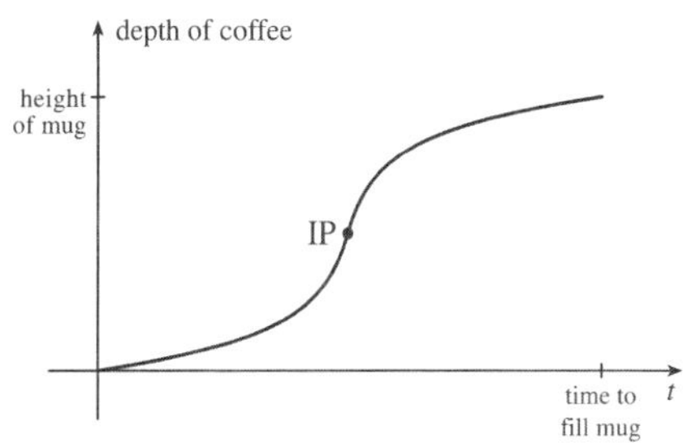


**11.** (a) $f$ is increasing where $f'$ is positive, that is, on $(0, 2)$, $(4, 6)$, and $(8, \infty)$; and decreasing where $f'$ is negative, that is, on $(2, 4)$ and $(6, 8)$.

(b) $f$ has local maxima where $f'$ changes from positive to negative, at $x = 2$ and at $x = 6$, and local minima where $f'$ changes from negative to positive, at $x = 4$ and at $x = 8$.

(c) $f$ is concave upward (CU) where $f'$ is increasing, that is, on $(3, 6)$ and $(6, \infty)$, and concave downward (CD) where $f'$ is decreasing, that is, on $(0, 3)$.

(d) There is a point of inflection where $f$ changes from being CD to being CU, that is, at $x = 3$.

(e)

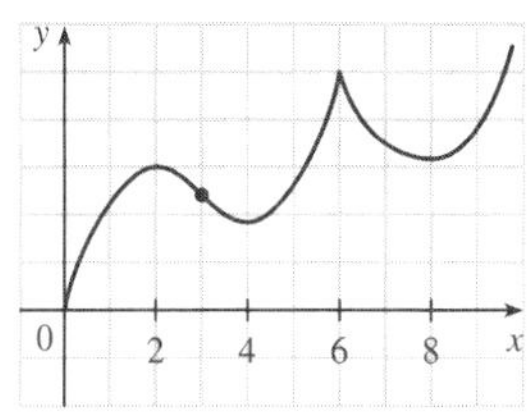


**12.** (a) $f$ is increasing where $f'$ is positive, on $(1, 6)$ and $(8, \infty)$, and decreasing where $f'$ is negative, on $(0, 1)$ and $(6, 8)$.

(b) $f$ has a local maximum where $f'$ changes from positive to negative, at $x = 6$, and local minima where $f'$ changes from negative to positive, at $x = 1$ and at $x = 8$.

(c) $f$ is concave upward where $f'$ is increasing, that is, on $(0, 2)$, $(3, 5)$, and $(7, \infty)$, and concave downward where $f'$ is decreasing, that is, on $(2, 3)$ and $(5, 7)$.

(d) There are points of inflection where $f$ changes its direction of concavity, at $x = 2$, $x = 3$, $x = 5$ and $x = 7$.

(e)

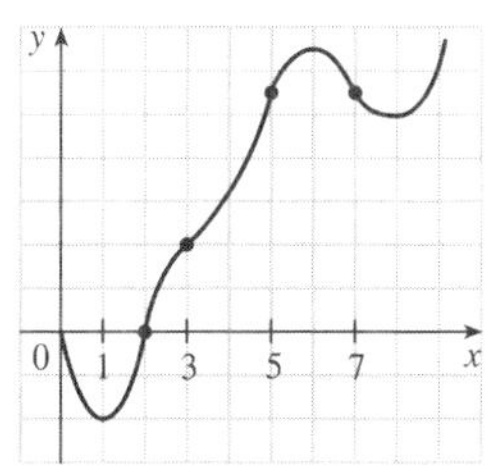

**13.** The function must be always decreasing (since the first derivative is always negative) and concave downward (since the second derivative is always negative).

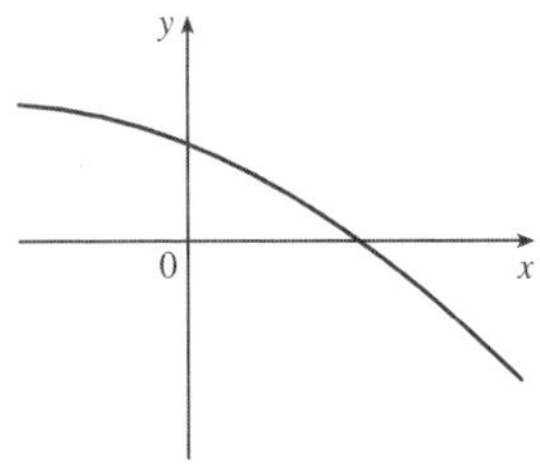


**14.** The function must be always decreasing and concave upward.

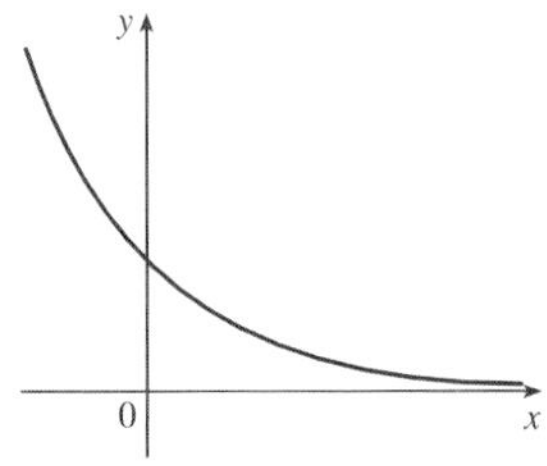


**15.** $f'(0) = f'(4) = 0 \Rightarrow$ horizontal tangents at $x = 0, 4$.
$f'(x) > 0$ if $x < 0 \Rightarrow f$ is increasing on $(-\infty, 0)$.
$f'(x) < 0$ if $0 < x < 4$ or if $x > 4 \Rightarrow f$ is decreasing on $(0, 4)$ and $(4, \infty)$.
$f''(x) > 0$ if $2 < x < 4 \Rightarrow f$ is concave upward on $(2, 4)$.
$f''(x) < 0$ if $x < 2$ or $x > 4 \Rightarrow f$ is concave downward on $(-\infty, 2)$ and $(4, \infty)$. There are inflection points when $x = 2$ and $4$.

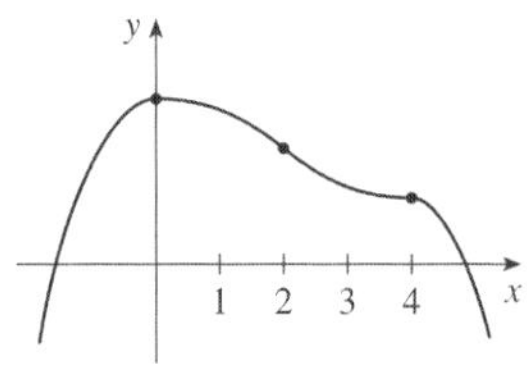


**16.** $f'(x) > 0$ for all $x \neq 1$ with vertical asymptote $x = 1$, so $f$ is increasing on $(-\infty, 1)$ and $(1, \infty)$. $f''(x) > 0$ if $x < 1$ or $x > 3$, and $f''(x) < 0$ if $1 < x < 3$, so $f$ is concave upward on $(-\infty, 1)$ and $(3, \infty)$, and concave downward on $(1, 3)$. There is an inflection point when $x = 3$.

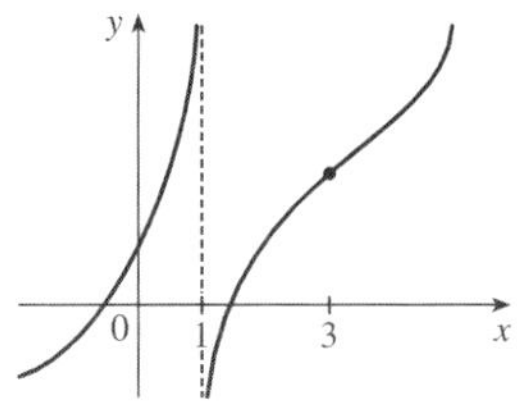


**17.** $f'(0) = f'(2) = f'(4) = 0 \Rightarrow$ horizontal tangents at $x = 0, 2, 4$.
$f'(x) > 0$ if $x < 0$ or $2 < x < 4 \Rightarrow f$ is increasing on $(-\infty, 0)$ and $(2, 4)$.
$f'(x) < 0$ if $0 < x < 2$ or $x > 4 \Rightarrow f$ is decreasing on $(0, 2)$ and $(4, \infty)$.
$f''(x) > 0$ if $1 < x < 3 \Rightarrow f$ is concave upward on $(1, 3)$.
$f''(x) < 0$ if $x < 1$ or $x > 3 \Rightarrow f$ is concave downward on $(-\infty, 1)$ and $(3, \infty)$. There are inflection points when $x = 1$ and $3$.

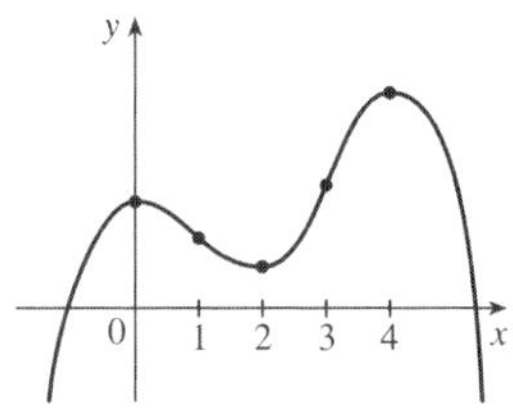


**18.** $f'(1) = f'(-1) = 0 \Rightarrow$ horizontal tangents at $x = \pm 1$. $f'(x) < 0$ if $|x| < 1$ $\Rightarrow f$ is decreasing on $(-1, 1)$. $f'(x) > 0$ if $1 < |x| < 2 \Rightarrow f$ is increasing on $(-2, -1)$ and $(1, 2)$. $f'(x) = -1$ if $|x| > 2 \Rightarrow$ the graph of $f$ has constant slope $-1$ on $(-\infty, -2)$ and $(2, \infty)$. $f''(x) < 0$ if $-2 < x < 0 \Rightarrow$ $f$ is concave downward on $(-2, 0)$. The point $(0, 1)$ is an inflection point.

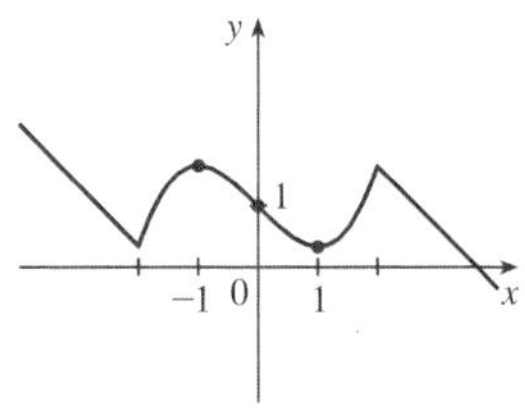


**19.** $f'(x) > 0$ if $|x| < 2 \Rightarrow f$ is increasing on $(-2, 2)$.
$f'(x) < 0$ if $|x| > 2 \Rightarrow f$ is decreasing on $(-\infty, -2)$ and $(2, \infty)$. $f'(-2) = 0 \Rightarrow$ horizontal tangent at $x = -2$. $\lim_{x \to 2} |f'(x)| = \infty \Rightarrow$ there is a vertical asymptote or vertical tangent (cusp) at $x = 2$. $f''(x) > 0$ if $x \neq 2 \Rightarrow f$ is concave upward on $(-\infty, 2)$ and $(2, \infty)$.

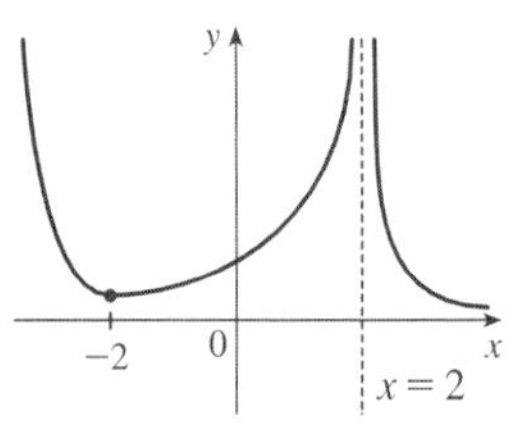


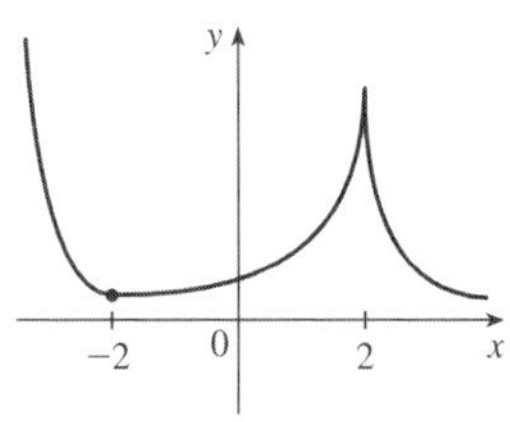

**20.** $f'(x) > 0$ if $|x| < 2 \Rightarrow f$ is increasing on $(-2, 2)$. $f'(x) < 0$ if $|x| > 2 \Rightarrow f$ is decreasing on $(-\infty, -2)$ and $(2, \infty)$. $f'(2) = 0$, so $f$ has a horizontal tangent (and local maximum) at $x = 2$. $\lim_{x\to\infty} f(x) = 1 \Rightarrow y = 1$ is a horizontal asymptote. $f(-x) = -f(x)$ $\Rightarrow f$ is an odd function (its graph is symmetric about the origin). Finally, $f''(x) < 0$ if $0 < x < 3$ and $f''(x) > 0$ if $x > 3$, so $f$ is CD on $(0, 3)$ and CU on $(3, \infty)$.

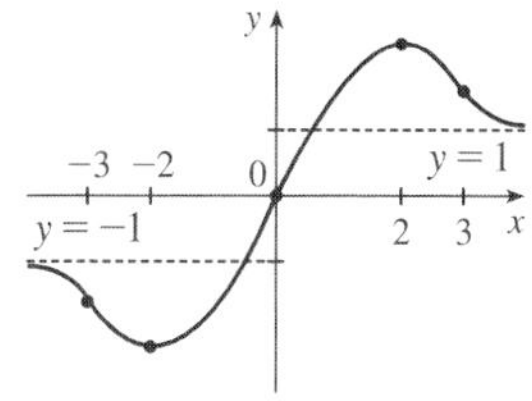


**21.** (a) Since $e^{-x^2}$ is positive for all $x$, $f'(x) = xe^{-x^2}$ is positive where $x > 0$ and negative where $x < 0$. Thus, $f$ is increasing on $(0, \infty)$ and decreasing on $(-\infty, 0)$.

(b) Since $f$ changes from decreasing to increasing at $x = 0$, $f$ has a minimum value there.

**22.** Since $f'(x) = e^{-x^2} > 0$ on $\mathbb{R}$, $f$ is increasing on $\mathbb{R}$.

**23.** (a) To find the intervals on which $f$ is increasing, we need to find the intervals on which $f'(x) = 3x^2 - 1$ is positive. $3x^2 - 1 > 0 \Leftrightarrow 3x^2 > 1 \Leftrightarrow x^2 > \frac{1}{3} \Leftrightarrow |x| > \sqrt{\frac{1}{3}}$, so $x \in \left(-\infty, -\sqrt{\frac{1}{3}}\right) \cup \left(\sqrt{\frac{1}{3}}, \infty\right)$. Thus, $f$ is increasing on $\left(-\infty, -\sqrt{\frac{1}{3}}\right)$ and on $\left(\sqrt{\frac{1}{3}}, \infty\right)$. In a similar fashion, $f$ is decreasing on $\left(-\sqrt{\frac{1}{3}}, \sqrt{\frac{1}{3}}\right)$.

(b) To find the intervals on which $f$ is concave upward, we need to find the intervals on which $f''(x) = 6x$ is positive. $6x > 0 \Leftrightarrow x > 0$. So $f$ is concave upward on $(0, \infty)$ and $f$ is concave downward on $(-\infty, 0)$.

(c) There is an inflection point at $(0, 0)$ since $f$ changes its direction of concavity at $x = 0$.

**24.** (a)
$$f'(x) = \lim_{h\to 0} \frac{f(x+h) - f(x)}{h} = \lim_{h\to 0} \frac{[(x+h)^4 - 2(x+h)^2] - (x^4 - 2x^2)}{h}$$
$$= \lim_{h\to 0} \frac{(x^4 + 4x^3h + 6x^2h^2 + 4xh^3 + h^4 - 2x^2 - 4xh - 2h^2) - (x^4 - 2x^2)}{h}$$
$$= \lim_{h\to 0} \frac{4x^3h + 6x^2h^2 + 4xh^3 + h^4 - 4xh - 2h^2}{h} = \lim_{h\to 0} (4x^3 + 6x^2h + 4xh^2 + h^3 - 4x - 2h) = 4x^3 - 4x$$

$$f''(x) = \lim_{h\to 0} \frac{f'(x+h) - f'(x)}{h} = \lim_{h\to 0} \frac{[4(x+h)^3 - 4(x+h)] - (4x^3 - 4x)}{h}$$
$$= \lim_{h\to 0} \frac{(4x^3 + 12x^2h + 12xh^2 + 4h^3 - 4x - 4h) - (4x^3 - 4x)}{h} = \lim_{h\to 0} \frac{12x^2h + 12xh^2 + 4h^3 - 4h}{h}$$
$$= \lim_{h\to 0} (12x^2 + 12xh + 4h^2 - 4) = 12x^2 - 4$$

(b) $f'(x) > 0 \Leftrightarrow 4x^3 - 4x > 0 \Leftrightarrow 4x(x^2 - 1) > 0 \Leftrightarrow 4x(x+1)(x-1) > 0$, so $f$ is increasing on $(-1, 0)$ and $(1, \infty)$ and $f$ is decreasing on $(-\infty, -1)$ and $(0, 1)$.

(c) $f''(x) > 0 \Leftrightarrow 12x^2 - 4 > 0 \Leftrightarrow 12x^2 > 4 \Leftrightarrow x^2 > \frac{1}{3} \Leftrightarrow |x| > \sqrt{\frac{1}{3}}$, so $f$ is CU on $\left(-\infty, -\sqrt{\frac{1}{3}}\right)$ and $\left(\sqrt{\frac{1}{3}}, \infty\right)$ and $f$ is CD on $\left(-\sqrt{\frac{1}{3}}, \sqrt{\frac{1}{3}}\right)$.

**25.** $b$ is the antiderivative of $f$. For small $x$, $f$ is negative, so the graph of its antiderivative must be decreasing. But both $a$ and $c$ are increasing for small $x$, so only $b$ can be $f$'s antiderivative. Also, $f$ is positive where $b$ is increasing, which supports our conclusion.

**26.** We know right away that $c$ cannot be $f$'s antiderivative, since the slope of $c$ is not zero at the $x$-value where $f = 0$. Now $f$ is positive when $a$ is increasing and negative when $a$ is decreasing, so $a$ is the antiderivative of $f$.

**27.** The graph of $F$ will have a minimum at 0 and a maximum at 2, since $f = F'$ goes from negative to positive at $x = 0$, and from positive to negative at $x = 2$.

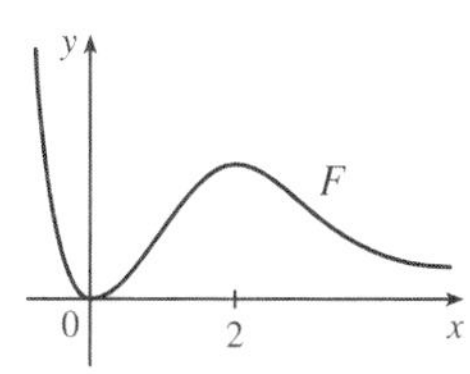

**28.** The position function is the antiderivative of the velocity function, so its graph has to be horizontal where the velocity function is equal to 0.

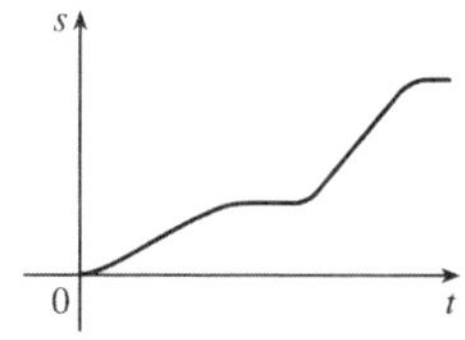


**29.**

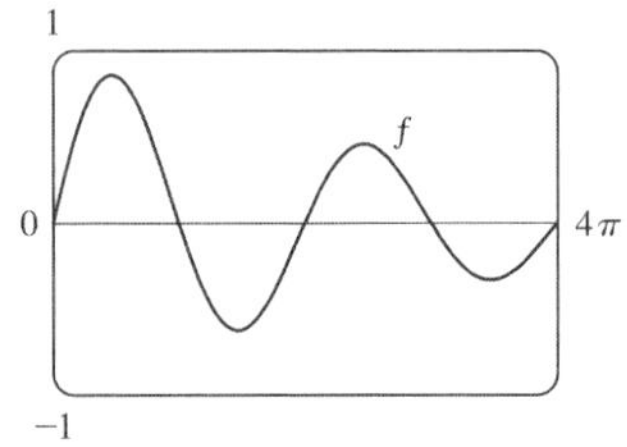


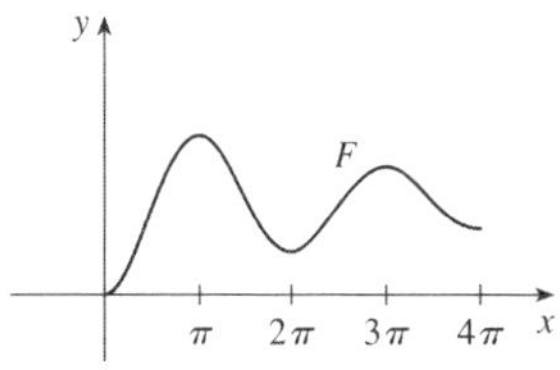


$f$ is positive on $(0, \pi)$ and $(2\pi, 3\pi)$, so $F$ is increasing on those intervals. $f$ is negative on $(\pi, 2\pi)$ and $(3\pi, 4\pi)$, so $F$ is decreasing on those intervals. $f$ is zero at $x = 0, \pi, 2\pi, 3\pi$, and $4\pi$, so $F$ has horizontal tangents at those points.

**30.**

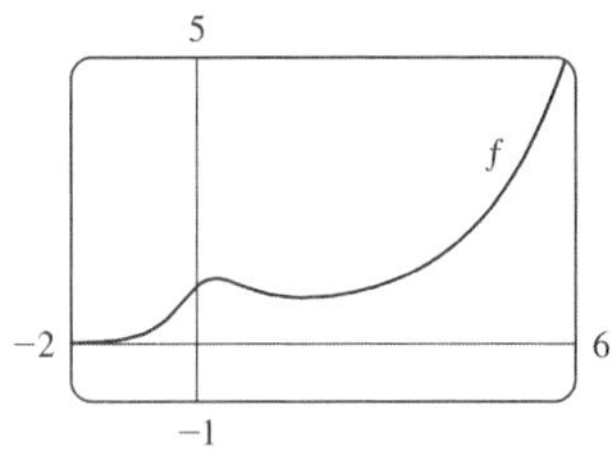


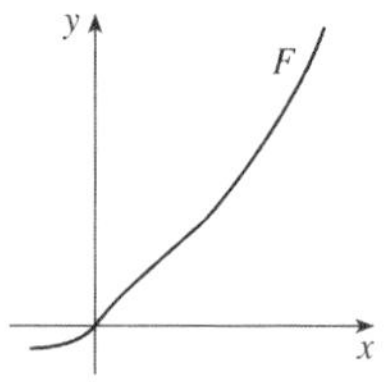


$f$ is always positive, so $F$ is always increasing.

## 2 Review

### CONCEPT CHECK

**1.** (a) $\lim_{x \to a} f(x) = L$: See Definition 2.2.1 and Figures 1 and 2 in Section 2.2.

(b) $\lim_{x \to a^+} f(x) = L$: See the paragraph after Definition 2.2.2 and Figure 9(b) in Section 2.2.

(c) $\lim_{x \to a^-} f(x) = L$: See Definition 2.2.2 and Figure 9(a) in Section 2.2.

(d) $\lim_{x \to a} f(x) = \infty$: See Definition 2.5.1 and Figure 2 in Section 2.5.

(e) $\lim_{x \to \infty} f(x) = L$: See Definition 2.5.4 and Figure 9 in Section 2.5.

**2.** In general, the limit of a function fails to exist when the function does not approach a fixed number. For each of the following functions, the limit fails to exist at $x = 2$.

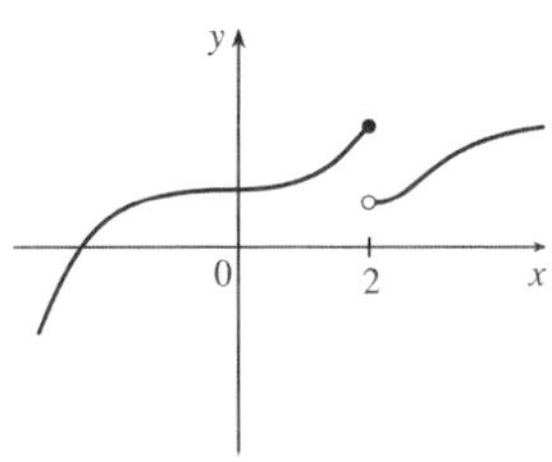


The left- and right-hand limits are not equal.

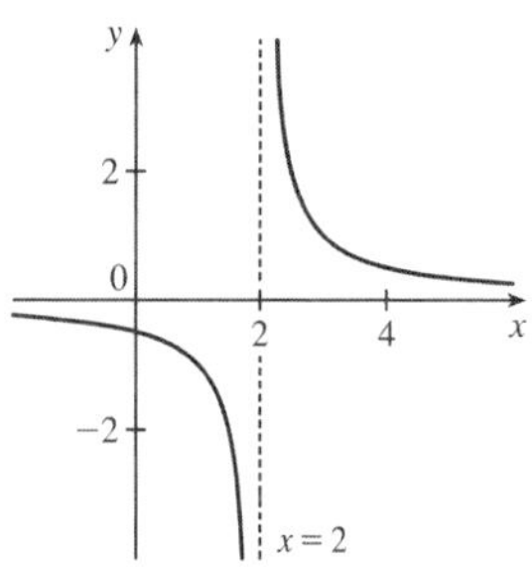


There is an infinite discontinuity.

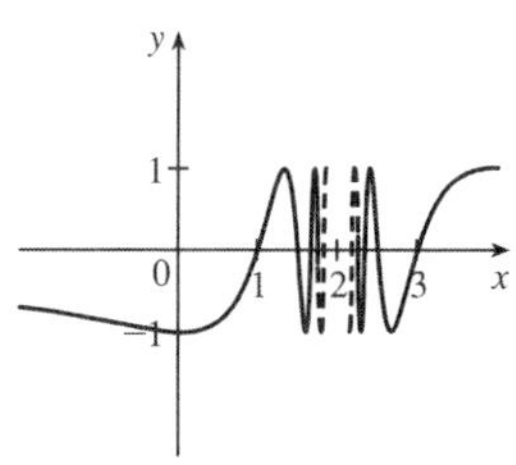


There are an infinite number of oscillations.

**3.** (a)–(g) See the statements of Limit Laws 1–6 and 11 in Section 2.3.

**4.** See Theorem 3 in Section 2.3.

**5.** (a) See Definition 2.5.2 and Figures 2 – 4 in Section 2.5.

(b) See Definition 2.5.5 and Figures 9 and 10 in Section 2.5.

**6.** (a) $y = x^4$: No asymptote

(b) $y = \sin x$: No asymptote

(c) $y = \tan x$: Vertical asymptotes $x = \frac{\pi}{2} + \pi n$, $n$ an integer

(d) $y = \tan^{-1} x$: Horizontal asymptotes $y = \pm\frac{\pi}{2}$

(e) $y = e^x$: Horizontal asymptote $y = 0$ $\left(\lim\limits_{x \to -\infty} e^x = 0\right)$

(f) $y = \ln x$: Vertical asymptote $x = 0$ $\left(\lim\limits_{x \to 0^+} \ln x = -\infty\right)$

(g) $y = 1/x$: Vertical asymptote $x = 0$, horizontal asymptote $y = 0$

(h) $y = \sqrt{x}$: No asymptote

**7.** (a) A function $f$ is continuous at a number $a$ if $f(x)$ approaches $f(a)$ as $x$ approaches $a$; that is, $\lim\limits_{x \to a} f(x) = f(a)$.

(b) A function $f$ is continuous on the interval $(-\infty, \infty)$ if $f$ is continuous at every real number $a$. The graph of such a function has no breaks and every vertical line crosses it.

**8.** See Theorem 2.4.10.

**9.** See Definition 2.6.1.

**10.** See the paragraph containing Formula 3 in Section 2.6.

**11.** (a) The average rate of change of $y$ with respect to $x$ over the interval $[x_1, x_2]$ is $\dfrac{f(x_2) - f(x_1)}{x_2 - x_1}$.

(b) The instantaneous rate of change of $y$ with respect to $x$ at $x = x_1$ is $\lim\limits_{x_2 \to x_1} \dfrac{f(x_2) - f(x_1)}{x_2 - x_1}$.

**12.** See Definition 2.7.2. The pages following the definition discuss interpretations of $f'(a)$ as the slope of a tangent line to the graph of $f$ at $x = a$ and as an instantaneous rate of change of $f(x)$ with respect to $x$ when $x = a$.

**13.** See the paragraphs before and after Example 7 in Section 2.8.

**14.** (a) A function $f$ is differentiable at a number $a$ if its derivative $f'$ exists at $x = a$; that is, if $f'(a)$ exists.

(b) See Theorem 2.8.4. This theorem also tells us that if $f$ is *not* continuous at $a$, then $f$ is *not* differentiable at $a$.

(c)

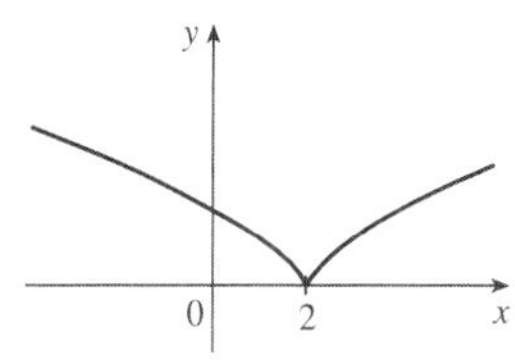


**15.** See the discussion and Figure 8 on page 162.

**16.** (a) See the first box in Section 2.9.

(b) See the second box in Section 2.9.

**17.** (a) An antiderivative of a function $f$ is a function $F$ such that $F' = f$.

(b) The antiderivative of a velocity function is a position function (the derivative of a position function is a velocity function). The antiderivative of an acceleration function is a velocity function (the derivative of a velocity function is an acceleration function).

## TRUE-FALSE QUIZ

**1.** False. Limit Law 2 applies only if the individual limits exist (these don't).

**2.** False. Limit Law 5 cannot be applied if the limit of the denominator is 0 (it is).

**3.** True. Limit Law 5 applies.

**4.** True. The limit doesn't exist since $f(x)/g(x)$ doesn't approach any real number as $x$ approaches 5. (The denominator approaches 0 and the numerator doesn't.)

**5.** False. Consider $\lim_{x\to5}\dfrac{x(x-5)}{x-5}$ or $\lim_{x\to5}\dfrac{\sin(x-5)}{x-5}$. The first limit exists and is equal to 5. By Example 3 in Section 2.2, we know that the latter limit exists (and it is equal to 1).

**6.** False. Consider $\lim_{x\to6}[f(x)g(x)]=\lim_{x\to6}\left[(x-6)\dfrac{1}{x-6}\right]$. It exists (its value is 1) but $f(6)=0$ and $g(6)$ does not exist, so $f(6)g(6)\neq1$.

**7.** True. A polynomial is continuous everywhere, so $\lim_{x\to b}p(x)$ exists and is equal to $p(b)$.

**8.** False. Consider $\lim_{x\to0}[f(x)-g(x)]=\lim_{x\to0}\left(\dfrac{1}{x^2}-\dfrac{1}{x^4}\right)$. This limit is $-\infty$ (not 0), but each of the individual functions approaches $\infty$.

**9.** True. See Figure 11 in Section 2.5.

**10.** False. Consider $f(x)=\sin x$ for $x\geq0$. $\lim_{x\to\infty}f(x)\neq\pm\infty$ and $f$ has no horizontal asymptote.

**11.** False. Consider $f(x)=\begin{cases}1/(x-1) & \text{if } x\neq1\\ 2 & \text{if } x=1\end{cases}$

**12.** False. The function $f$ must be *continuous* in order to use the Intermediate Value Theorem. For example, let $f(x)=\begin{cases}1 & \text{if } 0\leq x<3\\ -1 & \text{if } x=3\end{cases}$ There is no number $c\in[0,3]$ with $f(c)=0$.

**13.** True. Use Theorem 2.4.8 with $a=2$, $b=5$, and $g(x)=4x^2-11$. Note that $f(4)=3$ is not needed.

**14.** True. Use the Intermediate Value Theorem with $a=-1$, $b=1$, and $N=\pi$, since $3<\pi<4$.

**15.** False. See the note after Theorem 4 in Section 2.8.

**16.** True. $f'(r)$ exists $\Rightarrow$ $f$ is differentiable at $r$ $\Rightarrow$ $f$ is continuous at $r$ $\Rightarrow$ $\lim_{x\to r}f(x)=f(r)$.

**17.** False. $\dfrac{d^2y}{dx^2}$ is the second derivative while $\left(\dfrac{dy}{dx}\right)^2$ is the first derivative squared. For example, if $y=x$, then $\dfrac{d^2y}{dx^2}=0$, but $\left(\dfrac{dy}{dx}\right)^2=1$.

**18.** False. For example, let $f(x)=\begin{cases}x^2+1 & \text{if } x\neq0\\ 2 & \text{if } x=0\end{cases}$

Then $f(x)>1$ for all $x$, but $\lim_{x\to0}f(x)=\lim_{x\to0}(x^2+1)=1$.

## EXERCISES

**1.** (a) (i) $\lim_{x\to2^+}f(x)=3$ (ii) $\lim_{x\to-3^+}f(x)=0$

(iii) $\lim_{x\to-3}f(x)$ does not exist since the left and right limits are not equal. (The left limit is $-2$.)

(iv) $\lim_{x\to4}f(x)=2$

(v) $\lim_{x\to0}f(x)=\infty$ (vi) $\lim_{x\to2^-}f(x)=-\infty$

(vii) $\lim_{x\to\infty}f(x)=4$ (viii) $\lim_{x\to-\infty}f(x)=-1$

(b) The equations of the horizontal asymptotes are $y=-1$ and $y=4$.

(c) The equations of the vertical asymptotes are $x = 0$ and $x = 2$.

(d) $f$ is discontinuous at $x = -3$, 0, 2, and 4. The discontinuities are jump, infinite, infinite, and removable, respectively.

**2.** $\lim\limits_{x\to-\infty} f(x) = -2$, $\lim\limits_{x\to\infty} f(x) = 0$, $\lim\limits_{x\to-3} f(x) = \infty$,

$\lim\limits_{x\to3^-} f(x) = -\infty$, $\lim\limits_{x\to3^+} f(x) = 2$,

$f$ is continuous from the right at 3

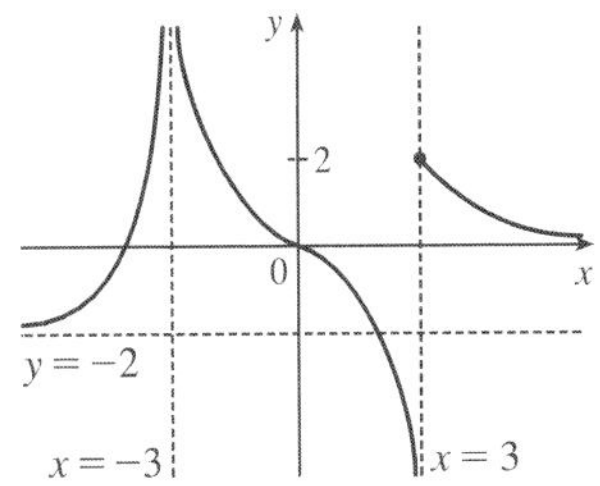


**3.** Since the exponential function is continuous, $\lim\limits_{x\to1} e^{x^3-x} = e^{1-1} = e^0 = 1$.

**4.** Since rational functions are continuous, $\lim\limits_{x\to3} \dfrac{x^2-9}{x^2+2x-3} = \dfrac{3^2-9}{3^2+2(3)-3} = \dfrac{0}{12} = 0$.

**5.** $$\lim_{x\to-3} \frac{x^2-9}{x^2+2x-3} = \lim_{x\to-3} \frac{(x+3)(x-3)}{(x+3)(x-1)} = \lim_{x\to-3} \frac{x-3}{x-1} = \frac{-3-3}{-3-1} = \frac{-6}{-4} = \frac{3}{2}$$

**6.** $\lim\limits_{x\to1^+} \dfrac{x^2-9}{x^2+2x-3} = -\infty$ since $x^2+2x-3 \to 0$ as $x \to 1^+$ and $\dfrac{x^2-9}{x^2+2x-3} < 0$ for $1 < x < 3$.

**7.** $$\lim_{h\to0} \frac{(h-1)^3+1}{h} = \lim_{h\to0} \frac{(h^3-3h^2+3h-1)+1}{h} = \lim_{h\to0} \frac{h^3-3h^2+3h}{h} = \lim_{h\to0} \left(h^2-3h+3\right) = 3$$

*Another solution:* Factor the numerator as a sum of two cubes and then simplify.

$$\lim_{h\to0} \frac{(h-1)^3+1}{h} = \lim_{h\to0} \frac{(h-1)^3+1^3}{h} = \lim_{h\to0} \frac{[(h-1)+1]\left[(h-1)^2-1(h-1)+1^2\right]}{h}$$
$$= \lim_{h\to0} \left[(h-1)^2-h+2\right] = 1-0+2 = 3$$

**8.** $$\lim_{t\to2} \frac{t^2-4}{t^3-8} = \lim_{t\to2} \frac{(t+2)(t-2)}{(t-2)(t^2+2t+4)} = \lim_{t\to2} \frac{t+2}{t^2+2t+4} = \frac{2+2}{4+4+4} = \frac{4}{12} = \frac{1}{3}$$

**9.** $\lim\limits_{r\to9} \dfrac{\sqrt{r}}{(r-9)^4} = \infty$ since $(r-9)^4 \to 0$ as $r \to 9$ and $\dfrac{\sqrt{r}}{(r-9)^4} > 0$ for $r \neq 9$.

**10.** $$\lim_{v\to4^+} \frac{4-v}{|4-v|} = \lim_{v\to4^+} \frac{4-v}{-(4-v)} = \lim_{v\to4^+} \frac{1}{-1} = -1$$

**11.** Let $t = \sin x$. Then as $x \to \pi^-$, $\sin x \to 0^+$, so $t \to 0^+$. Thus, $\lim\limits_{x\to\pi^-} \ln(\sin x) = \lim\limits_{t\to0^+} \ln t = -\infty$.

**12.** $$\lim_{x\to-\infty} \frac{1-2x^2-x^4}{5+x-3x^4} = \lim_{x\to-\infty} \frac{(1-2x^2-x^4)/x^4}{(5+x-3x^4)/x^4} = \lim_{x\to-\infty} \frac{1/x^4-2/x^2-1}{5/x^4+1/x^3-3} = \frac{0-0-1}{0+0-3} = \frac{-1}{-3} = \frac{1}{3}$$

**13.** $$\lim_{x\to\infty} \left(\sqrt{x^2+4x+1}-x\right) = \lim_{x\to\infty} \left[\frac{\sqrt{x^2+4x+1}-x}{1} \cdot \frac{\sqrt{x^2+4x+1}+x}{\sqrt{x^2+4x+1}+x}\right]$$
$$= \lim_{x\to\infty} \frac{(x^2+4x+1)-x^2}{\sqrt{x^2+4x+1}+x}$$
$$= \lim_{x\to\infty} \frac{(4x+1)/x}{(\sqrt{x^2+4x+1}+x)/x} \qquad \left[\text{divide by } x = \sqrt{x^2} \text{ for } x > 0\right]$$
$$= \lim_{x\to\infty} \frac{4+1/x}{\sqrt{1+4/x+1/x^2}+1} = \frac{4+0}{\sqrt{1+0+0}+1} = \frac{4}{2} = 2$$

**14.** Let $t = x - x^2 = x(1-x)$. Then as $x \to \infty$, $t \to -\infty$, and $\lim\limits_{x\to\infty} e^{x-x^2} = \lim\limits_{t\to-\infty} e^t = 0$.

**15.** Let $t = 1/x$. Then as $x \to 0^+$, $t \to \infty$, and $\lim\limits_{x\to0^+} \tan^{-1}(1/x) = \lim\limits_{t\to\infty} \tan^{-1} t = \dfrac{\pi}{2}$.

**16.** $\displaystyle\lim_{x\to 1}\left(\frac{1}{x-1}+\frac{1}{x^2-3x+2}\right)=\lim_{x\to 1}\left[\frac{1}{x-1}+\frac{1}{(x-1)(x-2)}\right]=\lim_{x\to 1}\left[\frac{x-2}{(x-1)(x-2)}+\frac{1}{(x-1)(x-2)}\right]$

$$=\lim_{x\to 1}\left[\frac{x-1}{(x-1)(x-2)}\right]=\lim_{x\to 1}\frac{1}{x-2}=\frac{1}{1-2}=-1$$

**17.** From the graph of $y=(\cos^2 x)/x^2$, it appears that $y=0$ is the horizontal asymptote and $x=0$ is the vertical asymptote. Now $0\le(\cos x)^2\le 1 \;\Rightarrow$ $\dfrac{0}{x^2}\le\dfrac{\cos^2 x}{x^2}\le\dfrac{1}{x^2} \;\Rightarrow\; 0\le\dfrac{\cos^2 x}{x^2}\le\dfrac{1}{x^2}$. But $\lim\limits_{x\to\pm\infty}0=0$ and $\lim\limits_{x\to\pm\infty}\dfrac{1}{x^2}=0$, so by the Squeeze Theorem, $\lim\limits_{x\to\pm\infty}\dfrac{\cos^2 x}{x^2}=0$.

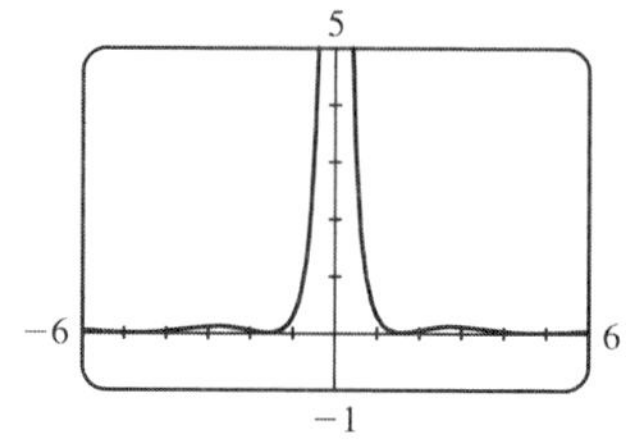


Thus, $y=0$ is the horizontal asymptote. $\lim\limits_{x\to 0}\dfrac{\cos^2 x}{x^2}=\infty$ because $\cos^2 x\to 1$ and $x^2\to 0$ as $x\to 0$, so $x=0$ is the vertical asymptote.

**18.** From the graph of $y=f(x)=\sqrt{x^2+x+1}-\sqrt{x^2-x}$, it appears that there are 2 horizontal asymptotes and possibly 2 vertical asymptotes. To obtain a different form for $f$, let's multiply and divide it by its conjugate.

$$f_1(x)=\left(\sqrt{x^2+x+1}-\sqrt{x^2-x}\right)\frac{\sqrt{x^2+x+1}+\sqrt{x^2-x}}{\sqrt{x^2+x+1}+\sqrt{x^2-x}}$$

$$=\frac{(x^2+x+1)-(x^2-x)}{\sqrt{x^2+x+1}+\sqrt{x^2-x}}=\frac{2x+1}{\sqrt{x^2+x+1}+\sqrt{x^2-x}}$$

Now

$$\lim_{x\to\infty}f_1(x)=\lim_{x\to\infty}\frac{2x+1}{\sqrt{x^2+x+1}+\sqrt{x^2-x}}$$

$$=\lim_{x\to\infty}\frac{2+(1/x)}{\sqrt{1+(1/x)+(1/x^2)}+\sqrt{1-(1/x)}}\qquad\left[\text{since }\sqrt{x^2}=x\text{ for }x>0\right]$$

$$=\frac{2}{1+1}=1,$$

so $y=1$ is a horizontal asymptote. For $x<0$, we have $\sqrt{x^2}=|x|=-x$, so when we divide the denominator by $x$, with $x<0$, we get

$$\frac{\sqrt{x^2+x+1}+\sqrt{x^2-x}}{x}=-\frac{\sqrt{x^2+x+1}+\sqrt{x^2-x}}{\sqrt{x^2}}=-\left[\sqrt{1+\frac{1}{x}+\frac{1}{x^2}}+\sqrt{1-\frac{1}{x}}\right]$$

Therefore,

$$\lim_{x\to-\infty}f_1(x)=\lim_{x\to-\infty}\frac{2x+1}{\sqrt{x^2+x+1}+\sqrt{x^2-x}}=\lim_{x\to\infty}\frac{2+(1/x)}{-\left[\sqrt{1+(1/x)+(1/x^2)}+\sqrt{1-(1/x)}\right]}$$

$$=\frac{2}{-(1+1)}=-1,$$

so $y=-1$ is a horizontal asymptote. As $x\to 0^-$, $f(x)\to 1$, so $x=0$ is *not* a vertical asymptote. As $x\to 1^+$, $f(x)\to\sqrt{3}$, so $x=1$ is *not* a vertical asymptote and hence there are no vertical asymptotes.

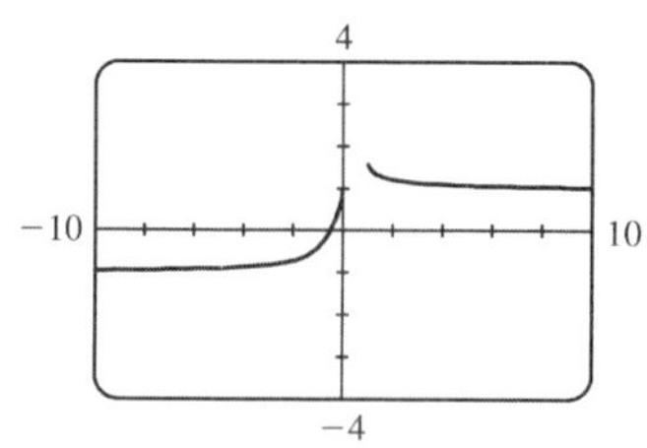

**19.** Since $2x - 1 \le f(x) \le x^2$ for $0 < x < 3$ and $\lim\limits_{x \to 1}(2x - 1) = 1 = \lim\limits_{x \to 1} x^2$, we have $\lim\limits_{x \to 1} f(x) = 1$ by the Squeeze Theorem.

**20.** Let $f(x) = -x^2$, $g(x) = x^2 \cos(1/x^2)$ and $h(x) = x^2$. Then since $|\cos(1/x^2)| \le 1$ for $x \ne 0$, we have $f(x) \le g(x) \le h(x)$ for $x \ne 0$, and so $\lim\limits_{x \to 0} f(x) = \lim\limits_{x \to 0} h(x) = 0 \;\Rightarrow\; \lim\limits_{x \to 0} g(x) = 0$ by the Squeeze Theorem.

**21.** (a) $f(x) = \sqrt{-x}$ if $x < 0$, $f(x) = 3 - x$ if $0 \le x < 3$, $f(x) = (x - 3)^2$ if $x > 3$.

(i) $\lim\limits_{x \to 0^+} f(x) = \lim\limits_{x \to 0^+} (3 - x) = 3$

(ii) $\lim\limits_{x \to 0^-} f(x) = \lim\limits_{x \to 0^-} \sqrt{-x} = 0$

(iii) Because of (i) and (ii), $\lim\limits_{x \to 0} f(x)$ does not exist.

(iv) $\lim\limits_{x \to 3^-} f(x) = \lim\limits_{x \to 3^-} (3 - x) = 0$

(v) $\lim\limits_{x \to 3^+} f(x) = \lim\limits_{x \to 3^+} (x - 3)^2 = 0$

(vi) Because of (iv) and (v), $\lim\limits_{x \to 3} f(x) = 0$.

(b) $f$ is discontinuous at 0 since $\lim\limits_{x \to 0} f(x)$ does not exist.

$f$ is discontinuous at 3 since $f(3)$ does not exist.

(c)

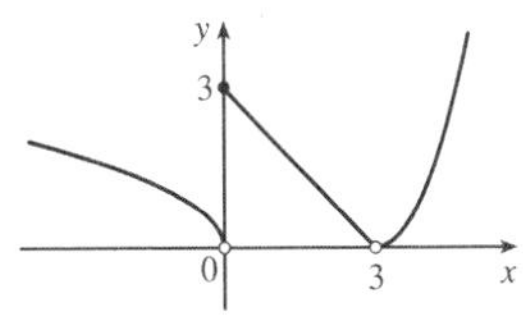


**22.** (a) $x^2 - 9$ is continuous on $\mathbb{R}$ since it is a polynomial and $\sqrt{x}$ is continuous on $[0, \infty)$, so the composition $\sqrt{x^2 - 9}$ is continuous on $\{x \mid x^2 - 9 \ge 0\} = (-\infty, -3] \cup [3, \infty)$. Note that $x^2 - 2 \ne 0$ on this set and so the quotient function $g(x) = \dfrac{\sqrt{x^2 - 9}}{x^2 - 2}$ is continuous on its domain, $(-\infty, -3] \cup [3, \infty)$.

(b) $f$ is discontinuous at 0 since $\lim\limits_{x \to 0} f(x)$ does not exist. $f$ is discontinuous at 3 since $f(3)$ does not exist.

**23.** $f(x) = 2x^3 + x^2 + 2$ is a polynomial, so it is continuous on $[-2, -1]$ and $f(-2) = -10 < 0 < 1 = f(-1)$. So by the Intermediate Value Theorem there is a number $c$ in $(-2, -1)$ such that $f(c) = 0$, that is, the equation $2x^3 + x^2 + 2 = 0$ has a root in $(-2, -1)$.

**24.** $f(x) = e^{-x^2} - x$ is continuous on $\mathbb{R}$ so it is continuous on $[0, 1]$. $f(0) = 1 > 0 > 1/e - 1 = f(1)$. So by the Intermediate Value Theorem, there is a number $c$ in $(0, 1)$ such that $f(c) = 0$. Thus, $e^{-x^2} - x = 0$, or $e^{-x^2} = x$, has a root in $(0, 1)$.

**25.** (a) $s = s(t) = 1 + 2t + t^2/4$. The average velocity over the time interval $[1, 1 + h]$ is

$$v_{\text{ave}} = \frac{s(1 + h) - s(1)}{(1 + h) - 1} = \frac{1 + 2(1 + h) + (1 + h)^2/4 - 13/4}{h} = \frac{10h + h^2}{4h} = \frac{10 + h}{4}.$$

So for the following intervals the average velocities are:

(i) $[1, 3]$: $h = 2$, $v_{\text{ave}} = (10 + 2)/4 = 3\text{ m/s}$

(ii) $[1, 2]$: $h = 1$, $v_{\text{ave}} = (10 + 1)/4 = 2.75\text{ m/s}$

(iii) $[1, 1.5]$: $h = 0.5$, $v_{\text{ave}} = (10 + 0.5)/4 = 2.625\text{ m/s}$

(iv) $[1, 1.1]$: $h = 0.1$, $v_{\text{ave}} = (10 + 0.1)/4 = 2.525\text{ m/s}$

(b) When $t = 1$, the instantaneous velocity is $\lim\limits_{h \to 0} \dfrac{s(1 + h) - s(1)}{h} = \lim\limits_{h \to 0} \dfrac{10 + h}{4} = \dfrac{10}{4} = 2.5\text{ m/s}$.

**26.** (a) When $V$ increases from $200\text{ in}^3$ to $250\text{ in}^3$, we have $\Delta V = 250 - 200 = 50\text{ in}^3$, and since $P = 800/V$,

$\Delta P = P(250) - P(200) = \dfrac{800}{250} - \dfrac{800}{200} = 3.2 - 4 = -0.8\text{ lb/in}^2$. So the average rate of change is $\dfrac{\Delta P}{\Delta V} = \dfrac{-0.8}{50} = -0.016\,\dfrac{\text{lb/in}^2}{\text{in}^3}$.

(b) Since $V = 800/P$, the instantaneous rate of change of $V$ with respect to $P$ is

$$\begin{aligned}\lim_{h \to 0} \frac{\Delta V}{\Delta P} &= \lim_{h \to 0} \frac{V(P + h) - V(P)}{h} = \lim_{h \to 0} \frac{800/(P + h) - 800/P}{h} \\ &= \lim_{h \to 0} \frac{800\,[P - (P + h)]}{h(P + h)P} = \lim_{h \to 0} \frac{-800}{(P + h)P} = -\frac{800}{P^2}\end{aligned}$$

which is inversely proportional to the square of $P$.

**27.** Estimating the slopes of the tangent lines at $x = 2$, 3, and 5, we obtain approximate values 0.4, 2, and 0.1. Since the graph is concave downward at $x = 5$, $f''(5)$ is negative. Arranging the numbers in increasing order, we have: $f''(5) < 0 < f'(5) < f'(2) < 1 < f'(3)$.

**28.** (a) $f'(2) = \lim_{x\to 2} \dfrac{f(x) - f(2)}{x - 2} = \lim_{x\to 2} \dfrac{x^3 - 2x - 4}{x - 2}$

$= \lim_{x\to 2} \dfrac{(x-2)(x^2 + 2x + 2)}{x - 2} = \lim_{x\to 2} (x^2 + 2x + 2) = 10$

(b) $y - 4 = 10(x - 2)$ or $y = 10x - 16$

(c)

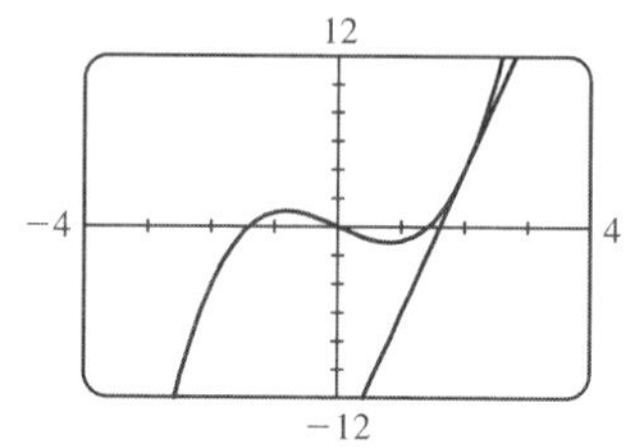


**29.** (a) Estimating $f'(1)$ from the triangle in the graph,

we get $\dfrac{\Delta y}{\Delta x} \approx \dfrac{-0.37}{0.50} = -0.74$.

To estimate $f'(1)$ numerically, we have

$$f'(1) = \lim_{h\to 0} \frac{f(1+h) - f(1)}{h} = \lim_{h\to 0} \frac{e^{-(1+h)^2} - e^{-1}}{h} = y$$

From the table, we have $f'(1) \approx -0.736$.

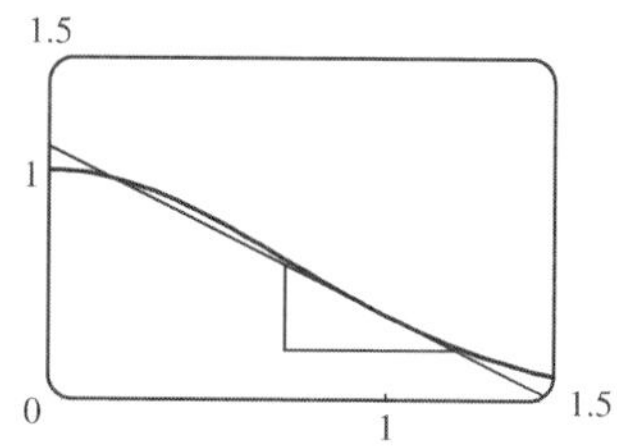


| $h$ | $y$ |
|---|---|
| 0.01 | −0.732 |
| 0.001 | −0.735 |
| 0.0001 | −0.736 |
| −0.01 | −0.739 |
| −0.001 | −0.736 |
| −0.0001 | −0.736 |

(b) $y - e^{-1} \approx -0.736(x - 1)$ or $y \approx -0.736x + 1.104$

(c) See the graph in part (a).

**30.** $2^6 = 64$, so $f(x) = x^6$ and $a = 2$.

**31.** (a) $f'(r)$ is the rate at which the total cost changes with respect to the interest rate. Its units are dollars/(percent per year).

(b) The total cost of paying off the loan is increasing by \$1200/(percent per year) as the interest rate reaches 10%. So if the interest rate goes up from 10% to 11%, the cost goes up approximately \$1200.

(c) As $r$ increases, $C$ increases. So $f'(r)$ will always be positive.

**32.**

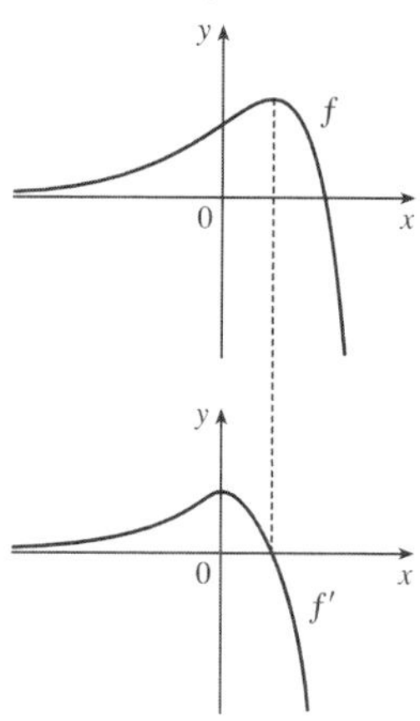


**33.**

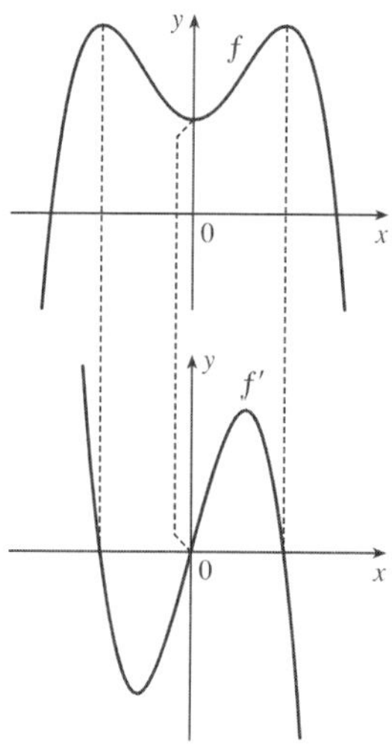


**34.**

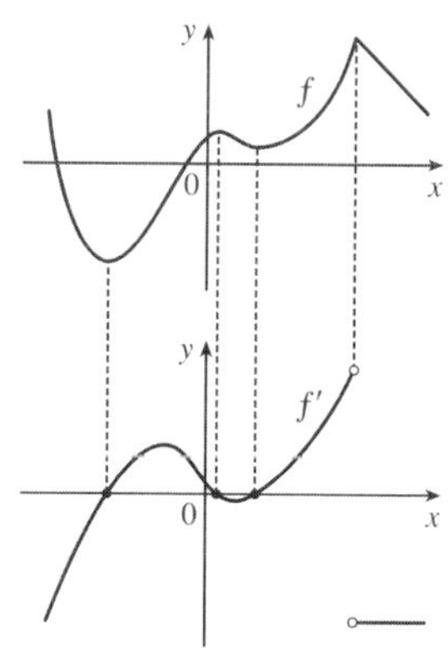

**35.** (a) $f'(x) = \lim_{h\to 0} \dfrac{f(x+h)-f(x)}{h} = \lim_{h\to 0} \dfrac{\sqrt{3-5(x+h)}-\sqrt{3-5x}}{h} \dfrac{\sqrt{3-5(x+h)}+\sqrt{3-5x}}{\sqrt{3-5(x+h)}+\sqrt{3-5x}}$

$$= \lim_{h\to 0} \frac{[3-5(x+h)]-(3-5x)}{h\left(\sqrt{3-5(x+h)}+\sqrt{3-5x}\right)} = \lim_{h\to 0} \frac{-5}{\sqrt{3-5(x+h)}+\sqrt{3-5x}} = \frac{-5}{2\sqrt{3-5x}}$$

(b) Domain of $f$: (the radicand must be nonnegative) $3-5x \geq 0 \;\Rightarrow$

$5x \leq 3 \;\Rightarrow\; x \in \left(-\infty, \frac{3}{5}\right]$

Domain of $f'$: exclude $\frac{3}{5}$ because it makes the denominator zero;

$x \in \left(-\infty, \frac{3}{5}\right)$

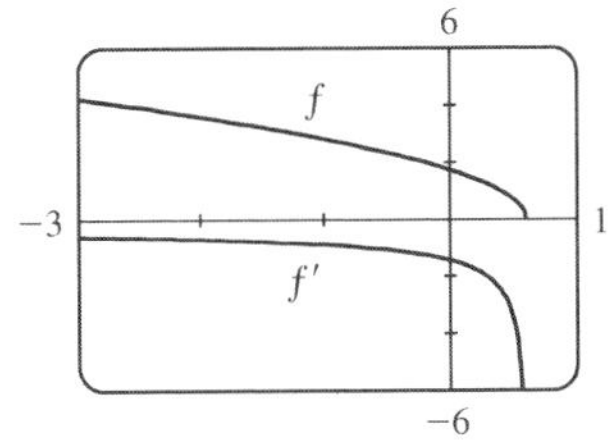


(c) Our answer to part (a) is reasonable because $f'(x)$ is always negative and $f$ is always decreasing.

**36.** (a) As $x \to \pm\infty$, $f(x) = (4-x)/(3+x) \to -1$, so there is a horizontal asymptote at $y=-1$. As $x \to -3^+$, $f(x) \to \infty$, and as $x \to -3^-$, $f(x) \to -\infty$. Thus, there is a vertical asymptote at $x=-3$.

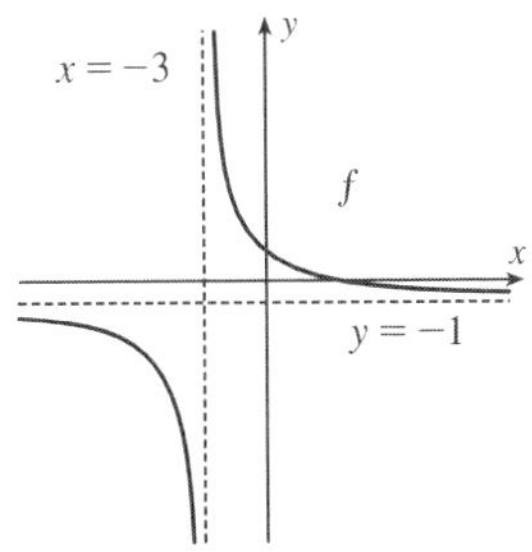


(b) Note that $f$ is decreasing on $(-\infty,-3)$ and $(-3,\infty)$, so $f'$ is negative on those intervals. As $x \to \pm\infty$, $f' \to 0$. As $x \to -3^-$ and as $x \to -3^+$, $f' \to -\infty$.

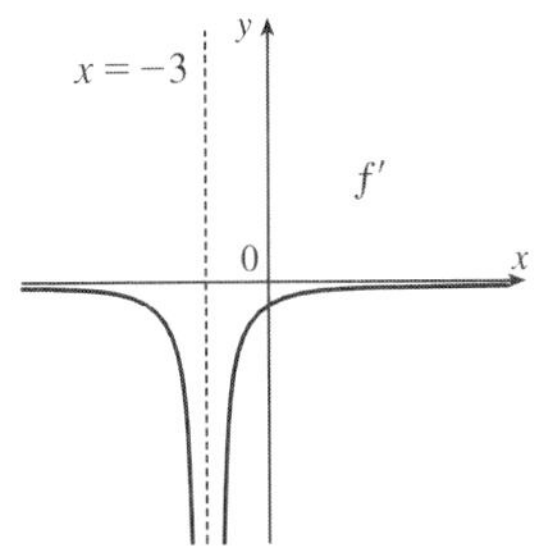


(c) $f'(x) = \lim_{h\to 0} \dfrac{f(x+h)-f(x)}{h} = \lim_{h\to 0} \dfrac{\dfrac{4-(x+h)}{3+(x+h)} - \dfrac{4-x}{3+x}}{h}$

$$= \lim_{h\to 0} \frac{(3+x)[4-(x+h)]-(4-x)[3+(x+h)]}{h[3+(x+h)](3+x)}$$

$$= \lim_{h\to 0} \frac{(12-3x-3h+4x-x^2-hx)-(12+4x+4h-3x-x^2-hx)}{h[3+(x+h)](3+x)}$$

$$= \lim_{h\to 0} \frac{-7h}{h[3+(x+h)](3+x)} = \lim_{h\to 0} \frac{-7}{[3+(x+h)](3+x)} = -\frac{7}{(3+x)^2}$$

(d) The graphing device confirms our graph in part (b).

**37.** $f$ is not differentiable: at $x=-4$ because $f$ is not continuous, at $x=-1$ because $f$ has a corner, at $x=2$ because $f$ is not continuous, and at $x=5$ because $f$ has a vertical tangent.

**38.** The graph of $a$ has tangent lines with positive slope for $x<0$ and negative slope for $x>0$, and the values of $c$ fit this pattern, so $c$ must be the graph of the derivative of the function for $a$. The graph of $c$ has horizontal tangent lines to the left and right of the $x$-axis and $b$ has zeros at these points. Hence, $b$ is the graph of the derivative of the function for $c$. Therefore, $a$ is the graph of $f$, $c$ is the graph of $f'$, and $b$ is the graph of $f''$.

**39.** $C'(1990)$ is the rate at which the total value of U.S. currency in circulation is changing in billions of dollars per year. To estimate the value of $C'(1990)$, we will average the difference quotients obtained using the times $t = 1985$ and $t = 1995$.

Let $A = \dfrac{C(1985) - C(1990)}{1985 - 1990} = \dfrac{187.3 - 271.9}{-5} = \dfrac{-84.6}{-5} = 16.92$ and

$B = \dfrac{C(1995) - C(1990)}{1995 - 1990} = \dfrac{409.3 - 271.9}{5} = \dfrac{137.4}{5} = 27.48$. Then

$C'(1990) = \displaystyle\lim_{t \to 1990} \frac{C(t) - C(1990)}{t - 1990} \approx \frac{A + B}{2} = \frac{16.92 + 27.48}{2} = \frac{44.4}{2} = 22.2$ billion dollars/year.

**40.** Let $C(t)$ be the function that denotes the cost of living in terms of time $t$. $C(t)$ is an increasing function, so $C'(t) > 0$. Since the cost of living is rising at a slower rate, the slopes of the tangent lines are positive but decreasing as $t$ increases. Hence, $C''(t) < 0$.

**41.** (a) $f'(x) > 0$ on $(-2, 0)$ and $(2, \infty)$ $\Rightarrow$ $f$ is increasing on those intervals. $f'(x) < 0$ on $(-\infty, -2)$ and $(0, 2)$ $\Rightarrow$ $f$ is decreasing on those intervals.

(b) $f'(x) = 0$ at $x = -2$, 0, and 2, so these are where local maxima or minima will occur. At $x = \pm 2$, $f'$ changes from negative to positive, so $f$ has local minima at those values. At $x = 0$, $f'$ changes from positive to negative, so $f$ has a local maximum there.

(c) $f'$ is increasing on $(-\infty, -1)$ and $(1, \infty)$ $\Rightarrow$ $f'' > 0$ and $f$ is concave upward on those intervals. $f'$ is decreasing on $(-1, 1)$ $\Rightarrow$ $f'' < 0$ and $f$ is concave downward on this interval.

(d)

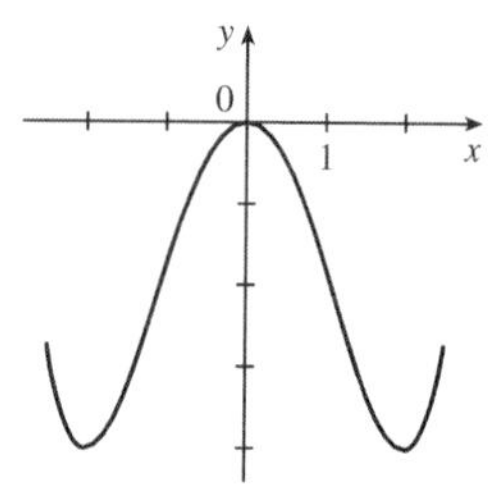

**42.** (a)

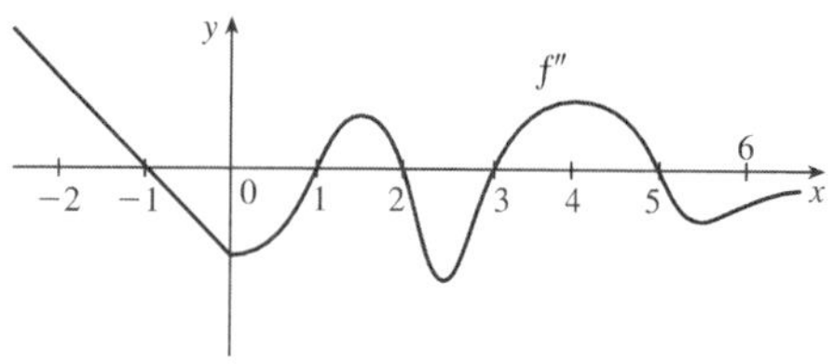


(b)

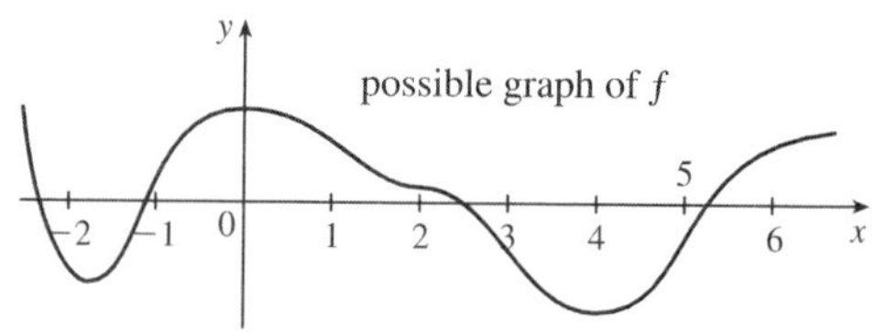


**43.** $f(0) = 0$, $f'(-2) = f'(1) = f'(9) = 0$, $\displaystyle\lim_{x \to \infty} f(x) = 0$, $\displaystyle\lim_{x \to 6} f(x) = -\infty$,

$f'(x) < 0$ on $(-\infty, -2)$, $(1, 6)$, and $(9, \infty)$, $f'(x) > 0$ on $(-2, 1)$ and $(6, 9)$,

$f''(x) > 0$ on $(-\infty, 0)$ and $(12, \infty)$, $f''(x) < 0$ on $(0, 6)$ and $(6, 12)$

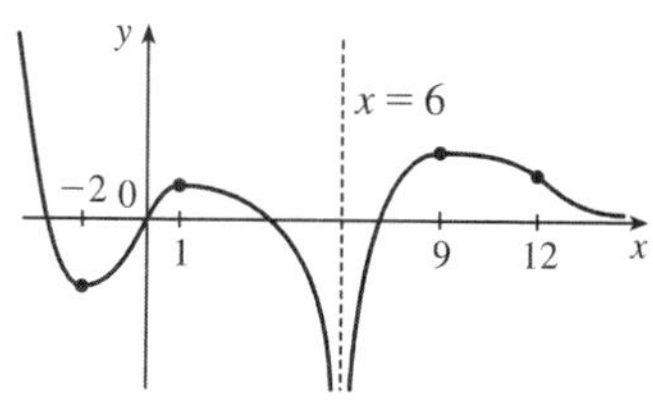


**44.** (a) Drawing slope triangles, we obtain the following estimates: $F'(1950) \approx \frac{1.1}{10} = 0.11$, $F'(1965) \approx \frac{-1.6}{10} = -0.16$, and $F'(1987) \approx \frac{0.2}{10} = 0.02$.

(b) The rate of change of the average number of children born to each woman was increasing by 0.11 in 1950, decreasing by 0.16 in 1965, and increasing by 0.02 in 1987.

(c) There are many possible reasons:

- In the baby-boom era (post-WWII), there was optimism about the economy and family size was rising.
- In the baby-bust era, there was less economic optimism, and it was considered less socially responsible to have a large family.
- In the baby-boomlet era, there was increased economic optimism and a return to more conservative attitudes.

**45.** (a) Using the data closest to $t = 6$, we have $\dfrac{s(8) - s(6)}{8 - 6} = \dfrac{180 - 95}{2} = 42.5$ and $\dfrac{s(4) - s(6)}{4 - 6} = \dfrac{40 - 95}{-2} = 27.5$.

Averaging these two values gives us $\dfrac{42.5 + 27.5}{2} = 35$ ft/s as an estimate for the speed of the car after 6 seconds.

(b) From the graph, it appears that the inflection point is at $(8, 180)$.

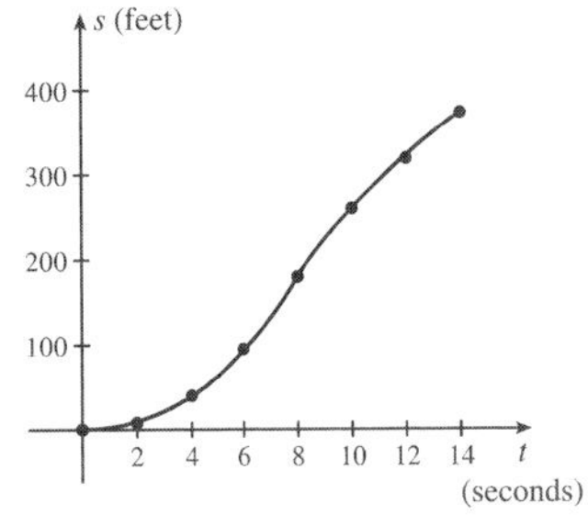


(c) The velocity of the car is at a maximum at the inflection point.

**46.** Let $f$ be the function shown. Since $f$ is negative for $x < 0$ and positive for $x > 0$, $F$ is decreasing for $x < 0$ and increasing for $x > 0$. $f$ is increasing on $(-a, a)$ (from the low point to the high point) so its derivative $f'$ (the second derivative of $F$) is positive, making $F$ concave upward on $(-a, a)$. $f$ is decreasing elsewhere, so its derivative $f'$ is negative and $F$ is concave downward on $(-\infty, -a)$ and $(a, \infty)$.

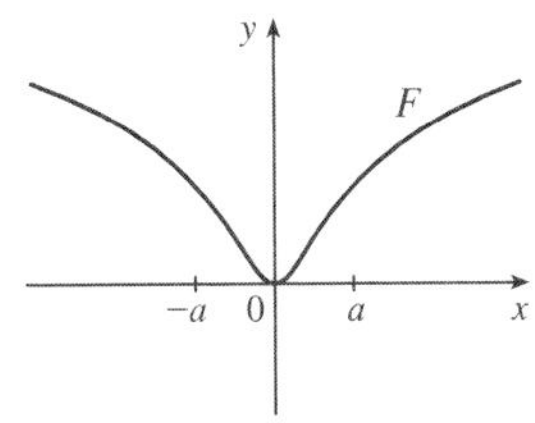

## FOCUS ON PROBLEM SOLVING

1. Let $t = \sqrt[6]{x}$, so $x = t^6$. Then $t \to 1$ as $x \to 1$, so

$$\lim_{x\to1}\frac{\sqrt[3]{x}-1}{\sqrt{x}-1} = \lim_{t\to1}\frac{t^2-1}{t^3-1} = \lim_{t\to1}\frac{(t-1)(t+1)}{(t-1)(t^2+t+1)} = \lim_{t\to1}\frac{t+1}{t^2+t+1} = \frac{1+1}{1^2+1+1} = \frac{2}{3}.$$

*Another method:* Multiply both the numerator and the denominator by $(\sqrt{x}+1)\left(\sqrt[3]{x^2}+\sqrt[3]{x}+1\right)$.

2. First rationalize the numerator: $\lim\limits_{x\to0}\dfrac{\sqrt{ax+b}-2}{x}\cdot\dfrac{\sqrt{ax+b}+2}{\sqrt{ax+b}+2} = \lim\limits_{x\to0}\dfrac{ax+b-4}{x\left(\sqrt{ax+b}+2\right)}$. Now since the denominator approaches 0 as $x \to 0$, the limit will exist only if the numerator also approaches 0 as $x \to 0$. So we require that $a(0)+b-4=0 \Rightarrow b=4$. So the equation becomes $\lim\limits_{x\to0}\dfrac{a}{\sqrt{ax+4}+2} = 1 \Rightarrow \dfrac{a}{\sqrt{4}+2} = 1 \Rightarrow a = 4$.

Therefore, $a = b = 4$.

3. For $-\frac{1}{2} < x < \frac{1}{2}$, we have $2x-1<0$ and $2x+1>0$, so $|2x-1| = -(2x-1)$ and $|2x+1| = 2x+1$.

Therefore, $\lim\limits_{x\to0}\dfrac{|2x-1|-|2x+1|}{x} = \lim\limits_{x\to0}\dfrac{-(2x-1)-(2x+1)}{x} = \lim\limits_{x\to0}\dfrac{-4x}{x} = \lim\limits_{x\to0}(-4) = -4$.

4. Let $R$ be the midpoint of $OP$, so the coordinates of $R$ are $\left(\frac{1}{2}x, \frac{1}{2}x^2\right)$ since the coordinates of $P$ are $(x, x^2)$. Let $Q = (0, a)$.

Since the slope $m_{OP} = \dfrac{x^2}{x} = x$, $m_{QR} = -\dfrac{1}{x}$ (negative reciprocal). But $m_{QR} = \dfrac{\frac{1}{2}x^2-a}{\frac{1}{2}x-0} = \dfrac{x^2-2a}{x}$, so we conclude that $-1 = x^2-2a \Rightarrow 2a = x^2+1 \Rightarrow a = \frac{1}{2}x^2+\frac{1}{2}$. As $x\to0$, $a\to\frac{1}{2}$, and the limiting position of $Q$ is $\left(0,\frac{1}{2}\right)$.

5. Since $[\![x]\!] \le x < [\![x]\!]+1$, we have $\dfrac{[\![x]\!]}{[\![x]\!]} \le \dfrac{x}{[\![x]\!]} < \dfrac{[\![x]\!]+1}{[\![x]\!]} \Rightarrow 1 \le \dfrac{x}{[\![x]\!]} < 1+\dfrac{1}{[\![x]\!]}$ for $x \ge 1$. As $x\to\infty$, $[\![x]\!]\to\infty$, so $\dfrac{1}{[\![x]\!]}\to0$ and $1+\dfrac{1}{[\![x]\!]}\to1$. Thus, $\lim\limits_{x\to\infty}\dfrac{x}{[\![x]\!]} = 1$ by the Squeeze Theorem.

6. (a) $[\![x]\!]^2+[\![y]\!]^2 = 1$. Since $[\![x]\!]^2$ and $[\![y]\!]^2$ are positive integers or 0, there are only 4 cases:

*Case (i):* $[\![x]\!]=1$, $[\![y]\!]=0 \Rightarrow 1 \le x < 2$ and $0 \le y < 1$

*Case (ii):* $[\![x]\!]=-1$, $[\![y]\!]=0 \Rightarrow -1 \le x < 0$ and $0 \le y < 1$

*Case (iii):* $[\![x]\!]=0$, $[\![y]\!]=1 \Rightarrow 0 \le x < 1$ and $1 \le y < 2$

*Case (iv):* $[\![x]\!]=0$, $[\![y]\!]=-1 \Rightarrow 0 \le x < 1$ and $-1 \le y < 0$

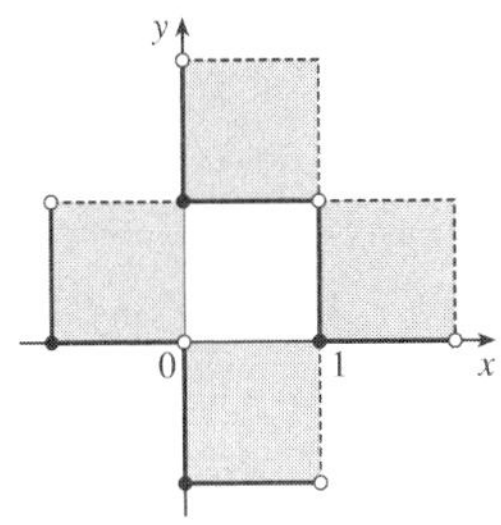


(b) $[\![x]\!]^2-[\![y]\!]^2 = 3$. The only integral solution of $n^2-m^2=3$ is $n=\pm2$ and $m=\pm1$. So the graph is

$$\{(x,y)\mid [\![x]\!]=\pm2,\ [\![y]\!]=\pm1\} = \left\{(x,y)\ \middle|\ \begin{array}{l} 2\le x\le 3 \text{ or } -2\le x<1, \\ 1\le y<2 \text{ or } -1\le y<0 \end{array}\right\}.$$

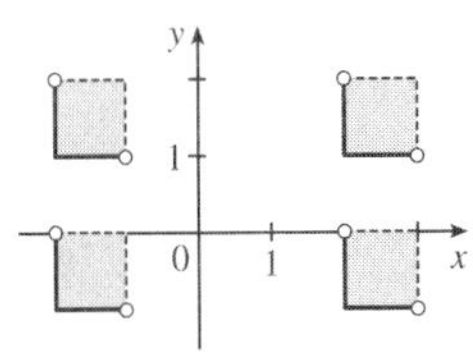


(c) $[\![x+y]\!]^2 = 1 \Rightarrow [\![x+y]\!] = \pm1 \Rightarrow 1 \le x+y < 2$ or $-1 \le x+y < 0$

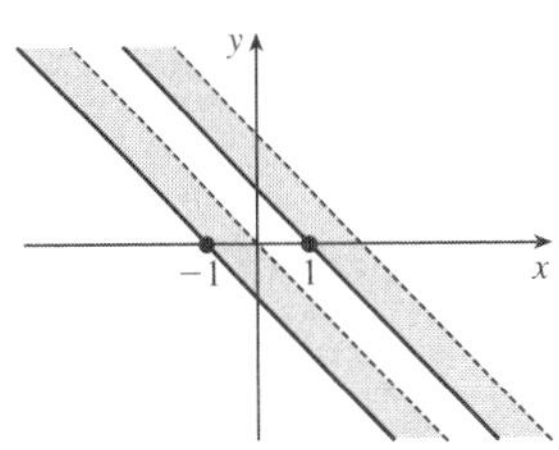

(d) For $n \le x < n+1$, $[\![x]\!] = n$. Then $[\![x]\!] + [\![y]\!] = 1 \;\Rightarrow\; [\![y]\!] = 1-n \;\Rightarrow$ $1-n \le y < 2-n$. Choosing integer values for $n$ produces the graph.

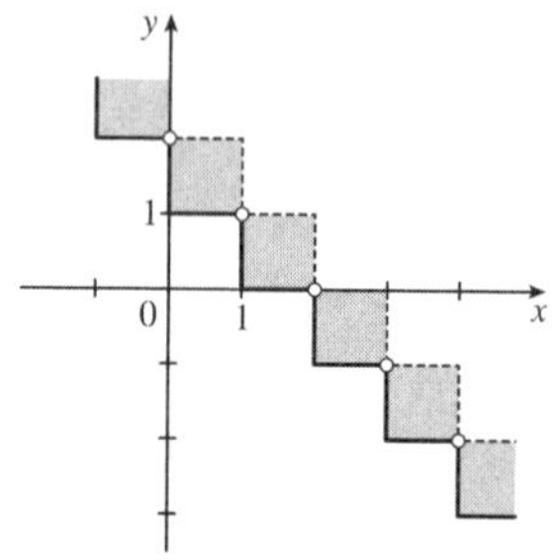


**7.** $f$ is continuous on $(-\infty, a)$ and $(a, \infty)$. To make $f$ continuous on $\mathbb{R}$, we must have continuity at $a$. Thus,
$\lim\limits_{x\to a^+} f(x) = \lim\limits_{x\to a^-} f(x) \;\Rightarrow\; \lim\limits_{x\to a^+} x^2 = \lim\limits_{x\to a^-} (x+1) \;\Rightarrow\; a^2 = a+1 \;\Rightarrow\; a^2 - a - 1 = 0 \;\Rightarrow$
[by the quadratic formula] $a = \left(1 \pm \sqrt{5}\right)/2 \approx 1.618$ or $-0.618$.

**8.** (a) Here are a few possibilities:

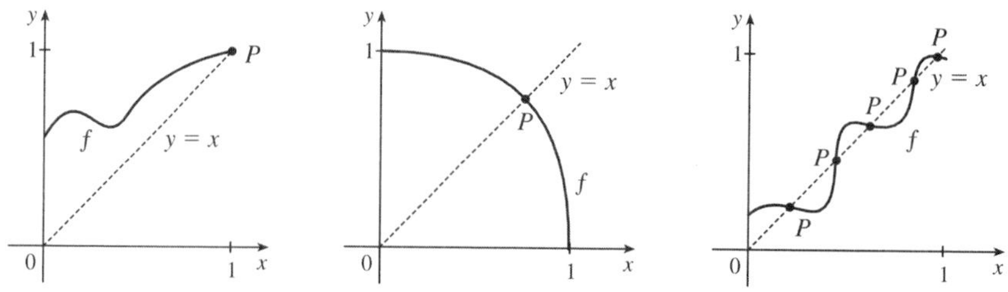


(b) The "obstacle" is the line $x = y$ (see diagram). Any intersection of the graph of $f$ with the line $y = x$ constitutes a fixed point, and if the graph of the function does not cross the line somewhere in $(0, 1)$, then it must either start at $(0, 0)$ (in which case 0 is a fixed point) or finish at $(1, 1)$ (in which case 1 is a fixed point).

(c) Consider the function $F(x) = f(x) - x$, where $f$ is any continuous function with domain $[0, 1]$ and range in $[0, 1]$. We shall prove that $f$ has a fixed point. Now if $f(0) = 0$ then we are done: $f$ has a fixed point (the number 0), which is what we are trying to prove. So assume $f(0) \ne 0$. For the same reason we can assume that $f(1) \ne 1$. Then $F(0) = f(0) > 0$ and $F(1) = f(1) - 1 < 0$. So by the Intermediate Value Theorem, there exists some number $c$ in the interval $(0, 1)$ such that $F(c) = f(c) - c = 0$. So $f(c) = c$, and therefore $f$ has a fixed point.

**9.** (a) Consider $G(x) = T(x + 180°) - T(x)$. Fix any number $a$. If $G(a) = 0$, we are done: Temperature at $a =$ Temperature at $a + 180°$. If $G(a) > 0$, then $G(a + 180°) = T(a + 360°) - T(a + 180°) = T(a) - T(a + 180°) = -G(a) < 0$. Also, $G$ is continuous since temperature varies continuously. So, by the Intermediate Value Theorem, $G$ has a zero on the interval $[a, a + 180°]$. If $G(a) < 0$, then a similar argument applies.

(b) Yes. The same argument applies.

(c) The same argument applies for quantities that vary continuously, such as barometric pressure. But one could argue that altitude above sea level is sometimes discontinuous, so the result might not always hold for that quantity.

**10.** (a) *Solution 1:* We introduce a coordinate system and drop a perpendicular from $P$, as shown. We see from $\angle NCP$ that $\tan 2\theta = \dfrac{y}{1-x}$, and from $\angle NBP$ that $\tan\theta = y/x$. Using the double-angle formula for tangents, we get $\dfrac{y}{1-x} = \tan 2\theta = \dfrac{2\tan\theta}{1-\tan^2\theta} = \dfrac{2(y/x)}{1-(y/x)^2}$. After a bit of simplification, this becomes $\dfrac{1}{1-x} = \dfrac{2x}{x^2-y^2} \;\Leftrightarrow\; y^2 = x\,(3x-2)$.

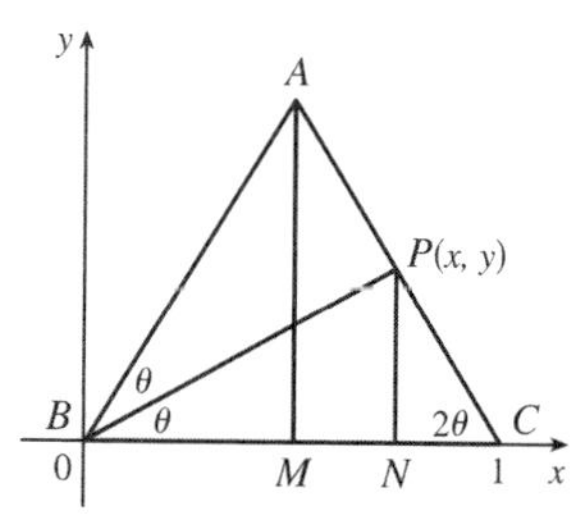


As the altitude $AM$ decreases in length, the point $P$ will approach the $x$-axis, that is, $y \to 0$, so the limiting location of $P$ must be one of the roots of the equation $x(3x-2) = 0$. Obviously it is not $x = 0$ (the point $P$ can never be to the left of the altitude $AM$, which it would have to be in order to approach 0) so it must be $3x - 2 = 0$, that is, $x = \frac{2}{3}$.

*Solution 2:* We add a few lines to the original diagram, as shown. Now note that $\angle BPQ = \angle PBC$ (alternate angles; $QP \parallel BC$ by symmetry) and similarly $\angle CQP = \angle QCB$. So $\Delta BPQ$ and $\Delta CQP$ are isosceles, and the line segments $BQ$, $QP$ and $PC$ are all of equal length. As $|AM| \to 0$, $P$ and $Q$ approach points on the base, and the point $P$ is seen to approach a position two-thirds of the way between $B$ and $C$, as above.

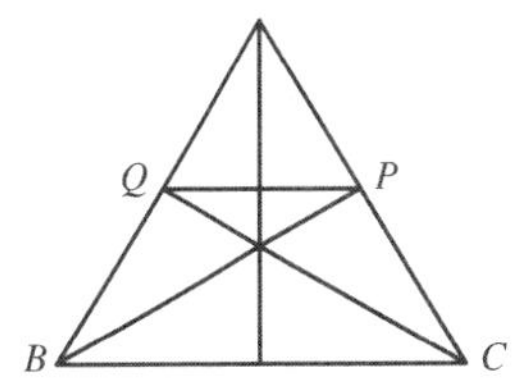


(b) The equation $y^2 = x(3x - 2)$ calculated in part (a) is the equation of the curve traced out by $P$. Now as $|AM| \to \infty$, $2\theta \to \frac{\pi}{2}$, $\theta \to \frac{\pi}{4}$, $x \to 1$, and since $\tan\theta = y/x$, $y \to 1$. Thus, $P$ only traces out the part of the curve with $0 \le y < 1$.

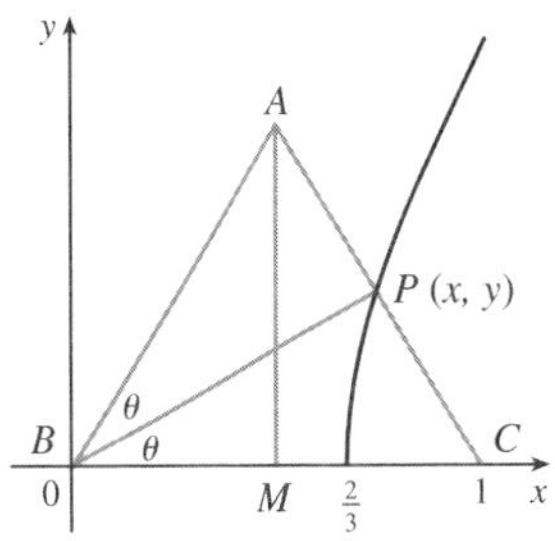


**11.** Let $a$ be the $x$-coordinate of $Q$. Since the derivative of $y = 1 - x^2$ is $y' = -2x$, the slope at $Q$ is $-2a$. But since the triangle is equilateral, $\overline{AO}/\overline{OC} = \sqrt{3}/1$, so the slope at $Q$ is $-\sqrt{3}$. Therefore, we must have that $-2a = -\sqrt{3} \;\Rightarrow\; a = \frac{\sqrt{3}}{2}$. Thus, the point $Q$ has coordinates $\left(\frac{\sqrt{3}}{2}, 1 - \left(\frac{\sqrt{3}}{2}\right)^2\right) = \left(\frac{\sqrt{3}}{2}, \frac{1}{4}\right)$ and by symmetry, $P$ has coordinates $\left(-\frac{\sqrt{3}}{2}, \frac{1}{4}\right)$.

**12.** (a) $V'(t)$ is the rate of change of the volume of the water with respect to time. $H'(t)$ is the rate of change of the height of the water with respect to time. Since the volume and the height are increasing, $V'(t)$ and $H'(t)$ are positive.

(b) $V'(t)$ is constant, so $V''(t)$ is zero (the slope of a constant function is 0).

(c) At first, the height $H$ of the water increases quickly because the tank is narrow. But as the sphere widens, the rate of increase of the height slows down, reaching a minimum at $t = t_2$. Thus, the height is increasing at a decreasing rate on $(0, t_2)$, so its graph is concave downward and $H''(t_1) < 0$. As the sphere narrows for $t > t_2$, the rate of increase of the height begins to increase, and the graph of $H$ is concave upward. Therefore, $H''(t_2) = 0$ and $H''(t_3) > 0$.

**13.** (a) Put $x = 0$ and $y = 0$ in the equation: $f(0 + 0) = f(0) + f(0) + 0^2 \cdot 0 + 0 \cdot 0^2 \;\Rightarrow\; f(0) = 2f(0)$. Subtracting $f(0)$ from each side of this equation gives $f(0) = 0$.

(b) $$f'(0) = \lim_{h\to 0} \frac{f(0+h) - f(0)}{h} = \lim_{h\to 0} \frac{\left[f(0) + f(h) + 0^2 h + 0h^2\right] - f(0)}{h} = \lim_{h\to 0} \frac{f(h)}{h} = \lim_{x\to 0} \frac{f(x)}{x} = 1$$

(c) $$f'(x) = \lim_{h\to 0} \frac{f(x+h) - f(x)}{h} = \lim_{h\to 0} \frac{\left[f(x) + f(h) + x^2 h + xh^2\right] - f(x)}{h} = \lim_{h\to 0} \frac{f(h) + x^2 h + xh^2}{h}$$
$$= \lim_{h\to 0} \left[\frac{f(h)}{h} + x^2 + xh\right] = 1 + x^2$$

**14.** We find the equation of the parabola by substituting the point $(-100, 100)$, at which the car is situated, into the general equation $y = ax^2$: $100 = a(-100)^2 \;\Rightarrow\; a = \frac{1}{100}$. Now we find the equation of a tangent to the parabola at the point $(x_0, y_0)$. We can show that $y' = a(2x) = \frac{1}{100}(2x) = \frac{1}{50}x$, so an equation of the tangent is $y - y_0 = \frac{1}{50}x_0(x - x_0)$. Since the point $(x_0, y_0)$ is on the parabola, we must have $y_0 = \frac{1}{100}x_0^2$, so our equation of the tangent can be simplified to $y = \frac{1}{100}x_0^2 + \frac{1}{50}x_0(x - x_0)$. We want the statue to be located on the tangent line, so we substitute its coordinates $(100, 50)$ into this equation: $50 = \frac{1}{100}x_0^2 + \frac{1}{50}x_0(100 - x_0) \;\Rightarrow\; x_0^2 - 200x_0 + 5000 = 0 \;\Rightarrow$

$x_0 = \frac{1}{2}\left[200 \pm \sqrt{200^2 - 4\,(5000)}\right] \;\Rightarrow\; x_0 = 100 \pm 50\sqrt{2}$. But $x_0 < 100$, so the car's headlights illuminate the statue when it is located at the point $\left(100 - 50\sqrt{2}, 150 - 100\sqrt{2}\right) \approx (29.3, 8.6)$, that is, about 29.3 m east and 8.6 m north of the origin.

**15.** $\lim\limits_{x\to a} f(x) = \lim\limits_{x\to a} \left(\frac{1}{2}[f(x)+g(x)] + \frac{1}{2}[f(x)-g(x)]\right) = \frac{1}{2}\lim\limits_{x\to a}[f(x)+g(x)] + \frac{1}{2}\lim\limits_{x\to a}[f(x)-g(x)]$

$= \frac{1}{2}\cdot 2 + \frac{1}{2}\cdot 1 = \frac{3}{2}$,

and $\lim\limits_{x\to a} g(x) = \lim\limits_{x\to a}\Big([f(x)+g(x)] - f(x)\Big) = \lim\limits_{x\to a}[f(x)+g(x)] - \lim\limits_{x\to a} f(x) = 2 - \frac{3}{2} = \frac{1}{2}$.

So $\lim\limits_{x\to a}[f(x)g(x)] = \left[\lim\limits_{x\to a} f(x)\right]\left[\lim\limits_{x\to a} g(x)\right] = \frac{3}{2}\cdot\frac{1}{2} = \frac{3}{4}$.

*Another solution:* Since $\lim\limits_{x\to a}[f(x)+g(x)]$ and $\lim\limits_{x\to a}[f(x)-g(x)]$ exist, we must have

$\lim\limits_{x\to a}[f(x)+g(x)]^2 = \left(\lim\limits_{x\to a}[f(x)+g(x)]\right)^2$ and $\lim\limits_{x\to a}[f(x)-g(x)]^2 = \left(\lim\limits_{x\to a}[f(x)-g(x)]\right)^2$, so

$\lim\limits_{x\to a}[f(x)\,g(x)] = \lim\limits_{x\to a}\frac{1}{4}\left([f(x)+g(x)]^2 - [f(x)-g(x)]^2\right)$ [because all of the $f^2$ and $g^2$ cancel]

$= \frac{1}{4}\left(\lim\limits_{x\to a}[f(x)+g(x)]^2 - \lim\limits_{x\to a}[f(x)-g(x)]^2\right) = \frac{1}{4}(2^2 - 1^2) = \frac{3}{4}$.

**16.** $g'(x) = \lim\limits_{h\to 0}\dfrac{g(x+h)-g(x)}{h} = \lim\limits_{h\to 0}\dfrac{(x+h)f(x+h) - xf(x)}{h} = \lim\limits_{h\to 0}\left[\dfrac{xf(x+h) - xf(x)}{h} + \dfrac{hf(x+h)}{h}\right]$

$= x\lim\limits_{h\to 0}\dfrac{f(x+h)-f(x)}{h} + \lim\limits_{h\to 0} f(x+h) = xf'(x) + f(x)$

because $f$ is differentiable and therefore continuous.

**17.** We are given that $|f(x)| \le x^2$ for all $x$. In particular, $|f(0)| \le 0$, but $|a| \ge 0$ for all $a$. The only conclusion is

that $f(0) = 0$. Now $\left|\dfrac{f(x)-f(0)}{x-0}\right| = \left|\dfrac{f(x)}{x}\right| = \dfrac{|f(x)|}{|x|} \le \dfrac{x^2}{|x|} = \dfrac{|x^2|}{|x|} = |x| \;\Rightarrow\; -|x| \le \dfrac{f(x)-f(0)}{x-0} \le |x|$.

But $\lim\limits_{x\to 0}(-|x|) = 0 = \lim\limits_{x\to 0}|x|$, so by the Squeeze Theorem, $\lim\limits_{x\to 0}\dfrac{f(x)-f(0)}{x-0} = 0$. So by the definition of a derivative,

$f$ is differentiable at 0 and, furthermore, $f'(0) = 0$.

# 3 □ DIFFERENTIATION RULES

## 3.1 Derivatives of Polynomials and Exponential Functions

**1.** (a) $e$ is the number such that $\lim\limits_{h\to 0}\dfrac{e^h-1}{h}=1$.

(b)

| $x$ | $\dfrac{2.7^x-1}{x}$ |
|---|---|
| $-0.001$ | $0.9928$ |
| $-0.0001$ | $0.9932$ |
| $0.001$ | $0.9937$ |
| $0.0001$ | $0.9933$ |

| $x$ | $\dfrac{2.8^x-1}{x}$ |
|---|---|
| $-0.001$ | $1.0291$ |
| $-0.0001$ | $1.0296$ |
| $0.001$ | $1.0301$ |
| $0.0001$ | $1.0297$ |

From the tables (to two decimal places),

$\lim\limits_{h\to 0}\dfrac{2.7^h-1}{h}=0.99$ and $\lim\limits_{h\to 0}\dfrac{2.8^h-1}{h}=1.03$.

Since $0.99<1<1.03$, $2.7<e<2.8$.

**2.** (a)

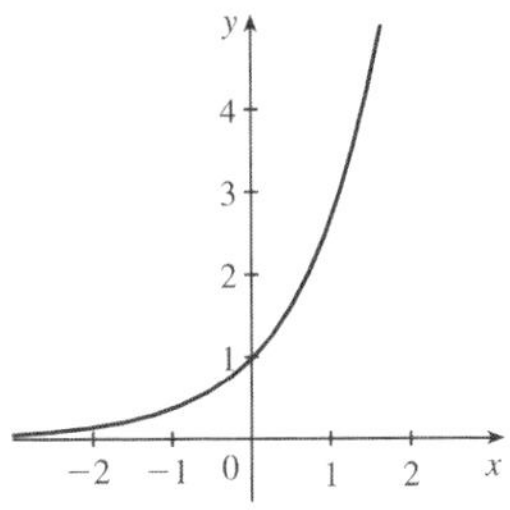


The function value at $x=0$ is 1 and the slope at $x=0$ is 1.

(b) $f(x)=e^x$ is an exponential function and $g(x)=x^e$ is a power function. $\dfrac{d}{dx}(e^x)=e^x$ and $\dfrac{d}{dx}(x^e)=ex^{e-1}$.

(c) $s=f(t)=t^3-12t^2+36t,\ t\geq 0\ \Rightarrow\ v(t)=f'(t)=3t^2-24t+36$. $a(t)=v'(t)=6t-24$.
$a(3)=6(3)-24=-6\ (\text{m/s})/\text{s}$ or $\text{m/s}^2$.

**3.** $f(x)=186.5$ is a constant function, so its derivative is 0, that is, $f'(x)=0$.

**4.** $f(x)=\sqrt{30}$ is a constant function, so its derivative is 0, that is, $f'(x)=0$.

**5.** $f(x)=x^3-4x+6\ \Rightarrow\ f'(x)=3x^2-4(1)+0=3x^2-4$

**6.** $f(t)=\frac{1}{2}t^6-3t^4+t\ \Rightarrow\ f'(t)=\frac{1}{2}(6t^5)-3(4t^3)+1=3t^5-12t^3+1$

**7.** $f(t)=\frac{1}{4}(t^4+8)\ \Rightarrow\ f'(t)=\frac{1}{4}(t^4+8)'=\frac{1}{4}(4t^{4-1}+0)=t^3$

**8.** $h(x)=(x-2)(2x+3)=2x^2-x-6\ \Rightarrow\ h'(x)=2(2x)-1-0=4x-1$

**9.** $y=x^{-2/5}\ \Rightarrow\ y'=-\frac{2}{5}x^{(-2/5)-1}=-\frac{2}{5}x^{-7/5}=-\dfrac{2}{5x^{7/5}}$

**10.** $y=5e^x+3\ \Rightarrow\ y'=5(e^x)+0=5e^x$

**11.** $G(x)=\sqrt{x}-2e^x=x^{1/2}-2e^x\ \Rightarrow\ G'(x)=\frac{1}{2}x^{-1/2}-2e^x=\dfrac{1}{2\sqrt{x}}-2e^x$

**12.** $R(t)=5t^{-3/5}\ \Rightarrow\ R'(t)=5\left[-\frac{3}{5}t^{(-3/5)-1}\right]=-3t^{-8/5}$

**13.** $V(r)=\frac{4}{3}\pi r^3\ \Rightarrow\ V'(r)=\frac{4}{3}\pi(3r^2)=4\pi r^2$

**14.** $R(x)=\dfrac{\sqrt{10}}{x^7}=\sqrt{10}\,x^{-7}\ \Rightarrow\ R'(x)=-7\sqrt{10}\,x^{-8}=-\dfrac{7\sqrt{10}}{x^8}$

**15.** $F(x)=(\frac{1}{2}x)^5=\left(\frac{1}{2}\right)^5x^5=\frac{1}{32}x^5\ \Rightarrow\ F'(x)=\frac{1}{32}(5x^4)=\frac{5}{32}x^4$

**16.** $y = \sqrt{x}\,(x-1) = x^{3/2} - x^{1/2} \;\Rightarrow\; y' = \frac{3}{2}x^{1/2} - \frac{1}{2}x^{-1/2} = \frac{1}{2}x^{-1/2}(3x-1)$ [factor out $\frac{1}{2}x^{-1/2}$]
or $y' = \dfrac{3x-1}{2\sqrt{x}}$.

**17.** $y = 4\pi^2 \;\Rightarrow\; y' = 0$ since $4\pi^2$ is a constant.

**18.** $g(u) = \sqrt{2}\,u + \sqrt{3u} = \sqrt{2}\,u + \sqrt{3}\,\sqrt{u} \;\Rightarrow\; g'(u) = \sqrt{2}\,(1) + \sqrt{3}\left(\frac{1}{2}u^{-1/2}\right) = \sqrt{2} + \sqrt{3}/(2\sqrt{u})$

**19.** $y = \dfrac{x^2+4x+3}{\sqrt{x}} = x^{3/2} + 4x^{1/2} + 3x^{-1/2} \;\Rightarrow$

$y' = \frac{3}{2}x^{1/2} + 4\left(\frac{1}{2}\right)x^{-1/2} + 3\left(-\frac{1}{2}\right)x^{-3/2} = \frac{3}{2}\sqrt{x} + \dfrac{2}{\sqrt{x}} - \dfrac{3}{2x\sqrt{x}}$

[note that $x^{3/2} = x^{2/2}\cdot x^{1/2} = x\sqrt{x}$]

**20.** $y = \dfrac{x^2 - 2\sqrt{x}}{x} = x - 2x^{-1/2} \;\Rightarrow\; y' = 1 - 2\left(-\frac{1}{2}\right)x^{-3/2} = 1 + 1/(x\sqrt{x})$

**21.** $v = t^2 - \dfrac{1}{\sqrt[4]{t^3}} = t^2 - t^{-3/4} \;\Rightarrow\; v' = 2t - \left(-\frac{3}{4}\right)t^{-7/4} = 2t + \dfrac{3}{4t^{7/4}} = 2t + \dfrac{3}{4t\,\sqrt[4]{t^3}}$

**22.** $y = ae^v + \dfrac{b}{v} + \dfrac{c}{v^2} = ae^v + bv^{-1} + cv^{-2} \;\Rightarrow\; y' = ae^v - bv^{-2} - 2cv^{-3} = ae^v - \dfrac{b}{v^2} - \dfrac{2c}{v^3}$

**23.** $z = \dfrac{A}{y^{10}} + Be^y = Ay^{-10} + Be^y \;\Rightarrow\; z' = -10Ay^{-11} + Be^y = -\dfrac{10A}{y^{11}} + Be^y$

**24.** $u = \sqrt[3]{t^2} + 2\sqrt{t^3} = t^{2/3} + 2t^{3/2} \;\Rightarrow\; u' = \frac{2}{3}t^{-1/3} + 2\left(\frac{3}{2}\right)t^{1/2} = \dfrac{2}{3\sqrt[3]{t}} + 3\sqrt{t}$

**25.** $y = x^4 + 2e^x \;\Rightarrow\; y' = 4x^3 + 2e^x$. At $(0,2)$, $y' = 2$ and an equation of the tangent line is $y - 2 = 2(x-0)$ or $y = 2x + 2$. The slope of the normal line is $-\frac{1}{2}$ (the negative reciprocal of 2) and an equation of the normal line is $y - 2 = -\frac{1}{2}(x-0)$ or $y = -\frac{1}{2}x + 2$.

**26.** $y = (1+2x)^2 = 1 + 4x + 4x^2 \;\Rightarrow\; y' = 4 + 8x$. At $(1,9)$, $y' = 12$ and an equation of the tangent line is $y - 9 = 12(x-1)$ or $y = 12x - 3$. The slope of the normal line is $-\frac{1}{12}$ (the negative reciprocal of 12) and an equation of the normal line is $y - 9 = -\frac{1}{12}(x-1)$ or $y = -\frac{1}{12}x + \frac{109}{12}$.

**27.** $y = f(x) = x + \sqrt{x} \;\Rightarrow\; f'(x) = 1 + \frac{1}{2}x^{-1/2}$.

So the slope of the tangent line at $(1,2)$ is $f'(1) = 1 + \frac{1}{2}(1) = \frac{3}{2}$ and its equation is $y - 2 = \frac{3}{2}(x-1)$ or $y = \frac{3}{2}x + \frac{1}{2}$.

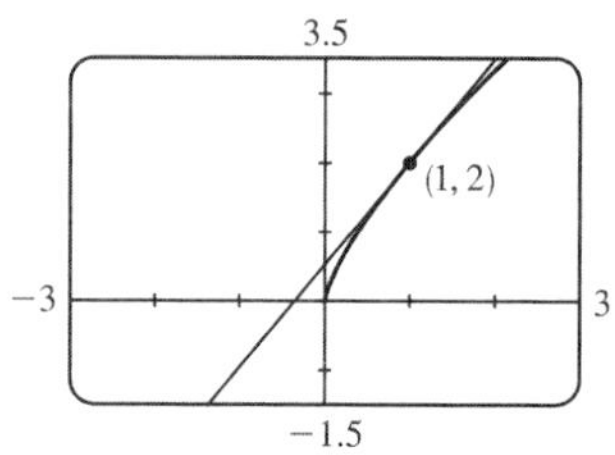


**28.** $y = 3x^2 - x^3 \;\Rightarrow\; y' = 6x - 3x^2$. At $(1,2)$, $y' = 6 - 3 = 3$, so an equation of the tangent line is $y - 2 = 3(x-1)$, or $y = 3x - 1$.

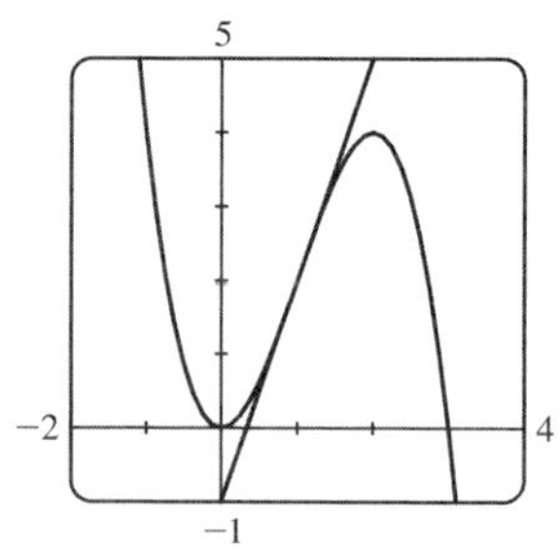

**29.** $f(x) = e^x - 5x \Rightarrow f'(x) = e^x - 5.$

Notice that $f'(x) = 0$ when $f$ has a horizontal tangent, $f'$ is positive when $f$ is increasing, and $f'$ is negative when $f$ is decreasing.

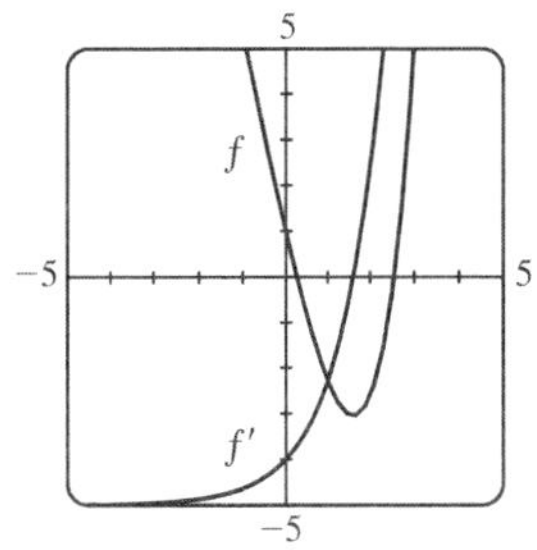


**30.** $f(x) = 3x^5 - 20x^3 + 50x \Rightarrow f'(x) = 15x^4 - 60x^2 + 50.$

Notice that $f'(x) = 0$ when $f$ has a horizontal tangent and that $f'$ is an even function while $f$ is an odd function.

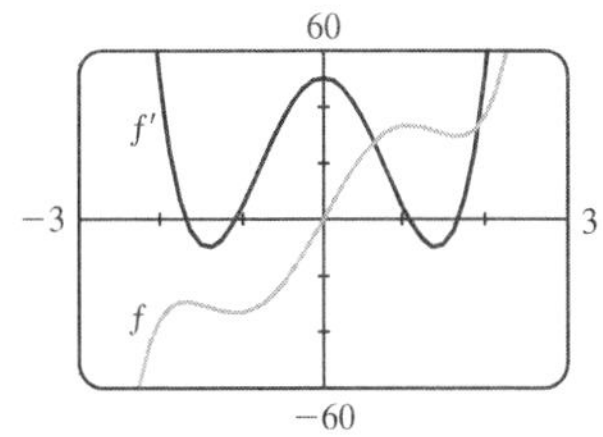


**31.** $f(x) = x - 3x^{1/3} \Rightarrow f'(x) = 1 - x^{-2/3} = 1 - 1/x^{2/3}.$

Note that $f'(x) = 0$ when $f$ has a horizontal tangent, $f'$ is positive when $f$ is increasing, and $f'$ is negative when $f$ is decreasing.

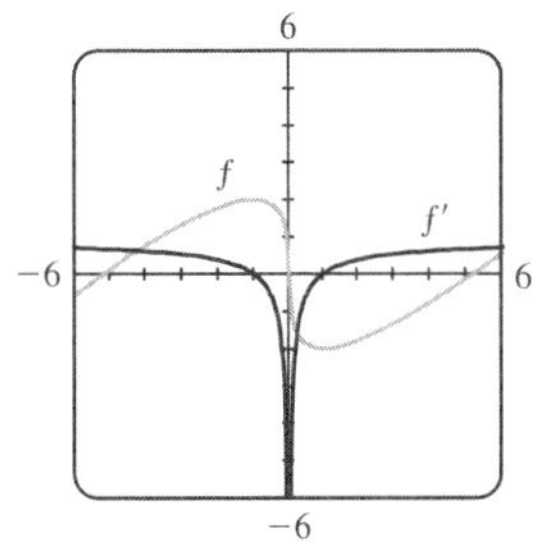


**32.** $f(x) = x + 1/x = x + x^{-1} \Rightarrow f'(x) = 1 - x^{-2} = 1 - 1/x^2.$

Notice that $f'(x) = 0$ when $f$ has a horizontal tangent, $f'$ is positive when $f$ is increasing, and $f'$ is negative when $f$ is decreasing.

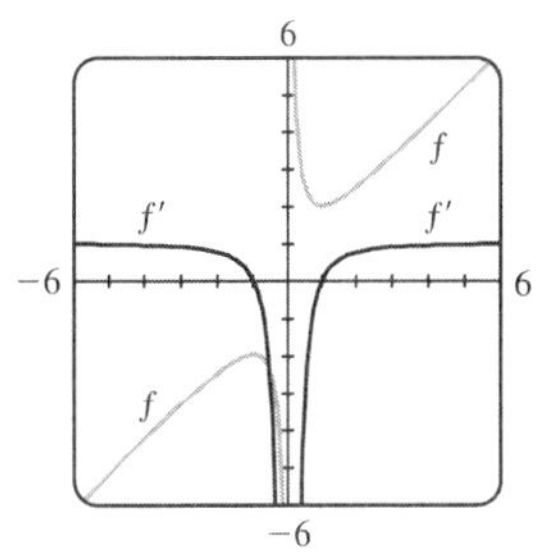


**33.** To graphically estimate the value of $f'(1)$ for $f(x) = 3x^2 - x^3$, we'll graph $f$ in the viewing rectangle $[1 - 0.1, 1 + 0.1]$ by $[f(0.9), f(1.1)]$, as shown in the figure. [When assigning values to the window variables, it is convenient to use $Y_1(0.9)$ for $Y_{\text{min}}$ and $Y_1(1.1)$ for $Y_{\text{max}}$.] If we have sufficiently zoomed in on the graph of $f$, we should obtain a graph that looks like a diagonal line; if not, graph again with $1 - 0.01$ and $1 + 0.01$, etc.

**Estimated value:**

$$f'(1) \approx \frac{2.299 - 1.701}{1.1 - 0.9} = \frac{0.589}{0.2} = 2.99.$$

**Exact value:** $f(x) = 3x^2 - x^3 \Rightarrow f'(x) = 6x - 3x^2$, so $f'(1) = 6 - 3 = 3.$

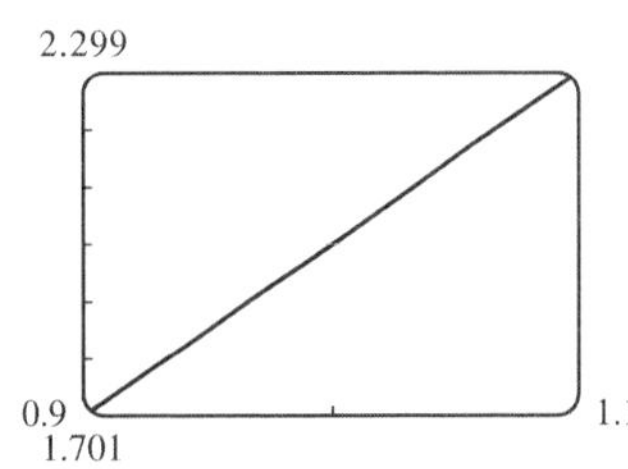


**34.** See the previous exercise. Since $f$ is a decreasing function, assign $Y_1(3.9)$ to $Y_{\text{max}}$ and $Y_1(4.1)$ to $Y_{\text{min}}$.

**Estimated value:** $f'(4) \approx \dfrac{0.49386 - 0.50637}{4.1 - 3.9} = \dfrac{-0.01251}{0.2} = -0.06255.$

**Exact value:** $f(x) = x^{-1/2} \Rightarrow f'(x) = -\frac{1}{2}x^{-3/2}$, so $f'(4) = -\frac{1}{2}(4^{-3/2}) = -\frac{1}{2}(\frac{1}{8}) = -\frac{1}{16} = -0.0625.$

**35.** (a)

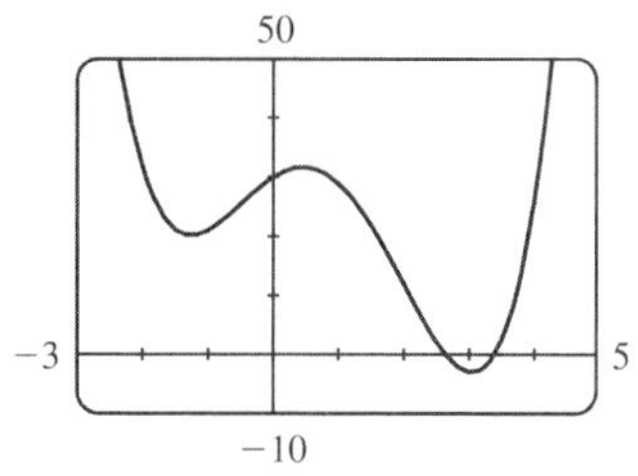


(b) From the graph in part (a), it appears that $f'$ is zero at $x_1 \approx -1.25$, $x_2 \approx 0.5$, and $x_3 \approx 3$. The slopes are negative (so $f'$ is negative) on $(-\infty, x_1)$ and $(x_2, x_3)$. The slopes are positive (so $f'$ is positive) on $(x_1, x_2)$ and $(x_3, \infty)$.

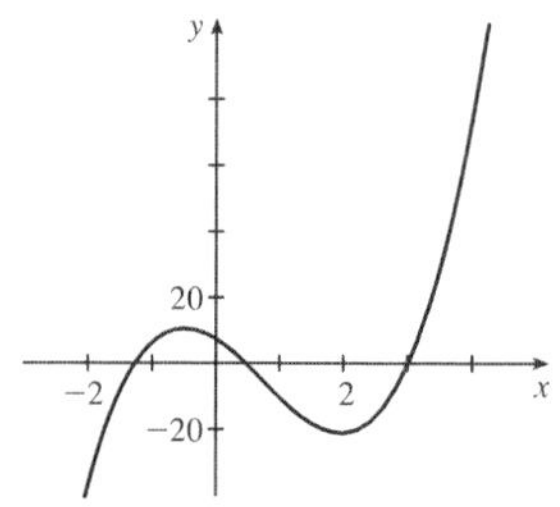


(c) $f(x) = x^4 - 3x^3 - 6x^2 + 7x + 30 \quad \Rightarrow$

$f'(x) = 4x^3 - 9x^2 - 12x + 7$

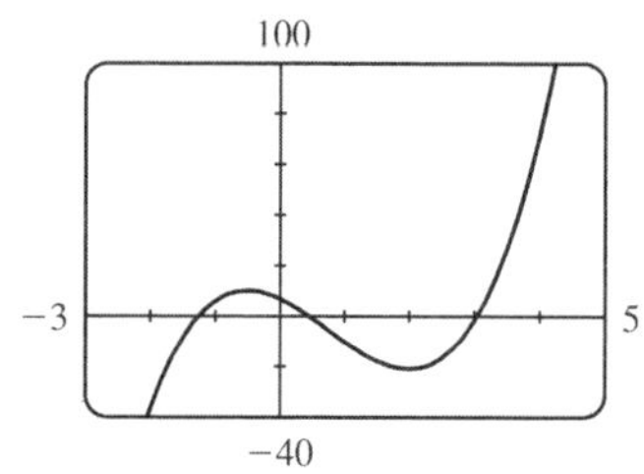


**36.** (a)

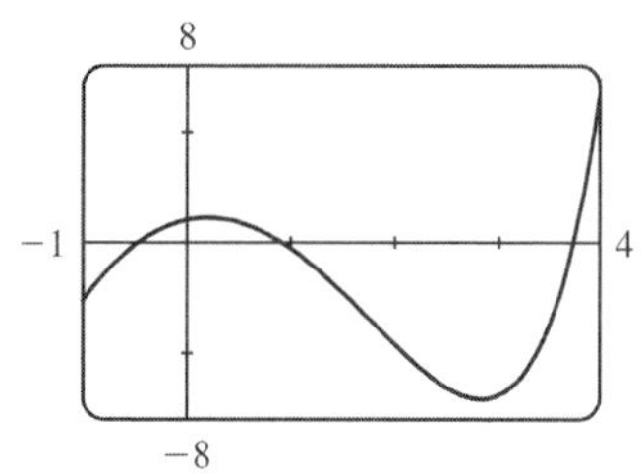


(b) From the graph in part (a), it appears that $f'$ is zero at $x_1 \approx 0.2$ and $x_2 \approx 2.8$. The slopes are positive (so $f'$ is positive) on $(-\infty, x_1)$ and $(x_2, \infty)$. The slopes are negative (so $f'$ is negative) on $(x_1, x_2)$.

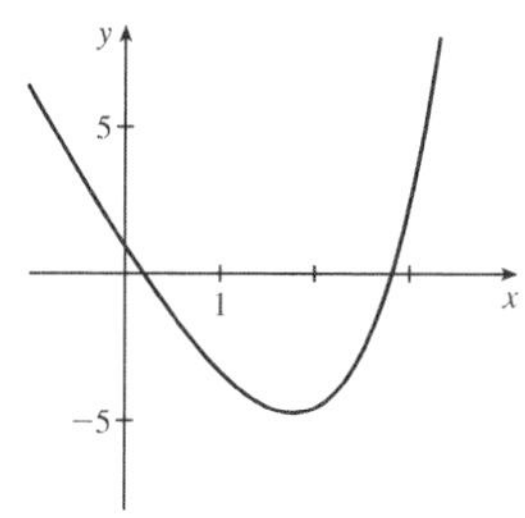


(c) $g(x) = e^x - 3x^2 \quad \Rightarrow \quad g'(x) = e^x - 6x$

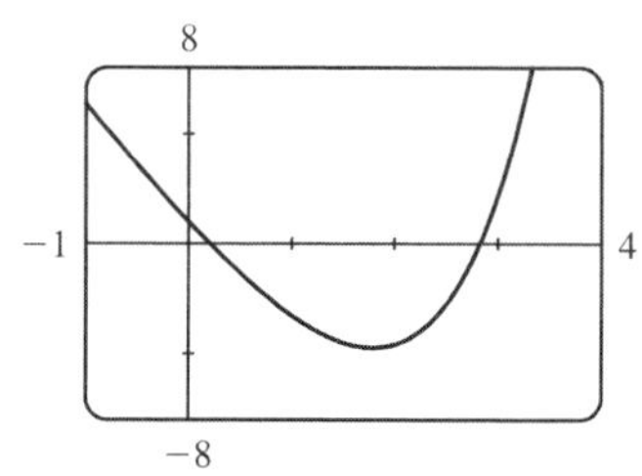


**37.** $f(x) = x^4 - 3x^3 + 16x \quad \Rightarrow \quad f'(x) = 4x^3 - 9x^2 + 16 \quad \Rightarrow \quad f''(x) = 12x^2 - 18x$

**38.** $G(r) = \sqrt{r} + \sqrt[3]{r} \quad \Rightarrow \quad G'(r) = \frac{1}{2}r^{-1/2} + \frac{1}{3}r^{-2/3} \quad \Rightarrow \quad G''(r) = -\frac{1}{4}r^{-3/2} - \frac{2}{9}r^{-5/3}$

**39.** $f(x) = 2x - 5x^{3/4} \quad \Rightarrow \quad f'(x) = 2 - \frac{15}{4}x^{-1/4} \quad \Rightarrow \quad f''(x) = \frac{15}{16}x^{-5/4}$

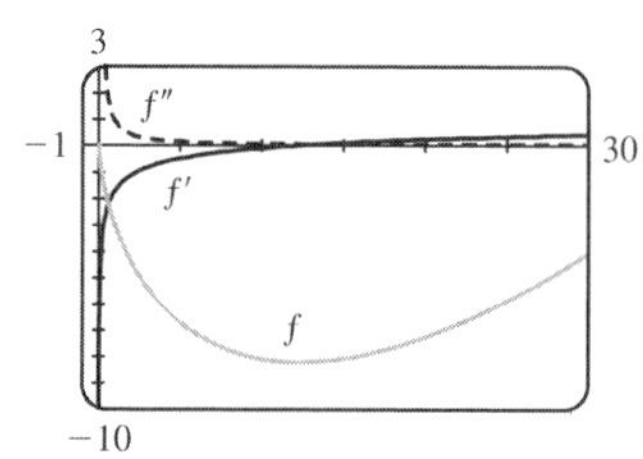


Note that $f'$ is negative when $f$ is decreasing and positive when $f$ is increasing. $f''$ is always positive since $f'$ is always increasing.

**40.** $f(x) = e^x - x^3 \Rightarrow f'(x) = e^x - 3x^2 \Rightarrow f''(x) = e^x - 6x$

Note that $f'(x) = 0$ when $f$ has a horizontal tangent and that $f''(x) = 0$ when $f'$ has a horizontal tangent.

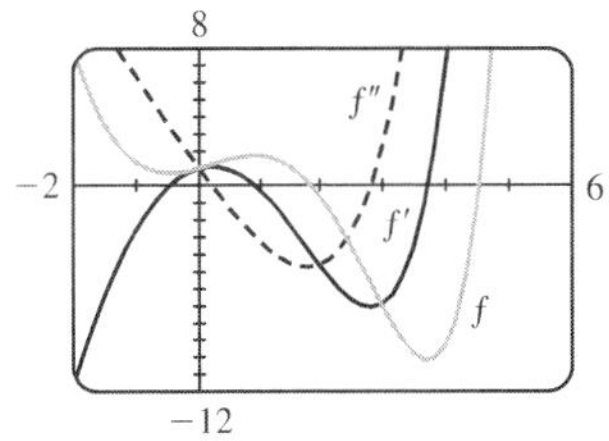


**41.** (a) $s = t^3 - 3t \Rightarrow v(t) = s'(t) = 3t^2 - 3 \Rightarrow a(t) = v'(t) = 6t$

(b) $a(2) = 6(2) = 12\text{ m/s}^2$

(c) $v(t) = 3t^2 - 3 = 0$ when $t^2 = 1$, that is, $t = 1$ and $a(1) = 6\text{ m/s}^2$.

**42.** (a) $s = 2t^3 - 7t^2 + 4t + 1 \Rightarrow v(t) = s'(t) = 6t^2 - 14t + 4 \Rightarrow a(t) = v'(t) = 12t - 14$

(b) $a(1) = 12 - 14 = -2\text{ m/s}^2$

(c)

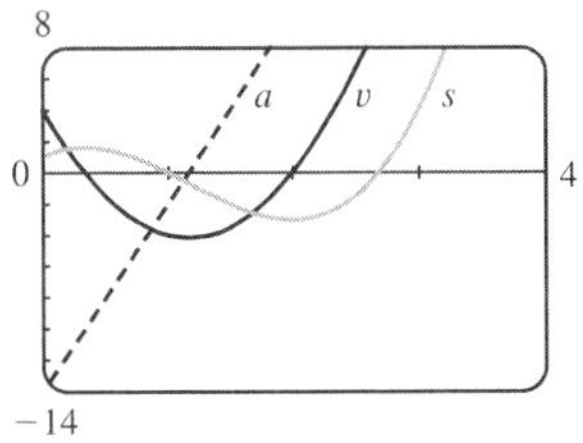


**43.** $f(x) = 1 + 2e^x - 3x \Rightarrow f'(x) = 2e^x - 3$. $f'(x) > 0 \Rightarrow 2e^x - 3 > 0 \Rightarrow 2e^x > 3 \Rightarrow e^x > 1.5 \Rightarrow x > \ln 1.5 \approx 0.41$. $f$ is increasing when $f'$ is positive; that is, on $(\ln 1.5, \infty)$.

**44.** $f(x) = x^3 - 4x^2 + 5x \Rightarrow f'(x) = 3x^2 - 8x + 5 \Rightarrow f''(x) = 6x - 8$.
$f''(x) > 0 \Rightarrow 6x - 8 > 0 \Rightarrow x > \frac{4}{3}$. $f$ is concave upward when $f''(x) > 0$; that is, on $\left(\frac{4}{3}, \infty\right)$.

**45.** The curve $y = 2x^3 + 3x^2 - 12x + 1$ has a horizontal tangent when $y' = 6x^2 + 6x - 12 = 0 \Leftrightarrow 6(x^2 + x - 2) = 0 \Leftrightarrow 6(x + 2)(x - 1) = 0 \Leftrightarrow x = -2$ or $x = 1$. The points on the curve are $(-2, 21)$ and $(1, -6)$.

**46.** $f(x) = x^3 + 3x^2 + x + 3$ has a horizontal tangent when $f'(x) = 3x^2 + 6x + 1 = 0 \Leftrightarrow$
$x = \dfrac{-6 \pm \sqrt{36 - 12}}{6} = -1 \pm \frac{1}{3}\sqrt{6}$.

**47.** $y = 6x^3 + 5x - 3 \Rightarrow m = y' = 18x^2 + 5$, but $x^2 \geq 0$ for all $x$, so $m \geq 5$ for all $x$.

**48.** $y = x\sqrt{x} = x^{3/2} \Rightarrow y' = \frac{3}{2}x^{1/2}$. The slope of the line $y = 1 + 3x$ is 3, so the slope of any line parallel to it is also 3. Thus, $y' = 3 \Rightarrow \frac{3}{2}x^{1/2} = 3 \Rightarrow \sqrt{x} = 2 \Rightarrow x = 4$, which is the $x$-coordinate of the point on the curve at which the slope is 3. The $y$-coordinate is $y = 4\sqrt{4} = 8$, so an equation of the tangent line is $y - 8 = 3(x - 4)$ or $y = 3x - 4$.

**49.** The slope of the line $12x - y = 1$ (or $y = 12x - 1$) is 12, so the slope of both lines tangent to the curve is 12. $y = 1 + x^3 \Rightarrow y' = 3x^2$. Thus, $3x^2 = 12 \Rightarrow x^2 = 4 \Rightarrow x = \pm 2$, which are the $x$-coordinates at which the tangent lines have slope 12. The points on the curve are $(2, 9)$ and $(-2, -7)$, so the tangent line equations are $y - 9 = 12(x - 2)$ or $y = 12x - 15$ and $y + 7 = 12(x + 2)$ or $y = 12x + 17$.

**50.** The slope of $y = 1 + 2e^x - 3x$ is given by $m = y' = 2e^x - 3$.

The slope of $3x - y = 5 \Leftrightarrow y = 3x - 5$ is 3.

$m = 3 \Rightarrow 2e^x - 3 = 3 \Rightarrow e^x = 3 \Rightarrow x = \ln 3$.

This occurs at the point $(\ln 3, 7 - 3\ln 3) \approx (1.1, 3.7)$.

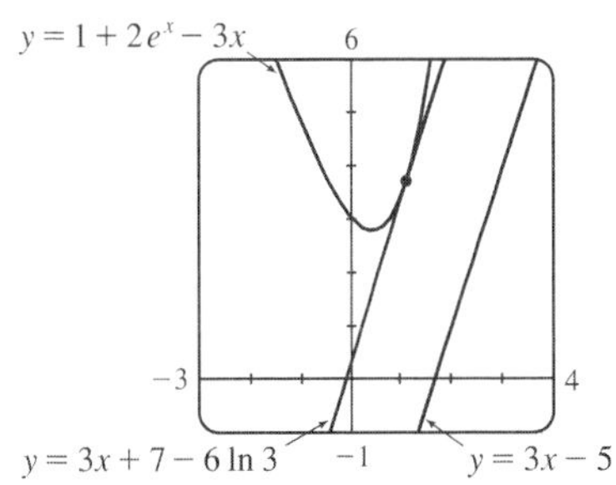

**51.** The slope of $y = x^2 - 5x + 4$ is given by $m = y' = 2x - 5$. The slope of $x - 3y = 5 \quad \Leftrightarrow \quad y = \frac{1}{3}x - \frac{5}{3}$ is $\frac{1}{3}$, so the desired normal line must have slope $\frac{1}{3}$, and hence, the tangent line to the parabola must have slope $-3$. This occurs if $2x - 5 = -3 \quad \Rightarrow \quad 2x = 2 \quad \Rightarrow \quad x = 1$. When $x = 1$, $y = 1^2 - 5(1) + 4 = 0$, and an equation of the normal line is $y - 0 = \frac{1}{3}(x - 1)$ or $y = \frac{1}{3}x - \frac{1}{3}$.

**52.** $y = f(x) = x - x^2 \quad \Rightarrow \quad f'(x) = 1 - 2x$. So $f'(1) = -1$, and the slope of the normal line is the negative reciprocal of that of the tangent line, that is, $-1/(-1) = 1$. So the equation of the normal line at $(1, 0)$ is $y - 0 = 1(x - 1)$ $\Leftrightarrow \quad y = x - 1$. Substituting this into the equation of the parabola, we obtain $x - 1 = x - x^2 \quad \Leftrightarrow \quad x = \pm 1$. The solution $x = -1$ is the one we require. Substituting $x = -1$ into the equation of the parabola to find the $y$-coordinate, we have $y = -2$. So the point of intersection is $(-1, -2)$, as shown in the sketch.

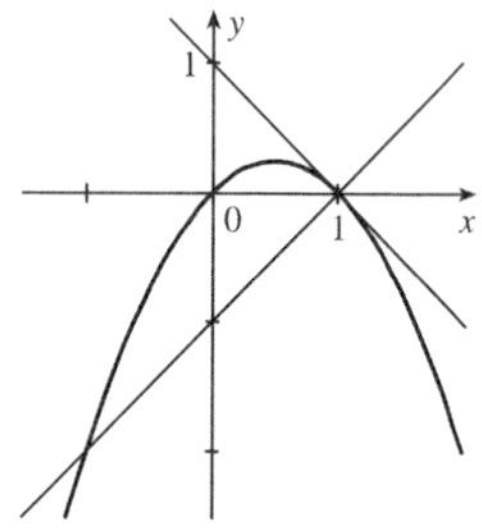


**53.**

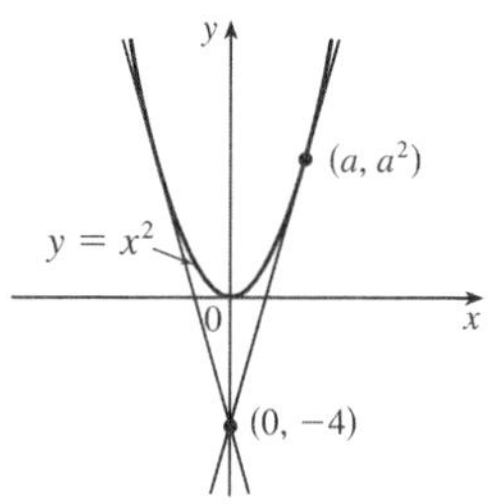


Let $(a, a^2)$ be a point on the parabola at which the tangent line passes through the point $(0, -4)$. The tangent line has slope $2a$ and equation $y - (-4) = 2a(x - 0)$ $\Leftrightarrow \quad y = 2ax - 4$. Since $(a, a^2)$ also lies on the line, $a^2 = 2a(a) - 4$, or $a^2 = 4$. So $a = \pm 2$ and the points are $(2, 4)$ and $(-2, 4)$.

**54.** (a) If $y = x^2 + x$, then $y' = 2x + 1$. If the point at which a tangent meets the parabola is $(a, a^2 + a)$, then the slope of the tangent is $2a + 1$. But since it passes through $(2, -3)$, the slope must also be $\dfrac{\Delta y}{\Delta x} = \dfrac{a^2 + a + 3}{a - 2}$.

Therefore, $2a + 1 = \dfrac{a^2 + a + 3}{a - 2}$. Solving this equation for $a$ we get $a^2 + a + 3 = 2a^2 - 3a - 2 \quad \Leftrightarrow$ $a^2 - 4a - 5 = (a - 5)(a + 1) = 0 \quad \Leftrightarrow \quad a = 5$ or $-1$. If $a = -1$, the point is $(-1, 0)$ and the slope is $-1$, so the equation is $y - 0 = (-1)(x + 1)$ or $y = -x - 1$. If $a = 5$, the point is $(5, 30)$ and the slope is $11$, so the equation is $y - 30 = 11(x - 5)$ or $y = 11x - 25$.

(b) As in part (a), but using the point $(2, 7)$, we get the equation $2a + 1 = \dfrac{a^2 + a - 7}{a - 2}$ $\Rightarrow \quad 2a^2 - 3a - 2 = a^2 + a - 7 \quad \Leftrightarrow \quad a^2 - 4a + 5 = 0$. The last equation has no real solution (discriminant $= -16 < 0$), so there is no line through the point $(2, 7)$ that is tangent to the parabola. The diagram shows that the point $(2, 7)$ is "inside" the parabola, but tangent lines to the parabola do not pass through points inside the parabola.

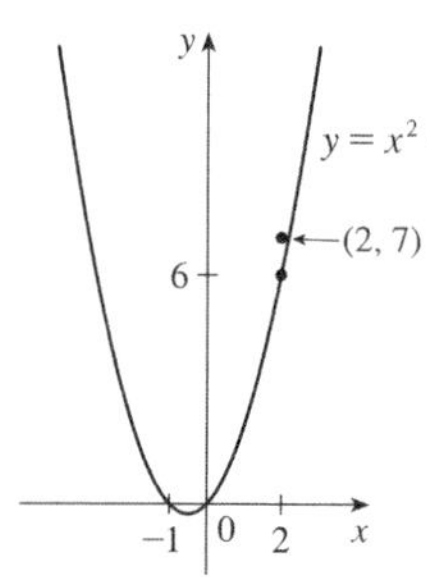


**55.** $f'(x) = \lim\limits_{h \to 0} \dfrac{f(x + h) - f(x)}{h} = \lim\limits_{h \to 0} \dfrac{\dfrac{1}{x + h} - \dfrac{1}{x}}{h} = \lim\limits_{h \to 0} \dfrac{x - (x + h)}{hx(x + h)} = \lim\limits_{h \to 0} \dfrac{-h}{hx(x + h)} = \lim\limits_{h \to 0} \dfrac{-1}{x(x + h)} = -\dfrac{1}{x^2}$

**56.** (a) $f(x) = x^n \quad \Rightarrow \quad f'(x) = nx^{n-1} \quad \Rightarrow \quad f''(x) = n(n-1)x^{n-2} \quad \Rightarrow \quad \cdots \quad \Rightarrow$ $f^{(n)}(x) = n(n-1)(n-2) \cdots 2 \cdot 1x^{n-n} = n!$

(b) $f(x) = x^{-1} \quad \Rightarrow \quad f'(x) = (-1)x^{-2} \quad \Rightarrow \quad f''(x) = (-1)(-2)x^{-3} \quad \Rightarrow \quad \cdots \quad \Rightarrow$ $f^{(n)}(x) = (-1)(-2)(-3) \cdots (-n)x^{-(n+1)} = (-1)^n n! x^{-(n+1)}$ or $\dfrac{(-1)^n \, n!}{x^{n+1}}$

**57.** Let $P(x) = ax^2 + bx + c$. Then $P'(x) = 2ax + b$ and $P''(x) = 2a$. $P''(2) = 2 \Rightarrow 2a = 2 \Rightarrow a = 1$.
$P'(2) = 3 \Rightarrow 2(1)(2) + b = 3 \Rightarrow 4 + b = 3 \Rightarrow b = -1$.
$P(2) = 5 \Rightarrow 1(2)^2 + (-1)(2) + c = 5 \Rightarrow 2 + c = 5 \Rightarrow c = 3$. So $P(x) = x^2 - x + 3$.

**58.** $y = Ax^2 + Bx + C \Rightarrow y' = 2Ax + B \Rightarrow y'' = 2A$. We substitute these expressions into the equation $y'' + y' - 2y = x^2$ to get

$$(2A) + (2Ax + B) - 2(Ax^2 + Bx + C) = x^2$$
$$2A + 2Ax + B - 2Ax^2 - 2Bx - 2C = x^2$$
$$(-2A)x^2 + (2A - 2B)x + (2A + B - 2C) = (1)x^2 + (0)x + (0)$$

The coefficients of $x^2$ on each side must be equal, so $-2A = 1 \Rightarrow A = -\frac{1}{2}$. Similarly, $2A - 2B = 0 \Rightarrow A = B = -\frac{1}{2}$ and $2A + B - 2C = 0 \Rightarrow -1 - \frac{1}{2} - 2C = 0 \Rightarrow C = -\frac{3}{4}$.

**59.** (a) At this stage, we would guess that an antiderivative of $x^2$ must have $x^3$ in it. Differentiating $x^3$ gives us $3x^2$, so we know that we must divide $x^3$ by 3. That gives us $F(x) = \frac{1}{3}x^3$. Checking, we have $F'(x) = \frac{1}{3}(3x^2) = x^2 = f(x)$. Because we can add an arbitrary constant $C$ to $F$ without changing its derivative, we have an infinite number of antiderivatives of the form $F(x) = \frac{1}{3}x^3 + C$.

(b) As in part (a), antiderivatives of $f(x) = x^3$ and $f(x) = x^4$ are $F(x) = \frac{1}{4}x^4 + C$ and $F(x) = \frac{1}{5}x^5 + C$.

(c) Similarly, an antiderivative for $f(x) = x^n$ is $F(x) = \dfrac{1}{n+1}x^{n+1} + C$, since then

$F'(x) = \dfrac{1}{n+1}[(n+1)x^n] = x^n = f(x)$ for $n \neq -1$.

**60.** (a) $f(x) = \sqrt{x} = x^{1/2} \Rightarrow F(x) = \frac{1}{(1/2)+1}x^{(1/2)+1} + C = \frac{2}{3}x^{3/2} + C$

(b) $f(x) = e^x + 8x^3 \Rightarrow F(x) = e^x + 8 \cdot \frac{1}{3+1}x^{3+1} + C = e^x + 2x^4 + C$

**61.** $y = f(x) = ax^2 \Rightarrow f'(x) = 2ax$. So the slope of the tangent to the parabola at $x = 2$ is $m = 2a(2) = 4a$. The slope of the given line, $2x + y = b \Leftrightarrow y = -2x + b$, is seen to be $-2$, so we must have $4a = -2 \Leftrightarrow a = -\frac{1}{2}$. So when $x = 2$, the point in question has $y$-coordinate $-\frac{1}{2} \cdot 2^2 = -2$. Now we simply require that the given line, whose equation is $2x + y = b$, pass through the point $(2, -2)$: $2(2) + (-2) = b \Leftrightarrow b = 2$. So we must have $a = -\frac{1}{2}$ and $b = 2$.

**62.** $y = ax^2 + bx + c \Rightarrow y'(x) = 2ax + b$. The parabola has slope 4 at $x = 1$ and slope $-8$ at $x = -1$, so $y'(1) = 4 \Rightarrow 2a + b = 4$ **(1)** and $y'(-1) = -8 \Rightarrow -2a + b = -8$ **(2)**. Adding **(1)** and **(2)** gives us $2b = -4 \Leftrightarrow b = -2$. From **(1)**, $2a - 2 = 4 \Leftrightarrow a = 3$. Thus, the equation of the parabola is $y = 3x^2 - 2x + c$. Since it passes through the point $(2, 15)$, we have $15 = 3(2)^2 - 2(2) + c \Rightarrow c = 7$, so the equation is $y = 3x^2 - 2x + 7$.

**63.** $y = f(x) = ax^3 + bx^2 + cx + d \Rightarrow f'(x) = 3ax^2 + 2bx + c$. The point $(-2, 6)$ is on $f$, so $f(-2) = 6 \Rightarrow -8a + 4b - 2c + d = 6$ **(1)**. The point $(2, 0)$ is on $f$, so $f(2) = 0 \Rightarrow 8a + 4b + 2c + d = 0$ **(2)**. Since there are horizontal tangents at $(-2, 6)$ and $(2, 0)$, $f'(\pm 2) = 0$. $f'(-2) = 0 \Rightarrow 12a - 4b + c = 0$ **(3)** and $f'(2) = 0 \Rightarrow 12a + 4b + c = 0$ **(4)**. Subtracting equation **(3)** from **(4)** gives $8b = 0 \Rightarrow b = 0$. Adding **(1)** and **(2)** gives $8b + 2d = 6$, so $d = 3$ since $b = 0$. From **(3)** we have $c = -12a$, so **(2)** becomes $8a + 4(0) + 2(-12a) + 3 = 0 \Rightarrow 3 = 16a \Rightarrow a = \frac{3}{16}$. Now $c = -12a = -12\left(\frac{3}{16}\right) = -\frac{9}{4}$ and the desired cubic function is $y = \frac{3}{16}x^3 - \frac{9}{4}x + 3$.

**64.** (a) $xy = c \Rightarrow y = \dfrac{c}{x}$. Let $P = \left(a, \dfrac{c}{a}\right)$. The slope of the tangent line at $x = a$ is $y'(a) = -\dfrac{c}{a^2}$. Its equation is $y - \dfrac{c}{a} = -\dfrac{c}{a^2}(x - a)$ or $y = -\dfrac{c}{a^2}x + \dfrac{2c}{a}$, so its $y$-intercept is $\dfrac{2c}{a}$. Setting $y = 0$ gives $x = 2a$, so the $x$-intercept is $2a$. The midpoint of the line segment joining $\left(0, \dfrac{2c}{a}\right)$ and $(2a, 0)$ is $\left(a, \dfrac{c}{a}\right) = P$.

(b) We know the $x$- and $y$-intercepts of the tangent line from part (a), so the area of the triangle bounded by the axes and the tangent is $\frac{1}{2}(\text{base})(\text{height}) = \frac{1}{2}xy = \frac{1}{2}(2a)(2c/a) = 2c$, a constant.

**65.** *Solution 1:* Let $f(x) = x^{1000}$. Then, by the definition of a derivative, $f'(1) = \lim_{x\to 1} \dfrac{f(x) - f(1)}{x - 1} = \lim_{x\to 1} \dfrac{x^{1000} - 1}{x - 1}$. But this is just the limit we want to find, and we know (from the Power Rule) that $f'(x) = 1000x^{999}$, so $f'(1) = 1000(1)^{999} = 1000$. So $\lim_{x\to 1} \dfrac{x^{1000} - 1}{x - 1} = 1000$.

*Solution 2:* Note that $(x^{1000} - 1) = (x - 1)(x^{999} + x^{998} + x^{997} + \cdots + x^2 + x + 1)$. So

$$\lim_{x\to 1} \frac{x^{1000} - 1}{x - 1} = \lim_{x\to 1} \frac{(x - 1)(x^{999} + x^{998} + x^{997} + \cdots + x^2 + x + 1)}{x - 1} = \lim_{x\to 1} (x^{999} + x^{998} + x^{997} + \cdots + x^2 + x + 1)$$
$$= \underbrace{1 + 1 + 1 + \cdots + 1 + 1 + 1}_{1000 \text{ ones}} = 1000, \text{ as above.}$$

**66.**

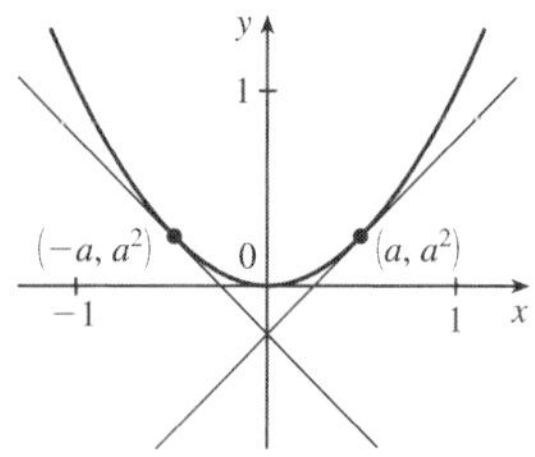


In order for the two tangents to intersect on the $y$-axis, the points of tangency must be at equal distances from the $y$-axis, since the parabola $y = x^2$ is symmetric about the $y$-axis. Say the points of tangency are $(a, a^2)$ and $(-a, a^2)$, for some $a > 0$. Then since the derivative of $y = x^2$ is $dy/dx = 2x$, the left-hand tangent has slope $-2a$ and equation $y - a^2 = -2a(x + a)$, or $y = -2ax - a^2$, and similarly the right-hand tangent line has equation $y - a^2 = 2a(x - a)$, or $y = 2ax - a^2$. So the two lines intersect at $(0, -a^2)$. Now if the lines are perpendicular, then the product of their slopes is $-1$, so $(-2a)(2a) = -1 \Leftrightarrow a^2 = \frac{1}{4} \Leftrightarrow a = \frac{1}{2}$. So the lines intersect at $\left(0, -\frac{1}{4}\right)$.

**67.** $y = x^2 \Rightarrow y' = 2x$, so the slope of a tangent line at the point $(a, a^2)$ is $y' = 2a$ and the slope of a normal line is $-1/(2a)$, for $a \neq 0$. The slope of the normal line through the points $(a, a^2)$ and $(0, c)$ is $\dfrac{a^2 - c}{a - 0}$, so $\dfrac{a^2 - c}{a} = -\dfrac{1}{2a} \Rightarrow a^2 - c = -\frac{1}{2} \Rightarrow a^2 = c - \frac{1}{2}$. The last equation has two solutions if $c > \frac{1}{2}$, one solution if $c = \frac{1}{2}$, and no solution if $c < \frac{1}{2}$. Since the $y$-axis is normal to $y = x^2$ regardless of the value of $c$ (this is the case for $a = 0$), we have three normal lines if $c > \frac{1}{2}$ and one normal line if $c \leq \frac{1}{2}$.

**68.**

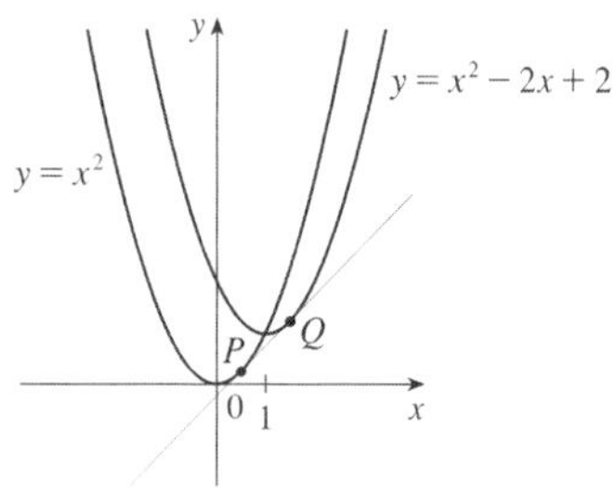


From the sketch, it appears that there may be a line that is tangent to both curves. The slope of the line through the points $P(a, a^2)$ and $Q(b, b^2 - 2b + 2)$ is $\dfrac{b^2 - 2b + 2 - a^2}{b - a}$. The slope of the tangent line at $P$ is $2a$ $[y' = 2x]$ and at $Q$ is $2b - 2$ $[y' = 2x - 2]$. All three slopes are equal, so $2a = 2b - 2 \Leftrightarrow a = b - 1$.

Also, $2b - 2 = \dfrac{b^2 - 2b + 2 - a^2}{b - a} \Rightarrow 2b - 2 = \dfrac{b^2 - 2b + 2 - (b - 1)^2}{b - (b - 1)} \Rightarrow 2b - 2 = b^2 - 2b + 2 - b^2 + 2b - 1 \Rightarrow$ $2b = 3 \Rightarrow b = \frac{3}{2}$ and $a = \frac{3}{2} - 1 = \frac{1}{2}$. Thus, an equation of the tangent line at $P$ is $y - \left(\frac{1}{2}\right)^2 = 2\left(\frac{1}{2}\right)\left(x - \frac{1}{2}\right)$ or $y = x - \frac{1}{4}$.

## APPLIED PROJECT Building a Better Roller Coaster

**1.** (a) $f(x) = ax^2 + bx + c \quad \Rightarrow \quad f'(x) = 2ax + b.$

The origin is at $P$: $f(0) = 0 \quad \Rightarrow \quad c = 0$

The slope of the ascent is $0.8$: $f'(0) = 0.8 \quad \Rightarrow \quad b = 0.8$

The slope of the drop is $-1.6$: $f'(100) = -1.6 \quad \Rightarrow \quad 200a + b = -1.6$

(b) $b = 0.8$, so $200a + b = -1.6 \quad \Rightarrow \quad 200a + 0.8 = -1.6 \quad \Rightarrow \quad 200a = -2.4 \quad \Rightarrow \quad a = -\dfrac{2.4}{200} = -0.012.$

Thus, $f(x) = -0.012x^2 + 0.8x$.

(c) Since $L_1$ passes through the origin with slope $0.8$, it has equation $y = 0.8x$.

The horizontal distance between $P$ and $Q$ is 100, so the $y$-coordinate at $Q$ is $f(100) = -0.012(100)^2 + 0.8(100) = -40$. Since $L_2$ passes through the point $(100, -40)$ and has slope $-1.6$, it has equation $y + 40 = -1.6(x - 100)$ or $y = -1.6x + 120$.

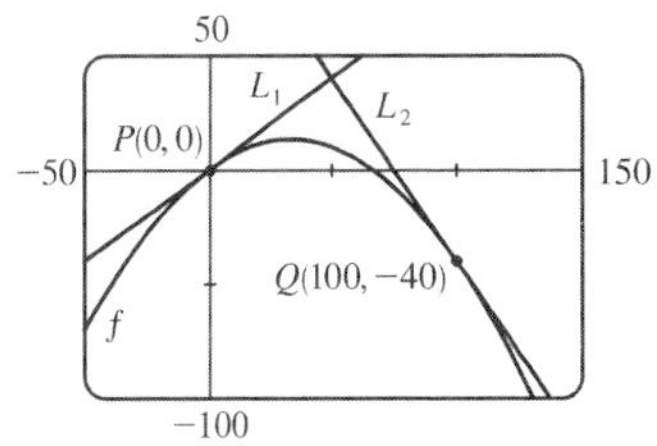


(d) The difference in elevation between $P(0, 0)$ and $Q(100, -40)$ is $0 - (-40) = 40$ feet.

**2.** (a)

| Interval | Function | First Derivative | Second Derivative |
|---|---|---|---|
| $(-\infty, 0)$ | $L_1(x) = 0.8x$ | $L_1'(x) = 0.8$ | $L_1''(x) = 0$ |
| $[0, 10)$ | $g(x) = kx^3 + lx^2 + mx + n$ | $g'(x) = 3kx^2 + 2lx + m$ | $g''(x) = 6kx + 2l$ |
| $[10, 90]$ | $q(x) = ax^2 + bx + c$ | $q'(x) = 2ax + b$ | $q''(x) = 2a$ |
| $(90, 100]$ | $h(x) = px^3 + qx^2 + rx + s$ | $h'(x) = 3px^2 + 2qx + r$ | $h''(x) = 6px + 2q$ |
| $(100, \infty)$ | $L_2(x) = -1.6x + 120$ | $L_2'(x) = -1.6$ | $L_2''(x) = 0$ |

There are 4 values of $x$ (0, 10, 90, and 100) for which we must make sure the function values are equal, the first derivative values are equal, and the second derivative values are equal. The third column in the following table contains the value of each side of the condition—these are found after solving the system in part (b).

| At $x =$ | Condition | Value | Resulting Equation |
|---|---|---|---|
| 0 | $g(0) = L_1(0)$ | $0$ | $n = 0$ |
| | $g'(0) = L_1'(0)$ | $\frac{4}{5}$ | $m = 0.8$ |
| | $g''(0) = L_1''(0)$ | $0$ | $2l = 0$ |
| 10 | $g(10) = q(10)$ | $\frac{68}{9}$ | $1000k + 100l + 10m + n = 100a + 10b + c$ |
| | $g'(10) = q'(10)$ | $\frac{2}{3}$ | $300k + 20l + m = 20a + b$ |
| | $g''(10) = q''(10)$ | $-\frac{2}{75}$ | $60k + 2l = 2a$ |
| 90 | $h(90) = q(90)$ | $-\frac{220}{9}$ | $729{,}000p + 8100q + 90r + s = 8100a + 90b + c$ |
| | $h'(90) = q'(90)$ | $-\frac{22}{15}$ | $24{,}300p + 180q + r = 180a + b$ |
| | $h''(90) = q''(90)$ | $-\frac{2}{75}$ | $540p + 2q = 2a$ |
| 100 | $h(100) = L_2(100)$ | $-40$ | $1{,}000{,}000p + 10{,}000q + 100r + s = -40$ |
| | $h'(100) = L_2'(100)$ | $-\frac{8}{5}$ | $30{,}000p + 200q + r = -1.6$ |
| | $h''(100) = L_2''(100)$ | $0$ | $600p + 2q = 0$ |

(b) We can arrange our work in a $12 \times 12$ matrix as follows.

| $a$ | $b$ | $c$ | $k$ | $l$ | $m$ | $n$ | $p$ | $q$ | $r$ | $s$ | constant |
|---|---|---|---|---|---|---|---|---|---|---|---|
| 0 | 0 | 0 | 0 | 0 | 0 | 1 | 0 | 0 | 0 | 0 | 0 |
| 0 | 0 | 0 | 0 | 0 | 1 | 0 | 0 | 0 | 0 | 0 | 0.8 |
| 0 | 0 | 0 | 0 | 2 | 0 | 0 | 0 | 0 | 0 | 0 | 0 |
| −100 | −10 | −1 | 1000 | 100 | 10 | 1 | 0 | 0 | 0 | 0 | 0 |
| −20 | −1 | 0 | 300 | 20 | 1 | 0 | 0 | 0 | 0 | 0 | 0 |
| −2 | 0 | 0 | 60 | 2 | 0 | 0 | 0 | 0 | 0 | 0 | 0 |
| −8100 | −90 | −1 | 0 | 0 | 0 | 0 | 729,000 | 8100 | 90 | 1 | 0 |
| −180 | −1 | 0 | 0 | 0 | 0 | 0 | 24,300 | 180 | 1 | 0 | 0 |
| −2 | 0 | 0 | 0 | 0 | 0 | 0 | 540 | 2 | 0 | 0 | 0 |
| 0 | 0 | 0 | 0 | 0 | 0 | 0 | 1,000,000 | 10,000 | 100 | 1 | −40 |
| 0 | 0 | 0 | 0 | 0 | 0 | 0 | 30,000 | 200 | 1 | 0 | −1.6 |
| 0 | 0 | 0 | 0 | 0 | 0 | 0 | 600 | 2 | 0 | 0 | 0 |

Solving the system gives us the formulas for $q$, $g$, and $h$.

$$\left.\begin{aligned} a &= -0.01\overline{3} = -\tfrac{1}{75} \\ b &= 0.9\overline{3} = \tfrac{14}{15} \\ c &= -0.\overline{4} = -\tfrac{4}{9} \end{aligned}\right\} q(x) = -\tfrac{1}{75}x^2 + \tfrac{14}{15}x - \tfrac{4}{9}$$

$$\left.\begin{aligned} k &= -0.000\overline{4} = -\tfrac{1}{2250} \\ l &= 0 \\ m &= 0.8 = \tfrac{4}{5} \\ n &= 0 \end{aligned}\right\} g(x) = -\tfrac{1}{2250}x^3 + \tfrac{4}{5}x$$

$$\left.\begin{aligned} p &= 0.000\overline{4} = \tfrac{1}{2250} \\ q &= -0.1\overline{3} = -\tfrac{2}{15} \\ r &= 11.7\overline{3} = \tfrac{176}{15} \\ s &= -324.\overline{4} = -\tfrac{2920}{9} \end{aligned}\right\} h(x) = \tfrac{1}{2250}x^3 - \tfrac{2}{15}x^2 + \tfrac{176}{15}x - \tfrac{2920}{9}$$

(c) Graph of $L_1$, $q$, $g$, $h$, and $L_2$:

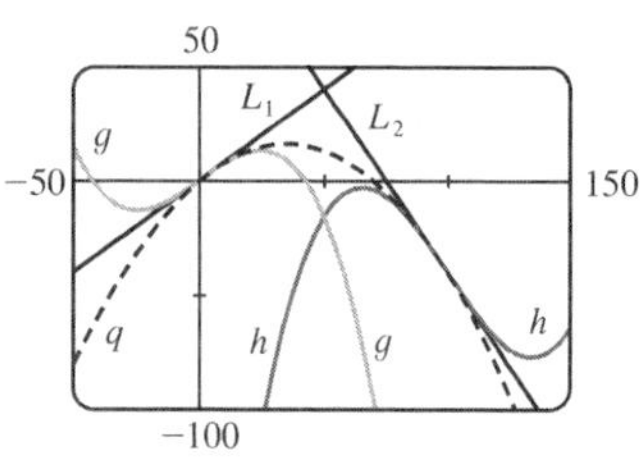


The graph of the five functions as a piecewise-defined function:

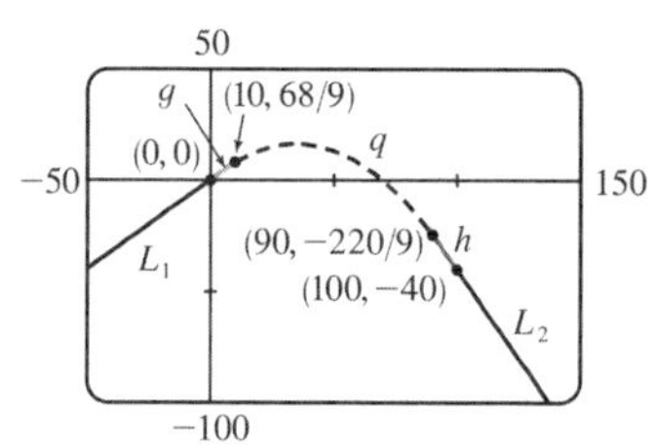


This is the piecewise-defined function assignment on a TI-83 Plus calculator, where $Y_2 = L_1$, $Y_6 = g$, $Y_5 = q$, $Y_7 = h$, and $Y_3 = L_2$.

```
Plot1 Plot2 Plot3
\Y8=Y2*(X<0)+Y6*
(X≥0 and X<10)+Y
5*(X≥10 and X≤90
)+Y7*(X>90 and X
≤100)+Y3*(X>100)

\Y9=
```

A comparison of the graphs in part 1(c) and part 2(c):

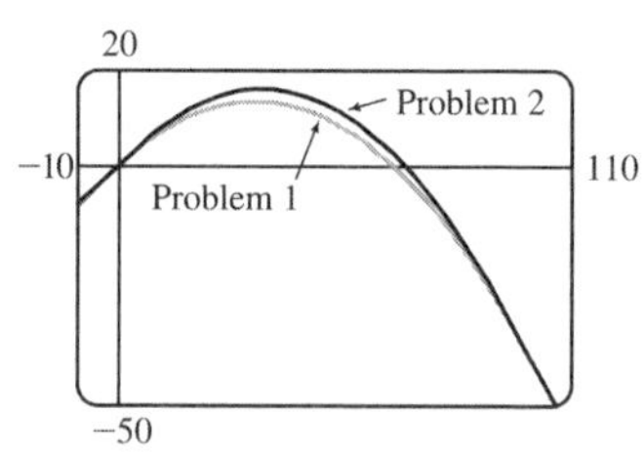

## 3.2 The Product and Quotient Rules

**1.** Product Rule: $y = (x^2+1)(x^3+1) \;\Rightarrow$

$$y' = (x^2+1)(3x^2) + (x^3+1)(2x) = 3x^4 + 3x^2 + 2x^4 + 2x = 5x^4 + 3x^2 + 2x.$$

Multiplying first: $y = (x^2+1)(x^3+1) = x^5 + x^3 + x^2 + 1 \;\Rightarrow\; y' = 5x^4 + 3x^2 + 2x$ (equivalent).

**2.** Quotient Rule: $F(x) = \dfrac{x - 3x\sqrt{x}}{\sqrt{x}} = \dfrac{x - 3x^{3/2}}{x^{1/2}} \;\Rightarrow$

$$F'(x) = \frac{x^{1/2}\left(1 - \frac{9}{2}x^{1/2}\right) - \left(x - 3x^{3/2}\right)\left(\frac{1}{2}x^{-1/2}\right)}{(x^{1/2})^2} = \frac{x^{1/2} - \frac{9}{2}x - \frac{1}{2}x^{1/2} + \frac{3}{2}x}{x} = \frac{\frac{1}{2}x^{1/2} - 3x}{x} = \tfrac{1}{2}x^{-1/2} - 3$$

Simplifying first: $F(x) = \dfrac{x - 3x\sqrt{x}}{\sqrt{x}} = \sqrt{x} - 3x = x^{1/2} - 3x \;\Rightarrow\; F'(x) = \frac{1}{2}x^{-1/2} - 3$ (equivalent).

For this problem, simplifying first seems to be the better method.

**3.** By the Product Rule, $f(x) = x^2e^x \;\Rightarrow\; f'(x) = x^2\dfrac{d}{dx}(e^x) + e^x\dfrac{d}{dx}(x^2) = x^2e^x + e^x(2x) = xe^x(x+2)$.

**4.** By the Product Rule, $g(x) = \sqrt{x}\,e^x = x^{1/2}e^x \;\Rightarrow\; g'(x) = x^{1/2}(e^x) + e^x\left(\frac{1}{2}x^{-1/2}\right) = \frac{1}{2}x^{-1/2}e^x(2x+1)$.

**5.** By the Quotient Rule, $y = \dfrac{e^x}{x^2} \;\Rightarrow$

$$y' = \frac{x^2\dfrac{d}{dx}(e^x) - e^x\dfrac{d}{dx}(x^2)}{(x^2)^2} = \frac{x^2(e^x) - e^x(2x)}{x^4} = \frac{xe^x(x-2)}{x^4} = \frac{e^x(x-2)}{x^3}.$$

**6.** By the Quotient Rule, $y = \dfrac{e^x}{1+x} \;\Rightarrow\; y' = \dfrac{(1+x)e^x - e^x(1)}{(1+x)^2} = \dfrac{e^x + xe^x - e^x}{(x+1)^2} = \dfrac{xe^x}{(x+1)^2}$.

The notations $\overset{\text{PR}}{\Rightarrow}$ and $\overset{\text{QR}}{\Rightarrow}$ indicate the use of the Product and Quotient Rules, respectively.

**7.** $g(x) = \dfrac{3x-1}{2x+1} \;\overset{\text{QR}}{\Rightarrow}\; g'(x) = \dfrac{(2x+1)(3) - (3x-1)(2)}{(2x+1)^2} = \dfrac{6x+3-6x+2}{(2x+1)^2} = \dfrac{5}{(2x+1)^2}$

**8.** $f(t) = \dfrac{2t}{4+t^2} \;\overset{\text{QR}}{\Rightarrow}\; f'(t) = \dfrac{(4+t^2)(2) - (2t)(2t)}{(4+t^2)^2} = \dfrac{8+2t^2-4t^2}{(4+t^2)^2} = \dfrac{8-2t^2}{(4+t^2)^2}$

**9.** $F(y) = \left(\dfrac{1}{y^2} - \dfrac{3}{y^4}\right)(y+5y^3) = (y^{-2} - 3y^{-4})(y+5y^3) \;\overset{\text{PR}}{\Rightarrow}$

$$\begin{aligned} F'(y) &= (y^{-2} - 3y^{-4})(1+15y^2) + (y+5y^3)(-2y^{-3} + 12y^{-5}) \\ &= (y^{-2} + 15 - 3y^{-4} - 45y^{-2}) + (-2y^{-2} + 12y^{-4} - 10 + 60y^{-2}) \\ &= 5 + 14y^{-2} + 9y^{-4} \text{ or } 5 + 14/y^2 + 9/y^4 \end{aligned}$$

**10.** $R(t) = (t+e^t)(3-\sqrt{t}) \;\overset{\text{PR}}{\Rightarrow}$

$$\begin{aligned} R'(t) &= (t+e^t)\left(-\tfrac{1}{2}t^{-1/2}\right) + (3-\sqrt{t})(1+e^t) \\ &= \left(-\tfrac{1}{2}t^{1/2} - \tfrac{1}{2}t^{-1/2}e^t\right) + (3 + 3e^t - \sqrt{t} - \sqrt{t}\,e^t) = 3 + 3e^t - \tfrac{3}{2}\sqrt{t} - \sqrt{t}\,e^t - e^t/(2\sqrt{t}) \end{aligned}$$

**11.** $y = \dfrac{t^2}{3t^2 - 2t + 1} \;\overset{\text{QR}}{\Rightarrow}$

$$y' = \frac{(3t^2-2t+1)(2t) - t^2(6t-2)}{(3t^2-2t+1)^2} = \frac{2t[3t^2-2t+1-t(3t-1)]}{(3t^2-2t+1)^2} = \frac{2t(3t^2-2t+1-3t^2+t)}{(3t^2-2t+1)^2} = \frac{2t(1-t)}{(3t^2-2t+1)^2}$$

**12.** $y = \dfrac{t^3+t}{t^4-2} \overset{\text{QR}}{\Rightarrow} y' = \dfrac{(t^4-2)(3t^2+1)-(t^3+t)(4t^3)}{(t^4-2)^2} = \dfrac{(3t^6+t^4-6t^2-2)-(4t^6+4t^4)}{(t^4-2)^2}$

$$= \frac{-t^6-3t^4-6t^2-2}{(t^4-2)^2} = -\frac{t^6+3t^4+6t^2+2}{(t^4-2)^2}$$

**13.** $y = (r^2-2r)e^r \overset{\text{PR}}{\Rightarrow} y' = (r^2-2r)(e^r)+e^r(2r-2) = e^r(r^2-2r+2r-2) = e^r(r^2-2)$

**14.** $y = \dfrac{1}{s+ke^s} \overset{\text{QR}}{\Rightarrow} y' = \dfrac{(s+ke^s)(0)-(1)(1+ke^s)}{(s+ke^s)^2} = -\dfrac{1+ke^s}{(s+ke^s)^2}$

**15.** $y = \dfrac{v^3-2v\sqrt{v}}{v} = v^2-2\sqrt{v} = v^2-2v^{1/2} \Rightarrow y' = 2v-2\left(\frac{1}{2}\right)v^{-1/2} = 2v-v^{-1/2}$.

We can change the form of the answer as follows: $2v-v^{-1/2} = 2v-\dfrac{1}{\sqrt{v}} = \dfrac{2v\sqrt{v}-1}{\sqrt{v}} = \dfrac{2v^{3/2}-1}{\sqrt{v}}$

**16.** $z = w^{3/2}(w+ce^w) = w^{5/2}+cw^{3/2}e^w \Rightarrow z' = \frac{5}{2}w^{3/2}+c\left(w^{3/2}\cdot e^w+e^w\cdot\frac{3}{2}w^{1/2}\right) = \frac{5}{2}w^{3/2}+\frac{1}{2}cw^{1/2}e^w(2w+3)$

**17.** $f(x) = \dfrac{A}{B+Ce^x} \overset{\text{QR}}{\Rightarrow} f'(x) = \dfrac{(B+Ce^x)\cdot 0-A(Ce^x)}{(B+Ce^x)^2} = -\dfrac{ACe^x}{(B+Ce^x)^2}$

**18.** $f(x) = \dfrac{1-xe^x}{x+e^x} \overset{\text{QR}}{\Rightarrow} f'(x) = \dfrac{(x+e^x)(-xe^x)'-(1-xe^x)(1+e^x)}{(x+e^x)^2}$

$$\overset{\text{PR}}{\Rightarrow} f'(x) = \frac{(x+e^x)[-(xe^x+e^x\cdot 1)]-(1+e^x-xe^x-xe^{2x})}{(x+e^x)^2}$$

$$= \frac{-x^2e^x-xe^x-xe^{2x}-e^{2x}-1-e^x+xe^x+xe^{2x}}{(x+e^x)^2} = \frac{-x^2e^x-e^{2x}-e^x-1}{(x+e^x)^2}$$

**19.** $f(x) = \dfrac{x}{x+c/x} \Rightarrow f'(x) = \dfrac{(x+c/x)(1)-x(1-c/x^2)}{\left(x+\frac{c}{x}\right)^2} = \dfrac{x+c/x-x+c/x}{\left(\frac{x^2+c}{x}\right)^2} = \dfrac{2c/x}{\frac{(x^2+c)^2}{x^2}}\cdot\dfrac{x^2}{x^2} = \dfrac{2cx}{(x^2+c)^2}$

**20.** $f(x) = \dfrac{ax+b}{cx+d} \Rightarrow f'(x) = \dfrac{(cx+d)(a)-(ax+b)(c)}{(cx+d)^2} = \dfrac{acx+ad-acx-bc}{(cx+d)^2} = \dfrac{ad-bc}{(cx+d)^2}$

**21.** $y = 2xe^x \Rightarrow y' = 2(x\cdot e^x+e^x\cdot 1) = 2e^x((x+1)$. At $(0,0)$, $y' = 2e^0(0+1) = 2\cdot 1\cdot 1 = 2$, and an equation of the tangent line is $y-0 = 2(x-0)$, or $y = 2x$. The slope of the normal line is $-\frac{1}{2}$, so an equation of the normal line is $y-0 = -\frac{1}{2}(x-0)$, or $y = -\frac{1}{2}x$.

**22.** $y = \dfrac{\sqrt{x}}{x+1} \Rightarrow y' = \dfrac{(x+1)\left(\frac{1}{2\sqrt{x}}\right)-\sqrt{x}\,(1)}{(x+1)^2} = \dfrac{(x+1)-(2x)}{2\sqrt{x}\,(x+1)^2} = \dfrac{1-x}{2\sqrt{x}\,(x+1)^2}$. At $(4,0.4)$, $y' = \frac{-3}{100} = -0.03$, and an equation of the tangent line is $y-0.4 = -0.03(x-4)$, or $y = -0.03x+0.52$. The slope of the normal line is $\frac{100}{3}$, so an equation of the normal line is $y-0.4 = \frac{100}{3}(x-4) \Leftrightarrow y = \frac{100}{3}x-\frac{400}{3}+\frac{2}{5} \Leftrightarrow y = \frac{100}{3}x-\frac{1994}{15}$.

**23.** (a) $y = f(x) = \dfrac{1}{1+x^2} \Rightarrow$

$f'(x) = \dfrac{(1+x^2)(0)-1(2x)}{(1+x^2)^2} = \dfrac{-2x}{(1+x^2)^2}$. So the slope of the tangent line at the point $\left(-1,\frac{1}{2}\right)$ is $f'(-1) = \dfrac{2}{2^2} = \frac{1}{2}$ and its equation is $y-\frac{1}{2} = \frac{1}{2}(x+1)$ or $y = \frac{1}{2}x+1$.

(b)

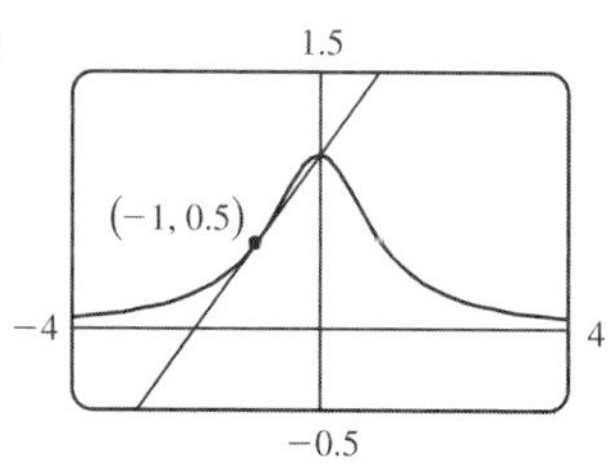

**24.** (a) $y = f(x) = \dfrac{x}{1+x^2} \Rightarrow$

$f'(x) = \dfrac{(1+x^2)1 - x(2x)}{(1+x^2)^2} = \dfrac{1-x^2}{(1+x^2)^2}$. So the slope of the tangent line at the point $(3, 0.3)$ is $f'(3) = \frac{-8}{100}$ and its equation is $y - 0.3 = -0.08(x-3)$ or $y = -0.08x + 0.54$.

(b)

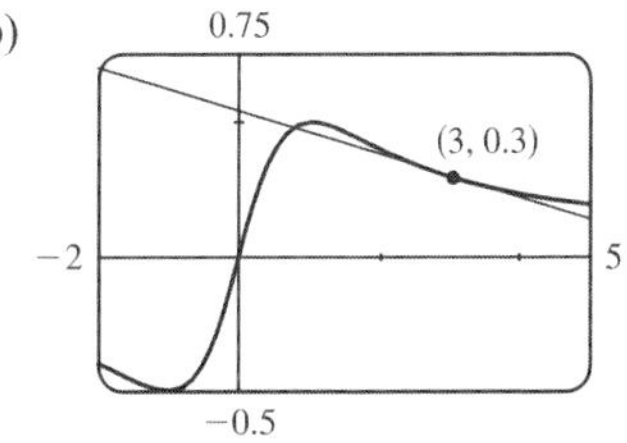


**25.** (a) $f(x) = \dfrac{e^x}{x^3} \Rightarrow f'(x) = \dfrac{x^3(e^x) - e^x(3x^2)}{(x^3)^2} = \dfrac{x^2e^x(x-3)}{x^6} = \dfrac{e^x(x-3)}{x^4}$

(b)

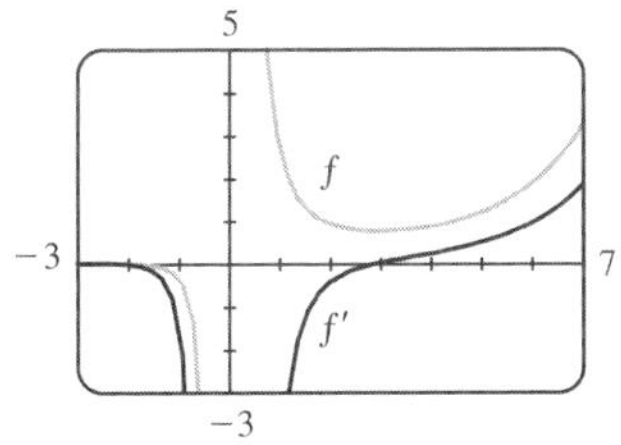


$f' = 0$ when $f$ has a horizontal tangent line, $f'$ is negative when $f$ is decreasing, and $f'$ is positive when $f$ is increasing.

**26.** (a) $f(x) = \dfrac{x}{x^2-1} \Rightarrow f'(x) = \dfrac{(x^2-1)1 - x(2x)}{(x^2-1)^2} = \dfrac{-x^2-1}{(x^2-1)^2}$

(b)

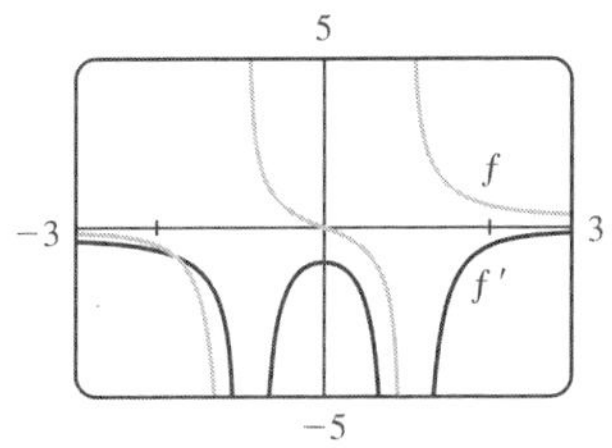


Notice that the slopes of all tangents to $f$ are negative and $f'(x) < 0$ always.

**27.** (a) $f(x) = (x-1)e^x \Rightarrow f'(x) = (x-1)e^x + e^x(1) = e^x(x-1+1) = xe^x$.

$f''(x) = x(e^x) + e^x(1) = e^x(x+1)$

(b)

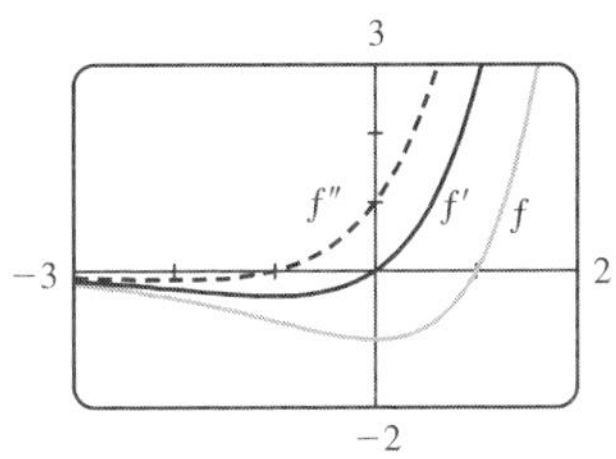


$f' = 0$ when $f$ has a horizontal tangent and $f'' = 0$ when $f'$ has a horizontal tangent. $f'$ is negative when $f$ is decreasing and positive when $f$ is increasing. $f''$ is negative when $f'$ is decreasing and positive when $f'$ is increasing. $f''$ is negative when $f$ is concave down and positive when $f$ is concave up.

**28.** (a) $f(x) = \dfrac{x}{x^2+1} \Rightarrow f'(x) = \dfrac{(x^2+1)(1) - x(2x)}{(x^2+1)^2} = \dfrac{1-x^2}{(x^2+1)^2} = \dfrac{1-x^2}{x^4+2x^2+1} \Rightarrow$

$f''(x) = \dfrac{(x^2+1)^2(-2x) - (1-x^2)(4x^3+4x)}{(x^2+1)^4} = \dfrac{(2x)(x^2+1)[-(x^2+1) - (1-x^2)(2)]}{(x^2+1)^4} = \dfrac{2x(x^2-3)}{(x^2+1)^3}$

(b)

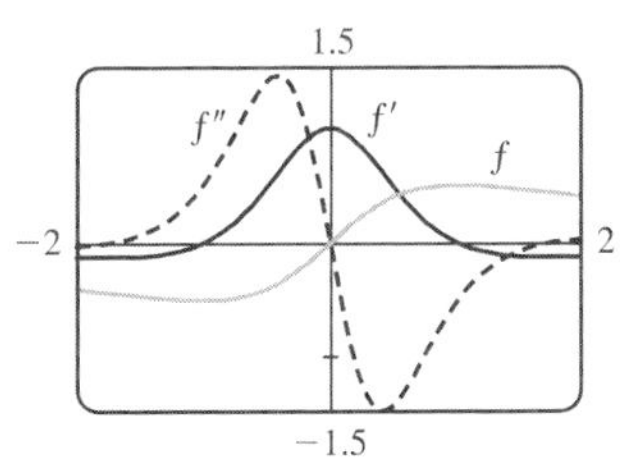


We can see that our answers are plausible, since $f$ has horizontal tangents where $f'(x) = 0$, and $f'$ has horizontal tangents where $f''(x) = 0$.

**29.** $f(x) = \dfrac{x^2}{1+x} \Rightarrow f'(x) = \dfrac{(1+x)(2x) - x^2(1)}{(1+x)^2} = \dfrac{2x + 2x^2 - x^2}{(1+x)^2} = \dfrac{x^2 + 2x}{x^2 + 2x + 1} \Rightarrow$

$$f''(x) = \frac{(x^2 + 2x + 1)(2x + 2) - (x^2 + 2x)(2x + 2)}{(x^2 + 2x + 1)^2} = \frac{(2x + 2)(x^2 + 2x + 1 - x^2 - 2x)}{[(x+1)^2]^2}$$

$$= \frac{2(x+1)(1)}{(x+1)^4} = \frac{2}{(x+1)^3},$$

so $f''(1) = \dfrac{2}{(1+1)^3} = \dfrac{2}{8} = \dfrac{1}{4}$.

**30.** $g(x) = \dfrac{x}{e^x} \Rightarrow g'(x) = \dfrac{e^x \cdot 1 - x \cdot e^x}{(e^x)^2} = \dfrac{e^x(x-1)}{(e^x)^2} = \dfrac{x-1}{e^x} \Rightarrow$

$$g''(x) = \frac{e^x \cdot 1 - (x-1)e^x}{(e^x)^2} = \frac{e^x[1 - (x-1)]}{(e^x)^2} = \frac{2-x}{e^x} \Rightarrow$$

$$g'''(x) = \frac{e^x \cdot (-1) - (2-x)e^x}{(e^x)^2} = \frac{e^x[-1 - (2-x)]}{(e^x)^2} = \frac{x-3}{e^x} \Rightarrow$$

$$g^{(4)}(x) = \frac{e^x \cdot 1 - (x-3)e^x}{(e^x)^2} = \frac{e^x[1 - (x-3)]}{(e^x)^2} = \frac{4-x}{e^x}.$$

The pattern suggests that $g^{(n)}(x) = \dfrac{(x-n)(-1)^n}{e^x}$. (We could use mathematical induction to prove this formula.)

**31.** We are given that $f(5) = 1$, $f'(5) = 6$, $g(5) = -3$, and $g'(5) = 2$.

(a) $(fg)'(5) = f(5)g'(5) + g(5)f'(5) = (1)(2) + (-3)(6) = 2 - 18 = -16$

(b) $\left(\dfrac{f}{g}\right)'(5) = \dfrac{g(5)f'(5) - f(5)g'(5)}{[g(5)]^2} = \dfrac{(-3)(6) - (1)(2)}{(-3)^2} = -\dfrac{20}{9}$

(c) $\left(\dfrac{g}{f}\right)'(5) = \dfrac{f(5)g'(5) - g(5)f'(5)}{[f(5)]^2} = \dfrac{(1)(2) - (-3)(6)}{(1)^2} = 20$

**32.** We are given that $f(3) = 4$, $g(3) = 2$, $f'(3) = -6$, and $g'(3) = 5$.

(a) $(f+g)'(3) = f'(3) + g'(3) = -6 + 5 = -1$

(b) $(fg)'(3) = f(3)g'(3) + g(3)f'(3) = (4)(5) + (2)(-6) = 20 - 12 = 8$

(c) $\left(\dfrac{f}{g}\right)'(3) = \dfrac{g(3)f'(3) - f(3)g'(3)}{[g(3)]^2} = \dfrac{(2)(-6) - (4)(5)}{(2)^2} = \dfrac{-32}{4} = -8$

(d) $\left(\dfrac{f}{f-g}\right)'(3) = \dfrac{[f(3) - g(3)]f'(3) - f(3)[f'(3) - g'(3)]}{[f(3) - g(3)]^2} = \dfrac{(4-2)(-6) - 4(-6-5)}{(4-2)^2} = \dfrac{-12 + 44}{2^2} = 8$

**33.** $f(x) = e^x g(x) \Rightarrow f'(x) = e^x g'(x) + g(x)e^x = e^x[g'(x) + g(x)]$. $f'(0) = e^0[g'(0) + g(0)] = 1(5+2) = 7$

**34.** $\dfrac{d}{dx}\left[\dfrac{h(x)}{x}\right] = \dfrac{xh'(x) - h(x) \cdot 1}{x^2} \Rightarrow \dfrac{d}{dx}\left[\dfrac{h(x)}{x}\right]_{x=2} = \dfrac{2h'(2) - h(2)}{2^2} = \dfrac{2(-3) - (4)}{4} = \dfrac{-10}{4} = -2.5$

**35.** (a) From the graphs of $f$ and $g$, we obtain the following values: $f(1) = 2$ since the point $(1, 2)$ is on the graph of $f$; $g(1) = 1$ since the point $(1, 1)$ is on the graph of $g$; $f'(1) = 2$ since the slope of the line segment between $(0, 0)$ and $(2, 4)$ is $\dfrac{4-0}{2-0} = 2$; $g'(1) = -1$ since the slope of the line segment between $(-2, 4)$ and $(2, 0)$ is $\dfrac{0-4}{2-(-2)} = -1$.

Now $u(x) = f(x)g(x)$, so $u'(1) = f(1)g'(1) + g(1)\,f'(1) = 2 \cdot (-1) + 1 \cdot 2 = 0$.

(b) $v(x) = f(x)/g(x)$, so $v'(5) = \dfrac{g(5)f'(5) - f(5)g'(5)}{[g(5)]^2} = \dfrac{2\left(-\frac{1}{3}\right) - 3 \cdot \frac{2}{3}}{2^2} = \dfrac{-\frac{8}{3}}{4} = -\dfrac{2}{3}$

**36.** (a) $P(x) = F(x)\,G(x)$, so $P'(2) = F(2)\,G'(2) + G(2)\,F'(2) = 3 \cdot \frac{2}{4} + 2 \cdot 0 = \frac{3}{2}$.

(b) $Q(x) = F(x)/G(x)$, so $Q'(7) = \dfrac{G(7)\,F'(7) - F(7)\,G'(7)}{[G(7)]^2} = \dfrac{1 \cdot \frac{1}{4} - 5 \cdot \left(-\frac{2}{3}\right)}{1^2} = \dfrac{1}{4} + \dfrac{10}{3} = \dfrac{43}{12}$

**37.** (a) $y = xg(x) \quad \Rightarrow \quad y' = xg'(x) + g(x) \cdot 1 = xg'(x) + g(x)$

(b) $y = \dfrac{x}{g(x)} \quad \Rightarrow \quad y' = \dfrac{g(x) \cdot 1 - xg'(x)}{[g(x)]^2} = \dfrac{g(x) - xg'(x)}{[g(x)]^2}$

(c) $y = \dfrac{g(x)}{x} \quad \Rightarrow \quad y' = \dfrac{xg'(x) - g(x) \cdot 1}{(x)^2} = \dfrac{xg'(x) - g(x)}{x^2}$

**38.** (a) $y = x^2 f(x) \quad \Rightarrow \quad y' = x^2 f'(x) + f(x)(2x)$

(b) $y = \dfrac{f(x)}{x^2} \quad \Rightarrow \quad y' = \dfrac{x^2 f'(x) - f(x)(2x)}{(x^2)^2} = \dfrac{xf'(x) - 2f(x)}{x^3}$

(c) $y = \dfrac{x^2}{f(x)} \quad \Rightarrow \quad y' = \dfrac{f(x)(2x) - x^2 f'(x)}{[f(x)]^2}$

(d) $y = \dfrac{1 + xf(x)}{\sqrt{x}} \quad \Rightarrow$

$$y' = \frac{\sqrt{x}\,[xf'(x) + f(x)] - [1 + xf(x)]\,\dfrac{1}{2\sqrt{x}}}{\left(\sqrt{x}\,\right)^2}$$

$$= \frac{x^{3/2} f'(x) + x^{1/2} f(x) - \frac{1}{2}x^{-1/2} - \frac{1}{2}x^{1/2} f(x)}{x} \cdot \frac{2x^{1/2}}{2x^{1/2}} = \frac{xf(x) + 2x^2 f'(x) - 1}{2x^{3/2}}$$

**39.** If $P(t)$ denotes the population at time $t$ and $A(t)$ the average annual income, then $T(t) = P(t)A(t)$ is the total personal income. The rate at which $T(t)$ is rising is given by $T'(t) = P(t)A'(t) + A(t)P'(t) \quad \Rightarrow$

$$T'(1999) = P(1999)A'(1999) + A(1999)P'(1999) = (961{,}400)(\$1400/\text{yr}) + (\$30{,}593)(9200/\text{yr})$$
$$= \$1{,}345{,}960{,}000/\text{yr} + \$281{,}455{,}600/\text{yr} = \$1{,}627{,}415{,}600/\text{yr}$$

So the total personal income was rising by about \$1.627 billion per year in 1999.

The term $P(t)A'(t) \approx \$1.346$ billion represents the portion of the rate of change of total income due to the existing population's increasing income. The term $A(t)P'(t) \approx \$281$ million represents the portion of the rate of change of total income due to increasing population.

**40.** (a) $f(20) = 10{,}000$ means that when the price of the fabric is \$20/yard, 10,000 yards will be sold.

$f'(20) = -350$ means that as the price of the fabric increases past \$20/yard, the amount of fabric which will be sold is decreasing at a rate of 350 yards per (dollar per yard).

(b) $R(p) = pf(p) \quad \Rightarrow \quad R'(p) = pf'(p) + f(p) \cdot 1 \quad \Rightarrow \quad R'(20) = 20f'(20) + f(20) \cdot 1 = 20(-350) + 10{,}000 = 3000$.
This means that as the price of the fabric increases past \$20/yard, the total revenue is increasing at \$3000/(\$/yard). Note that the Product Rule indicates that we will lose \$7000/(\$/yard) due to selling less fabric, but this loss is more than made up for by the additional revenue due to the increase in price.

**41.** $f$ is increasing when $f'$ is positive. $f(x) = x^3 e^x \quad \Rightarrow \quad f'(x) = x^3 e^x + e^x(3x^2) = x^2 e^x(x + 3)$. Now $x^2 \geq 0$ and $e^x > 0$ for all $x$, so $f'(x) > 0$ when $x + 3 > 0$ and $x \neq 0$; that is, when $x \in (-3, 0) \cup (0, \infty)$. So $f$ is increasing on $(-3, \infty)$.

**42.** $f$ is concave downward when $f''$ is negative. $f(x) = x^2 e^x \quad \Rightarrow \quad f'(x) = x^2 e^x + e^x(2x) \quad \Rightarrow$
$f''(x) = x^2 e^x + e^x(2x) + e^x(2) + (2x)e^x = e^x(x^2 + 2x + 2 + 2x) = e^x(x^2 + 4x + 2)$. Note that $e^x > 0$ for all $x$ and $f''(x) = 0 \quad \Leftrightarrow \quad x = -2 \pm \sqrt{2}$. $f''(x) < 0$ when $x \in \left(-2 - \sqrt{2}, -2 + \sqrt{2}\,\right)$.

**43.** If $y = f(x) = \dfrac{x}{x+1}$, then $f'(x) = \dfrac{(x+1)(1) - x(1)}{(x+1)^2} = \dfrac{1}{(x+1)^2}$. When $x = a$, the equation of the tangent line is

$y - \dfrac{a}{a+1} = \dfrac{1}{(a+1)^2}(x-a)$. This line passes through $(1, 2)$ when $2 - \dfrac{a}{a+1} = \dfrac{1}{(a+1)^2}(1-a) \quad \Leftrightarrow$

$2(a+1)^2 - a(a+1) = 1 - a \quad \Leftrightarrow \quad 2a^2 + 4a + 2 - a^2 - a - 1 + a = 0 \quad \Leftrightarrow \quad a^2 + 4a + 1 = 0.$

The quadratic formula gives the roots of this equation as $a = \dfrac{-4 \pm \sqrt{4^2 - 4(1)(1)}}{2(1)} = \dfrac{-4 \pm \sqrt{12}}{2} = -2 \pm \sqrt{3}$,

so there are two such tangent lines. Since

$$f\left(-2 \pm \sqrt{3}\right) = \frac{-2 \pm \sqrt{3}}{-2 \pm \sqrt{3} + 1} = \frac{-2 \pm \sqrt{3}}{-1 \pm \sqrt{3}} \cdot \frac{-1 \mp \sqrt{3}}{-1 \mp \sqrt{3}}$$

$$= \frac{2 \pm 2\sqrt{3} \mp \sqrt{3} - 3}{1 - 3} = \frac{-1 \pm \sqrt{3}}{-2} = \frac{1 \mp \sqrt{3}}{2},$$

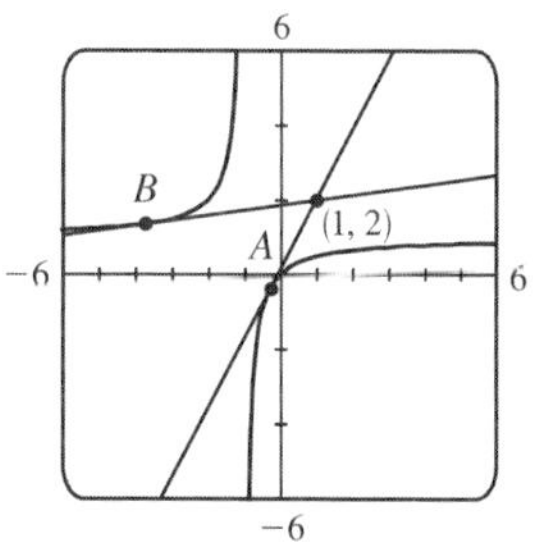


the lines touch the curve at $A\left(-2 + \sqrt{3}, \frac{1 - \sqrt{3}}{2}\right) \approx (-0.27, -0.37)$

and $B\left(-2 - \sqrt{3}, \frac{1 + \sqrt{3}}{2}\right) \approx (-3.73, 1.37)$.

**44.** $y = \dfrac{x-1}{x+1} \quad \Rightarrow \quad y' = \dfrac{(x+1)(1) - (x-1)(1)}{(x+1)^2} = \dfrac{2}{(x+1)^2}$. If the tangent intersects

the curve when $x = a$, then its slope is $2/(a+1)^2$. But if the tangent is parallel to

$x - 2y = 2$, that is, $y = \frac{1}{2}x - 1$, then its slope is $\frac{1}{2}$. Thus, $\dfrac{2}{(a+1)^2} = \dfrac{1}{2} \quad \Rightarrow$

$(a+1)^2 = 4 \quad \Rightarrow \quad a + 1 = \pm 2 \quad \Rightarrow \quad a = 1$ or $-3$. When $a = 1$, $y = 0$ and the

equation of the tangent is $y - 0 = \frac{1}{2}(x - 1)$ or $y = \frac{1}{2}x - \frac{1}{2}$. When $a = -3$, $y = 2$ and

the equation of the tangent is $y - 2 = \frac{1}{2}(x + 3)$ or $y = \frac{1}{2}x + \frac{7}{2}$.

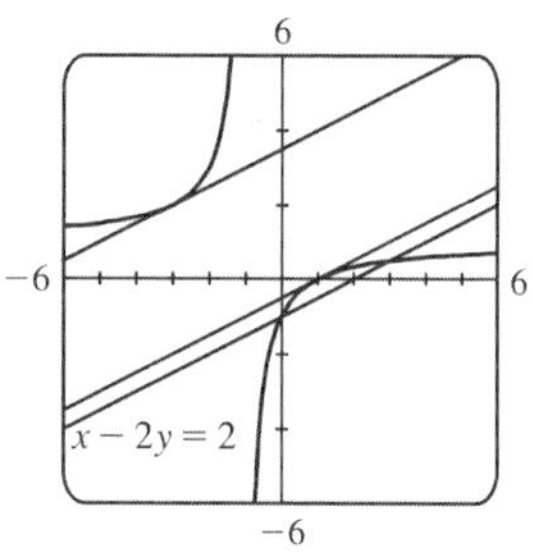


We will sometimes use the form $f'g + fg'$ rather than the form $fg' + gf'$ for the Product Rule.

**45.** (a) $(fgh)' = [(fg)h]' = (fg)'h + (fg)h' = (f'g + fg')h + (fg)h' = f'gh + fg'h + fgh'$

(b) Putting $f = g = h$ in part (a), we have $\dfrac{d}{dx}[f(x)]^3 = (fff)' = f'ff + ff'f + fff' = 3fff' = 3[f(x)]^2 f'(x)$.

(c) $\dfrac{d}{dx}\left(e^{3x}\right) = \dfrac{d}{dx}\left(e^x\right)^3 = 3(e^x)^2 e^x = 3e^{2x}e^x = 3e^{3x}$

**46.** (a) We use the Product Rule repeatedly: $F = fg \quad \Rightarrow \quad F' = f'g + fg' \quad \Rightarrow$

$F'' = (f''g + f'g') + (f'g' + fg'') = f''g + 2f'g' + fg''$.

(b) $F''' = f'''g + f''g' + 2\left(f''g' + f'g''\right) + f'g'' + fg''' = f'''g + 3f''g' + 3f'g'' + fg''' \quad \Rightarrow$

$F^{(4)} = f^{(4)}g + f'''g' + 3\left(f'''g' + f''g''\right) + 3\left(f''g'' + f'g'''\right) + f'g''' + fg^{(4)}$

$= f^{(4)}g + 4f'''g' + 6f''g'' + 4f'g''' + fg^{(4)}$

(c) By analogy with the Binomial Theorem, we make the guess:

$$F^{(n)} = f^{(n)}g + nf^{(n-1)}g' + \binom{n}{2}f^{(n-2)}g'' + \cdots + \binom{n}{k}f^{(n-k)}g^{(k)} + \cdots + nf'g^{(n-1)} + fg^{(n)},$$

where $\dbinom{n}{k} = \dfrac{n!}{k!\,(n-k)!} = \dfrac{n(n-1)(n-2)\cdots(n-k+1)}{k!}$.

**47.** For $f(x) = x^2e^x$, $f'(x) = x^2e^x + e^x(2x) = e^x(x^2 + 2x)$. Similarly, we have

$$f''(x) = e^x(x^2 + 4x + 2)$$
$$f'''(x) = e^x(x^2 + 6x + 6)$$
$$f^{(4)}(x) = e^x(x^2 + 8x + 12)$$
$$f^{(5)}(x) = e^x(x^2 + 10x + 20)$$

It appears that the coefficient of $x$ in the quadratic term increases by 2 with each differentiation. The pattern for the constant terms seems to be $0 = 1 \cdot 0$, $2 = 2 \cdot 1$, $6 = 3 \cdot 2$, $12 = 4 \cdot 3$, $20 = 5 \cdot 4$. So a reasonable guess is that $f^{(n)}(x) = e^x[x^2 + 2nx + n(n-1)]$.

*Proof:* Let $S_n$ be the statement that $f^{(n)}(x) = e^x[x^2 + 2nx + n(n-1)]$.

1. $S_1$ is true because $f'(x) = e^x(x^2 + 2x)$.
2. Assume that $S_k$ is true; that is, $f^{(k)}(x) = e^x[x^2 + 2kx + k(k-1)]$. Then

$$\begin{aligned} f^{(k+1)}(x) &= \frac{d}{dx}\left[f^{(k)}(x)\right] = e^x(2x + 2k) + [x^2 + 2kx + k(k-1)]e^x \\ &= e^x[x^2 + (2k+2)x + (k^2+k)] = e^x[x^2 + 2(k+1)x + (k+1)k] \end{aligned}$$

   This shows that $S_{k+1}$ is true.

3. Therefore, by mathematical induction, $S_n$ is true for all $n$; that is, $f^{(n)}(x) = e^x[x^2 + 2nx + n(n-1)]$ for every positive integer $n$.

**48.** (a) $\displaystyle \frac{d}{dx}\left(\frac{1}{g(x)}\right) = \frac{g(x) \cdot \frac{d}{dx}(1) - 1 \cdot \frac{d}{dx}[g(x)]}{[g(x)]^2}$ [Quotient Rule] $\displaystyle = \frac{g(x) \cdot 0 - 1 \cdot g'(x)}{[g(x)]^2} = \frac{0 - g'(x)}{[g(x)]^2} = -\frac{g'(x)}{[g(x)]^2}$

(b) $\displaystyle y = \frac{1}{s + ke^s} \quad\Rightarrow\quad y' = -\frac{1 + ke^s}{(s + ke^s)^2}$

(c) $\displaystyle \frac{d}{dx}\left(x^{-n}\right) = \frac{d}{dx}\left(\frac{1}{x^n}\right) = -\frac{(x^n)'}{(x^n)^2}$ [by the Reciprocal Rule] $\displaystyle = -\frac{nx^{n-1}}{x^{2n}} = -nx^{n-1-2n} = -nx^{-n-1}$

## 3.3 Rates of Change in the Natural and Social Sciences

**1.** (a) $s = f(t) = t^3 - 12t^2 + 36t \quad\Rightarrow\quad v(t) = f'(t) = 3t^2 - 24t + 36$

(b) $v(3) = 27 - 72 + 36 = -9$ m/s

(c) The particle is at rest when $v(t) = 0$. $3t^2 - 24t + 36 = 0 \quad\Rightarrow\quad 3(t-2)(t-6) = 0 \quad\Rightarrow\quad t = 2, 6$.

(d) The particle is moving in the positive direction when $v(t) > 0$. $3(t-2)(t-6) > 0 \quad\Leftrightarrow\quad 0 \le t < 2$ or $t > 6$.

(e) Since the particle is moving forward and backward, we need to calculate the distance traveled in the intervals $[0, 2]$, $[2, 6]$, and $[6, 8]$ separately.

$$|f(2) - f(0)| = |32 - 0| = 32.$$
$$|f(6) - f(2)| = |0 - 32| = 32.$$
$$|f(8) - f(6)| = |32 - 0| = 32.$$

The total distance is $32 + 32 + 32 = 96$ m.

(f)

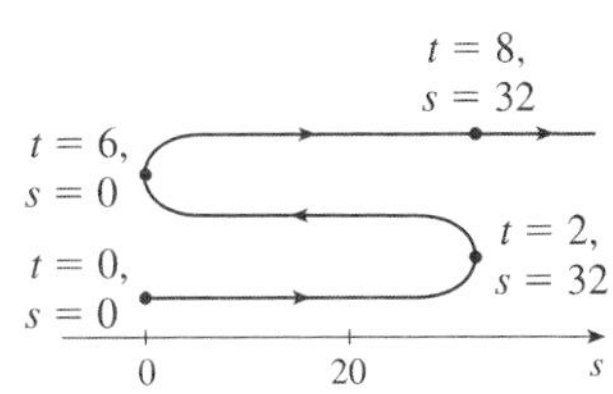

(g) $s = f(t) = t^3 - 12t^2 + 36t, t \geq 0 \;\Rightarrow\; v(t) = f'(t) = 3t^2 - 24t + 36.$ $a(t) = v'(t) = 6t - 24.$
$a(3) = 6(3) - 24 = -6\ (\text{m/s})/\text{s}$ or $\text{m/s}^2$.

(h)

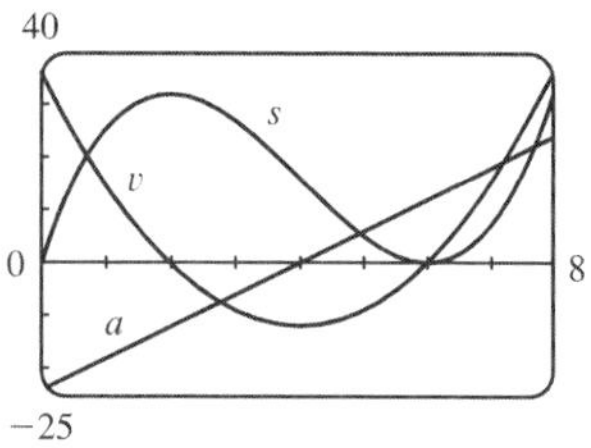


(i) The particle is speeding up when $v$ and $a$ have the same sign. This occurs when $2 < t < 4$ and when $t > 6$. It is slowing down when $v$ and $a$ have opposite signs; that is, when $0 \leq t < 2$ and when $4 < t < 6$.

**2.** (a) $x(t) = \dfrac{t}{1+t^2} \;\Rightarrow\; v(t) = x'(t) = \dfrac{(1+t^2)(1) - t(2t)}{(1+t^2)^2} = \dfrac{1-t^2}{(1+t^2)^2}$

(b) Right: $v(t) > 0 \;\Rightarrow\; 1 - t^2 > 0 \;\Rightarrow\; t^2 < 1 \;\Rightarrow\; |t| < 1 \;\Rightarrow\; 0 \leq t < 1$
Left: $v(t) < 0 \;\Rightarrow\; 1 - t^2 < 0 \;\Rightarrow\; t > 1$

(c) $|x(1) - x(0)| + |x(4) - x(1)| = \left|\frac{1}{2} - 0\right| + \left|\frac{4}{17} - \frac{1}{2}\right| = \frac{1}{2} + \frac{9}{34} = \frac{13}{17}$

(d) $x(t) = \dfrac{t}{1+t^2} \;\Rightarrow\; v(t) = x'(t) = \dfrac{(1+t^2)(1) - t(2t)}{(1+t^2)^2} = \dfrac{1-t^2}{(1+t^2)^2}.$ $a(t) = v'(t) = \dfrac{2t(t^2-3)}{(1+t^2)^3}.$
$a(t) = 0 \;\Rightarrow\; 2t(t^2 - 3) = 0 \;\Rightarrow\; t = 0$ or $\sqrt{3}$

(e)

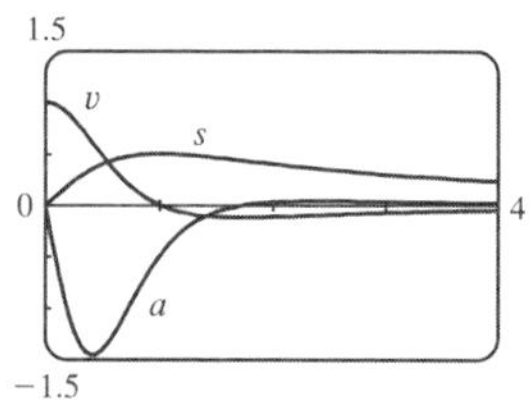


(f) $v$ and $a$ have the same sign and the particle is speeding up when $1 < t < \sqrt{3}$. The particle is slowing down and $v$ and $a$ have opposite signs when $0 < t < 1$ and when $t > \sqrt{3}$.

**3.** (a) From the figure, the velocity $v$ is positive on the interval $(0, 2)$ and negative on the interval $(2, 3)$. The acceleration $a$ is positive (negative) when the slope of the tangent line is positive (negative), so the acceleration is positive on the interval $(0, 1)$, and negative on the interval $(1, 3)$. The particle is speeding up when $v$ and $a$ have the same sign, that is, on the interval $(0, 1)$ when $v > 0$ and $a > 0$, and on the interval $(2, 3)$ when $v < 0$ and $a < 0$. The particle is slowing down when $v$ and $a$ have opposite signs, that is, on the interval $(1, 2)$ when $v > 0$ and $a < 0$.

(b) $v > 0$ on $(0, 3)$ and $v < 0$ on $(3, 4)$. $a > 0$ on $(1, 2)$ and $a < 0$ on $(0, 1)$ and $(2, 4)$. The particle is speeding up on $(1, 2)$ $[v > 0, a > 0]$ and on $(3, 4)$ $[v < 0, a < 0]$. The particle is slowing down on $(0, 1)$ and $(2, 3)$ $[v > 0, a < 0]$.

**4.** (a) The velocity $v$ is positive when $s$ is increasing, that is, on the intervals $(0, 1)$ and $(3, 4)$; and it is negative when $s$ is decreasing, that is, on the interval $(1, 3)$. The acceleration $a$ is positive when the graph of $s$ is concave upward ($v$ is increasing), that is, on the interval $(2, 4)$; and it is negative when the graph of $s$ is concave downward ($v$ is decreasing), that is, on the interval $(0, 2)$. The particle is speeding up on the interval $(1, 2)$ $[v < 0, a > 0]$ and on $(3, 4)$ $[v > 0, a > 0]$. The particle is slowing down on the interval $(0, 1)$ $[v > 0, a < 0]$ and on $(2, 3)$ $[v < 0, a > 0]$.

(b) The velocity $v$ is positive on $(3, 4)$ and negative on $(0, 3)$. The acceleration $a$ is positive on $(0, 1)$ and $(2, 4)$ and negative on $(1, 2)$. The particle is speeding up on the interval $(1, 2)$ $[v < 0, a < 0]$ and on $(3, 4)$ $[v > 0, a > 0]$. The particle is slowing down on the interval $(0, 1)$ $[v < 0, a > 0]$ and on $(2, 3)$ $[v < 0, a > 0]$.

**5.** (a) $s(t) = t^3 - 4.5t^2 - 7t \;\Rightarrow\; v(t) = s'(t) = 3t^2 - 9t - 7 = 5 \;\Leftrightarrow\; 3t^2 - 9t - 12 = 0 \;\Leftrightarrow\;$
$3(t-4)(t+1) = 0 \;\Leftrightarrow\; t = 4$ or $-1$. Since $t \geq 0$, the particle reaches a velocity of 5 m/s at $t = 4$ s.

(b) $a(t) = v'(t) = 6t - 9 = 0 \;\Leftrightarrow\; t = 1.5$. The acceleration changes from negative to positive, so the velocity changes from decreasing to increasing. Thus, at $t = 1.5$ s, the velocity has its minimum value.

**6.** (a) $s = 5t + 3t^2 \;\Rightarrow\; v(t) = \dfrac{ds}{dt} = 5 + 6t$, so $v(2) = 5 + 6(2) = 17$ m/s.

(b) $v(t) = 35 \;\Rightarrow\; 5 + 6t = 35 \;\Rightarrow\; 6t = 30 \;\Rightarrow\; t = 5$ s.

**7.** (a) $h = 10t - 0.83t^2 \Rightarrow v(t) = \dfrac{dh}{dt} = 10 - 1.66t$, so $v(3) = 10 - 1.66(3) = 5.02$ m/s.

(b) $h = 25 \Rightarrow 10t - 0.83t^2 = 25 \Rightarrow 0.83t^2 - 10t + 25 = 0 \Rightarrow t = \dfrac{10 \pm \sqrt{17}}{1.66} \approx 3.54$ or $8.51$.

The value $t_1 = \left(10 - \sqrt{17}\right)/1.66$ corresponds to the time it takes for the stone to rise 25 m and $t_2 = \left(10 + \sqrt{17}\right)/1.66$ corresponds to the time when the stone is 25 m high on the way down. Thus, $v(t_1) = 10 - 1.66\left[\left(10 - \sqrt{17}\right)/1.66\right] = \sqrt{17} \approx 4.12$ m/s.

**8.** (a) At maximum height the velocity of the ball is 0 ft/s. $v(t) = s'(t) = 80 - 32t = 0 \Leftrightarrow 32t = 80 \Leftrightarrow t = \frac{5}{2}$. So the maximum height is $s\left(\frac{5}{2}\right) = 80\left(\frac{5}{2}\right) - 16\left(\frac{5}{2}\right)^2 = 200 - 100 = 100$ ft.

(b) $s(t) = 80t - 16t^2 = 96 \Leftrightarrow 16t^2 - 80t + 96 = 0 \Leftrightarrow 16(t^2 - 5t + 6) = 0 \Leftrightarrow 16(t-3)(t-2) = 0$. So the ball has a height of 96 ft on the way up at $t = 2$ and on the way down at $t = 3$. At these times the velocities are $v(2) = 80 - 32(2) = 16$ ft/s and $v(3) = 80 - 32(3) = -16$ ft/s, respectively.

**9.** (a) $A(x) = x^2 \Rightarrow A'(x) = 2x$. $A'(15) = 30\ \text{mm}^2/\text{mm}$ is the rate at which the area is increasing with respect to the side length as $x$ reaches 15 mm.

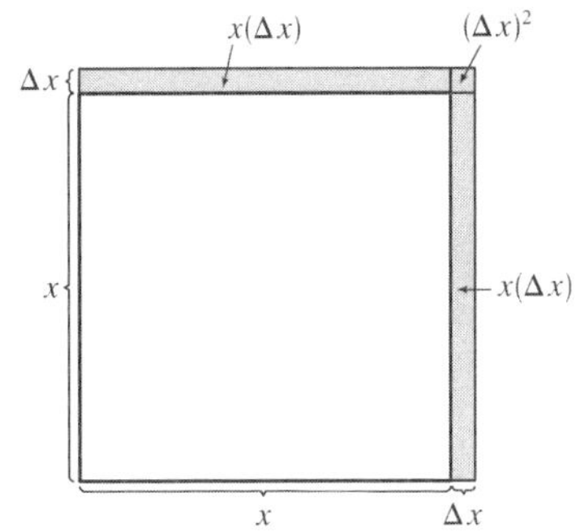


(b) The perimeter is $P(x) = 4x$, so $A'(x) = 2x = \frac{1}{2}(4x) = \frac{1}{2}P(x)$. The figure suggests that if $\Delta x$ is small, then the change in the area of the square is approximately half of its perimeter (2 of the 4 sides) times $\Delta x$. From the figure, $\Delta A = 2x\,(\Delta x) + (\Delta x)^2$. If $\Delta x$ is small, then $\Delta A \approx 2x\,(\Delta x)$ and so $\Delta A/\Delta x \approx 2x$.

**10.** (a) $V(x) = x^3 \Rightarrow \dfrac{dV}{dx} = 3x^2$. $\left.\dfrac{dV}{dx}\right|_{x=3} = 3(3)^2 = 27\ \text{mm}^3/\text{mm}$ is the rate at which the volume is increasing as $x$ increases past 3 mm.

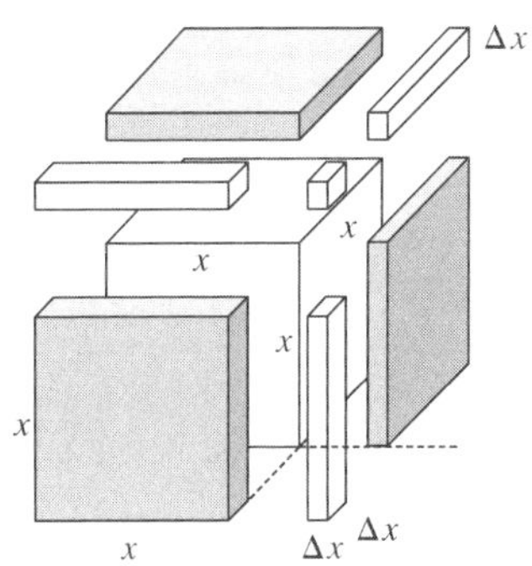


(b) The surface area is $S(x) = 6x^2$, so $V'(x) = 3x^2 = \frac{1}{2}(6x^2) = \frac{1}{2}S(x)$. The figure suggests that if $\Delta x$ is small, then the change in the volume of the cube is approximately half of its surface area (the area of 3 of the 6 faces) times $\Delta x$. From the figure, $\Delta V = 3x^2(\Delta x) + 3x(\Delta x)^2 + (\Delta x)^3$. If $\Delta x$ is small, then $\Delta V \approx 3x^2(\Delta x)$ and so $\Delta V/\Delta x \approx 3x^2$.

**11.** (a) Using $A(r) = \pi r^2$, we find that the average rate of change is:

(i) $\dfrac{A(3) - A(2)}{3 - 2} = \dfrac{9\pi - 4\pi}{1} = 5\pi$ (ii) $\dfrac{A(2.5) - A(2)}{2.5 - 2} = \dfrac{6.25\pi - 4\pi}{0.5} = 4.5\pi$

(iii) $\dfrac{A(2.1) - A(2)}{2.1 - 2} = \dfrac{4.41\pi - 4\pi}{0.1} = 4.1\pi$

(b) $A(r) = \pi r^2 \Rightarrow A'(r) = 2\pi r$, so $A'(2) = 4\pi$.

(c) The circumference is $C(r) = 2\pi r = A'(r)$. The figure suggests that if $\Delta r$ is small, then the change in the area of the circle (a ring around the outside) is approximately equal to its circumference times $\Delta r$. Straightening out this ring gives us a shape that is approximately rectangular with length $2\pi r$ and width $\Delta r$, so $\Delta A \approx 2\pi r(\Delta r)$. Algebraically, $\Delta A = A(r + \Delta r) - A(r) = \pi(r + \Delta r)^2 - \pi r^2 = 2\pi r(\Delta r) + \pi(\Delta r)^2$. So we see that if $\Delta r$ is small, then $\Delta A \approx 2\pi r(\Delta r)$ and therefore, $\Delta A/\Delta r \approx 2\pi r$.

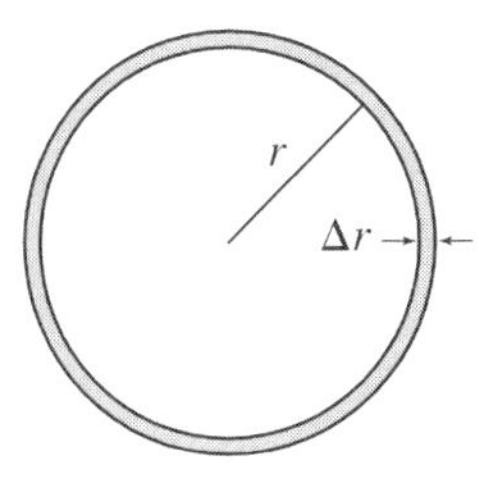

**12.** After $t$ seconds the radius is $r = 60t$, so the area is $A(t) = \pi(60t)^2 = 3600\pi t^2 \Rightarrow A'(t) = 7200\pi t \Rightarrow$

(a) $A'(1) = 7200\pi \text{ cm}^2/\text{s}$ (b) $A'(3) = 21,600\pi \text{ cm}^2/\text{s}$ (c) $A'(5) = 36{,}000\pi \text{ cm}^2/\text{s}$

As time goes by, the area grows at an increasing rate. In fact, the rate of change is linear with respect to time.

**13.** $S(r) = 4\pi r^2 \Rightarrow S'(r) = 8\pi r \Rightarrow$

(a) $S'(1) = 8\pi \text{ ft}^2/\text{ft}$ (b) $S'(2) = 16\pi \text{ ft}^2/\text{ft}$ (c) $S'(3) = 24\pi \text{ ft}^2/\text{ft}$

As the radius increases, the surface area grows at an increasing rate. In fact, the rate of change is linear with respect to the radius.

**14.** (a) Using $V(r) = \frac{4}{3}\pi r^3$, we find that the average rate of change is:

(i) $\dfrac{V(8) - V(5)}{8 - 5} = \dfrac{\frac{4}{3}\pi(512) - \frac{4}{3}\pi(125)}{3} = 172\pi \ \mu\text{m}^3/\mu\text{m}$

(ii) $\dfrac{V(6) - V(5)}{6 - 5} = \dfrac{\frac{4}{3}\pi(216) - \frac{4}{3}\pi(125)}{1} = 121.\overline{3}\pi \ \mu\text{m}^3/\mu\text{m}$

(iii) $\dfrac{V(5.1) - V(5)}{5.1 - 5} = \dfrac{\frac{4}{3}\pi(5.1)^3 - \frac{4}{3}\pi(5)^3}{0.1} = 102.01\overline{3}\pi \ \mu\text{m}^3/\mu\text{m}$

(b) $V'(r) = 4\pi r^2$, so $V'(5) = 100\pi \ \mu\text{m}^3/\mu\text{m}$.

(c) $V(r) = \frac{4}{3}\pi r^3 \Rightarrow V'(r) = 4\pi r^2 = S(r)$. By analogy with Exercise 11(c), we can say that the change in the volume of the spherical shell, $\Delta V$, is approximately equal to its thickness, $\Delta r$, times the surface area of the inner sphere. Thus, $\Delta V \approx 4\pi r^2(\Delta r)$ and so $\Delta V/\Delta r \approx 4\pi r^2$.

**15.** The mass is $f(x) = 3x^2$, so the linear density at $x$ is $\rho(x) = f'(x) = 6x$.

(a) $\rho(1) = 6$ kg/m (b) $\rho(2) = 12$ kg/m (c) $\rho(3) = 18$ kg/m

Since $\rho$ is an increasing function, the density will be the highest at the right end of the rod and lowest at the left end.

**16.** $V(t) = 5000\left(1 - \frac{1}{40}t\right)^2 = 5000\left(1 - \frac{1}{20}t + \frac{1}{1600}t^2\right) \Rightarrow V'(t) = 5000\left(-\frac{1}{20} + \frac{1}{800}t\right) = -250\left(1 - \frac{1}{40}t\right)$

(a) $V'(5) = -250\left(1 - \frac{5}{40}\right) = -218.75$ gal/min (b) $V'(10) = -250\left(1 - \frac{10}{40}\right) = -187.5$ gal/min

(c) $V'(20) = -250\left(1 - \frac{20}{40}\right) = -125$ gal/min (d) $V'(40) = -250\left(1 - \frac{40}{40}\right) = 0$ gal/min

The water is flowing out the fastest at the beginning—when $t = 0$, $V'(t) = -250$ gal/min. The water is flowing out the slowest at the end—when $t = 40$, $V'(t) = 0$. As the tank empties, the water flows out more slowly.

**17.** The quantity of charge is $Q(t) = t^3 - 2t^2 + 6t + 2$, so the current is $Q'(t) = 3t^2 - 4t + 6$.

(a) $Q'(0.5) = 3(0.5)^2 - 4(0.5) + 6 = 4.75$ A (b) $Q'(1) = 3(1)^2 - 4(1) + 6 = 5$ A

The current is lowest when $Q'$ has a minimum. $Q''(t) = 6t - 4 < 0$ when $t < \frac{2}{3}$. So the current decreases when $t < \frac{2}{3}$ and increases when $t > \frac{2}{3}$. Thus, the current is lowest at $t = \frac{2}{3}$ s.

**18.** (a) $F = \dfrac{GmM}{r^2} = (GmM)r^{-2} \Rightarrow \dfrac{dF}{dr} = -2(GmM)r^{-3} = -\dfrac{2GmM}{r^3}$, which is the rate of change of the force with respect to the distance between the bodies. The minus sign indicates that as the distance $r$ between the bodies increases, the magnitude of the force $F$ exerted by the body of mass $m$ on the body of mass $M$ is decreasing.

(b) Given $F'(20{,}000) = -2$, find $F'(10{,}000)$. $-2 = -\dfrac{2GmM}{20{,}000^3} \Rightarrow GmM = 20{,}000^3$.

$$F'(10,000) = -\frac{2(20{,}000^3)}{10{,}000^3} = -2 \cdot 2^3 = -16 \text{ N/km}$$

**19.** (a) To find the rate of change of volume with respect to pressure, we first solve for $V$ in terms of $P$.

$$PV = C \Rightarrow V = \frac{C}{P} \Rightarrow \frac{dV}{dP} = -\frac{C}{P^2}.$$

(b) From the formula for $dV/dP$ in part (a), we see that as $P$ increases, the absolute value of $dV/dP$ decreases. Thus, the volume is decreasing more rapidly at the beginning.

(c) $\beta = -\dfrac{1}{V}\dfrac{dV}{dP} = -\dfrac{1}{V}\left(-\dfrac{C}{P^2}\right) = \dfrac{C}{(PV)P} = \dfrac{C}{CP} = \dfrac{1}{P}$

**20.** (a) (i) $\dfrac{C(6)-C(2)}{6-2} = \dfrac{0.0295-0.0570}{4} = -0.006875$ (moles/L)/min

(ii) $\dfrac{C(4)-C(2)}{4-2} = \dfrac{0.0408-0.0570}{2} = -0.008$ (moles/L)/min

(iii) $\dfrac{C(2)-C(0)}{2-0} = \dfrac{0.0570-0.0800}{2} = -0.0115$ (moles/L)/min

(b) Slope $= \dfrac{\Delta C}{\Delta t} \approx -\dfrac{0.077}{7.8} \approx -0.01$ (moles/L)/min

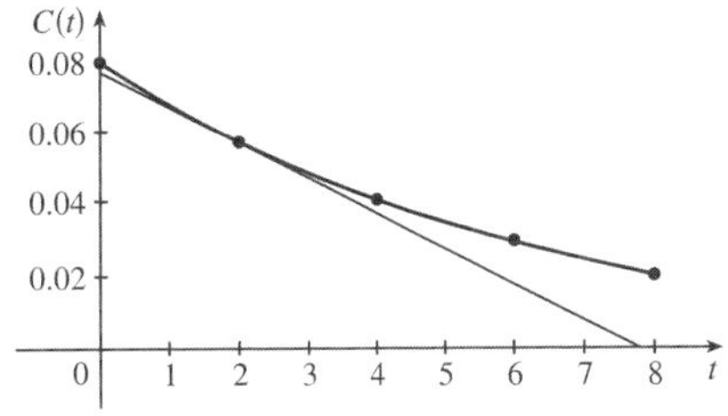


**21.** (a) **1920:** $m_1 = \dfrac{1860 - 1750}{1920 - 1910} = \dfrac{110}{10} = 11$, $m_2 = \dfrac{2070 - 1860}{1930 - 1920} = \dfrac{210}{10} = 21$,

$(m_1 + m_2)/2 = (11 + 21)/2 = 16$ million/year

**1980:** $m_1 = \dfrac{4450 - 3710}{1980 - 1970} = \dfrac{740}{10} = 74$, $m_2 = \dfrac{5280 - 4450}{1990 - 1980} = \dfrac{830}{10} = 83$,

$(m_1 + m_2)/2 = (74 + 83)/2 = 78.5$ million/year

(b) $P(t) = at^3 + bt^2 + ct + d$ (in millions of people), where $a \approx 0.0012937063$, $b \approx -7.061421911$, $c \approx 12{,}822.97902$, and $d \approx -7{,}743{,}770.396$.

(c) $P(t) = at^3 + bt^2 + ct + d \;\Rightarrow\; P'(t) = 3at^2 + 2bt + c$ (in millions of people per year)

(d) $P'(1920) = 3(0.0012937063)(1920)^2 + 2(-7.061421911)(1920) + 12{,}822.97902$

$\approx 14.48$ million/year [smaller than the answer in part (a), but close to it]

$P'(1980) \approx 75.29$ million/year (smaller, but close)

(e) $P'(1985) \approx 81.62$ million/year, so the rate of growth in 1985 was about 81.62 million/year.

**22.** (a) $A(t) = at^4 + bt^3 + ct^2 + dt + e$, where $a \approx -3.076923 \times 10^{-6}$, $b \approx 0.0243620824$, $c \approx -72.33147086$, $d \approx 95442.50365$, and $e \approx -47{,}224{,}986.6$.

(b) $A(t) = at^4 + bt^3 + ct^2 + dt + e \;\Rightarrow\; A'(t) = 4at^3 + 3bt^2 + 2ct + d$.

(c) Part (b) gives $A'(1990) \approx 0.11$ years of age per year.

(d)

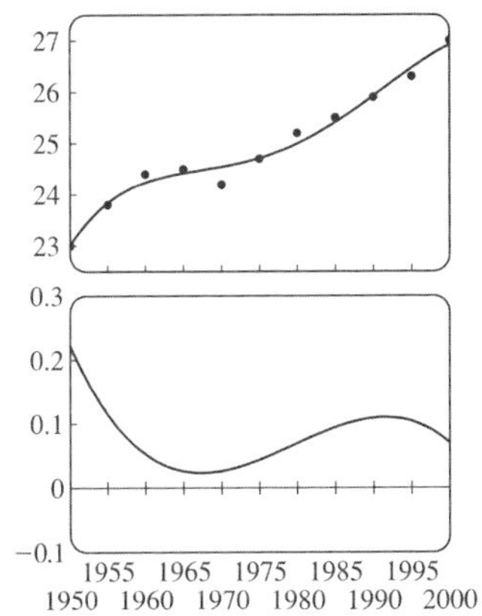


**23.** (a) $[\mathrm{C}] = \dfrac{a^2kt}{akt+1} \;\Rightarrow\;$ rate of reaction $= \dfrac{d[\mathrm{C}]}{dt} = \dfrac{(akt+1)(a^2k) - (a^2kt)(ak)}{(akt+1)^2} = \dfrac{a^2k(akt+1-akt)}{(akt+1)^2} = \dfrac{a^2k}{(akt+1)^2}$

(b) If $x = [\mathrm{C}]$, then $a - x = a - \dfrac{a^2kt}{akt+1} = \dfrac{a^2kt + a - a^2kt}{akt+1} = \dfrac{a}{akt+1}$.

So $k(a-x)^2 = k\left(\dfrac{a}{akt+1}\right)^2 = \dfrac{a^2k}{(akt+1)^2} = \dfrac{d[\mathrm{C}]}{dt}$ [from part (a)] $= \dfrac{dx}{dt}$.

(c) As $t \to \infty$, $[\mathrm{C}] = \dfrac{a^2kt}{akt+1} = \dfrac{(a^2kt)/t}{(akt+1)/t} = \dfrac{a^2k}{ak+(1/t)} \to \dfrac{a^2k}{ak} = a$ moles/L.

(d) As $t \to \infty$, $\dfrac{d[\mathrm{C}]}{dt} = \dfrac{a^2k}{(akt+1)^2} \to 0$.

(e) As $t$ increases, nearly all of the reactants A and B are converted into product C. In practical terms, the reaction virtually stops.

**24.** (a) After an hour the population is $n(1) = 3 \cdot 500$; after two hours it is $n(2) = 3(3 \cdot 500) = 3^2 \cdot 500$; after three hours, $n(3) = 3(3^2 \cdot 500) = 3^3 \cdot 500$; after four hours, $n(4) = 3^4 \cdot 500$. From this pattern, we see that the population after $t$ hours is $n(t) = 3^t \cdot 500 = 500 \cdot 3^t$.

(b) From (5) in Section 3.1, we have $\dfrac{d}{dx}(3^x) \approx (1.10)3^x$. Thus, for $n(t) = 500 \cdot 3^t$, $\dfrac{dn}{dt} = 500\,\dfrac{d}{dt}(3^t) \approx 500(1.10)3^t \;\Rightarrow$

$\left.\dfrac{dn}{dt}\right|_{t=6} \approx 500(1.10)3^6 \approx 400{,}950$ bacteria/hour.

**25.** (a) Using $v = \dfrac{P}{4\eta l}(R^2 - r^2)$ with $R = 0.01$, $l = 3$, $P = 3000$, and $\eta = 0.027$, we have $v$ as a function of $r$:

$v(r) = \dfrac{3000}{4(0.027)3}(0.01^2 - r^2)$. $v(0) = 0.\overline{925}$ cm/s, $v(0.005) = 0.69\overline{4}$ cm/s, $v(0.01) = 0$.

(b) $v(r) = \dfrac{P}{4\eta l}(R^2 - r^2) \;\Rightarrow\; v'(r) = \dfrac{P}{4\eta l}(-2r) = -\dfrac{Pr}{2\eta l}$. When $l = 3$, $P = 3000$, and $\eta = 0.027$, we have

$v'(r) = -\dfrac{3000r}{2(0.027)3}$. $v'(0) = 0$, $v'(0.005) = -92.\overline{592}$ (cm/s)/cm, and $v'(0.01) = -185.\overline{185}$ (cm/s)/cm.

(c) The velocity is greatest where $r = 0$ (at the center) and the velocity is changing most where $r = R = 0.01$ cm (at the edge).

**26.** (a) (i) $f = \dfrac{1}{2L}\sqrt{\dfrac{T}{\rho}} = \left(\dfrac{1}{2}\sqrt{\dfrac{T}{\rho}}\right)L^{-1} \;\Rightarrow\; \dfrac{df}{dL} = -\left(\dfrac{1}{2}\sqrt{\dfrac{T}{\rho}}\right)L^{-2} = -\dfrac{1}{2L^2}\sqrt{\dfrac{T}{\rho}}$

(ii) $f = \dfrac{1}{2L}\sqrt{\dfrac{T}{\rho}} = \left(\dfrac{1}{2L\sqrt{\rho}}\right)T^{1/2} \;\Rightarrow\; \dfrac{df}{dT} = \dfrac{1}{2}\left(\dfrac{1}{2L\sqrt{\rho}}\right)T^{-1/2} = \dfrac{1}{4L\sqrt{T\rho}}$

(iii) $f = \dfrac{1}{2L}\sqrt{\dfrac{T}{\rho}} = \left(\dfrac{\sqrt{T}}{2L}\right)\rho^{-1/2} \;\Rightarrow\; \dfrac{df}{d\rho} = -\dfrac{1}{2}\left(\dfrac{\sqrt{T}}{2L}\right)\rho^{-3/2} = -\dfrac{\sqrt{T}}{4L\rho^{3/2}}$

(b) *Note:* Illustrating tangent lines on the generic figures may help to explain the results.

(i) $\dfrac{df}{dL} < 0$ and $L$ is decreasing $\;\Rightarrow\;$ $f$ is increasing $\;\Rightarrow\;$ higher note

(ii) $\dfrac{df}{dT} > 0$ and $T$ is increasing $\;\Rightarrow\;$ $f$ is increasing $\;\Rightarrow\;$ higher note

(iii) $\dfrac{df}{d\rho} < 0$ and $\rho$ is increasing $\;\Rightarrow\;$ $f$ is decreasing $\;\Rightarrow\;$ lower note

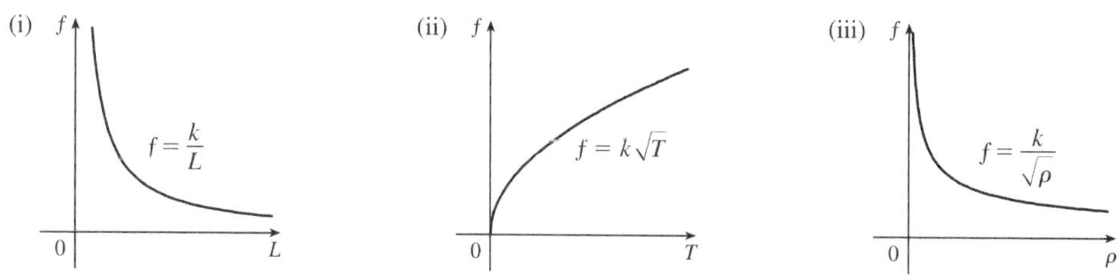


**27.** (a) $C(x) = 2000 + 3x + 0.01x^2 + 0.0002x^3 \;\Rightarrow\; C'(x) = 3 + 0.02x + 0.0006x^2$

(b) $C'(100) = 3 + 0.02(100) + 0.0006(10{,}000) = 3 + 2 + 6 = \$11$/pair. $C'(100)$ is the rate at which the cost is increasing as the 100th pair of jeans is produced. It predicts the cost of the 101st pair.

(c) The cost of manufacturing the 101st pair of jeans is
$C(101) - C(100) = (2000 + 303 + 102.01 + 206.0602) - (2000 + 300 + 100 + 200) = 11.0702 \approx \$11.07$.

**28.** (a) $C(x) = 84 + 0.16x - 0.0006x^2 + 0.000003x^3 \Rightarrow C'(x) = 0.16 - 0.0012x + 0.000009x^2 \Rightarrow C'(100) = 0.13$. This is the rate at which the cost is increasing as the 100th item is produced.

(b) $C(101) - C(100) = 97.13030299 - 97 \approx \$0.13$.

**29.** (a) $A(x) = \dfrac{p(x)}{x} \Rightarrow A'(x) = \dfrac{xp'(x) - p(x) \cdot 1}{x^2} = \dfrac{xp'(x) - p(x)}{x^2}$. $A'(x) > 0 \Rightarrow A(x)$ is increasing; that is, the average productivity increases as the size of the workforce increases.

(b) $p'(x)$ is greater than the average productivity $\Rightarrow p'(x) > A(x) \Rightarrow p'(x) > \dfrac{p(x)}{x} \Rightarrow xp'(x) > p(x) \Rightarrow$

$xp'(x) - p(x) > 0 \Rightarrow \dfrac{xp'(x) - p(x)}{x^2} > 0 \Rightarrow A'(x) > 0$.

**30.** (a) $S = \dfrac{dR}{dx} = \dfrac{(1 + 4x^{0.4})(9.6x^{-0.6}) - (40 + 24x^{0.4})(1.6x^{-0.6})}{(1 + 4x^{0.4})^2}$

$$= \frac{9.6x^{-0.6} + 38.4x^{-0.2} - 64x^{-0.6} - 38.4x^{-0.2}}{(1 + 4x^{0.4})^2} = -\frac{54.4x^{-0.6}}{(1 + 4x^{0.4})^2}$$

(b)

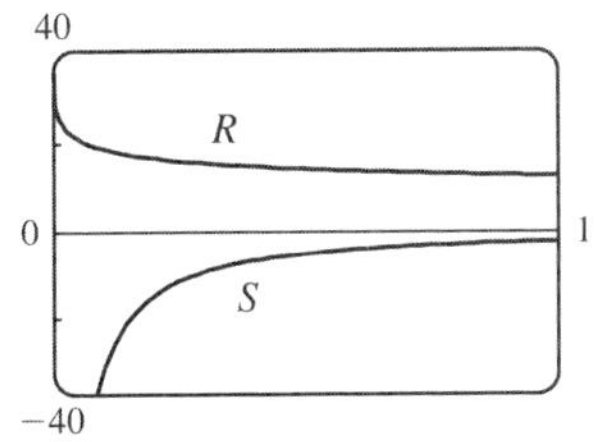


At low levels of brightness, $R$ is quite large $[R(0) = 40]$ and is quickly decreasing, that is, $S$ is negative with large absolute value. This is to be expected: at low levels of brightness, the eye is more sensitive to slight changes than it is at higher levels of brightness.

**31.** $PV = nRT \Rightarrow T = \dfrac{PV}{nR} = \dfrac{PV}{(10)(0.0821)} = \dfrac{1}{0.821}(PV)$. Using the Product Rule, we have

$\dfrac{dT}{dt} = \dfrac{1}{0.821}[P(t)V'(t) + V(t)P'(t)] = \dfrac{1}{0.821}[(8)(-0.15) + (10)(0.10)] \approx -0.2436$ K/min.

**32.** (a) If $dP/dt = 0$, the population is stable (it is constant).

(b) $\dfrac{dP}{dt} = 0 \Rightarrow \beta P = r_0\left(1 - \dfrac{P}{P_c}\right)P \Rightarrow \dfrac{\beta}{r_0} = 1 - \dfrac{P}{P_c} \Rightarrow \dfrac{P}{P_c} = 1 - \dfrac{\beta}{r_0} \Rightarrow P = P_c\left(1 - \dfrac{\beta}{r_0}\right)$.

If $P_c = 10{,}000$, $r_0 = 5\% = 0.05$, and $\beta = 4\% = 0.04$, then $P = 10{,}000\left(1 - \frac{4}{5}\right) = 2000$.

(c) If $\beta = 0.05$, then $P = 10{,}000\left(1 - \frac{5}{5}\right) = 0$. There is no stable population.

**33.** (a) If the populations are stable, then the growth rates are neither positive nor negative; that is, $\dfrac{dC}{dt} = 0$ and $\dfrac{dW}{dt} = 0$.

(b) "The caribou go extinct" means that the population is zero, or mathematically, $C = 0$.

(c) We have the equations $\dfrac{dC}{dt} = aC - bCW$ and $\dfrac{dW}{dt} = -cW + dCW$. Let $dC/dt = dW/dt = 0$, $a = 0.05$, $b = 0.001$, $c = 0.05$, and $d = 0.0001$ to obtain $0.05C - 0.001CW = 0$ **(1)** and $-0.05W + 0.0001CW = 0$ **(2)**. Adding 10 times **(2)** to **(1)** eliminates the $CW$-terms and gives us $0.05C - 0.5W = 0 \Rightarrow C = 10W$. Substituting $C = 10W$ into **(1)** results in $0.05(10W) - 0.001(10W)W = 0 \Leftrightarrow 0.5W - 0.01W^2 = 0 \Leftrightarrow 50W - W^2 = 0 \Leftrightarrow$ $W(50 - W) = 0 \Leftrightarrow W = 0$ or $50$. Since $C = 10W$, $C = 0$ or $500$. Thus, the population pairs $(C, W)$ that lead to stable populations are $(0, 0)$ and $(500, 50)$. So it is possible for the two species to live in harmony.

## 3.4 Derivatives of Trigonometric Functions

**1.** $f(x) = \sin x + \frac{1}{2}\cot x \Rightarrow f'(x) = \cos x - \frac{1}{2}\csc^2 x$

**2.** $y = 2\csc x + 5\cos x \Rightarrow y' = -2\csc x \cot x - 5\sin x$

**3.** $g(t) = t^3 \cos t \;\Rightarrow\; g'(t) = t^3(-\sin t) + (\cos t) \cdot 3t^2 = 3t^2 \cos t - t^3 \sin t$ or $t^2(3\cos t - t\sin t)$

**4.** $f(x) = \sqrt{x} \sin x \;\Rightarrow\; f'(x) = \sqrt{x} \cos x + \sin x \left(\frac{1}{2}x^{-1/2}\right) = \sqrt{x}\cos x + \dfrac{\sin x}{2\sqrt{x}}$

**5.** $y = \sec\theta \tan\theta \;\Rightarrow\; y' = \sec\theta\,(\sec^2\theta) + \tan\theta\,(\sec\theta\tan\theta) = \sec\theta\,(\sec^2\theta + \tan^2\theta)$. Using the identity $1 + \tan^2\theta = \sec^2\theta$, we can write alternative forms of the answer as $\sec\theta\,(1 + 2\tan^2\theta)$ or $\sec\theta\,(2\sec^2\theta - 1)$.

**6.** $y = e^u(\cos u + cu) \;\Rightarrow\; y' = e^u(-\sin u + c) + (\cos u + cu)e^u = e^u(\cos u - \sin u + cu + c)$

**7.** $y = \dfrac{x}{\cos x} \;\Rightarrow\; y' = \dfrac{(\cos x)(1) - (x)(-\sin x)}{(\cos x)^2} = \dfrac{\cos x + x\sin x}{\cos^2 x}$

**8.** $y = \dfrac{1+\sin x}{x + \cos x} \;\Rightarrow$

$$y' = \frac{(x+\cos x)(\cos x) - (1+\sin x)(1-\sin x)}{(x+\cos x)^2} = \frac{x\cos x + \cos^2 x - (1-\sin^2 x)}{(x+\cos x)^2}$$
$$= \frac{x\cos x + \cos^2 x - (\cos^2 x)}{(x+\cos x)^2} = \frac{x\cos x}{(x+\cos x)^2}$$

**9.** $f(\theta) = \dfrac{\sec\theta}{1+\sec\theta} \;\Rightarrow$

$$f'(\theta) = \frac{(1+\sec\theta)(\sec\theta\tan\theta) - (\sec\theta)(\sec\theta\tan\theta)}{(1+\sec\theta)^2} = \frac{(\sec\theta\tan\theta)[(1+\sec\theta) - \sec\theta]}{(1+\sec\theta)^2} = \frac{\sec\theta\tan\theta}{(1+\sec\theta)^2}$$

**10.** $y = \dfrac{1-\sec x}{\tan x} \;\Rightarrow$

$$y' = \frac{\tan x\,(-\sec x\tan x) - (1-\sec x)(\sec^2 x)}{(\tan x)^2} = \frac{\sec x\left(-\tan^2 x - \sec x + \sec^2 x\right)}{\tan^2 x} = \frac{\sec x\,(1-\sec x)}{\tan^2 x}$$

**11.** Using Exercise 3.2.45(a), $f(x) = xe^x \csc x \;\Rightarrow$

$$f'(x) = (x)'e^x\csc x + x(e^x)'\csc x + xe^x(\csc x)' = 1e^x\csc x + xe^x\csc x + xe^x(-\cot x\csc x)$$
$$= e^x\csc x(1 + x - x\cot x)$$

**12.** $y = \csc\theta\,(\theta + \cot\theta) \;\Rightarrow$

$$y' = \csc\theta\,(1 - \csc^2\theta) + (\theta + \cot\theta)(-\csc\theta\cot\theta) = \csc\theta\,(1 - \csc^2\theta - \theta\cot\theta - \cot^2\theta)$$
$$= \csc\theta\,(-\cot^2\theta - \theta\cot\theta - \cot^2\theta) \qquad \left[1 + \cot^2\theta = \csc^2\theta\right]$$
$$= \csc\theta\,(-\theta\cot\theta - 2\cot^2\theta) = -\csc\theta\cot\theta\,(\theta + 2\cot\theta)$$

**13.** $\dfrac{d}{dx}(\csc x) = \dfrac{d}{dx}\left(\dfrac{1}{\sin x}\right) = \dfrac{(\sin x)(0) - 1(\cos x)}{\sin^2 x} = \dfrac{-\cos x}{\sin^2 x} = -\dfrac{1}{\sin x}\cdot\dfrac{\cos x}{\sin x} = -\csc x\cot x$

**14.** $\dfrac{d}{dx}(\sec x) = \dfrac{d}{dx}\left(\dfrac{1}{\cos x}\right) = \dfrac{(\cos x)(0) - 1(-\sin x)}{\cos^2 x} = \dfrac{\sin x}{\cos^2 x} = \dfrac{1}{\cos x}\cdot\dfrac{\sin x}{\cos x} = \sec x\tan x$

**15.** $\dfrac{d}{dx}(\cot x) = \dfrac{d}{dx}\left(\dfrac{\cos x}{\sin x}\right) = \dfrac{(\sin x)(-\sin x) - (\cos x)(\cos x)}{\sin^2 x} = -\dfrac{\sin^2 x + \cos^2 x}{\sin^2 x} = -\dfrac{1}{\sin^2 x} = -\csc^2 x$

**16.** $f(x) = \cos x \;\Rightarrow$

$$f'(x) = \lim_{h\to 0}\frac{f(x+h) - f(x)}{h} = \lim_{h\to 0}\frac{\cos(x+h) - \cos x}{h} = \lim_{h\to 0}\frac{\cos x\cos h - \sin x\sin h - \cos x}{h}$$
$$= \lim_{h\to 0}\left(\cos x\,\frac{\cos h - 1}{h} - \sin x\,\frac{\sin h}{h}\right) = \cos x\lim_{h\to 0}\frac{\cos h - 1}{h} - \sin x\lim_{h\to 0}\frac{\sin h}{h}$$
$$= (\cos x)(0) - (\sin x)(1) = -\sin x$$

**17.** $y = \sec x \;\Rightarrow\; y' = \sec x\tan x$, so $y'\left(\frac{\pi}{3}\right) = \sec\frac{\pi}{3}\tan\frac{\pi}{3} = 2\sqrt{3}$. An equation of the tangent line to the curve $y = \sec x$ at the point $\left(\frac{\pi}{3}, 2\right)$ is $y - 2 = 2\sqrt{3}\left(x - \frac{\pi}{3}\right)$ or $y = 2\sqrt{3}\,x + 2 - \frac{2}{3}\sqrt{3}\,\pi$.

**18.** $y = e^x \cos x \quad \Rightarrow \quad y' = e^x(-\sin x) + (\cos x)e^x = e^x(\cos x - \sin x) \quad \Rightarrow$ the slope of the tangent line at $(0, 1)$ is $e^0(\cos 0 - \sin 0) = 1(1 - 0) = 1$ and an equation is $y - 1 = 1(x - 0)$ or $y = x + 1$.

**19.** (a) $y = x \cos x \quad \Rightarrow \quad y' = x(-\sin x) + \cos x(1) = \cos x - x \sin x$.

So the slope of the tangent at the point $(\pi, -\pi)$ is

$\cos \pi - \pi \sin \pi = -1 - \pi(0) = -1$, and an equation is

$y + \pi = -(x - \pi)$ or $y = -x$.

(b)

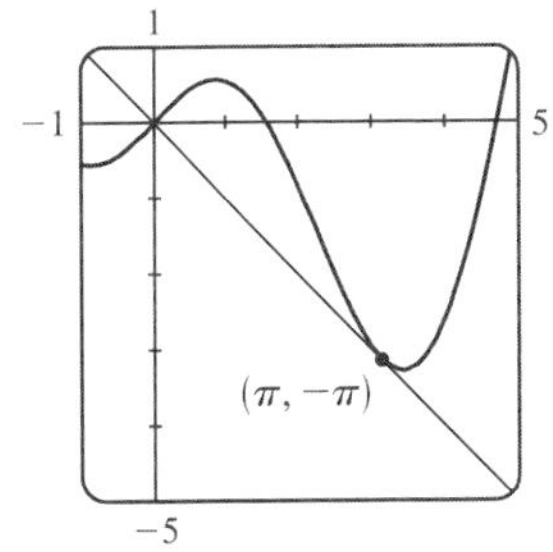


**20.** (a) $y = \sec x - 2 \cos x \quad \Rightarrow \quad y' = \sec x \tan x + 2 \sin x \quad \Rightarrow$

the slope of the tangent line at $\left(\frac{\pi}{3}, 1\right)$ is

$\sec \frac{\pi}{3} \tan \frac{\pi}{3} + 2 \sin \frac{\pi}{3} = 2 \cdot \sqrt{3} + 2 \cdot \frac{\sqrt{3}}{2} = 3\sqrt{3}$ and an equation

is $y - 1 = 3\sqrt{3}\left(x - \frac{\pi}{3}\right)$ or $y = 3\sqrt{3}x + 1 - \pi\sqrt{3}$.

(b)

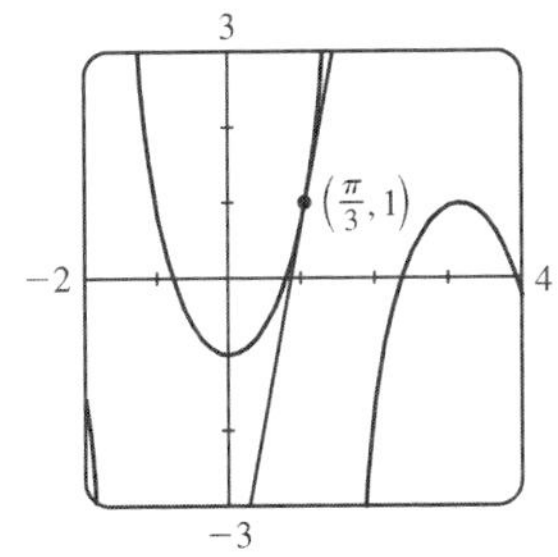


**21.** (a) $f(x) = 2x + \cot x \quad \Rightarrow \quad f'(x) = 2 - \csc^2 x$

(b)

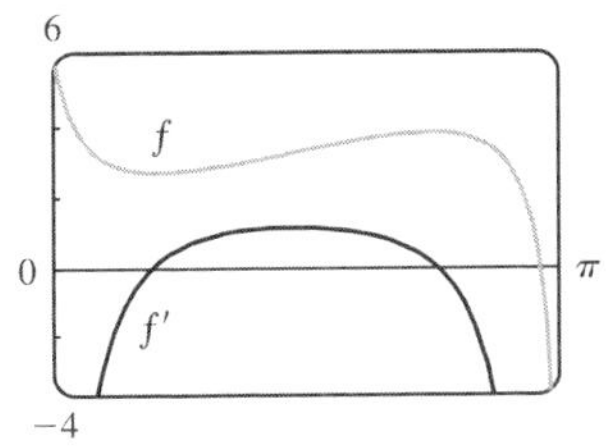


Notice that $f'(x) = 0$ when $f$ has a horizontal tangent.

$f'$ is positive when $f$ is increasing and $f'$ is negative when $f$ is decreasing.

Also, $f'(x)$ is large negative when the graph of $f$ is steep.

**22.** (a) $f(x) = e^x \cos x \quad \Rightarrow f'(x) = e^x(-\sin x) + (\cos x)\, e^x = e^x(\cos x - \sin x) \quad \Rightarrow$

$f''(x) = e^x(-\sin x - \cos x) + (\cos x - \sin x)\, e^x = e^x(-\sin x - \cos x + \cos x - \sin x) = -2e^x \sin x$

(b)

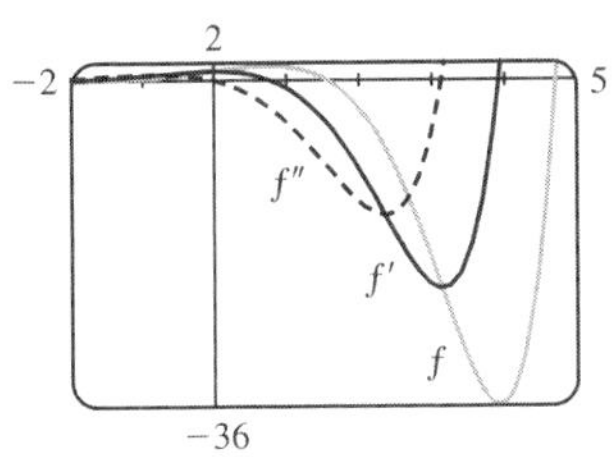


Note that $f' = 0$ where $f$ has a minimum and $f'' = 0$ where $f'$ has a minimum. Also note that $f'$ is negative when $f$ is decreasing and $f''$ is negative when $f'$ is decreasing.

**23.** $H(\theta) = \theta \sin \theta \quad \Rightarrow \quad H'(\theta) = \theta\,(\cos \theta) + (\sin \theta) \cdot 1 = \theta \cos \theta + \sin \theta \quad \Rightarrow$
$H''(\theta) = \theta\,(-\sin \theta) + (\cos \theta) \cdot 1 + \cos \theta = -\theta \sin \theta + 2 \cos \theta$

**24.** $f(x) = \sec x \quad \Rightarrow \quad f'(x) = \sec x \tan x \quad \Rightarrow \quad f''(x) = \sec x\,(\sec^2 x) + \tan x\,(\sec x \tan x) = \sec x\,(\sec^2 x + \tan^2 x)$.
$f''\left(\frac{\pi}{4}\right) = \sqrt{2}\left[\left(\sqrt{2}\right)^2 + 1^2\right] = \sqrt{2}\,(2 + 1) = 3\sqrt{2}$

**25.** (a) $f(x) = \dfrac{\tan x - 1}{\sec x} \quad \Rightarrow$

$$f'(x) = \frac{\sec x(\sec^2 x) - (\tan x - 1)(\sec x \tan x)}{(\sec x)^2} = \frac{\sec x(\sec^2 x - \tan^2 x + \tan x)}{\sec^2 x} = \frac{1 + \tan x}{\sec x}$$

(b) $f(x) = \dfrac{\tan x - 1}{\sec x} = \dfrac{\dfrac{\sin x}{\cos x} - 1}{\dfrac{1}{\cos x}} = \dfrac{\dfrac{\sin x - \cos x}{\cos x}}{\dfrac{1}{\cos x}} = \sin x - \cos x \quad \Rightarrow \quad f'(x) = \cos x - (-\sin x) = \cos x + \sin x$

(c) From part (a), $f'(x) = \dfrac{1 + \tan x}{\sec x} = \dfrac{1}{\sec x} + \dfrac{\tan x}{\sec x} = \cos x + \sin x$, which is the expression for $f'(x)$ in part (b).

**26.** (a) $g(x) = f(x)\sin x \;\Rightarrow\; g'(x) = f(x)\cos x + \sin x \cdot f'(x)$, so

$g'(\frac{\pi}{3}) = f(\frac{\pi}{3})\cos\frac{\pi}{3} + \sin\frac{\pi}{3} \cdot f'(\frac{\pi}{3}) = 4 \cdot \frac{1}{2} + \frac{\sqrt{3}}{2} \cdot (-2) = 2 - \sqrt{3}$

(b) $h(x) = \dfrac{\cos x}{f(x)} \;\Rightarrow\; h'(x) = \dfrac{f(x) \cdot (-\sin x) - \cos x \cdot f'(x)}{[f(x)]^2}$, so

$$h'(\tfrac{\pi}{3}) = \frac{f(\frac{\pi}{3}) \cdot (-\sin\frac{\pi}{3}) - \cos\frac{\pi}{3} \cdot f'(\frac{\pi}{3})}{[f(\frac{\pi}{3})]^2} = \frac{4\left(-\frac{\sqrt{3}}{2}\right) - \left(\frac{1}{2}\right)(-2)}{4^2} = \frac{-2\sqrt{3}+1}{16} = \frac{1-2\sqrt{3}}{16}$$

**27.** $f(x) = x + 2\sin x$ has a horizontal tangent when $f'(x) = 0 \;\Leftrightarrow\; 1 + 2\cos x = 0 \;\Leftrightarrow\; \cos x = -\frac{1}{2} \;\Leftrightarrow$ $x = \frac{2\pi}{3} + 2\pi n$ or $\frac{4\pi}{3} + 2\pi n$, where $n$ is an integer. Note that $\frac{4\pi}{3}$ and $\frac{2\pi}{3}$ are $\pm\frac{\pi}{3}$ units from $\pi$. This allows us to write the solutions in the more compact equivalent form $(2n+1)\pi \pm \frac{\pi}{3}$, $n$ an integer.

**28.** $y = \dfrac{\cos x}{2 + \sin x} \;\Rightarrow\; y' = \dfrac{(2+\sin x)(-\sin x) - \cos x\cos x}{(2+\sin x)^2} = \dfrac{-2\sin x - \sin^2 x - \cos^2 x}{(2+\sin x)^2} = \dfrac{-2\sin x - 1}{(2+\sin x)^2} = 0$ when $-2\sin x - 1 = 0 \;\Leftrightarrow\; \sin x = -\frac{1}{2} \;\Leftrightarrow\; x = \frac{11\pi}{6} + 2\pi n$ or $x = \frac{7\pi}{6} + 2\pi n$, $n$ an integer. So $y = \frac{1}{\sqrt{3}}$ or $y = -\frac{1}{\sqrt{3}}$ and the points on the curve with horizontal tangents are: $\left(\frac{11\pi}{6} + 2\pi n, \frac{1}{\sqrt{3}}\right)$, $\left(\frac{7\pi}{6} + 2\pi n, -\frac{1}{\sqrt{3}}\right)$, $n$ an integer.

**29.** $f(x) = x - 2\sin x$, $0 \le x \le 2\pi$. $f'(x) = 1 - 2\cos x$. So $f'(x) > 0 \;\Leftrightarrow\; 1 - 2\cos x > 0 \;\Leftrightarrow\; -2\cos x > -1 \;\Leftrightarrow$ $\cos x < \frac{1}{2} \;\Leftrightarrow\; \frac{\pi}{3} < x < \frac{5\pi}{3} \;\Rightarrow\; f$ is increasing on $(\frac{\pi}{3}, \frac{5\pi}{3})$.

**30.** $f(x) = x - \sin x$, $0 \le x \le 2\pi$. $f'(x) = 1 - \cos x \;\Rightarrow\; f''(x) = \sin x$. Since $\sin x > 0$ on $(0, \pi)$, $f$ is concave upward on $(0, \pi)$.

**31.** (a) $x(t) = 8\sin t \;\Rightarrow\; v(t) = x'(t) = 8\cos t \;\Rightarrow\; a(t) = x''(t) = -8\sin t$

(b) The mass at time $t = \frac{2\pi}{3}$ has position $x(\frac{2\pi}{3}) = 8\sin\frac{2\pi}{3} = 8\left(\frac{\sqrt{3}}{2}\right) = 4\sqrt{3}$, velocity $v(\frac{2\pi}{3}) = 8\cos\frac{2\pi}{3} = 8(-\frac{1}{2}) = -4$, and acceleration $a(\frac{2\pi}{3}) = -8\sin\frac{2\pi}{3} = -8\left(\frac{\sqrt{3}}{2}\right) = -4\sqrt{3}$. Since $v(\frac{2\pi}{3}) < 0$, the particle is moving to the left. Because $v$ and $a$ have the same sign, the particle is speeding up.

**32.** (a) $s(t) = 2\cos t + 3\sin t \;\Rightarrow$

$v(t) = -2\sin t + 3\cos t \;\Rightarrow$

$a(t) = -2\cos t - 3\sin t$

(b) 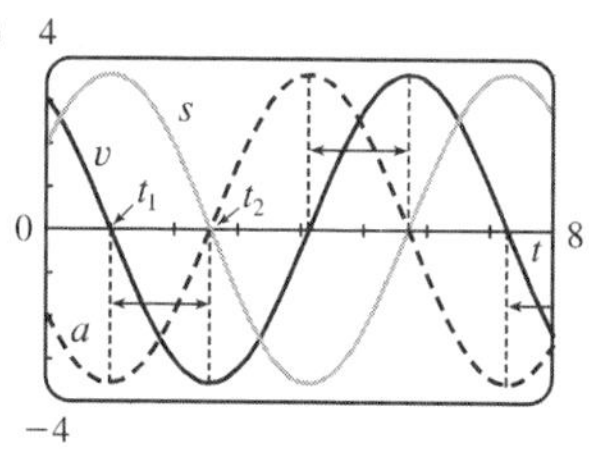


(c) $s = 0 \;\Rightarrow\; t_2 \approx 2.55$. So the mass passes through the equilibrium position for the first time when $t \approx 2.55$ s.

(d) $v = 0 \;\Rightarrow\; t_1 \approx 0.98$, $s(t_1) \approx 3.61$ cm. So the mass travels a maximum of about 3.6 cm (upward and downward) from its equilibrium position.

(e) The speed $|v|$ is greatest when $s = 0$; that is, when $t = t_2 + n\pi$, $n$ a positive integer. The mass is speeding up when $v$ and $a$ have the same sign. From the figure, we see that this is the case on the intervals $(t_1 + n\pi, t_2 + n\pi)$ where $n$ is a whole number.

**33.** 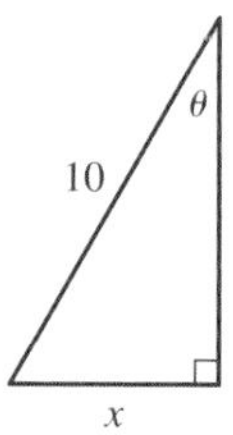


From the diagram we can see that $\sin\theta = x/10 \;\Leftrightarrow\; x = 10\sin\theta$. We want to find the rate of change of $x$ with respect to $\theta$; that is, $dx/d\theta$. Taking the derivative of the above expression, $dx/d\theta = 10(\cos\theta)$. So when $\theta = \frac{\pi}{3}$, $\frac{dx}{d\theta} = 10\cos\frac{\pi}{3} = 10(\frac{1}{2}) = 5$ ft/rad.

**34.** (a) $F = \dfrac{\mu W}{\mu \sin\theta + \cos\theta} \quad\Rightarrow\quad \dfrac{dF}{d\theta} = \dfrac{(\mu\sin\theta + \cos\theta)(0) - \mu W(\mu\cos\theta - \sin\theta)}{(\mu\sin\theta+\cos\theta)^2} = \dfrac{\mu W(\sin\theta - \mu\cos\theta)}{(\mu\sin\theta + \cos\theta)^2}$

(b) $\dfrac{dF}{d\theta} = 0 \quad\Rightarrow\quad \mu W(\sin\theta - \mu\cos\theta) = 0 \quad\Rightarrow\quad \sin\theta = \mu\cos\theta \quad\Rightarrow\quad \tan\theta = \mu \quad\Rightarrow\quad \theta = \tan^{-1}\mu$

(c)

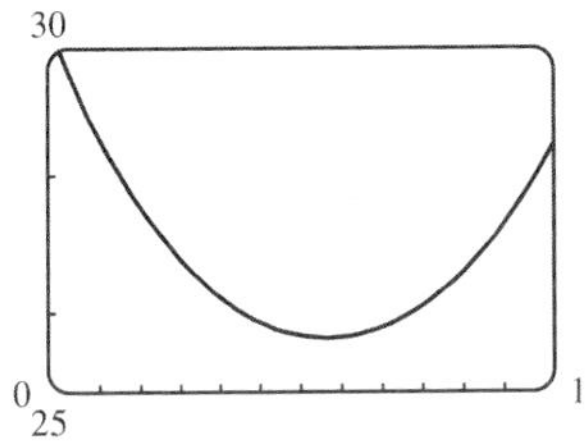


From the graph of $F = \dfrac{0.6(50)}{0.6\sin\theta + \cos\theta}$ for $0 \le \theta \le 1$, we see that $\dfrac{dF}{d\theta} = 0 \quad\Rightarrow\quad \theta \approx 0.54$. Checking this with part (b) and $\mu = 0.6$, we calculate $\theta = \tan^{-1} 0.6 \approx 0.54$. So the value from the graph is consistent with the value in part (b).

**35.** $\dfrac{d}{dx}(\sin x) = \cos x \quad\Rightarrow\quad \dfrac{d^2}{dx^2}(\sin x) = -\sin x \quad\Rightarrow\quad \dfrac{d^3}{dx^3}(\sin x) = -\cos x \quad\Rightarrow\quad \dfrac{d^4}{dx^4}(\sin x) = \sin x.$

The derivatives of $\sin x$ occur in a cycle of four. Since $99 = 4(24) + 3$, we have $\dfrac{d^{99}}{dx^{99}}(\sin x) = \dfrac{d^3}{dx^3}(\sin x) = -\cos x.$

**36.** Let $f(x) = x\sin x$ and $h(x) = \sin x$, so $f(x) = xh(x)$. Then $f'(x) = h(x) + xh'(x)$,
$f''(x) = h'(x) + h'(x) + xh''(x) = 2h'(x) + xh''(x)$,
$f'''(x) = 2h''(x) + h''(x) + xh'''(x) = 3h''(x) + xh'''(x), \cdots, f^{(n)}(x) = nh^{(n-1)}(x) + xh^{(n)}(x)$.

Since $34 = 4(8) + 2$, we have $h^{(34)}(x) = h^{(2)}(x) = \dfrac{d^2}{dx^2}(\sin x) = -\sin x$ and $h^{(35)}(x) = -\cos x$.

Thus, $\dfrac{d^{35}}{dx^{35}}(x\sin x) = 35h^{(34)}(x) + xh^{(35)}(x) = -35\sin x - x\cos x$.

**37.** $y = A\sin x + B\cos x \quad\Rightarrow\quad y' = A\cos x - B\sin x \quad\Rightarrow\quad y'' = -A\sin x - B\cos x$. Substituting these expressions for $y$, $y'$, and $y''$ into the given differential equation $y'' + y' - 2y = \sin x$ gives us
$(-A\sin x - B\cos x) + (A\cos x - B\sin x) - 2(A\sin x + B\cos x) = \sin x \quad\Leftrightarrow$
$-3A\sin x - B\sin x + A\cos x - 3B\cos x = \sin x \quad\Leftrightarrow\quad (-3A - B)\sin x + (A - 3B)\cos x = 1\sin x$, so we must have $-3A - B = 1$ and $A - 3B = 0$ (since 0 is the coefficient of $\cos x$ on the right side). Solving for $A$ and $B$, we add the first equation to three times the second to get $B = -\frac{1}{10}$ and $A = -\frac{3}{10}$.

**38.** (a) If $x \to 0$, then $5x \to 0$, and hence $\theta\ [= 5x] \to 0$. $\displaystyle\lim_{x\to 0}\frac{\sin 5x}{x} = \lim_{\theta\to 0}\frac{\sin\theta}{\frac{1}{5}\theta} = 5\lim_{\theta\to 0}\frac{\sin\theta}{\theta} = 5(1) = 5.$

(b) Let $f(x) = \sin 5x$. Then

$$\begin{aligned}\frac{d}{dx}(\sin 5x) = f'(x) &= \lim_{h\to 0}\frac{f(x+h) - f(x)}{h} = \lim_{h\to 0}\frac{\sin[5(x+h)] - \sin 5x}{h} = \lim_{h\to 0}\frac{\sin(5x+5h) - \sin 5x}{h}\\ &= \lim_{h\to 0}\frac{\sin 5x\cos 5h + \cos 5x\sin 5h - \sin 5x}{h} = \lim_{h\to 0}\left(\sin 5x\cdot\frac{\cos 5h - 1}{h} + \cos 5x\,\frac{\sin 5h}{h}\right)\\ &= \sin 5x\lim_{h\to 0}\frac{\cos 5h - 1}{h} + \cos 5x\lim_{h\to 0}\frac{\sin 5h}{h}\end{aligned}$$

The first limit can be shown to equal 0 in a manner similar to part (a) and Equation 3. The second limit is equal to 5, by part (a). Thus, we have $f'(x) = (\sin 5x)(0) + (\cos 5x)(5) = 5\cos 5x$.

**39.**
$$\begin{aligned}\lim_{t\to 0}\frac{\tan 6t}{\sin 2t} &= \lim_{t\to 0}\left(\frac{\sin 6t}{t}\cdot\frac{1}{\cos 6t}\cdot\frac{t}{\sin 2t}\right) = \lim_{t\to 0}\frac{6\sin 6t}{6t}\cdot\lim_{t\to 0}\frac{1}{\cos 6t}\cdot\lim_{t\to 0}\frac{2t}{2\sin 2t}\\ &= 6\lim_{t\to 0}\frac{\sin 6t}{6t}\cdot\lim_{t\to 0}\frac{1}{\cos 6t}\cdot\frac{1}{2}\lim_{t\to 0}\frac{2t}{\sin 2t} = 6(1)\cdot\frac{1}{1}\cdot\frac{1}{2}(1) = 3\end{aligned}$$

**40.** $$\lim_{x\to 0} x\cot x = \lim_{x\to 0} x\cdot\frac{\cos x}{\sin x} = \lim_{x\to 0}\frac{x\cos x}{\sin x} = \lim_{x\to 0}\frac{\dfrac{x\cos x}{x}}{\dfrac{\sin x}{x}} = \lim_{x\to 0}\frac{\dfrac{\cos x}{1}}{\dfrac{\sin x}{x}} = \frac{\displaystyle\lim_{x\to 0}\cos x}{\displaystyle\lim_{x\to 0}\frac{\sin x}{x}} = \frac{1}{1} = 1$$

**41.** $\displaystyle\lim_{\theta\to 0}\frac{\sin\theta}{\theta+\tan\theta}=\frac{\displaystyle\lim_{\theta\to 0}\frac{\sin\theta}{\theta}}{\displaystyle\lim_{\theta\to 0}\frac{\theta+\tan\theta}{\theta}}=\frac{1}{\displaystyle\lim_{\theta\to 0}\left(1+\frac{\sin\theta}{\theta}\cdot\frac{1}{\cos\theta}\right)}=\frac{1}{1+1\cdot 1}=\frac{1}{2}$

**42.** (a) Let $\theta=\dfrac{1}{x}$. Then as $x\to\infty$, $\theta\to 0$, and $\displaystyle\lim_{x\to\infty}x\sin\frac{1}{x}=\lim_{\theta\to 0}\frac{1}{\theta}\sin\theta=\lim_{\theta\to 0}\frac{\sin\theta}{\theta}=1.$

(b) Since $-1\le\sin(1/x)\le 1$, we have (as illustrated in the figure) $-|x|\le x\sin(1/x)\le|x|$. We know that $\displaystyle\lim_{x\to 0}(|x|)=0$ and $\displaystyle\lim_{x\to 0}(-|x|)=0$; so by the Squeeze Theorem, $\displaystyle\lim_{x\to 0}x\sin(1/x)=0.$

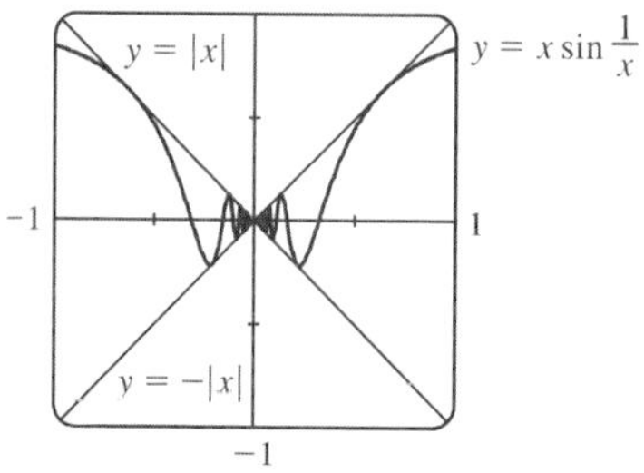


(c)

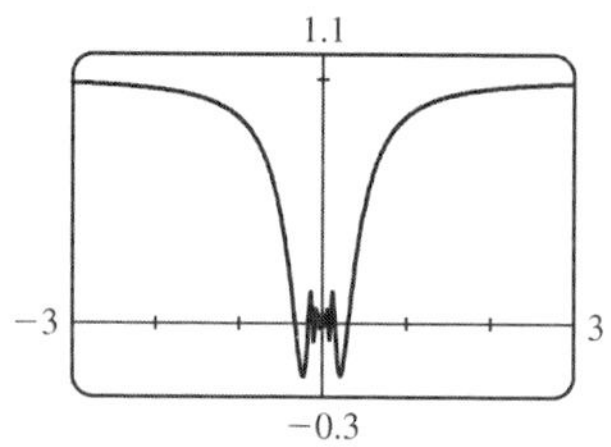


**43.** By the definition of radian measure, $s=r\theta$, where $r$ is the radius of the circle.By drawing the bisector of the angle $\theta$, we can see that $\sin\dfrac{\theta}{2}=\dfrac{d/2}{r}\Rightarrow d=2r\sin\dfrac{\theta}{2}$. So $\displaystyle\lim_{\theta\to 0^+}\frac{s}{d}=\lim_{\theta\to 0^+}\frac{r\theta}{2r\sin(\theta/2)}=\lim_{\theta\to 0^+}\frac{2\cdot(\theta/2)}{2\sin(\theta/2)}=\lim_{\theta\to 0}\frac{\theta/2}{\sin(\theta/2)}=1.$ [This is just the reciprocal of the limit $\lim_{x\to 0}\frac{\sin x}{x}=1$ combined with the fact that as $\theta\to 0$, $\frac{\theta}{2}\to 0$ also.]

**44.** Let $|PR|=x$. Then we get the following formulas for $r$ and $h$ in terms of $\theta$ and $x$:

$$\sin\frac{\theta}{2}=\frac{r}{x}\quad\Rightarrow\quad r=x\sin\frac{\theta}{2}\textit{ and }\cos\frac{\theta}{2}=\frac{h}{x}\quad\Rightarrow\quad h=x\cos\frac{\theta}{2}$$

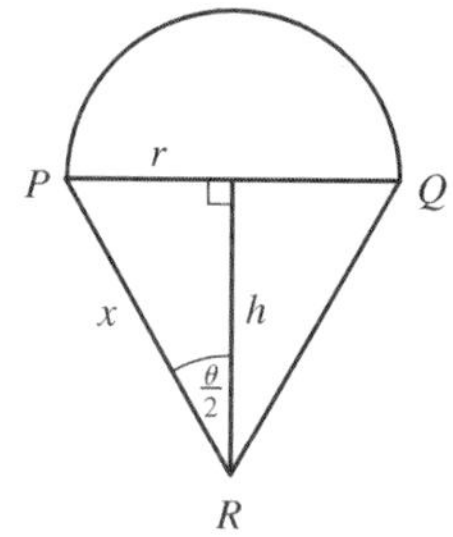


Now $A(\theta)=\frac{1}{2}\pi r^2$ and $B(\theta)=\frac{1}{2}(2r)h=rh$. So

$$\lim_{\theta\to 0^+}\frac{A(\theta)}{B(\theta)}=\lim_{\theta\to 0^+}\frac{\frac{1}{2}\pi r^2}{rh}=\tfrac{1}{2}\pi\lim_{\theta\to 0^+}\frac{r}{h}=\tfrac{1}{2}\pi\lim_{\theta\to 0^+}\frac{x\sin(\theta/2)}{x\cos(\theta/2)}$$
$$=\tfrac{1}{2}\pi\lim_{\theta\to 0^+}\tan(\theta/2)=0$$

## 3.5 The Chain Rule

**1.** Let $u=g(x)=4x$ and $y=f(u)=\sin u$. Then $\dfrac{dy}{dx}=\dfrac{dy}{du}\dfrac{du}{dx}=(\cos u)(4)=4\cos 4x.$

**2.** Let $u=g(x)=4+3x$ and $y=f(u)=\sqrt{u}=u^{1/2}$. Then $\dfrac{dy}{dx}=\dfrac{dy}{du}\dfrac{du}{dx}=\frac{1}{2}u^{-1/2}(3)=\dfrac{3}{2\sqrt{u}}=\dfrac{3}{2\sqrt{4+3x}}.$

**3.** Let $u=g(x)=1-x^2$ and $y=f(u)=u^{10}$. Then $\dfrac{dy}{dx}=\dfrac{dy}{du}\dfrac{du}{dx}=(10u^9)(-2x)=-20x(1-x^2)^9.$

**4.** Let $u=g(x)=\sin x$ and $y=f(u)=\tan u$. Then $\dfrac{dy}{dx}=\dfrac{dy}{du}\dfrac{du}{dx}=(\sec^2 u)(\cos x)=\sec^2(\sin x)\cdot\cos x,$ or equivalently, $[\sec(\sin x)]^2\cos x$.

**5.** Let $u=g(x)=\sqrt{x}$ and $y=f(u)=e^u$. Then $\dfrac{dy}{dx}=\dfrac{dy}{du}\dfrac{du}{dx}=(e^u)\left(\frac{1}{2}x^{-1/2}\right)=e^{\sqrt{x}}\cdot\dfrac{1}{2\sqrt{x}}=\dfrac{e^{\sqrt{x}}}{2\sqrt{x}}.$

**6.** Let $u=g(x)=e^x$ and $y=f(u)=\sin u$. Then $\dfrac{dy}{dx}=\dfrac{dy}{du}\dfrac{du}{dx}=(\cos u)(e^x)=e^x\cos e^x.$

**7.** $F(x) = \sqrt[4]{1+2x+x^3} = (1+2x+x^3)^{1/4} \quad \Rightarrow$

$$F'(x) = \tfrac{1}{4}(1+2x+x^3)^{-3/4} \cdot \frac{d}{dx}\,(1+2x+x^3) = \frac{1}{4(1+2x+x^3)^{3/4}} \cdot (2+3x^2)$$

$$= \frac{2+3x^2}{4(1+2x+x^3)^{3/4}} = \frac{2+3x^2}{4\,\sqrt[4]{(1+2x+x^3)^3}}$$

**8.** $F(x) = (x^2-x+1)^3 \quad \Rightarrow \quad F'(x) = 3(x^2-x+1)^2(2x-1)$

**9.** $g(t) = \dfrac{1}{(t^4+1)^3} = (t^4+1)^{-3} \quad \Rightarrow \quad g'(t) = -3(t^4+1)^{-4}(4t^3) = -12t^3(t^4+1)^{-4} = \dfrac{-12t^3}{(t^4+1)^4}$

**10.** $f(t) = \sqrt[3]{1+\tan t} = (1+\tan t)^{1/3} \quad \Rightarrow \quad f'(t) = \frac{1}{3}(1+\tan t)^{-2/3}\sec^2 t = \dfrac{\sec^2 t}{3\,\sqrt[3]{(1+\tan t)^2}}$

**11.** $y = \cos(a^3+x^3) \quad \Rightarrow \quad y' = -\sin(a^3+x^3)\cdot 3x^2 \quad [a^3 \text{ is just a constant}] \quad = -3x^2\sin(a^3+x^3)$

**12.** $y = a^3+\cos^3 x \quad \Rightarrow \quad y' = 3(\cos x)^2(-\sin x) \quad [a^3 \text{ is just a constant}] \quad = -3\sin x\cos^2 x$

**13.** $h(t) = t^3 - 3^t \quad \Rightarrow \quad h'(t) = 3t^2 - 3^t \ln 3 \qquad$ [by Formula 5]

**14.** $y = 3\cot(n\theta) \quad \Rightarrow \quad y' = 3[-\csc^2(n\theta)\cdot n] = -3n\csc^2(n\theta)$

**15.** $y = xe^{-x^2} \quad \Rightarrow \quad y' = xe^{-x^2}(-2x) + e^{-x^2}\cdot 1 = e^{-x^2}(-2x^2+1) = e^{-x^2}(1-2x^2)$

**16.** $y = e^{-5x}\cos 3x \quad \Rightarrow \quad y' = e^{-5x}(-3\sin 3x) + (\cos 3x)(-5e^{-5x}) = -e^{-5x}(3\sin 3x + 5\cos 3x)$

**17.** $g(x) = (1+4x)^5(3+x-x^2)^8 \quad \Rightarrow$

$$\begin{aligned} g'(x) &= (1+4x)^5\cdot 8(3+x-x^2)^7(1-2x) + (3+x-x^2)^8\cdot 5(1+4x)^4\cdot 4 \\ &= 4(1+4x)^4(3+x-x^2)^7[2(1+4x)(1-2x)+5(3+x-x^2)] \\ &= 4(1+4x)^4(3+x-x^2)^7[(2+4x-16x^2)+(15+5x-5x^2)] = 4(1+4x)^4(3+x-x^2)^7(17+9x-21x^2) \end{aligned}$$

**18.** $h(t) = (t^4-1)^3(t^3+1)^4 \quad \Rightarrow$

$$\begin{aligned} h'(t) &= (t^4-1)^3\cdot 4(t^3+1)^3(3t^2) + (t^3+1)^4\cdot 3(t^4-1)^2(4t^3) \\ &= 12t^2(t^4-1)^2(t^3+1)^3[(t^4-1)+t(t^3+1)] = 12t^2(t^4-1)^2(t^3+1)^3(2t^4+t-1) \end{aligned}$$

**19.** $y = e^{x\cos x} \quad \Rightarrow \quad y' = e^{x\cos x}\cdot \dfrac{d}{dx}\,(x\cos x) = e^{x\cos x}\,[x(-\sin x) + (\cos x)\cdot 1] = e^{x\cos x}(\cos x - x\sin x)$

**20.** Using Formula 5 and the Chain Rule, $y = 10^{1-x^2} \quad \Rightarrow \quad y' = 10^{1-x^2}(\ln 10)\cdot \dfrac{d}{dx}\,(1-x^2) = -2x(\ln 10)10^{1-x^2}$.

**21.** $F(z) = \sqrt{\dfrac{z-1}{z+1}} = \left(\dfrac{z-1}{z+1}\right)^{1/2} \quad \Rightarrow$

$$\begin{aligned} F'(z) &= \frac{1}{2}\left(\frac{z-1}{z+1}\right)^{-1/2}\cdot \frac{d}{dz}\left(\frac{z-1}{z+1}\right) = \frac{1}{2}\left(\frac{z+1}{z-1}\right)^{1/2}\cdot \frac{(z+1)(1)-(z-1)(1)}{(z+1)^2} \\ &= \frac{1}{2}\frac{(z+1)^{1/2}}{(z-1)^{1/2}}\cdot \frac{z+1-z+1}{(z+1)^2} = \frac{1}{2}\frac{(z+1)^{1/2}}{(z-1)^{1/2}}\cdot \frac{2}{(z+1)^2} = \frac{1}{(z-1)^{1/2}(z+1)^{3/2}} \end{aligned}$$

**22.** $G(y) = \left(\dfrac{y^2}{y+1}\right)^5 \quad \Rightarrow \quad G'(y) = 5\left(\dfrac{y^2}{y+1}\right)^4\cdot \dfrac{(y+1)(2y)-y^2(1)}{(y+1)^2} = 5\cdot \dfrac{y^8}{(y+1)^4}\cdot \dfrac{y(2y+2-y)}{(y+1)^2} = \dfrac{5y^9(y+2)}{(y+1)^6}$

**23.** $y = \sec^2 x + \tan^2 x = (\sec x)^2 + (\tan x)^2 \quad \Rightarrow$

$$y' = 2(\sec x)(\sec x\tan x) + 2(\tan x)(\sec^2 x) = 2\sec^2 x\tan x + 2\sec^2 x\tan x = 4\sec^2 x\tan x$$

**24.** $y = e^{k\tan\sqrt{x}} \quad \Rightarrow \quad y' = e^{k\tan\sqrt{x}}\cdot \dfrac{d}{dx}\,(k\tan\sqrt{x}\,) = e^{k\tan\sqrt{x}}\left(k\sec^2\sqrt{x}\cdot \tfrac{1}{2}x^{-1/2}\right) = \dfrac{k\sec^2\sqrt{x}}{2\sqrt{x}}\,e^{k\tan\sqrt{x}}$

**25.** $y = \dfrac{r}{\sqrt{r^2+1}} \quad \Rightarrow$

$$y' = \frac{\sqrt{r^2+1}\,(1) - r\cdot\frac{1}{2}(r^2+1)^{-1/2}(2r)}{\left(\sqrt{r^2+1}\right)^2} = \frac{\sqrt{r^2+1} - \dfrac{r^2}{\sqrt{r^2+1}}}{\left(\sqrt{r^2+1}\right)^2} = \frac{\dfrac{\sqrt{r^2+1}\,\sqrt{r^2+1} - r^2}{\sqrt{r^2+1}}}{\left(\sqrt{r^2+1}\right)^2}$$

$$= \frac{(r^2+1) - r^2}{\left(\sqrt{r^2+1}\right)^3} = \frac{1}{(r^2+1)^{3/2}} \text{ or } (r^2+1)^{-3/2}$$

*Another solution:* Write $y$ as a product and make use of the Product Rule. $y = r(r^2+1)^{-1/2} \quad \Rightarrow$
$y' = r\cdot -\frac{1}{2}(r^2+1)^{-3/2}(2r) + (r^2+1)^{-1/2}\cdot 1 = (r^2+1)^{-3/2}[-r^2 + (r^2+1)^1] = (r^2+1)^{-3/2}(1) = (r^2+1)^{-3/2}$.
The step that students usually have trouble with is factoring out $(r^2+1)^{-3/2}$. But this is no different than factoring out $x^2$ from $x^2 + x^5$; that is, we are just factoring out a factor with the *smallest* exponent that appears on it. In this case, $-\frac{3}{2}$ is smaller than $-\frac{1}{2}$.

**26.** $y = \dfrac{e^u - e^{-u}}{e^u + e^{-u}} \quad \Rightarrow$

$$y' = \frac{(e^u + e^{-u})(e^u - (-e^u)) - (e^u - e^{-u})(e^u + (-e^{-u}))}{(e^u + e^{-u})^2}$$

$$= \frac{e^{2u} + e^0 + e^0 + e^{-2u} - (e^{2u} - e^0 - e^0 + e^{-2u})}{(e^u + e^{-u})^2} = \frac{4e^0}{(e^u + e^{-u})^2} = \frac{4}{(e^u + e^{-u})^2}$$

**27.** Using Formula 5 and the Chain Rule, $y = 2^{\sin \pi x} \quad \Rightarrow$
$y' = 2^{\sin \pi x}(\ln 2)\cdot \dfrac{d}{dx}(\sin \pi x) = 2^{\sin \pi x}(\ln 2)\cdot \cos \pi x \cdot \pi = 2^{\sin \pi x}(\pi \ln 2)\cos \pi x$

**28.** $y = \tan^2(3\theta) = (\tan 3\theta)^2 \quad \Rightarrow \quad y' = 2(\tan 3\theta)\cdot \dfrac{d}{d\theta}(\tan 3\theta) = 2\tan 3\theta \cdot \sec^2 3\theta \cdot 3 = 6\tan 3\theta \sec^2 3\theta$

**29.** $y = \cot^2(\sin\theta) = [\cot(\sin\theta)]^2 \quad \Rightarrow$
$y' = 2[\cot(\sin\theta)]\cdot \dfrac{d}{d\theta}[\cot(\sin\theta)] = 2\cot(\sin\theta)\cdot[-\csc^2(\sin\theta)\cdot\cos\theta] = -2\cos\theta\,\cot(\sin\theta)\,\csc^2(\sin\theta)$

**30.** $y = \sin(\sin(\sin x)) \quad \Rightarrow \quad y' = \cos(\sin(\sin x))\,\dfrac{d}{dx}(\sin(\sin x)) = \cos(\sin(\sin x))\,\cos(\sin x)\,\cos x$

**31.** $y = \sin\left(\tan\sqrt{\sin x}\right) \quad \Rightarrow$

$$y' = \cos\left(\tan\sqrt{\sin x}\right)\cdot\frac{d}{dx}\left(\tan\sqrt{\sin x}\right) = \cos\left(\tan\sqrt{\sin x}\right)\sec^2\sqrt{\sin x}\cdot\frac{d}{dx}(\sin x)^{1/2}$$

$$= \cos\left(\tan\sqrt{\sin x}\right)\sec^2\sqrt{\sin x}\cdot\tfrac{1}{2}(\sin x)^{-1/2}\cdot\cos x = \cos\left(\tan\sqrt{\sin x}\right)\left(\sec^2\sqrt{\sin x}\right)\left(\frac{1}{2\sqrt{\sin x}}\right)(\cos x)$$

**32.** $y = \sqrt{x + \sqrt{x + \sqrt{x}}} \quad \Rightarrow \quad y' = \frac{1}{2}\left(x + \sqrt{x + \sqrt{x}}\right)^{-1/2}\left[1 + \frac{1}{2}\left(x + \sqrt{x}\right)^{-1/2}\left(1 + \frac{1}{2}x^{-1/2}\right)\right]$

**33.** $y = e^{\alpha x}\sin\beta x \quad \Rightarrow \quad y' = e^{\alpha x}\cdot\beta\cos\beta x + \sin\beta x\cdot\alpha e^{\alpha x} = e^{\alpha x}(\beta\cos\beta x + \alpha\sin\beta x) \quad \Rightarrow$
$y'' = e^{\alpha x}(-\beta^2\sin\beta x + \alpha\beta\cos\beta x) + (\beta\cos\beta x + \alpha\sin\beta x)\cdot\alpha e^{\alpha x}$

$$= e^{\alpha x}(-\beta^2\sin\beta x + \alpha\beta\cos\beta x + \alpha\beta\cos\beta x + \alpha^2\sin\beta x) = e^{\alpha x}(\alpha^2\sin\beta x - \beta^2\sin\beta x + 2\alpha\beta\cos\beta x)$$

$$= e^{\alpha x}\left[(\alpha^2 - \beta^2)\sin\beta x + 2\alpha\beta\cos\beta x\right]$$

**34.** $y = e^{e^x} \quad \Rightarrow \quad y' = e^{e^x}\cdot(e^x)' = e^{e^x}\cdot e^x \quad \Rightarrow$
$y'' = e^{e^x}\cdot(e^x)' + e^x\cdot\left(e^{e^x}\right)' = e^{e^x}\cdot e^x + e^x\cdot e^{e^x}\cdot e^x = e^{e^x}\cdot e^x(1 + e^x) \quad \text{or} \quad e^{e^x + x}(1 + e^x)$

**35.** $y = (1 + 2x)^{10} \quad \Rightarrow \quad y' = 10(1 + 2x)^9\cdot 2 = 20(1 + 2x)^9$. At $(0, 1)$, $y' = 20(1 + 0)^9 = 20$, and an equation of the tangent line is $y - 1 = 20(x - 0)$, or $y = 20x + 1$.

**36.** $y = \sin x + \sin^2 x \quad \Rightarrow \quad y' = \cos x + 2\sin x\cos x$. At $(0, 0)$, $y' = 1$, and an equation of the tangent line is $y - 0 = 1(x - 0)$, or $y = x$.

**37.** (a) $y = \dfrac{2}{1 + e^{-x}} \Rightarrow y' = \dfrac{(1 + e^{-x})(0) - 2(-e^{-x})}{(1 + e^{-x})^2} = \dfrac{2e^{-x}}{(1 + e^{-x})^2}$.

At $(0, 1)$, $y' = \dfrac{2e^0}{(1 + e^0)^2} = \dfrac{2(1)}{(1 + 1)^2} = \dfrac{2}{2^2} = \dfrac{1}{2}$. So an equation of the tangent line is $y - 1 = \frac{1}{2}(x - 0)$ or $y = \frac{1}{2}x + 1$.

(b)

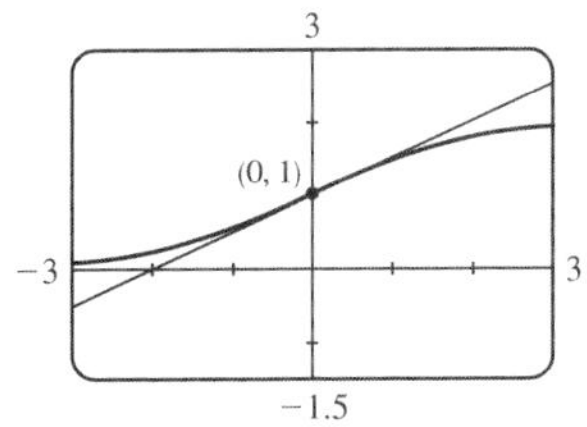


**38.** (a) For $x > 0$, $|x| = x$, and $y = f(x) = \dfrac{x}{\sqrt{2 - x^2}} \Rightarrow$

$$f'(x) = \frac{\sqrt{2 - x^2}\,(1) - x\left(\frac{1}{2}\right)(2 - x^2)^{-1/2}(-2x)}{\left(\sqrt{2 - x^2}\,\right)^2} \cdot \frac{(2 - x^2)^{1/2}}{(2 - x^2)^{1/2}}$$

$$= \frac{(2 - x^2) + x^2}{(2 - x^2)^{3/2}} = \frac{2}{(2 - x^2)^{3/2}}$$

So at $(1, 1)$, the slope of the tangent line is $f'(1) = 2$ and its equation is $y - 1 = 2(x - 1)$ or $y = 2x - 1$.

(b)

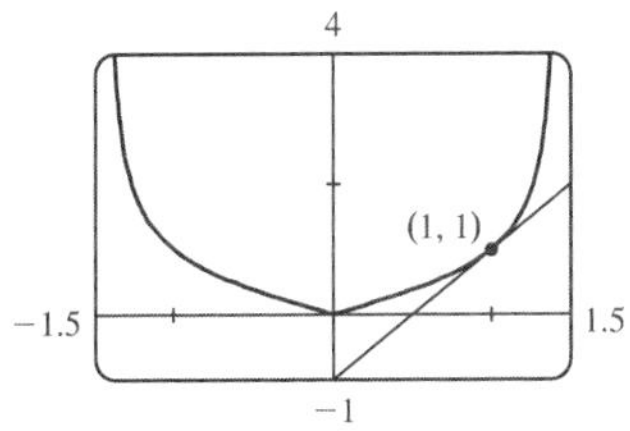


**39.** (a) $f(x) = x\sqrt{2 - x^2} = x(2 - x^2)^{1/2} \Rightarrow$

$$f'(x) = x \cdot \tfrac{1}{2}(2 - x^2)^{-1/2}(-2x) + (2 - x^2)^{1/2} \cdot 1 = (2 - x^2)^{-1/2}[-x^2 + (2 - x^2)] = \frac{2 - 2x^2}{\sqrt{2 - x^2}}$$

(b)

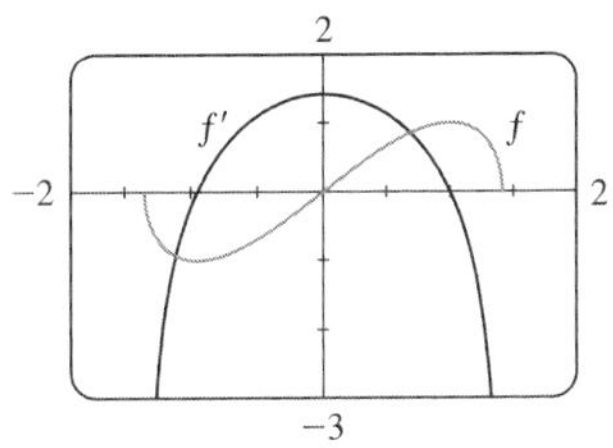


$f' = 0$ when $f$ has a horizontal tangent line, $f'$ is negative when $f$ is decreasing, and $f'$ is positive when $f$ is increasing.

**40.** (a)

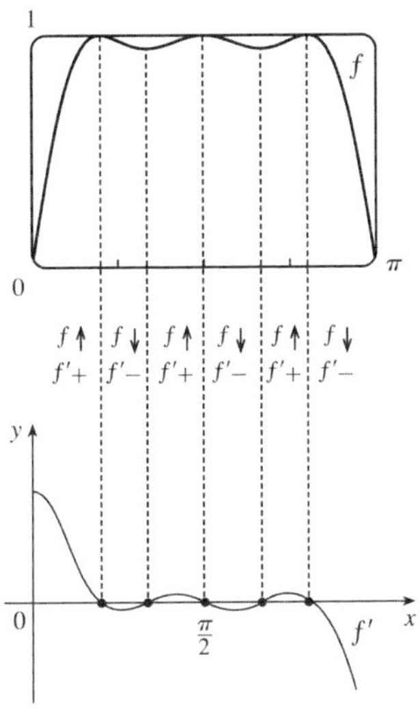


From the graph of $f$, we see that there are 5 horizontal tangents, so there must be 5 zeros on the graph of $f'$. From the symmetry of the graph of $f$, we must have the graph of $f'$ as high at $x = 0$ as it is low at $x = \pi$. The intervals of increase and decrease as well as the signs of $f'$ are indicated in the figure.

(b) $f(x) = \sin(x + \sin 2x) \Rightarrow$

$$f'(x) = \cos(x + \sin 2x) \cdot \frac{d}{dx}(x + \sin 2x) = \cos(x + \sin 2x)(1 + 2\cos 2x)$$

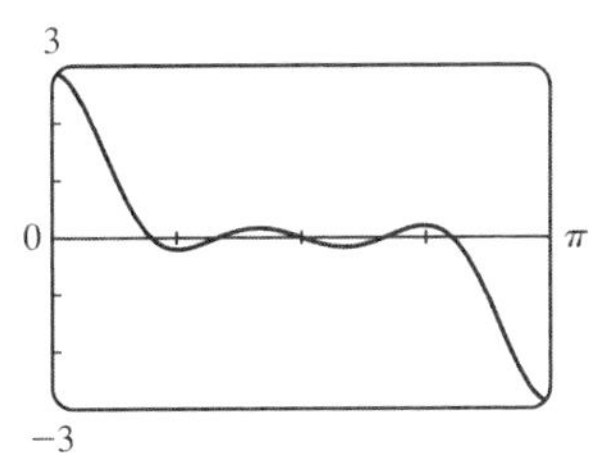

**41.** $F(x) = f(g(x)) \;\Rightarrow\; F'(x) = f'(g(x)) \cdot g'(x)$, so $F'(5) = f'(g(5)) \cdot g'(5) = f'(-2) \cdot 6 = 4 \cdot 6 = 24$

**42.** $h(x) = \sqrt{4 + 3f(x)} \;\Rightarrow\; h'(x) = \frac{1}{2}(4 + 3f(x))^{-1/2} \cdot 3f'(x)$, so
$h'(1) = \frac{1}{2}(4 + 3f(1))^{-1/2} \cdot 3f'(1) = \frac{1}{2}(4 + 3 \cdot 7)^{-1/2} \cdot 3 \cdot 4 = \frac{6}{\sqrt{25}} = \frac{6}{5}$

**43.** (a) $h(x) = f(g(x)) \;\Rightarrow\; h'(x) = f'(g(x)) \cdot g'(x)$, so $h'(1) = f'(g(1)) \cdot g'(1) = f'(2) \cdot 6 = 5 \cdot 6 = 30$.

(b) $H(x) = g(f(x)) \;\Rightarrow\; H'(x) = g'(f(x)) \cdot f'(x)$, so $H'(1) = g'(f(1)) \cdot f'(1) = g'(3) \cdot 4 = 9 \cdot 4 = 36$.

**44.** (a) $F(x) = f(f(x)) \;\Rightarrow\; F'(x) = f'(f(x)) \cdot f'(x)$, so $F'(2) = f'(f(2)) \cdot f'(2) = f'(1) \cdot 5 = 4 \cdot 5 = 20$.

(b) $G(x) = g(g(x)) \;\Rightarrow\; G'(x) = g'(g(x)) \cdot g'(x)$, so $G'(3) = g'(g(3)) \cdot g'(3) = g'(2) \cdot 9 = 7 \cdot 9 = 63$.

**45.** (a) $u(x) = f(g(x)) \;\Rightarrow\; u'(x) = f'(g(x))g'(x)$. So $u'(1) = f'(g(1))g'(1) = f'(3)g'(1)$. To find $f'(3)$, note that $f$ is linear from $(2,4)$ to $(6,3)$, so its slope is $\dfrac{3-4}{6-2} = -\dfrac{1}{4}$. To find $g'(1)$, note that $g$ is linear from $(0,6)$ to $(2,0)$, so its slope is $\dfrac{0-6}{2-0} = -3$. Thus, $f'(3)g'(1) = \left(-\frac{1}{4}\right)(-3) = \frac{3}{4}$.

(b) $v(x) = g(f(x)) \;\Rightarrow\; v'(x) = g'(f(x))f'(x)$. So $v'(1) = g'(f(1))f'(1) = g'(2)f'(1)$, which does not exist since $g'(2)$ does not exist.

(c) $w(x) = g(g(x)) \;\Rightarrow\; w'(x) = g'(g(x))g'(x)$. So $w'(1) = g'(g(1))g'(1) = g'(3)g'(1)$. To find $g'(3)$, note that $g$ is linear from $(2,0)$ to $(5,2)$, so its slope is $\dfrac{2-0}{5-2} = \dfrac{2}{3}$. Thus, $g'(3)g'(1) = \left(\frac{2}{3}\right)(-3) = -2$.

**46.** (a) $h(x) = f(f(x)) \;\Rightarrow\; h'(x) = f'(f(x))f'(x)$. So $h'(2) = f'(f(2))f'(2) = f'(1)f'(2) \approx (-1)(-1) = 1$.

(b) $g(x) = f(x^2) \;\Rightarrow\; g'(x) = f'(x^2) \cdot \dfrac{d}{dx}(x^2) = f'(x^2)(2x)$. So $g'(2) = f'(2^2)(2 \cdot 2) = 4f'(4) \approx 4(1.5) = 6$.

**47.** $h(x) = f(g(x)) \;\Rightarrow\; h'(x) = f'(g(x))g'(x)$. So $h'(0.5) = f'(g(0.5))g'(0.5) = f'(0.1)g'(0.5)$.
We can estimate the derivatives by taking the average of two secant slopes.

For $f'(0.1)$: $m_1 = \dfrac{14.8 - 12.6}{0.1 - 0} = 22$, $m_2 = \dfrac{18.4 - 14.8}{0.2 - 0.1} = 36$. So $f'(0.1) \approx \dfrac{m_1 + m_2}{2} = \dfrac{22 + 36}{2} = 29$.

For $g'(0.5)$: $m_1 = \dfrac{0.10 - 0.17}{0.5 - 0.4} = -0.7$, $m_2 = \dfrac{0.05 - 0.10}{0.6 - 0.5} = -0.5$. So $g'(0.5) \approx \dfrac{m_1 + m_2}{2} = -0.6$.

Hence, $h'(0.5) = f'(0.1)g'(0.5) \approx (29)(-0.6) = -17.4$.

**48.** $g(x) = f(f(x)) \;\Rightarrow\; g'(x) = f'(f(x))f'(x)$. So $g'(1) = f'(f(1))f'(1) = f'(2)f'(1)$.

For $f'(2)$: $m_1 = \dfrac{3.1 - 2.4}{2.0 - 1.5} = 1.4$, $m_2 = \dfrac{4.4 - 3.1}{2.5 - 2.0} = 2.6$. So $f'(2) \approx \dfrac{m_1 + m_2}{2} = 2$.

For $f'(1)$: $m_1 = \dfrac{2.0 - 1.8}{1.0 - 0.5} = 0.4$, $m_2 = \dfrac{2.4 - 2.0}{1.5 - 1.0} = 0.8$. So $f'(1) \approx \dfrac{m_1 + m_2}{2} = 0.6$.

Hence, $g'(1) = f'(2)f'(1) \approx (2)(0.6) = 1.2$.

**49.** (a) $F(x) = f(e^x) \;\Rightarrow\; F'(x) = f'(e^x)\dfrac{d}{dx}(e^x) = f'(e^x)\,e^x$

(b) $G(x) = e^{f(x)} \;\Rightarrow\; G'(x) = e^{f(x)}\dfrac{d}{dx}f(x) = e^{f(x)}f'(x)$

**50.** (a) $F(x) = f(x^\alpha) \;\Rightarrow\; F'(x) = f'(x^\alpha)\dfrac{d}{dx}(x^\alpha) = f'(x^\alpha)\alpha x^{\alpha-1}$

(b) $G(x) = [f(x)]^\alpha \;\Rightarrow\; G'(x) = \alpha\,[f(x)]^{\alpha-1}\,f'(x)$

**51.** $r(x) = f(g(h(x))) \;\Rightarrow\; r'(x) = f'(g(h(x))) \cdot g'(h(x)) \cdot h'(x)$, so
$r'(1) = f'(g(h(1))) \cdot g'(h(1)) \cdot h'(1) = f'(g(2)) \cdot g'(2) \cdot 4 = f'(3) \cdot 5 \cdot 4 = 6 \cdot 5 \cdot 4 = 120$

**52.** $f(x) = xg(x^2) \;\Rightarrow\; f'(x) = xg'(x^2)\,2x + g(x^2) \cdot 1 = 2x^2g'(x^2) + g(x^2) \;\Rightarrow$
$f''(x) = 2x^2g''(x^2)\,2x + g'(x^2)\,4x + g'(x^2)\,2x = 4x^3g''(x^2) + 4xg'(x^2) + 2xg'(x^2) = 6xg'(x^2) + 4x^3g''(x^2)$

**53.** For the tangent line to be horizontal, $f'(x) = 0$. $f(x) = 2\sin x + \sin^2 x \quad \Rightarrow \quad f'(x) = 2\cos x + 2\sin x\cos x = 0 \quad \Leftrightarrow$ $2\cos x\,(1 + \sin x) = 0 \quad \Leftrightarrow \quad \cos x = 0$ or $\sin x = -1$, so $x = \frac{\pi}{2} + 2n\pi$ or $\frac{3\pi}{2} + 2n\pi$, where $n$ is any integer. Now $f\left(\frac{\pi}{2}\right) = 3$ and $f\left(\frac{3\pi}{2}\right) = -1$, so the points on the curve with a horizontal tangent are $\left(\frac{\pi}{2} + 2n\pi, 3\right)$ and $\left(\frac{3\pi}{2} + 2n\pi, -1\right)$, where $n$ is any integer.

**54.** $y = e^{-x^2} \quad \Rightarrow \quad y' = e^{-x^2}(-2x) \quad \Rightarrow \quad y'' = e^{-x^2}(-2) + (-2x)e^{-x^2}(-2x) = 2e^{-x^2}(2x^2 - 1)$. $\quad y'' = 0 \quad \Leftrightarrow$ $2x^2 - 1 = 0 \quad \Leftrightarrow \quad x = \pm\frac{\sqrt{2}}{2}$. The curve is concave downward when $y'' < 0$. This is the case on the interval $\left(-\frac{\sqrt{2}}{2}, \frac{\sqrt{2}}{2}\right)$.

**55.** $y = Ae^{-x} + Bxe^{-x} \quad \Rightarrow \quad y' = A(-e^{-x}) + B[x(-e^{-x}) + e^{-x}\cdot 1]$

$$= -Ae^{-x} + Be^{-x} - Bxe^{-x} = (B - A)e^{-x} - Bxe^{-x}$$

$$\Rightarrow \quad y'' = (B - A)(-e^{-x}) - B[x(-e^{-x}) + e^{-x}\cdot 1]$$

$$= (A - B)e^{-x} - Be^{-x} + Bxe^{-x} = (A - 2B)e^{-x} + Bxe^{-x},$$

so $\quad y'' + 2y' + y = (A - 2B)e^{-x} + Bxe^{-x} + 2[(B - A)e^{-x} - Bxe^{-x}] + Ae^{-x} + Bxe^{-x}$

$$= [(A - 2B) + 2(B - A) + A]e^{-x} + [B - 2B + B]xe^{-x} = 0.$$

**56.** $y = e^{rx} \quad \Rightarrow \quad y' = re^{rx} \quad \Rightarrow \quad y'' = r^2e^{rx}$, so
$y'' + 5y' - 6y = r^2e^{rx} + 5re^{rx} - 6e^{rx} = e^{rx}(r^2 + 5r - 6) = e^{rx}(r + 6)(r - 1) = 0 \quad \Rightarrow$
$(r + 6)(r - 1) = 0 \quad \Rightarrow \quad r = 1$ or $-6$.

**57.** The use of $D, D^2, \ldots, D^n$ is just a derivative notation (see text page 159). In general, $Df(2x) = 2f'(2x)$, $D^2f(2x) = 4f''(2x), \ldots, D^nf(2x) = 2^nf^{(n)}(2x)$. Since $f(x) = \cos x$ and $50 = 4(12) + 2$, we have $f^{(50)}(x) = f^{(2)}(x) = -\cos x$, so $D^{50}\cos 2x = -2^{50}\cos 2x$.

**58.** $f(x) = xe^{-x}$, $f'(x) = e^{-x} - xe^{-x} = (1 - x)e^{-x}$, $f''(x) = -e^{-x} + (1 - x)(-e^{-x}) = (x - 2)e^{-x}$. Similarly, $f'''(x) = (3 - x)e^{-x}$, $f^{(4)}(x) = (x - 4)e^{-x}, \ldots, f^{(1000)}(x) = (x - 1000)e^{-x}$.

**59.** $s(t) = 10 + \frac{1}{4}\sin(10\pi t) \quad \Rightarrow$ the velocity after $t$ seconds is $v(t) = s'(t) = \frac{1}{4}\cos(10\pi t)(10\pi) = \frac{5\pi}{2}\cos(10\pi t)$ cm/s.

**60.** (a) $s = A\cos(\omega t + \delta) \quad \Rightarrow \quad$ velocity $= s' = -\omega A\sin(\omega t + \delta)$.

(b) If $A \neq 0$ and $\omega \neq 0$, then $s' = 0 \quad \Leftrightarrow \quad \sin(\omega t + \delta) = 0 \quad \Leftrightarrow \quad \omega t + \delta = n\pi \quad \Leftrightarrow \quad t = \dfrac{n\pi - \delta}{\omega}$, $n$ an integer.

**61.** (a) $B(t) = 4.0 + 0.35\sin\dfrac{2\pi t}{5.4} \quad \Rightarrow \quad \dfrac{dB}{dt} = \left(0.35\cos\dfrac{2\pi t}{5.4}\right)\left(\dfrac{2\pi}{5.4}\right) = \dfrac{0.7\pi}{5.4}\cos\dfrac{2\pi t}{5.4} = \dfrac{7\pi}{54}\cos\dfrac{2\pi t}{5.4}$

(b) At $t = 1$, $\dfrac{dB}{dt} = \dfrac{7\pi}{54}\cos\dfrac{2\pi}{5.4} \approx 0.16$.

**62.** $L(t) = 12 + 2.8\sin\left(\frac{2\pi}{365}(t - 80)\right) \quad \Rightarrow \quad L'(t) = 2.8\cos\left(\frac{2\pi}{365}(t - 80)\right)\left(\frac{2\pi}{365}\right)$.
On March 21, $t = 80$, and $L'(80) \approx 0.0482$ hours per day. On May 21, $t = 141$, and $L'(141) \approx 0.02398$, which is approximately one-half of $L'(80)$.

**63.** $s(t) = 2e^{-1.5t}\sin 2\pi t \quad \Rightarrow$
$v(t) = s'(t) = 2[e^{-1.5t}(\cos 2\pi t)(2\pi) + (\sin 2\pi t)e^{-1.5t}(-1.5)] = 2e^{-1.5t}(2\pi\cos 2\pi t - 1.5\sin 2\pi t)$

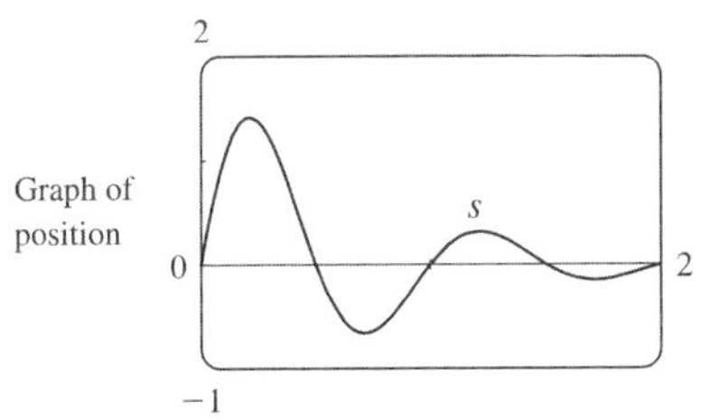


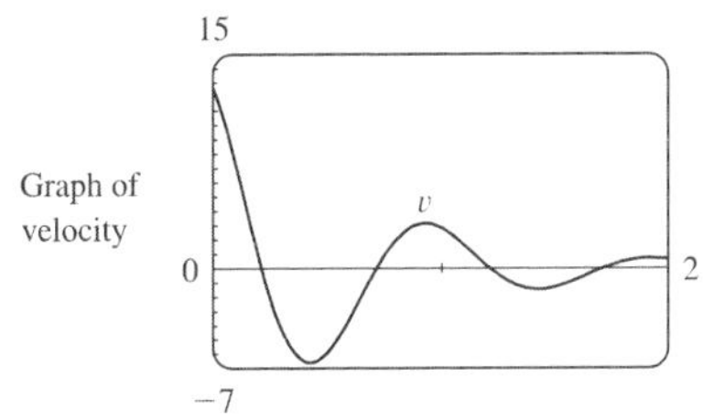


**64.** (a) $\displaystyle\lim_{t\to\infty} p(t) = \lim_{t\to\infty}\frac{1}{1 + ae^{-kt}} = \frac{1}{1 + a\cdot 0} = 1$, since $k > 0 \quad \Rightarrow \quad -kt \to -\infty \quad \Rightarrow \quad e^{-kt} \to 0$.

(b) $p(t) = (1 + ae^{-kt})^{-1} \Rightarrow \dfrac{dp}{dt} = -(1 + ae^{-kt})^{-2}(-kae^{-kt}) = \dfrac{kae^{-kt}}{(1 + ae^{-kt})^2}$

(c)

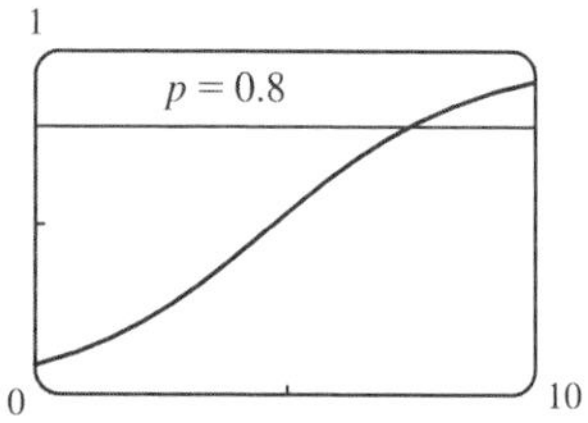


From the graph of $p(t) = (1 + 10e^{-0.5t})^{-1}$, it seems that $p(t) = 0.8$ (indicating that 80% of the population has heard the rumor) when $t \approx 7.4$ hours.

**65.** By the Chain Rule, $a(t) = \dfrac{dv}{dt} = \dfrac{dv}{ds}\dfrac{ds}{dt} = \dfrac{dv}{ds}v(t) = v(t)\dfrac{dv}{ds}$. The derivative $dv/dt$ is the rate of change of the velocity with respect to time (in other words, the acceleration) whereas the derivative $dv/ds$ is the rate of change of the velocity with respect to the displacement.

**66.** (a) The derivative $dV/dr$ represents the rate of change of the volume with respect to the radius and the derivative $dV/dt$ represents the rate of change of the volume with respect to time.

(b) Since $V = \frac{4}{3}\pi r^3$, $\dfrac{dV}{dt} = \dfrac{dV}{dr}\dfrac{dr}{dt} = 4\pi r^2 \dfrac{dr}{dt}$.

**67.** (a) Using a calculator or CAS, we obtain the model $Q = ab^t$ with $a \approx 100.0124369$ and $b \approx 0.000045145933$. We can change this model to one with base $e$ and exponent $\ln b$ [$b^t = e^{t \ln b}$ from precalculus mathematics]: $Q = ae^{t \ln b} \approx 100.012437e^{-10.005531t}$.

(b) Use $Q'(t) = ab^t \ln b$ or the calculator command `nDeriv(Y₁,X,.04)` with `Y₁=ab^x` to get $Q'(0.04) \approx -670.63\ \mu$A. The result of Example 2 in Section 2.1 was $-670\ \mu$A.

**68.** (a) $P = ab^t$ with $a = 4.502714 \times 10^{-20}$ and $b = 1.029953851$, where $P$ is measured in thousands of people. The fit appears to be very good.

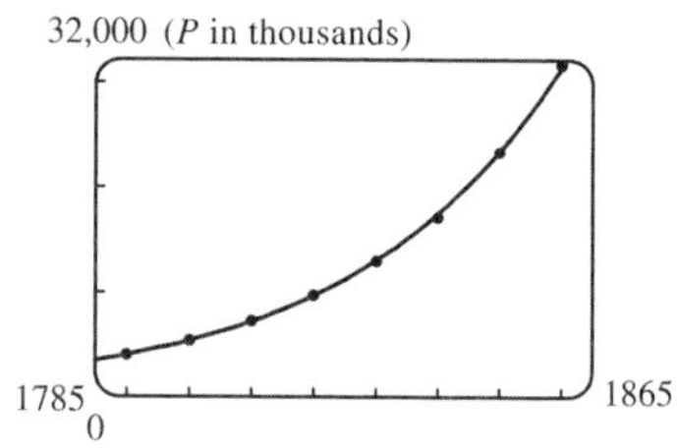


(b) **For 1800:** $m_1 = \dfrac{5308 - 3929}{1800 - 1790} = 137.9$, $m_2 = \dfrac{7240 - 5308}{1810 - 1800} = 193.2$.
So $P'(1800) \approx (m_1 + m_2)/2 = 165.55$ thousand people/year.

**For 1850:** $m_1 = \dfrac{23{,}192 - 17{,}063}{1850 - 1840} = 612.9$, $m_2 = \dfrac{31{,}443 - 23{,}192}{1860 - 1850} = 825.1$.
So $P'(1850) \approx (m_1 + m_2)/2 = 719$ thousand people/year.

(c) Use the calculator command `nDeriv(Y₁,X,year)` with `Y₁=ab^x` to get $P'(1800) \approx 156.85$ and $P'(1850) \approx 686.07$. These estimates are somewhat less than the ones in part (b).

(d) $P(1870) \approx 41{,}946.56$. The difference of 3.4 million people is most likely due to the Civil War (1861–1865).

**69.** $x = t^4 + 1$, $y = t^3 + t$; $t = -1$. $\dfrac{dy}{dt} = 3t^2 + 1$, $\dfrac{dx}{dt} = 4t^3$, and $\dfrac{dy}{dx} = \dfrac{dy/dt}{dx/dt} = \dfrac{3t^2 + 1}{4t^3}$. When $t = -1$, $(x, y) = (2, -2)$ and $dy/dx = \frac{4}{-4} = -1$, so an equation of the tangent to the curve at the point corresponding to $t = -1$ is $y - (-2) = (-1)(x - 2)$, or $y = -x$.

**70.** $x = \cos\theta + \sin 2\theta$, $y = \sin\theta + \cos 2\theta$; $\theta = 0$. $\dfrac{dy}{dx} = \dfrac{dy/d\theta}{dx/d\theta} = \dfrac{\cos\theta - 2\sin 2\theta}{-\sin\theta + 2\cos 2\theta}$. When $\theta = 0$, $(x, y) = (1, 1)$ and $dy/dx = \frac{1}{2}$, so an equation of the tangent to the curve is $y - 1 = \frac{1}{2}(x - 1)$, or $y = \frac{1}{2}x + \frac{1}{2}$.

**71.** $x = e^{\sqrt{t}}$, $y = t - t^2$; $t = 1$. $\dfrac{dy}{dt} = 1 - 2t$, $\dfrac{dx}{dt} = e^{\sqrt{t}} \cdot \dfrac{1}{2}t^{-1/2}$, and $\dfrac{dy}{dx} = \dfrac{dy/dt}{dx/dt} = \dfrac{1 - 2t}{e^{\sqrt{t}}/(2\sqrt{t})}$.

When $t = 1$, $(x, y) = (e, 0)$ and $\dfrac{dy}{dx} = \dfrac{-1}{e/2} = -\dfrac{2}{e}$, so an equation of the tangent to the curve at the point corresponding to $t = 1$ is $y - 0 = -\dfrac{2}{e}(x - e)$, or $y = -\dfrac{2}{e}x + 2$.

**72.** $x = \sin t$, $y = \sin(t + \sin t)$; $(0, 0)$.

$$\frac{dy}{dx} = \frac{dy/dt}{dx/dt} = \frac{\cos(t+\sin t)(1+\cos t)}{\cos t} = \cos(t+\sin t)\,\frac{1+\cos t}{\cos t} = \cos(t+\sin t)\left(\frac{1}{\cos t}+1\right)$$
$$= (\sec t + 1)\cos(t + \sin t).$$

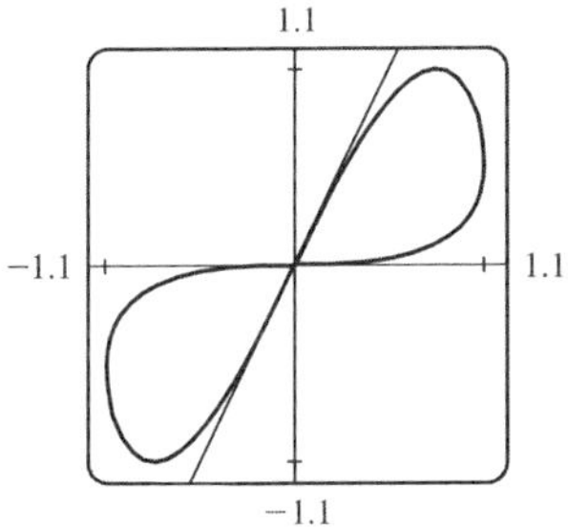


Now $x = \sin t$ is 0 when $t = 0$ and $t = \pi$, so there are two tangents at the point $(0, 0)$ since both $t = 0$ and $t = \pi$ correspond to the origin. The tangent corresponding to $t = 0$ has slope $(\sec 0 + 1)\cos(0 + \sin 0) = 2\cos 0 = 2$, and its equation is $y = 2x$. The tangent corresponding to $t = \pi$ has slope $(\sec\pi + 1)\cos(\pi + \sin\pi) = 0$, so it is the $x$-axis; that is, $y = 0$.

**73.** (a) $x = t^2$, $y = t^3 - 3t \;\Rightarrow\; \dfrac{dy}{dx} = \dfrac{dy/dt}{dx/dt} = \dfrac{3t^2-3}{2t}$. At the point $(3, 0)$, $x = 3 \;\Rightarrow\; t^2 = 3 \;\Rightarrow\; t = \pm\sqrt{3} \;\Rightarrow$

$$\frac{dy}{dx} = \frac{3(\pm\sqrt{3})^2 - 3}{2(\pm\sqrt{3})} = \frac{6}{2(\pm\sqrt{3})} = \pm\frac{3}{\sqrt{3}} = \pm\sqrt{3}.$$ When $t = \sqrt{3}$, an equation of the tangent line is $y - 0 = \sqrt{3}\,(x - 3)$ or $y = \sqrt{3}\,x - 3\sqrt{3}$. When $t = -\sqrt{3}$, an equation of the tangent line is $y - 0 = -\sqrt{3}\,(x - 3)$ or $y = -\sqrt{3}\,x + 3\sqrt{3}$.

(b) Horizontal tangent: $dy/dx = 0 \;\Leftrightarrow\; 3t^2 - 3 = 0 \;\Leftrightarrow\; 3(t^2 - 1) = 0 \;\Leftrightarrow\; t^2 = 1 \;\Leftrightarrow\; t = \pm 1$. $t = 1$ corresponds to the point $(x, y) = (t^2, t^3 - 3t) = (1^2, 1^3 - 3\cdot 1) = (1, -2)$ and $t = -1$ to $(1, 2)$.

Vertical tangent: $dy/dx$ is undefined $\;\Leftrightarrow\; 2t = 0 \;\Leftrightarrow\; t = 0$. The value $t = 0$ corresponds to the origin; that is, $(0, 0)$.

(c)

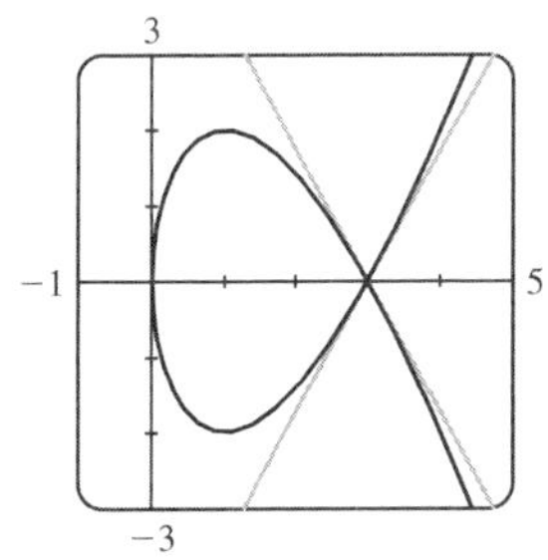


**74.** (a) $x = r(\theta - \sin\theta)$, $y = r(1 - \cos\theta) \;\Rightarrow\; \dfrac{dy}{dx} = \dfrac{dy/d\theta}{dx/d\theta} = \dfrac{r(\sin\theta)}{r(1-\cos\theta)} = \dfrac{\sin\theta}{1-\cos\theta}$. When $\theta = \frac{\pi}{3}$,

$$\frac{dy}{dx} = \frac{\frac{\sqrt{3}}{2}}{1-\frac{1}{2}} = \sqrt{3},\; (x, y) = \left(r\left(\tfrac{\pi}{3} - \tfrac{\sqrt{3}}{2}\right), \tfrac{1}{2}r\right),$$ and the tangent is $y - \frac{1}{2}r = \sqrt{3}\left[x - r\left(\frac{\pi}{3} - \frac{\sqrt{3}}{2}\right)\right]$.

(b) Horizontal tangent: $dy/dx = 0 \;\Leftrightarrow\; \sin\theta = 0$ (and $\cos\theta \neq 1$) $\;\Leftrightarrow\; \theta = (2n + 1)\pi$. The corresponding points are $((2n + 1)\pi r, 2r)$.

Vertical tangent: $dy/dx$ is undefined $\;\Leftrightarrow\; 1 - \cos\theta = 0 \;\Leftrightarrow\; \cos\theta = 1 \;\Leftrightarrow\; \theta = 2n\pi$. The corresponding points are $(2n\pi r, 0)$.

(c)

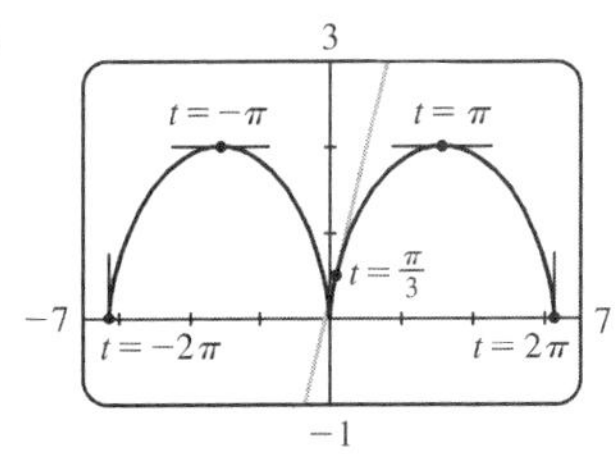


**75.** (a) Derive gives $g'(t) = \dfrac{45(t-2)^8}{(2t+1)^{10}}$ without simplifying. With either Maple or Mathematica, we first get $g'(t) = 9\dfrac{(t-2)^8}{(2t+1)^9} - 18\dfrac{(t-2)^9}{(2t+1)^{10}}$, and the simplification command results in the above expression.

(b) Derive gives $y' = 2(x^3 - x + 1)^3(2x + 1)^4(17x^3 + 6x^2 - 9x + 3)$ without simplifying. With either Maple or Mathematica, we first get $y' = 10(2x + 1)^4(x^3 - x + 1)^4 + 4(2x + 1)^5(x^3 - x + 1)^3(3x^2 - 1)$. If we use

Mathematica's `Factor` or `Simplify`, or Maple's `factor`, we get the above expression, but Maple's `simplify` gives the polynomial expansion instead. For locating horizontal tangents, the factored form is the most helpful.

**76.** (a) $f(x) = \left(\dfrac{x^4 - x + 1}{x^4 + x + 1}\right)^{1/2}$. Derive gives $f'(x) = \dfrac{(3x^4 - 1)\sqrt{\dfrac{x^4 - x + 1}{x^4 + x + 1}}}{(x^4 + x + 1)(x^4 - x + 1)}$ whereas either Maple or Mathematica give $f'(x) = \dfrac{3x^4 - 1}{\sqrt{\dfrac{x^4 - x + 1}{x^4 + x + 1}}\,(x^4 + x + 1)^2}$ after simplification.

(b) $f'(x) = 0 \quad \Leftrightarrow \quad 3x^4 - 1 = 0 \quad \Leftrightarrow \quad x = \pm\sqrt[4]{\frac{1}{3}} \approx \pm 0.7598$.

(c) $f'(x) = 0$ where $f$ has horizontal tangents. $f'$ has two maxima and one minimum where $f$ has inflection points.

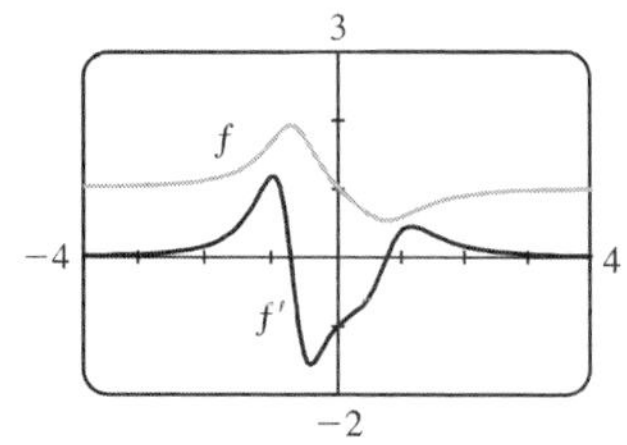


**77.** (a)
$$\begin{aligned}\frac{d}{dx}(\sin^n x \cos nx) &= n\sin^{n-1}x\cos x\cos nx + \sin^n x\,(-n\sin nx) && \text{[Product Rule]}\\ &= n\sin^{n-1}x\,(\cos nx\cos x - \sin nx\sin x) && [\text{factor out } n\sin^{n-1}x]\\ &= n\sin^{n-1}x\cos(nx + x) && \text{[Addition Formula for cosine]}\\ &= n\sin^{n-1}x\cos[(n+1)x] && [\text{factor out } x]\end{aligned}$$

(b)
$$\begin{aligned}\frac{d}{dx}(\cos^n x \cos nx) &= n\cos^{n-1}x\,(-\sin x)\cos nx + \cos^n x\,(-n\sin nx) && \text{[Product Rule]}\\ &= -n\cos^{n-1}x\,(\cos nx\sin x + \sin nx\cos x) && [\text{factor out } -n\cos^{n-1}x]\\ &= -n\cos^{n-1}x\sin(nx + x) && \text{[Addition Formula for sine]}\\ &= -n\cos^{n-1}x\sin[(n+1)x] && [\text{factor out } x]\end{aligned}$$

**78.** "The rate of change of $y^5$ with respect to $x$ is eighty times the rate of change of $y$ with respect to $x$" $\Leftrightarrow$ $\frac{d}{dx}y^5 = 80\frac{dy}{dx} \quad \Leftrightarrow \quad 5y^4\frac{dy}{dx} = 80\frac{dy}{dx} \quad \Leftrightarrow \quad 5y^4 = 80$ (Note that $dy/dx \neq 0$ since the curve never has a horizontal tangent) $\Leftrightarrow \quad y^4 = 16 \quad \Leftrightarrow \quad y = 2$ (since $y > 0$ for all $x$)

**79.** Since $\theta° = \left(\frac{\pi}{180}\right)\theta$ rad, we have $\frac{d}{d\theta}(\sin\theta°) = \frac{d}{d\theta}\left(\sin\frac{\pi}{180}\theta\right) = \frac{\pi}{180}\cos\frac{\pi}{180}\theta = \frac{\pi}{180}\cos\theta°$.

**80.** (a) $f(x) = |x| = \sqrt{x^2} = (x^2)^{1/2} \quad \Rightarrow \quad f'(x) = \frac{1}{2}(x^2)^{-1/2}(2x) = x/\sqrt{x^2} = x/|x|$ for $x \neq 0$.
$f$ is not differentiable at $x = 0$.

(b) $f(x) = |\sin x| = \sqrt{\sin^2 x} \quad \Rightarrow$

$$f'(x) = \tfrac{1}{2}(\sin^2 x)^{-1/2}2\sin x\cos x = \frac{\sin x}{|\sin x|}\cos x = \begin{cases}\cos x & \text{if } \sin x > 0\\ -\cos x & \text{if } \sin x < 0\end{cases}$$

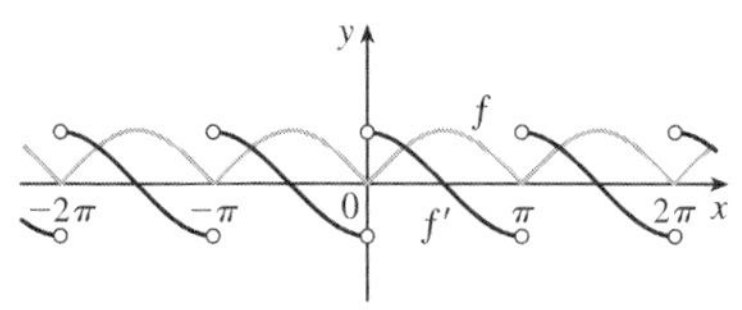


$f$ is not differentiable when $x = n\pi$, $n$ an integer.

(c) $g(x) = \sin|x| = \sin\sqrt{x^2} \quad \Rightarrow$

$$g'(x) = \cos|x| \cdot \frac{x}{|x|} = \frac{x}{|x|}\cos x = \begin{cases}\cos x & \text{if } x > 0\\ -\cos x & \text{if } x < 0\end{cases}$$

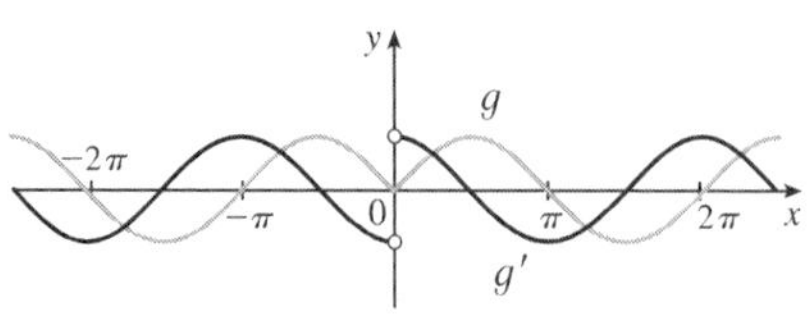


$g$ is not differentiable at 0.

**81.** $\dfrac{d^2y}{dx^2} = \dfrac{d}{dx}\left(\dfrac{dy}{dx}\right)$ [Leibniz notation for the second derivative]

$= \dfrac{d}{dx}\left(\dfrac{dy}{du}\dfrac{du}{dx}\right)$ [Chain Rule]

$= \dfrac{dy}{du}\cdot\dfrac{d}{dx}\left(\dfrac{du}{dx}\right) + \dfrac{du}{dx}\cdot\dfrac{d}{dx}\left(\dfrac{dy}{du}\right)$ [Product Rule]

$= \dfrac{dy}{du}\cdot\dfrac{d^2u}{dx^2} + \dfrac{du}{dx}\cdot\dfrac{d}{du}\left(\dfrac{dy}{du}\right)\cdot\dfrac{du}{dx}$ [$dy/du$ is a function of $u$]

$= \dfrac{dy}{du}\dfrac{d^2u}{dx^2} + \dfrac{d^2y}{du^2}\left(\dfrac{du}{dx}\right)^2$

*Or:* Using function notation for $y = f(u)$ and $u = g(x)$, we have $y = f(g(x))$, so
$y' = f'(g(x))\cdot g'(x)$ [by the Chain Rule] $\Rightarrow$
$(y')' = [f'(g(x))\cdot g'(x)]' = f'(g(x))\cdot g''(x) + g'(x)\cdot f''(g(x))\cdot g'(x) = f'(g(x))\cdot g''(x) + f''(g(x))\cdot [g'(x)]^2$.

**82.** $V = \frac{4}{3}\pi r^3 \Leftrightarrow \dfrac{dV}{dt} = 4\pi r^2\dfrac{dr}{dt}$. But $\dfrac{dV}{dt}$ is proportional to the surface area, so $\dfrac{dV}{dt} = k\cdot 4\pi r^2$ for some constant $k$.
Therefore, $4\pi r^2\dfrac{dr}{dt} = k\cdot 4\pi r^2 \Leftrightarrow \dfrac{dr}{dt} = k =$ constant. An antiderivative of $k$ with respect to $t$ is $kt$, so $r = kt + C$.
When $t = 0$, the radius $r$ must equal the original radius $r_0$, so $C = r_0$, and $r = kt + r_0$. To find $k$ we use the fact that when $t = 3$, $r = 3k + r_0$ and $V = \frac{1}{2}V_0 \Rightarrow \frac{4}{3}\pi(3k + r_0)^3 = \frac{1}{2}\cdot\frac{4}{3}\pi r_0^3 \Rightarrow (3k + r_0)^3 = \frac{1}{2}r_0^3 \Rightarrow 3k + r_0 = \frac{1}{\sqrt[3]{2}}r_0 \Rightarrow$
$k = \frac{1}{3}r_0\left(\frac{1}{\sqrt[3]{2}} - 1\right)$. Since $r = kt + r_0$, $r = \frac{1}{3}r_0\left(\frac{1}{\sqrt[3]{2}} - 1\right)t + r_0$. When the snowball has melted completely we have
$r = 0 \Rightarrow \frac{1}{3}r_0\left(\frac{1}{\sqrt[3]{2}} - 1\right)t + r_0 = 0$ which gives $t = \dfrac{3\sqrt[3]{2}}{\sqrt[3]{2} - 1}$. Hence, it takes $\dfrac{3\sqrt[3]{2}}{\sqrt[3]{2} - 1} - 3 = \dfrac{3}{\sqrt[3]{2} - 1} \approx 11$ h 33 min longer.

## LABORATORY PROJECT Bezier Curves

**1.** The parametric equations for a cubic Bézier curve are

$$x = x_0(1-t)^3 + 3x_1t(1-t)^2 + 3x_2t^2(1-t) + x_3t^3$$
$$y = y_0(1-t)^3 + 3y_1t(1-t)^2 + 3y_2t^2(1-t) + y_3t^3$$

where $0 \le t \le 1$. We are given the points $P_0(x_0, y_0) = (4, 1)$, $P_1(x_1, y_1) = (28, 48)$, $P_2(x_2, y_2) = (50, 42)$, and $P_3(x_3, y_3) = (40, 5)$. The curve is then given by

$$x(t) = 4(1-t)^3 + 3\cdot 28t(1-t)^2 + 3\cdot 50t^2(1-t) + 40t^3$$
$$y(t) = 1(1-t)^3 + 3\cdot 48t(1-t)^2 + 3\cdot 42t^2(1-t) + 5t^3$$

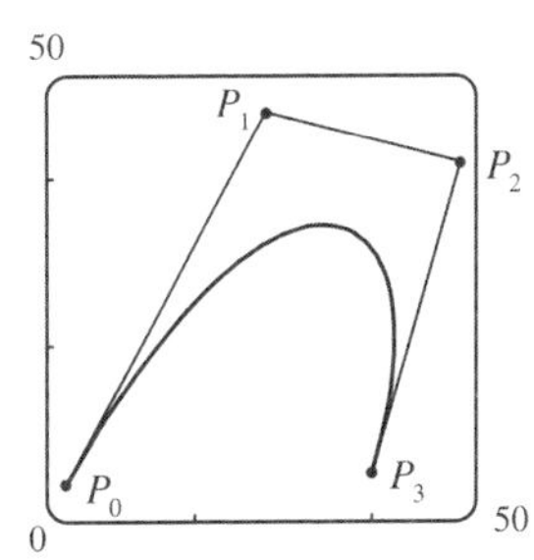


where $0 \le t \le 1$. The line segments are of the form $x = x_0 + (x_1 - x_0)t$, $y = y_0 + (y_1 - y_0)t$:

| | | |
|---|---|---|
| $P_0P_1$ | $x = 4 + 24t,$ | $y = 1 + 47t$ |
| $P_1P_2$ | $x = 28 + 22t,$ | $y = 48 - 6t$ |
| $P_2P_3$ | $x = 50 - 10t,$ | $y = 42 - 37t$ |

2. It suffices to show that the slope of the tangent at $P_0$ is the same as that of line segment $P_0P_1$, namely $\dfrac{y_1 - y_0}{x_1 - x_0}$. We calculate the slope of the tangent to the Bézier curve:

$$\frac{dy/dt}{dx/dt} = \frac{-3y_0(1-t)^2 + 3y_1\left[-2t(1-t) + (1-t)^2\right] + 3y_2\left[-t^2 + (2t)(1-t)\right] + 3y_3t^2}{-3x_0^2(1-t) + 3x_1[-2t(1-t) + (1-t)^2] + 3x_2[-t^2 + (2t)(1-t)] + 3x_3t^2}$$

At point $P_0$, $t = 0$, so the slope of the tangent is $\dfrac{-3y_0 + 3y_1}{-3x_0 + 3x_1} = \dfrac{y_1 - y_0}{x_1 - x_0}$. So the tangent to the curve at $P_0$ passes through $P_1$. Similarly, the slope of the tangent at point $P_3$ (where $t = 1$) is $\dfrac{-3y_2 + 3y_3}{-3x_2 + 3x_3} = \dfrac{y_3 - y_2}{x_3 - x_2}$, which is also the slope of line $P_2P_3$.

3. It seems that if $P_1$ were to the right of $P_2$, a loop would appear. We try setting $P_1 = (110, 30)$, and the resulting curve does indeed have a loop.

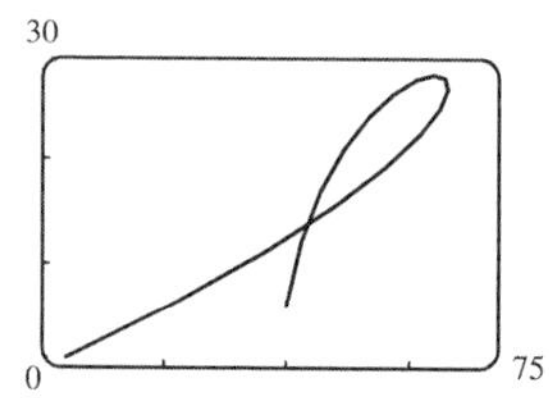


4. Based on the behavior of the Bézier curve in Problems 1–3, we suspect that the four control points should be in an exaggerated C shape. We try $P_0(10, 12)$, $P_1(4, 15)$, $P_2(4, 5)$, and $P_3(10, 8)$, and these produce a decent C. If you are using a CAS, it may be necessary to instruct it to make the $x$- and $y$-scales the same so as not to distort the figure (this is called a "constrained projection" in Maple.)

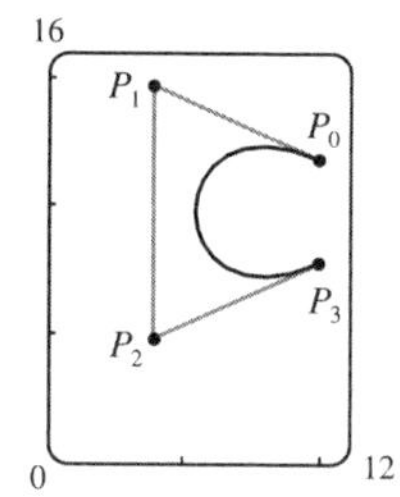


5. We use the same $P_0$ and $P_1$ as in Problem 4, and use part of our C as the top of an S. To prevent the center line from slanting up too much, we move $P_2$ up to $(4, 6)$ and $P_3$ down and to the left, to $(8, 7)$. In order to have a smooth joint between the top and bottom halves of the S (and a symmetric S), we determine points $P_4$, $P_5$, and $P_6$ by rotating points $P_2$, $P_1$, and $P_0$ about the center of the letter (point $P_3$). The points are therefore $P_4(12, 8)$, $P_5(12, -1)$, and $P_6(6, 2)$.

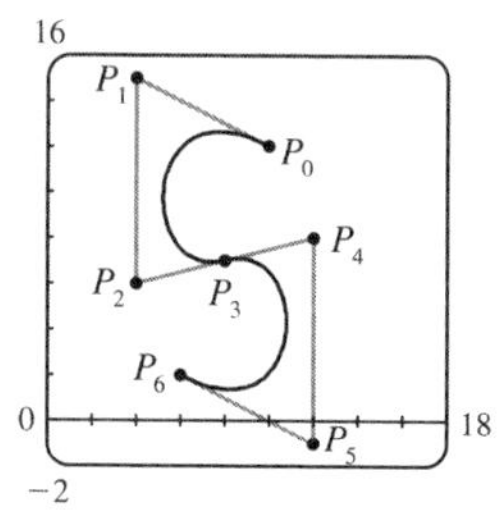


## APPLIED PROJECT Where Should a Pilot Start Descent?

1. Condition (i) will hold if and only if all of the following four conditions hold:

   ($\alpha$) $P(0) = 0$

   ($\beta$) $P'(0) = 0$ (for a smooth landing)

   ($\gamma$) $P'(\ell) = 0$ (since the plane is cruising horizontally when it begins its descent)

   ($\delta$) $P(\ell) = h$.

   First of all, condition $\alpha$ implies that $P(0) = d = 0$, so $P(x) = ax^3 + bx^2 + cx \;\Rightarrow\; P'(x) = 3ax^2 + 2bx + c$. But

$P'(0) = c = 0$ by condition $\beta$. So $P'(\ell) = 3a\ell^2 + 2b\ell = \ell(3a\ell + 2b)$. Now by condition $\gamma$, $3a\ell + 2b = 0 \Rightarrow a = -\dfrac{2b}{3\ell}$.

Therefore, $P(x) = -\dfrac{2b}{3\ell}x^3 + bx^2$. Setting $P(\ell) = h$ for condition $\delta$, we get $P(\ell) = -\dfrac{2b}{3\ell}\ell^3 + b\ell^2 = h \Rightarrow$

$-\dfrac{2}{3}b\ell^2 + b\ell^2 = h \Rightarrow \dfrac{1}{3}b\ell^2 = h \Rightarrow b = \dfrac{3h}{\ell^2} \Rightarrow a = -\dfrac{2h}{\ell^3}$. So $y = P(x) = -\dfrac{2h}{\ell^3}x^3 + \dfrac{3h}{\ell^2}x^2$.

**2.** By condition (ii), $\dfrac{dx}{dt} = -v$ for all $t$, so $x(t) = \ell - vt$. Condition (iii) states that $\left|\dfrac{d^2y}{dt^2}\right| \le k$. By the Chain Rule, we

have $\dfrac{dy}{dt} = \dfrac{dy}{dx}\dfrac{dx}{dt} = -\dfrac{2h}{\ell^3}\left(3x^2\right)\dfrac{dx}{dt} + \dfrac{3h}{\ell^2}(2x)\dfrac{dx}{dt} = \dfrac{6hx^2v}{\ell^3} - \dfrac{6hxv}{\ell^2}$ (for $x \le \ell$) $\Rightarrow$

$\dfrac{d^2y}{dt^2} = \dfrac{6hv}{\ell^3}(2x)\dfrac{dx}{dt} - \dfrac{6hv}{\ell^2}\dfrac{dx}{dt} = -\dfrac{12hv^2}{\ell^3}x + \dfrac{6hv^2}{\ell^2}$. In particular, when $t = 0$, $x = \ell$ and so

$\left.\dfrac{d^2y}{dt^2}\right|_{t=0} = -\dfrac{12hv^2}{\ell^3}\ell + \dfrac{6hv^2}{\ell^2} = -\dfrac{6hv^2}{\ell^2}$. Thus, $\left|\dfrac{d^2y}{dt^2}\right|_{t=0} = \dfrac{6hv^2}{\ell^2} \le k$. (This condition also follows from taking $x = 0$.)

**3.** We substitute $k = 860\ \text{mi/h}^2$, $h = 35{,}000\ \text{ft} \times \dfrac{1\ \text{mi}}{5280\ \text{ft}}$, and $v = 300\ \text{mi/h}$ into the result of part (b):

$\dfrac{6\left(35{,}000 \cdot \frac{1}{5280}\right)(300)^2}{\ell^2} \le 860 \Rightarrow \ell \ge 300\sqrt{6 \cdot \dfrac{35{,}000}{5280 \cdot 860}} \approx 64.5$ miles.

**4.** Substituting the values of $h$ and $\ell$ in Problem 3 into

$P(x) = -\dfrac{2h}{\ell^3}x^3 + \dfrac{3h}{\ell^2}x^2$ gives us $P(x) = ax^3 + bx^2$,

where $a \approx 4.937 \times 10^{-5}$ and $b \approx 4.78 \times 10^{-3}$.

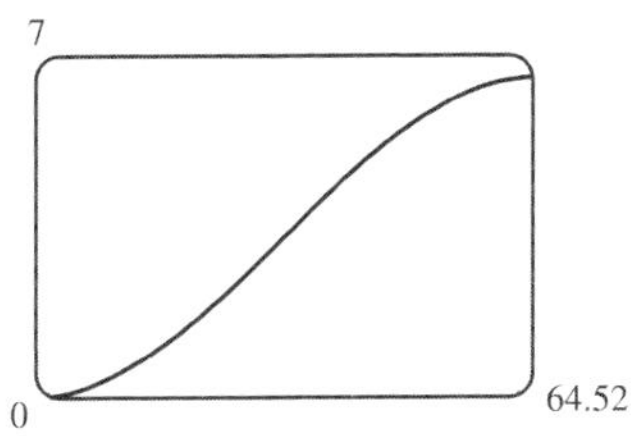


## 3.6 Implicit Differentiation

**1.** (a) $\dfrac{d}{dx}\left(xy + 2x + 3x^2\right) = \dfrac{d}{dx}(4) \Rightarrow (x \cdot y' + y \cdot 1) + 2 + 6x = 0 \Rightarrow xy' = -y - 2 - 6x \Rightarrow$

$y' = \dfrac{-y - 2 - 6x}{x}$ or $y' = -6 - \dfrac{y+2}{x}$.

(b) $xy + 2x + 3x^2 = 4 \Rightarrow xy = 4 - 2x - 3x^2 \Rightarrow y = \dfrac{4 - 2x - 3x^2}{x} = \dfrac{4}{x} - 2 - 3x$, so $y' = -\dfrac{4}{x^2} - 3$.

(c) From part (a), $y' = \dfrac{-y - 2 - 6x}{x} = \dfrac{-(4/x - 2 - 3x) - 2 - 6x}{x} = \dfrac{-4/x - 3x}{x} = -\dfrac{4}{x^2} - 3$.

**2.** (a) $\dfrac{d}{dx}\left(4x^2 + 9y^2\right) = \dfrac{d}{dx}(36) \Rightarrow 8x + 18y \cdot y' = 0 \Rightarrow y' = -\dfrac{8x}{18y} = -\dfrac{4x}{9y}$

(b) $4x^2 + 9y^2 = 36 \Rightarrow 9y^2 = 36 - 4x^2 \Rightarrow y^2 = \frac{4}{9}(9 - x^2) \Rightarrow y = \pm\frac{2}{3}\sqrt{9 - x^2}$, so

$y' = \pm\frac{2}{3} \cdot \frac{1}{2}(9 - x^2)^{-1/2}(-2x) = \mp\dfrac{2x}{3\sqrt{9 - x^2}}$

(c) From part (a), $y' = -\dfrac{4x}{9y} = -\dfrac{4x}{9\left(\pm\frac{2}{3}\sqrt{9 - x^2}\right)} = \mp\dfrac{2x}{3\sqrt{9 - x^2}}$.

**3.** $\dfrac{d}{dx}\left(x^3+x^2y+4y^2\right)=\dfrac{d}{dx}(6) \;\Rightarrow\; 3x^2+(x^2y'+y\cdot 2x)+8yy'=0 \;\Rightarrow\; x^2y'+8yy'=-3x^2-2xy \;\Rightarrow$

$(x^2+8y)y'=-3x^2-2xy \;\Rightarrow\; y'=-\dfrac{3x^2+2xy}{x^2+8y}=-\dfrac{x(3x+2y)}{x^2+8y}$

**4.** $\dfrac{d}{dx}\left(x^2-2xy+y^3\right)=\dfrac{d}{dx}(c) \;\Rightarrow\; 2x-2(xy'+y\cdot 1)+3y^2y'=0 \;\Rightarrow\; 2x-2y=2xy'-3y^2y' \;\Rightarrow$

$2x-2y=y'(2x-3y^2) \;\Rightarrow\; y'=\dfrac{2x-2y}{2x-3y^2}$

**5.** $\dfrac{d}{dx}\left(x^2y+xy^2\right)=\dfrac{d}{dx}(3x) \;\Rightarrow\; (x^2y'+y\cdot 2x)+(x\cdot 2yy'+y^2\cdot 1)=3 \;\Rightarrow\; x^2y'+2xyy'=3-2xy-y^2 \;\Rightarrow$

$y'(x^2+2xy)=3-2xy-y^2 \;\Rightarrow\; y'=\dfrac{3-2xy-y^2}{x^2+2xy}$

**6.** $\dfrac{d}{dx}\left(y^5+x^2y^3\right)=\dfrac{d}{dx}\left(1+ye^{x^2}\right) \;\Rightarrow\; 5y^4y'+(x^2\cdot 3y^2y'+y^3\cdot 2x)=0+y\cdot e^{x^2}\cdot 2x+e^{x^2}\cdot y' \;\Rightarrow$

$y'\left(5y^4+3x^2y^2-e^{x^2}\right)=2xye^{x^2}-2xy^3 \;\Rightarrow\; y'=\dfrac{2xy\left(e^{x^2}-y^2\right)}{5y^4+3x^2y^2-e^{x^2}}$

**7.** $\dfrac{d}{dx}\left(x^2y^2+x\sin y\right)=\dfrac{d}{dx}(4) \;\Rightarrow\; x^2\cdot 2yy'+y^2\cdot 2x+x\cos y\cdot y'+\sin y\cdot 1=0 \;\Rightarrow$

$2x^2yy'+x\cos y\cdot y'=-2xy^2-\sin y \;\Rightarrow\; (2x^2y+x\cos y)y'=-2xy^2-\sin y \;\Rightarrow\; y'=\dfrac{-2xy^2-\sin y}{2x^2y+x\cos y}$

**8.** $\dfrac{d}{dx}(1+x)=\dfrac{d}{dx}\left[\sin(xy^2)\right] \;\Rightarrow\; 1=[\cos(xy^2)](x\cdot 2yy'+y^2\cdot 1) \;\Rightarrow\; 1=2xy\cos(xy^2)y'+y^2\cos(xy^2) \;\Rightarrow$

$1-y^2\cos(xy^2)=2xy\cos(xy^2)y' \;\Rightarrow\; y'=\dfrac{1-y^2\cos(xy^2)}{2xy\cos(xy^2)}$

**9.** $\dfrac{d}{dx}(4\cos x\sin y)=\dfrac{d}{dx}(1) \;\Rightarrow\; 4\left[\cos x\cdot\cos y\cdot y'+\sin y\cdot(-\sin x)\right]=0 \;\Rightarrow$

$y'(4\cos x\cos y)=4\sin x\sin y \;\Rightarrow\; y'=\dfrac{4\sin x\sin y}{4\cos x\cos y}=\tan x\tan y$

**10.** $\dfrac{d}{dx}\left[y\sin(x^2)\right]=\dfrac{d}{dx}\left[x\sin(y^2)\right] \;\Rightarrow\; y\cos(x^2)\cdot 2x+\sin(x^2)\cdot y'=x\cos(y^2)\cdot 2yy'+\sin(y^2)\cdot 1 \;\Rightarrow$

$y'\left[\sin(x^2)-2xy\cos(y^2)\right]=\sin(y^2)-2xy\cos(x^2) \;\Rightarrow\; y'=\dfrac{\sin(y^2)-2xy\cos(x^2)}{\sin(x^2)-2xy\cos(y^2)}$

**11.** $\dfrac{d}{dx}\left(e^{x/y}\right)=\dfrac{d}{dx}(x-y) \;\Rightarrow\; e^{x/y}\cdot\dfrac{d}{dx}\left(\dfrac{x}{y}\right)=1-y' \;\Rightarrow$

$e^{x/y}\cdot\dfrac{y\cdot 1-x\cdot y'}{y^2}=1-y' \;\Rightarrow\; e^{x/y}\cdot\dfrac{1}{y}-\dfrac{xe^{x/y}}{y^2}\cdot y'=1-y' \;\Rightarrow\; y'-\dfrac{xe^{x/y}}{y^2}\cdot y'=1-\dfrac{e^{x/y}}{y} \;\Rightarrow$

$y'\left(1-\dfrac{xe^{x/y}}{y^2}\right)=\dfrac{y-e^{x/y}}{y} \;\Rightarrow\; y'=\dfrac{\dfrac{y-e^{x/y}}{y}}{\dfrac{y^2-xe^{x/y}}{y^2}}=\dfrac{y(y-e^{x/y})}{y^2-xe^{x/y}}$

**12.** $\sin x+\cos y=\sin x\cos y \;\Rightarrow\; \cos x-\sin y\cdot y'=\sin x\,(-\sin y\cdot y')+\cos y\cos x \;\Rightarrow$

$(\sin x\sin y-\sin y)\,y'=\cos x\cos y-\cos x \;\Rightarrow\; y'=\dfrac{\cos x\,(\cos y-1)}{\sin y\,(\sin x-1)}$

**13.** $\dfrac{d}{dx}\left\{f(x)+x^2[f(x)]^3\right\}=\dfrac{d}{dx}(10) \;\Rightarrow\; f'(x)+x^2\cdot 3[f(x)]^2\cdot f'(x)+[f(x)]^3\cdot 2x=0$. If $x=1$, we have

$f'(1)+1^2\cdot 3[f(1)]^2\cdot f'(1)+[f(1)]^3\cdot 2(1)=0 \;\Rightarrow\; f'(1)+1\cdot 3\cdot 2^2\cdot f'(1)+2^3\cdot 2=0 \;\Rightarrow$

$f'(1)+12f'(1)=-16 \;\Rightarrow\; 13f'(1)=-16 \;\Rightarrow\; f'(1)=-\frac{16}{13}$.

**14.** $\dfrac{d}{dx}\left[g(x) + x\sin g(x)\right] = \dfrac{d}{dx}\left(x^2\right) \;\Rightarrow\; g'(x) + x\cos g(x)\cdot g'(x) + \sin g(x)\cdot 1 = 2x$. If $x = 0$, we have $g'(0) + 0 + \sin g(0) = 2(0) \;\Rightarrow\; g'(0) + \sin 0 = 0 \;\Rightarrow\; g'(0) + 0 = 0 \;\Rightarrow\; g'(0) = 0$.

**15.** $x^2 + xy + y^2 = 3 \;\Rightarrow\; 2x + xy' + y\cdot 1 + 2yy' = 0 \;\Rightarrow\; xy' + 2yy' = -2x - y \;\Rightarrow\; y'(x + 2y) = -2x - y \;\Rightarrow$ $y' = \dfrac{-2x - y}{x + 2y}$. When $x = 1$ and $y = 1$, we have $y' = \dfrac{-2 - 1}{1 + 2\cdot 1} = \dfrac{-3}{3} = -1$, so an equation of the tangent line is $y - 1 = -1(x - 1)$ or $y = -x + 2$.

**16.** $x^2 + 2xy - y^2 + x = 2 \;\Rightarrow\; 2x + 2(xy' + y\cdot 1) - 2yy' + 1 = 0 \;\Rightarrow\; 2xy' - 2yy' = -2x - 2y - 1 \;\Rightarrow$ $y'(2x - 2y) = -2x - 2y - 1 \;\Rightarrow\; y' = \dfrac{-2x - 2y - 1}{2x - 2y}$. When $x = 1$ and $y = 2$, we have $y' = \dfrac{-2 - 4 - 1}{2 - 4} = \dfrac{-7}{-2} = \dfrac{7}{2}$, so an equation of the tangent line is $y - 2 = \frac{7}{2}(x - 1)$ or $y = \frac{7}{2}x - \frac{3}{2}$.

**17.** $x^2 + y^2 = (2x^2 + 2y^2 - x)^2 \;\Rightarrow\; 2x + 2yy' = 2(2x^2 + 2y^2 - x)(4x + 4yy' - 1)$. When $x = 0$ and $y = \frac{1}{2}$, we have $0 + y' = 2(\frac{1}{2})(2y' - 1) \;\Rightarrow\; y' = 2y' - 1 \;\Rightarrow\; y' = 1$, so an equation of the tangent line is $y - \frac{1}{2} = 1(x - 0)$ or $y = x + \frac{1}{2}$.

**18.** $x^{2/3} + y^{2/3} = 4 \;\Rightarrow\; \frac{2}{3}x^{-1/3} + \frac{2}{3}y^{-1/3}y' = 0 \;\Rightarrow\; \dfrac{1}{\sqrt[3]{x}} + \dfrac{y'}{\sqrt[3]{y}} = 0 \;\Rightarrow\; y' = -\dfrac{\sqrt[3]{y}}{\sqrt[3]{x}}$. When $x = -3\sqrt{3}$ and $y = 1$, we have $y' = -\dfrac{1}{\left(-3\sqrt{3}\right)^{1/3}} = -\dfrac{\left(-3\sqrt{3}\right)^{2/3}}{-3\sqrt{3}} = \dfrac{3}{3\sqrt{3}} = \dfrac{1}{\sqrt{3}}$, so an equation of the tangent line is $y - 1 = \frac{1}{\sqrt{3}}\left(x + 3\sqrt{3}\right)$ or $y = \frac{1}{\sqrt{3}}x + 4$.

**19.** $2(x^2 + y^2)^2 = 25(x^2 - y^2) \;\Rightarrow\; 4(x^2 + y^2)(2x + 2yy') = 25(2x - 2yy') \;\Rightarrow$ $4(x + yy')(x^2 + y^2) = 25(x - yy') \;\Rightarrow\; 4yy'(x^2 + y^2) + 25yy' = 25x - 4x(x^2 + y^2) \;\Rightarrow$ $y' = \dfrac{25x - 4x(x^2 + y^2)}{25y + 4y(x^2 + y^2)}$. When $x = 3$ and $y = 1$, we have $y' = \frac{75 - 120}{25 + 40} = -\frac{45}{65} = -\frac{9}{13}$, so an equation of the tangent line is $y - 1 = -\frac{9}{13}(x - 3)$ or $y = -\frac{9}{13}x + \frac{40}{13}$.

**20.** $y^2(y^2 - 4) = x^2(x^2 - 5) \;\Rightarrow\; y^4 - 4y^2 = x^4 - 5x^2 \;\Rightarrow\; 4y^3y' - 8yy' = 4x^3 - 10x$. When $x = 0$ and $y = -2$, we have $-32y' + 16y' = 0 \;\Rightarrow\; -16y' = 0 \;\Rightarrow\; y' = 0$, so an equation of the tangent line is $y + 2 = 0(x - 0)$ or $y = -2$.

**21.** (a) $y^2 = 5x^4 - x^2 \;\Rightarrow\; 2yy' = 5(4x^3) - 2x \;\Rightarrow\; y' = \dfrac{10x^3 - x}{y}$.

So at the point $(1, 2)$ we have $y' = \dfrac{10(1)^3 - 1}{2} = \dfrac{9}{2}$, and an equation of the tangent line is $y - 2 = \frac{9}{2}(x - 1)$ or $y = \frac{9}{2}x - \frac{5}{2}$.

(b)

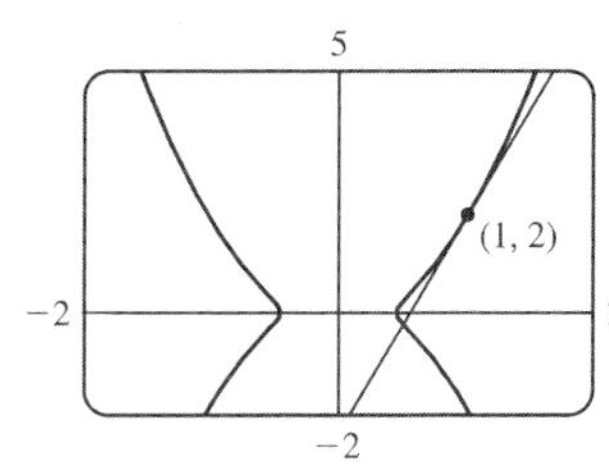


**22.** (a) $y^2 = x^3 + 3x^2 \;\Rightarrow\; 2yy' = 3x^2 + 3(2x) \;\Rightarrow\; y' = \dfrac{3x^2 + 6x}{2y}$. So at the point $(1, -2)$ we have $y' = \dfrac{3(1)^2 + 6(1)}{2(-2)} = -\dfrac{9}{4}$, and an equation of the tangent line is $y + 2 = -\frac{9}{4}(x - 1)$ or $y = -\frac{9}{4}x + \frac{1}{4}$.

(b) The curve has a horizontal tangent where $y' = 0 \;\Leftrightarrow$

$3x^2 + 6x = 0 \;\Leftrightarrow\; 3x(x + 2) = 0 \;\Leftrightarrow\; x = 0$ or $x = -2$.

But note that at $x = 0$, $y = 0$ also, so the derivative does not exist.

At $x = -2$, $y^2 = (-2)^3 + 3(-2)^2 = -8 + 12 = 4$, so $y = \pm 2$.

So the two points at which the curve has a horizontal tangent are $(-2, -2)$ and $(-2, 2)$.

(c)

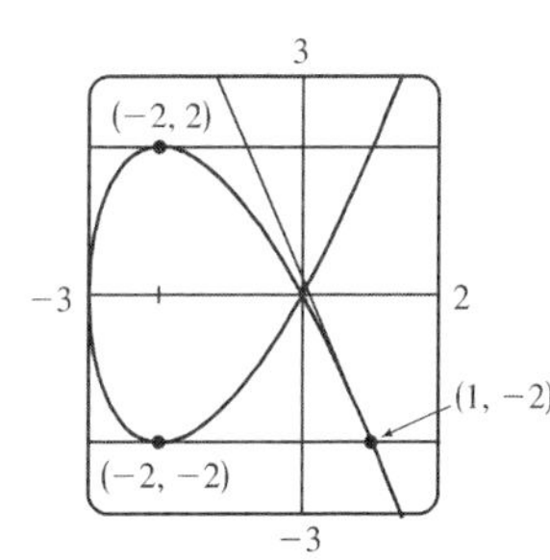

**23.** (a) There are eight points with horizontal tangents: four at $x \approx 1.57735$ and four at $x \approx 0.42265$.

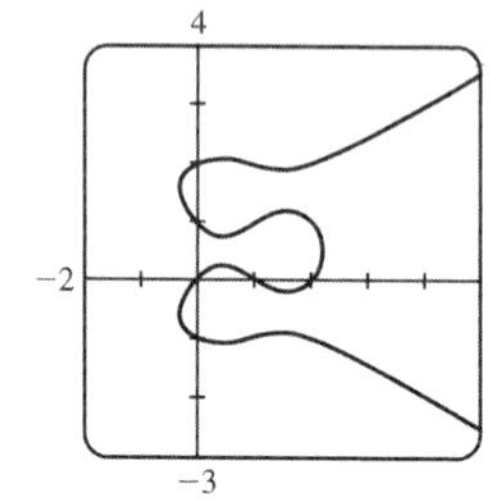

(b) $y' = \dfrac{3x^2 - 6x + 2}{2(2y^3 - 3y^2 - y + 1)} \quad \Rightarrow \quad y' = -1$ at $(0, 1)$ and $y' = \frac{1}{3}$ at $(0, 2)$.

Equations of the tangent lines are $y = -x + 1$ and $y = \frac{1}{3}x + 2$.

(c) $y' = 0 \quad \Rightarrow \quad 3x^2 - 6x + 2 = 0 \quad \Rightarrow \quad x = 1 \pm \frac{1}{3}\sqrt{3}$

(d) By multiplying the right side of the equation by $x - 3$, we obtain the first graph.By modifying the equation in other ways, we can generate the other graphs.

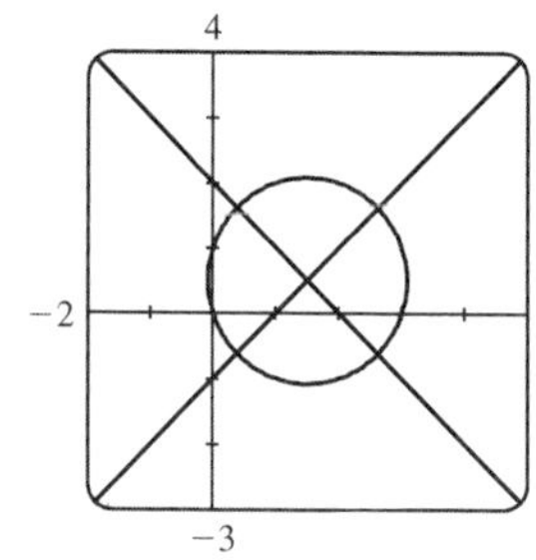

$y(y^2 - 1)(y - 2)$
$= x(x - 1)(x - 2)(x - 3)$

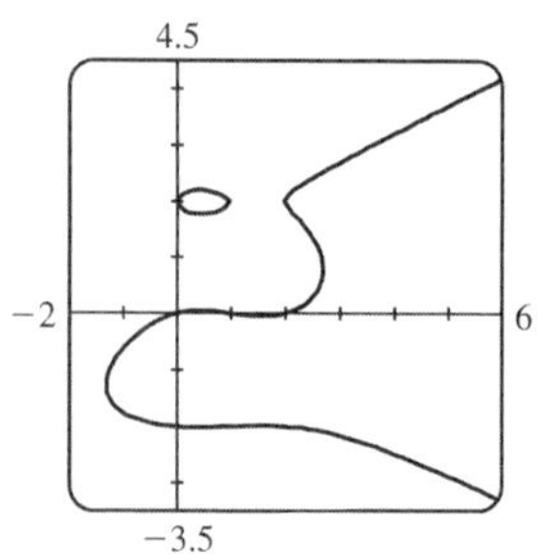

$y(y^2 - 4)(y - 2)$
$= x(x - 1)(x - 2)$

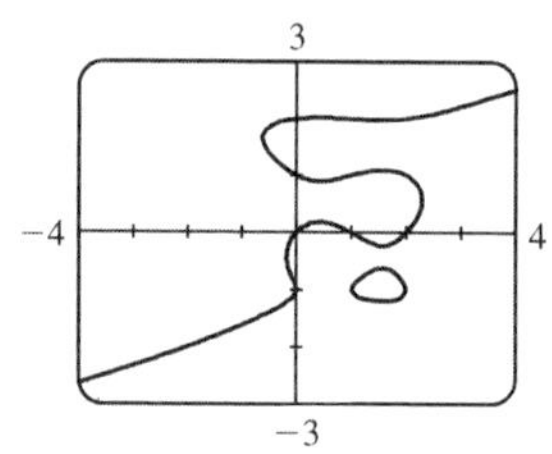

$y(y + 1)(y^2 - 1)(y - 2)$
$= x(x - 1)(x - 2)$

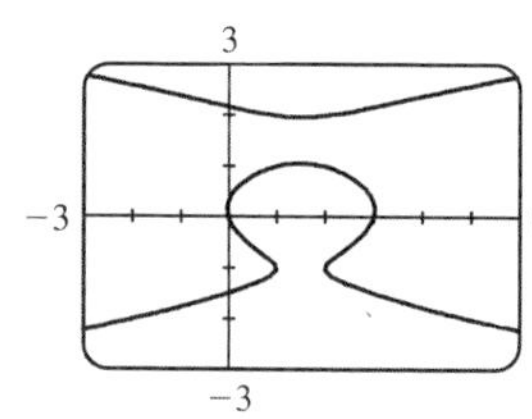

$(y + 1)(y^2 - 1)(y - 2)$
$= (x - 1)(x - 2)$

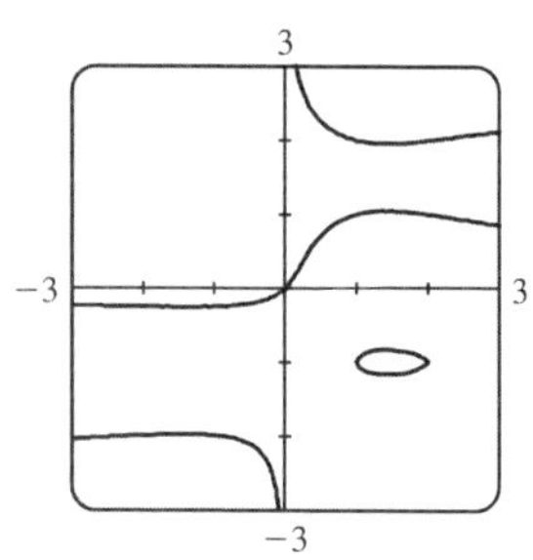

$x(y + 1)(y^2 - 1)(y - 2)$
$= y(x - 1)(x - 2)$

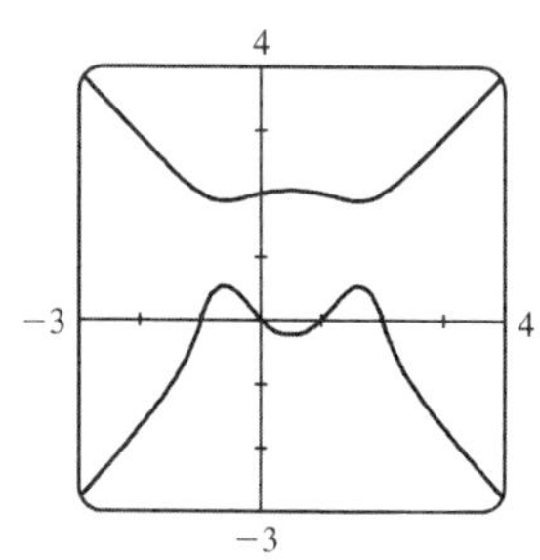

$y(y^2 + 1)(y - 2)$
$= x(x^2 - 1)(x - 2)$

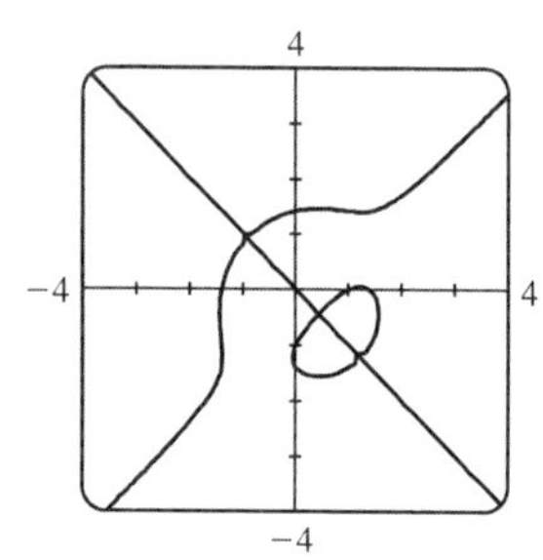

$y(y + 1)(y^2 - 2)$
$= x(x - 1)(x^2 - 2)$

**24.** (a)

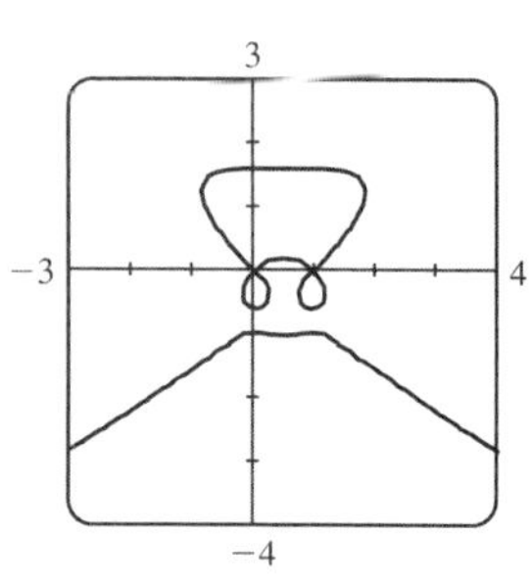

(b) There are 9 points with horizontal tangents: 3 at $x = 0$, 3 at $x = \frac{1}{2}$, and 3 at $x = 1$. The three horizontal tangents along the top of the wagon are hard to find, but by limiting the $y$-range of the graph (to $[1.6, 1.7]$, for example) they are distinguishable.

**25.** From Exercise 19, a tangent to the lemniscate will be horizontal if $y' = 0 \Rightarrow 25x - 4x(x^2 + y^2) = 0 \Rightarrow$ $x[25 - 4(x^2 + y^2)] = 0 \Rightarrow x^2 + y^2 = \frac{25}{4}$ **(1)**. (Note that when $x$ is 0, $y$ is also 0, and there is no horizontal tangent at the origin.) Substituting $\frac{25}{4}$ for $x^2 + y^2$ in the equation of the lemniscate, $2(x^2 + y^2)^2 = 25(x^2 - y^2)$, we get $x^2 - y^2 = \frac{25}{8}$ **(2)**. Solving **(1)** and **(2)**, we have $x^2 = \frac{75}{16}$ and $y^2 = \frac{25}{16}$, so the four points are $\left(\pm\frac{5\sqrt{3}}{4}, \pm\frac{5}{4}\right)$.

**26.** $\dfrac{x^2}{a^2} + \dfrac{y^2}{b^2} = 1 \Rightarrow \dfrac{2x}{a^2} + \dfrac{2yy'}{b^2} = 0 \Rightarrow y' = -\dfrac{b^2x}{a^2y} \Rightarrow$ an equation of the tangent line at $(x_0, y_0)$ is $y - y_0 = \dfrac{-b^2x_0}{a^2y_0}(x - x_0)$. Multiplying both sides by $\dfrac{y_0}{b^2}$ gives $\dfrac{y_0y}{b^2} - \dfrac{y_0^2}{b^2} = -\dfrac{x_0x}{a^2} + \dfrac{x_0^2}{a^2}$. Since $(x_0, y_0)$ lies on the ellipse, we have $\dfrac{x_0x}{a^2} + \dfrac{y_0y}{b^2} = \dfrac{x_0^2}{a^2} + \dfrac{y_0^2}{b^2} = 1$.

**27.** (a) $x^4 + y^4 = 16 \Rightarrow 4x^3 + 4y^3y' = 0 \Rightarrow y^3y' = -x^3 \Rightarrow y' = -x^3/y^3$

(b) $y'' = -\dfrac{y^3(3x^2) - (x^3)(3y^2y')}{(y^3)^2} = -\dfrac{3x^2y^3 - 3x^3y^2(-x^3/y^3)}{y^6} \cdot \dfrac{y}{y} = -\dfrac{3x^2y^4 + 3x^6}{y^7}$

(c) $y'' = -\dfrac{3x^2(y^4 + x^4)}{y^7} = -\dfrac{3x^2(16)}{y^7} = -48\dfrac{x^2}{y^7}$

**28.** $x^2 + 6xy + y^2 = 8 \Rightarrow 2x + 6xy' + 6y + 2yy' = 0 \Rightarrow (3x + y)y' = -x - 3y \Rightarrow y' = -\dfrac{x + 3y}{3x + y} \Rightarrow$

$$y'' = -\frac{(3x + y)(1 + 3y') - (x + 3y)(3 + y')}{(3x + y)^2} = -\frac{-8y + 8xy'}{(3x + y)^2} = \frac{8(y - xy')}{(3x + y)^2}$$
$$= \frac{8[y - x(-x - 3y)/(3x + y)]}{(3x + y)^2} \cdot \frac{3x + y}{3x + y} = \frac{8[y(3x + y) + x(x + 3y)]}{(3x + y)^3} = \frac{8(x^2 + 6xy + y^2)}{(3x + y)^3} = \frac{64}{(3x + y)^3}$$

At the last step, we used the fact that $x$ and $y$ must satisfy the original equation, $x^2 + 6xy + y^2 = 8$.

**29.** $y = \tan^{-1}\sqrt{x} \Rightarrow y' = \dfrac{1}{1 + (\sqrt{x})^2} \cdot \dfrac{d}{dx}(\sqrt{x}) = \dfrac{1}{1 + x}\left(\frac{1}{2}x^{-1/2}\right) = \dfrac{1}{2\sqrt{x}(1 + x)}$

**30.** $y = \sqrt{\tan^{-1} x} = (\tan^{-1} x)^{1/2} \Rightarrow$

$y' = \frac{1}{2}(\tan^{-1} x)^{-1/2} \cdot \dfrac{d}{dx}(\tan^{-1} x) = \dfrac{1}{2\sqrt{\tan^{-1} x}} \cdot \dfrac{1}{1 + x^2} = \dfrac{1}{2\sqrt{\tan^{-1} x}\,(1 + x^2)}$

**31.** $y = \sin^{-1}(2x + 1) \Rightarrow$

$y' = \dfrac{1}{\sqrt{1 - (2x + 1)^2}} \cdot \dfrac{d}{dx}(2x + 1) = \dfrac{1}{\sqrt{1 - (4x^2 + 4x + 1)}} \cdot 2 = \dfrac{2}{\sqrt{-4x^2 - 4x}} = \dfrac{1}{\sqrt{-x^2 - x}}$

**32.** $h(x) = \sqrt{1 - x^2} \arcsin x \Rightarrow h'(x) = \sqrt{1 - x^2} \cdot \dfrac{1}{\sqrt{1 - x^2}} + \arcsin x\left[\frac{1}{2}(1 - x^2)^{-1/2}(-2x)\right] = 1 - \dfrac{x \arcsin x}{\sqrt{1 - x^2}}$

**33.** $H(x) = (1 + x^2)\arctan x \Rightarrow H'(x) = (1 + x^2)\dfrac{1}{1 + x^2} + (\arctan x)(2x) = 1 + 2x\arctan x$

**34.** $y = \tan^{-1}\left(x - \sqrt{x^2 + 1}\right) \Rightarrow$

$$y' = \frac{1}{1 + \left(x - \sqrt{x^2 + 1}\right)^2}\left(1 - \frac{x}{\sqrt{x^2 + 1}}\right) = \frac{1}{1 + x^2 - 2x\sqrt{x^2 + 1} + x^2 + 1}\left(\frac{\sqrt{x^2 + 1} - x}{\sqrt{x^2 + 1}}\right)$$
$$= \frac{\sqrt{x^2 + 1} - x}{2\left(1 + x^2 - x\sqrt{x^2 + 1}\right)\sqrt{x^2 + 1}} = \frac{\sqrt{x^2 + 1} - x}{2\left[\sqrt{x^2 + 1}\,(1 + x^2) - x(x^2 + 1)\right]} = \frac{\sqrt{x^2 + 1} - x}{2\left[(1 + x^2)\left(\sqrt{x^2 + 1} - x\right)\right]}$$
$$= \frac{1}{2(1 + x^2)}$$

**35.** $y = \arcsin(\tan\theta) \Rightarrow y' = \dfrac{1}{\sqrt{1 - (\tan\theta)^2}} \cdot \dfrac{d}{d\theta}(\tan\theta) = \dfrac{\sec^2\theta}{\sqrt{1 - \tan^2\theta}}$

**36.** (a) Let $y = \cos^{-1} x$. Then $\cos y = x$ and $0 \le y \le \pi \;\Rightarrow\; \dfrac{d}{dx}(\cos y) = \dfrac{d}{dx}(x) \;\Rightarrow\; -\sin y \dfrac{dy}{dx} = 1 \;\Rightarrow$

$$\frac{dy}{dx} = -\frac{1}{\sin y} = -\frac{1}{\sqrt{1-\cos^2 y}} = -\frac{1}{\sqrt{1-x^2}}$$ (Note that $\sin y \ge 0$ for $0 \le y \le \pi$.)

(b) $y = x\cos^{-1} x - \sqrt{1-x^2} \;\Rightarrow\; y' = \cos^{-1} x - \dfrac{x}{\sqrt{1-x^2}} + \dfrac{x}{\sqrt{1-x^2}} = \cos^{-1} x$

**37.** $f(x) = e^x - x^2 \arctan x \;\Rightarrow$

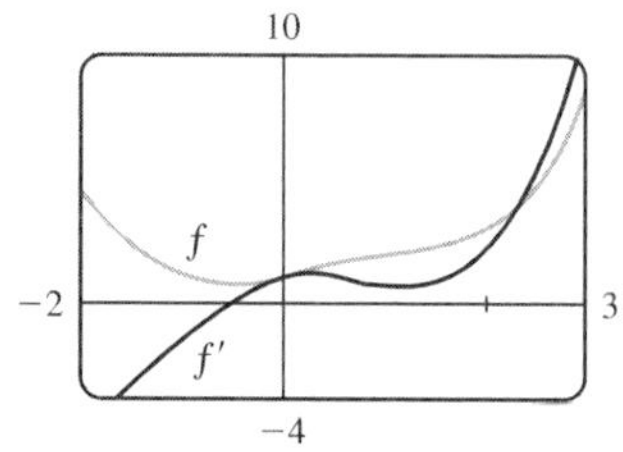


$$f'(x) = e^x - \left[x^2\left(\frac{1}{1+x^2}\right) + (\arctan x)(2x)\right]$$
$$= e^x - \frac{x^2}{1+x^2} - 2x\arctan x$$

This is reasonable because the graphs show that $f$ is increasing when $f'$ is positive, and $f'$ is zero when $f$ has a minimum.

**38.** $f(x) = x\arcsin(1-x^2) \;\Rightarrow$

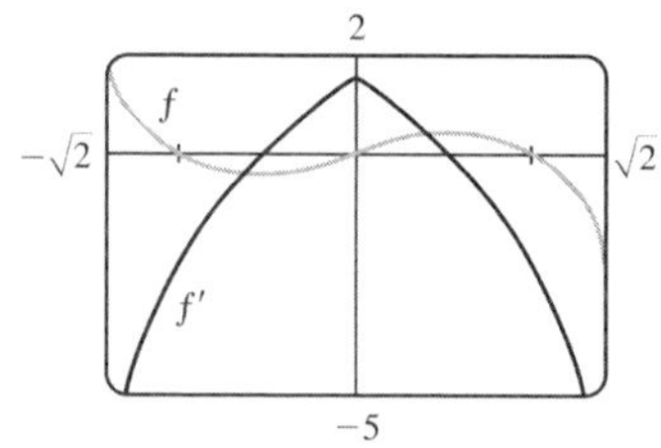


$$f'(x) = x\left[\frac{-2x}{\sqrt{1-(1-x^2)^2}}\right] + \arcsin(1-x^2)\cdot 1$$
$$= \arcsin(1-x^2) - \frac{2x^2}{\sqrt{2x^2-x^4}}$$

This is reasonable because the graphs show that $f$ is increasing when $f'$ is positive, and that $f$ has an inflection point when $f'$ changes from increasing to decreasing.

**39.** $2x^2 + y^2 = 3$ and $x = y^2$ intersect when $2x^2 + x - 3 = 0 \;\Leftrightarrow\; (2x+3)(x-1) = 0 \;\Leftrightarrow\; x = -\frac{3}{2}$ or 1, but $-\frac{3}{2}$ is extraneous since $x = y^2$ is nonnegative. When $x = 1$, $1 = y^2 \;\Rightarrow\; y = \pm 1$, so there are two points of intersection: $(1, \pm 1)$. $2x^2 + y^2 = 3 \;\Rightarrow\; 4x + 2yy' = 0 \;\Rightarrow\; y' = -2x/y$, and $x = y^2 \;\Rightarrow\; 1 = 2yy' \;\Rightarrow\; y' = 1/(2y)$. At $(1,1)$, the slopes are $m_1 = -2(1)/1 = -2$ and $m_2 = 1/(2\cdot 1) = \frac{1}{2}$, so the curves are orthogonal (since $m_1$ and $m_2$ are negative reciprocals of each other). By symmetry, the curves are also orthogonal at $(1, -1)$.

**40.** $x^2 - y^2 = 5$ and $4x^2 + 9y^2 = 72$ intersect when $4x^2 + 9(x^2 - 5) = 72 \;\Leftrightarrow\; 13x^2 = 117 \;\Leftrightarrow\; x = \pm 3$, so there are four points of intersection: $(\pm 3, \pm 2)$. $x^2 - y^2 = 5 \;\Rightarrow\; 2x - 2yy' = 0 \;\Rightarrow\; y' = x/y$, and $4x^2 + 9y^2 = 72 \;\Rightarrow$ $8x + 18yy' = 0 \;\Leftrightarrow\; y' = -4x/9y$. At $(3, 2)$, the slopes are $m_1 = \frac{3}{2}$ and $m_2 = -\frac{2}{3}$, so the curves are orthogonal. By symmetry, the curves are also orthogonal at $(3, -2)$, $(-3, 2)$ and $(-3, -2)$.

**41.**

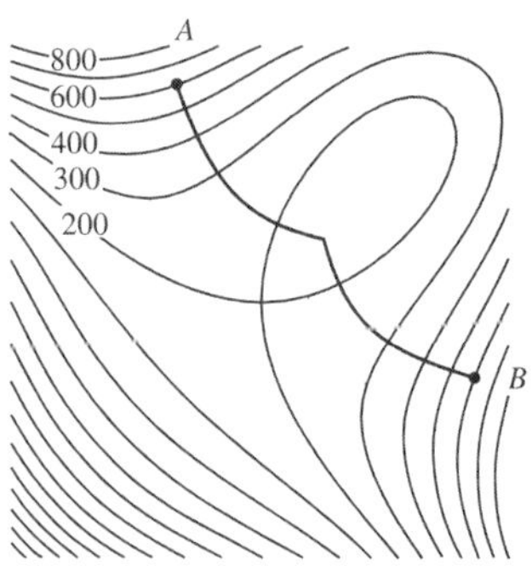


**42.** The orthogonal family represents the direction of the wind.

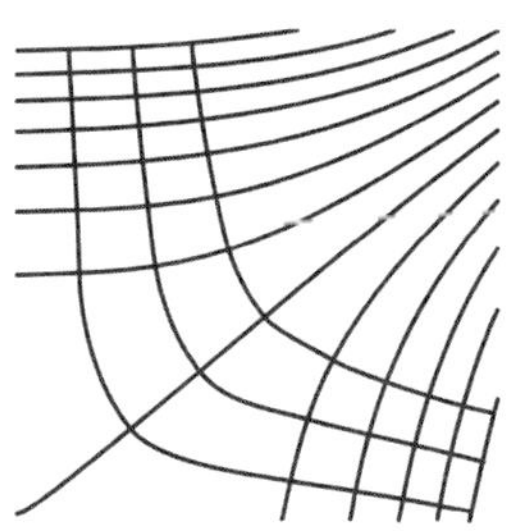

**43.** $x^2 + y^2 = r^2$ is a circle with center $O$ and $ax + by = 0$ is a line through $O$.

$x^2 + y^2 = r^2 \Rightarrow 2x + 2yy' = 0 \Rightarrow y' = -x/y$, so the slope of the tangent line at $P_0\,(x_0, y_0)$ is $-x_0/y_0$. The slope of the line $OP_0$ is $y_0/x_0$, which is the negative reciprocal of $-x_0/y_0$. Hence, the curves are orthogonal, and the families of curves are orthogonal trajectories of each other.

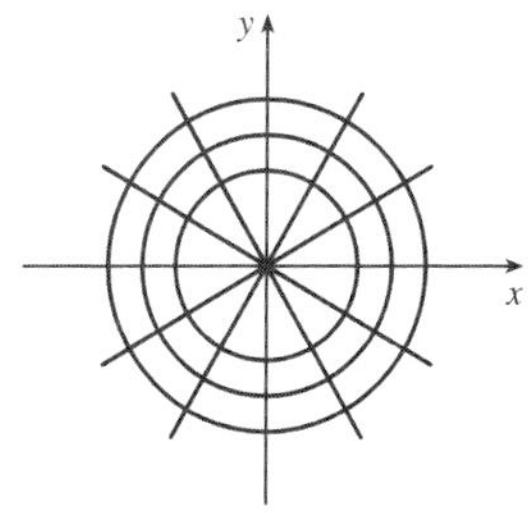


**44.** The circles $x^2 + y^2 = ax$ and $x^2 + y^2 = by$ intersect at the origin where the tangents are vertical and horizontal. If $(x_0, y_0)$ is the other point of intersection, then $x_0^2 + y_0^2 = ax_0$ **(1)** and $x_0^2 + y_0^2 = by_0$ **(2)**.

Now $x^2 + y^2 = ax \Rightarrow 2x + 2yy' = a \Rightarrow y' = \dfrac{a - 2x}{2y}$ and $x^2 + y^2 = by \Rightarrow$

$2x + 2yy' = by' \Rightarrow y' = \dfrac{2x}{b - 2y}$. Thus, the curves are orthogonal at $(x_0, y_0) \Leftrightarrow$

$\dfrac{a - 2x_0}{2y_0} = -\dfrac{b - 2y_0}{2x_0} \Leftrightarrow 2ax_0 - 4x_0^2 = 4y_0^2 - 2by_0 \Leftrightarrow ax_0 + by_0 = 2(x_0^2 + y_0^2)$,

which is true by **(1)** and **(2)**.

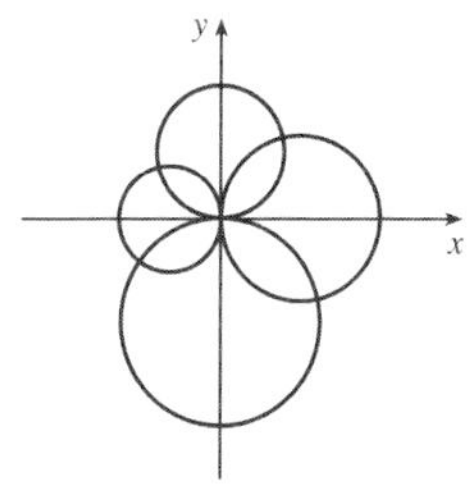


**45.** $y = cx^2 \Rightarrow y' = 2cx$ and $x^2 + 2y^2 = k \Rightarrow 2x + 4yy' = 0 \Rightarrow$

$2yy' = -x \Rightarrow y' = -\dfrac{x}{2(y)} = -\dfrac{x}{2(cx^2)} = -\dfrac{1}{2cx}$, so the curves are orthogonal.

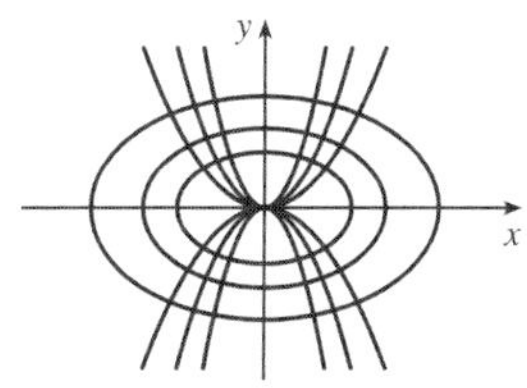


**46.** $y = ax^3 \Rightarrow y' = 3ax^2$ and $x^2 + 3y^2 = b \Rightarrow 2x + 6yy' = 0 \Rightarrow$

$3yy' = -x \Rightarrow y' = -\dfrac{x}{3(y)} = -\dfrac{x}{3(ax^3)} = -\dfrac{1}{3ax^2}$, so the curves are orthogonal.

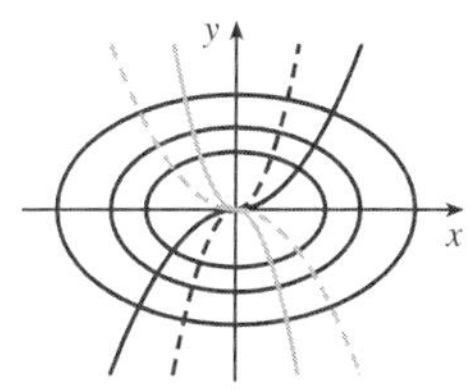


**47.** If the circle has radius $r$, its equation is $x^2 + y^2 = r^2 \Rightarrow 2x + 2yy' = 0 \Rightarrow y' = -\dfrac{x}{y}$, so the slope of the tangent line at $P(x_0, y_0)$ is $-\dfrac{x_0}{y_0}$. The negative reciprocal of that slope is $\dfrac{-1}{-x_0/y_0} = \dfrac{y_0}{x_0}$, which is the slope of $OP$, so the tangent line at $P$ is perpendicular to the radius $OP$.

**48.** $\sqrt{x} + \sqrt{y} = \sqrt{c} \Rightarrow \dfrac{1}{2\sqrt{x}} + \dfrac{y'}{2\sqrt{y}} = 0 \Rightarrow y' = -\dfrac{\sqrt{y}}{\sqrt{x}} \Rightarrow$ an equation of the tangent line at $(x_0, y_0)$ is $y - y_0 = -\dfrac{\sqrt{y_0}}{\sqrt{x_0}}(x - x_0)$. Now $x = 0 \Rightarrow y = y_0 - \dfrac{\sqrt{y_0}}{\sqrt{x_0}}(-x_0) = y_0 + \sqrt{x_0}\sqrt{y_0}$, so the $y$-intercept is $y_0 + \sqrt{x_0}\sqrt{y_0}$. And $y = 0 \Rightarrow -y_0 = -\dfrac{\sqrt{y_0}}{\sqrt{x_0}}(x - x_0) \Rightarrow x - x_0 = \dfrac{y_0\sqrt{x_0}}{\sqrt{y_0}} \Rightarrow$

$x = x_0 + \sqrt{x_0}\sqrt{y_0}$, so the $x$-intercept is $x_0 + \sqrt{x_0}\sqrt{y_0}$. The sum of the intercepts is

$\left(y_0 + \sqrt{x_0}\sqrt{y_0}\right) + \left(x_0 + \sqrt{x_0}\sqrt{y_0}\right) = x_0 + 2\sqrt{x_0}\sqrt{y_0} + y_0 = \left(\sqrt{x_0} + \sqrt{y_0}\right)^2 = \left(\sqrt{c}\right)^2 = c$.

**49.** To find the points at which the ellipse $x^2 - xy + y^2 = 3$ crosses the $x$-axis, let $y = 0$ and solve for $x$.
$y = 0 \quad\Rightarrow\quad x^2 - x(0) + 0^2 = 3 \quad\Leftrightarrow\quad x = \pm\sqrt{3}$. So the graph of the ellipse crosses the $x$-axis at the points $(\pm\sqrt{3}, 0)$.
Using implicit differentiation to find $y'$, we get $2x - xy' - y + 2yy' = 0 \quad\Rightarrow\quad y'(2y - x) = y - 2x \quad\Leftrightarrow\quad y' = \dfrac{y - 2x}{2y - x}$.

So $y'$ at $(\sqrt{3}, 0)$ is $\dfrac{0 - 2\sqrt{3}}{2(0) - \sqrt{3}} = 2$ and $y'$ at $(-\sqrt{3}, 0)$ is $\dfrac{0 + 2\sqrt{3}}{2(0) + \sqrt{3}} = 2$. Thus, the tangent lines at these points are parallel.

**50.** (a) We use implicit differentiation to find $y' = \dfrac{y - 2x}{2y - x}$ as in Exercise 49. The slope of the tangent line at $(-1, 1)$ is $m = \dfrac{1 - 2(-1)}{2(1) - (-1)} = \dfrac{3}{3} = 1$, so the slope of the normal line is $-\dfrac{1}{m} = -1$, and its equation is $y - 1 = -1(x + 1) \quad\Leftrightarrow$ $y = -x$. Substituting $-x$ for $y$ in the equation of the ellipse, we get $x^2 - x(-x) + (-x)^2 = 3 \quad\Rightarrow\quad 3x^2 = 3 \quad\Leftrightarrow\quad x = \pm 1$. So the normal line must intersect the ellipse again at $x = 1$, and since the equation of the line is $y = -x$, the other point of intersection must be $(1, -1)$.

(b)

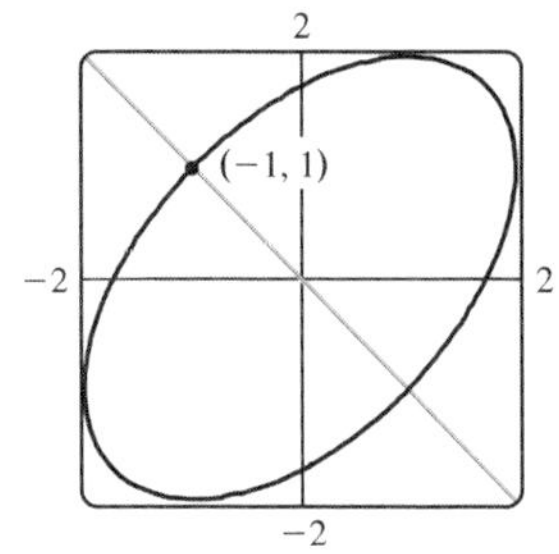


**51.** $x^2y^2 + xy = 2 \quad\Rightarrow\quad x^2 \cdot 2yy' + y^2 \cdot 2x + x \cdot y' + y \cdot 1 = 0 \quad\Leftrightarrow\quad y'(2x^2y + x) = -2xy^2 - y \quad\Leftrightarrow$
$y' = -\dfrac{2xy^2 + y}{2x^2y + x}$. So $-\dfrac{2xy^2 + y}{2x^2y + x} = -1 \quad\Leftrightarrow\quad 2xy^2 + y = 2x^2y + x \quad\Leftrightarrow\quad y(2xy + 1) = x(2xy + 1) \quad\Leftrightarrow$
$y(2xy + 1) - x(2xy + 1) = 0 \quad\Leftrightarrow\quad (2xy + 1)(y - x) = 0 \quad\Leftrightarrow\quad xy = -\frac{1}{2}$ or $y = x$. But $xy = -\frac{1}{2} \quad\Rightarrow$
$x^2y^2 + xy = \frac{1}{4} - \frac{1}{2} \neq 2$, so we must have $x = y$. Then $x^2y^2 + xy = 2 \quad\Rightarrow\quad x^4 + x^2 = 2 \quad\Leftrightarrow\quad x^4 + x^2 - 2 = 0 \quad\Leftrightarrow$
$(x^2 + 2)(x^2 - 1) = 0$. So $x^2 = -2$, which is impossible, or $x^2 = 1 \quad\Leftrightarrow\quad x = \pm 1$. Since $x = y$, the points on the curve where the tangent line has a slope of $-1$ are $(-1, -1)$ and $(1, 1)$.

**52.** $x^2 + 4y^2 = 36 \quad\Rightarrow\quad 2x + 8yy' = 0 \quad\Rightarrow\quad y' = -\dfrac{x}{4y}$. Let $(a, b)$ be a point on $x^2 + 4y^2 = 36$ whose tangent line passes through $(12, 3)$. The tangent line is then $y - 3 = -\dfrac{a}{4b}(x - 12)$, so $b - 3 = -\dfrac{a}{4b}(a - 12)$. Multiplying both sides by $4b$ gives $4b^2 - 12b = -a^2 + 12a$, so $4b^2 + a^2 = 12(a + b)$. But $4b^2 + a^2 = 36$, so $36 = 12(a + b) \quad\Rightarrow\quad a + b = 3 \quad\Rightarrow$
$b = 3 - a$. Substituting $3 - a$ for $b$ into $a^2 + 4b^2 = 36$ gives $a^2 + 4(3 - a)^2 = 36 \quad\Leftrightarrow\quad a^2 + 36 - 24a + 4a^2 = 36 \quad\Leftrightarrow$
$5a^2 - 24a = 0 \quad\Leftrightarrow\quad a(5a - 24) = 0$, so $a = 0$ or $a = \frac{24}{5}$. If $a = 0$, $b = 3 - 0 = 3$, and if $a = \frac{24}{5}$, $b = 3 - \frac{24}{5} = -\frac{9}{5}$.
So the two points on the ellipse are $(0, 3)$ and $\left(\frac{24}{5}, -\frac{9}{5}\right)$. Using
$y - 3 = -\dfrac{a}{4b}(x - 12)$ with $(a, b) = (0, 3)$ gives us the tangent line
$y - 3 = 0$ or $y = 3$. With $(a, b) = \left(\frac{24}{5}, -\frac{9}{5}\right)$, we have
$y - 3 = -\frac{24/5}{4(-9/5)}(x - 12) \quad\Leftrightarrow\quad y - 3 = \frac{2}{3}(x - 12) \quad\Leftrightarrow\quad y = \frac{2}{3}x - 5$.
A graph of the ellipse and the tangent lines confirms our results.

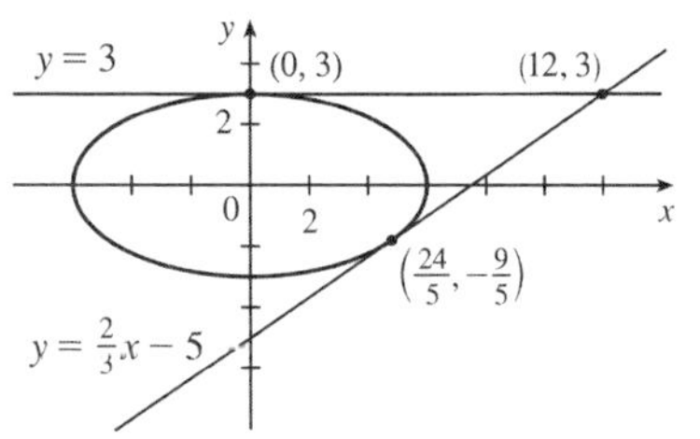


**53.** (a) If $y = f^{-1}(x)$, then $f(y) = x$. Differentiating implicitly with respect to $x$ and remembering that $y$ is a function of $x$, we get $f'(y)\dfrac{dy}{dx} = 1$, so $\dfrac{dy}{dx} = \dfrac{1}{f'(y)} \quad\Rightarrow\quad \left(f^{-1}\right)'(x) = \dfrac{1}{f'(f^{-1}(x))}$.

(b) $f(4) = 5 \quad\Rightarrow\quad f^{-1}(5) = 4$. By part (a), $\left(f^{-1}\right)'(5) = 1/f'\left(f^{-1}(5)\right) = 1/f'(4) = 1\big/\left(\frac{2}{3}\right) = \frac{3}{2}$.

**54.** (a) $f(x) = 2x + \cos x \quad \Rightarrow \quad f'(x) = 2 - \sin x > 0$ for all $x$. Thus, $f$ is increasing for all $x$ and is therefore one-to-one.

(b) Since $f$ is one-to-one, $f^{-1}(1) = k \quad \Leftrightarrow \quad f(k) = 1$. By inspection, we see that $f(0) = 2(0) + \cos 0 = 1$, so $k = f^{-1}(1) = 0$.

(c) $(f^{-1})'(1) = 1/f'(f^{-1}(1)) = 1/f'(0) = 1/(2 - \sin 0) = \frac{1}{2}$

**55.** (a) $y = J(x)$ and $xy'' + y' + xy = 0 \quad \Rightarrow \quad xJ''(x) + J'(x) + xJ(x) = 0$. If $x = 0$, we have $0 + J'(0) + 0 = 0$, so $J'(0) = 0$.

(b) Differentiating $xy'' + y' + xy = 0$ implicitly, we get $xy''' + y'' \cdot 1 + y'' + xy' + y \cdot 1 = 0 \quad \Rightarrow$ $xy''' + 2y'' + xy' + y = 0$, so $xJ'''(x) + 2J''(x) + xJ'(x) + J(x) = 0$. If $x = 0$, we have $0 + 2J''(0) + 0 + 1 \quad [J(0) = 1 \text{ is given}] \quad = 0 \quad \Rightarrow \quad 2J''(0) = -1 \quad \Rightarrow \quad J''(0) = -\frac{1}{2}$.

**56.** $x^2 + 4y^2 = 5 \quad \Rightarrow \quad 2x + 4(2yy') = 0 \quad \Rightarrow \quad y' = -\dfrac{x}{4y}$. Now let $h$ be the height of the lamp, and let $(a, b)$ be the point of tangency of the line passing through the points $(3, h)$ and $(-5, 0)$. This line has slope $(h - 0)/[3 - (-5)] = \frac{1}{8}h$. But the slope of the tangent line through the point $(a, b)$ can be expressed as $y' = -\dfrac{a}{4b}$, or as $\dfrac{b - 0}{a - (-5)} = \dfrac{b}{a + 5}$ [since the line passes through $(-5, 0)$ and $(a, b)$], so $-\dfrac{a}{4b} = \dfrac{b}{a + 5} \quad \Leftrightarrow \quad 4b^2 = -a^2 - 5a \quad \Leftrightarrow \quad a^2 + 4b^2 = -5a$. But $a^2 + 4b^2 = 5$ [since $(a, b)$ is on the ellipse], so $5 = -5a \quad \Leftrightarrow \quad a = -1$. Then $4b^2 = -a^2 - 5a = -1 - 5(-1) = 4 \quad \Rightarrow \quad b = 1$, since the point is on the top half of the ellipse. So $\dfrac{h}{8} = \dfrac{b}{a + 5} = \dfrac{1}{-1 + 5} = \dfrac{1}{4} \quad \Rightarrow \quad h = 2$. So the lamp is located 2 units above the $x$-axis.

## 3.7 Derivatives of Logarithmic Functions

**1.** The differentiation formula for logarithmic functions, $\dfrac{d}{dx}(\log_a x) = \dfrac{1}{x \ln a}$, is simplest when $a = e$ because $\ln e = 1$.

**2.** $f(x) = \ln(x^2 + 10) \quad \Rightarrow \quad f'(x) = \dfrac{1}{x^2 + 10} \dfrac{d}{dx}(x^2 + 10) = \dfrac{2x}{x^2 + 10}$

**3.** $f(\theta) = \ln(\cos\theta) \quad \Rightarrow \quad f'(\theta) = \dfrac{1}{\cos\theta} \dfrac{d}{d\theta}(\cos\theta) = \dfrac{-\sin\theta}{\cos\theta} = -\tan\theta$

**4.** $f(x) = \cos(\ln x) \quad \Rightarrow \quad f'(x) = -\sin(\ln x) \cdot \dfrac{1}{x} = \dfrac{-\sin(\ln x)}{x}$

**5.** $f(x) = \sqrt[5]{\ln x} = (\ln x)^{1/5} \quad \Rightarrow \quad f'(x) = \frac{1}{5}(\ln x)^{-4/5} \dfrac{d}{dx}(\ln x) = \dfrac{1}{5(\ln x)^{4/5}} \cdot \dfrac{1}{x} = \dfrac{1}{5x\sqrt[5]{(\ln x)^4}}$

**6.** $f(x) = \ln\sqrt[5]{x} = \ln x^{1/5} = \frac{1}{5}\ln x \quad \Rightarrow \quad f'(x) = \dfrac{1}{5} \cdot \dfrac{1}{x} = \dfrac{1}{5x}$

**7.** $f(x) = \log_2(1 - 3x) \quad \Rightarrow \quad f'(x) = \dfrac{1}{(1 - 3x)\ln 2} \dfrac{d}{dx}(1 - 3x) = \dfrac{-3}{(1 - 3x)\ln 2}$ or $\dfrac{3}{(3x - 1)\ln 2}$

**8.** $f(x) = \log_5(xe^x) \quad \Rightarrow \quad f'(x) = \dfrac{1}{xe^x \ln 5} \dfrac{d}{dx}(xe^x) = \dfrac{1}{xe^x \ln 5}(xe^x + e^x \cdot 1) = \dfrac{e^x(x + 1)}{xe^x \ln 5} = \dfrac{x + 1}{x \ln 5}$

*Another solution:* We can change the form of the function by first using logarithm properties.

$f(x) = \log_5(xe^x) = \log_5 x + \log_5 e^x \quad \Rightarrow \quad f'(x) = \dfrac{1}{x \ln 5} + \dfrac{1}{e^x \ln 5} \cdot e^x = \dfrac{1}{x \ln 5} + \dfrac{1}{\ln 5}$ or $\dfrac{1 + x}{x \ln 5}$

**9.** $f(x) = \sin x \ln(5x) \quad \Rightarrow \quad f'(x) = \sin x \cdot \dfrac{1}{5x} \cdot \dfrac{d}{dx}(5x) + \ln(5x) \cdot \cos x$

$$= \frac{\sin x \cdot 5}{5x} + \cos x \ln(5x) = \frac{\sin x}{x} + \cos x \ln(5x)$$

**10.** $f(t) = \dfrac{1+\ln t}{1-\ln t} \Rightarrow f'(t) = \dfrac{(1-\ln t)(1/t)-(1+\ln t)(-1/t)}{(1-\ln t)^2} = \dfrac{(1/t)[(1-\ln t)+(1+\ln t)]}{(1-\ln t)^2} = \dfrac{2}{t(1-\ln t)^2}$

**11.** $F(t) = \ln\dfrac{(2t+1)^3}{(3t-1)^4} = \ln(2t+1)^3 - \ln(3t-1)^4 = 3\ln(2t+1) - 4\ln(3t-1) \Rightarrow$

$F'(t) = 3\cdot\dfrac{1}{2t+1}\cdot 2 - 4\cdot\dfrac{1}{3t-1}\cdot 3 = \dfrac{6}{2t+1} - \dfrac{12}{3t-1}$, or combined, $\dfrac{-6(t+3)}{(2t+1)(3t-1)}$.

**12.** $F(y) = y\ln(1+e^y) \Rightarrow F'(y) = y\cdot\dfrac{1}{1+e^y}\cdot e^y + \ln(1+e^y)\cdot 1 = \dfrac{ye^y}{1+e^y} + \ln(1+e^y)$

**13.** $f(u) = \dfrac{\ln u}{1+\ln(2u)} \Rightarrow$

$f'(u) = \dfrac{[1+\ln(2u)]\cdot\frac{1}{u} - \ln u\cdot\frac{1}{2u}\cdot 2}{[1+\ln(2u)]^2} = \dfrac{\frac{1}{u}[1+\ln(2u)-\ln u]}{[1+\ln(2u)]^2} = \dfrac{1+(\ln 2+\ln u)-\ln u}{u[1+\ln(2u)]^2} = \dfrac{1+\ln 2}{u[1+\ln(2u)]^2}$

**14.** $y = \ln(x^4\sin^2 x) = \ln x^4 + \ln(\sin x)^2 = 4\ln x + 2\ln\sin x \Rightarrow y' = 4\cdot\dfrac{1}{x} + 2\cdot\dfrac{1}{\sin x}\cdot\cos x = \dfrac{4}{x} + 2\cot x$

**15.** $y = \ln\left|2-x-5x^2\right| \Rightarrow y' = \dfrac{1}{2-x-5x^2}\cdot(-1-10x) = \dfrac{-10x-1}{2-x-5x^2}$ or $\dfrac{10x+1}{5x^2+x-2}$

**16.** $G(u) = \ln\sqrt{\dfrac{3u+2}{3u-2}} = \frac{1}{2}[\ln(3u+2)-\ln(3u-2)] \Rightarrow G'(u) = \dfrac{1}{2}\left(\dfrac{3}{3u+2} - \dfrac{3}{3u-2}\right) = \dfrac{-6}{9u^2-4}$

**17.** $y = \ln(e^{-x}+xe^{-x}) = \ln(e^{-x}(1+x)) = \ln(e^{-x}) + \ln(1+x) = -x + \ln(1+x) \Rightarrow$

$y' = -1 + \dfrac{1}{1+x} = \dfrac{-1-x+1}{1+x} = -\dfrac{x}{1+x}$

**18.** $y = [\ln(1+e^x)]^2 \Rightarrow y' = 2[\ln(1+e^x)]\cdot\dfrac{1}{1+e^x}\cdot e^x = \dfrac{2e^x\ln(1+e^x)}{1+e^x}$

**19.** $y = e^x\ln x \Rightarrow y' = e^x\cdot\dfrac{1}{x} + (\ln x)\cdot e^x = e^x\left(\dfrac{1}{x}+\ln x\right) \Rightarrow$

$y'' = e^x\left(-\dfrac{1}{x^2}+\dfrac{1}{x}\right) + \left(\dfrac{1}{x}+\ln x\right)e^x = e^x\left(-\dfrac{1}{x^2}+\dfrac{1}{x}+\dfrac{1}{x}+\ln x\right) = e^x\left(\ln x + \dfrac{2}{x} - \dfrac{1}{x^2}\right)$

**20.** $y = \dfrac{\ln x}{x^2} \Rightarrow y' = \dfrac{x^2(1/x)-(\ln x)(2x)}{(x^2)^2} = \dfrac{x(1-2\ln x)}{x^4} = \dfrac{1-2\ln x}{x^3} \Rightarrow$

$y'' = \dfrac{x^3(-2/x)-(1-2\ln x)(3x^2)}{(x^3)^2} = \dfrac{x^2(-2-3+6\ln x)}{x^6} = \dfrac{6\ln x - 5}{x^4}$

**21.** $f(x) = \dfrac{x}{1-\ln(x-1)} \Rightarrow$

$f'(x) = \dfrac{[1-\ln(x-1)]\cdot 1 - x\cdot\dfrac{-1}{x-1}}{[1-\ln(x-1)]^2} = \dfrac{\dfrac{(x-1)[1-\ln(x-1)]+x}{x-1}}{[1-\ln(x-1)]^2} = \dfrac{x-1-(x-1)\ln(x-1)+x}{(x-1)[1-\ln(x-1)]^2}$

$= \dfrac{2x-1-(x-1)\ln(x-1)}{(x-1)[1-\ln(x-1)]^2}$

$\text{Dom}(f) = \{x \mid x-1>0 \text{ and } 1-\ln(x-1)\neq 0\} = \{x \mid x>1 \text{ and } \ln(x-1)\neq 1\}$

$= \{x \mid x>1 \text{ and } x-1\neq e^1\} = \{x \mid x>1 \text{ and } x\neq 1+e\} = (1,1+e)\cup(1+e,\infty)$

**22.** $f(x) = \ln\ln\ln x \Rightarrow f'(x) = \dfrac{1}{\ln\ln x}\cdot\dfrac{1}{\ln x}\cdot\dfrac{1}{x}$.

$\text{Dom}(f) = \{x \mid \ln\ln x > 0\} = \{x \mid \ln x > 1\} = \{x \mid x > e\} = (e,\infty)$.

**23.** $y = \ln(x^2-3) \Rightarrow y' = \dfrac{1}{x^2-3}\cdot 2x = \dfrac{2x}{x^2-3}$.

$y'(2) = \dfrac{2(2)}{2^2-3} = 4$, so an equation of the tangent line at $(2,0)$ is $y-0 = 4(x-2)$ or $y = 4x-8$.

**24.** $y = \dfrac{\ln x}{x} \Rightarrow y' = \dfrac{x(1/x) - \ln x}{x^2} = \dfrac{1 - \ln x}{x^2}$.

$y'(1) = \dfrac{1-0}{1^2} = 1$ and $y'(e) = \dfrac{1-1}{e^2} = 0 \Rightarrow$ equations of tangent lines are $y - 0 = 1(x-1)$ or $y = x - 1$ and $y - 1/e = 0(x - e)$ or $y = 1/e$.

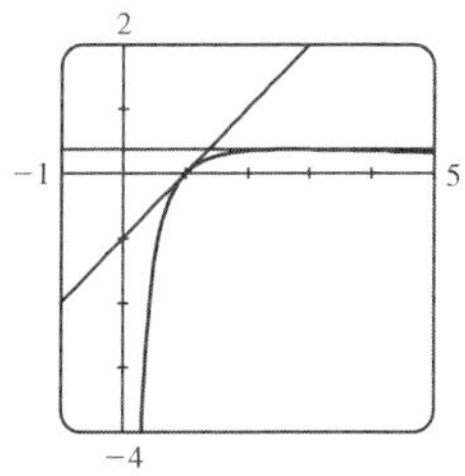


**25.** (a) The domain of $f(x) = x \ln x$ is $(0, \infty)$. $f'(x) = x(1/x) + (\ln x) \cdot 1 = 1 + \ln x$. So $f'(x) < 0$ when $1 + \ln x < 0 \Leftrightarrow \ln x < -1 \Leftrightarrow x < e^{-1}$. Therefore, $f$ is decreasing on $(0, 1/e)$.

(b) $f'(x) = 1 + \ln x \Rightarrow f''(x) = 1/x > 0$ for $x > 0$. So the curve is concave upward on $(0, \infty)$.

**26.** $f(x) = \sin x + \ln x \Rightarrow f'(x) = \cos x + 1/x$. This is reasonable, because the graph shows that $f$ increases when $f'$ is positive, and $f'(x) = 0$ when $f$ has a horizontal tangent.

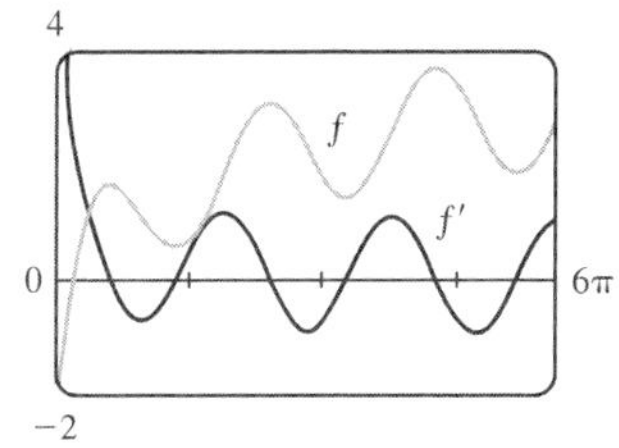


**27.** $y = (2x+1)^5(x^4-3)^6 \Rightarrow \ln y = \ln\left((2x+1)^5(x^4-3)^6\right) \Rightarrow \ln y = 5\ln(2x+1) + 6\ln(x^4-3) \Rightarrow$

$\dfrac{1}{y}y' = 5 \cdot \dfrac{1}{2x+1} \cdot 2 + 6 \cdot \dfrac{1}{x^4-3} \cdot 4x^3 \Rightarrow$

$y' = y\left(\dfrac{10}{2x+1} + \dfrac{24x^3}{x^4-3}\right) = (2x+1)^5(x^4-3)^6\left(\dfrac{10}{2x+1} + \dfrac{24x^3}{x^4-3}\right)$.

[The answer could be simplified to $y' = 2(2x+1)^4(x^4-3)^5(29x^4+12x^3-15)$, but this is unnecessary.]

**28.** $y = \sqrt{x}\, e^{x^2}(x^2+1)^{10} \Rightarrow \ln y = \ln\sqrt{x} + \ln e^{x^2} + \ln(x^2+1)^{10} \Rightarrow \ln y = \frac{1}{2}\ln x + x^2 + 10\ln(x^2+1) \Rightarrow$

$\dfrac{1}{y}y' = \dfrac{1}{2} \cdot \dfrac{1}{x} + 2x + 10 \cdot \dfrac{1}{x^2+1} \cdot 2x \Rightarrow y' = \sqrt{x}\, e^{x^2}(x^2+1)^{10}\left(\dfrac{1}{2x} + 2x + \dfrac{20x}{x^2+1}\right)$

**29.** $y = \dfrac{\sin^2 x \tan^4 x}{(x^2+1)^2} \Rightarrow \ln y = \ln(\sin^2 x \tan^4 x) - \ln(x^2+1)^2 \Rightarrow$

$\ln y = \ln(\sin x)^2 + \ln(\tan x)^4 - \ln(x^2+1)^2 \Rightarrow \ln y = 2\ln|\sin x| + 4\ln|\tan x| - 2\ln(x^2+1) \Rightarrow$

$\dfrac{1}{y}y' = 2 \cdot \dfrac{1}{\sin x} \cdot \cos x + 4 \cdot \dfrac{1}{\tan x} \cdot \sec^2 x - 2 \cdot \dfrac{1}{x^2+1} \cdot 2x \Rightarrow y' = \dfrac{\sin^2 x \tan^4 x}{(x^2+1)^2}\left(2\cot x + \dfrac{4\sec^2 x}{\tan x} - \dfrac{4x}{x^2+1}\right)$

**30.** $y = \sqrt[4]{\dfrac{x^2+1}{x^2-1}} \Rightarrow \ln y = \frac{1}{4}\ln(x^2+1) - \frac{1}{4}\ln(x^2-1) \Rightarrow \dfrac{1}{y}y' = \dfrac{1}{4} \cdot \dfrac{1}{x^2+1} \cdot 2x - \dfrac{1}{4} \cdot \dfrac{1}{x^2-1} \cdot 2x \Rightarrow$

$y' = \sqrt[4]{\dfrac{x^2+1}{x^2-1}} \cdot \dfrac{1}{2}\left(\dfrac{x}{x^2+1} - \dfrac{x}{x^2-1}\right) = \dfrac{1}{2}\sqrt[4]{\dfrac{x^2+1}{x^2-1}}\left(\dfrac{-2x}{x^4-1}\right) = \dfrac{x}{1-x^4}\sqrt[4]{\dfrac{x^2+1}{x^2-1}}$

**31.** $y = x^x \Rightarrow \ln y = \ln x^x \Rightarrow \ln y = x\ln x \Rightarrow y'/y = x(1/x) + (\ln x) \cdot 1 \Rightarrow y' = y(1 + \ln x) \Rightarrow$

$y' = x^x(1 + \ln x)$

**32.** $y = x^{\cos x} \Rightarrow \ln y = \ln x^{\cos x} \Rightarrow \ln y = \cos x \ln x \Rightarrow \dfrac{1}{y}y' = \cos x \cdot \dfrac{1}{x} + \ln x \cdot (-\sin x) \Rightarrow$

$y' = y\left(\dfrac{\cos x}{x} - \ln x \sin x\right) \Rightarrow y' = x^{\cos x}\left(\dfrac{\cos x}{x} - \ln x \sin x\right)$

**33.** $y = (\cos x)^x \Rightarrow \ln y = \ln(\cos x)^x \Rightarrow \ln y = x\ln\cos x \Rightarrow \dfrac{1}{y}y' = x \cdot \dfrac{1}{\cos x} \cdot (-\sin x) + \ln\cos x \cdot 1 \Rightarrow$

$y' = y\left(\ln\cos x - \dfrac{x\sin x}{\cos x}\right) \Rightarrow y' = (\cos x)^x(\ln\cos x - x\tan x)$

**34.** $y = \sqrt{x}^{\,x} \Rightarrow \ln y = \ln \sqrt{x}^{\,x} \Rightarrow \ln y = x \ln x^{1/2} \Rightarrow \ln y = \frac{1}{2} x \ln x \Rightarrow \frac{1}{y} y' = \frac{1}{2} x \cdot \frac{1}{x} + \ln x \cdot \frac{1}{2} \Rightarrow$
$y' = y\left(\frac{1}{2} + \frac{1}{2}\ln x\right) \Rightarrow y' = \frac{1}{2}\sqrt{x}^{\,x}(1 + \ln x)$

**35.** $y = (\tan x)^{1/x} \Rightarrow \ln y = \ln(\tan x)^{1/x} \Rightarrow \ln y = \frac{1}{x} \ln \tan x \Rightarrow$
$\frac{1}{y} y' = \frac{1}{x} \cdot \frac{1}{\tan x} \cdot \sec^2 x + \ln \tan x \cdot \left(-\frac{1}{x^2}\right) \Rightarrow y' = y\left(\frac{\sec^2 x}{x \tan x} - \frac{\ln \tan x}{x^2}\right) \Rightarrow$
$y' = (\tan x)^{1/x}\left(\frac{\sec^2 x}{x \tan x} - \frac{\ln \tan x}{x^2}\right)$ or $y' = (\tan x)^{1/x} \cdot \frac{1}{x}\left(\csc x \sec x - \frac{\ln \tan x}{x}\right)$

**36.** $y = (\sin x)^{\ln x} \Rightarrow \ln y = \ln(\sin x)^{\ln x} \Rightarrow \ln y = \ln x \cdot \ln \sin x \Rightarrow$
$\frac{1}{y} y' = \ln x \cdot \frac{1}{\sin x} \cdot \cos x + \ln \sin x \cdot \frac{1}{x} \Rightarrow y' = y\left(\ln x \cdot \frac{\cos x}{\sin x} + \frac{\ln \sin x}{x}\right) \Rightarrow$
$y' = (\sin x)^{\ln x}\left(\ln x \cot x + \frac{\ln \sin x}{x}\right)$

**37.** $y = \ln(x^2 + y^2) \Rightarrow y' = \frac{1}{x^2 + y^2} \frac{d}{dx}(x^2 + y^2) \Rightarrow y' = \frac{2x + 2yy'}{x^2 + y^2} \Rightarrow x^2 y' + y^2 y' = 2x + 2yy' \Rightarrow$
$x^2 y' + y^2 y' - 2yy' = 2x \Rightarrow (x^2 + y^2 - 2y)y' = 2x \Rightarrow y' = \frac{2x}{x^2 + y^2 - 2y}$

**38.** $x^y = y^x \Rightarrow y \ln x = x \ln y \Rightarrow y \cdot \frac{1}{x} + (\ln x) \cdot y' = x \cdot \frac{1}{y} \cdot y' + \ln y \Rightarrow y' \ln x - \frac{x}{y} y' = \ln y - \frac{y}{x} \Rightarrow$
$y' = \frac{\ln y - y/x}{\ln x - x/y}$

**39.** $f(x) = \ln(x-1) \Rightarrow f'(x) = 1/(x-1) = (x-1)^{-1} \Rightarrow f''(x) = -(x-1)^{-2} \Rightarrow f'''(x) = 2(x-1)^{-3} \Rightarrow$
$f^{(4)}(x) = -2 \cdot 3(x-1)^{-4} \Rightarrow \cdots \Rightarrow f^{(n)}(x) = (-1)^{n-1} \cdot 2 \cdot 3 \cdot 4 \cdot \cdots \cdot (n-1)(x-1)^{-n} = (-1)^{n-1}\frac{(n-1)!}{(x-1)^n}$

**40.** $y = x^8 \ln x$, so $D^9 y = D^8 y' = D^8\left(8x^7 \ln x + x^7\right)$. But the eighth derivative of $x^7$ is 0, so we now have
$D^8\left(8x^7 \ln x\right) = D^7\left(8 \cdot 7x^6 \ln x + 8x^6\right) = D^7\left(8 \cdot 7x^6 \ln x\right) = D^6\left(8 \cdot 7 \cdot 6x^5 \ln x\right) = \cdots = D\left(8!\, x^0 \ln x\right) = 8!/x$.

**41.** If $f(x) = \ln(1+x)$, then $f'(x) = \frac{1}{1+x}$, so $f'(0) = 1$.
Thus, $\lim\limits_{x \to 0} \frac{\ln(1+x)}{x} = \lim\limits_{x \to 0} \frac{f(x)}{x} = \lim\limits_{x \to 0} \frac{f(x) - f(0)}{x - 0} = f'(0) = 1$.

**42.** Let $m = n/x$. Then $n = xm$, and as $n \to \infty$, $m \to \infty$.
Therefore, $\lim\limits_{n \to \infty}\left(1 + \frac{x}{n}\right)^n = \lim\limits_{m \to \infty}\left(1 + \frac{1}{m}\right)^{mx} = \left[\lim\limits_{m \to \infty}\left(1 + \frac{1}{m}\right)^m\right]^x = e^x$ by Equation 6.

## DISCOVERY PROJECT Hyperbolic Functions

**1.** (a)

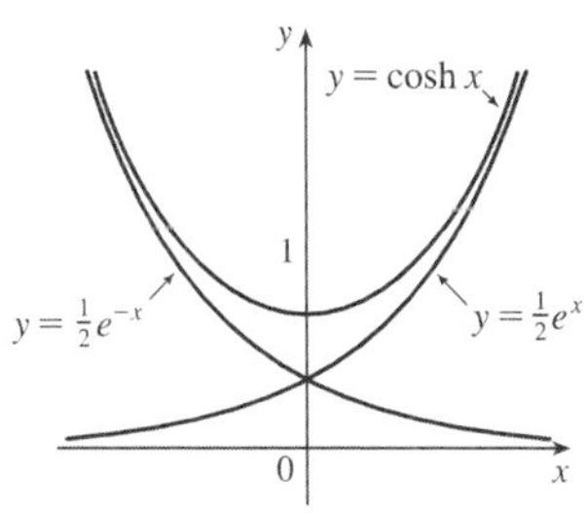


(b) The domain of $y = \cosh x$ is $\mathbb{R}$.

The range is $y \geq 1$.

**2.** 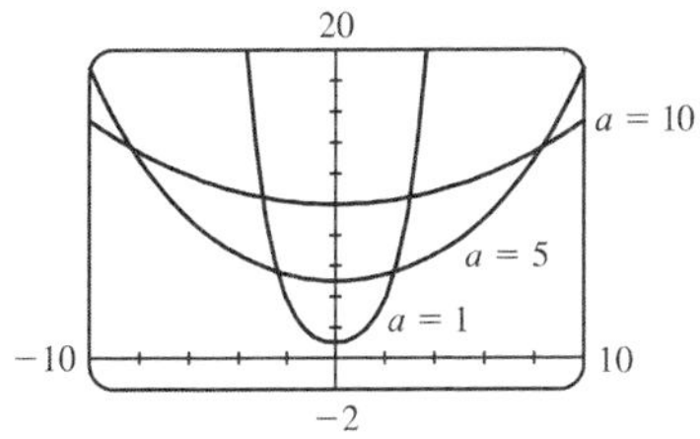


The value of $a$ is the $y$-intercept. As $a$ increases, the graph flattens out.

**3.** 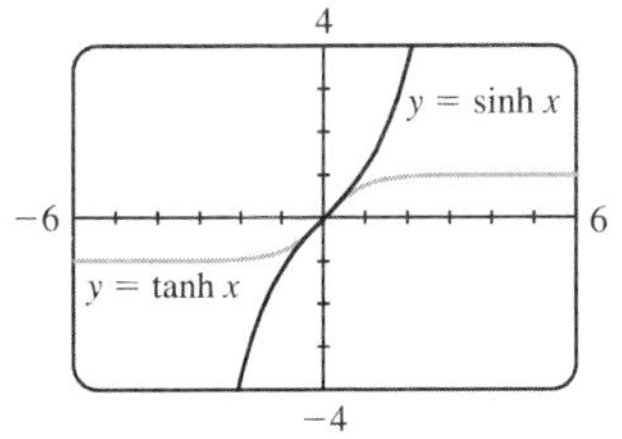


From the graphs, it appears that cosh is even and that sinh and tanh are odd.

$$f(x) = \cosh x = \frac{e^x + e^{-x}}{2} \quad \Rightarrow$$

$$f(-x) = \frac{e^{-x} + e^{-(-x)}}{2} = \frac{e^{-x} + e^x}{2} = f(x), \text{ so cosh is even.}$$

$$f(x) = \sinh x = \frac{e^x - e^{-x}}{2} \quad \Rightarrow \quad f(-x) = \frac{e^{-x} - e^{-(-x)}}{2} = \frac{e^{-x} - e^x}{2} = -\frac{e^x - e^{-x}}{2} = -f(x), \text{ so sinh is odd.}$$

$$f(x) = \tanh x = \frac{\sinh x}{\cosh x} \quad \Rightarrow \quad f(-x) = \frac{\sinh(-x)}{\cosh(-x)} = \frac{-\sinh x}{\cosh x} = -f(x), \text{ so tanh is odd.}$$

**4.** $\cosh^2 x - \sinh^2 x = \left(\dfrac{e^x + e^{-x}}{2}\right)^2 - \left(\dfrac{e^x - e^{-x}}{2}\right)^2 = \dfrac{e^{2x} + 2 + e^{-2x}}{4} - \dfrac{e^{2x} - 2 + e^{-2x}}{4} = \dfrac{4}{4} = 1.$

**5.** The curve looks like one branch of a hyperbola. We can confirm this by using the identity proved in Problem 4; $x^2 - y^2 = \cosh^2 t - \sinh^2 t = 1$. So the curve is the right branch of the hyperbola $x^2 - y^2 = 1$.

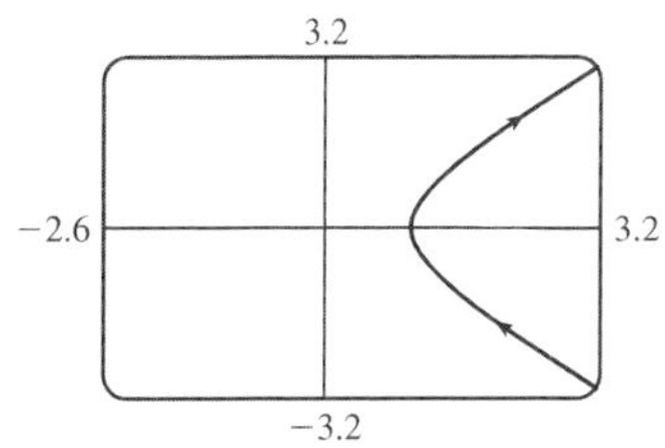


**6.**
$$\begin{aligned}
\sinh x \cosh y + \cosh x \sinh y &= \left[\tfrac{1}{2}(e^x - e^{-x})\right]\left[\tfrac{1}{2}(e^y + e^{-y})\right] + \left[\tfrac{1}{2}(e^x + e^{-x})\right]\left[\tfrac{1}{2}(e^y - e^{-y})\right] \\
&= \tfrac{1}{4}[(e^{x+y} + e^{x-y} - e^{-x+y} - e^{-x-y}) + (e^{x+y} - e^{x-y} + e^{-x+y} - e^{-x-y})] \\
&= \tfrac{1}{4}(2e^{x+y} - 2e^{-x-y}) = \tfrac{1}{2}[e^{x+y} - e^{-(x+y)}] = \sinh(x + y)
\end{aligned}$$

**7.** Substituting $-y$ for $y$ in Problem 6 and using the results from Problem 3 gives us $\sinh(x - y) = \sinh x \, \cosh y - \cosh x \, \sinh y$. By analogy with the trigonometric identity $\sin 2x = 2 \sin x \, \cos x$, we might guess that $\sinh 2x = 2 \sinh x \, \cosh x$ and we can prove it by putting $y = x$ in the identity in Problem 6. In view of the trigonometric identity $\cos(x + y) = \cos x \, \cos y - \sin x \, \sin y$, we expect that there is a similar hyperbolic identity. By a calculation similar to that in Problem 6, we can prove that $\cosh(x + y) = \cosh x \, \cosh y + \sinh x \, \sinh y$. Then, by putting $y = x$, we get $\cosh 2x = \cosh^2 x + \sinh^2 x$. By substituting $-y$ for $y$ gives us $\cosh(x - y) = \cosh x \, \cosh y - \sinh x \, \sinh y$. Also, using the identities for $\sinh(x \pm y)$ and $\cosh(x \pm y)$, we get

$$\tanh(x \pm y) = \frac{\sinh(x \pm y)}{\cosh(x \pm y)} = \frac{(\sinh x \, \cosh y \pm \cosh x \, \sinh y)/(\cosh x \, \cosh y)}{(\cosh x \, \cosh y \pm \sinh x \, \sinh y)/(\cosh x \, \cosh y)} = \frac{\tanh x \pm \tanh y}{1 \pm \tanh x \, \tanh y}.$$

**8.** (a) $\dfrac{d}{dx}(\sinh x) = \dfrac{d}{dx}\left(\dfrac{e^x - e^{-x}}{2}\right) = \dfrac{e^x + e^{-x}}{2} = \cosh x$

(b) $\dfrac{d}{dx}(\cosh x) = \dfrac{d}{dx}\left(\dfrac{e^x + e^{-x}}{2}\right) = \dfrac{e^x - e^{-x}}{2} = \sinh x$

$$\frac{d}{dx}(\tanh x) = \frac{d}{dx}\left(\frac{\sinh x}{\cosh x}\right) = \frac{\cosh x\,\cosh x - \sinh x\,\sinh x}{(\cosh x)^2} = \frac{\cosh^2 x - \sinh^2 x}{\cosh^2 x} = \frac{1}{\cosh^2 x} = \operatorname{sech}^2 x$$

**9.** (a) sinh is an increasing function on its domain, so it passes the Horizontal Line Test, and is one-to-one.

(b) Let $y = \sinh^{-1} x$. Then $\sinh y = x$. If we differentiate this equation implicitly with respect to $x$, we get $\cosh y\,\dfrac{dy}{dx} = 1$. Since $\cosh^2 y - \sinh^2 y = 1$ and $\cosh y \geq 0$, we have $\cosh y = \sqrt{1 + \sinh^2 y}$, so $\dfrac{dy}{dx} = \dfrac{1}{\cosh y} = \dfrac{1}{\sqrt{1 + \sinh^2 y}} = \dfrac{1}{\sqrt{1 + x^2}}$.

(c) Let $y = \sinh^{-1} x$. Then $x = \sinh y = \dfrac{e^y - e^{-y}}{2}$, so $e^y - 2x - e^{-y} = 0$ or, multiplying by $e^y$, $e^{2y} - 2xe^y - 1 = 0$. This is really a quadratic equation in $e^y$: $(e^y)^2 - 2x(e^y) - 1 = 0$. Solving by the quadratic formula, we get $e^y = \dfrac{2x \pm \sqrt{4x^2 + 4}}{2} = x \pm \sqrt{x^2 + 1}$. Note that $e^y > 0$, but $x - \sqrt{x^2 + 1} < 0$ $\left[\text{because } x < \sqrt{x^2 + 1}\right]$. Thus, the minus sign is inadmissible and we have $e^y = x + \sqrt{x^2 + 1}$. Therefore, $y = \ln(e^y) = \ln\left(x + \sqrt{x^2 + 1}\right)$.

(d)
$$\begin{aligned}\frac{d}{dx}\left[\ln\left(x + \sqrt{x^2 + 1}\right)\right] &= \frac{1}{x + \sqrt{x^2 + 1}}\left[1 + \tfrac{1}{2}(x^2 + 1)^{-1/2}(2x)\right] = \frac{1}{x + \sqrt{x^2 + 1}}\left(1 + \frac{x}{\sqrt{x^2 + 1}}\right) \\ &= \frac{1}{x + \sqrt{x^2 + 1}} \cdot \frac{\sqrt{x^2 + 1} + x}{\sqrt{x^2 + 1}} = \frac{1}{\sqrt{x^2 + 1}}\end{aligned}$$

**10.** (a) From the graph in Problem 3, tanh is an increasing function on its domain, so it passes the Horizontal Line Test, and is one-to-one.

(b) Let $y = \tanh^{-1} x$. Then $\tanh y = x \Rightarrow \operatorname{sech}^2 y \cdot y' = 1 \Rightarrow y' = \dfrac{1}{\operatorname{sech}^2 y} = \dfrac{1}{1 - \tanh^2 y} = \dfrac{1}{1 - x^2}$. To show that $\operatorname{sech}^2 y = 1 - \tanh^2 y$, divide the equation $\cosh^2 y - \sinh^2 y = 1$ by $\cosh^2 y$.

(c) Let $y = \tanh^{-1} x$. Then $x = \tanh y = \dfrac{\sinh y}{\cosh y} = \dfrac{e^y - e^{-y}}{e^y + e^{-y}} \cdot \dfrac{e^y}{e^y} = \dfrac{e^{2y} - 1}{e^{2y} + 1} \Rightarrow xe^{2y} + x = e^{2y} - 1 \Rightarrow$ $1 + x = e^{2y} - xe^{2y} \Rightarrow e^{2y} = \dfrac{1 + x}{1 - x} \Rightarrow 2y = \ln\left(\dfrac{1 + x}{1 - x}\right) \Rightarrow y = \dfrac{1}{2}\ln\left(\dfrac{1 + x}{1 - x}\right)$.

(d)
$$\begin{aligned}\frac{d}{dx}\left[\frac{1}{2}\ln\left(\frac{1 + x}{1 - x}\right)\right] &= \frac{d}{dx}\left[\tfrac{1}{2}\ln(1 + x) - \tfrac{1}{2}\ln(1 - x)\right] = \frac{1}{2}\frac{1}{1 + x} - \frac{1}{2}\frac{1}{1 - x}(-1) \\ &= \frac{(1 - x) + (1 + x)}{2(1 + x)(1 - x)} = \frac{2}{2(1 - x^2)} = \frac{1}{1 - x^2}\end{aligned}$$

**11.** The tangent to $y = \cosh x$ has slope 1 when $y' = \sinh x = 1 \Rightarrow x = \sinh^{-1} 1 = \ln\left(1 + \sqrt{2}\right)$, by Problem 9(c). Since $\sinh x = 1$ and $y = \cosh x = \sqrt{1 + \sinh^2 x}$, we have $\cosh x = \sqrt{2}$. The point is $\left(\ln\left(1 + \sqrt{2}\right), \sqrt{2}\right)$.

## 3.8 Linear Approximations and Differentials

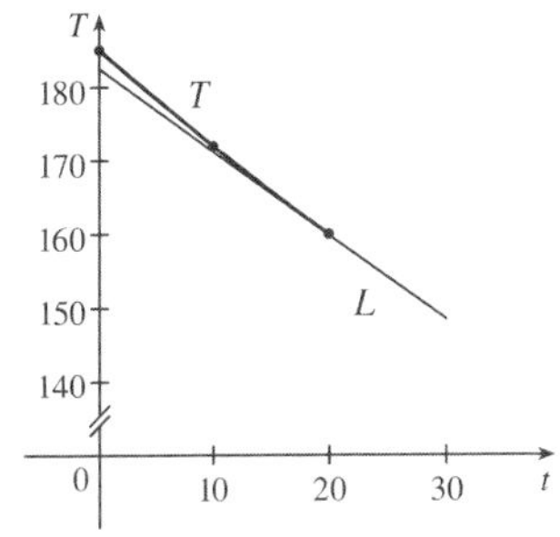


**1.** As in Example 1, $T(0) = 185$, $T(10) = 172$, $T(20) = 160$, and

$$T'(20) \approx \frac{T(10) - T(20)}{10 - 20} = \frac{172 - 160}{-10} = -1.2^\circ\text{F/min}.$$

$T(30) \approx T(20) + T'(20)(30 - 20) \approx 160 - 1.2(10) = 148^\circ\text{F}$.

We would expect the temperature of the turkey to get closer to $75^\circ$F as time increases. Since the temperature decreased $13^\circ$F in the first 10 minutes and $12^\circ$F in the second 10 minutes, we can assume that the slopes of the tangent line are increasing through negative values: $-1.3, -1.2, \ldots$. Hence, the tangent lines are under the curve and $148^\circ$F is an underestimate. From the figure, we estimate the slope of the tangent line at $t = 20$ to be $\frac{184 - 147}{0 - 30} = -\frac{37}{30}$. Then the linear approximation becomes $T(30) \approx T(20) + T'(20) \cdot 10 \approx 160 - \frac{37}{30}(10) = 147\frac{2}{3} \approx 147.7$.

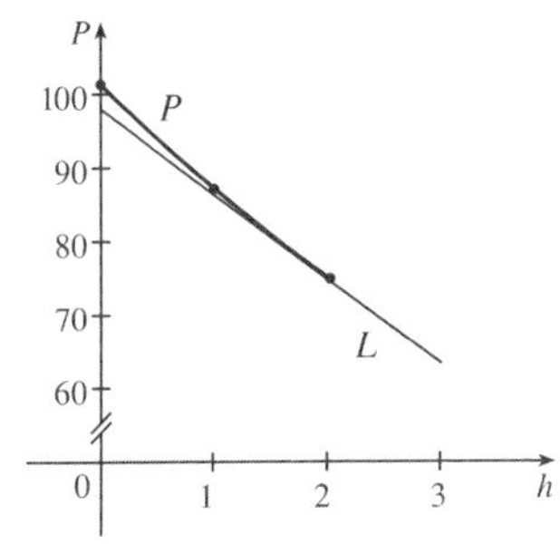


**2.** $P'(2) \approx \dfrac{P(1) - P(2)}{1 - 2} = \dfrac{87.1 - 74.9}{-1} = -12.2$ kilopascals/km.

$P(3) \approx P(2) + P'(2)(3 - 2) \approx 74.9 - 12.2(1) = 62.7$ kPa.

From the figure, we estimate the slope of the tangent line at $h = 2$ to be $\dfrac{98 - 63}{0 - 3} = -\dfrac{35}{3}$. Then the linear approximation becomes

$P(3) \approx P(2) + P'(2) \cdot 1 \approx 74.9 - \frac{35}{3} \approx 63.23$ kPa.

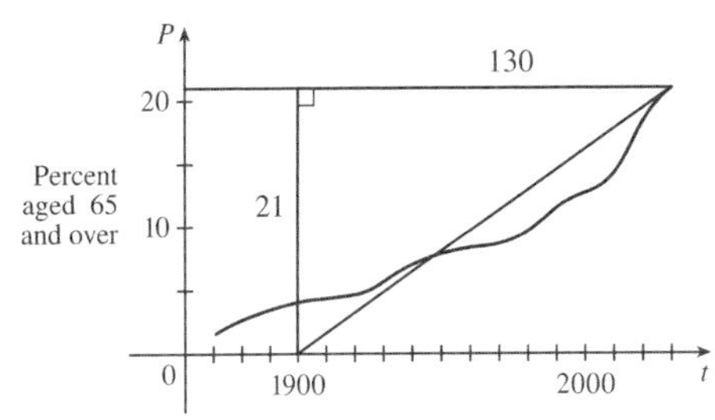


**3.** Extend the tangent line at the point $(2030, 21)$ to the $t$-axis. Answers will vary based on this approximation—we'll use $t = 1900$ as our $t$-intercept. The linearization is then

$$P(t) \approx P(2030) + P'(2030)(t - 2030) \approx 21 + \tfrac{21}{130}(t - 2030)$$

$P(2040) = 21 + \frac{21}{130}(2040 - 2030) \approx 22.6\%$

$P(2050) = 21 + \frac{21}{130}(2050 - 2030) \approx 24.2\%$

These predictions are probably too high since the tangent line lies above the graph at $t = 2030$.

**4.** Let $A = \dfrac{N(1980) - N(1985)}{1980 - 1985} = \dfrac{15.0 - 17.0}{-5} = 0.4$ and $B = \dfrac{N(1990) - N(1985)}{1990 - 1985} = \dfrac{19.3 - 17.0}{5} = 0.46$. Then

$$N'(1985) = \lim_{t \to 1985} \frac{N(t) - N(1985)}{t - 1985} \approx \frac{A + B}{2} = 0.43 \text{ million/year. So}$$

$N(1984) \approx N(1985) + N'(1985)(1984 - 1985) \approx 17.0 + 0.43(-1) = 16.57$ million.

$$N'(2000) \approx \frac{N(1995) - N(2000)}{1995 - 2000} = \frac{22.0 - 24.9}{-5} = 0.58 \text{ million/year.}$$

$N(2006) \approx N(2000) + N'(2000)(2006 - 2000) \approx 24.9 + 0.58(6) = 28.38$ million.

**5.** $f(x) = x^4 + 3x^2 \;\Rightarrow\; f'(x) = 4x^3 + 6x$, so $f(-1) = 4$ and $f'(-1) = -10$.

Thus, $L(x) = f(-1) + f'(-1)(x - (-1)) = 4 + (-10)(x + 1) = -10x - 6$.

**6.** $f(x) = \ln x \;\Rightarrow\; f'(x) = 1/x$, so $f(1) = 0$ and $f'(1) = 1$. Thus, $L(x) = f(1) + f'(1)(x-1) = 0 + 1(x-1) = x - 1$.

**7.** $f(x) = \cos x \;\Rightarrow\; f'(x) = -\sin x$, so $f\left(\frac{\pi}{2}\right) = 0$ and $f'\left(\frac{\pi}{2}\right) = -1$.

Thus, $L(x) = f\left(\frac{\pi}{2}\right) + f'\left(\frac{\pi}{2}\right)\left(x - \frac{\pi}{2}\right) = 0 - 1\left(x - \frac{\pi}{2}\right) = -x + \frac{\pi}{2}$.

**8.** $f(x) = x^{3/4} \;\Rightarrow\; f'(x) = \frac{3}{4}x^{-1/4}$, so $f(16) = 8$ and $f'(16) = \frac{3}{8}$.

Thus, $L(x) = f(16) + f'(16)(x - 16) = 8 + \frac{3}{8}(x - 16) = \frac{3}{8}x + 2$.

**9.** $f(x) = \sqrt{1-x} \;\Rightarrow\; f'(x) = \dfrac{-1}{2\sqrt{1-x}}$, so $f(0) = 1$ and $f'(0) = -\frac{1}{2}$.

Therefore,

$\sqrt{1-x} = f(x) \approx f(0) + f'(0)(x-0) = 1 + \left(-\frac{1}{2}\right)(x-0) = 1 - \frac{1}{2}x$.

So $\sqrt{0.9} = \sqrt{1-0.1} \approx 1 - \frac{1}{2}(0.1) = 0.95$

and $\sqrt{0.99} = \sqrt{1-0.01} \approx 1 - \frac{1}{2}(0.01) = 0.995$.

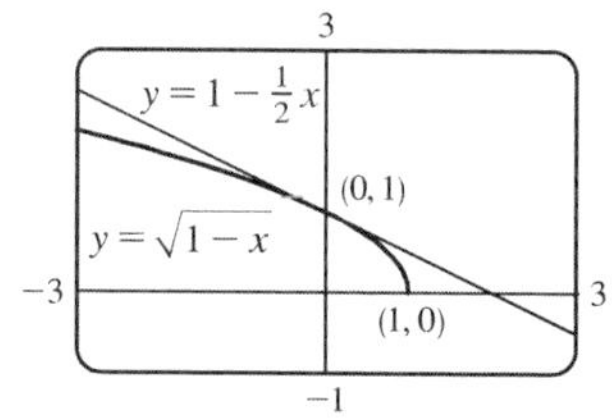


**10.** $g(x) = \sqrt[3]{1+x} = (1+x)^{1/3} \;\Rightarrow\; g'(x) = \frac{1}{3}(1+x)^{-2/3}$, so $g(0) = 1$ and $g'(0) = \frac{1}{3}$. Therefore, $\sqrt[3]{1+x} = g(x) \approx g(0) + g'(0)(x-0) = 1 + \frac{1}{3}x$.

So $\sqrt[3]{0.95} = \sqrt[3]{1+(-0.05)} \approx 1 + \frac{1}{3}(-0.05) = 0.98\overline{3}$,

and $\sqrt[3]{1.1} = \sqrt[3]{1+0.1} \approx 1 + \frac{1}{3}(0.1) = 1.0\overline{3}$.

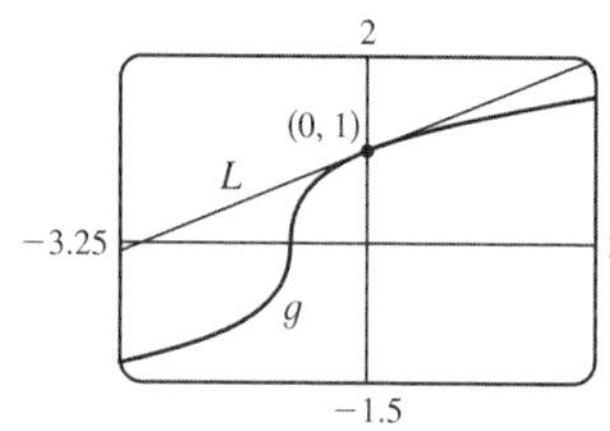


**11.** $f(x) = \sqrt[3]{1-x} = (1-x)^{1/3} \;\Rightarrow\; f'(x) = -\frac{1}{3}(1-x)^{-2/3}$, so $f(0) = 1$ and $f'(0) = -\frac{1}{3}$. Thus, $f(x) \approx f(0) + f'(0)(x-0) = 1 - \frac{1}{3}x$. We need $\sqrt[3]{1-x} - 0.1 < 1 - \frac{1}{3}x < \sqrt[3]{1-x} + 0.1$, which is true when $-1.204 < x < 0.706$.

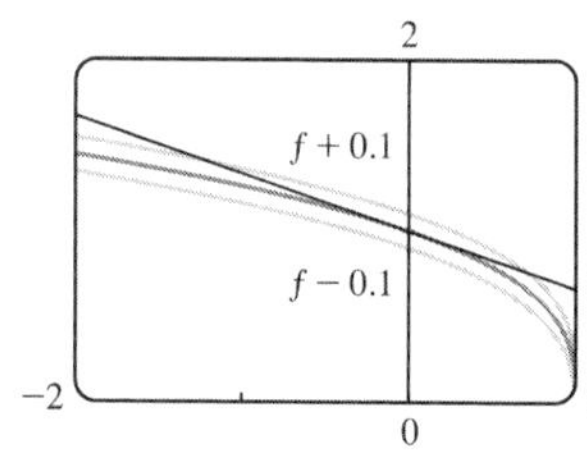


**12.** $f(x) = \tan x \;\Rightarrow\; f'(x) = \sec^2 x$, so $f(0) = 0$ and $f'(0) = 1$.

Thus, $f(x) \approx f(0) + f'(0)(x-0) = 0 + 1(x-0) = x$.

We need $\tan x - 0.1 < x < \tan x + 0.1$, which is true when $-0.63 < x < 0.63$.

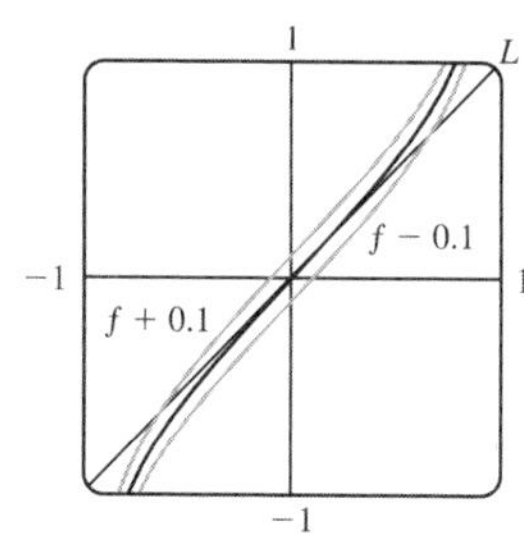


**13.** $f(x) = \dfrac{1}{(1+2x)^4} = (1+2x)^{-4} \;\Rightarrow$

$f'(x) = -4(1+2x)^{-5}(2) = \dfrac{-8}{(1+2x)^5}$, so $f(0) = 1$ and $f'(0) = -8$.

Thus, $f(x) \approx f(0) + f'(0)(x-0) = 1 + (-8)(x-0) = 1 - 8x$.

We need $1/(1+2x)^4 - 0.1 < 1 - 8x < 1/(1+2x)^4 + 0.1$, which is true when $-0.045 < x < 0.055$.

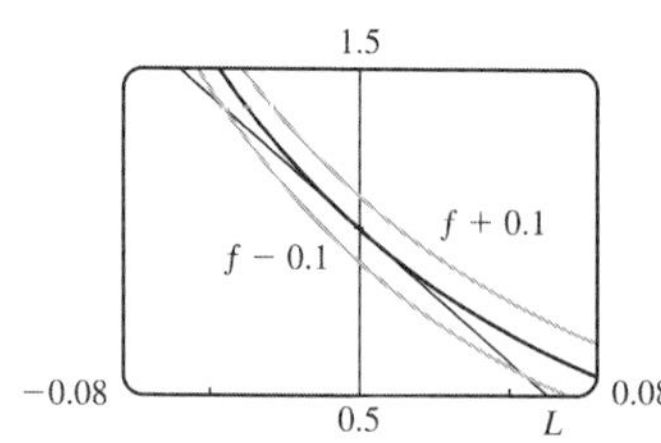

**14.** $f(x) = e^x \Rightarrow f'(x) = e^x$, so $f(0) = 1$ and $f'(0) = 1$.

Thus, $f(x) \approx f(0) + f'(0)(x - 0) = 1 + 1(x - 0) = 1 + x$.

We need $e^x - 0.1 < 1 + x < e^x + 0.1$, which is true when $-0.483 < x < 0.416$.

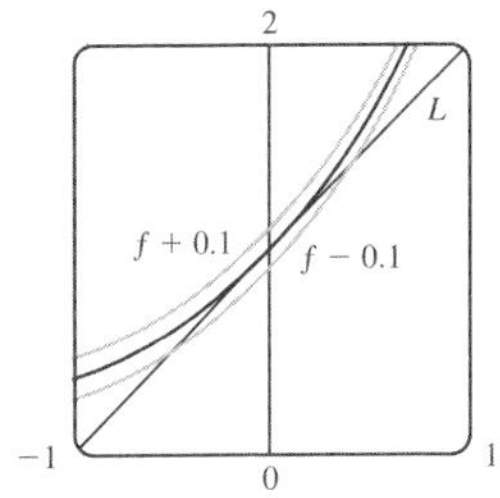


**15.** To estimate $(2.001)^5$, we'll find the linearization of $f(x) = x^5$ at $a = 2$. Since $f'(x) = 5x^4$, $f(2) = 32$, and $f'(2) = 80$, we have $L(x) = 32 + 80(x - 2) = 80x - 128$. Thus, $x^5 \approx 80x - 128$ when $x$ is near 2 , so $(2.001)^5 \approx 80(2.001) - 128 = 160.08 - 128 = 32.08$.

**16.** To estimate $e^{-0.015}$, we'll find the linearization of $f(x) = e^x$ at $a = 0$. Since $f'(x) = e^x$, $f(0) = 1$, and $f'(0) = 1$, we have $L(x) = 1 + 1(x - 0) = x + 1$. Thus, $e^x \approx x + 1$ when $x$ is near 0, so $e^{-0.015} \approx -0.015 + 1 = 0.985$.

**17.** To estimate $(8.06)^{2/3}$, we'll find the linearization of $f(x) = x^{2/3}$ at $a = 8$. Since $f'(x) = \frac{2}{3}x^{-1/3} = 2/(3\sqrt[3]{x})$, $f(8) = 4$, and $f'(8) = \frac{1}{3}$, we have $L(x) = 4 + \frac{1}{3}(x - 8) = \frac{1}{3}x + \frac{4}{3}$. Thus, $x^{2/3} \approx \frac{1}{3}x + \frac{4}{3}$ when $x$ is near 8, so $(8.06)^{2/3} \approx \frac{1}{3}(8.06) + \frac{4}{3} = \frac{12.06}{3} = 4.02$.

**18.** To estimate $1/1002$, we'll find the linearization of $f(x) = 1/x$ at $a = 1000$. Since $f'(x) = -1/x^2$, $f(1000) = 0.001$, and $f'(1000) = -0.000001$, we have $L(x) = 0.001 - 0.000001(x - 1000) = -0.000001x + 0.002$. Thus, $1/x \approx -0.000001x + 0.002$ when $x$ is near 1000, so $1/1002 \approx -0.000001(1002) + 0.002 = 0.000998$.

**19.** $y = f(x) = \sec x \Rightarrow f'(x) = \sec x \tan x$, so $f(0) = 1$ and $f'(0) = 1 \cdot 0 = 0$. The linear approximation of $f$ at 0 is $f(0) + f'(0)(x - 0) = 1 + 0(x) = 1$. Since 0.08 is close to 0, approximating $\sec 0.08$ with 1 is reasonable.

**20.** If $y = x^6$, $y' = 6x^5$ and the tangent line approximation at $(1, 1)$ has slope 6. If the change in $x$ is 0.01, the change in $y$ on the tangent line is 0.06, and approximating $(1.01)^6$ with 1.06 is reasonable.

**21.** $y = f(x) = \ln x \Rightarrow f'(x) = 1/x$, so $f(1) = 0$ and $f'(1) = 1$. The linear approximation of $f$ at 1 is $f(1) + f'(1)(x - 1) = 0 + 1(x - 1) = x - 1$. Now $f(1.05) = \ln 1.05 \approx 1.05 - 1 = 0.05$, so the approximation is reasonable.

**22.** (a) $f(x) = (x - 1)^2 \Rightarrow f'(x) = 2(x - 1)$, so $f(0) = 1$ and $f'(0) = -2$.
Thus, $f(x) \approx L_f(x) = f(0) + f'(0)(x - 0) = 1 - 2x$.
$g(x) = e^{-2x} \Rightarrow g'(x) = -2e^{-2x}$, so $g(0) = 1$ and $g'(0) = -2$.
Thus, $g(x) \approx L_g(x) = g(0) + g'(0)(x - 0) = 1 - 2x$.
$h(x) = 1 + \ln(1 - 2x) \Rightarrow h'(x) = \dfrac{-2}{1 - 2x}$, so $h(0) = 1$ and $h'(0) = -2$.
Thus, $h(x) \approx L_h(x) = h(0) + h'(0)(x - 0) = 1 - 2x$.
Notice that $L_f = L_g = L_h$. This happens because $f$, $g$, and $h$ have the same function values and the same derivative values at $a = 0$.

(b) The linear approximation appears to be the best for the function $f$ since it is closer to $f$ for a larger domain than it is to $g$ and $h$. The approximation looks worst for $h$ since $h$ moves away from $L$ faster than $f$ and $g$ do.

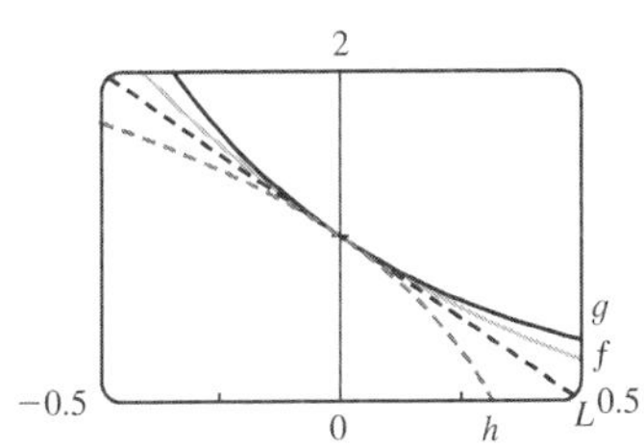

**23.** (a) The differential $dy$ is defined in terms of $dx$ by the equation $dy = f'(x)\,dx$. For $y = f(x) = x^2 \sin 2x$, $f'(x) = x^2 \cos 2x \cdot 2 + \sin 2x \cdot 2x = 2x(x\cos 2x + \sin 2x)$, so $dy = 2x(x\cos 2x + \sin 2x)\,dx$.

(b) For $y = f(t) = \ln\sqrt{1+t^2} = \frac{1}{2}\ln(1+t^2)$, $f'(t) = \dfrac{1}{2}\cdot\dfrac{1}{1+t^2}\cdot 2t = \dfrac{t}{1+t^2}$, so $dy = \dfrac{t}{1+t^2}\,dt$.

**24.** (a) For $y = f(s) = \dfrac{s}{1+2s}$, $f'(s) = \dfrac{(1+2s)(1) - s(2)}{(1+2s)^2} = \dfrac{1}{(1+2s)^2}$, so $dy = \dfrac{1}{(1+2s)^2}\,ds$.

(b) For $y = f(u) = e^{-u}\cos u$, $f'(u) = e^{-u}(-\sin u) + \cos u(-e^{-u}) = -e^{-u}(\sin u + \cos u)$, so $dy = -e^{-u}(\sin u + \cos u)\,du$.

**25.** (a) $y = e^{x/10} \quad\Rightarrow\quad dy = e^{x/10}\cdot\frac{1}{10}\,dx = \frac{1}{10}e^{x/10}dx$

(b) $x = 0$ and $dx = 0.1 \quad\Rightarrow\quad dy = \frac{1}{10}e^{0/10}(0.1) = 0.01$.

$\Delta y = f(x+\Delta x) - f(x) = e^{(x+\Delta x)/10} - e^x = e^{(0+0.1)/10} - e^0 = e^{1/100} - 1 \approx 0.0101$

**26.** (a) $y = \sqrt{x} \quad\Rightarrow\quad dy = \frac{1}{2}x^{-1/2}dx = \dfrac{1}{2\sqrt{x}}\,dx$

(b) $x = 1$ and $dx = 1 \quad\Rightarrow\quad dy = \frac{1}{2(1)}(1) = \frac{1}{2}$. $\Delta y = f(x+\Delta x) - f(x) = \sqrt{1+1} - \sqrt{1} = \sqrt{2} - 1 \approx 0.414$.

(c)

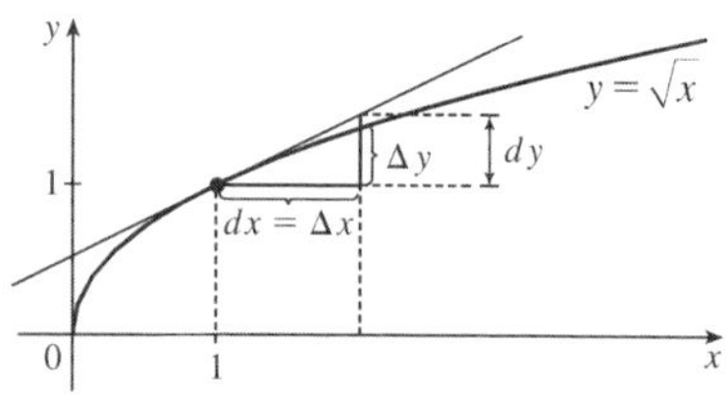


Remember, $\Delta y$ represents the amount that the curve $y = f(x)$ rises or falls when $x$ changes by an amount $dx$, whereas $dy$ represents the amount that the tangent line rises or falls (the change in the linearization).

**27.** (a) If $x$ is the edge length, then $V = x^3 \quad\Rightarrow\quad dV = 3x^2\,dx$. When $x = 30$ and $dx = 0.1$, $dV = 3(30)^2(0.1) = 270$, so the maximum possible error in computing the volume of the cube is about 270 $\text{cm}^3$. The relative error is calculated by dividing the change in $V$, $\Delta V$, by $V$. We approximate $\Delta V$ with $dV$.

Relative error $= \dfrac{\Delta V}{V} \approx \dfrac{dV}{V} = \dfrac{3x^2\,dx}{x^3} = 3\dfrac{dx}{x} = 3\left(\dfrac{0.1}{30}\right) = 0.01$.

Percentage error = relative error × 100% = 0.01 × 100% = 1%.

(b) $S = 6x^2 \quad\Rightarrow\quad dS = 12x\,dx$. When $x = 30$ and $dx = 0.1$, $dS = 12(30)(0.1) = 36$, so the maximum possible error in computing the surface area of the cube is about 36 $\text{cm}^2$.

Relative error $= \dfrac{\Delta S}{S} \approx \dfrac{dS}{S} = \dfrac{12x\,dx}{6x^2} = 2\dfrac{dx}{x} = 2\left(\dfrac{0.1}{30}\right) = 0.00\overline{6}$.

Percentage error = relative error ×100% = $0.00\overline{6}$ × 100% = $0.\overline{6}$%.

**28.** (a) $A = \pi r^2 \quad\Rightarrow\quad dA = 2\pi r\,dr$. When $r = 24$ and $dr = 0.2$, $dA = 2\pi(24)(0.2) = 9.6\pi$, so the maximum possible error in the calculated area of the disk is about $9.6\pi \approx 30\ \text{cm}^2$.

(b) Relative error $= \dfrac{\Delta A}{A} \approx \dfrac{dA}{A} = \dfrac{2\pi r\,dr}{\pi r^2} = \dfrac{2\,dr}{r} = \dfrac{2(0.2)}{24} = \dfrac{0.2}{12} = \dfrac{1}{60} = 0.01\overline{6}$.

Percentage error = relative error ×100% = $0.01\overline{6}$ × 100% = $1.\overline{6}$%.

**29.** (a) For a sphere of radius $r$, the circumference is $C = 2\pi r$ and the surface area is $S = 4\pi r^2$, so

$r = \dfrac{C}{2\pi} \quad\Rightarrow\quad S = 4\pi\left(\dfrac{C}{2\pi}\right)^2 = \dfrac{C^2}{\pi} \quad\Rightarrow\quad dS = \dfrac{2}{\pi}C\,dC$. When $C = 84$ and $dC = 0.5$, $dS = \dfrac{2}{\pi}(84)(0.5) = \dfrac{84}{\pi}$,

so the maximum error is about $\dfrac{84}{\pi} \approx 27\ \text{cm}^2$. Relative error $\approx \dfrac{dS}{S} = \dfrac{84/\pi}{84^2/\pi} = \dfrac{1}{84} \approx 0.012$

(b) $V = \frac{4}{3}\pi r^3 = \frac{4}{3}\pi\left(\frac{C}{2\pi}\right)^3 = \frac{C^3}{6\pi^2} \Rightarrow dV = \frac{1}{2\pi^2}C^2\,dC$. When $C = 84$ and $dC = 0.5$,

$dV = \frac{1}{2\pi^2}(84)^2(0.5) = \frac{1764}{\pi^2}$, so the maximum error is about $\frac{1764}{\pi^2} \approx 179\text{ cm}^3$.

The relative error is approximately $\frac{dV}{V} = \frac{1764/\pi^2}{(84)^3/(6\pi^2)} = \frac{1}{56} \approx 0.018$.

**30.** For a hemispherical dome, $V = \frac{2}{3}\pi r^3 \Rightarrow dV = 2\pi r^2\,dr$. When $r = \frac{1}{2}(50) = 25$ m and $dr = 0.05$ cm $= 0.0005$ m, $dV = 2\pi(25)^2(0.0005) = \frac{5\pi}{8}$, so the amount of paint needed is about $\frac{5\pi}{8} \approx 2\text{ m}^3$.

**31.** $F = kR^4 \Rightarrow dF = 4kR^3\,dR \Rightarrow \frac{dF}{F} = \frac{4kR^3\,dR}{kR^4} = 4\left(\frac{dR}{R}\right)$. Thus, the relative change in $F$ is about 4 times the relative change in $R$. So a 5% increase in the radius corresponds to a 20% increase in blood flow.

**32.** (a) $f(x) = \sin x \Rightarrow f'(x) = \cos x$, so $f(0) = 0$ and $f'(0) = 1$. Thus, $f(x) \approx f(0) + f'(0)(x - 0) = 0 + 1(x - 0) = x$.

(b)

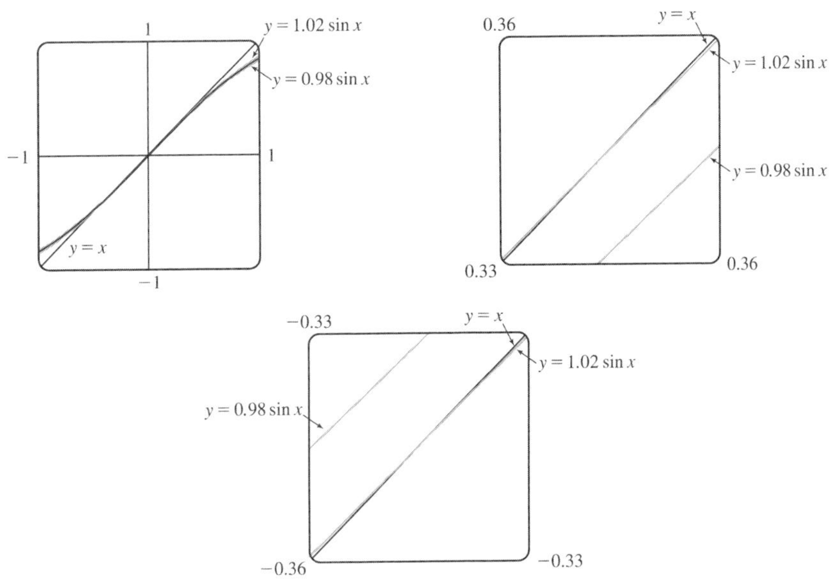


We want to know the values of $x$ for which $y = x$ approximates $y = \sin x$ with less than a 2% difference; that is, the values of $x$ for which

$$\left|\frac{x - \sin x}{\sin x}\right| < 0.02 \quad\Leftrightarrow\quad -0.02 < \frac{x - \sin x}{\sin x} < 0.02 \quad\Leftrightarrow$$

$$\begin{cases} -0.02\sin x < x - \sin x < 0.02\sin x & \text{if } \sin x > 0 \\ -0.02\sin x > x - \sin x > 0.02\sin x & \text{if } \sin x < 0 \end{cases} \quad\Leftrightarrow\quad \begin{cases} 0.98\sin x < x < 1.02\sin x & \text{if } \sin x > 0 \\ 1.02\sin x < x < 0.98\sin x & \text{if } \sin x < 0 \end{cases}$$

In the first figure, we see that the graphs are very close to each other near $x = 0$. Changing the viewing rectangle and using an intersect feature (see the second figure) we find that $y = x$ intersects $y = 1.02\sin x$ at $x \approx 0.344$. By symmetry, they also intersect at $x \approx -0.344$ (see the third figure). Converting 0.344 radians to degrees, we get $0.344\left(\frac{180^\circ}{\pi}\right) \approx 19.7^\circ \approx 20^\circ$, which verifies the statement.

**33.** (a) The graph shows that $f'(1) = 2$, so $L(x) = f(1) + f'(1)(x - 1) = 5 + 2(x - 1) = 2x + 3$. $f(0.9) \approx L(0.9) = 4.8$ and $f(1.1) \approx L(1.1) = 5.2$.

(b) From the graph, we see that $f'(x)$ is positive and decreasing. This means that the slopes of the tangent lines are positive, but the tangents are becoming less steep. So the tangent lines lie *above* the curve. Thus, the estimates in part (a) are too large.

**34.** (a) $g'(x) = \sqrt{x^2+5} \;\Rightarrow\; g'(2) = \sqrt{9} = 3$. $g(1.95) \approx g(2) + g'(2)(1.95-2) = -4 + 3(-0.05) = -4.15$.
$g(2.05) \approx g(2) + g'(2)(2.05-2) = -4 + 3(0.05) = -3.85$.

(b) The formula $g'(x) = \sqrt{x^2+5}$ shows that $g'(x)$ is positive and increasing. This means that the slopes of the tangent lines are positive and the tangents are getting steeper. So the tangent lines lie *below* the graph of $g$. Hence, the estimates in part (a) are too small.

## LABORATORY PROJECT Taylor Polynomials

**1.** We first write the functions described in conditions (i), (ii), and (iii):

$$\begin{aligned} P(x) &= A + Bx + Cx^2 & f(x) &= \cos x \\ P'(x) &= B + 2Cx & f'(x) &= -\sin x \\ P''(x) &= 2C & f''(x) &= -\cos x \end{aligned}$$

So, taking $a = 0$, our three conditions become

$$\begin{aligned} P(0) &= f(0): & A &= \cos 0 = 1 \\ P'(0) &= f'(0): & B &= -\sin 0 = 0 \\ P''(0) &= f''(0): & 2C &= -\cos 0 = -1 \;\Rightarrow\; C = -\tfrac{1}{2} \end{aligned}$$

The desired quadratic function is $P(x) = 1 - \frac{1}{2}x^2$, so the quadratic approximation is $\cos x \approx 1 - \frac{1}{2}x^2$.

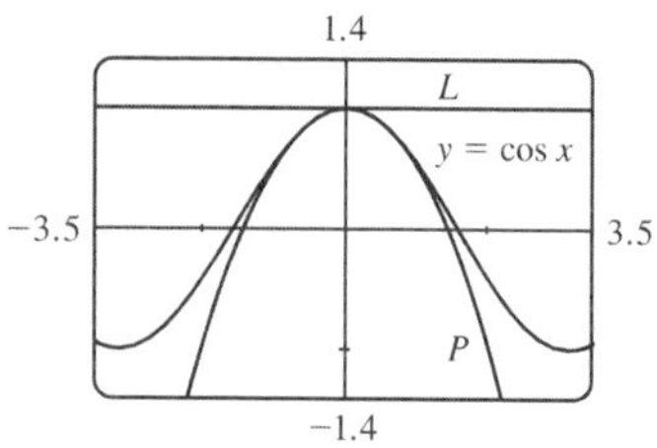


The figure shows a graph of the cosine function together with its linear approximation $L(x) = 1$ and quadratic approximation $P(x) = 1 - \frac{1}{2}x^2$ near 0. You can see that the quadratic approximation is much better than the linear one.

**2.** Accuracy to within 0.1 means that $\left|\cos x - \left(1 - \frac{1}{2}x^2\right)\right| < 0.1 \;\Leftrightarrow\; -0.1 < \cos x - \left(1 - \frac{1}{2}x^2\right) < 0.1$
$\Leftrightarrow\; 0.1 > \left(1 - \frac{1}{2}x^2\right) - \cos x > -0.1 \;\Leftrightarrow\; \cos x + 0.1 > 1 - \frac{1}{2}x^2 > \cos x - 0.1 \;\Leftrightarrow$
$\cos x - 0.1 < 1 - \frac{1}{2}x^2 < \cos x + 0.1$.

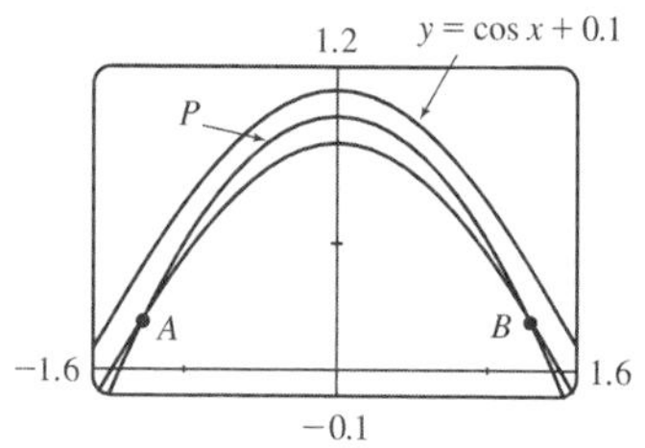


From the figure we see that this is true between $A$ and $B$. Zooming in or using an intersect feature, we find that the $x$-coordinates of $B$ and $A$ are about $\pm 1.26$. Thus, the approximation $\cos x \approx 1 - \frac{1}{2}x^2$ is accurate to within 0.1 when $-1.26 < x < 1.26$.

**3.** If $P(x) = A + B(x-a) + C(x-a)^2$, then $P'(x) = B + 2C(x-a)$ and $P''(x) = 2C$. Applying the conditions (i), (ii), and (iii), we get

$$\begin{aligned} P(a) &= f(a): & A &= f(a) \\ P'(a) &= f'(a): & B &= f'(a) \\ P''(a) &= f''(a): & 2C &= f''(a) \;\Rightarrow\; C = \tfrac{1}{2}f''(a) \end{aligned}$$

Thus, $P(x) = A + B(x-a) + C(x-a)^2$ can be written in the form $P(x) = f(a) + f'(a)(x-a) + \frac{1}{2}f''(a)(x-a)^2$.

4. From Example 2 in Section 3.8, we have $f(1) = 2$, $f'(1) = \frac{1}{4}$, and $f'(x) = \frac{1}{2}(x+3)^{-1/2}$. So $f''(x) = -\frac{1}{4}(x+3)^{-3/2} \Rightarrow$ $f''(1) = -\frac{1}{32}$. From Problem 3, the quadratic approximation $P(x)$ is $\sqrt{x+3} \approx f(1) + f'(1)(x-1) + \frac{1}{2}f''(1)(x-1)^2 = 2 + \frac{1}{4}(x-1) - \frac{1}{64}(x-1)^2$.

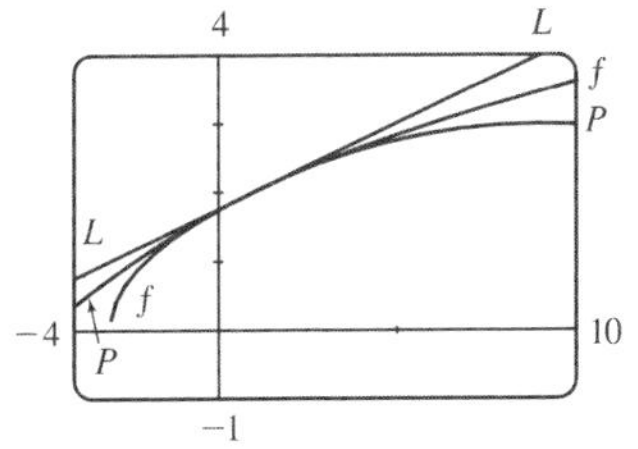


The figure shows the function $f(x) = \sqrt{x+3}$ together with its linear approximation $L(x) = \frac{1}{4}x + \frac{7}{4}$ and its quadratic approximation $P(x)$. You can see that $P(x)$ is a better approximation than $L(x)$ and this is borne out by the numerical values in the following chart.

| | from $L(x)$ | actual value | from $P(x)$ |
|---|---|---|---|
| $\sqrt{3.98}$ | 1.9950 | 1.99499373... | 1.99499375 |
| $\sqrt{4.05}$ | 2.0125 | 2.01246118... | 2.01246094 |
| $\sqrt{4.2}$ | 2.0500 | 2.04939015... | 2.04937500 |

5. $T_n(x) = c_0 + c_1(x-a) + c_2(x-a)^2 + c_3(x-a)^3 + \cdots + c_n(x-a)^n$. If we put $x = a$ in this equation, then all terms after the first are 0 and we get $T_n(a) = c_0$. Now we differentiate $T_n(x)$ and obtain $T_n'(x) = c_1 + 2c_2(x-a) + 3c_3(x-a)^2 + 4c_4(x-a)^3 + \cdots + nc_n(x-a)^{n-1}$. Substituting $x = a$ gives $T_n'(a) = c_1$. Differentiating again, we have $T_n''(x) = 2c_2 + 2\cdot 3c_3(x-a) + 3\cdot 4c_4(x-a^2) + \cdots + (n-1)nc_n(x-a)^{n-2}$ and so $T_n''(a) = 2c_2$. Continuing in this manner, we get $T_n'''(x) = 2\cdot 3c_3 + 2\cdot 3\cdot 4c_4(x-a) + \cdots + (n-2)(n-1)nc_n(x-a)^{n-3}$ and $T_n'''(a) = 2\cdot 3c_3$. By now we see the pattern. If we continue to differentiate and substitute $x = a$, we obtain $T_n^{(4)}(a) = 2\cdot 3\cdot 4c_4$ and in general, for any integer $k$ between 1 and $n$, $T_n^{(k)}(a) = 2\cdot 3\cdot 4\cdot 5\cdot\cdots\cdot kc_k = k!\,c_k \Rightarrow$ $c_k = \dfrac{T_n^{(k)}(a)}{k!}$. Because we want $T_n$ and $f$ to have the same derivatives at $a$, we require that $c_k = \dfrac{f^{(k)}(a)}{k!}$ for $k = 1, 2, \ldots, n$.

6. $T_n(x) = f(a) + f'(a)(x-a) + \dfrac{f''(a)}{2!}(x-a)^2 + \cdots + \dfrac{f^{(n)}(a)}{n!}(x-a)^n$. To compute the coefficients in this equation we need to calculate the derivatives of $f$ at 0:

$$
\begin{aligned}
f(x) &= \cos x & f(0) &= \cos 0 = 1 \\
f'(x) &= -\sin x & f'(0) &= -\sin 0 = 0 \\
f''(x) &= -\cos x & f''(0) &= -1 \\
f'''(x) &= \sin x & f'''(0) &= 0 \\
f^{(4)}(x) &= \cos x & f^{(4)}(0) &= 1
\end{aligned}
$$

We see that the derivatives repeat in a cycle of length 4, so $f^{(5)}(0) = 0$, $f^{(6)}(0) = -1$, $f^{(7)}(0) = 0$, and $f^{(8)}(0) = 1$. From the original expression for $T_n(x)$, with $n = 8$ and $a = 0$, we have

$$
\begin{aligned}
T_8(x) &= f(0) + f'(0)(x-0) + \frac{f''(0)}{2!}(x-0)^2 + \frac{f'''(0)}{3!}(x-0)^3 + \cdots + \frac{f^{(8)}(0)}{8!}(x-0)^8 \\
&= 1 + 0\cdot x + \frac{-1}{2!}x^2 + 0\cdot x^3 + \frac{1}{4!}x^4 + 0\cdot x^5 + \frac{-1}{6!}x^6 + 0\cdot x^7 + \frac{1}{8!}x^8 \\
&= 1 - \frac{x^2}{2!} + \frac{x^4}{4!} - \frac{x^6}{6!} + \frac{x^8}{8!}
\end{aligned}
$$

and the desired approximation is $\cos x \approx 1 - \frac{x^2}{2!} + \frac{x^4}{4!} - \frac{x^6}{6!} + \frac{x^8}{8!}$. The Taylor polynomials $T_2$, $T_4$, and $T_6$ consist of the initial terms of $T_8$ up through degree 2, 4, and 6, respectively. Therefore, $T_2(x) = 1 - \frac{x^2}{2!}$, $T_4(x) = 1 - \frac{x^2}{2!} + \frac{x^4}{4!}$, and $T_6(x) = 1 - \frac{x^2}{2!} + \frac{x^4}{4!} - \frac{x^6}{6!}$. We graph $T_2$, $T_4$, $T_6$, $T_8$, and $f$:

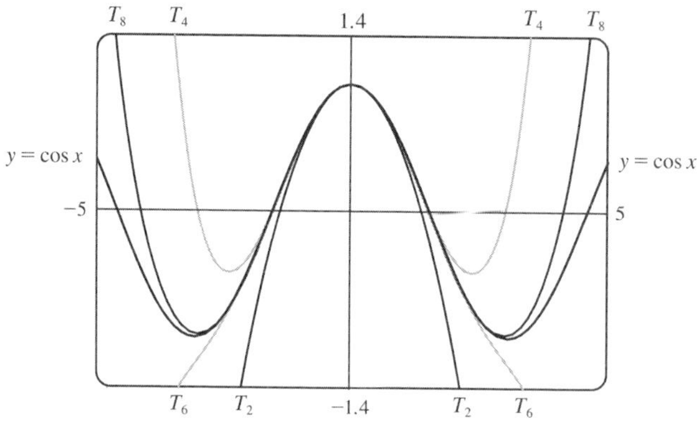


Notice that $T_2(x)$ is a good approximation to $\cos x$ near 0, $T_4(x)$ is a good approximation on a larger interval, $T_6(x)$ is a better approximation, and $T_8(x)$ is better still. Each successive Taylor polynomial is a good approximation on a larger interval than the previous one.

## 3 Review

### CONCEPT CHECK

**1.** (a) The Power Rule: If $n$ is any real number, then $\frac{d}{dx}(x^n) = nx^{n-1}$. The derivative of a variable base raised to a constant power is the power times the base raised to the power minus one.

(b) The Constant Multiple Rule: If $c$ is a constant and $f$ is a differentiable function, then $\frac{d}{dx}[cf(x)] = c\frac{d}{dx}f(x)$. The derivative of a constant times a function is the constant times the derivative of the function.

(c) The Sum Rule: If $f$ and $g$ are both differentiable, then $\frac{d}{dx}[f(x) + g(x)] = \frac{d}{dx}f(x) + \frac{d}{dx}g(x)$. The derivative of a sum of functions is the sum of the derivatives.

(d) The Difference Rule: If $f$ and $g$ are both differentiable, then $\frac{d}{dx}[f(x) - g(x)] = \frac{d}{dx}f(x) - \frac{d}{dx}g(x)$. The derivative of a difference of functions is the difference of the derivatives.

(e) The Product Rule: If $f$ and $g$ are both differentiable, then $\frac{d}{dx}[f(x)g(x)] = f(x)\frac{d}{dx}g(x) + g(x)\frac{d}{dx}f(x)$. The derivative of a product of two functions is the first function times the derivative of the second function plus the second function times the derivative of the first function.

(f) The Quotient Rule: If $f$ and $g$ are both differentiable, then $\frac{d}{dx}\left[\frac{f(x)}{g(x)}\right] = \frac{g(x)\frac{d}{dx}f(x) - f(x)\frac{d}{dx}g(x)}{[g(x)]^2}$.
The derivative of a quotient of functions is the denominator times the derivative of the numerator minus the numerator times the derivative of the denominator, all divided by the square of the denominator.

(g) The Chain Rule: If $f$ and $g$ are both differentiable and $F = f \circ g$ is the composite function defined by $F(x) = f(g(x))$, then $F$ is differentiable and $F'$ is given by the product $F'(x) = f'(g(x))g'(x)$. The derivative of a composite function is the derivative of the outer function evaluated at the inner function times the derivative of the inner function.

2. (a) $y = x^n \Rightarrow y' = nx^{n-1}$
(b) $y = e^x \Rightarrow y' = e^x$
(c) $y = a^x \Rightarrow y' = a^x \ln a$
(d) $y = \ln x \Rightarrow y' = 1/x$
(e) $y = \log_a x \Rightarrow y' = 1/(x \ln a)$
(f) $y = \sin x \Rightarrow y' = \cos x$
(g) $y = \cos x \Rightarrow y' = -\sin x$
(h) $y = \tan x \Rightarrow y' = \sec^2 x$
(i) $y = \csc x \Rightarrow y' = -\csc x \cot x$
(j) $y = \sec x \Rightarrow y' = \sec x \tan x$
(k) $y = \cot x \Rightarrow y' = -\csc^2 x$
(l) $y = \sin^{-1} x \Rightarrow y' = 1/\sqrt{1 - x^2}$
(m) $y = \tan^{-1} x \Rightarrow y' = 1/(1 + x^2)$

3. (a) $e$ is the number such that $\lim\limits_{h \to 0} \dfrac{e^h - 1}{h} = 1$.
(b) $e = \lim\limits_{x \to 0} (1 + x)^{1/x}$
(c) The differentiation formula for $y = a^x$ $[y' = a^x \ln a]$ is simplest when $a = e$ because $\ln e = 1$.
(d) The differentiation formula for $y = \log_a x$ $[y' = 1/(x \ln a)]$ is simplest when $a = e$ because $\ln e = 1$.

4. (a) Implicit differentiation consists of differentiating both sides of an equation involving $x$ and $y$ with respect to $x$, and then solving the resulting equation for $y'$.
(b) Logarithmic differentiation consists of taking natural logarithms of both sides of an equation $y = f(x)$, simplifying, differentiating implicitly with respect to $x$, and then solving the resulting equation for $y'$.

5. The linearization $L$ of $f$ at $x = a$ is $L(x) = f(a) + f'(a)(x - a)$.

## TRUE-FALSE QUIZ

1. True. This is the Sum Rule.

2. False. See the warning before the Product Rule.

3. True. This is the Chain Rule.

4. True by the Chain Rule.

5. False. $\dfrac{d}{dx} f(\sqrt{x}) = \dfrac{f'(\sqrt{x})}{2\sqrt{x}}$ by the Chain Rule.

6. False. $e^2$ is a constant, so $y' = 0$.

7. False. $\dfrac{d}{dx} 10^x = 10^x \ln 10$

8. False. $\ln 10$ is a constant, so its derivative is 0.

9. True. $\dfrac{d}{dx}(\tan^2 x) = 2 \tan x \sec^2 x$, and $\dfrac{d}{dx}(\sec^2 x) = 2 \sec x (\sec x \tan x) = 2 \tan x \sec^2 x$.

10. False. $f(x) = |x^2 + x| = x^2 + x$ for $x \geq 0$ or $x \leq -1$ and $|x^2 + x| = -(x^2 + x)$ for $-1 < x < 0$.
So $f'(x) = 2x + 1$ for $x > 0$ or $x < -1$ and $f'(x) = -(2x + 1)$ for $-1 < x < 0$. But $|2x + 1| = 2x + 1$ for $x \geq -\frac{1}{2}$ and $|2x + 1| = -2x - 1$ for $x < -\frac{1}{2}$.

11. True. $g(x) = x^5 \Rightarrow g'(x) = 5x^4 \Rightarrow g'(2) = 5(2)^4 = 80$, and by the definition of the derivative,
$\lim\limits_{x \to 2} \dfrac{g(x) - g(2)}{x - 2} = g'(2) = 80$.

12. False. A tangent line to the parabola $y = x^2$ has slope $dy/dx = 2x$, so at $(-2, 4)$ the slope of the tangent is $2(-2) = -4$ and an equation of the tangent line is $y - 4 = -4(x + 2)$. [The given equation, $y - 4 = 2x(x + 2)$, is not even linear!]

## EXERCISES

**1.** $y = (x^4 - 3x^2 + 5)^3 \quad \Rightarrow$

$y' = 3(x^4 - 3x^2 + 5)^2 \dfrac{d}{dx}(x^4 - 3x^2 + 5) = 3(x^4 - 3x^2 + 5)^2(4x^3 - 6x) = 6x(x^4 - 3x^2 + 5)^2(2x^2 - 3)$

**2.** $y = \cos(\tan x) \quad \Rightarrow \quad y' = -\sin(\tan x)\dfrac{d}{dx}(\tan x) = -\sin(\tan x)(\sec^2 x)$

**3.** $y = \sqrt{x} + \dfrac{1}{\sqrt[3]{x^4}} = x^{1/2} + x^{-4/3} \quad \Rightarrow \quad y' = \frac{1}{2}x^{-1/2} - \frac{4}{3}x^{-7/3} = \dfrac{1}{2\sqrt{x}} - \dfrac{4}{3\sqrt[3]{x^7}}$

**4.** $y = \dfrac{3x - 2}{\sqrt{2x + 1}} \quad \Rightarrow$

$y' = \dfrac{\sqrt{2x+1}\,(3) - (3x - 2)\frac{1}{2}(2x + 1)^{-1/2}(2)}{\left(\sqrt{2x+1}\right)^2} \cdot \dfrac{(2x + 1)^{1/2}}{(2x + 1)^{1/2}} = \dfrac{3(2x + 1) - (3x - 2)}{(2x + 1)^{3/2}} = \dfrac{3x + 5}{(2x + 1)^{3/2}}$

**5.** $y = 2x\sqrt{x^2 + 1} \quad \Rightarrow$

$y' = 2x \cdot \frac{1}{2}(x^2 + 1)^{-1/2}(2x) + \sqrt{x^2 + 1}\,(2) = \dfrac{2x^2}{\sqrt{x^2 + 1}} + 2\sqrt{x^2 + 1} = \dfrac{2x^2 + 2(x^2 + 1)}{\sqrt{x^2 + 1}} = \dfrac{2(2x^2 + 1)}{\sqrt{x^2 + 1}}$

**6.** $y = \dfrac{e^x}{1 + x^2} \quad \Rightarrow \quad y' = \dfrac{(1 + x^2)e^x - e^x(2x)}{(1 + x^2)^2} = \dfrac{e^x(x^2 - 2x + 1)}{(1 + x^2)^2} = \dfrac{e^x(x - 1)^2}{(1 + x^2)^2}$

**7.** $y = e^{\sin 2\theta} \quad \Rightarrow \quad y' = e^{\sin 2\theta}\dfrac{d}{d\theta}(\sin 2\theta) = e^{\sin 2\theta}(\cos 2\theta)(2) = 2\cos 2\theta\, e^{\sin 2\theta}$

**8.** $y = e^{-t}(t^2 - 2t + 2) \quad \Rightarrow \quad y' = e^{-t}(2t - 2) + (t^2 - 2t + 2)(-e^{-t}) = e^{-t}(2t - 2 - t^2 + 2t - 2) = e^{-t}(-t^2 + 4t - 4)$

**9.** $y = \dfrac{t}{1 - t^2} \quad \Rightarrow \quad y' = \dfrac{(1 - t^2)(1) - t(-2t)}{(1 - t^2)^2} = \dfrac{1 - t^2 + 2t^2}{(1 - t^2)^2} = \dfrac{t^2 + 1}{(1 - t^2)^2}$

**10.** $y = \sin^{-1}(e^x) \quad \Rightarrow \quad y' = \dfrac{1}{\sqrt{1 - (e^x)^2}} \cdot e^x = e^x/\sqrt{1 - e^{2x}}$

**11.** $y = xe^{-1/x} \quad \Rightarrow \quad y' = xe^{-1/x}(1/x^2) + e^{-1/x} \cdot 1 = e^{-1/x}(1/x + 1)$

**12.** $y = x^r e^{sx} \quad \Rightarrow \quad y' = x^r(se^{sx}) + e^{sx}(rx^{r-1}) = e^{sx}x^{r-1}(sx + r)$

**13.** $\dfrac{d}{dx}(xy^4 + x^2y) = \dfrac{d}{dx}(x + 3y) \quad \Rightarrow \quad x \cdot 4y^3y' + y^4 \cdot 1 + x^2 \cdot y' + y \cdot 2x = 1 + 3y' \quad \Rightarrow$

$y'(4xy^3 + x^2 - 3) = 1 - y^4 - 2xy \quad \Rightarrow \quad y' = \dfrac{1 - y^4 - 2xy}{4xy^3 + x^2 - 3}$

**14.** $y = \ln(\csc 5x) \quad \Rightarrow \quad y' = \dfrac{1}{\csc 5x}(-\csc 5x\, \cot 5x)(5) = -5\cot 5x$

**15.** $y = \dfrac{\sec 2\theta}{1 + \tan 2\theta} \quad \Rightarrow$

$y' = \dfrac{(1 + \tan 2\theta)(\sec 2\theta\, \tan 2\theta \cdot 2) - (\sec 2\theta)(\sec^2 2\theta \cdot 2)}{(1 + \tan 2\theta)^2} = \dfrac{2\sec 2\theta\,[(1 + \tan 2\theta)\tan 2\theta - \sec^2 2\theta]}{(1 + \tan 2\theta)^2}$

$= \dfrac{2\sec 2\theta\,(\tan 2\theta + \tan^2 2\theta - \sec^2 2\theta)}{(1 + \tan 2\theta)^2} = \dfrac{2\sec 2\theta\,(\tan 2\theta - 1)}{(1 + \tan 2\theta)^2} \quad [1 + \tan^2 x = \sec^2 x]$

**16.** $\dfrac{d}{dx}(x^2\cos y + \sin 2y) = \dfrac{d}{dx}(xy) \quad \Rightarrow \quad x^2(-\sin y \cdot y') + (\cos y)(2x) + \cos 2y \cdot 2y' = x \cdot y' + y \cdot 1 \quad \Rightarrow$

$y'(-x^2\sin y + 2\cos 2y - x) = y - 2x\cos y \quad \Rightarrow \quad y' = \dfrac{y - 2x\cos y}{2\cos 2y - x^2\sin y - x}$

**17.** $y = e^{cx}(c\sin x - \cos x) \quad \Rightarrow$

$y' = e^{cx}(c\cos x + \sin x) + ce^{cx}(c\sin x - \cos x)$

$= e^{cx}(c^2\sin x - c\cos x + c\cos x + \sin x) = e^{cx}(c^2\sin x + \sin x) = e^{cx}\sin x\,(c^2 + 1)$

**18.** $y = \ln(x^2 e^x) = \ln x^2 + \ln e^x = 2\ln x + x \quad\Rightarrow\quad y' = 2/x + 1$

**19.** $y = \log_5(1+2x) \quad\Rightarrow\quad y' = \dfrac{1}{(1+2x)\ln 5}\dfrac{d}{dx}(1+2x) = \dfrac{2}{(1+2x)\ln 5}$

**20.** $y = (\ln x)^{\cos x} \quad\Rightarrow\quad \ln y = \cos x\,\ln(\ln x) \quad\Rightarrow\quad \dfrac{y'}{y} = \cos x \cdot \dfrac{1}{\ln x}\cdot\dfrac{1}{x} + (\ln\ln x)(-\sin x) \quad\Rightarrow$

$y' = (\ln x)^{\cos x}\left(\dfrac{\cos x}{x\ln x} - \sin x\,\ln\ln x\right)$

**21.** $\sin(xy) = x^2 - y \quad\Rightarrow\quad \cos(xy)(xy' + y\cdot 1) = 2x - y' \quad\Rightarrow\quad x\cos(xy)y' + y' = 2x - y\cos(xy) \quad\Rightarrow$

$y'[x\cos(xy)+1] = 2x - y\cos(xy) \quad\Rightarrow\quad y' = \dfrac{2x - y\cos(xy)}{x\cos(xy)+1}$

**22.** $y = \sqrt{t\ln(t^4)} \quad\Rightarrow$

$y' = \frac{1}{2}[t\ln(t^4)]^{-1/2}\dfrac{d}{dt}[t\ln(t^4)] = \dfrac{1}{2\sqrt{t\ln(t^4)}}\cdot\left[1\cdot\ln(t^4) + t\cdot\dfrac{1}{t^4}\cdot 4t^3\right] = \dfrac{1}{2\sqrt{t\ln(t^4)}}\cdot[\ln(t^4)+4] = \dfrac{\ln(t^4)+4}{2\sqrt{t\ln(t^4)}}$

*Or:* Since $y$ is only defined for $t > 0$, we can write $y = \sqrt{t\cdot 4\ln t} = 2\sqrt{t\ln t}$. Then

$y' = 2\cdot\dfrac{1}{2\sqrt{t\ln t}}\cdot\left(1\cdot\ln t + t\cdot\dfrac{1}{t}\right) = \dfrac{\ln t + 1}{\sqrt{t\ln t}}$. This agrees with our first answer since

$\dfrac{\ln(t^4)+4}{2\sqrt{t\ln(t^4)}} = \dfrac{4\ln t + 4}{2\sqrt{t\cdot 4\ln t}} = \dfrac{4(\ln t + 1)}{2\cdot 2\sqrt{t\ln t}} = \dfrac{\ln t + 1}{\sqrt{t\ln t}}$.

**23.** $y = 3^{x\ln x} \quad\Rightarrow\quad y' = 3^{x\ln x}\cdot\ln 3\cdot\dfrac{d}{dx}(x\ln x) = 3^{x\ln x}\cdot\ln 3\left(x\cdot\dfrac{1}{x} + \ln x\cdot 1\right) = 3^{x\ln x}\cdot\ln 3(1+\ln x)$

**24.** $xe^y = y - 1 \quad\Rightarrow\quad xe^y y' + e^y = y' \quad\Rightarrow\quad e^y = y' - xe^y y' \quad\Rightarrow\quad y' = e^y/(1 - xe^y)$

**25.** $y = \ln\sin x - \frac{1}{2}\sin^2 x \quad\Rightarrow\quad y' = \dfrac{1}{\sin x}\cdot\cos x - \frac{1}{2}\cdot 2\sin x\cdot\cos x = \cot x - \sin x\cos x$

**26.** $y = \dfrac{(x^2+1)^4}{(2x+1)^3(3x-1)^5} \quad\Rightarrow$

$\ln y = \ln\dfrac{(x^2+1)^4}{(2x+1)^3(3x-1)^5} = \ln(x^2+1)^4 - \ln[(2x+1)^3(3x-1)^5]$

$= 4\ln(x^2+1) - [\ln(2x+1)^3 + \ln(3x-1)^5] = 4\ln(x^2+1) - 3\ln(2x+1) - 5\ln(3x-1) \quad\Rightarrow$

$\dfrac{y'}{y} = 4\cdot\dfrac{1}{x^2+1}\cdot 2x - 3\cdot\dfrac{1}{2x+1}\cdot 2 - 5\cdot\dfrac{1}{3x-1}\cdot 3 \quad\Rightarrow\quad y' = \dfrac{(x^2+1)^4}{(2x+1)^3(3x-1)^5}\left(\dfrac{8x}{x^2+1} - \dfrac{6}{2x+1} - \dfrac{15}{3x-1}\right).$

[The answer could be simplified to $y' = -\dfrac{(x^2+56x+9)(x^2+1)^3}{(2x+1)^4(3x-1)^6}$, but this is unnecessary.]

**27.** $y = x\tan^{-1}(4x) \quad\Rightarrow\quad y' = x\cdot\dfrac{1}{1+(4x)^2}\cdot 4 + \tan^{-1}(4x)\cdot 1 = \dfrac{4x}{1+16x^2} + \tan^{-1}(4x)$

**28.** $y = e^{\cos x} + \cos(e^x) \quad\Rightarrow\quad y' = e^{\cos x}(-\sin x) + [-\sin(e^x)\cdot e^x] = -\sin x\,e^{\cos x} - e^x\sin(e^x)$

**29.** $y = \ln|\sec 5x + \tan 5x| \quad\Rightarrow$

$y' = \dfrac{1}{\sec 5x + \tan 5x}(\sec 5x\,\tan 5x\cdot 5 + \sec^2 5x\cdot 5) = \dfrac{5\sec 5x\,(\tan 5x + \sec 5x)}{\sec 5x + \tan 5x} = 5\sec 5x$

**30.** $y = 10^{\tan\pi\theta} \quad\Rightarrow\quad y' = 10^{\tan\pi\theta}\cdot\ln 10\cdot\sec^2\pi\theta\cdot\pi = \pi(\ln 10)10^{\tan\pi\theta}\sec^2\pi\theta$

**31.** $y = \tan^2(\sin\theta) = [\tan(\sin\theta)]^2 \quad\Rightarrow\quad y' = 2[\tan(\sin\theta)]\cdot\sec^2(\sin\theta)\cdot\cos\theta$

**32.** $y = \ln\left|\dfrac{x^2-4}{2x+5}\right| = \ln|x^2-4| - \ln|2x+5| \quad\Rightarrow\quad y' = \dfrac{2x}{x^2-4} - \dfrac{2}{2x+5}$ or $\dfrac{2(x+1)(x+4)}{(x+2)(x-2)(2x+5)}$

**33.** $y = \sin\left(\tan\sqrt{1+x^3}\right) \quad\Rightarrow\quad y' = \cos\left(\tan\sqrt{1+x^3}\right)\left(\sec^2\sqrt{1+x^3}\right)\left[3x^2/\left(2\sqrt{1+x^3}\right)\right]$

**34.** $y = \arctan(\arcsin\sqrt{x}) \quad\Rightarrow\quad y' = \dfrac{1}{1+(\arcsin\sqrt{x})^2}\cdot\dfrac{1}{\sqrt{1-x}}\cdot\dfrac{1}{2\sqrt{x}}$

**35.** $f(t) = \sqrt{4t+1} \;\Rightarrow\; f'(t) = \frac{1}{2}(4t+1)^{-1/2} \cdot 4 = 2(4t+1)^{-1/2} \;\Rightarrow$
$f''(t) = 2(-\frac{1}{2})(4t+1)^{-3/2} \cdot 4 = -4/(4t+1)^{3/2}$, so $f''(2) = -4/9^{3/2} = -\frac{4}{27}$.

**36.** $g(\theta) = \theta\sin\theta \;\Rightarrow\; g'(\theta) = \theta\cos\theta + \sin\theta \cdot 1 \;\Rightarrow\; g''(\theta) = \theta(-\sin\theta) + \cos\theta \cdot 1 + \cos\theta = 2\cos\theta - \theta\sin\theta$,
so $g''(\pi/6) = 2\cos(\pi/6) - (\pi/6)\sin(\pi/6) = 2\left(\sqrt{3}/2\right) - (\pi/6)(1/2) = \sqrt{3} - \pi/12$.

**37.** $f(x) = 2^x \;\Rightarrow\; f'(x) = 2^x \ln 2 \;\Rightarrow\; f''(x) = (2^x \ln 2)\ln 2 = 2^x(\ln 2)^2 \;\Rightarrow$
$f'''(x) = (2^x \ln 2)(\ln 2)^2 = 2^x(\ln 2)^3 \;\Rightarrow\; \cdots \;\Rightarrow\; f^{(n)}(x) = (2^x \ln 2)(\ln 2)^{n-1} = 2^x(\ln 2)^n$

**38.** $x^6 + y^6 = 1 \;\Rightarrow\; 6x^5 + 6y^5y' = 0 \;\Rightarrow\; y' = -x^5/y^5 \;\Rightarrow$

$$y'' = -\frac{y^5(5x^4) - x^5(5y^4y')}{(y^5)^2} = -\frac{5x^4y^4[y - x(-x^5/y^5)]}{y^{10}} = -\frac{5x^4[(y^6+x^6)/y^5]}{y^6} = -\frac{5x^4}{y^{11}}$$

**39.** $y = 4\sin^2 x \;\Rightarrow\; y' = 4 \cdot 2\sin x\cos x$. At $\left(\frac{\pi}{6}, 1\right)$, $y' = 8 \cdot \frac{1}{2} \cdot \frac{\sqrt{3}}{2} = 2\sqrt{3}$, so an equation of the tangent line is $y - 1 = 2\sqrt{3}\left(x - \frac{\pi}{6}\right)$, or $y = 2\sqrt{3}\,x + 1 - \pi\sqrt{3}/3$.

**40.** $y = \dfrac{x^2-1}{x^2+1} \;\Rightarrow\; y' = \dfrac{(x^2+1)(2x) - (x^2-1)(2x)}{(x^2+1)^2} = \dfrac{4x}{(x^2+1)^2}$.

At $(0, -1)$, $y' = 0$, so an equation of the tangent line is $y + 1 = 0(x - 0)$, or $y = -1$.

**41.** $y = (2+x)e^{-x} \;\Rightarrow\; y' = (2+x)(-e^{-x}) + e^{-x} \cdot 1 = e^{-x}[-(2+x)+1] = e^{-x}(-x-1)$.
At $(0, 2)$, $y' = 1(-1) = -1$, so an equation of the tangent line is $y - 2 = -1(x - 0)$, or $y = -x + 2$.
The slope of the normal line is 1, so an equation of the normal line is $y - 2 = 1(x - 0)$, or $y = x + 2$.

**42.** $x^2 + 4xy + y^2 = 13 \;\Rightarrow\; 2x + 4(xy' + y \cdot 1) + 2yy' = 0 \;\Rightarrow\; x + 2xy' + 2y + yy' = 0 \;\Rightarrow$

$2xy' + yy' = -x - 2y \;\Rightarrow\; y'(2x+y) = -x - 2y \;\Rightarrow\; y' = \dfrac{-x-2y}{2x+y}$.

At $(2, 1)$, $y' = \dfrac{-2-2}{4+1} = -\dfrac{4}{5}$, so an equation of the tangent line is $y - 1 = -\frac{4}{5}(x-2)$, or $y = -\frac{4}{5}x + \frac{13}{5}$.

The slope of the normal line is $\frac{5}{4}$, so an equation of the normal line is $y - 1 = \frac{5}{4}(x-2)$, or $y = \frac{5}{4}x - \frac{3}{2}$.

**43.** (a) $f(x) = x\sqrt{5-x} \;\Rightarrow$

$$f'(x) = x\left[\tfrac{1}{2}(5-x)^{-1/2}(-1)\right] + \sqrt{5-x} = \frac{-x}{2\sqrt{5-x}} + \sqrt{5-x} \cdot \frac{2\sqrt{5-x}}{2\sqrt{5-x}}$$

$$= \frac{-x}{2\sqrt{5-x}} + \frac{2(5-x)}{2\sqrt{5-x}} = \frac{-x+10-2x}{2\sqrt{5-x}} = \frac{10-3x}{2\sqrt{5-x}}$$

(b) At $(1, 2)$: $f'(1) = \frac{7}{4}$. So an equation of the tangent line is $y - 2 = \frac{7}{4}(x-1)$ or $y = \frac{7}{4}x + \frac{1}{4}$.
At $(4, 4)$: $f'(4) = -\frac{2}{2} = -1$. So an equation of the tangent line is $y - 4 = -1(x-4)$ or $y = -x + 8$.

(c)

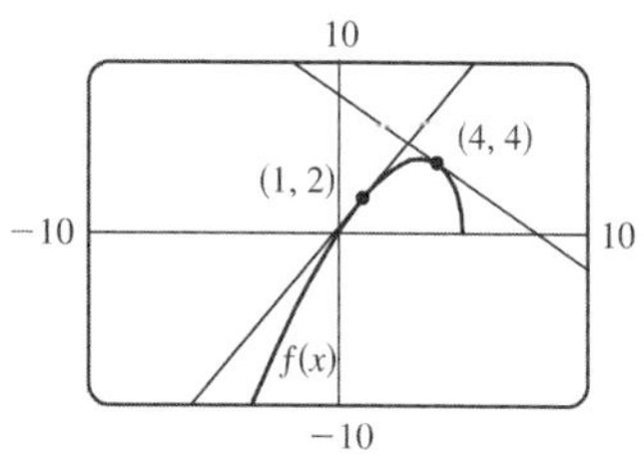


(d)

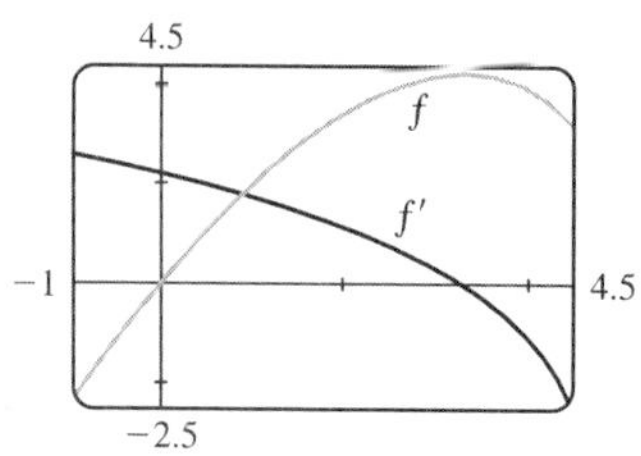


The graphs look reasonable, since $f'$ is positive where $f$ has tangents with positive slope, and $f'$ is negative where $f$ has tangents with negative slope.

**44.** (a) $f(x) = 4x - \tan x \quad \Rightarrow \quad f'(x) = 4 - \sec^2 x \quad \Rightarrow \quad f''(x) = -2\sec x\,(\sec x\,\tan x) = -2\sec^2 x\,\tan x.$

(b)

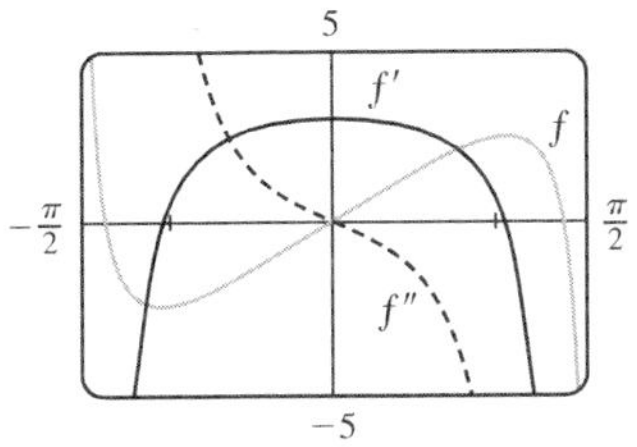


We can see that our answers are reasonable, since the graph of $f'$ is 0 where $f$ has a horizontal tangent, and the graph of $f'$ is positive where $f$ has tangents with positive slope and negative where $f$ has tangents with negative slope. The same correspondence holds between the graphs of $f'$ and $f''$.

**45.** $f(x) = xe^{\sin x} \quad \Rightarrow \quad f'(x) = x[e^{\sin x}(\cos x)] + e^{\sin x}(1) = e^{\sin x}(x\cos x + 1)$. As a check on our work, we notice from the graphs that $f'(x) > 0$ when $f$ is increasing. Also, we see in the larger viewing rectangle a certain similarity in the graphs of $f$ and $f'$: the sizes of the oscillations of $f$ and $f'$ are linked.

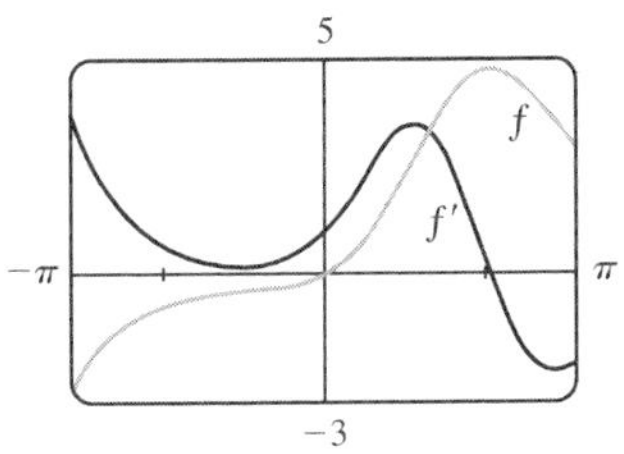


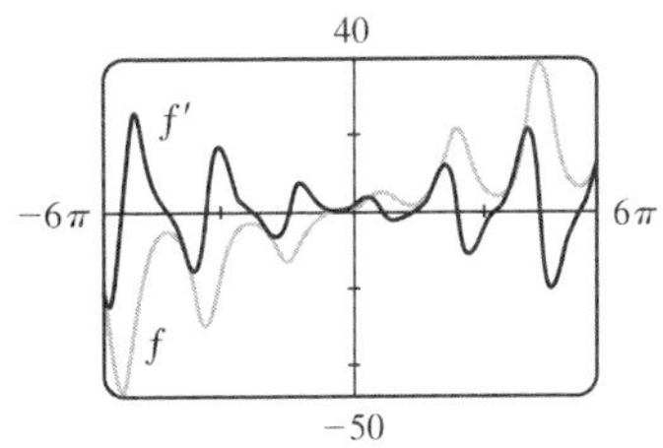


**46.** (a)

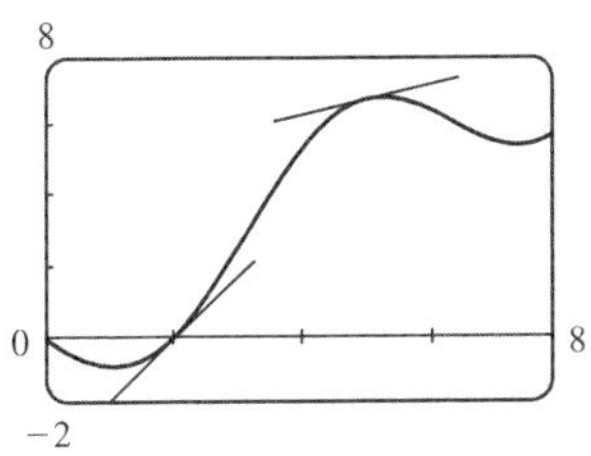


(b) The average rate of change is larger on $[2, 3]$.

(c) The instantaneous rate of change (the slope of the tangent) is larger at $x = 2$.

(d) $f(x) = x - 2\sin x \quad \Rightarrow \quad f'(x) = 1 - 2\cos x$, so $f'(2) = 1 - 2\cos 2 \approx 1.8323$ and $f'(5) = 1 - 2\cos 5 \approx 0.4327$. So $f'(2) > f'(5)$, as predicted in part (c).

**47.** (a) $h(x) = f(x)g(x) \quad \Rightarrow \quad h'(x) = f(x)g'(x) + g(x)f'(x) \quad \Rightarrow$
$h'(2) = f(2)g'(2) + g(2)f'(2) = (3)(4) + (5)(-2) = 12 - 10 = 2$

(b) $F(x) = f(g(x)) \quad \Rightarrow \quad F'(x) = f'(g(x))g'(x) \quad \Rightarrow \quad F'(2) = f'(g(2))g'(2) = f'(5)(4) = 11 \cdot 4 = 44$

**48.** (a) $P(x) = f(x)g(x) \quad \Rightarrow \quad P'(x) = f(x)g'(x) + g(x)f'(x) \quad \Rightarrow$
$P'(2) = f(2)g'(2) + g(2)f'(2) = (1)\left(\frac{6-0}{3-0}\right) + (4)\left(\frac{0-3}{3-0}\right) = (1)(2) + (4)(-1) = 2 - 4 = -2$

(b) $Q(x) = \dfrac{f(x)}{g(x)} \quad \Rightarrow \quad Q'(x) = \dfrac{g(x)f'(x) - f(x)g'(x)}{[g(x)]^2} \quad \Rightarrow$
$$Q'(2) = \frac{g(2)f'(2) - f(2)g'(2)}{[g(2)]^2} = \frac{(4)(-1) - (1)(2)}{4^2} = \frac{-6}{16} = -\frac{3}{8}$$

(c) $C(x) = f(g(x)) \quad \Rightarrow \quad C'(x) = f'(g(x))g'(x) \quad \Rightarrow$
$C'(2) = f'(g(2))g'(2) = f'(4)g'(2) = \left(\frac{6-0}{5-3}\right)(2) = (3)(2) = 6$

**49.** $f(x) = x^2 g(x) \quad \Rightarrow \quad f'(x) = x^2 g'(x) + g(x)(2x) = x[xg'(x) + 2g(x)]$

**50.** $f(x) = g(x^2) \quad \Rightarrow \quad f'(x) = g'(x^2)(2x) = 2xg'(x^2)$

**51.** $f(x) = [g(x)]^2 \quad \Rightarrow \quad f'(x) = 2[g(x)]^1 \cdot g'(x) = 2g(x)g'(x)$

**52.** $f(x) = g(g(x)) \quad \Rightarrow \quad f'(x) = g'(g(x))g'(x)$

**53.** $f(x) = g(e^x) \quad \Rightarrow \quad f'(x) = g'(e^x)e^x$

**54.** $f(x) = e^{g(x)} \quad \Rightarrow \quad f'(x) = e^{g(x)} g'(x)$

**55.** $f(x) = \ln |g(x)| \quad \Rightarrow \quad f'(x) = \dfrac{1}{g(x)} g'(x) = \dfrac{g'(x)}{g(x)}$

**56.** $f(x) = g(\ln x) \quad \Rightarrow \quad f'(x) = g'(\ln x) \cdot \dfrac{1}{x} = \dfrac{g'(\ln x)}{x}$

**57.** $h(x) = \dfrac{f(x)g(x)}{f(x) + g(x)} \quad \Rightarrow$

$$h'(x) = \frac{[f(x)+g(x)]\,[f(x)g'(x)+g(x)f'(x)] - f(x)g(x)\,[f'(x)+g'(x)]}{[f(x)+g(x)]^2}$$

$$= \frac{[f(x)]^2 g'(x) + f(x)g(x)f'(x) + f(x)g(x)g'(x) + [g(x)]^2 f'(x) - f(x)g(x)f'(x) - f(x)g(x)g'(x)}{[f(x)+g(x)]^2}$$

$$= \frac{f'(x)[g(x)]^2 + g'(x)[f(x)]^2}{[f(x)+g(x)]^2}$$

**58.** Using the Chain Rule repeatedly, $h(x) = f(g(\sin 4x)) \quad \Rightarrow$

$h'(x) = f'(g(\sin 4x)) \cdot \dfrac{d}{dx}(g(\sin 4x)) = f'(g(\sin 4x)) \cdot g'(\sin 4x) \cdot \dfrac{d}{dx}(\sin 4x) = f'(g(\sin 4x))g'(\sin 4x)(\cos 4x)(4).$

**59.** $y = [\ln(x+4)]^2 \quad \Rightarrow \quad y' = 2[\ln(x+4)]^1 \cdot \dfrac{1}{x+4} \cdot 1 = 2\dfrac{\ln(x+4)}{x+4}$ and $y' = 0 \quad \Leftrightarrow \quad \ln(x+4) = 0 \quad \Leftrightarrow$

$x + 4 = e^0 \quad \Rightarrow \quad x + 4 = 1 \quad \Leftrightarrow \quad x = -3$, so the tangent is horizontal at the point $(-3, 0)$.

**60.** (a) The line $x - 4y = 1$ has slope $\frac{1}{4}$. A tangent to $y = e^x$ has slope $\frac{1}{4}$ when $y' = e^x = \frac{1}{4} \quad \Rightarrow \quad x = \ln \frac{1}{4} = -\ln 4$. Since $y = e^x$, the $y$-coordinate is $\frac{1}{4}$ and the point of tangency is $\left(-\ln 4, \frac{1}{4}\right)$. Thus, an equation of the tangent line is $y - \frac{1}{4} = \frac{1}{4}(x + \ln 4)$ or $y = \frac{1}{4}x + \frac{1}{4}(\ln 4 + 1)$.

(b) The slope of the tangent at the point $(a, e^a)$ is $\left.\dfrac{d}{dx} e^x\right|_{x=a} = e^a$. Thus, an equation of the tangent line is $y - e^a = e^a(x - a)$. We substitute $x = 0$, $y = 0$ into this equation, since we want the line to pass through the origin: $0 - e^a = e^a(0 - a) \quad \Leftrightarrow \quad -e^a = e^a(-a) \quad \Leftrightarrow \quad a = 1$. So an equation of the tangent line at the point $(a, e^a) = (1, e)$ is $y - e = e(x - 1)$ or $y = ex$.

**61.** $x^2 + 2y^2 = 1 \quad \Rightarrow \quad 2x + 4yy' = 0 \quad \Rightarrow \quad y' = -x/(2y) = 1 \quad \Leftrightarrow \quad x = -2y$. Since the points lie on the ellipse, we have $(-2y)^2 + 2y^2 = 1 \quad \Rightarrow \quad 6y^2 = 1 \quad \Rightarrow \quad y = \pm\frac{1}{\sqrt{6}}$. The points are $\left(-\frac{2}{\sqrt{6}}, \frac{1}{\sqrt{6}}\right)$ and $\left(\frac{2}{\sqrt{6}}, -\frac{1}{\sqrt{6}}\right)$.

**62.** (a) $f(x) = \dfrac{\ln x}{x} \quad \Rightarrow \quad f'(x) = \dfrac{x(1/x) - (\ln x) \cdot 1}{x^2} = \dfrac{1 - \ln x}{x^2}$. $f'(x) > 0 \quad \Rightarrow \quad 1 - \ln x > 0 \quad \Rightarrow$ $\ln x < 1 \quad \Rightarrow \quad x < e$. Since the domain of $f$ is $(0, \infty)$, $f$ is increasing on $(0, e)$.

(b) $f''(x) = \dfrac{x^2(-1/x) - (1 - \ln x)(2x)}{(x^2)^2} = \dfrac{x[-1 - 2(1 - \ln x)]}{x^4} = \dfrac{2\ln x - 3}{x^3}$.

$f''(x) > 0 \quad \Rightarrow \quad 2\ln x - 3 > 0 \quad \Rightarrow \quad \ln x > \frac{3}{2} \quad \Rightarrow \quad x > e^{3/2} \approx 4.48$. $f$ is concave upward on $\left(e^{3/2}, \infty\right)$.

**63.** $y = f(x) = ax^2 + bx + c \quad \Rightarrow \quad f'(x) = 2ax + b$. We know that $f'(-1) = 6$ and $f'(5) = -2$, so $-2a + b = 6$ and $10a + b = -2$. Subtracting the first equation from the second gives $12a = -8 \quad \Rightarrow \quad a = -\frac{2}{3}$. Substituting $-\frac{2}{3}$ for $a$ in the first equation gives $b = \frac{14}{3}$. Now $f(1) = 4 \quad \Rightarrow \quad 4 = a + b + c$, so $c = 4 + \frac{2}{3} - \frac{14}{3} = 0$ and hence, $f(x) = -\frac{2}{3}x^2 + \frac{14}{3}x$.

**64.** (a) $y = t^3 - 12t + 3 \Rightarrow v(t) = y' = 3t^2 - 12 \Rightarrow a(t) = v'(t) = 6t$

(b) $v(t) = 3(t^2 - 4) > 0$ when $t > 2$, so it moves upward when $t > 2$ and downward when $0 \le t < 2$.

(c) Distance upward $= y(3) - y(2) = -6 - (-13) = 7$,
Distance downward $= y(0) - y(2) = 3 - (-13) = 16$. Total distance $= 7 + 16 = 23$.

(d)

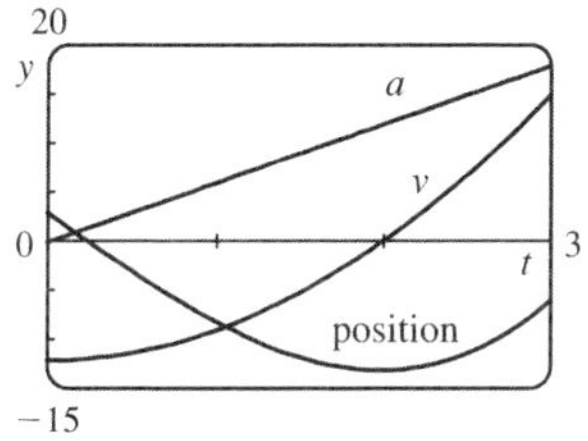


(e) The particle is speeding up when $v$ and $a$ have the same sign, that is, when $t > 2$. The particle is slowing down when $v$ and $a$ have opposite signs; that is, when $0 < t < 2$.

**65.** $s(t) = Ae^{-ct}\cos(\omega t + \delta) \Rightarrow$

$$\begin{aligned} v(t) = s'(t) &= A\{e^{-ct}[-\omega\sin(\omega t + \delta)] + \cos(\omega t + \delta)(-ce^{-ct})\} \\ &= -Ae^{-ct}[\omega\sin(\omega t + \delta) + c\cos(\omega t + \delta)] \quad \Rightarrow \end{aligned}$$

$$\begin{aligned} a(t) = v'(t) &= -A\{e^{-ct}[\omega^2\cos(\omega t + \delta) - c\omega\sin(\omega t + \delta)] + [\omega\sin(\omega t + \delta) + c\cos(\omega t + \delta)](-ce^{-ct})\} \\ &= -Ae^{-ct}[\omega^2\cos(\omega t + \delta) - c\omega\sin(\omega t + \delta) - c\omega\sin(\omega t + \delta) - c^2\cos(\omega t + \delta)] \\ &= -Ae^{-ct}[(\omega^2 - c^2)\cos(\omega t + \delta) - 2c\omega\sin(\omega t + \delta)] \\ &= Ae^{-ct}[(c^2 - \omega^2)\cos(\omega t + \delta) + 2c\omega\sin(\omega t + \delta)] \end{aligned}$$

**66.** (a) $x = \sqrt{b^2 + c^2t^2} \Rightarrow v(t) = x' = \left[1/\left(2\sqrt{b^2 + c^2t^2}\right)\right]2c^2t = c^2t/\sqrt{b^2 + c^2t^2} \Rightarrow$

$$a(t) = v'(t) = \frac{c^2\sqrt{b^2 + c^2t^2} - c^2t\left(c^2t/\sqrt{b^2 + c^2t^2}\right)}{b^2 + c^2t^2} = \frac{b^2c^2}{(b^2 + c^2t^2)^{3/2}}$$

(b) $v(t) > 0$ for $t > 0$, so the particle always moves in the positive direction.

**67.** The linear density $\rho$ is the rate of change of mass $m$ with respect to length $x$. $m = x(1 + \sqrt{x}) = x + x^{3/2} \Rightarrow$ $\rho = dm/dx = 1 + \frac{3}{2}\sqrt{x}$, so the linear density when $x = 4$ is $1 + \frac{3}{2}\sqrt{4} = 4$ kg/m.

**68.** (a) $V = \frac{1}{3}\pi r^2 h \Rightarrow dV/dh = \frac{1}{3}\pi r^2$ [$r$ constant]

(b) $V = \frac{1}{3}\pi r^2 h \Rightarrow dV/dr = \frac{2}{3}\pi rh$ [$h$ constant]

**69.** (a) $C(x) = 920 + 2x - 0.02x^2 + 0.00007x^3 \Rightarrow C'(x) = 2 - 0.04x + 0.00021x^2$

(b) $C'(100) = 2 - 4 + 2.1 = \$0.10/\text{unit}$. This value represents the rate at which costs are increasing as the hundredth unit is produced, and is the approximate cost of producing the 101st unit.

(c) The cost of producing the 101st item is $C(101) - C(100) = 990.10107 - 990 = \$0.10107$, slightly larger than $C'(100)$.

(d) $C''(x) = -0.04 + 0.00042x = 0 \Rightarrow x = \frac{0.04}{0.00042} \approx 95.24$ and $C''$ changes from negative to positive at this value of $x$. This is the value of $x$ at which the marginal cost is minimized.

**70.** (a) $\lim\limits_{t\to\infty} C(t) = \lim\limits_{t\to\infty}[K(e^{-at} - e^{-bt})] = K\lim\limits_{t\to\infty}(e^{-at} - e^{-bt}) = K(0 - 0) = 0$ because $-at \to -\infty$ and $-bt \to -\infty$ as $t \to \infty$.

(b) $C(t) = K(e^{-at} - e^{-bt}) \Rightarrow C'(t) = K(e^{-at}(-a) - e^{-bt}(-b)) = K(-ae^{-at} + be^{-bt})$

(c) $C'(t) = 0 \Rightarrow be^{-bt} = ae^{-at} \Rightarrow \dfrac{b}{a} = e^{(-a+b)t} \Rightarrow \ln\dfrac{b}{a} = (b - a)t \Rightarrow t = \dfrac{\ln(b/a)}{b - a}$

**71.** (a) $f(x) = \sqrt[3]{1 + 3x} = (1 + 3x)^{1/3} \Rightarrow f'(x) = (1 + 3x)^{-2/3}$, so the linearization of $f$ at $a = 0$ is $L(x) = f(0) + f'(0)(x - 0) = 1^{1/3} + 1^{-2/3}x = 1 + x$. Thus, $\sqrt[3]{1 + 3x} \approx 1 + x \Rightarrow$ $\sqrt[3]{1.03} = \sqrt[3]{1 + 3(0.01)} \approx 1 + (0.01) = 1.01$.

(b) The linear approximation is $\sqrt[3]{1+3x} \approx 1+x$, so for the required accuracy we want $\sqrt[3]{1+3x} - 0.1 < 1 + x < \sqrt[3]{1+3x} + 0.1$. From the graph, it appears that this is true when $-0.23 < x < 0.40$.

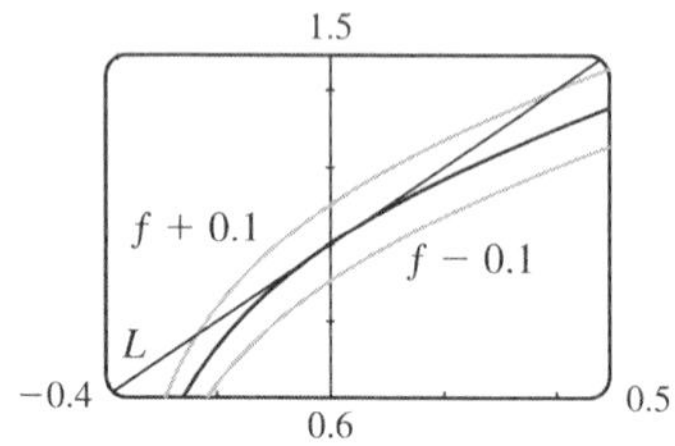


**72.** $A = x^2 + \frac{1}{2}\pi\left(\frac{1}{2}x\right)^2 = \left(1 + \frac{\pi}{8}\right)x^2 \Rightarrow dA = \left(2 + \frac{\pi}{4}\right)x\,dx$. When $x = 60$ and $dx = 0.1$, $dA = \left(2 + \frac{\pi}{4}\right)60(0.1) = 12 + \frac{3\pi}{2}$, so the maximum error is approximately $12 + \frac{3\pi}{2} \approx 16.7\text{ cm}^2$.

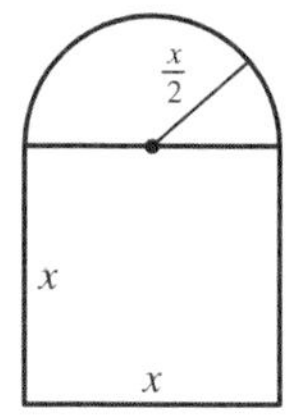


**73.** $\displaystyle \lim_{\theta \to \pi/3} \frac{\cos\theta - 0.5}{\theta - \pi/3} = \left[\frac{d}{d\theta}\cos\theta\right]_{\theta = \pi/3} = -\sin\frac{\pi}{3} = -\frac{\sqrt{3}}{2}$

**74.** $\dfrac{d}{dx}[f(2x)] = x^2 \Rightarrow f'(2x)\cdot 2 = x^2 \Rightarrow f'(2x) = \frac{1}{2}x^2$. Let $t = 2x$. Then $f'(t) = \frac{1}{2}\left(\frac{1}{2}t\right)^2 = \frac{1}{8}t^2$, so $f'(x) = \frac{1}{8}x^2$.

**75.**

$$\begin{aligned}
\lim_{x\to 0} \frac{\sqrt{1+\tan x} - \sqrt{1+\sin x}}{x^3} &= \lim_{x\to 0} \frac{\left(\sqrt{1+\tan x} - \sqrt{1+\sin x}\right)\left(\sqrt{1+\tan x} + \sqrt{1+\sin x}\right)}{x^3\left(\sqrt{1+\tan x} + \sqrt{1+\sin x}\right)} \\
&= \lim_{x\to 0} \frac{(1+\tan x) - (1+\sin x)}{x^3\left(\sqrt{1+\tan x} + \sqrt{1+\sin x}\right)} = \lim_{x\to 0} \frac{\sin x\,(1/\cos x - 1)}{x^3\left(\sqrt{1+\tan x} + \sqrt{1+\sin x}\right)} \cdot \frac{\cos x}{\cos x} \\
&= \lim_{x\to 0} \frac{\sin x\,(1-\cos x)}{x^3\left(\sqrt{1+\tan x} + \sqrt{1+\sin x}\right)\cos x} \cdot \frac{1+\cos x}{1+\cos x} \\
&= \lim_{x\to 0} \frac{\sin x \cdot \sin^2 x}{x^3\left(\sqrt{1+\tan x} + \sqrt{1+\sin x}\right)\cos x\,(1+\cos x)} \\
&= \left(\lim_{x\to 0}\frac{\sin x}{x}\right)^3 \lim_{x\to 0} \frac{1}{\left(\sqrt{1+\tan x} + \sqrt{1+\sin x}\right)\cos x\,(1+\cos x)} \\
&= 1^3 \cdot \frac{1}{\left(\sqrt{1}+\sqrt{1}\right)\cdot 1 \cdot (1+1)} = \frac{1}{4}
\end{aligned}$$

**76.** Let $(b, c)$ be on the curve, that is, $b^{2/3} + c^{2/3} = a^{2/3}$. Now $x^{2/3} + y^{2/3} = a^{2/3} \Rightarrow \frac{2}{3}x^{-1/3} + \frac{2}{3}y^{-1/3}\dfrac{dy}{dx} = 0$, so $\dfrac{dy}{dx} = -\dfrac{y^{1/3}}{x^{1/3}} = -\left(\dfrac{y}{x}\right)^{1/3}$, so at $(b, c)$ the slope of the tangent line is $-(c/b)^{1/3}$ and an equation of the tangent line is $y - c = -(c/b)^{1/3}(x - b)$ or $y = -(c/b)^{1/3}x + \left(c + b^{2/3}c^{1/3}\right)$. Setting $y = 0$, we find that the $x$-intercept is $b^{1/3}c^{2/3} + b = b^{1/3}\left(c^{2/3} + b^{2/3}\right)$ and setting $x = 0$ we find that the $y$-intercept is $c + b^{2/3}c^{1/3} = c^{1/3}\left(c^{2/3} + b^{2/3}\right)$. So the length of the tangent line between these two points is

$$\begin{aligned}
\sqrt{\left[b^{1/3}\left(c^{2/3} + b^{2/3}\right)\right]^2 + \left[c^{1/3}\left(c^{2/3} + b^{2/3}\right)\right]^2} &= \sqrt{b^{2/3}\left(a^{2/3}\right)^2 + c^{2/3}\left(a^{2/3}\right)^2} = \sqrt{\left(b^{2/3} + c^{2/3}\right)a^{4/3}} \\
&= \sqrt{a^{2/3}a^{4/3}} = \sqrt{a^2} = a = \text{constant}
\end{aligned}$$

## FOCUS ON PROBLEM SOLVING

**1.** We must find a value $x_0$ such that the normal lines to the parabola $y = x^2$ at $x = \pm x_0$ intersect at a point one unit from the points $(\pm x_0, x_0^2)$. The normals to $y = x^2$ at $x = \pm x_0$ have slopes $-\dfrac{1}{\pm 2x_0}$ and pass through $(\pm x_0, x_0^2)$ respectively, so the normals have the equations $y - x_0^2 = -\dfrac{1}{2x_0}(x - x_0)$ and $y - x_0^2 = \dfrac{1}{2x_0}(x + x_0)$. The common $y$-intercept is $x_0^2 + \frac{1}{2}$. We want to find the value of $x_0$ for which the distance from $(0, x_0^2 + \frac{1}{2})$ to $(x_0, x_0^2)$ equals 1. The square of the distance is $(x_0 - 0)^2 + \left[x_0^2 - (x_0^2 + \frac{1}{2})\right]^2 = x_0^2 + \frac{1}{4} = 1 \quad \Leftrightarrow \quad x_0 = \pm\frac{\sqrt{3}}{2}$. For these values of $x_0$, the $y$-intercept is $x_0^2 + \frac{1}{2} = \frac{5}{4}$, so the center of the circle is at $(0, \frac{5}{4})$.

*Another solution:* Let the center of the circle be $(0, a)$. Then the equation of the circle is $x^2 + (y - a)^2 = 1$. Solving with the equation of the parabola, $y = x^2$, we get $x^2 + (x^2 - a)^2 = 1 \quad \Leftrightarrow \quad x^2 + x^4 - 2ax^2 + a^2 = 1 \quad \Leftrightarrow$ $x^4 + (1 - 2a)x^2 + a^2 - 1 = 0$. The parabola and the circle will be tangent to each other when this quadratic equation in $x^2$ has equal roots; that is, when the discriminant is 0. Thus, $(1 - 2a)^2 - 4(a^2 - 1) = 0 \quad \Leftrightarrow$ $1 - 4a + 4a^2 - 4a^2 + 4 = 0 \quad \Leftrightarrow \quad 4a = 5$, so $a = \frac{5}{4}$. The center of the circle is $(0, \frac{5}{4})$.

**2.** $y = x^3 - 3x + 4 \quad \Rightarrow \quad y' = 3x^2 - 3$, and $y = 3(x^2 - x) \quad \Rightarrow$ $y' = 6x - 3$. The slopes of the tangents of the two curves are equal when $3x^2 - 3 = 6x - 3$; that is, when $x = 0$ or 2. At $x = 0$, both tangents have slope $-3$, but the curves do not intersect. At $x = 2$, both tangents have slope 9 and the curves intersect at $(2, 6)$. So there is a common tangent line at $(2, 6)$, $y = 9x - 12$.

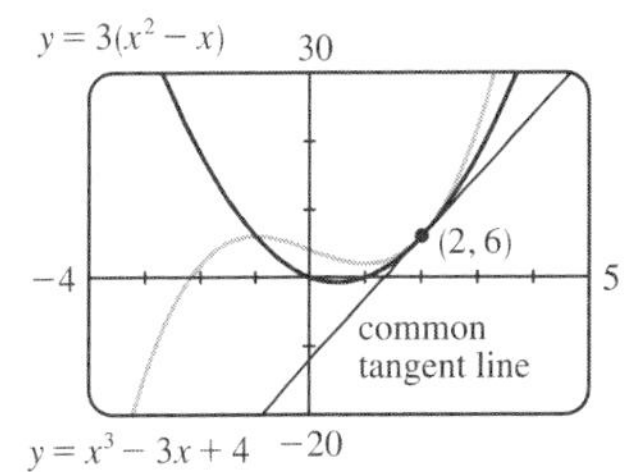


**3.**

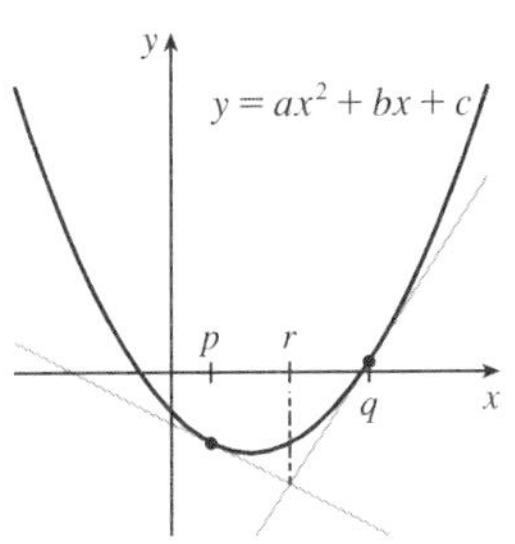


We must show that $r$ (in the figure) is halfway between $p$ and $q$, that is, $r = (p + q)/2$. For the parabola $y = ax^2 + bx + c$, the slope of the tangent line is given by $y' = 2ax + b$. An equation of the tangent line at $x = p$ is $y - (ap^2 + bp + c) = (2ap + b)(x - p)$. Solving for $y$ gives us

$$y = (2ap + b)x - 2ap^2 - bp + (ap^2 + bp + c)$$

$$\text{or} \quad y = (2ap + b)x + c - ap^2 \qquad \textbf{(1)}$$

Similarly, an equation of the tangent line at $x = q$ is

$$y = (2aq + b)x + c - aq^2 \qquad \textbf{(2)}$$

We can eliminate $y$ and solve for $x$ by subtracting equation **(1)** from equation **(2)**.

$$\begin{aligned} [(2aq + b) - (2ap + b)]x - aq^2 + ap^2 &= 0 \\ (2aq - 2ap)x &= aq^2 - ap^2 \\ 2a(q - p)x &= a(q^2 - p^2) \\ x &= \frac{a(q + p)(q - p)}{2a(q - p)} = \frac{p + q}{2} \end{aligned}$$

Thus, the $x$-coordinate of the point of intersection of the two tangent lines, namely $r$, is $(p + q)/2$.

**4.** $\lim_{x\to a}\dfrac{f(x)-f(a)}{\sqrt{x}-\sqrt{a}} = \lim_{x\to a}\left[\dfrac{f(x)-f(a)}{\sqrt{x}-\sqrt{a}}\cdot\dfrac{\sqrt{x}+\sqrt{a}}{\sqrt{x}+\sqrt{a}}\right] = \lim_{x\to a}\left[\dfrac{f(x)-f(a)}{x-a}\cdot(\sqrt{x}+\sqrt{a})\right]$

$= \lim_{x\to a}\dfrac{f(x)-f(a)}{x-a}\cdot\lim_{x\to a}(\sqrt{x}+\sqrt{a}) = f'(a)\cdot(\sqrt{a}+\sqrt{a}) = 2\sqrt{a}\,f'(a)$

**5.** We can assume without loss of generality that $\theta = 0$ at time $t = 0$, so that $\theta = 12\pi t$ rad. [The angular velocity of the wheel is 360 rpm $= 360\cdot(2\pi\text{ rad})/(60\text{ s}) = 12\pi$ rad/s.] Then the position of $A$ as a function of time is

$A = (40\cos\theta, 40\sin\theta) = (40\cos 12\pi t, 40\sin 12\pi t)$, so $\sin\alpha = \dfrac{y}{1.2\text{ m}} = \dfrac{40\sin\theta}{120} = \dfrac{\sin\theta}{3} = \frac{1}{3}\sin 12\pi t$.

(a) Differentiating the expression for $\sin\alpha$, we get $\cos\alpha\cdot\dfrac{d\alpha}{dt} = \frac{1}{3}\cdot 12\pi\cdot\cos 12\pi t = 4\pi\cos\theta$. When $\theta = \frac{\pi}{3}$, we have

$\sin\alpha = \frac{1}{3}\sin\theta = \frac{\sqrt{3}}{6}$, so $\cos\alpha = \sqrt{1-\left(\frac{\sqrt{3}}{6}\right)^2} = \sqrt{\frac{11}{12}}$ and $\dfrac{d\alpha}{dt} = \dfrac{4\pi\cos\frac{\pi}{3}}{\cos\alpha} = \dfrac{2\pi}{\sqrt{11/12}} = \dfrac{4\pi\sqrt{3}}{\sqrt{11}} \approx 6.56$ rad/s.

(b) By the Law of Cosines, $|AP|^2 = |OA|^2 + |OP|^2 - 2\,|OA|\,|OP|\cos\theta \;\Rightarrow$

$120^2 = 40^2 + |OP|^2 - 2\cdot 40\,|OP|\cos\theta \;\Rightarrow\; |OP|^2 - (80\cos\theta)\,|OP| - 12{,}800 = 0 \;\Rightarrow$

$|OP| = \frac{1}{2}\left(80\cos\theta \pm \sqrt{6400\cos^2\theta + 51{,}200}\right) = 40\cos\theta \pm 40\sqrt{\cos^2\theta + 8} = 40\left(\cos\theta + \sqrt{8+\cos^2\theta}\right)$ cm

[since $|OP| > 0$]. As a check, note that $|OP| = 160$ cm when $\theta = 0$ and $|OP| = 80\sqrt{2}$ cm when $\theta = \frac{\pi}{2}$.

(c) By part (b), the $x$-coordinate of $P$ is given by $x = 40\left(\cos\theta + \sqrt{8+\cos^2\theta}\right)$, so

$\dfrac{dx}{dt} = \dfrac{dx}{d\theta}\dfrac{d\theta}{dt} = 40\left(-\sin\theta - \dfrac{2\cos\theta\sin\theta}{2\sqrt{8+\cos^2\theta}}\right)\cdot 12\pi = -480\pi\sin\theta\left(1 + \dfrac{\cos\theta}{\sqrt{8+\cos^2\theta}}\right)$ cm/s.

In particular, $dx/dt = 0$ cm/s when $\theta = 0$ and $dx/dt = -480\pi$ cm/s when $\theta = \frac{\pi}{2}$.

**6.** The equation of $T_1$ is $y - x_1^2 = 2x_1(x - x_1) = 2x_1x - 2x_1^2$ or $y = 2x_1x - x_1^2$.

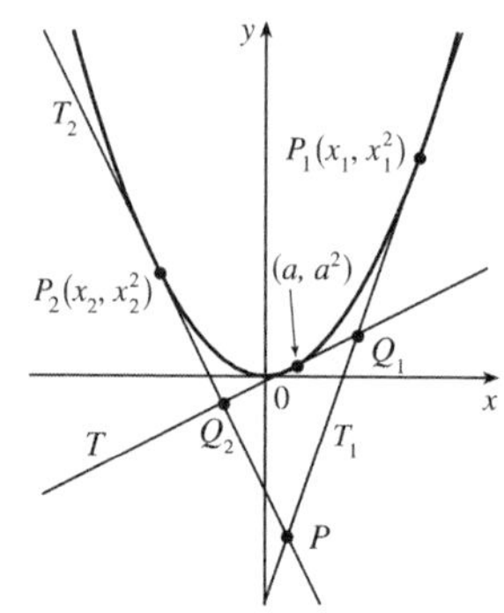


The equation of $T_2$ is $y = 2x_2x - x_2^2$. Solving for the point of intersection, we get $2x(x_1 - x_2) = x_1^2 - x_2^2 \;\Rightarrow\; x = \frac{1}{2}(x_1 + x_2)$. Therefore, the coordinates of $P$ are $\left(\frac{1}{2}(x_1+x_2), x_1x_2\right)$. So if the point of contact of $T$ is $(a, a^2)$, then $Q_1$ is $\left(\frac{1}{2}(a+x_1), ax_1\right)$ and $Q_2$ is $\left(\frac{1}{2}(a+x_2), ax_2\right)$. Therefore,

$|PQ_1|^2 = \frac{1}{4}(a-x_2)^2 + x_1^2(a-x_2)^2 = (a-x_2)^2\left(\frac{1}{4}+x_1^2\right)$ and

$|PP_1|^2 = \frac{1}{4}(x_1-x_2)^2 + x_1^2(x_1-x_2)^2 = (x_1-x_2)^2\left(\frac{1}{4}+x_1^2\right)$.

So $\dfrac{|PQ_1|^2}{|PP_1|^2} = \dfrac{(a-x_2)^2}{(x_1-x_2)^2}$, and similarly $\dfrac{|PQ_2|^2}{|PP_2|^2} = \dfrac{(x_1-a)^2}{(x_1-x_2)^2}$. Finally, $\dfrac{|PQ_1|}{|PP_1|} + \dfrac{|PQ_2|}{|PP_2|} = \dfrac{a-x_2}{x_1-x_2} + \dfrac{x_1-a}{x_1-x_2} = 1$.

**7.** Consider the statement that $\dfrac{d^n}{dx^n}(e^{ax}\sin bx) = r^ne^{ax}\sin(bx+n\theta)$. For $n = 1$,

$\dfrac{d}{dx}(e^{ax}\sin bx) = ae^{ax}\sin bx + be^{ax}\cos bx$, and

$$re^{ax}\sin(bx+\theta) = re^{ax}[\sin bx\cos\theta + \cos bx\sin\theta] = re^{ax}\left(\frac{a}{r}\sin bx + \frac{b}{r}\cos bx\right)$$
$$= ae^{ax}\sin bx + be^{ax}\cos bx$$

since $\tan\theta = \dfrac{b}{a} \;\Rightarrow\; \sin\theta = \dfrac{b}{r}$ and $\cos\theta = \dfrac{a}{r}$. So the statement is true for $n = 1$.

Assume it is true for $n = k$. Then

$$\frac{d^{k+1}}{dx^{k+1}}(e^{ax}\sin bx) = \frac{d}{dx}\left[r^ke^{ax}\sin(bx+k\theta)\right] = r^kae^{ax}\sin(bx+k\theta) + r^ke^{ax}b\cos(bx+k\theta)$$
$$= r^ke^{ax}[a\sin(bx+k\theta) + b\cos(bx+k\theta)]$$

But

$$\sin[bx+(k+1)\theta]=\sin[(bx+k\theta)+\theta]=\sin(bx+k\theta)\cos\theta+\sin\theta\cos(bx+k\theta)$$
$$=\frac{a}{r}\sin(bx+k\theta)+\frac{b}{r}\cos(bx+k\theta)$$

Hence, $a\sin(bx+k\theta)+b\cos(bx+k\theta)=r\sin[bx+(k+1)\theta]$. So

$$\frac{d^{k+1}}{dx^{k+1}}\left(e^{ax}\sin bx\right)=r^k e^{ax}[a\sin(bx+k\theta)+b\cos(bx+k\theta)]=r^k e^{ax}[r\sin(bx+(k+1)\theta)]$$
$$=r^{k+1}e^{ax}[\sin(bx+(k+1)\theta)]$$

Therefore, the statement is true for all $n$ by mathematical induction.

**8.** We recognize this limit as the definition of the derivative of the function $f(x)=e^{\sin x}$ at $x=\pi$, since it is of the form $\lim\limits_{x\to\pi}\dfrac{f(x)-f(\pi)}{x-\pi}$. Therefore, the limit is equal to $f'(\pi)=(\cos\pi)e^{\sin\pi}=-1\cdot e^0=-1$.

**9.** It seems from the figure that as $P$ approaches the point $(0,2)$ from the right, $x_T\to\infty$ and $y_T\to 2^+$. As $P$ approaches the point $(3,0)$ from the left, it appears that $x_T\to 3^+$ and $y_T\to\infty$. So we guess that $x_T\in(3,\infty)$ and $y_T\in(2,\infty)$. It is more difficult to estimate the range of values for $x_N$ and $y_N$. We might perhaps guess that $x_N\in(0,3)$, and $y_N\in(-\infty,0)$ or $(-2,0)$.

In order to actually solve the problem, we implicitly differentiate the equation of the ellipse to find the equation of the tangent line: $\dfrac{x^2}{9}+\dfrac{y^2}{4}=1 \;\Rightarrow\; \dfrac{2x}{9}+\dfrac{2y}{4}y'=0$, so $y'=-\dfrac{4}{9}\dfrac{x}{y}$. So at the point $(x_0,y_0)$ on the ellipse, an equation of the tangent line is $y-y_0=-\dfrac{4}{9}\dfrac{x_0}{y_0}(x-x_0)$ or $4x_0x+9y_0y=4x_0^2+9y_0^2$. This can be written as $\dfrac{x_0x}{9}+\dfrac{y_0y}{4}=\dfrac{x_0^2}{9}+\dfrac{y_0^2}{4}=1$, because $(x_0,y_0)$ lies on the ellipse. So an equation of the tangent line is $\dfrac{x_0x}{9}+\dfrac{y_0y}{4}=1$.

Therefore, the $x$-intercept $x_T$ for the tangent line is given by $\dfrac{x_0x_T}{9}=1 \;\Leftrightarrow\; x_T=\dfrac{9}{x_0}$, and the $y$-intercept $y_T$ is given by $\dfrac{y_0y_T}{4}=1 \;\Leftrightarrow\; y_T=\dfrac{4}{y_0}$.

So as $x_0$ takes on all values in $(0,3)$, $x_T$ takes on all values in $(3,\infty)$, and as $y_0$ takes on all values in $(0,2)$, $y_T$ takes on all values in $(2,\infty)$. At the point $(x_0,y_0)$ on the ellipse, the slope of the normal line is $-\dfrac{1}{y'(x_0,y_0)}=\dfrac{9}{4}\dfrac{y_0}{x_0}$, and its equation is $y-y_0=\dfrac{9}{4}\dfrac{y_0}{x_0}(x-x_0)$. So the $x$-intercept $x_N$ for the normal line is given by $0-y_0=\dfrac{9}{4}\dfrac{y_0}{x_0}(x_N-x_0) \;\Rightarrow\; x_N=-\dfrac{4x_0}{9}+x_0=\dfrac{5x_0}{9}$, and the $y$-intercept $y_N$ is given by $y_N-y_0=\dfrac{9}{4}\dfrac{y_0}{x_0}(0-x_0) \;\Rightarrow\; y_N=-\dfrac{9y_0}{4}+y_0=-\dfrac{5y_0}{4}$.

So as $x_0$ takes on all values in $(0,3)$, $x_N$ takes on all values in $\left(0,\frac{5}{3}\right)$, and as $y_0$ takes on all values in $(0,2)$, $y_N$ takes on all values in $\left(-\frac{5}{2},0\right)$.

**10.** $$\lim_{x\to 0}\frac{f(x)}{g(x)}=\lim_{x\to 0}\frac{f(x)-0}{g(x)-0}=\lim_{x\to 0}\frac{f(x)-f(0)}{g(x)-g(0)}=\lim_{x\to 0}\frac{\dfrac{f(x)-f(0)}{x-0}}{\dfrac{g(x)-g(0)}{x-0}}=\frac{\lim\limits_{x\to 0}\dfrac{f(x)-f(0)}{x-0}}{\lim\limits_{x\to 0}\dfrac{g(x)-g(0)}{x-0}}=\frac{f'(0)}{g'(0)}$$

**11.** $y = \dfrac{x}{\sqrt{a^2-1}} - \dfrac{2}{\sqrt{a^2-1}} \arctan \dfrac{\sin x}{a + \sqrt{a^2-1} + \cos x}$. Let $k = a + \sqrt{a^2-1}$. Then

$$\begin{aligned} y' &= \frac{1}{\sqrt{a^2-1}} - \frac{2}{\sqrt{a^2-1}} \cdot \frac{1}{1+\sin^2 x/(k+\cos x)^2} \cdot \frac{\cos x(k+\cos x)+\sin^2 x}{(k+\cos x)^2} \\ &= \frac{1}{\sqrt{a^2-1}} - \frac{2}{\sqrt{a^2-1}} \cdot \frac{k\cos x + \cos^2 x + \sin^2 x}{(k+\cos x)^2 + \sin^2 x} = \frac{1}{\sqrt{a^2-1}} - \frac{2}{\sqrt{a^2-1}} \cdot \frac{k\cos x + 1}{k^2 + 2k\cos x + 1} \\ &= \frac{k^2 + 2k\cos x + 1 - 2k\cos x - 2}{\sqrt{a^2-1}\,(k^2+2k\cos x+1)} = \frac{k^2-1}{\sqrt{a^2-1}\,(k^2+2k\cos x+1)} \end{aligned}$$

But $k^2 = 2a^2 + 2a\sqrt{a^2-1} - 1 = 2a\left(a+\sqrt{a^2-1}\right) - 1 = 2ak - 1$, so $k^2 + 1 = 2ak$, and $k^2 - 1 = 2(ak - 1)$.
So $y' = \dfrac{2(ak-1)}{\sqrt{a^2-1}\,(2ak + 2k\cos x)} = \dfrac{ak-1}{\sqrt{a^2-1}k\,(a+\cos x)}$. But $ak - 1 = a^2 + a\sqrt{a^2-1} - 1 = k\sqrt{a^2-1}$,
so $y' = 1/(a + \cos x)$.

**12.** We see that at $x = 0$, $f(x) = a^x = 1 + x = 1$, so if $y = a^x$ is to lie above $y = 1 + x$, the two curves must just touch at $(0, 1)$, that is, we must have $f'(0) = 1$. [To see this analytically, note that $a^x \geq 1 + x \;\Rightarrow\; a^x - 1 \geq x \;\Rightarrow$ $\dfrac{a^x - 1}{x} \geq 1$ for $x > 0$, so $f'(0) = \lim\limits_{x\to 0^+} \dfrac{a^x-1}{x} \geq 1$. Similarly, for $x < 0$, $a^x - 1 \geq x \;\Rightarrow\; \dfrac{a^x-1}{x} \leq 1$, so $f'(0) = \lim\limits_{x\to 0^-} \dfrac{a^x-1}{x} \leq 1$.

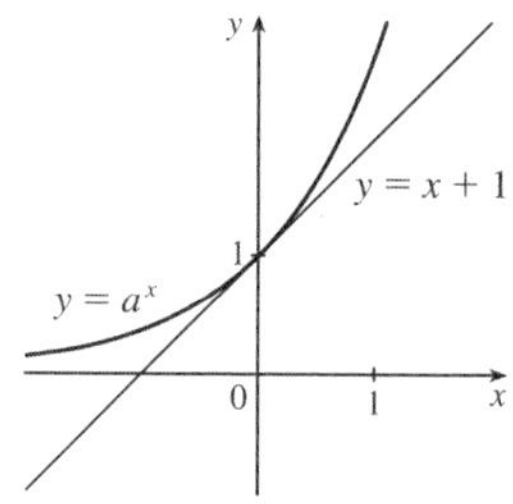


Since $1 \leq f'(0) \leq 1$, we must have $f'(0) = 1$.] But $f'(x) = a^x \ln a \;\Rightarrow\; f'(0) = \ln a$, so we have $\ln a = 1 \;\Leftrightarrow\; a = e$.

*Another method:* The inequality certainly holds for $x \leq -1$, so consider $x > -1$, $x \neq 0$. Then $a^x \geq 1 + x \;\Rightarrow$ $a \geq (1+x)^{1/x}$ for $x > 0 \;\Rightarrow\; a \geq \lim\limits_{x\to 0^+} (1+x)^{1/x} = e$, by Equation 3.7.5. Also, $a^x \geq 1 + x \;\Rightarrow\; a \leq (1+x)^{1/x}$ for $x < 0 \;\Rightarrow\; a \leq \lim\limits_{x\to 0^-} (1+x)^{1/x} = e$. So since $e \leq a \leq e$, we must have $a = e$.

**13.**

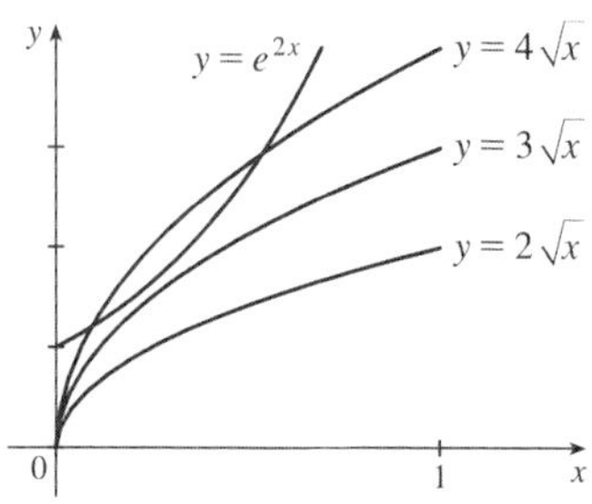


Let $f(x) = e^{2x}$ and $g(x) = k\sqrt{x}$ $(k > 0)$. From the graphs of $f$ and $g$, we see that $f$ will intersect $g$ exactly once when $f$ and $g$ share a tangent line. Thus, we must have $f = g$ and $f' = g'$ at $x = a$. $f(a) = g(a) \;\Rightarrow$ $e^{2a} = k\sqrt{a}$ $(\star)$ and $f'(a) = g'(a) \;\Rightarrow\; 2e^{2a} = \dfrac{k}{2\sqrt{a}} \;\Rightarrow$ $e^{2a} = \dfrac{k}{4\sqrt{a}}$. So we must have $k\sqrt{a} = \dfrac{k}{4\sqrt{a}} \;\Rightarrow\; \left(\sqrt{a}\right)^2 = \dfrac{k}{4k} \;\Rightarrow$ $a = \frac{1}{4}$. From $(\star)$, $e^{2(1/4)} = k\sqrt{1/4} \;\Rightarrow\; k = 2e^{1/2} = 2\sqrt{e} \approx 3.297$.

**14.** (a) $f(x) = x(x-2)(x-6) = x^3 - 8x^2 + 12x \;\Rightarrow$ $f'(x) = 3x^2 - 16x + 12$. The average of the first pair of zeros is $(0+2)/2 = 1$. At $x = 1$, the slope of the tangent line is $f'(1) = -1$, so an equation of the tangent line has the form $y = -1x + b$. Since $f(1) = 5$, we have $5 = -1 + b \;\Rightarrow\; b = 6$ and the tangent has equation $y = -x + 6$.

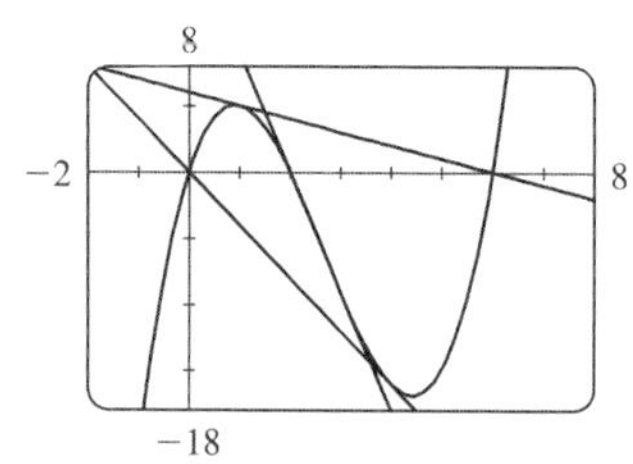


Similarly, at $x = \dfrac{0+6}{2} = 3$, $y = -9x + 18$; at $x = \dfrac{2+6}{2} = 4$, $y = -4x$. From the graph, we see that each tangent line drawn at the average of two zeros intersects the graph of $f$ at the third zero.

(b) A CAS gives $f'(x) = (x - b)(x - c) + (x - a)(x - c) + (x - a)(x - b)$ or $f'(x) = 3x^2 - 2(a + b + c)x + ab + ac + bc$. Using the `Simplify` command, we get $f'\left(\dfrac{a+b}{2}\right) = -\dfrac{(a-b)^2}{4}$ and $f\left(\dfrac{a+b}{2}\right) = -\dfrac{(a-b)^2}{8}(a + b - 2c)$, so an equation of the tangent line at $x = \dfrac{a+b}{2}$ is $y = -\dfrac{(a-b)^2}{4}\left(x - \dfrac{a+b}{2}\right) - \dfrac{(a-b)^2}{8}(a + b - 2c)$. To find the $x$-intercept, let $y = 0$ and use the `Solve` command. The result is $x = c$.

Using Derive, we can begin by authoring the expression $(x - a)(x - b)(x - c)$. Now load the utility file Dif_apps. Next we author tangent $(\#1, x, (a + b)/2)$—this is the command to find an equation of the tangent line of the function in #1 whose independent variable is $x$ at the $x$-value $(a + b)/2$. We then simplify that expression and obtain the equation $y = \#3$. The form in expression #3 makes it easy to see that the $x$-intercept is the third zero, namely $c$. In a similar fashion we see that $b$ is the $x$-intercept for the tangent line at $(a + c)/2$ and $a$ is the $x$-intercept for the tangent line at $(b + c)/2$.

```
#1:  (x - a)·(x - b)·(x - c)                          Author the function y=

                                        a + b
#2:  TANGENT((x - a)·(x - b)·(x - c), x, -----)      Tangent(#1, x,(a+b)/2)
                                          2

        2              2
      (a  - 2·a·b + b )·(c - x)
#3:  --------------------------                       0.0s Simp(#2)
                  4
```

**15.** (a) If the two lines $L_1$ and $L_2$ have slopes $m_1$ and $m_2$ and angles of inclination $\phi_1$ and $\phi_2$, then $m_1 = \tan\phi_1$ and $m_2 = \tan\phi_2$. The triangle in the figure shows that $\phi_1 + \alpha + (180^\circ - \phi_2) = 180^\circ$ and so $\alpha = \phi_2 - \phi_1$. Therefore, using the identity for $\tan(x - y)$, we have

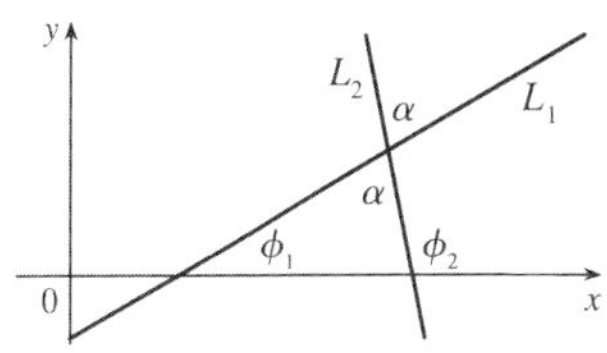


$\tan\alpha = \tan(\phi_2 - \phi_1) = \dfrac{\tan\phi_2 - \tan\phi_1}{1 + \tan\phi_2\tan\phi_1}$ and so $\tan\alpha = \dfrac{m_2 - m_1}{1 + m_1 m_2}$.

(b) (i) The parabolas intersect when $x^2 = (x - 2)^2 \Rightarrow x = 1$. If $y = x^2$, then $y' = 2x$, so the slope of the tangent to $y = x^2$ at $(1, 1)$ is $m_1 = 2(1) = 2$. If $y = (x - 2)^2$, then $y' = 2(x - 2)$, so the slope of the tangent to $y = (x - 2)^2$ at $(1, 1)$ is $m_2 = 2(1 - 2) = -2$. Therefore, $\tan\alpha = \dfrac{m_2 - m_1}{1 + m_1 m_2} = \dfrac{-2 - 2}{1 + 2(-2)} = \frac{4}{3}$ and so $\alpha = \tan^{-1}\left(\frac{4}{3}\right) \approx 53^\circ$ [or $127^\circ$].

(ii) $x^2 - y^2 = 3$ and $x^2 - 4x + y^2 + 3 = 0$ intersect when $x^2 - 4x + (x^2 - 3) + 3 = 0 \Leftrightarrow 2x(x - 2) = 0 \Rightarrow x = 0$ or 2, but 0 is extraneous. If $x = 2$, then $y = \pm 1$. If $x^2 - y^2 = 3$ then $2x - 2yy' = 0 \Rightarrow y' = x/y$ and $x^2 - 4x + y^2 + 3 = 0 \Rightarrow 2x - 4 + 2yy' = 0 \Rightarrow y' = \dfrac{2 - x}{y}$. At $(2, 1)$ the slopes are $m_1 = 2$ and $m_2 = 0$, so $\tan\alpha = \frac{0 - 2}{1 + 2 \cdot 0} = -2 \Rightarrow \alpha \approx 117^\circ$. At $(2, -1)$ the slopes are $m_1 = -2$ and $m_2 = 0$, so $\tan\alpha = \dfrac{0 - (-2)}{1 + (-2)(0)} = 2 \Rightarrow \alpha \approx 63^\circ$ [or $117^\circ$].

**16.** $y^2 = 4px \Rightarrow 2yy' = 4p \Rightarrow y' = 2p/y \Rightarrow$ slope of tangent at $P(x_1, y_1)$ is $m_1 = 2p/y_1$. The slope of $FP$ is $m_2 = \dfrac{y_1}{x_1 - p}$, so by the formula from Problem 15(a),

$$\tan\alpha = \frac{y_1/(x_1 - p) - 2p/y_1}{1 + (2p/y_1)[y_1/(x_1 - p)]} \cdot \frac{y_1(x_1 - p)}{y_1(x_1 - p)} = \frac{y_1^2 - 2p(x_1 - p)}{y_1(x_1 - p) + 2py_1}$$

$$= \frac{4px_1 - 2px_1 + 2p^2}{x_1 y_1 - py_1 + 2py_1} = \frac{2p(p + x_1)}{y_1(p + x_1)} = \frac{2p}{y_1} = \text{slope of tangent at } P = \tan\beta$$

Since $0 \le \alpha, \beta \le \frac{\pi}{2}$, this proves that $\alpha = \beta$.

**17.** Since $\angle ROQ = \angle OQP = \theta$, the triangle $QOR$ is isosceles, so $|QR| = |RO| = x$. By the Law of Cosines, $x^2 = x^2 + r^2 - 2rx\cos\theta$. Hence, $2rx\cos\theta = r^2$, so $x = \dfrac{r^2}{2r\cos\theta} = \dfrac{r}{2\cos\theta}$. Note that as $y \to 0^+$, $\theta \to 0^+$ (since $\sin\theta = y/r$), and hence $x \to \dfrac{r}{2\cos 0} = \dfrac{r}{2}$. Thus, as $P$ is taken closer and closer to the $x$-axis, the point $R$ approaches the midpoint of the radius $AO$.

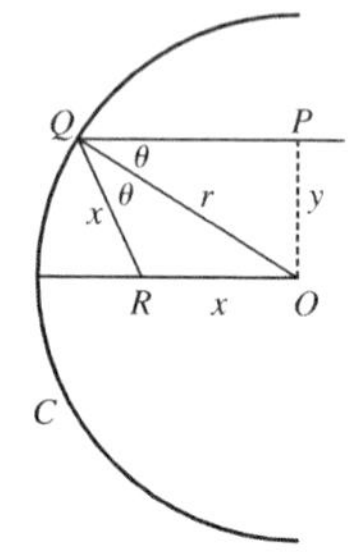


**18.** Suppose that $y = mx + c$ is a tangent line to the ellipse. Then it intersects the ellipse at only one point, so the discriminant of the equation $\dfrac{x^2}{a^2} + \dfrac{(mx+c)^2}{b^2} = 1 \iff (b^2 + a^2m^2)x^2 + 2mca^2x + a^2c^2 - a^2b^2 = 0$ must be 0; that is,

$$\begin{aligned} 0 &= (2mca^2)^2 - 4(b^2 + a^2m^2)(a^2c^2 - a^2b^2) = 4a^4c^2m^2 - 4a^2b^2c^2 + 4a^2b^4 - 4a^4c^2m^2 + 4a^4b^2m^2 \\ &= 4a^2b^2(a^2m^2 + b^2 - c^2) \end{aligned}$$

Therefore, $a^2m^2 + b^2 - c^2 = 0$.

Now if a point $(\alpha, \beta)$ lies on the line $y = mx + c$, then $c = \beta - m\alpha$, so from above,

$$0 = a^2m^2 + b^2 - (\beta - m\alpha)^2 = (a^2 - \alpha^2)m^2 + 2\alpha\beta m + b^2 - \beta^2 \iff m^2 + \frac{2\alpha\beta}{a^2 - \alpha^2}m + \frac{b^2 - \beta^2}{a^2 - \alpha^2} = 0.$$

(a) Suppose that the two tangent lines from the point $(\alpha, \beta)$ to the ellipse have slopes $m$ and $\dfrac{1}{m}$. Then $m$ and $\dfrac{1}{m}$ are roots of the equation $z^2 + \dfrac{2\alpha\beta}{a^2 - \alpha^2}z + \dfrac{b^2 - \beta^2}{a^2 - \alpha^2} = 0$. This implies that $(z - m)\left(z - \dfrac{1}{m}\right) = 0 \iff z^2 - \left(m + \dfrac{1}{m}\right)z + m\left(\dfrac{1}{m}\right) = 0$, so equating the constant terms in the two quadratic equations, we get $\dfrac{b^2 - \beta^2}{a^2 - \alpha^2} = m\left(\dfrac{1}{m}\right) = 1$, and hence $b^2 - \beta^2 = a^2 - \alpha^2$. So $(\alpha, \beta)$ lies on the hyperbola $x^2 - y^2 = a^2 - b^2$.

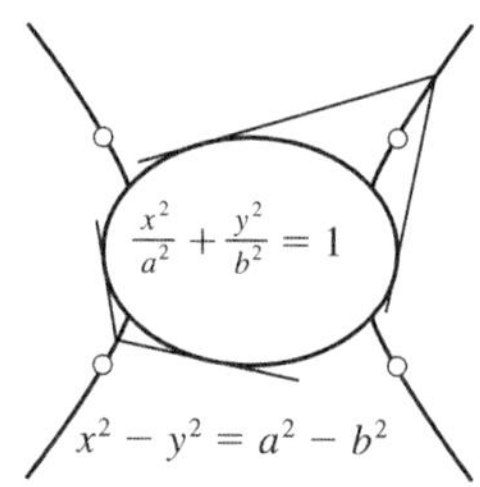


(b) If the two tangent lines from the point $(\alpha, \beta)$ to the ellipse have slopes $m$ and $-\dfrac{1}{m}$, then $m$ and $-\dfrac{1}{m}$ are roots of the quadratic equation, and so $(z - m)\left(z + \dfrac{1}{m}\right) = 0$, and equating the constant terms as in part (a), we get $\dfrac{b^2 - \beta^2}{a^2 - \alpha^2} = -1$, and hence $b^2 - \beta^2 = \alpha^2 - a^2$. So the point $(\alpha, \beta)$ lies on the circle $x^2 + y^2 = a^2 + b^2$.

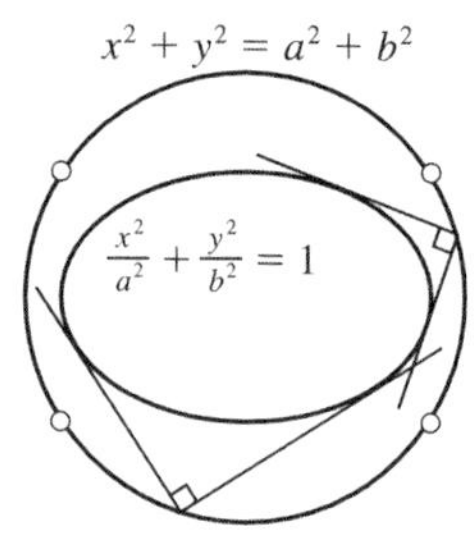


**19.** $y = x^4 - 2x^2 - x \Rightarrow y' = 4x^3 - 4x - 1$. The equation of the tangent line at $x = a$ is $y - (a^4 - 2a^2 - a) = (4a^3 - 4a - 1)(x - a)$ or $y = (4a^3 - 4a - 1)x + (-3a^4 + 2a^2)$ and similarly for $x = b$. So if at $x = a$ and $x = b$ we have the same tangent line, then $4a^3 - 4a - 1 = 4b^3 - 4b - 1$ and $-3a^4 + 2a^2 = -3b^4 + 2b^2$. The first equation gives $a^3 - b^3 = a - b \Rightarrow (a - b)(a^2 + ab + b^2) = (a - b)$. Assuming $a \neq b$, we have $1 = a^2 + ab + b^2$. The second equation gives $3(a^4 - b^4) = 2(a^2 - b^2) \Rightarrow 3(a^2 - b^2)(a^2 + b^2) = 2(a^2 - b^2)$ which is true if $a = -b$.

Substituting into $1 = a^2 + ab + b^2$ gives $1 = a^2 - a^2 + a^2 \;\Rightarrow\; a = \pm 1$ so that $a = 1$ and $b = -1$ or vice versa. Thus, the points $(1, -2)$ and $(-1, 0)$ have a common tangent line.

As long as there are only two such points, we are done. So we show that these are in fact the only two such points. Suppose that $a^2 - b^2 \neq 0$. Then $3(a^2 - b^2)(a^2 + b^2) = 2(a^2 - b^2)$ gives $3(a^2 + b^2) = 2$ or $a^2 + b^2 = \frac{2}{3}$. Thus, $ab = (a^2 + ab + b^2) - (a^2 + b^2) = 1 - \frac{2}{3} = \frac{1}{3}$, so $b = \dfrac{1}{3a}$. Hence, $a^2 + \dfrac{1}{9a^2} = \dfrac{2}{3}$, so $9a^4 + 1 = 6a^2 \;\Rightarrow$ $0 = 9a^4 - 6a^2 + 1 = (3a^2 - 1)^2$. So $3a^2 - 1 = 0 \;\Rightarrow\; a^2 = \frac{1}{3} \;\Rightarrow\; b^2 = \dfrac{1}{9a^2} = \frac{1}{3} = a^2$, contradicting our assumption that $a^2 \neq b^2$.

**20.** Suppose that the normal lines at the three points $(a_1, a_1^2)$, $(a_2, a_2^2)$, and $(a_3, a_3^2)$ intersect at a common point. Now if one of the $a_i$ is 0 (suppose $a_1 = 0$) then by symmetry $a_2 = -a_3$, so $a_1 + a_2 + a_3 = 0$. So we can assume that none of the $a_i$ is 0.

The slope of the tangent line at $(a_i, a_i^2)$ is $2a_i$, so the slope of the normal line is $-\dfrac{1}{2a_i}$ and its equation is $y - a_i^2 = -\dfrac{1}{2a_i}(x - a_i)$. We solve for the $x$-coordinate of the intersection of the normal lines from $(a_1, a_1^2)$ and $(a_2, a_2^2)$: $y = a_1^2 - \dfrac{1}{2a_1}(x - a_1) = a_2^2 - \dfrac{1}{2a_2}(x - a_2) \;\Rightarrow\; x\left(\dfrac{1}{2a_2} - \dfrac{1}{2a_1}\right) = a_2^2 - a_1^2 \;\Rightarrow$ $x\left(\dfrac{a_1 - a_2}{2a_1 a_2}\right) = (-a_1 - a_2)(a_1 + a_2) \;\Leftrightarrow\; x = -2a_1 a_2(a_1 + a_2)$ **(1)**. Similarly, solving for the $x$-coordinate of the intersections of the normal lines from $(a_1, a_1^2)$ and $(a_3, a_3^2)$ gives $x = -2a_1 a_3(a_1 + a_3)$ **(2)**.

Equating **(1)** and **(2)** gives $a_2(a_1 + a_2) = a_3(a_1 + a_3) \;\Leftrightarrow\; a_1(a_2 - a_3) = a_3^2 - a_2^2 = -(a_2 + a_3)(a_2 - a_3) \;\Leftrightarrow$ $a_1 = -(a_2 + a_3) \;\Leftrightarrow\; a_1 + a_2 + a_3 = 0$.

**21.** Because of the periodic nature of the lattice points, it suffices to consider the points in the $5 \times 2$ grid shown. We can see that the minimum value of $r$ occurs when there is a line with slope $\frac{2}{5}$ which touches the circle centered at $(3, 1)$ and the circles centered at $(0, 0)$ and $(5, 2)$.

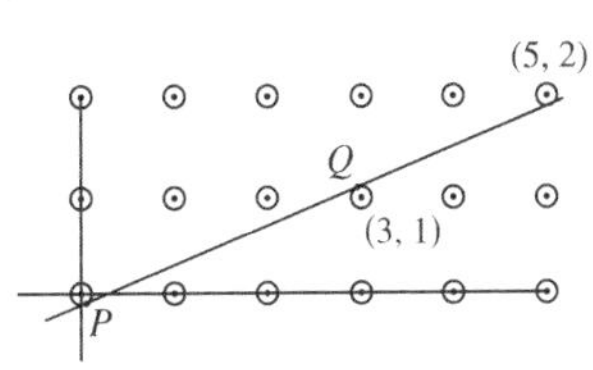


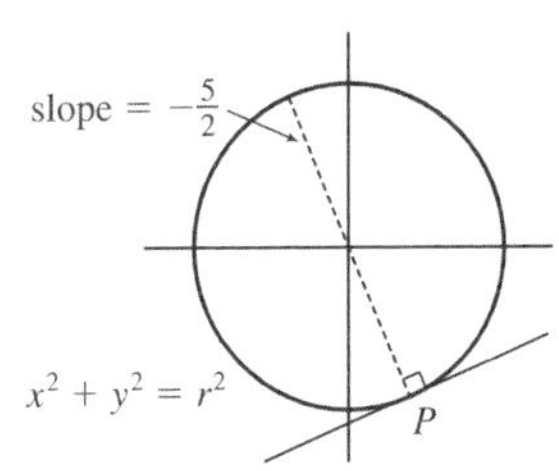


To find $P$, the point at which the line is tangent to the circle at $(0, 0)$, we simultaneously solve $x^2 + y^2 = r^2$ and $y = -\frac{5}{2}x \;\Rightarrow\; x^2 + \frac{25}{4}x^2 = r^2 \;\Rightarrow\; x^2 = \frac{4}{29}r^2 \;\Rightarrow\; x = \frac{2}{\sqrt{29}}r$, $y = -\frac{5}{\sqrt{29}}r$. To find $Q$, we either use symmetry or solve $(x - 3)^2 + (y - 1)^2 = r^2$ and $y - 1 = -\frac{5}{2}(x - 3)$. As above, we get $x = 3 - \frac{2}{\sqrt{29}}r$, $y = 1 + \frac{5}{\sqrt{29}}r$. Now the slope of the line $PQ$ is $\frac{2}{5}$, so $m_{PQ} = \dfrac{1 + \frac{5}{\sqrt{29}}r - \left(-\frac{5}{\sqrt{29}}r\right)}{3 - \frac{2}{\sqrt{29}}r - \frac{2}{\sqrt{29}}r} = \dfrac{1 + \frac{10}{\sqrt{29}}r}{3 - \frac{4}{\sqrt{29}}r} = \dfrac{\sqrt{29} + 10r}{3\sqrt{29} - 4r} = \dfrac{2}{5} \;\Rightarrow$ $5\sqrt{29} + 50r = 6\sqrt{29} - 8r \;\Leftrightarrow\; 58r = \sqrt{29} \;\Leftrightarrow\; r = \frac{\sqrt{29}}{58}$. So the minimum value of $r$ for which any line with slope $\frac{2}{5}$ intersects circles with radius $r$ centered at the lattice points on the plane is $r = \frac{\sqrt{29}}{58} \approx 0.093$.

# 4 ☐ APPLICATIONS OF DIFFERENTIATION

## 4.1 Related Rates

**1.** $V = x^3 \;\Rightarrow\; \dfrac{dV}{dt} = \dfrac{dV}{dx}\dfrac{dx}{dt} = 3x^2\,\dfrac{dx}{dt}$

**2.** (a) $A = \pi r^2 \;\Rightarrow\; \dfrac{dA}{dt} = \dfrac{dA}{dr}\dfrac{dr}{dt} = 2\pi r\,\dfrac{dr}{dt}$

(b) $\dfrac{dA}{dt} = 2\pi r\,\dfrac{dr}{dt} = 2\pi(30\text{ m})(1\text{ m/s}) = 60\pi\text{ m}^2/\text{s}$

**3.** Let $s$ denote the side of a square. The square's area $A$ is given by $A = s^2$. Differentiating with respect to $t$ gives us $\dfrac{dA}{dt} = 2s\,\dfrac{ds}{dt}$. When $A = 16$, $s = 4$. Substitution 4 for $s$ and 6 for $\dfrac{ds}{dt}$ gives us $\dfrac{dA}{dt} = 2(4)(6) = 48\text{ cm}^2/\text{s}$.

**4.** $A = \ell w \;\Rightarrow\; \dfrac{dA}{dt} = \ell\cdot\dfrac{dw}{dt} + w\cdot\dfrac{d\ell}{dt} = 20(3) + 10(8) = 140\text{ cm}^2/\text{s}$.

**5.** $y = x^3 + 2x \;\Rightarrow\; \dfrac{dy}{dt} = \dfrac{dy}{dx}\dfrac{dx}{dt} = (3x^2+2)(5) = 5(3x^2+2)$. When $x = 2$, $\dfrac{dy}{dt} = 5(14) = 70$.

**6.** $x^2 + y^2 = 25 \;\Rightarrow\; 2x\,\dfrac{dx}{dt} + 2y\,\dfrac{dy}{dt} = 0 \;\Rightarrow\; x\,\dfrac{dx}{dt} = -y\,\dfrac{dy}{dt} \;\Rightarrow\; \dfrac{dx}{dt} = -\dfrac{y}{x}\dfrac{dy}{dt}$.

When $y = 4$, $x^2 + 4^2 = 25 \;\Rightarrow\; x = \pm 3$. For $\dfrac{dy}{dt} = 6$, $\dfrac{dx}{dt} = -\dfrac{4}{\pm 3}(6) = \mp 8$.

**7.** $z^2 = x^2 + y^2 \;\Rightarrow\; 2z\,\dfrac{dz}{dt} = 2x\,\dfrac{dx}{dt} + 2y\,\dfrac{dy}{dt} \;\Rightarrow\; \dfrac{dz}{dt} = \dfrac{1}{z}\left(x\,\dfrac{dx}{dt} + y\,\dfrac{dy}{dt}\right)$. When $x = 5$ and $y = 12$,

$z^2 = 5^2 + 12^2 \;\Rightarrow\; z^2 = 169 \;\Rightarrow\; z = \pm 13$. For $\dfrac{dx}{dt} = 2$ and $\dfrac{dy}{dt} = 3$, $\dfrac{dz}{dt} = \dfrac{1}{\pm 13}(5\cdot 2 + 12\cdot 3) = \pm\dfrac{46}{13}$.

**8.** $y = \sqrt{1+x^3} \;\Rightarrow\; \dfrac{dy}{dt} = \dfrac{dy}{dx}\dfrac{dx}{dt} = \frac{1}{2}(1+x^3)^{-1/2}(3x^2)\,\dfrac{dx}{dt} = \dfrac{3x^2}{2\sqrt{1+x^3}}\,\dfrac{dx}{dt}$. With $\dfrac{dy}{dt} = 4$ when $x = 2$ and $y = 3$, we have $4 = \dfrac{3(4)}{2(3)}\dfrac{dx}{dt} \;\Rightarrow\; \dfrac{dx}{dt} = 2\text{ cm/s}$.

**9.** (a) Given: the rate of decrease of the surface area is $1\text{ cm}^2/\text{min}$. If we let $t$ be time (in minutes) and $S$ be the surface area (in $\text{cm}^2$), then we are given that $dS/dt = -1\text{ cm}^2/\text{s}$.

(b) Unknown: the rate of decrease of the diameter when the diameter is 10 cm. If we let $x$ be the diameter, then we want to find $dx/dt$ when $x = 10$ cm.

(c)

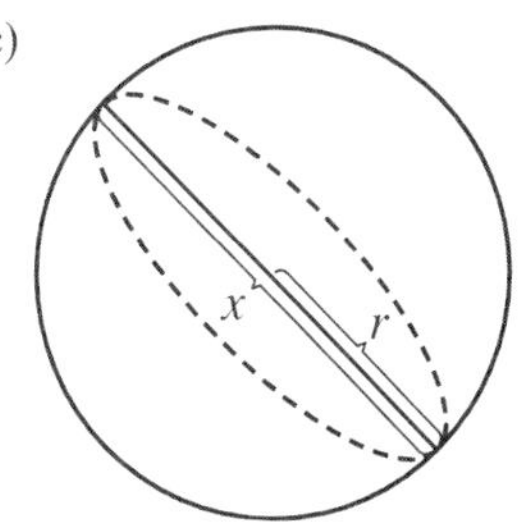


(d) If the radius is $r$ and the diameter $x = 2r$, then $r = \frac{1}{2}x$ and

$S = 4\pi r^2 = 4\pi\left(\frac{1}{2}x\right)^2 = \pi x^2 \;\Rightarrow\; \dfrac{dS}{dt} = \dfrac{dS}{dx}\dfrac{dx}{dt} = 2\pi x\,\dfrac{dx}{dt}$.

(e) $-1 = \dfrac{dS}{dt} = 2\pi x\,\dfrac{dx}{dt} \;\Rightarrow\; \dfrac{dx}{dt} = -\dfrac{1}{2\pi x}$. When $x = 10$, $\dfrac{dx}{dt} = -\dfrac{1}{20\pi}$. So the rate of decrease is $\frac{1}{20\pi}$ cm/min.

**10.** (a) Given: at noon, ship A is 150 km west of ship B; ship A is sailing east at 35 km/h, and ship B is sailing north at 25 km/h. If we let $t$ be time (in hours), $x$ be the distance traveled by ship A (in km), and $y$ be the distance traveled by ship B (in km), then we are given that $dx/dt = 35$ km/h and $dy/dt = 25$ km/h.

(b) Unknown: the rate at which the distance between the ships is changing at 4:00 P.M. If we let $z$ be the distance between the ships, then we want to find $dz/dt$ when $t = 4$ h.

(c)

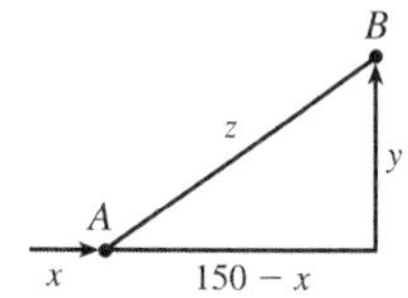


(d) $z^2 = (150 - x)^2 + y^2 \quad\Rightarrow\quad 2z\,\dfrac{dz}{dt} = 2(150 - x)\left(-\dfrac{dx}{dt}\right) + 2y\,\dfrac{dy}{dt}$

(e) At 4:00 P.M., $x = 4(35) = 140$ and $y = 4(25) = 100 \quad\Rightarrow\quad z = \sqrt{(150 - 140)^2 + 100^2} = \sqrt{10{,}100}$.

So $\dfrac{dz}{dt} = \dfrac{1}{z}\left[(x - 150)\,\dfrac{dx}{dt} + y\,\dfrac{dy}{dt}\right] = \dfrac{-10(35) + 100(25)}{\sqrt{10{,}100}} = \dfrac{215}{\sqrt{101}} \approx 21.4$ km/h.

**11.** (a) Given: a plane flying horizontally at an altitude of 1 mi and a speed of 500 mi/h passes directly over a radar station. If we let $t$ be time (in hours) and $x$ be the horizontal distance traveled by the plane (in mi), then we are given that $dx/dt = 500$ mi/h.

(b) Unknown: the rate at which the distance from the plane to the station is increasing when it is 2 mi from the station. If we let $y$ be the distance from the plane to the station, then we want to find $dy/dt$ when $y = 2$ mi.

(c)

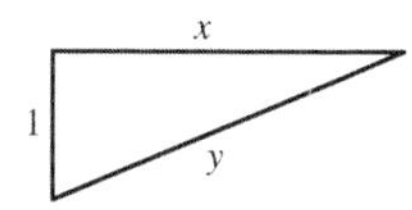


(d) By the Pythagorean Theorem, $y^2 = x^2 + 1 \quad\Rightarrow\quad 2y\,(dy/dt) = 2x\,(dx/dt)$.

(e) $\dfrac{dy}{dt} = \dfrac{x}{y}\dfrac{dx}{dt} = \dfrac{x}{y}(500)$. Since $y^2 = x^2 + 1$, when $y = 2$, $x = \sqrt{3}$, so $\dfrac{dy}{dt} = \frac{\sqrt{3}}{2}(500) = 250\sqrt{3} \approx 433$ mi/h.

**12.** (a) Given: a man 6 ft tall walks away from a street light mounted on a 15-ft-tall pole at a rate of 5 ft/s. If we let $t$ be time (in s) and $x$ be the distance from the pole to the man (in ft), then we are given that $dx/dt = 5$ ft/s.

(b) Unknown: the rate at which the tip of his shadow is moving when he is 40 ft from the pole. If we let $y$ be the distance from the man to the tip of his shadow (in ft), then we want to find $\dfrac{d}{dt}(x + y)$ when $x = 40$ ft.

(c)

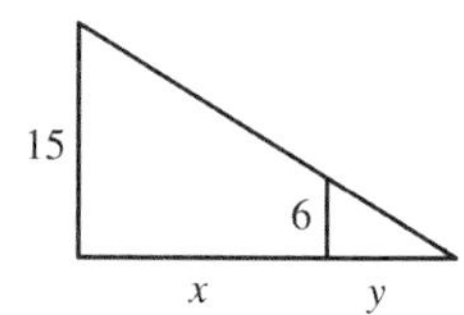


(d) By similar triangles, $\dfrac{15}{6} = \dfrac{x + y}{y} \quad\Rightarrow\quad 15y = 6x + 6y \quad\Rightarrow\quad 9y = 6x \quad\Rightarrow\quad y = \frac{2}{3}x$.

(e) The tip of the shadow moves at a rate of $\dfrac{d}{dt}(x + y) = \dfrac{d}{dt}\left(x + \frac{2}{3}x\right) = \dfrac{5}{3}\dfrac{dx}{dt} = \frac{5}{3}(5) = \frac{25}{3}$ ft/s.

**13.**

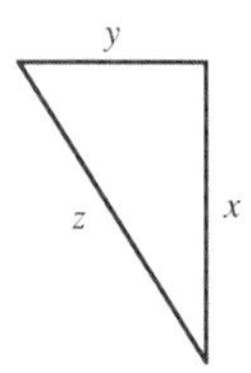


We are given that $\dfrac{dx}{dt} = 60$ mi/h and $\dfrac{dy}{dt} = 25$ mi/h. $z^2 = x^2 + y^2 \quad\Rightarrow$

$2z\,\dfrac{dz}{dt} = 2x\,\dfrac{dx}{dt} + 2y\,\dfrac{dy}{dt} \quad\Rightarrow\quad z\,\dfrac{dz}{dt} = x\,\dfrac{dx}{dt} + y\,\dfrac{dy}{dt} \quad\Rightarrow\quad \dfrac{dz}{dt} = \dfrac{1}{z}\left(x\,\dfrac{dx}{dt} + y\,\dfrac{dy}{dt}\right)$.

After 2 hours, $x = 2\,(60) = 120$ and $y = 2\,(25) = 50 \quad\Rightarrow\quad z = \sqrt{120^2 + 50^2} = 130$,

so $\dfrac{dz}{dt} = \dfrac{1}{z}\left(x\,\dfrac{dx}{dt} + y\,\dfrac{dy}{dt}\right) = \dfrac{120(60) + 50(25)}{130} = 65$ mi/h.

**14.**

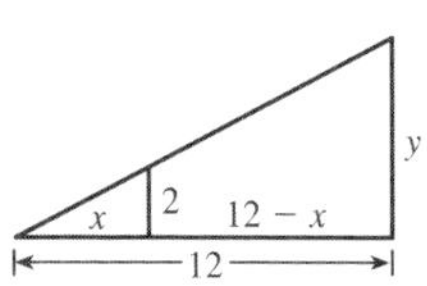


We are given that $\dfrac{dx}{dt} = 1.6$ m/s. By similar triangles, $\dfrac{y}{12} = \dfrac{2}{x} \quad\Rightarrow\quad y = \dfrac{24}{x} \quad\Rightarrow$

$\dfrac{dy}{dt} = -\dfrac{24}{x^2}\dfrac{dx}{dt} = -\dfrac{24}{x^2}(1.6)$. When $x = 8$, $\dfrac{dy}{dt} = -\dfrac{24(1.6)}{64} = -0.6$ m/s, so the shadow is decreasing at a rate of 0.6 m/s

**15.** 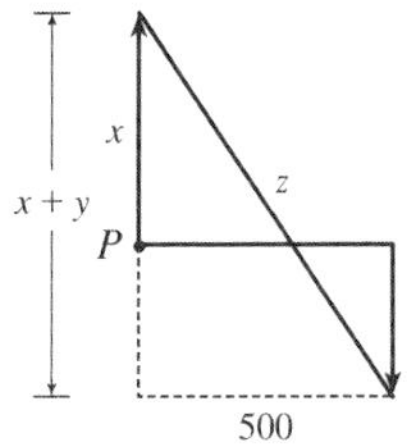


We are given that $\dfrac{dx}{dt} = 4$ ft/s and $\dfrac{dy}{dt} = 5$ ft/s. $z^2 = (x+y)^2 + 500^2 \Rightarrow$ $2z\,\dfrac{dz}{dt} = 2(x+y)\left(\dfrac{dx}{dt} + \dfrac{dy}{dt}\right)$. 15 minutes after the woman starts, we have $x = (4\text{ ft/s})(20\text{ min})(60\text{ s/min}) = 4800$ ft and $y = 5 \cdot 15 \cdot 60 = 4500 \Rightarrow$ $z = \sqrt{(4800+4500)^2 + 500^2} = \sqrt{86{,}740{,}000}$, so

$$\frac{dz}{dt} = \frac{x+y}{z}\left(\frac{dx}{dt} + \frac{dy}{dt}\right) = \frac{4800+4500}{\sqrt{86{,}740{,}000}}(4+5) = \frac{837}{\sqrt{8674}} \approx 8.99 \text{ ft/s}.$$

**16.** We are given that $\dfrac{dx}{dt} = 24$ ft/s.

(a) 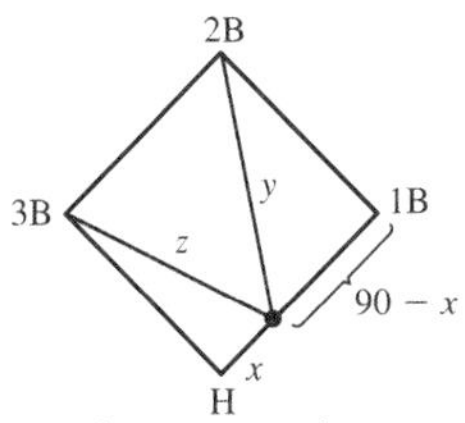


$y^2 = (90-x)^2 + 90^2 \Rightarrow 2y\,\dfrac{dy}{dt} = 2(90-x)\left(-\dfrac{dx}{dt}\right)$. When $x = 45$, $y = \sqrt{45^2 + 90^2} = 45\sqrt{5}$, so $\dfrac{dy}{dt} = \dfrac{90-x}{y}\left(-\dfrac{dx}{dt}\right) = \dfrac{45}{45\sqrt{5}}(-24) = -\dfrac{24}{\sqrt{5}}$, so the distance from second base is decreasing at a rate of $\frac{24}{\sqrt{5}} \approx 10.7$ ft/s.

(b) Due to the symmetric nature of the problem in part (a), we expect to get the same answer—and we do.

$z^2 = x^2 + 90^2 \Rightarrow 2z\,\dfrac{dz}{dt} = 2x\,\dfrac{dx}{dt}$. When $x = 45$, $z = 45\sqrt{5}$, so $\dfrac{dz}{dt} = \dfrac{45}{45\sqrt{5}}(24) = \dfrac{24}{\sqrt{5}} \approx 10.7$ ft/s.

**17.** $A = \frac{1}{2}bh$, where $b$ is the base and $h$ is the altitude. We are given that $\dfrac{dh}{dt} = 1$ cm/min and $\dfrac{dA}{dt} = 2$ cm$^2$/min. Using the Product Rule, we have $\dfrac{dA}{dt} = \dfrac{1}{2}\left(b\,\dfrac{dh}{dt} + h\,\dfrac{db}{dt}\right)$. When $h = 10$ and $A = 100$, we have $100 = \frac{1}{2}b(10) \Rightarrow \frac{1}{2}b = 10 \Rightarrow$ $b = 20$, so $2 = \dfrac{1}{2}\left(20 \cdot 1 + 10\,\dfrac{db}{dt}\right) \Rightarrow 4 = 20 + 10\,\dfrac{db}{dt} \Rightarrow \dfrac{db}{dt} = \dfrac{4-20}{10} = -1.6$ cm/min.

**18.** 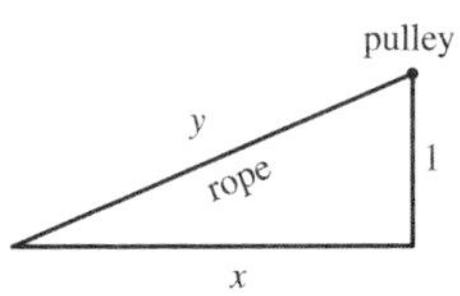


Given $\dfrac{dy}{dt} = -1$ m/s, find $\dfrac{dx}{dt}$ when $x = 8$ m. $y^2 = x^2 + 1 \Rightarrow 2y\,\dfrac{dy}{dt} = 2x\,\dfrac{dx}{dt} \Rightarrow$ $\dfrac{dx}{dt} = \dfrac{y}{x}\dfrac{dy}{dt} = -\dfrac{y}{x}$. When $x = 8$, $y = \sqrt{65}$, so $\dfrac{dx}{dt} = -\dfrac{\sqrt{65}}{8}$. Thus, the boat approaches the dock at $\dfrac{\sqrt{65}}{8} \approx 1.01$ m/s.

**19.** 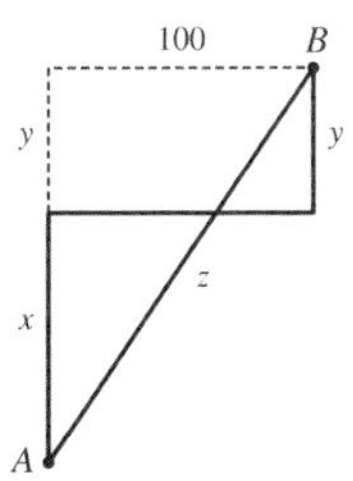


We are given that $\dfrac{dx}{dt} = 35$ km/h and $\dfrac{dy}{dt} = 25$ km/h. $z^2 = (x+y)^2 + 100^2 \Rightarrow$ $2z\,\dfrac{dz}{dt} = 2(x+y)\left(\dfrac{dx}{dt} + \dfrac{dy}{dt}\right)$. At 4:00 P.M., $x = 4(35) = 140$ and $y = 4(25) = 100 \Rightarrow z = \sqrt{(140+100)^2 + 100^2} = \sqrt{67{,}600} = 260$, so

$$\frac{dz}{dt} = \frac{x+y}{z}\left(\frac{dx}{dt} + \frac{dy}{dt}\right) = \frac{140+100}{260}(35+25) = \frac{720}{13} \approx 55.4 \text{ km/h}.$$

**20.** Let $D$ denote the distance from the origin $(0,0)$ to the point on the curve $y = \sqrt{x}$.

$$D = \sqrt{(x-0)^2 + (y-0)^2} = \sqrt{x^2 + (\sqrt{x})^2} = \sqrt{x^2 + x} \Rightarrow \frac{dD}{dt} = \tfrac{1}{2}(x^2+x)^{-1/2}(2x+1)\frac{dx}{dt} = \frac{2x+1}{2\sqrt{x^2+x}}\frac{dx}{dt}.$$

With $\dfrac{dx}{dt} = 3$ when $x = 4$, $\dfrac{dD}{dt} = \dfrac{9}{2\sqrt{20}}(3) = \dfrac{27}{4\sqrt{5}} \approx 3.02$ cm/s.

**21.** Using $Q$ for the origin, we are given $\dfrac{dx}{dt} = -2$ ft/s and need to find $\dfrac{dy}{dt}$ when $x = -5$.

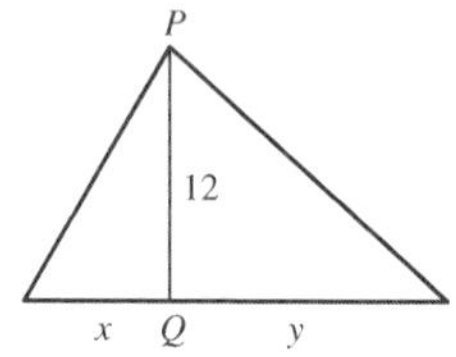


Using the Pythagorean Theorem twice, we have $\sqrt{x^2+12^2} + \sqrt{y^2+12^2} = 39$, the total length of the rope. Differentiating with respect to $t$, we get

$$\frac{x}{\sqrt{x^2+12^2}}\frac{dx}{dt} + \frac{y}{\sqrt{y^2+12^2}}\frac{dy}{dt} = 0, \text{ so } \frac{dy}{dt} = -\frac{x\sqrt{y^2+12^2}}{y\sqrt{x^2+12^2}}\frac{dx}{dt}.$$

Now when $x = -5$, $39 = \sqrt{(-5)^2+12^2} + \sqrt{y^2+12^2} = 13 + \sqrt{y^2+12^2} \Leftrightarrow \sqrt{y^2+12^2} = 26$, and

$y = \sqrt{26^2-12^2} = \sqrt{532}$. So when $x = -5$, $\dfrac{dy}{dt} = -\dfrac{(-5)(26)}{\sqrt{532}\,(13)}(-2) = -\dfrac{10}{\sqrt{133}} \approx -0.87$ ft/s. So cart $B$ is moving towards $Q$ at about 0.87 ft/s.

**22.** If $C =$ the rate at which water is pumped in, then $\dfrac{dV}{dt} = C - 10{,}000$, where

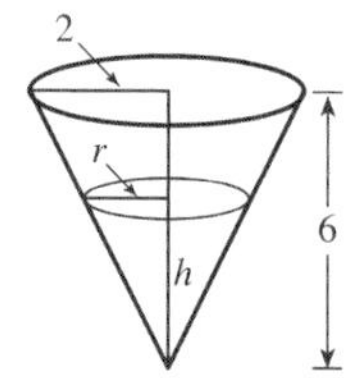


$V = \frac{1}{3}\pi r^2 h$ is the volume at time $t$. By similar triangles, $\dfrac{r}{2} = \dfrac{h}{6} \Rightarrow r = \frac{1}{3}h \Rightarrow$

$V = \frac{1}{3}\pi\left(\frac{1}{3}h\right)^2 h = \frac{\pi}{27}h^3 \Rightarrow \dfrac{dV}{dt} = \frac{\pi}{9}h^2\dfrac{dh}{dt}$. When $h = 200$ cm,

$\dfrac{dh}{dt} = 20$ cm/min, so $C - 10{,}000 = \frac{\pi}{9}(200)^2(20) \Rightarrow C = 10{,}000 + \frac{800{,}000}{9}\pi \approx 289{,}253$ cm$^3$/min.

**23.** By similar triangles, $\dfrac{3}{1} = \dfrac{b}{h}$, so $b = 3h$. The trough has volume

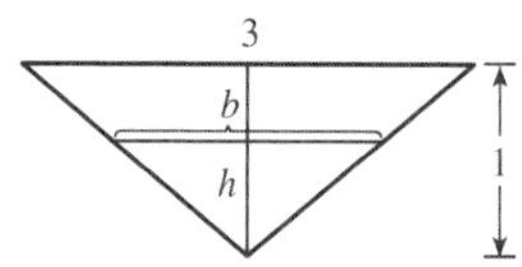


$V = \frac{1}{2}bh(10) = 5(3h)h = 15h^2 \Rightarrow 12 = \dfrac{dV}{dt} = 30h\dfrac{dh}{dt} \Rightarrow \dfrac{dh}{dt} = \dfrac{2}{5h}$.

When $h = \frac{1}{2}$, $\dfrac{dh}{dt} = \dfrac{2}{5 \cdot \frac{1}{2}} = \dfrac{4}{5}$ ft/min.

**24.** The figure is drawn without the top 3 feet.

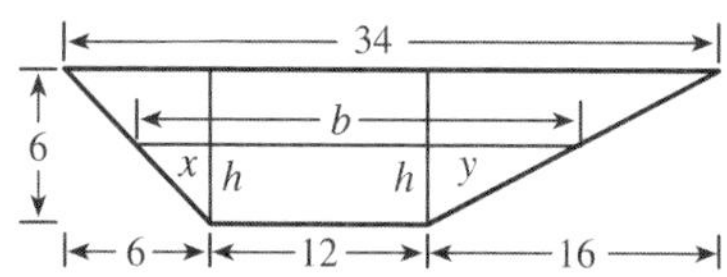


$V = \frac{1}{2}(b+12)h(20) = 10(b+12)h$ and, from similar triangles,

$\dfrac{x}{h} = \dfrac{6}{6}$ and $\dfrac{y}{h} = \dfrac{16}{6} = \dfrac{8}{3}$, so $b = x + 12 + y = h + 12 + \dfrac{8h}{3} = 12 + \dfrac{11h}{3}$.

Thus, $V = 10\left(24 + \dfrac{11h}{3}\right)h = 240h + \dfrac{110h^2}{3}$ and so $0.8 = \dfrac{dV}{dt} = \left(240 + \dfrac{220}{3}h\right)\dfrac{dh}{dt}$. When $h = 5$,

$\dfrac{dh}{dt} = \dfrac{0.8}{240 + 5(220/3)} = \dfrac{3}{2275} \approx 0.00132$ ft/min.

**25.** We are given that $\dfrac{dV}{dt} = 30$ ft$^3$/min. $V = \frac{1}{3}\pi r^2 h = \frac{1}{3}\pi\left(\dfrac{h}{2}\right)^2 h = \dfrac{\pi h^3}{12} \Rightarrow$

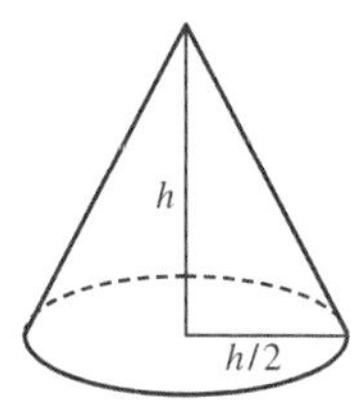


$\dfrac{dV}{dt} = \dfrac{dV}{dh}\dfrac{dh}{dt} \Rightarrow 30 = \dfrac{\pi h^2}{4}\dfrac{dh}{dt} \Rightarrow \dfrac{dh}{dt} = \dfrac{120}{\pi h^2}$. When $h = 10$ ft,

$\dfrac{dh}{dt} = \dfrac{120}{10^2\pi} = \dfrac{6}{5\pi} \approx 0.38$ ft/min.

**26.** We are given $dx/dt = 8$ ft/s. $\cot\theta = \dfrac{x}{100} \Rightarrow x = 100\cot\theta \Rightarrow$

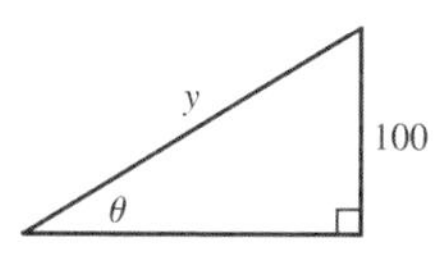


$\dfrac{dx}{dt} = -100\csc^2\theta\,\dfrac{d\theta}{dt} \Rightarrow \dfrac{d\theta}{dt} = -\dfrac{\sin^2\theta}{100} \cdot 8$. When $y = 200$, $\sin\theta = \dfrac{100}{200} = \dfrac{1}{2} \Rightarrow$

$\dfrac{d\theta}{dt} = -\dfrac{(1/2)^2}{100} \cdot 8 = -\dfrac{1}{50}$ rad/s. The angle is decreasing at a rate of $\frac{1}{50}$ rad/s.

**27.** $A = \frac{1}{2}bh$, but $b = 5$ m and $\sin\theta = \dfrac{h}{4} \quad \Rightarrow \quad h = 4\sin\theta$, so $A = \frac{1}{2}(5)(4\sin\theta) = 10\sin\theta$.

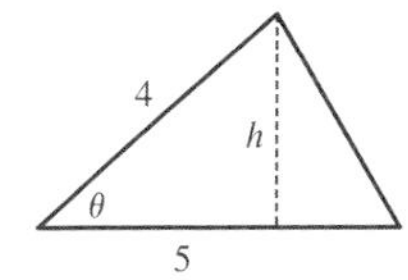


We are given $\dfrac{d\theta}{dt} = 0.06$ rad/s, so $\dfrac{dA}{dt} = \dfrac{dA}{d\theta}\dfrac{d\theta}{dt} = (10\cos\theta)(0.06) = 0.6\cos\theta$.

When $\theta = \frac{\pi}{3}$, $\dfrac{dA}{dt} = 0.6\left(\cos\frac{\pi}{3}\right) = (0.6)\left(\frac{1}{2}\right) = 0.3\ \text{m}^2/\text{s}$.

**28.** We are given $d\theta/dt = 2°/\text{min} = \frac{\pi}{90}$ rad/min. By the Law of Cosines,

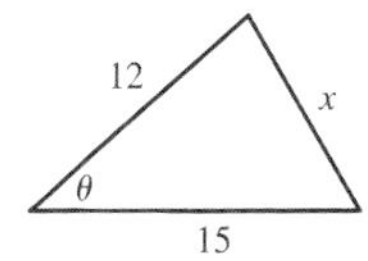


$x^2 = 12^2 + 15^2 - 2(12)(15)\cos\theta = 369 - 360\cos\theta \quad \Rightarrow$

$2x\dfrac{dx}{dt} = 360\sin\theta\dfrac{d\theta}{dt} \quad \Rightarrow \quad \dfrac{dx}{dt} = \dfrac{180\sin\theta}{x}\dfrac{d\theta}{dt}$. When $\theta = 60°$,

$x = \sqrt{369 - 360\cos 60°} = \sqrt{189} = 3\sqrt{21}$, so $\dfrac{dx}{dt} = \dfrac{180\sin 60°}{3\sqrt{21}}\dfrac{\pi}{90} = \dfrac{\pi\sqrt{3}}{3\sqrt{21}} = \dfrac{\sqrt{7}\,\pi}{21} \approx 0.396$ m/min.

**29.** Differentiating both sides of $PV = C$ with respect to $t$ and using the Product Rule gives us $P\dfrac{dV}{dt} + V\dfrac{dP}{dt} = 0 \quad \Rightarrow$

$\dfrac{dV}{dt} = -\dfrac{V}{P}\dfrac{dP}{dt}$. When $V = 600$, $P = 150$ and $\dfrac{dP}{dt} = 20$, so we have $\dfrac{dV}{dt} = -\dfrac{600}{150}(20) = -80$. Thus, the volume is decreasing at a rate of $80\ \text{cm}^3/\text{min}$.

**30.** $PV^{1.4} = C \quad \Rightarrow \quad P \cdot 1.4V^{0.4}\dfrac{dV}{dt} + V^{1.4}\dfrac{dP}{dt} = 0 \quad \Rightarrow \quad \dfrac{dV}{dt} = -\dfrac{V^{1.4}}{P \cdot 1.4V^{0.4}}\dfrac{dP}{dt} = -\dfrac{V}{1.4P}\dfrac{dP}{dt}$. When $V = 400$, $P = 80$ and $\dfrac{dP}{dt} = -10$, so we have $\dfrac{dV}{dt} = -\dfrac{400}{1.4(80)}(-10) = \dfrac{250}{7}$. Thus, the volume is increasing at a rate of $\frac{250}{7} \approx 36\ \text{cm}^3/\text{min}$.

**31.** With $R_1 = 80$ and $R_2 = 100$, $\dfrac{1}{R} = \dfrac{1}{R_1} + \dfrac{1}{R_2} = \dfrac{1}{80} + \dfrac{1}{100} = \dfrac{180}{8000} = \dfrac{9}{400}$, so $R = \dfrac{400}{9}$. Differentiating $\dfrac{1}{R} = \dfrac{1}{R_1} + \dfrac{1}{R_2}$ with respect to $t$, we have $-\dfrac{1}{R^2}\dfrac{dR}{dt} = -\dfrac{1}{R_1^2}\dfrac{dR_1}{dt} - \dfrac{1}{R_2^2}\dfrac{dR_2}{dt} \quad \Rightarrow \quad \dfrac{dR}{dt} = R^2\left(\dfrac{1}{R_1^2}\dfrac{dR_1}{dt} + \dfrac{1}{R_2^2}\dfrac{dR_2}{dt}\right)$. When $R_1 = 80$ and $R_2 = 100$, $\dfrac{dR}{dt} = \dfrac{400^2}{9^2}\left[\dfrac{1}{80^2}(0.3) + \dfrac{1}{100^2}(0.2)\right] = \dfrac{107}{810} \approx 0.132\ \Omega/\text{s}$.

**32.** We want to find $\dfrac{dB}{dt}$ when $L = 18$ using $B = 0.007W^{2/3}$ and $W = 0.12L^{2.53}$.

$$\frac{dB}{dt} = \frac{dB}{dW}\frac{dW}{dL}\frac{dL}{dt} = \left(0.007 \cdot \tfrac{2}{3}W^{-1/3}\right)\left(0.12 \cdot 2.53 \cdot L^{1.53}\right)\left(\frac{20 - 15}{10{,}000{,}000}\right)$$

$$= \left[0.007 \cdot \tfrac{2}{3}\left(0.12 \cdot 18^{2.53}\right)^{-1/3}\right]\left(0.12 \cdot 2.53 \cdot 18^{1.53}\right)\left(\frac{5}{10^7}\right) \approx 1.045 \times 10^{-8}\ \text{g/yr}$$

**33.** (a) By the Pythagorean Theorem, $4000^2 + y^2 = \ell^2$. Differentiating with respect to $t$, we obtain

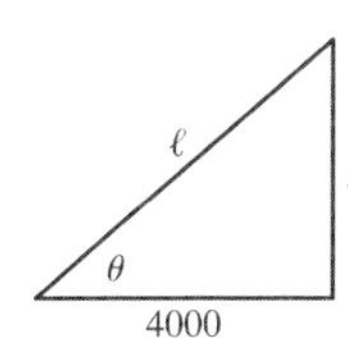


$2y\dfrac{dy}{dt} = 2\ell\dfrac{d\ell}{dt}$. We know that $\dfrac{dy}{dt} = 600$ ft/s, so when $y = 3000$ ft,

$\ell = \sqrt{4000^2 + 3000^2} = \sqrt{25{,}000{,}000} = 5000$ ft

and $\dfrac{d\ell}{dt} = \dfrac{y}{\ell}\dfrac{dy}{dt} = \dfrac{3000}{5000}(600) = \dfrac{1800}{5} = 360$ ft/s.

(b) Here $\tan\theta = \dfrac{y}{4000} \quad \Rightarrow \quad \dfrac{d}{dt}(\tan\theta) = \dfrac{d}{dt}\left(\dfrac{y}{4000}\right) \quad \Rightarrow \quad \sec^2\theta\,\dfrac{d\theta}{dt} = \dfrac{1}{4000}\dfrac{dy}{dt} \quad \Rightarrow \quad \dfrac{d\theta}{dt} = \dfrac{\cos^2\theta}{4000}\dfrac{dy}{dt}$. When $y = 3000$ ft, $\dfrac{dy}{dt} = 600$ ft/s, $\ell = 5000$ and $\cos\theta = \dfrac{4000}{\ell} = \dfrac{4000}{5000} = \dfrac{4}{5}$, so $\dfrac{d\theta}{dt} = \dfrac{(4/5)^2}{4000}(600) = 0.096$ rad/s.

**34.** We are given that $\dfrac{d\theta}{dt} = 4(2\pi) = 8\pi$ rad/min. $x = 3\tan\theta \;\Rightarrow$

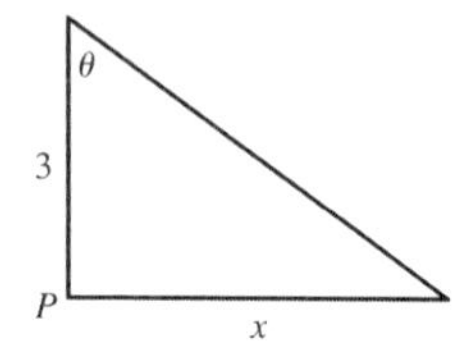


$\dfrac{dx}{dt} = 3\sec^2\theta\,\dfrac{d\theta}{dt}$. When $x = 1$, $\tan\theta = \frac{1}{3}$, so $\sec^2\theta = 1 + \left(\frac{1}{3}\right)^2 = \frac{10}{9}$

and $\dfrac{dx}{dt} = 3\left(\frac{10}{9}\right)(8\pi) = \frac{80\pi}{3} \approx 83.8$ km/min.

**35.** We are given that $\dfrac{dx}{dt} = 300$ km/h. By the Law of Cosines,

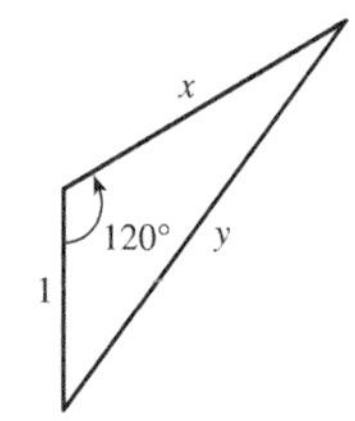


$y^2 = x^2 + 1^2 - 2(1)(x)\cos 120° = x^2 + 1 - 2x\left(-\frac{1}{2}\right) = x^2 + x + 1$, so

$2y\,\dfrac{dy}{dt} = 2x\,\dfrac{dx}{dt} + \dfrac{dx}{dt} \;\Rightarrow\; \dfrac{dy}{dt} = \dfrac{2x+1}{2y}\,\dfrac{dx}{dt}$. After 1 minute, $x = \frac{300}{60} = 5$ km $\;\Rightarrow$

$y = \sqrt{5^2 + 5 + 1} = \sqrt{31}$ km $\;\Rightarrow\; \dfrac{dy}{dt} = \dfrac{2(5)+1}{2\sqrt{31}}(300) = \dfrac{1650}{\sqrt{31}} \approx 296$ km/h.

**36.** We are given that $\dfrac{dx}{dt} = 3$ mi/h and $\dfrac{dy}{dt} = 2$ mi/h. By the Law of Cosines,

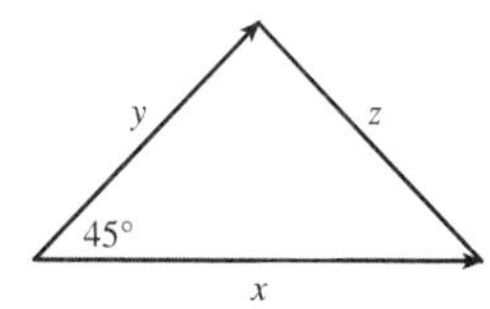


$z^2 = x^2 + y^2 - 2xy\cos 45° = x^2 + y^2 - \sqrt{2}\,xy \;\Rightarrow$

$2z\dfrac{dz}{dt} = 2x\,\dfrac{dx}{dt} + 2y\,\dfrac{dy}{dt} - \sqrt{2}\,x\,\dfrac{dy}{dt} - \sqrt{2}\,y\,\dfrac{dx}{dt}$. After 15 minutes $\left[= \frac{1}{4}\text{ h}\right]$,

we have $x = \frac{3}{4}$ and $y = \frac{2}{4} = \frac{1}{2} \;\Rightarrow\; z^2 = \left(\frac{3}{4}\right)^2 + \left(\frac{2}{4}\right)^2 - \sqrt{2}\left(\frac{3}{4}\right)\left(\frac{2}{4}\right) \;\Rightarrow\; z = \dfrac{\sqrt{13 - 6\sqrt{2}}}{4}$ and

$$\begin{aligned}\frac{dz}{dt} &= \frac{2}{\sqrt{13 - 6\sqrt{2}}}\left[2\left(\tfrac{3}{4}\right)3 + 2\left(\tfrac{1}{2}\right)2 - \sqrt{2}\left(\tfrac{3}{4}\right)2 - \sqrt{2}\left(\tfrac{1}{2}\right)3\right] \\ &= \frac{2}{\sqrt{13 - 6\sqrt{2}}}\,\frac{13 - 6\sqrt{2}}{2} = \sqrt{13 - 6\sqrt{2}} \approx 2.125 \text{ mi/h}.\end{aligned}$$

**37.** Let the distance between the runner and the friend be $\ell$. Then by the Law of Cosines,

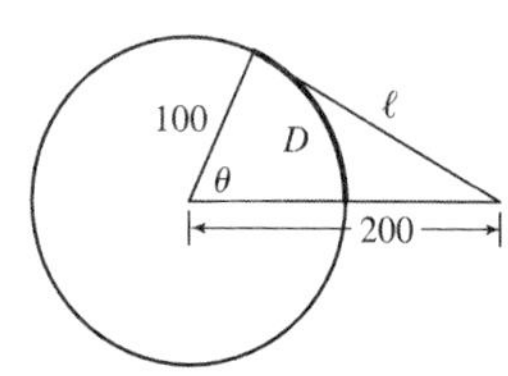


$\ell^2 = 200^2 + 100^2 - 2\cdot 200\cdot 100\cdot\cos\theta = 50{,}000 - 40{,}000\cos\theta$ $(\star)$. Differentiating implicitly with respect to $t$, we obtain $2\ell\,\dfrac{d\ell}{dt} = -40{,}000(-\sin\theta)\,\dfrac{d\theta}{dt}$. Now if $D$ is the distance run when the angle is $\theta$ radians, then by the formula for the length of an arc on a circle, $s = r\theta$, we have $D = 100\theta$, so $\theta = \dfrac{1}{100}D \;\Rightarrow\; \dfrac{d\theta}{dt} = \dfrac{1}{100}\,\dfrac{dD}{dt} = \dfrac{7}{100}$. To substitute into the expression for $\dfrac{d\ell}{dt}$, we must know $\sin\theta$ at the time when $\ell = 200$, which we find from $(\star)$: $200^2 = 50{,}000 - 40{,}000\cos\theta \;\Leftrightarrow$

$\cos\theta = \dfrac{1}{4} \;\Rightarrow\; \sin\theta = \sqrt{1 - \left(\frac{1}{4}\right)^2} = \dfrac{\sqrt{15}}{4}$. Substituting, we get $2(200)\,\dfrac{d\ell}{dt} = 40{,}000\dfrac{\sqrt{15}}{4}\left(\dfrac{7}{100}\right) \;\Rightarrow$

$d\ell/dt = \frac{7\sqrt{15}}{4} \approx 6.78$ m/s. Whether the distance between them is increasing or decreasing depends on the direction in which the runner is running.

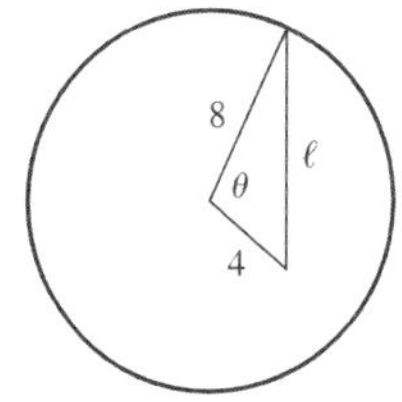


**38.** The hour hand of a clock goes around once every 12 hours or, in radians per hour, $\frac{2\pi}{12} = \frac{\pi}{6}$ rad/h. The minute hand goes around once an hour, or at the rate of $2\pi$ rad/h. So the angle $\theta$ between them (measuring clockwise from the minute hand to the hour hand) is changing at the rate of $d\theta/dt = \frac{\pi}{6} - 2\pi = -\frac{11\pi}{6}$ rad/h. Now, to relate $\theta$ to $\ell$, we use the Law of Cosines: $\ell^2 = 4^2 + 8^2 - 2 \cdot 4 \cdot 8 \cdot \cos\theta = 80 - 64\cos\theta$ $(\star)$. Differentiating implicitly with respect to $t$, we get $2\ell\,\frac{d\ell}{dt} = -64(-\sin\theta)\frac{d\theta}{dt}$. At 1:00, the angle between the two hands is one-twelfth of the circle, that is, $\frac{2\pi}{12} = \frac{\pi}{6}$ radians. We use $(\star)$ to find $\ell$ at 1:00: $\ell = \sqrt{80 - 64\cos\frac{\pi}{6}} = \sqrt{80 - 32\sqrt{3}}$. Substituting, we get $2\ell\,\frac{d\ell}{dt} = 64\sin\frac{\pi}{6}\left(-\frac{11\pi}{6}\right) \Rightarrow \frac{d\ell}{dt} = \frac{64\left(\frac{1}{2}\right)\left(-\frac{11\pi}{6}\right)}{2\sqrt{80 - 32\sqrt{3}}} = -\frac{88\pi}{3\sqrt{80 - 32\sqrt{3}}} \approx -18.6$.

So at 1:00, the distance between the tips of the hands is decreasing at a rate of 18.6 mm/h $\approx 0.005$ mm/s.

## 4.2 Maximum and Minimum Values

**1.** A function $f$ has an **absolute minimum** at $x = c$ if $f(c)$ is the smallest function value on the entire domain of $f$, whereas $f$ has a **local minimum** at $c$ if $f(c)$ is the smallest function value when $x$ is near $c$.

**2.** (a) The Extreme Value Theorem

(b) See the Closed Interval Method.

**3.** Absolute maximum at $b$; absolute minimum at $d$; local maxima at $b$ and $e$; local minima at $d$ and $s$; neither a maximum nor a minimum at $a$, $c$, $r$, and $t$.

**4.** Absolute maximum at $e$; absolute minimum at $t$; local maxima at $c$, $e$, and $s$; local minima at $b$, $c$, $d$, and $r$; neither a maximum nor a minimum at $a$.

**5.** Absolute maximum value is $f(4) = 4$; absolute minimum value is $f(7) = 0$; local maximum values are $f(4) = 4$ and $f(6) = 3$; local minimum values are $f(2) = 1$ and $f(5) = 2$.

**6.** Absolute maximum value is $f(7) = 5$; absolute minimum value is $f(1) = 0$; local maximum values are $f(0) = 2$, $f(3) = 4$, and $f(5) = 3$; local minimum values are $f(1) = 0$, $f(4) = 2$, and $f(6) = 1$.

**7.** Absolute minimum at 2, absolute maximum at 3, local minimum at 4

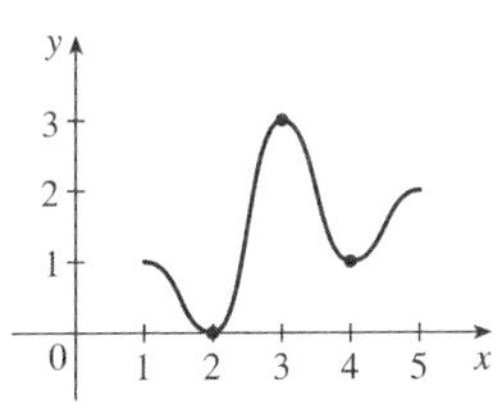


**8.** Absolute minimum at 1, absolute maximum at 5, local maximum at 2, local minimum at 4

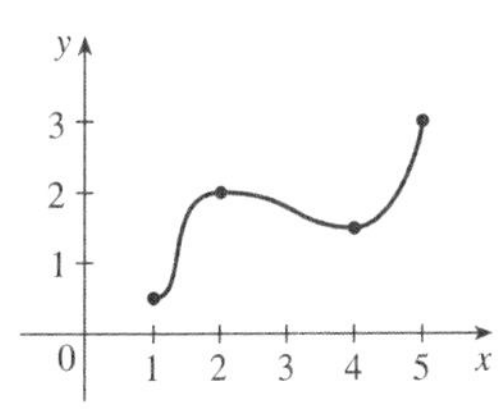

**9.** Absolute maximum at 5, absolute minimum at 2, local maximum at 3, local minima at 2 and 4

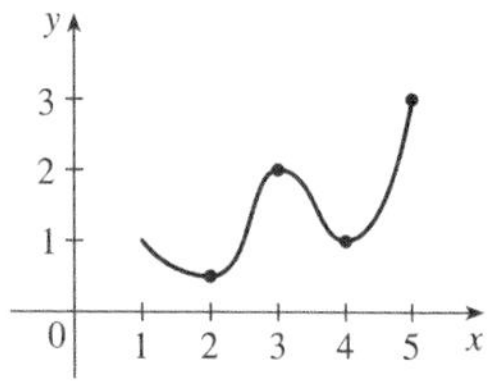

**10.** $f$ has no local maximum or minimum, but 2 and 4 are critical numbers

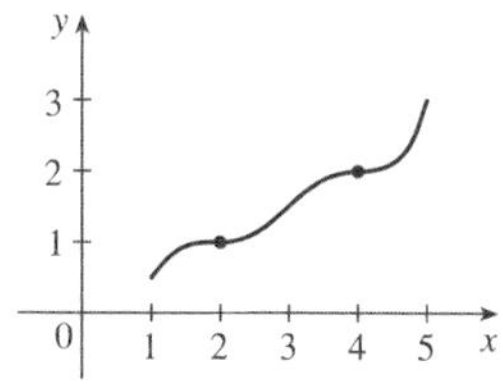

**11.** (a)

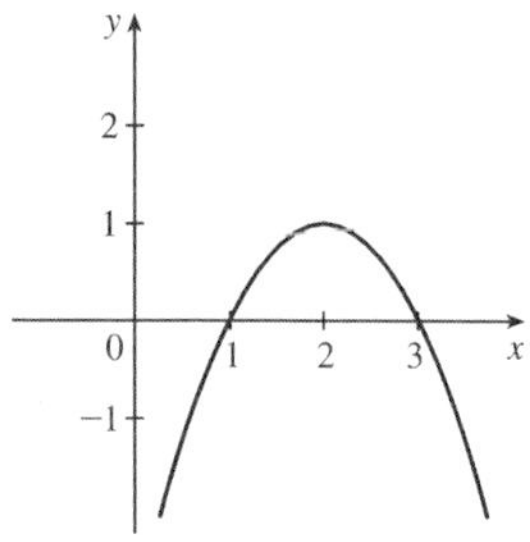

(b)

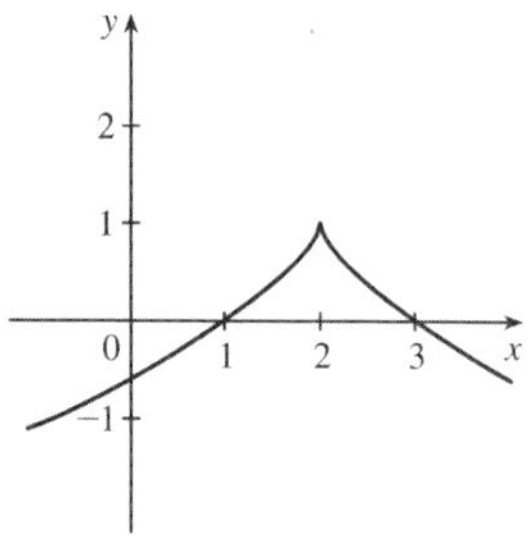

(c)

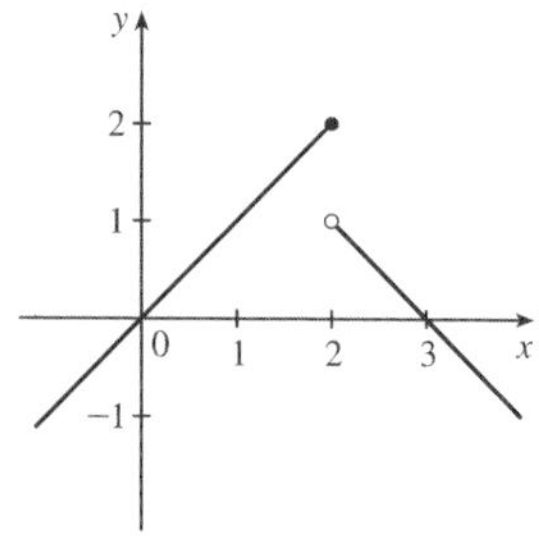

**12.** (a) Note that a local maximum cannot occur at an endpoint.

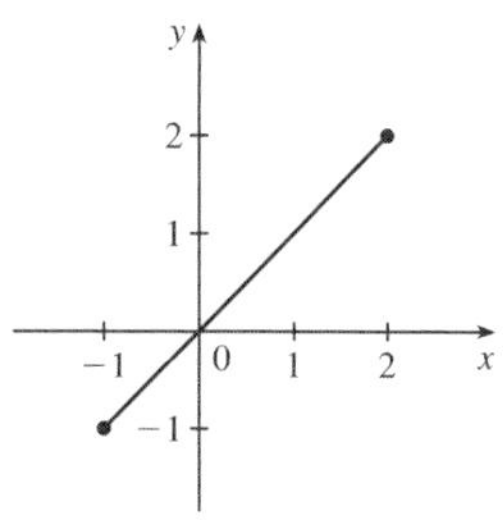

(b)

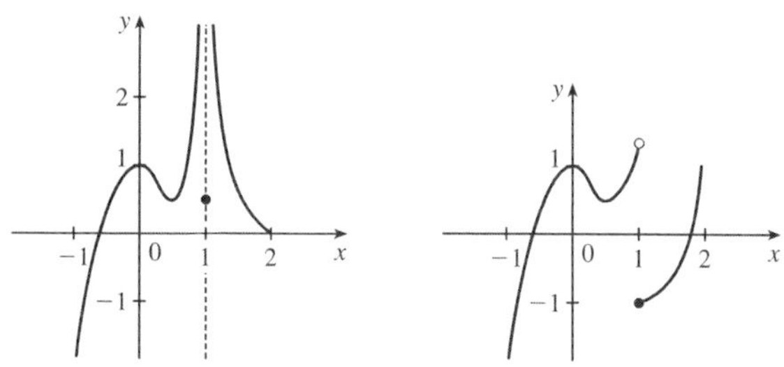

*Note:* By the Extreme Value Theorem, $f$ must *not* be continuous.

**13.** (a) *Note:* By the Extreme Value Theorem, $f$ must *not* be continuous; because if it were, it would attain an absolute minimum.

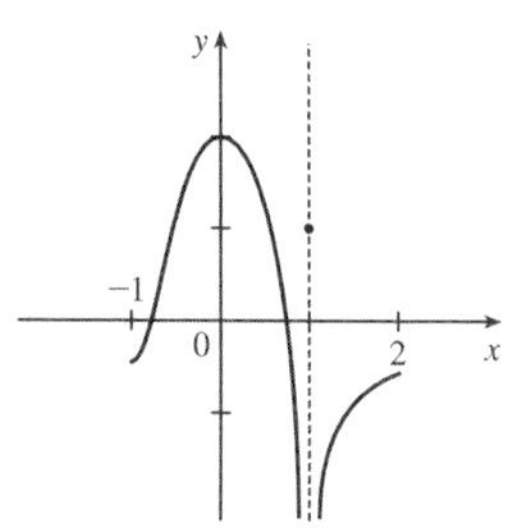

(b)

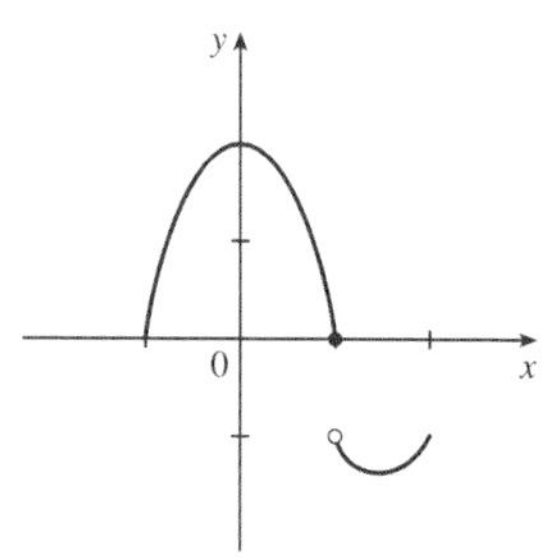

**14.** (a)

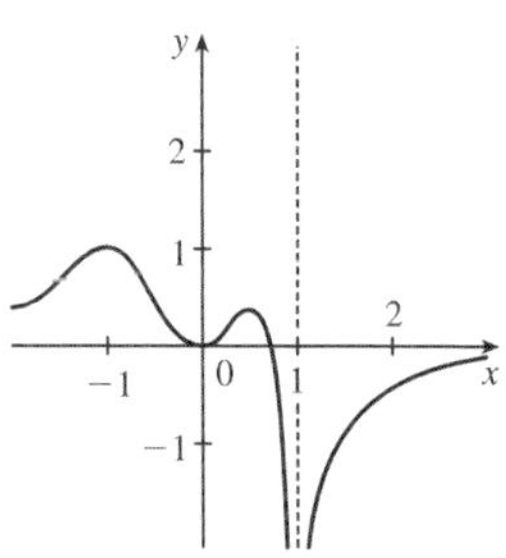

(b)

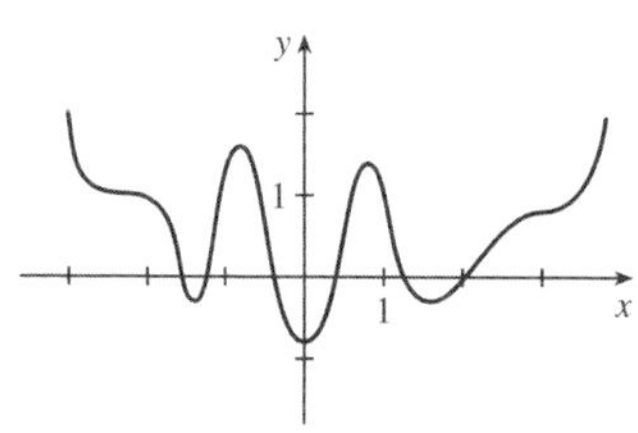

**15.** $f(x) = 8 - 3x$, $x \geq 1$. Absolute maximum $f(1) = 5$; no local maximum. No absolute or local minimum.

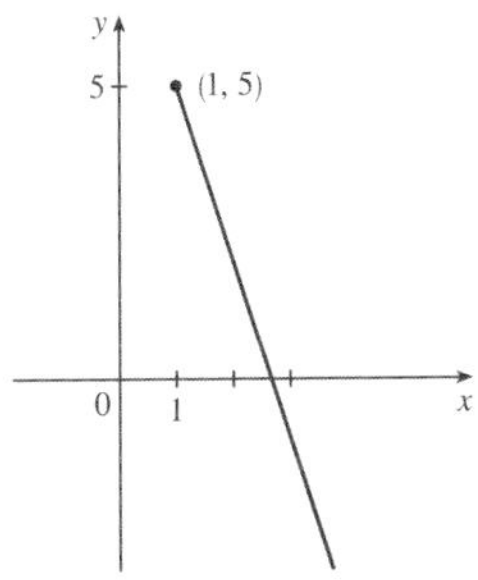


**16.** $f(x) = 3 - 2x$, $x \leq 5$. Absolute minimum $f(5) = -7$; no local minimum. No absolute or local maximum.

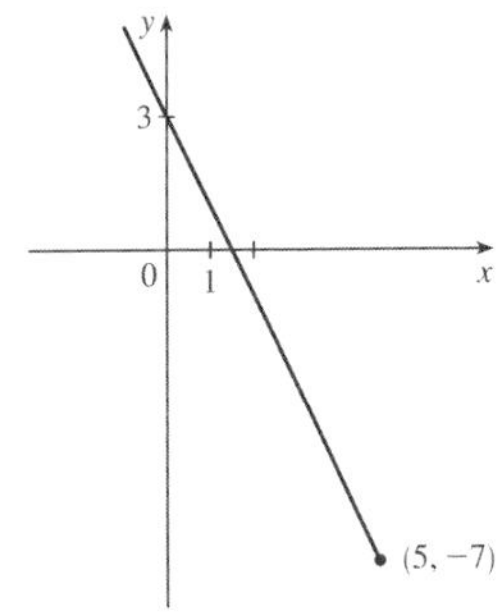


**17.** $f(x) = x^2$, $0 < x < 2$. No absolute or local maximum or minimum value.

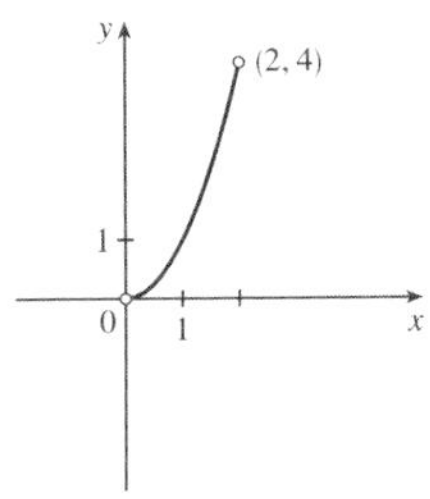


**18.** $f(x) = e^x$. No absolute or local maximum or minimum value.

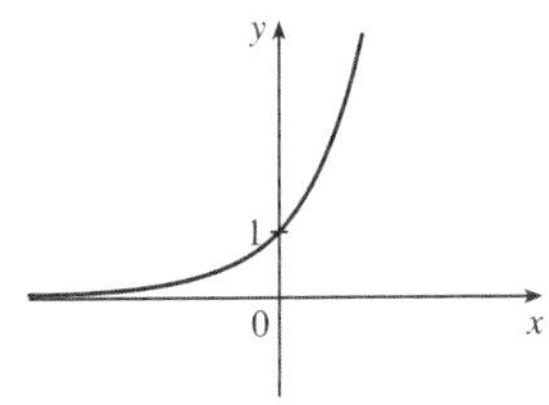


**19.** $f(\theta) = \sin\theta$, $-2\pi \leq \theta \leq 2\pi$. Absolute and local maxima $f\left(-\frac{3\pi}{2}\right) = f\left(\frac{\pi}{2}\right) = 1$. Absolute and local minima $f\left(-\frac{\pi}{2}\right) = f\left(\frac{3\pi}{2}\right) = -1$.

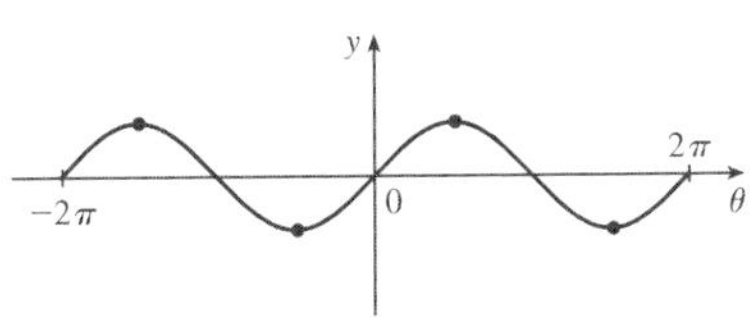


**20.** $f(\theta) = \tan\theta$, $-\frac{\pi}{4} \leq \theta < \frac{\pi}{2}$. Absolute minimum $f\left(-\frac{\pi}{4}\right) = -1$; no local minimum. No absolute or local maximum.

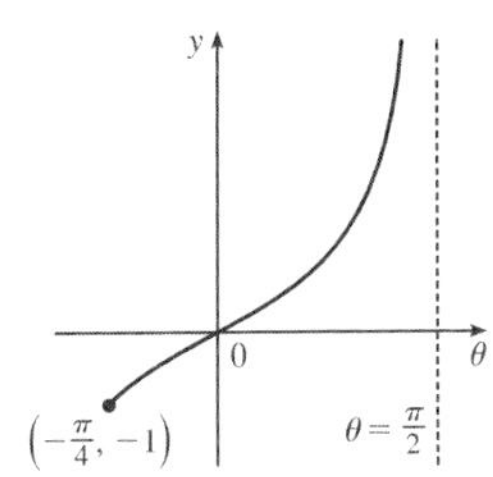


**21.** $f(x) = 1 - \sqrt{x}$. Absolute maximum $f(0) = 1$; no local maximum. No absolute or local minimum.

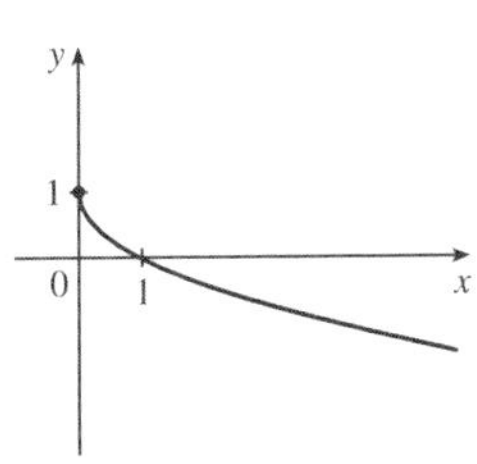


**22.** $f(x) = \begin{cases} 4 - x^2 & \text{if } -2 \leq x < 0 \\ 2x - 1 & \text{if } 0 \leq x \leq 2 \end{cases}$

Abolute minimum $f(0) = -1$; no local minimim. No absolute or local maximum.

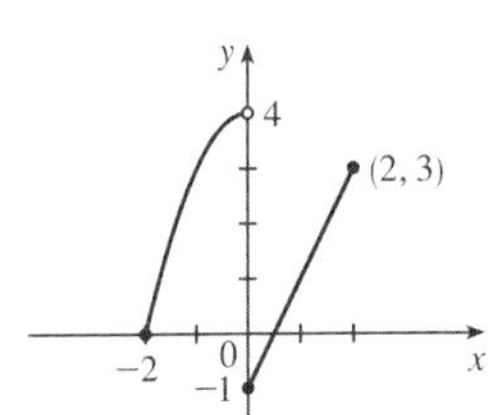

**23.** $f(x) = 5x^2 + 4x \Rightarrow f'(x) = 10x + 4$. $f'(x) = 0 \Rightarrow x = -\frac{2}{5}$, so $-\frac{2}{5}$ is the only critical number.

**24.** $f(x) = x^3 + x^2 - x \Rightarrow f'(x) = 3x^2 + 2x - 1$.
$f'(x) = 0 \Rightarrow (x+1)(3x-1) = 0 \Rightarrow x = -1, \frac{1}{3}$. These are the only critical numbers.

**25.** $f(x) = x^3 + 3x^2 - 24x \Rightarrow f'(x) = 3x^2 + 6x - 24 = 3(x^2 + 2x - 8)$.
$f'(x) = 0 \Rightarrow 3(x+4)(x-2) = 0 \Rightarrow x = -4, 2$. These are the only critical numbers.

**26.** $f(x) = x^3 + x^2 + x \Rightarrow f'(x) = 3x^2 + 2x + 1$. $f'(x) = 0 \Rightarrow 3x^2 + 2x + 1 = 0 \Rightarrow x = \dfrac{-2 \pm \sqrt{4-12}}{6}$.
Neither of these is a real number. Thus, there are no critical numbers.

**27.** $s(t) = 3t^4 + 4t^3 - 6t^2 \Rightarrow s'(t) = 12t^3 + 12t^2 - 12t$. $s'(t) = 0 \Rightarrow 12t(t^2 + t - 1) \Rightarrow$
$t = 0$ or $t^2 + t - 1 = 0$. Using the quadratic formula to solve the latter equation gives us
$t = \dfrac{-1 \pm \sqrt{1^2 - 4(1)(-1)}}{2(1)} = \dfrac{-1 \pm \sqrt{5}}{2} \approx 0.618, -1.618$. The three critical numbers are 0, $\dfrac{-1 \pm \sqrt{5}}{2}$.

**28.** $g(t) = |3t - 4| = \begin{cases} 3t - 4 & \text{if } 3t - 4 \geq 0 \\ -(3t-4) & \text{if } 3t - 4 < 0 \end{cases} = \begin{cases} 3t - 4 & \text{if } t \geq \frac{4}{3} \\ 4 - 3t & \text{if } t < \frac{4}{3} \end{cases}$

$g'(t) = \begin{cases} 3 & \text{if } t > \frac{4}{3} \\ -3 & \text{if } t < \frac{4}{3} \end{cases}$ and $g'(t)$ does not exist at $t = \frac{4}{3}$, so $t = \frac{4}{3}$ is a critical number.

**29.** $g(y) = \dfrac{y-1}{y^2 - y + 1} \Rightarrow$

$$g'(y) = \frac{(y^2 - y + 1)(1) - (y-1)(2y-1)}{(y^2 - y + 1)^2} = \frac{y^2 - y + 1 - (2y^2 - 3y + 1)}{(y^2 - y + 1)^2} = \frac{-y^2 + 2y}{(y^2 - y + 1)^2} = \frac{y(2-y)}{(y^2 - y + 1)^2}.$$

$g'(y) = 0 \Rightarrow y = 0, 2$. The expression $y^2 - y + 1$ is never equal to 0, so $g'(y)$ exists for all real numbers. The critical numbers are 0 and 2.

**30.** $h(p) = \dfrac{p-1}{p^2+4} \Rightarrow h'(p) = \dfrac{(p^2+4)(1) - (p-1)(2p)}{(p^2+4)^2} = \dfrac{p^2 + 4 - 2p^2 + 2p}{(p^2+4)^2} = \dfrac{-p^2 + 2p + 4}{(p^2+4)^2}$.

$h'(p) = 0 \Rightarrow p = \dfrac{-2 \pm \sqrt{4 + 16}}{-2} = 1 \pm \sqrt{5}$. The critical numbers are $1 \pm \sqrt{5}$. [$h'(p)$ exists for all real numbers.]

**31.** $F(x) = x^{4/5}(x-4)^2 \Rightarrow$

$$F'(x) = x^{4/5} \cdot 2(x-4) + (x-4)^2 \cdot \tfrac{4}{5}x^{-1/5} = \tfrac{1}{5}x^{-1/5}(x-4)\left[5 \cdot x \cdot 2 + (x-4) \cdot 4\right]$$
$$= \frac{(x-4)(14x-16)}{5x^{1/5}} = \frac{2(x-4)(7x-8)}{5x^{1/5}}$$

$F'(x) = 0 \Rightarrow x = 4, \frac{8}{7}$. $F'(0)$ does not exist. Thus, the three critical numbers are 0, $\frac{8}{7}$, and 4.

**32.** $G(x) = \sqrt[3]{x^2 - x} \Rightarrow G'(x) = \frac{1}{3}(x^2 - x)^{-2/3}(2x - 1)$. $G'(x)$ does not exist when $x^2 - x = 0$, that is, when $x = 0$ or 1. $G'(x) = 0 \Leftrightarrow 2x - 1 = 0 \Leftrightarrow x = \frac{1}{2}$. So the critical numbers are $x = 0, \frac{1}{2}, 1$.

**33.** $f(\theta) = 2\cos\theta + \sin^2\theta \Rightarrow f'(\theta) = -2\sin\theta + 2\sin\theta\cos\theta$. $f'(\theta) = 0 \Rightarrow 2\sin\theta(\cos\theta - 1) = 0 \Rightarrow \sin\theta = 0$ or $\cos\theta = 1 \Rightarrow \theta = n\pi$ ($n$ an integer) or $\theta = 2n\pi$. The solutions $\theta = n\pi$ include the solutions $\theta = 2n\pi$, so the critical numbers are $\theta = n\pi$.

**34.** $g(\theta) = 4\theta - \tan\theta \Rightarrow g'(\theta) = 4 - \sec^2\theta$. $g'(\theta) = 0 \Rightarrow \sec^2\theta = 4 \Rightarrow \sec\theta = \pm 2 \Rightarrow \cos\theta = \pm\frac{1}{2} \Rightarrow$
$\theta = \frac{\pi}{3} + 2n\pi$, $\frac{5\pi}{3} + 2n\pi$, $\frac{2\pi}{3} + 2n\pi$, and $\frac{4\pi}{3} + 2n\pi$ are critical numbers.
*Note:* The values of $\theta$ that make $g'(\theta)$ undefined are not in the domain of $g$.

**35.** $f(x) = x\ln x \quad \Rightarrow \quad f'(x) = x(1/x) + (\ln x)\cdot 1 = \ln x + 1.\quad f'(x) = 0 \quad \Leftrightarrow \quad \ln x = -1 \quad \Leftrightarrow \quad x = e^{-1} = 1/e.$ Therefore, the only critical number is $x = 1/e$.

**36.** $f(x) = xe^{2x} \quad \Rightarrow \quad f'(x) = x(2e^{2x}) + e^{2x} = e^{2x}(2x+1)$. Since $e^{2x}$ is never 0, we have $f'(x) = 0$ only when $2x + 1 = 0 \quad \Leftrightarrow \quad x = -\frac{1}{2}$. So $-\frac{1}{2}$ is the only critical number.

**37.** $f(x) = 3x^2 - 12x + 5,\ [0,3].\quad f'(x) = 6x - 12 = 0 \quad \Leftrightarrow \quad x = 2$. Applying the Closed Interval Method, we find that $f(0) = 5$, $f(2) = -7$, and $f(3) = -4$. So $f(0) = 5$ is the absolute maximum value and $f(2) = -7$ is the absolute minimum value.

**38.** $f(x) = x^3 - 3x + 1,\ [0,3].\quad f'(x) = 3x^2 - 3 = 0 \quad \Leftrightarrow \quad x = \pm 1$, but $-1$ is not in $[0,3]$. $f(0) = 1$, $f(1) = -1$, and $f(3) = 19$. So $f(3) = 19$ is the absolute maximum value and $f(1) = -1$ is the absolute minimum value.

**39.** $f(x) = 2x^3 - 3x^2 - 12x + 1,\ [-2,3].\quad f'(x) = 6x^2 - 6x - 12 = 6(x^2 - x - 2) = 6(x-2)(x+1) = 0 \quad \Leftrightarrow$ $x = 2, -1.\quad f(-2) = -3$, $f(-1) = 8$, $f(2) = -19$, and $f(3) = -8$. So $f(-1) = 8$ is the absolute maximum value and $f(2) = -19$ is the absolute minimum value.

**40.** $f(x) = x^3 - 6x^2 + 9x + 2,\ [-1,4].\quad f'(x) = 3x^2 - 12x + 9 = 3(x^2 - 4x + 3) = 3(x-1)(x-3) = 0 \quad \Leftrightarrow \quad x = 1, 3.$ $f(-1) = -14$, $f(1) = 6$, $f(3) = 2$, and $f(4) = 6$. So $f(1) = f(4) = 6$ is the absolute maximum value and $f(-1) = -14$ is the absolute minimum value.

**41.** $f(x) = x^4 - 2x^2 + 3,\ [-2,3].\quad f'(x) = 4x^3 - 4x = 4x(x^2-1) = 4x(x+1)(x-1) = 0 \quad \Leftrightarrow \quad x = -1, 0, 1.$ $f(-2) = 11$, $f(-1) = 2$, $f(0) = 3$, $f(1) = 2$, $f(3) = 66$. So $f(3) = 66$ is the absolute maximum value and $f(\pm 1) = 2$ is the absolute minimum value.

**42.** $f(x) = (x^2-1)^3,\ [-1,2].\quad f'(x) = 3(x^2-1)^2(2x) = 6x(x+1)^2(x-1)^2 = 0 \quad \Leftrightarrow \quad x = -1, 0, 1.\quad f(\pm 1) = 0$, $f(0) = -1$, and $f(2) = 27$. So $f(2) = 27$ is the absolute maximum value and $f(0) = -1$ is the absolute minimum value.

**43.** $f(t) = t\sqrt{4-t^2},\ [-1,2].$

$$f'(t) = t\cdot\tfrac{1}{2}\left(4-t^2\right)^{-1/2}(-2t) + \left(4-t^2\right)^{1/2}\cdot 1 = \frac{-t^2}{\sqrt{4-t^2}} + \sqrt{4-t^2} = \frac{-t^2 + \left(4-t^2\right)}{\sqrt{4-t^2}} = \frac{4-2t^2}{\sqrt{4-t^2}}.$$

$f'(t) = 0 \quad \Rightarrow \quad 4 - 2t^2 = 0 \quad \Rightarrow \quad t^2 = 2 \quad \Rightarrow \quad t = \pm\sqrt{2}$, but $t = -\sqrt{2}$ is not in the given interval, $[-1,2]$. $f'(t)$ does not exist if $4 - t^2 = 0 \quad \Rightarrow \quad t = \pm 2$, but $-2$ is not in the given interval. $f(-1) = -\sqrt{3}$, $f\left(\sqrt{2}\right) = 2$, and $f(2) = 0$. So $f\left(\sqrt{2}\right) = 2$ is the absolute maximum value and $f(-1) = -\sqrt{3}$ is the absolute minimum value.

**44.** $f(x) = \dfrac{x}{x^2+4},\ [0,3].\quad f'(x) = \dfrac{(x^2+4)1 - x(2x)}{(x^2+4)^2} = \dfrac{4-x^2}{(x^2+4)^2} = 0 \quad \Leftrightarrow \quad x = \pm 2$, but $-2$ is not in the interval $[0,3]$. $f(0) = 0$, $f(2) = \frac{1}{4} = 0.25$, $f(3) = \frac{3}{13} \approx 0.23$. So $f(2) = \frac{1}{4}$ is the absolute maximum and $f(0) = 0$ is the absolute minimum.

**45.** $f(x) = \sin x + \cos x,\ \left[0, \frac{\pi}{3}\right].\quad f'(x) = \cos x - \sin x = 0 \quad \Leftrightarrow \quad \sin x = \cos x \quad \Rightarrow \quad \dfrac{\sin x}{\cos x} = 1 \quad \Rightarrow \quad \tan x = 1 \quad \Rightarrow$ $x = \frac{\pi}{4}.\quad f(0) = 1$, $f\left(\frac{\pi}{4}\right) = \sqrt{2} \approx 1.41$, $f\left(\frac{\pi}{3}\right) = \frac{\sqrt{3}+1}{2} \approx 1.37$. So $f\left(\frac{\pi}{4}\right) = \sqrt{2}$ is the absolute maximum value and $f(0) = 1$ is the absolute minimum value.

**46.** $f(x) = x - 2\cos x,\ [-\pi, \pi].\quad f'(x) = 1 + 2\sin x = 0 \quad \Leftrightarrow \quad \sin x = -\frac{1}{2} \quad \Leftrightarrow \quad x = -\frac{5\pi}{6}, -\frac{\pi}{6}.$ $f(-\pi) = 2 - \pi \approx -1.14$, $f\left(-\frac{5\pi}{6}\right) = \sqrt{3} - \frac{5\pi}{6} \approx -0.886$, $f\left(-\frac{\pi}{6}\right) = -\frac{\pi}{6} - \sqrt{3} \approx -2.26$, $f(\pi) = \pi + 2 \approx 5.14$. So $f(\pi) = \pi + 2$ is the absolute maximum value and $f\left(-\frac{\pi}{6}\right) = -\frac{\pi}{6} - \sqrt{3}$ is the absolute minimum value.

**47.** $f(x) = xe^{-x^2/8}$, $[-1, 4]$. $f'(x) = x \cdot e^{-x^2/8} \cdot (-\frac{x}{4}) + e^{-x^2/8} \cdot 1 = e^{-x^2/8}(-\frac{x^2}{4} + 1)$. Since $e^{-x^2/8}$ is never 0, $f'(x) = 0 \Rightarrow -x^2/4 + 1 = 0 \Rightarrow 1 = x^2/4 \Rightarrow x^2 = 4 \Rightarrow x = \pm 2$, but $-2$ is not in the given interval, $[-1, 4]$. $f(-1) = -e^{-1/8} \approx -0.88$, $f(2) = 2e^{-1/2} \approx 1.21$, and $f(4) = 4e^{-2} \approx 0.54$. So $f(2) = 2e^{-1/2}$ is the absolute maximum value and $f(-1) = -e^{-1/8}$ is the absolute minimum value.

**48.** $f(x) = x - \ln x$, $[\frac{1}{2}, 2]$. $f'(x) = 1 - \frac{1}{x} = \frac{x-1}{x}$. $f'(x) = 0 \Rightarrow x = 1$. (Note that 0 is not in the domain of $f$.) $f(\frac{1}{2}) = \frac{1}{2} - \ln \frac{1}{2} \approx 1.19$, $f(1) = 1$, and $f(2) = 2 - \ln 2 \approx 1.31$. So $f(2) = 2 - \ln 2$ is the absolute maximum value and $f(1) = 1$ is the absolute minimum value.

**49.** $f(x) = x^a(1-x)^b$, $0 \le x \le 1$, $a > 0$, $b > 0$.

$$\begin{aligned} f'(x) &= x^a \cdot b(1-x)^{b-1}(-1) + (1-x)^b \cdot ax^{a-1} = x^{a-1}(1-x)^{b-1}\left[x \cdot b(-1) + (1-x) \cdot a\right] \\ &= x^{a-1}(1-x)^{b-1}(a - ax - bx) \end{aligned}$$

At the endpoints, we have $f(0) = f(1) = 0$ [the minimum value of $f$]. In the interval $(0, 1)$, $f'(x) = 0 \Leftrightarrow x = \dfrac{a}{a+b}$.

$$f\left(\frac{a}{a+b}\right) = \left(\frac{a}{a+b}\right)^a \left(1 - \frac{a}{a+b}\right)^b = \frac{a^a}{(a+b)^a}\left(\frac{a+b-a}{a+b}\right)^b = \frac{a^a}{(a+b)^a} \cdot \frac{b^b}{(a+b)^b} = \frac{a^a b^b}{(a+b)^{a+b}}.$$

So $f\left(\dfrac{a}{a+b}\right) = \dfrac{a^a b^b}{(a+b)^{a+b}}$ is the absolute maximum value.

**50.**

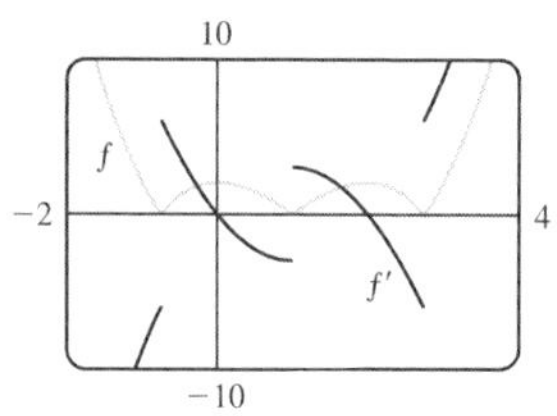


We see that $f'(x) = 0$ at about $x = 0.0$ and $2.0$, and that $f'(x)$ does not exist at about $x = -0.7$, $1.0$, and $2.7$, so the critical numbers of $f$ are about $-0.7$, $0.0$, $1.0$, $2.0$, and $2.7$.

**51.** (a)

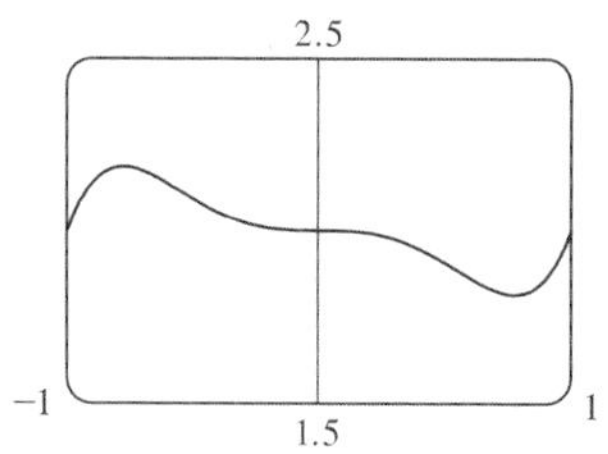


From the graph, it appears that the absolute maximum value is about $f(-0.77) = 2.19$, and the absolute minimum value is about $f(0.77) = 1.81$.

(b) $f(x) = x^5 - x^3 + 2 \Rightarrow f'(x) = 5x^4 - 3x^2 = x^2(5x^2 - 3)$. So $f'(x) = 0 \Rightarrow x = 0, \pm\sqrt{\frac{3}{5}}$.

$$f\left(-\sqrt{\tfrac{3}{5}}\right) = \left(-\sqrt{\tfrac{3}{5}}\right)^5 - \left(-\sqrt{\tfrac{3}{5}}\right)^3 + 2 = -\left(\tfrac{3}{5}\right)^2\sqrt{\tfrac{3}{5}} + \tfrac{3}{5}\sqrt{\tfrac{3}{5}} + 2 = \left(\tfrac{3}{5} - \tfrac{9}{25}\right)\sqrt{\tfrac{3}{5}} + 2 = \tfrac{6}{25}\sqrt{\tfrac{3}{5}} + 2 \text{ (maximum)}$$

and similarly, $f\left(\sqrt{\frac{3}{5}}\right) = -\frac{6}{25}\sqrt{\frac{3}{5}} + 2$ (minimum).

**52.** (a)

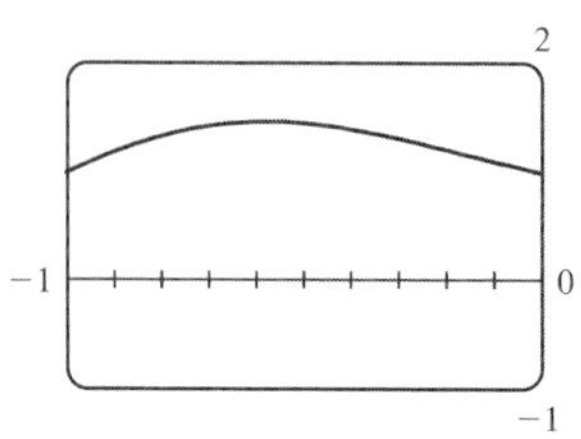


From the graph, it appears that the absolute maximum value is about $f(-0.58) = 1.47$, and the absolute minimum value is about $f(-1) = f(0) = 1.00$; that is, at both endpoints.

(b) $f(x) = e^{x^3 - x} \Rightarrow f'(x) = e^{x^3 - x}(3x^2 - 1)$. So $f'(x) = 0$ on $[-1, 0] \Rightarrow x = -\sqrt{1/3}$.
$f(-1) = f(0) = 1$ (minima) and $f\left(-\sqrt{1/3}\right) = e^{-\sqrt{3}/9 + \sqrt{3}/3} = e^{2\sqrt{3}/9}$ (maximum).

**53.** (a)

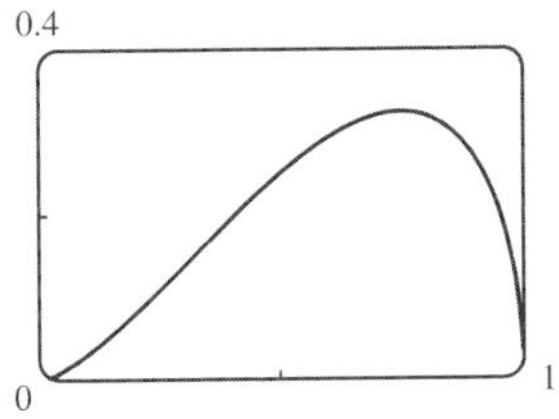


From the graph, it appears that the absolute maximum value is about $f(0.75) = 0.32$, and the absolute minimum value is $f(0) = f(1) = 0$; that is, at both endpoints.

(b) $f(x) = x\sqrt{x - x^2} \quad \Rightarrow \quad f'(x) = x \cdot \dfrac{1 - 2x}{2\sqrt{x - x^2}} + \sqrt{x - x^2} = \dfrac{(x - 2x^2) + (2x - 2x^2)}{2\sqrt{x - x^2}} = \dfrac{3x - 4x^2}{2\sqrt{x - x^2}}$.

So $f'(x) = 0 \quad \Rightarrow \quad 3x - 4x^2 = 0 \quad \Rightarrow \quad x(3 - 4x) = 0 \quad \Rightarrow \quad x = 0$ or $\frac{3}{4}$.

$f(0) = f(1) = 0$ (minimum), and $f\left(\frac{3}{4}\right) = \frac{3}{4}\sqrt{\frac{3}{4} - \left(\frac{3}{4}\right)^2} = \frac{3}{4}\sqrt{\frac{3}{16}} = \frac{3\sqrt{3}}{16}$ (maximum).

**54.** (a)

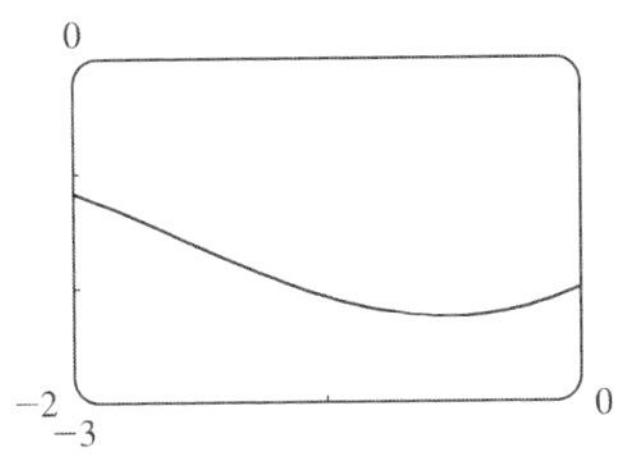


From the graph , it appears that the absolute maximum value is about $f(-2) = -1.17$, and the absolute minimum value is about $f(-0.52) = -2.26$.

(b) $f(x) = x - 2\cos x \quad \Rightarrow \quad f'(x) = 1 + 2\sin x$. So $f'(x) = 0 \quad \Rightarrow \quad \sin x = -\frac{1}{2} \quad \Rightarrow \quad x = -\frac{\pi}{6}$ on $[-2, 0]$.

$f(-2) = -2 - 2\cos(-2)$ (maximum) and $f\left(-\frac{\pi}{6}\right) = -\frac{\pi}{6} - 2\cos\left(-\frac{\pi}{6}\right) = -\frac{\pi}{6} - 2\left(\frac{\sqrt{3}}{2}\right) = -\frac{\pi}{6} - \sqrt{3}$ (minimum).

**55.** The density is defined as $\rho = \dfrac{\text{mass}}{\text{volume}} = \dfrac{1000}{V(T)}$ (in g/cm$^3$). But a critical point of $\rho$ will also be a critical point of $V$ $\left[\text{since } \dfrac{d\rho}{dT} = -1000V^{-2}\dfrac{dV}{dT} \text{ and } V \text{ is never } 0\right]$, and $V$ is easier to differentiate than $\rho$.

$V(T) = 999.87 - 0.06426T + 0.0085043T^2 - 0.0000679T^3 \quad \Rightarrow \quad V'(T) = -0.06426 + 0.0170086T - 0.0002037T^2$.

Setting this equal to 0 and using the quadratic formula to find $T$, we get

$T = \dfrac{-0.0170086 \pm \sqrt{0.0170086^2 - 4 \cdot 0.0002037 \cdot 0.06426}}{2(-0.0002037)} \approx 3.9665°\text{C}$ or $79.5318°\text{C}$. Since we are only interested in the region $0°\text{C} \le T \le 30°\text{C}$, we check the density $\rho$ at the endpoints and at 3.9665°C: $\rho(0) \approx \dfrac{1000}{999.87} \approx 1.00013$;

$\rho(30) \approx \dfrac{1000}{1003.7628} \approx 0.99625$; $\rho(3.9665) \approx \dfrac{1000}{999.7447} \approx 1.000255$. So water has its maximum density at about 3.9665°C.

**56.** $F = \dfrac{\mu W}{\mu\sin\theta + \cos\theta} \quad \Rightarrow \quad \dfrac{dF}{d\theta} = \dfrac{(\mu\sin\theta + \cos\theta)(0) - \mu W(\mu\cos\theta - \sin\theta)}{(\mu\sin\theta + \cos\theta)^2} = \dfrac{-\mu W(\mu\cos\theta - \sin\theta)}{(\mu\sin\theta + \cos\theta)^2}$.

So $\dfrac{dF}{d\theta} = 0 \quad \Rightarrow \quad \mu\cos\theta - \sin\theta = 0 \quad \Rightarrow \quad \mu = \dfrac{\sin\theta}{\cos\theta} = \tan\theta$. Substituting $\tan\theta$ for $\mu$ in $F$ gives us

$$F = \frac{(\tan\theta)W}{(\tan\theta)\sin\theta + \cos\theta} = \frac{W\tan\theta}{\dfrac{\sin^2\theta}{\cos\theta} + \cos\theta} = \frac{W\tan\theta\cos\theta}{\sin^2\theta + \cos^2\theta} = \frac{W\sin\theta}{1} = W\sin\theta.$$

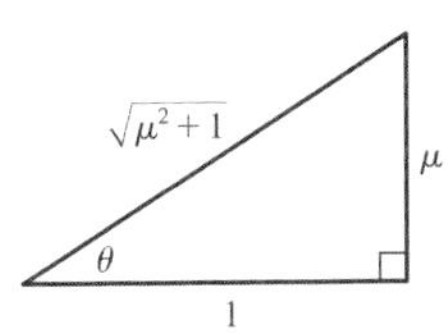


If $\tan\theta = \mu$, then $\sin\theta = \dfrac{\mu}{\sqrt{\mu^2 + 1}}$ (see the figure), so $F = \dfrac{\mu}{\sqrt{\mu^2 + 1}}W$.

We compare this with the value of $F$ at the endpoints: $F(0) = \mu W$ and $F\left(\frac{\pi}{2}\right) = W$.

Now because $\dfrac{\mu}{\sqrt{\mu^2 + 1}} \le 1$ and $\dfrac{\mu}{\sqrt{\mu^2 + 1}} \le \mu$, we have that $\dfrac{\mu}{\sqrt{\mu^2 + 1}}W$ is less than or equal to each of $F(0)$ and $F\left(\frac{\pi}{2}\right)$.

Hence, $\dfrac{\mu}{\sqrt{\mu^2 + 1}}W$ is the absolute minimum value of $F(\theta)$, and it occurs when $\tan\theta = \mu$.

**57.** We apply the Closed Interval Method to the continuous function

$I(t) = 0.00009045t^5 + 0.001438t^4 - 0.06561t^3 + 0.4598t^2 - 0.6270t + 99.33$ on $[0, 10]$. Its derivative is

$I'(t) = 0.00045225t^4 + 0.005752t^3 - 0.19683t^2 + 0.9196t - 0.6270$. Since $I'$ exists for all $t$, the only critical numbers of $I$ occur when $I'(t) = 0$. We use a root-finder on a computer algebra system (or a graphing device) to find that $I'(t) = 0$ when $t \approx -29.7186, 0.8231, 5.1309$, or $11.0459$, but only the second and third roots lie in the interval $[0, 10]$. The values of $I$ at these critical numbers are $I(0.8231) \approx 99.09$ and $I(5.1309) \approx 100.67$. The values of $I$ at the endpoints of the interval are $I(0) = 99.33$ and $I(10) \approx 96.86$. Comparing these four numbers, we see that food was most expensive at $t \approx 5.1309$ (corresponding roughly to August, 1989) and cheapest at $t = 10$ (midyear 1994).

**58.** (a) The equation of the graph in the figure is

$v(t) = 0.00146t^3 - 0.11553t^2 + 24.98169t - 21.26872.$

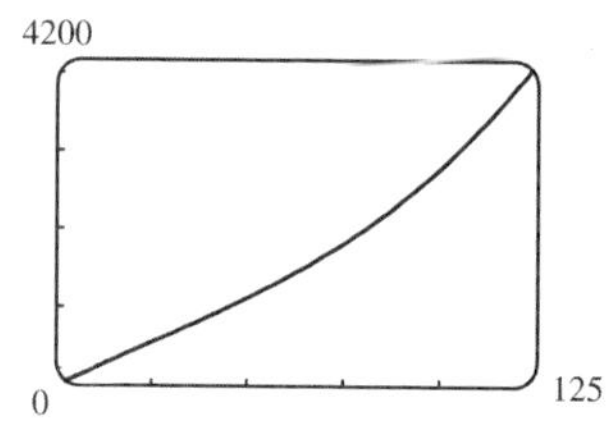


(b) $a(t) = v'(t) = 0.00438t^2 - 0.23106t + 24.98169 \Rightarrow$

$a'(t) = 0.00876t - 0.23106$. $a'(t) = 0 \Rightarrow t_1 = \frac{0.23106}{0.00876} \approx 26.4$.

$a(0) \approx 24.98$, $a(t_1) \approx 21.93$, and $a(125) \approx 64.54$.

The maximum acceleration is about $64.5 \text{ ft/s}^2$ and the minimum acceleration is about $21.93 \text{ ft/s}^2$.

**59.** (a) $v(r) = k(r_0 - r)r^2 = kr_0r^2 - kr^3 \Rightarrow v'(r) = 2kr_0r - 3kr^2$. $v'(r) = 0 \Rightarrow kr(2r_0 - 3r) = 0 \Rightarrow$ $r = 0$ or $\frac{2}{3}r_0$ (but 0 is not in the interval). Evaluating $v$ at $\frac{1}{2}r_0$, $\frac{2}{3}r_0$, and $r_0$, we get $v\left(\frac{1}{2}r_0\right) = \frac{1}{8}kr_0^3$, $v\left(\frac{2}{3}r_0\right) = \frac{4}{27}kr_0^3$, and $v(r_0) = 0$. Since $\frac{4}{27} > \frac{1}{8}$, $v$ attains its maximum value at $r = \frac{2}{3}r_0$. This supports the statement in the text.

(b) From part (a), the maximum value of $v$ is $\frac{4}{27}kr_0^3$.

(c)

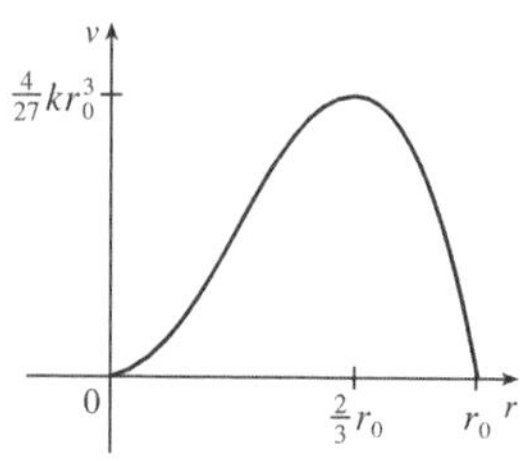


**60.** (a) $f(x) = ax^3 + bx^2 + cx + d$, $a \neq 0$. So $f'(x) = 3ax^2 + 2bx + c$ is a quadratic and hence has either 2, 1, or 0 real roots, so $f(x)$ has either 2, 1 or 0 critical numbers.

**Case (i)** [2 critical numbers]:

$f(x) = x^3 - 3x \Rightarrow$

$f'(x) = 3x^2 - 3$, so $x = -1, 1$ are critical numbers.

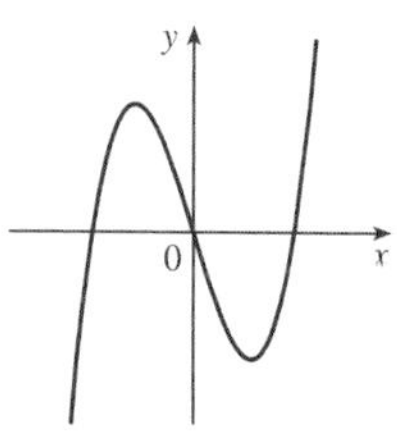


**Case (ii)** [1 critical number]:

$f(x) = x^3 \Rightarrow$

$f'(x) = 3x^2$, so $x = 0$ is the only critical number.

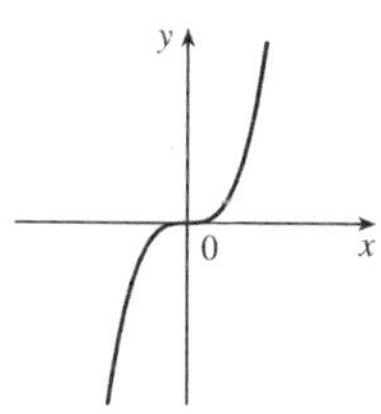


**Case (iii)** [no critical number]:

$f(x) = x^3 + 3x \Rightarrow$

$f'(x) = 3x^2 + 3$, so there is no critical number.

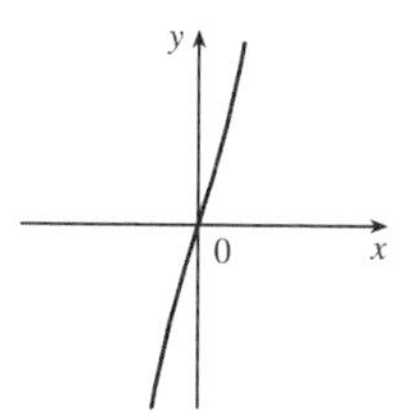


(b) Since there are at most two critical numbers, it can have at most two local extreme values and by (i) this can occur. By (iii) it can have no local extreme value. However, if there is only one critical number, then there is no local extreme value.

## APPLIED PROJECT The Calculus of Rainbows

**1.** From Snell's Law, we have $\sin\alpha = k\sin\beta \approx \frac{4}{3}\sin\beta \quad\Leftrightarrow\quad \beta \approx \arcsin\left(\frac{3}{4}\sin\alpha\right)$. We substitute this into $D(\alpha) = \pi + 2\alpha - 4\beta = \pi + 2\alpha - 4\arcsin\left(\frac{3}{4}\sin\alpha\right)$, and then differentiate to find the minimum:
$D'(\alpha) = 2 - 4\left[1 - \left(\frac{3}{4}\sin\alpha\right)^2\right]^{-1/2}\left(\frac{3}{4}\cos\alpha\right) = 2 - \dfrac{3\cos\alpha}{\sqrt{1 - \frac{9}{16}\sin^2\alpha}}$. This is 0 when $\dfrac{3\cos\alpha}{\sqrt{1 - \frac{9}{16}\sin^2\alpha}} = 2 \quad\Leftrightarrow$

$\frac{9}{4}\cos^2\alpha = 1 - \frac{9}{16}\sin^2\alpha \quad\Leftrightarrow\quad \frac{9}{4}\cos^2\alpha = 1 - \frac{9}{16}(1 - \cos^2\alpha) \quad\Leftrightarrow\quad \frac{27}{16}\cos^2\alpha = \frac{7}{16} \quad\Leftrightarrow\quad \cos\alpha = \sqrt{\frac{7}{27}} \quad\Leftrightarrow$

$\alpha = \arccos\sqrt{\frac{7}{27}} \approx 59.4^\circ$, and so the local minimum is $D(59.4^\circ) \approx 2.4$ radians $\approx 138^\circ$.
To see that this is an absolute minimum, we check the endpoints, which in this case are $\alpha = 0$ and $\alpha = \frac{\pi}{2}$:
$D(0) = \pi$ radians $= 180^\circ$, and $D\left(\frac{\pi}{2}\right) \approx 166^\circ$.
*Another method:* We first calculate $\dfrac{d\beta}{d\alpha}$: $\sin\alpha = \frac{4}{3}\sin\beta \quad\Leftrightarrow\quad \cos\alpha = \frac{4}{3}\cos\beta\,\dfrac{d\beta}{d\alpha} \quad\Leftrightarrow\quad \dfrac{d\beta}{d\alpha} = \dfrac{3\cos\alpha}{4\cos\beta}$, so since
$D'(\alpha) = 2 - 4\dfrac{d\beta}{d\alpha} = 0 \quad\Leftrightarrow\quad \dfrac{d\beta}{d\alpha} = \dfrac{1}{2}$, the minimum occurs when $3\cos\alpha = 2\cos\beta$. Now we square both sides and substitute $\sin\alpha = \frac{4}{3}\sin\beta$, leading to the same result.

**2.** If we repeat Problem 1 with $k$ in place of $\frac{4}{3}$, we get $D(\alpha) = \pi + 2\alpha - 4\arcsin\left(\dfrac{1}{k}\sin\alpha\right) \quad\Rightarrow$

$D'(\alpha) = 2 - \dfrac{4\cos\alpha}{k\sqrt{1 - [(\sin\alpha)/k]^2}}$, which is 0 when $\dfrac{2\cos\alpha}{k} = \sqrt{1 - \left(\dfrac{\sin\alpha}{k}\right)^2} \quad\Leftrightarrow\quad \left(\dfrac{2\cos\alpha}{k}\right)^2 = 1 - \left(\dfrac{\sin\alpha}{k}\right)^2$

$\Leftrightarrow\quad 4\cos^2\alpha = k^2 - \sin^2\alpha \quad\Leftrightarrow\quad 3\cos^2\alpha = k^2 - 1 \quad\Leftrightarrow\quad \alpha = \arccos\sqrt{\dfrac{k^2-1}{3}}$. So for $k \approx 1.3318$ (red light) the minimum occurs at $\alpha_1 \approx 1.038$ radians, and so the rainbow angle is about $\pi - D(\alpha_1) \approx 42.3^\circ$. For $k \approx 1.3435$ (violet light) the minimum occurs at $\alpha_2 \approx 1.026$ radians, and so the rainbow angle is about $\pi - D(\alpha_2) \approx 40.6^\circ$.
*Another method:* As in Problem 1, we can instead find $D'(\alpha)$ in terms of $\dfrac{d\beta}{d\alpha}$, and then substitute $\dfrac{d\beta}{d\alpha} = \dfrac{\cos\alpha}{k\cos\beta}$.

**3.** At each reflection or refraction, the light is bent in a counterclockwise direction: the bend at $A$ is $\alpha - \beta$, the bend at $B$ is $\pi - 2\beta$, the bend at $C$ is again $\pi - 2\beta$, and the bend at $D$ is $\alpha - \beta$. So the total bend is
$D(\alpha) = 2(\alpha - \beta) + 2(\pi - 2\beta) = 2\alpha - 6\beta + 2\pi$, as required. We substitute $\beta = \arcsin\left(\dfrac{\sin\alpha}{k}\right)$ and differentiate, to get

$D'(\alpha) = 2 - \dfrac{6\cos\alpha}{k\sqrt{1 - [(\sin\alpha)/k]^2}}$, which is 0 when $\dfrac{3\cos\alpha}{k} = \sqrt{1 - \left(\dfrac{\sin\alpha}{k}\right)^2} \quad\Leftrightarrow\quad 9\cos^2\alpha = k^2 - \sin^2\alpha \quad\Leftrightarrow$

$8\cos^2\alpha = k^2 - 1 \quad\Leftrightarrow\quad \cos\alpha = \sqrt{\frac{1}{8}(k^2-1)}$. If $k = \frac{4}{3}$, then the minimum occurs at

$\alpha_1 = \arccos\sqrt{\dfrac{(4/3)^2 - 1}{8}} \approx 1.254$ radians. Thus, the minimum *counterclockwise* rotation is $D(\alpha_1) \approx 231^\circ$, which is equivalent to a *clockwise* rotation of $360^\circ - 231^\circ = 129^\circ$ (see the figure). So the rainbow angle for the secondary rainbow is about $180^\circ - 129^\circ = 51^\circ$, as required.
In general, the rainbow angle for the secondary rainbow is $\pi - [2\pi - D(\alpha)] = D(\alpha) - \pi$.

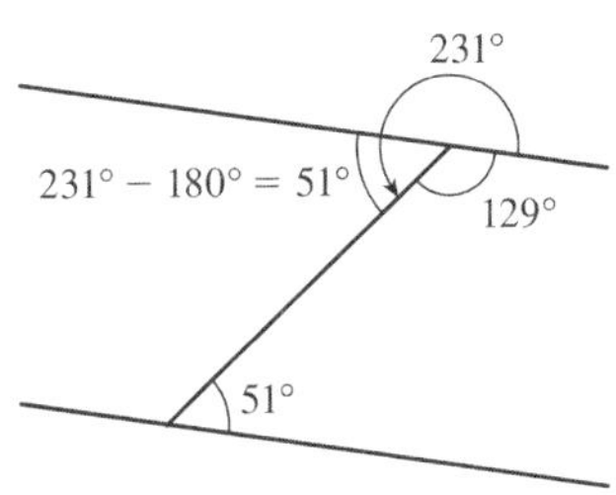

**4.** In the primary rainbow, the rainbow angle gets smaller as $k$ gets larger, as we found in Problem 2, so the colors appear from top to bottom in order of increasing $k$. But in the secondary rainbow, the rainbow angle gets larger as $k$ gets larger. To see this, we find the minimum deviations for red light and for violet light in the secondary rainbow. For $k \approx 1.3318$ (red light) the minimum occurs at $\alpha_1 \approx \arccos\sqrt{\dfrac{1.3318^2 - 1}{8}} \approx 1.255$ radians, and so the rainbow angle is $D(\alpha_1) - \pi \approx 50.6°$. For $k \approx 1.3435$ (violet light) the minimum occurs at $\alpha_2 \approx \arccos\sqrt{\dfrac{1.3435^2 - 1}{8}} \approx 1.248$ radians, and so the rainbow angle is $D(\alpha_2) - \pi \approx 53.6°$. Consequently, the rainbow angle is larger for colors with higher indices of refraction, and the colors appear from bottom to top in order of increasing $k$, the reverse of their order in the primary rainbow.

Note that our calculations above also explain why the secondary rainbow is more spread out than the primary rainbow: in the primary rainbow, the difference between rainbow angles for red and violet light is about $1.7°$, whereas in the secondary rainbow it is about $3°$.

## 4.3 Derivatives and the Shapes of Curves

**1.** $\dfrac{f(8) - f(0)}{8 - 0} = \dfrac{6 - 4}{8} = \dfrac{1}{4}$. The values of $c$ which satisfy $f'(c) = \frac{1}{4}$ seem to be about $c = 0.8, 3.2, 4.4$, and $6.1$.

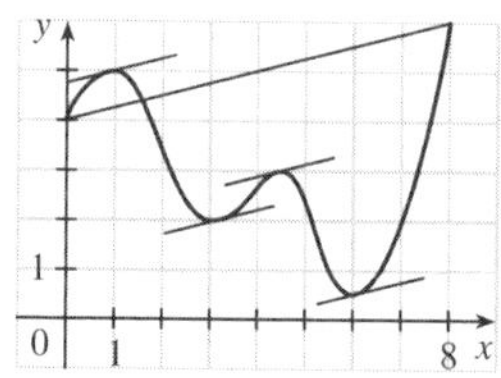


**2.** (a) $g$ is concave upward on $(-1, 2)$ and $(7, 8)$.

(b) $g$ is concave downward on $(2, 4)$ and $(4, 7)$.

(c) The only point of inflection is $(2, 2)$. Note that 7 is not in the domain of this function.

**3.** (a) Use the Increasing/Decreasing (I/D) Test.

(b) Use the Concavity Test.

(c) At any value of $x$ where the concavity changes, we have an inflection point at $(x, f(x))$.

**4.** (a) See the First Derivative Test.

(b) See the Second Derivative Test and the note that precedes Example 5.

**5.** There is an inflection point at $x = 1$ because $f''(x)$ changes from negative to positive there, and so the graph of $f$ changes from concave downward to concave upward. There is an inflection point at $x = 7$ because $f''(x)$ changes from positive to negative there, and so the graph of $f$ changes from concave upward to concave downward.

**6.** (a) $f$ is increasing on the intervals where $f'(x) > 0$, namely, $(2, 4)$ and $(6, 9)$.

(b) $f$ has a local maximum where it changes from increasing to decreasing, that is, where $f'$ changes from positive to negative (at $x = 4$). Similarly, where $f'$ changes from negative to positive, $f$ has a local minimum (at $x = 2$ and at $x = 6$).

(c) When $f'$ is increasing, its derivative $f''$ is positive and hence, $f$ is concave upward. This happens on $(1, 3)$, $(5, 7)$, and $(8, 9)$. Similarly, $f$ is concave downward when $f'$ is decreasing—that is, on $(0, 1)$, $(3, 5)$, and $(7, 8)$.

(d) $f$ has inflection points at $x = 1, 3, 5, 7$, and $8$, since the direction of concavity changes at each of these values.

**7.** (a) $f(x) = x^3 - 12x + 1 \Rightarrow f'(x) = 3x^2 - 12 = 3(x+2)(x-2)$.
We don't need to include "3" in the chart to determine the sign of $f'(x)$.

| Interval | $x+2$ | $x-2$ | $f'(x)$ | $f$ |
|---|---|---|---|---|
| $x < -2$ | $-$ | $-$ | $+$ | increasing on $(-\infty, -2)$ |
| $-2 < x < 2$ | $+$ | $-$ | $-$ | decreasing on $(-2, 2)$ |
| $x > 2$ | $+$ | $+$ | $+$ | increasing on $(2, \infty)$ |

So $f$ is increasing on $(-\infty, -2)$ and $(2, \infty)$ and $f$ is decreasing on $(-2, 2)$.

(b) $f$ changes from increasing to decreasing at $x = -2$ and from decreasing to increasing at $x = 2$. Thus, $f(-2) = 17$ is a local maximum value and $f(2) = -15$ is a local minimum value.

(c) $f''(x) = 6x$. $f''(x) > 0 \Leftrightarrow x > 0$ and $f''(x) < 0 \Leftrightarrow x < 0$. Thus, $f$ is concave upward on $(0, \infty)$ and concave downward on $(-\infty, 0)$. There is an inflection point where the concavity changes, at $(0, f(0)) = (0, 1)$.

**8.** (a) $f(x) = x^4 - 4x - 1 \Rightarrow f'(x) = 4x^3 - 4 = 4(x^3 - 1) = 4(x-1)(x^2 + x + 1)$. So $f'(x) > 0 \Leftrightarrow$ $x - 1 > 0$ $[4(x^2 + x + 1) > 0] \Leftrightarrow x > 1$. Thus, $f$ is increasing on $(1, \infty)$ and decreasing on $(-\infty, 1)$.

(b) $f$ changes from decreasing to increasing at its only critical number, $x = 1$. Thus, $f(1) = -4$ is a local minimum value.

(c) $f'(x) = 4x^3 - 4 \Rightarrow f''(x) = 12x^2$. $f''(x) > 0$ for all $x$ except $x = 0$. Thus, $f$ is concave upward on $(-\infty, 0)$ and $(0, \infty)$. Moreover, since $f'$ is increasing on $(-\infty, \infty)$, $f$ is concave upward on $(-\infty, \infty)$. There are no inflection points.

**9.** (a) $f(x) = x - 2\sin x$ on $(0, 3\pi) \Rightarrow f'(x) = 1 - 2\cos x$. $f'(x) > 0 \Leftrightarrow 1 - 2\cos x > 0 \Leftrightarrow \cos x < \frac{1}{2} \Leftrightarrow$ $\frac{\pi}{3} < x < \frac{5\pi}{3}$ or $\frac{7\pi}{3} < x < 3\pi$. $f'(x) < 0 \Leftrightarrow \cos x > \frac{1}{2} \Leftrightarrow 0 < x < \frac{\pi}{3}$ or $\frac{5\pi}{3} < x < \frac{7\pi}{3}$. So $f$ is increasing on $\left(\frac{\pi}{3}, \frac{5\pi}{3}\right)$ and $\left(\frac{7\pi}{3}, 3\pi\right)$, and $f$ is decreasing on $\left(0, \frac{\pi}{3}\right)$ and $\left(\frac{5\pi}{3}, \frac{7\pi}{3}\right)$.

(b) $f$ changes from increasing to decreasing at $x = \frac{5\pi}{3}$, and from decreasing to increasing at $x = \frac{\pi}{3}$ and at $x = \frac{7\pi}{3}$. Thus, $f\left(\frac{5\pi}{3}\right) = \frac{5\pi}{3} + \sqrt{3} \approx 6.97$ is a local maximum value and $f\left(\frac{\pi}{3}\right) = \frac{\pi}{3} - \sqrt{3} \approx -0.68$ and $f\left(\frac{7\pi}{3}\right) = \frac{7\pi}{3} - \sqrt{3} \approx 5.60$ are local minimum values.

(c) $f''(x) = 2\sin x > 0 \Leftrightarrow 0 < x < \pi$ and $2\pi < x < 3\pi$, $f''(x) < 0 \Leftrightarrow \pi < x < 2\pi$. Thus, $f$ is concave upward on $(0, \pi)$ and $(2\pi, 3\pi)$, and $f$ is concave downward on $(\pi, 2\pi)$. There are inflection points at $(\pi, \pi)$ and $(2\pi, 2\pi)$.

**10.** (a) $f(x) = \dfrac{x^2}{x^2 + 3} \Rightarrow f'(x) = \dfrac{(x^2+3)(2x) - x^2(2x)}{(x^2+3)^2} = \dfrac{6x}{(x^2+3)^2}$. The denominator is positive so the sign of $f'(x)$ is determined by the sign of $x$. Thus, $f'(x) > 0 \Leftrightarrow x > 0$ and $f'(x) < 0 \Leftrightarrow x < 0$. So $f$ is increasing on $(0, \infty)$ and $f$ is decreasing on $(-\infty, 0)$.

(b) $f$ changes from decreasing to increasing at $x = 0$. Thus, $f(0) = 0$ is a local minimum value.

(c) $$f''(x) = \frac{(x^2+3)^2(6) - 6x \cdot 2(x^2+3)(2x)}{[(x^2+3)^2]^2} = \frac{6(x^2+3)\left[x^2 + 3 - 4x^2\right]}{(x^2+3)^4} = \frac{6(3-3x^2)}{(x^2+3)^3} = \frac{-18(x+1)(x-1)}{(x^2+3)^3}.$$

$f''(x) > 0 \Leftrightarrow -1 < x < 1$ and $f''(x) < 0 \Leftrightarrow x < -1$ or $x > 1$. Thus, $f$ is concave upward on $(-1, 1)$ and concave downward on $(-\infty, -1)$ and $(1, \infty)$. There are inflection points at $\left(\pm 1, \frac{1}{4}\right)$.

**11.** (a) $y = f(x) = xe^x \Rightarrow f'(x) = xe^x + e^x = e^x(x+1)$. So $f'(x) > 0 \Leftrightarrow x + 1 > 0 \Leftrightarrow x > -1$. Thus, $f$ is increasing on $(-1, \infty)$ and decreasing on $(-\infty, -1)$.

(b) $f$ changes from decreasing to increasing at its only critical number, $x = -1$. Thus, $f(-1) = -e^{-1}$ is a local minimum value.

(c) $f'(x) = e^x(x+1) \Rightarrow f''(x) = e^x(1) + (x+1)e^x = e^x(x+2)$. So $f''(x) > 0 \Leftrightarrow x + 2 > 0 \Leftrightarrow x > -2$. Thus, $f$ is concave upward on $(-2, \infty)$ and concave downward on $(-\infty, -2)$. Since the concavity changes direction at $x = -2$, the point $\left(-2, -2e^{-2}\right)$ is an inflection point.

**12.** (a) $y = f(x) = x^2e^x \Rightarrow f'(x) = x^2e^x + 2xe^x = x(x+2)e^x$. So $f'(x) > 0 \Leftrightarrow x(x+2) > 0 \Leftrightarrow$ either $x < -2$ or $x > 0$. Therefore $f$ is increasing on $(-\infty, -2)$ and $(0, \infty)$, and decreasing on $(-2, 0)$.

(b) $f$ changes from increasing to decreasing at $x = -2$, so $f(-2) = 4e^{-2}$ is a local maximum value. $f$ changes from decreasing to increasing at $x = 0$, so $f(0) = 0$ is a local minimum value.

(c) $f'(x) = (x^2 + 2x)e^x \Rightarrow f''(x) = (x^2 + 2x)e^x + e^x(2x + 2) = e^x(x^2 + 4x + 2)$. $f''(x) = 0 \Leftrightarrow$ $x^2 + 4x + 2 = 0 \Leftrightarrow x = -2 \pm \sqrt{2}$. $f''(x) < 0 \Leftrightarrow -2 - \sqrt{2} < x < -2 + \sqrt{2}$, so $f$ is concave downward on $(-2-\sqrt{2}, -2+\sqrt{2})$ and concave upward on $(-\infty, -2-\sqrt{2})$ and $(-2+\sqrt{2}, \infty)$. There are inflection points at $(-2-\sqrt{2}, f(-2-\sqrt{2})) \approx (-3.41, 0.38)$ and $(-2+\sqrt{2}, f(-2+\sqrt{2})) \approx (-0.59, 0.19)$.

**13.** (a) $y = f(x) = \dfrac{\ln x}{\sqrt{x}}$. (Note that $f$ is only defined for $x > 0$.)

$$f'(x) = \frac{\sqrt{x}\,(1/x) - \ln x\left(\frac{1}{2}x^{-1/2}\right)}{x} = \frac{\dfrac{1}{\sqrt{x}} - \dfrac{\ln x}{2\sqrt{x}}}{x} \cdot \frac{2\sqrt{x}}{2\sqrt{x}} = \frac{2 - \ln x}{2x^{3/2}} > 0 \Leftrightarrow 2 - \ln x > 0 \Leftrightarrow$$

$\ln x < 2 \Leftrightarrow x < e^2$. Therefore $f$ is increasing on $(0, e^2)$ and decreasing on $(e^2, \infty)$.

(b) $f$ changes from increasing to decreasing at $x = e^2$, so $f(e^2) = \dfrac{\ln e^2}{\sqrt{e^2}} = \dfrac{2}{e}$ is a local maximum value.

(c) 
$$f''(x) = \frac{2x^{3/2}(-1/x) - (2 - \ln x)(3x^{1/2})}{(2x^{3/2})^2} = \frac{-2x^{1/2} + 3x^{1/2}(\ln x - 2)}{4x^3}$$
$$= \frac{x^{1/2}(-2 + 3\ln x - 6)}{4x^3} = \frac{3\ln x - 8}{4x^{5/2}}$$

$f''(x) = 0 \Leftrightarrow \ln x = \frac{8}{3} \Leftrightarrow x = e^{8/3}$. $f''(x) > 0 \Leftrightarrow x > e^{8/3}$, so $f$ is concave upward on $(e^{8/3}, \infty)$ and concave downward on $(0, e^{8/3})$. There is an inflection point at $\left(e^{8/3}, \frac{8}{3}e^{-4/3}\right) \approx (14.39, 0.70)$.

**14.** (a) $y = f(x) = x\ln x$. (Note that $f$ is only defined for $x > 0$.)

$f'(x) = x(1/x) + \ln x = 1 + \ln x$. $f'(x) > 0 \Leftrightarrow \ln x + 1 > 0 \Leftrightarrow \ln x > -1 \Leftrightarrow x > e^{-1}$. Therefore $f$ is increasing on $(1/e, \infty)$ and decreasing on $(0, 1/e)$.

(b) $f$ changes from decreasing to increasing at $x = 1/e$, so $f(1/e) = -1/e$ is a local minimum value.

(c) $f''(x) = 1/x > 0$ for $x > 0$. So $f$ is concave upward on its entire domain, and has no inflection point.

**15.** $f(x) = x + \sqrt{1-x} \Rightarrow f'(x) = 1 + \frac{1}{2}(1-x)^{-1/2}(-1) = 1 - \dfrac{1}{2\sqrt{1-x}}$. Note that $f$ is defined for $1 - x \geq 0$; that is, for $x \leq 1$. $f'(x) = 0 \Rightarrow 2\sqrt{1-x} = 1 \Rightarrow \sqrt{1-x} = \frac{1}{2} \Rightarrow 1 - x = \frac{1}{4} \Rightarrow x = \frac{3}{4}$. $f'$ does not exist at $x = 1$, but we can't have a local maximum or minimum at an endpoint.

*First Derivative Test:* $f'(x) > 0 \Rightarrow x < \frac{3}{4}$ and $f'(x) < 0 \Rightarrow \frac{3}{4} < x < 1$. Since $f'$ changes from positive to negative at $x = \frac{3}{4}$, $f\left(\frac{3}{4}\right) = \frac{5}{4}$ is a local maximum value.

*Second Derivative Test:* $f''(x) = -\frac{1}{2}\left(-\frac{1}{2}\right)(1-x)^{-3/2}(-1) = -\dfrac{1}{4\left(\sqrt{1-x}\right)^3}$. $f''\left(\frac{3}{4}\right) = -2 < 0 \Rightarrow$ $f\left(\frac{3}{4}\right) = \frac{5}{4}$ is a local maximum value.

*Preference:* The First Derivative Test may be slightly easier to apply in this case.

**16.** $f(x) = \dfrac{x}{x^2+4} \quad \Rightarrow \quad f'(x) = \dfrac{(x^2+4)\cdot 1 - x(2x)}{(x^2+4)^2} = \dfrac{4-x^2}{(x^2+4)^2} = \dfrac{(2+x)(2-x)}{(x^2+4)^2}.$

*First Derivative Test:* $f'(x) > 0 \quad \Rightarrow \quad -2 < x < 2$ and $f'(x) < 0 \quad \Rightarrow \quad x > 2$ or $x < -2$. Since $f'$ changes from positive to negative at $x = 2$, $f(2) = \frac{1}{4}$ is a local maximum value; and since $f'$ changes from negative to positive at $x = -2$, $f(-2) = -\frac{1}{4}$ is a local minimum value.

*Second Derivative Test:*

$$\begin{aligned} f''(x) &= \frac{(x^2+4)^2(-2x) - (4-x^2)\cdot 2(x^2+4)(2x)}{[(x^2+4)^2]^2} \\ &= \frac{-2x(x^2+4)\left[(x^2+4) + 2(4-x^2)\right]}{(x^2+4)^4} = \frac{-2x(12-x^2)}{(x^2+4)^3} \end{aligned}$$

$f'(x) = 0 \quad \Leftrightarrow \quad x = \pm 2$. $f''(-2) = \frac{1}{16} > 0 \quad \Rightarrow \quad f(-2) = -\frac{1}{4}$ is a local minimum value.

$f''(2) = -\frac{1}{16} < 0 \quad \Rightarrow \quad f(2) = \frac{1}{4}$ is a local maximum value.

*Preference:* Since calculating the second derivative is fairly difficult, the First Derivative Test is easier to use for this function.

**17.** (a) By the Second Derivative Test, if $f'(2) = 0$ and $f''(2) = -5 < 0$, $f$ has a local maximum at $x = 2$.

(b) If $f'(6) = 0$, we know that $f$ has a horizontal tangent at $x = 6$. Knowing that $f''(6) = 0$ does not provide any additional information since the Second Derivative Test fails. For example, the first and second derivatives of $y = (x-6)^4$, $y = -(x-6)^4$, and $y = (x-6)^3$ all equal zero for $x = 6$, but the first has a local minimum at $x = 6$, the second has a local maximum at $x = 6$, and the third has an inflection point at $x = 6$.

**18.** (a) $f(x) = x^4(x-1)^3 \quad \Rightarrow \quad f'(x) = x^4\cdot 3(x-1)^2 + (x-1)^3\cdot 4x^3 = x^3(x-1)^2\,[3x + 4(x-1)] = x^3(x-1)^2(7x-4)$

The critical numbers are 0, 1, and $\frac{4}{7}$.

(b) $$\begin{aligned} f''(x) &= 3x^2(x-1)^2(7x-4) + x^3\cdot 2(x-1)(7x-4) + x^3(x-1)^2\cdot 7 \\ &= x^2(x-1)\,[3(x-1)(7x-4) + 2x(7x-4) + 7x(x-1)] \end{aligned}$$

Now $f''(0) = f''(1) = 0$, so the Second Derivative Test gives no information for $x = 0$ or $x = 1$.

$f''\left(\frac{4}{7}\right) = \left(\frac{4}{7}\right)^2\left(\frac{4}{7}-1\right)\left[0 + 0 + 7\left(\frac{4}{7}\right)\left(\frac{4}{7}-1\right)\right] = \left(\frac{4}{7}\right)^2\left(-\frac{3}{7}\right)(4)\left(-\frac{3}{7}\right) > 0$, so there is a local minimum at $x = \frac{4}{7}$.

(c) $f'$ is positive on $(-\infty, 0)$, negative on $\left(0, \frac{4}{7}\right)$, positive on $\left(\frac{4}{7}, 1\right)$, and positive on $(1, \infty)$. So $f$ has a local maximum at $x = 0$, a local minimum at $x = \frac{4}{7}$, and no local maximum or minimum at $x = 1$.

**19.** (a) $f(x) = 2x^3 - 3x^2 - 12x \quad \Rightarrow \quad f'(x) = 6x^2 - 6x - 12 = 6(x^2 - x - 2) = 6(x-2)(x+1)$.

$f'(x) > 0 \quad \Leftrightarrow \quad x < -1$ or $x > 2$ and $f'(x) < 0 \quad \Leftrightarrow \quad -1 < x < 2$. So $f$ is increasing on $(-\infty, -1)$ and $(2, \infty)$, and $f$ is decreasing on $(-1, 2)$.

(b) Since $f$ changes from increasing to decreasing at $x = -1$, $f(-1) = 7$ is a local maximum value. Since $f$ changes from decreasing to increasing at $x = 2$, $f(2) = -20$ is a local minimum value.

(c) $f''(x) = 6(2x-1) \quad \Rightarrow \quad f''(x) > 0$ on $\left(\frac{1}{2}, \infty\right)$ and $f''(x) < 0$ on $\left(-\infty, \frac{1}{2}\right)$. So $f$ is concave upward on $\left(\frac{1}{2}, \infty\right)$ and concave downward on $\left(-\infty, \frac{1}{2}\right)$. There is a change in concavity at $x = \frac{1}{2}$, and we have an inflection point at $\left(\frac{1}{2}, -\frac{13}{2}\right)$.

(d)

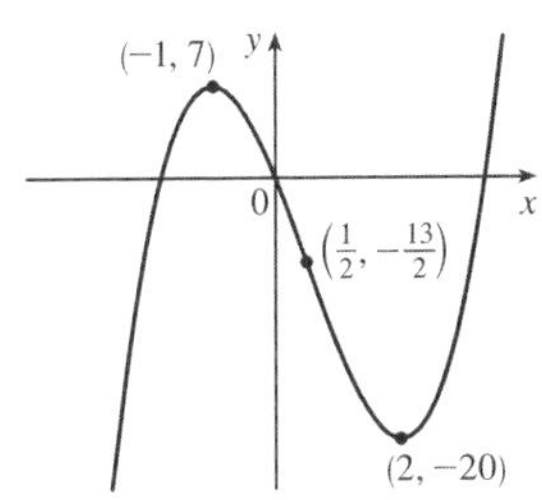

**20.** (a) $g(x) = 200 + 8x^3 + x^4 \quad \Rightarrow \quad g'(x) = 24x^2 + 4x^3 = 4x^2(6 + x) = 0$ when $x = -6$ and when $x = 0$. $g'(x) > 0 \quad \Leftrightarrow$ $x > -6 \ (x \neq 0)$ and $g'(x) < 0 \quad \Leftrightarrow \quad x < -6$, so $g$ is decreasing on $(-\infty, -6)$ and $g$ is increasing on $(-6, \infty)$, with a horizontal tangent at $x = 0$.

(b) $g(-6) = -232$ is a local minimum value.

There is no local maximum value.

(c) $g''(x) = 48x + 12x^2 = 12x(4 + x) = 0$ when $x = -4$ and when $x = 0$.

$g''(x) > 0 \quad \Leftrightarrow \quad x < -4$ or $x > 0$ and $g''(x) < 0 \quad \Leftrightarrow \quad -4 < x < 0$, so $g$ is CU on $(-\infty, -4)$ and $(0, \infty)$, and $g$ is CD on $(-4, 0)$. Inflection points at $(-4, -56)$ and $(0, 200)$

(d)

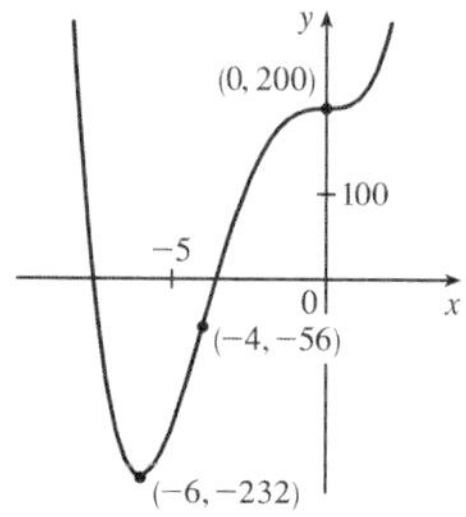


**21.** (a) $h(x) = 3x^5 - 5x^3 + 3 \quad \Rightarrow \quad h'(x) = 15x^4 - 15x^2 = 15x^2(x^2 - 1) = 0$ when $x = 0, \pm 1$. Since $15x^2$ is nonnegative, $h'(x) > 0 \quad \Leftrightarrow \quad x^2 > 1 \quad \Leftrightarrow \quad |x| > 1 \quad \Leftrightarrow \quad x > 1$ or $x < -1$, so $h$ is increasing on $(-\infty, -1)$ and $(1, \infty)$ and decreasing on $(-1, 1)$, with a horizontal tangent at $x = 0$.

(b) Local maximum value $h(-1) = 5$, local minimum value $h(1) = 1$

(c) $h''(x) = 60x^3 - 30x = 30x(2x^2 - 1)$

$$= 60x\left(x + \tfrac{1}{\sqrt{2}}\right)\left(x - \tfrac{1}{\sqrt{2}}\right) \quad \Rightarrow$$

$h''(x) > 0$ when $x > \frac{1}{\sqrt{2}}$ or $-\frac{1}{\sqrt{2}} < x < 0$, so $h$ is CU on $\left(-\frac{1}{\sqrt{2}}, 0\right)$ and $\left(\frac{1}{\sqrt{2}}, \infty\right)$ and CD on $\left(-\infty, -\frac{1}{\sqrt{2}}\right)$ and $\left(0, \frac{1}{\sqrt{2}}\right)$. Inflection points at $(0, 3)$ and $\left(\pm\frac{1}{\sqrt{2}}, 3 \mp \frac{7}{8}\sqrt{2}\right)$ [about $(-0.71, 4.24)$ and $(0.71, 1.76)$].

(d)

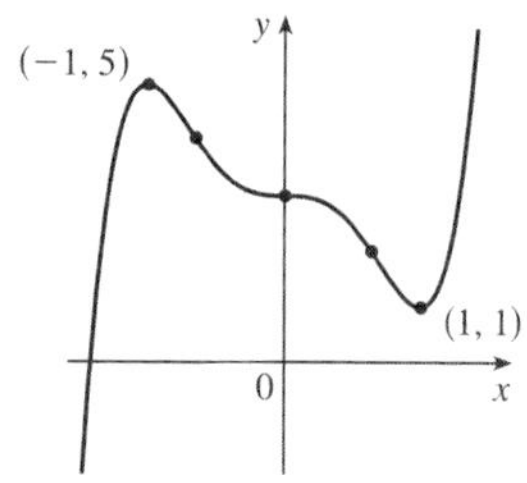


**22.** (a) $h(x) = (x^2 - 1)^3 \quad \Rightarrow \quad h'(x) = 6x(x^2 - 1)^2 \geq 0 \quad \Leftrightarrow$

$x > 0 \ (x \neq 1)$, so $h$ is increasing on $(0, \infty)$ and decreasing on $(-\infty, 0)$.

(b) $h(0) = -1$ is a local minimum value.

(c) $h''(x) = 6(x^2 - 1)^2 + 24x^2(x^2 - 1) = 6(x^2 - 1)(5x^2 - 1)$. The roots $\pm 1$ and $\pm\frac{1}{\sqrt{5}}$ divide $\mathbb{R}$ into five intervals.

| Interval | $x^2 - 1$ | $5x^2 - 1$ | $h''(x)$ | Concavity |
|---|---|---|---|---|
| $x < -1$ | $+$ | $+$ | $+$ | upward |
| $-1 < x < -\frac{1}{\sqrt{5}}$ | $-$ | $+$ | $-$ | downward |
| $-\frac{1}{\sqrt{5}} < x < \frac{1}{\sqrt{5}}$ | $-$ | $-$ | $+$ | upward |
| $\frac{1}{\sqrt{5}} < x < 1$ | $-$ | $+$ | $-$ | downward |
| $x > 1$ | $+$ | $+$ | $+$ | upward |

From the table, we see that $h$ is CU on $(-\infty, -1)$, $\left(-\frac{1}{\sqrt{5}}, \frac{1}{\sqrt{5}}\right)$ and $(1, \infty)$, and CD on $\left(-1, -\frac{1}{\sqrt{5}}\right)$ and $\left(\frac{1}{\sqrt{5}}, 1\right)$. Inflection points at $(\pm 1, 0)$ and $\left(\pm\frac{1}{\sqrt{5}}, -\frac{64}{125}\right)$

(d)

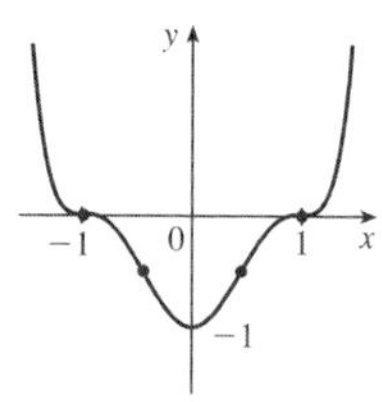

**23.** (a) $A(x) = x\sqrt{x+3} \quad \Rightarrow \quad A'(x) = x \cdot \frac{1}{2}(x+3)^{-1/2} + \sqrt{x+3} \cdot 1 = \dfrac{x}{2\sqrt{x+3}} + \sqrt{x+3} = \dfrac{x + 2(x+3)}{2\sqrt{x+3}} = \dfrac{3x+6}{2\sqrt{x+3}}.$
The domain of $A$ is $[-3, \infty)$. $A'(x) > 0$ for $x > -2$ and $A'(x) < 0$ for $-3 < x < -2$, so $A$ is increasing on $(-2, \infty)$ and decreasing on $(-3, -2)$.

(b) $A(-2) = -2$ is a local minimum value.

(c) $$A''(x) = \frac{2\sqrt{x+3} \cdot 3 - (3x+6) \cdot \dfrac{1}{\sqrt{x+3}}}{\left(2\sqrt{x+3}\right)^2} = \frac{6(x+3) - (3x+6)}{4(x+3)^{3/2}} = \frac{3x+12}{4(x+3)^{3/2}} = \frac{3(x+4)}{4(x+3)^{3/2}}$$

$A''(x) > 0$ for all $x > -3$, so $A$ is concave upward on $(-3, \infty)$. There is no inflection point.

(d)

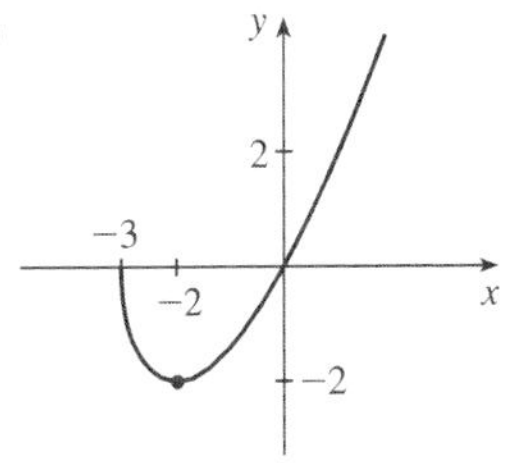


**24.** (a) $B(x) = 3x^{2/3} - x \quad \Rightarrow \quad B'(x) = 2x^{-1/3} - 1 = \dfrac{2}{\sqrt[3]{x}} - 1 = \dfrac{2 - \sqrt[3]{x}}{\sqrt[3]{x}}$. $B'(x) > 0$ if $0 < x < 8$ and $B'(x) < 0$ if $x < 0$ or $x > 8$, so $B$ is decreasing on $(-\infty, 0)$ and $(8, \infty)$, and $B$ is increasing on $(0, 8)$.

(b) $B(0) = 0$ is a local minimum value.
$B(8) = 4$ is a local maximum value.

(c) $B''(x) = -\frac{2}{3}x^{-4/3} = \dfrac{-2}{3x^{4/3}}$, so $B''(x) < 0$ for all $x \neq 0$. $B$ is concave downward on $(-\infty, 0)$ and $(0, \infty)$. There is no inflection point.

(d)

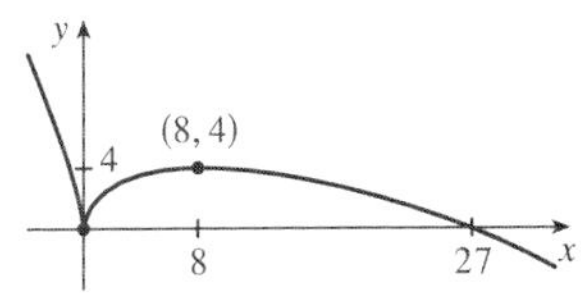


**25.** (a) $C(x) = x^{1/3}(x+4) = x^{4/3} + 4x^{1/3} \quad \Rightarrow \quad C'(x) = \frac{4}{3}x^{1/3} + \frac{4}{3}x^{-2/3} = \frac{4}{3}x^{-2/3}(x+1) = \dfrac{4(x+1)}{3\sqrt[3]{x^2}}$. $C'(x) > 0$ if $-1 < x < 0$ or $x > 0$ and $C'(x) < 0$ for $x < -1$, so $C$ is increasing on $(-1, \infty)$ and $C$ is decreasing on $(-\infty, -1)$.

(b) $C(-1) = -3$ is a local minimum value.

(c) $C''(x) = \frac{4}{9}x^{-2/3} - \frac{8}{9}x^{-5/3} = \frac{4}{9}x^{-5/3}(x-2) = \dfrac{4(x-2)}{9\sqrt[3]{x^5}}$. $C''(x) < 0$ for $0 < x < 2$ and $C''(x) > 0$ for $x < 0$ and $x > 2$, so $C$ is concave downward on $(0, 2)$ and concave upward on $(-\infty, 0)$ and $(2, \infty)$. There are inflection points at $(0, 0)$ and $\left(2, 6\sqrt[3]{2}\right) \approx (2, 7.56)$.

(d)

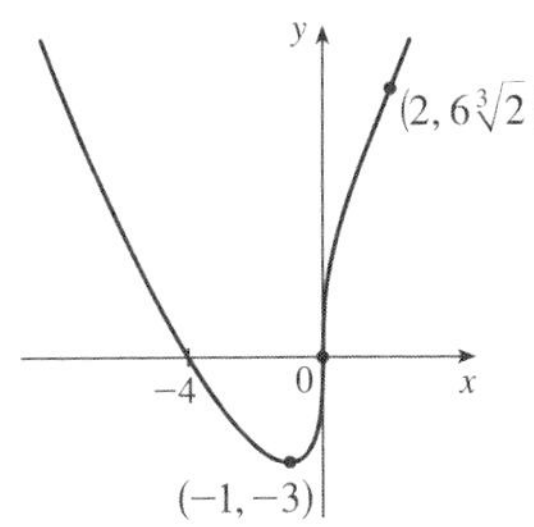


**26.** (a) $f(x) = \ln(x^4 + 27) \quad \Rightarrow \quad f'(x) = \dfrac{4x^3}{x^4 + 27}$. $f'(x) > 0$ if $x > 0$ and $f'(x) < 0$ if $x < 0$, so $f$ is increasing on $(0, \infty)$ and $f$ is decreasing on $(-\infty, 0)$.

(b) $f(0) = \ln 27 \approx 3.3$ is a local minimum value.

(c) $$f''(x) = \frac{(x^4+27)(12x^2) - 4x^3(4x^3)}{(x^4+27)^2} = \frac{4x^2\left[3(x^4+27) - 4x^4\right]}{(x^4+27)^2} = \frac{4x^2(81 - x^4)}{(x^4+27)^2} = \frac{-4x^2(x^2+9)(x+3)(x-3)}{(x^4+27)^2}$$

$f''(x) > 0$ if $-3 < x < 0$ and $0 < x < 3$, and $f''(x) < 0$ if $x < -3$ or $x > 3$. Thus, $f$ is concave upward on $(-3, 0)$ and $(0, 3)$ [hence on $(-3, 3)$] and $f$ is concave downward on $(-\infty, -3)$ and $(3, \infty)$. There are inflection points at $(\pm 3, \ln 108) \approx (\pm 3, 4.68)$.

(d)

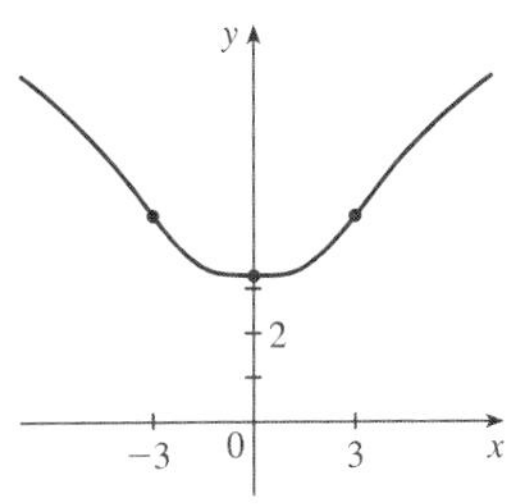

**27.** (a) $f(\theta) = 2\cos\theta - \cos 2\theta,\ \ 0 \le \theta \le 2\pi$.
$f'(\theta) = -2\sin\theta + 2\sin 2\theta = -2\sin\theta + 2(2\sin\theta\cos\theta) = 2\sin\theta\,(2\cos\theta - 1)$.

| Interval | $\sin\theta$ | $2\cos\theta - 1$ | $f'(\theta)$ | $f$ |
|---|---|---|---|---|
| $0 < \theta < \frac{\pi}{3}$ | + | + | + | increasing on $(0, \frac{\pi}{3})$ |
| $\frac{\pi}{3} < \theta < \pi$ | + | − | − | decreasing on $(\frac{\pi}{3}, \pi)$ |
| $\pi < \theta < \frac{5\pi}{3}$ | − | − | + | increasing on $(\pi, \frac{5\pi}{3})$ |
| $\frac{5\pi}{3} < \theta < 2\pi$ | − | + | − | decreasing on $(\frac{5\pi}{3}, 2\pi)$ |

(b) $f(\frac{\pi}{3}) = \frac{3}{2}$ and $f(\frac{5\pi}{3}) = \frac{3}{2}$ are local maximum values and $f(\pi) = -3$ is a local minimum value.

(c) $f'(\theta) = -2\sin\theta + 2\sin 2\theta \ \Rightarrow$

$$f''(\theta) = -2\cos\theta + 4\cos 2\theta = -2\cos\theta + 4(2\cos^2\theta - 1) = 2(4\cos^2\theta - \cos\theta - 2)$$

$$f''(\theta) = 0 \ \Leftrightarrow\ \cos\theta = \frac{1 \pm \sqrt{33}}{8} \ \Leftrightarrow\ \theta = \cos^{-1}\left(\frac{1 \pm \sqrt{33}}{8}\right)$$

$$\Leftrightarrow\ \theta = \cos^{-1}\left(\frac{1 + \sqrt{33}}{8}\right) \approx 0.5678,$$

(d) 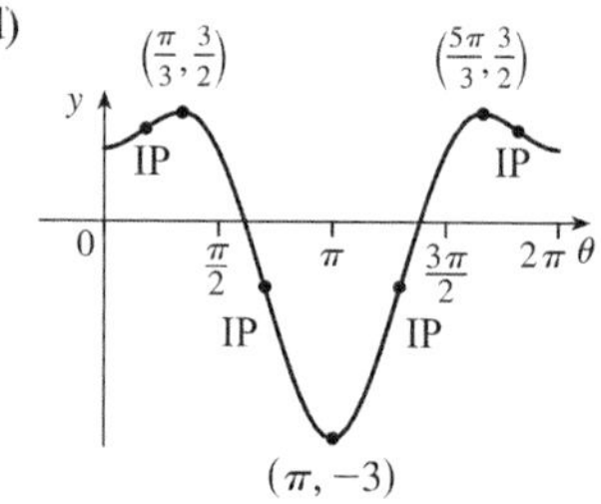


$2\pi - \cos^{-1}\left(\frac{1 + \sqrt{33}}{8}\right) \approx 5.7154$, $\cos^{-1}\left(\frac{1 - \sqrt{33}}{8}\right) \approx 2.2057$, or $2\pi - \cos^{-1}\left(\frac{1 - \sqrt{33}}{8}\right) \approx 4.0775$. Denote these four values of $\theta$ by $\theta_1$, $\theta_4$, $\theta_2$, and $\theta_3$, respectively. Then $f$ is CU on $(0, \theta_1)$, CD on $(\theta_1, \theta_2)$, CU on $(\theta_2, \theta_3)$, CD on $(\theta_3, \theta_4)$, and CU on $(\theta_4, 2\pi)$. To find the *exact* $y$-coordinate for $\theta = \theta_1$, we have

$$f(\theta_1) = 2\cos\theta_1 - \cos 2\theta_1 = 2\cos\theta_1 - (2\cos^2\theta_1 - 1) = 2\left(\frac{1 + \sqrt{33}}{8}\right) - 2\left(\frac{1 + \sqrt{33}}{8}\right)^2 + 1$$
$$= \tfrac{1}{4} + \tfrac{1}{4}\sqrt{33} - \tfrac{1}{32} - \tfrac{1}{16}\sqrt{33} - \tfrac{33}{32} + 1 = \tfrac{3}{16} + \tfrac{3}{16}\sqrt{33} = \tfrac{3}{16}(1 + \sqrt{33}) = y_1 \approx 1.26.$$

Similarly, $f(\theta_2) = \frac{3}{16}(1 - \sqrt{33}) = y_2 \approx -0.89$. So $f$ has inflection points at $(\theta_1, y_1)$, $(\theta_2, y_2)$, $(\theta_3, y_2)$, and $(\theta_4, y_1)$.

**28.** (a) $f(t) = t + \cos t$, $-2\pi \le t \le 2\pi \ \Rightarrow\ f'(t) = 1 - \sin t \ge 0$ for all $t$ and $f'(t) = 0$ when $\sin t = 1 \ \Leftrightarrow$ $t = -\frac{3\pi}{2}$ or $\frac{\pi}{2}$, so $f$ is increasing on $(-2\pi, 2\pi)$.

(b) No maximum or minimum

(c) $f''(t) = -\cos t > 0 \ \Leftrightarrow\ t \in \left(-\frac{3\pi}{2}, -\frac{\pi}{2}\right) \cup \left(\frac{\pi}{2}, \frac{3\pi}{2}\right)$, so $f$ is CU on these intervals and CD on $\left(-2\pi, -\frac{3\pi}{2}\right)$, $\left(-\frac{\pi}{2}, \frac{\pi}{2}\right)$, and $\left(\frac{3\pi}{2}, 2\pi\right)$. Points of inflection at $\left(\pm\frac{3\pi}{2}, \pm\frac{3\pi}{2}\right)$ and $\left(\pm\frac{\pi}{2}, \pm\frac{\pi}{2}\right)$

(d) 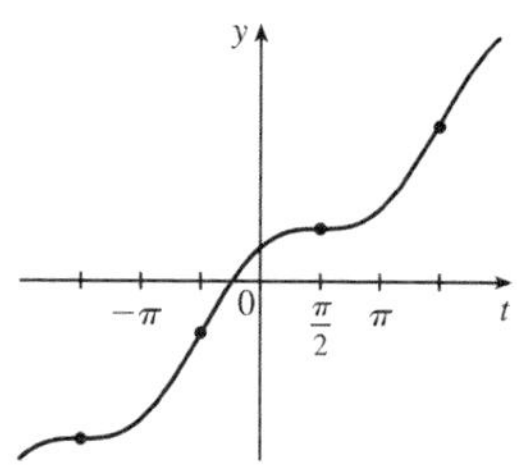


**29.** $f(x) = \dfrac{x^2}{x^2 - 1} = \dfrac{x^2}{(x+1)(x-1)}$ has domain $(-\infty, -1) \cup (-1, 1) \cup (1, \infty)$.

(a) $\displaystyle\lim_{x\to\pm\infty} f(x) = \lim_{x\to\pm\infty} \frac{x^2/x^2}{(x^2 - 1)/x^2} = \lim_{x\to\pm\infty} \frac{1}{1 - 1/x^2} = \frac{1}{1 - 0} = 1$, so $y = 1$ is a HA.

$\displaystyle\lim_{x\to -1^-} \frac{x^2}{x^2 - 1} = \infty$ since $x^2 \to 1$ and $(x^2 - 1) \to 0^+$ as $x \to -1^-$, so $x = -1$ is a VA.

$\displaystyle\lim_{x\to 1^+} \frac{x^2}{x^2 - 1} = \infty$ since $x^2 \to 1$ and $(x^2 - 1) \to 0^+$ as $x \to 1^+$, so $x = 1$ is a VA.

(b) $f(x) = \dfrac{x^2}{x^2-1} \Rightarrow f'(x) = \dfrac{(x^2-1)(2x) - x^2(2x)}{(x^2-1)^2} = \dfrac{2x[(x^2-1)-x^2]}{(x^2-1)^2} = \dfrac{-2x}{(x^2-1)^2}$. Since $(x^2-1)^2$ is positive for all $x$ in the domain of $f$, the sign of the derivative is determined by the sign of $-2x$. Thus, $f'(x) > 0$ if $x < 0$ $(x \neq -1)$ and $f'(x) < 0$ if $x > 0$ $(x \neq 1)$. So $f$ is increasing on $(-\infty, -1)$ and $(-1, 0)$, and $f$ is decreasing on $(0, 1)$ and $(1, \infty)$.

(c) $f'(x) = 0 \Rightarrow x = 0$ and $f(0) = 0$ is a local maximum value.

(d) $f''(x) = \dfrac{(x^2-1)^2(-2) - (-2x)\cdot 2(x^2-1)(2x)}{[(x^2-1)^2]^2}$

$= \dfrac{2(x^2-1)[-(x^2-1)+4x^2]}{(x^2-1)^4} = \dfrac{2(3x^2+1)}{(x^2-1)^3}$.

The sign of $f''(x)$ is determined by the denominator; that is, $f''(x) > 0$ if $|x| > 1$ and $f''(x) < 0$ if $|x| < 1$. Thus, $f$ is CU on $(-\infty, -1)$ and $(1, \infty)$, and $f$ is CD on $(-1, 1)$. There are no inflection points.

(e)

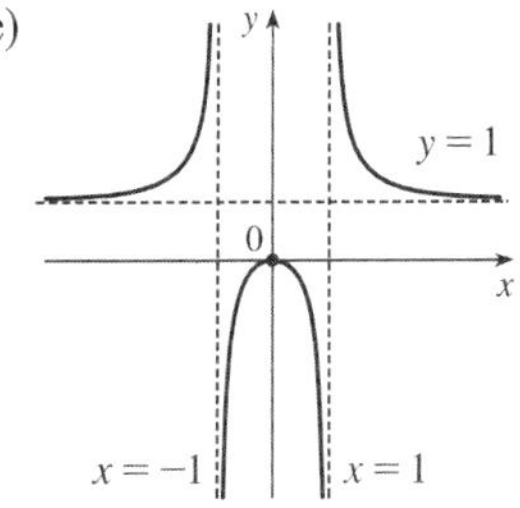


**30.** $f(x) = \dfrac{x^2}{(x-2)^2}$ has domain $(-\infty, 2) \cup (2, \infty)$.

(a) $\displaystyle\lim_{x\to\pm\infty} \frac{x^2}{x^2-4x+4} = \lim_{x\to\pm\infty} \frac{x^2/x^2}{(x^2-4x+4)/x^2} = \lim_{x\to\pm\infty} \frac{1}{1-4/x+4/x^2} = \frac{1}{1-0+0} = 1$,

so $y = 1$ is a HA. $\displaystyle\lim_{x\to 2^+} \frac{x^2}{(x-2)^2} = \infty$ since $x^2 \to 4$ and $(x-2)^2 \to 0^+$ as $x \to 2^+$, so $x = 2$ is a VA.

(b) $f(x) = \dfrac{x^2}{(x-2)^2} \Rightarrow f'(x) = \dfrac{(x-2)^2(2x) - x^2 \cdot 2(x-2)}{[(x-2)^2]^2} = \dfrac{2x(x-2)[(x-2)-x]}{(x-2)^4} = \dfrac{-4x}{(x-2)^3}$. $f'(x) > 0$ if $0 < x < 2$ and $f'(x) < 0$ if $x < 0$ or $x > 2$, so $f$ is increasing on $(0, 2)$ and $f$ is decreasing on $(-\infty, 0)$ and $(2, \infty)$.

(c) $f(0) = 0$ is a local minimum value.

(d) $f''(x) = \dfrac{(x-2)^3(-4) - (-4x)\cdot 3(x-2)^2}{[(x-2)^3]^2}$

$= \dfrac{4(x-2)^2[-(x-2)+3x]}{(x-2)^6} = \dfrac{8(x+1)}{(x-2)^4}$

$f''(x) > 0$ if $x > -1$ $(x \neq 2)$ and $f''(x) < 0$ if $x < -1$. Thus, $f$ is CU on $(-1, 2)$ and $(2, \infty)$, and $f$ is CD on $(-\infty, -1)$. There is an inflection point at $\left(-1, \frac{1}{9}\right)$.

(e)

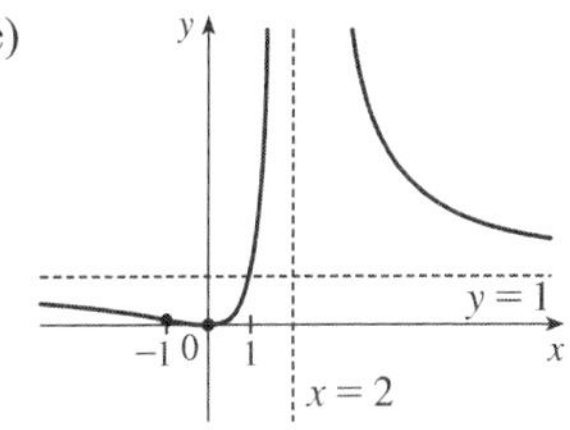


**31.** (a) $\displaystyle\lim_{x\to-\infty} \left(\sqrt{x^2+1} - x\right) = \infty$ and

$\displaystyle\lim_{x\to\infty} \left(\sqrt{x^2+1} - x\right) = \lim_{x\to\infty} \left(\sqrt{x^2+1} - x\right) \frac{\sqrt{x^2+1}+x}{\sqrt{x^2+1}+x} = \lim_{x\to\infty} \frac{1}{\sqrt{x^2+1}+x} = 0$, so $y = 0$ is a HA.

(b) $f(x) = \sqrt{x^2+1} - x \Rightarrow f'(x) = \dfrac{x}{\sqrt{x^2+1}} - 1$. Since $\dfrac{x}{\sqrt{x^2+1}} < 1$ for all $x$, $f'(x) < 0$, so $f$ is decreasing on $\mathbb{R}$.

(c) No minimum or maximum

(d) $f''(x) = \dfrac{(x^2+1)^{1/2}(1) - x \cdot \frac{1}{2}(x^2+1)^{-1/2}(2x)}{\left(\sqrt{x^2+1}\right)^2}$

$= \dfrac{(x^2+1)^{1/2} - \dfrac{x^2}{(x^2+1)^{1/2}}}{x^2+1} = \dfrac{(x^2+1)-x^2}{(x^2+1)^{3/2}}$

$= \dfrac{1}{(x^2+1)^{3/2}} > 0$, so $f$ is CU on $\mathbb{R}$. No IP

(e)

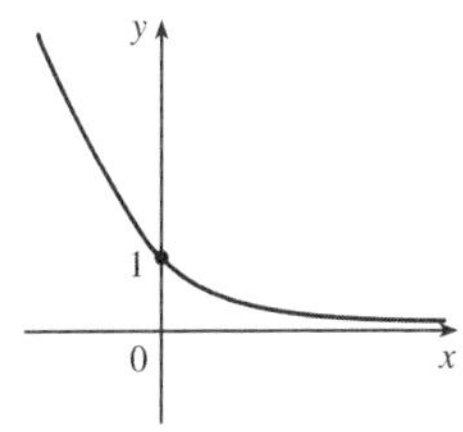

**32.** (a) $\lim\limits_{x\to\pi/2^-} x\tan x = \infty$ and $\lim\limits_{x\to-\pi/2^+} x\tan x = \infty$, so $x = \frac{\pi}{2}$ and $x = -\frac{\pi}{2}$ are VA.

(b) $f(x) = x\tan x$, $-\frac{\pi}{2} < x < \frac{\pi}{2}$. $f'(x) = x\sec^2 x + \tan x > 0 \iff$

$0 < x < \frac{\pi}{2}$, so $f$ increases on $\left(0, \frac{\pi}{2}\right)$ and decreases on $\left(-\frac{\pi}{2}, 0\right)$.

(c) $f(0) = 0$ is a local minimum value.

(d) $f''(x) = 2\sec^2 x + 2x\,\tan x\,\sec^2 x > 0$ for $-\frac{\pi}{2} < x < \frac{\pi}{2}$, so $f$ is CU on $\left(-\frac{\pi}{2}, \frac{\pi}{2}\right)$. No IP

(e)

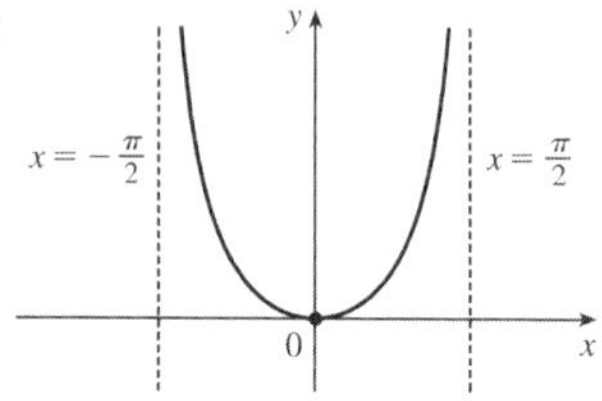


**33.** $f(x) = \ln(1 - \ln x)$ is defined when $x > 0$ (so that $\ln x$ is defined) and $1 - \ln x > 0$ [so that $\ln(1 - \ln x)$ is defined]. The second condition is equivalent to $1 > \ln x \iff x < e$, so $f$ has domain $(0, e)$.

(a) As $x \to 0^+$, $\ln x \to -\infty$, so $1 - \ln x \to \infty$ and $f(x) \to \infty$. As $x \to e^-$, $\ln x \to 1^-$, so $1 - \ln x \to 0^+$ and $f(x) \to -\infty$. Thus, $x = 0$ and $x = e$ are vertical asymptotes. There is no horizontal asymptote.

(b) $f'(x) = \dfrac{1}{1 - \ln x}\left(-\dfrac{1}{x}\right) = -\dfrac{1}{x(1 - \ln x)} < 0$ on $(0, e)$. Thus, $f$ is decreasing on its domain, $(0, e)$.

(c) $f'(x) \neq 0$ on $(0, e)$, so $f$ has no local maximum or minimum value.

(d) $f''(x) = -\dfrac{-[x(1 - \ln x)]'}{[x(1 - \ln x)]^2} = \dfrac{x(-1/x) + (1 - \ln x)}{x^2(1 - \ln x)^2}$

$= -\dfrac{\ln x}{x^2(1 - \ln x)^2}$

so $f''(x) > 0 \iff \ln x < 0 \iff 0 < x < 1$. Thus, $f$ is CU on $(0, 1)$ and CD on $(1, e)$. There is an inflection point at $(1, 0)$.

(e)

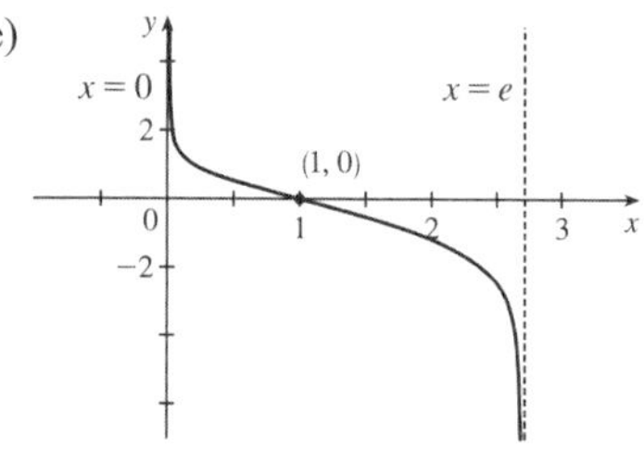


**34.** $f(x) = \dfrac{e^x}{1 + e^x}$ has domain $\mathbb{R}$.

(a) $\lim\limits_{x\to\infty} f(x) = \lim\limits_{x\to\infty} \dfrac{e^x/e^x}{(1 + e^x)/e^x} = \lim\limits_{x\to\infty} \dfrac{1}{e^{-x} + 1} = \dfrac{1}{0 + 1} = 1$, so $y = 1$ is a HA.

$\lim\limits_{x\to-\infty} f(x) = \lim\limits_{x\to-\infty} \dfrac{e^x}{1 + e^x} = \dfrac{0}{1 + 0} = 0$, so $y = 0$ is a HA. No VA.

(b) $f'(x) = \dfrac{(1 + e^x)e^x - e^x \cdot e^x}{(1 + e^x)^2} = \dfrac{e^x}{(1 + e^x)^2} > 0$ for all $x$. Thus, $f$ is increasing on $\mathbb{R}$.

(c) There is no local maximum or minimum.

(d) $f''(x) = \dfrac{(1 + e^x)^2 e^x - e^x \cdot 2(1 + e^x)\,e^x}{\left[(1 + e^x)^2\right]^2}$

$= \dfrac{e^x(1 + e^x)[(1 + e^x) - 2e^x]}{(1 + e^x)^4} = \dfrac{e^x(1 - e^x)}{(1 + e^x)^3}$

$f''(x) > 0 \iff 1 - e^x > 0 \iff x < 0$, so $f$ is CU on $(-\infty, 0)$ and CD on $(0, \infty)$. There is an inflection point at $\left(0, \frac{1}{2}\right)$.

(e)

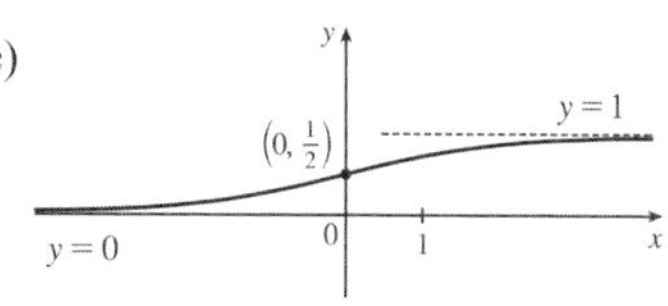


**35.** (a) $\lim\limits_{x\to\pm\infty} e^{-1/(x+1)} = 1$ since $-1/(x + 1) \to 0$, so $y = 1$ is a HA. $\lim\limits_{x\to-1^+} e^{-1/(x+1)} = 0$ since $-1/(x + 1) \to -\infty$,

$\lim\limits_{x\to-1^-} e^{-1/(x+1)} = \infty$ since $-1/(x + 1) \to \infty$, so $x = -1$ is a VA.

(b) $f(x) = e^{-1/(x+1)} \Rightarrow f'(x) = e^{-1/(x+1)}\left[-(-1)\dfrac{1}{(x + 1)^2}\right]$ [Reciprocal Rule] $= e^{-1/(x+1)}/(x + 1)^2 \Rightarrow$

$f'(x) > 0$ for all $x$ except $-1$, so $f$ is increasing on $(-\infty, -1)$ and $(-1, \infty)$.

(c) No local maximum or minimum

(d) $f''(x) = \dfrac{(x+1)^2 e^{-1/(x+1)}\left[1/(x+1)^2\right] - e^{-1/(x+1)}\left[2(x+1)\right]}{\left[(x+1)^2\right]^2}$

$= \dfrac{e^{-1/(x+1)}\left[1-(2x+2)\right]}{(x+1)^4} = -\dfrac{e^{-1/(x+1)}(2x+1)}{(x+1)^4} \Rightarrow$

$f''(x) > 0 \Leftrightarrow 2x+1<0 \Leftrightarrow x < -\frac{1}{2}$, so $f$ is CU on $(-\infty,-1)$ and $\left(-1,-\frac{1}{2}\right)$, and CD on $\left(-\frac{1}{2},\infty\right)$. $f$ has an IP at $\left(-\frac{1}{2},e^{-2}\right)$.

(e)

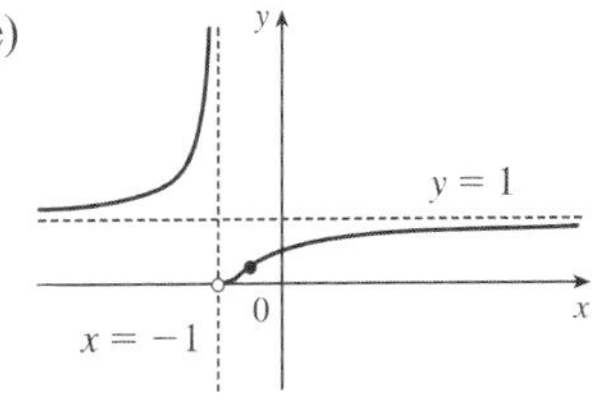


**36.** (a) $f$ is periodic with period $\pi$, so we consider only $-\frac{\pi}{2} < x < \frac{\pi}{2}$. $\lim\limits_{x\to 0}\ln(\tan^2 x) = -\infty$, $\lim\limits_{x\to(\pi/2)^-}\ln(\tan^2 x) = \infty$, and $\lim\limits_{x\to(-\pi/2)^+}\ln(\tan^2 x) = \infty$, so $x=0$, $x=\pm\frac{\pi}{2}$ are VA.

(b) $f(x) = \ln(\tan^2 x) \Rightarrow f'(x) = \dfrac{2\tan x\sec^2 x}{\tan^2 x} = 2\dfrac{\sec^2 x}{\tan x} > 0 \Leftrightarrow \tan x > 0 \Leftrightarrow 0 < x < \frac{\pi}{2}$, so $f$ is increasing on $\left(0,\frac{\pi}{2}\right)$ and decreasing on $\left(-\frac{\pi}{2},0\right)$.

(c) No maximum or minimum

(d) $f'(x) = \dfrac{2}{\sin x\cos x} = \dfrac{4}{\sin 2x} \Rightarrow f''(x) = \dfrac{-8\cos 2x}{\sin^2 2x} < 0$

$\Leftrightarrow \cos 2x > 0 \Leftrightarrow -\frac{\pi}{4} < x < \frac{\pi}{4}$, so $f$ is CD on $\left(-\frac{\pi}{4},0\right)$ and $\left(0,\frac{\pi}{4}\right)$, and CU on $\left(-\frac{\pi}{2},-\frac{\pi}{4}\right)$ and $\left(\frac{\pi}{4},\frac{\pi}{2}\right)$. IP are $\left(\pm\frac{\pi}{4},0\right)$.

(e)

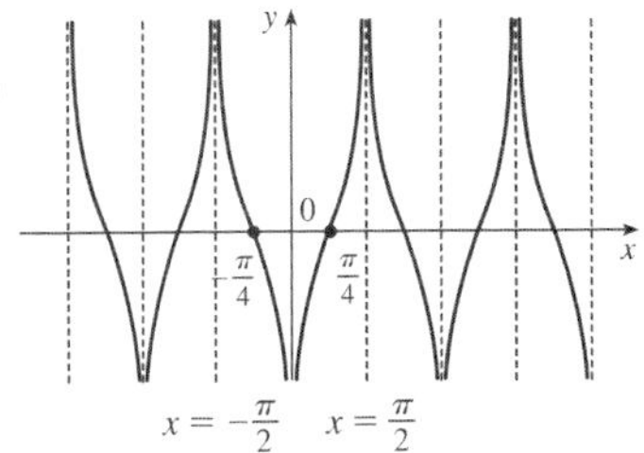


**37.** The nonnegative factors $(x+1)^2$ and $(x-6)^4$ do not affect the sign of $f'(x) = (x+1)^2(x-3)^5(x-6)^4$. So $f'(x) > 0 \Rightarrow (x-3)^5 > 0 \Rightarrow x-3>0 \Rightarrow x>3$. Thus, $f$ is increasing on the interval $(3,\infty)$.

**38.** $y = f(x) = x^3 - 3a^2x + 2a^3$, $a>0$. The $y$-intercept is $f(0) = 2a^3$. $y' = 3x^2 - 3a^2 = 3(x^2-a^2) = 3(x+a)(x-a)$. The critical numbers are $-a$ and $a$. $f'<0$ on $(-a,a)$, so $f$ is decreasing on $(-a,a)$ and $f$ is increasing on $(-\infty,-a)$ and $(a,\infty)$. $f(-a) = 4a^3$ is a local maximum value and $f(a)=0$ is a local minimum value. Since $f(a)=0$, $a$ is an $x$-intercept, and $x-a$ is a factor of $f$. Synthetically dividing $y = x^3-3a^2x+2a^3$ by $x-a$ gives us the following result: $y = x^3 - 3a^2x + 2a^3 = (x-a)(x^2+ax-2a^2) = (x-a)(x-a)(x+2a) = (x-a)^2(x+2a)$, which tells us

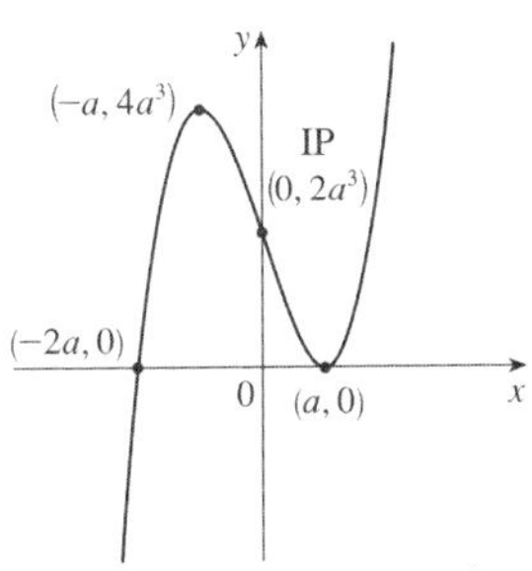


that the only $x$-intercepts are $-2a$ and $a$. $y' = 3x^2 - 3a^2 \Rightarrow y'' = 6x$, so $y''>0$ on $(0,\infty)$ and $y''<0$ on $(-\infty,0)$. This tells us that $f$ is CU on $(0,\infty)$ and CD on $(-\infty,0)$. There is an inflection point at $(0,2a^3)$. The graph illustrates these features.

What the curves in the family have in common is that they are all CD on $(-\infty,0)$, CU on $(0,\infty)$, and have the same basic shape. But as $a$ increases, the four key points shown in the figure move further away from the origin.

**39.** (a)

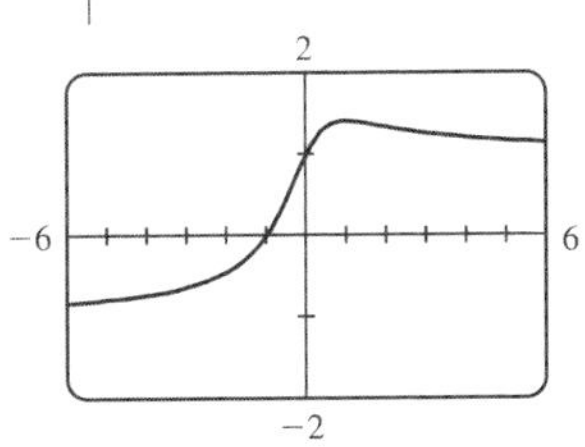


From the graph, we get an estimate of $f(1)\approx 1.41$ as a local maximum value, and no local minimum value.

$f(x) = \dfrac{x+1}{\sqrt{x^2+1}} \Rightarrow f'(x) = \dfrac{1-x}{(x^2+1)^{3/2}}$.

$f'(x) = 0 \Leftrightarrow x = 1$. $f(1) = \frac{2}{\sqrt{2}} = \sqrt{2}$ is the exact value.

(b) From the graph in part (a), $f$ increases most rapidly somewhere between $x=-\frac{1}{2}$ and $x=-\frac{1}{4}$. To find the exact value, we need to find the maximum value of $f'$, which we can do by finding the critical numbers of $f'$.

$f''(x) = \dfrac{2x^2-3x-1}{(x^2+1)^{5/2}} = 0 \Leftrightarrow x = \dfrac{3\pm\sqrt{17}}{4}$. $x = \dfrac{3+\sqrt{17}}{4}$ corresponds to the *minimum* value of $f'$. The maximum value of $f'$ is at $\left(\frac{3-\sqrt{17}}{4}, \sqrt{\frac{7}{6}-\frac{\sqrt{17}}{6}}\right) \approx (-0.28, 0.69)$.

**40.** (a)

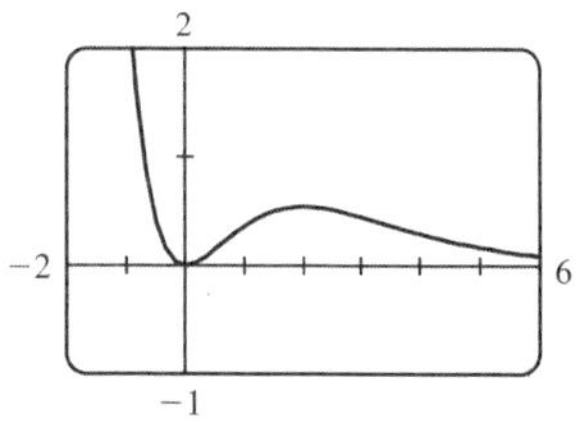


Tracing the graph gives us estimates of $f(0) = 0$ for a local minimum value and $f(2) = 0.54$ for a local maximum value. $f(x) = x^2e^{-x} \;\Rightarrow\; f'(x) = xe^{-x}(2 - x)$. $f'(x) = 0 \;\Leftrightarrow\; x = 0$ or $2$. $f(0) = 0$ and $f(2) = 4e^{-2}$ are the exact values.

(b) From the graph in part (a), $f$ increases most rapidly around $x = \frac{3}{4}$. To find the exact value, we need to find the maximum value of $f'$, which we can do by finding the critical numbers of $f'$. $f''(x) = e^{-x}\left(x^2 - 4x + 2\right) = 0 \;\Rightarrow\; x = 2 \pm \sqrt{2}$. $x = 2 + \sqrt{2}$ corresponds to the *minimum* value of $f'$. The maximum value of $f'$ is at $\left(2 - \sqrt{2}, \left(2 - \sqrt{2}\right)^2 e^{-2+\sqrt{2}}\right) \approx (0.59, 0.19)$.

**41.** $f(x) = \cos x + \frac{1}{2}\cos 2x \;\Rightarrow\; f'(x) = -\sin x - \sin 2x \;\Rightarrow\; f''(x) = -\cos x - 2\cos 2x$

(a)

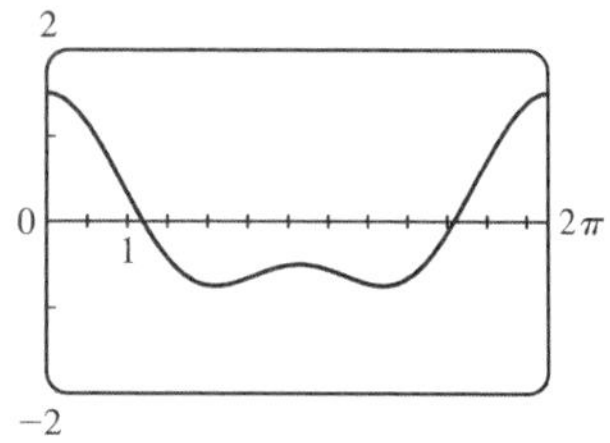


From the graph of $f$, it seems that $f$ is CD on $(0, 1)$, CU on $(1, 2.5)$, CD on $(2.5, 3.7)$, CU on $(3.7, 5.3)$, and CD on $(5.3, 2\pi)$. The points of inflection appear to be at $(1, 0.4)$, $(2.5, -0.6)$, $(3.7, -0.6)$, and $(5.3, 0.4)$.

(b)

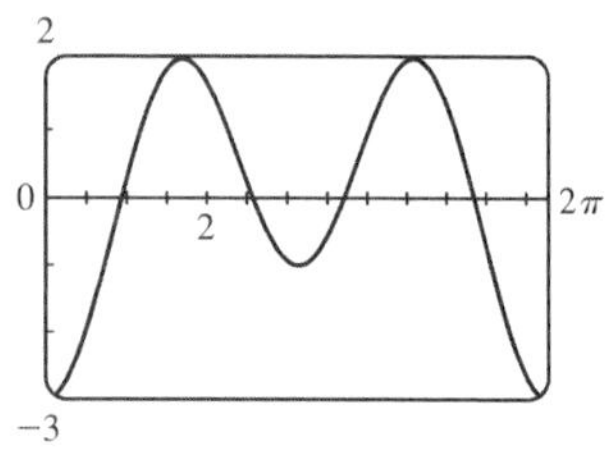


From the graph of $f''$ (and zooming in near the zeros), it seems that $f$ is CD on $(0, 0.94)$, CU on $(0.94, 2.57)$, CD on $(2.57, 3.71)$, CU on $(3.71, 5.35)$, and CD on $(5.35, 2\pi)$. Refined estimates of the inflection points are $(0.94, 0.44)$, $(2.57, -0.63)$, $(3.71, -0.63)$, and $(5.35, 0.44)$.

**42.** $f(x) = x^3(x - 2)^4 \;\Rightarrow$

$$f'(x) = x^3 \cdot 4(x - 2)^3 + (x - 2)^4 \cdot 3x^2 = x^2(x - 2)^3[4x + 3(x - 2)] = x^2(x - 2)^3(7x - 6) \;\Rightarrow$$

$$\begin{aligned} f''(x) &= (2x)(x - 2)^3(7x - 6) + x^2 \cdot 3(x - 2)^2(7x - 6) + x^2(x - 2)^3(7) \\ &= x(x - 2)^2[2(x - 2)(7x - 6) + 3x(7x - 6) + 7x(x - 2)] \\ &= x(x - 2)^2[42x^2 - 72x + 24] = 6x(x - 2)^2(7x^2 - 12x + 4) \end{aligned}$$

(a)

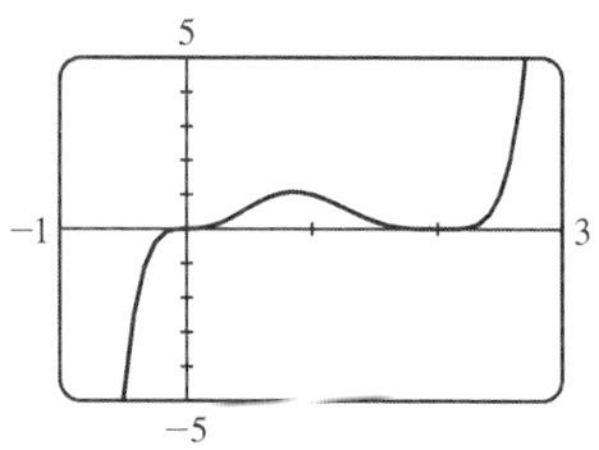


From the graph of $f$, it seems that $f$ is CD on $(-\infty, 0)$, CU on $(0, 0.5)$, CD on $(0.5, 1.3)$, and CU on $(1.3, \infty)$. The points of inflection appear to be at $(0, 0)$, $(0.5, 0.5)$, and $(1.3, 0.6)$.

(b)

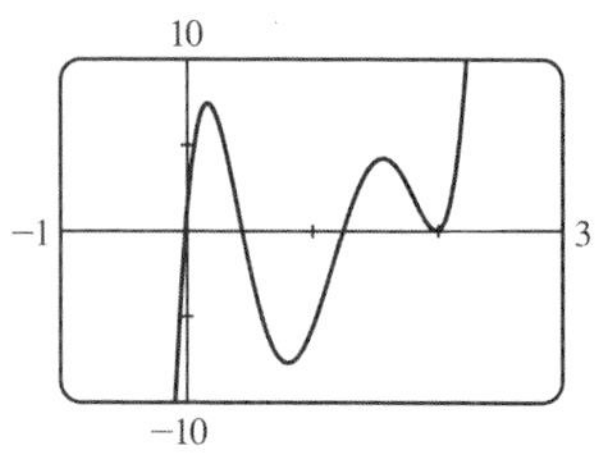


From the graph of $f''$ (and zooming in near the zeros), it seems that $f$ is CD on $(-\infty, 0)$, CU on $(0, 0.45)$, CD on $(0.45, 1.26)$, and CU on $(1.26, \infty)$. Refined estimates of the inflection points are $(0, 0)$, $(0.45, 0.53)$, and $(1.26, 0.60)$.

**43.** In Maple, we define $f$ and then use the command `plot(diff(diff(f,x),x),x=-3..3);`. In Mathematica, we define $f$ and then use `Plot[Dt[Dt[f,x],x],{x,-3,3}]`.We see that $f'' > 0$ for $x > 0.1$ and $f'' < 0$ for $x < 0.1$. So $f$ is concave up on $(0.1, \infty)$ and concave down on $(-\infty, 0.1)$.

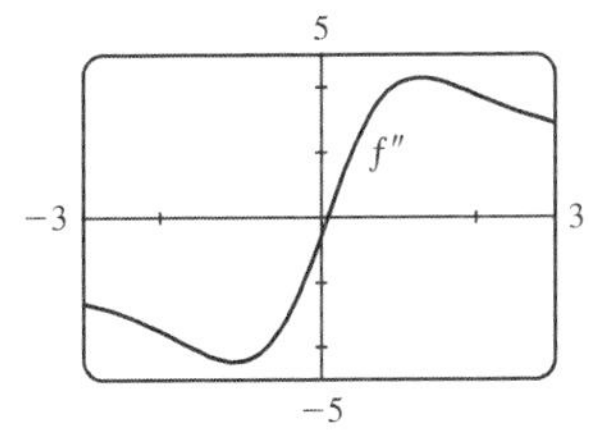


**44.** It appears that $f''$ is positive (and thus $f$ is concave up) on $(-1.8, 0.3)$ and $(1.5, \infty)$ and negative (so $f$ is concave down) on $(-\infty, -1.8)$ and $(0.3, 1.5)$.

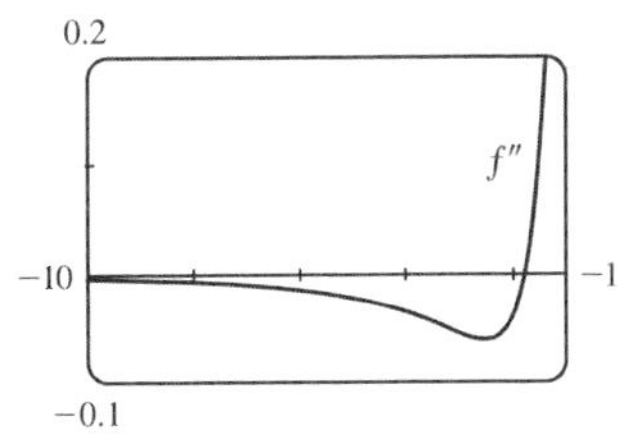


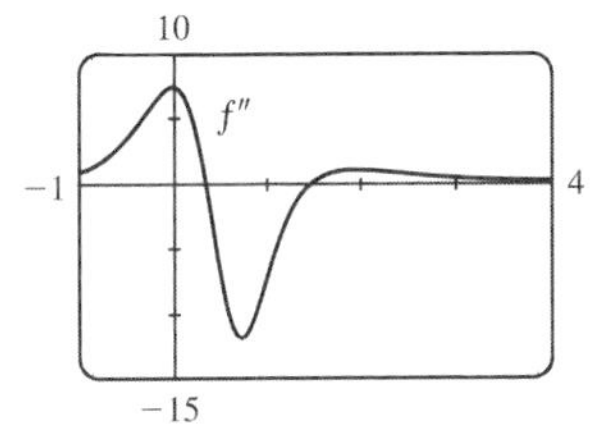


**45.** $y = -\dfrac{W}{24EI}x^4 + \dfrac{WL}{12EI}x^3 - \dfrac{WL^2}{24EI}x^2 = -\dfrac{W}{24EI}x^2(x^2 - 2Lx + L^2)$

$= \dfrac{-W}{24EI}x^2(x - L)^2 = cx^2(x - L)^2$

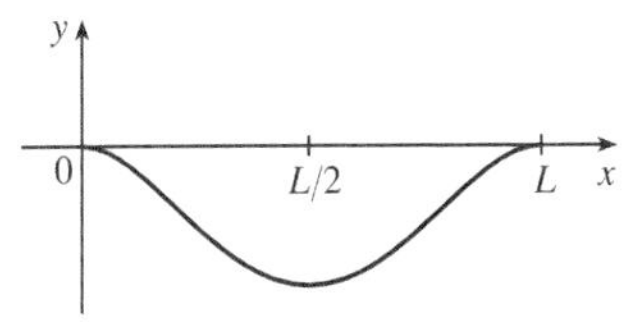


where $c = -\dfrac{W}{24EI}$ is a negative constant and $0 \le x \le L$. We sketch $f(x) = cx^2(x - L)^2$ for $c = -1$. $f(0) = f(L) = 0$.

$f'(x) = cx^2[2(x - L)] + (x - L)^2(2cx) = 2cx(x - L)\,[x + (x - L)] = 2cx(x - L)(2x - L)$. So for $0 < x < L$, $f'(x) > 0 \;\Leftrightarrow\; x(x - L)(2x - L) < 0$ [since $c < 0$] $\Leftrightarrow\; L/2 < x < L$ and $f'(x) < 0 \;\Leftrightarrow\; 0 < x < L/2$. Thus, $f$ is increasing on $(L/2, L)$ and decreasing on $(0, L/2)$, and there is a local and absolute minimum at the point $(L/2, f(L/2)) = \left(L/2, cL^4/16\right)$.

$f'(x) = 2c\,[x(x - L)(2x - L)] \;\Rightarrow$

$f''(x) = 2c\,[1(x - L)(2x - L) + x(1)(2x - L) + x(x - L)(2)] = 2c\left(6x^2 - 6Lx + L^2\right) = 0 \;\Leftrightarrow$

$x = \dfrac{6L \pm \sqrt{12L^2}}{12} = \frac{1}{2}L \pm \frac{\sqrt{3}}{6}L$, and these are the $x$-coordinates of the two inflection points.

**46.** $F(x) = -\dfrac{k}{x^2} + \dfrac{k}{(x-2)^2}$, where $k > 0$ and $0 < x < 2$. For $0 < x < 2$, $x - 2 < 0$, so $F'(x) = \dfrac{2k}{x^3} - \dfrac{2k}{(x-2)^3} > 0$ and $F$ is increasing. $\lim\limits_{x\to 0^+} F(x) = -\infty$ and $\lim\limits_{x\to 2^-} F(x) = \infty$, so $x = 0$ and $x = 2$ are vertical asymptotes. Notice that when the middle particle is at $x = 1$, the net force acting on it is 0. When $x > 1$, the net force is positive, meaning that it acts to the right. And if the particle approaches $x = 2$, the force on it rapidly becomes very large. When $x < 1$, the net force is negative, so it acts to the left. If the particle approaches 0, the force becomes very large to the left.

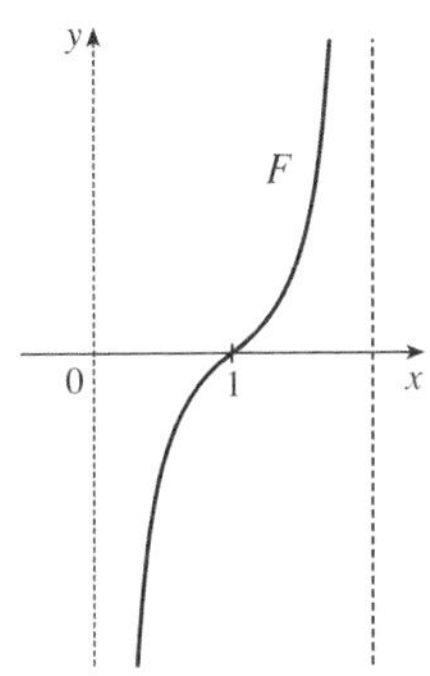

**47.** 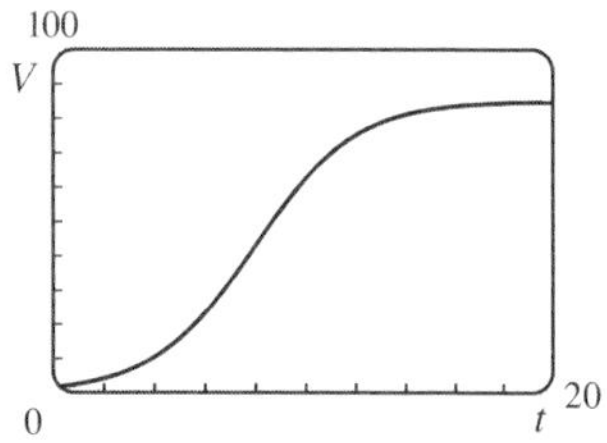


From the graph, we estimate that the most rapid increase in the percentage of households in the United States with at least one VCR occurs at about $t = 8$. To maximize the first derivative, we need to determine the values for which the second derivative is 0. We'll use $V(t) = \dfrac{a}{1 + be^{ct}}$, and substitute $a = 85$, $b = 53$, and $c = -0.5$ later.

$V'(t) = -\dfrac{a(bce^{ct})}{(1 + be^{ct})^2}$ [by the Reciprocal Rule] and

$$V''(t) = -abc \cdot \frac{(1 + be^{ct})^2 \cdot ce^{ct} - e^{ct} \cdot 2(1 + be^{ct}) \cdot bce^{ct}}{[(1 + be^{ct})^2]^2}$$
$$= \frac{-abc \cdot ce^{ct}(1 + be^{ct})[(1 + be^{ct}) - 2be^{ct}]}{(1 + be^{ct})^4} = \frac{-abc^2e^{ct}(1 - be^{ct})}{(1 + be^{ct})^3}$$

So $V''(t) = 0 \;\Leftrightarrow\; 1 = be^{ct} \;\Leftrightarrow\; e^{ct} = 1/b \;\Leftrightarrow\; ct = \ln(1/b) \;\Leftrightarrow\; t = (1/c)\ln(1/b) = -2\ln\frac{1}{53} \approx 7.94$ years, which corresponds to roughly midyear 1988.

**48.** (a) As $|x| \to \infty$, $t = -x^2/(2\sigma^2) \to -\infty$, and $e^t \to 0$. The HA is $y = 0$. Since $t$ takes on its maximum value at $x = 0$, so does $e^t$. Showing this result using derivatives, we have $f(x) = e^{-x^2/(2\sigma^2)} \;\Rightarrow\; f'(x) = e^{-x^2/(2\sigma^2)}\left(-x/\sigma^2\right)$. $f'(x) = 0 \;\Leftrightarrow\; x = 0$. Because $f'$ changes from positive to negative at $x = 0$, $f(0) = 1$ is a local maximum. For inflection points, we find

$f''(x) = -\dfrac{1}{\sigma^2}\left[e^{-x^2/(2\sigma^2)} \cdot 1 + xe^{-x^2/(2\sigma^2)}\left(-x/\sigma^2\right)\right] = \dfrac{-1}{\sigma^2}e^{-x^2/(2\sigma^2)}\left(1 - x^2/\sigma^2\right)$.

$f''(x) = 0 \;\Leftrightarrow\; x^2 = \sigma^2 \;\Leftrightarrow\; x = \pm\sigma$. $f''(x) < 0 \;\Leftrightarrow\; x^2 < \sigma^2 \;\Leftrightarrow\; -\sigma < x < \sigma$. So $f$ is CD on $(-\sigma, \sigma)$ and CU on $(-\infty, -\sigma)$ and $(\sigma, \infty)$. IP at $\left(\pm\sigma, e^{-1/2}\right)$.

(b) Since we have IP at $x = \pm\sigma$, the inflection points move away from the $y$-axis as $\sigma$ increases.

(c) 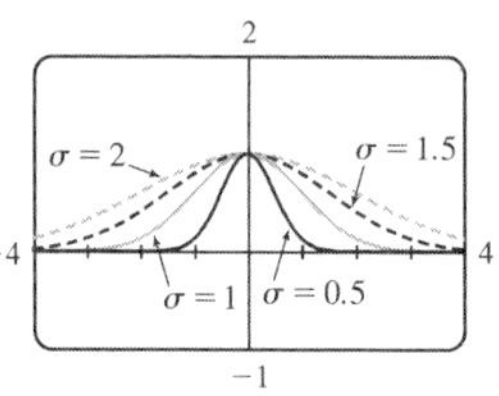


From the graph, we see that as $\sigma$ increases, the graph tends to spread out and there is more area between the curve and the $x$-axis.

**49.** $f(x) = ax^3 + bx^2 + cx + d \;\Rightarrow\; f'(x) = 3ax^2 + 2bx + c$. We are given that $f(1) = 0$ and $f(-2) = 3$, so $f(1) = a + b + c + d = 0$ and $f(-2) = -8a + 4b - 2c + d = 3$. Also $f'(1) = 3a + 2b + c = 0$ and $f'(-2) = 12a - 4b + c = 0$ by Fermat's Theorem. Solving these four equations, we get $a = \frac{2}{9}$, $b = \frac{1}{3}$, $c = -\frac{4}{3}$, $d = \frac{7}{9}$, so the function is $f(x) = \frac{1}{9}\left(2x^3 + 3x^2 - 12x + 7\right)$.

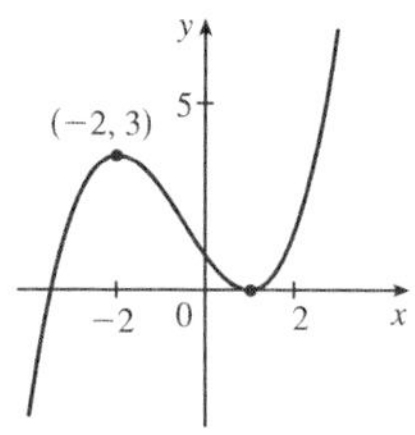


**50.** $f(x) = axe^{bx^2} \;\Rightarrow\; f'(x) = a\left[xe^{bx^2} \cdot 2bx + e^{bx^2} \cdot 1\right] = ae^{bx^2}\left(2bx^2 + 1\right)$. For $f(2) = 1$ to be a maximum value, we must have $f'(2) = 0$. $f(2) = 1 \;\Rightarrow\; 1 = 2ae^{4b}$ and $f'(2) = 0 \;\Rightarrow\; 0 = (8b + 1)ae^{4b}$. So $8b + 1 = 0$ $[a \neq 0]$ $\Rightarrow$ $b = -\frac{1}{8}$ and now $1 = 2ae^{-1/2} \;\Rightarrow\; a = \sqrt{e}/2$.

**51.** $f(x) = \tan x - x \;\Rightarrow\; f'(x) = \sec^2 x - 1 > 0$ for $0 < x < \frac{\pi}{2}$ since $\sec^2 x > 1$ for $0 < x < \frac{\pi}{2}$. So $f$ is increasing on $\left(0, \frac{\pi}{2}\right)$. Thus, $f(x) > f(0) = 0$ for $0 < x < \frac{\pi}{2} \;\Rightarrow\; \tan x - x > 0 \;\Rightarrow\; \tan x > x$ for $0 < x < \frac{\pi}{2}$.

**52.** (a) Let $f(x) = e^x - 1 - x$. Now $f(0) = e^0 - 1 = 0$, and for $x \geq 0$, we have $f'(x) = e^x - 1 \geq 0$. Now, since $f(0) = 0$ and $f$ is increasing on $[0, \infty)$, $f(x) \geq 0$ for $x \geq 0 \quad \Rightarrow \quad e^x - 1 - x \geq 0 \quad \Rightarrow \quad e^x \geq 1 + x$.

(b) Let $f(x) = e^x - 1 - x - \frac{1}{2}x^2$. Thus, $f'(x) = e^x - 1 - x$, which is positive for $x \geq 0$ by part (a). Thus, $f(x)$ is increasing on $(0, \infty)$, so on that interval, $0 = f(0) \leq f(x) = e^x - 1 - x - \frac{1}{2}x^2 \quad \Rightarrow \quad e^x \geq 1 + x + \frac{1}{2}x^2$.

(c) By part (a), the result holds for $n = 1$. Suppose that $e^x \geq 1 + x + \dfrac{x^2}{2!} + \cdots + \dfrac{x^k}{k!}$ for $x \geq 0$.

Let $f(x) = e^x - 1 - x - \dfrac{x^2}{2!} - \cdots - \dfrac{x^k}{k!} - \dfrac{x^{k+1}}{(k+1)!}$. Then $f'(x) = e^x - 1 - x - \cdots - \dfrac{x^k}{k!} \geq 0$ by assumption. Hence, $f(x)$ is increasing on $(0, \infty)$. So $0 \leq x$ implies that $0 = f(0) \leq f(x) = e^x - 1 - x - \cdots - \dfrac{x^k}{k!} - \dfrac{x^{k+1}}{(k+1)!}$, and hence $e^x \geq 1 + x + \cdots + \dfrac{x^k}{k!} + \dfrac{x^{k+1}}{(k+1)!}$ for $x \geq 0$. Therefore, for $x \geq 0$, $e^x \geq 1 + x + \dfrac{x^2}{2!} + \cdots + \dfrac{x^n}{n!}$ for every positive integer $n$, by mathematical induction.

**53.** We are given that $f$ is differentiable (and therefore continuous) everywhere. In particular, we can apply the Mean Value Theorem on the interval $[0, 4]$. There exists a number $c$ in $(0, 4)$ such that $f(4) - f(0) = f'(c)(4 - 0)$, so $f(4) = f(0) + 4f'(c) = -3 + 4f'(c)$. We are given that $f'(x) \leq 5$ for all $x$, so in particular we know that $f'(c) \leq 5$. Multiplying both sides of this inequality by 4, we have $4f'(c) \leq 20$, so $f(4) = -3 + 4f'(c) \leq -3 + 20 = 17$. The largest possible value for $f(4)$ is 17.

**54.** If $3 \leq f'(x) \leq 5$ for all $x$, then by the Mean Value Theorem, $f(8) - f(2) = f'(c) \cdot (8 - 2)$ for some $c$ in $[2, 8]$. ($f$ is differentiable for all $x$, so, in particular, $f$ is differentiable on $(2, 8)$ and continuous on $[2, 8]$. Thus, the hypotheses of the Mean Value Theorem are satisfied.) Since $f(8) - f(2) = 6f'(c)$ and $3 \leq f'(c) \leq 5$, it follows that $6 \cdot 3 \leq 6f'(c) \leq 6 \cdot 5 \quad \Rightarrow \quad 18 \leq f(8) - f(2) \leq 30$.

**55.** Let $g(t)$ and $h(t)$ be the position functions of the two runners and let $f(t) = g(t) - h(t)$. By hypothesis, $f(0) = g(0) - h(0) = 0$ and $f(b) = g(b) - h(b) = 0$, where $b$ is the finishing time. Then by the Mean Value Theorem, there is a time $c$, with $0 < c < b$, such that $f'(c) = \dfrac{f(b) - f(0)}{b - 0}$. But $f(b) = f(0) = 0$, so $f'(c) = 0$. Since $f'(c) = g'(c) - h'(c) = 0$, we have $g'(c) = h'(c)$. So at time $c$, both runners have the same speed $g'(c) = h'(c)$.

**56.** Let $v(t)$ be the velocity of the car $t$ hours after 2:00 P.M. Then $\dfrac{v(1/6) - v(0)}{1/6 - 0} = \dfrac{50 - 30}{1/6} = 120$. By the Mean Value Theorem, there is a number $c$ such that $0 < c < \frac{1}{6}$ with $v'(c) = 120$. Since $v'(t)$ is the acceleration at time $t$, the acceleration $c$ hours after 2:00 P.M. is exactly 120 mi/h$^2$.

**57.** Let the cubic function be $f(x) = ax^3 + bx^2 + cx + d \quad \Rightarrow \quad f'(x) = 3ax^2 + 2bx + c \quad \Rightarrow \quad f''(x) = 6ax + 2b$. So $f$ is CU when $6ax + 2b > 0 \quad \Leftrightarrow \quad x > -b/(3a)$, CD when $x < -b/(3a)$, and so the only point of inflection occurs when $x = -b/(3a)$. If the graph has three $x$-intercepts $x_1$, $x_2$ and $x_3$, then the expression for $f(x)$ must factor as $f(x) = a(x - x_1)(x - x_2)(x - x_3)$. Multiplying these factors together gives us $f(x) = a\left[x^3 - (x_1 + x_2 + x_3)x^2 + (x_1x_2 + x_1x_3 + x_2x_3)x - x_1x_2x_3\right]$. Equating the coefficients of the $x^2$-terms for the two forms of $f$ gives us $b = -a(x_1 + x_2 + x_3)$. Hence, the $x$-coordinate of the point of inflection is

$$-\frac{b}{3a} = -\frac{-a(x_1 + x_2 + x_3)}{3a} = \frac{x_1 + x_2 + x_3}{3}.$$

**58.** $P(x) = x^4 + cx^3 + x^2 \;\Rightarrow\; P'(x) = 4x^3 + 3cx^2 + 2x \;\Rightarrow\; P''(x) = 12x^2 + 6cx + 2$. The graph of $P''(x)$ is a parabola. If $P''(x)$ has two roots, then it changes sign twice and so has two inflection points. This happens when the discriminant of $P''(x)$ is positive, that is, $(6c)^2 - 4 \cdot 12 \cdot 2 > 0 \;\Leftrightarrow\; 36c^2 - 96 > 0 \;\Leftrightarrow\; |c| > \frac{2\sqrt{6}}{3} \approx 1.63$. If $36c^2 - 96 = 0 \;\Leftrightarrow\; c = \pm\frac{2\sqrt{6}}{3}$, $P''(x)$ is 0 at one point, but there is still no inflection point since $P''(x)$ never changes sign, and if $36c^2 - 96 < 0 \;\Leftrightarrow\; |c| < \frac{2\sqrt{6}}{3}$, then $P''(x)$ never changes sign, and so there is no inflection point.

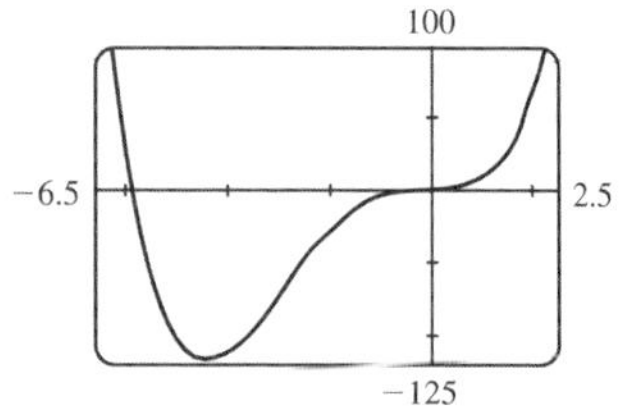


$c = 6$

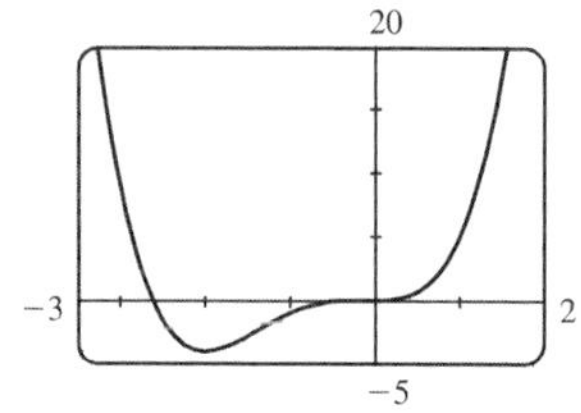


$c = 3$

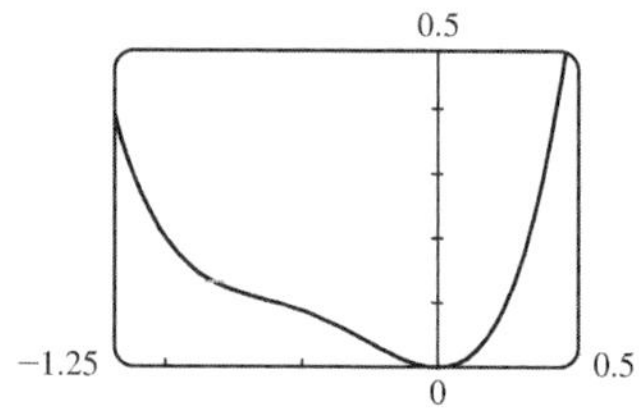


$c = 1.8$

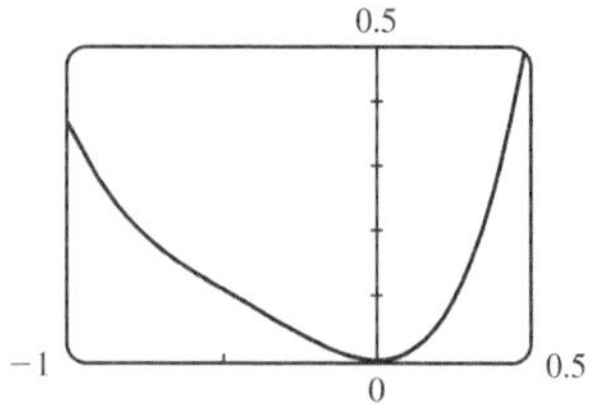


$c = \frac{2\sqrt{6}}{3}$

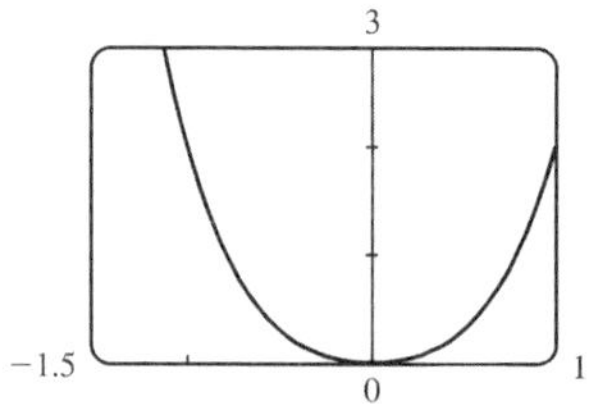


$c = 0$

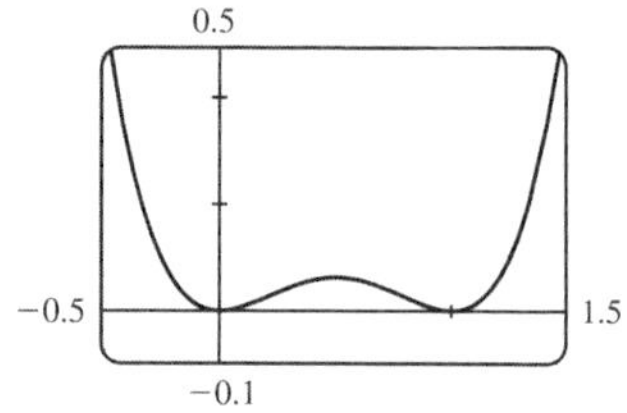


$c = -2$

For large positive $c$, the graph of $f$ has two inflection points and a large dip to the left of the $y$-axis. As $c$ decreases, the graph of $f$ becomes flatter for $x < 0$, and eventually the dip rises above the $x$-axis, and then disappears entirely, along with the inflection points. As $c$ continues to decrease, the dip and the inflection points reappear, to the right of the origin.

## 4.4 Graphing with Calculus and Calculators

**1.** $f(x) = 4x^4 - 32x^3 + 89x^2 - 95x + 29 \;\Rightarrow\; f'(x) = 16x^3 - 96x^2 + 178x - 95 \;\Rightarrow$ $f''(x) = 48x^2 - 192x + 178$. $f(x) = 0 \;\Leftrightarrow\; x \approx 0.5,\ 1.60$; $f'(x) = 0 \;\Leftrightarrow\; x \approx 0.92,\ 2.5,\ 2.58$ and $f''(x) = 0 \;\Leftrightarrow\; x \approx 1.46,\ 2.54$.

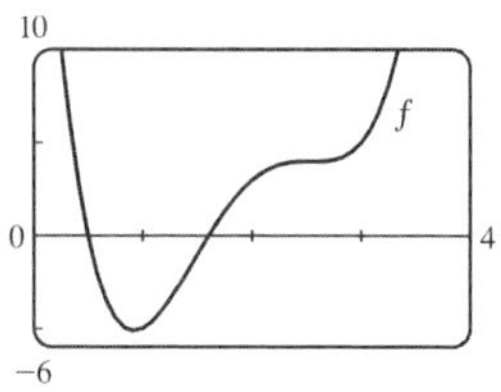


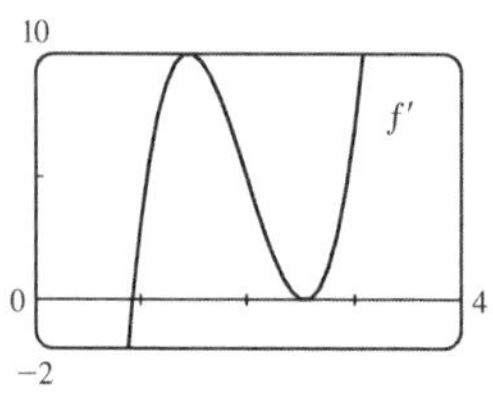


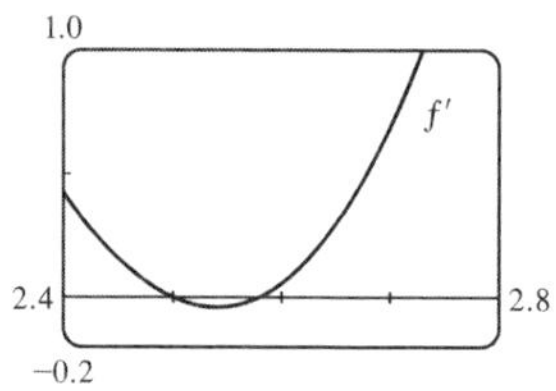


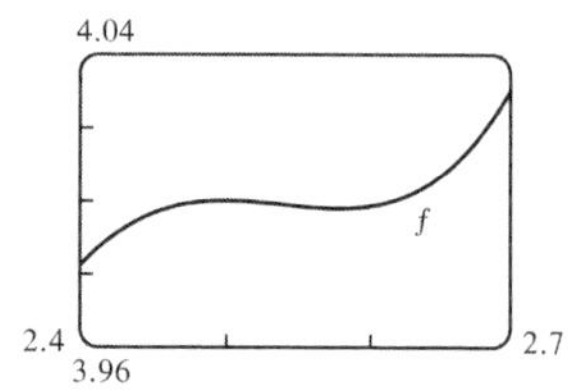


From the graphs of $f'$, we estimate that $f' < 0$ and that $f$ is decreasing on $(-\infty, 0.92)$ and $(2.5, 2.58)$, and that $f' > 0$ and $f$ is increasing on $(0.92, 2.5)$ and $(2.58, \infty)$ with local minimum values $f(0.92) \approx -5.12$ and $f(2.58) \approx 3.998$ and local maximum value $f(2.5) = 4$. The graphs of $f'$ make it clear that $f$ has a maximum and a minimum near $x = 2.5$, shown more clearly in the fourth graph.

From the graph of $f''$, we estimate that $f'' > 0$ and that $f$ is CU on $(-\infty, 1.46)$ and $(2.54, \infty)$, and that $f'' < 0$ and $f$ is CD on $(1.46, 2.54)$. There are inflection points at about $(1.46, -1.40)$ and $(2.54, 3.999)$.

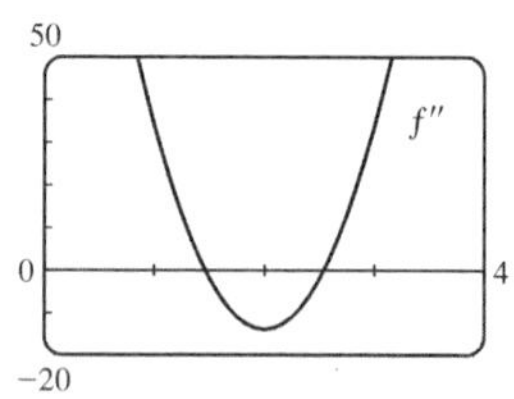

**2.** $f(x) = x^6 - 15x^5 + 75x^4 - 125x^3 - x \;\Rightarrow$

$f'(x) = 6x^5 - 75x^4 + 300x^3 - 375x^2 - 1 \;\Rightarrow\; f''(x) = 30x^4 - 300x^3 + 900x^2 - 750x.$

$f(x) = 0 \;\Leftrightarrow\; x = 0$ or $x \approx 5.33$; $f'(x) = 0 \;\Leftrightarrow\; x \approx 2.50, 4.95,$ or $5.05$;

$f''(x) = 0 \;\Leftrightarrow\; x = 0, 5$ or $x \approx 1.38, 3.62$.

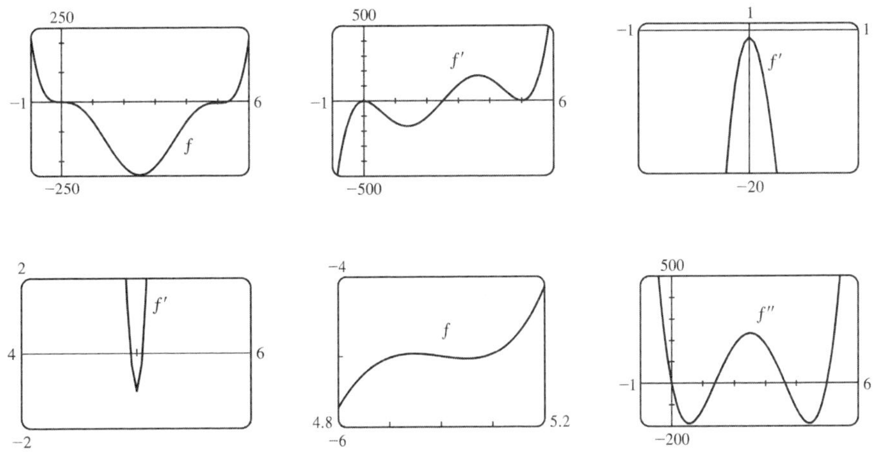

From the graphs of $f'$, we estimate that $f$ is decreasing on $(-\infty, 2.50)$, increasing on $(2.50, 4.95)$, decreasing on $(4.95, 5.05)$, and increasing on $(5.05, \infty)$, with local minimum values $f(2.50) \approx -246.6$ and $f(5.05) \approx -5.03$, and local maximum value $f(4.95) \approx -4.965$ (notice the second graph of $f$). From the graph of $f''$, we estimate that $f$ is CU on $(-\infty, 0)$, CD on $(0, 1.38)$, CU on $(1.38, 3.62)$, CD on $(3.62, 5)$, and CU on $(5, \infty)$. There are inflection points at $(0, 0)$ and $(5, -5)$, and at about $(1.38, -126.38)$ and $(3.62, -128.62)$.

**3.** $f(x) = \sqrt[3]{x^2 - 3x - 5} \;\Rightarrow\; f'(x) = \dfrac{1}{3}\dfrac{2x - 3}{(x^2 - 3x - 5)^{2/3}} \;\Rightarrow\; f''(x) = -\dfrac{2}{9}\dfrac{x^2 - 3x + 24}{(x^2 - 3x - 5)^{5/3}}$

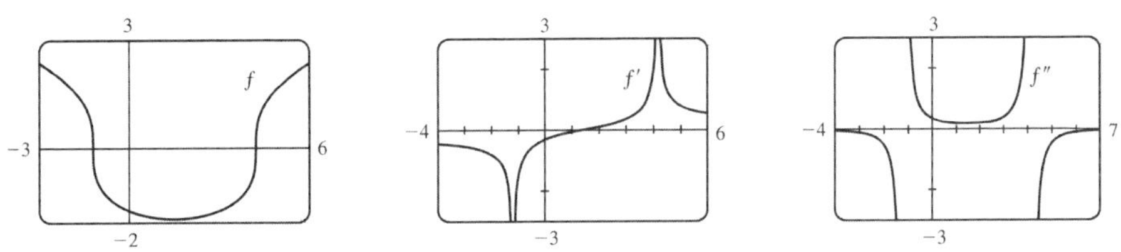

*Note:* With some CAS's, including Maple, it is necessary to define $f(x) = \dfrac{x^2 - 3x - 5}{|x^2 - 3x - 5|}\left|x^2 - 3x - 5\right|^{1/3}$, since the CAS does not compute real cube roots of negative numbers. We estimate from the graph of $f'$ that $f$ is increasing on $(1.5, \infty)$, and decreasing on $(-\infty, 1.5)$. $f$ has no maximum. Minimum value: $f(1.5) \approx -1.9$.

From the graph of $f''$, we estimate that $f$ is CU on $(-1.2, 4.2)$ and CD on $(-\infty, -1.2)$ and $(4.2, \infty)$. IP at $(-1.2, 0)$ and $(4.2, 0)$.

**4.** $f(x) = \dfrac{x^4 + x^3 - 2x^2 + 2}{x^2 + x - 2} \;\Rightarrow\; f'(x) = 2\,\dfrac{x^5 + 2x^4 - 3x^3 - 4x^2 + 2x - 1}{(x^2 + x - 2)^2} \;\Rightarrow$

$f''(x) = 2\,\dfrac{x^6 + 3x^5 - 3x^4 - 11x^3 + 12x^2 + 18x - 2}{(x^2 + x - 2)^3}$

[continued]

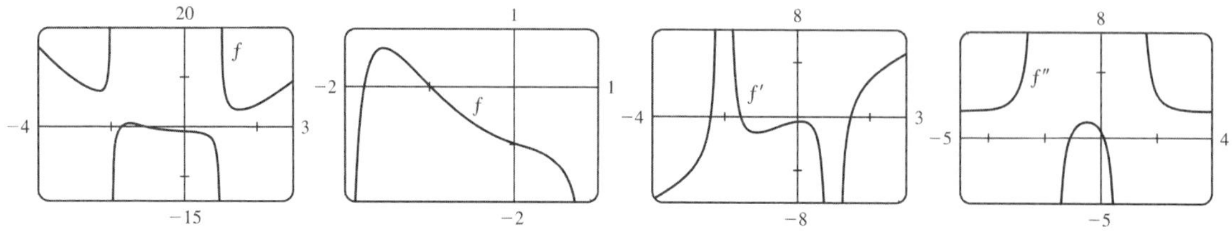


We estimate from the graph of $f'$ that $f$ is increasing on $(-2.4, -2)$, $(-2, -1.5)$ and $(1.5, \infty)$ and decreasing on $(-\infty, -2.4)$, $(-1.5, 1)$ and $(1, 1.5)$. Local maximum value: $f(-1.5) \approx 0.7$.

Local minimum values: $f(-2.4) \approx 7.2$, $f(1.5) \approx 3.4$. From the graph of $f''$, we estimate that $f$ is CU on $(-\infty, -2)$, $(-1.1, 0.1)$ and $(1, \infty)$ and CD on $(-2, -1.1)$ and $(0.1, 1)$. $f$ has IP at $(-1.1, 0.2)$ and $(0.1, -1.1)$.

**5.** $f(x) = \dfrac{x}{x^3 - x^2 - 4x + 1} \Rightarrow f'(x) = \dfrac{-2x^3 + x^2 + 1}{(x^3 - x^2 - 4x + 1)^2} \Rightarrow f''(x) = \dfrac{2(3x^5 - 3x^4 + 5x^3 - 6x^2 + 3x + 4)}{(x^3 - x^2 - 4x + 1)^3}$

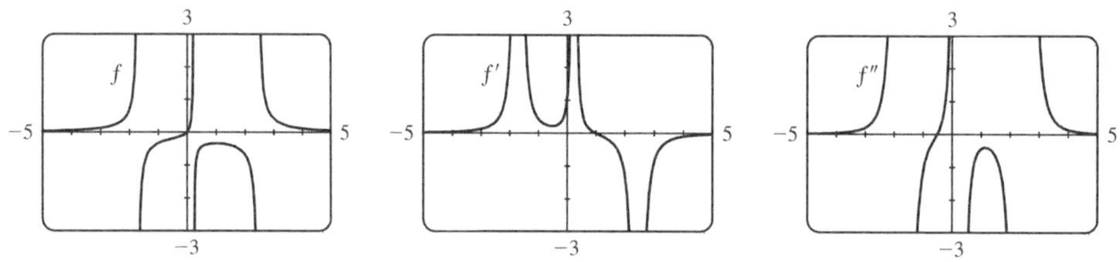


We estimate from the graph of $f$ that $y = 0$ is a horizontal asymptote, and that there are vertical asymptotes at $x = -1.7$, $x = 0.24$, and $x = 2.46$. From the graph of $f'$, we estimate that $f$ is increasing on $(-\infty, -1.7)$, $(-1.7, 0.24)$, and $(0.24, 1)$, and that $f$ is decreasing on $(1, 2.46)$ and $(2.46, \infty)$. There is a local maximum value at $f(1) = -\frac{1}{3}$. From the graph of $f''$, we estimate that $f$ is CU on $(-\infty, -1.7)$, $(-0.506, 0.24)$, and $(2.46, \infty)$, and that $f$ is CD on $(-1.7, -0.506)$ and $(0.24, 2.46)$. There is an inflection point at $(-0.506, -0.192)$.

**6.** $f(x) = \tan x + 5\cos x \Rightarrow f'(x) = \sec^2 x - 5\sin x \Rightarrow f''(x) = 2\sec^2 x\,\tan x - 5\,\cos x$. Since $f$ is periodic with period $2\pi$, and defined for all $x$ except odd multiples of $\frac{\pi}{2}$, we graph $f$ and its derivatives on $\left[-\frac{\pi}{2}, \frac{3\pi}{2}\right]$.

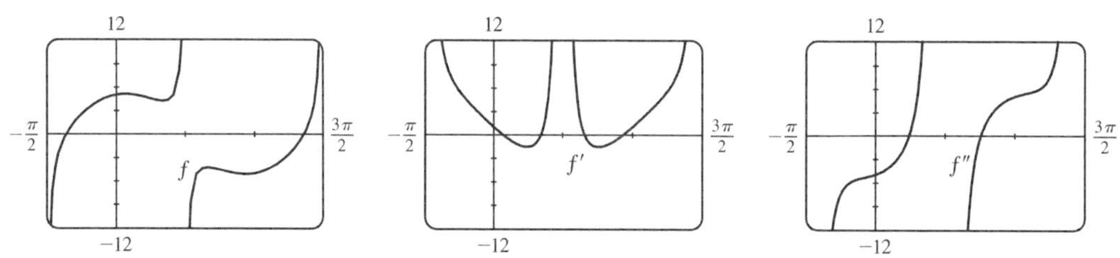


We estimate from the graph of $f'$ that $f$ is increasing on $\left(-\frac{\pi}{2}, 0.21\right)$, $\left(1.07, \frac{\pi}{2}\right)$, $\left(\frac{\pi}{2}, 2.07\right)$, and $\left(2.93, \frac{3\pi}{2}\right)$, and decreasing on $(0.21, 1.07)$ and $(2.07, 2.93)$. Local minimum values: $f(1.07) \approx 4.23$, $f(2.93) \approx -5.10$. Local maximum values: $f(0.21) \approx 5.10$, $f(2.07) \approx -4.23$.

From the graph of $f''$, we estimate that $f$ is CU on $\left(0.76, \frac{\pi}{2}\right)$ and $\left(2.38, \frac{3\pi}{2}\right)$, and CD on $\left(-\frac{\pi}{2}, 0.76\right)$ and $\left(\frac{\pi}{2}, 2.38\right)$. $f$ has IP at $(0.76, 4.57)$ and $(2.38, -4.57)$.

**7.** $f(x) = x^2 - 4x + 7\cos x,\ \ -4 \leq x \leq 4.\ \ f'(x) = 2x - 4 - 7\sin x \ \ \Rightarrow \ \ f''(x) = 2 - 7\cos x.$

$f(x) = 0 \ \Leftrightarrow \ x \approx 1.10$; $f'(x) = 0 \ \Leftrightarrow \ x \approx -1.49, -1.07$, or $2.89$; $f''(x) = 0 \ \Leftrightarrow \ x = \pm\cos^{-1}\left(\frac{2}{7}\right) \approx \pm 1.28.$

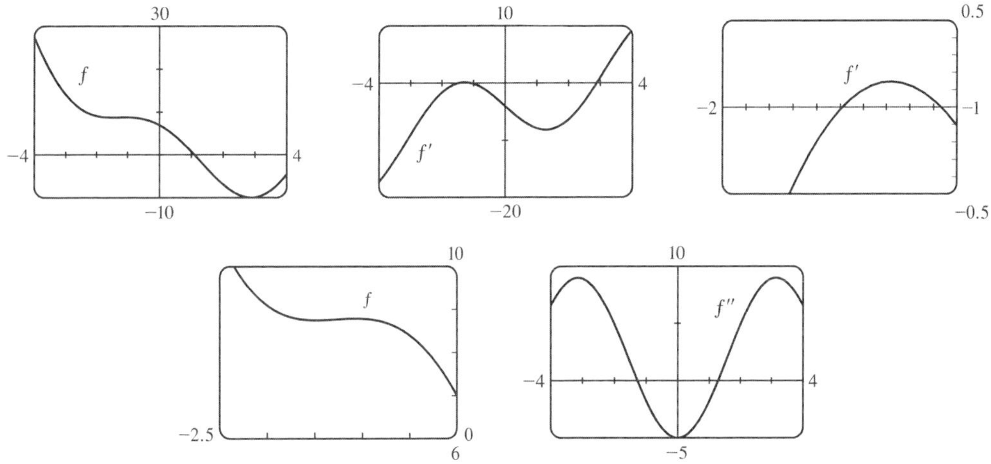


From the graphs of $f'$, we estimate that $f$ is decreasing ($f' < 0$) on $(-4, -1.49)$, increasing on $(-1.49, -1.07)$, decreasing on $(-1.07, 2.89)$, and increasing on $(2.89, 4)$, with local minimum values $f(-1.49) \approx 8.75$ and $f(2.89) \approx -9.99$ and local maximum value $f(-1.07) \approx 8.79$ (notice the second graph of $f$). From the graph of $f''$, we estimate that $f$ is CU ($f'' > 0$) on $(-4, -1.28)$, CD on $(-1.28, 1.28)$, and CU on $(1.28, 4)$. There are inflection points at about $(-1.28, 8.77)$ and $(1.28, -1.48)$.

**8.** $f(x) = \dfrac{e^x}{x^2 - 9} \ \ \Rightarrow \ \ f'(x) = \dfrac{e^x(x^2 - 2x - 9)}{(x^2 - 9)^2} \ \ \Rightarrow \ \ f''(x) = \dfrac{e^x(x^4 - 4x^3 - 12x^2 + 36x + 99)}{(x^2 - 9)^3}$

There are vertical asymptotes at $x = \pm 3$. It is difficult to show all the important features in one viewing rectangle, so we'll show $f$, $f'$, and $f''$ for $x < 3$ and also for $x > 3$.

**For $x < 3$:**

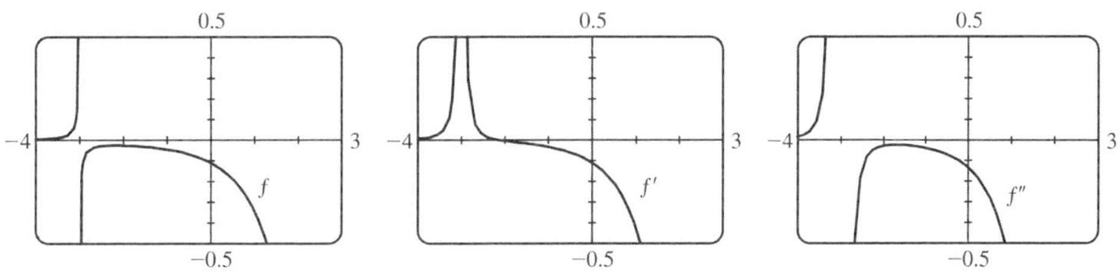


**For $x > 3$:**

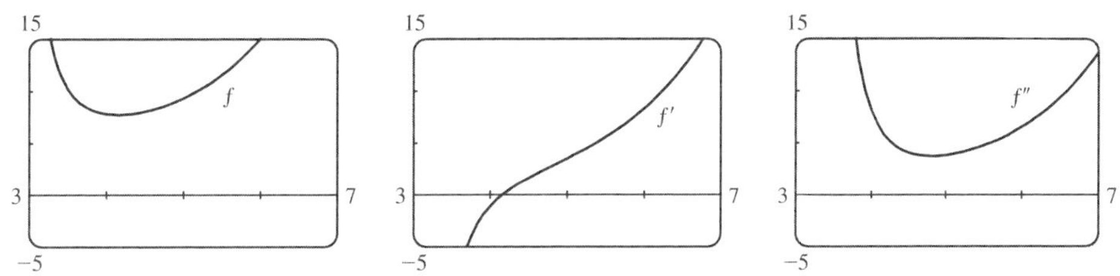


We estimate from the graphs of $f'$ and $f$ that $f$ is increasing on $(-\infty, -3)$, $(-3, -2.16)$, and $(4.16, \infty)$ and decreasing on $(-2.16, 3)$ and $(3, 4.16)$. There is a local maximum value of $f(-2.16) \approx -0.03$ and a local minimum value of $f(4.16) \approx 7.71$. From the graphs of $f''$, we see that $f$ is CU on $(-\infty, -3)$ and $(3, \infty)$ and CD on $(-3, 3)$. There is no inflection point.

**9.** $f(x) = 1 + \dfrac{1}{x} + \dfrac{8}{x^2} + \dfrac{1}{x^3} \;\Rightarrow\; f'(x) = -\dfrac{1}{x^2} - \dfrac{16}{x^3} - \dfrac{3}{x^4} = -\dfrac{1}{x^4}(x^2 + 16x + 3) \;\Rightarrow$

$f''(x) = \dfrac{2}{x^3} + \dfrac{48}{x^4} + \dfrac{12}{x^5} = \dfrac{2}{x^5}(x^2 + 24x + 6).$

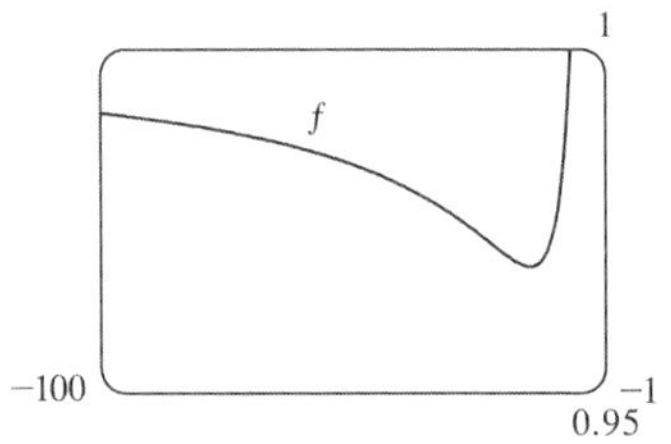


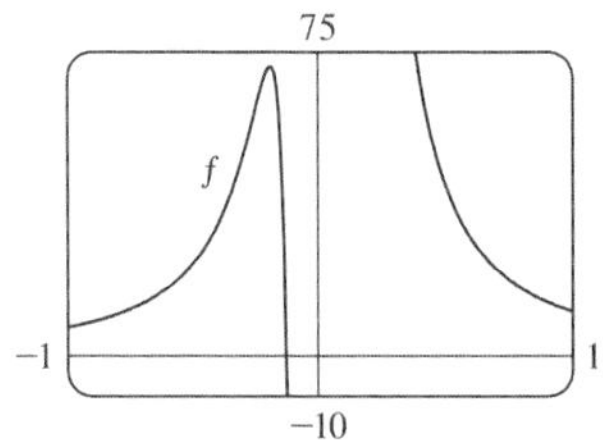


From the graphs, it appears that $f$ increases on $(-15.8, -0.2)$ and decreases on $(-\infty, -15.8)$, $(-0.2, 0)$, and $(0, \infty)$; that $f$ has a local minimum value of $f(-15.8) \approx 0.97$ and a local maximum value of $f(-0.2) \approx 72$; that $f$ is CD on $(-\infty, -24)$ and $(-0.25, 0)$ and is CU on $(-24, -0.25)$ and $(0, \infty)$; and that $f$ has IPs at $(-24, 0.97)$ and $(-0.25, 60)$.

To find the exact values, note that $f' = 0 \;\Rightarrow\; x = \dfrac{-16 \pm \sqrt{256 - 12}}{2} = -8 \pm \sqrt{61}$ $[\approx -0.19 \text{ and } -15.81]$. $f'$ is positive ($f$ is increasing) on $\left(-8 - \sqrt{61}, -8 + \sqrt{61}\right)$ and $f'$ is negative ($f$ is decreasing) on $\left(-\infty, -8 - \sqrt{61}\right)$, $\left(-8 + \sqrt{61}, 0\right)$, and $(0, \infty)$. $f'' = 0 \;\Rightarrow\; x = \dfrac{-24 \pm \sqrt{576 - 24}}{2} = -12 \pm \sqrt{138}$ $[\approx -0.25 \text{ and } -23.75]$. $f''$ is positive ($f$ is CU) on $\left(-12 - \sqrt{138}, -12 + \sqrt{138}\right)$ and $(0, \infty)$ and $f''$ is negative ($f$ is CD) on $\left(-\infty, -12 - \sqrt{138}\right)$ and $\left(-12 + \sqrt{138}, 0\right)$.

**10.** $f(x) = \dfrac{1}{x^8} - \dfrac{c}{x^4}$ $[c = 2 \times 10^8]$ $\Rightarrow\; f'(x) = -\dfrac{8}{x^9} + \dfrac{4c}{x^5} = -\dfrac{4}{x^9}(2 - cx^4) \;\Rightarrow$

$f''(x) = \dfrac{72}{x^{10}} - \dfrac{20c}{x^6} = \dfrac{4}{x^{10}}(18 - 5cx^4).$

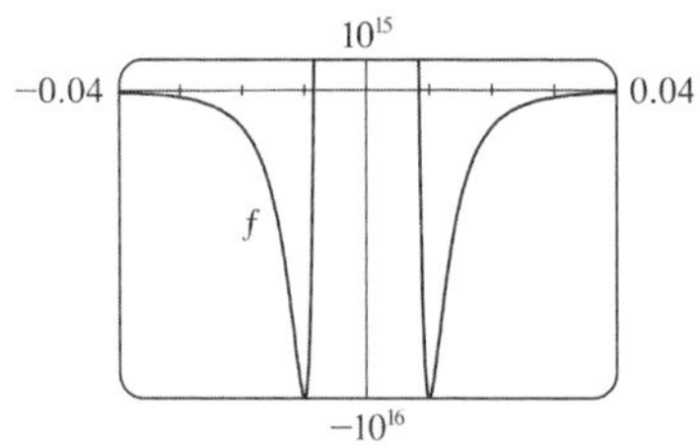


From the graph, it appears that $f$ increases on $(-0.01, 0)$ and $(0.01, \infty)$ and decreases on $(-\infty, -0.01)$ and $(0, 0.01)$; that $f$ has a local minimum value of $f(\pm 0.01) = -10^{16}$; and that $f$ is CU on $(-0.012, 0)$ and $(0, 0.012)$ and $f$ is CD on $(-\infty, -0.012)$ and $(0.012, \infty)$.

To find the exact values, note that $f' = 0 \;\Rightarrow\; x^4 = \dfrac{2}{c} \;\Rightarrow\; x \pm \sqrt[4]{\dfrac{2}{c}} = \pm\dfrac{1}{100}$ $[c = 2 \times 10^8]$. $f'$ is positive ($f$ is increasing) on $(-0.01, 0)$ and $(0.01, \infty)$ and $f'$ is negative ($f$ is decreasing) on $(-\infty, -0.01)$ and $(0, 0.01)$.

$f'' = 0 \;\Rightarrow\; x^4 = \dfrac{18}{5c} \;\Rightarrow\; x = \pm\sqrt[4]{\dfrac{18}{5c}} = \pm\dfrac{1}{100}\sqrt[4]{1.8}$ $[\approx \pm 0.0116]$. $f''$ is positive ($f$ is CU) on $\left(-\frac{1}{100}\sqrt[4]{1.8}, 0\right)$ and $\left(0, \frac{1}{100}\sqrt[4]{1.8}\right)$ and $f''$ is negative ($f$ is CD) on $\left(-\infty, -\frac{1}{100}\sqrt[4]{1.8}\right)$ and $\left(\frac{1}{100}\sqrt[4]{1.8}, \infty\right)$.

**11.** 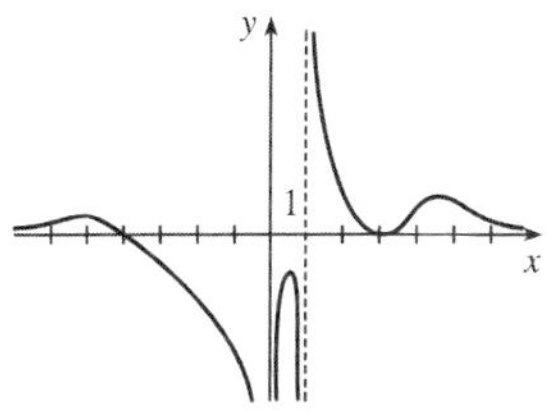


$f(x) = \dfrac{(x+4)(x-3)^2}{x^4(x-1)}$ has VA at $x = 0$ and at $x = 1$ since $\lim\limits_{x \to 0} f(x) = -\infty$, $\lim\limits_{x \to 1^-} f(x) = -\infty$ and $\lim\limits_{x \to 1^+} f(x) = \infty$.

$$f(x) = \frac{\dfrac{x+4}{x} \cdot \dfrac{(x-3)^2}{x^2}}{\dfrac{x^4}{x^3} \cdot (x-1)} \qquad \text{[dividing numerator and denominator by } x^3\text{]}$$

$$= \frac{(1+4/x)(1-3/x)^2}{x(x-1)} \to 0 \text{ as } x \to \pm\infty, \text{ so } f \text{ is asymptotic to the } x\text{-axis.}$$

Since $f$ is undefined at $x = 0$, it has no $y$-intercept. $f(x) = 0 \;\Rightarrow\; (x+4)(x-3)^2 = 0 \;\Rightarrow\; x = -4$ or $x = 3$, so $f$ has $x$-intercepts $-4$ and 3. Note, however, that the graph of $f$ is only tangent to the $x$-axis and does not cross it at $x = 3$, since $f$ is positive as $x \to 3^-$ and as $x \to 3^+$.

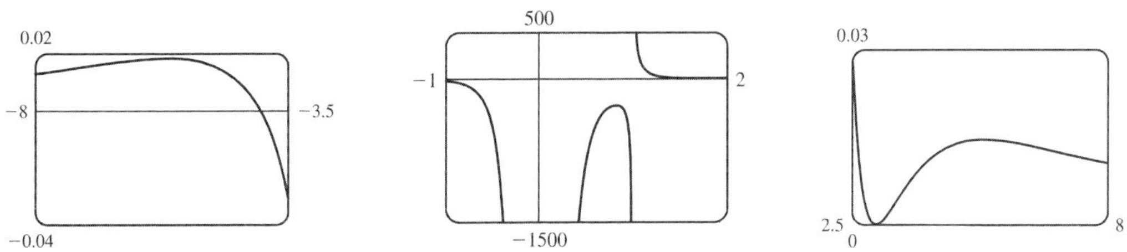


From these graphs, it appears that $f$ has three maximum values and one minimum value. The maximum values are approximately $f(-5.6) = 0.0182$, $f(0.82) = -281.5$ and $f(5.2) = 0.0145$ and we know (since the graph is tangent to the $x$-axis at $x = 3$) that the minimum value is $f(3) = 0$.

**12.** 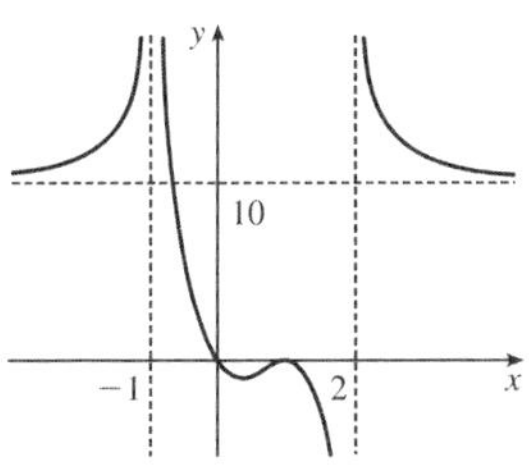


$f(x) = \dfrac{10x(x-1)^4}{(x-2)^3(x+1)^2}$ has VA at $x = -1$ and at $x = 2$ since $\lim\limits_{x \to -1} f(x) = \infty$, $\lim\limits_{x \to 2^-} f(x) = -\infty$ and $\lim\limits_{x \to 2^+} f(x) = \infty$.

$f(x) = \dfrac{10(1-1/x)^4}{(1-2/x)^3(1+1/x)^2} \to 10$ as $x \to \pm\infty$, so $f$ is asymptotic to the line $y = 10$. $f(0) = 0$, so $f$ has a $y$-intercept at 0. $f(x) = 0 \;\Rightarrow\; 10x(x-1)^4 = 0 \;\Rightarrow\; x = 0$ or $x = 1$. So $f$ has $x$-intercepts 0 and 1.

Note, however, that $f$ does not change sign at $x = 1$, so the graph is tangent to the $x$-axis and does not cross it. We know (since the graph is tangent to the $x$-axis at $x = 1$) that the maximum value is $f(1) = 0$. From the graphs it appears that the minimum value is about $f(0.2) = -0.1$.

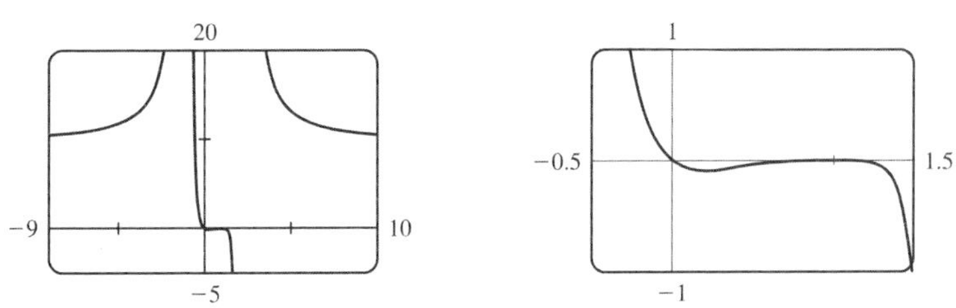

**13.** $f(x) = \dfrac{x^2(x+1)^3}{(x-2)^2(x-4)^4} \quad \Rightarrow \quad f'(x) = -\dfrac{x(x+1)^2(x^3+18x^2-44x-16)}{(x-2)^3(x-4)^5}$ (from CAS).

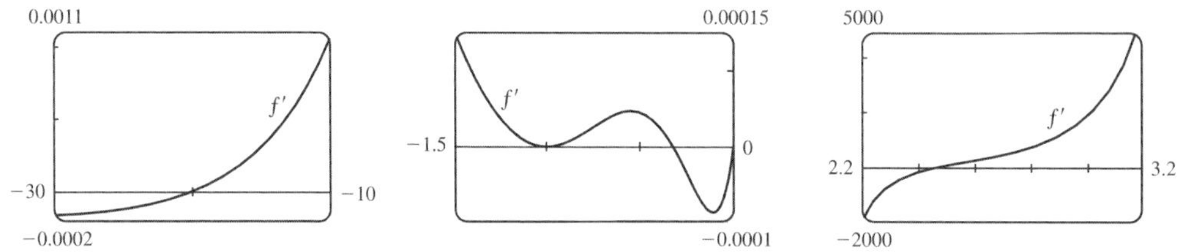


From the graphs of $f'$, it seems that the critical points which indicate extrema occur at $x \approx -20$, $-0.3$, and $2.5$, as estimated in Example 3. (There is another critical point at $x = -1$, but the sign of $f'$ does not change there.) We differentiate again, obtaining $f''(x) = 2\dfrac{(x+1)(x^6+36x^5+6x^4-628x^3+684x^2+672x+64)}{(x-2)^4(x-4)^6}$.

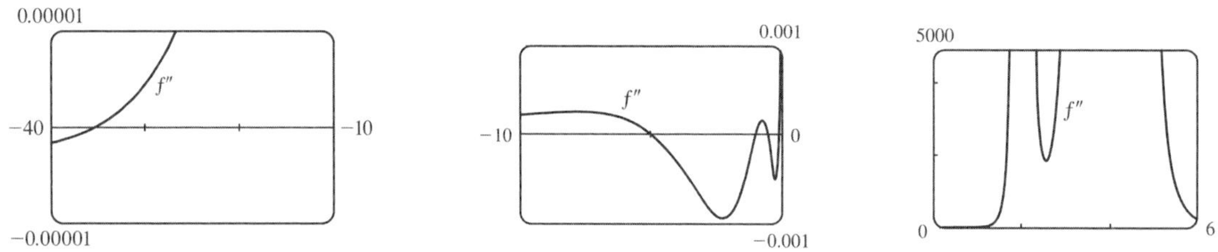


From the graphs of $f''$, it appears that $f$ is CU on $(-35.3, -5.0)$, $(-1, -0.5)$, $(-0.1, 2)$, $(2, 4)$ and $(4, \infty)$ and CD on $(-\infty, -35.3)$, $(-5.0, -1)$ and $(-0.5, -0.1)$. We check back on the graphs of $f$ to find the $y$-coordinates of the inflection points, and find that these points are approximately $(-35.3, -0.015)$, $(-5.0, -0.005)$, $(-1, 0)$, $(-0.5, 0.00001)$, and $(-0.1, 0.0000066)$.

**14.** $f(x) = \dfrac{10x(x-1)^4}{(x-2)^3(x+1)^2} \quad \Rightarrow \quad f'(x) = -20\dfrac{(x-1)^3(5x-1)}{(x-2)^4(x+1)^3}$ (from CAS).

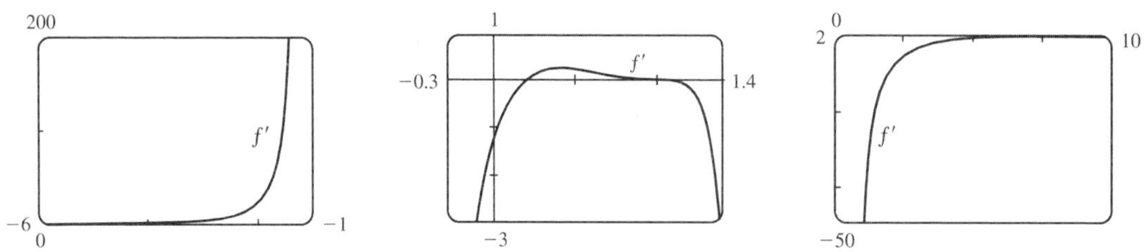


From the graphs of $f'$, we estimate that $f$ is increasing on $(-\infty, -1)$ and $(0.2, 1)$ and decreasing on $(-1, 0.2)$, $(1, 2)$ and $(2, \infty)$. Differentiating $f'(x)$, we get $f''(x) = 60\dfrac{(x-1)^2(5x^3-8x^2+17x-6)}{(x-2)^5(x+1)^4}$.

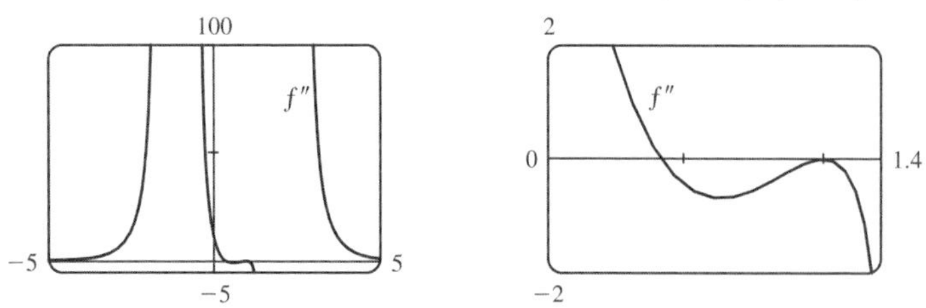


From the graphs of $f''$, it seems that $f$ is CU on $(-\infty, -1.0)$, $(-1.0, 0.4)$ and $(2.0, \infty)$, and CD on $(0.4, 2)$. There is an inflection point at about $(0.4, -0.06)$.

**15.** $y = f(x) = \dfrac{\sin^2 x}{\sqrt{x^2+1}}$ with $0 \le x \le 3\pi$. From a CAS, $y' = \dfrac{\sin x\left[2(x^2+1)\cos x - x\sin x\right]}{(x^2+1)^{3/2}}$ and

$$y'' = \frac{(4x^4+6x^2+5)\cos^2 x - 4x(x^2+1)\sin x\cos x - 2x^4 - 2x^2 - 3}{(x^2+1)^{5/2}}.$$

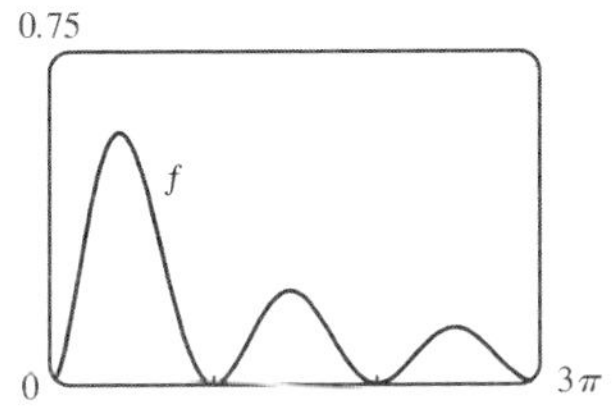


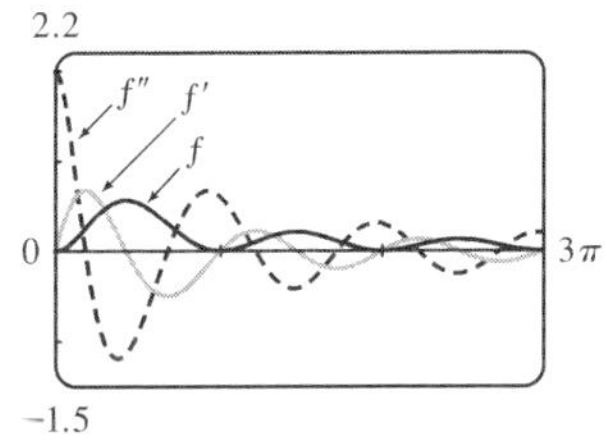


From the graph of $f'$ and the formula for $y'$, we determine that $y' = 0$ when $x = \pi, 2\pi, 3\pi$, or $x \approx 1.3, 4.6$, or $7.8$. So $f$ is increasing on $(0, 1.3)$, $(\pi, 4.6)$, and $(2\pi, 7.8)$. $f$ is decreasing on $(1.3, \pi)$, $(4.6, 2\pi)$, and $(7.8, 3\pi)$. Local maximum values: $f(1.3) \approx 0.6$, $f(4.6) \approx 0.21$, and $f(7.8) \approx 0.13$. Local minimum values: $f(\pi) = f(2\pi) = 0$. From the graph of $f''$, we see that $y'' = 0 \quad\Leftrightarrow\quad x \approx 0.6, 2.1, 3.8, 5.4, 7.0$, or $8.6$. So $f$ is CU on $(0, 0.6)$, $(2.1, 3.8)$, $(5.4, 7.0)$, and $(8.6, 3\pi)$. $f$ is CD on $(0.6, 2.1)$, $(3.8, 5.4)$, and $(7.0, 8.6)$. There are IP at $(0.6, 0.25)$, $(2.1, 0.31)$, $(3.8, 0.10)$, $(5.4, 0.11)$, $(7.0, 0.061)$, and $(8.6, 0.065)$.

**16.** $f(x) = \dfrac{2x-1}{\sqrt[4]{x^4+x+1}} \quad\Rightarrow\quad f'(x) = \dfrac{4x^3+6x+9}{4(x^4+x+1)^{5/4}} \quad\Rightarrow\quad f''(x) = -\dfrac{32x^6+96x^4+152x^3-48x^2+6x+21}{16(x^4+x+1)^{9/4}}$

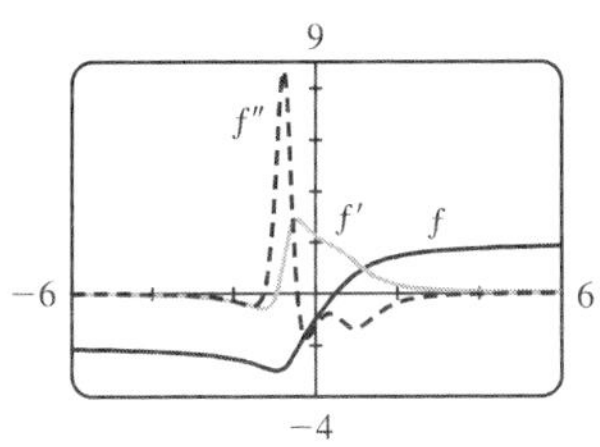


From the graph of $f'$, $f$ appears to be decreasing on $(-\infty, -0.94)$ and increasing on $(-0.94, \infty)$. There is a local minimum value of $f(-0.94) \approx -3.01$. From the graph of $f''$, $f$ appears to be CU on $(-1.25, -0.44)$ and CD on $(-\infty, -1.25)$ and $(-0.44, \infty)$. There are inflection points at $(-1.25, -2.87)$ and $(-0.44, -2.14)$.

**17.** $y = f(x) = \dfrac{1-e^{1/x}}{1+e^{1/x}}$. From a CAS, $y' = \dfrac{2e^{1/x}}{x^2(1+e^{1/x})^2}$ and $y'' = \dfrac{-2e^{1/x}(1-e^{1/x}+2x+2xe^{1/x})}{x^4(1+e^{1/x})^3}$.

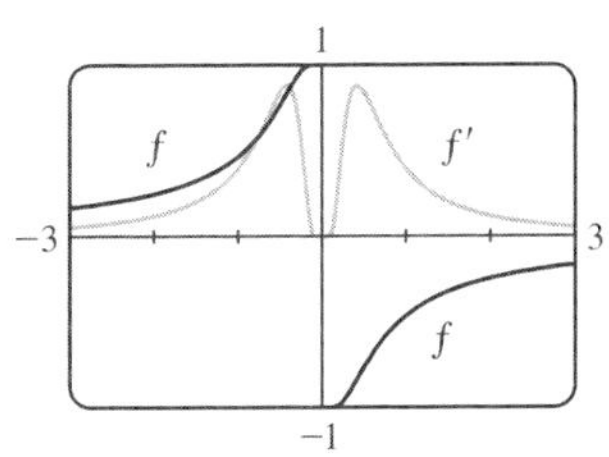


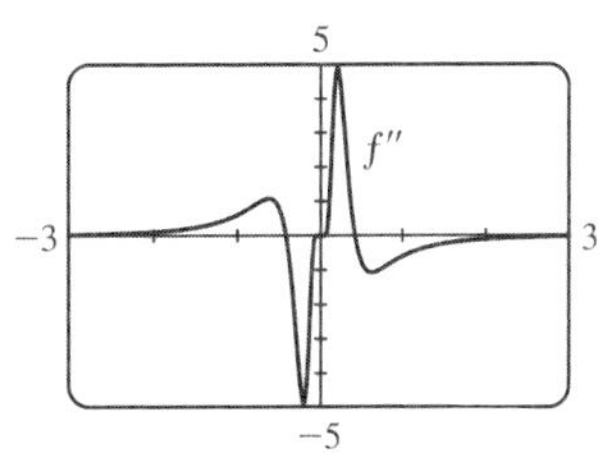


$f$ is an odd function defined on $(-\infty, 0) \cup (0, \infty)$. Its graph has no $x$- or $y$-intercepts. Since $\lim_{x\to\pm\infty} f(x) = 0$, the $x$-axis is a HA. $f'(x) > 0$ for $x \ne 0$, so $f$ is increasing on $(-\infty, 0)$ and $(0, \infty)$. It has no local extreme values. $f''(x) = 0$ for $x \approx \pm 0.417$, so $f$ is CU on $(-\infty, -0.417)$, CD on $(-0.417, 0)$, CU on $(0, 0.417)$, and CD on $(0.417, \infty)$. $f$ has IPs at $(-0.417, 0.834)$ and $(0.417, -0.834)$.

**18.** $y = f(x) = \dfrac{1}{1 + e^{\tan x}}$. From a CAS, $y' = -\dfrac{e^{\tan x}}{\cos^2 x\,(1 + e^{\tan x})^2}$ and

$y'' = -\dfrac{e^{\tan x}[e^{\tan x}(2\sin x\cos x - 1) + 2\sin x\cos x + 1]}{\cos^4 x\,(1 + e^{\tan x})^3}$. $f$ is a periodic function with period $\pi$ that has positive values throughout its domain, which consists of all real numbers except odd multiples of $\frac{\pi}{2}$ (that is, $\pm\frac{\pi}{2}, \pm\frac{3\pi}{2}, \pm\frac{5\pi}{2}$, and so on). $f$ has $y$-intercept $\frac{1}{2}$, but no $x$-intercepts. We graph $f$, $f'$, and $f''$ on one period, $\left(-\frac{\pi}{2}, \frac{\pi}{2}\right)$.

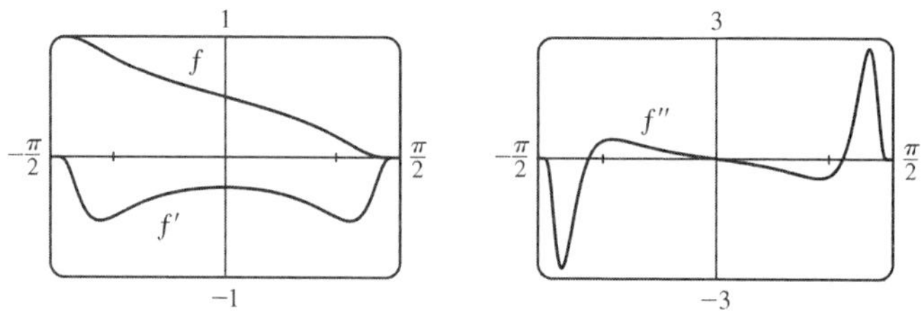


Since $f'(x) < 0$ for all $x$ in the domain of $f$, $f$ is decreasing on the intervals between odd multiples of $\frac{\pi}{2}$. $f''(x) = 0$ for $x = 0 + n\pi$ and for $x \approx \pm 1.124 + n\pi$, so $f$ is CD on $\left(-\frac{\pi}{2}, -1.124\right)$, CU on $(-1.124, 0)$, CD on $(0, 1.124)$, and CU on $\left(1.124, \frac{\pi}{2}\right)$. Since $f$ is periodic, this behavior repeats on every interval of length $\pi$. $f$ has IPs at $(-1.124 + n\pi, 0.890)$, $\left(n\pi, \frac{1}{2}\right)$, and $(1.124 + n\pi, 0.110)$.

**19.**

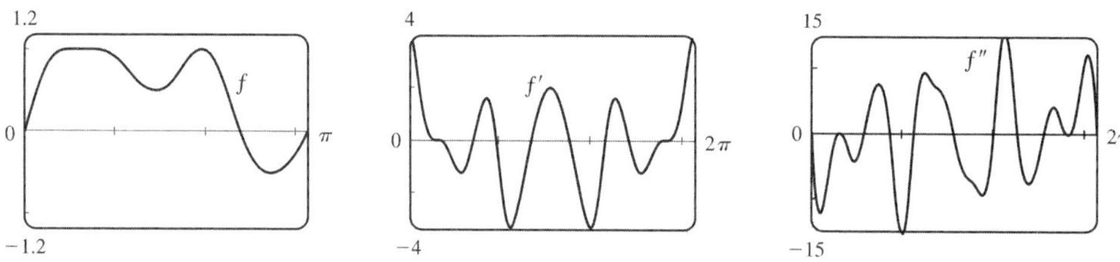


From the graph of $f(x) = \sin(x + \sin 3x)$ in the viewing rectangle $[0, \pi]$ by $[-1.2, 1.2]$, it looks like $f$ has two maxima and two minima. If we calculate and graph $f'(x) = [\cos(x + \sin 3x)]\,(1 + 3\cos 3x)$ on $[0, 2\pi]$, we see that the graph of $f'$ appears to be almost tangent to the $x$-axis at about $x = 0.7$. The graph of $f'' = -[\sin(x + \sin 3x)]\,(1 + 3\cos 3x)^2 + \cos(x + \sin 3x)(-9\sin 3x)$ is even more interesting near this $x$-value: it seems to just touch the $x$-axis.

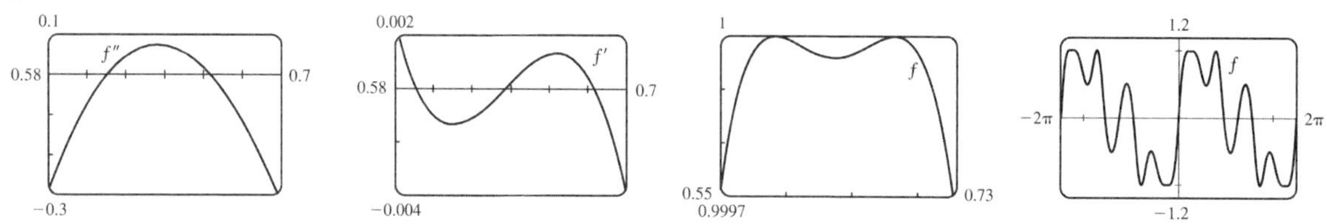


If we zoom in on this place on the graph of $f''$, we see that $f''$ actually does cross the axis twice near $x = 0.65$, indicating a change in concavity for a very short interval. If we look at the graph of $f'$ on the same interval, we see that it changes sign three times near $x = 0.65$, indicating that what we had thought was a broad extremum at about $x = 0.7$ actually consists of three extrema (two maxima and a minimum). These maximum values are roughly $f(0.59) = 1$ and $f(0.68) = 1$, and the minimum value is roughly $f(0.64) = 0.99996$. There are also a maximum value of about $f(1.96) = 1$ and minimum values of about $f(1.46) = 0.49$ and $f(2.73) = -0.51$. The points of inflection on $(0, \pi)$ are about $(0.61, 0.99998)$, $(0.66, 0.99998)$, $(1.17, 0.72)$, $(1.75, 0.77)$, and $(2.28, 0.34)$. On $(\pi, 2\pi)$, they are about $(4.01, -0.34)$, $(4.54, -0.77)$, $(5.11, -0.72)$, $(5.62, -0.99998)$, and $(5.67, -0.99998)$. There are also IP at $(0, 0)$ and $(\pi, 0)$. Note that the function is odd and periodic with period $2\pi$, and it is also rotationally symmetric about all points of the form $((2n + 1)\pi, 0)$, $n$ an integer.

**20.** From the graph, it appears that the leftmost point on the curve $x = t^4 - t^2$, $y = t + \ln t$ is about $(-0.25, 0.36)$. To find the exact coordinates, we find the value of $t$ for which the graph has a vertical tangent; that is, $dx/dt = 0$ and $dy/dt \neq 0$. $dx/dt = 0 \;\Leftrightarrow\; 4t^3 - 2t = 0 \;\Leftrightarrow\; 2t(2t^2 - 1) = 0$ $\Leftrightarrow\; 2t(\sqrt{2}t + 1)(\sqrt{2}t - 1) = 0 \;\Leftrightarrow\; t = 0$ or $\pm\frac{1}{\sqrt{2}}$. The negative and 0 roots are inadmissible since $y(t)$ is only defined for $t > 0$, so the leftmost point must be $\left(x\left(\frac{1}{\sqrt{2}}\right), y\left(\frac{1}{\sqrt{2}}\right)\right) = \left(\left(\frac{1}{\sqrt{2}}\right)^4 - \left(\frac{1}{\sqrt{2}}\right)^2, \frac{1}{\sqrt{2}} + \ln\frac{1}{\sqrt{2}}\right) = \left(-\frac{1}{4}, \frac{1}{\sqrt{2}} - \frac{1}{2}\ln 2\right)$.

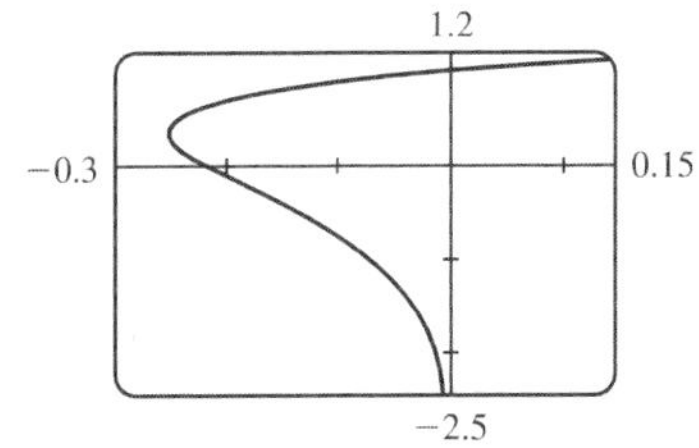


**21.**

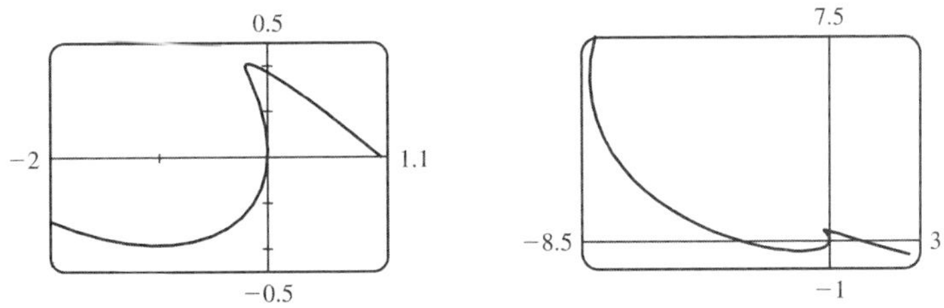


We graph the curve $x = t^4 - 2t^3 - 2t^2$, $y = t^3 - t$ in the viewing rectangle $[-2, 1.1]$ by $[-0.5, 0.5]$. This rectangle corresponds approximately to $t \in [-1, 0.8]$.We estimate that the curve has horizontal tangents at about $(-1, -0.4)$ and $(-0.17, 0.39)$ and vertical tangents at about $(0, 0)$ and $(-0.19, 0.37)$. We calculate $\dfrac{dy}{dx} = \dfrac{dy/dt}{dx/dt} = \dfrac{3t^2 - 1}{4t^3 - 6t^2 - 4t}$. The horizontal tangents occur when $dy/dt = 3t^2 - 1 = 0 \;\Leftrightarrow\; t = \pm\frac{1}{\sqrt{3}}$, so both horizontal tangents are shown in our graph. $t = \frac{1}{\sqrt{3}}$ corresponds to the point $\left(\frac{-2\sqrt{3}-5}{9}, \frac{-2\sqrt{3}}{9}\right) \approx (-0.94, -0.38)$ and $t = -\frac{1}{\sqrt{3}}$ corresponds to $\left(\frac{2\sqrt{3}-5}{9}, \frac{2\sqrt{3}}{9}\right) \approx (-0.17, 0.38)$. The vertical tangents occur when $dx/dt = 2t(2t^2 - 3t - 2) = 0 \;\Leftrightarrow$ $2t(2t + 1)(t - 2) = 0 \;\Leftrightarrow\; t = 0, -\frac{1}{2}$ or 2. It seems that we have missed one vertical tangent, and indeed if we plot the curve on the $t$-interval $[-1.2, 2.2]$ we see that there is another vertical tangent at $(-8, 6)$. The $t$-values and points at which there are vertical tangents are $t = 0$, $(0, 0)$; $t = -\frac{1}{2}$, $\left(-\frac{3}{16}, \frac{3}{8}\right)$; and $t = 2$, $(-8, 6)$.

**22.**

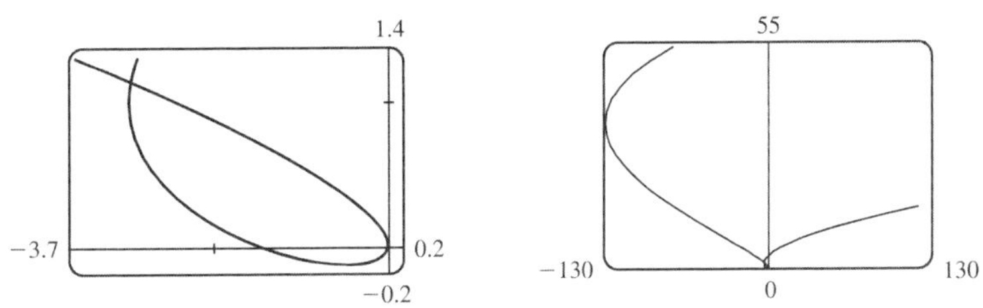


We graph the curve $x = t^4 + 4t^3 - 8t^2$, $y = 2t^2 - t$ in the viewing rectangle $[-3.7, 0.2]$ by $[-0.2, 1.4]$. It appears that there is a horizontal tangent at about $(-0.4, -0.1)$, and vertical tangents at about $(-3, 1)$ and $(0, 0)$. We calculate $\dfrac{dy}{dx} = \dfrac{dy/dt}{dx/dt} = \dfrac{4t - 1}{4t^3 + 12t^2 - 16t}$, so there is a horizontal tangent where $dy/dt = 4t - 1 = 0 \;\Leftrightarrow\; t = \frac{1}{4}$. This point (the lowest point) is shown in the first graph. Its coordinates are $\left(-\frac{111}{256}, -\frac{1}{8}\right)$. There are vertical tangents where $dx/dt = 4t^3 + 12t^2 - 16t = 0 \;\Leftrightarrow\; 4t(t^2 + 3t - 4) = 0 \;\Leftrightarrow\; 4t(t + 4)(t - 1) = 0$. We have missed one vertical tangent corresponding to $t = -4$, and if we plot the graph for $t \in [-5, 3]$, we see that the curve has another vertical tangent line at approximately $(-128, 36)$. The $t$-values and points at which there are vertical tangents are $t = 0$, $(0, 0)$; $t = -4$, $(-128, 36)$; and $t = 1$, $(-3, 1)$.

**23.** $x = t^3 - ct$, $y = t^2$. For $c = 0$, there is a cusp at $(0, 0)$. For $c < 0$, there is a local minimum at $(0, 0)$. For $c > 0$, there is a loop whose size increases as $c$ increases ($c = \frac{1}{2}$ and $c = 1$ are shown in the figure). The curve intersects itself on the $y$-axis; that is, when $x = 0 \quad\Leftrightarrow\quad t^3 - ct = 0 \quad\Leftrightarrow\quad t(t^2 - c) = 0 \quad\Leftrightarrow\quad t = 0, \pm\sqrt{c}$. Substituting $\pm\sqrt{c}$ for $t$ gives us $y = c$, so the point of intersection is $(0, c)$.

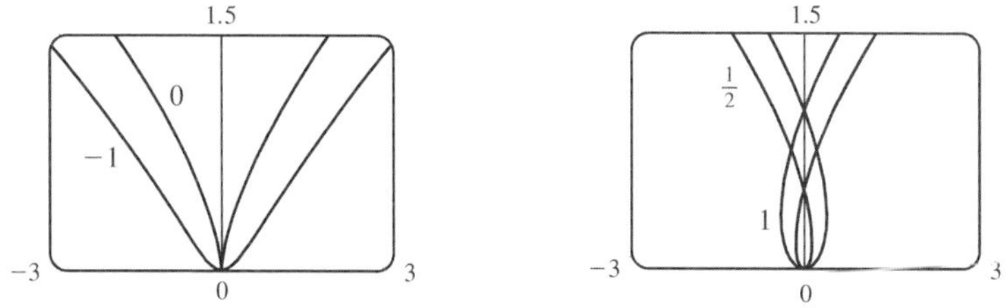


From the second figure, we see that the left- and rightmost points of the loop occur when there are vertical tangent lines. $dx/dt = 0 \quad\Rightarrow\quad 3t^2 - c = 0 \quad\Rightarrow\quad t = \pm\sqrt{c/3}$. The rightmost point occurs when $t = -\sqrt{c/3}$ and has coordinates $\left(\dfrac{2c\sqrt{3c}}{9}, \dfrac{c}{3}\right)$. The leftmost point occurs when $t = \sqrt{c/3}$ and has coordinates $\left(-\dfrac{2c\sqrt{3c}}{9}, \dfrac{c}{3}\right)$.

**24.** For $f(t) = C(e^{-at} - e^{-bt})$, $C$ affects only vertical stretching, so we let $C = 1$. From the first figure, we notice that the graphs all pass through the origin, approach the $t$-axis as $t$ increases, and approach $-\infty$ as $t \to -\infty$. Next we let $a = 2$ and produce the second figure.

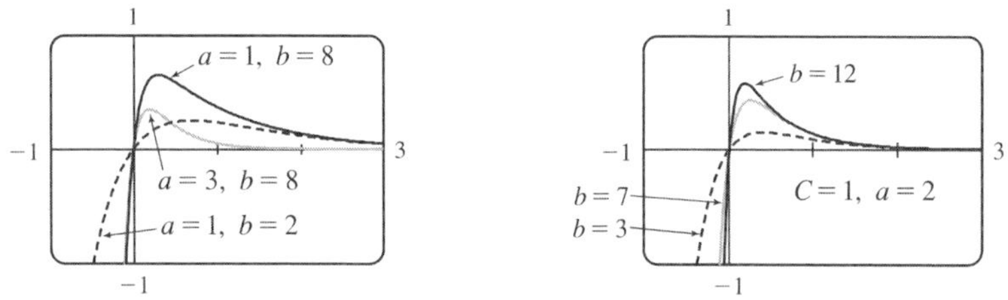


Here, as $b$ increases, the slope of the tangent at the origin increases and the local maximum value increases. $f(t) = e^{-2t} - e^{-bt} \quad\Rightarrow\quad f'(t) = be^{-bt} - 2e^{-2t}$. $f'(0) = b - 2$, which increases as $b$ increases.

$f'(t) = 0 \quad\Rightarrow\quad be^{-bt} = 2e^{-2t} \quad\Rightarrow\quad \dfrac{b}{2} = e^{(b-2)t} \quad\Rightarrow\quad \ln\dfrac{b}{2} = (b-2)t \quad\Rightarrow\quad t = t_1 = \dfrac{\ln b - \ln 2}{b - 2}$, which decreases as $b$ increases (the maximum is getting closer to the $y$-axis). $f(t_1) = \dfrac{(b-2)2^{2/(b-2)}}{b^{1+2/(b-2)}}$. We can show that this value increases as $b$ increases by considering it to be a function of $b$ and graphing its derivative with respect to $b$, which is always positive.

**25.** $f(x) = x^4 + cx^2 = x^2(x^2 + c)$. Note that $f$ is an even function. For $c \geq 0$, the only $x$-intercept is the point $(0, 0)$. We calculate $f'(x) = 4x^3 + 2cx = 4x(x^2 + \frac{1}{2}c) \quad\Rightarrow\quad f''(x) = 12x^2 + 2c$. If $c \geq 0$, $x = 0$ is the only critical point and there is no inflection point. As we can see from the examples, there is no change in the basic shape of the graph for $c > 0$; it merely becomes steeper as $c$ increases. For $c = 0$, the graph is the simple curve $y = x^4$. For $c < 0$, there are $x$-intercepts at 0 and at $\pm\sqrt{-c}$. Also, there is a maximum at $(0, 0)$, and there are minima at $\left(\pm\sqrt{-\frac{1}{2}c}, -\frac{1}{4}c^2\right)$. As $c \to -\infty$, the $x$-coordinates of these minima get larger in absolute value, and the minimum points move downward. There are inflection points at $\left(\pm\sqrt{-\frac{1}{6}c}, -\frac{5}{36}c^2\right)$, which also move away from the origin as $c \to -\infty$.

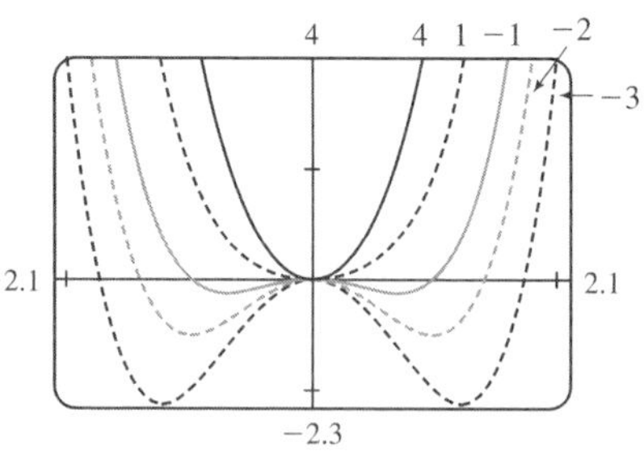

**26.** $f(x) = x^3 + cx = x(x^2 + c) \quad \Rightarrow \quad f'(x) = 3x^2 + c \quad \Rightarrow \quad f''(x) = 6x$

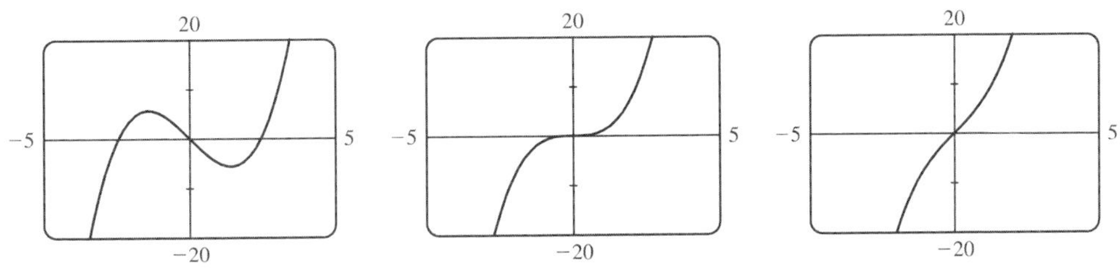


$c = -6$ $\qquad$ $c = 0$ $\qquad$ $c = 6$

$x$-intercepts: When $c \geq 0$, 0 is the only $x$-intercept. When $c < 0$, the $x$-intercepts are 0 and $\pm\sqrt{-c}$.

$y$-intercept $= f(0) = 0$. $f$ is odd, so the graph is symmetric with respect to the origin. $f''(x) < 0$ for $x < 0$ and $f''(x) > 0$ for $x > 0$, so $f$ is CD on $(-\infty, 0)$ and CU on $(0, \infty)$. The origin is the only inflection point.

If $c > 0$, then $f'(x) > 0$ for all $x$, so $f$ is increasing and has no local maximum or minimum.

If $c = 0$, then $f'(x) \geq 0$ with equality at $x = 0$, so again $f$ is increasing and has no local maximum or minimum.

If $c < 0$, then $f'(x) = 3[x^2 - (-c/3)] = 3\left(x + \sqrt{-c/3}\right)\left(x - \sqrt{-c/3}\right)$,

so $f'(x) > 0$ on $\left(-\infty, -\sqrt{-c/3}\right)$ and $\left(\sqrt{-c/3}, \infty\right)$; $f'(x) < 0$ on $\left(-\sqrt{-c/3}, \sqrt{-c/3}\right)$. It follows that $f\left(-\sqrt{-c/3}\right) = -\frac{2}{3}c\sqrt{-c/3}$ is a local maximum value and $f\left(\sqrt{-c/3}\right) = \frac{2}{3}c\sqrt{-c/3}$ is a local minimum value. As $c$ decreases (toward more negative values), the local maximum and minimum move further apart.

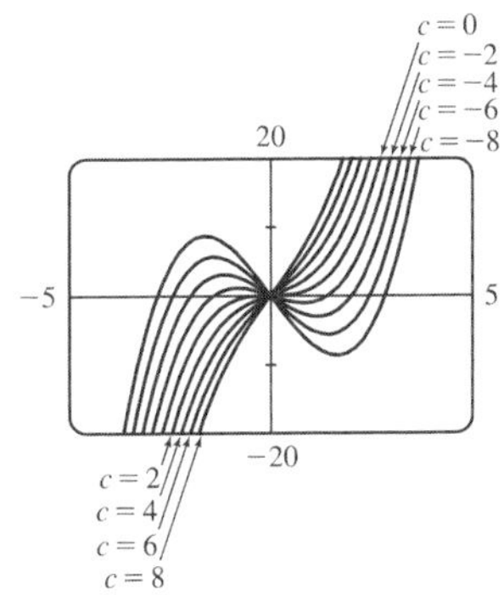


There is no absolute maximum or minimum value. The only transitional value of $c$ corresponding to a change in character of the graph is $c = 0$.

**27.**

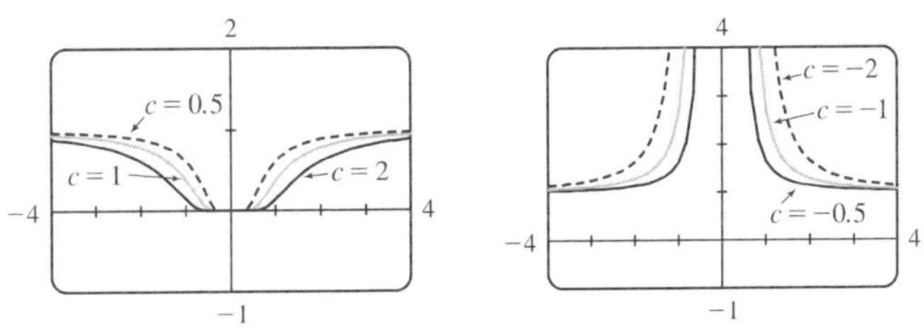


$c = 0$ is a transitional value—we get the graph of $y = 1$. For $c > 0$, we see that there is a HA at $y = 1$, and that the graph spreads out as $c$ increases. At first glance there appears to be a minimum at $(0, 0)$, but $f(0)$ is undefined, so there is no minimum or maximum. For $c < 0$, we still have the HA at $y = 1$, but the range is $(1, \infty)$ rather than $(0, 1)$. We also have a VA at $x = 0$. $f(x) = e^{-c/x^2} \quad \Rightarrow \quad f'(x) = e^{-c/x^2}\left(\dfrac{2c}{x^3}\right) \quad \Rightarrow \quad f''(x) = \dfrac{2c(2c - 3x^2)}{x^6 e^{c/x^2}}$. $f'(x) \neq 0$ and $f'(x)$ exists for all $x \neq 0$ (and 0 is not in the domain of $f$), so there are no maxima or minima. $f''(x) = 0 \quad \Rightarrow \quad x = \pm\sqrt{2c/3}$, so if $c > 0$, the inflection points spread out as $c$ increases, and if $c < 0$, there are no IP. For $c > 0$, there are IP at $\left(\pm\sqrt{2c/3}, e^{-3/2}\right)$. Note that the $y$-coordinate of the IP is constant.

**28.** We see that if $c \le 0$, $f(x) = \ln(x^2 + c)$ is only defined for $x^2 > -c \quad \Rightarrow \quad |x| > \sqrt{-c}$, and $\lim\limits_{x \to \sqrt{-c}^+} f(x) = \lim\limits_{x \to -\sqrt{-c}^-} f(x) = -\infty$, since $\ln y \to -\infty$ as $y \to 0$. Thus, for $c < 0$, there are vertical asymptotes at $x = \pm\sqrt{c}$, and as $c$ decreases (that is, $|c|$ increases), the asymptotes get further apart. For $c = 0$, $\lim\limits_{x \to 0} f(x) = -\infty$, so there is a vertical asymptote at $x = 0$. If $c > 0$, there are no asymptotes. To find the extrema and inflection points, we differentiate: $f(x) = \ln(x^2 + c) \quad \Rightarrow \quad f'(x) = \dfrac{1}{x^2 + c}(2x)$, so by the First Derivative Test there is a local and absolute minimum at $x = 0$. Differentiating again, we get

$$f''(x) = \frac{1}{x^2 + c}(2) + 2x\left[-\left(x^2 + c\right)^{-2}(2x)\right] = \frac{2(c - x^2)}{(x^2 + c)^2}.$$

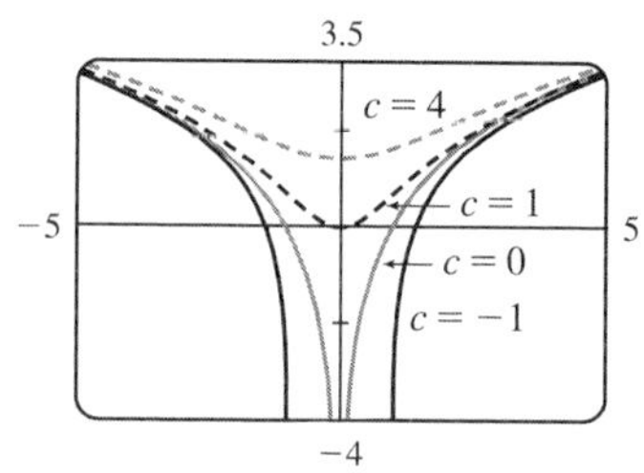


Now if $c \le 0$, $f''$ is always negative, so $f$ is concave down on both of the intervals on which it is defined. If $c > 0$, then $f''$ changes sign when $c = x^2 \quad \Leftrightarrow \quad x = \pm\sqrt{c}$. So for $c > 0$ there are inflection points at $x = \pm\sqrt{c}$, and as $c$ increases, the inflection points get further apart.

**29.** Note that $c = 0$ is a transitional value at which the graph consists of the $x$-axis. Also, we can see that if we substitute $-c$ for $c$, the function $f(x) = \dfrac{cx}{1 + c^2x^2}$ will be reflected in the $x$-axis, so we investigate only positive values of $c$ (except $c = -1$, as a demonstration of this reflective property). Also, $f$ is an odd function. $\lim\limits_{x \to \pm\infty} f(x) = 0$, so $y = 0$ is a horizontal asymptote for all $c$. We calculate $f'(x) = \dfrac{(1 + c^2x^2)c - cx(2c^2x)}{(1 + c^2x^2)^2} = -\dfrac{c(c^2x^2 - 1)}{(1 + c^2x^2)^2}$. $f'(x) = 0 \quad \Leftrightarrow \quad c^2x^2 - 1 = 0 \quad \Leftrightarrow \quad x = \pm 1/c$.

So there is an absolute maximum value of $f(1/c) = \frac{1}{2}$ and an absolute minimum value of $f(-1/c) = -\frac{1}{2}$. These extrema have the same value regardless of $c$, but the maximum points move closer to the $y$-axis as $c$ increases.

$$f''(x) = \frac{(-2c^3x)(1 + c^2x^2)^2 - (-c^3x^2 + c)[2(1 + c^2x^2)(2c^2x)]}{(1 + c^2x^2)^4}$$

$$= \frac{(-2c^3x)(1 + c^2x^2) + (c^3x^2 - c)(4c^2x)}{(1 + c^2x^2)^3} = \frac{2c^3x(c^2x^2 - 3)}{(1 + c^2x^2)^3}$$

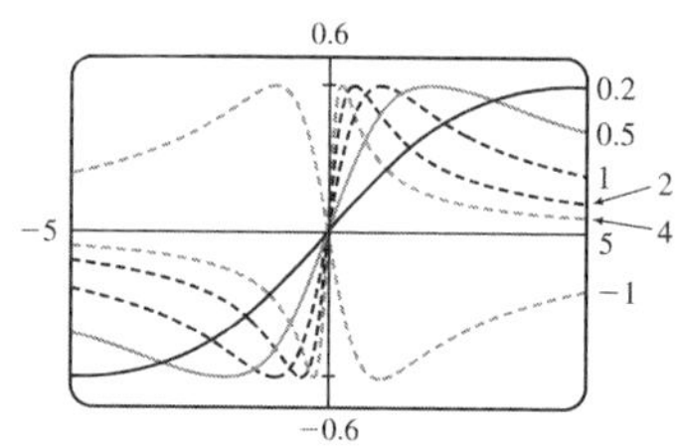


$f''(x) = 0 \quad \Leftrightarrow \quad x = 0$ or $\pm\sqrt{3}/c$, so there are inflection points at $(0, 0)$ and at $(\pm\sqrt{3}/c, \pm\sqrt{3}/4)$.

Again, the $y$-coordinate of the inflection points does not depend on $c$, but as $c$ increases, both inflection points approach the $y$-axis.

**30.** Note that $f(x) = \dfrac{1}{(1 - x^2)^2 + cx^2}$ is an even function, and also that $\lim\limits_{x \to \pm\infty} f(x) = 0$ for any value of $c$, so $y = 0$ is a horizontal asymptote. We calculate the derivatives:

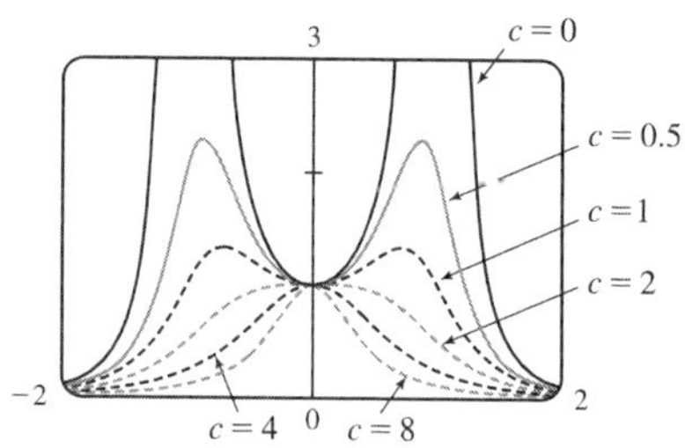


$$f'(x) = \frac{-4(1 - x^2)x + 2cx}{\left[(1 - x^2)^2 + cx^2\right]^2} = \frac{4x\left[x^2 + \left(\frac{1}{2}c - 1\right)\right]}{\left[(1 - x^2)^2 + cx^2\right]^2}, \text{ and}$$

$$f''(x) = 2\frac{10x^6 + (9c - 18)x^4 + (3c^2 - 12c + 6)x^2 + 2 - c}{\left[x^4 + (c - 2)x^2 + 1\right]^3}.$$

We first consider the case $c > 0$. Then the denominator of $f'$ is positive, that is, $(1 - x^2)^2 + cx^2 > 0$ for all $x$, so $f$ has domain $\mathbb{R}$ and also $f > 0$. If $\frac{1}{2}c - 1 \geq 0$; that is, $c \geq 2$, then the only critical point is $f(0) = 1$, a maximum. Graphing a few examples for $c \geq 2$ shows that there are two IP which approach the $y$-axis as $c \to \infty$.

$c = 2$ and $c = 0$ are transitional values of $c$ at which the shape of the curve changes. For $0 < c < 2$, there are three critical points: $f(0) = 1$, a minimum value, and $f\left(\pm\sqrt{1 - \frac{1}{2}c}\right) = \dfrac{1}{c(1 - c/4)}$, both maximum values. As $c$ decreases from 2 to 0, the maximum values get larger and larger, and the $x$-values at which they occur go from 0 to $\pm 1$. Graphs show that there are four inflection points for $0 < c < 2$, and that they get farther away from the origin, both vertically and horizontally, as $c \to 0^+$. For $c = 0$, the function is simply asymptotic to the $x$-axis and to the lines $x = \pm 1$, approaching $+\infty$ from both sides of each. The $y$-intercept is 1, and $(0, 1)$ is a local minimum. There are no inflection points. Now if $c < 0$, we can write

$f(x) = \dfrac{1}{(1 - x^2)^2 + cx^2} = \dfrac{1}{(1 - x^2)^2 - \left(\sqrt{-c}\,x\right)^2} = \dfrac{1}{\left(x^2 - \sqrt{-c}\,x - 1\right)\left(x^2 + \sqrt{-c}\,x - 1\right)}$. So $f$ has vertical asymptotes where $x^2 \pm \sqrt{-c}\,x - 1 = 0 \quad\Leftrightarrow\quad x = \left(-\sqrt{-c} \pm \sqrt{4 - c}\right)/\,2$ or $x = \left(\sqrt{-c} \pm \sqrt{4 - c}\right)/\,2$. As $c$ decreases, the two exterior asymptotes move away from the origin, while the two interior ones move toward it. We graph a few examples to see the behavior of the graph near the asymptotes, and the nature of the critical points $x = 0$ and $x = \pm\sqrt{1 - \frac{1}{2}c}$:

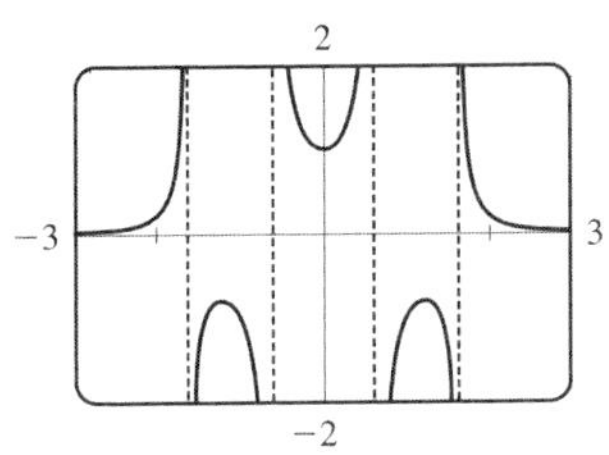

$c = -1$

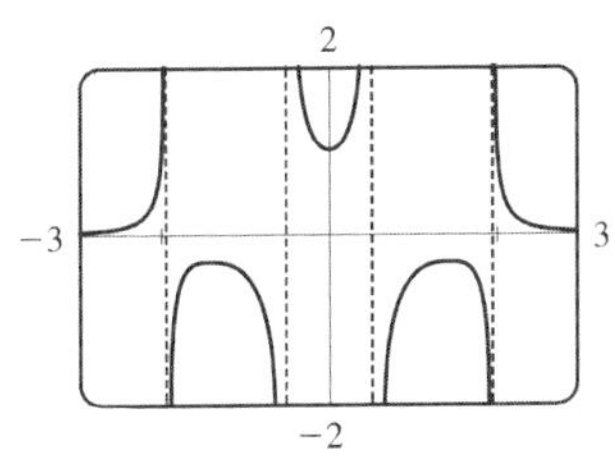

$c = -2$

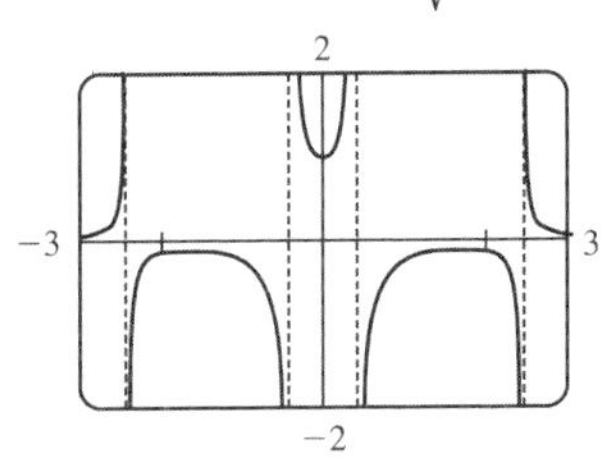

$c = -4$

We see that there is one local minimum value, $f(0) = 1$, and there are two local maximum values, $f\left(\pm\sqrt{1 - \frac{1}{2}c}\right) = \dfrac{1}{c(1 - c/4)}$ as before. As $c$ decreases, the $x$-values at which these maxima occur get larger, and the maximum values themselves approach 0, though they are always negative.

**31.** $f(x) = cx + \sin x \quad\Rightarrow\quad f'(x) = c + \cos x \quad\Rightarrow\quad f''(x) = -\sin x$

$f(-x) = -f(x)$, so $f$ is an odd function and its graph is symmetric with respect to the origin.

$f(x) = 0 \quad\Leftrightarrow\quad \sin x = -cx$, so 0 is always an $x$-intercept.

$f'(x) = 0 \quad\Leftrightarrow\quad \cos x = -c$, so there is no critical number when $|c| > 1$. If $|c| \leq 1$, then there are infinitely many critical numbers. If $x_1$ is the unique solution of $\cos x = -c$ in the interval $[0, \pi]$, then the critical numbers are $2n\pi \pm x_1$, where $n$ ranges over the integers. (Special cases: When $c = 1$, $x_1 = 0$; when $c = 0$, $x = \frac{\pi}{2}$; and when $c = -1$, $x_1 = \pi$.)

$f''(x) < 0 \quad\Leftrightarrow\quad \sin x > 0$, so $f$ is CD on intervals of the form $(2n\pi, (2n + 1)\pi)$. $f$ is CU on intervals of the form $((2n - 1)\pi, 2n\pi)$. The inflection points of $f$ are the points $(2n\pi, 2n\pi c)$, where $n$ is an integer.

If $c \geq 1$, then $f'(x) \geq 0$ for all $x$, so $f$ is increasing and has no extremum. If $c \leq -1$, then $f'(x) \leq 0$ for all $x$, so $f$ is decreasing and has no extremum. If $|c| < 1$, then $f'(x) > 0 \quad\Leftrightarrow\quad \cos x > -c \quad\Leftrightarrow\quad x$ is in an interval of the form $(2n\pi - x_1, 2n\pi + x_1)$ for some integer $n$. These are the intervals on which $f$ is increasing. Similarly, we find that $f$ is decreasing on the intervals of the form $(2n\pi + x_1, 2(n + 1)\pi - x_1)$. Thus, $f$ has local maxima at the points $2n\pi + x_1$, where $f$ has the values $c(2n\pi + x_1) + \sin x_1 = c(2n\pi + x_1) + \sqrt{1 - c^2}$, and $f$ has local minima at the points

$2n\pi - x_1$, where we have $f(2n\pi - x_1) = c(2n\pi - x_1) - \sin x_1 = c(2n\pi - x_1) - \sqrt{1 - c^2}$.

The transitional values of $c$ are $-1$ and $1$. The inflection points move vertically, but not horizontally, when $c$ changes. When $|c| \geq 1$, there is no extremum. For $|c| < 1$, the maxima are spaced $2\pi$ apart horizontally, as are the minima. The horizontal spacing between maxima and adjacent minima is regular (and equals $\pi$) when $c = 0$, but the horizontal space between a local maximum and the nearest local minimum shrinks as $|c|$ approaches 1.

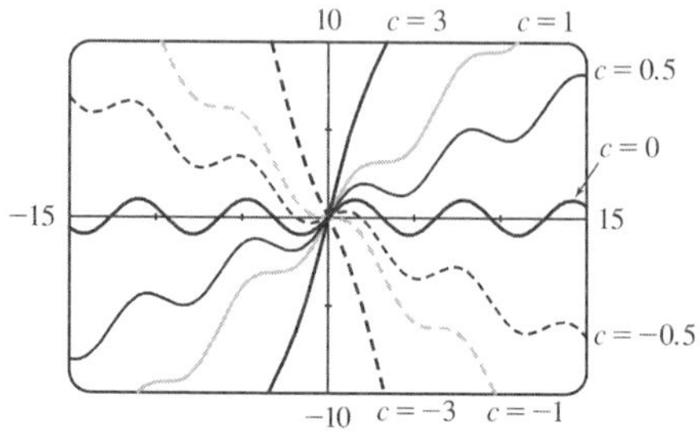


**32.** For $c = 0$, there is no inflection point; the curve is CU everywhere. If $c$ increases, the curve simply becomes steeper, and there are still no inflection points. If $c$ starts at 0 and decreases, a slight upward bulge appears near $x = 0$, so that there are two inflection points for any $c < 0$. This can be seen algebraically by calculating the second derivative: $f(x) = x^4 + cx^2 + x$ $\Rightarrow$ $f'(x) = 4x^3 + 2cx + 1$ $\Rightarrow$ $f''(x) = 12x^2 + 2c$. Thus, $f''(x) > 0$ when $c > 0$. For $c < 0$, there are inflection points when $x = \pm\sqrt{-\frac{1}{6}c}$. For $c = 0$, the graph has one critical number, at the absolute minimum somewhere around $x = -0.6$. As $c$ increases, the number of critical points does not change. If $c$ instead decreases from 0, we see that the graph eventually sprouts another local minimum, to the right of the origin, somewhere between $x = 1$ and $x = 2$. Consequently, there is also a maximum near $x = 0$.

After a bit of experimentation, we find that at $c = -1.5$, there appear to be two critical numbers: the absolute minimum at about $x = -1$, and a horizontal tangent with no extremum at about $x = 0.5$. For any $c$ smaller than this there will be 3 critical points, as shown in the graphs with $c = -3$ and with $c = -5$. To prove this algebraically, we calculate $f'(x) = 4x^3 + 2cx + 1$. Now if we substitute our value of $c = -1.5$, the formula for $f'(x)$ becomes $4x^3 - 3x + 1 = (x + 1)(2x - 1)^2$. This has a double root at $x = \frac{1}{2}$, indicating that the function has two critical points: $x = -1$ and $x = \frac{1}{2}$, just as we had guessed from the graph.

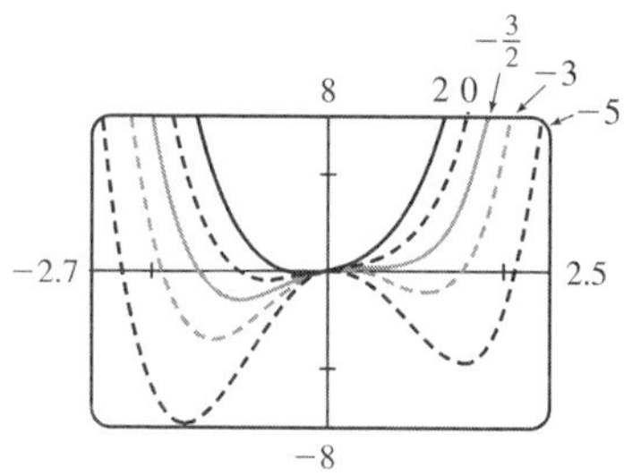


**33.** (a) $f(x) = cx^4 - 2x^2 + 1$. For $c = 0$, $f(x) = -2x^2 + 1$, a parabola whose vertex, $(0, 1)$, is the absolute maximum. For $c > 0$, $f(x) = cx^4 - 2x^2 + 1$ opens upward with two minimum points. As $c \to 0$, the minimum points spread apart and move downward; they are below the $x$-axis for $0 < c < 1$ and above for $c > 1$. For $c < 0$, the graph opens downward, and has an absolute maximum at $x = 0$ and no local minimum.

(b) $f'(x) = 4cx^3 - 4x = 4cx(x^2 - 1/c)$ $(c \neq 0)$. If $c \leq 0$, 0 is the only critical number. $f''(x) = 12cx^2 - 4$, so $f''(0) = -4$ and there is a local maximum at $(0, f(0)) = (0, 1)$, which lies on $y = 1 - x^2$. If $c > 0$, the critical numbers are 0 and $\pm 1/\sqrt{c}$. As before, there is a local maximum at $(0, f(0)) = (0, 1)$, which lies on $y = 1 - x^2$. $f''(\pm 1/\sqrt{c}) = 12 - 4 = 8 > 0$, so there is a local minimum at $x = \pm 1/\sqrt{c}$. Here $f(\pm 1/\sqrt{c}) = c(1/c^2) - 2/c + 1 = -1/c + 1$. But $(\pm 1/\sqrt{c}, -1/c + 1)$ lies on $y = 1 - x^2$ since $1 - (\pm 1/\sqrt{c})^2 = 1 - 1/c$.

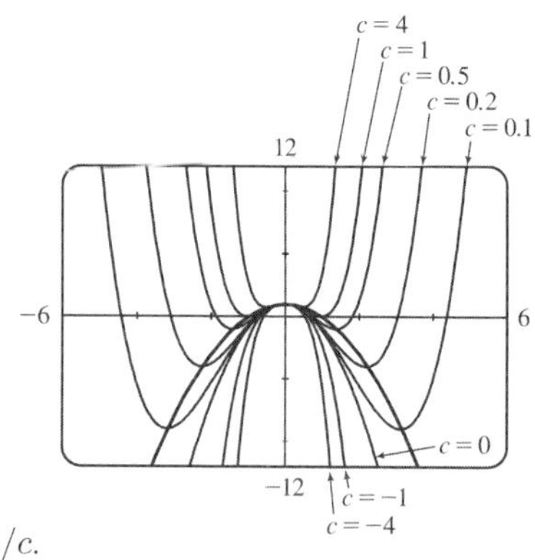

**34.** (a) $f(x) = 2x^3 + cx^2 + 2x \Rightarrow f'(x) = 6x^2 + 2cx + 2 = 2(3x^2 + cx + 1)$. $f'(x) = 0 \Leftrightarrow x = \dfrac{-c \pm \sqrt{c^2 - 12}}{6}$. So $f$ has critical points $\Leftrightarrow c^2 - 12 \geq 0 \Leftrightarrow |c| \geq 2\sqrt{3}$. For $c = \pm 2\sqrt{3}$, $f'(x) \geq 0$ on $(-\infty, \infty)$, so $f'$ does not change signs at $-c/6$, and there is no extremum. If $c^2 - 12 > 0$, then $f'$ changes from positive to negative at $x = \dfrac{-c - \sqrt{c^2 - 12}}{6}$ and from negative to positive at $x = \dfrac{-c + \sqrt{c^2 - 12}}{6}$. So $f$ has a local maximum at $x = \dfrac{-c - \sqrt{c^2 - 12}}{6}$ and a local minimum at $x = \dfrac{-c + \sqrt{c^2 - 12}}{6}$.

(b) Let $x_0$ be a critical number for $f(x)$. Then $f'(x_0) = 0 \Rightarrow$ $3x_0^2 + cx_0 + 1 = 0 \Leftrightarrow c = \dfrac{-1 - 3x_0^2}{x_0}$. Now

$$f(x_0) = 2x_0^3 + cx_0^2 + 2x_0 = 2x_0^3 + x_0^2\left(\frac{-1 - 3x_0^2}{x_0}\right) + 2x_0$$

$$= 2x_0^3 - x_0 - 3x_0^3 + 2x_0 = x_0 - x_0^3$$

So the point is $(x_0, y_0) = (x_0, x_0 - x_0^3)$; that is, the point lies on the curve $y = x - x^3$.

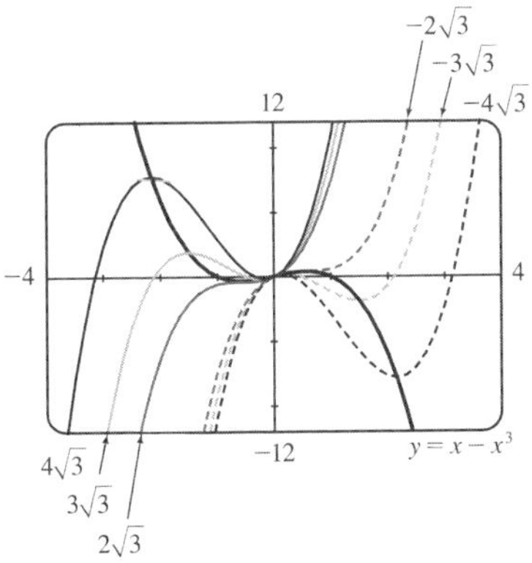


## 4.5 Indeterminate Forms and l'Hospital's Rule

Note: The use of l'Hospital's Rule is indicated by an H above the equal sign: $\overset{\text{H}}{=}$

**1.** (a) $\displaystyle\lim_{x \to a} \frac{f(x)}{g(x)}$ is an indeterminate form of type $\dfrac{0}{0}$.

(b) $\displaystyle\lim_{x \to a} \frac{f(x)}{p(x)} = 0$ because the numerator approaches 0 while the denominator becomes large.

(c) $\displaystyle\lim_{x \to a} \frac{h(x)}{p(x)} = 0$ because the numerator approaches a finite number while the denominator becomes large.

(d) If $\displaystyle\lim_{x \to a} p(x) = \infty$ and $f(x) \to 0$ through positive values, then $\displaystyle\lim_{x \to a} \frac{p(x)}{f(x)} = \infty$. [For example, take $a = 0$, $p(x) = 1/x^2$, and $f(x) = x^2$.] If $f(x) \to 0$ through negative values, then $\displaystyle\lim_{x \to a} \frac{p(x)}{f(x)} = -\infty$. [For example, take $a = 0$, $p(x) = 1/x^2$, and $f(x) = -x^2$.] If $f(x) \to 0$ through both positive and negative values, then the limit might not exist. [For example, take $a = 0$, $p(x) = 1/x^2$, and $f(x) = x$.]

(e) $\displaystyle\lim_{x \to a} \frac{p(x)}{q(x)}$ is an indeterminate form of type $\dfrac{\infty}{\infty}$.

**2.** (a) $\displaystyle\lim_{x \to a} [f(x)p(x)]$ is an indeterminate form of type $0 \cdot \infty$.

(b) When $x$ is near $a$, $p(x)$ is large and $h(x)$ is near 1, so $h(x)p(x)$ is large. Thus, $\displaystyle\lim_{x \to a} [h(x)p(x)] = \infty$.

(c) When $x$ is near $a$, $p(x)$ and $q(x)$ are both large, so $p(x)q(x)$ is large. Thus, $\displaystyle\lim_{x \to a} [p(x)q(x)] = \infty$.

**3.** (a) When $x$ is near $a$, $f(x)$ is near 0 and $p(x)$ is large, so $f(x) - p(x)$ is large negative. Thus, $\displaystyle\lim_{x \to a} [f(x) - p(x)] = -\infty$.

(b) $\displaystyle\lim_{x \to a} [p(x) - q(x)]$ is an indeterminate form of type $\infty - \infty$.

(c) When $x$ is near $a$, $p(x)$ and $q(x)$ are both large, so $p(x) + q(x)$ is large. Thus, $\displaystyle\lim_{x \to a} [p(x) + q(x)] = \infty$.

4. (a) $\lim_{x\to a} [f(x)]^{g(x)}$ is an indeterminate form of type $0^0$.

(b) If $y = [f(x)]^{p(x)}$, then $\ln y = p(x)\ln f(x)$. When $x$ is near $a$, $p(x) \to \infty$ and $\ln f(x) \to -\infty$, so $\ln y \to -\infty$. Therefore, $\lim_{x\to a} [f(x)]^{p(x)} = \lim_{x\to a} y = \lim_{x\to a} e^{\ln y} = 0$, provided $f^p$ is defined.

(c) $\lim_{x\to a} [h(x)]^{p(x)}$ is an indeterminate form of type $1^\infty$.

(d) $\lim_{x\to a} [p(x)]^{f(x)}$ is an indeterminate form of type $\infty^0$.

(e) If $y = [p(x)]^{q(x)}$, then $\ln y = q(x)\ln p(x)$. When $x$ is near $a$, $q(x) \to \infty$ and $\ln p(x) \to \infty$, so $\ln y \to \infty$. Therefore, $\lim_{x\to a} [p(x)]^{q(x)} = \lim_{x\to a} y = \lim_{x\to a} e^{\ln y} = \infty$.

(f) $\lim_{x\to a} \sqrt[q(x)]{p(x)} = \lim_{x\to a} [p(x)]^{1/q(x)}$ is an indeterminate form of type $\infty^0$.

5. This limit has the form $\frac{0}{0}$. We can simply factor the numerator to evaluate this limit.

$$\lim_{x\to -1} \frac{x^2-1}{x+1} = \lim_{x\to -1} \frac{(x+1)(x-1)}{x+1} = \lim_{x\to -1} (x-1) = -2$$

6. $\lim_{x\to 1} \dfrac{x^a - 1}{x^b - 1} \overset{\text{H}}{=} \lim_{x\to 1} \dfrac{ax^{a-1}}{bx^{b-1}} = \dfrac{a}{b}$

7. This limit has the form $\frac{0}{0}$. $\lim_{x\to(\pi/2)^+} \dfrac{\cos x}{1-\sin x} \overset{\text{H}}{=} \lim_{x\to(\pi/2)^+} \dfrac{-\sin x}{-\cos x} = \lim_{x\to(\pi/2)^+} \tan x = -\infty$.

8. $\lim_{x\to 0} \dfrac{x + \tan x}{\sin x} \overset{\text{H}}{=} \lim_{x\to 0} \dfrac{1+\sec^2 x}{\cos x} = \dfrac{1+1^2}{1} = 2$

9. This limit has the form $\frac{0}{0}$. $\lim_{t\to 0} \dfrac{e^t - 1}{t^3} \overset{\text{H}}{=} \lim_{t\to 0} \dfrac{e^t}{3t^2} = \infty$ since $e^t \to 1$ and $3t^2 \to 0^+$ as $t \to 0$.

10. $\lim_{t\to 0} \dfrac{e^{3t}-1}{t} \overset{\text{H}}{=} \lim_{t\to 0} \dfrac{3e^{3t}}{1} = 3$

11. This limit has the form $\frac{0}{0}$. $\lim_{x\to 0} \dfrac{\tan px}{\tan qx} \overset{\text{H}}{=} \lim_{x\to 0} \dfrac{p\sec^2 px}{q\sec^2 qx} = \dfrac{p(1)^2}{q(1)^2} = \dfrac{p}{q}$

12. $\lim_{\theta\to\pi/2} \dfrac{1-\sin\theta}{\csc\theta} = \dfrac{0}{1} = 0$. L'Hospital's Rule does not apply.

13. $\lim_{x\to 0^+} [(\ln x)/x] = -\infty$ since $\ln x \to -\infty$ as $x \to 0^+$ and dividing by small values of $x$ just increases the magnitude of the quotient $(\ln x)/x$. L'Hospital's Rule does not apply.

14. $\lim_{x\to\infty} \dfrac{\ln\ln x}{x} \overset{\text{H}}{=} \lim_{x\to\infty} \dfrac{\frac{1}{\ln x}\cdot\frac{1}{x}}{1} = \lim_{x\to\infty} \dfrac{1}{x\ln x} = 0$

15. This limit has the form $\frac{0}{0}$. $\lim_{t\to 0} \dfrac{5^t - 3^t}{t} \overset{\text{H}}{=} \lim_{t\to 0} \dfrac{5^t \ln 5 - 3^t \ln 3}{1} = \ln 5 - \ln 3 = \ln \frac{5}{3}$

16. $\lim_{x\to 1} \dfrac{\ln x}{\sin \pi x} \overset{\text{H}}{=} \lim_{x\to 1} \dfrac{1/x}{\pi\cos\pi x} = \dfrac{1}{\pi(-1)} = -\dfrac{1}{\pi}$

17. This limit has the form $\frac{0}{0}$. $\lim_{x\to 0} \dfrac{e^x - 1 - x}{x^2} \overset{\text{H}}{=} \lim_{x\to 0} \dfrac{e^x - 1}{2x} \overset{\text{H}}{=} \lim_{x\to 0} \dfrac{e^x}{2} = \dfrac{1}{2}$

18. This limit has the form $\frac{\infty}{\infty}$. $\lim_{x\to\infty} \dfrac{e^x}{x^3} \overset{\text{H}}{=} \lim_{x\to\infty} \dfrac{e^x}{3x^2} \overset{\text{H}}{=} \lim_{x\to\infty} \dfrac{e^x}{6x} \overset{\text{H}}{=} \lim_{x\to\infty} \dfrac{e^x}{6} = \infty$

19. This limit has the form $\frac{\infty}{\infty}$. $\lim_{x\to\infty} \dfrac{x}{\ln(1+2e^x)} \overset{\text{H}}{=} \lim_{x\to\infty} \dfrac{1}{\dfrac{1}{1+2e^x}\cdot 2e^x} = \lim_{x\to\infty} \dfrac{1+2e^x}{2e^x} \overset{\text{H}}{=} \lim_{x\to\infty} \dfrac{2e^x}{2e^x} = 1$

**20.** $\lim\limits_{x\to 0} \dfrac{\cos mx - \cos nx}{x^2} \overset{\text{H}}{=} \lim\limits_{x\to 0} \dfrac{-m\sin mx + n\sin nx}{2x} \overset{\text{H}}{=} \lim\limits_{x\to 0} \dfrac{-m^2\cos mx + n^2\cos nx}{2} = \frac{1}{2}(n^2 - m^2)$

**21.** This limit has the form $\frac{0}{0}$. $\lim\limits_{x\to 1} \dfrac{1 - x + \ln x}{1 + \cos \pi x} \overset{\text{H}}{=} \lim\limits_{x\to 1} \dfrac{-1 + 1/x}{-\pi\sin \pi x} \overset{\text{H}}{=} \lim\limits_{x\to 1} \dfrac{-1/x^2}{-\pi^2\cos \pi x} = \dfrac{-1}{-\pi^2(-1)} = -\dfrac{1}{\pi^2}$

**22.** $\lim\limits_{x\to 0} \dfrac{x}{\tan^{-1}(4x)} \overset{\text{H}}{=} \lim\limits_{x\to 0} \dfrac{1}{\dfrac{1}{1+(4x)^2}\cdot 4} = \lim\limits_{x\to 0} \dfrac{1+16x^2}{4} = \dfrac{1}{4}$

**23.** This limit has the form $\frac{0}{0}$. $\lim\limits_{x\to 1} \dfrac{x^a - ax + a - 1}{(x-1)^2} \overset{\text{H}}{=} \lim\limits_{x\to 1} \dfrac{ax^{a-1} - a}{2(x-1)} \overset{\text{H}}{=} \lim\limits_{x\to 1} \dfrac{a(a-1)x^{a-2}}{2} = \dfrac{a(a-1)}{2}$

**24.** $\lim\limits_{x\to 0} \dfrac{1 - e^{-2x}}{\sec x} = \dfrac{1-1}{1} = 0$. L'Hospital's Rule does not apply.

**25.** This limit has the form $0 \cdot (-\infty)$. We need to write this product as a quotient, but keep in mind that we will have to differentiate both the numerator and the denominator. If we differentiate $\dfrac{1}{\ln x}$, we get a complicated expression that results in a more difficult limit. Instead we write the quotient as $\dfrac{\ln x}{x^{-1/2}}$.

$$\lim_{x\to 0^+} \sqrt{x}\ln x = \lim_{x\to 0^+} \frac{\ln x}{x^{-1/2}} \overset{\text{H}}{=} \lim_{x\to 0^+} \frac{1/x}{-\frac{1}{2}x^{-3/2}} \cdot \frac{-2x^{3/2}}{-2x^{3/2}} = \lim_{x\to 0^+}(-2\sqrt{x}) = 0$$

**26.** $\lim\limits_{x\to -\infty} x^2 e^x = \lim\limits_{x\to -\infty} \dfrac{x^2}{e^{-x}} \overset{\text{H}}{=} \lim\limits_{x\to -\infty} \dfrac{2x}{-e^{-x}} \overset{\text{H}}{=} \lim\limits_{x\to -\infty} \dfrac{2}{e^{-x}} = \lim\limits_{x\to -\infty} 2e^x = 0$

**27.** This limit has the form $\infty \cdot 0$. We'll change it to the form $\frac{0}{0}$.

$$\lim_{x\to 0} \cot 2x \sin 6x = \lim_{x\to 0} \frac{\sin 6x}{\tan 2x} \overset{\text{H}}{=} \lim_{x\to 0} \frac{6\cos 6x}{2\sec^2 2x} = \frac{6(1)}{2(1)^2} = 3$$

**28.** $\lim\limits_{x\to 0^+} \sin x \ln x = \lim\limits_{x\to 0^+} \dfrac{\ln x}{\csc x} \overset{\text{H}}{=} \lim\limits_{x\to 0^+} \dfrac{1/x}{-\csc x \cot x} = -\lim\limits_{x\to 0^+}\left(\dfrac{\sin x}{x}\cdot \tan x\right)$

$$= -\left(\lim_{x\to 0^+} \frac{\sin x}{x}\right)\left(\lim_{x\to 0^+} \tan x\right) = -1 \cdot 0 = 0$$

**29.** This limit has the form $\infty \cdot 0$. $\lim\limits_{x\to\infty} x^3 e^{-x^2} = \lim\limits_{x\to\infty} \dfrac{x^3}{e^{x^2}} \overset{\text{H}}{=} \lim\limits_{x\to\infty} \dfrac{3x^2}{2xe^{x^2}} = \lim\limits_{x\to\infty} \dfrac{3x}{2e^{x^2}} \overset{\text{H}}{=} \lim\limits_{x\to\infty} \dfrac{3}{4xe^{x^2}} = 0$

**30.** $\lim\limits_{x\to\infty} x\tan(1/x) = \lim\limits_{x\to\infty} \dfrac{\tan(1/x)}{1/x} \overset{\text{H}}{=} \lim\limits_{x\to\infty} \dfrac{\sec^2(1/x)(-1/x^2)}{-1/x^2} = \lim\limits_{x\to\infty} \sec^2(1/x) = 1^2 = 1$

**31.** As $x \to \infty$, $1/x \to 0$, and $e^{1/x} \to 1$. So the limit has the form $\infty - \infty$ and we will change the form to a product by factoring out $x$.

$$\lim_{x\to\infty}\left(xe^{1/x} - x\right) = \lim_{x\to\infty} x\left(e^{1/x} - 1\right) = \lim_{x\to\infty} \frac{e^{1/x} - 1}{1/x} \overset{\text{H}}{=} \lim_{x\to\infty} \frac{e^{1/x}(-1/x^2)}{-1/x^2} = \lim_{x\to\infty} e^{1/x} = e^0 = 1$$

**32.** $\lim\limits_{x\to 0}(\csc x - \cot x) = \lim\limits_{x\to 0}\left(\dfrac{1}{\sin x} - \dfrac{\cos x}{\sin x}\right) = \lim\limits_{x\to 0} \dfrac{1 - \cos x}{\sin x} \overset{\text{H}}{=} \lim\limits_{x\to 0} \dfrac{\sin x}{\cos x} = 0$

**33.** The limit has the form $\infty - \infty$ and we will change the form to a product by factoring out $x$.

$$\lim_{x\to\infty}(x - \ln x) = \lim_{x\to\infty} x\left(1 - \frac{\ln x}{x}\right) = \infty \text{ since } \lim_{x\to\infty} \frac{\ln x}{x} \overset{\text{H}}{=} \lim_{x\to\infty} \frac{1/x}{1} = 0.$$

**34.** $\lim\limits_{x\to 1}\left(\dfrac{1}{\ln x} - \dfrac{1}{x-1}\right) = \lim\limits_{x\to 1} \dfrac{x - 1 - \ln x}{(x-1)\ln x} \overset{\text{H}}{=} \lim\limits_{x\to 1} \dfrac{1 - 1/x}{(x-1)(1/x) + \ln x} \cdot \dfrac{x}{x}$

$$= \lim_{x\to 1} \frac{x-1}{x - 1 + x\ln x} \overset{\text{H}}{=} \lim_{x\to 1} \frac{1}{1 + 1 + \ln x} = \frac{1}{2+0} = \frac{1}{2}$$

**35.** $y = x^{x^2} \quad \Rightarrow \quad \ln y = x^2 \ln x$, so $\lim\limits_{x\to 0^+} \ln y = \lim\limits_{x\to 0^+} x^2 \ln x = \lim\limits_{x\to 0^+} \dfrac{\ln x}{1/x^2} \overset{\text{H}}{=} \lim\limits_{x\to 0^+} \dfrac{1/x}{-2/x^3} = \lim\limits_{x\to 0^+} \left(-\dfrac{1}{2}x^2\right) = 0 \quad \Rightarrow$

$\lim\limits_{x\to 0^+} x^{x^2} = \lim\limits_{x\to 0^+} e^{\ln y} = e^0 = 1.$

**36.** $y = (\tan 2x)^x \quad \Rightarrow \quad \ln y = x \cdot \ln \tan 2x$, so

$$\lim_{x\to 0^+} \ln y = \lim_{x\to 0^+} x \cdot \ln \tan 2x = \lim_{x\to 0^+} \frac{\ln \tan 2x}{1/x}$$

$$\overset{\text{H}}{=} \lim_{x\to 0^+} \frac{(1/\tan 2x)(2\sec^2 2x)}{-1/x^2} = \lim_{x\to 0^+} \frac{-2x^2 \cos 2x}{\sin 2x \cos^2 2x} = \lim_{x\to 0^+} \frac{2x}{\sin 2x} \cdot \lim_{x\to 0^+} \frac{-x}{\cos 2x} = 1 \cdot 0 = 0 \quad \Rightarrow$$

$\lim\limits_{x\to 0^+} (\tan 2x)^x = \lim\limits_{x\to 0^+} e^{\ln y} = e^0 = 1.$

**37.** $y = (1-2x)^{1/x} \quad \Rightarrow \quad \ln y = \dfrac{1}{x}\ln(1-2x)$, so $\lim\limits_{x\to 0} \ln y = \lim\limits_{x\to 0} \dfrac{\ln(1-2x)}{x} \overset{\text{H}}{=} \lim\limits_{x\to 0} \dfrac{-2/(1-2x)}{1} = -2 \quad \Rightarrow$

$\lim\limits_{x\to 0} (1-2x)^{1/x} = \lim\limits_{x\to 0} e^{\ln y} = e^{-2}.$

**38.** $y = \left(1 + \dfrac{a}{x}\right)^{bx} \quad \Rightarrow \quad \ln y = bx \ln\left(1 + \dfrac{a}{x}\right)$, so

$$\lim_{x\to\infty} \ln y = \lim_{x\to\infty} \frac{b\ln(1+a/x)}{1/x} \overset{\text{H}}{=} \lim_{x\to\infty} \frac{b\left(\dfrac{1}{1+a/x}\right)\left(-\dfrac{a}{x^2}\right)}{-1/x^2} = \lim_{x\to\infty} \frac{ab}{1+a/x} = ab \quad \Rightarrow$$

$\lim\limits_{x\to\infty} \left(1 + \dfrac{a}{x}\right)^{bx} = \lim\limits_{x\to\infty} e^{\ln y} = e^{ab}.$

**39.** $y = (\cos x)^{1/x^2} \quad \Rightarrow \quad \ln y = \dfrac{1}{x^2} \ln \cos x \quad \Rightarrow$

$$\lim_{x\to 0^+} \ln y = \lim_{x\to 0^+} \frac{\ln \cos x}{x^2} \overset{\text{H}}{=} \lim_{x\to 0^+} \frac{-\tan x}{2x} \overset{\text{H}}{=} \lim_{x\to 0^+} \frac{-\sec^2 x}{2} = -\frac{1}{2} \quad \Rightarrow$$

$\lim\limits_{x\to 0^+} (\cos x)^{1/x^2} = \lim\limits_{x\to 0^+} e^{\ln y} = e^{-1/2} = 1/\sqrt{e}$

**40.** $y = x^{(\ln 2)/(1+\ln x)} \quad \Rightarrow \quad \ln y = \dfrac{\ln 2}{1 + \ln x} \ln x \quad \Rightarrow$

$\lim\limits_{x\to\infty} \ln y = \lim\limits_{x\to\infty} \dfrac{(\ln 2)(\ln x)}{1 + \ln x} \overset{\text{H}}{=} \lim\limits_{x\to\infty} \dfrac{(\ln 2)(1/x)}{1/x} = \lim\limits_{x\to\infty} \ln 2 = \ln 2$, so $\lim\limits_{x\to\infty} x^{(\ln 2)/(1+\ln x)} = \lim\limits_{x\to\infty} e^{\ln y} = e^{\ln 2} = 2.$

**41.** From the graph, it appears that $\lim\limits_{x\to\infty} x\,[\ln(x+5) - \ln x] = 5$. To prove this, we first note that

$\ln(x+5) - \ln x = \ln \dfrac{x+5}{x} = \ln\left(1 + \dfrac{5}{x}\right) \to \ln 1 = 0$ as $x \to \infty$. Thus,

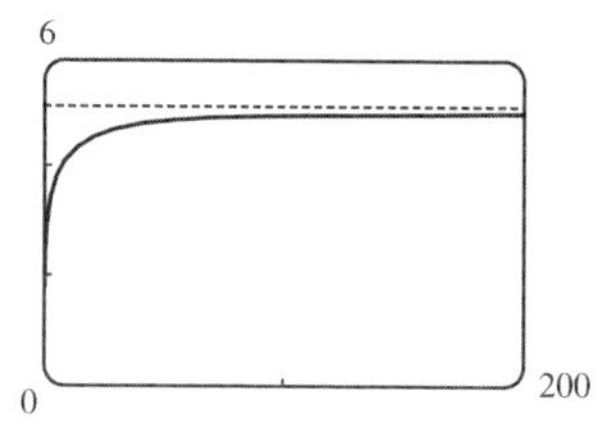


$$\lim_{x\to\infty} x\,[\ln(x+5) - \ln x] = \lim_{x\to\infty} \frac{\ln(x+5) - \ln x}{1/x} \overset{\text{H}}{=} \lim_{x\to\infty} \frac{\dfrac{1}{x+5} - \dfrac{1}{x}}{-1/x^2}$$

$$= \lim_{x\to\infty} \left[\frac{x - (x+5)}{x(x+5)} \cdot \frac{-x^2}{1}\right] = \lim_{x\to\infty} \frac{5x^2}{x^2 + 5x} = 5$$

**42.** From the graph, it appears that $\lim\limits_{x\to\pi/4} (\tan x)^{\tan 2x} \approx 0.368$. The limit has the form $1^\infty$.

Now $y = (\tan x)^{\tan 2x} \quad \Rightarrow \quad \ln y = \tan 2x \, \ln(\tan x)$, so

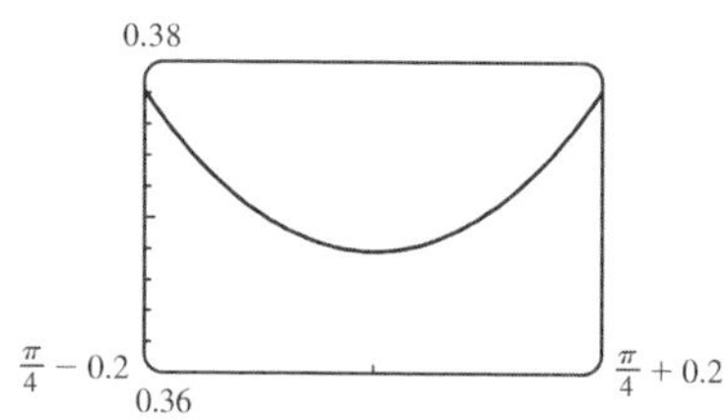


$$\lim_{x\to\pi/4} \ln y = \lim_{x\to\pi/4} \frac{\ln(\tan x)}{\cot 2x} \overset{\text{H}}{=} \lim_{x\to\pi/4} \frac{\sec^2 x/\tan x}{-2\csc^2 2x} = \frac{2/1}{-2(1)} = -1 \quad \Rightarrow$$

$\lim\limits_{x\to\pi/4} (\tan x)^{\tan 2x} = \lim\limits_{x\to\pi/4} e^{\ln y} = e^{-1} = 1/e \approx 0.3679.$

**43.** 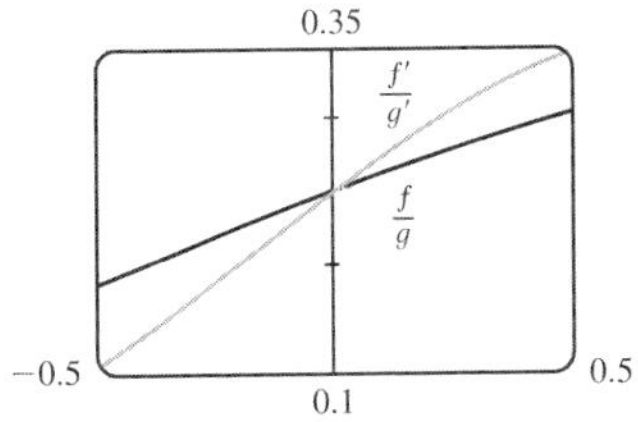


From the graph, it appears that $\lim\limits_{x\to 0}\dfrac{f(x)}{g(x)}=\lim\limits_{x\to 0}\dfrac{f'(x)}{g'(x)}=0.25$. We calculate $\lim\limits_{x\to 0}\dfrac{f(x)}{g(x)}=\lim\limits_{x\to 0}\dfrac{e^x-1}{x^3+4x}\overset{\text{H}}{=}\lim\limits_{x\to 0}\dfrac{e^x}{3x^2+4}=\dfrac{1}{4}$.

**44.** 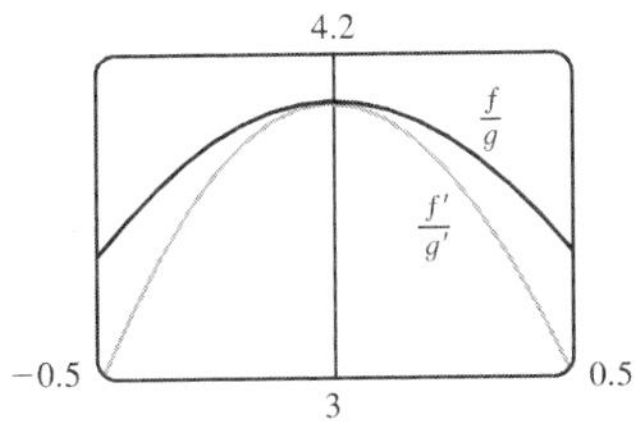


From the graph, it appears that $\lim\limits_{x\to 0}\dfrac{f(x)}{g(x)}=\lim\limits_{x\to 0}\dfrac{f'(x)}{g'(x)}=4$. We calculate

$$\lim_{x\to 0}\frac{f(x)}{g(x)}=\lim_{x\to 0}\frac{2x\sin x}{\sec x-1}\overset{\text{H}}{=}\lim_{x\to 0}\frac{2(x\cos x+\sin x)}{\sec x\,\tan x}$$

$$\overset{\text{H}}{=}\lim_{x\to 0}\frac{2(-x\sin x+\cos x+\cos x)}{\sec x(\sec^2 x)+\tan x(\sec x\,\tan x)}=\frac{4}{1}=4$$

**45.** $\lim\limits_{x\to\infty}xe^{-x}=\lim\limits_{x\to\infty}(x/e^x)\overset{\text{H}}{=}\lim\limits_{x\to\infty}(1/e^x)=0$, so $y=0$ is a HA. $\lim\limits_{x\to-\infty}xe^{-x}=-\infty$.

$f(x)=xe^{-x}\quad\Rightarrow\quad f'(x)=x(-e^{-x})+e^{-x}\cdot 1=e^{-x}(1-x)>0\quad\Leftrightarrow\quad 1-x>0\quad\Leftrightarrow\quad x<1$, so $f$ is increasing on $(-\infty,1)$ and decreasing on $(1,\infty)$. By the FDT, $f(1)=1/e$ is a local maximum.

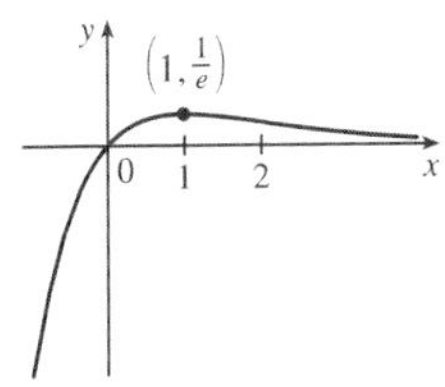


$f''(x)=e^{-x}(-1)+(1-x)(-e^{-x})=e^{-x}(-1-1+x)=e^{-x}(x-2)>0\quad\Leftrightarrow$

$x-2>0\quad\Leftrightarrow\quad x>2$, so $f$ is CU on $(2,\infty)$ and CD on $(-\infty,2)$. IP is $\left(2,2/e^2\right)$.

**46.** $\lim\limits_{x\to\infty}\dfrac{e^x}{x}\overset{\text{H}}{=}\lim\limits_{x\to\infty}\dfrac{e^x}{1}=\infty$, $\lim\limits_{x\to-\infty}\dfrac{e^x}{x}=0$, so $y=0$ is a HA. $\lim\limits_{x\to 0^+}\dfrac{e^x}{x}=\infty$, $\lim\limits_{x\to 0^-}\dfrac{e^x}{x}=-\infty$, so $x=0$ is a VA.

$f(x)=\dfrac{e^x}{x}\quad\Rightarrow\quad f'(x)=\dfrac{xe^x-e^x}{x^2}>0\quad\Leftrightarrow\quad (x-1)e^x>0\quad\Leftrightarrow\quad x>1$, so $f$ is increasing on $(1,\infty)$, and decreasing on $(-\infty,0)$ and $(0,1)$. By the FDT, $f(1)=e$ is a local minimum.

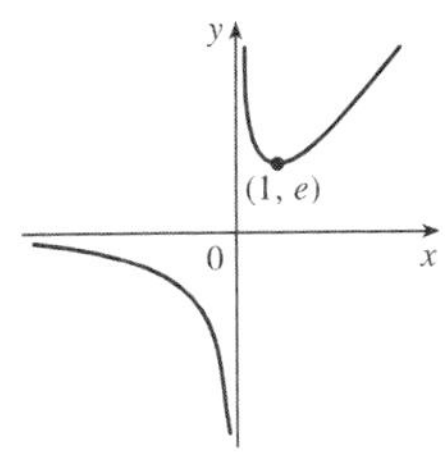


$$f''(x)=\frac{x^2(xe^x)-(xe^x-e^x)(2x)}{(x^2)^2}=\frac{xe^x\left(x^2-2x+2\right)}{x^4}$$

$$=\frac{e^x\left(x^2-2x+2\right)}{x^3}>0\quad\Leftrightarrow\quad x>0$$

since $x^2-2x+2>0$ for all $x$. So $f$ is CU on $(0,\infty)$ and CD on $(-\infty,0)$. No IP

**47.** 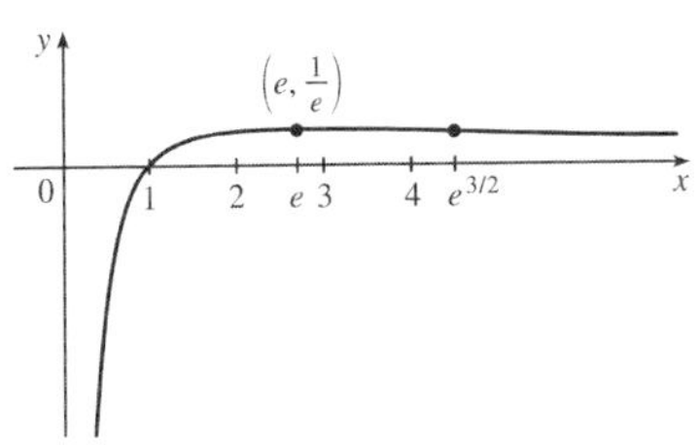


$\lim\limits_{x\to\infty}\dfrac{\ln x}{x}\overset{\text{H}}{=}\lim\limits_{x\to\infty}\dfrac{1/x}{1}=0$, so $y=0$ is a HA. Also $\lim\limits_{x\to 0^+}\dfrac{\ln x}{x}=-\infty$ since $\ln x\to-\infty$ and $x\to 0^+$, so $x=0$ is a VA.

$f(x)=\dfrac{\ln x}{x}\quad\Rightarrow\quad f'(x)=\dfrac{x(1/x)-(\ln x)(1)}{x^2}=\dfrac{1-\ln x}{x^2}=0$ when $\ln x=1\quad\Leftrightarrow\quad x=e$.

$f'(x)>0\quad\Leftrightarrow\quad 1-\ln x>0\quad\Leftrightarrow\quad \ln x<1\quad\Leftrightarrow\quad 0<x<e$. $f'(x)<0\quad\Leftrightarrow\quad x>e$.
So $f$ is increasing on $(0,e)$ and decreasing on $(e,\infty)$. By the FDT, $f(e)=1/e$ is a local maximum.

$f''(x)=\dfrac{x^2(-1/x)-(1-\ln x)(2x)}{(x^2)^2}=\dfrac{x(-1-2+2\ln x)}{x^4}=\dfrac{2\ln x-3}{x^3}$, so $f''(x)>0\quad\Leftrightarrow\quad 2\ln x-3>0\quad\Leftrightarrow$

$\ln x>\frac{3}{2}\quad\Leftrightarrow\quad x>e^{3/2}$. $f''(x)<0\quad\Leftrightarrow\quad 0<x<e^{3/2}$. So $f$ is CU on $\left(e^{3/2},\infty\right)$ and CD on $\left(0,e^{3/2}\right)$. There is an inflection point at $\left(e^{3/2},\frac{3}{2}e^{-3/2}\right)$.

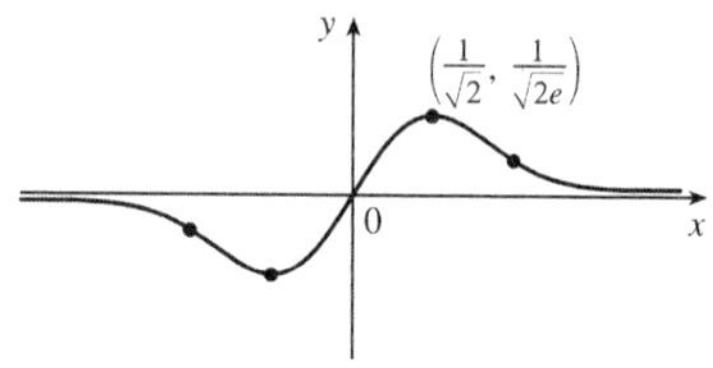


**48.** $\lim\limits_{x\to\pm\infty} xe^{-x^2} = \lim\limits_{x\to\pm\infty} \dfrac{x}{e^{x^2}} \overset{\text{H}}{=} \lim\limits_{x\to\pm\infty} \dfrac{1}{2xe^{x^2}} = 0$, so $y = 0$ is a HA.

$f(x) = xe^{-x^2} \Rightarrow f'(x) = xe^{-x^2}(-2x) + e^{-x^2} \cdot 1 = e^{-x^2}(1 - 2x^2) > 0 \Leftrightarrow$ $x^2 < \frac{1}{2} \Leftrightarrow |x| < \frac{1}{\sqrt{2}}$, so $f$ is increasing on $\left(-\frac{1}{\sqrt{2}}, \frac{1}{\sqrt{2}}\right)$ and decreasing on $\left(-\infty, -\frac{1}{\sqrt{2}}\right)$ and $\left(\frac{1}{\sqrt{2}}, \infty\right)$. By the FDT, $f\left(\frac{1}{\sqrt{2}}\right) = 1/\sqrt{2e}$ is a local maximum and $f\left(-\frac{1}{\sqrt{2}}\right) = -1/\sqrt{2e}$ is a local minimum.

$f''(x) = e^{-x^2}(-4x) + (1 - 2x^2)e^{-x^2}(-2x) = 2xe^{-x^2}(-2 - 1 + 2x^2) = 2xe^{-x^2}(2x^2 - 3) > 0 \Leftrightarrow x > \sqrt{\frac{3}{2}}$ or $-\sqrt{\frac{3}{2}} < x < 0$, so $f$ is CU on $\left(\sqrt{\frac{3}{2}}, \infty\right)$ and $\left(-\sqrt{\frac{3}{2}}, 0\right)$ and CD on $\left(-\infty, -\sqrt{\frac{3}{2}}\right)$ and $\left(0, \sqrt{\frac{3}{2}}\right)$. IP are $(0, 0)$ and $\left(\pm\sqrt{\frac{3}{2}}, \pm\sqrt{\frac{3}{2}}e^{-3/2}\right)$.

**49.** (a) $f(x) = x^2 \ln x$. The domain of $f$ is $(0, \infty)$.

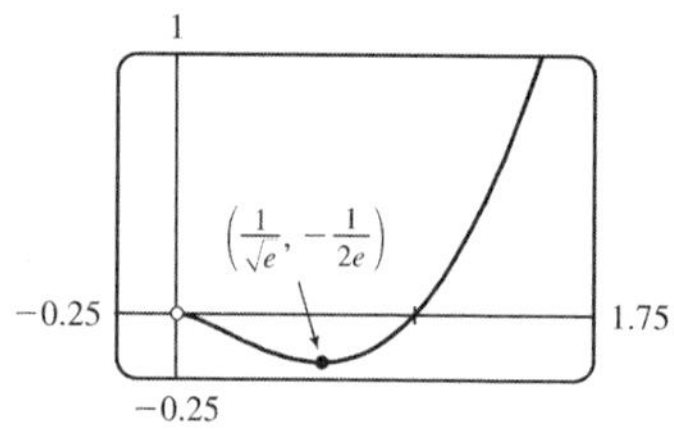


(b) $\lim\limits_{x\to 0^+} x^2 \ln x = \lim\limits_{x\to 0^+} \dfrac{\ln x}{1/x^2} \overset{\text{H}}{=} \lim\limits_{x\to 0^+} \dfrac{1/x}{-2/x^3} = \lim\limits_{x\to 0^+}\left(-\dfrac{x^2}{2}\right) = 0$. There is a hole at $(0, 0)$.

(c) It appears that there is an IP at about $(0.2, -0.06)$ and a local minimum at $(0.6, -0.18)$. $f(x) = x^2 \ln x \Rightarrow$ $f'(x) = x^2(1/x) + (\ln x)(2x) = x(2\ln x + 1) > 0 \Leftrightarrow \ln x > -\frac{1}{2} \Leftrightarrow x > e^{-1/2}$, so $f$ is increasing on $(1/\sqrt{e}, \infty)$, decreasing on $(0, 1/\sqrt{e})$. By the FDT, $f(1/\sqrt{e}) = -1/(2e)$ is a local minimum value. This point is approximately $(0.6065, -0.1839)$, which agrees with our estimate.

$f''(x) = x(2/x) + (2\ln x + 1) = 2\ln x + 3 > 0 \Leftrightarrow \ln x > -\frac{3}{2} \Leftrightarrow x > e^{-3/2}$, so $f$ is CU on $\left(e^{-3/2}, \infty\right)$ and CD on $\left(0, e^{-3/2}\right)$. IP is $\left(e^{-3/2}, -3/(2e^3)\right) \approx (0.2231, -0.0747)$.

**50.** (a) $f(x) = xe^{1/x}$. The domain of $f$ is $(-\infty, 0) \cup (0, \infty)$.

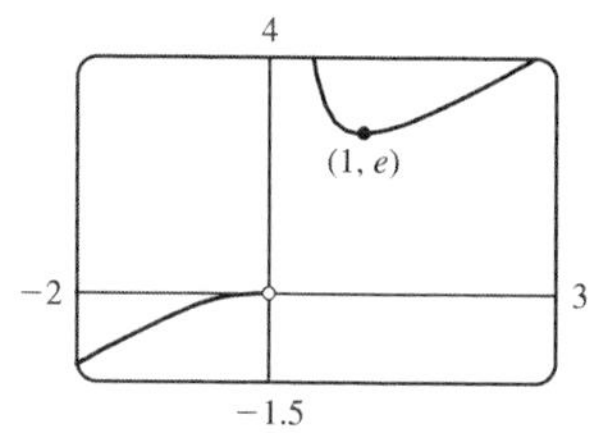


(b) $\lim\limits_{x\to 0^+} xe^{1/x} = \lim\limits_{x\to 0^+} \dfrac{e^{1/x}}{1/x} \overset{\text{H}}{=} \lim\limits_{x\to 0^+} \dfrac{e^{1/x}(-1/x^2)}{-1/x^2} = \lim\limits_{x\to 0^+} e^{1/x} = \infty$, so $x = 0$ is a VA.

Also $\lim\limits_{x\to 0^-} xe^{1/x} = 0$ since $1/x \to -\infty \Rightarrow e^{1/x} \to 0$.

(c) It appears that there is a local minimum at $(1, 2.7)$. There are no IP and $f$ is CD on $(-\infty, 0)$ and CU on $(0, \infty)$.

$f(x) = xe^{1/x} \Rightarrow f'(x) = xe^{1/x}\left(-\dfrac{1}{x^2}\right) + e^{1/x} = e^{1/x}\left(1 - \dfrac{1}{x}\right) > 0 \Leftrightarrow \dfrac{1}{x} < 1 \Leftrightarrow x < 0$ or $x > 1$, so $f$ is increasing on $(-\infty, 0)$ and $(1, \infty)$, and decreasing on $(0, 1)$. By the FDT, $f(1) = e$ is a local minimum value, which agrees with our estimate.

$f''(x) = e^{1/x}(1/x^2) + (1 - 1/x)e^{1/x}(-1/x^2) = \left(e^{1/x}/x^2\right)(1 - 1 + 1/x) = e^{1/x}/x^3 > 0 \Leftrightarrow x > 0$, so $f$ is CU on $(0, \infty)$ and CD on $(-\infty, 0)$. No IP.

**51.** (a) $f(x) = x^{1/x}$

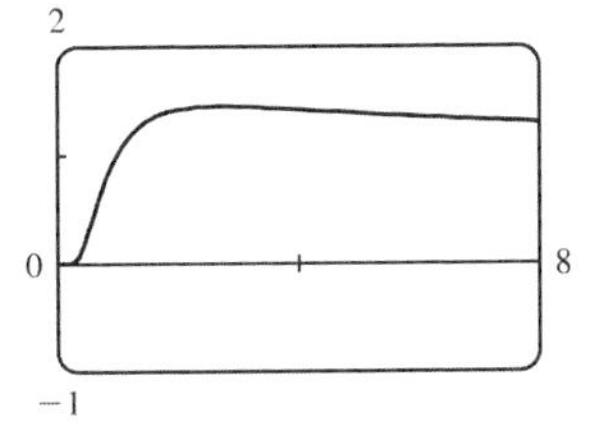


(b) Recall that $a^b = e^{b \ln a}$. $\lim\limits_{x \to 0^+} x^{1/x} = \lim\limits_{x \to 0^+} e^{(1/x)\ln x}$. As $x \to 0^+$, $\dfrac{\ln x}{x} \to -\infty$, so $x^{1/x} = e^{(1/x)\ln x} \to 0$. This indicates that there is a hole at $(0, 0)$. As $x \to \infty$, we have the indeterminate form $\infty^0$. $\lim\limits_{x \to \infty} x^{1/x} = \lim\limits_{x \to \infty} e^{(1/x)\ln x}$, but $\lim\limits_{x \to \infty} \dfrac{\ln x}{x} \overset{\text{H}}{=} \lim\limits_{x \to \infty} \dfrac{1/x}{1} = 0$, so $\lim\limits_{x \to \infty} x^{1/x} = e^0 = 1$. This indicates that $y = 1$ is a HA.

(c) Estimated maximum: $(2.72, 1.45)$. No estimated minimum. We use logarithmic differentiation to find any critical numbers. $y = x^{1/x} \quad \Rightarrow \quad \ln y = \dfrac{1}{x} \ln x \quad \Rightarrow \quad \dfrac{y'}{y} = \dfrac{1}{x} \cdot \dfrac{1}{x} + (\ln x)\left(-\dfrac{1}{x^2}\right) \quad \Rightarrow$

$y' = x^{1/x}\left(\dfrac{1 - \ln x}{x^2}\right) = 0 \quad \Rightarrow \quad \ln x = 1 \quad \Rightarrow \quad x = e$. For $0 < x < e$, $y' > 0$ and for $x > e$, $y' < 0$, so $f(e) = e^{1/e}$ is a local maximum value. This point is approximately $(2.7183, 1.4447)$, which agrees with our estimate.

(d)

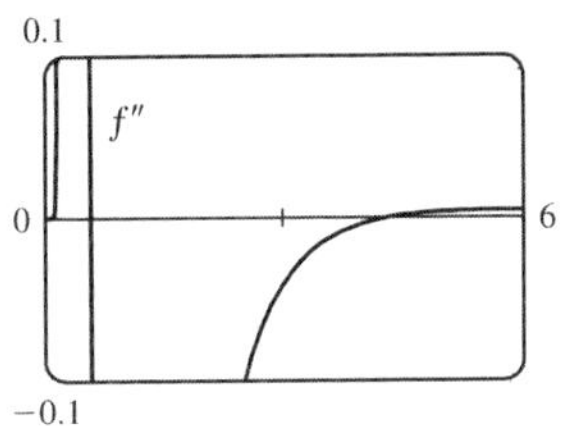


From the graph, we see that $f''(x) = 0$ at $x \approx 0.58$ and $x \approx 4.37$. Since $f''$ changes sign at these values, they are $x$-coordinates of inflection points.

**52.** (a) $f(x) = (\sin x)^{\sin x}$ is continuous where $\sin x > 0$, that is, on intervals of the form $(2n\pi, (2n + 1)\pi)$, so we have graphed $f$ on $(0, \pi)$.

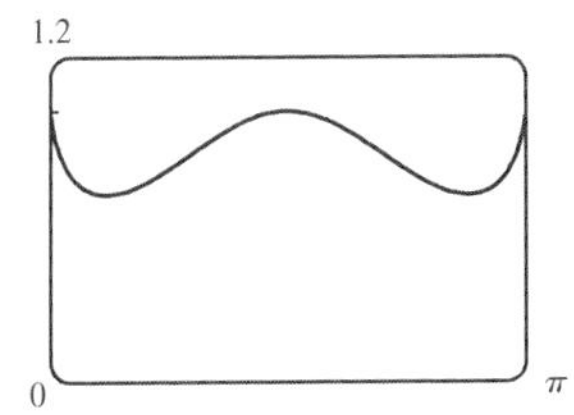


(b) $y = (\sin x)^{\sin x} \quad \Rightarrow \quad \ln y = \sin x \ln \sin x$, so

$$\lim_{x \to 0^+} \ln y = \lim_{x \to 0^+} \sin x \ln \sin x = \lim_{x \to 0^+} \frac{\ln \sin x}{\csc x} \overset{\text{H}}{=} \lim_{x \to 0^+} \frac{\cot x}{-\csc x \cot x} = \lim_{x \to 0^+} (-\sin x) = 0 \quad \Rightarrow$$

$\lim\limits_{x \to 0^+} y = e^0 = 1$.

(c) It appears that we have a local maximum at $(1.57, 1)$ and local minima at $(0.38, 0.69)$ and $(2.76, 0.69)$.

$y = (\sin x)^{\sin x} \quad \Rightarrow \quad \ln y = \sin x \ln \sin x \quad \Rightarrow \quad \dfrac{y'}{y} = (\sin x)\left(\dfrac{\cos x}{\sin x}\right) + (\ln \sin x) \cos x = \cos x (1 + \ln \sin x) \quad \Rightarrow$

$y' = (\sin x)^{\sin x}(\cos x)(1 + \ln \sin x)$. $y' = 0 \quad \Rightarrow \quad \cos x = 0$ or $\ln \sin x = -1 \quad \Rightarrow \quad x_2 = \frac{\pi}{2}$ or $\sin x = e^{-1}$. On $(0, \pi)$, $\sin x = e^{-1} \quad \Rightarrow \quad x_1 = \sin^{-1}(e^{-1})$ and $x_3 = \pi - \sin^{-1}(e^{-1})$. Approximating these points gives us $(x_1, f(x_1)) \approx (0.3767, 0.6922)$, $(x_2, f(x_2)) \approx (1.5708, 1)$, and $(x_3, f(x_3)) \approx (2.7649, 0.6922)$. The approximations confirm our estimates.

(d)

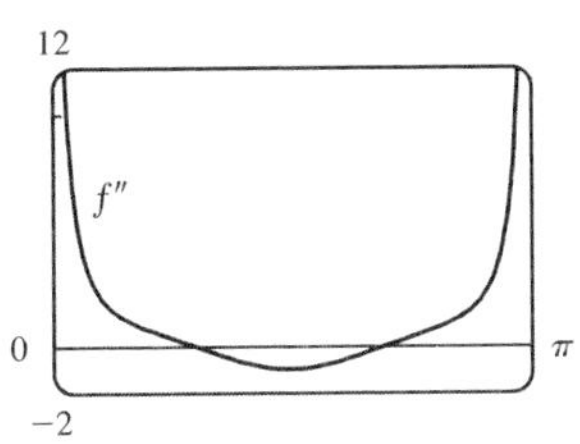


From the graph, we see that $f''(x) = 0$ at $x \approx 0.94$ and $x \approx 2.20$. Since $f''$ changes sign at these values, they are $x$-coordinates of inflection points.

**53.** If $c < 0$, then $\lim\limits_{x\to-\infty} f(x) = \lim\limits_{x\to-\infty} xe^{-cx} = \lim\limits_{x\to-\infty} \dfrac{x}{e^{cx}} \overset{\text{H}}{=} \lim\limits_{x\to-\infty} \dfrac{1}{ce^{cx}} = 0$, and $\lim\limits_{x\to\infty} f(x) = \infty$.

If $c > 0$, then $\lim\limits_{x\to-\infty} f(x) = -\infty$, and $\lim\limits_{x\to\infty} f(x) \overset{\text{H}}{=} \lim\limits_{x\to\infty} \dfrac{1}{ce^{cx}} = 0$.

If $c = 0$, then $f(x) = x$, so $\lim\limits_{x\to\pm\infty} f(x) = \pm\infty$, respectively.

So we see that $c = 0$ is a transitional value. We now exclude the case $c = 0$, since we know how the function behaves in that case. To find the maxima and minima of $f$, we differentiate: $f(x) = xe^{-cx} \Rightarrow$ $f'(x) = x(-ce^{-cx}) + e^{-cx} = (1 - cx)e^{-cx}$. This is 0 when $1 - cx = 0 \Leftrightarrow x = 1/c$. If $c < 0$ then this represents a minimum value of $f(1/c) = 1/(ce)$, since $f'(x)$ changes from negative to positive at $x = 1/c$; and if $c > 0$, it represents a maximum value. As $|c|$ increases, the maximum or minimum point gets closer to the origin. To find the inflection points, we differentiate again: $f'(x) = e^{-cx}(1 - cx) \Rightarrow$ $f''(x) = e^{-cx}(-c) + (1 - cx)(-ce^{-cx}) = (cx - 2)ce^{-cx}$. This changes sign when $cx - 2 = 0 \Leftrightarrow x = 2/c$. So as $|c|$ increases, the points of inflection get closer to the origin.

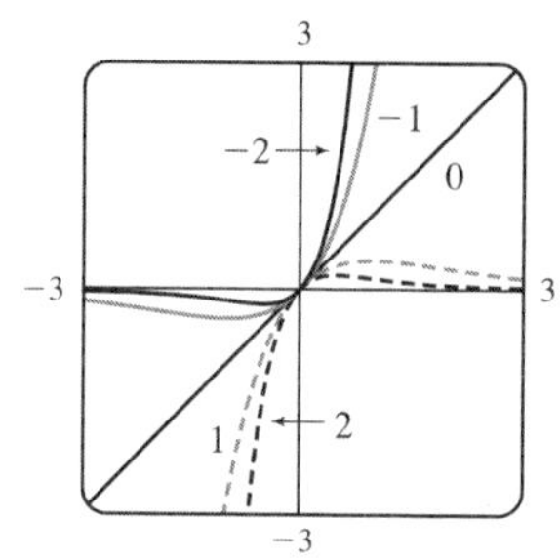


**54.**

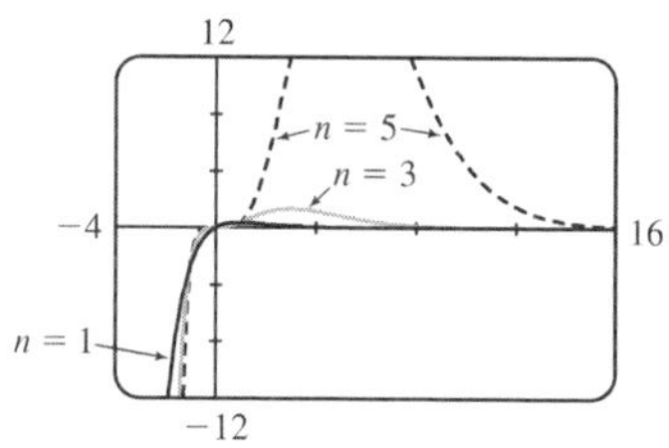


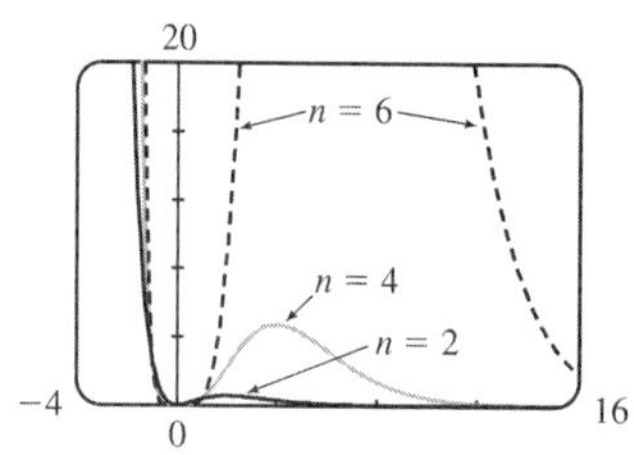


The first figure shows representative examples of $f(x) = x^n e^{-x}$ with $n$ odd. $n$ is even in the second figure. All curves pass through the origin and approach $y = 0$ as $x \to \infty$.

$f'(x) = \dfrac{x^n(n - x)}{xe^x} = 0 \Leftrightarrow x = n$ or $x = 0$ (the latter for $n > 1$). At $x = 0$, we have a local minimum for $n$ even. At $x = n$, we have a local maximum for all $n$. As $n$ increases, $(n, f(n))$ gets farther away from the origin.

$f''(x) = \dfrac{x^n(x^2 - 2nx + n^2 - n)}{x^2e^x} = 0 \Leftrightarrow x = n \pm \sqrt{n}$ or $x = 0$ (the latter for $n > 2$). As $n$ increases, the IP move farther away from the origin—they are symmetric about the line $x = n$.

**55.** $\lim\limits_{x\to\infty} \dfrac{e^x}{x^n} \overset{\text{H}}{=} \lim\limits_{x\to\infty} \dfrac{e^x}{nx^{n-1}} \overset{\text{H}}{=} \lim\limits_{x\to\infty} \dfrac{e^x}{n(n-1)x^{n-2}} \overset{\text{H}}{=} \cdots \overset{\text{H}}{=} \lim\limits_{x\to\infty} \dfrac{e^x}{n!} = \infty$

**56.** $\lim\limits_{x\to\infty} \dfrac{\ln x}{x^p} \overset{\text{H}}{=} \lim\limits_{x\to\infty} \dfrac{1/x}{px^{p-1}} = \lim\limits_{x\to\infty} \dfrac{1}{px^p} = 0$ since $p > 0$.

**57.** First we will find $\lim\limits_{n\to\infty} \left(1 + \dfrac{r}{n}\right)^{nt}$, which is of the form $1^\infty$. $y = \left(1 + \dfrac{r}{n}\right)^{nt} \Rightarrow \ln y = nt \ln\left(1 + \dfrac{r}{n}\right)$, so

$\lim\limits_{n\to\infty} \ln y = \lim\limits_{n\to\infty} nt \ln\left(1 + \dfrac{r}{n}\right) = t \lim\limits_{n\to\infty} \dfrac{\ln(1 + r/n)}{1/n} \overset{\text{H}}{=} t \lim\limits_{n\to\infty} \dfrac{(-r/n^2)}{(1 + r/n)(-1/n^2)} = t \lim\limits_{n\to\infty} \dfrac{r}{1 + i/n} = tr \Rightarrow$

$\lim\limits_{n\to\infty} y = e^{rt}$. Thus, as $n \to \infty$, $A = A_0\left(1 + \dfrac{r}{n}\right)^{nt} \to A_0e^{rt}$.

**58.** (a) $\lim\limits_{t\to\infty} v = \lim\limits_{t\to\infty} \dfrac{mg}{c}\left(1 - e^{-ct/m}\right) = \dfrac{mg}{c}\lim\limits_{t\to\infty}\left(1 - e^{-ct/m}\right) = \dfrac{mg}{c}(1-0)$ [because $-ct/m \to -\infty$ as $t\to\infty$]

$= \dfrac{mg}{c}$, which is the speed the object approaches as time goes on, the so-called limiting velocity.

(b) $\lim\limits_{c\to 0^+} v = \lim\limits_{c\to 0^+} \dfrac{mg}{c}\left(1 - e^{-ct/m}\right) = mg \lim\limits_{c\to 0^+} \dfrac{1-e^{-ct/m}}{c}$ [form is $\frac{0}{0}$]

$\overset{\text{H}}{=} mg \lim\limits_{c\to 0^+} \dfrac{\left(-e^{-ct/m}\right)\cdot(-t/m)}{1} = \dfrac{mgt}{m}\lim\limits_{c\to 0^+} e^{-ct/m} = gt(1) = gt$

The velocity of a falling object in a vacuum is directly proportional to the amount of time it falls.

**59.** $\lim\limits_{E\to 0^+} P(E) = \lim\limits_{E\to 0^+}\left(\dfrac{e^E + e^{-E}}{e^E - e^{-E}} - \dfrac{1}{E}\right)$

$= \lim\limits_{E\to 0^+} \dfrac{E(e^E + e^{-E}) - 1(e^E - e^{-E})}{(e^E - e^{-E})E} = \lim\limits_{E\to 0^+}\dfrac{Ee^E + Ee^{-E} - e^E + e^{-E}}{Ee^E - Ee^{-E}}$ [form is $\frac{0}{0}$]

$\overset{\text{H}}{=} \lim\limits_{E\to 0^+} \dfrac{Ee^E + e^E\cdot 1 + E(-e^{-E}) + e^{-E}\cdot 1 - e^E + (-e^{-E})}{Ee^E + e^E\cdot 1 - [E(-e^{-E}) + e^{-E}\cdot 1]}$

$= \lim\limits_{E\to 0^+} \dfrac{Ee^E - Ee^{-E}}{Ee^E + e^E + Ee^{-E} - e^{-E}} = \lim\limits_{E\to 0^+} \dfrac{e^E - e^{-E}}{e^E + \dfrac{e^E}{E} + e^{-E} - \dfrac{e^{-E}}{E}}$ [divide by $E$]

$= \dfrac{0}{2+L}$, where $L = \lim\limits_{E\to 0^+}\dfrac{e^E - e^{-E}}{E}$ [form is $\frac{0}{0}$]

$\overset{\text{H}}{=} \lim\limits_{E\to 0^+}\dfrac{e^E + e^{-E}}{1} = \dfrac{1+1}{1} = 2$

Thus, $\lim\limits_{E\to 0^+} P(E) = \dfrac{0}{2+2} = 0.$

**60.** (a) $\lim\limits_{R\to r^+} v = \lim\limits_{R\to r^+}\left[-c\left(\dfrac{r}{R}\right)^2 \ln\left(\dfrac{r}{R}\right)\right] = -cr^2 \lim\limits_{R\to r^+}\left[\left(\dfrac{1}{R}\right)^2\ln\left(\dfrac{r}{R}\right)\right] = -cr^2\cdot\dfrac{1}{r^2}\cdot\ln 1 = -c\cdot 0 = 0$

As the insulation of a metal cable becomes thinner, the velocity of an electrical impulse in the cable approaches zero.

(b) $\lim\limits_{r\to 0^+} v = \lim\limits_{r\to 0^+}\left[-c\left(\dfrac{r}{R}\right)^2\ln\left(\dfrac{r}{R}\right)\right] = -\dfrac{c}{R^2}\lim\limits_{r\to 0^+}\left[r^2\ln\left(\dfrac{r}{R}\right)\right]$ [form is $0\cdot\infty$]

$= -\dfrac{c}{R^2}\lim\limits_{r\to 0^+}\dfrac{\ln\left(\dfrac{r}{R}\right)}{\dfrac{1}{r^2}}$ [form is $\infty/\infty$]

$\overset{\text{H}}{=} -\dfrac{c}{R^2}\lim\limits_{r\to 0^+}\dfrac{\dfrac{R}{r}\cdot\dfrac{1}{R}}{\dfrac{-2}{r^3}} = -\dfrac{c}{R^2}\lim\limits_{r\to 0^+}\left(-\dfrac{r^2}{2}\right) = 0$

As the radius of the metal cable approaches zero, the velocity of an electrical impulse in the cable approaches zero.

**61.** We see that both numerator and denominator approach 0, so we can use l'Hospital's Rule:

$$\lim_{x\to a}\frac{\sqrt{2a^3x - x^4} - a\sqrt[3]{aax}}{a - \sqrt[4]{ax^3}} \overset{\text{H}}{=} \lim_{x\to a}\frac{\frac{1}{2}(2a^3x - x^4)^{-1/2}(2a^3 - 4x^3) - a\left(\frac{1}{3}\right)(aax)^{-2/3}a^2}{-\frac{1}{4}(ax^3)^{-3/4}(3ax^2)}$$

$$= \frac{\frac{1}{2}(2a^3a - a^4)^{-1/2}(2a^3 - 4a^3) - \frac{1}{3}a^3(a^2a)^{-2/3}}{-\frac{1}{4}(aa^3)^{-3/4}(3aa^2)}$$

$$= \frac{(a^4)^{-1/2}(-a^3) - \frac{1}{3}a^3(a^3)^{-2/3}}{-\frac{3}{4}a^3(a^4)^{-3/4}} = \frac{-a - \frac{1}{3}a}{-\frac{3}{4}} = \frac{4}{3}\left(\frac{4}{3}a\right) = \frac{16}{9}a$$

**62.** Let the radius of the circle be $r$. We see that $A(\theta)$ is the area of the whole figure (a sector of the circle with radius 1), minus the area of $\triangle OPR$. But the area of the sector of the circle is $\frac{1}{2}r^2\theta$ (see Reference Page 1), and the area of the triangle is $\frac{1}{2}r\,|PQ| = \frac{1}{2}r(r\sin\theta) = \frac{1}{2}r^2\sin\theta$. So we have $A(\theta) = \frac{1}{2}r^2\theta - \frac{1}{2}r^2\sin\theta = \frac{1}{2}r^2(\theta - \sin\theta)$. Now by elementary trigonometry, $B(\theta) = \frac{1}{2}\,|QR|\,|PQ| = \frac{1}{2}(r - |OQ|)\,|PQ| = \frac{1}{2}(r - r\cos\theta)(r\sin\theta) = \frac{1}{2}r^2(1-\cos\theta)\sin\theta$.
So the limit we want is

$$\begin{aligned}\lim_{\theta\to 0^+}\frac{A(\theta)}{B(\theta)} &= \lim_{\theta\to 0^+}\frac{\frac{1}{2}r^2(\theta-\sin\theta)}{\frac{1}{2}r^2(1-\cos\theta)\sin\theta} \overset{\text{H}}{=} \lim_{\theta\to 0^+}\frac{1-\cos\theta}{(1-\cos\theta)\cos\theta+\sin\theta\,(\sin\theta)}\\ &= \lim_{\theta\to 0^+}\frac{1-\cos\theta}{\cos\theta-\cos^2\theta+\sin^2\theta} \overset{\text{H}}{=} \lim_{\theta\to 0^+}\frac{\sin\theta}{-\sin\theta-2\cos\theta\,(-\sin\theta)+2\sin\theta\,(\cos\theta)}\\ &= \lim_{\theta\to 0^+}\frac{\sin\theta}{-\sin\theta+4\sin\theta\,\cos\theta} = \lim_{\theta\to 0^+}\frac{1}{-1+4\cos\theta} = \frac{1}{-1+4\cos 0} = \frac{1}{3}\end{aligned}$$

**63.** Since $f(2)=0$, the given limit has the form $\frac{0}{0}$.

$$\lim_{x\to 0}\frac{f(2+3x)+f(2+5x)}{x} \overset{\text{H}}{=} \lim_{x\to 0}\frac{f'(2+3x)\cdot 3+f'(2+5x)\cdot 5}{1} = f'(2)\cdot 3+f'(2)\cdot 5 = 8f'(2) = 8\cdot 7 = 56$$

**64.** $L = \lim\limits_{x\to 0}\left(\dfrac{\sin 2x}{x^3}+a+\dfrac{b}{x^2}\right) = \lim\limits_{x\to 0}\dfrac{\sin 2x+ax^3+bx}{x^3} \overset{\text{H}}{=} \lim\limits_{x\to 0}\dfrac{2\cos 2x+3ax^2+b}{3x^2}$. As $x\to 0$, $3x^2\to 0$, and $\left(2\cos 2x+3ax^2+b\right)\to b+2$, so the last limit exists only if $b+2=0$, that is, $b=-2$. Thus,

$\lim\limits_{x\to 0}\dfrac{2\cos 2x+3ax^2-2}{3x^2} \overset{\text{H}}{=} \lim\limits_{x\to 0}\dfrac{-4\sin 2x+6ax}{6x} \overset{\text{H}}{=} \lim\limits_{x\to 0}\dfrac{-8\cos 2x+6a}{6} = \dfrac{6a-8}{6}$, which is equal to 0 if and only if $a=\frac{4}{3}$. Hence, $L=0$ if and only if $b=-2$ and $a=\frac{4}{3}$.

**65.** Since $\lim\limits_{h\to 0}[f(x+h)-f(x-h)] = f(x)-f(x) = 0$ ($f$ is differentiable and hence continuous) and $\lim\limits_{h\to 0}2h = 0$, we use l'Hospital's Rule:

$$\lim_{h\to 0}\frac{f(x+h)-f(x-h)}{2h} \overset{\text{H}}{=} \lim_{h\to 0}\frac{f'(x+h)(1)-f'(x-h)(-1)}{2} = \frac{f'(x)+f'(x)}{2} = \frac{2f'(x)}{2} = f'(x)$$

$\dfrac{f(x+h)-f(x-h)}{2h}$ is the slope of the secant line between $(x-h, f(x-h))$ and $(x+h, f(x+h))$. As $h\to 0$, this line gets closer to the tangent line and its slope approaches $f'(x)$.

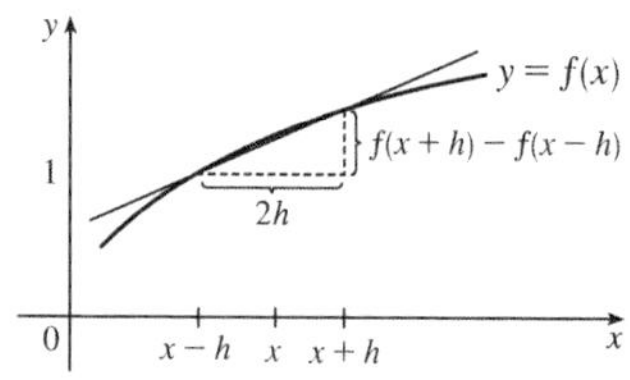


**66.** (a) For $f$ to be continuous, we need $\lim\limits_{x\to 0}f(x) = f(0) = 1$. We note that for $x\neq 0$, $\ln f(x) = \ln|x|^x = x\ln|x|$. So

$\lim\limits_{x\to 0}\ln f(x) = \lim\limits_{x\to 0}x\ln|x| = \lim\limits_{x\to 0}\dfrac{\ln|x|}{1/x} \overset{\text{H}}{=} \lim\limits_{x\to 0}\dfrac{1/x}{-1/x^2} = 0$. Therefore, $\lim\limits_{x\to 0}f(x) = \lim\limits_{x\to 0}e^{\ln f(x)} = e^0 = 1$. So $f$ is continuous at 0.

(b) From the graphs, it appears that $f$ is differentiable at 0.

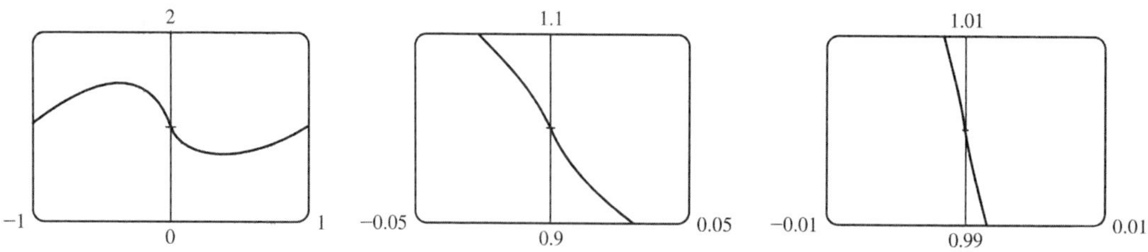

(c) To find $f'$, we use logarithmic differentiation: $\ln f(x) = x \ln |x| \quad \Rightarrow \quad \dfrac{f'(x)}{f(x)} = x\left(\dfrac{1}{x}\right) + \ln |x| \quad \Rightarrow$

$f'(x) = f(x)(1 + \ln |x|) = |x|^x (1 + \ln |x|)$, $x \neq 0$. Now $f'(x) \to -\infty$ as $x \to 0$ [since $|x|^x \to 1$ and $(1 + \ln |x|) \to -\infty$], so the curve has a vertical tangent at $(0, 1)$ and is therefore not differentiable there. The fact cannot be seen in the graphs in part (b) because $\ln |x| \to -\infty$ very slowly as $x \to 0$.

## 4.6 Optimization Problems

**1.** (a)

| First Number | Second Number | Product |
|---|---|---|
| 1 | 22 | 22 |
| 2 | 21 | 42 |
| 3 | 20 | 60 |
| 4 | 19 | 76 |
| 5 | 18 | 90 |
| 6 | 17 | 102 |
| 7 | 16 | 112 |
| 8 | 15 | 120 |
| 9 | 14 | 126 |
| 10 | 13 | 130 |
| 11 | 12 | 132 |

We needn't consider pairs where the first number is larger than the second, since we can just interchange the numbers in such cases. The answer appears to be 11 and 12, but we have considered only integers in the table.

(b) Call the two numbers $x$ and $y$. Then $x + y = 23$, so $y = 23 - x$. Call the product $P$. Then $P = xy = x(23 - x) = 23x - x^2$, so we wish to maximize the function $P(x) = 23x - x^2$. Since $P'(x) = 23 - 2x$, we see that $P'(x) = 0 \quad \Leftrightarrow \quad x = \frac{23}{2} = 11.5$. Thus, the maximum value of $P$ is $P(11.5) = (11.5)^2 = 132.25$ and it occurs when $x = y = 11.5$.

*Or:* Note that $P''(x) = -2 < 0$ for all $x$, so $P$ is everywhere concave downward and the local maximum at $x = 11.5$ must be an absolute maximum.

**2.** The two numbers are $x + 100$ and $x$. Minimize $f(x) = (x + 100)x = x^2 + 100x$. $f'(x) = 2x + 100 = 0 \quad \Rightarrow \quad x = -50$. Since $f''(x) = 2 > 0$, there is an absolute minimum at $x = -50$. The two numbers are 50 and $-50$.

**3.** The two numbers are $x$ and $\dfrac{100}{x}$, where $x > 0$. Minimize $f(x) = x + \dfrac{100}{x}$. $f'(x) = 1 - \dfrac{100}{x^2} = \dfrac{x^2 - 100}{x^2}$. The critical number is $x = 10$. Since $f'(x) < 0$ for $0 < x < 10$ and $f'(x) > 0$ for $x > 10$, there is an absolute minimum at $x = 10$. The numbers are 10 and 10.

**4.** Let $x > 0$ and let $f(x) = x + 1/x$. We wish to minimize $f(x)$. Now $f'(x) = 1 - \dfrac{1}{x^2} = \dfrac{1}{x^2}(x^2 - 1) = \dfrac{1}{x^2}(x + 1)(x - 1)$, so the only critical number in $(0, \infty)$ is 1.

$f'(x) < 0$ for $0 < x < 1$ and $f'(x) > 0$ for $x > 1$, so $f$ has an absolute minimum at $x = 1$, and $f(1) = 2$.

*Or:* $f''(x) = 2/x^3 > 0$ for all $x > 0$, so $f$ is concave upward everywhere and the critical point $(1, 2)$ must correspond to a local minimum for $f$.

**5.** If the rectangle has dimensions $x$ and $y$, then its perimeter is $2x + 2y = 100$ m, so $y = 50 - x$. Thus, the area is $A = xy = x(50 - x)$. We wish to maximize the function $A(x) = x(50 - x) = 50x - x^2$, where $0 < x < 50$. Since $A'(x) = 50 - 2x = -2(x - 25)$, $A'(x) > 0$ for $0 < x < 25$ and $A'(x) < 0$ for $25 < x < 50$. Thus, $A$ has an absolute maximum at $x = 25$, and $A(25) = 25^2 = 625 \text{ m}^2$. The dimensions of the rectangle that maximize its area are $x = y = 25$ m. (The rectangle is a square.)

**6.** If the rectangle has dimensions $x$ and $y$, then its area is $xy = 1000\text{ m}^2$, so $y = 1000/x$. The perimeter $P = 2x + 2y = 2x + 2000/x$. We wish to minimize the function $P(x) = 2x + 2000/x$ for $x > 0$. $P'(x) = 2 - 2000/x^2 = (2/x^2)(x^2 - 1000)$, so the only critical number in the domain of $P$ is $x = \sqrt{1000}$. $P''(x) = 4000/x^3 > 0$, so $P$ is concave upward throughout its domain and $P(\sqrt{1000}) = 4\sqrt{1000}$ is an absolute minimum value. The dimensions of the rectangle with minimal perimeter are $x = y = \sqrt{1000} = 10\sqrt{10}$ m. (The rectangle is a square.)

**7.** (a)

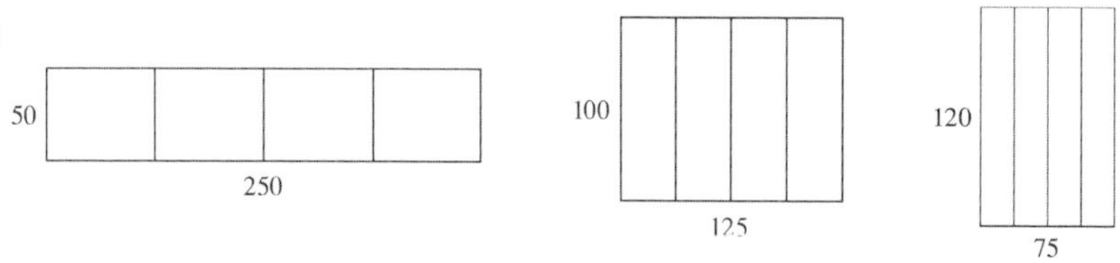


The areas of the three figures are 12,500, 12,500, and 9000 $\text{ft}^2$. There appears to be a maximum area of at least 12,500 $\text{ft}^2$.

(b) Let $x$ denote the length of each of two sides and three dividers.

Let $y$ denote the length of the other two sides.

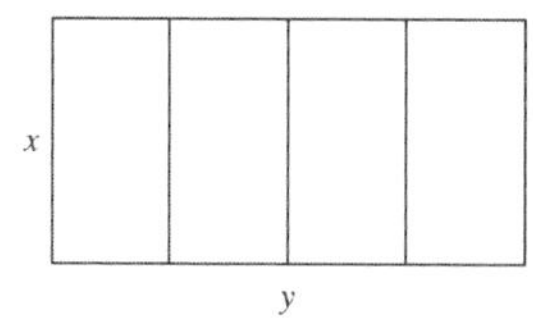


(c) Area $A = \text{length} \times \text{width} = y \cdot x$

(d) Length of fencing $= 750 \quad\Rightarrow\quad 5x + 2y = 750$

(e) $5x + 2y = 750 \quad\Rightarrow\quad y = 375 - \frac{5}{2}x \quad\Rightarrow\quad A(x) = \left(375 - \frac{5}{2}x\right)x = 375x - \frac{5}{2}x^2$

(f) $A'(x) = 375 - 5x = 0 \quad\Rightarrow\quad x = 75$. Since $A''(x) = -5 < 0$ there is an absolute maximum when $x = 75$. Then $y = \frac{375}{2} = 187.5$. The largest area is $75\left(\frac{375}{2}\right) = 14{,}062.5\text{ ft}^2$. These values of $x$ and $y$ are between the values in the first and second figures in part (a). Our original estimate was low.

**8.** (a)

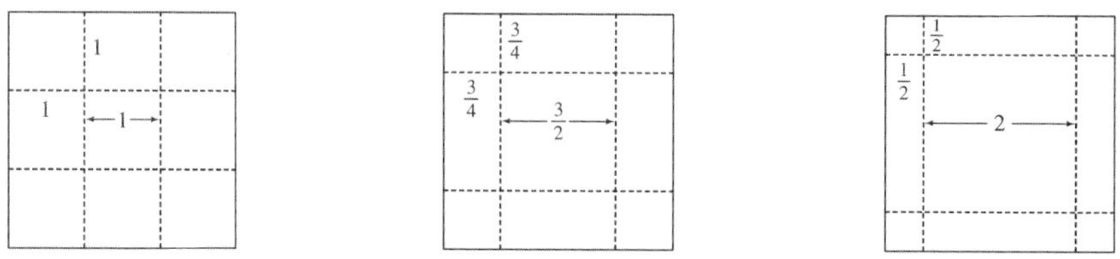


The volumes of the resulting boxes are 1, 1.6875, and 2 $\text{ft}^3$. There appears to be a maximum volume of at least 2 $\text{ft}^3$.

(b) Let $x$ denote the length of the side of the square being cut out. Let $y$ denote the length of the base.

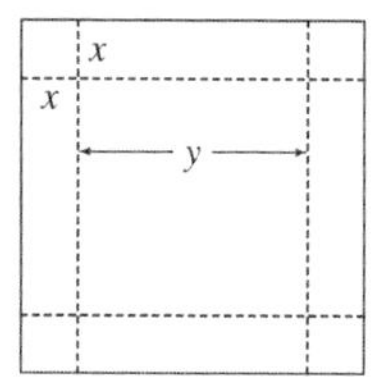


(c) Volume $V = \text{length} \times \text{width} \times \text{height} \quad\Rightarrow\quad V = y \cdot y \cdot x = xy^2$

(d) Length of cardboard $= 3 \quad\Rightarrow\quad x + y + x = 3 \quad\Rightarrow\quad y + 2x = 3$

(e) $y + 2x = 3 \quad\Rightarrow\quad y = 3 - 2x \quad\Rightarrow\quad V(x) = x(3 - 2x)^2$

(f) $V(x) = x(3 - 2x)^2 \quad\Rightarrow$

$V'(x) = x \cdot 2(3 - 2x)(-2) + (3 - 2x)^2 \cdot 1 = (3 - 2x)[-4x + (3 - 2x)] = (3 - 2x)(-6x + 3)$, so the critical numbers are $x = \frac{3}{2}$ and $x = \frac{1}{2}$. Now $0 \le x \le \frac{3}{2}$ and $V(0) = V\left(\frac{3}{2}\right) = 0$, so the maximum is $V\left(\frac{1}{2}\right) = \left(\frac{1}{2}\right)(2)^2 = 2\text{ ft}^3$, which is the value found from our third figure in part (a).

**9.** Let $b$ be the length of the base of the box and $h$ the height. The surface area is $1200 = b^2 + 4hb \quad\Rightarrow$ $h = (1200 - b^2)/(4b)$. The volume is $V = b^2h = b^2(1200 - b^2)/4b = 300b - b^3/4 \quad\Rightarrow\quad V'(b) = 300 - \frac{3}{4}b^2$. $V'(b) = 0 \quad\Rightarrow\quad 300 = \frac{3}{4}b^2 \quad\Rightarrow\quad b^2 = 400 \quad\Rightarrow\quad b = \sqrt{400} = 20$. Since $V'(b) > 0$ for $0 < b < 20$ and $V'(b) < 0$ for $b > 20$, there is an absolute maximum when $b = 20$ by the First Derivative Test for Absolute Extreme Values (see page 308). If $b = 20$, then $h = (1200 - 20^2)/(4 \cdot 20) = 10$, so the largest possible volume is $b^2h = (20)^2(10) = 4000\text{ cm}^3$.

**10.** Let $b$ be the length of the base of the box and $h$ the height. The volume is $32{,}000 = b^2h \Rightarrow h = 32{,}000/b^2$. The surface area of the open box is $S = b^2 + 4hb = b^2 + 4(32{,}000/b^2)b = b^2 + 4(32{,}000)/b$. So $S'(b) = 2b - 4(32{,}000)/b^2 = 2(b^3 - 64{,}000)/b^2 = 0 \Leftrightarrow b = \sqrt[3]{64{,}000} = 40$. This gives an absolute minimum since $S'(b) < 0$ if $0 < b < 40$ and $S'(b) > 0$ if $b > 40$. The box should be $40 \times 40 \times 20$.

**11.** (a) Let the rectangle have sides $x$ and $y$ and area $A$, so $A = xy$ or $y = A/x$. The problem is to minimize the perimeter $= 2x + 2y = 2x + 2A/x = P(x)$. Now $P'(x) = 2 - 2A/x^2 = 2(x^2 - A)/x^2$. So the critical number is $x = \sqrt{A}$. Since $P'(x) < 0$ for $0 < x < \sqrt{A}$ and $P'(x) > 0$ for $x > \sqrt{A}$, there is an absolute minimum at $x = \sqrt{A}$. The sides of the rectangle are $\sqrt{A}$ and $A/\sqrt{A} = \sqrt{A}$, so the rectangle is a square.

(b) Let $p$ be the perimeter and $x$ and $y$ the lengths of the sides, so $p = 2x + 2y \Rightarrow 2y = p - 2x \Rightarrow y = \frac{1}{2}p - x$. The area is $A(x) = x\left(\frac{1}{2}p - x\right) = \frac{1}{2}px - x^2$. Now $A'(x) = 0 \Rightarrow \frac{1}{2}p - 2x = 0 \Rightarrow 2x = \frac{1}{2}p \Rightarrow x = \frac{1}{4}p$. Since $A''(x) = -2 < 0$, there is an absolute maximum for $A$ when $x = \frac{1}{4}p$ by the Second Derivative Test. The sides of the rectangle are $\frac{1}{4}p$ and $\frac{1}{2}p - \frac{1}{4}p = \frac{1}{4}p$, so the rectangle is a square.

**12.**

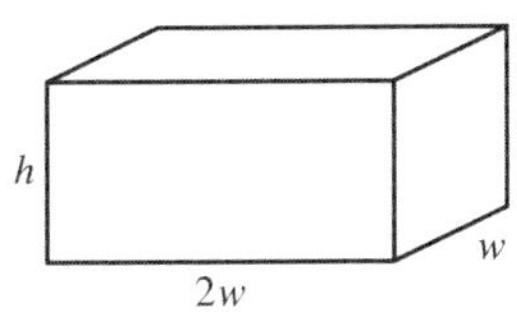


$V = lwh \Rightarrow 10 = (2w)(w)h = 2w^2h$, so $h = 5/w^2$. The cost is $10(2w^2) + 6[2(2wh) + 2(hw)] = 20w^2 + 36wh$, so $C(w) = 20w^2 + 36w(5/w^2) = 20w^2 + 180/w$. $C'(w) = 40w - 180/w^2 = 40\left(w^3 - \frac{9}{2}\right)/w^2 \Rightarrow w = \sqrt[3]{\frac{9}{2}}$ is the critical number. There is an absolute minimum for $C$ when $w = \sqrt[3]{\frac{9}{2}}$ since $C'(w) < 0$ for $0 < w < \sqrt[3]{\frac{9}{2}}$ and $C'(w) > 0$ for $w > \sqrt[3]{\frac{9}{2}}$. $C\left(\sqrt[3]{\frac{9}{2}}\right) = 20\left(\sqrt[3]{\frac{9}{2}}\right)^2 + \frac{180}{\sqrt[3]{9/2}} \approx \$163.54$.

**13.**

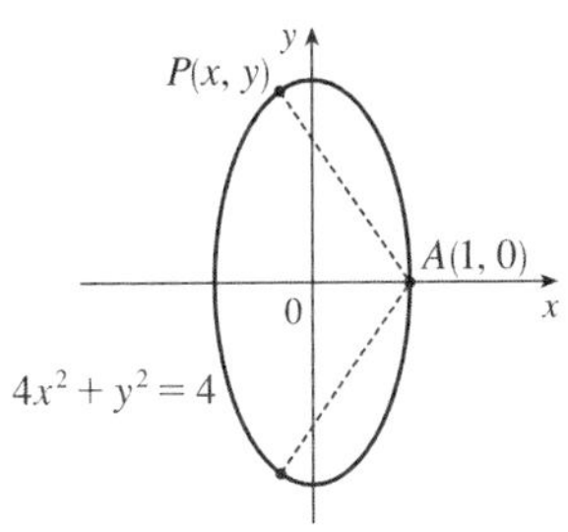


From the figure, we see that there are two points that are farthest away from $A(1, 0)$. The distance $d$ from $A$ to an arbitrary point $P(x, y)$ on the ellipse is $d = \sqrt{(x-1)^2 + (y-0)^2}$ and the square of the distance is $S = d^2 = x^2 - 2x + 1 + y^2 = x^2 - 2x + 1 + (4 - 4x^2) = -3x^2 - 2x + 5$. $S' = -6x - 2$ and $S' = 0 \Rightarrow x = -\frac{1}{3}$. Now $S'' = -6 < 0$, so we know that $S$ has a maximum at $x = -\frac{1}{3}$. Since $-1 \le x \le 1$, $S(-1) = 4$, $S\left(-\frac{1}{3}\right) = \frac{16}{3}$, and $S(1) = 0$, we see that the maximum distance is $\sqrt{\frac{16}{3}}$. The corresponding $y$-values are $y = \pm\sqrt{4 - 4\left(-\frac{1}{3}\right)^2} = \pm\sqrt{\frac{32}{9}} = \pm\frac{4}{3}\sqrt{2} \approx \pm 1.89$. The points are $\left(-\frac{1}{3}, \pm\frac{4}{3}\sqrt{2}\right)$.

**14.**

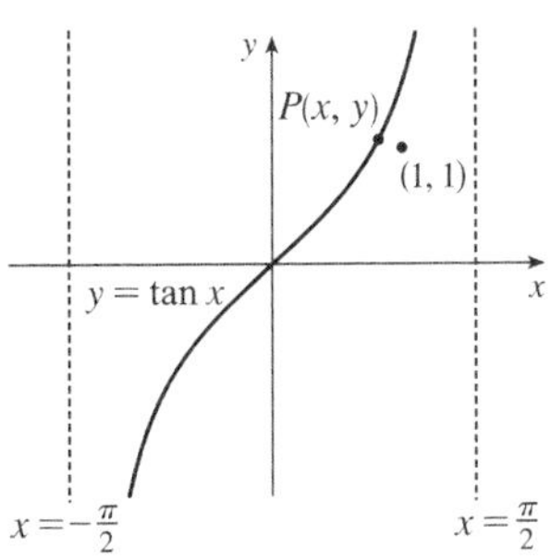


The distance $d$ from $(1, 1)$ to an arbitrary point $P(x, y)$ on the curve $y = \tan x$ is $d = \sqrt{(x-1)^2 + (y-1)^2}$ and the square of the distance is $S = d^2 = (x-1)^2 + (\tan x - 1)^2$. $S' = 2(x-1) + 2(\tan x - 1)\sec^2 x$. Graphing $S'$ on $\left(-\frac{\pi}{2}, \frac{\pi}{2}\right)$ gives us a zero at $x \approx 0.82$, and so $\tan x \approx 1.08$. The point on $y = \tan x$ that is closest to $(1, 1)$ is approximately $(0.82, 1.08)$.

**15.**

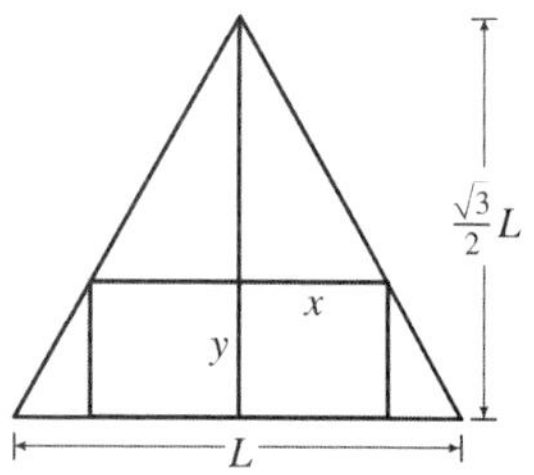


The height $h$ of the equilateral triangle with sides of length $L$ is $\frac{\sqrt{3}}{2}L$, since $h^2 + (L/2)^2 = L^2 \Rightarrow h^2 = L^2 - \frac{1}{4}L^2 = \frac{3}{4}L^2 \Rightarrow$ $h = \frac{\sqrt{3}}{2}L$. Using similar triangles, $\dfrac{\frac{\sqrt{3}}{2}L - y}{x} = \dfrac{\frac{\sqrt{3}}{2}L}{L/2} = \sqrt{3} \Rightarrow$

$$\sqrt{3}\,x = \frac{\sqrt{3}}{2}L - y \Rightarrow y = \frac{\sqrt{3}}{2}L - \sqrt{3}\,x \Rightarrow y = \frac{\sqrt{3}}{2}(L - 2x).$$

The area of the inscribed rectangle is $A(x) = (2x)y = \sqrt{3}\,x(L - 2x) = \sqrt{3}\,Lx - 2\sqrt{3}\,x^2$, where $0 \le x \le L/2$. Now $0 = A'(x) = \sqrt{3}\,L - 4\sqrt{3}\,x \Rightarrow x = \sqrt{3}\,L/(4\sqrt{3}) = L/4$. Since $A(0) = A(L/2) = 0$, the maximum occurs when $x = L/4$, and $y = \frac{\sqrt{3}}{2}L - \frac{\sqrt{3}}{4}L = \frac{\sqrt{3}}{4}L$, so the dimensions are $L/2$ and $\frac{\sqrt{3}}{4}L$.

**16.**

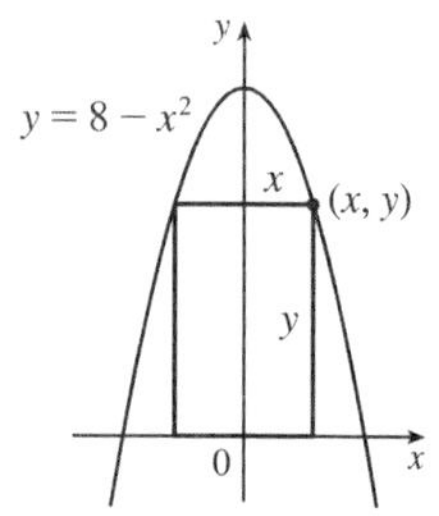


The rectangle has area $A(x) = 2xy = 2x(8 - x^2) = 16x - 2x^3$, where $0 \le x \le 2\sqrt{2}$. Now $A'(x) = 16 - 6x^2 = 0 \Rightarrow x = 2\sqrt{\frac{2}{3}}$. Since $A(0) = A(2\sqrt{2}) = 0$, there is a maximum when $x = 2\sqrt{\frac{2}{3}}$. Then $y = \frac{16}{3}$, so the rectangle has dimensions $4\sqrt{\frac{2}{3}}$ and $\frac{16}{3}$.

**17.**

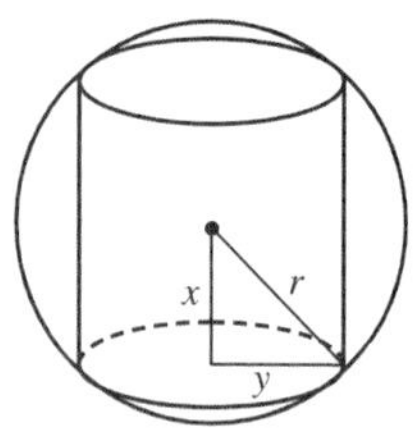


The cylinder has volume $V = \pi y^2(2x)$. Also $x^2 + y^2 = r^2 \Rightarrow$ $y^2 = r^2 - x^2$, so $V(x) = \pi(r^2 - x^2)(2x) = 2\pi(r^2 x - x^3)$, where $0 \le x \le r$. $V'(x) = 2\pi(r^2 - 3x^2) = 0 \Rightarrow x = r/\sqrt{3}$. Now $V(0) = V(r) = 0$, so there is a maximum when $x = r/\sqrt{3}$ and $V(r/\sqrt{3}) = \pi(r^2 - r^2/3)(2r/\sqrt{3}) = 4\pi r^3/(3\sqrt{3})$.

**18.**

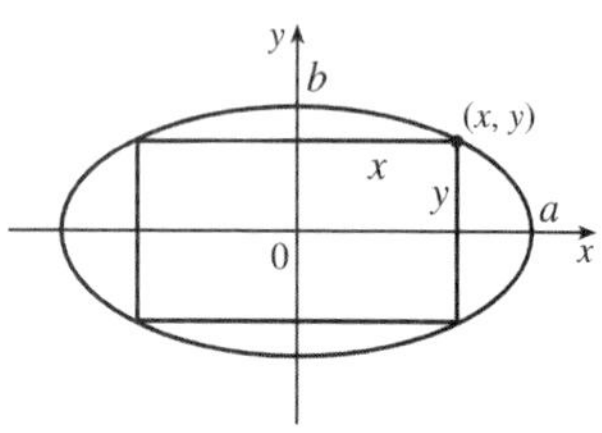


The area of the rectangle is $(2x)(2y) = 4xy$. Now $\dfrac{x^2}{a^2} + \dfrac{y^2}{b^2} = 1$ gives $y = \dfrac{b}{a}\sqrt{a^2 - x^2}$, so we maximize $A(x) = 4\dfrac{b}{a}x\sqrt{a^2 - x^2}$.

$$A'(x) = \frac{4b}{a}\left[x \cdot \tfrac{1}{2}(a^2 - x^2)^{-1/2}(-2x) + (a^2 - x^2)^{1/2} \cdot 1\right]$$

$$= \frac{4b}{a}(a^2 - x^2)^{-1/2}\left[-x^2 + a^2 - x^2\right] = \frac{4b}{a\sqrt{a^2 - x^2}}\left[a^2 - 2x^2\right]$$

So the critical number is $x = \frac{1}{\sqrt{2}}a$, and this clearly gives a maximum. Then $y = \frac{1}{\sqrt{2}}b$, so the maximum area is $4\left(\frac{1}{\sqrt{2}}a\right)\left(\frac{1}{\sqrt{2}}b\right) = 2ab$.

**19.**

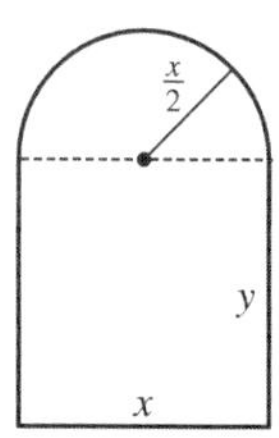


Perimeter $= 30 \Rightarrow 2y + x + \pi\left(\dfrac{x}{2}\right) = 30 \Rightarrow$ $y = \dfrac{1}{2}\left(30 - x - \dfrac{\pi x}{2}\right) = 15 - \dfrac{x}{2} - \dfrac{\pi x}{4}$. The area is the area of the rectangle plus the area of the semicircle, or $xy + \frac{1}{2}\pi\left(\dfrac{x}{2}\right)^2$, so $A(x) = x\left(15 - \dfrac{x}{2} - \dfrac{\pi x}{4}\right) + \frac{1}{8}\pi x^2 = 15x - \frac{1}{2}x^2 - \frac{\pi}{8}x^2$.

$A'(x) = 15 - \left(1 + \frac{\pi}{4}\right)x = 0 \Rightarrow x = \dfrac{15}{1 + \pi/4} = \dfrac{60}{4 + \pi}$. $A''(x) = -\left(1 + \dfrac{\pi}{4}\right) < 0$, so this gives a maximum. The dimensions are $x = \dfrac{60}{4 + \pi}$ ft and $y = 15 - \dfrac{30}{4 + \pi} - \dfrac{15\pi}{4 + \pi} = \dfrac{60 + 15\pi - 30 - 15\pi}{4 + \pi} = \dfrac{30}{4 + \pi}$ ft, so the height of the rectangle is half the base.

**20.** 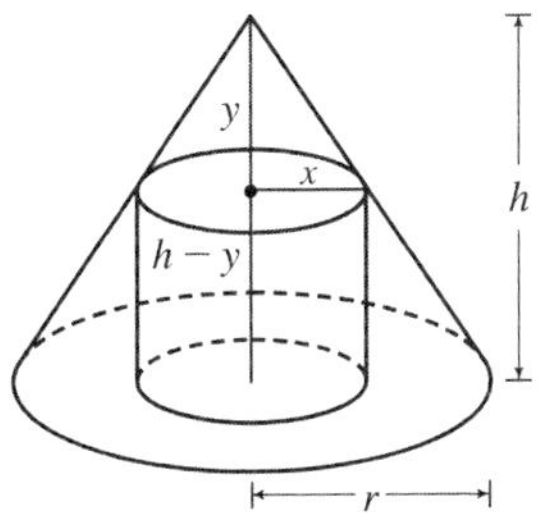


By similar triangles, $y/x = h/r$, so $y = hx/r$. The volume of the cylinder is $\pi x^2(h - y) = \pi hx^2 - (\pi h/r)x^3 = V(x)$. Now $V'(x) = 2\pi hx - (3\pi h/r)x^2 = \pi hx(2 - 3x/r)$.
So $V'(x) = 0 \quad \Rightarrow \quad x = 0$ or $x = \frac{2}{3}r$. The maximum clearly occurs when $x = \frac{2}{3}r$ and then the volume is
$\pi hx^2 - (\pi h/r)x^3 = \pi hx^2(1 - x/r) = \pi\left(\frac{2}{3}r\right)^2 h\left(1 - \frac{2}{3}\right) = \frac{4}{27}\pi r^2 h.$

**21.** 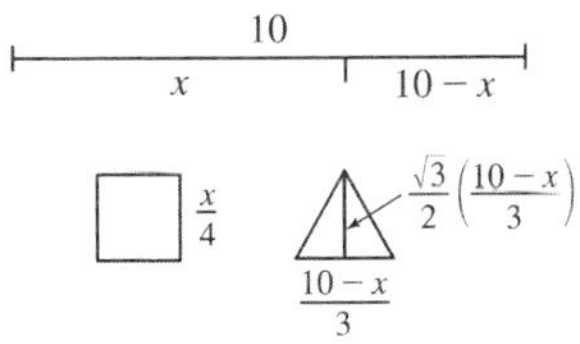


Let $x$ be the length of the wire used for the square. The total area is

$$A(x) = \left(\frac{x}{4}\right)^2 + \frac{1}{2}\left(\frac{10 - x}{3}\right)\frac{\sqrt{3}}{2}\left(\frac{10 - x}{3}\right)$$
$$= \tfrac{1}{16}x^2 + \tfrac{\sqrt{3}}{36}(10 - x)^2, \quad 0 \le x \le 10$$

$A'(x) = \frac{1}{8}x - \frac{\sqrt{3}}{18}(10 - x) = 0 \quad \Leftrightarrow \quad \frac{9}{72}x + \frac{4\sqrt{3}}{72}x - \frac{40\sqrt{3}}{72} = 0 \quad \Leftrightarrow \quad x = \frac{40\sqrt{3}}{9 + 4\sqrt{3}}$. Now $A(0) = \left(\frac{\sqrt{3}}{36}\right)100 \approx 4.81$,
$A(10) = \frac{100}{16} = 6.25$ and $A\left(\frac{40\sqrt{3}}{9 + 4\sqrt{3}}\right) \approx 2.72$, so
(a) The maximum area occurs when $x = 10$ m, and all the wire is used for the square.
(b) The minimum area occurs when $x = \frac{40\sqrt{3}}{9 + 4\sqrt{3}} \approx 4.35$ m.

**22.** 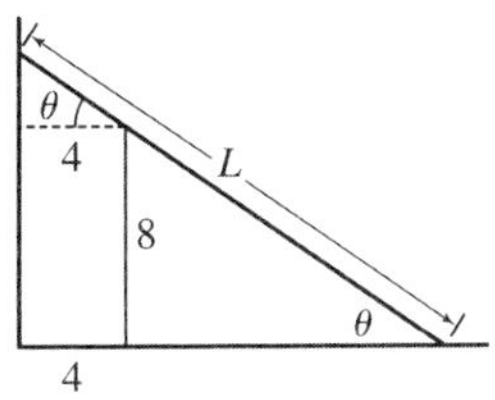


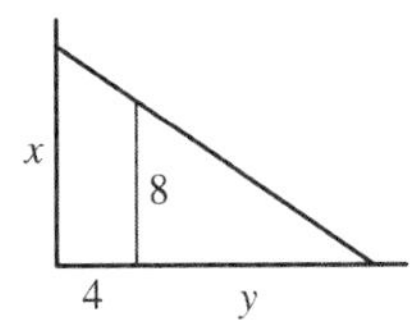


$L = 8\csc\theta + 4\sec\theta$, $0 < \theta < \frac{\pi}{2}$,
$\dfrac{dL}{d\theta} = -8\csc\theta\,\cot\theta + 4\sec\theta\,\tan\theta = 0$ when
$\sec\theta\,\tan\theta = 2\csc\theta\,\cot\theta \quad \Leftrightarrow \quad \tan^3\theta = 2 \quad \Leftrightarrow \quad \tan\theta = \sqrt[3]{2} \quad \Leftrightarrow$
$\theta = \tan^{-1}\sqrt[3]{2}$.
$dL/d\theta < 0$ when $0 < \theta < \tan^{-1}\sqrt[3]{2}$, $dL/d\theta > 0$ when $\tan^{-1}\sqrt[3]{2} < \theta < \frac{\pi}{2}$, so $L$ has an absolute minimum when $\theta = \tan^{-1}\sqrt[3]{2}$, and the shortest ladder has length
$$L = 8\frac{\sqrt{1 + 2^{2/3}}}{2^{1/3}} + 4\sqrt{1 + 2^{2/3}} \approx 16.65 \text{ ft.}$$
*Another method:* Minimize $L^2 = x^2 + (4 + y)^2$, where $\dfrac{x}{4 + y} = \dfrac{8}{y}$.

**23.** 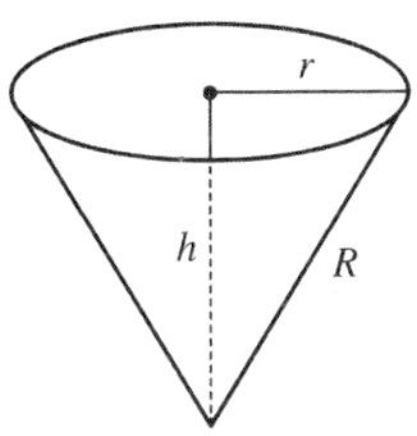


$h^2 + r^2 = R^2 \quad \Rightarrow \quad V = \frac{\pi}{3}r^2 h = \frac{\pi}{3}(R^2 - h^2)h = \frac{\pi}{3}(R^2 h - h^3)$.
$V'(h) = \frac{\pi}{3}(R^2 - 3h^2) = 0$ when $h = \frac{1}{\sqrt{3}}R$. This gives an absolute maximum, since $V'(h) > 0$ for $0 < h < \frac{1}{\sqrt{3}}R$ and $V'(h) < 0$ for $h > \frac{1}{\sqrt{3}}R$. The maximum volume is
$$V\left(\tfrac{1}{\sqrt{3}}R\right) = \tfrac{\pi}{3}\left(\tfrac{1}{\sqrt{3}}R^3 - \tfrac{1}{3\sqrt{3}}R^3\right) = \tfrac{2}{9\sqrt{3}}\pi R^3.$$

**24.** The volume and surface area of a cone with radius $r$ and height $h$ are given by $V = \frac{1}{3}\pi r^2 h$ and $S = \pi r\sqrt{r^2 + h^2}$.
We'll minimize $A = S^2$ subject to $V = 27$. $\;V = 27 \quad \Rightarrow \quad \frac{1}{3}\pi r^2 h = 27 \quad \Rightarrow \quad r^2 = \dfrac{81}{\pi h}$ **(1)**.

$A = \pi^2 r^2(r^2 + h^2) = \pi^2\left(\dfrac{81}{\pi h}\right)\left(\dfrac{81}{\pi h} + h^2\right) = \dfrac{81^2}{h^2} + 81\pi h$, so $A' = 0 \quad \Rightarrow \quad \dfrac{-2 \cdot 81^2}{h^3} + 81\pi = 0 \quad \Rightarrow$

$81\pi = \dfrac{2 \cdot 81^2}{h^3} \quad \Rightarrow \quad h^3 = \dfrac{162}{\pi} \quad \Rightarrow \quad h = \sqrt[3]{\dfrac{162}{\pi}} = 3\sqrt[3]{\dfrac{6}{\pi}} \approx 3.722$. From **(1)**, $r^2 = \dfrac{81}{\pi h} = \dfrac{81}{\pi \cdot 3\sqrt[3]{6/\pi}} = \dfrac{27}{\sqrt[3]{6\pi^2}} \quad \Rightarrow$

$r = \dfrac{3\sqrt{3}}{\sqrt[6]{6\pi^2}} \approx 2.632$. $A'' = 6 \cdot 81^2/h^4 > 0$, so $A$ and hence $S$ has an absolute minimum at these values of $r$ and $h$.

**25.**

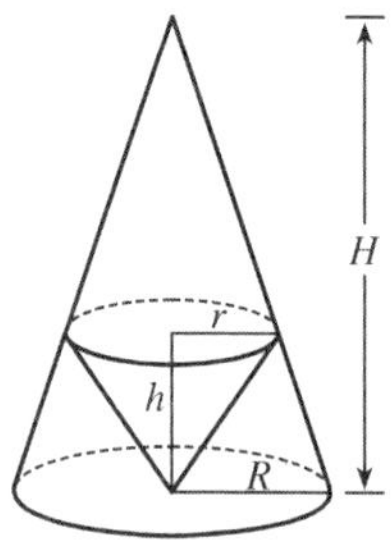


By similar triangles, $\dfrac{H}{R} = \dfrac{H-h}{r}$ **(1)**. The volume of the inner cone is $V = \frac{1}{3}\pi r^2 h$, so we'll solve **(1)** for $h$. $\dfrac{Hr}{R} = H - h \;\Rightarrow$

$$h = H - \frac{Hr}{R} = \frac{HR - Hr}{R} = \frac{H}{R}(R-r) \quad \textbf{(2)}.$$

Thus, $V(r) = \dfrac{\pi}{3}r^2 \cdot \dfrac{H}{R}(R-r) = \dfrac{\pi H}{3R}(Rr^2 - r^3) \;\Rightarrow$

$$V'(r) = \frac{\pi H}{3R}(2Rr - 3r^2) = \frac{\pi H}{3R}r(2R - 3r).$$

$V'(r) = 0 \;\Rightarrow\; r = 0$ or $2R = 3r \;\Rightarrow\; r = \frac{2}{3}R$ and from **(2)**, $h = \dfrac{H}{R}\left(R - \frac{2}{3}R\right) = \dfrac{H}{R}\left(\frac{1}{3}R\right) = \frac{1}{3}H$.

$V'(r)$ changes from positive to negative at $r = \frac{2}{3}R$, so the inner conc has a maximum volume of

$V = \frac{1}{3}\pi r^2 h = \frac{1}{3}\pi\left(\frac{2}{3}R\right)^2\left(\frac{1}{3}H\right) = \frac{4}{27}\cdot\frac{1}{3}\pi R^2 H$, which is approximately 15% of the volume of the larger cone.

**26.** We note that since $c$ is the consumption in gallons per hour, and $v$ is the velocity in miles per hour, then $\dfrac{c}{v} = \dfrac{\text{gallons/hour}}{\text{miles/hour}} = \dfrac{\text{gallons}}{\text{mile}}$ gives us the consumption in gallons per mile, that is, the quantity $G$. To find the minimum, we calculate $\dfrac{dG}{dv} = \dfrac{d}{dv}\left(\dfrac{c}{v}\right) = \dfrac{v\dfrac{dc}{dv} - c\dfrac{dv}{dv}}{v^2} = \dfrac{v\dfrac{dc}{dv} - c}{v^2}$.

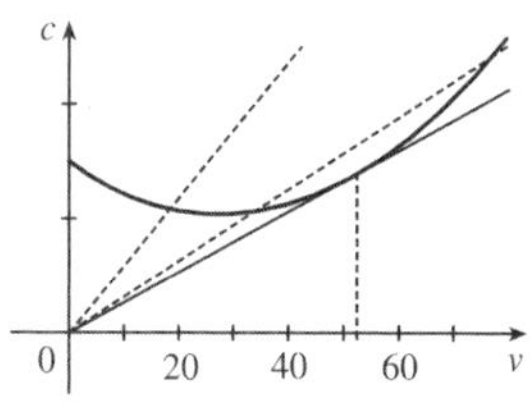


This is 0 when $v\dfrac{dc}{dv} - c = 0 \;\Leftrightarrow\; \dfrac{dc}{dv} = \dfrac{c}{v}$. This implies that the tangent line of $c(v)$ passes through the origin, and this occurs when $v \approx 53$ mi/h. Note that the slope of the secant line through the origin and a point $(v, c(v))$ on the graph is equal to $G(v)$, and it is intuitively clear that $G$ is minimized in the case where the secant is in fact a tangent.

**27.** $P(R) = \dfrac{E^2 R}{(R+r)^2} \;\Rightarrow$

$$P'(R) = \frac{(R+r)^2\cdot E^2 - E^2R\cdot 2(R+r)}{[(R+r)^2]^2} = \frac{(R^2 + 2Rr + r^2)\,E^2 - 2E^2R^2 - 2E^2Rr}{(R+r)^4}$$
$$= \frac{E^2r^2 - E^2R^2}{(R+r)^4} = \frac{E^2(r^2 - R^2)}{(R+r)^4} = \frac{E^2(r+R)(r-R)}{(R+r)^4} = \frac{E^2(r-R)}{(R+r)^3}$$

$P'(R) = 0 \;\Rightarrow\; R = r \;\Rightarrow\; P(r) = \dfrac{E^2 r}{(r+r)^2} = \dfrac{E^2 r}{4r^2} = \dfrac{E^2}{4r}$.

The expression for $P'(R)$ shows that $P'(R) > 0$ for $R < r$ and $P'(R) < 0$ for $R > r$. Thus, the maximum value of the power is $E^2/(4r)$, and this occurs when $R = r$.

**28.** (a) $E(v) = \dfrac{aLv^3}{v-u} \;\Rightarrow\; E'(v) = aL\dfrac{(v-u)3v^2 - v^3}{(v-u)^2} = 0$ when

$2v^3 = 3uv^2 \;\Rightarrow\; 2v = 3u \;\Rightarrow\; v = \frac{3}{2}u$.

The First Derivative Test shows that this value of $v$ gives the minimum value of $E$.

(b)

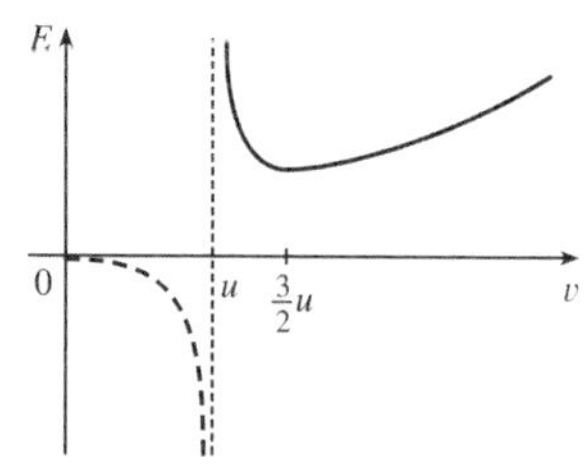

**29.** $S = 6sh - \frac{3}{2}s^2 \cot\theta + 3s^2 \frac{\sqrt{3}}{2} \csc\theta$

(a) $\dfrac{dS}{d\theta} = \frac{3}{2}s^2 \csc^2\theta - 3s^2 \frac{\sqrt{3}}{2} \csc\theta \cot\theta$ or $\frac{3}{2}s^2 \csc\theta \left(\csc\theta - \sqrt{3}\cot\theta\right)$.

(b) $\dfrac{dS}{d\theta} = 0$ when $\csc\theta - \sqrt{3}\cot\theta = 0 \;\Rightarrow\; \dfrac{1}{\sin\theta} - \sqrt{3}\dfrac{\cos\theta}{\sin\theta} = 0 \;\Rightarrow\; \cos\theta = \frac{1}{\sqrt{3}}$. The First Derivative Test shows that the minimum surface area occurs when $\theta = \cos^{-1}\left(\frac{1}{\sqrt{3}}\right) \approx 55^\circ$.

(c)

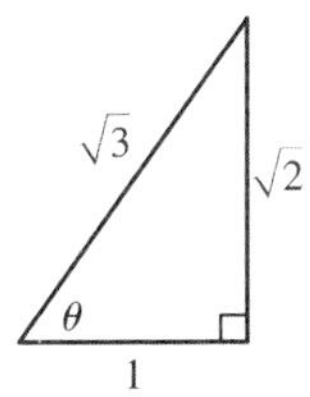


If $\cos\theta = \frac{1}{\sqrt{3}}$, then $\cot\theta = \frac{1}{\sqrt{2}}$ and $\csc\theta = \frac{\sqrt{3}}{\sqrt{2}}$, so the surface area is

$$S = 6sh - \tfrac{3}{2}s^2 \tfrac{1}{\sqrt{2}} + 3s^2 \tfrac{\sqrt{3}}{2}\tfrac{\sqrt{3}}{\sqrt{2}} = 6sh - \tfrac{3}{2\sqrt{2}}s^2 + \tfrac{9}{2\sqrt{2}}s^2$$

$$= 6sh + \tfrac{6}{2\sqrt{2}}s^2 = 6s\left(h + \tfrac{1}{2\sqrt{2}}s\right)$$

**30.**

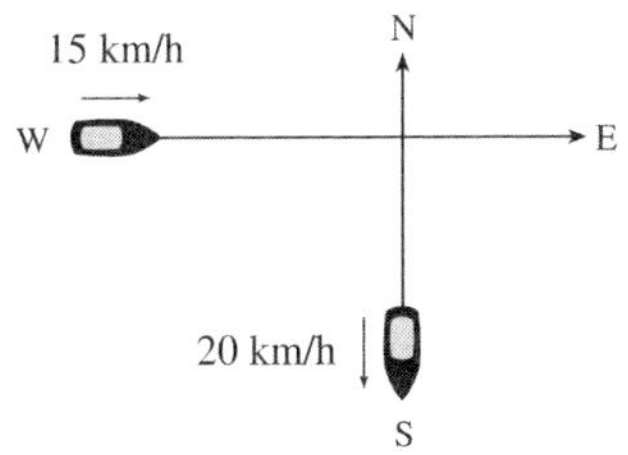


Let $t$ be the time, in hours, after 2:00 P.M. The position of the boat heading south at time $t$ is $(0, -20t)$. The position of the boat heading east at time $t$ is $(-15 + 15t, 0)$. If $D(t)$ is the distance between the boats at time $t$, we minimize $f(t) = [D(t)]^2 = 20^2t^2 + 15^2(t-1)^2$.

$f'(t) = 800t + 450(t-1) = 1250t - 450 = 0$ when $t = \frac{450}{1250} = 0.36$ h.

$0.36 \text{ h} \times \frac{60 \text{ min}}{\text{h}} = 21.6 \text{ min} = 21 \text{ min } 36 \text{ s}$. Since $f''(t) > 0$, this gives a minimum, so the boats are closest together at 2:21:36 P.M.

**31.**

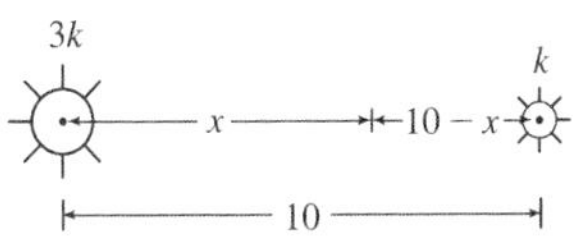


The total illumination is $I(x) = \dfrac{3k}{x^2} + \dfrac{k}{(10-x)^2}$, $0 < x < 10$. Then

$$I'(x) = \frac{-6k}{x^3} + \frac{2k}{(10-x)^3} = 0 \;\Rightarrow\; 6k(10-x)^3 = 2kx^3 \;\Rightarrow$$

$3(10-x)^3 = x^3 \;\Rightarrow\; \sqrt[3]{3}\,(10-x) = x \;\Rightarrow\; 10\sqrt[3]{3} - \sqrt[3]{3}\,x = x \;\Rightarrow\; 10\sqrt[3]{3} = x + \sqrt[3]{3}\,x \;\Rightarrow$

$10\sqrt[3]{3} = \left(1 + \sqrt[3]{3}\right)x \;\Rightarrow\; x = \dfrac{10\sqrt[3]{3}}{1 + \sqrt[3]{3}} \approx 5.9$ ft. This gives a minimum since $I''(x) > 0$ for $0 < x < 10$.

**32.**

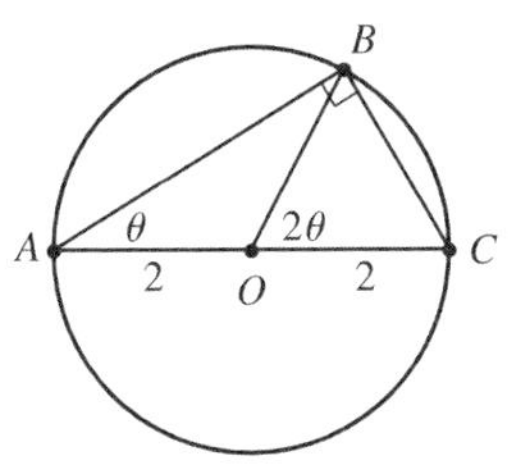


In isosceles triangle $AOB$, $\angle O = 180^\circ - \theta - \theta$, so $\angle BOC = 2\theta$. The distance rowed is $4\cos\theta$ while the distance walked is the length of arc $BC = 2(2\theta) = 4\theta$. The time taken is given by

$$T(\theta) = \frac{4\cos\theta}{2} + \frac{4\theta}{4} = 2\cos\theta + \theta, \quad 0 \le \theta \le \tfrac{\pi}{2}.$$

$$T'(\theta) = -2\sin\theta + 1 = 0 \;\Leftrightarrow\; \sin\theta = \tfrac{1}{2} \;\Rightarrow\; \theta = \tfrac{\pi}{6}.$$

Check the value of $T$ at $\theta = \frac{\pi}{6}$ and at the endpoints of the domain of $T$; that is, $\theta = 0$ and $\theta = \frac{\pi}{2}$. $T(0) = 2$, $T\left(\frac{\pi}{6}\right) = \sqrt{3} + \frac{\pi}{6} \approx 2.26$, and $T\left(\frac{\pi}{2}\right) = \frac{\pi}{2} \approx 1.57$. Therefore, the minimum value of $T$ is $\frac{\pi}{2}$ when $\theta = \frac{\pi}{2}$; that is, the woman should walk all the way. Note that $T''(\theta) = -2\cos\theta < 0$ for $0 \le \theta < \frac{\pi}{2}$, so $\theta = \frac{\pi}{6}$ gives a maximum time.

**33.** 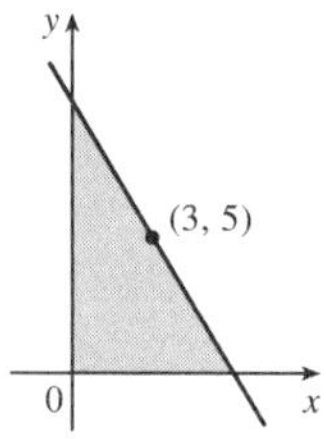


The line with slope $m$ (where $m < 0$) through $(3, 5)$ has equation $y - 5 = m(x - 3)$ or $y = mx + (5 - 3m)$. The $y$-intercept is $5 - 3m$ and the $x$-intercept is $-5/m + 3$. So the triangle has area $A(m) = \frac{1}{2}(5 - 3m)(-5/m + 3) = 15 - 25/(2m) - \frac{9}{2}m$. Now $A'(m) = \dfrac{25}{2m^2} - \dfrac{9}{2} = 0 \quad \Leftrightarrow \quad m^2 = \frac{25}{9} \quad \Rightarrow \quad m = -\frac{5}{3}$ (since $m < 0$).

$A''(m) = -\dfrac{25}{m^3} > 0$, so there is an absolute minimum when $m = -\frac{5}{3}$. Thus, an equation of the line is $y - 5 = -\frac{5}{3}(x - 3)$ or $y = -\frac{5}{3}x + 10$.

**34.** $y = 1 + 40x^3 - 3x^5 \quad \Rightarrow \quad y' = 120x^2 - 15x^4$, so the tangent line to the curve at $x = a$ has slope $m(a) = 120a^2 - 15a^4$. Now $m'(a) = 240a - 60a^3 = -60a(a^2 - 4) = -60a(a + 2)(a - 2)$, so $m'(a) > 0$ for $a < -2$, and $0 < a < 2$, and $m'(a) < 0$ for $-2 < a < 0$ and $a > 2$. Thus, $m$ is increasing on $(-\infty, -2)$, decreasing on $(-2, 0)$, increasing on $(0, 2)$, and decreasing on $(2, \infty)$. Clearly, $m(a) \to -\infty$ as $a \to \pm\infty$, so the maximum value of $m(a)$ must be one of the two local maxima, $m(-2)$ or $m(2)$. But both $m(-2)$ and $m(2)$ equal $120 \cdot 2^2 - 15 \cdot 2^4 = 480 - 240 = 240$. So 240 is the largest slope, and it occurs at the points $(-2, -223)$ and $(2, 225)$. *Note:* $a = 0$ corresponds to a local *minimum* of $m$.

**35.** 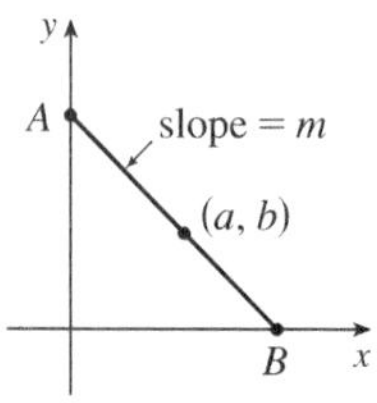


Every line segment in the first quadrant passing through $(a, b)$ with endpoints on the $x$- and $y$-axes satisfies an equation of the form $y - b = m(x - a)$, where $m < 0$. By setting $x = 0$ and then $y = 0$, we find its endpoints, $A(0, b - am)$ and $B\left(a - \frac{b}{m}, 0\right)$. The distance $d$ from $A$ to $B$ is given by $d = \sqrt{\left[\left(a - \frac{b}{m}\right) - 0\right]^2 + [0 - (b - am)]^2}$.

It follows that the square of the length of the line segment, as a function of $m$, is given by

$S(m) = \left(a - \dfrac{b}{m}\right)^2 + (am - b)^2 = a^2 - \dfrac{2ab}{m} + \dfrac{b^2}{m^2} + a^2m^2 - 2abm + b^2$. Thus,

$$\begin{aligned} S'(m) &= \frac{2ab}{m^2} - \frac{2b^2}{m^3} + 2a^2m - 2ab = \frac{2}{m^3}(abm - b^2 + a^2m^4 - abm^3) \\ &= \frac{2}{m^3}[b(am - b) + am^3(am - b)] = \frac{2}{m^3}(am - b)(b + am^3) \end{aligned}$$

Thus, $S'(m) = 0 \quad \Leftrightarrow \quad m = b/a$ or $m = -\sqrt[3]{\frac{b}{a}}$. Since $b/a > 0$ and $m < 0$, $m$ must equal $-\sqrt[3]{\frac{b}{a}}$. Since $\dfrac{2}{m^3} < 0$, we see that $S'(m) < 0$ for $m < -\sqrt[3]{\frac{b}{a}}$ and $S'(m) > 0$ for $m > -\sqrt[3]{\frac{b}{a}}$. Thus, $S$ has its absolute minimum value when $m = -\sqrt[3]{\frac{b}{a}}$. That value is

$$\begin{aligned} S\left(-\sqrt[3]{\tfrac{b}{a}}\right) &= \left(a + b\sqrt[3]{\tfrac{a}{b}}\right)^2 + \left(-a\sqrt[3]{\tfrac{b}{a}} - b\right)^2 = \left(a + \sqrt[3]{ab^2}\right)^2 + \left(\sqrt[3]{a^2b} + b\right)^2 \\ &= a^2 + 2a^{4/3}b^{2/3} + a^{2/3}b^{4/3} + a^{4/3}b^{2/3} + 2a^{2/3}b^{4/3} + b^2 = a^2 + 3a^{4/3}b^{2/3} + 3a^{2/3}b^{4/3} + b^2 \end{aligned}$$

The last expression is of the form $x^3 + 3x^2y + 3xy^2 + y^3 \quad \left[= (x + y)^3\right]$ with $x = a^{2/3}$ and $y = b^{2/3}$, so we can write it as $(a^{2/3} + b^{2/3})^3$ and the shortest such line segment has length $\sqrt{S} = (a^{2/3} + b^{2/3})^{3/2}$.

**36.** See the figure. The area is given by

$A(x) = \frac{1}{2}\left(2\sqrt{a^2-x^2}\right)x + \frac{1}{2}\left(2\sqrt{a^2-x^2}\right)\left(\sqrt{x^2+b^2-a^2}\right) = \sqrt{a^2-x^2}\left(x+\sqrt{x^2+b^2-a^2}\right)$ for $0 \le x \le a$. Now

$$A'(x) = \sqrt{a^2-x^2}\left(1+\frac{x}{\sqrt{x^2+b^2-a^2}}\right) + \left(x+\sqrt{x^2+b^2-a^2}\right)\frac{-x}{\sqrt{a^2-x^2}} = 0 \quad \Leftrightarrow$$

$$\frac{x}{\sqrt{a^2-x^2}}\left(x+\sqrt{x^2+b^2-a^2}\right) = \sqrt{a^2-x^2}\left(\frac{x+\sqrt{x^2+b^2-a^2}}{\sqrt{x^2+b^2-a^2}}\right).$$

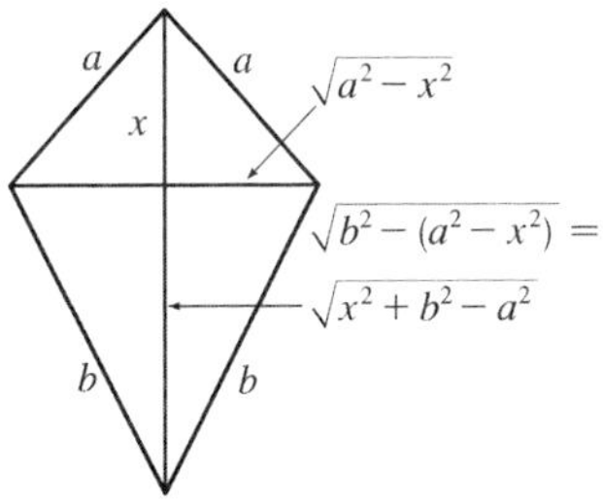


Except for the trivial case where $x = 0$, $a = b$ and $A(x) = 0$, we have $x+\sqrt{x^2+b^2-a^2} > 0$. Hence, cancelling this factor gives

$$\frac{x}{\sqrt{a^2-x^2}} = \frac{\sqrt{a^2-x^2}}{\sqrt{x^2+b^2-a^2}} \quad\Rightarrow\quad x\sqrt{x^2+b^2-a^2} = a^2-x^2 \quad\Rightarrow$$

$$x^2\left(x^2+b^2-a^2\right) = a^4 - 2a^2x^2 + x^4 \quad\Rightarrow\quad x^2\left(b^2-a^2\right) = a^4-2a^2x^2$$

$$\Rightarrow\quad x^2\left(b^2+a^2\right) = a^4 \quad\Rightarrow\quad x = \frac{a^2}{\sqrt{a^2+b^2}}.$$

Now we must check the value of $A$ at this point as well as at the endpoints of the domain to see which gives the maximum value. $A(0) = a\sqrt{b^2-a^2}$, $A(a) = 0$ and

$$A\left(\frac{a^2}{\sqrt{a^2+b^2}}\right) = \sqrt{a^2-\left(\frac{a^2}{\sqrt{a^2+b^2}}\right)^2}\left[\frac{a^2}{\sqrt{a^2+b^2}}+\sqrt{\left(\frac{a^2}{\sqrt{a^2+b^2}}\right)^2+b^2-a^2}\right]$$

$$= \frac{ab}{\sqrt{a^2+b^2}}\left[\frac{a^2}{\sqrt{a^2+b^2}}+\frac{b^2}{\sqrt{a^2+b^2}}\right] = \frac{ab\left(a^2+b^2\right)}{a^2+b^2} = ab$$

Since $b \ge \sqrt{b^2-a^2}$, $A\left(a^2/\sqrt{a^2+b^2}\right) \ge A(0)$. So there is an absolute maximum when $x = \dfrac{a^2}{\sqrt{a^2+b^2}}$. In this case the horizontal piece should be $\dfrac{2ab}{\sqrt{a^2+b^2}}$ and the vertical piece should be $\dfrac{a^2+b^2}{\sqrt{a^2+b^2}} = \sqrt{a^2+b^2}$.

**37.**

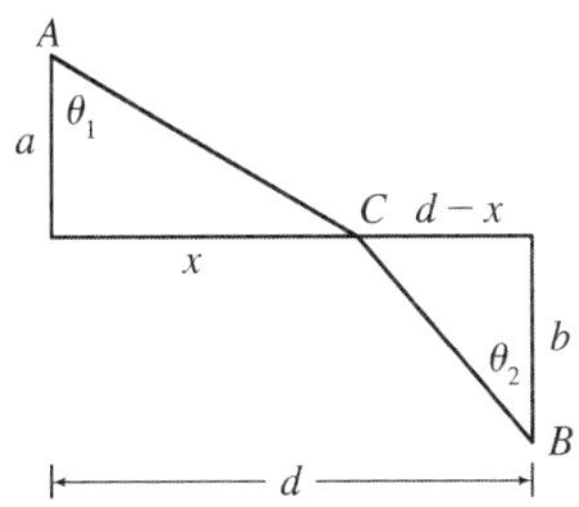


The total time is

$$T(x) = (\text{time from } A \text{ to } C) + (\text{time from } C \text{ to } B)$$

$$= \frac{\sqrt{a^2+x^2}}{v_1} + \frac{\sqrt{b^2+(d-x)^2}}{v_2}, \quad 0 < x < d$$

$$T'(x) = \frac{x}{v_1\sqrt{a^2+x^2}} - \frac{d-x}{v_2\sqrt{b^2+(d-x)^2}} = \frac{\sin\theta_1}{v_1} - \frac{\sin\theta_2}{v_2}$$

The minimum occurs when $T'(x) = 0 \;\Rightarrow\; \dfrac{\sin\theta_1}{v_1} = \dfrac{\sin\theta_2}{v_2}$.

[*Note:* $T''(x) > 0$]

**38.**

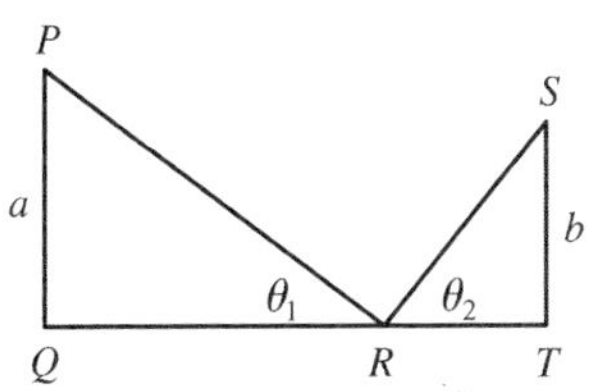


If $d = |QT|$, we minimize $f(\theta_1) = |PR| + |RS| = a\csc\theta_1 + b\csc\theta_2$.

Differentiating with respect to $\theta_1$, and setting $\dfrac{df}{d\theta_1}$ equal to 0, we get

$$\frac{df}{d\theta_1} = 0 = -a\csc\theta_1\cot\theta_1 - b\csc\theta_2\cot\theta_2\,\frac{d\theta_2}{d\theta_1}.$$

So we need to find an expression for $\dfrac{d\theta_2}{d\theta_1}$. We can do this by observing that $|QT| = \text{constant} = a\cot\theta_1 + b\cot\theta_2$.

Differentiating this equation implicitly with respect to $\theta_1$, we get $-a\csc^2\theta_1 - b\csc^2\theta_2\,\dfrac{d\theta_2}{d\theta_1} = 0 \quad\Rightarrow$

$\dfrac{d\theta_2}{d\theta_1} = -\dfrac{a\csc^2\theta_1}{b\csc^2\theta_2}$. We substitute this into the expression for $\dfrac{df}{d\theta_1}$ to get

$-a\csc\theta_1\cot\theta_1 - b\csc\theta_2\cot\theta_2\left(-\dfrac{a\csc^2\theta_1}{b\csc^2\theta_2}\right) = 0 \quad\Leftrightarrow\quad -a\csc\theta_1\cot\theta_1 + a\dfrac{\csc^2\theta_1\cot\theta_2}{\csc\theta_2} = 0 \quad\Leftrightarrow$

$\cot\theta_1\csc\theta_2 = \csc\theta_1\cot\theta_2 \quad\Leftrightarrow\quad \dfrac{\cot\theta_1}{\csc\theta_1} = \dfrac{\cot\theta_2}{\csc\theta_2} \quad\Leftrightarrow\quad \cos\theta_1 = \cos\theta_2$. Since $\theta_1$ and $\theta_2$ are both acute, we have $\theta_1 = \theta_2$.

**39.**

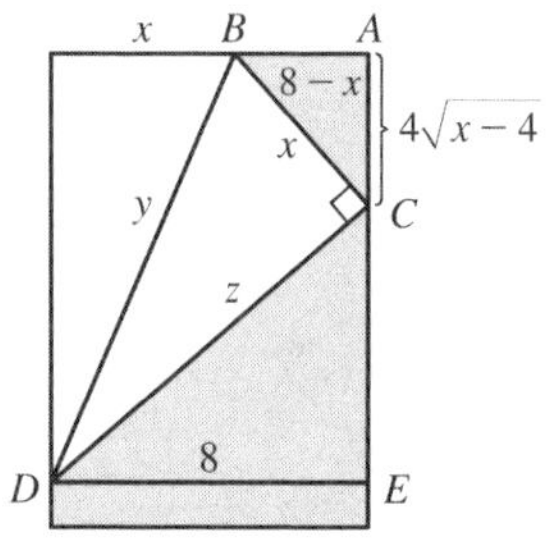


$y^2 = x^2 + z^2$, but triangles $CDE$ and $BCA$ are similar, so $z/8 = x/\left(4\sqrt{x-4}\right) \;\Rightarrow\; z = 2x/\sqrt{x-4}$. Thus, we minimize

$f(x) = y^2 = x^2 + 4x^2/(x-4) = x^3/(x-4), \quad 4 < x \le 8.$

$f'(x) = \dfrac{(x-4)(3x^2) - x^3}{(x-4)^2} = \dfrac{x^2[3(x-4) - x]}{(x-4)^2} = \dfrac{2x^2(x-6)}{(x-4)^2} = 0$ when $x = 6$. $f'(x) < 0$ when $x < 6$, $f'(x) > 0$ when $x > 6$, so the minimum occurs when $x = 6$ in.

**40.**

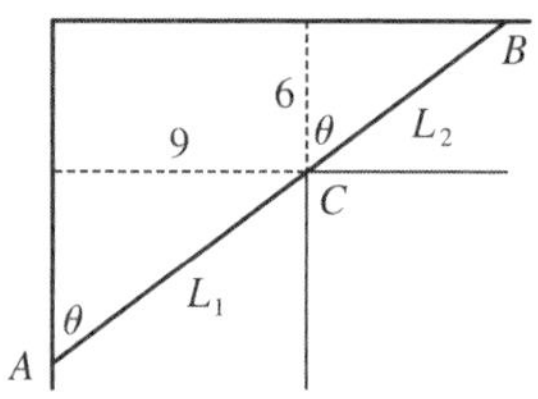


Paradoxically, we solve this maximum problem by solving a minimum problem. Let $L$ be the length of the line $ACB$ going from wall to wall touching the inner corner $C$. As $\theta \to 0$ or $\theta \to \frac{\pi}{2}$, we have $L \to \infty$ and there will be an angle that makes $L$ a minimum. A pipe of this length will just fit around the corner.

From the diagram, $L = L_1 + L_2 = 9\csc\theta + 6\sec\theta \;\Rightarrow\; dL/d\theta = -9\csc\theta\,\cot\theta + 6\sec\theta\,\tan\theta = 0$ when $6\sec\theta\,\tan\theta = 9\csc\theta\,\cot\theta \;\Leftrightarrow\; \tan^3\theta = \frac{9}{6} = 1.5 \;\Leftrightarrow\; \tan\theta = \sqrt[3]{1.5}$. Then $\sec^2\theta = 1 + \left(\frac{3}{2}\right)^{2/3}$ and $\csc^2\theta = 1 + \left(\frac{3}{2}\right)^{-2/3}$, so the longest pipe has length $L = 9\left[1 + \left(\frac{3}{2}\right)^{-2/3}\right]^{1/2} + 6\left[1 + \left(\frac{3}{2}\right)^{2/3}\right]^{1/2} \approx 21.07$ ft.

Or, use $\theta = \tan^{-1}\left(\sqrt[3]{1.5}\right) \approx 0.852 \;\Rightarrow\; L = 9\csc\theta + 6\sec\theta \approx 21.07$ ft.

**41.**

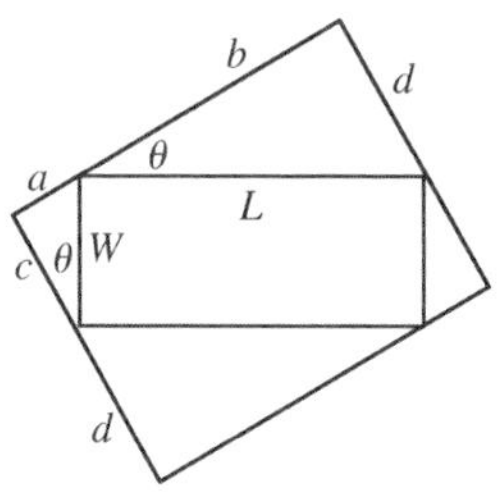


In the small triangle with sides $a$ and $c$ and hypotenuse $W$, $\sin\theta = \dfrac{a}{W}$ and $\cos\theta = \dfrac{c}{W}$. In the triangle with sides $b$ and $d$ and hypotenuse $L$, $\sin\theta = \dfrac{d}{L}$ and $\cos\theta = \dfrac{b}{L}$. Thus, $a = W\sin\theta$, $c = W\cos\theta$, $d = L\sin\theta$, and $b = L\cos\theta$, so the area of the circumscribed rectangle is

$$\begin{aligned} A(\theta) &= (a+b)(c+d) = (W\sin\theta + L\cos\theta)(W\cos\theta + L\sin\theta) \\ &= W^2\sin\theta\,\cos\theta + WL\sin^2\theta + LW\cos^2\theta + L^2\sin\theta\,\cos\theta \\ &= LW\sin^2\theta + LW\cos^2\theta + (L^2 + W^2)\sin\theta\,\cos\theta \\ &= LW(\sin^2\theta + \cos^2\theta) + (L^2 + W^2)\cdot\tfrac{1}{2}\cdot 2\sin\theta\,\cos\theta \\ &= LW + \tfrac{1}{2}(L^2 + W^2)\sin 2\theta, \quad 0 \le \theta \le \tfrac{\pi}{2} \end{aligned}$$

This expression shows, without calculus, that the maximum value of $A(\theta)$ occurs when $\sin 2\theta = 1 \;\Leftrightarrow\; 2\theta = \frac{\pi}{2} \;\Rightarrow\; \theta = \frac{\pi}{4}$. So the maximum area is $A\left(\frac{\pi}{4}\right) = LW + \frac{1}{2}(L^2 + W^2) = \frac{1}{2}(L^2 + 2LW + W^2) = \frac{1}{2}(L+W)^2$.

**42.** We maximize the cross-sectional area

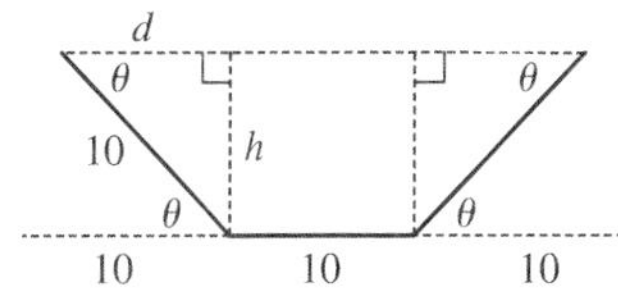


$$A(\theta) = 10h + 2\left(\tfrac{1}{2}dh\right) = 10h + dh = 10(10\sin\theta) + (10\cos\theta)(10\sin\theta)$$
$$= 100(\sin\theta + \sin\theta\,\cos\theta), \quad 0 \le \theta \le \tfrac{\pi}{2}$$

$$A'(\theta) = 100\left(\cos\theta + \cos^2\theta - \sin^2\theta\right) = 100\left(\cos\theta + 2\cos^2\theta - 1\right)$$
$$= 100(2\cos\theta - 1)(\cos\theta + 1) = 0 \text{ when } \cos\theta = \tfrac{1}{2} \quad \Leftrightarrow \quad \theta = \tfrac{\pi}{3}. \quad [\cos\theta \ne -1 \text{ since } 0 \le \theta \le \tfrac{\pi}{2}.]$$

Now $A(0) = 0$, $A\left(\frac{\pi}{2}\right) = 100$ and $A\left(\frac{\pi}{3}\right) = 75\sqrt{3} \approx 129.9$, so the maximum occurs when $\theta = \frac{\pi}{3}$.

**43.**

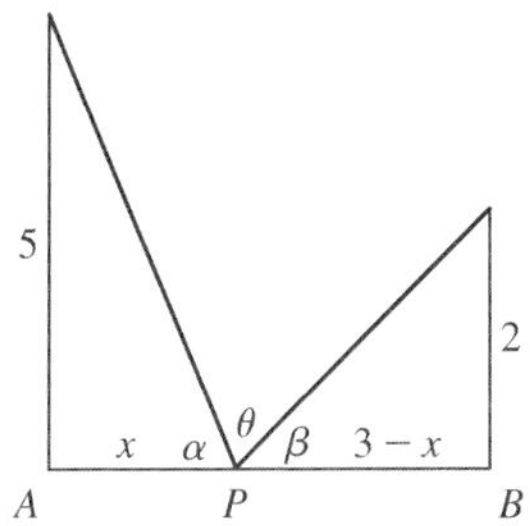


From the figure, $\tan\alpha = \dfrac{5}{x}$ and $\tan\beta = \dfrac{2}{3-x}$. Since

$$\alpha + \beta + \theta = 180^\circ = \pi,\ \theta = \pi - \tan^{-1}\left(\frac{5}{x}\right) - \tan^{-1}\left(\frac{2}{3-x}\right) \quad \Rightarrow$$

$$\frac{d\theta}{dx} = -\frac{1}{1+\left(\dfrac{5}{x}\right)^2}\left(-\frac{5}{x^2}\right) - \frac{1}{1+\left(\dfrac{2}{3-x}\right)^2}\left[\frac{2}{(3-x)^2}\right]$$
$$= \frac{x^2}{x^2+25}\cdot\frac{5}{x^2} - \frac{(3-x)^2}{(3-x)^2+4}\cdot\frac{2}{(3-x)^2}. \text{ Now}$$

$$\frac{d\theta}{dx} = 0 \quad \Rightarrow \quad \frac{5}{x^2+25} = \frac{2}{x^2-6x+13} \quad \Rightarrow \quad 2x^2 + 50 = 5x^2 - 30x + 65 \quad \Rightarrow$$

$3x^2 - 30x + 15 = 0 \quad \Rightarrow \quad x^2 - 10x + 5 = 0 \quad \Rightarrow \quad x = 5 \pm 2\sqrt{5}$. We reject the root with the $+$ sign, since it is larger than 3. $d\theta/dx > 0$ for $x < 5 - 2\sqrt{5}$ and $d\theta/dx < 0$ for $x > 5 - 2\sqrt{5}$, so $\theta$ is maximized when $|AP| = x = 5 - 2\sqrt{5} \approx 0.53$.

**44.** Let $x$ be the distance from the observer to the wall. Then, from the given figure,

$$\theta = \tan^{-1}\left(\frac{h+d}{x}\right) - \tan^{-1}\left(\frac{d}{x}\right),\ x > 0 \quad \Rightarrow$$

$$\frac{d\theta}{dx} = \frac{1}{1+[(h+d)/x]^2}\left[-\frac{h+d}{x^2}\right] - \frac{1}{1+(d/x)^2}\left[-\frac{d}{x^2}\right] = -\frac{h+d}{x^2+(h+d)^2} + \frac{d}{x^2+d^2}$$
$$= \frac{d\left[x^2+(h+d)^2\right] - (h+d)\left(x^2+d^2\right)}{\left[x^2+(h+d)^2\right]\left(x^2+d^2\right)} = \frac{h^2d + hd^2 - hx^2}{\left[x^2+(h+d)^2\right]\left(x^2+d^2\right)} = 0 \quad \Leftrightarrow$$

$hx^2 = h^2d + hd^2 \quad \Leftrightarrow \quad x^2 = hd + d^2 \quad \Leftrightarrow \quad x = \sqrt{d(h+d)}$. Since $d\theta/dx > 0$ for all $x < \sqrt{d(h+d)}$ and $d\theta/dx < 0$ for all $x > \sqrt{d(h+d)}$, the absolute maximum occurs when $x = \sqrt{d(h+d)}$.

**45.** (a)

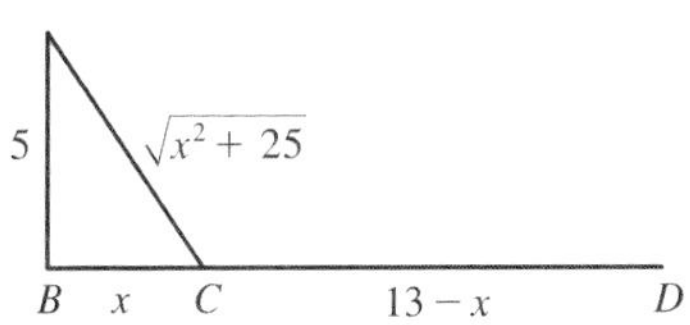


If $k$ = energy/km over land, then energy/km over water $= 1.4k$.

So the total energy is $E = 1.4k\sqrt{25+x^2} + k(13-x)$, $0 \le x \le 13$,

and so $\dfrac{dE}{dx} = \dfrac{1.4kx}{(25+x^2)^{1/2}} - k$.

Set $\dfrac{dE}{dx} = 0$: $1.4kx = k\left(25+x^2\right)^{1/2} \quad \Rightarrow \quad 1.96x^2 = x^2 + 25 \quad \Rightarrow \quad 0.96x^2 = 25 \quad \Rightarrow \quad x = \frac{5}{\sqrt{0.96}} \approx 5.1$. Testing against the value of $E$ at the endpoints: $E(0) = 1.4k(5) + 13k = 20k$, $E(5.1) \approx 17.9k$, $E(13) \approx 19.5k$. Thus, to minimize energy, the bird should fly to a point about 5.1 km from $B$.

(b) If $W/L$ is large, the bird would fly to a point $C$ that is closer to $B$ than to $D$ to minimize the energy used flying over water. If $W/L$ is small, the bird would fly to a point $C$ that is closer to $D$ than to $B$ to minimize the distance of the flight.

$E = W\sqrt{25+x^2} + L(13-x) \Rightarrow \dfrac{dE}{dx} = \dfrac{Wx}{\sqrt{25+x^2}} - L = 0$ when $\dfrac{W}{L} = \dfrac{\sqrt{25+x^2}}{x}$. By the same sort of argument as in part (a), this ratio will give the minimal expenditure of energy if the bird heads for the point $x$ km from $B$.

(c) For flight direct to $D$, $x = 13$, so from part (b), $W/L = \frac{\sqrt{25+13^2}}{13} \approx 1.07$. There is no value of $W/L$ for which the bird should fly directly to $B$. But note that $\lim\limits_{x\to 0^+} (W/L) = \infty$, so if the point at which $E$ is a minimum is close to $B$, then $W/L$ is large.

(d) Assuming that the birds instinctively choose the path that minimizes the energy expenditure, we can use the equation for $dE/dx = 0$ from part (a) with $1.4k = c$, $x = 4$, and $k = 1$: $c(4) = 1 \cdot \left(25+4^2\right)^{1/2} \Rightarrow c = \sqrt{41}/4 \approx 1.6$.

**46.** (a) Let $D$ be the point such that $a = |AD|$. From the figure, $\sin\theta = \dfrac{b}{|BC|} \Rightarrow |BC| = b\csc\theta$ and

$\cos\theta = \dfrac{|BD|}{|BC|} = \dfrac{a-|AB|}{|BC|} \Rightarrow |BC| = (a-|AB|)\sec\theta$. Eliminating $|BC|$ gives

$(a-|AB|)\sec\theta = b\csc\theta \Rightarrow b\cot\theta = a - |AB| \Rightarrow |AB| = a - b\cot\theta$. The total resistance is

$$R(\theta) = C\frac{|AB|}{r_1^4} + C\frac{|BC|}{r_2^4} = C\left(\frac{a-b\cot\theta}{r_1^4} + \frac{b\csc\theta}{r_2^4}\right).$$

(b) $R'(\theta) = C\left(\dfrac{b\csc^2\theta}{r_1^4} - \dfrac{b\csc\theta\cot\theta}{r_2^4}\right) = bC\csc\theta\left(\dfrac{\csc\theta}{r_1^4} - \dfrac{\cot\theta}{r_2^4}\right)$.

$R'(\theta) = 0 \Leftrightarrow \dfrac{\csc\theta}{r_1^4} = \dfrac{\cot\theta}{r_2^4} \Leftrightarrow \dfrac{r_2^4}{r_1^4} = \dfrac{\cot\theta}{\csc\theta} = \cos\theta$.

$R'(\theta) > 0 \Leftrightarrow \dfrac{\csc\theta}{r_1^4} > \dfrac{\cot\theta}{r_2^4} \Rightarrow \cos\theta < \dfrac{r_2^4}{r_1^4}$ and $R'(\theta) < 0$ when $\cos\theta > \dfrac{r_2^4}{r_1^4}$, so there is an absolute minimum when $\cos\theta = r_2^4/r_1^4$.

(c) When $r_2 = \frac{2}{3}r_1$, we have $\cos\theta = \left(\frac{2}{3}\right)^4$, so $\theta = \cos^{-1}\left(\frac{2}{3}\right)^4 \approx 79^\circ$.

**47.** (a) Distance = rate × time, so time = distance/rate. $T_1 = \dfrac{D}{c_1}$, $T_2 = \dfrac{2|PR|}{c_1} + \dfrac{|RS|}{c_2} = \dfrac{2h\sec\theta}{c_1} + \dfrac{D-2h\tan\theta}{c_2}$,

$T_3 = \dfrac{2\sqrt{h^2+D^2/4}}{c_1} = \dfrac{\sqrt{4h^2+D^2}}{c_1}$.

(b) $\dfrac{dT_2}{d\theta} = \dfrac{2h}{c_1}\cdot\sec\theta\tan\theta - \dfrac{2h}{c_2}\sec^2\theta = 0$ when $2h\sec\theta\left(\dfrac{1}{c_1}\tan\theta - \dfrac{1}{c_2}\sec\theta\right) = 0 \Rightarrow \dfrac{1}{c_1}\dfrac{\sin\theta}{\cos\theta} - \dfrac{1}{c_2}\dfrac{1}{\cos\theta} = 0$

$\Rightarrow \dfrac{\sin\theta}{c_1\cos\theta} = \dfrac{1}{c_2\cos\theta} \Rightarrow \sin\theta = \dfrac{c_1}{c_2}$. The First Derivative Test shows that this gives a minimum.

(c) Using part (a) with $D = 1$ and $T_1 = 0.26$, we have $T_1 = \dfrac{D}{c_1} \Rightarrow c_1 = \frac{1}{0.26} \approx 3.85$ km/s. $T_3 = \dfrac{\sqrt{4h^2+D^2}}{c_1} \Rightarrow$

$4h^2 + D^2 = T_3^2c_1^2 \Rightarrow h = \frac{1}{2}\sqrt{T_3^2c_1^2 - D^2} = \frac{1}{2}\sqrt{(0.34)^2(1/0.26)^2 - 1^2} \approx 0.42$ km. To find $c_2$, we use $\sin\theta = \dfrac{c_1}{c_2}$ from part (b) and $T_2 = \dfrac{2h\sec\theta}{c_1} + \dfrac{D-2h\tan\theta}{c_2}$ from part (a). From the figure,

$\sin\theta = \dfrac{c_1}{c_2} \Rightarrow \sec\theta = \dfrac{c_2}{\sqrt{c_2^2-c_1^2}}$ and $\tan\theta = \dfrac{c_1}{\sqrt{c_2^2-c_1^2}}$, so

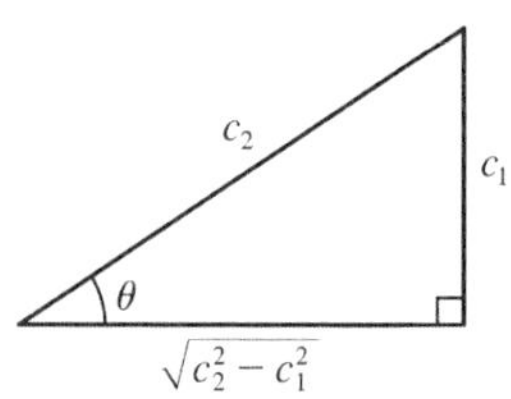


$T_2 = \dfrac{2hc_2}{c_1\sqrt{c_2^2-c_1^2}} + \dfrac{D\sqrt{c_2^2-c_1^2} - 2hc_1}{c_2\sqrt{c_2^2-c_1^2}}$. Using the values for $T_2$ [given as 0.32],

$h$, $c_1$, and $D$, we can graph $Y_1 = T_2$ and $Y_2 = \dfrac{2hc_2}{c_1\sqrt{c_2^2 - c_1^2}} + \dfrac{D\sqrt{c_2^2 - c_1^2} - 2hc_1}{c_2\sqrt{c_2^2 - c_1^2}}$ and find their intersection points. Doing so gives us $c_2 \approx 4.10$ and $7.66$, but if $c_2 = 4.10$, then $\theta = \arcsin(c_1/c_2) \approx 69.6°$, which implies that point $S$ is to the left of point $R$ in the diagram. So $c_2 = 7.66$ km/s.

**48.** (a) $I(x) \propto \dfrac{\text{strength of source}}{(\text{distance from source})^2}$. Adding the intensities from the left and right lightbulbs,

$$I(x) = \frac{k}{x^2 + d^2} + \frac{k}{(10 - x)^2 + d^2} = \frac{k}{x^2 + d^2} + \frac{k}{x^2 - 20x + 100 + d^2}.$$

(b) The magnitude of the constant $k$ won't affect the location of the point of maximum intensity, so for convenience we take $k = 1$. $I'(x) = -\dfrac{2x}{(x^2 + d^2)^2} - \dfrac{2\,(x - 10)}{(x^2 - 20x + 100 + d^2)^2}$.

Substituting $d = 5$ into the equations for $I(x)$and $I'(x)$, we get

$$I_5(x) = \frac{1}{x^2 + 25} + \frac{1}{x^2 - 20x + 125} \qquad \text{and} \qquad I_5'(x) = -\frac{2x}{(x^2 + 25)^2} - \frac{2(x - 10)}{(x^2 - 20x + 125)^2}$$

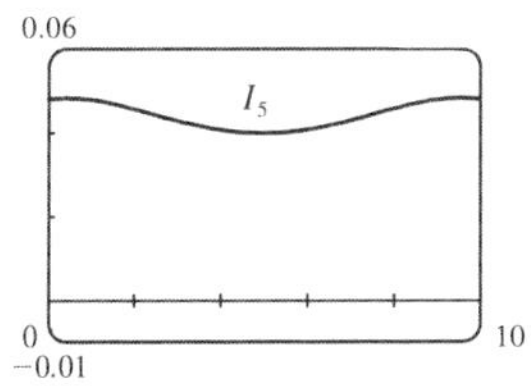


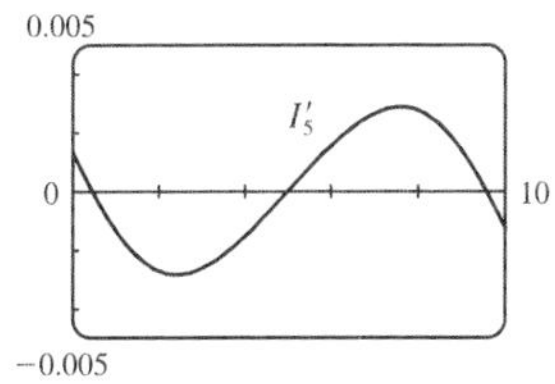


From the graphs, it appears that $I_5(x)$ has a minimum at $x = 5$ m.

(c) Substituting $d = 10$ into the equations for $I(x)$ and $I'(x)$ gives $I_{10}(x) = \dfrac{1}{x^2 + 100} + \dfrac{1}{x^2 - 20x + 200}$ and

$$I_{10}'(x) = -\frac{2x}{(x^2 + 100)^2} - \frac{2\,(x - 10)}{(x^2 - 20x + 200)^2}.$$

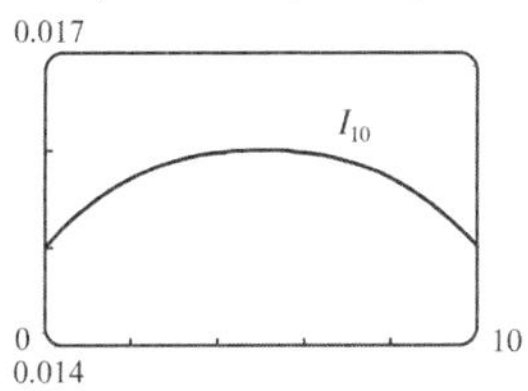


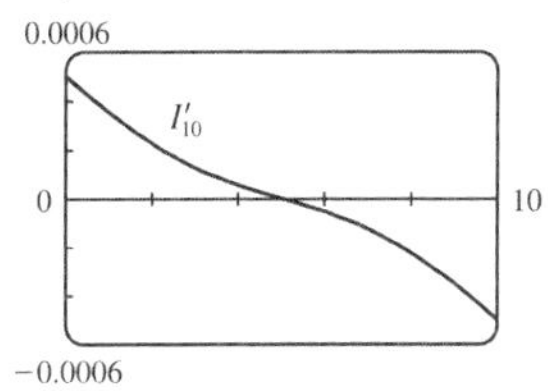


From the graphs, it seems that for $d = 10$, the intensity is minimized at the endpoints, that is, $x = 0$ and $x = 10$. The midpoint is now the most brightly lit point!

(d) From the first figures in parts (b) and (c), we see that the minimal illumination changes from the midpoint ($x = 5$ with $d = 5$) to the endpoints ($x = 0$ and $x = 10$ with $d = 10$).

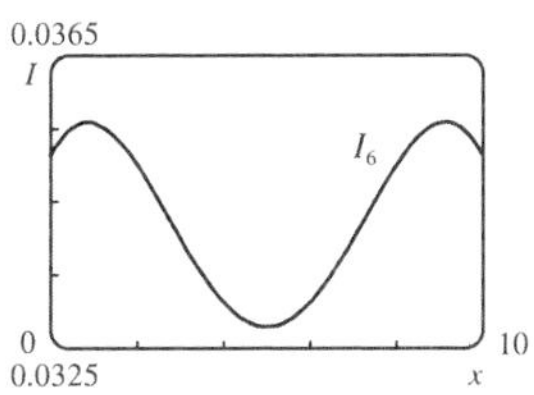


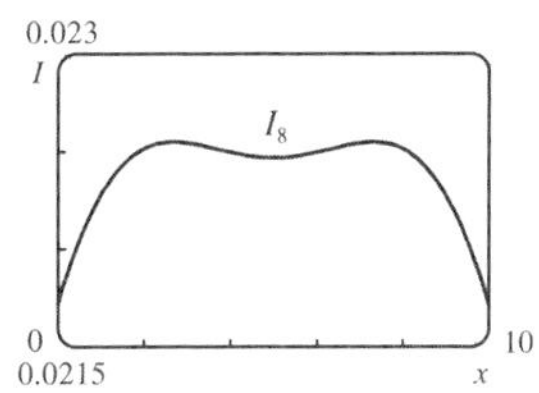


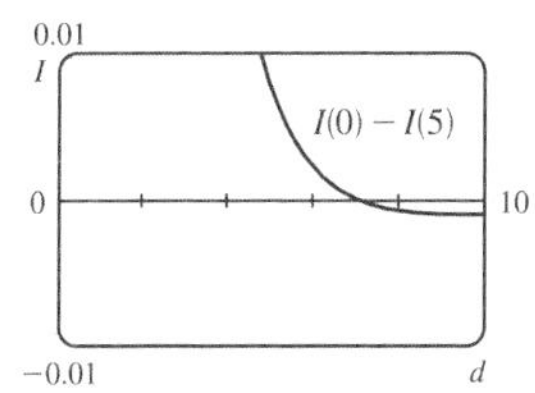


So we try $d = 6$ (see the first figure) and we see that the minimum value still occurs at $x = 5$. Next, we let $d = 8$ (see the second figure) and we see that the minimum value occurs at the endpoints. It appears that for some value of $d$ between 6 and 8, we must have minima at both the midpoint and the endpoints, that is, $I(5)$ must equal $I(0)$. To find this value of $d$, we solve $I(0) = I(5)$ (with $k = 1$):

$$\frac{1}{d^2} + \frac{1}{100 + d^2} = \frac{1}{25 + d^2} + \frac{1}{25 + d^2} = \frac{2}{25 + d^2} \quad \Rightarrow$$

$(25+d^2)(100+d^2)+d^2(25+d^2)=2d^2(100+d^2) \Rightarrow 2500+125d^2+d^4+25d^2+d^4=200d^2+2d^4 \Rightarrow$ $2500=50d^2 \Rightarrow d^2=50 \Rightarrow d=5\sqrt{2}\approx 7.071$ (for $0\le d\le 10$). The third figure, a graph of $I(0)-I(5)$ with $d$ independent, confirms that $I(0)-I(5)=0$, that is, $I(0)=I(5)$, when $d=5\sqrt{2}$. Thus, the point of minimal illumination changes abruptly from the midpoint to the endpoints when $d=5\sqrt{2}$.

## APPLIED PROJECT The Shape of a Can

1. In this case, the amount of metal used in the making of each top or bottom is $(2r)^2=4r^2$. So the quantity we want to minimize is $A=2\pi rh+2(4r^2)$. But $V=\pi r^2h \Leftrightarrow h=V/\pi r^2$. Substituting this expression for $h$ in $A$ gives $A=2V/r+8r^2$. Differentiating $A$ with respect to $r$, we get $dA/dr=-2V/r^2+16r=0 \Rightarrow$ $16r^3=2V=2\pi r^2h \Leftrightarrow \dfrac{h}{r}=\dfrac{8}{\pi}\approx 2.55$. This gives a minimum because $\dfrac{d^2A}{dr^2}=16+\dfrac{4V}{r^3}>0$.

2. 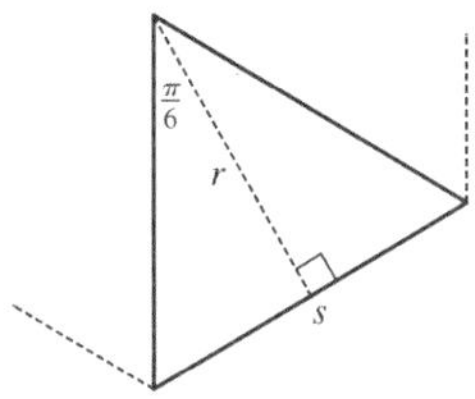


We need to find the area of metal used up by each end, that is, the area of each hexagon. We subdivide the hexagon into six congruent triangles, each sharing one side ($s$ in the diagram) with the hexagon. We calculate the length of $s=2r\tan\frac{\pi}{6}=\frac{2}{\sqrt{3}}r$, so the area of each triangle is $\frac{1}{2}sr=\frac{1}{\sqrt{3}}r^2$, and the total area of the hexagon is $6\cdot\frac{1}{\sqrt{3}}r^2=2\sqrt{3}\,r^2$. So the quantity we want to minimize is $A=2\pi rh+2\cdot 2\sqrt{3}\,r^2$.

Substituting for $h$ as in Problem 1 and differentiating, we get $\dfrac{dA}{dr}=-\dfrac{2V}{r^2}+8\sqrt{3}\,r$. Setting this equal to 0, we get $8\sqrt{3}\,r^3=2V=2\pi r^2h \Rightarrow \dfrac{h}{r}=\dfrac{4\sqrt{3}}{\pi}\approx 2.21$. Again this minimizes $A$ because $\dfrac{d^2A}{dr^2}=8\sqrt{3}+\dfrac{4V}{r^3}>0$.

3. Let $C=4\sqrt{3}\,r^2+2\pi rh+k\,(4\pi r+h)=4\sqrt{3}\,r^2+2\pi r\left(\dfrac{V}{\pi r^2}\right)+k\left(4\pi r+\dfrac{V}{\pi r^2}\right)$. Then $\dfrac{dC}{dr}=8\sqrt{3}\,r-\dfrac{2V}{r^2}+4k\pi-\dfrac{2kV}{\pi r^3}$. Setting this equal to 0, dividing by 2 and substituting $\dfrac{V}{r^2}=\pi h$ and $\dfrac{V}{\pi r^3}=\dfrac{h}{r}$ in the second and fourth terms respectively, we get $0=4\sqrt{3}\,r-\pi h+2k\pi-\dfrac{kh}{r} \Leftrightarrow$ $k\left(2\pi-\dfrac{h}{r}\right)=\pi h-4\sqrt{3}\,r \Rightarrow \dfrac{k}{r}\,\dfrac{2\pi-h/r}{\pi h/r-4\sqrt{3}}=1$. We now multiply by $\dfrac{\sqrt[3]{V}}{k}$, noting that $\dfrac{\sqrt[3]{V}}{k}\,\dfrac{k}{r}=\sqrt[3]{\dfrac{V}{r^3}}=\sqrt[3]{\dfrac{\pi h}{r}}$, and get $\dfrac{\sqrt[3]{V}}{k}=\sqrt[3]{\dfrac{\pi h}{r}}\cdot\dfrac{2\pi-h/r}{\pi h/r-4\sqrt{3}}$.

4. 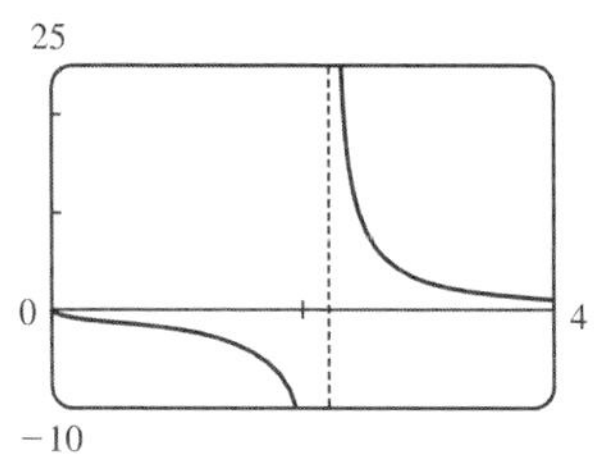


Let $\sqrt[3]{V}/k=T$ and $h/r=x$ so that $T(x)=\sqrt[3]{\pi x}\cdot\dfrac{2\pi-x}{\pi x-4\sqrt{3}}$. We see from the graph of $T$ that when the ratio $\sqrt[3]{V}/k$ is large; that is, either the volume of the can is large or the cost of joining (proportional to $k$) is small, the optimum value of $h/r$ is about 2.21, but when $\sqrt[3]{V}/k$ is small, indicating small volume or expensive joining, the optimum value of $h/r$ is larger. (The part of the graph for $\sqrt[3]{V}/k<0$ has no physical meaning, but confirms the location of the asymptote.)

5. Our conclusion is usually true in practice. But there are exceptions, such as cans of tuna, which may have to do with the shape of a reasonable slice of tuna. And for a comfortable grip on a soda or beer can, the geometry of the human hand is a restriction on the radius. Other possible considerations are packaging, transportation and stocking constraints, aesthetic appeal and other marketing concerns. Also, there may be better models than ours which prescribe a differently shaped can in special circumstances.

## 4.7 Applications to Business and Economics

1. (a) $C(0)$ represents the fixed costs of production, such as rent, utilities, machinery etc., which are incurred even when nothing is produced.

(b) The inflection point is the point at which $C''(x)$ changes from negative to positive; that is, the marginal cost $C'(x)$ changes from decreasing to increasing. Thus, the marginal cost is minimized at the inflection point.

(c) The marginal cost function is $C'(x)$. We graph it as in Example 1 in Section 2.8.

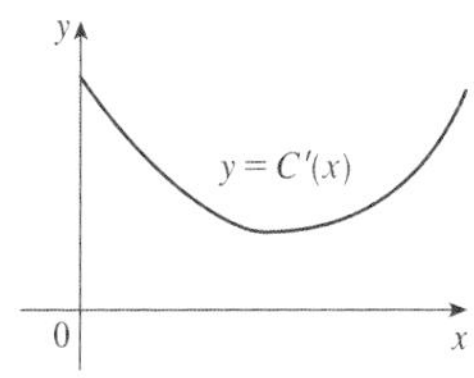


2. (a) We graph $C'$ as in Example 1 in Section 2.8.

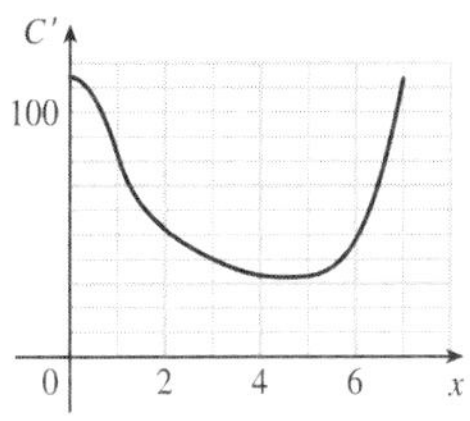


(b) By reading values of $C(x)$ from its graph, we can plot $c(x) = C(x)/x$.

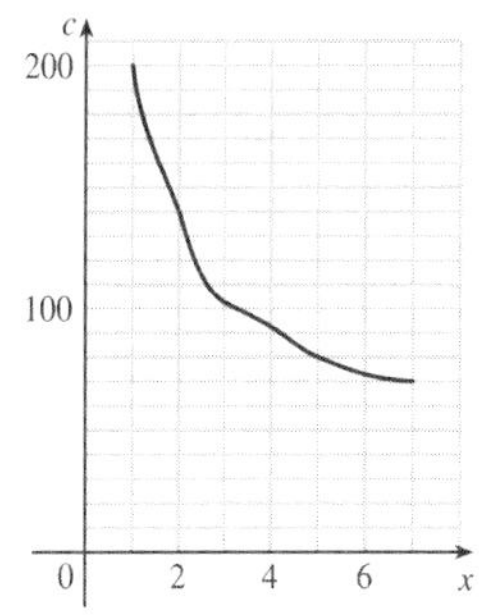


(c) Since the graph in part (b) is decreasing, we estimate that the minimum value of $c(x)$ occurs at $x = 7$. The average cost and the marginal cost are equal at that value. See the box preceding Example 1.

3. $c(x) = 21.4 - 0.002x$ and $c(x) = C(x)/x \Rightarrow C(x) = 21.4x - 0.002x^2$. $C'(x) = 21.4 - 0.004x$ and $C'(1000) = 17.4$. This means that the cost of producing the 1001st unit is about \$17.40.

4. (a) Profit is maximized when the marginal revenue is equal to the marginal cost; that is, when $R$ and $C$ have equal slopes. See the box preceding Example 2.

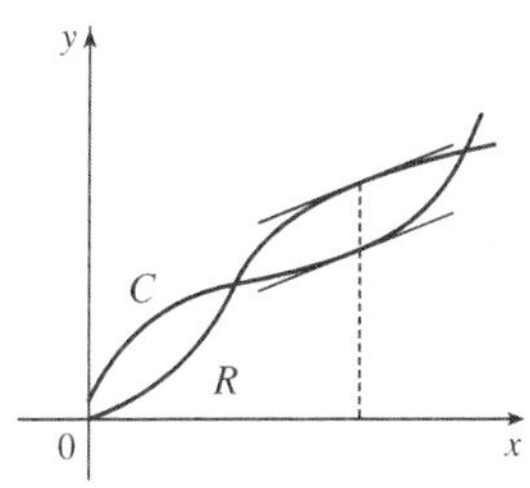

(b) $P(x) = R(x) - C(x)$ is sketched.

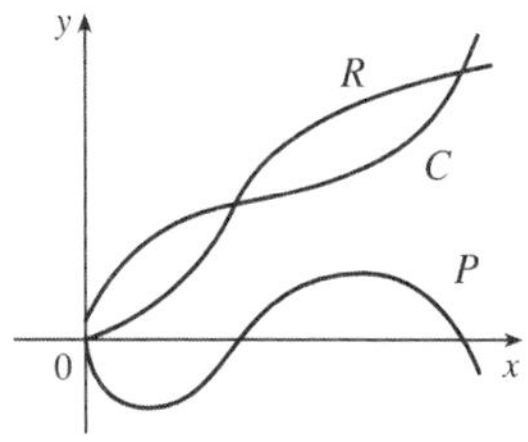


(c) The marginal profit function is defined as $P'(x)$.

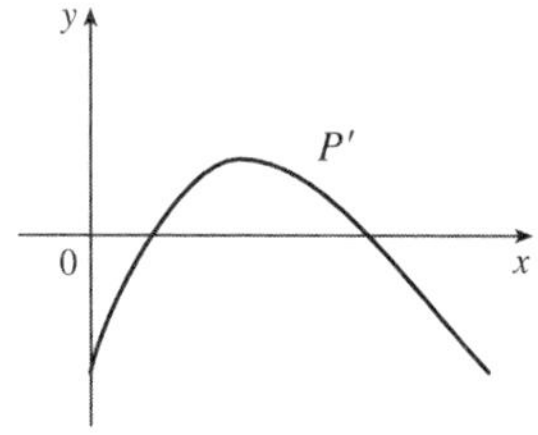


**5.** (a) $C(x) = 16{,}000 + 200x + 4x^{3/2}$, $C(1000) = 16{,}000 + 200{,}000 + 40{,}000\sqrt{10} \approx 216{,}000 + 126{,}491$, so $C(1000) \approx \$342{,}491$. $c(x) = C(x)/x = \dfrac{16{,}000}{x} + 200 + 4x^{1/2}$, $c(1000) \approx \$342.49/\text{unit}$. $C'(x) = 200 + 6x^{1/2}$, $C'(1000) = 200 + 60\sqrt{10} \approx \$389.74/\text{unit}$.

(b) We must have $C'(x) = c(x) \;\Leftrightarrow\; 200 + 6x^{1/2} = \dfrac{16{,}000}{x} + 200 + 4x^{1/2} \;\Leftrightarrow\; 2x^{3/2} = 16{,}000 \;\Leftrightarrow$ $x = (8{,}000)^{2/3} = 400$ units. To check that this is a minimum, we calculate $c'(x) = \dfrac{-16{,}000}{x^2} + \dfrac{2}{\sqrt{x}} = \dfrac{2}{x^2}(x^{3/2} - 8000)$. This is negative for $x < (8000)^{2/3} = 400$, zero at $x = 400$, and positive for $x > 400$, so $c$ is decreasing on $(0, 400)$ and increasing on $(400, \infty)$. Thus, $c$ has an absolute minimum at $x = 400$. [*Note:* $c''(x)$ is *not* positive for all $x > 0$.]

(c) The minimum average cost is $c(400) = 40 + 200 + 80 = \$320/\text{unit}$.

**6.** (a) $C(x) = 10{,}000 + 340x - 0.3x^2 + 0.0001x^3$, $C(1000) = \$150{,}000$.
$c(x) = C(x)/x = \dfrac{10{,}000}{x} + 340 - 0.3x + 0.0001x^2$, $c(1000) = \$150/\text{unit}$. $C'(x) = 340 - 0.6x + 0.0003x^2$, $C'(1000) = \$40/\text{unit}$.

(b) We must have $C'(x) = c(x) \;\Leftrightarrow\; 340 - 0.6x + 0.0003x^2 = \dfrac{10{,}000}{x} + 340 - 0.3x + 0.0001x^2 \;\Leftrightarrow$ $0.0002x^2 = \dfrac{10{,}000}{x} + 0.3x \;\Leftrightarrow\; 0.0002x^3 - 0.3x^2 - 10{,}000 = 0 \;\Leftrightarrow\; x^3 - 1500x^2 - 50{,}000{,}000 = 0 \;\Rightarrow$ $x \approx 1521.60 \approx 1522$ units. This gives a minimum since $c''(x) = \dfrac{20{,}000}{x^3} + 0.0002 > 0$.

(c) The minimum average cost is about $c(1521.60) \approx \$121.62/\text{unit}$.

**7.** (a) $C(x) = 3700 + 5x - 0.04x^2 + 0.0003x^3 \;\Rightarrow\; C'(x) = 5 - 0.08x + 0.0009x^2$ (marginal cost).
$c(x) = \dfrac{C(x)}{x} = \dfrac{3700}{x} + 5 - 0.04x + 0.0003x^2$ (average cost).

(b)

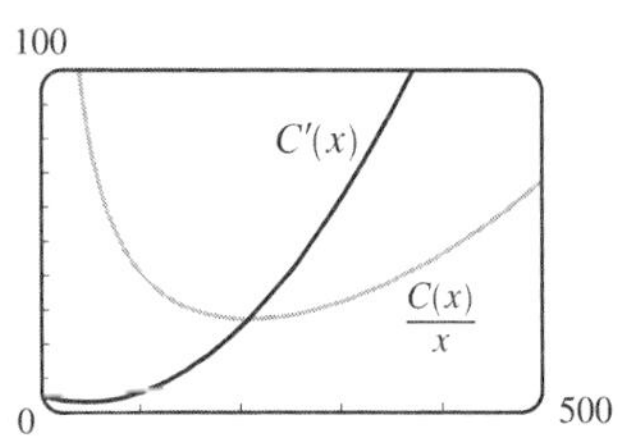


The graphs intersect at $(208.51, 27.45)$, so the production level that minimizes average cost is about 209 units.

(c) $c'(x) = -\dfrac{3700}{x^2} - 0.04 + 0.0006x = 0 \;\Rightarrow\; 3700 + 0.04x^2 - 0.0006x^3 = 0 \;\Rightarrow\; x_1 \approx 208.51$.
$c(x_1) \approx \$27.45/\text{unit}$.

(d) The marginal cost is given by $C'(x)$, so to find its minimum value we'll find the derivative of $C'$; that is, $C''$.
$C''(x) = -0.08 + 0.0018x = 0 \;\Rightarrow\; x_1 = \frac{800}{18} = 44.\overline{4}$. $C'(x_1) = \$3.22/\text{unit}$.
$C'''(x) = 0.0018 > 0$ for all $x$, so this is the minimum marginal cost. $C'''$ is the second derivative of $C'$.

**8.** (a) $C(x) = 339 + 25x - 0.09x^2 + 0.0004x^3 \quad \Rightarrow \quad C'(x) = 25 - 0.18x + 0.0012x^2$ (marginal cost).

$c(x) = \dfrac{C(x)}{x} = \dfrac{339}{x} + 25 - 0.09x + 0.0004x^2$ (average cost).

(b)

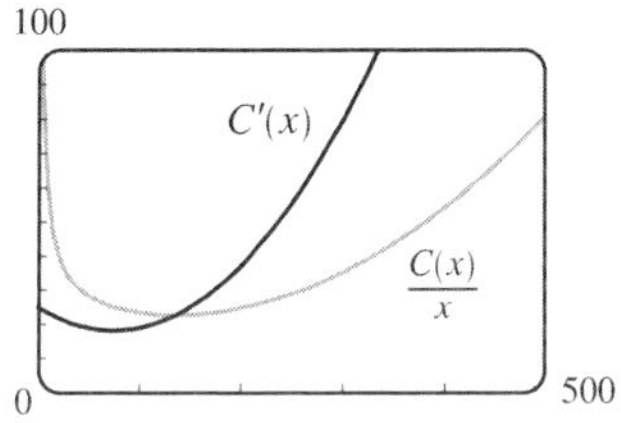


The graphs intersect at $(135.56, 22.65)$, so the production level that minimizes average cost is about 136 units.

(c) $c'(x) = -\dfrac{339}{x^2} - 0.09 + 0.0008x = 0 \quad \Rightarrow \quad x_1 \approx 135.56. \quad c(x_1) \approx \$22.65/\text{unit}.$

(d) $C''(x) = -0.18 + 0.0024x = 0 \quad \Rightarrow \quad x = \frac{1800}{24} = 75. \quad C'(75) = \$18.25/\text{unit}.$

$C'''(x) = 0.0024 > 0$ for all $x$, so this is the minimum marginal cost.

**9.** $C(x) = 680 + 4x + 0.01x^2$, $p(x) = 12 - x/500$. Then $R(x) = xp(x) = 12x - x^2/500$. If the profit is maximum, then $R'(x) = C'(x)$ [See the box preceding Example 2.] $\Leftrightarrow \quad 12 - x/250 = 4 + 0.02x \quad \Leftrightarrow \quad 8 = 0.024x \quad \Leftrightarrow$ $x = 8/0.024 = \frac{1000}{3}$. The profit is maximized if $P''(x) < 0$, but since $P''(x) = R''(x) - C''(x)$, we can just check the condition $R''(x) < C''(x)$. Now $R''(x) = -\frac{1}{250} < 0.02 = C''(x)$, so $x = \frac{1000}{3}$ gives a maximum.

**10.** $C(x) = 16{,}000 + 500x - 1.6x^2 + 0.004x^3$, $p(x) = 1700 - 7x$. Then $R(x) = xp(x) = 1700x - 7x^2$. If the profit is maximum, then $R'(x) = C'(x) \quad \Leftrightarrow \quad 1700 - 14x = 500 - 3.2x + 0.012x^2 \quad \Leftrightarrow \quad 0.012x^2 + 10.8x - 1200 = 0 \quad \Leftrightarrow$ $x^2 + 900x - 100{,}000 = 0 \quad \Leftrightarrow \quad (x + 1000)(x - 100) = 0 \quad \Leftrightarrow \quad x = 100$ (since $x > 0$). The profit is maximized if $P''(x) < 0$, but since $P''(x) = R''(x) - C''(x)$, we can just check the condition $R''(x) < C''(x)$. Now $R''(x) = -14 < -3.2 + 0.024x = C''(x)$ for $x > 0$, so there is a maximum at $x = 100$.

**11.** The marginal cost function, $C'(x)$, starts to increase when $C''(x)$ becomes positive.

$C(x) = 0.0008x^3 - 0.72x^2 + 325.3x + 78{,}000 \quad \Rightarrow \quad C'(x) = 0.0024x^2 - 1.44x + 325.3 \quad \Rightarrow$

$C''(x) = 0.0048x - 1.44. \quad C'' > 0 \quad \Rightarrow \quad 0.0048x > 1.44 \quad \Rightarrow \quad x > \dfrac{1.44}{0.0048} = 300.$ Thus, the marginal cost function starts to increase when the production level is 300.

**12.** The marginal cost function, $C'(x)$, starts to increase when $C''(x)$ becomes positive.

$C(x) = 0.001x^3 - 0.84x^2 + 416x + 55{,}000 \quad \Rightarrow \quad C'(x) = 0.003x^2 - 1.68x + 416 \quad \Rightarrow \quad C''(x) = 0.006x - 1.68.$

$C'' > 0 \quad \Rightarrow \quad 0.006x > 1.68 \quad \Rightarrow \quad x > \dfrac{1.68}{0.006} = 280.$ Thus, the marginal cost function starts to increase when the production level is 280.

**13.** (a) $C(x) = 1200 + 12x - 0.1x^2 + 0.0005x^3$.

$R(x) = xp(x) = 29x - 0.00021x^2$.

Since the profit is maximized when $R'(x) = C'(x)$, we examine the curves $R$ and $C$ in the figure, looking for $x$-values at which the slopes of the tangent lines are equal. It appears that $x = 200$ is a good estimate.

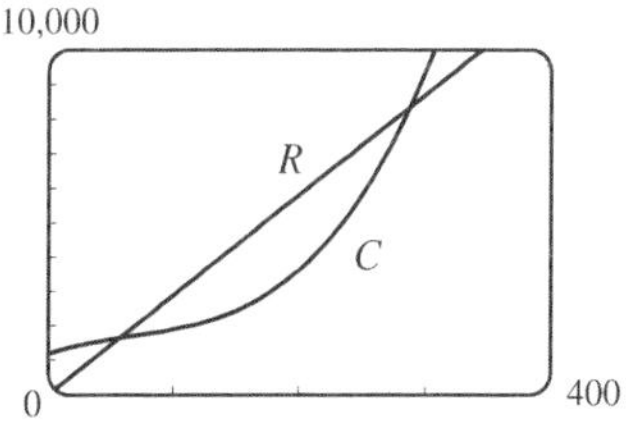


(b) $R'(x) = C'(x) \quad \Rightarrow \quad 29 - 0.00042x = 12 - 0.2x + 0.0015x^2 \quad \Rightarrow \quad 0.0015x^2 - 0.19958x - 17 = 0 \quad \Rightarrow$ $x \approx 192.06$ (for $x > 0$). As in Exercise 9, $R''(x) < C''(x) \quad \Rightarrow \quad -0.00042 < -0.2 + 0.003x \quad \Leftrightarrow$ $0.003x > 0.19958 \quad \Leftrightarrow \quad x > 66.5$. Our value of 192 is in this range, so we have a maximum profit when we produce 192 yards of fabric.

**14.** (a) Cost = setup cost + manufacturing cost $\Rightarrow$ $C(x) = 500 + m(x) = 500 + 20x - 5x^{3/4} + 0.01x^2$.
We can solve $x(p) = 320 - 7.7p$ for $p$ in terms of $x$ to find the demand (or price) function.
$x = 320 - 7.7p \Rightarrow 7.7p = 320 - x \Rightarrow p(x) = \dfrac{320 - x}{7.7}$. $R(x) = xp(x) = \dfrac{320x - x^2}{7.7}$.

(b) $C'(x) = R'(x) \Rightarrow 20 - \frac{15}{4}x^{-1/4} + 0.02x = \dfrac{320 - 2x}{7.7} \Rightarrow x \approx 81.53$ planes, and $p(x) = \$30.97$ million. The maximum profit associated with these values is about \$463.59 million.

**15.** (a) We are given that the demand function $p$ is linear and $p(27{,}000) = 10$, $p(33{,}000) = 8$, so the slope is $\frac{10 - 8}{27{,}000 - 33{,}000} = -\frac{1}{3000}$ and an equation of the line is $y - 10 = \left(-\frac{1}{3000}\right)(x - 27{,}000) \Rightarrow$ $y = p(x) = -\frac{1}{3000}x + 19 = 19 - (x/3000)$.

(b) The revenue is $R(x) = xp(x) = 19x - (x^2/3000) \Rightarrow R'(x) = 19 - (x/1500) = 0$ when $x = 28{,}500$. Since $R''(x) = -1/1500 < 0$, the maximum revenue occurs when $x = 28{,}500 \Rightarrow$ the price is $p(28{,}500) = \$9.50$.

**16.** (a) Let $p(x)$ be the demand function. Then $p(x)$ is linear and $y = p(x)$ passes through $(20, 10)$ and $(18, 11)$, so the slope is $-\frac{1}{2}$ and an equation of the line is $y - 10 = -\frac{1}{2}(x - 20) \Leftrightarrow y = -\frac{1}{2}x + 20$. Thus, the demand is $p(x) = -\frac{1}{2}x + 20$ and the revenue is $R(x) = xp(x) = -\frac{1}{2}x^2 + 20x$.

(b) The cost is $C(x) = 6x$, so the profit is $P(x) = R(x) - C(x) = -\frac{1}{2}x^2 + 14x$. Then $0 = P'(x) = -x + 14 \Rightarrow$ $x = 14$. Since $P''(x) = -1 < 0$, the selling price for maximum profit is $p(14) = -\frac{1}{2}(14) + 20 = \$13$.

**17.** (a) As in Example 3, we see that the demand function $p$ is linear. We are given that $p(1000) = 450$ and deduce that $p(1100) = 440$, since a \$10 reduction in price increases sales by 100 per week. The slope for $p$ is $\frac{440 - 450}{1100 - 1000} = -\frac{1}{10}$, so an equation is $p - 450 = -\frac{1}{10}(x - 1000)$ or $p(x) = -\frac{1}{10}x + 550$.

(b) $R(x) = xp(x) = -\frac{1}{10}x^2 + 550x$. $R'(x) = -\frac{1}{5}x + 550 = 0$ when $x = 5(550) = 2750$.
$p(2750) = 275$, so the rebate should be $450 - 275 = \$175$.

(c) $C(x) = 68{,}000 + 150x \Rightarrow P(x) = R(x) - C(x) = -\frac{1}{10}x^2 + 550x - 68{,}000 - 150x = -\frac{1}{10}x^2 + 400x - 68{,}000$,
$P'(x) = -\frac{1}{5}x + 400 = 0$ when $x = 2000$. $p(2000) = 350$. Therefore, the rebate to maximize profits should be $450 - 350 = \$100$.

**18.** Let $x$ denote the number of \$10 increases in rent. Then the price is $p(x) = 800 + 10x$, and the number of units occupied is $100 - x$. Now the revenue is

$$\begin{aligned} R(x) &= (\text{rental price per unit}) \times (\text{number of units rented}) \\ &= (800 + 10x)(100 - x) = -10x^2 + 200x + 80{,}000 \text{ for } 0 \le x \le 100 \quad \Rightarrow \end{aligned}$$

$R'(x) = -20x + 200 = 0 \Leftrightarrow x = 10$. This is a maximum since $R''(x) = -20 < 0$ for all $x$. Now we must check the value of $R(x) = (800 + 10x)(100 - x)$ at $x = 10$ and at the endpoints of the domain to see which value of $x$ gives the maximum value of $R$. $R(0) = 80{,}000$, $R(10) = (900)(90) = 81{,}000$, and $R(100) = (1800)(0) = 0$. Thus, the maximum revenue of \$81,000/week occurs when 90 units are occupied at a rent of \$900/week.

**19.** If the reorder quantity is $x$, then the manager places $\dfrac{800}{x}$ orders per year. Storage costs for the year are $\frac{1}{2}x \cdot 4 = 2x$ dollars. Handling costs are \$100 per delivery, for a total of $\dfrac{800}{x} \cdot 100 = \dfrac{80{,}000}{x}$ dollars. The total costs for the year are $C(x) = 2x + \dfrac{80{,}000}{x}$. To minimize $C(x)$, we calculate $C'(x) = 2 - \dfrac{80{,}000}{x^2} = \dfrac{2}{x^2}(x^2 - 40{,}000)$. This is negative when $x < 200$, zero when $x = 200$, and positive when $x > 200$, so $C$ is decreasing on $(0, 200)$ and increasing on $(200, \infty)$, reaching its absolute minimum when $x = 200$. Thus, the optimal reorder quantity is 200 cases. The manager will place 4 orders per year for a total cost of $C(200) = \$800$.

**20.** She will have $A/n$ dollars after each withdrawal and 0 dollars just before the next withdrawal, so her average cash balance at any given time is $\frac{1}{2}(A/n + 0) = A/(2n)$. The transaction costs for $n$ withdrawals are $nT$. The lost interest cost on the average cash balance is $[A/(2n)] \cdot R$. Thus, the total cost for $n$ transactions is $C(n) = nT + \dfrac{AR}{2n}$. Now $C'(n) = T - \dfrac{AR}{2n^2}$ and $C'(n) = 0 \;\Rightarrow\; \dfrac{AR}{2n^2} = T \;\Rightarrow\; n^2 = \dfrac{AR}{2T} \;\Rightarrow\; n = \sqrt{\dfrac{AR}{2T}}$, the value of $n$ that minimizes total costs since $C''(n) = -\dfrac{AR}{n^3} < 0$. Thus, the optimal average cash balance is $\dfrac{A}{2n} = \dfrac{A\sqrt{2T}}{2\sqrt{AR}} = \dfrac{\sqrt{AT}}{\sqrt{2R}} = \sqrt{\dfrac{AT}{2R}}$.

## 4.8 Newton's Method

**1.** (a)

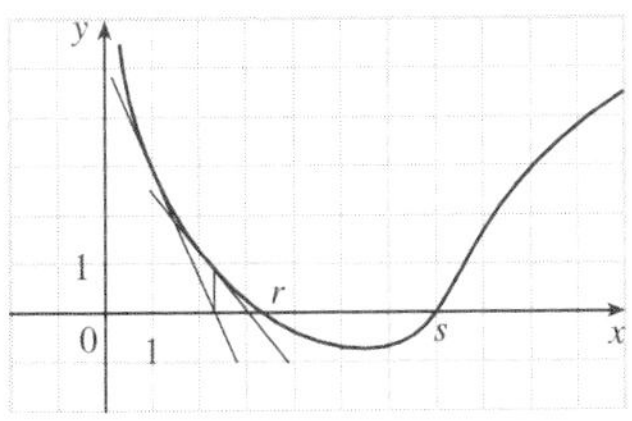


The tangent line at $x = 1$ intersects the $x$-axis at $x \approx 2.3$, so $x_2 \approx 2.3$. The tangent line at $x = 2.3$ intersects the $x$-axis at $x \approx 3$, so $x_3 \approx 3.0$.

(b) $x_1 = 5$ would *not* be a better first approximation than $x_1 = 1$ since the tangent line is nearly horizontal. In fact, the second approximation for $x_1 = 5$ appears to be to the left of $x = 1$.

**2.**

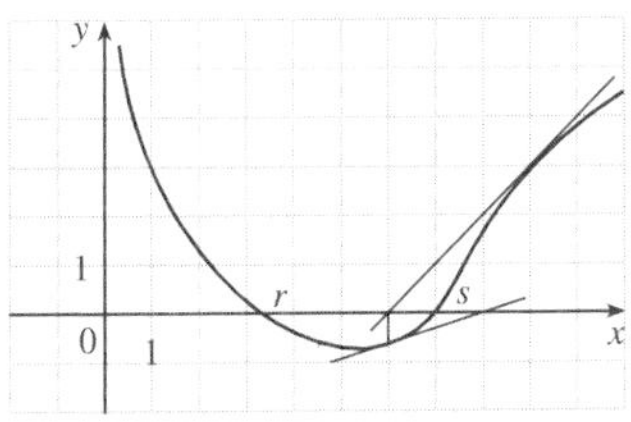


The tangent line at $x = 9$ intersects the $x$-axis at $x \approx 6.0$, so $x_2 \approx 6.0$. The tangent line at $x = 6.0$ intersects the $x$-axis at $x \approx 8.0$, so $x_3 \approx 8.0$.

**3.** Since $x_1 = 3$ and $y = 5x - 4$ is tangent to $y = f(x)$ at $x = 3$, we simply need to find where the tangent line intersects the $x$-axis. $y = 0 \;\Rightarrow\; 5x_2 - 4 = 0 \;\Rightarrow\; x_2 = \frac{4}{5}$.

**4.** (a)

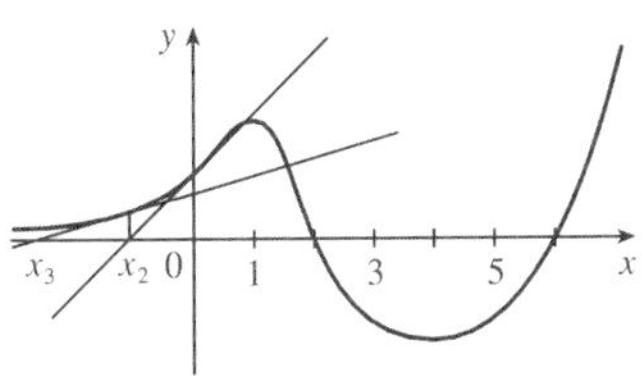


If $x_1 = 0$, then $x_2$ is negative, and $x_3$ is even more negative. The sequence of approximations does not converge, that is, Newton's method fails.

(b)

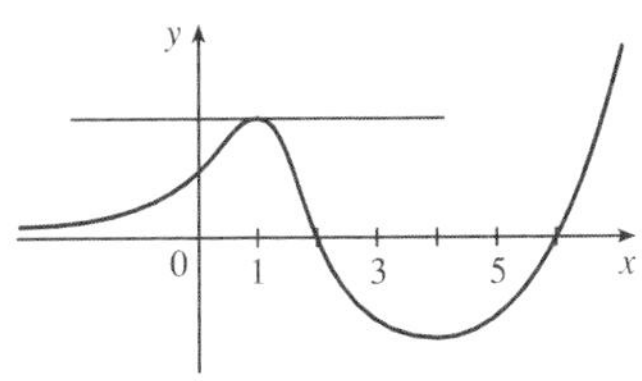


If $x_1 = 1$, the tangent line is horizontal and Newton's method fails.

(c)

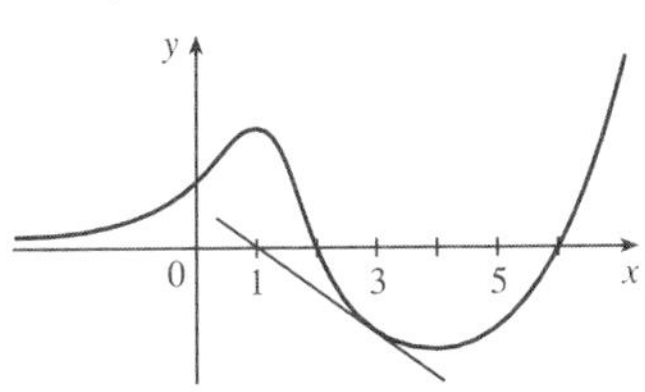


If $x_1 = 3$, then $x_2 = 1$ and we have the same situation as in part (b). Newton's method fails again.

(d)

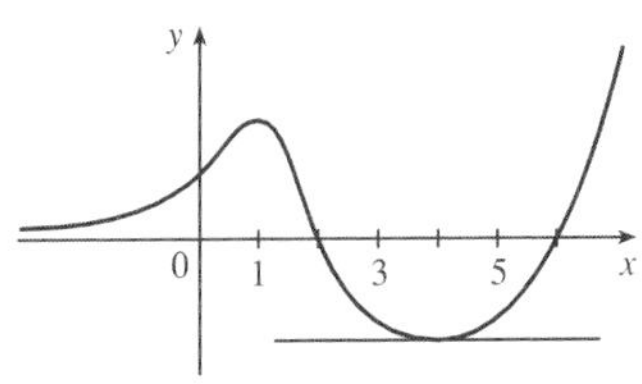


If $x_1 = 4$, the tangent line is horizontal and Newton's method fails.

(e)

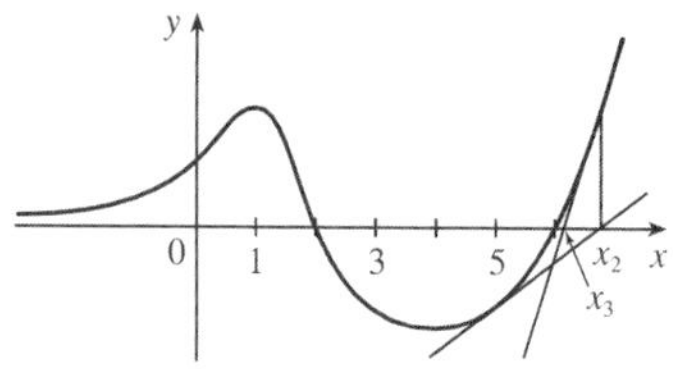


If $x_1 = 5$, then $x_2$ is greater than 6, $x_3$ gets closer to 6, and the sequence of approximations converges to 6. Newton's method succeeds!

**5.** $f(x) = x^3 + 2x - 4 \Rightarrow f'(x) = 3x^2 + 2$, so $x_{n+1} = x_n - \dfrac{x_n^3 + 2x_n - 4}{3x_n^2 + 2}$. Now $x_1 = 1 \Rightarrow$

$x_2 = 1 - \dfrac{1 + 2 - 4}{3 \cdot 1^2 + 2} = 1 - \dfrac{-1}{5} = 1.2 \Rightarrow x_3 = 1.2 - \dfrac{(1.2)^3 + 2(1.2) - 4}{3(1.2)^2 + 2} \approx 1.1797.$

**6.** $f(x) = x^5 + 2 \Rightarrow f'(x) = 5x^4$, so $x_{n+1} = x_n - \dfrac{x_n^5 + 2}{5x_n^4}$. Now $x_1 = -1 \Rightarrow$

$x_2 = -1 - \dfrac{(-1)^5 + 2}{5 \cdot (-1)^4} = -1 - \dfrac{1}{5} = -1.2 \Rightarrow x_3 = -1.2 - \dfrac{(-1.2)^5 + 2}{5(-1.2)^4} \approx -1.1529.$

**7.** $f(x) = x^3 + x + 3 \Rightarrow f'(x) = 3x^2 + 1$, so $x_{n+1} = x_n - \dfrac{x_n^3 + x_n + 3}{3x_n^2 + 1}$.

Now $x_1 = -1 \Rightarrow$

$x_2 = -1 - \dfrac{(-1)^3 + (-1) + 3}{3(-1)^2 + 1} = -1 - \dfrac{-1 - 1 + 3}{3 + 1} = -1 - \dfrac{1}{4} = -1.25.$

Newton's method follows the tangent line at $(-1, 1)$ down to its intersection with the $x$-axis at $(-1.25, 0)$, giving the second approximation $x_2 = -1.25$.

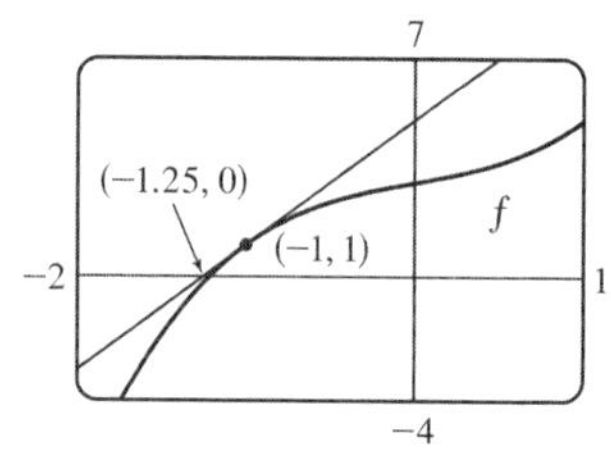


**8.** $f(x) = x^4 - x - 1 \Rightarrow f'(x) = 4x^3 - 1$, so $x_{n+1} = x_n - \dfrac{x_n^4 - x_n - 1}{4x_n^3 - 1}$.

Now $x_1 = 1 \Rightarrow x_2 = 1 - \dfrac{1^4 - 1 - 1}{4 \cdot 1^3 - 1} = 1 - \dfrac{-1}{3} = \dfrac{4}{3}$. Newton's method follows the tangent line at $(1, -1)$ up to its intersection with the $x$-axis at $\left(\frac{4}{3}, 0\right)$, giving the second approximation $x_2 = \frac{4}{3}$.

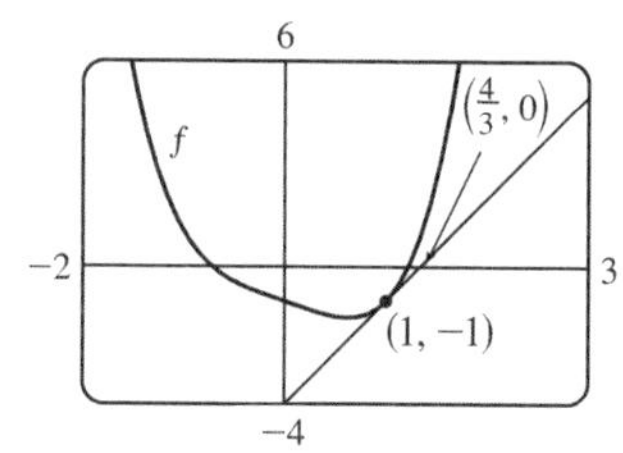


**9.** To approximate $x = \sqrt[3]{30}$ (so that $x^3 = 30$), we can take $f(x) = x^3 - 30$. So $f'(x) = 3x^2$, and thus, $x_{n+1} = x_n - \dfrac{x_n^3 - 30}{3x_n^2}$. Since $\sqrt[3]{27} = 3$ and 27 is close to 30, we'll use $x_1 = 3$. We need to find approximations until they agree to eight decimal places. $x_1 = 3 \Rightarrow x_2 \approx 3.11111111$, $x_3 \approx 3.10723734$, $x_4 \approx 3.10723251 \approx x_5$. So $\sqrt[3]{30} \approx 3.10723251$, to eight decimal places.

Here is a quick and easy method for finding the iterations for Newton's method on a programmable calculator. (The screens shown are from the TI-83 Plus, but the method is similar on other calculators.) Assign $f(x) = x^3 - 30$ to $\text{Y}_1$, and $f'(x) = 3x^2$ to $\text{Y}_2$. Now store $x_1 = 3$ in X and then enter $\text{X} - \text{Y}_1/\text{Y}_2 \to \text{X}$ to get $x_2 = 3.\overline{1}$. By successively pressing the ENTER key, you get the approximations $x_3, x_4, \ldots$.

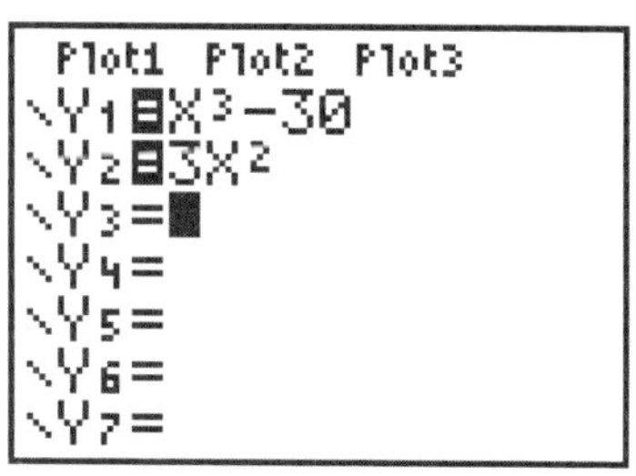


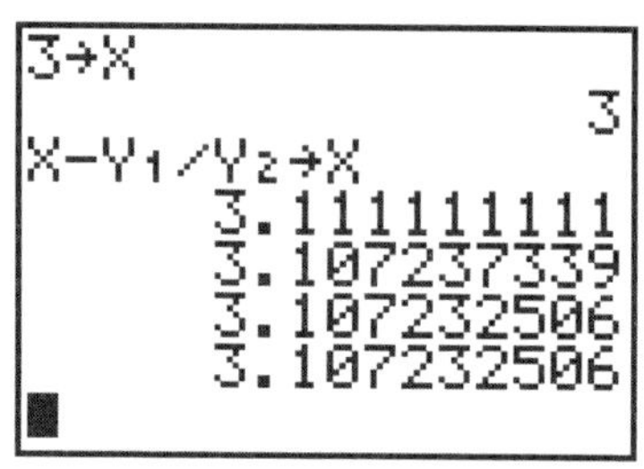


In Derive, load the utility file `SOLVE`. Enter `NEWTON(x^3-30,x,3)` and then `APPROXIMATE` to get

[3, 3.11111111, 3.10723733, 3.10723250, 3.10723250]. You can request a specific iteration by adding a fourth argument. For example, NEWTON(x^3-30,x,3,2) gives [3, 3.11111111, 3.10723733].

In Maple, make the assignments $f := x \to x\hat{\ }3 - 30;$, $g := x \to x - f(x)/D(f)(x);$, and $x := 3.;$. Repeatedly execute the command $x := g(x);$ to generate successive approximations.

In Mathematica, make the assignments $f[x\_] := x\hat{\ }3 - 30$, $g[x\_] := x - f[x]/f'[x]$, and $x = 3$. Repeatedly execute the command $x = g[x]$ to generate successive approximations.

**10.** $f(x) = x^7 - 1000 \;\Rightarrow\; f'(x) = 7x^6$, so $x_{n+1} = x_n - \dfrac{x_n^7 - 1000}{7x_n^6}$. We need to find approximations until they agree to eight decimal places. $x_1 = 3 \;\Rightarrow\; x_2 \approx 2.76739173$, $x_3 \approx 2.69008741$, $x_4 \approx 2.68275645$, $x_5 \approx 2.68269580 \approx x_6$. So $\sqrt[7]{1000} \approx 2.68269580$, to eight decimal places.

**11.** $\sin x = x^2$, so $f(x) = \sin x - x^2 \;\Rightarrow\; f'(x) = \cos x - 2x \;\Rightarrow$

$x_{n+1} = x_n - \dfrac{\sin x_n - x_n^2}{\cos x_n - 2x_n}$. From the figure, the positive root of $\sin x = x^2$ is near 1. $x_1 = 1 \;\Rightarrow\; x_2 \approx 0.891396$, $x_3 \approx 0.876985$, $x_4 \approx 0.876726 \approx x_5$. So the positive root is 0.876726, to six decimal places.

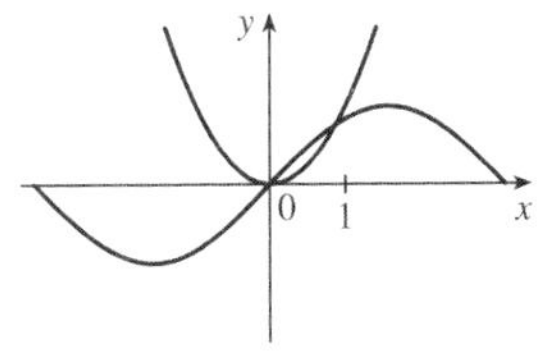


**12.** $2\cos x = x^4$, so $f(x) = 2\cos x - x^4 \;\Rightarrow\; f'(x) = -2\sin x - 4x^3 \;\Rightarrow$

$x_{n+1} = x_n - \dfrac{2\cos x_n - x_n^4}{-2\sin x_n - 4x_n^3}$. From the figure, the positive root of $2\cos x = x^4$ is near 1. $x_1 = 1 \;\Rightarrow\; x_2 \approx 1.014184$, $x_3 \approx 1.013958 \approx x_4$. So the positive root is 1.013958, to six decimal places.

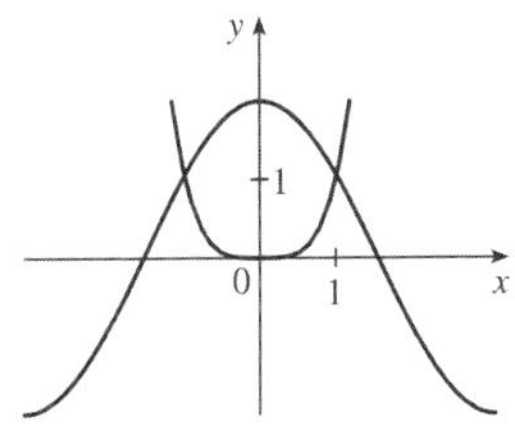


**13.**

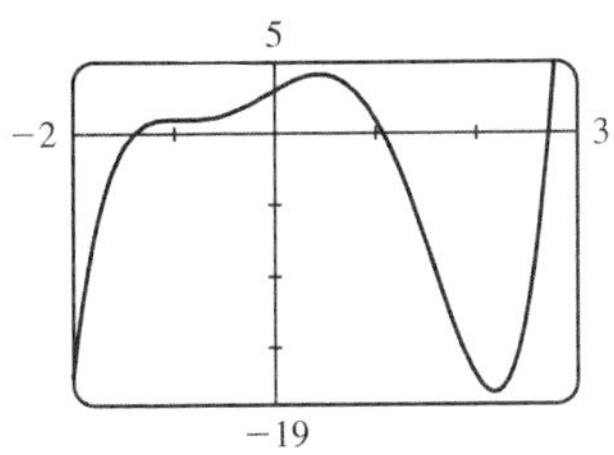


$f(x) = x^5 - x^4 - 5x^3 - x^2 + 4x + 3 \;\Rightarrow$

$f'(x) = 5x^4 - 4x^3 - 15x^2 - 2x + 4 \;\Rightarrow$

$x_{n+1} = x_n - \dfrac{x_n^5 - x_n^4 - 5x_n^3 - x_n^2 + 4x_n + 3}{5x_n^4 - 4x_n^3 - 15x_n^2 - 2x_n + 4}$. From the graph of $f$, there appear to be roots near $-1.4$, 1.1, and 2.7.

| | | |
|---|---|---|
| $x_1 = -1.4$ | $x_1 = 1.1$ | $x_1 = 2.7$ |
| $x_2 \approx -1.39210970$ | $x_2 \approx 1.07780402$ | $x_2 \approx 2.72046250$ |
| $x_3 \approx -1.39194698$ | $x_3 \approx 1.07739442$ | $x_3 \approx 2.71987870$ |
| $x_4 \approx -1.39194691 \approx x_5$ | $x_4 \approx 1.07739428 \approx x_5$ | $x_4 \approx 2.71987822 \approx x_5$ |

To eight decimal places, the roots of the equation are $-1.39194691$, 1.07739428, and 2.71987822.

**14.**

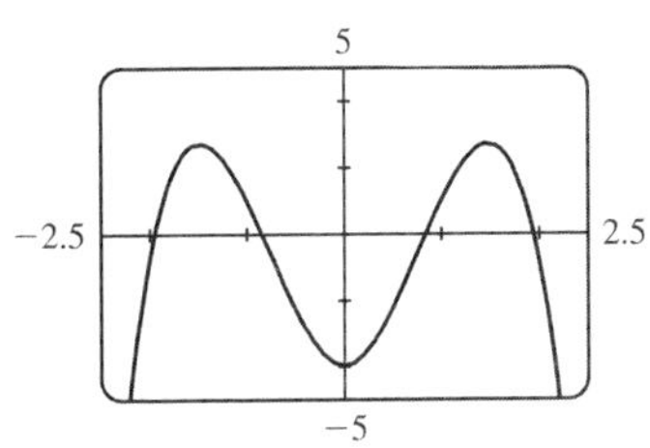


Solving $x^2(4 - x^2) = \dfrac{4}{x^2 + 1}$ is the same as solving

$f(x) = 4x^2 - x^4 - \dfrac{4}{x^2 + 1} = 0$. $f'(x) = 8x - 4x^3 + \dfrac{8x}{(x^2 + 1)^2} \;\Rightarrow$

$x_{n+1} = x_n - \dfrac{4x_n^2 - x_n^4 - 4/(x_n^2 + 1)}{8x_n - 4x_n^3 + 8x_n/(x_n^2 + 1)^2}$. From the graph of $f(x)$, there appear to be roots near $x = \pm 1.9$ and $x = \pm 0.8$. Since $f$ is even, we only need to find the positive roots.

| | |
|---|---|
| $x_1 = 0.8$ | $x_1 = 1.9$ |
| $x_2 \approx 0.84287645$ | $x_2 \approx 1.94689103$ |
| $x_3 \approx 0.84310820$ | $x_3 \approx 1.94383891$ |
| $x_4 \approx 0.84310821 \approx x_5$ | $x_4 \approx 1.94382538 \approx x_5$ |

To eight decimal places, the roots of the equation are $\pm 0.84310821$ and $\pm 1.94382538$.

15. 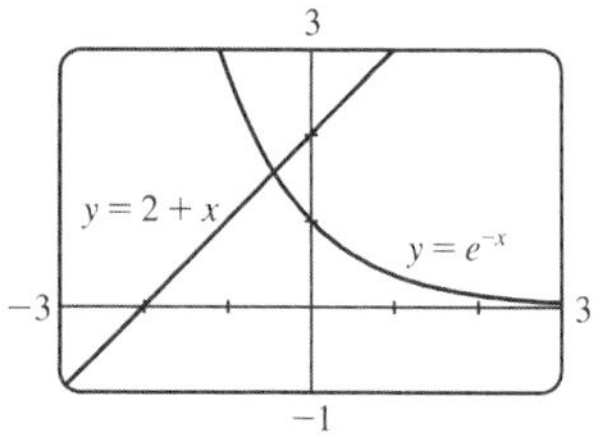


Solving $e^{-x} = 2 + x$ is the same as solving $f(x) = e^{-x} - x - 2 = 0$. $f'(x) = -e^{-x} - 1 \quad\Rightarrow\quad x_{n+1} = x_n - \dfrac{e^{-x_n} - x_n - 2}{-e^{-x_n} - 1}$. From the graph of $y = e^{-x}$ and $y = 2 + x$, there appears to be a root near $x = -0.5$. Now $x_1 = -0.5 \quad\Rightarrow\quad x_2 \approx -0.44385167$, $x_3 \approx -0.44285470$, $x_4 \approx -0.44285440 \approx x_5$. To eight decimal places, the root of the equation is $-0.44285440$.

16. 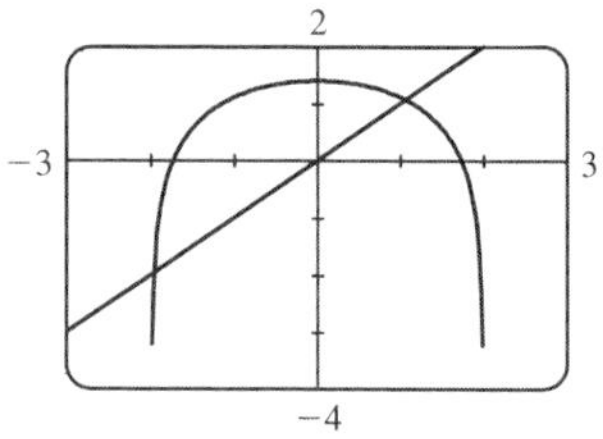


From the graph, $y = \ln(4 - x^2)$ and $y = x$ intersect twice, at $x \approx -2$ and at $x \approx 1$. $f(x) = \ln(4 - x^2) - x \quad\Rightarrow\quad f'(x) = \dfrac{-2x}{4 - x^2} - 1$, so $x_{n+1} = x_n - \dfrac{\ln(4 - x_n^2) - x_n}{[-2x_n/(4 - x_n^2)] - 1}$. Trying $x_1 = -2$ won't work because it's not in the domain of $y = \ln(4 - x^2)$. Trying $x_1 = -1.9$ also fails after one iteration because the approximation $x_2$ is less than $-2$. We try $x_1 = -1.99$.

| | |
|---|---|
| $x_1 = -1.99$ | $x_1 = 1.1$ |
| $x_2 \approx -1.97753026$ | $x_2 \approx 1.05864851$ |
| $x_3 \approx -1.96741777$ | $x_3 \approx 1.05800655$ |
| $x_4 \approx -1.96475281$ | $x_4 \approx 1.05800640 \approx x_5$ |
| $x_5 \approx -1.96463580$ | |
| $x_6 \approx -1.96463560 \approx x_7$ | |

To eight decimal places, the roots of the equation are $-1.96463560$ and $1.05800640$.

17. 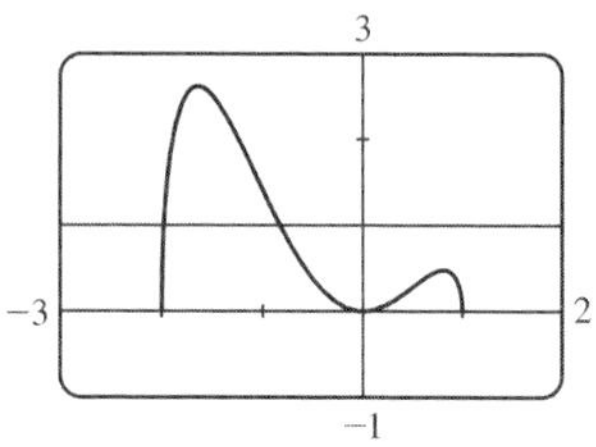


From the graph, $y = x^2\sqrt{2 - x - x^2}$ and $y = 1$ intersect twice, at $x \approx -2$ and at $x \approx -1$. $f(x) = x^2\sqrt{2 - x - x^2} - 1 \quad\Rightarrow$

$$\begin{aligned} f'(x) &= x^2 \cdot \tfrac{1}{2}(2 - x - x^2)^{-1/2}(-1 - 2x) + (2 - x - x^2)^{1/2} \cdot 2x \\ &= \tfrac{1}{2}x(2 - x - x^2)^{-1/2}\left[x(-1 - 2x) + 4(2 - x - x^2)\right] \\ &= \frac{x(8 - 5x - 6x^2)}{2\sqrt{(2 + x)(1 - x)}}, \end{aligned}$$

so $x_{n+1} = x_n - \dfrac{x_n^2\sqrt{2 - x_n - x_n^2} - 1}{\dfrac{x_n(8 - 5x_n - 6x_n^2)}{2\sqrt{(2 + x_n)(1 - x_n)}}}$. Trying $x_1 = -2$ won't work because $f'(-2)$ is undefined, so we'll try $x_1 = -1.95$.

| | |
|---|---|
| $x_1 = -1.95$ | $x_1 = -0.8$ |
| $x_2 \approx -1.98580357$ | $x_2 \approx -0.82674444$ |
| $x_3 \approx -1.97899778$ | $x_3 \approx -0.82646236$ |
| $x_4 \approx -1.97807848$ | $x_4 \approx -0.82646233 \approx x_5$ |
| $x_5 \approx -1.97806682$ | |
| $x_6 \approx -1.97806681 \approx x_7$ | |

To eight decimal places, the roots of the equation are $-1.97806681$ and $-0.82646233$.

**18.** 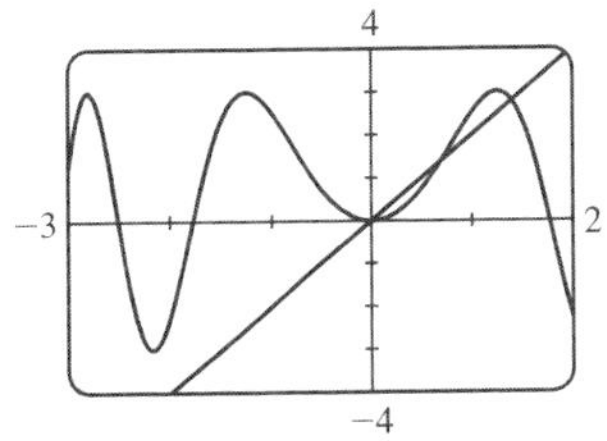


From the equations $y = 3\sin(x^2)$ and $y = 2x$ and the graph, we deduce that one root of the equation $3\sin(x^2) = 2x$ is $x = 0$. We also see that the graphs intersect at approximately $x = 0.7$ and $x = 1.4$.

$f(x) = 3\sin(x^2) - 2x \quad \Rightarrow \quad f'(x) = 3\cos(x^2)\cdot 2x - 2$, so

$$x_{n+1} = x_n - \frac{3\sin(x_n^2) - 2x_n}{6x_n\cos(x_n^2) - 2}.$$

| | |
|---|---|
| $x_1 = 0.7$ | $x_1 = 1.4$ |
| $x_2 \approx 0.69303689$ | $x_2 \approx 1.39530295$ |
| $x_3 \approx 0.69299996 \approx x_4$ | $x_3 \approx 1.39525078$ |
| | $x_4 \approx 1.39525077 \approx x_5$ |

To eight decimal places, the roots of the equation are $0.69299996$ and $1.39525077$.

**19.** 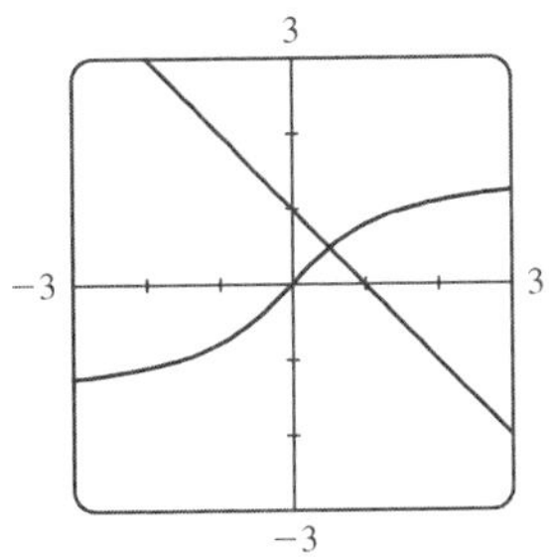


From the graph of $y = \tan^{-1} x$ and $y = 1 - x$, there appears to be a point of intersection near $x = 0.5$. Solving $\tan^{-1} x = 1 - x$ is the same as solving $f(x) = \tan^{-1} x + x - 1 = 0$. $f(x) = \tan^{-1} x + x - 1 \quad \Rightarrow$

$$f'(x) = \frac{1}{1 + x^2} + 1, \text{ so } x_{n+1} = x_n - \frac{\tan^{-1} x_n + x_n - 1}{1/(1 + x_n^2) + 1}.$$

Now $x_1 = 0.5 \quad \Rightarrow \quad x_2 \approx 0.52019577$, $x_3 \approx 0.52026899 \approx x_4$.
To eight decimal places, the root of the equation is $0.52026899$.

**20.** 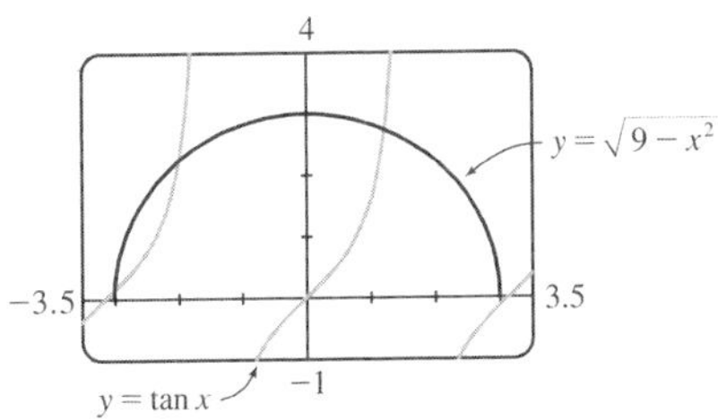


From the graph, $y = \tan x$ and $y = \sqrt{9 - x^2}$ intersect three times, at $x \approx -3$, $x \approx -2$, and $x \approx 1.2$. $f(x) = \tan x - \sqrt{9 - x^2} \quad \Rightarrow$

$$f'(x) = \sec^2 x + \frac{x}{\sqrt{9 - x^2}}, \text{ so } x_{n+1} = x_n - \frac{\tan x_n - \sqrt{9 - x_n^2}}{\sec^2 x_n + x_n/\sqrt{9 - x_n^2}}.$$

Trying $x_1 = -3$ won't work because it's not in the domain of $f'(x)$. Trying $x_1 = -2.9$ also fails after one iteration because the approximation $x_2$ is less than $-3$. We try $-2.99$.

| | | |
|---|---|---|
| $x_1 = -2.99$ | $x_1 = -2$ | $x_1 = 1.2$ |
| $x_2 \approx -2.99821730$ | $x_2 \approx -1.98954336$ | $x_2 \approx 1.22202993$ |
| $x_3 \approx -2.99675310$ | $x_3 \approx -1.98982579$ | $x_3 \approx 1.22089613$ |
| $x_4 \approx -2.99644398$ | $x_4 \approx -1.98982601 \approx x_5$ | $x_4 \approx 1.22089273 \approx x_5$ |
| $x_5 \approx -2.99643623 \approx x_6$ | | |

To eight decimal places, the roots of the equation are $-2.99643623$, $-1.98982601$, and $1.22089273$.

**21.** (a) $f(x) = x^2 - a \quad \Rightarrow \quad f'(x) = 2x$, so Newton's method gives

$$x_{n+1} = x_n - \frac{x_n^2 - a}{2x_n} = x_n - \frac{1}{2}x_n + \frac{a}{2x_n} = \frac{1}{2}x_n + \frac{a}{2x_n} = \frac{1}{2}\left(x_n + \frac{a}{x_n}\right).$$

(b) Using (a) with $a = 1000$ and $x_1 = \sqrt{900} = 30$, we get $x_2 \approx 31.666667$, $x_3 \approx 31.622807$, and $x_4 \approx 31.622777 \approx x_5$. So $\sqrt{1000} \approx 31.622777$.

**22.** (a) $f(x) = \dfrac{1}{x} - a \quad \Rightarrow \quad f'(x) = -\dfrac{1}{x^2}$, so $x_{n+1} = x_n - \dfrac{1/x_n - a}{-1/x_n^2} = x_n + x_n - ax_n^2 = 2x_n - ax_n^2$.

(b) Using (a) with $a = 1.6894$ and $x_1 = \frac{1}{2} = 0.5$, we get $x_2 = 0.5754$, $x_3 \approx 0.588485$, and $x_4 \approx 0.588789 \approx x_5$. So $1/1.6984 \approx 0.588789$.

**23.** $f(x) = x^3 - 3x + 6 \quad \Rightarrow \quad f'(x) = 3x^2 - 3$. If $x_1 = 1$, then $f'(x_1) = 0$ and the tangent line used for approximating $x_2$ is horizontal. Attempting to find $x_2$ results in trying to divide by zero.

**24.** $x^3 - x = 1 \quad \Leftrightarrow \quad x^3 - x - 1 = 0$. $f(x) = x^3 - x - 1 \quad \Rightarrow \quad f'(x) = 3x^2 - 1$, so $x_{n+1} = x_n - \dfrac{x_n^3 - x_n - 1}{3x_n^2 - 1}$.

(a) $x_1 = 1$, $x_2 = 1.5$, $x_3 \approx 1.347826$, $x_4 \approx 1.325200$, $x_5 \approx 1.324718 \approx x_6$

(b) $x_1 = 0.6$, $x_2 = 17.9$, $x_3 \approx 11.946802$, $x_4 \approx 7.985520$, $x_5 \approx 5.356909$, $x_6 \approx 3.624996$, $x_7 \approx 2.505589$, $x_8 \approx 1.820129$, $x_9 \approx 1.461044$, $x_{10} \approx 1.339323$, $x_{11} \approx 1.324913$, $x_{12} \approx 1.324718 \approx x_{13}$

(c) $x_1 = 0.57$, $x_2 \approx -54.165455$, $x_3 \approx -36.114293$, $x_4 \approx -24.082094$, $x_5 \approx -16.063387$, $x_6 \approx -10.721483$, $x_7 \approx -7.165534$, $x_8 \approx -4.801704$, $x_9 \approx -3.233425$, $x_{10} \approx -2.193674$, $x_{11} \approx -1.496867$, $x_{12} \approx -0.997546$, $x_{13} \approx -0.496305$, $x_{14} \approx -2.894162$, $x_{15} \approx -1.967962$, $x_{16} \approx -1.341355$, $x_{17} \approx -0.870187$, $x_{18} \approx -0.249949$, $x_{19} \approx -1.192219$, $x_{20} \approx -0.731952$, $x_{21} \approx 0.355213$, $x_{22} \approx -1.753322$, $x_{23} \approx -1.189420$, $x_{24} \approx -0.729123$, $x_{25} \approx 0.377844$, $x_{26} \approx -1.937872$, $x_{27} \approx -1.320350$, $x_{28} \approx -0.851919$, $x_{29} \approx -0.200959$, $x_{30} \approx -1.119386$, $x_{31} \approx -0.654291$, $x_{32} \approx 1.547010$, $x_{33} \approx 1.360051$, $x_{34} \approx 1.325828$, $x_{35} \approx 1.324719$, $x_{36} \approx 1.324718 \approx x_{37}$.

(d)

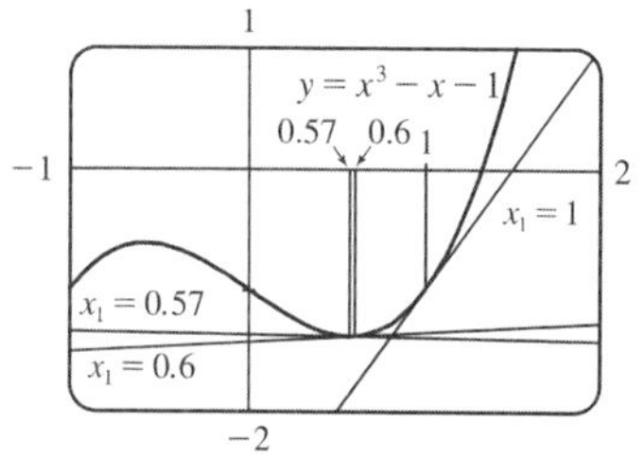


From the figure, we see that the tangent line corresponding to $x_1 = 1$ results in a sequence of approximations that converges quite quickly ($x_5 \approx x_6$). The tangent line corresponding to $x_1 = 0.6$ is close to being horizontal, so $x_2$ is quite far from the root. But the sequence still converges — just a little more slowly ($x_{12} \approx x_{13}$). Lastly, the tangent line corresponding to $x_1 = 0.57$ is very nearly horizontal, $x_2$ is farther away from the root, and the sequence takes more iterations to converge ($x_{36} \approx x_{37}$).

**25.** For $f(x) = x^{1/3}$, $f'(x) = \frac{1}{3}x^{-2/3}$ and

$$x_{n+1} = x_n - \frac{f(x_n)}{f'(x_n)} = x_n - \frac{x_n^{1/3}}{\frac{1}{3}x_n^{-2/3}} = x_n - 3x_n = -2x_n.$$

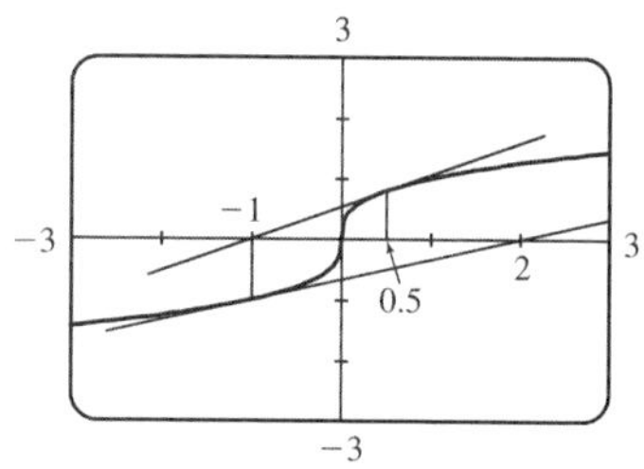


Therefore, each successive approximation becomes twice as large as the previous one in absolute value, so the sequence of approximations fails to converge to the root, which is 0. In the figure, we have $x_1 = 0.5$, $x_2 = -2(0.5) = -1$, and $x_3 = -2(-1) = 2$.

**26.** $f(x) = x^2 + \sin x \quad \Rightarrow \quad f'(x) = 2x + \cos x$. $f'(x)$ exists for all $x$, so to find the minimum of $f$, we can examine the zeros of $f'$. From the graph of $f'$, we see that a good choice for $x_1$ is $x_1 = -0.5$. Use $g(x) = 2x + \cos x$ and $g'(x) = 2 - \sin x$ to obtain $x_2 \approx -0.450627$, $x_3 \approx -0.450184 \approx x_4$. Since $f''(x) = 2 - \sin x > 0$ for all $x$, $f(-0.450184) \approx -0.232466$ is the absolute minimum.

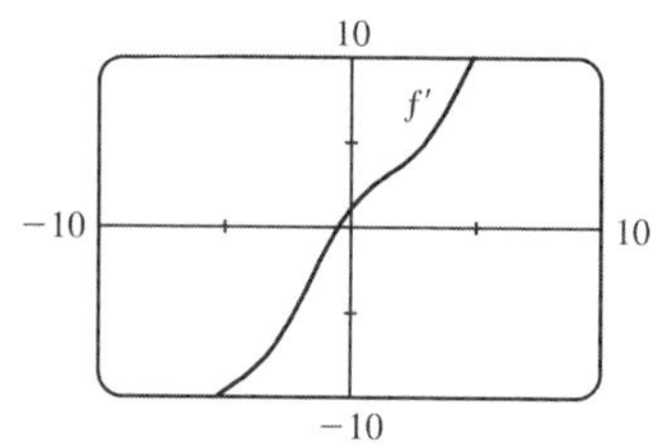


**27.**

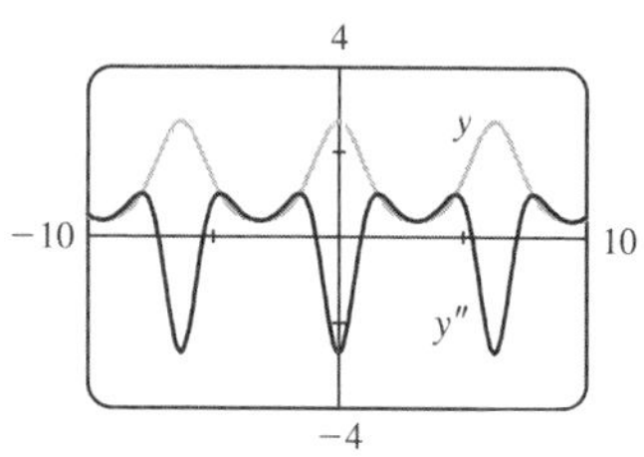


From the figure, we see that $y = f(x) = e^{\cos x}$ is periodic with period $2\pi$. To find the $x$-coordinates of the IP, we only need to approximate the zeros of $y''$ on $[0, \pi]$. $f'(x) = -e^{\cos x} \sin x \quad \Rightarrow \quad f''(x) = e^{\cos x}(\sin^2 x - \cos x)$. Since $e^{\cos x} \neq 0$, we will use Newton's method with $g(x) = \sin^2 x - \cos x$, $g'(x) = 2 \sin x \cos x + \sin x$, and $x_1 = 1$. $x_2 \approx 0.904173$, $x_3 \approx 0.904557 \approx x_4$. Thus, $(0.904557, 1.855277)$ is the IP.

**28.** 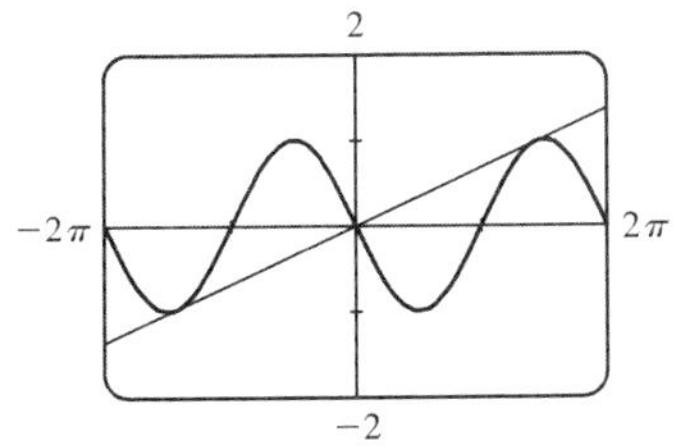


$f(x) = -\sin x \quad \Rightarrow \quad f'(x) = -\cos x$. At $x = a$, the slope of the tangent line is $f'(a) = -\cos a$. The line through the origin and $(a, f(a))$ is $y = \dfrac{-\sin a - 0}{a - 0}x$. If this line is to be tangent to $f$ at $x = a$, then its slope must equal $f'(a)$. Thus, $\dfrac{-\sin a}{a} = -\cos a \quad \Rightarrow \quad \tan a = a$.

To solve this equation using Newton's method, let $g(x) = \tan x - x$, $g'(x) = \sec^2 x - 1$, and $x_{n+1} = x_n - \dfrac{\tan x_n - x_n}{\sec^2 x_n - 1}$ with $x_1 = 4.5$ (estimated from the figure). $x_2 \approx 4.493614$, $x_3 \approx 4.493410$, $x_4 \approx 4.493409 \approx x_5$. Thus, the slope of the line that has the largest slope is $f'(x_5) \approx 0.217234$.

**29.** 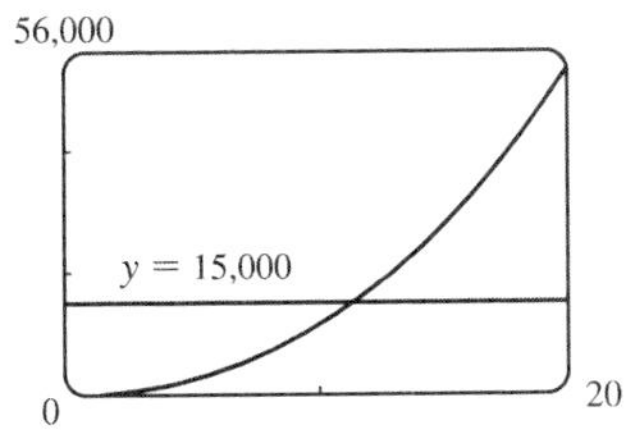


The volume of the silo, in terms of its radius, is $V(r) = \pi r^2(30) + \frac{1}{2}\left(\frac{4}{3}\pi r^3\right) = 30\pi r^2 + \frac{2}{3}\pi r^3$. From a graph of $V$, we see that $V(r) = 15{,}000$ at $r \approx 11$ ft. Now we use Newton's method to solve the equation $V(r) - 15{,}000 = 0$. $\dfrac{dV}{dr} = 60\pi r + 2\pi r^2$, so $r_{n+1} = r_n - \dfrac{30\pi r_n^2 + \frac{2}{3}\pi r_n^3 - 15{,}000}{60\pi r_n + 2\pi r_n^2}$.

Taking $r_1 = 11$, we get $r_2 \approx 11.2853$, $r_3 \approx 11.2807 \approx r_4$. So in order for the silo to hold 15,000 ft$^3$ of grain, its radius must be about 11.2807 ft.

**30.** Let the radius of the circle be $r$. Using $s = r\theta$, we have $5 = r\theta$ and so $r = 5/\theta$. From the Law of Cosines we get $4^2 = r^2 + r^2 - 2 \cdot r \cdot r \cdot \cos\theta \quad \Leftrightarrow \quad 16 = 2r^2(1 - \cos\theta) = 2(5/\theta)^2(1 - \cos\theta)$.
Multiplying by $\theta^2$ gives $16\theta^2 = 50(1 - \cos\theta)$, so we take $f(\theta) = 16\theta^2 + 50\cos\theta - 50$ and $f'(\theta) = 32\theta - 50\sin\theta$. The formula for Newton's method is $\theta_{n+1} = \theta_n - \dfrac{16\theta_n^2 + 50\cos\theta_n - 50}{32\theta_n - 50\sin\theta_n}$. From the graph of $f$, we can use $\theta_1 = 2.2$, giving us $\theta_2 \approx 2.2662$, $\theta_3 \approx 2.2622 \approx \theta_4$. So correct to four decimal places, the angle is 2.2622 radians $\approx 130^\circ$.

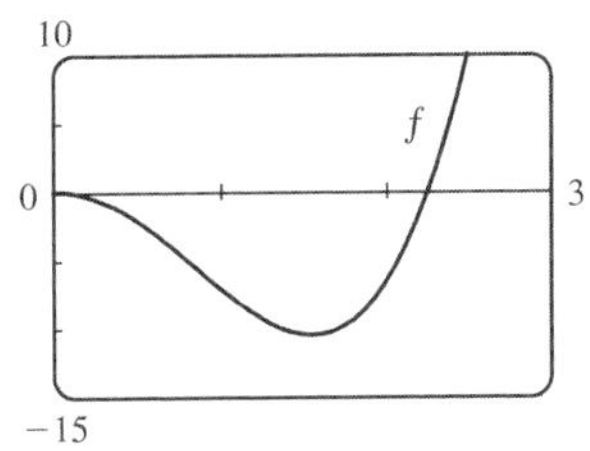


**31.** In this case, $A = 18{,}000$, $R = 375$, and $n = 5(12) = 60$. So the formula $A = \dfrac{R}{i}\left[1 - (1+i)^{-n}\right]$ becomes

$18{,}000 = \dfrac{375}{x}\left[1 - (1+x)^{-60}\right] \quad \Leftrightarrow \quad 48x = 1 - (1+x)^{-60}$ [multiply each term by $(1+x)^{60}$] $\Leftrightarrow$

$48x(1+x)^{60} - (1+x)^{60} + 1 = 0$. Let the LHS be called $f(x)$, so that

$$\begin{aligned} f'(x) &= 48x(60)(1+x)^{59} + 48(1+x)^{60} - 60(1+x)^{59} \\ &= 12(1+x)^{59}\left[4x(60) + 4(1+x) - 5\right] = 12(1+x)^{59}(244x - 1) \end{aligned}$$

$x_{n+1} = x_n - \dfrac{48x_n(1+x_n)^{60} - (1+x_n)^{60} + 1}{12(1+x_n)^{59}(244x_n - 1)}$. An interest rate of 1% per month seems like a reasonable estimate for $x = i$. So let $x_1 = 1\% = 0.01$, and we get $x_2 \approx 0.0082202$, $x_3 \approx 0.0076802$, $x_4 \approx 0.0076291$, $x_5 \approx 0.0076286 \approx x_6$. Thus, the dealer is charging a monthly interest rate of 0.76286% (or 9.55% per year, compounded monthly).

**32.** (a) $p(x) = x^5 - (2+r)x^4 + (1+2r)x^3 - (1-r)x^2 + 2(1-r)x + r - 1 \quad \Rightarrow$

$p'(x) = 5x^4 - 4(2+r)x^3 + 3(1+2r)x^2 - 2(1-r)x + 2(1-r)$. So we use

$$x_{n+1} = x_n - \frac{x_n^5 - (2+r)x_n^4 + (1+2r)x_n^3 - (1-r)x_n^2 + 2(1-r)x_n + r - 1}{5x_n^4 - 4(2+r)x_n^3 + 3(1+2r)x_n^2 - 2(1-r)x_n + 2(1-r)}.$$

We substitute in the value $r \approx 3.04042 \times 10^{-6}$ in order to evaluate the approximations numerically. The libration point $L_1$ is slightly less than 1 AU from the Sun, so we take $x_1 = 0.95$ as our first approximation, and get $x_2 \approx 0.96682$, $x_3 \approx 0.97770$, $x_4 \approx 0.98451$, $x_5 \approx 0.98830$, $x_6 \approx 0.98976$, $x_7 \approx 0.98998$, $x_8 \approx 0.98999 \approx x_9$. So, to five decimal places, $L_1$ is located 0.98999 AU from the Sun (or 0.01001 AU from Earth).

(b) In this case we use Newton's method with the function

$p(x) - 2rx^2 = x^5 - (2+r)x^4 + (1+2r)x^3 - (1+r)x^2 + 2(1-r)x + r - 1 \quad \Rightarrow$

$\left[p(x) - 2rx^2\right]' = 5x^4 - 4(2+r)x^3 + 3(1+2r)x^2 - 2(1+r)x + 2(1-r)$. So

$$x_{n+1} = x_n - \frac{x_n^5 - (2+r)x_n^4 + (1+2r)x_n^3 - (1+r)x_n^2 + 2(1-r)x_n + r - 1}{5x_n^4 - 4(2+r)x_n^3 + 3(1+2r)x_n^2 - 2(1+r)x_n + 2(1-r)}.$$ Again, we substitute

$r \approx 3.04042 \times 10^{-6}$. $L_2$ is slightly more than 1 AU from the Sun and, judging from the result of part (a), probably less than 0.02 AU from Earth. So we take $x_1 = 1.02$ and get $x_2 \approx 1.01422$, $x_3 \approx 1.01118$, $x_4 \approx 1.01018$, $x_5 \approx 1.01008 \approx x_6$. So, to five decimal places, $L_2$ is located 1.01008 AU from the Sun (or 0.01008 AU from Earth).

## 4.9 Antiderivatives

**1.** $f(x) = 6x^2 - 8x + 3 \quad \Rightarrow \quad F(x) = 6\dfrac{x^{2+1}}{2+1} - 8\dfrac{x^{1+1}}{1+1} + 3x + C = 2x^3 - 4x^2 + 3x + C$

*Check:* $F'(x) = 2 \cdot 3x^2 - 4 \cdot 2x + 3 + 0 = 6x^2 - 8x + 3 = f(x)$

**2.** $f(x) = 1 - x^3 + 12x^5 \quad \Rightarrow \quad F(x) = x - \dfrac{x^{3+1}}{3+1} + 12\dfrac{x^{5+1}}{5+1} + C = x - \frac{1}{4}x^4 + 2x^6 + C$

**3.** $f(x) = 5x^{1/4} - 7x^{3/4} \quad \Rightarrow \quad F(x) = 5\dfrac{x^{1/4+1}}{\frac{1}{4}+1} - 7\dfrac{x^{3/4+1}}{\frac{3}{4}+1} + C = 5\dfrac{x^{5/4}}{5/4} - 7\dfrac{x^{7/4}}{7/4} + C = 4x^{5/4} - 4x^{7/4} + C$

**4.** $f(x) = 2x + 3x^{1.7} \quad \Rightarrow \quad F(x) = x^2 + \frac{3}{2.7}x^{2.7} + C = x^2 + \frac{10}{9}x^{2.7} + C$

**5.** $f(x) = 3\sqrt{x} + \dfrac{5}{x^6} = 3x^{1/2} + 5x^{-6}$ has domain $(-\infty, 0) \cup (0, \infty)$, so

$$F(x) = \begin{cases} 3\dfrac{x^{1/2+1}}{\frac{1}{2}+1} + 5\dfrac{x^{-6+1}}{-6+1} + C_1 = 2x^{3/2} - x^{-5} + C_1 & \text{if } x < 0 \\ 2x^{3/2} - x^{-5} + C_2 & \text{if } x > 0 \end{cases}$$

See Example 1(b) for a similar problem.

**6.** $f(x) = \sqrt[4]{x^3} + \sqrt[3]{x^4} = x^{3/4} + x^{4/3} \quad \Rightarrow \quad F(x) = \dfrac{x^{7/4}}{7/4} + \dfrac{x^{7/3}}{7/3} + C = \frac{4}{7}x^{7/4} + \frac{3}{7}x^{7/3} + C$

**7.** $f(u) = \dfrac{u^4 + 3\sqrt{u}}{u^2} = \dfrac{u^4}{u^2} + \dfrac{3u^{1/2}}{u^2} = u^2 + 3u^{-3/2} \quad \Rightarrow$

$$F(u) = \frac{u^3}{3} + 3\,\frac{u^{-3/2+1}}{-3/2+1} + C = \frac{1}{3}u^3 + 3\frac{u^{-1/2}}{-1/2} + C = \frac{1}{3}u^3 - \frac{6}{\sqrt{u}} + C$$

**8.** $g(x) = \dfrac{5 - 4x^3 + 2x^6}{x^6} = 5x^{-6} - 4x^{-3} + 2$ has domain $(-\infty, 0) \cup (0, \infty)$, so

$$G(x) = \begin{cases} 5\,\dfrac{x^{-5}}{-5} - 4\,\dfrac{x^{-2}}{-2} + 2x + C_1 = -\dfrac{1}{x^5} + \dfrac{2}{x^2} + 2x + C_1 & \text{if } x < 0 \\ -\dfrac{1}{x^5} + \dfrac{2}{x^2} + 2x + C_2 & \text{if } x > 0 \end{cases}$$

**9.** $g(\theta) = \cos\theta - 5\sin\theta \quad\Rightarrow\quad G(\theta) = \sin\theta - 5(-\cos\theta) + C = \sin\theta + 5\cos\theta + C$

**10.** $f(x) = 3e^x + 7\sec^2 x \quad\Rightarrow\quad F(x) = 3e^x + 7\tan x + C_n$ on the interval $\left(n\pi - \frac{\pi}{2}, n\pi + \frac{\pi}{2}\right)$.

**11.** $f(x) = 2x + 5\left(1-x^2\right)^{-1/2} = 2x + \dfrac{5}{\sqrt{1-x^2}} \quad\Rightarrow\quad F(x) = x^2 + 5\sin^{-1} x + C$

**12.** $f(x) = \dfrac{x^2+x+1}{x} = x + 1 + \dfrac{1}{x} \quad\Rightarrow\quad F(x) = \begin{cases} \frac{1}{2}x^2 + x + \ln|x| + C_1 & \text{if } x < 0 \\ \frac{1}{2}x^2 + x + \ln|x| + C_2 & \text{if } x > 0 \end{cases}$

**13.** $f(x) = 5x^4 - 2x^5 \quad\Rightarrow\quad F(x) = 5\cdot\dfrac{x^5}{5} - 2\cdot\dfrac{x^6}{6} + C = x^5 - \frac{1}{3}x^6 + C.$

$F(0) = 4 \quad\Rightarrow\quad 0^5 - \frac{1}{3}\cdot 0^6 + C = 4 \quad\Rightarrow\quad C = 4$, so $F(x) = x^5 - \frac{1}{3}x^6 + 4.$

The graph confirms our answer since $f(x) = 0$ when $F$ has a local maximum, $f$ is positive when $F$ is increasing, and $f$ is negative when $F$ is decreasing.

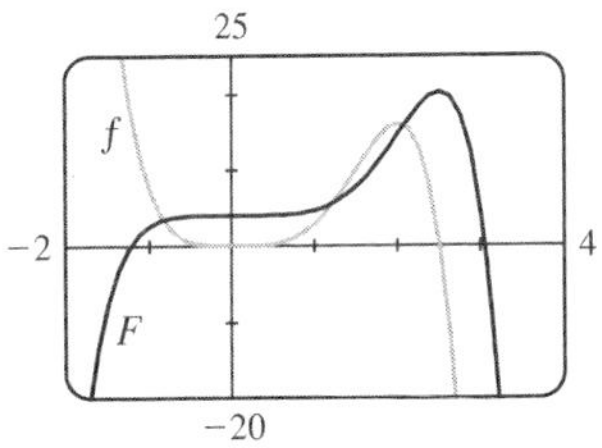


**14.** $f(x) = 4 - 3\left(1+x^2\right)^{-1} = 4 - \dfrac{3}{1+x^2} \quad\Rightarrow\quad F(x) = 4x - 3\tan^{-1} x + C.$

$F(1) = 0 \quad\Rightarrow\quad 4 - 3\left(\frac{\pi}{4}\right) + C = 0 \quad\Rightarrow\quad C = \frac{3\pi}{4} - 4$, so

$F(x) = 4x - 3\tan^{-1} x + \frac{3\pi}{4} - 4$. Note that $f$ is positive and $F$ is increasing on $\mathbb{R}$. Also, $f$ has smaller values where the slopes of the tangent lines of $F$ are smaller.

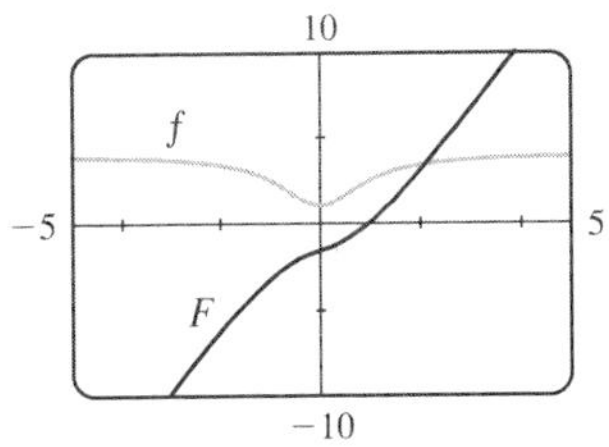


**15.** $f''(x) = 6x + 12x^2 \quad\Rightarrow\quad f'(x) = 6\cdot\dfrac{x^2}{2} + 12\cdot\dfrac{x^3}{3} + C = 3x^2 + 4x^3 + C \quad\Rightarrow$

$f(x) = 3\cdot\dfrac{x^3}{3} + 4\cdot\dfrac{x^4}{4} + Cx + D = x^3 + x^4 + Cx + D$ [$C$ and $D$ are just arbitrary constants]

**16.** $f''(x) = 2 + x^3 + x^6 \quad\Rightarrow\quad f'(x) = 2x + \frac{1}{4}x^4 + \frac{1}{7}x^7 + C \quad\Rightarrow\quad f(x) = x^2 + \frac{1}{20}x^5 + \frac{1}{56}x^8 + Cx + D$

**17.** $f''(x) = 1 + x^{4/5} \quad\Rightarrow\quad f'(x) = x + \frac{5}{9}x^{9/5} + C \quad\Rightarrow$

$f(x) = \frac{1}{2}x^2 + \frac{5}{9}\cdot\frac{5}{14}x^{14/5} + Cx + D = \frac{1}{2}x^2 + \frac{25}{126}x^{14/5} + Cx + D$

**18.** $f''(x) = \cos x \quad\Rightarrow\quad f'(x) = \sin x + C \quad\Rightarrow\quad f(x) = -\cos x + Cx + D$

**19.** $f'(x) = \sqrt{x}(6 + 5x) = 6x^{1/2} + 5x^{3/2} \quad\Rightarrow\quad f(x) = 4x^{3/2} + 2x^{5/2} + C.$

$f(1) = 6 + C$ and $f(1) = 10 \quad\Rightarrow\quad C = 4$, so $f(x) = 4x^{3/2} + 2x^{5/2} + 4.$

**20.** $f'(x) = 2x - 3/x^4 = 2x - 3x^{-4} \quad\Rightarrow\quad f(x) = x^2 + x^{-3} + C$ because we're given that $x > 0$.

$f(1) = 2 + C$ and $f(1) = 3 \quad\Rightarrow\quad C = 1$, so $f(x) = x^2 + 1/x^3 + 1.$

**21.** $f'(t) = 2\cos t + \sec^2 t \quad\Rightarrow\quad f(t) = 2\sin t + \tan t + C$ because $-\pi/2 < t < \pi/2$.

$f\left(\frac{\pi}{3}\right) = 2\left(\sqrt{3}/2\right) + \sqrt{3} + C = 2\sqrt{3} + C$ and $f\left(\frac{\pi}{3}\right) = 4 \quad\Rightarrow\quad C = 4 - 2\sqrt{3}$, so $f(t) = 2\sin t + \tan t + 4 - 2\sqrt{3}.$

**22.** $f'(x) = 4/\sqrt{1-x^2} \quad\Rightarrow\quad f(x) = 4\sin^{-1} x + C.\quad f\left(\frac{1}{2}\right) = 4\sin^{-1}\left(\frac{1}{2}\right) + C = 4\cdot\frac{\pi}{6} + C$ and $f\left(\frac{1}{2}\right) = 1 \quad\Rightarrow$

$\frac{2\pi}{3} + C = 1 \quad\Rightarrow\quad C = 1 - \frac{2\pi}{3}$, so $f(x) = 4\sin^{-1} x + 1 - \frac{2\pi}{3}.$

**23.** $f''(x) = 24x^2 + 2x + 10 \quad\Rightarrow\quad f'(x) = 8x^3 + x^2 + 10x + C.\ f'(1) = 8 + 1 + 10 + C$ and $f'(1) = -3 \quad\Rightarrow$

$19 + C = -3 \quad\Rightarrow\quad C = -22$, so $f'(x) = 8x^3 + x^2 + 10x - 22$ and hence, $f(x) = 2x^4 + \frac{1}{3}x^3 + 5x^2 - 22x + D.$

$f(1) = 2 + \frac{1}{3} + 5 - 22 + D$ and $f(1) = 5 \quad\Rightarrow\quad D = 22 - \frac{7}{3} = \frac{59}{3}$, so $f(x) = 2x^4 + \frac{1}{3}x^3 + 5x^2 - 22x + \frac{59}{3}.$

**24.** $f''(x) = 4 - 6x - 40x^3 \quad\Rightarrow\quad f'(x) = 4x - 3x^2 - 10x^4 + C.\ f'(0) = C$ and $f'(0) = 1 \quad\Rightarrow\quad C = 1$, so

$f'(x) = 4x - 3x^2 - 10x^4 + 1$ and hence, $f(x) = 2x^2 - x^3 - 2x^5 + x + D.\ f(0) = D$ and $f(0) = 2 \quad\Rightarrow\quad D = 2$, so

$f(x) = 2x^2 - x^3 - 2x^5 + x + 2.$

**25.** $f''(\theta) = \sin\theta + \cos\theta \quad\Rightarrow\quad f'(\theta) = -\cos\theta + \sin\theta + C$. $f'(0) = -1 + C$ and $f'(0) = 4 \quad\Rightarrow\quad C = 5$, so $f'(\theta) = -\cos\theta + \sin\theta + 5$ and hence, $f(\theta) = -\sin\theta - \cos\theta + 5\theta + D$. $f(0) = -1 + D$ and $f(0) = 3 \quad\Rightarrow\quad D = 4$, so $f(\theta) = -\sin\theta - \cos\theta + 5\theta + 4$.

**26.** $f''(t) = 3/\sqrt{t} = 3t^{-1/2} \quad\Rightarrow\quad f'(t) = 6t^{1/2} + C$. $f'(4) = 12 + C$ and $f'(4) = 7 \quad\Rightarrow\quad C = -5$, so $f'(t) = 6t^{1/2} - 5$ and hence, $f(t) = 4t^{3/2} - 5t + D$. $f(4) = 32 - 20 + D$ and $f(4) = 20 \quad\Rightarrow\quad D = 8$, so $f(t) = 4t^{3/2} - 5t + 8$.

**27.** $f''(x) = x^{-2}, x > 0 \quad\Rightarrow\quad f'(x) = -1/x + C \quad\Rightarrow\quad f(x) = -\ln|x| + Cx + D = -\ln x + Cx + D$ (since $x > 0$). $f(1) = 0 \quad\Rightarrow\quad C + D = 0$ and $f(2) = 0 \quad\Rightarrow\quad -\ln 2 + 2C + D = 0 \quad\Rightarrow$ $-\ln 2 + 2C - C = 0$ [since $D = -C$] $\quad\Rightarrow\quad -\ln 2 + C = 0 \quad\Rightarrow\quad C = \ln 2$ and $D = -\ln 2$. So $f(x) = -\ln x + (\ln 2)x - \ln 2$.

**28.** $f''(t) = 2e^t + 3\sin t \quad\Rightarrow\quad f'(t) = 2e^t - 3\cos t + C \quad\Rightarrow\quad f(t) = 2e^t - 3\sin t + Ct + D$. $f(0) = 2 + D$ and $f(0) = 0 \quad\Rightarrow\quad D = -2$. $f(\pi) = 2e^\pi + \pi C - 2$ and $f(\pi) = 0 \quad\Rightarrow\quad \pi C = 2 - 2e^\pi \quad\Rightarrow\quad C = \dfrac{2 - 2e^\pi}{\pi}$, so $f(t) = 2e^t - 3\sin t + \dfrac{2 - 2e^\pi}{\pi}t - 2$.

**29.** Given $f'(x) = 2x + 1$, we have $f(x) = x^2 + x + C$. Since $f$ passes through $(1, 6)$, $f(1) = 6 \quad\Rightarrow\quad 1^2 + 1 + C = 6 \quad\Rightarrow$ $C = 4$. Therefore, $f(x) = x^2 + x + 4$ and $f(2) = 2^2 + 2 + 4 = 10$.

**30.** $f'(x) = x^3 \quad\Rightarrow\quad f(x) = \frac{1}{4}x^4 + C$. $x + y = 0 \quad\Rightarrow\quad y = -x \quad\Rightarrow\quad m = -1$. Now $m = f'(x) \quad\Rightarrow\quad -1 = x^3 \quad\Rightarrow$ $x = -1 \quad\Rightarrow\quad y = 1$ (from the equation of the tangent line), so $(-1, 1)$ is a point on the graph of $f$. From $f$, $1 = \frac{1}{4}(-1)^4 + C \quad\Rightarrow\quad C = \frac{3}{4}$. Therefore, the function is $f(x) = \frac{1}{4}x^4 + \frac{3}{4}$.

**31.**

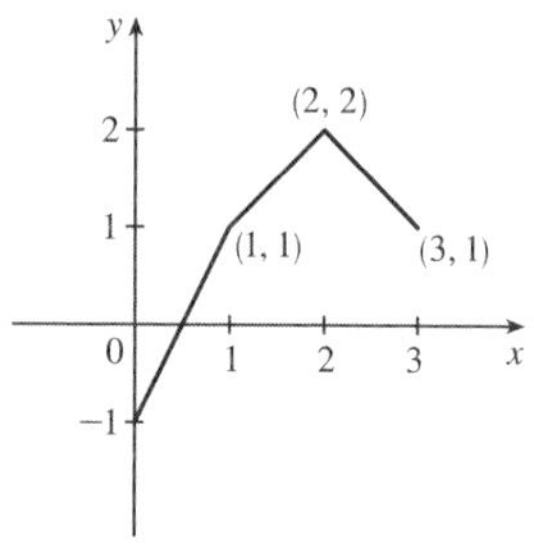


$$f'(x) = \begin{cases} 2 & \text{if } 0 \le x < 1 \\ 1 & \text{if } 1 < x < 2 \\ -1 & \text{if } 2 < x \le 3 \end{cases} \quad\Rightarrow\quad f(x) = \begin{cases} 2x + C & \text{if } 0 \le x < 1 \\ x + D & \text{if } 1 < x < 2 \\ -x + E & \text{if } 2 < x \le 3 \end{cases}$$

$f(0) = -1 \quad\Rightarrow\quad 2(0) + C = -1 \quad\Rightarrow\quad C = -1$. Starting at the point $(0, -1)$ and moving to the right on a line with slope 2 gets us to the point $(1, 1)$. The slope for $1 < x < 2$ is 1, so we get to the point $(2, 2)$. Here we have used the fact that $f$ is continuous. We can include the point $x = 1$ on either the first or the second part of $f$. The line connecting $(1, 1)$ to $(2, 2)$ is $y = x$, so $D = 0$. The slope for $2 < x \le 3$ is $-1$, so we get to $(3, 1)$. $f(3) = 1 \quad\Rightarrow\quad -3 + E = 1 \quad\Rightarrow\quad E = 4$. Thus,

$$f(x) = \begin{cases} 2x - 1 & \text{if } 0 \le x \le 1 \\ x & \text{if } 1 < x < 2 \\ -x + 4 & \text{if } 2 \le x \le 3 \end{cases}$$

Note that $f'(x)$ does not exist at $x = 1$ or at $x = 2$.

**32.** (a)

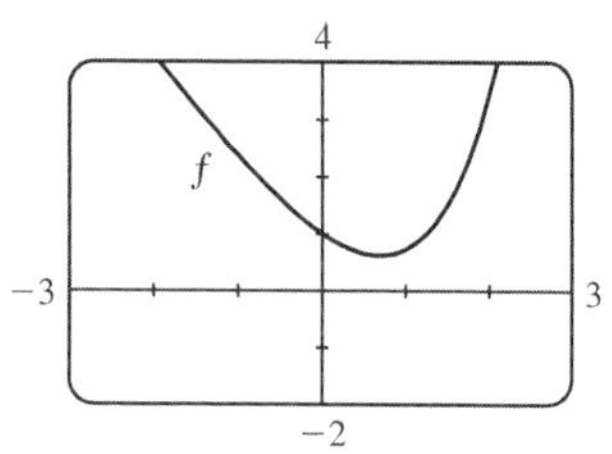


(b)

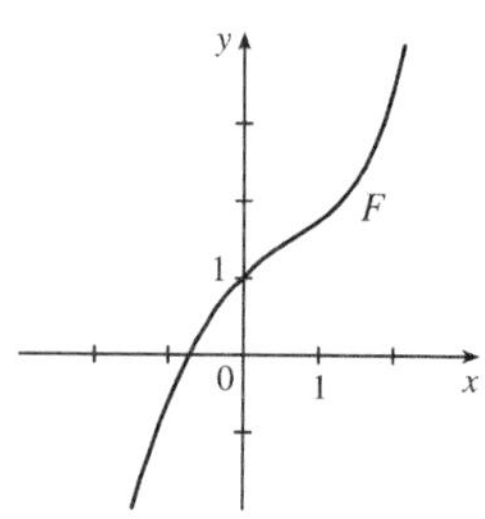


Since $f$ is always positive, $F$ is always increasing.

(c) $f(x) = e^x - 2x \quad \Rightarrow \quad F(x) = e^x - x^2 + C.\ F(0) = 1 \quad \Rightarrow \quad 1 - 0 + C = 1 \quad \Rightarrow \quad C = 0,$
so $F(x) = e^x - x^2$.

(d) If we use a graphing device to graph $F(x) = e^x - x^2$, we see that the graph looks similar to the one in part (b).

**33.**

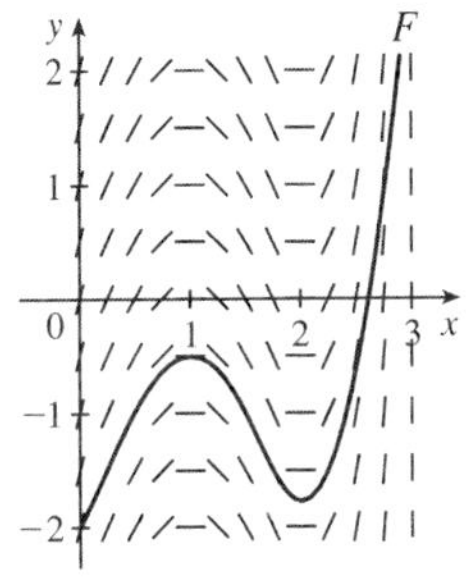


**34.**

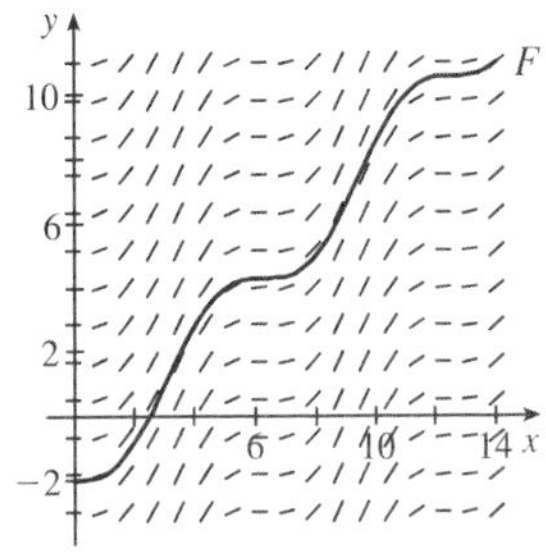


**35.**

| $x$ | $f(x)$ |
|---|---|
| 0 | 1 |
| 0.5 | 0.959 |
| 1.0 | 0.841 |
| 1.5 | 0.665 |
| 2.0 | 0.455 |
| 2.5 | 0.239 |
| 3.0 | 0.047 |

| $x$ | $f(x)$ |
|---|---|
| 3.5 | −0.100 |
| 4.0 | −0.189 |
| 4.5 | −0.217 |
| 5.0 | −0.192 |
| 5.5 | −0.128 |
| 6.0 | −0.047 |

We compute slopes [values of $f(x) = (\sin x)/x$ for $0 < x < 2\pi$] as in the table [$\lim_{x \to 0^+} f(x) = 1$] and draw a direction field as in Example 6. Then we use the direction field to graph $F$ starting at $(0, 0)$.

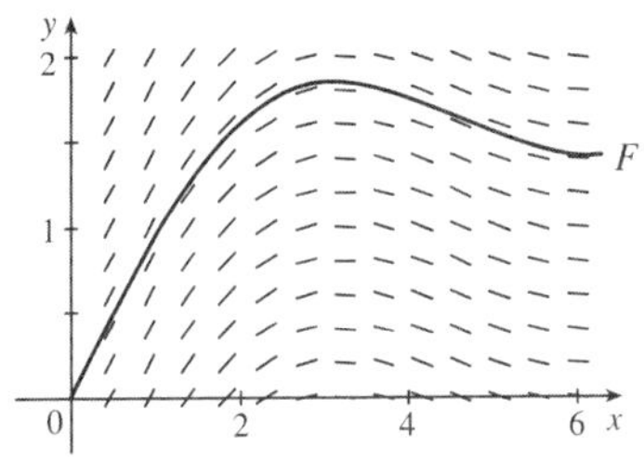


**36.**

| $x$ | $f(x)$ |
|---|---|
| 0 | 0 |
| $\pm 0.2$ | 0.041 |
| $\pm 0.4$ | 0.169 |
| $\pm 0.6$ | 0.410 |
| $\pm 0.8$ | 0.824 |
| $\pm 1.0$ | 1.557 |
| $\pm 1.2$ | 3.087 |
| $\pm 1.4$ | 8.117 |
| $\pm 1.5$ | 21.152 |

We compute slopes [values of $f(x) = x \tan x$ for $-\pi/2 < x < \pi/2$] as in the table and draw a direction field as in Example 6. Then we use the direction field to graph $F$ starting at $(0, 0)$ and extending in both directions. Note that if $f$ is an even function, then the antiderivative $F$ that passes through the origin is an odd function.

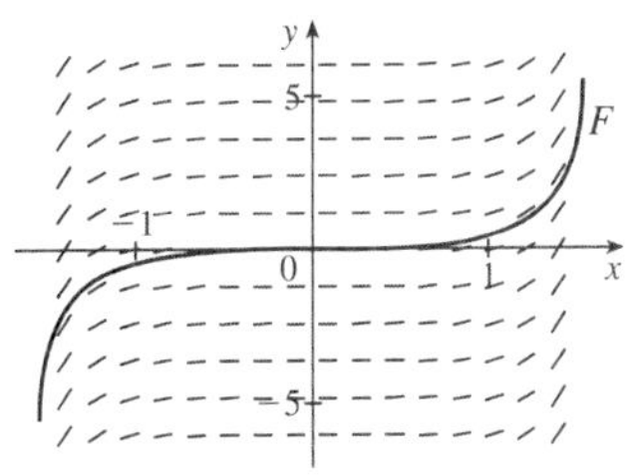


**37.**

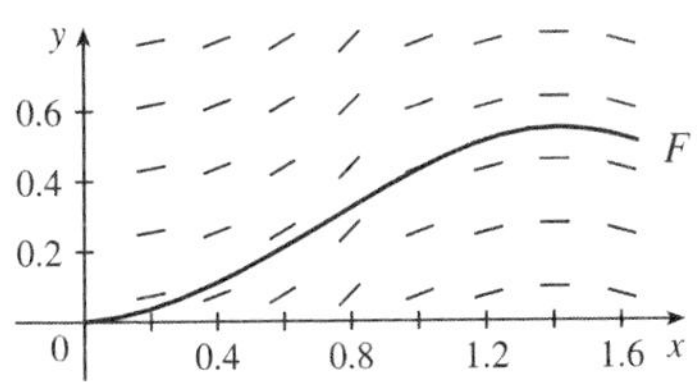


Remember that the given table values of $f$ are the slopes of $F$ at any $x$. For example, at $x = 1.4$, the slope of $F$ is $f(1.4) = 0$.

**38.** (a)

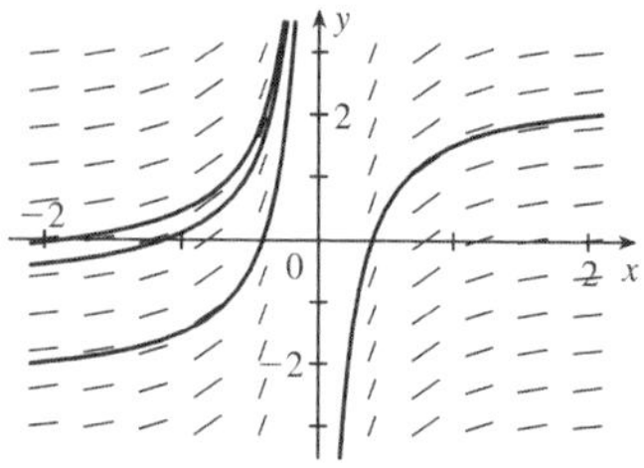

(b) The general antiderivative of $f(x) = x^{-2}$ is

$$F(x) = \begin{cases} -1/x + C_1 & \text{if } x < 0 \\ -1/x + C_2 & \text{if } x > 0 \end{cases}$$ since $f(x)$ is not defined at $x = 0$.

The graph of the general antiderivatives of $f(x)$ looks like the graph in part (a), as expected.

**39.** $v(t) = s'(t) = \sin t - \cos t \;\Rightarrow\; s(t) = -\cos t - \sin t + C$. $s(0) = -1 + C$ and $s(0) = 0 \;\Rightarrow\; C = 1$, so $s(t) = -\cos t - \sin t + 1$.

**40.** $a(t) = v'(t) = 5 + 4t - 2t^2 \;\Rightarrow\; v(t) = 5t + 2t^2 - \frac{2}{3}t^3 + C$. $v(0) = 3 \;\Rightarrow\; C = 3$, so $v(t) = 5t + 2t^2 - \frac{2}{3}t^3 + 3$. $v(t) = s'(t) \;\Rightarrow\; s(t) = \frac{5}{2}t^2 + \frac{2}{3}t^3 - \frac{1}{6}t^4 + 3t + D$. $s(0) = 10 \;\Rightarrow\; D = 10$, so the particle's position after $t$ seconds is given by $s(t) = \frac{5}{2}t^2 + \frac{2}{3}t^3 - \frac{1}{6}t^4 + 3t + 10$.

**41.** (a) We first observe that since the stone is dropped 450 m above the ground, $v(0) = 0$ and $s(0) = 450$. $v'(t) = a(t) = -9.8 \;\Rightarrow\; v(t) = -9.8t + C$. Now $v(0) = 0 \;\Rightarrow\; C = 0$, so $v(t) = -9.8t \;\Rightarrow\; s(t) = -4.9t^2 + D$. Last, $s(0) = 450 \;\Rightarrow\; D = 450 \;\Rightarrow\; s(t) = 450 - 4.9t^2$.

(b) The stone reaches the ground when $s(t) = 0$. $450 - 4.9t^2 = 0 \;\Rightarrow\; t^2 = 450/4.9 \;\Rightarrow\; t_1 = \sqrt{450/4.9} \approx 9.58$ s.

(c) The velocity with which the stone strikes the ground is $v(t_1) = -9.8\sqrt{450/4.9} \approx -93.9$ m/s.

(d) This is just reworking parts (a) and (b) with $v(0) = -5$. Using $v(t) = -9.8t + C$, $v(0) = -5 \;\Rightarrow\; 0 + C = -5 \;\Rightarrow\; v(t) = -9.8t - 5$. So $s(t) = -4.9t^2 - 5t + D$ and $s(0) = 450 \;\Rightarrow\; D = 450 \;\Rightarrow\; s(t) = -4.9t^2 - 5t + 450$. Solving $s(t) = 0$ by using the quadratic formula gives us $t = \left(5 \pm \sqrt{8845}\,\right)/(-9.8) \;\Rightarrow\; t_1 \approx 9.09$ s.

**42.** $v'(t) = a(t) = a \;\Rightarrow\; v(t) = at + C$ and $v_0 = v(0) = C \;\Rightarrow\; v(t) = at + v_0 \;\Rightarrow\; s(t) = \frac{1}{2}at^2 + v_0 t + D \;\Rightarrow\; s_0 = s(0) = D \;\Rightarrow\; s(t) = \frac{1}{2}at^2 + v_0 t + s_0$

**43.** By Exercise 42 with $a = -9.8$, $s(t) = -4.9t^2 + v_0 t + s_0$ and $v(t) = s'(t) = -9.8t + v_0$. So $[v(t)]^2 = (-9.8t + v_0)^2 = (9.8)^2 t^2 - 19.6 v_0 t + v_0^2 = v_0^2 + 96.04t^2 - 19.6 v_0 t = v_0^2 - 19.6\left(-4.9t^2 + v_0 t\right)$. But $-4.9t^2 + v_0 t$ is just $s(t)$ without the $s_0$ term; that is, $s(t) - s_0$. Thus, $[v(t)]^2 = v_0^2 - 19.6\,[s(t) - s_0]$.

**44.** For the first ball, $s_1(t) = -16t^2 + 48t + 432$ from Example 7. For the second ball, $a(t) = -32 \;\Rightarrow\; v(t) = -32t + C$, but $v(1) = -32(1) + C = 24 \;\Rightarrow\; C = 56$, so $v(t) = -32t + 56 \;\Rightarrow\; s(t) = -16t^2 + 56t + D$, but $s(1) = -16(1)^2 + 56(1) + D = 432 \;\Rightarrow\; D = 392$, and $s_2(t) = -16t^2 + 56t + 392$. The balls pass each other when $s_1(t) = s_2(t) \;\Rightarrow\; -16t^2 + 48t + 432 = -16t^2 + 56t + 392 \;\Leftrightarrow\; 8t = 40 \;\Leftrightarrow\; t = 5$ s.

*Another solution:* From Exercise 42, we have $s_1(t) = -16t^2 + 48t + 432$ and $s_2(t) = -16t^2 + 24t + 432$. We now want to solve $s_1(t) = s_2(t - 1) \;\Rightarrow\; -16t^2 + 48t + 432 = -16(t-1)^2 + 24(t-1) + 432 \;\Rightarrow\; 48t = 32t - 16 + 24t - 24 \;\Rightarrow\; 40 = 8t \;\Rightarrow\; t = 5$ s.

**45.** Marginal cost $= 1.92 - 0.002x = C'(x) \;\Rightarrow\; C(x) = 1.92x - 0.001x^2 + K$. But $C(1) = 1.92 - 0.001 + K = 562 \;\Rightarrow\; K = 560.081$. Therefore, $C(x) = 1.92x - 0.001x^2 + 560.081 \;\Rightarrow\; C(100) = 742.081$, so the cost of producing 100 items is \$742.08.

**46.** Let the mass, measured from one end, be $m(x)$. Then $m(0) = 0$ and $\rho = \dfrac{dm}{dx} = x^{-1/2} \Rightarrow m(x) = 2x^{1/2} + C$ and $m(0) = C = 0$, so $m(x) = 2\sqrt{x}$. Thus, the mass of the 100-centimeter rod is $m(100) = 2\sqrt{100} = 20$ g.

**47.** Using Exercise 42 with $a = -32$, $v_0 = 0$, and $s_0 = h$ (the height of the cliff), we know that the height at time $t$ is $s(t) = -16t^2 + h$. $v(t) = s'(t) = -32t$ and $v(t) = -120 \Rightarrow -32t = -120 \Rightarrow t = 3.75$, so $0 = s(3.75) = -16(3.75)^2 + h \Rightarrow h = 16(3.75)^2 = 225$ ft.

**48.** $v'(t) = a(t) = -22$. The initial velocity is 50 mi/h $= \frac{50 \cdot 5280}{3600} = \frac{220}{3}$ ft/s, so $v(t) = -22t + \frac{220}{3}$. The car stops when $v(t) = 0 \Leftrightarrow t = \frac{220}{3 \cdot 22} = \frac{10}{3}$. Since $s(t) = -11t^2 + \frac{220}{3}t$, the distance covered is $s\left(\frac{10}{3}\right) = -11\left(\frac{10}{3}\right)^2 + \frac{220}{3} \cdot \frac{10}{3} = \frac{1100}{9} = 122.\overline{2}$ ft.

**49.** $a(t) = k$, the initial velocity is 30 mi/h $= 30 \cdot \frac{5280}{3600} = 44$ ft/s, and the final velocity (after 5 seconds) is 50 mi/h $= 50 \cdot \frac{5280}{3600} = \frac{220}{3}$ ft/s. So $v(t) = kt + C$ and $v(0) = 44 \Rightarrow C = 44$. Thus, $v(t) = kt + 44 \Rightarrow v(5) = 5k + 44$. But $v(5) = \frac{220}{3}$, so $5k + 44 = \frac{220}{3} \Rightarrow 5k = \frac{88}{3} \Rightarrow k = \frac{88}{15} \approx 5.87$ ft/s$^2$.

**50.** $a(t) = -16 \Rightarrow v(t) = -16t + v_0$ where $v_0$ is the car's speed (in ft/s) when the brakes were applied. The car stops when $-16t + v_0 = 0 \Leftrightarrow t = \frac{1}{16}v_0$. Now $s(t) = \frac{1}{2}(-16)t^2 + v_0 t = -8t^2 + v_0 t$. The car travels 200 ft in the time that it takes to stop, so $s\left(\frac{1}{16}v_0\right) = 200 \Rightarrow 200 = -8\left(\frac{1}{16}v_0\right)^2 + v_0\left(\frac{1}{16}v_0\right) = \frac{1}{32}v_0^2 \Rightarrow v_0^2 = 32 \cdot 200 = 6400 \Rightarrow v_0 = 80$ ft/s ($54.\overline{54}$ mi/h).

**51.** Let the acceleration be $a(t) = k$ km/h$^2$. We have $v(0) = 100$ km/h and we can take the initial position $s(0)$ to be 0. We want the time $t_f$ for which $v(t) = 0$ to satisfy $s(t) < 0.08$ km. In general, $v'(t) = a(t) = k$, so $v(t) = kt + C$, where $C = v(0) = 100$. Now $s'(t) = v(t) = kt + 100$, so $s(t) = \frac{1}{2}kt^2 + 100t + D$, where $D = s(0) = 0$. Thus, $s(t) = \frac{1}{2}kt^2 + 100t$. Since $v(t_f) = 0$, we have $kt_f + 100 = 0$ or $t_f = -100/k$, so

$s(t_f) = \dfrac{1}{2}k\left(-\dfrac{100}{k}\right)^2 + 100\left(-\dfrac{100}{k}\right) = 10{,}000\left(\dfrac{1}{2k} - \dfrac{1}{k}\right) = -\dfrac{5{,}000}{k}$. The condition $s(t_f)$ must satisfy is

$-\dfrac{5{,}000}{k} < 0.08 \Rightarrow -\dfrac{5{,}000}{0.08} > k$ [$k$ is negative] $\Rightarrow k < -62{,}500$ km/h$^2$, or equivalently, $k < -\frac{3125}{648} \approx -4.82$ m/s$^2$.

**52.** (a) $EIy'' = mg(L-x) + \frac{1}{2}\rho g(L-x)^2 \Rightarrow EIy' = -\frac{1}{2}mg(L-x)^2 - \frac{1}{6}\rho g(L-x)^3 + C \Rightarrow$
$EIy = \frac{1}{6}mg(L-x)^3 + \frac{1}{24}\rho g(L-x)^4 + Cx + D$. Since the left end of the board is fixed, we must have $y = y' = 0$ when $x = 0$. Thus, $0 = -\frac{1}{2}mgL^2 - \frac{1}{6}\rho gL^3 + C$ and $0 = \frac{1}{6}mgL^3 + \frac{1}{24}\rho gL^4 + D$. It follows that
$EIy = \frac{1}{6}mg(L-x)^3 + \frac{1}{24}\rho g(L-x)^4 + \left(\frac{1}{2}mgL^2 + \frac{1}{6}\rho gL^3\right)x - \left(\frac{1}{6}mgL^3 + \frac{1}{24}\rho gL^4\right)$ and

$$f(x) = y = \frac{1}{EI}\left[\tfrac{1}{6}mg(L-x)^3 + \tfrac{1}{24}\rho g(L-x)^4 + \left(\tfrac{1}{2}mgL^2 + \tfrac{1}{6}\rho gL^3\right)x - \left(\tfrac{1}{6}mgL^3 + \tfrac{1}{24}\rho gL^4\right)\right]$$

(b) $f(L) < 0$, so the end of the board is a *distance* approximately $-f(L)$ below the horizontal. From our result in (a), we calculate

$$-f(L) = \frac{-1}{EI}\left[\tfrac{1}{2}mgL^3 + \tfrac{1}{6}\rho gL^4 - \tfrac{1}{6}mgL^3 - \tfrac{1}{24}\rho gL^4\right] = \frac{-1}{EI}\left(\tfrac{1}{3}mgL^3 + \tfrac{1}{8}\rho gL^4\right) = -\frac{gL^3}{EI}\left(\frac{m}{3} + \frac{\rho L}{8}\right)$$

*Note:* This is positive because $g$ is negative.

**53.** (a) The Mean Value Theorem says that there exists a number $c$ in the interval $(x_1, x_2)$ such that $H'(c) = \dfrac{H(x_2) - H(x_1)}{x_2 - x_1}$.

Since $H = G - F$ and $G$ and $F$ are antiderivatives of $f$, $H'(c) = G'(c) - F'(c) = f(c) - f(c) = 0$. So now $\dfrac{H(x_2) - H(x_1)}{x_2 - x_1} = 0 \quad \Rightarrow \quad H(x_2) - H(x_1) = 0 \; (x_2 \neq x_1) \quad \Rightarrow \quad H(x_2) = H(x_1)$. Since this is true for any $x_1 < x_2$ in $I$, $H$ must be a constant function.

(b) We have $H = G - F$ and $H(x) = C$, so $C = G - F \quad \Rightarrow \quad G(x) = F(x) + C$. Thus, any antiderivative $G$ can be expressed as $F(x) + C$.

**54.** Taking the upward direction to be positive we have that for $0 \leq t \leq 10$ (using the subscript 1 to refer to $0 \leq t \leq 10$), $a_1(t) = -(9 - 0.9t) = v_1'(t) \quad \Rightarrow \quad v_1(t) = -9t + 0.45t^2 + v_0$, but $v_1(0) = v_0 = -10 \quad \Rightarrow$ $v_1(t) = -9t + 0.45t^2 - 10 = s_1'(t) \quad \Rightarrow \quad s_1(t) = -\frac{9}{2}t^2 + 0.15t^3 - 10t + s_0$. But $s_1(0) = 500 = s_0 \quad \Rightarrow$ $s_1(t) = -\frac{9}{2}t^2 + 0.15t^3 - 10t + 500$. $s_1(10) = -450 + 150 - 100 + 500 = 100$, so it takes more than 10 seconds for the raindrop to fall. Now for $t > 10$, $a(t) = 0 = v'(t) \quad \Rightarrow$ $v(t) = \text{constant} = v_1(10) = -9(10) + 0.45(10)^2 - 10 = -55 \quad \Rightarrow \quad v(t) = -55$.

At 55 ft/s, it will take $100/55 \approx 1.8$ s to fall the last 100 ft. Hence, the total time is $10 + \frac{100}{55} = \frac{130}{11} \approx 11.8$ s.

**55.** (a) First note that $90 \text{ mi/h} = 90 \times \frac{5280}{3600} \text{ ft/s} = 132 \text{ ft/s}$. Then $a(t) = 4 \text{ ft/s}^2 \quad \Rightarrow \quad v(t) = 4t + C$, but $v(0) = 0 \quad \Rightarrow$ $C = 0$. Now $4t = 132$ when $t = \frac{132}{4} = 33$ s, so it takes 33 s to reach 132 ft/s. Therefore, taking $s(0) = 0$, we have $s(t) = 2t^2$, $0 \leq t \leq 33$. So $s(33) = 2178$ ft. 15 minutes $= 15(60) = 900$ s, so for $33 < t \leq 933$ we have $v(t) = 132 \text{ ft/s} \quad \Rightarrow \quad s(933) = 132(900) + 2178 = 120{,}978 \text{ ft} = 22.9125 \text{ mi}$.

(b) As in part (a), the train accelerates for 33 s and travels 2178 ft while doing so. Similarly, it decelerates for 33 s and travels 2178 ft at the end of its trip. During the remaining $900 - 66 = 834$ s it travels at 132 ft/s, so the distance traveled is $132 \cdot 834 = 110{,}088$ ft. Thus, the total distance is $2178 + 110{,}088 + 2178 = 114{,}444 \text{ ft} = 21.675$ mi.

(c) $45 \text{ mi} = 45(5280) = 237{,}600$ ft. Subtract $2(2178)$ to take care of the speeding up and slowing down, and we have 233,244 ft at 132 ft/s for a trip of $233{,}244/132 = 1767$ s at 90 mi/h. The total time is $1767 + 2(33) = 1833 \text{ s} = 30 \text{ min } 33 \text{ s} = 30.55$ min.

(d) $37.5(60) = 2250$ s. $2250 - 2(33) = 2184$ s at maximum speed. $2184(132) + 2(2178) = 292{,}644$ total feet or $292{,}644/5280 = 55.425$ mi.

**56.** (a) For $0 \leq t \leq 3$ we have $a(t) = 60t \quad \Rightarrow \quad v(t) = 30t^2 + C \quad \Rightarrow \quad v(0) = 0 = C \quad \Rightarrow \quad v(t) = 30t^2$, so $s(t) = 10t^3 + C \quad \Rightarrow \quad s(0) = 0 = C \quad \Rightarrow \quad s(t) = 10t^3$. Note that $v(3) = 270$ and $s(3) = 270$.

For $3 < t \leq 17$: $a(t) = -g = -32 \text{ ft/s} \quad \Rightarrow \quad v(t) = -32(t - 3) + C \quad \Rightarrow \quad v(3) = 270 = C \quad \Rightarrow$ $v(t) = -32(t - 3) + 270 \quad \Rightarrow \quad s(t) = -16(t - 3)^2 + 270(t - 3) + C \quad \Rightarrow \quad s(3) = 270 = C \quad \Rightarrow$ $s(t) = -16(t - 3)^2 + 270(t - 3) + 270$. Note that $v(17) = -178$ and $s(17) = 914$.

For $17 < t \leq 22$: The velocity increases linearly from $-178$ ft/s to $-18$ ft/s during this period, so $\dfrac{\Delta v}{\Delta t} = \dfrac{-18 - (-178)}{22 - 17} = \dfrac{160}{5} = 32$. Thus, $v(t) = 32(t - 17) - 178 \quad \Rightarrow \quad s(t) = 16(t - 17)^2 - 178(t - 17) + 914$ and $s(22) = 424$ ft.

For $t > 22$: $v(t) = -18 \quad \Rightarrow \quad s(t) = -18(t - 22) + C$. But $s(22) = 424 = C \quad \Rightarrow \quad s(t) = -18(t - 22) + 424$.

Therefore, until the rocket lands, we have

$$v(t) = \begin{cases} 30t^2 & \text{if } 0 \le t \le 3 \\ -32\,(t-3) + 270 & \text{if } 3 < t \le 17 \\ 32(t-17) - 178 & \text{if } 17 < t \le 22 \\ -18 & \text{if } t > 22 \end{cases}$$

and

$$s(t) = \begin{cases} 10t^3 & \text{if } 0 \le t \le 3 \\ -16(t-3)^2 + 270(t-3) + 270 & \text{if } 3 < t \le 17 \\ 16(t-17)^2 - 178\,(t-17) + 914 & \text{if } 17 < t \le 22 \\ -18(t-22) + 424 & \text{if } t > 22 \end{cases}$$

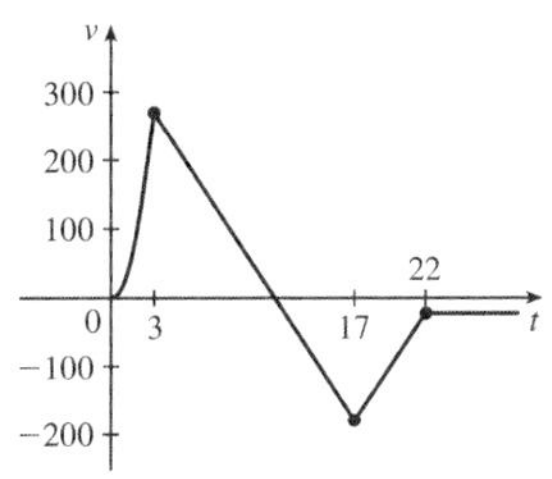


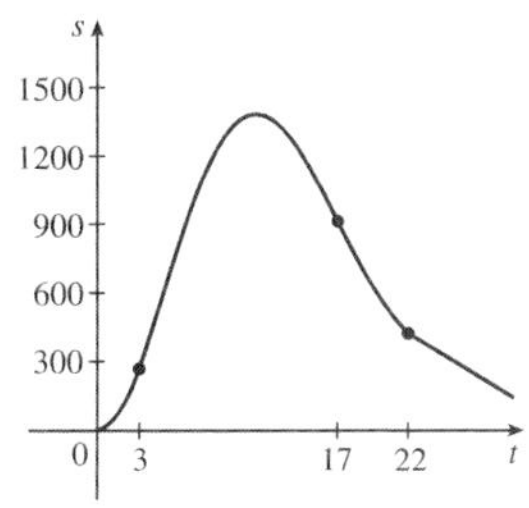


(b) To find the maximum height, set $v(t)$ on $3 < t \le 17$ equal to 0. $-32(t-3) + 270 = 0 \;\Rightarrow\; t_1 = 11.4375$ s and the maximum height is $s(t_1) = -16(t_1 - 3)^2 + 270(t_1 - 3) + 270 = 1409.0625$ ft.

(c) To find the time to land, set $s(t) = -18(t-22) + 424 = 0$. Then $t - 22 = \frac{424}{18} = 23.\overline{5}$, so $t \approx 45.6$ s.

## 4 Review

### CONCEPT CHECK

1. A function $f$ has an **absolute maximum** at $x = c$ if $f(c)$ is the largest function value on the entire domain of $f$, whereas $f$ has a **local maximum** at $c$ if $f(c)$ is the largest function value when $x$ is near $c$. See Figure 4 in Section 4.2.

2. (a) See Theorem 4.2.3.

   (b) See the Closed Interval Method before Example 6 in Section 4.2.

3. (a) See Theorem 4.2.4.

   (b) See Definition 4.2.5.

4. See the Mean Value Theorem in Section 4.3. Geometric interpretation—there is some point $P$ on the graph of a function $f$ [on the interval $(a, b)$] where the tangent line is parallel to the secant line that connects $(a, f(a))$ and $(b, f(b))$.

5. (a) See the I/D Test before Example 2 in Section 4.3.

   (b) A function $f$ is concave upward on an interval $I$ if $f'$ is an increasing function on $I$ (or, equivalently, the graph of $f$ lies above all of its tangent lines on $I$).

   (c) See the Concavity Test before Example 4 in Section 4.3.

   (d) An inflection point is a point where a curve changes its direction of concavity. They can be found by determining the points at which the second derivative changes sign.

6. (a) See the First Derivative Test after Example 2 in Section 4.3.

   (b) See the Second Derivative Test before Example 4 in Section 4.3.

   (c) See the note before Example 5 in Section 4.3.

**7.** (a) See l'Hospital's Rule and the three notes that follow it in Section 4.5.

(b) Write $fg$ as $\dfrac{f}{1/g}$ or $\dfrac{g}{1/f}$.

(c) Convert the difference into a quotient using a common denominator, rationalizing, factoring, or some other method.

(d) Convert the power to a product by taking the natural logarithm of both sides of $y = f^g$ or by writing $f^g$ as $e^{g \ln f}$.

**8.** Without calculus you could get misleading graphs that fail to show the most interesting features of a function. See the discussion following Figure 3 in Section 4.5 and the first paragraph in Section 4.4.

**9.** (a) See Figure 3 in Section 4.8.

(b) $x_2 = x_1 - \dfrac{f(x_1)}{f'(x_1)}$

(c) $x_{n+1} = x_n - \dfrac{f(x_n)}{f'(x_n)}$

(d) Newton's method is likely to fail or to work very slowly when $f'(x_1)$ is close to 0.

**10.** (a) See the definition at the beginning of Section 4.9.

(b) If $F_1$ and $F_2$ are both antiderivatives of $f$ on an interval $I$, then they differ by a constant.

## TRUE-FALSE QUIZ

**1.** False. For example, take $f(x) = x^3$, then $f'(x) = 3x^2$ and $f'(0) = 0$, but $f(0) = 0$ is not a maximum or minimum; $(0, 0)$ is an inflection point.

**2.** False. For example, $f(x) = |x|$ has an absolute minimum at 0, but $f'(0)$ does not exist.

**3.** False. For example, $f(x) = x$ is continuous on $(0, 1)$ but attains neither a maximum nor a minimum value on $(0, 1)$. Don't confuse this with $f$ being continuous on the *closed* interval $[a, b]$, which would make the statement true.

**4.** True. By the Mean Value Theorem, $f'(c) = \dfrac{f(1) - f(-1)}{1 - (-1)} = \dfrac{0}{2} = 0$. Note that $|c| < 1 \;\Leftrightarrow\; c \in (-1, 1)$.

**5.** True. This is an example of part (b) of the I/D Test.

**6.** False. For example, the curve $y = f(x) = 1$ has no inflection points but $f''(c) = 0$ for all $c$.

**7.** False. $f'(x) = g'(x) \;\Rightarrow\; f(x) = g(x) + C$. For example, if $f(x) = x + 2$ and $g(x) = x + 1$, then $f'(x) = g'(x) = 1$, but $f(x) \neq g(x)$.

**8.** False. Assume there is a function $f$ such that $f(1) = -2$ and $f(3) = 0$. Then by the Mean Value Theorem there exists a number $c \in (1, 3)$ such that $f'(c) = \dfrac{f(3) - f(1)}{3 - 1} = \dfrac{0 - (-2)}{2} = 1$. But $f'(x) > 1$ for all $x$, a contradiction.

**9.** True. The graph of one such function is sketched.

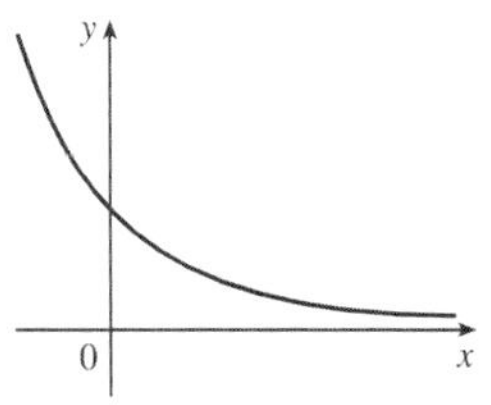


**10.** False. At any point $(a, f(a))$, we know that $f'(a) < 0$. So since the tangent line at $(a, f(a))$ is not horizontal, it must cross the $x$-axis—at $x = b$, say. But since $f''(x) > 0$ for all $x$, the graph of $f$ must lie above all of its tangents; in particular, $f(b) > 0$. But this is a contradiction, since we are given that $f(x) < 0$ for all $x$.

**11.** True. By the Mean Value Theorem, there exists a number $c$ in $(0, 1)$ such that $f(1) - f(0) = f'(c)(1 - 0) = f'(c)$. Since $f'(c)$ is nonzero, $f(1) - f(0) \neq 0$, so $f(1) \neq f(0)$.

**12.** False. The most general antiderivative is $F(x) = -1/x + C_1$ for $x < 0$ and $F(x) = -1/x + C_2$ for $x > 0$ (see Example 1 in Section 4.9).

**13.** False. $\lim\limits_{x\to0} \dfrac{x}{e^x} = \dfrac{\lim\limits_{x\to0} x}{\lim\limits_{x\to0} e^x} = \dfrac{0}{1} = 0$, not 1.

## EXERCISES

**1.** $f(x) = 10 + 27x - x^3$, $0 \le x \le 4$. $f'(x) = 27 - 3x^2 = -3(x^2 - 9) = -3(x+3)(x-3) = 0$ only when $x = 3$ (since $-3$ is not in the domain). $f'(x) > 0$ for $x < 3$ and $f'(x) < 0$ for $x > 3$, so $f(3) = 64$ is a local maximum value. Checking the endpoints, we find $f(0) = 10$ and $f(4) = 54$. Thus, $f(0) = 10$ is the absolute minimum value and $f(3) = 64$ is the absolute maximum value.

**2.** $f(x) = x - \sqrt{x}$, $0 \le x \le 4$. $f'(x) = 1 - 1/(2\sqrt{x}) = 0 \iff 2\sqrt{x} = 1 \Rightarrow x = \frac{1}{4}$. $f'(x)$ does not exist $\iff x = 0$. $f'(x) < 0$ for $0 < x < \frac{1}{4}$ and $f'(x) > 0$ for $\frac{1}{4} < x < 4$, so $f\left(\frac{1}{4}\right) = -\frac{1}{4}$ is a local and absolute minimum value. $f(0) = 0$ and $f(4) = 2$, so $f(4) = 2$ is the absolute maximum value.

**3.** $f(x) = \dfrac{x}{x^2 + x + 1}$, $-2 \le x \le 0$. $f'(x) = \dfrac{(x^2 + x + 1)(1) - x(2x + 1)}{(x^2 + x + 1)^2} = \dfrac{1 - x^2}{(x^2 + x + 1)^2} = 0 \iff x = -1$ (since 1 is not in the domain). $f'(x) < 0$ for $-2 < x < -1$ and $f'(x) > 0$ for $-1 < x < 0$, so $f(-1) = -1$ is a local and absolute minimum value. $f(-2) = -\frac{2}{3}$ and $f(0) = 0$, so $f(0) = 0$ is an absolute maximum value.

**4.** $f(x) = \dfrac{\ln x}{x^2}$, $[1, 3]$. $f'(x) = \dfrac{x^2 \cdot \frac{1}{x} - (\ln x)(2x)}{(x^2)^2} = \dfrac{x - 2x \ln x}{x^4} = \dfrac{1 - 2\ln x}{x^3} = 0 \iff \ln x = \frac{1}{2} \iff x = e^{1/2} = \sqrt{e} \approx 1.65$. $f'(x) > 0$ for $x < \sqrt{e}$ and $f'(x) < 0$ for $x > \sqrt{e}$, so $f$ is increasing on $(1, \sqrt{e})$ and decreasing on $(\sqrt{e}, 3)$. Hence, $f(\sqrt{e}) = \dfrac{1}{2e}$ is a local maximum value. $f(1) = 0$ and $f(3) = \dfrac{\ln 3}{9} \approx 0.12$. Since $\dfrac{1}{2e} \approx 0.18$, $f(\sqrt{e}) = \dfrac{1}{2e}$ is the absolute maximum value and $f(1) = 0$ is the absolute minimum value.

**5.** (a) $f(x) = 2 - 2x - x^3$ is a polynomial, so there is no asymptote.

(b) $f'(x) = -2 - 3x^2 = -1(3x^2 + 2) < 0$, so $f$ is decreasing on $\mathbb{R}$.

(c) No local extrema

(d) $f''(x) = -6x < 0$ on $(0, \infty)$ and $f''(x) > 0$ on $(-\infty, 0)$, so $f$ is CD on $(0, \infty)$ and CU on $(-\infty, 0)$. There is an IP at $(0, 2)$.

(e)

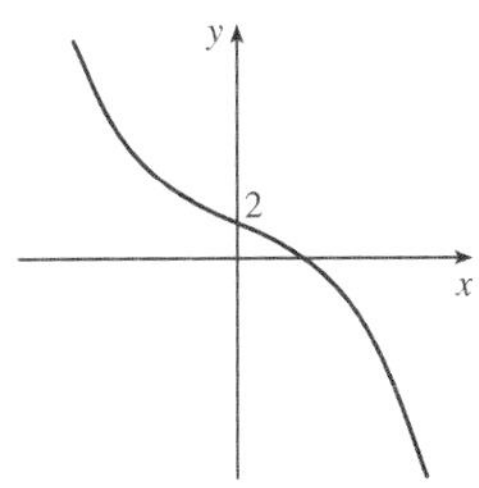


**6.** (a) $f(x) = x^4 + 4x^3$ is a polynomial, so there is no asymptote.

(b) $f'(x) = 4x^3 + 12x^2 = 4x^2(x + 3) > 0 \iff x > -3$, so $f$ is increasing on $(-3, \infty)$ and decreasing on $(-\infty, -3)$.

(c) Local minimum $f(-3) = -27$, no local maximum

(d) $f''(x) = 12x^2 + 24x = 12x(x + 2) < 0 \iff -2 < x < 0$, so $f$ is CD on $(-2, 0)$ and CU on $(-\infty, -2)$ and $(0, \infty)$. IP at $(0, 0)$ and $(-2, -16)$.

(e)

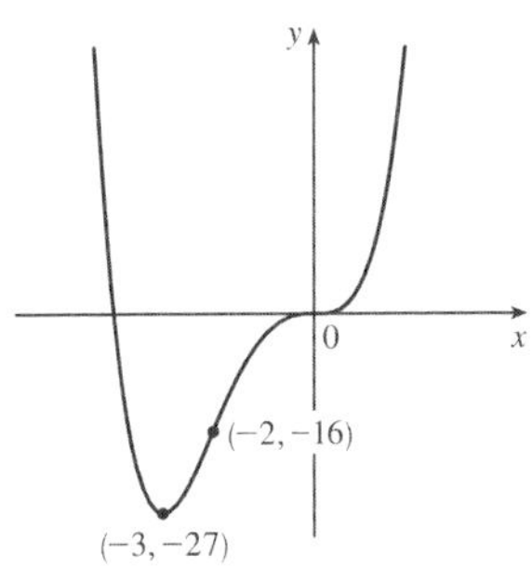

**7.** (a) $f(x) = x + \sqrt{1-x}$ has no asymptote.

(b) $f'(x) = 1 - 1/\left(2\sqrt{1-x}\right) = 0 \quad \Leftrightarrow \quad 2\sqrt{1-x} = 1 \quad \Leftrightarrow$ $1 - x = \frac{1}{4} \quad \Leftrightarrow \quad x = \frac{3}{4}$ and $f'(x) > 0 \quad \Leftrightarrow \quad x < \frac{3}{4}$, so $f$ is increasing on $\left(-\infty, \frac{3}{4}\right)$ and decreasing on $\left(\frac{3}{4}, 1\right)$.

(c) $f\left(\frac{3}{4}\right) = \frac{3}{4} + \sqrt{1 - \frac{3}{4}} = \frac{3}{4} + \sqrt{\frac{1}{4}} = \frac{3}{4} + \frac{1}{2} = \frac{5}{4}$ is a local maximum.

(d) $f''(x) = -\dfrac{1}{4(1-x)^{3/2}} < 0$ on the domain of $f$, so $f$ is CD on $(-\infty, 1)$. No IP

(e)

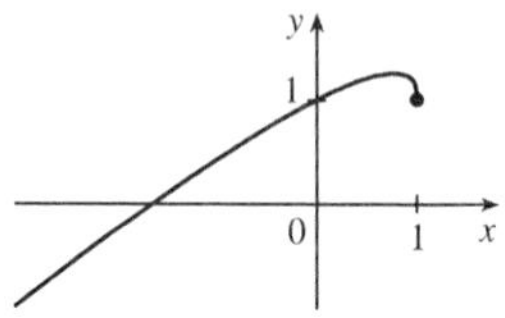


**8.** (a) $f(x) = \dfrac{1}{1-x^2} = \dfrac{1}{(1+x)(1-x)}$ has vertical asymptotes $x = \pm 1$. $\lim\limits_{x \to \pm\infty} f(x) = 0$, so $f$ has a horizontal asymptote of $y = 0$.

(b) $f'(x) = \dfrac{2x}{(1-x^2)^2} = 0 \quad \Leftrightarrow \quad x = 0$, so $f$ is decreasing on $(-\infty, -1)$ and $(-1, 0)$, and increasing on $(0, 1)$ and $(1, \infty)$.

(c) Local minimum $f(0) = 1$; no local maximum

(d) $f''(x) = \dfrac{(1-x^2)^2 \cdot 2 - 2x \cdot 2(1-x^2)(-2x)}{(1-x^2)^4} = \dfrac{2(1-x^2)^2 + 8x^2}{(1-x^2)^3} =$ $\dfrac{6x^2+2}{(1-x^2)^3} < 0 \quad \Rightarrow \quad x^2 > 1$, so $f$ is CD on $(-\infty, -1)$ and $(1, \infty)$, and CU on $(-1, 1)$. No IP

(e)

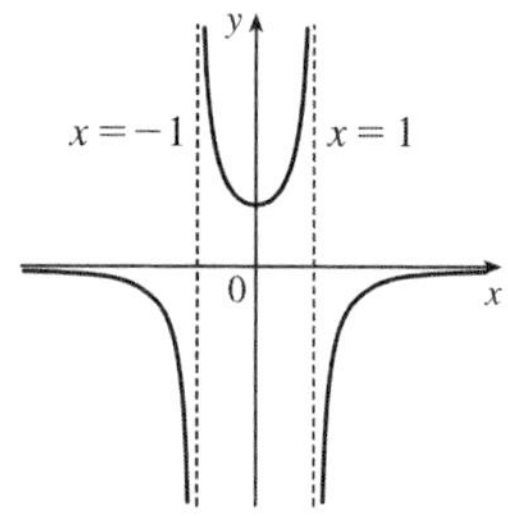


**9.** (a) $y = f(x) = \sin^2 x - 2\cos x$ has no asymptote.

(b) $y' = 2\sin x\,\cos x + 2\sin x = 2\sin x\,(\cos x + 1)$. $y' = 0 \quad \Leftrightarrow \quad \sin x = 0$ or $\cos x = -1 \quad \Leftrightarrow \quad x = n\pi$ or $x = (2n+1)\pi$. $y' > 0$ when $\sin x > 0$, since $\cos x + 1 \geq 0$ for all $x$. Therefore, $y' > 0$ (and so $f$ is increasing) on $(2n\pi, (2n+1)\pi)$; $y' < 0$ (and so $f$ is decreasing) on $((2n-1)\pi, 2n\pi)$ or equivalently, $((2n+1)\pi, (2n+2)\pi)$.

(c) Local maxima are $f((2n+1)\pi) = 2$; local minima are $f(2n\pi) = -2$.

(d) $y' = \sin 2x + 2\sin x \quad \Rightarrow$

$$y'' = 2\cos 2x + 2\cos x = 2\left(2\cos^2 x - 1\right) + 2\cos x = 4\cos^2 x + 2\cos x - 2$$
$$= 2\left(2\cos^2 x + \cos x - 1\right) = 2(2\cos x - 1)(\cos x + 1)$$

$y'' = 0 \quad \Leftrightarrow \quad \cos x = \frac{1}{2}$ or $-1 \quad \Leftrightarrow \quad x = 2n\pi \pm \frac{\pi}{3}$ or $x = (2n+1)\pi$. $y'' > 0$ (and so $f$ is CU) on $\left(2n\pi - \frac{\pi}{3}, 2n\pi + \frac{\pi}{3}\right)$; $y'' \leq 0$ (and so $f$ is CD) on $\left(2n\pi + \frac{\pi}{3}, 2n\pi + \frac{5\pi}{3}\right)$. There are inflection points at $\left(2n\pi \pm \frac{\pi}{3}, -\frac{1}{4}\right)$.

(e)

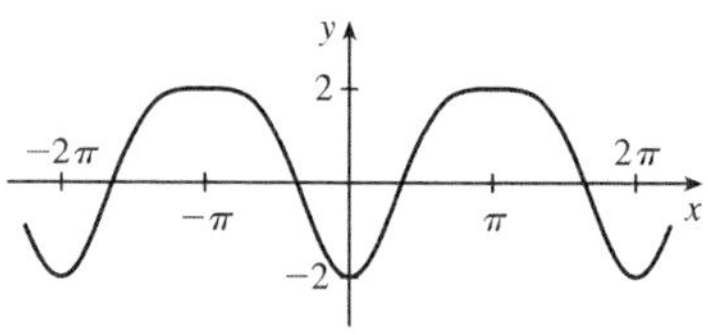


**10.** (a) $\lim\limits_{x \to \pm\infty} e^{2x - x^2} = 0$, so $y = 0$ is a HA.

(b) $y = f(x) = e^{2x-x^2} \quad \Rightarrow \quad f'(x) = 2(1-x)e^{2x-x^2} > 0 \quad \Leftrightarrow \quad x < 1$, so $f$ is increasing on $(-\infty, 1)$ and decreasing on $(1, \infty)$.

(c) $f(1) = e$ is a local and absolute maximum.

(d) $f''(x) = 2\left(2x^2 - 4x + 1\right)e^{2x-x^2} = 0 \quad \Leftrightarrow \quad x = 1 \pm \frac{\sqrt{2}}{2}$. $f''(x) > 0$ $\Leftrightarrow \quad x < 1 - \frac{\sqrt{2}}{2}$ or $x > 1 + \frac{\sqrt{2}}{2}$, so $f$ is CU on $\left(-\infty, 1 - \frac{\sqrt{2}}{2}\right)$ and $\left(1 + \frac{\sqrt{2}}{2}, \infty\right)$, and CD on $\left(1 - \frac{\sqrt{2}}{2}, 1 + \frac{\sqrt{2}}{2}\right)$. IP at $\left(1 \pm \frac{\sqrt{2}}{2}, \sqrt{e}\right)$.

(e)

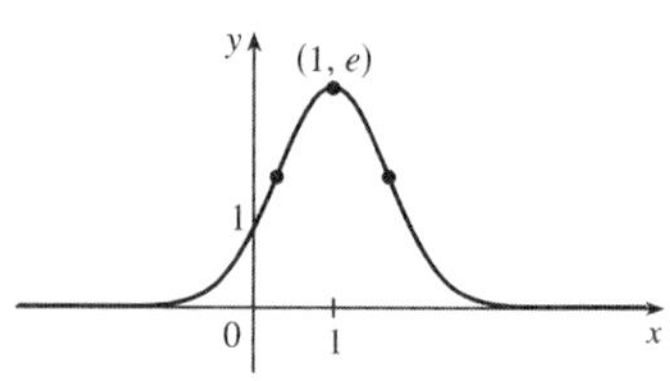

**11.** (a) $\lim\limits_{x\to\pm\infty}\left(e^x+e^{-3x}\right)=\infty$, no asymptote.

(b) $y=f(x)=e^x+e^{-3x}\quad\Rightarrow\quad f'(x)=e^x-3e^{-3x}=e^{-3x}\left(e^{4x}-3\right)>0\quad\Leftrightarrow$

$e^{4x}>3\quad\Leftrightarrow\quad 4x>\ln 3\quad\Leftrightarrow\quad x>\frac{1}{4}\ln 3$, so $f$ is increasing on $\left(\frac{1}{4}\ln 3,\infty\right)$ and decreasing on $\left(-\infty,\frac{1}{4}\ln 3\right)$.

(c) $f\left(\frac{1}{4}\ln 3\right)=3^{1/4}+3^{-3/4}\approx 1.75$ is a local and absolute minimum.

(d) $f''(x)=e^x+9e^{-3x}>0$, so $f$ is CU on $(-\infty,\infty)$. No IP

(e)

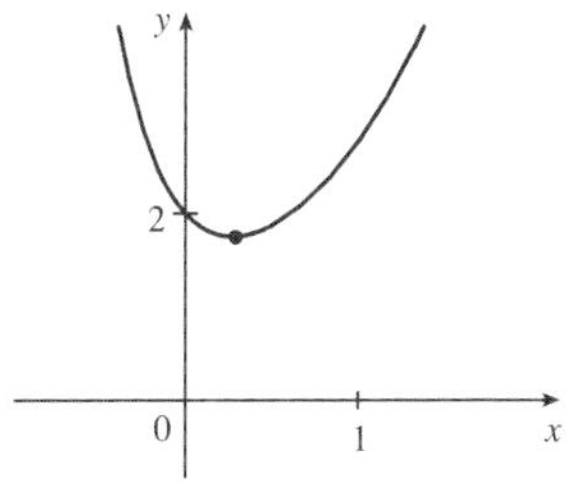


**12.** (a) $\lim\limits_{x\to\pm\infty}\ln\left(x^2-1\right)=\infty$, $\lim\limits_{x\to 1^+}\ln\left(x^2-1\right)=-\infty$, $\lim\limits_{x\to -1^-}\ln\left(x^2-1\right)=-\infty$, so $x=1$ and $x=-1$ are VA.

(b) $y=f(x)=\ln\left(x^2-1\right)\quad\Rightarrow\quad f'(x)=\dfrac{2x}{x^2-1}>0$ for $x>1$ and $f'(x)<0$ for $x<-1$, so $f$ is increasing on $(1,\infty)$ and decreasing on $(-\infty,-1)$. Note that the domain of $f$ is $|x|>1$.

(c) No maximum or minimum.

(d) $f''(x)=-2\dfrac{x^2+1}{\left(x^2-1\right)^2}<0$, so $f$ is CD on $(-\infty,-1)$ and $(1,\infty)$. No IP

(e)

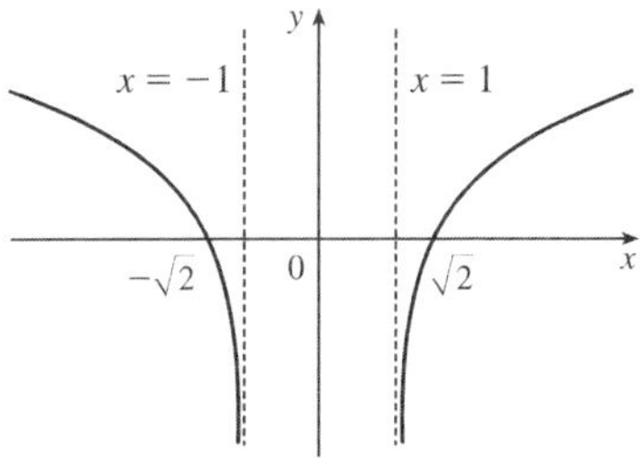


**13.** $f(x)=\dfrac{x^2-1}{x^3}\quad\Rightarrow\quad f'(x)=\dfrac{x^3(2x)-\left(x^2-1\right)3x^2}{x^6}=\dfrac{3-x^2}{x^4}\quad\Rightarrow$

$f''(x)=\dfrac{x^4(-2x)-\left(3-x^2\right)4x^3}{x^8}=\dfrac{2x^2-12}{x^5}$

*Estimates:* From the graphs of $f'$ and $f''$, it appears that $f$ is increasing on $(-1.73,0)$ and $(0,1.73)$ and decreasing on $(-\infty,-1.73)$ and $(1.73,\infty)$; $f$ has a local maximum of about $f(1.73)=0.38$ and a local minimum of about $f(-1.7)=-0.38$; $f$ is CU on $(-2.45,0)$ and $(2.45,\infty)$, and CD on $(-\infty,-2.45)$ and $(0,2.45)$; and $f$ has inflection points at about $(-2.45,-0.34)$ and $(2.45,0.34)$.

*Exact:* Now $f'(x)=\dfrac{3-x^2}{x^4}$ is positive for $0<x^2<3$, that is, $f$ is increasing on $\left(-\sqrt{3},0\right)$ and $\left(0,\sqrt{3}\right)$; and $f'(x)$ is negative (and so $f$ is decreasing) on $\left(-\infty,-\sqrt{3}\right)$ and $\left(\sqrt{3},\infty\right)$. $f'(x)=0$ when $x=\pm\sqrt{3}$.

$f'$ goes from positive to negative at $x=\sqrt{3}$, so $f$ has a local maximum of $f\left(\sqrt{3}\right)=\frac{\left(\sqrt{3}\right)^2-1}{\left(\sqrt{3}\right)^3}=\frac{2\sqrt{3}}{9}$; and since $f$ is odd, we know that maxima on the interval $(0,\infty)$ correspond to minima on $(-\infty,0)$, so $f$ has a local minimum of $f\left(-\sqrt{3}\right)=-\frac{2\sqrt{3}}{9}$. Also, $f''(x)=\dfrac{2x^2-12}{x^5}$ is positive (so $f$ is CU) on $\left(-\sqrt{6},0\right)$ and $\left(\sqrt{6},\infty\right)$, and negative (so $f$ is CD) on $\left(-\infty,-\sqrt{6}\right)$ and $\left(0,\sqrt{6}\right)$. There are IP at $\left(\sqrt{6},\frac{5\sqrt{6}}{36}\right)$ and $\left(-\sqrt{6},-\frac{5\sqrt{6}}{36}\right)$.

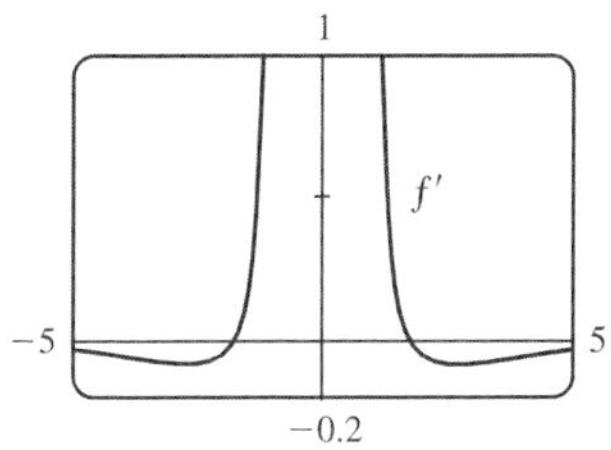


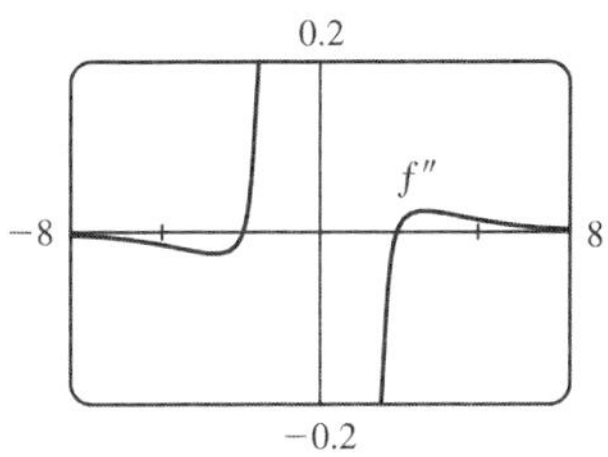


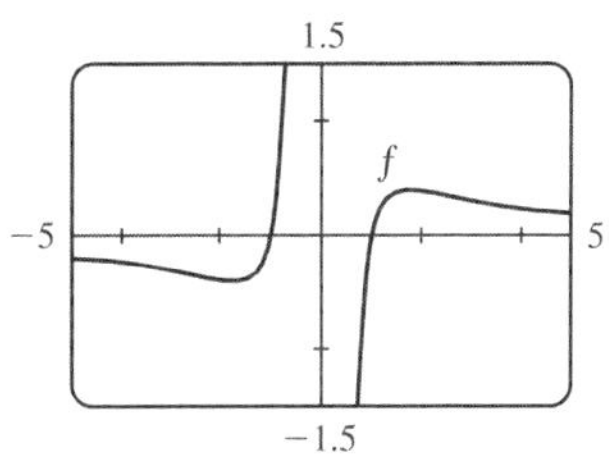


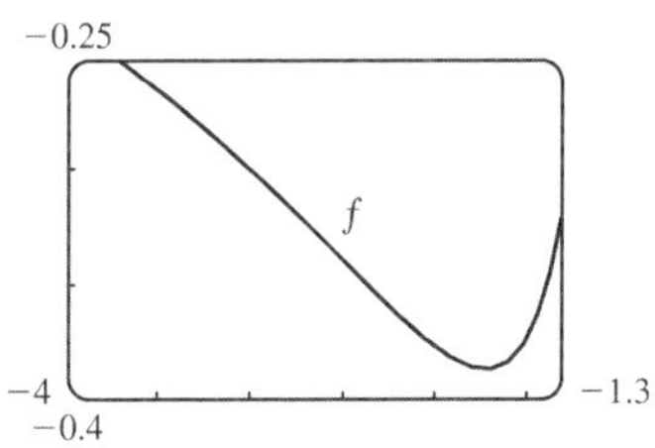

**14.** $f(x) = \dfrac{\sqrt[3]{x}}{1-x} = x^{1/3}(1-x)^{-1} \;\Rightarrow\; f'(x) = x^{1/3}(-1)(1-x)^{-2}(-1) + (1-x)^{-1}\left(\frac{1}{3}\right)x^{-2/3} = \dfrac{x^{-2/3}}{3}\dfrac{1+2x}{(x-1)^2} \;\Rightarrow$

$$f''(x) = \frac{x^{-2/3}}{3}\frac{(x-1)^2(2)-(1+2x)(2)(x-1)}{(x-1)^4} + \frac{1+2x}{(x-1)^2}\left(\frac{-2x^{-5/3}}{9}\right) = -\frac{2x^{-5/3}}{9}\frac{5x^2+5x-1}{(x-1)^3}$$

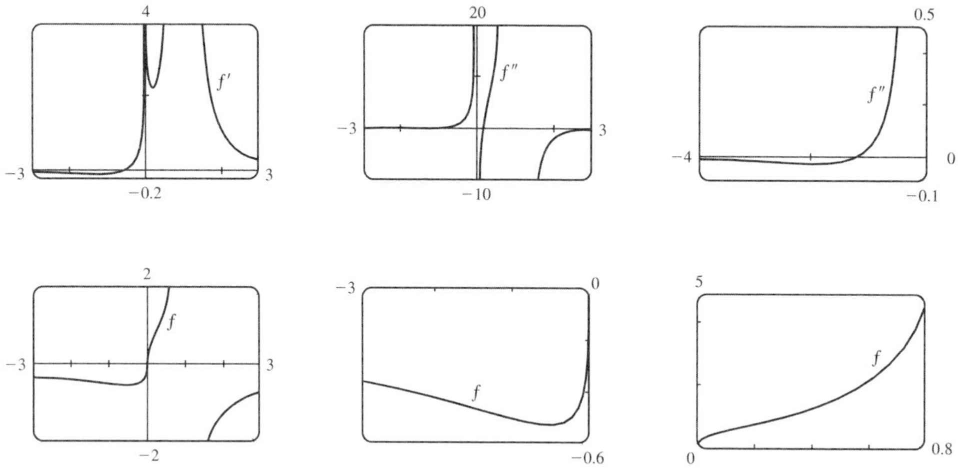


From the graphs, it appears that $f$ is increasing on $(-0.50, 1)$ and $(1, \infty)$, with a vertical asymptote at $x = 1$, and decreasing on $(-\infty, -0.50)$; $f$ has no local maximum, but a local minimum of about $f(-0.50) = -0.53$; $f$ is CU on $(-1.17, 0)$ and $(0.17, 1)$ and CD on $(-\infty, -1.17)$, $(0, 0.17)$ and $(1, \infty)$; and $f$ has inflection points at about $(-1.17, -0.49)$, $(0, 0)$ and $(0.17, 0.67)$. Note also that $\lim\limits_{x\to\pm\infty} f(x) = 0$, so $y = 0$ is a horizontal asymptote.

**15.** $f(x) = 3x^6 - 5x^5 + x^4 - 5x^3 - 2x^2 + 2 \;\Rightarrow\; f'(x) = 18x^5 - 25x^4 + 4x^3 - 15x^2 - 4x \;\Rightarrow$

$f''(x) = 90x^4 - 100x^3 + 12x^2 - 30x - 4$

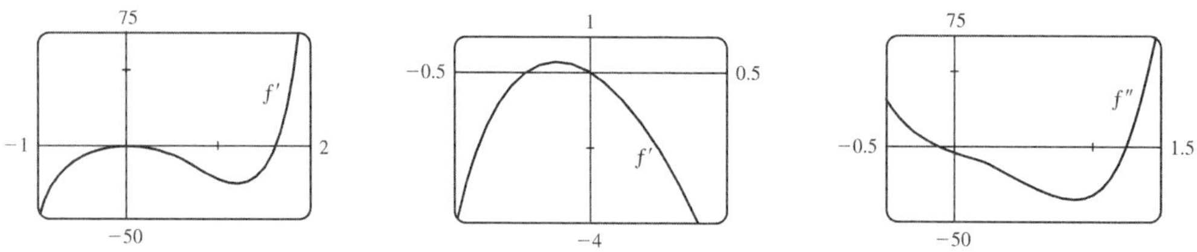


From the graphs of $f'$ and $f''$, it appears that $f$ is increasing on $(-0.23, 0)$ and $(1.62, \infty)$ and decreasing on $(-\infty, -0.23)$ and $(0, 1.62)$; $f$ has a local maximum of about $f(0) = 2$ and local minima of about $f(-0.23) = 1.96$ and $f(1.62) = -19.2$; $f$ is CU on $(-\infty, -0.12)$ and $(1.24, \infty)$ and CD on $(-0.12, 1.24)$; and $f$ has inflection points at about $(-0.12, 1.98)$ and $(1.24, -12.1)$.

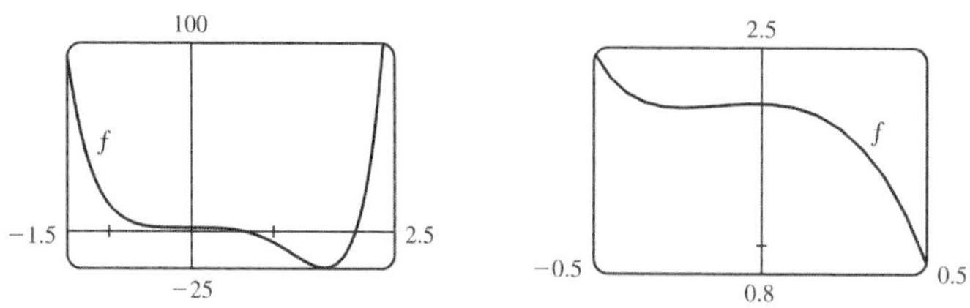

**16.** $f(x) = \sin x \cos^2 x \quad \Rightarrow \quad f'(x) = \cos^3 x - 2\sin^2 x \cos x \quad \Rightarrow \quad f''(x) = -7\sin x \cos^2 x + 2\sin^3 x$

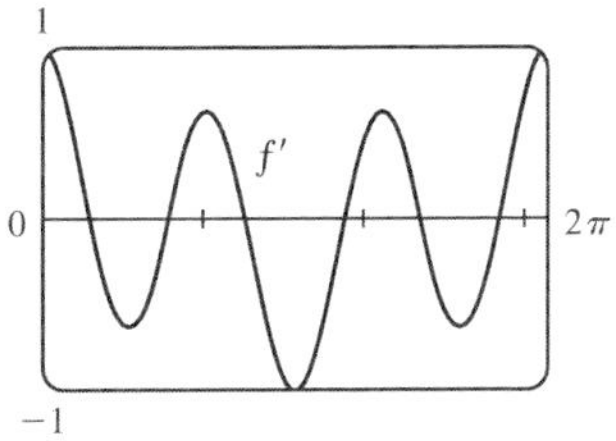
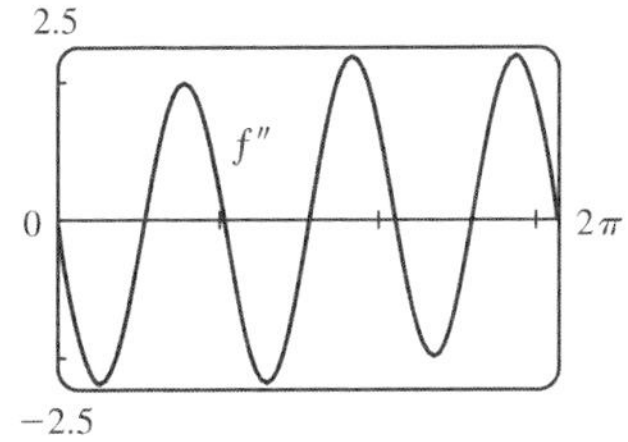
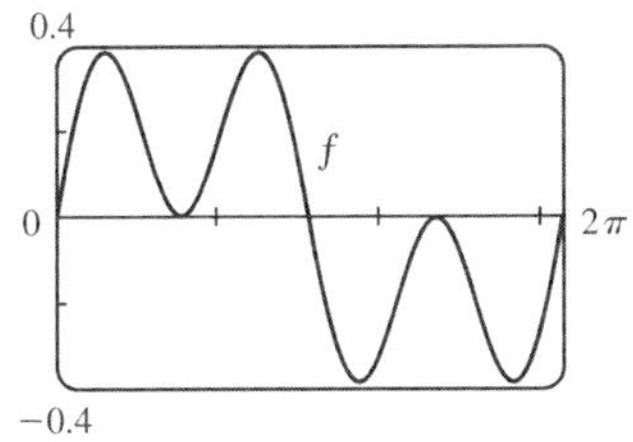

From the graphs of $f'$ and $f''$, it appears that $f$ is increasing on $(0, 0.62)$, $(1.57, 2.53)$, $(3.76, 4.71)$ and $(5.67, 2\pi)$ and decreasing on $(0.62, 1.57)$, $(2.53, 3.76)$ and $(4.71, 5.67)$; $f$ has local maxima of about $f(0.62) = f(2.53) = 0.38$ and $f(4.71) = 0$ and local minima of about $f(1.57) = 0$ and $f(3.76) = f(5.67) = -0.38$; $f$ is CU on $(1.08, 2.06)$, $(3.14, 4.22)$ and $(5.20, 2\pi)$ and CD on $(0, 1.08)$, $(2.06, 3.14)$ and $(4.22, 5.20)$; and $f$ has inflection points at about $(0, 0)$, $(1.08, 0.20)$, $(2.06, 0.20)$, $(3.14, 0)$, $(4.22, -0.20)$, $(5.20, -0.20)$ and $(2\pi, 0)$.

**17.**

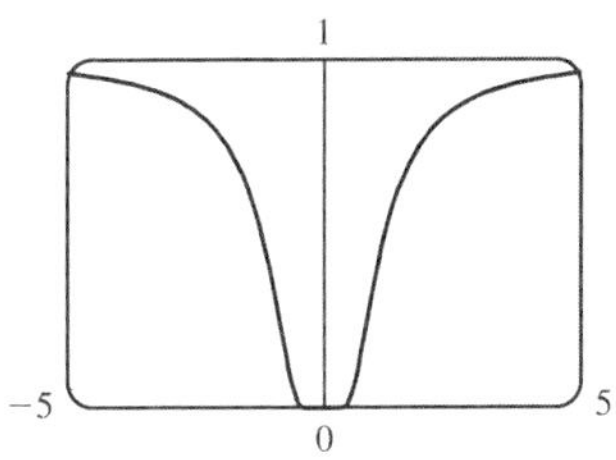

From the graph, we estimate the points of inflection to be about $(\pm 0.82, 0.22)$.

$f(x) = e^{-1/x^2} \quad \Rightarrow \quad f'(x) = 2x^{-3}e^{-1/x^2} \quad \Rightarrow$

$f''(x) = 2\left[x^{-3}(2x^{-3})e^{-1/x^2} + e^{-1/x^2}(-3x^{-4})\right] = 2x^{-6}e^{-1/x^2}(2 - 3x^2)$.

This is 0 when $2 - 3x^2 = 0 \quad \Leftrightarrow \quad x = \pm\sqrt{\frac{2}{3}}$, so the inflection points are $\left(\pm\sqrt{\frac{2}{3}}, e^{-3/2}\right)$.

**18.** (a)

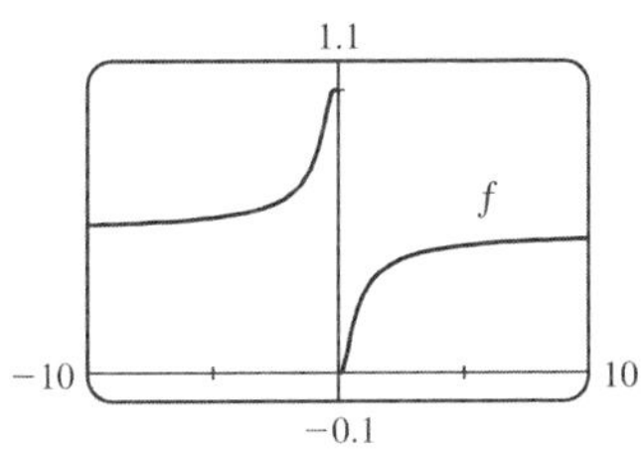

(b) $f(x) = \dfrac{1}{1 + e^{1/x}}$. $\displaystyle\lim_{x\to\infty} f(x) = \frac{1}{1+1} = \frac{1}{2}$,

$\displaystyle\lim_{x\to-\infty} f(x) = \frac{1}{1+1} = \frac{1}{2}$,

$\displaystyle\lim_{x\to 0^+} f(x) = \frac{1}{1+\infty} = 0$, $\displaystyle\lim_{x\to 0^-} f(x) = \frac{1}{1+0} = 1$

(c) From the graph of $f$, estimates for the IP are $(-0.4, 0.9)$ and $(0.4, 0.08)$.

(d) $f''(x) = -\dfrac{e^{1/x}\left[e^{1/x}(2x-1) + 2x + 1\right]}{x^4(e^{1/x}+1)^3}$

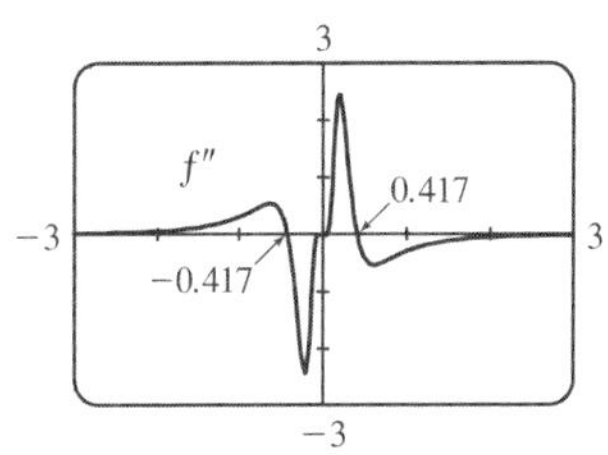

(e) From the graph, we see that $f''$ changes sign at $x = \pm 0.417$ ($x = 0$ is not in the domain of $f$). IP are approximately $(0.417, 0.083)$ and $(-0.417, 0.917)$.

**19.** $f(x) = \arctan(\cos(3 \arcsin x))$. We use a CAS to compute $f'$ and $f''$, and to graph $f$, $f'$, and $f''$:

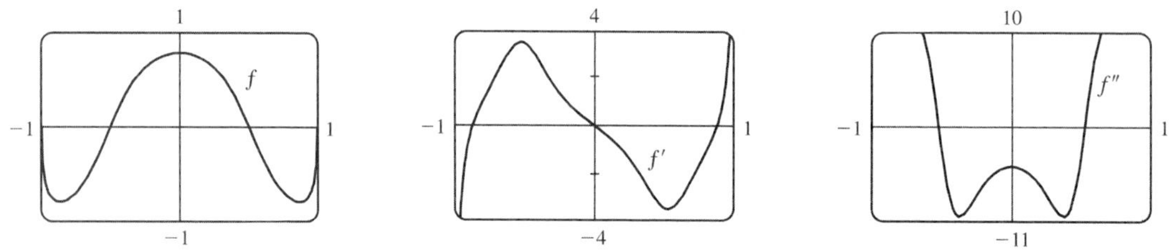


From the graph of $f'$, it appears that the only maximum occurs at $x = 0$ and there are minima at $x = \pm 0.87$.

From the graph of $f''$, it appears that there are inflection points at $x = \pm 0.52$.

**20.** $f(x) = \ln(2x + x \sin x)$. We use the CAS to calculate

$$f'(x) = \frac{2 + \sin x + x \cos x}{2x + x \sin x} \quad \text{and}$$

$$f''(x) = \frac{2x^2 \sin x + 4 \sin x - \cos^2 x + x^2 + 5}{x^2(\cos^2 x - 4 \sin x - 5)}.$$

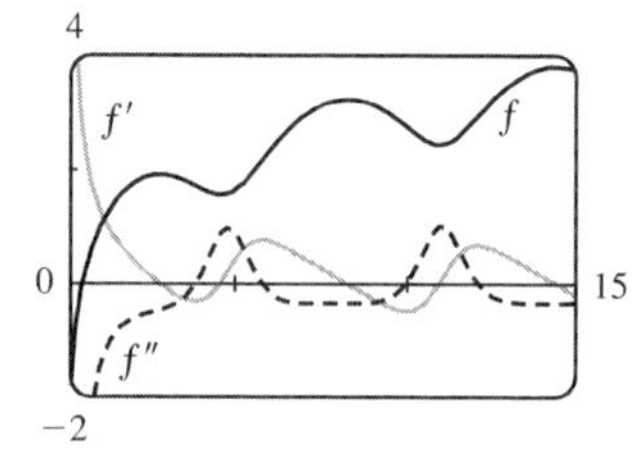


From the graphs, it seems that $f' > 0$ (and so $f$ is increasing) on approximately the intervals $(0, 2.7)$, $(4.5, 8.2)$ and $(10.9, 14.3)$. It seems that $f''$ changes sign (indicating inflection points) at $x \approx 3.8, 5.7, 10.0$ and $12.0$. Looking back at the graph of $f$, this implies that the inflection points have approximate coordinates $(3.8, 1.7)$, $(5.7, 2.1)$, $(10.0, 2.7)$, and $(12.0, 2.9)$.

**21.** The family of functions $f(x) = \ln(\sin x + C)$ all have the same period and all have maximum values at $x = \frac{\pi}{2} + 2\pi n$. Since the domain of ln is $(0, \infty)$, $f$ has a graph only if $\sin x + C > 0$ somewhere. Since $-1 \leq \sin x \leq 1$, this happens if $C > -1$, that is, $f$ has no graph if $C \leq -1$. Similarly, if $C > 1$, then $\sin x + C > 0$ and $f$ is continuous on $(-\infty, \infty)$. As $C$ increases, the graph of $f$ is shifted vertically upward and flattens out.

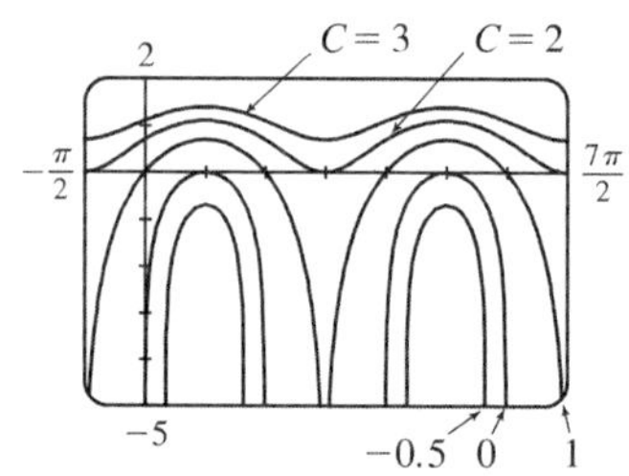


If $-1 < C \leq 1$, $f$ is defined where $\sin x + C > 0 \quad \Leftrightarrow \quad \sin x > -C \quad \Leftrightarrow \quad \sin^{-1}(-C) < x < \pi - \sin^{-1}(-C)$. Since the period is $2\pi$, the domain of $f$ is $\left(2n\pi + \sin^{-1}(-C), (2n+1)\pi - \sin^{-1}(-C)\right)$, $n$ an integer.

**22.** We exclude the case $c = 0$, since in that case $f(x) = 0$ for all $x$. To find the maxima and minima, we differentiate:

$$f(x) = cxe^{-cx^2} \quad \Rightarrow \quad f'(x) = c\left[xe^{-cx^2}(-2cx) + e^{-cx^2}(1)\right] = ce^{-cx^2}(-2cx^2 + 1).$$

This is 0 where $-2cx^2 + 1 = 0 \quad \Leftrightarrow \quad x = \pm 1/\sqrt{2c}$. So if $c > 0$, there are two maxima or minima, whose $x$-coordinates approach 0 as $c$ increases. The negative root gives a minimum and the positive root gives a maximum, by the First Derivative Test. By substituting back into the equation, we see that $f\left(\pm 1/\sqrt{2c}\right) = c\left(\pm 1/\sqrt{2c}\right) e^{-c\left(\pm 1/\sqrt{2c}\right)^2} = \pm\sqrt{c/2e}$. So as $c$ increases, the extreme points become more pronounced. Note that if $c > 0$, then $\lim\limits_{x \to \pm\infty} f(x) = 0$. If $c < 0$, then there are no extreme values, and $\lim\limits_{x \to \pm\infty} f(x) = \mp\infty$.

To find the points of inflection, we differentiate again: $f'(x) = ce^{-cx^2}(-2cx^2+1) \Rightarrow$ $f''(x) = c\left[e^{-cx^2}(-4cx) + (-2cx^2+1)\left(-2cxe^{-cx^2}\right)\right] = -2c^2xe^{-cx^2}(3-2cx^2)$. This is 0 at $x=0$ and where $3-2cx^2=0 \Leftrightarrow x = \pm\sqrt{3/(2c)} \Rightarrow$ IP at $\left(\pm\sqrt{3/(2c)}, \pm\sqrt{3c/2}\,e^{-3/2}\right)$. If $c>0$ there are three inflection points, and as $c$ increases, the $x$-coordinates of the nonzero inflection points approach 0. If $c<0$, there is only one inflection point, the origin.

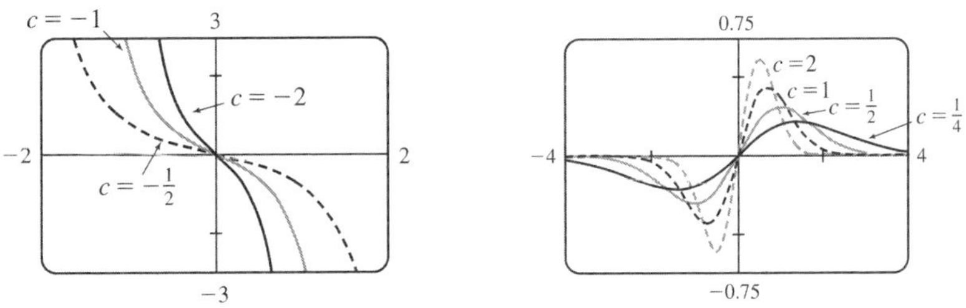


**23.** For $(1,6)$ to be on the curve $y = x^3+ax^2+bx+1$, we have that $6 = 1+a+b+1 \Rightarrow b = 4-a$. Now $y' = 3x^2+2ax+b$ and $y'' = 6x+2a$. Also, for $(1,6)$ to be an inflection point it must be true that $y''(1) = 6(1)+2a = 0 \Rightarrow a=-3 \Rightarrow b = 4-(-3) = 7$. Note that with $a=-3$, we have $y'' = 6x-6 = 6(x-1)$, so $y''$ changes sign at $x=1$, proving that $(1,6)$ is a point of inflection. [This does not follow from the fact that $y''(1)=0$.]

**24.** (a) $g(x) = f(x^2) \Rightarrow g'(x) = 2xf'(x^2)$ by the Chain Rule. Since $f'(x)>0$ for all $x \neq 0$, we must have $f'(x^2)>0$ for $x\neq 0$, so $g'(x)=0 \Leftrightarrow x=0$. Now $g'(x)$ changes sign (from negative to positive) at $x=0$, since one of its factors, $f'(x^2)$, is positive for all $x$, and its other factor, $2x$, changes from negative to positive at this point, so by the First Derivative Test, $f$ has a local and absolute minimum at $x=0$.

(b) $g'(x) = 2xf'(x^2) \Rightarrow g''(x) = 2\left[xf''(x^2)(2x) + f'(x^2)\right] = 4x^2f''(x^2) + 2f'(x^2)$ by the Product Rule and the Chain Rule. But $x^2>0$ for all $x\neq 0$, $f''(x^2)>0$ (since $f$ is CU for $x>0$), and $f'(x^2)>0$ for all $x \neq 0$, so since all of its factors are positive, $g''(x)>0$ for $x\neq 0$. Whether $g''(0)$ is positive or 0 doesn't matter (since the sign of $g''$ does not change there); $g$ is concave upward on $\mathbb{R}$.

**25.** $\displaystyle\lim_{x\to 0}\frac{\tan \pi x}{\ln(1+x)} \overset{\text{H}}{=} \lim_{x\to 0}\frac{\pi\sec^2 \pi x}{1/(1+x)} = \frac{\pi\cdot 1^2}{1/1} = \pi$

**26.** $\displaystyle\lim_{x\to 0}\frac{1-\cos x}{x^2+x} \overset{\text{H}}{=} \lim_{x\to 0}\frac{\sin x}{2x+1} = \frac{0}{1} = 0$

**27.** $\displaystyle\lim_{x\to 0}\frac{e^{4x}-1-4x}{x^2} \overset{\text{H}}{=} \lim_{x\to 0}\frac{4e^{4x}-4}{2x} \overset{\text{H}}{=} \lim_{x\to 0}\frac{16e^{4x}}{2} = \lim_{x\to 0} 8e^{4x} = 8\cdot 1 = 8$

**28.** $\displaystyle\lim_{x\to \infty}\frac{e^{4x}-1-4x}{x^2} \overset{\text{H}}{=} \lim_{x\to \infty}\frac{4e^{4x}-4}{2x} \overset{\text{H}}{=} \lim_{x\to \infty}\frac{16e^{4x}}{2} = \lim_{x\to \infty} 8e^{4x} = \infty$

**29.** $\displaystyle\lim_{x\to\infty} x^3e^{-x} = \lim_{x\to\infty}\frac{x^3}{e^x} \overset{\text{H}}{=} \lim_{x\to\infty}\frac{3x^2}{e^x} \overset{\text{H}}{=} \lim_{x\to\infty}\frac{6x}{e^x} \overset{\text{H}}{=} \lim_{x\to\infty}\frac{6}{e^x} = 0$

**30.** $\displaystyle\lim_{x\to 0^+} x^2\ln x = \lim_{x\to 0^+}\frac{\ln x}{1/x^2} \overset{\text{H}}{=} \lim_{x\to 0^+}\frac{1/x}{-2/x^3} = \lim_{x\to 0^+}\left(-\tfrac{1}{2}x^2\right) = 0$

**31.** $\lim_{x\to 1^+}\left(\frac{x}{x-1}-\frac{1}{\ln x}\right)=\lim_{x\to 1^+}\left(\frac{x\ln x-x+1}{(x-1)\ln x}\right)\overset{\text{H}}{=}\lim_{x\to 1^+}\frac{x\cdot(1/x)+\ln x-1}{(x-1)\cdot(1/x)+\ln x}$

$=\lim_{x\to 1^+}\frac{\ln x}{1-1/x+\ln x}\overset{\text{H}}{=}\lim_{x\to 1^+}\frac{1/x}{1/x^2+1/x}=\frac{1}{1+1}=\frac{1}{2}$

**32.** $y=(\tan x)^{\cos x}\ \Rightarrow\ \ln y=\cos x\ln\tan x$, so

$$\lim_{x\to(\pi/2)^-}\ln y=\lim_{x\to(\pi/2)^-}\frac{\ln\tan x}{\sec x}\overset{\text{H}}{=}\lim_{x\to(\pi/2)^-}\frac{(1/\tan x)\sec^2 x}{\sec x\tan x}=\lim_{x\to(\pi/2)^-}\frac{\sec x}{\tan^2 x}$$

$$=\lim_{x\to(\pi/2)^-}\frac{\cos x}{\sin^2 x}=\frac{0}{1^2}=0,$$

so $\lim_{x\to(\pi/2)^-}(\tan x)^{\cos x}=\lim_{x\to(\pi/2)^-}e^{\ln y}=e^0=1.$

**33.** We are given $d\theta/dt=-0.25$ rad/h. $\tan\theta=400/x\ \Rightarrow$ $x=400\cot\theta\ \Rightarrow\ \frac{dx}{dt}=-400\csc^2\theta\,\frac{d\theta}{dt}$. When $\theta=\frac{\pi}{6}$, $\frac{dx}{dt}=-400(2)^2(-0.25)=400$ ft/h.

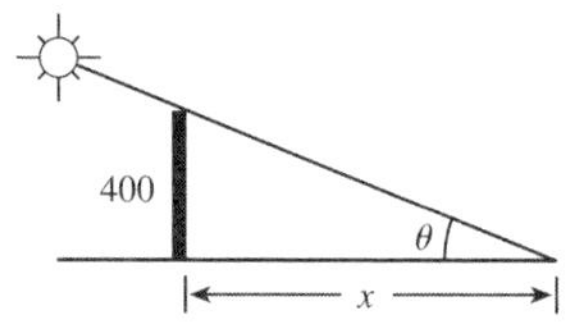


**34.** Given $dV/dt=2$, find $dh/dt$ when $h=5$. $V=\frac{1}{3}\pi r^2h$ and, from similar triangles, $\frac{r}{h}=\frac{3}{10}\ \Rightarrow\ V=\frac{\pi}{3}\left(\frac{3h}{10}\right)^2h=\frac{3\pi}{100}h^3$, so

$2=\frac{dV}{dt}=\frac{9\pi}{100}h^2\frac{dh}{dt}\ \Rightarrow\ \frac{dh}{dt}=\frac{200}{9\pi h^2}=\frac{200}{9\pi\,(5)^2}=\frac{8}{9\pi}$ cm/s

when $h=5$.

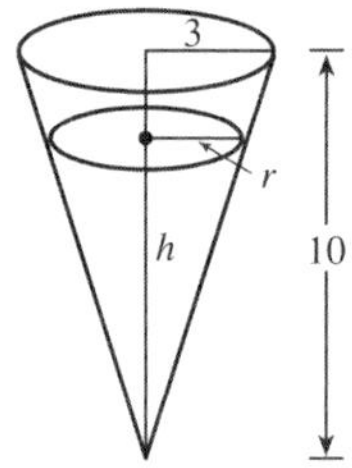


**35.** Given $dh/dt=5$ and $dx/dt=15$, find $dz/dt$. $z^2=x^2+h^2\ \Rightarrow$ $2z\,\frac{dz}{dt}=2x\,\frac{dx}{dt}+2h\,\frac{dh}{dt}\ \Rightarrow\ \frac{dz}{dt}=\frac{1}{z}(15x+5h)$. When $t=3$, $h=45+3(5)=60$ and $x=15(3)=45\ \Rightarrow\ z=\sqrt{45^2+60^2}=75$, so $\frac{dz}{dt}=\frac{1}{75}\,[15(45)+5(60)]=13$ ft/s.

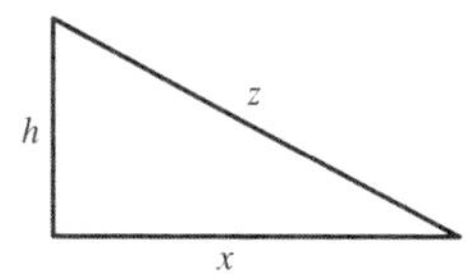


**36.** We are given $dz/dt=30$ ft/s. By similar triangles, $\frac{y}{z}=\frac{4}{\sqrt{241}}\ \Rightarrow$ $y=\frac{4}{\sqrt{241}}z$, so $\frac{dy}{dt}=\frac{4}{\sqrt{241}}\frac{dz}{dt}=\frac{120}{\sqrt{241}}\approx 7.7$ ft/s.

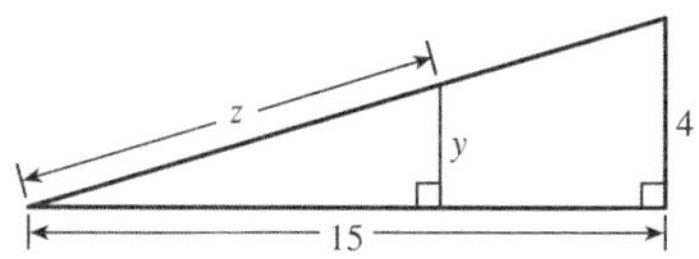


**37.** Call the two integers $x$ and $y$. Then $x+4y=1000$, so $x=1000-4y$. Their product is $P=xy=(1000-4y)y$, so our problem is to maximize the function $P(y)=1000y-4y^2$, where $0<y<250$ and $y$ is an integer. $P'(y)=1000-8y$, so $P'(y)=0\ \Leftrightarrow\ y=125$. $P''(y)=-8<0$, so $P(125)=62{,}500$ is an absolute maximum. Since the optimal $y$ turned out to be an integer, we have found the desired pair of numbers, namely $x=1000-4(125)=500$ and $y=125$.

**38.** On the hyperbola $xy=8$, if $d(x)$ is the distance from the point $(x,y)=(x,8/x)$ to the point $(3,0)$, then $[d(x)]^2=(x-3)^2+64/x^2=f(x)$. $f'(x)=2(x-3)-128/x^3=0\ \Rightarrow\ x^4-3x^3-64=0\ \Rightarrow$ $(x-4)(x^3+x^2+4x+16)=0\ \Rightarrow\ x=4$ since the solution must have $x>0$. Then $y=\frac{8}{4}=2$, so the point is $(4,2)$.

**39.**

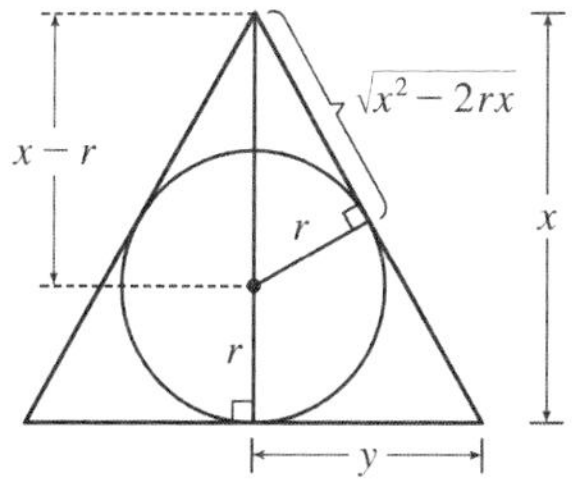


By similar triangles, $\dfrac{y}{x} = \dfrac{r}{\sqrt{x^2-2rx}}$, so the area of the triangle is

$$A(x) = \tfrac{1}{2}(2y)x = xy = \frac{rx^2}{\sqrt{x^2-2rx}} \quad \Rightarrow$$

$$A'(x) = \frac{2rx\sqrt{x^2-2rx} - rx^2(x-r)/\sqrt{x^2-2rx}}{x^2-2rx} = \frac{rx^2(x-3r)}{(x^2-2rx)^{3/2}} = 0 \text{ when } x = 3r.$$

$A'(x) < 0$ when $2r < x < 3r$, $A'(x) > 0$ when $x > 3r$. So $x = 3r$ gives a minimum and $A(3r) = r(9r^2)/(\sqrt{3}\,r) = 3\sqrt{3}\,r^2$.

**40.**

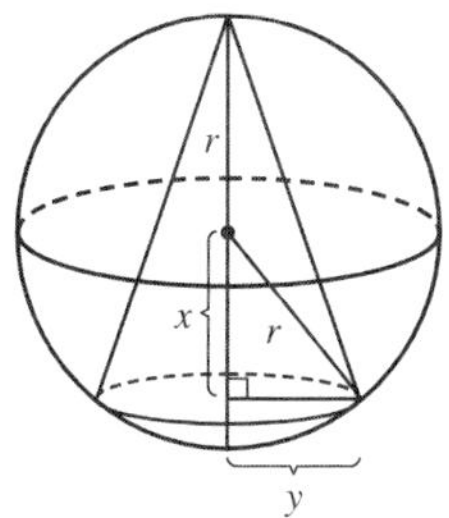


The volume of the cone is $V = \frac{1}{3}\pi y^2(r+x) = \frac{1}{3}\pi(r^2-x^2)(r+x)$, $-r \le x \le r$.

$$V'(x) = \tfrac{\pi}{3}\left[(r^2-x^2)(1) + (r+x)(-2x)\right] = \tfrac{\pi}{3}\left[(r+x)(r-x-2x)\right] = \tfrac{\pi}{3}(r+x)(r-3x) = 0 \text{ when } x = -r \text{ or } x = r/3.$$

Now $V(r) = 0 = V(-r)$, so the maximum occurs at $x = r/3$ and the volume is $V\left(\dfrac{r}{3}\right) = \dfrac{\pi}{3}\left(r^2 - \dfrac{r^2}{9}\right)\left(\dfrac{4r}{3}\right) = \dfrac{32\pi r^3}{81}$.

**41.**

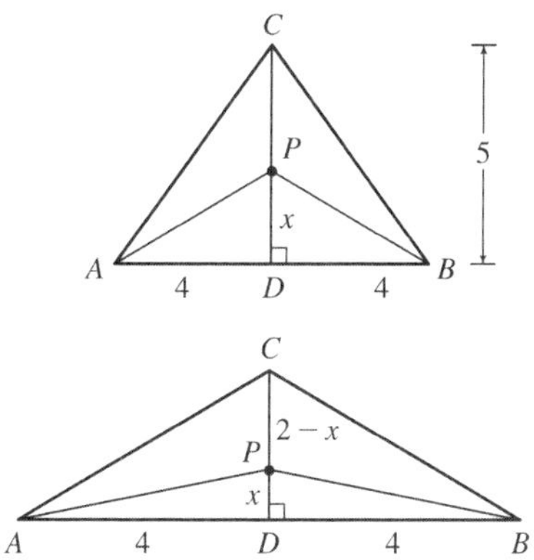


We minimize $L(x) = |PA| + |PB| + |PC| = 2\sqrt{x^2+16} + (5-x)$, $0 \le x \le 5$. $L'(x) = 2x/\sqrt{x^2+16} - 1 = 0 \Leftrightarrow 2x = \sqrt{x^2+16} \Leftrightarrow 4x^2 = x^2+16 \Leftrightarrow x = \frac{4}{\sqrt{3}}$. $L(0) = 13$, $L\left(\frac{4}{\sqrt{3}}\right) \approx 11.9$, $L(5) \approx 12.8$, so the minimum occurs when $x = \frac{4}{\sqrt{3}} \approx 2.3$.

If $|CD| = 2$, $L(x)$ changes from $(5-x)$ to $(2-x)$ with $0 \le x \le 2$. But we still get $L'(x) = 0 \Leftrightarrow x = \frac{4}{\sqrt{3}}$, which isn't in the interval $[0, 2]$. Now $L(0) = 10$ and $L(2) = 2\sqrt{20} = 4\sqrt{5} \approx 8.9$. The minimum occurs when $P = C$.

**42.**

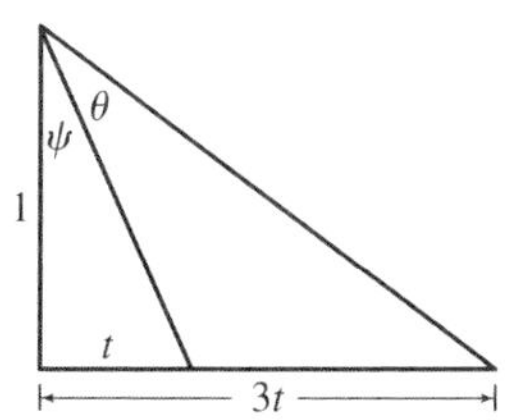


It suffices to maximize $\tan\theta$. Now

$$\frac{3t}{1} = \tan(\psi+\theta) = \frac{\tan\psi + \tan\theta}{1 - \tan\psi\,\tan\theta} = \frac{t+\tan\theta}{1 - t\tan\theta}.$$

So $3t(1 - t\tan\theta) = t + \tan\theta \Rightarrow 2t = (1+3t^2)\tan\theta \Rightarrow \tan\theta = \dfrac{2t}{1+3t^2}$. Let $f(t) = \tan\theta = \dfrac{2t}{1+3t^2} \Rightarrow$

$f'(t) = \dfrac{2(1+3t^2) - 2t(6t)}{(1+3t^2)^2} = \dfrac{2(1-3t^2)}{(1+3t^2)^2} = 0 \Leftrightarrow 1 - 3t^2 = 0 \Leftrightarrow t = \frac{1}{\sqrt{3}}$ since $t \ge 0$. Now $f'(t) > 0$ for

$0 \le t < \frac{1}{\sqrt{3}}$ and $f'(t) < 0$ for $t > \frac{1}{\sqrt{3}}$, so $f$ has an absolute maximum when $t = \frac{1}{\sqrt{3}}$ and $\tan\theta = \dfrac{2(1/\sqrt{3})}{1 + 3(1/\sqrt{3})^2} = \dfrac{1}{\sqrt{3}}$

$\Rightarrow \quad \theta = \frac{\pi}{6}$. Substituting for $t$ and $\theta$ in $3t = \tan(\psi + \theta)$ gives us $\sqrt{3} = \tan(\psi + \frac{\pi}{6}) \quad \Rightarrow \quad \psi = \frac{\pi}{6}$.

**43.** $v = K\sqrt{\dfrac{L}{C} + \dfrac{C}{L}} \quad \Rightarrow \quad \dfrac{dv}{dL} = \dfrac{K}{2\sqrt{(L/C) + (C/L)}}\left(\dfrac{1}{C} - \dfrac{C}{L^2}\right) = 0 \quad \Leftrightarrow \quad \dfrac{1}{C} = \dfrac{C}{L^2} \quad \Leftrightarrow \quad L^2 = C^2 \quad \Leftrightarrow \quad L = C.$

This gives the minimum velocity since $v' < 0$ for $0 < L < C$ and $v' > 0$ for $L > C$.

**44.**

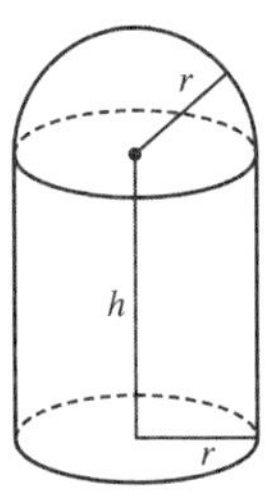


We minimize the surface area $S = \pi r^2 + 2\pi rh + \frac{1}{2}(4\pi r^2) = 3\pi r^2 + 2\pi rh$.

Solving $V = \pi r^2 h + \frac{2}{3}\pi r^3$ for $h$, we get $h = \dfrac{V - \frac{2}{3}\pi r^3}{\pi r^2} = \dfrac{V}{\pi r^2} - \frac{2}{3}r$, so

$$S(r) = 3\pi r^2 + 2\pi r\left[\frac{V}{\pi r^2} - \tfrac{2}{3}r\right] = \tfrac{5}{3}\pi r^2 + \frac{2V}{r}.$$

$$S'(r) = -\frac{2V}{r^2} + \tfrac{10}{3}\pi r = \frac{\frac{10}{3}\pi r^3 - 2V}{r^2} = 0 \quad \Leftrightarrow \quad \tfrac{10}{3}\pi r^3 = 2V \quad \Leftrightarrow$$

$r^3 = \dfrac{3V}{5\pi} \quad \Leftrightarrow \quad r = \sqrt[3]{\dfrac{3V}{5\pi}}$. This gives an absolute minimum since $S'(r) < 0$ for $0 < r < \sqrt[3]{\dfrac{3V}{5\pi}}$ and $S'(r) > 0$

for $r > \sqrt[3]{\dfrac{3V}{5\pi}}$. Thus, $h = \dfrac{V - \frac{2}{3}\pi \cdot \frac{3V}{5\pi}}{\pi\sqrt[3]{\frac{(3V)^2}{(5\pi)^2}}} = \dfrac{(V - \frac{2}{5}V)\sqrt[3]{(5\pi)^2}}{\pi\sqrt[3]{(3V)^2}} = \dfrac{3V\sqrt[3]{(5\pi)^2}}{5\pi\sqrt[3]{(3V)^2}} = \sqrt[3]{\dfrac{3V}{5\pi}} = r.$

**45.** Let $x$ denote the number of \$1 decreases in ticket price. Then the ticket price is $\$12 - \$1(x)$, and the average attendance is $11{,}000 + 1000(x)$. Now the revenue per game is

$$R(x) = (\text{price per person}) \times (\text{number of people per game})$$
$$= (12 - x)(11{,}000 + 1000x) = -1000x^2 + 1000x + 132{,}000$$

for $0 \le x \le 4$ (since the seating capacity is 15,000) $\Rightarrow \quad R'(x) = -2000x + 1000 = 0 \quad \Leftrightarrow \quad x = 0.5$. This is a maximum since $R''(x) = -2000 < 0$ for all $x$. Now we must check the value of $R(x) = (12 - x)(11{,}000 + 1000x)$ at $x = 0.5$ and at the endpoints of the domain to see which value of $x$ gives the maximum value of $R$. $R(0) = (12)(11{,}000) = 132{,}000$, $R(0.5) = (11.5)(11{,}500) = 132{,}250$, and $R(4) = (8)(15{,}000) = 120{,}000$. Thus, the maximum revenue of \$132,250 per game occurs when the average attendance is 11,500 and the ticket price is \$11.50.

**46.** (a) $C(x) = 1800 + 25x - 0.2x^2 + 0.001x^3$ and $R(x) = xp(x) = 48.2x - 0.03x^2$. The profit is maximized when $C'(x) = R'(x)$.

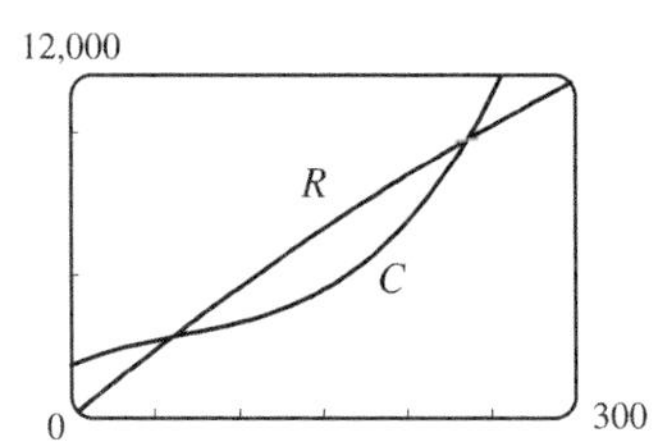


From the figure, we estimate that the tangents are parallel when $x \approx 160$.

(b) $C'(x) = 25 - 0.4x + 0.003x^2$ and $R'(x) = 48.2 - 0.06x$. $C'(x) = R'(x) \quad \Rightarrow \quad 0.003x^2 - 0.34x - 23.2 = 0 \quad \Rightarrow$ $x_1 \approx 161.3$ $(x > 0)$. $R''(x) = -0.06$ and $C''(x) = -0.4 + 0.006x$, so $R''(x_1) = -0.06 < C''(x_1) \approx 0.57 \quad \Rightarrow$ profit is maximized by producing 161 units.

(c) $c(x) = \frac{C(x)}{x} = \frac{1800}{x} + 25 - 0.2x + 0.001x^2$ is the average cost. Since the average cost is minimized when the marginal cost equals the average cost, we graph $c(x)$ and $C'(x)$ and estimate the point of intersection. From the figure, $C'(x) = c(x) \Leftrightarrow x \approx 144$.

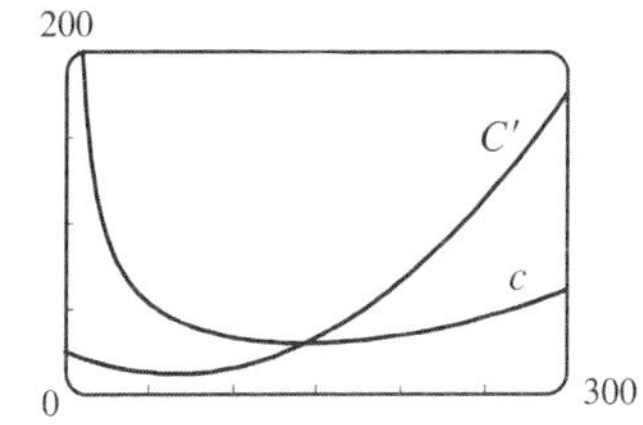


**47.** $f(t) = \cos t + t - t^2 \Rightarrow f'(t) = -\sin t + 1 - 2t$. $f'(t)$ exists for all $t$, so to find the maximum of $f$, we can examine the zeros of $f'$. From the graph of $f'$, we see that a good choice for $t_1$ is $t_1 = 0.3$. Use $g(t) = -\sin t + 1 - 2t$ and $g'(t) = -\cos t - 2$ to obtain $t_2 \approx 0.33535293$, $t_3 \approx 0.33541803 \approx t_4$. Since $f''(t) = -\cos t - 2 < 0$ for all $t$, $f(0.33541803) \approx 1.16718557$ is the absolute maximum.

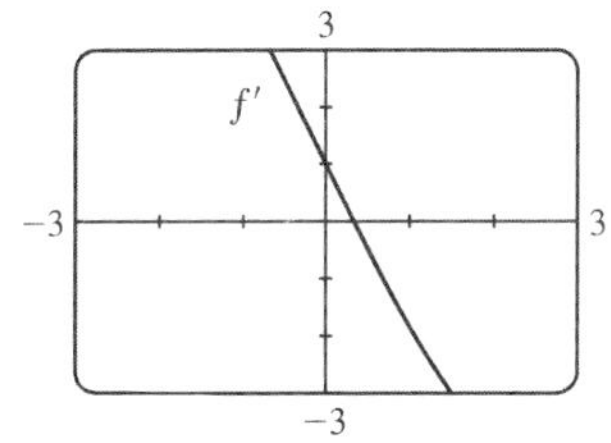


**48.** Graphing $y = \sin x$ and $y = x^2 - 3x + 1$ shows that there are two roots, one about 0.3 and the other about 2.8. $f(x) = \sin x - x^2 + 3x - 1 \Rightarrow$

$f'(x) = \cos x - 2x + 3 \Rightarrow x_{n+1} = x_n - \frac{\sin x_n - x_n^2 + 3x_n - 1}{\cos x_n - 2x_n + 3}$.

Now $x_1 = 0.3 \Rightarrow x_2 \approx 0.268552 \Rightarrow x_3 \approx 0.268881 \approx x_4$ and $x_1 = 2.8 \Rightarrow x_2 \approx 2.770354 \Rightarrow x_3 \approx 2.770058 \approx x_4$, so to six decimal places, the roots are 0.268881 and 2.770058.

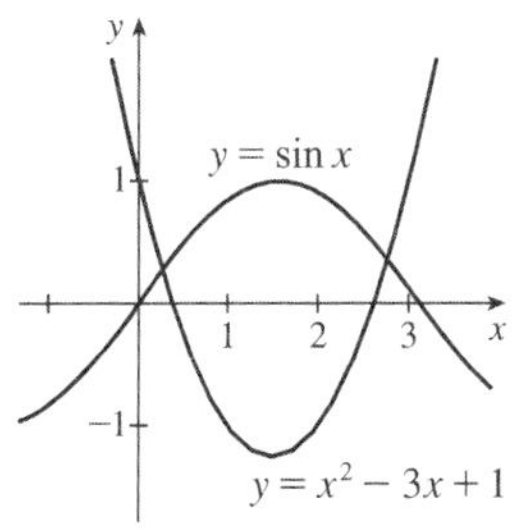


**49.** $f(x) = e^x - (2/\sqrt{x}) = e^x - 2x^{-1/2} \Rightarrow F(x) = e^x - 2\frac{x^{-1/2+1}}{-1/2+1} + C = e^x - 2\frac{x^{1/2}}{1/2} + C = e^x - 4\sqrt{x} + C$

**50.** $g(t) = \frac{1+t}{\sqrt{t}} = \frac{1}{\sqrt{t}} + \frac{t}{\sqrt{t}} = t^{-1/2} + t^{1/2} \Rightarrow G(t) = 2t^{1/2} + \frac{2}{3}t^{3/2} + C$

**51.** $f'(x) = 2/(1 + x^2) \Rightarrow f(x) = 2\arctan x + C$. $f(0) = 2\arctan 0 + C = 0 + C = C$ and $f(0) = -1 \Rightarrow C = -1$. Therefore, $f(x) = 2\arctan x - 1$.

**52.** $f'(u) = \frac{u^2 + \sqrt{u}}{u} = u + u^{-1/2} \Rightarrow f(u) = \frac{1}{2}u^2 + 2u^{1/2} + C$.

$f(1) = \frac{1}{2} + 2 + C$ and $f(1) = 3 \Rightarrow C = \frac{1}{2}$, so $f(u) = \frac{1}{2}u^2 + 2\sqrt{u} + \frac{1}{2}$.

**53.** $f''(x) = 1 - 6x + 48x^2 \Rightarrow f'(x) = x - 3x^2 + 16x^3 + C$. $f'(0) = C$ and $f'(0) = 2 \Rightarrow C = 2$, so $f'(x) = x - 3x^2 + 16x^3 + 2$ and hence, $f(x) = \frac{1}{2}x^2 - x^3 + 4x^4 + 2x + D$.

$f(0) = D$ and $f(0) = 1 \Rightarrow D = 1$, so $f(x) = \frac{1}{2}x^2 - x^3 + 4x^4 + 2x + 1$.

**54.** $f''(x) = 2x^3 + 3x^2 - 4x + 5 \Rightarrow f'(x) = \frac{1}{2}x^4 + x^3 - 2x^2 + 5x + C \Rightarrow$

$f(x) = \frac{1}{10}x^5 + \frac{1}{4}x^4 - \frac{2}{3}x^3 + \frac{5}{2}x^2 + Cx + D$. $f(0) = D$ and $f(0) = 2 \Rightarrow D = 2$.

$f(1) = \frac{1}{10} + \frac{1}{4} - \frac{2}{3} + \frac{5}{2} + C + 2$ and $f(1) = 0 \Rightarrow C = -\frac{6}{60} - \frac{15}{60} + \frac{40}{60} - \frac{150}{60} - \frac{120}{60} = -\frac{251}{60}$, so

$f(x) = \frac{1}{10}x^5 + \frac{1}{4}x^4 - \frac{2}{3}x^3 + \frac{5}{2}x^2 - \frac{251}{60}x + 2$.

**55.** (a) Since $f$ is 0 just to the left of the $y$-axis, we must have a minimum of $F$ at the same place since we are increasing through $(0, 0)$ on $F$. There must be a local maximum to the left of $x = -3$, since $f$ changes from positive to negative there.

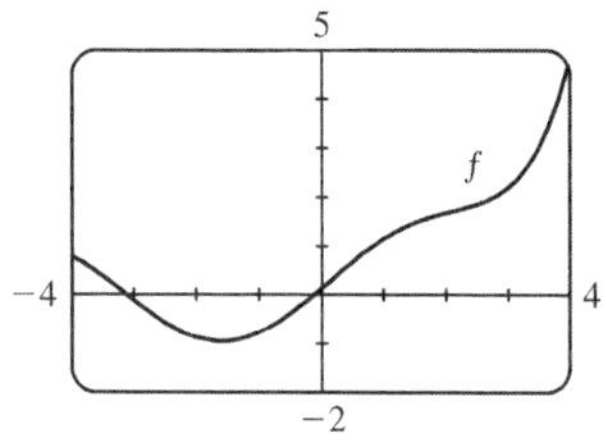


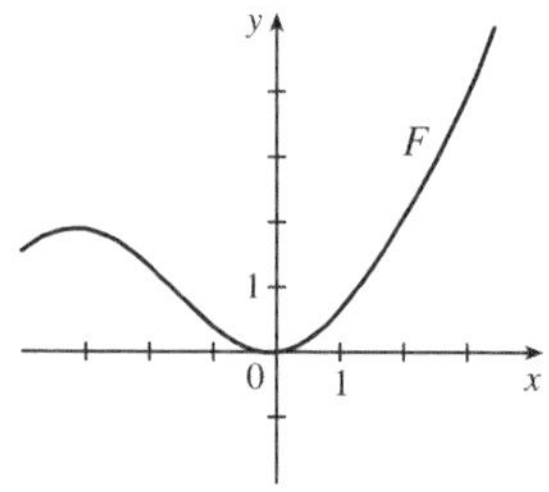


(b) $f(x) = 0.1e^x + \sin x \quad \Rightarrow$

$F(x) = 0.1e^x - \cos x + C$. $F(0) = 0 \quad \Rightarrow$

$0.1 - 1 + C = 0 \quad \Rightarrow \quad C = 0.9$, so

$F(x) = 0.1e^x - \cos x + 0.9$.

(c)

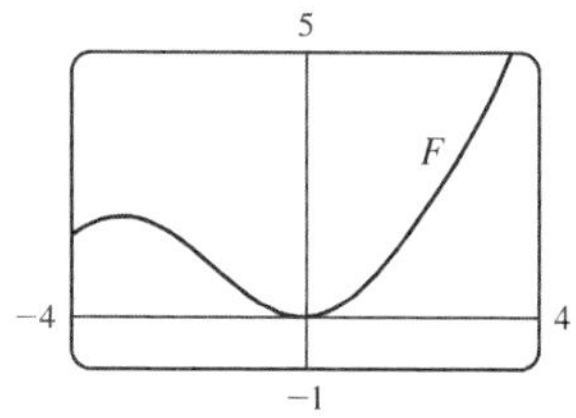


**56.** On $(0, 1)$, $f(x) = x^2$ since $f'(x) = 2x$ and $f(0) = 0$. On $(1, 3)$, $f$ is linear with slope $-1$ since $f'(x) = -1$. If $x > 3$, $f$ is linear with slope 1 since $f'(x) = 1$. Because $f$ is an even function, we can just reflect this graph through the $y$-axis to get the complete graph.

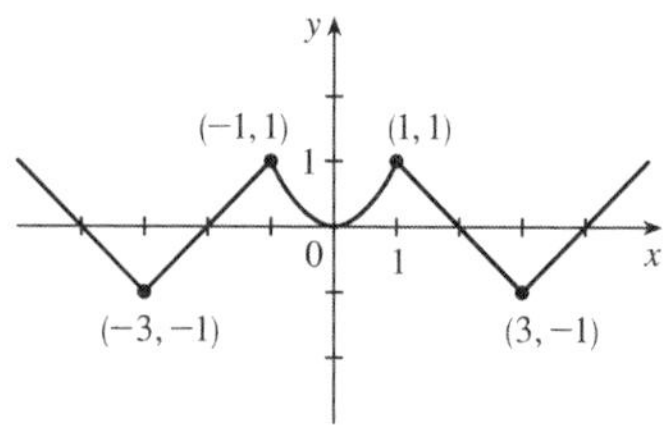


**57.** Choosing the positive direction to be upward, we have $a(t) = -9.8 \quad \Rightarrow \quad v(t) = -9.8t + v_0$, but $v(0) = 0 = v_0 \quad \Rightarrow$ $v(t) = -9.8t = s'(t) \quad \Rightarrow \quad s(t) = -4.9t^2 + s_0$, but $s(0) = s_0 = 500 \quad \Rightarrow \quad s(t) = -4.9t^2 + 500$. When $s = 0$, $-4.9t^2 + 500 = 0 \quad \Rightarrow \quad t_1 = \sqrt{\frac{500}{4.9}} \approx 10.1 \quad \Rightarrow \quad v(t_1) = -9.8\sqrt{\frac{500}{4.9}} \approx -98.995$ m/s. Since the canister has been designed to withstand an impact velocity of 100 m/s, the canister will *not burst*.

**58.** $f(x) = x^4 + x^3 + cx^2 \quad \Rightarrow \quad f'(x) = 4x^3 + 3x^2 + 2cx$. This is 0 when $x(4x^2 + 3x + 2c) = 0 \quad \Leftrightarrow \quad x = 0$ or $4x^2 + 3x + 2c = 0$. Using the quadratic formula, we find that the roots of this last equation are $x = \dfrac{-3 \pm \sqrt{9 - 32c}}{8}$. Now if $9 - 32c < 0 \quad \Leftrightarrow \quad c > \frac{9}{32}$, then $(0, 0)$ is the only critical point, a minimum. If $c = \frac{9}{32}$, then there are two critical points (a minimum at $x = 0$, and a horizontal tangent with no maximum or minimum at $x = -\frac{3}{8}$) and if $c < \frac{9}{32}$, then there are three critical points except when $c = 0$, in which case the root with the $+$ sign coincides with the critical point at $x = 0$. For $0 < c < \frac{9}{32}$, there is a minimum at $x = -\dfrac{3}{8} - \dfrac{\sqrt{9 - 32c}}{8}$, a maximum at $x = -\dfrac{3}{8} + \dfrac{\sqrt{9 - 32c}}{8}$, and a minimum at $x = 0$. For $c = 0$, there is a minimum at $x = -\frac{3}{4}$ and a horizontal tangent with no extremum at $x = 0$, and for $c < 0$, there is a maximum at $x = 0$, and there are minima at $x = -\dfrac{3}{8} \pm \dfrac{\sqrt{9 - 32c}}{8}$. Now we calculate $f''(x) = 12x^2 + 6x + 2c$.

The roots of this equation are $x = \dfrac{-6 \pm \sqrt{36 - 4 \cdot 12 \cdot 2c}}{24}$. So if $36 - 96c \le 0 \;\Leftrightarrow\; c \ge \frac{3}{8}$, then there is no inflection point. If $c < \frac{3}{8}$, then there are two inflection points at $x = -\dfrac{1}{4} \pm \dfrac{\sqrt{9 - 24c}}{12}$.

| Value of $c$ | No. of CP | No. of IP |
|---|---|---|
| $c < 0$ | 3 | 2 |
| $c = 0$ | 2 | 2 |
| $0 < c < \frac{9}{32}$ | 3 | 2 |
| $c = \frac{9}{32}$ | 2 | 2 |
| $\frac{9}{32} < c < \frac{3}{8}$ | 1 | 2 |
| $c \ge \frac{3}{8}$ | 1 | 0 |

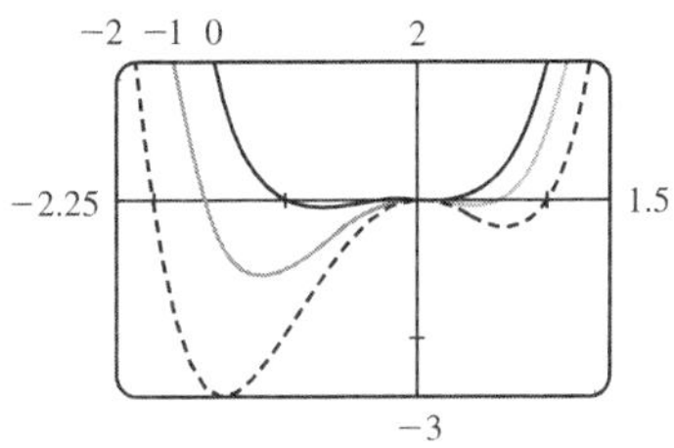


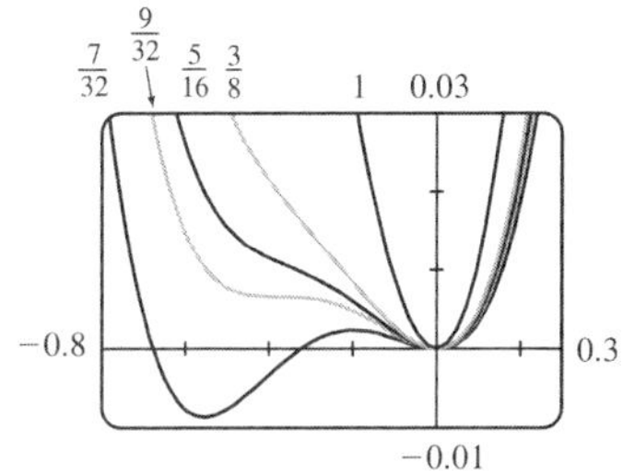


**59.** (a)

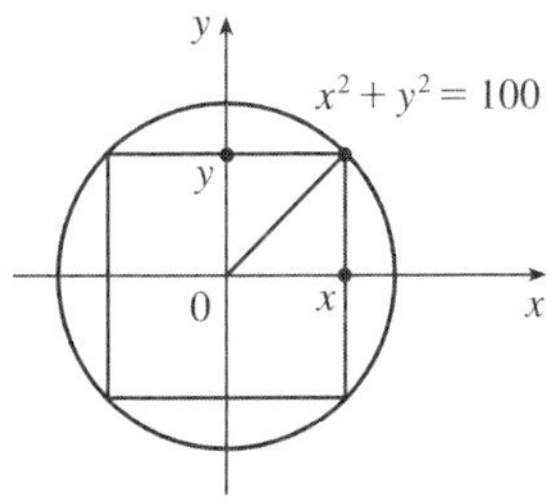


The cross-sectional area of the rectangular beam is

$A = 2x \cdot 2y = 4xy = 4x\sqrt{100 - x^2}$, $0 \le x \le 10$, so

$$\frac{dA}{dx} = 4x\left(\tfrac{1}{2}\right)\left(100 - x^2\right)^{-1/2}(-2x) + \left(100 - x^2\right)^{1/2} \cdot 4$$

$$= \frac{-4x^2}{\left(100 - x^2\right)^{1/2}} + 4\left(100 - x^2\right)^{1/2} = \frac{4\left[-x^2 + \left(100 - x^2\right)\right]}{\left(100 - x^2\right)^{1/2}}.$$

$\dfrac{dA}{dx} = 0$ when $-x^2 + \left(100 - x^2\right) = 0 \;\Rightarrow\; x^2 = 50 \;\Rightarrow\; x = \sqrt{50} \approx 7.07 \;\Rightarrow\; y = \sqrt{100 - \left(\sqrt{50}\right)^2} = \sqrt{50}.$

Since $A(0) = A(10) = 0$, the rectangle of maximum area is a square.

(b)

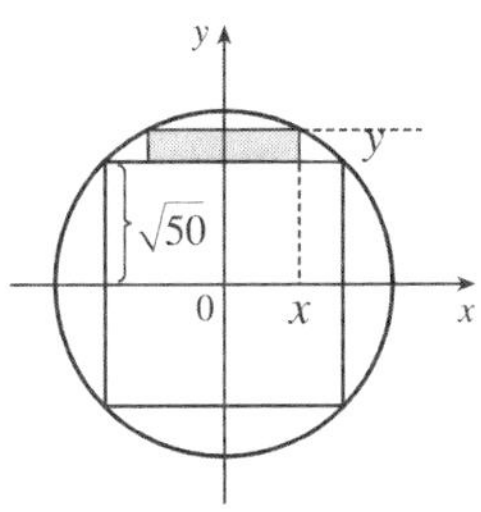


The cross-sectional area of each rectangular plank (shaded in the figure) is

$A = 2x\left(y - \sqrt{50}\right) = 2x\left[\sqrt{100 - x^2} - \sqrt{50}\right]$, $0 \le x \le \sqrt{50}$, so

$$\frac{dA}{dx} = 2\left(\sqrt{100 - x^2} - \sqrt{50}\right) + 2x\left(\tfrac{1}{2}\right)(100 - x^2)^{-1/2}(-2x)$$

$$= 2(100 - x^2)^{1/2} - 2\sqrt{50} - \frac{2x^2}{(100 - x^2)^{1/2}}$$

Set $\dfrac{dA}{dx} = 0$: $(100 - x^2) - \sqrt{50}\,(100 - x^2)^{1/2} - x^2 = 0 \;\Rightarrow\; 100 - 2x^2 = \sqrt{50}\,(100 - x^2)^{1/2} \;\Rightarrow$

$10{,}000 - 400x^2 + 4x^4 = 50(100 - x^2) \;\Rightarrow\; 4x^4 - 350x^2 + 5000 = 0 \;\Rightarrow$

$2x^4 - 175x^2 + 2500 = 0 \;\Rightarrow\; x^2 = \dfrac{175 \pm \sqrt{10{,}625}}{4} \approx 69.52$ or $17.98 \;\Rightarrow\; x \approx 8.34$ or $4.24$.

But $8.34 > \sqrt{50}$, so $x_1 \approx 4.24 \;\Rightarrow\; y - \sqrt{50} = \sqrt{100 - x_1^2} - \sqrt{50} \approx 1.99$. Each plank should have dimensions about $8\frac{1}{2}$ inches by 2 inches.

(c) From the figure in part (a), the width is $2x$ and the depth is $2y$, so the strength is

$S = k(2x)(2y)^2 = 8kxy^2 = 8kx\left(100 - x^2\right) = 800kx - 8kx^3$, $0 \le x \le 10$. $dS/dx = 800k - 24kx^2 = 0$ when

$24kx^2 = 800k \;\Rightarrow\; x^2 = \frac{100}{3} \;\Rightarrow\; x = \frac{10}{\sqrt{3}} \;\Rightarrow\; y = \sqrt{\frac{200}{3}} = \frac{10\sqrt{2}}{\sqrt{3}} = \sqrt{2}\,x$. Since $S(0) = S(10) = 0$, the maximum strength occurs when $x = \frac{10}{\sqrt{3}}$. The dimensions should be $\frac{20}{\sqrt{3}} \approx 11.55$ inches by $\frac{20\sqrt{2}}{\sqrt{3}} \approx 16.33$ inches.

**60.** (a)

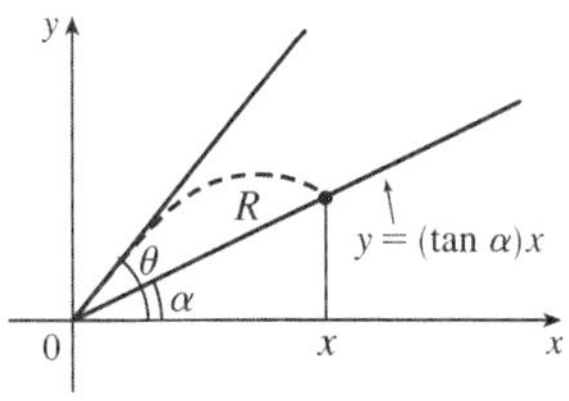


$y = (\tan\theta)x - \dfrac{g}{2v^2\cos^2\theta}x^2$. The parabola intersects the line when

$$(\tan\alpha)x = (\tan\theta)x - \frac{g}{2v^2\cos^2\theta}x^2 \Rightarrow x = \frac{(\tan\theta - \tan\alpha)2v^2\cos^2\theta}{g} \Rightarrow$$

$$R(\theta) = \frac{x}{\cos\alpha} = \left(\frac{\sin\theta}{\cos\theta} - \frac{\sin\alpha}{\cos\alpha}\right)\frac{2v^2\cos^2\theta}{g\cos\alpha} = \left(\frac{\sin\theta}{\cos\theta} - \frac{\sin\alpha}{\cos\alpha}\right)(\cos\theta\,\cos\alpha)\frac{2v^2\cos\theta}{g\cos^2\alpha}$$

$$= (\sin\theta\,\cos\alpha - \sin\alpha\,\cos\theta)\frac{2v^2\cos\theta}{g\cos^2\alpha} = \sin(\theta - \alpha)\frac{2v^2\cos\theta}{g\cos^2\alpha}$$

(b) $$R'(\theta) = \frac{2v^2}{g\cos^2\alpha}[\cos\theta\cdot\cos(\theta-\alpha) + \sin(\theta-\alpha)(-\sin\theta)] = \frac{2v^2}{g\cos^2\alpha}\cos[\theta + (\theta - \alpha)]$$

$$= \frac{2v^2}{g\cos^2\alpha}\cos(2\theta - \alpha) = 0$$

when $\cos(2\theta - \alpha) = 0 \Rightarrow 2\theta - \alpha = \frac{\pi}{2} \Rightarrow \theta = \dfrac{\pi/2 + \alpha}{2} = \dfrac{\pi}{4} + \dfrac{\alpha}{2}$. The First Derivative Test shows that this gives a maximum value for $R(\theta)$. [This could be done without calculus by applying the formula for $\sin x\cos y$ to $R(\theta)$.]

(c)

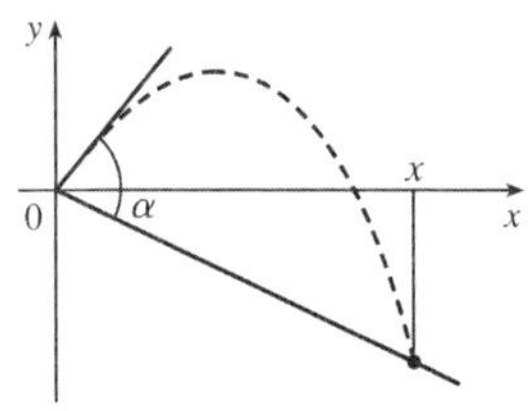


Replacing $\alpha$ by $-\alpha$ in part (a), we get $R(\theta) = \dfrac{2v^2\cos\theta\,\sin(\theta + \alpha)}{g\cos^2\alpha}$.

Proceeding as in part (b), or simply by replacing $\alpha$ by $-\alpha$ in the result of part (b), we see that $R(\theta)$ is maximized when $\theta = \dfrac{\pi}{4} - \dfrac{\alpha}{2}$.

**61.** (a) $$I = \frac{k\cos\theta}{d^2} = \frac{k(h/d)}{d^2} = k\frac{h}{d^3} = k\frac{h}{\left(\sqrt{40^2 + h^2}\right)^3} = k\frac{h}{(1600 + h^2)^{3/2}} \Rightarrow$$

$$\frac{dI}{dh} = k\frac{(1600 + h^2)^{3/2} - h\frac{3}{2}(1600 + h^2)^{1/2}\cdot 2h}{[(1600 + h^2)^{3/2}]^2} = \frac{k(1600 + h^2)^{1/2}(1600 + h^2 - 3h^2)}{(1600 + h^2)^{3/2}}$$

$$= \frac{k(1600 - 2h^2)}{(1600 + h^2)^{5/2}} \quad [k \text{ is the constant of proportionality}]$$

Set $dI/dh = 0$: $1600 - 2h^2 = 0 \Rightarrow h^2 = 800 \Rightarrow h = \sqrt{800} = 20\sqrt{2}$. By the First Derivative Test, $I$ has a local maximum at $h = 20\sqrt{2} \approx 28$ ft.

(b)

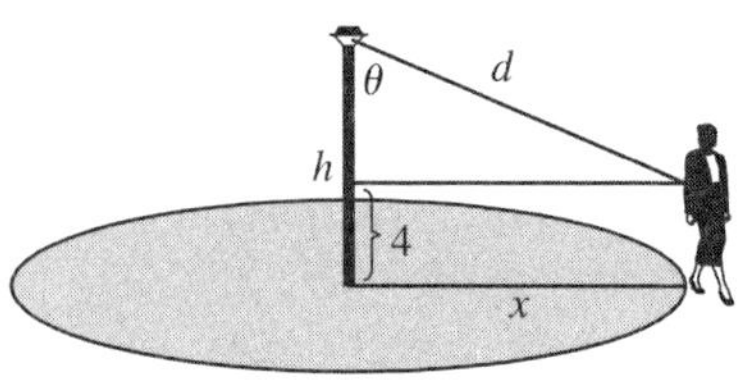


$\dfrac{dx}{dt} = 4$ ft/s

$$I = \frac{k\cos\theta}{d^2} = \frac{k[(h-4)/d]}{d^2} = \frac{k(h-4)}{d^3}$$

$$= \frac{k(h-4)}{[(h-4)^2 + x^2]^{3/2}} = k(h-4)\left[(h-4)^2 + x^2\right]^{-3/2}$$

$$\frac{dI}{dt} = \frac{dI}{dx}\cdot\frac{dx}{dt} = k(h-4)\left(-\tfrac{3}{2}\right)\left[(h-4)^2 + x^2\right]^{-5/2}\cdot 2x\cdot\frac{dx}{dt}$$

$$= k(h-4)(-3x)\left[(h-4)^2 + x^2\right]^{-5/2}\cdot 4 = \frac{-12xk(h-4)}{[(h-4)^2 + x^2]^{5/2}}$$

$$\left.\frac{dI}{dt}\right|_{x=40} = -\frac{480k(h-4)}{[(h-4)^2 + 1600]^{5/2}}$$

## □ FOCUS ON PROBLEM SOLVING

1. Let $y = f(x) = e^{-x^2}$. The area of the rectangle under the curve from $-x$ to $x$ is $A(x) = 2xe^{-x^2}$ where $x \geq 0$. We maximize $A(x)$: $A'(x) = 2e^{-x^2} - 4x^2e^{-x^2} = 2e^{-x^2}(1 - 2x^2) = 0 \quad \Rightarrow \quad x = \frac{1}{\sqrt{2}}$. This gives a maximum since $A'(x) > 0$ for $0 \leq x < \frac{1}{\sqrt{2}}$ and $A'(x) < 0$ for $x > \frac{1}{\sqrt{2}}$. We next determine the points of inflection of $f(x)$. Notice that $f'(x) = -2xe^{-x^2} = -A(x)$. So $f''(x) = -A'(x)$ and hence, $f''(x) < 0$ for $-\frac{1}{\sqrt{2}} < x < \frac{1}{\sqrt{2}}$ and $f''(x) > 0$ for $x < -\frac{1}{\sqrt{2}}$ and $x > \frac{1}{\sqrt{2}}$. So $f(x)$ changes concavity at $x = \pm\frac{1}{\sqrt{2}}$, and the two vertices of the rectangle of largest area are at the inflection points.

2. Let $f(x) = \sin x - \cos x$ on $[0, 2\pi]$ since $f$ has period $2\pi$. $f'(x) = \cos x + \sin x = 0 \quad \Leftrightarrow \quad \cos x = -\sin x \quad \Leftrightarrow$ $\tan x = -1 \quad \Leftrightarrow \quad x = \frac{3\pi}{4}$ or $\frac{7\pi}{4}$. Evaluating $f$ at its critical numbers and endpoints, we get $f(0) = -1$, $f\left(\frac{3\pi}{4}\right) = \sqrt{2}$, $f\left(\frac{7\pi}{4}\right) = -\sqrt{2}$, and $f(2\pi) = -1$. So $f$ has absolute maximum value $\sqrt{2}$ and absolute minimum value $-\sqrt{2}$. Thus, $-\sqrt{2} \leq \sin x - \cos x \leq \sqrt{2} \quad \Rightarrow \quad |\sin x - \cos x| \leq \sqrt{2}$.

3. First, we recognize some symmetry in the inequality: $\dfrac{e^{x+y}}{xy} \geq e^2 \quad \Leftrightarrow \quad \dfrac{e^x}{x} \cdot \dfrac{e^y}{y} \geq e \cdot e$. This suggests that we need to show that $\dfrac{e^x}{x} \geq e$ for $x > 0$. If we can do this, then the inequality $\dfrac{e^y}{y} \geq e$ is true, and the given inequality follows. $f(x) = \dfrac{e^x}{x} \quad \Rightarrow$ $f'(x) = \dfrac{xe^x - e^x}{x^2} = \dfrac{e^x(x-1)}{x^2} = 0 \quad \Rightarrow \quad x = 1$. By the First Derivative Test, we have a minimum of $f(1) = e$, so $e^x/x \geq e$ for all $x$.

4. $x^2y^2(4 - x^2)(4 - y^2) = x^2(4 - x^2)y^2(4 - y^2) = f(x)f(y)$, where $f(t) = t^2(4 - t^2)$. We will show that $0 \leq f(t) \leq 4$ for $|t| \leq 2$, which gives $0 \leq f(x)f(y) \leq 16$ for $|x| \leq 2$ and $|y| \leq 2$.

   $f(t) = 4t^2 - t^4 \quad \Rightarrow \quad f'(t) = 8t - 4t^3 = 4t(2 - t^2) = 0 \quad \Rightarrow \quad t = 0$ or $\pm\sqrt{2}$.

   $f(0) = 0$, $f(\pm\sqrt{2}) = 2(4 - 2) = 4$, and $f(2) = 0$. So 0 is the absolute minimum value of $f(t)$ on $[-2, 2]$ and 4 is the absolute maximum value of $f(t)$ on $[-2, 2]$. We conclude that $0 \leq f(t) \leq 4$ for $|t| \leq 2$ and hence, $0 \leq f(x)f(y) \leq 4^2$ or $0 \leq x^2(4 - x^2)y^2(4 - y^2) \leq 16$.

5. First we show that $x(1 - x) \leq \frac{1}{4}$ for all $x$. Let $f(x) = x(1 - x) = x - x^2$. Then $f'(x) = 1 - 2x$. This is 0 when $x = \frac{1}{2}$ and $f'(x) > 0$ for $x < \frac{1}{2}$, $f'(x) < 0$ for $x > \frac{1}{2}$, so the absolute maximum of $f$ is $f\left(\frac{1}{2}\right) = \frac{1}{4}$. Thus, $x(1 - x) \leq \frac{1}{4}$ for all $x$.

   Now suppose that the given assertion is false, that is, $a(1 - b) > \frac{1}{4}$ and $b(1 - a) > \frac{1}{4}$. Multiply these inequalities: $a(1 - b)b(1 - a) > \frac{1}{16} \quad \Rightarrow \quad [a(1 - a)]\,[b(1 - b)] > \frac{1}{16}$. But we know that $a(1 - a) \leq \frac{1}{4}$ and $b(1 - b) \leq \frac{1}{4} \quad \Rightarrow$ $[a(1 - a)]\,[b(1 - b)] \leq \frac{1}{16}$. Thus, we have a contradiction, so the given assertion is proved.

**6.** Let $P(a, 1-a^2)$ be the point of contact. The equation of the tangent line at $P$ is $y-(1-a^2)=(-2a)(x-a) \Rightarrow$ $y-1+a^2=-2ax+2a^2 \Rightarrow y=-2ax+a^2+1$. To find the $x$-intercept, put $y=0$: $2ax=a^2+1 \Rightarrow$ $x=\dfrac{a^2+1}{2a}$. To find the $y$-intercept, put $x=0$: $y=a^2+1$. Therefore, the area of the triangle is $\dfrac{1}{2}\left(\dfrac{a^2+1}{2a}\right)(a^2+1)=\dfrac{(a^2+1)^2}{4a}$. Therefore, we minimize the function $A(a)=\dfrac{(a^2+1)^2}{4a}$, $0<a\leq 1$.

$$A'(a)=\frac{(4a)2(a^2+1)(2a)-(a^2+1)^2(4)}{16a^2}=\frac{(a^2+1)[4a^2-(a^2+1)]}{4a^2}=\frac{(a^2+1)(3a^2-1)}{4a^2}.$$

$A'(a)=0$ when $3a^2-1=0 \Rightarrow a=\frac{1}{\sqrt{3}}$. $A'(a)<0$ for $a<\frac{1}{\sqrt{3}}$, $A'(a)>0$ for $a>\frac{1}{\sqrt{3}}$. So by the First Derivative Test, there is an absolute minimum when $a=\frac{1}{\sqrt{3}}$. The required point is $\left(\frac{1}{\sqrt{3}},\frac{2}{3}\right)$ and the corresponding minimum area is $A\left(\frac{1}{\sqrt{3}}\right)=\frac{4\sqrt{3}}{9}$.

**7.** Differentiating $x^2+xy+y^2=12$ implicitly with respect to $x$ gives $2x+y+x\dfrac{dy}{dx}+2y\dfrac{dy}{dx}=0$, so $\dfrac{dy}{dx}=-\dfrac{2x+y}{x+2y}$.

At a highest or lowest point, $\dfrac{dy}{dx}=0 \Leftrightarrow y=-2x$. Substituting $-2x$ for $y$ in the original equation gives $x^2+x(-2x)+(-2x)^2=12$, so $3x^2=12$ and $x=\pm 2$. If $x=2$, then $y=-2x=-4$, and if $x=-2$ then $y=4$. Thus, the highest and lowest points are $(-2,4)$ and $(2,-4)$.

**8.** Let the circle have radius $r$, so $|OP|=|OQ|=r$, where $O$ is the center of the circle. Now $\angle POR$ has measure $\frac{1}{2}\theta$, and $\angle OPR$ is a right angle, so $\tan\frac{1}{2}\theta=\dfrac{|PR|}{r}$ and the area of $\triangle OPR$ is $\frac{1}{2}|OP|\,|PR|=\frac{1}{2}r^2\tan\frac{1}{2}\theta$. The area of the sector cut by $OP$ and $OR$ is $\frac{1}{2}r^2\left(\frac{1}{2}\theta\right)=\frac{1}{4}r^2\theta$. Let $S$ be the intersection of $PQ$ and $OR$. Then $\sin\frac{1}{2}\theta=\dfrac{|PS|}{r}$ and $\cos\frac{1}{2}\theta=\dfrac{|OS|}{r}$, and the area of $\triangle OSP$ is $\frac{1}{2}|OS|\,|PS|=\frac{1}{2}\left(r\cos\frac{1}{2}\theta\right)\left(r\sin\frac{1}{2}\theta\right)=\frac{1}{2}r^2\sin\frac{1}{2}\theta\cos\frac{1}{2}\theta=\frac{1}{4}r^2\sin\theta$.

So $B(\theta)=2\left(\frac{1}{2}r^2\tan\frac{1}{2}\theta-\frac{1}{4}r^2\theta\right)=r^2\left(\tan\frac{1}{2}\theta-\frac{1}{2}\theta\right)$ and $A(\theta)=2\left(\frac{1}{4}r^2\theta-\frac{1}{4}r^2\sin\theta\right)=\frac{1}{2}r^2(\theta-\sin\theta)$. Thus,

$$\begin{aligned}\lim_{\theta\to 0^+}\frac{A(\theta)}{B(\theta)}&=\lim_{\theta\to 0^+}\frac{\frac{1}{2}r^2(\theta-\sin\theta)}{r^2\left(\tan\frac{1}{2}\theta-\frac{1}{2}\theta\right)}=\lim_{\theta\to 0^+}\frac{\theta-\sin\theta}{2\left(\tan\frac{1}{2}\theta-\frac{1}{2}\theta\right)}\overset{\text{H}}{=}\lim_{\theta\to 0^+}\frac{1-\cos\theta}{2\left(\frac{1}{2}\sec^2\frac{1}{2}\theta-\frac{1}{2}\right)}\\&=\lim_{\theta\to 0^+}\frac{1-\cos\theta}{\sec^2\frac{1}{2}\theta-1}=\lim_{\theta\to 0^+}\frac{1-\cos\theta}{\tan^2\frac{1}{2}\theta}\overset{\text{H}}{=}\lim_{\theta\to 0^+}\frac{\sin\theta}{2\left(\tan\frac{1}{2}\theta\right)\left(\sec^2\frac{1}{2}\theta\right)\frac{1}{2}}\\&=\lim_{\theta\to 0^+}\frac{\sin\theta\cos^3\frac{1}{2}\theta}{\sin\frac{1}{2}\theta}=\lim_{\theta\to 0^+}\frac{\left(2\sin\frac{1}{2}\theta\cos\frac{1}{2}\theta\right)\cos^3\frac{1}{2}\theta}{\sin\frac{1}{2}\theta}=2\lim_{\theta\to 0^+}\cos^4\left(\tfrac{1}{2}\theta\right)=2(1)^4=2\end{aligned}$$

**9.** Let $L=\displaystyle\lim_{x\to 0}\frac{ax^2+\sin bx+\sin cx+\sin dx}{3x^2+5x^4+7x^6}$. Now $L$ has the indeterminate form of type $\frac{0}{0}$, so we can apply l'Hospital's Rule. $L=\displaystyle\lim_{x\to 0}\frac{2ax+b\cos bx+c\cos cx+d\cos dx}{6x+20x^3+42x^5}$. The denominator approaches 0 as $x\to 0$, so the numerator must also approach 0 (because the limit exists). But the numerator approaches $0+b+c+d$, so $b+c+d=0$. Apply l'Hospital's Rule again. $L=\displaystyle\lim_{x\to 0}\frac{2a-b^2\sin bx-c^2\sin cx-d^2\sin dx}{6+60x^2+210x^4}=\frac{2a-0}{6+0}=\frac{2a}{6}$, which must equal 8.

$\dfrac{2a}{6}=8 \Rightarrow a=24$. Thus, $a+b+c+d=a+(b+c+d)=24+0=24$.

**10.** To sketch the region $\{(x,y) \mid 2xy \le |x-y| \le x^2+y^2\}$, we consider two cases.

*Case 1:* $x \ge y$ This is the case in which $(x,y)$ lies on or below the line $y=x$. The double inequality becomes $2xy \le x-y \le x^2+y^2$. The right-hand inequality holds if and only if $x^2-x+y^2+y \ge 0 \Leftrightarrow$ $\left(x-\frac{1}{2}\right)^2+\left(y+\frac{1}{2}\right)^2 \ge \frac{1}{2} \Leftrightarrow (x,y)$ lies on or outside the circle with radius $\frac{1}{\sqrt{2}}$ centered at $\left(\frac{1}{2},-\frac{1}{2}\right)$.

The left-hand inequality holds if and only if $2xy-x+y \le 0 \Leftrightarrow xy-\frac{1}{2}x+\frac{1}{2}y \le 0 \Leftrightarrow$ $\left(x+\frac{1}{2}\right)\left(y-\frac{1}{2}\right) \le -\frac{1}{4} \Leftrightarrow (x,y)$ lies on or below the hyperbola $\left(x+\frac{1}{2}\right)\left(y-\frac{1}{2}\right)=-\frac{1}{4}$, which passes through the origin and approaches the lines $y=\frac{1}{2}$ and $x=-\frac{1}{2}$ asymptotically.

*Case 2:* $y \ge x$ This is the case in which $(x,y)$ lies on or above the line $y=x$. The double inequality becomes $2xy \le y-x \le x^2+y^2$. The right-hand inequality holds if and only if $x^2+x+y^2-y \ge 0 \Leftrightarrow$ $\left(x+\frac{1}{2}\right)^2+\left(y-\frac{1}{2}\right)^2 \ge \frac{1}{2} \Leftrightarrow (x,y)$ lies on or outside the circle of radius $\frac{1}{\sqrt{2}}$ centered at $\left(-\frac{1}{2},\frac{1}{2}\right)$. The left-hand inequality holds if and only if $2xy+x-y \le 0 \Leftrightarrow xy+\frac{1}{2}x-\frac{1}{2}y \le 0 \Leftrightarrow \left(x-\frac{1}{2}\right)\left(y+\frac{1}{2}\right) \le -\frac{1}{4} \Leftrightarrow (x,y)$ lies on or above the left-hand branch of the hyperbola $\left(x-\frac{1}{2}\right)\left(y+\frac{1}{2}\right)=-\frac{1}{4}$, which passes through the origin and approaches the lines $y=-\frac{1}{2}$ and $x=\frac{1}{2}$ asymptotically. Therefore, the region of interest consists of the points on or above the left branch of the hyperbola $\left(x-\frac{1}{2}\right)\left(y+\frac{1}{2}\right)=-\frac{1}{4}$ that are on or outside the circle $\left(x+\frac{1}{2}\right)^2+\left(y-\frac{1}{2}\right)^2=\frac{1}{2}$, together with the points on or below the right branch of the hyperbola $\left(x+\frac{1}{2}\right)\left(y-\frac{1}{2}\right)=-\frac{1}{4}$ that are on or outside the circle $\left(x-\frac{1}{2}\right)^2+\left(y+\frac{1}{2}\right)^2=\frac{1}{2}$. Note that the inequalities are unchanged when $x$ and $y$ are interchanged, so the region is symmetric about the line $y=x$. So we need only have analyzed case 1 and then reflected that region about the line $y=x$, instead of considering case 2.

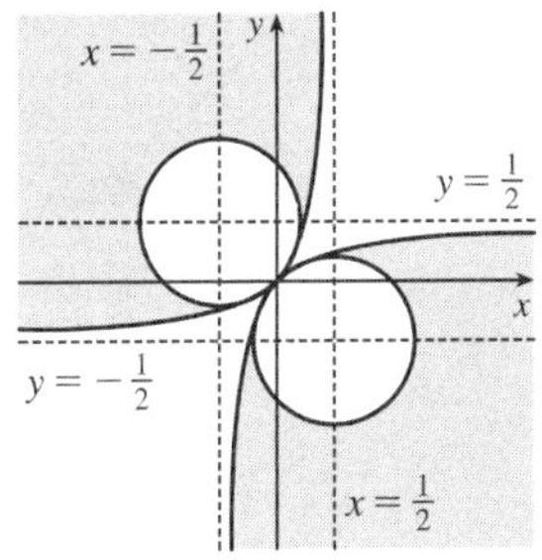


**11.** We first show that $\frac{x}{1+x^2} < \tan^{-1}x$ for $x>0$. Let $f(x)=\tan^{-1}x-\frac{x}{1+x^2}$. Then

$f'(x)=\frac{1}{1+x^2}-\frac{1(1+x^2)-x(2x)}{(1+x^2)^2}=\frac{(1+x^2)-(1-x^2)}{(1+x^2)^2}=\frac{2x^2}{(1+x^2)^2}>0$ for $x>0$. So $f(x)$ is increasing on $(0,\infty)$. Hence, $0<x \Rightarrow 0=f(0)<f(x)=\tan^{-1}x-\frac{x}{1+x^2}$. So $\frac{x}{1+x^2}<\tan^{-1}x$ for $0<x$. We next show that $\tan^{-1}x<x$ for $x>0$. Let $h(x)=x-\tan^{-1}x$. Then $h'(x)=1-\frac{1}{1+x^2}=\frac{x^2}{1+x^2}>0$. Hence, $h(x)$ is increasing on $(0,\infty)$. So for $0<x$, $0=h(0)<h(x)=x-\tan^{-1}x$. Hence, $\tan^{-1}x<x$ for $x>0$, and we conclude that $\frac{x}{1+x^2}<\tan^{-1}x<x$ for $x>0$.

**12.** If $L=\lim\limits_{x\to\infty}\left(\frac{x+a}{x-a}\right)^x$, then $L$ has the indeterminate form $1^\infty$, so

$$\ln L=\lim_{x\to\infty}\ln\left(\frac{x+a}{x-a}\right)^x=\lim_{x\to\infty}x\ln\left(\frac{x+a}{x-a}\right)=\lim_{x\to\infty}\frac{\ln(x+a)-\ln(x-a)}{1/x}\overset{\text{H}}{=}\lim_{x\to\infty}\frac{\frac{1}{x+a}-\frac{1}{x-a}}{-1/x^2}$$

$$=\lim_{x\to\infty}\left[\frac{(x-a)-(x+a)}{(x+a)(x-a)}\cdot\frac{-x^2}{1}\right]=\lim_{x\to\infty}\frac{2ax^2}{x^2-a^2}=\lim_{x\to\infty}\frac{2a}{1-a^2/x^2}=2a$$

Hence, $\ln L=2a$, so $L=e^{2a}$. From the original equation, we want $L=e^1 \Rightarrow 2a=1 \Rightarrow a=\frac{1}{2}$.

**13.** $f(x) = (a^2 + a - 6)\cos 2x + (a-2)x + \cos 1 \;\Rightarrow\; f'(x) = -(a^2 + a - 6)\sin 2x\,(2) + (a-2)$. The derivative exists for all $x$, so the only possible critical points will occur where $f'(x) = 0 \;\Leftrightarrow\; 2(a-2)(a+3)\sin 2x = a - 2 \;\Leftrightarrow$ either $a = 2$ or $2(a+3)\sin 2x = 1$, with the latter implying that $\sin 2x = \dfrac{1}{2(a+3)}$. Since the range of $\sin 2x$ is $[-1, 1]$, this equation has no solution whenever either $\dfrac{1}{2(a+3)} < -1$ or $\dfrac{1}{2(a+3)} > 1$. Solving these inequalities, we get $-\frac{7}{2} < a < -\frac{5}{2}$.

**14.** *Case (i) (first graph):* For $x + y \geq 0$, that is, $y \geq -x$, $|x+y| = x + y \leq e^x \;\Rightarrow\; y \leq e^x - x$.
Note that $y = e^x - x$ is always above the line $y = -x$ and that $y = -x$ is a slant asymptote.

*Case (ii) (second graph):* For $x + y < 0$, that is, $y < -x$, $|x+y| = -x - y \leq e^x \;\Rightarrow\; y \geq -x - e^x$.
Note that $-x - e^x$ is always below the line $y = -x$ and $y = -x$ is a slant asymptote.

Putting the two pieces together gives the third graph.

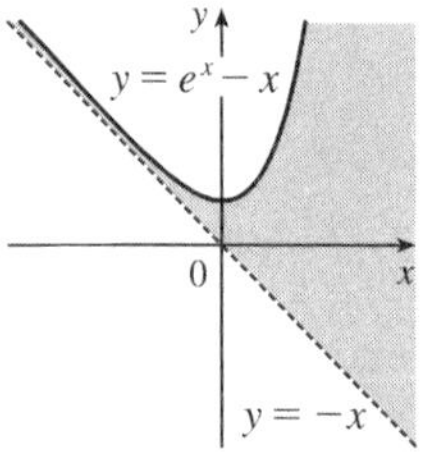


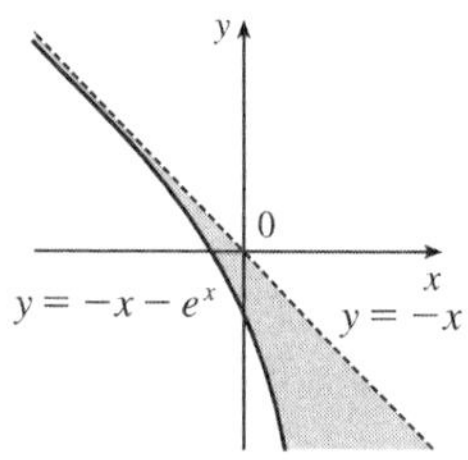


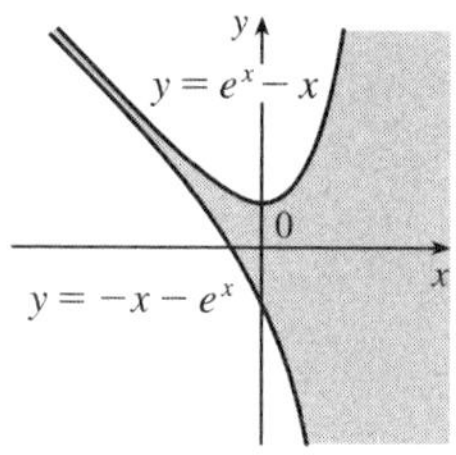


**15.** $y = x^2 \;\Rightarrow\; y' = 2x$, so the slope of the tangent line at $P(a, a^2)$ is $2a$ and the slope of the normal line is $-\dfrac{1}{2a}$ for $a \neq 0$.

An equation of the normal line is $y - a^2 = -\dfrac{1}{2a}(x - a)$. Substitute $x^2$ for $y$ to find the $x$-coordinates of the two points of intersection of the parabola and the normal line. $x^2 - a^2 = -\dfrac{x}{2a} + \dfrac{1}{2} \;\Rightarrow\; 2ax^2 + x - 2a^3 - a = 0 \;\Rightarrow$

$$x = \frac{-1 \pm \sqrt{1 - 4(2a)(-2a^3 - a)}}{2(2a)} = \frac{-1 \pm \sqrt{1 + 16a^4 + 8a^2}}{4a} = \frac{-1 \pm \sqrt{(4a^2+1)^2}}{4a} = \frac{-1 \pm (4a^2+1)}{4a}$$

$$= \frac{4a^2}{4a} \text{ or } \frac{-4a^2 - 2}{4a}, \text{ or equivalently, } a \text{ or } -a - \frac{1}{2a}.$$

So the point $Q$ has coordinates $\left(-a - \dfrac{1}{2a}, \left(-a - \dfrac{1}{2a}\right)^2\right)$. The square $S$ of the distance from $P$ to $Q$ is given by

$$S = \left(a - \frac{1}{2a} - a\right)^2 + \left[\left(-a - \frac{1}{2a}\right)^2 - a^2\right]^2 = \left(-2a - \frac{1}{2a}\right)^2 + \left[\left(a^2 + 1 + \frac{1}{4a^2}\right) - a^2\right]^2$$

$$= \left(4a^2 + 2 + \frac{1}{4a^2}\right) + \left(1 + \frac{1}{4a^2}\right)^2 = \left(4a^2 + 2 + \frac{1}{4a^2}\right) + 1 + \frac{2}{4a^2} + \frac{1}{16a^4} = 4a^2 + 3 + \frac{3}{4a^2} + \frac{1}{16a^4}$$

$S' = 8a - \dfrac{6}{4a^3} - \dfrac{4}{16a^5} = 8a - \dfrac{3}{2a^3} - \dfrac{1}{4a^5} = \dfrac{32a^6 - 6a^2 - 1}{4a^5}$. The only real positive zero of the equation $S' = 0$ is $a = \dfrac{1}{\sqrt{2}}$. Since $S'' = 8 + \dfrac{9}{2a^4} + \dfrac{5}{4a^6} > 0$, $a = \dfrac{1}{\sqrt{2}}$ corresponds to the shortest possible length of the line segment $PQ$.

**16.** A straight line intersects the curve $y = f(x) = x^4 + cx^3 + 12x^2 - 5x + 2$ in four distinct points if and only if the graph of $f$ has two inflection points. $f'(x) = 4x^3 + 3cx^2 + 24x - 5$ and $f''(x) = 12x^2 + 6cx + 24$.

$f''(x) = 0 \quad \Leftrightarrow \quad x = \dfrac{-6c \pm \sqrt{(6c)^2 - 4(12)(24)}}{2(12)}$. There are two distinct roots for $f''(x) = 0$ (and hence two inflection points) if and only if the discriminant is positive; that is, $36c^2 - 1152 > 0 \quad \Leftrightarrow \quad c^2 > 32 \quad \Leftrightarrow \quad |c| > \sqrt{32}$. Thus, the desired values of $c$ are $c < -4\sqrt{2}$ or $c > 4\sqrt{2}$.

**17.** (a) Let $y = |AD|$, $x = |AB|$, and $1/x = |AC|$, so that $|AB| \cdot |AC| = 1$.

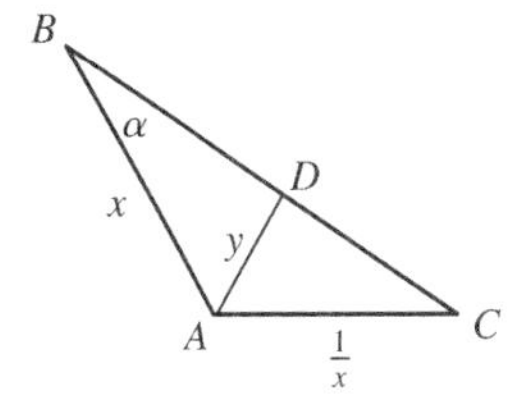


We compute the area $\mathcal{A}$ of $\triangle ABC$ in two ways.

First, $\mathcal{A} = \frac{1}{2}|AB||AC| \sin \frac{2\pi}{3} = \frac{1}{2} \cdot 1 \cdot \frac{\sqrt{3}}{2} = \frac{\sqrt{3}}{4}$.

Second,

$$\begin{aligned} \mathcal{A} &= (\text{area of } \triangle ABD) + (\text{area of } \triangle ACD) \\ &= \tfrac{1}{2}|AB||AD| \sin \tfrac{\pi}{3} + \tfrac{1}{2}|AD||AC| \sin \tfrac{\pi}{3} \\ &= \tfrac{1}{2}xy\tfrac{\sqrt{3}}{2} + \tfrac{1}{2}y(1/x)\tfrac{\sqrt{3}}{2} = \tfrac{\sqrt{3}}{4}y(x + 1/x) \end{aligned}$$

Equating the two expressions for the area, we get $\frac{\sqrt{3}}{4}y\left(x + \frac{1}{x}\right) = \frac{\sqrt{3}}{4} \quad \Leftrightarrow \quad y = \dfrac{1}{x + 1/x} = \dfrac{x}{x^2 + 1}$, $x > 0$.

*Another method:* Use the Law of Sines on the triangles $ABD$ and $ABC$. In $\triangle ABD$, we have $\angle A + \angle B + \angle D = 180° \quad \Leftrightarrow \quad 60° + \alpha + \angle D = 180° \quad \Leftrightarrow \quad \angle D = 120° - \alpha$. Thus,

$\dfrac{x}{y} = \dfrac{\sin(120° - \alpha)}{\sin \alpha} = \dfrac{\sin 120° \cos \alpha - \cos 120° \sin \alpha}{\sin \alpha} = \dfrac{\frac{\sqrt{3}}{2}\cos \alpha + \frac{1}{2}\sin \alpha}{\sin \alpha} \quad \Rightarrow \quad \dfrac{x}{y} = \frac{\sqrt{3}}{2}\cot \alpha + \frac{1}{2}$, and by a similar argument with $\triangle ABC$, $\frac{\sqrt{3}}{2}\cot \alpha = x^2 + \frac{1}{2}$. Eliminating $\cot \alpha$ gives $\dfrac{x}{y} = \left(x^2 + \frac{1}{2}\right) + \frac{1}{2} \quad \Rightarrow$

$y = \dfrac{x}{x^2 + 1}$, $x > 0$.

(b) We differentiate our expression for $y$ with respect to $x$ to find the maximum:

$\dfrac{dy}{dx} = \dfrac{(x^2 + 1) - x(2x)}{(x^2 + 1)^2} = \dfrac{1 - x^2}{(x^2 + 1)^2} = 0$ when $x = 1$. This indicates a maximum by the First Derivative Test, since $y'(x) > 0$ for $0 < x < 1$ and $y'(x) < 0$ for $x > 1$, so the maximum value of $y$ is $y(1) = \frac{1}{2}$.

**18.** Let $x = |AE|$, $y = |AF|$ as shown. The area $\mathcal{A}$ of the $\Delta AEF$ is $\mathcal{A} = \frac{1}{2}xy$. We need to find a relationship between $x$ and $y$, so that we can take the derivative $d\mathcal{A}/dx$ and then find the maximum and minimum areas. Now let $A'$ be the point on which $A$ ends up after the fold has been performed, and let $P$ be the intersection of $AA'$ and $EF$. Note that $AA'$ is perpendicular to $EF$ since we are reflecting $A$ through the line $EF$ to get to $A'$, and that $|AP| = |PA'|$ for the same reason.

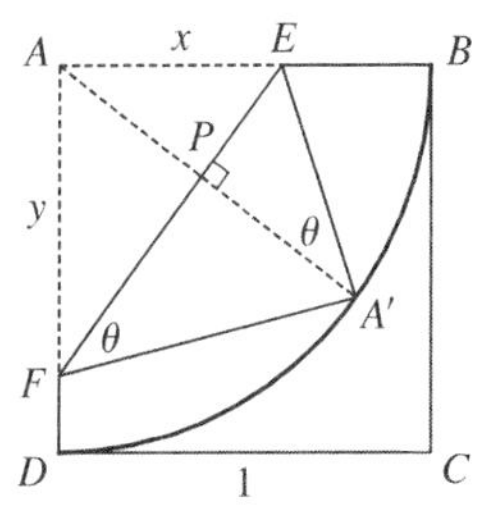


But $|AA'| = 1$, since $AA'$ is a radius of the circle. Since $|AP| + |PA'| = |AA'|$, we have $|AP| = \frac{1}{2}$. Another way to express the area of the triangle is $\mathcal{A} = \frac{1}{2}|EF||AP| = \frac{1}{2}\sqrt{x^2 + y^2}\left(\frac{1}{2}\right) = \frac{1}{4}\sqrt{x^2 + y^2}$. Equating the two expressions for $\mathcal{A}$, we get $\frac{1}{2}xy = \frac{1}{4}\sqrt{x^2 + y^2} \quad \Rightarrow \quad 4x^2y^2 = x^2 + y^2 \quad \Rightarrow \quad y^2(4x^2 - 1) = x^2 \quad \Rightarrow \quad y = x/\sqrt{4x^2 - 1}$.

(Note that we could also have derived this result from the similarity of $\triangle A'PE$ and $\triangle A'FE$; that is,

$\dfrac{|A'P|}{|PE|} = \dfrac{|A'F|}{|A'E|} \quad \Rightarrow \quad \dfrac{\frac{1}{2}}{\sqrt{x^2 - \left(\frac{1}{2}\right)^2}} = \dfrac{y}{x} \quad \Rightarrow \quad y = \dfrac{\frac{1}{2}x}{\sqrt{4x^2 - 1}/2} = \dfrac{x}{\sqrt{4x^2 - 1}}$.) Now we can substitute for $y$ and

calculate $\dfrac{d\mathcal{A}}{dx}$: $\mathcal{A} = \dfrac{1}{2}\dfrac{x^2}{\sqrt{4x^2 - 1}} \quad \Rightarrow \quad \dfrac{d\mathcal{A}}{dx} = \dfrac{1}{2}\left[\dfrac{\sqrt{4x^2 - 1}\,(2x) - x^2\left(\frac{1}{2}\right)\left(4x^2 - 1\right)^{-1/2}(8x)}{4x^2 - 1}\right]$. This is 0 when

$2x\sqrt{4x^2 - 1} - 4x^3\left(4x^2 - 1\right)^{-1/2} = 0 \quad \Leftrightarrow \quad 2x\left(4x^2 - 1\right)^{-1/2}\left[\left(4x^2 - 1\right) - 2x^2\right] = 0 \quad \Rightarrow \quad \left(4x^2 - 1\right) - 2x^2 = 0$

$(x > 0) \quad \Leftrightarrow \quad 2x^2 = 1 \quad \Rightarrow \quad x = \frac{1}{\sqrt{2}}$. So this is one possible value for an extremum. We must also test the endpoints of the interval over which $x$ ranges. The largest value that $x$ can attain is 1, and the smallest value of $x$ occurs when $y = 1 \quad \Leftrightarrow$ $1 = x/\sqrt{4x^2 - 1} \quad \Leftrightarrow \quad x^2 = 4x^2 - 1 \quad \Leftrightarrow \quad 3x^2 = 1 \quad \Leftrightarrow \quad x = \frac{1}{\sqrt{3}}$. This will give the same value of $\mathcal{A}$ as will $x = 1$, since the geometric situation is the same (reflected through the line $y = x$). We calculate

$\mathcal{A}\left(\frac{1}{\sqrt{2}}\right) = \dfrac{1}{2}\dfrac{\left(1/\sqrt{2}\right)^2}{\sqrt{4\left(1/\sqrt{2}\right)^2 - 1}} = \dfrac{1}{4} = 0.25$, and $\mathcal{A}(1) = \dfrac{1}{2}\dfrac{1^2}{\sqrt{4(1)^2 - 1}} = \dfrac{1}{2\sqrt{3}} \approx 0.29$. So the maximum area is

$\mathcal{A}(1) = \mathcal{A}\left(\frac{1}{\sqrt{3}}\right) = \frac{1}{2\sqrt{3}}$ and the minimum area is $\mathcal{A}\left(\frac{1}{\sqrt{2}}\right) = \frac{1}{4}$.

*Another method:* Use the angle $\theta$ (see diagram above) as a variable:

$\mathcal{A} = \frac{1}{2}xy = \frac{1}{2}\left(\frac{1}{2}\sec\theta\right)\left(\frac{1}{2}\csc\theta\right) = \dfrac{1}{8\sin\theta\cos\theta} = \dfrac{1}{4\sin 2\theta}$. $\mathcal{A}$ is minimized when $\sin 2\theta$ is maximal, that is, when

$\sin 2\theta = 1 \quad \Rightarrow \quad 2\theta = \frac{\pi}{2} \quad \Rightarrow \quad \theta = \frac{\pi}{4}$. Also note that $A'E = x = \frac{1}{2}\sec\theta \le 1 \quad \Rightarrow \quad \sec\theta \le 2 \quad \Rightarrow$

$\cos\theta \ge \frac{1}{2} \quad \Rightarrow \quad \theta \le \frac{\pi}{3}$, and similarly, $A'F = y = \frac{1}{2}\csc\theta \le 1 \quad \Rightarrow \quad \csc\theta \le 2 \quad \Rightarrow \quad \sin\theta \le \frac{1}{2} \quad \Rightarrow \quad \theta \ge \frac{\pi}{6}$.

As above, we find that $\mathcal{A}$ is maximized at these endpoints: $\mathcal{A}\left(\frac{\pi}{6}\right) = \dfrac{1}{4\sin\frac{\pi}{3}} = \dfrac{1}{2\sqrt{3}} = \dfrac{1}{4\sin\frac{2\pi}{3}} = \mathcal{A}\left(\frac{\pi}{3}\right)$; and minimized at

$\theta = \frac{\pi}{4}$: $\mathcal{A}\left(\frac{\pi}{4}\right) = \dfrac{1}{4\sin\frac{\pi}{2}} = \dfrac{1}{4}$.

**19.** Let $s_A(t)$ and $s_B(t)$ be the position functions for cars $A$ and $B$ and let $f(t) = s_A(t) - s_B(t)$. Since $A$ passed $B$ twice ($B$ passed $A$ once), there must be three values of $t$ such that $f(t) = 0$. Let these times be denoted $t_1$, $t_2$, and $t_3$. By the Mean Value Theorem, we know that for some number $c_1$ in $(t_1, t_2)$, $f'(c_1) = \dfrac{f(t_2) - f(t_1)}{t_2 - t_1}$, but $f(t_2) - f(t_1) = 0$, so $f'(c_1) = 0 \quad \Leftrightarrow \quad s'_A(c_1) - s'_B(c_1) = 0 \quad \Leftrightarrow \quad v_A(c_1) - v_B(c_1) = 0 \quad \Rightarrow \quad v_A(c_1) = v_B(c_1)$.

By a similar argument, there exists some number $c_2$ in $(t_2, t_3)$ such that $v_A(c_2) = v_B(c_2)$. Now let $g(t) = v_A(t) - v_B(t)$ and apply the Mean Value Theorem on $[c_1, c_2]$ with $c_1 < c < c_2$. $g'(c) = \dfrac{g(c_2) - g(c_1)}{c_2 - c_1}$,

but $g(c_2) = g(c_1) = 0$, so $g'(c) = 0 \quad \Rightarrow \quad v'_A(c) - v'_B(c) = 0 \quad \Rightarrow \quad a_A(c) - a_B(c) = 0 \quad \Rightarrow \quad a_A(c) = a_B(c)$; that is, $A$ and $B$ had equal accelerations at $t = c$.

**20.** By ignoring the bottom hemisphere of the initial spherical bubble, we can rephrase the problem as follows: Prove that the maximum height of a stack of $n$ hemispherical bubbles is $\sqrt{n}$ if the radius of the bottom hemisphere is 1. We proceed by induction. The case $n = 1$ is obvious since $\sqrt{1}$ is the height of the first hemisphere. Suppose the assertion is true for $n = k$ and let's suppose we have $k + 1$ hemispherical bubbles forming a stack of maximum height. Suppose the second hemisphere (counting from the bottom) has radius $r$. Then by our induction hypothesis (scaled to the setting of a bottom hemisphere of radius $r$), the height of the stack formed by the top $k$ bubbles is $\sqrt{k}\,r$. (If it were shorter, then the total stack of $k + 1$ bubbles wouldn't have maximum height.)

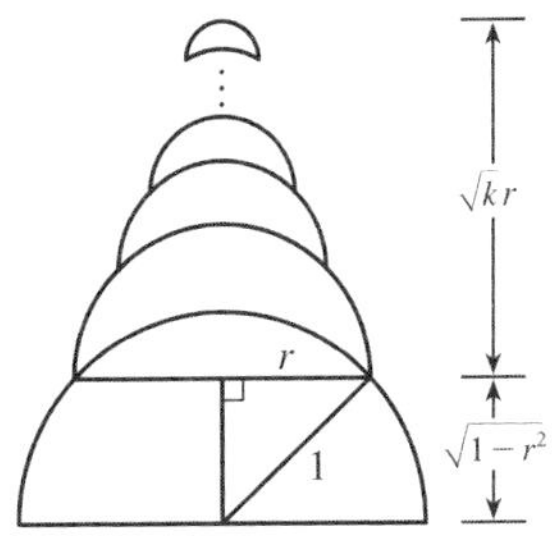


The height of the whole stack is $H(r) = \sqrt{k}\,r + \sqrt{1-r^2}$. (See the figure.) We want to choose $r$ so as to maximize $H(r)$. Note that $0 < r < 1$.

We calculate $H'(r) = \sqrt{k} - \dfrac{r}{\sqrt{1-r^2}}$ and $H''(r) = \dfrac{-1}{(1-r^2)^{3/2}}$.

$H'(r) = 0 \;\Leftrightarrow\; r^2 = k(1-r^2) \;\Leftrightarrow\; (k+1)r^2 = k \;\Leftrightarrow\; r = \sqrt{\dfrac{k}{k+1}}$.

This is the only critical number in $(0,1)$ and it represents a local maximum

(hence an absolute maximum) since $H''(r) < 0$ on $(0,1)$. When $r = \sqrt{\dfrac{k}{k+1}}$,

$H(r) = \sqrt{k}\dfrac{\sqrt{k}}{\sqrt{k+1}} + \sqrt{1 - \dfrac{k}{k+1}} = \dfrac{k}{\sqrt{k+1}} + \dfrac{1}{\sqrt{k+1}} = \sqrt{k+1}$. Thus, the assertion is true for $n = k+1$ when it is true for $n = k$. By induction, it is true for all positive integers $n$.

*Note:* In general, a maximally tall stack of $n$ hemispherical bubbles consists of bubbles with radii

$$1, \sqrt{\frac{n-1}{n}}, \sqrt{\frac{n-2}{n}}, \ldots, \sqrt{\frac{2}{n}}, \sqrt{\frac{1}{n}}.$$

**21.**

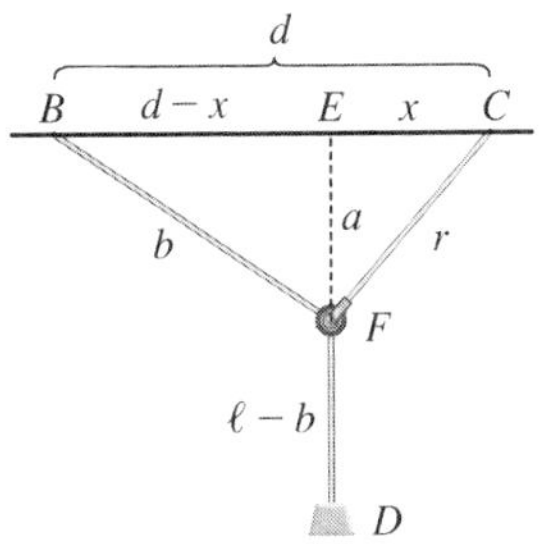


Let $a = |EF|$ and $b = |BF|$ as shown in the figure. Since $\ell = |BF| + |FD|$, $|FD| = \ell - b$. Now

$$\begin{aligned} |ED| &= |EF| + |FD| = a + \ell - b \\ &= \sqrt{r^2 - x^2} + \ell - \sqrt{(d-x)^2 + a^2} \\ &= \sqrt{r^2 - x^2} + \ell - \sqrt{(d-x)^2 + \left(\sqrt{r^2-x^2}\right)^2} \\ &= \sqrt{r^2 - x^2} + \ell - \sqrt{d^2 - 2dx + x^2 + r^2 - x^2} \end{aligned}$$

Let $f(x) = \sqrt{r^2 - x^2} + \ell - \sqrt{d^2 + r^2 - 2dx}$.

$f'(x) = \frac{1}{2}\left(r^2 - x^2\right)^{-1/2}(-2x) - \frac{1}{2}\left(d^2 + r^2 - 2dx\right)^{-1/2}(-2d) = \dfrac{-x}{\sqrt{r^2 - x^2}} + \dfrac{d}{\sqrt{d^2 + r^2 - 2dx}}$.

$f'(x) = 0 \;\Rightarrow\; \dfrac{x}{\sqrt{r^2 - x^2}} = \dfrac{d}{\sqrt{d^2 + r^2 - 2dx}} \;\Rightarrow\; \dfrac{x^2}{r^2 - x^2} = \dfrac{d^2}{d^2 + r^2 - 2dx} \;\Rightarrow$

$d^2x^2 + r^2x^2 - 2dx^3 = d^2r^2 - d^2x^2 \;\Rightarrow\; 0 = 2dx^3 - 2d^2x^2 - r^2x^2 + d^2r^2 \;\Rightarrow$

$0 = 2dx^2(x-d) - r^2\left(x^2 - d^2\right) \;\Rightarrow\; 0 = 2dx^2(x-d) - r^2(x+d)(x-d) \;\Rightarrow\; 0 = (x-d)\left[2dx^2 - r^2(x+d)\right]$

But $d > r > x$, so $x \neq d$. Thus, we solve $2dx^2 - r^2x - dr^2 = 0$ for $x$:

$x = \dfrac{-(-r^2) \pm \sqrt{(-r^2)^2 - 4(2d)(-dr^2)}}{2(2d)} = \dfrac{r^2 \pm \sqrt{r^4 + 8d^2r^2}}{4d}$. Because $\sqrt{r^4 + 8d^2r^2} > r^2$, the "negative" can be discarded. Thus, $x = \dfrac{r^2 + \sqrt{r^2}\sqrt{r^2 + 8d^2}}{4d} = \dfrac{r^2 + r\sqrt{r^2 + 8d^2}}{4d} \quad [r > 0] \quad = \dfrac{r}{4d}\left(r + \sqrt{r^2 + 8d^2}\right)$. The maximum value of $|ED|$ occurs at this value of $x$.

**22.**

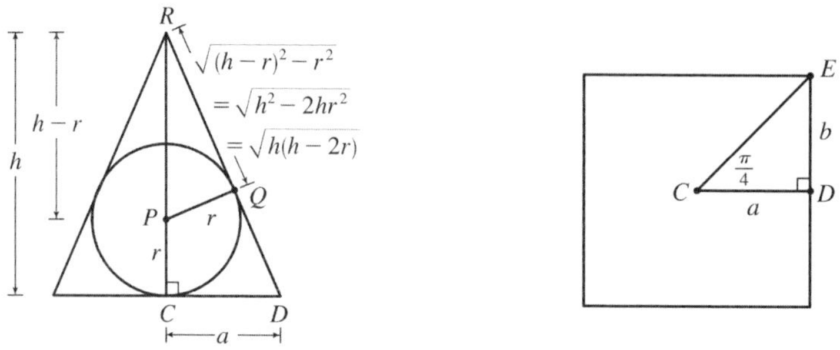


Let $a = \overline{CD}$ denote the distance from the center $C$ of the base to the midpoint $D$ of a side of the base.

Since $\Delta PQR$ is similar to $\Delta DCR$, $\dfrac{a}{h} = \dfrac{r}{\sqrt{h(h-2r)}} \quad \Rightarrow \quad a = \dfrac{rh}{\sqrt{h(h-2r)}} = r\dfrac{\sqrt{h}}{\sqrt{h-2r}}$.

Let $b$ denote one-half the length of a side of the base. The area $A$ of the base is

$A = 8(\text{area of } \Delta CDE) = 8\left(\frac{1}{2}ab\right) = 4a\left(a \tan \frac{\pi}{4}\right) = 4a^2$.

The volume of the pyramid is $V = \frac{1}{3}Ah = \frac{1}{3}\left(4a^2\right)h = \dfrac{4}{3}\left(r\dfrac{\sqrt{h}}{\sqrt{h-2r}}\right)^2 h = \frac{4}{3}r^2\dfrac{h^2}{h-2r}$, with domain $h > 2r$.

Now $\dfrac{dV}{dh} = \dfrac{4}{3}r^2 \cdot \dfrac{(h-2r)(2h) - h^2(1)}{(h-2r)^2} = \dfrac{4}{3}r^2\dfrac{h^2-4hr}{(h-2r)^2} = \dfrac{4}{3}r^2\dfrac{h(h-4r)}{(h-2r)^2}$

and
$$\begin{aligned}\frac{d^2V}{dh^2} &= \frac{4}{3}r^2 \cdot \frac{(h-2r)^2(2h-4r) - \left(h^2-4hr\right)(2)(h-2r)(1)}{[(h-2r)^2]^2} \\ &= \frac{4}{3}r^2 \cdot \frac{2(h-2r)\left[\left(h^2-4hr+4r^2\right) - \left(h^2-4hr\right)\right]}{(h-2r)^2} \\ &= \frac{8}{3}r^2 \cdot \frac{4r^2}{(h-2r)^3} = \frac{32}{3}r^4 \cdot \frac{1}{(h-2r)^3}.\end{aligned}$$

The first derivative is equal to zero for $h = 4r$ and the second derivative is positive for $h > 2r$, so the volume of the pyramid is minimized when $h = 4r$.

To extend our solution to a regular $n$-gon, we make the following changes:

(1) the number of sides of the base is $n$

(2) the number of triangles in the base is $2n$

(3) $\angle DCE = \dfrac{\pi}{n}$

(4) $b = a \tan \dfrac{\pi}{n}$

We then obtain the following results: $A = na^2 \tan \dfrac{\pi}{n}$, $V = \dfrac{nr^2}{3} \cdot \tan\left(\dfrac{\pi}{n}\right) \cdot \dfrac{h^2}{h-2r}$, $\dfrac{dV}{dh} = \dfrac{nr^2}{3} \cdot \tan\left(\dfrac{\pi}{n}\right) \cdot \dfrac{h(h-4r)}{(h-2r)^2}$,

and $\dfrac{d^2V}{dh^2} = \dfrac{8nr^4}{3} \cdot \tan\left(\dfrac{\pi}{n}\right) \cdot \dfrac{1}{(h-2r)^3}$. Notice that the answer, $h = 4r$, is independent of the number of sides of the base of the polygon!

**23.** 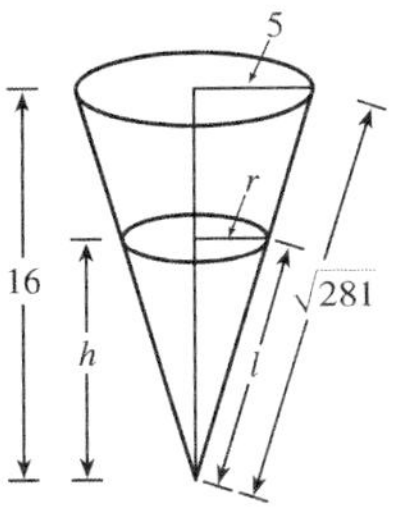


By similar triangles, $\frac{r}{5} = \frac{h}{16} \Rightarrow r = \frac{5h}{16}$. The volume of the cone is $V = \frac{1}{3}\pi r^2 h = \frac{1}{3}\pi\left(\frac{5h}{16}\right)^2 h = \frac{25\pi}{768}h^3$, so $\frac{dV}{dt} = \frac{25\pi}{256}h^2\frac{dh}{dt}$. Now the rate of change of the volume is also equal to the difference of what is being added ($2\text{ cm}^3/\text{min}$) and what is oozing out ($k\pi rl$, where $\pi rl$ is the area of the cone and $k$ is a proportionality constant). Thus, $\frac{dV}{dt} = 2 - k\pi rl$.

Equating the two expressions for $\frac{dV}{dt}$ and substituting $h = 10$, $\frac{dh}{dt} = -0.3$, $r = \frac{5(10)}{16} = \frac{25}{8}$, and $\frac{l}{\sqrt{281}} = \frac{10}{16} \Leftrightarrow$ $l = \frac{5}{8}\sqrt{281}$, we get $\frac{25\pi}{256}(10)^2(-0.3) = 2 - k\pi\frac{25}{8}\cdot\frac{5}{8}\sqrt{281} \Leftrightarrow \frac{125k\pi\sqrt{281}}{64} = 2 + \frac{750\pi}{256}$. Solving for $k$ gives us $k = \frac{256 + 375\pi}{250\pi\sqrt{281}}$. To maintain a certain height, the rate of oozing, $k\pi rl$, must equal the rate of the liquid being poured in; that is, $\frac{dV}{dt} = 0$. $k\pi rl = \frac{256 + 375\pi}{250\pi\sqrt{281}}\cdot\pi\cdot\frac{25}{8}\cdot\frac{5\sqrt{281}}{8} = \frac{256 + 375\pi}{128} \approx 11.204\text{ cm}^3/\text{min}$.

**24.** 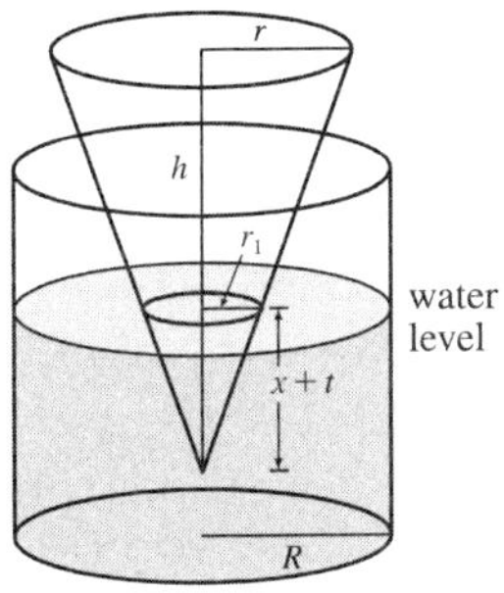


Assume the axes of the cone and the cylinder are parallel. Let $H$ denote the initial height of the water. When the cone has been dropping for $t$ seconds, the water level has risen $x$ meters, so the tip of the cone is $x + 1t$ meters below the water line. We want to find $dx/dt$ when $x + t = h$ (when the cone is completely submerged). Using similar triangles, $\frac{r_1}{x+t} = \frac{r}{h} \Rightarrow r_1 = \frac{r}{h}(x+t)$.

| volume of water and cone at time $t$ | = | original volume of water | + | volume of submerged part of cone |
|---|---|---|---|---|
| $\pi R^2(H + x)$ | = | $\pi R^2 H$ | + | $\frac{1}{3}\pi r_1^2(x+t)$ |
| $\pi R^2 H + \pi R^2 x$ | = | $\pi R^2 H$ | + | $\frac{1}{3}\pi\frac{r^2}{h^2}(x+t)^3$ |
| $3h^2R^2x$ | = | $r^2(x+t)^3$ | | |

Differentiating implicitly with respect to $t$ gives us

$$3h^2R^2\frac{dx}{dt} = r^2\left[3(x+t)^2\frac{dx}{dt} + 3(x+t)^2\frac{dt}{dt}\right] \Rightarrow \frac{dx}{dt} = \frac{r^2(x+t)^2}{h^2R^2 - r^2(x+t)^2} \Rightarrow$$

$$\left.\frac{dx}{dt}\right|_{x+t=h} = \frac{r^2h^2}{h^2R^2 - r^2h^2} = \frac{r^2}{R^2 - r^2}$$

Thus, the water level is rising at a rate of $\frac{r^2}{R^2 - r^2}$ cm/s at the instant the cone is completely submerged.

# 5 ☐ INTEGRALS

## 5.1 Areas and Distances

**1.** (a) Since $f$ is *increasing*, we can obtain a *lower* estimate by using *left* endpoints. We are instructed to use five rectangles, so $n = 5$.

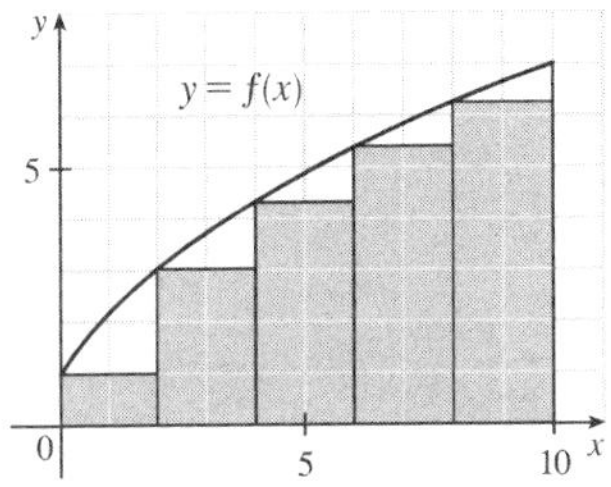


$$L_5 = \sum_{i=1}^{5} f(x_{i-1})\,\Delta x \quad [\Delta x = \tfrac{b-a}{n} = \tfrac{10-0}{5} = 2]$$
$$= f(x_0)\cdot 2 + f(x_1)\cdot 2 + f(x_2)\cdot 2 + f(x_3)\cdot 2 + f(x_4)\cdot 2$$
$$= 2\,[f(0) + f(2) + f(4) + f(6) + f(8)]$$
$$\approx 2(1 + 3 + 4.3 + 5.4 + 6.3) = 2(20) = 40$$

Since $f$ is *increasing*, we can obtain an *upper* estimate by using *right* endpoints.

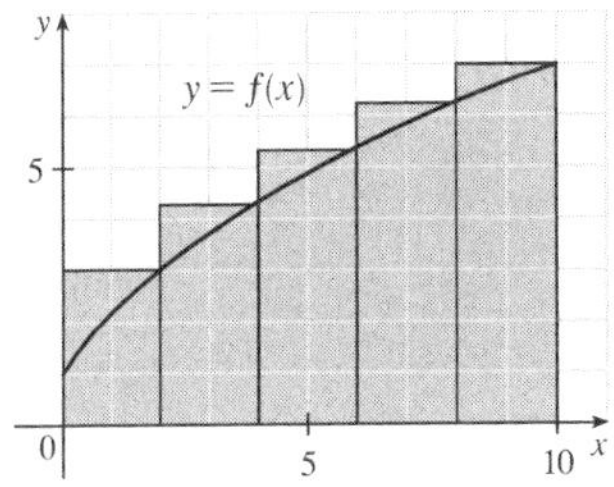


$$R_5 = \sum_{i=1}^{5} f(x_i)\,\Delta x$$
$$= 2\,[f(x_1) + f(x_2) + f(x_3) + f(x_4) + f(x_5)]$$
$$= 2\,[f(2) + f(4) + f(6) + f(8) + f(10)]$$
$$\approx 2(3 + 4.3 + 5.4 + 6.3 + 7) = 2(26) = 52$$

Comparing $R_5$ to $L_5$, we see that we have added the area of the rightmost upper rectangle, $f(10)\cdot 2$, to the sum and subtracted the area of the leftmost lower rectangle, $f(0)\cdot 2$, from the sum.

(b) $$L_{10} = \sum_{i=1}^{10} f(x_{i-1})\,\Delta x \quad [\Delta x = \tfrac{10-0}{10} = 1]$$
$$= 1\,[f(x_0) + f(x_1) + \cdots + f(x_9)]$$
$$= f(0) + f(1) + \cdots + f(9)$$
$$\approx 1 + 2.1 + 3 + 3.7 + 4.3 + 4.9 + 5.4 + 5.8 + 6.3 + 6.7$$
$$= 43.2$$

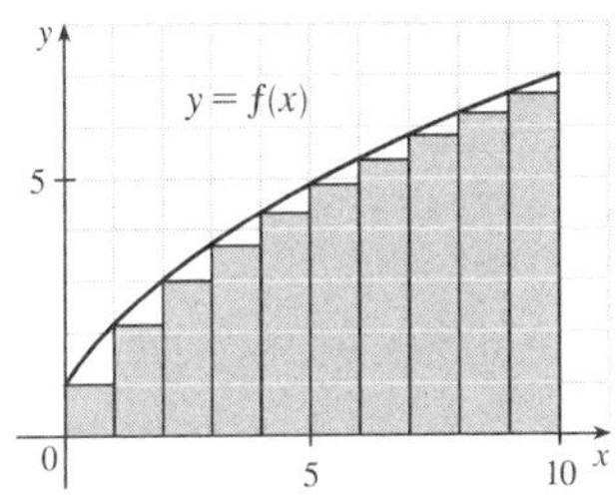


$$R_{10} = \sum_{i=1}^{10} f(x_i)\,\Delta x = f(1) + f(2) + \cdots + f(10)$$
$$= L_{10} + 1\cdot f(10) - 1\cdot f(0) \quad \left[\begin{array}{c}\text{add rightmost upper rectangle,}\\ \text{subtract leftmost lower rectangle}\end{array}\right]$$
$$= 43.2 + 7 - 1 = 49.2$$

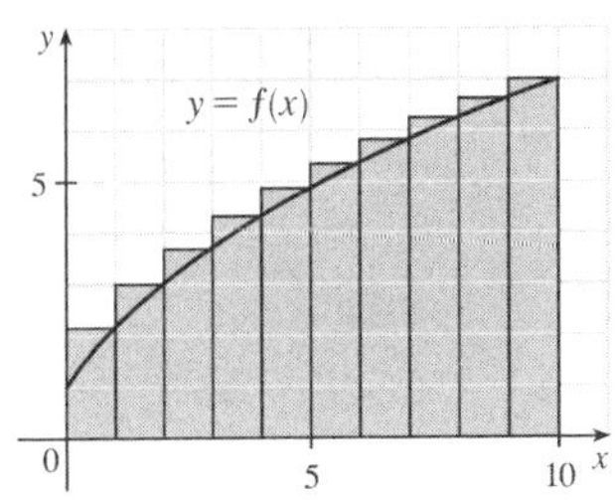

**2.** (a) (i) $L_6 = \sum_{i=1}^{6} f(x_{i-1})\Delta x \quad [\Delta x = \frac{12-0}{6} = 2]$

$= 2[f(x_0) + f(x_1) + f(x_2) + f(x_3) + f(x_4) + f(x_5)]$

$= 2[f(0) + f(2) + f(4) + f(6) + f(8) + f(10)]$

$\approx 2(9 + 8.8 + 8.2 + 7.3 + 5.9 + 4.1)$

$= 2(43.3) = 86.6$

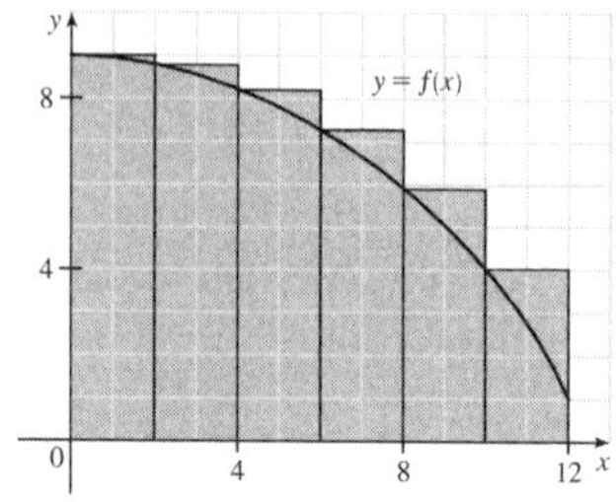


(ii) $R_6 = L_6 + 2 \cdot f(12) - 2 \cdot f(0)$

$\approx 86.6 + 2(1) - 2(9) = 70.6$

[Add area of rightmost lower rectangle and subtract area of leftmost upper rectangle.]

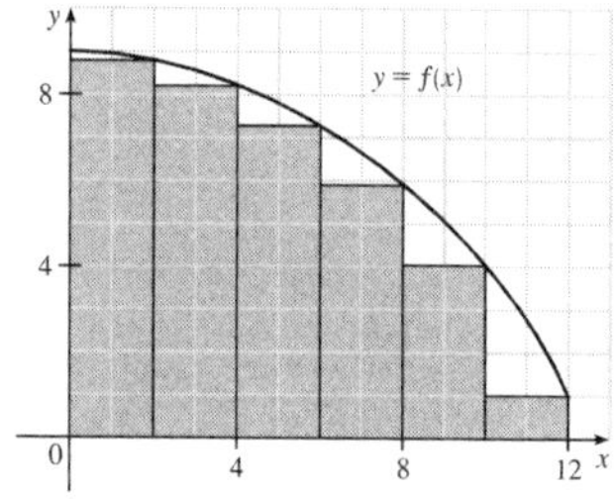


(iii) $M_6 = \sum_{i=1}^{6} f(x_i^*)\,\Delta x$

$= 2[f(1) + f(3) + f(5) + f(7) + f(9) + f(11)]$

$\approx 2(8.9 + 8.5 + 7.8 + 6.6 + 5.1 + 2.8)$

$= 2(39.7) = 79.4$

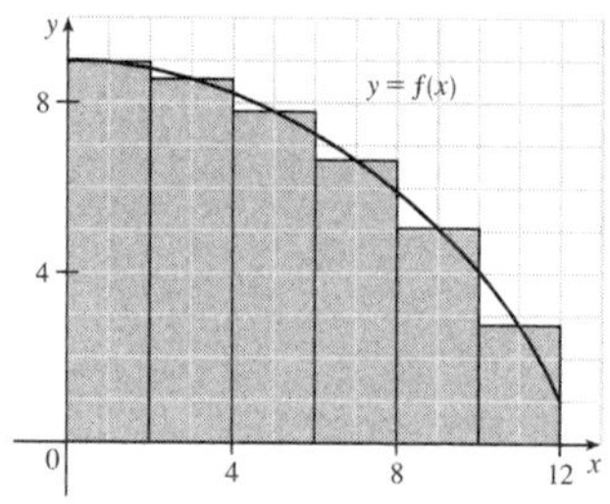


(b) Since $f$ is *decreasing*, we obtain an *overestimate* by using *left* endpoints; that is, $L_6$.

(c) Since $f$ is *decreasing*, we obtain an *underestimate* by using *right* endpoints; that is, $R_6$.

(d) $M_6$ gives the best estimate, since the area of each rectangle appears to be closer to the true area than the overestimates and underestimates in $L_6$ and $R_6$.

**3.** (a) $R_4 = \sum_{i=1}^{4} f(x_i)\,\Delta x \quad [\Delta x = \frac{5-1}{4} = 1]$

$= f(x_1) \cdot 1 + f(x_2) \cdot 1 + f(x_3) \cdot 1 + f(x_4) \cdot 1$

$= f(2) + f(3) + f(4) + f(5)$

$= \frac{1}{2} + \frac{1}{3} + \frac{1}{4} + \frac{1}{5} = \frac{77}{60} = 1.28\overline{3}$

Since $f$ is *decreasing* on $[1, 5]$, an *underestimate* is obtained by using the *right* endpoint approximation, $R_4$.

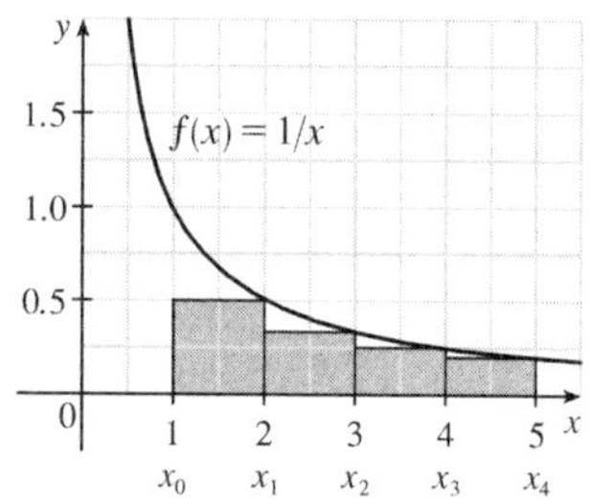


(b) $L_4 = \sum_{i=1}^{4} f(x_{i-1})\,\Delta x$

$= f(1) + f(2) + f(3) + f(4)$

$= 1 + \frac{1}{2} + \frac{1}{3} + \frac{1}{4} = \frac{25}{12} = 2.08\overline{3}$

$L_4$ is an overestimate. Alternatively, we could just add the area of the leftmost upper rectangle and subtract the area of the rightmost lower rectangle; that is, $L_4 = R_4 + f(1) \cdot 1 - f(5) \cdot 1$.

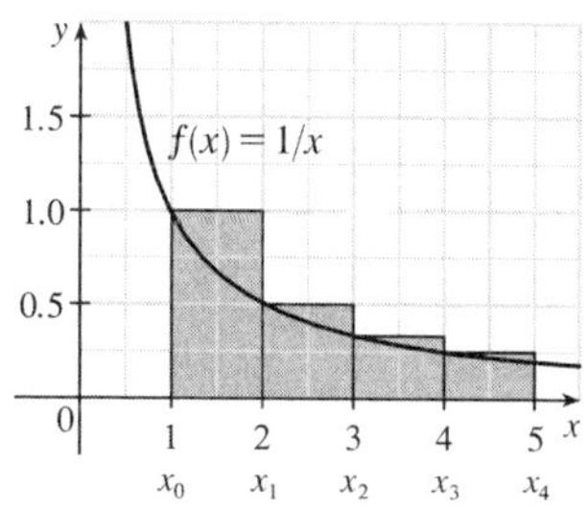

**4.** (a) $R_5 = \sum_{i=1}^{5} f(x_i)\,\Delta x \quad [\Delta x = \frac{5-0}{5} = 1]$

$= f(x_1)\cdot 1 + f(x_2)\cdot 1 + f(x_3)\cdot 1 + f(x_4)\cdot 1 + f(x_5)\cdot 1$

$= f(1) + f(2) + f(3) + f(4) + f(5)$

$= 24 + 21 + 16 + 9 + 0 = 70$

Since $f$ is decreasing on $[0, 5]$, $R_5$ is an underestimate.

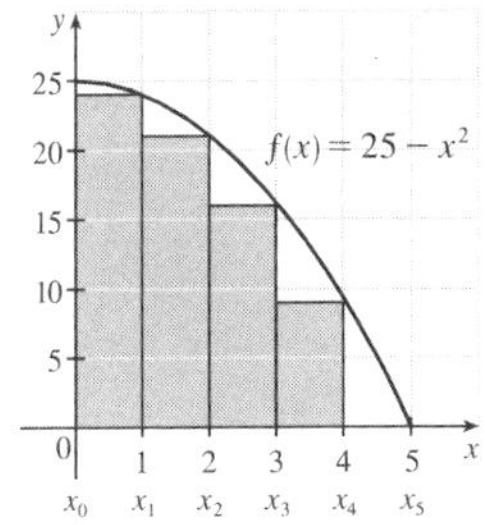


(b) $L_5 = \sum_{i=1}^{5} f(x_{i-1})\,\Delta x$

$= f(0) + f(1) + f(2) + f(3) + f(4)$

$= 25 + 24 + 21 + 16 + 9 = 95$

$L_5$ is an overestimate.

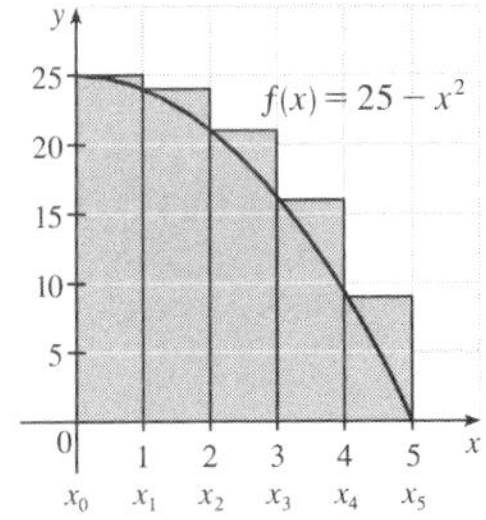


**5.** (a) $f(x) = 1 + x^2$ and $\Delta x = \dfrac{2-(-1)}{3} = 1 \quad\Rightarrow$

$R_3 = 1\cdot f(0) + 1\cdot f(1) + 1\cdot f(2) = 1\cdot 1 + 1\cdot 2 + 1\cdot 5 = 8.$

$\Delta x = \dfrac{2-(-1)}{6} = 0.5 \quad\Rightarrow$

$R_6 = 0.5[f(-0.5) + f(0) + f(0.5) + f(1) + f(1.5) + f(2)]$

$= 0.5(1.25 + 1 + 1.25 + 2 + 3.25 + 5)$

$= 0.5(13.75) = 6.875$

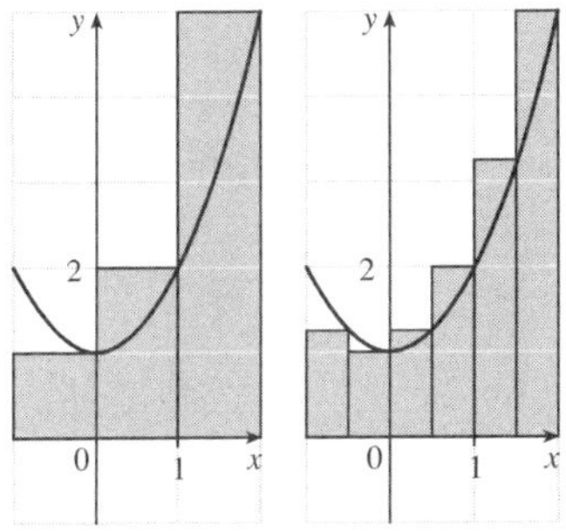


(b) $L_3 = 1\cdot f(-1) + 1\cdot f(0) + 1\cdot f(1) = 1\cdot 2 + 1\cdot 1 + 1\cdot 2 = 5$

$L_6 = 0.5[f(-1) + f(-0.5) + f(0) + f(0.5) + f(1) + f(1.5)]$

$= 0.5(2 + 1.25 + 1 + 1.25 + 2 + 3.25)$

$= 0.5(10.75) = 5.375$

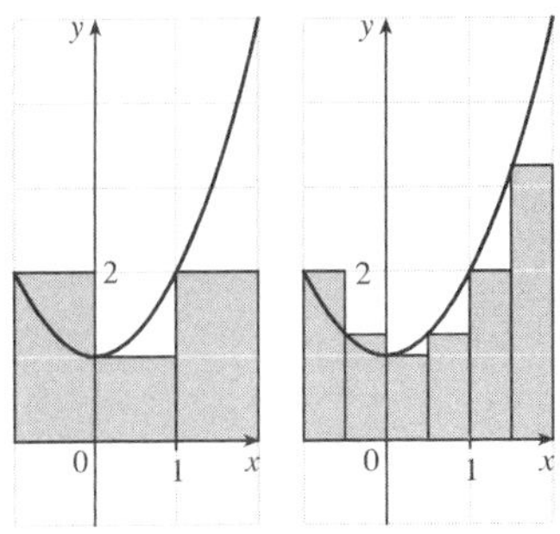


(c) $M_3 = 1\cdot f(-0.5) + 1\cdot f(0.5) + 1\cdot f(1.5)$

$= 1\cdot 1.25 + 1\cdot 1.25 + 1\cdot 3.25 = 5.75$

$M_6 = 0.5[f(-0.75) + f(-0.25) + f(0.25)$

$+ f(0.75) + f(1.25) + f(1.75)]$

$= 0.5(1.5625 + 1.0625 + 1.0625 + 1.5625 + 2.5625 + 4.0625)$

$= 0.5(11.875) = 5.9375$

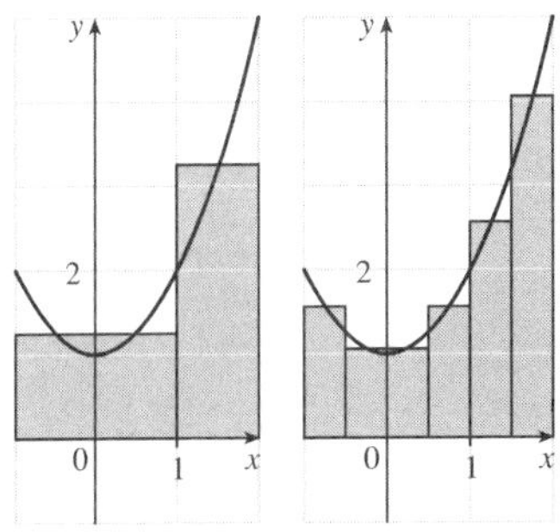


(d) $M_6$ appears to be the best estimate.

**6.** (a)

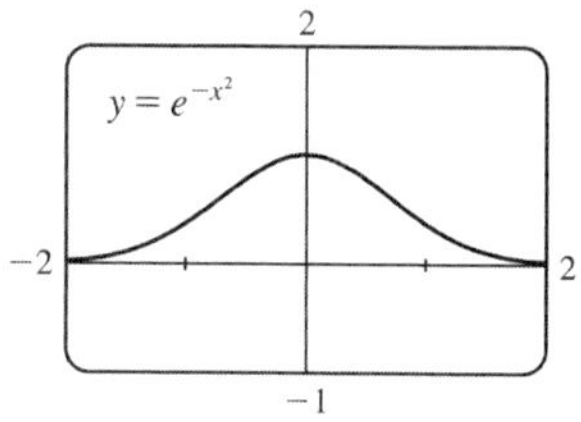


(b) $f(x) = e^{-x^2}$ and $\Delta x = \dfrac{2 - (-2)}{4} = 1 \;\Rightarrow$

(i) $R_4 = 1 \cdot f(-1) + 1 \cdot f(0) + 1 \cdot f(1) + 1 \cdot f(2)$
$= e^{-1} + 1 + e^{-1} + e^{-4}$
$\approx 1.754$

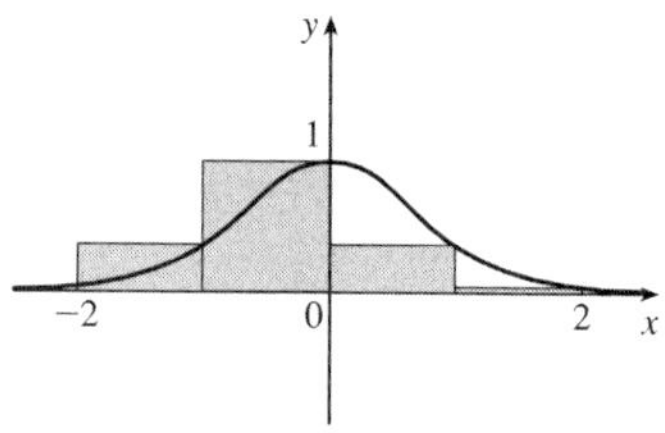


(ii) $M_4 = 1 \cdot f(-1.5) + 1 \cdot f(-0.5) + 1 \cdot f(0.5) + 1 \cdot f(1.5)$
$= e^{-2.25} + e^{-0.25} + e^{-0.25} + e^{-2.25}$
$\approx 1.768$

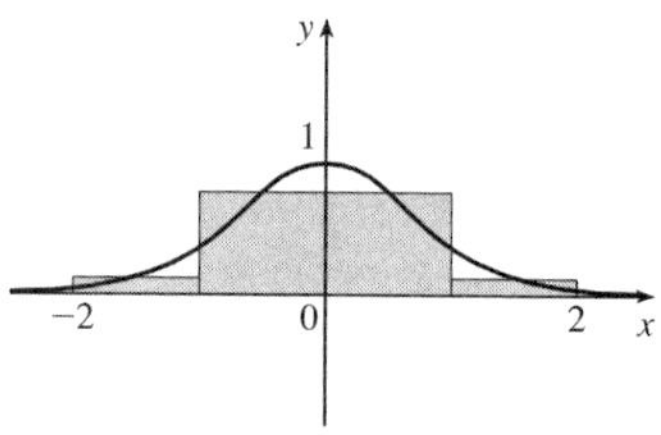


(c) (i) $R_8 = 0.5[f(-1.5) + f(-1) + f(-0.5) + f(0) + f(0.5) + f(1) + f(1.5) + f(2)]$
$= e^{-2.25} + e^{-1} + e^{-0.25} + 1 + e^{-0.25} + e^{-1} + e^{-2.25} + e^{-4}$
$\approx 1.761$

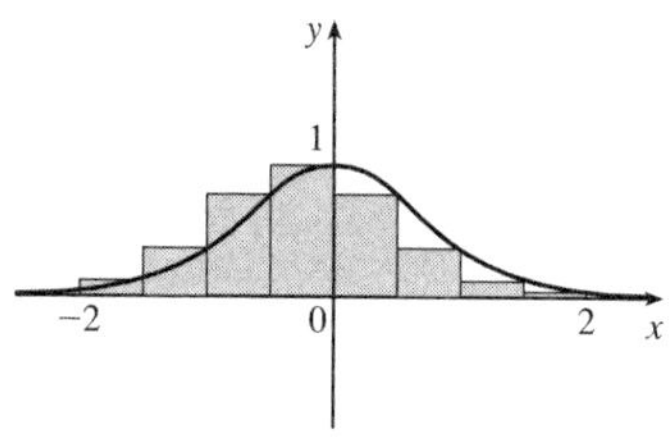


(ii) Due to the symmetry of the figure, we see that

$M_8 = (0.5)(2)[f(0.25) + f(0.75) + f(1.25) + f(1.75)]$
$= e^{-0.0625} + e^{-0.5625} + e^{-1.5625} + e^{-3.0625}$
$\approx 1.766$

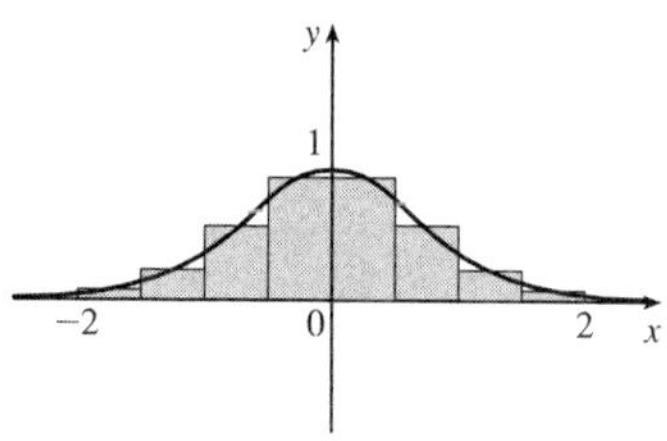


**7.** Here is one possible algorithm (ordered sequence of operations) for calculating the sums:

1 Let SUM = 0, X_MIN = 0, X_MAX = 1, N = 10 (depending on which sum we are calculating), DELTA_X = (X_MAX - X_MIN)/N, and RIGHT_ENDPOINT = X_MIN + DELTA_X.

2 Repeat steps 2a, 2b in sequence until RIGHT_ENDPOINT > X_MAX.

2a Add (RIGHT_ENDPOINT)^4 to SUM.

2b Add DELTA_X to RIGHT_ENDPOINT.

At the end of this procedure, (DELTA_X)·(SUM) is equal to the answer we are looking for. We find that

$R_{10} = \dfrac{1}{10}\sum\limits_{i=1}^{10}\left(\dfrac{i}{10}\right)^4 \approx 0.2533$, $R_{30} = \dfrac{1}{30}\sum\limits_{i=1}^{30}\left(\dfrac{i}{30}\right)^4 \approx 0.2170$, $R_{50} = \dfrac{1}{50}\sum\limits_{i=1}^{50}\left(\dfrac{i}{50}\right)^4 \approx 0.2101$, and

$R_{100} = \dfrac{1}{100}\sum\limits_{i=1}^{100}\left(\dfrac{i}{100}\right)^4 \approx 0.2050$. It appears that the exact area is 0.2.

Shown below is program SUMRIGHT and its output from a TI-83 Plus calculator. To generalize the program, we have input (rather than assign) values for Xmin, Xmax, and N. Also, the function, $x^4$, is assigned to $\text{Y}_1$, enabling us to evaluate any right sum merely by changing $\text{Y}_1$ and running the program.

```
PROGRAM:SUMRIGHT
:0→S
:Prompt Xmin
:Prompt Xmax
:Prompt N
:(Xmax-Xmin)/N→D
:Xmin+D→R
:For(I,1,N)
:S+Y1(R)→S
:R+D→R
:End
:D*S→Z
:Disp Z
```

```
prgmSUMRIGHT
Xmin=?0
Xmax=?1
N=?10
                .25333
                  Done
```

**8.** We can use the algorithm from Exercise 7 with X_MIN = 0, X_MAX = $\pi/2$, and cos(RIGHT_ENDPOINT) instead of (RIGHT_ENDPOINT)^4 in step 2a. We find that $R_{10} = \dfrac{\pi/2}{10}\sum\limits_{i=1}^{10}\cos\left(\dfrac{i\pi}{20}\right) \approx 0.9194$, $R_{30} = \dfrac{\pi/2}{30}\sum\limits_{i=1}^{30}\cos\left(\dfrac{i\pi}{60}\right) \approx 0.9736$,

and $R_{50} = \dfrac{\pi/2}{50}\sum\limits_{i=1}^{50}\cos\left(\dfrac{i\pi}{100}\right) \approx 0.9842$, and $R_{100} = \dfrac{\pi/2}{100}\sum\limits_{i=1}^{100}\cos\left(\dfrac{i\pi}{200}\right) \approx 0.9921$. It appears that the exact area is 1.

**9.** In Maple, we have to perform a number of steps before getting a numerical answer. After loading the student package [command: `with(student);`] we use the command `left_sum:=leftsum(1/(x^2+1),x=0..1,10` [or `30`, or `50`]`);` which gives us the expression in summation notation. To get a numerical approximation to the sum, we use `evalf(left_sum);`. Mathematica does not have a special command for these sums, so we must type them in manually. For example, the first left sum is given by `(1/10)*Sum[1/(((i-1)/10)^2+1)],{i,1,10}]`, and we use the `N` command on the resulting output to get a numerical approximation.

In Derive, we use the `LEFT_RIEMANN` command to get the left sums, but must define the right sums ourselves. (We can define a new function using `LEFT_RIEMANN` with $k$ ranging from 1 to $n$ instead of from 0 to $n-1$.)

(a) With $f(x) = \dfrac{1}{x^2+1}$, $0 \le x \le 1$, the left sums are of the form $L_n = \dfrac{1}{n}\sum\limits_{i=1}^{n}\dfrac{1}{\left(\dfrac{i-1}{n}\right)^2+1}$. Specifically, $L_{10} \approx 0.8100$,

$L_{30} \approx 0.7937$, and $L_{50} \approx 0.7904$. The right sums are of the form $R_n = \dfrac{1}{n}\sum\limits_{i=1}^{n}\dfrac{1}{\left(\dfrac{i}{n}\right)^2+1}$. Specifically, $R_{10} \approx 0.7600$,

$R_{30} \approx 0.7770$, and $R_{50} \approx 0.7804$.

(b) In Maple, we use the `leftbox` (with the same arguments as `left_sum`) and `rightbox` commands to generate the graphs.

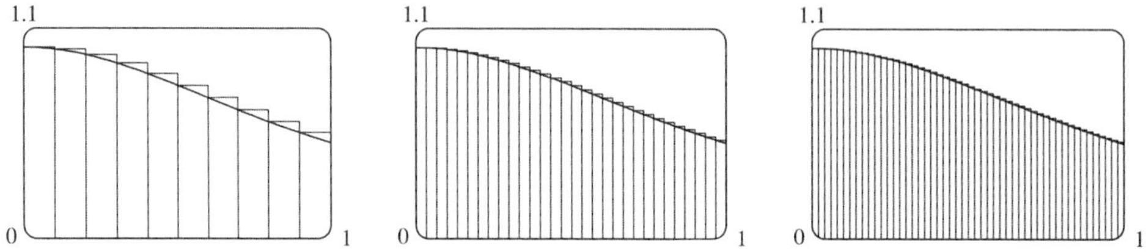

left endpoints, $n = 10$ left endpoints, $n = 30$ left endpoints, $n = 50$

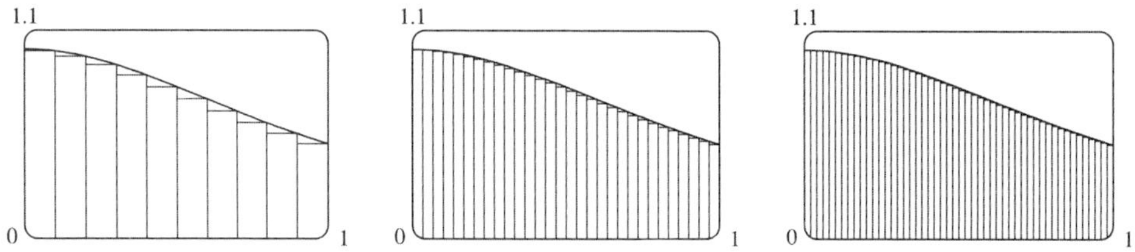

right endpoints, $n = 10$ right endpoints, $n = 30$ right endpoints, $n = 50$

(c) We know that since $y = 1/(x^2 + 1)$ is a decreasing function on $(0, 1)$, all of the left sums are larger than the actual area, and all of the right sums are smaller than the actual area. Since the left sum with $n = 50$ is about $0.7904 < 0.791$ and the right sum with $n = 50$ is about $0.7804 > 0.780$, we conclude that $0.780 < R_{50} <$ exact area $< L_{50} < 0.791$, so the exact area is between $0.780$ and $0.791$.

**10.** See the solution to Exercise 9 for the CAS commands for evaluating the sums.

(a) With $f(x) = \ln x$, $1 \le x \le 4$, the left sums are of the form $L_n = \frac{3}{n}\sum_{i=1}^{n} \ln\left(1 + \frac{3(i-1)}{n}\right)$. In particular, $L_{10} \approx 2.3316$, $L_{30} \approx 2.4752$, and $L_{50} \approx 2.5034$. The right sums are of the form $R_n = \frac{3}{n}\sum_{i=1}^{n} \ln\left(1 + \frac{3i}{n}\right)$. In particular, $R_{10} \approx 2.7475$, $R_{30} \approx 2.6139$, and $R_{50} \approx 2.5865$.

(b) In Maple, we use the `leftbox` (with the same arguments as `left_sum`) and `rightbox` commands to generate the graphs.

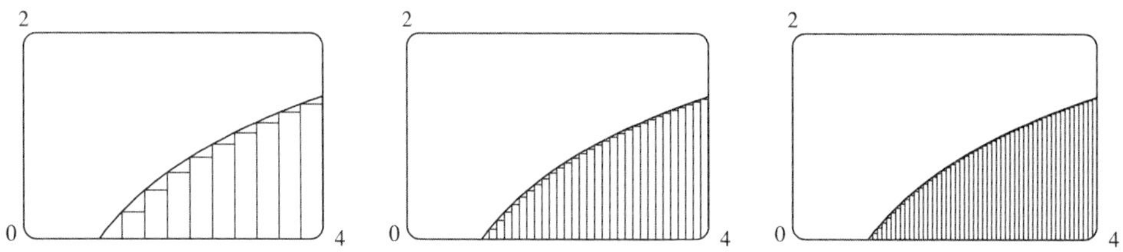

left endpoints, $n = 10$ left endpoints, $n = 30$ left endpoints, $n = 50$

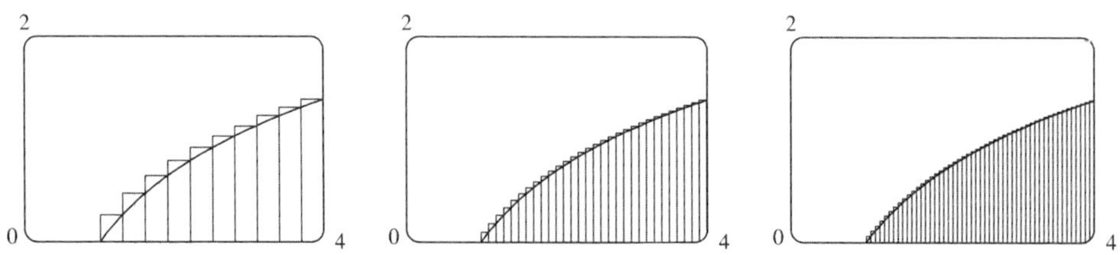

right endpoints, $n = 10$ right endpoints, $n = 30$ right endpoints, $n = 50$

(c) We know that since $y = \ln x$ is an increasing function on $(1, 4)$, all of the left sums are smaller than the actual area, and all of the right sums are larger than the actual area. Since the left sum with $n = 50$ is about $2.503 > 2.50$ and the right sum with $n = 50$ is about $2.587 < 2.59$, we conclude that $2.50 < L_{50} <$ exact area $< R_{50} < 2.59$, so the exact area is between 2.50 and 2.59.

**11.** Since $v$ is an increasing function, $L_6$ will give us a lower estimate and $R_6$ will give us an upper estimate.

$$L_6 = (0 \text{ ft/s})(0.5 \text{ s}) + (6.2)(0.5) + (10.8)(0.5) + (14.9)(0.5) + (18.1)(0.5) + (19.4)(0.5)$$
$$= 0.5(69.4) = 34.7 \text{ ft}$$

$$R_6 = 0.5(6.2 + 10.8 + 14.9 + 18.1 + 19.4 + 20.2) = 0.5(89.6) = 44.8 \text{ ft}$$

**12.** (a) $d \approx L_5 = (30 \text{ ft/s})(12 \text{ s}) + 28 \cdot 12 + 25 \cdot 12 + 22 \cdot 12 + 24 \cdot 12$
$$= (30 + 28 + 25 + 22 + 24) \cdot 12 = 129 \cdot 12 = 1548 \text{ ft}$$

(b) $d \approx R_5 = (28 + 25 + 22 + 24 + 27) \cdot 12 = 126 \cdot 12 = 1512 \text{ ft}$

(c) The estimates are neither lower nor upper estimates since $v$ is neither an increasing nor a decreasing function of $t$.

**13.** Lower estimate for oil leakage: $R_5 = (7.6 + 6.8 + 6.2 + 5.7 + 5.3)(2) = (31.6)(2) = 63.2$ L.

Upper estimate for oil leakage: $L_5 = (8.7 + 7.6 + 6.8 + 6.2 + 5.7)(2) = (35)(2) = 70$ L.

**14.** We can find an upper estimate by using the final velocity for each time interval. Thus, the distance $d$ traveled after 62 seconds can be approximated by

$$d = \sum_{i=1}^{6} v(t_i)\Delta t_i = (185 \text{ ft/s})(10 \text{ s}) + 319 \cdot 5 + 447 \cdot 5 + 742 \cdot 12 + 1325 \cdot 27 + 1445 \cdot 3 = 54{,}694 \text{ ft}$$

**15.** For a decreasing function, using left endpoints gives us an overestimate and using right endpoints results in an underestimate. We will use $M_6$ to get an estimate. $\Delta t = 1$, so

$$M_6 = 1[v(0.5) + v(1.5) + v(2.5) + v(3.5) + v(4.5) + v(5.5)]$$
$$\approx 55 + 40 + 28 + 18 + 10 + 4 = 155 \text{ ft}$$

For a very rough check on the above calculation, we can draw a line from $(0, 70)$ to $(6, 0)$ and calculate the area of the triangle: $\frac{1}{2}(70)(6) = 210$. This is clearly an overestimate, so our midpoint estimate of 155 is reasonable.

**16.** For an increasing function, using left endpoints gives us an underestimate and using right endpoints results in an overestimate. We will use $M_6$ to get an estimate. $\Delta t = \frac{30-0}{6} = 5 \text{ s} = \frac{5}{3600} \text{ h} = \frac{1}{720}$ h.

$$M_6 = \tfrac{1}{720}[v(2.5) + v(7.5) + v(12.5) + v(17.5) + v(22.5) + v(27.5)]$$
$$= \tfrac{1}{720}(31.25 + 66 + 88 + 103.5 + 113.75 + 119.25) = \tfrac{1}{720}(521.75) \approx 0.725 \text{ km}$$

For a very rough check on the above calculation, we can draw a line from $(0, 0)$ to $(30, 120)$ and calculate the area of the triangle: $\frac{1}{2}(30)(120) = 1800$. Divide by 3600 to get 0.5, which is clearly an underestimate, making our midpoint estimate of 0.725 seem reasonable. Of course, answers will vary due to different readings of the graph.

**17.** $f(x) = \sqrt[4]{x}$, $1 \le x \le 16$. $\Delta x = (16 - 1)/n = 15/n$ and $x_i = 1 + i\,\Delta x = 1 + 15i/n$.

$$A = \lim_{n\to\infty} R_n = \lim_{n\to\infty} \sum_{i=1}^{n} f(x_i)\,\Delta x = \lim_{n\to\infty} \sum_{i=1}^{n} \sqrt[4]{1 + \tfrac{15i}{n}} \cdot \tfrac{15}{n}.$$

**18.** $f(x) = \dfrac{\ln x}{x}$, $3 \le x \le 10$. $\Delta x = (10 - 3)/n = 7/n$ and $x_i = 3 + i\,\Delta x = 3 + 7i/n$.

$$A = \lim_{n\to\infty} R_n = \lim_{n\to\infty} \sum_{i=1}^{n} f(x_i)\,\Delta x = \lim_{n\to\infty} \sum_{i=1}^{n} \frac{\ln(3 + 7i/n)}{3 + 7i/n} \cdot \frac{7}{n}.$$

**19.** $\displaystyle\lim_{n\to\infty} \sum_{i=1}^{n} \frac{\pi}{4n} \tan\frac{i\pi}{4n}$ can be interpreted as the area of the region lying under the graph of $y = \tan x$ on the interval $\left[0, \frac{\pi}{4}\right]$,

since for $y = \tan x$ on $\left[0, \frac{\pi}{4}\right]$ with $\Delta x = \dfrac{\pi/4 - 0}{n} = \dfrac{\pi}{4n}$, $x_i = 0 + i\,\Delta x = \dfrac{i\pi}{4n}$, and $x_i^* = x_i$, the expression for the area is

$A = \lim_{n\to\infty} \sum_{i=1}^{n} f(x_i^*)\,\Delta x = \lim_{n\to\infty} \sum_{i=1}^{n} \tan\left(\frac{i\pi}{4n}\right)\frac{\pi}{4n}$. Note that this answer is not unique, since the expression for the area is the same for the function $y = \tan(x - k\pi)$ on the interval $\left[k\pi, k\pi + \frac{\pi}{4}\right]$, where $k$ is any integer.

**20.** (a) $\Delta x = \dfrac{1-0}{n} = \dfrac{1}{n}$ and $x_i = 0 + i\,\Delta x = \dfrac{i}{n}$. $A = \lim_{n\to\infty} R_n = \lim_{n\to\infty} \sum_{i=1}^{n} f(x_i)\,\Delta x = \lim_{n\to\infty} \sum_{i=1}^{n} \left(\dfrac{i}{n}\right)^3 \cdot \dfrac{1}{n}$.

(b) $$\lim_{n\to\infty} \sum_{i=1}^{n} \frac{i^3}{n^3}\cdot\frac{1}{n} = \lim_{n\to\infty} \frac{1}{n^4}\sum_{i=1}^{n} i^3 = \lim_{n\to\infty} \frac{1}{n^4}\left[\frac{n(n+1)}{2}\right]^2 = \lim_{n\to\infty} \frac{(n+1)^2}{4n^2} = \frac{1}{4}\lim_{n\to\infty}\left(1+\frac{1}{n}\right)^2 = \frac{1}{4}$$

**21.** (a) $y = f(x) = x^5$. $\Delta x = \dfrac{2-0}{n} = \dfrac{2}{n}$ and $x_i = 0 + i\,\Delta x = \dfrac{2i}{n}$.

$$A = \lim_{n\to\infty} R_n = \lim_{n\to\infty} \sum_{i=1}^{n} f(x_i)\,\Delta x = \lim_{n\to\infty} \sum_{i=1}^{n} \left(\frac{2i}{n}\right)^5 \cdot \frac{2}{n} = \lim_{n\to\infty} \sum_{i=1}^{n} \frac{32i^5}{n^5}\cdot\frac{2}{n} = \lim_{n\to\infty} \frac{64}{n^6}\sum_{i=1}^{n} i^5.$$

(b) $$\sum_{i=1}^{n} i^5 \overset{\text{CAS}}{=} \frac{n^2(n+1)^2(2n^2+2n-1)}{12}$$

(c) $$\lim_{n\to\infty} \frac{64}{n^6}\cdot\frac{n^2(n+1)^2(2n^2+2n-1)}{12} = \frac{64}{12}\lim_{n\to\infty}\frac{(n^2+2n+1)(2n^2+2n-1)}{n^2\cdot n^2}$$
$$= \frac{16}{3}\lim_{n\to\infty}\left(1+\frac{2}{n}+\frac{1}{n^2}\right)\left(2+\frac{2}{n}-\frac{1}{n^2}\right) = \tfrac{16}{3}\cdot 1\cdot 2 = \tfrac{32}{3}$$

**22.** From Example 3(a), we have $A = \lim_{n\to\infty} \dfrac{2}{n}\sum_{i=1}^{n} e^{-2i/n}$. Using a CAS, $\sum_{i=1}^{n} e^{-2i/n} = \dfrac{e^{-2}(e^2-1)}{e^{2/n}-1}$ and

$\lim_{n\to\infty} \dfrac{2}{n}\cdot\dfrac{e^{-2}(e^2-1)}{e^{2/n}-1} = e^{-2}(e^2-1) \approx 0.8647$, whereas the estimate from Example 3(b) using $M_{10}$ was 0.8632.

**23.** $y = f(x) = \cos x$. $\Delta x = \dfrac{b-0}{n} = \dfrac{b}{n}$ and $x_i = 0 + i\,\Delta x = \dfrac{bi}{n}$.

$$A = \lim_{n\to\infty} R_n = \lim_{n\to\infty} \sum_{i=1}^{n} f(x_i)\,\Delta x = \lim_{n\to\infty} \sum_{i=1}^{n} \cos\left(\frac{bi}{n}\right)\cdot\frac{b}{n} \overset{\text{CAS}}{=} \lim_{n\to\infty}\left[\frac{b\sin\left(b\left(\frac{1}{2n}+1\right)\right)}{2n\sin\left(\frac{b}{2n}\right)} - \frac{b}{2n}\right] \overset{\text{CAS}}{=} \sin b$$

If $b = \frac{\pi}{2}$, then $A = \sin\frac{\pi}{2} = 1$.

**24.** (a)

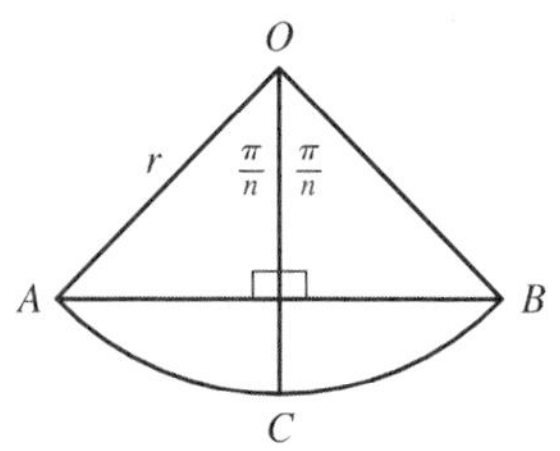


The diagram shows one of the $n$ congruent triangles, $\triangle AOB$, with central angle $2\pi/n$. $O$ is the center of the circle and $AB$ is one of the sides of the polygon. Radius $OC$ is drawn so as to bisect $\angle AOB$. It follows that $OC$ intersects $AB$ at right angles and bisects $AB$. Thus, $\triangle AOB$ is divided into two right triangles with legs of length $\frac{1}{2}(AB) = r\sin(\pi/n)$ and $r\cos(\pi/n)$. $\triangle AOB$ has area $2\cdot\frac{1}{2}[r\sin(\pi/n)][r\cos(\pi/n)] = r^2\sin(\pi/n)\cos(\pi/n) = \frac{1}{2}r^2\sin(2\pi/n)$, so $A_n = n\cdot\text{area}(\triangle AOB) = \frac{1}{2}nr^2\sin(2\pi/n)$.

(b) To use Equation 3.4.2, $\lim_{\theta\to 0}\dfrac{\sin\theta}{\theta} = 1$, we need to have the same expression in the denominator as we have in the argument of the sine function—in this case, $2\pi/n$.

$\lim_{n\to\infty} A_n = \lim_{n\to\infty} \frac{1}{2}nr^2\sin(2\pi/n) = \lim_{n\to\infty} \frac{1}{2}nr^2\dfrac{\sin(2\pi/n)}{2\pi/n}\cdot\dfrac{2\pi}{n} = \lim_{n\to\infty}\dfrac{\sin(2\pi/n)}{2\pi/n}\pi r^2$. Let $\theta = \dfrac{2\pi}{n}$.

Then as $n\to\infty$, $\theta\to 0$, so $\lim_{n\to\infty}\dfrac{\sin(2\pi/n)}{2\pi/n}\pi r^2 = \lim_{\theta\to 0}\dfrac{\sin\theta}{\theta}\pi r^2 = (1)\,\pi r^2 = \pi r^2$.

## 5.2 The Definite Integral

**1.** $R_4 = \sum_{i=1}^{4} f(x_i)\,\Delta x \quad [x_i^* = x_i \text{ is a right endpoint and } \Delta x = 0.5]$

$= 0.5\,[f(0.5) + f(1) + f(1.5) + f(2)] \quad [f(x) = 2 - x^2]$

$= 0.5\,[1.75 + 1 + (-0.25) + (-2)]$

$= 0.5(0.5) = 0.25$

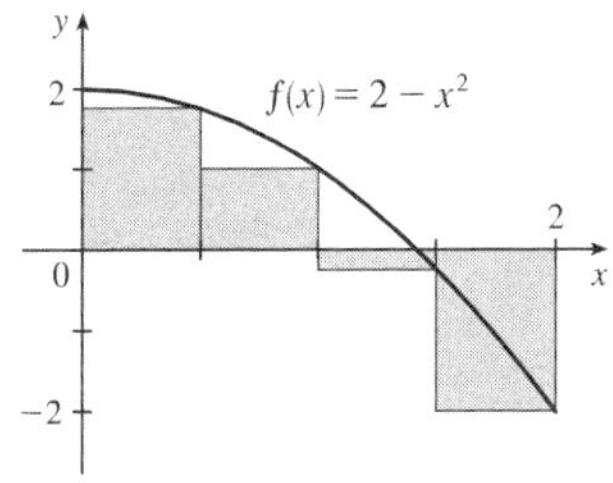


The Riemann sum represents the sum of the areas of the two rectangles above the $x$-axis minus the sum of the areas of the two rectangles below the $x$-axis; that is, the *net area* of the rectangles with respect to the $x$-axis.

**2.** $L_6 = \sum_{i=1}^{6} f(x_{i-1})\Delta x \quad [x_i^* = x_{i-1} \text{ is a left endpoint and } \Delta x = 0.5]$

$= 0.5\,[f(1) + f(1.5) + f(2) + f(2.5) + f(3) + f(3.5)] \quad [f(x) = \ln x - 1]$

$\approx 0.5(-1 - 0.5945349 - 0.3068528 - 0.0837093 + 0.0986123 + 0.2527630)$

$= 0.5(-1.6337217) \approx -0.816861$

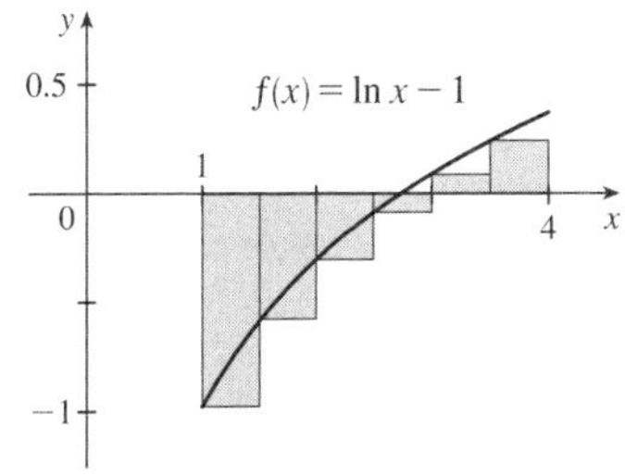


The Riemann sum represents the sum of the areas of the two rectangles above the $x$-axis minus the sum of the areas of the four rectangles below the $x$-axis; that is, the *net area* of the rectangles with respect to the $x$-axis.

**3.** $M_5 = \sum_{i=1}^{5} f(\overline{x}_i)\,\Delta x \quad [x_i^* = \overline{x}_i = \frac{1}{2}(x_{i-1} + x_i) \text{ is a midpoint and } \Delta x = 1]$

$= 1\,[f(1.5) + f(2.5) + f(3.5) + f(4.5) + f(5.5)] \quad [f(x) = \sqrt{x} - 2]$

$\approx -0.856759$

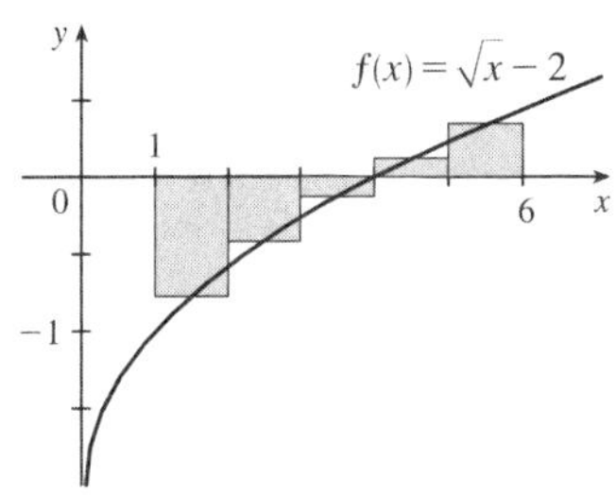


The Riemann sum represents the sum of the areas of the two rectangles above the $x$-axis minus the sum of the areas of the three rectangles below the $x$-axis.

**4.** (a) $R_6 = \sum_{i=1}^{6} f(x_i)\,\Delta x \quad [x_i^* = x_i \text{ is a right endpoint and } \Delta x = 0.5]$

$= 0.5\,[f(0.5) + f(1) + f(1.5) + f(2) + f(2.5) + f(3)] \quad [f(x) = x - 2\sin 2x]$

$\approx 5.353254$

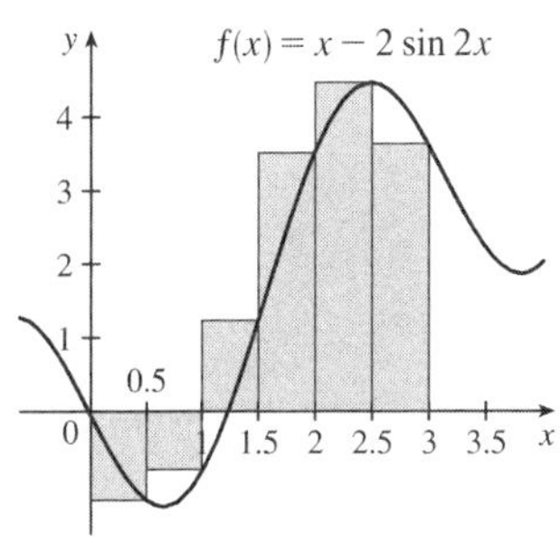


The Riemann sum represents the sum of the areas of the four rectangles above the $x$-axis minus the sum of the areas of the two rectangles below the $x$-axis.

(b) $M_6 = \sum_{i=1}^{6} f(\overline{x}_i)\,\Delta x \quad [x_i^* = \overline{x}_i \text{ is a midpoint and } \Delta x = 0.5]$

$= 0.5[f(0.25) + f(0.75) + f(1.25) + f(1.75)$

$\quad + f(2.25) + f(2.75)] \quad [f(x) = x - 2\sin 2x]$

$\approx 4.458461$

The Riemann sum represents the sum of the areas of the four rectangles above the $x$-axis minus the sum of the areas of the two rectangles below the $x$-axis.

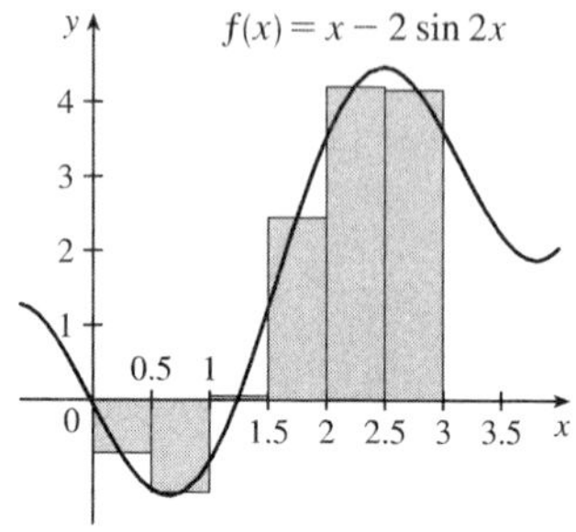


**5.** $\Delta x = (b-a)/n = (8-0)/4 = 8/4 = 2$.

(a) Using the right endpoints to approximate $\int_0^8 f(x)\,dx$, we have

$$\sum_{i=1}^{4} f(x_i)\,\Delta x = 2[f(2) + f(4) + f(6) + f(8)] \approx 2[1 + 2 + (-2) + 1] = 4.$$

(b) Using the left endpoints to approximate $\int_0^8 f(x)\,dx$, we have

$$\sum_{i=1}^{4} f(x_{i-1})\,\Delta x = 2[f(0) + f(2) + f(4) + f(6)] \approx 2[2 + 1 + 2 + (-2)] = 6.$$

(c) Using the midpoint of each subinterval to approximate $\int_0^8 f(x)\,dx$, we have

$$\sum_{i=1}^{4} f(\overline{x}_i)\,\Delta x = 2[f(1) + f(3) + f(5) + f(7)] \approx 2[3 + 2 + 1 + (-1)] = 10.$$

**6.** (a) Using the right endpoints to approximate $\int_{-3}^{3} g(x)\,dx$, we have

$$\sum_{i=1}^{6} g(x_i)\,\Delta x = 1[g(-2) + g(-1) + g(0) + g(1) + g(2) + g(3)]$$

$$\approx 1 - 0.5 - 1.5 - 1.5 - 0.5 + 2.5 = -0.5$$

(b) Using the left endpoints to approximate $\int_{-3}^{3} g(x)\,dx$, we have

$$\sum_{i=1}^{6} g(x_{i-1})\,\Delta x = 1[g(-3) + g(-2) + g(-1) + g(0) + g(1) + g(2)]$$

$$\approx 2 + 1 - 0.5 - 1.5 - 1.5 - 0.5 = -1$$

(c) Using the midpoint of each subinterval to approximate $\int_{-3}^{3} g(x)\,dx$, we have

$$\sum_{i=1}^{6} g(\overline{x}_i)\,\Delta x = 1[g(-2.5) + g(-1.5) + g(-0.5) + g(0.5) + g(1.5) + g(2.5)]$$

$$\approx 1.5 + 0 - 1 - 1.75 - 1 + 0.5 = -1.75$$

**7.** Since $f$ is increasing, $L_5 \le \int_0^{25} f(x)\,dx \le R_5$.

$$\text{Lower estimate} = L_5 = \sum_{i=1}^{5} f(x_{i-1})\,\Delta x = 5[f(0) + f(5) + f(10) + f(15) + f(20)]$$

$$= 5(-42 - 37 - 25 - 6 + 15) = 5(-95) = -475$$

$$\text{Upper estimate} = R_5 = \sum_{i=1}^{5} f(x_i)\,\Delta x = 5[f(5) + f(10) + f(15) + f(20) + f(25)]$$

$$= 5(-37 - 25 - 6 + 15 + 36) = 5(-17) = -85$$

**8.** (a) Using the right endpoints to approximate $\int_0^6 f(x)\,dx$, we have

$$\sum_{i=1}^{3} f(x_i)\,\Delta x = 2[f(2) + f(4) + f(6)] = 2(8.3 + 2.3 - 10.5) = 0.2$$

(b) Using the left endpoints to approximate $\int_0^6 f(x)\,dx$, we have

$$\sum_{i=1}^{3} f(x_{i-1})\,\Delta x = 2[f(0)+f(2)+f(4)] = 2(9.3+8.3+2.3) = 39.8$$

(c) Using the midpoint of each interval to approximate $\int_0^6 f(x)\,dx$, we have

$$\sum_{i=1}^{3} f(\overline{x}_i)\,\Delta x = 2[f(1)+f(3)+f(5)] = 2(9.0+6.5-7.6) = 15.8.$$

The estimate using the right endpoints must be less than $\int_0^6 f(x)\,dx$, since if we take $x_i^*$ to be the right endpoint $x_i$ of each interval, then $f(x_i) \le f(x)$ for all $x$ on $[x_{i-1}, x_i]$, which implies that $f(x_i)\,\Delta x \le \int_{x_{i-1}}^{x_i} f(x)\,dx$, and so the sum $\sum_{i=1}^{3}[f(x_i)\,\Delta x] \le \sum_{i=1}^{3}\left[\int_{x_{i-1}}^{x_i} f(x)\,dx\right] = \int_0^6 f(x)\,dx$. Similarly, if we take $x_i^*$ to be the left endpoint $x_{i-1}$ of each interval, then $f(x_{i-1}) \ge f(x)$ for all $x$ on $[x_{i-1}, x_i]$, and so $\sum_{i=1}^{3}[f(x_{i-1})\,\Delta x] \ge \int_0^6 f(x)\,dx$. We cannot say anything about the midpoint estimate.

**9.** $\Delta x = (10-2)/4 = 2$, so the endpoints are 2, 4, 6, 8, and 10, and the midpoints are 3, 5, 7, and 9. The Midpoint Rule gives

$$\int_2^{10} \sqrt{x^3+1}\,dx \approx \sum_{i=1}^{4} f(\overline{x}_i)\,\Delta x = 2\left(\sqrt{3^3+1}+\sqrt{5^3+1}+\sqrt{7^3+1}+\sqrt{9^3+1}\right) \approx 124.1644.$$

**10.** $\Delta x = (\pi-0)/6 = \frac{\pi}{6}$, so the endpoints are $0, \frac{\pi}{6}, \frac{2\pi}{6}, \frac{3\pi}{6}, \frac{4\pi}{6}, \frac{5\pi}{6}$, and $\frac{6\pi}{6}$, and the midpoints are $\frac{\pi}{12}, \frac{3\pi}{12}, \frac{5\pi}{12}, \frac{7\pi}{12}, \frac{9\pi}{12}$, and $\frac{11\pi}{12}$. The Midpoint Rule gives

$$\int_0^{\pi} \sec(x/3)\,dx \approx \sum_{i=1}^{6} f(\overline{x}_i)\,\Delta x = \frac{\pi}{6}\left(\sec\frac{\pi}{36}+\sec\frac{3\pi}{36}+\sec\frac{5\pi}{36}+\sec\frac{7\pi}{36}+\sec\frac{9\pi}{36}+\sec\frac{11\pi}{36}\right) \approx 3.9379.$$

**11.** $\Delta x = (1-0)/5 = 0.2$, so the endpoints are 0, 0.2, 0.4, 0.6, 0.8, and 1, and the midpoints are 0.1, 0.3, 0.5, 0.7, and 0.9. The Midpoint Rule gives

$$\int_0^1 \sin(x^2)\,dx \approx \sum_{i=1}^{5} f(\overline{x}_i)\,\Delta x = 0.2\left[\sin(0.1)^2+\sin(0.3)^2+\sin(0.5)^2+\sin(0.7)^2+\sin(0.9)^2\right] \approx 0.3084.$$

**12.** $\Delta x = (5-1)/4 = 1$, so the endpoints are 1, 2, 3, 4, and 5, and the midpoints are 1.5, 2.5, 3.5, and 4.5. The Midpoint Rule gives

$$\int_1^5 x^2 e^{-x}\,dx \approx \sum_{n=1}^{4} f(\overline{x}_i)\,\Delta x = 1\left[(1.5)^2 e^{-1.5}+(2.5)^2 e^{-2.5}+(3.5)^2 e^{-3.5}+(4.5)^2 e^{-4.5}\right] \approx 1.6099.$$

**13.** In Maple, we use the command `with(student);` to load the sum and box commands, then `m:=middlesum(sin(x^2),x=0..1,5);` which gives us the sum in summation notation, then `M:=evalf(m);` which gives $M_5 \approx 0.30843908$, confirming the result of Exercise 11. The command `middlebox(sin(x^2),x=0..1,5)` generates the graph. Repeating for $n = 10$ and $n = 20$ gives $M_{10} \approx 0.30981629$ and $M_{20} \approx 0.31015563$.

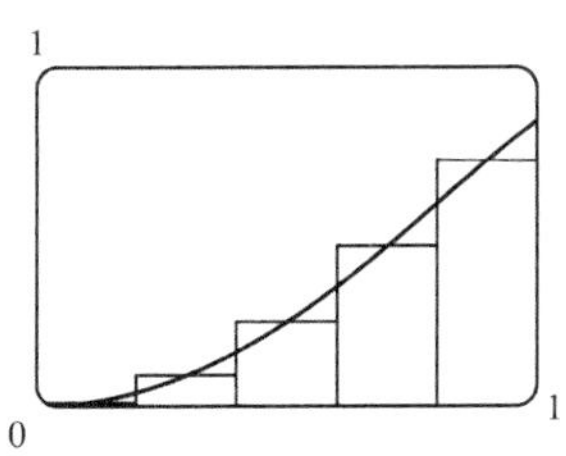


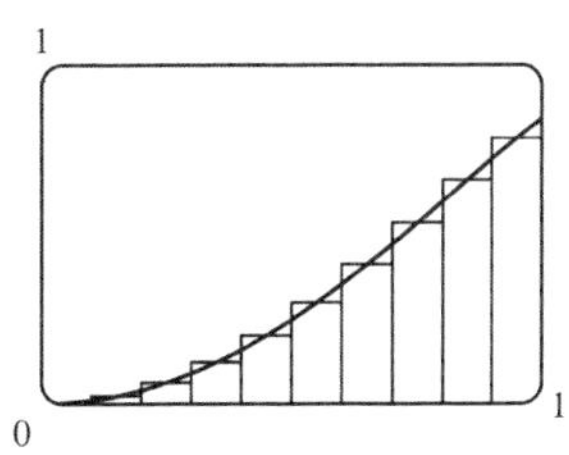


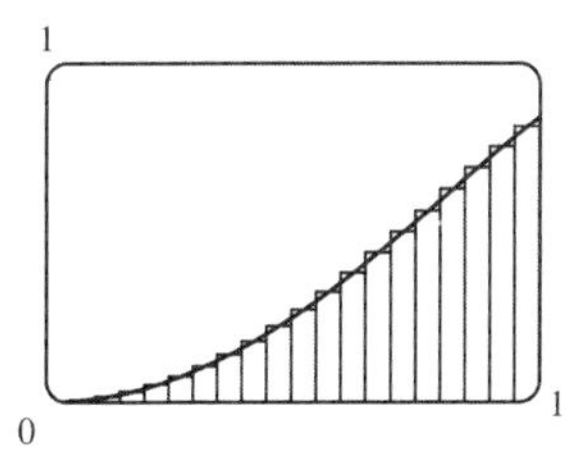

**14.** See the solution to Exercise 5.1.7 for a possible algorithm to calculate the sums. With $\Delta x = (1-0)/100 = 0.01$ and subinterval endpoints 1, 1.01, 1.02, ..., 1.99, 2, we calculate that the left Riemann sum is

$L_{100} = \sum_{i=1}^{100} \sin(x_{i-1}^2)\,\Delta x \approx 0.30607$, and the right Riemann sum is $R_{100} = \sum_{i=1}^{100} \sin(x_i^2)\,\Delta x \approx 0.31448$.

Since $f(x) = \sin(x^2)$ is an increasing function, we must have $L_{100} \le \int_0^1 \sin(x^2)\,dx \le R_{100}$, so

$0.306 < L_{100} \le \int_0^1 \sin(x^2)\,dx \le R_{100} < 0.315$. Therefore, the approximate value $0.3084 \approx 0.31$ in Exercise 11 must be accurate to two decimal places.

**15.** We'll create the table of values to approximate $\int_0^\pi \sin x\,dx$ by using the program in the solution to Exercise 5.1.7 with $Y_1 = \sin x$, Xmin $= 0$, Xmax $= \pi$, and $n = 5$, 10, 50, and 100.

| $n$ | $R_n$ |
|---|---|
| 5 | 1.933766 |
| 10 | 1.983524 |
| 50 | 1.999342 |
| 100 | 1.999836 |

The values of $R_n$ appear to be approaching 2.

**16.** $\int_0^2 e^{-x^2}\,dx$ with $n = 5$, 10, 50, and 100.

| $n$ | $L_n$ | $R_n$ |
|---|---|---|
| 5 | 1.077467 | 0.684794 |
| 10 | 0.980007 | 0.783670 |
| 50 | 0.901705 | 0.862438 |
| 100 | 0.891896 | 0.872262 |

The value of the integral lies between 0.872 and 0.892. Note that $f(x) = e^{-x^2}$ is decreasing on $(0,2)$. We cannot make a similar statement for $\int_{-1}^2 e^{-x^2}\,dx$ since $f$ is increasing on $(-1,0)$.

**17.** On $[0,\pi]$, $\lim\limits_{n\to\infty} \sum_{i=1}^n x_i \sin x_i\,\Delta x = \int_0^\pi x\sin x\,dx$.

**18.** On $[1,5]$, $\lim\limits_{n\to\infty} \sum_{i=1}^n \dfrac{e^{x_i}}{1+x_i}\,\Delta x = \displaystyle\int_1^5 \frac{e^x}{1+x}\,dx$.

**19.** On $[1,8]$, $\lim\limits_{n\to\infty} \sum_{i=1}^n \sqrt{2x_i^* + (x_i^*)^2}\,\Delta x = \int_1^8 \sqrt{2x+x^2}\,dx$.

**20.** On $[0,2]$, $\lim\limits_{n\to\infty} \sum_{i=1}^n \left[4 - 3(x_i^*)^2 + 6(x_i^*)^5\right]\Delta x = \int_0^2 (4 - 3x^2 + 6x^5)\,dx$.

**21.** Note that $\Delta x = \dfrac{5-(-1)}{n} = \dfrac{6}{n}$ and $x_i = -1 + i\,\Delta x = -1 + \dfrac{6i}{n}$.

$$\begin{aligned}
\int_{-1}^5 (1+3x)\,dx &= \lim_{n\to\infty} \sum_{i=1}^n f(x_i)\,\Delta x = \lim_{n\to\infty} \sum_{i=1}^n \left[1 + 3\left(-1+\frac{6i}{n}\right)\right]\frac{6}{n} \\
&= \lim_{n\to\infty} \frac{6}{n}\sum_{i=1}^n \left[-2 + \frac{18i}{n}\right] = \lim_{n\to\infty} \frac{6}{n}\left[\sum_{i=1}^n (-2) + \sum_{i=1}^n \frac{18i}{n}\right] \\
&= \lim_{n\to\infty} \frac{6}{n}\left[-2n + \frac{18}{n}\sum_{i=1}^n i\right] = \lim_{n\to\infty} \frac{6}{n}\left[-2n + \frac{18}{n}\cdot\frac{n(n+1)}{2}\right] \\
&= \lim_{n\to\infty} \left[-12 + \frac{108}{n^2}\cdot\frac{n(n+1)}{2}\right] = \lim_{n\to\infty} \left[-12 + 54\frac{n+1}{n}\right] \\
&= \lim_{n\to\infty} \left[-12 + 54\left(1+\frac{1}{n}\right)\right] = -12 + 54\cdot 1 = 42
\end{aligned}$$

**22.** $\displaystyle\int_1^4 (x^2 + 2x - 5)\,dx = \lim_{n\to\infty} \sum_{i=1}^n f(x_i)\Delta x \qquad [\Delta x = 3/n \text{ and } x_i = 1 + 3i/n]$

$$= \lim_{n\to\infty} \sum_{i=1}^n \left[\left(1 + \frac{3i}{n}\right)^2 + 2\left(1 + \frac{3i}{n}\right) - 5\right]\left(\frac{3}{n}\right)$$

$$= \lim_{n\to\infty} \frac{3}{n}\left[\sum_{i=1}^n \left(1 + \frac{6i}{n} + \frac{9i^2}{n^2} + 2 + \frac{6i}{n} - 5\right)\right]$$

$$= \lim_{n\to\infty} \frac{3}{n}\left[\sum_{i=1}^n \left(\frac{9}{n^2}\cdot i^2 + \frac{12}{n}\cdot i - 2\right)\right]$$

$$= \lim_{n\to\infty} \frac{3}{n}\left[\frac{9}{n^2}\sum_{i=1}^n i^2 + \frac{12}{n}\sum_{i=1}^n i - \sum_{i=1}^n 2\right]$$

$$= \lim_{n\to\infty} \left(\frac{27}{n^3}\cdot\frac{n(n+1)(2n+1)}{6} + \frac{36}{n^2}\cdot\frac{n(n+1)}{2} - \frac{6}{n}\cdot n\right)$$

$$= \lim_{n\to\infty} \left(\frac{9}{2}\cdot\frac{n+1}{n}\cdot\frac{2n+1}{n} + 18\cdot\frac{n+1}{n} - 6\right)$$

$$= \lim_{n\to\infty} \left[\frac{9}{2}\left(1 + \frac{1}{n}\right)\left(2 + \frac{1}{n}\right) + 18\left(1 + \frac{1}{n}\right) - 6\right] = \tfrac{9}{2}\cdot 1\cdot 2 + 18\cdot 1 - 6 = 21$$

**23.** Note that $\Delta x = \dfrac{2-0}{n} = \dfrac{2}{n}$ and $x_i = 0 + i\,\Delta x = \dfrac{2i}{n}$.

$$\int_0^2 (2 - x^2)\,dx = \lim_{n\to\infty} \sum_{i=1}^n f(x_i)\,\Delta x = \lim_{n\to\infty} \sum_{i=1}^n \left(2 - \frac{4i^2}{n^2}\right)\left(\frac{2}{n}\right)$$

$$= \lim_{n\to\infty} \frac{2}{n}\left[\sum_{i=1}^n 2 - \frac{4}{n^2}\sum_{i=1}^n i^2\right] = \lim_{n\to\infty} \frac{2}{n}\left(2n - \frac{4}{n^2}\sum_{i=1}^n i^2\right)$$

$$= \lim_{n\to\infty} \left[4 - \frac{8}{n^3}\cdot\frac{n(n+1)(2n+1)}{6}\right] = \lim_{n\to\infty} \left(4 - \frac{4}{3}\cdot\frac{n+1}{n}\cdot\frac{2n+1}{n}\right)$$

$$= \lim_{n\to\infty} \left[4 - \frac{4}{3}\left(1 + \frac{1}{n}\right)\left(2 + \frac{1}{n}\right)\right] = 4 - \tfrac{4}{3}\cdot 1\cdot 2 = \tfrac{4}{3}$$

**24.** $\displaystyle\int_0^5 (1 + 2x^3)\,dx = \lim_{n\to\infty} \sum_{i=1}^n f(x_i)\,\Delta x \quad [\Delta x = 5/n \text{ and } x_i = 5i/n]$

$$= \lim_{n\to\infty} \sum_{i=1}^n \left(1 + 2\cdot\frac{125i^3}{n^3}\right)\left(\frac{5}{n}\right) = \lim_{n\to\infty} \frac{5}{n}\left[\sum_{i=1}^n 1 + \frac{250}{n^3}\sum_{i=1}^n i^3\right]$$

$$= \lim_{n\to\infty} \frac{5}{n}\left(1\cdot n + \frac{250}{n^3}\sum_{i=1}^n i^3\right) = \lim_{n\to\infty} \left[5 + \frac{1250}{n^4}\cdot\frac{n^2(n+1)^2}{4}\right]$$

$$= \lim_{n\to\infty} \left[5 + 312.5\cdot\frac{(n+1)^2}{n^2}\right] = \lim_{n\to\infty} \left[5 + 312.5\left(1 + \frac{1}{n}\right)^2\right]$$

$$= 5 + 312.5 = 317.5$$

**25.** Note that $\Delta x = \dfrac{2-1}{n} = \dfrac{1}{n}$ and $x_i = 1 + i\,\Delta x = 1 + i(1/n) = 1 + i/n$.

$$\begin{aligned}
\int_1^2 x^3\,dx &= \lim_{n\to\infty}\sum_{i=1}^n f(x_i)\,\Delta x = \lim_{n\to\infty}\sum_{i=1}^n\left(1+\frac{i}{n}\right)^3\left(\frac{1}{n}\right) = \lim_{n\to\infty}\frac{1}{n}\sum_{i=1}^n\left(\frac{n+i}{n}\right)^3 \\
&= \lim_{n\to\infty}\frac{1}{n^4}\sum_{i=1}^n\left(n^3+3n^2i+3ni^2+i^3\right) = \lim_{n\to\infty}\frac{1}{n^4}\left[\sum_{i=1}^n n^3+\sum_{i=1}^n 3n^2i+\sum_{i=1}^n 3ni^2+\sum_{i=1}^n i^3\right] \\
&= \lim_{n\to\infty}\frac{1}{n^4}\left[n\cdot n^3+3n^2\sum_{i=1}^n i+3n\sum_{i=1}^n i^2+\sum_{i=1}^n i^3\right] \\
&= \lim_{n\to\infty}\left[1+\frac{3}{n^2}\cdot\frac{n(n+1)}{2}+\frac{3}{n^3}\cdot\frac{n(n+1)(2n+1)}{6}+\frac{1}{n^4}\cdot\frac{n^2(n+1)^2}{4}\right] \\
&= \lim_{n\to\infty}\left[1+\frac{3}{2}\cdot\frac{n+1}{n}+\frac{1}{2}\cdot\frac{n+1}{n}\cdot\frac{2n+1}{n}+\frac{1}{4}\cdot\frac{(n+1)^2}{n^2}\right] \\
&= \lim_{n\to\infty}\left[1+\frac{3}{2}\left(1+\frac{1}{n}\right)+\frac{1}{2}\left(1+\frac{1}{n}\right)\left(2+\frac{1}{n}\right)+\frac{1}{4}\left(1+\frac{1}{n}\right)^2\right] = 1+\frac{3}{2}+\tfrac{1}{2}\cdot 2+\tfrac{1}{4} = 3.75
\end{aligned}$$

**26.** (a) $\Delta x = (4-0)/8 = 0.5$ and $x_i^* = x_i = 0.5i$.

$$\begin{aligned}
\textstyle\int_0^4 (x^2-3x)\,dx &\approx \sum_{i=1}^8 f(x_i^*)\,\Delta x \\
&= 0.5\{[0.5^2-3(0.5)]+[1.0^2-3(1.0)]+\cdots \\
&\qquad +[3.5^2-3(3.5)]+[4.0^2-3(4.0)]\} \\
&= \tfrac{1}{2}\left(-\tfrac{5}{4}-2-\tfrac{9}{4}-2-\tfrac{5}{4}+0+\tfrac{7}{4}+4\right) = -1.5
\end{aligned}$$

(b)

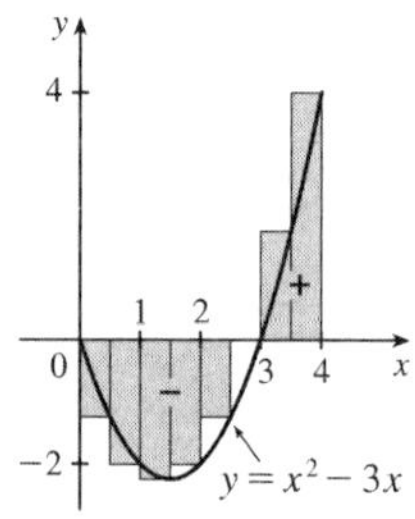


(c)
$$\begin{aligned}
\int_0^4 (x^2-3x)\,dx &= \lim_{n\to\infty}\sum_{i=1}^n\left[\left(\frac{4i}{n}\right)^2-3\left(\frac{4i}{n}\right)\right]\left(\frac{4}{n}\right) \\
&= \lim_{n\to\infty}\frac{4}{n}\left[\frac{16}{n^2}\sum_{i=1}^n i^2-\frac{12}{n}\sum_{i=1}^n i\right] \\
&= \lim_{n\to\infty}\left[\frac{64}{n^3}\cdot\frac{n(n+1)(2n+1)}{6}-\frac{48}{n^2}\cdot\frac{n(n+1)}{2}\right] \\
&= \lim_{n\to\infty}\left[\frac{32}{3}\left(1+\frac{1}{n}\right)\left(2+\frac{1}{n}\right)-24\left(1+\frac{1}{n}\right)\right] \\
&= \tfrac{32}{3}\cdot 2-24 = -\tfrac{8}{3}
\end{aligned}$$

(d) $\int_0^4 (x^2-3x)\,dx = A_1 - A_2$, where $A_1$ is the area marked $+$ and $A_2$ is the area marked $-$.

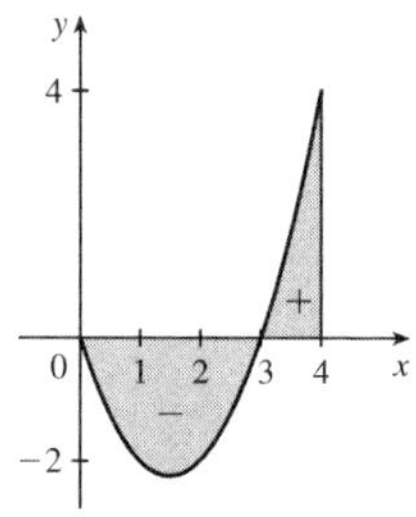


**27.** $f(x) = \dfrac{x}{1+x^5}$, $a = 2$, $b = 6$, and $\Delta x = \dfrac{6-2}{n} = \dfrac{4}{n}$. Using Equation 3, we get $x_i^* = x_i = 2 + i\,\Delta x = 2 + \dfrac{4i}{n}$, so

$$\int_2^6 \frac{x}{1+x^5}\,dx = \lim_{n\to\infty} R_n = \lim_{n\to\infty}\sum_{i=1}^n \frac{2+\dfrac{4i}{n}}{1+\left(2+\dfrac{4i}{n}\right)^5}\cdot\frac{4}{n}.$$

**28.** $\Delta x = \dfrac{10-1}{n} = \dfrac{9}{n}$ and $x_i = 1 + i\,\Delta x = 1 + \dfrac{9i}{n}$, so

$$\int_1^{10} (x - 4\ln x)\,dx = \lim_{n\to\infty} R_n = \lim_{n\to\infty} \sum_{i=1}^{n} \left[\left(1+\frac{9i}{n}\right) - 4\ln\left(1+\frac{9i}{n}\right)\right] \cdot \frac{9}{n}.$$

**29.** $\Delta x = (\pi - 0)/n = \pi/n$ and $x_i^* = x_i = \pi i/n$.

$$\int_0^{\pi} \sin 5x\,dx = \lim_{n\to\infty} \sum_{i=1}^{n} (\sin 5x_i)\left(\frac{\pi}{n}\right) = \lim_{n\to\infty} \sum_{i=1}^{n} \left(\sin\frac{5\pi i}{n}\right)\frac{\pi}{n} \overset{\text{CAS}}{=} \pi \lim_{n\to\infty} \frac{1}{n}\cot\left(\frac{5\pi}{2n}\right) \overset{\text{CAS}}{=} \pi\left(\frac{2}{5\pi}\right) = \frac{2}{5}$$

**30.** $\Delta x = (10-2)/n = 8/n$ and $x_i^* = x_i = 2 + 8i/n$.

$$\begin{aligned}\int_2^{10} x^6\,dx &= \lim_{n\to\infty} \sum_{i=1}^{n} \left(2+\frac{8i}{n}\right)^6 \left(\frac{8}{n}\right) = 8 \lim_{n\to\infty} \frac{1}{n} \sum_{i=1}^{n} \left(2+\frac{8i}{n}\right)^6 \\ &\overset{\text{CAS}}{=} 8 \lim_{n\to\infty} \frac{1}{n} \cdot \frac{64(58{,}593n^6 + 164{,}052n^5 + 131{,}208n^4 - 27{,}776n^2 + 2048)}{21n^5} \\ &\overset{\text{CAS}}{=} 8\left(\frac{1{,}249{,}984}{7}\right) = \frac{9{,}999{,}872}{7} \approx 1{,}428{,}553.1\end{aligned}$$

**31.** (a) Think of $\int_0^2 f(x)\,dx$ as the area of a trapezoid with bases 1 and 3 and height 2. The area of a trapezoid is $A = \frac{1}{2}(b+B)h$, so $\int_0^2 f(x)\,dx = \frac{1}{2}(1+3)2 = 4$.

(b) $\int_0^5 f(x)\,dx = \underset{\text{trapezoid}}{\int_0^2 f(x)\,dx} + \underset{\text{rectangle}}{\int_2^3 f(x)\,dx} + \underset{\text{triangle}}{\int_3^5 f(x)\,dx}$

$= \frac{1}{2}(1+3)2 + 3\cdot 1 + \frac{1}{2}\cdot 2\cdot 3 = 4 + 3 + 3 = 10$

(c) $\int_5^7 f(x)\,dx$ is the negative of the area of the triangle with base 2 and height 3. $\int_5^7 f(x)\,dx = -\frac{1}{2}\cdot 2\cdot 3 = -3$.

(d) $\int_7^9 f(x)\,dx$ is the negative of the area of a trapezoid with bases 3 and 2 and height 2, so it equals $-\frac{1}{2}(B+b)h = -\frac{1}{2}(3+2)2 = -5$. Thus,
$\int_0^9 f(x)\,dx = \int_0^5 f(x)\,dx + \int_5^7 f(x)\,dx + \int_7^9 f(x)\,dx = 10 + (-3) + (-5) = 2$.

**32.** (a) $\int_0^2 g(x)\,dx = \frac{1}{2}\cdot 4\cdot 2 = 4$ (area of a triangle)

(b) $\int_2^6 g(x)\,dx = -\frac{1}{2}\pi(2)^2 = -2\pi$ (negative of the area of a semicircle)

(c) $\int_6^7 g(x)\,dx = \frac{1}{2}\cdot 1\cdot 1 = \frac{1}{2}$ (area of a triangle)
$\int_0^7 g(x)\,dx = \int_0^2 g(x)\,dx + \int_2^6 g(x)\,dx + \int_6^7 g(x)\,dx = 4 - 2\pi + \frac{1}{2} = 4.5 - 2\pi$

**33.** $\int_0^3 \left(\frac{1}{2}x - 1\right)dx$ can be interpreted as the area of the triangle above the $x$-axis minus the area of the triangle below the $x$-axis; that is,

$\frac{1}{2}(1)\left(\frac{1}{2}\right) - \frac{1}{2}(2)(1) = \frac{1}{4} - 1 = -\frac{3}{4}$.

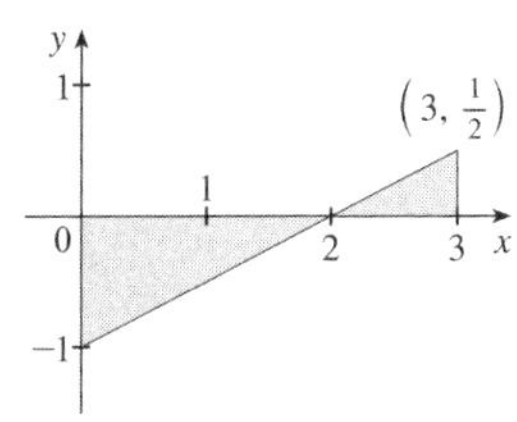


**34.** $\int_{-2}^{2} \sqrt{4-x^2}\,dx$ can be interpreted as the area under the graph of $f(x) = \sqrt{4-x^2}$ between $x = -2$ and $x = 2$. This is equal to half the area of the circle with radius 2, so $\int_{-2}^{2}\sqrt{4-x^2}\,dx = \frac{1}{2}\pi\cdot 2^2 = 2\pi$.

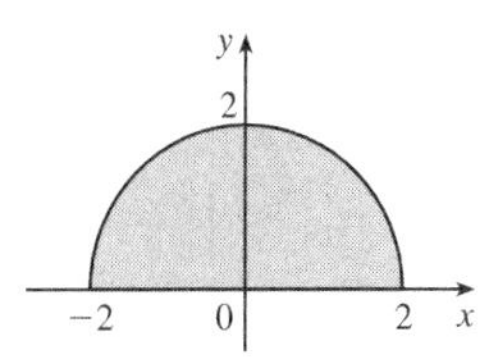

**35.** $\int_{-3}^{0}\left(1+\sqrt{9-x^2}\right)dx$ can be interpreted as the area under the graph of $f(x)=1+\sqrt{9-x^2}$ between $x=-3$ and $x=0$. This is equal to one-quarter the area of the circle with radius 3, plus the area of the rectangle, so $\int_{-3}^{0}\left(1+\sqrt{9-x^2}\right)dx=\frac{1}{4}\pi\cdot 3^2+1\cdot 3=3+\frac{9}{4}\pi.$

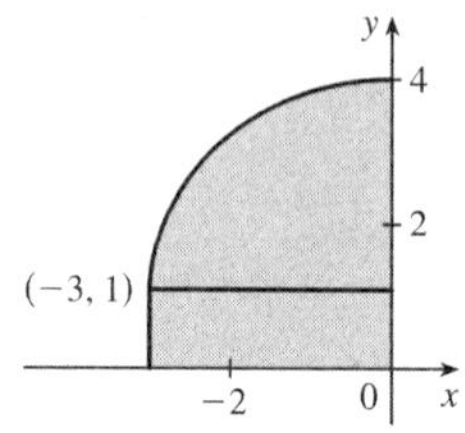


**36.** $\int_{-1}^{3}(3-2x)\,dx$ can be interpreted as the area of the triangle above the $x$-axis minus the area of the triangle below the $x$-axis; that is,

$\frac{1}{2}\left(\frac{5}{2}\right)(5)-\frac{1}{2}\left(\frac{3}{2}\right)(3)=\frac{25}{4}-\frac{9}{4}=4.$

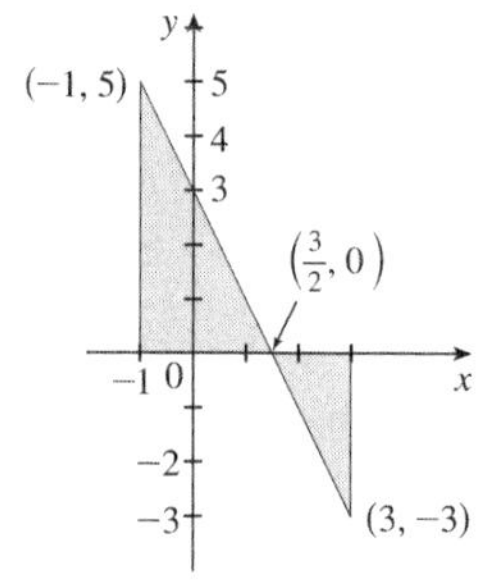


**37.** $\int_{-1}^{2}|x|\,dx$ can be interpreted as the sum of the areas of the two shaded triangles; that is, $\frac{1}{2}(1)(1)+\frac{1}{2}(2)(2)=\frac{1}{2}+\frac{4}{2}=\frac{5}{2}$.

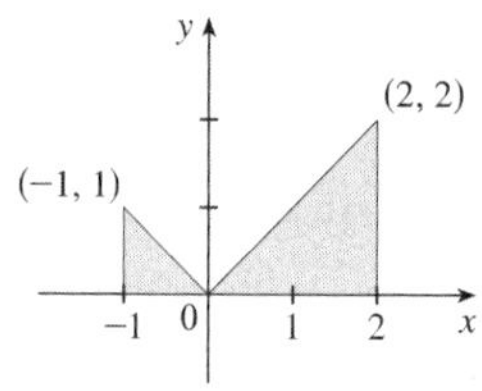


**38.** $\int_{0}^{10}|x-5|\,dx$ can be interpreted as the sum of the areas of the two shaded triangles; that is, $2\left(\frac{1}{2}\right)(5)(5)=25$.

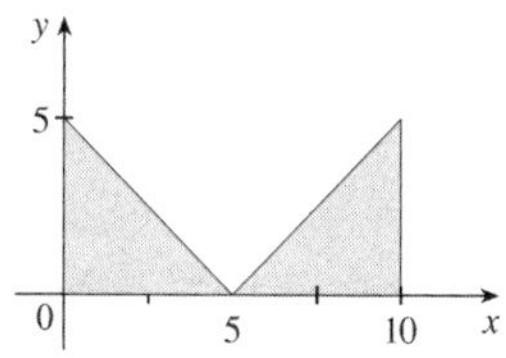


**39.** $\int_{9}^{4}\sqrt{t}\,dt=-\int_{4}^{9}\sqrt{t}\,dt$ [because we reversed the limits of integration]

$=-\int_{4}^{9}\sqrt{x}\,dx$ [we can use any letter without changing the value of the integral]

$=-\frac{38}{3}$

**40.** $\int_{1}^{1}x^2\cos x\,dx=0$ since the limits of integration are equal.

**41.** $\int_{-2}^{2}f(x)\,dx+\int_{2}^{5}f(x)\,dx-\int_{-2}^{-1}f(x)\,dx=\int_{-2}^{5}f(x)\,dx+\int_{-1}^{-2}f(x)\,dx$ [by Property 5 and reversing limits]

$=\int_{-1}^{5}f(x)\,dx$ [Property 5]

**42.** $\int_{1}^{4}f(x)\,dx=\int_{1}^{5}f(x)\,dx-\int_{4}^{5}f(x)\,dx=12-3.6=8.4$

**43.** $\int_{0}^{9}[2f(x)+3g(x)]\,dx=2\int_{0}^{9}f(x)\,dx+3\int_{0}^{9}g(x)\,dx=2(37)+3(16)=122$

**44.** If $f(x)=\begin{cases}3 & \text{for } x<3\\ x & \text{for } x\geq 3\end{cases}$, then $\int_{0}^{5}f(x)\,dx$ can be interpreted as the area of the shaded region, which consists of a 5-by-3 rectangle surmounted by an isosceles right triangle whose legs have length 2. Thus,

$\int_{0}^{5}f(x)\,dx=5(3)+\frac{1}{2}(2)(2)=17.$

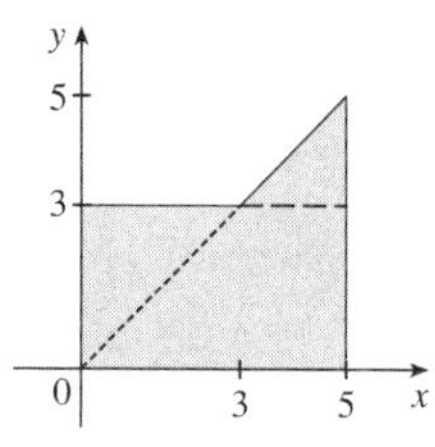


**45.** $\int_{0}^{1}\left(5-6x^2\right)dx=\int_{0}^{1}5\,dx-6\int_{0}^{1}x^2\,dx=5(1-0)-6\left(\frac{1}{3}\right)=5-2=3$

**46.** $\int_{1}^{3}\left(2e^x-1\right)dx=2\int_{1}^{3}e^x\,dx-\int_{1}^{3}1\,dx=2\left(e^3-e\right)-1\,(3-1)=2e^3-2e-2$

**47.** $\int_1^3 e^{x+2}\,dx = \int_1^3 e^x \cdot e^2\,dx = e^2 \int_1^3 e^x\,dx = e^2(e^3 - e) = e^5 - e^3$

**48.** Using Integral Comparison Property 8,

$m \le f(x) \le M \quad \Rightarrow \quad m(2-0) \le \int_0^2 f(x)\,dx \le M(2-0) \quad \Rightarrow \quad 2m \le \int_0^2 f(x)\,dx \le 2M.$

**49.** On $[1,3]$, $\ln 1 \le \ln x \le \ln 3 \quad \Rightarrow \quad 0(3-1) \le \int_1^3 \ln x\,dx \le (\ln 3)(3-1) \quad \Rightarrow \quad 0 \le \int_1^3 \ln x\,dx \le 2\ln 3.$

**50.** If $0 \le x \le 2$, then $0 \le x^3 \le 8$, so $1 \le x^3 + 1 \le 9$ and $1 \le \sqrt{x^3+1} \le 3$.

Thus, $1(2-0) \le \int_0^2 \sqrt{x^3+1}\,dx \le 3(2-0)$; that is, $2 \le \int_0^2 \sqrt{x^3+1}\,dx \le 6$.

**51.** $\lim\limits_{n\to\infty} \sum\limits_{i=1}^{n} \dfrac{i^4}{n^5} = \lim\limits_{n\to\infty} \sum\limits_{i=1}^{n} \dfrac{i^4}{n^4} \cdot \dfrac{1}{n} = \lim\limits_{n\to\infty} \sum\limits_{i=1}^{n} \left(\dfrac{i}{n}\right)^4 \dfrac{1}{n}$. At this point, we need to recognize the limit as being of the form $\lim\limits_{n\to\infty} \sum\limits_{i=1}^{n} f(x_i)\,\Delta x$, where $\Delta x = (1-0)/n = 1/n$, $x_i = 0 + i\,\Delta x = i/n$, and $f(x) = x^4$. Thus, the definite integral is $\int_0^1 x^4\,dx$.

## 5.3 Evaluating Definite Integrals

**1.** $\displaystyle\int_{-1}^{3} x^5\,dx = \left[\frac{x^6}{6}\right]_{-1}^{3} = \frac{3^6}{6} - \frac{(-1)^6}{6} = \frac{729-1}{6} = \frac{364}{3}$

**2.** $\int_1^3 (1 + 2x - 4x^3)\,dx = \left[x + 2\cdot\frac12 x^2 - 4\cdot\frac14 x^4\right]_1^3 = \left[x + x^2 - x^4\right]_1^3$
$= (3 + 9 - 81) - (1 + 1 - 1) = -69 - 1 = -70$

**3.** $\int_0^2 (6x^2 - 4x + 5)\,dx = \left[6\cdot\frac13 x^3 - 4\cdot\frac12 x^2 + 5x\right]_0^2 = \left[2x^3 - 2x^2 + 5x\right]_0^2 = (16 - 8 + 10) - 0 = 18$

**4.** $\int_{-2}^{0} (u^5 - u^3 + u^2)\,du = \left[\frac16 u^6 - \frac14 u^4 + \frac13 u^3\right]_{-2}^{0} = 0 - \left(\frac{32}{3} - 4 - \frac83\right) = -4$

**5.** $\int_0^1 x^{4/5}\,dx = \left[\frac59 x^{9/5}\right]_0^1 = \frac59 - 0 = \frac59$

**6.** $\int_1^8 \sqrt[3]{x}\,dx = \int_1^8 x^{1/3}\,dx = \left[\frac34 x^{4/3}\right]_1^8 = \frac34(8^{4/3} - 1^{4/3}) = \frac34(2^4 - 1) = \frac34(16-1) = \frac34(15) = \frac{45}{4}$

**7.** $\int_{-1}^{0} (2x - e^x)\,dx = \left[x^2 - e^x\right]_{-1}^{0} = (0 - 1) - (1 - e^{-1}) = -2 + 1/e$

**8.** $\int_\pi^{2\pi} \cos\theta\,d\theta = \left[\sin\theta\right]_\pi^{2\pi} = \sin 2\pi - \sin\pi = 0 - 0 = 0$

**9.** $\int_{-2}^{2} (3u+1)^2\,du = \int_{-2}^{2} (9u^2 + 6u + 1)\,du = \left[9\cdot\frac13 u^3 + 6\cdot\frac12 u^2 + u\right]_{-2}^{2} = \left[3u^3 + 3u^2 + u\right]_{-2}^{2}$
$= (24 + 12 + 2) - (-24 + 12 - 2) = 38 - (-14) = 52$

**10.** $\int_0^4 (2v+5)(3v-1)\,dv = \int_0^4 (6v^2 + 13v - 5)\,dv = \left[6\cdot\frac13 v^3 + 13\cdot\frac12 v^2 - 5v\right]_0^4 = \left[2v^3 + \frac{13}{2}v^2 - 5v\right]_0^4$
$= (128 + 104 - 20) - 0 = 212$

**11.** $\displaystyle\int_{-2}^{-1} \left(4y^3 + \frac{2}{y^3}\right) dy = \left[4\cdot\tfrac14 y^4 + 2\cdot\frac{1}{-2}y^{-2}\right]_{-2}^{-1} = \left[y^4 - \frac{1}{y^2}\right]_{-2}^{-1} = (1-1) - \left(16 - \tfrac14\right) = -\tfrac{63}{4}$

**12.** $\displaystyle\int_1^2 \frac{y + 5y^7}{y^3}\,dy = \int_1^2 (y^{-2} + 5y^4)\,dy = \left[-y^{-1} + 5\cdot\tfrac15 y^5\right]_1^2 = \left[-\frac1y + y^5\right]_1^2 = \left(-\tfrac12 + 32\right) - (-1 + 1) = \tfrac{63}{2}$

**13.** $\int_0^1 x(\sqrt[3]{x} + \sqrt[4]{x})\,dx = \int_0^1 (x^{4/3} + x^{5/4})\,dx = \left[\frac37 x^{7/3} + \frac49 x^{9/4}\right]_0^1 = \left(\frac37 + \frac49\right) - 0 = \frac{55}{63}$

**14.** $\displaystyle\int_1^9 \frac{3x - 2}{\sqrt{x}}\,dx = \int_1^9 (3x^{1/2} - 2x^{-1/2})\,dx = \left[3\cdot\tfrac23 x^{3/2} - 2\cdot 2x^{1/2}\right]_1^9 = \left[2x^{3/2} - 4x^{1/2}\right]_1^9$
$= (54 - 12) - (2 - 4) = 44$

**15.** $\int_0^{\pi/4} \sec^2 t\,dt = \left[\tan t\right]_0^{\pi/4} = \tan\frac{\pi}{4} - \tan 0 = 1 - 0 = 1$

**16.** $\int_0^5 (2e^x + 4\cos x)\,dx = \left[2e^x + 4\sin x\right]_0^5 = \left(2e^5 + 4\sin 5\right) - \left(2e^0 + 4\sin 0\right) = 2e^5 + 4\sin 5 - 2 \approx 290.99$

**17.** $\displaystyle\int_1^9 \frac{1}{2x}\,dx = \frac{1}{2}\int_1^9 \frac{1}{x}\,dx = \frac{1}{2}\Big[\ln|x|\Big]_1^9 = \frac{1}{2}(\ln 9 - \ln 1) = \frac{1}{2}\ln 9 - 0 = \ln 9^{1/2} = \ln 3$

**18.** $\displaystyle\int_0^1 10^x\,dx = \left[\frac{10^x}{\ln 10}\right]_0^1 = \frac{10}{\ln 10} - \frac{1}{\ln 10} = \frac{9}{\ln 10}$

**19.** $\displaystyle\int_{1/2}^{\sqrt{3}/2} \frac{6}{\sqrt{1-t^2}}\,dt = 6\int_{1/2}^{\sqrt{3}/2} \frac{1}{\sqrt{1-t^2}}\,dt = 6\left[\sin^{-1} t\right]_{1/2}^{\sqrt{3}/2} = 6\left[\sin^{-1}\left(\tfrac{\sqrt{3}}{2}\right) - \sin^{-1}\left(\tfrac{1}{2}\right)\right]$

$= 6\left(\frac{\pi}{3} - \frac{\pi}{6}\right) = 6\left(\frac{\pi}{6}\right) = \pi$

**20.** $\displaystyle\int_0^1 \frac{4}{t^2+1}\,dt = 4\int_0^1 \frac{1}{1+t^2}\,dt = 4\left[\tan^{-1} t\right]_0^1 = 4\left(\tan^{-1} 1 - \tan^{-1} 0\right) = 4\left(\frac{\pi}{4} - 0\right) = \pi$

**21.** $\displaystyle\int_1^{64} \frac{1+\sqrt[3]{x}}{\sqrt{x}}\,dx = \int_1^{64}\left(\frac{1}{x^{1/2}} + \frac{x^{1/3}}{x^{1/2}}\right)dx = \int_1^{64}\left(x^{-1/2} + x^{(1/3)-(1/2)}\right)dx = \int_1^{64}\left(x^{-1/2} + x^{-1/6}\right)dx$

$= \left[2x^{1/2} + \frac{6}{5}x^{5/6}\right]_1^{64} = \left(16 + \frac{192}{5}\right) - \left(2 + \frac{6}{5}\right) = 14 + \frac{186}{5} = \frac{256}{5}$

**22.** $\int_{\pi/4}^{\pi/3} \sec\theta\tan\theta\,d\theta = \left[\sec\theta\right]_{\pi/4}^{\pi/3} = \sec\frac{\pi}{3} - \sec\frac{\pi}{4} = 2 - \sqrt{2}$

**23.** $\int_{-1}^{1} e^{u+1}\,du = \left[e^{u+1}\right]_{-1}^{1} = e^2 - e^0 = e^2 - 1$ [or start with $e^{u+1} = e^u e^1$]

**24.** $\int_0^1 (1+x^2)^3\,dx = \int_0^1 (1 + 3x^2 + 3x^4 + x^6)\,dx = \left[x + x^3 + \frac{3}{5}x^5 + \frac{1}{7}x^7\right]_0^1 = \left(1 + 1 + \frac{3}{5} + \frac{1}{7}\right) - 0 = \frac{96}{35}$

**25.** $\displaystyle\int_0^{\pi/4} \frac{1+\cos^2\theta}{\cos^2\theta}\,d\theta = \int_0^{\pi/4}\left(\frac{1}{\cos^2\theta} + \frac{\cos^2\theta}{\cos^2\theta}\right)d\theta = \int_0^{\pi/4}\left(\sec^2\theta + 1\right)d\theta$

$= \left[\tan\theta + \theta\right]_0^{\pi/4} = \left(\tan\frac{\pi}{4} + \frac{\pi}{4}\right) - (0 + 0) = 1 + \frac{\pi}{4}$

**26.** $\displaystyle\int_0^{\pi/3} \frac{\sin\theta + \sin\theta\tan^2\theta}{\sec^2\theta}\,d\theta = \int_0^{\pi/3} \frac{\sin\theta\,(1+\tan^2\theta)}{\sec^2\theta}\,d\theta = \int_0^{\pi/3} \frac{\sin\theta\sec^2\theta}{\sec^2\theta}\,d\theta = \int_0^{\pi/3} \sin\theta\,d\theta$

$= \left[-\cos\theta\right]_0^{\pi/3} = -\frac{1}{2} - (-1) = \frac{1}{2}$

**27.** $\displaystyle\int_1^e \frac{x^2+x+1}{x}\,dx = \int_1^e \left(x + 1 + \frac{1}{x}\right)dx = \left[\tfrac{1}{2}x^2 + x + \ln|x|\right]_1^e$

$= \left(\frac{1}{2}e^2 + e + \ln e\right) - \left(\frac{1}{2} + 1 + \ln 1\right) = \frac{1}{2}e^2 + e - \frac{1}{2}$

**28.** $\int_0^{3\pi/2} |\sin x|\,dx = \int_0^{\pi} \sin x\,dx + \int_{\pi}^{3\pi/2} (-\sin x)\,dx = \left[-\cos x\right]_0^{\pi} + \left[\cos x\right]_{\pi}^{3\pi/2}$

$= [1 - (-1)] + [0 - (-1)] = 2 + 1 = 3$

**29.** $f(x) = 1/x^2$ is not continuous on the interval $[-1, 3]$, so the Evaluation Theorem does not apply. In fact, $f$ has an infinite discontinuity at $x = 0$, so $\int_{-1}^{3} (1/x^2)\,dx$ does not exist.

**30.** $f(x) = \sec^2 x$ is not continuous on the interval $[0, \pi]$, so the Evaluation Theorem does not apply. In fact, $f$ has an infinite discontinuity at $x = \pi/2$, so $\int_0^{\pi} \sec^2 x\,dx$ does not exist.

**31.** It appears that the area under the graph is about $\frac{2}{3}$ of the area of the viewing rectangle, or about $\frac{2}{3}\pi \approx 2.1$. The actual area is $\int_0^{\pi} \sin x\,dx = \left[-\cos x\right]_0^{\pi} = (-\cos\pi) - (-\cos 0) = -(-1) + 1 = 2$.

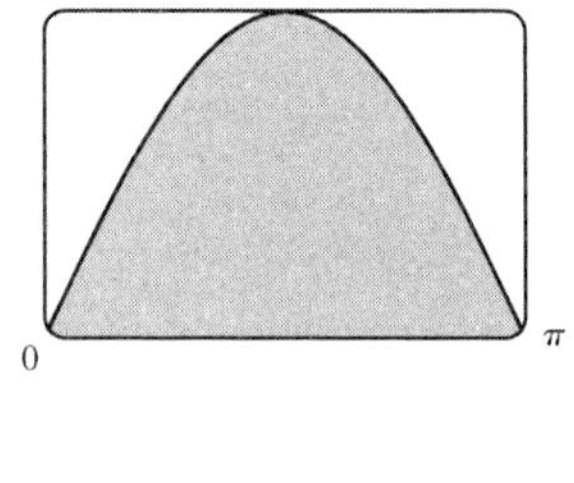

**32.** Splitting up the region as shown, we estimate that the area under the graph is $\frac{\pi}{3} + \frac{1}{4}\left(3 \cdot \frac{\pi}{3}\right) \approx 1.8$. The actual area is $\int_0^{\pi/3} \sec^2 x\,dx = [\tan x]_0^{\pi/3} = \sqrt{3} - 0 = \sqrt{3} \approx 1.73$.

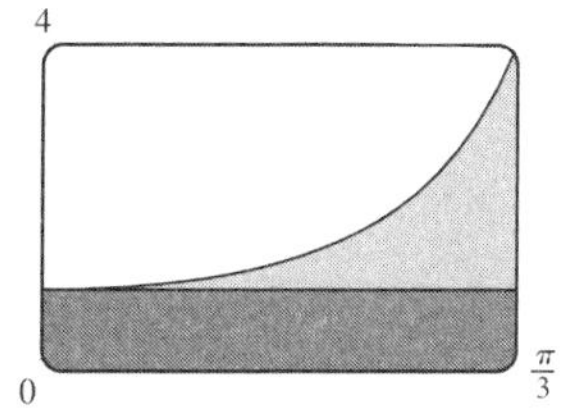


**33.** The graph shows that $y = x + x^2 - x^4$ has $x$-intercepts at $x = 0$ and at $x = a \approx 1.32$. So the area of the region that lies under the curve and above the $x$-axis is

$$\begin{aligned}\int_0^a (x + x^2 - x^4)\,dx &= \left[\tfrac{1}{2}x^2 + \tfrac{1}{3}x^3 - \tfrac{1}{5}x^5\right]_0^a \\ &= \left(\tfrac{1}{2}a^2 + \tfrac{1}{3}a^3 - \tfrac{1}{5}a^5\right) - 0 \\ &\approx 0.84\end{aligned}$$

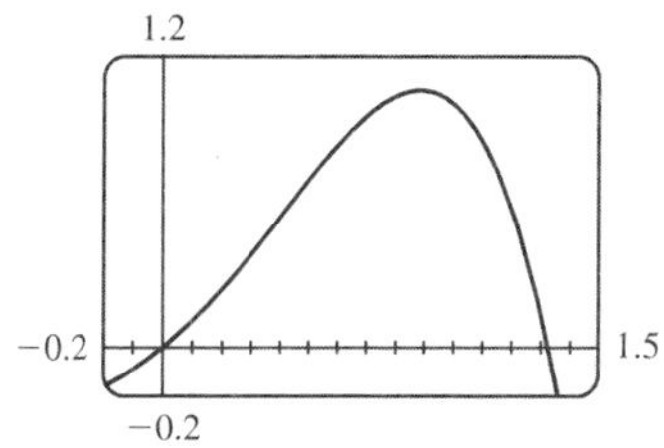


**34.** The graph shows that $y = 2x + 3x^4 - 2x^6$ has $x$-intercepts at $x = 0$ and at $x = a \approx 1.37$. So the area of the region that lies under the curve and above the $x$-axis is

$$\begin{aligned}\int_0^a (2x + 3x^4 - 2x^6)\,dx &= \left[x^2 + \tfrac{3}{5}x^5 - \tfrac{2}{7}x^7\right]_0^a \\ &= \left(a^2 + \tfrac{3}{5}a^5 - \tfrac{2}{7}a^7\right) - 0 \\ &\approx 2.18\end{aligned}$$

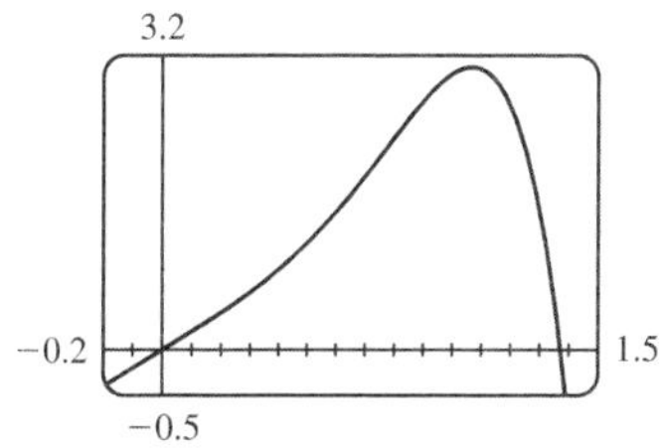


**35.** $\int_{-1}^{2} x^3\,dx = \left[\frac{1}{4}x^4\right]_{-1}^{2} = 4 - \frac{1}{4} = \frac{15}{4} = 3.75$

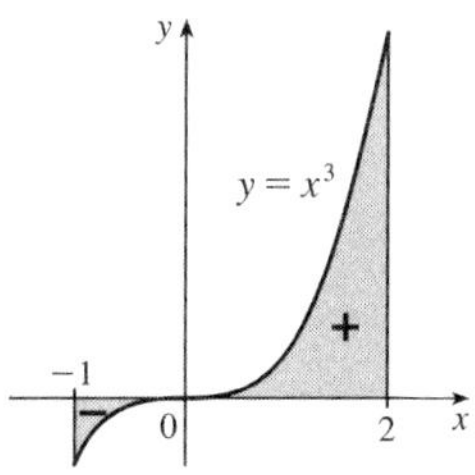


**36.** $\int_{\pi/4}^{5\pi/2} \sin x\,dx = [-\cos x]_{\pi/4}^{5\pi/2} = 0 + \frac{\sqrt{2}}{2} = \frac{\sqrt{2}}{2}$

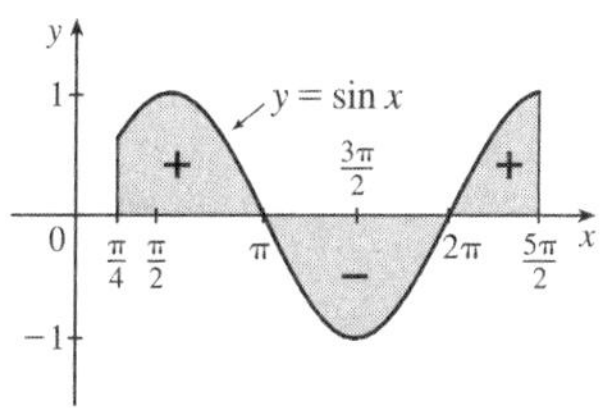


**37.** $\dfrac{d}{dx}\left[\sqrt{x^2+1} + C\right] = \dfrac{d}{dx}\left[\left(x^2+1\right)^{1/2} + C\right] = \frac{1}{2}\left(x^2+1\right)^{-1/2} \cdot 2x = \dfrac{x}{\sqrt{x^2+1}}$

**38.** $\dfrac{d}{dx}\left[x \sin x + \cos x + C\right] = x\cos x + (\sin x)\cdot 1 - \sin x = x\cos x$

**39.** $\int x\sqrt{x}\,dx = \int x^{3/2}\,dx = \frac{2}{5}x^{5/2} + C$.

The members of the family in the figure correspond to $C = 5, 3, 0, -2$, and $-4$.

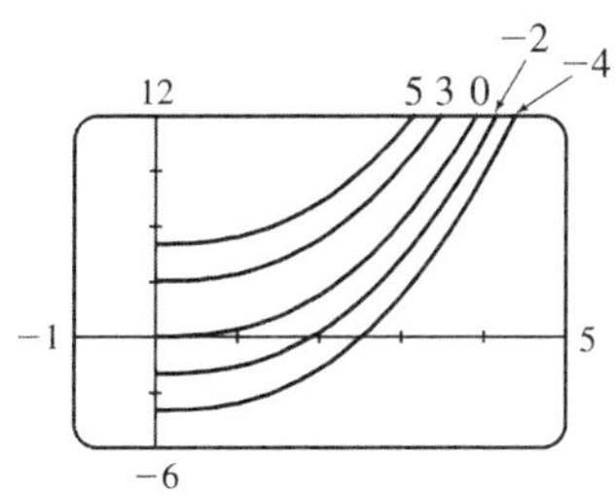


**40.** $\int(\cos x - 2\sin x)\,dx = \sin x + 2\cos x + C$.

The members of the family in the figure correspond to $C = 5, 3, 0, -2$, and $-4$.

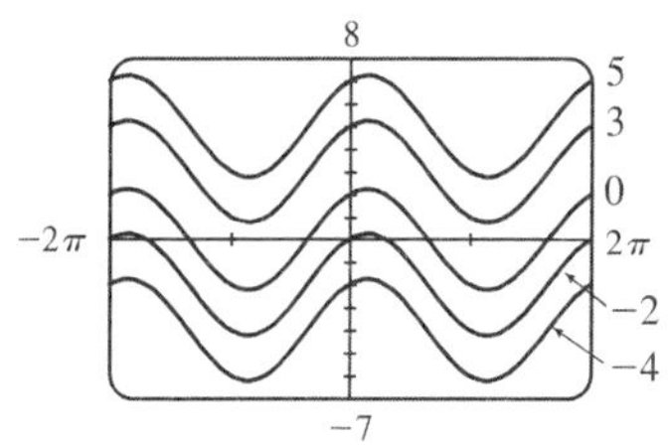


**41.** $\displaystyle\int (1-t)(2+t^2)\,dt = \int \left(2 - 2t + t^2 - t^3\right)dt = 2t - 2\frac{t^2}{2} + \frac{t^3}{3} - \frac{t^4}{4} + C = 2t - t^2 + \tfrac{1}{3}t^3 - \tfrac{1}{4}t^4 + C$

**42.** $\displaystyle\int x(1+2x^4)\,dx = \int (x+2x^5)\,dx = \frac{x^2}{2} + 2\frac{x^6}{6} + C = \tfrac{1}{2}x^2 + \tfrac{1}{3}x^6 + C$

**43.** $\displaystyle\int \frac{\sin x}{1-\sin^2 x}\,dx = \int \frac{\sin x}{\cos^2 x}\,dx = \int \frac{1}{\cos x}\cdot\frac{\sin x}{\cos x}\,dx = \int \sec x\,\tan x\,dx = \sec x + C$

**44.** $\displaystyle\int \frac{\sin 2x}{\sin x}\,dx = \int \frac{2\sin x\cos x}{\sin x}\,dx = \int 2\cos x\,dx = 2\sin x + C$

**45.** $A = \int_0^2 (2y - y^2)\,dy = \left[y^2 - \frac{1}{3}y^3\right]_0^2 = \left(4 - \frac{8}{3}\right) - 0 = \frac{4}{3}$

**46.** $y = \sqrt[4]{x} \;\Rightarrow\; x = y^4$, so $A = \int_0^1 y^4\,dy = \left[\frac{1}{5}y^5\right]_0^1 = \frac{1}{5}$.

**47.** If $w'(t)$ is the rate of change of weight in pounds per year, then $w(t)$ represents the weight in pounds of the child at age $t$. We know from the Net Change Theorem that $\int_5^{10} w'(t)\,dt = w(10) - w(5)$, so the integral represents the increase in the child's weight (in pounds) between the ages of 5 and 10.

**48.** $\int_a^b I(t)\,dt = \int_a^b Q'(t)\,dt = Q(b) - Q(a)$ by the Net Change Theorem, so it represents the change in the charge $Q$ from time $t = a$ to $t = b$.

**49.** Since $r(t)$ is the rate at which oil leaks, we can write $r(t) = -V'(t)$, where $V(t)$ is the volume of oil at time $t$. [Note that the minus sign is needed because $V$ is decreasing, so $V'(t)$ is negative, but $r(t)$ is positive.] Thus, by the Net Change Theorem, $\int_0^{120} r(t)\,dt = -\int_0^{120} V'(t)\,dt = -[V(120) - V(0)] = V(0) - V(120)$, which is the number of gallons of oil that leaked from the tank in the first two hours (120 minutes).

**50.** By the Net Change Theorem, $\int_0^{15} n'(t)\,dt = n(15) - n(0) = n(15) - 100$ represents the increase in the bee population in 15 weeks. So $100 + \int_0^{15} n'(t)\,dt = n(15)$ represents the total bee population after 15 weeks.

**51.** By the Net Change Theorem, $\int_{1000}^{5000} R'(x)\,dx = R(5000) - R(1000)$, so it represents the increase in revenue when production is increased from 1000 units to 5000 units.

**52.** The slope of the trail is the rate of change of the elevation $E$, so $f(x) = E'(x)$. By the Net Change Theorem, $\int_3^5 f(x)\,dx = \int_3^5 E'(x)\,dx = E(5) - E(3)$ is the change in the elevation $E$ between $x = 3$ miles and $x = 5$ miles from the start of the trail.

**53.** In general, the unit of measurement for $\int_a^b f(x)\,dx$ is the product of the unit for $f(x)$ and the unit for $x$. Since $f(x)$ is measured in newtons and $x$ is measured in meters, the units for $\int_0^{100} f(x)\,dx$ are newton-meters. (A newton-meter is abbreviated N·m and is called a joule.)

**54.** The units for $a(x)$ are pounds per foot and the units for $x$ are feet, so the units for $da/dx$ are pounds per foot per foot, denoted (lb/ft)/ft. The unit of measurement for $\int_2^8 a(x)\,dx$ is the product of pounds per foot and feet; that is, pounds.

**55.** (a) Displacement $= \int_0^3 (3t-5)\,dt = \left[\frac{3}{2}t^2 - 5t\right]_0^3 = \frac{27}{2} - 15 = -\frac{3}{2}$ m

(b) Distance traveled $= \int_0^3 |3t-5|\,dt = \int_0^{5/3} (5-3t)\,dt + \int_{5/3}^3 (3t-5)\,dt$

$= \left[5t - \frac{3}{2}t^2\right]_0^{5/3} + \left[\frac{3}{2}t^2 - 5t\right]_{5/3}^3 = \frac{25}{3} - \frac{3}{2}\cdot\frac{25}{9} + \frac{27}{2} - 15 - \left(\frac{3}{2}\cdot\frac{25}{9} - \frac{25}{3}\right) = \frac{41}{6}$ m

**56.** (a) Displacement $= \int_1^6 (t^2 - 2t - 8)\,dt = \left[\frac{1}{3}t^3 - t^2 - 8t\right]_1^6 = (72 - 36 - 48) - \left(\frac{1}{3} - 1 - 8\right) = -\frac{10}{3}$ m

(b) Distance traveled $= \int_1^6 |t^2 - 2t - 8|\,dt = \int_1^6 |(t-4)(t+2)|\,dt$

$= \int_1^4 (-t^2 + 2t + 8)\,dt + \int_4^6 (t^2 - 2t - 8)\,dt = \left[-\frac{1}{3}t^3 + t^2 + 8t\right]_1^4 + \left[\frac{1}{3}t^3 - t^2 - 8t\right]_4^6$

$= \left(-\frac{64}{3} + 16 + 32\right) - \left(-\frac{1}{3} + 1 + 8\right) + (72 - 36 - 48) - \left(\frac{64}{3} - 16 - 32\right) = \frac{98}{3}$ m

**57.** (a) $v'(t) = a(t) = t + 4 \;\Rightarrow\; v(t) = \frac{1}{2}t^2 + 4t + C \;\Rightarrow\; v(0) = C = 5 \;\Rightarrow\; v(t) = \frac{1}{2}t^2 + 4t + 5$ m/s

(b) Distance traveled $= \int_0^{10} |v(t)|\,dt = \int_0^{10} \left|\frac{1}{2}t^2 + 4t + 5\right|dt = \int_0^{10} \left(\frac{1}{2}t^2 + 4t + 5\right)dt$

$= \left[\frac{1}{6}t^3 + 2t^2 + 5t\right]_0^{10} = \frac{500}{3} + 200 + 50 = 416\frac{2}{3}$ m

**58.** (a) $v'(t) = a(t) = 2t + 3 \;\Rightarrow\; v(t) = t^2 + 3t + C \;\Rightarrow\; v(0) = C = -4 \;\Rightarrow\; v(t) = t^2 + 3t - 4$

(b) Distance traveled $= \int_0^3 \left|t^2 + 3t - 4\right| dt = \int_0^3 |(t+4)(t-1)|\, dt$

$$= \int_0^1 (-t^2 - 3t + 4)\, dt + \int_1^3 (t^2 + 3t - 4)\, dt$$
$$= \left[-\tfrac{1}{3}t^3 - \tfrac{3}{2}t^2 + 4t\right]_0^1 + \left[\tfrac{1}{3}t^3 + \tfrac{3}{2}t^2 - 4t\right]_1^3$$
$$= \left(-\tfrac{1}{3} - \tfrac{3}{2} + 4\right) + \left(9 + \tfrac{27}{2} - 12\right) - \left(\tfrac{1}{3} + \tfrac{3}{2} - 4\right) = \tfrac{89}{6} \text{ m}$$

**59.** Since $m'(x) = \rho(x)$, $m = \int_0^4 \rho(x)\, dx = \int_0^4 (9 + 2\sqrt{x})\, dx = \left[9x + \frac{4}{3}x^{3/2}\right]_0^4 = 36 + \frac{32}{3} - 0 = \frac{140}{3} = 46\frac{2}{3}$ kg.

**60.** By the Net Change Theorem, the amount of water that flows from the tank during the first 10 minutes is $\int_0^{10} r(t)\, dt = \int_0^{10} (200 - 4t)\, dt = \left[200t - 2t^2\right]_0^{10} = (2000 - 200) - 0 = 1800$ liters.

**61.** Let $s$ be the position of the car. We know from Equation 2 that $s(100) - s(0) = \int_0^{100} v(t)\, dt$. We use the Midpoint Rule for $0 \le t \le 100$ with $n = 5$. Note that the length of each of the five time intervals is 20 seconds $= \frac{20}{3600}$ hour $= \frac{1}{180}$ hour. So the distance traveled is

$$\int_0^{100} v(t)\, dt \approx \tfrac{1}{180}[v(10) + v(30) + v(50) + v(70) + v(90)]$$
$$= \tfrac{1}{180}(38 + 58 + 51 + 53 + 47)$$
$$= \tfrac{247}{180} \approx 1.4 \text{ miles}$$

**62.** (a) By the Net Change Theorem, the total amount spewed into the atmosphere is $Q(6) - Q(0) = \int_0^6 r(t)\, dt = Q(6)$ since $Q(0) = 0$. The rate $r(t)$ is positive, so $Q$ is an increasing function. Thus, an upper estimate for $Q(6)$ is $R_6$ and a lower estimate for $Q(6)$ is $L_6$. $\Delta t = \dfrac{b-a}{n} = \dfrac{6-0}{6} = 1$.

$R_6 = \sum\limits_{i=1}^{6} r(t_i)\, \Delta t = 10 + 24 + 36 + 46 + 54 + 60 = 230$ tonnes.

$L_6 = \sum\limits_{i=1}^{6} r(t_{i-1})\, \Delta t = R_6 + r(0) - r(6) = 230 + 2 - 60 = 172$ tonnes.

(b) $\Delta t = \dfrac{b-a}{n} = \dfrac{6-0}{3} = 2$. $Q(6) \approx M_3 = 2[r(1) + r(3) + r(5)] = 2(10 + 36 + 54) = 2(100) = 200$ tonnes.

**63.** From the Net Change Theorem, the increase in cost if the production level is raised from 2000 yards to 4000 yards is $C(4000) - C(2000) = \int_{2000}^{4000} C'(x)\, dx$.

$$\int_{2000}^{4000} C'(x)\, dx = \int_{2000}^{4000} (3 - 0.01x + 0.000006x^2)\, dx$$
$$= \left[3x - 0.005x^2 + 0.000002x^3\right]_{2000}^{4000} = 60{,}000 - 2{,}000 = \$58{,}000$$

**64.** By the Net Change Theorem, the amount of water after four days is

$$25{,}000 + \int_0^4 r(t)\, dt \approx 25{,}000 + M_4$$
$$= 25{,}000 + \tfrac{4-0}{4}\,[r(0.5) + r(1.5) + r(2.5) + r(3.5)]$$
$$\approx 25{,}000 + [1500 + 1770 + 740 + (-690)] = 28{,}320 \text{ liters}$$

**65.** (a) We can find the area between the Lorenz curve and the line $y = x$ by subtracting the area under $y = L(x)$ from the area under $y = x$. Thus,

$$\text{coefficient of inequality} = \frac{\text{area between Lorenz curve and line } y = x}{\text{area under line } y = x} = \frac{\int_0^1 [x - L(x)]\, dx}{\int_0^1 x\, dx}$$
$$= \frac{\int_0^1 [x - L(x)]\, dx}{\left[x^2/2\right]_0^1} = \frac{\int_0^1 [x - L(x)]\, dx}{1/2} = 2\int_0^1 [x - L(x)]\, dx$$

(b) $L(x) = \frac{5}{12}x^2 + \frac{7}{12}x \;\Rightarrow\; L(50\%) = L\left(\frac{1}{2}\right) = \frac{5}{48} + \frac{7}{24} = \frac{19}{48} = 0.3958\overline{3}$, so the bottom 50% of the households receive at most about 40% of the income. Using the result in part (a),

$$\begin{aligned}\text{coefficient of inequality} &= 2\int_0^1 [x - L(x)]\,dx = 2\int_0^1 \left(x - \tfrac{5}{12}x^2 - \tfrac{7}{12}x\right)dx \\ &= 2\int_0^1 \left(\tfrac{5}{12}x - \tfrac{5}{12}x^2\right)dx = 2\int_0^1 \tfrac{5}{12}\left(x - x^2\right)dx \\ &= \tfrac{5}{6}\left[\tfrac{1}{2}x^2 - \tfrac{1}{3}x^3\right]_0^1 = \tfrac{5}{6}\left(\tfrac{1}{2} - \tfrac{1}{3}\right) = \tfrac{5}{6}\left(\tfrac{1}{6}\right) = \tfrac{5}{36}\end{aligned}$$

**66.** (a) From Exercise 4.2.58(a), $v(t) = 0.00146t^3 - 0.11553t^2 + 24.98169t - 21.26872$.

(b) $h(125) - h(0) = \int_0^{125} v(t)\,dt = \left[0.000365t^4 - 0.03851t^3 + 12.490845t^2 - 21.26872t\right]_0^{125} \approx 206{,}407$ ft

**67.** The second derivative is the derivative of the first derivative, so we'll apply the Net Change Theorem with $F = h'$. $\int_1^2 h''(u)\,du = \int_1^2 (h')'(u)\,du = h'(2) - h'(1) = 5 - 2 = 3$. The other information is unnecessary.

**68.** $B = 3A \;\Rightarrow\; \int_0^b e^x\,dx = 3\int_0^a e^x\,dx \;\Rightarrow\; \left[e^x\right]_0^b = 3\left[e^x\right]_0^a \;\Rightarrow\; e^b - 1 = 3(e^a - 1) \;\Rightarrow$ $e^b = 3e^a - 2 \;\Rightarrow\; b = \ln(3e^a - 2)$

## DISCOVERY PROJECT Area Functions

**1.** (a)

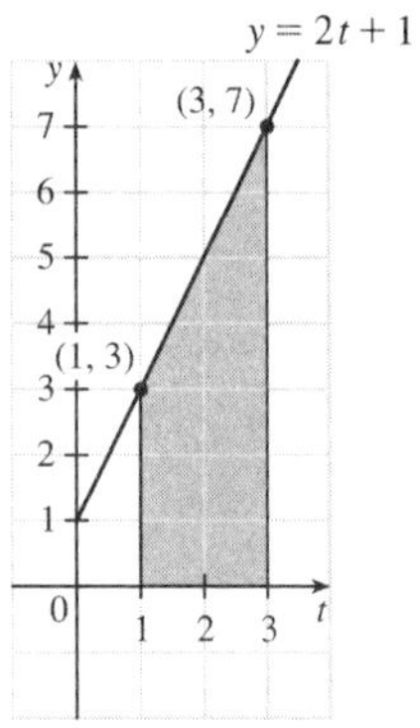


Area of trapezoid $= \frac{1}{2}(b_1 + b_2)h = \frac{1}{2}(3 + 7)2$
$= 10$ square units

*Or:*

Area of rectangle + area of triangle
$= b_r h_r + \frac{1}{2}b_t h_t$
$= (2)(3) + \frac{1}{2}(2)(4) = 10$ square units

(b)

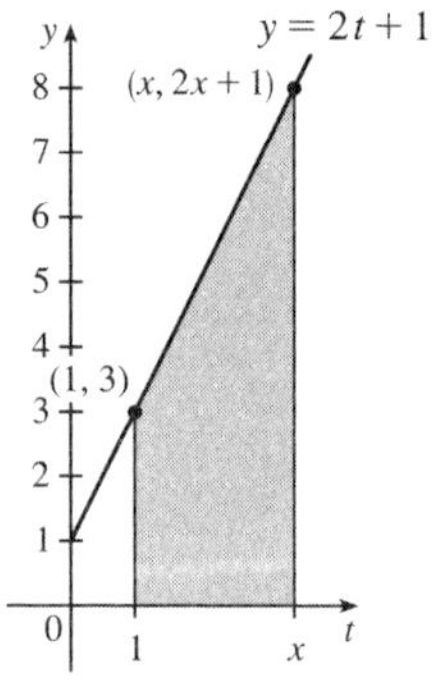


As in part (a),

$$\begin{aligned}A(x) &= \tfrac{1}{2}[3 + (2x + 1)](x - 1) \\ &= \tfrac{1}{2}(2x + 4)(x - 1) \\ &= (x + 2)(x - 1) \\ &= x^2 + x - 2 \text{ square units}\end{aligned}$$

(c) $A'(x) = 2x + 1$. This is the $y$-coordinate of the point $(x, 2x + 1)$ on the given line.

**2.** (a)

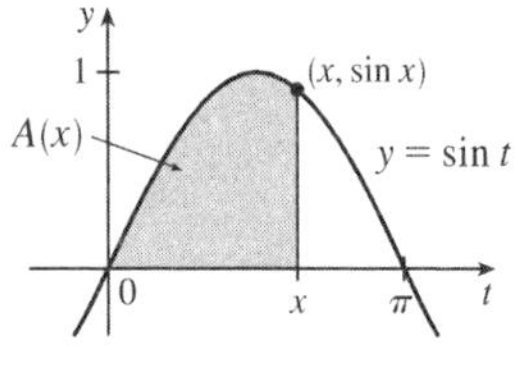


(b) $A(x) = \int_0^x \sin t\,dt = [-\cos t]_0^x$
$= -\cos x - (-1) = 1 - \cos x$

(c) $A'(x) = \sin x$. This is the $y$-coordinate of the point $(x, \sin x)$ on the given curve.

(d)

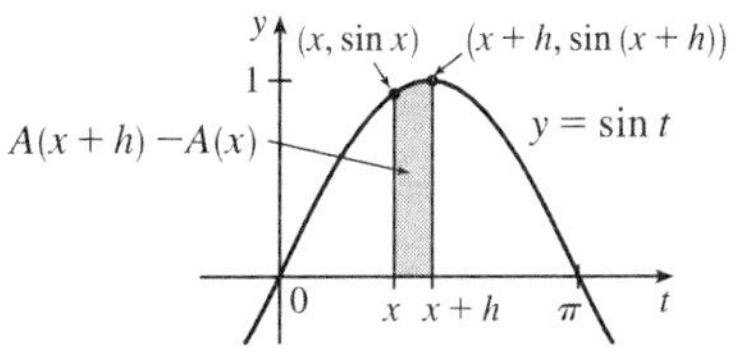


$A(x+h) - A(x)$ is the area under the curve $y = \sin t$ from $t = x$ to $t = x + h$.

(e)

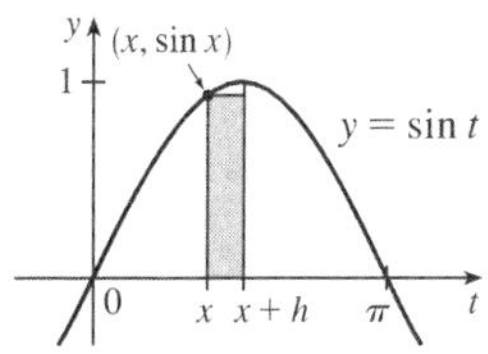


An approximating rectangle is shown in the figure. It has height $\sin x$, width $h$, and area $h \sin x$, so

$$A(x+h) - A(x) \approx h \sin x \quad \Rightarrow$$

$$\frac{A(x+h) - A(x)}{h} \approx \sin x.$$

(f) Part (e) says that the average rate of change of $A$ is approximately $\sin x$. As $h$ approaches 0, the quotient approaches the instantaneous rate of change—namely, $A'(x)$. So the result of part (c), $A'(x) = \sin x$, is geometrically plausible.

**3.** (a) $f(x) = \cos(x^2)$

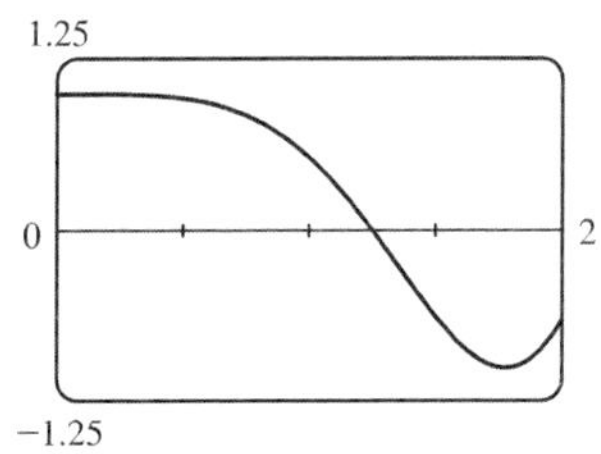


(b) $g(x)$ starts to decrease at that value of $x$ where $\cos(t^2)$ changes from positive to negative; that is, at about $x = 1.25$.

(c) $g(x) = \int_0^x \cos(t^2)\,dt$. Using an integration command, we find that $g(0) = 0$, $g(0.2) \approx 0.200$, $g(0.4) \approx 0.399$, $g(0.6) \approx 0.592$, $g(0.8) \approx 0.768$, $g(1.0) \approx 0.905$, $g(1.2) \approx 0.974$, $g(1.4) \approx 0.950$, $g(1.6) \approx 0.826$, $g(1.8) \approx 0.635$, and $g(2.0) \approx 0.461$.

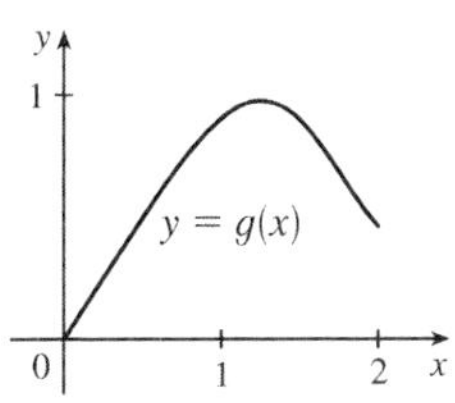


(d) We sketch the graph of $g'$ using the method of Example 1 in Section 2.8. The graphs of $g'(x)$ and $f(x)$ look alike, so we guess that $g'(x) = f(x)$.

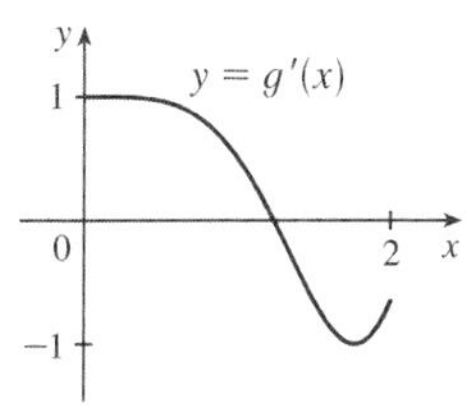


**4.** In Problems 1 and 2, we showed that if $g(x) = \int_a^x f(t)\,dt$, then $g'(x) = f(x)$, for the functions $f(t) = 2t + 1$ and $f(t) = \sin t$. In Problem 3 we guessed that the same is true for $f(t) = \cos(t^2)$, based on visual evidence. So we conjecture that $g'(x) = f(x)$ for any continuous function $f$. This turns out to be true and is proved in Section 5.4 (the Fundamental Theorem of Calculus).

## 5.4 The Fundamental Theorem of Calculus

1. The precise version of this statement is given by the Fundamental Theorem of Calculus. See the statement of this theorem and the paragraph that follows it on page 381.

2. (a) $g(x) = \int_0^x f(t)\,dt$, so $g(0) = \int_0^0 f(t)\,dt = 0$.

$g(1) = \int_0^1 f(t)\,dt = \frac{1}{2} \cdot 1 \cdot 1$ [area of triangle] $= \frac{1}{2}$.

$g(2) = \int_0^2 f(t)\,dt = \int_0^1 f(t)\,dt + \int_1^2 f(t)\,dt$ [below the $x$-axis]

$= \frac{1}{2} - \frac{1}{2} \cdot 1 \cdot 1 = 0.$

$g(3) = g(2) + \int_2^3 f(t)\,dt = 0 - \frac{1}{2} \cdot 1 \cdot 1 = -\frac{1}{2}.$

$g(4) = g(3) + \int_3^4 f(t)\,dt = -\frac{1}{2} + \frac{1}{2} \cdot 1 \cdot 1 = 0.$

$g(5) = g(4) + \int_4^5 f(t)\,dt = 0 + 1.5 = 1.5.$

$g(6) = g(5) + \int_5^6 f(t)\,dt = 1.5 + 2.5 = 4.$

(b) $g(7) = g(6) + \int_6^7 f(t)\,dt \approx 4 + 2.2$ [estimate from the graph] $= 6.2$.

(c) The answers from part (a) and part (b) indicate that $g$ has a minimum at $x = 3$ and a maximum at $x = 7$. This makes sense from the graph of $f$ since we are subtracting area on $1 < x < 3$ and adding area on $3 < x < 7$.

(d)

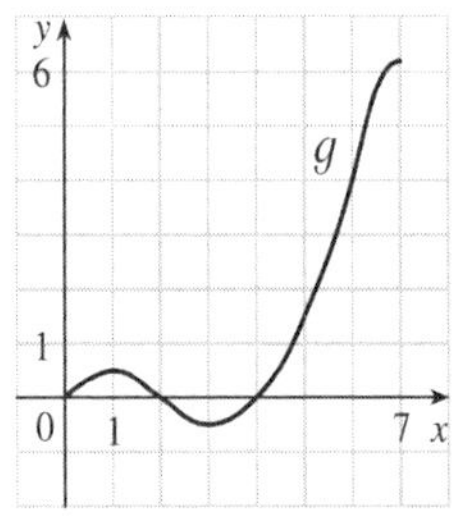


3. (a) $g(x) = \int_0^x f(t)\,dt$.

$g(0) = \int_0^0 f(t)\,dt = 0$

$g(1) = \int_0^1 f(t)\,dt = 1 \cdot 2 = 2$ [rectangle],

$g(2) = \int_0^2 f(t)\,dt = \int_0^1 f(t)\,dt + \int_1^2 f(t)\,dt = g(1) + \int_1^2 f(t)\,dt$

$= 2 + 1 \cdot 2 + \frac{1}{2} \cdot 1 \cdot 2 = 5$ [rectangle plus triangle],

$g(3) = \int_0^3 f(t)\,dt = g(2) + \int_2^3 f(t)\,dt = 5 + \frac{1}{2} \cdot 1 \cdot 4 = 7,$

$g(6) = g(3) + \int_3^6 f(t)\,dt$ [the integral is negative since $f$ lies under the $x$-axis]

$= 7 + \left[-\left(\frac{1}{2} \cdot 2 \cdot 2 + 1 \cdot 2\right)\right] = 7 - 4 = 3$

(b) $g$ is increasing on $(0, 3)$ because as $x$ increases from 0 to 3, we keep adding more area.

(c) $g$ has a maximum value when we start subtracting area; that is, at $x = 3$.

(d)

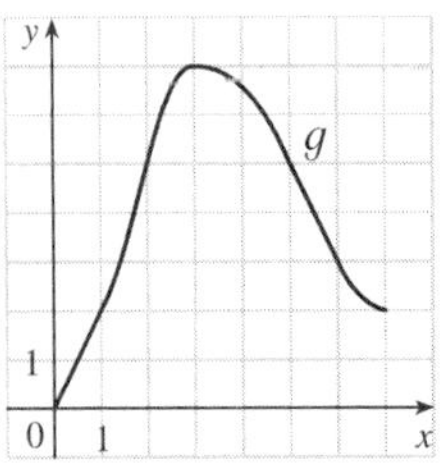

**4.** (a) $g(-3) = \int_{-3}^{-3} f(t)\,dt = 0$, $g(3) = \int_{-3}^{3} f(t)\,dt = \int_{-3}^{0} f(t)\,dt + \int_{0}^{3} f(t)\,dt = 0$ by symmetry, since the area above the $x$-axis is the same as the area below the axis.

(b) From the graph, it appears that to the nearest $\frac{1}{2}$, $g(-2) = \int_{-3}^{-2} f(t)\,dt \approx 1$, $g(-1) = \int_{-3}^{-1} f(t)\,dt \approx 3\frac{1}{2}$, and $g(0) = \int_{-3}^{0} f(t)\,dt \approx 5\frac{1}{2}$.

(c) $g$ is increasing on $(-3, 0)$ because as $x$ increases from $-3$ to 0, we keep adding more area.

(d) $g$ has a maximum value when we start subtracting area; that is, at $x = 0$.

(e)

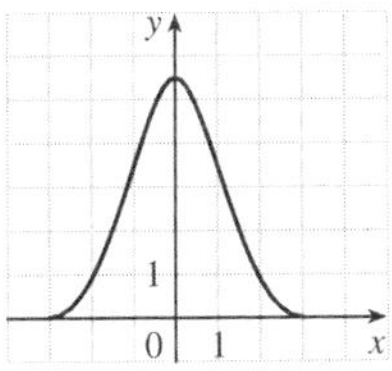


(f) The graph of $g'(x)$ is the same as that of $f(x)$, as indicated by FTC1.

**5.**

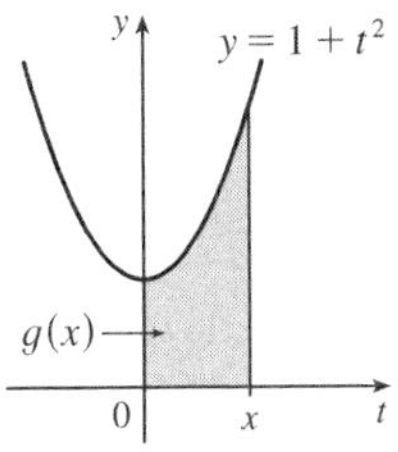


(a) By FTC1, $g(x) = \int_0^x (1+t^2)dt \quad\Rightarrow\quad g'(x) = f(x) = 1 + x^2$.
(b) By FTC2, $g(x) = \int_0^x (1+t^2)dt = \left[t + \frac{1}{3}t^3\right]_0^x = \left(x + \frac{1}{3}x^3\right) - 0 \quad\Rightarrow\quad g'(x) = 1 + x^2$.

**6.**

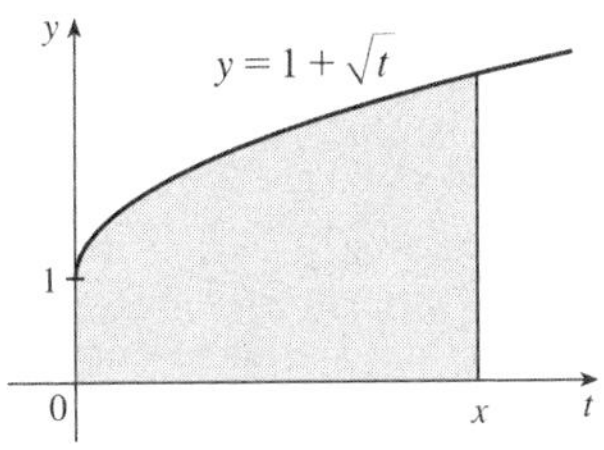


(a) By FTC1 with $f(t) = 1 + \sqrt{t}$ and $a = 0$, $g(x) = \int_0^x (1+\sqrt{t}\,)\,dt \quad\Rightarrow\quad g'(x) = f(x) = 1 + \sqrt{x}$.
(b) Using FTC2, $g(x) = \int_0^x (1+\sqrt{t}\,)\,dt = \left[t + \frac{2}{3}t^{3/2}\right]_0^x = x + \frac{2}{3}x^{3/2} \quad\Rightarrow\quad g'(x) = 1 + x^{1/2} = 1 + \sqrt{x}$.

**7.** $f(t) = \sqrt{1+2t}$ and $g(x) = \int_0^x \sqrt{1+2t}\,dt$, so by FTC1, $g'(x) = f(x) = \sqrt{1+2x}$.

**8.** $f(t) = \ln t$ and $g(x) = \int_1^x \ln t\,dt$, so by FTC1, $g'(x) = f(x) = \ln x$.

**9.** $f(t) = t^2 \sin t$ and $g(y) = \int_2^y t^2 \sin t\,dt$, so by FTC1, $g'(y) = f(y) = y^2 \sin y$.

**10.** $f(\theta) = \tan\theta$ and $F(x) = \int_x^{10} \tan\theta\,d\theta = -\int_{10}^x \tan\theta\,d\theta$, so by FTC1, $F'(x) = -f(x) = -\tan x$.

**11.** Let $u = \dfrac{1}{x}$. Then $\dfrac{du}{dx} = -\dfrac{1}{x^2}$. Also, $\dfrac{dh}{dx} = \dfrac{dh}{du}\dfrac{du}{dx}$, so

$$h'(x) = \frac{d}{dx}\int_2^{1/x} \arctan t\,dt = \frac{d}{du}\int_2^{u} \arctan t\,dt \cdot \frac{du}{dx} = \arctan u\,\frac{du}{dx} = -\frac{\arctan(1/x)}{x^2}.$$

**12.** Let $u = x^2$. Then $\dfrac{du}{dx} = 2x$. Also, $\dfrac{dh}{dx} = \dfrac{dh}{du}\dfrac{du}{dx}$, so

$$h'(x) = \frac{d}{dx}\int_0^{x^2} \sqrt{1+r^3}\,dr = \frac{d}{du}\int_0^{u} \sqrt{1+r^3}\,dr \cdot \frac{du}{dx} = \sqrt{1+u^3}(2x) = 2x\sqrt{1+(x^2)^3} = 2x\sqrt{1+x^6}.$$

**13.** Let $u = \sqrt{x}$. Then $\dfrac{du}{dx} = \dfrac{1}{2\sqrt{x}}$. Also, $\dfrac{dy}{dx} = \dfrac{dy}{du}\dfrac{du}{dx}$, so

$$y' = \frac{d}{dx}\int_3^{\sqrt{x}} \frac{\cos t}{t}\,dt = \frac{d}{du}\int_3^{u} \frac{\cos t}{t}\,dt \cdot \frac{du}{dx} = \frac{\cos u}{u}\cdot\frac{1}{2\sqrt{x}} = \frac{\cos\sqrt{x}}{\sqrt{x}}\cdot\frac{1}{2\sqrt{x}} = \frac{\cos\sqrt{x}}{2x}.$$

**14.** Let $u = e^x$. Then $\dfrac{du}{dx} = e^x$. Also, $\dfrac{dy}{dx} = \dfrac{dy}{du}\dfrac{du}{dx}$, so

$$y' = \frac{d}{dx}\int_{e^x}^{0} \sin^3 t\,dt = \frac{d}{du}\int_u^0 \sin^3 t\,dt \cdot \frac{du}{dx} = -\frac{d}{du}\int_0^u \sin^3 t\,dt \cdot \frac{du}{dx} = -\sin^3 u \cdot e^x = -e^x\sin^3(e^x).$$

**15.** $g(x) = \displaystyle\int_{2x}^{3x} \frac{u^2-1}{u^2+1}\,du = \int_{2x}^{0} \frac{u^2-1}{u^2+1}\,du + \int_0^{3x} \frac{u^2-1}{u^2+1}\,du = -\int_0^{2x} \frac{u^2-1}{u^2+1}\,du + \int_0^{3x} \frac{u^2-1}{u^2+1}\,du \;\Rightarrow$

$$g'(x) = -\frac{(2x)^2-1}{(2x)^2+1}\cdot\frac{d}{dx}(2x) + \frac{(3x)^2-1}{(3x)^2+1}\cdot\frac{d}{dx}(3x) = -2\cdot\frac{4x^2-1}{4x^2+1} + 3\cdot\frac{9x^2-1}{9x^2+1}$$

**16.** $y = \displaystyle\int_{\sin x}^{\cos x} (1+v^2)^{10}\,dv = \int_{\sin x}^{0} (1+v^2)^{10}\,dv + \int_0^{\cos x} (1+v^2)^{10}\,dv$

$$= -\int_0^{\sin x} (1+v^2)^{10}\,dv + \int_0^{\cos x} (1+v^2)^{10}\,dv \quad\Rightarrow$$

$$y' = -(1+(\sin x)^2)^{10}\cdot\frac{d}{dx}(\sin x) + (1+(\cos x)^2)^{10}\cdot\frac{d}{dx}(\cos x) = -(1+\sin^2 x)^{10}\cos x - (1+\cos^2 x)^{10}\sin x$$

**17.** $F(x) = \displaystyle\int_1^x f(t)\,dt \;\Rightarrow\; F'(x) = f(x) = \int_1^{x^2} \frac{\sqrt{1+u^4}}{u}\,du \left[\text{since } f(t) = \int_1^{t^2} \frac{\sqrt{1+u^4}}{u}\,du\right] \;\Rightarrow$

$$F''(x) = f'(x) = \frac{\sqrt{1+(x^2)^4}}{x^2}\cdot\frac{d}{dx}(x^2) = \frac{\sqrt{1+x^8}}{x^2}\cdot 2x = \frac{2\sqrt{1+x^8}}{x}.\text{ So } F''(2) = \sqrt{1+2^8} = \sqrt{257}.$$

**18.** For the curve to be concave upward, we must have $y'' > 0$. $y = \displaystyle\int_0^x \frac{1}{1+t+t^2}\,dt \;\Rightarrow\; y' = \frac{1}{1+x+x^2} \;\Rightarrow$

$y'' = \dfrac{-(1+2x)}{(1+x+x^2)^2}$. For this expression to be positive, we must have $(1+2x) < 0$, since $(1+x+x^2)^2 > 0$ for all $x$. $(1+2x) < 0 \;\Leftrightarrow\; x < -\frac{1}{2}$. Thus, the curve is concave upward on $\left(-\infty, -\frac{1}{2}\right)$.

**19.** (a) By FTC1, $g'(x) = f(x)$. So $g'(x) = f(x) = 0$ at $x = 1, 3, 5, 7$, and $9$. $g$ has local maxima at $x = 1$ and $5$ (since $f = g'$ changes from positive to negative there) and local minima at $x = 3$ and $7$. There is no local maximum or minimum at $x = 9$, since $f$ is not defined for $x > 9$.

(b) We can see from the graph that $\left|\int_0^1 f\,dt\right| < \left|\int_1^3 f\,dt\right| < \left|\int_3^5 f\,dt\right| < \left|\int_5^7 f\,dt\right| < \left|\int_7^9 f\,dt\right|$. So $g(1) = \left|\int_0^1 f\,dt\right|$, $g(5) = \int_0^5 f\,dt = g(1) - \left|\int_1^3 f\,dt\right| + \left|\int_3^5 f\,dt\right|$, and $g(9) = \int_0^9 f\,dt = g(5) - \left|\int_5^7 f\,dt\right| + \left|\int_7^9 f\,dt\right|$. Thus, $g(1) < g(5) < g(9)$, and so the absolute maximum of $g(x)$ occurs at $x = 9$.

(c) $g$ is concave downward on those intervals where $g'' < 0$. But $g'(x) = f(x)$, so $g''(x) = f'(x)$, which is negative on (approximately) $\left(\frac{1}{2}, 2\right)$, $(4, 6)$ and $(8, 9)$. So $g$ is concave downward on these intervals.

(d)

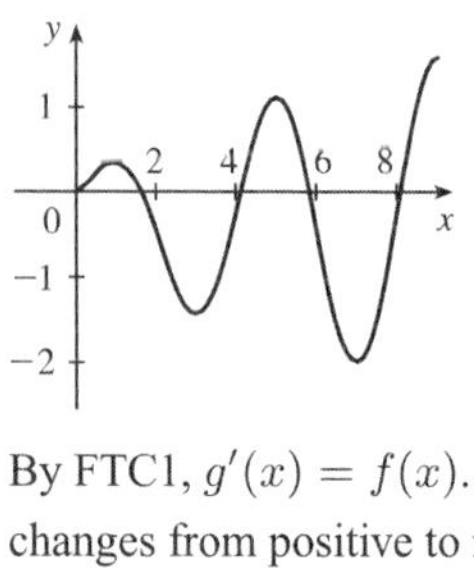


**20.** (a) By FTC1, $g'(x) = f(x)$. So $g'(x) = f(x) = 0$ at $x = 2, 4, 6, 8$, and $10$. $g$ has local maxima at $x = 2$ and $6$ (since $f = g'$ changes from positive to negative there) and local minima at $x = 4$ and $8$. There is no local maximum or minimum at $x = 10$, since $f$ is not defined for $x > 10$.

(b) We can see from the graph that $\left|\int_0^2 f\,dt\right| > \left|\int_2^4 f\,dt\right| > \left|\int_4^6 f\,dt\right| > \left|\int_6^8 f\,dt\right| > \left|\int_8^{10} f\,dt\right|$. So $g(2) = \left|\int_0^2 f\,dt\right|$, $g(6) = \int_0^6 f\,dt = g(2) - \left|\int_2^4 f\,dt\right| + \left|\int_4^6 f\,dt\right|$, and $g(10) = \int_0^{10} f\,dt = g(6) - \left|\int_6^8 f\,dt\right| + \left|\int_8^{10} f\,dt\right|$. Thus, $g(2) > g(6) > g(10)$, and so the absolute maximum of $g(x)$ occurs at $x = 2$.

(c) $g$ is concave downward on those intervals where $g'' < 0$. But $g'(x) = f(x)$, so $g''(x) = f'(x)$, which is negative on $(1, 3)$, $(5, 7)$ and $(9, 10)$. So $g$ is concave downward on these intervals.

(d)

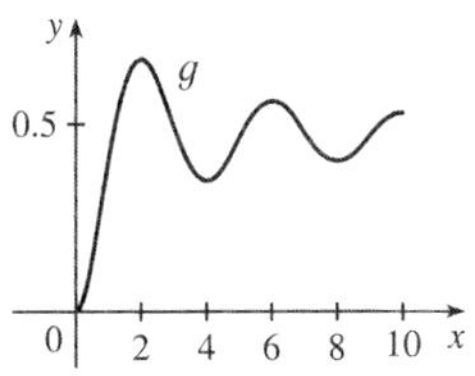


**21.** By FTC2, $\int_1^4 f'(x)\,dx = f(4) - f(1)$, so $17 = f(4) - 12 \quad\Rightarrow\quad f(4) = 17 + 12 = 29$.

**22.** (a) $\operatorname{erf}(x) = \dfrac{2}{\sqrt{\pi}}\displaystyle\int_0^x e^{-t^2}\,dt \quad\Rightarrow\quad \int_0^x e^{-t^2}\,dt = \dfrac{\sqrt{\pi}}{2}\operatorname{erf}(x)$. By Property 5 of definite integrals in Section 5.2, $\int_0^b e^{-t^2}\,dt = \int_0^a e^{-t^2}\,dt + \int_a^b e^{-t^2}\,dt$, so

$$\int_a^b e^{-t^2}\,dt = \int_0^b e^{-t^2}\,dt - \int_0^a e^{-t^2}\,dt = \frac{\sqrt{\pi}}{2}\operatorname{erf}(b) - \frac{\sqrt{\pi}}{2}\operatorname{erf}(a) = \tfrac{1}{2}\sqrt{\pi}\,[\operatorname{erf}(b) - \operatorname{erf}(a)].$$

(b) $y = e^{x^2}\operatorname{erf}(x) \quad\Rightarrow$

$y' = 2xe^{x^2}\operatorname{erf}(x) + e^{x^2}\operatorname{erf}'(x) = 2xy + e^{x^2}\cdot\dfrac{2}{\sqrt{\pi}}e^{-x^2}$ [by FTC1] $= 2xy + 2/\sqrt{\pi}$.

**23.** (a) The Fresnel function $S(x) = \int_0^x \sin\left(\frac{\pi}{2}t^2\right)dt$ has local maximum values where $0 = S'(x) = \sin\left(\frac{\pi}{2}x^2\right)$ and $S'$ changes from positive to negative. For $x > 0$, this happens when $\frac{\pi}{2}x^2 = (2n-1)\pi$ [odd multiples of $\pi$] $\Leftrightarrow\quad x^2 = 2(2n-1)$ $\Leftrightarrow\quad x = \sqrt{4n-2}$, $n$ any positive integer. For $x < 0$, $S'$ changes from positive to negative where $\frac{\pi}{2}x^2 = 2n\pi$ [even multiples of $\pi$] $\Leftrightarrow\quad x^2 = 4n \quad\Leftrightarrow\quad x = -2\sqrt{n}$. $S'$ does not change sign at $x = 0$.

(b) $S$ is concave upward on those intervals where $S''(x) > 0$. Differentiating our expression for $S'(x)$, we get $S''(x) = \cos\left(\frac{\pi}{2}x^2\right)\left(2\frac{\pi}{2}x\right) = \pi x\cos\left(\frac{\pi}{2}x^2\right)$. For $x > 0$, $S''(x) > 0$ where $\cos\left(\frac{\pi}{2}x^2\right) > 0 \quad\Leftrightarrow\quad 0 < \frac{\pi}{2}x^2 < \frac{\pi}{2}$ or $\left(2n - \frac{1}{2}\right)\pi < \frac{\pi}{2}x^2 < \left(2n + \frac{1}{2}\right)\pi$, $n$ any integer $\quad\Leftrightarrow\quad 0 < x < 1$ or $\sqrt{4n-1} < x < \sqrt{4n+1}$, $n$ any positive integer. For $x < 0$, $S''(x) > 0$ where $\cos\left(\frac{\pi}{2}x^2\right) < 0 \quad\Leftrightarrow\quad \left(2n - \frac{3}{2}\right)\pi < \frac{\pi}{2}x^2 < \left(2n - \frac{1}{2}\right)\pi$, $n$ any integer $\quad\Leftrightarrow$ $4n - 3 < x^2 < 4n - 1 \quad\Leftrightarrow\quad \sqrt{4n-3} < |x| < \sqrt{4n-1} \quad\Rightarrow\quad \sqrt{4n-3} < -x < \sqrt{4n-1} \quad\Rightarrow$ $-\sqrt{4n-3} > x > -\sqrt{4n-1}$, so the intervals of upward concavity for $x < 0$ are $\left(-\sqrt{4n-1}, -\sqrt{4n-3}\right)$, $n$ any positive integer. To summarize: $S$ is concave upward on the intervals $(0, 1)$, $\left(-\sqrt{3}, -1\right)$, $\left(\sqrt{3}, \sqrt{5}\right)$, $\left(-\sqrt{7}, -\sqrt{5}\right)$, $\left(\sqrt{7}, 3\right)$, ….

(c) In Maple, we use `plot({int(sin(Pi*t^2/2),t=0..x),0.2},x=0..2);`. Note that Maple recognizes the Fresnel function, calling it `FresnelS(x)`. In Mathematica, we use `Plot[{Integrate[Sin[Pi*t^2/2],{t,0,x}],0.2},{x,0,2}]`. In Derive, we load the utility file `FRESNEL` and plot `FRESNEL_SIN(x)`. From the graphs, we see that $\int_0^x \sin\left(\frac{\pi}{2}t^2\right)dt = 0.2$ at $x \approx 0.74$.

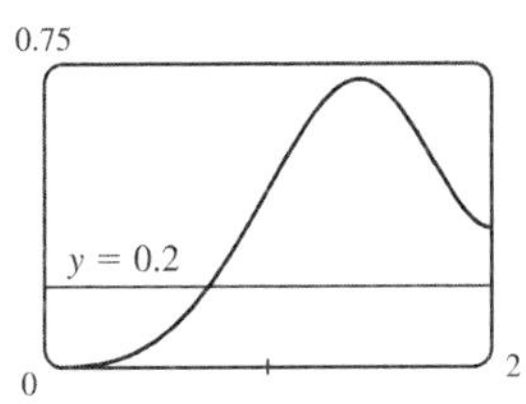


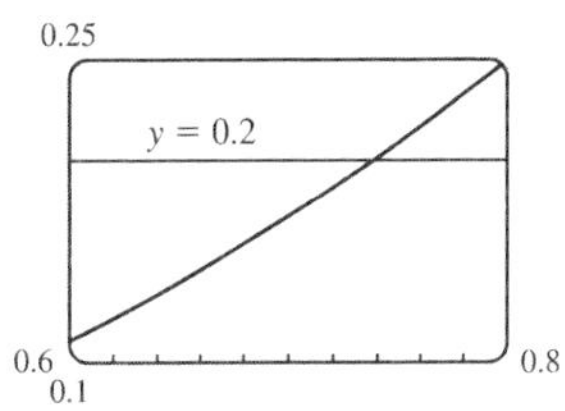

**24.** (a) In Maple, we should start by setting `si:=int(sin(t)/t,t=0..x);`. In Mathematica, the command is `si=Integrate[Sin[t]/t,{t,0,x}]`. Note that both systems recognize this function; Maple calls it `Si(x)` and Mathematica calls it `SinIntegral[x]`. In Maple, the command to generate the graph is `plot(si,x=-4*Pi..4*Pi);`. In Mathematica, it is `Plot[si,{x,-4*Pi,4*Pi}]`. In Derive, we load the utility file `EXP_INT` and plot `SI(x)`.

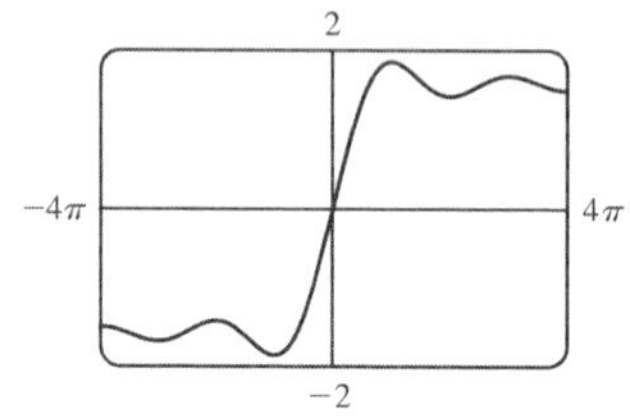


(b) $\mathrm{Si}(x)$ has local maximum values where $\mathrm{Si}'(x)$ changes from positive to negative, passing through 0. From the Fundamental Theorem we know that $\mathrm{Si}'(x) = \dfrac{d}{dx}\displaystyle\int_0^x \frac{\sin t}{t}\,dt = \frac{\sin x}{x}$, so we must have $\sin x = 0$ for a maximum, and for $x > 0$ we must have $x = (2n-1)\pi$, $n$ any positive integer, for $\mathrm{Si}'$ to be changing from positive to negative at $x$. For $x < 0$, we must have $x = 2n\pi$, $n$ any positive integer, for a maximum, since the denominator of $\mathrm{Si}'(x)$ is negative for $x < 0$. Thus, the local maxima occur at $x = \pi, -2\pi, 3\pi, -4\pi, 5\pi, -6\pi, \ldots$.

(c) To find the first inflection point, we solve $\mathrm{Si}''(x) = \dfrac{\cos x}{x} - \dfrac{\sin x}{x^2} = 0$. We can see from the graph that the first inflection point lies somewhere between $x = 3$ and $x = 5$. Using a root finder gives the value $x \approx 4.4934$. To find the $y$-coordinate of the inflection point, we evaluate $\mathrm{Si}(4.4934) \approx 1.6556$. So the coordinates of the first inflection point to the right of the origin are about $(4.4934, 1.6556)$. Alternatively, we could graph $S''(x)$ and estimate the first positive $x$-value at which it changes sign.

(d) It seems from the graph that the function has horizontal asymptotes at $y \approx 1.5$, with $\lim\limits_{x\to\pm\infty} \mathrm{Si}(x) \approx \pm 1.5$ respectively. Using the limit command, we get $\lim\limits_{x\to\infty} \mathrm{Si}(x) = \frac{\pi}{2}$. Since $\mathrm{Si}(x)$ is an odd function, $\lim\limits_{x\to-\infty} \mathrm{Si}(x) = -\frac{\pi}{2}$. So $\mathrm{Si}(x)$ has the horizontal asymptotes $y = \pm\frac{\pi}{2}$.

(e) We use the `fsolve` command in Maple (or `FindRoot` in Mathematica) to find that the solution is $x \approx 1.1$. Or, as in Exercise 23(c), we graph $y = \mathrm{Si}(x)$ and $y = 1$ on the same screen to see where they intersect.

**25.** By FTC2, $\int_1^x f'(t)\,dt = f(x) - f(1) \quad\Rightarrow\quad f(x) = f(1) + \int_1^x f'(t)\,dt = f(1) + \int_1^x (2^t/t)\,dt$. This integral cannot be expressed in a simpler form. Since we want $f(1) = 0$, we have $f(x) = \int_1^x (2^t/t)\,dt$.

**26.** (a) If $x < 0$, then $g(x) = \int_0^x f(t)\,dt = \int_0^x 0\,dt = 0$.

If $0 \le x \le 1$, then $g(x) = \int_0^x f(t)\,dt = \int_0^x t\,dt = \left[\frac{1}{2}t^2\right]_0^x = \frac{1}{2}x^2$.

If $1 < x \le 2$, then

$$\begin{aligned} g(x) &= \int_0^x f(t)\,dt = \int_0^1 f(t)\,dt + \int_1^x f(t)\,dt \\ &= g(1) + \int_1^x (2-t)\,dt = \tfrac{1}{2}(1)^2 + \left[2t - \tfrac{1}{2}t^2\right]_1^x \\ &= \tfrac{1}{2} + \left(2x - \tfrac{1}{2}x^2\right) - \left(2 - \tfrac{1}{2}\right) = 2x - \tfrac{1}{2}x^2 - 1. \end{aligned}$$

If $x > 2$, then $g(x) = \int_0^x f(t)\,dt = g(2) + \int_2^x 0\,dt = 1 + 0 = 1$. So

$$g(x) = \begin{cases} 0 & \text{if } x < 0 \\ \frac{1}{2}x^2 & \text{if } 0 \le x \le 1 \\ 2x - \frac{1}{2}x^2 - 1 & \text{if } 1 < x \le 2 \\ 1 & \text{if } x > 2 \end{cases}$$

(b)

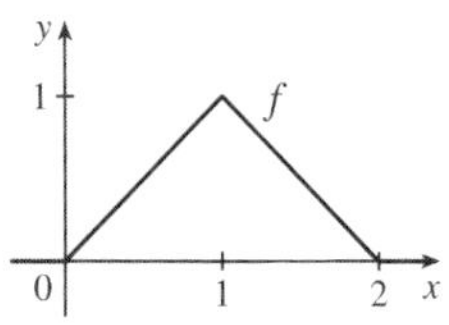


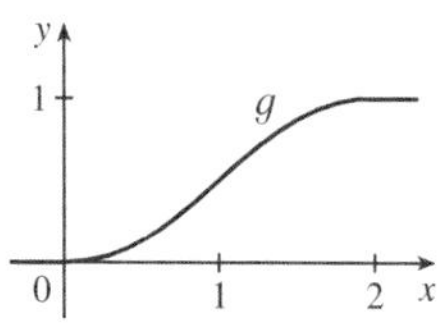


(c) $f$ is not differentiable at its corners at $x = 0$, 1, and 2. $f$ is differentiable on $(-\infty, 0)$, $(0, 1)$, $(1, 2)$ and $(2, \infty)$. $g$ is differentiable on $(-\infty, \infty)$.

**27.** Using FTC1, we differentiate both sides of $6 + \int_a^x \frac{f(t)}{t^2}\,dt = 2\sqrt{x}$ to get $\frac{f(x)}{x^2} = 2\,\frac{1}{2\sqrt{x}} \Rightarrow f(x) = x^{3/2}$. To find $a$, we substitute $x = a$ in the original equation to obtain $6 + \int_a^a \frac{f(t)}{t^2}\,dt = 2\sqrt{a} \Rightarrow 6 + 0 = 2\sqrt{a} \Rightarrow 3 = \sqrt{a} \Rightarrow a = 9$.

**28.** (a) $C(t) = \frac{1}{t}\int_0^t [f(s) + g(s)]\,ds$. Using FTC1 and the Product Rule, we have

$C'(t) = \frac{1}{t}[f(t) + g(t)] - \frac{1}{t^2}\int_0^t [f(s) + g(s)]\,ds$. Set $C'(t) = 0$: $\frac{1}{t}[f(t) + g(t)] - \frac{1}{t^2}\int_0^t [f(s) + g(s)]\,ds = 0 \Rightarrow$

$[f(t) + g(t)] - \frac{1}{t}\int_0^t [f(s) + g(s)]\,ds = 0 \Rightarrow [f(t) + g(t)] - C(t) = 0 \Rightarrow C(t) = f(t) + g(t)$.

(b) For $0 \le t \le 30$, we have $D(t) = \int_0^t \left(\frac{V}{15} - \frac{V}{450}s\right) ds = \left[\frac{V}{15}s - \frac{V}{900}s^2\right]_0^t = \frac{V}{15}t - \frac{V}{900}t^2$.

So $D(t) = V \Rightarrow \frac{V}{15}t - \frac{V}{900}t^2 = V \Rightarrow 60t - t^2 = 900 \Rightarrow t^2 - 60t + 900 = 0 \Rightarrow$

$(t - 30)^2 = 0 \Rightarrow t = 30$. So the length of time $T$ is 30 months.

(c) $C(t) = \frac{1}{t}\int_0^t \left(\frac{V}{15} - \frac{V}{450}s + \frac{V}{12{,}900}s^2\right) ds = \frac{1}{t}\left[\frac{V}{15}s - \frac{V}{900}s^2 + \frac{V}{38{,}700}s^3\right]_0^t$

$= \frac{1}{t}\left(\frac{V}{15}t - \frac{V}{900}t^2 + \frac{V}{38{,}700}t^3\right) = \frac{V}{15} - \frac{V}{900}t + \frac{V}{38{,}700}t^2 \Rightarrow$

$C'(t) = -\frac{V}{900} + \frac{V}{19{,}350}t = 0$ when $\frac{1}{19{,}350}t = \frac{1}{900} \Rightarrow t = 21.5$.

$C(21.5) = \frac{V}{15} - \frac{V}{900}(21.5) + \frac{V}{38{,}700}(21.5)^2 \approx 0.05472V$, $C(0) = \frac{V}{15} \approx 0.06667V$, and

$C(30) = \frac{V}{15} - \frac{V}{900}(30) + \frac{V}{38{,}700}(30)^2 \approx 0.05659V$, so the absolute minimum is $C(21.5) \approx 0.05472V$.

(d) As in part (c), we have $C(t) = \frac{V}{15} - \frac{V}{900}t + \frac{V}{38{,}700}t^2$, so $C(t) = f(t) + g(t)$

$\Leftrightarrow \frac{V}{15} - \frac{V}{900}t + \frac{V}{38{,}700}t^2 = \frac{V}{15} - \frac{V}{450}t + \frac{V}{12{,}900}t^2 \Leftrightarrow$

$t^2\left(\frac{1}{12{,}900} - \frac{1}{38{,}700}\right) = t\left(\frac{1}{450} - \frac{1}{900}\right) \Leftrightarrow$

$t = \frac{1/900}{2/38{,}700} = \frac{43}{2} = 21.5$. This is the value of $t$ that we obtained as the critical number of $C$ in part (c), so we have verified the result of (a) in this case.

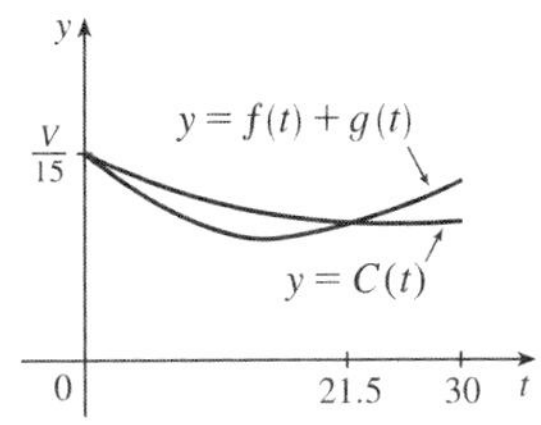


**29.** (a) Let $F(t) = \int_0^t f(s)\,ds$. Then, by FTC1, $F'(t) = f(t) =$ rate of depreciation, so $F(t)$ represents the loss in value over the interval $[0, t]$.

(b) $C(t) = \frac{1}{t}\left[A + \int_0^t f(s)\,ds\right] = \frac{A + F(t)}{t}$ represents the average expenditure per unit of $t$ during the interval $[0, t]$, assuming that there has been only one overhaul during that time period. The company wants to minimize average expenditure.

(c) $C(t) = \frac{1}{t}\left[A + \int_0^t f(s)\,ds\right]$. Using FTC1, we have $C'(t) = -\frac{1}{t^2}\left[A + \int_0^t f(s)\,ds\right] + \frac{1}{t}f(t)$.

$C'(t) = 0 \Rightarrow t\,f(t) = A + \int_0^t f(s)\,ds \Rightarrow f(t) = \frac{1}{t}\left[A + \int_0^t f(s)\,ds\right] = C(t)$.

## 5.5 The Substitution Rule

**1.** Let $u = 3x$. Then $du = 3\,dx$, so $dx = \frac{1}{3}du$. Thus, $\int \cos 3x\,dx = \int \cos u\left(\frac{1}{3}\,du\right) = \frac{1}{3}\int \cos u\,du = \frac{1}{3}\sin u + C = \frac{1}{3}\sin 3x + C$. Don't forget that it is often very easy to check an indefinite integration by differentiating your answer. In this case, $\frac{d}{dx}\left(\frac{1}{3}\sin 3x + C\right) = \frac{1}{3}(\cos 3x)\cdot 3 = \cos 3x$, the desired result.

**2.** Let $u = 4 + x^2$. Then $du = 2x\,dx$ and $x\,dx = \frac{1}{2}\,du$, so
$\int x(4+x^2)^{10}\,dx = \int u^{10}\left(\frac{1}{2}\,du\right) = \frac{1}{2}\cdot\frac{1}{11}u^{11} + C = \frac{1}{22}(4+x^2)^{11} + C$.

**3.** Let $u = x^3 + 1$. Then $du = 3x^2\,dx$ and $x^2\,dx = \frac{1}{3}\,du$, so

$$\int x^2\sqrt{x^3+1}\,dx = \int \sqrt{u}\left(\frac{1}{3}\,du\right) = \frac{1}{3}\frac{u^{3/2}}{3/2} + C = \frac{1}{3}\cdot\frac{2}{3}u^{3/2} + C = \frac{2}{9}(x^3+1)^{3/2} + C.$$

**4.** Let $u = \sqrt{x}$. Then $du = \dfrac{1}{2\sqrt{x}}\,dx$ and $\dfrac{1}{\sqrt{x}}\,dx = 2\,du$, so

$$\int \frac{\sin\sqrt{x}}{\sqrt{x}}\,dx = \int \sin u\,(2\,du) = 2(-\cos u) + C = -2\cos\sqrt{x} + C.$$

**5.** Let $u = 1 + 2x$. Then $du = 2\,dx$ and $dx = \frac{1}{2}\,du$, so

$$\int \frac{4}{(1+2x)^3}\,dx = 4\int u^{-3}\left(\tfrac{1}{2}\,du\right) = 2\frac{u^{-2}}{-2} + C = -\frac{1}{u^2} + C = -\frac{1}{(1+2x)^2} + C.$$

**6.** Let $u = \sin\theta$. Then $du = \cos\theta\,d\theta$, so $\int e^{\sin\theta}\cos\theta\,d\theta = \int e^u\,du = e^u + C = e^{\sin\theta} + C$.

**7.** Let $u = x^2 + 3$. Then $du = 2x\,dx$, so $\int 2x(x^2+3)^4\,dx = \int u^4\,du = \frac{1}{5}u^5 + C = \frac{1}{5}(x^2+3)^5 + C$.

**8.** Let $u = x^3 + 5$. Then $du = 3x^2dx$ and $x^2dx = \frac{1}{3}\,du$, so
$\int x^2(x^3+5)^9\,dx = \int u^9\left(\frac{1}{3}\,du\right) = \frac{1}{3}\cdot\frac{1}{10}u^{10} + C = \frac{1}{30}(x^3+5)^{10} + C$.

**9.** Let $u = 3x - 2$. Then $du = 3\,dx$ and $dx = \frac{1}{3}\,du$, so $\int(3x-2)^{20}\,dx = \int u^{20}\left(\frac{1}{3}\,du\right) = \frac{1}{3}\cdot\frac{1}{21}u^{21} + C = \frac{1}{63}(3x-2)^{21} + C$.

**10.** Let $u = x^2$. Then $du = 2x\,dx$, so $\int xe^{x^2}\,dx = \int e^u\left(\frac{1}{2}\,du\right) = \frac{1}{2}e^u + C = \frac{1}{2}e^{x^2} + C$.

**11.** Let $u = \ln x$. Then $du = \dfrac{dx}{x}$, so $\displaystyle\int \frac{(\ln x)^2}{x}\,dx = \int u^2\,du = \frac{1}{3}u^3 + C = \frac{1}{3}(\ln x)^3 + C$.

**12.** Let $u = 2 - x$. Then $du = -dx$ and $dx = -du$, so $\int(2-x)^6\,dx = \int u^6(-du) = -\frac{1}{7}u^7 + C = -\frac{1}{7}(2-x)^7 + C$.

**13.** Let $u = 5 - 3x$. Then $du = -3\,dx$ and $dx = -\frac{1}{3}\,du$, so

$$\int \frac{dx}{5-3x} = \int \frac{1}{u}\left(-\tfrac{1}{3}\,du\right) = -\tfrac{1}{3}\ln|u| + C = -\tfrac{1}{3}\ln|5-3x| + C.$$

**14.** Let $u = x^2 + 1$. Then $du = 2x\,dx$ and $x\,dx = \frac{1}{2}\,du$, so

$$\int \frac{x}{(x^2+1)^2}\,dx = \int u^{-2}\left(\tfrac{1}{2}\,du\right) = \tfrac{1}{2}\cdot\frac{-1}{u} + C = \frac{-1}{2u} + C = \frac{-1}{2(x^2+1)} + C.$$

**15.** Let $u = 3ax + bx^3$. Then $du = (3a + 3bx^2)\,dx = 3(a + bx^2)\,dx$, so

$$\int \frac{a+bx^2}{\sqrt{3ax+bx^3}}\,dx = \int \frac{\frac{1}{3}\,du}{u^{1/2}} = \frac{1}{3}\int u^{-1/2}\,du = \tfrac{1}{3}\cdot 2u^{1/2} + C = \tfrac{2}{3}\sqrt{3ax+bx^3} + C.$$

**16.** Let $u = 5t + 4$. Then $du = 5\,dt$ and $dt = \frac{1}{5}\,du$, so

$$\int \frac{1}{(5t+4)^{2.7}}\,dt = \int u^{-2.7}\left(\tfrac{1}{5}\,du\right) = \frac{1}{5}\cdot\frac{1}{-1.7}u^{-1.7} + C = \frac{-1}{8.5}u^{-1.7} + C = \frac{-2}{17(5t+4)^{1.7}} + C.$$

**17.** Let $u = \pi t$. Then $du = \pi\,dt$ and $dt = \frac{1}{\pi}\,du$, so $\int \sin\pi t\,dt = \int \sin u\left(\frac{1}{\pi}\,du\right) = \frac{1}{\pi}(-\cos u) + C = -\frac{1}{\pi}\cos\pi t + C$.

**18.** Let $u = 2y^4 - 1$. Then $du = 8y^3\,dy$ and $y^3\,dy = \frac{1}{8}\,du$, so
$\int y^3\sqrt{2y^4-1}\,dy = \int u^{1/2}\left(\frac{1}{8}\,du\right) = \frac{1}{8}\cdot\frac{2}{3}u^{3/2} + C = \frac{1}{12}(2y^4-1)^{3/2} + C.$

**19.** Let $u = 1 + e^x$. Then $du = e^x\,dx$, so $\int e^x\sqrt{1+e^x}\,dx = \int \sqrt{u}\,du = \frac{2}{3}u^{3/2} + C = \frac{2}{3}(1+e^x)^{3/2} + C.$
*Or:* Let $u = \sqrt{1+e^x}$. Then $u^2 = 1 + e^x$ and $2u\,du = e^x\,dx$, so
$\int e^x\sqrt{1+e^x}\,dx = \int u\cdot 2u\,du = \frac{2}{3}u^3 + C = \frac{2}{3}(1+e^x)^{3/2} + C.$

**20.** Let $u = 2\theta$. Then $du = 2\,d\theta$ and $d\theta = \frac{1}{2}\,du$, so $\int \sec 2\theta\,\tan 2\theta\,d\theta = \int \sec u \tan u\left(\frac{1}{2}\,du\right) = \frac{1}{2}\sec u + C = \frac{1}{2}\sec 2\theta + C.$

**21.** Let $u = \sin\theta$. Then $du = \cos\theta\,d\theta$, so $\int \cos\theta\,\sin^6\theta\,d\theta = \int u^6\,du = \frac{1}{7}u^7 + C = \frac{1}{7}\sin^7\theta + C.$

**22.** Let $u = x^2 + 1$. Then $du = 2x\,dx$ and $x\,dx = \frac{1}{2}\,du$, so

$$\int \frac{x}{x^2+1}\,dx = \int \frac{\frac{1}{2}\,du}{u} = \tfrac{1}{2}\ln|u| + C = \tfrac{1}{2}\ln\left|x^2+1\right| + C = \tfrac{1}{2}\ln\left(x^2+1\right) + C \quad [\text{since } x^2+1>0]$$

or $\ln(x^2+1)^{1/2} + C = \ln\sqrt{x^2+1} + C.$

**23.** Let $u = \cot x$. Then $du = -\csc^2 x\,dx$ and $\csc^2 x\,dx = -du$, so

$$\int \sqrt{\cot x}\,\csc^2 x\,dx = \int \sqrt{u}\,(-du) = -\frac{u^{3/2}}{3/2} + C = -\tfrac{2}{3}(\cot x)^{3/2} + C.$$

**24.** Let $u = \tan^{-1}x$. Then $du = \dfrac{dx}{1+x^2}$, so $\displaystyle\int \frac{\tan^{-1}x}{1+x^2}\,dx = \int u\,du = \frac{u^2}{2} + C = \frac{\left(\tan^{-1}x\right)^2}{2} + C.$

**25.** Let $u = \sin^{-1}x$. Then $du = \dfrac{1}{\sqrt{1-x^2}}\,dx$, so $\displaystyle\int \frac{dx}{\sqrt{1-x^2}\,\sin^{-1}x} = \int \frac{1}{u}\,du = \ln|u| + C = \ln\left|\sin^{-1}x\right| + C.$

**26.** Let $u = \dfrac{\pi}{x}$. Then $du = -\dfrac{\pi}{x^2}\,dx$ and $\dfrac{1}{x^2}\,dx = -\dfrac{1}{\pi}\,du$, so

$$\int \frac{\cos(\pi/x)}{x^2}\,dx = \int \cos u\left(-\frac{1}{\pi}\,du\right) = -\frac{1}{\pi}\sin u + C = -\frac{1}{\pi}\sin\frac{\pi}{x} + C.$$

**27.** Let $u = \sec x$. Then $du = \sec x\,\tan x\,dx$, so
$\int \sec^3 x\,\tan x\,dx = \int \sec^2 x\,(\sec x\,\tan x)\,dx = \int u^2\,du = \frac{1}{3}u^3 + C = \frac{1}{3}\sec^3 x + C.$

**28.** Let $u = \cos t$. Then $du = -\sin t\,dt$ and $\sin t\,dt = -du$, so
$\int \sin t\,\sec^2(\cos t)\,dt = \int \sec^2 u\cdot(-du) = -\tan u + C = -\tan(\cos t) + C.$

**29.** $\displaystyle\int \frac{e^x+1}{e^x}\,dx = \int\left(\frac{e^x}{e^x} + \frac{1}{e^x}\right)dx = \int\left(1 + e^{-x}\right)dx = x - e^{-x} + C$ [Substitute $u = -x$.]

**30.** Let $u = e^x + 1$. Then $du = e^x\,dx$, so $\displaystyle\int \frac{e^x}{e^x+1}\,dx = \int \frac{du}{u} = \ln|u| + C = \ln(e^x+1) + C.$

**31.** $\displaystyle\int \frac{\sin 2x}{1+\cos^2 x}\,dx = 2\int \frac{\sin x\cos x}{1+\cos^2 x}\,dx = 2I$. Let $u = \cos x$. Then $du = -\sin x\,dx$, so

$$2I = -2\int \frac{u\,du}{1+u^2} = -2\cdot\tfrac{1}{2}\ln(1+u^2) + C = -\ln(1+u^2) + C = -\ln(1+\cos^2 x) + C.$$

*Or:* Let $u = 1 + \cos^2 x$.

**32.** Let $u = \cos x$. Then $du = -\sin x\,dx$ and $\sin x\,dx = -du$, so

$$\int \frac{\sin x}{1+\cos^2 x}\,dx = \int \frac{-du}{1+u^2} = -\tan^{-1}u + C = -\tan^{-1}(\cos x) + C.$$

**33.** Let $u = 1 + x^2$. Then $du = 2x\,dx$, so

$$\begin{aligned}\int \frac{1+x}{1+x^2}\,dx &= \int \frac{1}{1+x^2}\,dx + \int \frac{x}{1+x^2}\,dx = \tan^{-1}x + \int \frac{\frac{1}{2}du}{u} = \tan^{-1}x + \tfrac{1}{2}\ln|u| + C \\ &= \tan^{-1}x + \tfrac{1}{2}\ln\left|1+x^2\right| + C = \tan^{-1}x + \tfrac{1}{2}\ln\left(1+x^2\right) + C \ [\text{since } 1+x^2>0].\end{aligned}$$

**34.** Let $u = x^2$. Then $du = 2x\,dx$, so $\displaystyle\int \frac{x}{1+x^4}\,dx = \int \frac{\frac{1}{2}\,du}{1+u^2} = \tfrac{1}{2}\tan^{-1}u + C = \tfrac{1}{2}\tan^{-1}\left(x^2\right) + C.$

**35.** $f(x) = \dfrac{3x-1}{(3x^2-2x+1)^4}$.

$u = 3x^2 - 2x + 1 \quad \Rightarrow \quad du = (6x-2)\,dx = 2(3x-1)\,dx$, so

$$\int \frac{3x-1}{(3x^2-2x+1)^4}\,dx = \int \frac{1}{u^4}\left(\frac{1}{2}\,du\right) = \frac{1}{2}\int u^{-4}\,du$$

$$= -\frac{1}{6}u^{-3} + C = -\frac{1}{6\,(3x^2-2x+1)^3} + C$$

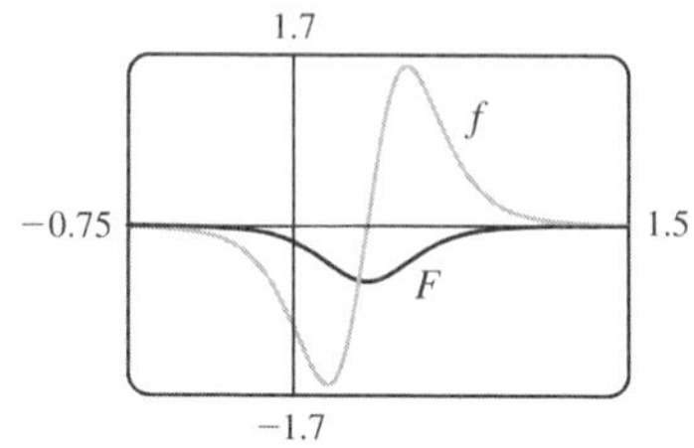


Notice that at $x = \frac{1}{3}$, $f$ changes from negative to positive, and $F$ has a local minimum.

**36.** $f(x) = \dfrac{x}{\sqrt{x^2+1}}$. $u = x^2 + 1 \quad \Rightarrow \quad du = 2x\,dx$, so

$$\int \frac{x}{\sqrt{x^2+1}}\,dx = \int \frac{1}{\sqrt{u}}\left(\frac{1}{2}\,du\right) = \frac{1}{2}\int u^{-1/2}\,du$$

$$= u^{1/2} + C = \sqrt{x^2+1} + C.$$

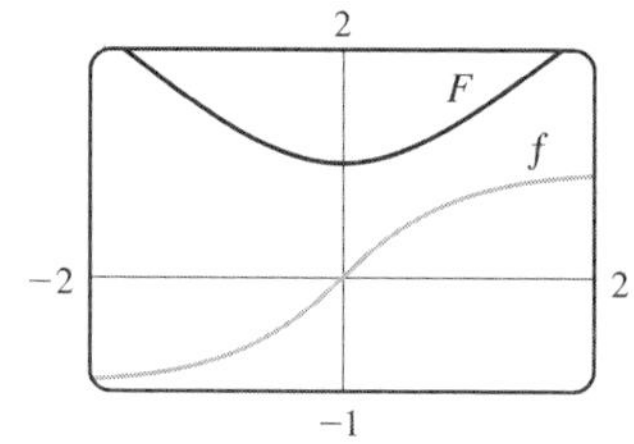


Note that at $x = 0$, $f$ changes from negative to positive and $F$ has a local minimum.

**37.** $f(x) = \sin^3 x \cos x$. $u = \sin x \quad \Rightarrow \quad du = \cos x\,dx$, so

$$\textstyle\int \sin^3 x \cos x\,dx = \int u^3\,du = \frac{1}{4}u^4 + C = \frac{1}{4}\sin^4 x + C$$

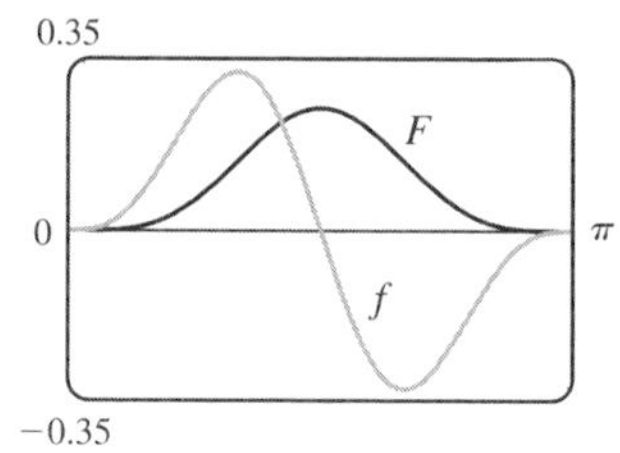


Note that at $x = \frac{\pi}{2}$, $f$ changes from positive to negative and $F$ has a local maximum. Also, both $f$ and $F$ are periodic with period $\pi$, so at $x = 0$ and at $x = \pi$, $f$ changes from negative to positive and $F$ has local minima.

**38.** $f(\theta) = \tan^2\theta \sec^2\theta$. $u = \tan\theta \quad \Rightarrow \quad du = \sec^2\theta\,d\theta$, so

$$\textstyle\int \tan^2\theta \sec^2\theta\,d\theta = \int u^2\,du = \frac{1}{3}u^3 + C = \frac{1}{3}\tan^3\theta + C$$

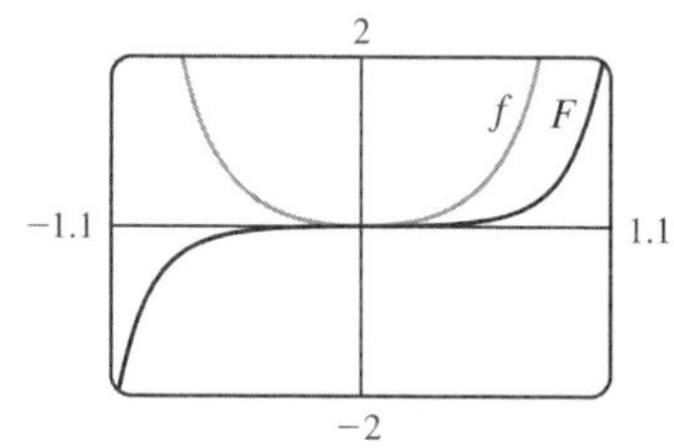


Note that $f$ is positive and $F$ is increasing. At $x = 0$, $f = 0$ and $F$ has a horizontal tangent.

**39.** Let $u = x - 1$, so $du = dx$. When $x = 0$, $u = -1$; when $x = 2$, $u = 1$. Thus, $\int_0^2 (x-1)^{25}\,dx = \int_{-1}^1 u^{25}\,du = 0$ by Theorem 6(b), since $f(u) = u^{25}$ is an odd function.

**40.** Let $u = 4 + 3x$, so $du = 3\,dx$. When $x = 0$, $u = 4$; when $x = 7$, $u = 25$. Thus,

$$\int_0^7 \sqrt{4+3x}\,dx = \int_4^{25} \sqrt{u}\left(\tfrac{1}{3}\,du\right) = \frac{1}{3}\left[\frac{u^{3/2}}{3/2}\right]_4^{25} = \frac{2}{9}(25^{3/2} - 4^{3/2}) = \frac{2}{9}(125-8) = \frac{234}{9} = 26$$

**41.** Let $u = 1 + 2x^3$, so $du = 6x^2\,dx$. When $x = 0$, $u = 1$; when $x = 1$, $u = 3$. Thus,

$$\int_0^1 x^2\left(1+2x^3\right)^5\,dx = \int_1^3 u^5\left(\tfrac{1}{6}\,du\right) = \tfrac{1}{6}\left[\tfrac{1}{6}u^6\right]_1^3 = \tfrac{1}{36}(3^6 - 1^6) = \tfrac{1}{36}(729-1) = \tfrac{728}{36} = \tfrac{182}{9}$$

**42.** Let $u = x^2$, so $du = 2x\,dx$. When $x = 0$, $u = 0$; when $x = \sqrt{\pi}$, $u = \pi$. Thus,

$\int_0^{\sqrt{\pi}} x\cos(x^2)\,dx = \int_0^{\pi} \cos u\left(\frac{1}{2}\,du\right) = \frac{1}{2}[\sin u]_0^{\pi} = \frac{1}{2}(\sin\pi - \sin 0) = \frac{1}{2}(0-0) = 0.$

**43.** Let $u = t/4$, so $du = \frac{1}{4}\,dt$. When $t = 0$, $u = 0$; when $t = \pi$, $u = \pi/4$. Thus,

$\int_0^{\pi} \sec^2(t/4)\,dt = \int_0^{\pi/4} \sec^2 u\,(4\,du) = 4\big[\tan u\big]_0^{\pi/4} = 4\big(\tan\frac{\pi}{4} - \tan 0\big) = 4(1-0) = 4.$

**44.** Let $u = \pi t$, so $du = \pi\,dt$. When $t = \frac{1}{6}$, $u = \frac{\pi}{6}$; when $t = \frac{1}{2}$, $u = \frac{\pi}{2}$. Thus,
$\int_{1/6}^{1/2} \csc \pi t \cot \pi t\,dt = \int_{\pi/6}^{\pi/2} \csc u \cot u \left(\frac{1}{\pi}\,du\right) = \frac{1}{\pi}\big[-\csc u\big]_{\pi/6}^{\pi/2} = -\frac{1}{\pi}(1-2) = \frac{1}{\pi}$.

**45.** Let $u = \sqrt{x}$, so $du = \dfrac{1}{2\sqrt{x}}\,dx$. When $x = 1$, $u = 1$; when $x = 4$, $u = 2$. Thus,

$$\int_1^4 \frac{e^{\sqrt{x}}}{\sqrt{x}}\,dx = \int_1^2 e^u\,(2\,du) = 2\left[e^u\right]_1^2 = 2(e^2 - e)$$

**46.** Let $u = \sin x$, so $du = \cos x\,dx$. When $x = 0$, $u = 0$; when $x = \frac{\pi}{2}$, $u = 1$. Thus,
$\int_0^{\pi/2} \cos x \sin(\sin x)\,dx = \int_0^1 \sin u\,du = \big[-\cos u\big]_0^1 = -(\cos 1 - 1) = 1 - \cos 1$.

**47.** Let $u = x - 1$, so $u + 1 = x$ and $du = dx$. When $x = 1$, $u = 0$; when $x = 2$, $u = 1$. Thus,
$\int_1^2 x\sqrt{x-1}\,dx = \int_0^1 (u+1)\sqrt{u}\,du = \int_0^1 (u^{3/2} + u^{1/2})\,du = \left[\frac{2}{5}u^{5/2} + \frac{2}{3}u^{3/2}\right]_0^1 = \frac{2}{5} + \frac{2}{3} = \frac{16}{15}$.

**48.** $\displaystyle\int_{-\pi/2}^{\pi/2} \frac{x^2 \sin x}{1+x^6}\,dx = 0$ by Theorem 6(b), since $f(x) = \dfrac{x^2 \sin x}{1+x^6}$ is an odd function.

**49.** Let $u = e^z + z$, so $du = (e^z + 1)\,dz$. When $z = 0$, $u - 1$; when $z = 1$, $u = e + 1$. Thus,
$\displaystyle\int_0^1 \frac{e^z+1}{e^z+z}\,dz = \int_1^{e+1} \frac{1}{u}\,du = \Big[\ln|u|\Big]_1^{e+1} = \ln|e+1| - \ln|1| = \ln(e+1)$.

**50.** Let $u = 1 + 2x$, so $x = \frac{1}{2}(u-1)$ and $du = 2\,dx$. When $x = 0$, $u = 1$; when $x = 4$, $u = 9$. Thus,

$$\int_0^4 \frac{x\,dx}{\sqrt{1+2x}} = \int_1^9 \frac{\frac{1}{2}(u-1)}{\sqrt{u}}\frac{du}{2} = \frac{1}{4}\int_1^9 (u^{1/2} - u^{-1/2})\,du = \frac{1}{4}\left[\frac{2}{3}u^{3/2} - 2u^{1/2}\right]_1^9$$
$$= \frac{1}{4}\cdot\frac{2}{3}\left[u^{3/2} - 3u^{1/2}\right]_1^9 = \frac{1}{6}[(27-9)-(1-3)] = \frac{20}{6} = \frac{10}{3}$$

**51.** $\int_{-\pi/6}^{\pi/6} \tan^3\theta\,d\theta = 0$ by Theorem 6(b), since $f(\theta) = \tan^3\theta$ is an odd function.

**52.** Assume $a > 0$. Let $u = a^2 - x^2$, so $du = -2x\,dx$. When $x = 0$, $u = a^2$; when $x = a$, $u = 0$. Thus,
$\int_0^a x\sqrt{a^2-x^2}\,dx = \int_{a^2}^0 u^{1/2}\left(-\frac{1}{2}\,du\right) = \frac{1}{2}\int_0^{a^2} u^{1/2}\,du = \frac{1}{2}\cdot\left[\frac{2}{3}u^{3/2}\right]_0^{a^2} = \frac{1}{3}a^3$.

**53.** Let $u = \ln x$, so $du = \dfrac{dx}{x}$. When $x = e$, $u = 1$; when $x = e^4$; $u = 4$. Thus,
$\displaystyle\int_e^{e^4} \frac{dx}{x\sqrt{\ln x}} = \int_1^4 u^{-1/2}\,du = 2\Big[u^{1/2}\Big]_1^4 = 2(2-1) = 2$.

**54.** Let $u = \sin^{-1} x$, so $du = \dfrac{dx}{\sqrt{1-x^2}}$. When $x = 0$, $u = 0$; when $x = \frac{1}{2}$, $u = \frac{\pi}{6}$. Thus,
$\displaystyle\int_0^{1/2} \frac{\sin^{-1} x}{\sqrt{1-x^2}}\,dx = \int_0^{\pi/6} u\,du = \left[\frac{u^2}{2}\right]_0^{\pi/6} = \frac{\pi^2}{72}$.

**55.** From the graph, it appears that the area under the curve is about $1 + \left(\text{a little more than } \frac{1}{2}\cdot 1\cdot 0.7\right)$, or about 1.4. The exact area is given by $A = \int_0^1 \sqrt{2x+1}\,dx$. Let $u = 2x + 1$, so $du = 2\,dx$. The limits change to $2\cdot 0 + 1 = 1$ and $2\cdot 1 + 1 = 3$, and
$A = \int_1^3 \sqrt{u}\left(\frac{1}{2}\,du\right) = \frac{1}{2}\left[\frac{2}{3}u^{3/2}\right]_1^3 = \frac{1}{3}(3\sqrt{3} - 1) = \sqrt{3} - \frac{1}{3} \approx 1.399$.

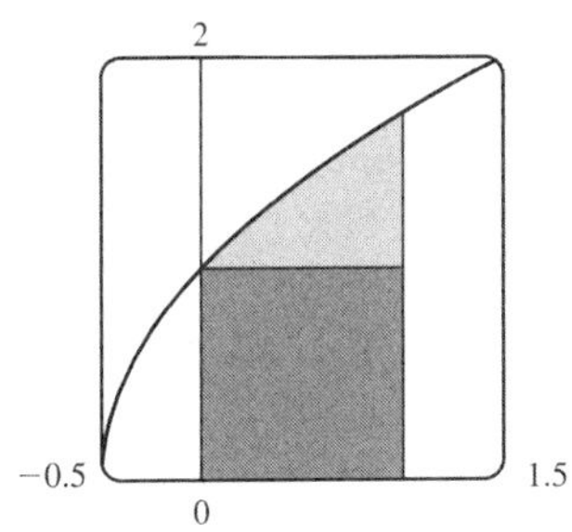

**56.** From the graph, it appears that the area under the curve is almost $\frac{1}{2} \cdot \pi \cdot 2.6$, or about 4. The exact area is given by

$$A = \int_0^\pi (2\sin x - \sin 2x)dx = -2\left[\cos x\right]_0^\pi - \int_0^\pi \sin 2x\,dx$$
$$= -2(-1-1) - 0 = 4$$

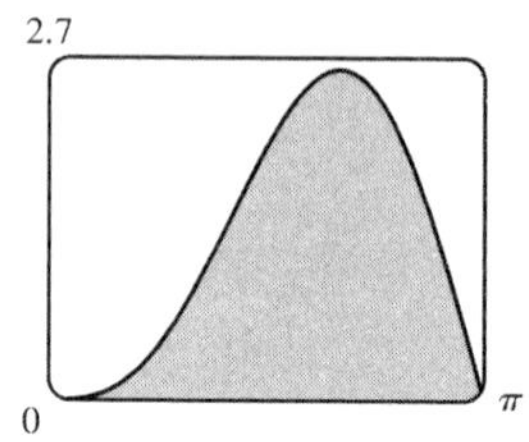


*Note:* $\int_0^\pi \sin 2x\,dx = 0$ since it is clear from the graph of $y = \sin 2x$ that $\int_{\pi/2}^\pi \sin 2x\,dx = -\int_0^{\pi/2} \sin 2x\,dx$.

**57.** First write the integral as a sum of two integrals:
$I = \int_{-2}^2 (x+3)\sqrt{4-x^2}\,dx = I_1 + I_2 = \int_{-2}^2 x\sqrt{4-x^2}\,dx + \int_{-2}^2 3\sqrt{4-x^2}\,dx$. $I_1 = 0$ by Theorem 6(b), since $f(x) = x\sqrt{4-x^2}$ is an odd function and we are integrating from $x = -2$ to $x = 2$. We interpret $I_2$ as three times the area of a semicircle with radius 2, so $I = 0 + 3 \cdot \frac{1}{2}\left(\pi \cdot 2^2\right) = 6\pi$.

**58.** Let $u = x^2$. Then $du = 2x\,dx$ and the limits are unchanged ($0^2 = 0$ and $1^2 = 1$), so
$I = \int_0^1 x\sqrt{1-x^4}\,dx = \frac{1}{2}\int_0^1 \sqrt{1-u^2}\,du$. But this integral can be interpreted as the area of a quarter-circle with radius 1. So $I = \frac{1}{2} \cdot \frac{1}{4}\left(\pi \cdot 1^2\right) = \frac{1}{8}\pi$.

**59.** **First Figure** Let $u = \sqrt{x}$, so $x = u^2$ and $dx = 2u\,du$. When $x = 0$, $u = 0$; when $x = 1$, $u = 1$. Thus,

$$A_1 = \int_0^1 e^{\sqrt{x}}\,dx = \int_0^1 e^u(2u\,du) = 2\int_0^1 ue^u\,du.$$

**Second Figure** $A_2 = \int_0^1 2xe^x\,dx = 2\int_0^1 ue^u\,du$.

**Third Figure** Let $u = \sin x$, so $du = \cos x\,dx$. When $x = 0$, $u = 0$; when $x = \frac{\pi}{2}$, $u = 1$. Thus,

$$A_3 = \int_0^{\pi/2} e^{\sin x}\sin 2x\,dx = \int_0^{\pi/2} e^{\sin x}(2\sin x\,\cos x)\,dx = \int_0^1 e^u(2u\,du) = 2\int_0^1 ue^u\,du.$$

Since $A_1 = A_2 = A_3$, all three areas are equal.

**60.** Let $r(t) = ae^{bt}$ with $a = 450.268$ and $b = 1.12567$, and $n(t) =$ population after $t$ hours. Since $r(t) = n'(t)$, $\int_0^3 r(t)\,dt = n(3) - n(0)$ is the total change in the population after three hours. Since we start with 400 bacteria, the population will be

$$n(3) = 400 + \int_0^3 r(t)\,dt = 400 + \int_0^3 ae^{bt}\,dt = 400 + \frac{a}{b}\left[e^{bt}\right]_0^3 = 400 + \frac{a}{b}\left(e^{3b} - 1\right)$$
$$\approx 400 + 11{,}313 = 11{,}713 \text{ bacteria}$$

**61.** The volume of inhaled air in the lungs at time $t$ is

$$V(t) = \int_0^t f(u)\,du = \int_0^t \frac{1}{2}\sin\left(\tfrac{2\pi}{5}u\right)du = \int_0^{2\pi t/5} \tfrac{1}{2}\sin v\left(\tfrac{5}{2\pi}\,dv\right) \quad \left[\text{substitute } v = \tfrac{2\pi}{5}u,\ dv = \tfrac{2\pi}{5}\,du\right]$$
$$= \tfrac{5}{4\pi}\left[-\cos v\right]_0^{2\pi t/5} = \tfrac{5}{4\pi}\left[-\cos\left(\tfrac{2\pi}{5}t\right) + 1\right] = \tfrac{5}{4\pi}\left[1 - \cos\left(\tfrac{2\pi}{5}t\right)\right] \text{ liters}$$

**62.** Number of calculators $= x(4) - x(2) = \int_2^4 5000\left[1 - 100(t+10)^{-2}\right]dt$

$$= 5000\left[t + 100(t+10)^{-1}\right]_2^4 = 5000\left[\left(4 + \tfrac{100}{14}\right) - \left(2 + \tfrac{100}{12}\right)\right] \approx 4048$$

**63.** Let $u = 2x$. Then $du = 2\,dx$, so $\int_0^2 f(2x)\,dx = \int_0^4 f(u)\left(\frac{1}{2}\,du\right) = \frac{1}{2}\int_0^4 f(u)\,du = \frac{1}{2}(10) = 5$.

**64.** Let $u = x^2$. Then $du = 2x\,dx$, so $\int_0^3 xf\left(x^2\right)dx = \int_0^9 f(u)\left(\frac{1}{2}\,du\right) = \frac{1}{2}\int_0^9 f(u)\,du = \frac{1}{2}(4) = 2$.

**65.** Let $u = -x$. Then $du = -dx$, so

$$\int_a^b f(-x)\,dx = \int_{-a}^{-b} f(u)(-du) = \int_{-b}^{-a} f(u)\,du = \int_{-b}^{-a} f(x)\,dx$$

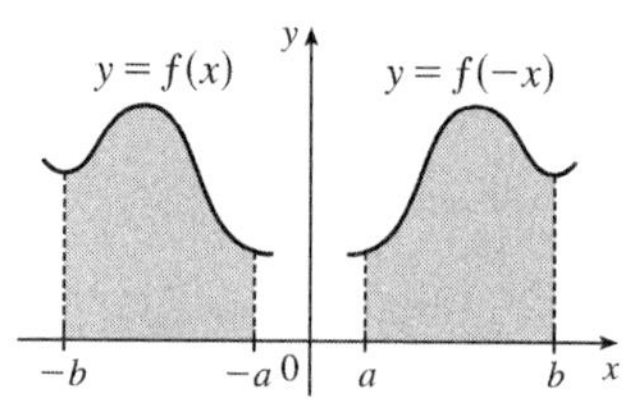


From the diagram, we see that the equality follows from the fact that we are reflecting the graph of $f$, and the limits of integration, about the $y$-axis.

**66.** Let $u = x + c$. Then $du = dx$, so

$\int_a^b f(x+c)\,dx = \int_{a+c}^{b+c} f(u)\,du = \int_{a+c}^{b+c} f(x)\,dx$

From the diagram, we see that the equality follows from the fact that we are translating the graph of $f$, and the limits of integration, by a distance $c$.

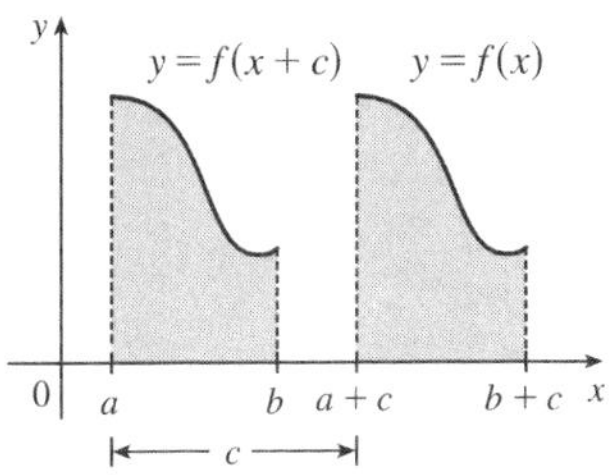


**67.** Let $u = 1 - x$. Then $x = 1 - u$ and $dx = -du$, so
$\int_0^1 x^a(1-x)^b\,dx = \int_1^0 (1-u)^a\,u^b(-du) = \int_0^1 u^b(1-u)^a\,du = \int_0^1 x^b(1-x)^a\,dx$.

## 5.6 Integration by Parts

**1.** Let $u = \ln x$, $dv = x\,dx \;\Rightarrow\; du = dx/x$, $v = \frac{1}{2}x^2$. Then by Equation 2, $\int u\,dv = uv - \int v\,du$,

$$\begin{aligned}\int x\ln x\,dx &= \tfrac{1}{2}x^2\ln x - \int \tfrac{1}{2}x^2(dx/x) = \tfrac{1}{2}x^2\ln x - \tfrac{1}{2}\int x\,dx = \tfrac{1}{2}x^2\ln x - \tfrac{1}{2}\cdot\tfrac{1}{2}x^2 + C\\ &= \tfrac{1}{2}x^2\ln x - \tfrac{1}{4}x^2 + C\end{aligned}$$

**2.** Let $u = \theta$, $dv = \cos\theta\,d\theta \;\Rightarrow\; du = d\theta$, $v = \sin\theta$. Then by Equation 2,
$\int \theta\cos\theta\,d\theta = \theta\sin\theta - \int \sin\theta\,d\theta = \theta\sin\theta + \cos\theta + C$.

Note: A mnemonic device which is helpful for selecting $u$ when using integration by parts is the LIATE principle of precedence for $u$:

Logarithmic
Inverse trigonometric
Algebraic
Trigonometric
Exponential

If the integrand has several factors, then we try to choose among them a $u$ which appears as high as possible on the list. For example, in $\int xe^{2x}\,dx$ the integrand is $xe^{2x}$, which is the product of an algebraic function ($x$) and an exponential function ($e^{2x}$). Since Algebraic appears before Exponential, we choose $u = x$. Sometimes the integration turns out to be similar regardless of the selection of $u$ and $dv$, but it is advisable to refer to LIATE when in doubt.

**3.** Let $u = x$, $dv = \cos 5x\,dx \;\Rightarrow\; du = dx$, $v = \frac{1}{5}\sin 5x$. Then by Equation 2,
$\int x\cos 5x\,dx = \frac{1}{5}x\sin 5x - \int \frac{1}{5}\sin 5x\,dx = \frac{1}{5}x\sin 5x + \frac{1}{25}\cos 5x + C$.

**4.** Let $u = x$, $dv = e^{-x}\,dx \;\Rightarrow\; du = dx$, $v = -e^{-x}$. Then $\int xe^{-x}\,dx = -xe^{-x} + \int e^{-x}\,dx = -xe^{-x} - e^{-x} + C$.

**5.** Let $u = r$, $dv = e^{r/2}\,dr \;\Rightarrow\; du = dr$, $v = 2e^{r/2}$. Then $\int re^{r/2}\,dr = 2re^{r/2} - \int 2e^{r/2}\,dr = 2re^{r/2} - 4e^{r/2} + C$.

**6.** Let $u = t$, $dv = \sin 2t\,dt \;\Rightarrow\; du = dt$, $v = -\frac{1}{2}\cos 2t$. Then
$\int t\sin 2t\,dt = -\frac{1}{2}t\cos 2t + \frac{1}{2}\int \cos 2t\,dt = -\frac{1}{2}t\cos 2t + \frac{1}{4}\sin 2t + C$.

**7.** Let $u = x^2$, $dv = \sin\pi x\,dx \;\Rightarrow\; du = 2x\,dx$ and $v = -\frac{1}{\pi}\cos\pi x$. Then
$I = \int x^2\sin\pi x\,dx = -\frac{1}{\pi}x^2\cos\pi x + \frac{2}{\pi}\int x\cos\pi x\,dx$ $(*)$.
Next let $U = x$, $dV = \cos\pi x\,dx \;\Rightarrow\; dU = dx$, $V = \frac{1}{\pi}\sin\pi x$, so
$\int x\cos\pi x\,dx = \frac{1}{\pi}x\sin\pi x - \frac{1}{\pi}\int \sin\pi x\,dx = \frac{1}{\pi}x\sin\pi x + \frac{1}{\pi^2}\cos\pi x + C_1$. Substituting for $\int x\cos\pi x\,dx$ in $(*)$, we get
$I = -\frac{1}{\pi}x^2\cos\pi x + \frac{2}{\pi}\left(\frac{1}{\pi}x\sin\pi x + \frac{1}{\pi^2}\cos\pi x + C_1\right) = -\frac{1}{\pi}x^2\cos\pi x + \frac{2}{\pi^2}x\sin\pi x + \frac{2}{\pi^3}\cos\pi x + C$, where $C = \frac{2}{\pi}C_1$.

**8.** Let $u = x^2$, $dv = \cos mx\,dx \;\Rightarrow\; du = 2x\,dx$, $v = \frac{1}{m}\sin mx$. Then

$I = \displaystyle\int x^2 \cos mx\,dx = \frac{1}{m}x^2 \sin mx - \frac{2}{m}\int x \sin mx\,dx$ (*). Next let

$U = x$, $dV = \sin mx\,dx \;\Rightarrow\; dU = dx$, $V = -\frac{1}{m}\cos mx$, so

$\int x \sin mx\,dx = -\frac{1}{m}x\cos mx + \frac{1}{m}\int \cos mx\,dx = -\frac{1}{m}x\cos mx + \frac{1}{m^2}\sin mx + C_1$. Substituting for $\int x\sin mx\,dx$ in (*), we get $I = \frac{1}{m}x^2\sin mx - \frac{2}{m}\left(-\frac{1}{m}x\cos mx + \frac{1}{m^2}\sin mx + C_1\right) = \frac{1}{m}x^2\sin mx + \frac{2}{m^2}x\cos mx - \frac{2}{m^3}\sin mx + C$, where $C = -\frac{2}{m}C_1$.

**9.** Let $u = \ln(2x+1)$, $dv = dx \;\Rightarrow\; du = \dfrac{2}{2x+1}\,dx$, $v = x$. Then

$$\begin{aligned}\int \ln(2x+1)\,dx &= x\ln(2x+1) - \int \frac{2x}{2x+1}\,dx = x\ln(2x+1) - \int \frac{(2x+1)-1}{2x+1}\,dx \\ &= x\ln(2x+1) - \int\left(1 - \frac{1}{2x+1}\right)dx = x\ln(2x+1) - x + \tfrac{1}{2}\ln(2x+1) + C \\ &= \tfrac{1}{2}(2x+1)\,\ln(2x+1) - x + C\end{aligned}$$

**10.** Let $u = \ln p$, $dv = p^5 dp \;\Rightarrow\; du = \dfrac{1}{p}\,dp$, $v = \frac{1}{6}p^6$. Then $\int p^5 \ln p\,dp = \frac{1}{6}p^6\ln p - \frac{1}{6}\int p^5 dp = \frac{1}{6}p^6 \ln p - \frac{1}{36}p^6 + C$.

**11.** Let $u = \arctan 4t$, $dv = dt \;\Rightarrow\; du = \dfrac{4}{1+(4t)^2}\,dt = \dfrac{4}{1+16t^2}\,dt$, $v = t$. Then

$$\int \arctan 4t\,dt = t\arctan 4t - \int \frac{4t}{1+16t^2}\,dt = t\arctan 4t - \frac{1}{8}\int \frac{32t}{1+16t^2}\,dt = t\arctan 4t - \tfrac{1}{8}\ln(1+16t^2) + C.$$

**12.** Let $u = t^3$, $dv = e^t\,dt \;\Rightarrow\; du = 3t^2\,dt$, $v = e^t$. Then $I = \int t^3 e^t\,dt = t^3e^t - \int 3t^2e^t\,dt$. Integrate by parts twice more with $dv = e^t\,dt$.

$$\begin{aligned}I &= t^3e^t - \left(3t^2e^t - \textstyle\int 6te^t\,dt\right) = t^3e^t - 3t^2e^t + 6te^t - \textstyle\int 6e^t\,dt \\ &= t^3e^t - 3t^2e^t + 6te^t - 6e^t + C = (t^3 - 3t^2 + 6t - 6)e^t + C\end{aligned}$$

More generally, if $p(t)$ is a polynomial of degree $n$ in $t$, then repeated integration by parts shows that

$\int p(t)\,e^t\,dt = \left[p(t) - p'(t) + p''(t) - p'''(t) + \cdots + (-1)^n\,p^{(n)}(t)\right]e^t + C.$

**13.** First let $u = \sin 3\theta$, $dv = e^{2\theta}\,d\theta \;\Rightarrow\; du = 3\cos 3\theta\,d\theta$, $v = \frac{1}{2}e^{2\theta}$. Then

$I = \int e^{2\theta}\sin 3\theta\,d\theta = \frac{1}{2}e^{2\theta}\sin 3\theta - \frac{3}{2}\int e^{2\theta}\cos 3\theta\,d\theta$. Next let $U = \cos 3\theta$, $dV = e^{2\theta}\,d\theta \;\Rightarrow\; dU = -3\sin 3\theta\,d\theta$, $V = \frac{1}{2}e^{2\theta}$ to get $\int e^{2\theta}\cos 3\theta\,d\theta = \frac{1}{2}e^{2\theta}\cos 3\theta + \frac{3}{2}\int e^{2\theta}\sin 3\theta\,d\theta$. Substituting in the previous formula gives

$I = \frac{1}{2}e^{2\theta}\sin 3\theta - \frac{3}{4}e^{2\theta}\cos 3\theta - \frac{9}{4}\int e^{2\theta}\sin 3\theta\,d\theta = \frac{1}{2}e^{2\theta}\sin 3\theta - \frac{3}{4}e^{2\theta}\cos 3\theta - \frac{9}{4}I \;\Rightarrow$

$\frac{13}{4}I = \frac{1}{2}e^{2\theta}\sin 3\theta - \frac{3}{4}e^{2\theta}\cos 3\theta + C_1$. Hence, $I = \frac{1}{13}e^{2\theta}(2\sin 3\theta - 3\cos 3\theta) + C$, where $C = \frac{4}{13}C_1$.

**14.** First let $u = e^{-\theta}$, $dv = \cos 2\theta\,d\theta \;\Rightarrow\; du = -e^{-\theta}\,d\theta$, $v = \frac{1}{2}\sin 2\theta$. Then

$I = \int e^{-\theta}\cos 2\theta\,d\theta = \frac{1}{2}e^{-\theta}\sin 2\theta - \int \frac{1}{2}\sin 2\theta\,(-e^{-\theta}\,d\theta) = \frac{1}{2}e^{-\theta}\sin 2\theta + \frac{1}{2}\int e^{-\theta}\sin 2\theta\,d\theta.$

Next let $U = e^{-\theta}$, $dV = \sin 2\theta\,d\theta \;\Rightarrow\; dU = -e^{-\theta}\,d\theta$, $V = -\frac{1}{2}\cos 2\theta$, so

$\int e^{-\theta}\sin 2\theta\,d\theta = -\frac{1}{2}e^{-\theta}\cos 2\theta - \int\left(-\frac{1}{2}\right)\cos 2\theta(-e^{-\theta}\,d\theta) = -\frac{1}{2}e^{-\theta}\cos 2\theta - \frac{1}{2}\int e^{-\theta}\cos 2\theta\,d\theta.$

So $I = \frac{1}{2}e^{-\theta}\sin 2\theta + \frac{1}{2}\left[\left(-\frac{1}{2}e^{-\theta}\cos 2\theta\right) - \frac{1}{2}I\right] = \frac{1}{2}e^{-\theta}\sin 2\theta - \frac{1}{4}e^{-\theta}\cos 2\theta - \frac{1}{4}I \;\Rightarrow$

$\frac{5}{4}I = \frac{1}{2}e^{-\theta}\sin 2\theta - \frac{1}{4}e^{-\theta}\cos 2\theta + C_1 \;\Rightarrow\; I = \frac{4}{5}\left(\frac{1}{2}e^{-\theta}\sin 2\theta - \frac{1}{4}e^{-\theta}\cos 2\theta + C_1\right) = \frac{2}{5}e^{-\theta}\sin 2\theta - \frac{1}{5}e^{-\theta}\cos 2\theta + C.$

**15.** Let $u = t$, $dv = \sin 3t\,dt \;\Rightarrow\; du = dt$, $v = -\frac{1}{3}\cos 3t$. Then

$\int_0^\pi t\sin 3t\,dt = \left[-\frac{1}{3}t\cos 3t\right]_0^\pi + \frac{1}{3}\int_0^\pi \cos 3t\,dt = \left(\frac{1}{3}\pi - 0\right) + \frac{1}{9}\left[\sin 3t\right]_0^\pi = \frac{\pi}{3}.$

**16.** First let $u = x^2 + 1$, $dv = e^{-x}\,dx \;\Rightarrow\; du = 2x\,dx$, $v = -e^{-x}$. By (6),

$\int_0^1 (x^2+1)e^{-x}\,dx = \left[-(x^2+1)e^{-x}\right]_0^1 + \int_0^1 2xe^{-x}\,dx = -2e^{-1} + 1 + 2\int_0^1 xe^{-x}\,dx$. Next

let $U = x$, $dV = e^{-x}\,dx \;\Rightarrow\; dU = dx$, $V = -e^{-x}$. By (6) again,

$\int_0^1 xe^{-x}\,dx = \left[-xe^{-x}\right]_0^1 + \int_0^1 e^{-x}\,dx = -e^{-1} + \left[-e^{-x}\right]_0^1 = -e^{-1} - e^{-1} + 1 = -2e^{-1} + 1$. So

$\int_0^1 (x^2+1)e^{-x}\,dx = -2e^{-1} + 1 + 2(-2e^{-1}+1) = -2e^{-1} + 1 - 4e^{-1} + 2 = -6e^{-1} + 3$.

**17.** Let $u = \ln x$, $dv = x^{-2}\,dx \;\Rightarrow\; du = \dfrac{1}{x}\,dx$, $v = -x^{-1}$. By (6),

$$\int_1^2 \frac{\ln x}{x^2}\,dx = \left[-\frac{\ln x}{x}\right]_1^2 + \int_1^2 x^{-2}\,dx = -\tfrac{1}{2}\ln 2 + \ln 1 + \left[-\frac{1}{x}\right]_1^2 = -\tfrac{1}{2}\ln 2 + 0 - \tfrac{1}{2} + 1 = \tfrac{1}{2} - \tfrac{1}{2}\ln 2.$$

**18.** Let $u = \ln t$, $dv = \sqrt{t}\,dt \;\Rightarrow\; du = dt/t$, $v = \tfrac{2}{3}t^{3/2}$. By Formula 6,

$$\int_1^4 \sqrt{t}\ln t\,dt = \left[\tfrac{2}{3}t^{3/2}\ln t\right]_1^4 - \tfrac{2}{3}\int_1^4 \sqrt{t}\,dt = \tfrac{2}{3}\cdot 8\cdot\ln 4 - 0 - \left[\tfrac{2}{3}\cdot\tfrac{2}{3}t^{3/2}\right]_1^4 = \tfrac{16}{3}\ln 4 - \tfrac{4}{9}(8-1) = \tfrac{16}{3}\ln 4 - \tfrac{28}{9}.$$

**19.** Let $u = y$, $dv = \dfrac{dy}{e^{2y}} = e^{-2y}dy \;\Rightarrow\; du = dy$, $v = -\dfrac{1}{2}e^{-2y}$. Then

$$\int_0^1 \frac{y}{e^{2y}}\,dy = \left[-\tfrac{1}{2}ye^{-2y}\right]_0^1 + \tfrac{1}{2}\int_0^1 e^{-2y}dy = \left(-\tfrac{1}{2}e^{-2} + 0\right) - \tfrac{1}{4}\left[e^{-2y}\right]_0^1 = -\tfrac{1}{2}e^{-2} - \tfrac{1}{4}e^{-2} + \tfrac{1}{4} = \tfrac{1}{4} - \tfrac{3}{4}e^{-2}.$$

**20.** Let $u = \arctan(1/x)$, $dv = dx \;\Rightarrow\; du = \dfrac{1}{1+(1/x)^2}\cdot\dfrac{-1}{x^2}\,dx = \dfrac{-dx}{x^2+1}$, $v = x$. Then

$$\begin{aligned}\int_1^{\sqrt{3}} \arctan(1/x)\,dx &= \left[x\arctan\left(\frac{1}{x}\right)\right]_1^{\sqrt{3}} + \int_1^{\sqrt{3}} \frac{x\,dx}{x^2+1} = \sqrt{3}\,\frac{\pi}{6} - 1\cdot\frac{\pi}{4} + \frac{1}{2}\left[\ln(x^2+1)\right]_1^{\sqrt{3}} \\ &= \frac{\pi\sqrt{3}}{6} - \frac{\pi}{4} + \frac{1}{2}(\ln 4 - \ln 2) = \frac{\pi\sqrt{3}}{6} - \frac{\pi}{2} + \frac{1}{2}\ln\frac{4}{2} = \frac{\pi\sqrt{3}}{6} - \frac{\pi}{2} + \frac{1}{2}\ln 2\end{aligned}$$

**21.** Let $u = \sin^{-1}x$, $dv = dx \;\Rightarrow\; du = \dfrac{dx}{\sqrt{1-x^2}}$, $v = x$. By Formula 6,

$I = \displaystyle\int_0^{1/2} \sin^{-1}x\,dx = \left[x\sin^{-1}x\right]_0^{1/2} - \int_0^{1/2} \frac{x\,dx}{\sqrt{1-x^2}} = \frac{1}{2}\cdot\frac{\pi}{6} - \int_0^{1/2}\frac{x\,dx}{\sqrt{1-x^2}}$. To evaluate the last integral, let

$t = 1 - x^2$, so $dt = -2x\,dx$ and $x\,dx = -\tfrac{1}{2}\,dt$. When $x = 0$, $t = 1$; when $x = \tfrac{1}{2}$, $t = \tfrac{3}{4}$. So

$$\int_0^{1/2} \frac{x\,dx}{\sqrt{1-x^2}} = \int_1^{3/4} \frac{1}{\sqrt{t}}\left(-\frac{1}{2}\,dt\right) = \frac{1}{2}\int_{3/4}^1 t^{-1/2}dt = \tfrac{1}{2}\left[2t^{1/2}\right]_{3/4}^1 = \sqrt{1} - \sqrt{\tfrac{3}{4}} = 1 - \tfrac{\sqrt{3}}{2}.$$

Thus, $I = \frac{\pi}{12} - \left(1 - \frac{\sqrt{3}}{2}\right) = \frac{\pi}{12} - 1 + \frac{\sqrt{3}}{2} = \frac{1}{12}\left(\pi - 12 + 6\sqrt{3}\right)$.

**22.** Let $u = r^2$, $dv = \dfrac{r}{\sqrt{4+r^2}}\,dr \;\Rightarrow\; du = 2r\,dr$, $v = \sqrt{4+r^2}$. By (6),

$$\begin{aligned}\int_0^1 \frac{r^3}{\sqrt{4+r^2}}\,dr &= \left[r^2\sqrt{4+r^2}\right]_0^1 - 2\int_0^1 r\sqrt{4+r^2}\,dr = \sqrt{5} - \tfrac{2}{3}\left[(4+r^2)^{3/2}\right]_0^1 \\ &= \sqrt{5} - \tfrac{2}{3}(5)^{3/2} + \tfrac{2}{3}(8) = \sqrt{5}\left(1 - \tfrac{10}{3}\right) + \tfrac{16}{3} = \tfrac{16}{3} - \tfrac{7}{3}\sqrt{5}\end{aligned}$$

**23.** Let $u = (\ln x)^2$, $dv = dx \;\Rightarrow\; du = \dfrac{2}{x}\ln x\,dx$, $v = x$. By Formula 6, $I = \int_1^2 (\ln x)^2\,dx = \left[x(\ln x)^2\right]_1^2 - 2\int_1^2 \ln x\,dx$.

To evaluate the last integral, let $U = \ln x$, $dV = dx \;\Rightarrow\; dU = \dfrac{1}{x}\,dx$, $V = x$. Thus,

$$\begin{aligned}I &= \left[x(\ln x)\right]_1^2 - 2\left(\left[x\ln x\right]_1^2 - \textstyle\int_1^2 dx\right) = \left[x(\ln x) - 2x\ln x + 2x\right]_1^2 \\ &= \left(2(\ln 2)^2 - 4\ln 2 + 4\right) - (0 - 0 + 2) = 2(\ln 2)^2 - 4\ln 2 + 2\end{aligned}$$

**24.** Let $u = \sin(t-s)$, $dv = e^s\,ds \Rightarrow du = -\cos(t-s)\,ds$, $v = e^s$. Then
$I = \int_0^t e^s \sin(t-s)\,ds = [e^s\sin(t-s)]_0^t + \int_0^t e^s\cos(t-s)\,ds = e^t\sin 0 - e^0 \sin t + I_1$. For $I_1$, let $U = \cos(t-s)$, $dV = e^s\,ds \Rightarrow dU = \sin(t-s)\,ds$, $V = e^s$. So $I_1 = [e^s\cos(t-s)]_0^t - \int_0^t e^s\sin(t-s)ds = e^t\cos 0 - e^0\cos t - I$.
Thus, $I = -\sin t + e^t - \cos t - I \Rightarrow 2I = e^t - \cos t - \sin t \Rightarrow I = \frac{1}{2}(e^t - \cos t - \sin t)$.

**25.** Let $w = \sqrt{x}$, so that $x = w^2$ and $dx = 2w\,dw$. Thus, $\int \sin\sqrt{x}\,dx = \int 2w\sin w\,dw$. Now use parts with $u = 2w$, $dv = \sin w\,dw$, $du = 2\,dw$, $v = -\cos w$ to get

$$\begin{aligned}\int 2w\sin w\,dw &= -2w\cos w + \int 2\cos w\,dw = -2w\cos w + 2\sin w + C \\ &= -2\sqrt{x}\cos\sqrt{x} + 2\sin\sqrt{x} + C = 2\left(\sin\sqrt{x} - \sqrt{x}\cos\sqrt{x}\right) + C\end{aligned}$$

**26.** Let $t = x^3$, so that $dt = 3x^2\,dx$. Thus, $\int x^5\cos(x^3)\,dx = \frac{1}{3}\int x^3\cos(x^3)\cdot 3x^2\,dx = \frac{1}{3}\int t\cos t\,dt$.
Now use parts with $u = t$, $dv = \cos t\,dt$, $du = dt$, $v = \sin t$ to get
$\frac{1}{3}\int t\cos t\,dt = \frac{1}{3}\left(t\sin t - \int \sin t\,dt\right) = \frac{1}{3}t\sin t + \frac{1}{3}\cos t + C = \frac{1}{3}x^3\sin(x^3) + \frac{1}{3}\cos(x^3) + C$.

**27.** Let $x = \theta^2$, so that $dx = 2\theta\,d\theta$. Thus, $\int_{\sqrt{\pi/2}}^{\sqrt{\pi}} \theta^3\cos(\theta^2)\,d\theta = \int_{\sqrt{\pi/2}}^{\sqrt{\pi}} \theta^2\cos(\theta^2)\cdot\frac{1}{2}(2\theta\,d\theta) = \frac{1}{2}\int_{\pi/2}^{\pi} x\cos x\,dx$. Now use parts with $u = x$, $dv = \cos x\,dx$, $du = dx$, $v = \sin x$ to get

$$\begin{aligned}\frac{1}{2}\int_{\pi/2}^{\pi} x\cos x\,dx &= \frac{1}{2}\left([x\sin x]_{\pi/2}^{\pi} - \int_{\pi/2}^{\pi}\sin x\,dx\right) = \frac{1}{2}[x\sin x + \cos x]_{\pi/2}^{\pi} \\ &= \frac{1}{2}(\pi\sin\pi + \cos\pi) - \frac{1}{2}\left(\frac{\pi}{2}\sin\frac{\pi}{2} + \cos\frac{\pi}{2}\right) = \frac{1}{2}(\pi\cdot 0 - 1) - \frac{1}{2}\left(\frac{\pi}{2}\cdot 1 + 0\right) = -\frac{1}{2} - \frac{\pi}{4}\end{aligned}$$

**28.** Let $w = \sqrt{x}$, so that $x = w^2$ and $dx = 2w\,dw$. Thus, $\int_1^4 e^{\sqrt{x}}\,dx = \int_1^2 e^w 2w\,dw$. Now use parts with $u = 2w$, $dv = e^w\,dw$, $du = 2\,dw$, $v = e^w$ to get $\int_1^2 e^w 2w\,dw = [2we^w]_1^2 - 2\int_1^2 e^w\,dw = 4e^2 - 2e - 2(e^2 - e) = 2e^2$.

**29.** Let $u = x$, $dv = \cos\pi x\,dx \Rightarrow du = dx$, $v = (\sin\pi x)/\pi$. Then

$$\int x\cos\pi x\,dx = x\cdot\frac{\sin\pi x}{\pi} - \int\frac{\sin\pi x}{\pi}\,dx = \frac{x\sin\pi x}{\pi} + \frac{\cos\pi x}{\pi^2} + C.$$

We see from the graph that this is reasonable, since $F$ has extreme values where $f$ is 0.

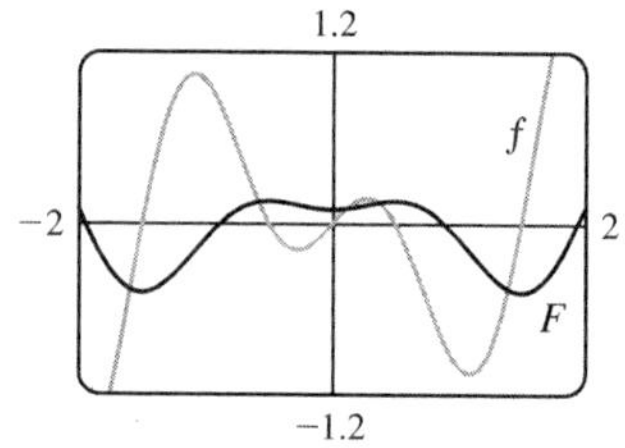


**30.** Let $u = \ln x$, $dv = x^{3/2}\,dx \Rightarrow du = \frac{1}{x}\,dx$, $v = \frac{2}{5}x^{5/2}$. Then

$$\begin{aligned}\int x^{3/2}\ln x\,dx &= \tfrac{2}{5}x^{5/2}\ln x - \tfrac{2}{5}\int x^{3/2}\,dx = \tfrac{2}{5}x^{5/2}\ln x - \left(\tfrac{2}{5}\right)^2 x^{5/2} + C \\ &= \tfrac{2}{5}x^{5/2}\ln x - \tfrac{4}{25}x^{5/2} + C.\end{aligned}$$

We see from the graph that this is reasonable, since $F$ has a minimum where $f$ changes from negative to positive.

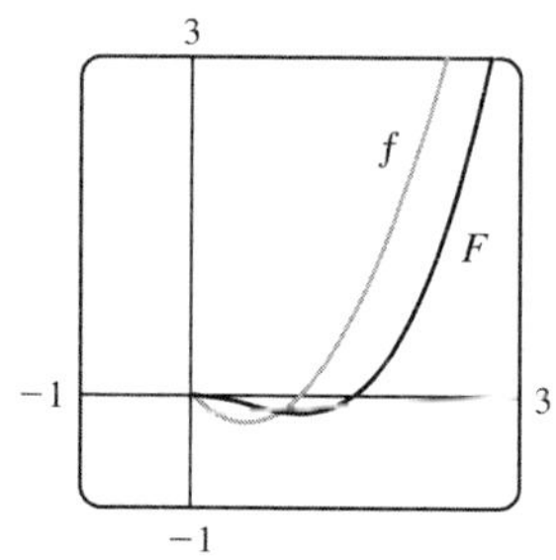


**31.** Let $u = 2x + 3$, $dv = e^x\,dx \Rightarrow du = 2\,dx$, $v = e^x$. Then
$\int(2x+3)e^x\,dx = (2x+3)e^x - 2\int e^x\,dx = (2x+3)e^x - 2e^x + C =$
$(2x+1)\,e^x + C$. We see from the graph that this is reasonable, since $F$ has a minimum where $f$ changes from negative to positive.

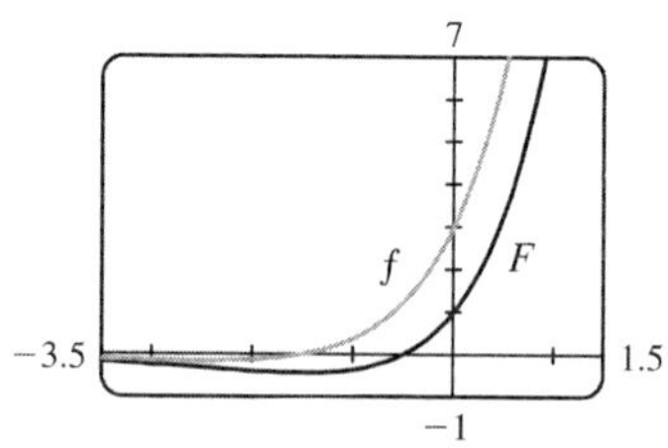

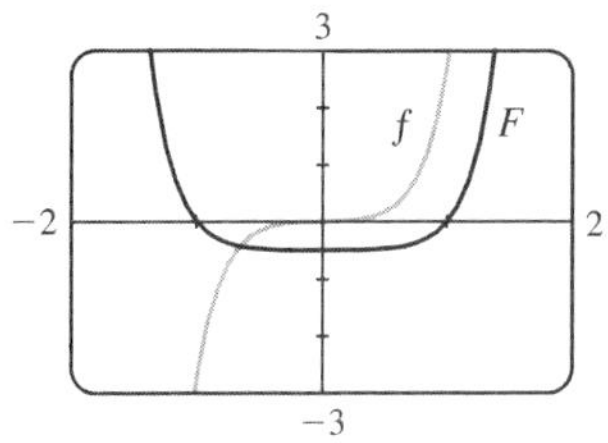


**32.** $\int x^3 e^{x^2}\,dx = \int x^2 \cdot xe^{x^2}\,dx = I$.

Let $u = x^2$, $dv = xe^{x^2}\,dx \;\Rightarrow\; du = 2x\,dx$, $v = \frac{1}{2}e^{x^2}$. Then

$I = \frac{1}{2}x^2 e^{x^2} - \int xe^{x^2}\,dx = \frac{1}{2}x^2 e^{x^2} - \frac{1}{2}e^{x^2} + C = \frac{1}{2}e^{x^2}(x^2-1) + C$.

We see from the graph that this is reasonable, since $F$ has a minimum where $f$ changes from negative to positive.

**33.** (a) Take $n = 2$ in Example 6 to get $\displaystyle\int \sin^2 x\,dx = -\frac{1}{2}\cos x\sin x + \frac{1}{2}\int 1\,dx = \frac{x}{2} - \frac{\sin 2x}{4} + C$.

(b) $\int \sin^4 x\,dx = -\frac{1}{4}\cos x\sin^3 x + \frac{3}{4}\int \sin^2 x\,dx = -\frac{1}{4}\cos x\sin^3 x + \frac{3}{8}x - \frac{3}{16}\sin 2x + C$.

**34.** (a) Let $u = \cos^{n-1} x$, $dv = \cos x\,dx \;\Rightarrow\; du = -(n-1)\cos^{n-2} x\sin x\,dx$, $v = \sin x$ in (2):

$$\begin{aligned}\int \cos^n x\,dx &= \cos^{n-1} x\sin x + (n-1)\int \cos^{n-2} x\sin^2 x\,dx \\ &= \cos^{n-1} x\sin x + (n-1)\int \cos^{n-2} x(1-\cos^2 x)\,dx \\ &= \cos^{n-1} x\sin x + (n-1)\int \cos^{n-2} x\,dx - (n-1)\int \cos^n x\,dx\end{aligned}$$

Rearranging terms gives $n\int \cos^n x\,dx = \cos^{n-1} x\sin x + (n-1)\int \cos^{n-2} x\,dx$ or

$$\int \cos^n x\,dx = \frac{1}{n}\cos^{n-1} x\sin x + \frac{n-1}{n}\int \cos^{n-2} x\,dx$$

(b) Take $n = 2$ in part (a) to get $\int \cos^2 x\,dx = \frac{1}{2}\cos x\sin x + \frac{1}{2}\int 1\,dx = \dfrac{x}{2} + \dfrac{\sin 2x}{4} + C$.

(c) $\int \cos^4 x\,dx = \frac{1}{4}\cos^3 x\sin x + \frac{3}{4}\int \cos^2 x\,dx = \frac{1}{4}\cos^3 x\sin x + \frac{3}{8}x + \frac{3}{16}\sin 2x + C$

**35.** (a) From Example 6, $\displaystyle\int \sin^n x\,dx = -\frac{1}{n}\cos x\sin^{n-1} x + \frac{n-1}{n}\int \sin^{n-2} x\,dx$. Using (6),

$$\begin{aligned}\int_0^{\pi/2} \sin^n x\,dx &= \left[-\frac{\cos x\sin^{n-1} x}{n}\right]_0^{\pi/2} + \frac{n-1}{n}\int_0^{\pi/2} \sin^{n-2} x\,dx \\ &= (0-0) + \frac{n-1}{n}\int_0^{\pi/2} \sin^{n-2} x\,dx = \frac{n-1}{n}\int_0^{\pi/2} \sin^{n-2} x\,dx\end{aligned}$$

(b) Using $n = 3$ in part (a), we have $\int_0^{\pi/2} \sin^3 x\,dx = \frac{2}{3}\int_0^{\pi/2} \sin x\,dx = \left[-\frac{2}{3}\cos x\right]_0^{\pi/2} = \frac{2}{3}$.

Using $n = 5$ in part (a), we have $\int_0^{\pi/2} \sin^5 x\,dx = \frac{4}{5}\int_0^{\pi/2} \sin^3 x\,dx = \frac{4}{5}\cdot\frac{2}{3} = \frac{8}{15}$.

(c) The formula holds for $n = 1$ (that is, $2n+1 = 3$) by (b). Assume it holds for some $k \geq 1$. Then

$\displaystyle\int_0^{\pi/2} \sin^{2k+1} x\,dx = \frac{2\cdot 4\cdot 6\cdot\cdots\cdot(2k)}{3\cdot 5\cdot 7\cdot\cdots\cdot(2k+1)}$. By Example 6,

$$\begin{aligned}\int_0^{\pi/2} \sin^{2k+3} x\,dx &= \frac{2k+2}{2k+3}\int_0^{\pi/2} \sin^{2k+1} x\,dx = \frac{2k+2}{2k+3}\cdot\frac{2\cdot 4\cdot 6\cdot\cdots\cdot(2k)}{3\cdot 5\cdot 7\cdot\cdots\cdot(2k+1)} \\ &= \frac{2\cdot 4\cdot 6\cdot\cdots\cdot(2k)[2(k+1)]}{3\cdot 5\cdot 7\cdot\cdots\cdot(2k+1)[2(k+1)+1]},\end{aligned}$$

so the formula holds for $n = k+1$. By induction, the formula holds for all $n \geq 1$.

**36.** Using Exercise 35(a), we see that the formula holds for $n = 1$, because $\int_0^{\pi/2} \sin^2 x\,dx = \frac{1}{2}\int_0^{\pi/2} 1\,dx = \frac{1}{2}\left[x\right]_0^{\pi/2} = \frac{1}{2}\cdot\frac{\pi}{2}$.

Now assume it holds for some $k \geq 1$. Then $\displaystyle\int_0^{\pi/2} \sin^{2k} x\,dx = \frac{1\cdot 3\cdot 5\cdot\cdots\cdot(2k-1)}{2\cdot 4\cdot 6\cdot\cdots\cdot(2k)}\frac{\pi}{2}$. By Exercise 35(a),

$$\begin{aligned}\int_0^{\pi/2} \sin^{2(k+1)} x\,dx &= \frac{2k+1}{2k+2}\int_0^{\pi/2} \sin^{2k} x\,dx = \frac{2k+1}{2k+2}\cdot\frac{1\cdot 3\cdot 5\cdot\cdots\cdot(2k-1)}{2\cdot 4\cdot 6\cdot\cdots\cdot(2k)}\frac{\pi}{2} \\ &= \frac{1\cdot 3\cdot 5\cdot\cdots\cdot(2k-1)(2k+1)}{2\cdot 4\cdot 6\cdot\cdots\cdot(2k)(2k+2)}\cdot\frac{\pi}{2},\end{aligned}$$

so the formula holds for $n = k+1$. By induction, the formula holds for all $n \geq 1$.

**37.** Let $u = (\ln x)^n$, $dv = dx \;\Rightarrow\; du = n(\ln x)^{n-1}(dx/x)$, $v = x$. By Equation 2,
$\int (\ln x)^n\,dx = x(\ln x)^n - \int nx(\ln x)^{n-1}(dx/x) = x(\ln x)^n - n\int (\ln x)^{n-1}\,dx$.

**38.** Let $u = x^n$, $dv = e^x\,dx \;\Rightarrow\; du = nx^{n-1}\,dx$, $v = e^x$. By Equation 2, $\int x^n e^x\,dx = x^n e^x - n\int x^{n-1}e^x\,dx$.

**39.** Take $n = 3$ in Exercise 37 to get $\int (\ln x)^3\,dx = x\,(\ln x)^3 - 3\int (\ln x)^2\,dx = x(\ln x)^3 - 3x(\ln x)^2 + 6x\ln x - 6x + C$ [by Exercise 23].
*Or*: Instead of using Exercise 23, apply Exercise 37 again with $n = 2$.

**40.** Take $n = 4$ in Exercise 38 to get

$$\begin{aligned}\int x^4 e^x\,dx &= x^4 e^x - 4\int x^3 e^x\,dx = x^4 e^x - 4\left(x^3 - 3x^2 + 6x - 6\right)e^x + C \quad \text{[by Exercise 12]}\\ &= e^x\left(x^4 - 4x^3 + 12x^2 - 24x + 24\right) + C\end{aligned}$$

*Or:* Instead of using Exercise 12, apply Exercise 38 with $n = 3$, then $n = 2$, then $n = 1$.

**41.** Since $v(t) > 0$ for all $t$, the desired distance is $s(t) = \int_0^t v(w)dw = \int_0^t w^2 e^{-w}\,dw$.

First let $u = w^2$, $dv = e^{-w}\,dw \;\Rightarrow\; du = 2w\,dw$, $v = -e^{-w}$. Then $s(t) = \left[-w^2e^{-w}\right]_0^t + 2\int_0^t we^{-w}\,dw$.

Next let $U = w$, $dV = e^{-w}\,dw \;\Rightarrow\; dU = dw$, $V = -e^{-w}$. Then

$$\begin{aligned}s(t) &= -t^2e^{-t} + 2\Big(\left[-we^{-w}\right]_0^t + \int_0^t e^{-w}\,dw\Big) = -t^2e^{-t} + 2\Big(-te^{-t} + 0 + \left[-e^{-w}\right]_0^t\Big)\\ &= -t^2e^{-t} + 2\left(-te^{-t} - e^{-t} + 1\right) = -t^2e^{-t} - 2te^{-t} - 2e^{-t} + 2\\ &= 2 - e^{-t}\left(t^2 + 2t + 2\right) \text{ meters}\end{aligned}$$

**42.** The rocket will have height $H = \int_0^{60} v(t)\,dt$ after 60 seconds.

$$\begin{aligned}H &= \int_0^{60}\left[-gt - v_e\ln\!\left(\frac{m-rt}{m}\right)\right]dt = -g\left[\tfrac{1}{2}t^2\right]_0^{60} - v_e\left[\int_0^{60}\ln(m-rt)\,dt - \int_0^{60}\ln m\,dt\right]\\ &= -g(1800) + v_e(\ln m)(60) - v_e\int_0^{60}\ln(m-rt)\,dt\end{aligned}$$

Let $u = \ln(m-rt)$, $dv = dt \;\Rightarrow\; du = \dfrac{1}{m-rt}(-r)\,dt$, $v = t$. Then

$$\begin{aligned}\int_0^{60}\ln(m-rt)\,dt &= \left[t\ln(m-rt)\right]_0^{60} + \int_0^{60}\frac{rt}{m-rt}\,dt = 60\ln(m-60r) + \int_0^{60}\left(-1 + \frac{m}{m-rt}\right)dt\\ &= 60\ln(m-60r) + \left[-t - \frac{m}{r}\ln(m-rt)\right]_0^{60}\\ &= 60\ln(m-60r) - 60 - \frac{m}{r}\ln(m-60r) + \frac{m}{r}\ln m\end{aligned}$$

So $H = -1800g + 60v_e\ln m - 60v_e\ln(m-60r) + 60v_e + \dfrac{m}{r}v_e\ln(m-60r) - \dfrac{m}{r}v_e\ln m$. Substituting $g = 9.8$, $m = 30{,}000$, $r = 160$, and $v_e = 3000$ gives us $H \approx 14{,}844$ m.

**43.** For $I = \int_1^4 xf''(x)\,dx$, let $u = x$, $dv = f''(x)\,dx \;\Rightarrow\; du = dx$, $v = f'(x)$. Then
$I = \left[xf'(x)\right]_1^4 - \int_1^4 f'(x)\,dx = 4f'(4) - 1\cdot f'(1) - [f(4) - f(1)] = 4\cdot 3 - 1\cdot 5 - (7-2) = 12 - 5 - 5 = 2$.
We used the fact that $f''$ is continuous to guarantee that $I$ exists.

**44.** (a) Take $g(x) = x$ and $g'(x) = 1$ in Equation 1.

(b) By part (a), $\int_a^b f(x)\,dx = bf(b) - a\,f(a) - \int_a^b x\,f'(x)\,dx$. Now let $y = f(x)$, so that $x = g(y)$ and $dy = f'(x)\,dx$. Then $\int_a^b x\,f'(x)\,dx = \int_{f(a)}^{f(b)} g(y)\,dy$. The result follows.

(c) Part (b) says that the area of region $ABFC$ is

$$= \quad bf(b) \quad - \quad af(a) \quad - \quad \int_{f(a)}^{f(b)} g(y)\,dy$$

$$= (\text{area of rectangle } OBFE) - (\text{area of rectangle } OACD) - (\text{area of region } DCFE)$$

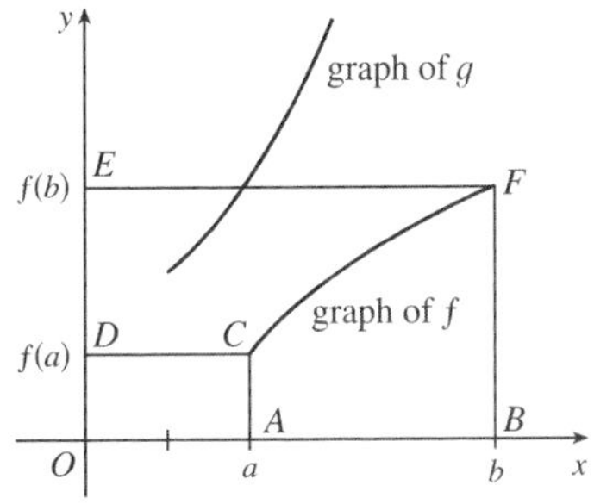


(d) We have $f(x) = \ln x$, so $f^{-1}(x) = e^x$, and since $g = f^{-1}$, we have $g(y) = e^y$. By part (b),

$\int_1^e \ln x\,dx = e\ln e - 1\ln 1 - \int_{\ln 1}^{\ln e} e^y\,dy = e - \int_0^1 e^y\,dy = e - \left[e^y\right]_0^1 = e - (e-1) = 1.$

**45.** Suppose $f(0) = g(0) = 0$ and let $u = f(x)$, $dv = g''(x)\,dx \Rightarrow du = f'(x)\,dx$, $v = g'(x)$.

Then $\int_0^a f(x)g''(x)\,dx = \left[f(x)g'(x)\right]_0^a - \int_0^a f'(x)g'(x)\,dx = f(a)g'(a) - \int_0^a f'(x)g'(x)\,dx.$

Now let $U = f'(x)$, $dV = g'(x)\,dx \Rightarrow dU = f''(x)\,dx$ and $V = g(x)$, so

$\int_0^a f'(x)g'(x)\,dx = \left[f'(x)g(x)\right]_0^a - \int_0^a f''(x)g(x)\,dx = f'(a)g(a) - \int_0^a f''(x)g(x)\,dx.$

Combining the two results, we get $\int_0^a f(x)g''(x)\,dx = f(a)g'(a) - f'(a)g(a) + \int_0^a f''(x)g(x)\,dx.$

**46.** (a) We note that for $0 \le x \le \frac{\pi}{2}$, $0 \le \sin x \le 1$, so $\sin^{2n+2} x \le \sin^{2n+1} x \le \sin^{2n} x$. So by the second Comparison Property of the Integral, $I_{2n+2} \le I_{2n+1} \le I_{2n}$.

(b) Substituting directly into the result from Exercise 36, we get

$$\frac{I_{2n+2}}{I_{2n}} = \frac{\dfrac{1\cdot 3\cdot 5\cdot\cdots\cdot[2(n+1)-1]}{2\cdot 4\cdot 6\cdot\cdots\cdot[2(n+1)]}\dfrac{\pi}{2}}{\dfrac{1\cdot 3\cdot 5\cdot\cdots\cdot(2n-1)}{2\cdot 4\cdot 6\cdot\cdots\cdot(2n)}\dfrac{\pi}{2}} = \frac{2(n+1)-1}{2(n+1)} = \frac{2n+1}{2n+2}$$

(c) We divide the result from part (a) by $I_{2n}$. The inequalities are preserved since $I_{2n}$ is positive: $\dfrac{I_{2n+2}}{I_{2n}} \le \dfrac{I_{2n+1}}{I_{2n}} \le \dfrac{I_{2n}}{I_{2n}}$.

Now from part (b), the left term is equal to $\dfrac{2n+1}{2n+2}$, so the expression becomes $\dfrac{2n+1}{2n+2} \le \dfrac{I_{2n+1}}{I_{2n}} \le 1$. Now

$\lim\limits_{n\to\infty}\dfrac{2n+1}{2n+2} = \lim\limits_{n\to\infty} 1 = 1$, so by the Squeeze Theorem, $\lim\limits_{n\to\infty}\dfrac{I_{2n+1}}{I_{2n}} = 1$.

(d) We substitute the results from Exercises 35 and 36 into the result from part (c):

$$1 = \lim_{n\to\infty}\frac{I_{2n+1}}{I_{2n}} = \lim_{n\to\infty}\frac{\dfrac{2\cdot 4\cdot 6\cdot\cdots\cdot(2n)}{3\cdot 5\cdot 7\cdot\cdots\cdot(2n+1)}}{\dfrac{1\cdot 3\cdot 5\cdot\cdots\cdot(2n-1)}{2\cdot 4\cdot 6\cdot\cdots\cdot(2n)}\dfrac{\pi}{2}}$$

$$= \lim_{n\to\infty}\left[\frac{2\cdot 4\cdot 6\cdot\cdots\cdot(2n)}{3\cdot 5\cdot 7\cdot\cdots\cdot(2n+1)}\right]\left[\frac{2\cdot 4\cdot 6\cdot\cdots\cdot(2n)}{1\cdot 3\cdot 5\cdot\cdots\cdot(2n-1)}\left(\frac{2}{\pi}\right)\right]$$

$$= \lim_{n\to\infty}\frac{2}{1}\cdot\frac{2}{3}\cdot\frac{4}{3}\cdot\frac{4}{5}\cdot\frac{6}{5}\cdot\frac{6}{7}\cdot\cdots\cdot\frac{2n}{2n-1}\cdot\frac{2n}{2n+1}\cdot\frac{2}{\pi} \qquad \text{[rearrange terms]}$$

Multiplying both sides by $\frac{\pi}{2}$ gives us the *Wallis product*:

$$\frac{\pi}{2} = \frac{2}{1}\cdot\frac{2}{3}\cdot\frac{4}{3}\cdot\frac{4}{5}\cdot\frac{6}{5}\cdot\frac{6}{7}\cdot\cdots$$

(e) The area of the $k$th rectangle is $k$. At the $2n$th step, the area is increased from $2n-1$ to $2n$ by multiplying the width by $\dfrac{2n}{2n-1}$, and at the $(2n+1)$th step, the area is increased from $2n$ to $2n+1$ by multiplying the height by $\dfrac{2n+1}{2n}$. These two steps multiply the ratio of width to height by $\dfrac{2n}{2n-1}$ and $\dfrac{1}{(2n+1)/(2n)} = \dfrac{2n}{2n+1}$ respectively. So, by part (d), the limiting ratio is $\dfrac{2}{1}\cdot\dfrac{2}{3}\cdot\dfrac{4}{3}\cdot\dfrac{4}{5}\cdot\dfrac{6}{5}\cdot\dfrac{6}{7}\cdot\cdots = \dfrac{\pi}{2}$.

## 5.7 Additional Techniques of Integration

The symbols $\overset{s}{=}$ and $\overset{c}{=}$ indicate the use of the substitutions $\{u=\sin x,\, du=\cos x\,dx\}$ and $\{u=\cos x,\, du=-\sin x\,dx\}$, respectively.

**1.** $\int \sin^3 x\,\cos^2 x\,dx = \int \sin^2 x\,\cos^2 x\,\sin x\,dx = \int (1-\cos^2 x)\cos^2 x\,\sin x\,dx \overset{c}{=} \int (1-u^2)u^2(-du)$
$= \int (u^2-1)u^2\,du = \int (u^4-u^2)\,du = \frac{1}{5}u^5 - \frac{1}{3}u^3 + C = \frac{1}{5}\cos^5 x - \frac{1}{3}\cos^3 x + C$

**2.** $\int_0^{\pi/2} \cos^5 x\,dx = \int_0^{\pi/2} (\cos^2 x)^2 \cos x\,dx = \int_0^{\pi/2} (1-\sin^2 x)^2 \cos x\,dx \overset{s}{=} \int_0^1 (1-u^2)^2\,du$
$= \int_0^1 (1-2u^2+u^4)\,du = \left[u - \frac{2}{3}u^3 + \frac{1}{5}u^5\right]_0^1 = \left(1-\frac{2}{3}+\frac{1}{5}\right) - 0 = \frac{8}{15}$

**3.** $\int_{\pi/2}^{3\pi/4} \sin^5 x\,\cos^3 x\,dx = \int_{\pi/2}^{3\pi/4} \sin^5 x\,\cos^2 x\,\cos x\,dx = \int_{\pi/2}^{3\pi/4} \sin^5 x\,(1-\sin^2 x)\cos x\,dx$
$\overset{s}{=} \int_1^{\sqrt{2}/2} u^5(1-u^2)\,du = \int_1^{\sqrt{2}/2} (u^5-u^7)\,du = \left[\frac{1}{6}u^6 - \frac{1}{8}u^8\right]_1^{\sqrt{2}/2}$
$= \left(\frac{1/8}{6} - \frac{1/16}{8}\right) - \left(\frac{1}{6}-\frac{1}{8}\right) = -\frac{11}{384}$

**4.** $\int \sin^3(mx)\,dx = \int (1-\cos^2 mx)\sin mx\,dx = -\frac{1}{m}\int (1-u^2)\,du \quad [u=\cos mx,\, du=-m\sin mx\,dx]$
$= -\frac{1}{m}\left(u-\frac{1}{3}u^3\right) + C = -\frac{1}{m}\left(\cos mx - \frac{1}{3}\cos^3 mx\right) + C$
$= \frac{1}{3m}\cos^3 mx - \frac{1}{m}\cos mx + C$

**5.** $\int_0^{2\pi} \cos^2(6\theta)\,d\theta = \frac{1}{2}\int_0^{2\pi} [1+\cos(12\theta)]\,d\theta = \frac{1}{2}\left[\theta + \frac{1}{12}\sin(12\theta)\right]_0^{2\pi} = \frac{1}{2}[(2\pi+0)-(0+0)] = \pi$

**6.** $\int_0^{\pi/2} \sin^2 x\,\cos^2 x\,dx = \int_0^{\pi/2} \frac{1}{4}(4\sin^2 x\,\cos^2 x)\,dx = \int_0^{\pi/2} \frac{1}{4}(2\sin x\,\cos x)^2 dx = \frac{1}{4}\int_0^{\pi/2} \sin^2 2x\,dx$
$= \frac{1}{4}\int_0^{\pi/2} \frac{1}{2}(1-\cos 4x)\,dx = \frac{1}{8}\int_0^{\pi/2} (1-\cos 4x)\,dx = \frac{1}{8}\left[x - \frac{1}{4}\sin 4x\right]_0^{\pi/2}$
$= \frac{1}{8}\left(\frac{\pi}{2}\right) = \frac{\pi}{16}$

**7.** Let $u=\sec x$. Then $du = \sec x\,\tan x\,dx$, so

$$\int \tan^3 x\,\sec x\,dx = \int (\tan^2 x)(\tan x\,\sec x)\,dx = \int (\sec^2 x - 1)(\sec x\,\tan x\,dx)$$
$$= \int (u^2-1)\,du = \tfrac{1}{3}u^3 - u + C = \tfrac{1}{3}\sec^3 x - \sec x + C$$

**8.** Let $u = \tan x$. Then $du = \sec^2 x\,dx$, so

$$\int_0^{\pi/4} \tan^2 x\,\sec^4 x\,dx = \int_0^{\pi/4} \tan^2 x\,\sec^2 x(\sec^2 x\,dx) = \int_0^{\pi/4} \tan^2 x(1+\tan^2 x)(\sec^2 x\,dx)$$
$$= \int_0^1 u^2(1+u^2)\,du = \int_0^1 (u^2+u^4)\,du$$
$$= \left[\tfrac{1}{3}u^3 + \tfrac{1}{5}u^5\right]_0^1 = \tfrac{1}{3} + \tfrac{1}{5} = \tfrac{8}{15}$$

**9.** $x = 3\sin\theta$, where $-\pi/2 \le \theta \le \pi/2$. Then $dx = 3\cos\theta\, d\theta$ and $\sqrt{9 - x^2} = \sqrt{9 - 9\sin^2\theta} = \sqrt{9\cos^2\theta} = 3\,|\cos\theta| = 3\cos\theta$. (Note that $\cos\theta \ge 0$ because $-\pi/2 \le \theta \le \pi/2$.) Thus, substitution gives

$$\int \frac{\sqrt{9-x^2}}{x^2}\,dx = \int \frac{3\cos\theta}{9\sin^2\theta}\,3\cos\theta\,d\theta = \int \frac{\cos^2\theta}{\sin^2\theta}\,d\theta = \int \cot^2\theta\,d\theta$$

$$= \int (\csc^2\theta - 1)\,d\theta = -\cot\theta - \theta + C$$

Since this is an indefinite integral, we must return to the original variable $x$. This can be done either by using trigonometric identities to express $\cot\theta$ in terms of $\sin\theta = x/3$ or by drawing a diagram, as shown, where $\theta$ is interpreted as an angle of a right triangle.

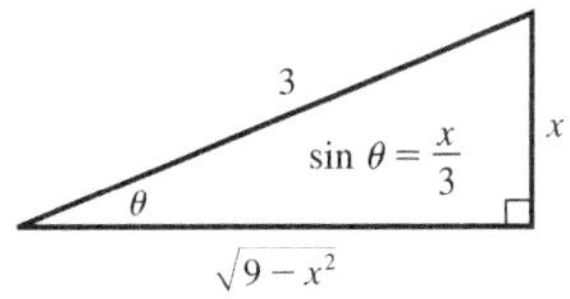


Since $\sin\theta = x/3$, we label the opposite side and the hypotenuse as having lengths $x$ and 3. Then the Pythagorean Theorem gives the length of the adjacent side as $\sqrt{9 - x^2}$, so we can simply read the value of $\cot\theta$ from the figure: $\cot\theta = \dfrac{\sqrt{9-x^2}}{x}$. (Although $\theta > 0$ in the diagram, this expression for $\cot\theta$ is valid even when $\theta < 0$.) Since $\sin\theta = x/3$, we have $\theta = \sin^{-1}(x/3)$ and so $\displaystyle\int \frac{\sqrt{9-x^2}}{x^2}\,dx = -\frac{\sqrt{9-x^2}}{x} - \sin^{-1}\!\left(\frac{x}{3}\right) + C.$

**10.** $x = \sec\theta$, where $0 \le \theta < \pi/2$ or $\pi \le \theta < 3\pi/2$. Then $dx = \sec\theta\,\tan\theta\,d\theta$ and $\sqrt{x^2 - 1} = \sqrt{\sec^2\theta - 1} = \sqrt{\tan^2\theta} = |\tan\theta| = \tan\theta$ (since $\tan\theta \ge 0$ for the specified values of $\theta$). Thus, substitution gives

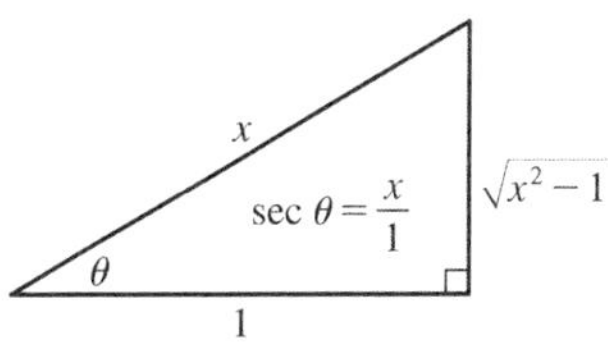


$$\int \frac{\sqrt{x^2-1}}{x^4}\,dx = \int \frac{\tan\theta}{\sec^4\theta}\sec\theta\,\tan\theta\,d\theta = \int \frac{\tan^2\theta}{\sec^3\theta}\,d\theta = \int \frac{\sin^2\theta}{\cos^2\theta}\cdot\cos^3\theta\,d\theta$$

$$= \int \sin^2\theta\,\cos\theta\,d\theta = \int u^2\,du \quad [u = \sin\theta,\ du = \cos\theta\,d\theta]$$

$$= \frac{1}{3}u^3 + C = \frac{1}{3}\sin^3\theta + C = \frac{1}{3}\left(\frac{\sqrt{x^2-1}}{x}\right)^3 + C = \frac{(x^2-1)^{3/2}}{3x^3} + C$$

**11.** $x = 2\tan\theta$, where $-\pi/2 < \theta < \pi/2$. Then $dx = 2\sec^2\theta\,d\theta$ and

$$\sqrt{x^2+4} = \sqrt{(2\tan\theta)^2 + 4} = \sqrt{4\tan^2\theta + 4}$$
$$= \sqrt{4(\tan^2\theta + 1)} = 2\sqrt{\sec^2\theta} = 2\,|\sec\theta|$$
$$= 2\sec\theta \quad (\text{since } \sec\theta \ge 0 \text{ for } -\pi/2 < \theta < \pi/2).$$

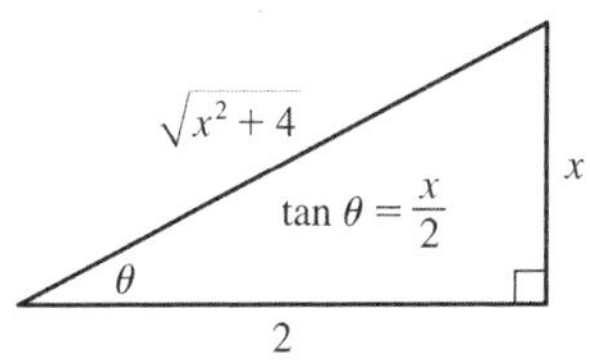


Thus, substitution gives

$$\int \frac{1}{x^2\sqrt{x^2+4}}\,dx = \int \frac{1}{4\tan^2\theta\,(2\sec\theta)}\left(2\sec^2\theta\,d\theta\right) = \frac{1}{4}\int \frac{\sec\theta}{\tan^2\theta}\,d\theta = \frac{1}{4}\int \frac{1}{\cos\theta}\cdot\frac{\cos^2\theta}{\sin^2\theta}\,d\theta$$

$$= \frac{1}{4}\int \frac{\cos\theta}{\sin^2\theta}\,d\theta = \frac{1}{4}\int \frac{1}{u^2}\,du \quad [u = \sin\theta,\ du = \cos\theta\,d\theta]$$

$$= \frac{1}{4}\left(-\frac{1}{u}\right) + C = -\frac{1}{4}\,\frac{1}{\sin\theta} + C = -\frac{1}{4}\cdot\frac{\sqrt{x^2+4}}{x} + C = -\frac{\sqrt{x^2+4}}{4x} + C$$

**12.** (a) $\dfrac{d}{d\theta}\left[\frac{1}{2}(\sec\theta\,\tan\theta + \ln|\sec\theta + \tan\theta|) + C\right]$

$$= \frac{1}{2}\left(\sec\theta\cdot\sec^2\theta + \tan\theta\cdot\sec\theta\,\tan\theta + \frac{1}{\sec\theta + \tan\theta}\cdot\sec\theta\,\tan\theta + \sec^2\theta\right) + 0$$
$$= \frac{1}{2}\left[\sec\theta\left(\sec^2\theta + \tan^2\theta\right) + \frac{\sec\theta\,(\tan\theta + \sec\theta)}{\sec\theta + \tan\theta}\right]$$
$$= \tfrac{1}{2}\left[\sec\theta\left(\sec^2\theta + \sec^2\theta - 1\right) + \sec\theta\right] = \tfrac{1}{2}\sec\theta\left[\left(2\sec^2\theta - 1\right) + 1\right]$$
$$= \tfrac{1}{2}\sec\theta\left(2\sec^2\theta\right) = \sec^3\theta.$$

Thus, $\int \sec^3\theta\,d\theta = \frac{1}{2}(\sec\theta\,\tan\theta + \ln|\sec\theta + \tan\theta|) + C$.

(b) As in Exercise 11, we use the substitution $x = \tan\theta$, where $-\pi/2 < \theta < \pi/2$. Then $dx = \sec^2\theta\,d\theta$ and $\sqrt{x^2+1} = \sqrt{\tan^2\theta + 1} = \sqrt{\sec^2\theta} = |\sec\theta| = \sec\theta$ (since $\sec\theta \geq 0$ for $-\pi/2 < \theta < \pi/2$). When $x = 0$, $\tan\theta = 0 \Rightarrow \theta = 0$, and when $x = 1$, $\tan\theta = 1 \Rightarrow \theta = \frac{\pi}{4}$. Thus, substitution gives

$$\int_0^1 \sqrt{x^2+1}\,dx = \int_0^{\pi/4} \sec\theta\left(\sec^2\theta\,d\theta\right) = \int_0^{\pi/4} \sec^3\theta\,d\theta = \left[\tfrac{1}{2}\left(\sec\theta\,\tan\theta + \ln|\sec\theta + \tan\theta|\right)\right]_0^{\pi/4}$$
$$= \tfrac{1}{2}\left[\left(\sqrt{2}\cdot 1 + \ln\left|\sqrt{2} + 1\right|\right) - \left(1\cdot 0 + \ln|1 + 0|\right)\right]$$
$$= \tfrac{1}{2}\left[\sqrt{2} + \ln\left(\sqrt{2} + 1\right)\right]$$

**13.** Let $t = \sec\theta$, so $dt = \sec\theta\tan\theta\,d\theta$, $t = \sqrt{2} \Rightarrow \theta = \frac{\pi}{4}$, and $t = 2 \Rightarrow \theta = \frac{\pi}{3}$. Then

$$\int_{\sqrt{2}}^{2} \frac{1}{t^3\sqrt{t^2-1}}\,dt = \int_{\pi/4}^{\pi/3} \frac{1}{\sec^3\theta\tan\theta}\sec\theta\tan\theta\,d\theta = \int_{\pi/4}^{\pi/3} \frac{1}{\sec^2\theta}\,d\theta = \int_{\pi/4}^{\pi/3} \cos^2\theta\,d\theta$$
$$= \int_{\pi/4}^{\pi/3} \tfrac{1}{2}(1 + \cos 2\theta)\,d\theta = \tfrac{1}{2}\left[\theta + \tfrac{1}{2}\sin 2\theta\right]_{\pi/4}^{\pi/3}$$
$$= \tfrac{1}{2}\left[\left(\tfrac{\pi}{3} + \tfrac{1}{2}\tfrac{\sqrt{3}}{2}\right) - \left(\tfrac{\pi}{4} + \tfrac{1}{2}\cdot 1\right)\right] = \tfrac{1}{2}\left(\tfrac{\pi}{12} + \tfrac{\sqrt{3}}{4} - \tfrac{1}{2}\right) = \tfrac{\pi}{24} + \tfrac{\sqrt{3}}{8} - \tfrac{1}{4}$$

**14.** Let $x = 4\sin\theta$, where $-\pi/2 \leq \theta \leq \pi/2$. Then $dx = 4\cos\theta\,d\theta$ and $\sqrt{16 - x^2} = \sqrt{16 - 16\sin^2\theta} = \sqrt{16\cos^2\theta} = 4|\cos\theta| = 4\cos\theta$. When $x = 0$, $4\sin\theta = 0 \Rightarrow \theta = 0$, and when $x = 2\sqrt{3}$, $4\sin\theta = 2\sqrt{3} \Rightarrow \sin\theta = \frac{\sqrt{3}}{2} \Rightarrow \theta = \frac{\pi}{3}$. Thus, substitution gives

$$\int_0^{2\sqrt{3}} \frac{x^3}{\sqrt{16 - x^2}}\,dx = \int_0^{\pi/3} \frac{4^3\sin^3\theta}{4\cos\theta} 4\cos\theta\,d\theta = 4^3\int_0^{\pi/3} \sin^3\theta\,d\theta$$
$$= 4^3\int_0^{\pi/3} \left(1 - \cos^2\theta\right)\sin\theta\,d\theta$$
$$\overset{c}{=} -4^3\int_1^{1/2} \left(1 - u^2\right)du = -64\left[u - \tfrac{1}{3}u^3\right]_1^{1/2}$$
$$= -64\left[\left(\tfrac{1}{2} - \tfrac{1}{24}\right) - \left(1 - \tfrac{1}{3}\right)\right] = -64\left(-\tfrac{5}{24}\right) = \tfrac{40}{3}$$

*Or:* Let $u = 16 - x^2$, $x^2 = 16 - u$, $du = -2x\,dx$.

**15.** (a) $\dfrac{2}{x^2 + 3x - 4} = \dfrac{2}{(x+4)(x-1)} = \dfrac{A}{x+4} + \dfrac{B}{x-1}$

(b) $x^2 + x + 1$ is irreducible, so $\dfrac{x^2}{(x-1)(x^2+x+1)} = \dfrac{A}{x-1} + \dfrac{Bx + C}{x^2 + x + 1}$.

**16.** (a) $\dfrac{x-1}{x^3 + x^2} = \dfrac{x-1}{x^2(x+1)} = \dfrac{A}{x} + \dfrac{B}{x^2} + \dfrac{C}{x+1}$

(b) $\dfrac{x-1}{x^3 + x} = \dfrac{x-1}{x(x^2+1)} = \dfrac{A}{x} + \dfrac{Bx + C}{x^2 + 1}$

**17.** $\dfrac{x-9}{(x+5)(x-2)} = \dfrac{A}{x+5} + \dfrac{B}{x-2}$. Multiply both sides by $(x+5)(x-2)$ to get $x-9 = A(x-2) + B(x+5)\,(*)$, or equivalently, $x - 9 = (A+B)x - 2A + 5B$. Equating coefficients of $x$ on each side of the equation gives us $1 = A + B$ **(1)** and equating constants gives us $-9 = -2A + 5B$ **(2)**. Adding two times **(1)** to **(2)** gives us $-7 = 7B \Leftrightarrow B = -1$ and hence, $A = 2$. [Alternatively, to find the coefficients $A$ and $B$, we may use substitution as follows: substitute 2 for $x$ in $(*)$ to get $-7 = 7B \Leftrightarrow B = -1$, then substitute $-5$ for $x$ in $(*)$ to get $-14 = -7A \Leftrightarrow A = 2$.] Thus,

$$\int \frac{x-9}{(x+5)(x-2)}\,dx = \int \left(\frac{2}{x+5} + \frac{-1}{x-2}\right) dx = 2\ln|x+5| - \ln|x-2| + C.$$

To find the constants in problems involving partial fractions, we may use the coefficient comparison method or the substitution method (as in the solution for Exercise 17) or a combination of both methods.

**18.** $\dfrac{x-1}{x^2+3x+2} = \dfrac{A}{x+1} + \dfrac{B}{x+2}$. Multiply both sides by $(x+1)(x+2)$ to get $x - 1 = A(x+2) + B(x+1)$. Substituting $-2$ for $x$ gives $-3 = -B \Leftrightarrow B = 3$. Substituting $-1$ for $x$ gives $-2 = A$. Thus,

$$\begin{aligned}\int_0^1 \frac{x-1}{x^2+3x+2}\,dx &= \int_0^1 \left(\frac{-2}{x+1} + \frac{3}{x+2}\right) dx = \big[-2\ln|x+1| + 3\ln|x+2|\big]_0^1 \\ &= (-2\ln 2 + 3\ln 3) - (-2\ln 1 + 3\ln 2) = 3\ln 3 - 5\ln 2 \quad \left[\text{or } \ln\tfrac{27}{32}\right]\end{aligned}$$

**19.** $\dfrac{1}{x^2-1} = \dfrac{1}{(x+1)(x-1)} = \dfrac{A}{x+1} + \dfrac{B}{x-1}$. Multiply both sides by $(x+1)(x-1)$ to get $1 = A(x-1) + B(x+1)$. Substituting 1 for $x$ gives $1 = 2B \Leftrightarrow B = \frac{1}{2}$. Substituting $-1$ for $x$ gives $1 = -2A \Leftrightarrow A = -\frac{1}{2}$. Thus,

$$\begin{aligned}\int_2^3 \frac{1}{x^2-1}\,dx &= \int_2^3 \left(\frac{-1/2}{x+1} + \frac{1/2}{x-1}\right) dx = \left[-\tfrac{1}{2}\ln|x+1| + \tfrac{1}{2}\ln|x-1|\right]_2^3 \\ &= \left(-\tfrac{1}{2}\ln 4 + \tfrac{1}{2}\ln 2\right) - \left(-\tfrac{1}{2}\ln 3 + \tfrac{1}{2}\ln 1\right) = \tfrac{1}{2}(\ln 2 + \ln 3 - \ln 4) \quad \left[\text{or } \tfrac{1}{2}\ln\tfrac{3}{2}\right]\end{aligned}$$

**20.** $\dfrac{x^2+2x-1}{x^3-x} = \dfrac{x^2+2x-1}{x(x+1)(x-1)} = \dfrac{A}{x} + \dfrac{B}{x+1} + \dfrac{C}{x-1}$. Multiply both sides by $x(x+1)(x-1)$ to get $x^2 + 2x - 1 = A(x+1)(x-1) + Bx(x-1) + Cx(x+1)$. Substituting 0 for $x$ gives $-1 = -A \Leftrightarrow A = 1$. Substituting $-1$ for $x$ gives $-2 = 2B \Leftrightarrow B = -1$. Substituting 1 for $x$ gives $2 = 2C \Leftrightarrow C = 1$. Thus,

$$\int \frac{x^2+2x-1}{x^3-x}\,dx = \int \left(\frac{1}{x} - \frac{1}{x+1} + \frac{1}{x-1}\right) dx = \ln|x| - \ln|x+1| + \ln|x-1| + C = \ln\left|\frac{x(x-1)}{x+1}\right| + C.$$

**21.** $\dfrac{10}{(x-1)(x^2+9)} = \dfrac{A}{x-1} + \dfrac{Bx+C}{x^2+9}$. Multiply both sides by $(x-1)(x^2+9)$ to get $10 = A(x^2+9) + (Bx+C)(x-1)$ $(*)$. Substituting 1 for $x$ gives $10 = 10A \Leftrightarrow A = 1$. Substituting 0 for $x$ gives $10 = 9A - C \Rightarrow C = 9(1) - 10 = -1$. The coefficients of the $x^2$-terms in $(*)$ must be equal, so $0 = A + B \Rightarrow B = -1$. Thus,

$$\begin{aligned}\int \frac{10}{(x-1)(x^2+9)}\,dx &= \int \left(\frac{1}{x-1} + \frac{-x-1}{x^2+9}\right) dx = \int \left(\frac{1}{x-1} - \frac{x}{x^2+9} - \frac{1}{x^2+9}\right) dx \\ &= \ln|x-1| - \tfrac{1}{2}\ln(x^2+9)\ \ [\text{let } u = x^2+9] \ - \tfrac{1}{3}\tan^{-1}\left(\tfrac{x}{3}\right)\ [\text{Formula 10}] + C\end{aligned}$$

**22.** $\dfrac{2x^2+5}{(x^2+1)(x^2+4)} = \dfrac{Ax+B}{x^2+1} + \dfrac{Cx+D}{x^2+4}$. Multiply both sides by $(x^2+1)(x^2+4)$ to get

$2x^2 + 5 = (Ax+B)(x^2+4) + (Cx+D)(x^2+1) \Leftrightarrow$

$2x^2 + 5 = (Ax^3 + Bx^2 + 4Ax + 4B) + (Cx^3 + Dx^2 + Cx + D) \Leftrightarrow$

$2x^2 + 5 = (A+C)x^3 + (B+D)x^2 + (4A+C)x + (4B+D)$. Comparing coefficients gives us $A + C = 0$, $B + D = 2$, $4A + C = 0$, and $4B + D = 5$. Solving gives us $A = C = 0$ and $B = D = 1$. Thus,

$$\int \frac{2x^2+5}{(x^2+1)(x^2+4)}\,dx = \int \left(\frac{1}{x^2+1} + \frac{1}{x^2+4}\right) dx = \tan^{-1}x + \tfrac{1}{2}\tan^{-1}\left(\tfrac{x}{2}\right) + C.$$

**23.** $\dfrac{x^3+x^2+2x+1}{(x^2+1)(x^2+2)} = \dfrac{Ax+B}{x^2+1} + \dfrac{Cx+D}{x^2+2}$. Multiply both sides by $(x^2+1)(x^2+2)$ to get

$x^3+x^2+2x+1 = (Ax+B)(x^2+2)+(Cx+D)(x^2+1) \quad \Leftrightarrow$

$x^3+x^2+2x+1 = (Ax^3+Bx^2+2Ax+2B)+(Cx^3+Dx^2+Cx+D) \quad \Leftrightarrow$

$x^3+x^2+2x+1 = (A+C)x^3+(B+D)x^2+(2A+C)x+(2B+D)$. Comparing coefficients gives us the following system of equations:

$$A+C=1 \quad \textbf{(1)} \qquad B+D=1 \quad \textbf{(2)}$$
$$2A+C=2 \quad \textbf{(3)} \qquad 2B+D=1 \quad \textbf{(4)}$$

Subtracting equation **(1)** from equation **(3)** gives us $A=1$, so $C=0$. Subtracting equation **(2)** from equation **(4)** gives us $B=0$, so $D=1$. Thus, $I=\displaystyle\int\frac{x^3+x^2+2x+1}{(x^2+1)(x^2+2)}\,dx=\int\left(\frac{x}{x^2+1}+\frac{1}{x^2+2}\right)dx$. For $\displaystyle\int\frac{x}{x^2+1}\,dx$, let $u=x^2+1$ so $du=2x\,dx$ and then $\displaystyle\int\frac{x}{x^2+1}\,dx=\frac{1}{2}\int\frac{1}{u}\,du=\frac{1}{2}\ln|u|+C=\frac{1}{2}\ln(x^2+1)+C$. For $\displaystyle\int\frac{1}{x^2+2}\,dx$, use Formula 10 with $a=\sqrt{2}$. So $\displaystyle\int\frac{1}{x^2+2}\,dx=\int\frac{1}{x^2+(\sqrt{2})^2}\,dx=\frac{1}{\sqrt{2}}\tan^{-1}\frac{x}{\sqrt{2}}+C$.

Thus, $I=\dfrac{1}{2}\ln(x^2+1)+\dfrac{1}{\sqrt{2}}\tan^{-1}\dfrac{x}{\sqrt{2}}+C$.

**24.** $\dfrac{x^2-x+6}{x^3+3x}=\dfrac{x^2-x+6}{x(x^2+3)}=\dfrac{A}{x}+\dfrac{Bx+C}{x^2+3}$. Multiply by $x(x^2+3)$ to get $x^2-x+6=A(x^2+3)+(Bx+C)x$.

Substituting 0 for $x$ gives $6=3A \quad \Leftrightarrow \quad A=2$. The coefficients of the $x^2$-terms must be equal, so $1=A+B \quad \Rightarrow$ $B=1-2=-1$. The coefficients of the $x$-terms must be equal, so $-1=C$. Thus,

$$\int\frac{x^2-x+6}{x^3+3x}\,dx=\int\left(\frac{2}{x}+\frac{-x-1}{x^2+3}\right)dx=\int\left(\frac{2}{x}-\frac{x}{x^2+3}-\frac{1}{x^2+3}\right)dx$$
$$=2\ln|x|-\frac{1}{2}\ln(x^2+3)-\frac{1}{\sqrt{3}}\tan^{-1}\frac{x}{\sqrt{3}}+C$$

**25.** $\displaystyle\int\frac{x}{x-6}\,dx=\int\frac{(x-6)+6}{x-6}dx=\int\left(1+\frac{6}{x-6}\right)dx=x+6\ln|x-6|+C$

**26.** $\displaystyle\int\frac{r^2}{r+4}\,dr=\int\left(\frac{r^2-16}{r+4}+\frac{16}{r+4}\right)dr=\int\left(r-4+\frac{16}{r+4}\right)dr$ [or use long division]

$=\frac{1}{2}r^2-4r+16\ln|r+4|+C$

**27.**

$$\begin{array}{r|l} & x \\ \hline x^2+4 & x^3+0x^2+0x+4 \\ & x^3 \qquad\quad +4x \\ \hline & \qquad -4x+4 \end{array}$$

By long division, $\dfrac{x^3+4}{x^2+4}=x+\dfrac{-4x+4}{x^2+4}$. Thus,

$$\int\frac{x^3+4}{x^2+4}\,dx=\int\left(x+\frac{-4x+4}{x^2+4}\right)dx$$
$$=\int\left(x-\frac{4x}{x^2+4}+\frac{4}{x^2+2^2}\right)dx=\frac{1}{2}x^2-4\cdot\frac{1}{2}\ln|x^2+4|+4\cdot\frac{1}{2}\tan^{-1}\left(\frac{x}{2}\right)+C$$
$$=\frac{1}{2}x^2-2\ln(x^2+4)+2\tan^{-1}\left(\frac{x}{2}\right)+C$$

**28.** $\dfrac{x^3-4x-10}{x^2-x-6}=x+1+\dfrac{3x-4}{(x-3)(x+2)}$. Write $\dfrac{3x-4}{(x-3)(x+2)}=\dfrac{A}{x-3}+\dfrac{B}{x+2}$. Then $3x-4=A(x+2)+B(x-3)$. Taking $x=3$ and $x=-2$, we get $5=5A \Leftrightarrow A=1$ and $-10=-5B \Leftrightarrow B=2$, so

$$\int_0^1 \frac{x^3-4x-10}{x^2-x-6}\,dx=\int_0^1\left(x+1+\frac{1}{x-3}+\frac{2}{x+2}\right)dx=\left[\frac{1}{2}x^2+x+\ln|x-3|+2\ln(x+2)\right]_0^1$$
$$=\left(\tfrac{1}{2}+1+\ln 2+2\ln 3\right)-(0+0+\ln 3+2\ln 2)=\tfrac{3}{2}+\ln 3-\ln 2=\tfrac{3}{2}+\ln\tfrac{3}{2}$$

**29.** Let $u=\sqrt{x}$, so $u^2=x$ and $dx=2u\,du$. Thus,

$$\int_9^{16}\frac{\sqrt{x}}{x-4}\,dx=\int_3^4\frac{u}{u^2-4}2u\,du=2\int_3^4\frac{u^2}{u^2-4}\,du=2\int_3^4\left(1+\frac{4}{u^2-4}\right)du \qquad \text{[by long division]}$$
$$=2+8\int_3^4\frac{du}{(u+2)(u-2)}. \quad (*)$$

Multiply $\dfrac{1}{(u+2)(u-2)}=\dfrac{A}{u+2}+\dfrac{B}{u-2}$ by $(u+2)(u-2)$ to get $1=A(u-2)+B(u+2)$. Equating coefficients we get $A+B=0$ and $-2A+2B=1$. Solving gives us $B=\frac{1}{4}$ and $A=-\frac{1}{4}$, so $\dfrac{1}{(u+2)(u-2)}=\dfrac{-1/4}{u+2}+\dfrac{1/4}{u-2}$ and $(*)$ is

$$2+8\int_3^4\left(\frac{-1/4}{u+2}+\frac{1/4}{u-2}\right)du=2+8\left[-\tfrac{1}{4}\ln|u+2|+\tfrac{1}{4}\ln|u-2|\right]_3^4$$
$$=2+\Big[2\ln|u-2|-2\ln|u+2|\Big]_3^4=2+2\left[\ln\left|\frac{u-2}{u+2}\right|\right]_3^4$$
$$=2+2\left(\ln\tfrac{2}{6}-\ln\tfrac{1}{5}\right)=2+2\ln\tfrac{2/6}{1/5}$$
$$=2+2\ln\tfrac{5}{3}\ \text{ or }\ 2+\ln\left(\tfrac{5}{3}\right)^2=2+\ln\tfrac{25}{9}$$

**30.** Let $u=\sqrt{x+2}$. Then $x=u^2-2$, $dx=2u\,du \Rightarrow I=\displaystyle\int\frac{dx}{x-\sqrt{x+2}}=\int\frac{2u\,du}{u^2-2-u}=2\int\frac{u\,du}{u^2-u-2}$ and

$\dfrac{u}{u^2-u-2}=\dfrac{A}{u-2}+\dfrac{B}{u+1} \Rightarrow u=A(u+1)+B(u-2)$. Substituting $-1$ for $u$ gives $-1=-3B \Leftrightarrow B=\frac{1}{3}$ and substituting 2 for $u$ gives $2=3A \Leftrightarrow A=\frac{2}{3}$. Thus,

$$I=\frac{2}{3}\int\left[\frac{2}{u-2}+\frac{1}{u+1}\right]du=\tfrac{2}{3}(2\ln|u-2|+\ln|u+1|)+C$$
$$=\tfrac{2}{3}\left[2\ln\left|\sqrt{x+2}-2\right|+\ln\left(\sqrt{x+2}+1\right)\right]+C$$

**31.** $x^2+x+1=x^2+x+\frac{1}{4}+1-\frac{1}{4}$ [add and subtract the square of one-half the coefficient of $x$ to complete the square]

$$=x^2+x+\tfrac{1}{4}+\tfrac{3}{4}=\left(x+\tfrac{1}{2}\right)^2+\left(\tfrac{\sqrt{3}}{2}\right)^2$$

So $I=\displaystyle\int\frac{dx}{x^2+x+1}=\int\frac{1}{\left(x+\frac{1}{2}\right)^2+\left(\frac{\sqrt{3}}{2}\right)^2}\,dx$. Now let $u=x+\frac{1}{2} \Rightarrow du=dx$ and

$$I=\int\frac{1}{u^2+\left(\frac{\sqrt{3}}{2}\right)^2}\,du=\frac{1}{\frac{\sqrt{3}}{2}}\tan^{-1}\frac{u}{\frac{\sqrt{3}}{2}}+C=\frac{2}{\sqrt{3}}\tan^{-1}\frac{2\left(x+\frac{1}{2}\right)}{\sqrt{3}}+C=\frac{2}{\sqrt{3}}\tan^{-1}\frac{2x+1}{\sqrt{3}}+C.$$

**32.** $3 - 2x - x^2 = 3 - (x^2 + 2x) = 3 - (x^2 + 2x + 1 - 1) = 4 - (x + 1)^2.$

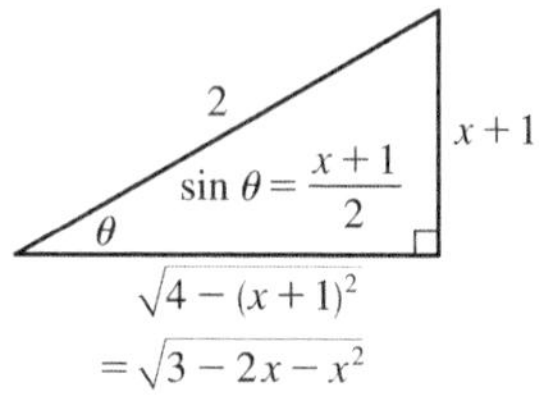


$$I = \int \frac{x}{\sqrt{3 - 2x - x^2}}\,dx = \int \frac{x}{\sqrt{4 - (x+1)^2}}\,dx.$$

Let $x + 1 = 2\sin\theta$, where $-\pi/2 \le \theta \le \pi/2$, so $dx = 2\cos\theta\,d\theta$ and

$$\sqrt{4 - (x+1)^2} = \sqrt{4 - 4\sin^2\theta} = 2\sqrt{\cos^2\theta} = 2\,|\cos\theta| = 2\cos\theta.$$

Thus,

$$\begin{aligned} I &= \int \frac{2\sin\theta - 1}{2\cos\theta}(2\cos\theta\,d\theta) = \int (2\sin\theta - 1)\,d\theta = -2\cos\theta - \theta + C \\ &= -2\,\frac{\sqrt{3 - 2x - x^2}}{2} - \sin^{-1}\frac{x+1}{2} + C = -\sqrt{3 - 2x - x^2} - \sin^{-1}\frac{x+1}{2} + C \end{aligned}$$

## 5.8 Integration Using Tables and Computer Algebra Systems

Keep in mind that there are several ways to approach many of these exercises, and different methods can lead to different forms of the answer.

**1.** Using long division, $\dfrac{x^3 - x^2 + x - 1}{x^2 + 9} = x - 1 - 8\,\dfrac{x - 1}{x^2 + 9}$.

$$I = \int \left(x - 1 - 8\,\frac{x-1}{x^2+9}\right)dx = \int (x-1)dx - 8\left[\int \frac{x}{x^2+9}\,dx - \frac{1}{x^2+9}\,dx\right]$$

Using Formula 17 with $a = 3$, we have

$I \overset{17}{=} \frac{1}{2}x^2 - x - 8 \cdot \frac{1}{2}\ln(x^2 + 9) + 8 \cdot \frac{1}{3}\tan^{-1}(x/3) + C = \frac{1}{2}x^2 - x - 4\ln(x^2 + 9) + \frac{8}{3}\tan^{-1}(x/3) + C.$

**2.** $\displaystyle\int e^{2\theta}\sin 3\theta\,d\theta \overset{98}{=} \frac{e^{2\theta}}{2^2 + 3^2}(2\sin 3\theta - 3\cos 3\theta) + C = \tfrac{2}{13}e^{2\theta}\sin 3\theta - \tfrac{3}{13}e^{2\theta}\cos 3\theta + C$

**3.** Let $u = \pi x \;\Rightarrow\; du = \pi\,dx$, so

$$\begin{aligned} \textstyle\int \sec^3(\pi x)\,dx &= \tfrac{1}{\pi}\textstyle\int \sec^3 u\,du \overset{71}{=} \tfrac{1}{\pi}\left(\tfrac{1}{2}\sec u\tan u + \tfrac{1}{2}\ln|\sec u + \tan u|\right) + C \\ &= \tfrac{1}{2\pi}\sec\pi x\tan\pi x + \tfrac{1}{2\pi}\ln|\sec\pi x + \tan\pi x| + C \end{aligned}$$

**4.** $$\begin{aligned} \int_2^3 \frac{1}{x^2\sqrt{4x^2 - 7}}\,dx &= \int_4^6 \frac{1}{\left(\frac{1}{2}u\right)^2\sqrt{u^2 - 7}}\left(\frac{1}{2}\,du\right) \qquad [u = 2x,\ du = 2\,dx] \\ &= 2\int_4^6 \frac{du}{u^2\sqrt{u^2 - 7}} \overset{45}{=} 2\left[\frac{\sqrt{u^2 - 7}}{7u}\right]_4^6 \\ &= 2\left(\tfrac{\sqrt{29}}{42} - \tfrac{3}{28}\right) = \tfrac{\sqrt{29}}{21} - \tfrac{3}{14} \end{aligned}$$

**5.** Let $u = 2x$ and $a = 3$. Then $du = 2\,dx$ and

$$\begin{aligned} \int \frac{dx}{x^2\sqrt{4x^2 + 9}} &= \int \frac{\frac{1}{2}\,du}{\frac{u^2}{4}\sqrt{u^2 + a^2}} = 2\int \frac{du}{u^2\sqrt{a^2 + u^2}} \overset{28}{=} -2\,\frac{\sqrt{a^2 + u^2}}{a^2 u} + C \\ &= -2\,\frac{\sqrt{4x^2 + 9}}{9 \cdot 2x} + C = -\frac{\sqrt{4x^2 + 9}}{9x} + C \end{aligned}$$

**6.** Let $u = \sqrt{2}\,y$ and $a = \sqrt{3}$. Then $du = \sqrt{2}\,dy$ and

$$\begin{aligned}\int \frac{\sqrt{2y^2-3}}{y^2}\,dy &= \int \frac{\sqrt{u^2-a^2}}{\frac{1}{2}u^2}\,\frac{du}{\sqrt{2}} = \sqrt{2}\int \frac{\sqrt{u^2-a^2}}{u^2}\,du \\ &\overset{42}{=} \sqrt{2}\left(-\frac{\sqrt{u^2-a^2}}{u} + \ln\left|u+\sqrt{u^2-a^2}\right|\right) + C \\ &= \sqrt{2}\left(-\frac{\sqrt{2y^2-3}}{\sqrt{2}\,y} + \ln\left|\sqrt{2}\,y+\sqrt{2y^2-3}\right|\right) + C \\ &= -\frac{\sqrt{2y^2-3}}{y} + \sqrt{2}\ln\left|\sqrt{2}\,y+\sqrt{2y^2-3}\right| + C\end{aligned}$$

**7.** $\int x^3 \sin x\,dx \overset{84}{=} -x^3\cos x + 3\int x^2\cos x\,dx$, $\int x^2\cos x\,dx \overset{85}{=} x^2\sin x - 2\int x\sin x\,dx$, and $\int x\sin x\,dx \overset{82}{=} \sin x - x\cos x + C$. Substituting, we get

$\int x^3\sin x\,dx = -x^3\cos x + 3\left[x^2\sin x - 2(\sin x - x\cos x)\right] + C = -x^3\cos x + 3x^2\sin x - 6\sin x + 6x\cos x + C$.

So $\int_0^\pi x^3\sin x\,dx = \left[-x^3\cos x + 3x^2\sin x - 6\sin x + 6x\cos x\right]_0^\pi = \left(-\pi^3\cdot -1 + 6\pi\cdot -1\right) - (0) = \pi^3 - 6\pi$.

**8.** Let $u = e^x$. Then $du = e^x\,dx$, so

$$\int \frac{e^{2x}}{\sqrt{2+e^x}}\,dx = \int \frac{e^x}{\sqrt{2+e^x}}(e^x\,dx) = \int \frac{u}{\sqrt{2+u}}\,du \overset{55}{=} \tfrac{2}{3}(u-2\cdot 2)\sqrt{2+u} + C = \tfrac{2}{3}(e^x-4)\sqrt{2+e^x} + C.$$

*Another method*: Let $u = 2 + e^x$. Then $du = e^x\,dx$, so

$$\begin{aligned}\int \frac{e^{2x}}{\sqrt{2+e^x}}\,dx &= \int \frac{e^x}{\sqrt{2+e^x}}(e^x\,dx) = \int \frac{u-2}{\sqrt{u}}\,du = \int \left(u^{1/2} - 2u^{-1/2}\right)du = \tfrac{2}{3}u^{3/2} - 2\left(2u^{1/2}\right) + C \\ &= \tfrac{2}{3}(2+e^x)^{3/2} - 4\,(2+e^x)^{1/2} + C.\end{aligned}$$

**9.** $$\begin{aligned}\int \frac{\tan^3(1/z)}{z^2}\,dz \quad \begin{bmatrix} u = 1/z, \\ du = -dz/z^2\end{bmatrix} \quad &= -\int \tan^3 u\,du \overset{69}{=} -\frac{1}{2}\tan^2 u - \ln|\cos u| + C \\ &= -\frac{1}{2}\tan^2\left(\frac{1}{z}\right) - \ln\left|\cos\left(\frac{1}{z}\right)\right| + C\end{aligned}$$

**10.** Let $u = \sqrt{x}$. Then $u^2 = x$ and $2u\,du = dx$, so

$$\begin{aligned}\int \sin^{-1}\sqrt{x}\,dx &= 2\int u\sin^{-1}u\,du \overset{90}{=} \frac{2u^2-1}{2}\sin^{-1}u + \frac{u\sqrt{1-u^2}}{2} + C \\ &= \frac{2x-1}{2}\sin^{-1}\sqrt{x} + \frac{\sqrt{x(1-x)}}{2} + C\end{aligned}$$

**11.** Let $z = 6 + 4y - 4y^2 = 6 - (4y^2 - 4y + 1) + 1 = 7 - (2y-1)^2$, $u = 2y - 1$, and $a = \sqrt{7}$. Then $z = a^2 - u^2$, $du = 2\,dy$, and

$$\begin{aligned}\int y\sqrt{6+4y-4y^2}\,dy &= \int y\sqrt{z}\,dy = \int \tfrac{1}{2}(u+1)\sqrt{a^2-u^2}\,\tfrac{1}{2}\,du \\ &= \tfrac{1}{4}\int u\sqrt{a^2-u^2}\,du + \tfrac{1}{4}\int\sqrt{a^2-u^2}\,du \\ &= \tfrac{1}{4}\int\sqrt{a^2-u^2}\,du - \tfrac{1}{8}\int(-2u)\sqrt{a^2-u^2}\,du \\ &\overset{30}{=} \frac{u}{8}\sqrt{a^2-u^2} + \frac{a^2}{8}\sin^{-1}\left(\frac{u}{a}\right) - \frac{1}{8}\int\sqrt{w}\,dw \qquad \begin{bmatrix} w = a^2 - u^2, \\ dw = -2u\,du\end{bmatrix} \\ &= \frac{2y-1}{8}\sqrt{6+4y-4y^2} + \frac{7}{8}\sin^{-1}\frac{2y-1}{\sqrt{7}} - \frac{1}{8}\cdot\frac{2}{3}w^{3/2} + C \\ &= \frac{2y-1}{8}\sqrt{6+4y-4y^2} + \frac{7}{8}\sin^{-1}\frac{2y-1}{\sqrt{7}} - \frac{1}{12}(6+4y-4y^2)^{3/2} + C.\end{aligned}$$

[continued]

This can be rewritten as

$$\sqrt{6+4y-4y^2}\left[\frac{1}{8}(2y-1)-\frac{1}{12}(6+4y-4y^2)\right]+\frac{7}{8}\sin^{-1}\frac{2y-1}{\sqrt{7}}+C$$
$$=\left(\frac{1}{3}y^2-\frac{1}{12}y-\frac{5}{8}\right)\sqrt{6+4y-4y^2}+\frac{7}{8}\sin^{-1}\left(\frac{2y-1}{\sqrt{7}}\right)+C$$
$$=\frac{1}{24}(8y^2-2y-15)\sqrt{6+4y-4y^2}+\frac{7}{8}\sin^{-1}\left(\frac{2y-1}{\sqrt{7}}\right)+C$$

**12.** Let $u=x^2$, so that $du=2x\,dx$. Then

$$\int x\sin(x^2)\cos(3x^2)\,dx=\frac{1}{2}\int\sin u\cos 3u\,du\overset{81}{=}-\frac{1}{2}\frac{\cos(1-3)u}{2(1-3)}-\frac{1}{2}\frac{\cos(1+3)u}{2(1+3)}+C$$
$$=\tfrac{1}{8}\cos 2u-\tfrac{1}{16}\cos 4u+C=\tfrac{1}{8}\cos(2x^2)-\tfrac{1}{16}\cos(4x^2)+C$$

**13.** Let $u=\sin x$. Then $du=\cos x\,dx$, so

$$\int\sin^2x\,\cos x\,\ln(\sin x)\,dx=\int u^2\ln u\,du\overset{101}{=}\frac{u^{2+1}}{(2+1)^2}\left[(2+1)\ln u-1\right]+C=\tfrac{1}{9}u^3(3\ln u-1)+C$$
$$=\tfrac{1}{9}\sin^3x\,[3\ln(\sin x)-1]+C$$

**14.** Let $u=3\theta$. Then $du=3\,d\theta$, so $I=\int_0^\pi\cos^4(3\theta)\,d\theta=\frac{1}{3}\int_0^{3\pi}\cos^4u\,du$. Now $\int\cos^4u\,du\overset{74}{=}\frac{1}{4}\cos^3u\,\sin u+\frac{3}{4}\int\cos^2u\,du$ and $\int\cos^2u\,du\overset{64}{=}\frac{1}{2}u+\frac{1}{4}\sin 2u+C$. So

$$I=\tfrac{1}{3}\left[\tfrac{1}{4}\cos^3u\,\sin u+\tfrac{3}{4}\left(\tfrac{1}{2}u+\tfrac{1}{4}\sin 2u\right)\right]_0^{3\pi}$$
$$=\left[\tfrac{1}{12}\cos^3u\,\sin u+\tfrac{1}{8}u+\tfrac{1}{16}\sin 2u\right]_0^{3\pi}=\tfrac{3\pi}{8}-0=\tfrac{3\pi}{8}.$$

**15.** Let $u=e^x$ and $a=\sqrt{3}$. Then $du=e^x\,dx$ and

$$\int\frac{e^x}{3-e^{2x}}\,dx=\int\frac{du}{a^2-u^2}\overset{19}{=}\frac{1}{2a}\ln\left|\frac{u+a}{u-a}\right|+C=\frac{1}{2\sqrt{3}}\ln\left|\frac{e^x+\sqrt{3}}{e^x-\sqrt{3}}\right|+C.$$

**16.** Let $u=x^2$ and $a=2$. Then $du=2x\,dx$ and

$$\int_0^2x^3\sqrt{4x^2-x^4}\,dx=\frac{1}{2}\int_0^2x^2\sqrt{2\cdot2\cdot x^2-(x^2)^2}\cdot2x\,dx=\frac{1}{2}\int_0^4u\sqrt{2au-u^2}\,du$$
$$\overset{114}{=}\left[\frac{2u^2-au-3a^2}{12}\sqrt{2au-u^2}+\frac{a^3}{4}\cos^{-1}\left(\frac{a-u}{a}\right)\right]_0^4$$
$$=\left[\frac{2u^2-2u-12}{12}\sqrt{4u-u^2}+\frac{8}{4}\cos^{-1}\left(\frac{2-u}{2}\right)\right]_0^4$$
$$=\left[\frac{u^2-u-6}{6}\sqrt{4u-u^2}+2\cos^{-1}\left(\frac{2-u}{2}\right)\right]_0^4$$
$$=[0+2\cos^{-1}(-1)]-(0+2\cos^{-1}1)=2\cdot\pi-2\cdot0=2\pi$$

**17.** $$\int\frac{x^4\,dx}{\sqrt{x^{10}-2}}=\int\frac{x^4\,dx}{\sqrt{(x^5)^2-2}}=\frac{1}{5}\int\frac{du}{\sqrt{u^2-2}}\qquad[u=x^5,\ du=5x^4\,dx]$$
$$\overset{43}{=}\tfrac{1}{5}\ln\left|u+\sqrt{u^2-2}\right|+C=\tfrac{1}{5}\ln\left|x^5+\sqrt{x^{10}-2}\right|+C$$

**18.** $\int x^4e^{-x}\,dx\overset{97}{=}-x^4e^{-x}+4\int x^3e^{-x}\,dx\overset{97}{=}-x^4e^{-x}+4\left(-x^3e^{-x}+3\int x^2e^{-x}\,dx\right)$

$$\overset{97}{=}-(x^4+4x^3)e^{-x}+12\left(-x^2e^{-x}+2\int xe^{-x}\,dx\right)$$
$$\overset{96}{=}-(x^4+4x^3+12x^2)e^{-x}+24\left[(-x-1)e^{-x}\right]+C=-(x^4+4x^3+12x^2+24x+24)e^{-x}+C$$

So $\int_0^1x^4e^{-x}\,dx=\left[-(x^4+4x^3+12x^2+24x+24)e^{-x}\right]_0^1=-(1+4+12+24+24)e^{-1}+24e^0=24-65e^{-1}$

**19.** Let $u = \ln x$ and $a = 2$. Then $du = \dfrac{dx}{x}$ and

$$\int \frac{\sqrt{4+(\ln x)^2}}{x}\,dx = \int \sqrt{a^2+u^2}\,du \overset{21}{=} \frac{u}{2}\sqrt{a^2+u^2} + \frac{a^2}{2}\ln\left(u+\sqrt{a^2+u^2}\right) + C$$
$$= \tfrac{1}{2}(\ln x)\sqrt{4+(\ln x)^2} + 2\ln\left[\ln x + \sqrt{4+(\ln x)^2}\,\right] + C$$

**20.** Let $u = \tan\theta$ and $a = 3$. Then $du = \sec^2\theta\,d\theta$ and

$$\int \frac{\sec^2\theta\,\tan^2\theta}{\sqrt{9-\tan^2\theta}}\,d\theta = \int \frac{u^2}{\sqrt{a^2-u^2}}\,du \overset{34}{=} -\frac{u}{2}\sqrt{a^2-u^2} + \frac{a^2}{2}\sin^{-1}\left(\frac{u}{a}\right) + C$$
$$= -\frac{1}{2}\tan\theta\sqrt{9-\tan^2\theta} + \frac{9}{2}\sin^{-1}\left(\frac{\tan\theta}{3}\right) + C$$

**21.** Let $u = e^x$. Then $x = \ln u$, $dx = du/u$, so

$$\int \sqrt{e^{2x}-1}\,dx = \int \frac{\sqrt{u^2-1}}{u}\,du \overset{41}{=} \sqrt{u^2-1} - \cos^{-1}(1/u) + C = \sqrt{e^{2x}-1} - \cos^{-1}(e^{-x}) + C.$$

**22.** Let $u = \alpha t - 3$ and assume that $\alpha \neq 0$. Then $du = \alpha\,dt$ and

$$\int e^t \sin(\alpha t - 3)\,dt = \tfrac{1}{\alpha}\int e^{(u+3)/\alpha}\sin u\,du = \tfrac{1}{\alpha}e^{3/\alpha}\int e^{(1/\alpha)u}\sin u\,du$$
$$\overset{98}{=} \frac{1}{\alpha}e^{3/\alpha}\frac{e^{(1/\alpha)u}}{(1/\alpha)^2+1^2}\left(\frac{1}{\alpha}\sin u - \cos u\right) + C$$
$$= \frac{1}{\alpha}e^{3/\alpha}e^{(1/\alpha)u}\frac{\alpha^2}{1+\alpha^2}\left(\frac{1}{\alpha}\sin u - \cos u\right) + C$$
$$= \frac{1}{1+\alpha^2}e^{(u+3)/\alpha}(\sin u - \alpha\cos u) + C$$
$$= \frac{1}{1+\alpha^2}e^t\left[\sin(\alpha t-3) - \alpha\cos(\alpha t-3)\right] + C$$

**23.** (a) $$\frac{d}{du}\left[\frac{1}{b^3}\left(a+bu-\frac{a^2}{a+bu}-2a\ln|a+bu|\right)+C\right] = \frac{1}{b^3}\left[b+\frac{ba^2}{(a+bu)^2}-\frac{2ab}{(a+bu)}\right]$$
$$= \frac{1}{b^3}\left[\frac{b(a+bu)^2+ba^2-(a+bu)2ab}{(a+bu)^2}\right] = \frac{1}{b^3}\left[\frac{b^3u^2}{(a+bu)^2}\right] = \frac{u^2}{(a+bu)^2}$$

(b) Let $t = a + bu \;\Rightarrow\; dt = b\,du$. Note that $u = \dfrac{t-a}{b}$ and $du = \dfrac{1}{b}\,dt$.

$$\int \frac{u^2\,du}{(a+bu)^2} = \frac{1}{b^3}\int\frac{(t-a)^2}{t^2}\,dt = \frac{1}{b^3}\int\frac{t^2-2at+a^2}{t^2}\,dt$$
$$= \frac{1}{b^3}\int\left(1-\frac{2a}{t}+\frac{a^2}{t^2}\right)dt = \frac{1}{b^3}\left(t-2a\ln|t|-\frac{a^2}{t}\right)+C$$
$$= \frac{1}{b^3}\left(a+bu-\frac{a^2}{a+bu}-2a\ln|a+bu|\right)+C$$

**24.** (a) $$\frac{d}{du}\left[\frac{u}{8}(2u^2-a^2)\sqrt{a^2-u^2}+\frac{a^4}{8}\sin^{-1}\frac{u}{a}+C\right]$$
$$= \frac{u}{8}(2u^2-a^2)\frac{-u}{\sqrt{a^2-u^2}} + \sqrt{a^2-u^2}\left[\frac{u}{8}(4u)+(2u^2-a^2)\tfrac{1}{8}\right] + \frac{a^4}{8}\frac{1/a}{\sqrt{1-u^2/a^2}}$$
$$= -\frac{u^2(2u^2-a^2)}{8\sqrt{a^2-u^2}} + \sqrt{a^2-u^2}\left[\frac{u^2}{2}+\frac{2u^2-a^2}{8}\right] + \frac{a^4}{8\sqrt{a^2-u^2}}$$
$$= \tfrac{1}{2}(a^2-u^2)^{-1/2}\left[-\frac{u^2}{4}(2u^2-a^2)+u^2(a^2-u^2)+\tfrac{1}{4}(a^2-u^2)(2u^2-a^2)+\frac{a^4}{4}\right]$$
$$= \tfrac{1}{2}(a^2-u^2)^{-1/2}\left[2u^2a^2-2u^4\right] = \frac{u^2(a^2-u^2)}{\sqrt{a^2-u^2}} = u^2\sqrt{a^2-u^2}$$

(b) Let $u = a\sin\theta \Rightarrow du = a\cos\theta\,d\theta$. Then

$$\begin{aligned}
\int u^2\sqrt{a^2-u^2}\,du &= \int a^2\sin^2\theta\,a\sqrt{1-\sin^2\theta}\,a\cos\theta\,d\theta = a^4\int\sin^2\theta\cos^2\theta\,d\theta \\
&= a^4\int\tfrac12(1+\cos 2\theta)\tfrac12(1-\cos 2\theta)d\theta = \tfrac14 a^4\int(1-\cos^2 2\theta)\,d\theta \\
&= \tfrac14 a^4\int\left[1-\tfrac12(1+\cos 4\theta)\right]d\theta = \tfrac14 a^4\left(\tfrac12\theta-\tfrac18\sin 4\theta\right)+C \\
&= \tfrac14 a^4\left(\tfrac12\theta-\tfrac18 2\sin 2\theta\cos 2\theta\right)+C = \tfrac14 a^4\left[\tfrac12\theta-\tfrac12\sin\theta\cos\theta\left(1-2\sin^2\theta\right)\right]+C \\
&= \frac{a^4}{8}\left[\sin^{-1}\frac{u}{a}-\frac{u}{a}\frac{\sqrt{a^2-u^2}}{a}\left(1-\frac{2u^2}{a^2}\right)\right]+C \\
&= \frac{a^4}{8}\left[\sin^{-1}\frac{u}{a}-\frac{u}{a}\frac{\sqrt{a^2-u^2}}{a}\frac{a^2-2u^2}{a^2}\right]+C \\
&= \frac{u}{8}\left(2u^2-a^2\right)\sqrt{a^2-u^2}+\frac{a^4}{8}\sin^{-1}\frac{u}{a}+C
\end{aligned}$$

**25.** Maple, Mathematica and Derive all give $\int x^2\sqrt{5-x^2}\,dx = -\frac14 x\left(5-x^2\right)^{3/2}+\frac58 x\sqrt{5-x^2}+\frac{25}{8}\sin^{-1}\left(\frac{1}{\sqrt5}x\right)$. Using Formula 31, we get $\int x^2\sqrt{5-x^2}\,dx = \frac18 x\left(2x^2-5\right)\sqrt{5-x^2}+\frac18\left(5^2\right)\sin^{-1}\left(\frac{1}{\sqrt5}x\right)+C$. But $-\frac14 x\left(5-x^2\right)^{3/2}+\frac58 x\sqrt{5-x^2} = \frac18 x\sqrt{5-x^2}\left[5-2\left(5-x^2\right)\right] = \frac18 x\left(2x^2-5\right)\sqrt{5-x^2}$, and the $\sin^{-1}$ terms are the same in each expression, so the answers are equivalent.

**26.** Maple and Mathematica both give $\int x^2\left(1+x^3\right)^4dx = \frac{1}{15}x^{15}+\frac13 x^{12}+\frac23 x^9+\frac23 x^6+\frac13 x^3$, while Derive gives $\int x^2\left(1+x^3\right)^4dx = \frac{1}{15}\left(x^3+1\right)^5$. Using the substitution $u = 1+x^3 \Rightarrow du = 3x^2\,dx$, we get $\int x^2\left(1+x^3\right)^4dx = \int u^4\left(\frac13\,du\right) = \frac{1}{15}u^5+C = \frac{1}{15}\left(1+x^3\right)^5+C$. We can use the Binomial Theorem or a CAS to expand this expression, and we get $\frac{1}{15}\left(1+x^3\right)^5+C = \frac{1}{15}+\frac13 x^3+\frac23 x^6+\frac23 x^9+\frac13 x^{12}+\frac{1}{15}x^{15}+C$.

**27.** Maple and Derive both give $\int\sin^3x\,\cos^2x\,dx = -\frac15\sin^2x\,\cos^3x-\frac{2}{15}\cos^3x$ (although Derive factors the expression), and Mathematica gives $\int\sin^3x\,\cos^2x\,dx = -\frac18\cos x-\frac{1}{48}\cos 3x+\frac{1}{80}\cos 5x$. We can use a CAS to show that both of these expressions are equal to $-\frac13\cos^3x+\frac15\cos^5x$. Using Formula 86, we write

$$\begin{aligned}
\int\sin^3x\,\cos^2x\,dx &= -\tfrac15\sin^2x\,\cos^3x+\tfrac25\int\sin x\,\cos^2x\,dx = -\tfrac15\sin^2x\,\cos^3x+\tfrac25\left(-\tfrac13\cos^3x\right)+C \\
&= -\tfrac15\sin^2x\,\cos^3x-\tfrac{2}{15}\cos^3x+C
\end{aligned}$$

**28.** Maple gives $\int\tan^2x\,\sec^4\,dx = \dfrac15\dfrac{\sin^3x}{\cos^5x}+\dfrac{2}{15}\dfrac{\sin^3x}{\cos^3x}$, Mathematica gives $\int\tan^2x\,\sec^4\,dx = -\frac{1}{120}\sec^5x\left(-20\sin x+5\sin 3x+\sin 5x\right)$, and Derive gives $\int\tan^2x\,\sec^4\,dx = -\frac{2}{15}\tan x-\dfrac{\sin x}{15\cos^3x}+\dfrac{\sin x}{5\cos^5x}$. All of these expressions can be "simplified" to $-\dfrac{1}{15}\dfrac{\sin x\left(\cos^2x-2\cos^4x-3\right)}{\cos^5x}$ using Maple. Using the identity $1+\tan^2x = \sec^2x$, we write $\int\tan^2x\,\sec^4x\,dx = \int\tan^2x\left(1+\tan^2x\right)\sec^2x\,dx = \int\left(\tan^2x+\tan^4x\right)\sec^2x\,dx$. Now we substitute $u = \tan x \Rightarrow du = \sec^2x\,dx$, and the integral becomes $\int\left(u^2+u^4\right)du = \frac13u^3+\frac15u^5+C = \frac13\tan^3x+\frac15\tan^5x+C$. If we write $\sin^5x = \sin^3x\left(1-\cos^2x\right)$ and substitute into the numerator of the $\tan^5x$ term, this becomes $\dfrac13\dfrac{\sin^3x}{\cos^3x}+\dfrac15\dfrac{\sin^3x\left(1-\cos^2x\right)}{\cos^5x}+C = \dfrac15\dfrac{\sin^3x}{\cos^5x}+\left(\dfrac13-\dfrac15\right)\dfrac{\sin^3x}{\cos^3x}+C = \dfrac15\dfrac{\sin^3x}{\cos^5x}+\dfrac{2}{15}\dfrac{\sin^3x}{\cos^3x}+C$, which is the same as Maple's expression.

**29.** Maple gives $\int x\sqrt{1+2x}\,dx = \frac{1}{10}(1+2x)^{5/2} - \frac{1}{6}(1+2x)^{3/2}$, Mathematica gives $\sqrt{1+2x}\left(\frac{2}{5}x^2 + \frac{1}{15}x - \frac{1}{15}\right)$, and Derive gives $\frac{1}{15}(1+2x)^{3/2}(3x-1)$. The first two expressions can be simplified to Derive's result. If we use Formula 54, we get

$$\begin{aligned}\int x\sqrt{1+2x}\,dx &= \frac{2}{15(2)^2}(3\cdot 2x - 2\cdot 1)(1+2x)^{3/2} + C = \frac{1}{30}(6x-2)(1+2x)^{3/2} + C \\ &= \frac{1}{15}(3x-1)(1+2x)^{3/2}\end{aligned}$$

**30.** Maple and Derive both give $\int \sin^4 x\,dx = -\frac{1}{4}\sin^3 x\cos x - \frac{3}{8}\cos x\sin x + \frac{3}{8}x$, while Mathematica gives $\frac{1}{32}(12x - 8\sin 2x + \sin 4x)$, which can be expanded and simplified to give the other expression. Now

$$\begin{aligned}\int \sin^4 x\,dx &\overset{73}{=} -\frac{1}{4}\sin^3 x\cos x + \frac{3}{4}\int \sin^2 x\,dx \overset{63}{=} -\frac{1}{4}\sin^3 x\cos x + \frac{3}{4}\left(\frac{1}{2}x - \frac{1}{4}\sin 2x\right) + C \\ &= -\frac{1}{4}\sin^3 x\cos x - \frac{3}{8}\sin x\cos x + \frac{3}{8}x + C \text{ since } \sin 2x = 2\sin x\cos x\end{aligned}$$

**31.** Maple gives $\int \tan^5 x\,dx = \frac{1}{4}\tan^4 x - \frac{1}{2}\tan^2 x + \frac{1}{2}\ln(1+\tan^2 x)$, Mathematica gives $\int \tan^5 x\,dx = \frac{1}{4}[-1 - 2\cos(2x)]\sec^4 x - \ln(\cos x)$, and Derive gives $\int \tan^5 x\,dx = \frac{1}{4}\tan^4 x - \frac{1}{2}\tan^2 x - \ln(\cos x)$. These expressions are equivalent, and none includes absolute value bars or a constant of integration. Note that Mathematica's and Derive's expressions suggest that the integral is undefined where $\cos x < 0$, which is not the case.

Using Formula 75, $\int \tan^5 x\,dx = \frac{1}{5-1}\tan^{5-1} x - \int \tan^{5-2} x\,dx = \frac{1}{4}\tan^4 x - \int \tan^3 x\,dx$. Using Formula 69, $\int \tan^3 x\,dx = \frac{1}{2}\tan^2 x + \ln|\cos x| + C$, so $\int \tan^5 x\,dx = \frac{1}{4}\tan^4 x - \frac{1}{2}\tan^2 x - \ln|\cos x| + C$.

**32.** Maple gives $\int x^5\sqrt{x^2+1}\,dx = \frac{1}{35}x^4\sqrt{1+x^2} - \frac{4}{105}x^2\sqrt{1+x^2} + \frac{8}{105}\sqrt{1+x^2} + \frac{1}{7}x^6\sqrt{1+x^2}$. When we use the `factor` command on this expression, it becomes $\frac{1}{105}(1+x^2)^{3/2}(15x^4 - 12x^2 + 8)$. Mathematica gives $\sqrt{1+x^2}\left(\frac{8}{105} - \frac{4}{105}x^2 + \frac{1}{35}x^4 + \frac{1}{7}x^6\right)$, which again factors to give the above expression, and Derive gives the factored form immediately. If we substitute $u = \sqrt{x^2+1} \;\Rightarrow\; x^4 = (u^2-1)^2$, $x\,dx = u\,du$, then the integral becomes

$$\begin{aligned}\int (u^2-1)^2 u(u\,du) &= \int (u^4 - 2u^2 + 1)u^2\,du = \frac{1}{7}u^7 - \frac{2}{5}u^5 + \frac{1}{3}u^3 + C \\ &= (x^2+1)^{3/2}\left[\frac{1}{7}(x^2+1)^2 - \frac{2}{5}(x^2+1) + \frac{1}{3}\right] + C \\ &= \frac{1}{105}(x^2+1)^{3/2}\left[15(x^2+1)^2 - 42(x^2+1) + 35\right] + C \\ &= \frac{1}{105}(x^2+1)^{3/2}(15x^4 - 12x^2 + 8) + C\end{aligned}$$

**33.** Derive gives $I = \displaystyle\int 2^x\sqrt{4^x - 1}\,dx = \frac{2^{x-1}\sqrt{2^{2x}-1}}{\ln 2} - \frac{\ln\left(\sqrt{2^{2x}-1} + 2^x\right)}{2\ln 2}$ immediately. Neither Maple nor Mathematica is able to evaluate $I$ in its given form. However, if we instead write $I$ as $\int 2^x\sqrt{(2^x)^2 - 1}\,dx$, both systems give the same answer as Derive (after minor simplification). Our trick works because the CAS now recognizes $2^x$ as a promising substitution.

**34.** None of Maple, Mathematica and Derive is able to evaluate $\int (1+\ln x)\sqrt{1+(x\ln x)^2}\,dx$. However, if we let $u = x\ln x$, then $du = (1+\ln x)dx$ and the integral is simply $\int \sqrt{1+u^2}\,du$, which any CAS can evaluate. The antiderivative is $\frac{1}{2}\ln\left(x\ln x + \sqrt{1+(x\ln x)^2}\right) + \frac{1}{2}x\ln x\sqrt{1+(x\ln x)^2} + C$.

## DISCOVERY PROJECT Patterns in Integrals

**1.** (a) The CAS results are listed. Note that the absolute value symbols are missing, as is the familiar "$+\,C$".

(i) $\displaystyle\int \frac{1}{(x+2)(x+3)}\,dx = \ln(x+2) - \ln(x+3)$

(ii) $\displaystyle\int \frac{1}{(x+1)(x+5)}\,dx = \frac{\ln(x+1)}{4} - \frac{\ln(x+5)}{4}$

(iii) $\displaystyle\int \frac{1}{(x+2)(x-5)}\,dx = \frac{\ln(x-5)}{7} - \frac{\ln(x+2)}{7}$

(iv) $\displaystyle\int \frac{1}{(x+2)^2}\,dx = -\frac{1}{x+2}$

(b) If $a \neq b$, it appears that $\ln(x+a)$ is divided by $b-a$ and $\ln(x+b)$ is divided by $a-b$, so we guess that

$$\int \frac{1}{(x+a)(x+b)}\,dx = \frac{\ln(x+a)}{b-a} + \frac{\ln(x+b)}{a-b} + C$$

If $a = b$, as in part (a)(iv), it appears that

$$\int \frac{1}{(x+a)^2}\,dx = -\frac{1}{x+a} + C$$

(c) The CAS verifies our guesses. Now

$$\frac{1}{(x+a)(x+b)} = \frac{A}{x+a} + \frac{B}{x+b} \quad\Rightarrow\quad 1 = A(x+b) + B(x+a)$$

Setting $x = -b$ gives $B = 1/(a-b)$ and setting $x = -a$ gives $A = 1/(b-a)$. So

$$\int \frac{1}{(x+a)(x+b)}\,dx = \int \left[\frac{1/(b-a)}{x+a} + \frac{1/(a-b)}{x+b}\right] dx = \frac{\ln|x+a|}{b-a} + \frac{\ln|x+b|}{a-b} + C$$

and our guess for $a \neq b$ is correct.

If $a = b$, then $\dfrac{1}{(x+a)(x+b)} = \dfrac{1}{(x+a)^2} = (x+a)^{-2}$. Letting $u = x+a \;\Rightarrow\; du = dx$, we have

$\displaystyle\int (x+a)^{-2}\,dx = \int u^{-2}\,du = -\frac{1}{u} + C = -\frac{1}{x+a} + C$, and our guess for $a = b$ is also correct.

**2.** (a) (i) $\int \sin x \,\cos 2x\,dx = \dfrac{\cos x}{2} - \dfrac{\cos 3x}{6}$

(ii) $\int \sin 3x \,\cos 7x\,dx = \dfrac{\cos 4x}{8} - \dfrac{\cos 10x}{20}$

(iii) $\int \sin 8x \,\cos 3x\,dx = -\dfrac{\cos 11x}{22} - \dfrac{\cos 5x}{10}$

(b) Looking at the sums and differences of $a$ and $b$ in part (a), we guess that

$$\int \sin ax \,\cos bx\,dx = \frac{\cos((a-b)x)}{2(b-a)} - \frac{\cos((a+b)x)}{2(a+b)} + C$$

Note that $\cos((a-b)x) = \cos((b-a)x)$.

(c) The CAS verifies our guess. Again, we can prove that the guess is correct by differentiating:

$$\frac{d}{dx}\left[\frac{\cos((a-b)x)}{2(b-a)} - \frac{\cos((a+b)x)}{2(a+b)}\right]$$
$$= \frac{1}{2(b-a)}\left[-\sin((a-b)x)\right](a-b) - \frac{1}{2(a+b)}\left[-\sin((a+b)x)\right](a+b)$$
$$= \tfrac{1}{2}\sin(ax-bx) + \tfrac{1}{2}\sin(ax+bx)$$
$$= \tfrac{1}{2}(\sin ax\,\cos bx - \cos ax\,\sin bx) + \tfrac{1}{2}(\sin ax\,\cos bx + \cos ax\,\sin bx)$$
$$= \sin ax\,\cos bx$$

Our formula is valid for $a \neq b$.

**3.** (a) (i) $\int \ln x\,dx = x\ln x - x$

(ii) $\int x\ln x\,dx = \frac{1}{2}x^2\ln x - \frac{1}{4}x^2$

(iii) $\int x^2\ln x\,dx = \frac{1}{3}x^3\ln x - \frac{1}{9}x^3$

(iv) $\int x^3\ln x\,dx = \frac{1}{4}x^4\ln x - \frac{1}{16}x^4$

(v) $\int x^7\ln x\,dx = \frac{1}{8}x^8\ln x - \frac{1}{64}x^8$

(b) We guess that $\int x^n\ln x\,dx = \dfrac{1}{n+1}x^{n+1}\ln x - \dfrac{1}{(n+1)^2}x^{n+1}$.

(c) Let $u = \ln x$, $dv = x^n\,dx \;\Rightarrow\; du = \dfrac{dx}{x}$, $v = \dfrac{1}{n+1}x^{n+1}$.

Then
$$\int x^n\ln x\,dx = \frac{1}{n+1}x^{n+1}\ln x - \frac{1}{n+1}\int x^n\,dx$$
$$= \frac{1}{n+1}x^{n+1}\ln x - \frac{1}{n+1}\cdot\frac{1}{n+1}x^{n+1},$$

which verifies our guess. We must have $n+1 \neq 0 \;\Leftrightarrow\; n \neq -1$.

**4.** (a) (i) $\int xe^x\,dx = e^x(x-1)$

(ii) $\int x^2e^x\,dx = e^x(x^2-2x+2)$

(iii) $\int x^3e^x\,dx = e^x(x^3-3x^2+6x-6)$

(iv) $\int x^4e^x\,dx = e^x(x^4-4x^3+12x^2-24x+24)$

(v) $\int x^5e^x\,dx = e^x(x^5-5x^4+20x^3-60x^2+120x-120)$

(b) Notice from part (a) that we can write

$$\int x^4e^x\,dx = e^x\left(x^4 - 4x^3 + 4\cdot 3x^2 - 4\cdot 3\cdot 2x + 4\cdot 3\cdot 2\cdot 1\right)$$

and

$$\int x^5e^x\,dx = e^x\left(x^5 - 5x^4 + 5\cdot 4x^3 - 5\cdot 4\cdot 3x^2 + 5\cdot 4\cdot 3\cdot 2x - 5\cdot 4\cdot 3\cdot 2\cdot 1\right)$$

So we guess that

$$\int x^6e^x\,dx = e^x\left(x^6 - 6x^5 + 6\cdot 5x^4 - 6\cdot 5\cdot 4x^3 + 6\cdot 5\cdot 4\cdot 3x^2 - 6\cdot 5\cdot 4\cdot 3\cdot 2x + 6\cdot 5\cdot 4\cdot 3\cdot 2\cdot 1\right)$$
$$= e^x\left(x^6 - 6x^5 + 30x^4 - 120x^3 + 360x^2 - 720x + 720\right)$$

The CAS verifies our guess.

(c) From the results in part (a), as well as our prediction in part (b), we speculate that

$$\int x^ne^x\,dx = e^x\left[x^n - nx^{n-1} + n(n-1)x^{n-2} - n(n-1)(n-2)x^{n-3} + \cdots \pm n!x \mp n!\right]$$
$$= e^x\sum_{i=0}^{n}(-1)^{n-i}\frac{n!}{i!}x^i$$

(We have reversed the order of the polynomial's terms.)

(d) Let $S_n$ be the statement that $\int x^n e^x\,dx = e^x \sum_{i=0}^{n} (-1)^{n-i} \frac{n!}{i!} x^i$.

$S_1$ is true by part (a)(i). Suppose $S_k$ is true for some $k$, and consider $S_{k+1}$. Integrating by parts with $u = x^{k+1}$, $dv = e^x\,dx \quad \Rightarrow \quad du = (k+1)x^k\,dx$, $v = e^x$, we get

$$\begin{aligned}\int x^{k+1} e^x\,dx &= x^{k+1}e^x - (k+1)\int x^k e^x\,dx \\ &= x^{k+1}e^x - (k+1)\left[e^x \sum_{i=0}^{k} (-1)^{k-i}\frac{k!}{i!}x^i\right] = e^x\left[x^{k+1} - (k+1)\sum_{i=0}^{k}(-1)^{k-i}\frac{k!}{i!}x^i\right] \\ &= e^x\left[x^{k+1} + \sum_{i=0}^{k}(-1)^{k-i+1}\frac{(k+1)k!}{i!}x^i\right] = e^x \sum_{i=0}^{k+1}(-1)^{(k+1)-i}\frac{(k+1)!}{i!}x^i\end{aligned}$$

This verifies $S_n$ for $n = k+1$. Thus, by mathematical induction, $S_n$ is true for all $n$, where $n$ is a positive integer.

## 5.9 Approximate Integration

**1.** (a) $\Delta x = (b-a)/n = (4-0)/2 = 2$

$$L_2 = \sum_{i=1}^{2} f(x_{i-1})\,\Delta x = f(x_0)\cdot 2 + f(x_1)\cdot 2 = 2\,[f(0) + f(2)] = 2(0.5 + 2.5) = 6$$

$$R_2 = \sum_{i=1}^{2} f(x_i)\,\Delta x = f(x_1)\cdot 2 + f(x_2)\cdot 2 = 2\,[f(2) + f(4)] = 2(2.5 + 3.5) = 12$$

$$M_2 = \sum_{i=1}^{2} f(\overline{x}_i)\Delta x = f(\overline{x}_1)\cdot 2 + f(\overline{x}_2)\cdot 2 = 2\,[f(1) + f(3)] \approx 2(1.6 + 3.2) = 9.6$$

(b)

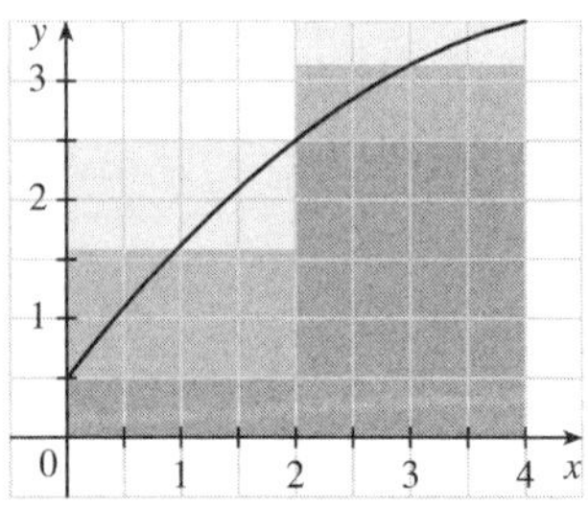


$L_2$ is an underestimate, since the area under the small rectangles is less than the area under the curve, and $R_2$ is an overestimate, since the area under the large rectangles is greater than the area under the curve. It appears that $M_2$ is an overestimate, though it is fairly close to $I$. See the solution to Exercise 37 for a proof of the fact that if $f$ is concave down on $[a,b]$, then the Midpoint Rule is an overestimate of $\int_a^b f(x)\,dx$.

(c) $T_2 = \left(\frac{1}{2}\,\Delta x\right)[f(x_0) + 2f(x_1) + f(x_2)] = \frac{2}{2}[f(0) + 2f(2) + f(4)] = 0.5 + 2(2.5) + 3.5 = 9.$
This approximation is an underestimate, since the graph is concave down. Thus, $T_2 = 9 < I$. See the solution to Exercise 37 for a general proof of this conclusion.

(d) For any $n$, we will have $L_n < T_n < I < M_n < R_n$.

**2.**

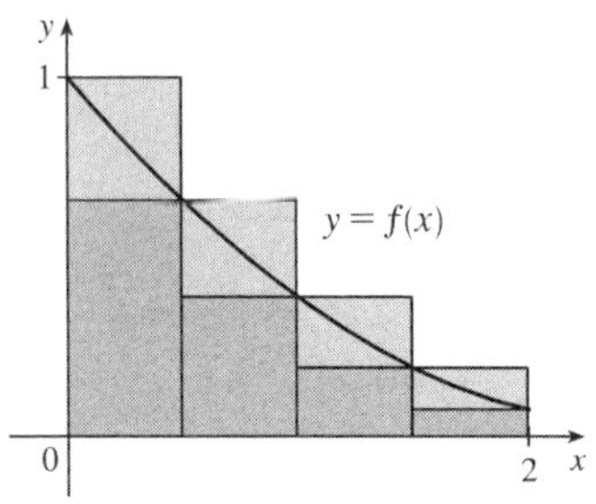


The diagram shows that $L_4 > T_4 > \int_0^2 f(x)\,dx > R_4$, and it appears that $M_4$ is a bit less than $\int_0^2 f(x)\,dx$. In fact, for any function that is concave upward, it can be shown that

$L_n > T_n > \int_0^2 f(x)\,dx > M_n > R_n$.

(a) Since $0.9540 > 0.8675 > 0.8632 > 0.7811$, it follows that $L_n = 0.9540$, $T_n = 0.8675$, $M_n = 0.8632$, and $R_n = 0.7811$.

(b) Since $M_n < \int_0^2 f(x)\,dx < T_n$, we have $0.8632 < \int_0^2 f(x)\,dx < 0.8675$.

3. $f(x) = \cos(x^2)$, $\Delta x = \frac{1-0}{4} = \frac{1}{4}$

(a) $T_4 = \frac{1}{4 \cdot 2}\left[f(0) + 2f\left(\frac{1}{4}\right) + 2f\left(\frac{2}{4}\right) + 2f\left(\frac{3}{4}\right) + f(1)\right] \approx 0.895759$

(b) $M_4 = \frac{1}{4}\left[f\left(\frac{1}{8}\right) + f\left(\frac{3}{8}\right) + f\left(\frac{5}{8}\right) + f\left(\frac{7}{8}\right)\right] \approx 0.908907$

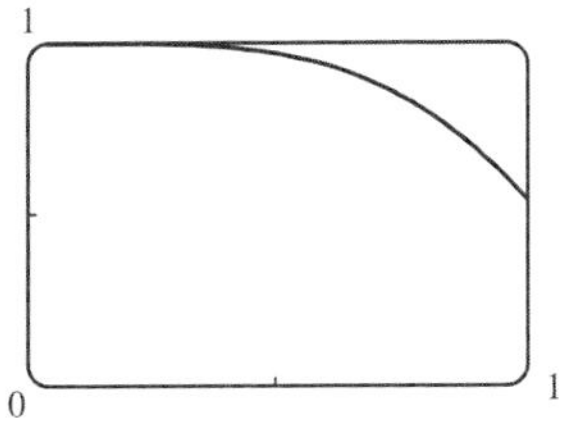


The graph shows that $f$ is concave down on $[0, 1]$. So $T_4$ is an underestimate and $M_4$ is an overestimate. We can conclude that $0.895759 < \int_0^1 \cos(x^2)\,dx < 0.908907$.

4. 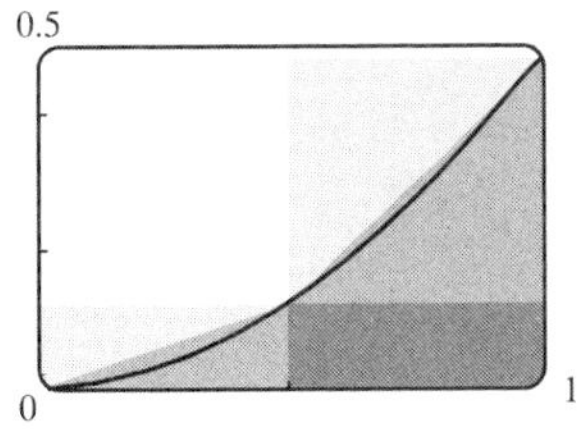


(a) Since $f$ is increasing on $[0, 1]$, $L_2$ will underestimate $I$ (since the area of the darkest rectangle is less than the area under the curve), and $R_2$ will overestimate $I$. Since $f$ is concave upward on $[0, 1]$, $M_2$ will underestimate $I$ and $T_2$ will overestimate $I$ (the area under the straight line segments is greater than the area under the curve).

(b) For any $n$, we will have $L_n < M_n < I < T_n < R_n$.

(c) $L_5 = \sum_{i=1}^{5} f(x_{i-1})\,\Delta x = \frac{1}{5}[f(0.0) + f(0.2) + f(0.4) + f(0.6) + f(0.8)] \approx 0.1187$

$R_5 = \sum_{i=1}^{5} f(x_i)\,\Delta x = \frac{1}{5}[f(0.2) + f(0.4) + f(0.6) + f(0.8) + f(1)] \approx 0.2146$

$M_5 = \sum_{i=1}^{5} f(\overline{x}_i)\,\Delta x = \frac{1}{5}[f(0.1) + f(0.3) + f(0.5) + f(0.7) + f(0.9)] \approx 0.1622$

$T_5 = \left(\frac{1}{2}\,\Delta x\right)[f(0) + 2f(0.2) + 2f(0.4) + 2f(0.6) + 2f(0.8) + f(1)] \approx 0.1666$

From the graph, it appears that the Midpoint Rule gives the best approximation. (This is in fact the case, since $I \approx 0.16371405$.)

5. $f(x) = x^2 \sin x$, $\Delta x = \dfrac{b-a}{n} = \dfrac{\pi - 0}{8} = \dfrac{\pi}{8}$

(a) $M_8 = \frac{\pi}{8}\left[f\left(\frac{\pi}{16}\right) + f\left(\frac{3\pi}{16}\right) + f\left(\frac{5\pi}{16}\right) + \cdots + f\left(\frac{15\pi}{16}\right)\right] \approx 5.932957$

(b) $S_8 = \frac{\pi}{8 \cdot 3}\left[f(0) + 4f\left(\frac{\pi}{8}\right) + 2f\left(\frac{2\pi}{8}\right) + 4f\left(\frac{3\pi}{8}\right) + 2f\left(\frac{4\pi}{8}\right) + 4f\left(\frac{5\pi}{8}\right) + 2f\left(\frac{6\pi}{8}\right) + 4f\left(\frac{7\pi}{8}\right) + f(\pi)\right]$

$\approx 5.869247$

*Actual:* $\int_0^\pi x^2 \sin x\,dx \overset{84}{=} \left[-x^2 \cos x\right]_0^\pi + 2\int_0^\pi x \cos x\,dx \overset{83}{=} \left[-\pi^2(-1) - 0\right] + 2[\cos x + x \sin x]_0^\pi$

$= \pi^2 + 2[(-1+0) - (1+0)] = \pi^2 - 4 \approx 5.869604$

*Errors:* $E_M = \text{actual} - M_8 = \int_0^\pi x^2 \sin x\,dx - M_8 \approx -0.063353$

$E_S = \text{actual} - S_8 = \int_0^\pi x^2 \sin x\,dx - S_8 \approx 0.000357$

6. $f(x) = e^{-\sqrt{x}}$, $\Delta x = \dfrac{b-a}{n} = \dfrac{1-0}{6} = \dfrac{1}{6}$

(a) $M_6 = \frac{1}{6}\left[f\left(\frac{1}{12}\right) + f\left(\frac{3}{12}\right) + f\left(\frac{5}{12}\right) + f\left(\frac{7}{12}\right) + f\left(\frac{9}{12}\right) + f\left(\frac{11}{12}\right)\right] \approx 0.525100$

(b) $S_6 = \frac{1}{6 \cdot 3}\left[f(0) + 4f\left(\frac{1}{6}\right) + 2f\left(\frac{2}{6}\right) + 4f\left(\frac{3}{6}\right) + 2f\left(\frac{4}{6}\right) + 4f\left(\frac{5}{6}\right) + f(1)\right] \approx 0.533979$

*Actual:* $\int_0^1 e^{-\sqrt{x}}\,dx = \int_0^{-1} e^u 2u\,du \quad [u = -\sqrt{x},\ u^2 = x,\ 2u\,du = dx]$

$\overset{96}{=} 2[(u-1)e^u]_0^{-1} = 2\left[-2e^{-1} - (-1e^0)\right] = 2 - 4e^{-1} \approx 0.528482$

*Errors:* $E_M = \text{actual} - M_6 = \int_0^1 e^{-\sqrt{x}}\,dx - M_6 \approx 0.003382$

$E_S = \text{actual} - S_6 = \int_0^1 e^{-\sqrt{x}}\,dx - S_6 \approx -0.005497$

7. $f(x) = \sqrt[4]{1+x^2}$, $\Delta x = \dfrac{2-0}{8} = \dfrac{1}{4}$

(a) $T_8 = \frac{1}{4 \cdot 2}\left[f(0) + 2f\left(\frac{1}{4}\right) + 2f\left(\frac{1}{2}\right) + \cdots + 2f\left(\frac{3}{2}\right) + 2f\left(\frac{7}{4}\right) + f(2)\right] \approx 2.413790$

(b) $M_8 = \frac{1}{4}\left[f\left(\frac{1}{8}\right) + f\left(\frac{3}{8}\right) + \cdots + f\left(\frac{13}{8}\right) + f\left(\frac{15}{8}\right)\right] \approx 2.411453$

(c) $S_8 = \frac{1}{4 \cdot 3}\left[f(0) + 4f\left(\frac{1}{4}\right) + 2f\left(\frac{1}{2}\right) + 4f\left(\frac{3}{4}\right) + 2f(1) + 4f\left(\frac{5}{4}\right) + 2f\left(\frac{3}{2}\right) + 4f\left(\frac{7}{4}\right) + f(2)\right] \approx 2.412232$

8. $f(x) = \sin(x^2)$, $\Delta x = \dfrac{\frac{1}{2}-0}{4} = \dfrac{1}{8}$

(a) $T_4 = \frac{1}{8 \cdot 2}\left[f(0) + 2f\left(\frac{1}{8}\right) + 2f\left(\frac{2}{8}\right) + 2f\left(\frac{3}{8}\right) + f\left(\frac{1}{2}\right)\right] \approx 0.042743$

(b) $M_4 = \frac{1}{8}\left[f\left(\frac{1}{16}\right) + f\left(\frac{3}{16}\right) + f\left(\frac{5}{16}\right) + f\left(\frac{7}{16}\right)\right] \approx 0.040850$

(c) $S_4 = \frac{1}{8 \cdot 3}\left[f(0) + 4f\left(\frac{1}{8}\right) + 2f\left(\frac{2}{8}\right) + 4f\left(\frac{3}{8}\right) + f\left(\frac{1}{2}\right)\right] \approx 0.041478$

9. $f(x) = \dfrac{\ln x}{1+x}$, $\Delta x = \dfrac{2-1}{10} = \dfrac{1}{10}$

(a) $T_{10} = \frac{1}{10 \cdot 2}[f(1) + 2f(1.1) + 2f(1.2) + \cdots + 2f(1.8) + 2f(1.9) + f(2)] \approx 0.146879$

(b) $M_{10} = \frac{1}{10}[f(1.05) + f(1.15) + \cdots + f(1.85) + f(1.95)] \approx 0.147391$

(c) $S_{10} = \frac{1}{10 \cdot 3}[f(1) + 4f(1.1) + 2f(1.2) + 4f(1.3) + 2f(1.4) + 4f(1.5) + 2f(1.6) + 4f(1.7)$
$+ 2f(1.8) + 4f(1.9) + f(2)]$

$\approx 0.147219$

10. $f(t) = \dfrac{1}{1+t^2+t^4}$, $\Delta t = \dfrac{3-0}{6} = \dfrac{1}{2}$

(a) $T_6 = \frac{1}{2 \cdot 2}\left[f(0) + 2f\left(\frac{1}{2}\right) + 2f(1) + 2f\left(\frac{3}{2}\right) + 2f(2) + 2f\left(\frac{5}{2}\right) + f(3)\right] \approx 0.895122$

(b) $M_6 = \frac{1}{2}\left[f\left(\frac{1}{4}\right) + f\left(\frac{3}{4}\right) + f\left(\frac{5}{4}\right) + f\left(\frac{7}{4}\right) + f\left(\frac{9}{4}\right) + f\left(\frac{11}{4}\right)\right] \approx 0.895478$

(c) $S_6 = \frac{1}{2 \cdot 3}\left[f(0) + 4f\left(\frac{1}{2}\right) + 2f(1) + 4f\left(\frac{3}{2}\right) + 2f(2) + 4f\left(\frac{5}{2}\right) + f(3)\right] \approx 0.898014$

11. $f(t) = e^{\sqrt{t}} \sin t$, $\Delta t = \dfrac{4-0}{8} = \dfrac{1}{2}$

(a) $T_8 = \frac{1}{2 \cdot 2}\left[f(0) + 2f\left(\frac{1}{2}\right) + 2f(1) + 2f\left(\frac{3}{2}\right) + 2f(2) + 2f\left(\frac{5}{2}\right) + 2f(3) + 2f\left(\frac{7}{2}\right) + f(4)\right] \approx 4.513618$

(b) $M_8 = \frac{1}{2}\left[f\left(\frac{1}{4}\right) + f\left(\frac{3}{4}\right) + f\left(\frac{5}{4}\right) + f\left(\frac{7}{4}\right) + f\left(\frac{9}{4}\right) + f\left(\frac{11}{4}\right) + f\left(\frac{13}{4}\right) + f\left(\frac{15}{4}\right)\right] \approx 4.748256$

(c) $S_8 = \frac{1}{2 \cdot 3}\left[f(0) + 4f\left(\frac{1}{2}\right) + 2f(1) + 4f\left(\frac{3}{2}\right) + 2f(2) + 4f\left(\frac{5}{2}\right) + 2f(3) + 4f\left(\frac{7}{2}\right) + f(4)\right] \approx 4.675111$

12. $f(x) = \sqrt{1+\sqrt{x}}$, $\Delta x = \dfrac{4-0}{8} = \dfrac{1}{2}$

(a) $T_8 = \frac{1}{2 \cdot 2}\left[f(0) + 2f\left(\frac{1}{2}\right) + 2f(1) + \cdots + 2f(3) + 2f\left(\frac{7}{2}\right) + f(4)\right] \approx 6.042985$

(b) $M_8 = \frac{1}{2}\left[f\left(\frac{1}{4}\right) + f\left(\frac{3}{4}\right) + \cdots + f\left(\frac{13}{4}\right) + f\left(\frac{15}{4}\right)\right] \approx 6.084778$

(c) $S_8 = \frac{1}{2 \cdot 3}\left[f(0) + 4f\left(\frac{1}{2}\right) + 2f(1) + 4f\left(\frac{3}{2}\right) + 2f(2) + 4f\left(\frac{5}{2}\right) + 2f(3) + 4f\left(\frac{7}{2}\right) + f(4)\right] \approx 6.061678$

13. $f(x) = \dfrac{\cos x}{x}$, $\Delta x = \dfrac{5-1}{8} = \dfrac{1}{2}$

(a) $T_8 = \frac{1}{2 \cdot 2}\left[f(1) + 2f\left(\frac{3}{2}\right) + 2f(2) + \cdots + 2f(4) + 2f\left(\frac{9}{2}\right) + f(5)\right] \approx -0.495333$

(b) $M_8 = \frac{1}{2}\left[f\left(\frac{5}{4}\right) + f\left(\frac{7}{4}\right) + f\left(\frac{9}{4}\right) + f\left(\frac{11}{4}\right) + f\left(\frac{13}{4}\right) + f\left(\frac{15}{4}\right) + f\left(\frac{17}{4}\right) + f\left(\frac{19}{4}\right)\right] \approx -0.543321$

(c) $S_8 = \frac{1}{2 \cdot 3}\left[f(1) + 4f\left(\frac{3}{2}\right) + 2f(2) + 4f\left(\frac{5}{2}\right) + 2f(3) + 4f\left(\frac{7}{2}\right) + 2f(4) + 4f\left(\frac{9}{2}\right) + f(5)\right] \approx -0.526123$

14. $f(x) = \ln(x^3+2)$, $\Delta x = \dfrac{6-4}{10} = \dfrac{1}{5}$

(a) $T_{10} = \frac{1}{5 \cdot 2}[f(4) + 2f(4.2) + 2f(4.4) + \cdots + 2f(5.6) + 2f(5.8) + f(6)] \approx 9.649753$

(b) $M_{10} = \frac{1}{5}[f(4.1) + f(4.3) + \cdots + f(5.7) + f(5.9)] \approx 9.650912$

(c) $S_{10} = \frac{1}{5 \cdot 3}[f(4) + 4f(4.2) + 2f(4.4) + 4f(4.6) + 2f(4.8) + 4f(5) + 2f(5.2) + 4f(5.4)$
$+2f(5.6) + 4f(5.8) + f(6)]$

$\approx 9.650526$

**15.** $f(y) = \dfrac{1}{1+y^5}$, $\Delta y = \dfrac{3-0}{6} = \dfrac{1}{2}$

(a) $T_6 = \frac{1}{2\cdot 2}\left[f(0) + 2f\left(\frac{1}{2}\right) + 2f\left(\frac{2}{2}\right) + 2f\left(\frac{3}{2}\right) + 2f\left(\frac{4}{2}\right) + 2f\left(\frac{5}{2}\right) + f(3)\right] \approx 1.064275$

(b) $M_6 = \frac{1}{2}\left[f\left(\frac{1}{4}\right) + f\left(\frac{3}{4}\right) + f\left(\frac{5}{4}\right) + f\left(\frac{7}{4}\right) + f\left(\frac{9}{4}\right) + f\left(\frac{11}{4}\right)\right] \approx 1.067416$

(c) $S_6 = \frac{1}{2\cdot 3}\left[f(0) + 4f\left(\frac{1}{2}\right) + 2f\left(\frac{2}{2}\right) + 4f\left(\frac{3}{2}\right) + 2f\left(\frac{4}{2}\right) + 4f\left(\frac{5}{2}\right) + f(3)\right] \approx 1.074915$

**16.** $f(z) = \sqrt{z}e^{-z}$, $\Delta z = \dfrac{1-0}{10} = \dfrac{1}{10}$

(a) $T_{10} = \frac{1}{10\cdot 2}\left\{f(0) + 2\left[f(0.1) + f(0.2) + \cdots + f(0.9)\right] + f(1)\right\} \approx 0.372299$

(b) $M_{10} = \frac{1}{10}\left[f(0.05) + f(0.15) + f(0.25) + \cdots + f(0.95)\right] \approx 0.380894$

(c) $S_{10} = \frac{1}{10\cdot 3}[f(0) + 4f(0.1) + 2f(0.2) + 4f(0.3) + 2f(0.4) + 4f(0.5) + 2f(0.6)$
$+ 4f(0.7) + 2f(0.8) + 4f(0.9) + f(1)]$
$\approx 0.376330$

**17.** (a) $f(x) = e^{-x^2}$, $\Delta x = \dfrac{2-0}{10} = \dfrac{1}{5}$

$T_{10} = \frac{1}{5\cdot 2}\{f(0) + 2[f(0.2) + f(0.4) + \cdots + f(1.8)] + f(2)\} \approx 0.881839$

$M_{10} = \frac{1}{5}[f(0.1) + f(0.3) + f(0.5) + \cdots + f(1.7) + f(1.9)] \approx 0.882202$

(b) $f(x) = e^{-x^2}$, $f'(x) = -2xe^{-x^2}$, $f''(x) = \left(4x^2 - 2\right)e^{-x^2}$, $f'''(x) = 4x\left(3 - 2x^2\right)e^{-x^2}$.

$f'''(x) = 0 \quad\Leftrightarrow\quad x = 0$ or $x = \pm\sqrt{\frac{3}{2}}$. So to find the maximum value of $|f''(x)|$ on $[0, 2]$, we need only consider its values at $x = 0$, $x = 2$, and $x = \sqrt{\frac{3}{2}}$. $|f''(0)| = 2$, $|f''(2)| \approx 0.2564$ and $\left|f''\left(\sqrt{\frac{3}{2}}\right)\right| \approx 0.8925$. Thus, taking $K = 2$, $a = 0$, $b = 2$, and $n = 10$ in Theorem 3, we get $|E_T| \le 2 \cdot 2^3/\left(12 \cdot 10^2\right) = \frac{1}{75} = 0.01\overline{3}$, and $|E_M| \le |E_T|/2 \le 0.00\overline{6}$.

(c) Take $K = 2$ [as in part (b)] in Theorem 3. $|E_T| \le \dfrac{K(b-a)^3}{12n^2} \le 10^{-5} \quad\Leftrightarrow\quad \dfrac{2(2-0)^3}{12n^2} \le 10^{-5} \quad\Leftrightarrow\quad \frac{3}{4}n^2 \ge 10^5 \quad\Leftrightarrow$ $n \ge 365.1\ldots \quad\Leftrightarrow\quad n \ge 366$. Take $n = 366$ for $T_n$. For $E_M$, again take $K = 2$ in Theorem 3 to get $|E_M| \le 10^{-5} \quad\Leftrightarrow$ $\frac{3}{2}n^2 \ge 10^5 \quad\Leftrightarrow\quad n \ge 258.2 \quad\Rightarrow\quad n \ge 259$. Take $n = 259$ for $M_n$.

**18.** (a) $T_8 = \frac{1}{8\cdot 2}\left\{f(0) + 2\left[f\left(\frac{1}{8}\right) + f\left(\frac{2}{8}\right) + \cdots + f\left(\frac{7}{8}\right)\right] + f(1)\right\} \approx 0.902333$

$M_8 = \frac{1}{8}\left[f\left(\frac{1}{16}\right) + f\left(\frac{3}{16}\right) + f\left(\frac{5}{16}\right) + \cdots + f\left(\frac{15}{16}\right)\right] = 0.905620$

(b) $f(x) = \cos(x^2)$, $f'(x) = -2x\sin(x^2)$, $f''(x) = -2\sin(x^2) - 4x^2\cos(x^2)$. For $0 \le x \le 1$, sin and cos are positive, so $|f''(x)| = 2\sin(x^2) + 4x^2\cos(x^2) \le 2 \cdot 1 + 4 \cdot 1 \cdot 1 = 6$ since $\sin(x^2) \le 1$ and $\cos(x^2) \le 1$ for all $x$, and $x^2 \le 1$ for $0 \le x \le 1$. So for $n = 8$, we take $K = 6$, $a = 0$, and $b = 1$ in Theorem 3, to get $|E_T| \le 6 \cdot 1^3/\left(12 \cdot 8^2\right) = \frac{1}{128} = 0.0078125$ and $|E_M| \le \frac{1}{256} = 0.00390625$. [A better estimate is obtained by noting from a graph of $f''$ that $|f''(x)| \le 4$ for $0 \le x \le 1$.]

(c) Using $K = 6$ as in part (b), we have $|E_T| \le 6 \cdot 1^3/\left(12n^2\right) = 1/\left(2n^2\right) \le 10^{-5} \quad\Rightarrow\quad 2n^2 \ge 10^5 \quad\Rightarrow\quad n \ge \sqrt{\frac{1}{2} \cdot 10^5}$ or $n \ge 224$. To guarantee that $|E_M| \le 0.00001$, we need $6 \cdot 1^3/\left(24n^2\right) \le 10^{-5} \quad\Rightarrow\quad 4n^2 \ge 10^5 \quad\Rightarrow\quad n \ge \sqrt{\frac{1}{4} \cdot 10^5}$ or $n \ge 159$.

**19.** (a) $T_{10} = \frac{1}{10\cdot 2}\{f(0) + 2[f(0.1) + f(0.2) + \cdots + f(0.9)] + f(1)\} \approx 1.71971349$

$S_{10} = \frac{1}{10\cdot 3}[f(0) + 4f(0.1) + 2f(0.2) + 4f(0.3) + \cdots + 4f(0.9) + f(1)] \approx 1.71828278$

Since $I = \int_0^1 e^x\,dx = \left[e^x\right]_0^1 = e - 1 \approx 1.71828183$, $E_T = I - T_{10} \approx -0.00143166$ and $E_S = I - S_{10} \approx -0.00000095$.

(b) $f(x) = e^x \quad\Rightarrow\quad f''(x) = e^x \le e$ for $0 \le x \le 1$. Taking $K = e$, $a = 0$, $b = 1$, and $n = 10$ in Theorem 3, we get $|E_T| \le e(1)^3/\left(12 \cdot 10^2\right) \approx 0.002265 > 0.00143166$ [actual $|E_T|$ from (a)]. $f^{(4)}(x) = e^x < e$ for $0 \le x \le 1$. Using Theorem 4, we have $|E_S| \le e(1)^5/\left(180 \cdot 10^4\right) \approx 0.0000015 > 0.00000095$ [actual $|E_S|$ from (a)]. We see that the actual errors are about two-thirds the size of the error estimates.

(c) From part (b), we take $K = e$ to get $|E_T| \le \dfrac{K(b-a)^3}{12n^2} \le 0.00001 \Rightarrow n^2 \ge \dfrac{e(1^3)}{12(0.00001)} \Rightarrow n \ge 150.5$. Take $n = 151$ for $T_n$. Now $|E_M| \le \dfrac{K(b-a)^3}{24n^2} \le 0.00001 \Rightarrow n \ge 106.4$. Take $n = 107$ for $M_n$. Finally, $|E_S| \le \dfrac{K(b-a)^5}{180n^4} \le 0.00001 \Rightarrow n^4 \ge \dfrac{e(1^5)}{180(0.00001)} \Rightarrow n \ge 6.23$. Take $n = 8$ for $S_n$ (since $n$ has to be even for Simpson's Rule).

**20.** From Example 7(b), we take $K = 76e$ to get $|E_S| \le 76e(1)^5/(180n^4) \le 0.00001 \Rightarrow n^4 \ge 76e/[180(0.00001)] \Rightarrow n \ge 18.4$. Take $n = 20$ (since $n$ must be even).

**21.** (a) Using a CAS, we differentiate $f(x) = e^{\cos x}$ twice, and find that $f''(x) = e^{\cos x}(\sin^2 x - \cos x)$. From the graph, we see that the maximum value of $|f''(x)|$ occurs at the endpoints of the interval $[0, 2\pi]$. Since $f''(0) = -e$, we can use $K = e$ or $K = 2.8$.

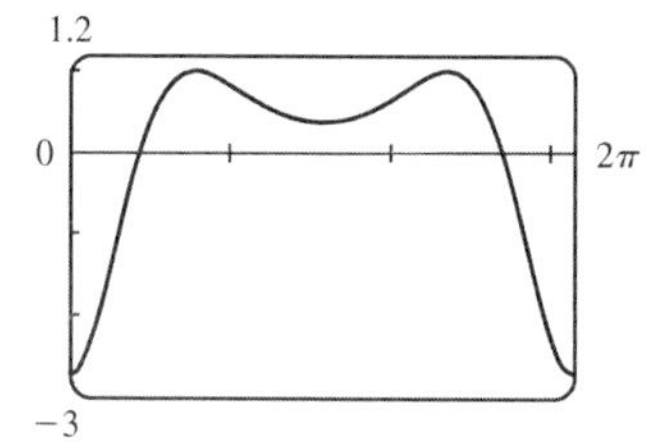


(b) A CAS gives $M_{10} \approx 7.954926518$. (In Maple, use `student[middlesum]`.)

(c) Using Theorem 3 for the Midpoint Rule, with $K = e$, we get $|E_M| \le \dfrac{e(2\pi - 0)^3}{24 \cdot 10^2} \approx 0.280945995$. With $K = 2.8$, we get $|E_M| \le \dfrac{2.8(2\pi - 0)^3}{24 \cdot 10^2} = 0.289391916$.

(d) A CAS gives $I \approx 7.954926521$.

(e) The actual error is only about $3 \times 10^{-9}$, much less than the estimate in part (c).

(f) We use the CAS to differentiate twice more, and then graph $f^{(4)}(x) = e^{\cos x}(\sin^4 x - 6\sin^2 x \cos x + 3 - 7\sin^2 x + \cos x)$. From the graph, we see that the maximum value of $\left|f^{(4)}(x)\right|$ occurs at the endpoints of the interval $[0, 2\pi]$. Since $f^{(4)}(0) = 4e$, we can use $K = 4e$ or $K = 10.9$.

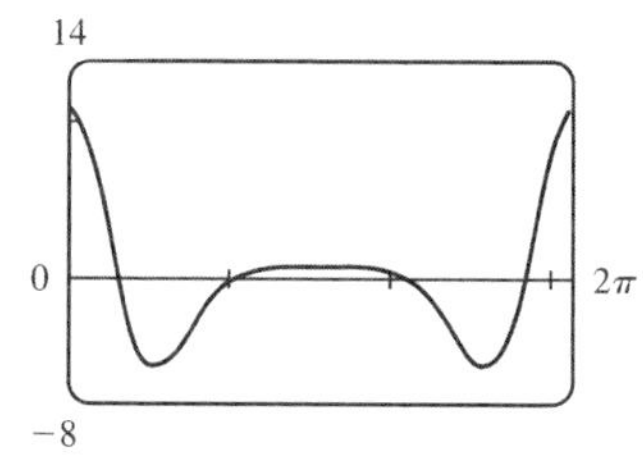


(g) A CAS gives $S_{10} \approx 7.953789422$. (In Maple, use `student[simpson]`.)

(h) Using Theorem 4 with $K = 4e$, we get $|E_S| \le \dfrac{4e(2\pi - 0)^5}{180 \cdot 10^4} \approx 0.059153618$. With $K = 10.9$, we get $|E_S| \le \dfrac{10.9(2\pi - 0)^5}{180 \cdot 10^4} \approx 0.059299814$.

(i) The actual error is about $7.954926521 - 7.953789422 \approx 0.00114$. This is quite a bit smaller than the estimate in part (h), though the difference is not nearly as great as it was in the case of the Midpoint Rule.

(j) To ensure that $|E_S| \le 0.0001$, we use Theorem 4: $|E_S| \le \dfrac{4e(2\pi)^5}{180 \cdot n^4} \le 0.0001 \Rightarrow \dfrac{4e(2\pi)^5}{180 \cdot 0.0001} \le n^4 \Rightarrow n^4 \ge 5{,}915{,}362 \Leftrightarrow n \ge 49.3$. So we must take $n \ge 50$ to ensure that $|I - S_n| \le 0.0001$. ($K = 10.9$ leads to the same value of $n$.)

**22.** (a) Using the CAS, we differentiate $f(x) = \sqrt{4 - x^3}$ twice, and find that $f''(x) = -\dfrac{9x^4}{4(4-x^3)^{3/2}} - \dfrac{3x}{(4-x^3)^{1/2}}$.

From the graph, we see that $|f''(x)| < 2.2$ on $[-1, 1]$.

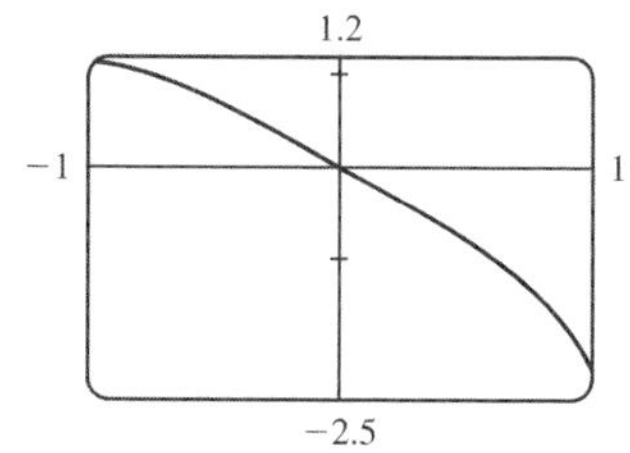


(b) A CAS gives $M_{10} \approx 3.995804152$. (In Maple, use `student[middlesum]`.)

(c) Using Theorem 3 for the Midpoint Rule, with $K = 2.2$, we get $|E_M| \le \dfrac{2.2\,[1-(-1)]^3}{24 \cdot 10^2} \approx 0.00733$.

(d) A CAS gives $I \approx 3.995487677$.

(e) The actual error is about $-0.0003165$, much less than the estimate in part (c).

(f) We use the CAS to differentiate twice more, and then graph

$$f^{(4)}(x) = \frac{9}{16}\frac{x^2(x^6 - 224x^3 - 1280)}{(4-x^3)^{7/2}}.$$

From the graph, we see that $\left|f^{(4)}(x)\right| < 18.1$ on $[-1, 1]$.

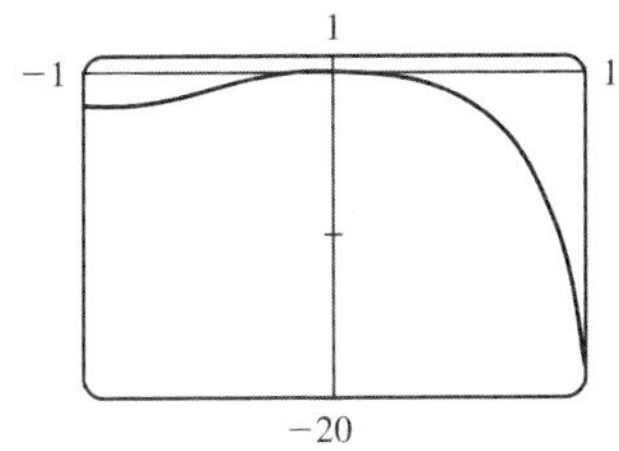


(g) A CAS gives $S_{10} \approx 3.995449790$. (In Maple, use `student[simpson]`.)

(h) Using Theorem 4 with $K = 18.1$, we get $|E_S| \le \dfrac{18.1\,[1-(-1)]^5}{180 \cdot 10^4} \approx 0.000322$.

(i) The actual error is about $3.995487677 - 3.995449790 \approx 0.0000379$. This is quite a bit smaller than the estimate in part (h).

(j) To ensure that $|E_S| \le 0.0001$,we use Theorem 4: $|E_S| \le \dfrac{18.1(2)^5}{180 \cdot n^4} \le 0.0001 \;\Rightarrow\; \dfrac{18.1(2)^5}{180 \cdot 0.0001} \le n^4 \;\Rightarrow$ $n^4 \ge 32{,}178 \;\Rightarrow n \ge 13.4$. So we must take $n \ge 14$ to ensure that $|I - S_n| \le 0.0001$.

**23.** $I = \int_0^1 x^3 dx = \left[\frac{1}{4}x^4\right]_0^1 = 0.25$. $f(x) = x^3$.

$n = 4$:

$$L_4 = \tfrac{1}{4}\left[0^3 + \left(\tfrac{1}{4}\right)^3 + \left(\tfrac{2}{4}\right)^3 + \left(\tfrac{3}{4}\right)^3\right] = 0.140625$$

$$R_4 = \tfrac{1}{4}\left[\left(\tfrac{1}{4}\right)^3 + \left(\tfrac{2}{4}\right)^3 + \left(\tfrac{3}{4}\right)^3 + 1^3\right] = 0.390625$$

$$T_4 = \tfrac{1}{4 \cdot 2}\left[0^3 + 2\left(\tfrac{1}{4}\right)^3 + 2\left(\tfrac{2}{4}\right)^3 + 2\left(\tfrac{3}{4}\right)^3 + 1^3\right] = 0.265625,$$

$$M_4 = \tfrac{1}{4}\left[\left(\tfrac{1}{8}\right)^3 + \left(\tfrac{3}{8}\right)^3 + \left(\tfrac{5}{8}\right)^3 + \left(\tfrac{7}{8}\right)^3\right] = 0.2421875,$$

$$E_L = I - L_4 = \tfrac{1}{4} - 0.140625 = 0.109375,\; E_R = \tfrac{1}{4} - 0.390625 = -0.140625,$$

$$E_T = \tfrac{1}{4} - 0.265625 = -0.015625,\; E_M = \tfrac{1}{4} - 0.2421875 = 0.0078125$$

$n = 8$:

$$L_8 = \tfrac{1}{8}\left[f(0) + f\left(\tfrac{1}{8}\right) + f\left(\tfrac{2}{8}\right) + \cdots + f\left(\tfrac{7}{8}\right)\right] \approx 0.191406$$

$$R_8 = \tfrac{1}{8}\left[f\left(\tfrac{1}{8}\right) + f\left(\tfrac{2}{8}\right) + \cdots + f\left(\tfrac{7}{8}\right) + f(1)\right] \approx 0.316406$$

$$T_8 = \tfrac{1}{8 \cdot 2}\left\{f(0) + 2\left[f\left(\tfrac{1}{8}\right) + f\left(\tfrac{2}{8}\right) + \cdots + f\left(\tfrac{7}{8}\right)\right] + f(1)\right\} \approx 0.253906$$

$$M_8 = \tfrac{1}{8}\left[f\left(\tfrac{1}{16}\right) + f\left(\tfrac{3}{16}\right) + \cdots + f\left(\tfrac{13}{16}\right) + f\left(\tfrac{15}{16}\right)\right] = 0.248047$$

$$E_L \approx \tfrac{1}{4} - 0.191406 \approx 0.058594,\; E_R \approx \tfrac{1}{4} - 0.316406 \approx -0.066406,$$

$$E_T \approx \tfrac{1}{4} - 0.253906 \approx -0.003906,\; E_M \approx \tfrac{1}{4} - 0.248047 \approx 0.001953.$$

$n = 16$:

$$L_{16} = \tfrac{1}{16}\left[f(0) + f\left(\tfrac{1}{16}\right) + f\left(\tfrac{2}{16}\right) + \cdots + f\left(\tfrac{15}{16}\right)\right] \approx 0.219727$$

$$R_{16} = \tfrac{1}{16}\left[f\left(\tfrac{1}{16}\right) + f\left(\tfrac{2}{16}\right) + \cdots + f\left(\tfrac{15}{16}\right) + f(1)\right] \approx 0.282227$$

$$T_{16} = \tfrac{1}{16 \cdot 2}\left\{f(0) + 2\left[f\left(\tfrac{1}{16}\right) + f\left(\tfrac{2}{16}\right) + \cdots + f\left(\tfrac{15}{16}\right)\right] + f(1)\right\} \approx 0.250977$$

$$M_{16} = \tfrac{1}{16}\left[f\left(\tfrac{1}{32}\right) + f\left(\tfrac{3}{32}\right) + \cdots + f\left(\tfrac{31}{32}\right)\right] \approx 0.249512$$

$$E_L \approx \tfrac{1}{4} - 0.219727 \approx 0.030273,\; E_R \approx \tfrac{1}{4} - 0.282227 \approx -0.032227,$$

$$E_T \approx \tfrac{1}{4} - 0.250977 \approx -0.000977,\; E_M \approx \tfrac{1}{4} - 0.249512 \approx 0.000488.$$

[continued]

| $n$ | $L_n$ | $R_n$ | $T_n$ | $M_n$ |
|---|---|---|---|---|
| 4 | 0.140625 | 0.390625 | 0.265625 | 0.242188 |
| 8 | 0.191406 | 0.316406 | 0.253906 | 0.248047 |
| 16 | 0.219727 | 0.282227 | 0.250977 | 0.249512 |

| $n$ | $E_L$ | $E_R$ | $E_T$ | $E_M$ |
|---|---|---|---|---|
| 4 | 0.109375 | −0.140625 | −0.015625 | 0.007813 |
| 8 | 0.058594 | −0.066406 | −0.003906 | 0.001953 |
| 16 | 0.030273 | −0.032227 | −0.000977 | 0.000488 |

*Observations:*

1. $E_L$ and $E_R$ are always opposite in sign, as are $E_T$ and $E_M$.
2. As $n$ is doubled, $E_L$ and $E_R$ are decreased by about a factor of 2, and $E_T$ and $E_M$ are decreased by a factor of about 4.
3. The Midpoint approximation is about twice as accurate as the Trapezoidal approximation.
4. All the approximations become more accurate as the value of $n$ increases.
5. The Midpoint and Trapezoidal approximations are much more accurate than the endpoint approximations.

**24.** $I = \int_{-1}^{2} xe^x\,dx = \left[xe^x - e^x\right]_{-1}^{2} = e^2 + 2/e \approx 8.124815.$ $f(x) = xe^x$.

$n = 6$: $\Delta x = [2-(-1)]/6 = \frac{1}{2}$

$T_6 = \frac{1}{2\cdot 2}\{f(-1) + 2[f(-0.5) + f(0) + \cdots + f(1.5)] + f(2)\} \approx 8.583514$

$M_6 = \frac{1}{2}[f(-0.75) + f(-0.25) + \cdots + f(1.75)] \approx 7.896632$

$S_6 = \frac{1}{2\cdot 3}[f(-1) + 4f(-0.5) + 2f(0) + 4f(0.5) + 2f(1) + 4f(1.5) + f(2)] \approx 8.136885$

$E_T \approx I - 8.583514 \approx -0.458699$, $E_M \approx I - 7.896632 \approx 0.228183$,

$E_S \approx I - 8.136885 \approx -0.012070$.

$n = 12$: $\Delta x = [2-(-1)]/12 = \frac{1}{4}$

$T_{12} = \frac{1}{4\cdot 2}\{f(-1) + 2[f(-0.75) + f(-0.5) + \cdots + f(1.75)] + f(2)\} \approx 8.240073$

$M_{12} = \frac{1}{4}\left[f\left(-\frac{7}{8}\right) + f\left(-\frac{5}{8}\right) + \cdots + f\left(\frac{13}{8}\right) + f\left(\frac{15}{8}\right)\right] \approx 8.067259$

$S_{12} = \frac{1}{4\cdot 3}[f(-1) + 4f(-0.75) + 2f(-0.5) + \cdots + 2f(1.5) + 4f(1.75) + f(2)] \approx 8.125593$

$E_T \approx I - 8.240073 \approx -0.115258$, $E_M \approx I - 8.067259 \approx 0.057556$,

$E_S \approx I - 8.125593 \approx -0.000778$

| $n$ | $T_n$ | $M_n$ | $S_n$ |
|---|---|---|---|
| 6 | 8.583514 | 7.896632 | 8.136885 |
| 12 | 8.240073 | 8.067259 | 8.125593 |

| $n$ | $E_T$ | $E_M$ | $E_S$ |
|---|---|---|---|
| 6 | −0.458699 | 0.228183 | −0.012070 |
| 12 | −0.115258 | 0.057556 | −0.000778 |

*Observations:*

1. $E_T$ and $E_M$ are opposite in sign and decrease by a factor of about 4 as $n$ is doubled.
2. The Simpson's approximation is much more accurate than the Midpoint and Trapezoidal approximations, and seems to decrease by a factor of about 16 as $n$ is doubled.

**25.** $\Delta x = (4-0)/4 = 1$

(a) $T_4 = \frac{1}{2}[f(0) + 2f(1) + 2f(2) + 2f(3) + f(4)] \approx \frac{1}{2}[0 + 2(3) + 2(5) + 2(3) + 1] = 11.5$

(b) $M_4 = 1\cdot[f(0.5) + f(1.5) + f(2.5) + f(3.5)] \approx 1 + 4.5 + 4.5 + 2 = 12$

(c) $S_4 = \frac{1}{3}[f(0) + 4f(1) + 2f(2) + 4f(3) + f(4)] \approx \frac{1}{3}[0 + 4(3) + 2(5) + 4(3) + 1] = 11.\overline{6}$

**26.** We use Simpson's Rule with $n = 10$ and $\Delta x = \frac{1}{2}$:

$$\begin{aligned}\text{distance} = \int_0^5 v(t)\,dt \approx S_{10} &= \tfrac{1}{2\cdot 3}[f(0) + 4f(0.5) + 2f(1) + \cdots + 4f(4.5) + f(5)] \\ &= \tfrac{1}{6}[0 + 4(4.67) + 2(7.34) + 4(8.86) + 2(9.73) + 4(10.22) \\ &\qquad + 2(10.51) + 4(10.67) + 2(10.76) + 4(10.81) + 10.81] \\ &= \tfrac{1}{6}(268.41) = 44.735 \text{ m}\end{aligned}$$

**27.** By the Net Change Theorem, the increase in velocity is equal to $\int_0^6 a(t)\,dt$. We use Simpson's Rule with $n = 6$ and $\Delta t = (6 - 0)/6 = 1$ to estimate this integral:

$$\begin{aligned}\int_0^6 a(t)\,dt \approx S_6 &= \tfrac{1}{3}[a(0) + 4a(1) + 2a(2) + 4a(3) + 2a(4) + 4a(5) + a(6)] \\ &\approx \tfrac{1}{3}[0 + 4(0.5) + 2(4.1) + 4(9.8) + 2(12.9) + 4(9.5) + 0] = \tfrac{1}{3}(113.2) = 37.7\overline{3} \text{ ft/s}\end{aligned}$$

**28.** By the Net Change Theorem, the total amount of water that leaked out during the first six hours is equal to $\int_0^6 r(t)\,dt$. We use Simpson's Rule with $n = 6$ and $\Delta t = \frac{6-0}{6} = 1$ to estimate this integral:

$$\begin{aligned}\int_0^6 r(t)\,dt \approx S_6 &= \tfrac{1}{3}[r(0) + 4r(1) + 2r(2) + 4r(3) + 2r(4) + 4r(5) + r(6)] \\ &\approx \tfrac{1}{3}[4 + 4(3) + 2(2.4) + 4(1.9) + 2(1.4) + 4(1.1) + 1] \\ &= \tfrac{1}{3}(36.6) = 12.2 \text{ liters}\end{aligned}$$

The function values were obtained from a high-resolution graph.

**29.** By the Net Change Theorem, the energy used is equal to $\int_0^6 P(t)\,dt$. We use Simpson's Rule with $n = 12$ and $\Delta t = (6 - 0)/12 = \frac{1}{2}$ to estimate this integral:

$$\begin{aligned}\int_0^6 P(t)\,dt \approx S_{12} &= \tfrac{1/2}{3}[P(0) + 4P(0.5) + 2P(1) + 4P(1.5) + 2P(2) + 4P(2.5) \\ &\qquad + 2P(3) + 4P(3.5) + 2P(4) + 4P(4.5) + 2P(5) + 4P(5.5) + P(6)] \\ &= \tfrac{1}{6}[1814 + 4(1735) + 2(1686) + 4(1646) + 2(1637) + 4(1609) + 2(1604) \\ &\qquad + 4(1611) + 2(1621) + 4(1666) + 2(1745) + 4(1886) + 2052] \\ &= \tfrac{1}{6}(61{,}064) = 10{,}177.\overline{3} \text{ megawatt-hours.}\end{aligned}$$

**30.** By the Net Change Theorem, the total amount of data transmitted is equal to $\int_0^8 D(t)\,dt \times 3600$ [since $D(t)$ is measured in megabits per second and $t$ is in hours]. We use Simpson's Rule with $n = 8$ and $\Delta t = (8 - 0)/8 = 1$ to estimate this integral:

$$\begin{aligned}\int_0^8 D(t)\,dt \approx S_8 &= \tfrac{1}{3}[D(0) + 4D(1) + 2D(2) + 4D(3) + 2D(4) + 4D(5) + 2D(6) + 4D(7) + D(8)] \\ &\approx \tfrac{1}{3}[0.35 + 4(0.32) + 2(0.41) + 4(0.50) + 2(0.51) + 4(0.56) + 2(0.56) + 4(0.83) + 0.88] \\ &= \tfrac{1}{3}(13.03) = 4.34\overline{3}\end{aligned}$$

Now multiply by 3600 to obtain 15,636 megabits.

**31.** (a) We are given the function values at the endpoints of 8 intervals of length 0.4, so we'll use the Midpoint Rule with $n = 8/2 = 4$ and $\Delta x = (3.2 - 0)/4 = 0.8$.

$$\begin{aligned}\int_0^{3.2} f(x)\,dx \approx M_4 &= 0.8[f(0.4) + f(1.2) + f(2.0) + f(2.8)] \\ &= 0.8[6.5 + 6.4 + 7.6 + 8.8] \\ &= 0.8(29.3) = 23.44\end{aligned}$$

(b) $-4 \le f''(x) \le 1 \;\Rightarrow\; |f''(x)| \le 4$, so use $K = 4$, $a = 0$, $b = 3.2$, and $n = 4$ in Theorem 3. So

$$|E_M| \le \frac{4(3.2 - 0)^3}{24(4)^2} = \frac{128}{375} = 0.341\overline{3}.$$

**32.** Using Simpson's Rule with $n = 10$, $\Delta x = \frac{\pi/2}{10}$, $L = 1$, $\theta_0 = \frac{42\pi}{180}$ radians, $g = 9.8\text{ m/s}^2$, $k^2 = \sin^2\left(\frac{1}{2}\theta_0\right)$, and $f(x) = 1/\sqrt{1 - k^2\sin^2 x}$, we get

$$T = 4\sqrt{\frac{L}{g}}\int_0^{\pi/2}\frac{dx}{\sqrt{1-k^2\sin^2 x}} \approx 4\sqrt{\frac{L}{g}}\,S_{10}$$

$$= 4\sqrt{\tfrac{1}{9.8}}\left(\tfrac{\pi/2}{10\cdot 3}\right)\left[f(0) + 4f\left(\tfrac{\pi}{20}\right) + 2f\left(\tfrac{2\pi}{20}\right) + \cdots + 4f\left(\tfrac{9\pi}{20}\right) + f\left(\tfrac{\pi}{2}\right)\right] \approx 2.07665$$

**33.** $I(\theta) = \dfrac{N^2\sin^2 k}{k^2}$, where $k = \dfrac{\pi N d\sin\theta}{\lambda}$, $N = 10{,}000$, $d = 10^{-4}$, and $\lambda = 632.8\times 10^{-9}$. So $I(\theta) = \dfrac{(10^4)^2\sin^2 k}{k^2}$, where $k = \dfrac{\pi(10^4)(10^{-4})\sin\theta}{632.8\times 10^{-9}}$. Now $n = 10$ and $\Delta\theta = \dfrac{10^{-6} - (-10^{-6})}{10} = 2\times 10^{-7}$, so $M_{10} = 2\times 10^{-7}[I(-0.0000009) + I(-0.0000007) + \cdots + I(0.0000009)] \approx 59.4$.

**34.** Consider the function $f(x) = |x - 1|$, $0 \le x \le 2$. The area $\int_0^2 f(x)\,dx$ is exactly 1. So is the right endpoint approximation: $R_2 = f(1)\,\Delta x + f(2)\,\Delta x = 0\cdot 1 + 1\cdot 1 = 1$. But Simpson's Rule approximates $f$ with the parabola $y = (x-1)^2$, shown dashed, and $S_2 = \dfrac{\Delta x}{3}\,[f(0) + 4f(1) + f(2)] = \dfrac{1}{3}\,[1 + 4\cdot 0 + 1] = \dfrac{2}{3}$.

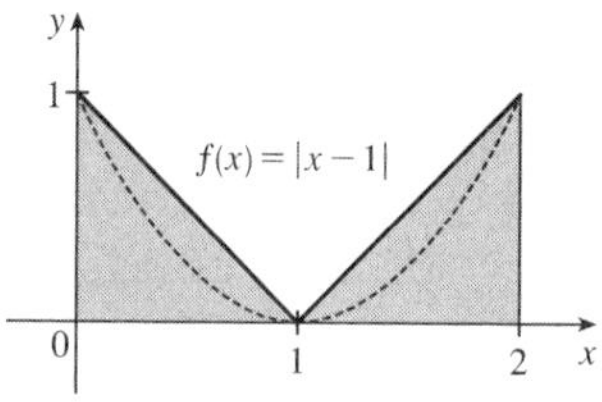


**35.** Consider the function $f$ whose graph is shown. The area $\int_0^2 f(x)\,dx$ is close to 2. The Trapezoidal Rule gives

$T_2 = \frac{2-0}{2\cdot 2}\,[f(0) + 2f(1) + f(2)] = \frac{1}{2}\,[1 + 2\cdot 1 + 1] = 2.$

The Midpoint Rule gives

$M_2 = \frac{2-0}{2}\,[f(0.5) + f(1.5)] = 1[0 + 0] = 0,$

so the Trapezoidal Rule is more accurate.

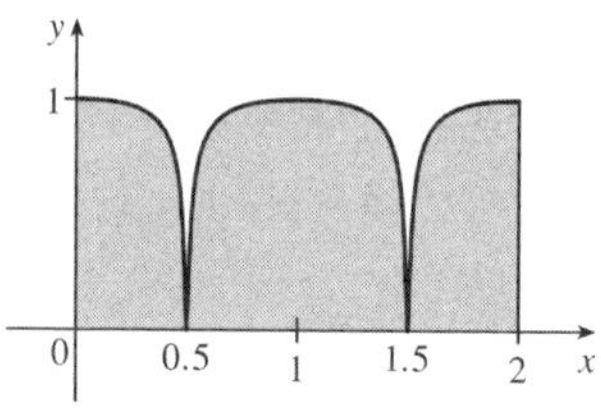


**36.** $f(x) = \cos(\pi x)$, $\Delta x = \frac{20-0}{10} = 2 \;\Rightarrow$

$$\begin{aligned} T_{10} &= \tfrac{2}{2}\{f(0) + 2[f(2) + f(4) + \cdots + f(18)] + f(20)\} \\ &= 1[\cos 0 + 2(\cos 2\pi + \cos 4\pi + \cdots + \cos 18\pi) + \cos 20\pi] \\ &= 1 + 2(1+1+1+1+1+1+1+1+1) + 1 = 20 \end{aligned}$$

The actual value is $\int_0^{20}\cos(\pi x)\,dx = \frac{1}{\pi}[\sin\pi x]_0^{20} = \frac{1}{\pi}(\sin 20\pi - \sin 0) = 0$. The discrepancy is due to the fact that the function is sampled only at points of the form $2n$, where its value is $f(2n) = \cos(2n\pi) = 1$.

**37.** Since the Trapezoidal and Midpoint approximations on the interval $[a,b]$ are the sums of the Trapezoidal and Midpoint approximations on the subintervals $[x_{i-1}, x_i]$, $i = 1, 2, \ldots, n$, we can focus our attention on one such interval. The condition $f''(x) < 0$ for $a \le x \le b$ means that the graph of $f$ is concave down as in Figure 5. In that figure, $T_n$ is the area of the trapezoid $AQRD$, $\int_a^b f(x)\,dx$ is the area of the region $AQPRD$, and $M_n$ is the area of the trapezoid $ABCD$, so $T_n < \int_a^b f(x)\,dx < M_n$. In general, the condition $f'' < 0$ implies that the graph of $f$ on $[a,b]$ lies above the chord joining the points $(a, f(a))$ and $(b, f(b))$. Thus, $\int_a^b f(x)\,dx > T_n$. Since $M_n$ is the area under a tangent to the graph, and since $f'' < 0$ implies that the tangent lies above the graph, we also have $M_n > \int_a^b f(x)\,dx$. Thus, $T_n < \int_a^b f(x)\,dx < M_n$.

**38.** Let $f$ be a polynomial of degree $\le 3$; say $f(x) = Ax^3 + Bx^2 + Cx + D$. It will suffice to show that Simpson's estimate is exact when there are two subintervals ($n = 2$), because for a larger even number of subintervals the sum of exact estimates is exact. As in the derivation of Simpson's Rule, we can assume that $x_0 = -h$, $x_1 = 0$, and $x_2 = h$. Then Simpson's approximation is

$$\begin{aligned}\int_{-h}^{h} f(x)\,dx &\approx \tfrac{1}{3}h[f(-h) + 4f(0) + f(h)] \\ &= \tfrac{1}{3}h\big[\big(-Ah^3 + Bh^2 - Ch + D\big) + 4D + \big(Ah^3 + Bh^2 + Ch + D\big)\big] \\ &= \tfrac{1}{3}h\big[2Bh^2 + 6D\big] = \tfrac{2}{3}Bh^3 + 2Dh\end{aligned}$$

The exact value of the integral is

$$\begin{aligned}\int_{-h}^{h} \big(Ax^3 + Bx^2 + Cx + D\big)dx &= 2\int_0^h \big(Bx^2 + D\big)\,dx \quad \text{[by Theorem 5.5.6(a) and (b)]} \\ &= 2\big[\tfrac{1}{3}Bx^3 + Dx\big]_0^h = \tfrac{2}{3}Bh^3 + 2Dh\end{aligned}$$

Thus, Simpson's Rule is exact.

**39.** $T_n = \frac{1}{2}\,\Delta x\,[f(x_0) + 2f(x_1) + \cdots + 2f(x_{n-1}) + f(x_n)]$ and
$M_n = \Delta x\,[f(\overline{x}_1) + f(\overline{x}_2) + \cdots + f(\overline{x}_{n-1}) + f(\overline{x}_n)]$, where $\overline{x}_i = \frac{1}{2}(x_{i-1} + x_i)$. Now

$$\begin{aligned}T_{2n} = \tfrac{1}{2}\big(\tfrac{1}{2}\Delta x\big)[f(x_0) &+ 2f(\overline{x}_1) + 2f(x_1) + 2f(\overline{x}_2) + 2f(x_2) + \cdots \\ &+ 2f(\overline{x}_{n-1}) + 2f(x_{n-1}) + 2f(\overline{x}_n) + f(x_n)]\end{aligned}$$

so

$$\begin{aligned}\tfrac{1}{2}(T_n + M_n) &= \tfrac{1}{2}T_n + \tfrac{1}{2}M_n \\ &= \tfrac{1}{4}\,\Delta x\,[f(x_0) + 2f(x_1) + \cdots + 2f(x_{n-1}) + f(x_n)] \\ &\qquad + \tfrac{1}{4}\Delta x\,[2f(\overline{x}_1) + 2f(\overline{x}_2) + \cdots + 2f(\overline{x}_{n-1}) + 2f(\overline{x}_n)] \\ &= T_{2n}\end{aligned}$$

**40.** $T_n = \dfrac{\Delta x}{2}\left[f(x_0) + 2\displaystyle\sum_{i=1}^{n-1} f(x_i) + f(x_n)\right]$ and $M_n = \Delta x \displaystyle\sum_{i=1}^{n} f\!\left(x_i - \frac{\Delta x}{2}\right)$, so

$$\tfrac{1}{3}T_n + \tfrac{2}{3}M_n = \tfrac{1}{3}(T_n + 2M_n) = \frac{\Delta x}{3\cdot 2}\left[f(x_0) + 2\sum_{i=1}^{n-1} f(x_i) + f(x_n) + 4\sum_{i=1}^{n} f\!\left(x_i - \frac{\Delta x}{2}\right)\right]$$

where $\Delta x = \dfrac{b-a}{n}$. Let $\delta x = \dfrac{b-a}{2n}$. Then $\Delta x = 2\delta x$, so

$$\begin{aligned}\tfrac{1}{3}T_n + \tfrac{2}{3}M_n &= \frac{\delta x}{3}\left[f(x_0) + 2\sum_{i=1}^{n-1} f(x_i) + f(x_n) + 4\sum_{i=1}^{n} f(x_i - \delta x)\right] \\ &= \tfrac{1}{3}\delta x[f(x_0) + 4f(x_1 - \delta x) + 2f(x_1) + 4f(x_2 - \delta x) \\ &\qquad + 2f(x_2) + \cdots + 2f(x_{n-1}) + 4f(x_n - \delta x) + f(x_n)]\end{aligned}$$

Since $x_0, x_1 - \delta x, x_1, x_2 - \delta x, x_2, \ldots, x_{n-1}, x_n - \delta x, x_n$ are the subinterval endpoints for $S_{2n}$, and since $\delta x = \dfrac{b-a}{2n}$ is the width of the subintervals for $S_{2n}$, the last expression for $\frac{1}{3}T_n + \frac{2}{3}M_n$ is the usual expression for $S_{2n}$. Therefore, $\frac{1}{3}T_n + \frac{2}{3}M_n = S_{2n}$.

## 5.10 Improper Integrals

1. (a) Since $\int_1^\infty x^4 e^{-x^4}\,dx$ has an infinite interval of integration, it is an improper integral of Type I.

(b) Since $y = \sec x$ has an infinite discontinuity at $x = \frac{\pi}{2}$, $\int_0^{\pi/2} \sec x\,dx$ is a Type II improper integral.

(c) Since $y = \dfrac{x}{(x-2)(x-3)}$ has an infinite discontinuity at $x = 2$, $\displaystyle\int_0^2 \frac{x}{x^2-5x+6}\,dx$ is a Type II improper integral.

(d) Since $\displaystyle\int_{-\infty}^0 \frac{1}{x^2+5}\,dx$ has an infinite interval of integration, it is an improper integral of Type I.

2. (a) Since $y = \dfrac{1}{2x-1}$ is defined and continuous on $[1, 2]$, $\displaystyle\int_1^2 \frac{1}{2x-1}\,dx$ is proper.

(b) Since $y = \dfrac{1}{2x-1}$ has an infinite discontinuity at $x = \frac{1}{2}$, $\displaystyle\int_0^1 \frac{1}{2x-1}\,dx$ is a Type II improper integral.

(c) Since $\displaystyle\int_{-\infty}^\infty \frac{\sin x}{1+x^2}\,dx$ has an infinite interval of integration, it is an improper integral of Type I.

(d) Since $y = \ln(x-1)$ has an infinite discontinuity at $x = 1$, $\int_1^2 \ln(x-1)dx$ is a Type II improper integral.

3. The area under the graph of $y = 1/x^3 = x^{-3}$ between $x = 1$ and $x = t$ is

$A(t) = \int_1^t x^{-3}\,dx = \left[-\frac{1}{2}x^{-2}\right]_1^t = -\frac{1}{2}t^{-2} - \left(-\frac{1}{2}\right) = \frac{1}{2} - 1/(2t^2)$. So the area for $1 \le x \le 10$ is $A(10) = 0.5 - 0.005 = 0.495$, the area for $1 \le x \le 100$ is $A(100) = 0.5 - 0.00005 = 0.49995$, and the area for $1 \le x \le 1000$ is $A(1000) = 0.5 - 0.0000005 = 0.4999995$. The total area under the curve for $x \ge 1$ is $\lim\limits_{t\to\infty} A(t) = \lim\limits_{t\to\infty}\left[\frac{1}{2} - 1/(2t^2)\right] = \frac{1}{2}$.

4. (a)

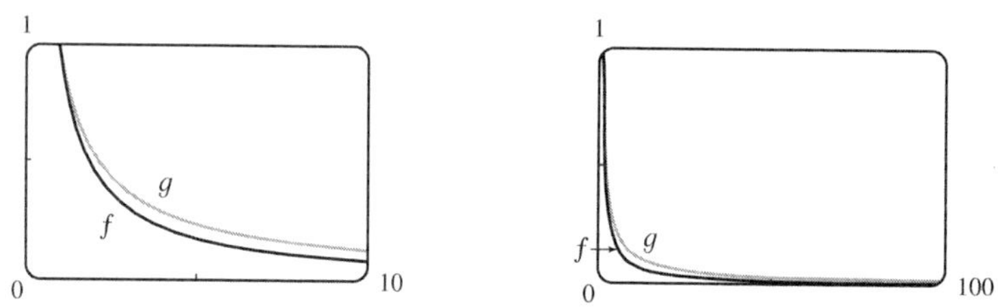


(b) The area under the graph of $f$ from $x = 1$ to $x = t$ is

$$F(t) = \int_1^t f(x)\,dx = \int_1^t x^{-1.1}dx = \left[-\tfrac{1}{0.1}x^{-0.1}\right]_1^t$$
$$= -10(t^{-0.1} - 1) = 10(1 - t^{-0.1})$$

and the area under the graph of $g$ is

$$G(t) = \int_1^t g(x)dx = \int_1^t x^{-0.9}dx = \left[\tfrac{1}{0.1}x^{0.1}\right]_1^t$$
$$= 10(t^{0.1} - 1)$$

| $t$ | $F(t)$ | $G(t)$ |
|---|---|---|
| 10 | 2.06 | 2.59 |
| 100 | 3.69 | 5.85 |
| $10^4$ | 6.02 | 15.12 |
| $10^6$ | 7.49 | 29.81 |
| $10^{10}$ | 9 | 90 |
| $10^{20}$ | 9.9 | 990 |

(c) The total area under the graph of $f$ is $\lim\limits_{t\to\infty} F(t) = \lim\limits_{t\to\infty} 10(1 - t^{-0.1}) = 10$.

The total area under the graph of $g$ does not exist, since $\lim\limits_{t\to\infty} G(t) = \lim\limits_{t\to\infty} 10(t^{0.1} - 1) = \infty$.

**5.** $I = \int_1^{\infty} \frac{1}{(3x+1)^2}\,dx = \lim_{t\to\infty} \int_1^t \frac{1}{(3x+1)^2}\,dx$. Now

$\int \frac{1}{(3x+1)^2}\,dx = \frac{1}{3}\int \frac{1}{u^2}\,du \quad [u = 3x+1,\ du = 3\,dx] \quad = -\frac{1}{3u} + C = -\frac{1}{3(3x+1)} + C,$

so $I = \lim_{t\to\infty}\left[-\frac{1}{3(3x+1)}\right]_1^t = \lim_{t\to\infty}\left[-\frac{1}{3(3t+1)} + \frac{1}{12}\right] = 0 + \frac{1}{12} = \frac{1}{12}$. Convergent

**6.** $\int_{-\infty}^{0} \frac{1}{2x-5}\,dx = \lim_{t\to-\infty} \int_t^0 \frac{1}{2x-5}\,dx = \lim_{t\to-\infty}\left[\frac{1}{2}\ln|2x-5|\right]_t^0 = \lim_{t\to\infty}\left[\frac{1}{2}\ln 5 - \frac{1}{2}\ln|2t-5|\right] = -\infty$. Divergent

**7.** $\int_{-\infty}^{-1} \frac{1}{\sqrt{2-w}}\,dw = \lim_{t\to-\infty}\int_t^{-1} \frac{1}{\sqrt{2-w}}\,dw = \lim_{t\to-\infty}\left[-2\sqrt{2-w}\,\right]_t^{-1} \quad [u = 2-w,\ du = -dw]$

$= \lim_{t\to-\infty}\left[-2\sqrt{3} + 2\sqrt{2-t}\,\right] = \infty$. Divergent

**8.** $\int_0^{\infty} \frac{x}{(x^2+2)^2}\,dx = \lim_{t\to\infty}\int_0^t \frac{x}{(x^2+2)^2}\,dx = \lim_{t\to\infty}\frac{1}{2}\left[\frac{-1}{x^2+2}\right]_0^t = \frac{1}{2}\lim_{t\to\infty}\left(\frac{-1}{t^2+2} + \frac{1}{2}\right)$

$= \frac{1}{2}\left(0 + \frac{1}{2}\right) = \frac{1}{4}$. Convergent

**9.** $\int_4^{\infty} e^{-y/2}\,dy = \lim_{t\to\infty}\int_4^t e^{-y/2}\,dy = \lim_{t\to\infty}\left[-2e^{-y/2}\right]_4^t = \lim_{t\to\infty}(-2e^{-t/2} + 2e^{-2}) = 0 + 2e^{-2} = 2e^{-2}$.

Convergent

**10.** $\int_{-\infty}^{-1} e^{-2t}\,dt = \lim_{x\to-\infty}\int_x^{-1} e^{-2t}\,dt = \lim_{x\to-\infty}\left[-\frac{1}{2}e^{-2t}\right]_x^{-1} = \lim_{x\to-\infty}\left[-\frac{1}{2}e^2 + \frac{1}{2}e^{-2x}\right] = \infty$. Divergent

**11.** $\int_{2\pi}^{\infty} \sin\theta\,d\theta = \lim_{t\to\infty}\int_{2\pi}^t \sin\theta\,d\theta = \lim_{t\to\infty}\left[-\cos\theta\right]_{2\pi}^t = \lim_{t\to\infty}(-\cos t + 1)$. This limit does not exist, so the integral is divergent. Divergent

**12.** $I = \int_{-\infty}^{\infty}(2-v^4)\,dv = I_1 + I_2 = \int_{-\infty}^{0}(2-v^4)\,dv + \int_0^{\infty}(2-v^4)\,dv$, but

$I_1 = \lim_{t\to-\infty}\left[2v - \frac{1}{5}v^5\right]_t^0 = \lim_{t\to-\infty}\left(-2t + \frac{1}{5}t^5\right) = -\infty$. Since $I_1$ is divergent, $I$ is divergent, and there is no need to evaluate $I_2$. Divergent

**13.** $\int_{-\infty}^{\infty} xe^{-x^2}\,dx = \int_{-\infty}^{0} xe^{-x^2}\,dx + \int_0^{\infty} xe^{-x^2}\,dx$.

$\int_{-\infty}^{0} xe^{-x^2}\,dx = \lim_{t\to-\infty}\left(-\frac{1}{2}\right)\left[e^{-x^2}\right]_t^0 = \lim_{t\to-\infty}\left(-\frac{1}{2}\right)\left(1 - e^{-t^2}\right) = -\frac{1}{2}\cdot 1 = -\frac{1}{2}$, and

$\int_0^{\infty} xe^{-x^2}\,dx = \lim_{t\to\infty}\left(-\frac{1}{2}\right)\left[e^{-x^2}\right]_0^t = \lim_{t\to\infty}\left(-\frac{1}{2}\right)\left(e^{-t^2} - 1\right) = -\frac{1}{2}\cdot(-1) = \frac{1}{2}$.

Therefore, $\int_{-\infty}^{\infty} xe^{-x^2}\,dx = -\frac{1}{2} + \frac{1}{2} = 0$. Convergent

**14.** $\int_{-\infty}^{\infty} x^2e^{-x^3}\,dx = \int_{-\infty}^{0} x^2e^{-x^3}\,dx + \int_0^{\infty} x^2e^{-x^3}\,dx$, and

$\int_{-\infty}^{0} x^2e^{-x^3}\,dx = \lim_{t\to-\infty}\left[-\frac{1}{3}e^{-x^3}\right]_t^0 = -\frac{1}{3} + \frac{1}{3}\left(\lim_{t\to-\infty} e^{-t^3}\right) = \infty$. Divergent

**15.** $\int_0^{\infty} se^{-5s}\,ds = \lim_{t\to\infty}\int_0^t se^{-5s}\,ds = \lim_{t\to\infty}\left[-\frac{1}{5}se^{-5s} - \frac{1}{25}e^{-5s}\right]_0^t \quad$ [by integration by parts with $u = s$]

$= \lim_{t\to\infty}\left(-\frac{1}{5}te^{-5t} - \frac{1}{25}e^{-5t} + \frac{1}{25}\right) = 0 - 0 + \frac{1}{25}$ [by l'Hospital's Rule]

$= \frac{1}{25}$. Convergent

**16.** $I = \int_{-\infty}^{\infty}\cos\pi t\,dt = I_1 + I_2 = \int_{-\infty}^{0}\cos\pi t\,dt + \int_0^{\infty}\cos\pi t\,dt$, but $I_1 = \lim_{s\to-\infty}\left[\frac{1}{\pi}\sin\pi t\right]_s^0 = \lim_{s\to-\infty}\left(-\frac{1}{\pi}\sin\pi t\right)$ and this limit does not exist. Since $I_1$ is divergent, $I$ is divergent, and there is no need to evaluate $I_2$. Divergent

**17.** $\displaystyle\int_1^\infty \frac{\ln x}{x}\,dx = \lim_{t\to\infty}\left[\frac{(\ln x)^2}{2}\right]_1^t$ (by substitution with $u = \ln x$, $du = dx/x$) $\displaystyle= \lim_{t\to\infty}\frac{(\ln t)^2}{2} = \infty.$ Divergent

**18.** $\displaystyle\int_{-\infty}^6 re^{r/3}\,dr = \lim_{t\to-\infty}\int_t^6 re^{r/3}\,dr = \lim_{t\to-\infty}\left[3re^{r/3} - 9e^{r/3}\right]_t^6$ $\left[\begin{array}{l}\text{by integration by}\\ \text{parts with } u = r\end{array}\right]$

$$= \lim_{t\to-\infty}\left(18e^2 - 9e^2 - 3te^{t/3} + 9e^{t/3}\right) = 9e^2 - 0 + 0 \quad \text{[by l'Hospital's Rule]}$$

$= 9e^2.$ Convergent

**19.** Integrate by parts with $u = \ln x$, $dv = dx/x^2$ $\Rightarrow$ $du = dx/x$, $v = -1/x$.

$$\int_1^\infty \frac{\ln x}{x^2}\,dx = \lim_{t\to\infty}\int_1^t \frac{\ln x}{x^2}\,dx = \lim_{t\to\infty}\left[-\frac{\ln x}{x} - \frac{1}{x}\right]_1^t = \lim_{t\to\infty}\left(-\frac{\ln t}{t} - \frac{1}{t} + 0 + 1\right)$$

$$= -0 - 0 + 0 + 1 = 1$$

since $\displaystyle\lim_{t\to\infty}\frac{\ln t}{t} \overset{\text{H}}{=} \lim_{t\to\infty}\frac{1/t}{1} = 0.$ Convergent

**20.** Integrate by parts with $u = \ln x$, $dv = dx/x^3$ $\Rightarrow$ $du = dx/x$, $v = -1/(2x^2)$.

$$\int_1^\infty \frac{\ln x}{x^3}\,dx = \lim_{t\to\infty}\int_1^t \frac{\ln x}{x^3}\,dx = \lim_{t\to\infty}\left(\left[-\frac{1}{2x^2}\ln x\right]_1^t + \frac{1}{2}\int_1^t \frac{1}{x^3}\,dx\right)$$

$$= \lim_{t\to\infty}\left(-\frac{1}{2}\frac{\ln t}{t^2} + 0 - \frac{1}{4t^2} + \frac{1}{4}\right) = \frac{1}{4}$$

since $\displaystyle\lim_{t\to\infty}\frac{\ln t}{t^2} \overset{\text{H}}{=} \lim_{t\to\infty}\frac{1/t}{2t} = \lim_{t\to\infty}\frac{1}{2t^2} = 0.$ Convergent

**21.** $\displaystyle\int_{-\infty}^\infty \frac{x^2}{9+x^6}\,dx = \int_{-\infty}^0 \frac{x^2}{9+x^6}\,dx + \int_0^\infty \frac{x^2}{9+x^6}\,dx = 2\int_0^\infty \frac{x^2}{9+x^6}\,dx$ [since the integrand is even].

Now $\displaystyle\int \frac{x^2\,dx}{9+x^6}$ $\left[\begin{array}{l}u = x^3\\ du = 3x^2\,dx\end{array}\right]$ $\displaystyle= \int \frac{\frac{1}{3}\,du}{9+u^2}$ $\left[\begin{array}{l}u = 3v\\ du = 3\,dv\end{array}\right]$ $\displaystyle= \int \frac{\frac{1}{3}(3\,dv)}{9+9v^2} = \frac{1}{9}\int \frac{dv}{1+v^2}$

$$= \frac{1}{9}\tan^{-1}v + C = \frac{1}{9}\tan^{-1}\left(\frac{u}{3}\right) + C = \frac{1}{9}\tan^{-1}\left(\frac{x^3}{3}\right) + C,$$

so $\displaystyle 2\int_0^\infty \frac{x^2}{9+x^6}\,dx = 2\lim_{t\to\infty}\int_0^t \frac{x^2}{9+x^6}\,dx = 2\lim_{t\to\infty}\left[\frac{1}{9}\tan^{-1}\left(\frac{x^3}{3}\right)\right]_0^t$

$\displaystyle= 2\lim_{t\to\infty}\frac{1}{9}\tan^{-1}\left(\frac{t^3}{3}\right) = \frac{2}{9}\cdot\frac{\pi}{2} = \frac{\pi}{9}.$ Convergent

**22.** $\displaystyle\int_0^\infty \frac{e^x}{e^{2x}+3}\,dx = \lim_{t\to\infty}\int_0^t \frac{e^x}{(e^x)^2 + (\sqrt{3})^2}\,dx = \lim_{t\to\infty}\left[\frac{1}{\sqrt{3}}\arctan\frac{e^x}{\sqrt{3}}\right]_0^t$

$$= \frac{1}{\sqrt{3}}\lim_{t\to\infty}\left(\arctan\frac{e^t}{\sqrt{3}} - \arctan\frac{1}{\sqrt{3}}\right) = \frac{1}{\sqrt{3}}\left(\frac{\pi}{2} - \frac{\pi}{6}\right)$$

$\displaystyle= \frac{1}{\sqrt{3}}\left(\frac{\pi}{3}\right) = \frac{\pi\sqrt{3}}{9}.$ Convergent

**23.** $\displaystyle\int_0^1 \frac{3}{x^5}\,dx = \lim_{t\to0^+}\int_t^1 3x^{-5}\,dx = \lim_{t\to0^+}\left[-\frac{3}{4x^4}\right]_t^1 = -\frac{3}{4}\lim_{t\to0^+}\left(1 - \frac{1}{t^4}\right) = \infty.$ Divergent

**24.** $\displaystyle\int_2^3 \frac{1}{\sqrt{3-x}}\,dx = \lim_{t\to3^-}\int_2^t (3-x)^{-1/2}\,dx = \lim_{t\to3^-}\left[-2(3-x)^{1/2}\right]_2^t = -2\lim_{t\to3^-}\left(\sqrt{3-t} - \sqrt{1}\right)$

$= -2(0 - 1) = 2.$ Convergent

**25.** $\displaystyle\int_{-2}^{14} \frac{dx}{\sqrt[4]{x+2}} = \lim_{t\to -2^+} \int_t^{14} (x+2)^{-1/4}\,dx = \lim_{t\to -2^+} \left[\frac{4}{3}(x+2)^{3/4}\right]_t^{14} = \frac{4}{3}\lim_{t\to -2^+}\left[16^{3/4} - (t+2)^{3/4}\right]$

$= \frac{4}{3}(8-0) = \frac{32}{3}.$ Convergent

**26.** $\displaystyle\int_6^8 \frac{4}{(x-6)^3}\,dx = \lim_{t\to 6^+}\int_t^8 4(x-6)^{-3}\,dx = \lim_{t\to 6^+}\left[-2(x-6)^{-2}\right]_t^8 = -2\lim_{t\to 6^+}\left[\frac{1}{2^2} - \frac{1}{(t-6)^2}\right] = \infty.$ Divergent

**27.** There is an infinite discontinuity at $x=1$. $\int_0^{33}(x-1)^{-1/5}\,dx = \int_0^1 (x-1)^{-1/5}\,dx + \int_1^{33}(x-1)^{-1/5}\,dx$. Here

$\int_0^1 (x-1)^{-1/5}\,dx = \lim_{t\to 1^-}\int_0^t (x-1)^{-1/5}\,dx = \lim_{t\to 1^-}\left[\frac{5}{4}(x-1)^{4/5}\right]_0^t = \lim_{t\to 1^-}\left[\frac{5}{4}(t-1)^{4/5} - \frac{5}{4}\right] = -\frac{5}{4}$ and

$\int_1^{33}(x-1)^{-1/5}\,dx = \lim_{t\to 1^+}\int_t^{33}(x-1)^{-1/5}\,dx = \lim_{t\to 1^+}\left[\frac{5}{4}(x-1)^{4/5}\right]_t^{33} = \lim_{t\to 1^+}\left[\frac{5}{4}\cdot 16 - \frac{5}{4}(t-1)^{4/5}\right] = 20$. Thus,

$\int_0^{33}(x-1)^{-1/5}\,dx = -\frac{5}{4} + 20 = \frac{75}{4}.$ Convergent

**28.** $f(y) = 1/(4y-1)$ has an infinite discontinuity at $y = \frac{1}{4}$.

$$\int_{1/4}^1 \frac{1}{4y-1}\,dy = \lim_{t\to(1/4)^+}\int_t^1 \frac{1}{4y-1}\,dy = \lim_{t\to(1/4)^+}\left[\tfrac{1}{4}\ln|4y-1|\right]_t^1$$
$$= \lim_{t\to(1/4)^+}\left[\tfrac{1}{4}\ln 3 - \tfrac{1}{4}\ln(4t-1)\right] = \infty$$

so $\displaystyle\int_{1/4}^1 \frac{1}{4y-1}\,dy$ diverges, and hence, $\displaystyle\int_0^1 \frac{1}{4y-1}\,dy$ diverges. Divergent

**29.** There is an infinite discontinuity at $x=0$. $\displaystyle\int_{-1}^1 \frac{e^x}{e^x-1}\,dx = \int_{-1}^0 \frac{e^x}{e^x-1}\,dx + \int_0^1 \frac{e^x}{e^x-1}\,dx.$

$\displaystyle\int_{-1}^0 \frac{e^x}{e^x-1}\,dx = \lim_{t\to 0^-}\int_{-1}^t \frac{e^x}{e^x-1}\,dx = \lim_{t\to 0^-}\Big[\ln|e^x-1|\Big]_{-1}^t = \lim_{t\to 0^-}\Big[\ln|e^t-1| - \ln|e^{-1}-1|\Big] = -\infty,$

so $\displaystyle\int_{-1}^1 \frac{e^x}{e^x-1}\,dx$ is divergent. The integral $\displaystyle\int_0^1 \frac{e^x}{e^x-1}\,dx$ also diverges since

$\displaystyle\int_0^1 \frac{e^x}{e^x-1}\,dx = \lim_{t\to 0^+}\int_t^1 \frac{e^x}{e^x-1}\,dx = \lim_{t\to 0^+}\Big[\ln|e^x-1|\Big]_t^1 = \lim_{t\to 0^+}\Big[\ln|e-1| - \ln|e^t-1|\Big] = \infty.$

Divergent

**30.** $\displaystyle\int_0^1 \frac{dx}{\sqrt{1-x^2}} = \lim_{t\to 1^-}\int_0^t \frac{dx}{\sqrt{1-x^2}} = \lim_{t\to 1^-}\left[\sin^{-1}x\right]_0^t = \lim_{t\to 1^-}\sin^{-1}t = \frac{\pi}{2}.$ Convergent

**31.** $I = \int_0^2 z^2\ln z\,dz = \lim_{t\to 0^+}\int_t^2 z^2\ln z\,dz \overset{101}{=} \lim_{t\to 0^+}\left[\frac{z^3}{3^2}(3\ln z - 1)\right]_t^2$

$= \lim_{t\to 0^+}\left[\frac{8}{9}(3\ln 2 - 1) - \frac{1}{9}t^3(3\ln t - 1)\right] = \frac{8}{3}\ln 2 - \frac{8}{9} - \frac{1}{9}\lim_{t\to 0^+}\left[t^3(3\ln t - 1)\right] = \frac{8}{3}\ln 2 - \frac{8}{9} - \frac{1}{9}L.$

Now $L = \displaystyle\lim_{t\to 0^+}\left[t^3(3\ln t - 1)\right] = \lim_{t\to 0^+}\frac{3\ln t - 1}{t^{-3}} \overset{\text{H}}{=} \lim_{t\to 0^+}\frac{3/t}{-3/t^4} = \lim_{t\to 0^+}(-t^3) = 0.$

Thus, $L = 0$ and $I = \frac{8}{3}\ln 2 - \frac{8}{9}.$ Convergent

**32.** Integrate by parts with $u = \ln x$, $dv = dx/\sqrt{x}$ $\Rightarrow$ $du = dx/x$, $v = 2\sqrt{x}$.

$$\int_0^1 \frac{\ln x}{\sqrt{x}}\,dx = \lim_{t\to 0^+}\int_t^1 \frac{\ln x}{\sqrt{x}}\,dx = \lim_{t\to 0^+}\left(\left[2\sqrt{x}\ln x\right]_t^1 - 2\int_t^1 \frac{dx}{\sqrt{x}}\right) = \lim_{t\to 0^+}\left(-2\sqrt{t}\ln t - 4\left[\sqrt{x}\right]_t^1\right)$$
$$= \lim_{t\to 0^+}\left(-2\sqrt{t}\ln t - 4 + 4\sqrt{t}\right) = -4$$

since $\displaystyle\lim_{t\to 0^+}\sqrt{t}\ln t = \lim_{t\to 0^+}\frac{\ln t}{t^{-1/2}} \overset{\text{H}}{=} \lim_{t\to 0^+}\frac{1/t}{-t^{-3/2}/2} = \lim_{t\to 0^+}\left(-2\sqrt{t}\right) = 0.$ Convergent

**33.**

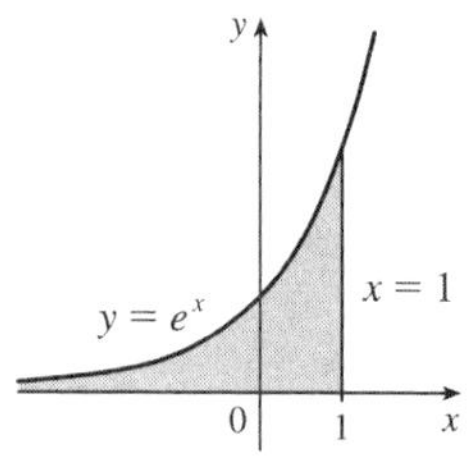


$$\text{Area} = \int_{-\infty}^{1} e^x\,dx = \lim_{t\to-\infty} \left[e^x\right]_t^1$$
$$= e - \lim_{t\to-\infty} e^t = e$$

**34.**

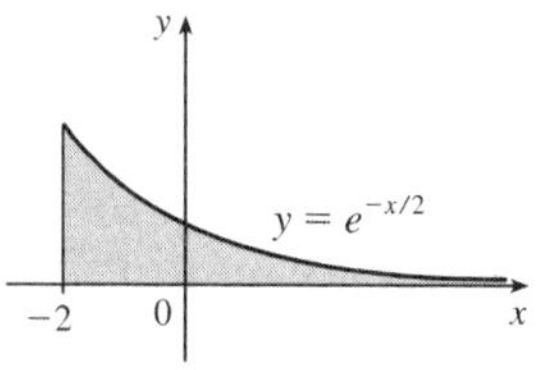


$$\text{Area} = \int_{-2}^{\infty} e^{-x/2}\,dx = -2 \lim_{t\to\infty} \left[e^{-x/2}\right]_{-2}^{t}$$
$$= -2 \lim_{t\to\infty} e^{-t/2} + 2e = 2e$$

**35.**

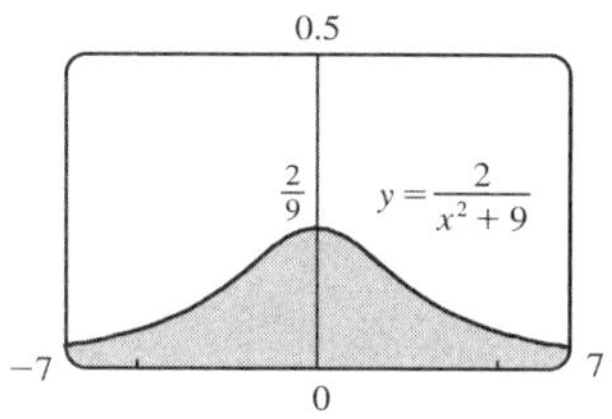


$$\text{Area} = \int_{-\infty}^{\infty} \frac{2}{x^2+9}\,dx = 2\cdot 2\int_0^{\infty} \frac{1}{x^2+9}\,dx$$
$$= 4 \lim_{t\to\infty} \int_0^t \frac{1}{x^2+9}\,dx = 4 \lim_{t\to\infty} \left[\frac{1}{3}\tan^{-1}\frac{x}{3}\right]_0^t$$
$$= \frac{4}{3} \lim_{t\to\infty} \left[\tan^{-1}\frac{t}{3} - 0\right] = \frac{4}{3}\cdot\frac{\pi}{2} = \frac{2\pi}{3}$$

**36.**

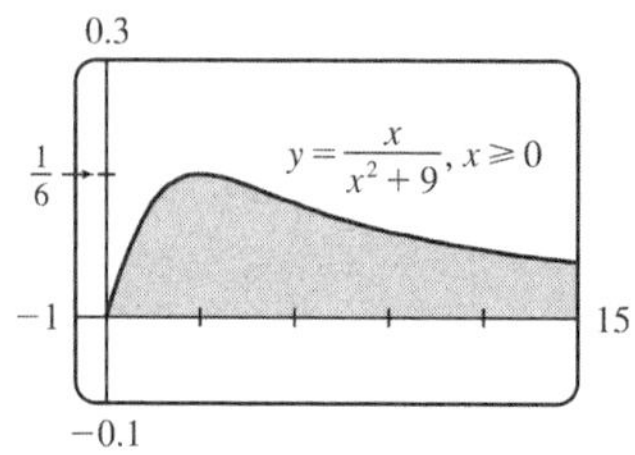


$$\text{Area} = \int_0^{\infty} \frac{x}{x^2+9}\,dx = \lim_{t\to\infty} \int_0^t \frac{x}{x^2+9}\,dx$$
$$= \lim_{t\to\infty} \left[\tfrac{1}{2}\ln(x^2+9)\right]_0^t$$
$$= \tfrac{1}{2} \lim_{t\to\infty} \left[\ln(t^2+9) - \ln 9\right] = \infty$$

Infinite area

**37.**

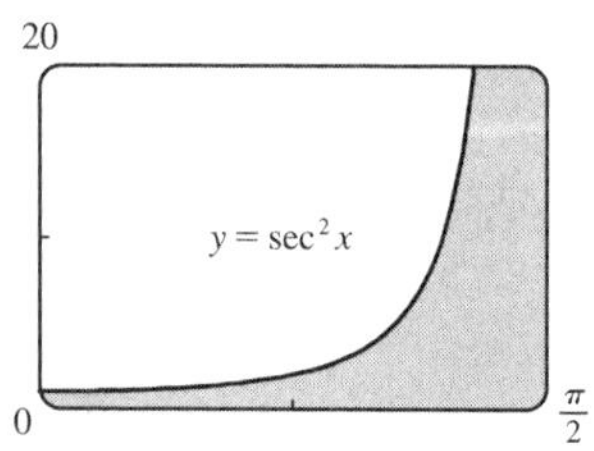


$$\text{Area} = \int_0^{\pi/2} \sec^2 x\,dx = \lim_{t\to(\pi/2)^-} \int_0^t \sec^2 x\,dx$$
$$= \lim_{t\to(\pi/2)^-} \left[\tan x\right]_0^t = \lim_{t\to(\pi/2)^-} (\tan t - 0)$$
$$= \infty$$

Infinite area

**38.**

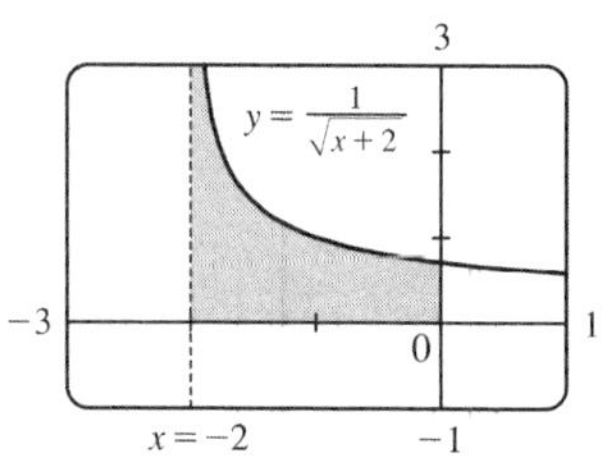


$$\text{Area} = \int_{-2}^{0} \frac{1}{\sqrt{x+2}}\,dx = \lim_{t\to-2^+} \int_t^0 \frac{1}{\sqrt{x+2}}\,dx$$
$$= \lim_{t\to-2^+} \left[2\sqrt{x+2}\right]_t^0 = \lim_{t\to-2^+} \left(2\sqrt{2} - 2\sqrt{t+2}\right)$$
$$= 2\sqrt{2} - 0 = 2\sqrt{2}$$

**39.** (a)

| $t$ | $\int_1^t g(x)\,dx$ |
|---|---|
| 2 | 0.447453 |
| 5 | 0.577101 |
| 10 | 0.621306 |
| 100 | 0.668479 |
| 1000 | 0.672957 |
| 10,000 | 0.673407 |

$g(x) = \dfrac{\sin^2 x}{x^2}$.

It appears that the integral is convergent.

(b) $-1 \le \sin x \le 1 \;\Rightarrow\; 0 \le \sin^2 x \le 1 \;\Rightarrow\; 0 \le \dfrac{\sin^2 x}{x^2} \le \dfrac{1}{x^2}$. Since $\displaystyle\int_1^\infty \frac{1}{x^2}\,dx$ is convergent (Equation 2 with $p = 2 > 1$), $\displaystyle\int_1^\infty \frac{\sin^2 x}{x^2}\,dx$ is convergent by the Comparison Theorem.

(c)

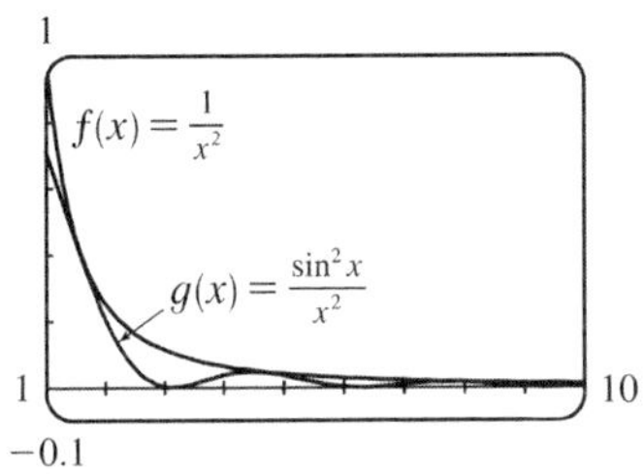


Since $\int_1^\infty f(x)\,dx$ is finite and the area under $g(x)$ is less than the area under $f(x)$ on any interval $[1, t]$, $\int_1^\infty g(x)\,dx$ must be finite; that is, the integral is convergent.

**40.** (a)

| $t$ | $\int_2^t g(x)dx$ |
|---|---|
| 5 | 3.830327 |
| 10 | 6.801200 |
| 100 | 23.328769 |
| 1000 | 69.023361 |
| 10,000 | 208.124560 |

$g(x) = \dfrac{1}{\sqrt{x} - 1}$.

It appears that the integral is divergent.

(b) For $x \ge 2$, $\sqrt{x} > \sqrt{x} - 1 \;\Rightarrow\; \dfrac{1}{\sqrt{x}} < \dfrac{1}{\sqrt{x} - 1}$. Since $\displaystyle\int_2^\infty \frac{1}{\sqrt{x}}\,dx$ is divergent (Equation 2 with $p = \frac{1}{2} \le 1$), $\displaystyle\int_2^\infty \frac{1}{\sqrt{x} - 1}\,dx$ is divergent by the Comparison Theorem.

(c)

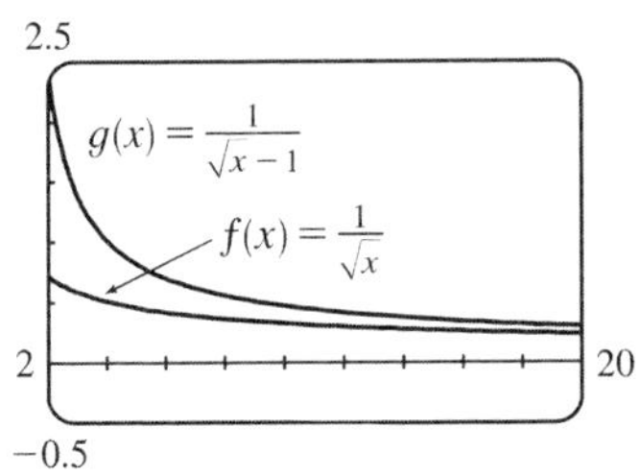


Since $\int_2^\infty f(x)\,dx$ is infinite and the area under $g(x)$ is greater than the area under $f(x)$ on any interval $[2, t]$, $\int_2^\infty g(x)\,dx$ must be infinite; that is, the integral is divergent.

**41.** For $x \ge 1$, $\dfrac{\cos^2 x}{1 + x^2} \le \dfrac{1}{1 + x^2} < \dfrac{1}{x^2}$. $\displaystyle\int_1^\infty \frac{1}{x^2}\,dx$ is convergent by Equation 2 with $p = 2 > 1$, so $\displaystyle\int_1^\infty \frac{\cos^2 x}{1 + x^2}\,dx$ is convergent by the Comparison Theorem.

**42.** For $x \geq 1$, $\dfrac{2+e^{-x}}{x} > \dfrac{2}{x}$ [since $e^{-x} > 0$] $> \dfrac{1}{x}$. $\displaystyle\int_1^\infty \frac{1}{x}\,dx$ is divergent by Equation 2 with $p = 1 \leq 1$, so

$\displaystyle\int_1^\infty \frac{2+e^{-x}}{x}\,dx$ is divergent by the Comparison Theorem.

**43.** For $x \geq 1$, $x + e^{2x} > e^{2x} > 0 \;\Rightarrow\; \dfrac{1}{x+e^{2x}} \leq \dfrac{1}{e^{2x}} = e^{-2x}$ on $[1, \infty)$.

$$\int_1^\infty e^{-2x}\,dx = \lim_{t\to\infty}\left[-\frac{1}{2}e^{-2x}\right]_1^t = \lim_{t\to\infty}\left[-\frac{1}{2}e^{-2t} + \frac{1}{2}e^{-2}\right] = \frac{1}{2}e^{-2}.$$

Therefore, $\displaystyle\int_1^\infty e^{-2x}\,dx$ is convergent, and by the Comparison Theorem, $\displaystyle\int_1^\infty \frac{dx}{x+e^{2x}}$ is also convergent.

**44.** For $x \geq 1$, $0 < \dfrac{x}{\sqrt{1+x^6}} < \dfrac{x}{\sqrt{x^6}} = \dfrac{x}{x^3} = \dfrac{1}{x^2}$. $\displaystyle\int_1^\infty \frac{1}{x^2}\,dx$ is convergent by Equation 2 with $p = 2 > 1$, so

$\displaystyle\int_1^\infty \frac{x}{\sqrt{1+x^6}}\,dx$ is convergent by the Comparison Theorem.

**45.** $\dfrac{1}{x\sin x} \geq \dfrac{1}{x}$ on $\left(0, \frac{\pi}{2}\right]$ since $0 \leq \sin x \leq 1$. $\displaystyle\int_0^{\pi/2} \frac{dx}{x} = \lim_{t\to 0^+}\int_t^{\pi/2} \frac{dx}{x} = \lim_{t\to 0^+}\left[\ln x\right]_t^{\pi/2}$.

But $\ln t \to -\infty$ as $t \to 0^+$, so $\displaystyle\int_0^{\pi/2} \frac{dx}{x}$ is divergent, and by the Comparison Theorem, $\displaystyle\int_0^{\pi/2} \frac{dx}{x\sin x}$ is also divergent.

**46.** For $0 \leq x \leq 1$, $e^{-x} \leq 1 \;\Rightarrow\; \dfrac{e^{-x}}{\sqrt{x}} \leq \dfrac{1}{\sqrt{x}}$.

$$\int_0^1 \frac{1}{\sqrt{x}}\,dx = \lim_{t\to 0^+}\int_t^1 \frac{1}{\sqrt{x}}\,dx = \lim_{t\to 0^+}\left[2\sqrt{x}\right]_t^1 = \lim_{t\to 0^+}\left(2 - 2\sqrt{t}\right) = 2 \text{ is convergent.}$$

Therefore, $\displaystyle\int_0^1 \frac{e^{-x}}{\sqrt{x}}\,dx$ is convergent by the Comparison Theorem.

**47.** $\displaystyle\int_0^\infty \frac{dx}{\sqrt{x}\,(1+x)} = \int_0^1 \frac{dx}{\sqrt{x}\,(1+x)} + \int_1^\infty \frac{dx}{\sqrt{x}\,(1+x)} = \lim_{t\to 0^+}\int_t^1 \frac{dx}{\sqrt{x}\,(1+x)} + \lim_{t\to\infty}\int_1^t \frac{dx}{\sqrt{x}\,(1+x)}$. Now

$$\begin{aligned}\int \frac{dx}{\sqrt{x}\,(1+x)} &= \int \frac{2u\,du}{u(1+u^2)} \quad [u = \sqrt{x}, x = u^2, dx = 2u\,du]\\ &= 2\int \frac{du}{1+u^2} = 2\tan^{-1}u + C = 2\tan^{-1}\sqrt{x} + C,\end{aligned}$$

so
$$\begin{aligned}\int_0^\infty \frac{dx}{\sqrt{x}\,(1+x)} &= \lim_{t\to 0^+}\left[2\tan^{-1}\sqrt{x}\right]_t^1 + \lim_{t\to\infty}\left[2\tan^{-1}\sqrt{x}\right]_1^t\\ &= \lim_{t\to 0^+}\left[2\left(\tfrac{\pi}{4}\right) - 2\tan^{-1}\sqrt{t}\right] + \lim_{t\to\infty}\left[2\tan^{-1}\sqrt{t} - 2\left(\tfrac{\pi}{4}\right)\right] = \tfrac{\pi}{2} - 0 + 2\left(\tfrac{\pi}{2}\right) - \tfrac{\pi}{2} = \pi.\end{aligned}$$

**48.** Let $u = \ln x$. Then $du = dx/x \;\Rightarrow\; \displaystyle\int_e^\infty \frac{dx}{x\,(\ln x)^p} = \int_1^\infty \frac{du}{u^p}$. By Example 4, this converges to $\dfrac{1}{p-1}$ if $p > 1$ and diverges otherwise.

**49.** If $p = 1$, then $\displaystyle\int_0^1 \frac{dx}{x^p} = \lim_{t\to 0^+} \int_t^1 \frac{dx}{x} = \lim_{t\to 0^+} [\ln x]_t^1 = \infty$. Divergent.

If $p \neq 1$, then $\displaystyle\int_0^1 \frac{dx}{x^p} = \lim_{t\to 0^+} \int_t^1 \frac{dx}{x^p}$ (note that the integral is not improper if $p < 0$)

$$= \lim_{t\to 0^+} \left[\frac{x^{-p+1}}{-p+1}\right]_t^1 = \lim_{t\to 0^+} \frac{1}{1-p}\left[1 - \frac{1}{t^{p-1}}\right]$$

If $p > 1$, then $p - 1 > 0$, so $\dfrac{1}{t^{p-1}} \to \infty$ as $t \to 0^+$, and the integral diverges.

If $p < 1$, then $p - 1 < 0$, so $\dfrac{1}{t^{p-1}} \to 0$ as $t \to 0^+$ and $\displaystyle\int_0^1 \frac{dx}{x^p} = \frac{1}{1-p}\left[\lim_{t\to 0^+} \left(1 - t^{1-p}\right)\right] = \frac{1}{1-p}$.

Thus, the integral converges if and only if $p < 1$, and in that case its value is $\dfrac{1}{1-p}$.

**50.** (a) $\boldsymbol{n = 0}$: $\displaystyle\int_0^\infty x^n e^{-x}\,dx = \lim_{t\to\infty} \int_0^t e^{-x}\,dx = \lim_{t\to\infty} \left[-e^{-x}\right]_0^t$

$$= \lim_{t\to\infty} \left[-e^{-t} + 1\right] = 0 + 1 = 1$$

$\boldsymbol{n = 1}$: $\displaystyle\int_0^\infty x^n e^{-x}\,dx = \lim_{t\to\infty} \int_0^t xe^{-x}\,dx$. To evaluate $\displaystyle\int xe^{-x}\,dx$, we'll use integration by parts with $u = x$, $dv = e^{-x}\,dx \Rightarrow du = dx$, $v = -e^{-x}$.

So $\displaystyle\int xe^{-x}\,dx = -xe^{-x} - \int -e^{-x}\,dx = -xe^{-x} - e^{-x} + C = (-x-1)e^{-x} + C$ and

$$\lim_{t\to\infty} \int_0^t xe^{-x}\,dx = \lim_{t\to\infty} \left[(-x-1)e^{-x}\right]_0^t$$
$$= \lim_{t\to\infty} \left[(-t-1)e^{-t} + 1\right] = \lim_{t\to\infty} \left[-te^{-t} - e^{-t} + 1\right]$$
$$= 0 - 0 + 1 \quad \text{[use l'Hospital's Rule]} \quad = 1$$

$\boldsymbol{n = 2}$: $\displaystyle\int_0^\infty x^n e^{-x}\,dx = \lim_{t\to\infty} \int_0^t x^2 e^{-x}\,dx$. To evaluate $\displaystyle\int x^2 e^{-x}\,dx$, we could use integration by parts again or Formula 97. Thus,

$$\lim_{t\to\infty} \int_0^t x^2 e^{-x}\,dx = \lim_{t\to\infty} \left[-x^2 e^{-x}\right]_0^t + 2\lim_{t\to\infty} \int_0^t xe^{-x}\,dx$$
$$= 0 + 0 + 2(1) \quad \text{[use l'Hospital's Rule and the result for } n = 1\text{]} \quad = 2$$

$\boldsymbol{n = 3}$: $\displaystyle\int_0^\infty x^n e^{-x}\,dx = \lim_{t\to\infty} \int_0^t x^3 e^{-x}\,dx \overset{97}{=} \lim_{t\to\infty} \left[-x^3 e^{-x}\right]_0^t + 3\lim_{t\to\infty} \int_0^t x^2 e^{-x}\,dx$

$$= 0 + 0 + 3(2) \quad \text{[use l'Hospital's Rule and the result for } n = 2\text{]} \quad = 6$$

(b) For $n = 1$, 2, and 3, we have $\displaystyle\int_0^\infty x^n e^{-x}\,dx = 1$, 2, and 6. The values for the integral are equal to the factorials for $n$, so we guess $\displaystyle\int_0^\infty x^n e^{-x}\,dx = n!$.

(c) Suppose that $\int_0^\infty x^k e^{-x}\,dx = k!$ for some positive integer $k$. Then $\int_0^\infty x^{k+1}e^{-x}\,dx = \lim\limits_{t\to\infty}\int_0^t x^{k+1}e^{-x}\,dx$. To evaluate $\int x^{k+1}e^{-x}\,dx$, we use parts with $u = x^{k+1}$, $dv = e^{-x}\,dx \Rightarrow du = (k+1)x^k\,dx$, $v = -e^{-x}$.

So $\int x^{k+1}e^{-x}\,dx = -x^{k+1}e^{-x} - \int -(k+1)x^k e^{-x}\,dx = -x^{k+1}e^{-x} + (k+1)\int x^k e^{-x}\,dx$ and

$$\begin{aligned}\lim_{t\to\infty}\int_0^t x^{k+1}e^{-x}\,dx &= \lim_{t\to\infty}\left[-x^{k+1}e^{-x}\right]_0^t + (k+1)\lim_{t\to\infty}\int_0^t x^k e^{-x}\,dx \\ &= \lim_{t\to\infty}\left[-t^{k+1}e^{-t}+0\right] + (k+1)k! = 0+0+(k+1)! = (k+1)!,\end{aligned}$$

so the formula holds for $k+1$. By induction, the formula holds for all positive integers. (Since $0! = 1$, the formula holds for $n = 0$, too.)

**51.** (a) $I = \int_{-\infty}^\infty x\,dx = \int_{-\infty}^0 x\,dx + \int_0^\infty x\,dx$, and $\int_0^\infty x\,dx = \lim\limits_{t\to\infty}\int_0^t x\,dx = \lim\limits_{t\to\infty}\left[\frac12 x^2\right]_0^t = \lim\limits_{t\to\infty}\left[\frac12 t^2 - 0\right] = \infty$, so $I$ is divergent.

(b) $\int_{-t}^t x\,dx = \left[\frac12 x^2\right]_{-t}^t = \frac12 t^2 - \frac12 t^2 = 0$, so $\lim\limits_{t\to\infty}\int_{-t}^t x\,dx = 0$. Therefore, $\int_{-\infty}^\infty x\,dx \neq \lim\limits_{t\to\infty}\int_{-t}^t x\,dx$.

**52.** Assume without loss of generality that $a < b$. Then

$$\begin{aligned}\int_{-\infty}^a f(x)\,dx + \int_a^\infty f(x)\,dx &= \lim_{t\to-\infty}\int_t^a f(x)\,dx + \lim_{u\to\infty}\int_a^u f(x)\,dx \\ &= \lim_{t\to-\infty}\int_t^a f(x)\,dx + \lim_{u\to\infty}\left[\int_a^b f(x)\,dx + \int_b^u f(x)\,dx\right] \\ &= \lim_{t\to-\infty}\int_t^a f(x)\,dx + \int_a^b f(x)\,dx + \lim_{u\to\infty}\int_b^u f(x)\,dx \\ &= \lim_{t\to-\infty}\left[\int_t^a f(x)\,dx + \int_a^b f(x)\,dx\right] + \int_b^\infty f(x)\,dx \\ &= \lim_{t\to-\infty}\int_t^b f(x)\,dx + \int_b^\infty f(x)\,dx \\ &= \int_{-\infty}^b f(x)\,dx + \int_b^\infty f(x)\,dx\end{aligned}$$

**53.** We would expect a small percentage of bulbs to burn out in the first few hundred hours, most of the bulbs to burn out after close to 700 hours, and a few overachievers to burn on and on.

(a)

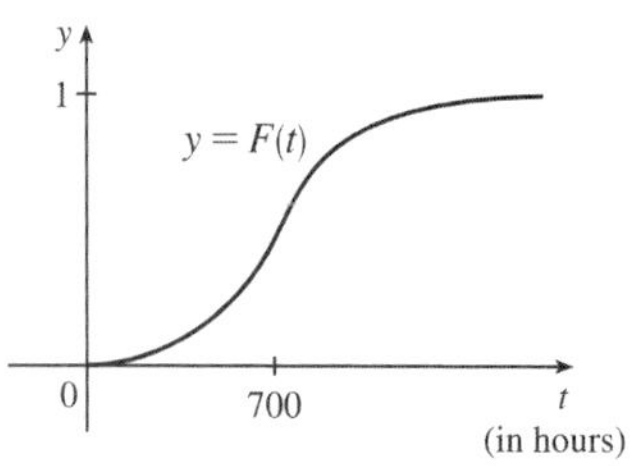


(b) $r(t) = F'(t)$ is the rate at which the fraction $F(t)$ of burnt-out bulbs increases as $t$ increases. This could be interpreted as a fractional burnout rate.

(c) $\int_0^\infty r(t)\,dt = \lim\limits_{x\to\infty} F(x) = 1$, since all of the bulbs will eventually burn out.

**54.** Let $k = \dfrac{M}{2RT}$ so that $\overline{v} = \dfrac{4}{\sqrt{\pi}}\, k^{3/2} \displaystyle\int_0^\infty v^3 e^{-kv^2}\, dv$. Let $I$ denote the integral and use parts to integrate $I$. Let $\alpha = v^2$, $d\beta = ve^{-kv^2}\, dv \quad\Rightarrow\quad d\alpha = 2v\, dv$, $\beta = -\dfrac{1}{2k} e^{-kv^2}$:

$$\begin{aligned} I &= \lim_{t\to\infty}\left[-\frac{1}{2k}v^2 e^{-kv^2}\right]_0^t + \frac{1}{k}\int_0^\infty ve^{-kv^2}\, dv_0^t \\ &= -\frac{1}{2k}\lim_{t\to\infty}\left(t^2 e^{-kt^2}\right) + \frac{1}{k}\lim_{t\to\infty}\left[-\frac{1}{2k}e^{-kv^2}\right] \\ &\overset{\text{H}}{=} -\frac{1}{2k}\cdot 0 - \frac{1}{2k^2}(0-1) = \frac{1}{2k^2} \end{aligned}$$

Thus, $\overline{v} = \dfrac{4}{\sqrt{\pi}} k^{3/2} \cdot \dfrac{1}{2k^2} = \dfrac{2}{(k\pi)^{1/2}} = \dfrac{2}{[\pi M/(2RT)]^{1/2}} = \dfrac{2\sqrt{2}\sqrt{RT}}{\sqrt{\pi M}} = \sqrt{\dfrac{8RT}{\pi M}}$.

**55.** $I = \displaystyle\int_0^\infty te^{kt}\, dt = \lim_{s\to\infty}\left[\frac{1}{k^2}(kt-1)\,e^{kt}\right]_0^s$ [Formula 96, or parts]

$= \displaystyle\lim_{s\to\infty}\left[\left(\frac{1}{k}se^{ks} - \frac{1}{k^2}e^{ks}\right) - \left(-\frac{1}{k^2}\right)\right]$.

Since $k < 0$ the first two terms approach 0 (you can verify that the first term does so with l'Hospital's Rule), so the limit is equal to $1/k^2$. Thus, $M = -kI = -k(1/k^2) = -1/k = -1/(-0.000121) \approx 8264.5$ years.

**56.** $y(s) = \displaystyle\int_s^R \frac{2r}{\sqrt{r^2-s^2}}\, x(r)\, dr$ and $x(r) = \frac{1}{2}(R-r)^2 \quad\Rightarrow$

$$\begin{aligned} y(s) &= \lim_{t\to s^+}\int_t^R \frac{r(R-r)^2}{\sqrt{r^2-s^2}}\, dr = \lim_{t\to s^+}\int_t^R \frac{r^3 - 2Rr^2 + R^2 r}{\sqrt{r^2-s^2}}\, dr \\ &= \lim_{t\to s^+}\left[\int_t^R \frac{r^3\, dr}{\sqrt{r^2-s^2}} - 2R\int_t^R \frac{r^2\, dr}{\sqrt{r^2-s^2}} + R^2\int_t^R \frac{r\, dr}{\sqrt{r^2-s^2}}\right] \\ &= \lim_{t\to s^+}\left(I_1 - 2RI_2 + R^2 I_3\right) = L \end{aligned}$$

For $I_1$: Let $u = \sqrt{r^2-s^2} \quad\Rightarrow\quad u^2 = r^2 - s^2$, $r^2 = u^2 + s^2$, $2r\, dr = 2u\, du$, so, omitting limits and constant of integration,

$$\begin{aligned} I_1 &= \int \frac{(u^2+s^2)u}{u}\, du = \int (u^2+s^2)\, du = \tfrac{1}{3}u^3 + s^2 u = \tfrac{1}{3}u(u^2+3s^2) \\ &= \tfrac{1}{3}\sqrt{r^2-s^2}\,(r^2 - s^2 + 3s^2) = \tfrac{1}{3}\sqrt{r^2-s^2}\,(r^2+2s^2) \end{aligned}$$

For $I_2$: Using Formula 44, $I_2 = \dfrac{r}{2}\sqrt{r^2-s^2} + \dfrac{s^2}{2}\ln\left|r + \sqrt{r^2-s^2}\,\right|$.

For $I_3$: Let $u = r^2 - s^2 \quad\Rightarrow\quad du = 2r\, dr$. Then $I_3 = \dfrac{1}{2}\displaystyle\int \frac{du}{\sqrt{u}} = \frac{1}{2}\cdot 2\sqrt{u} = \sqrt{r^2-s^2}$.

[continued]

Thus,

$$L = \lim_{t \to s^+} \left[ \tfrac{1}{3}\sqrt{r^2 - s^2}\,(r^2 + 2s^2) - 2R\left( \frac{r}{2}\sqrt{r^2 - s^2} + \frac{s^2}{2}\ln\left| r + \sqrt{r^2 - s^2} \right| \right) + R^2\sqrt{r^2 - s^2} \right]_t^R$$

$$= \lim_{t \to s^+} \left[ \tfrac{1}{3}\sqrt{R^2 - s^2}\,(R^2 + 2s^2) - 2R\left( \frac{R}{2}\sqrt{R^2 - s^2} + \frac{s^2}{2}\ln\left| R + \sqrt{R^2 - s^2} \right| \right) + R^2\sqrt{R^2 - s^2} \right]$$

$$- \lim_{t \to s^+} \left[ \tfrac{1}{3}\sqrt{t^2 - s^2}\,(t^2 + 2s^2) - 2R\left( \frac{t}{2}\sqrt{t^2 - s^2} + \frac{s^2}{2}\ln\left| t + \sqrt{t^2 - s^2} \right| \right) + R^2\sqrt{t^2 - s^2} \right]$$

$$= \left[ \tfrac{1}{3}\sqrt{R^2 - s^2}\,(R^2 + 2s^2) - Rs^2\ln\left| R + \sqrt{R^2 - s^2} \right| \right] - \left[ -Rs^2\ln|s| \right]$$

$$= \tfrac{1}{3}\sqrt{R^2 - s^2}\,(R^2 + 2s^2) - Rs^2\ln\left( \frac{R + \sqrt{R^2 - s^2}}{s} \right)$$

**57.** $I = \displaystyle\int_a^\infty \frac{1}{x^2+1}\,dx = \lim_{t\to\infty} \int_a^t \frac{1}{x^2+1}\,dx = \lim_{t\to\infty} \left[\tan^{-1} x\right]_a^t = \lim_{t\to\infty} \left(\tan^{-1} t - \tan^{-1} a\right) = \tfrac{\pi}{2} - \tan^{-1} a.$

$I < 0.001 \;\Rightarrow\; \tfrac{\pi}{2} - \tan^{-1} a < 0.001 \;\Rightarrow\; \tan^{-1} a > \tfrac{\pi}{2} - 0.001 \;\Rightarrow\; a > \tan\left(\tfrac{\pi}{2} - 0.001\right) \approx 1000.$

**58.** $f(x) = e^{-x^2}$ and $\Delta x = \frac{4-0}{8} = \frac{1}{2}$.

$$\textstyle\int_0^4 f(x)\,dx \approx S_8] = \frac{1}{2\cdot 3}[f(0) + 4f(0.5) + 2f(1) + \cdots + 2f(3) + 4f(3.5) + f(4)$$
$$\approx \tfrac{1}{6}(5.31717808) \approx 0.8862$$

Now $x > 4 \;\Rightarrow\; -x\cdot x < -x \cdot 4 \;\Rightarrow\; e^{-x^2} < e^{-4x} \;\Rightarrow\; \int_4^\infty e^{-x^2}\,dx < \int_4^\infty e^{-4x}\,dx.$

$\int_4^\infty e^{-4x}\,dx = \lim_{t\to\infty} \left[-\frac{1}{4}e^{-4x}\right]_4^t = -\frac{1}{4}\left(0 - e^{-16}\right) = 1/(4e^{16}) \approx 0.0000000281 < 0.0000001$, as desired.

**59.** We use integration by parts: let $u = x$, $dv = xe^{-x^2}\,dx \;\Rightarrow\; du = dx$, $v = -\frac{1}{2}e^{-x^2}$. So

$$\textstyle\int_0^\infty x^2 e^{-x^2}\,dx = \lim_{t\to\infty} \left[-\frac{1}{2}xe^{-x^2}\right]_0^t + \frac{1}{2}\int_0^\infty e^{-x^2}\,dx$$
$$\textstyle = \lim_{t\to\infty} \left[-t\big/\left(2e^{t^2}\right)\right] + \frac{1}{2}\int_0^\infty e^{-x^2}\,dx = \frac{1}{2}\int_0^\infty e^{-x^2}\,dx$$

(The limit is 0 by l'Hospital's Rule.)

**60.** $\int_0^\infty e^{-x^2}\,dx$ is the area under the curve $y = e^{-x^2}$ for $0 \le x < \infty$ and $0 < y \le 1$. Solving $y = e^{-x^2}$ for $x$, we get $y = e^{-x^2} \;\Rightarrow\; \ln y = -x^2 \;\Rightarrow\; -\ln y = x^2 \;\Rightarrow\; x = \pm\sqrt{-\ln y}$. Since $x$ is positive, choose $x = \sqrt{-\ln y}$, and the area is represented by $\int_0^1 \sqrt{-\ln y}\,dy$. Therefore, each integral represents the same area, so the integrals are equal.

**61.** For the first part of the integral, let $x = 2\tan\theta \;\Rightarrow\; dx = 2\sec^2\theta\,d\theta$.

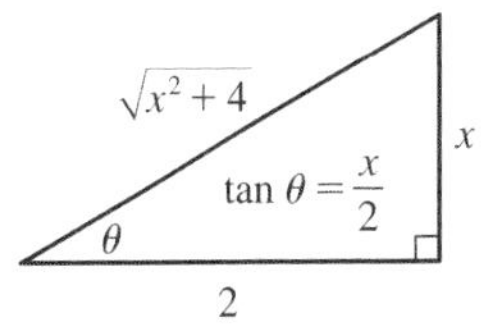


$$\int \frac{1}{\sqrt{x^2+4}}\,dx = \int \frac{2\sec^2\theta}{2\sec\theta}\,d\theta = \int \sec\theta\,d\theta = \ln|\sec\theta + \tan\theta|.$$

From the figure, $\tan\theta = \dfrac{x}{2}$, and $\sec\theta = \dfrac{\sqrt{x^2+4}}{2}$. So

$$\begin{aligned} I = \int_0^\infty \left(\tfrac{1}{\sqrt{x^2+4}} - \tfrac{C}{x+2}\right)dx &= \lim_{t\to\infty}\left[\ln\left|\tfrac{\sqrt{x^2+4}}{2} + \tfrac{x}{2}\right| - C\ln|x+2|\right]_0^t \\ &= \lim_{t\to\infty}\left[\ln\frac{\sqrt{t^2+4}+t}{2} - C\ln(t+2) - (\ln 1 - C\ln 2)\right] \\ &= \lim_{t\to\infty}\left[\ln\left(\tfrac{\sqrt{t^2+4}+t}{2(t+2)^C}\right) + \ln 2^C\right] \\ &= \ln\left(\lim_{t\to\infty}\tfrac{t+\sqrt{t^2+4}}{(t+2)^C}\right) + \ln 2^{C-1} \end{aligned}$$

Now $L = \displaystyle\lim_{t\to\infty}\frac{t+\sqrt{t^2+4}}{(t+2)^C} \overset{\text{H}}{=} \lim_{t\to\infty}\frac{1+t/\sqrt{t^2+4}}{C\,(t+2)^{C-1}} = \frac{2}{C\lim\limits_{t\to\infty}(t+2)^{C-1}}$.

If $C < 1$, $L = \infty$ and $I$ diverges.

If $C = 1$, $L = 2$ and $I$ converges to $\ln 2 + \ln 2^0 = \ln 2$.

If $C > 1$, $L = 0$ and $I$ diverges to $-\infty$.

**62.**
$$\begin{aligned} I = \int_0^\infty\left(\frac{x}{x^2+1} - \frac{C}{3x+1}\right)dx &= \lim_{t\to\infty}\left[\tfrac{1}{2}\ln(x^2+1) - \tfrac{1}{3}C\ln(3x+1)\right]_0^t \\ &= \lim_{t\to\infty}\left[\ln(t^2+1)^{1/2} - \ln(3t+1)^{C/3}\right] \\ &= \lim_{t\to\infty}\left(\ln\frac{(t^2+1)^{1/2}}{(3t+1)^{C/3}}\right) = \ln\left(\lim_{t\to\infty}\frac{\sqrt{t^2+1}}{(3t+1)^{C/3}}\right) \end{aligned}$$

For $C \le 0$, the integral diverges. For $C > 0$, we have

$$L = \lim_{t\to\infty}\frac{\sqrt{t^2+1}}{(3t+1)^{C/3}} \overset{\text{H}}{=} \lim_{t\to\infty}\frac{t/\sqrt{t^2+1}}{C\,(3t+1)^{(C/3)-1}} = \frac{1}{C}\lim_{t\to\infty}\frac{1}{(3t+1)^{(C/3)-1}}.$$

For $C/3 < 1 \;\Leftrightarrow\; C < 3$, $L = \infty$ and $I$ diverges.

For $C = 3$, $L = \frac{1}{3}$ and $I = \ln\frac{1}{3}$.

For $C > 3$, $L = 0$ and $I$ diverges to $-\infty$.

## 5 Review

### CONCEPT CHECK

**1.** (a) $\sum_{i=1}^{n} f(x_i^*)\,\Delta x$ is an expression for a Riemann sum of a function $f$.
$x_i^*$ is a point in the $i$th subinterval $[x_{i-1}, x_i]$ and $\Delta x$ is the length of the subintervals.

(b) See Figure 1 in Section 5.2.

(c) In Section 5.2, see Figure 3 and the paragraph beside it.

**2.** (a) See Definition 5.2.2.

(b) See Figure 2 in Section 5.2.

(c) In Section 5.2, see Figure 4 and the paragraph above it.

**3.** (a) See the Evaluation Theorem at the beginning of Section 5.3.

(b) See the Net Change Theorem after Example 6 in Section 5.3.

**4.** $\int_{t_1}^{t_2} r(t)\,dt$ represents the change in the amount of water in the reservoir between time $t_1$ and time $t_2$.

**5.** (a) $\int_{60}^{120} v(t)\,dt$ represents the change in position of the particle from $t = 60$ to $t = 120$ seconds.

(b) $\int_{60}^{120} |v(t)|\,dt$ represents the total distance traveled by the particle from $t = 60$ to 120 seconds.

(c) $\int_{60}^{120} a(t)\,dt$ represents the change in the velocity of the particle from $t = 60$ to $t = 120$ seconds.

**6.** (a) $\int f(x)\,dx$ is the family of functions $\{F \mid F' = f\}$. Any two such functions differ by a constant.

(b) The connection is given by the Evaluation Theorem: $\int_a^b f(x)\,dx = \left[\int f(x)\,dx\right]_a^b$ if $f$ is continuous.

**7.** See the Fundamental Theorem of Calculus after Example 5 in Section 5.4.

**8.** (a) See the Substitution Rule (5.5.4). This says that it is permissible to operate with the $dx$ after an integral sign as if it were a differential.

(b) See Formula 5.6.1 or 5.6.2. We try to choose $u = f(x)$ to be a function that becomes simpler when differentiated (or at least not more complicated) as long as $dv = g'(x)\,dx$ can be readily integrated to give $v$.

**9.** See the Midpoint Rule, the Trapezoidal Rule, and Simpson's Rule, as well as their associated error bounds, all in Section 5.9. We would expect the best estimate to be given by Simpson's Rule.

**10.** See Definitions 1(a), (b), and (c) in Section 5.10.

**11.** See Definitions 3(b), (a), and (c) in Section 5.10.

**12.** See the Comparison Theorem after Example 8 in Section 5.10.

**13.** The precise version of this statement is given by the Fundamental Theorem of Calculus. See the statement of this theorem and the paragraph that follows it in Section 5.4.

## TRUE-FALSE QUIZ

**1.** True by Property 2 of the Integral in Section 5.2.

**2.** False. Try $a = 0$, $b = 2$, $f(x) = g(x) = 1$ as a counterexample.

**3.** True by Property 3 of the Integral in Section 5.2.

**4.** False. You can't take a variable outside the integral sign. For example, using $f(x) = 1$ on $[0, 1]$,

$$\int_0^1 x\,f(x)\,dx = \int_0^1 x\,dx = \left[\tfrac{1}{2}x^2\right]_0^1 = \tfrac{1}{2} \text{ (a constant) while } x\int_0^1 1\,dx = x\,[x]_0^1 = x \cdot 1 = x \text{ (a variable).}$$

**5.** False. For example, let $f(x) = x^2$. Then $\int_0^1 \sqrt{x^2}\,dx = \int_0^1 x\,dx = \frac{1}{2}$, but $\sqrt{\int_0^1 x^2\,dx} = \sqrt{\frac{1}{3}} = \frac{1}{\sqrt{3}}$.

**6.** True by the Net Change Theorem.

**7.** True by Comparison Property 7 of the Integral in Section 5.2.

**8.** False. For example, let $a = 0$, $b = 1$, $f(x) = 3$, $g(x) = x$. $f(x) > g(x)$ for each $x$ in $(0, 1)$, but $f'(x) = 0 < 1 = g'(x)$ for $x \in (0, 1)$.

**9.** True. The integrand is an odd function that is continuous on $[-1, 1]$, so the result follows from Theorem 5.5.6(b).

**10.** True.

$$\begin{aligned}\int_{-5}^{5}(ax^2 + bx + c)\,dx &= \int_{-5}^{5}(ax^2 + c)\,dx + \int_{-5}^{5} bx\,dx \\ &= 2\int_0^5 (ax^2 + c)\,dx \;\; [\text{by 5.5.6(a)}] \; + 0 \;\; [\text{by 5.5.6(b)}]\end{aligned}$$

**11.** False. This is an improper integral, since the denominator vanishes at $x = 1$.

$$\int_0^4 \frac{x}{x^2-1}\,dx = \int_0^1 \frac{x}{x^2-1}\,dx + \int_1^4 \frac{x}{x^2-1}\,dx \text{ and}$$

$$\int_0^1 \frac{x}{x^2-1}\,dx = \lim_{t\to1^-}\int_0^t \frac{x}{x^2-1}\,dx = \lim_{t\to1^-}\left[\tfrac{1}{2}\ln\left|x^2-1\right|\right]_0^t = \lim_{t\to1^-}\tfrac{1}{2}\ln\left|t^2-1\right| = \infty$$

So the integral diverges.

**12.** True by Theorem 5.10.2 with $p = \sqrt{2} > 1$.

**13.** False. See the remarks and Figure 4 before Example 1 in Section 5.2, and notice that $y = x - x^3 < 0$ for $1 < x \le 2$.

**14.** True by FTC1.

**15.** False. For example, the function $y = |x|$ is continuous on $\mathbb{R}$, but has no derivative at $x = 0$.

**16.** False. For example, with $n = 1$ the Trapezoidal Rule is much more accurate than the Midpoint Rule for the function in the diagram.

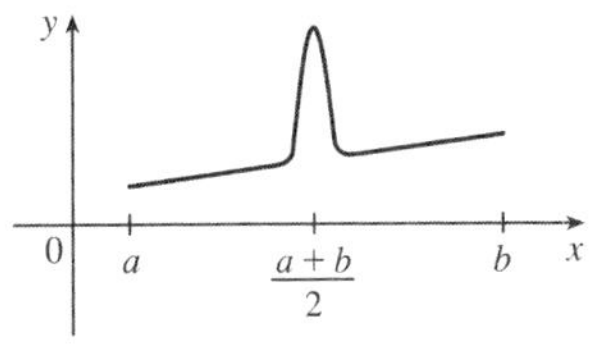


**17.** False. See Exercise 51 in Section 5.10.

**18.** True. If $f$ is continuous on $[0,\infty)$, then $\int_0^1 f(x)\,dx$ is finite. Since $\int_1^\infty f(x)\,dx$ is finite, so is $\int_0^\infty f(x)\,dx = \int_0^1 f(x)\,dx + \int_1^\infty f(x)\,dx$.

**19.** False. If $f(x) = 1/x$, then $f$ is continuous and decreasing on $[1,\infty)$ with $\lim_{x\to\infty} f(x) = 0$, but $\int_1^\infty f(x)\,dx$ is divergent.

**20.** True.

$$\begin{aligned}\int_a^\infty [f(x)+g(x)]\,dx &= \lim_{t\to\infty}\int_a^t [f(x)+g(x)]\,dx = \lim_{t\to\infty}\left(\int_a^t f(x)\,dx + \int_a^t g(x)\,dx\right)\\ &= \lim_{t\to\infty}\int_a^t f(x)\,dx + \lim_{t\to\infty}\int_a^t g(x)\,dx \quad \left[\begin{matrix}\text{since both limits}\\ \text{in the sum exist}\end{matrix}\right]\\ &= \int_a^\infty f(x)\,dx + \int_a^\infty g(x)\,dx\end{aligned}$$

Since the two integrals are finite, so is their sum.

**21.** False. Take $f(x) = 1$ for all $x$ and $g(x) = -1$ for all $x$. Then $\int_a^\infty f(x)\,dx = \infty$ [divergent] and $\int_a^\infty g(x)\,dx = -\infty$ [divergent], but $\int_a^\infty [f(x)+g(x)]\,dx = 0$ [convergent].

**22.** False. $\int_0^\infty f(x)\,dx$ could converge or diverge. For example, if $g(x) = 1$, then $\int_0^\infty f(x)\,dx$ diverges if $f(x) = 1$ and converges if $f(x) = 0$.

## EXERCISES

**1.** (a)

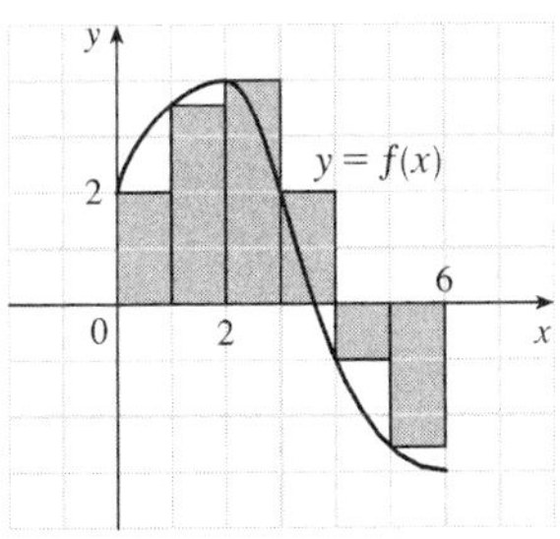


$$L_6 = \sum_{i=1}^{6} f(x_{i-1})\,\Delta x \quad \left[\Delta x = \tfrac{6-0}{6} = 1\right]$$

$$\begin{aligned}&= f(x_0)\cdot 1 + f(x_1)\cdot 1 + f(x_2)\cdot 1\\ &\quad + f(x_3)\cdot 1 + f(x_4)\cdot 1 + f(x_5)\cdot 1\\ &\approx 2 + 3.5 + 4 + 2 + (-1) + (-2.5) = 8\end{aligned}$$

The Riemann sum represents the sum of the areas of the four rectangles above the $x$-axis minus the sum of the areas of the two rectangles below the $x$-axis.

(b)

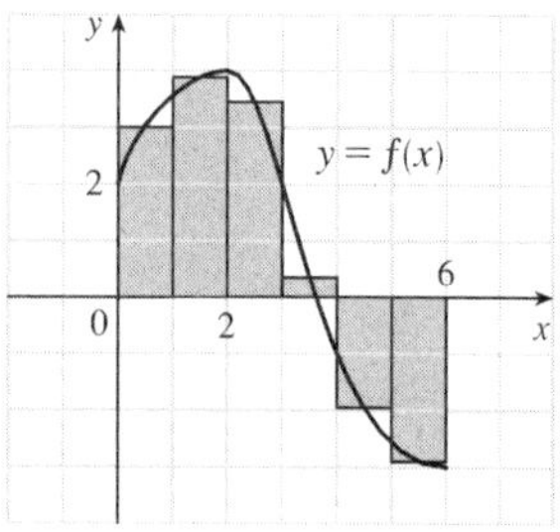


$$M_6 = \sum_{i=1}^{6} f(\overline{x}_i)\,\Delta x \quad [\Delta x = \tfrac{6-0}{6} = 1]$$

$$= f(\overline{x}_1)\cdot 1 + f(\overline{x}_2)\cdot 1 + f(\overline{x}_3)\cdot 1 + f(\overline{x}_4)\cdot 1 + f(\overline{x}_5)\cdot 1 + f(\overline{x}_6)\cdot 1$$

$$= f(0.5) + f(1.5) + f(2.5) + f(3.5) + f(4.5) + f(5.5)$$

$$\approx 3 + 3.9 + 3.4 + 0.3 + (-2) + (-2.9) = 5.7$$

The Riemann sum represents the sum of the areas of the four rectangles above the $x$-axis minus the sum of the areas of the two rectangles below the $x$-axis.

**2.** (a)

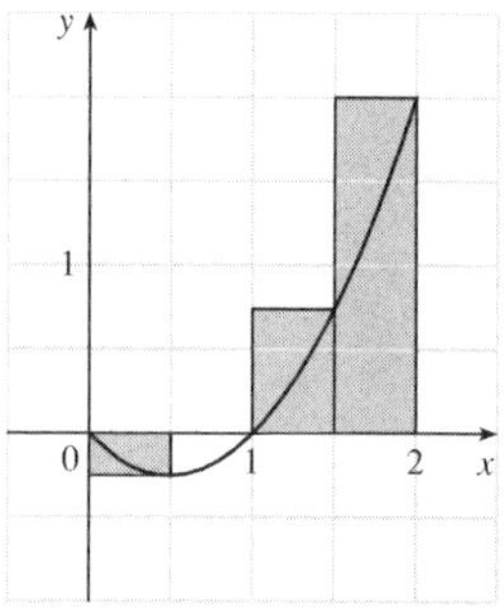


$f(x) = x^2 - x$ and $\Delta x = \frac{2-0}{4} = 0.5 \;\Rightarrow$

$$R_4 = 0.5f(0.5) + 0.5f(1) + 0.5f(1.5) + 0.5f(2)$$

$$= 0.5(-0.25 + 0 + 0.75 + 2) = 1.25$$

The Riemann sum represents the sum of the areas of the two rectangles above the $x$-axis minus the area of the rectangle below the $x$-axis. (The second rectangle vanishes.)

(b) $\int_0^2 (x^2 - x)\,dx = \lim\limits_{n\to\infty} \sum\limits_{i=1}^{n} f(x_i)\,\Delta x \quad [\Delta x = 2/n \text{ and } x_i = 2i/n]$

$$= \lim_{n\to\infty} \sum_{i=1}^{n} \left(\frac{4i^2}{n^2} - \frac{2i}{n}\right)\left(\frac{2}{n}\right) = \lim_{n\to\infty} \frac{2}{n}\left[\frac{4}{n^2}\sum_{i=1}^{n} i^2 - \frac{2}{n}\sum_{i=1}^{n} i\right]$$

$$= \lim_{n\to\infty}\left[\frac{8}{n^3}\cdot\frac{n(n+1)(2n+1)}{6} - \frac{4}{n^2}\cdot\frac{n(n+1)}{2}\right] = \lim_{n\to\infty}\left[\frac{4}{3}\cdot\frac{n+1}{n}\cdot\frac{2n+1}{n} - 2\cdot\frac{n+1}{n}\right]$$

$$= \lim_{n\to\infty}\left[\frac{4}{3}\left(1+\frac{1}{n}\right)\left(2+\frac{1}{n}\right) - 2\left(1+\frac{1}{n}\right)\right] = \tfrac{4}{3}\cdot 1\cdot 2 - 2\cdot 1 = \tfrac{2}{3}$$

(c) $\int_0^2 (x^2 - x)\,dx = \left[\frac{1}{3}x^3 - \frac{1}{2}x^2\right]_0^2 = \left(\frac{8}{3} - 2\right) = \frac{2}{3}$

(d)

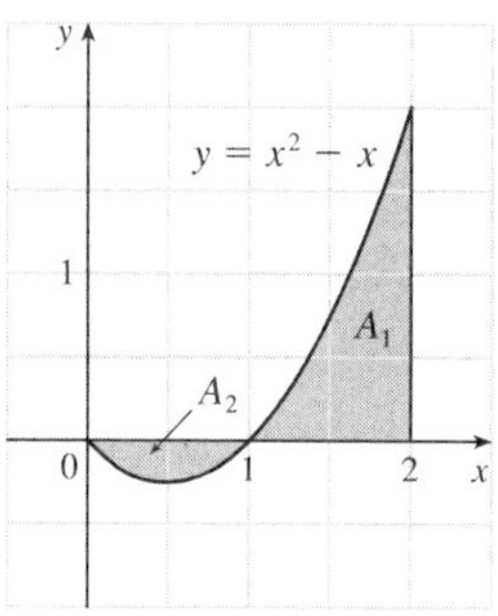


$\int_0^2 (x^2 - x)\,dx = A_1 - A_2$, where $A_1$ and $A_2$ are the areas shown in the diagram.

**3.** $\int_0^1 \left(x + \sqrt{1-x^2}\,\right)dx = \int_0^1 x\,dx + \int_0^1 \sqrt{1-x^2}\,dx = I_1 + I_2.$

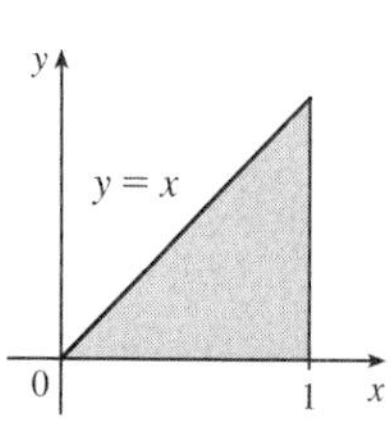


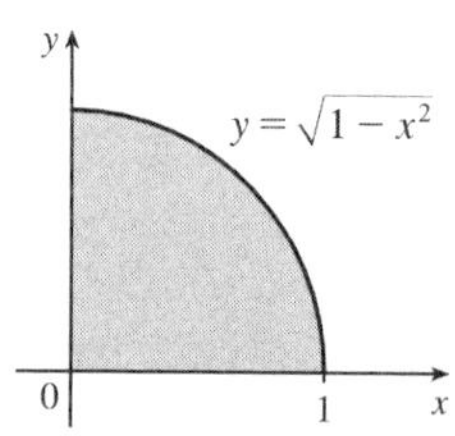


$I_1$ can be interpreted as the area of the triangle shown in the figure and $I_2$ can be interpreted as the area of the quarter-circle. Area $= \frac{1}{2}(1)(1) + \frac{1}{4}(\pi)(1)^2 = \frac{1}{2} + \frac{\pi}{4}$.

**4.** On $[0, \pi]$, $\lim\limits_{n\to\infty} \sum\limits_{i=1}^{n} \sin x_i \, \Delta x = \int_0^\pi \sin x \, dx = [-\cos x]_0^\pi = -(-1) - (-1) = 2.$

**5.** $\int_0^6 f(x)\,dx = \int_0^4 f(x)\,dx + \int_4^6 f(x)\,dx \;\Rightarrow\; 10 = 7 + \int_4^6 f(x)\,dx \;\Rightarrow\; \int_4^6 f(x)\,dx = 10 - 7 = 3$

**6.** (a) $f(x) = e^{3x}$, $\Delta x = (2-0)/n = 2/n$, and $x_i = 2i/n \;\Rightarrow$

$$\int_0^2 e^{3x}\,dx = \lim_{n\to\infty} \sum_{i=1}^{n} f\left(\frac{2i}{n}\right)\left(\frac{2}{n}\right) = \lim_{n\to\infty} \sum_{i=1}^{n} e^{3(2i/n)}\left(\frac{2}{n}\right) \overset{\text{CAS}}{=} \lim_{n\to\infty} \frac{2e^{6/n}(e^6-1)}{n(e^{6/n}-1)} \overset{\text{CAS}}{=} \frac{e^6-1}{3} \approx 134.14.$$

(b) $\int_0^2 e^{3x}\,dx = \left[\frac{1}{3}e^{3x}\right]_0^2 = \frac{1}{3}(e^6 - 1)$, as in part (a).

**7.** First note that either $a$ or $b$ must be the graph of $\int_0^x f(t)\,dt$, since $\int_0^0 f(t)\,dt = 0$, and $c(0) \neq 0$. Now notice that $b > 0$ when $c$ is increasing, and that $c > 0$ when $a$ is increasing. It follows that $c$ is the graph of $f(x)$, $b$ is the graph of $f'(x)$, and $a$ is the graph of $\int_0^x f(t)\,dt$.

**8.** (a) By the Evaluation Theorem (FTC2), $\displaystyle\int_0^1 \frac{d}{dx}\left(e^{\arctan x}\right) dx = \left[e^{\arctan x}\right]_0^1 = e^{\pi/4} - 1$

(b) $\displaystyle\frac{d}{dx}\int_0^1 e^{\arctan x}\,dx = 0$ since this is the derivative of a constant.

(c) By FTC1, $\displaystyle\frac{d}{dx}\int_0^x e^{\arctan t}\,dt = e^{\arctan x}$.

**9.** $\int_1^2 (8x^3 + 3x^2)\,dx = \left[8\cdot\frac{1}{4}x^4 + 3\cdot\frac{1}{3}x^3\right]_1^2 = \left[2x^4 + x^3\right]_1^2 = (2\cdot 2^4 + 2^3) - (2+1) = 40 - 3 = 37$

**10.** $\displaystyle\int_0^T (x^4 - 8x + 7)\,dx = \left[\frac{1}{5}x^5 - 4x^2 + 7x\right]_0^T = \left(\frac{1}{5}T^5 - 4T^2 + 7T\right) - 0 = \frac{1}{5}T^5 - 4T^2 + 7T$

**11.** $\int_0^1 (1 - x^9)\,dx = \left[x - \frac{1}{10}x^{10}\right]_0^1 = \left(1 - \frac{1}{10}\right) - 0 = \frac{9}{10}$

**12.** Let $u = 1 - x$, so $du = -dx$ and $dx = -du$. When $x = 0$, $u = 1$; when $x = 1$, $u = 0$. Thus,
$\int_0^1 (1-x)^9\,dx = \int_1^0 u^9(-du) = \int_0^1 u^9\,du = \frac{1}{10}\left[u^{10}\right]_0^1 = \frac{1}{10}(1-0) = \frac{1}{10}.$

**13.** $\displaystyle\int_1^9 \frac{\sqrt{u} - 2u^2}{u}\,du = \int_1^9 (u^{-1/2} - 2u)\,du = \left[2u^{1/2} - u^2\right]_1^9 = (6 - 81) - (2 - 1) = -76$

**14.** $\int_0^1 \left(\sqrt[4]{u} + 1\right)^2 du = \int_0^1 (u^{1/2} + 2u^{1/4} + 1)\,du = \left[\frac{2}{3}u^{3/2} + \frac{8}{5}u^{5/4} + u\right]_0^1 = \left(\frac{2}{3} + \frac{8}{5} + 1\right) - 0 = \frac{49}{15}$

**15.** $u = x^2 + 1$, $du = 2x\,dx$, so $\displaystyle\int_0^1 \frac{x}{x^2+1}\,dx = \int_1^2 \frac{1}{u}\left(\frac{1}{2}\,du\right) = \frac{1}{2}\left[\ln u\right]_1^2 = \frac{1}{2}\ln 2.$

**16.** $\displaystyle\int_0^1 \frac{1}{x^2+1}\,dx = \left[\tan^{-1} x\right]_0^1 = \tan^{-1} 1 - \tan^{-1} 0 = \frac{\pi}{4} - 0 = \frac{\pi}{4}$

**17.** Let $u = v^3$, so $du = 3v^2\,dv$. When $v = 0$, $u = 0$; when $v = 1$, $u = 1$. Thus,
$\int_0^1 v^2\cos(v^3)\,dv = \int_0^1 \cos u\left(\frac{1}{3}\,du\right) = \frac{1}{3}\left[\sin u\right]_0^1 = \frac{1}{3}(\sin 1 - 0) = \frac{1}{3}\sin 1.$

**18.** Let $u = 3\pi t$, so $du = 3\pi\,dt$. When $t = 0$, $u = 1$; when $t = 1$, $u = 3\pi$. Thus,
$$\int_0^1 \sin(3\pi t)\,dt = \int_0^{3\pi} \sin u\left(\frac{1}{3\pi}\,du\right) = \frac{1}{3\pi}\left[-\cos u\right]_0^{3\pi} = -\frac{1}{3\pi}(-1-1) = \frac{2}{3\pi}.$$

**19.** $\int_0^1 e^{\pi t}\,dt = \left[\frac{1}{\pi}e^{\pi t}\right]_0^1 = \frac{1}{\pi}(e^\pi - 1)$

**20.** Let $u = 2 - 3x$, so $du = -3\,dx$. When $x = 1$, $u = -1$; when $x = 2$, $u = -4$. Thus,
$$\int_1^2 \frac{1}{2-3x}\,dx = \int_{-1}^{-4} \frac{1}{u}\left(-\frac{1}{3}\,du\right) = -\frac{1}{3}\left[\ln|u|\right]_{-1}^{-4} = -\frac{1}{3}(\ln 4 - \ln 1) = -\frac{1}{3}\ln 4.$$

**21.** Let $u = x^2 + 4x$. Then $du = (2x+4)\,dx = 2(x+2)\,dx$, so
$$\int \frac{x+2}{\sqrt{x^2+4x}}\,dx = \int u^{-1/2}\left(\frac{1}{2}\,du\right) = \frac{1}{2}\cdot 2u^{1/2} + C = \sqrt{u} + C = \sqrt{x^2+4x} + C.$$

**22.** Integrate by parts with $u = \ln x$, $dv = x^3\,dx \Rightarrow du = dx/x$, $v = x^4/4$:

$\int_1^2 x^3 \ln x\,dx = \left[\frac{1}{4}x^4 \ln x\right]_1^2 - \frac{1}{4}\int_1^2 x^3\,dx = 4\ln 2 - \frac{1}{16}\left[x^4\right]_1^2 = 4\ln 2 - \frac{15}{16}$.

**23.** $$\int_0^5 \frac{x}{x+10}\,dx = \int_0^5 \left(1 - \frac{10}{x+10}\right)dx = \Big[x - 10\ln(x+10)\Big]_0^5$$
$$= 5 - 10\ln 15 + 10\ln 10 = 5 + 10\ln\tfrac{10}{15} = 5 + 10\ln\tfrac{2}{3}$$

**24.** $$\int_0^5 ye^{-0.6y}\,dy \quad \begin{bmatrix} u = y, & dv = e^{-0.6y}\,dy, \\ du = dy & v = -\frac{5}{3}e^{-0.6y} \end{bmatrix} = \left[-\tfrac{5}{3}ye^{-0.6y}\right]_0^5 - \int_0^5 \left(-\tfrac{5}{3}e^{-0.6y}\right)dy$$
$$= -\tfrac{25}{3}e^{-3} - \tfrac{25}{9}\left[e^{-0.6y}\right]_0^5 = -\tfrac{25}{3}e^{-3} - \tfrac{25}{9}\left(e^{-3} - 1\right)$$
$$= -\tfrac{25}{3}e^{-3} - \tfrac{25}{9}e^{-3} + \tfrac{25}{9} = \tfrac{25}{9} - \tfrac{100}{9}e^{-3}$$

**25.** $$\int_0^{\pi/2} \frac{\cos\theta}{1+\sin\theta}\,d\theta = \Big[\ln(1+\sin\theta)\Big]_0^{\pi/2} = \ln 2 - \ln 1 = \ln 2$$

**26.** $$\int_1^4 \frac{dt}{(2t+1)^3} \quad \begin{bmatrix} u = 2t+1, \\ du = 2\,dt \end{bmatrix} = \int_3^9 \frac{\frac{1}{2}\,du}{u^3} = \frac{-1}{4}\left[\frac{1}{u^2}\right]_3^9 = -\frac{1}{4}\left(\frac{1}{81} - \frac{1}{9}\right) = -\frac{1}{4}\left(-\frac{8}{81}\right) = \frac{2}{81}$$

**27.** $$\int_1^4 x^{3/2}\ln x\,dx \quad \begin{bmatrix} u = \ln x, & dv = x^{3/2}\,dx, \\ du = dx/x & v = \frac{2}{5}x^{5/2} \end{bmatrix} = \frac{2}{5}\left[x^{5/2}\ln x\right]_1^4 - \frac{2}{5}\int_1^4 x^{3/2}\,dx$$
$$= \tfrac{2}{5}(32\ln 4 - \ln 1) - \tfrac{2}{5}\left[\tfrac{2}{5}x^{5/2}\right]_1^4$$
$$= \tfrac{2}{5}(64\ln 2) - \tfrac{4}{25}(32 - 1)$$
$$= \tfrac{128}{5}\ln 2 - \tfrac{124}{25} \quad \left(\text{or } \tfrac{64}{5}\ln 4 - \tfrac{124}{25}\right)$$

**28.** Let $u = \cos x$. Then $du = -\sin x\,dx$, so $\int \sin x\,\cos(\cos x)\,dx = -\int \cos u\,du = -\sin u + C = -\sin(\cos x) + C$.

**29.** $\dfrac{1}{t^2+6t+8} = \dfrac{1}{(t+2)(t+4)} = \dfrac{A}{t+2} + \dfrac{B}{t+4}$. Multiply both sides by $(t+2)(t+4)$ to get $1 = A(t+4) + B(t+2)$. Substituting $-4$ for $t$ gives $1 = -2B \Leftrightarrow B = -\frac{1}{2}$. Substituting $-2$ for $t$ gives $1 = 2A \Leftrightarrow A = \frac{1}{2}$. Thus,

$$\int \frac{dt}{t^2+6t+8} = \int\left(\frac{1/2}{t+2} - \frac{1/2}{t+4}\right)dt = \tfrac{1}{2}\ln|t+2| - \tfrac{1}{2}\ln|t+4| + C = \tfrac{1}{2}\ln\left|\frac{t+2}{t+4}\right| + C.$$

**30.** Let $u = x^2$. Then $du = 2x\,dx$, so $\displaystyle\int \frac{x}{\sqrt{1-x^4}}\,dx = \frac{1}{2}\int \frac{du}{\sqrt{1-u^2}} = \tfrac{1}{2}\sin^{-1}u + C = \tfrac{1}{2}\sin^{-1}(x^2) + C$.

**31.** Let $w = \sqrt[3]{x}$. Then $w^3 = x$ and $3w^2\,dw = dx$, so $\int e^{\sqrt[3]{x}}\,dx = \int e^w \cdot 3w^2\,dw = 3I$. To evaluate $I$, let $u = w^2$, $dv = e^w\,dw \Rightarrow du = 2w\,dw$, $v = e^w$, so $I = \int w^2 e^w\,dw = w^2e^w - \int 2we^w\,dw$. Now let $U = w$, $dV = e^w\,dw \Rightarrow dU = dw$, $V = e^w$. Thus, $I = w^2e^w - 2\left[we^w - \int e^w\,dw\right] = w^2e^w - 2we^w + 2e^w + C_1$, and hence $3I = 3e^w(w^2 - 2w + 2) + C = 3e^{\sqrt[3]{x}}(x^{2/3} - 2x^{1/3} + 2) + C$.

**32.** Let $u = \tan^{-1}x$, $dv = dx \Rightarrow du = \dfrac{1}{1+x^2}\,dx$, $v = x$:

$$\int \tan^{-1}x\,dx = x\tan^{-1}x - \int \frac{x}{1+x^2}\,dx = x\tan^{-1}x - \tfrac{1}{2}\ln(1+x^2) + C.$$

**33.** Let $u = 1 + \sec\theta$. Then $du = \sec\theta\,\tan\theta\,d\theta$, so

$$\int \frac{\sec\theta\,\tan\theta}{1+\sec\theta}\,d\theta = \int \frac{1}{1+\sec\theta}\,(\sec\theta\,\tan\theta\,d\theta) = \int \frac{1}{u}\,du = \ln|u| + C = \ln|1+\sec\theta| + C.$$

**34.** $\displaystyle\int_{-1}^{1} \frac{\sin x}{1+x^2}\,dx = 0$ by Theorem 5.5.6(b), since $f(x) = \dfrac{\sin x}{1+x^2}$ is an odd function.

**35.** Let $u = 1 + \sin x$. Then $du = \cos x\,dx$, so

$$\int \frac{\cos x\,dx}{\sqrt{1+\sin x}} = \int u^{-1/2}\,du = 2u^{1/2} + C = 2\sqrt{1+\sin x} + C.$$

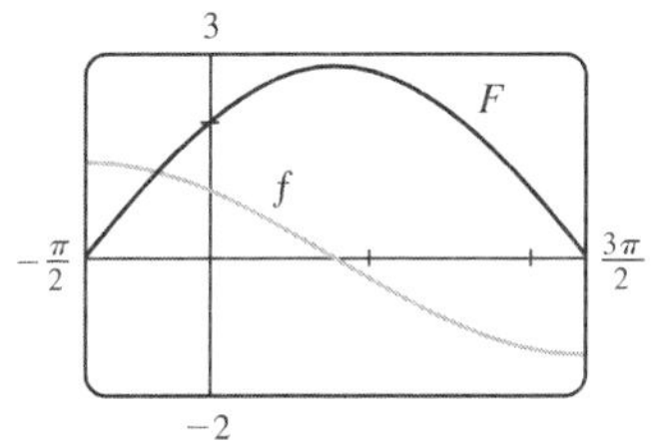


**36.** Let $u = x^2 + 1$. Then $x^2 = u - 1$ and $x\,dx = \frac{1}{2}\,du$, so

$$\begin{aligned}\int \frac{x^3}{\sqrt{x^2+1}}\,dx &= \int \frac{(u-1)}{\sqrt{u}}\left(\tfrac{1}{2}\,du\right) = \frac{1}{2}\int \left(u^{1/2} - u^{-1/2}\right)du \\ &= \tfrac{1}{2}\left(\tfrac{2}{3}u^{3/2} - 2u^{1/2}\right) + C \\ &= \tfrac{1}{3}\left(x^2+1\right)^{3/2} - \left(x^2+1\right)^{1/2} + C \\ &= \tfrac{1}{3}\left(x^2+1\right)^{1/2}\left[\left(x^2+1\right) - 3\right] + C = \tfrac{1}{3}\sqrt{x^2+1}\left(x^2-2\right) + C\end{aligned}$$

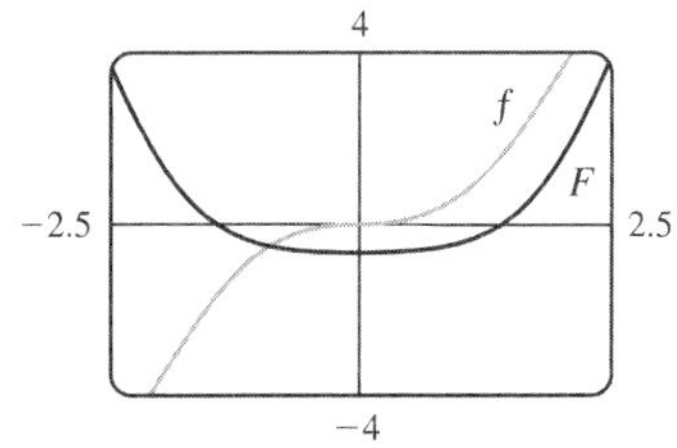


**37.** From the graph, it appears that the area under the curve $y = x\sqrt{x}$ between $x = 0$ and $x = 4$ is somewhat less than half the area of an $8 \times 4$ rectangle, so perhaps about 13 or 14. To find the exact value, we evaluate

$$\int_0^4 x\sqrt{x}\,dx = \int_0^4 x^{3/2}\,dx = \left[\tfrac{2}{5}x^{5/2}\right]_0^4 = \tfrac{2}{5}(4)^{5/2} = \tfrac{64}{5} = 12.8.$$

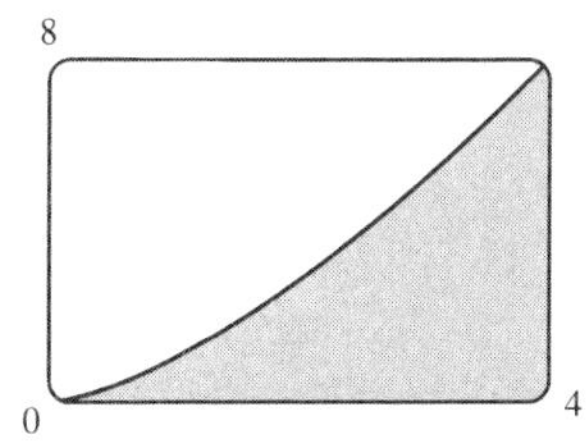


**38.** From the graph, it seems as though $\int_0^{2\pi} \cos^2 x\,\sin^3 x\,dx$ is equal to 0.

To evaluate the integral, we write the integral as

$I = \int_0^{2\pi} \cos^2 x\left(1 - \cos^2 x\right)\sin x\,dx$ and let $u = \cos x \quad\Rightarrow$

$du = -\sin x\,dx$. Thus, $I = \int_1^1 u^2\left(1-u^2\right)(-du) = 0$.

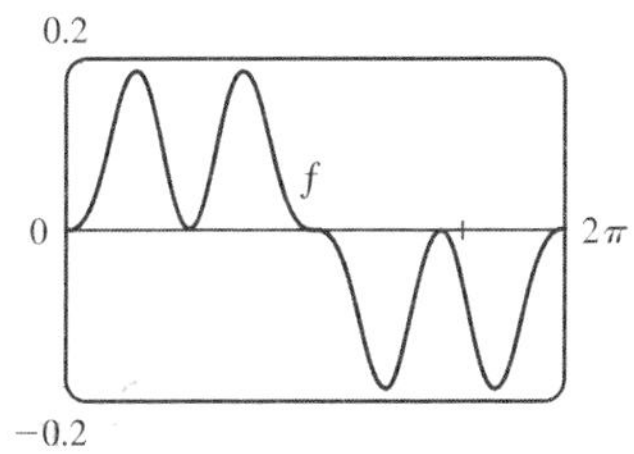


**39.** By FTC1, $F(x) = \int_1^x \sqrt{1+t^4}\,dt \quad\Rightarrow\quad F'(x) = \sqrt{1+x^4}$.

**40.** Let $u = \cos x$. Then $\dfrac{du}{dx} = -\sin x$. Also, $\dfrac{dg}{dx} = \dfrac{dg}{du}\dfrac{du}{dx}$, so

$$\begin{aligned}\frac{d}{dx}\int_1^{\cos x} \sqrt[3]{1-t^2}\,dt &= \frac{d}{du}\int_1^{u} \sqrt[3]{1-t^2}\,dt \cdot \frac{du}{dx} = \sqrt[3]{1-u^2}\,(-\sin x) \\ &= \sqrt[3]{1-\cos^2 x}\,(-\sin x) = -\sin x\,\sqrt[3]{\sin^2 x} = -(\sin x)^{5/3}\end{aligned}$$

**41.** $y = \int_{\sqrt{x}}^{x} \frac{e^t}{t}\,dt = \int_{\sqrt{x}}^{1} \frac{e^t}{t}\,dt + \int_{1}^{x} \frac{e^t}{t}\,dt = -\int_{1}^{\sqrt{x}} \frac{e^t}{t}\,dt + \int_{1}^{x} \frac{e^t}{t}\,dt \Rightarrow$

$\frac{dy}{dx} = -\frac{d}{dx}\left(\int_{1}^{\sqrt{x}} \frac{e^t}{t}\,dt\right) + \frac{d}{dx}\left(\int_{1}^{x} \frac{e^t}{t}\,dt\right)$. Let $u = \sqrt{x}$. Then

$$\frac{d}{dx}\int_{1}^{\sqrt{x}} \frac{e^t}{t}\,dt = \frac{d}{dx}\int_{1}^{u} \frac{e^t}{t}\,dt = \frac{d}{du}\left(\int_{1}^{u} \frac{e^t}{t}\,dt\right)\frac{du}{dx} = \frac{e^u}{u}\cdot\frac{1}{2\sqrt{x}} = \frac{e^{\sqrt{x}}}{\sqrt{x}}\cdot\frac{1}{2\sqrt{x}} = \frac{e^{\sqrt{x}}}{2x},$$

so $\frac{dy}{dx} = -\frac{e^{\sqrt{x}}}{2x} + \frac{e^x}{x}$.

**42.** $y = \int_{2x}^{3x+1} \sin(t^4)\,dt = \int_{2x}^{0} \sin(t^4)\,dt + \int_{0}^{3x+1} \sin(t^4)\,dt = \int_{0}^{3x+1} \sin(t^4)\,dt - \int_{0}^{2x} \sin(t^4)\,dt \Rightarrow$

$y' = \sin\left[(3x+1)^4\right]\cdot\frac{d}{dx}(3x+1) - \sin\left[(2x)^4\right]\cdot\frac{d}{dx}(2x) = 3\sin\left[(3x+1)^4\right] - 2\sin\left[(2x)^4\right]$

**43.** $u = e^x \Rightarrow du = e^x\,dx$, so

$$\int e^x\sqrt{1-e^{2x}}\,dx = \int\sqrt{1-u^2}\,du \overset{30}{=} \tfrac{1}{2}u\sqrt{1-u^2} + \tfrac{1}{2}\sin^{-1}u + C = \tfrac{1}{2}\left[e^x\sqrt{1-e^{2x}} + \sin^{-1}(e^x)\right] + C$$

**44.** $\int \csc^5 t\,dt \overset{78}{=} -\frac{1}{4}\cot t\,\csc^3 t + \frac{3}{4}\int \csc^3 t\,dt$

$\overset{72}{=} -\frac{1}{4}\cot t\,\csc^3 t + \frac{3}{4}\left[-\frac{1}{2}\csc t\,\cot t + \frac{1}{2}\ln|\csc t - \cot t|\right] + C$

$= -\frac{1}{4}\cot t\,\csc^3 t - \frac{3}{8}\csc t\,\cot t + \frac{3}{8}\ln|\csc t - \cot t| + C$

**45.** $\int\sqrt{x^2+x+1}\,dx = \int\sqrt{x^2+x+\frac{1}{4}+\frac{3}{4}}\,dx = \int\sqrt{\left(x+\frac{1}{2}\right)^2+\frac{3}{4}}\,dx$

$= \int\sqrt{u^2+\left(\frac{\sqrt{3}}{2}\right)^2}\,du \quad [u = x+\frac{1}{2},\ du = dx]$

$\overset{21}{=} \frac{1}{2}u\sqrt{u^2+\frac{3}{4}} + \frac{3}{8}\ln\left(u+\sqrt{u^2+\frac{3}{4}}\right) + C$

$= \frac{2x+1}{4}\sqrt{x^2+x+1} + \frac{3}{8}\ln\left(x+\frac{1}{2}+\sqrt{x^2+x+1}\right) + C$

**46.** Let $u = \sin x$. Then $du = \cos x\,dx$, so

$$\int\frac{\cot x\,dx}{\sqrt{1+2\sin x}} = \int\frac{du}{u\sqrt{1+2u}} \overset{57\text{ with } a=1,\,b=2}{=} \ln\left|\frac{\sqrt{1+2u}-1}{\sqrt{1+2u}+1}\right| + C = \ln\left|\frac{\sqrt{1+2\sin x}-1}{\sqrt{1+2\sin x}+1}\right| + C$$

**47.** $f(x) = \sqrt{1+x^4}$, $\Delta x = \frac{b-a}{n} = \frac{1-0}{10} = \frac{1}{10}$.

(a) $T_{10} = \frac{1}{10\cdot 2}\{f(0) + 2\left[f(0.1) + f(0.2) + \cdots + f(0.9)\right] + f(1)\} \approx 1.090608$

(b) $M_{10} = \frac{1}{10}\left[f\left(\frac{1}{20}\right) + f\left(\frac{3}{20}\right) + f\left(\frac{5}{20}\right) + \cdots + f\left(\frac{19}{20}\right)\right] \approx 1.088840$

(c) $S_{10} = \frac{1}{10\cdot 3}[f(0) + 4f(0.1) + 2f(0.2) + \cdots + 4f(0.9) + f(1)] \approx 1.089429$

$f$ is concave upward, so the Trapezoidal Rule gives us an overestimate, the Midpoint Rule gives an underestimate, and we cannot tell whether Simpson's Rule gives us an overestimate or an underestimate.

**48.** $f(x) = \sqrt{\sin x}$, $\Delta x = \frac{\frac{\pi}{2}-0}{10} = \frac{\pi}{20}$.

(a) $T_{10} = \frac{\pi}{20\cdot 2}\left\{f(0) + 2\left[f\left(\frac{\pi}{20}\right) + f\left(\frac{2\pi}{20}\right) + \cdots + f\left(\frac{9\pi}{20}\right)\right] + f\left(\frac{\pi}{2}\right)\right\} \approx 1.185197$

(b) $M_{10} = \frac{\pi}{20}\left[f\left(\frac{\pi}{40}\right) + f\left(\frac{3\pi}{40}\right) + f\left(\frac{5\pi}{40}\right) + \cdots + f\left(\frac{17\pi}{40}\right) + f\left(\frac{19\pi}{40}\right)\right] \approx 1.201932$

(c) $S_{10} = \frac{\pi}{20 \cdot 3}\left[f(0) + 4f\left(\frac{\pi}{20}\right) + 2f\left(\frac{2\pi}{20}\right) + \cdots + 4f\left(\frac{9\pi}{20}\right) + f\left(\frac{\pi}{2}\right)\right] \approx 1.193089$

$f$ is concave downward, so the Trapezoidal Rule gives us an underestimate, the Midpoint Rule gives an overestimate, and we cannot tell whether Simpson's Rule gives us an overestimate or an underestimate.

**49.** $f(x) = (1+x^4)^{1/2}$, $f'(x) = \frac{1}{2}(1+x^4)^{-1/2}(4x^3) = 2x^3(1+x^4)^{-1/2}$, $f''(x) = (2x^6+6x^2)(1+x^4)^{-3/2}$.
A graph of $f''$ on $[0,1]$ shows that it has its maximum at $x=1$, so $|f''(x)| \le f''(1) = \sqrt{8}$ on $[0,1]$. By taking $K=\sqrt{8}$, we find that the error in Exercise 47(a) is bounded by $\dfrac{K(b-a)^3}{12n^2} = \dfrac{\sqrt{8}}{1200} \approx 0.0024$, and in (b) by about $\frac{1}{2}(0.0024) = 0.0012$.

*Note:* Another way to estimate $K$ is to let $x=1$ in the factor $2x^6+6x^2$ (maximizing the numerator) and let $x=0$ in the factor $(1+x^4)^{-3/2}$ (minimizing the denominator). Doing so gives us $K=8$ and errors of $0.00\overline{6}$ and $0.00\overline{3}$.

Using $K=8$ for the Trapezoidal Rule, we have $|E_T| \le \dfrac{K(b-a)^3}{12n^2} \le 0.00001 \quad\Leftrightarrow\quad \dfrac{8(1-0)^3}{12n^2} \le \dfrac{1}{100{,}000} \quad\Leftrightarrow$

$n^2 \ge \dfrac{800{,}000}{12} \quad\Leftrightarrow\quad n \gtrsim 258.2$, so we should take $n = 259$.

For the Midpoint Rule, $|E_M| \le \dfrac{K(b-a)^3}{24n^2} \le 0.00001 \quad\Leftrightarrow\quad n^2 \ge \dfrac{800{,}000}{24} \quad\Leftrightarrow\quad n \gtrsim 182.6$, so we should take $n = 183$.

**50.** $\displaystyle\int_1^4 \frac{e^x}{x}\,dx \approx S_6 = \frac{(4-1)/6}{3}\left[f(1) + 4f(1.5) + 2f(2) + 4f(2.5) + 2f(3) + 4f(3.5) + f(4)\right] \approx 17.739438$

**51.** (a) $f(x) = \sin(\sin x)$. A CAS gives

$$f^{(4)}(x) = \sin(\sin x)\left[\cos^4 x + 7\cos^2 x - 3\right] + \cos(\sin x)\left[6\cos^2 x \sin x + \sin x\right]$$

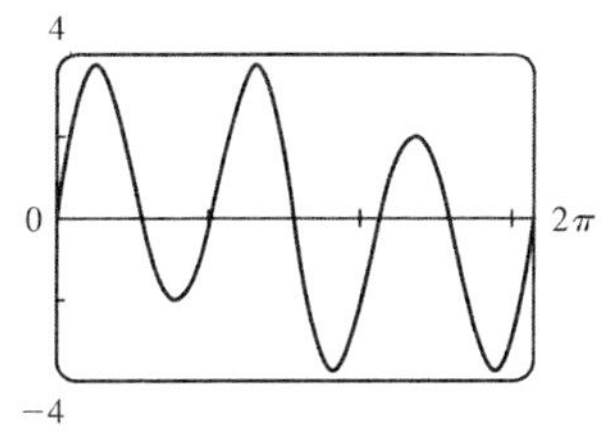


From the graph, we see that $\left|f^{(4)}(x)\right| < 3.8$ for $x \in [0,\pi]$.

(b) We use Simpson's Rule with $f(x) = \sin(\sin x)$ and $\Delta x = \frac{\pi}{10}$:

$$\int_0^\pi f(x)\,dx \approx \frac{\pi}{10 \cdot 3}\left[f(0) + 4f\left(\frac{\pi}{10}\right) + 2f\left(\frac{2\pi}{10}\right) + \cdots + 4f\left(\frac{9\pi}{10}\right) + f(\pi)\right] \approx 1.786721$$

From part (a), we know that $\left|f^{(4)}(x)\right| < 3.8$ on $[0,\pi]$, so we use Theorem 5.9.4 with $K=3.8$, and estimate the error as $|E_S| \le \dfrac{3.8(\pi-0)^5}{180(10)^4} \approx 0.000646$.

(c) If we want the error to be less than 0.00001, we must have $|E_S| \le \frac{3.8\pi^5}{180n^4} \le 0.00001$, so
$n^4 \ge \frac{3.8\pi^5}{180(0.00001)} \approx 646{,}041.6 \quad\Rightarrow\quad n \ge 28.35$. Since $n$ must be even for Simpson's Rule, we must have $n \ge 30$ to ensure the desired accuracy.

**52.** (a) To evaluate $\int x^5 e^{-2x}\,dx$ by hand, we would integrate by parts repeatedly, always taking $dv = e^{-2x}$ and starting with $u = x^5$. Each time we would reduce the degree of the $x$-factor by 1.

(b) To evaluate the integral using tables, we would use Formula 97 (which is proved using integration by parts) until the exponent of $x$ was reduced to 1, and then we would use Formula 96.

(c) $\int x^5 e^{-2x}\,dx = -\frac{1}{8}e^{-2x}\left(4x^5 + 10x^4 + 20x^3 + 30x^2 + 30x + 15\right) + C$

(d)

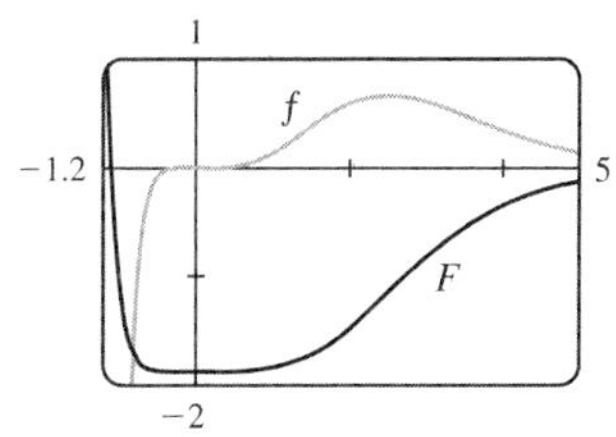

**53.** If $1 \le x \le 3$, then $\sqrt{1^2+3} \le \sqrt{x^2+3} \le \sqrt{3^2+3} \quad \Rightarrow \quad 2 \le \sqrt{x^2+3} \le 2\sqrt{3}$, so $2(3-1) \le \int_1^3 \sqrt{x^2+3}\,dx \le 2\sqrt{3}(3-1)$; that is, $4 \le \int_1^3 \sqrt{x^2+3}\,dx \le 4\sqrt{3}$.

**54.** On $[0,1]$, $x^4 \ge x^4 \cos x$ (since $0 \le \cos x \le 1$), so by Property 7, $\int_0^1 x^4\,dx \ge \int_0^1 x^4 \cos x\,dx$. Also, $x^4 \cos x \ge 0$, so by Property 6, $\int_0^1 x^4 \cos x\,dx \ge 0$. But $\int_0^1 x^4\,dx = \left[\frac{1}{5}x^5\right]_0^1 = \frac{1}{5} = 0.2$, so $0 \le \int_0^1 x^4 \cos x\,dx \le 0.2$.

**55.** $$\int_1^\infty \frac{1}{(2x+1)^3}\,dx = \lim_{t\to\infty} \int_1^t \frac{1}{(2x+1)^3}\,dx = \lim_{t\to\infty} \int_1^t \tfrac{1}{2}(2x+1)^{-3}\,2\,dx$$
$$= \lim_{t\to\infty} \left[-\frac{1}{4(2x+1)^2}\right]_1^t = -\frac{1}{4} \lim_{t\to\infty} \left[\frac{1}{(2t+1)^2} - \frac{1}{9}\right] = -\frac{1}{4}\left(0 - \frac{1}{9}\right) = \frac{1}{36}$$

**56.** $I = \displaystyle\int_0^\infty \frac{\ln x}{x^4}\,dx = \int_0^1 \frac{\ln x}{x^4}\,dx + \int_1^\infty \frac{\ln x}{x^4}\,dx = I_1 + I_2$. Integrate by parts with $u = \ln x$, $dv = dx/x^4 \quad \Rightarrow$ $du = dx/x$, $v = -1/\left(3x^3\right)$:

$$\int \frac{\ln x}{x^4}\,dx = -\frac{\ln x}{3x^3} + \frac{1}{3}\int \frac{1}{x^4}\,dx = -\frac{\ln x}{3x^3} + \frac{1}{3}\left(-\frac{1}{3x^3}\right) + C = -\frac{1}{9}\cdot\frac{3\ln x + 1}{x^3} + C$$

$$I_1 = \lim_{t\to 0^+} \int_t^1 \frac{\ln x}{x^4}\,dx = -\frac{1}{9}\lim_{t\to 0^+}\left[\frac{3\ln x + 1}{x^3}\right]_t^1 = -\frac{1}{9}\lim_{t\to 0^+}\left[1 - \frac{3\ln t + 1}{t^3}\right] = -\infty$$

So $I_1$ diverges and hence, $I$ diverges. Divergent

**57.** $\int_{-\infty}^0 e^{-2x}\,dx = \lim\limits_{t\to-\infty} \int_t^0 e^{-2x}\,dx = \lim\limits_{t\to-\infty} \left[-\frac{1}{2}e^{-2x}\right]_t^0 = \lim\limits_{t\to-\infty} \left(-\frac{1}{2} + \frac{1}{2}e^{-2t}\right) = \infty$. Divergent

**58.** Note that $f(x) = 1/(2-3x)$ has an infinite discontinuity at $x = \frac{2}{3}$. Now

$$\int_0^{2/3} \frac{1}{2-3x}\,dx = \lim_{t\to(2/3)^-} \int_0^t \frac{1}{2-3x}\,dx = \lim_{t\to(2/3)^-} \left[-\tfrac{1}{3}\ln|2-3x|\right]_0^t$$
$$= -\tfrac{1}{3} \lim_{t\to(2/3)^-} \Big[\ln|2-3t| - \ln 2\Big] = \infty$$

Since $\displaystyle\int_0^{2/3} \frac{1}{2-3x}\,dx$ diverges, so does $\displaystyle\int_0^1 \frac{1}{2-3x}\,dx$.

**59.** Let $u = \ln x$. Then $du = dx/x$, so $\displaystyle\int \frac{dx}{x\sqrt{\ln x}} = \int \frac{du}{\sqrt{u}} = 2\sqrt{u} + C = 2\sqrt{\ln x} + C$.

Thus, $\displaystyle\int_1^e \frac{dx}{x\sqrt{\ln x}} = \lim_{t\to1^+} \int_t^e \frac{dx}{x\sqrt{\ln x}} = \lim_{t\to1^+} \left[2\sqrt{\ln x}\right]_t^e = \lim_{t\to1^+} \left(2\sqrt{\ln e} - 2\sqrt{\ln t}\right) = 2\cdot1 - 2\cdot0 = 2.$

**60.** Let $u = \sqrt{y-2}$. Then $y = u^2 + 2$ and $dy = 2u\,du$, so

$$\int \frac{y\,dy}{\sqrt{y-2}} = \int \frac{(u^2+2)2u\,du}{u} = 2\int \left(u^2+2\right)du = 2\left[\tfrac{1}{3}u^3 + 2u\right] + C$$

Thus, $$\int_2^6 \frac{y\,dy}{\sqrt{y-2}} = \lim_{t\to2^+} \int_t^6 \frac{y\,dy}{\sqrt{y-2}} = \lim_{t\to2^+} \left[\tfrac{2}{3}(y-2)^{3/2} + 4\sqrt{y-2}\right]_t^6$$
$$= \lim_{t\to2^+} \left[\tfrac{16}{3} + 8 - \tfrac{2}{3}(t-2)^{3/2} - 4\sqrt{t-2}\right] = \tfrac{40}{3}.$$

**61.** $\dfrac{x^3}{x^5+2} \le \dfrac{x^3}{x^5} = \dfrac{1}{x^2}$ for $x$ in $[1,\infty)$. $\displaystyle\int_1^\infty \frac{1}{x^2}\,dx$ is convergent by (5.10.2) with $p = 2 > 1$. Therefore, $\displaystyle\int_1^\infty \frac{x^3}{x^5+2}\,dx$ is convergent by the Comparison Theorem.

**62.** $I = \int_0^\infty e^{ax}\cos x\,dx = \lim\limits_{t\to\infty}\int_0^t e^{ax}\cos x\,dx \overset{\substack{99\text{ with}\\ b=1}}{=} \lim\limits_{t\to\infty}\left[\frac{e^{ax}}{a^2+1}(a\cos x + \sin x)\right]_0^t$

$= \lim\limits_{t\to\infty}\left[\frac{e^{at}}{a^2+1}(a\cos t+\sin t) - \frac{1}{a^2+1}(a)\right] = \frac{1}{a^2+1}\lim\limits_{t\to\infty}\left[e^{at}(a\cos t+\sin t) - a\right].$

For $a \geq 0$, the limit does not exist due to oscillation. For $a < 0$, $\lim\limits_{t\to\infty}\left[e^{at}(a\cos t + \sin t)\right] = 0$ by the Squeeze Theorem, because $\left|e^{at}(a\cos t+\sin t)\right| \leq e^{at}(|a|+1)$, so $I = \frac{1}{a^2+1}(-a) = -\frac{a}{a^2+1}$.

**63.** (a) Displacement $= \int_0^5 (t^2 - t)\,dt = \left[\frac{1}{3}t^3 - \frac{1}{2}t^2\right]_0^5 = \frac{125}{3} - \frac{25}{2} = \frac{175}{6} = 29.1\overline{6}$ meters

(b) Distance traveled $= \int_0^5 |t^2 - t|\,dt = \int_0^5 |t(t-1)|\,dt = \int_0^1 (t - t^2)\,dt + \int_1^5 (t^2 - t)\,dt$

$= \left[\frac{1}{2}t^2 - \frac{1}{3}t^3\right]_0^1 + \left[\frac{1}{3}t^3 - \frac{1}{2}t^2\right]_1^5$

$= \frac{1}{2} - \frac{1}{3} - 0 + \left(\frac{125}{3} - \frac{25}{2}\right) - \left(\frac{1}{3} - \frac{1}{2}\right) = \frac{177}{6} = 29.5$ meters

**64.** $\Delta t = \left(\frac{10}{60} - 0\right)/\,10 = \frac{1}{60}$.

Distance traveled $= \int_0^{10} v\,dt \approx S_{10}$

$= \frac{1}{60\cdot 3}[40 + 4(42) + 2(45) + 4(49) + 2(52) + 4(54) + 2(56) + 4(57) + 2(57) + 4(55) + 56]$

$= \frac{1}{180}(1544) = 8.5\overline{7}$ mi

**65.** Note that $r(t) = b'(t)$, where $b(t) =$ the number of barrels of oil consumed up to time $t$. So, by the Net Change Theorem, $\int_0^3 r(t)\,dt = b(3) - b(0)$ represents the number of barrels of oil consumed from Jan. 1, 2000, through Jan. 1, 2003.

**66.** We use Simpson's Rule with $n = 6$ and $\Delta t = \frac{24-0}{6} = 4$:

Increase in bee population $= \int_0^{24} r(t)\,dt \approx S_6$

$= \frac{4}{3}[r(0) + 4r(4) + 2r(8) + 4r(12) + 2r(16) + 4r(20) + r(24)]$

$= \frac{4}{3}[0 + 4(300) + 2(3000) + 4(11{,}000) + 2(4000) + 4(400) + 0]$

$= \frac{4}{3}(60{,}800) \approx 81{,}067$ bees

**67.** Both numerator and denominator approach 0 as $a \to 0$, so we use l'Hospital's Rule. (Note that we are differentiating *with respect to* $a$, since that is the quantity which is changing.) We also use FTC1:

$$\lim_{a\to 0} T(x,t) = \lim_{a\to 0}\frac{C\int_0^a e^{-(x-u)^2/(4kt)}\,du}{a\sqrt{4\pi kt}} \overset{\text{H}}{=} \lim_{a\to 0}\frac{Ce^{-(x-a)^2/(4kt)}}{\sqrt{4\pi kt}} = \frac{Ce^{-x^2/(4kt)}}{\sqrt{4\pi kt}}$$

**68.** (a) $C$ is increasing on those intervals where $C'$ is positive. By the Fundamental Theorem of Calculus,

$C'(x) = \frac{d}{dx}\left[\int_0^x \cos\left(\frac{\pi}{2}t^2\right)dt\right] = \cos\left(\frac{\pi}{2}x^2\right)$. This is positive when $\frac{\pi}{2}x^2$ is in the interval $\left(\left(2n - \frac{1}{2}\right)\pi, \left(2n + \frac{1}{2}\right)\pi\right)$, $n$ any integer. This implies that $\left(2n - \frac{1}{2}\right)\pi < \frac{\pi}{2}x^2 < \left(2n+\frac{1}{2}\right)\pi \iff 0 \leq |x| \leq 1$ or $\sqrt{4n-1} < |x| < \sqrt{4n+1}$, $n$ any positive integer. So $C$ is increasing on the intervals $[-1,1]$, $\left[\sqrt{3},\sqrt{5}\right]$, $\left[-\sqrt{5},-\sqrt{3}\right]$, $\left[\sqrt{7},3\right]$, $\left[-3,-\sqrt{7}\right]$, ....

(b) $C$ is concave upward on those intervals where $C'' > 0$. We differentiate $C'$ to find $C''$: $C'(x) = \cos\left(\frac{\pi}{2}x^2\right) \Rightarrow$ $C''(x) = -\sin\left(\frac{\pi}{2}x^2\right)\left(\frac{\pi}{2}\cdot 2x\right) = -\pi x\sin\left(\frac{\pi}{2}x^2\right)$. For $x > 0$, this is positive where $(2n-1)\pi < \frac{\pi}{2}x^2 < 2n\pi$, $n$ any positive integer $\iff \sqrt{2(2n-1)} < x < 2\sqrt{n}$, $n$ any positive integer. Since there is a factor of $-x$ in $C''$, the intervals of upward concavity for $x < 0$ are $\left(-\sqrt{2(2n+1)}, -2\sqrt{n}\right)$, $n$ any nonnegative integer. That is, $C$ is concave upward on $\left(-\sqrt{2},0\right)$, $\left(\sqrt{2},2\right)$, $\left(-\sqrt{6},-2\right)$, $\left(\sqrt{6},2\sqrt{2}\right)$, ....

(c) 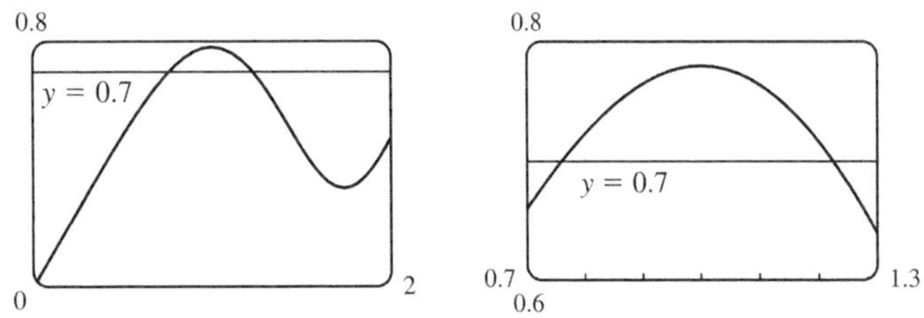


From the graphs, we can determine that $\int_0^x \cos\left(\frac{\pi}{2}t^2\right) dt = 0.7$ at $x \approx 0.76$ and $x \approx 1.22$.

(d) 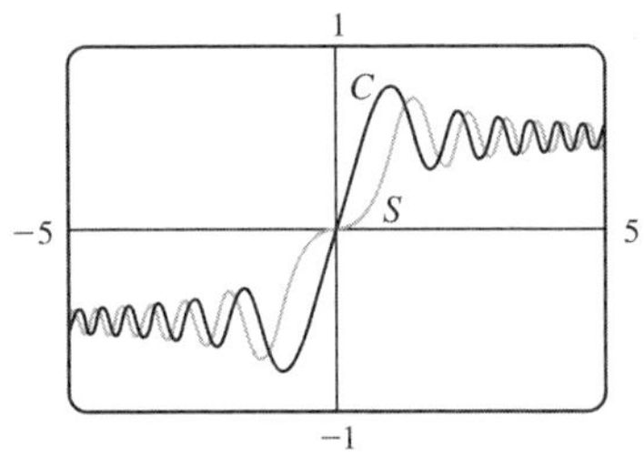


The graphs of $S(x)$ and $C(x)$ have similar shapes, except that $S$'s flattens out near the origin, while $C$'s does not. Note that for $x > 0$, $C$ is increasing where $S$ is concave up, and $C$ is decreasing where $S$ is concave down. Similarly, $S$ is increasing where $C$ is concave down, and $S$ is decreasing where $C$ is concave up. For $x < 0$, these relationships are reversed; that is, $C$ is increasing where $S$ is concave down, and $S$ is increasing where $C$ is concave up. See Example 5.4.4 and Exercise 5.4.23 for a discussion of $S(x)$.

**69.** Using FTC1, we differentiate both sides of the given equation, $\int_0^x f(t)\,dt = xe^{2x} + \int_0^x e^{-t} f(t)\,dt$, and get

$$f(x) = e^{2x} + 2xe^{2x} + e^{-x} f(x) \quad\Rightarrow\quad f(x)\left(1 - e^{-x}\right) = e^{2x} + 2xe^{2x} \quad\Rightarrow\quad f(x) = \frac{e^{2x}(1+2x)}{1 - e^{-x}}.$$

**70.** $2\int_a^x f(t)\,dt = 2\sin x - 1 \quad\Rightarrow\quad \int_a^x f(t)\,dt = \sin x - \frac{1}{2}$. Differentiating both sides using FTC1 gives $f(x) = \cos x$. We put $x = a$ into the last equation to get $0 = \sin a - \frac{1}{2}$, so $a = \frac{\pi}{6}$ satisfies the given equation.

**71.** Let $u = f(x)$ and $du = f'(x)\,dx$. So $2\int_a^b f(x)f'(x)\,dx = 2\int_{f(a)}^{f(b)} u\,du = \left[u^2\right]_{f(a)}^{f(b)} = [f(b)]^2 - [f(a)]^2$.

**72.** Integrate by parts with $u = (\ln x)^n$, $dv = dx \quad\Rightarrow\quad du = n(\ln x)^{n-1} \cdot \frac{1}{x}\,dx$, $v = x$:

$\int (\ln x)^n\,dx = x(\ln x)^n - \int x \cdot n(\ln x)^{n-1}(dx/x) = x(\ln x)^n - n\int (\ln x)^{n-1}\,dx$. Thus,

$$\begin{aligned}\int_0^1 (\ln x)^n\,dx &= \lim_{t\to 0^+} \int_t^1 (\ln x)^n\,dx = \lim_{t\to 0^+} \left[x(\ln x)^n\right]_t^1 - n \lim_{t\to 0^+} \int_t^1 (\ln x)^{n-1}\,dx \\ &= -\lim_{t\to 0^+} \frac{(\ln t)^n}{1/t} - n\int_0^1 (\ln x)^{n-1}\,dx = -n\int_0^1 (\ln x)^{n-1}\,dx,\end{aligned}$$

by repeated application of l'Hospital's Rule. We want to prove that $\int_0^1 (\ln x)^n\,dx = (-1)^n n!$ for every positive integer $n$. For $n = 1$, we have $\int_0^1 (\ln x)^1\,dx = (-1)\int_0^1 (\ln x)^0\,dx = -\int_0^1 dx = -1$ (or $\int_0^1 \ln x\,dx = \lim_{t\to 0^+} \left[x\ln x - x\right]_t^1 = -1$). Assuming that the formula holds for $n$, we find that

$\int_0^1 (\ln x)^{n+1}\,dx = -(n+1)\int_0^1 (\ln x)^n\,dx = -(n+1)(-1)^n n! = (-1)^{n+1}(n+1)!$.

This is the formula for $n + 1$. Thus, the formula holds for all positive integers $n$ by induction.

**73.** By the Fundamental Theorem of Calculus,

$$\int_0^\infty f'(x)\,dx = \lim_{t\to\infty} \int_0^t f'(x)\,dx = \lim_{t\to\infty} [f(t) - f(0)] = \lim_{t\to\infty} f(t) - f(0) = 0 - f(0) = -f(0).$$

**74.** The area $A(t) = \int_0^t \sin(x^2)\,dx$, and the area $B(t) = \frac{1}{2}t\sin(t^2)$. Since $\lim_{t\to 0^+} A(t) = 0 = \lim_{t\to 0^+} B(t)$, we can use l'Hospital's Rule:

$$\lim_{t\to 0^+} \frac{A(t)}{B(t)} \overset{H}{=} \lim_{t\to 0^+} \frac{\sin(t^2)}{\frac{1}{2}\sin(t^2) + \frac{1}{2}t[2t\cos(t^2)]} \quad \text{[by FTC1 and the Product Rule]}$$

$$\overset{H}{=} \lim_{t\to 0^+} \frac{2t\cos(t^2)}{t\cos(t^2) - 2t^3\sin(t^2) + 2t\cos(t^2)} = \lim_{t\to 0^+} \frac{2\cos(t^2)}{3\cos(t^2) - 2t^2\sin(t^2)} = \frac{2}{3 - 0} = \frac{2}{3}$$

# FOCUS ON PROBLEM SOLVING

**1.**

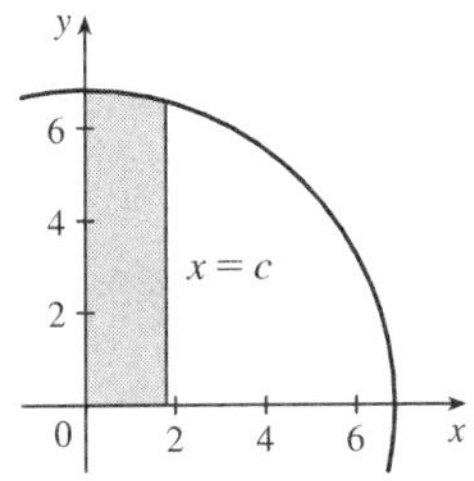


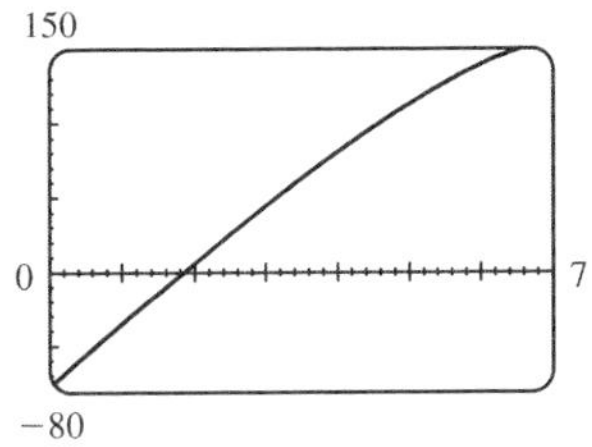


By symmetry, the problem can be reduced to finding the line $x = c$ such that the shaded area is one-third of the area of the quarter-circle. The equation of the circle is $y = \sqrt{49 - x^2}$, so we require that $\int_0^c \sqrt{49 - x^2}\,dx = \frac{1}{3} \cdot \frac{1}{4}\pi(7)^2 \quad \Leftrightarrow$ $\left[\frac{1}{2}x\sqrt{49 - x^2} + \frac{49}{2}\sin^{-1}(x/7)\right]_0^c = \frac{49}{12}\pi$ [by Formula 30] $\Leftrightarrow \quad \frac{1}{2}c\sqrt{49 - c^2} + \frac{49}{2}\sin^{-1}(c/7) = \frac{49}{12}\pi$.

This equation would be difficult to solve exactly, so we plot the left-hand side as a function of $c$, and find that the equation holds for $c \approx 1.85$. So the cuts should be made at distances of about 1.85 inches from the center of the pizza.

**2.** (a)

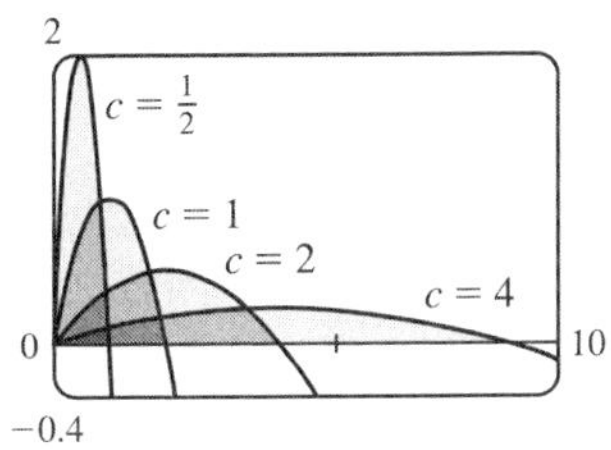


From the graph of $f(x) = \dfrac{2cx - x^2}{c^3}$, it appears that the areas are equal; that is, the area enclosed is independent of $c$.

(b) We first find the $x$-intercepts of the curve, to determine the limits of integration: $y = 0 \quad \Leftrightarrow \quad 2cx - x^2 = 0 \quad \Leftrightarrow \quad x = 0$ or $x = 2c$. Now we integrate the function between these limits to find the enclosed area:

$$A = \int_0^{2c} \frac{2cx - x^2}{c^3}\,dx = \frac{1}{c^3}\left[cx^2 - \tfrac{1}{3}x^3\right]_0^{2c} = \frac{1}{c^3}\left[c(2c)^2 - \tfrac{1}{3}(2c)^3\right] = \frac{1}{c^3}\left[4c^3 - \tfrac{8}{3}c^3\right] = \tfrac{4}{3}, \text{ a constant.}$$

(c)

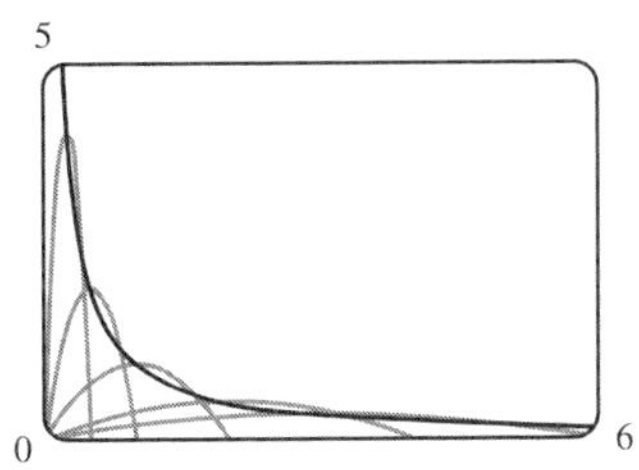


The vertices of the family of parabolas seem to determine a branch of a hyperbola.

(d) For a particular $c$, the vertex is the point where the maximum occurs. We have seen that the $x$-intercepts are 0 and $2c$, so by symmetry, the maximum occurs at $x = c$, and its value is $\dfrac{2c(c) - c^2}{c^3} = \dfrac{1}{c}$. So we are interested in the curve consisting of all points of the form $\left(c, \dfrac{1}{c}\right)$, $c > 0$. This is the part of the hyperbola $y = 1/x$ lying in the first quadrant.

**3.** Differentiating both sides of the equation $x \sin \pi x = \int_0^{x^2} f(t)\,dt$ (using FTC1 and the Chain Rule for the right side) gives $\sin \pi x + \pi x \cos \pi x = 2x f(x^2)$. Letting $x = 2$ so that $f(x^2) = f(4)$, we obtain $\sin 2\pi + 2\pi \cos 2\pi = 4f(4)$, so $f(4) = \frac{1}{4}(0 + 2\pi \cdot 1) = \frac{\pi}{2}$.

**4.** If $f(x) = \int_0^x x^2 \sin(t^2)\,dt = x^2 \int_0^x \sin(t^2)\,dt$, then $f'(x) = x^2 \sin(x^2) + 2x \int_0^x \sin(t^2)\,dt$, by the Product Rule and FTC1.

**5.** By FTC2, $\int_0^1 f'(x)\,dx = f(1) - f(0) = 1 - 0 = 1$.

**6.** $n$ is a positive integer, so

$$\int (\ln x)^n\,dx = x(\ln x)^n - \int x \cdot n(\ln x)^{n-1}\,(dx/x) \quad \text{[by parts]} \quad = x(\ln x)^n - n\int (\ln x)^{n-1}\,dx$$

Thus,

$$\int_0^1 (\ln x)^n\,dx = \lim_{t\to 0^+} \int_t^1 (\ln x)^n\,dx = \lim_{t\to 0^+} [x(\ln x)^n]_t^1 - n \lim_{t\to 0^+} \int_t^1 (\ln x)^{n-1}\,dx$$

$$= -\lim_{t\to 0^+} \frac{(\ln t)^n}{1/t} - n\int_0^1 (\ln x)^{n-1}\,dx = -n\int_0^1 (\ln x)^{n-1}\,dx$$

by repeated application of l'Hospital's Rule. We want to prove that $\int_0^1 (\ln x)^n\,dx = (-1)^n n!$ for every positive integer $n$. For $n = 1$, we have

$$\int_0^1 (\ln x)^1\,dx = (-1)\int_0^1 (\ln x)^0\,dx = -\int_0^1 dx = -1 \qquad \left[\text{or } \int_0^1 \ln x\,dx = \lim_{t\to 0^+} [x\ln x - x]_t^1 = -1\right]$$

Assuming that the formula holds for $n$, we find that

$$\int_0^1 (\ln x)^{n+1}\,dx = -(n+1)\int_0^1 (\ln x)^n\,dx = -(n+1)(-1)^n n! = (-1)^{n+1}(n+1)!$$

This is the formula for $n+1$. Thus, the formula holds for all positive integers $n$ by induction.

**7.** Differentiating the given equation, $\int_0^x f(t)\,dt = [f(x)]^2$, using FTC1 gives $f(x) = 2f(x)f'(x) \Rightarrow$ $f(x)\,[2f'(x) - 1] = 0$, so $f(x) = 0$ or $f'(x) = \frac{1}{2}$. Since $f(x)$ is never 0, we must have $f'(x) = \frac{1}{2}$ and $f'(x) = \frac{1}{2} \Rightarrow$ $f(x) = \frac{1}{2}x + C$. To find $C$, we substitute into the given equation to get $\int_0^x \left(\frac{1}{2}t + C\right)dt = \left(\frac{1}{2}x + C\right)^2 \Leftrightarrow$ $\frac{1}{4}x^2 + Cx = \frac{1}{4}x^2 + Cx + C^2$. It follows that $C^2 = 0$, so $C = 0$, and $f(x) = \frac{1}{2}x$.

**8.**

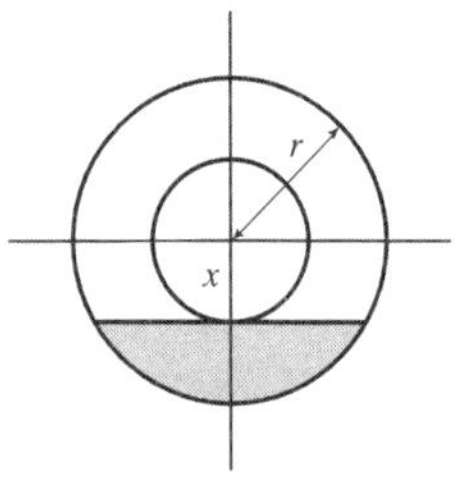


Let $x$ be the distance between the center of the disk and the surface of the liquid. The wetted circular region has area $\pi r^2 - \pi x^2$ while the unexposed wetted region (shaded in the diagram) has area $2\int_x^r \sqrt{r^2 - t^2}\,dt$, so the exposed wetted region has area $A(x) = \pi r^2 - \pi x^2 - 2\int_x^r \sqrt{r^2 - t^2}\,dt$, $0 \le x \le r$. By FTC1, we have $A'(x) = -2\pi x + 2\sqrt{r^2 - x^2}$.

Now $A'(x) > 0 \Rightarrow -2\pi x + 2\sqrt{r^2 - x^2} > 0 \Rightarrow \sqrt{r^2 - x^2} > \pi x \Rightarrow r^2 - x^2 > \pi^2 x^2 \Rightarrow$ $r^2 > \pi^2 x^2 + x^2 \Rightarrow r^2 > x^2(\pi^2 + 1) \Rightarrow x^2 < \dfrac{r^2}{\pi^2 + 1} \Rightarrow x < \dfrac{r}{\sqrt{\pi^2 + 1}}$, and we'll call this value $x^*$.

Since $A'(x) > 0$ for $0 < x < x^*$ and $A'(x) < 0$ for $x^* < x < r$, we have an absolute maximum when $x = x^*$.

**9.** By l'Hospital's Rule and the Fundamental Theorem, using the notation $\exp(y) = e^y$,

$$\lim_{x\to 0} \frac{\int_0^x (1 - \tan 2t)^{1/t}\,dt}{x} \overset{\text{H}}{=} \lim_{x\to 0} \frac{(1 - \tan 2x)^{1/x}}{1} = \exp\left(\lim_{x\to 0} \frac{\ln(1 - \tan 2x)}{x}\right)$$

$$\overset{\text{H}}{=} \exp\left(\lim_{x\to 0} \frac{-2\sec^2 2x}{1 - \tan 2x}\right) = \exp\left(\frac{-2 \cdot 1^2}{1 - 0}\right) = e^{-2}.$$

**10.** $f(x) = \int_0^{g(x)} \frac{1}{\sqrt{1+t^3}}\,dt$, where $g(x) = \int_0^{\cos x} \left[1 + \sin(t^2)\right] dt$. Using FTC1 and the Chain Rule (twice) we have

$f'(x) = \frac{1}{\sqrt{1+[g(x)]^3}} g'(x) = \frac{1}{\sqrt{1+[g(x)]^3}} \left[1 + \sin(\cos^2 x)\right](-\sin x)$. Now $g\left(\frac{\pi}{2}\right) = \int_0^0 \left[1 + \sin(t^2)\right] dt = 0$, so

$f'\left(\frac{\pi}{2}\right) = \frac{1}{\sqrt{1+0}}(1 + \sin 0)(-1) = 1 \cdot 1 \cdot (-1) = -1$.

**11.** Such a function cannot exist. $f'(x) > 3$ for all $x$ means that $f$ is differentiable (and hence continuous) for all $x$. So by FTC2, $\int_1^4 f'(x)\,dx = f(4) - f(1) = 7 - (-1) = 8$. However, if $f'(x) > 3$ for all $x$, then $\int_1^4 f'(x)\,dx \geq 3 \cdot (4-1) = 9$ by Comparison Property 8 in Section 5.2.

*Another solution:* By the Mean Value Theorem, there exists a number $c \in (1, 4)$ such that

$f'(c) = \frac{f(4) - f(1)}{4-1} = \frac{7-(-1)}{3} = \frac{8}{3} \quad \Rightarrow \quad 8 = 3f'(c)$. But $f'(x) > 3 \quad \Rightarrow \quad 3f'(c) > 9$, so such a function cannot exist.

**12.**

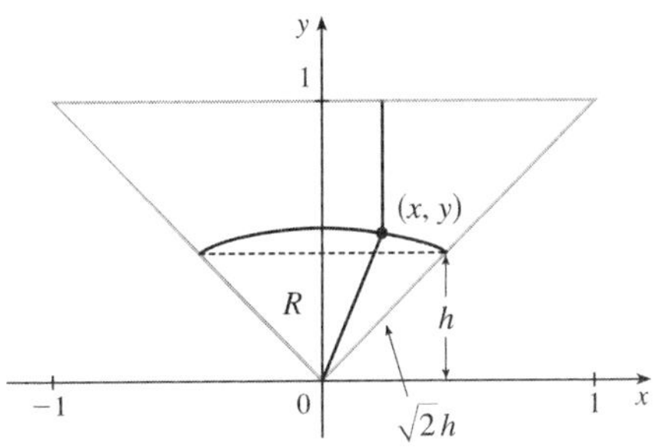


We restrict our attention to the triangle shown. A point in this triangle is closer to the side shown than to any other side, so if we find the area of the region $R$ consisting of all points in the triangle that are closer to the center than to that side, we can multiply this area by 4 to find the total area. We find the equation of the set of points which are equidistant from the center and the side: the distance of the point $(x, y)$ from the side is $1 - y$, and its distance from the center is $\sqrt{x^2 + y^2}$.

So the distances are equal if $\sqrt{x^2 + y^2} = 1 - y \quad \Leftrightarrow \quad x^2 + y^2 = 1 - 2y + y^2 \quad \Leftrightarrow \quad y = \frac{1}{2}(1 - x^2)$. Note that the area we are interested in is equal to the area of a triangle plus a crescent-shaped area. To find these areas, we have to find the $y$-coordinate $h$ of the horizontal line separating them. From the diagram, $1 - h = \sqrt{2}h \quad \Leftrightarrow \quad h = \frac{1}{1+\sqrt{2}} = \sqrt{2} - 1$. We calculate the areas in terms of $h$, and substitute afterward.

The area of the triangle is $\frac{1}{2}(2h)(h) = h^2$, and the area of the crescent-shaped section is

$$\int_{-h}^{h} \left[\tfrac{1}{2}(1 - x^2) - h\right] dx = 2\int_0^h \left(\tfrac{1}{2} - h - \tfrac{1}{2}x^2\right) dx = 2\left[\left(\tfrac{1}{2} - h\right)x - \tfrac{1}{6}x^3\right]_0^h = h - 2h^2 - \tfrac{1}{3}h^3.$$

So the area of the whole region is

$$4\left[\left(h - 2h^2 - \tfrac{1}{3}h^3\right) + h^2\right] = 4h\left(1 - h - \tfrac{1}{3}h^2\right) = 4\left(\sqrt{2} - 1\right)\left[1 - \left(\sqrt{2} - 1\right) - \tfrac{1}{3}\left(\sqrt{2} - 1\right)^2\right]$$

$$= 4\left(\sqrt{2} - 1\right)\left(1 - \tfrac{1}{3}\sqrt{2}\right) = \tfrac{4}{3}\left(4\sqrt{2} - 5\right)$$

**13.** $f(x) = 2 + x - x^2 = (-x + 2)(x + 1) = 0 \quad \Leftrightarrow \quad x = 2$ or $x = -1$. $f(x) \geq 0$ for $x \in [-1, 2]$ and $f(x) < 0$ everywhere else. The integral $\int_a^b (2 + x - x^2)\,dx$ has a maximum on the interval where the integrand is positive, which is $[-1, 2]$. So $a = -1$, $b = 2$. (Any larger interval gives a smaller integral since $f(x) < 0$ outside $[-1, 2]$. Any smaller interval also gives a smaller integral since $f(x) \geq 0$ in $[-1, 2]$.)

**14.** The shaded region has area $\int_0^1 f(x)\,dx = \frac{1}{3}$. The integral $\int_0^1 f^{-1}(y)\,dy$ gives the area of the unshaded region, which we know to be $1 - \frac{1}{3} = \frac{2}{3}$. So $\int_0^1 f^{-1}(y)\,dy = \frac{2}{3}$.

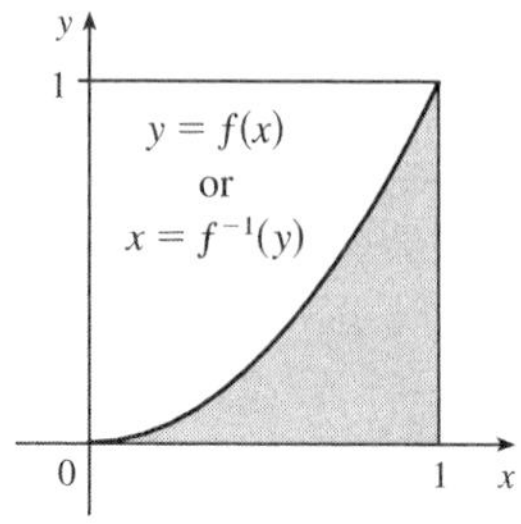


**15.** By FTC1, $\frac{d}{dx}\int_0^x \left(\int_1^{\sin t} \sqrt{1+u^4}\,du\right)dt = \int_1^{\sin x} \sqrt{1+u^4}\,du$. Again using FTC1,

$$\frac{d^2}{dx^2}\int_0^x \left(\int_1^{\sin t} \sqrt{1+u^4}\,du\right)dt = \frac{d}{dx}\int_1^{\sin x} \sqrt{1+u^4}\,du = \sqrt{1+\sin^4 x}\cos x.$$

**16.** This sum can be interpreted as a Riemann sum, with the right endpoints of the subintervals as sample points and with $a = 0$, $b = 10{,}000$, and $f(x) = \sqrt{x}$. So we approximate

$$\sum_{i=1}^{10{,}000} \sqrt{i} \approx \lim_{n\to\infty} \frac{10{,}000}{n} \sum_{i=1}^{n} \sqrt{\frac{10{,}000i}{n}} = \int_0^{10{,}000} \sqrt{x}\,dx = \left[\tfrac{2}{3}x^{3/2}\right]_0^{10{,}000} = \tfrac{2}{3}(1{,}000{,}000) \approx 666{,}667.$$

*Alternate method:* We can use graphical methods as follows:

From the figure we see that $\int_{i-1}^{i} \sqrt{x}\,dx < \sqrt{i} < \int_i^{i+1} \sqrt{x}\,dx$, so

$\int_0^{10{,}000} \sqrt{x}\,dx < \sum_{i=1}^{10{,}000} \sqrt{i} < \int_1^{10{,}001} \sqrt{x}\,dx$. Since

$\int \sqrt{x}\,dx = \frac{2}{3}x^{3/2} + C$, we get $\int_0^{10{,}000} \sqrt{x}\,dx = 666{,}666.\overline{6}$ and

$\int_1^{10{,}001} \sqrt{x}\,dx = \frac{2}{3}\left[(10{,}001)^{3/2} - 1\right] \approx 666{,}766.$

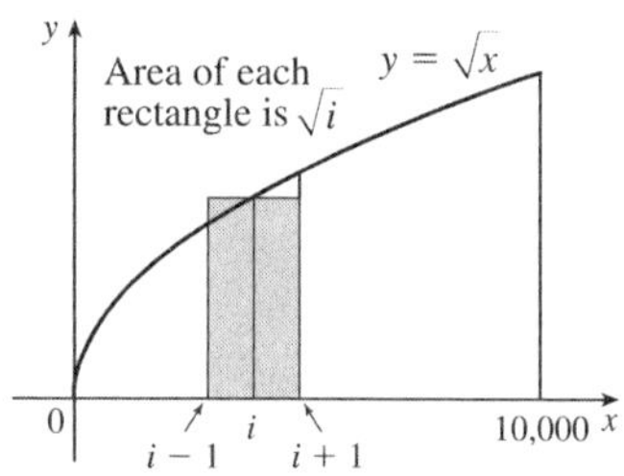


Hence, $666{,}666.\overline{6} < \sum_{i=1}^{10{,}000} \sqrt{i} < 666{,}766$. We can estimate the sum by averaging these bounds:

$$\sum_{i=1}^{10{,}000} \approx \frac{666{,}666.\overline{6} + 666{,}766}{2} \approx 666{,}716.$$ The actual value is about 666,716.46.

**17.** The given integral represents the difference of the shaded areas, which appears to be 0. It can be calculated by integrating with respect to either $x$ or $y$, so we find $x$ in terms of $y$ for each curve: $y = \sqrt[3]{1-x^7} \;\Rightarrow\; x = \sqrt[7]{1-y^3}$ and $y = \sqrt[7]{1-x^3} \;\Rightarrow\; x = \sqrt[3]{1-y^7}$, so

$\int_0^1 \left(\sqrt[3]{1-y^7} - \sqrt[7]{1-y^3}\right)dy = \int_0^1 \left(\sqrt[7]{1-x^3} - \sqrt[3]{1-x^7}\right)dx$. But this equation is of the form $z = -z$. So $\int_0^1 \left(\sqrt[3]{1-x^7} - \sqrt[7]{1-x^3}\right)dx = 0$.

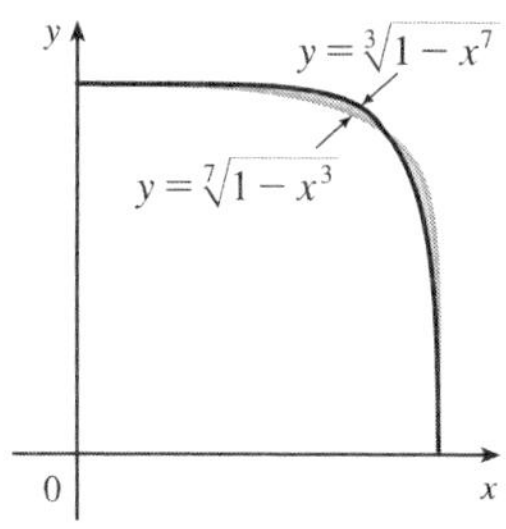


**18.** The problem can be reduced to finding the line which minimizes the shaded area in the diagram. The equation of the circle in the first quadrant is $y = \sqrt{1-x^2}$, so if the equation of the line is $y = h$, then the circle and the line intersect where $h = \sqrt{1-x^2} \;\Rightarrow\; x = \sqrt{1-h^2}$.

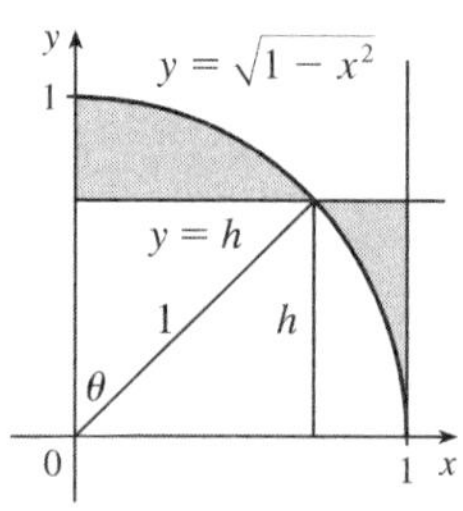

So the shaded area is

$$A = \int_0^{\sqrt{1-h^2}} \left(\sqrt{1-x^2} - h\right) dx + \int_{\sqrt{1-h^2}}^{1} \left(h - \sqrt{1-x^2}\right) dx$$

$$\overset{\star}{=} \left[-hx\right]_0^{\sqrt{1-h^2}} + \left[hx\right]_{\sqrt{1-h^2}}^{1} + \int_0^{\sqrt{1-h^2}} \sqrt{1-x^2}\,dx + \int_1^{\sqrt{1-h^2}} \sqrt{1-x^2}\,dx$$

$$= -h\sqrt{1-h^2} + h - h\sqrt{1-h^2} + \int_0^{\sqrt{1-h^2}} \sqrt{1-x^2}\,dx + \int_1^{\sqrt{1-h^2}} \sqrt{1-x^2}\,dx$$

$$= h\left(1 - 2\sqrt{1-h^2}\,\right) + \int_0^{\sqrt{1-h^2}} \sqrt{1-x^2}\,dx + \int_1^{\sqrt{1-h^2}} \sqrt{1-x^2}\,dx$$

Note that at ($\star$), we reversed the limits of integration and changed the sign in the last integral. We are interested in the minimum of $A(h) = h\left(1 - 2\sqrt{1-h^2}\,\right) + \int_0^{\sqrt{1-h^2}} \sqrt{1-x^2}\,dx + \int_1^{\sqrt{1-h^2}} \sqrt{1-x^2}\,dx$, so we find $dA/dh$ using FTC1 and the Chain Rule:

$$\frac{dA}{dh} = h\left(-2\frac{-h}{\sqrt{1-h^2}}\right) + \left(1 - 2\sqrt{1-h^2}\,\right) + 2\left[\sqrt{1 - \left(\sqrt{1-h^2}\,\right)^2}\,\right] \tfrac{d}{dh}\left(\sqrt{1-h^2}\,\right)$$

$$= \frac{1}{\sqrt{1-h^2}}\left[2h^2 + \sqrt{1-h^2} - 2\left(1-h^2\right)\right] + 2h\,\frac{-h}{\sqrt{1-h^2}} = \frac{1}{\sqrt{1-h^2}}\left[\sqrt{1-h^2} - 2\left(1-h^2\right)\right]$$

This is 0 when $\sqrt{1-h^2} - 2\left(1-h^2\right) = 0 \;\Leftrightarrow\; u - 2u^2 = 0$ (where $u = \sqrt{1-h^2}\,$) $\;\Leftrightarrow\; u = 0$ or $\frac{1}{2} \;\Leftrightarrow\; h = 1$ or $\frac{\sqrt{3}}{2}$. By the First Derivative Test, $h = \frac{\sqrt{3}}{2}$ represents a minimum for $A(h)$, since $A'(h) = 1 - \dfrac{2}{\sqrt{1-h^2}}$ goes from negative to positive at $h = \frac{\sqrt{3}}{2}$.

*Another method:* Use FTC2 to evaluate all of the integrals before differentiating.

*Note:* Another strategy is to use the angle $\theta$ as the variable (see diagram above) and show that $A = \theta + \cos\theta - \frac{\pi}{4} - \frac{1}{2}\sin 2\theta$, which is minimized when $\theta = \frac{\pi}{6}$.

**19.** In accordance with the hint, we let $I_k = \int_0^1 \left(1-x^2\right)^k dx$, and we find an expression for $I_{k+1}$ in terms of $I_k$. We integrate $I_{k+1}$ by parts with $u = \left(1-x^2\right)^{k+1}$, $dv = dx \;\Rightarrow\; du = (k+1)\left(1-x^2\right)^k(-2x)$, $v = x$, and then split the remaining integral into identifiable quantities:

$$I_{k+1} = \left[x\left(1-x^2\right)^{k+1}\right]_0^1 + 2(k+1)\int_0^1 x^2\left(1-x^2\right)^k dx = (2k+2)\int_0^1 \left(1-x^2\right)^k\left[1 - \left(1-x^2\right)\right] dx$$

$$= (2k+2)(I_k - I_{k+1})$$

So $I_{k+1}\left[1 + (2k+2)\right] = (2k+2)I_k \Rightarrow I_{k+1} = \dfrac{2k+2}{2k+3}I_k$.

Now to complete the proof, we use induction: $I_0 = 1 = \dfrac{2^0(0!)^2}{1!}$, so the formula holds for $n = 0$. Now suppose it holds for $n = k$. Then

$$I_{k+1} = \frac{2k+2}{2k+3}I_k = \frac{2k+2}{2k+3}\left[\frac{2^{2k}(k!)^2}{(2k+1)!}\right] = \frac{2(k+1)2^{2k}(k!)^2}{(2k+3)(2k+1)!} = \frac{2(k+1)2^{2k}(k!)^2}{(2k+3)(2k+1)!}\cdot\frac{2(k+1)}{2k+2}$$

$$= \frac{\left[2(k+1)\right]^2 2^{2k}(k!)^2}{(2k+3)(2k+2)(2k+1)!} = \frac{2^{2(k+1)}\left[(k+1)!\right]^2}{\left[2(k+1)+1\right]!}$$

So by induction, the formula holds for all integers $n \geq 0$.

**20.** 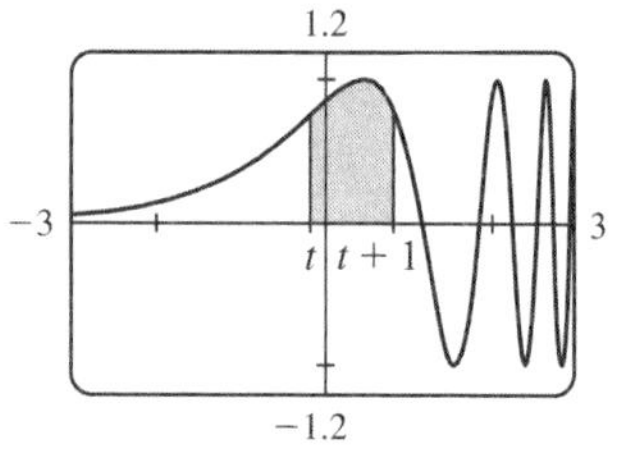


From the graph, it appears that the area under the graph of $f(x) = \sin(e^x)$ on the interval $[t, t+1]$ is greatest when $t \approx -0.2$. To find the exact value, we write the integral as $I = \int_t^{t+1} f(x)\,dx = \int_0^{t+1} f(x)\,dx - \int_0^t f(x)\,dx$, and use FTC1 to find $dI/dt = f(t+1) - f(t) = \sin(e^{t+1}) - \sin(e^t) = 0$ when $\sin(e^{t+1}) = \sin(e^t)$.

Now we have $\sin x = \sin y$ whenever $x - y = 2k\pi$ and also whenever $x$ and $y$ are the same distance from $\left(k + \frac{1}{2}\right)\pi$, $k$ any integer, since $\sin x$ is symmetric about the line $x = \left(k + \frac{1}{2}\right)\pi$. The first possibility is the more obvious one, but if we calculate $e^{t+1} - e^t = 2k\pi$, we get $t = \ln(2k\pi/(e-1))$, which is about 1.3 for $k = 1$ (the least possible value of $k$). From the graph, this looks unlikely to give the maximum we are looking for. So instead we set $e^{t+1} - \left(k + \frac{1}{2}\right)\pi = \left(k + \frac{1}{2}\right)\pi - e^t \;\Leftrightarrow$ $e^{t+1} + e^t = (2k+1)\pi \;\Leftrightarrow\; e^t(e+1) = (2k+1)\pi \;\Leftrightarrow\; t = \ln((2k+1)\pi/(e+1))$. Now $k = 0 \;\Rightarrow$ $t = \ln(\pi/(e+1)) \approx -0.16853$, which does give the maximum value, as we have seen from the graph of $f$.

**21.** (a) The tangent to the curve $y = f(x)$ at $x = x_0$ has the equation $y - f(x_0) = f'(x_0)(x - x_0)$. The $y$-intercept of this tangent line is $f(x_0) - f'(x_0)x_0$. Thus, $L$ is the distance from the point $(0, f(x_0) - f'(x_0)x_0)$ to the point $(x_0, f(x_0))$; that is, $L^2 = x_0^2 + [f'(x_0)]^2 x_0^2$, so $[f'(x_0)]^2 = \dfrac{L^2 - x_0^2}{x_0^2}$ and $f'(x_0) = -\dfrac{\sqrt{L^2 - x_0^2}}{x_0}$ for $0 < x_0 < L$.

(b) $\dfrac{dy}{dx} = -\dfrac{\sqrt{L^2 - x^2}}{x} \;\Rightarrow\; y = \displaystyle\int\left(-\frac{\sqrt{L^2 - x^2}}{x}\right)dx.$

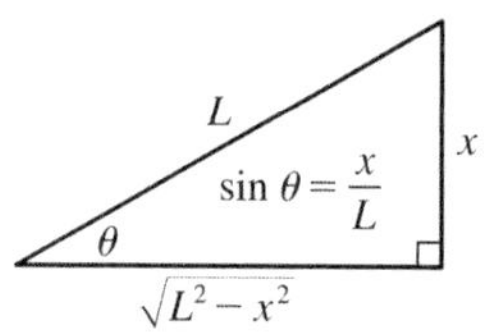


Let $x = L\sin\theta$. Then $dx = L\cos\theta\,d\theta$ and

$$y = \int \frac{-L\cos\theta\, L\cos\theta\,d\theta}{L\sin\theta} = L\int\frac{\sin^2\theta - 1}{\sin\theta}\,d\theta = L\int(\sin\theta - \csc\theta)\,d\theta$$

$$= -L\cos\theta - L\ln|\csc\theta - \cot\theta| + C = -\sqrt{L^2 - x^2} - L\ln\left(\frac{L}{x} - \frac{\sqrt{L^2 - x^2}}{x}\right) + C$$

When $x = L$, $y = 0$, and $0 = -0 - L\ln(1 - 0) + C$, so $C = 0$. Therefore, $y = -\sqrt{L^2 - x^2} - L\ln\left(\dfrac{L - \sqrt{L^2 - x^2}}{x}\right)$.

**22.** Note that the graphs of $(x - c)^2$ and $[(x - c) - 2]^2$ intersect when $|x - c| = |x - c - 2| \;\Leftrightarrow\; c - x = x - c - 2 \;\Leftrightarrow$ $x = c + 1$. The integration will proceed differently depending on the value of $c$.

*Case 1:* $-2 \le c < -1$ In this case, $f_c(x) = (x - c - 2)^2$ for $x \in [0, 1]$, so

$$g(c) = \int_0^1 (x - c - 2)^2\,dx = \tfrac{1}{3}\left[(x - c - 2)^3\right]_0^1 = \tfrac{1}{3}\left[(-c-1)^3 - (-c-2)^3\right]$$

$$= \tfrac{1}{3}(3c^2 + 9c + 7) = c^2 + 3c + \frac{7}{3} = \left(c + \tfrac{3}{2}\right)^2 + \tfrac{1}{12}$$

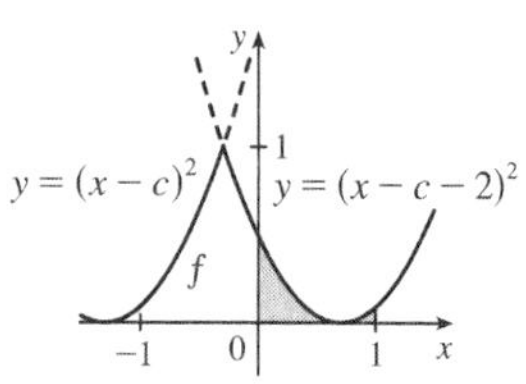


This is a parabola; its maximum value for $-2 \le c < -1$ is $g(-2) = \frac{1}{3}$, and its minimum value is $g\left(-\frac{3}{2}\right) = \frac{1}{12}$.

*Case 2:* $-1 \le c < 0$ In this case, $f_c(x) = \begin{cases} (x-c)^2 & \text{if } 0 \le x \le c+1 \\ (x-c-2)^2 & \text{if } c+1 < x \le 1 \end{cases}$

Therefore,

$$\begin{aligned} g(c) &= \int_0^1 f_c(x)\,dx = \int_0^{c+1} (x-c)^2\,dx + \int_{c+1}^1 (x-c-2)^2\,dx \\ &= \tfrac{1}{3}\left[(x-c)^3\right]_0^{c+1} + \tfrac{1}{3}\left[(x-c-2)^3\right]_{c+1}^1 = \tfrac{1}{3}\left[1 + c^3 + (-c-1)^3 - (-1)\right] \\ &= -c^2 - c + \tfrac{1}{3} = -\left(c + \tfrac{1}{2}\right)^2 + \tfrac{7}{12} \end{aligned}$$

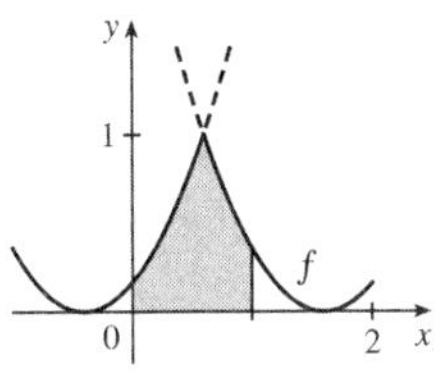


Again, this is a parabola, whose maximum value for $-1 \le c < 0$ is $g\left(-\frac{1}{2}\right) = \frac{7}{12}$, and whose minimum value on this $c$-interval is $g(-1) = \frac{1}{3}$.

*Case 3:* $0 \le c \le 2$ In this case, $f_c(x) = (x-c)^2$ for $x \in [0,1]$, so

$g(c) = \int_0^1 (x-c)^2\,dx = \frac{1}{3}\left[(x-c)^3\right]_0^1 = \frac{1}{3}\left[(1-c)^3 - (-c)^3\right] = c^2 - c + \frac{1}{3} = \left(c - \frac{1}{2}\right)^2 + \frac{1}{12}$.

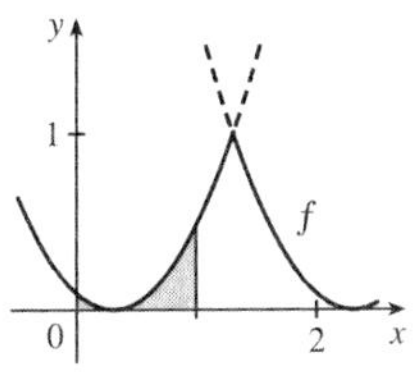


This parabola has a maximum value of $g(2) = \frac{7}{3}$ and a minimum value of $g\left(\frac{1}{2}\right) = \frac{1}{12}$.

We conclude that $g(c)$ has an absolute maximum value of $g(2) = \frac{7}{3}$, and absolute minimum values of $g\left(-\frac{3}{2}\right) = g\left(\frac{1}{2}\right) = \frac{1}{12}$.

# 6 □ APPLICATIONS OF INTEGRATION

## 6.1 More about Areas

**1.** $A = \int_{x=0}^{x=4} (y_T - y_B)\,dx = \int_0^4 \left[(5x - x^2) - x\right] dx = \int_0^4 (4x - x^2)\,dx$
$= \left[2x^2 - \frac{1}{3}x^3\right]_0^4 = \left(32 - \frac{64}{3}\right) - (0) = \frac{32}{3}$

**2.** $A = \int_0^2 \left(\sqrt{x+2} - \frac{1}{x+1}\right) dx = \left[\frac{2}{3}(x+2)^{3/2} - \ln(x+1)\right]_0^2$
$= \left[\frac{2}{3}(4)^{3/2} - \ln 3\right] - \left[\frac{2}{3}(2)^{3/2} - \ln 1\right] = \frac{16}{3} - \ln 3 - \frac{4}{3}\sqrt{2}$

**3.** $A = \int_{y=-1}^{y=1} (x_R - x_L)\,dy = \int_{-1}^1 \left[e^y - (y^2 - 2)\right] dy$
$= \int_{-1}^1 (e^y - y^2 + 2)\,dy = \left[e^y - \frac{1}{3}y^3 + 2y\right]_{-1}^1 = \left(e^1 - \frac{1}{3} + 2\right) - \left(e^{-1} + \frac{1}{3} - 2\right) = e - \frac{1}{e} + \frac{10}{3}$

**4.** $A = \int_0^3 \left[(2y - y^2) - (y^2 - 4y)\right] dy = \int_0^3 (-2y^2 + 6y)\,dy$
$= \left[-\frac{2}{3}y^3 + 3y^2\right]_0^3 = (-18 + 27) - 0 = 9$

**5.** $A = \int_{-1}^2 \left[(9 - x^2) - (x+1)\right] dx$
$= \int_{-1}^2 (8 - x - x^2)\,dx$
$= \left[8x - \frac{x^2}{2} - \frac{x^3}{3}\right]_{-1}^2$
$= \left(16 - 2 - \frac{8}{3}\right) - \left(-8 - \frac{1}{2} + \frac{1}{3}\right)$
$= 22 - 3 + \frac{1}{2} = \frac{39}{2}$

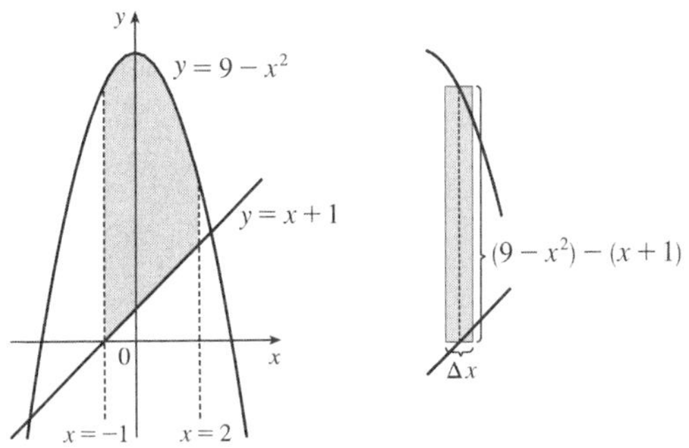


**6.** $A = \int_0^{\pi/2} (e^x - \sin x)\,dx$
$= \left[e^x + \cos x\right]_0^{\pi/2}$
$= \left(e^{\pi/2} + 0\right) - (1 + 1)$
$= e^{\pi/2} - 2$

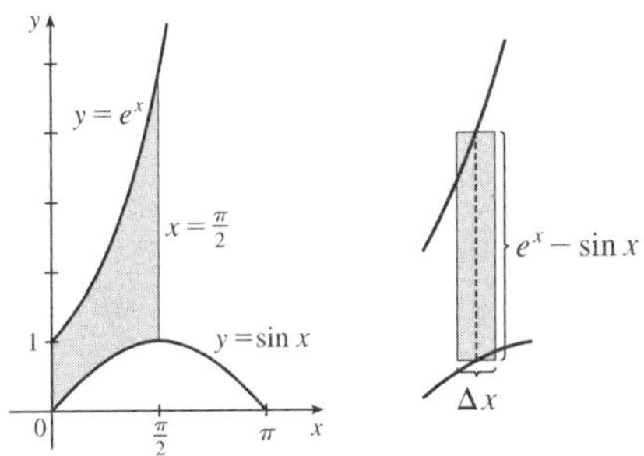


**7.** The curves intersect when $x = x^2 \;\Rightarrow\; x^2 - x = 0 \;\Leftrightarrow\; x(x-1) = 0 \;\Leftrightarrow\; x = 0, 1.$

$A = \int_0^1 (x - x^2)\,dx = \left[\frac{1}{2}x^2 - \frac{1}{3}x^3\right]_0^1$
$= \frac{1}{2} - \frac{1}{3} = \frac{1}{6}$

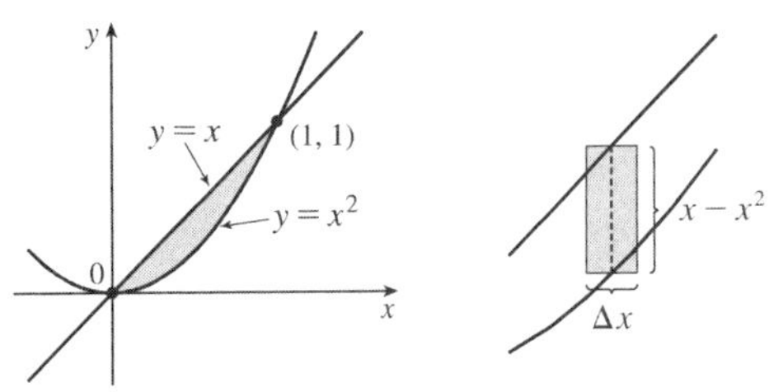

**8.** $1+\sqrt{x}=\dfrac{3+x}{3}=1+\dfrac{x}{3} \;\Rightarrow\; \sqrt{x}=\dfrac{x}{3} \;\Rightarrow\; x=\dfrac{x^2}{9} \;\Rightarrow\; 9x-x^2=0 \;\Rightarrow\; x(9-x)=0 \;\Rightarrow\; x=0$ or $9$, so

$$A=\int_0^9\left[\left(1+\sqrt{x}\right)-\left(\frac{3+x}{3}\right)\right]dx=\int_0^9\left[\left(1+\sqrt{x}\right)-\left(1+\frac{x}{3}\right)\right]dx$$

$$=\int_0^9\left(\sqrt{x}-\tfrac{1}{3}x\right)dx=\left[\tfrac{2}{3}x^{3/2}-\tfrac{1}{6}x^2\right]_0^9=18-\tfrac{27}{2}=\tfrac{9}{2}$$

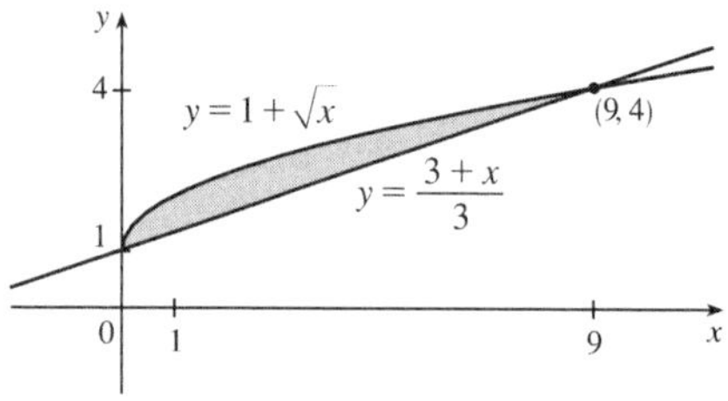


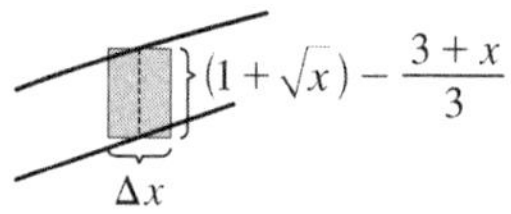


**9.** $12-x^2=x^2-6 \;\Leftrightarrow\; 2x^2=18 \;\Leftrightarrow\; x^2=9 \;\Leftrightarrow\; x=\pm3$, so

$$A=\int_{-3}^{3}\left[\left(12-x^2\right)-\left(x^2-6\right)\right]dx=2\int_0^3\left(18-2x^2\right)dx \qquad \text{[by symmetry]}$$

$$=2\left[18x-\tfrac{2}{3}x^3\right]_0^3=2\left[(54-18)-0\right]=2(36)=72$$

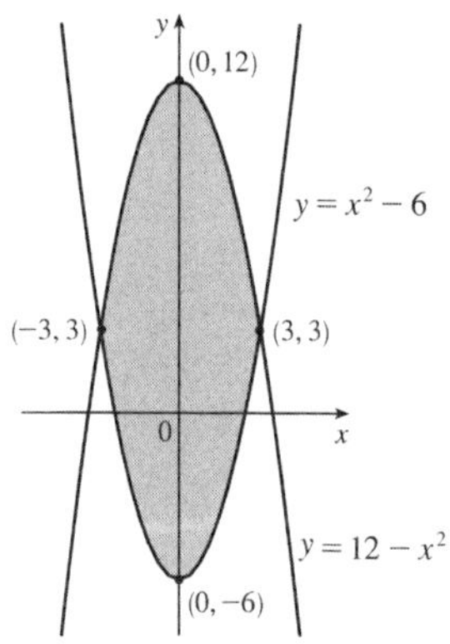


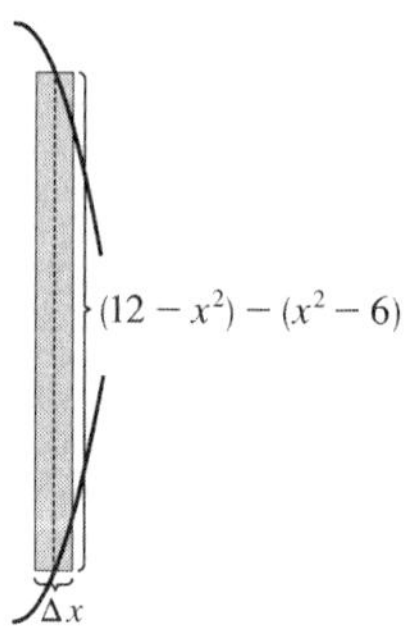


**10.** $x^2=4x-x^2 \;\Leftrightarrow\; 2x^2-4x=0 \;\Leftrightarrow\; 2x(x-2)=0 \;\Leftrightarrow\; x=0$ or $2$, so

$$A=\int_0^2\left[\left(4x-x^2\right)-x^2\right]dx=\int_0^2\left(4x-2x^2\right)dx=\left[2x^2-\tfrac{2}{3}x^3\right]_0^2=8-\tfrac{16}{3}=\tfrac{8}{3}$$

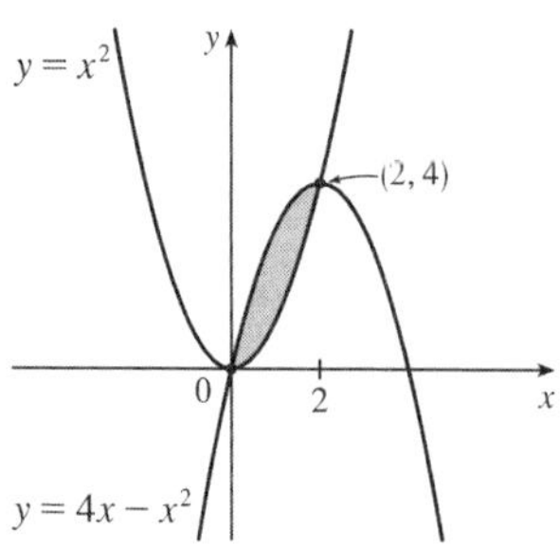


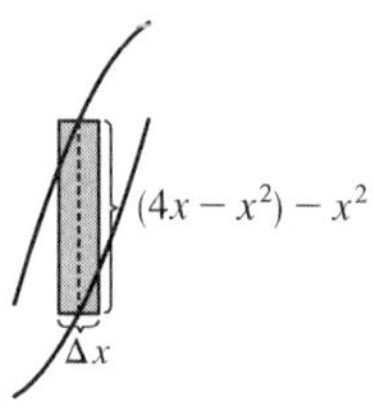

**11.** $2y^2 = 1 - y \quad \Leftrightarrow \quad 2y^2 + y - 1 = 0 \quad \Leftrightarrow \quad (2y-1)(y+1) = 0 \quad \Leftrightarrow \quad y = \frac{1}{2}$ or $-1$, so $x = \frac{1}{2}$ or 2 and

$$A = \int_{-1}^{1/2} \left[(1-y) - 2y^2\right] dy = \int_{-1}^{1/2} (1 - y - 2y^2)\, dy = \left[y - \tfrac{1}{2}y^2 - \tfrac{2}{3}y^3\right]_{-1}^{1/2}$$

$$= \left(\tfrac{1}{2} - \tfrac{1}{8} - \tfrac{1}{12}\right) - \left(-1 - \tfrac{1}{2} + \tfrac{2}{3}\right) = \tfrac{7}{24} - \left(-\tfrac{5}{6}\right) = \tfrac{7}{24} + \tfrac{20}{24} = \tfrac{27}{24} = \tfrac{9}{8}$$

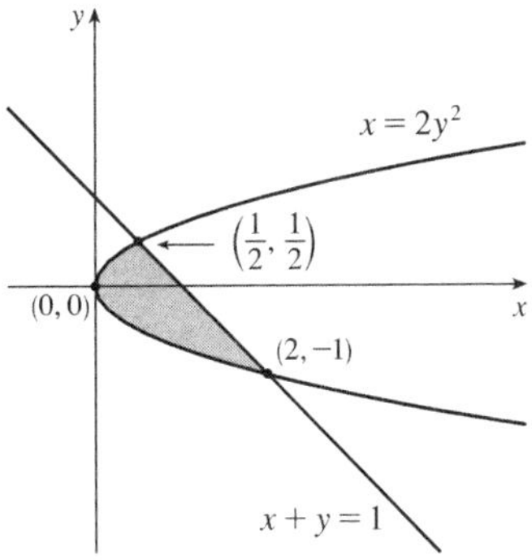


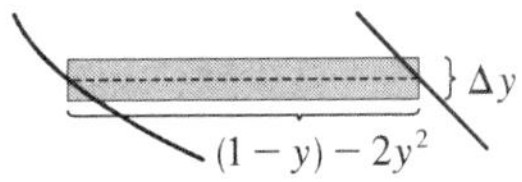


**12.** $4x + x^2 = 12 \quad \Leftrightarrow \quad (x+6)(x-2) = 0 \quad \Leftrightarrow \quad x = -6$ or $x = 2$, so $y = -6$ or $y = 2$ and

$$A = \int_{-6}^{2} \left[\left(-\tfrac{1}{4}y^2 + 3\right) - y\right] dy = \left[-\tfrac{1}{12}y^3 - \tfrac{1}{2}y^2 + 3y\right]_{-6}^{2}$$

$$= \left(-\tfrac{2}{3} - 2 + 6\right) - (18 - 18 - 18) = 22 - \tfrac{2}{3} = \tfrac{64}{3}$$

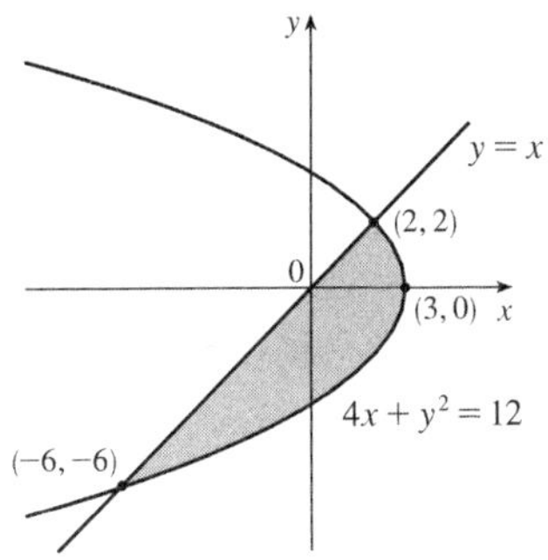


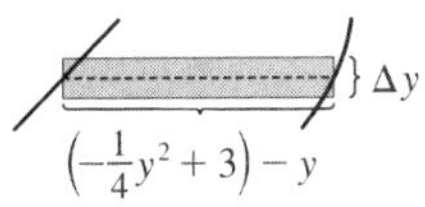


**13.** $2y^2 = 4 + y^2 \quad \Leftrightarrow \quad y^2 = 4 \quad \Leftrightarrow \quad y = \pm 2$, so

$$A = \int_{-2}^{2} [(4 + y^2) - 2y^2]\, dy = 2\int_{0}^{2} (4 - y^2)\, dy \qquad \text{[by symmetry]}$$

$$= 2\left[4y - \tfrac{1}{3}y^3\right]_0^2 = 2\left(8 - \tfrac{8}{3}\right) = \tfrac{32}{3}$$

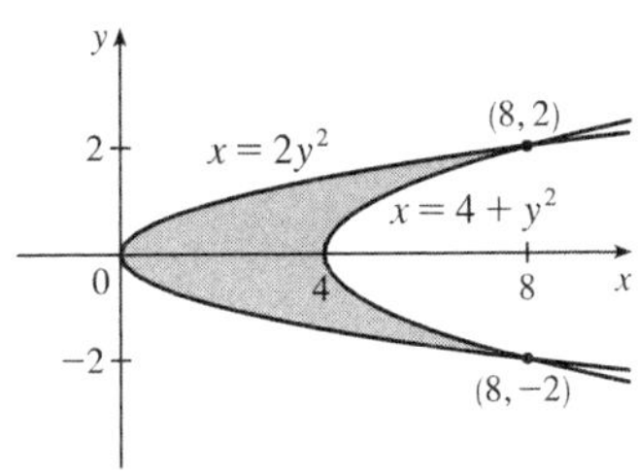


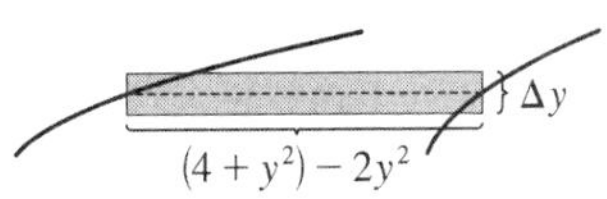

**14.** By observation, $y = \sin x$ and $y = 2x/\pi$ intersect at $(0,0)$ and $(\pi/2, 1)$ for $x \geq 0$.

$$A = \int_0^{\pi/2} \left(\sin x - \frac{2x}{\pi}\right) dx = \left[-\cos x - \frac{1}{\pi}x^2\right]_0^{\pi/2} = \left(0 - \frac{\pi}{4}\right) - (-1) = 1 - \frac{\pi}{4}$$

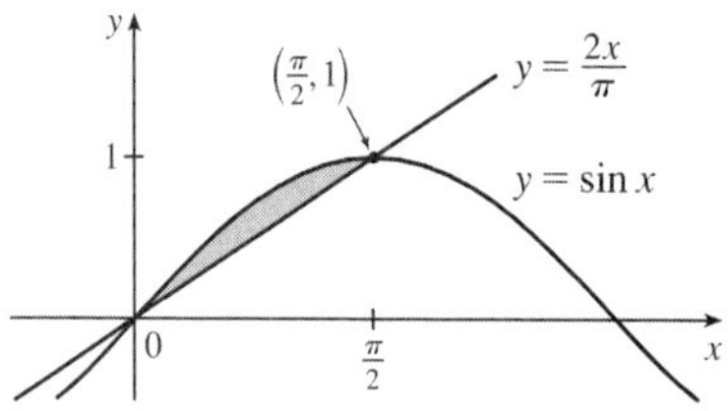


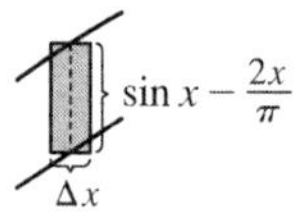


**15.** $1/x = x \quad \Leftrightarrow \quad 1 = x^2 \quad \Leftrightarrow \quad x = \pm 1$ and $1/x = \frac{1}{4}x \quad \Leftrightarrow \quad 4 = x^2 \quad \Leftrightarrow \quad x = \pm 2$, so for $x > 0$,

$$A = \int_0^1 \left(x - \frac{1}{4}x\right) dx + \int_1^2 \left(\frac{1}{x} - \frac{1}{4}x\right) dx = \int_0^1 \left(\frac{3}{4}x\right) dx + \int_1^2 \left(\frac{1}{x} - \frac{1}{4}x\right) dx = \left[\frac{3}{8}x^2\right]_0^1 + \left[\ln|x| - \frac{1}{8}x^2\right]_1^2$$
$$= \tfrac{3}{8} + \left(\ln 2 - \tfrac{1}{2}\right) - \left(0 - \tfrac{1}{8}\right) = \ln 2$$

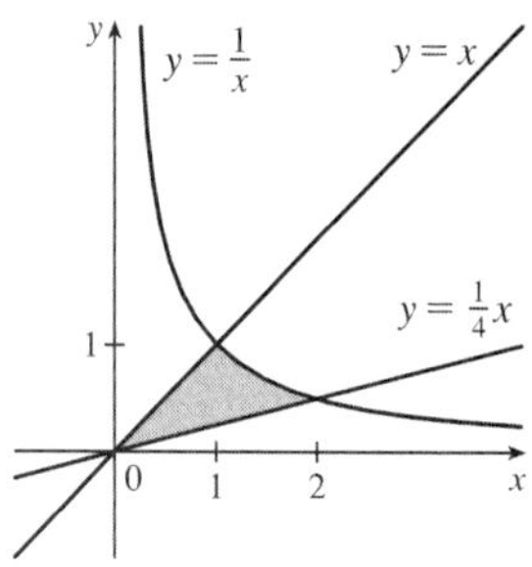


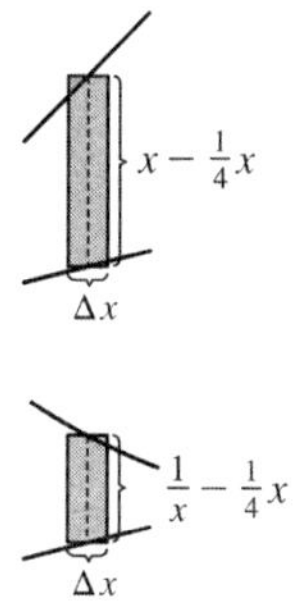


**16.** For $x > 0$, $x = x^2 - 2 \quad \Rightarrow \quad 0 = x^2 - x - 2 \quad \Rightarrow \quad 0 = (x-2)(x+1) \quad \Rightarrow \quad x = 2$. By symmetry,

$$\int_{-2}^{2} d\left[|x| - (x^2 - 2)\right] x = 2\int_0^2 \left[x - (x^2 - 2)\right] dx = 2\int_0^2 (x - x^2 + 2)\, dx = 2\left[\tfrac{1}{2}x^2 - \tfrac{1}{3}x^3 + 2x\right]_0^2$$
$$= 2\left(2 - \tfrac{8}{3} + 4\right) = \tfrac{20}{3}$$

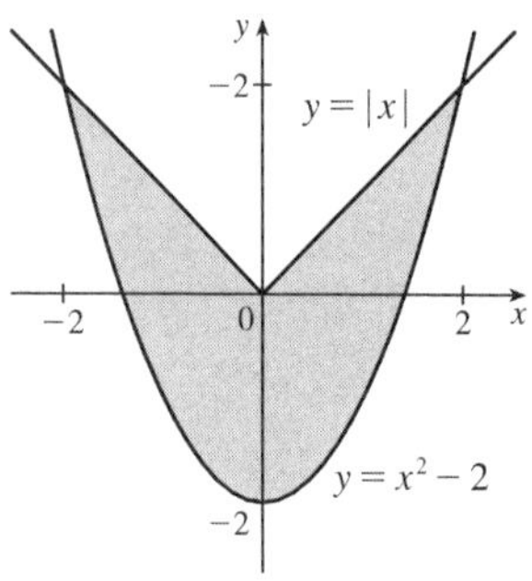


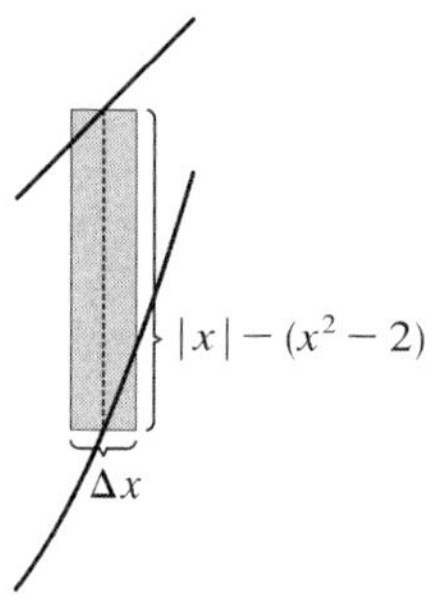


**17.**

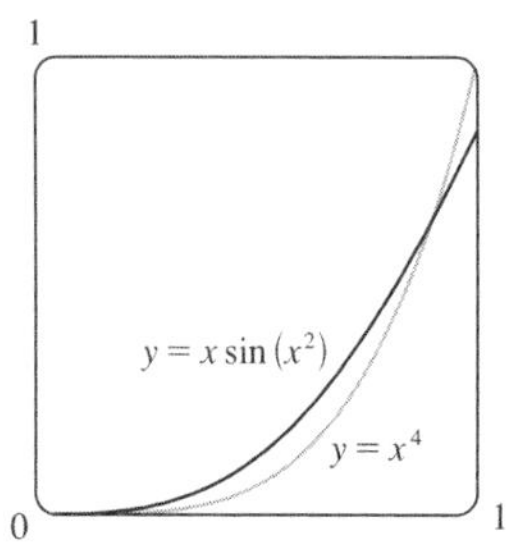


From the graph, we see that the curves intersect at $x = 0$ and $x = a \approx 0.896$, with $x\sin(x^2) > x^4$ on $(0, a)$. So the area $A$ of the region bounded by the curves is

$$A = \int_0^a \left[x\sin(x^2) - x^4\right] dx = \left[-\tfrac{1}{2}\cos(x^2) - \tfrac{1}{5}x^5\right]_0^a$$
$$= -\tfrac{1}{2}\cos(a^2) - \tfrac{1}{5}a^5 + \tfrac{1}{2} \approx 0.037$$

**18.** 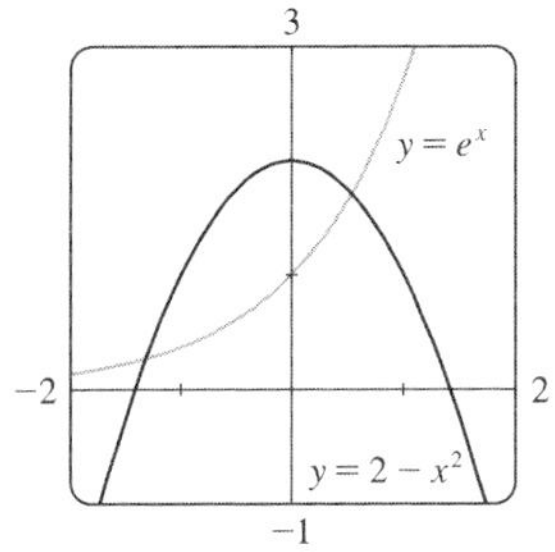


From the graph, we see that the curves intersect at $x = a \approx -1.32$ and $x = b \approx 0.54$, with $2 - x^2 > e^x$ on $(a, b)$. So the area $A$ of the region bounded by the curves is

$$A = \int_a^b \left[(2 - x^2) - e^x\right] dx = \left[2x - \tfrac{1}{3}x^3 - e^x\right]_a^b \approx 1.45$$

**19.** 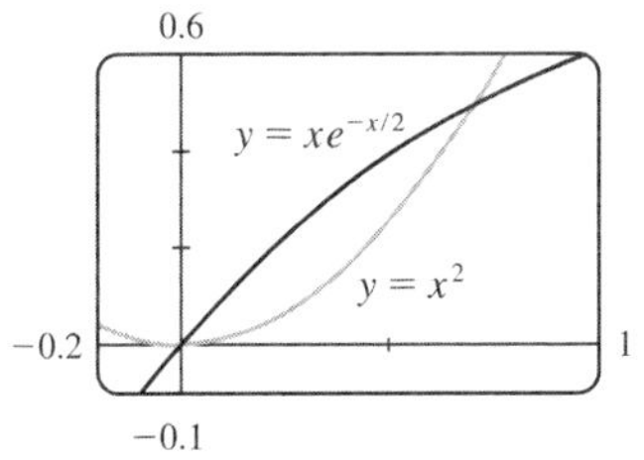


From the graph, we see that the curves intersect at $x = 0$ and $x = a \approx 0.70$, with $xe^{-x/2} > x^2$ on $(0, a)$. So the area $A$ of the region bounded by the curves is

$$\begin{aligned} A &= \int_0^a (xe^{-x/2} - x^2)\, dx \\ &= \left[4(-\tfrac{1}{2}x - 1)e^{-x/2} - \tfrac{1}{3}x^3\right]_0^a \quad \text{[Formula 96 with } a = -\tfrac{1}{2}\text{]} \\ &\approx 0.08 \end{aligned}$$

**20.** 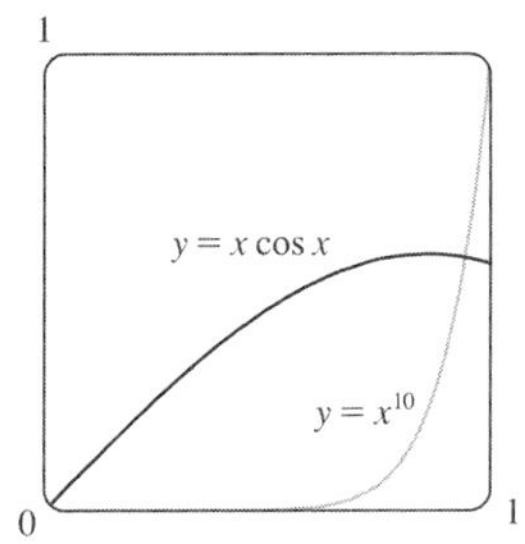


From the graph, we see that the curves intersect at $x = 0$ and $x = a \approx 0.94$, with $x \cos x > x^{10}$ on $(0, a)$. So the area $A$ of the region bounded by the curves is

$$\begin{aligned} A &= \int_0^a (x \cos x - x^{10})\, dx \\ &= \left[x \sin x + \cos x - \tfrac{1}{11}x^{11}\right]_0^a \quad \begin{bmatrix} u = x, & dv = \cos x\, dx \\ du = dx, & v = \sin x \end{bmatrix} \\ &\approx 0.30 \end{aligned}$$

**21.** As in Example 4, we approximate the distance between the two cars after ten seconds using Simpson's Rule with $\Delta t = 1 \text{ s} = \frac{1}{3600}$ h.

$$\begin{aligned} \text{distance}_{\text{Kelly}} - \text{distance}_{\text{Chris}} &= \int_0^{10} v_K\, dt - \int_0^{10} v_C\, dt = \int_0^{10} (v_K - v_C)\, dt \approx S_{10} \\ &= \frac{1}{3 \cdot 3600}[(0 - 0) + 4(22 - 20) + 2(37 - 32) + 4(52 - 46) + 2(61 - 54) + 4(71 - 62) \\ &\qquad + 2(80 - 69) + 4(86 - 75) + 2(93 - 81) + 4(98 - 86) + (102 - 90)] \\ &= \frac{1}{10{,}800}(242) = \frac{121}{5400} \text{ mi} \end{aligned}$$

So after 10 seconds, Kelly's car is about $\dfrac{121}{5400} \text{ mi}\left(5280\dfrac{\text{ft}}{\text{mi}}\right) \approx 118$ ft ahead of Chris's.

**22.** We know that the area under curve $A$ between $t = 0$ and $t = x$ is $\int_0^x v_A(t)\, dt = s_A(x)$, where $v_A(t)$ is the velocity of car $A$ and $s_A$ is its displacement. Similarly, the area under curve $B$ between $t = 0$ and $t = x$ is $\int_0^x v_B(t)\, dt = s_B(x)$.

(a) After one minute, the area under curve $A$ is greater than the area under curve $B$. So car $A$ is ahead after one minute.

(b) The area of the shaded region has numerical value $s_A(1) - s_B(1)$, which is the distance by which $A$ is ahead of $B$ after 1 minute.

(c) After two minutes, car $B$ is traveling faster than car $A$ and has gained some ground, but the area under curve $A$ from $t = 0$ to $t = 2$ is still greater than the corresponding area for curve $B$, so car $A$ is still ahead.

(d) From the graph, it appears that the area between curves $A$ and $B$ for $0 \le t \le 1$ (when car $A$ is going faster), which corresponds to the distance by which car $A$ is ahead, seems to be about 3 squares. Therefore, the cars will be side by side at the time $x$ where the area between the curves for $1 \le t \le x$ (when car $B$ is going faster) is the same as the area for $0 \le t \le 1$. From the graph, it appears that this time is $x \approx 2.2$. So the cars are side by side when $t \approx 2.2$ minutes.

**23.** If $x$ = distance from left end of pool and $w = w(x)$ = width at $x$, then Simpson's Rule with $n = 8$ and $\Delta x = 2$ gives

Area $= \int_0^{16} w\,dx \approx \frac{2}{3}\,[0 + 4(6.2) + 2(7.2) + 4(6.8) + 2(5.6) + 4(5.0) + 2(4.8) + 4(4.8) + 0] = \frac{2}{3}(126.4) \approx 84\ \text{m}^2$.

**24.** Let $h(x)$ denote the height of the wing at $x$ cm from the left end.

$$\begin{aligned} A \approx S_{10} &= \frac{200-0}{3(10)}\,[h(0) + 4h(20) + 2h(40) + \cdots + 4h(180) + h(200)] \\ &= \tfrac{20}{3}\,[5.8 + 4(20.3) + 2(26.7) + 4(29.0) + 2(27.6) \\ &\qquad + 4(27.3) + 2(23.8) + 4(20.5) + 2(15.1) + 4(8.7) + 2.8] \\ &= \tfrac{20}{3}(618.2) \approx 4121\ \text{cm}^2 \end{aligned}$$

**25.** For $0 \le t \le 10$, $b(t) > d(t)$, so the area between the curves is given by

$$\begin{aligned} \int_0^{10} [b(t) - d(t)]\,dt &= \int_0^{10} (2200e^{0.024t} - 1460e^{0.018t})\,dt = \left[\frac{2200}{0.024}e^{0.024t} - \frac{1460}{0.018}e^{0.018t}\right]_0^{10} \\ &= \left(\frac{275{,}000}{3}e^{0.24} - \frac{730{,}000}{9}e^{0.18}\right) - \left(\frac{275{,}000}{3} - \frac{730{,}000}{9}\right) \approx 8868 \text{ people} \end{aligned}$$

This area represents the increase in population over a 10-year period.

**26.** The area under $R'(x)$ from $x = 50$ to $x = 100$ represents the change in revenue, and the area under $C'(x)$ from $x = 50$ to $x = 100$ represents the change in cost. The shaded region represents the difference between these two values; that is, the increase in profit as the production level increases from 50 units to 100 units. We use the Midpoint Rule with $n = 5$ and $\Delta x = 10$:

$$M_5 = \Delta x\{[R'(55) - C'(55)] + [R'(65) - C'(65)] + [R'(75) - C'(75)] + [R'(85) - C'(85)] + [R'(95) - C'(95)]\}$$

$$\approx 10(2.40 - 0.85 + 2.20 - 0.90 + 2.00 - 1.00 + 1.80 - 1.10 + 1.70 - 1.20)$$

$$= 10(5.05) = 50.5 \text{ thousand dollars}$$

Using $M_1$ would give us $50(2 - 1) = 50$ thousand dollars.

**27.** $\cos x = \sin 2x = 2\sin x\,\cos x \quad\Leftrightarrow\quad 2\sin x\,\cos x - \cos x = 0 \quad\Leftrightarrow\quad \cos x\,(2\sin x - 1) = 0 \quad\Leftrightarrow$
$2\sin x = 1$ or $\cos x = 0 \quad\Leftrightarrow\quad x = \frac{\pi}{6}$ or $\frac{\pi}{2}$.

$$\begin{aligned} A &= \int_0^{\pi/6}(\cos x - \sin 2x)\,dx + \int_{\pi/6}^{\pi/2}(\sin 2x - \cos x)\,dx \\ &= \left[\sin x + \tfrac{1}{2}\cos 2x\right]_0^{\pi/6} + \left[-\tfrac{1}{2}\cos 2x - \sin x\right]_{\pi/6}^{\pi/2} \\ &= \left(\tfrac{1}{2} + \tfrac{1}{2}\cdot\tfrac{1}{2}\right) - \left(0 + \tfrac{1}{2}\cdot 1\right) + \left[-\tfrac{1}{2}\cdot(-1) - 1\right] - \left(-\tfrac{1}{2}\cdot\tfrac{1}{2} - \tfrac{1}{2}\right) \\ &= \tfrac{3}{4} - \tfrac{1}{2} - \tfrac{1}{2} + \tfrac{3}{4} = \tfrac{1}{2} \end{aligned}$$

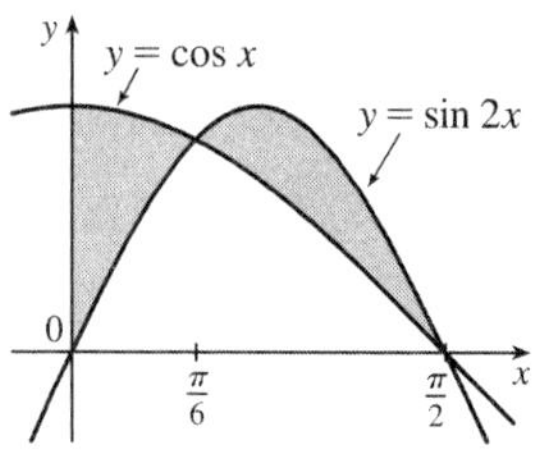


**28.**
$$\begin{aligned} A &= \int_0^1 \left[(x^3 - 4x^2 + 3x) - (x^2 - x)\right]dx \\ &\qquad + \int_1^4 \left[(x^2 - x) - (x^3 - 4x^2 + 3x)\right]dx \\ &= \int_0^1 (x^3 - 5x^2 + 4x)\,dx + \int_1^4 (-x^3 + 5x^2 - 4x)\,dx \\ &= \left[\tfrac{1}{4}x^4 - \tfrac{5}{3}x^3 + 2x^2\right]_0^1 + \left[-\tfrac{1}{4}x^4 + \tfrac{5}{3}x^3 - 2x^2\right]_1^4 \\ &= \left(\tfrac{1}{4} - \tfrac{5}{3} + 2\right) - 0 + \left(-64 + \tfrac{320}{3} - 32\right) - \left(-\tfrac{1}{4} + \tfrac{5}{3} - 2\right) = \tfrac{71}{6} \end{aligned}$$

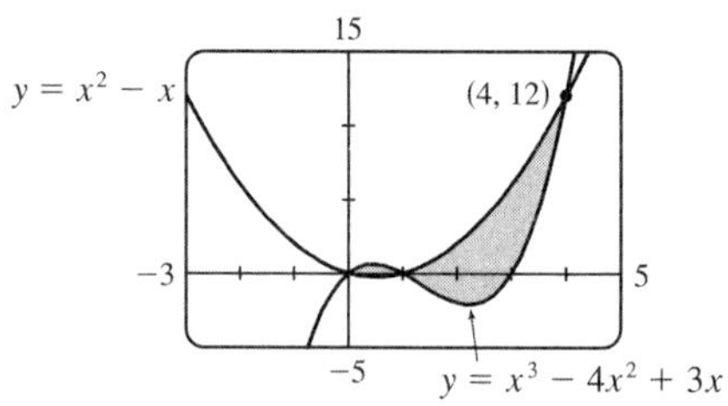

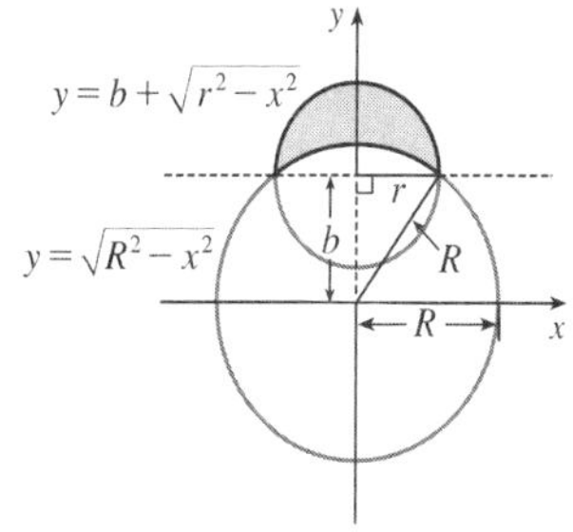


**29.** Let the equation of the large circle be $x^2+y^2=R^2$. Then the equation of the small circle is $x^2+(y-b)^2=r^2$, where $b=\sqrt{R^2-r^2}$ is the distance between the centers of the circles. The desired area is

$$\begin{aligned} A &= \int_{-r}^{r}\left[\left(b+\sqrt{r^2-x^2}\right)-\sqrt{R^2-x^2}\right]dx \\ &= 2\int_0^r\left(b+\sqrt{r^2-x^2}-\sqrt{R^2-x^2}\right)dx \\ &= 2\int_0^r b\,dx+2\int_0^r\sqrt{r^2-x^2}\,dx-2\int_0^r\sqrt{R^2-x^2}\,dx \end{aligned}$$

The first integral is just $2br=2r\sqrt{R^2-r^2}$. The second integral represents the area of a quarter-circle of radius $r$, so its value is $\frac{1}{4}\pi r^2$. To evaluate the other integral, note that

$$\begin{aligned} \int\sqrt{a^2-x^2}\,dx &= \int a^2\cos^2\theta\,d\theta \quad [x=a\sin\theta,\ dx=a\cos\theta\,d\theta] \quad = \left(\tfrac{1}{2}a^2\right)\int(1+\cos 2\theta)\,d\theta \\ &= \tfrac{1}{2}a^2\left(\theta+\tfrac{1}{2}\sin 2\theta\right)+C=\tfrac{1}{2}a^2(\theta+\sin\theta\,\cos\theta)+C \\ &= \tfrac{a^2}{2}\arcsin\left(\tfrac{x}{a}\right)+\tfrac{a^2}{2}\left(\tfrac{x}{a}\right)\tfrac{\sqrt{a^2-x^2}}{a}+C=\tfrac{a^2}{2}\arcsin\left(\tfrac{x}{a}\right)+\tfrac{x}{2}\sqrt{a^2-x^2}+C \end{aligned}$$

Thus, the desired area is

$$\begin{aligned} A &= 2r\sqrt{R^2-r^2}+2\left(\tfrac{1}{4}\pi r^2\right)-\left[R^2\arcsin(x/R)+x\sqrt{R^2-x^2}\right]_0^r \\ &= 2r\sqrt{R^2-r^2}+\tfrac{1}{2}\pi r^2-\left[R^2\arcsin(r/R)+r\sqrt{R^2-r^2}\right]=r\sqrt{R^2-r^2}+\tfrac{\pi}{2}r^2-R^2\arcsin(r/R) \end{aligned}$$

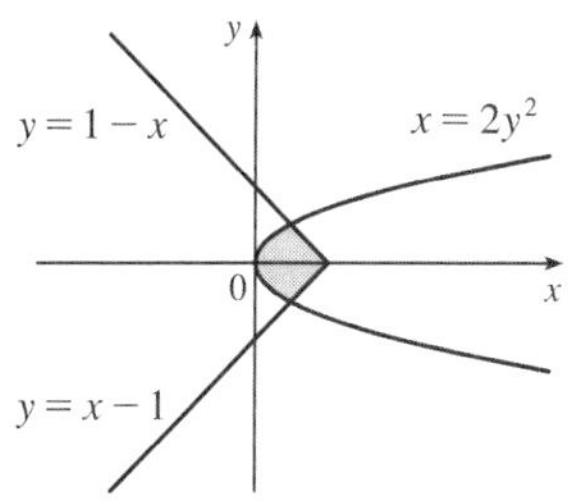


**30.** The inequality $x\ge 2y^2$ describes the region that lies on, or to the right of, the parabola $x=2y^2$. The inequality $x\le 1-|y|$ describes the region that lies on, or to the left of, the curve $x=1-|y|=\begin{cases}1-y & \text{if } y\ge 0\\ 1+y & \text{if } y<0\end{cases}$.

So the given region is the shaded region that lies between the curves. The graphs of $x=1-y$ and $x=2y^2$ intersect when $1-y=2y^2 \Leftrightarrow 2y^2+y-1=0 \Leftrightarrow (2y-1)(y+1)=0 \Rightarrow y=\frac{1}{2}$ (for $y\ge 0$).

By symmetry, $A=2\displaystyle\int_0^{1/2}\left[(1-y)-2y^2\right]dy=2\left[-\tfrac{2}{3}y^3-\tfrac{1}{2}y^2+y\right]_0^{1/2}=2\left[\left(-\tfrac{1}{12}-\tfrac{1}{8}+\tfrac{1}{2}\right)-0\right]=2\left(\tfrac{7}{24}\right)=\tfrac{7}{12}$.

**31.** By symmetry of the ellipse about the $x$- and $y$-axes,

$$\begin{aligned} A &= 4\int_0^a y\,dx=4\int_{\pi/2}^{0}b\sin\theta\,(-a\sin\theta)\,d\theta \qquad \left[\begin{matrix} x=a\cos\theta=0 \ \Rightarrow\ \theta=\frac{\pi}{2} \text{ and} \\ x=a\cos\theta=a \ \Rightarrow\ \theta=0 \end{matrix}\right] \\ &= 4ab\int_0^{\pi/2}\sin^2\theta\,d\theta=4ab\int_0^{\pi/2}\tfrac{1}{2}(1-\cos 2\theta)\,d\theta \\ &= 2ab\left[\theta-\tfrac{1}{2}\sin 2\theta\right]_0^{\pi/2}=2ab\left(\tfrac{\pi}{2}\right)=\pi ab \end{aligned}$$

Note that the formula for the area of a circle, $A=\pi r^2$, is just a special case of this formula with $a=b=r$.

**32.** $y = 2.5 \;\Rightarrow\; t + 1/t = 2.5 \;\Leftrightarrow\; t^2 - 2.5t + 1 = 0 \;\Rightarrow\; t = \frac{1}{2}$ or 2, and for $\frac{1}{2} < t < 2$, we have $t + 1/t < 2.5$. $x = -\frac{3}{2}$ when $t = \frac{1}{2}$ and $x = \frac{3}{2}$ when $t = 2$.

$$\begin{aligned} A &= \int_{-3/2}^{3/2} (2.5 - y)\, dx \\ &= \int_{1/2}^{2} \left(\frac{5}{2} - t - 1/t\right)\left(1 + 1/t^2\right) dt \quad [x = t - 1/t \;\Rightarrow\; dx = \left(1 + 1/t^2\right) dt] \\ &= \int_{1/2}^{2} \left(-t + \frac{5}{2} - 2t^{-1} + \frac{5}{2}t^{-2} - t^{-3}\right) dt = \left[\frac{-t^2}{2} + \frac{5t}{2} - 2\ln|t| - \frac{5}{2t} + \frac{1}{2t^2}\right]_{1/2}^{2} \\ &= \left(-2 + 5 - 2\ln 2 - \tfrac{5}{4} + \tfrac{1}{8}\right) - \left(-\tfrac{1}{8} + \tfrac{5}{4} + 2\ln 2 - 5 + 2\right) \quad \left[\ln\tfrac{1}{2} = \ln 2^{-1} = -\ln 2\right] \\ &= \tfrac{15}{4} - 4\ln 2 \end{aligned}$$

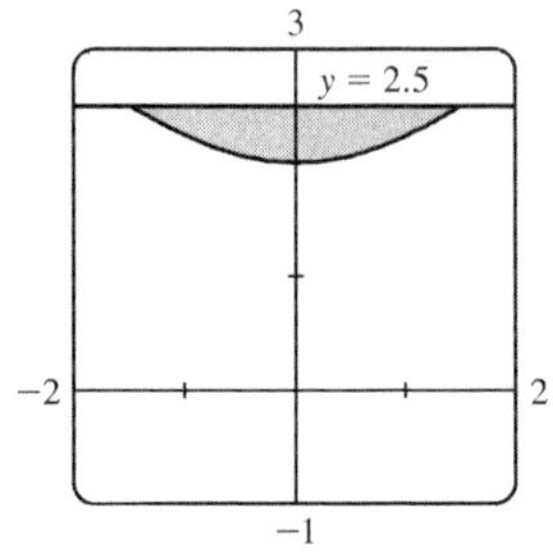


**33.**

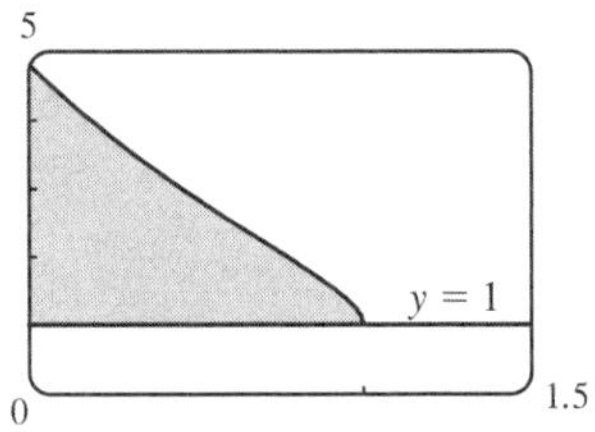


$x = \cos t,\ y = e^t,\ 0 \le t \le \frac{\pi}{2}$. When $x = 0$, $t = \pi/2$; when $x = 1$, $t = 0$.

$$\begin{aligned} A &= \textstyle\int_0^1 (y - 1)\, dx = \int_{\pi/2}^{0} (e^t - 1)(-\sin t)\, dt \\ &= \textstyle\int_0^{\pi/2} (e^t \sin t - \sin t)\, dt \end{aligned}$$

By Example 4 in Section 5.6 or Integration Formula 98, $\int e^x \sin x\, dx = \frac{1}{2}e^x(\sin x - \cos x) + C$, so

$$\begin{aligned} \textstyle\int_0^{\pi/2} (e^t \sin t - \sin t)\, dt &= \left[\tfrac{1}{2}e^t(\sin t - \cos t) + \cos t\right]_0^{\pi/2} = \left[\tfrac{1}{2}e^{\pi/2}(1 - 0) + 0\right] - \left[\tfrac{1}{2}e^0(0 - 1) + 1\right] \\ &= \tfrac{1}{2}e^{\pi/2} - \tfrac{1}{2} = \tfrac{1}{2}(e^{\pi/2} - 1) \end{aligned}$$

**34.**

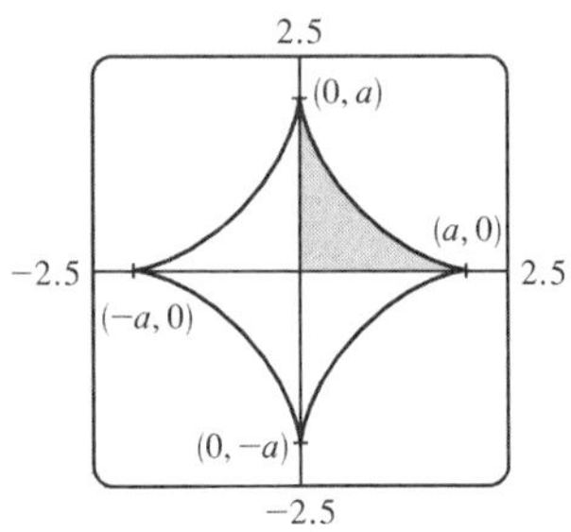


$x = a\cos^3\theta$, $y = a\sin^3\theta$. (In the diagram, $a = 2$.) By symmetry,

$$\begin{aligned} A &= 4\int_0^a y\, dx = 4\int_{\pi/2}^{0} a\sin^3\theta\left(-3a\cos^2\theta\sin\theta\right) d\theta \\ &= 12a^2 \int_0^{\pi/2} \sin^4\theta\cos^2\theta\, d\theta \overset{\text{CAS}}{=} 12a^2\left(\frac{\pi}{32}\right) = \frac{3}{8}\pi a^2 \end{aligned}$$

**35.** By symmetry, the area of the region enclosed by the loop is twice the area above the $x$-axis inside the loop. $y = 0 \;\Leftrightarrow\; t^3 - 3t = 0 \;\Leftrightarrow\; t(t^2 - 3) = 0 \;\Leftrightarrow\; t = 0, \pm\sqrt{3}$. The top half of the loop is described by $x = t^2$, $y = t^3 - 3t$, $-\sqrt{3} \le t \le 0$, so, using the Substitution Rule with $y = t^3 - 3t$ and $dx = 2t\, dt$, we find that

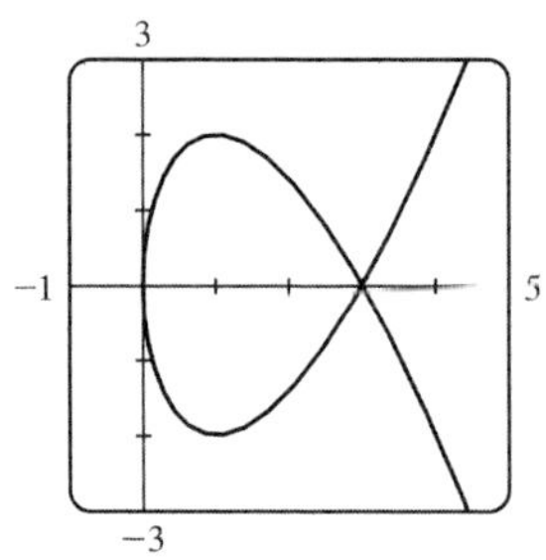


$$\begin{aligned} \text{area} &= 2\textstyle\int_0^3 y\, dx = 2\int_0^{-\sqrt{3}} (t^3 - 3t)2t\, dt = 4\int_0^{-\sqrt{3}} (t^4 - 3t^2)\, dt \\ &= 4\left[\tfrac{1}{5}t^5 - t^3\right]_0^{-\sqrt{3}} = 4\left[\tfrac{1}{5}\left(-3^{1/2}\right)^5 - \left(-3^{1/2}\right)^3\right] \\ &= 4\left[\tfrac{1}{5}(-9\sqrt{3}) - (-3\sqrt{3})\right] = \tfrac{24}{5}\sqrt{3} \approx 8.31. \end{aligned}$$

**36.** We plot the curve $x = t^3 - 12t$, $y = 3t^2 + 2t + 5$ in the parameter interval $t \in [-4, 3.5]$. In order to find the area of the loop, we need to estimate the two $t$-values corresponding to the point at which the curve crosses itself. By zooming in, we estimate the $y$-coordinate of the point of intersection to be 39.667, and so the two $t$-values at the point of intersection are approximately the two solutions of the equation $y = 3t^2 + 2t + 5 = 39.667$, which are

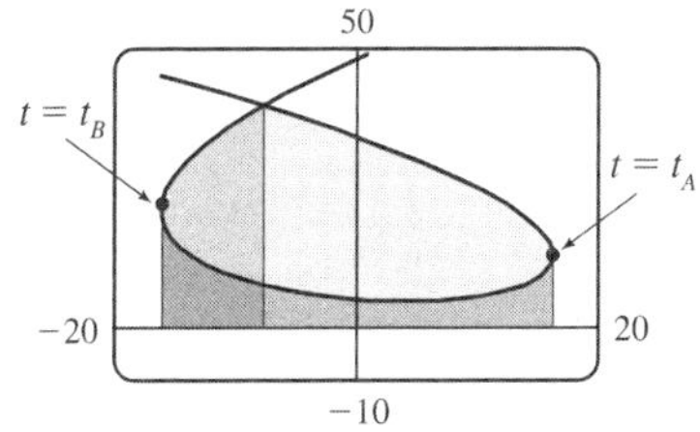


$t = -\frac{1}{3} \pm \frac{\sqrt{420.004}}{6} \approx -3.7490$ or $3.0823$. We can evaluate the area of the loop simply by integrating $y\,dx$ between these two $t$-values, since this integral represents the area under the upper part of the loop for $t$ between the first $t$-value and $t_A$, minus the area under the bottom part between $t_A$ and $t_B$, plus the area under the top part between $t_B$ and the final $t$-value. So since $dx = (3t^2 - 12)\,dt$, the area of the loop is $A \approx \int_{-3.7490}^{3.0823} (3t^2 + 2t + 5)(3t^2 - 12)\,dt \approx 743.85$.

**37.** We first assume that $c > 0$, since $c$ can be replaced by $-c$ in both equations without changing the graphs, and if $c = 0$ the curves do not enclose a region. We see from the graph that the enclosed area $A$ lies between $x = -c$ and $x = c$, and by symmetry, it is equal to four times the area in the first quadrant. The enclosed area is

$$A = 4\int_0^c (c^2 - x^2)\,dx = 4\left[c^2x - \tfrac{1}{3}x^3\right]_0^c$$

$$= 4\left(c^3 - \tfrac{1}{3}c^3\right) = 4\left(\tfrac{2}{3}c^3\right) = \tfrac{8}{3}c^3$$

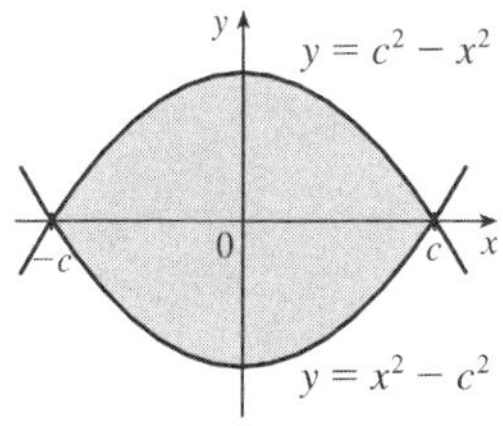


So $A = 576 \iff \frac{8}{3}c^3 = 576 \iff c^3 = 216 \iff c = \sqrt[3]{216} = 6$.

Note that $c = -6$ is another solution, since the graphs are the same.

**38.**

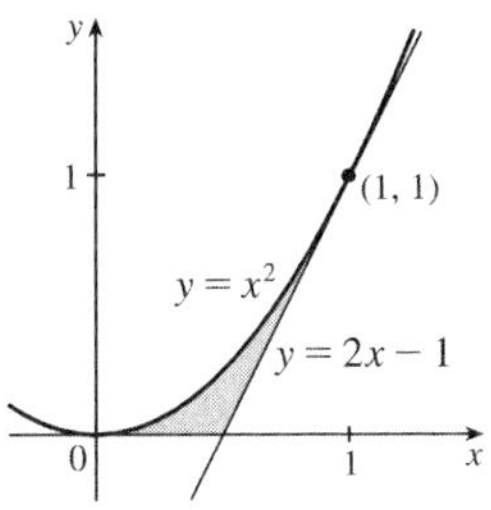


We start by finding the equation of the tangent line to $y = x^2$ at the point $(1, 1)$: $y' = 2x$, so the slope of the tangent is $2(1) = 2$, and its equation is $y - 1 = 2(x - 1)$, or $y = 2x - 1$. We would need two integrals to integrate with respect to $x$, but only one to integrate with respect to $y$.

$$A = \int_0^1 \left[\tfrac{1}{2}(y + 1) - \sqrt{y}\right] dy = \left[\tfrac{1}{4}y^2 + \tfrac{1}{2}y - \tfrac{2}{3}y^{3/2}\right]_0^1$$

$$= \tfrac{1}{4} + \tfrac{1}{2} - \tfrac{2}{3} = \tfrac{1}{12}$$

**39.**

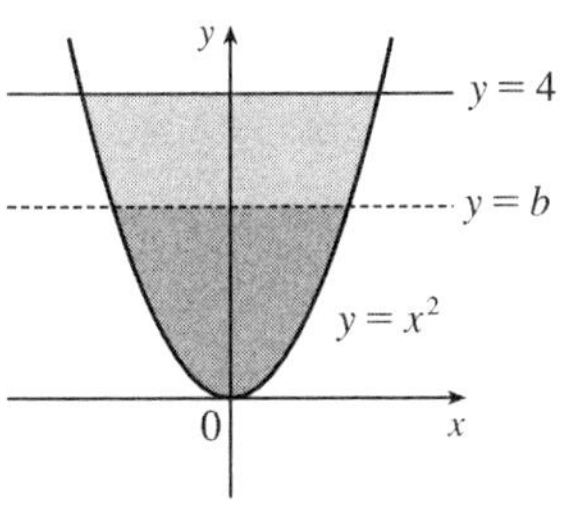


By the symmetry of the problem, we consider only the first quadrant, where $y = x^2 \Rightarrow x = \sqrt{y}$. We are looking for a number $b$ such that

$$\int_0^b \sqrt{y}\,dy = \int_b^4 \sqrt{y}\,dy \Rightarrow \tfrac{2}{3}\left[y^{3/2}\right]_0^b = \tfrac{2}{3}\left[y^{3/2}\right]_b^4 \Rightarrow$$

$$b^{3/2} = 4^{3/2} - b^{3/2} \Rightarrow 2b^{3/2} = 8 \Rightarrow b^{3/2} = 4 \Rightarrow b = 4^{2/3} \approx 2.52.$$

**40.** (a) We want to choose $a$ so that

$$\int_1^a \frac{1}{x^2}\,dx = \int_a^4 \frac{1}{x^2}\,dx \Rightarrow \left[\frac{-1}{x}\right]_1^a = \left[\frac{-1}{x}\right]_a^4 \Rightarrow -\frac{1}{a} + 1 = -\frac{1}{4} + \frac{1}{a} \Rightarrow \frac{5}{4} = \frac{2}{a} \Rightarrow a = \frac{8}{5}.$$

(b) The area under the curve $y = 1/x^2$ from $x = 1$ to $x = 4$ is $\frac{3}{4}$ [take $a = 4$ in the first integral in part (a)]. Now the line $y = b$ must intersect the curve $x = 1/\sqrt{y}$ and not the line $x = 4$, since the area under the line $y = 1/4^2$ from $x = 1$ to $x = 4$ is only $\frac{3}{16}$, which is less than half of $\frac{3}{4}$. We want to choose $b$ so that the upper area in the diagram is half of the total area under the curve $y = 1/x^2$ from $x = 1$ to $x = 4$. This implies that

$\int_b^1 (1/\sqrt{y} - 1)\,dy = \frac{1}{2}\cdot\frac{3}{4} \;\Rightarrow\; \left[2\sqrt{y} - y\right]_b^1 = \frac{3}{8} \;\Rightarrow\; 1 - 2\sqrt{b} + b = \frac{3}{8} \;\Rightarrow$
$b - 2\sqrt{b} + \frac{5}{8} = 0$. Letting $c = \sqrt{b}$, we get $c^2 - 2c + \frac{5}{8} = 0 \;\Rightarrow$
$8c^2 - 16c + 5 = 0$. Thus, $c = \frac{16 \pm \sqrt{256 - 160}}{16} = 1 \pm \frac{\sqrt{6}}{4}$. But $c = \sqrt{b} < 1 \;\Rightarrow$
$c = 1 - \frac{\sqrt{6}}{4} \;\Rightarrow\; b = c^2 = 1 + \frac{3}{8} - \frac{\sqrt{6}}{2} = \frac{1}{8}\left(11 - 4\sqrt{6}\right) \approx 0.1503$.

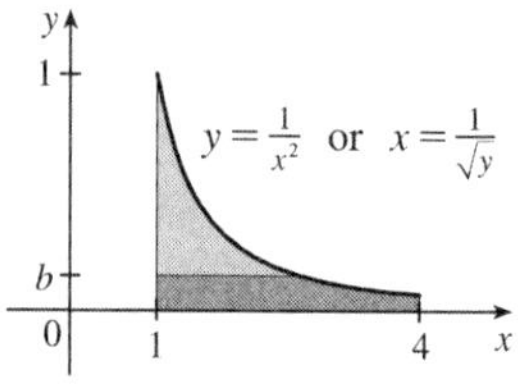


**41.** The area under the graph of $f$ from 0 to $t$ is equal to $\int_0^t f(x)\,dx$, so the requirement is that $\int_0^t f(x)\,dx = t^3$ for all $t$. We differentiate both sides of this equation with respect to $t$ (with the help of FTC1) to get $f(t) = 3t^2$. This function is positive and continuous, as required.

**42.**

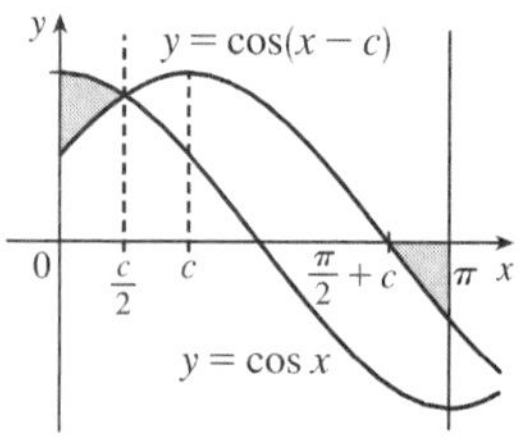


It appears from the diagram that the curves $y = \cos x$ and $y = \cos(x - c)$ intersect halfway between 0 and $c$, namely, when $x = c/2$. We can verify that this is indeed true by noting that $\cos(c/2 - c) = \cos(-c/2) = \cos(c/2)$. The point where $\cos(x - c)$ crosses the $x$-axis is $x = \frac{\pi}{2} + c$. So we require that

$\int_0^{c/2} [\cos x - \cos(x - c)]\,dx = -\int_{\pi/2+c}^{\pi} \cos(x - c)\,dx$ [the negative sign on the RHS is needed since the second area is beneath the $x$-axis] $\;\Leftrightarrow\; [\sin x - \sin(x - c)]_0^{c/2} = -[\sin(x - c)]_{\pi/2+c}^{\pi} \;\Rightarrow$
$[\sin(c/2) - \sin(-c/2)] - [-\sin(-c)] = -\sin(\pi - c) + \sin\left[\left(\frac{\pi}{2} + c\right) - c\right] \;\Leftrightarrow\; 2\sin(c/2) - \sin c = -\sin c + 1$.
[Here we have used the oddness of the sine function, and the fact that $\sin(\pi - c) = \sin c$]. So $2\sin(c/2) = 1 \;\Leftrightarrow$
$\sin(c/2) = \frac{1}{2} \;\Leftrightarrow\; c/2 = \frac{\pi}{6} \;\Leftrightarrow\; c = \frac{\pi}{3}$.

**43.** The curve and the line will determine a region when they intersect at two or more points. So we solve the equation $x/(x^2 + 1) = mx \;\Rightarrow\; x = x(mx^2 + m) \;\Rightarrow$
$x(mx^2 + m) - x = 0 \;\Rightarrow\; x(mx^2 + m - 1) = 0 \;\Rightarrow$
$x = 0$ or $mx^2 + m - 1 = 0 \;\Rightarrow$

$$x = 0 \text{ or } x^2 = \frac{1 - m}{m} \;\Rightarrow\; x = 0 \text{ or } x = \pm\sqrt{\frac{1}{m} - 1}.$$

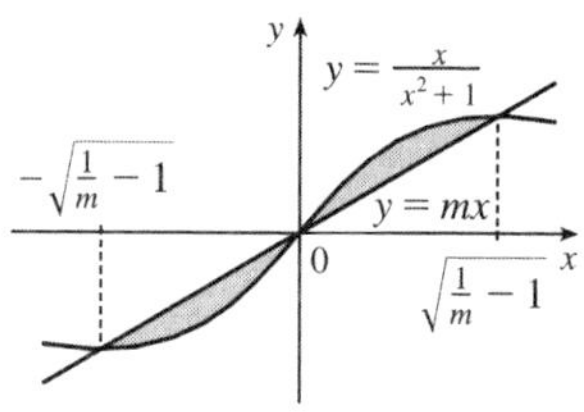


Note that if $m = 1$, this has only the solution $x = 0$, and no region is determined. But if $1/m - 1 > 0 \;\Leftrightarrow\; 1/m > 1 \;\Leftrightarrow$ $0 < m < 1$, then there are two solutions. [Another way of seeing this is to observe that the slope of the tangent to $y = x/(x^2 + 1)$ at the origin is $y' = 1$ and therefore we must have $0 < m < 1$.] Note that we cannot just integrate between the positive and negative roots, since the curve and the line cross at the origin. Since $mx$ and $x/(x^2 + 1)$ are both odd functions, the total area is twice the area between the curves on the interval $\left[0, \sqrt{1/m - 1}\right]$. So the total area enclosed is

$$\begin{aligned} 2\int_0^{\sqrt{1/m-1}} \left[\frac{x}{x^2 + 1} - mx\right] dx &= 2\left[\frac{1}{2}\ln(x^2 + 1) - \frac{1}{2}mx^2\right]_0^{\sqrt{1/m-1}} \\ &= [\ln(1/m - 1 + 1) - m(1/m - 1)] - (\ln 1 - 0) \\ &= \ln(1/m) - 1 + m = m - \ln m - 1 \end{aligned}$$

## 6.2 Volumes

**1.** A cross-section is a disk with radius $1/x$, so its area is $A(x) = \pi(1/x)^2$.

$$V = \int_1^2 A(x)\,dx = \int_1^2 \pi\left(\frac{1}{x}\right)^2 dx$$
$$= \pi \int_1^2 \frac{1}{x^2}\,dx = \pi\left[-\frac{1}{x}\right]_1^2$$
$$= \pi\left[-\tfrac{1}{2} - (-1)\right] = \frac{\pi}{2}$$

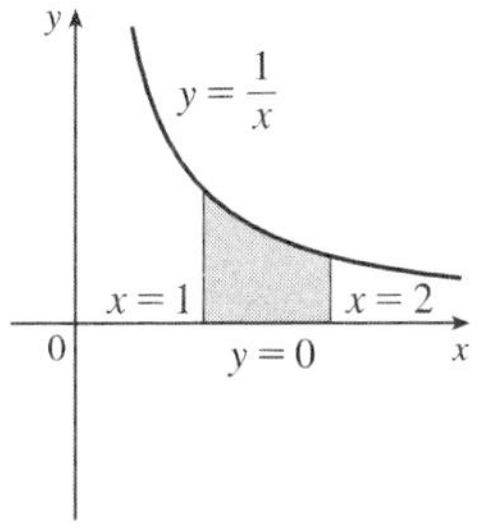


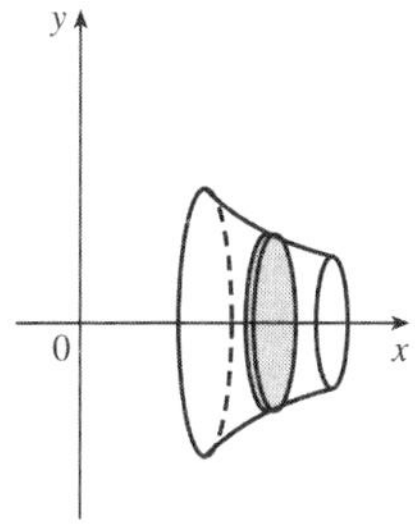


**2.** A cross-section is a disk with radius $1 - x^2$, so its area is $A(x) = \pi(1 - x^2)^2$.

$$V = \int_{-1}^1 A(x)\,dx = \int_{-1}^1 \pi(1 - x^2)^2\,dx$$
$$= 2\pi \int_0^1 (1 - 2x^2 + x^4)\,dx$$
$$= 2\pi\left[x - \tfrac{2}{3}x^3 + \tfrac{1}{5}x^5\right]_0^1 = 2\pi\left(1 - \tfrac{2}{3} + \tfrac{1}{5}\right)$$
$$= 2\pi\left(\tfrac{8}{15}\right) = \tfrac{16\pi}{15}$$

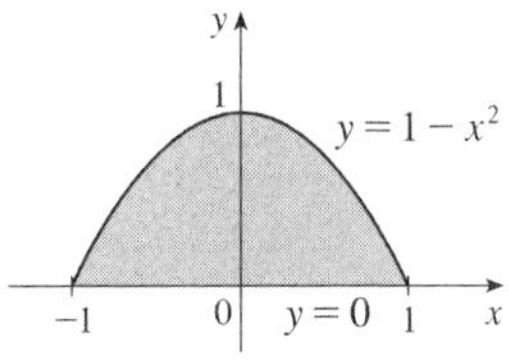


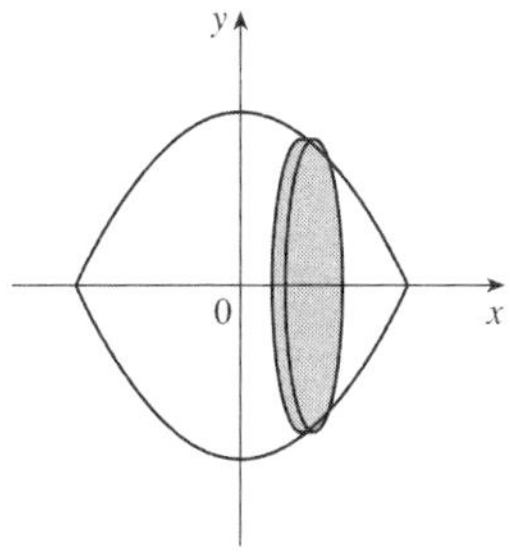


**3.** A cross-section is a disk with radius $2\sqrt{y}$, so its area is $A(y) = \pi\left(2\sqrt{y}\right)^2$.

$$V = \int_0^9 A(y)\,dy = \int_0^9 \pi\left(2\sqrt{y}\right)^2 dy$$
$$= 4\pi \int_0^9 y\,dy = 4\pi\left[\tfrac{1}{2}y^2\right]_0^9$$
$$= 2\pi(81) = 162\pi$$

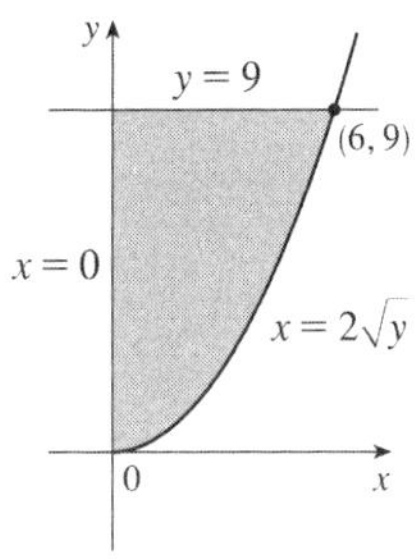


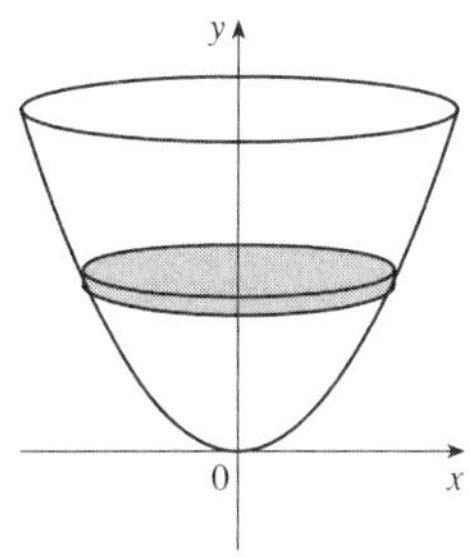


**4.** A cross-section is a disk with radius $e^y$, so its area is $A(y) = \pi(e^y)^2$.

$$V = \int_1^2 \pi(e^y)^2\,dy = \pi \int_1^2 e^{2y}\,dy = \pi\left[\frac{1}{2}e^{2y}\right]_1^2 = \frac{\pi}{2}\left(e^4 - e^2\right)$$

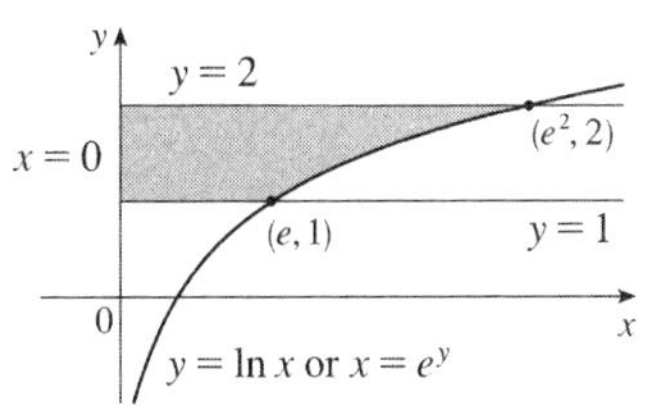


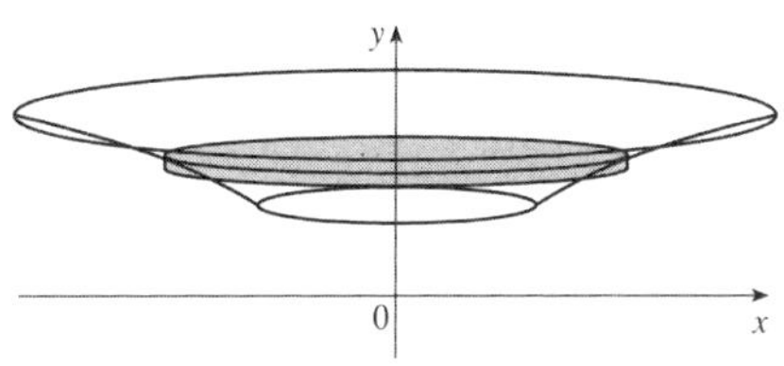

**5.** A cross-section is a washer (annulus) with inner radius $x^3$ and outer radius $x$, so its area is $A(x) = \pi(x)^2 - \pi(x^3)^2 = \pi(x^2 - x^6)$.

$$V = \int_0^1 A(x)\,dx = \int_0^1 \pi(x^2 - x^6)\,dx$$
$$= \pi\left[\tfrac{1}{3}x^3 - \tfrac{1}{7}x^7\right]_0^1 = \pi\left(\tfrac{1}{3} - \tfrac{1}{7}\right) = \tfrac{4\pi}{21}$$

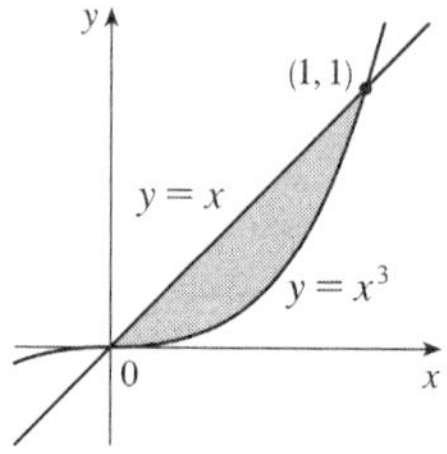


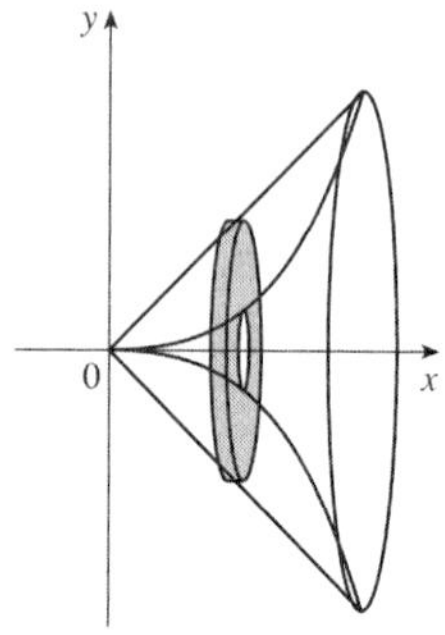


**6.** A cross-section is a washer with inner radius $\frac{1}{4}x^2$ and outer radius $5 - x^2$, so its area is

$$A(x) = \pi(5 - x^2)^2 - \pi\left(\tfrac{1}{4}x^2\right)^2$$
$$= \pi\left(25 - 10x^2 + x^4 - \tfrac{1}{16}x^4\right).$$

$$V = \int_{-2}^{2} A(x)\,dx = \int_{-2}^{2} \pi\left(25 - 10x^2 + \tfrac{15}{16}x^4\right)dx$$
$$= 2\pi\int_0^2 \left(25 - 10x^2 + \tfrac{15}{16}x^4\right)dx$$
$$= 2\pi\left[25x - \tfrac{10}{3}x^3 + \tfrac{3}{16}x^5\right]_0^2 = 2\pi\left(50 - \tfrac{80}{3} + 6\right) = \tfrac{176\pi}{3}$$

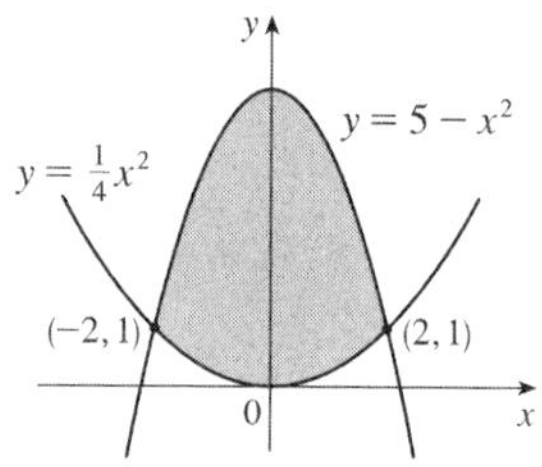


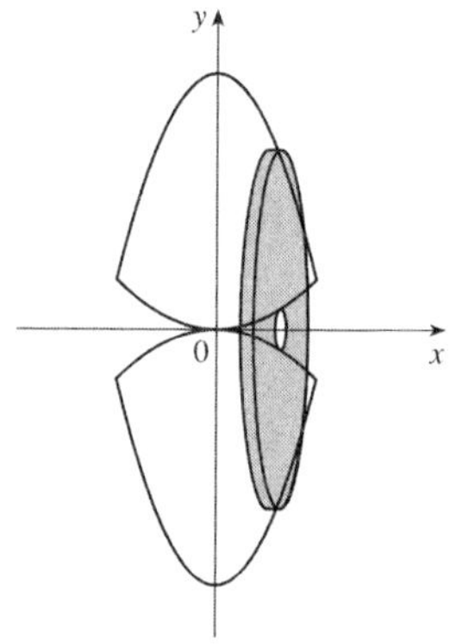


**7.** A cross-section is a washer with inner radius $y^2$ and outer radius $2y$, so its area is $A(y) = \pi(2y)^2 - \pi(y^2)^2 = \pi(4y^2 - y^4)$.

$$V = \int_0^2 A(y)\,dy = \pi\int_0^2 (4y^2 - y^4)\,dy$$
$$= \pi\left[\tfrac{4}{3}y^3 - \tfrac{1}{5}y^5\right]_0^2 = \pi\left(\tfrac{32}{3} - \tfrac{32}{5}\right) = \tfrac{64\pi}{15}$$

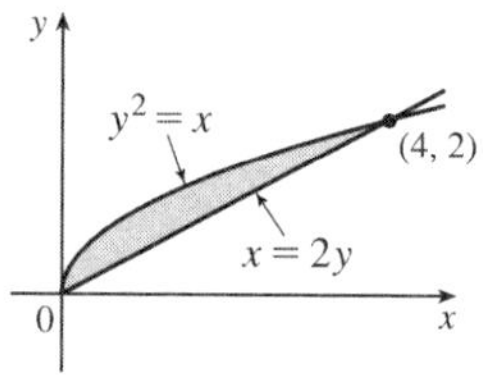


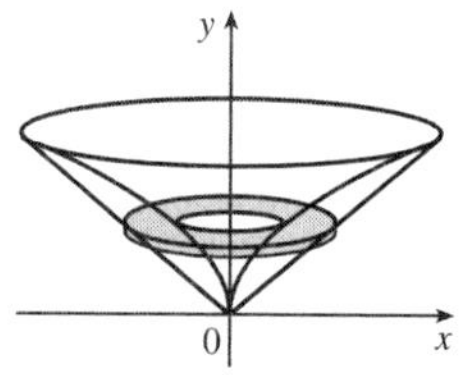


**8.** $y = x^{2/3} \Leftrightarrow x = y^{3/2}$, so a cross-section is a washer with inner radius $y^{3/2}$ and outer radius 1, and its area is $A(y) = \pi(1)^2 - \pi(y^{3/2})^2 = \pi(1 - y^3)$.

$$V = \int_0^1 A(y)\,dy = \pi\int_0^1 (1 - y^3)\,dy$$
$$= \pi\left[y - \tfrac{1}{4}y^4\right]_0^1 = \tfrac{3}{4}\pi$$

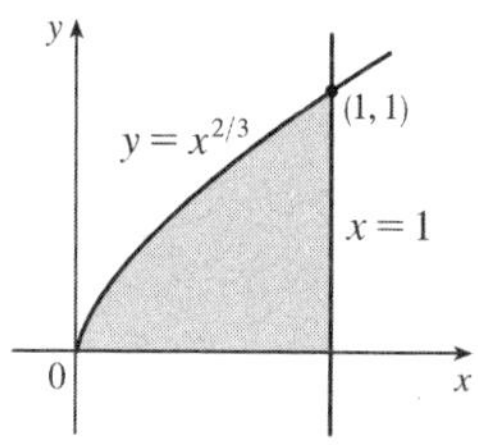


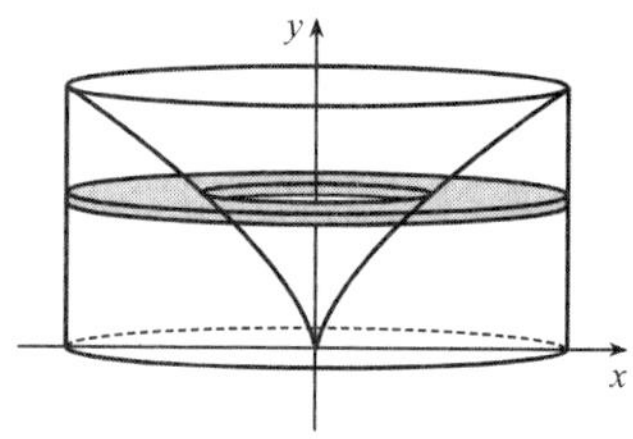


**9.** A cross-section is a washer with inner radius $1 - \sqrt{x}$ and outer radius $1 - x$, so its area is

$$A(x) = \pi(1 - x)^2 - \pi\left(1 - \sqrt{x}\right)^2$$
$$= \pi\left[(1 - 2x + x^2) - (1 - 2\sqrt{x} + x)\right]$$
$$= \pi\left(-3x + x^2 + 2\sqrt{x}\right).$$

$$V = \int_0^1 A(x)\,dx = \pi\int_0^1 \left(-3x + x^2 + 2\sqrt{x}\right)dx$$
$$= \pi\left[-\tfrac{3}{2}x^2 + \tfrac{1}{3}x^3 + \tfrac{4}{3}x^{3/2}\right]_0^1 = \pi\left(-\tfrac{3}{2} + \tfrac{5}{3}\right) = \tfrac{\pi}{6}$$

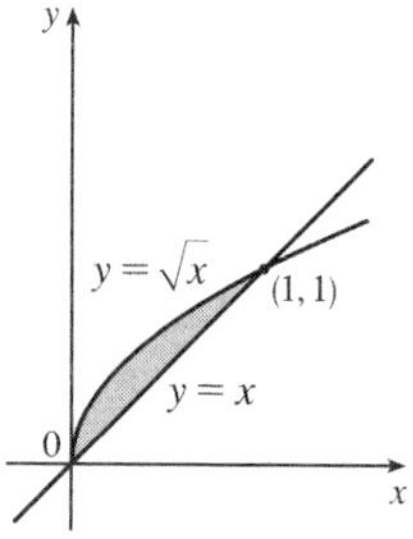


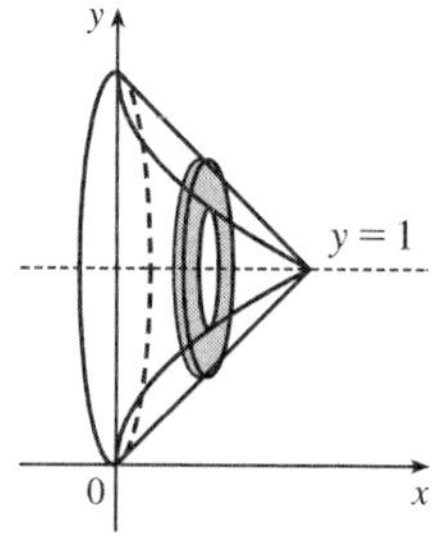

**10.** $V = \int_1^3 \pi \left\{ \left[ \frac{1}{x} - (-1) \right]^2 - [0 - (-1)]^2 \right\} dx = \pi \int_1^3 \left[ \left( \frac{1}{x} + 1 \right)^2 - 1^2 \right] dx = \pi \int_1^3 \left( \frac{1}{x^2} + \frac{2}{x} \right) dx$

$$= \pi \left[ -\frac{1}{x} + 2 \ln x \right]_1^3 = \pi \left[ \left( -\tfrac{1}{3} + 2 \ln 3 \right) - (-1 + 0) \right] = \pi \left( 2 \ln 3 + \tfrac{2}{3} \right) = 2\pi \left( \ln 3 + \tfrac{1}{3} \right)$$

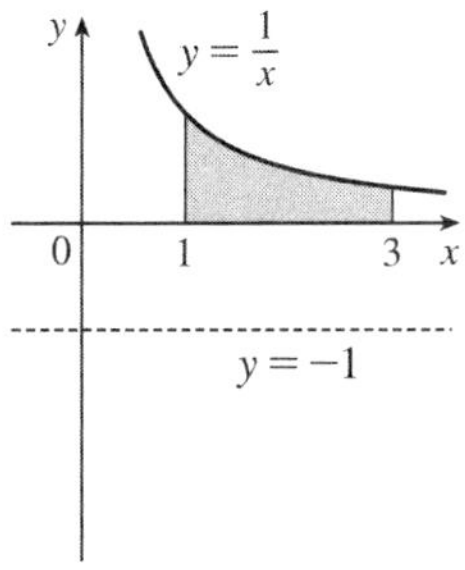


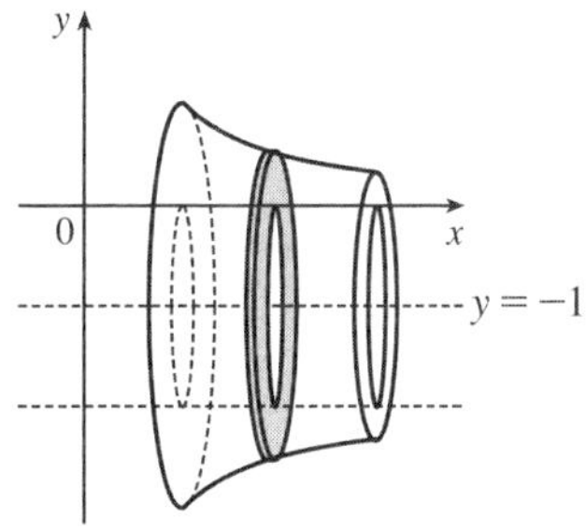


**11.** $y = x^2 \;\Rightarrow\; x = \sqrt{y}$ for $x \geq 0$. The outer radius is the distance from $x = -1$ to $x = \sqrt{y}$ and the inner radius is the distance from $x = -1$ to $x = y^2$.

$$V = \int_0^1 \pi \left\{ \left[ \sqrt{y} - (-1) \right]^2 - \left[ y^2 - (-1) \right]^2 \right\} dy = \pi \int_0^1 \left[ (\sqrt{y} + 1)^2 - (y^2 + 1)^2 \right] dy$$

$$= \pi \int_0^1 \left( y + 2\sqrt{y} + 1 - y^4 - 2y^2 - 1 \right) dy = \pi \int_0^1 \left( y + 2\sqrt{y} - y^4 - 2y^2 \right) dy$$

$$= \pi \left[ \tfrac{1}{2} y^2 + \tfrac{4}{3} y^{3/2} - \tfrac{1}{5} y^5 - \tfrac{2}{3} y^3 \right]_0^1 = \pi \left( \tfrac{1}{2} + \tfrac{4}{3} - \tfrac{1}{5} - \tfrac{2}{3} \right) = \tfrac{29}{30} \pi$$

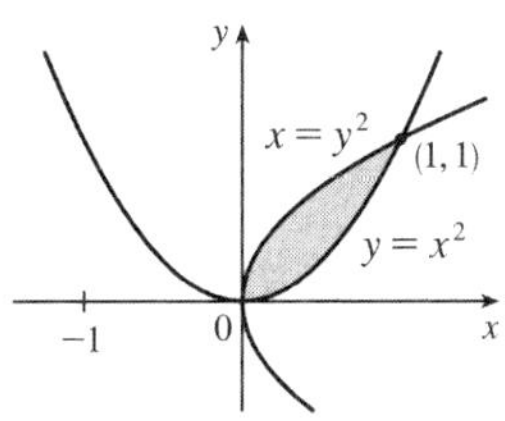


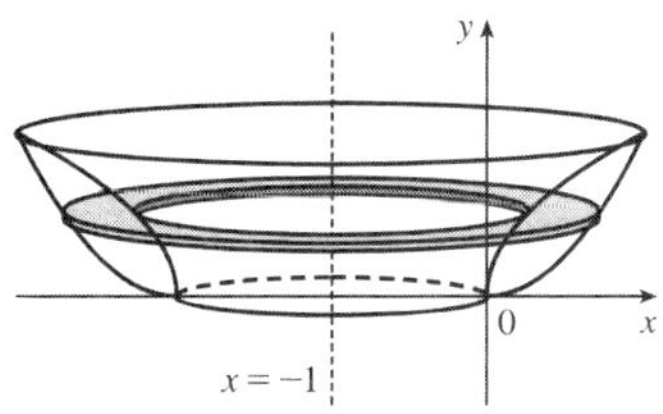


**12.** $y = \sqrt{x} \;\Rightarrow\; x = y^2$, so the outer radius is $2 - y^2$.

$$V = \int_0^1 \pi \left[ (2 - y^2)^2 - (2 - y)^2 \right] dy = \pi \int_0^1 \left[ (4 - 4y^2 + y^4) - (4 - 4y + y^2) \right] dy$$

$$= \pi \int_0^1 \left( y^4 - 5y^2 + 4y \right) dy = \pi \left[ \tfrac{1}{5} y^5 - \tfrac{5}{3} y^3 + 2y^2 \right]_0^1 = \pi \left( \tfrac{1}{5} - \tfrac{5}{3} + 2 \right) = \tfrac{8}{15} \pi$$

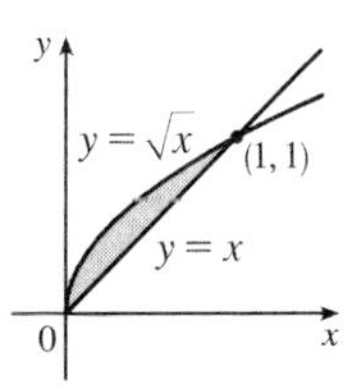


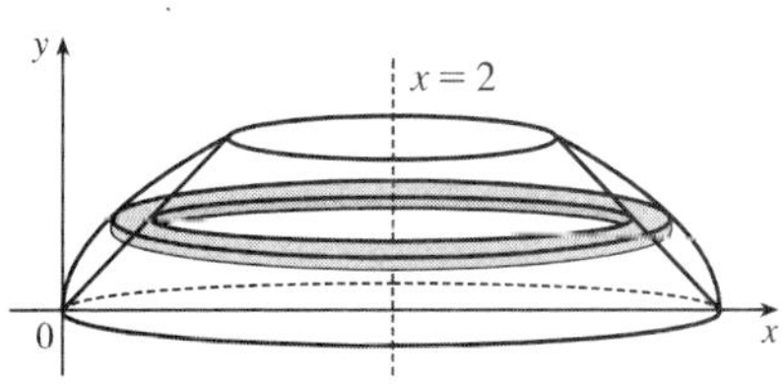

**13.** $y = \sqrt{x} \;\Rightarrow\; x = y^2$ and $y = x^3 \;\Rightarrow\; x = \sqrt[3]{y}$. A cross-section is a washer with inner radius $1 - \sqrt[3]{y}$ and outer radius $1 - y^2$, so its area is

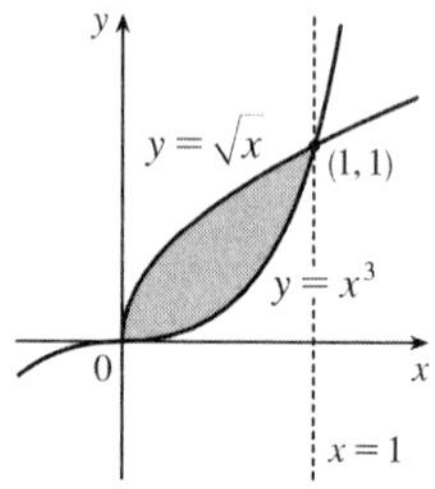


$A(y) = \pi(1 - y^2)^2 - \pi\left(1 - \sqrt[3]{y}\right)^2$.

$$\begin{aligned} V &= \int_0^1 A(y)\,dy = \int_0^1 \left[\pi(1-y^2)^2 - \pi\left(1 - \sqrt[3]{y}\right)^2\right] dy \\ &= \pi \int_0^1 [(1 - 2y^2 + y^4) - (1 - 2y^{1/3} + y^{2/3})]\,dy \\ &= \pi \int_0^1 (-2y^2 + y^4 + 2y^{1/3} - y^{2/3})\,dy = \pi\left[-\tfrac{2}{3}y^3 + \tfrac{1}{5}y^5 + \tfrac{3}{2}y^{4/3} - \tfrac{3}{5}y^{5/3}\right]_0^1 \\ &= \pi\left(-\tfrac{2}{3} + \tfrac{1}{5} + \tfrac{3}{2} - \tfrac{3}{5}\right) = \tfrac{13\pi}{30} \end{aligned}$$

**14.** A cross-section is a washer with inner radius $1 - \sqrt{x}$ and outer radius $1 - x^3$, so its area is $A(x) = \pi(1 - x^3)^2 - \pi(1 - \sqrt{x})^2$.

$$\begin{aligned} V &= \int_0^1 A(x)\,dx = \int_0^1 \left[\pi(1-x^3)^2 - \pi\left(1 - \sqrt{x}\right)^2\right] dx = \pi \int_0^1 [(1 - 2x^3 + x^6) - (1 - 2x^{1/2} + x)]\,dx \\ &= \pi \int_0^1 (-2x^3 + x^6 + 2x^{1/2} - x)\,dx = \pi\left[-\tfrac{1}{2}x^4 + \tfrac{1}{7}x^7 + \tfrac{4}{3}x^{3/2} - \tfrac{1}{2}x^2\right]_0^1 = \pi\left(-\tfrac{1}{2} + \tfrac{1}{7} + \tfrac{4}{3} - \tfrac{1}{2}\right) = \tfrac{10\pi}{21} \end{aligned}$$

**15.** $V = \pi \displaystyle\int_0^{\pi/4} (1 - \tan^3 x)^2\,dx$

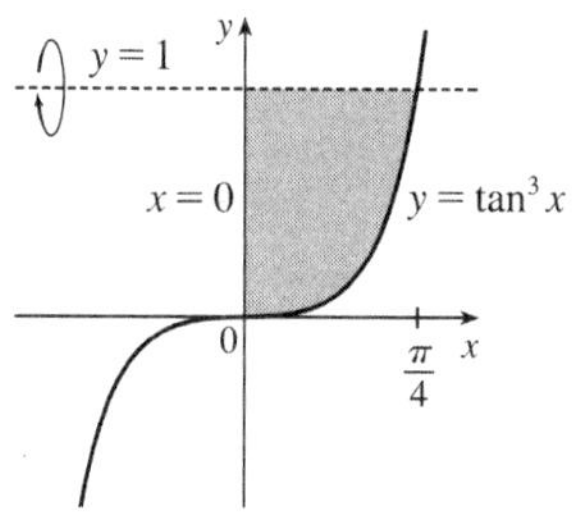


**16.** $y = (x-2)^4$ and $8x - y = 16$ intersect when

$(x-2)^4 = 8x - 16 = 8(x-2) \;\Leftrightarrow$

$(x-2)^4 - 8(x-2) = 0 \;\Leftrightarrow\; (x-2)\left[(x-2)^3 - 8\right] = 0 \;\Leftrightarrow$

$x - 2 = 0$ or $x - 2 = 2 \;\Leftrightarrow\; x = 2$ or $4$. $y = (x-2)^4 \;\Rightarrow$

$x - 2 = \pm\sqrt[4]{y} \;\Rightarrow\; x = 2 + \sqrt[4]{y}$ [since $x \geq 2$].

$8x - y = 16 \;\Rightarrow\; 8x = y + 16 \;\Rightarrow\; x = \tfrac{1}{8}y + 2$.

$V = \pi \displaystyle\int_0^{16} \left\{\left[10 - \left(\tfrac{1}{8}y + 2\right)\right]^2 - \left[10 - (2 + \sqrt[4]{y})\right]^2\right\} dy$

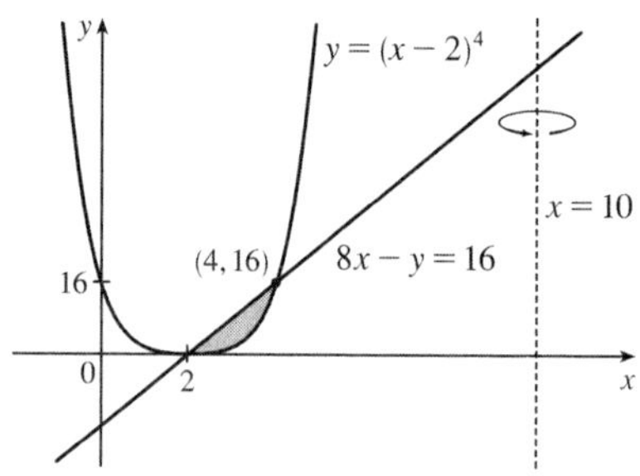


**17.**

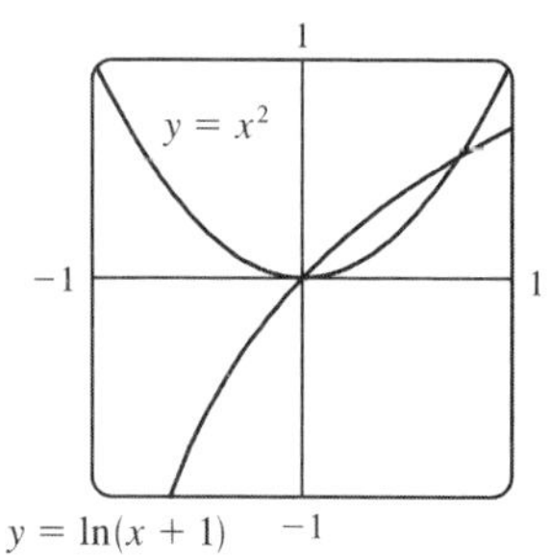


$y = x^2$ and $y = \ln(x+1)$ intersect at $x = 0$ and at $x = a \approx 0.747$.

$V = \pi \displaystyle\int_0^a \left\{[\ln(x+1)]^2 - \left(x^2\right)^2\right\} dx \approx 0.132$

**18.**

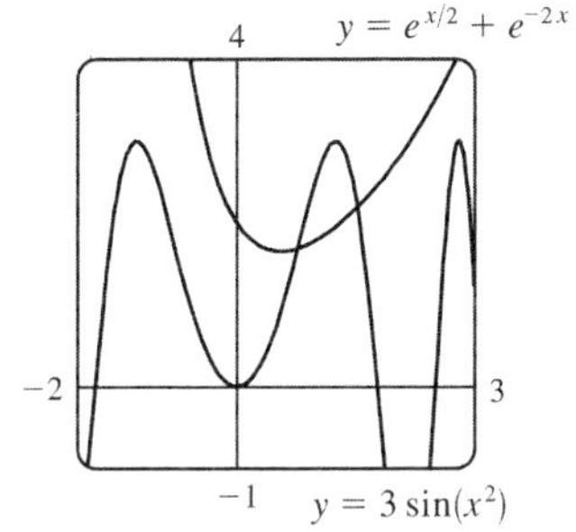


$y = 3\sin(x^2)$ and $y = e^{x/2} + e^{-2x}$ intersect at $x = a \approx 0.772$ and at $x = b \approx 1.524$.

$$V = \pi \int_a^b \left\{ \left[3\sin(x^2)\right]^2 - \left(e^{x/2} + e^{-2x}\right)^2 \right\} dx \approx 7.519$$

**19.** $V = \pi \int_0^{\pi} \left\{ \left[\sin^2 x - (-1)\right]^2 - [0 - (-1)]^2 \right\} dx \overset{\text{CAS}}{=} \frac{11}{8}\pi^2$

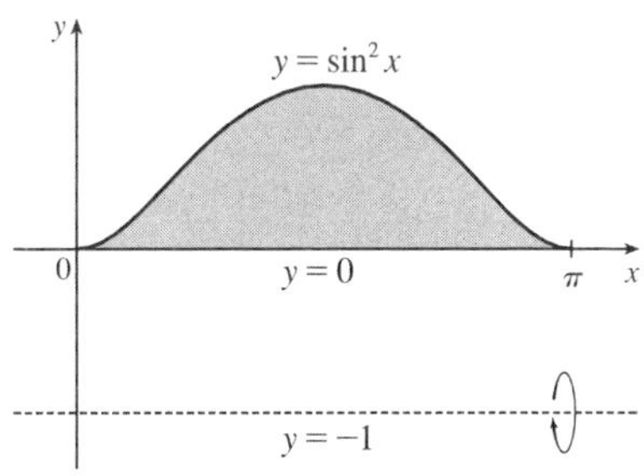


**20.** $V = \pi \int_0^2 \left[ (3 - x)^2 - (3 - xe^{1-x/2})^2 \right] dx \overset{\text{CAS}}{=} \pi\left(-2e^2 + 24e - \frac{142}{3}\right)$

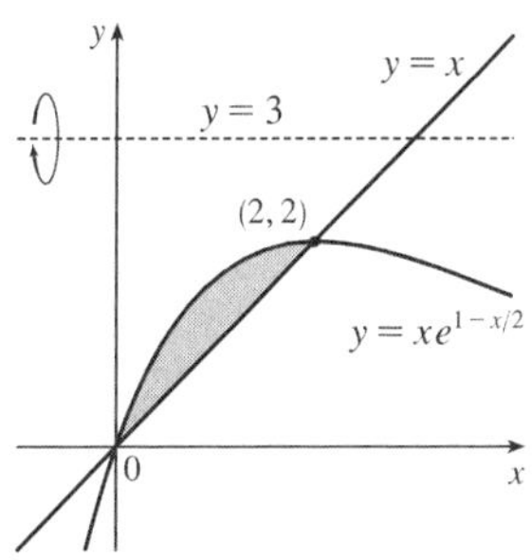


**21.** (a) $\pi \int_0^{\pi/2} \cos^2 x\, dx$ describes the volume of the solid obtained by rotating the region $\mathcal{R} = \left\{(x, y) \mid 0 \le x \le \frac{\pi}{2}, 0 \le y \le \cos x\right\}$ of the $xy$-plane about the $x$-axis.

(b) $\pi \int_0^1 (y^4 - y^8)\, dy = \pi \int_0^1 \left[ (y^2)^2 - (y^4)^2 \right] dy$ describes the volume of the solid obtained by rotating the region $\mathcal{R} = \left\{(x, y) \mid 0 \le y \le 1, y^4 \le x \le y^2\right\}$ of the $xy$-plane about the $y$-axis.

**22.** (a) $\pi \int_2^5 y\, dy = \pi \int_2^5 \left(\sqrt{y}\right)^2 dy$ describes the volume of the solid obtained by rotating the region $\mathcal{R} = \left\{(x, y) \mid 2 \le y \le 5, 0 \le x \le \sqrt{y}\right\}$ of the $xy$-plane about the $y$-axis.

(b) $\pi \int_0^{\pi/2} \left[ (1 + \cos x)^2 - 1^2 \right] dx$ describes the volume of the solid obtained by rotating the region $\mathcal{R} = \left\{(x, y) \mid 0 \le x \le \frac{\pi}{2}, 1 \le y \le 1 + \cos x\right\}$ of the $xy$-plane about the $x$-axis.
*Or:* The solid could be obtained by rotating the region $\mathcal{R}' = \left\{(x, y) \mid 0 \le x \le \frac{\pi}{2}, 0 \le y \le \cos x\right\}$ about the line $y = -1$.

**23.** There are 10 subintervals over the 15-cm length, so we'll use $n = 10/2 = 5$ for the Midpoint Rule.

$$V = \int_0^{15} A(x)\, dx \approx M_5 = \frac{15-0}{5}[A(1.5) + A(4.5) + A(7.5) + A(10.5) + A(13.5)]$$
$$= 3(18 + 79 + 106 + 128 + 39) = 3 \cdot 370 = 1110 \text{ cm}^3$$

**24.** $$V = \int_0^{10} A(x)\, dx \approx M_5 = \frac{10-0}{5}[A(1) + A(3) + A(5) + A(7) + A(9)]$$
$$= 2(0.65 + 0.61 + 0.59 + 0.55 + 0.50) = 2(2.90) = 5.80 \text{ m}^3$$

**25.** We'll form a right circular cone with height $h$ and base radius $r$ by revolving the line $y = \frac{r}{h}x$ about the $x$-axis.

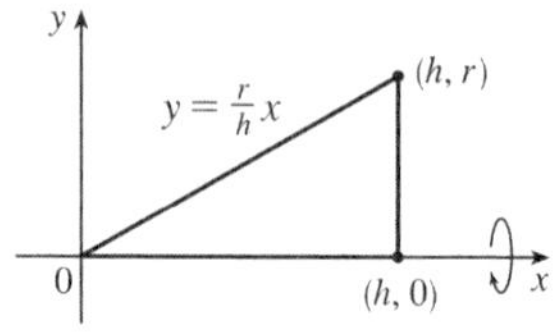


$$V = \pi \int_0^h \left(\frac{r}{h}x\right)^2 dx = \pi \int_0^h \frac{r^2}{h^2}x^2 dx = \pi \frac{r^2}{h^2}\left[\frac{1}{3}x^3\right]_0^h$$

$$= \pi \frac{r^2}{h^2}\left(\frac{1}{3}h^3\right) = \frac{1}{3}\pi r^2 h$$

*Another solution:* Revolve $x = -\dfrac{r}{h}y + r$ about the $y$-axis.

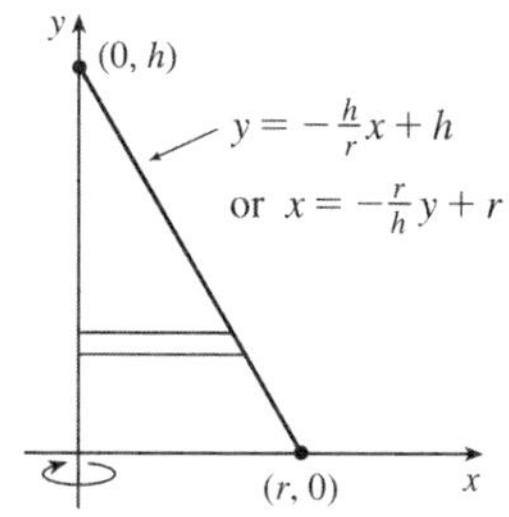


$$V = \pi \int_0^h \left(-\frac{r}{h}y + r\right)^2 dy \stackrel{*}{=} \pi \int_0^h \left[\frac{r^2}{h^2}y^2 - \frac{2r^2}{h}y + r^2\right] dy$$

$$= \pi\left[\frac{r^2}{3h^2}y^3 - \frac{r^2}{h}y^2 + r^2 y\right]_0^h = \pi\left(\tfrac{1}{3}r^2h - r^2h + r^2h\right) = \tfrac{1}{3}\pi r^2 h$$

* Or use substitution with $u = r - \dfrac{r}{h}y$ and $du = -\dfrac{r}{h}\,dy$ to get

$$\pi \int_r^0 u^2\left(-\frac{h}{r}\,du\right) = -\pi\frac{h}{r}\left[\frac{1}{3}u^3\right]_r^0 = -\pi\frac{h}{r}\left(-\frac{1}{3}r^3\right) = \frac{1}{3}\pi r^2 h.$$

**26.** $V = \pi \displaystyle\int_0^h \left(R - \frac{R-r}{h}y\right)^2 dy$

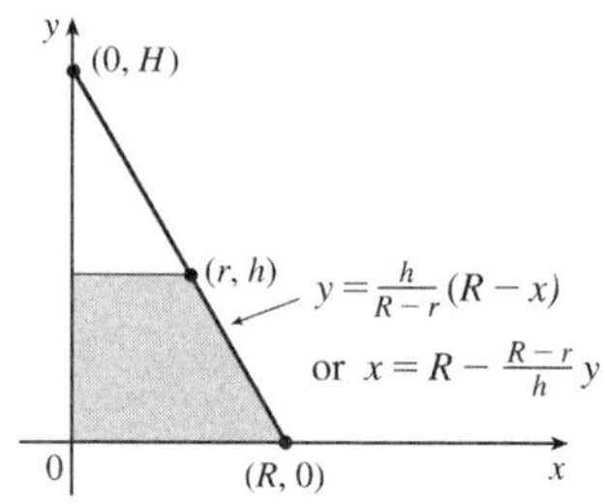


$$= \pi \int_0^h \left[R^2 - \frac{2R(R-r)}{h}y + \left(\frac{R-r}{h}\right)^2 y^2\right] dy$$

$$= \pi\left[R^2 y - \frac{R(R-r)}{h}y^2 + \frac{1}{3}\left(\frac{R-r}{h}\right)^2 y^3\right]_0^h$$

$$= \pi\left[R^2h - R(R-r)h + \tfrac{1}{3}(R-r)^2h\right]$$

$$= \tfrac{1}{3}\pi h\left[3Rr + (R^2 - 2Rr + r^2)\right] = \tfrac{1}{3}\pi h(R^2 + Rr + r^2)$$

*Another solution:* $\dfrac{H}{R} = \dfrac{H-h}{r}$ by similar triangles. Therefore,

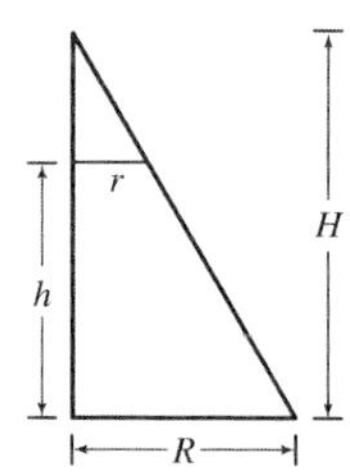


$Hr = HR - hR \;\Rightarrow\; hR = H(R-r) \;\Rightarrow\; H = \dfrac{hR}{R-r}$. Now

$V = \tfrac{1}{3}\pi R^2 H - \tfrac{1}{3}\pi r^2(H-h)$ [by Exercise 25]

$$= \tfrac{1}{3}\pi R^2\frac{hR}{R-r} - \tfrac{1}{3}\pi r^2\frac{rh}{R-r} \qquad \left[H - h = \frac{rH}{R} = \frac{rhR}{R(R-r)}\right]$$

$$= \frac{1}{3}\pi h\frac{R^3 - r^3}{R-r} = \tfrac{1}{3}\pi h(R^2 + Rr + r^2)$$

$$= \tfrac{1}{3}\left[\pi R^2 + \pi r^2 + \sqrt{(\pi R^2)(\pi r^2)}\right]h = \tfrac{1}{3}\left(A_1 + A_2 + \sqrt{A_1A_2}\right)h$$

where $A_1$ and $A_2$ are the areas of the bases of the frustum. (See Exercise 28 for a related result.)

**27.** $x^2 + y^2 = r^2 \quad \Leftrightarrow \quad x^2 = r^2 - y^2$

$$V = \pi \int_{r-h}^{r} (r^2 - y^2)\,dy = \pi \left[ r^2 y - \frac{y^3}{3} \right]_{r-h}^{r}$$

$$= \pi \left\{ \left[ r^3 - \frac{r^3}{3} \right] - \left[ r^2(r-h) - \frac{(r-h)^3}{3} \right] \right\}$$

$$= \pi \left\{ \tfrac{2}{3} r^3 - \tfrac{1}{3}(r-h)\left[3r^2 - (r-h)^2\right] \right\}$$

$$= \tfrac{1}{3}\pi \left\{ 2r^3 - (r-h)\left[3r^2 - (r^2 - 2rh + h^2)\right] \right\}$$

$$= \tfrac{1}{3}\pi \left\{ 2r^3 - (r-h)\left[2r^2 + 2rh - h^2\right] \right\}$$

$$= \tfrac{1}{3}\pi \left( 2r^3 - 2r^3 - 2r^2h + rh^2 + 2r^2h + 2rh^2 - h^3 \right)$$

$$= \tfrac{1}{3}\pi \left( 3rh^2 - h^3 \right) = \tfrac{1}{3}\pi h^2 (3r - h), \text{ or, equivalently, } \pi h^2 \left( r - \frac{h}{3} \right)$$

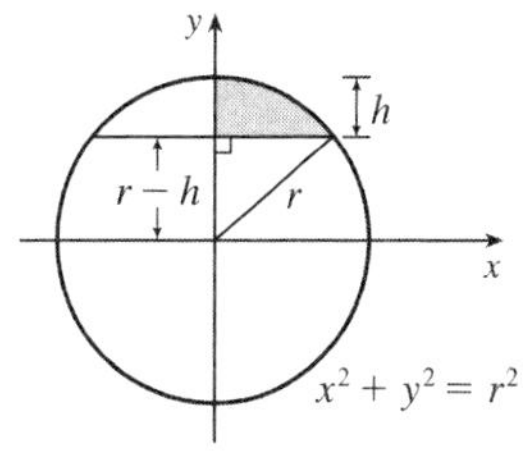


**28.** An equation of the line is $x = \dfrac{\Delta x}{\Delta y}\,y + (x\text{-intercept}) = \dfrac{a/2 - b/2}{h - 0}\,y + \dfrac{b}{2} = \dfrac{a-b}{2h}\,y + \dfrac{b}{2}$.

$$V = \int_0^h A(y)\,dy = \int_0^h (2x)^2\,dy$$

$$= \int_0^h \left[ 2\left( \frac{a-b}{2h}y + \frac{b}{2} \right) \right]^2 dy = \int_0^h \left[ \frac{a-b}{h}y + b \right]^2 dy$$

$$= \int_0^h \left[ \frac{(a-b)^2}{h^2}y^2 + \frac{2b(a-b)}{h}y + b^2 \right] dy$$

$$= \left[ \frac{(a-b)^2}{3h^2}y^3 + \frac{b(a-b)}{h}y^2 + b^2 y \right]_0^h$$

$$= \tfrac{1}{3}(a-b)^2 h + b(a-b)h + b^2 h = \tfrac{1}{3}(a^2 - 2ab + b^2 + 3ab)h$$

$$= \tfrac{1}{3}(a^2 + ab + b^2)h$$

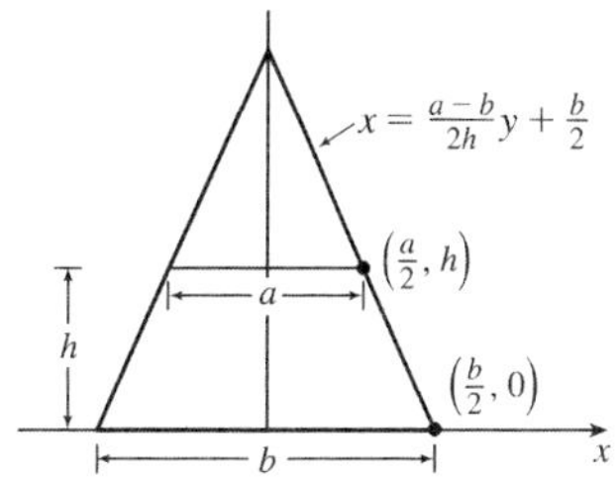


[Note that this can be written as $\frac{1}{3}\left(A_1 + A_2 + \sqrt{A_1 A_2}\,\right)h$, as in Exercise 26.]

If $a = b$, we get a rectangular solid with volume $b^2 h$. If $a = 0$, we get a square pyramid with volume $\frac{1}{3}b^2 h$.

**29.** For a cross-section at height $y$, we see from similar triangles that $\dfrac{\alpha/2}{b/2} = \dfrac{h-y}{h}$, so $\alpha = b\left(1 - \frac{y}{h}\right)$.

Similarly, for cross-sections having $2b$ as their base and $\beta$ replacing $\alpha$, $\beta = 2b\left(1 - \frac{y}{h}\right)$. So

$$V = \int_0^h A(y)\,dy = \int_0^h \left[ b\left(1 - \tfrac{y}{h}\right) \right] \left[ 2b\left(1 - \tfrac{y}{h}\right) \right] dy$$

$$= \int_0^h 2b^2 \left(1 - \tfrac{y}{h}\right)^2 dy = 2b^2 \int_0^h \left( 1 - \frac{2y}{h} + \frac{y^2}{h^2} \right) dy$$

$$= 2b^2 \left[ y - \frac{y^2}{h} + \frac{y^3}{3h^2} \right]_0^h = 2b^2 \left[ h - h + \tfrac{1}{3}h \right]$$

$$= \tfrac{2}{3}b^2 h \quad \left[ = \tfrac{1}{3}Bh \text{ where } B \text{ is the area of the base, as with any pyramid.} \right]$$

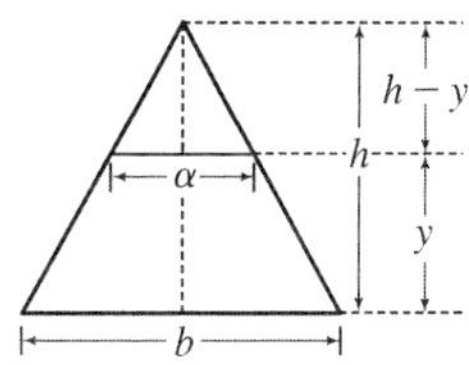

**30.** Consider the triangle consisting of two vertices of the base and the center of the base. This triangle is similar to the corresponding triangle at a height $y$, so $a/b = \alpha/\beta \;\Rightarrow\; \alpha = a\beta/b$. Also by similar triangles, $b/h = \beta/(h-y)$ $\Rightarrow\; \beta = b(h-y)/h$. These two equations imply that $\alpha = a(1-y/h)$, and since the cross-section is an equilateral triangle, it has area

$$A(y) = \frac{1}{2}\cdot\alpha\cdot\frac{\sqrt{3}}{2}\alpha = \frac{a^2(1-y/h)^2}{4}\sqrt{3}, \text{ so}$$

$$V = \int_0^h A(y)\,dy = \frac{a^2\sqrt{3}}{4}\int_0^h\left(1-\frac{y}{h}\right)^2 dy$$

$$= \frac{a^2\sqrt{3}}{4}\left[-\frac{h}{3}\left(1-\frac{y}{h}\right)^3\right]_0^h = -\frac{\sqrt{3}}{12}a^2h(-1) = \frac{\sqrt{3}}{12}a^2h$$

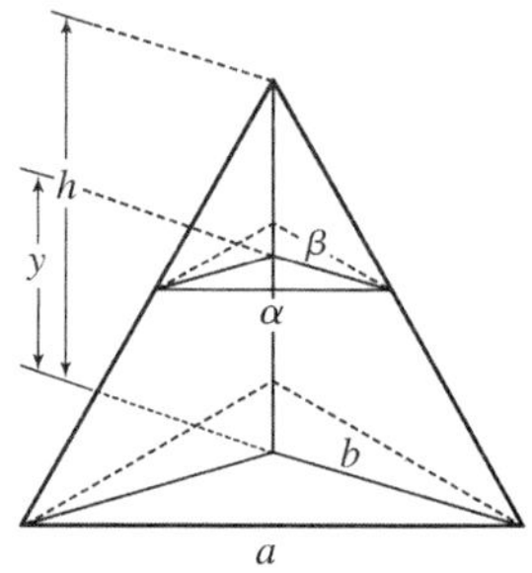


**31.** A cross-section at height $z$ is a triangle similar to the base, so we'll multiply the legs of the base triangle, 3 and 4, by a proportionality factor of $(5-z)/5$. Thus, the triangle at height $z$ has area

$$A(z) = \frac{1}{2}\cdot 3\left(\frac{5-z}{5}\right)\cdot 4\left(\frac{5-z}{5}\right) = 6\left(1-\frac{z}{5}\right)^2, \text{ so}$$

$$V = \textstyle\int_0^5 A(z)\,dz = 6\int_0^5 (1-z/5)^2\,dz$$

$$= 6\textstyle\int_1^0 u^2(-5\,du) \quad \left[u = 1 - z/5,\ du = -\frac{1}{5}dz\right]$$

$$= -30\left[\tfrac{1}{3}u^3\right]_1^0 = -30\left(-\tfrac{1}{3}\right) = 10\text{ cm}^3$$

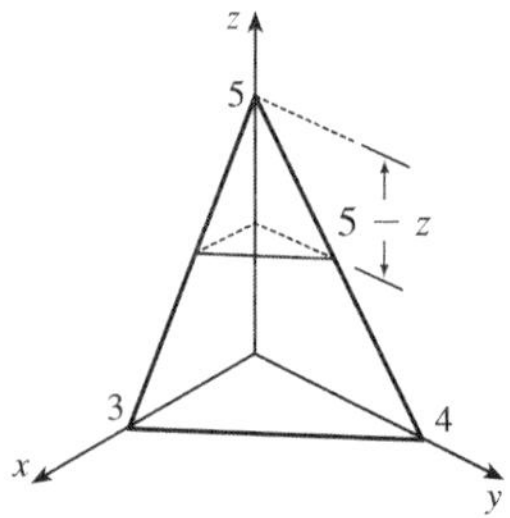


**32.** A cross-section is shaded in the diagram.

$$A(x) = (2y)^2 = \left(2\sqrt{r^2-x^2}\right)^2, \text{ so}$$

$$V = \textstyle\int_{-r}^r A(x)\,dx = 2\int_0^r 4(r^2-x^2)\,dx$$

$$= 8\left[r^2x - \tfrac{1}{3}x^3\right]_0^r$$

$$= 8\left(\tfrac{2}{3}r^3\right) = \tfrac{16}{3}r^3$$

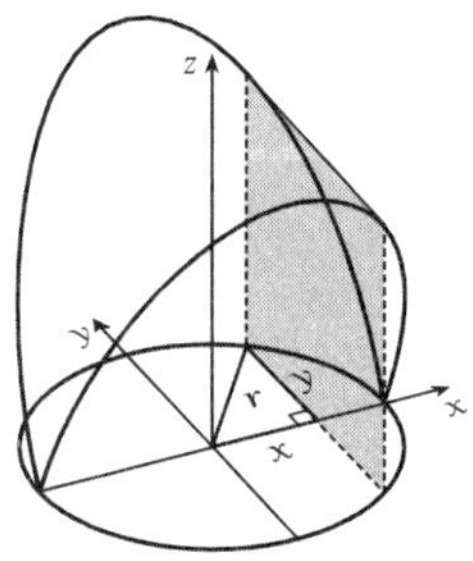


**33.** If $l$ is a leg of the isosceles right triangle and $2y$ is the hypotenuse, then $l^2 + l^2 = (2y)^2 \;\Rightarrow\; 2l^2 = 4y^2 \;\Rightarrow\; l^2 = 2y^2$.

$$V = \textstyle\int_{-2}^2 A(x)\,dx = 2\int_0^2 A(x)\,dx = 2\int_0^2 \frac{1}{2}(l)(l)\,dx = 2\int_0^2 y^2\,dx$$

$$= 2\textstyle\int_0^2 \frac{1}{4}(36-9x^2)\,dx = \frac{9}{2}\int_0^2 (4-x^2)\,dx$$

$$= \tfrac{9}{2}\left[4x - \tfrac{1}{3}x^3\right]_0^2 = \tfrac{9}{2}\left(8 - \tfrac{8}{3}\right) = 24$$

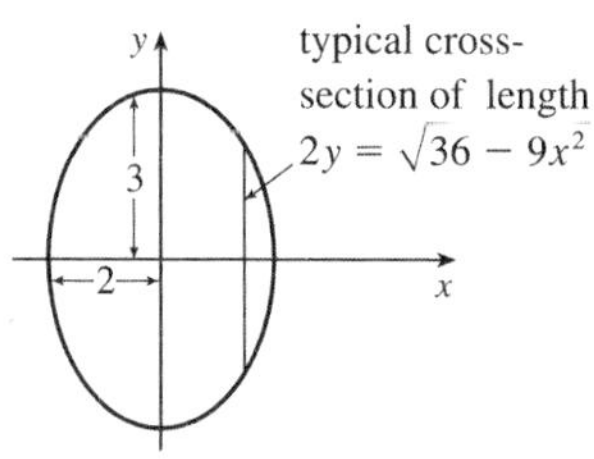

**34.** The cross-section of the base corresponding to the coordinate $y$ has length $2x = 2\sqrt{y}$. The corresponding equilateral triangle with side $s$ has area $A(y) = s^2\left(\frac{\sqrt{3}}{4}\right) = (2x)^2\left(\frac{\sqrt{3}}{4}\right) = \left(2\sqrt{y}\right)^2\left(\frac{\sqrt{3}}{4}\right) = y\sqrt{3}$. Therefore, $V = \int_0^1 A(y)\,dy = \int_0^1 y\sqrt{3}\,dy = \sqrt{3}\left[\frac{1}{2}y^2\right]_0^1 = \frac{\sqrt{3}}{2}$.

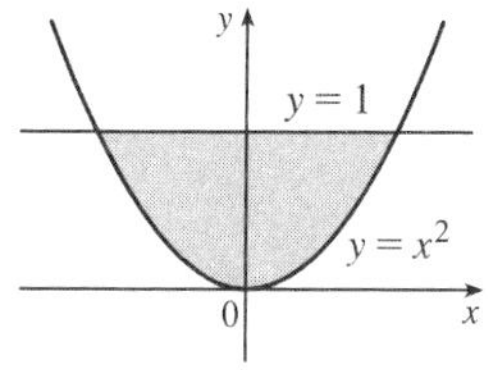


**35.** The cross-section of the base corresponding to the coordinate $y$ has length $2x = 2\sqrt{y}$. The square has area $A(y) = \left(2\sqrt{y}\right)^2 = 4y$, so $V = \int_0^1 A(y)\,dy = \int_0^1 4y\,dy = \left[2y^2\right]_0^1 = 2$.

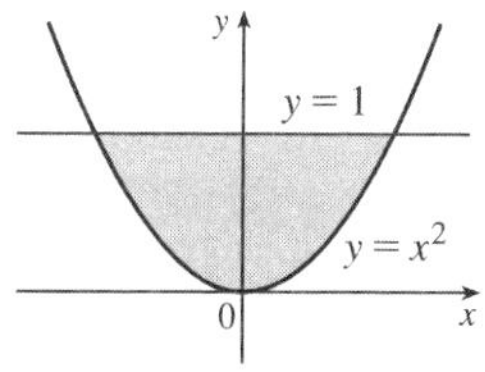


**36.** A typical cross-section perpendicular to the $y$-axis in the base has length $\ell(y) = 3 - \frac{3}{2}y$. This length is the diameter of a cross-sectional semicircle in $S$, so

$$V = \int_0^2 A(y)\,dy = \int_0^2 \frac{\pi}{2}\left[\frac{\ell(y)}{2}\right]^2 dy = \frac{\pi}{8}\int_0^2 \left(3 - \tfrac{3}{2}y\right)^2 dy$$

$$= \tfrac{\pi}{8}\int_3^0 u^2\left(-\tfrac{2}{3}\,du\right) \quad \left[u = 3 - \tfrac{3}{2}y,\ du = -\tfrac{3}{2}\,dy\right]$$

$$= -\tfrac{\pi}{12}\left[\tfrac{1}{3}u^3\right]_3^0 = -\tfrac{\pi}{12}(-9) = \tfrac{3\pi}{4}$$

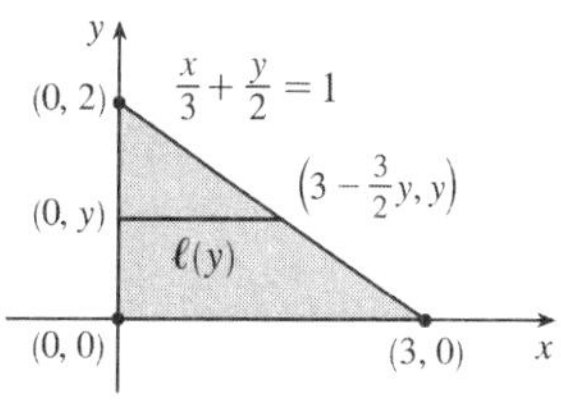


**37.** A typical cross-section perpendicular to the $y$-axis in the base has length $\ell(y) = 3 - \frac{3}{2}y$. This length is the leg of an isosceles right triangle, so

$$A(y) = \tfrac{1}{2}\left[\ell(y)\right]^2 \quad \left[\tfrac{1}{2}bh \text{ with base} = \text{height}\right]$$

$$= \tfrac{1}{2}\left[3\left(1 - \tfrac{1}{2}y\right)\right]^2 = \tfrac{9}{2}\left(1 - \tfrac{1}{2}y\right)^2$$

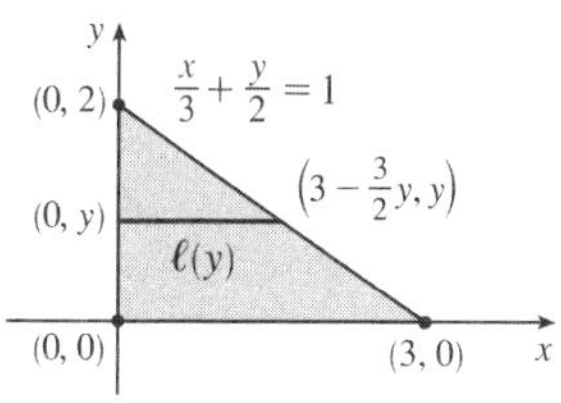


Thus,

$$V = \int_0^2 A(y)\,dy = \tfrac{9}{2}\int_1^0 u^2(-2\,du) \qquad \left[u = 1 - \tfrac{1}{2}y,\ du = -\tfrac{1}{2}\,dy\right]$$

$$= -9\left[\tfrac{1}{3}u^3\right]_1^0 = -9\left(-\tfrac{1}{3}\right) = 3$$

**38.** (a) $$V = \int_{-r}^{r} A(x)\,dx = 2\int_0^r A(x)\,dx = 2\int_0^r \tfrac{1}{2}h\left(2\sqrt{r^2 - x^2}\right)dx$$

$$= 2h\int_0^r \sqrt{r^2 - x^2}\,dx$$

(b) Observe that the integral represents one quarter of the area of a circle of radius $r$, so $V = 2h \cdot \frac{1}{4}\pi r^2 = \frac{1}{2}\pi h r^2$.

**39.** (a) The torus is obtained by rotating the circle $(x - R)^2 + y^2 = r^2$ about the $y$-axis. Solving for $x$, we see that the right half of the circle is given by $x = R + \sqrt{r^2 - y^2} = f(y)$ and the left half by $x = R - \sqrt{r^2 - y^2} = g(y)$. So

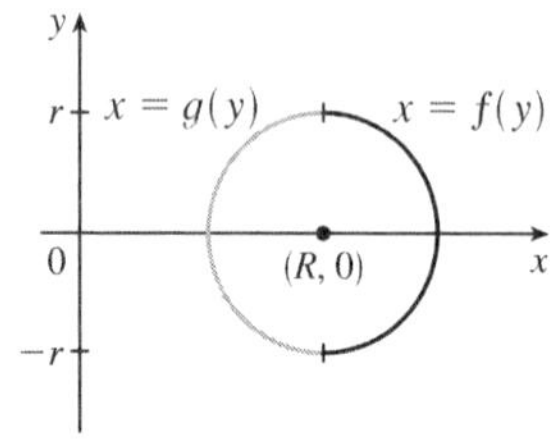


$$V = \pi \int_{-r}^{r} \{[f(y)]^2 - [g(y)]^2\}\, dy$$
$$= 2\pi \int_0^r \left[\left(R^2 + 2R\sqrt{r^2 - y^2} + r^2 - y^2\right) - \left(R^2 - 2R\sqrt{r^2 - y^2} + r^2 - y^2\right)\right] dy$$
$$= 2\pi \int_0^r 4R\sqrt{r^2 - y^2}\, dy = 8\pi R \int_0^r \sqrt{r^2 - y^2}\, dy$$

(b) Observe that the integral represents a quarter of the area of a circle with radius $r$, so

$$8\pi R \int_0^r \sqrt{r^2 - y^2}\, dy = 8\pi R \cdot \tfrac{1}{4}\pi r^2 = 2\pi^2 r^2 R.$$

**40.**

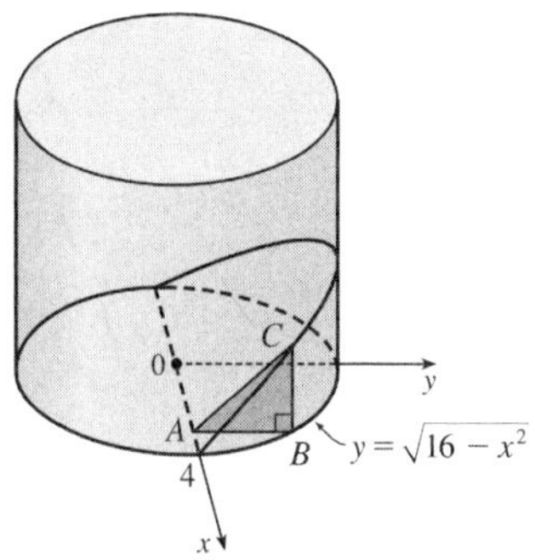


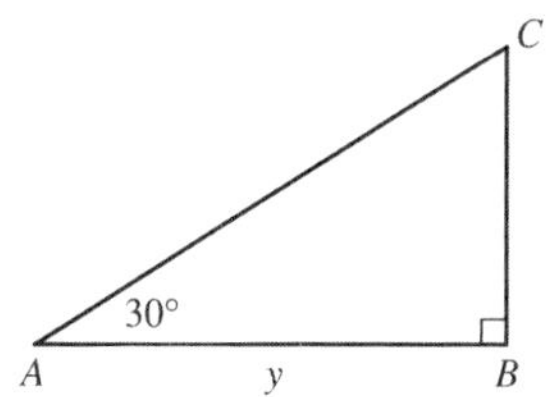


If we place the $x$-axis along the diameter where the planes meet, then the base of the solid is a semicircle with equation $y = \sqrt{16 - x^2}$, $-4 \le x \le 4$. A cross-section perpendicular to the $x$-axis at a distance $x$ from the origin is a triangle $ABC$, as shown in the figure, whose base is $y = \sqrt{16 - x^2}$ and whose height is $|BC| = y \tan 30^\circ = \sqrt{16 - x^2}/\sqrt{3}$. Thus, the cross-sectional area is $A(x) = \frac{1}{2}\sqrt{16 - x^2} \cdot \dfrac{1}{\sqrt{3}}\sqrt{16 - x^2} = \dfrac{16 - x^2}{2\sqrt{3}}$ and the volume is

$$V = \int_{-4}^{4} A(x)\, dx = \int_{-4}^{4} \frac{16 - x^2}{2\sqrt{3}}\, dx = \frac{1}{\sqrt{3}} \int_0^4 (16 - x^2)\, dx = \frac{1}{\sqrt{3}} \left[16x - \frac{1}{3}x^3\right]_0^4 = \frac{128}{3\sqrt{3}}$$

*Another method:* The cross-sections perpendicular to the $y$-axis in the figure are rectangles. The rectangle corresponding to the coordinate $y$ has a base of length $2\sqrt{16 - y^2}$ in the $xy$-plane and a height of $\frac{1}{\sqrt{3}}y$, since $\angle BAC = 30^\circ$ and $|BC| = \frac{1}{\sqrt{3}}|AB|$. Thus, $A(y) = \frac{2}{\sqrt{3}}\, y\sqrt{16 - y^2}$ and

$$V = \int_0^4 A(y)\, dy = \tfrac{2}{\sqrt{3}} \int_0^4 \sqrt{16 - y^2}\, y\, dy = \tfrac{2}{\sqrt{3}} \int_{16}^0 u^{1/2}\left(-\tfrac{1}{2}\, du\right) \qquad \begin{bmatrix} u = 16 - y^2, \\ du = -2y\, dy \end{bmatrix}$$
$$= \tfrac{1}{\sqrt{3}} \int_0^{16} u^{1/2}\, du = \tfrac{1}{\sqrt{3}}\tfrac{2}{3}\left[u^{3/2}\right]_0^{16} = \tfrac{2}{3\sqrt{3}}(64) = \tfrac{128}{3\sqrt{3}}$$

**41.** (a) Volume$(S_1) = \int_0^h A(z)\,dz =$ Volume$(S_2)$ since the cross-sectional area $A(z)$ at height $z$ is the same for both solids.

(b) By Cavalieri's Principle, the volume of the cylinder in the figure is the same as that of a right circular cylinder with radius $r$ and height $h$, that is, $\pi r^2 h$.

**42.** Each cross-section of the solid $S$ in a plane perpendicular to the $x$-axis is a square (since the edges of the cut lie on the cylinders, which are perpendicular). One-quarter of this square and one-eighth of $S$ are shown. The area of this quarter-square is $|PQ|^2 = r^2 - x^2$. Therefore, $A(x) = 4(r^2 - x^2)$ and the volume of $S$ is

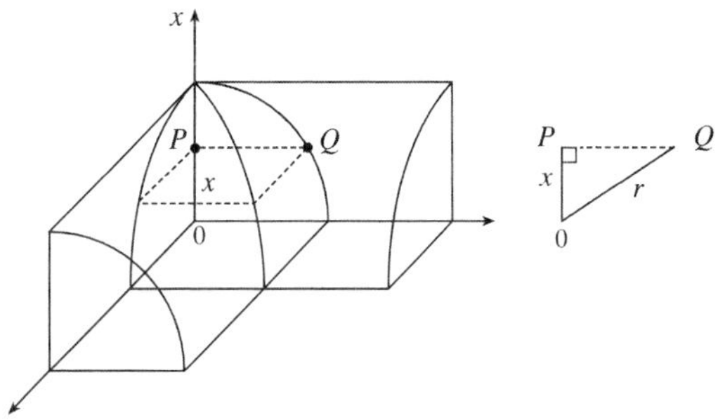


$$V = \int_{-r}^{r} A(x)\,dx = 4\int_{-r}^{r}(r^2 - x^2)\,dx$$
$$= 8\int_0^r (r^2 - x^2)\,dx = 8\left[r^2x - \tfrac{1}{3}x^3\right]_0^r = \tfrac{16}{3}r^3$$

**43.** The volume is obtained by rotating the area common to two circles of radius $r$, as shown. The volume of the right half is

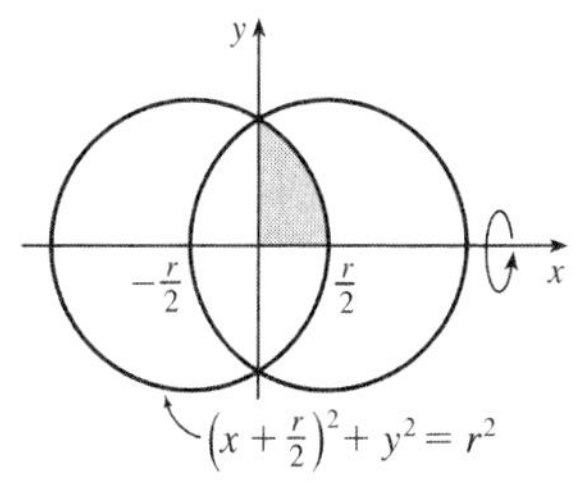


$$V_{\text{right}} = \pi\int_0^{r/2} y^2\,dx = \pi\int_0^{r/2}\left[r^2 - \left(\tfrac{1}{2}r + x\right)^2\right]dx$$
$$= \pi\left[r^2x - \tfrac{1}{3}\left(\tfrac{1}{2}r + x\right)^3\right]_0^{r/2} = \pi\left[\left(\tfrac{1}{2}r^3 - \tfrac{1}{3}r^3\right) - \left(0 - \tfrac{1}{24}r^3\right)\right] = \tfrac{5}{24}\pi r^3$$

So by symmetry, the total volume is twice this, or $\frac{5}{12}\pi r^3$.

*Another solution:* We observe that the volume is the twice the volume of a cap of a sphere, so we can use the formula from Exercise 27 with $h = \frac{1}{2}r$: $V = 2\cdot\frac{1}{3}\pi h^2(3r - h) = \frac{2}{3}\pi\left(\frac{1}{2}r\right)^2\left(3r - \frac{1}{2}r\right) = \frac{5}{12}\pi r^3$.

**44.** We consider two cases: one in which the ball is not completely submerged and the other in which it is.

*Case 1:* $0 \le h \le 10$ The ball will not be completely submerged, and so a cross-section of the water parallel to the surface will be the shaded area shown in the first diagram. We can find the area of the cross-section at height $x$ above the bottom of the bowl by using the Pythagorean Theorem: $R^2 = 15^2 - (15 - x)^2$ and $r^2 = 5^2 - (x - 5)^2$, so $A(x) = \pi(R^2 - r^2) = 20\pi x$. The volume of water when it has depth $h$ is then $V(h) = \int_0^h A(x)\,dx = \int_0^h 20\pi x\,dx = \left[10\pi x^2\right]_0^h = 10\pi h^2$ cm$^3$, $0 \le h \le 10$.

*Case 2:* $10 < h \le 15$ In this case we can find the volume by simply subtracting the volume displaced by the ball from the total volume inside the bowl underneath the surface of the water. The total volume underneath the surface is just the volume of a cap of the bowl, so we use the formula from Exercise 27: $V_{\text{cap}}(h) = \frac{1}{3}\pi h^2(45 - h)$. The volume of the small sphere is $V_{\text{ball}} = \frac{4}{3}\pi(5)^3 = \frac{500}{3}\pi$, so the total volume is $V_{\text{cap}} - V_{\text{ball}} = \frac{1}{3}\pi(45h^2 - h^3 - 500)$ cm$^3$.

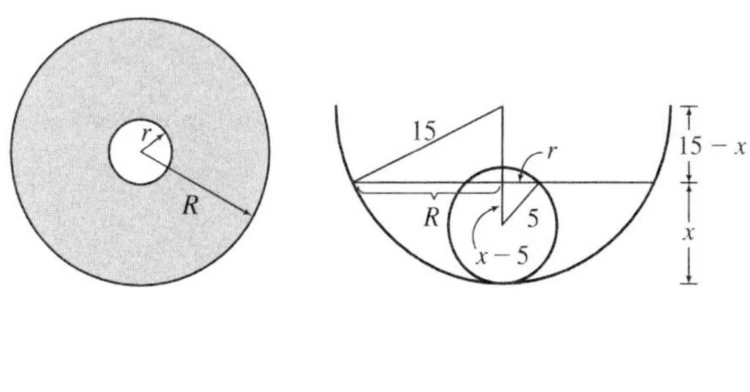

**45.** Take the $x$-axis to be the axis of the cylindrical hole of radius $r$.

A quarter of the cross-section through $y$, perpendicular to the $y$-axis, is the rectangle shown. Using the Pythagorean Theorem twice, we see that the dimensions of this rectangle are

$x = \sqrt{R^2 - y^2}$ and $z = \sqrt{r^2 - y^2}$, so

$\frac{1}{4}A(y) = xz = \sqrt{r^2 - y^2}\,\sqrt{R^2 - y^2}$, and

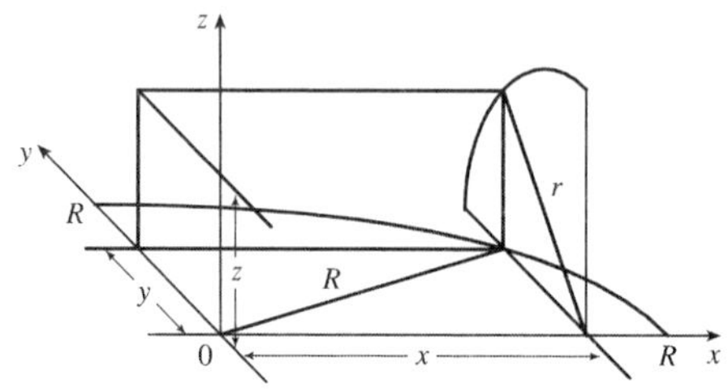


$$V = \int_{-r}^{r} A(y)\,dy = \int_{-r}^{r} 4\sqrt{r^2 - y^2}\,\sqrt{R^2 - y^2}\,dy = 8\int_0^r \sqrt{r^2 - y^2}\,\sqrt{R^2 - y^2}\,dy$$

**46.** The line $y = r$ intersects the semicircle $y = \sqrt{R^2 - x^2}$ when $r = \sqrt{R^2 - x^2} \;\Rightarrow\; r^2 = R^2 - x^2 \;\Rightarrow$ $x^2 = R^2 - r^2 \;\Rightarrow\; x = \pm\sqrt{R^2 - r^2}$. Rotating the shaded region about the $x$-axis gives us

$$V = \int_{-\sqrt{R^2-r^2}}^{\sqrt{R^2-r^2}} \pi\left[\left(\sqrt{R^2 - x^2}\right)^2 - r^2\right] dx = 2\pi\int_0^{\sqrt{R^2-r^2}} \left(R^2 - x^2 - r^2\right) dx \qquad \text{[by symmetry]}$$

$$= 2\pi\int_0^{\sqrt{R^2-r^2}} \left[\left(R^2 - r^2\right) - x^2\right] dx = 2\pi\left[\left(R^2 - r^2\right)x - \tfrac{1}{3}x^3\right]_0^{\sqrt{R^2-r^2}}$$

$$= 2\pi\left[\left(R^2 - r^2\right)^{3/2} - \tfrac{1}{3}\left(R^2 - r^2\right)^{3/2}\right] = 2\pi\cdot\tfrac{2}{3}\left(R^2 - r^2\right)^{3/2} = \tfrac{4\pi}{3}\left(R^2 - r^2\right)^{3/2}$$

Our answer makes sense in limiting cases. As $r \to 0$, $V \to \frac{4}{3}\pi R^3$, which is the volume of the full sphere. As $r \to R$, $V \to 0$, which makes sense because the hole's radius is approaching that of the sphere

.

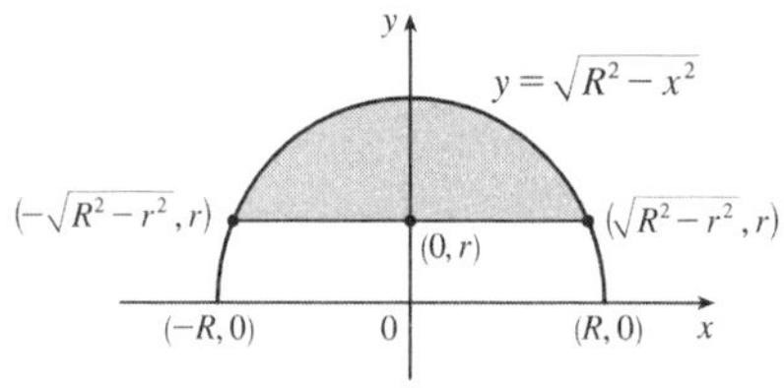


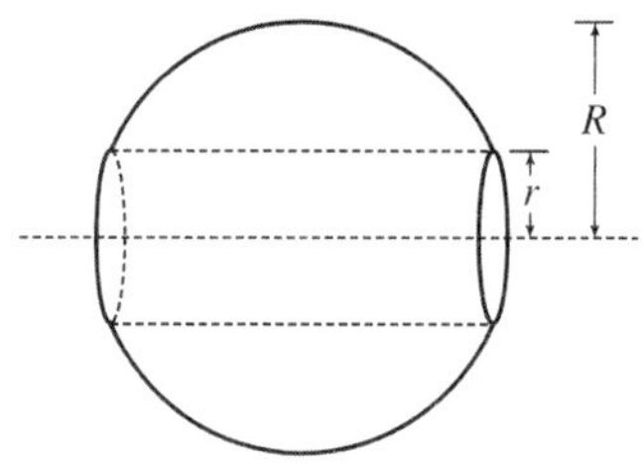


**47.** (a) The radius of the barrel is the same at each end by symmetry, since the function $y = R - cx^2$ is even. Since the barrel is obtained by rotating the graph of the function $y$ about the $x$-axis, this radius is equal to the value of $y$ at $x = \frac{1}{2}h$, which is $R - c\left(\frac{1}{2}h\right)^2 = R - d = r$.

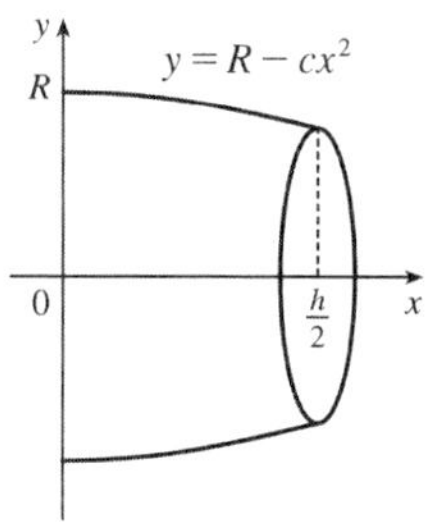


(b) The barrel is symmetric about the $y$-axis, so its volume is twice the volume of that part of the barrel for $x > 0$. Also, the barrel is a volume of rotation, so

$$V = 2\int_0^{h/2} \pi y^2\,dx = 2\pi\int_0^{h/2} \left(R - cx^2\right)^2 dx = 2\pi\left[R^2x - \tfrac{2}{3}Rcx^3 + \tfrac{1}{5}c^2x^5\right]_0^{h/2}$$

$$= 2\pi\left(\tfrac{1}{2}R^2h - \tfrac{1}{12}Rch^3 + \tfrac{1}{160}c^2h^5\right)$$

Trying to make this look more like the expression we want, we rewrite it as $V = \frac{1}{3}\pi h\left[2R^2 + \left(R^2 - \frac{1}{2}Rch^2 + \frac{3}{80}c^2h^4\right)\right]$.

But $R^2 - \frac{1}{2}Rch^2 + \frac{3}{80}c^2h^4 = \left(R - \frac{1}{4}ch^2\right)^2 - \frac{1}{40}c^2h^4 = (R - d)^2 - \frac{2}{5}\left(\frac{1}{4}ch^2\right)^2 = r^2 - \frac{2}{5}d^2$.

Substituting this back into $V$, we see that $V = \frac{1}{3}\pi h\left(2R^2 + r^2 - \frac{2}{5}d^2\right)$, as required.

**48.** (a) $V = \int_{-1}^{1} \pi \left[ (ax^3 + bx^2 + cx + d) \sqrt{1 - x^2} \right]^2 dx \overset{\text{CAS}}{=} \dfrac{4\{5a^2 + 18ac + 3[3b^2 + 14bd + 7(c^2 + 5d^2)]\}\,\pi}{315}$

(b) $y = (-0.06x^3 + 0.04x^2 + 0.1x + 0.54)\sqrt{1 - x^2}$ is graphed in the figure. Substitute $a = -0.06$, $b = 0.04$, $c = 0.1$, and $d = 0.54$ in the answer for part (a) to get $V \overset{\text{CAS}}{=} \dfrac{3769\pi}{9375} \approx 1.263$.

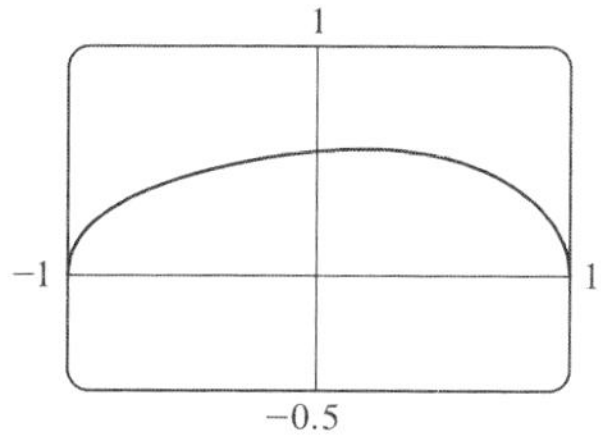


**49.**

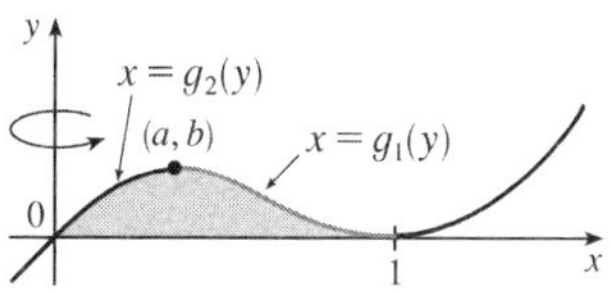


If we were to use the "washer method," we would first have to locate the local maximum point $(a, b)$ of $y = x(x - 1)^2$ using the methods of Chapter 4. Then we would have to solve the equation $y = x(x - 1)^2$ for $x$ in terms of $y$ to obtain the functions $x = g_1(y)$ and $x = g_2(y)$ shown in the figure above. This step would be difficult because it involves the cubic formula. Finally we would find the volume using $V = \pi \int_0^b \{[g_1(y)]^2 - [g_2(y)]^2\}\, dy$.

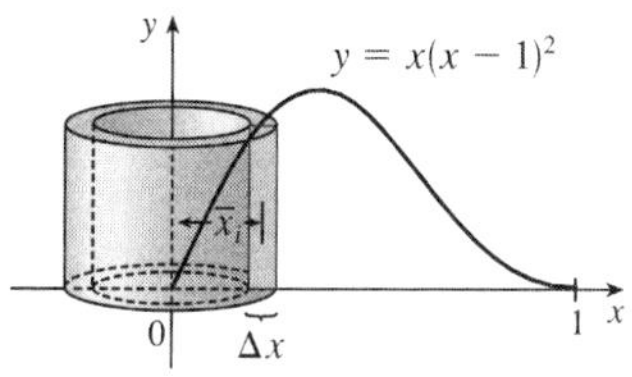


Instead, we use cylindrical shells. As in Example 9, we rotate an approximating rectangle with width $\Delta x$ about the $y$-axis, to get a cylindrical shell whose average radius is $\overline{x}_i$ and whose volume is $2\pi \overline{x}_i \left[ \overline{x}_i (\overline{x}_i - 1)^2 \right] \Delta x$. So the total volume is

$$V = \lim_{n \to \infty} \sum_{i=1}^{n} 2\pi\, \overline{x}_i \left[ \overline{x}_i (\overline{x}_i - 1)^2 \right] \Delta x = \int_0^1 2\pi x \left[ x(x - 1)^2 \right] dx$$

$$= 2\pi \int_0^1 (x^4 - 2x^3 + x^2)\, dx = 2\pi \left[ \frac{x^5}{5} - 2\frac{x^4}{4} + \frac{x^3}{3} \right]_0^1$$

$$= 2\pi \left( \frac{1}{5} - \frac{1}{2} + \frac{1}{3} \right) = 2\pi \left( \frac{1}{30} \right) = \frac{\pi}{15}$$

**50.**

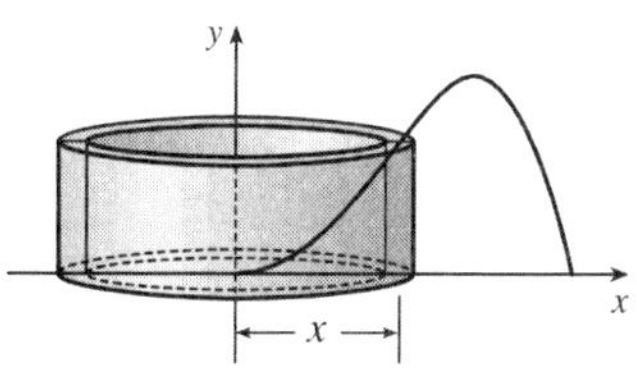


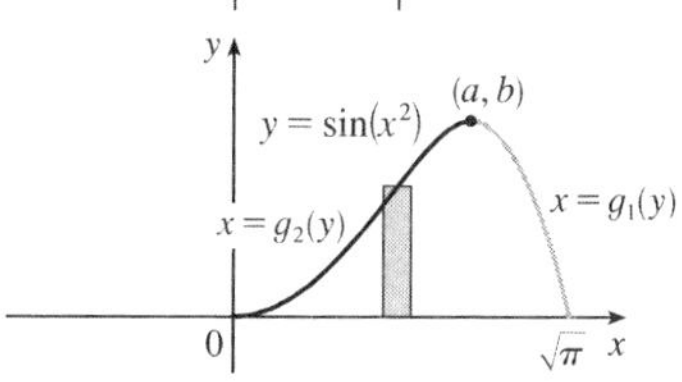


A typical cylindrical shell has circumference $2\pi x$ and height $\sin(x^2)$. $V = \int_0^{\sqrt{\pi}} 2\pi x \sin(x^2)\, dx$. Let $u = x^2$. Then $du = 2x\, dx$, so $V = \pi \int_0^{\pi} \sin u\, du = \pi[-\cos u]_0^{\pi} = \pi[1 - (-1)] = 2\pi$.

For slicing, we would first have to locate the local maximum point $(a, b)$ of $y = \sin(x^2)$ using the methods of Chapter 4. Then we would have to solve the equation $y = \sin(x^2)$ for $x$ in terms of $y$ to obtain the functions $x = g_1(y)$ and $x = g_2(y)$ shown in the second figure. Finally we would find the volume using $V = \pi \int_0^b \{[g_1(y)]^2 - [g_2(y)]^2\}\, dy$. Using shells is definitely preferable to slicing.

**51.** (a) Let $y = f(x)$ denote the curve. Using cylindrical shells, $V = \int_2^{10} 2\pi x f(x)\,dx = 2\pi \int_2^{10} x f(x)\,dx = 2\pi I_1$. Now use Simpson's Rule to approximate $I_1$:

$$\begin{aligned} I_1 \approx S_8 &= \tfrac{10-2}{3(8)}\,[2f(2) + 4\cdot 3f(3) + 2\cdot 4f(4) + 4\cdot 5f(5) + 2\cdot 6f(6) \\ &\qquad + 4\cdot 7f(7) + 2\cdot 8f(8) + 4\cdot 9f(9) + 10f(10)] \\ &\approx \tfrac{1}{3}[2(0) + 12(1.5) + 8(1.9) + 20(2.2) + 12(3.0) + 28(3.8) + 16(4.0) + 36(3.1) + 10(0)] \\ &= \tfrac{1}{3}(395.2) \end{aligned}$$

Thus, $V \approx 2\pi \cdot \frac{1}{3}(395.2) \approx 827.7$ or 828 cubic units.

(b) Using disks, $V = \int_2^{10} \pi[f(x)]^2\,dx = \pi \int_2^{10} [f(x)]^2\,dx = \pi I_2$. Now use Simpson's Rule to approximate $I_2$:

$$\begin{aligned} I_2 \approx S_8 &= \tfrac{10-2}{3(8)}\,\{[f(2)]^2 + 4\,[f(3)]^2 + 2\,[f(4)]^2 + 4\,[f(5)]^2 + 2\,[f(6)]^2 \\ &\qquad + 4\,[f(7)]^2 + 2\,[f(8)]^2 + 4\,[f(9)]^2 + [f(10)]^2\} \\ &\approx \tfrac{1}{3}\,[(0)^2 + 4(1.5)^2 + 2(1.9)^2 + 4(2.2)^2 + 2(3.0)^2 + 4(3.8)^2 + 2(4.0)^2 + 4(3.1)^2 + (0)^2] \\ &= \tfrac{1}{3}(181.78) \end{aligned}$$

Thus, $V \approx \pi \cdot \frac{1}{3}(181.78) \approx 190.4$ or 190 cubic units.

**52.**

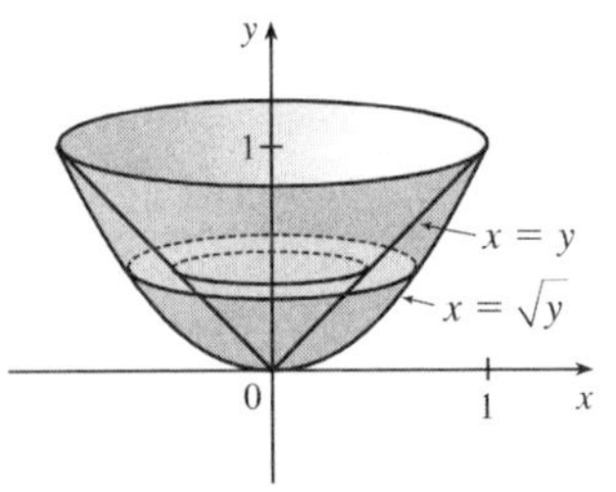


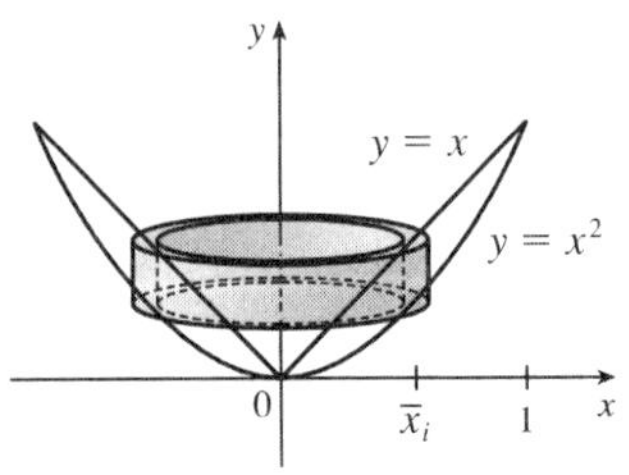


The first figure shows a cross-section perpendicular to the $y$-axis. It is a washer with inner radius $y$ and outer radius $\sqrt{y}$, so the volume by slicing is

$$V = \int_0^1 A(y)\,dy = \pi \int_0^1 \left[(\sqrt{y}\,)^2 - y^2\right] dy$$

$$= \pi \int_0^1 (y - y^2)\,dy = \pi\left[\frac{y^2}{2} - \frac{y^3}{3}\right]_0^1 = \frac{\pi}{6}$$

Next, we use cylindrical shells to find the volume. The height of a shell is $\overline{x}_i - \overline{x}_i^2$, so its volume is $2\pi\,\overline{x}_i\left(\overline{x}_i - \overline{x}_i^2\right)\Delta x$. Thus,

$$V = \lim_{n\to\infty} \sum_{i=1}^{n} 2\pi\,\overline{x}_i\left(\overline{x}_i - \overline{x}_i^2\right)\Delta x = \int_0^1 2\pi x\left(x - x^2\right)dx$$

$$= 2\pi \int_0^1 \left(x^2 - x^3\right)dx = 2\pi\left[\frac{x^3}{3} - \frac{x^4}{4}\right]_0^1 = 2\pi\left(\frac{1}{12}\right) = \frac{\pi}{6}$$

**53.**

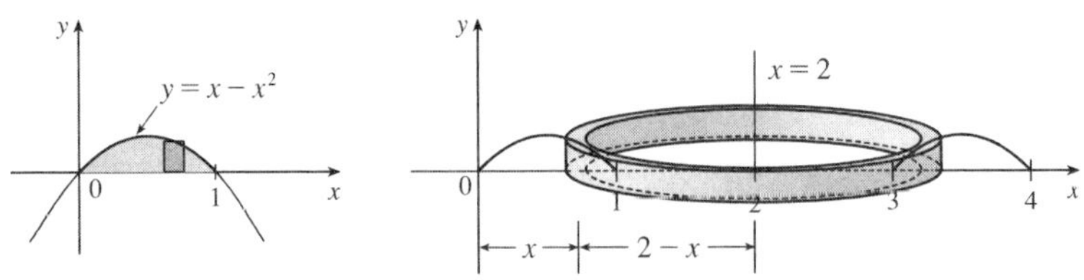


$$V = \int_0^1 (\text{circumference})\,(\text{height})\,(\text{thickness}) = \int_0^1 \left[2\pi(2 - x)\right]\left(x - x^2\right)dx$$

$$= 2\pi \int_0^1 \left(x^3 - 3x^2 + 2x\right)dx = 2\pi\left[\tfrac{1}{4}x^4 - x^3 + x^2\right]_0^1 = 2\pi\left(\tfrac{1}{4}\right) = \tfrac{\pi}{2}$$

See the solution for Exercise 49 as to why the method of cylindrical shells is preferable to slicing.

**54.** By symmetry, the volume of a napkin ring obtained by drilling a hole of radius $r$ through a sphere with radius $R$ is twice the volume obtained by rotating the area above the $x$-axis and below the curve $y = \sqrt{R^2 - x^2}$ (the equation of the top half of the cross-section of the sphere), between $x = r$ and $x = R$, about the $y$-axis. This volume is equal to

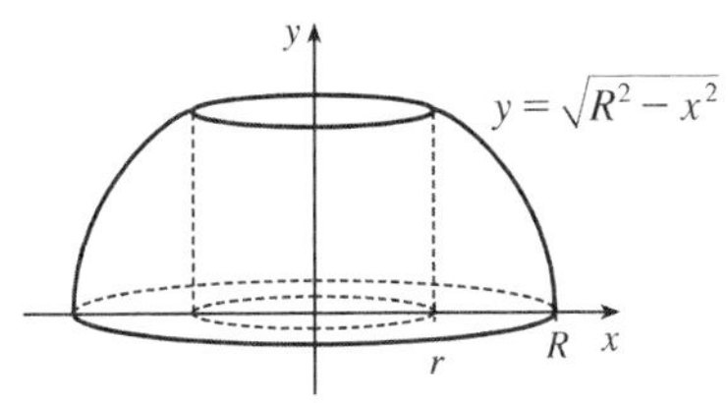


$$2\int_{\text{inner radius}}^{\text{outer radius}} 2\pi r h\,dx = 2\cdot 2\pi\int_r^R x\sqrt{R^2-x^2}\,dx = 4\pi\left[-\tfrac{1}{3}\left(R^2-x^2\right)^{3/2}\right]_r^R = \tfrac{4}{3}\pi\left(R^2-r^2\right)^{3/2}$$

But by the Pythagorean Theorem, $R^2 - r^2 = \left(\frac{1}{2}h\right)^2$, so the volume of the napkin ring is $\frac{4}{3}\pi\left(\frac{1}{2}h\right)^3 = \frac{1}{6}\pi h^3$, which is independent of both $R$ and $r$; that is, the amount of wood in a napkin ring of height $h$ is the same regardless of the size of the sphere used. Note that most of this calculation has been done already, but with more difficulty, in Exercise 46.

*Another solution:* The height of the missing cap is the radius of the sphere minus half the height of the cut-out cylinder, that is, $R - \frac{1}{2}h$. Using Exercise 27,

$$V_{\text{napkin ring}} = V_{\text{sphere}} - V_{\text{cylinder}} - 2V_{\text{cap}} = \tfrac{4}{3}\pi R^3 - \pi r^2 h - 2\cdot\tfrac{\pi}{3}\left(R-\tfrac{1}{2}h\right)^2\left[3R - \left(R-\tfrac{1}{2}h\right)\right] = \tfrac{1}{6}\pi h^3$$

## DISCOVERY PROJECT Rotating on a Slant

**1.**

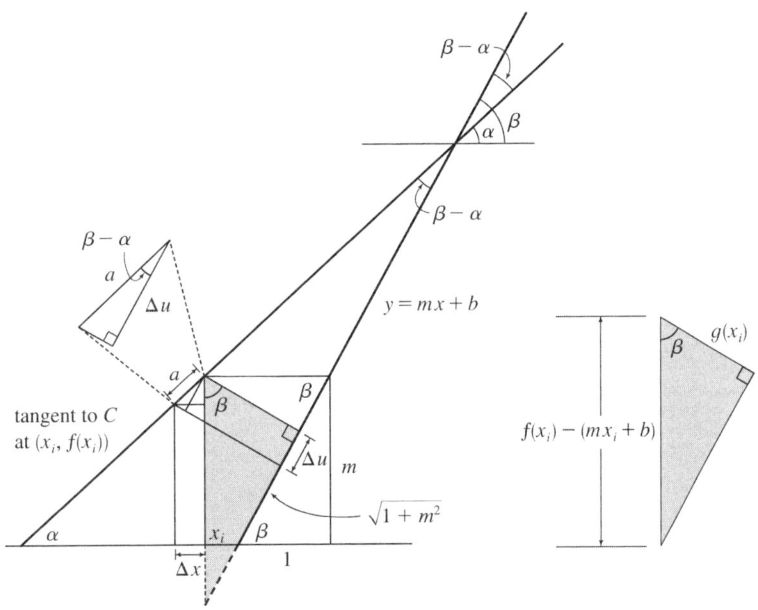


In the figure, the segment $a$ lying above the interval $[x_i - \Delta x, x_i]$ along the tangent to $C$ has length $\Delta x \sec\alpha = \Delta x\sqrt{1+\tan^2\alpha} = \sqrt{1+[f'(x_i)]^2}\,\Delta x$. The segment from $(x_i, f(x_i))$ drawn perpendicular to the line

$y = mx + b$ has length

$$g(x_i) = [f(x_i) - mx_i - b]\cos\beta = \frac{f(x_i) - mx_i - b}{\sec\beta} = \frac{f(x_i) - mx_i - b}{\sqrt{1+\tan^2\beta}} = \frac{f(x_i) - mx_i - b}{\sqrt{1+m^2}}$$

Also, $\cos(\beta - \alpha) = \dfrac{\Delta u}{\Delta x\ \sec\alpha} \quad \Rightarrow$

$$\Delta u = \Delta x\ \sec\alpha\ \cos(\beta-\alpha) = \Delta x\,\frac{\cos\beta\ \cos\alpha + \sin\beta\ \sin\alpha}{\cos\alpha} = \Delta x(\cos\beta + \sin\beta\ \tan\alpha)$$

$$= \Delta x\left[\frac{1}{\sqrt{1+m^2}} + \frac{m}{\sqrt{1+m^2}}f'(x_i)\right] = \frac{1+mf'(x_i)}{\sqrt{1+m^2}}\,\Delta x$$

Thus,

$$\text{Area}(\mathfrak{R}) = \lim_{n\to\infty}\sum_{i=1}^{n} g(x_i)\,\Delta u = \lim_{n\to\infty}\sum_{i=1}^{n}\frac{f(x_i)-mx_i-b}{\sqrt{1+m^2}}\cdot\frac{1+mf'(x_i)}{\sqrt{1+m^2}}\,\Delta x$$

$$= \frac{1}{1+m^2}\int_p^q [f(x) - mx - b]\,[1+mf'(x)]\,dx$$

**2.** From Problem 1 with $m = 1$, $f(x) = x + \sin x$, $mx + b = x - 2$, $p = 0$, and $q = 2\pi$,

$$\text{Area} = \tfrac{1}{1+1^2}\int_0^{2\pi}[x + \sin x - (x-2)]\,[1 + 1(1+\cos x)]\,dx = \tfrac{1}{2}\int_0^{2\pi}(\sin x + 2)(2+\cos x)\,dx$$

$$= \tfrac{1}{2}\int_0^{2\pi}(2\sin x + \sin x\cos x + 4 + 2\cos x)\,dx = \tfrac{1}{2}\left[-2\cos x + \tfrac{1}{2}\sin^2 x + 4x + 2\sin x\right]_0^{2\pi}$$

$$= \tfrac{1}{2}[(-2+0+8\pi+0) - (-2+0+0+0)] = \tfrac{1}{2}(8\pi) = 4\pi$$

**3.** $V = \displaystyle\lim_{n\to\infty}\sum_{i=1}^{n}\pi\,[g(x_i)]^2\,\Delta u = \lim_{n\to\infty}\sum_{i=1}^{n}\pi\left[\frac{f(x_i)-mx_i-b}{\sqrt{1+m^2}}\right]^2\frac{1+mf'(x_i)}{\sqrt{1+m^2}}\,\Delta x$

$$= \frac{\pi}{(1+m^2)^{3/2}}\int_p^q [f(x)-mx-b]^2\,[1+mf'(x)]\,dx$$

**4.** $V = \dfrac{\pi}{(1+1^2)^{3/2}}\displaystyle\int_0^{2\pi}(x+\sin x - x + 2)^2(1+1+\cos x)\,dx$

$$= \frac{\pi}{2\sqrt{2}}\int_0^{2\pi}(\sin x + 2)^2(\cos x + 2)\,dx$$

$$= \tfrac{\pi}{2\sqrt{2}}\int_0^{2\pi}(\sin^2 x + 4\sin x + 4)(\cos x + 2)\,dx$$

$$= \tfrac{\pi}{2\sqrt{2}}\int_0^{2\pi}(\sin^2 x\ \cos x + 4\sin x\ \cos x + 4\cos x + 2\sin^2 x + 8\sin x + 8)\,dx$$

$$= \tfrac{\pi}{2\sqrt{2}}\left[\tfrac{1}{3}\sin^3 x + 2\sin^2 x + 4\sin x + x - \tfrac{1}{2}\sin 2x - 8\cos x + 8x\right]_0^{2\pi} \qquad [\text{since } 2\sin^2 x = 1 - \cos 2x]$$

$$= \tfrac{\pi}{2\sqrt{2}}[(2\pi - 8 + 16\pi) - (-8)] = \tfrac{9\sqrt{2}}{2}\,\pi^2$$

## 6.3 Arc Length

**1.** $y = 2 - 3x \quad \Rightarrow \quad L = \int_{-2}^{1} \sqrt{1 + (dy/dx)^2}\, dx = \int_{-2}^{1} \sqrt{1 + (-3)^2}\, dx = \sqrt{10}\,[1 - (-2)] = 3\sqrt{10}.$

The arc length can be calculated using the distance formula, since the curve is a line segment, so

$$L = [\text{distance from } (-2, 8) \text{ to } (1, -1)] = \sqrt{[1 - (-2)]^2 + [(-1) - 8]^2} = \sqrt{90} = 3\sqrt{10}$$

**2.** (a) $x = \cos t,\ y = \sin t,\ 0 \le t \le 2\pi.\ \left(\dfrac{dx}{dt}\right)^2 + \left(\dfrac{dy}{dt}\right)^2 = (-\sin t)^2 + (\cos t)^2 = \sin^2 t + \cos^2 t = 1.$ So by formula (1),

$L = \int_0^{2\pi} \sqrt{1}\, dt = \big[\, t \,\big]_0^{2\pi} = 2\pi$, as expected.

(b) $x = \sin 2t,\ y = \cos 2t,\ 0 \le t \le 2\pi.\ \left(\dfrac{dx}{dt}\right)^2 + \left(\dfrac{dy}{dt}\right)^2 = (2\cos 2t)^2 + (-2\sin 2t)^2 = 4\cos^2 2t + 4\sin^2 2t = 4.$

$L = \int_0^{2\pi} \sqrt{4}\, dt = 2\big[\, t \,\big]_0^{2\pi} = 2(2\pi) = 4\pi$. The discrepancy results from the fact that the unit circle is traversed twice with this parametrization.

**3.** $y = \cos x \quad \Rightarrow \quad dy/dx = -\sin x \quad \Rightarrow \quad 1 + (dy/dx)^2 = 1 + \sin^2 x.$ So $L = \int_0^{2\pi} \sqrt{1 + \sin^2 x}\, dx.$

**4.** $x = \ln t,\ y = \sqrt{t + 1},\ \ 1 \le t \le 5.\ \ \dfrac{dx}{dt} = \dfrac{1}{t}$ and $\dfrac{dy}{dt} = \dfrac{1}{2\sqrt{t+1}}$, so $\left(\dfrac{dx}{dt}\right)^2 + \left(\dfrac{dy}{dt}\right)^2 = \dfrac{1}{t^2} + \dfrac{1}{4\,(t+1)} = \dfrac{t^2 + 4t + 4}{4t^2\,(t+1)}.$

Thus, $L = \displaystyle\int_a^b \sqrt{\left(\frac{dx}{dt}\right)^2 + \left(\frac{dy}{dt}\right)^2}\, dt = \int_1^5 \sqrt{\frac{t^2 + 4t + 4}{4t^2\,(t+1)}}\, dt = \int_1^5 \sqrt{\frac{(t+2)^2}{(2t)^2\,(t+1)}}\, dt = \int_1^5 \frac{t+2}{2t\sqrt{t+1}}\, dt.$

**5.**

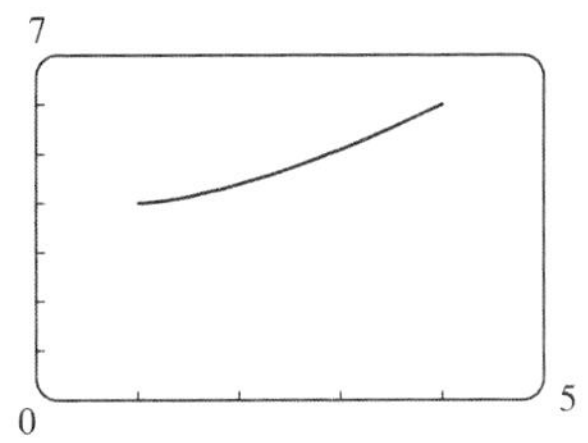


$x = 1 + 3t^2,\quad y = 4 + 2t^3,\quad 0 \le t \le 1.$

$dx/dt = 6t$ and $dy/dt = 6t^2$, so $(dx/dt)^2 + (dy/dt)^2 = 36t^2 + 36t^4.$

Thus, $\quad L = \int_0^1 \sqrt{36t^2 + 36t^4}\, dt$

$= \int_0^1 6t\sqrt{1 + t^2}\, dt = 6\int_1^2 \sqrt{u}\,\left(\tfrac{1}{2}du\right) \quad [u = 1 + t^2,\ du = 2t\,dt]$

$= 3\left[\tfrac{2}{3}u^{3/2}\right]_1^2 = 2(2^{3/2} - 1) = 2\left(2\sqrt{2} - 1\right)$

**6.**

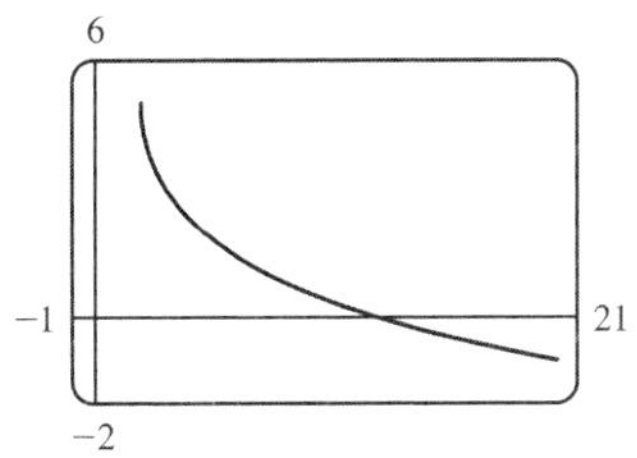


$x = e^t + e^{-t},\ y = 5 - 2t,\ 0 \le t \le 3.$

$dx/dt = e^t - e^{-t}$ and $dy/dt = -2$, so

$\left(\dfrac{dx}{dt}\right)^2 + \left(\dfrac{dy}{dt}\right)^2 = e^{2t} - 2 + e^{-2t} + 4 = e^{2t} + 2 + e^{-2t} = \left(e^t + e^{-t}\right)^2$

and

$L = \int_0^3 \left(e^t + e^{-t}\right) dt = \left[e^t - e^{-t}\right]_0^3 = e^3 - e^{-3} - (1 - 1) = e^3 - e^{-3}.$

**7.**

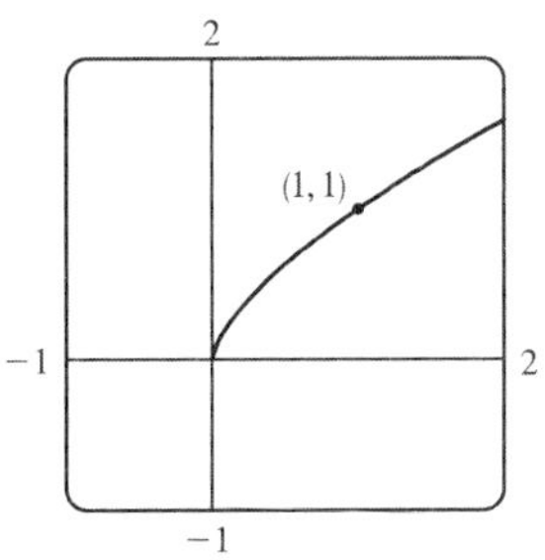


$x = y^{3/2} \quad \Rightarrow \quad 1 + (dx/dy)^2 = 1 + \left(\tfrac{3}{2}y^{1/2}\right)^2 = 1 + \tfrac{9}{4}y.$

$L = \int_0^1 \sqrt{1 + \tfrac{9}{4}y}\, dy = \int_1^{13/4} \sqrt{u}\,\left(\tfrac{4}{9}\,du\right) \quad \left[u = 1 + \tfrac{9}{4}\,y,\ du = \tfrac{9}{4}\,dy\right]$

$= \tfrac{4}{9} \cdot \tfrac{2}{3}\left[u^{3/2}\right]_1^{13/4} = \tfrac{8}{27}\left(\tfrac{13\sqrt{13}}{8} - 1\right) = \tfrac{13\sqrt{13} - 8}{27}.$

**8.** 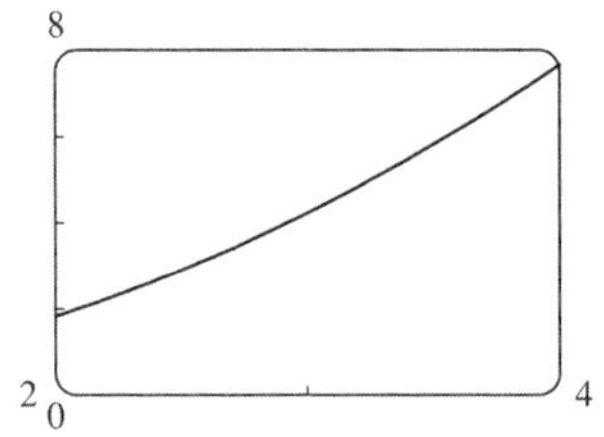


$$y=\frac{x^2}{2}-\frac{\ln x}{4} \Rightarrow \frac{dy}{dx}=x-\frac{1}{4x} \Rightarrow$$

$$1+\left(\frac{dy}{dx}\right)^2=x^2+\frac{1}{2}+\frac{1}{16x^2}. \text{ So}$$

$$L=\int_2^4\left(x+\frac{1}{4x}\right)dx=\left[\frac{x^2}{2}+\frac{\ln x}{4}\right]_2^4=\left(8+\frac{2\ln 2}{4}\right)-\left(2+\frac{\ln 2}{4}\right)$$

$$=6+\frac{\ln 2}{4}$$

**9.** 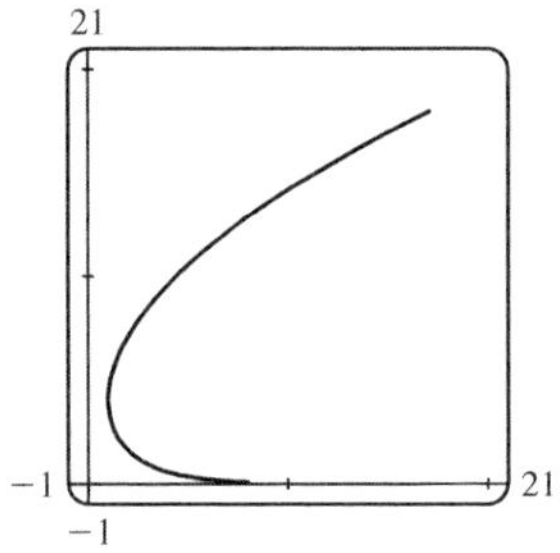


$x=e^t-t,\ y=4e^{t/2},\ -8\le t\le 3$

$$(dx/dt)^2+(dy/dt)^2=(e^t-1)^2+(2e^{t/2})^2=e^{2t}-2e^t+1+4e^t$$

$$=e^{2t}+2e^t+1=(e^t+1)^2$$

$$L=\textstyle\int_{-8}^3\sqrt{(e^t+1)^2}\,dt=\int_{-8}^3(e^t+1)\,dt=\left[e^t+t\right]_{-8}^{3t}$$

$$=(e^3+3)-(e^{-8}-8)=e^3-e^{-8}+11$$

**10.** 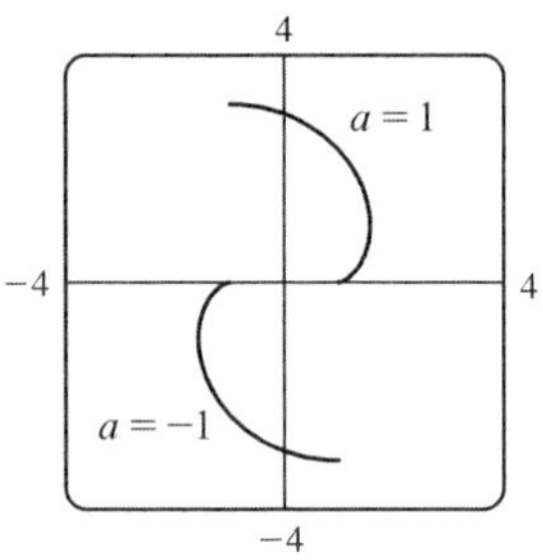


$x=a(\cos\theta+\theta\sin\theta),\ y=a(\sin\theta-\theta\cos\theta),\ 0\le\theta\le\pi.$

$$\left(\frac{dx}{d\theta}\right)^2+\left(\frac{dy}{d\theta}\right)^2$$

$$=a^2[(-\sin\theta+\theta\cos\theta+\sin\theta)^2+(\cos\theta+\theta\sin\theta-\cos\theta)^2]$$

$$=a^2\theta^2(\cos^2\theta+\sin^2\theta)=(a\theta)^2$$

$$L=\textstyle\int_0^\pi a\theta\,d\theta=a\left[\frac{1}{2}\theta^2\right]_0^\pi=\frac{1}{2}\pi^2 a$$

**11.** $y=xe^{-x} \Rightarrow dy/dx=e^{-x}-xe^{-x}=e^{-x}(1-x) \Rightarrow 1+(dy/dx)^2=1+e^{-2x}(1-x)^2$. Let $f(x)=\sqrt{1+(dy/dx)^2}=\sqrt{1+e^{-2x}(1-x)^2}$. Then $L=\int_0^5 f(x)\,dx$. Since $n=10$, $\Delta x=\frac{5-0}{10}=\frac{1}{2}$. Now

$$L\approx S_{10}=\tfrac{1/2}{3}[f(0)+4f(\tfrac{1}{2})+2f(1)+4f(\tfrac{3}{2})+2f(2)+4f(\tfrac{5}{2})+2f(3)$$

$$+4f(\tfrac{7}{2})+2f(4)+4f(\tfrac{9}{2})+f(5)]\approx 5.115840$$

The value of the integral produced by a calculator is 5.113568 (to six decimal places).

**12.** $x=y+\sqrt{y} \Rightarrow dx/dy=1+\dfrac{1}{2\sqrt{y}} \Rightarrow 1+(dx/dy)^2=1+\left(1+\dfrac{1}{2\sqrt{y}}\right)^2=2+\dfrac{1}{\sqrt{y}}+\dfrac{1}{4y}$.

Let $g(y)=\sqrt{1+(dx/dy)^2}$. Then $L=\int_1^2 g(y)\,dy$. Since $n=10$, $\Delta y=\frac{2-1}{10}=\frac{1}{10}$. Now

$$L\approx S_{10}=\tfrac{1/10}{3}[g(1)+4g(1.1)+2g(1.2)+4g(1.3)+2g(1.4)+4g(1.5)+2g(1.6)$$

$$+4g(1.7)+2g(1.8)+4g(1.9)+g(2)]\approx 1.732215,$$

which is the same value of the integral produced by a calculator to six decimal places.

**13.** $x = \sin t,\ y = t^2 \quad \Rightarrow \quad (dx/dt)^2 + (dy/dt)^2 = (\cos t)^2 + (2t)^2 = \cos^2 t + 4t^2 \quad \Rightarrow \quad L = \int_0^{2\pi} \sqrt{\cos^2 t + 4t^2}\, dt.$

Using Simpson's Rule with $n = 10$, $\Delta t = \dfrac{2\pi - 0}{10} = \dfrac{\pi}{5}$, and $f(t) = \sqrt{\cos^2 t + 4t^2}\, dt$, we get

$$L \approx S_{10} = \tfrac{2\pi - 0}{3(10)} \left[ f(0) + 4f(\tfrac{\pi}{5}) + 2f(\tfrac{2\pi}{5}) + 4f(\tfrac{3\pi}{5}) + 2f(\tfrac{4\pi}{5}) + 4f(\pi) + 2f(\tfrac{6\pi}{5}) + 4f(\tfrac{7\pi}{5}) + 2f(\tfrac{8\pi}{5}) + 4f(\tfrac{9\pi}{5}) + f(2\pi) \right]$$

$$\approx 40.056222$$

The value of the integral produced by a calculator is 40.051156 (to six decimal places).

**14.** $x = 3t - t^3$, $y = 3t^2$. $dx/dt = 3 - 3t^2$ and $dy/dt = 6t$, so

$$\left(\frac{dx}{dt}\right)^2 + \left(\frac{dy}{dt}\right)^2 = (3 - 3t^2)^2 + (6t)^2 = (3 + 3t^2)^2$$

and the length of the loop is given by

$$L = \int_{-\sqrt{3}}^{\sqrt{3}} (3 + 3t^2)\, dt = 2 \int_0^{\sqrt{3}} (3 + 3t^2)\, dt = 2\left[3t + t^3\right]_0^{\sqrt{3}}$$
$$= 2\left(3\sqrt{3} + 3\sqrt{3}\right) = 12\sqrt{3}$$

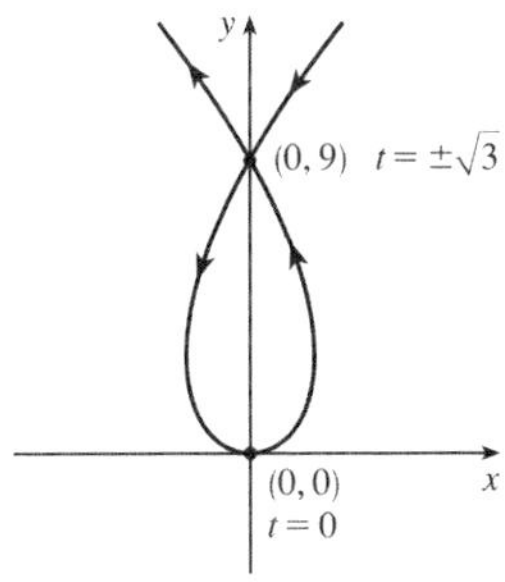


**15.** (a)

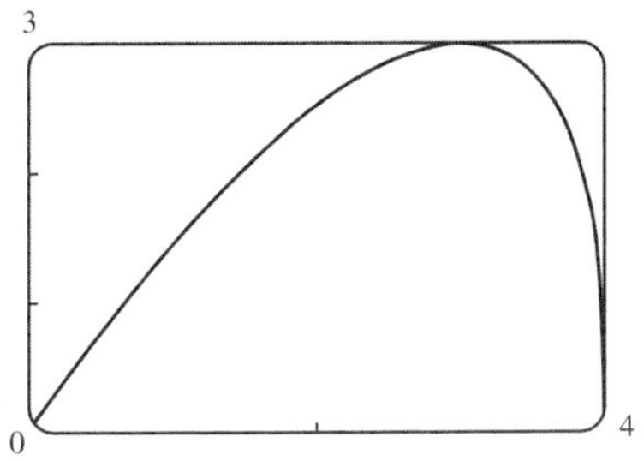


(b)

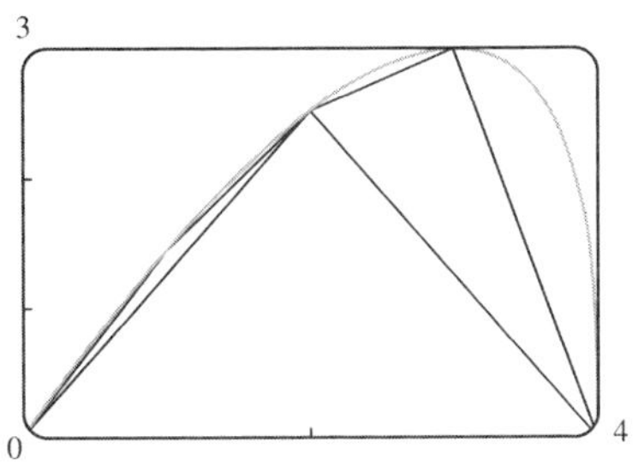


Let $f(x) = y = x\sqrt[3]{4 - x}$. The polygon with one side is just the line segment joining the points $(0, f(0)) = (0, 0)$ and $(4, f(4)) = (4, 0)$, and its length is 4. The polygon with two sides joins the points $(0, 0)$, $(2, f(2)) = \left(2, 2\sqrt[3]{2}\right)$ and $(4, 0)$.

Its length is

$$\sqrt{(2 - 0)^2 + \left(2\sqrt[3]{2} - 0\right)^2} + \sqrt{(4 - 2)^2 + \left(0 - 2\sqrt[3]{2}\right)^2} = 2\sqrt{4 + 2^{8/3}} \approx 6.43$$

Similarly, the inscribed polygon with four sides joins the points $(0, 0)$, $\left(1, \sqrt[3]{3}\right)$, $\left(2, 2\sqrt[3]{2}\right)$, $(3, 3)$, and $(4, 0)$, so its length is

$$\sqrt{1 + \left(\sqrt[3]{3}\right)^2} + \sqrt{1 + \left(2\sqrt[3]{2} - \sqrt[3]{3}\right)^2} + \sqrt{1 + \left(3 - 2\sqrt[3]{2}\right)^2} + \sqrt{1 + 9} \approx 7.50$$

(c) Using the arc length formula with $\dfrac{dy}{dx} = x\left[\tfrac{1}{3}(4 - x)^{-2/3}(-1)\right] + \sqrt[3]{4 - x} = \dfrac{12 - 4x}{3(4 - x)^{2/3}}$, the length of the curve is

$$L = \int_0^4 \sqrt{1 + \left(\frac{dy}{dx}\right)^2}\, dx = \int_0^4 \sqrt{1 + \left[\frac{12 - 4x}{3(4 - x)^{2/3}}\right]^2}\, dx.$$

(d) According to a CAS, the length of the curve is $L \approx 7.7988$. The actual value is larger than any of the approximations in part (b). This is always true, since any approximating straight line between two points on the curve is shorter than the length of the curve between the two points.

**16.** (a) Let $f(x) = y = x + \sin x$ with $0 \le x \le 2\pi$.

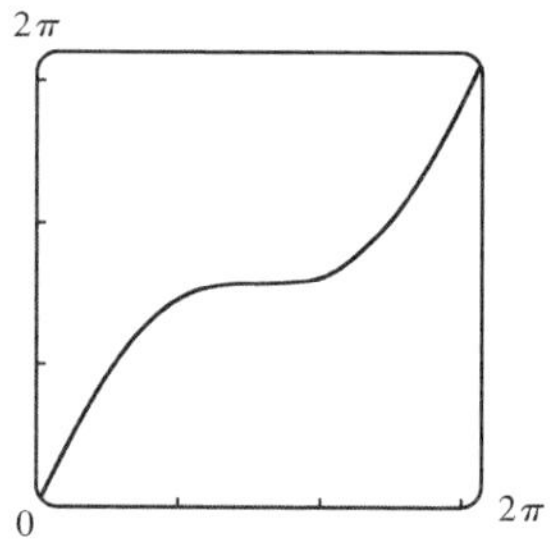


(b) The polygon with one side is just the line segment joining the points $(0, f(0)) = (0, 0)$ and $(2\pi, f(2\pi)) = (2\pi, 2\pi)$, and its length is $\sqrt{(2\pi - 0)^2 + (2\pi - 0)^2} = 2\sqrt{2}\,\pi \approx 8.9$.

The polygon with two sides joins the points $(0, 0)$, $(\pi, f(\pi)) = (\pi, \pi)$, and $(2\pi, 2\pi)$. Its length is

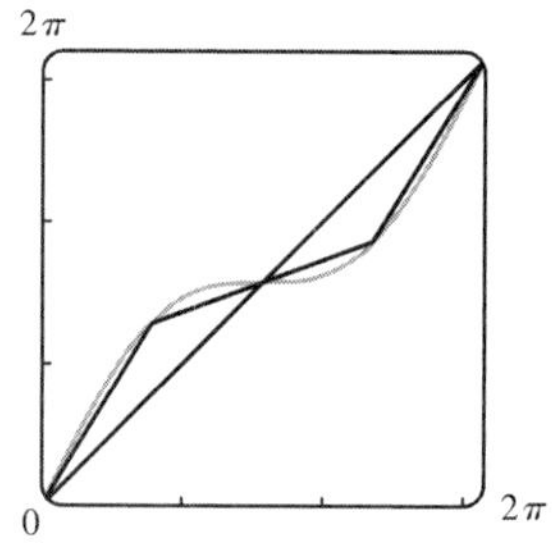


$$\sqrt{(\pi - 0)^2 + (\pi - 0)^2} + \sqrt{(2\pi - \pi)^2 + (2\pi - \pi)^2} = \sqrt{2}\,\pi + \sqrt{2}\,\pi$$
$$= 2\sqrt{2}\,\pi \approx 8.9$$

Note from the diagram that the two approximations are the same because the sides of the 2-sided polygon are in fact on the same line, since $f(\pi) = \pi = \frac{1}{2} f(2\pi)$.

The four-sided polygon joins the points $(0, 0)$, $\left(\frac{\pi}{2}, \frac{\pi}{2} + 1\right)$, $(\pi, \pi)$, $\left(\frac{3\pi}{2}, \frac{3\pi}{2} - 1\right)$, and $(2\pi, 2\pi)$, so its length is

$$\sqrt{\left(\tfrac{\pi}{2}\right)^2 + \left(\tfrac{\pi}{2} + 1\right)^2} + \sqrt{\left(\tfrac{\pi}{2}\right)^2 + \left(\tfrac{\pi}{2} - 1\right)^2} + \sqrt{\left(\tfrac{\pi}{2}\right)^2 + \left(\tfrac{\pi}{2} - 1\right)^2} + \sqrt{\left(\tfrac{\pi}{2}\right)^2 + \left(\tfrac{\pi}{2} + 1\right)^2} \approx 9.4$$

(c) Using the arc length formula with $dy/dx = 1 + \cos x$, the length of the curve is

$$L = \int_0^{2\pi} \sqrt{1 + (1 + \cos x)^2}\, dx = \int_0^{2\pi} \sqrt{2 + 2\cos x + \cos^2 x}\, dx$$

(d) The CAS approximates the integral as 9.5076. The actual length is larger than the approximations in part (b).

**17.** $x = t^3 \quad \Rightarrow \quad dx/dt = 3t^2$ and $y = t^4 \quad \Rightarrow \quad dy/dt = 4t^3$. So

$L = \int_0^1 \sqrt{9t^4 + 16t^6}\, dt = \int_0^1 \sqrt{t^4(9 + 16t^2)}\, dt = \int_0^1 t^2\sqrt{9 + 16t^2}\, dt.$

Now use Formula 22 from the table of integrals to evaluate $L$.

$$L = \int_0^4 \left(\tfrac{1}{4}u\right)^2 \sqrt{a^2 + u^2}\left(\tfrac{1}{4}u\right) \qquad [a = 3,\ u = 4t,\ du = 4\,dt]$$
$$= \tfrac{1}{64}\int_0^4 u^2\sqrt{a^2 + u^2}\, du = \tfrac{1}{64}\left[\tfrac{u}{8}(9 + 2u^2)\sqrt{9 + u^2} - \tfrac{81}{8}\ln\left(u + \sqrt{9 + u^2}\right)\right]_0^4$$
$$= \tfrac{1}{64}\left\{\left[\tfrac{1}{2}\cdot 41 \cdot 5 - \tfrac{81}{8}\ln(4 + 5)\right] - \left[0 - \tfrac{81}{8}\ln 3\right]\right\}$$
$$= \tfrac{1}{64}\left[\tfrac{205}{2} - \tfrac{81}{8}(2\ln 3) + \tfrac{81}{8}\ln 3\right] \qquad [\ln 9 = \ln 3^2 = 2\ln 3]$$
$$= \tfrac{1}{64}\left(\tfrac{205}{2} - \tfrac{81}{8}\ln 3\right) = \tfrac{205}{128} - \tfrac{81\ln 3}{512} \approx 1.428.$$

**18.** $y^2 = 4x$, $x = \frac{1}{4}y^2 \;\Rightarrow\; \frac{dx}{dy} = \frac{1}{2}y \;\Rightarrow\; 1 + \left(\frac{dx}{dy}\right)^2 = 1 + \frac{1}{4}y^2$. So

$$L = \int_0^2 \sqrt{1 + \tfrac{1}{4}y^2}\,dy = \int_0^1 \sqrt{1+u^2}\cdot 2\,du \qquad \left[u = \tfrac{1}{2}y,\ dy = 2\,du\right]$$

$$\overset{21}{=} \left[u\sqrt{1+u^2} + \ln\left|u + \sqrt{1+u^2}\right|\right]_0^1 = \sqrt{2} + \ln\left(1+\sqrt{2}\right)$$

**19.** $y = \ln(\cos x) \;\Rightarrow\; y' = \dfrac{1}{\cos x}(-\sin x) = -\tan x \;\Rightarrow\; 1 + (y')^2 = 1 + \tan^2 x = \sec^2 x$.

So $L = \int_0^{\pi/4} \sec x\,dx \overset{14}{=} \Big[\ln|\sec x + \tan x|\Big]_0^{\pi/4} = \ln\left(\sqrt{2}+1\right) - \ln(1+0) = \ln\left(\sqrt{2}+1\right) \approx 0.881$.

**20.** $y = \ln x \;\Rightarrow\; y' = \dfrac{1}{x} \;\Rightarrow\; 1 + (y')^2 = 1 + \dfrac{1}{x^2} = \dfrac{x^2+1}{x^2}$. So

$$L = \int_1^{\sqrt{3}} \sqrt{\frac{x^2+1}{x^2}}\,dx = \int_1^{\sqrt{3}} \frac{\sqrt{x^2+1}}{x}\,dx \overset{23}{=} \left[\sqrt{x^2+1} - \ln\left|\frac{1+\sqrt{x^2+1}}{x}\right|\right]_1^{\sqrt{3}}$$

$$= \left(2 - \ln\sqrt{3}\right) - \left(\sqrt{2} - \ln\left(1+\sqrt{2}\right)\right) = 2 - \sqrt{2} + \ln\left(1+\sqrt{2}\right) - \ln\sqrt{3}$$

**21.** The prey hits the ground when $y = 0 \;\Leftrightarrow\; 180 - \frac{1}{45}x^2 = 0 \;\Leftrightarrow\; x^2 = 45\cdot 180 \;\Rightarrow\; x = \sqrt{8100} = 90$, since $x$ must be positive. $y' = -\frac{2}{45}x \;\Rightarrow\; 1 + (y')^2 = 1 + \frac{4}{45^2}x^2$, so the distance traveled by the prey is

$$L = \int_0^{90} \sqrt{1 + \tfrac{4}{45^2}x^2}\,dx = \int_0^4 \sqrt{1+u^2}\left(\tfrac{45}{2}\,du\right) \qquad \left[u = \tfrac{2}{45}x,\ du = \tfrac{2}{45}\,dx\right]$$

$$\overset{21}{=} \tfrac{45}{2}\left[\tfrac{1}{2}u\sqrt{1+u^2} + \tfrac{1}{2}\ln\left(u+\sqrt{1+u^2}\right)\right]_0^4 = \tfrac{45}{2}\left[2\sqrt{17} + \tfrac{1}{2}\ln\left(4+\sqrt{17}\right)\right] = 45\sqrt{17} + \tfrac{45}{4}\ln\left(4+\sqrt{17}\right) \approx 209.1 \text{ m}$$

**22.** $y = 150 - \frac{1}{40}(x-50)^2 \;\Rightarrow\; y' = -\frac{1}{20}(x-50) \;\Rightarrow\; 1 + (y')^2 = 1 + \frac{1}{20^2}(x-50)^2$, so the distance traveled by the kite is

$$L = \int_0^{80} \sqrt{1 + \tfrac{1}{20^2}(x-50)^2}\,dx = \int_{-5/2}^{3/2} \sqrt{1+u^2}\,(20\,du) \qquad \left[u = \tfrac{1}{20}(x-50),\ du = \tfrac{1}{20}\,dx\right]$$

$$\overset{21}{=} 20\left[\tfrac{1}{2}u\sqrt{1+u^2} + \tfrac{1}{2}\ln\left(u+\sqrt{1+u^2}\right)\right]_{-5/2}^{3/2} = 10\left[\tfrac{3}{2}\sqrt{\tfrac{13}{4}} + \ln\left(\tfrac{3}{2}+\sqrt{\tfrac{13}{4}}\right) + \tfrac{5}{2}\sqrt{\tfrac{29}{4}} - \ln\left(-\tfrac{5}{2}+\sqrt{\tfrac{29}{4}}\right)\right]$$

$$= \tfrac{15}{2}\sqrt{13} + \tfrac{25}{2}\sqrt{29} + 10\ln\left(\frac{3+\sqrt{13}}{-5+\sqrt{29}}\right) \approx 122.8 \text{ ft}$$

**23.** The sine wave has amplitude 1 and period 14, since it goes through two periods in a distance of 28 in., so its equation is $y = 1\sin\left(\frac{2\pi}{14}x\right) = \sin\left(\frac{\pi}{7}x\right)$. The width $w$ of the flat metal sheet needed to make the panel is the arc length of the sine curve from $x = 0$ to $x = 28$. We set up the integral to evaluate $w$ using the arc length formula with $\frac{dy}{dx} = \frac{\pi}{7}\cos\left(\frac{\pi}{7}x\right)$:

$L = \int_0^{28} \sqrt{1 + \left[\frac{\pi}{7}\cos\left(\frac{\pi}{7}x\right)\right]^2}\,dx = 2\int_0^{14} \sqrt{1 + \left[\frac{\pi}{7}\cos\left(\frac{\pi}{7}x\right)\right]^2}\,dx$. This integral would be very difficult to evaluate exactly, so we use a CAS, and find that $L \approx 29.36$ inches.

**24.** $x = a\cos^3\theta$, $y = a\sin^3\theta$.

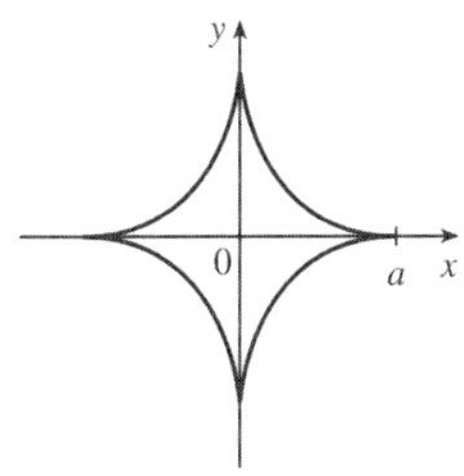


$$(dx/d\theta)^2 + (dy/d\theta)^2 = \left(-3a\cos^2\theta\,\sin\theta\right)^2 + \left(3a\sin^2\theta\,\cos\theta\right)^2$$
$$= 9a^2\cos^4\theta\,\sin^2\theta + 9a^2\sin^4\theta\,\cos^2\theta$$
$$= 9a^2\sin^2\theta\,\cos^2\theta\left(\cos^2\theta + \sin^2\theta\right) = 9a^2\sin^2\theta\,\cos^2\theta.$$

The graph has four-fold symmetry and the curve in the first quadrant corresponds to $0 \le \theta \le \pi/2$. Thus,

$$L = 4\int_0^{\pi/2} 3a\sin\theta\,\cos\theta\,d\theta \qquad [\text{since } a > 0 \text{ and } \sin\theta \text{ and } \cos\theta \text{ are positive for } 0 \le \theta \le \pi/2]$$

$$= 12a\left[\tfrac{1}{2}\sin^2\theta\right]_0^{\pi/2} = 12a\left(\tfrac{1}{2} - 0\right) = 6a$$

**25.** $x = a\sin\theta,\ y = b\cos\theta,\ 0 \le \theta \le 2\pi$.

$$\left(\frac{dx}{d\theta}\right)^2 + \left(\frac{dy}{d\theta}\right)^2 = (a\cos\theta)^2 + (-b\sin\theta)^2 = a^2\cos^2\theta + b^2\sin^2\theta = a^2(1-\sin^2\theta) + b^2\sin^2\theta$$

$$= a^2 - (a^2 - b^2)\sin^2\theta = a^2 - c^2\sin^2\theta = a^2\left(1 - \tfrac{c^2}{a^2}\sin^2\theta\right) = a^2(1 - e^2\sin^2\theta)$$

So $L = 4\int_0^{\pi/2}\sqrt{a^2(1-e^2\sin^2\theta)}\,d\theta$ [by symmetry] $= 4a\int_0^{\pi/2}\sqrt{1-e^2\sin^2\theta}\,d\theta$.

**26.** By symmetry, the length of the curve in each quadrant is the same, so we'll find the length in the first quadrant and multiply by 4.

$x^{2k} + y^{2k} = 1 \ \Rightarrow\ y^{2k} = 1 - x^{2k} \ \Rightarrow\ y = (1 - x^{2k})^{1/(2k)}$

(in the first quadrant), so we use the arc length formula with

$$\frac{dy}{dx} = \frac{1}{2k}(1 - x^{2k})^{1/(2k)-1}(-2kx^{2k-1})$$

$$= -x^{2k-1}(1-x^{2k})^{1/(2k)-1}$$

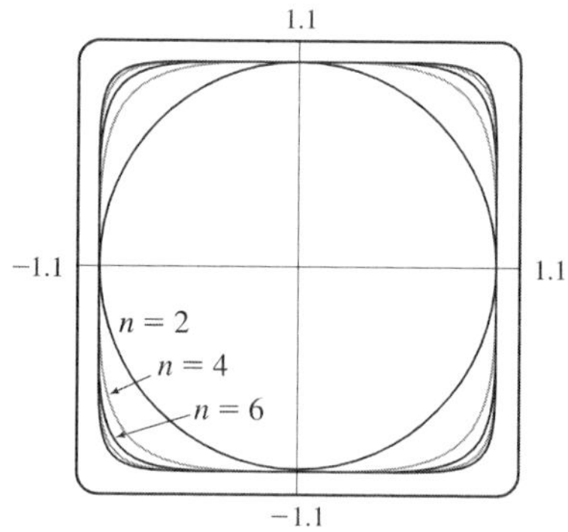


The total length is therefore

$$L_{2k} = 4\int_0^1 \sqrt{1 + \left[-x^{2k-1}(1-x^{2k})^{1/(2k)-1}\right]^2}\,dx = 4\int_0^1 \sqrt{1 + x^{2(2k-1)}(1-x^{2k})^{1/k-2}}\,dx$$

Now from the graph, we see that as $k$ increases, the "corners" of these fat circles get closer to the points $(\pm1, \pm1)$ and $(\pm1, \mp1)$, and the "edges" of the fat circles approach the lines joining these four points. It seems plausible that as $k \to \infty$, the total length of the fat circle with $n = 2k$ will approach the length of the perimeter of the square with sides of length 2. This is supported by taking the limit as $k \to \infty$ of the equation of the fat circle in the first quadrant: $\lim\limits_{k\to\infty}(1 - x^{2k})^{1/(2k)} = 1$ for $0 \le x < 1$. So we guess that $\lim\limits_{k\to\infty} L_{2k} = 4 \cdot 2 = 8$.

**27.** (a) Notice that $0 \le t \le 2\pi$ does not give the complete curve because $x(0) \ne x(2\pi)$. In fact, we must take $t \in [0, 4\pi]$ in order to obtain the complete curve, since the first term in each of the parametric equations has period $2\pi$ and the second has period $\frac{2\pi}{11/2} = \frac{4\pi}{11}$, and the least common integer multiple of these two numbers is $4\pi$.

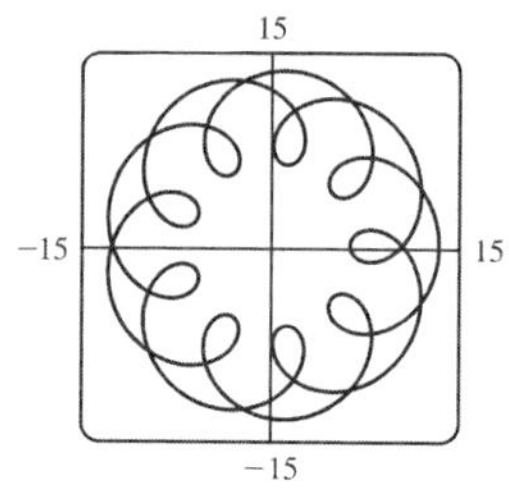


(b) We use the CAS to find the derivatives $dx/dt$ and $dy/dt$, and then use Formula 1 to find the arc length. Recent versions of Maple express the integral $\int_0^{4\pi}\sqrt{(dx/dt)^2 + (dy/dt)^2}\,dt$ as $88E(2\sqrt{2}\,i)$, where $E(x)$ is the elliptic integral $\displaystyle\int_0^1 \frac{\sqrt{1-x^2t^2}}{\sqrt{1-t^2}}\,dt$ and $i$ is the imaginary number $\sqrt{-1}$. Some earlier versions of Maple (as well as Mathematica) cannot do the integral exactly, so we use the command

`evalf(Int(sqrt(diff(x,t)^2+diff(y,t)^2),t=0..4*Pi));` to estimate the length, and find that the arc length is approximately 294.03. Derive's `Para_arc_length` function in the utility file `Int_apps` simplifies the integral to $11\int_0^{4\pi}\sqrt{-4\cos t\,\cos\left(\frac{11t}{2}\right)-4\sin t\,\sin\left(\frac{11t}{2}\right)+5}\,dt$.

**28.** (a) It appears that as $t\to\infty$, $(x,y)\to\left(\frac{1}{2},\frac{1}{2}\right)$, and as $t\to-\infty$, $(x,y)\to\left(-\frac{1}{2},-\frac{1}{2}\right)$.

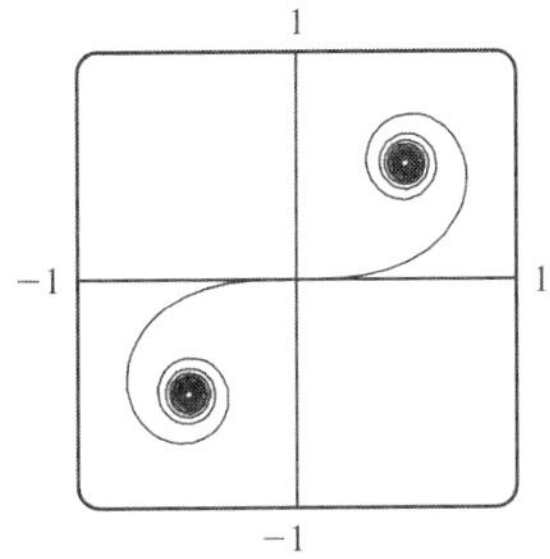


(b) By the Fundamental Theorem of Calculus, $dx/dt=\cos\left(\frac{\pi}{2}t^2\right)$ and $dy/dt=\sin\left(\frac{\pi}{2}t^2\right)$, so by Formula 1, the length of the curve from the origin to the point with parameter value $t$ is

$$L=\int_0^t\sqrt{(dx/du)^2+(dy/du)^2}\,du=\int_0^t\sqrt{\cos^2\left(\tfrac{\pi}{2}u^2\right)+\sin^2\left(\tfrac{\pi}{2}u^2\right)}\,du$$

$$=\int_0^t 1\,du=t\qquad[\text{or }-t\text{ if }t<0]$$

We have used $u$ as the dummy variable so as not to confuse it with the upper limit of integration.

## DISCOVERY PROJECT Arc Length Contest

For advice on how to run the contest and a list of student entries, see the article "Arc Length Contest" by Larry Riddle in *The College Mathematics Journal*, Volume 29, No. 4, September 1998, pages 314–320.

## 6.4 Average Value of a Function

**1.** $g_{\text{ave}}=\dfrac{1}{\frac{\pi}{2}-0}\int_0^{\pi/2}\cos x\,dx=\frac{2}{\pi}\left[\sin x\right]_0^{\pi/2}=\frac{2}{\pi}(1-0)=\frac{2}{\pi}$

**2.** $g_{\text{ave}}=\frac{1}{2-0}\int_0^2 x^2\sqrt{1+x^3}\,dx=\frac{1}{2}\int_1^9\sqrt{u}\cdot\frac{1}{3}\,du\quad[u=1+x^3,\ du=3x^2\,dx]\ =\frac{1}{6}\left[\frac{2}{3}u^{3/2}\right]_1^9=\frac{1}{9}(27-1)=\frac{26}{9}$

**3.** $f_{\text{ave}}=\frac{1}{3-0}\int_0^3 te^{t/2}\,dt$. Using integration by parts,

$$\int te^{t/2}dt=2te^{t/2}-\int 2e^{t/2}\,dt\quad\begin{bmatrix}u=t, & dv=e^{t/2}\,dt\\ du=dt, & v=2e^{t/2}\end{bmatrix}\quad=2te^{t/2}-4e^{t/2}.$$

Thus, $f_{\text{ave}}=\frac{1}{3}\left[2te^{t/2}-4e^{t/2}\right]_0^3=\frac{1}{3}[(6e^{3/2}-4e^{3/2})-(0-4)]=\frac{1}{3}(2e^{3/2}+4)$.

**4.** $h_{\text{ave}}=\dfrac{1}{6-1}\displaystyle\int_1^6\frac{3}{(1+r)^2}\,dr=\frac{1}{5}\int_2^7 3u^{-2}\,du\qquad[u=1+r,\ du=dr]$

$=-\frac{3}{5}\left[u^{-1}\right]_2^7=-\frac{3}{5}\left(\frac{1}{7}-\frac{1}{2}\right)=\frac{3}{5}\left(\frac{1}{2}-\frac{1}{7}\right)=\frac{3}{5}\cdot\frac{5}{14}=\frac{3}{14}$

**5.** (a) $f_{\text{ave}}=\dfrac{1}{5-2}\displaystyle\int_2^5(x-3)^2\,dx=\frac{1}{3}\left[\frac{1}{3}(x-3)^3\right]_2^5$

$=\frac{1}{9}\left[2^3-(-1)^3\right]=\frac{1}{9}(8+1)=1$

(b) $f(c)=f_{\text{ave}}\ \Leftrightarrow\ (c-3)^2=1\ \Leftrightarrow$
$c-3=\pm1\ \Leftrightarrow\ c=2\text{ or }4$

(c)

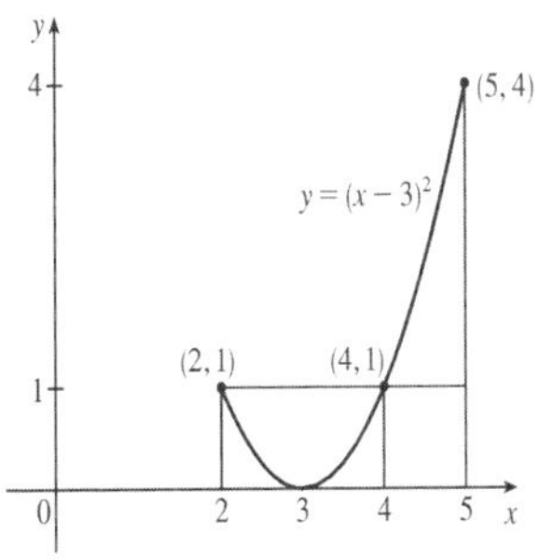

**6.** (a) $f_{\text{ave}} = \frac{1}{3-1}\int_1^3 \ln x\,dx = \frac{1}{2}[x\ln x - x]_1^3$ [by parts]

$= \frac{1}{2}[(3\ln 3 - 3) - (\ln 1 - 1)]$

$= \frac{1}{2}(3\ln 3 - 2) = \frac{3}{2}\ln 3 - 1$

(b) $f_{\text{ave}} = f(c) \;\Leftrightarrow\; \frac{3}{2}\ln 3 - 1 = \ln c \;\Leftrightarrow$

$c = e^{(3/2)\ln 3 - 1}$ or $c = 3\sqrt{3}/e \approx 1.91$

(c)

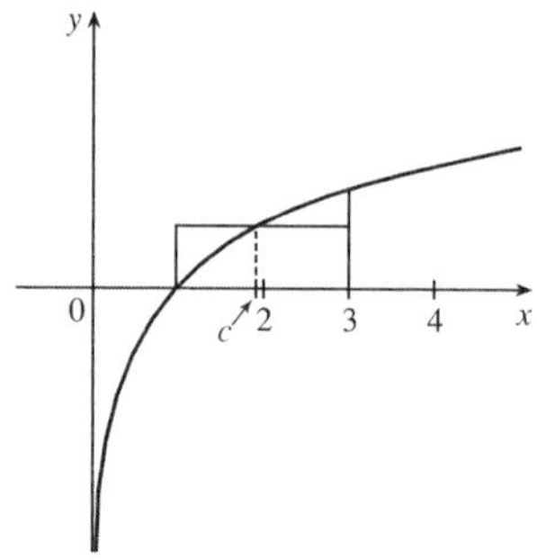


**7.** (a) $f_{\text{ave}} = \dfrac{1}{\pi - 0}\displaystyle\int_0^\pi (2\sin x - \sin 2x)\,dx$

$= \frac{1}{\pi}\left[-2\cos x + \frac{1}{2}\cos 2x\right]_0^\pi$

$= \frac{1}{\pi}\left[\left(2 + \frac{1}{2}\right) - \left(-2 + \frac{1}{2}\right)\right] = \frac{4}{\pi}$

(b) $f(c) = f_{\text{ave}} \;\Leftrightarrow\; 2\sin c - \sin 2c = \frac{4}{\pi} \;\Leftrightarrow$

$c_1 \approx 1.238$ or $c_2 \approx 2.808$

(c)

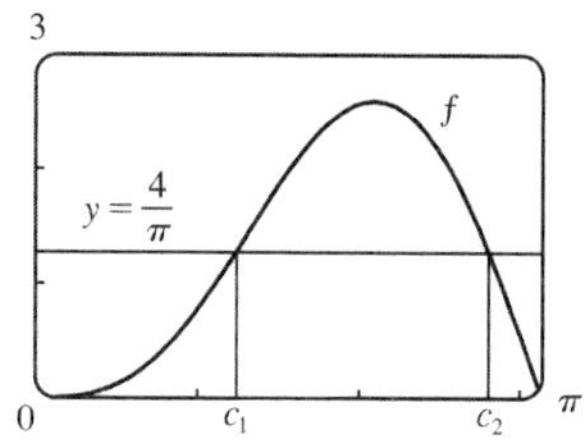


**8.** (a) $f_{\text{ave}} = \dfrac{1}{2-0}\displaystyle\int_0^2 \frac{2x}{(1+x^2)^2}\,dx$

$= \dfrac{1}{2}\displaystyle\int_1^5 \frac{1}{u^2}\,du$ $\quad [u = 1 + x^2,\ du = 2x\,dx]$

$= \dfrac{1}{2}\left[-\dfrac{1}{u}\right]_1^5 = -\dfrac{1}{2}\left(\dfrac{1}{5} - 1\right) = \dfrac{2}{5}$

(b) $f(c) = f_{\text{ave}} \;\Leftrightarrow\; \dfrac{2c}{(1+c^2)^2} = \dfrac{2}{5} \;\Leftrightarrow\; 5c = (1+c^2)^2 \;\Leftrightarrow$

$c_1 \approx 0.220$ or $c_2 \approx 1.207$

(c)

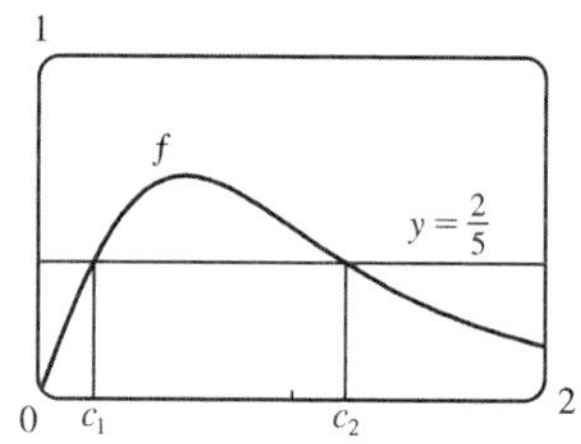


**9.** $f$ is continuous on $[1, 3]$, so by the Mean Value Theorem for Integrals there exists a number $c$ in $[1, 3]$ such that $\int_1^3 f(x)\,dx = f(c)(3-1) \;\Rightarrow\; 8 = 2f(c)$; that is, there is a number $c$ such that $f(c) = \frac{8}{2} = 4$.

**10.** The requirement is that $\dfrac{1}{b-0}\displaystyle\int_0^b f(x)\,dx = 3$. The LHS of this equation is equal to

$\dfrac{1}{b}\displaystyle\int_0^b (2 + 6x - 3x^2)\,dx = \dfrac{1}{b}\left[2x + 3x^2 - x^3\right]_0^b = 2 + 3b - b^2$, so we solve the equation $2 + 3b - b^2 = 3 \;\Leftrightarrow$

$b^2 - 3b + 1 = 0 \;\Leftrightarrow\; b = \dfrac{3 \pm \sqrt{(-3)^2 - 4\cdot 1\cdot 1}}{2\cdot 1} = \dfrac{3 \pm \sqrt{5}}{2}$. Both roots are valid since they are positive.

**11.** $f_{\text{ave}} = \dfrac{1}{50-20}\displaystyle\int_{20}^{50} f(x)\,dx \approx \frac{1}{30}M_3 = \dfrac{1}{30}\cdot\dfrac{50-20}{3}[f(25) + f(35) + f(45)] = \frac{1}{3}(38 + 29 + 48) = \frac{115}{3} = 38\frac{1}{3}$

**12.** (a) $v_{\text{ave}} = \frac{1}{12-0}\int_0^{12} v(t)\,dt = \frac{1}{12}I$. Use the Midpoint Rule with $n = 3$ and $\Delta t = \frac{12-0}{3} = 4$ to estimate $I$.

$I \approx M_3 = 4[v(2) + v(6) + v(10)] = 4[21 + 50 + 66] = 4(137) = 548$. Thus, $v_{\text{ave}} \approx \frac{1}{12}(548) = 45\frac{2}{3}$ km/h.

(b) Estimating from the graph, $v(t) = 45\frac{2}{3}$ when $t \approx 5.2$ s.

**13.** Let $t = 0$ and $t = 12$ correspond to 9 A.M. and 9 P.M., respectively.

$$T_{\text{ave}} = \tfrac{1}{12-0}\int_0^{12}\left[50 + 14\sin\tfrac{1}{12}\pi t\right]dt = \tfrac{1}{12}\left[50t - 14\cdot\tfrac{12}{\pi}\cos\tfrac{1}{12}\pi t\right]_0^{12}$$

$$= \tfrac{1}{12}\left[50\cdot 12 + 14\cdot\tfrac{12}{\pi} + 14\cdot\tfrac{12}{\pi}\right] = \left(50 + \tfrac{28}{\pi}\right)\,^\circ\text{F} \approx 59\,^\circ\text{F}$$

**14.** $T_{\text{ave}} = \dfrac{1}{30-0}\displaystyle\int_0^{30} (20 + 75e^{-t/50})\, dt = \dfrac{1}{30}\Big[20t - 50 \cdot 75e^{-t/50}\Big]_0^{30} = \dfrac{1}{30}[(600 - 3750e^{-3/5}) - (-3750)]$

$= \frac{1}{30}(4350 - 3750e^{-3/5}) = 145 - 125e^{-3/5} \approx 76.4\,^\circ\text{C}$

**15.** (a) We want to calculate the square root of the average value of $[E(t)]^2 = [155\sin(120\pi t)]^2 = 155^2\sin^2(120\pi t)$. First, we calculate the average value itself, by integrating $[E(t)]^2$ over one cycle (between $t = 0$ and $t = \frac{1}{60}$, since there are 60 cycles per second) and dividing by $\left(\frac{1}{60} - 0\right)$:

$$[E(t)]^2_{\text{ave}} = \tfrac{1}{1/60}\textstyle\int_0^{1/60} \left[155^2\sin^2(120\pi t)\right] dt = 60 \cdot 155^2 \int_0^{1/60} \frac{1}{2}[1 - \cos(240\pi t)]\, dt$$

$$= 60 \cdot 155^2\left(\tfrac{1}{2}\right)\left[t - \tfrac{1}{240\pi}\sin(240\pi t)\right]_0^{1/60} = 60 \cdot 155^2\left(\tfrac{1}{2}\right)\left[\left(\tfrac{1}{60} - 0\right) - (0 - 0)\right] = \tfrac{155^2}{2}$$

The RMS value is just the square root of this quantity, which is $\frac{155}{\sqrt{2}} \approx 110$ V.

(b) $220 = \sqrt{[E(t)]^2_{\text{ave}}} \Rightarrow$

$$220^2 = [E(t)]^2_{\text{ave}} = \tfrac{1}{1/60}\textstyle\int_0^{1/60} A^2\sin^2(120\pi t)\, dt = 60A^2\int_0^{1/60} \frac{1}{2}[1 - \cos(240\pi t)]\, dt$$

$$= 30A^2\left[t - \tfrac{1}{240\pi}\sin(240\pi t)\right]_0^{1/60} = 30A^2\left[\left(\tfrac{1}{60} - 0\right) - (0 - 0)\right] = \tfrac{1}{2}A^2$$

Thus, $220^2 = \frac{1}{2}A^2 \Rightarrow A = 220\sqrt{2} \approx 311$ V.

**16.** $s = \frac{1}{2}gt^2 \Rightarrow t = \sqrt{2s/g}$ [since $t \geq 0$]. Now $v = \dfrac{ds}{dt} = gt = g\sqrt{2s/g} = \sqrt{2gs} \Rightarrow v^2 = 2gs \Rightarrow s = \dfrac{v^2}{2g}$. We see that $v$ can be regarded as a function of $t$ or of $s$: $v = F(t) = gt$ and $v = G(s) = \sqrt{2gs}$. Note that $v_T = F(T) = gT$. Displacement can be viewed as a function of $t$: $s = s(t) = \frac{1}{2}gt^2$; also $s(t) = \dfrac{v^2}{2g} = \dfrac{[F(t)]^2}{2g}$. When $t = T$, these two formulas for $s(t)$ imply that

$$\sqrt{2gs(T)} = F(T) = v_T = gT = 2\left(\tfrac{1}{2}gT^2\right)/T = 2s(T)/T \qquad (*)$$

The average of the velocities with respect to time $t$ during the interval $[0, T]$ is

$$v_{t\text{-ave}} = F_{\text{ave}} = \frac{1}{T-0}\int_0^T F(t)\, dt = \frac{1}{T}[s(T) - s(0)] \qquad \text{[by FTC]}$$

$$= \frac{s(T)}{T} \quad [\text{since } s(0) = 0] \quad = \frac{1}{2}v_T \qquad [\text{by } (*)]$$

But the average of the velocities with respect to displacement $s$ during the corresponding displacement interval $[s(0), s(T)] = [0, s(T)]$ is

$$v_{s\text{-ave}} = G_{\text{ave}} = \frac{1}{s(T) - 0}\int_0^{s(T)} G(s)\, ds = \frac{1}{s(T)}\int_0^{s(T)} \sqrt{2gs}\, ds = \frac{\sqrt{2g}}{s(T)}\int_0^{s(T)} s^{1/2}\, ds$$

$$= \frac{\sqrt{2g}}{s(T)} \cdot \frac{2}{3}\left[s^{3/2}\right]_0^{s(T)} = \frac{2}{3} \cdot \frac{\sqrt{2g}}{s(T)} \cdot \left[s(T)\right]^{3/2} = \frac{2}{3}\sqrt{2gs(T)} = \frac{2}{3}v_T \qquad [\text{by } (*)]$$

**17.** $V_{\text{ave}} = \frac{1}{5}\int_0^5 V(t)\, dt = \frac{1}{5}\int_0^5 \frac{5}{4\pi}\left[1 - \cos\left(\frac{2}{5}\pi t\right)\right] dt = \frac{1}{4\pi}\int_0^5 \left[1 - \cos\left(\frac{2}{5}\pi t\right)\right] dt$

$= \frac{1}{4\pi}\left[t - \frac{5}{2\pi}\sin\left(\frac{2}{5}\pi t\right)\right]_0^5 = \frac{1}{4\pi}[(5 - 0) - 0] = \frac{5}{4\pi} \approx 0.4$ L

**18.** $v_{\text{ave}} = \dfrac{1}{R-0}\displaystyle\int_0^R v(r)\, dr = \dfrac{1}{R}\int_0^R \dfrac{P}{4\eta l}(R^2 - r^2)\, dr = \dfrac{P}{4\eta lR}\left[R^2 r - \tfrac{1}{3}r^3\right]_0^R = \dfrac{P}{4\eta lR}\left(\tfrac{2}{3}\right)R^3 = \dfrac{PR^2}{6\eta l}$.

Since $v(r)$ is decreasing on $(0, R]$, $v_{\max} = v(0) = \dfrac{PR^2}{4\eta l}$. Thus, $v_{\text{ave}} = \frac{2}{3}v_{\max}$.

**19.** Let $F(x) = \int_a^x f(t)\,dt$ for $x$ in $[a, b]$. Then $F$ is continuous on $[a, b]$ and differentiable on $(a, b)$, so by the Mean Value Theorem there is a number $c$ in $(a, b)$ such that $F(b) - F(a) = F'(c)(b - a)$. But $F'(x) = f(x)$ by the Fundamental Theorem of Calculus. Therefore, $\int_a^b f(t)\,dt - 0 = f(c)(b - a)$.

**20.** $$f_{\text{ave}}\,[a,b] = \frac{1}{b-a}\int_a^b f(x)\,dx = \frac{1}{b-a}\int_a^c f(x)\,dx + \frac{1}{b-a}\int_c^b f(x)\,dx$$
$$= \frac{c-a}{b-a}\left[\frac{1}{c-a}\int_a^c f(x)\,dx\right] + \frac{b-c}{b-a}\left[\frac{1}{b-c}\int_c^b f(x)\,dx\right] = \frac{c-a}{b-a}\,f_{\text{ave}}\,[a,c] + \frac{b-c}{b-a}\,f_{\text{ave}}\,[c,b]$$

## APPLIED PROJECT Where to Sit at the Movies

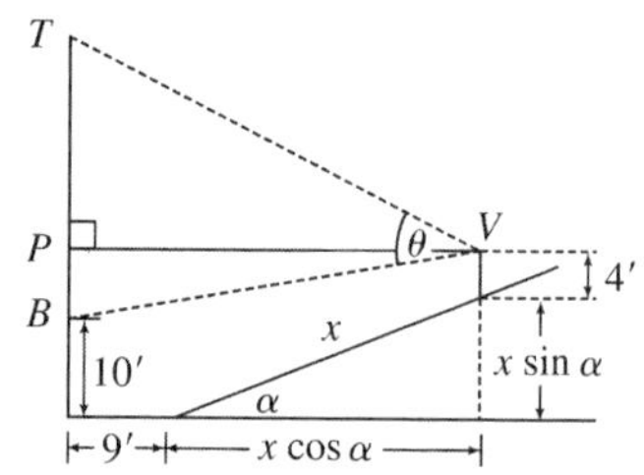


**1.** $|VP| = 9 + x\cos\alpha$, $|PT| = 35 - (4 + x\sin\alpha) = 31 - x\sin\alpha$, and $|PB| = (4 + x\sin\alpha) - 10 = x\sin\alpha - 6$. So using the Pythagorean Theorem, we have $|VT| = \sqrt{|VP|^2 + |PT|^2} = \sqrt{(9 + x\cos\alpha)^2 + (31 - x\sin\alpha)^2} = a$, and $|VB| = \sqrt{|VP|^2 + |PB|^2} = \sqrt{(9 + x\cos\alpha)^2 + (x\sin\alpha - 6)^2} = b$.

Using the Law of Cosines on $\triangle VBT$, we get $25^2 = a^2 + b^2 - 2ab\cos\theta \;\Leftrightarrow$

$\cos\theta = \dfrac{a^2 + b^2 - 625}{2ab} \;\Leftrightarrow\; \theta = \arccos\left(\dfrac{a^2 + b^2 - 625}{2ab}\right)$, as required.

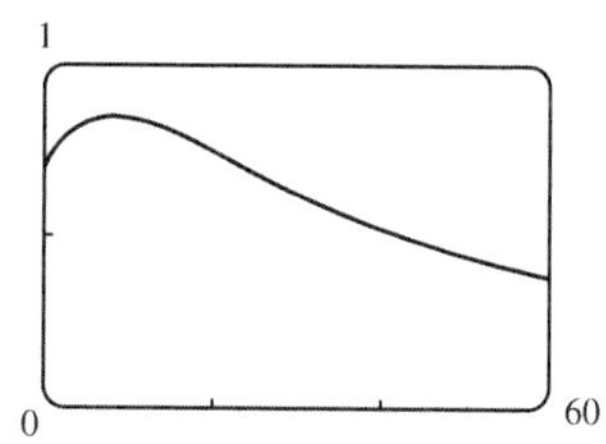


**2.** From the graph of $\theta$, it appears that the value of $x$ which maximizes $\theta$ is $x \approx 8.25$ ft. Assuming that the first row is at $x = 0$, the row closest to this value of $x$ is the fourth row, at $x = 9$ ft, and from the graph, the viewing angle in this row seems to be about $0.85$ radians, or about $49°$.

**3.** With a CAS, we type in the definition of $\theta$, substitute in the proper values of $a$ and $b$ in terms of $x$ and $\alpha = 20° = \frac{\pi}{9}$ radians, and then use the differentiation command to find the derivative. We use a numerical root finder and find that the root of the equation $d\theta/dx = 0$ is $x \approx 8.253062$, as approximated in Problem 2.

**4.** From the graph in Problem 2, it seems that the average value of the function on the interval $[0, 60]$ is about $0.6$. We can use a CAS to approximate $\frac{1}{60}\int_0^{60} \theta(x)\,dx \approx 0.625 \approx 36°$. (The calculation is much faster if we reduce the number of digits of accuracy required.) The minimum value is $\theta(60) \approx 0.38$ and, from Problem 2, the maximum value is about $0.85$.

## 6.5 Applications to Physics and Engineering

**1.** $W = \displaystyle\int_a^b f(x)\,dx = \int_0^9 \frac{10}{(1+x)^2}\,dx = 10\int_1^{10} \frac{1}{u^2}\,du \quad [u = 1 + x,\ du = dx] \;= 10\left[-\frac{1}{u}\right]_1^{10} = 10\left(-\tfrac{1}{10} + 1\right) = 9$ ft-lb

**2.** $W = \int_1^2 \cos\left(\frac{1}{3}\pi x\right) dx = \frac{3}{\pi}\left[\sin\left(\frac{1}{3}\pi x\right)\right]_1^2 = \frac{3}{\pi}\left(\frac{\sqrt{3}}{2} - \frac{\sqrt{3}}{2}\right) = 0$ N·m $= 0$ J.

*Interpretation:* From $x = 1$ to $x = \frac{3}{2}$, the force does work equal to $\int_1^{3/2} \cos\left(\frac{1}{3}\pi x\right) dx = \frac{3}{\pi}\left(1 - \frac{\sqrt{3}}{2}\right)$ J in accelerating the particle and increasing its kinetic energy. From $x = \frac{3}{2}$ to $x = 2$, the force opposes the motion of the particle, decreasing its kinetic energy. This is negative work, equal in magnitude but opposite in sign to the work done from $x = 1$ to $x = \frac{3}{2}$.

**3.** The force function is given by $F(x)$ (in newtons) and the work (in joules) is the area under the curve, given by

$\int_0^8 F(x)\,dx = \int_0^4 F(x)\,dx + \int_4^8 F(x)\,dx = \frac{1}{2}(4)(30) + (4)(30) = 180$ J.

**4.** Work $= \int_0^{18} f(x)\,dx \approx S_6 = \frac{18-0}{6\cdot 3}\,[f(0) + 4f(3) + 2f(6) + 4f(9) + 2f(12) + 4f(15) + f(18)]$

$= 1 \cdot [9.8 + 4(9.1) + 2(8.5) + 4(8.0) + 2(7.7) + 4(7.5) + 7.4] = 148$ joules

**5.** $10 = f(x) = kx = \frac{1}{3}k$ [4 inches $= \frac{1}{3}$ foot], so $k = 30$ lb/ft and $f(x) = 30x$. Now 6 inches $= \frac{1}{2}$ foot, so

$W = \int_0^{1/2} 30x\,dx = \left[15x^2\right]_0^{1/2} = \frac{15}{4}$ ft-lb.

**6.** $25 = f(x) = kx = k(0.1)$ [10 cm $= 0.1$ m], so $k = 250$ N/m and $f(x) = 250x$. Now 5 cm $= 0.05$ m, so

$W = \int_0^{0.05} 250x\,dx = \left[125x^2\right]_0^{0.05} = 125(0.0025) = 0.3125 \approx 0.31$ J.

**7.** (a) If $\int_0^{0.12} kx\,dx = 2$ J, then $2 = \left[\frac{1}{2}kx^2\right]_0^{0.12} = \frac{1}{2}k(0.0144) = 0.0072k$ and $k = \frac{2}{0.0072} = \frac{2500}{9} \approx 277.78$ N/m.

Thus, the work needed to stretch the spring from 35 cm to 40 cm is

$\int_{0.05}^{0.10} \frac{2500}{9}x\,dx = \left[\frac{1250}{9}x^2\right]_{1/20}^{1/10} = \frac{1250}{9}\left(\frac{1}{100} - \frac{1}{400}\right) = \frac{25}{24} \approx 1.04$ J.

(b) $f(x) = kx$, so $30 = \frac{2500}{9}x$ and $x = \frac{270}{2500}$ m $= 10.8$ cm

**8.** Let $L$ be the natural length of the spring in meters. Then

$6 = \int_{0.10-L}^{0.12-L} kx\,dx = \left[\frac{1}{2}kx^2\right]_{0.10-L}^{0.12-L} = \frac{1}{2}k\left[(0.12-L)^2 - (0.10-L)^2\right]$ and

$10 = \int_{0.12-L}^{0.14-L} kx\,dx = \left[\frac{1}{2}kx^2\right]_{0.12-L}^{0.14-L} = \frac{1}{2}k\left[(0.14-L)^2 - (0.12-L)^2\right]$. Simplifying gives us $12 = k(0.0044 - 0.04L)$ and $20 = k(0.0052 - 0.04L)$. Subtracting the first equation from the second gives $8 = 0.0008k$, so $k = 10{,}000$. Now the second equation becomes $20 = 52 - 400L$, so $L = \frac{32}{400}$ m $= 8$ cm.

In Exercises 9–16, $n$ is the number of subintervals of length $\Delta x$, and $x_i^*$ is a sample point in the $i$th subinterval $[x_{i-1}, x_i]$.

**9.** (a) The portion of the rope from $x$ ft to $(x + \Delta x)$ ft below the top of the building weighs $\frac{1}{2}\,\Delta x$ lb and must be lifted $x_i^*$ ft, so its contribution to the total work is $\frac{1}{2}x_i^*\,\Delta x$ ft-lb. The total work is

$$W = \lim_{n\to\infty} \sum_{i=1}^{n} \tfrac{1}{2}x_i^*\,\Delta x = \int_0^{50} \tfrac{1}{2}x\,dx = \left[\tfrac{1}{4}x^2\right]_0^{50} = \tfrac{2500}{4} = 625 \text{ ft-lb}$$

Notice that the exact height of the building does not matter (as long as it is more than 50 ft).

(b) When half the rope is pulled to the top of the building, the work to lift the top half of the rope is $W_1 = \int_0^{25} \frac{1}{2}x\,dx = \left[\frac{1}{4}x^2\right]_0^{25} = \frac{625}{4}$ ft-lb. The bottom half of the rope is lifted 25 ft and the work needed to accomplish that is $W_2 = \int_{25}^{50} \frac{1}{2} \cdot 25\,dx = \frac{25}{2}\left[x\right]_{25}^{50} = \frac{625}{2}$ ft-lb. The total work done in pulling half the rope to the top of the building is $W = W_1 + W_2 = \frac{625}{2} + \frac{625}{4} = \frac{3}{4} \cdot 625 = \frac{1875}{4}$ ft-lb.

**10.** *Assumptions*:

1. After lifting, the chain is L-shaped, with 4 m of the chain lying along the ground.
2. The chain slides effortlessly and without friction along the ground while its end is lifted.
3. The weight density of the chain is constant throughout its length and therefore equals $(8 \text{ kg/m})(9.8 \text{ m/s}^2) = 78.4$ N/m.

The part of the chain $x$ m from the lifted end is raised $6 - x$ m if $0 \le x \le 6$ m, and it is lifted 0 m if $x > 6$ m.

Thus, the work needed is

$$W = \lim_{n\to\infty} \sum_{i=1}^{n} (6 - x_i^*) \cdot 78.4\,\Delta x = \int_0^6 (6 - x)78.4\,dx = 78.4\left[6x - \tfrac{1}{2}x^2\right]_0^6 = (78.4)(18) = 1411.2 \text{ J}$$

**11.** The work needed to lift the cable is $\lim_{n\to\infty} \sum_{i=1}^{n} 2x_i^*\,\Delta x = \int_0^{500} 2x\,dx = \left[x^2\right]_0^{500} = 250{,}000$ ft-lb. The work needed to lift the coal is 800 lb $\cdot$ 500 ft $= 400{,}000$ ft-lb. Thus, the total work required is $250{,}000 + 400{,}000 = 650{,}000$ ft-lb.

**12.** The work needed to lift the bucket itself is 4 lb · 80 ft = 320 ft-lb. At time $t$ (in seconds) the bucket is $x_i^* = 2t$ ft above its original 80 ft depth, but it now holds only $(40 - 0.2t)$ lb of water. In terms of distance, the bucket holds $\left[40 - 0.2\left(\frac{1}{2}x_i^*\right)\right]$ lb of water when it is $x_i^*$ ft above its original 80 ft depth. Moving this amount of water a distance $\Delta x$ requires $\left(40 - \frac{1}{10}x_i^*\right)\Delta x$ ft-lb of work. Thus, the work needed to lift the water is

$$W = \lim_{n\to\infty}\sum_{i=1}^{n}\left(40 - \tfrac{1}{10}x_i^*\right)\Delta x = \int_0^{80}\left(40 - \tfrac{1}{10}x\right)dx = \left[40x - \tfrac{1}{20}x^2\right]_0^{80} = (3200 - 320) \text{ ft-lb}$$

Adding the work of lifting the bucket gives a total of 3200 ft-lb of work.

**13.** At a height of $x$ meters $(0 \le x \le 12)$, the mass of the rope is $(0.8 \text{ kg/m})(12 - x \text{ m}) = (9.6 - 0.8x)$ kg and the mass of the water is $\left(\frac{36}{12} \text{ kg/m}\right)(12 - x \text{ m}) = (36 - 3x)$ kg. The mass of the bucket is 10 kg, so the total mass is $(9.6 - 0.8x) + (36 - 3x) + 10 = (55.6 - 3.8x)$ kg, and hence, the total force is $9.8(55.6 - 3.8x)$ N. The work needed to lift the bucket $\Delta x$ m through the $i$th subinterval of $[0, 12]$ is $9.8(55.6 - 3.8x_i^*)\Delta x$, so the total work is

$$W = \lim_{n\to\infty}\sum_{i=1}^{n} 9.8(55.6 - 3.8x_i^*)\,\Delta x = \int_0^{12}(9.8)(55.6 - 3.8x)\,dx = 9.8\Big[55.6x - 1.9x^2\Big]_0^{12} = 9.8(393.6) \approx 3857 \text{ J}$$

**14.** The chain's weight density is $\dfrac{25 \text{ lb}}{10 \text{ ft}} = 2.5$ lb/ft. The part of the chain $x$ ft below the ceiling (for $5 \le x \le 10$) has to be lifted $2(x - 5)$ ft, so the work needed to lift the $i$th subinterval of the chain is $2(x_i^* - 5)(2.5\,\Delta x)$. The total work needed is

$$\begin{aligned} W &= \lim_{n\to\infty}\sum_{i=1}^{n} 2(x_i^* - 5)(2.5)\,\Delta x = \int_5^{10}[2(x - 5)(2.5)]\,dx = 5\int_5^{10}(x - 5)\,dx \\ &= 5\left[\tfrac{1}{2}x^2 - 5x\right]_5^{10} = 5\left[(50 - 50) - \left(\tfrac{25}{2} - 25\right)\right] = 5\left(\tfrac{25}{2}\right) = 62.5 \text{ ft-lb} \end{aligned}$$

**15.** A "slice" of water $\Delta x$ m thick and lying at a depth of $x_i^*$ m (where $0 \le x_i^* \le \frac{1}{2}$) has volume $(2 \times 1 \times \Delta x)$ m$^3$, a mass of $2000\,\Delta x$ kg, weighs about $(9.8)(2000\,\Delta x) = 19{,}600\,\Delta x$ N, and thus requires about $19{,}600x_i^*\,\Delta x$ J of work for its removal. So $W = \lim\limits_{n\to\infty}\sum\limits_{i=1}^{n} 19{,}600x_i^*\,\Delta x = \int_0^{1/2} 19{,}600x\,dx = \left[9800x^2\right]_0^{1/2} = 2450$ J.

**16.** A horizontal cylindrical slice of water $\Delta x$ ft thick has a volume of $\pi r^2 h = \pi \cdot 12^2 \cdot \Delta x$ ft$^3$ and weighs about $\left(62.5 \text{ lb/ft}^3\right)\left(144\pi\,\Delta x \text{ ft}^3\right) = 9000\pi\,\Delta x$ lb. If the slice lies $x_i^*$ ft below the edge of the pool (where $1 \le x_i^* \le 5$), then the work needed to pump it out is about $9000\pi x_i^*\,\Delta x$. Thus,

$$W = \lim_{n\to\infty}\sum_{i=1}^{n} 9000\pi x_i^*\,\Delta x = \int_1^5 9000\pi x\,dx = \left[4500\pi x^2\right]_1^5 = 4500\pi(25 - 1) = 108{,}000\pi \text{ ft-lb}$$

**17.** (a) A rectangular "slice" of water $\Delta x$ m thick and lying $x$ ft above the bottom has width $x$ ft and volume $8x\,\Delta x$ m$^3$. It weighs about $(9.8 \times 1000)(8x\,\Delta x)$ N, and must be lifted $(5 - x)$ m by the pump, so the work needed is about $(9.8 \times 10^3)(5 - x)(8x\,\Delta x)$J. The total work required is

$$\begin{aligned} W &\approx \int_0^3 (9.8 \times 10^3)(5 - x)8x\,dx = (9.8 \times 10^3)\int_0^3 (40x - 8x^2)\,dx - (9.8 \times 10^3)\left[20x^2 - \tfrac{8}{3}x^3\right]_0^3 \\ &= (9.8 \times 10^3)(180 - 72) = (9.8 \times 10^3)(108) = 1058.4 \times 10^3 \approx 1.06 \times 10^6 \text{ J} \end{aligned}$$

(b) If only $4.7 \times 10^5$ J of work is done, then only the water above a certain level (call it $h$) will be pumped out. So we use the same formula as in part (a), except that the work is fixed, and we are trying to find the lower limit of integration:

$4.7 \times 10^5 \approx \int_h^3 (9.8 \times 10^3)(5 - x)8x\,dx = (9.8 \times 10^3)\left[20x^2 - \frac{8}{3}x^3\right]_h^3 \quad \Leftrightarrow$

$\frac{4.7}{9.8} \times 10^2 \approx 48 = \left(20 \cdot 3^2 - \frac{8}{3} \cdot 3^3\right) - \left(20h^2 - \frac{8}{3}h^3\right) \quad \Leftrightarrow$

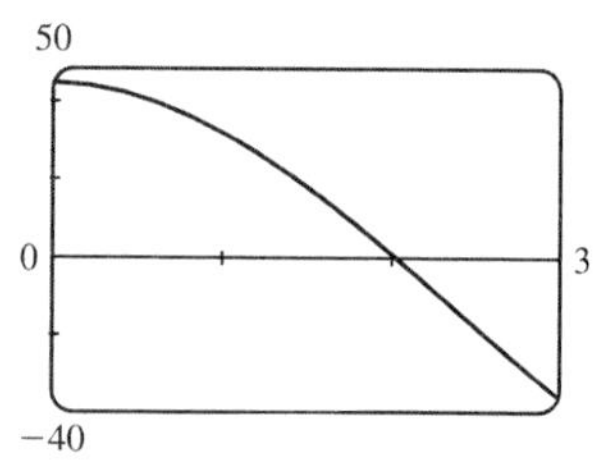

$2h^3 - 15h^2 + 45 = 0$. To find the solution of this equation, we plot $2h^3 - 15h^2 + 45$ between $h = 0$ and $h = 3$. We see that the equation is satisfied for $h \approx 2.0$. So the depth of water remaining in the tank is about 2.0 m.

**18.** Let $x$ be depth in feet, so that $0 \le x \le 5$. Then $\Delta W = (62.5)\pi\left(\sqrt{5^2 - x^2}\right)^2 \Delta x \cdot x$ ft-lb and

$$W \approx 62.5\pi \int_0^5 x(25 - x^2)\,dx = 62.5\pi\left[\tfrac{25}{2}x^2 - \tfrac{1}{4}x^4\right]_0^5 = 62.5\pi\left(\tfrac{625}{2} - \tfrac{625}{4}\right) = 62.5\pi\left(\tfrac{625}{4}\right) \approx 3.07 \times 10^4 \text{ ft-lb}$$

**19.** $V = \pi r^2 x$, so $V$ is a function of $x$ and $P$ can also be regarded as a function of $x$. If $V_1 = \pi r^2 x_1$ and $V_2 = \pi r^2 x_2$, then

$$W = \int_{x_1}^{x_2} F(x)\,dx = \int_{x_1}^{x_2} \pi r^2 P(V(x))\,dx = \int_{x_1}^{x_2} P(V(x))\,dV(x) \qquad [\text{Let } V(x) = \pi r^2 x, \text{ so } dV(x) = \pi r^2\,dx.]$$

$$= \int_{V_1}^{V_2} P(V)\,dV \quad \text{by the Substitution Rule.}$$

**20.** $160 \text{ lb/in}^2 = 160 \cdot 144 \text{ lb/ft}^2$, $100 \text{ in}^3 = \frac{100}{1728} \text{ ft}^3$, and $800 \text{ in}^3 = \frac{800}{1728} \text{ ft}^3$.

$k = PV^{1.4} = (160 \cdot 144)\left(\frac{100}{1728}\right)^{1.4} = 23{,}040\left(\frac{25}{432}\right)^{1.4} \approx 426.5$. Therefore, $P \approx 426.5V^{-1.4}$ and

$$W = \int_{100/1728}^{800/1728} 426.5V^{-1.4}\,dV = 426.5\left[\tfrac{1}{-0.4}V^{-0.4}\right]_{25/432}^{25/54} = (426.5)(2.5)\left[\left(\tfrac{432}{25}\right)^{0.4} - \left(\tfrac{54}{25}\right)^{0.4}\right] \approx 1.88 \times 10^3 \text{ ft-lb.}$$

**21.** (a) $W = \displaystyle\int_a^b F(r)\,dr = \int_a^b G\frac{m_1 m_2}{r^2}\,dr = Gm_1m_2\left[\frac{-1}{r}\right]_a^b = Gm_1m_2\left(\frac{1}{a} - \frac{1}{b}\right)$

(b) By part (a), $W = GMm\left(\dfrac{1}{R} - \dfrac{1}{R + 1{,}000{,}000}\right)$ where $M$ = mass of Earth in kg, $R$ = radius of Earth in m, and $m$ = mass of satellite in kg. (Note that 1000 km = 1,000,000 m.) Thus,

$$W = (6.67 \times 10^{-11})(5.98 \times 10^{24})(1000) \times \left(\frac{1}{6.37 \times 10^6} - \frac{1}{7.37 \times 10^6}\right) \approx 8.50 \times 10^9 \text{ J}$$

**22.** (a) $W = \displaystyle\int_R^\infty \frac{GMm}{r^2}\,dr = \lim_{t\to\infty}\int_R^t \frac{GMm}{r^2}\,dr = \lim_{t\to\infty} GMm\left[\frac{-1}{r}\right]_R^t = GMm\lim_{t\to\infty}\left(\frac{-1}{t} + \frac{1}{R}\right) = \frac{GMm}{R}$,

where $M$ = mass of Earth $= 5.98 \times 10^{24}$ kg, $m$ = mass of satellite $= 10^3$ kg,
$R$ = radius of Earth $= 6.37 \times 10^6$ m, and $G$ = gravitational constant $= 6.67 \times 10^{-11}$ N·m²/kg².

Therefore, work $= \dfrac{6.67 \times 10^{-11} \cdot 5.98 \times 10^{24} \cdot 10^3}{6.37 \times 10^6} \approx 6.26 \times 10^{10}$ J.

(b) From part (a), $W = \dfrac{GMm}{R}$. The initial kinetic energy supplies the needed work, so $\dfrac{1}{2}mv_0^2 = \dfrac{GMm}{R} \Rightarrow$ $v_0 = \sqrt{\dfrac{2GM}{R}}$.

**23.** Since an equation for the shape is $x^2 + y^2 = 10^2$ ($x \ge 0$), we have $y = \sqrt{100 - x^2}$. Thus, the area of the $i$th strip is $2\sqrt{100 - (x_i^*)^2}\,\Delta x$ and the pressure on the strip is $\rho g x_i^*$, so the hydrostatic force on the strip is $\rho g x_i^* \cdot 2\sqrt{100 - (x_i^*)^2}\,\Delta x$ and the total force on the plate $\approx \sum_{i=1}^{n} \rho g x_i^* \cdot 2\sqrt{100 - (x_i^*)^2}\,\Delta x$. The total force

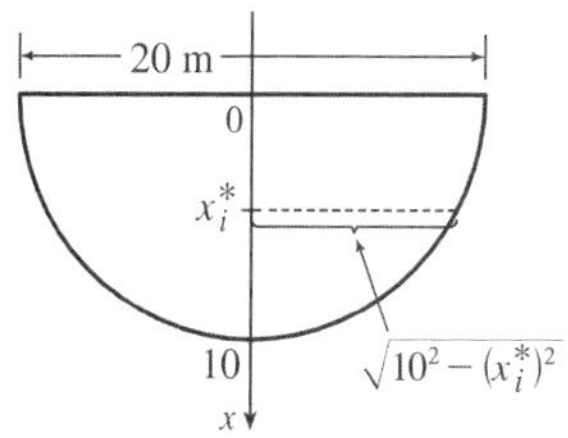


$$F = \lim_{n\to\infty}\sum_{i=1}^{n} \rho g x_i^* \cdot 2\sqrt{100 - (x_i^*)^2}\,\Delta x = \int_0^{10} 2\rho g x\sqrt{100 - x^2}\,dx$$

$$= -\rho g\int_0^{10}(100 - x^2)^{1/2}(-2x)\,dx = -\rho g\left[\tfrac{2}{3}(100 - x^2)^{3/2}\right]_0^{10} = -\tfrac{2}{3}\rho g(0 - 1000)$$

$$= \tfrac{2000}{3}\rho g \approx \tfrac{2000}{3} \cdot 1000 \cdot 9.8 \approx 6.5 \times 10^6 \text{ N} \qquad \left[\rho \approx 1000 \text{ kg/m}^3 \text{ and } g \approx 9.8 \text{ m/s}^2\right]$$

**24.** Set up a vertical $x$-axis as shown. Then the area of the $i$th rectangular strip is

$\frac{4}{3}(4 - x_i^*)\,\Delta x$. [By similar triangles, $\dfrac{w_i}{4 - x_i^*} = \dfrac{4}{3}$, so $w_i = \frac{4}{3}(4 - x_i^*)$.]

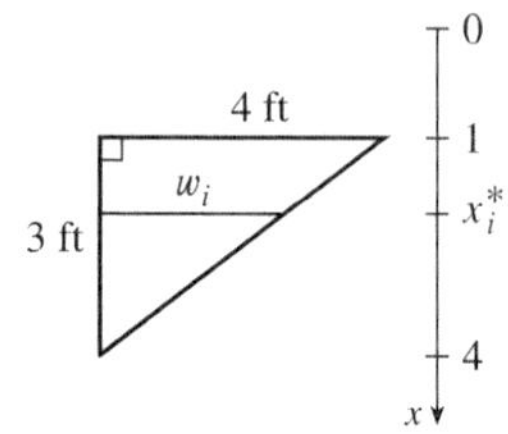


The pressure on the strip is $\delta x_i^*$, so the hydrostatic force on the strip is

$\delta x_i^* \cdot \frac{4}{3}(4 - x_i^*)\,\Delta x$ and the total force on the plate $\approx \sum\limits_{i=1}^{n} \delta x_i^* \cdot \frac{4}{3}(4 - x_i^*)\,\Delta x$.

The total force

$$F = \lim_{n\to\infty} \sum_{i=1}^{n} \delta x_i^* \cdot \tfrac{4}{3}(4 - x_i^*)\,\Delta x = \int_1^4 \delta x \cdot \tfrac{4}{3}(4 - x)\,dx = \tfrac{4}{3}\delta \int_1^4 (4x - x^2)\,dx$$

$$= \tfrac{4}{3}\delta\left[2x^2 - \tfrac{1}{3}x^3\right]_1^4 = \tfrac{4}{3}\delta\left[\left(32 - \tfrac{64}{3}\right) - \left(2 - \tfrac{1}{3}\right)\right] = \tfrac{4}{3}\delta(9) = 12\delta \approx 750 \text{ lb}$$

**25.** Using similar triangles, $\dfrac{4 \text{ ft wide}}{8 \text{ ft high}} = \dfrac{a \text{ ft wide}}{x_i^* \text{ ft high}}$, so $a = \frac{1}{2}x_i^*$ and the width of the $i$th rectangular strip is $12 + 2a = 12 + x_i^*$. The area of the strip is $(12 + x_i^*)\,\Delta x$. The pressure on the strip is $\delta x_i^*$.

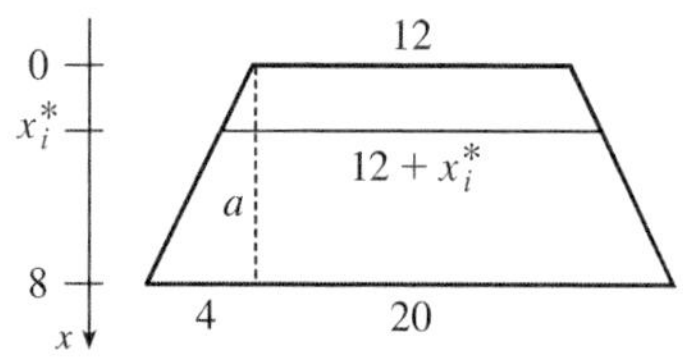


$$F = \lim_{n\to\infty} \sum_{i=1}^{n} \delta x_i^*(12 + x_i^*)\,\Delta x = \int_0^8 \delta x \cdot (12 + x)\,dx = \delta \int_0^8 (12x + x^2)\,dx$$

$$= \delta\left[6x^2 + \tfrac{x^3}{3}\right]_0^8 = \delta\left(384 + \tfrac{512}{3}\right) = (62.5)\tfrac{1664}{3} \approx 3.47 \times 10^4 \text{ lb}$$

**26.** By similar triangles, $w_i/4 = x_i^*/5$, so $w_i = \frac{4}{5}x_i^*$ and the area of the $i$th strip is $\frac{4}{5}x_i^*\,\Delta x$. The pressure on the strip is $\rho g x_i^*$, so the hydrostatic force on the strip is $\rho g x_i^* \cdot \frac{4}{5}x_i^*\,\Delta x$ and the total force on the plate $\approx \sum\limits_{i=1}^{n} \rho g x_i^* \cdot \frac{4}{5}x_i^*\,\Delta x$. The total force

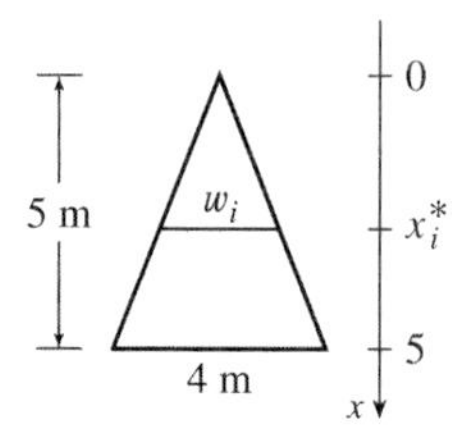


$$F = \lim_{n\to\infty} \sum_{i=1}^{n} \rho g x_i^* \cdot \tfrac{4}{5}x_i^*\,\Delta x = \int_0^5 \rho g x \cdot \tfrac{4}{5}x\,dx = \tfrac{4}{5}\rho g\left[\tfrac{1}{3}x^3\right]_0^5 = \tfrac{4}{5}\rho g \cdot \tfrac{125}{3} = \tfrac{100}{3}\rho g$$

$$\approx \tfrac{100}{3} \cdot 1000 \cdot 9.8 \approx 3.3 \times 10^5 \text{ N}$$

**27.** By similar triangles, $\dfrac{8}{4\sqrt{3}} = \dfrac{w_i}{x_i^*} \quad\Rightarrow\quad w_i = \dfrac{2x_i^*}{\sqrt{3}}$. The area of the $i$th rectangular strip is $\dfrac{2x_i^*}{\sqrt{3}}\,\Delta x$ and the pressure on it is $\rho g\left(4\sqrt{3} - x_i^*\right)$.

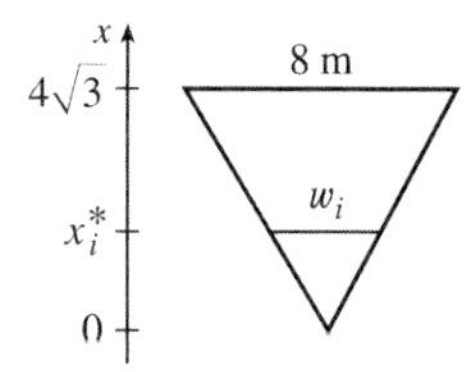


$$F = \int_0^{4\sqrt{3}} \rho g\left(4\sqrt{3} - x\right)\frac{2x}{\sqrt{3}}\,dx = 8\rho g \int_0^{4\sqrt{3}} x\,dx - \frac{2\rho g}{\sqrt{3}} \int_0^{4\sqrt{3}} x^2\,dx$$

$$= 4\rho g\left[x^2\right]_0^{4\sqrt{3}} - \frac{2\rho g}{3\sqrt{3}}\left[x^3\right]_0^{4\sqrt{3}} = 192\rho g - \frac{2\rho g}{3\sqrt{3}} 64 \cdot 3\sqrt{3} = 192\rho g - 128\rho g = 64\rho g$$

$$\approx 64(840)(9.8) \approx 5.27 \times 10^5 \text{ N}$$

**28.** The area of the $i$th rectangular strip is $2\sqrt{2y_i^*}\,\Delta y$ and the pressure on it is $\delta d_i = \delta(8 - y_i^*)$.

$$F = \int_0^8 \delta(8-y)2\sqrt{2y}\,dy = 42 \cdot 2 \cdot \sqrt{2}\int_0^8 (8-y)y^{1/2}\,dy$$

$$= 84\sqrt{2}\int_0^8 \left(8y^{1/2} - y^{3/2}\right)dy = 84\sqrt{2}\left[8 \cdot \tfrac{2}{3}y^{3/2} - \tfrac{2}{5}y^{5/2}\right]_0^8$$

$$= 84\sqrt{2}\left[8 \cdot \tfrac{2}{3} \cdot 16\sqrt{2} - \tfrac{2}{5} \cdot 128\sqrt{2}\right]$$

$$= 84\sqrt{2} \cdot 256\sqrt{2}\left(\tfrac{1}{3} - \tfrac{1}{5}\right) = 43{,}008 \cdot \tfrac{2}{15} = 5734.4 \text{ lb}$$

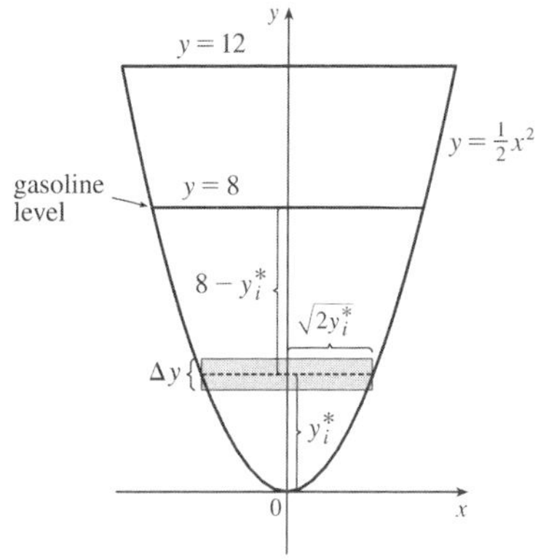


**29.** (a) The area of a strip is $20\,\Delta x$ and the pressure on it is $\delta x_i$.

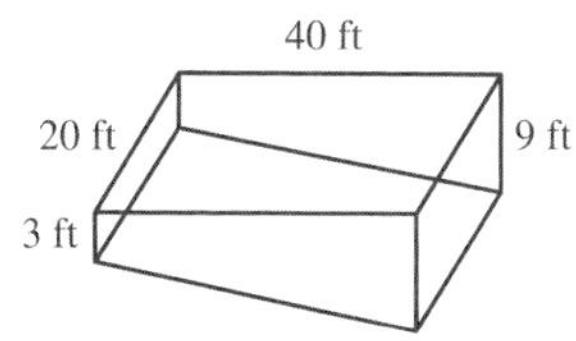


$$F = \int_0^3 \delta x 20\,dx = 20\delta\left[\tfrac{1}{2}x^2\right]_0^3 = 20\delta \cdot \tfrac{9}{2} = 90\delta$$

$$= 90(62.5) = 5625 \text{ lb} \approx 5.63 \times 10^3 \text{ lb}$$

(b) $F = \int_0^9 \delta x 20\,dx = 20\delta\left[\tfrac{1}{2}x^2\right]_0^9 = 20\delta \cdot \tfrac{81}{2} = 810\delta = 810(62.5) = 50{,}625 \text{ lb} \approx 5.06 \times 10^4$ lb.

(c) For the first 3 ft, the length of the side is constant at 40 ft. For $3 < x \leq 9$, we can use similar triangles to find the length $a$:

$$\frac{a}{40} = \frac{9-x}{6} \quad \Rightarrow \quad a = 40 \cdot \frac{9-x}{6}.$$

$$F = \int_0^3 \delta x 40\,dx + \int_3^9 \delta x(40)\tfrac{9-x}{6}\,dx = 40\delta\left[\tfrac{1}{2}x^2\right]_0^3 + \tfrac{20}{3}\delta\int_3^9 \left(9x - x^2\right)dx_3^9$$

$$= 180\delta + \tfrac{20}{3}\delta\left[\tfrac{9}{2}x^2 - \tfrac{1}{3}x^3\right] = 180\delta + \tfrac{20}{3}\delta\left[\left(\tfrac{729}{2} - 243\right) - \left(\tfrac{81}{2} - 9\right)\right]$$

$$= 180\delta + 600\delta = 780\delta = 780(62.5) = 48{,}750 \text{ lb} \approx 4.88 \times 10^4$$

(d) For any right triangle with hypotenuse on the bottom,

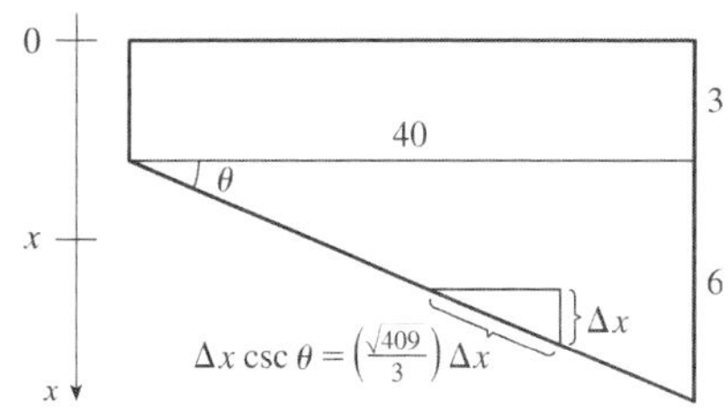


$$\csc\theta = \frac{\Delta x}{\text{hypotenuse}} \quad \Rightarrow$$

$$\text{hypotenuse} = \Delta x \csc\theta = \Delta x\,\frac{\sqrt{40^2 + 6^2}}{6} = \frac{\sqrt{409}}{3}\,\Delta x.$$

$$F = \int_3^9 \delta x 20\,\tfrac{\sqrt{409}}{3}\,dx = \tfrac{1}{3}\left(20\sqrt{409}\right)\delta\left[\tfrac{1}{2}x^2\right]_3^9$$

$$= \tfrac{1}{3} \cdot 10\sqrt{409}\,\delta(81 - 9) \approx 303{,}356 \text{ lb} \approx 3.03 \times 10^5 \text{ lb}$$

**30.** $F = \int_0^2 \rho g(10 - x)2\sqrt{4 - x^2}\,dx$

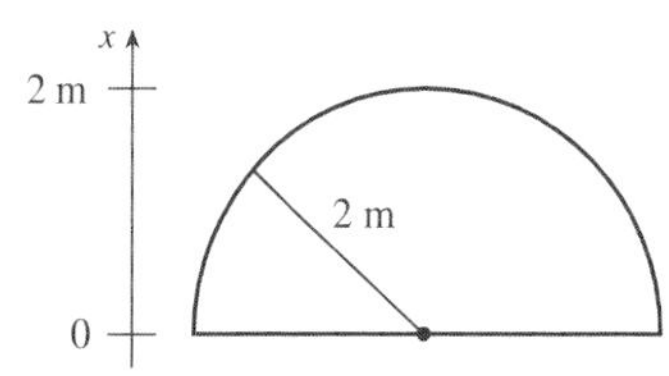


$$= 20\rho g\int_0^2 \sqrt{4-x^2}\,dx - \rho g\int_0^2 \sqrt{4-x^2}\,2x\,dx$$

$$= 20\rho g\tfrac{1}{4}\pi\left(2^2\right) - \rho g\int_0^4 u^{1/2}\,du \qquad \left[u = 4 - x^2,\ du = -2x\,dx\right]$$

$$= 20\pi\rho g - \tfrac{2}{3}\rho g\left[u^{3/2}\right]_0^4 = 20\pi\rho g - \tfrac{16}{3}\rho g = \rho g\left(20\pi - \tfrac{16}{3}\right)$$

$$= (1000)(9.8)\left(20\pi - \tfrac{16}{3}\right) \approx 5.63 \times 10^5 \text{ N}$$

**31.** $F = \int_2^5 \rho g x \cdot w(x)\,dx$, where $w(x)$ is the width of the plate at depth $x$. Since $n = 6$, $\Delta x = \frac{5-2}{6} = \frac{1}{2}$, and

$$\begin{aligned} F \approx S_6 &= \rho g \cdot \tfrac{1/2}{3}[2 \cdot w(2) + 4 \cdot 2.5 \cdot w(2.5) + 2 \cdot 3 \cdot w(3) + 4 \cdot 3.5 \cdot w(3.5) \\ &\qquad + 2 \cdot 4 \cdot w(4) + 4 \cdot 4.5 \cdot w(4.5) + 5 \cdot w(5)] \\ &= \tfrac{1}{6}\rho g(2 \cdot 0 + 10 \cdot 0.8 + 6 \cdot 1.7 + 14 \cdot 2.4 + 8 \cdot 2.9 + 18 \cdot 3.3 + 5 \cdot 3.6) \\ &= \tfrac{1}{6}(1000)(9.8)(152.4) \approx 2.5 \times 10^5 \text{ N} \end{aligned}$$

**32.** $M = m_1x_1 + m_2x_2 + m_3x_3 = 25(-2) + 20(3) + 10(7) = 80$; $\overline{x} = M/(m_1 + m_2 + m_3) = \frac{80}{55} = \frac{16}{11}$.

**33.** $m = \sum\limits_{i=1}^{3} m_i = 6 + 5 + 10 = 21$.

$M_x = \sum\limits_{i=1}^{3} m_iy_i = 6(5) + 5(-2) + 10(-1) = 10$; $M_y = \sum\limits_{i=1}^{3} m_ix_i = 6(1) + 5(3) + 10(-2) = 1$.

$\overline{x} = \dfrac{M_y}{m} = \dfrac{1}{21}$ and $\overline{y} = \dfrac{M_x}{m} = \dfrac{10}{21}$, so the center of mass of the system is $\left(\frac{1}{21}, \frac{10}{21}\right)$.

**34.** $M_x = \sum\limits_{i=1}^{4} m_iy_i = 6(-2) + 5(4) + 1(-7) + 4(-1) = -3$, $M_y = \sum\limits_{i=1}^{4} m_ix_i = 6(1) + 5(3) + 1(-3) + 4(6) = 42$,

and $m = \sum\limits_{i=1}^{4} m_i = 16$, so $\overline{x} = \dfrac{M_y}{m} = \dfrac{42}{16} = \dfrac{21}{8}$ and $\overline{y} = \dfrac{M_x}{m} = -\dfrac{3}{16}$; the center of mass is $(\overline{x}, \overline{y}) = \left(\frac{21}{8}, -\frac{3}{16}\right)$.

**35.** Since the region in the figure is symmetric about the $y$-axis, we know that $\overline{x} = 0$. The region is "bottom-heavy," so we know that $\overline{y} < 2$, and we might guess that $\overline{y} = 1.5$.

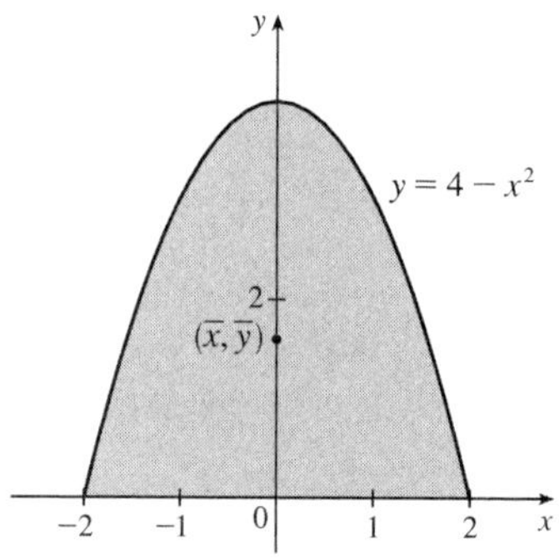


$$\begin{aligned} A &= \textstyle\int_{-2}^{2}(4 - x^2)\,dx = 2\int_0^2(4 - x^2)\,dx = 2\left[4x - \frac{1}{3}x^3\right]_0^2 \\ &= 2\left(8 - \tfrac{8}{3}\right) = \tfrac{32}{3} \end{aligned}$$

$\overline{x} = \frac{1}{A}\int_{-2}^{2} x(4 - x^2)\,dx = 0$ since $f(x) = x(4 - x^2)$ is an odd function (or since the region is symmetric about the $y$-axis).

$$\begin{aligned} \overline{y} &= \tfrac{1}{A}\textstyle\int_{-2}^{2} \frac{1}{2}(4 - x^2)^2\,dx = \frac{3}{32} \cdot \frac{1}{2} \cdot 2\int_0^2(16 - 8x^2 + x^4)\,dx = \frac{3}{32}\left[16x - \frac{8}{3}x^3 + \frac{1}{5}x^5\right]_0^2 \\ &= \tfrac{3}{32}\left(32 - \tfrac{64}{3} + \tfrac{32}{5}\right) = 3\left(1 - \tfrac{2}{3} + \tfrac{1}{5}\right) = 3\left(\tfrac{8}{15}\right) = \tfrac{8}{5} \end{aligned}$$

Thus, the centroid is $(\overline{x}, \overline{y}) = \left(0, \frac{8}{5}\right)$.

**36.** The region in the figure is "left-heavy" and "bottom-heavy," so we know $\overline{x} < 1$ and $\overline{y} < 1.5$, and we might guess that $\overline{x} = 0.7$ and $\overline{y} = 1.2$.

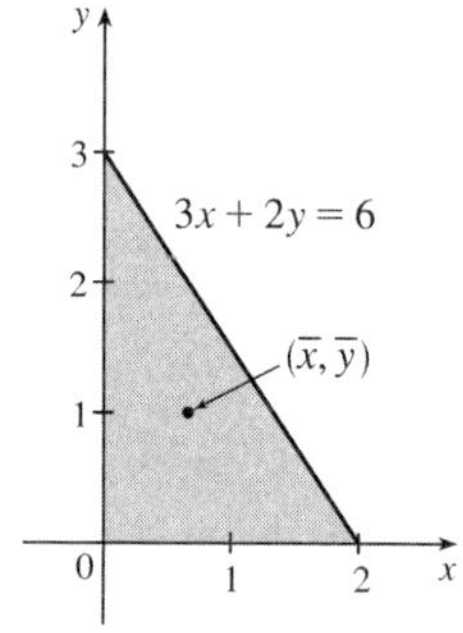


$3x + 2y = 6 \quad \Leftrightarrow \quad 2y = 6 - 3x \quad \Leftrightarrow \quad y = 3 - \frac{3}{2}x$.

$A = \int_0^2 \left(3 - \frac{3}{2}x\right)dx = \left[3x - \frac{3}{4}x^2\right]_0^2 = 6 - 3 = 3$.

$$\begin{aligned} \overline{x} &= \tfrac{1}{A}\textstyle\int_0^2 x\left(3 - \frac{3}{2}x\right)dx = \frac{1}{3}\int_0^2\left(3x - \frac{3}{2}x^2\right)dx = \frac{1}{3}\left[\frac{3}{2}x^2 - \frac{1}{2}x^3\right]_0^2 \\ &= \tfrac{1}{3}(6 - 4) = \tfrac{2}{3}; \end{aligned}$$

$$\overline{y} = \tfrac{1}{A}\textstyle\int_0^2 \frac{1}{2}\left(3 - \frac{3}{2}x\right)^2 dx = \frac{1}{3} \cdot \frac{1}{2}\int_0^2\left(9 - 9x + \frac{9}{4}x^2\right)dx = \frac{1}{6}\left[9x - \frac{9}{2}x^2 + \frac{3}{4}x^3\right]_0^2 = \frac{1}{6}(18 - 18 + 6) = 1.$$

Thus, the centroid is $(\overline{x}, \overline{y}) = \left(\frac{2}{3}, 1\right)$.

**37.** The region in the figure is "right-heavy" and "bottom-heavy," so we know $\overline{x} > 0.5$ and $\overline{y} < 1$, and we might guess that $\overline{x} = 0.6$ and $\overline{y} = 0.9$.

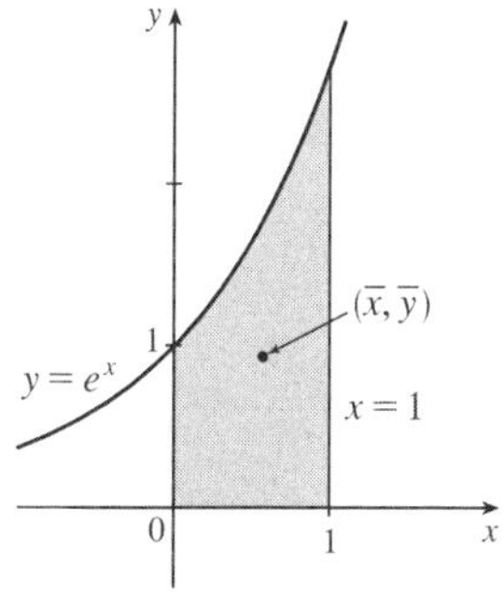


$A = \int_0^1 e^x\,dx = [e^x]_0^1 = e - 1$,

$\overline{x} = \frac{1}{A}\int_0^1 xe^x\,dx = \frac{1}{e-1}[xe^x - e^x]_0^1$ [by parts]

$= \frac{1}{e-1}[0-(-1)] = \frac{1}{e-1}$,

$\overline{y} = \frac{1}{A}\int_0^1 \frac{1}{2}(e^x)^2\,dx = \frac{1}{e-1}\cdot\frac{1}{4}\left[e^{2x}\right]_0^1 = \frac{1}{4(e-1)}\left(e^2-1\right) = \frac{e+1}{4}$.

Thus, the centroid is $(\overline{x},\overline{y}) = \left(\frac{1}{e-1}, \frac{e+1}{4}\right) \approx (0.58, 0.93)$.

**38.** The region in the figure is "left-heavy" and "bottom-heavy," so we know $\overline{x} < 1.5$ and $\overline{y} < 0.5$, and we might guess that $\overline{x} = 1.4$ and $\overline{y} = 0.4$.

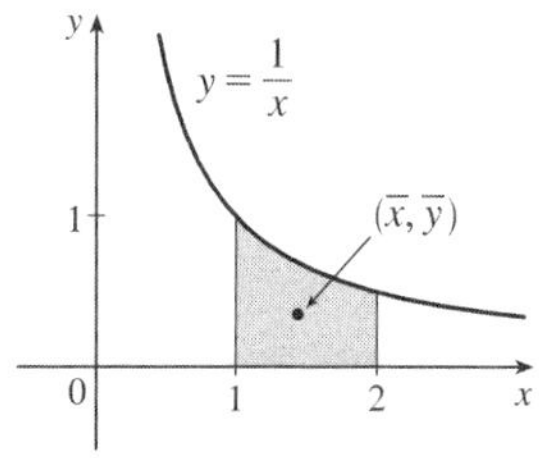


$A = \int_1^2 \frac{1}{x}\,dx = [\ln x]_1^2 = \ln 2$, $\overline{x} = \frac{1}{A}\int_1^2 x\cdot\frac{1}{x}\,dx = \frac{1}{A}[x]_1^2 = \frac{1}{A} = \frac{1}{\ln 2}$,

$\overline{y} = \frac{1}{A}\int_1^2 \frac{1}{2}\left(\frac{1}{x}\right)^2 dx = \frac{1}{2A}\int_1^2 x^{-2}\,dx = \frac{1}{2A}\left[-\frac{1}{x}\right]_1^2$

$= \frac{1}{2\ln 2}\left(-\frac{1}{2}+1\right) = \frac{1}{4\ln 2}$.

Thus, the centroid is $(\overline{x},\overline{y}) = \left(\frac{1}{\ln 2}, \frac{1}{4\ln 2}\right) \approx (1.44, 0.36)$.

**39.** By symmetry, $M_y = 0$ and $\overline{x} = 0$. $A = \frac{1}{2}bh = \frac{1}{2}\cdot 2\cdot 2 = 2$.

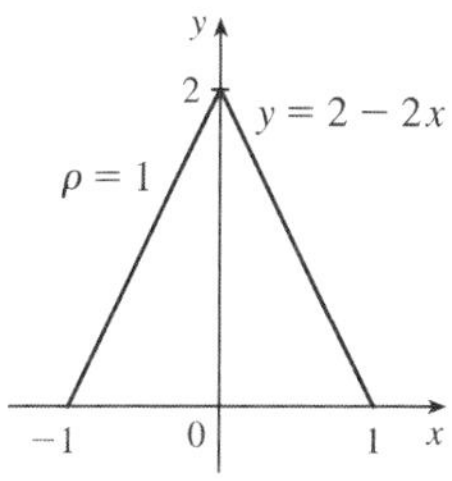


$M_x = \rho\int_{-1}^1 \frac{1}{2}(2-2x)^2\,dx = 2\rho\int_0^1 \frac{1}{2}(2-2x)^2\,dx$

$= \left(2\cdot 1\cdot\frac{1}{2}\cdot 2^2\right)\int_0^1 (1-x)^2\,dx = 4\int_1^0 u^2(-du)$ $[u = 1-x,\ du = -dx]$

$= -4\left[\frac{1}{3}u^3\right]_1^0 = -4\left(-\frac{1}{3}\right) = \frac{4}{3}$

$\overline{y} = \frac{1}{m}M_x = \frac{1}{\rho A}M_x = \frac{1}{1\cdot 2}\cdot\frac{4}{3} = \frac{2}{3}$. Thus, the centroid is $(\overline{x},\overline{y}) = \left(0, \frac{2}{3}\right)$.

**40.** By symmetry about the line $y = x$, we expect that $\overline{x} = \overline{y}$. $A = \frac{1}{4}\pi r^2$, so $m = \rho A = 2A = \frac{1}{2}\pi r^2$.

$M_x = \rho\int_0^r \frac{1}{2}\left(\sqrt{r^2-x^2}\right)^2 dx = 2\cdot\frac{1}{2}\int_0^r \left(r^2-x^2\right)dx = \left[r^2x - \frac{1}{3}x^3\right]_0^r = \frac{2}{3}r^3$.

$M_y = \rho\int_0^r x\sqrt{r^2-x^2}\,dx = \int_0^r \left(r^2-x^2\right)^{1/2}2x\,dx = \int_0^{r^2} u^{1/2}\,du$ $[u = r^2-x^2]$ $= \left[\frac{2}{3}u^{3/2}\right]_0^{r^2} = \frac{2}{3}r^3$.

$\overline{x} = \dfrac{1}{m}M_y = \dfrac{2}{\pi r^2}\left(\frac{2}{3}r^3\right) = \frac{4}{3\pi}r$; $\overline{y} = \dfrac{1}{m}M_x = \dfrac{2}{\pi r^2}\left(\frac{2}{3}r^3\right) = \frac{4}{3\pi}r$. Thus, the centroid is $(\overline{x},\overline{y}) = \left(\frac{4}{3\pi}r, \frac{4}{3\pi}r\right)$.

**41.** (a)

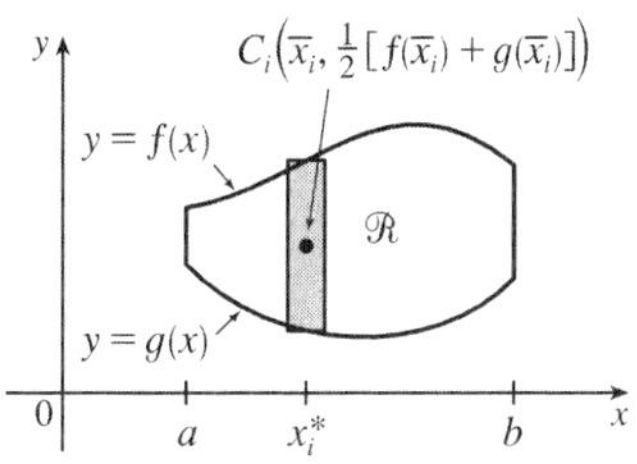


Suppose the region lies between two curves $y = f(x)$ and $y = g(x)$ where $f(x) \geq g(x)$, as illustrated in the figure. Use $n$ subintervals determined by points $x_i$ with $a = x_0 < x_1 < \cdots < x_n = b$ and choose $x_i^* = \overline{x}_i$ to be the midpoint of the $i$th subinterval; that is, $\overline{x}_i = \frac{1}{2}(x_{i-1}+x_i)$. Then the centroid of the $i$th approximating rectangle $R_i$ is its center $C_i = \left(\overline{x}_i, \frac{1}{2}[f(\overline{x}_i)+g(\overline{x}_i)]\right)$.

Its area is $[f(\overline{x}_i) - g(\overline{x}_i)]\,\Delta x$, so its mass is $\rho[f(\overline{x}_i) - g(\overline{x}_i)]\,\Delta x$.

Thus, $M_y(R_i) = \rho[f(\overline{x}_i) - g(\overline{x}_i)]\,\Delta x\cdot\overline{x}_i = \rho\overline{x}_i[f(\overline{x}_i) - g(\overline{x}_i)]\,\Delta x$ and

$M_x(R_i) = \rho[f(\overline{x}_i) - g(\overline{x}_i)]\,\Delta x \cdot \frac{1}{2}[f(\overline{x}_i) + g(\overline{x}_i)] = \rho \cdot \frac{1}{2}\left\{[f(\overline{x}_i)]^2 - [g(\overline{x}_i)]^2\right\}\Delta x$. Summing over $i$ and taking the limit as $n \to \infty$, we get $M_y = \lim_{n\to\infty} \sum_{i=1}^{n} \rho\overline{x}_i[f(\overline{x}_i) - g(\overline{x}_i)]\,\Delta x = \rho\int_a^b x[f(x) - g(x)]\,dx$ and

$M_x = \lim_{n\to\infty} \sum_{i=1}^{n} \rho \cdot \frac{1}{2}\left[f(\overline{x}_i)^2 - g(\overline{x}_i)^2\right]\Delta x = \rho\int_a^b \frac{1}{2}\{[f(x)]^2 - [g(x)]^2\}\,dx$. Thus,

$$\overline{x} = \frac{M_y}{m} = \frac{M_y}{\rho A} = \frac{1}{A}\int_a^b x[f(x) - g(x)]\,dx \text{ and } \overline{y} = \frac{M_x}{m} = \frac{M_x}{\rho A} = \frac{1}{A}\int_a^b \tfrac{1}{2}\{[f(x)]^2 - [g(x)]^2\}\,dx.$$

(b) 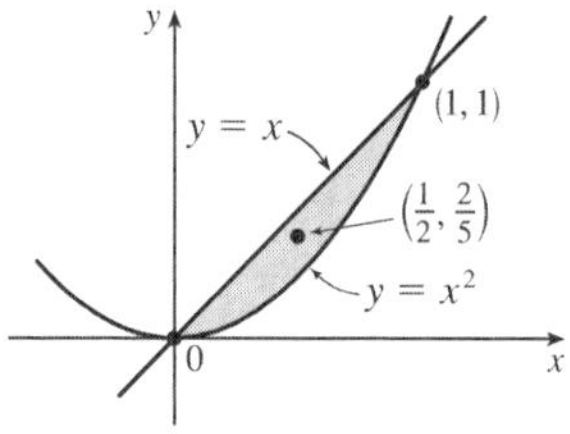


The region is sketched in the figure. We take $f(x) = x$, $g(x) = x^2$, $a = 0$, and $b = 1$ in the formulas in part (a). First we note that the area of the region is $A = \int_0^1 (x - x^2)\,dx = \left[\frac{1}{2}x^2 - \frac{1}{3}x^3\right]_0^1 = \frac{1}{6}$.

Therefore, $\overline{x} = \frac{1}{A}\int_0^1 x[f(x) - g(x)]\,dx = \frac{1}{1/6}\int_0^1 x(x - x^2)\,dx = 6\int_0^1 (x^2 - x^3)\,dx = 6\left[\frac{1}{3}x^3 - \frac{1}{4}x^4\right]_0^1 = \frac{1}{2}$

and $\overline{y} = \frac{1}{A}\int_0^1 \frac{1}{2}\left\{[f(x)]^2 - [g(x)]^2\right\}dx = \frac{1}{1/6}\int_0^1 \frac{1}{2}(x^2 - x^4)dx = 3\left[\frac{1}{3}x^3 - \frac{1}{5}x^5\right]_0^1 = \frac{2}{5}$.

The centroid is $\left(\frac{1}{2}, \frac{2}{5}\right)$.

**42.** (a) Let $0 \le x \le 1$. If $n < m$, then $x^n > x^m$; that is, raising $x$ to a larger power produces a smaller number.

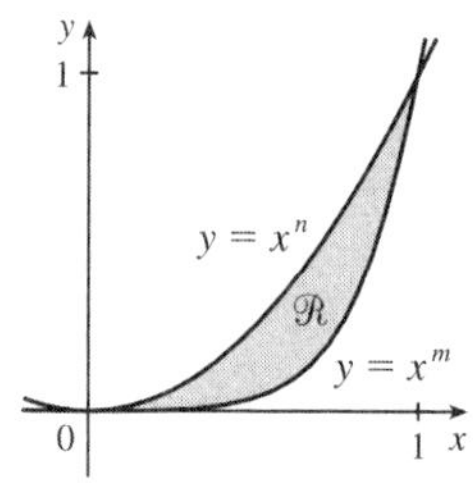


(b) Using Formulas 9 and the fact that the area of $\mathcal{R}$ is

$A = \int_0^1 (x^n - x^m)\,dx = \frac{1}{n+1} - \frac{1}{m+1} = \frac{m-n}{(n+1)(m+1)}$, we get

$$\overline{x} = \frac{(n+1)(m+1)}{m-n}\int_0^1 x[x^n - x^m]\,dx = \frac{(n+1)(m+1)}{m-n}\int_0^1 \left(x^{n+1} - x^{m+1}\right)dx$$
$$= \frac{(n+1)(m+1)}{m-n}\left[\frac{1}{n+2} - \frac{1}{m+2}\right] = \frac{(n+1)(m+1)}{(n+2)(m+2)}$$

and

$$\overline{y} = \frac{(n+1)(m+1)}{m-n}\int_0^1 \frac{1}{2}\left[(x^n)^2 - (x^m)^2\right]dx = \frac{(n+1)(m+1)}{2(m-n)}\int_0^1 \left(x^{2n} - x^{2m}\right)dx$$
$$= \frac{(n+1)(m+1)}{2(m-n)}\left[\frac{1}{2n+1} - \frac{1}{2m+1}\right] = \frac{(n+1)(m+1)}{(2n+1)(2m+1)}$$

(c) If we take $n = 3$ and $m = 4$, then

$$(\overline{x}, \overline{y}) = \left(\frac{4\cdot 5}{5\cdot 6}, \frac{4\cdot 5}{7\cdot 9}\right) = \left(\frac{2}{3}, \frac{20}{63}\right)$$

which lies outside $\mathcal{R}$ since $\left(\frac{2}{3}\right)^3 = \frac{8}{27} < \frac{20}{63}$. This is the simplest of many possibilities.

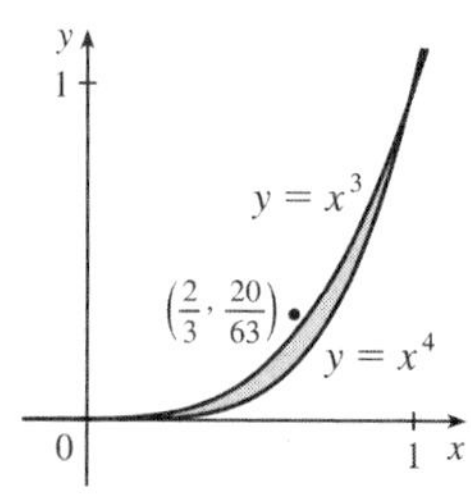

## 6.6 Applications to Economics and Biology

**1.** By the Net Change Theorem, $C(2000) - C(0) = \int_0^{2000} C'(x)\,dx \quad \Rightarrow$

$$\begin{aligned} C(2000) &= 20{,}000 + \int_0^{2000} (5 - 0.008x + 0.000009x^2)\,dx = 20{,}000 + \left[5x - 0.004x^2 + 0.000003x^3\right]_0^{2000} \\ &= 20{,}000 + 10{,}000 - 0.004(4{,}000{,}000) + 0.000003(8{,}000{,}000{,}000) = 30{,}000 - 16{,}000 + 24{,}000 \\ &= \$38{,}000 \end{aligned}$$

**2.** By the Net Change Theorem, $R(5000) - R(1000) = \int_{1000}^{5000} R'(x)\,dx \quad \Rightarrow$

$$\begin{aligned} R(5000) &= 12{,}400 + \int_{1000}^{5000} (12 - 0.0004x)\,dx = 12{,}400 + \left[12x - 0.0002x^2\right]_{1000}^{5000} \\ &= 12{,}400 + (60{,}000 - 5{,}000) - (12{,}000 - 200) = \$55{,}600 \end{aligned}$$

**3.** If the production level is raised from 1200 units to 1600 units, then the increase in cost is

$$\begin{aligned} C(1600) - C(1200) &= \int_{1200}^{1600} C'(x)\,dx = \int_{1200}^{1600} (74 + 1.1x - 0.002x^2 + 0.00004x^3)\,dx \\ &= \left[74x + 0.55x^2 - \tfrac{0.002}{3}x^3 + 0.00001x^4\right]_{1200}^{1600} \\ &= 64{,}331{,}733.33 - 20{,}464{,}800 = \$43{,}866{,}933.33 \end{aligned}$$

**4.** Consumer surplus $= \int_0^{30} [p(x) - p(30)]\,dx$

$$\begin{aligned} &= \int_0^{30} \left[5 - \tfrac{1}{10}x - \left(5 - \tfrac{30}{10}\right)\right] dx \\ &= \left[3x - \tfrac{1}{20}x^2\right]_0^{30} = 90 - 45 = \$45 \end{aligned}$$

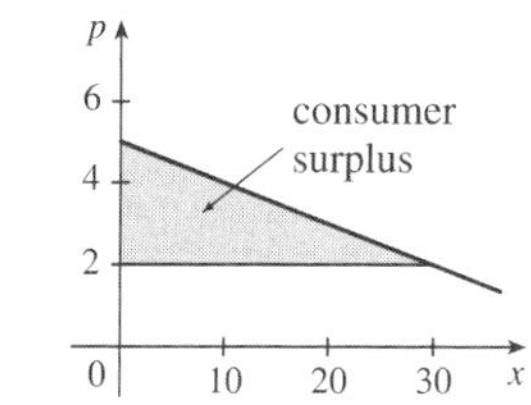


**5.** $p(x) = 10 \quad \Rightarrow \quad \dfrac{450}{x+8} = 10 \quad \Rightarrow \quad x + 8 = 45 \quad \Rightarrow \quad x = 37.$

Consumer surplus $= \displaystyle\int_0^{37} [p(x) - 10]\,dx = \int_0^{37} \left(\frac{450}{x+8} - 10\right) dx$

$$\begin{aligned} &= \left[450 \ln(x+8) - 10x\right]_0^{37} = (450 \ln 45 - 370) - 450 \ln 8 \\ &= 450 \ln\left(\tfrac{45}{8}\right) - 370 \approx \$407.25 \end{aligned}$$

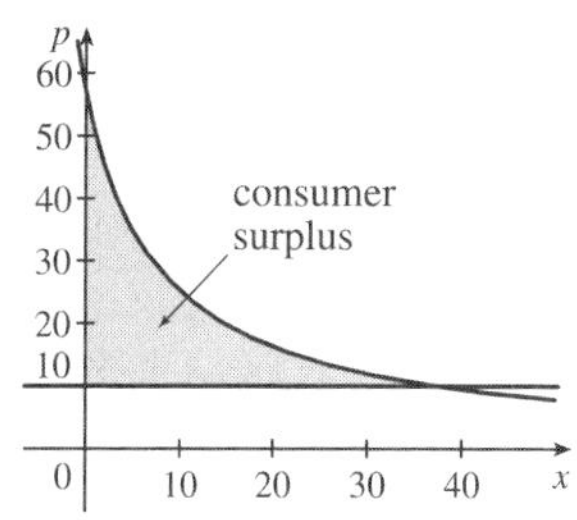


**6.** $p_S(x) = 3 + 0.01x^2$. $P = p_S(10) = 3 + 1 = 4$.

Producer surplus $= \int_0^{10} [P - p_S(x)]\,dx = \int_0^{10} \left[4 - 3 - 0.01x^2\right] dx$

$$= \left[x - \tfrac{0.01}{3}x^3\right]_0^{10} \approx 10 - 3.33 = \$6.67$$

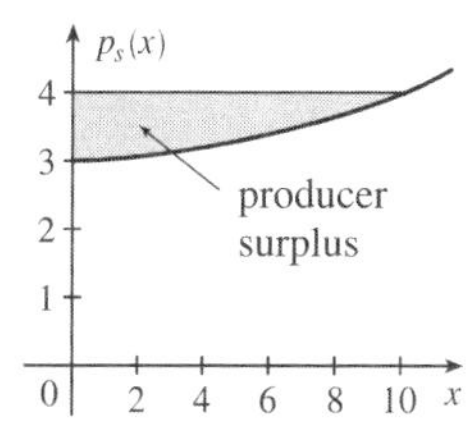


**7.** $P = p_S(x) \quad \Rightarrow \quad 400 = 200 + 0.2x^{3/2} \quad \Rightarrow \quad 200 = 0.2x^{3/2} \quad \Rightarrow \quad 1000 = x^{3/2} \quad \Rightarrow \quad x = 1000^{2/3} = 100.$

Producer surplus $= \int_0^{100} [P - p_S(x)]\,dx = \int_0^{100} \left[400 - (200 + 0.2x^{3/2})\right] dx = \int_0^{100} \left(200 - \tfrac{1}{5}x^{3/2}\right) dx$

$$= \left[200x - \tfrac{2}{25}x^{5/2}\right]_0^{100} = 20{,}000 - 8{,}000 = \$12{,}000$$

**8.** $p = 50 - \frac{1}{20}x$ and $p = 20 + \frac{1}{10}x$ intersect at $p = 40$ and $x = 200$.

Consumer surplus $= \int_0^{200} \left(50 - \frac{1}{20}x - 40\right) dx = \left[10x - \frac{1}{40}x^2\right]_0^{200} = \$1000$

Producer surplus $= \int_0^{200} \left(40 - 20 - \frac{1}{10}x\right) dx = \left[20x - \frac{1}{20}x^2\right]_0^{200} = \$2000$

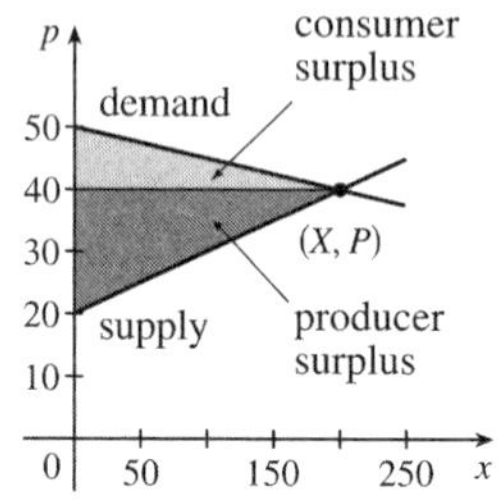


**9.** $p(x) = \dfrac{800{,}000e^{-x/5000}}{x + 20{,}000} = 16 \;\Rightarrow\; x = x_1 \approx 3727.04$.

Consumer surplus $= \int_0^{x_1} [p(x) - 16]\, dx \approx \$37{,}753$

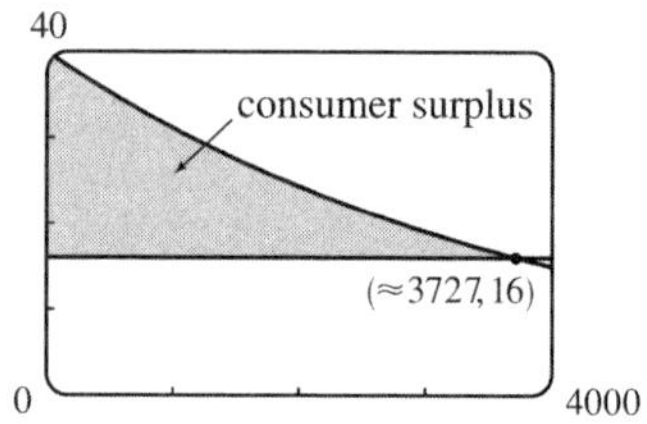


**10.** The demand function is linear with slope $\frac{-0.5}{35} = -\frac{1}{70}$ and $p(400) = 7.5$, so an equation is $p - 7.5 = -\frac{1}{70}(x - 400)$ or $p = -\frac{1}{70}x + \frac{185}{14}$. A selling price of \$6 implies that $6 = -\frac{1}{70}x + \frac{185}{14} \;\Rightarrow\; \frac{1}{70}x = \frac{185}{14} - \frac{84}{14} = \frac{101}{14} \;\Rightarrow\; x = 505$.

Consumer surplus $= \int_0^{505} \left(-\frac{1}{70}x + \frac{185}{14} - 6\right) dx = \left[-\frac{1}{140}x^2 + \frac{101}{14}x\right]_0^{505} \approx \$1821.61$.

**11.** $f(8) - f(4) = \int_4^8 f'(t)\, dt = \int_4^8 \sqrt{t}\, dt = \left[\frac{2}{3}t^{3/2}\right]_4^8 = \frac{2}{3}\left(16\sqrt{2} - 8\right) \approx \$9.75$ million

**12.** $n(9) - n(5) = \displaystyle\int_5^9 \left(2200 + 10e^{0.8t}\right) dt = \left[2200t + \frac{10e^{0.8t}}{0.8}\right]_5^9 = [2200t]_5^9 + \frac{25}{2}\left[e^{0.8t}\right]_5^9$

$= 2200(9 - 5) + 12.5\left(e^{7.2} - e^4\right) \approx 24{,}860$

**13.** $F = \dfrac{\pi P R^4}{8\eta l} = \dfrac{\pi(4000)(0.008)^4}{8(0.027)(2)} \approx 1.19 \times 10^{-4}\ \text{cm}^3/\text{s}$

**14.** If the flux remains constant, then $\dfrac{\pi P_0 R_0^4}{8\eta l} = \dfrac{\pi P R^4}{8\eta l} \;\Rightarrow\; P_0 R_0^4 = P R^4 \;\Rightarrow\; \dfrac{P}{P_0} = \left(\dfrac{R_0}{R}\right)^4$.

$R = \frac{3}{4}R_0 \;\Rightarrow\; \dfrac{P}{P_0} = \left(\dfrac{R_0}{\frac{3}{4}R_0}\right)^4 \;\Rightarrow\; P = P_0\left(\frac{4}{3}\right)^4 \approx 3.1605P_0 > 3P_0$; that is, the blood pressure is more than tripled.

**15.** $\int_0^{12} c(t)\, dt = \int_0^{12} \frac{1}{4}t(12 - t)\, dt = \int_0^{12} \left(3t - \frac{1}{4}t^2\right) dt = \left[\frac{3}{2}t^2 - \frac{1}{12}t^3\right]_0^{12} = (216 - 144) = 72\ \text{mg}\cdot\text{s/L}$.

Thus, the cardiac output is $F = \dfrac{A}{\int_0^{12} c(t)\, dt} = \dfrac{8\ \text{mg}}{72\ \text{mg}\cdot\text{s/L}} = \dfrac{1}{9}\ \text{L/s} = \dfrac{60}{9}\ \text{L/min}$.

**16.** As in Example 2, we will estimate the cardiac output using Simpson's Rule with $\Delta t = 2$.

$\int_0^{20} c(t)\, dt \approx \frac{2}{3}[1(0) + 4(2.4) + 2(5.1) + 4(7.8) + 2(7.6) + 4(5.4) + 2(3.9) + 4(2.3) + 2(1.6) + 4(0.7) + 1(0)]$

$= \frac{2}{3}(110.8) \approx 73.87\ \text{mg}\cdot\text{s/L}$

Therefore, $F \approx \dfrac{A}{73.87} = \dfrac{8}{73.87} \approx 0.1083\ \text{L/s}$ or $6.498\ \text{L/min}$.

## 6.7 Probability

**1.** (a) $\int_{30,000}^{40,000} f(x)\,dx$ is the probability that a randomly chosen tire will have a lifetime between 30,000 and 40,000 miles.

(b) $\int_{25,000}^{\infty} f(x)\,dx$ is the probability that a randomly chosen tire will have a lifetime of at least 25,000 miles.

**2.** (a) The probability that you drive to school in less than 15 minutes is $\int_0^{15} f(t)\,dt$.

(b) The probability that it takes you more than half an hour to get to school is $\int_{30}^{\infty} f(t)\,dt$.

**3.** (a) In general, we must satisfy the two conditions that are mentioned before Example 1—namely, (1) $f(x) \geq 0$ for all $x$, and (2) $\int_{-\infty}^{\infty} f(x)\,dx = 1$. For $0 \leq x \leq 4$, we have $f(x) = \frac{3}{64}x\sqrt{16 - x^2} \geq 0$, so $f(x) \geq 0$ for all $x$. Also,

$$\int_{-\infty}^{\infty} f(x)\,dx = \int_0^4 \tfrac{3}{64}x\sqrt{16-x^2}\,dx = -\tfrac{3}{128}\int_0^4 (16-x^2)^{1/2}(-2x)\,dx = -\tfrac{3}{128}\left[\tfrac{2}{3}(16-x^2)^{3/2}\right]_0^4$$
$$= -\tfrac{1}{64}\left[(16-x^2)^{3/2}\right]_0^4 = -\tfrac{1}{64}(0-64) = 1.$$

Therefore, $f$ is a probability density function.

(b) $P(X < 2) = \int_{-\infty}^{2} f(x)\,dx = \int_0^2 \frac{3}{64}x\sqrt{16-x^2}\,dx = -\frac{3}{128}\int_0^2 (16-x^2)^{1/2}(-2x)\,dx$

$$= -\tfrac{3}{128}\left[\tfrac{2}{3}(16-x^2)^{3/2}\right]_0^2 = -\tfrac{1}{64}\left[(16-x^2)^{3/2}\right]_0^2 = -\tfrac{1}{64}(12^{3/2} - 16^{3/2})$$
$$= \tfrac{1}{64}(64 - 12\sqrt{12}) = \tfrac{1}{64}(64 - 24\sqrt{3}) = 1 - \tfrac{3}{8}\sqrt{3} \approx 0.350481$$

**4.** (a) For $0 \leq x \leq 1$, we have $f(x) = kx^2(1-x)$, which is nonnegative if and only if $k \geq 0$. Also,

$\int_{-\infty}^{\infty} f(x)\,dx = \int_0^1 kx^2(1-x)\,dx = k\int_0^1 (x^2 - x^3)\,dx = k\left[\frac{1}{3}x^3 - \frac{1}{4}x^4\right]_0^1 = k/12$. Now $k/12 = 1 \Leftrightarrow k = 12$.

Therefore, $f$ is a probability density function if and only if $k = 12$.

(b) Let $k = 12$.

$$P\left(X \geq \tfrac{1}{2}\right) = \int_{1/2}^{\infty} f(x)\,dx = \int_{1/2}^{1} 12x^2(1-x)\,dx = \int_{1/2}^{1}(12x^2 - 12x^3)\,dx = \left[4x^3 - 3x^4\right]_{1/2}^{1}$$
$$= (4-3) - \left(\tfrac{1}{2} - \tfrac{3}{16}\right) = 1 - \tfrac{5}{16} = \tfrac{11}{16}$$

(c) The mean

$$\mu = \int_{-\infty}^{\infty} xf(x)\,dx = \int_0^1 x\cdot 12x^2(1-x)\,dx = 12\int_0^1 (x^3 - x^4)\,dx = 12\left[\tfrac{1}{4}x^4 - \tfrac{1}{5}x^5\right]_0^1$$
$$= 12\left(\tfrac{1}{4} - \tfrac{1}{5}\right) = \tfrac{12}{20} = \tfrac{3}{5}$$

**5.** (a) In general, we must satisfy the two conditions that are mentioned before Example 1—namely, (1) $f(x) \geq 0$ for all $x$, and (2) $\int_{-\infty}^{\infty} f(x)\,dx = 1$. Since $f(x) = 0$ or $f(x) = 0.1$, condition (1) is satisfied. For condition (2), we see that $\int_{-\infty}^{\infty} f(x)\,dx = \int_0^{10} 0.1\,dx = \left[\frac{1}{10}x\right]_0^{10} = 1$. Thus, $f(x)$ is a probability density function for the spinner's values.

(b) Since all the numbers between 0 and 10 are equally likely to be selected, we expect the mean to be halfway between the endpoints of the interval; that is, $x = 5$.

$$\mu = \int_{-\infty}^{\infty} xf(x)\,dx = \int_0^{10} x(0.1)\,dx = \left[\tfrac{1}{20}x^2\right]_0^{10} = \tfrac{100}{20} = 5, \quad \text{as expected.}$$

**6.** (a) As in the preceding exercise, (1) $f(x) \geq 0$ and (2) $\int_{-\infty}^{\infty} f(x)\,dx = \int_0^{10} f(x)\,dx = \frac{1}{2}(10)(0.2)$ [area of a triangle] $= 1$. So $f(x)$ is a probability density function.

(b) (i) $P(X < 3) = \int_0^3 f(x)\,dx = \frac{1}{2}(3)(0.1) = \frac{3}{20} = 0.15$

(ii) We first compute $P(X > 8)$ and then subtract that value and our answer in (i) from 1 (the total probability).

$P(X > 8) = \int_8^{10} f(x)\,dx = \frac{1}{2}(2)(0.1) = \frac{2}{20} = 0.10$. So $P(3 \leq X \leq 8) = 1 - 0.15 - 0.10 = 0.75$.

(c) We find equations of the lines from $(0, 0)$ to $(6, 0.2)$ and from $(6, 0.2)$ to $(10, 0)$, and find that

$$f(x) = \begin{cases} \frac{1}{30}x & \text{if } 0 \leq x < 6 \\ -\frac{1}{20}x + \frac{1}{2} & \text{if } 6 \leq x < 10 \\ 0 & \text{otherwise} \end{cases}$$

$$\mu = \int_{-\infty}^{\infty} xf(x)\,dx = \int_0^6 x\left(\tfrac{1}{30}x\right)dx + \int_6^{10} x\left(-\tfrac{1}{20}x + \tfrac{1}{2}\right)dx = \left[\tfrac{1}{90}x^3\right]_0^6 + \left[-\tfrac{1}{60}x^3 + \tfrac{1}{4}x^2\right]_6^{10}$$

$$= \tfrac{216}{90} + \left(-\tfrac{1000}{60} + \tfrac{100}{4}\right) - \left(-\tfrac{216}{60} + \tfrac{36}{4}\right) = \tfrac{16}{3} = 5.\overline{3}$$

**7.** We need to find $m$ so that $\int_m^\infty f(t)\,dt = \frac{1}{2}$ $\Rightarrow$ $\lim\limits_{x\to\infty} \int_m^x \frac{1}{5}e^{-t/5}\,dt = \frac{1}{2}$ $\Rightarrow$ $\lim\limits_{x\to\infty} \left[\frac{1}{5}(-5)e^{-t/5}\right]_m^x = \frac{1}{2}$ $\Rightarrow$ $(-1)(0 - e^{-m/5}) = \frac{1}{2}$ $\Rightarrow$ $e^{-m/5} = \frac{1}{2}$ $\Rightarrow$ $-m/5 = \ln\frac{1}{2}$ $\Rightarrow$ $m = -5\ln\frac{1}{2} = 5\ln 2 \approx 3.47$ min.

**8.** (a) $\mu = 1000$ $\Rightarrow$ $f(t) = \begin{cases} 0 & \text{if } t < 0 \\ \frac{1}{1000}e^{-t/1000} & \text{if } t \geq 0 \end{cases}$

(i) $P(0 \leq X \leq 200) = \int_0^{200} \frac{1}{1000}e^{-t/1000}\,dt = \left[-e^{-t/1000}\right]_0^{200} = -e^{-1/5} + 1 \approx 0.181$

(ii) $P(X > 800) = \int_{800}^\infty \frac{1}{1000}e^{-t/1000}\,dt = \lim\limits_{x\to\infty}\left[-e^{-t/1000}\right]_{800}^x = 0 + e^{-4/5} \approx 0.449$

(b) We need to find $m$ so that $\int_m^\infty f(t)\,dt = \frac{1}{2}$ $\Rightarrow$ $\lim\limits_{x\to\infty}\int_m^x \frac{1}{1000}e^{-t/1000}\,dt = \frac{1}{2}$ $\Rightarrow$ $\lim\limits_{x\to\infty}\left[-e^{-t/1000}\right]_m^x = \frac{1}{2}$ $\Rightarrow$ $0 + e^{-m/1000} = \frac{1}{2}$ $\Rightarrow$ $-m/1000 = \ln\frac{1}{2}$ $\Rightarrow$ $m = -1000\ln\frac{1}{2} = 1000\ln 2 \approx 693.1$ h.

**9.** We use an exponential density function with $\mu = 2.5$ min.

(a) $P(X > 4) = \int_4^\infty f(t)\,dt = \lim\limits_{x\to\infty}\int_4^x \frac{1}{2.5}e^{-t/2.5}\,dt = \lim\limits_{x\to\infty}\left[-e^{-t/2.5}\right]_4^x = 0 + e^{-4/2.5} \approx 0.202$

(b) $P(0 \leq X \leq 2) = \int_0^2 f(t)\,dt = \left[-e^{-t/2.5}\right]_0^2 = -e^{-2/2.5} + 1 \approx 0.551$

(c) We need to find a value $a$ so that $P(X \geq a) = 0.02$, or, equivalently, $P(0 \leq X \leq a) = 0.98$ $\Leftrightarrow$ $\int_0^a f(t)\,dt = 0.98$ $\Leftrightarrow$ $\left[-e^{-t/2.5}\right]_0^a = 0.98$ $\Leftrightarrow$ $-e^{-a/2.5} + 1 = 0.98$ $\Leftrightarrow$ $e^{-a/2.5} = 0.02$ $\Leftrightarrow$ $-a/2.5 = \ln 0.02$ $\Leftrightarrow$ $a = -2.5\ln\frac{1}{50} = 2.5\ln 50 \approx 9.78 \text{ min} \approx 10$ min. The ad should say that if you aren't served within 10 minutes, you get a free hamburger.

**10.** (a) With $\mu = 69$ and $\sigma = 2.8$, we have $P(65 \leq X \leq 73) = \displaystyle\int_{65}^{73} \frac{1}{2.8\sqrt{2\pi}} \exp\left(-\frac{(x-69)^2}{2\cdot 2.8^2}\right)dx \approx 0.847$ (using a calculator or computer to estimate the integral).

(b) $P(X > 6 \text{ feet}) = P(X > 72 \text{ inches}) = 1 - P(0 \leq X \leq 72) \approx 1 - 0.858 = 0.142$, so 14.2% of the adult male population is more than 6 feet tall.

**11.** $P(X \geq 10) = \displaystyle\int_{10}^\infty \frac{1}{4.2\sqrt{2\pi}} \exp\left(-\frac{(x-9.4)^2}{2\cdot 4.2^2}\right)dx$. To avoid the improper integral we approximate it by the integral from 10 to 100. Thus, $P(X \geq 10) \approx \displaystyle\int_{10}^{100} \frac{1}{4.2\sqrt{2\pi}} \exp\left(-\frac{(x-9.4)^2}{2\cdot 4.2^2}\right)dx \approx 0.443$ (using a calculator or computer to estimate the integral), so about 44 percent of the households throw out at least 10 lb of paper a week.

*Note*: We can't evaluate $1 - P(0 \leq X \leq 10)$ for this problem since a significant amount of area lies to the left of $X = 0$.

**12.** (a) $P(0 \leq X \leq 480) = \displaystyle\int_0^{480} \frac{1}{12\sqrt{2\pi}} \exp\left(-\frac{(x-500)^2}{2\cdot 12^2}\right)dx \approx 0.0478$ (using a calculator or computer to estimate the integral), so there is about a 4.78% chance that a particular box contains less than 480 g of cereal.

(b) We need to find $\mu$ so that $P(0 \le X < 500) = 0.05$. Using our calculator or computer to find $P(0 \le X \le 500)$ for various values of $\mu$, we find that if $\mu = 519.73$, $P = 0.05007$; and if $\mu = 519.74$, $P = 0.04998$. So a good target weight is at least $519.74$ g.

**13.** $P(\mu - 2\sigma \le X \le \mu + 2\sigma) = \int_{\mu-2\sigma}^{\mu+2\sigma} \frac{1}{\sigma\sqrt{2\pi}} \exp\left(-\frac{(x-\mu)^2}{2\sigma^2}\right) dx$. Substituting $t = \frac{x-\mu}{\sigma}$ and $dt = \frac{1}{\sigma}\,dx$ gives us $\int_{-2}^{2} \frac{1}{\sigma\sqrt{2\pi}} e^{-t^2/2}(\sigma\,dt) = \frac{1}{\sqrt{2\pi}} \int_{-2}^{2} e^{-t^2/2}\,dt \approx 0.9545$.

**14.** Let $f(x) = \begin{cases} 0 & \text{if } x < 0 \\ ce^{-cx} & \text{if } x \ge 0 \end{cases}$ where $c = 1/\mu$. By using parts, tables, or a CAS, we find that

(1): $\int xe^{bx}\,dx = \left(e^{bx}/b^2\right)(bx - 1)$

(2): $\int x^2e^{bx}\,dx = \left(e^{bx}/b^3\right)(b^2x^2 - 2bx + 2)$

Now

$$\begin{aligned} \sigma^2 &= \textstyle\int_{-\infty}^{\infty} (x-\mu)^2 f(x)\,dx = \int_{-\infty}^{0} (x-\mu)^2 f(x)\,dx + \int_0^{\infty} (x-\mu)^2 f(x)\,dx \\ &= 0 + \lim_{t\to\infty} c \textstyle\int_0^t (x-\mu)^2 e^{-cx}\,dx = c \cdot \lim_{t\to\infty} \int_0^t \left(x^2e^{-cx} - 2x\mu e^{-cx} + \mu^2 e^{-cx}\right) dx \end{aligned}$$

Next we use (2) and (1) with $b = -c$ to get

$$\sigma^2 = c \lim_{t\to\infty} \left[ -\frac{e^{-cx}}{c^3}(c^2x^2 + 2cx + 2) - 2\mu\frac{e^{-cx}}{c^2}(-cx - 1) + \mu^2 \frac{e^{-cx}}{-c} \right]_0^t$$

Using l'Hospital's Rule several times, along with the fact that $\mu = 1/c$, we get

$$\sigma^2 = c\left[0 - \left(-\frac{2}{c^3} + \frac{2}{c}\cdot\frac{1}{c^2} + \frac{1}{c^2}\cdot\frac{1}{-c}\right)\right] = c\left(\frac{1}{c^3}\right) = \frac{1}{c^2} \quad\Rightarrow\quad \sigma = \frac{1}{c} = \mu$$

## 6 Review

### CONCEPT CHECK

**1.** (a) See Section 6.1, Figure 2 and Equations 6.1.1 and 6.1.2.

(b) Instead of using "top minus bottom" and integrating from left to right, we use "right minus left" and integrate from bottom to top. See Figures 9 and 10 in Section 6.1.

**2.** The numerical value of the area represents the number of meters by which Sue is ahead of Kathy after 1 minute.

**3.** (a) See the discussion in Section 6.2, near Figures 2 and 3, ending in the Definition of Volume.

(b) See the discussion between Examples 5 and 6 in Section 6.2. If the cross-section is a disk, find the radius in terms of $x$ or $y$ and use $A = \pi(\text{radius})^2$. If the cross-section is a washer, find the inner radius $r_{\text{in}}$ and outer radius $r_{\text{out}}$ and use $A = \pi\left(r_{\text{out}}^2\right) - \pi\left(r_{\text{in}}^2\right)$.

**4.** (a) The length of a curve is defined to be the limit of the lengths of the inscribed polygons, as described near Figure 3 in Section 6.3.

(b) See Equation 6.3.1.

(c) See Equations 6.3.2 and 6.3.3.

**5.** (a) The average value of a function $f$ on an interval $[a, b]$ is $f_{\text{ave}} = \dfrac{1}{b-a}\displaystyle\int_a^b f(x)\,dx$.

(b) The Mean Value Theorem for Integrals says that there is a number $c$ at which the value of $f$ is exactly equal to the average value of the function, that is, $f(c) = f_{\text{ave}}$. For a geometric interpretation of the Mean Value Theorem for Integrals, see Figure 2 in Section 6.4 and the discussion that accompanies it.

**6.** $\int_0^6 f(x)\,dx$ represents the amount of work done. Its units are newton-meters, or joules.

**7.** Let $c(x)$ be the cross-sectional length of the wall (measured parallel to the surface of the fluid) at depth $x$. Then the hydrostatic force against the wall is given by $F = \int_a^b \delta x c(x)\,dx$, where $a$ and $b$ are the lower and upper limits for $x$ at points of the wall and $\delta$ is the weight density of the fluid.

**8.** (a) The center of mass is the point at which the plate balances horizontally.

(b) See Equations 6.5.12.

**9.** See Figure 3 in Section 6.6, and the discussion which precedes it.

**10.** (a) See the definition in the first paragraph of the subsection *Cardiac Output* in Section 6.6.

(b) See the discussion in the second paragraph of the subsection *Cardiac Output* in Section 6.6.

**11.** A probability density function $f$ is a function on the domain of a continuous random variable $X$ such that $\int_a^b f(x)\,dx$ measures the probability that $X$ lies between $a$ and $b$. Such a function $f$ has nonnegative values and satisfies the relation $\int_D f(x)\,dx = 1$, where $D$ is the domain of the corresponding random variable $X$. If $D = \mathbb{R}$, or if we define $f(x) = 0$ for real numbers $x \notin D$, then $\int_{-\infty}^{\infty} f(x)\,dx = 1$. (Of course, to work with $f$ in this way, we must assume that the integrals of $f$ exist.)

**12.** (a) $\int_0^{130} f(x)\,dx$ represents the probability that the weight of a randomly chosen female college student is less than 130 pounds.

(b) $\mu = \int_{-\infty}^{\infty} xf(x)\,dx = \int_0^{\infty} xf(x)\,dx$

(c) The median of $f$ is the number $m$ such that $\int_m^{\infty} f(x)\,dx = \frac{1}{2}$.

**13.** See the discussion near Equation 3 in Section 6.7.

## EXERCISES

**1.** $A = \int_0^1 \left[(e^x - 1) - (x^2 - x)\right] dx$

$= \int_0^1 (e^x - 1 - x^2 + x)\,dx = \left[e^x - x - \frac{1}{3}x^3 + \frac{1}{2}x^2\right]_0^1$

$= \left(e - 1 - \frac{1}{3} + \frac{1}{2}\right) - (1 - 0 - 0 + 0) = e - \frac{11}{6}$

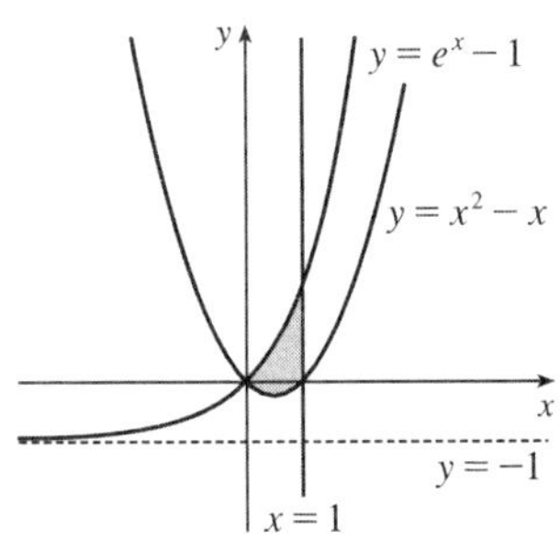


**2.** $y^2 + 3y = -y \;\Leftrightarrow\; y^2 + 4y = 0 \;\Leftrightarrow\; y(y+4) = 0 \;\Leftrightarrow$ $y = 0$ or $-4$.

$A = \int_{-4}^0 \left[-y - (y^2 + 3y)\right] dy = \int_{-4}^0 (-y^2 - 4y)\,dy$

$= \left[-\frac{1}{3}y^3 - 2y^2\right]_{-4}^0 = 0 - \left(\frac{64}{3} - 32\right) = \frac{32}{3}$

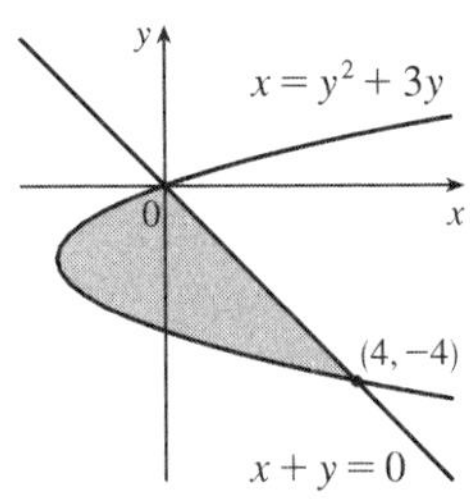

**3.** $x = 2\theta - \sin\theta \Rightarrow dx = (2 - \cos\theta)\,d\theta$

$$A = \int_0^{2\pi} y\,dx = \int_0^{2\pi} [(2-\cos\theta)(2-\cos\theta)]\,d\theta$$
$$= \int_0^{2\pi} (4 - 4\cos\theta + \cos^2\theta)\,d\theta = \int_0^{2\pi}\left(4 - 4\cos\theta + \tfrac{1}{2} + \tfrac{1}{2}\cos 2\theta\right) d\theta$$
$$= \left[4\theta - 4\sin\theta + \tfrac{1}{2}\theta + \tfrac{1}{4}\sin 2\theta\right]_0^{2\pi} = (8\pi - 0 + \pi + 0) - (0) = 9\pi$$

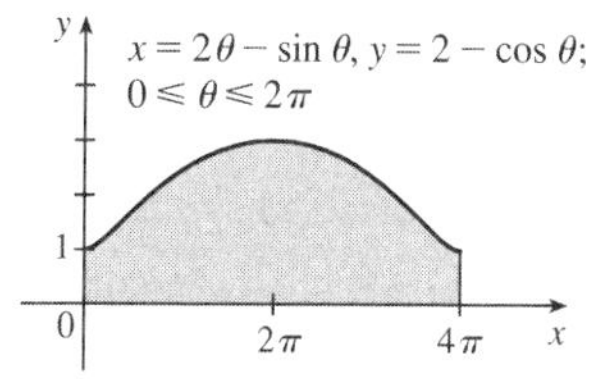


**4.** $V = \int_0^1 \pi\left[(1+x)^2 - \left(e^{-2x}\right)^2\right] dx = \pi\int_0^1 \left(1 + 2x + x^2 - e^{-4x}\right) dx$

$$= \pi\left[x + x^2 + \tfrac{1}{3}x^3 + \tfrac{1}{4}e^{-4x}\right]_0^1 = \pi\left(1 + 1 + \tfrac{1}{3} + \tfrac{1}{4}e^{-4} - \tfrac{1}{4}\right)$$
$$= \pi\left(\frac{25}{12} + \frac{1}{4e^4}\right)$$

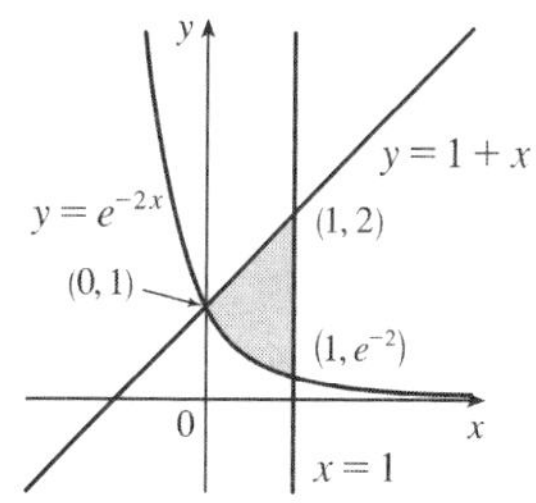


**5.** (a) Using the Midpoint Rule on $[0, 1]$ with $f(x) = \tan(x^2)$ and $n = 4$, we estimate

$$A = \int_0^1 \tan(x^2)\,dx \approx \tfrac{1}{4}\left[\tan\left(\left(\tfrac{1}{8}\right)^2\right) + \tan\left(\left(\tfrac{3}{8}\right)^2\right) + \tan\left(\left(\tfrac{5}{8}\right)^2\right) + \tan\left(\left(\tfrac{7}{8}\right)^2\right)\right] \approx \tfrac{1}{4}(1.53) \approx 0.38$$

(b) Using the Midpoint Rule on $[0, 1]$ with $f(x) = \pi\tan^2(x^2)$ (for disks) and $n = 4$, we estimate

$$V = \int_0^1 f(x)\,dx \approx \tfrac{1}{4}\pi\left[\tan^2\left(\left(\tfrac{1}{8}\right)^2\right) + \tan^2\left(\left(\tfrac{3}{8}\right)^2\right) + \tan^2\left(\left(\tfrac{5}{8}\right)^2\right) + \tan^2\left(\left(\tfrac{7}{8}\right)^2\right)\right] \approx \tfrac{\pi}{4}(1.114) \approx 0.87$$

**6.** (a) $A = \int_0^1 (2x - x^2 - x^3)\,dx = \left[x^2 - \tfrac{1}{3}x^3 - \tfrac{1}{4}x^4\right]_0^1 = 1 - \tfrac{1}{3} - \tfrac{1}{4} = \tfrac{5}{12}$

(b) A cross-section is a washer with inner radius $x^3$ and outer radius $2x - x^2$, so its area is $\pi(2x - x^2)^2 - \pi(x^3)^2$.

$$V = \int_0^1 A(x)\,dx = \int_0^1 \pi\left[(2x - x^2)^2 - (x^3)^2\right] dx = \int_0^1 \pi(4x^2 - 4x^3 + x^4 - x^6)\,dx$$
$$= \pi\left[\tfrac{4}{3}x^3 - x^4 + \tfrac{1}{5}x^5 - \tfrac{1}{7}x^7\right]_0^1 = \pi\left(\tfrac{4}{3} - 1 + \tfrac{1}{5} - \tfrac{1}{7}\right) = \tfrac{41\pi}{105}$$

(c) Using the method of cylindrical shells,

$V = \int_0^1 2\pi x(2x - x^2 - x^3)\,dx = \int_0^1 2\pi(2x^2 - x^3 - x^4)\,dx = 2\pi\left[\tfrac{2}{3}x^3 - \tfrac{1}{4}x^4 - \tfrac{1}{5}x^5\right]_0^1 = 2\pi\left(\tfrac{2}{3} - \tfrac{1}{4} - \tfrac{1}{5}\right) = \tfrac{13\pi}{30}$.

**7.** (a) A cross-section is a washer with inner radius $x^2$ and outer radius $x$.

$V = \int_0^1 \pi\left[(x)^2 - (x^2)^2\right] dx = \int_0^1 \pi(x^2 - x^4)\,dx = \pi\left[\tfrac{1}{3}x^3 - \tfrac{1}{5}x^5\right]_0^1 = \pi\left[\tfrac{1}{3} - \tfrac{1}{5}\right] = \tfrac{2\pi}{15}$

(b) A cross-section is a washer with inner radius $y$ and outer radius $\sqrt{y}$.

$V = \int_0^1 \pi\left[\left(\sqrt{y}\right)^2 - y^2\right] dy = \int_0^1 \pi(y - y^2)\,dy = \pi\left[\tfrac{1}{2}y^2 - \tfrac{1}{3}y^3\right]_0^1 = \pi\left[\tfrac{1}{2} - \tfrac{1}{3}\right] = \tfrac{\pi}{6}$

(c) A cross-section is a washer with inner radius $2 - x$ and outer radius $2 - x^2$.

$V = \int_0^1 \pi\left[(2 - x^2)^2 - (2 - x)^2\right] dx = \int_0^1 \pi(x^4 - 5x^2 + 4x)\,dx = \pi\left[\tfrac{1}{5}x^5 - \tfrac{5}{3}x^3 + 2x^2\right]_0^1 = \pi\left[\tfrac{1}{5} - \tfrac{5}{3} + 2\right] = \tfrac{8\pi}{15}$

**8.** (a)

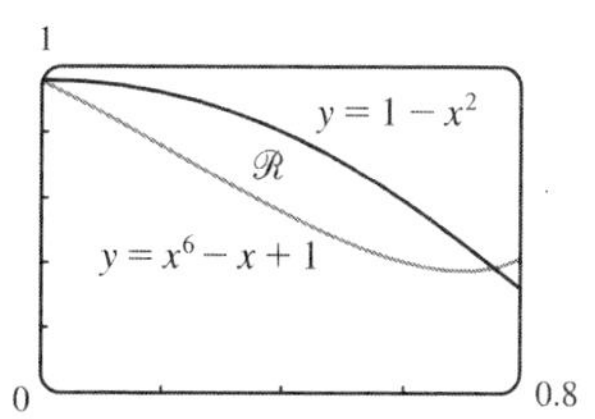


From the graph, we see that the curves intersect at $x = 0$ and at $x = a \approx 0.75$, with $1 - x^2 > x^6 - x + 1$ on $(0, a)$.

(b) The area of $\mathcal{R}$ is $A = \int_0^a \left[(1 - x^2) - (x^6 - x + 1)\right] dx = \left[-\tfrac{1}{3}x^3 - \tfrac{1}{7}x^7 + \tfrac{1}{2}x^2\right]_0^a \approx 0.12$.

(c) Using washers, the volume generated when $\mathcal{R}$ is rotated about the $x$-axis is

$$V = \pi\int_0^a \left[(1 - x^2)^2 - (x^6 - x + 1)^2\right] dx = \pi\int_0^a (-x^{12} + 2x^7 - 2x^6 + x^4 - 3x^2 + 2x)\,dx$$
$$= \pi\left[-\tfrac{1}{13}x^{13} + \tfrac{1}{4}x^8 - \tfrac{2}{7}x^7 + \tfrac{1}{5}x^5 - x^3 + x^2\right]_0^a \approx 0.54$$

(d) Using shells, the volume generated when $\mathcal{R}$ is rotated about the $y$-axis is
$V = \int_0^a 2\pi x[(1-x^2) - (x^6 - x + 1)]\,dx = 2\pi\int_0^a (-x^3 - x^7 + x^2)\,dx = 2\pi\left[-\frac{1}{4}x^4 - \frac{1}{8}x^8 + \frac{1}{3}x^3\right]_0^a \approx 0.31.$

**9.** (a) The solid is obtained by rotating the region $\mathcal{R} = \left\{(x,y) \mid 0 \le x \le \frac{\pi}{2}, 0 \le y \le \sqrt{2}\cos x\right\}$ about the $x$-axis.

(b) The solid is obtained by rotating the region $\mathcal{R} = \left\{(x,y) \mid 0 \le x \le 1, 2 - \sqrt{x} \le y \le 2 - x^2\right\}$ about the $x$-axis.
*Or*: The solid is obtained by rotating the region $\mathcal{R} = \left\{(x,y) \mid 0 \le x \le 1, x^2 \le y \le \sqrt{x}\right\}$ about the line $y = 2$.

**10.** With an $x$-axis in the normal position, at $x = 7$ we have $C = 2\pi r = 45 \Rightarrow r(7) = \frac{2\pi}{45}$.
Using Simpson's Rule with $n = 4$ and $\Delta x = 7$, we have
$V = \int_0^{28} \pi[r(x)]^2\,dx \approx S_4 = \frac{7}{3}\left[0 + 4\pi\left(\frac{45}{2\pi}\right)^2 + 2\pi\left(\frac{53}{2\pi}\right)^2 + 4\pi\left(\frac{45}{2\pi}\right)^2 + 0\right] = \frac{7}{3}\left(\frac{21{,}818}{4\pi}\right) \approx 4051\text{ cm}^3.$

**11.** Take the base to be the disk $x^2 + y^2 \le 9$. Then $V = \int_{-3}^3 A(x)\,dx$, where $A(x_0)$ is the area of the isosceles right triangle whose hypotenuse lies along the line $x = x_0$ in the $xy$-plane. The length of the hypotenuse is $2\sqrt{9 - x^2}$ and the length of each leg is $\sqrt{2}\sqrt{9-x^2}$. $A(x) = \frac{1}{2}\left(\sqrt{2}\sqrt{9-x^2}\right)^2 = 9 - x^2$, so

$$V = 2\int_0^3 A(x)\,dx = 2\int_0^3 (9 - x^2)\,dx = 2\left[9x - \frac{1}{3}x^3\right]_0^3 = 2(27 - 9) = 36$$

**12.** $V = \int_{-1}^1 A(x)\,dx = 2\int_0^1 A(x)\,dx = 2\int_0^1 \left[(2 - x^2) - x^2\right]^2 dx = 2\int_0^1 \left[2(1 - x^2)\right]^2 dx$
$= 8\int_0^1 (1 - 2x^2 + x^4)\,dx = 8\left[x - \frac{2}{3}x^3 + \frac{1}{5}x^5\right]_0^1 = 8\left(1 - \frac{2}{3} + \frac{1}{5}\right) = \frac{64}{15}$

**13.** Equilateral triangles with sides measuring $\frac{1}{4}x$ meters have height $\frac{1}{4}x\sin 60^\circ = \frac{\sqrt{3}}{8}x$. Therefore,
$A(x) = \frac{1}{2}\cdot\frac{1}{4}x\cdot\frac{\sqrt{3}}{8}x = \frac{\sqrt{3}}{64}x^2$. $V = \int_0^{20} A(x)\,dx = \frac{\sqrt{3}}{64}\int_0^{20} x^2\,dx = \frac{\sqrt{3}}{64}\left[\frac{1}{3}x^3\right]_0^{20} = \frac{8000\sqrt{3}}{64\cdot 3} = \frac{125\sqrt{3}}{3}\text{ m}^3.$

**14.** (a) By the symmetry of the problem, we consider only the solid to the right of the origin. The semicircular cross-sections perpendicular to the $x$-axis have radius $1 - x$, so $A(x) = \frac{1}{2}\pi(1-x)^2$. Now we can calculate
$V = 2\int_0^1 A(x)\,dx = 2\int_0^1 \frac{1}{2}\pi(1-x)^2\,dx = \int_0^1 \pi(1-x)^2\,dx = -\frac{\pi}{3}\left[(1-x)^3\right]_0^1 = \frac{\pi}{3}.$

(b) Cut the solid with a plane perpendicular to the $x$-axis and passing through the $y$-axis. Fold the half of the solid in the region $x \le 0$ under the $xy$-plane so that the point $(-1, 0)$ comes around and touches the point $(1, 0)$. The resulting solid is a right circular cone of radius 1 with vertex at $(x,y,z) = (1,0,0)$ and with its base in the $yz$-plane, centered at the origin. The volume of this cone is $\frac{1}{3}\pi r^2 h = \frac{1}{3}\pi\cdot 1^2\cdot 1 = \frac{\pi}{3}$.

**15.** $x = 3t^2, y = 2t^3, 0 \le t \le 2.$
$L = \int_0^2 \sqrt{(dx/dt)^2 + (dy/dt)^2}\,dt = \int_0^2 \sqrt{(6t)^2 + (6t^2)^2}\,dt = \int_0^2 \sqrt{36t^2 + 36t^4}\,dt = 6\int_0^2 t\sqrt{1+t^2}\,dt$
$= 6\int_1^5 \sqrt{u}\left(\frac{1}{2}du\right)\ \left[u = 1 + t^2, du = 2t\,dt\right] = 3\left[\frac{2}{3}u^{3/2}\right]_1^5 = 2(5\sqrt{5} - 1)$

**16.** $y = \dfrac{1}{x^2}, 1 \le x \le 2.$ $\dfrac{dy}{dx} = -\dfrac{2}{x^3}$, so $1 + \left(\dfrac{dy}{dx}\right)^2 = 1 + \dfrac{4}{x^6} \Rightarrow f(x) = \sqrt{1 + 4/x^6}$ and $L = \int_1^2 \sqrt{1 + 4/x^6}\,dx$. By Simpson's Rule with $n = 10$, $L \approx \frac{1/10}{3}[f(1) + 4f(1.1) + 2f(1.2) + 4f(1.3) + \cdots + 2f(1.8) + 4f(1.9) + f(2)] \approx 1.297$.

**17.** $y = \frac{1}{6}(x^2+4)^{3/2} \Rightarrow dy/dx = \frac{1}{4}(x^2+4)^{1/2}(2x) \Rightarrow$
$1 + (dy/dx)^2 = 1 + \left[\frac{1}{2}x(x^2+4)^{1/2}\right]^2 = 1 + \frac{1}{4}x^2(x^2+4) = \frac{1}{4}x^4 + x^2 + 1 = \left(\frac{1}{2}x^2 + 1\right)^2.$
Thus, $L = \int_0^3 \sqrt{\left(\frac{1}{2}x^2+1\right)^2}\,dx = \int_0^3 \left(\frac{1}{2}x^2 + 1\right)dx = \left[\frac{1}{6}x^3 + x\right]_0^3 = \frac{15}{2}.$

**18.** $y = \int_1^x \sqrt{\sqrt{t} - 1}\,dt \Rightarrow dy/dx = \sqrt{\sqrt{x} - 1} \Rightarrow 1 + (dy/dx)^2 = 1 + (\sqrt{x} - 1) = \sqrt{x}.$
Thus, $L = \int_1^{16} \sqrt{\sqrt{x}}\,dx = \int_1^{16} x^{1/4}\,dx = \frac{4}{5}\left[x^{5/4}\right]_1^{16} = \frac{4}{5}(32 - 1) = \frac{124}{5}.$

**19.** $f(x) = kx \quad \Rightarrow \quad 30 \text{ N} = k(15 - 12) \text{ cm} \quad \Rightarrow \quad k = 10 \text{ N/cm} = 1000 \text{ N/m}.\quad 20 \text{ cm} - 12 \text{ cm} = 0.08 \text{ m} \quad \Rightarrow$

$W = \int_0^{0.08} kx\,dx = 1000 \int_0^{0.08} x\,dx = 500\big[x^2\big]_0^{0.08} = 500(0.08)^2 = 3.2 \text{ N-m} = 3.2 \text{ J}.$

**20.** The work needed to raise the elevator alone is 1600 lb $\times$ 30 ft = 48,000 ft-lb. The work needed to raise the bottom 170 ft of cable is 170 ft $\times$ 10 lb/ft $\times$ 30 ft = 51,000 ft-lb. The work needed to raise the top 30 ft of cable is $\int_0^{30} 10x\,dx = \big[5x^2\big]_0^{30} = 5 \cdot 900 = 4500$ ft-lb. Adding these, we see that the total work needed is $48{,}000 + 51{,}000 + 4{,}500 = 103{,}500$ ft-lb.

**21.** (a) The parabola has equation $y = ax^2$ with vertex at the origin and passing through $(4, 4)$. $4 = a \cdot 4^2 \quad \Rightarrow \quad a = \frac{1}{4} \quad \Rightarrow \quad y = \frac{1}{4}x^2 \quad \Rightarrow \quad x^2 = 4y \quad \Rightarrow \quad x = 2\sqrt{y}$. Each circular disk has radius $2\sqrt{y}$ and is moved $4 - y$ ft.

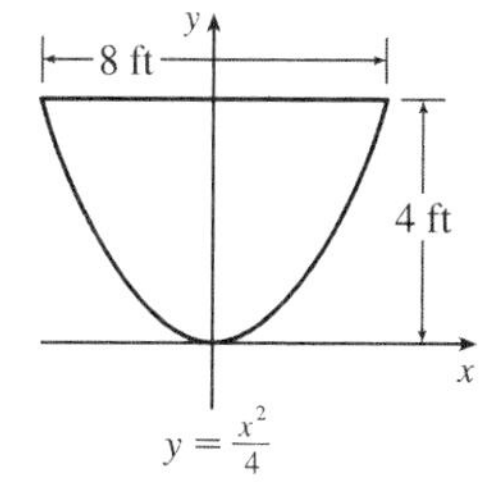


$$W = \int_0^4 \pi\left(2\sqrt{y}\right)^2 62.5\,(4 - y)\;dy = 250\pi \int_0^4 y(4 - y)\,dy$$

$$= 250\pi\left[2y^2 - \tfrac{1}{3}y^3\right]_0^4 = 250\pi\left(32 - \tfrac{64}{3}\right) = \tfrac{8000\pi}{3} \approx 8378 \text{ ft-lb}$$

(b) In part (a) we knew the final water level (0) but not the amount of work done. Here we use the same equation, except with the work fixed, and the lower limit of integration (that is, the final water level — call it $h$) unknown: $W = 4000 \quad \Leftrightarrow$ $250\pi\left[2y^2 - \frac{1}{3}y^3\right]_h^4 = 4000 \quad \Leftrightarrow \quad \frac{16}{\pi} = \left[\left(32 - \frac{64}{3}\right) - \left(2h^2 - \frac{1}{3}h^3\right)\right] \quad \Leftrightarrow$ $h^3 - 6h^2 + 32 - \frac{48}{\pi} = 0$. We graph the function $f(h) = h^3 - 6h^2 + 32 - \frac{48}{\pi}$ on the interval $[0, 4]$ to see where it is 0. From the graph, $f(h) = 0$ for $h \approx 2.1$. So the depth of water remaining is about 2.1 ft.

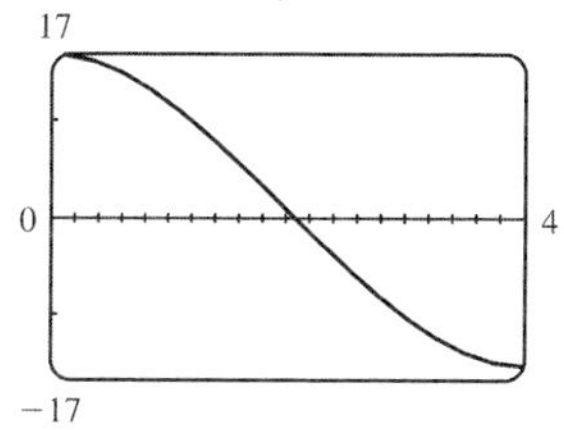


**22.** $F = \int_0^4 \delta(4 - y)2\left(2\sqrt{y}\right)dy = 4\delta \int_0^4 \left(4y^{1/2} - y^{3/2}\right)dy$

$= 4\delta\left[\frac{8}{3}y^{3/2} - \frac{2}{5}y^{5/2}\right]_0^4 = 4\delta\left(\frac{64}{3} - \frac{64}{5}\right) = 256\delta\left(\frac{1}{3} - \frac{1}{5}\right)$

$= \frac{512}{15}\delta \approx 2133.3 \text{ lb} \qquad [\delta \approx 62.5 \text{ lb/ft}^3]$

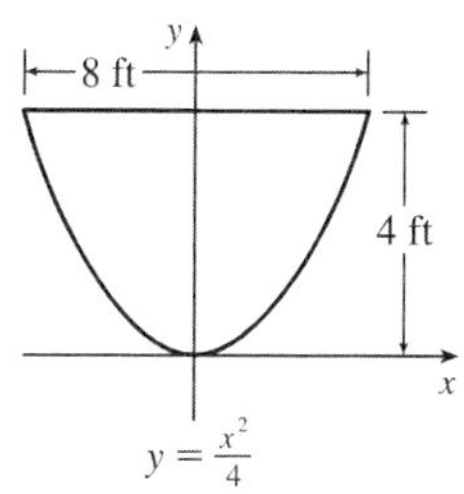


**23.** As in Example 4 of Section 6.5, $\dfrac{a}{2 - x} = \dfrac{1}{2} \quad \Rightarrow \quad 2a = 2 - x$ and $w = 2(1.5 + a) = 3 + 2a = 3 + 2 - x = 5 - x$.

Thus, $F = \int_0^2 \rho g x(5 - x)\,dx = \rho g\left[\frac{5}{2}x^2 - \frac{1}{3}x^3\right]_0^2 = \rho g\left(10 - \frac{8}{3}\right) = \frac{22}{3}\delta \quad [\rho g = \delta] \quad \approx \frac{22}{3} \cdot 62.5 \approx 458$ lb.

**24.** An equation of the line passing through $(0, 0)$ and $(3, 2)$ is $y = \frac{2}{3}x$. $A = \frac{1}{2} \cdot 3 \cdot 2 = 3$. Therefore, using Equations 6.5.12, $\overline{x} = \frac{1}{3}\int_0^3 x\left(\frac{2}{3}x\right)dx = \frac{2}{27}\left[x^3\right]_0^3 = 2$ and $\overline{y} = \frac{1}{3}\int_0^3 \frac{1}{2}\left(\frac{2}{3}x\right)^2 dx = \frac{2}{81}\left[x^3\right]_0^3 = \frac{2}{3}$. Thus, the centroid is $(\overline{x}, \overline{y}) = \left(2, \frac{2}{3}\right)$.

**25.** $x = 100 \quad \Rightarrow \quad P = 2000 - 0.1(100) - 0.01(100)^2 = 1890$

$$\text{Consumer surplus} = \int_0^{100} [p(x) - P]\,dx = \int_0^{100} \left(2000 - 0.1x - 0.01x^2 - 1890\right)dx$$

$$= \left[110x - 0.05x^2 - \tfrac{0.01}{3}x^3\right]_0^{100} = 11{,}000 - 500 - \tfrac{10{,}000}{3} \approx \$7166.67$$

**26.** $f_{\text{ave}} = \frac{1}{2-0}\int_0^2 x^2\sqrt{1+x^3}\,dx = \frac{1}{2}\cdot\frac{1}{3}\int_1^9 \sqrt{u}\,du \qquad [u = 1 + x^3,\, du = 3x^2\,dx]$

$= \frac{1}{6}\left[\frac{2}{3}u^{3/2}\right]_1^9 = (9^{3/2} - 1^{3/2}) = \frac{1}{9}(27-1) = \frac{26}{9}$

**27.** $\lim\limits_{h\to 0} f_{\text{ave}} = \lim\limits_{h\to 0}\frac{1}{(x+h)-x}\int_x^{x+h} f(t)\,dt = \lim\limits_{h\to 0}\frac{F(x+h)-F(x)}{h}$, where $F(x) = \int_a^x f(t)\,dt$. But we recognize this limit as being $F'(x)$ by the definition of a derivative. Therefore, $\lim\limits_{h\to 0} f_{\text{ave}} = F'(x) = f(x)$ by FTC1.

**28.** $\int_0^{24} c(t)\,dt \approx S_{12} = \frac{24-0}{12\cdot 3}[1(0) + 4(1.9) + 2(3.3) + 4(5.1) + 2(7.6) + 4(7.1) + 2(5.8)$

$+ 4(4.7) + 2(3.3) + 4(2.1) + 2(1.1) + 4(0.5) + 1(0)]$

$= \frac{2}{3}(127.8) = 85.2 \text{ mg}\cdot\text{s/L}$

Therefore, $F \approx A/85.2 = 6/85.2 \approx 0.0704$ L/s or 4.225 L/min.

**29.** $f(x) = \begin{cases} \frac{\pi}{20}\sin\left(\frac{\pi}{10}x\right) & \text{if } 0 \le x \le 10 \\ 0 & \text{if } x < 0 \text{ or } x > 10 \end{cases}$

(a) $f(x) \ge 0$ for all real numbers $x$ and

$$\int_{-\infty}^{\infty} f(x)\,dx = \int_0^{10}\frac{\pi}{20}\sin\left(\frac{\pi}{10}x\right)dx = \frac{\pi}{20}\cdot\frac{10}{\pi}\left[-\cos\left(\frac{\pi}{10}x\right)\right]_0^{10} = \frac{1}{2}(-\cos\pi + \cos 0) = \frac{1}{2}(1+1) = 1$$

Therefore, $f$ is a probability density function.

(b) $P(X<4) = \int_{-\infty}^4 f(x)\,dx = \int_0^4 \frac{\pi}{20}\sin\left(\frac{\pi}{10}x\right)dx = \frac{1}{2}\left[-\cos\left(\frac{\pi}{10}x\right)\right]_0^4 = \frac{1}{2}\left(-\cos\frac{2\pi}{5} + \cos 0\right)$

$\approx \frac{1}{2}(-0.309017 + 1) \approx 0.3455$

(c) $\mu = \int_{-\infty}^{\infty} xf(x)\,dx = \int_0^{10}\frac{\pi}{20}x\sin\left(\frac{\pi}{10}x\right)dx$

$= \int_0^{\pi}\frac{\pi}{20}\cdot\frac{10}{\pi}u(\sin u)\left(\frac{10}{\pi}\right)du \quad \left[u = \frac{\pi}{10}x,\, du = \frac{\pi}{10}\,dx\right]$

$= \frac{5}{\pi}\int_0^{\pi} u\sin u\,du \overset{82}{=} \frac{5}{\pi}[\sin u - u\cos u]_0^{\pi} = \frac{5}{\pi}[0 - \pi(-1)] = 5$

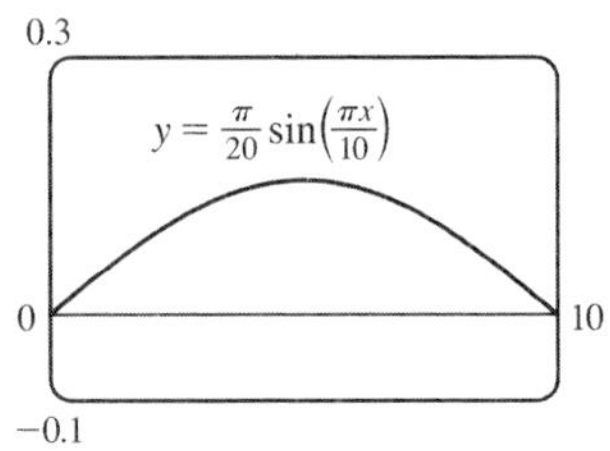


This answer is expected because the graph of $f$ is symmetric about the line $x = 5$.

**30.** $P(250 \le X \le 280) = \int_{250}^{280}\frac{1}{15\sqrt{2\pi}}\exp\left(\frac{-(x-268)^2}{2\cdot 15^2}\right)dx \approx 0.673$. Thus, the percentage of pregnancies that last between 250 and 280 days is about 67.3%.

**31.** (a) The probability density function is $f(t) = \begin{cases} 0 & \text{if } t < 0 \\ \frac{1}{8}e^{-t/8} & \text{if } t \ge 0 \end{cases}$

$P(0 \le X \le 3) = \int_0^3 \frac{1}{8}e^{-t/8}\,dt = \left[-e^{-t/8}\right]_0^3 = -e^{-3/8} + 1 \approx 0.3127$

(b) $P(X > 10) = \int_{10}^{\infty}\frac{1}{8}e^{-t/8}\,dt = \lim\limits_{x\to\infty}\left[-e^{-t/8}\right]_{10}^x = \lim\limits_{x\to\infty}\left(-e^{-x/8} + e^{-10/8}\right) = 0 + e^{-5/4} \approx 0.2865$

(c) We need to find $m$ such that $P(X \ge m) = \frac{1}{2} \;\Rightarrow\; \int_m^{\infty}\frac{1}{8}e^{-t/8}\,dt = \frac{1}{2} \;\Rightarrow\; \lim\limits_{x\to\infty}\left[-e^{-t/8}\right]_m^x = \frac{1}{2} \;\Rightarrow$

$\lim\limits_{x\to\infty}\left(-e^{-x/8} + e^{-m/8}\right) = \frac{1}{2} \;\Rightarrow\; e^{-m/8} = \frac{1}{2} \;\Rightarrow\; -m/8 = \ln\frac{1}{2} \;\Rightarrow\; m = -8\ln\frac{1}{2} = 8\ln 2 \approx 5.55$ minutes.

# FOCUS ON PROBLEM SOLVING

**1.** The volume generated from $x = 0$ to $x = b$ is $\int_0^b \pi[f(x)]^2\,dx$. Hence, we are given that $b^2 = \int_0^b \pi[f(x)]^2\,dx$ for all $b > 0$. Differentiating both sides of this equation with respect to $b$ using the Fundamental Theorem of Calculus gives $2b = \pi[f(b)]^2 \;\Rightarrow\; f(b) = \sqrt{2b/\pi}$, since $f$ is positive. Therefore, $f(x) = \sqrt{2x/\pi}$.

**2.** Let $a$ and $b$ be the $x$-coordinates of the points where the line intersects the curve. From the figure, $R_1 = R_2 \;\Rightarrow$

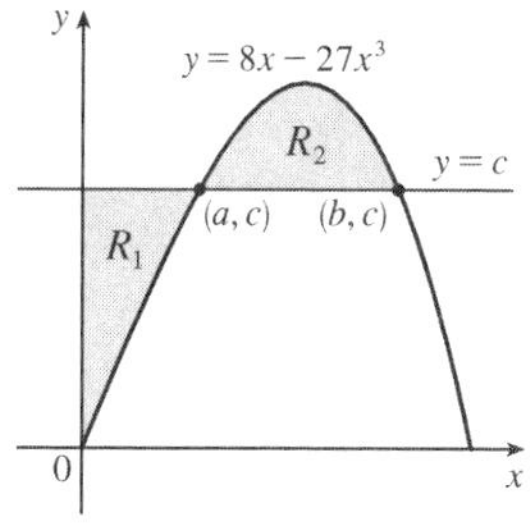


$$\int_0^a [c - (8x - 27x^3)]\,dx = \int_a^b [(8x - 27x^3) - c]\,dx$$

$$\left[cx - 4x^2 + \tfrac{27}{4}x^4\right]_0^a = \left[4x^2 - \tfrac{27}{4}x^4 - cx\right]_a^b$$

$$ac - 4a^2 + \tfrac{27}{4}a^4 = \left(4b^2 - \tfrac{27}{4}b^4 - bc\right) - \left(4a^2 - \tfrac{27}{4}a^4 - ac\right)$$

$$\begin{aligned} 0 &= 4b^2 - \tfrac{27}{4}b^4 - bc = 4b^2 - \tfrac{27}{4}b^4 - b(8b - 27b^3) \\ &= 4b^2 - \tfrac{27}{4}b^4 - 8b^2 + 27b^4 = \tfrac{81}{4}b^4 - 4b^2 \\ &= b^2\left(\tfrac{81}{4}b^2 - 4\right) \end{aligned}$$

So for $b > 0$, $b^2 = \frac{16}{81} \;\Rightarrow\; b = \frac{4}{9}$. Thus, $c = 8b - 27b^3 = 8\left(\frac{4}{9}\right) - 27\left(\frac{64}{729}\right) = \frac{32}{9} - \frac{64}{27} = \frac{32}{27}$.

**3.** (a) $V = \pi h^2(r - h/3) = \frac{1}{3}\pi h^2(3r - h)$. See the solution to Exercise 6.2.27.

(b) The smaller segment has height $h = 1 - x$ and so by part (a) its volume is $V = \frac{1}{3}\pi(1 - x)^2\,[3(1) - (1 - x)] = \frac{1}{3}\pi(x - 1)^2(x + 2)$. This volume must be $\frac{1}{3}$ of the total volume of the sphere, which is $\frac{4}{3}\pi(1)^3$. So $\frac{1}{3}\pi(x - 1)^2(x + 2) = \frac{1}{3}\left(\frac{4}{3}\pi\right) \;\Rightarrow\; (x^2 - 2x + 1)(x + 2) = \frac{4}{3} \;\Rightarrow\; x^3 - 3x + 2 = \frac{4}{3} \;\Rightarrow$ $3x^3 - 9x + 2 = 0$. Using Newton's method with $f(x) = 3x^3 - 9x + 2$, $f'(x) = 9x^2 - 9$, we get $x_{n+1} = x_n - \dfrac{3x_n^3 - 9x_n + 2}{9x_n^2 - 9}$. Taking $x_1 = 0$, we get $x_2 \approx 0.2222$, and $x_3 \approx 0.2261 \approx x_4$, so, correct to four decimal places, $x \approx 0.2261$.

(c) With $r = 0.5$ and $s = 0.75$, the equation $x^3 - 3rx^2 + 4r^3 s = 0$ becomes $x^3 - 3(0.5)x^2 + 4(0.5)^3(0.75) = 0 \;\Rightarrow$ $x^3 - \frac{3}{2}x^2 + 4\left(\frac{1}{8}\right)\frac{3}{4} = 0 \;\Rightarrow\; 8x^3 - 12x^2 + 3 = 0$. We use Newton's method with $f(x) = 8x^3 - 12x^2 + 3$, $f'(x) = 24x^2 - 24x$, so $x_{n+1} = x_n - \dfrac{8x_n^3 - 12x_n^2 + 3}{24x_n^2 - 24x_n}$. Take $x_1 = 0.5$. Then $x_2 \approx 0.6667$, and $x_3 \approx 0.6736 \approx x_4$. So to four decimal places the depth is 0.6736 m.

(d) (i) From part (a) with $r = 5$ in., the volume of water in the bowl is $V = \frac{1}{3}\pi h^2(3r - h) = \frac{1}{3}\pi h^2(15 - h) = 5\pi h^2 - \frac{1}{3}\pi h^3$. We are given that $\dfrac{dV}{dt} = 0.2\ \text{m}^3/\text{s}$ and we want to find $\dfrac{dh}{dt}$ when $h = 3$. Now $\dfrac{dV}{dt} = 10\pi h\dfrac{dh}{dt} - \pi h^2\dfrac{dh}{dt}$, so $\dfrac{dh}{dt} = \dfrac{0.2}{\pi(10h - h^2)}$. When $h = 3$, we have

$$\frac{dh}{dt} = \frac{0.2}{\pi(10 \cdot 3 - 3^2)} = \frac{1}{105\pi} \approx 0.003\ \text{in/s}.$$

(ii) From part (a), the volume of water required to fill the bowl from the instant that the water is 4 in. deep is $V = \frac{1}{2} \cdot \frac{4}{3}\pi(5)^3 - \frac{1}{3}\pi(4)^2(15 - 4) = \frac{2}{3} \cdot 125\pi - \frac{16}{3} \cdot 11\pi = \frac{74}{3}\pi$. To find the time required to fill the bowl we divide this volume by the rate: Time $= \frac{74\pi/3}{0.2} = \frac{370\pi}{3} \approx 387\ \text{s} \approx 6.5$ min.

**4.** (a) The volume above the surface is $\int_0^{L-h} A(y)\,dy = \int_{-h}^{L-h} A(y)\,dy - \int_{-h}^{0} A(y)\,dy$. So the proportion of volume above the surface is $\dfrac{\int_0^{L-h} A(y)\,dy}{\int_{-h}^{L-h} A(y)\,dy} = \dfrac{\int_{-h}^{L-h} A(y)\,dy - \int_{-h}^{0} A(y)\,dy}{\int_{-h}^{L-h} A(y)\,dy}$. Now by Archimedes' Principle, we have $F = W \Rightarrow$

$\rho_f g \int_{-h}^{0} A(y)\,dy = \rho_0 g \int_{-h}^{L-h} A(y)\,dy$, so $\int_{-h}^{0} A(y)\,dy = (\rho_0/\rho_f)\int_{-h}^{L-h} A(y)\,dy$. Therefore,

$\dfrac{\int_0^{L-h} A(y)\,dy}{\int_{-h}^{L-h} A(y)\,dy} = \dfrac{\int_{-h}^{L-h} A(y)\,dy - (\rho_0/\rho_f)\int_{-h}^{L-h} A(y)\,dy}{\int_{-h}^{L-h} A(y)\,dy} = \dfrac{\rho_f - \rho_0}{\rho_f}$, so the percentage of volume above the surface is $100\left(\dfrac{\rho_f - \rho_0}{\rho_f}\right)\%$.

(b) For an iceberg, the percentage of volume above the surface is $100\left(\frac{1030-917}{1030}\right)\% \approx 11\%$.

(c) No, the water does not overflow. Let $V_i$ be the volume of the ice cube, and let $V_w$ be the volume of the water which results from the melting. Then by the formula derived in part (a), the volume of ice above the surface of the water is $\left[(\rho_f - \rho_0)/\rho_f\right]V_i$, so the volume below the surface is $V_i - \left[(\rho_f - \rho_0)/\rho_f\right]V_i = (\rho_0/\rho_f)V_i$. Now the mass of the ice cube is the same as the mass of the water which is created when it melts, namely $m = \rho_0 V_i = \rho_f V_w \Rightarrow$ $V_w = (\rho_0/\rho_f)V_i$. So when the ice cube melts, the volume of the resulting water is the same as the underwater volume of the ice cube, and so the water does not overflow.

(d)

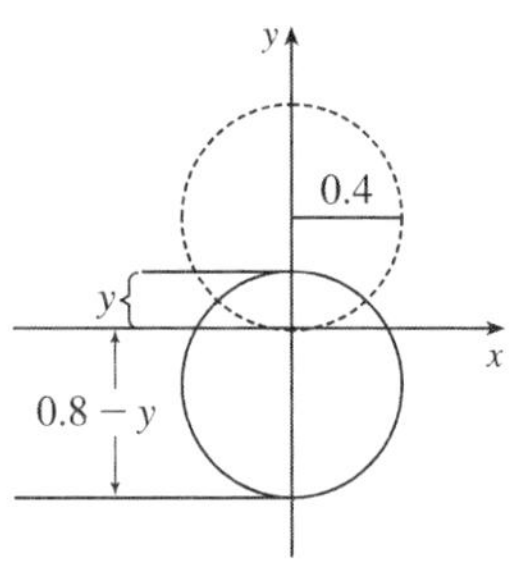


The figure shows the instant when the height of the exposed part of the ball is $y$. Using the formula in Problem 3(a) with $r = 0.4$ and $h = 0.8 - y$, we see that the volume of the submerged part of the sphere is $\frac{1}{3}\pi(0.8-y)^2[1.2-(0.8-y)]$, so its weight is $1000g \cdot \frac{1}{3}\pi s^2(1.2-s)$, where $s = 0.8 - y$. Then the work done to submerge the sphere is

$$\begin{aligned} W &= \int_0^{0.8} g\tfrac{1000}{3}\pi s^2(1.2-s)\,ds = g\tfrac{1000}{3}\pi \int_0^{0.8}(1.2s^2 - s^3)\,ds \\ &= g\tfrac{1000}{3}\pi\left[0.4s^3 - \tfrac{1}{4}s^4\right]_0^{0.8} = g\tfrac{1000}{3}\pi(0.2048 - 0.1024) \\ &= 9.8\,\tfrac{1000}{3}\,\pi(0.1024) \approx 1.05 \times 10^3 \text{ joules} \end{aligned}$$

**5.** We are given that the rate of change of the volume of water is $\dfrac{dV}{dt} = -kA(x)$, where $k$ is some positive constant and $A(x)$ is the area of the surface when the water has depth $x$. Now we are concerned with the rate of change of the depth of the water with respect to time, that is, $\dfrac{dx}{dt}$. But by the Chain Rule, $\dfrac{dV}{dt} = \dfrac{dV}{dx}\dfrac{dx}{dt}$, so the first equation can be written $\dfrac{dV}{dx}\dfrac{dx}{dt} = -kA(x)$ $(\star)$. Also, we know that the total volume of water up to a depth $x$ is $V(x) = \int_0^x A(s)\,ds$, where $A(s)$ is the area of a cross-section of the water at a depth $s$. Differentiating this equation with respect to $x$, we get $dV/dx = A(x)$. Substituting this into equation $\star$, we get $A(x)(dx/dt) = -kA(x) \Rightarrow dx/dt = -k$, a constant.

6. A typical sphere of radius $r$ is shown in the figure. We wish to maximize the shaded volume $V$, which can be thought of as the volume of a hemisphere of radius $r$ minus the volume of the spherical cap with height $h = 1 - \sqrt{1 - r^2}$ and radius 1.

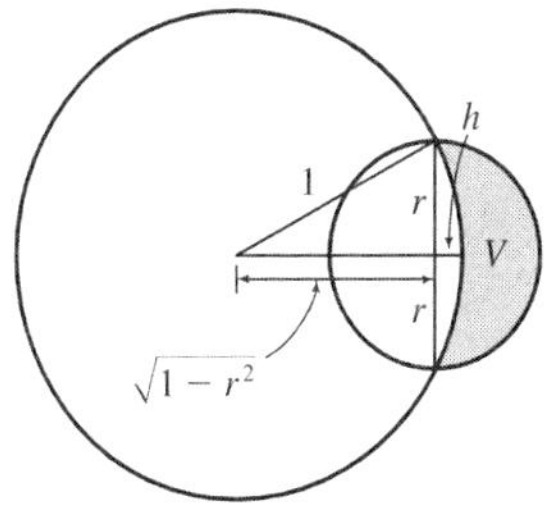


$$V = \tfrac{1}{2} \cdot \tfrac{4}{3}\pi r^3 - \tfrac{1}{3}\pi\left(1 - \sqrt{1 - r^2}\right)^2 \left[3(1) - \left(1 - \sqrt{1 - r^2}\right)\right] \quad \text{[by Problem 3(a)]}$$

$$= \tfrac{1}{3}\pi\left[2r^3 - \left(2 - 2\sqrt{1 - r^2} - r^2\right)\left(2 + \sqrt{1 - r^2}\right)\right]$$

$$= \tfrac{1}{3}\pi\left[2r^3 - 2 + \left(r^2 + 2\right)\sqrt{1 - r^2}\right]$$

$$V' = \tfrac{1}{3}\pi\left[6r^2 + \frac{(r^2 + 2)(-r)}{\sqrt{1 - r^2}} + \sqrt{1 - r^2}(2r)\right] = \tfrac{1}{3}\pi\left[\frac{6r^2\sqrt{1 - r^2} - r(r^2 + 2) + 2r(1 - r^2)}{\sqrt{1 - r^2}}\right]$$

$$= \tfrac{1}{3}\pi\left(\frac{6r^2\sqrt{1 - r^2} - 3r^3}{\sqrt{1 - r^2}}\right) = \frac{\pi r^2\left(2\sqrt{1 - r^2} - r\right)}{\sqrt{1 - r^2}}$$

$V'(r) = 0 \quad \Leftrightarrow \quad 2\sqrt{1 - r^2} = r \quad \Leftrightarrow \quad 4 - 4r^2 = r^2 \quad \Leftrightarrow \quad r^2 = \frac{4}{5} \quad \Leftrightarrow \quad r = \frac{2}{\sqrt{5}} \approx 0.89.$

Since $V'(r) > 0$ for $0 < r < \frac{2}{\sqrt{5}}$ and $V'(r) < 0$ for $\frac{2}{\sqrt{5}} < r < 1$, we know that $V$ attains a maximum at $r = \frac{2}{\sqrt{5}}$.

7. To find the height of the pyramid, we use similar triangles. The first figure shows a cross-section of the pyramid passing through the top and through two opposite corners of the square base. Now $|BD| = b$, since it is a radius of the sphere, which has diameter $2b$ since it is tangent to the opposite sides of the square base. Also, $|AD| = b$ since $\triangle ADB$ is isosceles. So the height is $|AB| = \sqrt{b^2 + b^2} = \sqrt{2}b$.

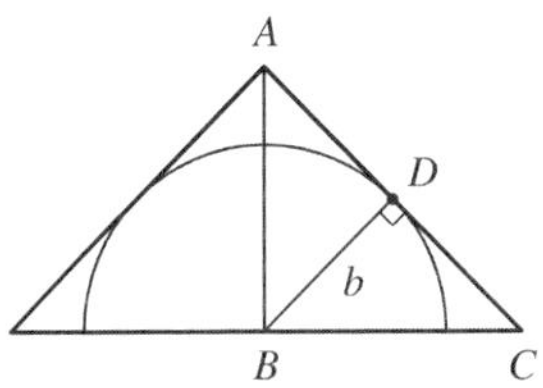


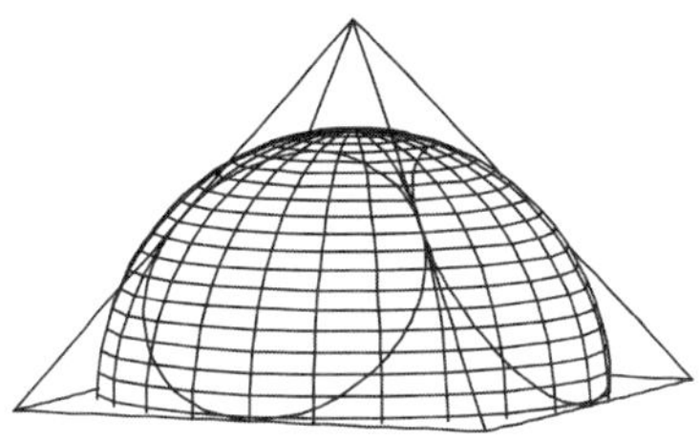

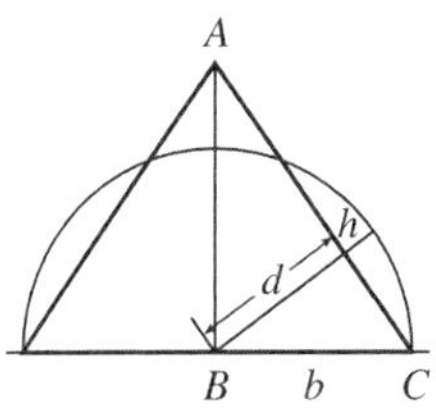


We first observe that the shared volume is equal to half the volume of the sphere, minus the sum of the four equal volumes (caps of the sphere) cut off by the triangular faces of the pyramid. See Exercise 6.2.27 for a derivation of the formula for the volume of a cap of a sphere. To use the formula, we need to find the perpendicular distance $h$ of each triangular face from the surface of the sphere. We first find the distance $d$ from the center of the sphere to one of the triangular faces. The third figure shows a cross-section of the pyramid through the top and through the midpoints of opposite sides of the square base. From similar triangles we find that

$$\frac{d}{b} = \frac{|AB|}{|AC|} = \frac{\sqrt{2}\,b}{\sqrt{b^2 + \left(\sqrt{2}\,b\right)^2}} \quad \Rightarrow \quad d = \frac{\sqrt{2}\,b^2}{\sqrt{3b^2}} = \frac{\sqrt{6}}{3}\,b$$

So $h = b - d = b - \frac{\sqrt{6}}{3}b = \frac{3 - \sqrt{6}}{3}b$. So, using the formula $V = \pi h^2(r - h/3)$ from Exercise 6.2.27 with $r = b$, we find that the volume of each of the caps is $\pi\left(\frac{3 - \sqrt{6}}{3}b\right)^2\left(b - \frac{3 - \sqrt{6}}{3 \cdot 3}b\right) = \frac{15 - 6\sqrt{6}}{9} \cdot \frac{6 + \sqrt{6}}{9}\pi b^3 = \left(\frac{2}{3} - \frac{7}{27}\sqrt{6}\right)\pi b^3$. So, using our first observation, the shared volume is $V = \frac{1}{2}\left(\frac{4}{3}\pi b^3\right) - 4\left(\frac{2}{3} - \frac{7}{27}\sqrt{6}\right)\pi b^3 = \left(\frac{28}{27}\sqrt{6} - 2\right)\pi b^3$.

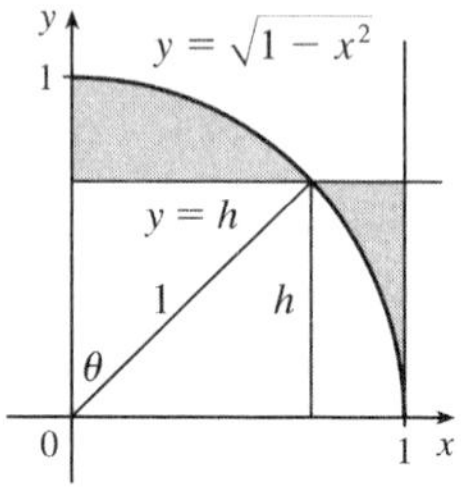


**8.** The problem can be reduced to finding the line which minimizes the shaded area in the diagram. The equation of the circle in the first quadrant is $y = \sqrt{1 - x^2}$, so if the equation of the line is $y = h$, then the circle and the line intersect where $h = \sqrt{1 - x^2} \Rightarrow x = \sqrt{1 - h^2}$. So the shaded area is

$$\begin{aligned} A &= \textstyle\int_0^{\sqrt{1-h^2}} \left(\sqrt{1 - x^2} - h\right) dx + \int_{\sqrt{1-h^2}}^1 \left(h - \sqrt{1 - x^2}\right) \\ &\overset{\star}{=} \left[-hx\right]_0^{\sqrt{1-h^2}} + \left[hx\right]_{\sqrt{1-h^2}}^1 + \textstyle\int_0^{\sqrt{1-h^2}} \sqrt{1 - x^2}\, dx + \int_1^{\sqrt{1-h^2}} \sqrt{1 - x^2}\, dx \\ &= -h\sqrt{1 - h^2} + h - h\sqrt{1 - h^2} + \textstyle\int_0^{\sqrt{1-h^2}} \sqrt{1 - x^2}\, dx + \int_1^{\sqrt{1-h^2}} \sqrt{1 - x^2}\, dx \\ &= h\left(1 - 2\sqrt{1 - h^2}\right) + \textstyle\int_0^{\sqrt{1-h^2}} \sqrt{1 - x^2}\, dx + \int_1^{\sqrt{1-h^2}} \sqrt{1 - x^2}\, dx \end{aligned}$$

Note that at $(\star)$, we reversed the limits of integration and changed the sign in the last integral.

We are interested in the minimum of $A\,(h) = h\left(1 - 2\sqrt{1 - h^2}\right) + \int_0^{\sqrt{1-h^2}} \sqrt{1 - x^2}\, dx + \int_1^{\sqrt{1-h^2}} \sqrt{1 - x^2}\, dx$, so we find $dA/dh$ using FTC1 and the Chain Rule:

$$\begin{aligned} \frac{dA}{dh} &= h\left(-2\frac{-h}{\sqrt{1 - h^2}}\right) + \left(1 - 2\sqrt{1 - h^2}\right) + 2\left[\sqrt{1 - \left(\sqrt{1 - h^2}\right)^2}\right] \tfrac{d}{dh}\left(\sqrt{1 - h^2}\right) \\ &= \frac{1}{\sqrt{1 - h^2}}\left[2h^2 + \sqrt{1 - h^2} - 2\left(1 - h^2\right)\right] + 2h\frac{-h}{\sqrt{1 - h^2}} \end{aligned}$$

This is 0 when $\sqrt{1 - h^2} - 2\left(1 - h^2\right) = 0 \Leftrightarrow u - 2u^2 = 0$ (where $u = \sqrt{1 - h^2}$) $\Leftrightarrow u = 0$ or $\frac{1}{2} \Leftrightarrow h = 1$ or $\frac{\sqrt{3}}{2}$. By the First Derivative Test, $h = \frac{\sqrt{3}}{2}$ represents a minimum for $A(h)$, since $A'(h) = 1 - \dfrac{2}{\sqrt{1 - h^2}}$ goes from negative to positive at $h = \frac{\sqrt{3}}{2}$.

*Another method:* Use FTC2 to evaluate all of the integrals before differentiating.

*Note:* Another strategy is to use the angle $\theta$ as the variable (see diagram above) and show that $A = \theta + \cos\theta - \frac{\pi}{4} - \frac{1}{2}\sin 2\theta$, which is minimized when $\theta = \frac{\pi}{6}$.

**9.** $x = \displaystyle\int_1^t \frac{\cos u}{u}\, du$, $y = \displaystyle\int_1^t \frac{\sin u}{u}\, du$, so by FTC1, we have $\dfrac{dx}{dt} = \dfrac{\cos t}{t}$ and $\dfrac{dy}{dt} = \dfrac{\sin t}{t}$. Vertical tangent lines occur when $dx/dt = 0 \Leftrightarrow \cos t = 0 \Leftrightarrow t = \frac{\pi}{2} + n\pi$. The parameter value corresponding to the origin, $(x, y) = (0, 0)$, is $t = 1$, so the nearest vertical tangent occurs when $t = \frac{\pi}{2}$. Therefore, the arc length between these points is

$$L = \int_1^{\pi/2} \sqrt{\left(\frac{dx}{dt}\right)^2 + \left(\frac{dy}{dt}\right)^2}\, dt = \int_1^{\pi/2} \sqrt{\frac{\cos^2 t}{t^2} + \frac{\sin^2 t}{t^2}}\, dt = \int_1^{\pi/2} \frac{dt}{t} = \left[\ln t\right]_1^{\pi/2} = \ln\frac{\pi}{2}$$

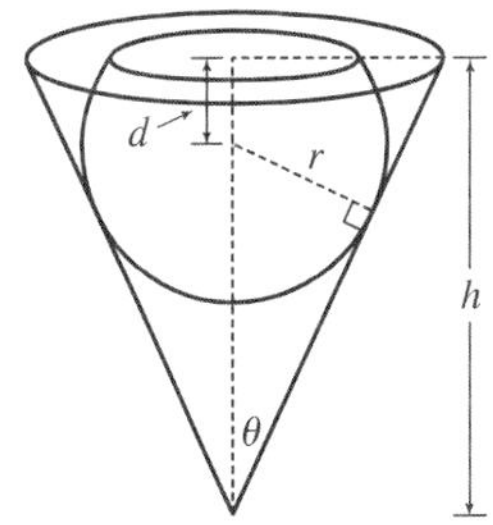

**10.** We want to find the volume of that part of the sphere which is below the surface of the water. As we can see from the diagram, this region is a cap of a sphere with radius $r$ and height $r + d$. If we can find an expression for $d$ in terms of $h$, $r$ and $\theta$, then we can determine the volume of the region [see Problem 3(a)], and then differentiate with respect to $r$ to find the maximum. We see that

$\sin\theta = \dfrac{r}{h-d} \quad\Leftrightarrow\quad h - d = \dfrac{r}{\sin\theta} \quad\Leftrightarrow\quad d = h - r\csc\theta.$

Now we can use the formula from Problem 3(a) to find the volume of water displaced:

$$V = \tfrac{1}{3}\pi h^2(3r-h) = \tfrac{1}{3}\pi(r+d)^2\,[3r-(r+d)] = \tfrac{1}{3}\pi(r+h-r\csc\theta)^2(2r-h+r\csc\theta)$$
$$= \tfrac{\pi}{3}[r(1-\csc\theta)+h]^2[r(2+\csc\theta)-h]$$

Now we differentiate with respect to $r$:

$$dV/dr = \tfrac{\pi}{3}\left([r(1-\csc\theta)+h]^2(2+\csc\theta) + 2[r(1-\csc\theta)+h](1-\csc\theta)[r(2+\csc\theta)-h]\right)$$
$$= \tfrac{\pi}{3}[r(1-\csc\theta)+h]([r(1-\csc\theta)+h](2+\csc\theta)+2(1-\csc\theta)[r(2+\csc\theta)-h])$$
$$= \tfrac{\pi}{3}[r(1-\csc\theta)+h](3(2+\csc\theta)(1-\csc\theta)r+[(2+\csc\theta)-2(1-\csc\theta)]h)$$
$$= \tfrac{\pi}{3}[r(1-\csc\theta)+h][3(2+\csc\theta)(1-\csc\theta)r+3h\csc\theta]$$

This is 0 when $r = \dfrac{h}{\csc\theta - 1}$ and when $r = \dfrac{h\csc\theta}{(\csc\theta+2)(\csc\theta-1)}$. Now since $V\left(\dfrac{h}{\csc\theta-1}\right) = 0$ (the first factor vanishes; this corresponds to $d = -r$), the maximum volume of water is displaced when $r = \dfrac{h\csc\theta}{(\csc\theta-1)(\csc\theta+2)}$.

(Our intuition tells us that a maximum value does exist, and it must occur at a critical number.) Multiplying numerator and denominator by $\sin^2\theta$, we get an alternative form of the answer: $r = \dfrac{h\sin\theta}{\sin\theta+\cos 2\theta}$.

**11.**

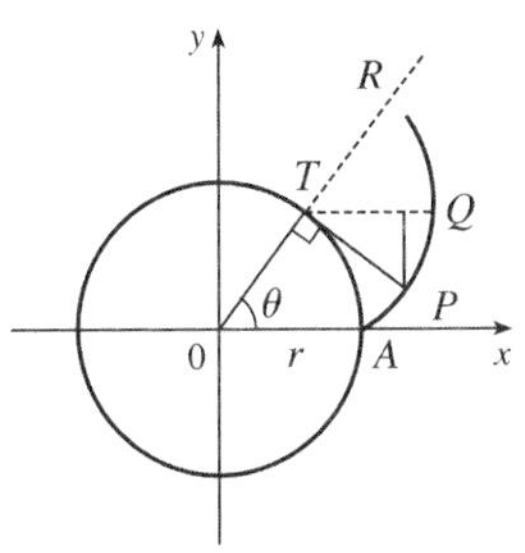


The coordinates of $T$ are $(x_1, y_1) = (r\cos\theta, r\sin\theta)$. Since $TP$ was unwound from arc $TA$, $TP$ has length $r\theta$. Also $\angle PTQ = \angle PTR - \angle QTR = \frac{1}{2}\pi - \theta$, so $P$ has coordinates

$x = x_1 + |TP|\cos\angle PTQ = r\cos\theta + r\theta\cos\left(\tfrac{1}{2}\pi - \theta\right) = r(\cos\theta + \theta\sin\theta),$

$y = y_1 + |TP|\sin\angle PTQ = r\sin\theta - r\theta\sin\left(\tfrac{1}{2}\pi - \theta\right) = r(\sin\theta - \theta\cos\theta).$

**12.** If the cow walks with the rope taut, it traces out the portion of the involute in Problem 11 corresponding to the range $0 \le \theta \le \pi$, arriving at the point $(-r, \pi r)$ when $\theta = \pi$. With the rope now fully extended, the cow walks in a semicircle of radius $\pi r$, arriving at $(-r, -\pi r)$. Finally, the cow traces out another portion of the involute, namely the reflection about the

$x$-axis of the initial involute path. (This corresponds to the range $-\pi \le \theta \le 0$.) Referring to the figure, we see that the total grazing area is $2(A_1 + A_3)$. $A_3$ is $\frac{1}{4}$ of the area of a circle of radius $\pi r$, so $A_3 = \frac{1}{4}\pi(\pi r)^2 = \frac{1}{4}\pi^3 r^2$.

We will compute $A_1 + A_2$ and then subtract $A_2 = \frac{1}{2}\pi r^2$ to obtain $A_1$.

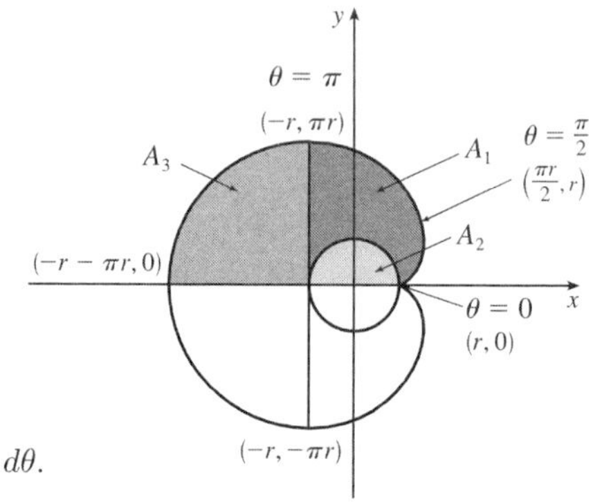


To find $A_1 + A_2$, first note that the rightmost point of the involute is $\left(\frac{\pi}{2}r, r\right)$. [To see this, note that $dx/d\theta = 0$ when $\theta = 0$ or $\frac{\pi}{2}$. $\theta = 0$ corresponds to the cusp at $(r, 0)$ and $\theta = \frac{\pi}{2}$ corresponds to $\left(\frac{\pi}{2}r, r\right)$.]

The leftmost point of the involute is $(-r, \pi r)$. Thus,

$A_1 + A_2 = \int_{\theta=\pi}^{\pi/2} y\,dx - \int_{\theta=0}^{\pi/2} y\,dx = \int_{\theta=\pi}^{0} y\,dx.$

Now $y\,dx = r(\sin\theta - \theta\cos\theta)\,r\theta\cos\theta\,d\theta = r^2\left(\theta\sin\theta\,\cos\theta - \theta^2\cos^2\theta\right)d\theta$.

Integrate: $(1/r^2)\int y\,dx = -\theta\cos^2\theta - \frac{1}{2}(\theta^2 - 1)\sin\theta\,\cos\theta - \frac{1}{6}\theta^3 + \frac{1}{2}\theta + C$. This enables us to compute

$$A_1 + A_2 = r^2\left[-\theta\cos^2\theta - (\theta^2 - 1)\sin\theta\,\cos\theta - \tfrac{1}{6}\theta^3 + \tfrac{1}{2}\theta\right]_\pi^0 = r^2\left[0 - \left(-\pi - \tfrac{\pi^3}{6} + \tfrac{\pi}{2}\right)\right] = r^2\left(\tfrac{\pi}{2} + \tfrac{\pi^3}{6}\right).$$

Therefore, $A_1 = (A_1 + A_2) - A_2 = \frac{1}{6}\pi^3 r^2$, so the grazing area is $2(A_1 + A_3) = 2\left(\frac{1}{6}\pi^3 r^2 + \frac{1}{4}\pi^3 r^2\right) = \frac{5}{6}\pi^3 r^2$.

**13.** We can assume that the cut is made along a vertical line $x = b > 0$, that the disk's boundary is the circle $x^2 + y^2 = 1$, and that the center of mass of the smaller piece (to the right of $x = b$) is $\left(\frac{1}{2}, 0\right)$. We wish to find $b$ to two decimal places. We have $\dfrac{1}{2} = \overline{x} = \dfrac{\int_b^1 x \cdot 2\sqrt{1 - x^2}\,dx}{\int_b^1 2\sqrt{1 - x^2}\,dx}$. Evaluating the numerator gives us

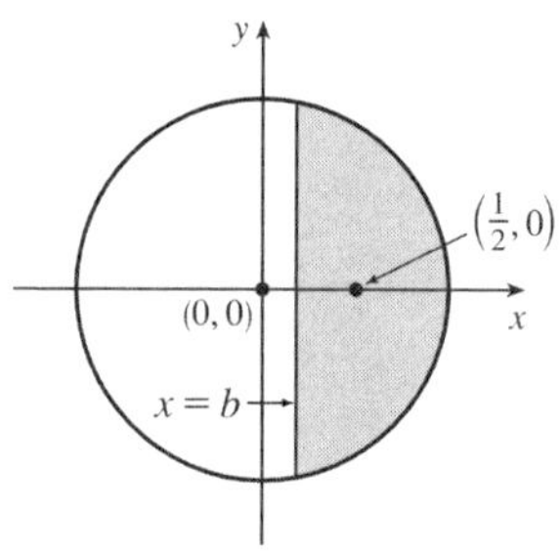


$$-\int_b^1 (1 - x^2)^{1/2}(-2x)\,dx = -\tfrac{2}{3}\left[\left(1 - x^2\right)^{3/2}\right]_b^1 = -\tfrac{2}{3}\left[0 - \left(1 - b^2\right)^{3/2}\right] = \tfrac{2}{3}(1 - b^2)^{3/2}$$

Using Formula 30 in the Table of Integrals, we find that the denominator is

$\left[x\sqrt{1 - x^2} + \sin^{-1}x\right]_b^1 = \left(0 + \frac{\pi}{2}\right) - \left(b\sqrt{1 - b^2} + \sin^{-1}b\right)$. Thus, we have $\dfrac{1}{2} = \overline{x} = \dfrac{\frac{2}{3}(1 - b^2)^{3/2}}{\frac{\pi}{2} - b\sqrt{1 - b^2} - \sin^{-1}b}$, or, equivalently, $\frac{2}{3}(1 - b^2)^{3/2} = \frac{\pi}{4} - \frac{1}{2}b\sqrt{1 - b^2} - \frac{1}{2}\sin^{-1}b$. Solving this equation numerically with a calculator or CAS, we obtain $b \approx 0.138173$, or $b = 0.14$ m to two decimal places.

**14.**

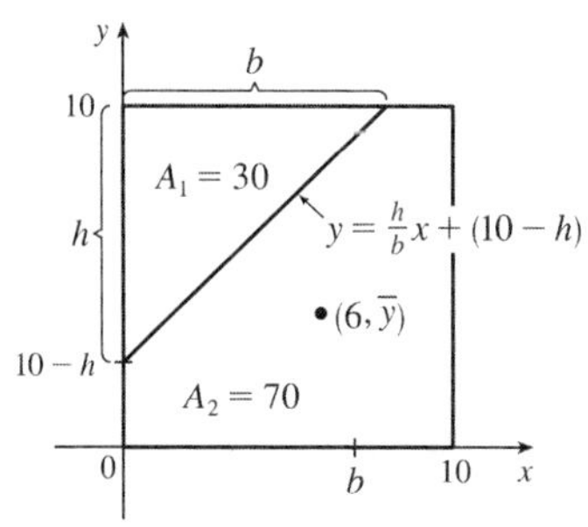


$A_1 = 30 \;\Rightarrow\; \frac{1}{2}bh = 30 \;\Rightarrow\; bh = 60.$

$\overline{x} = 6 \;\Rightarrow\; \dfrac{1}{A_2}\displaystyle\int_0^{10} x f(x)\,dx = 6 \;\Rightarrow$

$\displaystyle\int_0^b x\left(\frac{h}{b}x + 10 - h\right)dx + \int_b^{10} x(10)\,dx = 6(70) \;\Rightarrow$

$\displaystyle\int_0^b \left(\frac{h}{b}x^2 + 10x - hx\right)dx + 10 \cdot \tfrac{1}{2}\left[x^2\right]_b^{10} = 420 \;\Rightarrow$

$\left[\dfrac{h}{3b}x^3 + 5x^2 - \dfrac{h}{2}x^2\right]_0^b + 5(100 - b^2) = 420 \;\Rightarrow\; \frac{1}{3}hb^2 + 5b^2 - \frac{1}{2}hb^2 + 500 - 5b^2 = 420 \;\Rightarrow\; 80 = \frac{1}{6}hb^2 \;\Rightarrow$

$480 = (hb)b \;\Rightarrow\; 480 = 60b \;\Rightarrow\; b = 8.$

So $h = \dfrac{60}{8} = \dfrac{15}{2}$ and an equation of the line is $y = \dfrac{15/2}{8}x + \left(10 - \dfrac{15}{2}\right) = \dfrac{15}{16}x + \dfrac{5}{2}$. Now

$$\begin{aligned}\overline{y} &= \frac{1}{A_2}\int_0^{10} \frac{1}{2}[f(x)]^2\,dx = \frac{1}{70\cdot 2}\left[\int_0^8 \left(\frac{15}{16}x + \frac{5}{2}\right)^2 dx + \int_8^{10} (10)^2\,dx\right] \\ &= \tfrac{1}{140}\left[\int_0^8 \left(\tfrac{225}{256}x^2 + \tfrac{75}{16}x + \tfrac{25}{4}\right)dx + 100(10-8)\right] = \tfrac{1}{140}\left(\left[\tfrac{225}{768}x^3 + \tfrac{75}{32}x^2 + \tfrac{25}{4}x\right]_0^8 + 200\right) \\ &= \tfrac{1}{140}(150 + 150 + 50 + 200) = \tfrac{550}{140} = \tfrac{55}{14}\end{aligned}$$

Thus, the centroid is located $\frac{55}{14} \approx 3.93$ cm from the bottom of the square.

# 7 ☐ DIFFERENTIAL EQUATIONS

## 7.1 Modeling with Differential Equations

**1.** $y = x - x^{-1} \quad \Rightarrow \quad y' = 1 + x^{-2}$. To show that $y$ is a solution of the differential equation, we will substitute the expressions for $y$ and $y'$ in the left-hand side of the equation and show that the left-hand side is equal to the right-hand side.

$$\text{LHS} = xy' + y = x(1 + x^{-2}) + (x - x^{-1}) = x + x^{-1} + x - x^{-1} = 2x = \text{RHS}$$

**2.** $y = \sin x \cos x - \cos x \quad \Rightarrow \quad y' = \sin x\,(-\sin x) + \cos x\,(\cos x) - (-\sin x) = \cos^2 x - \sin^2 x + \sin x$.

$$\begin{aligned}\text{LHS} &= y' + (\tan x)y = \cos^2 x - \sin^2 x + \sin x + (\tan x)(\sin x \cos x - \cos x) \\ &= \cos^2 x - \sin^2 x + \sin x + \sin^2 x - \sin x = \cos^2 x = \text{RHS},\end{aligned}$$

so $y$ is a solution of the differential equation. Also, $y(0) = \sin 0 \cos 0 - \cos 0 = 0 \cdot 1 - 1 = -1$, so the initial condition is satisfied.

**3.** (a) $y = e^{rx} \quad \Rightarrow \quad y' = re^{rx} \quad \Rightarrow \quad y'' = r^2 e^{rx}$. Substituting these expressions into the differential equation $2y'' + y' - y = 0$, we get $2r^2 e^{rx} + re^{rx} - e^{rx} = 0 \quad \Rightarrow \quad (2r^2 + r - 1)e^{rx} = 0 \quad \Rightarrow$ $(2r - 1)(r + 1) = 0$ [since $e^{rx}$ is never zero] $\Rightarrow \quad r = \frac{1}{2}$ or $-1$.

(b) Let $r_1 = \frac{1}{2}$ and $r_2 = -1$, so we need to show that every member of the family of functions $y = ae^{x/2} + be^{-x}$ is a solution of the differential equation $2y'' + y' - y = 0$.

$y = ae^{x/2} + be^{-x} \quad \Rightarrow \quad y' = \frac{1}{2}ae^{x/2} - be^{-x} \quad \Rightarrow \quad y'' = \frac{1}{4}ae^{x/2} + be^{-x}$.

$$\begin{aligned}\text{LHS} = 2y'' + y' - y &= 2(\tfrac{1}{4}ae^{x/2} + be^{-x}) + (\tfrac{1}{2}ae^{x/2} - be^{-x}) - (ae^{x/2} + be^{-x}) \\ &= \tfrac{1}{2}ae^{x/2} + 2be^{-x} + \tfrac{1}{2}ae^{x/2} - be^{-x} - ae^{x/2} - be^{-x} \\ &= (\tfrac{1}{2}a + \tfrac{1}{2}a - a)e^{x/2} + (2b - b - b)e^{-x} \\ &= 0 = \text{RHS}\end{aligned}$$

**4.** (a) $y = \cos kt \quad \Rightarrow \quad y' = -k \sin kt \quad \Rightarrow \quad y'' = -k^2 \cos kt$. Substituting these expressions into the differential equation $4y'' = -25y$, we get $4(-k^2 \cos kt) = -25(\cos kt) \quad \Rightarrow \quad (25 - 4k^2)\cos kt = 0$ [for all $t$] $\Rightarrow$ $25 - 4k^2 = 0 \quad \Rightarrow \quad k^2 = \frac{25}{4} \quad \Rightarrow \quad k = \pm\frac{5}{2}$.

(b) $y = A \sin kt + B \cos kt \quad \Rightarrow \quad y' = Ak \cos kt - Bk \sin kt \quad \Rightarrow \quad y'' = -Ak^2 \sin kt - Bk^2 \cos kt$.
The given differential equation $4y'' = -25y$ is equivalent to $4y'' + 25y = 0$. Thus,

$$\begin{aligned}\text{LHS} = 4y'' + 25y &= 4(-Ak^2 \sin kt - Bk^2 \cos kt) + 25(A \sin kt + B \cos kt) \\ &= -4Ak^2 \sin kt - 4Bk^2 \cos kt + 25A \sin kt + 25B \cos kt \\ &= (25 - 4k^2)A \sin kt + (25 - 4k^2)B \cos kt \\ &= 0 \quad \text{since } k^2 = \tfrac{25}{4}.\end{aligned}$$

**5.** (a) $y = \sin x \quad \Rightarrow \quad y' = \cos x \quad \Rightarrow \quad y'' = -\sin x$.
$\text{LHS} = y'' + y = -\sin x + \sin x = 0 \neq \sin x$, so $y = \sin x$ **is not** a solution of the differential equation.

(b) $y = \cos x \quad \Rightarrow \quad y' = -\sin x \quad \Rightarrow \quad y'' = -\cos x$.
$\text{LHS} = y'' + y = -\cos x + \cos x = 0 \neq \sin x$, so $y = \cos x$ **is not** a solution of the differential equation.

(c) $y = \frac{1}{2}x \sin x \quad \Rightarrow \quad y' = \frac{1}{2}(x \cos x + \sin x) \quad \Rightarrow \quad y'' = \frac{1}{2}(-x \sin x + \cos x + \cos x)$.
$\text{LHS} = y'' + y = \frac{1}{2}(-x \sin x + 2\cos x) + \frac{1}{2}x \sin x = \cos x \neq \sin x$, so $y = \frac{1}{2}x \sin x$ **is not** a solution of the differential equation.

(d) $y = -\frac{1}{2}x\cos x \;\Rightarrow\; y' = -\frac{1}{2}(-x\sin x + \cos x) \;\Rightarrow\; y'' = -\frac{1}{2}(-x\cos x - \sin x - \sin x)$.
LHS $= y'' + y = -\frac{1}{2}(-x\cos x - 2\sin x) + \left(-\frac{1}{2}x\cos x\right) = \sin x =$ RHS, so $y = -\frac{1}{2}x\cos x$ **is** a solution of the differential equation.

**6.** (a) $y = \dfrac{\ln x + C}{x} \;\Rightarrow\; y' = \dfrac{x\cdot(1/x) - (\ln x + C)}{x^2} = \dfrac{1 - \ln x - C}{x^2}$.

$$\text{LHS} = x^2y' + xy = x^2\cdot\frac{1 - \ln x - C}{x^2} + x\cdot\frac{\ln x + C}{x}$$
$$= 1 - \ln x - C + \ln x + C = 1 = \text{RHS, so } y \text{ is a solution of the differential equation.}$$

(b)

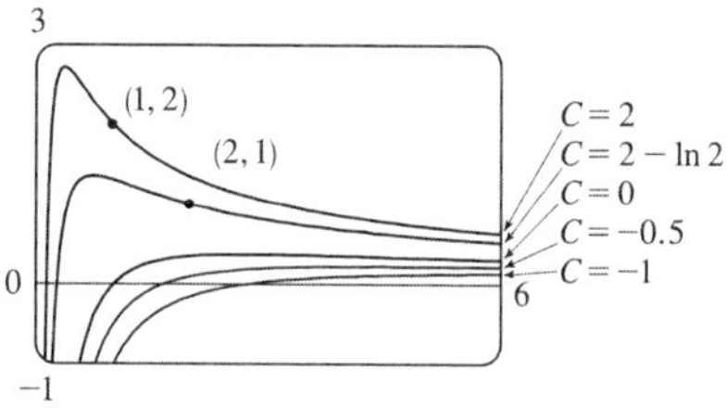


A few notes about the graph of $y = (\ln x + C)/x$:

(1) There is a vertical asymptote of $x = 0$.

(2) There is a horizontal asymptote of $y = 0$.

(3) $y = 0 \;\Rightarrow\; \ln x + C = 0 \;\Rightarrow\; x = e^{-C}$, so there is an $x$-intercept at $e^{-C}$.

(4) $y' = 0 \;\Rightarrow\; \ln x = 1 - C \;\Rightarrow\; x = e^{1-C}$, so there is a local maximum at $x = e^{1-C}$.

(c) $y(1) = 2 \;\Rightarrow\; 2 = \dfrac{\ln 1 + C}{1} \;\Rightarrow\; 2 = C$, so the solution is $y = \dfrac{\ln x + 2}{x}$ [shown in part (b)].

(d) $y(2) = 1 \;\Rightarrow\; 1 = \dfrac{\ln 2 + C}{2} \;\Rightarrow\; 2 + \ln 2 + C \;\Rightarrow\; C = 2 - \ln 2$, so the solution is $y = \dfrac{\ln x + 2 - \ln 2}{x}$ [shown in part (b)].

**7.** (a) Since the derivative $y' = -y^2$ is always negative (or 0 if $y = 0$), the function $y$ must be decreasing (or equal to 0) on any interval on which it is defined.

(b) $y = \dfrac{1}{x + C} \;\Rightarrow\; y' = -\dfrac{1}{(x + C)^2}$. LHS $= y' = -\dfrac{1}{(x + C)^2} = -\left(\dfrac{1}{x + C}\right)^2 = -y^2 =$ RHS

(c) $y = 0$ is a solution of $y' = -y^2$ that is not a member of the family in part (b).

(d) If $y(x) = \dfrac{1}{x + C}$, then $y(0) = \dfrac{1}{0 + C} = \dfrac{1}{C}$. Since $y(0) = 0.5$, $\dfrac{1}{C} = \dfrac{1}{2} \;\Rightarrow\; C = 2$, so $y = \dfrac{1}{x + 2}$.

**8.** (a) If $x$ is close to 0, then $xy^3$ is close to 0, and hence, $y'$ is close to 0. Thus, the graph of $y$ must have a tangent line that is nearly horizontal. If $x$ is large, then $xy^3$ is large, and the graph of $y$ must have a tangent line that is nearly vertical. (In both cases, we assume reasonable values for $y$.)

(b) $y = (c - x^2)^{-1/2} \;\Rightarrow\; y' = x(c - x^2)^{-3/2}$. RHS $= xy^3 = x[(c - x^2)^{-1/2}]^3 = x(c - x^2)^{-3/2} = y' =$ LHS

(c)

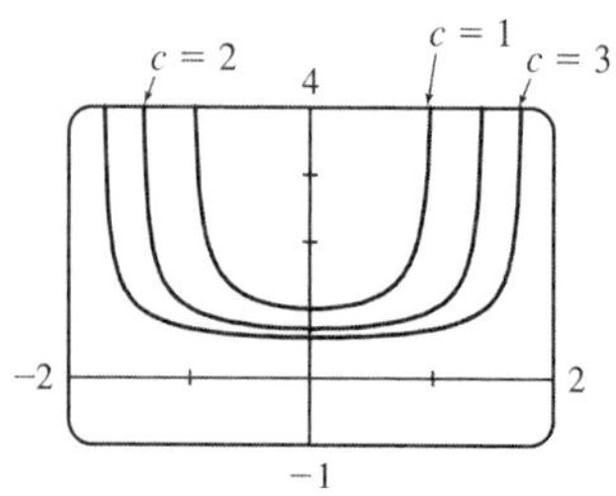


When $x$ is close to 0, $y'$ is also close to 0.
As $x$ gets larger, so does $|y'|$.

(d) $y(0) = (c - 0)^{-1/2} = 1/\sqrt{c}$ and $y(0) = 2 \;\Rightarrow\; \sqrt{c} = \frac{1}{2} \;\Rightarrow\; c = \frac{1}{4}$, so $y = \left(\frac{1}{4} - x^2\right)^{-1/2}$.

**9.** (a) $\dfrac{dP}{dt} = 1.2P\left(1 - \dfrac{P}{4200}\right)$. Now $\dfrac{dP}{dt} > 0 \;\Rightarrow\; 1 - \dfrac{P}{4200} > 0$ [assuming that $P > 0$] $\Rightarrow\; \dfrac{P}{4200} < 1 \;\Rightarrow$ $P < 4200 \;\Rightarrow$ the population is increasing for $0 < P < 4200$.

(b) $\dfrac{dP}{dt} < 0 \;\Rightarrow\; P > 4200$

(c) $\dfrac{dP}{dt} = 0 \;\Rightarrow\; P = 4200$ or $P = 0$

**10.** (a) $y = k \;\Rightarrow\; y' = 0$, so $\dfrac{dy}{dt} = y^4 - 6y^3 + 5y^2 \;\Leftrightarrow\; 0 = k^4 - 6k^3 + 5k^2 \;\Leftrightarrow\; k^2(k^2 - 6k + 5) = 0 \;\Leftrightarrow$ $k^2(k-1)(k-5) = 0 \;\Leftrightarrow\; k = 0, 1$, or $5$

(b) $y$ is increasing $\Leftrightarrow\; \dfrac{dy}{dt} > 0 \;\Leftrightarrow\; y^2(y-1)(y-5) > 0 \;\Leftrightarrow\; y \in (-\infty, 0) \cup (0, 1) \cup (5, \infty)$

(c) $y$ is decreasing $\Leftrightarrow\; \dfrac{dy}{dt} < 0 \;\Leftrightarrow\; y \in (1, 5)$

**11.** (a) This function is increasing *and* also decreasing. But $dy/dt = e^t(y-1)^2 \geq 0$ for all $t$, implying that the graph of the solution of the differential equation cannot be decreasing on any interval.

(b) When $y = 1$, $dy/dt = 0$, but the graph does not have a horizontal tangent line.

**12.** The graph for this exercise is shown in the figure at the right.

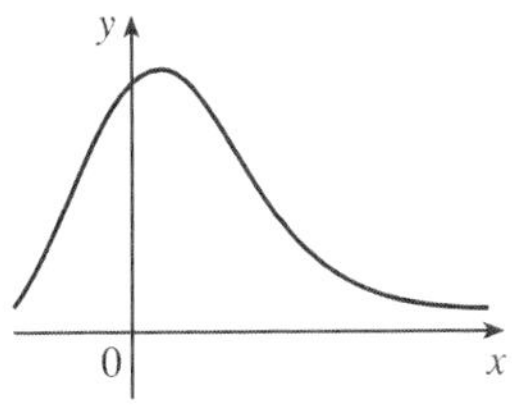


A. $y' = 1 + xy > 1$ for points in the first quadrant, but we can see that $y' < 0$ for some points in the first quadrant.

B. $y' = -2xy = 0$ when $x = 0$, but we can see that $y' > 0$ for $x = 0$.

Thus, equations A and B are incorrect, so the correct equation is C.

C. $y' = 1 - 2xy$ seems reasonable since:

(1) When $x = 0$, $y'$ could be 1.

(2) When $x < 0$, $y'$ could be greater than 1.

(3) Solving $y' = 1 - 2xy$ for $y$ gives us $y = \dfrac{1 - y'}{2x}$. If $y'$ takes on small negative values, then as $x \to \infty$, $y \to 0^+$, as shown in the figure.

**13.** (a) $P$ increases most rapidly at the beginning, since there are usually many simple, easily-learned sub-skills associated with learning a skill. As $t$ increases, we would expect $dP/dt$ to remain positive, but decrease. This is because as time progresses, the only points left to learn are the more difficult ones.

(b) $\dfrac{dP}{dt} = k(M - P)$ is always positive, so the level of performance $P$ is increasing. As $P$ gets close to $M$, $dP/dt$ gets close to 0; that is, the performance levels off, as explained in part (a).

(c)

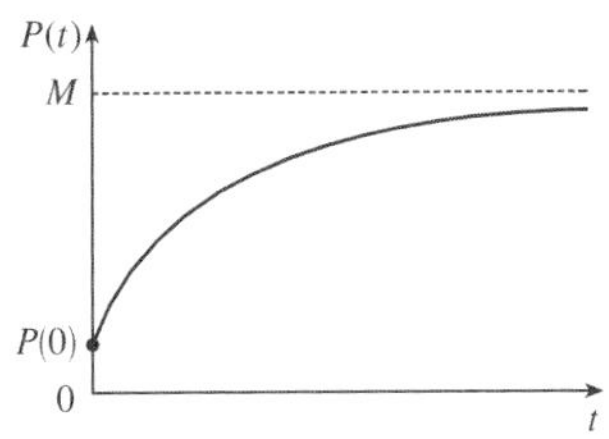


**14.** (a) The coffee cools most quickly as soon as it is removed from the heat source. The rate of cooling decreases toward 0 since the coffee approaches room temperature.

(b) $\dfrac{dy}{dt} = k(y - R)$, where $k$ is a proportionality constant, $y$ is the temperature of the coffee, and $R$ is the room temperature. The initial condition is $y(0) = 95\,°\text{C}$. The answer and the model support each other because as $y$ approaches $R$, $dy/dt$ approaches 0, so the model seems appropriate.

(c)

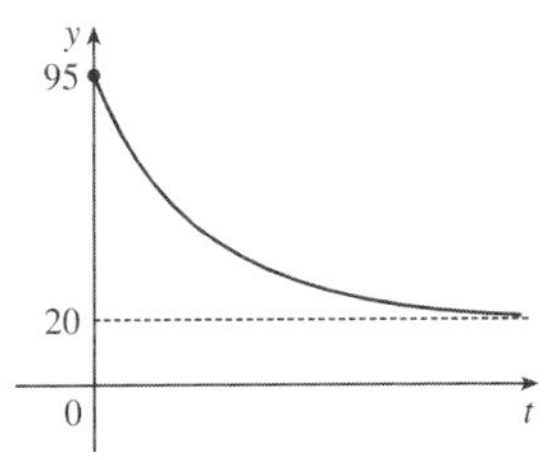

## 7.2 Direction Fields and Euler's Method

**1.** (a)

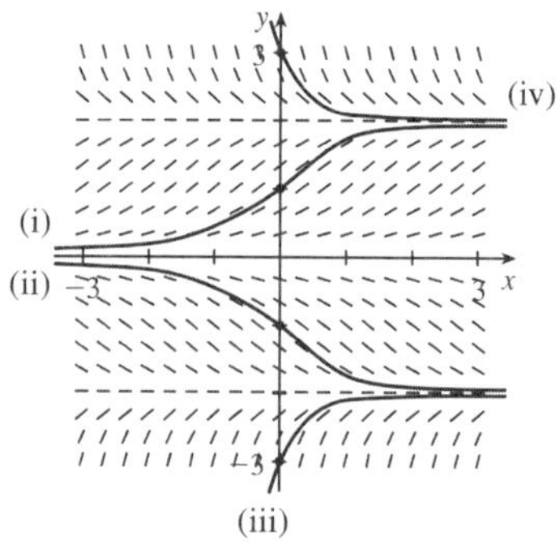


(b) It appears that the constant functions $y = 0$, $y = -2$, and $y = 2$ are equilibrium solutions. Note that these three values of $y$ satisfy the given differential equation $y' = y\left(1 - \frac{1}{4}y^2\right)$.

**2.** (a)

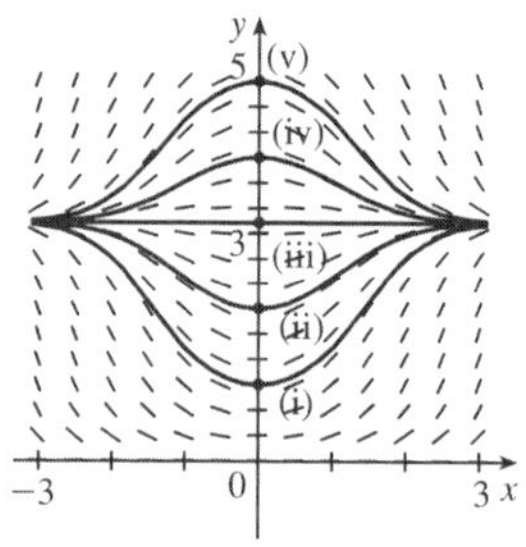


(b) From the figure, it appears that $y = \pi$ is an equilibrium solution. From the equation $y' = x \sin y$, we see that $y = n\pi$ ($n$ an integer) describes all the equilibrium solutions.

**3.** $y' = y - 2$. The slopes at each point are independent of $x$, so the slopes are the same along each line parallel to the $x$-axis. Thus, III is the direction field for this equation. Note that for $y = 2$, $y' = 0$.

**4.** $y' = x(2 - y) = 0$ on the lines $x = 0$ and $y = 2$. Direction field I satisfies these conditions.

**5.** $y' = x + y - 1 = 0$ on the line $y = -x + 1$. Direction field IV satisfies this condition. Notice also that on the line $y = -x$ we have $y' = -1$, which is true in IV.

**6.** $y' = \sin x \sin y = 0$ on the lines $x = 0$ and $y = 0$, and $y' > 0$ for $0 < x < \pi$, $0 < y < \pi$. Direction field II satisfies these conditions.

**7.** (a) $y(0) = 1$

(b) $y(0) = 2$

(c) $y(0) = -1$

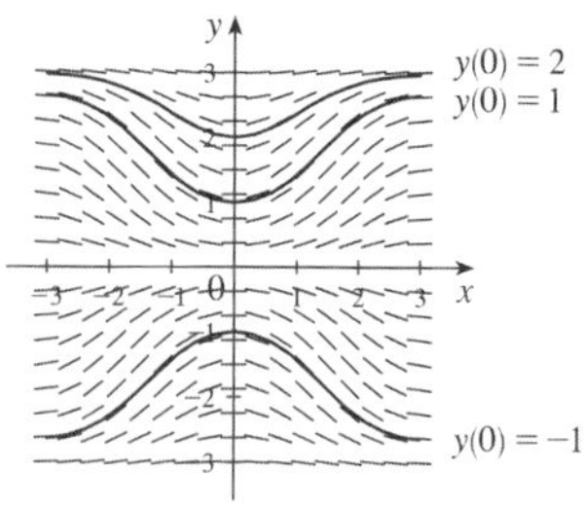


**8.** (a) $y(0) = -1$

(b) $y(0) = 0$

(c) $y(0) = 1$

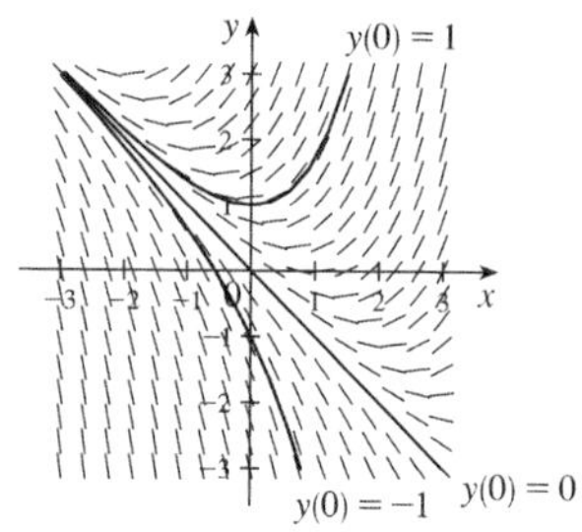


**9.**

| $x$ | $y$ | $y' = 1 + y$ |
|---|---|---|
| 0 | 0 | 1 |
| 0 | 1 | 2 |
| 0 | 2 | 3 |
| 0 | −3 | −2 |
| 0 | −2 | −1 |

Note that for $y = -1$, $y' = 0$. The three solution curves sketched go through $(0, 0)$, $(0, -1)$, and $(0, -2)$.

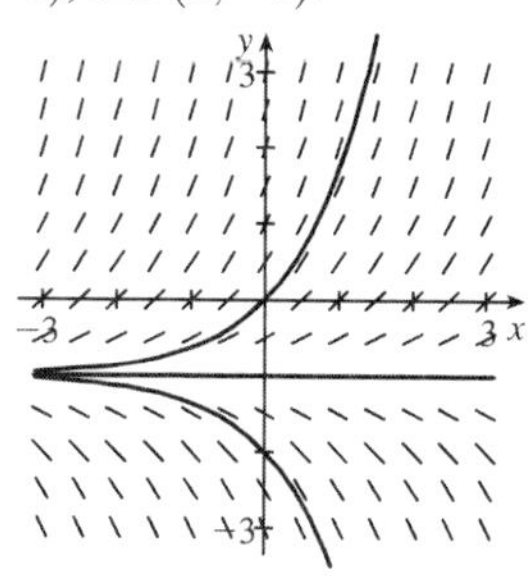

**10.**

| $x$ | $y$ | $y' = x^2 - y^2$ |
|---|---|---|
| $\pm 1$ | $\pm 3$ | $-8$ |
| $\pm 3$ | $\pm 1$ | $8$ |
| $\pm 1$ | $\pm 0.5$ | $0.75$ |
| $\pm 0.5$ | $\pm 1$ | $-0.75$ |

Note that $y' = 0$ for $y = \pm x$. If $|x| < |y|$, then $y' < 0$; that is, the slopes are negative for all points in quadrants I and II above both of the lines $y = x$ and $y = -x$, and all points in quadrants III and IV below both of the lines $y = -x$ and $y = x$. A similar statement holds for positive slopes.

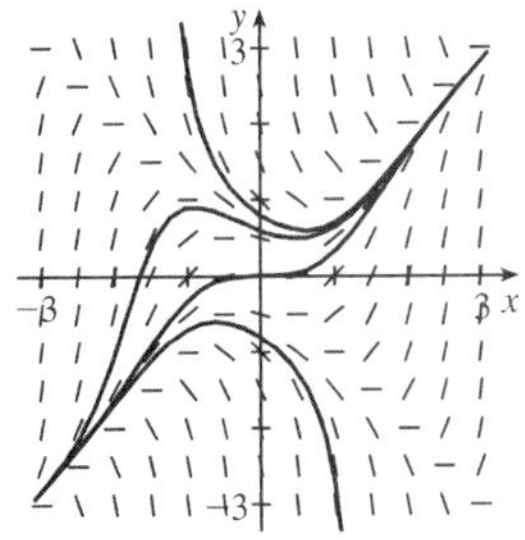


**11.**

| $x$ | $y$ | $y' = y - 2x$ |
|---|---|---|
| $-2$ | $-2$ | $2$ |
| $-2$ | $2$ | $6$ |
| $2$ | $2$ | $-2$ |
| $2$ | $-2$ | $-6$ |

Note that $y' = 0$ for any point on the line $y = 2x$. The slopes are positive to the left of the line and negative to the right of the line. The solution curve in the graph passes through $(1, 0)$.

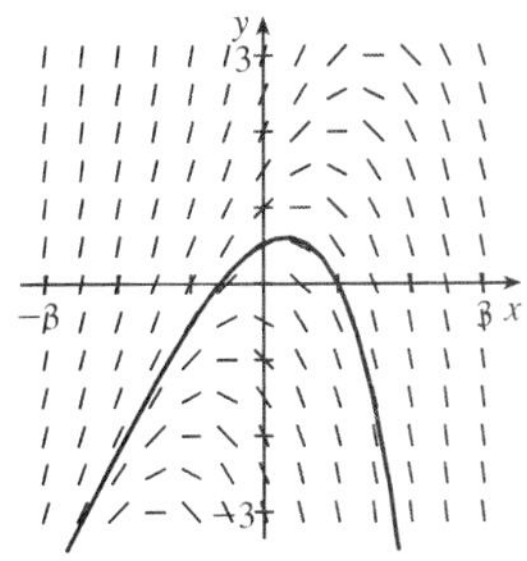


**12.**

| $x$ | $y$ | $y' = 1 - xy$ |
|---|---|---|
| $\pm 1$ | $\pm 1$ | $0$ |
| $\pm 2$ | $\pm 2$ | $-3$ |
| $\pm 2$ | $\mp 2$ | $5$ |

Note that $y' = 0$ for any point on the hyperbola $xy = 1$ (or $y = 1/x$). The slopes are negative at points "inside" the branches and positive at points everywhere else. The solution curve in the graph passes through $(0, 0)$.

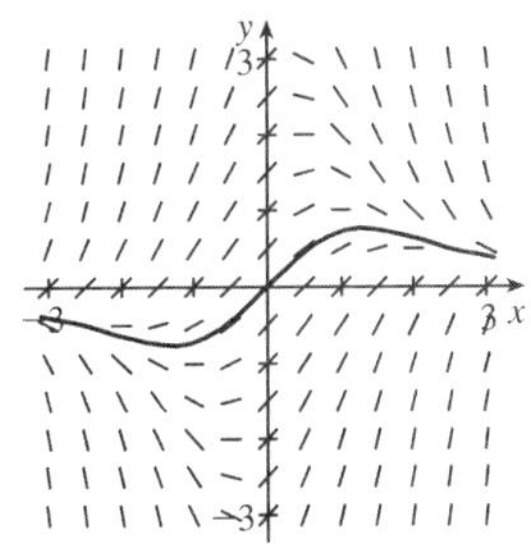

**13.**

| $x$ | $y$ | $y' = y + xy$ |
|---|---|---|
| 0 | $\pm 2$ | $\pm 2$ |
| 1 | $\pm 2$ | $\pm 4$ |
| $-3$ | $\pm 2$ | $\mp 4$ |

Note that $y' = y(x+1) = 0$ for any point on $y = 0$ or on $x = -1$. The slopes are positive when the factors $y$ and $x+1$ have the same sign and negative when they have opposite signs. The solution curve in the graph passes through $(0, 1)$.

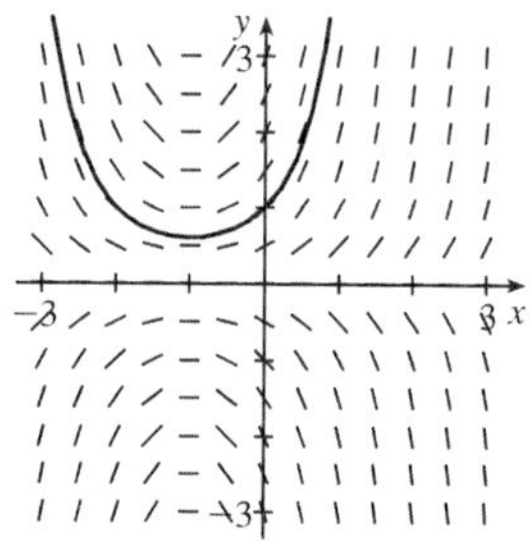


**14.**

| $x$ | $y$ | $y' = x - xy$ |
|---|---|---|
| $\pm 2$ | 0 | $\pm 2$ |
| $\pm 2$ | 3 | $\mp 4$ |
| $\pm 2$ | $-1$ | $\pm 4$ |

Note that $y' = x(1-y) = 0$ for any point on $x = 0$ or on $y = 1$. The slopes are positive when the factors $x$ and $1-y$ have the same sign and negative when they have opposite signs. The solution curve in the graph passes through $(1, 0)$.

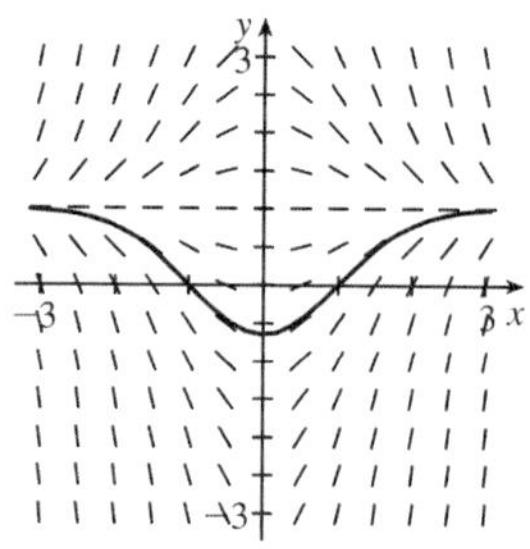


**15.** In Maple, we can use either `directionfield` (in Maple's share library) or `DEtools[DEplot]` to plot the direction field. To plot the solution, we can either use the initial-value option in `directionfield`, or actually solve the equation.

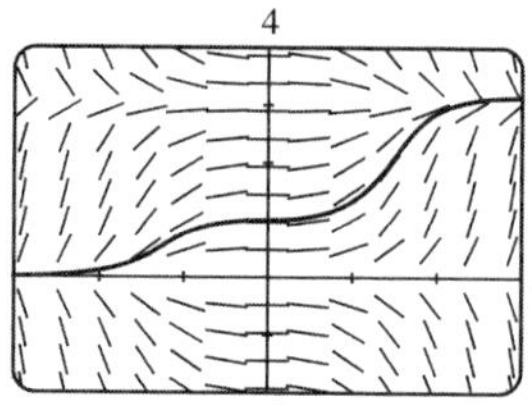


In Mathematica, we use `PlotVectorField` for the direction field, and the `Plot[Evaluate[...]]` construction to plot the solution, which is $y = 2\arctan\left(e^{x^3/3} \cdot \tan\frac{1}{2}\right)$.

In Derive, use `Direction_Field` (in utility file `ODE_APPR`) to plot the direction field. Then use `DSOLVE1(-x^2*SIN(y),1,x,y,0,1)` (in utility file `ODE1`) to solve the equation. Simplify each result.

**16.** See Exercise 15 for specific CAS directions. The exact solution is

$$y = \frac{2\left(3 - e^{2x^2}\right)}{e^{2x^2} + 3}$$

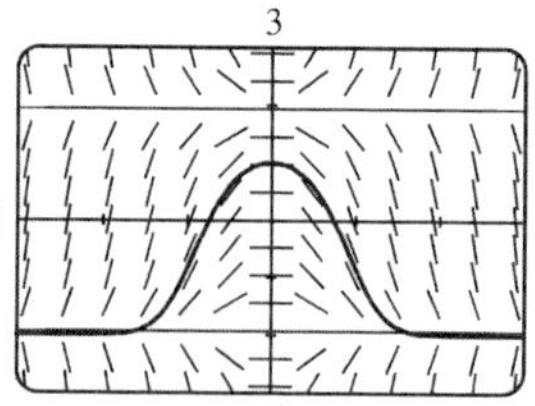

**17.**

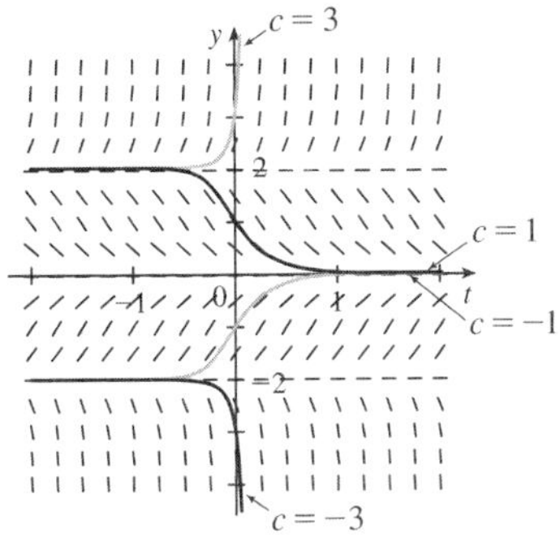


The direction field is for the differential equation $y' = y^3 - 4y$.

$L = \lim_{t\to\infty} y(t)$ exists for $-2 \le c \le 2$;

$L = \pm 2$ for $c = \pm 2$ and $L = 0$ for $-2 < c < 2$.

For other values of $c$, $L$ does not exist.

**18.**

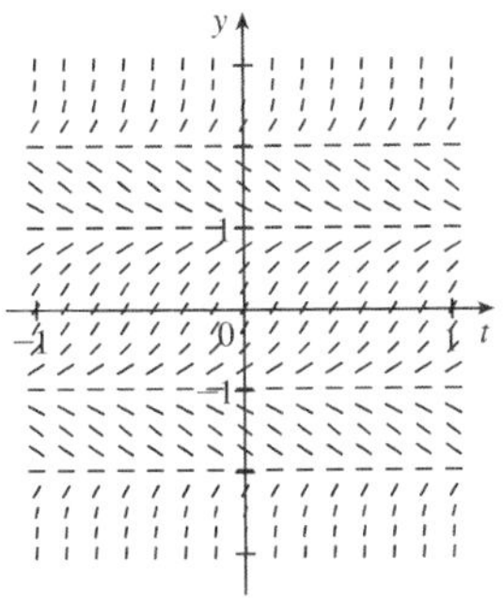

Note that when $f(y) = 0$ on the graph in the text, we have $y' = f(y) = 0$; so we get horizontal segments at $y = \pm 1, \pm 2$. We get segments with negative slopes only for $1 < |y| < 2$. All other segments have positive slope. For the limiting behavior of solutions:

- If $y(0) > 2$, then $\lim_{t\to\infty} y = \infty$ and $\lim_{t\to-\infty} y = 2$.
- If $1 < y(0) < 2$, then $\lim_{t\to\infty} y = 1$ and $\lim_{t\to-\infty} y = 2$.
- If $-1 < y(0) < 1$, then $\lim_{t\to\infty} y = 1$ and $\lim_{t\to-\infty} y = -1$.
- If $-2 < y(0) < -1$, then $\lim_{t\to\infty} y = -2$ and $\lim_{t\to-\infty} y = -1$.
- If $y < -2$, then $\lim_{t\to\infty} y = -2$ and $\lim_{t\to-\infty} y = -\infty$.

**19.** (a) $y' = F(x, y) = y$ and $y(0) = 1 \;\Rightarrow\; x_0 = 0,\, y_0 = 1$.

(i) $h = 0.4$ and $y_1 = y_0 + hF(x_0, y_0) \;\Rightarrow\; y_1 = 1 + 0.4 \cdot 1 = 1.4$. $x_1 = x_0 + h = 0 + 0.4 = 0.4$, so $y_1 = y\,(0.4) = 1.4$.

(ii) $h = 0.2 \;\Rightarrow\; x_1 = 0.2$ and $x_2 = 0.4$, so we need to find $y_2$.
$y_1 = y_0 + hF(x_0, y_0) = 1 + 0.2y_0 = 1 + 0.2 \cdot 1 = 1.2$,
$y_2 = y_1 + hF(x_1, y_1) = 1.2 + 0.2y_1 = 1.2 + 0.2 \cdot 1.2 = 1.44$.

(iii) $h = 0.1 \;\Rightarrow\; x_4 = 0.4$, so we need to find $y_4$. $y_1 = y_0 + hF(x_0, y_0) = 1 + 0.1y_0 = 1 + 0.1 \cdot 1 = 1.1$,
$y_2 = y_1 + hF(x_1, y_1) = 1.1 + 0.1y_1 = 1.1 + 0.1 \cdot 1.1 = 1.21$,
$y_3 = y_2 + hF(x_2, y_2) = 1.21 + 0.1y_2 = 1.21 + 0.1 \cdot 1.21 = 1.331$,
$y_4 = y_3 + hF(x_3, y_3) = 1.331 + 0.1y_3 = 1.331 + 0.1 \cdot 1.331 = 1.4641$.

(b)

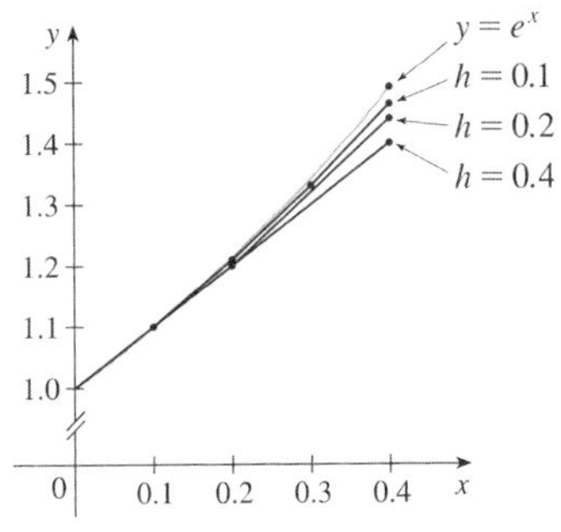


We see that the estimates are underestimates since they are all below the graph of $y = e^x$.

(c) (i) For $h = 0.4$: (exact value) − (approximate value) $= e^{0.4} - 1.4 \approx 0.0918$

(ii) For $h = 0.2$: (exact value) − (approximate value) $= e^{0.4} - 1.44 \approx 0.0518$

(iii) For $h = 0.1$: (exact value) − (approximate value) $= e^{0.4} - 1.4641 \approx 0.0277$

Each time the step size is halved, the error estimate also appears to be halved (approximately).

**20.**

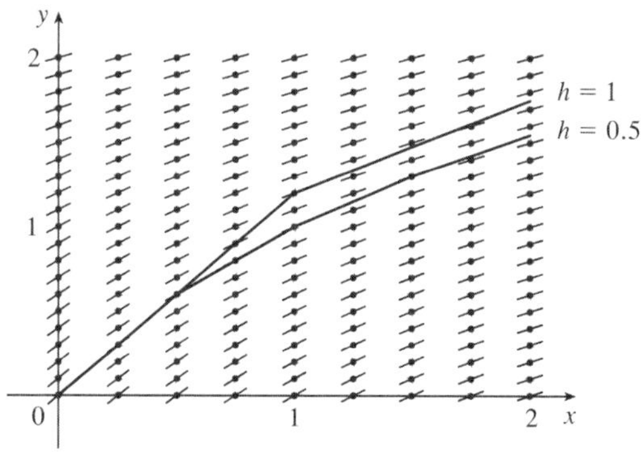


As $x$ increases, the slopes decrease and all of the estimates are above the true values. Thus, all of the estimates are overestimates.

**21.** $h = 0.5$, $x_0 = 1$, $y_0 = 0$, and $F(x, y) = y - 2x$.

Note that $x_1 = x_0 + h = 1 + 0.5 = 1.5$, $x_2 = 2$, and $x_3 = 2.5$.

$y_1 = y_0 + hF(x_0, y_0) = 0 + 0.5F(1, 0) = 0.5[0 - 2(1)] = -1.$

$y_2 = y_1 + hF(x_1, y_1) = -1 + 0.5F(1.5, -1) = -1 + 0.5[-1 - 2(1.5)] = -3.$

$y_3 = y_2 + hF(x_2, y_2) = -3 + 0.5F(2, -3) = -3 + 0.5[-3 - 2(2)] = -6.5.$

$y_4 = y_3 + hF(x_3, y_3) = -6.5 + 0.5F(2.5, -6.5) = -6.5 + 0.5[-6.5 - 2(2.5)] = -12.25.$

**22.** $h = 0.2$, $x_0 = 0$, $y_0 = 0$, and $F(x, y) = 1 - xy$.

Note that $x_1 = x_0 + h = 0 + 0.2 = 0.2$, $x_2 = 0.4$, $x_3 = 0.6$, and $x_4 = 0.8$.

$y_1 = y_0 + hF(x_0, y_0) = 0 + 0.2F(0, 0) = 0.2[1 - (0)(0)] = 0.2.$

$y_2 = y_1 + hF(x_1, y_1) = 0.2 + 0.2F(0.2, 0.2) = 0.2 + 0.2[1 - (0.2)(0.2)] = 0.392.$

$y_3 = y_2 + hF(x_2, y_2) = 0.392 + 0.2F(0.4, 0.392) = 0.392 + 0.2[1 - (0.4)(0.392)] = 0.56064.$

$y_4 = y_3 + hF(x_3, y_3) = 0.56064 + 0.2[1 - (0.6)(0.56064)] = 0.6933632.$

$y_5 = y_4 + hF(x_4, y_4) = 0.6933632 + 0.2[1 - (0.8)(0.6933632)] = 0.782425088.$

Thus, $y(1) \approx 0.7824$.

**23.** $h = 0.1$, $x_0 = 0$, $y_0 = 1$, and $F(x, y) = y + xy$.

Note that $x_1 = x_0 + h = 0 + 0.1 = 0.1$, $x_2 = 0.2$, $x_3 = 0.3$, and $x_4 = 0.4$.

$y_1 = y_0 + hF(x_0, y_0) = 1 + 0.1F(0, 1) = 1 + 0.1[1 + (0)(1)] = 1.1.$

$y_2 = y_1 + hF(x_1, y_1) = 1.1 + 0.1F(0.1, 1.1) = 1.1 + 0.1[1.1 + (0.1)(1.1)] = 1.221.$

$y_3 = y_2 + hF(x_2, y_2) = 1.221 + 0.1F(0.2, 1.221) = 1.221 + 0.1[1.221 + (0.2)(1.221)] = 1.36752.$

$$
\begin{aligned}
y_4 = y_3 + hF(x_3, y_3) &= 1.36752 + 0.1F(0.3, 1.36752) \\
&= 1.36752 + 0.1[1.36752 + (0.3)(1.36752)] = 1.5452976.
\end{aligned}
$$

$$
\begin{aligned}
y_5 = y_4 + hF(x_4, y_4) &= 1.5452976 + 0.1F(0.4, 1.5452976) \\
&= 1.5452976 + 0.1[1.5452976 + (0.4)(1.5452976)] = 1.761639264.
\end{aligned}
$$

Thus, $y(0.5) \approx 1.7616$.

**24.** (a) $h = 0.2$, $x_0 = 1$, $y_0 = 0$, and $F(x, y) = x - xy$.

We need to find $y_2$, because $x_1 = 1.2$ and $x_2 = 1.4$.

$y_1 = y_0 + hF(x_0, y_0) = 0 + 0.2F(1, 0) = 0.2[1 - (1)(0)] = 0.2.$

$y_2 = y_1 + hF(x_1, y_1) = 0.2 + 0.2F(1.2, 0.2) = 0.2 + 0.2[1.2 - (1.2)(0.2)] = 0.392 \approx y(1.4).$

(b) Now $h = 0.1$, so we need to find $y_4$.

$y_1 = 0 + 0.1[1 - (1)(0)] = 0.1,$

$y_2 = 0.1 + 0.1[1.1 - (1.1)(0.1)] = 0.199,$

$y_3 = 0.199 + 0.1[1.2 - (1.2)(0.199)] = 0.29512,$ and

$y_4 = 0.29512 + 0.1[1.3 - (1.3)(0.29512)] = 0.3867544 \approx y(1.4).$

**25.** (a) $dy/dx + 3x^2y = 6x^2 \Rightarrow y' = 6x^2 - 3x^2y$. Store this expression in $\text{Y}_1$ and use the following simple program to evaluate $y(1)$ for each part, using $\text{H} = h = 1$ and $\text{N} = 1$ for part (i), $\text{H} = 0.1$ and $\text{N} = 10$ for part (ii), and so forth.

$h \to$ H: $0 \to$ X: $3 \to$ Y:

For(I, 1, N): Y + H × $\text{Y}_1$ → Y: X + H → X:

End(loop):

Display Y. [To see all iterations, include this statement in the loop.]

(i) $\text{H} = 1, \text{N} = 1 \Rightarrow y(1) = 3$

(ii) $\text{H} = 0.1, \text{N} = 10 \Rightarrow y(1) \approx 2.3928$

(iii) $\text{H} = 0.01, \text{N} = 100 \Rightarrow y(1) \approx 2.3701$

(iv) $\text{H} = 0.001, \text{N} = 1000 \Rightarrow y(1) \approx 2.3681$

(b) $y = 2 + e^{-x^3} \Rightarrow y' = -3x^2e^{-x^3}$

$$\text{LHS} = y' + 3x^2y = -3x^2e^{-x^3} + 3x^2\left(2 + e^{-x^3}\right) = -3x^2e^{-x^3} + 6x^2 + 3x^2e^{-x^3} = 6x^2 = \text{RHS}$$

$y(0) = 2 + e^{-0} = 2 + 1 = 3$

(c) The exact value of $y(1)$ is $2 + e^{-1^3} = 2 + e^{-1}$.

(i) For $h = 1$: (exact value) − (approximate value) $= 2 + e^{-1} - 3 \approx -0.6321$

(ii) For $h = 0.1$: (exact value) − (approximate value) $= 2 + e^{-1} - 2.3928 \approx -0.0249$

(iii) For $h = 0.01$: (exact value) − (approximate value) $= 2 + e^{-1} - 2.3701 \approx -0.0022$

(iv) For $h = 0.001$: (exact value) − (approximate value) $= 2 + e^{-1} - 2.3681 \approx -0.0002$

In (ii)–(iv), it seems that when the step size is divided by 10, the error estimate is also divided by 10 (approximately).

**26.** (a) We use the program from the solution to Exercise 25 with $\text{Y}_1 = x^3 - y^3$, $\text{H} = 0.01$, and $\text{N} = \frac{2-0}{0.01} = 200$. With $(x_0, y_0) = (0, 1)$, we get $y(2) \approx 1.9000$.

(b)

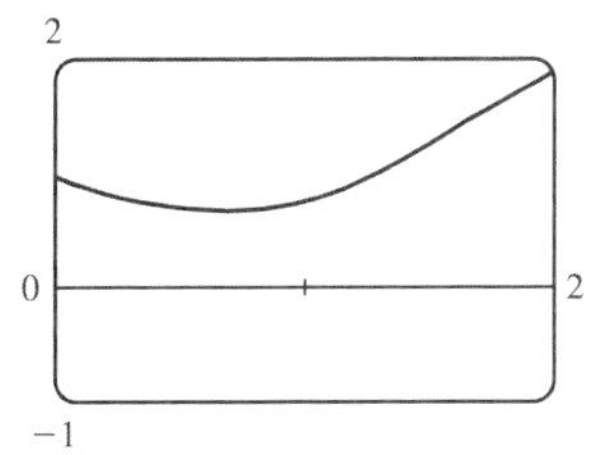


Notice from the graph that $y(2) \approx 1.9$, which serves as a check on our calculation in part (a).

**27.** (a) $R\dfrac{dQ}{dt} + \dfrac{1}{C}Q = E(t)$ becomes $5Q' + \dfrac{1}{0.05}Q = 60$ or $Q' + 4Q = 12$.

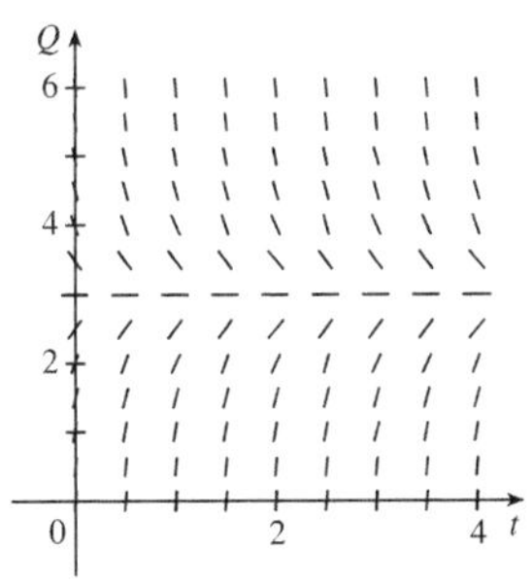


(b) From the graph, it appears that the limiting value of the charge $Q$ is about 3.

(c) If $Q' = 0$, then $4Q = 12 \Rightarrow Q = 3$ is an equilibrium solution.

(d)

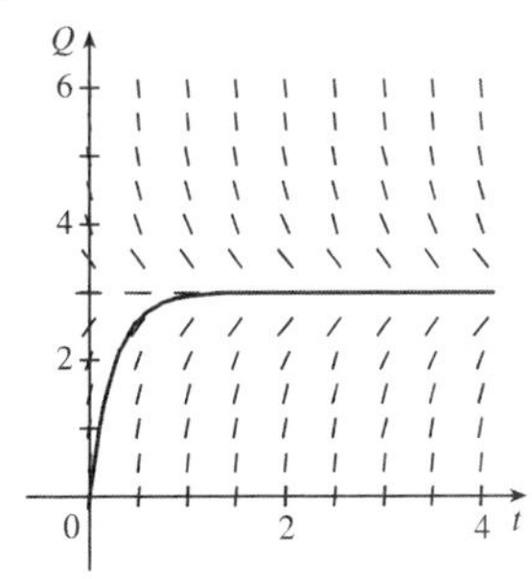

(e) $Q' + 4Q = 12 \Rightarrow Q' = 12 - 4Q$. Now $Q(0) = 0$, so $t_0 = 0$ and $Q_0 = 0$.

$$Q_1 = Q_0 + hF(t_0, Q_0) = 0 + 0.1(12 - 4 \cdot 0) = 1.2$$
$$Q_2 = Q_1 + hF(t_1, Q_1) = 1.2 + 0.1(12 - 4 \cdot 1.2) = 1.92$$
$$Q_3 = Q_2 + hF(t_2, Q_2) = 1.92 + 0.1(12 - 4 \cdot 1.92) = 2.352$$
$$Q_4 = Q_3 + hF(t_3, Q_3) = 2.352 + 0.1(12 - 4 \cdot 2.352) = 2.6112$$
$$Q_5 = Q_4 + hF(t_4, Q_4) = 2.6112 + 0.1(12 - 4 \cdot 2.6112) = 2.76672$$

Thus, $Q_5 = Q(0.5) \approx 2.77$ C.

**28.** (a) From Exercise 7.1.14, we have $dy/dt = k(y - R)$. We are given that $R = 20°\text{C}$ and $dy/dt = -1°\text{C/min}$ when $y = 70°\text{C}$. Thus, $-1 = k(70 - 20) \Rightarrow k = -\frac{1}{50}$ and the differential equation becomes $dy/dt = -\frac{1}{50}(y - 20)$.

(b)

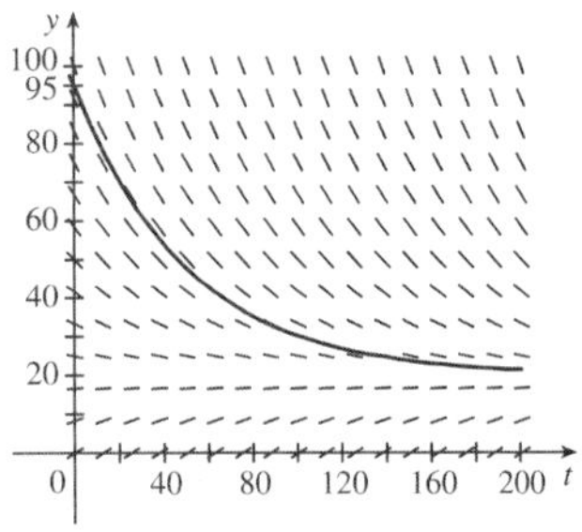


The limiting value of the temperature is 20°C; that is, the temperature of the room.

(c) From part (a), $dy/dt = -\frac{1}{50}(y - 20)$. With $t_0 = 0$, $y_0 = 95$, and $h = 2$ min, we get

$$y_1 = y_0 + hF(t_0, y_0) = 95 + 2\left[-\tfrac{1}{50}(95 - 20)\right] = 92$$
$$y_2 = y_1 + hF(t_1, y_1) = 92 + 2\left[-\tfrac{1}{50}(92 - 20)\right] = 89.12$$
$$y_3 = y_2 + hF(t_2, y_2) = 89.12 + 2\left[-\tfrac{1}{50}(89.12 - 20)\right] = 86.3552$$
$$y_4 = y_3 + hF(t_3, y_3) = 86.3552 + 2\left[-\tfrac{1}{50}(86.3552 - 20)\right] = 83.700992$$
$$y_5 = y_4 + hF(t_4, y_4) = 83.700992 + 2\left[-\tfrac{1}{50}(83.700992 - 20)\right] = 81.15295232$$

Thus, $y(10) \approx 81.15°\text{C}$.

## 7.3 Separable Equations

**1.** $\dfrac{dy}{dx} = \dfrac{y}{x} \Rightarrow \dfrac{dy}{y} = \dfrac{dx}{x}\ [y \neq 0] \Rightarrow \displaystyle\int \frac{dy}{y} = \int \frac{dx}{x} \Rightarrow \ln|y| = \ln|x| + C \Rightarrow$
$|y| = e^{\ln|x| + C} = e^{\ln|x|}e^C = e^C\,|x| \Rightarrow y = Kx$, where $K = \pm e^C$ is a constant. (In our derivation, $K$ was nonzero, but we can restore the excluded case $y = 0$ by allowing $K$ to be zero.)

**2.** $\dfrac{dy}{dx} = \dfrac{\sqrt{x}}{e^y} \Rightarrow e^y\,dy = \sqrt{x}\,dx \Rightarrow \int e^y\,dy = \int x^{1/2}\,dx \Rightarrow e^y = \frac{2}{3}x^{3/2} + C \Rightarrow y = \ln\left(\frac{2}{3}x^{3/2} + C\right)$

**3.** $(x^2 + 1)y' = xy \Rightarrow \dfrac{dy}{dx} = \dfrac{xy}{x^2 + 1} \Rightarrow \dfrac{dy}{y} = \dfrac{x\,dx}{x^2 + 1}\ [y \neq 0] \Rightarrow \displaystyle\int \frac{dy}{y} = \int \frac{x\,dx}{x^2 + 1} \Rightarrow$
$\ln|y| = \frac{1}{2}\ln(x^2 + 1) + C\ \ [u = x^2 + 1, du = 2x\,dx] = \ln(x^2 + 1)^{1/2} + \ln e^C = \ln\left(e^C\sqrt{x^2 + 1}\right) \Rightarrow$
$|y| = e^C\sqrt{x^2 + 1} \Rightarrow y = K\sqrt{x^2 + 1}$, where $K = \pm e^C$ is a constant. (In our derivation, $K$ was nonzero, but we can restore the excluded case $y = 0$ by allowing $K$ to be zero.)

**4.** $y' = y^2 \sin x \Rightarrow \dfrac{dy}{dx} = y^2 \sin x \Rightarrow \dfrac{dy}{y^2} = \sin x\,dx\ [y \neq 0] \Rightarrow \displaystyle\int \frac{dy}{y^2} = \int \sin x\,dx \Rightarrow$
$-\dfrac{1}{y} = -\cos x + C \Rightarrow \dfrac{1}{y} = \cos x - C \Rightarrow y = \dfrac{1}{\cos x + K}$, where $K = -C$. $y = 0$ is also a solution.

**5.** $(1 + \tan y)\, y' = x^2 + 1 \;\Rightarrow\; (1 + \tan y)\dfrac{dy}{dx} = x^2 + 1 \;\Rightarrow\; \left(1 + \dfrac{\sin y}{\cos y}\right) dy = (x^2 + 1)\, dx \;\Rightarrow$ $\displaystyle\int \left(1 - \frac{-\sin y}{\cos y}\right) dy = \int (x^2 + 1)\, dx \;\Rightarrow\; y - \ln|\cos y| = \tfrac{1}{3}x^3 + x + C$. Note: The left side is equivalent to $y + \ln|\sec y|$.

**6.** $\dfrac{dy}{d\theta} = \dfrac{e^y \sin^2\theta}{y \sec\theta} \;\Rightarrow\; \dfrac{y}{e^y}\, dy = \dfrac{\sin^2\theta}{\sec\theta}\, d\theta \;\Rightarrow\; \int y e^{-y}\, dy = \int \sin^2\theta \cos\theta\, d\theta$. Integrating the left side by parts with $u = y$, $dv = e^{-y}\, dy$ and the right side by the substitution $u = \sin\theta$, we obtain $-ye^{-y} - e^{-y} = \frac{1}{3}\sin^3\theta + C$. We cannot solve explicitly for $y$.

**7.** $\dfrac{du}{dt} = 2 + 2u + t + tu \;\Rightarrow\; \dfrac{du}{dt} = (1 + u)(2 + t) \;\Rightarrow\; \displaystyle\int \frac{du}{1 + u} = \int (2 + t) dt \;\; [u \neq -1] \;\Rightarrow$ $\ln|1 + u| = \frac{1}{2}t^2 + 2t + C \;\Rightarrow\; |1 + u| = e^{t^2/2 + 2t + C} = Ke^{t^2/2 + 2t}$, where $K = e^C \;\Rightarrow\; 1 + u = \pm Ke^{t^2/2 + 2t} \;\Rightarrow$ $u = -1 \pm Ke^{t^2/2 + 2t}$ where $K > 0$. $u = -1$ is also a solution, so $u = -1 + Ae^{t^2/2 + 2t}$, where $A$ is an arbitrary constant.

**8.** $\dfrac{dz}{dt} + e^{t + z} = 0 \;\Rightarrow\; \dfrac{dz}{dt} = -e^t e^z \;\Rightarrow\; \int e^{-z}\, dz = -\int e^t\, dt \;\Rightarrow\; -e^{-z} = -e^t + C \;\Rightarrow\; e^{-z} = e^t - C \;\Rightarrow$ $\dfrac{1}{e^z} = e^t - C \;\Rightarrow\; e^z = \dfrac{1}{e^t - C} \;\Rightarrow\; z = \ln\left(\dfrac{1}{e^t - C}\right) \;\Rightarrow\; z = -\ln(e^t - C)$

**9.** $\dfrac{du}{dt} = \dfrac{2t + \sec^2 t}{2u}$, $u(0) = -5$. $\int 2u\, du = \int (2t + \sec^2 t)\, dt \;\Rightarrow\; u^2 = t^2 + \tan t + C$, where $[u(0)]^2 = 0^2 + \tan 0 + C \;\Rightarrow\; C = (-5)^2 = 25$. Therefore, $u^2 = t^2 + \tan t + 25$, so $u = \pm\sqrt{t^2 + \tan t + 25}$. Since $u(0) = -5$, we must have $u = -\sqrt{t^2 + \tan t + 25}$.

**10.** $\dfrac{dy}{dx} = \dfrac{y \cos x}{1 + y^2}$, $y(0) = 1$. $(1 + y^2)\, dy = y \cos x\, dx \;\Rightarrow\; \dfrac{1 + y^2}{y} dy = \cos x\, dx \;\Rightarrow\; \int \left(\frac{1}{y} + y\right) dy = \int \cos x\, dx \;\Rightarrow$ $\ln|y| + \frac{1}{2}y^2 = \sin x + C$. $y(0) = 1 \;\Rightarrow\; \ln 1 + \frac{1}{2} = \sin 0 + C \;\Rightarrow\; C = \frac{1}{2}$, so $\ln|y| + \frac{1}{2}y^2 = \sin x + \frac{1}{2}$. We cannot solve explicitly for $y$.

**11.** $x \cos x = (2y + e^{3y})\, y' \;\Rightarrow\; x \cos x\, dx = (2y + e^{3y})\, dy \;\Rightarrow\; \int (2y + e^{3y})\, dy = \int x \cos x\, dx \;\Rightarrow$ $y^2 + \frac{1}{3}e^{3y} = x \sin x + \cos x + C$ [where the second integral is evaluated using integration by parts]. Now $y(0) = 0 \;\Rightarrow$ $0 + \frac{1}{3} = 0 + 1 + C \;\Rightarrow\; C = -\frac{2}{3}$. Thus, a solution is $y^2 + \frac{1}{3}e^{3y} = x \sin x + \cos x - \frac{2}{3}$. We cannot solve explicitly for $y$.

**12.** $\dfrac{dP}{dt} = \sqrt{Pt} \;\Rightarrow\; dP/\sqrt{P} = \sqrt{t}\, dt \;\Rightarrow\; \int P^{-1/2}\, dP = \int t^{1/2}\, dt \;\Rightarrow\; 2P^{1/2} = \frac{2}{3}t^{3/2} + C$. $P(1) = 2 \;\Rightarrow\; 2\sqrt{2} = \frac{2}{3} + C \;\Rightarrow\; C = 2\sqrt{2} - \frac{2}{3}$, so $2P^{1/2} = \frac{2}{3}t^{3/2} + 2\sqrt{2} - \frac{2}{3} \;\Rightarrow\; \sqrt{P} = \frac{1}{3}t^{3/2} + \sqrt{2} - \frac{1}{3} \;\Rightarrow$ $P = \left(\frac{1}{3}t^{3/2} + \sqrt{2} - \frac{1}{3}\right)^2$.

**13.** $y' \tan x = a + y$, $0 < x < \pi/2 \;\Rightarrow\; \dfrac{dy}{dx} = \dfrac{a + y}{\tan x} \;\Rightarrow\; \dfrac{dy}{a + y} = \cot x\, dx \;\; [a + y \neq 0] \;\Rightarrow$ $\displaystyle\int \frac{dy}{a + y} = \int \frac{\cos x}{\sin x}\, dx \;\Rightarrow\; \ln|a + y| = \ln|\sin x| + C \;\Rightarrow\; |a + y| = e^{\ln|\sin x| + C} = e^{\ln|\sin x|} \cdot e^C = e^C\,|\sin x| \;\Rightarrow$ $a + y = K \sin x$, where $K = \pm e^C$. (In our derivation, $K$ was nonzero, but we can restore the excluded case $y = -a$ by allowing $K$ to be zero.) $y(\pi/3) = a \;\Rightarrow$

$a + a = K \sin\left(\dfrac{\pi}{3}\right) \;\Rightarrow\; 2a = K\dfrac{\sqrt{3}}{2} \;\Rightarrow\; K = \dfrac{4a}{\sqrt{3}}$. Thus, $a + y = \dfrac{4a}{\sqrt{3}} \sin x$ and so $y = \dfrac{4a}{\sqrt{3}} \sin x - a$.

**14.** $\dfrac{dL}{dt} = kL^2 \ln t \;\Rightarrow\; \dfrac{dL}{L^2} = k \ln t\, dt \;\Rightarrow\; \displaystyle\int \frac{dL}{L^2} = \int k \ln t\, dt \;\Rightarrow\; -\frac{1}{L} = kt \ln t - \int k\, dt$ [by parts with $u = \ln t$, $dv = k\, dt$] $\Rightarrow\; -\dfrac{1}{L} = kt \ln t - kt + C \;\Rightarrow\; L = \dfrac{1}{kt - kt \ln t - C}$. $L(1) = -1 \;\Rightarrow\; -1 = \dfrac{1}{k - k \ln 1 - C} \;\Rightarrow\; C - k = 1 \;\Rightarrow\; C = k + 1$. Thus, $L = \dfrac{1}{kt - kt \ln t - k - 1}$.

**15.** $\dfrac{dy}{dx} = 4x^3y,\ y(0) = 7.\quad \dfrac{dy}{y} = 4x^3\,dx\ \ [\text{if } y \neq 0] \ \Rightarrow\ \displaystyle\int \frac{dy}{y} = \int 4x^3\,dx \ \Rightarrow\ \ln|y| = x^4 + C \ \Rightarrow$

$e^{\ln|y|} = e^{x^4+C} \ \Rightarrow\ |y| = e^{x^4}e^C \ \Rightarrow\ y = Ae^{x^4};\ y(0) = 7 \ \Rightarrow\ A = 7 \ \Rightarrow\ y = 7e^{x^4}.$

**16.** $\dfrac{dy}{dx} = \dfrac{y^2}{x^3},\ y(1) = 1.\quad \displaystyle\int \frac{dy}{y^2} = \int \frac{dx}{x^3} \ \Rightarrow\ -\frac{1}{y} = -\frac{1}{2x^2} + C.\ \ y(1) = 1 \ \Rightarrow\ -1 = -\tfrac{1}{2} + C \ \Rightarrow\ C = -\tfrac{1}{2}.$

So $\dfrac{1}{y} = \dfrac{1}{2x^2} + \dfrac{1}{2} = \dfrac{2 + 2x^2}{2 \cdot 2x^2} \ \Rightarrow\ y = \dfrac{2x^2}{x^2 + 1}.$

**17.** (a) $y' = 2x\sqrt{1 - y^2} \ \Rightarrow\ \dfrac{dy}{dx} = 2x\sqrt{1 - y^2} \ \Rightarrow\ \dfrac{dy}{\sqrt{1 - y^2}} = 2x\,dx \ \Rightarrow\ \displaystyle\int \frac{dy}{\sqrt{1 - y^2}} = \int 2x\,dx \ \Rightarrow$

$\sin^{-1} y = x^2 + C$ for $-\frac{\pi}{2} \leq x^2 + C \leq \frac{\pi}{2}$.

(b) $y(0) = 0 \ \Rightarrow\ \sin^{-1} 0 = 0^2 + C \ \Rightarrow$

$C = 0$, so $\sin^{-1} y = x^2$ and $y = \sin(x^2)$ for

$-\sqrt{\pi/2} \leq x \leq \sqrt{\pi/2}$.

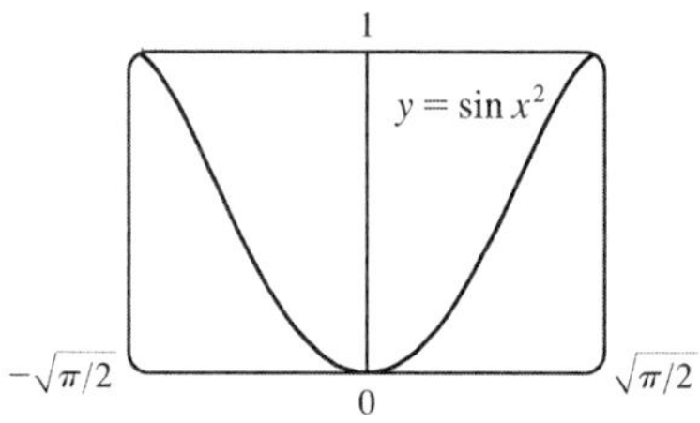


(c) For $\sqrt{1 - y^2}$ to be a real number, we must have $-1 \leq y \leq 1$; that is, $-1 \leq y(0) \leq 1$. Thus, the initial-value problem $y' = 2x\sqrt{1 - y^2},\ y(0) = 2$ does *not* have a solution.

**18.** $e^{-y}y' + \cos x = 0 \ \Leftrightarrow\ \int e^{-y}\,dy = -\int \cos x\,dx \ \Leftrightarrow\ -e^{-y} = -\sin x + C_1 \ \Leftrightarrow\ y = -\ln(\sin x + C)$. The solution is periodic, with period $2\pi$. Note that for $C > 1$, the domain of the solution is $\mathbb{R}$, but for $-1 < C \leq 1$ it is only defined on the intervals where $\sin x + C > 0$, and it is meaningless for $C \leq -1$, since then $\sin x + C \leq 0$, and the logarithm is undefined.

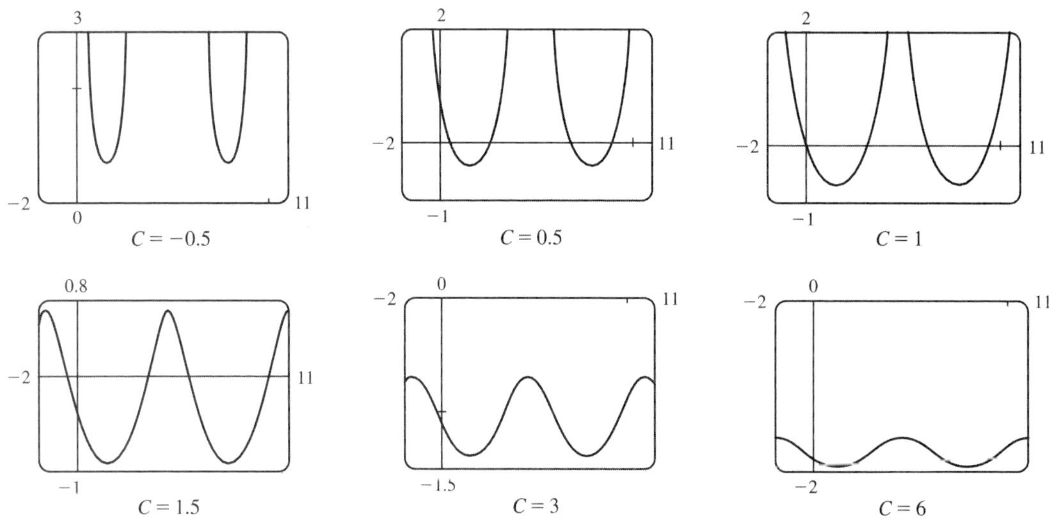


For $-1 < C < 1$, the solution curve consists of concave-up pieces separated by intervals on which the solution is not defined (where $\sin x + C \leq 0$). For $C = 1$, the solution curve consists of concave-up pieces separated by vertical asymptotes at the points where $\sin x + C = 0 \ \Leftrightarrow\ \sin x = -1$. For $C > 1$, the curve is continuous, and as $C$ increases, the graph moves downward, and the amplitude of the oscillations decreases.

**19.** $\dfrac{dy}{dx} = \dfrac{\sin x}{\sin y}$, $y(0) = \frac{\pi}{2}$. So $\int \sin y\, dy = \int \sin x\, dx \quad \Leftrightarrow \quad -\cos y = -\cos x + C \quad \Leftrightarrow$ $\cos y = \cos x - C$. From the initial condition, we need $\cos \frac{\pi}{2} = \cos 0 - C \quad \Rightarrow$ $0 = 1 - C \quad \Rightarrow \quad C = 1$, so the solution is $\cos y = \cos x - 1$. Note that we cannot take $\cos^{-1}$ of both sides, since that would unnecessarily restrict the solution to the case where $-1 \leq \cos x - 1 \quad \Leftrightarrow \quad 0 \leq \cos x$, as $\cos^{-1}$ is defined only on $[-1, 1]$. Instead we plot the graph using Maple's `plots[implicitplot]` or Mathematica's `Plot[Evaluate[···]]`.

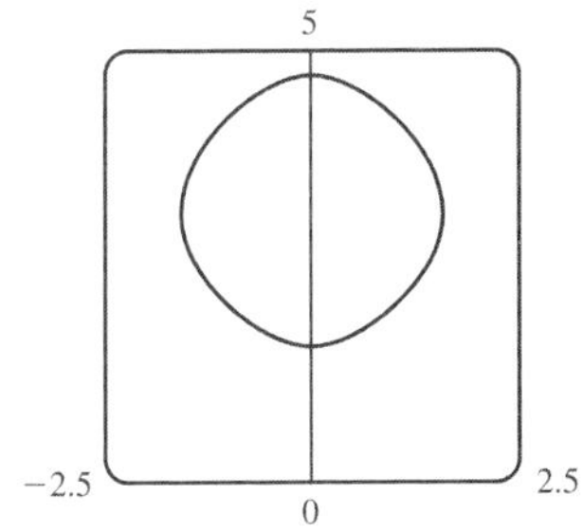


**20.** $\dfrac{dy}{dx} = \dfrac{x\sqrt{x^2+1}}{ye^y} \quad \Leftrightarrow \quad \int ye^y\, dy = \int x\sqrt{x^2+1}\, dx$. We use parts on the LHS with $u = y$, $dv = e^y\, dy$, and on the RHS we use the substitution $z = x^2 + 1$, so $dz = 2x\, dx$. The equation becomes $ye^y - \int e^y\, dy = \frac{1}{2}\int \sqrt{z}\, dz \quad \Leftrightarrow$ $e^y(y-1) = \frac{1}{3}(x^2+1)^{3/2} + C$, so we see that the curves are symmetric about the $y$-axis. Every point $(x, y)$ in the plane lies on one of the curves, namely the one for which $C = (y-1)e^y - \frac{1}{3}(x^2+1)^{3/2}$. For example, along the $y$-axis, $C = (y-1)e^y - \frac{1}{3}$, so the origin lies on the curve with $C = -\frac{4}{3}$. We use Maple's `plots[implicitplot]` command or `Plot[Evaluate[···]]` in Mathematica to plot the solution curves for various values of $C$.

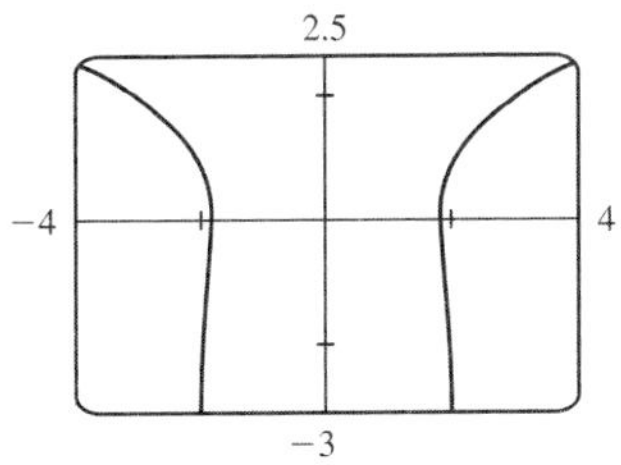

$C = -4$

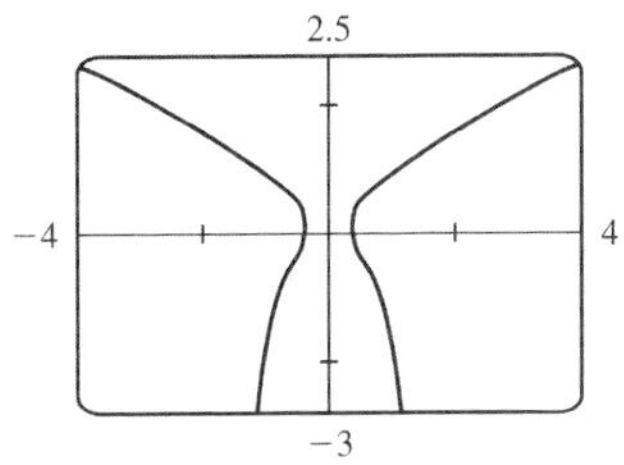

$C = -1.4$

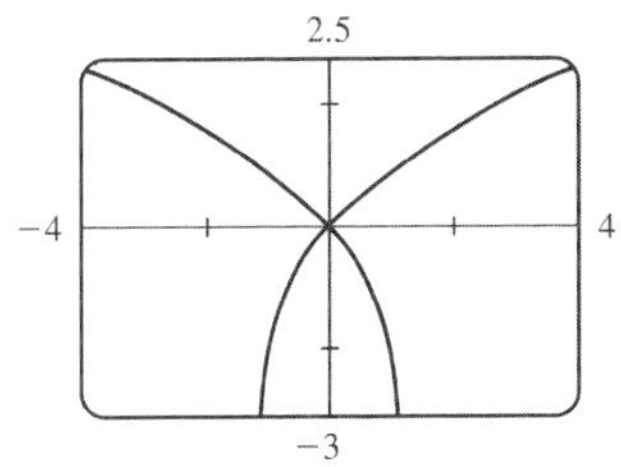

$C = -4/3$

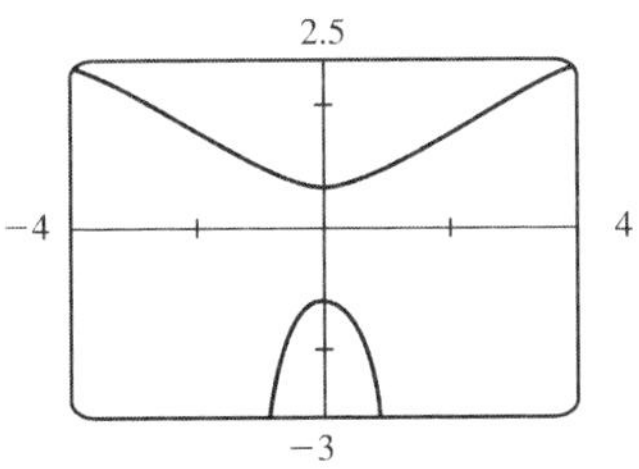

$C = -1$

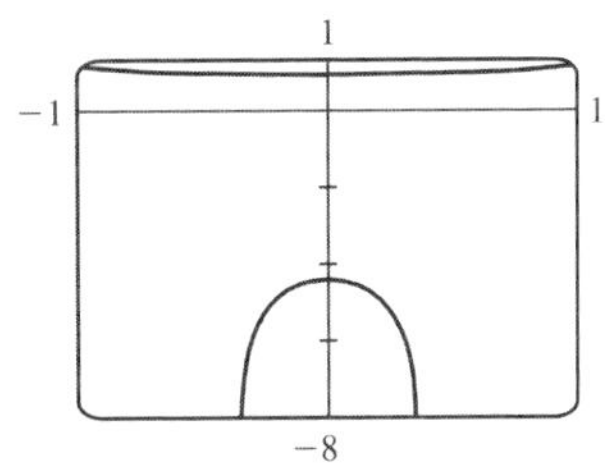

$C = -0.4$

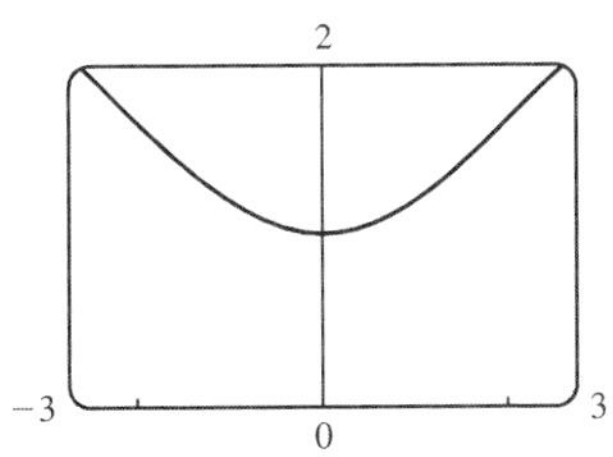

$C = -1/3$

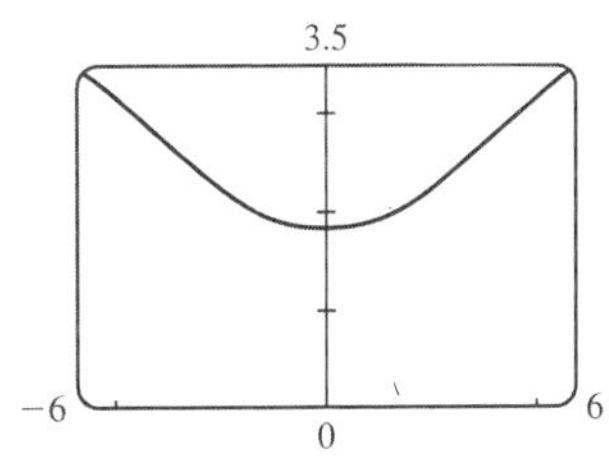

$C = 5$

It seems that the transitional values of $C$ are $-\frac{4}{3}$ and $-\frac{1}{3}$. For $C < -\frac{4}{3}$, the graph consists of left and right branches. At $C = -\frac{4}{3}$, the two branches become connected at the origin, and as $C$ increases, the graph splits into top and bottom branches. At $C = -\frac{1}{3}$, the bottom half disappears. As $C$ increases further, the graph moves upward, but doesn't change shape much.

**21.** (a)

| $x$ | $y$ | $y' = 1/y$ |
|---|---|---|
| 0 | 0.5 | 2 |
| 0 | −0.5 | −2 |
| 0 | 1 | 1 |
| 0 | −1 | −1 |
| 0 | 2 | 0.5 |

| $x$ | $y$ | $y' = 1/y$ |
|---|---|---|
| 0 | −2 | −0.5 |
| 0 | 4 | 0.25 |
| 0 | 3 | $0.\overline{3}$ |
| 0 | 0.25 | 4 |
| 0 | $0.\overline{3}$ | 3 |

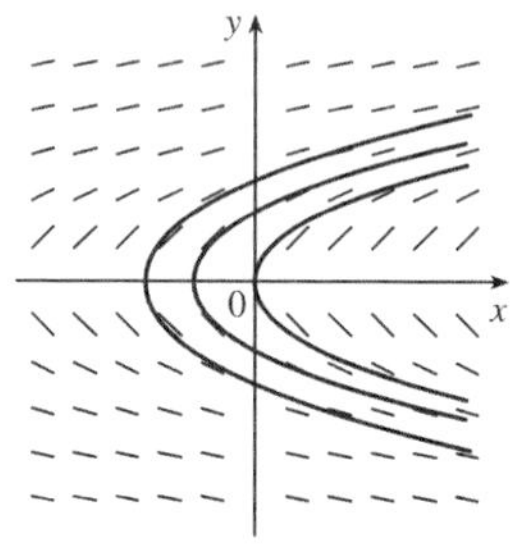


(b) $y' = 1/y \quad \Rightarrow \quad dy/dx = 1/y \quad \Rightarrow$

$y\,dy = dx \quad \Rightarrow \quad \int y\,dy = \int dx \quad \Rightarrow \quad \frac{1}{2}y^2 = x + C \quad \Rightarrow$

$y^2 = 2(x + C)$ or $y = \pm\sqrt{2(x + C)}$.

(c)

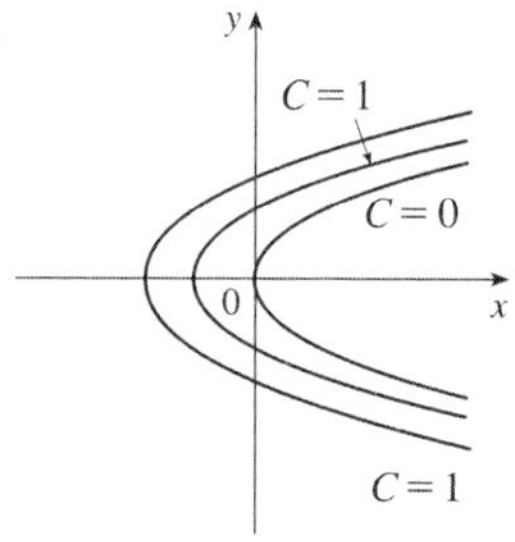


**22.** (a)

| $x$ | $y$ | $y' = x^2/y$ |
|---|---|---|
| 1 | 1 | 1 |
| −1 | 1 | 1 |
| −1 | −1 | −1 |
| 1 | −1 | −1 |
| 1 | 2 | 0.5 |
| 2 | 1 | 4 |
| 2 | 2 | 2 |
| 1 | 0.5 | 2 |
| 0.5 | 1 | 0.25 |
| 2 | 0.5 | 8 |

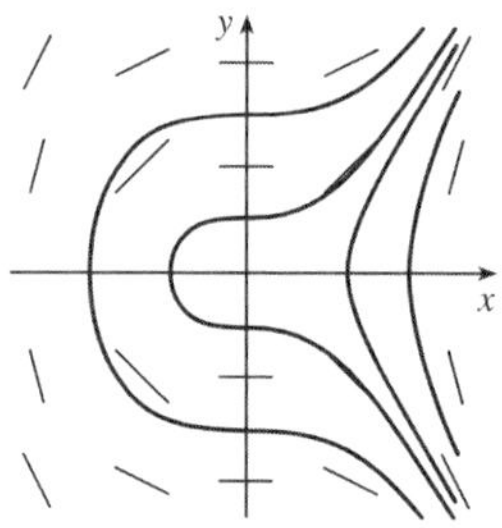


(b) $y' = x^2/y \quad \Rightarrow \quad y\,dy = x^2\,dx$, so

$\frac{1}{2}y^2 = \frac{1}{3}x^3 + C_1$, or

$y = \pm\left(\frac{2}{3}x^3 + C\right)^{1/2}$.

(c)

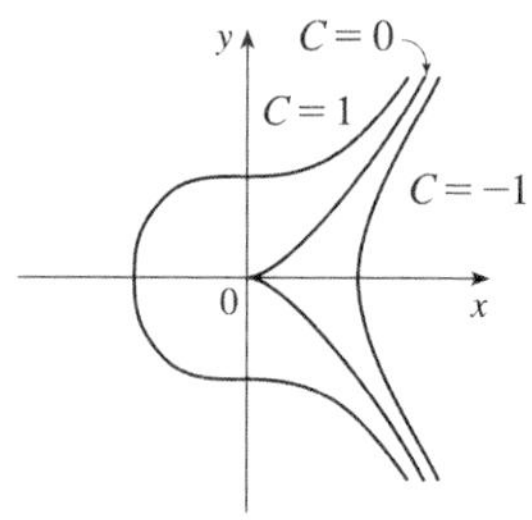


**23.** The curves $x^2 + 2y^2 = k^2$ form a family of ellipses with major axis on the $x$-axis. Differentiating gives

$\frac{d}{dx}\left(x^2 + 2y^2\right) = \frac{d}{dx}\left(k^2\right) \quad \Rightarrow \quad 2x + 4yy' = 0 \quad \Rightarrow \quad 4yy' = -2x \quad \Rightarrow \quad y' = \frac{-x}{2y}$. Thus, the slope of the tangent line at any point $(x, y)$ on one of the ellipses is $y' = \frac{-x}{2y}$, so the orthogonal trajectories must satisfy $y' = \frac{2y}{x} \quad \Leftrightarrow$

$\frac{dy}{dx} = \frac{2y}{x} \quad \Leftrightarrow \quad \frac{dy}{y} = 2 = \frac{dx}{x} \quad \Leftrightarrow \quad \int \frac{dy}{y} = 2\int \frac{dx}{x} \quad \Leftrightarrow \quad \ln|y| = 2\ln|x| + C_1 \quad \Leftrightarrow \quad \ln|y| = \ln|x|^2 + C_1 \quad \Leftrightarrow$

$|y| = e^{\ln x^2 + C_1} \quad \Leftrightarrow \quad y = \pm x^2 \cdot e^{C_1} = Cx^2$. This is a family of parabolas.

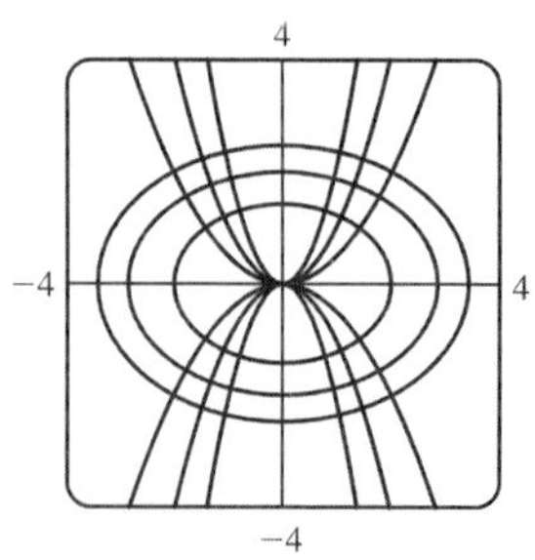

**24.** The curves $y^2 = kx^3$ form a family of power functions. Differentiating gives $\dfrac{d}{dx}(y^2) = \dfrac{d}{dx}(kx^3) \Rightarrow 2yy' = 3kx^2 \Rightarrow$ $y' = \dfrac{3kx^2}{2y} = \dfrac{3(y^2/x^3)x^2}{2y} = \dfrac{3y}{2x}$, the slope of the tangent line at $(x, y)$ on one of the curves. Thus, the orthogonal trajectories must satisfy $y' = -\dfrac{2x}{3y} \Leftrightarrow \dfrac{dy}{dx} = -\dfrac{2x}{3y} \Leftrightarrow 3y\,dy = -2x\,dx \Leftrightarrow \int 3y\,dy = \int -2x\,dx \Leftrightarrow$ $\frac{3}{2}y^2 = -x^2 + C_1 \Leftrightarrow 3y^2 = -2x^2 + C_2 \Leftrightarrow 2x^2 + 3y^2 = C$. This is a family of ellipses.

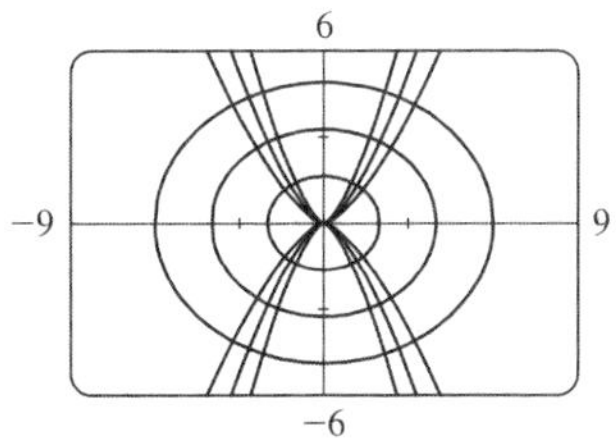

**25.** The curves $y = k/x$ form a family of hyperbolas with asymptotes $x = 0$ and $y = 0$. Differentiating gives $\dfrac{d}{dx}(y) = \dfrac{d}{dx}\left(\dfrac{k}{x}\right) \Rightarrow y' = -\dfrac{k}{x^2} \Rightarrow y' = -\dfrac{xy}{x^2}$ [since $y = k/x \Rightarrow xy = k$] $\Rightarrow y' = -\dfrac{y}{x}$. Thus, the slope of the tangent line at any point $(x, y)$ on one of the hyperbolas is $y' = -y/x$, so the orthogonal trajectories must satisfy $y' = x/y \Leftrightarrow \dfrac{dy}{dx} = \dfrac{x}{y} \Leftrightarrow y\,dy = x\,dx \Leftrightarrow \int y\,dy = \int x\,dx \Leftrightarrow \frac{1}{2}y^2 = \frac{1}{2}x^2 + C_1 \Leftrightarrow y^2 = x^2 + C_2 \Leftrightarrow$ $x^2 - y^2 = C$. This is a family of hyperbolas with asymptotes $y = \pm x$.

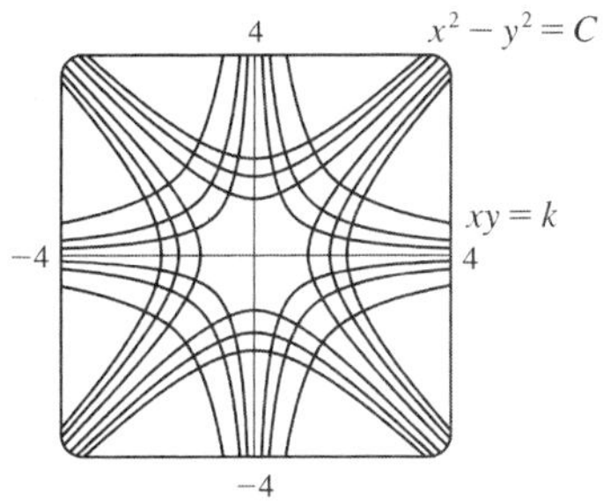


**26.** Differentiating $y = \dfrac{x}{1 + kx}$ gives $y' = \dfrac{1}{(1 + kx)^2}$, but $k = \dfrac{x - y}{xy}$, so $y' = \dfrac{1}{\left(1 + \dfrac{x - y}{y}\right)^2} = \dfrac{y^2}{x^2}$. Thus, the orthogonal trajectories must satisfy $y' = -x^2/y^2 \Leftrightarrow y^2\,dy = -x^2\,dx \Leftrightarrow \int y^2\,dy = -\int x^2\,dx \Leftrightarrow \frac{1}{3}y^3 = -\frac{1}{3}x^3 + C_1 \Leftrightarrow$ $y^3 = C - x^3 \Leftrightarrow y = \sqrt[3]{C - x^3}$.

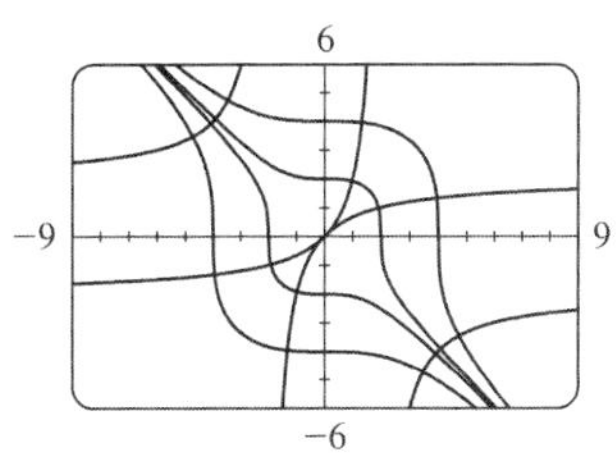

**27.** From Exercise 7.2.27, $\dfrac{dQ}{dt} = 12 - 4Q \Leftrightarrow \displaystyle\int \frac{dQ}{12 - 4Q} = \int dt \Leftrightarrow -\tfrac{1}{4}\ln|12 - 4Q| = t + C \Leftrightarrow$

$\ln|12 - 4Q| = -4t - 4C \Leftrightarrow |12 - 4Q| = e^{-4t-4C} \Leftrightarrow 12 - 4Q = Ke^{-4t}\ [K = \pm e^{-4C}] \Leftrightarrow$

$4Q = 12 - Ke^{-4t} \Leftrightarrow Q = 3 - Ae^{-4t}\ [A = K/4].\ Q(0) = 0 \Leftrightarrow 0 = 3 - A \Leftrightarrow A = 3 \Leftrightarrow$

$Q(t) = 3 - 3e^{-4t}$. As $t \to \infty$, $Q(t) \to 3 - 0 = 3$ (the limiting value).

**28.** From Exercise 7.2.28, $\dfrac{dy}{dt} = -\tfrac{1}{50}(y - 20) \Leftrightarrow \displaystyle\int \frac{dy}{y - 20} = \int \left(-\tfrac{1}{50}\right) dt \Leftrightarrow \ln|y - 20| = -\tfrac{1}{50}t + C \Leftrightarrow$

$y - 20 = Ke^{-t/50} \Leftrightarrow y(t) = Ke^{-t/50} + 20.\ y(0) = 95 \Leftrightarrow 95 = K + 20 \Leftrightarrow K = 75 \Leftrightarrow$

$y(t) = 75e^{-t/50} + 20.$

**29.** $\dfrac{dP}{dt} = k(M - P) \Leftrightarrow \displaystyle\int \frac{dP}{P - M} = \int (-k)\,dt \Leftrightarrow \ln|P - M| = -kt + C \Leftrightarrow |P - M| = e^{-kt+C} \Leftrightarrow$

$P - M = Ae^{-kt}\ [A = \pm e^{C}] \Leftrightarrow P = M + Ae^{-kt}$. If we assume that performance is at level 0 when $t = 0$, then

$P(0) = 0 \Leftrightarrow 0 = M + A \Leftrightarrow A = -M \Leftrightarrow P(t) = M - Me^{-kt}.\ \lim\limits_{t\to\infty} P(t) = M - M \cdot 0 = M.$

**30.** (a) $\dfrac{dx}{dt} = k\,(a - x)\,(b - x)$, $a \neq b$. Using partial fractions, $\dfrac{1}{(a - x)(b - x)} = \dfrac{1/(b - a)}{a - x} - \dfrac{1/(b - a)}{b - x}$, so

$\displaystyle\int \frac{dx}{(a - x)(b - x)} = \int k\,dt \Rightarrow \frac{1}{b - a}(-\ln|a - x| + \ln|b - x|) = kt + C \Rightarrow \ln\left|\frac{b - x}{a - x}\right| = (b - a)\,(kt + C).$

The concentrations [A] $= a - x$ and [B] $= b - x$ cannot be negative, so $\dfrac{b - x}{a - x} \geq 0$ and

$\left|\dfrac{b - x}{a - x}\right| = \dfrac{b - x}{a - x}$. We now have $\ln\left(\dfrac{b - x}{a - x}\right) = (b - a)(kt + C)$. Since $x(0) = 0$, we get

$\ln\left(\dfrac{b}{a}\right) = (b - a)C$. Hence, $\ln\left(\dfrac{b - x}{a - x}\right) = (b - a)kt + \ln\left(\dfrac{b}{a}\right) \Rightarrow \dfrac{b - x}{a - x} = \dfrac{b}{a}e^{(b-a)kt} \Rightarrow$

$x = \dfrac{b\left[e^{(b-a)kt} - 1\right]}{be^{(b-a)kt}/a - 1} = \dfrac{ab\left[e^{(b-a)kt} - 1\right]}{be^{(b-a)kt} - a}$ moles/L.

(b) If $b = a$, then $\dfrac{dx}{dt} = k(a - x)^2$, so $\displaystyle\int \frac{dx}{(a - x)^2} = \int k\,dt$ and $\dfrac{1}{a - x} = kt + C$. Since $x(0) = 0$, we get $C = \dfrac{1}{a}$. Thus,

$a - x = \dfrac{1}{kt + 1/a}$ and $x = a - \dfrac{a}{akt + 1} = \dfrac{a^2kt}{akt + 1}\dfrac{\text{moles}}{\text{L}}.$

Suppose $x = [\text{C}] = a/2$ when $t = 20$. Then $x(20) = a/2 \Rightarrow \dfrac{a}{2} = \dfrac{20a^2k}{20ak + 1} \Rightarrow 40a^2k = 20a^2k + a \Rightarrow$

$20a^2k = a \Rightarrow k = \dfrac{1}{20a}$, so $x = \dfrac{a^2t/(20a)}{1 + at/(20a)} = \dfrac{at/20}{1 + t/20} = \dfrac{at}{t + 20}\dfrac{\text{moles}}{\text{L}}.$

**31.** (a) If $a = b$, then $\dfrac{dx}{dt} = k(a - x)(b - x)^{1/2}$ becomes $\dfrac{dx}{dt} = k(a - x)^{3/2} \Rightarrow (a - x)^{-3/2}dx = k\,dt \Rightarrow$

$\int (a - x)^{-3/2}dx = \int k\,dt \Rightarrow 2(a - x)^{-1/2} = kt + C$ [by substitution] $\Rightarrow \dfrac{2}{kt + C} = \sqrt{a - x} \Rightarrow$

$\left(\dfrac{2}{kt + C}\right)^2 = a - x \Rightarrow x(t) = a - \dfrac{4}{(kt + C)^2}$. The initial concentration of HBr is 0, so $x(0) = 0 \Rightarrow$

$0 = a - \dfrac{4}{C^2} \Rightarrow \dfrac{4}{C^2} = a \Rightarrow C^2 = \dfrac{4}{a} \Rightarrow C = 2/\sqrt{a}$ ($C$ is positive since $kt + C = 2(a - x)^{-1/2} > 0$). Thus,

$x(t) = a - \dfrac{4}{(kt + 2/\sqrt{a})^2}.$

(b) $\frac{dx}{dt} = k(a-x)(b-x)^{1/2} \Rightarrow \frac{dx}{(a-x)\sqrt{b-x}} = k\,dt \Rightarrow \int \frac{dx}{(a-x)\sqrt{b-x}} = \int k\,dt$ $(\star)$. From the hint, $u = \sqrt{b-x} \Rightarrow u^2 = b-x \Rightarrow 2u\,du = -dx$, so

$$\int \frac{dx}{(a-x)\sqrt{b-x}} = \int \frac{-2u\,du}{[a-(b-u^2)]u} = -2\int \frac{du}{a-b+u^2} = -2\int \frac{du}{\left(\sqrt{a-b}\right)^2 + u^2}$$

$$\overset{17}{=} -2\left(\frac{1}{\sqrt{a-b}} \tan^{-1} \frac{u}{\sqrt{a-b}}\right)$$

So $(\star)$ becomes $\frac{-2}{\sqrt{a-b}} \tan^{-1} \frac{\sqrt{b-x}}{\sqrt{a-b}} = kt + C$. Now $x(0) = 0 \Rightarrow C = \frac{-2}{\sqrt{a-b}} \tan^{-1} \frac{\sqrt{b}}{\sqrt{a-b}}$ and we have

$\frac{-2}{\sqrt{a-b}} \tan^{-1} \frac{\sqrt{b-x}}{\sqrt{a-b}} = kt - \frac{2}{\sqrt{a-b}} \tan^{-1} \frac{\sqrt{b}}{\sqrt{a-b}} \Rightarrow \frac{2}{\sqrt{a-b}}\left(\tan^{-1}\sqrt{\frac{b}{a-b}} - \tan^{-1}\sqrt{\frac{b-x}{a-b}}\right) = kt \Rightarrow$

$t(x) = \frac{2}{k\sqrt{a-b}}\left(\tan^{-1}\sqrt{\frac{b}{a-b}} - \tan^{-1}\sqrt{\frac{b-x}{a-b}}\right)$.

**32.** If $S = \frac{dT}{dr}$, then $\frac{dS}{dr} = \frac{d^2T}{dr^2}$. The differential equation $\frac{d^2T}{dr^2} + \frac{2}{r}\frac{dT}{dr} = 0$ can be written as $\frac{dS}{dr} + \frac{2}{r}S = 0$. Thus,

$\frac{dS}{dr} = \frac{-2S}{r} \Rightarrow \frac{dS}{S} = -\frac{2}{r}\,dr \Rightarrow \int \frac{1}{S}\,dS = \int -\frac{2}{r}\,dr \Rightarrow \ln|S| = -2\ln|r| + C$. Assuming $S = dT/dr > 0$ and $r > 0$, we have $S = e^{-2\ln r + C} = e^{\ln r^{-2}} e^C = r^{-2}k$ $[k = e^C]$ $\Rightarrow S = \frac{1}{r^2}k \Rightarrow \frac{dT}{dr} = \frac{1}{r^2}k \Rightarrow$

$dT = \frac{1}{r^2}k\,dr \Rightarrow \int dT = \int \frac{1}{r^2}k\,dr \Rightarrow T(r) = -\frac{k}{r} + A.$

$T(1) = 15 \Rightarrow 15 = -k + A$ **(1)** and $T(2) = 25 \Rightarrow 25 = -\frac{1}{2}k + A$ **(2)**.

Now solve for $k$ and $A$: $-2$**(2)** + **(1)** $\Rightarrow -35 = -A$, so $A = 35$ and $k = 20$, and $T(r) = -20/r + 35$.

**33.** (a) $\frac{dC}{dt} = r - kC \Rightarrow \frac{dC}{dt} = -(kC - r) \Rightarrow \int \frac{dC}{kC - r} = \int -dt \Rightarrow (1/k)\ln|kC - r| = -t + M_1 \Rightarrow$

$\ln|kC - r| = -kt + M_2 \Rightarrow |kC - r| = e^{-kt + M_2} \Rightarrow kC - r = M_3 e^{-kt} \Rightarrow kC = M_3 e^{-kt} + r \Rightarrow$

$C(t) = M_4 e^{-kt} + r/k$. $C(0) = C_0 \Rightarrow C_0 = M_4 + r/k \Rightarrow M_4 = C_0 - r/k \Rightarrow$

$C(t) = (C_0 - r/k)e^{-kt} + r/k$.

(b) If $C_0 < r/k$, then $C_0 - r/k < 0$ and the formula for $C(t)$ shows that $C(t)$ increases and $\lim_{t\to\infty} C(t) = r/k$.

As $t$ increases, the formula for $C(t)$ shows how the role of $C_0$ steadily diminishes as that of $r/k$ increases.

**34.** (a) Use 1 billion dollars as the $x$-unit and 1 day as the $t$-unit. Initially, there is \$10 billion of old currency in circulation, so all of the \$50 million returned to the banks is old. At time $t$, the amount of new currency is $x(t)$ billion dollars, so $10 - x(t)$ billion dollars of currency is old. The fraction of circulating money that is old is $[10 - x(t)]/10$, and the amount of old currency being returned to the banks each day is $\frac{10 - x(t)}{10} 0.05$ billion dollars. This amount of new currency per day is introduced into circulation, so $\frac{dx}{dt} = \frac{10 - x}{10} \cdot 0.05 = 0.005(10 - x)$ billion dollars per day.

(b) $\frac{dx}{10 - x} = 0.005\,dt \Rightarrow \frac{-dx}{10 - x} = -0.005\,dt \Rightarrow \ln(10 - x) = -0.005t + c \Rightarrow 10 - x = Ce^{-0.005t}$, where $C = e^c \Rightarrow x(t) = 10 - Ce^{-0.005t}$. From $x(0) = 0$, we get $C = 10$, so $x(t) = 10(1 - e^{-0.005t})$.

(c) The new bills make up 90% of the circulating currency when $x(t) = 0.9 \cdot 10 = 9$ billion dollars.

$9 = 10(1 - e^{-0.005t}) \Rightarrow 0.9 = 1 - e^{-0.005t} \Rightarrow e^{-0.005t} = 0.1 \Rightarrow -0.005t = -\ln 10 \Rightarrow$

$t = 200\ln 10 \approx 460.517$ days $\approx 1.26$ years.

**35.** (a) Let $y(t)$ be the amount of salt (in kg) after $t$ minutes. Then $y(0) = 15$. The amount of liquid in the tank is 1000 L at all times, so the concentration at time $t$ (in minutes) is $y(t)/1000$ kg/L and $\frac{dy}{dt} = -\left[\frac{y(t)}{1000}\frac{\text{kg}}{\text{L}}\right]\left(10\,\frac{\text{L}}{\text{min}}\right) = -\frac{y(t)}{100}\frac{\text{kg}}{\text{min}}$.

$\int \frac{dy}{y} = -\frac{1}{100}\int dt \;\Rightarrow\; \ln y = -\frac{t}{100} + C$, and $y(0) = 15 \;\Rightarrow\; \ln 15 = C$, so $\ln y = \ln 15 - \frac{t}{100}$.

It follows that $\ln\left(\frac{y}{15}\right) = -\frac{t}{100}$ and $\frac{y}{15} = e^{-t/100}$, so $y = 15e^{-t/100}$ kg.

(b) After 20 minutes, $y = 15e^{-20/100} = 15e^{-0.2} \approx 12.3$ kg.

**36.** Let $y(t)$ be the amount of carbon dioxide in the room after $t$ minutes. Then $y(0) = 0.0015(180) = 0.27\text{ m}^3$. The amount of air in the room is $180\text{ m}^3$ at all times, so the percentage at time $t$ (in mimutes) is $y(t)/180 \times 100$, and the change in the amount of carbon dioxide with respect to time is

$$\frac{dy}{dt} = (0.0005)\left(2\,\frac{\text{m}^3}{\text{min}}\right) - \frac{y(t)}{180}\left(2\,\frac{\text{m}^3}{\text{min}}\right) = 0.001 - \frac{y}{90} = \frac{9-100y}{9000}\,\frac{\text{m}^3}{\text{min}}$$

Hence, $\int \frac{dy}{9-100y} = \int \frac{dt}{9000}$ and $-\frac{1}{100}\ln|9-100y| = \frac{1}{9000}t + C$. Because $y(0) = 0.27$, we have $-\frac{1}{100}\ln 18 = C$, so $-\frac{1}{100}\ln|9-100y| = \frac{1}{9000}t - \frac{1}{100}\ln 18 \;\Rightarrow\; \ln|9-100y| = -\frac{1}{90}t + \ln 18 \;\Rightarrow$ $\ln|9-100y| = \ln e^{-t/90} + \ln 18 \;\Rightarrow\; \ln|9-100y| = \ln(18e^{-t/90})$, and $|9-100y| = 18e^{-t/90}$. Since $y$ is continuous, $y(0) = 0.27$, and the right-hand side is never zero, we deduce that $9 - 100y$ is always negative. Thus, $|9-100y| = 100y - 9$ and we have $100y - 9 = 18e^{-t/90} \;\Rightarrow\; 100y = 9 + 18e^{-t/90} \;\Rightarrow\; y = 0.09 + 0.18e^{-t/90}$. The percentage of carbon dioxide in the room is

$$p(t) = \frac{y}{180}\times 100 = \frac{0.09+0.18e^{-t/90}}{180}\times 100 = (0.0005 + 0.001e^{-t/90})\times 100 = 0.05 + 0.1e^{-t/90}$$

In the long run, we have $\lim_{t\to\infty} p(t) = 0.05 + 0.1(0) = 0.05$; that is, the amount of carbon dioxide approaches 0.05% as time goes on.

**37.** Let $y(t)$ be the amount of alcohol in the vat after $t$ minutes. Then $y(0) = 0.04(500) = 20$ gal. The amount of beer in the vat is 500 gallons at all times, so the percentage at time $t$ (in minutes) is $y(t)/500 \times 100$, and the change in the amount of alcohol with respect to time $t$ is

$$\frac{dy}{dt} = \text{rate in} - \text{rate out} = 0.06\left(5\,\frac{\text{gal}}{\text{min}}\right) - \frac{y(t)}{500}\left(5\,\frac{\text{gal}}{\text{min}}\right) = 0.3 - \frac{y}{100} = \frac{30-y}{100}\,\frac{\text{gal}}{\text{min}}$$

Hence, $\int \frac{dy}{30-y} = \int \frac{dt}{100}$ and $-\ln|30-y| = \frac{1}{100}t + C$. Because $y(0) = 20$, we have $-\ln 10 = C$, so $-\ln|30-y| = \frac{1}{100}t - \ln 10 \;\Rightarrow\; \ln|30-y| = -t/100 + \ln 10 \;\Rightarrow\; \ln|30-y| = \ln e^{-t/100} + \ln 10 \;\Rightarrow$ $\ln|30-y| = \ln(10e^{-t/100}) \;\Rightarrow\; |30-y| = 10e^{-t/100}$. Since $y$ is continuous, $y(0) = 20$, and the right-hand side is never zero, we deduce that $30 - y$ is always positive. Thus, $30 - y = 10e^{-t/100} \;\Rightarrow\; y = 30 - 10e^{-t/100}$. The percentage of alcohol is $p(t) = y(t)/500 \times 100 = y(t)/5 = 6 - 2e^{-t/100}$. The percentage of alcohol after one hour is $p(60) = 6 - 2e^{-60/100} \approx 4.9$.

**38.** (a) If $y(t)$ is the amount of salt (in kg) after $t$ minutes, then $y(0)=0$ and the total amount of liquid in the tank remains constant at 1000 L.

$$\frac{dy}{dt}=\left(0.05\,\frac{\text{kg}}{\text{L}}\right)\left(5\,\frac{\text{L}}{\text{min}}\right)+\left(0.04\,\frac{\text{kg}}{\text{L}}\right)\left(10\,\frac{\text{L}}{\text{min}}\right)-\left(\frac{y(t)}{1000}\,\frac{\text{kg}}{\text{L}}\right)\left(15\,\frac{\text{L}}{\text{min}}\right)$$
$$=0.25+0.40-0.015y=0.65-0.015y=\frac{130-3y}{200}\,\frac{\text{kg}}{\text{min}}$$

Hence, $\displaystyle\int\frac{dy}{130-3y}=\int\frac{dt}{200}$ and $-\frac{1}{3}\ln|130-3y|=\frac{1}{200}t+C$. Because $y(0)=0$, we have $-\frac{1}{3}\ln 130=C$, so $-\frac{1}{3}\ln|130-3y|=\frac{1}{200}t-\frac{1}{3}\ln 130\;\Rightarrow\;\ln|130-3y|=-\frac{3}{200}t+\ln 130=\ln\left(130e^{-3t/200}\right)$, and $|130-3y|=130e^{-3t/200}$. Since $y$ is continuous, $y(0)=0$, and the right-hand side is never zero, we deduce that $130-3y$ is always positive. Thus, $130-3y=130e^{-3t/200}$ and $y=\frac{130}{3}\left(1-e^{-3t/200}\right)$ kg.

(b) After one hour, $y=\frac{130}{3}\left(1-e^{-3\cdot 60/200}\right)=\frac{130}{3}\left(1-e^{-0.9}\right)\approx 25.7$ kg.

*Note:* As $t\to\infty$, $y(t)\to\frac{130}{3}=43\frac{1}{3}$ kg.

**39.** Assume that the raindrop begins at rest, so that $v(0)=0$. $dm/dt=km$ and $(mv)'=gm\;\Rightarrow\;mv'+vm'=gm\;\Rightarrow$ $mv'+v(km)=gm\;\Rightarrow\;v'+vk=g\;\Rightarrow\;\dfrac{dv}{dt}=g-kv\;\Rightarrow\;\displaystyle\int\frac{dv}{g-kv}=\int dt\;\Rightarrow$ $-(1/k)\ln|g-kv|=t+C\;\Rightarrow\;\ln|g-kv|=-kt-kC\;\Rightarrow\;g-kv=Ae^{-kt}$. $v(0)=0\;\Rightarrow\;A=g$. So $kv=g-ge^{-kt}\;\Rightarrow\;v=(g/k)\left(1-e^{-kt}\right)$. Since $k>0$, as $t\to\infty$, $e^{-kt}\to 0$ and therefore, $\displaystyle\lim_{t\to\infty}v(t)=g/k$.

**40.** (a) $m\dfrac{dv}{dt}=-kv\;\Rightarrow\;\dfrac{dv}{v}=-\dfrac{k}{m}\,dt\;\Rightarrow\;\ln|v|=-\dfrac{k}{m}t+C$. Since $v(0)=v_0$, $\ln|v_0|=C$. Therefore, $\ln\left|\dfrac{v}{v_0}\right|=-\dfrac{k}{m}t\;\Rightarrow\;\left|\dfrac{v}{v_0}\right|=e^{-kt/m}\;\Rightarrow\;v(t)=\pm v_0e^{-kt/m}$. The sign is + when $t=0$, and we assume $v$ is continuous, so that the sign is + for all $t$. Thus, $v(t)=v_0e^{-kt/m}$. $ds/dt=v_0e^{-kt/m}\;\Rightarrow$ $s(t)=-\dfrac{mv_0}{k}e^{-kt/m}+C'$.

From $s(0)=s_0$, we get $s_0=-\dfrac{mv_0}{k}+C'$, so $C'=s_0+\dfrac{mv_0}{k}$ and $s(t)=s_0+\dfrac{mv_0}{k}\left(1-e^{-kt/m}\right)$.

The distance traveled from time 0 to time $t$ is $s(t)-s_0$, so the total distance traveled is $\displaystyle\lim_{t\to\infty}[s(t)-s_0]=\frac{mv_0}{k}$.

*Note:* In finding the limit, we use the fact that $k>0$ to conclude that $\displaystyle\lim_{t\to\infty}e^{-kt/m}=0$.

(b) $m\dfrac{dv}{dt}=-kv^2\;\Rightarrow\;\dfrac{dv}{v^2}=-\dfrac{k}{m}\,dt\;\Rightarrow\;\dfrac{-1}{v}=-\dfrac{kt}{m}+C\;\Rightarrow\;\dfrac{1}{v}=\dfrac{kt}{m}-C$. Since $v(0)=v_0$, $C=-\dfrac{1}{v_0}$ and $\dfrac{1}{v}=\dfrac{kt}{m}+\dfrac{1}{v_0}$. Therefore, $v(t)=\dfrac{1}{kt/m+1/v_0}=\dfrac{mv_0}{kv_0t+m}$. $\dfrac{ds}{dt}=\dfrac{mv_0}{kv_0t+m}\;\Rightarrow$

$s(t)=\dfrac{m}{k}\displaystyle\int\frac{kv_0\,dt}{kv_0t+m}=\dfrac{m}{k}\ln|kv_0t+m|+C'$. Since $s(0)=s_0$, we get $s_0=\dfrac{m}{k}\ln m+C'\;\Rightarrow$

$C'=s_0-\dfrac{m}{k}\ln m\;\Rightarrow\;s(t)=s_0+\dfrac{m}{k}(\ln|kv_0t+m|-\ln m)=s_0+\dfrac{m}{k}\ln\left|\dfrac{kv_0t+m}{m}\right|$.

We can rewrite the formulas for $v(t)$ and $s(t)$ as $v(t)=\dfrac{v_0}{1+(kv_0/m)t}$ and $s(t)=s_0+\dfrac{m}{k}\ln\left|1+\dfrac{kv_0}{m}t\right|$.

*Remarks:* This model of horizontal motion through a resistive medium was designed to handle the case in which $v_0>0$. Then the term $-kv^2$ representing the resisting force causes the object to decelerate. The absolute value in the expression for $s(t)$ is unnecessary (since $k$, $v_0$, and $m$ are all positive), and $\lim_{t\to\infty} s(t) = \infty$. In other words, the object travels infinitely far. However, $\lim_{t\to\infty} v(t) = 0$. When $v_0 < 0$, the term $-kv^2$ increases the magnitude of the object's negative velocity. According to the formula for $s(t)$, the position of the object approaches $-\infty$ as $t$ approaches $m/k(-v_0)$: $\lim_{t\to -m/(kv_0)} s(t) = -\infty$. Again the object travels infinitely far, but this time the feat is accomplished in a finite amount of time. Notice also that $\lim_{t\to -m/(kv_0)} v(t) = -\infty$ when $v_0 < 0$, showing that the speed of the object increases without limit.

**41.** (a) The rate of growth of the area is jointly proportional to $\sqrt{A(t)}$ and $M - A(t)$; that is, the rate is proportional to the product of those two quantities. So for some constant $k$, $dA/dt = k\sqrt{A}\,(M - A)$. We are interested in the maximum of the function $dA/dt$ (when the tissue grows the fastest), so we differentiate, using the Chain Rule and then substituting for $dA/dt$ from the differential equation:

$$\frac{d}{dt}\left(\frac{dA}{dt}\right) = k\left[\sqrt{A}\,(-1)\frac{dA}{dt} + (M - A)\cdot\tfrac{1}{2}A^{-1/2}\frac{dA}{dt}\right] = \tfrac{1}{2}kA^{-1/2}\frac{dA}{dt}\left[-2A + (M - A)\right]$$

$$= \tfrac{1}{2}kA^{-1/2}\left[k\sqrt{A}(M - A)\right][M - 3A] = \tfrac{1}{2}k^2(M - A)(M - 3A)$$

This is 0 when $M - A = 0$ [this situation never actually occurs, since the graph of $A(t)$ is asymptotic to the line $y = M$, as in the logistic model] and when $M - 3A = 0 \;\Leftrightarrow\; A(t) = M/3$. This represents a maximum by the First Derivative Test, since $\frac{d}{dt}\left(\frac{dA}{dt}\right)$ goes from positive to negative when $A(t) = M/3$.

(b) From the CAS, we get $A(t) = M\left(\dfrac{Ce^{\sqrt{M}kt} - 1}{Ce^{\sqrt{M}kt} + 1}\right)^2$. To get $C$ in terms of the initial area $A_0$ and the maximum area $M$, we substitute $t = 0$ and $A = A_0 = A(0)$: $A_0 = M\left(\dfrac{C - 1}{C + 1}\right)^2 \;\Leftrightarrow\; (C + 1)\sqrt{A_0} = (C - 1)\sqrt{M} \;\Leftrightarrow$

$C\sqrt{A_0} + \sqrt{A_0} = C\sqrt{M} - \sqrt{M} \;\Leftrightarrow\; \sqrt{M} + \sqrt{A_0} = C\sqrt{M} - C\sqrt{A_0} \;\Leftrightarrow$

$\sqrt{M} + \sqrt{A_0} = C\left(\sqrt{M} - \sqrt{A_0}\right) \;\Leftrightarrow\; C = \dfrac{\sqrt{M} + \sqrt{A_0}}{\sqrt{M} - \sqrt{A_0}}$. (Notice that if $A_0 = 0$, then $C = 1$.)

**42.** (a) According to the hint we use the Chain Rule: $m\dfrac{dv}{dt} = m\dfrac{dv}{dx}\cdot\dfrac{dx}{dt} = mv\dfrac{dv}{dx} = -\dfrac{mgR^2}{(x + R)^2} \;\Rightarrow$

$\displaystyle\int v\,dv = \int\frac{-gR^2\,dx}{(x + R)^2} \;\Rightarrow\; \frac{v^2}{2} = \frac{gR^2}{x + R} + C$. When $x = 0$, $v = v_0$, so $\dfrac{v_0^2}{2} = \dfrac{gR^2}{0 + R} + C \;\Rightarrow\; C = \tfrac{1}{2}v_0^2 - gR$

$\Rightarrow\; \tfrac{1}{2}v^2 - \tfrac{1}{2}v_0^2 = \dfrac{gR^2}{x + R} - gR$. Now at the top of its flight, the rocket's velocity will be 0, and its height will be $x = h$.

Solving for $v_0$: $-\tfrac{1}{2}v_0^2 = \dfrac{gR^2}{h + R} - gR \;\Rightarrow\; \dfrac{v_0^2}{2} = g\left[-\dfrac{R^2}{R + h} + \dfrac{R\,(R + h)}{R + h}\right] = \dfrac{gRh}{R + h} \;\Rightarrow\; v_0 = \sqrt{\dfrac{2gRh}{R + h}}$.

(b) $v_e = \displaystyle\lim_{h\to\infty} v_0 = \lim_{h\to\infty}\sqrt{\frac{2gRh}{R + h}} = \lim_{h\to\infty}\sqrt{\frac{2gR}{(R/h) + 1}} = \sqrt{2gR}$

(c) $v_e = \sqrt{2\cdot 32\text{ ft/s}^2\cdot 3960\text{ mi}\cdot 5280\text{ ft/mi}} \approx 36{,}581\text{ ft/s} \approx 6.93\text{ mi/s}$

## APPLIED PROJECT How Fast Does a Tank Drain?

**1.** (a) $V = \pi r^2 h \quad \Rightarrow \quad \dfrac{dV}{dt} = \pi r^2 \dfrac{dh}{dt}$ [implicit differentiation] $\Rightarrow$

$$\frac{dh}{dt} = \frac{1}{\pi r^2}\frac{dV}{dt} = \frac{1}{\pi r^2}\left(-a\sqrt{2gh}\right) = \frac{1}{\pi 2^2}\left[-\pi\left(\tfrac{1}{12}\right)^2\sqrt{2\cdot 32\sqrt{h}}\right] = -\tfrac{1}{72}\sqrt{h}$$

(b) $\dfrac{dh}{dt} = -\tfrac{1}{72}\sqrt{h} \quad \Rightarrow \quad h^{-1/2}\,dh = -\tfrac{1}{72}\,dt \quad \Rightarrow \quad 2\sqrt{h} = -\tfrac{1}{72}t + C.$

$h(0) = 6 \quad \Rightarrow \quad 2\sqrt{6} = 0 + C \quad \Rightarrow \quad C = 2\sqrt{6} \quad \Rightarrow \quad h(t) = \left(-\tfrac{1}{144}t + \sqrt{6}\right)^2.$

(c) We want to find $t$ when $h = 0$, so we set $h = 0 = \left(-\tfrac{1}{144}t + \sqrt{6}\right)^2 \quad \Rightarrow \quad t = 144\sqrt{6} \approx 5$ min 53 s.

**2.** (a) $\dfrac{dh}{dt} = k\sqrt{h} \quad \Rightarrow \quad h^{-1/2}\,dh = k\,dt \quad [h \neq 0] \quad \Rightarrow \quad 2\sqrt{h} = kt + C$

$\Rightarrow \quad h(t) = \tfrac{1}{4}(kt + C)^2$. Since $h(0) = 10$ cm, the relation $2\sqrt{h(t)} = kt + C$ gives us $2\sqrt{10} = C$. Also, $h(68) = 3$ cm, so $2\sqrt{3} = 68k + 2\sqrt{10}$ and $k = -\dfrac{\sqrt{10}-\sqrt{3}}{34}$. Thus,

$$h(t) = \frac{1}{4}\left(2\sqrt{10} - \frac{\sqrt{10}-\sqrt{3}}{34}t\right)^2 \approx 10 - 0.133t + 0.00044t^2.$$

Here is a table of values of $h(t)$ correct to one decimal place.

| $t$ (in s) | $h(t)$ (in cm) |
|---|---|
| 10 | 8.7 |
| 20 | 7.5 |
| 30 | 6.4 |
| 40 | 5.4 |
| 50 | 4.5 |
| 60 | 3.6 |

(b) The answers to this part are to be obtained experimentally. See the article by Tom Farmer and Fred Gass, *Physical Demonstrations in the Calculus Classroom,* College Mathematics Journal 1992, pp. 146–148.

**3.** $V(t) = \pi r^2 h(t) = 100\pi h(t) \quad \Rightarrow \quad \dfrac{dV}{dh} = 100\pi$ and $\dfrac{dV}{dt} = \dfrac{dV}{dh}\dfrac{dh}{dt} = 100\pi\dfrac{dh}{dt}.$

Diameter $= 2.5$ inches $\Rightarrow$ radius $= 1.25$ inches $= \tfrac{5}{4}\cdot\tfrac{1}{12}$ foot $= \tfrac{5}{48}$ foot. Thus,

$$\frac{dV}{dt} = -a\sqrt{2gh} \quad \Rightarrow \quad 100\pi\frac{dh}{dt} = -\pi\left(\tfrac{5}{48}\right)^2\sqrt{2\cdot 32h} = -\frac{25\pi}{288}\sqrt{h} \quad \Rightarrow \quad \frac{dh}{dt} = -\frac{\sqrt{h}}{1152} \quad \Rightarrow$$

$\int h^{-1/2}\,dh = \int -\tfrac{1}{1152}\,dt \quad \Rightarrow \quad 2\sqrt{h} = -\tfrac{1}{1152}t + C \quad \Rightarrow \quad \sqrt{h} = -\tfrac{1}{2304}t + k \quad \Rightarrow \quad h(t) = \left(-\tfrac{1}{2304}t + k\right)^2$. The water pressure after $t$ seconds is $62.5h(t)$ lb/ft$^2$, so the condition that the pressure be at least 2160 lb/ft$^2$ for 10 minutes (600 seconds) is the condition $62.5\cdot h(600) \geq 2160$; that is, $\left(k - \tfrac{600}{2304}\right)^2 \geq \tfrac{2160}{62.5} \quad \Rightarrow$

$\left|k - \tfrac{25}{96}\right| \geq \sqrt{34.56} \quad \Rightarrow \quad k \geq \tfrac{25}{96} + \sqrt{34.56}$. Now $h(0) = k^2$, so the height of the tank should be at least $\left(\tfrac{25}{96} + \sqrt{34.56}\right)^2 \approx 37.69$ ft.

**4.** (a) If the radius of the circular cross-section at height $h$ is $r$, then the Pythagorean Theorem gives $r^2 = 2^2 - (2-h)^2$ since the radius of the tank is 2 m. So $A(h) = \pi r^2 = \pi\left[4 - (2-h)^2\right] = \pi\left(4h - h^2\right)$. Thus, $A(h)\dfrac{dh}{dt} = -a\sqrt{2gh} \quad \Rightarrow$

$$\pi\left(4h - h^2\right)\frac{dh}{dt} = -\pi(0.01)^2\sqrt{2\cdot 10h} \quad \Rightarrow \quad \left(4h - h^2\right)\frac{dh}{dt} = -0.0001\sqrt{20h}.$$

(b) From part (a) we have $\left(4h^{1/2} - h^{3/2}\right)dh = \left(-0.0001\sqrt{20}\right)dt \quad \Rightarrow \quad \tfrac{8}{3}h^{3/2} - \tfrac{2}{5}h^{5/2} = \left(-0.0001\sqrt{20}\right)t + C.$

$h(0) = 2 \quad \Rightarrow \quad \tfrac{8}{3}(2)^{3/2} - \tfrac{2}{5}(2)^{5/2} = C \quad \Rightarrow \quad C = \left(\tfrac{16}{3} - \tfrac{8}{5}\right)\sqrt{2} = \tfrac{56}{15}\sqrt{2}$. To find out how long it will take to drain all the water we evaluate $t$ when $h = 0$: $0 = \left(-0.0001\sqrt{20}\right)t + C \quad \Rightarrow$

$$t = \frac{C}{0.0001\sqrt{20}} = \frac{56\sqrt{2}/15}{0.0001\sqrt{20}} = \frac{11{,}200\sqrt{10}}{3} \approx 11{,}806 \text{ s} \approx 3 \text{ h } 17 \text{ min}.$$

## APPLIED PROJECT Which is Faster, Going Up or Coming Down?

**1.** $mv' = -pv - mg \;\Rightarrow\; m\dfrac{dv}{dt} = -(pv + mg) \;\Rightarrow\; \displaystyle\int \frac{dv}{pv + mg} = \int -\frac{1}{m}\,dt \;\Rightarrow$

$\dfrac{1}{p}\ln(pv + mg) = -\dfrac{1}{m}t + C$ $[pv + mg > 0]$. At $t = 0$, $v = v_0$, so $C = \dfrac{1}{p}\ln(pv_0 + mg)$.

Thus, $\dfrac{1}{p}\ln(pv + mg) = -\dfrac{1}{m}t + \dfrac{1}{p}\ln(pv_0 + mg) \;\Rightarrow\; \ln(pv + mg) = -\dfrac{p}{m}t + \ln(pv_0 + mg) \;\Rightarrow$

$pv + mg = e^{-pt/m}(pv_0 + mg) \;\Rightarrow\; pv = (pv_0 + mg)e^{-pt/m} - mg \;\Rightarrow\; v(t) = \left(v_0 + \dfrac{mg}{p}\right)e^{-pt/m} - \dfrac{mg}{p}.$

**2.** $y(t) = \int v(t)\,dt = \displaystyle\int \left[\left(v_0 + \frac{mg}{p}\right)e^{-pt/m} - \frac{mg}{p}\right]dt = \left(v_0 + \frac{mg}{p}\right)e^{-pt/m}\left(-\frac{m}{p}\right) - \frac{mg}{p}t + C$

At $t = 0$, $y = 0$, so $C = \left(v_0 + \dfrac{mg}{p}\right)\dfrac{m}{p}$. Thus,

$$y(t) = \left(v_0 + \frac{mg}{p}\right)\frac{m}{p} - \left(v_0 + \frac{mg}{p}\right)\frac{m}{p}e^{-pt/m} - \frac{mgt}{p} = \left(v_0 + \frac{mg}{p}\right)\frac{m}{p}\left(1 - e^{-pt/m}\right) - \frac{mgt}{p}$$

**3.** $v(t) = 0 \;\Rightarrow\; \dfrac{mg}{p} = \left(v_0 + \dfrac{mg}{p}\right)e^{-pt/m} \;\Rightarrow\; e^{pt/m} = \dfrac{pv_0}{mg} + 1 \;\Rightarrow\; \dfrac{pt}{m} = \ln\left(\dfrac{pv_0}{mg} + 1\right) \;\Rightarrow$

$t_1 = \dfrac{m}{p}\ln\left(\dfrac{mg + pv_0}{mg}\right)$. With $m = 1$, $v_0 = 20$, $p = \frac{1}{10}$, and $g = 9.8$, we have $t_1 = 10\ln\left(\frac{11.8}{9.8}\right) \approx 1.86$ s.

**4.**

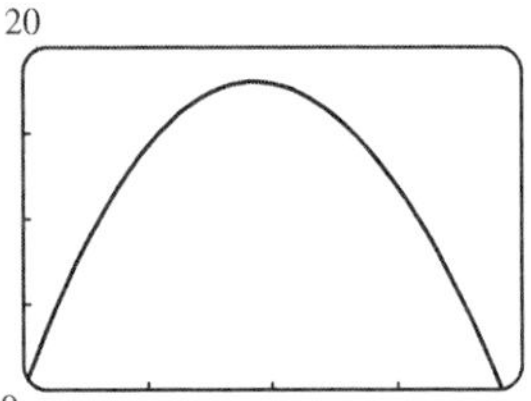


The figure shows the graph of $y = 1180\left(1 - e^{-0.1t}\right) - 98t$. The zeros are at $t = 0$ and $t_2 \approx 3.84$. Thus, $t_1 - 0 \approx 1.86$ and $t_2 - t_1 \approx 1.98$. So the time it takes to come down is about $0.12$ s longer than the time it takes to go up; hence, going up is faster.

**5.** $$y(2t_1) = \left(v_0 + \frac{mg}{p}\right)\frac{m}{p}\left(1 - e^{-2pt_1/m}\right) - \frac{mg}{p}\cdot 2t_1$$
$$= \left(\frac{pv_0 + mg}{p}\right)\frac{m}{p}\left[1 - \left(e^{pt_1/m}\right)^{-2}\right] - \frac{mg}{p}\cdot 2\frac{m}{p}\ln\left(\frac{pv_0 + mg}{mg}\right)$$

Substituting $x = e^{pt_1/m} = \dfrac{pv_0}{mg} + 1 = \dfrac{pv_0 + mg}{mg}$ (from Problem 3), we get

$y\,(2t_1) = \left(x \cdot \dfrac{mg}{p}\right)\dfrac{m}{p}\left(1 - x^{-2}\right) - \dfrac{m^2g}{p^2}\cdot 2\ln x = \dfrac{m^2g}{p^2}\left(x - \dfrac{1}{x} - 2\ln x\right)$. Now $p > 0$, $m > 0$, $t_1 > 0 \;\Rightarrow$

$x = e^{pt_1/m} > e^0 = 1$. $f(x) = x - \dfrac{1}{x} - 2\ln x \;\Rightarrow\; f'(x) = 1 + \dfrac{1}{x^2} - \dfrac{2}{x} = \dfrac{x^2 - 2x + 1}{x^2} = \dfrac{(x-1)^2}{x^2} > 0$ for

$x > 1 \;\Rightarrow\; f(x)$ is increasing for $x > 1$. Since $f(1) = 0$, it follows that $f(x) > 0$ for every $x > 1$. Therefore,

$y(2t_1) = \dfrac{m^2g}{p^2}f(x)$ is positive, which means that the ball has not yet reached the ground at time $2t_1$. This tells us that the time spent going up is always less than the time spent coming down, so *ascent is faster*.

## 7.4 Exponential Growth and Decay

**1.** The relative growth rate is $\frac{1}{P}\frac{dP}{dt} = 0.7944$, so $\frac{dP}{dt} = 0.7944P$ and, by Theorem 2, $P(t) = P(0)e^{0.7944t} = 2e^{0.7944t}$. Thus, $P(6) = 2e^{0.7944(6)} \approx 234.99$ or about 235 members.

**2.** (a) By Theorem 2, $P(t) = P(0)e^{kt} = 60e^{kt}$. In 20 minutes ($\frac{1}{3}$ hour), there are 120 cells, so $P\left(\frac{1}{3}\right) = 60e^{k/3} = 120 \Rightarrow e^{k/3} = 2 \Rightarrow k/3 = \ln 2 \Rightarrow k = 3\ln 2 = \ln(2^3) = \ln 8$.

(b) $P(t) = 60e^{(\ln 8)t} = 60 \cdot 8^t$

(c) $P(8) = 60 \cdot 8^8 = 60 \cdot 2^{24} = 1{,}006{,}632{,}960$

(d) $dP/dt = kP \Rightarrow P'(8) = kP(8) = (\ln 8)P(8) \approx 2.093$ billion cells/h

(e) $P(t) = 20{,}000 \Rightarrow 60 \cdot 8^t = 20{,}000 \Rightarrow 8^t = 1000/3 \Rightarrow t\ln 8 = \ln(1000/3) \Rightarrow$
$t = \dfrac{\ln(1000/3)}{\ln 8} \approx 2.79$ h

**3.** (a) By Theorem 2, $P(t) = P(0)e^{kt} = 100e^{kt}$. Now $P(1) = 100e^{k(1)} = 420 \Rightarrow e^k = \frac{420}{100} \Rightarrow k = \ln 4.2$.
So $P(t) = 100e^{(\ln 4.2)t} = 100(4.2)^t$.

(b) $P(3) = 100(4.2)^3 = 7408.8 \approx 7409$ bacteria

(c) $dP/dt = kP \Rightarrow P'(3) = k \cdot P(3) = (\ln 4.2)\left(100(4.2)^3\right)$ [from part (a)] $\approx 10{,}632$ bacteria/hour

(d) $P(t) = 100(4.2)^t = 10{,}000 \Rightarrow (4.2)^t = 100 \Rightarrow t = (\ln 100)/(\ln 4.2) \approx 3.2$ hours

**4.** (a) $y(t) = y(0)e^{kt} \Rightarrow y(2) = y(0)e^{2k} = 600$, $y(8) = y(0)e^{8k} = 75{,}000$. Dividing these equations, we get $e^{8k}/e^{2k} = 75{,}000/600 \Rightarrow e^{6k} = 125 \Rightarrow 6k = \ln 125 = \ln 5^3 = 3\ln 5 \Rightarrow k = \frac{3}{6}\ln 5 = \frac{1}{2}\ln 5$. Thus, $y(0) = 600/e^{2k} = 600/e^{\ln 5} = \frac{600}{5} = 120$.

(b) $y(t) = y(0)e^{kt} = 120e^{(\ln 5)t/2}$ or $y = 120 \cdot 5^{t/2}$

(c) $y(5) = 120 \cdot 5^{5/2} = 120 \cdot 25\sqrt{5} = 3000\sqrt{5} \approx 6708$ bacteria.

(d) $y(t) = 120 \cdot 5^{t/2} \Rightarrow y'(t) = 120 \cdot 5^{t/2} \cdot \ln 5 \cdot \frac{1}{2} = 60 \cdot \ln 5 \cdot 5^{t/2}$.
$y'(5) = 60 \cdot \ln 5 \cdot 5^{5/2} = 60 \cdot \ln 5 \cdot 25\sqrt{5} \approx 5398$ bacteria/hour.

(e) $y(t) = 200{,}000 \Leftrightarrow 120e^{(\ln 5)t/2} = 200{,}000 \Leftrightarrow e^{(\ln 5)t/2} = \frac{5000}{3} \Leftrightarrow (\ln 5)t/2 = \ln\frac{5000}{3} \Leftrightarrow$
$t = \left(2\ln\frac{5000}{3}\right)/\ln 5 \approx 9.2$ h.

**5.** (a) Let the population (in millions) in the year $t$ be $P(t)$. Since the initial time is the year 1750, we substitute $t - 1750$ for $t$ in Theorem 2, so the exponential model gives $P(t) = P(1750)e^{k(t-1750)}$. Then $P(1800) = 980 = 790e^{k(1800-1750)} \Rightarrow \frac{980}{790} = e^{k(50)} \Rightarrow \ln\frac{980}{790} = 50k \Rightarrow k = \frac{1}{50}\ln\frac{980}{790} \approx 0.0043104$. So with this model, we have $P(1900) = 790e^{k(1900-1750)} \approx 1508$ million, and $P(1950) = 790e^{k(1950-1750)} \approx 1871$ million. Both of these estimates are much too low.

(b) In this case, the exponential model gives $P(t) = P(1850)e^{k(t-1850)} \Rightarrow P(1900) = 1650 = 1260e^{k(1900-1850)} \Rightarrow \ln\frac{1650}{1260} = k(50) \Rightarrow k = \frac{1}{50}\ln\frac{1650}{1260} \approx 0.005393$. So with this model, we estimate $P(1950) = 1260e^{k(1950-1850)} \approx 2161$ million. This is still too low, but closer than the estimate of $P(1950)$ in part (a).

(c) The exponential model gives $P(t) = P(1900)e^{k(t-1900)} \Rightarrow P(1950) = 2560 = 1650e^{k(1950-1900)} \Rightarrow \ln\frac{2560}{1650} = k(50) \Rightarrow k = \frac{1}{50}\ln\frac{2560}{1650} \approx 0.008785$. With this model, we estimate $P(2000) = 1650e^{k(2000-1900)} \approx 3972$ million. This is much too low. The discrepancy is explained by the fact that the world birth rate (average yearly number of births per person) is about the same as always, whereas the mortality rate (especially the infant mortality rate) is much lower, owing mostly to advances in medical science and to the wars in the first part of the twentieth century. The exponential model assumes, among other things, that the birth and mortality rates will remain constant.

**6.** (a) Let $P(t)$ be the population (in millions) in the year $t$. Since the initial time is the year 1900, we substitute $t - 1900$ for $t$ in Theorem 2, and find that the exponential model gives $P(t) = P(1900)e^{k(t-1900)} \Rightarrow$ $P(1910) = 92 = 76e^{k(1910-1900)} \Rightarrow k = \frac{1}{10}\ln\frac{92}{76} \approx 0.0191$. With this model, we estimate $P(2000) = 76e^{k(2000-1900)} \approx 514$ million. This estimate is much too high. The discrepancy is explained by the fact that, between the years 1900 and 1910, an enormous number of immigrants (compared to the total population) came to the United States. Since that time, immigration (as a proportion of total population) has been much lower. Also, the birth rate in the United States has declined since the turn of the century. So our calculation of the constant $k$ was based partly on factors which no longer exist.

(b) Substituting $t - 1980$ for $t$ in Theorem 2, we find that the exponential model gives $P(t) = P(1980)e^{k(t-1980)} \Rightarrow$ $P(1990) = 250 = 227e^{k(1990-1980)} \Rightarrow k = \frac{1}{10}\ln\frac{250}{227} \approx 0.00965$. With this model, we estimate $P(2000) = 227e^{k(2000-1980)} \approx 275.3$ million. This is quite accurate. The further estimates are $P(2010) = 227e^{30k} \approx 303$ million and $P(2020) = 227e^{40k} \approx 334$ million.

(c)

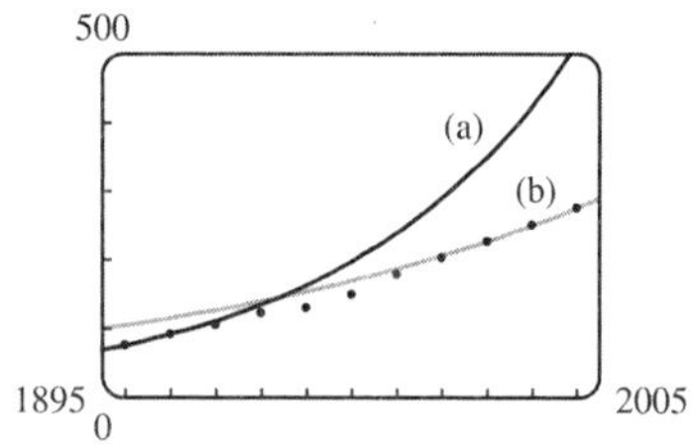


The model in part (a) is quite inaccurate after 1910 (off by 5 million in 1920 and 12 million in 1930). The model in part (b) is more accurate (which is not surprising, since it is based on more recent information).

**7.** (a) If $y = [N_2O_5]$ then by Theorem 2, $\dfrac{dy}{dt} = -0.0005y \Rightarrow y(t) = y(0)e^{-0.0005t} = Ce^{-0.0005t}$.

(b) $y(t) = Ce^{-0.0005t} = 0.9C \Rightarrow e^{-0.0005t} = 0.9 \Rightarrow -0.0005t = \ln 0.9 \Rightarrow t = -2000\ln 0.9 \approx 211$ s

**8.** (a) The mass remaining after $t$ days is $y(t) = y(0)e^{kt} = 800e^{kt}$. Since the half-life is 5.0 days, $y(5) = 800e^{5k} = 400 \Rightarrow$ $e^{5k} = \frac{1}{2} \Rightarrow 5k = \ln\frac{1}{2} \Rightarrow k = -(\ln 2)/5$, so $y(t) = 800e^{-(\ln 2)t/5} = 800 \cdot 2^{-t/5}$.

(b) $y(30) = 800 \cdot 2^{-30/5} = 12.5$ mg

(c) $800e^{-(\ln 2)t/5} = 1 \Leftrightarrow -(\ln 2)\frac{t}{5} = \ln\frac{1}{800} = -\ln 800 \Leftrightarrow$ $t = 5\,\frac{\ln 800}{\ln 2} \approx 48$ days

(d)

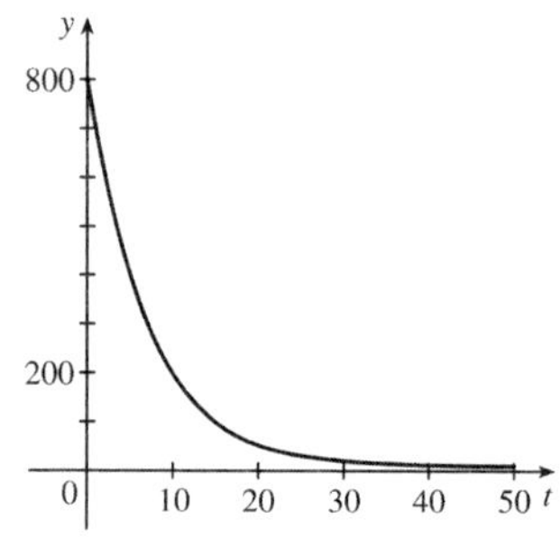


**9.** (a) If $y(t)$ is the mass (in mg) remaining after $t$ years, then $y(t) = y(0)e^{kt} = 100e^{kt}$. $y(30) = 100e^{30k} = \frac{1}{2}(100) \Rightarrow$ $e^{30k} = \frac{1}{2} \Rightarrow k = -(\ln 2)/30 \Rightarrow y(t) = 100e^{-(\ln 2)t/30} = 100 \cdot 2^{-t/30}$

(b) $y(100) = 100 \cdot 2^{-100/30} \approx 9.92$ mg

(c) $100e^{-(\ln 2)t/30} = 1 \Rightarrow -(\ln 2)t/30 = \ln\frac{1}{100} \Rightarrow t = -30\,\frac{\ln 0.01}{\ln 2} \approx 199.3$ years

**10.** (a) If $y(t)$ is the mass after $t$ days and $y(0) = A$, then $y(t) = Ae^{kt}$. $y(1) = Ae^{k} = 0.945A \Rightarrow$ $e^{k} = 0.945 \Rightarrow k = \ln 0.945$. Then $Ae^{(\ln 0.945)t} = \frac{1}{2}A \Leftrightarrow \ln e^{(\ln 0.945)t} = \ln\frac{1}{2} \Leftrightarrow (\ln 0.945)t = \ln\frac{1}{2} \Leftrightarrow$ $t = -\dfrac{\ln 2}{\ln 0.945} \approx 12.25$ years.

(b) $Ae^{(\ln 0.945)t} = 0.20A \Leftrightarrow (\ln 0.945)t = \ln\frac{1}{5} \Leftrightarrow t = -\dfrac{\ln 5}{\ln 0.945} \approx 28.45$ years

**11.** Let $y(t)$ be the level of radioactivity. Thus, $y(t) = y(0)e^{-kt}$ and $k$ is determined by using the half-life: $y(5730) = \frac{1}{2}y(0) \Rightarrow y(0)e^{-k(5730)} = \frac{1}{2}y(0) \Rightarrow e^{-5730k} = \frac{1}{2} \Rightarrow -5730k = \ln\frac{1}{2} \Rightarrow$ $k = -\dfrac{\ln\frac{1}{2}}{5730} = \dfrac{\ln 2}{5730}$. If 74% of the $^{14}$C remains, then we know that $y(t) = 0.74y(0) \Rightarrow 0.74 = e^{-t(\ln 2)/5730} \Rightarrow$ $\ln 0.74 = -\dfrac{t\ln 2}{5730} \Rightarrow t = -\dfrac{5730(\ln 0.74)}{\ln 2} \approx 2489 \approx 2500$ years.

**12.** From the information given, we know that $\dfrac{dy}{dx} = 2y \Rightarrow y = Ce^{2x}$ by Theorem 2. To calculate $C$ we use the point $(0,5)$: $5 = Ce^{2(0)} \Rightarrow C = 5$. Thus, the equation of the curve is $y = 5e^{2x}$.

**13.** (a) Using Newton's Law of Cooling, $\dfrac{dT}{dt} = k(T - T_s)$, we have $\dfrac{dT}{dt} = k(T - 75)$. Now let $y = T - 75$, so $y(0) = T(0) - 75 = 185 - 75 = 110$, so $y$ is a solution of the initial-value problem $dy/dt = ky$ with $y(0) = 110$ and by Theorem 2 we have $y(t) = y(0)e^{kt} = 110e^{kt}$.

$y(30) = 110e^{30k} = 150 - 75 \Rightarrow e^{30k} = \frac{75}{110} = \frac{15}{22} \Rightarrow k = \frac{1}{30}\ln\frac{15}{22}$, so $y(t) = 110e^{\frac{1}{30}t\ln\left(\frac{15}{22}\right)}$ and $y(45) = 110e^{\frac{45}{30}\ln\left(\frac{15}{22}\right)} \approx 62\,^\circ\text{F}$. Thus, $T(45) \approx 62 + 75 = 137\,^\circ\text{F}$.

(b) $T(t) = 100 \Rightarrow y(t) = 25$. $y(t) = 110e^{\frac{1}{30}t\ln\left(\frac{15}{22}\right)} = 25 \Rightarrow e^{\frac{1}{30}t\ln\left(\frac{15}{22}\right)} = \frac{25}{110} \Rightarrow \frac{1}{30}t\ln\frac{15}{22} = \ln\frac{25}{110} \Rightarrow$ $t = \dfrac{30\ln\frac{25}{110}}{\ln\frac{15}{22}} \approx 116$ min.

**14.** (a) Let $T(t) =$ temperature after $t$ minutes. Newton's Law of Cooling implies that $\dfrac{dT}{dt} = k(T - 5)$. Let $y(t) = T(t) - 5$. Then $\dfrac{dy}{dt} = ky$, so $y(t) = y(0)e^{kt} = 15e^{kt} \Rightarrow T(t) = 5 + 15e^{kt} \Rightarrow T(1) = 5 + 15e^{k} = 12 \Rightarrow e^{k} = \frac{7}{15} \Rightarrow$ $k = \ln\frac{7}{15}$, so $T(t) = 5 + 15e^{\ln(7/15)t}$ and $T(2) = 5 + 15e^{2\ln(7/15)} \approx 8.3°\text{C}$.

(b) $5 + 15e^{\ln(7/15)t} = 6$ when $e^{\ln(7/15)t} = \frac{1}{15} \Rightarrow \ln\left(\frac{7}{15}\right)t = \ln\frac{1}{15} \Rightarrow t = \dfrac{\ln\frac{1}{15}}{\ln\frac{7}{15}} \approx 3.6$ min.

**15.** $\dfrac{dT}{dt} = k(T - 20)$. Letting $y = T - 20$, we get $\dfrac{dy}{dt} = ky$, so $y(t) = y(0)e^{kt}$. $y(0) = T(0) - 20 = 5 - 20 = -15$, so $y(25) = y(0)e^{25k} = -15e^{25k}$, and $y(25) = T(25) - 20 = 10 - 20 = -10$, so $-15e^{25k} = -10 \Rightarrow e^{25k} = \frac{2}{3}$. Thus, $25k = \ln\left(\frac{2}{3}\right)$ and $k = \frac{1}{25}\ln\left(\frac{2}{3}\right)$, so $y(t) = y(0)e^{kt} = -15e^{(1/25)\ln(2/3)t}$. More simply, $e^{25k} = \frac{2}{3} \Rightarrow e^{k} = \left(\frac{2}{3}\right)^{1/25} \Rightarrow$ $e^{kt} = \left(\frac{2}{3}\right)^{t/25} \Rightarrow y(t) = -15\cdot\left(\frac{2}{3}\right)^{t/25}$.

(a) $T(50) = 20 + y(50) = 20 - 15\cdot\left(\frac{2}{3}\right)^{50/25} = 20 - 15\cdot\left(\frac{2}{3}\right)^{2} = 20 - \frac{20}{3} = 13.\bar{3}\,^\circ\text{C}$

(b) $15 = T(t) = 20 + y(t) = 20 - 15\cdot\left(\frac{2}{3}\right)^{t/25} \Rightarrow 15\cdot\left(\frac{2}{3}\right)^{t/25} = 5 \Rightarrow \left(\frac{2}{3}\right)^{t/25} = \frac{1}{3} \Rightarrow$ $(t/25)\ln\left(\frac{2}{3}\right) = \ln\left(\frac{1}{3}\right) \Rightarrow t = 25\ln\left(\frac{1}{3}\right)/\ln\left(\frac{2}{3}\right) \approx 67.74$ min.

**16.** $\dfrac{dT}{dt} = k(T - 20)$. Let $y = T - 20$. Then $\dfrac{dy}{dt} = ky$, so $y(t) = y(0)e^{kt}$. $y(0) = T(0) - 20 = 95 - 20 = 75$, so $y(t) = 75e^{kt}$. When $T(t) = 70$, $\dfrac{dT}{dt} = -1°\text{C/min}$. Equivalently, $\dfrac{dy}{dt} = -1$ when $y(t) = 50$. Thus, $-1 = \dfrac{dy}{dt} = ky(t) = 50k$ and $50 = y(t) = 75e^{kt}$. The first relation implies $k = -1/50$, so the second relation says $50 = 75e^{-t/50}$. Thus, $e^{-t/50} = \frac{2}{3} \Rightarrow -t/50 = \ln\left(\frac{2}{3}\right) \Rightarrow t = -50\ln\left(\frac{2}{3}\right) \approx 20.27$ min.

**17.** (a) Let $P(h)$ be the pressure at altitude $h$. Then $dP/dh = kP \Rightarrow P(h) = P(0)e^{kh} = 101.3e^{kh}$. $P(1000) = 101.3e^{1000k} = 87.14 \Rightarrow 1000k = \ln\left(\frac{87.14}{101.3}\right) \rightarrow k = \frac{1}{1000}\ln\left(\frac{87.14}{101.3}\right) \Rightarrow$ $P(h) = 101.3\,e^{\frac{1}{1000}h\ln\left(\frac{87.14}{101.3}\right)}$, so $P(3000) = 101.3e^{3\ln\left(\frac{87.14}{101.3}\right)} \approx 64.5$ kPa.

(b) $P(6187) = 101.3\,e^{\frac{6187}{1000}\ln\left(\frac{87.14}{101.3}\right)} \approx 39.9$ kPa

**18.** (a) Using $A = A_0\left(1 + \frac{r}{n}\right)^{nt}$ with $A_0 = 500$, $r = 0.14$, and $t = 2$, we have:

(i) Annually: $n = 1$; $A = 500\left(1 + \frac{0.14}{1}\right)^{1 \cdot 2} = \$649.80$

(ii) Quarterly: $n = 4$; $A = 500\left(1 + \frac{0.14}{4}\right)^{4 \cdot 2} = \$658.40$

(iii) Monthly: $n = 12$; $A = 500\left(1 + \frac{0.14}{12}\right)^{12 \cdot 2} = \$660.49$

(iv) Daily: $n = 365$; $A = 500\left(1 + \frac{0.14}{365}\right)^{365 \cdot 2} = \$661.53$

(v) Hourly: $n = 365 \cdot 24$; $A = 500\left(1 + \frac{0.14}{365 \cdot 24}\right)^{365 \cdot 24 \cdot 2} = \$661.56$

(vi) Continuously: $A = 500e^{(0.14)2} = \$661.56$

(b) 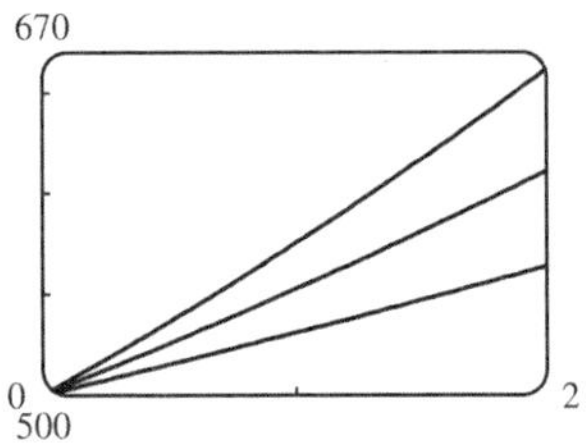


$A_{0.14}(2) = \$661.56$, $A_{0.10}(2) = \$610.70$, and $A_{0.06}(2) = \$563.75$.

**19.** (a) Using $A = A_0\left(1 + \frac{r}{n}\right)^{nt}$ with $A_0 = 3000$, $r = 0.05$, and $t = 5$, we have:

(i) Annually: $n = 1$; $A = 3000\left(1 + \frac{0.05}{1}\right)^{1 \cdot 5} = \$3828.84$

(ii) Semiannually: $n = 2$; $A = 3000\left(1 + \frac{0.05}{2}\right)^{2 \cdot 5} = \$3840.25$

(iii) Monthly: $n = 12$; $A = 3000\left(1 + \frac{0.05}{12}\right)^{12 \cdot 5} = \$3850.08$

(iv) Weekly: $n = 52$; $A = 3000\left(1 + \frac{0.05}{52}\right)^{52 \cdot 5} = \$3851.61$

(v) Daily: $n = 365$; $A = 3000\left(1 + \frac{0.05}{365}\right)^{365 \cdot 5} = \$3852.01$

(vi) Continuously: $A = 3000e^{(0.05)5} = \$3852.08$

(b) $dA/dt = 0.05A$ and $A(0) = 3000$.

**20.** (a) $A_0e^{0.06t} = 2A_0 \Leftrightarrow e^{0.06t} = 2 \Leftrightarrow 0.06t = \ln 2 \Leftrightarrow t = \frac{50}{3}\ln 2 \approx 11.55$, so the investment will double in about 11.55 years.

(b) The annual interest rate in $A = A_0(1 + r)^t$ is $r$. From part (a), we have $A = A_0e^{0.06t}$. These amounts must be equal, so $(1 + r)^t = e^{0.06t} \Rightarrow 1 + r = e^{0.06} \Rightarrow r = e^{0.06} - 1 \approx 0.0618 = 6.18\%$, which is the equivalent annual interest rate.

**21.** (a) $\dfrac{dP}{dt} = kP - m = k\left(P - \dfrac{m}{k}\right)$. Let $y = P - \dfrac{m}{k}$, so $\dfrac{dy}{dt} = \dfrac{dP}{dt}$ and the differential equation becomes $\dfrac{dy}{dt} = ky$.
The solution is $y = y_0e^{kt} \Rightarrow P - \dfrac{m}{k} = \left(P_0 - \dfrac{m}{k}\right)e^{kt} \Rightarrow P(t) = \dfrac{m}{k} + \left(P_0 - \dfrac{m}{k}\right)e^{kt}$.

(b) Since $k > 0$, there will be an exponential expansion $\Leftrightarrow P_0 - \dfrac{m}{k} > 0 \Leftrightarrow m < kP_0$.

(c) The population will be constant if $P_0 - \dfrac{m}{k} = 0 \Leftrightarrow m = kP_0$. It will decline if $P_0 - \dfrac{m}{k} < 0 \Leftrightarrow m > kP_0$.

(d) $P_0 = 8{,}000{,}000$, $k = \alpha - \beta = 0.016$, $m = 210{,}000 \Rightarrow m > kP_0$ (= 128,000), so by part (c), the population was declining.

**22.** (a) $\dfrac{dy}{dt} = ky^{1+c} \Rightarrow y^{-1-c}\,dy = k\,dt \Rightarrow \dfrac{y^{-c}}{-c} = kt + C$. Since $y(0) = y_0$, we have $C = \dfrac{y_0^{-c}}{-c}$. Thus,
$\dfrac{y^{-c}}{-c} = kt + \dfrac{y_0^{-c}}{-c}$, or $y^{-c} = y_0^{-c} - ckt$. So $y^c = \dfrac{1}{y_0^{-c} - ckt} = \dfrac{y_0^c}{1 - cy_0^ckt}$ and $y(t) = \dfrac{y_0}{(1 - cy_0^ckt)^{1/c}}$.

(b) $y(t) \to \infty$ as $1 - cy_0^c kt \to 0$, that is, as $t \to \dfrac{1}{cy_0^c k}$. Define $T = \dfrac{1}{cy_0^c k}$. Then $\lim\limits_{t \to T^-} y(t) = \infty$.

(c) According to the data given, we have $c = 0.01$, $y(0) = 2$, and $y(3) = 16$, where the time $t$ is given in months. Thus, $y_0 = 2$ and $16 = y(3) = \dfrac{y_0}{(1 - cy_0^c k \cdot 3)^{1/c}}$. Since $T = \dfrac{1}{cy_0^c k}$, we will solve for $cy_0^c k$. $16 = \dfrac{2}{(1 - 3cy_0^c k)^{100}} \Rightarrow$ $1 - 3cy_0^c k = \left(\frac{1}{8}\right)^{0.01} = 8^{-0.01} \Rightarrow cy_0^c k = \frac{1}{3}\left(1 - 8^{-0.01}\right)$. Thus, doomsday occurs when $t = T = \dfrac{1}{cy_0^c k} = \dfrac{3}{1 - 8^{-0.01}} \approx 145.77$ months or 12.15 years.

## APPLIED PROJECT Calculus and Baseball

**1.** (a) $F = ma = m\dfrac{dv}{dt}$, so by the Substitution Rule we have

$$\int_{t_0}^{t_1} F(t)\,dt = \int_{t_0}^{t_1} m\left(\frac{dv}{dt}\right)dt = m\int_{v_0}^{v_1} dv = [mv]_{v_0}^{v_1} = mv_1 - mv_0 = p(t_1) - p(t_0)$$

(b) (i) We have $v_1 = 110 \text{ mi/h} = \frac{110(5280)}{3600} \text{ ft/s} = 161.\overline{3} \text{ ft/s}$, $v_0 = -90 \text{ mi/h} = -132 \text{ ft/s}$, and the mass of the baseball is $m = \dfrac{w}{g} = \frac{5/16}{32} = \frac{5}{512}$. So the change in momentum is

$p(t_1) - p(t_0) = mv_1 - mv_0 = \frac{5}{512}\left[161.\overline{3} - (-132)\right] \approx 2.86$ slug-ft/s.

(ii) From part (a) and part (b)(i), we have $\int_0^{0.001} F(t)dt = p(0.001) - p(0) \approx 2.86$, so the average force over the interval $[0, 0.001]$ is $\frac{1}{0.001}\int_0^{0.001} F(t)\,dt \approx \frac{1}{0.001}(2.86) = 2860$ lb.

**2.** (a) $W = \displaystyle\int_{s_0}^{s_1} F(s)\,ds$, where $F(s) = m\dfrac{dv}{dt} = m\dfrac{dv}{ds}\dfrac{ds}{dt} = mv\dfrac{dv}{ds}$ and so, by the Substitution Rule,

$$W = \int_{s_0}^{s_1} F(s)\,ds = \int_{s_0}^{s_1} mv\frac{dv}{ds}\,ds = \int_{v(s_0)}^{v(s_1)} mv\,dv = \left[\tfrac{1}{2}mv^2\right]_{v_0}^{v_1} = \tfrac{1}{2}mv_1^2 - \tfrac{1}{2}mv_0^2$$

(b) From part (b)(i), 90 mi/h = 132 ft/s. Assume $v_0 = v(s_0) = 0$ and $v_1 = v(s_1) = 132$ ft/s (note that $s_1$ is the point of release of the baseball). $m = \frac{5}{512}$, so the work done is $W = \frac{1}{2}mv_1^2 - \frac{1}{2}mv_0^2 = \frac{1}{2} \cdot \frac{5}{512} \cdot (132)^2 \approx 85$ ft-lb.

**3.** (a) Here we have a differential equation of the form $dv/dt = kv$, so by Theorem 7.4.2, the solution is $v(t) = v(0)e^{kt}$. In this case $k = -\frac{1}{10}$ and $v(0) = 100$ ft/s, so $v(t) = 100e^{-t/10}$. We are interested in the time $t$ that the ball takes to travel 280 ft, so we find the distance function

$$s(t) = \int_0^t v(x)\,dx = \int_0^t 100e^{-x/10}\,dx = 100\Big[-10e^{-x/10}\Big]_0^t = -1000(e^{-t/10} - 1) = 1000(1 - e^{-t/10})$$

Now we set $s(t) = 280$ and solve for $t$: $280 = 1000(1 - e^{-t/10}) \Rightarrow 1 - e^{-t/10} = \frac{7}{25} \Rightarrow$

$-\frac{1}{10}t = \ln\left(1 - \frac{7}{25}\right) \Rightarrow t \approx 3.285$ seconds.

(b) Let $x$ be the distance of the shortstop from home plate. We calculate the time for the ball to reach home plate as a function of $x$, then differentiate with respect to $x$ to find the value of $x$ which corresponds to the minimum time. The total time that it takes the ball to reach home is the sum of the times of the two throws, plus the relay time ($\frac{1}{2}$ s). The distance from the fielder to the shortstop is $280 - x$, so to find the time $t_1$ taken by the first throw, we solve the equation

$s_1(t_1) = 280 - x \;\Leftrightarrow\; 1 - e^{-t_1/10} = \dfrac{280-x}{1000} \;\Leftrightarrow\; t_1 = -10\ln\dfrac{720+x}{1000}$. We find the time $t_2$ taken by the second throw if the shortstop throws with velocity $w$, since we see that this velocity varies in the rest of the problem. We use $v = we^{-t/10}$ and isolate $t_2$ in the equation $s(t_2) = 10w\left(1 - e^{-t_2/10}\right) = x \;\Leftrightarrow\; e^{-t_2/10} = 1 - \dfrac{x}{10w} \;\Leftrightarrow$ $t_2 = -10\ln\dfrac{10w-x}{10w}$, so the total time is $t_w(x) = \dfrac{1}{2} - 10\left[\ln\dfrac{720+x}{1000} + \ln\dfrac{10w-x}{10w}\right]$. To find the minimum, we differentiate: $\dfrac{dt_w}{dx} = -10\left[\dfrac{1}{720+x} - \dfrac{1}{10w-x}\right]$, which changes from negative to positive when $720 + x = 10w - x$ $\Leftrightarrow\; x = 5w - 360$. By the First Derivative Test, $t_w$ has a minimum at this distance from the shortstop to home plate. So if the shortstop throws at $w = 105$ ft/s from a point $x = 5(105) - 360 = 165$ ft from home plate, the minimum time is $t_{105}(165) = \frac{1}{2} - 10\left(\ln\frac{720+165}{1000} + \ln\frac{1050-165}{1050}\right) \approx 3.431$ seconds. This is longer than the time taken in part (a), so in this case the manager should encourage a direct throw. If $w = 115$ ft/s, then $x = 215$ ft from home, and the minimum time is $t_{115}(215) = \frac{1}{2} - 10\left(\ln\frac{720+215}{1000} + \ln\frac{1150-215}{1150}\right) \approx 3.242$ seconds. This is less than the time taken in part (a), so in this case, the manager should encourage a relayed throw.

(c) In general, the minimum time is

$$t_w(5w-360) = \frac{1}{2} - 10\left[\ln\frac{360+5w}{1000} + \ln\frac{360+5w}{10w}\right]$$
$$= \frac{1}{2} - 10\ln\frac{(w+72)^2}{400w}$$

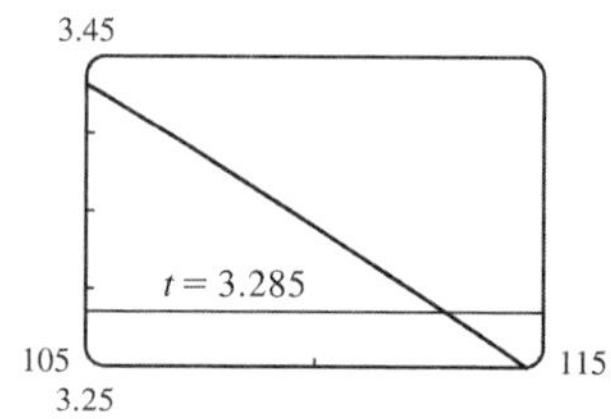


We want to find out when this is about 3.285 seconds, the same time as the direct throw. From the graph, we estimate that this is the case for $w \approx 112.8$ ft/s. So if the shortstop can throw the ball with this velocity, then a relayed throw takes the same time as a direct throw.

## 7.5 The Logistic Equation

1. (a) $dP/dt = 0.05P - 0.0005P^2 = 0.05P(1 - 0.01P) = 0.05P(1 - P/100)$. Comparing to Equation 1, $dP/dt = kP(1 - P/K)$, we see that the carrying capacity is $K = 100$ and the value of $k$ is 0.05.

(b) The slopes close to 0 occur where $P$ is near 0 or 100. The largest slopes appear to be on the line $P = 50$. The solutions are increasing for $0 < P_0 < 100$ and decreasing for $P_0 > 100$.

(c)

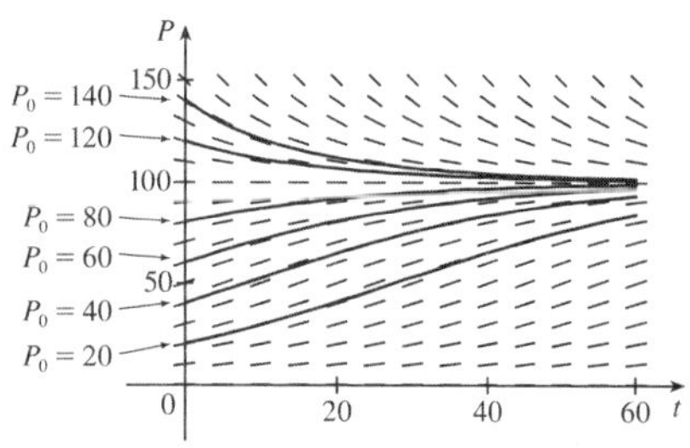


All of the solutions approach $P = 100$ as $t$ increases. As in part (b), the solutions differ since for $0 < P_0 < 100$ they are increasing, and for $P_0 > 100$ they are decreasing. Also, some have an IP and some don't. It appears that the solutions which have $P_0 = 20$ and $P_0 = 40$ have inflection points at $P = 50$.

(d) The equilibrium solutions are $P = 0$ (trivial solution) and $P = 100$. The increasing solutions move away from $P = 0$ and all nonzero solutions approach $P = 100$ as $t \to \infty$.

2. (a) $K = 6000$ and $k = 0.0015 \;\Rightarrow\; dP/dt = 0.0015P(1 - P/6000)$.

(b) 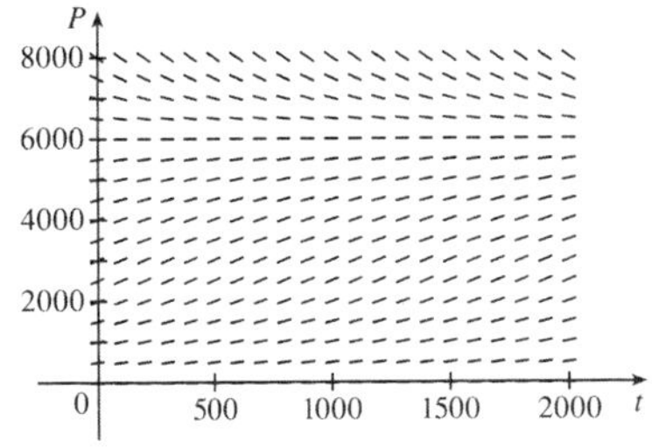


All of the solution curves approach 6000 as $t \to \infty$.

(c) 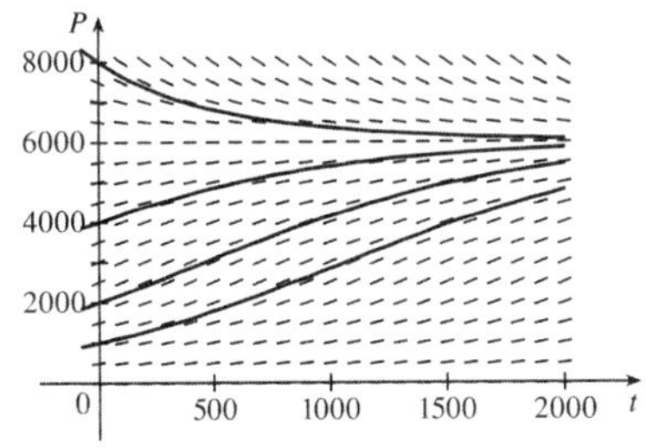


The curves with $P_0 = 1000$ and $P_0 = 2000$ appear to be concave upward at first and then concave downward. The curve with $P_0 = 4000$ appears to be concave downward everywhere. The curve with $P_0 = 8000$ appears to be concave upward everywhere. The inflection points are where the population grows the fastest.

(d) See the solution to Exercise 7.2.25 for a possible program to calculate $P(50)$. [In this case, we use $\mathrm{X} = 0$, $\mathrm{H} = 1$, $\mathrm{N} = 50$, $\mathrm{Y}_1 = 0.0015y(1 - y/6000)$, and $\mathrm{Y} = 1000$.] We find that $P(50) \approx 1064$.

(e) Using Equation 4 with $K = 6000$, $k = 0.0015$, and $P_0 = 1000$, we have $P(t) = \dfrac{K}{1 + Ae^{-kt}} = \dfrac{6000}{1 + Ae^{-0.0015t}}$, where $A = \dfrac{K - P_0}{P_0} = \dfrac{6000 - 1000}{1000} = 5$. Thus, $P(50) = \dfrac{6000}{1 + 5e^{-0.0015(50)}} \approx 1064.1$, which is extremely close to the estimate obtained in part (d).

(f) 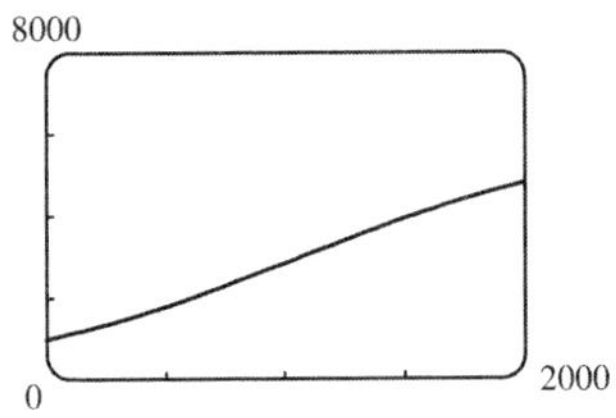


The curves are very similar.

**3.** (a) $\dfrac{dy}{dt} = ky\left(1 - \dfrac{y}{K}\right) \quad\Rightarrow\quad y(t) = \dfrac{K}{1 + Ae^{-kt}}$ with $A = \dfrac{K - y(0)}{y(0)}$. With $K = 8 \times 10^7$, $k = 0.71$, and $y(0) = 2 \times 10^7$, we get the model $y(t) = \dfrac{8 \times 10^7}{1 + 3e^{-0.71t}}$, so $y(1) = \dfrac{8 \times 10^7}{1 + 3e^{-0.71}} \approx 3.23 \times 10^7$ kg.

(b) $y(t) = 4 \times 10^7 \quad\Rightarrow\quad \dfrac{8 \times 10^7}{1 + 3e^{-0.71t}} = 4 \times 10^7 \quad\Rightarrow\quad 2 = 1 + 3e^{-0.71t} \quad\Rightarrow\quad e^{-0.71t} = \frac{1}{3} \quad\Rightarrow\quad -0.71t = \ln\frac{1}{3} \quad\Rightarrow$

$t = \dfrac{\ln 3}{0.71} \approx 1.55$ years

**4.** (a) 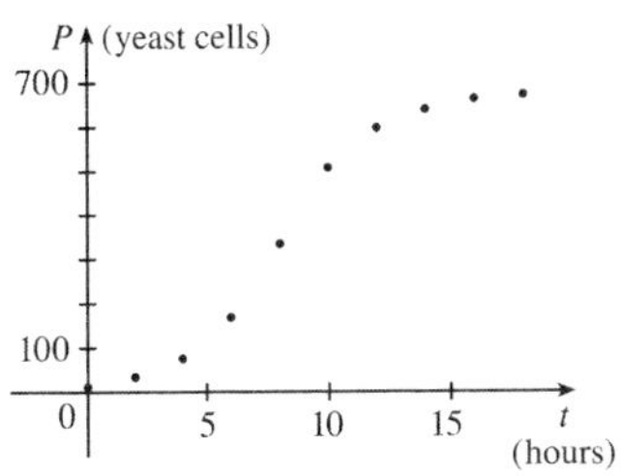


From the graph, we estimate the carrying capacity $K$ for the yeast population to be 680.

(b) An estimate of the initial relative growth rate is $\dfrac{1}{P_0}\dfrac{dP}{dt} = \dfrac{1}{18} \cdot \dfrac{39 - 18}{2 - 0} = \dfrac{7}{12} = 0.58\overline{3}$.

(c) An exponential model is $P(t) = 18e^{7t/12}$. A logistic model is $P(t) = \dfrac{680}{1 + Ae^{-7t/12}}$, where $A = \frac{680 - 18}{18} = \frac{331}{9}$.

(d)

| Time in Hours | Observed Values | Exponential Model | Logistic Model |
|---|---|---|---|
| 0 | 18 | 18 | 18 |
| 2 | 39 | 58 | 55 |
| 4 | 80 | 186 | 149 |
| 6 | 171 | 596 | 322 |
| 8 | 336 | 1914 | 505 |
| 10 | 509 | 6147 | 614 |
| 12 | 597 | 19,739 | 658 |
| 14 | 640 | 63,389 | 673 |
| 16 | 664 | 203,558 | 678 |
| 18 | 672 | 653,679 | 679 |

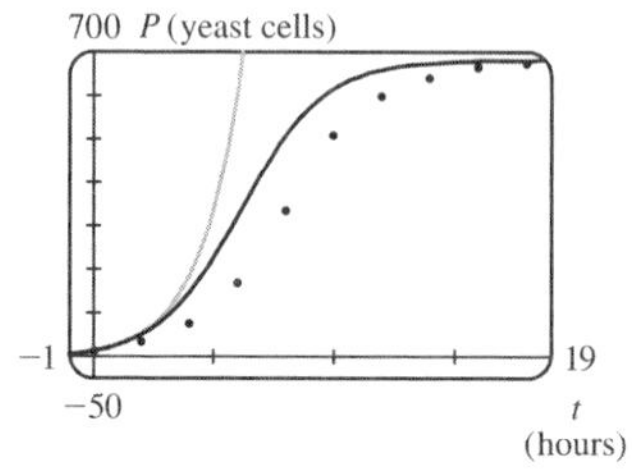


The exponential model is a poor fit for anything beyond the first two observed values. The logistic model varies more for the middle values than it does for the values at either end, but provides a good general fit, as shown in the figure.

(e) $P(7) = \dfrac{680}{1 + \frac{331}{9} e^{-7(7/12)}} \approx 420$ yeast cells

**5.** (a) We will assume that the difference in the birth and death rates is 20 million/year. Let $t = 0$ correspond to the year 1990 and use a unit of 1 billion for all calculations. $k \approx \dfrac{1}{P}\dfrac{dP}{dt} = \dfrac{1}{5.3}(0.02) = \dfrac{1}{265}$, so

$$\frac{dP}{dt} = kP\left(1 - \frac{P}{K}\right) = \frac{1}{265}P\left(1 - \frac{P}{100}\right), \qquad P \text{ in billions}$$

(b) $A = \dfrac{K - P_0}{P_0} = \dfrac{100 - 5.3}{5.3} = \dfrac{947}{53} \approx 17.8679$. $P(t) = \dfrac{K}{1 + Ae^{-kt}} = \dfrac{100}{1 + \frac{947}{53}e^{-(1/265)t}}$, so $P(10) \approx 5.49$ billion.

(c) $P(110) \approx 7.81$, and $P(510) \approx 27.72$. The predictions are 7.81 billion in the year 2100 and 27.72 billion in 2500.

(d) If $K = 50$, then $P(t) = \dfrac{50}{1 + \frac{447}{53}e^{-(1/265)t}}$. So $P(10) \approx 5.48$, $P(110) \approx 7.61$, and $P(510) \approx 22.41$. The predictions become 5.48 billion in the year 2000, 7.61 billion in 2100, and 22.41 billion in the year 2500.

**6.** (a) If we assume that the carrying capacity for the world population is 100 billion, it would seem reasonable that the carrying capacity for the U.S. is 3–5 billion by using current populations and simple proportions. We will use $K = 4$ billion or 4000 million. With $t = 0$ corresponding to 1980, we have $P(t) = \dfrac{4000}{1 + \left(\frac{4000 - 250}{250}\right)e^{-kt}} = \dfrac{4000}{1 + 15e^{-kt}}$.

(b) $P(10) = 275 \;\Rightarrow\; \dfrac{4000}{1 + 15e^{-10k}} = 275 \;\Rightarrow\; 1 + 15e^{-10k} = \frac{4000}{275} \;\Rightarrow\; e^{-10k} = \dfrac{\frac{160}{11} - 1}{15} \;\Rightarrow$
$-10k = \ln\frac{149}{165} \;\Rightarrow\; k = -\frac{1}{10}\ln\frac{149}{165} \approx 0.01019992$.

(c) $2100 - 1990 = 110$ and $P(110) \approx 680$ million.
$2200 - 1990 = 210$ and $P(210) \approx 1449$ million, or about 1.4 billion.

(d) $P(t) = 300 \;\Rightarrow\; \dfrac{4000}{1 + 15e^{-kt}} = 300 \;\Rightarrow\; 1 + 15e^{-kt} = \frac{40}{3} \;\Rightarrow\; e^{-kt} = \frac{37}{3} \cdot \frac{1}{15} \;\Rightarrow\; -kt = \ln\frac{37}{45} \;\Rightarrow$
$t = 10\dfrac{\ln\frac{37}{45}}{\ln\frac{149}{165}} \approx 19.19 \approx 19$. So we predict that the U.S. population will exceed 300 million in the year $1990 + 19 = 2009$.

**7.** (a) Our assumption is that $\dfrac{dy}{dt} = ky(1 - y)$, where $y$ is the fraction of the population that has heard the rumor.

(b) Using the logistic equation (1), $\dfrac{dP}{dt} = kP\left(1 - \dfrac{P}{K}\right)$, we substitute $y = \dfrac{P}{K}$, $P = Ky$, and $\dfrac{dP}{dt} = K\dfrac{dy}{dt}$, to obtain $K\dfrac{dy}{dt} = k(Ky)(1 - y) \;\Leftrightarrow\; \dfrac{dy}{dt} = ky(1 - y)$, our equation in part (a).

Now the solution to (1) is $P(t) = \dfrac{K}{1 + Ae^{-kt}}$, where $A = \dfrac{K - P_0}{P_0}$.

We use the same substitution to obtain $Ky = \dfrac{K}{1 + \dfrac{K - Ky_0}{Ky_0} e^{-kt}} \Rightarrow y = \dfrac{y_0}{y_0 + (1 - y_0)e^{-kt}}$.

Alternatively, we could use the same steps as outlined in "The Analytic Solution," following Example 2.

(c) Let $t$ be the number of hours since 8 A.M. Then $y_0 = y(0) = \frac{80}{1000} = 0.08$ and $y(4) = \frac{1}{2}$, so

$\frac{1}{2} = y(4) = \dfrac{0.08}{0.08 + 0.92e^{-4k}}$. Thus, $0.08 + 0.92e^{-4k} = 0.16$, $e^{-4k} = \frac{0.08}{0.92} = \frac{2}{23}$, and $e^{-k} = \left(\frac{2}{23}\right)^{1/4}$, so

$y = \dfrac{0.08}{0.08 + 0.92(2/23)^{t/4}} = \dfrac{2}{2 + 23(2/23)^{t/4}}$. Solving this equation for $t$, we get

$$2y + 23y\left(\frac{2}{23}\right)^{t/4} = 2 \Rightarrow \left(\frac{2}{23}\right)^{t/4} = \frac{2 - 2y}{23y} \Rightarrow \left(\frac{2}{23}\right)^{t/4} = \frac{2}{23} \cdot \frac{1 - y}{y} \Rightarrow \left(\frac{2}{23}\right)^{t/4 - 1} = \frac{1 - y}{y}.$$

It follows that $\dfrac{t}{4} - 1 = \dfrac{\ln[(1 - y)/y]}{\ln \frac{2}{23}}$, so $t = 4\left[1 + \dfrac{\ln((1 - y)/y)}{\ln \frac{2}{23}}\right]$.

When $y = 0.9$, $\dfrac{1 - y}{y} = \frac{1}{9}$, so $t = 4\left(1 - \dfrac{\ln 9}{\ln \frac{2}{23}}\right) \approx 7.6$ h or 7 h 36 min. Thus, 90% of the population will have heard the rumor by 3:36 P.M.

**8.** (a) $P(0) = P_0 = 400$, $P(1) = 1200$ and $K = 10{,}000$. From the solution to the logistic differential equation

$P(t) = \dfrac{P_0 K}{P_0 + (K - P_0)e^{-kt}}$, we get $P = \dfrac{400\,(10{,}000)}{400 + (9600)e^{-kt}} = \dfrac{10{,}000}{1 + 24e^{-kt}}$. $P(1) = 1200 \Rightarrow$

$1 + 24e^{-k} = \frac{100}{12} \Rightarrow e^k = \frac{288}{88} \Rightarrow k = \ln \frac{36}{11}$. So $P = \dfrac{10{,}000}{1 + 24e^{-t\ln(36/11)}} = \dfrac{10{,}000}{1 + 24 \cdot (11/36)^t}$.

(b) $5000 = \dfrac{10{,}000}{1 + 24(11/36)^t} \Rightarrow 24\left(\frac{11}{36}\right)^t = 1 \Rightarrow t \ln \frac{11}{36} = \ln \frac{1}{24} \Rightarrow t \approx 2.68$ years.

**9.** (a) $\dfrac{dP}{dt} = kP\left(1 - \dfrac{P}{K}\right) \Rightarrow \dfrac{d^2P}{dt^2} = k\left[P\left(-\dfrac{1}{K}\dfrac{dP}{dt}\right) + \left(1 - \dfrac{P}{K}\right)\dfrac{dP}{dt}\right] = k\dfrac{dP}{dt}\left(-\dfrac{P}{K} + 1 - \dfrac{P}{K}\right)$

$$= k\left[kP\left(1 - \frac{P}{K}\right)\right]\left(1 - \frac{2P}{K}\right) = k^2P\left(1 - \frac{P}{K}\right)\left(1 - \frac{2P}{K}\right)$$

(b) $P$ grows fastest when $P'$ has a maximum, that is, when $P'' = 0$. From part (a), $P'' = 0 \Leftrightarrow P = 0, P = K$, or $P = K/2$. Since $0 < P < K$, we see that $P'' = 0 \Leftrightarrow P = K/2$.

**10.** First we keep $k$ constant (at 0.1, say) and change $P_0$ in the function

$P = \dfrac{10P_0}{P_0 + (10 - P_0)e^{-0.1t}}$. (Notice that $P_0$ is the $P$-intercept.) If $P_0 = 0$, the function is 0 everywhere. For $0 < P_0 < 5$, the curve has an inflection point, which moves to the right as $P_0$ decreases. If $5 < P_0 < 10$, the graph is concave down everywhere. (We are considering only $t \geq 0$.) If $P_0 = 10$, the function is the constant function $P = 10$, and if $P_0 > 10$, the function decreases. For all $P_0 \neq 0$, $\lim_{t\to\infty} P = 10$.

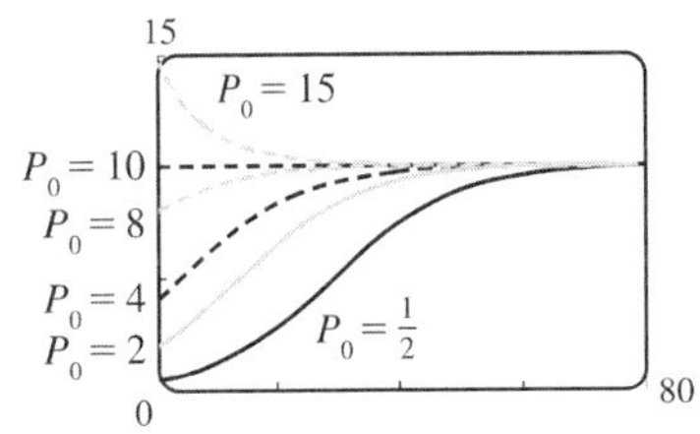


Now we instead keep $P_0$ constant (at $P_0 = 1$) and change $k$ in the function

$P = \dfrac{10}{1 + 9e^{-kt}}$. It seems that as $k$ increases, the graph approaches the line $P = 10$ more and more quickly. (Note that the only difference in the shape of the curves is in the horizontal scaling; if we choose suitable $x$-scales, the graphs all look the same.)

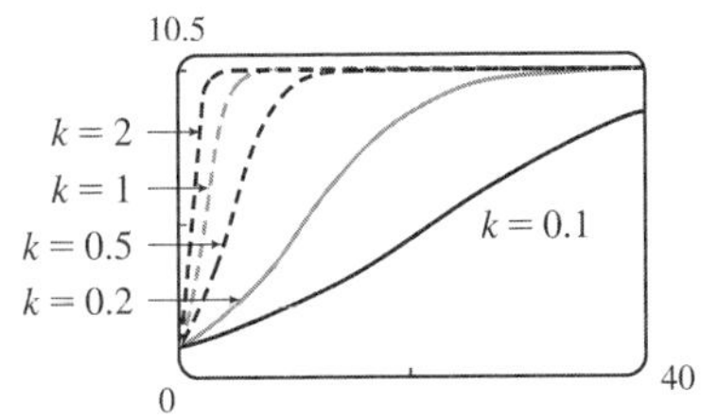

**11.** (a) The term $-15$ represents a harvesting of fish at a constant rate—in this case, 15 fish/week. This is the rate at which fish are caught.

(b)

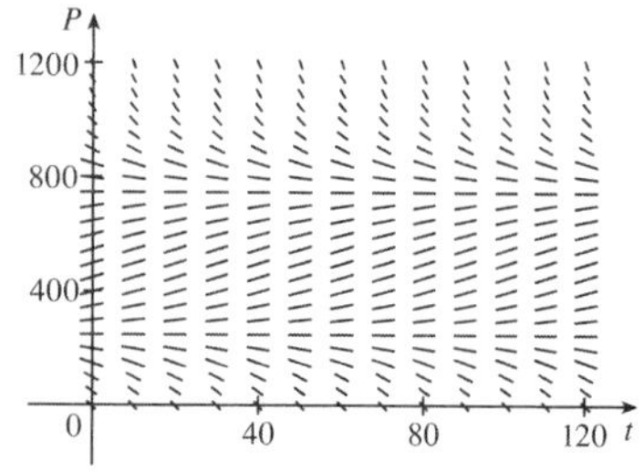


(c) From the graph in part (b), it appears that $P(t) = 250$ and $P(t) = 750$ are the equilibrium solutions. We confirm this analytically by solving the equation $dP/dt = 0$ as follows: $0.08P(1 - P/1000) - 15 = 0 \;\Rightarrow$

$0.08P - 0.00008P^2 - 15 = 0 \;\Rightarrow$

$-0.00008(P^2 - 1000P + 187{,}500) = 0 \;\Rightarrow$

$(P - 250)(P - 750) = 0 \;\Rightarrow\; P = 250$ or $750$.

(d)

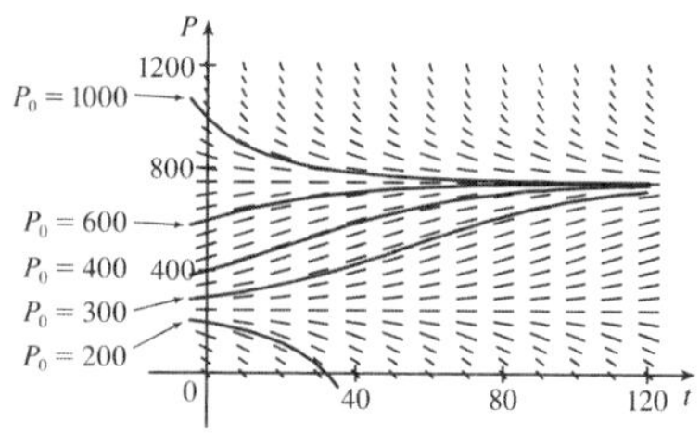


For $0 < P_0 < 250$, $P(t)$ decreases to 0. For $P_0 = 250$, $P(t)$ remains constant. For $250 < P_0 < 750$, $P(t)$ increases and approaches 750. For $P_0 = 750$, $P(t)$ remains constant. For $P_0 > 750$, $P(t)$ decreases and approaches 750.

(e) $\dfrac{dP}{dt} = 0.08P\left(1 - \dfrac{P}{1000}\right) - 15 \;\Leftrightarrow\; -\dfrac{100{,}000}{8} \cdot \dfrac{dP}{dt} = (0.08P - 0.00008P^2 - 15) \cdot \left(-\dfrac{100{,}000}{8}\right) \;\Leftrightarrow$

$-12{,}500\,\dfrac{dP}{dt} = P^2 - 1000P + 187{,}500 \;\Leftrightarrow\; \dfrac{dP}{(P-250)(P-750)} = -\dfrac{1}{12{,}500}\,dt \;\Leftrightarrow$

$\displaystyle\int\left(\frac{-1/500}{P-250} + \frac{1/500}{P-750}\right)dP = -\frac{1}{12{,}500}\,dt \;\Leftrightarrow\; \int\left(\frac{1}{P-250} - \frac{1}{P-750}\right)dP = \tfrac{1}{25}\,dt \;\Leftrightarrow$

$\ln|P - 250| - \ln|P - 750| = \frac{1}{25}t + C \;\Leftrightarrow\; \ln\left|\dfrac{P-250}{P-750}\right| = \frac{1}{25}t + C \;\Leftrightarrow\; \left|\dfrac{P-250}{P-750}\right| = e^{t/25+C} = ke^{t/25} \;\Leftrightarrow$

$\dfrac{P-250}{P-750} = ke^{t/25} \;\Leftrightarrow\; P - 250 = Pke^{t/25} - 750ke^{t/25} \;\Leftrightarrow\; P - Pke^{t/25} = 250 - 750ke^{t/25} \;\Leftrightarrow$

$P(t) = \dfrac{250 - 750ke^{t/25}}{1 - ke^{t/25}}$. If $t = 0$ and $P = 200$, then $200 = \dfrac{250 - 750k}{1-k} \;\Leftrightarrow\; 200 - 200k = 250 - 750k \;\Leftrightarrow$

$550k = 50 \;\Leftrightarrow\; k = \frac{1}{11}$. Similarly, if $t = 0$ and $P = 300$, then $k = -\frac{1}{9}$. Simplifying $P$ with these two values of $k$ gives us

$$P(t) = \frac{250\left(3e^{t/25} - 11\right)}{e^{t/25} - 11} \text{ and } P(t) = \frac{750\left(e^{t/25} + 3\right)}{e^{t/25} + 9}.$$

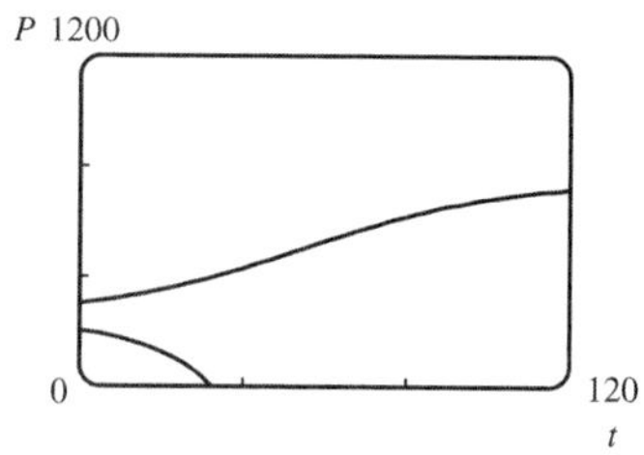


**12.** (a)

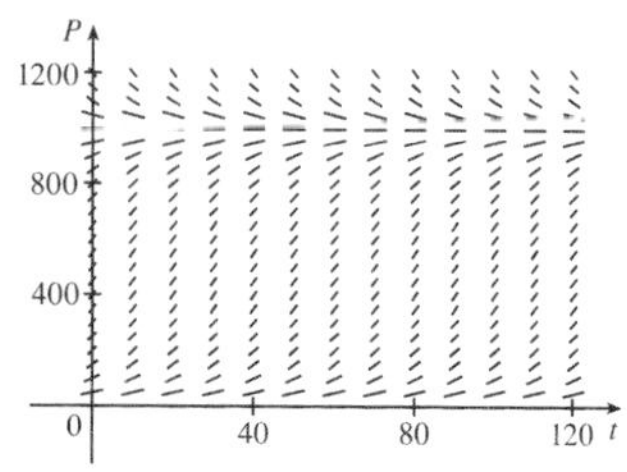


$c = 0$

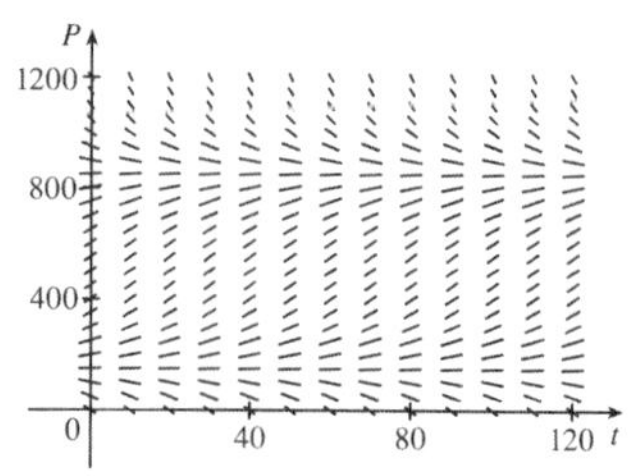


$c = 10$

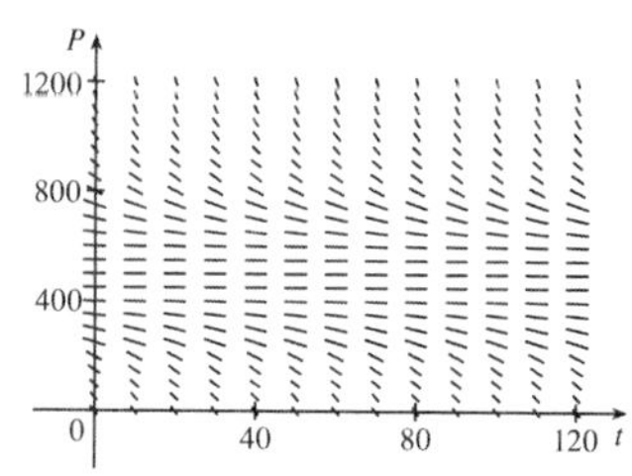


$c = 20$

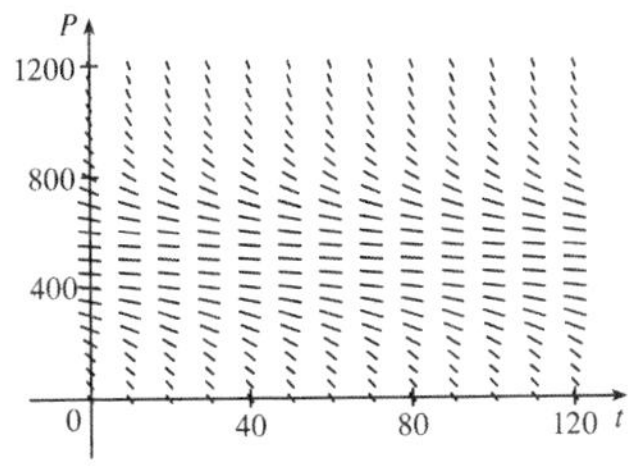


$c = 21$

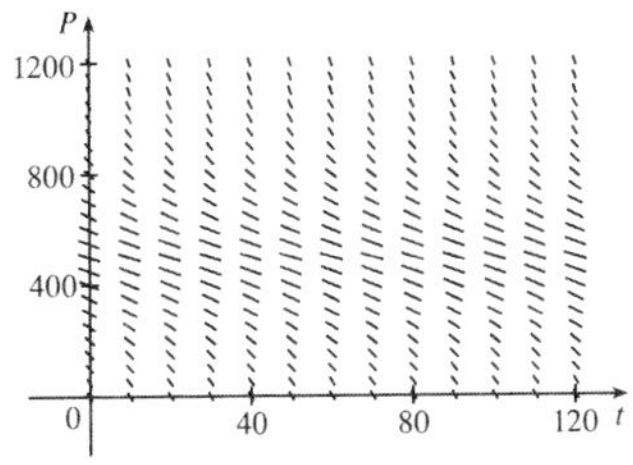


$c = 25$

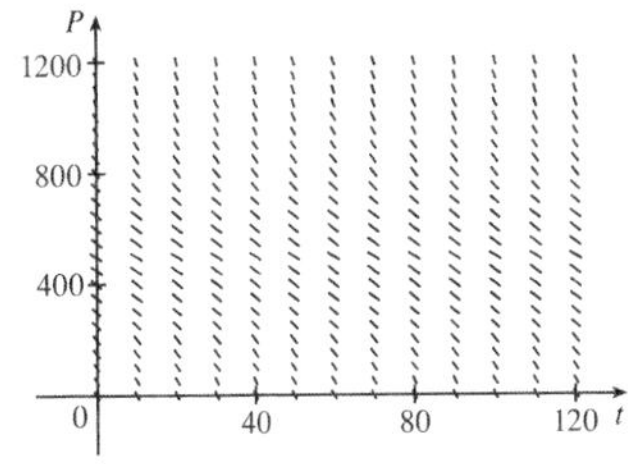


$c = 30$

(b) For $0 \leq c \leq 20$, there is at least one equilibrium solution. For $c > 20$, the population always dies out.

(c) $\dfrac{dP}{dt} = 0.08P - 0.00008P^2 - c$. $\dfrac{dP}{dt} = 0 \;\Leftrightarrow\; P = \dfrac{-0.08 \pm \sqrt{(0.08)^2 - 4(-0.00008)(-c)}}{2(-0.00008)}$, which has at least one solution when the discriminant is nonnegative $\Rightarrow$ $0.0064 - 0.00032c \geq 0 \;\Leftrightarrow\; c \leq 20$. For $0 \leq c \leq 20$, there is at least one value of $P$ such that $dP/dt = 0$ and hence, at least one equilibrium solution. For $c > 20$, $dP/dt < 0$ and the population always dies out.

(d) The weekly catch should be less than 20 fish per week.

**13.** (a) $\dfrac{dP}{dt} = (kP)\left(1 - \dfrac{P}{K}\right)\left(1 - \dfrac{m}{P}\right)$. If $m < P < K$, then $dP/dt = (+)(+)(+) = + \;\Rightarrow\;$ $P$ is increasing.
If $0 < P < m$, then $dP/dt = (+)(+)(-) = - \;\Rightarrow\;$ $P$ is decreasing.

(b)

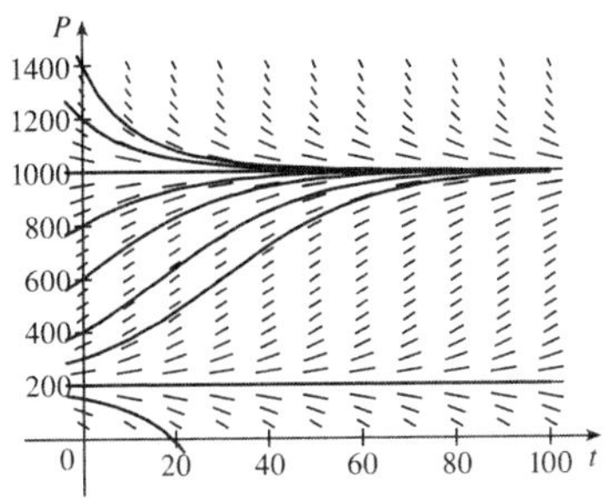


$k = 0.08$, $K = 1000$, and $m = 200 \;\Rightarrow$

$$\frac{dP}{dt} = 0.08P\left(1 - \frac{P}{1000}\right)\left(1 - \frac{200}{P}\right)$$

For $0 < P_0 < 200$, the population dies out. For $P_0 = 200$, the population is steady. For $200 < P_0 < 1000$, the population increases and approaches 1000. For $P_0 > 1000$, the population decreases and approaches 1000.

The equilibrium solutions are $P(t) = 200$ and $P(t) = 1000$.

(c) $\dfrac{dP}{dt} = kP\left(1 - \dfrac{P}{K}\right)\left(1 - \dfrac{m}{P}\right) = kP\left(\dfrac{K-P}{K}\right)\left(\dfrac{P-m}{P}\right) = \dfrac{k}{K}(K-P)(P-m) \;\Leftrightarrow$

$\displaystyle\int \frac{dP}{(K-P)(P-m)} = \int \frac{k}{K}\,dt.$

By partial fractions, $\dfrac{1}{(K-P)(P-m)} = \dfrac{A}{K-P} + \dfrac{B}{P-m}$, so $A(P-m) + B(K-P) = 1$.

If $P = m$, $B = \dfrac{1}{K-m}$; if $P = K$, $A = \dfrac{1}{K-m}$, so $\dfrac{1}{K-m}\displaystyle\int\left(\frac{1}{K-P} + \frac{1}{P-m}\right)dP = \int \frac{k}{K}\,dt \;\Rightarrow$

$\dfrac{1}{K-m}\left(-\ln|K-P| + \ln|P-m|\right) = \dfrac{k}{K}t + M \;\Rightarrow\; \dfrac{1}{K-m}\ln\left|\dfrac{P-m}{K-P}\right| = \dfrac{k}{K}t + M \;\Rightarrow$

$\ln\left|\dfrac{P-m}{K-P}\right| = (K-m)\dfrac{k}{K}t + M_1 \;\Leftrightarrow\; \dfrac{P-m}{K-P} = De^{(K-m)(k/K)t}$ $\quad [D = \pm e^{M_1}]$.

Let $t = 0$: $\dfrac{P_0 - m}{K - P_0} = D$. So $\dfrac{P-m}{K-P} = \dfrac{P_0 - m}{K - P_0}e^{(K-m)(k/K)t}$. Solving for $P$, we get

$$P(t) = \frac{m(K - P_0) + K(P_0 - m)e^{(K-m)(k/K)t}}{K - P_0 + (P_0 - m)e^{(K-m)(k/K)t}}.$$

(d) If $P_0 < m$, then $P_0 - m < 0$. Let $N(t)$ be the numerator of the expression for $P(t)$ in part (c). Then $N(0) = P_0(K - m) > 0$, and $P_0 - m < 0 \;\Leftrightarrow\; \lim\limits_{t\to\infty} K(P_0 - m)\,e^{(K-m)(k/K)t} = -\infty \;\Rightarrow\; \lim\limits_{t\to\infty} N(t) = -\infty$.
Since $N$ is continuous, there is a number $t$ such that $N(t) = 0$ and thus $P(t) = 0$. So the species will become extinct.

**14.** (a) $\dfrac{dP}{dt} = c\ln\left(\dfrac{K}{P}\right)P \;\Rightarrow\; \displaystyle\int \frac{dP}{P\ln(K/P)} = \int c\,dt$. Let $u = \ln\left(\dfrac{K}{P}\right) = \ln K - \ln P \;\Rightarrow$

$du = -\dfrac{dP}{P} \;\Rightarrow\; \displaystyle\int -\frac{du}{u} = ct + D \;\Rightarrow\; \ln|u| = -ct - D \;\Rightarrow\; |u| = e^{-(ct+D)} \;\Rightarrow$

$|\ln(K/P)| = e^{-(ct+D)} \;\Rightarrow\; \ln(K/P) = \pm e^{-(ct+D)}$. Letting $t = 0$, we get $\ln(K/P_0) = \pm e^{-D}$, so

$\ln(K/P) = \pm e^{-ct-D} = \pm e^{-ct}e^{-D} = \ln(K/P_0)\,e^{-ct} \;\Rightarrow\; K/P = e^{\ln(K/P_0)e^{-ct}} \;\Rightarrow$

$P(t) = Ke^{-\ln(K/P_0)e^{-ct}}$, $c \neq 0$.

(b) $\displaystyle\lim_{t\to\infty} P(t) = \lim_{t\to\infty} Ke^{-\ln(K/P_0)e^{-ct}} = Ke^{-\ln(K/P_0)\cdot 0} = Ke^0 = K$

(c)

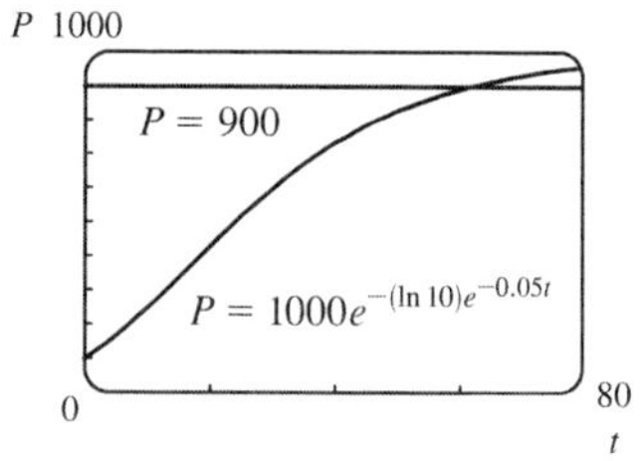


The graphs look very similar. For the Gompertz function, $P(40) \approx 732$, nearly the same as the logistic function. The Gompertz function reaches $P = 900$ at $t \approx 61.7$ and its value at $t = 80$ is about 959, so it doesn't increase quite as fast as the logistic curve.

(d) $\dfrac{dP}{dt} = c\ln\left(\dfrac{K}{P}\right)P = cP(\ln K - \ln P) \;\Rightarrow$

$$\frac{d^2P}{dt^2} = c\left[P\left(-\frac{1}{P}\frac{dP}{dt}\right) + (\ln K - \ln P)\frac{dP}{dt}\right] = c\,\frac{dP}{dt}\left[-1 + \ln\left(\frac{K}{P}\right)\right]$$
$$= c\,[c\ln(K/P)\,P]\,[\ln(K/P) - 1] = c^2P\ln(K/P)\,[\ln(K/P) - 1]$$

Since $0 < P < K$, $P'' = 0 \;\Leftrightarrow\; \ln(K/P) = 1 \;\Leftrightarrow\; K/P = e \;\Leftrightarrow\; P = K/e$. $P'' > 0$ for $0 < P < K/e$ and $P'' < 0$ for $K/e < P < K$, so $P'$ is a maximum (and $P$ grows fastest) when $P = K/e$.

*Note:* If $P > K$, then $\ln(K/P) < 0$, so $P''(t) > 0$.

**15.** (a) $dP/dt = kP\cos(rt - \phi) \;\Rightarrow\; (dP)/P = k\cos(rt - \phi)\,dt \;\Rightarrow\; \int (dP)/P = k\int \cos(rt - \phi)\,dt \;\Rightarrow$

$\ln P = (k/r)\sin(rt - \phi) + C$. (Since this is a growth model, $P > 0$ and we can write $\ln P$ instead of $\ln|P|$.) Since $P(0) = P_0$, we obtain $\ln P_0 = (k/r)\sin(-\phi) + C = -(k/r)\sin\phi + C \;\Rightarrow\; C = \ln P_0 + (k/r)\sin\phi$. Thus, $\ln P = (k/r)\sin(rt - \phi) + \ln P_0 + (k/r)\sin\phi$, which we can rewrite as $\ln(P/P_0) = (k/r)[\sin(rt - \phi) + \sin\phi]$ or, after exponentiation, $P(t) = P_0e^{(k/r)[\sin(rt-\phi)+\sin\phi]}$.

(b) As $k$ increases, the amplitude increases, but the minimum value stays the same.

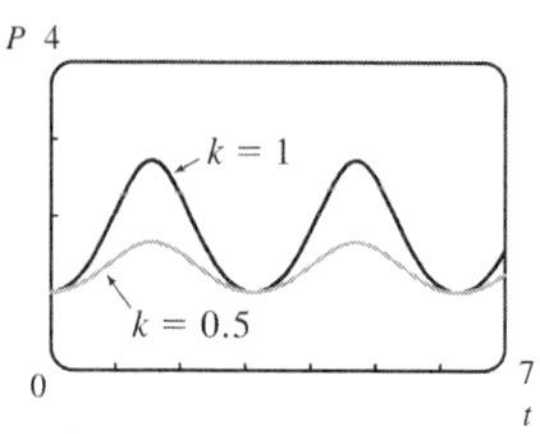


Comparing values of $k$ with $P_0 = 1$, $r = 2$, and $\phi = \pi/2$

As $r$ increases, the amplitude and the period decrease.

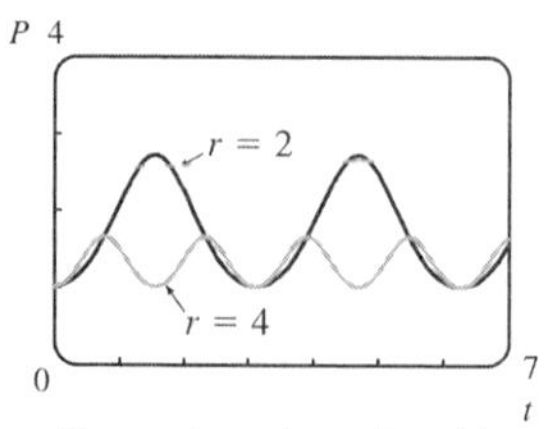


Comparing values of $r$ with $P_0 = 1$, $k = 1$, and $\phi = \pi/2$

A change in $\phi$ produces slight adjustments in the phase shift and amplitude.

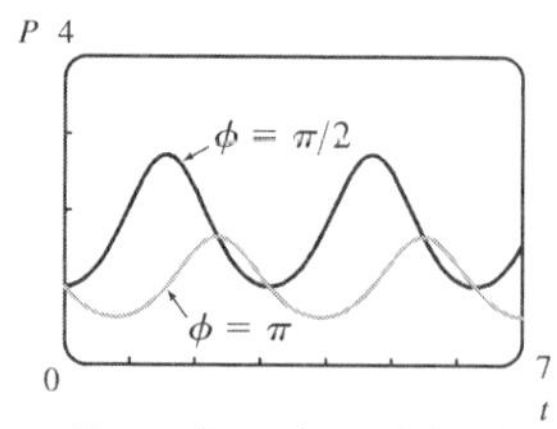


Comparing values of $\phi$ with $P_0 = 1$, $k = 1$, and $r = 2$

$P(t)$ oscillates between $P_0e^{(k/r)(1+\sin\phi)}$ and $P_0e^{(k/r)(-1+\sin\phi)}$ (the extreme values are attained when $rt - \phi$ is an odd multiple of $\frac{\pi}{2}$), so $\displaystyle\lim_{t\to\infty} P(t)$ does not exist.

**16.** (a) $dP/dt = kP\cos^2(rt - \phi) \;\Rightarrow\; (dP)/P = k\cos^2(rt-\phi)\,dt \;\Rightarrow\; \int(dP)/P = k\int\cos^2(rt-\phi)\,dt \;\Rightarrow$

$\ln P = k\displaystyle\int \frac{1+\cos(2(rt-\phi))}{2}\,dt = \frac{k}{2}t + \frac{k}{4r}\sin(2(rt-\phi)) + C$. From $P(0) = P_0$, we get

$\ln P_0 = \dfrac{k}{4r}\sin(-2\phi) + C = C - \dfrac{k}{4r}\sin 2\phi$, so $C = \ln P_0 + \dfrac{k}{4r}\sin 2\phi$ and

$\ln P = \dfrac{k}{2}t + \dfrac{k}{4r}\sin(2(rt-\phi)) + \ln P_0 + \dfrac{k}{4r}\sin 2\phi$. Simplifying, we get

$\ln\dfrac{P}{P_0} = \dfrac{k}{2}t + \dfrac{k}{4r}[\sin(2(rt-\phi)) + \sin 2\phi] = f(t)$, or $P(t) = P_0e^{f(t)}$.

(b) An increase in $k$ stretches the graph of $P$ vertically while maintaining $P(0) = P_0$.

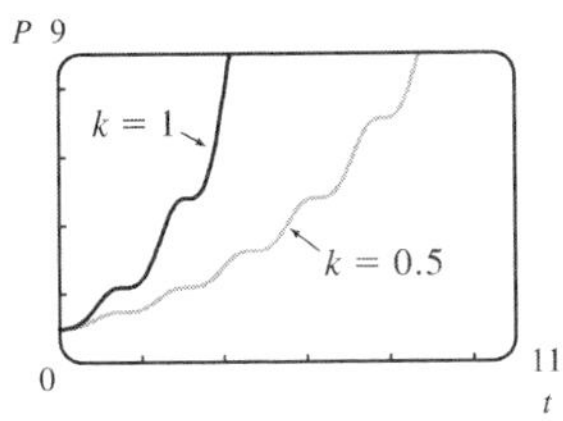


Comparing values of $k$ with $P_0 = 1$, $r = 2$, and $\phi = \pi/2$

An increase in $r$ compresses the graph of $P$ horizontally—similar to changing the period in Exercise 15.

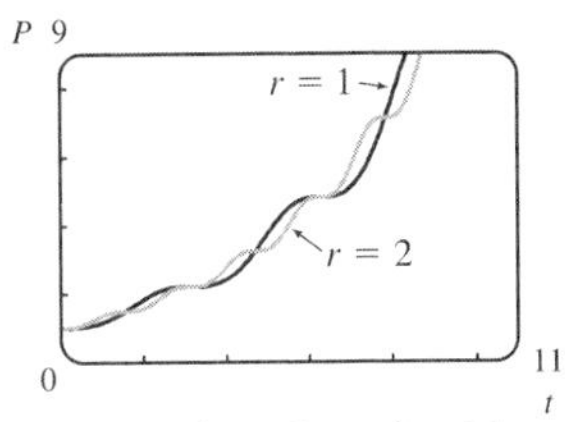


Comparing values of $r$ with $P_0 = 1$, $k = 0.5$, and $\phi = \pi/2$

As in Exercise 15, a change in $\phi$ only makes slight adjustments in the growth of $P$, as shown in the figure.

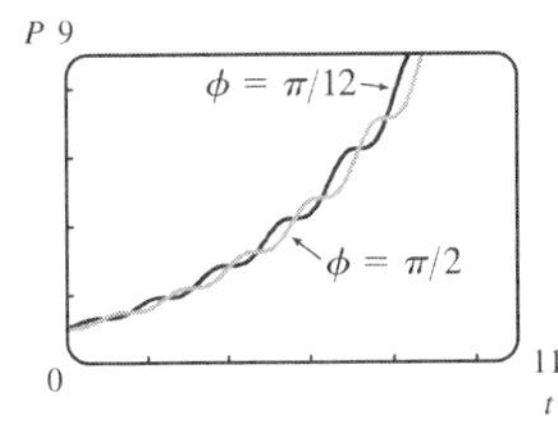


Comparing values of $\phi$ with $P_0 = 1$, $k = 0.5$, and $r = 2$

$f'(t) = k/2 + [k/(4r)][2r\cos(2(rt-\phi))] = (k/2)[1+\cos(2(rt-\phi))] \geq 0$. Since $P(t) = P_0e^{f(t)}$, we have $P'(t) = P_0f'(t)e^{f(t)} \geq 0$, with equality only when $\cos(2(rt-\phi)) = -1$; that is, when $rt - \phi$ is an odd multiple of $\frac{\pi}{2}$. Therefore, $P(t)$ is an increasing function on $(0, \infty)$. $P$ can also be written as $P(t) = P_0e^{kt/2}e^{(k/4r)[\sin(2(rt-\phi))+\sin 2\phi]}$. The second exponential oscillates between $e^{(k/4r)(1+\sin 2\phi)}$ and $e^{(k/4r)(-1+\sin 2\phi)}$, while the first one, $e^{kt/2}$, grows without bound. So $\lim\limits_{t\to\infty} P(t) = \infty$.

## 7.6 Predator-Prey Systems

**1.** (a) $dx/dt = -0.05x + 0.0001xy$. If $y = 0$, we have $dx/dt = -0.05x$, which indicates that in the absence of $y$, $x$ declines at a rate proportional to itself. So $x$ represents the predator population and $y$ represents the prey population. The growth of the prey population, $0.1y$ (from $dy/dt = 0.1y - 0.005xy$), is restricted only by encounters with predators (the term $-0.005xy$). The predator population increases only through the term $0.0001xy$; that is, by encounters with the prey and not through additional food sources.

(b) $dy/dt = -0.015y + 0.00008xy$. If $x = 0$, we have $dy/dt = -0.015y$, which indicates that in the absence of $x$, $y$ would decline at a rate proportional to itself. So $y$ represents the predator population and $x$ represents the prey population. The growth of the prey population, $0.2x$ (from $dx/dt = 0.2x - 0.0002x^2 - 0.006xy = 0.2x(1 - 0.001x) - 0.006xy$), is restricted by a carrying capacity of 1000 [from the term $1 - 0.001x = 1 - x/1000$] and by encounters with predators (the term $-0.006xy$). The predator population increases only through the term $0.00008xy$; that is, by encounters with the prey and not through additional food sources.

**2.** (a) $dx/dt = 0.12x - 0.0006x^2 + 0.00001xy$. $dy/dt = 0.08y + 0.00004xy$.

The $xy$ terms represent encounters between the two species $x$ and $y$. An increase in $y$ makes $dx/dt$ (the growth rate of $x$) larger due to the positive term $0.00001xy$. An increase in $x$ makes $dy/dt$ (the growth rate of $y$) larger due to the positive term $0.00004xy$. Hence, the system describes a cooperation model.

(b) $dx/dt = 0.15x - 0.0002x^2 - 0.0006xy = 0.15x(1 - x/750) - 0.0006xy.$

$dy/dt = 0.2y - 0.00008y^2 - 0.0002xy = 0.2y(1 - y/2500) - 0.0002xy.$

The system shows that $x$ and $y$ have carrying capacities of 750 and 2500. An increase in $x$ reduces the growth rate of $y$ due to the negative term $-0.0002xy$. An increase in $y$ reduces the growth rate of $x$ due to the negative term $-0.0006xy$. Hence, the system describes a competition model.

**3.** (a) At $t = 0$, there are about 300 rabbits and 100 foxes. At $t = t_1$, the number of foxes reaches a minimum of about 20 while the number of rabbits is about 1000. At $t = t_2$, the number of rabbits reaches a maximum of about 2400, while the number of foxes rebounds to 100. At $t = t_3$, the number of rabbits decreases to about 1000 and the number of foxes reaches a maximum of about 315. As $t$ increases, the number of foxes decreases greatly to 100, and the number of rabbits decreases to 300 (the initial populations), and the cycle starts again.

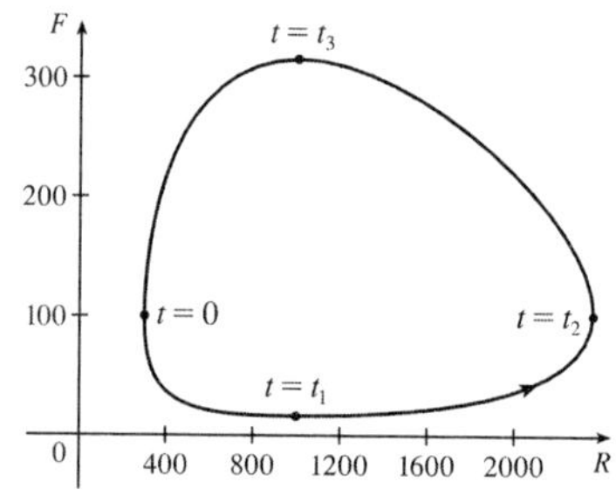


(b)

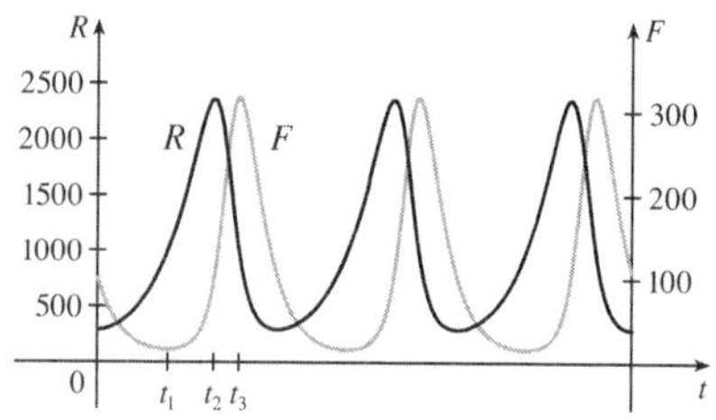


**4.** (a) At $t = 0$, there are about 600 rabbits and 160 foxes. At $t = t_1$, the number of rabbits reaches a minimum of about 80 and the number of foxes is also 80. At $t = t_2$, the number of foxes reaches a minimum of about 25 while the number of rabbits rebounds to 1000. At $t = t_3$, the number of foxes has increased to 40 and the rabbit population has reached a maximum of about 1750. The curve ends at $t = t_4$, where the number of foxes has increased to 65 and the number of rabbits has decreased to about 950.

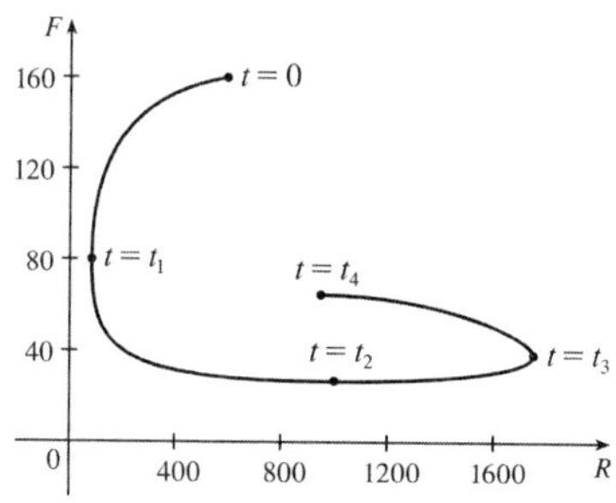


(b)

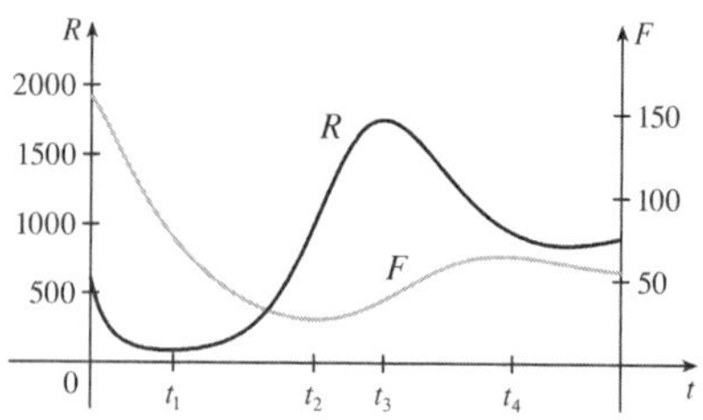


**5.**

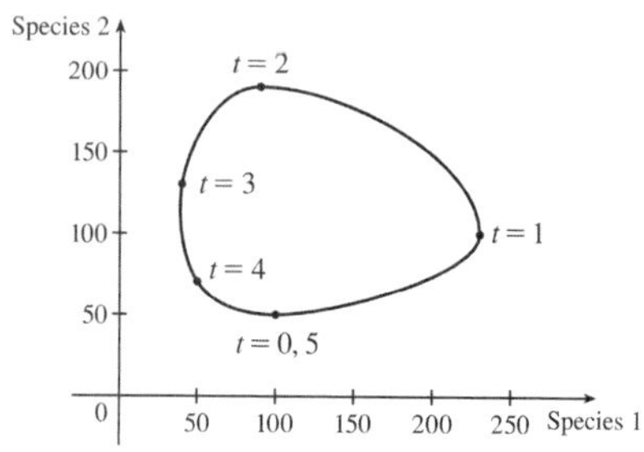


**6.**

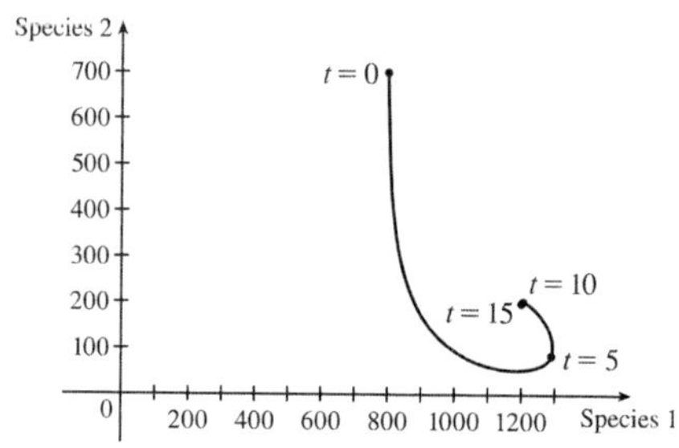

**7.** $\dfrac{dW}{dR} = \dfrac{-0.02W + 0.00002RW}{0.08R - 0.001RW} \Leftrightarrow (0.08 - 0.001W)R\,dW = (-0.02 + 0.00002R)W\,dR \Leftrightarrow$

$\dfrac{0.08 - 0.001W}{W}\,dW = \dfrac{-0.02 + 0.00002R}{R}\,dR \Leftrightarrow \displaystyle\int\left(\frac{0.08}{W} - 0.001\right)dW = \int\left(-\frac{0.02}{R} + 0.00002\right)dR \Leftrightarrow$

$0.08\ \ln|W| - 0.001W = -0.02\ \ln|R| + 0.00002R + K \Leftrightarrow 0.08\ln W + 0.02\ln R = 0.001W + 0.00002R + K \Leftrightarrow$

$\ln\left(W^{0.08}R^{0.02}\right) = 0.00002R + 0.001W + K \Leftrightarrow W^{0.08}R^{0.02} = e^{0.00002R + 0.001W + K} \Leftrightarrow$

$R^{0.02}W^{0.08} = Ce^{0.00002R}e^{0.001W} \Leftrightarrow \dfrac{R^{0.02}W^{0.08}}{e^{0.00002R}e^{0.001W}} = C$. In general, if $\dfrac{dy}{dx} = \dfrac{-ry + bxy}{kx - axy}$, then $C = \dfrac{x^r y^k}{e^{bx}e^{ay}}$.

**8.** (a) $A$ and $L$ are constant $\Rightarrow A' = 0$ and $L' = 0 \Rightarrow \begin{cases} 0 = 2A - 0.01AL \\ 0 = -0.5L + 0.0001AL \end{cases} \Rightarrow \begin{cases} 0 = A(2 - 0.01L) \\ 0 = L(-0.5 + 0.0001A) \end{cases}$

So either $A = L = 0$ or $L = \frac{2}{0.01} = 200$ and $A = \frac{0.5}{0.0001} = 5000$. The trivial solution $A = L = 0$ just says that if there aren't any aphids or ladybugs, then the populations will not change. The non-trivial solution, $L = 200$ and $A = 5000$, indicates the population sizes needed so that there are no changes in either the number of aphids or the number of ladybugs.

(b) $\dfrac{dL}{dA} = \dfrac{dL/dt}{dA/dt} = \dfrac{-0.5L + 0.0001AL}{2A - 0.01AL}$

(c) The solution curves (phase trajectories) are all closed curves that have the equilibrium point $(5000, 200)$ inside them.

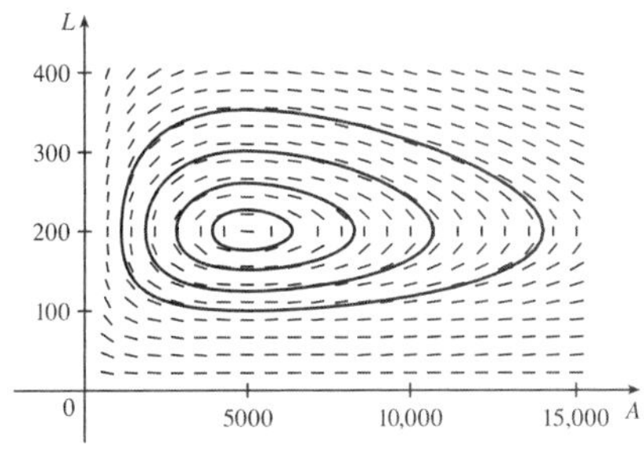


(d)

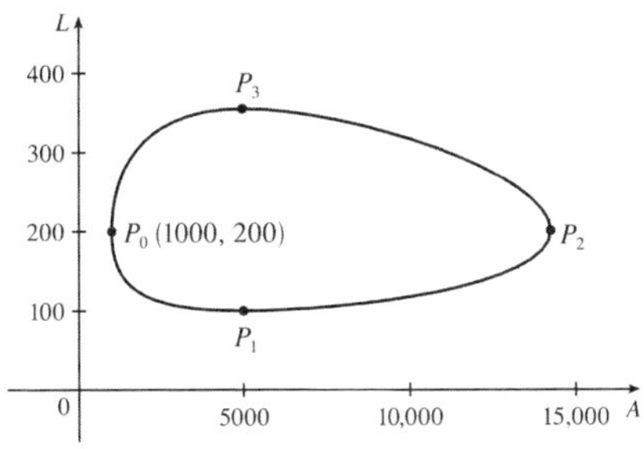


At $P_0(1000, 200)$, $dA/dt = 0$ and $dL/dt = -80 < 0$, so the number of ladybugs is decreasing and hence, we are proceeding in a counterclockwise direction. At $P_0$, there aren't enough aphids to support the ladybug population, so the number of ladybugs decreases and the number of aphids begins to increase. The ladybug population reaches a minimum at $P_1(5000, 100)$ while the aphid population increases in a dramatic way, reaching its maximum at $P_2(14{,}250, 200)$.

Meanwhile, the ladybug population is increasing from $P_1$ to $P_3(5000, 355)$, and as we pass through $P_2$, the increasing number of ladybugs starts to deplete the aphid population. At $P_3$ the ladybugs reach a maximum population, and start to decrease due to the reduced aphid population. Both populations then decrease until $P_0$, where the cycle starts over again.

(e) Both graphs have the same period and the graph of $L$ peaks about a quarter of a cycle after the graph of $A$.

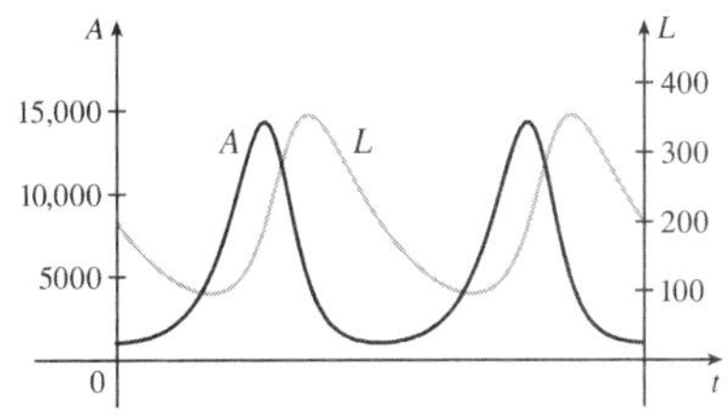


**9.** (a) Letting $W = 0$ gives us $dR/dt = 0.08R(1 - 0.0002R)$. $dR/dt = 0 \Leftrightarrow R = 0$ or 5000. Since $dR/dt > 0$ for $0 < R < 5000$, we would expect the rabbit population to *increase* to 5000 for these values of $R$. Since $dR/dt < 0$ for $R > 5000$, we would expect the rabbit population to *decrease* to 5000 for these values of $R$. Hence, in the absence of wolves, we would expect the rabbit population to stabilize at 5000.

(b) $R$ and $W$ are constant $\Rightarrow$ $R' = 0$ and $W' = 0$ $\Rightarrow$

$$\left.\begin{cases} 0 = 0.08R(1 - 0.0002R) - 0.001RW \\ 0 = -0.02W + 0.00002RW \end{cases}\right\} \Rightarrow \begin{cases} 0 = R[0.08(1 - 0.0002R) - 0.001W] \\ 0 = W(-0.02 + 0.00002R) \end{cases}$$

The second equation is true if $W = 0$ or $R = \frac{0.02}{0.00002} = 1000$. If $W = 0$ in the first equation, then either $R = 0$ or $R = \frac{1}{0.0002} = 5000$ [as in part (a)]. If $R = 1000$, then $0 = 1000[0.08(1 - 0.0002 \cdot 1000) - 0.001W]$ $\Leftrightarrow$ $0 = 80(1 - 0.2) - W$ $\Leftrightarrow$ $W = 64$.

*Case (i):* $W = 0$, $R = 0$: both populations are zero

*Case (ii):* $W = 0$, $R = 5000$: see part (a)

*Case (iii):* $R = 1000$, $W = 64$: the predator/prey interaction balances and the populations are stable.

(c) The populations of wolves and rabbits fluctuate around 64 and 1000, respectively, and eventually stabilize at those values.

(d)

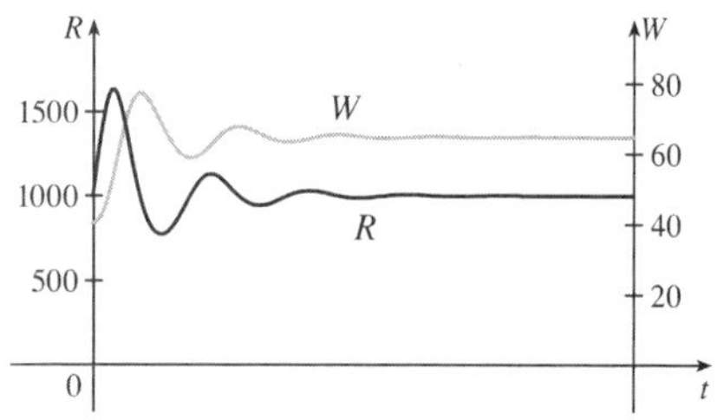


**10.** (a) If $L = 0$, $dA/dt = 2A\,(1 - 0.0001A)$, so $dA/dt = 0$ $\Leftrightarrow$ $A = 0$ or $A = \frac{1}{0.0001} = 10{,}000$. Since $dA/dt > 0$ for $0 < A < 10{,}000$, we expect the aphid population to *increase* to 10,000 for these values of $A$. Since $dA/dt < 0$ for $A > 10{,}000$, we expect the aphid population to *decrease* to 10,000 for these values of $A$. Hence, in the absence of ladybugs we expect the aphid population to stabilize at 10,000.

(b) $A$ and $L$ are constant $\Rightarrow$ $A' = 0$ and $L' = 0$ $\Rightarrow$

$$\left.\begin{cases} 0 = 2A(1 - 0.0001A) - 0.01AL \\ 0 = -0.5L + 0.0001AL \end{cases}\right\} \Rightarrow \begin{cases} 0 = A[2(1 - 0.0001A) - 0.01L] \\ 0 = L(-0.5 + 0.0001A) \end{cases}$$

The second equation is true if $L = 0$ or $A = \frac{0.5}{0.0001} = 5000$. If $L = 0$ in the first equation, then either $A = 0$ or $A = \frac{1}{0.0001} = 10{,}000$. If $A = 5000$, then $0 = 5000[2(1 - 0.0001 \cdot 5000) - 0.01L]$ $\Leftrightarrow$ $0 = 10{,}000(1 - 0.5) - 50L$ $\Leftrightarrow$ $50L = 5000$ $\Leftrightarrow$ $L = 100$.

The equilibrium solutions are: (i) $L = 0, A = 0$ (ii) $L = 0, A = 10{,}000$ (iii) $A = 5000, L = 100$

(c) $\dfrac{dL}{dA} = \dfrac{dL/dt}{dA/dt} = \dfrac{-0.5L + 0.0001AL}{2A(1 - 0.0001A) - 0.01AL}$

(d)

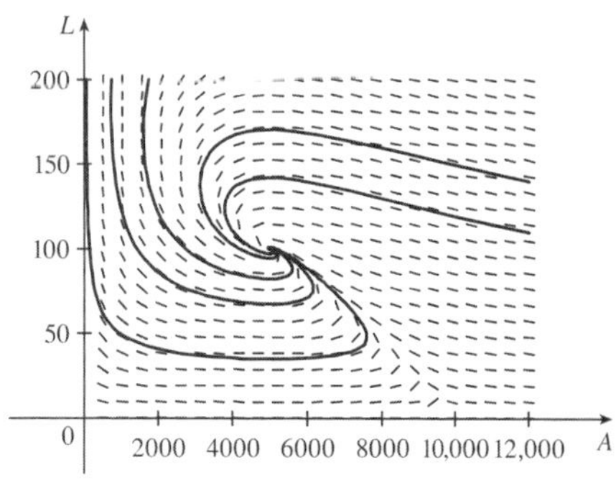


All of the phase trajectories spiral tightly around the equilibrium solution (5000, 100).

(e)

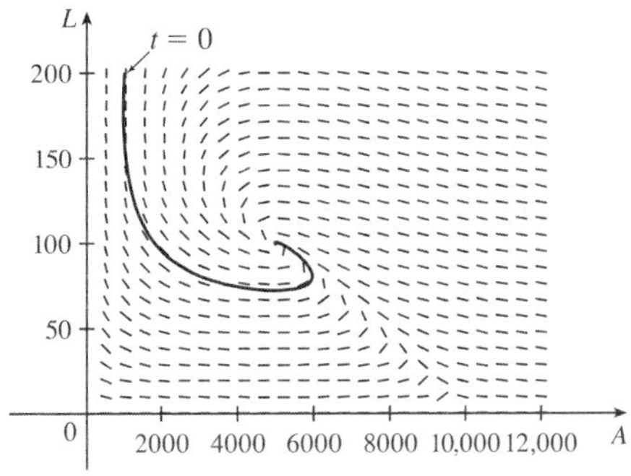


At $t = 0$, the ladybug population decreases rapidly and the aphid population decreases slightly before beginning to increase. As the aphid population continues to increase, the ladybug population reaches a minimum at about $(5000, 75)$. The ladybug population starts to increase and quickly stabilizes at 100, while the aphid population stabilizes at 5000.

(f)

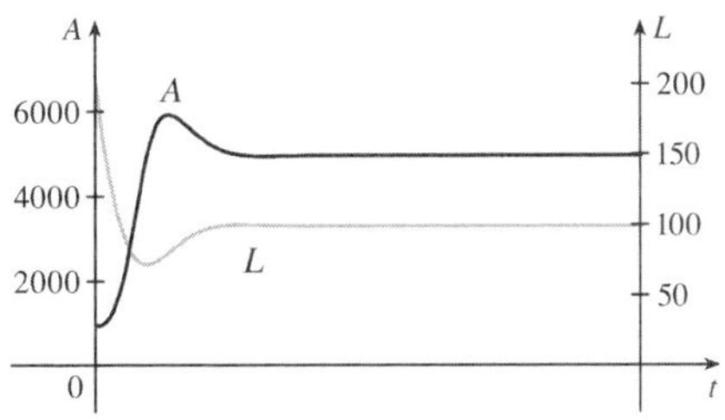


The graph of $A$ peaks just after the graph of $L$ has a minimum.

## 7 Review

CONCEPT CHECK

**1.** (a) A differential equation is an equation that contains an unknown function and one or more of its derivatives.

(b) The order of a differential equation is the order of the highest derivative that occurs in the equation.

(c) An initial condition is a condition of the form $y(t_0) = y_0$.

**2.** $y' = x^2 + y^2 \geq 0$ for all $x$ and $y$. $y' = 0$ only at the origin, so there is a horizontal tangent at $(0, 0)$, but nowhere else. The graph of the solution is increasing on every interval.

**3.** See the paragraph preceding Example 1 in Section 7.2.

**4.** See the paragraph after Figure 14 in Section 7.2.

**5.** A separable equation is a first-order differential equation in which the expression for $dy/dx$ can be factored as a function of $x$ times a function of $y$, that is, $dy/dx = g(x)f(y)$. We can solve the equation by integrating both sides of the equation $dy/f(y) = g(x)dx$ and solving for $y$.

**6.** (a) $\dfrac{dy}{dt} = ky$; the relative growth rate, $\dfrac{1}{y}\dfrac{dy}{dt}$, is constant.

(b) The equation in part (a) is an appropriate model for population growth, assuming that there is enough room and nutrition to support the growth.

(c) If $y(0) = y_0$, then the solution is $y(t) = y_0 e^{kt}$.

**7.** (a) $dP/dt = kP(1 - P/K)$, where $K$ is the carrying capacity.

(b) The equation in part (a) is an appropriate model for population growth, assuming that the population grows at a rate proportional to the size of the population in the beginning, but eventually levels off and approaches its carrying capacity because of limited resources.

**8.** (a) $dF/dt = kF - aFS$ and $dS/dt = -rS + bFS$.

(b) In the absence of sharks, an ample food supply would support exponential growth of the fish population, that is, $dF/dt = kF$, where $k$ is a positive constant. In the absence of fish, we assume that the shark population would decline at a rate proportional to itself, that is, $dS/dt = -rS$, where $r$ is a positive constant.

## TRUE-FALSE QUIZ

**1.** True. Since $y^4 \geq 0$, $y' = -1 - y^4 < 0$ and the solutions are decreasing functions.

**2.** True. $y = \dfrac{\ln x}{x} \;\Rightarrow\; y' = \dfrac{1 - \ln x}{x^2}$.

LHS $= x^2 y' + xy = x^2 \cdot \dfrac{1 - \ln x}{x^2} + x \cdot \dfrac{\ln x}{x} = (1 - \ln x) + \ln x = 1 =$ RHS, so $y = \dfrac{\ln x}{x}$ is a solution of $x^2 y' + xy = 1$.

**3.** False. $x + y$ cannot be written in the form $g(x)f(y)$.

**4.** True. $y' = 3y - 2x + 6xy - 1 = 6xy - 2x + 3y - 1 = 2x(3y - 1) + 1(3y - 1) = (2x + 1)(3y - 1)$, so $y'$ can be written in the form $g(x)f(y)$, and hence, is separable.

**5.** True. By comparing $\dfrac{dy}{dt} = 2y\left(1 - \dfrac{y}{5}\right)$ with the logistic differential equation (7.5.1), we see that the carrying capacity is 5; that is, $\lim\limits_{t \to \infty} y = 5$.

## EXERCISES

**1.** (a)

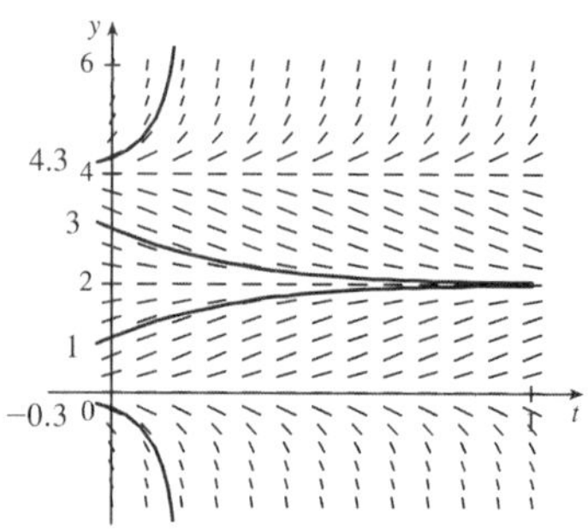


(b) $\lim\limits_{t \to \infty} y(t)$ appears to be finite for $0 \leq c \leq 4$. In fact $\lim\limits_{t \to \infty} y(t) = 4$ for $c = 4$, $\lim\limits_{t \to \infty} y(t) = 2$ for $0 < c < 4$, and $\lim\limits_{t \to \infty} y(t) = 0$ for $c = 0$. The equilibrium solutions are $y(t) = 0$, $y(t) = 2$, and $y(t) = 4$.

**2.** (a)

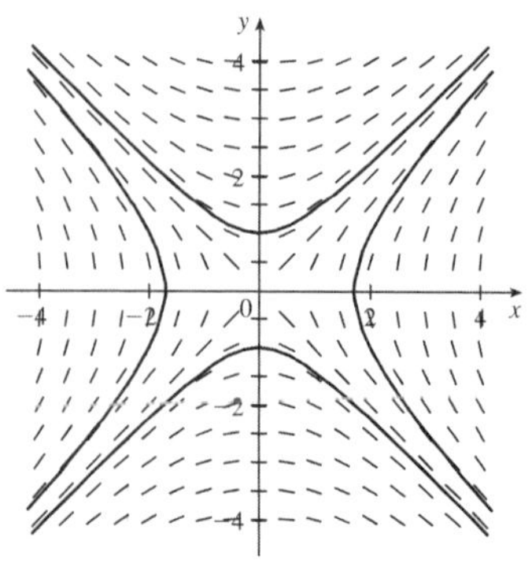


We sketch the direction field and four solution curves, as shown. Note that the slope $y' = x/y$ is not defined on the line $y = 0$.

(b) $y' = x/y \;\Leftrightarrow\; y\,dy = x\,dx \;\Leftrightarrow\; y^2 = x^2 + C$. For $C = 0$, this is the pair of lines $y = \pm x$. For $C \neq 0$, it is the hyperbola $x^2 - y^2 = -C$.

**3.** (a)

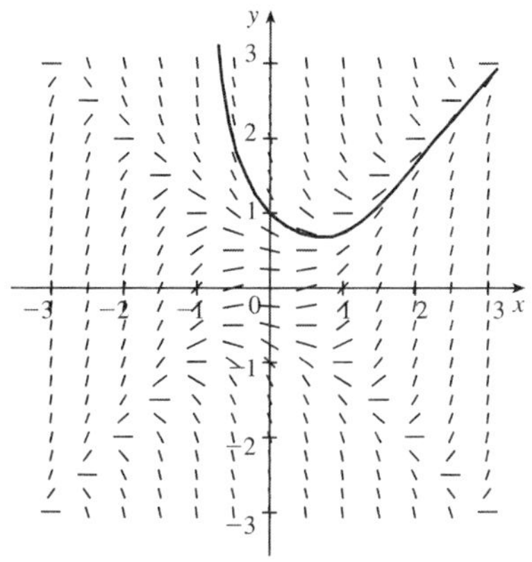

(b) $h = 0.1$, $x_0 = 0$, $y_0 = 1$ and $F(x, y) = x^2 - y^2$.

So $y_n = y_{n-1} + 0.1(x_{n-1}^2 - y_{n-1}^2)$. Thus,

$y_1 = 1 + 0.1(0^2 - 1^2) = 0.9$,

$y_2 = 0.9 + 0.1(0.1^2 - 0.9^2) = 0.82$,

$y_3 = 0.82 + 0.1(0.2^2 - 0.82^2) = 0.75676$. This is close to our graphical estimate of $y(0.3) \approx 0.8$.

We estimate that when $x = 0.3$, $y = 0.8$, so $y(0.3) \approx 0.8$.

(c) The centers of the horizontal line segments of the direction field are located on the lines $y = x$ and $y = -x$. When a solution curve crosses one of these lines, it has a local maximum or minimum.

**4.** (a) $h = 0.2$, $x_0 = 0$, $y_0 = 1$ and $F(x, y) = 2xy^2$. We need $y_2$.
$y_1 = 1 + 0.2(2 \cdot 0 \cdot 1^2) = 1$, $y_2 = 1 + 0.2(2 \cdot 0.2 \cdot 1^2) = 1.08 \approx y(0.4)$.

(b) $h = 0.1$ now, so $y_1 = 1 + 0.1(2 \cdot 0 \cdot 1^2) = 1$, $y_2 = 1 + 0.1(2 \cdot 0.1 \cdot 1^2) = 1.02$,
$y_3 = 1.02 + 0.1(2 \cdot 0.2 \cdot 1.02^2) \approx 1.06162$, $y_4 = 1.06162 + 0.1(2 \cdot 0.3 \cdot 1.06162^2) \approx 1.1292 \approx y(0.4)$.

(c) The equation is separable, so we write $\dfrac{dy}{y^2} = 2x\,dx \;\Rightarrow\; \displaystyle\int \frac{dy}{y^2} = \int 2x\,dx \;\Leftrightarrow\; -\frac{1}{y} = x^2 + C$, but $y(0) = 1$, so $C = -1$ and $y(x) = \dfrac{1}{1 - x^2} \;\Leftrightarrow\; y(0.4) = \dfrac{1}{1 - 0.16} \approx 1.1905$. From this we see that the approximation was greatly improved by increasing the number of steps, but the approximations were still far off.

**5.** $(3y^2 + 2y)\,y' = x\cos x \;\Rightarrow\; (3y^2 + 2y)\,dy = (x\cos x)\,dx \;\Rightarrow\; \int(3y^2 + 2y)\,dy = \int(x\cos x)\,dx \;\Rightarrow$
$y^3 + y^2 = \cos x + x\sin x + C$. For the last step, use integration by parts or Formula 83 in the Table of Integrals.

**6.** $\dfrac{dx}{dt} = 1 - t + x - tx = 1(1 - t) + x(1 - t) = (1 + x)(1 - t) \;\Rightarrow\; \dfrac{dx}{1 + x} = (1 - t)\,dt \;\Rightarrow$
$\displaystyle\int \frac{dx}{1 + x} = \int (1 - t)\,dt \;\Rightarrow\; \ln|1 + x| = t - \tfrac{1}{2}t^2 + C \;\Rightarrow\; |1 + x| = e^{t - t^2/2 + C} \;\Rightarrow$
$1 + x = \pm e^{t - t^2/2} \cdot e^C \;\Rightarrow\; x = -1 + Ke^{t - t^2/2}$, where $K$ is any nonzero constant.

**7.** $\dfrac{dr}{dt} + 2tr = r \;\Rightarrow\; \dfrac{dr}{dt} = r - 2tr = r(1 - 2t) \;\Rightarrow\; \displaystyle\int \frac{dr}{r} = \int (1 - 2t)\,dt \;\Rightarrow\; \ln|r| = t - t^2 + C \;\Rightarrow$
$|r| = e^{t - t^2 + C} = ke^{t - t^2}$. Since $r(0) = 5$, $5 = ke^0 = k$. Thus, $r(t) = 5e^{t - t^2}$.

**8.** $(1 + \cos x)y' = (1 + e^{-y})\sin x \;\Rightarrow\; \dfrac{dy}{1 + e^{-y}} = \dfrac{\sin x\,dx}{1 + \cos x} \;\Rightarrow\; \displaystyle\int \frac{dy}{1 + 1/e^y} = \int \frac{\sin x\,dx}{1 + \cos x} \;\Rightarrow$
$\displaystyle\int \frac{e^y\,dy}{1 + e^y} = \int \frac{\sin x\,dx}{1 + \cos x} \;\Rightarrow\; \ln|1 + e^y| = -\ln|1 + \cos x| + C \;\Rightarrow\; \ln(1 + e^y) = -\ln(1 + \cos x) + C \;\Rightarrow$
$1 + e^y = e^{-\ln(1 + \cos x)} \cdot e^C \;\Rightarrow\; e^y = ke^{-\ln(1 + \cos x)} - 1 \;\Rightarrow\; y = \ln[ke^{-\ln(1 + \cos x)} - 1]$. Since $y(0) = 0$,
$0 = \ln[ke^{-\ln 2} - 1] \;\Rightarrow\; e^0 = k(\tfrac{1}{2}) - 1 \;\Rightarrow\; k = 4$. Thus, $y(x) = \ln[4e^{-\ln(1 + \cos x)} - 1]$. An equavalent form is
$y(x) = \ln\dfrac{3 - \cos x}{1 + \cos x}$.

**9.** $\dfrac{d}{dx}(y) = \dfrac{d}{dx}(ke^x) \;\Rightarrow\; y' = ke^x = y$, so the orthogonal trajectories must have $y' = -\dfrac{1}{y} \;\Rightarrow\; \dfrac{dy}{dx} = -\dfrac{1}{y} \;\Rightarrow$
$y\,dy = -dx \;\Rightarrow\; \int y\,dy = -\int dx \;\Rightarrow\; \tfrac{1}{2}y^2 = -x + C \;\Rightarrow\; x = C - \tfrac{1}{2}y^2$, which are parabolas with a horizontal axis.

**10.** $\frac{d}{dx}(y) = \frac{d}{dx}(e^{kx}) \Rightarrow y' = ke^{kx} = ky = \frac{\ln y}{x} \cdot y$, so the orthogonal trajectories must have $y' = -\frac{x}{y \ln y} \Rightarrow$

$\frac{dy}{dx} = -\frac{x}{y\ln y} \Rightarrow y \ln y \, dy = -x\,dx \Rightarrow \int y \ln y \, dy = -\int x\,dx \Rightarrow \frac{1}{2}y^2 \ln y - \frac{1}{4}y^2$ [parts with $u = \ln y$, $dv = y\,dy$] $= -\frac{1}{2}x^2 + C_1 \Rightarrow 2y^2 \ln y - y^2 = C - 2x^2$.

**11.** (a) $y(t) = y(0)e^{kt} = 200e^{kt} \Rightarrow y(0.5) = 200e^{0.5k} = 360 \Rightarrow e^{0.5k} = 1.8 \Rightarrow 0.5k = \ln 1.8 \Rightarrow$ $k = 2\ln 1.8 = \ln(1.8)^2 = \ln 3.24 \Rightarrow y(t) = 200e^{(\ln 3.24)t} = 200(3.24)^t$

(b) $y(4) = 200(3.24)^4 \approx 22{,}040$ bacteria

(c) $y'(t) = 200(3.24)^t \cdot \ln 3.24$, so $y'(4) = 200(3.24)^4 \cdot \ln 3.24 \approx 25{,}910$ bacteria per hour

(d) $200(3.24)^t = 10{,}000 \Rightarrow (3.24)^t = 50 \Rightarrow t \ln 3.24 = \ln 50 \Rightarrow t = \ln 50/\ln 3.24 \approx 3.33$ hours

**12.** (a) If $y(t)$ is the mass remaining after $t$ years, then $y(t) = y(0)e^{kt} = 100e^{kt}$. $y(5.24) = 100e^{5.24k} = \frac{1}{2}\cdot 100 \Rightarrow$ $e^{5.24k} = \frac{1}{2} \Rightarrow 5.24k = -\ln 2 \Rightarrow k = -\frac{1}{5.24}\ln 2 \Rightarrow y(t) = 100e^{-(\ln 2)t/5.24} = 100 \cdot 2^{-t/5.24}$. Thus, $y(20) = 100 \cdot 2^{-20/5.24} \approx 7.1$ mg.

(b) $100 \cdot 2^{-t/5.24} = 1 \Rightarrow 2^{-t/5.24} = \frac{1}{100} \Rightarrow -\frac{t}{5.24}\ln 2 = \ln\frac{1}{100} \Rightarrow t = 5.24\frac{\ln 100}{\ln 2} \approx 34.8$ years

**13.** (a) $C'(t) = -kC(t) \Rightarrow C(t) = C(0)e^{-kt}$ by Theorem 10.4.2. But $C(0) = C_0$, so $C(t) = C_0 e^{-kt}$.

(b) $C(30) = \frac{1}{2}C_0$ since the concentration is reduced by half. Thus, $\frac{1}{2}C_0 = C_0 e^{-30k} \Rightarrow \ln\frac{1}{2} = -30k \Rightarrow$ $k = -\frac{1}{30}\ln\frac{1}{2} = \frac{1}{30}\ln 2$. Since 10% of the original concentration remains if 90% is eliminated, we want the value of $t$ such that $C(t) = \frac{1}{10}C_0$. Therefore, $\frac{1}{10}C_0 = C_0 e^{-t(\ln 2)/30} \Rightarrow \ln 0.1 = -t(\ln 2)/30 \Rightarrow t = -\frac{30}{\ln 2}\ln 0.1 \approx 100$ h.

**14.** (a) Let $t = 0$ correspond to 1990 so that $P(t) = 5.28e^{kt}$ is a starting point for the model. When $t = 10$, $P = 6.07$. So $6.07 = 5.28e^{10k} \Rightarrow 10k = \ln\frac{6.07}{5.28} \Rightarrow k = \frac{1}{10}\ln\frac{6.07}{5.28} \approx 0.01394$. For the year 2020, $t = 30$, and $P(30) = 5.28e^{30k} \approx 8.02$ billion.

(b) $P = 10 \Rightarrow 5.28e^{kt} = 10 \Rightarrow \frac{10}{5.28} = e^{kt} \Rightarrow kt = \ln\frac{10}{5.28} \Rightarrow t = 10\frac{\ln\frac{10}{5.28}}{\ln\frac{6.07}{5.28}} \approx 45.8$ years; that is, in $1990 + 45 = 2035$.

(c) $P(t) = \frac{K}{1 + Ae^{-kt}} = \frac{100}{1 + Ae^{-kt}}$, where $A = \frac{100 - 5.28}{5.28} \approx 17.94$. Using $k = \frac{1}{10}\ln\frac{6.07}{5.28}$ from part (a), a model is $P(t) \approx \frac{100}{1 + 17.94e^{-0.01394t}}$ and $P(30) \approx 7.81$ billion, slightly lower than our estimate of 8.02 billion in part (a).

(d) $P = 10 \Rightarrow 1 + Ae^{-kt} = \frac{100}{10} \Rightarrow Ae^{-kt} = 9 \Rightarrow e^{-kt} = 9/A \Rightarrow -kt = \ln(9/A) \Rightarrow$ $t = -\frac{1}{k}\ln\frac{9}{A} \approx 49.47$ years (that is, in 2039), which is later than the prediction of 2035 in part (b).

**15.** (a) $\frac{dL}{dt} \propto L_\infty - L \Rightarrow \frac{dL}{dt} = k(L_\infty - L) \Rightarrow \int\frac{dL}{L_\infty - L} = \int k\,dt \Rightarrow -\ln|L_\infty - L| = kt + C \Rightarrow$ $\ln|L_\infty - L| = -kt - C \Rightarrow |L_\infty - L| = e^{-kt-C} \Rightarrow L_\infty - L = Ae^{-kt} \Rightarrow L = L_\infty - Ae^{-kt}$. At $t = 0$, $L = L(0) = L_\infty - A \Rightarrow A = L_\infty - L(0) \Rightarrow L(t) = L_\infty - [L_\infty - L(0)]\,e^{-kt}$.

(b) $L_\infty = 53$ cm, $L(0) = 10$ cm, and $k = 0.2 \Rightarrow L(t) = 53 - (53 - 10)e^{-0.2t} = 53 - 43e^{-0.2t}$.

**16.** (a) If $y = u - 20$, $u(0) = 80 \Rightarrow y(0) = 80 - 20 = 60$, and the initial-value problem is $dy/dt = ky$ with $y(0) = 60$. So the solution is $y(t) = 60e^{kt}$. Now $y(0.5) = 60e^{k(0.5)} = 60 - 20 \Rightarrow e^{0.5k} = \frac{40}{60} = \frac{2}{3} \Rightarrow k = 2\ln\frac{2}{3} = \ln\frac{4}{9}$, so $y(t) = 60e^{(\ln 4/9)t} = 60(\frac{4}{9})^t$. Thus, $y(1) = 60(\frac{4}{9})^1 = \frac{80}{3} = 26\frac{2}{3}$ °C and $u(1) = 46\frac{2}{3}$ °C.

(b) $u(t) = 40 \Rightarrow y(t) = 20$. $y(t) = 60(\frac{4}{9})^t = 20 \Rightarrow (\frac{4}{9})^t = \frac{1}{3} \Rightarrow t\ln\frac{4}{9} = \ln\frac{1}{3} \Rightarrow t = \dfrac{\ln\frac{1}{3}}{\ln\frac{4}{9}} \approx 1.35$ h or 81.3 min.

**17.** Let $P$ represent the population and $I$ the number of infected people.

The rate of spread $dI/dt$ is jointly proportional to $I$ and to $P - I$, so for some constant $k$, $dI/dt = kI(P - I) \Rightarrow$

$$I = \frac{I_0 P}{I_0 + (P - I_0)e^{-kPt}} \quad \text{[from the discussion of logistic growth in Section 7.5].}$$

Now, measuring $t$ in days, we substitute $t = 7$, $P = 5000$, $I_0 = 160$ and $I(7) = 1200$ to find $k$:

$$1200 = \frac{160 \cdot 5000}{160 + (5000 - 160)e^{-5000 \cdot 7 \cdot k}} \Leftrightarrow 3 = \frac{2000}{160 + 4840e^{-35{,}000k}} \Leftrightarrow 480 + 14{,}520e^{-35{,}000k} = 2000 \Leftrightarrow$$

$$e^{-35{,}000k} = \frac{2000 - 480}{14{,}520} \Leftrightarrow -35{,}000k = \ln\frac{38}{363} \Leftrightarrow k = \frac{-1}{35{,}000}\ln\frac{38}{363} \approx 0.00006448.$$ Next, let

$I = 5000 \times 80\% = 4000$, and solve for $t$: $4000 = \dfrac{160 \cdot 5000}{160 + (5000 - 160)e^{-k \cdot 5000 \cdot t}} \Leftrightarrow 1 = \dfrac{200}{160 + 4840e^{-5000kt}} \Leftrightarrow$

$$160 + 4840e^{-5000kt} = 200 \Leftrightarrow e^{-5000kt} = \frac{200 - 160}{4840} \Leftrightarrow -5000kt = \ln\frac{1}{121} \Leftrightarrow$$

$t = \dfrac{-1}{5000k}\ln\dfrac{1}{121} = \dfrac{1}{\frac{1}{7}\ln\frac{38}{363}} \cdot \ln\dfrac{1}{121} = 7 \cdot \dfrac{\ln 121}{\ln\frac{363}{38}} \approx 14.875$. So it takes about 15 days for 80% of the population to be infected.

**18.** Denote the amount of salt in the tank (in kg) by $y$. $y(0) = 0$ since initially there is only water in the tank.

The rate at which $y$ increases is equal to the rate at which salt flows into the tank minus the rate at which it flows out.

That rate is $\dfrac{dy}{dt} = 0.1\dfrac{\text{kg}}{\text{L}} \times 10\dfrac{\text{L}}{\text{min}} - \dfrac{y}{100}\dfrac{\text{kg}}{\text{L}} \times 10\dfrac{\text{L}}{\text{min}} = 1 - \dfrac{y}{10}\dfrac{\text{kg}}{\text{min}} \Rightarrow \displaystyle\int\frac{dy}{10 - y} = \int\frac{1}{10}\,dt \Rightarrow$

$-\ln|10 - y| = \frac{1}{10}t + C \Rightarrow 10 - y = Ae^{-t/10}$. $y(0) = 0 \Rightarrow 10 = A \Rightarrow y = 10(1 - e^{-t/10})$.

At $t = 6$ minutes, $y = 10\left(1 - e^{-6/10}\right) \approx 4.512$ kg.

**19.** $\dfrac{dh}{dt} = -\dfrac{R}{V}\left(\dfrac{h}{k + h}\right) \Rightarrow \displaystyle\int\frac{k + h}{h}\,dh = \int\left(-\frac{R}{V}\right)dt \Rightarrow \int\left(1 + \frac{k}{h}\right)dh = -\frac{R}{V}\int 1\,dt \Rightarrow$

$h + k\ln h = -\dfrac{R}{V}t + C$. This equation gives a relationship between $h$ and $t$, but it is not possible to isolate $h$ and express it in terms of $t$.

**20.** $dx/dt = 0.4x - 0.002xy$, $dy/dt = -0.2y + 0.000008xy$

(a) The $xy$ terms represent encounters between the birds and the insects. Since the $y$-population increases from these terms and the $x$-population decreases, we expect $y$ to represent the birds and $x$ the insects.

(b) $x$ and $y$ are constant $\Rightarrow$ $x' = 0$ and $y' = 0$ $\Rightarrow$

$$\left.\begin{cases} 0 = 0.4x - 0.002xy \\ 0 = -0.2y + 0.000008xy \end{cases}\right\} \Rightarrow \begin{cases} 0 = 0.4x(1 - 0.005y) \\ 0 = -0.2y(1 - 0.00004x) \end{cases} \Rightarrow y = 0 \text{ and } x = 0 \text{ (zero populations)}$$

or $y = \frac{1}{0.005} = 200$ and $x = \frac{1}{0.00004} = 25{,}000$. The non-trivial solution represents the population sizes needed so that there are no changes in either the number of birds or the number of insects.

(c) $\dfrac{dy}{dx} = \dfrac{dy/dt}{dx/dt} = \dfrac{-0.2y + 0.000008xy}{0.4x - 0.002xy}$

(d)

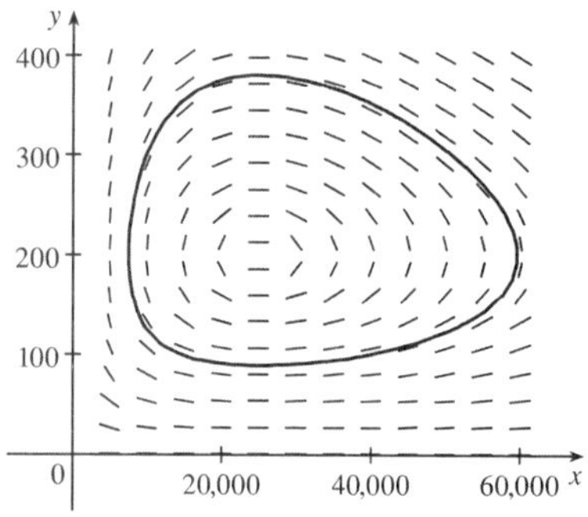


At $(x, y) = (40{,}000, 100)$, $dx/dt = 8000 > 0$, so as $t$ increases we are proceeding in a counterclockwise direction. The populations increase to approximately $(59{,}646, 200)$, at which point the insect population starts to decrease. The birds attain a maximum population of about 380 when the insect population is 25,000. The populations decrease to about $(7370, 200)$, at which point the insect population starts to increase. The birds attain a minimum population of about 88 when the insect population is 25,000, and then the cycle repeats.

(e)

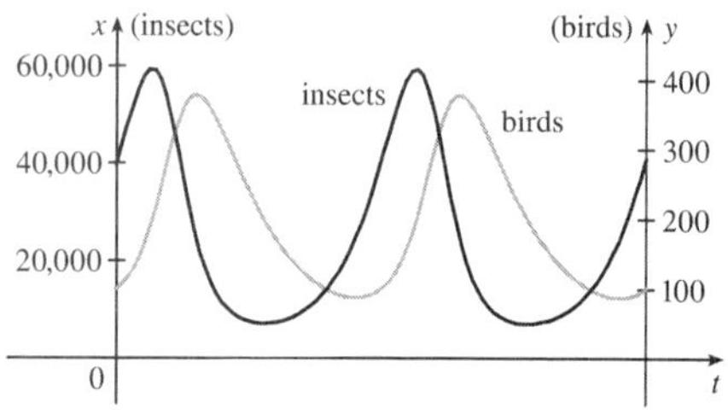


Both graphs have the same period and the bird population peaks about a quarter-cycle after the insect population.

**21.** (a) $dx/dt = 0.4x(1 - 0.000005x) - 0.002xy$, $dy/dt = -0.2y + 0.000008xy$. If $y = 0$, then $dx/dt = 0.4x(1 - 0.000005x)$, so $dx/dt = 0 \Leftrightarrow x = 0$ or $x = 200{,}000$, which shows that the insect population increases logistically with a carrying capacity of 200,000. Since $dx/dt > 0$ for $0 < x < 200{,}000$ and $dx/dt < 0$ for $x > 200{,}000$, we expect the insect population to stabilize at 200,000.

(b) $x$ and $y$ are constant $\Rightarrow$ $x' = 0$ and $y' = 0$ $\Rightarrow$

$$\left.\begin{cases} 0 = 0.4x(1 - 0.000005x) - 0.002xy \\ 0 = -0.2y + 0.000008xy \end{cases}\right\} \Rightarrow \begin{cases} 0 = 0.4x[(1 - 0.000005x) - 0.005y] \\ 0 = y(-0.2 + 0.000008x) \end{cases}$$

The second equation is true if $y = 0$ or $x = \frac{0.2}{0.000008} = 25{,}000$. If $y = 0$ in the first equation, then either $x = 0$ or $x = \frac{1}{0.000005} = 200{,}000$. If $x = 25{,}000$, then $0 = 0.4(25{,}000)[(1 - 0.000005 \cdot 25{,}000) - 0.005y] \Rightarrow$ $0 = 10{,}000[(1 - 0.125) - 0.005y] \Rightarrow 0 = 8750 - 50y \Rightarrow y = 175$.

*Case (i):* $y = 0$, $x = 0$: Zero populations

*Case (ii):* $y = 0$, $x = 200{,}000$: In the absence of birds, the insect population is always 200,000.

*Case (iii):* $x = 25{,}000$, $y = 175$: The predator/prey interaction balances and the populations are stable.

(c) The populations of the birds and insects fluctuate around 175 and 25,000, respectively, and eventually stabilize at those values.

(d)

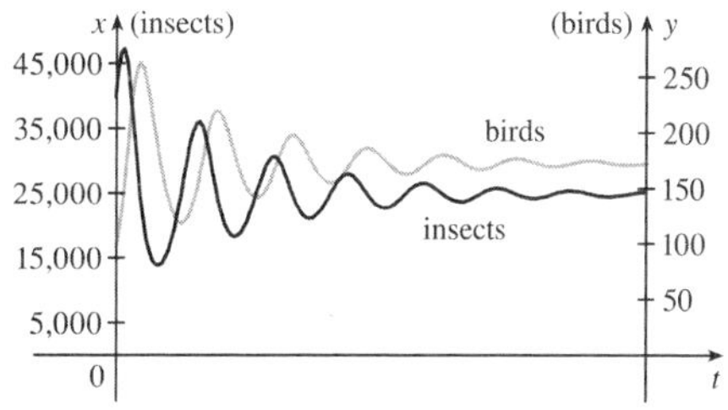


**22.** First note that, in this question, "weighs" is used in the informal sense, so what we really require is Barbara's mass $m$ in kg as a function of $t$. Barbara's net intake of calories per day at time $t$ (measured in days) is $c(t) = 1600 - 850 - 15m(t) = 750 - 15m(t)$, where $m(t)$ is her mass at time $t$. We are given that $m(0) = 60$ kg and $\dfrac{dm}{dt} = \dfrac{c(t)}{10{,}000}$, so $\dfrac{dm}{dt} = \dfrac{750 - 15m}{10{,}000} = \dfrac{150 - 3m}{2000} = \dfrac{-3(m - 50)}{2000}$ with $m(0) = 60$. From $\displaystyle\int \frac{dm}{m - 50} = \int \frac{-3\,dt}{2000}$, we get $\ln|m - 50| = -\frac{3}{2000}t + C$. Since $m(0) = 60$, $C = \ln 10$. Now $\ln \dfrac{|m - 50|}{10} = -\dfrac{3t}{2000}$, so $|m - 50| = 10e^{-3t/2000}$. The quantity $m - 50$ is continuous, initially positive, and the right-hand side is never zero. Thus, $m - 50$ is positive for all $t$, and $m(t) = 50 + 10e^{-3t/2000}$ kg. As $t \to \infty$, $m(t) \to 50$ kg. Thus, Barbara's mass gradually settles down to 50 kg.

# FOCUS ON PROBLEM SOLVING

**1.** We use the Fundamental Theorem of Calculus to differentiate the given equation:
$[f(x)]^2 = 100 + \int_0^x \left\{[f(t)]^2 + [f'(t)]^2\right\} dt \;\Rightarrow\; 2f(x)f'(x) = [f(x)]^2 + [f'(x)]^2 \;\Rightarrow$
$[f(x)]^2 + [f'(x)]^2 - 2f(x)f'(x) = 0 \;\Rightarrow\; [f(x) - f'(x)]^2 = 0 \;\Leftrightarrow\; f(x) = f'(x)$. We can solve this as a separable equation, or else use Theorem 7.4.2 with $k = 1$, which says that the solutions are $f(x) = Ce^x$. Now $[f(0)]^2 = 100$, so $f(0) = C = \pm 10$, and hence $f(x) = \pm 10e^x$ are the only functions satisfying the given equation.

**2.** $(fg)' = f'g'$, where $f(x) = e^{x^2} \;\Rightarrow\; \left(e^{x^2} g\right)' = 2xe^{x^2} g'$. Since the student's mistake did not affect the answer,
$\left(e^{x^2} g\right)' = e^{x^2} g' + 2xe^{x^2} g = 2xe^{x^2} g'$. So $(2x - 1)g' = 2xg$, or $\dfrac{g'}{g} = \dfrac{2x}{2x-1} = 1 + \dfrac{1}{2x-1} \;\Rightarrow$
$\ln|g(x)| = x + \frac{1}{2}\ln(2x-1) + C \;\Rightarrow\; g(x) = Ae^x\sqrt{2x-1}$.

**3.** $f'(x) = \lim\limits_{h\to 0} \dfrac{f(x+h) - f(x)}{h} = \lim\limits_{h\to 0} \dfrac{f(x)\,[f(h) - 1]}{h}$ [since $f(x+h) = f(x)f(h)$]

$$= f(x) \lim_{h\to 0} \frac{f(h) - 1}{h} = f(x) \lim_{h\to 0} \frac{f(h) - f(0)}{h - 0} = f(x)f'(0) = f(x)$$

Therefore, $f'(x) = f(x)$ for all $x$ and from Theorem 7.4.2 we get $f(x) = Ae^x$. Now $f(0) = 1 \;\Rightarrow\; A = 1 \;\Rightarrow$ $f(x) = e^x$.

**4.** $\left(\int f(x)\,dx\right)\left(\int \dfrac{dx}{f(x)}\right) = -1 \;\Rightarrow\; \int \dfrac{dx}{f(x)} = \dfrac{-1}{\int f(x)\,dx} \;\Rightarrow\; \dfrac{1}{f(x)} = \dfrac{f(x)}{\left[\int f(x)\,dx\right]^2}$ [after differentiating] $\Rightarrow$
$\int f(x)\,dx = \pm f(x)$ [after taking square roots] $\Rightarrow$ $f(x) = \pm f'(x)$ [after differentiating again] $\Rightarrow$ $y = Ae^x$ or $y = Ae^{-x}$ by Theorem 7.4.2. Therefore, $f(x) = Ae^x$ or $f(x) = Ae^{-x}$, for all nonzero constants $A$, are the functions satisfying the original equation.

**5.** Let $y(t)$ denote the temperature of the peach pie $t$ minutes after 5:00 P.M. and $R$ the temperature of the room. Newton's Law of Cooling gives us $dy/dt = k(y - R)$. Solving for $y$ we get $\dfrac{dy}{y - R} = k\,dt \;\Rightarrow\; \ln|y - R| = kt + C \;\Rightarrow$
$|y - R| = e^{kt+C} \;\Rightarrow\; y - R = \pm e^{kt} \cdot e^C \;\Rightarrow\; y = Me^{kt} + R$, where $M$ is a nonzero constant. We are given temperatures at three times.

$$\begin{aligned} y(0) &= 100 \;\Rightarrow\; 100 = M + R \;\Rightarrow\; R = 100 - M \\ y(10) &= 80 \;\Rightarrow\; 80 = Me^{10k} + R \quad \textbf{(1)} \\ y(20) &= 65 \;\Rightarrow\; 65 = Me^{20k} + R \quad \textbf{(2)} \end{aligned}$$

Substituting $100 - M$ for $R$ in **(1)** and **(2)** gives us

$$-20 = Me^{10k} - M \;\textbf{(3)} \quad \text{and} \quad -35 = Me^{20k} - M \;\textbf{(4)}$$

Dividing **(3)** by **(4)** gives us $\dfrac{-20}{-35} = \dfrac{M(e^{10k} - 1)}{M(e^{20k} - 1)} \;\Rightarrow\; \dfrac{4}{7} = \dfrac{e^{10k} - 1}{e^{20k} - 1} \;\Rightarrow\; 4e^{20k} - 4 = 7e^{10k} - 7 \;\Rightarrow$
$4e^{20k} - 7e^{10k} + 3 = 0$. This is a quadratic equation in $e^{10k}$. $\left(4e^{10k} - 3\right)\left(e^{10k} - 1\right) = 0 \;\Rightarrow\; e^{10k} = \frac{3}{4}$ or $1 \;\Rightarrow$
$10k = \ln\frac{3}{4}$ or $\ln 1 \;\Rightarrow\; k = \frac{1}{10}\ln\frac{3}{4}$ since $k$ is a nonzero constant of proportionality. Substituting $\frac{3}{4}$ for $e^{10k}$ in **(3)** gives us $-20 = M \cdot \frac{3}{4} - M \;\Rightarrow\; -20 = -\frac{1}{4}M \;\Rightarrow\; M = 80$. Now $R = 100 - M$ so $R = 20°\text{C}$.

**6.** Let $b$ be the number of hours before noon that it began to snow, $t$ the time measured in hours after noon, and $x = x(t) =$ distance traveled by the plow at time $t$. Then $dx/dt =$ speed of plow. Since the snow falls steadily, the height at time $t$ is $h(t) = k(t + b)$, where $k$ is a constant. We are given that the rate of removal is constant, say $R$ (in $\text{m}^3/\text{h}$). If the width of the path is $w$, then $R = \text{height} \times \text{width} \times \text{speed} = h(t) \times w \times \dfrac{dx}{dt} = k(t+b)w\,\dfrac{dx}{dt}$. Thus, $\dfrac{dx}{dt} = \dfrac{C}{t+b}$, where $C = \dfrac{R}{kw}$ is a constant. This is a separable equation. $\displaystyle\int dx = C\int \frac{dt}{t+b} \;\Rightarrow\; x(t) = C\ln(t+b) + K$.

Put $t = 0$: $0 = C\ln b + K \;\Rightarrow\; K = -C\ln b$, so $x(t) = C\ln(t+b) - C\ln b = C\ln(1 + t/b)$.

Put $t = 1$: $6000 = C\ln(1 + 1/b)$ $\quad[x = 6 \text{ km}]$.

Put $t = 2$: $9000 = C\ln(1 + 2/b)$ $\quad[x = (6+3) \text{ km}]$.

Solve for $b$: $\dfrac{\ln(1+1/b)}{6000} = \dfrac{\ln(1+2/b)}{9000} \;\Rightarrow\; 3\ln\left(1+\dfrac{1}{b}\right) = 2\ln\left(1+\dfrac{2}{b}\right) \;\Rightarrow\; \left(1+\dfrac{1}{b}\right)^3 = \left(1+\dfrac{2}{b}\right)^2 \;\Rightarrow$

$1 + \dfrac{3}{b} + \dfrac{3}{b^2} + \dfrac{1}{b^3} = 1 + \dfrac{4}{b} + \dfrac{4}{b^2} \;\Rightarrow\; \dfrac{1}{b} + \dfrac{1}{b^2} - \dfrac{1}{b^3} = 0 \;\Rightarrow\; b^2 + b - 1 = 0 \;\Rightarrow\; b = \frac{-1\pm\sqrt{5}}{2}$.

But $b > 0$, so $b = \frac{-1+\sqrt{5}}{2} \approx 0.618$ h $\approx 37$ min. The snow began to fall $\frac{\sqrt{5}-1}{2}$ hours before noon; that is, at about 11:23 A.M.

7. (a) While running from $(L, 0)$ to $(x, y)$, the dog travels a distance

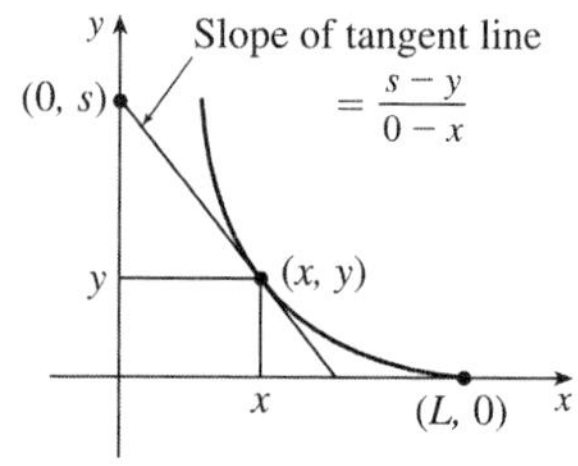


$$s = \int_x^L \sqrt{1 + (dy/dx)^2}\,dx = -\int_L^x \sqrt{1 + (dy/dx)^2}\,dx, \text{ so}$$

$\dfrac{ds}{dx} = -\sqrt{1 + (dy/dx)^2}$. The dog and rabbit run at the same speed, so the rabbit's position when the dog has traveled a distance $s$ is $(0, s)$.

Since the dog runs straight for the rabbit, $\dfrac{dy}{dx} = \dfrac{s-y}{0-x}$ (see the figure).

Thus, $s = y - x\,\dfrac{dy}{dx} \;\Rightarrow\; \dfrac{ds}{dx} = \dfrac{dy}{dx} - \left(x\,\dfrac{d^2y}{dx^2} + 1\,\dfrac{dy}{dx}\right) = -x\,\dfrac{d^2y}{dx^2}$. Equating the two expressions for $\dfrac{ds}{dx}$ gives us

$x\,\dfrac{d^2y}{dx^2} = \sqrt{1 + \left(\dfrac{dy}{dx}\right)^2}$, as claimed.

(b) Letting $z = \dfrac{dy}{dx}$, we obtain the differential equation $x\,\dfrac{dz}{dx} = \sqrt{1+z^2}$, or $\dfrac{dz}{\sqrt{1+z^2}} = \dfrac{dx}{x}$. Integrating:

$\ln x = \displaystyle\int \frac{dz}{\sqrt{1+z^2}} \overset{25}{=} \ln\left(z + \sqrt{1+z^2}\right) + C$. When $x = L$, $z = dy/dx = 0$, so $\ln L = \ln 1 + C$. Therefore, $C = \ln L$,

so $\ln x = \ln\left(\sqrt{1+z^2} + z\right) + \ln L = \ln\left[L\left(\sqrt{1+z^2} + z\right)\right] \;\Rightarrow\; x = L\left(\sqrt{1+z^2} + z\right) \;\Rightarrow\; \sqrt{1+z^2} = \dfrac{x}{L} - z \;\Rightarrow$

$1 + z^2 = \left(\dfrac{x}{L}\right)^2 - \dfrac{2xz}{L} + z^2 \;\Rightarrow\; \left(\dfrac{x}{L}\right)^2 - 2z\left(\dfrac{x}{L}\right) - 1 = 0 \;\Rightarrow\; z = \dfrac{(x/L)^2 - 1}{2(x/L)} = \dfrac{x^2 - L^2}{2Lx} = \dfrac{x}{2L} - \dfrac{L}{2}\dfrac{1}{x}$

[for $x > 0$]. Since $z = \dfrac{dy}{dx}$, $y = \dfrac{x^2}{4L} - \dfrac{L}{2}\ln x + C_1$. Since $y = 0$ when $x = L$, $0 = \dfrac{L}{4} - \dfrac{L}{2}\ln L + C_1 \;\Rightarrow$

$C_1 = \dfrac{L}{2}\ln L - \dfrac{L}{4}$. Thus, $y = \dfrac{x^2}{4L} - \dfrac{L}{2}\ln x + \dfrac{L}{2}\ln L - \dfrac{L}{4} = \dfrac{x^2 - L^2}{4L} - \dfrac{L}{2}\ln\left(\dfrac{x}{L}\right)$.

(c) As $x \to 0^+$, $y \to \infty$, so the dog never catches the rabbit.

8. (a) If the dog runs twice as fast as the rabbit, then the rabbit's position when the dog has traveled a distance $s$ is $(0, s/2)$.

Since the dog runs straight toward the rabbit, the tangent line to the dog's path has slope $\dfrac{dy}{dx} = \dfrac{s/2 - y}{0 - x}$.

Thus, $s = 2y - 2x\,\dfrac{dy}{dx} \;\Rightarrow\; \dfrac{ds}{dx} = 2\,\dfrac{dy}{dx} - \left(2x\,\dfrac{d^2y}{dx^2} + 2\,\dfrac{dy}{dx}\right) = -2x\,\dfrac{d^2y}{dx^2}$.

From Problem 7(a), $\dfrac{ds}{dx} = -\sqrt{1 + \left(\dfrac{dy}{dx}\right)^2}$, so $2x\,\dfrac{d^2y}{dx^2} = \sqrt{1 + \left(\dfrac{dy}{dx}\right)^2}$.

Letting $z = \dfrac{dy}{dx}$, we obtain the differential equation $2x\,\dfrac{dz}{dx} = \sqrt{1+z^2}$, or $\dfrac{2\,dz}{\sqrt{1+z^2}} = \dfrac{dx}{x}$.

Integrating, we get $\ln x = \int \frac{2\,dz}{\sqrt{1+z^2}} = 2\ln\left(\sqrt{1+z^2}+z\right)+C$. [See Problem 7(b).]

When $x = L$, $z = dy/dx = 0$, so $\ln L = 2\ln 1 + C = C$. Thus,

$\ln x = 2\ln\left(\sqrt{1+z^2}+z\right)+\ln L = \ln\left(L\left(\sqrt{1+z^2}+z\right)^2\right) \Rightarrow x = L\left(\sqrt{1+z^2}+z\right)^2 \Rightarrow$

$\sqrt{1+z^2} = \sqrt{\frac{x}{L}} - z \Rightarrow 1+z^2 = \frac{x}{L} - 2\sqrt{\frac{x}{L}}\,z + z^2 \Rightarrow 2\sqrt{\frac{x}{L}}\,z = \frac{x}{L} - 1 \Rightarrow$

$\frac{dy}{dx} = z = \frac{1}{2}\sqrt{\frac{x}{L}} - \frac{1}{2\sqrt{x/L}} = \frac{1}{2\sqrt{L}}x^{1/2} - \frac{\sqrt{L}}{2}x^{-1/2} \Rightarrow y = \frac{1}{3\sqrt{L}}x^{3/2} - \sqrt{L}\,x^{1/2} + C_1.$

When $x = L$, $y = 0$, so $0 = \frac{1}{3\sqrt{L}}L^{3/2} - \sqrt{L}\,L^{1/2} + C_1 = \frac{L}{3} - L + C_1 = C_1 - \frac{2}{3}L$. Therefore, $C_1 = \frac{2}{3}L$ and

$y = \frac{x^{3/2}}{3\sqrt{L}} - \sqrt{L}\,x^{1/2} + \frac{2}{3}L$. As $x \to 0$, $y \to \frac{2}{3}L$, so the dog catches the rabbit when the rabbit is at $\left(0, \frac{2}{3}L\right)$.

(At that point, the dog has traveled a distance of $\frac{4}{3}L$, twice as far as the rabbit has run.)

(b) As in the solutions to part (a) and Problem 7, we get $z = \frac{dy}{dx} = \frac{x^2}{2L^2} - \frac{L^2}{2x^2}$ and hence $y = \frac{x^3}{6L^2} + \frac{L^2}{2x} - \frac{2}{3}L$.

We want to minimize the distance $D$ from the dog at $(x, y)$ to the rabbit at $(0, 2s)$. Now $s = \frac{1}{2}y - \frac{1}{2}x\frac{dy}{dx} \Rightarrow$

$2s = y - xz \Rightarrow y - 2s = xz = x\left(\frac{x^2}{2L^2} - \frac{L^2}{2x^2}\right) = \frac{x^3}{2L^2} - \frac{L^2}{2x}$, so

$$D = \sqrt{(x-0)^2 + (y-2s)^2} = \sqrt{x^2 + \left(\frac{x^3}{2L^2} - \frac{L^2}{2x}\right)^2}$$

$$= \sqrt{\frac{x^6}{4L^4} + \frac{x^2}{2} + \frac{L^4}{4x^2}} = \sqrt{\left(\frac{x^3}{2L^2} + \frac{L^2}{2x}\right)^2} = \frac{x^3}{2L^2} + \frac{L^2}{2x}$$

$D' = 0 \Leftrightarrow \frac{3x^2}{2L^2} - \frac{L^2}{2x^2} = 0 \Leftrightarrow \frac{3x^2}{2L^2} = \frac{L^2}{2x^2} \Leftrightarrow x^4 = \frac{L^4}{3} \Leftrightarrow x = \frac{L}{\sqrt[4]{3}}$, $x > 0$, $L > 0$.

Since $D''(x) = \frac{3x}{L^2} + \frac{L^2}{x^3} > 0$ for all $x > 0$, we know that $D\left(\frac{L}{\sqrt[4]{3}}\right) = \frac{\left(L \cdot 3^{-1/4}\right)^3}{2L^2} + \frac{L^2}{2(L \cdot 3^{-1/4})} = \frac{2L}{3^{3/4}}$ is the minimum value of $D$, that is, the closest the dog gets to the rabbit. The positions at this distance are

Dog: $(x, y) = \left(\frac{L}{\sqrt[4]{3}}, \left(\frac{5}{3^{7/4}} - \frac{2}{3}\right)L\right) = \left(\frac{L}{\sqrt[4]{3}}, \frac{5\sqrt[4]{3}-6}{9}L\right)$

Rabbit: $(0, 2s) = \left(0, \frac{8\sqrt[4]{3}L}{9} - \frac{2L}{3}\right) = \left(0, \frac{8\sqrt[4]{3}-6}{9}L\right)$

**9.** (a) We are given that $V = \frac{1}{3}\pi r^2 h$, $dV/dt = 60{,}000\pi$ ft$^3$/h, and $r = 1.5h = \frac{3}{2}h$. So $V = \frac{1}{3}\pi\left(\frac{3}{2}h\right)^2 h = \frac{3}{4}\pi h^3 \Rightarrow$

$\frac{dV}{dt} = \frac{3}{4}\pi \cdot 3h^2\frac{dh}{dt} = \frac{9}{4}\pi h^2\frac{dh}{dt}$. Therefore, $\frac{dh}{dt} = \frac{4(dV/dt)}{9\pi h^2} = \frac{240{,}000\pi}{9\pi h^2} = \frac{80{,}000}{3h^2}$ $(\star) \Rightarrow$

$\int 3h^2\,dh = \int 80{,}000\,dt \Rightarrow h^3 = 80{,}000t + C$. When $t = 0$, $h = 60$. Thus, $C = 60^3 = 216{,}000$, so

$h^3 = 80{,}000t + 216{,}000$. Let $h = 100$. Then $100^3 = 1{,}000{,}000 = 80{,}000t + 216{,}000 \Rightarrow 80{,}000t = 784{,}000 \Rightarrow$

$t = 9.8$, so the time required is 9.8 hours.

(b) The floor area of the silo is $F = \pi \cdot 200^2 = 40{,}000\pi$ ft$^2$, and the area of the base of the pile is

$A = \pi r^2 = \pi\left(\frac{3}{2}h\right)^2 = \frac{9\pi}{4}h^2$. So the area of the floor which is not covered when $h = 60$ is

$F - A = 40{,}000\pi - 8100\pi = 31{,}900\pi \approx 100{,}217$ ft$^2$. Now $A = \frac{9\pi}{4}h^2 \Rightarrow dA/dt = \frac{9\pi}{4} \cdot 2h\,(dh/dt)$,

and from $(\star)$ in part (a) we know that when $h = 60$, $dh/dt = \frac{80{,}000}{3(60)^2} = \frac{200}{27}$ ft/h. Therefore,

$dA/dt = \frac{9\pi}{4}(2)(60)\left(\frac{200}{27}\right) = 2000\pi \approx 6283$ ft$^2$/h.

(c) At $h = 90$ ft, $dV/dt = 60{,}000\pi - 20{,}000\pi = 40{,}000\pi$ ft$^3$/h. From ($\star$) in part (a),
$\frac{dh}{dt} = \frac{4(dV/dt)}{9\pi h^2} = \frac{4(40{,}000\pi)}{9\pi h^2} = \frac{160{,}000}{9h^2} \Rightarrow \int 9h^2\,dh = \int 160{,}000\,dt \Rightarrow 3h^3 = 160{,}000t + C$. When $t = 0$, $h = 90$; therefore, $C = 3 \cdot 729{,}000 = 2{,}187{,}000$. So $3h^3 = 160{,}000t + 2{,}187{,}000$. At the top, $h = 100 \Rightarrow$ $3(100)^3 = 160{,}000t + 2{,}187{,}000 \Rightarrow t = \frac{813{,}000}{160{,}000} \approx 5.1$. The pile reaches the top after about 5.1 h.

**10.** Let $P(a, b)$ be any first-quadrant point on the curve $y = f(x)$. The tangent line at $P$ has equation $y - b = f'(a)(x - a)$, or equivalently, $y = mx + b - ma$, where $m = f'(a)$. If $Q(0, c)$ is the $y$-intercept, then $c = b - am$. If $R(k, 0)$ is the $x$-intercept, then $k = \frac{am - b}{m} = a - \frac{b}{m}$. Since the tangent line is bisected at $P$, we know that $|PQ| = |PR|$; that is, $\sqrt{(a-0)^2 + [b - (b - am)]^2} = \sqrt{[a - (a - b/m)]^2 + (b-0)^2}$. Squaring and simplifying gives us $a^2 + a^2m^2 = b^2/m^2 + b^2 \Rightarrow a^2m^2 + a^2m^4 = b^2 + b^2m^2 \Rightarrow a^2m^4 + (a^2 - b^2)m^2 - b^2 = 0 \Rightarrow$ $(a^2m^2 - b^2)(m^2 + 1) = 0 \Rightarrow m^2 = b^2/a^2$. Since $m$ is the slope of the line from a positive $y$-intercept to a positive $x$-intercept, $m$ must be negative. Since $a$ and $b$ are positive, we have $m = -b/a$, so we will solve the equivalent differential equation $\frac{dy}{dx} = -\frac{y}{x} \Rightarrow \frac{dy}{y} = -\frac{dx}{x} \Rightarrow \int \frac{dy}{y} = -\int \frac{dx}{x} \Rightarrow \ln y = -\ln x + C \; [x, y > 0] \Rightarrow$
$y = e^{-\ln x + C} = e^{\ln x^{-1}} \cdot e^C = x^{-1} \cdot A \Rightarrow y = A/x$. Since the point $(3, 2)$ is on the curve, $3 = A/2 \Rightarrow A = 6$ and the curve is $y = 6/x$ with $x > 0$.

**11.** Let $P(a, b)$ be any point on the curve. If $m$ is the slope of the tangent line at $P$, then $m = y'(a)$, and an equation of the normal line at $P$ is $y - b = -\frac{1}{m}(x - a)$, or equivalently, $y = -\frac{1}{m}x + b + \frac{a}{m}$. The $y$-intercept is always 6, so $b + \frac{a}{m} = 6 \Rightarrow \frac{a}{m} = 6 - b \Rightarrow m = \frac{a}{6-b}$. We will solve the equivalent differential equation $\frac{dy}{dx} = \frac{x}{6-y} \Rightarrow$
$(6-y)\,dy = x\,dx \Rightarrow \int (6-y)\,dy = \int x\,dx \Rightarrow 6y - \frac{1}{2}y^2 = \frac{1}{2}x^2 + C \Rightarrow 12y - y^2 = x^2 + K$.
Since $(3, 2)$ is on the curve, $12(2) - 2^2 = 3^2 + K \Rightarrow K = 11$. So the curve is given by $12y - y^2 = x^2 + 11 \Rightarrow$ $x^2 + y^2 - 12y + 36 = -11 + 36 \Rightarrow x^2 + (y-6)^2 = 25$, a circle with center $(0, 6)$ and radius 5.

**12.** Suppose $C$ is a curve with the required property and let $P = (x_0, y_0)$ be a point on $C$. The equation of the normal line to $C$ at $P$ is $y - y_0 = -\frac{1}{y_0'}(x - x_0)$, where $y_0'$ is the value of $\frac{dy}{dx}$ at $x = x_0$. This equation makes sense only if $y_0' \neq 0$. If $y_0' = 0$, then the normal line at $P$ is $x = x_0$, which does not intersect the $y$-axis at all unless $x_0 = 0$.

So let's assume that $y_0' \neq 0$. Then the normal line to $C$ at $P$ intersects the $x$-axis at $(x_0 + y_0y_0', 0)$, and it intersects the $y$-axis at $(0, y_0 + x_0/y_0')$. The condition on $C$ implies that

$$[\text{distance from } P(x_0, y_0) \text{ to } (0, y_0 + x_0/y_0')] = [\text{distance from } (0, y_0 + x_0/y_0') \text{ to } (x_0 + y_0y_0', 0)]$$

$$\sqrt{(0 - x_0)^2 + (y_0 + x_0/y_0' - y_0)^2} = \sqrt{(x_0 + y_0y_0' - 0)^2 + [0 - (y_0 + x_0/y_0')]^2}$$

Squaring both sides, we get $x_0^2 + x_0^2/(y_0')^2 = (x_0 + y_0y_0')^2 + (y_0 + x_0/y_0')^2$ or
$x_0^2 + \frac{x_0^2}{(y_0')^2} = x_0^2 + 2x_0y_0y_0' + y_0^2(y_0')^2 + y_0^2 + 2\frac{x_0y_0}{y_0'} + \frac{x_0^2}{(y_0')^2}$. Subtracting $x_0^2 + \frac{x_0^2}{(y_0')^2}$ from both sides and multiplying by $y_0'$, we get

$$0 = y_0^2y_0' + y_0^2(y_0')^3 + 2x_0y_0\left[1 + (y_0')^2\right] = y_0\left\{y_0y_0' + y_0(y_0')^3 + 2x_0\left[1 + (y_0')^2\right]\right\}$$
$$= y_0\left\{y_0y_0'\left[1 + (y_0')^2\right] + 2x_0\left[1 + (y_0')^2\right]\right\} = y_0(y_0y_0' + 2x_0)\left[1 + (y_0')^2\right]$$

Since $1 + (y_0')^2 \geq 1 > 0$, we conclude that $y_0(y_0y_0' + 2x_0) = 0$. Now $P$ is an arbitrary point on $C$ for which $y_0' \neq 0$. Thus, we have shown that $y(yy' + 2x) = 0$ for points $(x, y)$ along $C$ where $y' \neq 0$. One solution of this equation is $y = 0$, but that curve (the $x$-axis) doesn't satisfy the condition required of $C$, since its normal lines at points for $x \neq 0$ don't intersect the $y$-axis. Thus, we can focus our attention on points of $C$ where $y \neq 0$, and conclude that $yy' + 2x = 0$ at points of $C$ where $y \neq 0$ and $y' \neq 0$. Integrating both sides of $yy' + 2x = 0$, we get $\frac{1}{2}y^2 + x^2 = c$. Clearly $c > 0$ (since $y \neq 0$), so we can write $c = a^2$, where $a = \sqrt{c} > 0$. Thus, $\frac{1}{2}y^2 + x^2 = a^2$ and $x^2/a^2 + y^2/(\sqrt{2}a)^2 = 1$. This shows that $C$ is (part of) the ellipse centered at $(0, 0)$ with semimajor axis $\sqrt{2}\,a$ in the $y$-direction and semiminor axis $a$ in the $x$-direction.

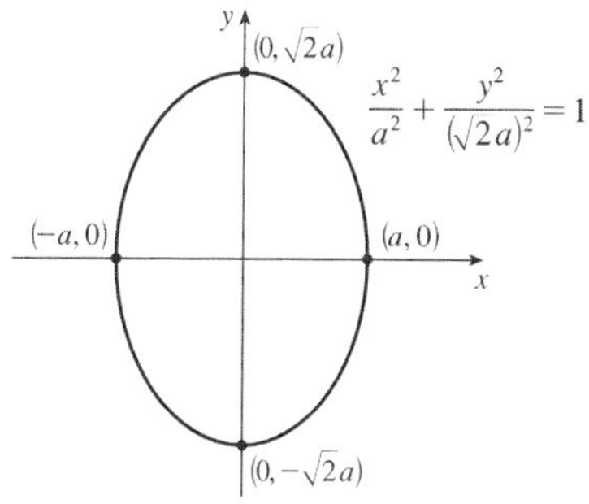


The points of $C$ where $y = 0$ or $y' = 0$ are the vertices $(0, \pm\sqrt{2}\,a)$ and $(\pm a, 0)$. At these points, the condition on $C$ is satisfied in a degenerate way. [When $P = (\pm a, 0)$, the normal line at $P$ *is* the $x$-axis, so *all* the points of the normal line can be viewed as points of intersection with the $x$-axis. The intersection with the $y$-axis at $(0, 0)$ is midway between $(a, 0)$ and $(-a, 0)$; one of these points is $P$, and the other can be regarded as an intersection of the normal line with the $x$-axis. Similarly, when $P = (0, \pm\sqrt{2}\,a)$, the normal line is the $y$-axis, and the point $(0, \pm\sqrt{2}\,a/2)$, which can be regarded as an intersection of the normal line with the $y$-axis, is midway between $P$ and $(0, 0)$, the intersection with the $x$-axis.]

Conversely, if $C$ is part of the ellipse $\dfrac{x^2}{a^2} + \dfrac{y^2}{2a^2} = 1$ for some $a > 0$, then the normal line at a point $(x_0, y_0)$ of $C$ (other than the four vertices) has equation $y - y_0 = \dfrac{y_0}{2x_0}(x - x_0)$. Its intersections with the coordinate axes are $\left(0, \dfrac{y_0}{2}\right)$ and $(-x_0, 0)$. $\left[\text{distance from } (x_0, y_0) \text{ to } \left(0, \dfrac{y_0}{2}\right)\right]^2 = x_0^2 + \dfrac{y_0^2}{4}$ and $\left[\text{distance from } \left(0, \dfrac{y_0}{2}\right) \text{ to } (-x_0, 0)\right]^2 = x_0^2 + \dfrac{y_0^2}{4}$, so the required condition is met at points other than the four vertices. As we have explained, if we are willing to interpret the condition broadly, then it can be viewed as holding even at the four vertices.

*Another method:* Let $P(x_0, y_0)$ be a point on the curve. Since the midpoint of the line segment determined by the normal line from $(x_0, y_0)$ to its intersection with the $x$-axis has $x$-coordinate 0, the $x$-coordinate of the point of intersection with the $x$-axis must be $-x_0$. Hence, the normal line has slope $\dfrac{y_0 - 0}{x_0 - (-x_0)} = \dfrac{y_0}{2x_0}$. So the tangent line has slope $-\dfrac{2x_0}{y_0}$. This gives the differential equation $y' = -\dfrac{2x}{y} \Rightarrow y\,dy = -2x\,dx \Rightarrow \int y\,dy = \int(-2x)\,dx \Rightarrow \frac{1}{2}y^2 = -x^2 + C \Rightarrow$ $x^2 + \frac{1}{2}y^2 = C$ $(C > 0)$.

# 8 □ INFINITE SEQUENCES AND SERIES

## 8.1 Sequences

**1.** (a) A sequence is an ordered list of numbers. It can also be defined as a function whose domain is the set of positive integers.

(b) The terms $a_n$ approach 8 as $n$ becomes large. In fact, we can make $a_n$ as close to 8 as we like by taking $n$ sufficiently large.

(c) The terms $a_n$ become large as $n$ becomes large. In fact, we can make $a_n$ as large as we like by taking $n$ sufficiently large.

**2.** (a) From Definition 1, a convergent sequence is a sequence for which $\lim_{n\to\infty} a_n$ exists. Examples: $\{1/n\}$, $\{1/2^n\}$

(b) A divergent sequence is a sequence for which $\lim_{n\to\infty} a_n$ *does not* exist. Examples: $\{n\}$, $\{\sin n\}$

**3.** The first six terms of $a_n = \dfrac{n}{2n+1}$ are $\dfrac{1}{3}, \dfrac{2}{5}, \dfrac{3}{7}, \dfrac{4}{9}, \dfrac{5}{11}, \dfrac{6}{13}$. It appears that the sequence is approaching $\dfrac{1}{2}$.

$\displaystyle\lim_{n\to\infty} \frac{n}{2n+1} = \lim_{n\to\infty} \frac{1}{2+1/n} = \frac{1}{2}$

**4.** $\{\cos(n\pi/3)\}_{n=1}^{9} = \left\{\frac{1}{2}, -\frac{1}{2}, -1, -\frac{1}{2}, \frac{1}{2}, 1, \frac{1}{2}, -\frac{1}{2}, -1\right\}$. The sequence does not appear to have a limit. The values will cycle through the first six numbers in the sequence—never approaching a particular number.

**5.** $\left\{1, -\frac{2}{3}, \frac{4}{9}, -\frac{8}{27}, \ldots\right\}$. Each term is $-\frac{2}{3}$ times the preceding one, so $a_n = \left(-\frac{2}{3}\right)^{n-1}$.

**6.** $\left\{-\frac{1}{4}, \frac{2}{9}, -\frac{3}{16}, \frac{4}{25}, \ldots\right\}$. The numerator of the $n$th term is $n$ and its denominator is $(n+1)^2$. Including the alternating signs, we get $a_n = (-1)^n \dfrac{n}{(n+1)^2}$.

**7.** $\{2, 7, 12, 17, \ldots\}$. Each term is larger than the preceding one by 5, so $a_n = a_1 + d(n-1) = 2 + 5(n-1) = 5n - 3$.

**8.** $\{5, 1, 5, 1, 5, 1, \ldots\}$. The average of 5 and 1 is 3, so we can think of the sequence as alternately adding 2 and $-2$ to 3. Thus, $a_n = 3 + (-1)^{n+1} \cdot 2$.

**9.** $a_n = \dfrac{3+5n^2}{n+n^2} = \dfrac{(3+5n^2)/n^2}{(n+n^2)/n^2} = \dfrac{5+3/n^2}{1+1/n}$, so $a_n \to \dfrac{5+0}{1+0} = 5$ as $n \to \infty$. Converges

**10.** $a_n = \dfrac{n+1}{3n-1} = \dfrac{1+1/n}{3-1/n}$, so $a_n \to \dfrac{1+0}{3-0} = \dfrac{1}{3}$ as $n \to \infty$. Converges

**11.** $a_n = \dfrac{2^n}{3^{n+1}} = \dfrac{1}{3}\left(\dfrac{2}{3}\right)^n$, so $\lim_{n\to\infty} a_n = \frac{1}{3} \lim_{n\to\infty} \left(\frac{2}{3}\right)^n = \frac{1}{3} \cdot 0 = 0$ by (8) with $r = \frac{2}{3}$. Converges

**12.** $a_n = \dfrac{\sqrt{n}}{1+\sqrt{n}} = \dfrac{1}{1/\sqrt{n}+1}$, so $a_n \to \dfrac{1}{0+1} = 1$ as $n \to \infty$. Converges

**13.** $a_n = \dfrac{(n+2)!}{n!} = \dfrac{(n+2)(n+1)n!}{n!} = (n+2)(n+1)$, so $a_n \to \infty$ as $n \to \infty$ and the sequence diverges.

**14.** $a_n = \dfrac{n}{1+\sqrt{n}} = \dfrac{\sqrt{n}}{1/\sqrt{n}+1}$. The numerator approaches $\infty$ and the denominator approaches $0 + 1 = 1$ as $n \to \infty$, so $a_n \to \infty$ as $n \to \infty$ and the sequence diverges.

**15.** $a_n = \dfrac{(-1)^{n-1} n}{n^2+1} = \dfrac{(-1)^{n-1}}{n+1/n}$, so $0 \le |a_n| = \dfrac{1}{n+1/n} \le \dfrac{1}{n} \to 0$ as $n \to \infty$, so $a_n \to 0$ by the Squeeze Theorem and Theorem 4. Converges

**16.** $a_n = \dfrac{(-1)^n n^3}{n^3 + 2n^2 + 1}$. Now $|a_n| = \dfrac{n^3}{n^3 + 2n^2 + 1} = \dfrac{1}{1 + \frac{2}{n} + \frac{1}{n^3}} \to 1$ as $n \to \infty$, but the terms of the sequence $\{a_n\}$ alternate in sign, so the sequence $a_1, a_3, a_5, \ldots$ converges to $-1$ and the sequence $a_2, a_4, a_6, \ldots$ converges to $+1$. This shows that the given sequence diverges since its terms don't approach a single real number.

**17.** $a_n = \dfrac{e^n + e^{-n}}{e^{2n} - 1} \cdot \dfrac{e^{-n}}{e^{-n}} = \dfrac{1 + e^{-2n}}{e^n - e^{-n}} \to \dfrac{1 + 0}{e^n - 0} \to 0$ as $n \to \infty$. Converges

**18.** $a_n = \cos(2/n)$. As $n \to \infty$, $2/n \to 0$, so $\cos(2/n) \to \cos 0 = 1$. Converges

**19.** $a_n = n^2 e^{-n} = \dfrac{n^2}{e^n}$. Since $\lim\limits_{x\to\infty} \dfrac{x^2}{e^x} \overset{H}{=} \lim\limits_{x\to\infty} \dfrac{2x}{e^x} \overset{H}{=} \lim\limits_{x\to\infty} \dfrac{2}{e^x} = 0$, it follows from Theorem 3 that $\lim\limits_{n\to\infty} a_n = 0$. Converges

**20.** $2n \to \infty$ as $n \to \infty$, so since $\lim\limits_{x\to\infty} \arctan x = \frac{\pi}{2}$, we have $\lim\limits_{n\to\infty} \arctan 2n = \frac{\pi}{2}$. Converges

**21.** $0 \le \dfrac{\cos^2 n}{2^n} \le \dfrac{1}{2^n}$ [since $0 \le \cos^2 n \le 1$], so since $\lim\limits_{n\to\infty} \dfrac{1}{2^n} = 0$, $\left\{\dfrac{\cos^2 n}{2^n}\right\}$ converges to 0 by the Squeeze Theorem.

**22.** $a_n = n \cos n\pi = n(-1)^n$. Since $|a_n| = n \to \infty$ as $n \to \infty$, the given sequence diverges.

**23.** $y = \left(1 + \dfrac{2}{x}\right)^x \Rightarrow \ln y = x \ln\left(1 + \dfrac{2}{x}\right)$, so

$$\lim_{x\to\infty} \ln y = \lim_{x\to\infty} \frac{\ln(1 + 2/x)}{1/x} \overset{H}{=} \lim_{x\to\infty} \frac{\left(\dfrac{1}{1 + 2/x}\right)\left(-\dfrac{2}{x^2}\right)}{-1/x^2} = \lim_{x\to\infty} \frac{2}{1 + 2/x} = 2 \Rightarrow$$

$\lim\limits_{x\to\infty}\left(1 + \dfrac{2}{x}\right)^x = \lim\limits_{x\to\infty} e^{\ln y} = e^2$, so by Theorem 2, $\lim\limits_{n\to\infty}\left(1 + \dfrac{2}{n}\right)^n = e^2$. Convergent

**24.** $a_n = \dfrac{\sin 2n}{1 + \sqrt{n}}$. $|a_n| \le \dfrac{1}{1 + \sqrt{n}}$ and $\lim\limits_{n\to\infty} \dfrac{1}{1 + \sqrt{n}} = 0$, so $\dfrac{-1}{1 + \sqrt{n}} \le a_n \le \dfrac{1}{1 + \sqrt{n}} \Rightarrow \lim\limits_{n\to\infty} a_n = 0$ by the Squeeze Theorem. Converges

**25.** $\{0, 1, 0, 0, 1, 0, 0, 0, 1, \ldots\}$ diverges since the sequence takes on only two values, 0 and 1, and never stays arbitrarily close to either one (or any other value) for $n$ sufficiently large.

**26.** $\lim\limits_{x\to\infty} \dfrac{(\ln x)^2}{x} \overset{H}{=} \lim\limits_{x\to\infty} \dfrac{2(\ln x)(1/x)}{1} = 2 \lim\limits_{x\to\infty} \dfrac{\ln x}{x} \overset{H}{=} 2 \lim\limits_{x\to\infty} \dfrac{1/x}{1} = 0$, so by Theorem 2, $\lim\limits_{n\to\infty} \dfrac{(\ln n)^2}{n} = 0$. Convergent

**27.** $a_n = \ln(2n^2 + 1) - \ln(n^2 + 1) = \ln\left(\dfrac{2n^2 + 1}{n^2 + 1}\right) = \ln\left(\dfrac{2 + 1/n^2}{1 + 1/n^2}\right) \to \ln 2$ as $n \to \infty$. Convergent

**28.** $0 < |a_n| = \dfrac{3^n}{n!} = \dfrac{3}{1} \cdot \dfrac{3}{2} \cdot \dfrac{3}{3} \cdot \cdots \cdot \dfrac{3}{(n-1)} \cdot \dfrac{3}{n} \le \dfrac{3}{1} \cdot \dfrac{3}{2} \cdot \dfrac{3}{n}$ [for $n > 2$] $= \dfrac{27}{2n} \to 0$ as $n \to \infty$, so by the Squeeze Theorem and Theorem 4, $\{(-3)^n/n\}$ converges to 0.

**29.**

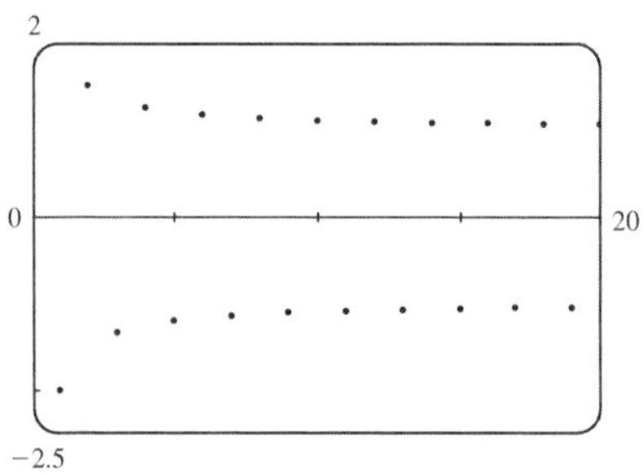


From the graph, we see that the sequence $\left\{(-1)^n \dfrac{n+1}{n}\right\}$ is divergent, since it oscillates between 1 and $-1$ (approximately).

**30.**

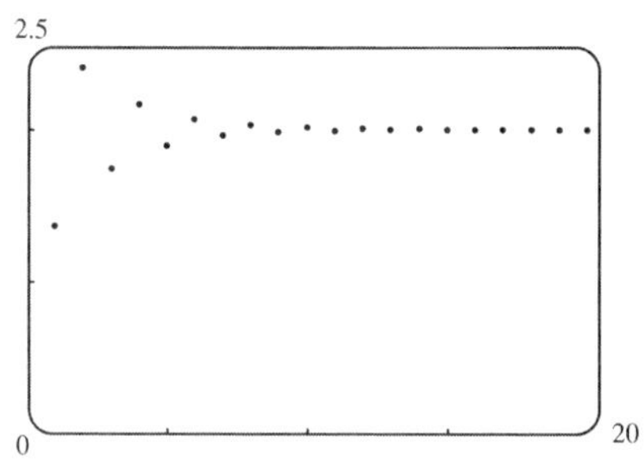


From the graph, it appears that the sequence converges to 2. $\left\{\left(-\frac{2}{\pi}\right)^n\right\}$ converges to 0 by (6), and hence $\left\{2 + \left(-\frac{2}{\pi}\right)^n\right\}$ converges to $2 + 0 = 2$.

**31.** 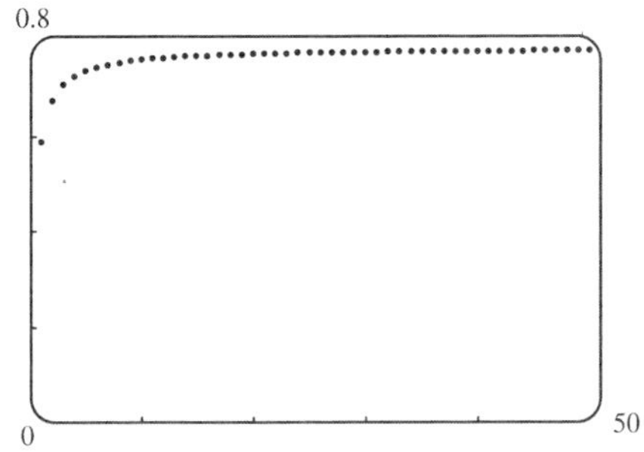


From the graph, it appears that the sequence converges to about 0.78.

$$\lim_{n\to\infty}\frac{2n}{2n+1}=\lim_{n\to\infty}\frac{2}{2+1/n}=1, \text{ so}$$

$$\lim_{n\to\infty}\arctan\left(\frac{2n}{2n+1}\right)=\arctan 1=\frac{\pi}{4}.$$

**32.** 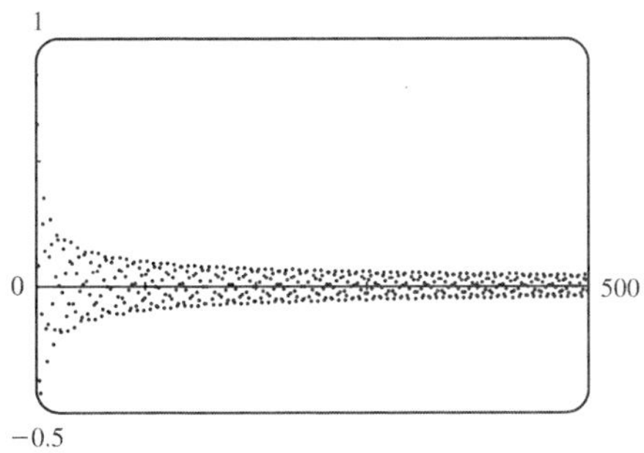


From the graph, it appears that the sequence converges (slowly) to 0.

$0\le\dfrac{|\sin n|}{\sqrt{n}}\le\dfrac{1}{\sqrt{n}}\to 0$ as $n\to\infty$, so by the Squeeze Theorem and Theorem 4, $\left\{\dfrac{\sin n}{\sqrt{n}}\right\}$ converges to 0.

**33.** 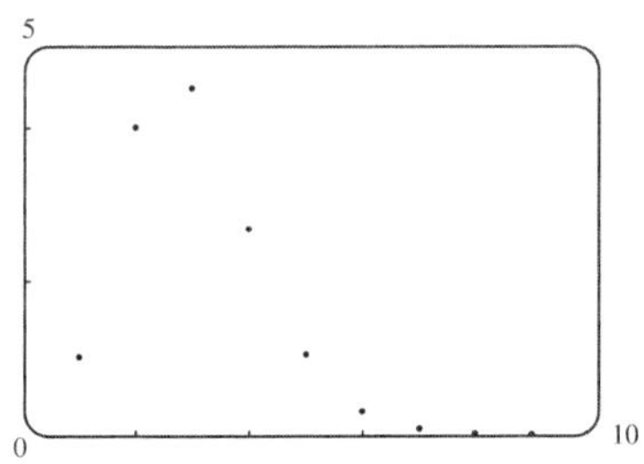


From the graph, it appears that the sequence converges to 0.

$$0<a_n=\frac{n^3}{n!}=\frac{n}{n}\cdot\frac{n}{(n-1)}\cdot\frac{n}{(n-2)}\cdot\frac{1}{(n-3)}\cdot\cdots\cdot\frac{1}{3}\cdot\frac{1}{2}\cdot\frac{1}{1}$$

$$\le\frac{n^2}{(n-1)(n-2)(n-3)}\quad[\text{for } n\ge 4]$$

$$=\frac{1/n}{(1-1/n)(1-2/n)(1-3/n)}\to 0 \text{ as } n\to\infty$$

So by the Squeeze Theorem, $\{n^3/n!\}$ converges to 0.

**34.** 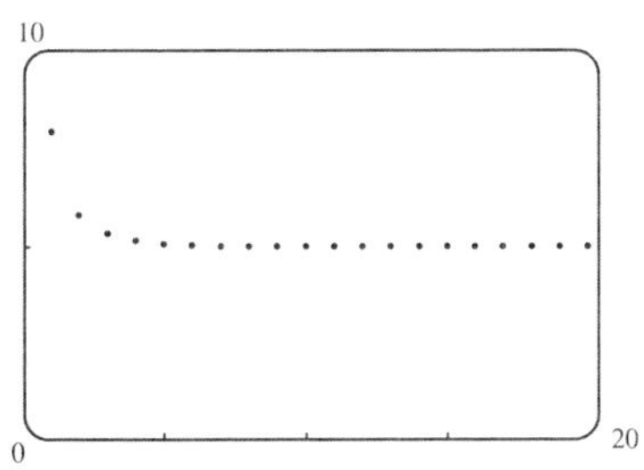


From the graph, it appears that the sequence converges to 5.

$$5=\sqrt[n]{5^n}\le\sqrt[n]{3^n+5^n}\le\sqrt[n]{5^n+5^n}=\sqrt[n]{2}\,\sqrt[n]{5^n}$$

$$=\sqrt[n]{2}\cdot 5\to 5 \text{ as } n\to\infty\qquad\left[\lim_{n\to\infty}2^{1/n}=2^0=1\right]$$

Hence, $a_n\to 5$ by the Squeeze Theorem.

*Alternate solution:* Let $y=(3^x+5^x)^{1/x}$. Then

$$\lim_{x\to\infty}\ln y=\lim_{x\to\infty}\frac{\ln(3^x+5^x)}{x}\overset{\text{H}}{=}\lim_{x\to\infty}\frac{3^x\ln 3+5^x\ln 5}{3^x+5^x}=\lim_{x\to\infty}\frac{\left(\frac{3}{5}\right)^x\ln 3+\ln 5}{\left(\frac{3}{5}\right)^x+1}=\ln 5,$$

so $\lim\limits_{x\to\infty}y=e^{\ln 5}=5$, and so $\left\{\sqrt[n]{3^n+5^n}\right\}$ converges to 5.

**35.** (a) $a_n=1000(1.06)^n \;\Rightarrow\; a_1=1060,\ a_2=1123.60,\ a_3=1191.02,\ a_4=1262.48$, and $a_5=1338.23$.

(b) $\lim\limits_{n\to\infty}a_n=1000\lim\limits_{n\to\infty}(1.06)^n$, so the sequence diverges by (6) with $r=1.06>1$.

**36.** $a_{n+1} = \begin{cases} \frac{1}{2}a_n & \text{if } a_n \text{ is an even number} \\ 3a_n + 1 & \text{if } a_n \text{ is an odd number} \end{cases}$ When $a_1 = 11$, the first 40 terms are 11, 34, 17, 52, 26, 13, 40, 20, 10, 5, 16, 8, 4, 2, 1, 4, 2, 1, 4, 2, 1, 4, 2, 1, 4, 2, 1, 4, 2, 1, 4, 2, 1, 4, 2, 1, 4, 2, 1, 4. When $a_1 = 25$, the first 40 terms are 25, 76, 38, 19, 58, 29, 88, 44, 22, 11, 34, 17, 52, 26, 13, 40, 20, 10, 5, 16, 8, 4, 2, 1, 4, 2, 1, 4, 2, 1, 4, 2, 1, 4, 2, 1, 4, 2, 1, 4. The famous Collatz conjecture is that this sequence always reaches 1, regardless of the starting point $a_1$.

**37.** (a) $a_1 = 1$, $a_2 = 4 - a_1 = 4 - 1 = 3$, $a_3 = 4 - a_2 = 4 - 3 = 1$, $a_4 = 4 - a_3 = 4 - 1 = 3$, $a_5 = 4 - a_4 = 4 - 3 = 1$. Since the terms of the sequence alternate between 1 and 3, the sequence is divergent.

(b) $a_1 = 2$, $a_2 = 4 - a_1 = 4 - 2 = 2$, $a_3 = 4 - a_2 = 4 - 2 = 2$. Since all of the terms are 2, $\lim_{n\to\infty} a_n = 2$ and hence, the sequence is convergent.

**38.** (a) Since $\lim_{n\to\infty} a_n = L$, the terms $a_n$ approach $L$ as $n$ becomes large. Because we can make $a_n$ as close to $L$ as we wish, $a_{n+1}$ will also be close, and so $\lim_{n\to\infty} a_{n+1} = L$.

(b) $a_1 = 1$, $a_2 = \dfrac{1}{1+a_1} = \dfrac{1}{1+1} = \dfrac{1}{2} = 0.5$, $a_3 = \dfrac{1}{1+a_2} = \dfrac{1}{1+\frac{1}{2}} = \dfrac{2}{3} \approx 0.66667$,

$a_4 = \dfrac{1}{1+a_3} = \dfrac{1}{1+\frac{2}{3}} = \dfrac{3}{5} = 0.6$, $a_5 = \dfrac{1}{1+a_4} = \dfrac{1}{1+\frac{3}{5}} = \dfrac{5}{8} = 0.625$,

$a_6 = \dfrac{1}{1+a_5} = \dfrac{1}{1+\frac{5}{8}} = \dfrac{8}{13} \approx 0.61538$, $a_7 = \dfrac{1}{1+a_6} = \dfrac{1}{1+\frac{8}{13}} = \dfrac{13}{21} \approx 0.61905$,

$a_8 = \dfrac{1}{1+a_7} = \dfrac{1}{1+\frac{13}{21}} = \dfrac{21}{34} \approx 0.61765$, $a_9 = \dfrac{1}{1+a_8} = \dfrac{1}{1+\frac{21}{34}} = \dfrac{34}{55} \approx 0.61818$,

$a_{10} = \dfrac{1}{1+a_9} = \dfrac{1}{1+\frac{34}{55}} = \dfrac{55}{89} \approx 0.61800$. It appears that $\lim_{n\to\infty} a_n \approx 0.618$; hence, the sequence is convergent.

(c) If $L = \lim_{n\to\infty} a_n$ then $\lim_{n\to\infty} a_{n+1} = L$ also, so $L$ must satisfy

$L = 1/(1+L) \quad\Rightarrow\quad L^2 + L - 1 = 0 \quad\Rightarrow\quad L = \frac{-1+\sqrt{5}}{2} \approx 0.618$ (since $L$ has to be non-negative if it exists).

**39.** (a) Let $a_n$ be the number of rabbit pairs in the $n$th month. Clearly $a_1 = 1 = a_2$. In the $n$th month, each pair that is 2 or more months old (that is, $a_{n-2}$ pairs) will produce a new pair to add to the $a_{n-1}$ pairs already present. Thus, $a_n = a_{n-1} + a_{n-2}$, so that $\{a_n\} = \{f_n\}$, the Fibonacci sequence.

(b) $a_n = \dfrac{f_{n+1}}{f_n} \quad\Rightarrow\quad a_{n-1} = \dfrac{f_n}{f_{n-1}} = \dfrac{f_{n-1}+f_{n-2}}{f_{n-1}} = 1 + \dfrac{f_{n-2}}{f_{n-1}} = 1 + \dfrac{1}{f_{n-1}/f_{n-2}} = 1 + \dfrac{1}{a_{n-2}}$. If $L = \lim_{n\to\infty} a_n$, then $L = \lim_{n\to\infty} a_{n-1}$ and $L = \lim_{n\to\infty} a_{n-2}$, so $L$ must satisfy $L = 1 + \dfrac{1}{L} \quad\Rightarrow\quad L^2 - L - 1 = 0 \quad\Rightarrow\quad L = \frac{1+\sqrt{5}}{2}$ (since $L$ must be positive).

**40.** $a_1 = 2^{1/2}$, $a_2 = 2^{3/4}$, $a_3 = 2^{7/8}$, ..., so $a_n = 2^{(2^n-1)/2^n} = 2^{1-(1/2^n)}$. $\lim_{n\to\infty} a_n = \lim_{n\to\infty} 2^{1-(1/2^n)} = 2^1 = 2$.

*Alternate solution*: Let $L = \lim_{n\to\infty} a_n$. (We could show the limit exists by showing that $\{a_n\}$ is bounded and increasing.) Then $L$ must satisfy $L = \sqrt{2 \cdot L} \quad\Rightarrow\quad L^2 = 2L \quad\Rightarrow\quad L(L-2) = 0$. $L \neq 0$ since the sequence increases, so $L = 2$.

**41.** $a_n = \dfrac{1}{2n+3}$ is decreasing since $a_{n+1} = \dfrac{1}{2(n+1)+3} = \dfrac{1}{2n+5} < \dfrac{1}{2n+3} = a_n$ for each $n \geq 1$. The sequence is bounded since $0 < a_n \leq \frac{1}{5}$ for all $n \geq 1$. Note that $a_1 = \frac{1}{5}$.

**42.** $a_n = \dfrac{2n-3}{3n+4}$ defines an increasing sequence since for $f(x) = \dfrac{2x-3}{3x+4}$,

$f'(x) = \dfrac{(3x+4)(2)-(2x-3)(3)}{(3x+4)^2} = \dfrac{17}{(3x+4)^2} > 0$. The sequence is bounded since $a_n \geq a_1 = -\frac{1}{7}$ for $n \geq 1$, and

$a_n < \dfrac{2n-3}{3n} < \dfrac{2n}{3n} = \dfrac{2}{3}$ for $n \geq 1$.

**43.** $a_n = \cos(n\pi/2)$ is not monotonic. The first few terms are $0, -1, 0, 1, 0, -1, 0, 1, \ldots$. In fact, the sequence consists of the terms $0, -1, 0, 1$ repeated over and over again in that order. The sequence is bounded since $|a_n| \le 1$ for all $n \ge 1$.

**44.** $a_n = n + \dfrac{1}{n}$ defines an increasing sequence since the function $g(x) = x + \dfrac{1}{x}$ is increasing for $x > 1$. $[g'(x) = 1 - 1/x^2 > 0$ for $x > 1$.] The sequence is unbounded since $a_n \to \infty$ as $n \to \infty$. (It is, however, bounded below by $a_1 = 2$.)

**45.** Since $\{a_n\}$ is a decreasing sequence, $a_n > a_{n+1}$ for all $n \ge 1$. Because all of its terms lie between 5 and 8, $\{a_n\}$ is a bounded sequence. By the Monotonic Sequence Theorem, $\{a_n\}$ is convergent; that is, $\{a_n\}$ has a limit $L$. $L$ must be less than 8 since $\{a_n\}$ is decreasing, so $5 \le L < 8$.

**46.** (a) Let $P_n$ be the statement that $a_{n+1} \ge a_n$ and $a_n \le 3$. $P_1$ is obviously true. We will assume that $P_n$ is true and then show that as a consequence $P_{n+1}$ must also be true. $a_{n+2} \ge a_{n+1} \Leftrightarrow \sqrt{2 + a_{n+1}} \ge \sqrt{2 + a_n} \Leftrightarrow 2 + a_{n+1} \ge 2 + a_n \Leftrightarrow a_{n+1} \ge a_n$, which is the induction hypothesis. $a_{n+1} \le 3 \Leftrightarrow \sqrt{2 + a_n} \le 3 \Leftrightarrow 2 + a_n \le 9 \Leftrightarrow a_n \le 7$, which is certainly true because we are assuming that $a_n \le 3$. So $P_n$ is true for all $n$, and so $a_1 \le a_n \le 3$ (showing that the sequence is bounded), and hence by the Monotonic Sequence Theorem, $\lim\limits_{n\to\infty} a_n$ exists.

(b) If $L = \lim\limits_{n\to\infty} a_n$, then $\lim\limits_{n\to\infty} a_{n+1} = L$ also, so $L = \sqrt{2 + L} \Rightarrow L^2 = 2 + L \Leftrightarrow L^2 - L - 2 = 0 \Leftrightarrow (L + 1)(L - 2) = 0 \Leftrightarrow L = 2$ (since $L$ can't be negative).

**47.** We show by induction that $\{a_n\}$ is increasing and bounded above by 3. Let $P_n$ be the proposition that $a_{n+1} > a_n$ and $0 < a_n < 3$. Clearly $P_1$ is true. Assume that $P_n$ is true.

Then $a_{n+1} > a_n \Rightarrow \dfrac{1}{a_{n+1}} < \dfrac{1}{a_n} \Rightarrow -\dfrac{1}{a_{n+1}} > -\dfrac{1}{a_n}$. Now $a_{n+2} = 3 - \dfrac{1}{a_{n+1}} > 3 - \dfrac{1}{a_n} = a_{n+1} \Leftrightarrow P_{n+1}$.

This proves that $\{a_n\}$ is increasing and bounded above by 3, so $1 = a_1 < a_n < 3$, that is, $\{a_n\}$ is bounded, and hence convergent by the Monotonic Sequence Theorem. If $L = \lim\limits_{n\to\infty} a_n$, then $\lim\limits_{n\to\infty} a_{n+1} = L$ also, so $L$ must satisfy

$L = 3 - 1/L \Rightarrow L^2 - 3L + 1 = 0 \Rightarrow L = \frac{3 \pm \sqrt{5}}{2}$. But $L > 1$, so $L = \frac{3 + \sqrt{5}}{2}$.

**48.** We use induction. Let $P_n$ be the statement that $0 < a_{n+1} \le a_n \le 2$. Clearly $P_1$ is true, since $a_2 = 1/(3 - 2) = 1$. Now assume that $P_n$ is true. Then $a_{n+1} \le a_n \Rightarrow -a_{n+1} \ge -a_n \Rightarrow 3 - a_{n+1} \ge 3 - a_n \Rightarrow$

$a_{n+2} = \dfrac{1}{3 - a_{n+1}} \le \dfrac{1}{3 - a_n} = a_{n+1}$. Also $a_{n+2} > 0$ (since $3 - a_{n+1}$ is positive) and $a_{n+1} \le 2$ by the induction hypothesis, so $P_{n+1}$ is true. To find the limit, we use the fact that $\lim\limits_{n\to\infty} a_n = \lim\limits_{n\to\infty} a_{n+1} \Rightarrow L = \frac{1}{3 - L} \Rightarrow$

$L^2 - 3L + 1 = 0 \Rightarrow L = \frac{3 \pm \sqrt{5}}{2}$. But $L \le 2$, so we must have $L = \frac{3 - \sqrt{5}}{2}$.

**49.** $(0.8)^n < 0.000001 \Rightarrow \ln(0.8)^n < \ln(0.000001) \Rightarrow n\ln(0.8) < \ln(0.000001) \Rightarrow$

$n > \dfrac{\ln(0.000001)}{\ln(0.8)} \Rightarrow n > 61.9$, so $n$ must be at least 62 to satisfy the given inequality.

**50.** (a) If $f$ is continuous, then $f(L) = f\left(\lim\limits_{n\to\infty} a_n\right) = \lim\limits_{n\to\infty} f(a_n) = \lim\limits_{n\to\infty} a_{n+1} = L$ by Exercise 38(a).

(b) By repeatedly pressing the cosine key on the calculator (that is, taking cosine of the previous answer) until the displayed value stabilizes, we see that $L \approx 0.73909$.

**51.** (a) Suppose $\{p_n\}$ converges to $p$. Then $p_{n+1} = \dfrac{bp_n}{a + p_n} \Rightarrow \lim\limits_{n\to\infty} p_{n+1} = \dfrac{b \lim\limits_{n\to\infty} p_n}{a + \lim\limits_{n\to\infty} p_n} \Rightarrow p = \dfrac{bp}{a + p} \Rightarrow$

$p^2 + ap = bp \Rightarrow p(p + a - b) = 0 \Rightarrow p = 0$ or $p = b - a$.

(b) $p_{n+1} = \dfrac{bp_n}{a + p_n} = \dfrac{\left(\dfrac{b}{a}\right)p_n}{1 + \dfrac{p_n}{a}} < \left(\dfrac{b}{a}\right)p_n$ since $1 + \dfrac{p_n}{a} > 1$.

(c) By part (b), $p_1 < \left(\frac{b}{a}\right)p_0$, $p_2 < \left(\frac{b}{a}\right)p_1 < \left(\frac{b}{a}\right)^2 p_0$, $p_3 < \left(\frac{b}{a}\right)p_2 < \left(\frac{b}{a}\right)^3 p_0$, etc. In general, $p_n < \left(\frac{b}{a}\right)^n p_0$,

so $\lim_{n\to\infty} p_n \le \lim_{n\to\infty}\left(\frac{b}{a}\right)^n \cdot p_0 = 0$ since $b < a$. $\left[\text{By result 8, } \lim_{n\to\infty} r^n = 0 \text{ if } -1 < r < 1. \text{ Here } r = \frac{b}{a} \in (0,1).\right]$

(d) Let $a < b$. We first show, by induction, that if $p_0 < b - a$, then $p_n < b - a$ and $p_{n+1} > p_n$.

For $n = 0$, we have $p_1 - p_0 = \dfrac{bp_0}{a + p_0} - p_0 = \dfrac{p_0(b - a - p_0)}{a + p_0} > 0$ since $p_0 < b - a$. So $p_1 > p_0$.

Now we suppose the assertion is true for $n = k$, that is, $p_k < b - a$ and $p_{k+1} > p_k$. Then

$b - a - p_{k+1} = b - a - \dfrac{bp_k}{a + p_k} = \dfrac{a(b-a) + bp_k - ap_k - bp_k}{a + p_k} = \dfrac{a(b - a - p_k)}{a + p_k} > 0$ because $p_k < b - a$. So

$p_{k+1} < b - a$. And $p_{k+2} - p_{k+1} = \dfrac{bp_{k+1}}{a + p_{k+1}} - p_{k+1} = \dfrac{p_{k+1}(b - a - p_{k+1})}{a + p_{k+1}} > 0$ since $p_{k+1} < b - a$. Therefore,

$p_{k+2} > p_{k+1}$. Thus, the assertion is true for $n = k + 1$. It is therefore true for all $n$ by mathematical induction. A similar proof by induction shows that if $p_0 > b - a$, then $p_n > b - a$ and $\{p_n\}$ is decreasing.

In either case the sequence $\{p_n\}$ is bounded and monotonic, so it is convergent by the Monotonic Sequence Theorem. It then follows from part (a) that $\lim_{n\to\infty} p_n = b - a$.

**52.** $a_1 = 1$, $a_2 = 1 + \frac{1}{1+1} = \frac{3}{2} = 1.5$, $a_3 = 1 + \frac{1}{5/2} = \frac{7}{5} = 1.4$, $a_4 = 1 + \frac{1}{12/5} = \frac{17}{12} = 1.41\overline{6}$,

$a_5 = 1 + \frac{1}{29/12} = \frac{41}{29} \approx 1.413793$, $a_6 = 1 + \frac{1}{70/29} = \frac{99}{70} \approx 1.414286$, $a_7 = 1 + \frac{1}{169/70} = \frac{239}{169} \approx 1.414201$,

$a_8 = 1 + \frac{1}{408/169} = \frac{577}{408} \approx 1.414216$. Notice that $a_1 < a_3 < a_5 < a_7$ and $a_2 > a_4 > a_6 > a_8$. It appears that the odd terms are increasing and the even terms are decreasing. Let's prove that $a_{2n-2} > a_{2n}$ and $a_{2n-1} < a_{2n+1}$ by mathematical induction. Suppose that $a_{2k-2} > a_{2k}$. Then $1 + a_{2k-2} > 1 + a_{2k} \quad\Rightarrow$

$$\frac{1}{1 + a_{2k-2}} < \frac{1}{1 + a_{2k}} \quad\Rightarrow\quad 1 + \frac{1}{1 + a_{2k-2}} < 1 + \frac{1}{1 + a_{2k}} \quad\Rightarrow\quad a_{2k-1} < a_{2k+1} \quad\Rightarrow$$

$$1 + a_{2k-1} < 1 + a_{2k+1} \quad\Rightarrow\quad \frac{1}{1 + a_{2k-1}} > \frac{1}{1 + a_{2k+1}} \quad\Rightarrow\quad 1 + \frac{1}{1 + a_{2k-1}} > 1 + \frac{1}{1 + a_{2k+1}} \quad\Rightarrow\quad a_{2k} > a_{2k+2}.$$

We have thus shown, by induction, that the odd terms are increasing and the even terms are decreasing. Also all terms lie between 1 and 2, so both $\{a_n\}$ and $\{b_n\}$ are bounded monotonic sequences and therefore convergent by the Monotonic Sequence Theorem. Let $\lim_{n\to\infty} a_{2n} = L$. Then $\lim_{n\to\infty} a_{2n+2} = L$ also. We have

$a_{n+2} = 1 + \dfrac{1}{1 + 1 + 1/(1 + a_n)} = 1 + \dfrac{1}{(3 + 2a_n)/(1 + a_n)} = \dfrac{4 + 3a_n}{3 + 2a_n}$, so $a_{2n+2} = \dfrac{4 + 3a_{2n}}{3 + 2a_{2n}}$. Taking limits of both

sides, we get $L = \dfrac{4 + 3L}{3 + 2L} \quad\Rightarrow\quad 3L + 2L^2 = 4 + 3L \quad\Rightarrow\quad L^2 = 2 \quad\Rightarrow\quad L = \sqrt{2}$ [since $L > 0$]. Thus,

$\lim_{n\to\infty} a_{2n} = \sqrt{2}$.

Similarly, we find that $\lim_{n\to\infty} a_{2n+1} = \sqrt{2}$. Since the even terms approach $\sqrt{2}$ and the odd terms also approach $\sqrt{2}$, it follows that the sequence as a whole approaches $\sqrt{2}$, that is, $\lim_{n\to\infty} a_n = \sqrt{2}$.

## LABORATORY PROJECT Logistic Sequences

1. To write such a program in Maple it is best to calculate all the points first and then graph them. One possible sequence of commands [taking $p_0 = \frac{1}{2}$ and $k = 1.5$ for the difference equation] is

```
t:='t';p(0):=1/2;k:=1.5;
for j from 1 to 20  do p(j):=k*p(j-1)*(1-p(j-1)) od;
plot([seq([t,p(t)] t=0..20)],t=0..20,p=0..0.5,style=point);
```

In Mathematica, we can use the following program:

```
p[0]=1/2
k=1.5
p[j_]:=k*p[j-1]*(1-p[j-1])
P=Table[p[t],{t,20}]
ListPlot[P]
```

With $p_0 = \frac{1}{2}$ and $k = 1.5$:

| $n$ | $p_n$ | $n$ | $p_n$ | $n$ | $p_n$ |
|---|---|---|---|---|---|
| 0 | 0.5 | 7 | 0.3338465076 | 14 | 0.3333373303 |
| 1 | 0.375 | 8 | 0.3335895255 | 15 | 0.3333353318 |
| 2 | 0.3515625 | 9 | 0.3334613309 | 16 | 0.3333343326 |
| 3 | 0.3419494629 | 10 | 0.3333973076 | 17 | 0.3333338329 |
| 4 | 0.3375300416 | 11 | 0.3333653143 | 18 | 0.3333335831 |
| 5 | 0.3354052689 | 12 | 0.3333493223 | 19 | 0.3333334582 |
| 6 | 0.3343628617 | 13 | 0.3333413274 | 20 | 0.3333333958 |

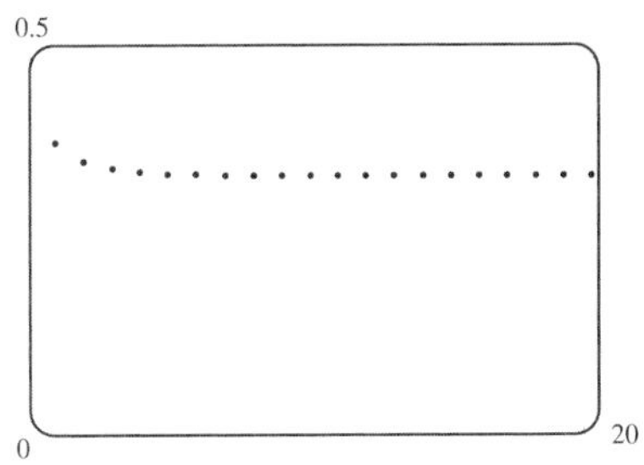


With $p_0 = \frac{1}{2}$ and $k = 2.5$:

| $n$ | $p_n$ | $n$ | $p_n$ | $n$ | $p_n$ |
|---|---|---|---|---|---|
| 0 | 0.5 | 7 | 0.6004164790 | 14 | 0.5999967417 |
| 1 | 0.625 | 8 | 0.5997913269 | 15 | 0.6000016291 |
| 2 | 0.5859375 | 9 | 0.6001042277 | 16 | 0.5999991854 |
| 3 | 0.6065368651 | 10 | 0.5999478590 | 17 | 0.6000004073 |
| 4 | 0.5966247409 | 11 | 0.6000260637 | 18 | 0.5999997964 |
| 5 | 0.6016591486 | 12 | 0.5999869664 | 19 | 0.6000001018 |
| 6 | 0.5991635437 | 13 | 0.6000065164 | 20 | 0.5999999491 |

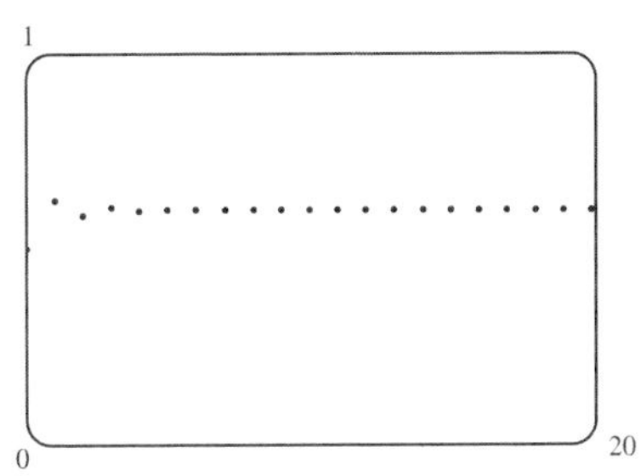


Both of these sequences seem to converge (the first to about $\frac{1}{3}$, the second to about 0.60).

With $p_0 = \frac{7}{8}$ and $k = 1.5$:

| $n$ | $p_n$ | $n$ | $p_n$ | $n$ | $p_n$ |
|---|---|---|---|---|---|
| 0 | 0.875 | 7 | 0.3239166554 | 14 | 0.3332554829 |
| 1 | 0.1640625 | 8 | 0.3284919837 | 15 | 0.3332943990 |
| 2 | 0.2057189941 | 9 | 0.3308775005 | 16 | 0.3333138639 |
| 3 | 0.2450980344 | 10 | 0.3320963702 | 17 | 0.3333235980 |
| 4 | 0.2775374819 | 11 | 0.3327125567 | 18 | 0.3333284655 |
| 5 | 0.3007656421 | 12 | 0.3330223670 | 19 | 0.3333308994 |
| 6 | 0.3154585059 | 13 | 0.3331777051 | 20 | 0.3333321164 |

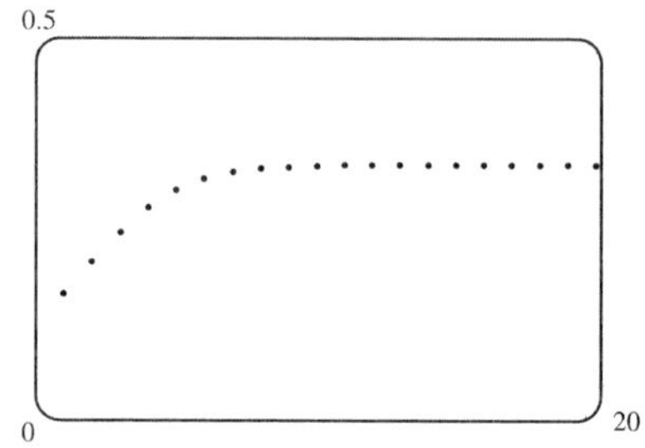


With $p_0 = \frac{7}{8}$ and $k = 2.5$:

| $n$ | $p_n$ | $n$ | $p_n$ | $n$ | $p_n$ |
|---|---|---|---|---|---|
| 0 | 0.875 | 7 | 0.6016572368 | 14 | 0.5999869815 |
| 1 | 0.2734375 | 8 | 0.5991645155 | 15 | 0.6000065088 |
| 2 | 0.4966735840 | 9 | 0.6004159972 | 16 | 0.5999967455 |
| 3 | 0.6249723374 | 10 | 0.5997915688 | 17 | 0.6000016272 |
| 4 | 0.5859547872 | 11 | 0.6001041070 | 18 | 0.5999991864 |
| 5 | 0.6065294364 | 12 | 0.5999479194 | 19 | 0.6000004068 |
| 6 | 0.5966286980 | 13 | 0.6000260335 | 20 | 0.5999997966 |

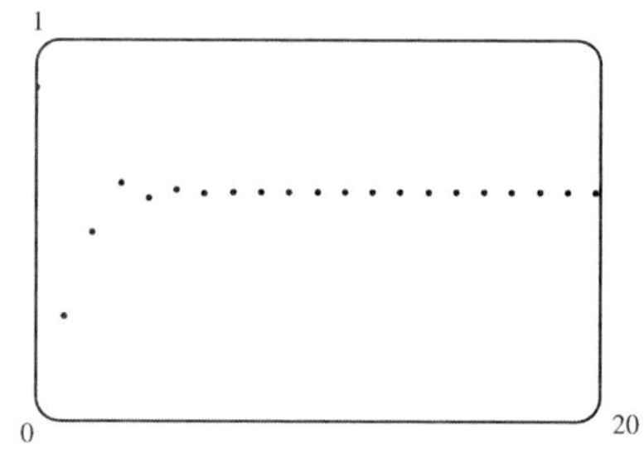


The limit of the sequence seems to depend on $k$, but not on $p_0$.

**2.** With $p_0 = \frac{7}{8}$ and $k = 3.2$:

| $n$ | $p_n$ | $n$ | $p_n$ | $n$ | $p_n$ |
|---|---|---|---|---|---|
| 0 | 0.875 | 7 | 0.5830728495 | 14 | 0.7990633827 |
| 1 | 0.35 | 8 | 0.7779164854 | 15 | 0.5137954979 |
| 2 | 0.728 | 9 | 0.5528397669 | 16 | 0.7993909896 |
| 3 | 0.6336512 | 10 | 0.7910654689 | 17 | 0.5131681132 |
| 4 | 0.7428395416 | 11 | 0.5288988570 | 18 | 0.7994451225 |
| 5 | 0.6112926626 | 12 | 0.7973275394 | 19 | 0.5130643795 |
| 6 | 0.7603646184 | 13 | 0.5171082698 | 20 | 0.7994538304 |

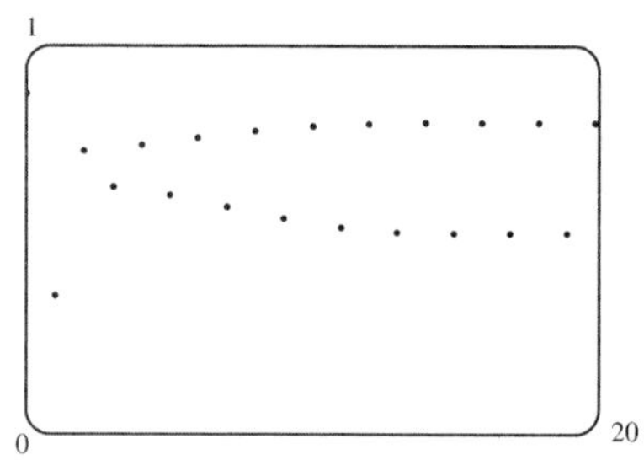


It seems that eventually the terms fluctuate between two values (about 0.5 and 0.8 in this case).

**3.** With $p_0 = \frac{7}{8}$ and $k = 3.42$:

| $n$ | $p_n$ | $n$ | $p_n$ | $n$ | $p_n$ |
|---|---|---|---|---|---|
| 0 | 0.875 | 7 | 0.4523028596 | 14 | 0.8442074951 |
| 1 | 0.3740625 | 8 | 0.8472194412 | 15 | 0.4498025048 |
| 2 | 0.8007579316 | 9 | 0.4426802161 | 16 | 0.8463823232 |
| 3 | 0.5456427596 | 10 | 0.8437633929 | 17 | 0.4446659586 |
| 4 | 0.8478752457 | 11 | 0.4508474156 | 18 | 0.8445284520 |
| 5 | 0.4411212220 | 12 | 0.8467373602 | 19 | 0.4490464985 |
| 6 | 0.8431438501 | 13 | 0.4438243545 | 20 | 0.8461207931 |

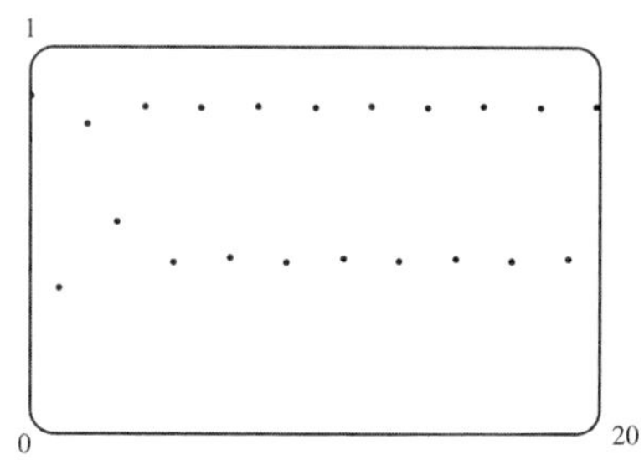

With $p_0 = \frac{7}{8}$ and $k = 3.45$:

| $n$ | $p_n$ | $n$ | $p_n$ | $n$ | $p_n$ |
|---|---|---|---|---|---|
| 0 | 0.875 | 7 | 0.4670259170 | 14 | 0.8403376122 |
| 1 | 0.37734375 | 8 | 0.8587488490 | 15 | 0.4628875685 |
| 2 | 0.8105962830 | 9 | 0.4184824586 | 16 | 0.8577482026 |
| 3 | 0.5296783241 | 10 | 0.8395743720 | 17 | 0.4209559716 |
| 4 | 0.8594612299 | 11 | 0.4646778983 | 18 | 0.8409445432 |
| 5 | 0.4167173034 | 12 | 0.8581956045 | 19 | 0.4614610237 |
| 6 | 0.8385707740 | 13 | 0.4198508858 | 20 | 0.8573758782 |

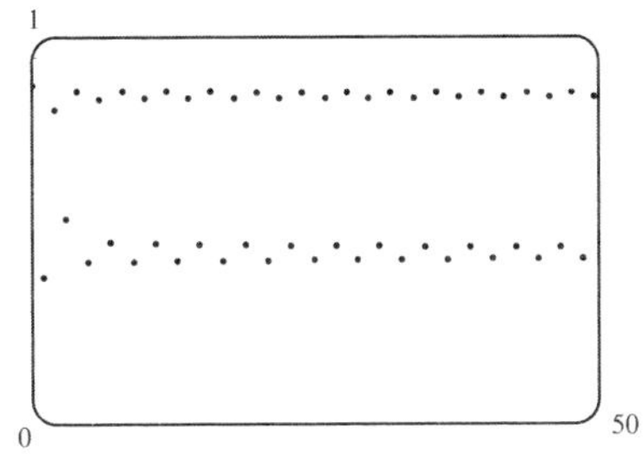


From the graphs above, it seems that for $k$ between 3.4 and 3.5, the terms eventually fluctuate between four values. In the graph below, the pattern followed by the terms is $0.395, 0.832, 0.487, 0.869, 0.395, \ldots$. Note that even for $k = 3.42$ (as in the first graph), there are four distinct "branches; even after 1000 terms, the first and third terms in the pattern differ by about $2 \times 10^{-9}$, while the first and fifth terms differ by only $2 \times 10^{-10}$. With $p_0 = \frac{7}{8}$ and $k = 3.48$:

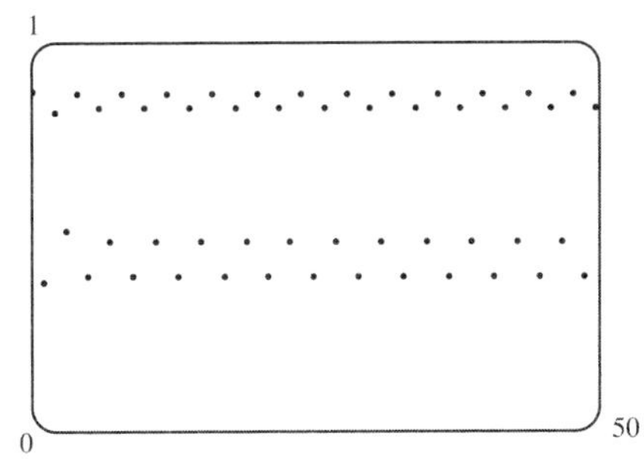


**4.**

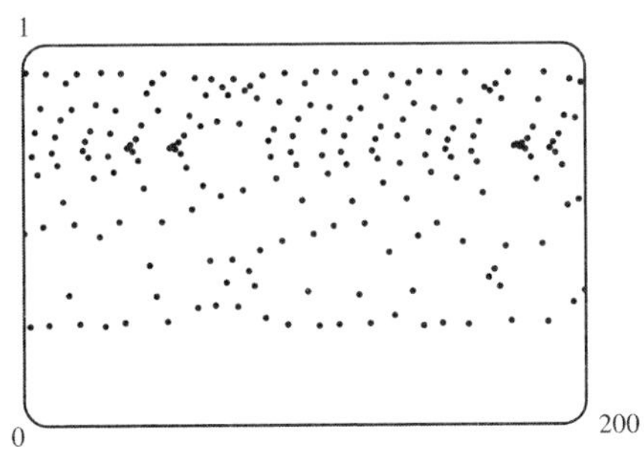


$p_0 = 0.5, k = 3.7$

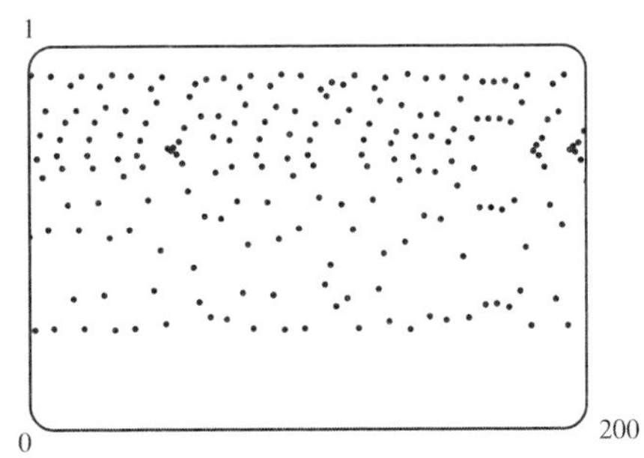


$p_0 = 0.501, k = 3.7$

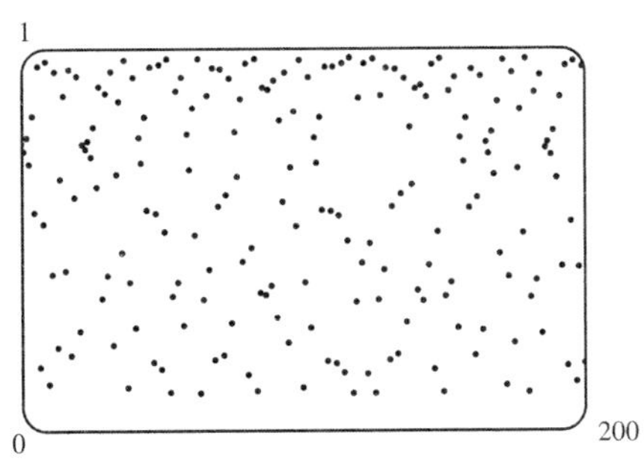


$p_0 = 0.75, k = 3.9$

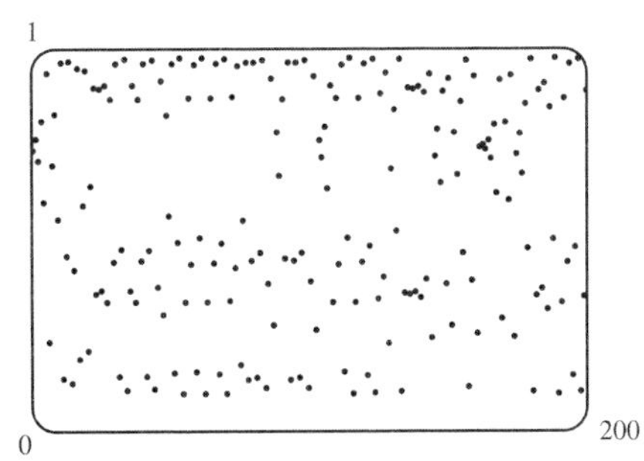


$p_0 = 0.749, k = 3.9$

[continued]

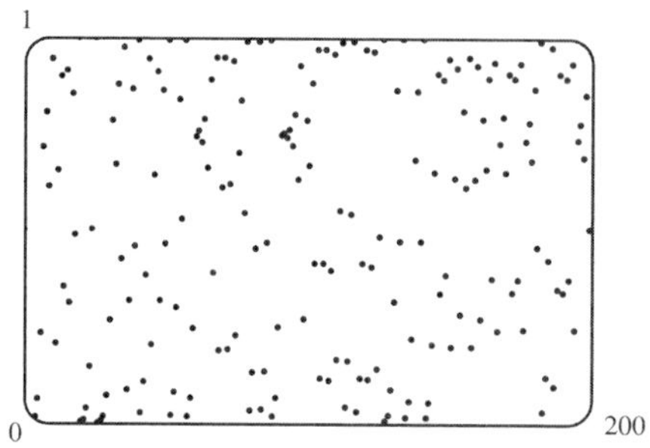


$p_0 = 0.5$, $k = 3.999$

From the graphs, it seems that if $p_0$ is changed by 0.001, the whole graph changes completely. (Note, however, that this might be partially due to accumulated round-off error in the CAS. These graphs were generated by Maple with 100-digit accuracy, and different degrees of accuracy give different graphs.) There seem to be some some fleeting patterns in these graphs, but on the whole they are certainly very chaotic. As $k$ increases, the graph spreads out vertically, with more extreme values close to 0 or 1.

## 8.2 Series

**1.** (a) A sequence is an ordered list of numbers whereas a series is the *sum* of a list of numbers.

(b) A series is convergent if the sequence of partial sums is a convergent sequence. A series is divergent if it is not convergent.

**2.** $\sum\limits_{n=1}^{\infty} a_n = 5$ means that by adding sufficiently many terms of the series we can get as close as we like to the number 5.

In other words, it means that $\lim_{n\to\infty} s_n = 5$, where $s_n$ is the $n$th partial sum, that is, $\sum\limits_{i=1}^{n} a_i$.

**3.**

| $n$ | $s_n$ |
|---|---|
| 1 | −2.40000 |
| 2 | −1.92000 |
| 3 | −2.01600 |
| 4 | −1.99680 |
| 5 | −2.00064 |
| 6 | −1.99987 |
| 7 | −2.00003 |
| 8 | −1.99999 |
| 9 | −2.00000 |
| 10 | −2.00000 |

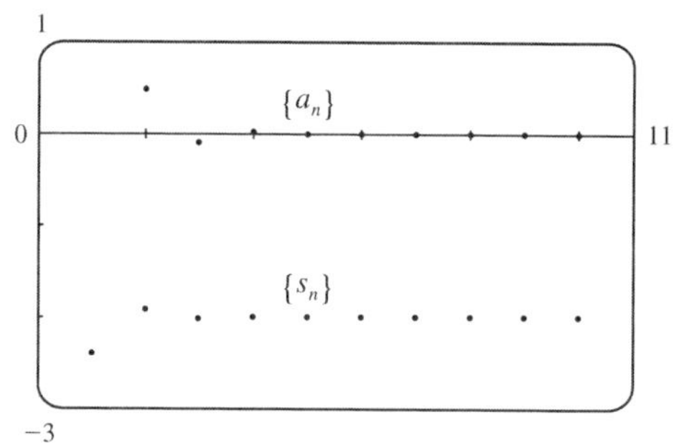


From the graph and the table, it seems that the series converges to $-2$. In fact, it is a geometric series with $a = -2.4$ and $r = -\frac{1}{5}$, so its sum is $\sum\limits_{n=1}^{\infty} \dfrac{12}{(-5)^n} = \dfrac{-2.4}{1-\left(-\frac{1}{5}\right)} = \dfrac{-2.4}{1.2} = -2.$

Note that the dot corresponding to $n = 1$ is part of both $\{a_n\}$ and $\{s_n\}$.

**TI-86 Note:** To graph $\{a_n\}$ and $\{s_n\}$, set your calculator to Param mode and DrawDot mode. (DrawDot is under GRAPH, MORE, FORMT (F3).) Now under `E(t)=` make the assignments: `xt1=t, yt1=12/(-5)^t, xt2=t, yt2=sum seq(yt1,t,1,t,1)`. (sum and seq are under LIST, OPS (F5), MORE.) Under WIND use `1,10,1,0,10,1,-3,1,1` to obtain a graph similar to the one above. Then use TRACE (F4) to see the values.

4.

| $n$ | $s_n$ |
|---|---|
| 1 | 0.50000 |
| 2 | 1.90000 |
| 3 | 3.60000 |
| 4 | 5.42353 |
| 5 | 7.30814 |
| 6 | 9.22706 |
| 7 | 11.16706 |
| 8 | 13.12091 |
| 9 | 15.08432 |
| 10 | 17.05462 |

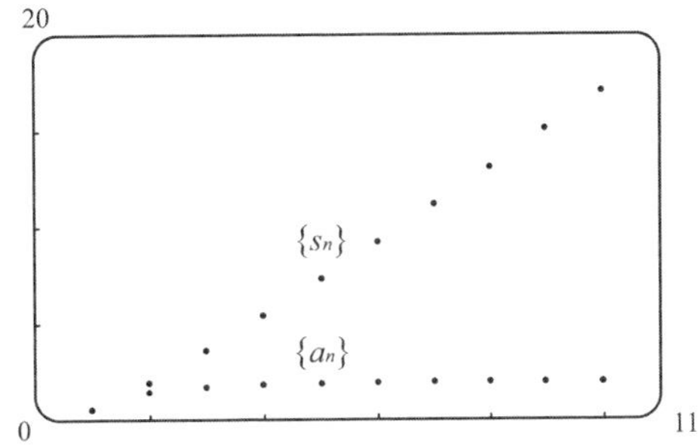


The series $\sum_{n=1}^{\infty} \frac{2n^2 - 1}{n^2 + 1}$ diverges, since its terms do not approach 0.

5.

| $n$ | $s_n$ |
|---|---|
| 1 | 1.55741 |
| 2 | −0.62763 |
| 3 | −0.77018 |
| 4 | 0.38764 |
| 5 | −2.99287 |
| 6 | −3.28388 |
| 7 | −2.41243 |
| 8 | −9.21214 |
| 9 | −9.66446 |
| 10 | −9.01610 |

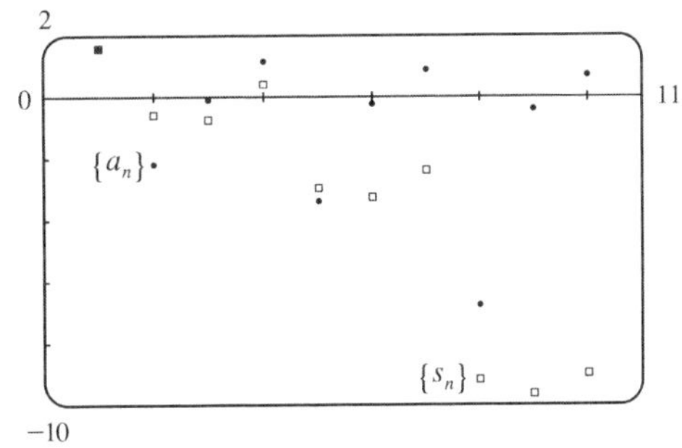


The series $\sum_{n=1}^{\infty} \tan n$ diverges, since its terms do not approach 0.

6.

| $n$ | $s_n$ |
|---|---|
| 1 | 1.00000 |
| 2 | 1.60000 |
| 3 | 1.96000 |
| 4 | 2.17600 |
| 5 | 2.30560 |
| 6 | 2.38336 |
| 7 | 2.43002 |
| 8 | 2.45801 |
| 9 | 2.47481 |
| 10 | 2.48488 |

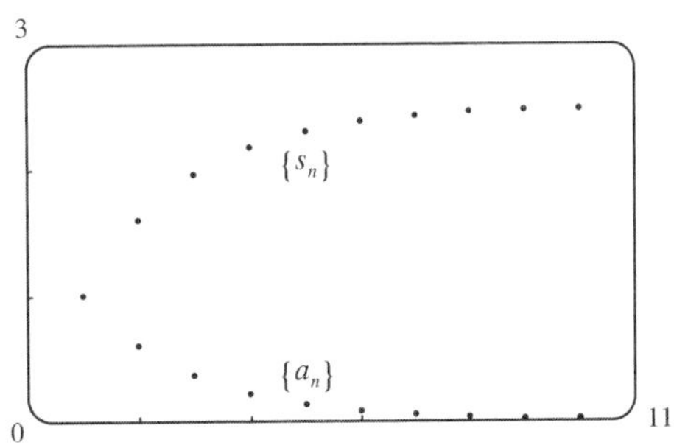


From the graph and the table, it seems that the series converges to 2.5. In fact, it is a geometric series with $a = 1$ and $r = 0.6$, so its sum is $\sum_{n=1}^{\infty} (0.6)^{n-1} = \frac{1}{1 - 0.6} = \frac{1}{2/5} = 2.5$.

7.

| $n$ | $s_n$ |
|---|---|
| 1 | 0.64645 |
| 2 | 0.80755 |
| 3 | 0.87500 |
| 4 | 0.91056 |
| 5 | 0.93196 |
| 6 | 0.94601 |
| 7 | 0.95581 |
| 8 | 0.96296 |
| 9 | 0.96838 |
| 10 | 0.97259 |

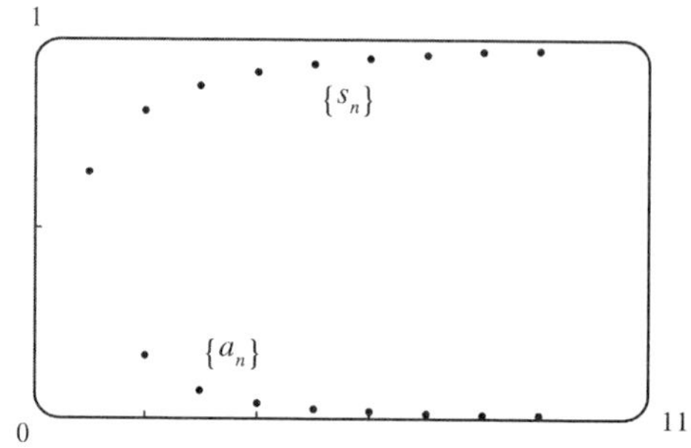


From the graph, it seems that the series converges to 1. To find the sum, we write

$$s_n = \sum_{i=1}^{n} \left( \frac{1}{i^{1.5}} - \frac{1}{(i+1)^{1.5}} \right)$$
$$= \left(1 - \frac{1}{2^{1.5}}\right) + \left(\frac{1}{2^{1.5}} - \frac{1}{3^{1.5}}\right) + \left(\frac{1}{3^{1.5}} - \frac{1}{4^{1.5}}\right) + \cdots + \left(\frac{1}{n^{1.5}} - \frac{1}{(n+1)^{1.5}}\right)$$
$$= 1 - \frac{1}{(n+1)^{1.5}}$$

So the sum is $\lim_{n\to\infty} s_n = 1 - 0 = 1$.

8.

| $n$ | $s_n$ |
|---|---|
| 2 | 0.50000 |
| 3 | 0.66667 |
| 4 | 0.75000 |
| 5 | 0.80000 |
| 6 | 0.83333 |
| 7 | 0.85714 |
| 8 | 0.87500 |
| 9 | 0.88889 |
| 10 | 0.90000 |
| 11 | 0.90909 |
| 100 | 0.99000 |

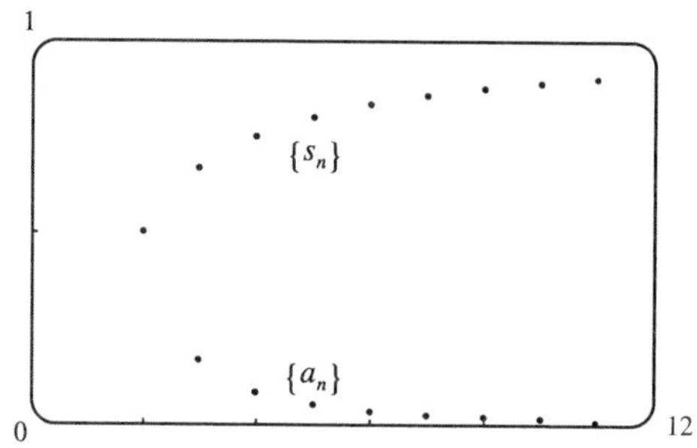


From the graph and the table, it seems that the series converges to 1. To find the sum, we write

$$s_n = \sum_{i=2}^{n} \frac{1}{i(i-1)} = \sum_{i=2}^{n} \left( \frac{1}{i-1} - \frac{1}{i} \right) \qquad \text{[partial fractions]}$$
$$= \left(1 - \frac{1}{2}\right) + \left(\frac{1}{2} - \frac{1}{3}\right) + \left(\frac{1}{3} - \frac{1}{4}\right) + \cdots + \left(\frac{1}{n-1} - \frac{1}{n}\right) = 1 - \frac{1}{n},$$

and so the sum is $\lim_{n\to\infty} s_n = 1 - 0 = 1$.

**9.** (a) $\lim_{n\to\infty} a_n = \lim_{n\to\infty} \frac{2n}{3n+1} = \frac{2}{3}$, so the *sequence* $\{a_n\}$ is convergent by (8.1.1).

(b) Since $\lim_{n\to\infty} a_n = \frac{2}{3} \neq 0$, the *series* $\sum_{n=1}^{\infty} a_n$ is divergent by the Test for Divergence (7).

**10.** (a) Both $\sum\limits_{i=1}^{n} a_i$ and $\sum\limits_{j=1}^{n} a_j$ represent the sum of the first $n$ terms of the sequence $\{a_n\}$, that is, the $n$th partial sum.

(b) $\sum\limits_{i=1}^{n} a_j = \underbrace{a_j + a_j + \cdots + a_j}_{n \text{ terms}} = na_j$, which, in general, is not the same as $\sum\limits_{i=1}^{n} a_i = a_1 + a_2 + \cdots + a_n$.

**11.** $5 - \frac{10}{3} + \frac{20}{9} - \frac{40}{27} + \cdots$ is a geometric series with $a = 5$ and $r = -\frac{2}{3}$. Since $|r| = \frac{2}{3} < 1$, the series converges to

$$\frac{a}{1-r} = \frac{5}{1-(-2/3)} = \frac{5}{5/3} = 3.$$

**12.** $1 + 0.4 + 0.16 + 0.064 + \cdots$ is a geometric series with ratio $0.4 = \frac{2}{5}$. Since $|r| = \frac{2}{5} < 1$, the series converges to

$$\frac{a}{1-r} = \frac{1}{1-2/5} = \frac{5}{3}.$$

**13.** $\sum\limits_{n=1}^{\infty} 5\left(\frac{2}{3}\right)^{n-1}$ is a geometric series with $a = 5$ and $r = \frac{2}{3}$. Since $|r| = \frac{2}{3} < 1$, the series converges to

$$\frac{a}{1-r} = \frac{5}{1-2/3} = \frac{5}{1/3} = 15.$$

**14.** $\sum\limits_{n=1}^{\infty} \frac{(-6)^{n-1}}{5^{n-1}}$ is a geometric series with $a = 1$ and $r = -\frac{6}{5}$. The series diverges since $|r| = \frac{6}{5} > 1$.

**15.** $\sum\limits_{n=0}^{\infty} \frac{\pi^n}{3^{n+1}} = \frac{1}{3}\sum\limits_{n=0}^{\infty}\left(\frac{\pi}{3}\right)^n$ is a geometric series with ratio $r = \frac{\pi}{3}$. Since $|r| > 1$, the series diverges.

**16.** $\sum\limits_{n=0}^{\infty} \frac{1}{(\sqrt{2})^n}$ is a geometric series with ratio $r = \frac{1}{\sqrt{2}}$. Since $|r| = \frac{1}{\sqrt{2}} < 1$, the series converges. Its sum is

$$\frac{1}{1-1/\sqrt{2}} = \frac{\sqrt{2}}{\sqrt{2}-1} = \frac{\sqrt{2}}{\sqrt{2}-1}\cdot\frac{\sqrt{2}+1}{\sqrt{2}+1} = \sqrt{2}\left(\sqrt{2}+1\right) = 2+\sqrt{2}.$$

**17.** $\sum\limits_{n=1}^{\infty} \frac{1}{2n} = \frac{1}{2}\sum\limits_{n=1}^{\infty}\frac{1}{n}$ diverges since each of its partial sums is $\frac{1}{2}$ times the corresponding partial sum of the harmonic series $\sum\limits_{n=1}^{\infty}\frac{1}{n}$, which diverges. $\left[\text{If } \sum\limits_{n=1}^{\infty}\frac{1}{2n} \text{ were to converge, then } \sum\limits_{n=1}^{\infty}\frac{1}{n} \text{ would also have to converge by Theorem 8(i).}\right]$ In general, constant multiples of divergent series are divergent.

**18.** $\sum\limits_{n=1}^{\infty} \frac{n+1}{2n-3}$ diverges since $\lim\limits_{n\to\infty} a_n = \lim\limits_{n\to\infty}\frac{n+1}{2n-3} = \frac{1}{2} \neq 0$. [Use (7), the Test for Divergence.]

**19.** $\sum\limits_{k=2}^{\infty} \frac{k^2}{k^2-1}$ diverges by the Test for Divergence since $\lim\limits_{k\to\infty} a_k = \lim\limits_{k\to\infty}\frac{k^2}{k^2-1} = 1 \neq 0$.

**20.** $\sum_{k=1}^{\infty} \frac{k(k+2)}{(k+3)^2}$ diverges by (7), the Test for Divergence, since $\lim_{k\to\infty} a_k = \lim_{k\to\infty} \frac{k(k+2)}{(k+3)^2} = \lim_{k\to\infty} \frac{1\cdot(1+2/k)}{(1+3/k)^2} = 1 \neq 0$.

**21.** Converges.

$$\sum_{n=1}^{\infty} \frac{1+2^n}{3^n} = \sum_{n=1}^{\infty}\left(\frac{1}{3^n}+\frac{2^n}{3^n}\right) = \sum_{n=1}^{\infty}\left[\left(\frac{1}{3}\right)^n+\left(\frac{2}{3}\right)^n\right] \quad \text{[sum of two convergent geometric series]}$$
$$= \frac{1/3}{1-1/3}+\frac{2/3}{1-2/3} = \frac{1}{2}+2 = \frac{5}{2}$$

**22.** $\sum_{n=1}^{\infty} \frac{1+3^n}{2^n} = \sum_{n=1}^{\infty}\left(\frac{1}{2^n}+\frac{3^n}{2^n}\right) = \sum_{n=1}^{\infty}\left[\left(\frac{1}{2}\right)^n+\left(\frac{3}{2}\right)^n\right] = \sum_{n=1}^{\infty}\left(\frac{1}{2}\right)^n + \sum_{n=1}^{\infty}\left(\frac{3}{2}\right)^n$. The first series is a convergent geometric series ($|r| = \frac{1}{2} < 1$), but the second series is a divergent geometric series ($|r| = \frac{3}{2} \geq 1$), so the original series is divergent.

**23.** $\sum_{n=1}^{\infty} \sqrt[n]{2} = 2 + \sqrt{2} + \sqrt[3]{2} + \sqrt[4]{2} + \cdots$ diverges by the Test for Divergence since

$\lim_{n\to\infty} a_n = \lim_{n\to\infty} \sqrt[n]{2} = \lim_{n\to\infty} 2^{1/n} = 2^0 = 1 \neq 0$.

**24.** $$\sum_{n=1}^{\infty}\left[(0.8)^{n-1}-(0.3)^n\right] = \sum_{n=1}^{\infty}(0.8)^{n-1} - \sum_{n=1}^{\infty}(0.3)^n \quad \text{[difference of two convergent geometric series]}$$
$$= \frac{1}{1-0.8} - \frac{0.3}{1-0.3} = 5 - \frac{3}{7} = \frac{32}{7}$$

**25.** $\lim_{n\to\infty} a_n = \lim_{n\to\infty} \arctan n = \frac{\pi}{2} \neq 0$, so the series diverges by the Test for Divergence.

**26.** $\sum_{k=1}^{\infty} (\cos 1)^k$ is a geometric series with ratio $r = \cos 1 \approx 0.540302$. It converges because $|r| < 1$. Its sum is

$\frac{\cos 1}{1-\cos 1} \approx 1.175343$.

**27.** Using partial fractions, the partial sums of the series $\sum_{n=2}^{\infty} \frac{2}{n^2-1}$ are

$$s_n = \sum_{i=2}^{n} \frac{2}{(i-1)(i+1)} = \sum_{i=2}^{n}\left(\frac{1}{i-1}-\frac{1}{i+1}\right)$$
$$= \left(1-\frac{1}{3}\right)+\left(\frac{1}{2}-\frac{1}{4}\right)+\left(\frac{1}{3}-\frac{1}{5}\right)+\cdots+\left(\frac{1}{n-3}-\frac{1}{n-1}\right)+\left(\frac{1}{n-2}-\frac{1}{n}\right)$$

This sum is a telescoping series and $s_n = 1 + \frac{1}{2} - \frac{1}{n-1} - \frac{1}{n}$.

Thus, $\sum_{n=2}^{\infty} \frac{2}{n^2-1} = \lim_{n\to\infty} s_n = \lim_{n\to\infty}\left(1+\frac{1}{2}-\frac{1}{n-1}-\frac{1}{n}\right) = \frac{3}{2}$.

**28.** For the series $\sum_{n=1}^{\infty} \frac{2}{n^2+4n+3}$, $s_n = \sum_{i=1}^{n} \frac{2}{i^2+4i+3} = \sum_{i=1}^{n}\left(\frac{1}{i+1} - \frac{1}{i+3}\right)$ [using partial fractions]. The latter sum is

$$\left(\tfrac{1}{2}-\tfrac{1}{4}\right)+\left(\tfrac{1}{3}-\tfrac{1}{5}\right)+\left(\tfrac{1}{4}-\tfrac{1}{6}\right)+\left(\tfrac{1}{5}-\tfrac{1}{7}\right)+\cdots+\left(\tfrac{1}{n}-\tfrac{1}{n+2}\right)+\left(\tfrac{1}{n+1}-\tfrac{1}{n+3}\right)$$

$$= \tfrac{1}{2}+\tfrac{1}{3}-\tfrac{1}{n+2}-\tfrac{1}{n+3} \quad \text{[telescoping series]}$$

Thus, $\sum_{n=1}^{\infty} \frac{2}{n^2+4n+3} = \lim_{n\to\infty} s_n = \lim_{n\to\infty}\left(\frac{1}{2}+\frac{1}{3}-\frac{1}{n+2}-\frac{1}{n+3}\right) = \frac{1}{2}+\frac{1}{3} = \frac{5}{6}$. Converges

**29.** For the series $\sum_{n=1}^{\infty} \frac{3}{n(n+3)}$, $s_n = \sum_{i=1}^{n} \frac{3}{i(i+3)} = \sum_{i=1}^{n}\left(\frac{1}{i} - \frac{1}{i+3}\right)$ [using partial fractions]. The latter sum is

$$\left(1-\tfrac{1}{4}\right)+\left(\tfrac{1}{2}-\tfrac{1}{5}\right)+\left(\tfrac{1}{3}-\tfrac{1}{6}\right)+\left(\tfrac{1}{4}-\tfrac{1}{7}\right)+\cdots+\left(\tfrac{1}{n-3}-\tfrac{1}{n}\right)+\left(\tfrac{1}{n-2}-\tfrac{1}{n+1}\right)+\left(\tfrac{1}{n-1}-\tfrac{1}{n+2}\right)+\left(\tfrac{1}{n}-\tfrac{1}{n+3}\right)$$

$$= 1+\tfrac{1}{2}+\tfrac{1}{3}-\tfrac{1}{n+1}-\tfrac{1}{n+2}-\tfrac{1}{n+3} \quad \text{[telescoping series]}$$

Thus, $\sum_{n=1}^{\infty} \frac{3}{n(n+3)} = \lim_{n\to\infty} s_n = \lim_{n\to\infty}\left(1+\tfrac{1}{2}+\tfrac{1}{3}-\tfrac{1}{n+1}-\tfrac{1}{n+2}-\tfrac{1}{n+3}\right) = 1+\tfrac{1}{2}+\tfrac{1}{3} = \tfrac{11}{6}$. Converges

**30.** For the series $\sum_{n=1}^{\infty} \ln\frac{n}{n+1}$, $s_n = (\ln 1 - \ln 2) + (\ln 2 - \ln 3) + (\ln 3 - \ln 4) + \cdots + [\ln n - \ln(n+1)]$
$= \ln 1 - \ln(n+1) = -\ln(n+1)$ [telescoping series]

Thus, $\lim_{n\to\infty} s_n = -\infty$, so the series is divergent.

**31.** $0.\overline{2} = \frac{2}{10} + \frac{2}{10^2} + \cdots$ is a geometric series with $a = \frac{2}{10}$ and $r = \frac{1}{10}$. It converges to $\frac{a}{1-r} = \frac{2/10}{1-1/10} = \frac{2}{9}$.

**32.** $0.\overline{73} = \frac{73}{10^2} + \frac{73}{10^4} + \cdots = \frac{73/10^2}{1-1/10^2} = \frac{73/100}{99/100} = \frac{73}{99}$

**33.** $3.\overline{417} = 3 + \frac{417}{10^3} + \frac{417}{10^6} + \cdots$. Now $\frac{417}{10^3} + \frac{417}{10^6} + \cdots$ is a geometric series with $a = \frac{417}{10^3}$ and $r = \frac{1}{10^3}$. It converges to

$\frac{a}{1-r} = \frac{417/10^3}{1-1/10^3} = \frac{417/10^3}{999/10^3} = \frac{417}{999}$. Thus, $3.\overline{417} = 3 + \frac{417}{999} = \frac{3414}{999} = \frac{1138}{333}$.

**34.** $6.2\overline{54} = 6.2 + \frac{54}{10^3} + \frac{54}{10^5} + \cdots = 6.2 + \frac{54/10^3}{1-1/10^2} = \frac{62}{10} + \frac{54}{990} = \frac{6192}{990} = \frac{344}{55}$

**35.** $\sum_{n=1}^{\infty} \frac{x^n}{3^n} = \sum_{n=1}^{\infty}\left(\frac{x}{3}\right)^n$ is a geometric series with $r = \frac{x}{3}$, so the series converges $\Leftrightarrow$ $|r| < 1$ $\Leftrightarrow$ $\frac{|x|}{3} < 1$ $\Leftrightarrow$ $|x| < 3$;

that is, $-3 < x < 3$. In that case, the sum of the series is $\frac{a}{1-r} = \frac{x/3}{1-x/3} = \frac{x/3}{1-x/3}\cdot\frac{3}{3} = \frac{x}{3-x}$.

**36.** $\sum_{n=0}^{\infty} 2^n (x+1)^n = \sum_{n=0}^{\infty} [2(x+1)]^n = \sum_{n=1}^{\infty} [2(x+1)]^{n-1}$ is a geometric series with $r = 2(x+1)$, so the series

converges $\Leftrightarrow$ $|r| < 1$ $\Leftrightarrow$ $|2(x+1)| < 1$ $\Leftrightarrow$ $|x+1| < \frac{1}{2}$ $\Leftrightarrow$ $-\frac{1}{2} < x+1 < \frac{1}{2}$ $\Leftrightarrow$ $-\frac{3}{2} < x < -\frac{1}{2}$.

In that case, the sum of the series is $\dfrac{a}{1-r} = \dfrac{1}{1-2(x+1)} = \dfrac{1}{-1-2x}$ or $\dfrac{-1}{2x+1}$.

**37.** $\sum_{n=0}^{\infty} \dfrac{\cos^n x}{2^n}$ is a geometric series with first term 1 and ratio $r = \dfrac{\cos x}{2}$, so it converges $\Leftrightarrow$ $|r| < 1$. But $|r| = \dfrac{|\cos x|}{2} \leq \dfrac{1}{2}$

for all $x$. Thus, the series converges for all real values of $x$ and the sum of the series is $\dfrac{1}{1-(\cos x)/2} = \dfrac{2}{2-\cos x}$.

**38.** Because $\dfrac{1}{n} \to 0$ and ln is continuous, we have $\lim_{n\to\infty} \ln\left(1+\dfrac{1}{n}\right) = \ln 1 = 0$.

We now show that the series $\sum_{n=1}^{\infty} \ln\left(1+\dfrac{1}{n}\right) = \sum_{n=1}^{\infty} \ln\left(\dfrac{n+1}{n}\right) = \sum_{n=1}^{\infty} [\ln(n+1) - \ln n]$ diverges.

$s_n = (\ln 2 - \ln 1) + (\ln 3 - \ln 2) + \cdots + (\ln(n+1) - \ln n) = \ln(n+1) - \ln 1 = \ln(n+1)$.

As $n \to \infty$, $s_n = \ln(n+1) \to \infty$, so the series diverges.

**39.** After defining $f$, We use `convert(f,parfrac);` in Maple, `Apart` in Mathematica, or `Expand Rational` and `Simplify` in Derive to find that the general term is $\dfrac{1}{(4n+1)(4n-3)} = -\dfrac{1/4}{4n+1} + \dfrac{1/4}{4n-3}$. So the $n$th partial sum is

$$s_n = \sum_{k=1}^{n} \left(-\frac{1/4}{4k+1} + \frac{1/4}{4k-3}\right) = \frac{1}{4}\sum_{k=1}^{n}\left(\frac{1}{4k-3} - \frac{1}{4k+1}\right)$$

$$= \frac{1}{4}\left[\left(1-\frac{1}{5}\right) + \left(\frac{1}{5}-\frac{1}{9}\right) + \left(\frac{1}{9}-\frac{1}{13}\right) + \cdots + \left(\frac{1}{4n-3} - \frac{1}{4n+1}\right)\right] = \frac{1}{4}\left(1 - \frac{1}{4n+1}\right)$$

The series converges to $\lim_{n\to\infty} s_n = \frac{1}{4}$. This can be confirmed by directly computing the sum using `sum(f,1..infinity);` (in Maple), `Sum[f,{n,1,Infinity}]` (in Mathematica), or `Calculus Sum` (from 1 to $\infty$) and `Simplify` (in Derive).

**40.** See Exercise 39 for specific CAS commands. $\dfrac{n^2+3n+1}{(n^2+n)^2} = \dfrac{1}{n^2} + \dfrac{1}{n} - \dfrac{1}{(n+1)^2} - \dfrac{1}{n+1}$. So the $n$th partial sum is

$$s_n = \sum_{k=1}^{n}\left(\frac{1}{k^2} + \frac{1}{k} - \frac{1}{(k+1)^2} - \frac{1}{k+1}\right)$$

$$= \left(1+1-\frac{1}{2^2}-\frac{1}{2}\right) + \left(\frac{1}{2^2}+\frac{1}{2}-\frac{1}{3^2}-\frac{1}{3}\right) + \cdots + \left(\frac{1}{n^2}+\frac{1}{n}-\frac{1}{(n+1)^2}-\frac{1}{n+1}\right)$$

$$= 1 + 1 - \frac{1}{(n+1)^2} - \frac{1}{n+1}$$

The series converges to $\lim_{n\to\infty} s_n = 2$.

**41.** For $n = 1$, $a_1 = 0$ since $s_1 = 0$. For $n > 1$,

$$a_n = s_n - s_{n-1} = \frac{n-1}{n+1} - \frac{(n-1)-1}{(n-1)+1} = \frac{(n-1)n-(n+1)(n-2)}{(n+1)n} = \frac{2}{n(n+1)}$$

Also, $\sum\limits_{n=1}^{\infty} a_n = \lim\limits_{n\to\infty} s_n = \lim\limits_{n\to\infty} \dfrac{1-1/n}{1+1/n} = 1.$

**42.** $a_1 = s_1 = 3 - \frac{1}{2} = \frac{5}{2}$. For $n \neq 1$,

$$a_n = s_n - s_{n-1} = \left(3 - n2^{-n}\right) - \left[3-(n-1)2^{-(n-1)}\right] = -\frac{n}{2^n} + \frac{n-1}{2^{n-1}}\cdot\frac{2}{2} = \frac{2(n-1)}{2^n} - \frac{n}{2^n} = \frac{n-2}{2^n}$$

Also, $\sum\limits_{n=1}^{\infty} a_n = \lim\limits_{n\to\infty} s_n = \lim\limits_{n\to\infty}\left(3 - \dfrac{n}{2^n}\right) = 3$ because $\lim\limits_{x\to\infty}\dfrac{x}{2^x} \overset{\text{H}}{=} \lim\limits_{x\to\infty}\dfrac{1}{2^x\ln 2} = 0.$

**43.** (a) The first step in the chain occurs when the local government spends $D$ dollars. The people who receive it spend a fraction $c$ of those $D$ dollars, that is, $Dc$ dollars. Those who receive the $Dc$ dollars spend a fraction $c$ of it, that is, $Dc^2$ dollars. Continuing in this way, we see that the total spending after $n$ transactions is

$S_n = D + Dc + Dc^2 + \cdots + Dc^{n-1} = \dfrac{D(1-c^n)}{1-c}$ by (3).

(b) $\lim\limits_{n\to\infty} S_n = \lim\limits_{n\to\infty}\dfrac{D(1-c^n)}{1-c} = \dfrac{D}{1-c}\lim\limits_{n\to\infty}(1-c^n) = \dfrac{D}{1-c}$ [since $0 < c < 1 \Rightarrow \lim\limits_{n\to\infty} c^n = 0$]

$= \dfrac{D}{s}$ [since $c + s = 1$] $= kD$ [since $k = 1/s$]

If $c = 0.8$, then $s = 1 - c = 0.2$ and the multiplier is $k = 1/s = 5$.

**44.** (a) Initially, the ball falls a distance $H$, then rebounds a distance $rH$, falls $rH$, rebounds $r^2H$, falls $r^2H$, etc. The total distance it travels is

$$H + 2rH + 2r^2H + 2r^3H + \cdots = H\left(1 + 2r + 2r^2 + 2r^3 + \cdots\right) = H\left[1 + 2r\left(1 + r + r^2 + \cdots\right)\right]$$
$$= H\left[1 + 2r\left(\frac{1}{1-r}\right)\right] = H\left(\frac{1+r}{1-r}\right) \text{ meters}$$

(b) From Example 3 in Section 2.1, we know that a ball falls $\frac{1}{2}gt^2$ meters in $t$ seconds, where $g$ is the gravitational acceleration. Thus, a ball falls $h$ meters in $t = \sqrt{2h/g}$ seconds. The total travel time in seconds is

$$\sqrt{\frac{2H}{g}} + 2\sqrt{\frac{2H}{g}r} + 2\sqrt{\frac{2H}{g}r^2} + 2\sqrt{\frac{2H}{g}r^3} + \cdots = \sqrt{\frac{2H}{g}}\left[1 + 2\sqrt{r} + 2\sqrt{r}^{\,2} + 2\sqrt{r}^{\,3} + \cdots\right]$$
$$= \sqrt{\frac{2H}{g}}\left(1 + 2\sqrt{r}\left[1 + \sqrt{r} + \sqrt{r}^{\,2} + \cdots\right]\right) = \sqrt{\frac{2H}{g}}\left[1 + 2\sqrt{r}\left(\frac{1}{1-\sqrt{r}}\right)\right] = \sqrt{\frac{2H}{g}}\,\frac{1+\sqrt{r}}{1-\sqrt{r}}$$

(c) It will help to make a chart of the time for each descent and each rebound of the ball, together with the velocity just before and just after each bounce. Recall that the time in seconds needed to fall $h$ meters is $\sqrt{2h/g}$. The ball hits the ground with velocity $-g\sqrt{2h/g} = -\sqrt{2hg}$ (taking the upward direction to be positive) and rebounds with velocity $kg\sqrt{2h/g} = k\sqrt{2hg}$, taking time $k\sqrt{2h/g}$ to reach the top of its bounce, where its velocity is 0. At that point, its height

is $k^2h$. All these results follow from the formulas for vertical motion with gravitational acceleration $-g$:
$\frac{d^2y}{dt^2} = -g \quad \Rightarrow \quad v = \frac{dy}{dt} = v_0 - gt \quad \Rightarrow \quad y = y_0 + v_0 t - \frac{1}{2}gt^2$.

| number of descent | time of descent | speed before bounce | speed after bounce | time of ascent | peak height |
|---|---|---|---|---|---|
| 1 | $\sqrt{2H/g}$ | $\sqrt{2Hg}$ | $k\sqrt{2Hg}$ | $k\sqrt{2H/g}$ | $k^2H$ |
| 2 | $\sqrt{2k^2H/g}$ | $\sqrt{2k^2Hg}$ | $k\sqrt{2k^2Hg}$ | $k\sqrt{2k^2H/g}$ | $k^4H$ |
| 3 | $\sqrt{2k^4H/g}$ | $\sqrt{2k^4Hg}$ | $k\sqrt{2k^4Hg}$ | $k\sqrt{2k^4H/g}$ | $k^6H$ |
| ⋯ | ⋯ | ⋯ | ⋯ | ⋯ | ⋯ |

The total travel time in seconds is

$$\sqrt{\frac{2H}{g}} + k\sqrt{\frac{2H}{g}} + k\sqrt{\frac{2H}{g}} + k^2\sqrt{\frac{2H}{g}} + k^2\sqrt{\frac{2H}{g}} + \cdots = \sqrt{\frac{2H}{g}}\left(1 + 2k + 2k^2 + 2k^3 + \cdots\right)$$
$$= \sqrt{\frac{2H}{g}}\left[1 + 2k(1 + k + k^2 + \cdots)\right] = \sqrt{\frac{2H}{g}}\left[1 + 2k\left(\frac{1}{1-k}\right)\right] = \sqrt{\frac{2H}{g}}\,\frac{1+k}{1-k}$$

*Another method:* We could use part (b). At the top of the bounce, the height is $k^2h = rh$, so $\sqrt{r} = k$ and the result follows from part (b).

**45.** $\sum_{n=2}^{\infty}(1+c)^{-n}$ is a geometric series with $a = (1+c)^{-2}$ and $r = (1+c)^{-1}$, so the series converges when $\left|(1+c)^{-1}\right| < 1$ $\Leftrightarrow$ $|1+c| > 1$ $\Leftrightarrow$ $1 + c > 1$ or $1 + c < -1$ $\Leftrightarrow$ $c > 0$ or $c < -2$. We calculate the sum of the series and set it equal to 2: $\frac{(1+c)^{-2}}{1-(1+c)^{-1}} = 2 \quad \Leftrightarrow \quad \left(\frac{1}{1+c}\right)^2 = 2 - 2\left(\frac{1}{1+c}\right) \quad \Leftrightarrow \quad 1 = 2(1+c)^2 - 2(1+c) \quad \Leftrightarrow \quad 2c^2 + 2c - 1 = 0$
$\Leftrightarrow \quad c = \frac{-2 \pm \sqrt{12}}{4} = \frac{\pm\sqrt{3}-1}{2}$. However, the negative root is inadmissible because $-2 < \frac{-\sqrt{3}-1}{2} < 0$. So $c = \frac{\sqrt{3}-1}{2}$.

**46.** The area between $y = x^{n-1}$ and $y = x^n$ for $0 \le x \le 1$ is

$$\int_0^1 \left(x^{n-1} - x^n\right) dx = \left[\frac{x^n}{n} - \frac{x^{n+1}}{n+1}\right]_0^1 = \frac{1}{n} - \frac{1}{n+1}$$
$$= \frac{(n+1) - n}{n(n+1)} = \frac{1}{n(n+1)}$$

We can see from the diagram that as $n \to \infty$, the sum of the areas between the successive curves approaches the area of the unit square, that is, 1. So $\sum_{n=1}^{\infty} \frac{1}{n(n+1)} = 1$.

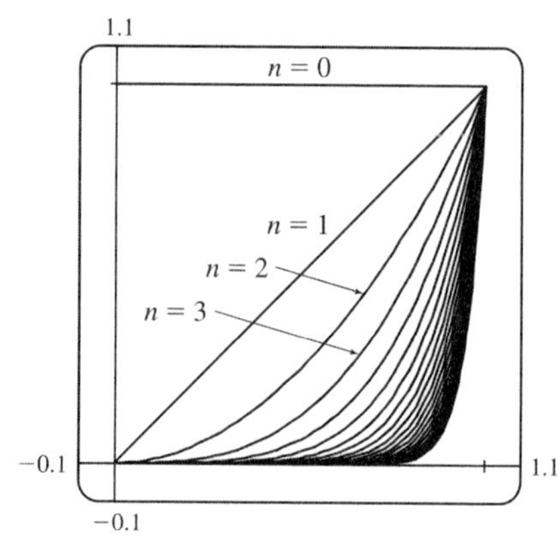

**47.** Let $d_n$ be the diameter of $C_n$. We draw lines from the centers of the $C_i$ to the center of $D$ (or $C$), and using the Pythagorean Theorem, we can write

$1^2 + \left(1 - \frac{1}{2}d_1\right)^2 = \left(1 + \frac{1}{2}d_1\right)^2 \Leftrightarrow$

$1 = \left(1 + \frac{1}{2}d_1\right)^2 - \left(1 - \frac{1}{2}d_1\right)^2 = 2d_1$ [difference of squares] $\Rightarrow d_1 = \frac{1}{2}$.

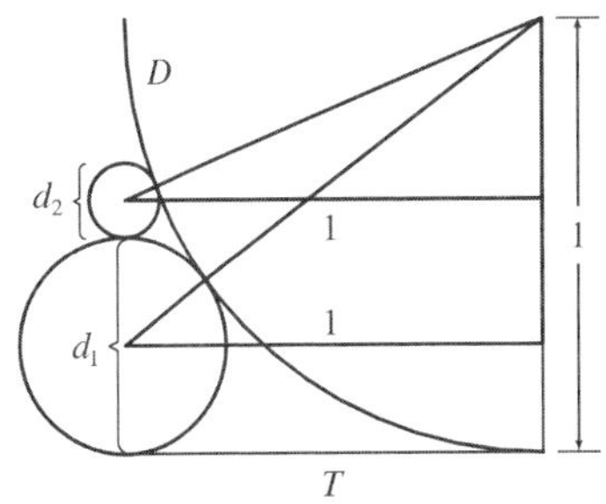


Similarly,

$1 = \left(1 + \frac{1}{2}d_2\right)^2 - \left(1 - d_1 - \frac{1}{2}d_2\right)^2 = 2d_2 + 2d_1 - d_1^2 - d_1 d_2$

$= (2 - d_1)(d_1 + d_2) \quad \Leftrightarrow$

$d_2 = \dfrac{1}{2 - d_1} - d_1 = \dfrac{(1 - d_1)^2}{2 - d_1}$, $1 = \left(1 + \frac{1}{2}d_3\right)^2 - \left(1 - d_1 - d_2 - \frac{1}{2}d_3\right)^2 \Leftrightarrow d_3 = \dfrac{[1 - (d_1 + d_2)]^2}{2 - (d_1 + d_2)}$, and in general,

$d_{n+1} = \dfrac{\left(1 - \sum_{i=1}^{n} d_i\right)^2}{2 - \sum_{i=1}^{n} d_i}$. If we actually calculate $d_2$ and $d_3$ from the formulas above, we find that they are $\dfrac{1}{6} = \dfrac{1}{2 \cdot 3}$ and

$\dfrac{1}{12} = \dfrac{1}{3 \cdot 4}$ respectively, so we suspect that in general, $d_n = \dfrac{1}{n(n+1)}$. To prove this, we use induction: Assume that for all

$k \le n$, $d_k = \dfrac{1}{k(k+1)} = \dfrac{1}{k} - \dfrac{1}{k+1}$. Then $\sum\limits_{i=1}^{n} d_i = 1 - \dfrac{1}{n+1} = \dfrac{n}{n+1}$ [telescoping sum]. Substituting this into our

formula for $d_{n+1}$, we get $d_{n+1} = \dfrac{\left[1 - \dfrac{n}{n+1}\right]^2}{2 - \left(\dfrac{n}{n+1}\right)} = \dfrac{\dfrac{1}{(n+1)^2}}{\dfrac{n+2}{n+1}} = \dfrac{1}{(n+1)(n+2)}$, and the induction is complete.

Now, we observe that the partial sums $\sum_{i=1}^{n} d_i$ of the diameters of the circles approach 1 as $n \to \infty$; that is,

$\sum\limits_{n=1}^{\infty} a_n = \sum\limits_{n=1}^{\infty} \dfrac{1}{n(n+1)} = 1$, which is what we wanted to prove.

**48.** $|CD| = b\sin\theta$, $|DE| = |CD|\sin\theta = b\sin^2\theta$, $|EF| = |DE|\sin\theta = b\sin^3\theta$, $\ldots$. Therefore,

$|CD| + |DE| + |EF| + |FG| + \cdots = b\sum\limits_{n=1}^{\infty} \sin^n\theta = b\left(\dfrac{\sin\theta}{1 - \sin\theta}\right)$ since this is a geometric series with $r = \sin\theta$ and

$|\sin\theta| < 1$ $\left[\text{because } 0 < \theta < \frac{\pi}{2}\right]$.

**49.** The series $1 - 1 + 1 - 1 + 1 - 1 + \cdots$ diverges (geometric series with $r = -1$) so we cannot say that

$0 = 1 - 1 + 1 - 1 + 1 - 1 + \cdots$.

**50.** If $\sum\limits_{n=1}^{\infty} a_n$ is convergent, then $\lim\limits_{n\to\infty} a_n = 0$ by Theorem 6, so $\lim\limits_{n\to\infty} \dfrac{1}{a_n} \ne 0$, and so $\sum\limits_{n=1}^{\infty} \dfrac{1}{a_n}$ is divergent by the Test for

Divergence.

**51.** Suppose on the contrary that $\sum(a_n + b_n)$ converges. Then $\sum(a_n + b_n)$ and $\sum a_n$ are convergent series. So by Theorem 8, $\sum[(a_n + b_n) - a_n]$ would also be convergent. But $\sum[(a_n + b_n) - a_n] = \sum b_n$, a contradiction, since $\sum b_n$ is given to be divergent.

**52.** No. For example, take $\sum a_n = \sum n$ and $\sum b_n = \sum(-n)$, which both diverge, yet $\sum(a_n + b_n) = \sum 0$, which converges with sum 0.

**53.** The partial sums $\{s_n\}$ form an increasing sequence, since $s_n - s_{n-1} = a_n > 0$ for all $n$. Also, the sequence $\{s_n\}$ is bounded since $s_n \le 1000$ for all $n$. So by Theorem 8.1.7, the sequence of partial sums converges, that is, the series $\sum a_n$ is convergent.

**54.** (a) $\text{RHS} = \dfrac{1}{f_{n-1}f_n} - \dfrac{1}{f_n f_{n+1}} = \dfrac{f_n f_{n+1} - f_n f_{n-1}}{f_n^2 f_{n-1} f_{n+1}} = \dfrac{f_{n+1} - f_{n-1}}{f_n f_{n-1} f_{n+1}} = \dfrac{(f_{n-1} + f_n) - f_{n-1}}{f_n f_{n-1} f_{n+1}} = \dfrac{1}{f_{n-1} f_{n+1}} = \text{LHS}$

(b) $$\sum_{n=2}^{\infty} \frac{1}{f_{n-1}f_{n+1}} = \sum_{n=2}^{\infty}\left(\frac{1}{f_{n-1}f_n} - \frac{1}{f_n f_{n+1}}\right) \quad \text{[from part (a)]}$$
$$= \lim_{n\to\infty}\left[\left(\frac{1}{f_1f_2} - \frac{1}{f_2f_3}\right) + \left(\frac{1}{f_2f_3} - \frac{1}{f_3f_4}\right) + \left(\frac{1}{f_3f_4} - \frac{1}{f_4f_5}\right) + \cdots + \left(\frac{1}{f_{n-1}f_n} - \frac{1}{f_nf_{n+1}}\right)\right]$$
$$= \lim_{n\to\infty}\left(\frac{1}{f_1f_2} - \frac{1}{f_nf_{n+1}}\right) = \frac{1}{f_1f_2} - 0 = \frac{1}{1\cdot 1} = 1 \quad \text{because } f_n \to \infty \text{ as } n \to \infty.$$

(c) $$\sum_{n=2}^{\infty} \frac{f_n}{f_{n-1}f_{n+1}} = \sum_{n=2}^{\infty}\left(\frac{f_n}{f_{n-1}f_n} - \frac{f_n}{f_n f_{n+1}}\right) \quad \text{[as above]}$$
$$= \sum_{n=2}^{\infty}\left(\frac{1}{f_{n-1}} - \frac{1}{f_{n+1}}\right)$$
$$= \lim_{n\to\infty}\left[\left(\frac{1}{f_1} - \frac{1}{f_3}\right) + \left(\frac{1}{f_2} - \frac{1}{f_4}\right) + \left(\frac{1}{f_3} - \frac{1}{f_5}\right) + \left(\frac{1}{f_4} - \frac{1}{f_6}\right) + \cdots + \left(\frac{1}{f_{n-1}} - \frac{1}{f_{n+1}}\right)\right]$$
$$= \lim_{n\to\infty}\left(\frac{1}{f_1} + \frac{1}{f_2} - \frac{1}{f_n} - \frac{1}{f_{n+1}}\right) = 1 + 1 - 0 - 0 = 2 \quad \text{because } f_n \to \infty \text{ as } n \to \infty.$$

**55.** (a) At the first step, only the interval $\left(\frac{1}{3}, \frac{2}{3}\right)$ (length $\frac{1}{3}$) is removed. At the second step, we remove the intervals $\left(\frac{1}{9}, \frac{2}{9}\right)$ and $\left(\frac{7}{9}, \frac{8}{9}\right)$, which have a total length of $2 \cdot \left(\frac{1}{3}\right)^2$. At the third step, we remove $2^2$ intervals, each of length $\left(\frac{1}{3}\right)^3$. In general, at the $n$th step we remove $2^{n-1}$ intervals, each of length $\left(\frac{1}{3}\right)^n$, for a length of $2^{n-1} \cdot \left(\frac{1}{3}\right)^n = \frac{1}{3}\left(\frac{2}{3}\right)^{n-1}$. Thus, the total length of all removed intervals is $\sum\limits_{n=1}^{\infty} \frac{1}{3}\left(\frac{2}{3}\right)^{n-1} = \frac{1/3}{1 - 2/3} = 1$ $\left[\text{geometric series with } a = \frac{1}{3} \text{ and } r = \frac{2}{3}\right]$. Notice that at the $n$th step, the leftmost interval that is removed is $\left(\left(\frac{1}{3}\right)^n, \left(\frac{2}{3}\right)^n\right)$, so we never remove 0, and 0 is in the Cantor set. Also, the rightmost interval removed is $\left(1 - \left(\frac{2}{3}\right)^n, 1 - \left(\frac{1}{3}\right)^n\right)$, so 1 is never removed. Some other numbers in the Cantor set are $\frac{1}{3}$, $\frac{2}{3}$, $\frac{1}{9}$, $\frac{2}{9}$, $\frac{7}{9}$, and $\frac{8}{9}$.

(b) The area removed at the first step is $\frac{1}{9}$; at the second step, $8 \cdot \left(\frac{1}{9}\right)^2$; at the third step, $(8)^2 \cdot \left(\frac{1}{9}\right)^3$. In general, the area removed at the $n$th step is $(8)^{n-1}\left(\dfrac{1}{9}\right)^n = \dfrac{1}{9}\left(\dfrac{8}{9}\right)^{n-1}$, so the total area of all removed squares is

$$\sum_{n=1}^{\infty} \frac{1}{9}\left(\frac{8}{9}\right)^{n-1} = \frac{1/9}{1 - 8/9} = 1.$$

**56.** (a)

| $a_1$ | 1 | 2 | 4 | 1 | 1 | 1000 |
|---|---|---|---|---|---|---|
| $a_2$ | 2 | 3 | 1 | 4 | 1000 | 1 |
| $a_3$ | 1.5 | 2.5 | 2.5 | 2.5 | 500.5 | 500.5 |
| $a_4$ | 1.75 | 2.75 | 1.75 | 3.25 | 750.25 | 250.75 |
| $a_5$ | 1.625 | 2.625 | 2.125 | 2.875 | 625.375 | 375.625 |
| $a_6$ | 1.6875 | 2.6875 | 1.9375 | 3.0625 | 687.813 | 313.188 |
| $a_7$ | 1.65625 | 2.65625 | 2.03125 | 2.96875 | 656.594 | 344.406 |
| $a_8$ | 1.67188 | 2.67188 | 1.98438 | 3.01563 | 672.203 | 328.797 |
| $a_9$ | 1.66406 | 2.66406 | 2.00781 | 2.99219 | 664.398 | 336.602 |
| $a_{10}$ | 1.66797 | 2.66797 | 1.99609 | 3.00391 | 668.301 | 332.699 |
| $a_{11}$ | 1.66602 | 2.66602 | 2.00195 | 2.99805 | 666.350 | 334.650 |
| $a_{12}$ | 1.66699 | 2.66699 | 1.99902 | 3.00098 | 667.325 | 333.675 |

The limits seem to be $\frac{5}{3}, \frac{8}{3}$, 2, 3, 667, and 334. Note that the limits appear to be "weighted" more toward $a_2$. In general, we guess that the limit is $\dfrac{a_1 + 2a_2}{3}$.

(b) $a_{n+1} - a_n = \frac{1}{2}(a_n + a_{n-1}) - a_n = -\frac{1}{2}(a_n - a_{n-1}) = -\frac{1}{2}\left[\frac{1}{2}(a_{n-1} + a_{n-2}) - a_{n-1}\right]$
$= -\frac{1}{2}\left[-\frac{1}{2}(a_{n-1} - a_{n-2})\right] = \cdots = \left(-\frac{1}{2}\right)^{n-1}(a_2 - a_1)$

Note that we have used the formula $a_k = \frac{1}{2}(a_{k-1} + a_{k-2})$ a total of $n-1$ times in this calculation, once for each $k$ between 3 and $n+1$. Now we can write

$$\begin{aligned} a_n &= a_1 + (a_2 - a_1) + (a_3 - a_2) + \cdots + (a_{n-1} - a_{n-2}) + (a_n - a_{n-1}) \\ &= a_1 + \textstyle\sum_{k=1}^{n-1}(a_{k+1} - a_k) = a_1 + \sum_{k=1}^{n-1}\left(-\frac{1}{2}\right)^{k-1}(a_2 - a_1) \end{aligned}$$

and so

$$\lim_{n\to\infty} a_n = a_1 + (a_2 - a_1)\textstyle\sum_{k=1}^{\infty}\left(-\frac{1}{2}\right)^{k-1} = a_1 + (a_2 - a_1)\left[\dfrac{1}{1-(-1/2)}\right] = a_1 + \frac{2}{3}(a_2 - a_1) = \dfrac{a_1 + 2a_2}{3}.$$

**57.** (a) For $\sum\limits_{n=1}^{\infty} \dfrac{n}{(n+1)!}$, $s_1 = \dfrac{1}{1\cdot 2} = \dfrac{1}{2}$, $s_2 = \dfrac{1}{2} + \dfrac{2}{1\cdot 2\cdot 3} = \dfrac{5}{6}$, $s_3 = \dfrac{5}{6} + \dfrac{3}{1\cdot 2\cdot 3\cdot 4} = \dfrac{23}{24}$,

$s_4 = \dfrac{23}{24} + \dfrac{4}{1\cdot 2\cdot 3\cdot 4\cdot 5} = \dfrac{119}{120}$. The denominators are $(n+1)!$, so a guess would be $s_n = \dfrac{(n+1)!-1}{(n+1)!}$.

(b) For $n = 1$, $s_1 = \dfrac{1}{2} = \dfrac{2!-1}{2!}$, so the formula holds for $n = 1$. Assume $s_k = \dfrac{(k+1)!-1}{(k+1)!}$. Then

$$\begin{aligned} s_{k+1} &= \frac{(k+1)!-1}{(k+1)!} + \frac{k+1}{(k+2)!} = \frac{(k+1)!-1}{(k+1)!} + \frac{k+1}{(k+1)!(k+2)} \\ &= \frac{(k+2)! - (k+2) + k + 1}{(k+2)!} = \frac{(k+2)!-1}{(k+2)!} \end{aligned}$$

Thus, the formula is true for $n = k+1$. So by induction, the guess is correct.

(c) $\lim\limits_{n\to\infty} s_n = \lim\limits_{n\to\infty} \dfrac{(n+1)!-1}{(n+1)!} = \lim\limits_{n\to\infty}\left[1 - \dfrac{1}{(n+1)!}\right] = 1$ and so $\sum\limits_{n=1}^{\infty} \dfrac{n}{(n+1)!} = 1$.

**58.**

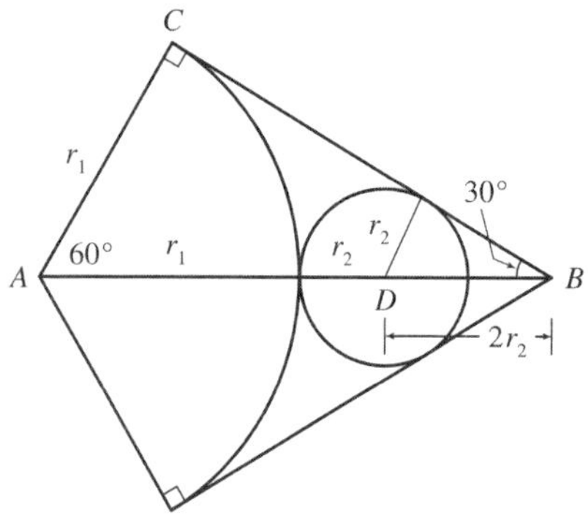


Let $r_1$ = radius of the large circle, $r_2$ = radius of next circle, and so on. From the figure we have $\angle BAC = 60°$ and $\cos 60° = r_1/|AB|$, so $|AB| = 2r_1$ and $|DB| = 2r_2$. Therefore, $2r_1 = r_1 + r_2 + 2r_2 \Rightarrow r_1 = 3r_2$. In general, we have $r_{n+1} = \frac{1}{3}r_n$, so the total area is

$$A = \pi r_1^2 + 3\pi r_2^2 + 3\pi r_3^2 + \cdots = \pi r_1^2 + 3\pi r_2^2\left(1 + \frac{1}{3^2} + \frac{1}{3^4} + \frac{1}{3^6} + \cdots\right)$$

$$= \pi r_1^2 + 3\pi r_2^2 \cdot \frac{1}{1 - 1/9} = \pi r_1^2 + \tfrac{27}{8}\pi r_2^2$$

Since the sides of the triangle have length 1, $|BC| = \frac{1}{2}$ and $\tan 30° = \dfrac{r_1}{1/2}$. Thus, $r_1 = \dfrac{\tan 30°}{2} = \dfrac{1}{2\sqrt{3}} \Rightarrow r_2 = \dfrac{1}{6\sqrt{3}}$,

so $A = \pi\left(\dfrac{1}{2\sqrt{3}}\right)^2 + \dfrac{27\pi}{8}\left(\dfrac{1}{6\sqrt{3}}\right)^2 = \dfrac{\pi}{12} + \dfrac{\pi}{32} = \dfrac{11\pi}{96}$. The area of the triangle is $\dfrac{\sqrt{3}}{4}$, so the circles occupy about 83.1% of the area of the triangle.

## 8.3 The Integral and Comparison Tests; Estimating Sums

**1.** The picture shows that $a_2 = \dfrac{1}{2^{1.3}} < \displaystyle\int_1^2 \frac{1}{x^{1.3}}\,dx$, $a_3 = \dfrac{1}{3^{1.3}} < \displaystyle\int_2^3 \frac{1}{x^{1.3}}\,dx$, and so on, so $\displaystyle\sum_{n=2}^{\infty} \frac{1}{n^{1.3}} < \int_1^{\infty} \frac{1}{x^{1.3}}\,dx$. The integral converges by (5.10.2) with $p = 1.3 > 1$, so the series converges.

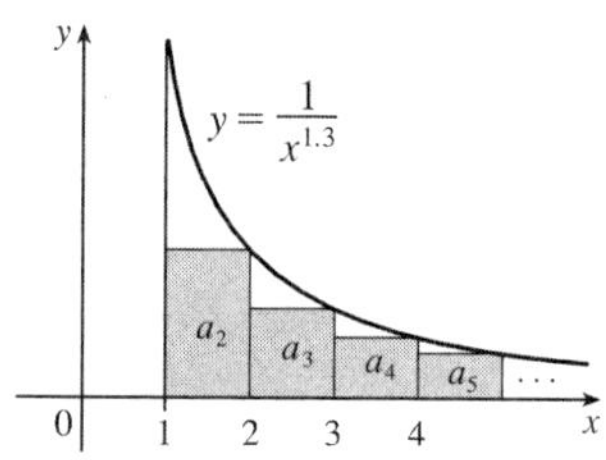


**2.** From the first figure, we see that $\int_1^6 f(x)dx < \sum_{i=1}^{5} a_i$. From the second figure, we see that $\sum_{i=2}^{6} a_i < \int_1^6 f(x)\,dx$. Thus, we have $\sum_{i=2}^{6} a_i < \int_1^6 f(x)\,dx < \sum_{i=1}^{5} a_i$.

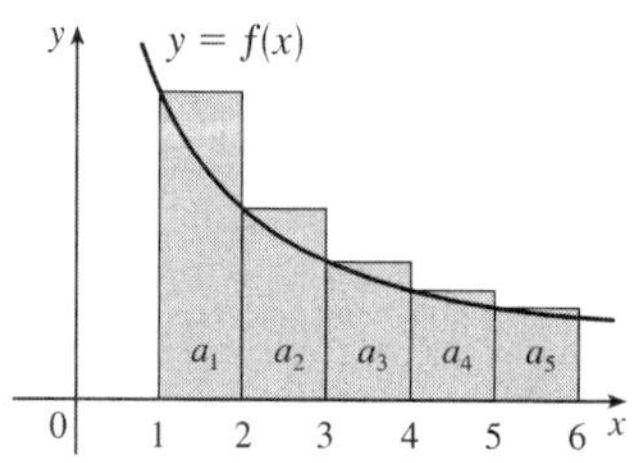


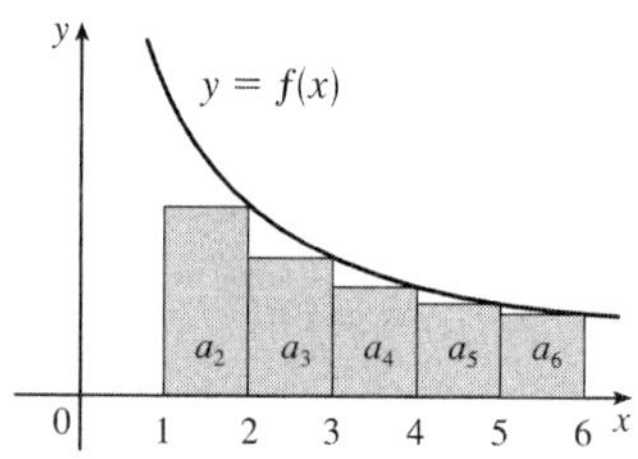


**3.** (a) We cannot say anything about $\sum a_n$. If $a_n > b_n$ for all $n$ and $\sum b_n$ is convergent, then $\sum a_n$ could be convergent or divergent. (See the note after Example 4.)

(b) If $a_n < b_n$ for all $n$, then $\sum a_n$ is convergent. [This is part (i) of the Comparison Test.]

**4.** (a) If $a_n > b_n$ for all $n$, then $\sum a_n$ is divergent. [This is part (ii) of the Comparison Test.]

(b) We cannot say anything about $\sum a_n$. If $a_n < b_n$ for all $n$ and $\sum b_n$ is divergent, then $\sum a_n$ could be convergent or divergent.

**5.** $\displaystyle\sum_{n=1}^{\infty} n^b$ is a $p$-series with $p = -b$. $\displaystyle\sum_{n=1}^{\infty} b^n$ is a geometric series. By (1), the $p$-series is convergent if $p > 1$. In this case, $\displaystyle\sum_{n=1}^{\infty} n^b = \sum_{n=1}^{\infty} (1/n^{-b})$, so $-b > 1 \Leftrightarrow b < -1$ are the values for which the series converge. A geometric series $\displaystyle\sum_{n=1}^{\infty} ar^{n-1}$ converges if $|r| < 1$, so $\displaystyle\sum_{n=1}^{\infty} b^n$ converges if $|b| < 1 \Leftrightarrow -1 < b < 1$.

**6.** The function $f(x) = 1/\sqrt[4]{x} = x^{-1/4}$ is continuous, positive, and decreasing on $[1, \infty)$, so the Integral Test applies.

$\int_1^\infty x^{-1/4}\,dx = \lim_{t\to\infty} \int_1^t x^{-1/4}\,dx = \lim_{t\to\infty} \left[\frac{4}{3}x^{3/4}\right]_1^t = \lim_{t\to\infty} \left(\frac{4}{3}t^{3/4} - \frac{4}{3}\right) = \infty$, so $\sum_{n=1}^\infty 1/\sqrt[4]{n}$ diverges.

**7.** The function $f(x) = 1/x^4$ is continuous, positive, and decreasing on $[1, \infty)$, so the Integral Test applies.

$\displaystyle\int_1^\infty \frac{1}{x^4}\,dx = \lim_{t\to\infty} \int_1^t x^{-4}\,dx = \lim_{t\to\infty} \left[\frac{x^{-3}}{-3}\right]_1^t = \lim_{t\to\infty} \left(-\frac{1}{3t^3} + \frac{1}{3}\right) = \frac{1}{3}$. Since this improper integral is convergent, the series $\displaystyle\sum_{n=1}^\infty \frac{1}{n^4}$ is also convergent by the Integral Test.

**8.** The function $f(x) = 1/(x^2+1)$ is continuous, positive, and decreasing on $[1, \infty)$, so the Integral Test applies.

$\displaystyle\int_1^\infty \frac{1}{x^2+1}\,dx = \lim_{t\to\infty} \int_1^t \frac{1}{x^2+1}\,dx = \lim_{t\to\infty} \left[\tan^{-1} x\right]_1^t = \lim_{t\to\infty} \left(\tan^{-1} t - \tan^{-1} 1\right) = \frac{\pi}{2} - \frac{\pi}{4} = \frac{\pi}{4}$, so $\displaystyle\sum_{n=1}^\infty \frac{1}{n^2+1}$ converges.

**9.** $\displaystyle\frac{1}{n^2+n+1} < \frac{1}{n^2}$ for all $n \geq 1$, so $\displaystyle\sum_{n=1}^\infty \frac{1}{n^2+n+1}$ converges by comparison with $\displaystyle\sum_{n=1}^\infty \frac{1}{n^2}$, which converges because it is a $p$-series with $p = 2 > 1$.

**10.** $\displaystyle\frac{\sqrt{n}}{n-1} > \frac{\sqrt{n}}{n} = \frac{1}{\sqrt{n}}$, so $\displaystyle\sum_{n=2}^\infty \frac{\sqrt{n}}{n-1}$ diverges by comparison with the divergent (partial) $p$-series $\displaystyle\sum_{n=2}^\infty \frac{1}{\sqrt{n}}$ $\quad (p = \frac{1}{2} \leq 1)$.

**11.** $\displaystyle 1 + \frac{1}{8} + \frac{1}{27} + \frac{1}{64} + \frac{1}{125} + \cdots = \sum_{n=1}^\infty \frac{1}{n^3}$. This is a $p$-series with $p = 3 > 1$, so it converges by (1).

**12.** $\displaystyle\sum_{n=1}^\infty \frac{1}{n^4}$ and $\displaystyle\sum_{n=1}^\infty \frac{1}{n^{3/2}}$ are convergent $p$-series with $p = 4 > 1$ and $p = \frac{3}{2} > 1$, respectively. Thus,

$\displaystyle\sum_{n=1}^\infty \left(\frac{5}{n^4} + \frac{4}{n\sqrt{n}}\right) = 5\sum_{n=1}^\infty \frac{1}{n^4} + 4\sum_{n=1}^\infty \frac{1}{n^{3/2}}$ is convergent by Theorems 8.2.8(i) and 8.2.8(ii).

**13.** $f(x) = xe^{-x}$ is continuous and positive on $[1, \infty)$. $f'(x) = -xe^{-x} + e^{-x} = e^{-x}(1-x) < 0$ for $x > 1$, so $f$ is decreasing on $[1, \infty)$. Thus, the Integral Test applies.

$$\begin{aligned}\int_1^\infty xe^{-x}\,dx &= \lim_{b\to\infty} \int_1^b xe^{-x}\,dx = \lim_{b\to\infty} \left[-xe^{-x} - e^{-x}\right]_1^b \quad \text{[by parts]}\\ &= \lim_{b\to\infty} \left[-be^{-b} - e^{-b} + e^{-1} + e^{-1}\right] = 2/e\end{aligned}$$

since $\lim_{b\to\infty} be^{-b} = \lim_{b\to\infty} (b/e^b) \overset{\text{H}}{=} \lim_{b\to\infty} (1/e^b) = 0$ and $\lim_{b\to\infty} e^{-b} = 0$. Thus, $\sum_{n=1}^\infty ne^{-n}$ converges.

**14.** $\displaystyle f(x) = \frac{x^2}{x^3+1}$ is continuous and positive on $[2, \infty)$, and also decreasing since $\displaystyle f'(x) = \frac{x(2-x^3)}{(x^3+1)^2} < 0$ for $x \geq 2$, so we can use the Integral Test [note that $f$ is *not* decreasing on $[1, \infty)$].

$\displaystyle\int_2^\infty \frac{x^2}{x^3+1}\,dx = \lim_{t\to\infty} \left[\frac{1}{3}\ln(x^3+1)\right]_2^t = \frac{1}{3}\lim_{t\to\infty} \left[\ln(t^3+1) - \ln 9\right] = \infty$, so the series $\displaystyle\sum_{n=2}^\infty \frac{n^2}{n^3+1}$ diverges, and so does the given series, $\displaystyle\sum_{n=1}^\infty \frac{n^2}{n^3+1}$.

*Another solution:* Use the Limit Comparison Test with $\displaystyle a_n = \frac{n^2}{n^3+1}$ and $\displaystyle b_n = \frac{1}{n}$:

$\displaystyle\lim_{n\to\infty} \frac{a_n}{b_n} = \lim_{n\to\infty} \frac{n^2 \cdot n}{n^3+1} = \lim_{n\to\infty} \frac{1}{1+1/n^3} = 1 > 0$. Since the harmonic series $\displaystyle\sum_{n=1}^\infty \frac{1}{n}$ diverges, so does $\displaystyle\sum_{n=1}^\infty \frac{n^2}{n^3+1}$.

**15.** $\displaystyle f(x) = \frac{1}{x\ln x}$ is continuous and positive on $[2, \infty)$, and also decreasing since $\displaystyle f'(x) = -\frac{1+\ln x}{x^2(\ln x)^2} < 0$ for $x > 2$, so we can use the Integral Test. $\displaystyle\int_2^\infty \frac{1}{x\ln x}\,dx = \lim_{t\to\infty} \left[\ln(\ln x)\right]_2^t = \lim_{t\to\infty} \left[\ln(\ln t) - \ln(\ln 2)\right] = \infty$, so the series $\displaystyle\sum_{n=2}^\infty \frac{1}{n\ln n}$ diverges.

**16.** $\dfrac{n^2-1}{3n^4+1} < \dfrac{n^2}{3n^4+1} < \dfrac{n^2}{3n^4} = \dfrac{1}{3}\dfrac{1}{n^2}$. $\sum\limits_{n=1}^{\infty} \dfrac{n^2-1}{3n^4+1}$ converges by comparison with $\sum\limits_{n=1}^{\infty} \dfrac{1}{3n^2}$, which converges because it is a constant multiple of a convergent $p$-series ($p = 2 > 1$). The terms of the given series are positive for $n > 1$, which is good enough.

**17.** $\dfrac{\cos^2 n}{n^2+1} \le \dfrac{1}{n^2+1} < \dfrac{1}{n^2}$, so the series $\sum\limits_{n=1}^{\infty} \dfrac{\cos^2 n}{n^2+1}$ converges by comparison with the $p$-series $\sum\limits_{n=1}^{\infty} \dfrac{1}{n^2}$ $(p = 2 > 1)$.

**18.** $\dfrac{4+3^n}{2^n} > \dfrac{3^n}{2^n} = \left(\dfrac{3}{2}\right)^n$ for all $n \ge 1$, so $\sum\limits_{n=1}^{\infty} \dfrac{4+3^n}{2^n}$ diverges by comparison with the divergent geometric series $\sum\limits_{n=1}^{\infty} \left(\frac{3}{2}\right)^n$.

**19.** $\dfrac{n-1}{n\,4^n}$ is positive for $n > 1$ and $\dfrac{n-1}{n\,4^n} < \dfrac{n}{n\,4^n} = \dfrac{1}{4^n} = \left(\dfrac{1}{4}\right)^n$, so $\sum\limits_{n=1}^{\infty} \dfrac{n-1}{n\,4^n}$ converges by comparison with the convergent geometric series $\sum\limits_{n=1}^{\infty} \left(\dfrac{1}{4}\right)^n$.

**20.** $\dfrac{1}{\sqrt{n^3+1}} < \dfrac{1}{\sqrt{n^3}} = \dfrac{1}{n^{3/2}}$, so $\sum\limits_{n=1}^{\infty} \dfrac{1}{\sqrt{n^3+1}}$ converges by comparison with the convergent $p$-series $\sum\limits_{n=1}^{\infty} \dfrac{1}{n^{3/2}}$ $\left(p = \frac{3}{2} > 1\right)$.

**21.** Use the Limit Comparison Test with $a_n = \dfrac{1}{\sqrt{n^2+1}}$ and $b_n = \dfrac{1}{n}$:

$\lim\limits_{n\to\infty} \dfrac{a_n}{b_n} = \lim\limits_{n\to\infty} \dfrac{n}{\sqrt{n^2+1}} = \lim\limits_{n\to\infty} \dfrac{1}{\sqrt{1+(1/n^2)}} = 1 > 0$. Since the harmonic series $\sum\limits_{n=1}^{\infty} \dfrac{1}{n}$ diverges, so does $\sum\limits_{n=1}^{\infty} \dfrac{1}{\sqrt{n^2+1}}$.

**22.** Use the Limit Comparison Test with $a_n = \dfrac{1}{2n+3}$ and $b_n = \dfrac{1}{n}$: $\lim\limits_{n\to\infty} \dfrac{a_n}{b_n} = \lim\limits_{n\to\infty} \dfrac{n}{2n+3} = \lim\limits_{n\to\infty} \dfrac{1}{2+(3/n)} = \dfrac{1}{2} > 0$. Since the harmonic series $\sum\limits_{n=1}^{\infty} \dfrac{1}{n}$ diverges, so does $\sum\limits_{n=1}^{\infty} \dfrac{1}{2n+3}$.

**23.** $\dfrac{2+(-1)^n}{n\sqrt{n}} \le \dfrac{3}{n\sqrt{n}}$, and $\sum\limits_{n=1}^{\infty} \dfrac{3}{n\sqrt{n}}$ converges because it is a constant multiple of the convergent $p$-series $\sum\limits_{n=1}^{\infty} \dfrac{1}{n\sqrt{n}}$ $\left(p = \frac{3}{2} > 1\right)$, so the given series converges by the Comparison Test.

**24.** $\dfrac{1+\sin n}{10^n} \le \dfrac{2}{10^n}$ and $\sum\limits_{n=0}^{\infty} \dfrac{2}{10^n} = 2\sum\limits_{n=0}^{\infty} \left(\dfrac{1}{10}\right)^n$, so the given series converges by comparison with a constant multiple of a convergent geometric series.

**25.** Use the Limit Comparison Test with $a_n = \sin\left(\dfrac{1}{n}\right)$ and $b_n = \dfrac{1}{n}$. Then $\sum a_n$ and $\sum b_n$ are series with positive terms and

$\lim\limits_{n\to\infty} \dfrac{a_n}{b_n} = \lim\limits_{n\to\infty} \dfrac{\sin(1/n)}{1/n} = \lim\limits_{\theta\to 0} \dfrac{\sin\theta}{\theta} = 1 > 0$. Since $\sum\limits_{n=1}^{\infty} b_n$ is the divergent harmonic series, $\sum\limits_{n=1}^{\infty} \sin(1/n)$ also diverges.

[Note that we could also use l'Hospital's Rule to evaluate the limit:

$$\lim_{x\to\infty} \frac{\sin(1/x)}{1/x} \overset{\text{H}}{=} \lim_{x\to\infty} \frac{\cos(1/x)\cdot(-1/x^2)}{-1/x^2} = \lim_{x\to\infty} \cos\frac{1}{x} = \cos 0 = 1.]$$

**26.** If $a_n = \dfrac{n+5}{\sqrt[3]{n^7+n^2}}$ and $b_n = \dfrac{n}{\sqrt[3]{n^7}} = \dfrac{n}{n^{7/3}} = \dfrac{1}{n^{4/3}}$, then

$$\lim_{n\to\infty} \frac{a_n}{b_n} = \lim_{n\to\infty} \frac{n^{7/3}+5n^{4/3}}{(n^7+n^2)^{1/3}} \cdot \frac{n^{-7/3}}{n^{-7/3}} = \lim_{n\to\infty} \frac{1+5/n}{\left[(n^7+n^2)/n^7\right]^{1/3}} = \lim_{n\to\infty} \frac{1+5/n}{(1+1/n^5)^{1/3}} = \frac{1+0}{(1+0)^{1/3}} = 1 > 0,$$

so $\sum\limits_{n=1}^{\infty} \dfrac{n+5}{\sqrt[3]{n^7+n^2}}$ converges by the Limit Comparison Test with the convergent $p$-series $\sum\limits_{n=1}^{\infty} \dfrac{1}{n^{4/3}}$.

**27.** We have already shown (in Exercise 15) that when $p = 1$ the series $\sum_{n=2}^{\infty} \frac{1}{n(\ln n)^p}$ diverges, so assume that $p \neq 1$.

$f(x) = \frac{1}{x(\ln x)^p}$ is continuous and positive on $[2, \infty)$, and $f'(x) = -\frac{p + \ln x}{x^2(\ln x)^{p+1}} < 0$ if $x > e^{-p}$, so that $f$ is eventually decreasing and we can use the Integral Test.

$$\int_2^{\infty} \frac{1}{x(\ln x)^p}\,dx = \lim_{t\to\infty}\left[\frac{(\ln x)^{1-p}}{1-p}\right]_2^t \quad [\text{for } p \neq 1] \quad = \lim_{t\to\infty}\left[\frac{(\ln t)^{1-p}}{1-p}\right] - \frac{(\ln 2)^{1-p}}{1-p}$$

This limit exists whenever $1 - p < 0 \iff p > 1$, so the series converges for $p > 1$.

**28.** (a) $f(x) = 1/x^4$ is positive and continuous and $f'(x) = -4/x^5$ is negative for $x > 0$, and so the Integral Test applies.

$\sum_{n=1}^{\infty} \frac{1}{n^4} \approx s_{10} = \frac{1}{1^4} + \frac{1}{2^4} + \frac{1}{3^4} + \cdots + \frac{1}{10^4} \approx 1.082037.$

$R_{10} \leq \int_{10}^{\infty} \frac{1}{x^4}\,dx = \lim_{t\to\infty}\left[\frac{1}{-3x^3}\right]_{10}^t = \lim_{t\to\infty}\left(-\frac{1}{3t^3} + \frac{1}{3\,(10)^3}\right) = \frac{1}{3000}$, so the error is at most $0.000\overline{3}$.

(b) $s_{10} + \int_{11}^{\infty} \frac{1}{x^4}\,dx \leq s \leq s_{10} + \int_{10}^{\infty} \frac{1}{x^4}\,dx \Rightarrow s_{10} + \frac{1}{3(11)^3} \leq s \leq s_{10} + \frac{1}{3(10)^3} \Rightarrow$ $1.082037 + 0.000250 = 1.082287 \leq s \leq 1.082037 + 0.000333 = 1.082370$, so we get $s \approx 1.08233$ with error $\leq 0.00005$.

(c) $R_n \leq \int_n^{\infty} \frac{1}{x^4}\,dx = \frac{1}{3n^3}$. So $R_n < 0.00001 \Rightarrow \frac{1}{3n^3} < \frac{1}{10^5} \Rightarrow 3n^3 > 10^5 \Rightarrow n > \sqrt[3]{(10)^5/3} \approx 32.2$, that is, for $n > 32$.

**29.** (a) $f(x) = \frac{1}{x^2}$ is positive and continuous and $f'(x) = -\frac{2}{x^3}$ is negative for $x > 0$, and so the Integral Test applies.

$\sum_{n=1}^{\infty} \frac{1}{n^2} \approx s_{10} = \frac{1}{1^2} + \frac{1}{2^2} + \frac{1}{3^2} + \cdots + \frac{1}{10^2} \approx 1.549768.$

$R_{10} \leq \int_{10}^{\infty} \frac{1}{x^2}\,dx = \lim_{t\to\infty}\left[\frac{-1}{x}\right]_{10}^t = \lim_{t\to\infty}\left(-\frac{1}{t} + \frac{1}{10}\right) = \frac{1}{10}$, so the error is at most $0.1$.

(b) $s_{10} + \int_{11}^{\infty} \frac{1}{x^2}\,dx \leq s \leq s_{10} + \int_{10}^{\infty} \frac{1}{x^2}\,dx \Rightarrow s_{10} + \frac{1}{11} \leq s \leq s_{10} + \frac{1}{10} \Rightarrow$ $1.549768 + 0.090909 = 1.640677 \leq s \leq 1.549768 + 0.1 = 1.649768$, so we get $s \approx 1.64522$ (the average of 1.640677 and 1.649768) with error $\leq 0.005$ (the maximum of $1.649768 - 1.64522$ and $1.64522 - 1.640677$, rounded up).

(c) $R_n \leq \int_n^{\infty} \frac{1}{x^2}\,dx = \frac{1}{n}$. So $R_n < 0.001$ if $\frac{1}{n} < \frac{1}{1000} \iff n > 1000$.

**30.** $f(x) = 1/x^5$ is positive and continuous and $f'(x) = -5/x^6$ is negative for $x > 0$, and so the Integral Test applies. Using (3), $R_n \leq \int_n^{\infty} x^{-5}\,dx = \lim_{t\to\infty}\left[\frac{-1}{4x^4}\right]_n^t = \frac{1}{4n^4}$. If we take $n = 5$, then $s_5 \approx 1.036662$ and $R_5 \leq 0.0004$. So $s \approx s_5 \approx 1.037$.

**31.** $f(x) = x^{-3/2}$ is positive and continuous and $f'(x) = -\frac{3}{2}x^{-5/2}$ is negative for $x > 0$, so the Integral Test applies. From the end of Example 7, we see that the error is at most half the length of the interval. From (4), the interval is $\left(s_n + \int_{n+1}^{\infty} f(x)\,dx,\, s_n + \int_n^{\infty} f(x)\,dx\right)$, so its length is $\int_n^{\infty} f(x)\,dx - \int_{n+1}^{\infty} f(x)\,dx = \int_n^{n+1} f(x)\,dx$. Thus, we need $n$ such that $0.01 > \frac{1}{2}\int_n^{n+1} x^{-3/2}\,dx = \frac{1}{2}\left[\frac{-2}{\sqrt{x}}\right]_n^{n+1} = \frac{1}{\sqrt{n}} - \frac{1}{\sqrt{n+1}} \iff n > 13.08$ (use a graphing calculator to solve $1/\sqrt{x} - 1/\sqrt{x+1} < 0.01$). Again from the end of Example 7, we approximate $s$ by the midpoint of this interval. In general, the midpoint is $\frac{1}{2}\left[\left(s_n + \int_{n+1}^{\infty} f(x)\,dx\right) + \left(s_n + \int_n^{\infty} f(x)\,dx\right)\right] = s_n + \frac{1}{2}\left(\int_{n+1}^{\infty} f(x)\,dx + \int_n^{\infty} f(x)\,dx\right)$. So using $n = 14$, we have $s \approx s_{14} + \frac{1}{2}\left(\int_{14}^{\infty} x^{-3/2}\,dx + \int_{15}^{\infty} x^{-3/2}\,dx\right) \approx 2.0872 + \frac{1}{\sqrt{14}} + \frac{1}{\sqrt{15}} \approx 2.6127 \approx 2.61$. Any larger value of $n$ will also work. For instance, $s \approx s_{30} + \frac{1}{\sqrt{30}} + \frac{1}{\sqrt{31}} \approx 2.6124$.

**32.** $f(x) = \dfrac{1}{x(\ln x)^2}$ is positive and continuous and $f'(x) = -\dfrac{\ln x + 2}{x^2(\ln x)^3}$ is negative for $x > 1$, so the Integral Test applies.

Using (3), we need $0.01 > \displaystyle\int_n^\infty \frac{dx}{x(\ln x)^2} = \lim_{t\to\infty}\left[\frac{-1}{\ln x}\right]_n^t = \frac{1}{\ln n}$. This is true for $n > e^{100}$, so we would have to take this many terms, which would be problematic because $e^{100} \approx 2.7 \times 10^{43}$.

**33.** $\displaystyle\sum_{n=1}^{10} \frac{1}{n^4+n^2} = \frac{1}{2} + \frac{1}{20} + \frac{1}{90} + \cdots + \frac{1}{10{,}100} \approx 0.567975$. Now $\dfrac{1}{n^4+n^2} < \dfrac{1}{n^4}$, so using the reasoning and notation of Example 8, the error is $R_{10} \le T_{10} = \displaystyle\sum_{n=11}^\infty \frac{1}{n^4} \le \int_{10}^\infty \frac{dx}{x^4} = \lim_{t\to\infty}\left[-\frac{x^{-3}}{3}\right]_{10}^t = \frac{1}{3000} = 0.000\overline{3}$.

**34.** $\displaystyle\sum_{n=1}^{10} \frac{n}{(n+1)3^n} = \frac{1}{6} + \frac{2}{27} + \frac{3}{108} + \cdots + \frac{10}{649{,}539} \approx 0.283597$. Now $\dfrac{n}{(n+1)3^n} < \dfrac{n}{n \cdot 3^n} = \dfrac{1}{3^n}$, so the error is $R_{10} \le T_{10} = \displaystyle\sum_{n=11}^\infty \frac{1}{3^n} = \frac{1/3^{11}}{1-1/3} \approx 0.0000085$.

**35.** (a) From the figure, $a_2 + a_3 + \cdots + a_n \le \int_1^n f(x)\,dx$, so with

$$f(x) = \frac{1}{x},\ \frac{1}{2} + \frac{1}{3} + \frac{1}{4} + \cdots + \frac{1}{n} \le \int_1^n \frac{1}{x}\,dx = \ln n. \text{ Thus,}$$

$$s_n = 1 + \frac{1}{2} + \frac{1}{3} + \frac{1}{4} + \cdots + \frac{1}{n} \le 1 + \ln n.$$

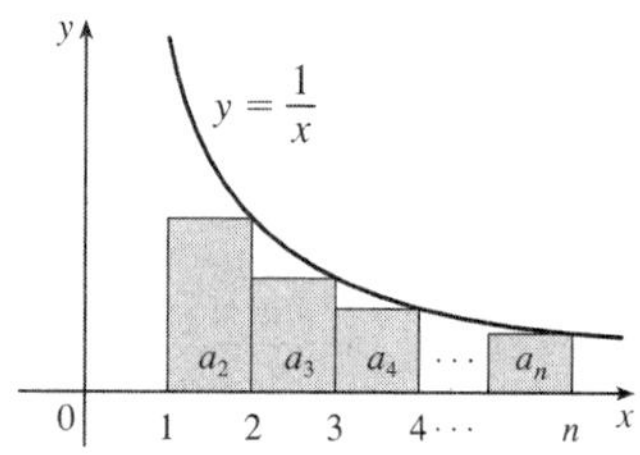


(b) By part (a), $s_{10^6} \le 1 + \ln 10^6 \approx 14.82 < 15$ and $s_{10^9} \le 1 + \ln 10^9 \approx 21.72 < 22$.

**36.** $\displaystyle\sum_{n=1}^\infty n^{-1.001} = \sum_{n=1}^\infty \frac{1}{n^{1.001}}$ is a convergent $p$-series with $p = 1.001 > 1$. Using (3), we get

$$R_n \le \int_n^\infty x^{-1.001}\,dx = \lim_{t\to\infty}\left[\frac{x^{-0.001}}{-0.001}\right]_n^t = -1000\lim_{t\to\infty}\left[\frac{1}{x^{0.001}}\right]_n^t = -1000\left(-\frac{1}{n^{0.001}}\right) = \frac{1000}{n^{0.001}}.$$

We want $R_n < 0.000\,000\,005 \Leftrightarrow \dfrac{1000}{n^{0.001}} < 5 \times 10^{-9} \Leftrightarrow n^{0.001} > \dfrac{1000}{5 \times 10^{-9}} \Leftrightarrow$

$n > (2 \times 10^{11})^{1000} = 2^{1000} \times 10^{11{,}000} \approx 1.07 \times 10^{301} \times 10^{11{,}000} = 1.07 \times 10^{11{,}301}$.

**37.** Since $\dfrac{d_n}{10^n} \le \dfrac{9}{10^n}$ for each $n$, and since $\displaystyle\sum_{n=1}^\infty \frac{9}{10^n}$ is a convergent geometric series $\left(|r| = \frac{1}{10} < 1\right)$, $0.d_1d_2d_3\ldots = \displaystyle\sum_{n=1}^\infty \frac{d_n}{10^n}$ will always converge by the Comparison Test.

**38.** $b^{\ln n} = \left(e^{\ln b}\right)^{\ln n} = \left(e^{\ln n}\right)^{\ln b} = n^{\ln b} = \dfrac{1}{n^{-\ln b}}$. This is a $p$-series, which converges for all $b$ such that $-\ln b > 1 \Leftrightarrow \ln b < -1 \Leftrightarrow b < e^{-1} \Leftrightarrow b < 1/e$ [with $b > 0$].

**39.** Yes. Since $\sum a_n$ is a convergent series with positive terms, $\lim\limits_{n\to\infty} a_n = 0$ by Theorem 8.2.6, and $\sum b_n = \sum \sin(a_n)$ is a series with positive terms (for large enough $n$). We have $\lim\limits_{n\to\infty} \dfrac{b_n}{a_n} = \lim\limits_{n\to\infty} \dfrac{\sin(a_n)}{a_n} = 1 > 0$ by Theorem 3.4.2. Thus, $\sum b_n$ is also convergent by the Limit Comparison Test.

**40.** First we observe that, by l'Hospital's Rule, $\lim\limits_{x\to 0} \dfrac{\ln(1+x)}{x} = \lim\limits_{x\to 0} \dfrac{1}{1+x} = 1$. Also, if $\sum a_n$ converges, then $\lim\limits_{n\to\infty} a_n = 0$ by Theorem 8.2.6. Therefore, $\lim\limits_{n\to\infty} \dfrac{\ln(1+a_n)}{a_n} = \lim\limits_{x\to 0} \dfrac{\ln(1+x)}{x} = 1 > 0$. We are given that $\sum a_n$ is convergent and $a_n > 0$. Thus, $\sum \ln(1+a_n)$ is convergent by the Limit Comparison Test.

## 8.4 Other Convergence Tests

**1.** (a) An alternating series is a series whose terms are alternately positive and negative.

(b) An alternating series $\sum_{n=1}^{\infty}(-1)^{n-1}b_n$ converges if $0 < b_{n+1} \le b_n$ for all $n$ and $\lim\limits_{n\to\infty} b_n = 0$. (This is the Alternating Series Test.)

(c) The error involved in using the partial sum $s_n$ as an approximation to the total sum $s$ is the remainder $R_n = s - s_n$ and the size of the error is smaller than $b_{n+1}$; that is, $|R_n| \le b_{n+1}$. (This is the Alternating Series Estimation Theorem.)

**2.** (a) Since $\lim\limits_{n\to\infty}\left|\dfrac{a_{n+1}}{a_n}\right| = 8 > 1$, part (b) of the Ratio Test tells us that the series $\sum a_n$ is divergent.

(b) Since $\lim\limits_{n\to\infty}\left|\dfrac{a_{n+1}}{a_n}\right| = 0.8 < 1$, part (a) of the Ratio Test tells us that the series $\sum a_n$ is absolutely convergent (and therefore convergent).

(c) Since $\lim\limits_{n\to\infty}\left|\dfrac{a_{n+1}}{a_n}\right| = 1$, the Ratio Test fails and the series $\sum a_n$ might converge or it might diverge.

**3.** $\dfrac{4}{7} - \dfrac{4}{8} + \dfrac{4}{9} - \dfrac{4}{10} + \dfrac{4}{11} - \cdots = \sum\limits_{n=1}^{\infty}(-1)^{n-1}\dfrac{4}{n+6}$. Now $b_n = \dfrac{4}{n+6} > 0$, $\{b_n\}$ is decreasing, and $\lim\limits_{n\to\infty} b_n = 0$, so the series converges by the Alternating Series Test.

**4.** $-\dfrac{1}{3} + \dfrac{2}{4} - \dfrac{3}{5} + \dfrac{4}{6} - \dfrac{5}{7} + \cdots = \sum\limits_{n=1}^{\infty}(-1)^n\dfrac{n}{n+2}$. Here $a_n = (-1)^n\dfrac{n}{n+2}$. Since $\lim\limits_{n\to\infty} a_n \ne 0$ (in fact the limit does not exist), the series diverges by the Test for Divergence.

**5.** $b_n = \dfrac{1}{\sqrt{n}} > 0$, $\{b_n\}$ is decreasing, and $\lim\limits_{n\to\infty} b_n = 0$, so the series $\sum\limits_{n=1}^{\infty}\dfrac{(-1)^{n-1}}{\sqrt{n}}$ converges by the Alternating Series Test.

**6.** $\sum\limits_{n=1}^{\infty} a_n = \sum\limits_{n=1}^{\infty}(-1)^n\dfrac{\sqrt{n}}{1+2\sqrt{n}} = \sum\limits_{n=1}^{\infty}(-1)^n b_n$. Now $\lim\limits_{n\to\infty} b_n = \lim\limits_{n\to\infty}\dfrac{1}{2+1/\sqrt{n}} = \dfrac{1}{2} \ne 0$. Since $\lim\limits_{n\to\infty} a_n \ne 0$ (in fact the limit does not exist), the series diverges by the Test for Divergence.

**7.** $\sum\limits_{n=1}^{\infty} a_n = \sum\limits_{n=1}^{\infty}(-1)^n\dfrac{3n-1}{2n+1} = \sum\limits_{n=1}^{\infty}(-1)^n b_n$. Now $\lim\limits_{n\to\infty} b_n = \lim\limits_{n\to\infty}\dfrac{3-1/n}{2+1/n} = \dfrac{3}{2} \ne 0$. Since $\lim\limits_{n\to\infty} a_n \ne 0$ (in fact the limit does not exist), the series diverges by the Test for Divergence.

**8.** $\sum\limits_{n=1}^{\infty}(-1)^{n-1}\left(\dfrac{\ln n}{n}\right) = 0 + \sum\limits_{n=2}^{\infty}(-1)^{n-1}\left(\dfrac{\ln n}{n}\right)$. $b_n = \dfrac{\ln n}{n} > 0$ for $n \ge 2$, and if $f(x) = \dfrac{\ln x}{x}$, then $f'(x) = \dfrac{1-\ln x}{x^2} < 0$ for $x > e$, so $\{b_n\}$ is eventually decreasing. Also,

$$\lim_{n\to\infty} b_n = \lim_{n\to\infty}\frac{\ln n}{n} = \lim_{x\to\infty}\frac{\ln x}{x} \overset{\text{H}}{=} \lim_{x\to\infty}\frac{1/x}{1} = 0,$$ so the series converges by the Alternating Series Test.

**9.** $\sum\limits_{n=1}^{\infty}\dfrac{(-1)^{n-1}}{n} = 1 - \dfrac{1}{2} + \dfrac{1}{3} - \dfrac{1}{4} + \cdots + \dfrac{1}{49} - \dfrac{1}{50} + \dfrac{1}{51} - \dfrac{1}{52} + \cdots$. The 50th partial sum of this series is an underestimate, since $\sum\limits_{n=1}^{\infty}\dfrac{(-1)^{n-1}}{n} = s_{50} + \left(\dfrac{1}{51} - \dfrac{1}{52}\right) + \left(\dfrac{1}{53} - \dfrac{1}{54}\right) + \cdots$, and the terms in parentheses are all positive. The result can be seen geometrically in Figure 1.

**10.**

| $n$ | $a_n$ | $s_n$ |
|---|---|---|
| 1 | 1 | 1 |
| 2 | $-0.125$ | 0.875 |
| 3 | 0.03704 | 0.91204 |
| 4 | $-0.01563$ | 0.89641 |
| 5 | 0.008 | 0.90441 |
| 6 | $-0.00463$ | 0.89978 |
| 7 | 0.00292 | 0.90270 |
| 8 | $-0.00195$ | 0.90074 |
| 9 | 0.00137 | 0.90212 |
| 10 | $-0.001$ | 0.90112 |

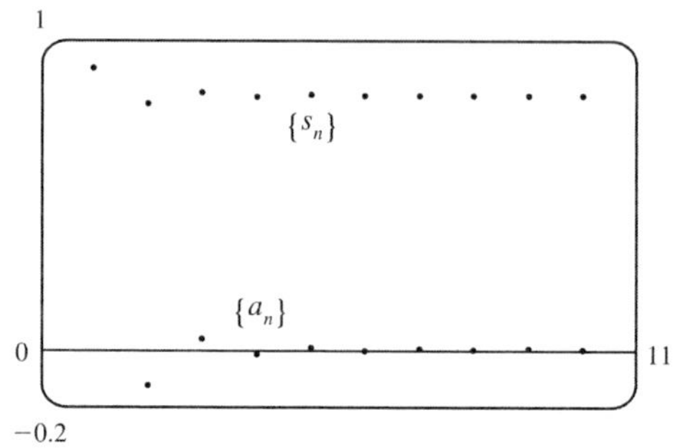


By the Alternating Series Estimation Theorem, the error in the approximation $\sum_{n=1}^{\infty} \frac{(-1)^{n-1}}{n^3} \approx 0.90112$ is $|s - s_{10}| \le b_{11} = 1/11^3 \approx 0.0007513$.

**11.** If $p > 0$, $\frac{1}{(n+1)^p} \le \frac{1}{n^p}$ ($\{1/n^p\}$ is decreasing) and $\lim_{n\to\infty} \frac{1}{n^p} = 0$, so the series converges by the Alternating Series Test. If $p \le 0$, $\lim_{n\to\infty} \frac{(-1)^{n-1}}{n^p}$ does not exist, so the series diverges by the Test for Divergence. Thus, $\sum_{n=1}^{\infty} \frac{(-1)^{n-1}}{n^p}$ converges $\Leftrightarrow$ $p > 0$.

**12.** The series $\sum_{n=1}^{\infty} \frac{(-1)^n}{n\,5^n}$ satisfies (i) of the Alternating Series Test because $\frac{1}{(n+1)5^{n+1}} < \frac{1}{n\,5^n}$ and (ii) $\lim_{n\to\infty} \frac{1}{n\,5^n} = 0$, so the series is convergent. Now $b_4 = \frac{1}{4 \cdot 5^4} = 0.0004 > 0.0001$ and $b_5 = \frac{1}{5 \cdot 5^5} = 0.000064 < 0.0001$, so by the Alternating Series Estimation Theorem, $n = 4$. (That is, since the 5th term is less than the desired error, we need to add the first 4 terms to get the sum to the desired accuracy.)

**13.** The series $\sum_{n=1}^{\infty} \frac{(-1)^{n+1}}{n^6}$ satisfies (i) of the Alternating Series Test because $\frac{1}{(n+1)^6} < \frac{1}{n^6}$ and (ii) $\lim_{n\to\infty} \frac{1}{n^6} = 0$, so the series is convergent. Now $b_5 = \frac{1}{5^6} = 0.000064 > 0.00005$ and $b_6 = \frac{1}{6^6} \approx 0.00002 < 0.00005$, so by the Alternating Series Estimation Theorem, $n = 5$. (That is, since the 6th term is less than the desired error, we need to add the first 5 terms to get the sum to the desired accuracy.)

**14.** Using the Ratio Test with the series $\sum_{n=1}^{\infty} (-1)^{n-1} n e^{-n} = \sum_{n=1}^{\infty} (-1)^{n-1} \frac{n}{e^n}$,

$$\lim_{n\to\infty} \left| \frac{a_{n+1}}{a_n} \right| = \lim_{n\to\infty} \left| \frac{(-1)^n (n+1)}{e^{n+1}} \cdot \frac{e^n}{(-1)^{n-1} n} \right| = \lim_{n\to\infty} \left| \frac{(-1)^1 (n+1)}{en} \right| = \frac{1}{e} \lim_{n\to\infty} \frac{n+1}{n} = \frac{1}{e}(1) = \frac{1}{e} < 1,$$

so the series is absolutely convergent (and therefore convergent). Now $b_6 = 6/e^6 \approx 0.015 > 0.01$ and $b_7 = 7/e^7 \approx 0.006 < 0.01$, so by the Alternating Series Estimation Theorem, $n = 6$. (That is, since the 7th term is less than the desired error, we need to add the first 6 terms to get the sum to the desired accuracy.)

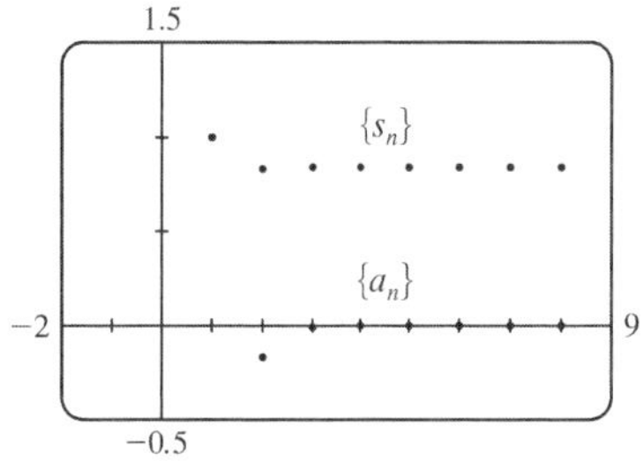


**15.** The graph gives us an estimate for the sum of the series $\sum_{n=1}^{\infty} \frac{(-1)^{n-1}}{(2n-1)!}$ of 0.84.

$$b_5 = \frac{1}{(2 \cdot 5 - 1)!} = \frac{1}{362{,}880} \approx 0.000\,003, \text{ so}$$

$$\sum_{n=1}^{\infty} \frac{(-1)^{n-1}}{(2n-1)!} \approx s_4 = \sum_{n=1}^{4} \frac{(-1)^{n-1}}{(2n-1)!} = 1 - \frac{1}{6} + \frac{1}{120} - \frac{1}{5040} \approx 0.841468.$$

Adding $b_5$ to $s_4$ does not change the fourth decimal place of $s_4$, so the sum of the series, correct to four decimal places, is 0.8415.

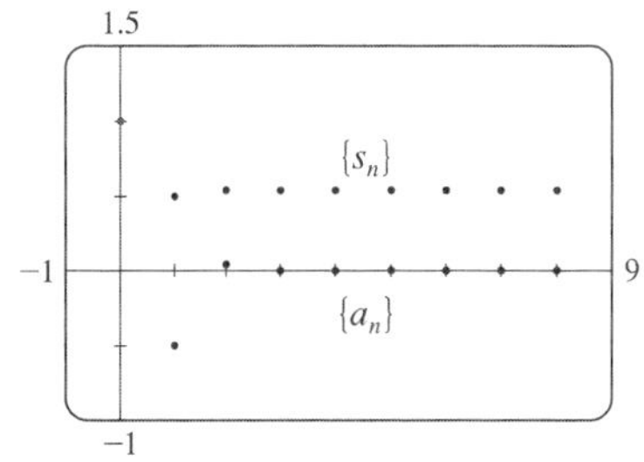


**16.** The graph gives us an estimate for the sum of the series $\sum_{n=0}^{\infty} \frac{(-1)^n}{(2n)!}$ of 0.54. $b_4 = \frac{1}{(2 \cdot 4)!} = \frac{1}{40{,}320} \approx 0.000\,025$, so

$$\sum_{n=0}^{\infty} \frac{(-1)^n}{(2n)!} \approx s_3 = \sum_{n=0}^{3} \frac{(-1)^n}{(2n)!} = 1 - \frac{1}{2} + \frac{1}{24} - \frac{1}{720} \approx 0.540278.$$

Adding $b_4$ to $s_3$ does not change the fourth decimal place of $s_3$, so the sum of the series, correct to four decimal places, is 0.5403.

**17.** $b_7 = \frac{7^2}{10^7} = 0.000\,004\,9$, so

$\sum_{n=1}^{\infty} \frac{(-1)^{n-1} n^2}{10^n} \approx s_6 = \sum_{n=1}^{6} \frac{(-1)^{n-1} n^2}{10^n} = \frac{1}{10} - \frac{4}{100} + \frac{9}{1000} - \frac{16}{10{,}000} + \frac{25}{100{,}000} - \frac{36}{1{,}000{,}000} = 0.067\,614$. Adding $b_7$ to $s_6$ does not change the fourth decimal place of $s_6$, so the sum of the series, correct to four decimal places, is 0.0676.

**18.** $b_6 = \frac{1}{3^6 \cdot 6!} = \frac{1}{524{,}880} \approx 0.000\,001\,9$, so

$\sum_{n=1}^{\infty} \frac{(-1)^n}{3^n n!} \approx s_5 = \sum_{n=1}^{5} \frac{(-1)^n}{3^n n!} = -\frac{1}{3} + \frac{1}{18} - \frac{1}{162} + \frac{1}{1944} - \frac{1}{29{,}160} \approx -0.283\,471$. Adding $b_6$ to $s_5$ does not change the fourth decimal place of $s_5$, so the sum of the series, correct to four decimal places, is $-0.2835$.

**19.** Using the Ratio Test, $\lim_{n\to\infty} \left| \frac{a_{n+1}}{a_n} \right| = \lim_{n\to\infty} \left| \frac{(-3)^{n+1}/(n+1)^3}{(-3)^n/n^3} \right| = \lim_{n\to\infty} \left| \frac{(-3)n^3}{(n+1)^3} \right| = 3 \lim_{n\to\infty} \left( \frac{n}{n+1} \right)^3 = 3 > 1$, so the series diverges.

**20.** The series $\sum_{n=1}^{\infty} \frac{n^2}{2^n}$ has positive terms and $\lim_{n\to\infty} \frac{a_{n+1}}{a_n} = \lim_{n\to\infty} \left[ \frac{(n+1)^2}{2^{n+1}} \cdot \frac{2^n}{n^2} \right] = \lim_{n\to\infty} \left( 1 + \frac{1}{n} \right)^2 \cdot \frac{1}{2} = \frac{1}{2} < 1$, so the series is absolutely convergent by the Ratio Test.

**21.** $\sum_{n=0}^{\infty} \frac{(-10)^n}{n!}$. Using the Ratio Test, $\lim_{n\to\infty} \left| \frac{a_{n+1}}{a_n} \right| = \lim_{n\to\infty} \left| \frac{(-10)^{n+1}}{(n+1)!} \cdot \frac{n!}{(-10)^n} \right| = \lim_{n\to\infty} \left| \frac{-10}{n+1} \right| = 0 < 1$, so the series is absolutely convergent.

**22.** $\sum_{n=1}^{\infty} \left| (-1)^n \frac{n}{n^2+1} \right| = \sum_{n=1}^{\infty} \frac{n}{n^2+1} = \sum_{n=1}^{\infty} a_n$. If $b_n = \frac{1}{n}$, then $\sum_{n=1}^{\infty} b_n$ is the divergent harmonic series. Applying the Limit Comparison Test, $\lim_{n\to\infty} \frac{a_n}{b_n} = \lim_{n\to\infty} \frac{n/(n^2+1)}{1/n} = \lim_{n\to\infty} \frac{n^2}{n^2+1} = 1 > 0$, so both series diverge and the given series is *not* absolutely convergent. (The Integral Test could also be used.)

**23.** Consider the series whose terms are the absolute values of the terms of the given series. $\sum_{n=1}^{\infty}\left|\frac{(-1)^{n-1}}{\sqrt{n}}\right| = \sum_{n=1}^{\infty}\frac{1}{n^{1/2}}$, which is a divergent $p$-series ($p = \frac{1}{2} \leq 1$). Thus, $\sum_{n=1}^{\infty}\frac{(-1)^{n-1}}{\sqrt{n}}$ is *not* absolutely convergent.

**24.** $\sum_{n=1}^{\infty}(-1)^{n-1}\frac{2^n}{n^4}$ diverges by the Test for Divergence. $\lim_{n\to\infty}\frac{2^n}{n^4} = \infty$, so $\lim_{n\to\infty}(-1)^{n-1}\frac{2^n}{n^4}$ does not exist.

**25.** $$\lim_{n\to\infty}\left|\frac{a_{n+1}}{a_n}\right| = \lim_{n\to\infty}\left[\frac{10^{n+1}}{(n+2)4^{2(n+1)+1}}\cdot\frac{(n+1)4^{2n+1}}{10^n}\right] = \lim_{n\to\infty}\left[\frac{10^{n+1}}{(n+2)4^{2n+3}}\cdot\frac{(n+1)4^{2n+1}}{10^n}\right]$$
$$= \lim_{n\to\infty}\left(\frac{10}{4^2}\cdot\frac{n+1}{n+2}\right) = \frac{5}{8} < 1,$$

so the series is absolutely convergent by the Ratio Test. Since the terms of this series are positive, absolute convergence is the same as convergence.

**26.** $\left|\frac{\sin 4n}{4^n}\right| \leq \frac{1}{4^n}$, so $\sum_{n=1}^{\infty}\left|\frac{\sin 4n}{4^n}\right|$ converges by comparison with the convergent geometric series $\sum_{n=1}^{\infty}\frac{1}{4^n}$ $\left(|r| = \frac{1}{4} < 1\right)$. Thus, $\sum_{n=1}^{\infty}\frac{\sin 4n}{4^n}$ is absolutely convergent.

**27.** $\left|\frac{(-1)^n \arctan n}{n^2}\right| < \frac{\pi/2}{n^2}$, so since $\sum_{n=1}^{\infty}\frac{\pi/2}{n^2} = \frac{\pi}{2}\sum_{n=1}^{\infty}\frac{1}{n^2}$ converges ($p = 2 > 1$), the given series $\sum_{n=1}^{\infty}\frac{(-1)^n \arctan n}{n^2}$ converges absolutely by the Comparison Test.

**28.** $\lim_{n\to\infty}\left|\frac{a_{n+1}}{a_n}\right| = \lim_{n\to\infty}\frac{5^n/\left[(n+2)^2\,4^{n+3}\right]}{5^{n-1}/\left[(n+1)^2\,4^{n+2}\right]} = \frac{5}{4}\lim_{n\to\infty}\left(\frac{n+1}{n+2}\right)^2 = \frac{5}{4} > 1$, so the series diverges by the Ratio Test.

**29.** Use the Ratio Test with the series

$$1 - \frac{1\cdot 3}{3!} + \frac{1\cdot 3\cdot 5}{5!} - \frac{1\cdot 3\cdot 5\cdot 7}{7!} + \cdots + (-1)^{n-1}\frac{1\cdot 3\cdot 5\cdot\cdots\cdot(2n-1)}{(2n-1)!} + \cdots = \sum_{n=1}^{\infty}(-1)^{n-1}\frac{1\cdot 3\cdot 5\cdot\cdots\cdot(2n-1)}{(2n-1)!}.$$

$$\lim_{n\to\infty}\left|\frac{a_{n+1}}{a_n}\right| = \lim_{n\to\infty}\left|\frac{(-1)^n\cdot 1\cdot 3\cdot 5\cdot\cdots\cdot(2n-1)[2(n+1)-1]}{[2(n+1)-1]!}\cdot\frac{(2n-1)!}{(-1)^{n-1}\cdot 1\cdot 3\cdot 5\cdot\cdots\cdot(2n-1)}\right|$$
$$= \lim_{n\to\infty}\left|\frac{(-1)(2n+1)(2n-1)!}{(2n+1)(2n)(2n-1)!}\right| = \lim_{n\to\infty}\frac{1}{2n} = 0 < 1,$$

so the given series is absolutely convergent and therefore convergent.

**30.** Use the Ratio Test with the series $\frac{2}{5} + \frac{2\cdot 6}{5\cdot 8} + \frac{2\cdot 6\cdot 10}{5\cdot 8\cdot 11} + \frac{2\cdot 6\cdot 10\cdot 14}{5\cdot 8\cdot 11\cdot 14} + \cdots = \sum_{n=1}^{\infty}\frac{2\cdot 6\cdot 10\cdot 14\cdot\cdots\cdot(4n-2)}{5\cdot 8\cdot 11\cdot 14\cdot\cdots\cdot(3n+2)}$.

$$\lim_{n\to\infty}\left|\frac{a_{n+1}}{a_n}\right| = \lim_{n\to\infty}\left|\frac{2\cdot 6\cdot 10\cdot\cdots\cdot(4n-2)[4(n+1)-2]}{5\cdot 8\cdot 11\cdot\cdots\cdot(3n+2)[3(n+1)+2]}\cdot\frac{5\cdot 8\cdot 11\cdot\cdots\cdot(3n+2)}{2\cdot 6\cdot 10\cdot\cdots\cdot(4n-2)}\right|$$
$$= \lim_{n\to\infty}\frac{4n+2}{3n+5} = \frac{4}{3} > 1,$$

so the given series is divergent.

**31.** By the recursive definition, $\lim_{n\to\infty}\left|\frac{a_{n+1}}{a_n}\right| = \lim_{n\to\infty}\left|\frac{5n+1}{4n+3}\right| = \frac{5}{4} > 1$, so the series diverges by the Ratio Test.

**32.** By the recursive definition, $\lim_{n\to\infty}\left|\frac{a_{n+1}}{a_n}\right| = \lim_{n\to\infty}\left|\frac{2+\cos n}{\sqrt{n}}\right| = 0 < 1$, so the series converges absolutely by the Ratio Test.

**33.** (a) $\lim_{n\to\infty}\left|\frac{1/(n+1)^3}{1/n^3}\right| = \lim_{n\to\infty}\frac{n^3}{(n+1)^3} = \lim_{n\to\infty}\frac{1}{(1+1/n)^3} = 1.$ Inconclusive

(b) $\lim_{n\to\infty}\left|\frac{(n+1)}{2^{n+1}}\cdot\frac{2^n}{n}\right| = \lim_{n\to\infty}\frac{n+1}{2n} = \lim_{n\to\infty}\left(\frac{1}{2}+\frac{1}{2n}\right) = \frac{1}{2}$. Conclusive (convergent)

(c) $\lim_{n\to\infty}\left|\frac{(-3)^n}{\sqrt{n+1}}\cdot\frac{\sqrt{n}}{(-3)^{n-1}}\right| = 3\lim_{n\to\infty}\sqrt{\frac{n}{n+1}} = 3\lim_{n\to\infty}\sqrt{\frac{1}{1+1/n}} = 3$. Conclusive (divergent)

(d) $\lim_{n\to\infty}\left|\frac{\sqrt{n+1}}{1+(n+1)^2}\cdot\frac{1+n^2}{\sqrt{n}}\right| = \lim_{n\to\infty}\left[\sqrt{1+\frac{1}{n}}\cdot\frac{1/n^2+1}{1/n^2+(1+1/n)^2}\right] = 1$. Inconclusive

**34.** We use the Ratio Test:

$$\lim_{n\to\infty}\left|\frac{a_{n+1}}{a_n}\right| = \lim_{n\to\infty}\left|\frac{[(n+1)!]^2/[k(n+1)]!}{(n!)^2/(kn)!}\right| = \lim_{n\to\infty}\left|\frac{(n+1)^2}{[k(n+1)]\,[k(n+1)-1]\cdots[kn+1]}\right|$$

Now if $k=1$, then this is equal to $\lim_{n\to\infty}\left|\frac{(n+1)^2}{(n+1)}\right| = \infty$, so the series diverges; if $k=2$, the limit is $\lim_{n\to\infty}\left|\frac{(n+1)^2}{(2n+2)(2n+1)}\right| = \frac{1}{4} < 1$, so the series converges, and if $k>2$, then the highest power of $n$ in the denominator is larger than 2, and so the limit is 0, indicating convergence. So the series converges for $k \geq 2$.

**35.** (a) $\lim_{n\to\infty}\left|\frac{a_{n+1}}{a_n}\right| = \lim_{n\to\infty}\left|\frac{x^{n+1}}{(n+1)!}\cdot\frac{n!}{x^n}\right| = \lim_{n\to\infty}\left|\frac{x}{n+1}\right| = |x|\lim_{n\to\infty}\frac{1}{n+1} = |x|\cdot 0 = 0 < 1$, so by the Ratio Test the series $\sum_{n=0}^{\infty}\frac{x^n}{n!}$ converges for all $x$.

(b) Since the series of part (a) always converges, we must have $\lim_{n\to\infty}\frac{x^n}{n!} = 0$ by Theorem 8.2.6.

**36.** (a) $\lim_{n\to\infty}\left|\frac{a_{n+1}}{a_n}\right| = \lim_{n\to\infty}\left|\frac{[4(n+1)]!\,[1103+26{,}390(n+1)]}{[(n+1)!]^4\,396^{4(n+1)}}\cdot\frac{(n!)^4\,396^{4n}}{(4n)!\,(1103+26{,}390n)}\right|$

$$= \lim_{n\to\infty}\frac{(4n+4)(4n+3)(4n+2)(4n+1)(26{,}390n+27{,}493)}{(n+1)^4\,396^4\,(26{,}390n+1103)}$$

$$= \frac{4^4}{396^4} = \frac{1}{99^4} < 1,$$

so by the Ratio Test, the series $\sum_{n=0}^{\infty}\frac{(4n)!\,(1103+26{,}390n)}{(n!)^4\,396^{4n}}$ converges.

(b) $\frac{1}{\pi} = \frac{2\sqrt{2}}{9801}\sum_{n=0}^{\infty}\frac{(4n)!\,(1103+26{,}390n)}{(n!)^4\,396^{4n}}$

With the first term ($n=0$), $\frac{1}{\pi}\approx\frac{2\sqrt{2}}{9801}\cdot\frac{1103}{1} \Rightarrow \pi\approx 3.141\,592\,73$, so we get 6 correct decimal places of $\pi$, which is 3.141 592 653 589 793 238 to 18 decimal places.

With the second term ($n=1$), $\frac{1}{\pi}\approx\frac{2\sqrt{2}}{9801}\left(\frac{1103}{1}+\frac{4!\,(1103+26{,}390)}{396^4}\right) \Rightarrow \pi\approx 3.141\,592\,653\,589\,793\,878$, so we get 15 correct decimal places of $\pi$.

## 8.5 Power Series

**1.** A power series is a series of the form $\sum_{n=0}^{\infty} c_n x^n = c_0 + c_1 x + c_2 x^2 + c_3 x^3 + \cdots$, where $x$ is a variable and the $c_n$'s are constants called the coefficients of the series.

More generally, a series of the form $\sum_{n=0}^{\infty} c_n (x-a)^n = c_0 + c_1(x-a) + c_2(x-a)^2 + \cdots$ is called a power series in $(x-a)$ or a power series centered at $a$ or a power series about $a$, where $a$ is a constant.

**2.** (a) Given the power series $\sum_{n=0}^{\infty} c_n(x-a)^n$, the radius of convergence is:

(i) 0 if the series converges only when $x = a$

(ii) $\infty$ if the series converges for all $x$, or

(iii) a positive number $R$ such that the series converges if $|x-a| < R$ and diverges if $|x-a| > R$.

In most cases, $R$ can be found by using the Ratio Test.

(b) The interval of convergence of a power series is the interval that consists of all values of $x$ for which the series converges. Corresponding to the cases in part (a), the interval of convergence is: (i) the single point $\{a\}$, (ii) all real numbers; that is, the real number line $(-\infty, \infty)$, or (iii) an interval with endpoints $a - R$ and $a + R$ which can contain neither, either, or both of the endpoints. In this case, we must test the series for convergence at each endpoint to determine the interval of convergence.

**3.** If $a_n = \dfrac{x^n}{\sqrt{n}}$, then $\lim\limits_{n\to\infty}\left|\dfrac{a_{n+1}}{a_n}\right| = \lim\limits_{n\to\infty}\left|\dfrac{x^{n+1}}{\sqrt{n+1}}\cdot\dfrac{\sqrt{n}}{x^n}\right| = \lim\limits_{n\to\infty}\left|\dfrac{x}{\sqrt{n+1}/\sqrt{n}}\right| = \lim\limits_{n\to\infty}\dfrac{|x|}{\sqrt{1+1/n}} = |x|$.

By the Ratio Test, the series $\sum\limits_{n=1}^{\infty}\dfrac{x^n}{\sqrt{n}}$ converges when $|x| < 1$, so the radius of convergence $R = 1$. Now we'll check the endpoints, that is, $x = \pm 1$. When $x = 1$, the series $\sum\limits_{n=1}^{\infty}\dfrac{1}{\sqrt{n}}$ diverges because it is a $p$-series with $p = \frac{1}{2} \le 1$. When $x = -1$, the series $\sum\limits_{n=1}^{\infty}\dfrac{(-1)^n}{\sqrt{n}}$ converges by the Alternating Series Test. Thus, the interval of convergence is $I = [-1, 1)$.

**4.** If $a_n = \dfrac{(-1)^n x^n}{n+1}$, then $\lim\limits_{n\to\infty}\left|\dfrac{a_{n+1}}{a_n}\right| = \lim\limits_{n\to\infty}\left|\dfrac{x^{n+1}}{n+2}\cdot\dfrac{n+1}{x^n}\right| = \lim\limits_{n\to\infty}\dfrac{|x|}{1+1/(n+1)} = |x|$.

By the Ratio Test, the series $\sum\limits_{n=0}^{\infty}\dfrac{(-1)^n x^n}{n+1}$ converges when $|x| < 1$, so $R = 1$. When $x = -1$, the series diverges because it is the harmonic series; when $x = 1$, it is the alternating harmonic series, which converges by the Alternating Series Test. Thus, $I = (-1, 1]$.

**5.** If $a_n = \dfrac{(-1)^{n-1}x^n}{n^3}$, then

$$\lim_{n\to\infty}\left|\frac{a_{n+1}}{a_n}\right| = \lim_{n\to\infty}\left|\frac{(-1)^n x^{n+1}}{(n+1)^3}\cdot\frac{n^3}{(-1)^{n-1}x^n}\right| = \lim_{n\to\infty}\left|\frac{(-1)xn^3}{(n+1)^3}\right|$$
$$= \lim_{n\to\infty}\left[\left(\frac{n}{n+1}\right)^3|x|\right] = 1^3\cdot|x| = |x|$$

By the Ratio Test, the series $\sum\limits_{n=1}^{\infty}\dfrac{(-1)^{n-1}x^n}{n^3}$ converges when $|x| < 1$, so the radius of convergence $R = 1$. Now we'll check the endpoints, that is, $x = \pm 1$. When $x = 1$, the series $\sum\limits_{n=1}^{\infty}\dfrac{(-1)^{n-1}}{n^3}$ converges by the Alternating Series Test. When $x = -1$, the series $\sum\limits_{n=1}^{\infty}\dfrac{(-1)^{n-1}(-1)^n}{n^3} = -\sum\limits_{n=1}^{\infty}\dfrac{1}{n^3}$ converges because it is a constant multiple of a convergent $p$-series ($p = 3 > 1$). Thus, the interval of convergence is $I = [-1, 1]$.

**6.** $a_n = \sqrt{n}\,x^n$, so we need $\lim\limits_{n\to\infty}\left|\dfrac{a_{n+1}}{a_n}\right| = \lim\limits_{n\to\infty}\dfrac{\sqrt{n+1}\,|x|^{n+1}}{\sqrt{n}\,|x|^n} = \lim\limits_{n\to\infty}\sqrt{1+\dfrac{1}{n}}\,|x| = |x| < 1$ for convergence (by the Ratio Test), so $R = 1$. When $x = \pm 1$, $\lim\limits_{n\to\infty}|a_n| = \lim\limits_{n\to\infty}\sqrt{n} = \infty$, so the series diverges by the Test for Divergence. Thus, $I = (-1, 1)$.

**7.** If $a_n = \dfrac{x^n}{n!}$, then $\lim\limits_{n\to\infty}\left|\dfrac{a_{n+1}}{a_n}\right| = \lim\limits_{n\to\infty}\left|\dfrac{x^{n+1}}{(n+1)!}\cdot\dfrac{n!}{x^n}\right| = \lim\limits_{n\to\infty}\left|\dfrac{x}{n+1}\right| = |x|\lim\limits_{n\to\infty}\dfrac{1}{n+1} = |x|\cdot 0 = 0 < 1$ for *all* real $x$.

So, by the Ratio Test, $R = \infty$, and $I = (-\infty, \infty)$.

**8.** If $a_n = \dfrac{x^n}{n3^n}$, then $\lim\limits_{n\to\infty}\left|\dfrac{a_{n+1}}{a_n}\right| = \lim\limits_{n\to\infty}\left|\dfrac{x^{n+1}}{(n+1)3^{n+1}}\cdot\dfrac{n3^n}{x^n}\right| = \lim\limits_{n\to\infty}\left|\dfrac{xn}{(n+1)3}\right| = \dfrac{|x|}{3}\lim\limits_{n\to\infty}\dfrac{n}{n+1} = \dfrac{|x|}{3}$. By the Ratio Test, the series converges when $\dfrac{|x|}{3} < 1 \;\Leftrightarrow\; |x| < 3$, so $R = 3$. When $x = -3$, the series is the alternating harmonic series, which converges by the Alternating Series Test. When $x = 3$, it is the harmonic series, which diverges. Thus, $I = [-3, 3)$.

**9.** $a_n = \dfrac{(-2)^n x^n}{\sqrt[4]{n}}$, so $\lim\limits_{n\to\infty}\left|\dfrac{a_{n+1}}{a_n}\right| = \lim\limits_{n\to\infty}\dfrac{2^{n+1}|x|^{n+1}}{\sqrt[4]{n+1}}\cdot\dfrac{\sqrt[4]{n}}{2^n|x|^n} = \lim\limits_{n\to\infty}2|x|\sqrt[4]{\dfrac{n}{n+1}} = 2|x|$, so by the Ratio Test, the series converges when $2|x| < 1 \;\Leftrightarrow\; |x| < \frac{1}{2}$, so $R = \frac{1}{2}$. When $x = -\frac{1}{2}$, we get the divergent $p$-series $\sum\limits_{n=1}^{\infty}\dfrac{1}{\sqrt[4]{n}}$ $\left(p = \frac{1}{4} \le 1\right)$. When $x = \frac{1}{2}$, we get the series $\sum\limits_{n=1}^{\infty}\dfrac{(-1)^n}{\sqrt[4]{n}}$, which converges by the Alternating Series Test. Thus, $I = \left(-\frac{1}{2}, \frac{1}{2}\right]$.

**10.** $a_n = \dfrac{x^n}{5^n n^5}$, so $\lim\limits_{n\to\infty}\left|\dfrac{a_{n+1}}{a_n}\right| = \lim\limits_{n\to\infty}\left|\dfrac{x^{n+1}}{5^{n+1}(n+1)^5}\cdot\dfrac{5^n n^5}{x^n}\right| = \lim\limits_{n\to\infty}\dfrac{|x|}{5}\left(\dfrac{n}{n+1}\right)^5 = \dfrac{|x|}{5}$. By the Ratio Test, the series $\sum\limits_{n=0}^{\infty}\dfrac{x^n}{5^n n^5}$ converges when $\dfrac{|x|}{5} < 1 \;\Leftrightarrow\; |x| < 5$, so $R = 5$. When $x = -5$, we get the series $\sum\limits_{n=1}^{\infty}\dfrac{(-1)^n}{n^5}$, which converges by the Alternating Series Test. When $x = 5$, we get the convergent $p$-series $\sum\limits_{n=1}^{\infty}\dfrac{1}{n^5}$ $(p = 5 > 1)$. Thus, $I = [-5, 5]$.

**11.** If $a_n = (-1)^n\dfrac{x^n}{4^n \ln n}$, then $\lim\limits_{n\to\infty}\left|\dfrac{a_{n+1}}{a_n}\right| = \lim\limits_{n\to\infty}\left|\dfrac{x^{n+1}}{4^{n+1}\ln(n+1)}\cdot\dfrac{4^n\ln n}{x^n}\right| = \dfrac{|x|}{4}\lim\limits_{n\to\infty}\dfrac{\ln n}{\ln(n+1)} = \dfrac{|x|}{4}\cdot 1$ [by l'Hospital's Rule] $= \dfrac{|x|}{4}$. By the Ratio Test, the series converges when $\dfrac{|x|}{4} < 1 \;\Leftrightarrow\; |x| < 4$, so $R = 4$. When $x = -4$, $\sum\limits_{n=2}^{\infty}(-1)^n\dfrac{x^n}{4^n\ln n} = \sum\limits_{n=2}^{\infty}\dfrac{[(-1)(-4)]^n}{4^n\ln n} = \sum\limits_{n=2}^{\infty}\dfrac{1}{\ln n}$. Since $\ln n < n$ for $n \ge 2$, $\dfrac{1}{\ln n} > \dfrac{1}{n}$ and $\sum\limits_{n=2}^{\infty}\dfrac{1}{n}$ is the divergent harmonic series (without the $n = 1$ term), $\sum\limits_{n=2}^{\infty}\dfrac{1}{\ln n}$ is divergent by the Comparison Test. When $x = 4$, $\sum\limits_{n=2}^{\infty}(-1)^n\dfrac{x^n}{4^n\ln n} = \sum\limits_{n=2}^{\infty}(-1)^n\dfrac{1}{\ln n}$, which converges by the Alternating Series Test. Thus, $I = (-4, 4]$.

**12.** $a_n = (-1)^n\dfrac{x^{2n}}{(2n)!}$, so $\lim\limits_{n\to\infty}\left|\dfrac{a_{n+1}}{a_n}\right| = \lim\limits_{n\to\infty}\dfrac{|x|^{2n+2}}{(2n+2)!}\cdot\dfrac{(2n)!}{|x|^{2n}} = \lim\limits_{n\to\infty}\dfrac{|x|^2}{(2n+1)(2n+2)} = 0$. Thus, by the Ratio Test, the series converges for *all* real $x$ and we have $R = \infty$ and $I = (-\infty, \infty)$.

**13.** If $a_n = (-1)^n\dfrac{(x+2)^n}{n\,2^n}$, then $\lim\limits_{n\to\infty}\left|\dfrac{a_{n+1}}{a_n}\right| = \lim\limits_{n\to\infty}\left[\dfrac{|x+2|^{n+1}}{(n+1)2^{n+1}}\cdot\dfrac{n2^n}{|x+2|^n}\right] = \lim\limits_{n\to\infty}\dfrac{n}{n+1}\cdot\dfrac{|x+2|}{2} = \dfrac{|x+2|}{2}$. By the Ratio Test, the series converges when $\dfrac{|x+2|}{2} < 1 \;\Leftrightarrow\; |x+2| < 2$ [so $R = 2$] $\Leftrightarrow\; -2 < x+2 < 2 \;\Leftrightarrow$ $-4 < x < 0$. When $x = -4$, the series becomes $\sum\limits_{n=1}^{\infty}(-1)^n\dfrac{(-2)^n}{n2^n} = \sum\limits_{n=1}^{\infty}\dfrac{2^n}{n\,2^n} = \sum\limits_{n=1}^{\infty}\dfrac{1}{n}$, which is the divergent harmonic series. When $x = 0$, the series is $\sum\limits_{n=1}^{\infty}\dfrac{(-1)^n}{n}$, the alternating harmonic series, which converges by the Alternating Series Test. Thus, $I = (-4, 0]$.

**14.** If $a_n = \dfrac{(-2)^n}{\sqrt{n}}(x+3)^n$, then

$$\lim_{n\to\infty}\left|\frac{a_{n+1}}{a_n}\right| = \lim_{n\to\infty}\left|\frac{(-2)^{n+1}(x+3)^{n+1}}{\sqrt{n+1}}\cdot\frac{\sqrt{n}}{(-2)^n(x+3)^n}\right| = \lim_{n\to\infty}\frac{2\,|x+3|}{\sqrt{1+1/n}} = 2\,|x+3| < 1 \quad\Leftrightarrow\quad |x+3| < \tfrac{1}{2}$$

$\left[\text{so } R = \frac{1}{2}\right] \quad\Leftrightarrow\quad -\frac{7}{2} < x < -\frac{5}{2}$. When $x = -\frac{7}{2}$, the series becomes $\sum\limits_{n=1}^{\infty}\dfrac{1}{\sqrt{n}}$, which diverges because it is a $p$-series with $p = \frac{1}{2} \le 1$. When $x = -\frac{5}{2}$, the series becomes $\sum\limits_{n=1}^{\infty}\dfrac{(-1)^n}{\sqrt{n}}$, which converges by the Alternating Series Test. Thus, $I = \left(-\frac{7}{2}, -\frac{5}{2}\right]$.

**15.** $a_n = \dfrac{n}{b^n}(x-a)^n$, where $b > 0$.

$$\lim_{n\to\infty}\left|\frac{a_{n+1}}{a_n}\right| = \lim_{n\to\infty}\frac{(n+1)\,|x-a|^{n+1}}{b^{n+1}}\cdot\frac{b^n}{n\,|x-a|^n} = \lim_{n\to\infty}\left(1+\frac{1}{n}\right)\frac{|x-a|}{b} = \frac{|x-a|}{b}.$$

By the Ratio Test, the series converges when $\dfrac{|x-a|}{b} < 1 \quad\Leftrightarrow\quad |x-a| < b \quad [\text{so } R = b] \quad\Leftrightarrow\quad -b < x-a < b \quad\Leftrightarrow$ $a-b < x < a+b$. When $|x-a| = b$, $\lim\limits_{n\to\infty}|a_n| = \lim\limits_{n\to\infty} n = \infty$, so the series diverges. Thus, $I = (a-b, a+b)$.

**16.** $a_n = \dfrac{n(x-4)^n}{n^3+1}$, so

$$\lim_{n\to\infty}\left|\frac{a_{n+1}}{a_n}\right| = \lim_{n\to\infty}\frac{(n+1)\,|x-4|^{n+1}}{(n+1)^3+1}\cdot\frac{n^3+1}{n\,|x-4|^n} = \lim_{n\to\infty}\left(1+\frac{1}{n}\right)\frac{n^3+1}{n^3+3n^2+3n+2}\,|x-4| = |x-4|.$$

By the Ratio Test, the series converges when $|x-4| < 1$ $[\text{so } R = 1]$ $\Leftrightarrow$ $-1 < x-4 < 1$ $\Leftrightarrow$ $3 < x < 5$. When $|x-4| = 1$, $\sum\limits_{n=1}^{\infty}|a_n| = \sum\limits_{n=1}^{\infty}\dfrac{n}{n^3+1}$, which converges by comparison with the convergent $p$-series $\sum\limits_{n=1}^{\infty}\dfrac{1}{n^2}$ $(p = 2 > 1)$. Thus, $I = [3, 5]$.

**17.** If $a_n = n!(2x-1)^n$, then $\lim\limits_{n\to\infty}\left|\dfrac{a_{n+1}}{a_n}\right| = \lim\limits_{n\to\infty}\left|\dfrac{(n+1)!(2x-1)^{n+1}}{n!(2x-1)^n}\right| = \lim\limits_{n\to\infty}(n+1)\,|2x-1| \to \infty$ as $n \to \infty$ for all $x \ne \frac{1}{2}$. Since the series diverges for all $x \ne \frac{1}{2}$, $R = 0$ and $I = \left\{\frac{1}{2}\right\}$.

**18.** $a_n = \dfrac{n^2x^n}{2\cdot 4\cdot 6\cdot\cdots\cdot(2n)} = \dfrac{n^2x^n}{2^n n!} = \dfrac{nx^n}{2^n(n-1)!}$, so

$\lim\limits_{n\to\infty}\left|\dfrac{a_{n+1}}{a_n}\right| = \lim\limits_{n\to\infty}\dfrac{(n+1)\,|x|^{n+1}}{2^{n+1}n!}\cdot\dfrac{2^n(n-1)!}{n\,|x|^n} = \lim\limits_{n\to\infty}\dfrac{n+1}{n^2}\dfrac{|x|}{2} = 0$. Thus, by the Ratio Test, the series converges for *all* real $x$ and we have $R = \infty$ and $I = (-\infty, \infty)$.

**19.** (a) We are given that the power series $\sum_{n=0}^{\infty} c_n x^n$ is convergent for $x = 4$. So by Theorem 3, it must converge for at least $-4 < x \le 4$. In particular, it converges when $x = -2$; that is, $\sum_{n=0}^{\infty} c_n(-2)^n$ is convergent.

(b) It does not follow that $\sum_{n=0}^{\infty} c_n(-4)^n$ is necessarily convergent. [See the comments after Theorem 3 about convergence at the endpoint of an interval. An example is $c_n = (-1)^n/(n4^n)$.]

**20.** We are given that the power series $\sum_{n=0}^{\infty} c_n x^n$ is convergent for $x = -4$ and divergent when $x = 6$. So by Theorem 3 it converges for at least $-4 \le x < 4$ and diverges for at least $x \ge 6$ and $x < -6$. Therefore:

(a) It converges when $x = 1$; that is, $\sum c_n$ is convergent.

(b) It diverges when $x = 8$; that is, $\sum c_n 8^n$ is divergent.

(c) It converges when $x = -3$; that is, $\sum c_n(-3^n)$ is convergent.

(d) It diverges when $x = -9$; that is, $\sum c_n(-9)^n = \sum(-1)^n c_n 9^n$ is divergent.

**21.** If $a_n = \dfrac{(n!)^k}{(kn)!}x^n$, then

$$\lim_{n\to\infty}\left|\frac{a_{n+1}}{a_n}\right| = \lim_{n\to\infty}\frac{[(n+1)!]^k\,(kn)!}{(n!)^k\,[k(n+1)]!}|x| = \lim_{n\to\infty}\frac{(n+1)^k}{(kn+k)(kn+k-1)\cdots(kn+2)(kn+1)}|x|$$

$$= \lim_{n\to\infty}\left[\frac{(n+1)}{(kn+1)}\frac{(n+1)}{(kn+2)}\cdots\frac{(n+1)}{(kn+k)}\right]|x|$$

$$= \lim_{n\to\infty}\left[\frac{n+1}{kn+1}\right]\lim_{n\to\infty}\left[\frac{n+1}{kn+2}\right]\cdots\lim_{n\to\infty}\left[\frac{n+1}{kn+k}\right]|x|$$

$$= \left(\frac{1}{k}\right)^k|x| < 1 \quad\Leftrightarrow\quad |x| < k^k \text{ for convergence, and the radius of convergence is } R = k^k.$$

**22.** The partial sums of the series $\sum_{n=0}^{\infty} x^n$ definitely do not converge to $f(x) = 1/(1-x)$ for $x \geq 1$, since $f$ is undefined at $x = 1$ and negative on $(1, \infty)$, while all the partial sums are positive on this interval. The partial sums also fail to converge to $f$ for $x \leq -1$, since $0 < f(x) < 1$ on this interval, while the partial sums are either larger than 1 or less than 0. The partial sums seem to converge to $f$ on $(-1, 1)$. This graphical evidence is consistent with what we know about geometric series: convergence for $|x| < 1$, divergence for $|x| \geq 1$ (see Example 8.2.5).

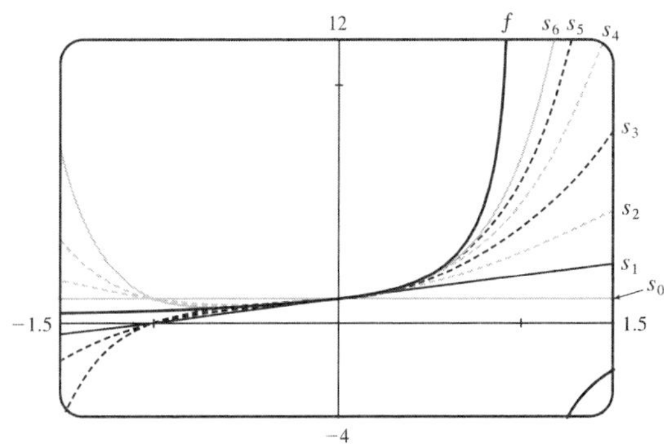


**23.** (a) If $a_n = \dfrac{(-1)^n\,x^{2n+1}}{n!(n+1)!\,2^{2n+1}}$, then

$$\lim_{n\to\infty}\left|\frac{a_{n+1}}{a_n}\right| = \lim_{n\to\infty}\left|\frac{x^{2n+3}}{(n+1)!(n+2)!\,2^{2n+3}}\cdot\frac{n!(n+1)!\,2^{2n+1}}{x^{2n+1}}\right| = \left(\frac{x}{2}\right)^2\lim_{n\to\infty}\frac{1}{(n+1)(n+2)} = 0 \text{ for all } x.$$

So $J_1(x)$ converges for all $x$ and its domain is $(-\infty, \infty)$.

(b), (c) The initial terms of $J_1(x)$ up to $n = 5$ are $a_0 = \dfrac{x}{2}$,

$a_1 = -\dfrac{x^3}{16}$, $a_2 = \dfrac{x^5}{384}$, $a_3 = -\dfrac{x^7}{18{,}432}$, $a_4 = \dfrac{x^9}{1{,}474{,}560}$,

and $a_5 = -\dfrac{x^{11}}{176{,}947{,}200}$. The partial sums seem to approximate $J_1(x)$ well near the origin, but as $|x|$ increases, we need to take a large number of terms to get a good approximation.

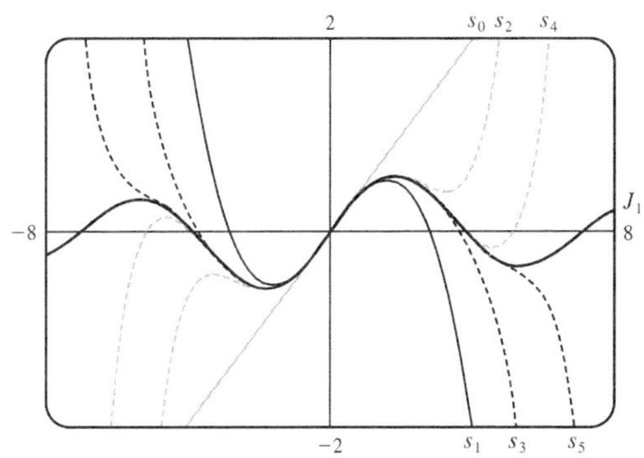


**24.** (a) $A(x) = 1 + \sum\limits_{n=1}^{\infty} a_n$, where $a_n = \dfrac{x^{3n}}{2\cdot 3\cdot 5\cdot 6\cdot\cdots\cdot(3n-1)(3n)}$, so $\lim\limits_{n\to\infty}\left|\dfrac{a_{n+1}}{a_n}\right| = |x|^3\lim\limits_{n\to\infty}\dfrac{1}{(3n+2)(3n+3)} = 0$ for all $x$, so the domain is $\mathbb{R}$.

(b), (c)

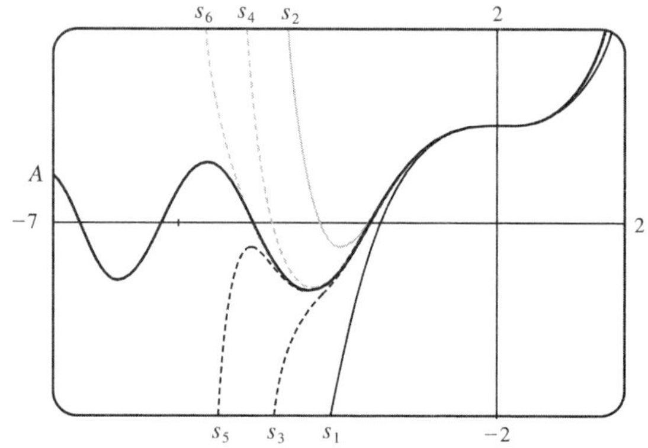


$s_0 = 1$ has been omitted from the graph. The partial sums seem to approximate $A(x)$ well near the origin, but as $|x|$ increases, we need to take a large number of terms to get a good approximation.

To plot $A$, we must first define $A(x)$ for the CAS. Note that for $n \geq 1$, the denominator of $a_n$ is

$2 \cdot 3 \cdot 5 \cdot 6 \cdot \cdots \cdot (3n-1) \cdot 3n = \dfrac{(3n)!}{1 \cdot 4 \cdot 7 \cdot \cdots \cdot (3n-2)} = \dfrac{(3n)!}{\prod_{k=1}^{n}(3k-2)}$, so $a_n = \dfrac{\prod_{k=1}^{n}(3k-2)}{(3n)!}x^{3n}$ and thus

$A(x) = 1 + \sum\limits_{n=1}^{\infty} \dfrac{\prod_{k=1}^{n}(3k-2)}{(3n)!}x^{3n}$. Both Maple and Mathematica are able to plot $A$ if we define it this way, and Derive is able to produce a similar graph using a suitable partial sum of $A(x)$.

Derive, Maple and Mathematica all have two initially known Airy functions, called `AI·SERIES(z,m)` and `BI·SERIES(z,m)` from `BESSEL.MTH` in Derive and `AiryAi` and `AiryBi` in Maple and Mathematica (just `Ai` and `Bi` in older versions of Maple). However, it is very difficult to solve for $A$ in terms of the CAS's Airy functions, although in fact $A(x) = \dfrac{\sqrt{3}\,\texttt{AiryAi}(x) + \texttt{AiryBi}(x)}{\sqrt{3}\,\texttt{AiryAi}(0) + \texttt{AiryBi}(0)}$.

**25.** $s_{2n-1} = 1 + 2x + x^2 + 2x^3 + x^4 + 2x^5 + \cdots + x^{2n-2} + 2x^{2n-1}$

$= 1(1+2x) + x^2(1+2x) + x^4(1+2x) + \cdots + x^{2n-2}(1+2x) = (1+2x)(1 + x^2 + x^4 + \cdots + x^{2n-2})$

$= (1+2x)\dfrac{1-x^{2n}}{1-x^2}$ [by (8.2.3)] with $r = x^2$] $\to \dfrac{1+2x}{1-x^2}$ as $n \to \infty$ [by (8.2.4)], when $|x| < 1$.

Also $s_{2n} = s_{2n-1} + x^{2n} \to \dfrac{1+2x}{1-x^2}$ since $x^{2n} \to 0$ for $|x| < 1$. Therefore, $s_n \to \dfrac{1+2x}{1-x^2}$ since $s_{2n}$ and $s_{2n-1}$ both approach $\dfrac{1+2x}{1-x^2}$ as $n \to \infty$. Thus, the interval of convergence is $(-1, 1)$ and $f(x) = \dfrac{1+2x}{1-x^2}$.

**26.** $s_{4n-1} = c_0 + c_1 x + c_2 x^2 + c_3 x^3 + c_0 x^4 + c_1 x^5 + c_2 x^6 + c_3 x^7 + \cdots + c_3 x^{4n-1}$

$= \left(c_0 + c_1 x + c_2 x^2 + c_3 x^3\right)\left(1 + x^4 + x^8 + \cdots + x^{4n-4}\right) \to \dfrac{c_0 + c_1 x + c_2 x^2 + c_3 x^3}{1-x^4}$ as $n \to \infty$

[by ( 8.2.4) with $r = x^4$] for $|x^4| < 1 \quad \Leftrightarrow \quad |x| < 1$. Also $s_{4n}, s_{4n+1}, s_{4n+2}$ have the same limits (for example, $s_{4n} = s_{4n-1} + c_0 x^{4n}$ and $x^{4n} \to 0$ for $|x| < 1$). So if at least one of $c_0$, $c_1$, $c_2$, and $c_3$ is nonzero, then the interval of convergence is $(-1, 1)$ and $f(x) = \dfrac{c_0 + c_1 x + c_2 x^2 + c_3 x^3}{1-x^4}$.

**27.** For $2 < x < 3$, $\sum c_n x^n$ diverges and $\sum d_n x^n$ converges. By Exercise 8.2.51, $\sum (c_n + d_n)\, x^n$ diverges. Since both series converge for $|x| < 2$, the radius of convergence of $\sum (c_n + d_n)\, x^n$ is 2.

**28.** Since $\sum c_n x^n$ converges whenever $|x| < R$, $\sum c_n x^{2n} = \sum c_n \left(x^2\right)^n$ converges whenever $|x^2| < R \quad \Leftrightarrow \quad |x| < \sqrt{R}$, so the second series has radius of convergence $\sqrt{R}$.

## 8.6 Representations of Functions as Power Series

**1.** If $f(x) = \sum_{n=0}^{\infty} c_n x^n$ has radius of convergence 10, then $f'(x) = \sum_{n=1}^{\infty} n c_n x^{n-1}$ also has radius of convergence 10 by Theorem 2.

**2.** If $f(x) = \sum_{n=0}^{\infty} b_n x^n$ converges on $(-2, 2)$, then $\int f(x)dx = C + \sum_{n=0}^{\infty} \frac{b_n}{n+1} x^{n+1}$ has the same radius of convergence (by Theorem 2), but may not have the same interval of convergence—it may happen that the integrated series converges at an endpoint (or both endpoints).

**3.** Our goal is to write the function in the form $\frac{1}{1-r}$, and then use Equation (1) to represent the function as a sum of a power series. $f(x) = \frac{1}{1+x} = \frac{1}{1-(-x)} = \sum_{n=0}^{\infty} (-x)^n = \sum_{n=0}^{\infty} (-1)^n x^n$ with $|-x| < 1 \iff |x| < 1$, so $R = 1$ and $I = (-1, 1)$.

**4.** $f(x) = \frac{3}{1-x^4} = 3\left(\frac{1}{1-x^4}\right) = 3(1 + x^4 + x^8 + x^{12} + \cdots) = 3\sum_{n=0}^{\infty} (x^4)^n = \sum_{n=0}^{\infty} 3x^{4n}$ with $|x^4| < 1 \iff |x| < 1$, so $R = 1$ and $I = (-1, 1)$.

[Note that $3\sum_{n=0}^{\infty} (x^4)^n$ converges $\iff$ $\sum_{n=0}^{\infty} (x^4)^n$ converges, so the appropriate condition [from equation (1)] is $|x^4| < 1$.]

**5.** Replacing $x$ with $x^3$ in (1) gives $f(x) = \frac{1}{1-x^3} = \sum_{n=0}^{\infty} (x^3)^n = \sum_{n=0}^{\infty} x^{3n}$. The series converges when $|x^3| < 1 \iff |x|^3 < 1 \iff |x| < \sqrt[3]{1} \iff |x| < 1$. Thus, $R = 1$ and $I = (-1, 1)$.

**6.** $f(x) = \frac{1}{1+9x^2} = \frac{1}{1-(-9x^2)} = \sum_{n=0}^{\infty} (-9x^2)^n = \sum_{n=0}^{\infty} (-1)^n 3^{2n} x^{2n}$. The series converges when $|-9x^2| < 1$; that is, when $|x| < \frac{1}{3}$, so $I = \left(-\frac{1}{3}, \frac{1}{3}\right)$.

**7.** $f(x) = \frac{1}{x-5} = -\frac{1}{5}\left(\frac{1}{1-x/5}\right) = -\frac{1}{5}\sum_{n=0}^{\infty}\left(\frac{x}{5}\right)^n$ or equivalently, $-\sum_{n=0}^{\infty} \frac{1}{5^{n+1}} x^n$. The series converges when $\left|\frac{x}{5}\right| < 1$; that is, when $|x| < 5$, so $I = (-5, 5)$.

**8.** $f(x) = \frac{x}{4x+1} = x \cdot \frac{1}{1-(-4x)} = x\sum_{n=0}^{\infty} (-4x)^n = \sum_{n=0}^{\infty} (-1)^n 2^{2n} x^{n+1}$. The series converges when $|-4x| < 1$; that is, when $|x| < \frac{1}{4}$, so $I = \left(-\frac{1}{4}, \frac{1}{4}\right)$.

**9.** $$f(x) = \frac{x}{9+x^2} = \frac{x}{9}\left[\frac{1}{1+(x/3)^2}\right] = \frac{x}{9}\left[\frac{1}{1-\{-(x/3)^2\}}\right] = \frac{x}{9}\sum_{n=0}^{\infty}\left[-\left(\frac{x}{3}\right)^2\right]^n = \frac{x}{9}\sum_{n=0}^{\infty}(-1)^n \frac{x^{2n}}{9^n}$$
$$= \sum_{n=0}^{\infty} (-1)^n \frac{x^{2n+1}}{9^{n+1}}$$

The geometric series $\sum_{n=0}^{\infty}\left[-\left(\frac{x}{3}\right)^2\right]^n$ converges when $\left|-\left(\frac{x}{3}\right)^2\right| < 1 \iff \frac{|x^2|}{9} < 1 \iff |x|^2 < 9 \iff |x| < 3$, so $R = 3$ and $I = (-3, 3)$.

**10.** $f(x) = \frac{x^2}{a^3 - x^3} = \frac{x^2}{a^3} \cdot \frac{1}{1 - x^3/a^3} = \frac{x^2}{a^3}\sum_{n=0}^{\infty}\left(\frac{x^3}{a^3}\right)^n = \sum_{n=0}^{\infty} \frac{x^{3n+2}}{a^{3n+3}}$. The series converges when $|x^3/a^3| < 1 \iff |x^3| < |a^3| \iff |x| < |a|$, so $R = |a|$ and $I = (-|a|, |a|)$.

**11.** (a) $f(x) = \dfrac{1}{(1+x)^2} = \dfrac{d}{dx}\left(\dfrac{-1}{1+x}\right) = -\dfrac{d}{dx}\left[\sum\limits_{n=0}^{\infty}(-1)^n x^n\right]$ [from Exercise 3]

$= \sum\limits_{n=1}^{\infty}(-1)^{n+1}nx^{n-1}$ [from Theorem 2(i)] $= \sum\limits_{n=0}^{\infty}(-1)^n(n+1)x^n$ with $R=1$.

In the last step, note that we *decreased* the initial value of the summation variable $n$ by 1, and then *increased* each occurrence of $n$ in the term by 1 [also note that $(-1)^{n+2}=(-1)^n$].

(b) $f(x) = \dfrac{1}{(1+x)^3} = -\dfrac{1}{2}\dfrac{d}{dx}\left[\dfrac{1}{(1+x)^2}\right] = -\dfrac{1}{2}\dfrac{d}{dx}\left[\sum\limits_{n=0}^{\infty}(-1)^n(n+1)x^n\right]$ [from part (a)]

$= -\frac{1}{2}\sum\limits_{n=1}^{\infty}(-1)^n(n+1)nx^{n-1} = \frac{1}{2}\sum\limits_{n=0}^{\infty}(-1)^n(n+2)(n+1)x^n$ with $R=1$.

(c) $f(x) = \dfrac{x^2}{(1+x)^3} = x^2\cdot\dfrac{1}{(1+x)^3} = x^2\cdot\dfrac{1}{2}\sum\limits_{n=0}^{\infty}(-1)^n(n+2)(n+1)x^n$ [from part (b)]

$= \dfrac{1}{2}\sum\limits_{n=0}^{\infty}(-1)^n(n+2)(n+1)x^{n+2}$

To write the power series with $x^n$ rather than $x^{n+2}$, we will *decrease* each occurrence of $n$ in the term by 2 and *increase* the initial value of the summation variable by 2. This gives us $\dfrac{1}{2}\sum\limits_{n=2}^{\infty}(-1)^n(n)(n-1)x^n$.

**12.** (a) $\dfrac{1}{1+x} = \dfrac{1}{1-(-x)} = \sum\limits_{n=0}^{\infty}(-1)^n x^n$ [geometric series with $R=1$], so

$f(x) = \ln(1+x) = \int\dfrac{dx}{1+x} = \int\left[\sum\limits_{n=0}^{\infty}(-1)^n x^n\right]dx = C + \sum\limits_{n=0}^{\infty}(-1)^n\dfrac{x^{n+1}}{n+1} = \sum\limits_{n=1}^{\infty}\dfrac{(-1)^{n-1}x^n}{n}$

[$C=0$ since $f(0)=\ln 1=0$], with $R=1$

(b) $f(x) = x\ln(1+x) = x\left[\sum\limits_{n=1}^{\infty}\dfrac{(-1)^{n-1}x^n}{n}\right]$ [by part (a)] $= \sum\limits_{n=1}^{\infty}\dfrac{(-1)^{n-1}x^{n+1}}{n} = \sum\limits_{n=2}^{\infty}\dfrac{(-1)^n x^n}{n-1}$ with $R=1$.

(c) $f(x) = \ln(x^2+1) = \sum\limits_{n=1}^{\infty}\dfrac{(-1)^{n-1}(x^2)^n}{n}$ [by part (a)] $= \sum\limits_{n=1}^{\infty}\dfrac{(-1)^{n-1}x^{2n}}{n}$ with $R=1$.

**13.** $f(x) = \ln(5-x) = -\displaystyle\int\frac{dx}{5-x} = -\frac{1}{5}\int\frac{dx}{1-x/5} = -\frac{1}{5}\int\left[\sum_{n=0}^{\infty}\left(\frac{x}{5}\right)^n\right]dx = C - \frac{1}{5}\sum_{n=0}^{\infty}\frac{x^{n+1}}{5^n(n+1)} = C - \sum_{n=1}^{\infty}\frac{x^n}{n\,5^n}$

Putting $x=0$, we get $C=\ln 5$. The series converges for $|x/5|<1 \Leftrightarrow |x|<5$, so $R=5$.

**14.** We know that $\dfrac{1}{1-2x} = \sum\limits_{n=0}^{\infty}(2x)^n$. Differentiating, we get $\dfrac{2}{(1-2x)^2} = \sum\limits_{n=1}^{\infty}2^n nx^{n-1} = \sum\limits_{n=0}^{\infty}2^{n+1}(n+1)x^n$, so

$f(x) = \dfrac{x^2}{(1-2x)^2} = \dfrac{x^2}{2}\cdot\dfrac{2}{(1-2x)^2} = \dfrac{x^2}{2}\sum\limits_{n=0}^{\infty}2^{n+1}(n+1)x^n = \sum\limits_{n=0}^{\infty}2^n(n+1)x^{n+2}$ or $\sum\limits_{n=2}^{\infty}2^{n-2}(n-1)x^n$,

with $R=\frac{1}{2}$.

**15.** $\dfrac{1}{2-x} = \dfrac{1}{2(1-x/2)} = \dfrac{1}{2}\sum\limits_{n=0}^{\infty}\left(\dfrac{x}{2}\right)^n = \sum\limits_{n=0}^{\infty}\dfrac{1}{2^{n+1}}x^n$ for $\left|\dfrac{x}{2}\right|<1 \Leftrightarrow |x|<2$. Now

$\dfrac{1}{(x-2)^2} = \dfrac{d}{dx}\left(\dfrac{1}{2-x}\right) = \dfrac{d}{dx}\left(\sum\limits_{n=0}^{\infty}\dfrac{1}{2^{n+1}}x^n\right) = \sum\limits_{n=1}^{\infty}\dfrac{n}{2^{n+1}}x^{n-1} = \sum\limits_{n=0}^{\infty}\dfrac{n+1}{2^{n+2}}x^n$. So

$f(x) = \dfrac{x^3}{(x-2)^2} = x^3\sum\limits_{n=0}^{\infty}\dfrac{n+1}{2^{n+2}}x^n = \sum\limits_{n=0}^{\infty}\dfrac{n+1}{2^{n+2}}x^{n+3}$ or $\sum\limits_{n=3}^{\infty}\dfrac{n-2}{2^{n-1}}x^n$ for $|x|<2$. Thus, $R=2$ and $I=(-2,2)$.

**16.** From Example 7, $g(x) = \arctan x = \sum\limits_{n=0}^{\infty}(-1)^n\dfrac{x^{2n+1}}{2n+1}$. Thus,

$f(x) = \arctan(x/3) = \sum\limits_{n=0}^{\infty}(-1)^n\dfrac{(x/3)^{2n+1}}{2n+1} = \sum\limits_{n=0}^{\infty}(-1)^n\dfrac{1}{3^{2n+1}(2n+1)}x^{2n+1}$ for $\left|\dfrac{x}{3}\right|<1 \Leftrightarrow |x|<3$, so $R=3$.

**17.** $f(x) = \ln(3+x) = \int \frac{dx}{3+x} = \frac{1}{3}\int \frac{dx}{1+x/3} = \frac{1}{3}\int \frac{dx}{1-(-x/3)} = \frac{1}{3}\int \sum_{n=0}^{\infty}\left(-\frac{x}{3}\right)^n dx$

$= C + \frac{1}{3}\sum_{n=0}^{\infty} \frac{(-1)^n}{(n+1)3^n} x^{n+1} = \ln 3 + \frac{1}{3}\sum_{n=1}^{\infty} \frac{(-1)^{n-1}}{n3^{n-1}} x^n \quad [C = f(0) = \ln 3]$

$= \ln 3 + \sum_{n=1}^{\infty} \frac{(-1)^{n-1}}{n\,3^n} x^n$. The series converges when $|-x/3| < 1 \;\Leftrightarrow\; |x| < 3$, so $R = 3$.

The terms of the series are $a_0 = \ln 3,\ a_1 = \frac{x}{3},\ a_2 = -\frac{x^2}{18},\ a_3 = \frac{x^3}{81},\ a_4 = -\frac{x^4}{324},\ a_5 = \frac{x^5}{1215}, \ldots$.

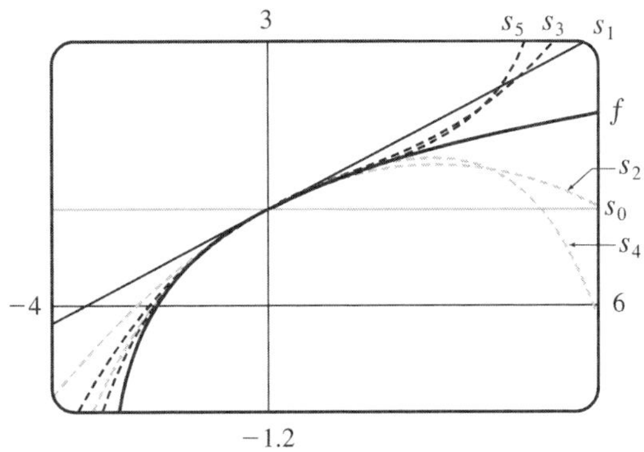


As $n$ increases, $s_n(x)$ approximates $f$ better on the interval of convergence, which is $(-3, 3)$.

**18.** $f(x) = \frac{1}{x^2+25} = \frac{1}{25}\left(\frac{1}{1+x^2/25}\right) = \frac{1}{25}\left(\frac{1}{1-(-x^2/25)}\right) = \frac{1}{25}\sum_{n=0}^{\infty}\left(-\frac{x^2}{25}\right)^n = \frac{1}{25}\sum_{n=0}^{\infty}(-1)^n\left(\frac{x}{5}\right)^{2n}$.

The series converges when $\left|-x^2/25\right| < 1 \;\Leftrightarrow\; x^2 < 25 \;\Leftrightarrow\; |x| < 5$, so $R = 5$.

The terms of the series are $a_0 = \frac{1}{25},\ a_1 = -\frac{x^2}{625},\ a_2 = \frac{x^4}{15{,}625}, \ldots$.

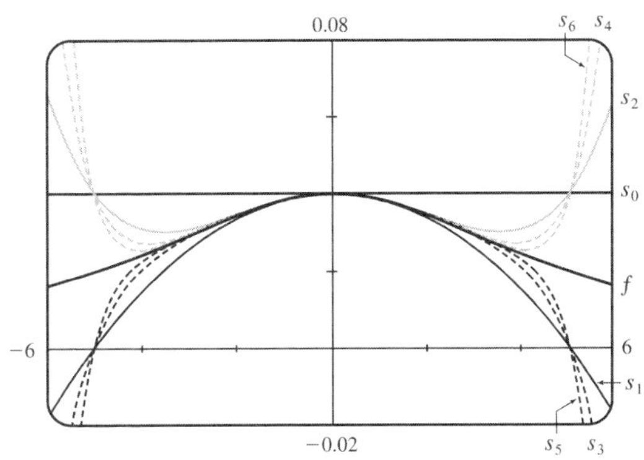


As $n$ increases, $s_n(x)$ approximates $f$ better on the interval of convergence, which is $(-5, 5)$.

**19.** $f(x) = \ln\left(\dfrac{1+x}{1-x}\right) = \ln(1+x) - \ln(1-x) = \displaystyle\int \frac{dx}{1+x} + \int \frac{dx}{1-x} = \int \frac{dx}{1-(-x)} + \int \frac{dx}{1-x}$

$$= \int \left[\sum_{n=0}^{\infty} (-1)^n x^n + \sum_{n=0}^{\infty} x^n\right] dx = \int \left[(1 - x + x^2 - x^3 + x^4 - \cdots) + (1 + x + x^2 + x^3 + x^4 + \cdots)\right] dx$$

$$= \int (2 + 2x^2 + 2x^4 + \cdots)\, dx = \int \sum_{n=0}^{\infty} 2x^{2n}\, dx = C + \sum_{n=0}^{\infty} \frac{2x^{2n+1}}{2n+1}$$

But $f(0) = \ln\frac{1}{1} = 0$, so $C = 0$ and we have $f(x) = \displaystyle\sum_{n=0}^{\infty} \frac{2x^{2n+1}}{2n+1}$ with $R = 1$. If $x = \pm 1$, then $f(x) = \pm 2 \displaystyle\sum_{n=0}^{\infty} \frac{1}{2n+1}$, which both diverge by the Limit Comparison Test with $b_n = \dfrac{1}{n}$.

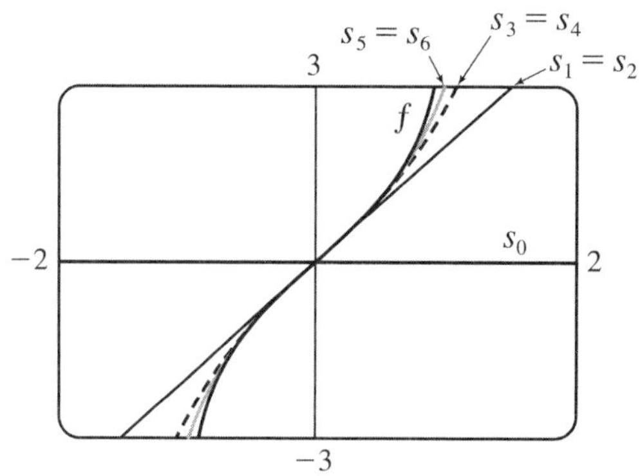


As $n$ increases, $s_n(x)$ approximates $f$ better on the interval of convergence, which is $(-1, 1)$.

**20.** $f(x) = \tan^{-1}(2x) = 2\displaystyle\int \frac{dx}{1+4x^2} = 2\int \sum_{n=0}^{\infty} (-1)^n \left(4x^2\right)^n dx = 2\int \sum_{n=0}^{\infty} (-1)^n 4^n x^{2n}\, dx$

$$= C + 2\sum_{n=0}^{\infty} \frac{(-1)^n 4^n x^{2n+1}}{2n+1} = \sum_{n=0}^{\infty} \frac{(-1)^n 2^{2n+1} x^{2n+1}}{2n+1} \quad [f(0) = \tan^{-1} 0 = 0, \text{ so } C = 0].$$

The series converges when $\left|4x^2\right| < 1 \;\Leftrightarrow\; |x| < \frac{1}{2}$, so $R = \frac{1}{2}$. If $x = \pm\frac{1}{2}$, then $f(x) = \displaystyle\sum_{n=0}^{\infty} (-1)^n \frac{1}{2n+1}$ and $f(x) = \displaystyle\sum_{n=0}^{\infty} (-1)^{n+1} \frac{1}{2n+1}$, respectively. Both series converge by the Alternating Series Test.

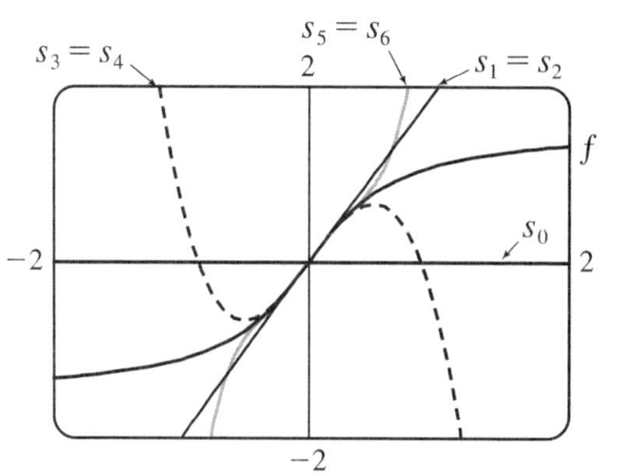


As $n$ increases, $s_n(x)$ approximates $f$ better on the interval of convergence, which is $\left[-\frac{1}{2}, \frac{1}{2}\right]$.

**21.** $\dfrac{t}{1-t^8} = t \cdot \dfrac{1}{1-t^8} = t \sum\limits_{n=0}^{\infty} (t^8)^n = \sum\limits_{n=0}^{\infty} t^{8n+1} \quad \Rightarrow \quad \displaystyle\int \frac{t}{1-t^8}\,dt = C + \sum_{n=0}^{\infty} \frac{t^{8n+2}}{8n+2}$. The series for $\dfrac{1}{1-t^8}$ converges when $|t^8| < 1 \quad \Leftrightarrow \quad |t| < 1$, so $R = 1$ for that series and also the series for $t/(1-t^8)$. By Theorem 2, the series for $\displaystyle\int \frac{t}{1-t^8}\,dt$ also has $R = 1$.

**22.** By Example 6, $\ln(1-t) = -\sum\limits_{n=1}^{\infty} \dfrac{t^n}{n}$ for $|t| < 1$, so $\dfrac{\ln(1-t)}{t} = -\sum\limits_{n=1}^{\infty} \dfrac{t^{n-1}}{n}$ and $\displaystyle\int \frac{\ln(1-t)}{t}\,dt = C - \sum_{n=1}^{\infty} \frac{t^n}{n^2}$.

By Theorem 2, $R = 1$.

**23.** By Example 7, $\tan^{-1} x = \sum\limits_{n=0}^{\infty} (-1)^n \dfrac{x^{2n+1}}{2n+1}$ with $R = 1$, so

$x - \tan^{-1} x = x - \left(x - \dfrac{x^3}{3} + \dfrac{x^5}{5} - \dfrac{x^7}{7} + \cdots\right) = \dfrac{x^3}{3} - \dfrac{x^5}{5} + \dfrac{x^7}{7} - \cdots = \sum\limits_{n=1}^{\infty} (-1)^{n+1} \dfrac{x^{2n+1}}{2n+1}$ and

$\dfrac{x - \tan^{-1} x}{x^3} = \sum\limits_{n=1}^{\infty} (-1)^{n+1} \dfrac{x^{2n-2}}{2n+1}$, so

$\displaystyle\int \frac{x - \tan^{-1} x}{x^3}\,dx = C + \sum_{n=1}^{\infty} (-1)^{n+1} \frac{x^{2n-1}}{(2n+1)(2n-1)} = C + \sum_{n=1}^{\infty} (-1)^{n+1} \frac{x^{2n-1}}{4n^2-1}$. By Theorem 2, $R = 1$.

**24.** By Example 7, $\displaystyle\int \tan^{-1}(x^2)\,dx = \int \sum_{n=0}^{\infty} (-1)^n \frac{(x^2)^{2n+1}}{2n+1}\,dx = C + \sum_{n=0}^{\infty} (-1)^n \frac{x^{4n+3}}{(2n+1)(4n+3)}$ with $R = 1$.

**25.** $\dfrac{1}{1+x^5} = \dfrac{1}{1-(-x^5)} = \sum\limits_{n=0}^{\infty} (-x^5)^n = \sum\limits_{n=0}^{\infty} (-1)^n x^{5n} \quad \Rightarrow$

$\displaystyle\int \frac{1}{1+x^5}\,dx = \int \sum_{n=0}^{\infty} (-1)^n x^{5n}\,dx = C + \sum_{n=0}^{\infty} (-1)^n \frac{x^{5n+1}}{5n+1}$. Thus,

$I = \displaystyle\int_0^{0.2} \frac{1}{1+x^5}\,dx = \left[x - \frac{x^6}{6} + \frac{x^{11}}{11} - \cdots\right]_0^{0.2} = 0.2 - \frac{(0.2)^6}{6} + \frac{(0.2)^{11}}{11} - \cdots$. The series is alternating, so if we use the first two terms, the error is at most $(0.2)^{11}/11 \approx 1.9 \times 10^{-9}$. So $I \approx 0.2 - (0.2)^6/6 \approx 0.199989$ to six decimal places.

**26.** From Example 6, we know $\ln(1-x) = -\sum\limits_{n=1}^{\infty} \dfrac{x^n}{n}$, so

$\ln(1+x^4) = \ln[1-(-x^4)] = -\sum\limits_{n=1}^{\infty} \dfrac{(-x^4)^n}{n} = \sum\limits_{n=1}^{\infty} (-1)^{n+1} \dfrac{x^{4n}}{n} \quad \Rightarrow$

$\displaystyle\int \ln(1+x^4)\,dx = \int \sum_{n=1}^{\infty} (-1)^{n+1} \frac{x^{4n}}{n}\,dx = C + \sum_{n=1}^{\infty} (-1)^{n+1} \frac{x^{4n+1}}{n(4n+1)}$. Thus,

$I = \displaystyle\int_0^{0.4} \ln(1+x^4)\,dx = \left[\frac{x^5}{5} - \frac{x^9}{18} + \frac{x^{13}}{39} - \frac{x^{17}}{68} + \cdots\right]_0^{0.4} = \frac{(0.4)^5}{5} - \frac{(0.4)^9}{18} + \frac{(0.4)^{13}}{39} - \frac{(0.4)^{17}}{68} + \cdots$.

The series is alternating, so if we use the first three terms, the error is at most $(0.4)^{17}/68 \approx 2.5 \times 10^{-9}$.

So $I \approx (0.4)^5/5 - (0.4)^9/18 + (0.9)^{13}/39 \approx 0.002034$ to six decimal places.

**27.** We substitute $3x$ for $x$ in Example 7, and find that

$$\int x\arctan(3x)\,dx = \int x\sum_{n=0}^{\infty}(-1)^n\frac{(3x)^{2n+1}}{2n+1}\,dx = \int\sum_{n=0}^{\infty}(-1)^n\frac{3^{2n+1}x^{2n+2}}{2n+1}\,dx$$
$$= C+\sum_{n=0}^{\infty}(-1)^n\frac{3^{2n+1}x^{2n+3}}{(2n+1)(2n+3)}$$

So $$\int_0^{0.1} x\arctan(3x)\,dx = \left[\frac{3x^3}{1\cdot 3}-\frac{3^3x^5}{3\cdot 5}+\frac{3^5x^7}{5\cdot 7}-\frac{3^7x^9}{7\cdot 9}+\cdots\right]_0^{0.1}$$
$$=\frac{1}{10^3}-\frac{9}{5\times 10^5}+\frac{243}{35\times 10^7}-\frac{2187}{63\times 10^9}+\cdots.$$

The series is alternating, so if we use three terms, the error is at most $\frac{2187}{63\times 10^9}\approx 3.5\times 10^{-8}$. So

$$\int_0^{0.1} x\arctan(3x)\,dx \approx \frac{1}{10^3}-\frac{9}{5\times 10^5}+\frac{243}{35\times 10^7}\approx 0.000\,983 \text{ to six decimal places.}$$

**28.** $$\int_0^{0.3}\frac{x^2}{1+x^4}\,dx = \int_0^{0.3}x^2\sum_{n=0}^{\infty}(-1)^n x^{4n}\,dx = \sum_{n=0}^{\infty}\left[\frac{(-1)^n x^{4n+3}}{4n+3}\right]_0^{0.3} = \sum_{n=0}^{\infty}\frac{(-1)^n 3^{4n+3}}{(4n+3)10^{4n+3}}$$
$$=\frac{3^3}{3\times 10^3}-\frac{3^7}{7\times 10^7}+\frac{3^{11}}{11\times 10^{11}}-\cdots$$

The series is alternating, so if we use only two terms, the error is at most $\frac{3^{11}}{11\times 10^{11}}\approx 0.000\,000\,16$. So, to six decimal places, $\int_0^{0.3}\frac{x^2}{1+x^4}\,dx \approx \frac{3^3}{3\times 10^3}-\frac{3^7}{7\times 10^7}\approx 0.008\,969$.

**29.** Using the result of Example 6, $\ln(1-x) = -\sum_{n=1}^{\infty}\frac{x^n}{n}$, with $x=-0.1$, we have

$\ln 1.1 = \ln[1-(-0.1)] = 0.1-\frac{0.01}{2}+\frac{0.001}{3}-\frac{0.0001}{4}+\frac{0.00001}{5}-\cdots$. The series is alternating, so if we use only the first four terms, the error is at most $\frac{0.00001}{5}=0.000002$. So $\ln 1.1\approx 0.1-\frac{0.01}{2}+\frac{0.001}{3}-\frac{0.0001}{4}\approx 0.09531$.

**30.** $f(x)=\sum_{n=0}^{\infty}\frac{(-1)^n x^{2n}}{(2n)!} \Rightarrow f'(x)=\sum_{n=1}^{\infty}\frac{(-1)^n 2nx^{2n-1}}{(2n)!}$ [the first term disappears], so

$$f''(x)=\sum_{n=1}^{\infty}\frac{(-1)^n(2n)(2n-1)x^{2n-2}}{(2n)!}=\sum_{n=1}^{\infty}\frac{(-1)^n x^{2(n-1)}}{[2(n-1)]!}=\sum_{n=0}^{\infty}\frac{(-1)^{n+1}x^{2n}}{(2n)!} \quad \text{[substituting } n+1 \text{ for } n\text{]}$$
$$=-\sum_{n=0}^{\infty}\frac{(-1)^n x^{2n}}{(2n)!}=-f(x) \Rightarrow f''(x)+f(x)=0.$$

**31.** (a) $J_0(x)=\sum_{n=0}^{\infty}\frac{(-1)^n x^{2n}}{2^{2n}(n!)^2}$, $J_0'(x)=\sum_{n=1}^{\infty}\frac{(-1)^n 2nx^{2n-1}}{2^{2n}(n!)^2}$, and $J_0''(x)=\sum_{n=1}^{\infty}\frac{(-1)^n 2n(2n-1)x^{2n-2}}{2^{2n}(n!)^2}$, so

$$x^2J_0''(x)+xJ_0'(x)+x^2J_0(x)=\sum_{n=1}^{\infty}\frac{(-1)^n 2n(2n-1)x^{2n}}{2^{2n}(n!)^2}+\sum_{n=1}^{\infty}\frac{(-1)^n 2nx^{2n}}{2^{2n}(n!)^2}+\sum_{n=0}^{\infty}\frac{(-1)^n x^{2n+2}}{2^{2n}(n!)^2}$$
$$=\sum_{n=1}^{\infty}\frac{(-1)^n 2n(2n-1)x^{2n}}{2^{2n}(n!)^2}+\sum_{n=1}^{\infty}\frac{(-1)^n 2nx^{2n}}{2^{2n}(n!)^2}+\sum_{n=1}^{\infty}\frac{(-1)^{n-1}x^{2n}}{2^{2n-2}[(n-1)!]^2}$$
$$=\sum_{n=1}^{\infty}\frac{(-1)^n 2n(2n-1)x^{2n}}{2^{2n}(n!)^2}+\sum_{n=1}^{\infty}\frac{(-1)^n 2nx^{2n}}{2^{2n}(n!)^2}+\sum_{n=1}^{\infty}\frac{(-1)^n(-1)^{-1}2^2n^2x^{2n}}{2^{2n}(n!)^2}$$
$$=\sum_{n=1}^{\infty}(-1)^n\left[\frac{2n(2n-1)+2n-2^2n^2}{2^{2n}(n!)^2}\right]x^{2n}$$
$$=\sum_{n=1}^{\infty}(-1)^n\left[\frac{4n^2-2n+2n-4n^2}{2^{2n}(n!)^2}\right]x^{2n}=0$$

(b) $\displaystyle\int_0^1 J_0(x)\,dx = \int_0^1 \left[\sum_{n=0}^{\infty} \frac{(-1)^n x^{2n}}{2^{2n}(n!)^2}\right] dx = \int_0^1 \left(1 - \frac{x^2}{4} + \frac{x^4}{64} - \frac{x^6}{2304} + \cdots\right) dx$

$$= \left[x - \frac{x^3}{3\cdot 4} + \frac{x^5}{5\cdot 64} - \frac{x^7}{7\cdot 2304} + \cdots\right]_0^1 = 1 - \frac{1}{12} + \frac{1}{320} - \frac{1}{16{,}128} + \cdots$$

Since $\frac{1}{16{,}128} \approx 0.000062$, it follows from The Alternating Series Estimation Theorem that, correct to three decimal places, $\int_0^1 J_0(x)\,dx \approx 1 - \frac{1}{12} + \frac{1}{320} \approx 0.920$.

**32.** (a) $J_1(x) = \displaystyle\sum_{n=0}^{\infty} \frac{(-1)^n x^{2n+1}}{n!\,(n+1)!\,2^{2n+1}}$, $J_1'(x) = \displaystyle\sum_{n=0}^{\infty} \frac{(-1)^n\,(2n+1)\,x^{2n}}{n!\,(n+1)!\,2^{2n+1}}$, and $J_1''(x) = \displaystyle\sum_{n=1}^{\infty} \frac{(-1)^n\,(2n+1)\,(2n)\,x^{2n-1}}{n!\,(n+1)!\,2^{2n+1}}$.

$$x^2 J_1''(x) + xJ_1'(x) + (x^2-1)J_1(x)$$
$$= \sum_{n=1}^{\infty} \frac{(-1)^n\,(2n+1)(2n)x^{2n+1}}{n!\,(n+1)!\,2^{2n+1}} + \sum_{n=0}^{\infty} \frac{(-1)^n\,(2n+1)x^{2n+1}}{n!\,(n+1)!\,2^{2n+1}} + \sum_{n=0}^{\infty} \frac{(-1)^n\,x^{2n+3}}{n!\,(n+1)!\,2^{2n+1}} - \sum_{n=0}^{\infty} \frac{(-1)^n\,x^{2n+1}}{n!\,(n+1)!\,2^{2n+1}}$$
$$= \sum_{n=1}^{\infty} \frac{(-1)^n\,(2n+1)(2n)x^{2n+1}}{n!\,(n+1)!\,2^{2n+1}} + \sum_{n=0}^{\infty} \frac{(-1)^n\,(2n+1)x^{2n+1}}{n!\,(n+1)!\,2^{2n+1}} - \sum_{n=1}^{\infty} \frac{(-1)^n\,x^{2n+1}}{(n-1)!\,n!\,2^{2n-1}} - \sum_{n=0}^{\infty} \frac{(-1)^n\,x^{2n+1}}{n!\,(n+1)!\,2^{2n+1}} \quad \left[\begin{array}{c}\text{Replace } n \text{ with } n-1\\ \text{in the third term}\end{array}\right]$$
$$= \frac{x}{2} - \frac{x}{2} + \sum_{n=1}^{\infty} (-1)^n \left[\frac{(2n+1)(2n) + (2n+1) - (n)(n+1)2^2 - 1}{n!\,(n+1)!\,2^{2n+1}}\right] x^{2n+1} = 0$$

(b) $J_0(x) = \displaystyle\sum_{n=0}^{\infty} \frac{(-1)^n\,x^{2n}}{2^{2n}\,(n!)^2} \Rightarrow$

$$J_0'(x) = \sum_{n=1}^{\infty} \frac{(-1)^n\,(2n)x^{2n-1}}{2^{2n}\,(n!)^2} = \sum_{n=0}^{\infty} \frac{(-1)^{n+1}\,2(n+1)x^{2n+1}}{2^{2n+2}\,[(n+1)!]^2} \quad [\text{Replace } n \text{ with } n+1]$$
$$= -\sum_{n=0}^{\infty} \frac{(-1)^n\,x^{2n+1}}{2^{2n+1}(n+1)!\,n!} \quad [\text{cancel 2 and } n+1\text{; take } -1 \text{ outside sum}] \quad = -J_1(x)$$

**33.** (a) $f(x) = \displaystyle\sum_{n=0}^{\infty} \frac{x^n}{n!} \Rightarrow f'(x) = \sum_{n=1}^{\infty} \frac{nx^{n-1}}{n!} = \sum_{n=1}^{\infty} \frac{x^{n-1}}{(n-1)!} = \sum_{n=0}^{\infty} \frac{x^n}{n!} = f(x)$

(b) By Theorem 7.4.2, the only solution to the differential equation $df(x)/dx = f(x)$ is $f(x) = Ke^x$, but $f(0) = 1$, so $K = 1$ and $f(x) = e^x$.

*Or:* We could solve the equation $df(x)/dx = f(x)$ as a separable differential equation.

**34.** $\dfrac{|\sin nx|}{n^2} \le \dfrac{1}{n^2}$, so $\displaystyle\sum_{n=1}^{\infty} \frac{\sin nx}{n^2}$ converges by the Comparison Test. $\dfrac{d}{dx}\left(\dfrac{\sin nx}{n^2}\right) = \dfrac{\cos nx}{n}$, so when $x = 2k\pi$ ($k$ an integer), $\displaystyle\sum_{n=1}^{\infty} f_n'(x) = \sum_{n=1}^{\infty} \frac{\cos(2kn\pi)}{n} = \sum_{n=1}^{\infty} \frac{1}{n}$, which diverges (harmonic series). $f_n''(x) = -\sin nx$, so $\displaystyle\sum_{n=1}^{\infty} f_n''(x) = -\sum_{n=1}^{\infty} \sin nx$, which converges only if $\sin nx = 0$, or $x = k\pi$ ($k$ an integer).

**35.** If $a_n = \dfrac{x^n}{n^2}$, then by the Ratio Test, $\lim\limits_{n\to\infty}\left|\dfrac{a_{n+1}}{a_n}\right| = \lim\limits_{n\to\infty}\left|\dfrac{x^{n+1}}{(n+1)^2}\cdot\dfrac{n^2}{x^n}\right| = |x|\lim\limits_{n\to\infty}\left(\dfrac{n}{n+1}\right)^2 = |x| < 1$ for convergence, so $R = 1$. When $x = \pm 1$, $\sum\limits_{n=1}^{\infty}\left|\dfrac{x^n}{n^2}\right| = \sum\limits_{n=1}^{\infty}\dfrac{1}{n^2}$ which is a convergent $p$-series ($p = 2 > 1$), so the interval of convergence for $f$ is $[-1,1]$. By Theorem 2, the radii of convergence of $f'$ and $f''$ are both 1, so we need only check the endpoints. $f(x) = \sum\limits_{n=1}^{\infty}\dfrac{x^n}{n^2} \Rightarrow f'(x) = \sum\limits_{n=1}^{\infty}\dfrac{nx^{n-1}}{n^2} = \sum\limits_{n=0}^{\infty}\dfrac{x^n}{n+1}$, and this series diverges for $x = 1$ (harmonic series) and converges for $x = -1$ (Alternating Series Test), so the interval of convergence is $[-1,1)$. $f''(x) = \sum\limits_{n=1}^{\infty}\dfrac{nx^{n-1}}{n+1}$ diverges at both 1 and $-1$ (Test for Divergence) since $\lim\limits_{n\to\infty}\dfrac{n}{n+1} = 1 \neq 0$, so its interval of convergence is $(-1,1)$.

**36.** (a) $\sum\limits_{n=1}^{\infty} nx^{n-1} = \sum\limits_{n=0}^{\infty}\dfrac{d}{dx}x^n = \dfrac{d}{dx}\left[\sum\limits_{n=0}^{\infty}x^n\right] = \dfrac{d}{dx}\left[\dfrac{1}{1-x}\right] = -\dfrac{1}{(1-x)^2}(-1) = \dfrac{1}{(1-x)^2}$, $|x| < 1$.

(b) (i) $\sum\limits_{n=1}^{\infty} nx^n = x\sum\limits_{n=1}^{\infty} nx^{n-1} = x\left[\dfrac{1}{(1-x)^2}\right]$ [from part (a)] $= \dfrac{x}{(1-x)^2}$ for $|x| < 1$.

(ii) Put $x = \frac{1}{2}$ in (i): $\sum\limits_{n=1}^{\infty}\dfrac{n}{2^n} = \sum\limits_{n=1}^{\infty} n\left(\frac{1}{2}\right)^n = \dfrac{1/2}{(1-1/2)^2} = 2.$

(c) (i) $\sum\limits_{n=2}^{\infty} n(n-1)x^n = x^2\sum\limits_{n=2}^{\infty} n(n-1)x^{n-2} = x^2\dfrac{d}{dx}\left[\sum\limits_{n=1}^{\infty} nx^{n-1}\right] = x^2\dfrac{d}{dx}\dfrac{1}{(1-x)^2}$

$= x^2\dfrac{2}{(1-x)^3} = \dfrac{2x^2}{(1-x)^3}$ for $|x| < 1$.

(ii) Put $x = \frac{1}{2}$ in (i): $\sum\limits_{n=2}^{\infty}\dfrac{n^2-n}{2^n} = \sum\limits_{n=2}^{\infty} n(n-1)\left(\frac{1}{2}\right)^n = \dfrac{2(1/2)^2}{(1-1/2)^3} = 4.$

(iii) From (b)(ii) and (c)(ii), we have $\sum\limits_{n=1}^{\infty}\dfrac{n^2}{2^n} = \sum\limits_{n=1}^{\infty}\dfrac{n^2-n}{2^n} + \sum\limits_{n=1}^{\infty}\dfrac{n}{2^n} = 4 + 2 = 6.$

**37.** By Example 7, $\tan^{-1}x = \sum\limits_{n=0}^{\infty}(-1)^n\dfrac{x^{2n+1}}{2n+1}$ for $|x| < 1$. In particular, for $x = \dfrac{1}{\sqrt{3}}$, we have $\dfrac{\pi}{6} = \tan^{-1}\left(\dfrac{1}{\sqrt{3}}\right) = \sum\limits_{n=0}^{\infty}(-1)^n\dfrac{\left(1/\sqrt{3}\right)^{2n+1}}{2n+1} = \sum\limits_{n=0}^{\infty}(-1)^n\left(\dfrac{1}{3}\right)^n\dfrac{1}{\sqrt{3}}\dfrac{1}{2n+1}$, so

$$\pi = \frac{6}{\sqrt{3}}\sum_{n=0}^{\infty}\frac{(-1)^n}{(2n+1)3^n} = 2\sqrt{3}\sum_{n=0}^{\infty}\frac{(-1)^n}{(2n+1)3^n}.$$

**38.** (a) $\displaystyle\int_0^{1/2}\frac{dx}{x^2-x+1} = \int_0^{1/2}\frac{dx}{(x-1/2)^2+3/4}$ $\left[\begin{array}{l} x - 1/2 = \left(\sqrt{3}/2\right)u,\ u = \left(2/\sqrt{3}\right)(x-1/2) \\ dx = \left(\sqrt{3}/2\right)du \end{array}\right]$

$$= \int_{-1/\sqrt{3}}^{0}\frac{\left(\sqrt{3}/2\right)du}{(3/4)(u^2+1)} = \frac{2\sqrt{3}}{3}\left[\tan^{-1}u\right]_{-1/\sqrt{3}}^{0} = \frac{2}{\sqrt{3}}\left[0 - \left(-\frac{\pi}{6}\right)\right] = \frac{\pi}{3\sqrt{3}}$$

(b) $\dfrac{1}{x^3+1} = \dfrac{1}{(x+1)(x^2-x+1)} \Rightarrow$

$$\frac{1}{x^2-x+1} = (x+1)\left(\frac{1}{1+x^3}\right) = (x+1)\frac{1}{1-(-x^3)} = (x+1)\sum_{n=0}^{\infty}(-1)^n x^{3n}$$

$$= \sum_{n=0}^{\infty}(-1)^n x^{3n+1} + \sum_{n=0}^{\infty}(-1)^n x^{3n} \text{ for } |x|<1 \Rightarrow$$

$$\int \frac{dx}{x^2-x+1} = C + \sum_{n=0}^{\infty}(-1)^n \frac{x^{3n+2}}{3n+2} + \sum_{n=0}^{\infty}(-1)^n \frac{x^{3n+1}}{3n+1} \text{ for } |x|<1 \Rightarrow$$

$$\int_0^{1/2} \frac{dx}{x^2-x+1} = \sum_{n=0}^{\infty}(-1)^n\left[\frac{1}{4\cdot 8^n(3n+2)} + \frac{1}{2\cdot 8^n(3n+1)}\right] = \frac{1}{4}\sum_{n=0}^{\infty}\frac{(-1)^n}{8^n}\left(\frac{2}{3n+1}+\frac{1}{3n+2}\right).$$

By part (a), this equals $\dfrac{\pi}{3\sqrt{3}}$, so $\pi = \dfrac{3\sqrt{3}}{4}\displaystyle\sum_{n=0}^{\infty}\frac{(-1)^n}{8^n}\left(\frac{2}{3n+1}+\frac{1}{3n+2}\right).$

## 8.7 Taylor and Maclaurin Series

**1.** Using Theorem 5 with $\sum_{n=0}^{\infty} b_n(x-5)^n$, $b_n = \dfrac{f^{(n)}(a)}{n!}$, so $b_8 = \dfrac{f^{(8)}(5)}{8!}$.

**2.** (a) Using Formula 6, a power series expansion of $f$ at 1 must have the form $f(1) + f'(1)(x-1) + \cdots$. Comparing to the given series, $1.6 - 0.8(x-1) + \cdots$, we must have $f'(1) = -0.8$. But from the graph, $f'(1)$ is positive. Hence, the given series is *not* the Taylor series of $f$ centered at 1.

(b) A power series expansion of $f$ at 2 must have the form $f(2) + f'(2)(x-2) + \frac{1}{2}f''(2)(x-2)^2 + \cdots$. Comparing to the given series, $2.8 + 0.5(x-2) + 1.5(x-2)^2 - 0.1(x-2)^3 + \cdots$, we must have $\frac{1}{2}f''(2) = 1.5$; that is, $f''(2)$ is positive. But from the graph, $f$ is concave downward near $x = 2$, so $f''(2)$ must be negative. Hence, the given series is *not* the Taylor series of $f$ centered at 2.

**3.** Since $f^{(n)}(0) = (n+1)!$, Equation 7 gives the Maclaurin series

$\displaystyle\sum_{n=0}^{\infty}\frac{f^{(n)}(0)}{n!}x^n = \sum_{n=0}^{\infty}\frac{(n+1)!}{n!}x^n = \sum_{n=0}^{\infty}(n+1)x^n$. Applying the Ratio Test with $a_n = (n+1)x^n$ gives us

$\displaystyle\lim_{n\to\infty}\left|\frac{a_{n+1}}{a_n}\right| = \lim_{n\to\infty}\left|\frac{(n+2)x^{n+1}}{(n+1)x^n}\right| = |x|\lim_{n\to\infty}\frac{n+2}{n+1} = |x|\cdot 1 = |x|$. For convergence, we must have $|x|<1$, so the radius of convergence $R = 1$.

**4.** Since $f^{(n)}(4) = \dfrac{(-1)^n\, n!}{3^n(n+1)}$, Equation 6 gives the Taylor series

$\displaystyle\sum_{n=0}^{\infty}\frac{f^{(n)}(4)}{n!}(x-4)^n = \sum_{n=0}^{\infty}\frac{(-1)^n\, n!}{3^n(n+1)\, n!}(x-4)^n = \sum_{n=0}^{\infty}\frac{(-1)^n}{3^n(n+1)}(x-4)^n$, which is the Taylor series for $f$ centered at 4. Apply the Ratio Test to find the radius of convergence $R$.

$$\lim_{n\to\infty}\left|\frac{a_{n+1}}{a_n}\right| = \lim_{n\to\infty}\left|\frac{(-1)^{n+1}(x-4)^{n+1}}{3^{n+1}(n+2)}\cdot\frac{3^n(n+1)}{(-1)^n(x-4)^n}\right| = \lim_{n\to\infty}\left|\frac{(-1)(x-4)(n+1)}{3(n+2)}\right|$$

$$= \frac{1}{3}|x-4|\lim_{n\to\infty}\frac{n+1}{n+2} = \frac{1}{3}|x-4|$$

For convergence, $\frac{1}{3}|x-4|<1 \Leftrightarrow |x-4|<3$, so $R = 3$.

5.

| $n$ | $f^{(n)}(x)$ | $f^{(n)}(0)$ |
|---|---|---|
| 0 | $\cos x$ | 1 |
| 1 | $-\sin x$ | 0 |
| 2 | $-\cos x$ | $-1$ |
| 3 | $\sin x$ | 0 |
| 4 | $\cos x$ | 1 |
| $\vdots$ | $\vdots$ | $\vdots$ |

We use Equation 7 with $f(x) = \cos x$.

$$\cos x = f(0) + f'(0)x + \frac{f''(0)}{2!}x^2 + \frac{f^{(3)}(0)}{3!}x^3 + \frac{f^{(4)}(0)}{4!}x^4 + \cdots$$

$$= 1 - \frac{x^2}{2!} + \frac{x^4}{4!} - \cdots = \sum_{n=0}^{\infty} \frac{(-1)^n x^{2n}}{(2n)!}$$

If $a_n = \dfrac{(-1)^n x^{2n}}{(2n)!}$, then

$$\lim_{n\to\infty}\left|\frac{a_{n+1}}{a_n}\right| = \lim_{n\to\infty}\left|\frac{x^{2n+2}}{(2n+2)!}\cdot\frac{(2n)!}{x^{2n}}\right| = x^2 \lim_{n\to\infty}\frac{1}{(2n+2)(2n+1)} = 0 < 1$$ for all $x$. So $R = \infty$ (Ratio Test).

6.

| $n$ | $f^{(n)}(x)$ | $f^{(n)}(0)$ |
|---|---|---|
| 0 | $\sin 2x$ | 0 |
| 1 | $2\cos 2x$ | 2 |
| 2 | $-2^2 \sin 2x$ | 0 |
| 3 | $-2^3 \cos 2x$ | $-2^3$ |
| 4 | $2^4 \sin 2x$ | 0 |
| $\vdots$ | $\vdots$ | $\vdots$ |

$f^{(n)}(0) = 0$ if $n$ is even and $f^{(2n+1)}(0) = (-1)^n 2^{2n+1}$, so

$$\sin 2x = \sum_{n=0}^{\infty}\frac{f^{(n)}(0)}{n!}x^n = \sum_{n=0}^{\infty}\frac{f^{(2n+1)}(0)}{(2n+1)!}x^{2n+1}$$

$$= \sum_{n=0}^{\infty}\frac{(-1)^n 2^{2n+1} x^{2n+1}}{(2n+1)!}$$

$$\lim_{n\to\infty}\left|\frac{a_{n+1}}{a_n}\right| = \lim_{n\to\infty}\frac{2^2|x|^2}{(2n+3)(2n+2)} = 0 < 1$$ for all $x$,

so $R = \infty$ (Ratio Test).

7.

| $n$ | $f^{(n)}(x)$ | $f^{(n)}(0)$ |
|---|---|---|
| 0 | $e^{5x}$ | 1 |
| 1 | $5e^{5x}$ | 5 |
| 2 | $5^2e^{5x}$ | 25 |
| 3 | $5^3e^{5x}$ | 125 |
| 4 | $5^4e^{5x}$ | 625 |
| $\vdots$ | $\vdots$ | $\vdots$ |

$$e^{5x} = \sum_{n=0}^{\infty}\frac{f^{(n)}(0)}{n!}x^n = \sum_{n=0}^{\infty}\frac{5^n}{n!}x^n.$$

$$\lim_{n\to\infty}\left|\frac{a_{n+1}}{a_n}\right| = \lim_{n\to\infty}\left[\frac{5^{n+1}|x|^{n+1}}{(n+1)!}\cdot\frac{n!}{5^n|x|^n}\right]$$

$$= \lim_{n\to\infty}\frac{5|x|}{n+1} = 0 < 1$$ for all $x$, so $R = \infty$.

8.

| $n$ | $f^{(n)}(x)$ | $f^{(n)}(0)$ |
|---|---|---|
| 0 | $xe^x$ | 0 |
| 1 | $(x+1)e^x$ | 1 |
| 2 | $(x+2)e^x$ | 2 |
| 3 | $(x+3)e^x$ | 3 |
| $\vdots$ | $\vdots$ | $\vdots$ |

$$xe^x = \sum_{n=0}^{\infty}\frac{f^{(n)}(0)}{n!}x^n = \sum_{n=0}^{\infty}\frac{n}{n!}x^n = \sum_{n=1}^{\infty}\frac{n}{n!}x^n = \sum_{n=1}^{\infty}\frac{x^n}{(n-1)!}.$$

$$\lim_{n\to\infty}\left|\frac{a_{n+1}}{a_n}\right| = \lim_{n\to\infty}\left[\frac{|x|^{n+1}}{n!}\cdot\frac{(n-1)!}{|x|^n}\right] = \lim_{n\to\infty}\frac{|x|}{n} = 0 < 1$$ for all $x$, so $R = \infty$.

**9.**

| $n$ | $f^{(n)}(x)$ | $f^{(n)}(2)$ |
|---|---|---|
| 0 | $1+x+x^2$ | 7 |
| 1 | $1+2x$ | 5 |
| 2 | 2 | 2 |
| 3 | 0 | 0 |
| 4 | 0 | 0 |
| $\vdots$ | $\vdots$ | $\vdots$ |

$$f(x) = 7 + 5(x-2) + \frac{2}{2!}(x-2)^2 + \sum_{n=3}^{\infty} \frac{0}{n!}(x-2)^n$$
$$= 7 + 5(x-2) + (x-2)^2$$

Since $a_n = 0$ for large $n$, $R = \infty$.

**10.**

| $n$ | $f^{(n)}(x)$ | $f^{(n)}(-1)$ |
|---|---|---|
| 0 | $x^3$ | $-1$ |
| 1 | $3x^2$ | 3 |
| 2 | $6x$ | $-6$ |
| 3 | 6 | 6 |
| 4 | 0 | 0 |
| 5 | 0 | 0 |
| $\vdots$ | $\vdots$ | $\vdots$ |

$$f(x) = -1 + 3(x+1) - \tfrac{6}{2!}(x+1)^2 + \tfrac{6}{3!}(x+1)^3$$
$$= -1 + 3(x+1) - 3(x+1)^2 + (x+1)^3$$

Since $a_n = 0$ for large $n$, $R = \infty$.

**11.** Clearly, $f^{(n)}(x) = e^x$, so $f^{(n)}(3) = e^3$ and $e^x = \sum\limits_{n=0}^{\infty} \dfrac{e^3}{n!}(x-3)^n$. If $a_n = \dfrac{e^3}{n!}(x-3)^n$, then

$$\lim_{n\to\infty}\left|\frac{a_{n+1}}{a_n}\right| = \lim_{n\to\infty}\left|\frac{e^3(x-3)^{n+1}}{(n+1)!}\cdot\frac{n!}{e^3(x-3)^n}\right| = \lim_{n\to\infty}\frac{|x-3|}{n+1} = 0 < 1 \text{ for all } x, \text{ so } R = \infty.$$

**12.**

| $n$ | $f^{(n)}(x)$ | $f^{(n)}(2)$ |
|---|---|---|
| 0 | $\ln x$ | $\ln 2$ |
| 1 | $x^{-1}$ | $\frac{1}{2}$ |
| 2 | $-x^{-2}$ | $-\frac{1}{4}$ |
| 3 | $2x^{-3}$ | $\frac{2}{8}$ |
| 4 | $-3\cdot 2x^{-4}$ | $-\frac{3\cdot 2}{16}$ |
| $\vdots$ | $\vdots$ | $\vdots$ |

$f^{(n)}(2) = \dfrac{(-1)^{n-1}(n-1)!}{2^n}$ for $n \geq 1$, so

$$\ln x = \ln 2 + \sum_{n=1}^{\infty} \frac{(-1)^{n-1}(x-2)^n}{n\cdot 2^n}.$$

$$\lim_{n\to\infty}\left|\frac{a_{n+1}}{a_n}\right| = \frac{|x-2|}{2}\lim_{n\to\infty}\frac{n}{n+1} = \frac{|x-2|}{2} < 1 \text{ for convergence, so}$$

$|x-2| < 2 \;\Rightarrow\; R = 2.$

**13.**

| $n$ | $f^{(n)}(x)$ | $f^{(n)}(\pi)$ |
|---|---|---|
| 0 | $\cos x$ | $-1$ |
| 1 | $-\sin x$ | 0 |
| 2 | $-\cos x$ | 1 |
| 3 | $\sin x$ | 0 |
| 4 | $\cos x$ | $-1$ |
| $\vdots$ | $\vdots$ | $\vdots$ |

$$\cos x = \sum_{k=0}^{\infty}\frac{f^{(k)}(\pi)}{k!}(x-\pi)^k$$
$$= -1 + \frac{(x-\pi)^2}{2!} - \frac{(x-\pi)^4}{4!} + \frac{(x-\pi)^6}{6!} - \cdots$$
$$= \sum_{n=0}^{\infty}(-1)^{n+1}\frac{(x-\pi)^{2n}}{(2n)!}$$

$$\lim_{n\to\infty}\left|\frac{a_{n+1}}{a_n}\right| = \lim_{n\to\infty}\left[\frac{|x-\pi|^{2n+2}}{(2n+2)!}\cdot\frac{(2n)!}{|x-\pi|^{2n}}\right] = \lim_{n\to\infty}\frac{|x-\pi|^2}{(2n+2)(2n+1)} = 0 < 1 \text{ for all } x, \text{ so } R = \infty.$$

**14.**

| $n$ | $f^{(n)}(x)$ | $f^{(n)}(\pi/2)$ |
|---|---|---|
| 0 | $\sin x$ | 1 |
| 1 | $\cos x$ | 0 |
| 2 | $-\sin x$ | $-1$ |
| 3 | $-\cos x$ | 0 |
| 4 | $\sin x$ | 1 |
| ⋮ | ⋮ | ⋮ |

$$\sin x = \sum_{k=0}^{\infty} \frac{f^{(k)}(\pi/2)}{k!}\left(x - \frac{\pi}{2}\right)^k = 1 - \frac{(x-\pi/2)^2}{2!} + \frac{(x-\pi/2)^4}{4!} - \frac{(x-\pi/2)^6}{6!} + \cdots = \sum_{n=0}^{\infty} (-1)^n \frac{(x-\pi/2)^{2n}}{(2n)!}$$

$$\lim_{n\to\infty}\left|\frac{a_{n+1}}{a_n}\right| = \lim_{n\to\infty}\left[\frac{|x-\pi/2|^{2n+2}}{(2n+2)!}\cdot\frac{(2n)!}{|x-\pi/2|^{2n}}\right] = \lim_{n\to\infty}\frac{|x-\pi/2|^2}{(2n+2)(2n+1)} = 0 < 1 \text{ for all } x, \text{ so } R = \infty.$$

**15.**

| $n$ | $f^{(n)}(x)$ | $f^{(n)}(9)$ |
|---|---|---|
| 0 | $x^{-1/2}$ | $\frac{1}{3}$ |
| 1 | $-\frac{1}{2}x^{-3/2}$ | $-\frac{1}{2}\cdot\frac{1}{3^3}$ |
| 2 | $\frac{3}{4}x^{-5/2}$ | $-\frac{1}{2}\cdot\left(-\frac{3}{2}\right)\cdot\frac{1}{3^5}$ |
| 3 | $-\frac{15}{8}x^{-7/2}$ | $-\frac{1}{2}\cdot\left(-\frac{3}{2}\right)\cdot\left(-\frac{5}{2}\right)\cdot\frac{1}{3^7}$ |
| ⋮ | ⋮ | ⋮ |

$$\frac{1}{\sqrt{x}} = \frac{1}{3} - \frac{1}{2\cdot 3^3}(x-9) + \frac{3}{2^2\cdot 3^5}\frac{(x-9)^2}{2!} - \frac{3\cdot 5}{2^3\cdot 3^7}\frac{(x-9)^3}{3!} + \cdots = \sum_{n=0}^{\infty}(-1)^n\frac{1\cdot 3\cdot 5\cdot\cdots\cdot(2n-1)}{2^n\cdot 3^{2n+1}\cdot n!}(x-9)^n.$$

$$\lim_{n\to\infty}\left|\frac{a_{n+1}}{a_n}\right| = \lim_{n\to\infty}\left[\frac{1\cdot 3\cdot 5\cdot\cdots\cdot(2n-1)[2(n+1)-1]\,|x-9|^{n+1}}{2^{n+1}\cdot 3^{[2(n+1)+1]}\cdot(n+1)!}\cdot\frac{2^n\cdot 3^{2n+1}\cdot n!}{1\cdot 3\cdot 5\cdot\cdots\cdot(2n-1)\,|x-9|^n}\right]$$

$$= \lim_{n\to\infty}\left[\frac{(2n+1)\,|x-9|}{2\cdot 3^2(n+1)}\right] = \frac{1}{9}|x-9| < 1$$

for convergence, so $|x-9| < 9$ and $R = 9$.

**16.**

| $n$ | $f^{(n)}(x)$ | $f^{(n)}(1)$ |
|---|---|---|
| 0 | $x^{-2}$ | 1 |
| 1 | $-2x^{-3}$ | $-2$ |
| 2 | $6x^{-4}$ | 6 |
| 3 | $-24x^{-5}$ | $-24$ |
| 4 | $120x^{-6}$ | 120 |
| ⋮ | ⋮ | ⋮ |

$$x^{-2} = 1 - 2(x-1) + 6\cdot\frac{(x-1)^2}{2!} - 24\cdot\frac{(x-1)^3}{3!} + 120\cdot\frac{(x-1)^4}{4!} - \cdots = 1 - 2(x-1) + 3(x-1)^2 - 4(x-1)^3 + 5(x-1)^4 - \cdots = \sum_{n=0}^{\infty}(-1)^n(n+1)(x-1)^n.$$

$$\lim_{n\to\infty}\left|\frac{a_{n+1}}{a_n}\right| = \lim_{n\to\infty}\frac{(n+2)\,|x-1|^{n+1}}{(n+1)\,|x-1|^n} = \lim_{n\to\infty}\left[\frac{n+2}{n+1}\cdot|x-1|\right] = |x-1| < 1 \text{ for convergence, so } R = 1.$$

**17.** If $f(x) = \cos x$, then $f^{(n+1)}(x) = \pm\sin x$ or $\pm\cos x$. In each case, $\left|f^{(n+1)}(x)\right| \le 1$, so by Formula 9 with $a = 0$ and $M = 1$, $|R_n(x)| \le \frac{1}{(n+1)!}|x|^{n+1}$. Thus, $|R_n(x)| \to 0$ as $n\to\infty$ by Equation 10. So $\lim_{n\to\infty} R_n(x) = 0$ and, by Theorem 8, the series in Exercise 5 represents $\cos x$ for all $x$.

**18.** If $f(x) = \sin x$, then $f^{(n+1)}(x) = \pm\sin x$ or $\pm\cos x$. In each case, $\left|f^{(n+1)}(x)\right| \le 1$, so by Formula 9 with $a = 0$ and $M = 1$, $|R_n(x)| \le \frac{1}{(n+1)!}\left|x - \frac{\pi}{2}\right|^{n+1}$. Thus, $|R_n(x)| \to 0$ as $n\to\infty$ by Equation 10. So $\lim_{n\to\infty} R_n(x) = 0$ and, by Theorem 8, the series in Exercise 14 represents $\sin x$ for all $x$.

**19.** $\cos x = \sum_{n=0}^{\infty} (-1)^n \frac{x^{2n}}{(2n)!} \Rightarrow f(x) = \cos(\pi x) = \sum_{n=0}^{\infty} \frac{(-1)^n (\pi x)^{2n}}{(2n)!} = \sum_{n=0}^{\infty} \frac{(-1)^n \pi^{2n} x^{2n}}{(2n)!}, \quad R = \infty$

**20.** $e^x = \sum_{n=0}^{\infty} \frac{x^n}{n!} \Rightarrow f(x) = e^{-x/2} = \sum_{n=0}^{\infty} \frac{(-x/2)^n}{n!} = \sum_{n=0}^{\infty} \frac{(-1)^n}{2^n\, n!} x^n, \quad R = \infty$

**21.** $\tan^{-1} x = \sum_{n=0}^{\infty} (-1)^n \frac{x^{2n+1}}{2n+1} \Rightarrow f(x) = x \tan^{-1} x = x \sum_{n=0}^{\infty} (-1)^n \frac{x^{2n+1}}{2n+1} = \sum_{n=0}^{\infty} (-1)^n \frac{x^{2n+2}}{2n+1}, \quad R = 1$

**22.** $\sin x = \sum_{n=0}^{\infty} (-1)^n \frac{x^{2n+1}}{(2n+1)!} \Rightarrow f(x) = \sin(x^4) = \sum_{n=0}^{\infty} (-1)^n \frac{(x^4)^{2n+1}}{(2n+1)!} = \sum_{n=0}^{\infty} \frac{(-1)^n}{(2n+1)!} x^{8n+4}, \quad R = \infty$

**23.** $e^x = \sum_{n=0}^{\infty} \frac{x^n}{n!} \Rightarrow f(x) = x^2 e^{-x} = x^2 \sum_{n=0}^{\infty} \frac{(-x)^n}{n!} = \sum_{n=0}^{\infty} \frac{(-1)^n\, x^{n+2}}{n!}, \quad R = \infty$

**24.** $\cos x = \sum_{n=0}^{\infty} (-1)^n \frac{x^{2n}}{(2n)!} \Rightarrow \cos 2x = \sum_{n=0}^{\infty} (-1)^n \frac{(2x)^{2n}}{(2n)!} = \sum_{n=0}^{\infty} \frac{(-1)^n 2^{2n}}{(2n)!} x^{2n} \Rightarrow$

$f(x) = x \cos 2x = \sum_{n=0}^{\infty} \frac{(-1)^n 2^{2n}}{(2n)!} x^{2n+1}, \quad R = \infty$

**25.** $\sin^2 x = \frac{1}{2}(1 - \cos 2x) = \frac{1}{2}\left[1 - \sum_{n=0}^{\infty} \frac{(-1)^n (2x)^{2n}}{(2n)!}\right] = \frac{1}{2}\left[1 - 1 - \sum_{n=1}^{\infty} \frac{(-1)^n (2x)^{2n}}{(2n)!}\right]$

$= \sum_{n=1}^{\infty} \frac{(-1)^{n+1} 2^{2n-1} x^{2n}}{(2n)!}, \quad R = \infty$

**26.** $\frac{x - \sin x}{x^3} = \frac{1}{x^3}\left[x - \sum_{n=0}^{\infty} \frac{(-1)^n x^{2n+1}}{(2n+1)!}\right] = \frac{1}{x^3}\left[x - x - \sum_{n=1}^{\infty} \frac{(-1)^n x^{2n+1}}{(2n+1)!}\right] = \frac{1}{x^3}\left[- \sum_{n=0}^{\infty} \frac{(-1)^{n+1} x^{2n+3}}{(2n+3)!}\right]$

$= \frac{1}{x^3} \sum_{n=0}^{\infty} \frac{(-1)^n x^{2n+3}}{(2n+3)!} = \sum_{n=0}^{\infty} \frac{(-1)^n x^{2n}}{(2n+3)!}$

and this series also gives the required value at $x = 0$ (namely $1/6$); $R = \infty$.

**27.**

| $n$ | $f^{(n)}(x)$ | $f^{(n)}(0)$ |
|---|---|---|
| 0 | $(1+x)^{1/2}$ | 1 |
| 1 | $\frac{1}{2}(1+x)^{-1/2}$ | $\frac{1}{2}$ |
| 2 | $-\frac{1}{4}(1+x)^{-3/2}$ | $-\frac{1}{4}$ |
| 3 | $\frac{3}{8}(1+x)^{-5/2}$ | $\frac{3}{8}$ |
| 4 | $-\frac{15}{16}(1+x)^{-7/2}$ | $-\frac{15}{16}$ |
| $\vdots$ | $\vdots$ | $\vdots$ |

So $f^{(n)}(0) = \frac{(-1)^{n-1}\, 1 \cdot 3 \cdot 5 \cdot \cdots \cdot (2n-3)}{2^n}$ for $n \geq 2$, and $\sqrt{1+x} = 1 + \frac{x}{2} + \sum_{n=2}^{\infty} \frac{(-1)^{n-1} 1 \cdot 3 \cdot 5 \cdot \cdots \cdot (2n-3)}{2^n n!} x^n$.

If $a_n = \frac{(-1)^{n-1} 1 \cdot 3 \cdot 5 \cdot \cdots \cdot (2n-3)}{2^n n!} x^n$, then

$$\lim_{n\to\infty} \left|\frac{a_{n+1}}{a_n}\right| = \lim_{n\to\infty} \left| \frac{1 \cdot 3 \cdot 5 \cdot \cdots \cdot (2n-3)(2n-1) x^{n+1}}{2^{n+1}(n+1)!} \cdot \frac{2^n n!}{1 \cdot 3 \cdot 5 \cdot \cdots \cdot (2n-3) x^n} \right|$$

$$= \frac{|x|}{2} \lim_{n\to\infty} \frac{2n-1}{n+1} = \frac{|x|}{2} \cdot 2 = |x| < 1 \text{ for convergence, so } R = 1.$$

[continued]

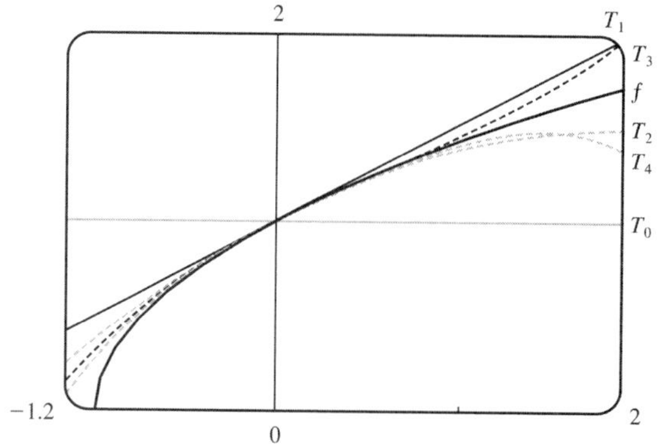


Notice that, as $n$ increases, $T_n(x)$ becomes a better approximation to $f(x)$ for $-1 < x < 1$.

**28.** $e^x \overset{(11)}{=} \sum\limits_{n=0}^{\infty} \dfrac{x^n}{n!}$, so $e^{-x^2} = \sum\limits_{n=0}^{\infty} \dfrac{(-x^2)^n}{n!} = \sum\limits_{n=0}^{\infty} (-1)^n \dfrac{x^{2n}}{n!}$.

Also, $\cos x \overset{(16)}{=} \sum\limits_{n=0}^{\infty} (-1)^n \dfrac{x^{2n}}{(2n)!}$, so

$$f(x) = e^{-x^2} + \cos x = \sum_{n=0}^{\infty} (-1)^n \left(\frac{1}{n!} + \frac{1}{(2n)!}\right) x^{2n}$$
$$= 2 - \frac{3}{2}x^2 + \frac{13}{24}x^4 - \frac{121}{720}x^6 + \cdots.$$

The series for $e^x$ and $\cos x$ converge for all $x$, so the same is true of the series for $f(x)$; that is, $R = \infty$. From the graphs of $f$ and the first few Taylor polynomials, we see that $T_n(x)$ provides a closer fit to $f(x)$ near 0 as $n$ increases.

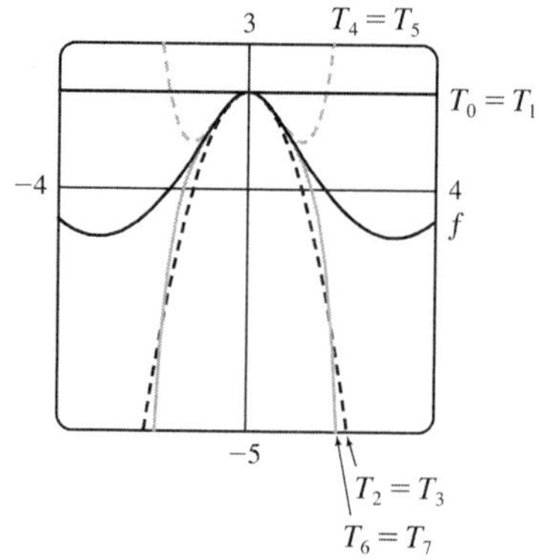


**29.** $\cos x = \sum\limits_{n=0}^{\infty} (-1)^n \dfrac{x^{2n}}{(2n)!} \Rightarrow f(x) = \cos(x^2) = \sum\limits_{n=0}^{\infty} \dfrac{(-1)^n (x^2)^{2n}}{(2n)!} = \sum\limits_{n=0}^{\infty} \dfrac{(-1)^n x^{4n}}{(2n)!}$, $R = \infty$

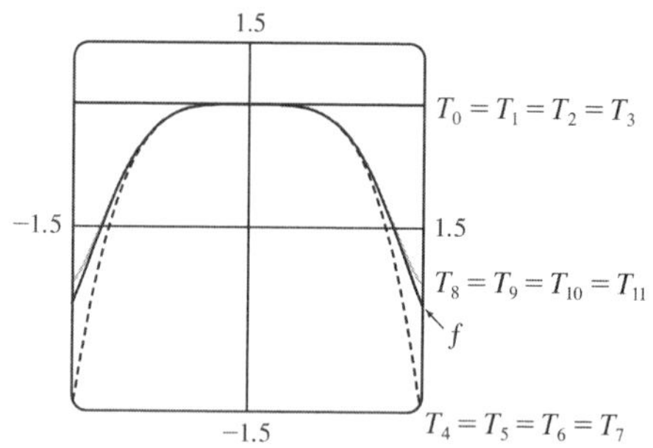


Notice that, as $n$ increases, $T_n(x)$ becomes a better approximation to $f(x)$.

**30.** $2^x = \left(e^{\ln 2}\right)^x = e^{x \ln 2} = \sum\limits_{n=0}^{\infty} \dfrac{(x \ln 2)^n}{n!}$

$= \sum\limits_{n=0}^{\infty} \dfrac{(\ln 2)^n x^n}{n!}$, $R = \infty$.

Notice that, as $n$ increases, $T_n(x)$ becomes a better approximation to $f(x)$.

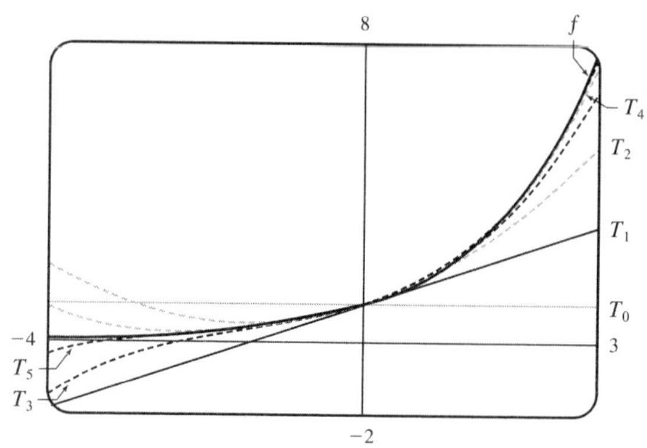

**31.** $e^x = \sum\limits_{n=0}^{\infty} \dfrac{x^n}{n!}$, so $e^{-0.2} = \sum\limits_{n=0}^{\infty} \dfrac{(-0.2)^n}{n!} = 1 - 0.2 + \dfrac{1}{2!}(0.2)^2 - \dfrac{1}{3!}(0.2)^3 + \dfrac{1}{4!}(0.2)^4 - \dfrac{1}{5!}(0.2)^5 + \dfrac{1}{6!}(0.2)^6 - \cdots$.

But $\dfrac{1}{6!}(0.2)^6 = 8.\overline{8} \times 10^{-8}$, so by the Alternating Series Estimation Theorem, $e^{-0.2} \approx \sum\limits_{n=0}^{5} \dfrac{(-0.2)^n}{n!} \approx 0.81873$, correct to five decimal places.

**32.** $3^\circ = \dfrac{\pi}{60}$ radians and $\sin x = \sum\limits_{n=0}^{\infty} \dfrac{(-1)^n\, x^{2n+1}}{(2n+1)!}$, so

$\sin \dfrac{\pi}{60} = \dfrac{\pi}{60} - \dfrac{\left(\frac{\pi}{60}\right)^3}{3!} + \dfrac{\left(\frac{\pi}{60}\right)^5}{5!} - \cdots = \dfrac{\pi}{60} - \dfrac{\pi^3}{1{,}296{,}000} + \dfrac{\pi^5}{93{,}312{,}000{,}000} - \cdots$. But $\dfrac{\pi^5}{93{,}312{,}000{,}000} < 10^{-8}$, so by the Alternating Series Estimation Theorem, $\sin \dfrac{\pi}{60} \approx \dfrac{\pi}{60} - \dfrac{\pi^3}{1{,}296{,}000} \approx 0.05234$.

**33.** $\cos x \overset{(16)}{=} \sum\limits_{n=0}^{\infty} (-1)^n \dfrac{x^{2n}}{(2n)!} \quad\Rightarrow\quad \cos(x^3) = \sum\limits_{n=0}^{\infty} (-1)^n \dfrac{(x^3)^{2n}}{(2n)!} = \sum\limits_{n=0}^{\infty} (-1)^n \dfrac{x^{6n}}{(2n)!} \quad\Rightarrow$

$x\cos(x^3) = \sum\limits_{n=0}^{\infty} (-1)^n \dfrac{x^{6n+1}}{(2n)!} \quad\Rightarrow\quad \displaystyle\int x\cos(x^3)\,dx = C + \sum_{n=0}^{\infty} (-1)^n \frac{x^{6n+2}}{(6n+2)(2n)!}$, with $R = \infty$.

**34.** $\dfrac{\sin x}{x} = \dfrac{1}{x}\sum\limits_{n=0}^{\infty} \dfrac{(-1)^n\, x^{2n+1}}{(2n+1)!} = \sum\limits_{n=0}^{\infty} \dfrac{(-1)^n\, x^{2n}}{(2n+1)!}$, so $\displaystyle\int \frac{\sin x}{x}\,dx = \int \sum_{n=0}^{\infty} \frac{(-1)^n\, x^{2n}}{(2n+1)!}\,dx = C + \sum_{n=0}^{\infty} \frac{(-1)^n\, x^{2n+1}}{(2n+1)(2n+1)!}$

**35.** Using the series from Exercise 27 and substituting $x^3$ for $x$, we get

$$\begin{aligned}\int \sqrt{x^3+1}\,dx &= \int \left[1 + \frac{x^3}{2} + \sum_{n=2}^{\infty} \frac{(-1)^{n-1}\, 1\cdot 3\cdot 5\cdot\cdots\cdot(2n-3)}{2^n n!} x^{3n}\right] dx \\ &= C + x + \frac{x^4}{8} + \sum_{n=2}^{\infty} \frac{(-1)^{n-1}\, 1\cdot 3\cdot 5\cdot\cdots\cdot(2n-3)}{2^n n!(3n+1)} x^{3n+1}\end{aligned}$$

**36.** $e^x \overset{(11)}{=} \sum\limits_{n=0}^{\infty} \dfrac{x^n}{n!} \quad\Rightarrow\quad e^x - 1 = \sum\limits_{n=1}^{\infty} \dfrac{x^n}{n!} \quad\Rightarrow\quad \dfrac{e^x-1}{x} = \sum\limits_{n=1}^{\infty} \dfrac{x^{n-1}}{n!} \quad\Rightarrow\quad \displaystyle\int \frac{e^x-1}{x}\,dx = C + \sum_{n=1}^{\infty} \frac{x^n}{n\cdot n!}$, with $R = \infty$.

**37.** By Exercise 33, $\displaystyle\int x\cos(x^3)\,dx = C + \sum_{n=0}^{\infty} (-1)^n \frac{x^{6n+2}}{(6n+2)(2n)!}$, so

$\displaystyle\int_0^1 x\cos(x^3)\,dx = \left[\sum_{n=0}^{\infty} (-1)^n \frac{x^{6n+2}}{(6n+2)(2n)!}\right]_0^1 = \sum_{n=0}^{\infty} \frac{(-1)^n}{(6n+2)(2n)!} = \frac{1}{2} - \frac{1}{8\cdot 2!} + \frac{1}{14\cdot 4!} - \frac{1}{20\cdot 6!} + \cdots$, but

$\dfrac{1}{20\cdot 6!} = \dfrac{1}{14{,}400} \approx 0.000\,069$, so $\displaystyle\int_0^1 x\cos(x^3)\,dx \approx \frac{1}{2} - \frac{1}{16} + \frac{1}{336} \approx 0.440$ (correct to three decimal places) by the Alternating Series Estimation Theorem.

**38.** From the table of Maclaurin series in this section, we see that

$\tan^{-1} x = \sum\limits_{n=0}^{\infty} (-1)^n \dfrac{x^{2n+1}}{2n+1}$ for $x$ in $[-1,1]$ and $\sin x = \sum\limits_{n=0}^{\infty} (-1)^n \dfrac{x^{2n+1}}{(2n+1)!}$ for all real numbers $x$, so

$\tan^{-1}(x^3) + \sin(x^3) = \sum\limits_{n=0}^{\infty} (-1)^n \dfrac{x^{6n+3}}{2n+1} + \sum\limits_{n=0}^{\infty} (-1)^n \dfrac{x^{6n+3}}{(2n+1)!}$ for $x^3$ in $[-1,1] \quad\Leftrightarrow\quad x$ in $[-1,1]$. Thus,

$$\begin{aligned}I &= \int_0^{0.2} [\tan^{-1}(x^3) + \sin(x^3)]\,dx = \int_0^{0.2} \sum_{n=0}^{\infty} (-1)^n x^{6n+3}\left(\frac{1}{2n+1} + \frac{1}{(2n+1)!}\right) dx \\ &= \left[\sum_{n=0}^{\infty} (-1)^n \frac{x^{6n+4}}{6n+4}\left(\frac{1}{2n+1} + \frac{1}{(2n+1)!}\right)\right]_0^{0.2} = \sum_{n=0}^{\infty} (-1)^n \frac{(0.2)^{6n+4}}{6n+4}\left(\frac{1}{2n+1} + \frac{1}{(2n+1)!}\right) \\ &= \frac{(0.2)^4}{4}(1+1) - \frac{(0.2)^{10}}{10}\left(\frac{1}{3} + \frac{1}{3!}\right) + \cdots\end{aligned}$$

[continued]

But $\dfrac{(0.2)^{10}}{10}\left(\dfrac{1}{3}+\dfrac{1}{3!}\right)=\dfrac{(0.2)^{10}}{20}=5.12\times 10^{-9}$, so by the Alternating Series Estimation Theorem,

$I\approx\dfrac{(0.2)^4}{2}=0.000\,80$ (correct to five decimal places). [Actually, the value is 0.000 800 0, correct to seven decimal places.]

**39.** We first find a series representation for $f(x)=(1+x)^{-1/2}$, and then substitute.

| $n$ | $f^{(n)}(x)$ | $f^{(n)}(0)$ |
|---|---|---|
| 0 | $(1+x)^{-1/2}$ | 1 |
| 1 | $-\frac{1}{2}(1+x)^{-3/2}$ | $-\frac{1}{2}$ |
| 2 | $\frac{3}{4}(1+x)^{-5/2}$ | $\frac{3}{4}$ |
| 3 | $-\frac{15}{8}(1+x)^{-7/2}$ | $-\frac{15}{8}$ |
| $\vdots$ | $\vdots$ | $\vdots$ |

$$\frac{1}{\sqrt{1+x}}=1-\frac{x}{2}+\frac{3}{4}\left(\frac{x^2}{2!}\right)-\frac{15}{8}\left(\frac{x^3}{3!}\right)+\cdots\quad\Rightarrow\quad\frac{1}{\sqrt{1+x^3}}=1-\frac{1}{2}x^3+\frac{3}{8}x^6-\frac{5}{16}x^9+\cdots\quad\Rightarrow$$

$\displaystyle\int_0^{0.1}\frac{dx}{\sqrt{1+x^3}}=\left[x-\frac{1}{8}x^4+\frac{3}{56}x^7-\frac{1}{32}x^{10}+\cdots\right]_0^{0.1}\approx(0.1)-\frac{1}{8}(0.1)^4$, by the Alternating Series Estimation Theorem, since $\frac{3}{56}(0.1)^7\approx 0.000\,000\,005\,4<10^{-8}$, which is the maximum desired error. Therefore,

$$\int_0^{0.1}\frac{dx}{\sqrt{1+x^3}}\approx 0.099\,987\,50.$$

**40.** $\displaystyle\int_0^{0.5}x^2e^{-x^2}\,dx=\int_0^{0.5}\sum_{n=0}^{\infty}\frac{(-1)^n x^{2n+2}}{n!}\,dx=\sum_{n=0}^{\infty}\left[\frac{(-1)^n x^{2n+3}}{n!(2n+3)}\right]_0^{1/2}=\sum_{n=0}^{\infty}\frac{(-1)^n}{n!(2n+3)2^{2n+3}}$ and since the term with $n=2$ is $\dfrac{1}{1792}<0.001$, we use $\displaystyle\sum_{n=0}^{1}\frac{(-1)^n}{n!(2n+3)2^{2n+3}}=\frac{1}{24}-\frac{1}{160}\approx 0.0354.$

**41.** 
$$\lim_{x\to 0}\frac{x-\tan^{-1}x}{x^3}=\lim_{x\to 0}\frac{x-\left(x-\frac{1}{3}x^3+\frac{1}{5}x^5-\frac{1}{7}x^7+\cdots\right)}{x^3}=\lim_{x\to 0}\frac{\frac{1}{3}x^3-\frac{1}{5}x^5+\frac{1}{7}x^7-\cdots}{x^3}$$
$$=\lim_{x\to 0}\left(\tfrac{1}{3}-\tfrac{1}{5}x^2+\tfrac{1}{7}x^4-\cdots\right)=\tfrac{1}{3}$$

since power series are continuous functions.

**42.** 
$$\lim_{x\to 0}\frac{1-\cos x}{1+x-e^x}=\lim_{x\to 0}\frac{1-\left(1-\frac{1}{2!}x^2+\frac{1}{4!}x^4-\frac{1}{6!}x^6+\cdots\right)}{1+x-\left(1+x+\frac{1}{2!}x^2+\frac{1}{3!}x^3+\frac{1}{4!}x^4+\frac{1}{5!}x^5+\frac{1}{6!}x^6+\cdots\right)}$$
$$=\lim_{x\to 0}\frac{\frac{1}{2!}x^2-\frac{1}{4!}x^4+\frac{1}{6!}x^6-\cdots}{-\frac{1}{2!}x^2-\frac{1}{3!}x^3-\frac{1}{4!}x^4-\frac{1}{5!}x^5-\frac{1}{6!}x^6-\cdots}$$
$$=\lim_{x\to 0}\frac{\frac{1}{2!}-\frac{1}{4!}x^2+\frac{1}{6!}x^4-\cdots}{-\frac{1}{2!}-\frac{1}{3!}x-\frac{1}{4!}x^2-\frac{1}{5!}x^3-\frac{1}{6!}x^4-\cdots}=\frac{\frac{1}{2}-0}{-\frac{1}{2}-0}=-1$$

since power series are continuous functions.

**43.** 
$$\lim_{x\to 0}\frac{\sin x-x+\frac{1}{6}x^3}{x^5}=\lim_{x\to 0}\frac{\left(x-\frac{1}{3!}x^3+\frac{1}{5!}x^5-\frac{1}{7!}x^7+\cdots\right)-x+\frac{1}{6}x^3}{x^5}$$
$$=\lim_{x\to 0}\frac{\frac{1}{5!}x^5-\frac{1}{7!}x^7+\cdots}{x^5}=\lim_{x\to 0}\left(\frac{1}{5!}-\frac{x^2}{7!}+\frac{x^4}{9!}-\cdots\right)=\frac{1}{5!}=\frac{1}{120}$$

since power series are continuous functions.

**44.** 
$$\lim_{x\to 0}\frac{\tan x-x}{x^3}=\lim_{x\to 0}\frac{\left(x+\frac{1}{3}x^3+\frac{2}{15}x^5+\cdots\right)-x}{x^3}=\lim_{x\to 0}\frac{\frac{1}{3}x^3+\frac{2}{15}x^5+\cdots}{x^3}=\lim_{x\to 0}\left(\tfrac{1}{3}+\tfrac{2}{15}x^2+\cdots\right)=\tfrac{1}{3}$$

since power series are continuous functions.

**45.** As in Example 8(a), we have $e^{-x^2} = 1 - \dfrac{x^2}{1!} + \dfrac{x^4}{2!} - \dfrac{x^6}{3!} + \cdots$ and we know that $\cos x = 1 - \dfrac{x^2}{2!} + \dfrac{x^4}{4!} - \cdots$ from Equation 16. Therefore, $e^{-x^2}\cos x = \left(1 - x^2 + \frac{1}{2}x^4 - \cdots\right)\left(1 - \frac{1}{2}x^2 + \frac{1}{24}x^4 - \cdots\right)$. Writing only the terms with degree $\leq 4$, we get $e^{-x^2}\cos x = 1 - \frac{1}{2}x^2 + \frac{1}{24}x^4 - x^2 + \frac{1}{2}x^4 + \frac{1}{2}x^4 + \cdots = 1 - \frac{3}{2}x^2 + \frac{25}{24}x^4 + \cdots$.

**46.** $\sec x = \dfrac{1}{\cos x} \overset{(16)}{=} \dfrac{1}{1 - \frac{1}{2}x^2 + \frac{1}{24}x^4 - \cdots}$.

$$
\begin{array}{r|l}
 & 1 + \frac{1}{2}x^2 + \frac{5}{24}x^4 + \cdots \\
\hline
1 - \frac{1}{2}x^2 + \frac{1}{24}x^4 - \cdots & 1 \\
 & 1 - \frac{1}{2}x^2 + \frac{1}{24}x^4 - \cdots \\
\hline
 & \quad \frac{1}{2}x^2 - \frac{1}{24}x^4 + \cdots \\
 & \quad \frac{1}{2}x^2 - \frac{1}{4}x^4 + \cdots \\
\hline
 & \qquad \frac{5}{24}x^4 + \cdots \\
 & \qquad \frac{5}{24}x^4 + \cdots \\
\hline
 & \qquad \cdots
\end{array}
$$

From the long division above, $\sec x = 1 + \frac{1}{2}x^2 + \frac{5}{24}x^4 + \cdots$.

**47.** $\dfrac{x}{\sin x} \overset{(15)}{=} \dfrac{x}{x - \frac{1}{6}x^3 + \frac{1}{120}x^5 - \cdots}$.

$$
\begin{array}{r|l}
 & 1 + \frac{1}{6}x^2 + \frac{7}{360}x^4 + \cdots \\
\hline
x - \frac{1}{6}x^3 + \frac{1}{120}x^5 - \cdots & x \\
 & x - \frac{1}{6}x^3 + \frac{1}{120}x^5 - \cdots \\
\hline
 & \quad \frac{1}{6}x^3 - \frac{1}{120}x^5 + \cdots \\
 & \quad \frac{1}{6}x^3 - \frac{1}{36}x^5 + \cdots \\
\hline
 & \qquad \frac{7}{360}x^5 + \cdots \\
 & \qquad \frac{7}{360}x^5 + \cdots \\
\hline
 & \qquad \cdots
\end{array}
$$

From the long division above, $\dfrac{x}{\sin x} = 1 + \frac{1}{6}x^2 + \frac{7}{360}x^4 + \cdots$.

**48.** From Example 6 in Section 8.6, we have $\ln(1-x) = -x - \frac{1}{2}x^2 - \frac{1}{3}x^3 - \cdots$, $|x| < 1$. Therefore,

$$
\begin{aligned}
e^x \ln(1-x) &= \left(1 + x + \tfrac{1}{2}x^2 + \cdots\right)\left(-x - \tfrac{1}{2}x^2 - \tfrac{1}{3}x^3 - \cdots\right) \\
&= -x - \tfrac{1}{2}x^2 - \tfrac{1}{3}x^3 - x^2 - \tfrac{1}{2}x^3 - \tfrac{1}{2}x^3 - \cdots = -x - \tfrac{3}{2}x^2 - \tfrac{4}{3}x^3 - \cdots,\ |x| < 1
\end{aligned}
$$

**49.** $\displaystyle\sum_{n=0}^{\infty} (-1)^n \frac{x^{4n}}{n!} = \sum_{n=0}^{\infty} \frac{\left(-x^4\right)^n}{n!} = e^{-x^4}$, by (11).

**50.** $\displaystyle\sum_{n=0}^{\infty} \frac{(-1)^n \pi^{2n}}{6^{2n}(2n)!} = \sum_{n=0}^{\infty} (-1)^n \frac{\left(\frac{\pi}{6}\right)^{2n}}{(2n)!} = \cos\frac{\pi}{6} = \frac{\sqrt{3}}{2}$, by (16).

**51.** $\displaystyle\sum_{n=0}^{\infty} \frac{(-1)^n \pi^{2n+1}}{4^{2n+1}(2n+1)!} = \sum_{n=0}^{\infty} \frac{(-1)^n \left(\frac{\pi}{4}\right)^{2n+1}}{(2n+1)!} = \sin\frac{\pi}{4} = \frac{1}{\sqrt{2}}$, by (15).

**52.** $\displaystyle\sum_{n=0}^{\infty} \frac{3^n}{5^n\, n!} = \sum_{n=0}^{\infty} \frac{(3/5)^n}{n!} = e^{3/5}$, by (11).

**53.** $3 + \dfrac{9}{2!} + \dfrac{27}{3!} + \dfrac{81}{4!} + \cdots = \dfrac{3^1}{1!} + \dfrac{3^2}{2!} + \dfrac{3^3}{3!} + \dfrac{3^4}{4!} + \cdots = \sum\limits_{n=1}^{\infty} \dfrac{3^n}{n!} = \sum\limits_{n=0}^{\infty} \dfrac{3^n}{n!} - 1 = e^3 - 1$, by (11).

**54.** $1 - \ln 2 + \dfrac{(\ln 2)^2}{2!} - \dfrac{(\ln 2)^3}{3!} + \cdots = \sum\limits_{n=0}^{\infty} \dfrac{(-\ln 2)^n}{n!} = e^{-\ln 2} = \left(e^{\ln 2}\right)^{-1} = 2^{-1} = \frac{1}{2}$, by (11).

**55.** Assume that $|f'''(x)| \le M$, so $f'''(x) \le M$ for $a \le x \le a + d$. Now $\int_a^x f'''(t)\,dt \le \int_a^x M\,dt \Rightarrow$
$f''(x) - f''(a) \le M(x - a) \Rightarrow f''(x) \le f''(a) + M(x - a)$. Thus, $\int_a^x f''(t)\,dt \le \int_a^x [f''(a) + M(t - a)]\,dt \Rightarrow$
$f'(x) - f'(a) \le f''(a)(x - a) + \frac{1}{2}M(x - a)^2 \Rightarrow f'(x) \le f'(a) + f''(a)(x - a) + \frac{1}{2}M(x - a)^2 \Rightarrow$
$\int_a^x f'(t)\,dt \le \int_a^x \left[f'(a) + f''(a)(t - a) + \frac{1}{2}M(t - a)^2\right] dt \Rightarrow$
$f(x) - f(a) \le f'(a)(x - a) + \frac{1}{2}f''(a)(x - a)^2 + \frac{1}{6}M(x - a)^3$. So
$f(x) - f(a) - f'(a)(x - a) - \frac{1}{2}f''(a)(x - a)^2 \le \frac{1}{6}M(x - a)^3$. But
$R_2(x) = f(x) - T_2(x) = f(x) - f(a) - f'(a)(x - a) - \frac{1}{2}f''(a)(x - a)^2$, so $R_2(x) \le \frac{1}{6}M(x - a)^3$.
A similar argument using $f'''(x) \ge -M$ shows that $R_2(x) \ge -\frac{1}{6}M(x - a)^3$. So $|R_2(x_2)| \le \frac{1}{6}M\,|x - a|^3$.
Although we have assumed that $x > a$, a similar calculation shows that this inequality is also true if $x < a$.

**56.** (a) $f(x) = \begin{cases} e^{-1/x^2} & \text{if } x \ne 0 \\ 0 & \text{if } x = 0 \end{cases}$ so $f'(0) = \lim\limits_{x \to 0} \dfrac{f(x) - f(0)}{x - 0} = \lim\limits_{x \to 0} \dfrac{e^{-1/x^2}}{x} = \lim\limits_{x \to 0} \dfrac{1/x}{e^{1/x^2}} = \lim\limits_{x \to 0} \dfrac{x}{2e^{1/x^2}} = 0$

(using l'Hospital's Rule and simplifying in the penultimate step). Similarly, we can use the definition of the derivative and l'Hospital's Rule to show that $f''(0) = 0$, $f^{(3)}(0) = 0, \ldots, f^{(n)}(0) = 0$, so that the Maclaurin series for $f$ consists entirely of zero terms. But since $f(x) \ne 0$ except for $x = 0$, we see that $f$ cannot equal its Maclaurin series except at $x = 0$.

(b)

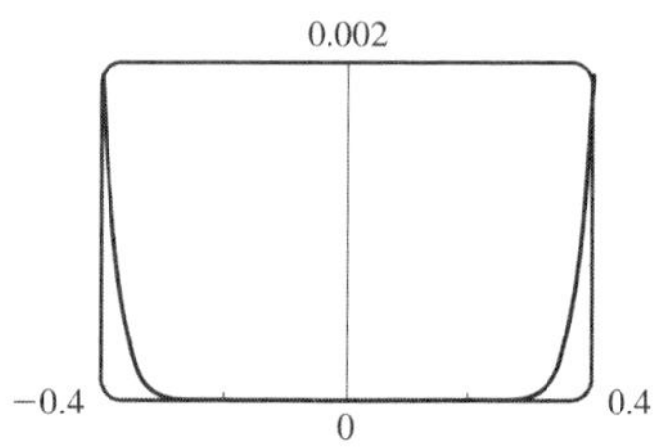


From the graph, it seems that the function is extremely flat at the origin. In fact, it could be said to be "infinitely flat" at $x = 0$, since all of its derivatives are 0 there.

## LABORATORY PROJECT An Elusive Limit

**1.** $f(x) = \dfrac{n(x)}{d(x)} = \dfrac{\sin(\tan x) - \tan(\sin x)}{\arcsin(\arctan x) - \arctan(\arcsin x)}$

| $x$ | $f(x)$ |
|---|---|
| 1 | 1.1838 |
| 0.1 | 0.9821 |
| 0.01 | 2.0000 |
| 0.001 | 3.3333 |
| 0.0001 | 3.3333 |

The table of function values were obtained using Maple with 10 digits of precision. The results of this project will vary depending on the CAS and precision level. It appears that as $x \to 0^+$, $f(x) \to \frac{10}{3}$. Since $f$ is an even function, we have $f(x) \to \frac{10}{3}$ as $x \to 0$.

2. The graph is inconclusive about the limit of $f$ as $x \to 0$.

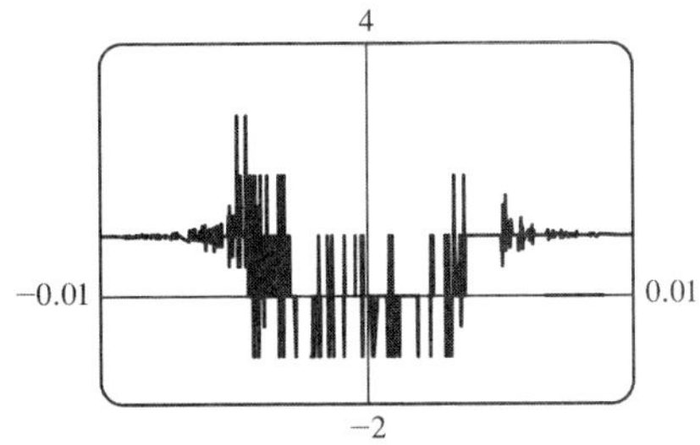


3. The limit has the indeterminate form $\frac{0}{0}$. Applying l'Hospital's Rule, we obtain the form $\frac{0}{0}$ six times. Finally, on the seventh application we obtain $\lim\limits_{x\to 0}\dfrac{n^{(7)}(x)}{d^{(7)}(x)} = \dfrac{-168}{-168} = 1$.

4. $$\lim_{x\to 0} f(x) = \lim_{x\to 0}\frac{n(x)}{d(x)} \overset{\text{CAS}}{=} \lim_{x\to 0}\frac{-\frac{1}{30}x^7 - \frac{29}{756}x^9 + \cdots}{-\frac{1}{30}x^7 + \frac{13}{756}x^9 + \cdots}$$
$$= \lim_{x\to 0}\frac{\left(-\frac{1}{30}x^7 - \frac{29}{756}x^9 + \cdots\right)/x^7}{\left(-\frac{1}{30}x^7 + \frac{13}{756}x^9 + \cdots\right)/x^7} = \lim_{x\to 0}\frac{-\frac{1}{30} - \frac{29}{756}x^2 + \cdots}{-\frac{1}{30} + \frac{13}{756}x^2 + \cdots} = \frac{-\frac{1}{30}}{-\frac{1}{30}} = 1$$

Note that $n^{(7)}(x) = d^{(7)}(x) = -\frac{7!}{30} = -\frac{5040}{30} = -168$, which agrees with the result in Problem 3.

5. The limit command gives the result that $\lim\limits_{x\to 0} f(x) = 1$.

6. The strange results (with only 10 digits of precision) must be due to the fact that the terms being subtracted in the numerator and denominator are very close in value when $|x|$ is small. Thus, the differences are imprecise (have few correct digits).

## 8.8 The Binomial Series

1. The general binomial series in (2) is

$$(1+x)^k = \sum_{n=0}^{\infty}\binom{k}{n}x^n = 1 + kx + \frac{k(k-1)}{2!}x^2 + \frac{k(k-1)(k-2)}{3!}x^3 + \cdots.$$

$$(1+x)^{1/2} = \sum_{n=0}^{\infty}\binom{\frac{1}{2}}{n}x^n = 1 + \left(\tfrac{1}{2}\right)x + \frac{\left(\frac{1}{2}\right)\left(-\frac{1}{2}\right)}{2!}x^2 + \frac{\left(\frac{1}{2}\right)\left(-\frac{1}{2}\right)\left(-\frac{3}{2}\right)}{3!}x^3 + \cdots$$
$$= 1 + \frac{x}{2} - \frac{x^2}{2^2\cdot 2!} + \frac{1\cdot 3\cdot x^3}{2^3\cdot 3!} - \frac{1\cdot 3\cdot 5\cdot x^4}{2^4\cdot 4!} + \cdots$$
$$= 1 + \frac{x}{2} + \sum_{n=2}^{\infty}\frac{(-1)^{n-1}\,1\cdot 3\cdot 5\cdot\cdots\cdot(2n-3)x^n}{2^n\cdot n!} \quad \text{for } |x|<1, \quad \text{so } R=1.$$

2. $\dfrac{1}{(1+x)^4} = (1+x)^{-4} = \sum\limits_{n=0}^{\infty}\dbinom{-4}{n}x^n$. The binomial coefficient is

$$\binom{-4}{n} = \frac{(-4)(-5)(-6)\cdot\cdots\cdot(-4-n+1)}{n!} = \frac{(-4)(-5)(-6)\cdot\cdots\cdot[-(n+3)]}{n!}$$
$$= \frac{(-1)^n\cdot 2\cdot 3\cdot 4\cdot 5\cdot 6\cdot\cdots\cdot(n+1)(n+2)(n+3)}{2\cdot 3\cdot n!} = \frac{(-1)^n(n+1)(n+2)(n+3)}{6}$$

Thus, $\dfrac{1}{(1+x)^4} = \sum\limits_{n=0}^{\infty}\dfrac{(-1)^n(n+1)(n+2)(n+3)}{6}x^n$ for $|x|<1$, so $R=1$.

**3.** $\dfrac{1}{(2+x)^3} = \dfrac{1}{[2(1+x/2)]^3} = \dfrac{1}{8}\left(1+\dfrac{x}{2}\right)^{-3} = \dfrac{1}{8}\sum\limits_{n=0}^{\infty}\binom{-3}{n}\left(\dfrac{x}{2}\right)^n$. The binomial coefficient is

$$\binom{-3}{n} = \frac{(-3)(-4)(-5)\cdot\cdots\cdot(-3-n+1)}{n!} = \frac{(-3)(-4)(-5)\cdot\cdots\cdot[-(n+2)]}{n!}$$

$$= \frac{(-1)^n\cdot 2\cdot 3\cdot 4\cdot 5\cdot\cdots\cdot(n+1)(n+2)}{2\cdot n!} = \frac{(-1)^n(n+1)(n+2)}{2}$$

Thus, $\dfrac{1}{(2+x)^3} = \dfrac{1}{8}\sum\limits_{n=0}^{\infty}\dfrac{(-1)^n(n+1)(n+2)}{2}\dfrac{x^n}{2^n} = \sum\limits_{n=0}^{\infty}\dfrac{(-1)^n(n+1)(n+2)x^n}{2^{n+4}}$ for $\left|\dfrac{x}{2}\right| < 1 \iff$ $|x| < 2$, so $R = 2$.

**4.** $(1-x)^{2/3} = \sum\limits_{n=0}^{\infty}\binom{\frac{2}{3}}{n}(-x)^n$

$$= 1 + \tfrac{2}{3}(-x) + \frac{\frac{2}{3}\left(-\frac{1}{3}\right)}{2!}(-x)^2 + \frac{\frac{2}{3}\left(-\frac{1}{3}\right)\left(-\frac{4}{3}\right)}{3!}(-x)^3 + \cdots$$

$$= 1 - \tfrac{2}{3}x + \sum_{n=2}^{\infty}\frac{(-1)^{n-1}(-1)^n\cdot 2\cdot[1\cdot 4\cdot 7\cdot\cdots\cdot(3n-5)]}{3^n\cdot n!}x^n$$

$$= 1 - \tfrac{2}{3}x - 2\sum_{n=2}^{\infty}\frac{1\cdot 4\cdot 7\cdot\cdots\cdot(3n-5)}{3^n\cdot n!}x^n$$

and $|-x| < 1 \iff |x| < 1$, so $R = 1$.

**5.** We must write the binomial in the form (1+ expression), so we'll factor out a 4.

$$\frac{x}{\sqrt{4+x^2}} = \frac{x}{\sqrt{4(1+x^2/4)}} = \frac{x}{2\sqrt{1+x^2/4}} = \frac{x}{2}\left(1+\frac{x^2}{4}\right)^{-1/2} = \frac{x}{2}\sum_{n=0}^{\infty}\binom{-\frac{1}{2}}{n}\left(\frac{x^2}{4}\right)^n$$

$$= \frac{x}{2}\left[1 + \left(-\tfrac{1}{2}\right)\frac{x^2}{4} + \frac{\left(-\frac{1}{2}\right)\left(-\frac{3}{2}\right)}{2!}\left(\frac{x^2}{4}\right)^2 + \frac{\left(-\frac{1}{2}\right)\left(-\frac{3}{2}\right)\left(-\frac{5}{2}\right)}{3!}\left(\frac{x^2}{4}\right)^3 + \cdots\right]$$

$$= \frac{x}{2} + \frac{x}{2}\sum_{n=1}^{\infty}(-1)^n\frac{1\cdot 3\cdot 5\cdot\cdots\cdot(2n-1)}{2^n\cdot 4^n\cdot n!}x^{2n}$$

$$= \frac{x}{2} + \sum_{n=1}^{\infty}(-1)^n\frac{1\cdot 3\cdot 5\cdot\cdots\cdot(2n-1)}{n!\,2^{3n+1}}x^{2n+1} \text{ and } \frac{x^2}{4} < 1 \iff \frac{|x|}{2} < 1 \iff |x| < 2, \quad \text{so } R = 2.$$

**6.** $\dfrac{x^2}{\sqrt{2+x}} = \dfrac{x^2}{\sqrt{2(1+x/2)}} = \dfrac{x^2}{\sqrt{2}}\left(1+\dfrac{x}{2}\right)^{-1/2} = \dfrac{x^2}{\sqrt{2}}\sum\limits_{n=0}^{\infty}\binom{-\frac{1}{2}}{n}\left(\dfrac{x}{2}\right)^n$

$$= \frac{x^2}{\sqrt{2}}\left[1 + \left(-\tfrac{1}{2}\right)\left(\frac{x}{2}\right) + \frac{\left(-\frac{1}{2}\right)\left(-\frac{3}{2}\right)}{2!}\left(\frac{x}{2}\right)^2 + \frac{\left(-\frac{1}{2}\right)\left(-\frac{3}{2}\right)\left(-\frac{5}{2}\right)}{3!}\left(\frac{x}{2}\right)^3 + \cdots\right]$$

$$= \frac{x^2}{\sqrt{2}} + \frac{x^2}{\sqrt{2}}\sum_{n=1}^{\infty}(-1)^n\frac{1\cdot 3\cdot 5\cdot\cdots\cdot(2n-1)}{n!\,2^{2n}}x^n$$

$$= \frac{x^2}{\sqrt{2}} + \sum_{n=1}^{\infty}(-1)^n\frac{1\cdot 3\cdot 5\cdot\cdots\cdot(2n-1)}{n!\,2^{2n+1/2}}x^{n+2} \text{ and } \left|\frac{x}{2}\right| < 1 \iff |x| < 2, \quad \text{so } R = 2.$$

**7.** $(1+2x)^{3/4} = 1 + \frac{3}{4}(2x) + \frac{\left(\frac{3}{4}\right)\left(-\frac{1}{4}\right)}{2!}(2x)^2 + \frac{\left(\frac{3}{4}\right)\left(-\frac{1}{4}\right)\left(-\frac{5}{4}\right)}{3!}(2x)^3 + \cdots$

$$= 1 + \frac{3}{2}x + 3\sum_{n=2}^{\infty}(-1)^{n+1}\frac{1\cdot 5\cdot 9\cdot\cdots\cdot(4n-7)}{4^n\cdot n!}\cdot 2^n x^n$$

$$= 1 + \frac{3}{2}x + 3\sum_{n=2}^{\infty}(-1)^{n+1}\frac{1\cdot 5\cdot 9\cdot\cdots\cdot(4n-7)}{2^n\cdot n!}x^n \text{ and } |2x| < 1 \iff |x| < \frac{1}{2}, \text{ so } R = \frac{1}{2}.$$

The three Taylor polynomials are $T_1(x) = 1 + \frac{3}{2}x$, $T_2(x) = 1 + \frac{3}{2}x - \frac{3}{8}x^2$, and $T_3(x) = 1 + \frac{3}{2}x - \frac{3}{8}x^2 + \frac{5}{16}x^3$.

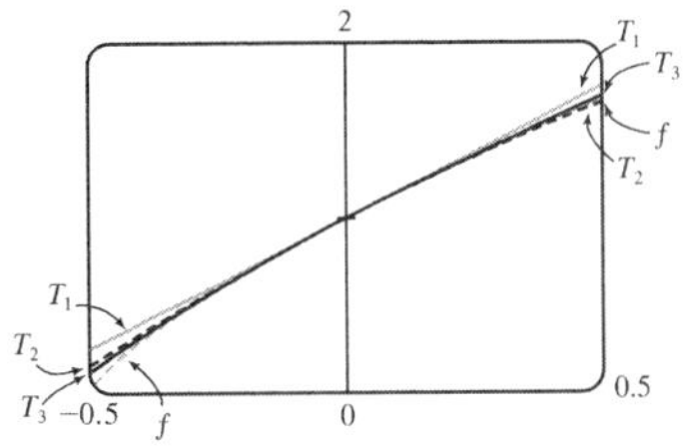


**8.** $\sqrt[3]{1+4x} = (1+4x)^{1/3}$

$$= 1 + \frac{1}{3}(4x) + \frac{\left(\frac{1}{3}\right)\left(-\frac{2}{3}\right)}{2!}(4x)^2 + \frac{\left(\frac{1}{3}\right)\left(-\frac{2}{3}\right)\left(-\frac{5}{3}\right)}{3!}(4x)^3 + \cdots$$

$$= 1 + \frac{4}{3}x + \sum_{n=2}^{\infty}(-1)^{n+1}\frac{2\cdot 5\cdot 8\cdot\cdots\cdot(3n-4)}{3^n\cdot n!}(4x)^n \text{ and } |4x| < 1 \iff |x| < \frac{1}{4}, \text{ so } R = \frac{1}{4}.$$

The three Taylor polynomials are $T_1(x) = 1 + \frac{4}{3}x$, $T_2(x) = 1 + \frac{4}{3}x - \frac{6}{9}x^2$, and $T_3(x) = 1 + \frac{4}{3}x - \frac{16}{9}x^2 + \frac{320}{81}x^3$.

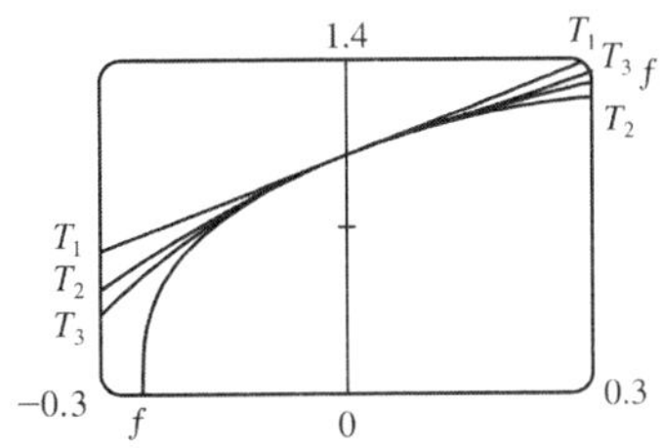


**9.** (a) $1/\sqrt{1-x^2} = \left[1 + (-x^2)\right]^{-1/2} = 1 + \left(-\frac{1}{2}\right)(-x^2) + \frac{\left(-\frac{1}{2}\right)\left(-\frac{3}{2}\right)}{2!}(-x^2)^2 + \frac{\left(-\frac{1}{2}\right)\left(-\frac{3}{2}\right)\left(-\frac{5}{2}\right)}{3!}(-x^2)^3 + \cdots$

$$= 1 + \sum_{n=1}^{\infty}\frac{1\cdot 3\cdot 5\cdot\cdots\cdot(2n-1)}{2^n\cdot n!}x^{2n}$$

(b) $\sin^{-1}x = \int \frac{1}{\sqrt{1-x^2}}\,dx = C + x + \sum_{n=1}^{\infty}\frac{1\cdot 3\cdot 5\cdot\cdots\cdot(2n-1)}{(2n+1)2^n\cdot n!}x^{2n+1}$

$$= x + \sum_{n=1}^{\infty}\frac{1\cdot 3\cdot 5\cdot\cdots\cdot(2n-1)}{(2n+1)2^n\cdot n!}x^{2n+1} \quad \text{since } 0 = \sin^{-1}0 = C.$$

**10.** (a) $1/\sqrt[4]{1+x} = (1+x)^{-1/4} = \sum_{n=0}^{\infty}\binom{-\frac{1}{4}}{n}x^n = 1 - \frac{1}{4}x + \frac{\left(-\frac{1}{4}\right)\left(-\frac{5}{4}\right)}{2!}x^2 + \frac{\left(-\frac{1}{4}\right)\left(-\frac{5}{4}\right)\left(-\frac{9}{4}\right)}{3!}x^3 + \cdots$

$$= 1 - \frac{1}{4}x + \sum_{n=2}^{\infty}(-1)^n\frac{1\cdot 5\cdot 9\cdot\cdots\cdot(4n-3)}{4^n\cdot n!}x^n$$

(b) $1/\sqrt[4]{1+x} = 1 - \frac{1}{4}x + \frac{5}{32}x^2 - \frac{15}{128}x^3 + \frac{195}{2048}x^4 - \cdots$. $1/\sqrt[4]{1.1} = 1/\sqrt[4]{1+0.1}$, so let $x = 0.1$. The sum of the first four terms is then $1 - \frac{1}{4}(0.1) + \frac{5}{32}(0.1)^2 - \frac{15}{128}(0.1)^3 \approx 0.976$. The fifth term is $\frac{195}{2048}(0.1)^4 \approx 0.000\,009\,5$, which does not affect the third decimal place of the sum, so we have $1/\sqrt[4]{1.1} \approx 0.976$. (Note that the third decimal place of the sum of the first three terms is affected by the fourth term, so we need to use more than three terms for the sum.)

**11.** (a) $[1+(-x)]^{-2} = 1 + (-2)(-x) + \frac{(-2)(-3)}{2!}(-x)^2 + \frac{(-2)(-3)(-4)}{3!}(-x)^3 + \cdots$

$$= 1 + 2x + 3x^2 + 4x^3 + \cdots = \sum_{n=0}^{\infty}(n+1)x^n,$$

so $\frac{x}{(1-x)^2} = x\sum_{n=0}^{\infty}(n+1)x^n = \sum_{n=0}^{\infty}(n+1)x^{n+1} = \sum_{n=1}^{\infty}nx^n.$

(b) With $x = \frac{1}{2}$ in part (a), we have $\sum_{n=1}^{\infty} n\left(\frac{1}{2}\right)^n = \sum_{n=1}^{\infty}\frac{n}{2^n} = \frac{\frac{1}{2}}{\left(1-\frac{1}{2}\right)^2} = \frac{\frac{1}{2}}{\frac{1}{4}} = 2.$

**12.** (a) $[1+(-x)]^{-3} = \sum_{n=0}^{\infty}\binom{-3}{n}(-x)^n = 1 + (-3)(-x) + \frac{(-3)(-4)}{2!}(-x)^2 + \frac{(-3)(-4)(-5)}{3!}(-x)^3 + \cdots$

$$= 1 + \sum_{n=1}^{\infty}\frac{3\cdot4\cdot5\cdot\cdots\cdot(n+2)}{n!}x^n = \sum_{n=0}^{\infty}\frac{2\cdot3\cdot4\cdot5\cdot\cdots\cdot(n+2)}{2\cdot n!}x^n$$

$$= \sum_{n=0}^{\infty}\frac{(n+1)(n+2)}{2}x^n \quad\Rightarrow$$

$$(x+x^2)[1+(-x)]^{-3} = x\,[1+(-x)]^{-3} + x^2\,[1+(-x)]^{-3}$$

$$= \sum_{n=0}^{\infty}\frac{(n+1)(n+2)}{2}x^{n+1} + \sum_{n=0}^{\infty}\frac{(n+1)(n+2)}{2}x^{n+2}$$

$$= \sum_{n=1}^{\infty}\frac{n(n+1)}{2}x^n + \sum_{n=1}^{\infty}\frac{n(n+1)}{2}x^{n+1}$$

$$= x + \sum_{n=2}^{\infty}\frac{n(n+1)}{2}x^n + \sum_{n=2}^{\infty}\frac{(n-1)n}{2}x^n = x + \sum_{n=2}^{\infty}\left[\frac{n(n+1)}{2} + \frac{(n-1)n}{2}\right]x^n$$

$$= x + \sum_{n=2}^{\infty}n^2x^n = \sum_{n=1}^{\infty}n^2x^n, \quad -1 < x < 1$$

(b) Setting $x = \frac{1}{2}$ in the last series above gives the required series, so $\sum_{n=1}^{\infty}\frac{n^2}{2^n} = \frac{\frac{1}{2}+\left(\frac{1}{2}\right)^2}{\left(1-\frac{1}{2}\right)^3} = \frac{\frac{3}{4}}{\frac{1}{8}} = 6.$

**13.** (a) $\left(1+x^2\right)^{1/2} = 1 + \left(\frac{1}{2}\right)x^2 + \frac{\left(\frac{1}{2}\right)\left(-\frac{1}{2}\right)}{2!}\left(x^2\right)^2 + \frac{\left(\frac{1}{2}\right)\left(-\frac{1}{2}\right)\left(-\frac{3}{2}\right)}{3!}\left(x^2\right)^3 + \cdots$

$$= 1 + \frac{x^2}{2} + \sum_{n=2}^{\infty}\frac{(-1)^{n-1}\,1\cdot3\cdot5\cdot\cdots\cdot(2n-3)}{2^n\cdot n!}x^{2n}$$

(b) The coefficient of $x^{10}$ (corresponding to $n=5$) in the above Maclaurin series is $\frac{f^{(10)}(0)}{10!}$, so

$$\frac{f^{(10)}(0)}{10!} = \frac{(-1)^4\cdot1\cdot3\cdot5\cdot7}{2^5\cdot5!} \quad\Rightarrow\quad f^{(10)}(0) = 10!\left(\frac{1\cdot3\cdot5\cdot7}{2^5\cdot5!}\right) = 99{,}225.$$

**14.** (a) $\left(1+x^3\right)^{-1/2} = \sum_{n=0}^{\infty}\binom{-\frac{1}{2}}{n}\left(x^3\right)^n = 1 + \left(-\frac{1}{2}\right)\left(x^3\right) + \frac{\left(-\frac{1}{2}\right)\left(-\frac{3}{2}\right)}{2!}\left(x^3\right)^2 + \frac{\left(-\frac{1}{2}\right)\left(-\frac{3}{2}\right)\left(-\frac{5}{2}\right)}{3!}\left(x^3\right)^3 + \cdots$

$$= 1 + \sum_{n=1}^{\infty}\frac{(-1)^n\,1\cdot3\cdot5\cdot\cdots\cdot(2n-1)\,x^{3n}}{2^n\cdot n!}$$

(b) The coefficient of $x^9$ (corresponding to $n=3$) in the preceding series is $\frac{f^{(9)}(0)}{9!}$, so

$$\frac{f^{(9)}(0)}{9!} = \frac{(-1)^3\,1\cdot3\cdot5}{2^3\cdot3!} \quad\Rightarrow\quad f^{(9)}(0) = -\frac{9!\cdot5}{8\cdot2} = -113{,}400.$$

**15.** (a) $g(x)=\sum\limits_{n=0}^{\infty}\binom{k}{n}x^n \Rightarrow g'(x)=\sum\limits_{n=1}^{\infty}\binom{k}{n}nx^{n-1}$, so

$$(1+x)g'(x)=(1+x)\sum_{n=1}^{\infty}\binom{k}{n}nx^{n-1}=\sum_{n=1}^{\infty}\binom{k}{n}nx^{n-1}+\sum_{n=1}^{\infty}\binom{k}{n}nx^n$$

$$=\sum_{n=0}^{\infty}\binom{k}{n+1}(n+1)x^n+\sum_{n=0}^{\infty}\binom{k}{n}nx^n \qquad \left[\begin{array}{c}\text{Replace } n \text{ with } n+1\\ \text{in the first series}\end{array}\right]$$

$$=\sum_{n=0}^{\infty}(n+1)\frac{k(k-1)(k-2)\cdots(k-n+1)(k-n)}{(n+1)!}x^n+\sum_{n=0}^{\infty}\left[(n)\frac{k(k-1)(k-2)\cdots(k-n+1)}{n!}\right]x^n$$

$$=\sum_{n=0}^{\infty}\frac{(n+1)k(k-1)(k-2)\cdots(k-n+1)}{(n+1)!}\left[(k-n)+n\right]x^n$$

$$=k\sum_{n=0}^{\infty}\frac{k(k-1)(k-2)\cdots(k-n+1)}{n!}x^n=k\sum_{n=0}^{\infty}\binom{k}{n}x^n=kg(x)$$

Thus, $g'(x)=\dfrac{kg(x)}{1+x}$.

(b) $h(x)=(1+x)^{-k}g(x) \Rightarrow$

$$h'(x)=-k(1+x)^{-k-1}g(x)+(1+x)^{-k}g'(x) \quad \text{[Product Rule]}$$

$$=-k(1+x)^{-k-1}g(x)+(1+x)^{-k}\frac{kg(x)}{1+x} \quad \text{[from part (a)]}$$

$$=-k(1+x)^{-k-1}g(x)+k(1+x)^{-k-1}g(x)=0$$

(c) From part (b) we see that $h(x)$ must be constant for $x\in(-1,1)$, so $h(x)=h(0)=1$ for $x\in(-1,1)$.
Thus, $h(x)=1=(1+x)^{-k}g(x) \Leftrightarrow g(x)=(1+x)^k$ for $x\in(-1,1)$.

**16.** By Exercise 1, $\sqrt{1+x}=(1+x)^{1/2}=1+\dfrac{x}{2}+\sum\limits_{n=2}^{\infty}\dfrac{(-1)^{n-1}1\cdot3\cdot5\cdot\cdots\cdot(2n-3)x^n}{2^n\cdot n!}$, so

$(1-x^2)^{1/2}=1-\dfrac{1}{2}x^2-\sum\limits_{n=2}^{\infty}\dfrac{1\cdot3\cdot5\cdot\cdots\cdot(2n-3)}{2^n\cdot n!}x^{2n}$ and

$\sqrt{1-e^2\sin^2\theta}=1-\dfrac{1}{2}e^2\sin^2\theta-\sum\limits_{n=2}^{\infty}\dfrac{1\cdot3\cdot5\cdot\cdots\cdot(2n-3)}{2^n\cdot n!}e^{2n}\sin^{2n}\theta$. Thus,

$$L=4a\int_0^{\pi/2}\sqrt{1-e^2\sin^2\theta}\,d\theta=4a\int_0^{\pi/2}\left(1-\frac{1}{2}e^2\sin^2\theta-\sum_{n=2}^{\infty}\frac{1\cdot3\cdot5\cdot\cdots\cdot(2n-3)}{2^n\cdot n!}e^{2n}\sin^{2n}\theta\right)d\theta$$

$$=4a\left[\frac{\pi}{2}-\frac{e^2}{2}S_1-\sum_{n=2}^{\infty}\frac{1\cdot3\cdot5\cdot\cdots\cdot(2n-3)}{n!}\left(\frac{e^2}{2}\right)^n S_n\right]$$

where $S_n=\displaystyle\int_0^{\pi/2}\sin^{2n}\theta\,d\theta=\frac{1\cdot3\cdot5\cdot\cdots\cdot(2n-1)}{2\cdot4\cdot6\cdot\cdots\cdot2n}\frac{\pi}{2}$ by Exercise 36 of 5.6.

[continued]

$$L = 4a\left(\frac{\pi}{2}\right)\left[1 - \frac{e^2}{2}\cdot\frac{1}{2} - \sum_{n=2}^{\infty}\frac{1\cdot 3\cdot 5\cdot\cdots\cdot(2n-3)}{n!}\left(\frac{e^2}{2}\right)^n\frac{1\cdot 3\cdot 5\cdot\cdots\cdot(2n-1)}{2\cdot 4\cdot 6\cdot\cdots\cdot 2n}\right]$$

$$= 2\pi a\left[1 - \frac{e^2}{4} - \sum_{n=2}^{\infty}\frac{e^{2n}}{2^n}\cdot\frac{1^2\cdot 3^2\cdot 5^2\cdot\cdots\cdot(2n-3)^2(2n-1)}{n!\cdot 2^n\cdot n!}\right]$$

$$= 2\pi a\left[1 - \frac{e^2}{4} - \sum_{n=2}^{\infty}\frac{e^{2n}}{4^n}\left(\frac{1\cdot 3\cdot\cdots\cdot(2n-3)}{n!}\right)^2(2n-1)\right]$$

$$= 2\pi a\left[1 - \frac{e^2}{4} - \frac{3e^4}{64} - \frac{5e^6}{256} - \cdots\right] = \frac{\pi a}{128}(256 - 64e^2 - 12e^4 - 5e^6 - \cdots)$$

## 8.9 Applications of Taylor Polynomials

**1.** (a)

| $n$ | $f^{(n)}(x)$ | $f^{(n)}(0)$ | $T_n(x)$ |
|---|---|---|---|
| 0 | $\cos x$ | 1 | 1 |
| 1 | $-\sin x$ | 0 | 1 |
| 2 | $-\cos x$ | $-1$ | $1 - \frac{1}{2}x^2$ |
| 3 | $\sin x$ | 0 | $1 - \frac{1}{2}x^2$ |
| 4 | $\cos x$ | 1 | $1 - \frac{1}{2}x^2 + \frac{1}{24}x^4$ |
| 5 | $-\sin x$ | 0 | $1 - \frac{1}{2}x^2 + \frac{1}{24}x^4$ |
| 6 | $-\cos x$ | $-1$ | $1 - \frac{1}{2}x^2 + \frac{1}{24}x^4 - \frac{1}{720}x^6$ |

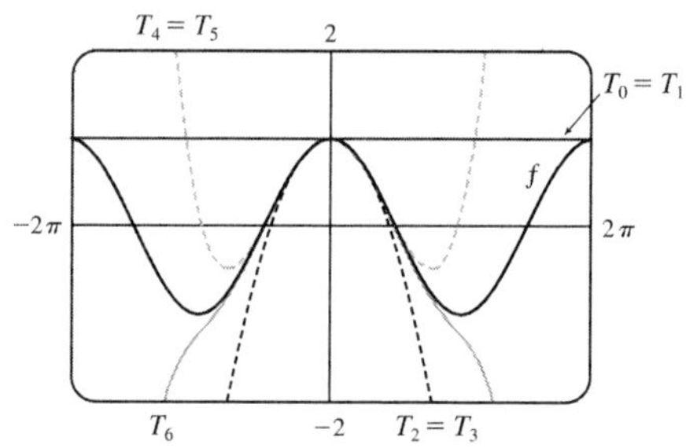


(b)

| $x$ | $f$ | $T_0 = T_1$ | $T_2 = T_3$ | $T_4 = T_5$ | $T_6$ |
|---|---|---|---|---|---|
| $\frac{\pi}{4}$ | 0.7071 | 1 | 0.6916 | 0.7074 | 0.7071 |
| $\frac{\pi}{2}$ | 0 | 1 | $-0.2337$ | 0.0200 | $-0.0009$ |
| $\pi$ | $-1$ | 1 | $-3.9348$ | 0.1239 | $-1.2114$ |

(c) As $n$ increases, $T_n(x)$ is a good approximation to $f(x)$ on a larger and larger interval.

**2.** (a)

| $n$ | $f^{(n)}(x)$ | $f^{(n)}(1)$ | $T_n(x)$ |
|---|---|---|---|
| 0 | $x^{-1}$ | 1 | 1 |
| 1 | $-x^{-2}$ | $-1$ | $1 - (x-1) = 2 - x$ |
| 2 | $2x^{-3}$ | 2 | $1 - (x-1) + (x-1)^2 = x^2 - 3x + 3$ |
| 3 | $6x^{-4}$ | $-6$ | $1 - (x-1) + (x-1)^2 - (x-1)^3 = -x^3 + 4x^2 - 6x + 4$ |

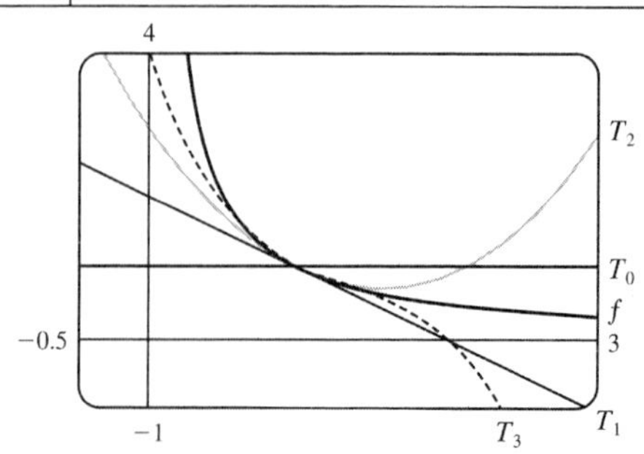

(b)

| $x$ | $f$ | $T_0$ | $T_1$ | $T_2$ | $T_3$ |
|---|---|---|---|---|---|
| 0.9 | $1.\overline{1}$ | 1 | 1.1 | 1.11 | 1.111 |
| 1.3 | 0.7692 | 1 | 0.7 | 0.79 | 0.763 |

(c) As $n$ increases, $T_n(x)$ is a good approximation to $f(x)$ on a larger and larger interval.

**3.**

| $n$ | $f^{(n)}(x)$ | $f^{(n)}(\frac{\pi}{6})$ |
|---|---|---|
| 0 | $\sin x$ | $\frac{1}{2}$ |
| 1 | $\cos x$ | $\frac{\sqrt{3}}{2}$ |
| 2 | $-\sin x$ | $-\frac{1}{2}$ |
| 3 | $-\cos x$ | $-\frac{\sqrt{3}}{2}$ |

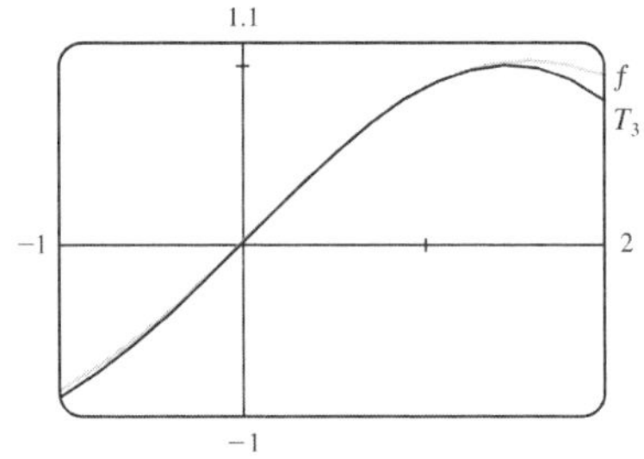


$$T_3(x) = \sum_{n=0}^{3} \frac{f^{(n)}(\frac{\pi}{6})}{n!}(x - \tfrac{\pi}{6})^n = \tfrac{1}{2} + \tfrac{\sqrt{3}}{2}(x - \tfrac{\pi}{6}) - \tfrac{1}{4}(x - \tfrac{\pi}{6})^2 - \tfrac{\sqrt{3}}{12}(x - \tfrac{\pi}{6})^3$$

**4.**

| $n$ | $f^{(n)}(x)$ | $f^{(n)}(2)$ |
|---|---|---|
| 0 | $e^x$ | $e^2$ |
| 1 | $e^x$ | $e^2$ |
| 2 | $e^x$ | $e^2$ |
| 3 | $e^x$ | $e^2$ |

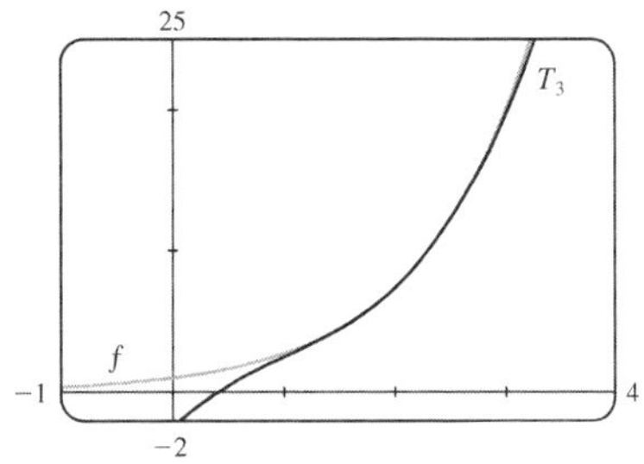


$$T_3(x) = \sum_{n=0}^{3} \frac{f^{(n)}(2)}{n!}(x-2)^n = e^2 + e^2(x-2) + \frac{e^2}{2}(x-2)^2 + \frac{e^2}{6}(x-2)^3$$

**5.**

| $n$ | $f^{(n)}(x)$ | $f^{(n)}(0)$ |
|---|---|---|
| 0 | $\arcsin x$ | 0 |
| 1 | $1/\sqrt{1-x^2}$ | 1 |
| 2 | $x/(1-x^2)^{3/2}$ | 0 |
| 3 | $(2x^2+1)/(1-x^2)^{5/2}$ | 1 |

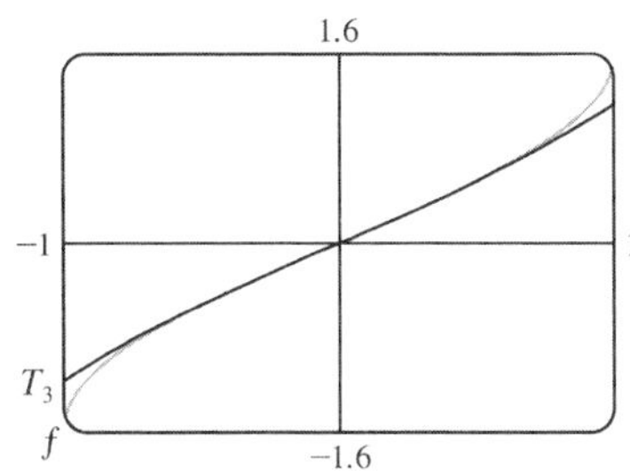


$$T_3(x) = \sum_{n=0}^{3} \frac{f^{(n)}(0)}{n!}x^n = x + \frac{x^3}{6}$$

**6.**

| $n$ | $f^{(n)}(x)$ | $f^{(n)}(1)$ |
|---|---|---|
| 0 | $(\ln x)/x$ | 0 |
| 1 | $(1-\ln x)/x^2$ | 1 |
| 2 | $(-3+2\ln x)/x^3$ | $-3$ |
| 3 | $(11-6\ln x)/x^4$ | 11 |

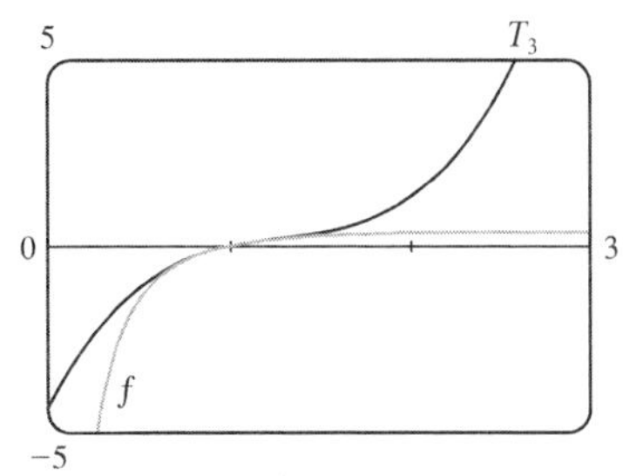


$$T_3(x) = \sum_{n=0}^{3} \frac{f^{(n)}(1)}{n!}(x-1)^n = (x-1) - \tfrac{3}{2}(x-1)^2 + \tfrac{11}{6}(x-1)^3$$

**7.**

| $n$ | $f^{(n)}(x)$ | $f^{(n)}(0)$ |
|---|---|---|
| 0 | $xe^{-2x}$ | 0 |
| 1 | $(1-2x)e^{-2x}$ | 1 |
| 2 | $4(x-1)e^{-2x}$ | $-4$ |
| 3 | $4(3-2x)e^{-2x}$ | 12 |

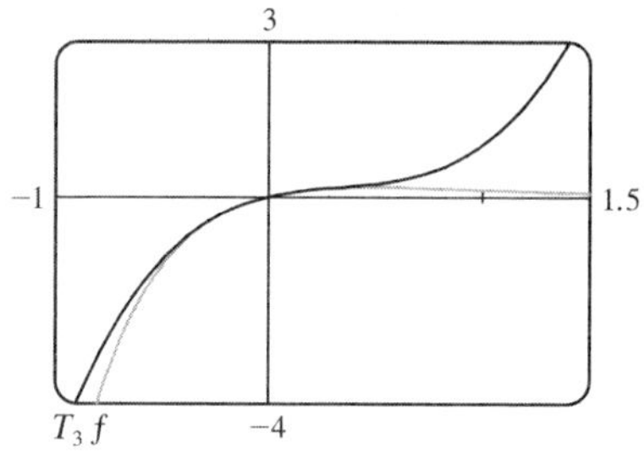


$$T_3(x) = \sum_{n=0}^{3} \frac{f^{(n)}(0)}{n!} x^n = \frac{0}{1} \cdot 1 + \frac{1}{1} x^1 + \frac{-4}{2} x^2 + \frac{12}{6} x^3 = x - 2x^2 + 2x^3$$

**8.**

| $n$ | $f^{(n)}(x)$ | $f^{(n)}(1)$ |
|---|---|---|
| 0 | $(3+x^2)^{1/2}$ | 2 |
| 1 | $x(3+x^2)^{-1/2}$ | $\frac{1}{2}$ |
| 2 | $3(3+x^2)^{-3/2}$ | $\frac{3}{8}$ |

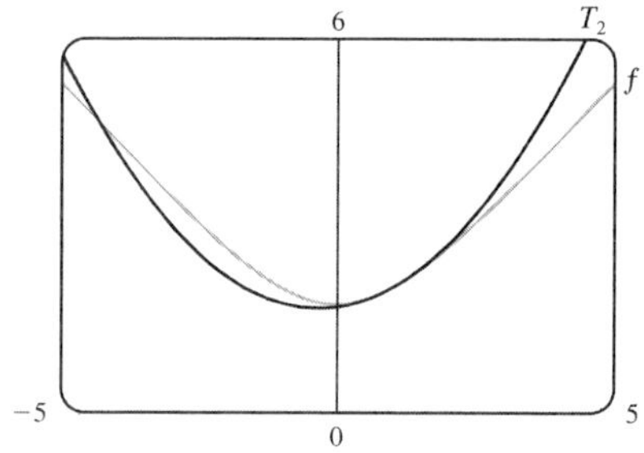


$$T_2(x) = \sum_{n=0}^{2} \frac{f^{(n)}(1)}{n!} (x-1)^n = 2 + \frac{1}{2}(x-1) + \frac{3/8}{2}(x-1)^2 = 2 + \frac{1}{2}(x-1) + \frac{3}{16}(x-1)^2$$

**9.** In Maple, we can find the Taylor polynomials by the following method: first define `f:=sec(x);` and then set `T2:=convert(taylor(f,x=0,3),polynom);`, `T4:=convert(taylor(f,x=0,5),polynom);`, etc. (The third argument in the `taylor` function is one more than the degree of the desired polynomial). We must `convert` to the type `polynom` because the output of the `taylor` function contains an error term which we do not want. In Mathematica, we use `Tn:=Normal[Series[f,{x,0,n}]]`, with `n=2, 4`, etc. Note that in Mathematica, the "degree" argument is the same as the degree of the desired polynomial. In Derive, author $\sec x$, then enter `Calculus, Taylor, 8, 0`; and then simplify the expression. The eighth Taylor polynomial is $T_8(x) = 1 + \frac{1}{2}x^2 + \frac{5}{24}x^4 + \frac{61}{720}x^6 + \frac{277}{8064}x^8$.

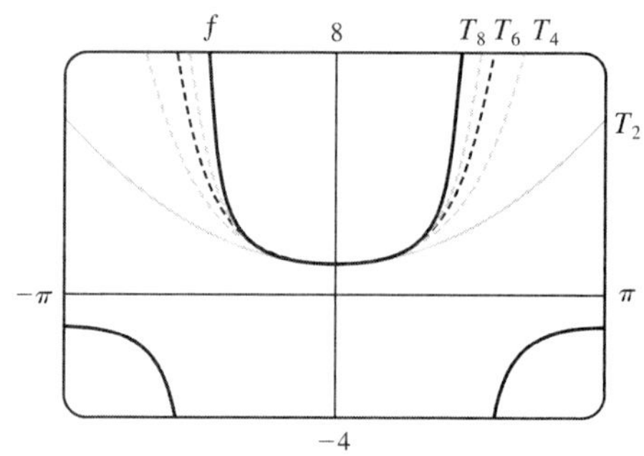


**10.** See Exercise 9 for the CAS commands used to generate the Taylor polynomials. The ninth Taylor polynomial for $\tan x$ is $T_9(x) = x + \frac{1}{3}x^3 + \frac{2}{15}x^5 + \frac{17}{315}x^7 + \frac{62}{2835}x^9$.

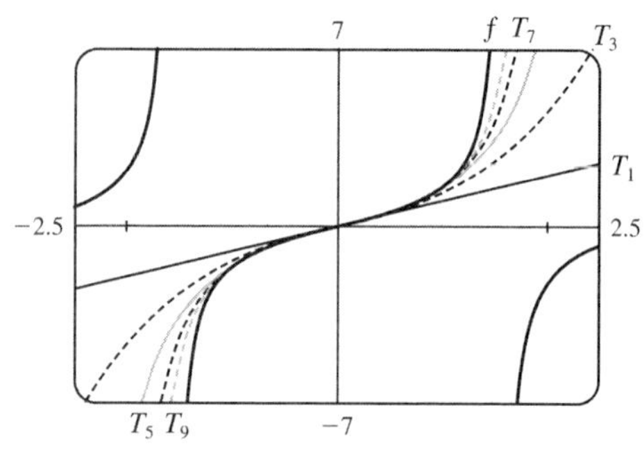

**11.**

| $n$ | $f^{(n)}(x)$ | $f^{(n)}(4)$ |
|---|---|---|
| 0 | $\sqrt{x}$ | 2 |
| 1 | $\frac{1}{2}x^{-1/2}$ | $\frac{1}{4}$ |
| 2 | $-\frac{1}{4}x^{-3/2}$ | $-\frac{1}{32}$ |
| 3 | $\frac{3}{8}x^{-5/2}$ | |

(a) $f(x) = \sqrt{x} \approx T_2(x) = 2 + \frac{1}{4}(x-4) - \frac{1/32}{2!}(x-4)^2 = 2 + \frac{1}{4}(x-4) - \frac{1}{64}(x-4)^2$

(b) $|R_2(x)| \le \dfrac{M}{3!}|x-4|^3$, where $|f'''(x)| \le M$. Now $4 \le x \le 4.2 \;\Rightarrow\; |x-4| \le 0.2 \;\Rightarrow\; |x-4|^3 \le 0.008$.

Since $f'''(x)$ is decreasing on $[4, 4.2]$, we can take $M = |f'''(4)| = \frac{3}{8}4^{-5/2} = \frac{3}{256}$, so

$|R_2(x)| \le \frac{3/256}{6}(0.008) = \frac{0.008}{512} = 0.000015625$.

(c)

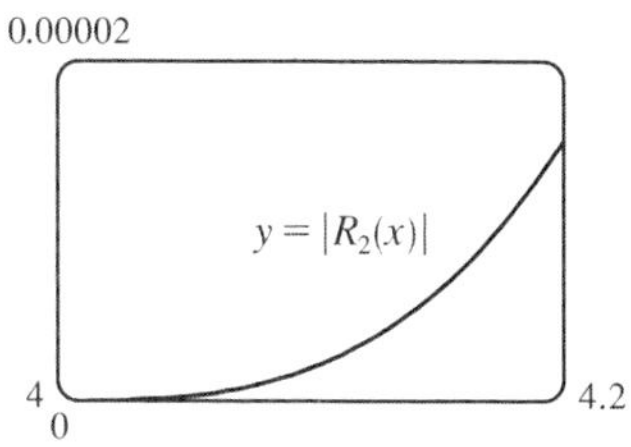


From the graph of $|R_2(x)| = |\sqrt{x} - T_2(x)|$, it seems that the error is less than $1.52 \times 10^{-5}$ on $[4, 4.2]$.

**12.**

| $n$ | $f^{(n)}(x)$ | $f^{(n)}(1)$ |
|---|---|---|
| 0 | $x^{-2}$ | 1 |
| 1 | $-2x^{-3}$ | $-2$ |
| 2 | $6x^{-4}$ | 6 |
| 3 | $-24x^{-5}$ | |

(a) $f(x) = x^{-2} \approx T_2(x) = 1 - 2(x-1) + \frac{6}{2!}(x-1)^2 = 1 - 2(x-1) + 3(x-1)^2$

(b) $|R_2(x)| \le \dfrac{M}{3!}|x-1|^3$, where $|f'''(x)| \le M$. Now $0.9 \le x \le 1.1 \;\Rightarrow\; |x-1| \le 0.1 \;\Rightarrow\; |x-1|^3 \le 0.001$. Since $f'''(x)$ is decreasing on $[0.9, 1.1]$, we can take $M = |f'''(0.9)| = \frac{24}{(0.9)^5}$, so

$$|R_2(x)| \le \frac{24/(0.9)^5}{6}(0.001) = \frac{0.004}{0.59049} \approx 0.00677404$$

(c)

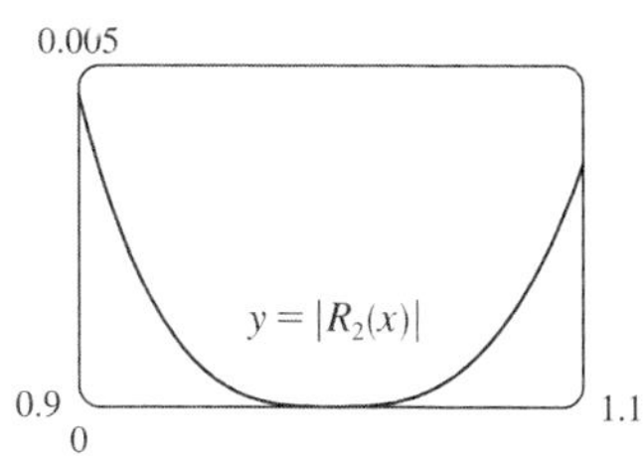


From the graph of $|R_2(x)| = \left|x^{-2} - T_2(x)\right|$, it seems that the error is less than $0.0046$ on $[0.9, 1.1]$.

**13.**

| $n$ | $f^{(n)}(x)$ | $f^{(n)}(1)$ |
|---|---|---|
| 0 | $x^{2/3}$ | 1 |
| 1 | $\frac{2}{3}x^{-1/3}$ | $\frac{2}{3}$ |
| 2 | $-\frac{2}{9}x^{-4/3}$ | $-\frac{2}{9}$ |
| 3 | $\frac{8}{27}x^{-7/3}$ | $\frac{8}{27}$ |
| 4 | $-\frac{56}{81}x^{-10/3}$ | |

(a) $f(x) = x^{2/3} \approx T_3(x) = 1 + \frac{2}{3}(x-1) - \frac{2/9}{2!}(x-1)^2 + \frac{8/27}{3!}(x-1)^3 = 1 + \frac{2}{3}(x-1) - \frac{1}{9}(x-1)^2 + \frac{4}{81}(x-1)^3$

(b) $|R_3(x)| \le \frac{M}{4!}|x-1|^4$, where $\left|f^{(4)}(x)\right| \le M$. Now $0.8 \le x \le 1.2 \;\Rightarrow\; |x-1| \le 0.2 \;\Rightarrow\; |x-1|^4 \le 0.0016$.

Since $\left|f^{(4)}(x)\right|$ is decreasing on $[0.8, 1.2]$, we can take $M = \left|f^{(4)}(0.8)\right| = \frac{56}{81}(0.8)^{-10/3}$, so

$|R_3(x)| \le \dfrac{\frac{56}{81}(0.8)^{-10/3}}{24}(0.0016) \approx 0.000\,096\,97$.

(c)

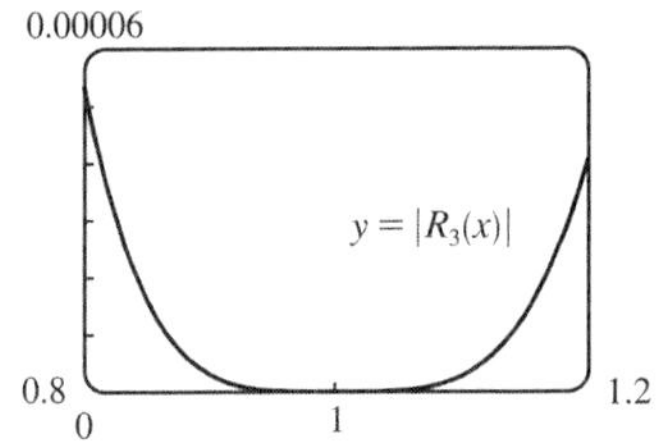


From the graph of $|R_3(x)| = \left|x^{2/3} - T_3(x)\right|$, it seems that the error is less than $0.000\,053\,3$ on $[0.8, 1.2]$.

**14.**

| $n$ | $f^{(n)}(x)$ | $f^{(n)}\left(\frac{\pi}{3}\right)$ |
|---|---|---|
| 0 | $\cos x$ | $\frac{1}{2}$ |
| 1 | $-\sin x$ | $-\frac{\sqrt{3}}{2}$ |
| 2 | $-\cos x$ | $-\frac{1}{2}$ |
| 3 | $\sin x$ | $\frac{\sqrt{3}}{2}$ |
| 4 | $\cos x$ | $\frac{1}{2}$ |
| 5 | $-\sin x$ | |

(a) $f(x) = \cos x \approx T_4(x) = \frac{1}{2} - \frac{\sqrt{3}}{2}\left(x - \frac{\pi}{3}\right) - \frac{1}{4}\left(x - \frac{\pi}{3}\right)^2 + \frac{\sqrt{3}}{12}\left(x - \frac{\pi}{3}\right)^3 + \frac{1}{48}\left(x - \frac{\pi}{3}\right)^4$

(b) $|R_4(x)| \le \dfrac{M}{5!}\left|x - \frac{\pi}{3}\right|^5$, where $\left|f^{(5)}(x)\right| \le M$. Now $0 \le x \le \frac{2\pi}{3} \;\Rightarrow\; \left(x - \frac{\pi}{3}\right)^5 \le \left(\frac{\pi}{3}\right)^5$, and letting $x = \frac{\pi}{2}$ gives

$M = 1$, so $|R_4(x)| \le \frac{1}{5!}\left(\frac{\pi}{3}\right)^5 \approx 0.0105$.

(c)

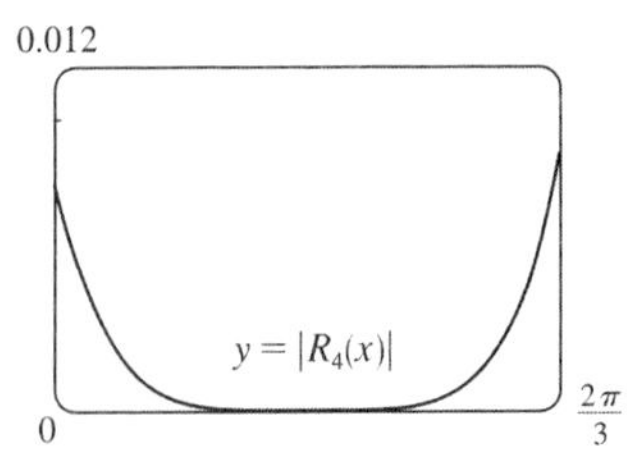


From the graph of $|R_4(x)| = |\cos x - T_4(x)|$, it seems that the error is less than 0.01 on $\left[0, \frac{2\pi}{3}\right]$.

**15.**

| $n$ | $f^{(n)}(x)$ | $f^{(n)}(0)$ |
|---|---|---|
| 0 | $e^{x^2}$ | 1 |
| 1 | $e^{x^2}(2x)$ | 0 |
| 2 | $e^{x^2}(2+4x^2)$ | 2 |
| 3 | $e^{x^2}(12x+8x^3)$ | 0 |
| 4 | $e^{x^2}(12+48x^2+16x^4)$ | |

(a) $f(x) = e^{x^2} \approx T_3(x) = 1 + \frac{2}{2!}x^2 = 1 + x^2$

(b) $|R_3(x)| \le \dfrac{M}{4!}|x|^4$, where $\left|f^{(4)}(x)\right| \le M$. Now $0 \le x \le 0.1 \;\Rightarrow\; x^4 \le (0.1)^4$, and letting $x = 0.1$ gives

$$|R_3(x)| \le \frac{e^{0.01}\,(12 + 0.48 + 0.0016)}{24}(0.1)^4 \approx 0.00006.$$

(c)

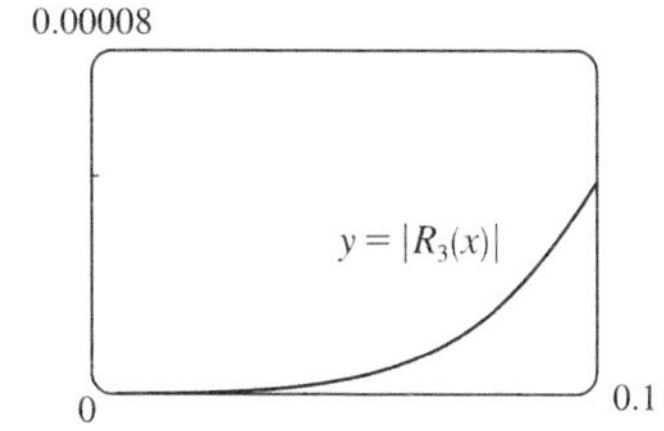


From the graph of $|R_3(x)| = \left|e^{x^2} - (1 + x^2)\right|$, it appears that the error is less than $0.000051$ on $[0, 0.1]$.

**16.**

| $n$ | $f^{(n)}(x)$ | $f^{(n)}(1)$ |
|---|---|---|
| 0 | $\ln(1+2x)$ | $\ln 3$ |
| 1 | $2/(1+2x)$ | $\frac{2}{3}$ |
| 2 | $-4/(1+2x)^2$ | $-\frac{4}{9}$ |
| 3 | $16/(1+2x)^3$ | $\frac{16}{27}$ |
| 4 | $-96/(1+2x)^4$ | |

(a) $f(x) = \ln(1+2x) \approx T_3(x) = \ln 3 + \frac{2}{3}(x-1) - \dfrac{4/9}{2!}(x-1)^2 + \dfrac{16/27}{3!}(x-1)^3$

(b) $|R_3(x)| \le \dfrac{M}{4!}|x-1|^4$, where $\left|f^{(4)}(x)\right| \le M$. Now $0.5 \le x \le 1.5 \;\Rightarrow\; -0.5 \le x - 1 \le 0.5 \;\Rightarrow$

$|x-1| \le 0.5 \;\Rightarrow\; |x-1|^4 \le \frac{1}{16}$, and letting $x = 0.5$ gives $M = 6$, so $|R_3(x)| \le \dfrac{6}{4!} \cdot \dfrac{1}{16} = \dfrac{1}{64} = 0.015625$.

(c)

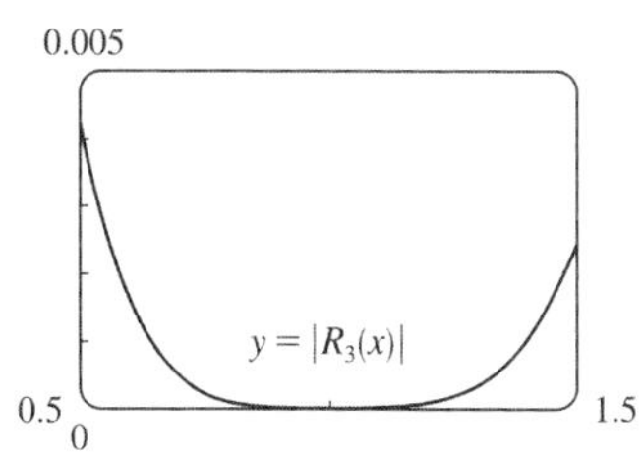


From the graph of $|R_3(x)| = |\ln(1+2x) - T_3(x)|$, it seems that the error is less than $0.005$ on $[0.5, 1.5]$.

**17.**

| $n$ | $f^{(n)}(x)$ | $f^{(n)}(0)$ |
|---|---|---|
| 0 | $x \sin x$ | 0 |
| 1 | $\sin x + x \cos x$ | 0 |
| 2 | $2 \cos x - x \sin x$ | 2 |
| 3 | $-3 \sin x - x \cos x$ | 0 |
| 4 | $-4 \cos x + x \sin x$ | $-4$ |
| 5 | $5 \sin x + x \cos x$ | |

(a) $f(x) = x \sin x \approx T_4(x) = \frac{2}{2!}(x-0)^2 + \frac{-4}{4!}(x-0)^4 = x^2 - \frac{1}{6}x^4$

(b) $|R_4(x)| \leq \frac{M}{5!}|x|^5$, where $\left|f^{(5)}(x)\right| \leq M$. Now $-1 \leq x \leq 1 \Rightarrow |x| \leq 1$, and a graph of $f^{(5)}(x)$ shows that $\left|f^{(5)}(x)\right| \leq 5$ for $-1 \leq x \leq 1$. Thus, we can take $M = 5$ and get $|R_4(x)| \leq \frac{5}{5!} \cdot 1^5 = \frac{1}{24} = 0.041\overline{6}$.

(c)

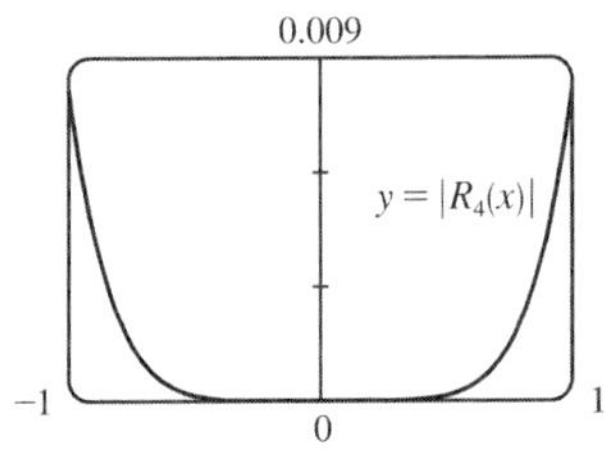


From the graph of $|R_4(x)| = |x \sin x - T_4(x)|$, it seems that the error is less than $0.0082$ on $[-1, 1]$.

**18.**

| $n$ | $f^{(n)}(x)$ | $f^{(n)}(1)$ |
|---|---|---|
| 0 | $x \ln x$ | 0 |
| 1 | $\ln x + 1$ | 1 |
| 2 | $1/x$ | 1 |
| 3 | $-1/x^2$ | $-1$ |
| 4 | $2/x^3$ | |

(a) $f(x) = x \ln x \approx T_3(x) = (x-1) + \frac{1}{2}(x-1)^2 - \frac{1}{6}(x-1)^3$

(b) $|R_3(x)| \leq \frac{M}{4!}|x-1|^4$, where $\left|f^{(4)}(x)\right| \leq M$. Now $0.5 \leq x \leq 1.5 \Rightarrow |x-1| \leq \frac{1}{2} \Rightarrow |x-1|^4 \leq \frac{1}{16}$.

Since $\left|f^{(4)}(x)\right|$ is decreasing on $[0.5, 1.5]$, we can take $M = \left|f^{(4)}(0.5)\right| = 2/(0.5)^3 = 16$, so $|R_3(x)| \leq \frac{16}{24}(1/16) = \frac{1}{24} = 0.041\overline{6}$.

(c)

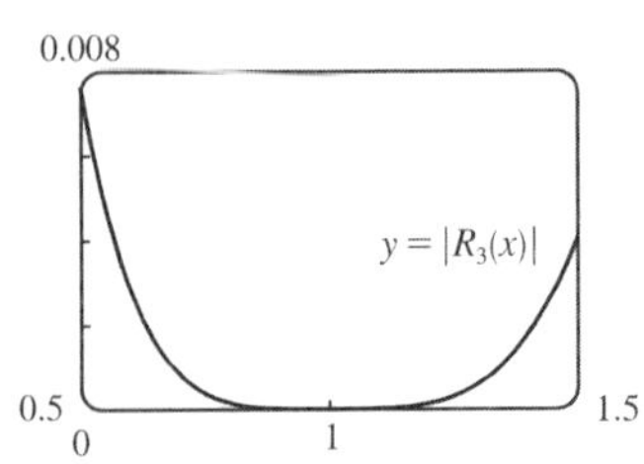


From the graph of $|R_3(x)| = |x \ln x - T_3(x)|$, it seems that the error is less than $0.0076$ on $[0.5, 1.5]$.

**19.** From Exercise 3, $\sin x = \frac{1}{2} + \frac{\sqrt{3}}{2}\left(x - \frac{\pi}{6}\right) - \frac{1}{4}\left(x - \frac{\pi}{6}\right)^2 - \frac{\sqrt{3}}{12}\left(x - \frac{\pi}{6}\right)^3 + R_3(x)$, where $|R_3(x)| \le \frac{M}{4!}\left|x - \frac{\pi}{6}\right|^4$ with

$\left|f^{(4)}(x)\right| = |\sin x| \le M = 1$. Now $x = 35^\circ = (30^\circ + 5^\circ) = \left(\frac{\pi}{6} + \frac{\pi}{36}\right)$ radians, so the error is

$\left|R_3\left(\frac{\pi}{36}\right)\right| \le \dfrac{\left(\frac{\pi}{36}\right)^4}{4!} < 0.000003$. Therefore, to five decimal places,

$\sin 35^\circ \approx \frac{1}{2} + \frac{\sqrt{3}}{2}\left(\frac{\pi}{36}\right) - \frac{1}{4}\left(\frac{\pi}{36}\right)^2 - \frac{\sqrt{3}}{12}\left(\frac{\pi}{36}\right)^3 \approx 0.57358$.

**20.** From Exercise 14, $\cos x = \frac{1}{2} - \frac{\sqrt{3}}{2}\left(x - \frac{\pi}{3}\right) - \frac{1}{4}\left(x - \frac{\pi}{3}\right)^2 + \frac{\sqrt{3}}{12}\left(x - \frac{\pi}{3}\right)^3 + \frac{1}{48}\left(x - \frac{\pi}{3}\right)^4 + R_4(x)$, where

$|R_4(x)| \le \frac{M}{5!}\left|x - \frac{\pi}{3}\right|^5$ with $\left|f^{(5)}(x)\right| = |-\sin x| \le M = 1$. Now $x = 69^\circ = (60^\circ + 9^\circ) = \left(\frac{\pi}{3} + \frac{\pi}{20}\right)$ radians, so the

error is $|R_4(x)| \le \dfrac{\left(\frac{\pi}{20}\right)^5}{5!} < 8 \times 10^{-7}$. Therefore, to five decimal places,

$\cos 69^\circ \approx \frac{1}{2} - \frac{\sqrt{3}}{2}\left(\frac{\pi}{20}\right) - \frac{1}{4}\left(\frac{\pi}{20}\right)^2 + \frac{\sqrt{3}}{12}\left(\frac{\pi}{20}\right)^3 + \frac{1}{48}\left(\frac{\pi}{20}\right)^4 \approx 0.35837$.

**21.** All derivatives of $e^x$ are $e^x$, so $|R_n(x)| \le \dfrac{e^x}{(n+1)!}|x|^{n+1}$, where $0 < x < 0.1$. Letting $x = 0.1$,

$R_n(0.1) \le \dfrac{e^{0.1}}{(n+1)!}(0.1)^{n+1} < 0.00001$, and by trial and error we find that $n = 3$ satisfies this inequality since

$R_3(0.1) < 0.0000046$. Thus, by adding the four terms of the Maclaurin series for $e^x$ corresponding to $n = 0$, 1, 2, and 3, we can estimate $e^{0.1}$ to within 0.00001. (In fact, this sum is $1.1051\overline{6}$ and $e^{0.1} \approx 1.10517$.)

**22.** $f(x) = \sum\limits_{n=0}^{\infty} \dfrac{f^{(n)}(4)}{n!}(x-4)^n = \sum\limits_{n=0}^{\infty} \dfrac{(-1)^n\, n!}{3^n(n+1)\,n!}(x-4)^n = \sum\limits_{n=0}^{\infty} \dfrac{(-1)^n}{3^n(n+1)}(x-4)^n$. Now

$f(5) = \sum\limits_{n=0}^{\infty} \dfrac{(-1)^n}{3^n(n+1)} = \sum\limits_{n=0}^{\infty} (-1)^n b_n$ is the sum of an alternating series that satisfies (i) $b_{n+1} \le b_n$ and

(ii) $\lim\limits_{n\to\infty} b_n = 0$, so by the Alternating Series Estimation Theorem, $|R_5(5)| = |f(5) - T_5(5)| \le b_6$, and

$b_6 = \dfrac{1}{3^6(7)} = \dfrac{1}{5103} \approx 0.000196 < 0.0002$; that is, the fifth-degree Taylor polynomial approximates $f(5)$ with error less than 0.0002.

**23.** $\sin x = x - \dfrac{1}{3!}x^3 + \dfrac{1}{5!}x^5 - \cdots$. By the Alternating Series Estimation Theorem, the error in the approximation $\sin x = x - \dfrac{1}{3!}x^3$ is less than

$\left|\dfrac{1}{5!}x^5\right| < 0.01 \quad\Leftrightarrow\quad \left|x^5\right| < 120(0.01) \quad\Leftrightarrow$

$|x| < (1.2)^{1/5} \approx 1.037$. The curves $y = x - \dfrac{1}{6}x^3$ and

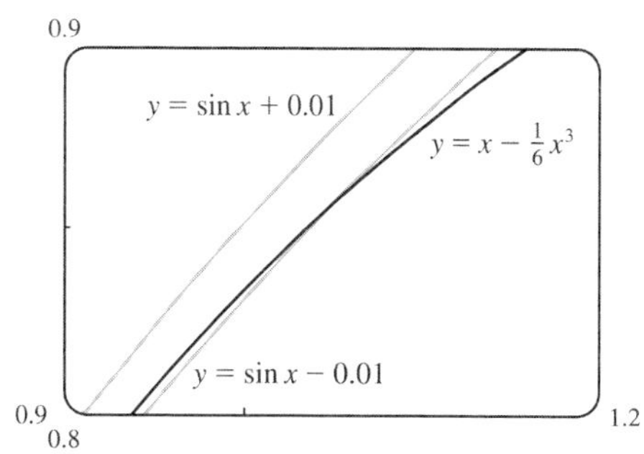


$y = \sin x - 0.01$ intersect at $x \approx 1.043$, so the graph confirms our estimate. Since both the sine function andthe given approximation are odd functions, we need to check the estimate only for $x > 0$. Thus, the desired range of values for $x$ is $-1.037 < x < 1.037$.

**24.** $\cos x = 1 - \frac{1}{2!}x^2 + \frac{1}{4!}x^4 - \frac{1}{6!}x^6 + \cdots$. By the Alternating Series Estimation Theorem, the error is less than $\left|-\frac{1}{6!}x^6\right| < 0.005 \Leftrightarrow x^6 < 720(0.005) \Leftrightarrow |x| < (3.6)^{1/6} \approx 1.238$. The curves $y = 1 - \frac{1}{2}x^2 + \frac{1}{24}x^4$ and $y = \cos x + 0.005$ intersect at $x \approx 1.244$, so the graph confirms our estimate. Since both the cosine function and the given approximation are even functions, we need to check the estimate only for $x > 0$. Thus, the desired range of values for $x$ is $-1.238 < x < 1.238$.

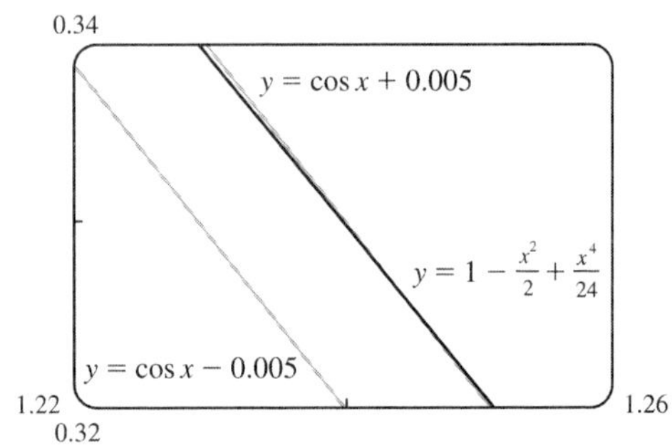


**25.** Let $s(t)$ be the position function of the car, and for convenience set $s(0) = 0$. The velocity of the car is $v(t) = s'(t)$ and the acceleration is $a(t) = s''(t)$, so the second degree Taylor polynomial is $T_2(t) = s(0) + v(0)t + \frac{a(0)}{2}t^2 = 20t + t^2$. We estimate the distance travelled during the next second to be $s(1) \approx T_2(1) = 20 + 1 = 21$ m. The function $T_2(t)$ would not be accurate over a full minute, since the car could not possibly maintain an acceleration of 2 m/s$^2$ for that long (if it did, its final speed would be 140 m/s $\approx 313$ mi/h!)

**26.** (a) $\frac{n_1}{\ell_o} + \frac{n_2}{\ell_i} = \frac{1}{R}\left(\frac{n_2 s_i}{\ell_i} - \frac{n_1 s_o}{\ell_o}\right)$ [Equation 1] where

$$\ell_o = \sqrt{R^2 + (s_o + R)^2 - 2R(s_o + R)\cos\phi} \quad \text{and} \quad \ell_i = \sqrt{R^2 + (s_i - R)^2 + 2R(s_i - R)\cos\phi} \quad (2)$$

Using $\cos\phi \approx 1$ gives

$$\ell_o = \sqrt{R^2 + (s_o + R)^2 - 2R(s_o + R)} = \sqrt{R^2 + s_o^2 + 2Rs_o + R^2 - 2Rs_o - 2R^2} = \sqrt{s_o^2} = s_o$$

and similarly, $\ell_i = s_i$. Thus, Equation 1 becomes $\frac{n_1}{s_o} + \frac{n_2}{s_i} = \frac{1}{R}\left(\frac{n_2 s_i}{s_i} - \frac{n_1 s_o}{s_o}\right) \Rightarrow \frac{n_1}{s_o} + \frac{n_2}{s_i} = \frac{n_2 - n_1}{R}$.

(b) Using $\cos\phi \approx 1 - \frac{1}{2}\phi^2$ in (2) gives us

$$\begin{aligned}\ell_o &= \sqrt{R^2 + (s_o + R)^2 - 2R(s_o + R)\left(1 - \tfrac{1}{2}\phi^2\right)} \\ &= \sqrt{R^2 + s_o^2 + 2Rs_o + R^2 - 2Rs_o + Rs_o\phi^2 - 2R^2 + R^2\phi^2} = \sqrt{s_o^2 + Rs_o\phi^2 + R^2\phi^2}\end{aligned}$$

Anticipating that we will use the binomial series expansion $(1 + x)^k \approx 1 + kx$, we can write the last expression for $\ell_o$ as $s_o\sqrt{1 + \phi^2\left(\frac{R}{s_o} + \frac{R^2}{s_o^2}\right)}$ and similarly, $\ell_i = s_i\sqrt{1 - \phi^2\left(\frac{R}{s_i} - \frac{R^2}{s_i^2}\right)}$. Thus, from Equation 1,

$$\frac{n_1}{\ell_o} + \frac{n_2}{\ell_i} = \frac{1}{R}\left(\frac{n_2 s_i}{\ell_i} - \frac{n_1 s_o}{\ell_o}\right) \Leftrightarrow n_1\ell_o^{-1} + n_2\ell_i^{-1} = \frac{n_2}{R}\cdot\frac{s_i}{\ell_i} - \frac{n_1}{R}\cdot\frac{s_o}{\ell_o} \Leftrightarrow$$

$$\begin{aligned}&\frac{n_1}{s_o}\left[1 + \phi^2\left(\frac{R}{s_o} + \frac{R^2}{s_o^2}\right)\right]^{-1/2} + \frac{n_2}{s_i}\left[1 - \phi^2\left(\frac{R}{s_i} - \frac{R^2}{s_i^2}\right)\right]^{-1/2} \\ &\qquad = \frac{n_2}{R}\left[1 - \phi^2\left(\frac{R}{s_i} - \frac{R^2}{s_i^2}\right)\right]^{-1/2} - \frac{n_1}{R}\left[1 + \phi^2\left(\frac{R}{s_o} + \frac{R^2}{s_o^2}\right)\right]^{-1/2}\end{aligned}$$

Approximating the expressions for $\ell_o^{-1}$ and $\ell_i^{-1}$ by the first two terms in their binomial series, we get

$$\begin{aligned}&\frac{n_1}{s_o}\left[1 - \tfrac{1}{2}\phi^2\left(\frac{R}{s_o} + \frac{R^2}{s_o^2}\right)\right] + \frac{n_2}{s_i}\left[1 + \tfrac{1}{2}\phi^2\left(\frac{R}{s_i} - \frac{R^2}{s_i^2}\right)\right] \\ &\qquad = \frac{n_2}{R}\left[1 + \tfrac{1}{2}\phi^2\left(\frac{R}{s_i} - \frac{R^2}{s_i^2}\right)\right] - \frac{n_1}{R}\left[1 - \tfrac{1}{2}\phi^2\left(\frac{R}{s_o} + \frac{R^2}{s_o^2}\right)\right] \Leftrightarrow\end{aligned}$$

$$\frac{n_1}{s_o} - \frac{n_1\phi^2}{2s_o}\left(\frac{R}{s_o} + \frac{R^2}{s_o^2}\right) + \frac{n_2}{s_i} + \frac{n_2\phi^2}{2s_i}\left(\frac{R}{s_i} - \frac{R^2}{s_i^2}\right) = \frac{n_2}{R} + \frac{n_2\phi^2}{2R}\left(\frac{R}{s_i} - \frac{R^2}{s_i^2}\right) - \frac{n_1}{R} + \frac{n_1\phi^2}{2R}\left(\frac{R}{s_o} + \frac{R^2}{s_o^2}\right) \Leftrightarrow$$

$$\begin{aligned}\frac{n_1}{s_o} + \frac{n_2}{s_i} &= \frac{n_2}{R} - \frac{n_1}{R} + \frac{n_1\phi^2}{2s_o}\left(\frac{R}{s_o} + \frac{R^2}{s_o^2}\right) + \frac{n_1\phi^2}{2R}\left(\frac{R}{s_o} + \frac{R^2}{s_o^2}\right) + \frac{n_2\phi^2}{2R}\left(\frac{R}{s_i} - \frac{R^2}{s_i^2}\right) - \frac{n_2\phi^2}{2s_i}\left(\frac{R}{s_i} - \frac{R^2}{s_i^2}\right) \\ &= \frac{n_2 - n_1}{R} + \frac{n_1\phi^2}{2}\left(\frac{R}{s_o} + \frac{R^2}{s_o^2}\right)\left(\frac{1}{s_o} + \frac{1}{R}\right) + \frac{n_2\phi^2}{2}\left(\frac{R}{s_i} - \frac{R^2}{s_i^2}\right)\left(\frac{1}{R} - \frac{1}{s_i}\right) \\ &= \frac{n_2 - n_1}{R} + \frac{n_1\phi^2R^2}{2s_o}\left(\frac{1}{R} + \frac{1}{s_o}\right)\left(\frac{1}{R} + \frac{1}{s_o}\right) + \frac{n_2\phi^2R^2}{2s_i}\left(\frac{1}{R} - \frac{1}{s_i}\right)\left(\frac{1}{R} - \frac{1}{s_i}\right) \\ &= \frac{n_2 - n_1}{R} + \phi^2R^2\left[\frac{n_1}{2s_o}\left(\frac{1}{R} + \frac{1}{s_o}\right)^2 + \frac{n_2}{2s_i}\left(\frac{1}{R} - \frac{1}{s_i}\right)^2\right]\end{aligned}$$

From Figure 8, we see that $\sin\phi = h/R$. So if we approximate $\sin\phi$ with $\phi$, we get $h = R\phi$ and $h^2 = \phi^2R^2$ and hence, Equation 4, as desired.

**27.** $E = \dfrac{q}{D^2} - \dfrac{q}{(D+d)^2} = \dfrac{q}{D^2} - \dfrac{q}{D^2(1+d/D)^2} = \dfrac{q}{D^2}\left[1 - \left(1 + \dfrac{d}{D}\right)^{-2}\right].$

We use the Binomial Series to expand $(1 + d/D)^{-2}$:

$$\begin{aligned}E &= \frac{q}{D^2}\left[1 - \left(1 - 2\left(\frac{d}{D}\right) + \frac{2\cdot 3}{2!}\left(\frac{d}{D}\right)^2 - \frac{2\cdot 3\cdot 4}{3!}\left(\frac{d}{D}\right)^3 + \cdots\right)\right] \\ &= \frac{q}{D^2}\left[2\left(\frac{d}{D}\right) - 3\left(\frac{d}{D}\right)^2 + 4\left(\frac{d}{D}\right)^3 - \cdots\right] \approx \frac{q}{D^2}\cdot 2\left(\frac{d}{D}\right) = 2qd\cdot\frac{1}{D^3}\end{aligned}$$

when $D$ is much larger than $d$; that is, when $P$ is far away from the dipole.

**28.** (a)

| $n$ | $\rho^{(n)}(t)$ | $\rho^{(n)}(20)$ |
|---|---|---|
| 0 | $\rho_{20}e^{\alpha(t-20)}$ | $\rho_{20}$ |
| 1 | $\alpha\rho_{20}e^{\alpha(t-20)}$ | $\alpha\rho_{20}$ |
| 2 | $\alpha^2\rho_{20}e^{\alpha(t-20)}$ | $\alpha^2\rho_{20}$ |

The linear approximation is $T_1(t) = \rho(20) + \rho'(20)(t-20) = \rho_{20}\left[1 + \alpha(t-20)\right]$. The quadratic approximation is $T_2(t) = \rho(20) + \rho'(20)(t-20) + \dfrac{\rho''(20)}{2}(t-20)^2 = \rho_{20}\left[1 + \alpha(t-20) + \frac{1}{2}\alpha^2(t-20)^2\right]$.

(b)

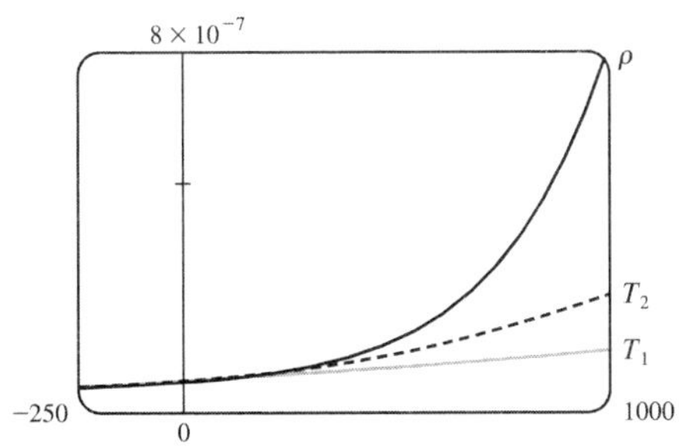

(c)

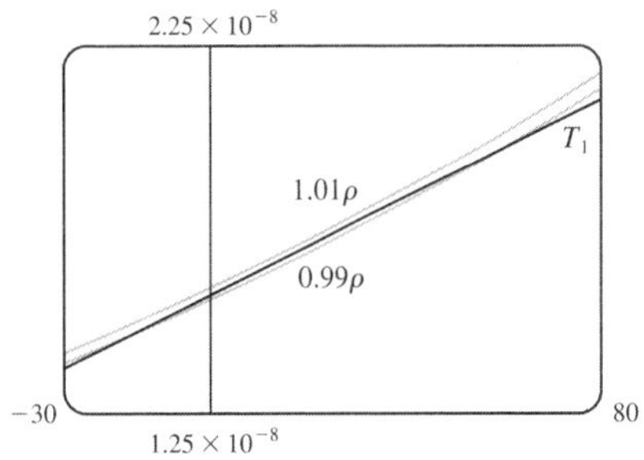


From the graph, it seems that $T_1(t)$ is within 1% of $\rho(t)$, that is, $0.99\rho(t) \leq T_1(t) \leq 1.01\rho(t)$, for $-14\ ^\circ\text{C} \leq t \leq 58\ ^\circ\text{C}$.

**29.** (a) $L$ is the length of the arc subtended by the angle $\theta$, so $L = R\theta \Rightarrow$ $\theta = L/R$. Now $\sec\theta = (R+C)/R \Rightarrow R\sec\theta = R + C \Rightarrow$ $C = R\sec\theta - R = R\sec(L/R) - R$.

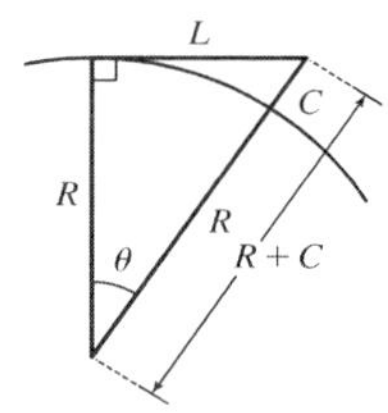


(b) From Exercise 9, $\sec x \approx T_4(x) = 1 + \frac{1}{2}x^2 + \frac{5}{24}x^4$. By part (a),

$$C \approx R\left[1 + \frac{1}{2}\left(\frac{L}{R}\right)^2 + \frac{5}{24}\left(\frac{L}{R}\right)^4\right] - R = R + \frac{1}{2}R \cdot \frac{L^2}{R^2} + \frac{5}{24}R \cdot \frac{L^4}{R^4} - R = \frac{L^2}{2R} + \frac{5L^4}{24R^3}.$$

(c) Taking $L = 100$ km and $R = 6370$ km, the formula in part (a) says that $C = R\sec(L/R) - R = 6370\sec(100/6370) - 6370 \approx 0.785\,009\,965\,44$ km. The formula in part (b) says that

$$C \approx \frac{L^2}{2R} + \frac{5L^4}{24R^3} = \frac{100^2}{2 \cdot 6370} + \frac{5 \cdot 100^4}{24 \cdot 6370^3} \approx 0.785\,009\,957\,36 \text{ km.}$$

The difference between these two results is only 0.000 000 008 08 km, or 0.000 008 08 m!

**30.** (a)

$$4\sqrt{\frac{L}{g}}\int_0^{\pi/2}\frac{dx}{\sqrt{1-k^2\sin^2 x}} = 4\sqrt{\frac{L}{g}}\int_0^{\pi/2}\left[1 + (-k^2\sin^2 x)\right]^{-1/2}dx$$

$$= 4\sqrt{\frac{L}{g}}\int_0^{\pi/2}\left[1 - \frac{1}{2}(-k^2\sin^2 x) + \frac{\frac{1}{2}\cdot\frac{3}{2}}{2!}(-k^2\sin^2 x)^2 - \frac{\frac{1}{2}\cdot\frac{3}{2}\cdot\frac{5}{2}}{3!}(-k^2\sin^2 x)^3 + \cdots\right]dx$$

$$= 4\sqrt{\frac{L}{g}}\int_0^{\pi/2}\left[1 + \left(\frac{1}{2}\right)k^2\sin^2 x + \left(\frac{1\cdot 3}{2\cdot 4}\right)k^4\sin^4 x + \left(\frac{1\cdot 3\cdot 5}{2\cdot 4\cdot 6}\right)k^6\sin^6 x + \cdots\right]dx$$

$$= 4\sqrt{\frac{L}{g}}\left[\frac{\pi}{2} + \left(\frac{1}{2}\right)\left(\frac{1}{2}\cdot\frac{\pi}{2}\right)k^2 + \left(\frac{1\cdot 3}{2\cdot 4}\right)\left(\frac{1\cdot 3}{2\cdot 4}\cdot\frac{\pi}{2}\right)k^4 + \left(\frac{1\cdot 3\cdot 5}{2\cdot 4\cdot 6}\right)\left(\frac{1\cdot 3\cdot 5}{2\cdot 4\cdot 6}\cdot\frac{\pi}{2}\right)k^6 + \cdots\right]$$

[split up the integral and use the result from Exercise 5.6.36]

$$= 2\pi\sqrt{\frac{L}{g}}\left[1 + \frac{1^2}{2^2}k^2 + \frac{1^2\cdot 3^2}{2^2\cdot 4^2}k^4 + \frac{1^2\cdot 3^2\cdot 5^2}{2^2\cdot 4^2\cdot 6^2}k^6 + \cdots\right]$$

(b) The first of the two inequalities is true because all of the terms in the series are positive. For the second,

$$T = 2\pi\sqrt{\frac{L}{g}}\left[1 + \frac{1^2}{2^2}k^2 + \frac{1^2\cdot 3^2}{2^2\cdot 4^2}k^4 + \frac{1^2\cdot 3^2\cdot 5^2}{2^2\cdot 4^2\cdot 6^2}k^6 + \frac{1^2\cdot 3^2\cdot 5^2\cdot 7^2}{2^2\cdot 4^2\cdot 6^2\cdot 8^2}k^8 + \cdots\right]$$

$$\leq 2\pi\sqrt{\frac{L}{g}}\left[1 + \frac{1}{4}k^2 + \frac{1}{4}k^4 + \frac{1}{4}k^6 + \frac{1}{4}k^8 + \cdots\right]$$

The terms in brackets (after the first) form a geometric series with $a = \frac{1}{4}k^2$ and $r = k^2 = \sin^2\left(\frac{1}{2}\theta_0\right) < 1$.

So $T \leq 2\pi\sqrt{\frac{L}{g}}\left[1 + \frac{k^2/4}{1-k^2}\right] = 2\pi\sqrt{\frac{L}{g}}\frac{4-3k^2}{4-4k^2}$.

(c) We substitute $L = 1$, $g = 9.8$, and $k = \sin(10^\circ/2) \approx 0.08716$, and the inequality from part (b) becomes $2.01090 \leq T \leq 2.01093$, so $T \approx 2.0109$. The estimate $T \approx 2\pi\sqrt{L/g} \approx 2.0071$ differs by about 0.2%. If $\theta_0 = 42^\circ$, then $k \approx 0.35837$ and the inequality becomes $2.07153 \leq T \leq 2.08103$, so $T \approx 2.0763$. The one-term estimate is the same, and the discrepancy between the two estimates increases to about 3.4%.

**31.** Using $f(x) = T_n(x) + R_n(x)$ with $n = 1$ and $x = r$, we have $f(r) = T_1(r) + R_1(r)$, where $T_1$ is the first-degree Taylor polynomial of $f$ at $a$. Because $a = x_n$, $f(r) = f(x_n) + f'(x_n)(r - x_n) + R_1(r)$. But $r$ is a root of $f$, so $f(r) = 0$ and we have $0 = f(x_n) + f'(x_n)(r - x_n) + R_1(r)$. Taking the first two terms to the left side gives us $f'(x_n)(x_n - r) - f(x_n) = R_1(r)$. Dividing by $f'(x_n)$, we get $x_n - r - \dfrac{f(x_n)}{f'(x_n)} = \dfrac{R_1(r)}{f'(x_n)}$. By the formula for Newton's method, the left side of the preceding equation is $x_{n+1} - r$, so $|x_{n+1} - r| = \left|\dfrac{R_1(r)}{f'(x_n)}\right|$. Taylor's Inequality gives us $|R_1(r)| \le \dfrac{|f''(r)|}{2!}|r - x_n|^2$. Combining this inequality with the facts $|f''(x)| \le M$ and $|f'(x)| \ge K$ gives us $|x_{n+1} - r| \le \dfrac{M}{2K}|x_n - r|^2$.

## APPLIED PROJECT Radiation from the Stars

**1.** If we write $f(\lambda) = \dfrac{8\pi hc\lambda^{-5}}{e^{hc/(\lambda kT)} - 1} = \dfrac{a\lambda^{-5}}{e^{b/(\lambda T)} - 1}$, then as $\lambda \to 0^+$, it is of the form $\infty/\infty$, and as $\lambda \to \infty$ it is of the form $0/0$, so in either case we can use l'Hospital's Rule. First of all,

$$\lim_{\lambda\to\infty} f(\lambda) \overset{\text{H}}{=} \lim_{\lambda\to\infty} \frac{a\left(-5\lambda^{-6}\right)}{-\dfrac{bT}{(\lambda T)^2}e^{b/(\lambda T)}} = 5\frac{aT}{b}\lim_{\lambda\to\infty}\frac{\lambda^2\lambda^{-6}}{e^{b/(\lambda T)}} = 5\frac{aT}{b}\lim_{\lambda\to\infty}\frac{\lambda^{-4}}{e^{b/(\lambda T)}} = 0$$

Also,

$$\lim_{\lambda\to0^+} f(\lambda) \overset{\text{H}}{=} 5\frac{aT}{b}\lim_{\lambda\to0^+}\frac{\lambda^{-4}}{e^{b/(\lambda T)}} \overset{\text{H}}{=} 5\frac{aT}{b}\lim_{\lambda\to0^+}\frac{-4\lambda^{-5}}{-\dfrac{bT}{(\lambda T)^2}e^{b/(\lambda T)}} = 20\frac{aT^2}{b^2}\lim_{\lambda\to0^+}\frac{\lambda^{-3}}{e^{b/(\lambda T)}}$$

This is still indeterminate, but note that each time we use l'Hospital's Rule, we gain a factor of $\lambda$ in the numerator, as well as a constant factor, and the denominator is unchanged. So if we use l'Hospital's Rule three more times, the exponent of $\lambda$ in the numerator will become 0. That is, for some $\{k_i\}$, all constant,

$$\lim_{\lambda\to0^+} f(\lambda) \overset{\text{H}}{=} k_1\lim_{\lambda\to0^+}\frac{\lambda^{-3}}{e^{b/(\lambda T)}} \overset{\text{H}}{=} k_2\lim_{\lambda\to0^+}\frac{\lambda^{-2}}{e^{b/(\lambda T)}} \overset{\text{H}}{=} k_3\lim_{\lambda\to0^+}\frac{\lambda^{-1}}{e^{b/(\lambda T)}} \overset{\text{H}}{=} k_4\lim_{\lambda\to0^+}\frac{1}{e^{b/(\lambda T)}} = 0$$

**2.** We expand the denominator of Planck's Law using the Taylor series $e^x = 1 + x + \dfrac{x^2}{2!} + \dfrac{x^3}{3!} + \cdots$ with $x = \dfrac{hc}{\lambda kT}$, and use the fact that if $\lambda$ is large, then all subsequent terms in the Taylor expansion are very small compared to the first one, so we can approximate using the Taylor polynomial $T_1$:

$$f(\lambda) = \frac{8\pi hc\lambda^{-5}}{e^{hc/(\lambda kT)} - 1} = \frac{8\pi hc\lambda^{-5}}{\left[1 + \dfrac{hc}{\lambda kT} + \dfrac{1}{2!}\left(\dfrac{hc}{\lambda kT}\right)^2 + \dfrac{1}{3!}\left(\dfrac{hc}{\lambda kT}\right)^3 + \cdots\right] - 1}$$

$$\approx \frac{8\pi hc\lambda^{-5}}{\left(1 + \dfrac{hc}{\lambda kT}\right) - 1} = \frac{8\pi kT}{\lambda^4}$$

which is the Rayleigh-Jeans Law.

**3.** To convert to $\mu$m, we substitute $\lambda/10^6$ for $\lambda$ in both laws. The first figure shows that the two laws are similar for large $\lambda$. The second figure shows that the two laws are very different for short wavelengths (Planck's Law gives a maximum at $\lambda \approx 0.51$ $\mu$m; the Rayleigh-Jeans Law gives no minimum or maximum.).

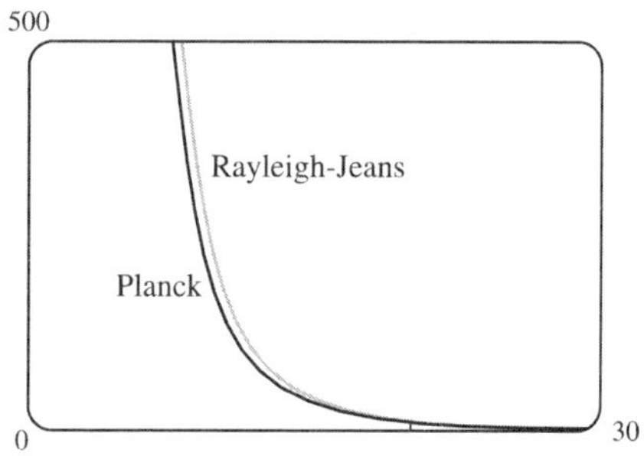


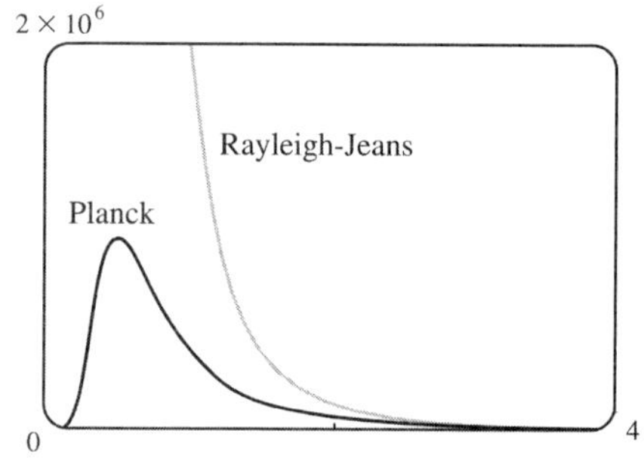


**4.** From the graph in Problem 3, $f(\lambda)$ has a maximum under Planck's Law at $\lambda \approx 0.51$ $\mu$m.

**5.**

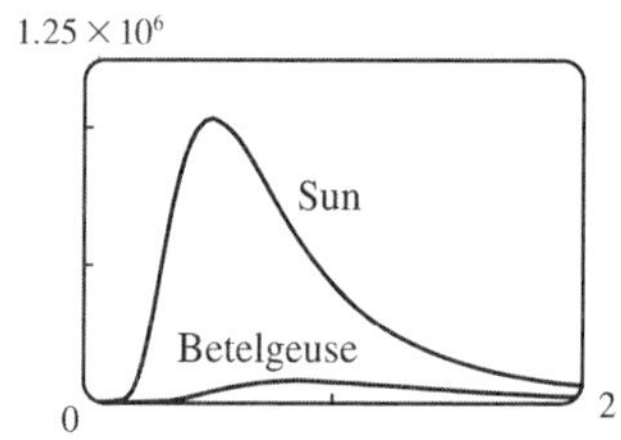


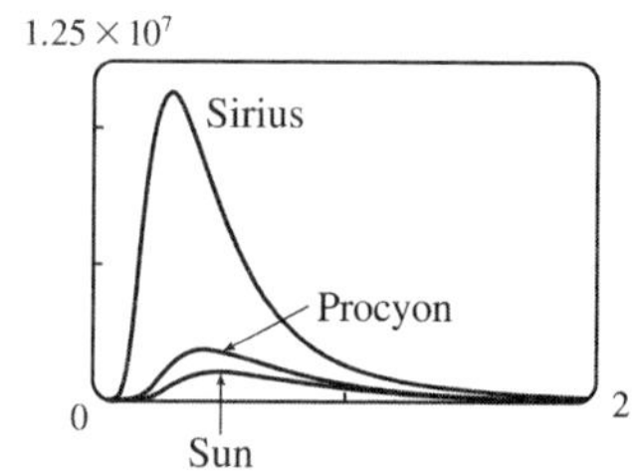


As $T$ gets larger, the total area under the curve increases, as we would expect: the hotter the star, the more energy it emits. Also, as $T$ increases, the $\lambda$-value of the maximum decreases, so the higher the temperature, the shorter the peak wavelength (and consequently the average wavelength) of light emitted. This is why Sirius is a blue star and Betelgeuse is a red star: most of Sirius's light is of a fairly short wavelength; that is, a higher frequency, toward the blue end of the spectrum, whereas most of Betelgeuse's light is of a lower frequency, toward the red end of the spectrum.

## 8 Review

### CONCEPT CHECK

**1.** (a) See Definition 8.1.1.

(b) See Definition 8.2.2.

(c) The terms of the sequence $\{a_n\}$ approach 3 as $n$ becomes large.

(d) By adding sufficiently many terms of the series, we can make the partial sums as close to 3 as we like.

**2.** (a) See the definition on page 563.

(b) A sequence is monotonic if it is either increasing or decreasing.

(c) By Theorem 8.1.7, every bounded, monotonic sequence is convergent.

**3.** (a) See (4) in Section 8.2.

(b) The $p$-series $\sum_{n=1}^{\infty} \dfrac{1}{n^p}$ is convergent if $p > 1$.

**4.** If $\sum a_n = 3$, then $\lim_{n\to\infty} a_n = 0$ and $\lim_{n\to\infty} s_n = 3$.

**5.** (a) See the Test for Divergence on page 572.

(b) See the Integral Test on page 578.

(c) See the Comparison Test on page 580.

(d) See the Limit Comparison Test on page 582.

(e) See the Alternating Series Test on page 587.

(f) See the Ratio Test on page 591.

**6.** (a) A series $\sum a_n$ is called *absolutely convergent* if the series of absolute values $\sum |a_n|$ is convergent.

(b) If a series $\sum a_n$ is absolutely convergent, then it is convergent.

**7.** (a) Use (4) in Section 8.3.

(b) See Example 8 in Section 8.3.

(c) By adding terms until you reach the desired accuracy given by the Alternating Series Estimation Theorem on page 588.

**8.** (a) $\sum_{n=0}^{\infty} c_n(x-a)^n$

(b) Given the power series $\sum_{n=0}^{\infty} c_n(x-a)^n$, the radius of convergence is:

(i) 0 if the series converges only when $x = a$

(ii) $\infty$ if the series converges for all $x$, or

(iii) a positive number $R$ such that the series converges if $|x-a| < R$ and diverges if $|x-a| > R$.

(c) The interval of convergence of a power series is the interval that consists of all values of $x$ for which the series converges. Corresponding to the cases in part (b), the interval of convergence is: (i) the single point $\{a\}$, (ii) all real numbers, that is, the real number line $(-\infty, \infty)$, or (iii) an interval with endpoints $a - R$ and $a + R$ which can contain neither, either, or both of the endpoints. In this case, we must test the series for convergence at each endpoint to determine the interval of convergence.

**9.** (a), (b) See Theorem 8.6.2.

**10.** (a) $T_n(x) = \sum\limits_{i=0}^{n} \dfrac{f^{(i)}(a)}{i!}(x-a)^i$

(b) $\sum\limits_{n=0}^{\infty} \dfrac{f^{(n)}(a)}{n!}(x-a)^n$

(c) $\sum\limits_{n=0}^{\infty} \dfrac{f^{(n)}(0)}{n!}x^n$ [$a = 0$ in part (b)]

(d) See Theorem 8.7.8.

(e) See Taylor's Inequality (8.7.9).

**11.** (a) – (e) See the table on page 612.

**12.** See the Binomial Series (8.8.2) for the expansion. The radius of convergence for the binomial series is 1.

## TRUE-FALSE QUIZ

**1.** False. See Note 2 after Theorem 8.2.6.

**2.** False. The series $\sum\limits_{n=1}^{\infty} n^{-\sin 1} = \sum\limits_{n=1}^{\infty} \dfrac{1}{n^{\sin 1}}$ is a $p$-series with $p = \sin 1 \approx 0.84 \le 1$, so the series diverges.

**3.** True. If $\lim\limits_{n\to\infty} a_n = L$, then given any $\varepsilon > 0$, we can find a positive integer $N$ such that $|a_n - L| < \varepsilon$ whenever $n > N$. If $n > N$, then $2n+1 > N$ and $|a_{2n+1} - L| < \varepsilon$. Thus, $\lim\limits_{n\to\infty} a_{2n+1} = L$.

**4.** True by Theorem 8.5.3.
*Or:* Use the Comparison Test to show that $\sum c_n(-2)^n$ converges absolutely.

**5.** False. For example, take $c_n = (-1)^n/(n6^n)$.

**6.** True by Theorem 8.5.3.

**7.** False, since $\lim\limits_{n\to\infty} \left|\dfrac{a_{n+1}}{a_n}\right| = \lim\limits_{n\to\infty} \left|\dfrac{1}{(n+1)^3} \cdot \dfrac{n^3}{1}\right| = \lim\limits_{n\to\infty} \left|\dfrac{n^3}{(n+1)^3} \cdot \dfrac{1/n^3}{1/n^3}\right| = \lim\limits_{n\to\infty} \dfrac{1}{(1+1/n)^3} = 1.$

**8.** True, since $\lim_{n\to\infty}\left|\dfrac{a_{n+1}}{a_n}\right| = \lim_{n\to\infty}\left|\dfrac{1}{(n+1)!}\cdot\dfrac{n!}{1}\right| = \lim_{n\to\infty}\dfrac{1}{n+1} = 0 < 1.$

**9.** False. See the note after Example 4 in Section 8.3.

**10.** True, since $\dfrac{1}{e} = e^{-1}$ and $e^x = \sum\limits_{n=0}^{\infty}\dfrac{x^n}{n!}$, so $e^{-1} = \sum\limits_{n=0}^{\infty}\dfrac{(-1)^n}{n!}$.

**11.** True. See (6) in Section 8.1.

**12.** True, because if $\sum|a_n|$ is convergent, then so is $\sum a_n$ by Theorem 8.4.1.

**13.** True. By Theorem 8.7.5 the coefficient of $x^3$ is $\dfrac{f'''(0)}{3!} = \dfrac{1}{3} \Rightarrow f'''(0) = 2.$

*Or:* Use Theorem 8.6.2 to differentiate $f$ three times.

**14.** False. Let $a_n = n$ and $b_n = -n$. Then $\{a_n\}$ and $\{b_n\}$ are divergent, but $a_n + b_n = 0$, so $\{a_n + b_n\}$ is convergent.

**15.** False. For example, let $a_n = b_n = (-1)^n$. Then $\{a_n\}$ and $\{b_n\}$ are divergent, but $a_n b_n = 1$, so $\{a_n b_n\}$ is convergent.

**16.** True by the Monotonic Sequence Theorem, since $\{a_n\}$ is decreasing and $0 < a_n \le a_1$ for all $n \Rightarrow \{a_n\}$ is bounded.

**17.** True by Theorem 8.4.1. [$\sum(-1)^n a_n$ is absolutely convergent and hence convergent.]

**18.** True. $\lim_{n\to\infty}\dfrac{a_{n+1}}{a_n} < 1 \Rightarrow \sum a_n$ converges (Ratio Test) $\Rightarrow \lim_{n\to\infty} a_n = 0$ [Theorem 8.2.6].

## EXERCISES

**1.** $\left\{\dfrac{2+n^3}{1+2n^3}\right\}$ converges since $\lim_{n\to\infty}\dfrac{2+n^3}{1+2n^3} = \lim_{n\to\infty}\dfrac{2/n^3+1}{1/n^3+2} = \dfrac{1}{2}.$

**2.** $a_n = \dfrac{9^{n+1}}{10^n} = 9\cdot\left(\frac{9}{10}\right)^n$, so $\lim_{n\to\infty} a_n = 9\lim_{n\to\infty}\left(\frac{9}{10}\right)^n = 9\cdot 0 = 0$ by (8.1.6).

**3.** $\lim_{n\to\infty} a_n = \lim_{n\to\infty}\dfrac{n^3}{1+n^2} = \lim_{n\to\infty}\dfrac{n}{1/n^2+1} = \infty$, so the sequence diverges.

**4.** $a_n = \cos(n\pi/2)$, so $a_n = 0$ if $n$ is odd and $a_n = \pm 1$ if $n$ is even. As $n$ increases, $a_n$ keeps cycling through the values 0, 1, 0, $-1$, so the sequence $\{a_n\}$ is divergent.

**5.** $|a_n| = \left|\dfrac{n\sin n}{n^2+1}\right| \le \dfrac{n}{n^2+1} < \dfrac{1}{n}$, so $|a_n| \to 0$ as $n\to\infty$. Thus, $\lim_{n\to\infty} a_n = 0$. The sequence $\{a_n\}$ is convergent.

**6.** $a_n = \dfrac{\ln n}{\sqrt{n}}$. Let $f(x) = \dfrac{\ln x}{\sqrt{x}}$ for $x > 0$. Then $\lim_{x\to\infty} f(x) = \lim_{x\to\infty}\dfrac{\ln x}{\sqrt{x}} \overset{\text{H}}{=} \lim_{x\to\infty}\dfrac{1/x}{1/(2\sqrt{x})} = \lim_{x\to\infty}\dfrac{2}{\sqrt{x}} = 0$. Thus, by Theorem 2 in Section 8.1, $\{a_n\}$ converges and $\lim_{n\to\infty} a_n = 0$.

**7.** $\left\{\left(1+\dfrac{3}{n}\right)^{4n}\right\}$ is convergent. Let $y = \left(1+\dfrac{3}{x}\right)^{4x}$. Then

$$\lim_{x\to\infty}\ln y = \lim_{x\to\infty} 4x\ln(1+3/x) = \lim_{x\to\infty}\frac{\ln(1+3/x)}{1/(4x)} \overset{\text{H}}{=} \lim_{x\to\infty}\frac{\dfrac{1}{1+3/x}\left(-\dfrac{3}{x^2}\right)}{-1/(4x^2)} = \lim_{x\to\infty}\frac{12}{1+3/x} = 12$$

so $\lim_{x\to\infty} y = \lim_{n\to\infty}\left(1+\dfrac{3}{n}\right)^{4n} = e^{12}.$

*Or:* Use Exercise 4.5.38.

**8.** We use induction, hypothesizing that $a_{n-1} < a_n < 2$. Note first that $1 < a_2 = \frac{1}{3}(1+4) = \frac{5}{3} < 2$, so the hypothesis holds for $n = 2$. Now assume that $a_{k-1} < a_k < 2$. Then $a_k = \frac{1}{3}(a_{k-1}+4) < \frac{1}{3}(a_k+4) < \frac{1}{3}(2+4) = 2$. So $a_k < a_{k+1} < 2$, and the induction is complete. To find the limit of the sequence, we note that $L = \lim\limits_{n\to\infty} a_n = \lim\limits_{n\to\infty} a_{n+1} \Rightarrow$

$L = \frac{1}{3}(L+4) \Rightarrow L = 2$.

**9.** $\dfrac{n}{n^3+1} < \dfrac{n}{n^3} = \dfrac{1}{n^2}$, so $\sum\limits_{n=1}^{\infty} \dfrac{n}{n^3+1}$ converges by the Comparison Test with the convergent $p$-series $\sum\limits_{n=1}^{\infty} \dfrac{1}{n^2}$ $(p = 2 > 1)$.

**10.** Let $a_n = \dfrac{n^2+1}{n^3+1}$ and $b_n = \dfrac{1}{n}$, so $\lim\limits_{n\to\infty} \dfrac{a_n}{b_n} = \lim\limits_{n\to\infty} \dfrac{n^3+n}{n^3+1} = \lim\limits_{n\to\infty} \dfrac{1+1/n^2}{1+1/n^3} = 1 > 0$. Since $\sum_{n=1}^{\infty} b_n$ is the divergent harmonic series, $\sum_{n=1}^{\infty} a_n$ also diverges by the Limit Comparison Test.

**11.** $\lim\limits_{n\to\infty} \left|\dfrac{a_{n+1}}{a_n}\right| = \lim\limits_{n\to\infty} \left[\dfrac{(n+1)^3}{5^{n+1}} \cdot \dfrac{5^n}{n^3}\right] = \lim\limits_{n\to\infty} \left(1+\dfrac{1}{n}\right)^3 \cdot \dfrac{1}{5} = \dfrac{1}{5} < 1$, so $\sum\limits_{n=1}^{\infty} \dfrac{n^3}{5^n}$ converges by the Ratio Test.

**12.** Let $b_n = \dfrac{1}{\sqrt{n+1}}$. Then $b_n$ is positive for $n \geq 1$, the sequence $\{b_n\}$ is decreasing, and $\lim\limits_{n\to\infty} b_n = 0$, so the series $\sum\limits_{n=1}^{\infty} \dfrac{(-1)^n}{\sqrt{n+1}}$ converges by the Alternating Series Test.

**13.** Let $f(x) = \dfrac{1}{x\sqrt{\ln x}}$. Then $f$ is continuous, positive, and decreasing on $[2, \infty)$, so the Integral Test applies.

$$\int_2^{\infty} f(x)\,dx = \lim_{t\to\infty} \int_2^t \frac{1}{x\sqrt{\ln x}}\,dx \quad \left[\begin{array}{c} u = \ln x, \\ du = \frac{1}{x}\,dx \end{array}\right] \quad = \lim_{t\to\infty} \int_{\ln 2}^{\ln t} u^{-1/2}\,du$$

$$= \lim_{t\to\infty} \left[2\sqrt{u}\right]_{\ln 2}^{\ln t} = \lim_{t\to\infty} \left(2\sqrt{\ln t} - 2\sqrt{\ln 2}\right) = \infty, \text{ so the series } \sum_{n=2}^{\infty} \frac{1}{n\sqrt{\ln n}} \text{ diverges.}$$

**14.** $\lim\limits_{n\to\infty} \dfrac{n}{3n+1} = \dfrac{1}{3}$, so $\lim\limits_{n\to\infty} \ln\left(\dfrac{n}{3n+1}\right) = \ln\frac{1}{3} \neq 0$. Thus, the series $\sum\limits_{n=1}^{\infty} \ln\left(\dfrac{n}{3n+1}\right)$ diverges by the Test for Divergence.

**15.** $b_n = \dfrac{\sqrt{n}}{n+1} > 0$, $\{b_n\}$ is decreasing, and $\lim\limits_{n\to\infty} b_n = 0$, so the series $\sum\limits_{n=1}^{\infty} (-1)^{n-1} \dfrac{\sqrt{n}}{n+1}$ converges by the Alternating Series Test.

**16.** $|a_n| = \left|\dfrac{\cos 3n}{1+(1.2)^n}\right| \leq \dfrac{1}{1+(1.2)^n} < \dfrac{1}{(1.2)^n} = \left(\dfrac{5}{6}\right)^n$, so $\sum\limits_{n=1}^{\infty} |a_n|$ converges by comparison with the convergent geometric series $\sum\limits_{n=1}^{\infty} \left(\dfrac{5}{6}\right)^n$ $\left(r = \frac{5}{6} < 1\right)$. It follows that $\sum\limits_{n=1}^{\infty} a_n$ converges (by Theorem 1 in Section 8.4).

**17.** $\lim\limits_{n\to\infty} \left|\dfrac{a_{n+1}}{a_n}\right| = \lim\limits_{n\to\infty} \dfrac{1\cdot 3\cdot 5\cdot\cdots\cdot(2n-1)(2n+1)}{5^{n+1}(n+1)!} \cdot \dfrac{5^n n!}{1\cdot 3\cdot 5\cdot\cdots\cdot(2n-1)} = \lim\limits_{n\to\infty} \dfrac{2n+1}{5(n+1)} = \dfrac{2}{5} < 1$, so the series converges by the Ratio Test.

**18.** $\sum\limits_{n=1}^{\infty} \dfrac{(-5)^{2n}}{n^2\,9^n} = \sum\limits_{n=1}^{\infty} \dfrac{1}{n^2}\left(\dfrac{25}{9}\right)^n$. Now $\lim\limits_{n\to\infty} \left|\dfrac{a_{n+1}}{a_n}\right| = \lim\limits_{n\to\infty} \dfrac{25^{n+1}}{(n+1)^2\cdot 9^{n+1}} \cdot \dfrac{n^2\cdot 9^n}{25^n} = \lim\limits_{n\to\infty} \dfrac{25n^2}{9(n+1)^2} = \dfrac{25}{9} > 1$, so the series diverges by the Ratio Test.

**19.** $\dfrac{2^{2n+1}}{5^n} = \dfrac{2^{2n}\cdot 2^1}{5^n} = \dfrac{(2^2)^n\cdot 2}{5^n} = 2\left(\dfrac{4}{5}\right)^n$, so $\sum\limits_{n=1}^{\infty} \dfrac{2^{2n+1}}{5^n} = 2\sum\limits_{n=1}^{\infty} \left(\dfrac{4}{5}\right)^n$ is a geometric series with $a = \dfrac{8}{5}$ and $r = \dfrac{4}{5}$. Since $|r| = \dfrac{4}{5} < 1$, the series converges to $\dfrac{a}{1-r} = \dfrac{8/5}{1-4/5} = \dfrac{8/5}{1/5} = 8$.

**20.** $\sum\limits_{n=0}^{\infty} \dfrac{(-1)^n \pi^n}{3^{2n}(2n)!} = \sum\limits_{n=0}^{\infty} (-1)^n \dfrac{1}{(2n)!} \cdot \dfrac{\pi^n}{3^{2n}} = \sum\limits_{n=0}^{\infty} (-1)^n \dfrac{1}{(2n)!} \cdot \left(\dfrac{\sqrt{\pi}}{3}\right)^{2n} = \cos\left(\dfrac{\sqrt{\pi}}{3}\right)$ since $\cos x = \sum\limits_{n=0}^{\infty} (-1)^n \dfrac{x^{2n}}{(2n)!}$ for all $x$.

**21.** $\sum_{n=1}^{\infty}\left[\tan^{-1}(n+1)-\tan^{-1}n\right] = \lim_{n\to\infty} s_n$

$$= \lim_{n\to\infty}\left[\left(\tan^{-1}2-\tan^{-1}1\right)+\left(\tan^{-1}3-\tan^{-1}2\right)+\cdots + \left(\tan^{-1}(n+1)-\tan^{-1}n\right)\right]$$

$$= \lim_{n\to\infty}\left[\tan^{-1}(n+1)-\tan^{-1}1\right] = \tfrac{\pi}{2}-\tfrac{\pi}{4} = \tfrac{\pi}{4}$$

**22.** $1-e+\dfrac{e^2}{2!}-\dfrac{e^3}{3!}+\dfrac{e^4}{4!}-\cdots = \sum_{n=0}^{\infty}(-1)^n\dfrac{e^n}{n!} = \sum_{n=0}^{\infty}\dfrac{(-e)^n}{n!} = e^{-e}$ since $e^x = \sum_{n=0}^{\infty}\dfrac{x^n}{n!}$ for all $x$.

**23.** $1.2345345345\ldots = 1.2 + 0.0\overline{345} = \dfrac{12}{10}+\dfrac{345/10{,}000}{1-1/1000} = \dfrac{12}{10}+\dfrac{345}{9990} = \dfrac{4111}{3330}$

**24.** $\sum_{n=1}^{\infty}(\ln x)^n$ is a geometric series which converges whenever $|\ln x|<1 \;\Rightarrow\; -1<\ln x<1 \;\Rightarrow\; e^{-1}<x<e$.

**25.** $\sum_{n=1}^{\infty}\dfrac{(-1)^{n+1}}{n^5} = 1-\dfrac{1}{32}+\dfrac{1}{243}-\dfrac{1}{1024}+\dfrac{1}{3125}-\dfrac{1}{7776}+\dfrac{1}{16{,}807}-\dfrac{1}{32{,}768}+\cdots$.

Since $b_8 = \dfrac{1}{8^5} = \dfrac{1}{32{,}768} < 0.000031$, $\sum_{n=1}^{\infty}\dfrac{(-1)^{n+1}}{n^5} \approx \sum_{n=1}^{7}\dfrac{(-1)^{n+1}}{n^5} \approx 0.9721$.

**26.** (a) $s_5 = \sum_{n=1}^{5}\dfrac{1}{n^6} = 1+\dfrac{1}{2^6}+\cdots+\dfrac{1}{5^6} \approx 1.017305$. The series $\sum_{n=1}^{\infty}\dfrac{1}{n^6}$ converges by the Integral Test, so we estimate the remainder $R_5$ with (8.3.3): $R_5 \le \int_5^{\infty}\dfrac{dx}{x^6} = \left[-\dfrac{x^{-5}}{5}\right]_5^{\infty} = \dfrac{5^{-5}}{5} = 0.000064$. So the error is at most 0.000064.

(b) In general, $R_n \le \int_n^{\infty}\dfrac{dx}{x^6} = \dfrac{1}{5n^5}$. If we take $n=9$, then $s_9 \approx 1.01734$ and $R_9 \le \dfrac{1}{5\cdot 9^5} \approx 3.4\times 10^{-6}$. So to five decimal places, $\sum_{n=1}^{\infty}\dfrac{1}{n^5} \approx \sum_{n=1}^{9}\dfrac{1}{n^5} \approx 1.01734$.

*Another method:* Use (8.3.4) instead of (8.3.3).

**27.** $\sum_{n=1}^{\infty}\dfrac{1}{2+5^n} \approx \sum_{n=1}^{8}\dfrac{1}{2+5^n} \approx 0.18976224$. To estimate the error, note that $\dfrac{1}{2+5^n} < \dfrac{1}{5^n}$, so the remainder term is

$R_8 = \sum_{n=9}^{\infty}\dfrac{1}{2+5^n} < \sum_{n=9}^{\infty}\dfrac{1}{5^n} = \dfrac{1/5^9}{1-1/5} = 6.4\times 10^{-7}$ [geometric series with $a=\frac{1}{5^9}$ and $r=\frac{1}{5}$].

**28.** (a) $\lim_{n\to\infty}\left|\dfrac{a_{n+1}}{a_n}\right| = \lim_{n\to\infty}\left|\dfrac{(n+1)^{n+1}}{[2(n+1)]!}\cdot\dfrac{(2n)!}{n^n}\right| = \lim_{n\to\infty}\dfrac{(n+1)^n(n+1)^1}{(2n+2)(2n+1)n^n} = \lim_{n\to\infty}\left(\dfrac{n+1}{n}\right)^n\dfrac{1}{2(2n+1)}$

$$= \lim_{n\to\infty}\left(1+\frac{1}{n}\right)^n\frac{1}{2(2n+1)} = e\cdot 0 = 0 < 1$$

so the series converges by the Ratio Test.

(b) The series in part (a) is convergent, so $\lim_{n\to\infty} a_n = 0$ by Theorem 8.2.6.

**29.** Use the Limit Comparison Test. $\lim_{n\to\infty}\left|\dfrac{\left(\frac{n+1}{n}\right)a_n}{a_n}\right| = \lim_{n\to\infty}\dfrac{n+1}{n} = \lim_{n\to\infty}\left(1+\dfrac{1}{n}\right) = 1 > 0$.

Since $\sum|a_n|$ is convergent, so is $\sum\left|\left(\dfrac{n+1}{n}\right)a_n\right|$, by the Limit Comparison Test.

**30.** $\lim_{n\to\infty}\left|\dfrac{a_{n+1}}{a_n}\right| = \lim_{n\to\infty}\left|\dfrac{x^{n+1}}{(n+1)^2\,5^{n+1}}\cdot\dfrac{n^2 5^n}{x^n}\right| = \lim_{n\to\infty}\dfrac{1}{(1+1/n)^2}\dfrac{|x|}{5} = \dfrac{|x|}{5}$, so by the Ratio Test, $\sum_{n=1}^{\infty}(-1)^n\dfrac{x^n}{n^2\,5^n}$ converges when $\dfrac{|x|}{5}<1 \;\Leftrightarrow\; |x|<5$, so $R=5$. When $x=-5$, the series becomes the convergent $p$-series $\sum_{n=1}^{\infty}\dfrac{1}{n^2}$ with $p=2>1$. When $x=5$, the series becomes $\sum_{n=1}^{\infty}\dfrac{(-1)^n}{n^2}$, which converges by the Alternating Series Test. Thus, $I=[-5,5]$.

**31.** $\lim_{n\to\infty}\left|\frac{a_{n+1}}{a_n}\right| = \lim_{n\to\infty}\left[\frac{|x+2|^{n+1}}{(n+1)\,4^{n+1}}\cdot\frac{n\,4^n}{|x+2|^n}\right] = \lim_{n\to\infty}\left[\frac{n}{n+1}\frac{|x+2|}{4}\right] = \frac{|x+2|}{4} < 1 \;\Leftrightarrow\; |x+2| < 4$, so $R = 4$.

$|x+2| < 4 \;\Leftrightarrow\; -4 < x+2 < 4 \;\Leftrightarrow\; -6 < x < 2$. If $x = -6$, then the series $\sum\limits_{n=1}^{\infty}\frac{(x+2)^n}{n4^n}$ becomes

$\sum\limits_{n=1}^{\infty}\frac{(-4)^n}{n4^n} = \sum\limits_{n=1}^{\infty}\frac{(-1)^n}{n}$, the alternating harmonic series, which converges by the Alternating Series Test. When $x = 2$, the

series becomes the harmonic series $\sum\limits_{n=1}^{\infty}\frac{1}{n}$, which diverges. Thus, $I = [-6, 2)$.

**32.** $\lim_{n\to\infty}\left|\frac{a_{n+1}}{a_n}\right| = \lim_{n\to\infty}\left|\frac{2^{n+1}\,(x-2)^{n+1}}{(n+3)!}\cdot\frac{(n+2)!}{2^n(x-2)^n}\right| = \lim_{n\to\infty}\frac{2}{n+3}\,|x-2| = 0 < 1$, so the series $\sum\limits_{n=1}^{\infty}\frac{2^n\,(x-2)^n}{(n+2)!}$

converges for all $x$. $R = \infty$ and $I = (-\infty, \infty)$.

**33.** $\lim_{n\to\infty}\left|\frac{a_{n+1}}{a_n}\right| = \lim_{n\to\infty}\left|\frac{2^{n+1}(x-3)^{n+1}}{\sqrt{n+4}}\cdot\frac{\sqrt{n+3}}{2^n(x-3)^n}\right| = 2\,|x-3|\lim_{n\to\infty}\sqrt{\frac{n+3}{n+4}} = 2\,|x-3| < 1 \;\Leftrightarrow\; |x-3| < \frac{1}{2}$, so

$R = \frac{1}{2}$. $|x-3| < \frac{1}{2} \;\Leftrightarrow\; -\frac{1}{2} < x-3 < \frac{1}{2} \;\Leftrightarrow\; \frac{5}{2} < x < \frac{7}{2}$. For $x = \frac{7}{2}$, the series $\sum\limits_{n=1}^{\infty}\frac{2^n(x-3)^n}{\sqrt{n+3}}$ becomes

$\sum\limits_{n=0}^{\infty}\frac{1}{\sqrt{n+3}} = \sum\limits_{n=3}^{\infty}\frac{1}{n^{1/2}}$, which diverges ($p = \frac{1}{2} \le 1$), but for $x = \frac{5}{2}$, we get $\sum\limits_{n=0}^{\infty}\frac{(-1)^n}{\sqrt{n+3}}$, which is a convergent

alternating series, so $I = \left[\frac{5}{2}, \frac{7}{2}\right)$.

**34.** $\lim_{n\to\infty}\left|\frac{a_{n+1}}{a_n}\right| = \lim_{n\to\infty}\left|\frac{(2n+2)!\,x^{n+1}}{[(n+1)!]^2}\cdot\frac{(n!)^2}{(2n)!\,x^n}\right| = \lim_{n\to\infty}\frac{(2n+2)(2n+1)}{(n+1)(n+1)}\,|x| = 4\,|x|$.

To converge, we must have $4\,|x| < 1 \;\Leftrightarrow\; |x| < \frac{1}{4}$, so $R = \frac{1}{4}$.

**35.**

| $n$ | $f^{(n)}(x)$ | $f^{(n)}\left(\frac{\pi}{6}\right)$ |
|---|---|---|
| 0 | $\sin x$ | $\frac{1}{2}$ |
| 1 | $\cos x$ | $\frac{\sqrt{3}}{2}$ |
| 2 | $-\sin x$ | $-\frac{1}{2}$ |
| 3 | $-\cos x$ | $-\frac{\sqrt{3}}{2}$ |
| 4 | $\sin x$ | $\frac{1}{2}$ |
| $\vdots$ | $\vdots$ | $\vdots$ |

$$\begin{aligned}\sin x &= f\left(\frac{\pi}{6}\right) + f'\left(\frac{\pi}{6}\right)\left(x-\frac{\pi}{6}\right) + \frac{f''\left(\frac{\pi}{6}\right)}{2!}\left(x-\frac{\pi}{6}\right)^2 + \frac{f^{(3)}\left(\frac{\pi}{6}\right)}{3!}\left(x-\frac{\pi}{6}\right)^3 + \frac{f^{(4)}\left(\frac{\pi}{6}\right)}{4!}\left(x-\frac{\pi}{6}\right)^4 + \cdots \\ &= \frac{1}{2}\left[1 - \frac{1}{2!}\left(x-\frac{\pi}{6}\right)^2 + \frac{1}{4!}\left(x-\frac{\pi}{6}\right)^4 - \cdots\right] + \frac{\sqrt{3}}{2}\left[\left(x-\frac{\pi}{6}\right) - \frac{1}{3!}\left(x-\frac{\pi}{6}\right)^3 + \cdots\right] \\ &= \frac{1}{2}\sum_{n=0}^{\infty}(-1)^n\frac{1}{(2n)!}\left(x-\frac{\pi}{6}\right)^{2n} + \frac{\sqrt{3}}{2}\sum_{n=0}^{\infty}(-1)^n\frac{1}{(2n+1)!}\left(x-\frac{\pi}{6}\right)^{2n+1}\end{aligned}$$

**36.**

| $n$ | $f^{(n)}(x)$ | $f^{(n)}\left(\frac{\pi}{3}\right)$ |
|---|---|---|
| 0 | $\cos x$ | $\frac{1}{2}$ |
| 1 | $-\sin x$ | $-\frac{\sqrt{3}}{2}$ |
| 2 | $-\cos x$ | $-\frac{1}{2}$ |
| 3 | $\sin x$ | $\frac{\sqrt{3}}{2}$ |
| 4 | $\cos x$ | $\frac{1}{2}$ |
| $\vdots$ | $\vdots$ | $\vdots$ |

$$\begin{aligned}\cos x &= f\left(\frac{\pi}{3}\right) + f'\left(\frac{\pi}{3}\right)\left(x-\frac{\pi}{3}\right) + \frac{f''\left(\frac{\pi}{3}\right)}{2!}\left(x-\frac{\pi}{3}\right)^2 + \frac{f^{(3)}\left(\frac{\pi}{3}\right)}{3!}\left(x-\frac{\pi}{3}\right)^3 + \frac{f^{(4)}\left(\frac{\pi}{3}\right)}{4!}\left(x-\frac{\pi}{3}\right)^4 + \cdots \\ &= \frac{1}{2}\left[1 - \frac{1}{2!}\left(x-\frac{\pi}{3}\right)^2 + \frac{1}{4!}\left(x-\frac{\pi}{3}\right)^4 - \cdots\right] + \frac{\sqrt{3}}{2}\left[-\left(x-\frac{\pi}{3}\right) + \frac{1}{3!}\left(x-\frac{\pi}{3}\right)^3 - \cdots\right] \\ &= \frac{1}{2}\sum_{n=0}^{\infty}(-1)^n\frac{1}{(2n)!}\left(x-\frac{\pi}{3}\right)^{2n} + \frac{\sqrt{3}}{2}\sum_{n=0}^{\infty}(-1)^{n+1}\frac{1}{(2n+1)!}\left(x-\frac{\pi}{3}\right)^{2n+1}\end{aligned}$$

**37.** $\dfrac{1}{1+x} = \dfrac{1}{1-(-x)} = \sum\limits_{n=0}^{\infty}(-x)^n = \sum\limits_{n=0}^{\infty}(-1)^n x^n$ for $|x|<1 \;\Rightarrow\; \dfrac{x^2}{1+x} = \sum\limits_{n=0}^{\infty}(-1)^n x^{n+2}$ with $R=1$.

**38.** $\tan^{-1}x = \sum\limits_{n=0}^{\infty}(-1)^n\dfrac{x^{2n+1}}{2n+1}$ with interval of convergence $[-1,1]$, so

$\tan^{-1}(x^2) = \sum\limits_{n=0}^{\infty}(-1)^n\dfrac{(x^2)^{2n+1}}{2n+1} = \sum\limits_{n=0}^{\infty}(-1)^n\dfrac{x^{4n+2}}{2n+1}$, which converges when $x^2\in[-1,1] \;\Leftrightarrow\; x\in[-1,1]$.

Therefore, $R=1$.

**39.** $\dfrac{1}{1-x} = \sum\limits_{n=0}^{\infty}x^n$ for $|x|<1 \;\Rightarrow\; \ln(1-x) = -\displaystyle\int\frac{dx}{1-x} = -\int\sum_{n=0}^{\infty}x^n\,dx = C - \sum_{n=0}^{\infty}\frac{x^{n+1}}{n+1}$.

$\ln(1-0) = C-0 \;\Rightarrow\; C=0 \;\Rightarrow\; \ln(1-x) = -\sum\limits_{n=0}^{\infty}\dfrac{x^{n+1}}{n+1} = \sum\limits_{n=1}^{\infty}\dfrac{-x^n}{n}$ with $R=1$.

**40.** $e^x = \sum\limits_{n=0}^{\infty}\dfrac{x^n}{n!} \;\Rightarrow\; e^{2x} = \sum\limits_{n=0}^{\infty}\dfrac{(2x)^n}{n!} \;\Rightarrow\; xe^{2x} = x\sum\limits_{n=0}^{\infty}\dfrac{2^nx^n}{n!} = \sum\limits_{n=0}^{\infty}\dfrac{2^nx^{n+1}}{n!}$, $R=\infty$

**41.** $\sin x = \sum\limits_{n=0}^{\infty}\dfrac{(-1)^nx^{2n+1}}{(2n+1)!} \;\Rightarrow\; \sin(x^4) = \sum\limits_{n=0}^{\infty}\dfrac{(-1)^n(x^4)^{2n+1}}{(2n+1)!} = \sum\limits_{n=0}^{\infty}\dfrac{(-1)^nx^{8n+4}}{(2n+1)!}$ for all $x$, so the radius of convergence is $\infty$.

**42.** $e^x = \sum\limits_{n=0}^{\infty}\dfrac{x^n}{n!} \;\Rightarrow\; 10^x = e^{(\ln 10)x} = \sum\limits_{n=0}^{\infty}\dfrac{[(\ln 10)x]^n}{n!} = \sum\limits_{n=0}^{\infty}\dfrac{(\ln 10)^nx^n}{n!}$, $R=\infty$

**43.** $f(x) = \dfrac{1}{\sqrt[4]{16-x}} = \dfrac{1}{\sqrt[4]{16(1-x/16)}} = \dfrac{1}{\sqrt[4]{16}\left(1-\frac{1}{16}x\right)^{1/4}} = \frac{1}{2}\left(1-\frac{1}{16}x\right)^{-1/4}$

$$\begin{aligned}&= \frac{1}{2}\left[1 + \left(-\frac{1}{4}\right)\left(-\frac{x}{16}\right) + \frac{\left(-\frac{1}{4}\right)\left(-\frac{5}{4}\right)}{2!}\left(-\frac{x}{16}\right)^2 + \frac{\left(-\frac{1}{4}\right)\left(-\frac{5}{4}\right)\left(-\frac{9}{4}\right)}{3!}\left(-\frac{x}{16}\right)^3 + \cdots\right] \\ &= \frac{1}{2} + \sum_{n=1}^{\infty}\frac{1\cdot5\cdot9\cdot\cdots\cdot(4n-3)}{2\cdot4^n\cdot n!\cdot16^n}x^n = \frac{1}{2} + \sum_{n=1}^{\infty}\frac{1\cdot5\cdot9\cdot\cdots\cdot(4n-3)}{2^{6n+1}\,n!}x^n\end{aligned}$$

for $\left|-\dfrac{x}{16}\right| < 1 \;\Leftrightarrow\; |x|<16$, so $R=16$.

**44.** $(1-3x)^{-5} = \sum\limits_{n=0}^{\infty} \binom{-5}{n} (-3x)^n = 1 + (-5)(-3x) + \dfrac{(-5)(-6)}{2!}(-3x)^2 + \dfrac{(-5)(-6)(-7)}{3!}(-3x)^3 + \cdots$

$= 1 + \sum\limits_{n=1}^{\infty} \dfrac{5 \cdot 6 \cdot 7 \cdot \cdots \cdot (n+4) \cdot 3^n x^n}{n!}$ for $|-3x| < 1 \quad \Leftrightarrow \quad |x| < \frac{1}{3}$, so $R = \frac{1}{3}$.

**45.** $e^x = \sum\limits_{n=0}^{\infty} \dfrac{x^n}{n!}$, so $\dfrac{e^x}{x} = \dfrac{1}{x} \sum\limits_{n=0}^{\infty} \dfrac{x^n}{n!} = \sum\limits_{n=0}^{\infty} \dfrac{x^{n-1}}{n!} = x^{-1} + \sum\limits_{n=1}^{\infty} \dfrac{x^{n-1}}{n!} = \dfrac{1}{x} + \sum\limits_{n=1}^{\infty} \dfrac{x^{n-1}}{n!}$ and

$\displaystyle\int \frac{e^x}{x}\,dx = C + \ln|x| + \sum_{n=1}^{\infty} \frac{x^n}{n \cdot n!}$.

**46.** $\left(1+x^4\right)^{1/2} = \sum\limits_{n=0}^{\infty} \binom{\frac{1}{2}}{n} (x^4)^n = 1 + \left(\frac{1}{2}\right)x^4 + \dfrac{\left(\frac{1}{2}\right)\left(-\frac{1}{2}\right)}{2!}(x^4)^2 + \dfrac{\left(\frac{1}{2}\right)\left(-\frac{1}{2}\right)\left(-\frac{3}{2}\right)}{3!}(x^4)^3 + \cdots$

$= 1 + \frac{1}{2}x^4 - \frac{1}{8}x^8 + \frac{1}{16}x^{12} - \cdots$

so $\int_0^1 \left(1+x^4\right)^{1/2} dx = \left[x + \frac{1}{10}x^5 - \frac{1}{72}x^9 + \frac{1}{208}x^{13} - \cdots\right]_0^1 = 1 + \frac{1}{10} - \frac{1}{72} + \frac{1}{208} - \cdots$.

This is an alternating series, so by the Alternating Series Test, the error in the approximation $\int_0^1 (1+x^4)^{1/2}\,dx \approx 1 + \frac{1}{10} - \frac{1}{72} \approx 1.086$ is less than $\frac{1}{208}$, sufficient for the desired accuracy.

Thus, correct to two decimal places, $\int_0^1 (1+x^4)^{1/2}\,dx \approx 1.09$.

**47.** (a)

| $n$ | $f^{(n)}(x)$ | $f^{(n)}(1)$ |
|---|---|---|
| 0 | $x^{1/2}$ | 1 |
| 1 | $\frac{1}{2}x^{-1/2}$ | $\frac{1}{2}$ |
| 2 | $-\frac{1}{4}x^{-3/2}$ | $-\frac{1}{4}$ |
| 3 | $\frac{3}{8}x^{-5/2}$ | $\frac{3}{8}$ |
| 4 | $-\frac{15}{16}x^{-7/2}$ | $-\frac{15}{16}$ |
| ⋮ | ⋮ | ⋮ |

$\sqrt{x} \approx T_3(x) = 1 + \dfrac{1/2}{1!}(x-1) - \dfrac{1/4}{2!}(x-1)^2 + \dfrac{3/8}{3!}(x-1)^3$

$= 1 + \frac{1}{2}(x-1) - \frac{1}{8}(x-1)^2 + \frac{1}{16}(x-1)^3$

(b)

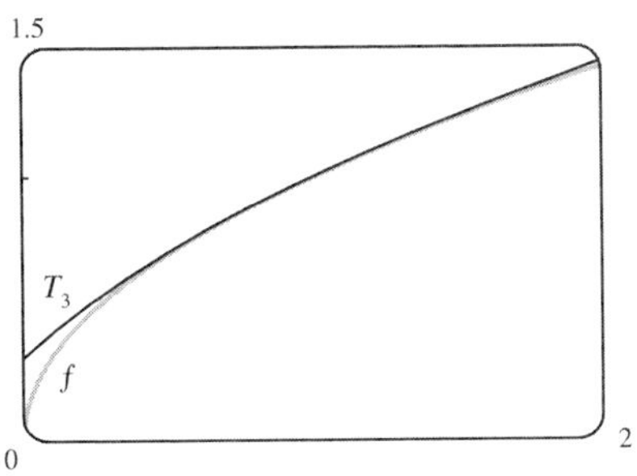


(c) $|R_3(x)| \le \dfrac{M}{4!}|x-1|^4$, where $\left|f^{(4)}(x)\right| \le M$ with

$f^{(4)}(x) = -\frac{15}{16}x^{-7/2}$. Now $0.9 \le x \le 1.1 \quad \Rightarrow$

$-0.1 \le x - 1 \le 0.1 \quad \Rightarrow \quad (x-1)^4 \le (0.1)^4$,

and letting $x = 0.9$ gives $M = \dfrac{15}{16(0.9)^{7/2}}$, so

$$|R_3(x)| \le \frac{15}{16(0.9)^{7/2}\,4!}(0.1)^4 \approx 0.000\,005\,648$$

$$\approx 0.000\,006 = 6 \times 10^{-6}$$

(d)

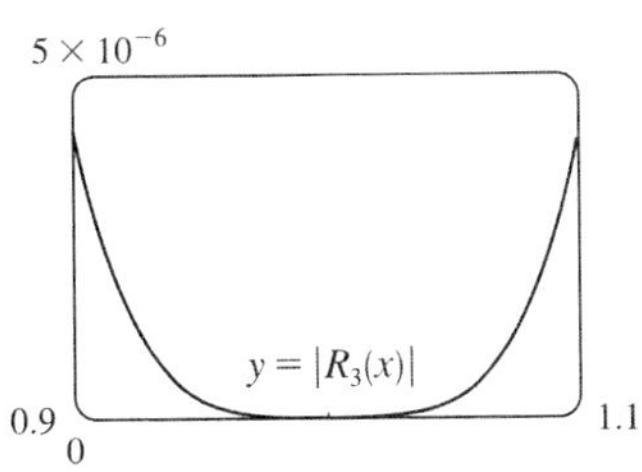


From the graph of $|R_3(x)| = |\sqrt{x} - T_3(x)|$, it appears that the error is less than $5 \times 10^{-6}$ on $[0.9, 1.1]$.

**48.** (a)

| $n$ | $f^{(n)}(x)$ | $f^{(n)}(0)$ |
|---|---|---|
| 0 | $\sec x$ | 1 |
| 1 | $\sec x \tan x$ | 0 |
| 2 | $\sec x \tan^2 x + \sec^3 x$ | 1 |
| 3 | $\sec x \tan^3 x + 5\sec^3 x \tan x$ | 0 |
| ⋮ | ⋮ | ⋮ |

$\sec x \approx T_2(x) = 1 + \frac{1}{2}x^2$

(b) 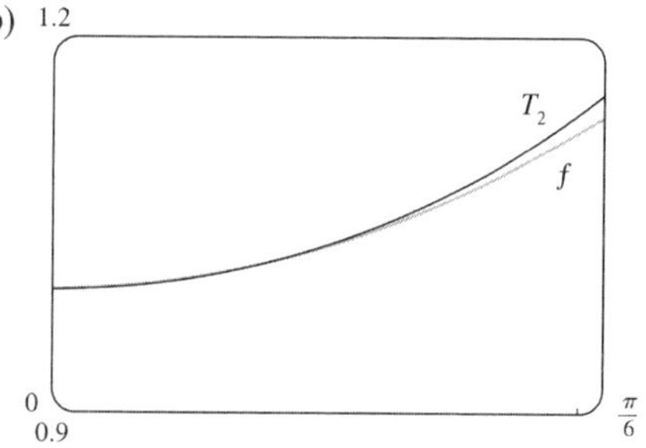


(c) $|R_2(x)| \le \frac{M}{3!}|x|^3$, where $\left|f^{(3)}(x)\right| \le M$ with $f^{(3)}(x) = \sec x \tan^3 x + 5\sec^3 x \tan x$. Now $0 \le x \le \frac{\pi}{6} \Rightarrow x^3 \le \left(\frac{\pi}{6}\right)^3$, and letting $x = \frac{\pi}{6}$ gives $M = \frac{14}{3}$, so $|R_2(x)| \le \frac{14}{3 \cdot 6}\left(\frac{\pi}{6}\right)^3 \approx 0.111648$.

(d) 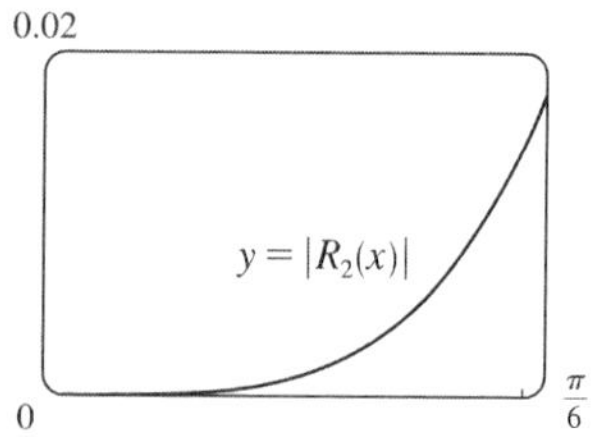


From the graph of $|R_2(x)| = |\sec x - T_2(x)|$, it appears that the error is less than 0.02 on $\left[0, \frac{\pi}{6}\right]$.

**49.** $\sin x = \sum_{n=0}^{\infty} (-1)^n \frac{x^{2n+1}}{(2n+1)!} = x - \frac{x^3}{3!} + \frac{x^5}{5!} - \frac{x^7}{7!} + \cdots$, so $\sin x - x = -\frac{x^3}{3!} + \frac{x^5}{5!} - \frac{x^7}{7!} + \cdots$ and $\frac{\sin x - x}{x^3} = -\frac{1}{3!} + \frac{x^2}{5!} - \frac{x^4}{7!} + \cdots$. Thus, $\lim_{x \to 0} \frac{\sin x - x}{x^3} = \lim_{x \to 0}\left(-\frac{1}{6} + \frac{x^2}{120} - \frac{x^4}{5040} + \cdots\right) = -\frac{1}{6}$.

**50.** (a) $F = \frac{mgR^2}{(R+h)^2} = \frac{mg}{(1+h/R)^2} = mg \sum_{n=0}^{\infty} \binom{-2}{n}\left(\frac{h}{R}\right)^n$ (Binomial Series)

(b) We expand $F = mg\left[1 - 2(h/R) + 3(h/R)^2 - \cdots\right]$. This is an alternating series, so by the Alternating Series Estimation Theorem, the error in the approximation $F = mg$ is less than $2mgh/R$, so for accuracy within 1% we want

$$\left|\frac{2mgh/R}{mgR^2/(R+h)^2}\right| < 0.01 \quad \Leftrightarrow \quad \frac{2h(R+h)^2}{R^3} < 0.01.$$

This inequality would be difficult to solve for $h$, so we substitute $R = 6{,}400$ km and plot both sides of the inequality. It appears that the approximation is accurate to within 1% for $h < 31$ km.

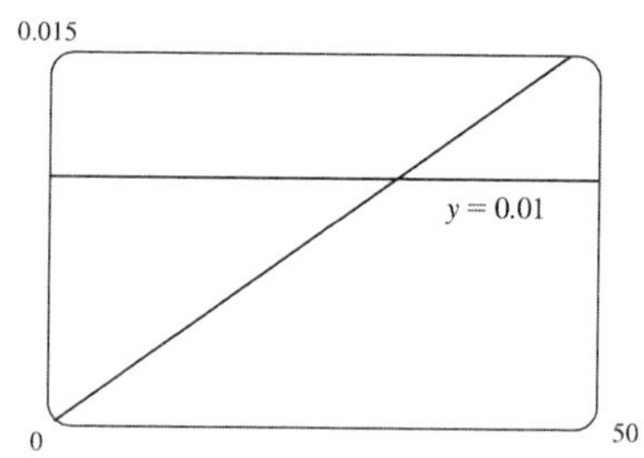


**51.** (a) From Formula 14a in Appendix C, with $x = y = \theta$, we get $\tan 2\theta = \frac{2\tan\theta}{1 - \tan^2\theta}$, so $\cot 2\theta = \frac{1 - \tan^2\theta}{2\tan\theta} \Rightarrow 2\cot 2\theta = \frac{1 - \tan^2\theta}{\tan\theta} = \cot\theta - \tan\theta$. Replacing $\theta$ by $\frac{1}{2}x$, we get $2\cot x = \cot\frac{1}{2}x - \tan\frac{1}{2}x$, or $\tan\frac{1}{2}x = \cot\frac{1}{2}x - 2\cot x$.

(b) From part (a) with $\frac{x}{2^{n-1}}$ in place of $x$, $\tan \frac{x}{2^n} = \cot \frac{x}{2^n} - 2\cot \frac{x}{2^{n-1}}$, so the $n$th partial sum of $\sum_{n=1}^{\infty} \frac{1}{2^n} \tan \frac{x}{2^n}$ is

$$\begin{aligned} s_n &= \frac{\tan(x/2)}{2} + \frac{\tan(x/4)}{4} + \frac{\tan(x/8)}{8} + \cdots + \frac{\tan(x/2^n)}{2^n} \\ &= \left[\frac{\cot(x/2)}{2} - \cot x\right] + \left[\frac{\cot(x/4)}{4} - \frac{\cot(x/2)}{2}\right] + \left[\frac{\cot(x/8)}{8} - \frac{\cot(x/4)}{4}\right] + \cdots \\ &\qquad + \left[\frac{\cot(x/2^n)}{2^n} - \frac{\cot(x/2^{n-1})}{2^{n-1}}\right] = -\cot x + \frac{\cot(x/2^n)}{2^n} \quad \text{[telescoping sum]} \end{aligned}$$

Now $\frac{\cot(x/2^n)}{2^n} = \frac{\cos(x/2^n)}{2^n \sin(x/2^n)} = \frac{\cos(x/2^n)}{x} \cdot \frac{x/2^n}{\sin(x/2^n)} \to \frac{1}{x} \cdot 1 = \frac{1}{x}$ as $n \to \infty$ since $x/2^n \to 0$ for $x \neq 0$. Therefore, if $x \neq 0$ and $x \neq k\pi$ where $k$ is any integer, then

$$\sum_{n=1}^{\infty} \frac{1}{2^n} \tan \frac{x}{2^n} = \lim_{n\to\infty} s_n = \lim_{n\to\infty}\left(-\cot x + \frac{1}{2^n}\cot\frac{x}{2^n}\right) = -\cot x + \frac{1}{x}$$

If $x = 0$, then all terms in the series are 0, so the sum is 0.

**52.** We use the problem-solving strategy of taking cases:

*Case (i):* If $|x| < 1$, then $0 \le x^2 < 1$, so $\lim_{n\to\infty} x^{2n} = 0$ (see Example 9 in Section 8.1)

and $f(x) = \lim_{n\to\infty} \frac{x^{2n} - 1}{x^{2n} + 1} = \frac{0-1}{0+1} = -1$.

*Case (ii):* If $|x| = 1$, that is, $x = \pm 1$, then $x^2 = 1$, so $f(x) = \lim_{n\to\infty} \frac{x^{2n}-1}{x^{2n}+1} = \lim_{n\to\infty} \frac{1-1}{1+1} = 0$.

*Case (iii):* If $|x| > 1$, then $x^2 > 1$, so $\lim_{n\to\infty} x^{2n} = \infty$ and

$$f(x) = \lim_{n\to\infty} \frac{x^{2n}-1}{x^{2n}+1} = \lim_{n\to\infty} \frac{1-(1/x^{2n})}{1+(1/x^{2n})} = \frac{1-0}{1+0} = 1.$$

Thus, $f(x) = \begin{cases} 1 & \text{if } x < -1 \\ 0 & \text{if } x = -1 \\ -1 & \text{if } -1 < x < 1 \\ 0 & \text{if } x = 1 \\ 1 & \text{if } x > 1 \end{cases}$

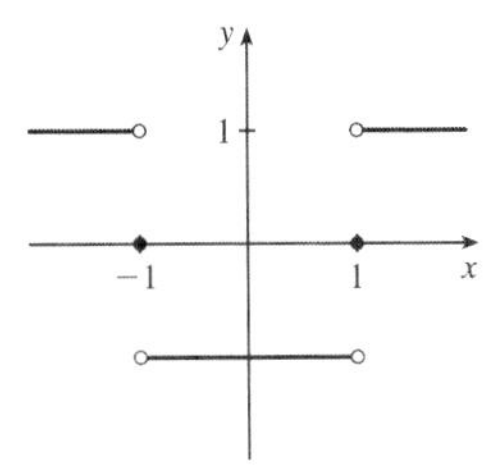


The graph shows that $f$ is continuous everywhere except at $x = \pm 1$.

# FOCUS ON PROBLEM SOLVING

**1.** It would be far too much work to compute 15 derivatives of $f$. The key idea is to remember that $f^{(n)}(0)$ occurs in the coefficient of $x^n$ in the Maclaurin series of $f$. We start with the Maclaurin series for sin: $\sin x = x - \frac{x^3}{3!} + \frac{x^5}{5!} - \cdots$. Then $\sin(x^3) = x^3 - \frac{x^9}{3!} + \frac{x^{15}}{5!} - \cdots$, and so the coefficient of $x^{15}$ is $\frac{f^{(15)}(0)}{15!} = \frac{1}{5!}$. Therefore, $f^{(15)}(0) = \frac{15!}{5!} = 6 \cdot 7 \cdot 8 \cdot 9 \cdot 10 \cdot 11 \cdot 12 \cdot 13 \cdot 14 \cdot 15 = 10{,}897{,}286{,}400$.

**2.** $|AP_2|^2 = 2$, $|AP_3|^2 = 2 + 2^2$, $|AP_4|^2 = 2 + 2^2 + \left(2^2\right)^2$, $|AP_5|^2 = 2 + 2^2 + \left(2^2\right)^2 + \left(2^3\right)^2, \ldots,$

$$|AP_n|^2 = 2 + 2^2 + \left(2^2\right)^2 + \cdots + \left(2^{n-2}\right)^2 \quad [\text{for } n \geq 3] \; = 2 + \left(4 + 4^2 + 4^3 + \cdots + 4^{n-2}\right)$$

$$= 2 + \frac{4(4^{n-2} - 1)}{4 - 1} \quad [\text{finite geometric sum with } a = 4, r = 4] \; = \frac{6}{3} + \frac{4^{n-1} - 4}{3} = \frac{2}{3} + \frac{4^{n-1}}{3}$$

So $\tan \angle P_n A P_{n+1} = \dfrac{|P_n P_{n+1}|}{|AP_n|} = \dfrac{2^{n-1}}{\sqrt{\frac{2}{3} + \frac{4^{n-1}}{3}}} = \dfrac{\sqrt{4^{n-1}}}{\sqrt{\frac{2}{3} + \frac{4^{n-1}}{3}}} = \dfrac{1}{\sqrt{\frac{2}{3 \cdot 4^{n-1}} + \frac{1}{3}}} \to \sqrt{3}$ as $n \to \infty$.

Thus, $\angle P_n A P_{n+1} \to \frac{\pi}{3}$ as $n \to \infty$.

**3.** (a) At each stage, each side is replaced by four shorter sides, each of length $\frac{1}{3}$ of the side length at the preceding stage. Writing $s_0$ and $\ell_0$ for the number of sides and the length of the side of the initial triangle, we generate the table at right. In general, we have $s_n = 3 \cdot 4^n$ and $\ell_n = \left(\frac{1}{3}\right)^n$, so the length of the perimeter at the $n$th stage of construction is $p_n = s_n \ell_n = 3 \cdot 4^n \cdot \left(\frac{1}{3}\right)^n = 3 \cdot \left(\frac{4}{3}\right)^n$.

| | |
|---|---|
| $s_0 = 3$ | $\ell_0 = 1$ |
| $s_1 = 3 \cdot 4$ | $\ell_1 = 1/3$ |
| $s_2 = 3 \cdot 4^2$ | $\ell_2 = 1/3^2$ |
| $s_3 = 3 \cdot 4^3$ | $\ell_3 = 1/3^3$ |
| $\vdots$ | $\vdots$ |

(b) $p_n = \dfrac{4^n}{3^{n-1}} = 4\left(\dfrac{4}{3}\right)^{n-1}$. Since $\frac{4}{3} > 1$, $p_n \to \infty$ as $n \to \infty$.

(c) The area of each of the small triangles added at a given stage is one-ninth of the area of the triangle added at the preceding stage. Let $a$ be the area of the original triangle. Then the area $a_n$ of each of the small triangles added at stage $n$ is $a_n = a \cdot \frac{1}{9^n} = \frac{a}{9^n}$. Since a small triangle is added to each side at every stage, it follows that the total area $A_n$ added to the figure at the $n$th stage is $A_n = s_{n-1} \cdot a_n = 3 \cdot 4^{n-1} \cdot \frac{a}{9^n} = a \cdot \frac{4^{n-1}}{3^{2n-1}}$. Then the total area enclosed by the snowflake curve is $A = a + A_1 + A_2 + A_3 + \cdots = a + a \cdot \frac{1}{3} + a \cdot \frac{4}{3^3} + a \cdot \frac{4^2}{3^5} + a \cdot \frac{4^3}{3^7} + \cdots$. After the first term, this is a geometric series with common ratio $\frac{4}{9}$, so $A = a + \frac{a/3}{1 - \frac{4}{9}} = a + \frac{a}{3} \cdot \frac{9}{5} = \frac{8a}{5}$. But the area of the original equilateral triangle with side 1 is $a = \frac{1}{2} \cdot 1 \cdot \sin \frac{\pi}{3} = \frac{\sqrt{3}}{4}$. So the area enclosed by the snowflake curve is $\frac{8}{5} \cdot \frac{\sqrt{3}}{4} = \frac{2\sqrt{3}}{5}$.

**4.** Let the series $S = 1 + \frac{1}{2} + \frac{1}{3} + \frac{1}{4} + \frac{1}{6} + \frac{1}{8} + \frac{1}{9} + \frac{1}{12} + \cdots$. Then every term in $S$ is of the form $\frac{1}{2^m 3^n}$, $m, n \geq 0$, and furthermore each term occurs only once. So we can write

$$S = \sum_{m=0}^{\infty} \sum_{n=0}^{\infty} \frac{1}{2^m 3^n} = \sum_{m=0}^{\infty} \sum_{n=0}^{\infty} \frac{1}{2^m} \frac{1}{3^n} = \sum_{m=0}^{\infty} \frac{1}{2^m} \sum_{n=0}^{\infty} \frac{1}{3^n} = \frac{1}{1 - \frac{1}{2}} \cdot \frac{1}{1 - \frac{1}{3}} = 2 \cdot \tfrac{3}{2} = 3$$

**5.** $\ln\left(1-\dfrac{1}{n^2}\right)=\ln\left(\dfrac{n^2-1}{n^2}\right)=\ln\dfrac{(n+1)(n-1)}{n^2}=\ln[(n+1)(n-1)]-\ln n^2$

$$=\ln(n+1)+\ln(n-1)-2\ln n=\ln(n-1)-\ln n-\ln n+\ln(n+1)$$

$$=\ln\frac{n-1}{n}-[\ln n-\ln(n+1)]=\ln\frac{n-1}{n}-\ln\frac{n}{n+1}.$$

Let $s_k=\sum\limits_{n=2}^{k}\ln\left(1-\dfrac{1}{n^2}\right)=\sum\limits_{n=2}^{k}\left(\ln\dfrac{n-1}{n}-\ln\dfrac{n}{n+1}\right)$ for $k\geq 2$. Then

$s_k=\left(\ln\dfrac{1}{2}-\ln\dfrac{2}{3}\right)+\left(\ln\dfrac{2}{3}-\ln\dfrac{3}{4}\right)+\cdots+\left(\ln\dfrac{k-1}{k}-\ln\dfrac{k}{k+1}\right)=\ln\dfrac{1}{2}-\ln\dfrac{k}{k+1}$, so

$\sum\limits_{n=2}^{\infty}\ln\left(1-\dfrac{1}{n^2}\right)=\lim\limits_{k\to\infty}s_k=\lim\limits_{k\to\infty}\left(\ln\dfrac{1}{2}-\ln\dfrac{k}{k+1}\right)=\ln\dfrac{1}{2}-\ln 1=\ln 1-\ln 2-\ln 1=-\ln 2.$

**6.** Place the $y$-axis as shown and let the length of each book be $L$. We want to show that the center of mass of the system of $n$ books lies above the table, that is, $\overline{x}<L$. The $x$-coordinates of the centers of mass of the books are

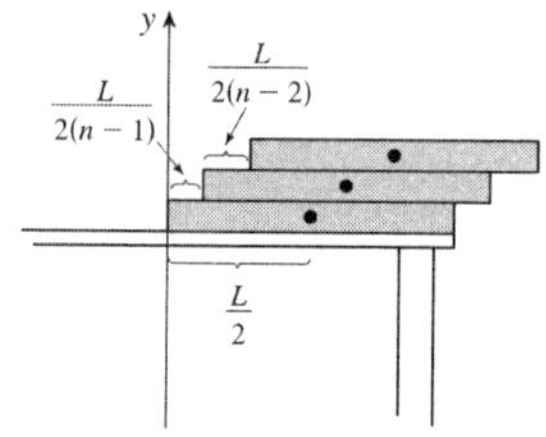


$x_1=\dfrac{L}{2}$, $x_2=\dfrac{L}{2(n-1)}+\dfrac{L}{2}$, $x_3=\dfrac{L}{2(n-1)}+\dfrac{L}{2(n-2)}+\dfrac{L}{2}$, and so on.

Each book has the same mass $m$, so if there are $n$ books, then

$$\overline{x}=\frac{mx_1+mx_2+\cdots+mx_n}{mn}=\frac{x_1+x_2+\cdots+x_n}{n}$$

$$=\frac{1}{n}\left[\frac{L}{2}+\left(\frac{L}{2(n-1)}+\frac{L}{2}\right)+\left(\frac{L}{2(n-1)}+\frac{L}{2(n-2)}+\frac{L}{2}\right)+\cdots\right.$$
$$\left.+\left(\frac{L}{2(n-1)}+\frac{L}{2(n-2)}+\cdots+\frac{L}{4}+\frac{L}{2}+\frac{L}{2}\right)\right]$$

$$=\frac{L}{n}\left[\frac{n-1}{2(n-1)}+\frac{n-2}{2(n-2)}+\cdots+\frac{2}{4}+\frac{1}{2}+\frac{n}{2}\right]=\frac{L}{n}\left[(n-1)\frac{1}{2}+\frac{n}{2}\right]=\frac{2n-1}{2n}L<L$$

This shows that, no matter how many books are added according to the given scheme, the center of mass lies above the table. It remains to observe that the series $\frac{1}{2}+\frac{1}{4}+\frac{1}{6}+\frac{1}{8}+\cdots=\frac{1}{2}\sum(1/n)$ is divergent (harmonic series), so we can make the top book extend as far as we like beyond the edge of the table if we add enough books.

**7.** $u=1+\dfrac{x^3}{3!}+\dfrac{x^6}{6!}+\dfrac{x^9}{9!}+\cdots$, $v=x+\dfrac{x^4}{4!}+\dfrac{x^7}{7!}+\dfrac{x^{10}}{10!}+\cdots$, $w=\dfrac{x^2}{2!}+\dfrac{x^5}{5!}+\dfrac{x^8}{8!}+\cdots$.

Use the Ratio Test to show that the series for $u$, $v$, and $w$ have positive radii of convergence ($\infty$ in each case), so Theorem 8.6.2 applies, and hence, we may differentiate each of these series:

$$\frac{du}{dx}=\frac{3x^2}{3!}+\frac{6x^5}{6!}+\frac{9x^8}{9!}+\cdots=\frac{x^2}{2!}+\frac{x^5}{5!}+\frac{x^8}{8!}+\cdots=w$$

Similarly, $\dfrac{dv}{dx}=1+\dfrac{x^3}{3!}+\dfrac{x^6}{6!}+\dfrac{x^9}{9!}+\cdots=u$, and $\dfrac{dw}{dx}=x+\dfrac{x^4}{4!}+\dfrac{x^7}{7!}+\dfrac{x^{10}}{10!}+\cdots=v$.

So $u'=w$, $v'=u$, and $w'=v$. Now differentiate the left hand side of the desired equation:

$$\frac{d}{dx}\left(u^3+v^3+w^3-3uvw\right)=3u^2u'+3v^2v'+3w^2w'-3(u'vw+uv'w+uvw')$$

$$=3u^2w+3v^2u+3w^2v-3(vw^2+u^2w+uv^2)=0\quad\Rightarrow$$

$u^3+v^3+w^3-3uvw=C$. To find the value of the constant $C$, we put $x=0$ in the last equation and get $1^3+0^3+0^3-3(1\cdot 0\cdot 0)=C\quad\Rightarrow\quad C=1$, so $u^3+v^3+w^3-3uvw=1$.

**8.** First notice that both series are absolutely convergent ($p$-series with $p > 1$.) Let the given expression be called $x$. Then

$$x = \frac{1 + \frac{1}{2^p} + \frac{1}{3^p} + \frac{1}{4^p} + \cdots}{1 - \frac{1}{2^p} + \frac{1}{3^p} - \frac{1}{4^p} + \cdots} = \frac{1 + \left(2 \cdot \frac{1}{2^p} - \frac{1}{2^p}\right) + \frac{1}{3^p} + \left(2 \cdot \frac{1}{4^p} - \frac{1}{4^p}\right) + \cdots}{1 - \frac{1}{2^p} + \frac{1}{3^p} - \frac{1}{4^p} + \cdots}$$

$$= \frac{\left(1 - \frac{1}{2^p} + \frac{1}{3^p} - \frac{1}{4^p} + \cdots\right) + \left(2 \cdot \frac{1}{2^p} + 2 \cdot \frac{1}{4^p} + 2 \cdot \frac{1}{6^p} + \cdots\right)}{1 - \frac{1}{2^p} + \frac{1}{3^p} - \frac{1}{4^p} + \cdots}$$

$$= 1 + \frac{2\left(\frac{1}{2^p} + \frac{1}{4^p} + \frac{1}{6^p} + \frac{1}{8^p} + \cdots\right)}{1 - \frac{1}{2^p} + \frac{1}{3^p} - \frac{1}{4^p} + \cdots} = 1 + \frac{\frac{1}{2^{p-1}}\left(1 + \frac{1}{2^p} + \frac{1}{3^p} + \frac{1}{4^p} + \cdots\right)}{1 - \frac{1}{2^p} + \frac{1}{3^p} - \frac{1}{4^p} + \cdots} = 1 + 2^{1-p}x$$

Therefore, $x = 1 + 2^{1-p}x \quad \Leftrightarrow \quad x - 2^{1-p}x = 1 \quad \Leftrightarrow \quad x(1 - 2^{1-p}) = 1 \quad \Leftrightarrow \quad x = \dfrac{1}{1 - 2^{1-p}}$.

**9.** If $L$ is the length of a side of the equilateral triangle, then the area is $A = \frac{1}{2}L \cdot \frac{\sqrt{3}}{2}L = \frac{\sqrt{3}}{4}L^2$ and so $L^2 = \frac{4}{\sqrt{3}}A$.

Let $r$ be the radius of one of the circles. When there are $n$ rows of circles, the figure shows that

$L = \sqrt{3}r + r + (n-2)(2r) + r + \sqrt{3}\,r = r\left(2n - 2 + 2\sqrt{3}\right)$, so $r = \dfrac{L}{2\left(n + \sqrt{3} - 1\right)}$.

The number of circles is $1 + 2 + \cdots + n = \dfrac{n(n+1)}{2}$, and so the total area of the circles is

$$A_n = \frac{n(n+1)}{2}\pi r^2 = \frac{n(n+1)}{2}\pi \frac{L^2}{4\left(n + \sqrt{3} - 1\right)^2} = \frac{n(n+1)}{2}\pi \frac{4A/\sqrt{3}}{4\left(n + \sqrt{3} - 1\right)^2}$$

$$= \frac{n(n+1)}{\left(n + \sqrt{3} - 1\right)^2} \frac{\pi A}{2\sqrt{3}} \quad \Rightarrow$$

$$\frac{A_n}{A} = \frac{n(n+1)}{\left(n + \sqrt{3} - 1\right)^2} \frac{\pi}{2\sqrt{3}}$$

$$= \frac{1 + 1/n}{\left[1 + \left(\sqrt{3} - 1\right)/n\right]^2} \frac{\pi}{2\sqrt{3}} \to \frac{\pi}{2\sqrt{3}} \text{ as } n \to \infty$$

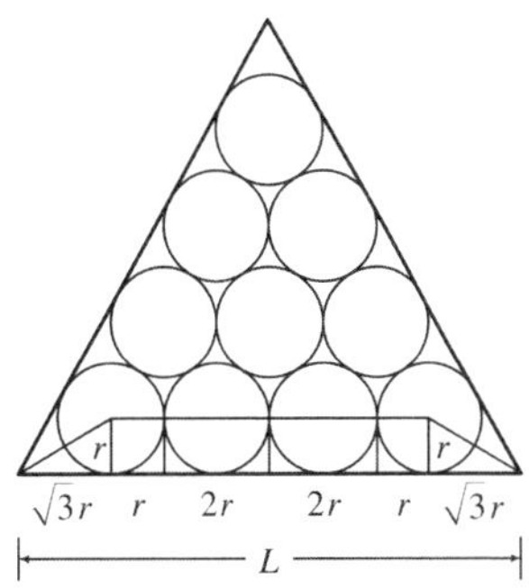


**10.** Given $a_0 = a_1 = 1$ and $a_n = \dfrac{(n-1)(n-2)a_{n-1} - (n-3)a_{n-2}}{n(n-1)}$, we calculate the next few terms of the sequence:

$$a_2 = \frac{1 \cdot 0 \cdot a_1 - (-1)a_0}{2 \cdot 1} = \frac{1}{2},\ a_3 = \frac{2 \cdot 1 \cdot a_2 - 0 \cdot a_1}{3 \cdot 2} = \frac{1}{6},\ a_4 = \frac{3 \cdot 2 \cdot a_3 - 1 \cdot a_2}{4 \cdot 3} = \frac{1}{24}.$$

It seems that $a_n = \dfrac{1}{n!}$, so we try to prove this by induction. The first step is done, so assume $a_k = \dfrac{1}{k!}$ and $a_{k-1} = \dfrac{1}{(k-1)!}$.

Then $a_{k+1} = \dfrac{k(k-1)a_k - (k-2)a_{k-1}}{(k+1)k} = \dfrac{\frac{k(k-1)}{k!} - \frac{k-2}{(k-1)!}}{(k+1)k} = \dfrac{(k-1) - (k-2)}{[(k+1)(k)]\,(k-1)!} = \dfrac{1}{(k+1)!}$ and the induction is complete. Therefore, $\sum\limits_{n=0}^{\infty} a_n = \sum\limits_{n=0}^{\infty} \dfrac{1}{n!} = e$.

**11.** As in Section 8.6 we have to integrate the function $x^x$ by integrating series. Writing $x^x = (e^{\ln x})^x = e^{x \ln x}$ and using the Maclaurin series for $e^x$, we have $x^x = (e^{\ln x})^x = e^{x \ln x} = \sum_{n=0}^{\infty} \frac{(x \ln x)^n}{n!} = \sum_{n=0}^{\infty} \frac{x^n (\ln x)^n}{n!}$. As with power series, we can integrate this series term-by-term: $\int_0^1 x^x\,dx = \sum_{n=0}^{\infty} \int_0^1 \frac{x^n (\ln x)^n}{n!}\,dx = \sum_{n=0}^{\infty} \frac{1}{n!} \int_0^1 x^n (\ln x)^n\,dx$. We integrate by parts with $u = (\ln x)^n$, $dv = x^n\,dx$, so $du = \frac{n(\ln x)^{n-1}}{x}\,dx$ and $v = \frac{x^{n+1}}{n+1}$:

$$\int_0^1 x^n (\ln x)^n\,dx = \lim_{t\to 0^+} \int_t^1 x^n (\ln x)^n\,dx = \lim_{t\to 0^+} \left[\frac{x^{n+1}}{n+1}(\ln x)^n\right]_t^1 - \lim_{t\to 0^+} \int_t^1 \frac{n}{n+1} x^n (\ln x)^{n-1}\,dx$$

$$= 0 - \frac{n}{n+1} \int_0^1 x^n (\ln x)^{n-1}\,dx$$

(where l'Hospital's Rule was used to help evaluate the first limit). Further integration by parts gives $\int_0^1 x^n (\ln x)^k\,dx = -\frac{k}{n+1} \int_0^1 x^n (\ln x)^{k-1}\,dx$ and, combining these steps, we get

$$\int_0^1 x^n (\ln x)^n\,dx = \frac{(-1)^n\, n!}{(n+1)^n} \int_0^1 x^n\,dx = \frac{(-1)^n\, n!}{(n+1)^{n+1}} \quad \Rightarrow$$

$$\int_0^1 x^x\,dx = \sum_{n=0}^{\infty} \frac{1}{n!} \int_0^1 x^n (\ln x)^n\,dx = \sum_{n=0}^{\infty} \frac{1}{n!} \frac{(-1)^n\, n!}{(n+1)^{n+1}} = \sum_{n=0}^{\infty} \frac{(-1)^n}{(n+1)^{n+1}} = \sum_{n=1}^{\infty} \frac{(-1)^{n-1}}{n^n}.$$

**12.** (a) Since $P_n$ is defined as the midpoint of $P_{n-4}P_{n-3}$, $x_n = \frac{1}{2}(x_{n-4} + x_{n-3})$ for $n \geq 5$. So we prove by induction that $\frac{1}{2}x_n + x_{n+1} + x_{n+2} + x_{n+3} = 2$. The case $n = 1$ is immediate, since $\frac{1}{2} \cdot 0 + 1 + 1 + 0 = 2$. Assume that the result holds for $n = k - 1$, that is, $\frac{1}{2}x_{k-1} + x_k + x_{k+1} + x_{k+2} = 2$. Then for $n = k$,

$$\tfrac{1}{2}x_k + x_{k+1} + x_{k+2} + x_{k+3} = \tfrac{1}{2}x_k + x_{k+1} + x_{k+2} + \tfrac{1}{2}(x_{k+3-4} + x_{k+3-3}) \text{ (by above)}$$

$$= \tfrac{1}{2}x_{k-1} + x_k + x_{k+1} + x_{k+2} = 2 \text{ (by the induction hypothesis)}$$

Similarly, for $n \geq 5$, $y_n = \frac{1}{2}(y_{n-4} + y_{n-3})$, so the same argument as above holds for $y$, with 2 replaced by $\frac{1}{2}y_1 + y_2 + y_3 + y_4 = \frac{1}{2} \cdot 1 + 1 + 0 + 0 = \frac{3}{2}$. So $\frac{1}{2}y_n + y_{n+1} + y_{n+2} + y_{n+3} = \frac{3}{2}$ for all $n$.

(b) $\lim_{n\to\infty} \left(\frac{1}{2}x_n + x_{n+1} + x_{n+2} + x_{n+3}\right) = \frac{1}{2} \lim_{n\to\infty} x_n + \lim_{n\to\infty} x_{n+1} + \lim_{n\to\infty} x_{n+2} + \lim_{n\to\infty} x_{n+3} = 2$. Since all the limits on the left hand side are the same, we get $\frac{7}{2} \lim_{n\to\infty} x_n = 2 \Rightarrow \lim_{n\to\infty} x_n = \frac{4}{7}$. In the same way, $\frac{7}{2} \lim_{n\to\infty} y_n = \frac{3}{2} \Rightarrow \lim_{n\to\infty} y_n = \frac{3}{7}$, so $P = \left(\frac{4}{7}, \frac{3}{7}\right)$.

**13.** Let $f(x) = \sum_{m=0}^{\infty} c_m x^m$ and $g(x) = e^{f(x)} = \sum_{n=0}^{\infty} d_n x^n$. Then $g'(x) = \sum_{n=0}^{\infty} n d_n x^{n-1}$, so $n d_n$ occurs as the coefficient of $x^{n-1}$. But also

$$g'(x) = e^{f(x)} f'(x) = \left(\sum_{n=0}^{\infty} d_n x^n\right)\left(\sum_{m=1}^{\infty} m c_m x^{m-1}\right)$$

$$= \left(d_0 + d_1 x + d_2 x^2 + \cdots + d_{n-1} x^{n-1} + \cdots\right)\left(c_1 + 2c_2 x + 3c_3 x^2 + \cdots + n c_n x^{n-1} + \cdots\right)$$

so the coefficient of $x^{n-1}$ is $c_1 d_{n-1} + 2c_2 d_{n-2} + 3c_3 d_{n-3} + \cdots + n c_n d_0 = \sum_{i=1}^{n} i c_i d_{n-i}$. Therefore, $n d_n = \sum_{i=1}^{n} i c_i d_{n-i}$.

**14.** By repeated use of the Pythagorean Theorem, we find that the base of the second right triangle has length $\sqrt{1^2+1^2}=\sqrt{2}$, the base of the third right triangle has length $\sqrt{\left(\sqrt{2}\right)^2+1^2}=\sqrt{3}$, and in general, the $n$th right triangle has base of length $\sqrt{\left(\sqrt{n-1}\right)^2+1^2}=\sqrt{n}$. Thus, $\theta_n=\tan^{-1}\dfrac{1}{\sqrt{n}}$ and $\sum\limits_{n=1}^{\infty}\theta_n=\sum\limits_{n=1}^{\infty}\tan^{-1}\dfrac{1}{\sqrt{n}}$. To show that the last series diverges, we use the Limit Comparison Test with the divergent $p$-series $\sum\limits_{n=1}^{\infty}\dfrac{1}{\sqrt{n}}$ $\left(p=\frac{1}{2}\leq 1\right)$.

$$\lim_{n\to\infty}\frac{a_n}{b_n}=\lim_{n\to\infty}\frac{\tan^{-1}(1/\sqrt{n})}{1/\sqrt{n}}\overset{\text{H}}{=}\lim_{n\to\infty}\frac{\dfrac{1}{1+(1/\sqrt{n})^2}\cdot\left(-\dfrac{1}{2n^{3/2}}\right)}{-\dfrac{1}{2n^{3/2}}}=\lim_{n\to\infty}\frac{1}{1+(1/n)}=1>0.$$ Thus,

$\sum\limits_{n=1}^{\infty}\tan^{-1}\dfrac{1}{\sqrt{n}}=\sum\limits_{n=1}^{\infty}\theta_n$ is a divergent series. It follows that there are infinitely many turns around $P$.

**15.** Call the series $S$. We group the terms according to the number of digits in their denominators:

$$S=\underbrace{\left(\tfrac{1}{1}+\tfrac{1}{2}+\cdots+\tfrac{1}{8}+\tfrac{1}{9}\right)}_{g_1}+\underbrace{\left(\tfrac{1}{11}+\cdots+\tfrac{1}{99}\right)}_{g_2}+\underbrace{\left(\tfrac{1}{111}+\cdots+\tfrac{1}{999}\right)}_{g_3}+\cdots$$

Now in the group $g_n$, since we have 9 choices for each of the $n$ digits in the denominator, there are $9^n$ terms. Furthermore, each term in $g_n$ is less than $\frac{1}{10^{n-1}}$ [except for the first term in $g_1$]. So $g_n<9^n\cdot\frac{1}{10^{n-1}}=9\left(\frac{9}{10}\right)^{n-1}$. Now $\sum_{n=1}^{\infty}9\left(\frac{9}{10}\right)^{n-1}$ is a geometric series with $a=9$ and $r=\frac{9}{10}<1$. Therefore, by the Comparison Test, $S=\sum_{n=1}^{\infty}g_n<\sum_{n=1}^{\infty}9\left(\frac{9}{10}\right)^{n-1}=\frac{9}{1-9/10}=90$.

**16.** (a) Let $f(x)=\dfrac{x}{1-x-x^2}=\sum\limits_{n=0}^{\infty}c_nx^n=c_0+c_1x+c_2x^2+c_3x^3+\cdots$. Then

$$x=\left(1-x-x^2\right)\left(c_0+c_1x+c_2x^2+c_3x^3+\cdots\right)$$

$$\begin{aligned}x=c_0+c_1x&+c_2x^2+c_3x^3+c_4x^4+c_5x^5+\cdots\\&-c_0x-c_1x^2-c_2x^3-c_3x^4-c_4x^5-\cdots\\&\qquad-c_0x^2-c_1x^3-c_2x^4-c_3x^5-\cdots\end{aligned}$$

$$x=c_0+(c_1-c_0)x+(c_2-c_1-c_0)x^2+(c_3-c_2-c_1)x^3+\cdots$$

Comparing coefficients of powers of $x$ gives us $c_0=0$ and

$$c_1-c_0=1\quad\Rightarrow\quad c_1=c_0+1=1$$

$$c_2-c_1-c_0=0\quad\Rightarrow\quad c_2=c_1+c_0=1+0=1$$

$$c_3-c_2-c_1=0\quad\Rightarrow\quad c_3=c_2+c_1=1+1=2$$

In general, we have $c_n=c_{n-1}+c_{n-2}$ for $n\geq 3$. Each $c_n$ is equal to the $n$th Fibonacci number; that is,

$$\sum_{n=0}^{\infty}c_nx^n=\sum_{n=1}^{\infty}c_nx^n=\sum_{n=1}^{\infty}f_nx^n.$$

(b) Completing the square on $x^2+x-1$ gives us

$$\left(x^2+x+\frac{1}{4}\right)-1-\frac{1}{4}=\left(x+\frac{1}{2}\right)^2-\frac{5}{4}=\left(x+\frac{1}{2}\right)^2-\left(\frac{\sqrt{5}}{2}\right)^2$$
$$=\left(x+\frac{1}{2}+\frac{\sqrt{5}}{2}\right)\left(x+\frac{1}{2}-\frac{\sqrt{5}}{2}\right)=\left(x+\frac{1+\sqrt{5}}{2}\right)\left(x+\frac{1-\sqrt{5}}{2}\right)$$

So $\dfrac{x}{1-x-x^2}=\dfrac{-x}{x^2+x-1}=\dfrac{-x}{\left(x+\frac{1+\sqrt{5}}{2}\right)\left(x+\frac{1-\sqrt{5}}{2}\right)}$. The factors in the denominator are linear, so the partial fraction decomposition is

$$\frac{-x}{\left(x+\frac{1+\sqrt{5}}{2}\right)\left(x+\frac{1-\sqrt{5}}{2}\right)}=\frac{A}{x+\frac{1+\sqrt{5}}{2}}+\frac{B}{x+\frac{1-\sqrt{5}}{2}}\quad -x=A\left(x+\tfrac{1-\sqrt{5}}{2}\right)+B\left(x+\tfrac{1+\sqrt{5}}{2}\right)$$

If $x=\frac{-1+\sqrt{5}}{2}$, then $-\frac{-1+\sqrt{5}}{2}=B\sqrt{5}\;\Rightarrow\;B=\frac{1-\sqrt{5}}{2\sqrt{5}}$.

If $x=\frac{-1-\sqrt{5}}{2}$, then $-\frac{-1-\sqrt{5}}{2}=A\left(-\sqrt{5}\right)\;\Rightarrow\;A=\frac{1+\sqrt{5}}{-2\sqrt{5}}$. Thus,

$$\frac{x}{1-x-x^2}=\frac{\frac{1+\sqrt{5}}{-2\sqrt{5}}}{x+\frac{1+\sqrt{5}}{2}}+\frac{\frac{1-\sqrt{5}}{2\sqrt{5}}}{x+\frac{1-\sqrt{5}}{2}}=\frac{\frac{1+\sqrt{5}}{-2\sqrt{5}}}{x+\frac{1+\sqrt{5}}{2}}\cdot\frac{\frac{2}{1+\sqrt{5}}}{\frac{2}{1+\sqrt{5}}}+\frac{\frac{1-\sqrt{5}}{2\sqrt{5}}}{x+\frac{1-\sqrt{5}}{2}}\cdot\frac{\frac{2}{1-\sqrt{5}}}{\frac{2}{1-\sqrt{5}}}$$
$$=\frac{-1/\sqrt{5}}{1+\frac{2}{1+\sqrt{5}}x}+\frac{1/\sqrt{5}}{1+\frac{2}{1-\sqrt{5}}x}=-\frac{1}{\sqrt{5}}\sum_{n=0}^{\infty}\left(-\frac{2}{1+\sqrt{5}}x\right)^n+\frac{1}{\sqrt{5}}\sum_{n=0}^{\infty}\left(-\frac{2}{1-\sqrt{5}}x\right)^n$$
$$=\frac{1}{\sqrt{5}}\sum_{n=0}^{\infty}\left[\left(\frac{-2}{1-\sqrt{5}}\right)^n-\left(\frac{-2}{1+\sqrt{5}}\right)^n\right]x^n$$
$$=\frac{1}{\sqrt{5}}\sum_{n=1}^{\infty}\left[\frac{(-2)^n\left(1+\sqrt{5}\right)^n-(-2)^n\left(1-\sqrt{5}\right)^n}{\left(1-\sqrt{5}\right)^n\left(1+\sqrt{5}\right)^n}\right]x^n\qquad[\text{the } n=0 \text{ term is } 0]$$
$$=\frac{1}{\sqrt{5}}\sum_{n=1}^{\infty}\left[\frac{(-2)^n\left(\left(1+\sqrt{5}\right)^n-\left(1-\sqrt{5}\right)^n\right)}{(1-5)^n}\right]x^n$$
$$=\frac{1}{\sqrt{5}}\sum_{n=1}^{\infty}\left[\frac{\left(1+\sqrt{5}\right)^n-\left(1-\sqrt{5}\right)^n}{2^n}\right]x^n\qquad[(-4)^n=(-2)^n\cdot 2^n]$$

From part (a), this series must equal $\sum\limits_{n=1}^{\infty} f_n x^n$, so $f_n=\dfrac{\left(1+\sqrt{5}\right)^n-\left(1-\sqrt{5}\right)^n}{2^n\sqrt{5}}$, which is an explicit formula for the $n$th Fibonacci number.

# APPENDIXES

## A INTERVALS, INEQUALITIES, AND ABSOLUTE VALUES

**1.** $|5-23| = |-18| = 18$

**2.** $|\pi - 2| = \pi - 2$ because $\pi - 2 > 0$.

**3.** $\left|\sqrt{5}-5\right| = -\left(\sqrt{5}-5\right) = 5-\sqrt{5}$ because $\sqrt{5}-5<0$.

**4.** $\left||-2|-|-3|\right| = |2-3| = |-1| = 1$

**5.** If $x<2$, $x-2<0$, so $|x-2| = -(x-2) = 2-x$.

**6.** If $x>2$, $x-2>0$, so $|x-2| = x-2$.

**7.** $|x+1| = \begin{cases} x+1 & \text{if } x+1\geq 0 \\ -(x+1) & \text{if } x+1<0 \end{cases} = \begin{cases} x+1 & \text{if } x\geq -1 \\ -x-1 & \text{if } x<-1 \end{cases}$

**8.** $|2x-1| = \begin{cases} 2x-1 & \text{if } 2x-1\geq 0 \\ -(2x-1) & \text{if } 2x-1<0 \end{cases} = \begin{cases} 2x-1 & \text{if } x\geq \frac{1}{2} \\ 1-2x & \text{if } x<\frac{1}{2} \end{cases}$

**9.** $\left|x^2+1\right| = x^2+1$ (since $x^2+1\geq 0$ for all $x$).

**10.** Determine when $1-2x^2<0 \;\Leftrightarrow\; 1<2x^2 \;\Leftrightarrow\; x^2>\frac{1}{2} \;\Leftrightarrow\; \sqrt{x^2}>\sqrt{\frac{1}{2}} \;\Leftrightarrow\; |x|>\sqrt{\frac{1}{2}} \;\Leftrightarrow$

$x<-\frac{1}{\sqrt{2}}$ or $x>\frac{1}{\sqrt{2}}$. Thus, $\left|1-2x^2\right| = \begin{cases} 1-2x^2 & \text{if } -\frac{1}{\sqrt{2}}\leq x\leq \frac{1}{\sqrt{2}} \\ 2x^2-1 & \text{if } x<-\frac{1}{\sqrt{2}} \text{ or } x>\frac{1}{\sqrt{2}} \end{cases}$

**11.** $2x+7>3 \;\Leftrightarrow\; 2x>-4 \;\Leftrightarrow\; x>-2$, so $x\in(-2,\infty)$.

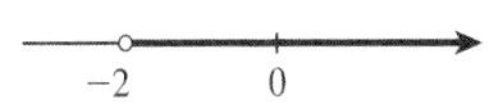


**12.** $4-3x\geq 6 \;\Leftrightarrow\; -3x\geq 2 \;\Leftrightarrow\; x\leq -\frac{2}{3}$, so $x\in\left(-\infty,-\frac{2}{3}\right]$.

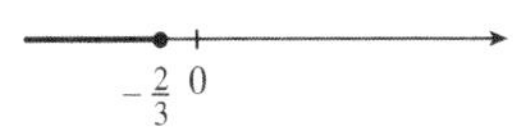


**13.** $1-x\leq 2 \;\Leftrightarrow\; -x\leq 1 \;\Leftrightarrow\; x\geq -1$, so $x\in[-1,\infty)$.

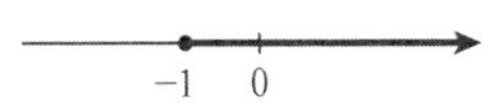


**14.** $1+5x>5-3x \;\Leftrightarrow\; 8x>4 \;\Leftrightarrow\; x>\frac{1}{2}$, so $x\in\left(\frac{1}{2},\infty\right)$.

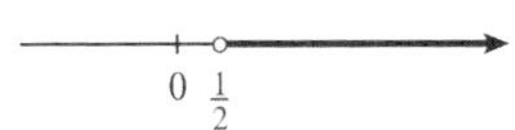


**15.** $0\leq 1-x<1 \;\Leftrightarrow\; -1\leq -x<0 \;\Leftrightarrow\; 1\geq x>0$, so $x\in(0,1]$.

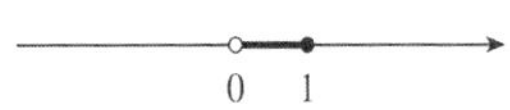


**16.** $1<3x+4\leq 16 \;\Leftrightarrow\; -3<3x\leq 12 \;\Leftrightarrow\; -1<x\leq 4$, so $x\in(-1,4]$.

**17.** $(x-1)(x-2)>0$. *Case 1:* (both factors are positive, so their product is positive)

$x-1>0 \;\Leftrightarrow\; x>1$, and $x-2>0 \;\Leftrightarrow\; x>2$, so $x\in(2,\infty)$.

*Case 2:* (both factors are negative, so their product is positive)

$x-1<0 \;\Leftrightarrow\; x<1$, and $x-2<0 \;\Leftrightarrow\; x<2$, so $x\in(-\infty,1)$.

Thus, the solution set is $(-\infty,1)\cup(2,\infty)$.

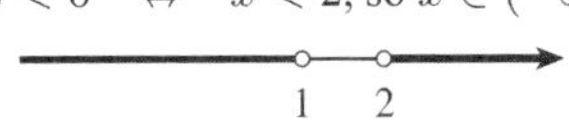

**18.** $x^2 < 2x + 8 \;\Leftrightarrow\; x^2 - 2x - 8 < 0 \;\Leftrightarrow\; (x-4)(x+2) < 0$. *Case 1:* $x > 4$ and $x < -2$, which is impossible. *Case 2:* $x < 4$ and $x > -2$. Thus, the solution set is $(-2, 4)$.

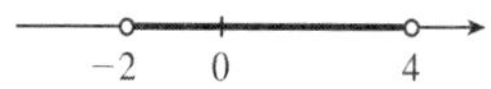


**19.** $x^2 < 3 \;\Leftrightarrow\; x^2 - 3 < 0 \;\Leftrightarrow\; \left(x - \sqrt{3}\right)\left(x + \sqrt{3}\right) < 0$. *Case 1:* $x > \sqrt{3}$ and $x < -\sqrt{3}$, which is impossible. *Case 2:* $x < \sqrt{3}$ and $x > -\sqrt{3}$. Thus, the solution set is $\left(-\sqrt{3}, \sqrt{3}\right)$.
*Another method:* $x^2 < 3 \;\Leftrightarrow\; |x| < \sqrt{3} \;\Leftrightarrow\; -\sqrt{3} < x < \sqrt{3}$.

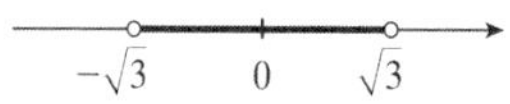


**20.** $x^2 \geq 5 \;\Leftrightarrow\; x^2 - 5 \geq 0 \;\Leftrightarrow\; \left(x - \sqrt{5}\right)\left(x + \sqrt{5}\right) \geq 0$. *Case 1:* $x \geq \sqrt{5}$ and $x \geq -\sqrt{5}$, so $x \in \left[\sqrt{5}, \infty\right)$. *Case 2:* $x \leq \sqrt{5}$ and $x \leq -\sqrt{5}$, so $x \in \left(-\infty, -\sqrt{5}\right]$. Thus, the solution set is $\left(-\infty, -\sqrt{5}\right] \cup \left[\sqrt{5}, \infty\right)$.
*Another method:* $x^2 \geq 5 \;\Leftrightarrow\; |x| \geq \sqrt{5} \;\Leftrightarrow\; x \geq \sqrt{5}$ or $x \leq -\sqrt{5}$.

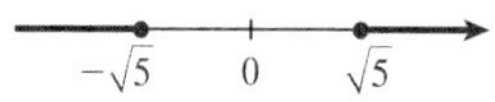


**21.** $x^3 - x^2 \leq 0 \;\Leftrightarrow\; x^2(x-1) \leq 0$. Since $x^2 \geq 0$ for all $x$, the inequality is satisfied when $x - 1 \leq 0 \;\Leftrightarrow\; x \leq 1$. Thus, the solution set is $(-\infty, 1]$.

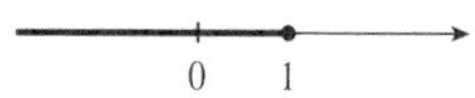


**22.** $(x+1)(x-2)(x+3) = 0 \;\Leftrightarrow\; x = -1, 2,$ or $-3$. Construct a chart:

| Interval | $x+1$ | $x-2$ | $x+3$ | $(x+1)(x-2)(x+3)$ |
|---|---|---|---|---|
| $x < -3$ | $-$ | $-$ | $-$ | $-$ |
| $-3 < x < -1$ | $-$ | $-$ | $+$ | $+$ |
| $-1 < x < 2$ | $+$ | $-$ | $+$ | $-$ |
| $x > 2$ | $+$ | $+$ | $+$ | $+$ |

Thus, $(x+1)(x-2)(x+3) \geq 0$ on $[-3, -1]$ and $[2, \infty)$, and the solution set is $[-3, -1] \cup [2, \infty)$.

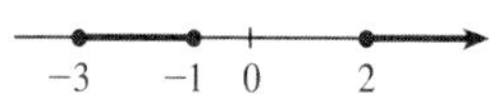


**23.** $x^3 > x \;\Leftrightarrow\; x^3 - x > 0 \;\Leftrightarrow\; x\left(x^2 - 1\right) > 0 \;\Leftrightarrow\; x(x-1)(x+1) > 0$. Construct a chart:

| Interval | $x$ | $x-1$ | $x+1$ | $x(x-1)(x+1)$ |
|---|---|---|---|---|
| $x < -1$ | $-$ | $-$ | $-$ | $-$ |
| $-1 < x < 0$ | $-$ | $-$ | $+$ | $+$ |
| $0 < x < 1$ | $+$ | $-$ | $+$ | $-$ |
| $x > 1$ | $+$ | $+$ | $+$ | $+$ |

Since $x^3 > x$ when the last column is positive, the solution set is $(-1, 0) \cup (1, \infty)$.

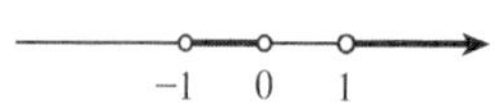

**24.** $x^3 + 3x < 4x^2 \quad\Leftrightarrow\quad x^3 - 4x^2 + 3x < 0 \quad\Leftrightarrow\quad x(x^2 - 4x + 3) < 0 \quad\Leftrightarrow\quad x(x-1)(x-3) < 0.$

| Interval | $x$ | $x-1$ | $x-3$ | $x(x-1)(x-3)$ |
|---|---|---|---|---|
| $x < 0$ | $-$ | $-$ | $-$ | $-$ |
| $0 < x < 1$ | $+$ | $-$ | $-$ | $+$ |
| $1 < x < 3$ | $+$ | $+$ | $-$ | $-$ |
| $x > 3$ | $+$ | $+$ | $+$ | $+$ |

Thus, the solution set is $(-\infty, 0) \cup (1, 3)$.

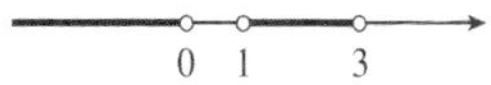


**25.** $1/x < 4$. This is clearly true for $x < 0$. So suppose $x > 0$. then $1/x < 4 \quad\Leftrightarrow\quad 1 < 4x \quad\Leftrightarrow\quad \frac{1}{4} < x$. Thus, the solution set is $(-\infty, 0) \cup \left(\frac{1}{4}, \infty\right)$.

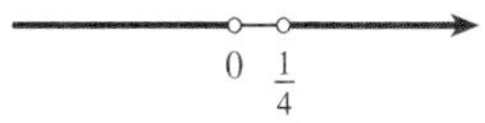


**26.** $-3 < 1/x \leq 1$. We solve the two inequalities separately and take the intersection of the solution sets. First, $-3 < 1/x$ is clearly true for $x > 0$. So suppose $x < 0$. Then $-3 < 1/x \quad\Leftrightarrow\quad -3x > 1 \quad\Leftrightarrow\quad x < -\frac{1}{3}$, so for this inequality, the solution set is $\left(-\infty, -\frac{1}{3}\right) \cup (0, \infty)$. Now $1/x \leq 1$ is clearly true if $x < 0$. So suppose $x > 0$. Then $1/x \leq 1 \quad\Leftrightarrow\quad 1 \leq x$, and the solution set here is $(-\infty, 0) \cup [1, \infty)$. Taking the intersection of the two solution sets gives the final solution set: $\left(-\infty, -\frac{1}{3}\right) \cup [1, \infty)$.

**27.** $C = \frac{5}{9}(F - 32) \quad\Rightarrow\quad F = \frac{9}{5}C + 32$. So $50 \leq F \leq 95 \quad\Rightarrow\quad 50 \leq \frac{9}{5}C + 32 \leq 95 \quad\Rightarrow\quad 18 \leq \frac{9}{5}C \leq 63 \quad\Rightarrow$ $10 \leq C \leq 35$. So the interval is $[10, 35]$.

**28.** Since $20 \leq C \leq 30$ and $C = \frac{5}{9}(F - 32)$, we have $20 \leq \frac{5}{9}(F - 32) \leq 30 \quad\Rightarrow\quad 36 \leq F - 32 \leq 54 \quad\Rightarrow\quad 68 \leq F \leq 86$. So the interval is $[68, 86]$.

**29.** (a) Let $T$ represent the temperature in degrees Celsius and $h$ the height in km. $T = 20$ when $h = 0$ and $T$ decreases by $10^\circ$C for every km ($1^\circ$C for each 100-m rise). Thus, $T = 20 - 10h$ when $0 \leq h \leq 12$.

(b) From part (a), $T = 20 - 10h \quad\Rightarrow\quad 10h = 20 - T \quad\Rightarrow\quad h = 2 - T/10$. So $0 \leq h \leq 5 \quad\Rightarrow\quad 0 \leq 2 - T/10 \leq 5 \quad\Rightarrow$ $-2 \leq -T/10 \leq 3 \quad\Rightarrow\quad -20 \leq -T \leq 30 \quad\Rightarrow\quad 20 \geq T \geq -30 \quad\Rightarrow\quad -30 \leq T \leq 20$. Thus, the range of temperatures (in $^\circ$C) to be expected is $[-30, 20]$.

**30.** The ball will be at least 32 ft above the ground if $h \geq 32 \quad\Leftrightarrow\quad 128 + 16t - 16t^2 \geq 32 \quad\Leftrightarrow\quad 16t^2 - 16t - 96 \leq 0 \quad\Leftrightarrow$ $16(t - 3)(t + 2) \leq 0$. $t = 3$ and $t = -2$ are endpoints of the interval we're looking for, and constructing a table gives $-2 \leq t \leq 3$. But $t \geq 0$, so the ball will be at least 32 ft above the ground in the time interval $[0, 3]$.

**31.** $|x + 3| = |2x + 1| \quad\Leftrightarrow\quad$ either $x + 3 = 2x + 1$ or $x + 3 = -(2x + 1)$. In the first case, $x = 2$, and in the second case, $x + 3 = -2x - 1 \quad\Leftrightarrow\quad 3x = -4 \quad\Leftrightarrow\quad x = -\frac{4}{3}$. So the solutions are $-\frac{4}{3}$ and 2.

**32.** $|3x + 5| = 1 \quad\Leftrightarrow\quad$ either $3x + 5 = 1$ or $-1$. In the first case, $3x = -4 \quad\Leftrightarrow\quad x = -\frac{4}{3}$, and in the second case, $3x = -6 \quad\Leftrightarrow\quad x = -2$. So the solutions are $-2$ and $-\frac{4}{3}$.

**33.** By Property 5 of absolute values, $|x| < 3 \quad\Leftrightarrow\quad -3 < x < 3$, so $x \in (-3, 3)$.

**34.** By Properties 4 and 6 of absolute values, $|x| \geq 3 \quad\Leftrightarrow\quad x \leq -3$ or $x \geq 3$, so $x \in (-\infty, -3] \cup [3, \infty)$.

**35.** $|x - 4| < 1 \quad\Leftrightarrow\quad -1 < x - 4 < 1 \quad\Leftrightarrow\quad 3 < x < 5$, so $x \in (3, 5)$.

**36.** $|x - 6| < 0.1 \quad\Leftrightarrow\quad -0.1 < x - 6 < 0.1 \quad\Leftrightarrow\quad 5.9 < x < 6.1$, so $x \in (5.9, 6.1)$.

**37.** $|x + 5| \geq 2 \quad\Leftrightarrow\quad x + 5 \geq 2$ or $x + 5 \leq -2 \quad\Leftrightarrow\quad x \geq -3$ or $x \leq -7$, so $x \in (-\infty, -7] \cup [-3, \infty)$.

**38.** $|x+1| \geq 3 \quad \Leftrightarrow \quad x+1 \geq 3$ or $x+1 \leq -3 \quad \Leftrightarrow \quad x \geq 2$ or $x \leq -4$, so $x \in (-\infty, -4] \cup [2, \infty)$.

**39.** $|2x-3| \leq 0.4 \quad \Leftrightarrow \quad -0.4 \leq 2x-3 \leq 0.4 \quad \Leftrightarrow \quad 2.6 \leq 2x \leq 3.4 \quad \Leftrightarrow \quad 1.3 \leq x \leq 1.7$, so $x \in [1.3, 1.7]$.

**40.** $|5x-2| < 6 \quad \Leftrightarrow \quad -6 < 5x-2 < 6 \quad \Leftrightarrow \quad -4 < 5x < 8 \quad \Leftrightarrow \quad -\frac{4}{5} < x < \frac{8}{5}$, so $x \in \left(-\frac{4}{5}, \frac{8}{5}\right)$.

**41.** $a(bx-c) \geq bc \quad \Leftrightarrow \quad bx - c \geq \dfrac{bc}{a} \quad \Leftrightarrow \quad bx \geq \dfrac{bc}{a} + c = \dfrac{bc+ac}{a} \quad \Leftrightarrow \quad x \geq \dfrac{bc+ac}{ab}$

**42.** $ax + b < c \quad \Leftrightarrow \quad ax < c - b \quad \Leftrightarrow \quad x > \dfrac{c-b}{a}$ (since $a < 0$)

**43.** $|ab| = \sqrt{(ab)^2} = \sqrt{a^2b^2} = \sqrt{a^2}\sqrt{b^2} = |a|\,|b|$

**44.** If $0 < a < b$, then $a \cdot a < a \cdot b$ and $a \cdot b < b \cdot b$ [using Rule 3 of Inequalities]. So $a^2 < ab < b^2$ and hence $a^2 < b^2$.

## B COORDINATE GEOMETRY

**1.** Use the distance formula with $P_1(x_1, y_1) = (1, 1)$ and $P_2(x_2, y_2) = (4, 5)$ to get

$$|P_1P_2| = \sqrt{(4-1)^2 + (5-1)^2} = \sqrt{3^2+4^2} = \sqrt{25} = 5$$

**2.** The distance from $(1, -3)$ to $(5, 7)$ is $\sqrt{(5-1)^2 + [7-(-3)]^2} = \sqrt{4^2+10^2} = \sqrt{116} = 2\sqrt{29}$.

**3.** With $P(-3, 3)$ and $Q(-1, -6)$, the slope $m$ of the line through $P$ and $Q$ is $m = \dfrac{-6-3}{-1-(-3)} = -\dfrac{9}{2}$.

**4.** $m = \dfrac{0-(-4)}{6-(-1)} = \dfrac{4}{7}$

**5.** Using $A(-2, 9)$, $B(4, 6)$, $C(1, 0)$, and $D(-5, 3)$, we have

$|AB| = \sqrt{[4-(-2)]^2 + (6-9)^2} = \sqrt{6^2+(-3)^2} = \sqrt{45} = \sqrt{9}\sqrt{5} = 3\sqrt{5}$,

$|BC| = \sqrt{(1-4)^2 + (0-6)^2} = \sqrt{(-3)^2+(-6)^2} = \sqrt{45} = \sqrt{9}\sqrt{5} = 3\sqrt{5}$,

$|CD| = \sqrt{(-5-1)^2 + (3-0)^2} = \sqrt{(-6)^2+3^2} = \sqrt{45} = \sqrt{9}\sqrt{5} = 3\sqrt{5}$, and

$|DA| = \sqrt{[-2-(-5)]^2 + (9-3)^2} = \sqrt{3^2+6^2} = \sqrt{45} = \sqrt{9}\sqrt{5} = 3\sqrt{5}$. So all sides are of equal length and we have a rhombus. Moreover, $m_{AB} = \dfrac{6-9}{4-(-2)} = -\dfrac{1}{2}$, $m_{BC} = \dfrac{0-6}{1-4} = 2$, $m_{CD} = \dfrac{3-0}{-5-1} = -\dfrac{1}{2}$, and

$m_{DA} = \dfrac{9-3}{-2-(-5)} = 2$, so the sides are perpendicular. Thus, $A$, $B$, $C$, and $D$ are vertices of a square.

**6.** (a) Using $A(-1, 3)$, $B(3, 11)$, and $C(5, 15)$, we have

$|AB| = \sqrt{[3-(-1)]^2 + (11-3)^2} = \sqrt{4^2+8^2} = \sqrt{80} = 4\sqrt{5}$,

$|BC| = \sqrt{(5-3)^2 + (15-11)^2} = \sqrt{2^2+4^2} = \sqrt{20} = 2\sqrt{5}$, and

$|AC| = \sqrt{[5-(-1)]^2 + (15-3)^2} = \sqrt{6^2+12^2} = \sqrt{180} = 6\sqrt{5}$. Thus, $|AC| = |AB| + |BC|$.

(b) $m_{AB} = \dfrac{11-3}{3-(-1)} = \dfrac{8}{4} = 2$ and $m_{AC} = \dfrac{15-3}{5-(-1)} = \dfrac{12}{6} = 2$. Since the segments $AB$ and $AC$ have the same slope, $A$, $B$ and $C$ must be collinear.

**7.** The graph of the equation $x = 3$ is a vertical line with $x$-intercept 3. The line does not have a slope.

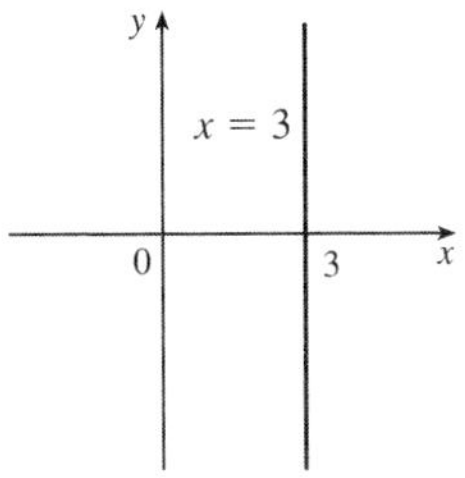


**8.** The graph of the equation $y = -2$ is a horizontal line with $y$-intercept $-2$. The line has slope 0.

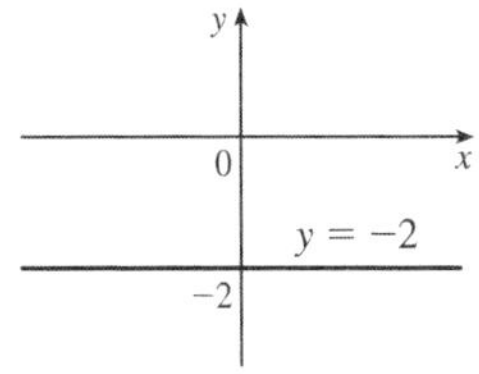


**9.** $xy = 0 \quad \Leftrightarrow \quad x = 0$ or $y = 0$. The graph consists of the coordinate axes.

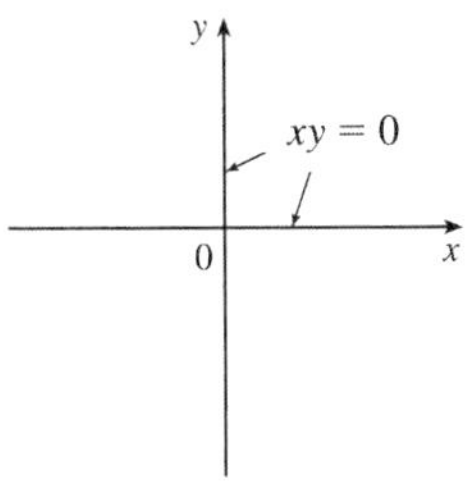


**10.** $|y| = 1 \quad \Leftrightarrow \quad y = 1$ or $y = -1$

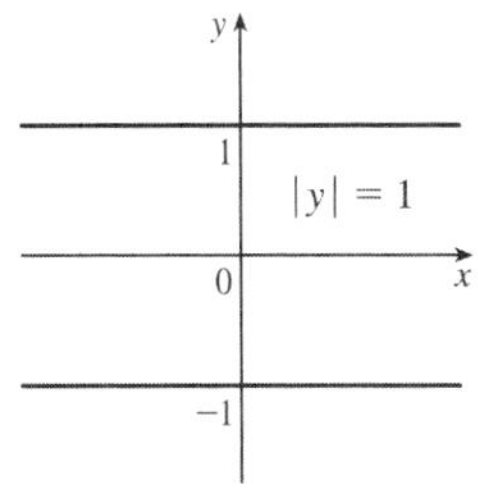


**11.** By the point-slope form of the equation of a line, an equation of the line through $(2, -3)$ with slope 6 is $y - (-3) = 6(x - 2)$ or $y = 6x - 15$.

**12.** $y - (-5) = -\frac{7}{2}\,[x - (-3)]$ or $y = -\frac{7}{2}x - \frac{31}{2}$

**13.** The slope of the line through $(2, 1)$ and $(1, 6)$ is $m = \dfrac{6 - 1}{1 - 2} = -5$, so an equation of the line is $y - 1 = -5(x - 2)$ or $y = -5x + 11$.

**14.** For $(-1, -2)$ and $(4, 3)$, $m = \dfrac{3 - (-2)}{4 - (-1)} = 1$. An equation of the line is $y - 3 = 1(x - 4)$ or $y = x - 1$.

**15.** By the slope-intercept form of the equation of a line, an equation of the line is $y = 3x - 2$.

**16.** By the slope-intercept form of the equation of a line, an equation of the line is $y = \frac{2}{5}x + 4$.

**17.** Since the line passes through $(1, 0)$ and $(0, -3)$, its slope is $m = \dfrac{-3 - 0}{0 - 1} = 3$, so an equation is $y = 3x - 3$.

*Another method:* From Exercise 46, $\dfrac{x}{1} + \dfrac{y}{-3} = 1 \quad \Rightarrow \quad -3x + y = -3 \quad \Rightarrow \quad y = 3x - 3$.

**18.** For $(-8, 0)$ and $(0, 6)$, $m = \dfrac{6 - 0}{0 - (-8)} = \dfrac{3}{4}$. So an equation is $y = \frac{3}{4}x + 6$.

*Another method:* From Exercise 46, $\dfrac{x}{-8} + \dfrac{y}{6} = 1 \quad \Rightarrow \quad -3x + 4y = 24 \quad \Rightarrow \quad y = \frac{3}{4}x + 6$.

**19.** The line is parallel to the $x$-axis, so it is horizontal and must have the form $y = k$. Since it goes through the point $(x, y) = (4, 5)$, the equation is $y = 5$.

**20.** The line is parallel to the $y$-axis, so it is vertical and must have the form $x = k$. Since it goes through the point $(x, y) = (4, 5)$, the equation is $x = 4$.

**21.** Putting the line $x + 2y = 6$ into its slope-intercept form gives us $y = -\frac{1}{2}x + 3$, so we see that this line has slope $-\frac{1}{2}$. Thus, we want the line of slope $-\frac{1}{2}$ that passes through the point $(1, -6)$: $y - (-6) = -\frac{1}{2}(x - 1) \quad \Leftrightarrow \quad y = -\frac{1}{2}x - \frac{11}{2}$.

**22.** $2x + 3y + 4 = 0 \quad \Leftrightarrow \quad y = -\frac{2}{3}x - \frac{4}{3}$, so $m = -\frac{2}{3}$ and the required line is $y = -\frac{2}{3}x + 6$.

**23.** $2x + 5y + 8 = 0 \quad \Leftrightarrow \quad y = -\frac{2}{5}x - \frac{8}{5}$. Since this line has slope $-\frac{2}{5}$, a line perpendicular to it would have slope $\frac{5}{2}$, so the required line is $y - (-2) = \frac{5}{2}\left[x - (-1)\right] \quad \Leftrightarrow \quad y = \frac{5}{2}x + \frac{1}{2}$.

**24.** $4x - 8y = 1 \quad \Leftrightarrow \quad y = \frac{1}{2}x - \frac{1}{8}$. Since this line has slope $\frac{1}{2}$, a line perpendicular to it would have slope $-2$, so the required line is $y - \left(-\frac{2}{3}\right) = -2\left(x - \frac{1}{2}\right) \quad \Leftrightarrow \quad y = -2x + \frac{1}{3}$.

**25.** $x + 3y = 0 \quad \Leftrightarrow \quad y = -\frac{1}{3}x$, so the slope is $-\frac{1}{3}$ and the $y$-intercept is 0.

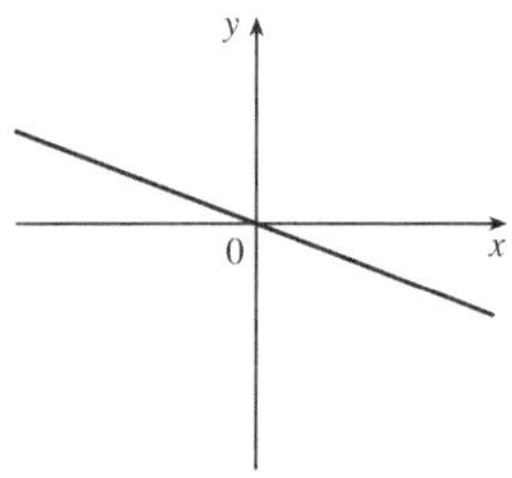


**26.** $2x - 3y + 6 = 0 \quad \Leftrightarrow$ $y = \frac{2}{3}x + 2$, so the slope is $\frac{2}{3}$ and the $y$-intercept is 2.

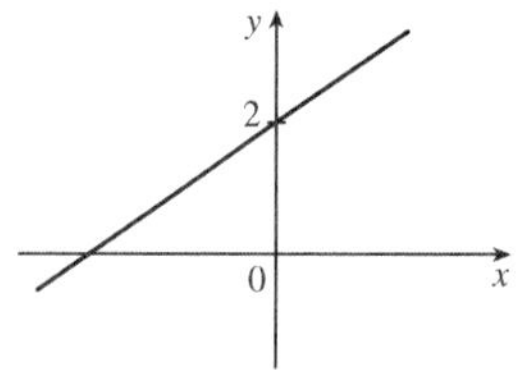


**27.** $3x - 4y = 12 \quad \Leftrightarrow$ $y = \frac{3}{4}x - 3$, so the slope is $\frac{3}{4}$ and the $y$-intercept is $-3$.

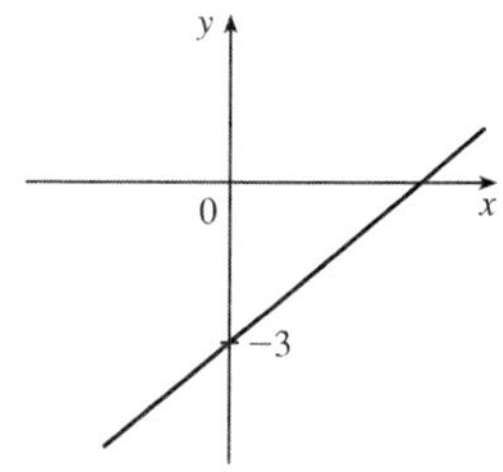


**28.** $4x + 5y = 10 \quad \Leftrightarrow$ $y = -\frac{4}{5}x + 2$, so the slope is $-\frac{4}{5}$ and the $y$-intercept is 2.

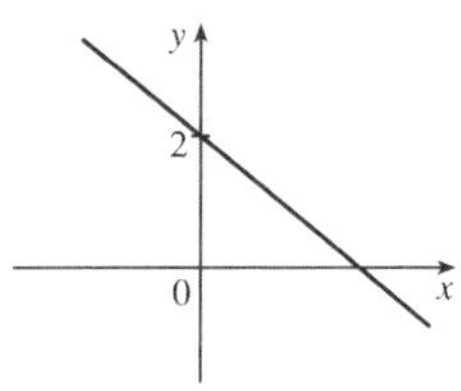


**29.** $\{(x, y) \mid x < 0\}$

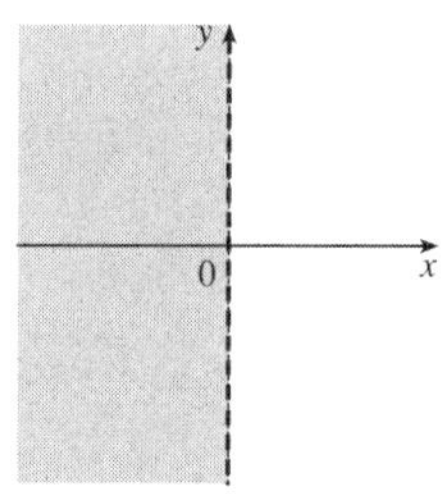


**30.** $\{(x, y) \mid x \geq 1 \text{ and } y < 3\}$

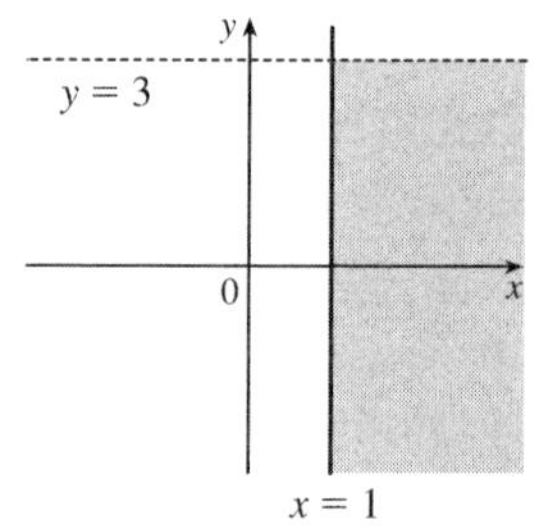


**31.** $\left\{(x, y) \,\middle|\, |x| \leq 2\right\} = \{(x, y) \mid -2 \leq x \leq 2\}$

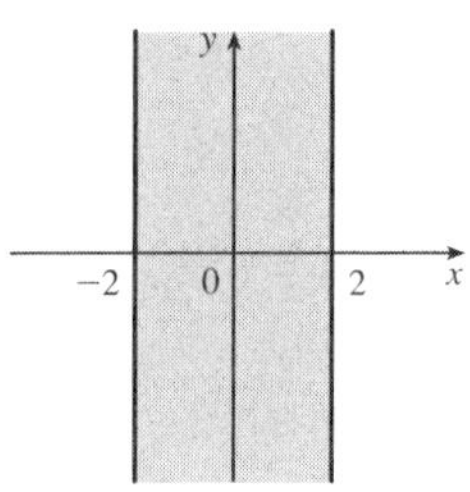


**32.** $\left\{(x, y) \,\middle|\, |x| < 3 \text{ and } |y| < 2\right\}$

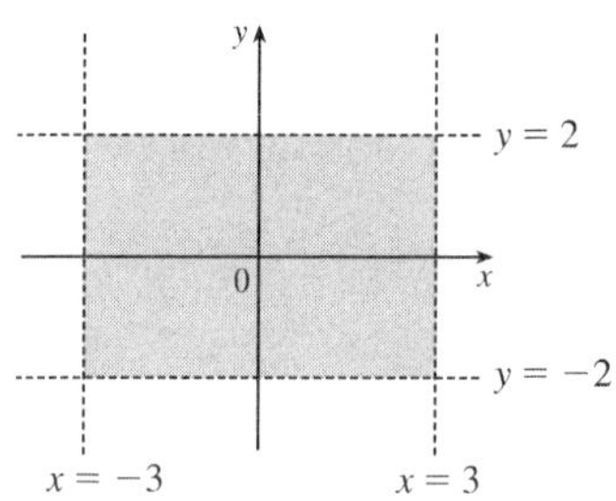


**33.** $\{(x, y) \mid 0 \leq y \leq 4, x \leq 2\}$

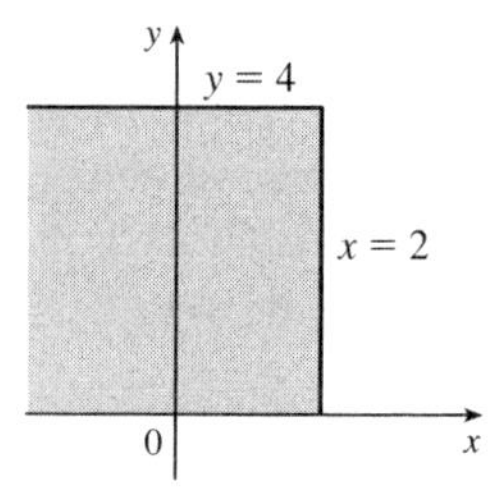


**34.** $\{(x, y) \mid y > 2x - 1\}$

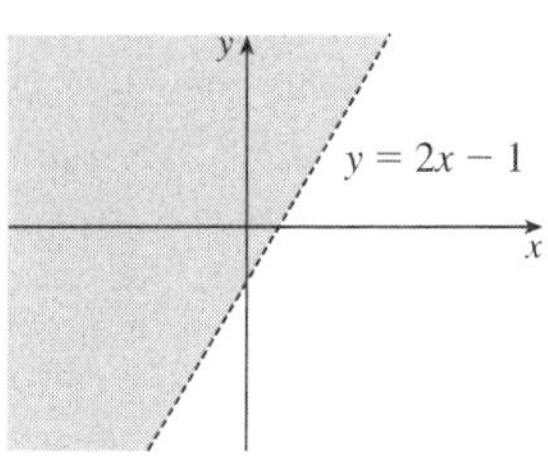


**35.** $\{(x, y) \mid 1 + x \leq y \leq 1 - 2x\}$

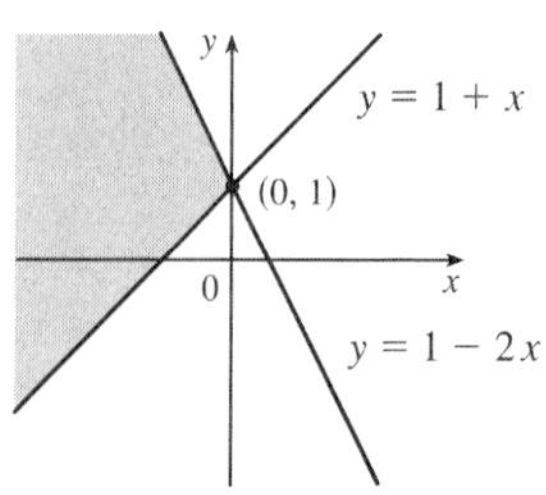


**36.** $\left\{(x, y) \,\middle|\, -x \leq y < \dfrac{x + 3}{2}\right\}$

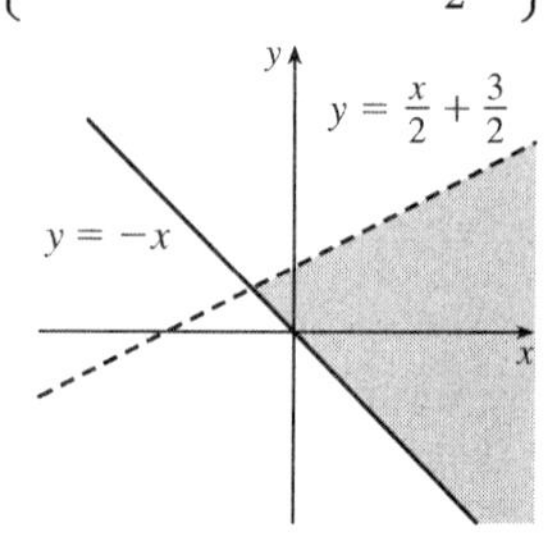

**37.** An equation of the circle with center $(3, -1)$ and radius 5 is $(x-3)^2 + (y+1)^2 = 5^2 = 25$.

**38.** The equation has the form $(x+1)^2 + (y-5)^2 = r^2$. Since $(-4, -6)$ lies on the circle, we have $r^2 = (-4+1)^2 + (-6-5)^2 = 130$. So an equation is $(x+1)^2 + (y-5)^2 = 130$.

**39.** $x^2 + y^2 - 4x + 10y + 13 = 0 \quad \Leftrightarrow \quad x^2 - 4x + y^2 + 10y = -13 \quad \Leftrightarrow$ $(x^2 - 4x + 4) + (y^2 + 10y + 25) = -13 + 4 + 25 = 16 \quad \Leftrightarrow \quad (x-2)^2 + (y+5)^2 = 4^2$. Thus, we have a circle with center $(2, -5)$ and radius 4.

**40.** $x^2 + y^2 + 6y + 2 = 0 \quad \Leftrightarrow \quad x^2 + (y^2 + 6y + 9) = -2 + 9 \quad \Leftrightarrow \quad x^2 + (y+3)^2 = 7$. Thus, we have a circle with center $(0, -3)$ and radius $\sqrt{7}$.

**41.** $2x - y = 4 \quad \Leftrightarrow \quad y = 2x - 4 \quad \Rightarrow \quad m_1 = 2$ and $6x - 2y = 10 \quad \Leftrightarrow \quad 2y = 6x - 10 \quad \Leftrightarrow \quad y = 3x - 5 \quad \Rightarrow \quad m_2 = 3$. Since $m_1 \neq m_2$, the two lines are not parallel. To find the point of intersection: $2x - 4 = 3x - 5 \quad \Leftrightarrow \quad x = 1 \quad \Rightarrow$ $y = -2$. Thus, the point of intersection is $(1, -2)$.

**42.** $3x - 5y + 19 = 0 \quad \Leftrightarrow \quad 5y = 3x + 19 \quad \Leftrightarrow \quad y = \frac{3}{5}x + \frac{19}{5} \quad \Rightarrow \quad m_1 = \frac{3}{5}$ and $10x + 6y - 50 = 0 \quad \Leftrightarrow$ $6y = -10x + 50 \quad \Leftrightarrow \quad y = -\frac{5}{3}x + \frac{25}{3} \quad \Rightarrow \quad m_2 = -\frac{5}{3}$. Since $m_1 m_2 = \frac{3}{5}\left(-\frac{5}{3}\right) = -1$, the two lines are perpendicular. To find the point of intersection: $\frac{3}{5}x + \frac{19}{5} = -\frac{5}{3}x + \frac{25}{3} \quad \Leftrightarrow \quad 9x + 57 = -25x + 125 \quad \Leftrightarrow \quad 34x = 68 \quad \Leftrightarrow \quad x = 2 \quad \Rightarrow$ $y = \frac{3}{5} \cdot 2 + \frac{19}{5} = \frac{25}{5} = 5$. Thus, the point of intersection is $(2, 5)$.

**43.** Let $M$ be the point $\left(\dfrac{x_1 + x_2}{2}, \dfrac{y_1 + y_2}{2}\right)$. Then

$|MP_1|^2 = \left(x_1 - \dfrac{x_1 + x_2}{2}\right)^2 + \left(y_1 - \dfrac{y_1 + y_2}{2}\right)^2 = \left(\dfrac{x_1 - x_2}{2}\right)^2 + \left(\dfrac{y_1 - y_2}{2}\right)^2$ and

$|MP_2|^2 = \left(x_2 - \dfrac{x_1 + x_2}{2}\right)^2 + \left(y_2 - \dfrac{y_1 + y_2}{2}\right)^2 = \left(\dfrac{x_2 - x_1}{2}\right)^2 + \left(\dfrac{y_2 - y_1}{2}\right)^2$. Hence, $|MP_1| = |MP_2|$; that is, $M$ is equidistant from $P_1$ and $P_2$.

**44.** Using the midpoint formula from Exercise 43 with $(1, 3)$ and $(7, 15)$, we get $\left(\frac{1+7}{2}, \frac{3+15}{2}\right) = (4, 9)$.

**45.** With $A(1, 4)$ and $B(7, -2)$, the slope of segment $AB$ is $\frac{-2-4}{7-1} = -1$, so its perpendicular bisector has slope 1. The midpoint of $AB$ is $\left(\frac{1+7}{2}, \frac{4+(-2)}{2}\right) = (4, 1)$, so an equation of the perpendicular bisector is $y - 1 = 1(x - 4)$ or $y = x - 3$.

**46.** (a) Since the $x$-intercept is $a$, the point $(a, 0)$ is on the line, and similarly since the $y$-intercept is $b$, $(0, b)$ is on the line. Hence, the slope of the line is $m = \dfrac{b - 0}{0 - a} = -\dfrac{b}{a}$. Substituting into $y = mx + b$ gives $y = -\dfrac{b}{a}x + b \quad \Leftrightarrow \quad \dfrac{b}{a}x + y = b \quad \Leftrightarrow$ $\dfrac{x}{a} + \dfrac{y}{b} = 1$.

(b) Letting $a = 6$ and $b = -8$ gives $\dfrac{x}{6} + \dfrac{y}{-8} = 1 \quad \Leftrightarrow \quad -8x + 6y = -48$ [multiply by $-48$] $\quad \Leftrightarrow \quad 6y = 8x - 48 \quad \Leftrightarrow$ $3y = 4x - 24 \quad \Leftrightarrow \quad y = \frac{4}{3}x - 8$.

**47.** If $P(x, y)$ is any point on the parabola, then the distance from $P$ to the focus is $|PF| = \sqrt{x^2 + (y-p)^2}$ and the distance from $P$ to the directrix is $|y + p|$. (Figure 14 in the text illustrates the case where $p > 0$.) The defining property of a parabola is that these distances are equal: $\sqrt{x^2 + (y-p)^2} = |y + p|$. We get an equivalent equation by squaring and simplifying: $x^2 + (y-p)^2 = |y+p|^2 = (y+p)^2 \quad \Leftrightarrow \quad x^2 + y^2 - 2py + p^2 = y^2 + 2py + p^2 \quad \Leftrightarrow \quad x^2 = 4py$. Thus, an equation of a parabola with focus $(0, p)$ and directrix $y = -p$ is $x^2 = 4py$.

**48.** From Exercise 47, we have $x^2 = 4py$ as a general equation of a parabola. Here, we have $x^2 = y$, so $4p = 1 \quad \Leftrightarrow \quad p = \frac{1}{4}$. Thus, the focus is $\left(0, \frac{1}{4}\right)$ and the directrix has equation $y = -\frac{1}{4}$.

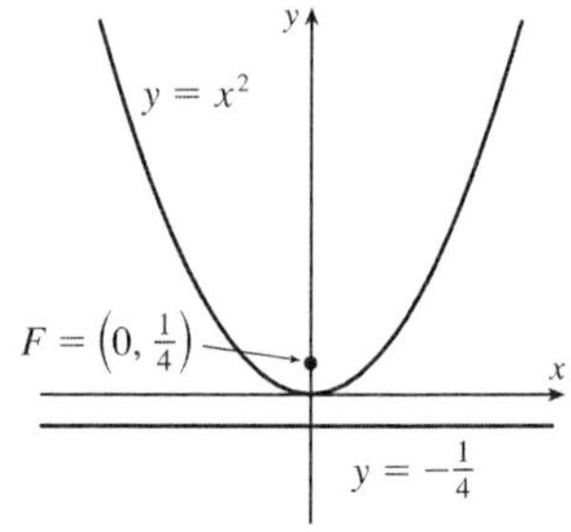


**49.** See Figure 20 in the text. $P(x, y)$ is a point on the ellipse when $|PF_1| + |PF_2| = 2a$; that is,

$\sqrt{(x+c)^2+y^2} + \sqrt{(x-c)^2+y^2} = 2a$ or $\sqrt{(x+c)^2+y^2} = 2a - \sqrt{(x+c)^2+y^2}$. Squaring both sides, we have $x^2 - 2cx + c^2 + y^2 = 4a^2 - 4a\sqrt{(x+c)^2+y^2} + x^2 + 2cx + c^2 + y^2$, which simplifies to $a\sqrt{(x+c)^2+y^2} = a^2 + cx$. We square again: $a^2\left(x^2 + 2cx + c^2 + y^2\right) = a^4 + 2a^2cx + c^2x^2$, which becomes $\left(a^2 - c^2\right)x^2 + a^2y^2 = a^2\left(a^2 - c^2\right)$. From triangle $F_1F_2P$ in Figure 20, we see that $2c < 2a$, so $c < a$ and, therefore, $a^2 - c^2 > 0$. For convenience, let $b^2 = a^2 - c^2$. Then the equation of the ellipse becomes $b^2x^2 + a^2y^2 = a^2b^2$ or, if both sides are divided by $a^2b^2$, $\dfrac{x^2}{a^2} + \dfrac{y^2}{b^2} = 1$.

**50.** $x^2 + 4y^2 = 4 \quad \Leftrightarrow \quad \frac{1}{4}x^2 + y^2 = 1 \quad \Rightarrow \quad a = 2, b = 1,$
$c = \sqrt{2^2 - 1^2} = \sqrt{3} \quad \Rightarrow$
center $(0, 0)$, vertices $(\pm 2, 0)$, foci $\left(\pm\sqrt{3}, 0\right)$

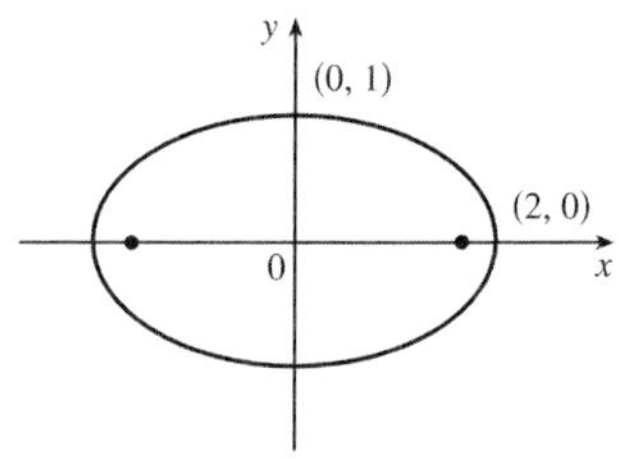


**51.** From Figure 23 in the text, $|PF_1| - |PF_2| = \pm 2a \quad \Leftrightarrow \quad \sqrt{(x+c)^2+y^2} - \sqrt{(x-c)^2+y^2} = \pm 2a \quad \Leftrightarrow$

$\sqrt{(x+c)^2+y^2} = \sqrt{(x-c)^2+y^2} \pm 2a \quad \Leftrightarrow$

$(x+c)^2 + y^2 = (x-c)^2 + y^2 + 4a^2 \pm 4a\sqrt{(x-c)^2+y^2} \quad \Leftrightarrow \quad 4cx - 4a^2 = \pm 4a\sqrt{(x-c)^2+y^2} \quad \Leftrightarrow$

$c^2x^2 - 2a^2cx + a^4 = a^2\left(x^2 - 2cx + c^2 + y^2\right) \quad \Leftrightarrow \quad \left(c^2 - a^2\right)x^2 - a^2y^2 = a^2\left(c^2 - a^2\right) \quad \Leftrightarrow$

$b^2x^2 - a^2y^2 = a^2b^2$ [where $b^2 = c^2 - a^2$] $\quad \Leftrightarrow \quad \dfrac{x^2}{a^2} - \dfrac{y^2}{b^2} = 1.$

**52.** (a) $x^2 - y^2 = 1 \quad \Rightarrow \quad a = b = 1,$
$c = \sqrt{1^2 + 1^2} = \sqrt{2} \quad \Rightarrow \quad$ center $(0, 0)$,
vertices $(\pm 1, 0)$, foci $\left(\pm\sqrt{2}, 0\right)$,
asymptotes $y = \pm x$

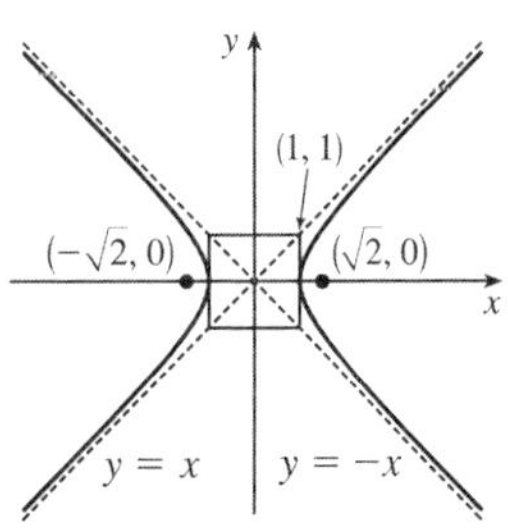


(b) $y^2 - x^2 = 1 \quad \Rightarrow \quad a = b = 1,$
$c = \sqrt{1^2 + 1^2} = \sqrt{2} \quad \Rightarrow \quad$ center $(0, 0)$,
vertices $(0, \pm 1)$, foci $\left(0, \pm\sqrt{2}\right)$,
asymptotes $y = \pm x$

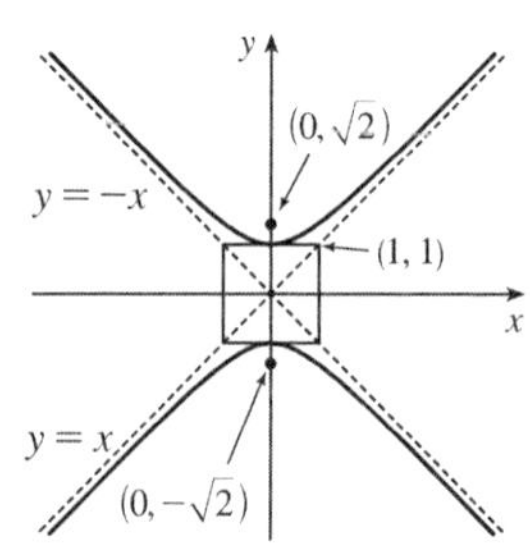

**53.**

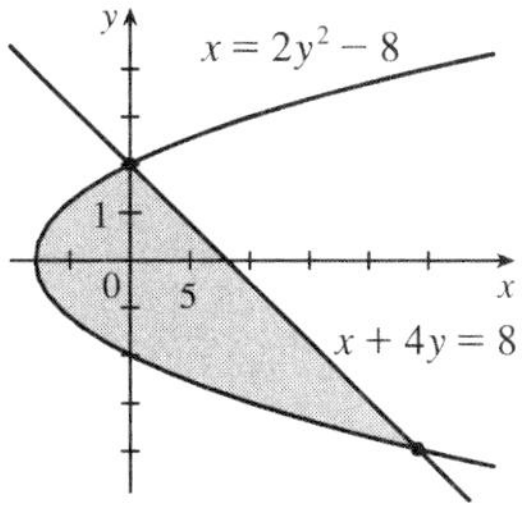


$x + 4y = 8$, $x = 2y^2 - 8$. Substitute $x$ from the second equation into the first: $(2y^2 - 8) + 4y = 8 \quad \Leftrightarrow \quad 2y^2 + 4y - 16 = 0$ $\Leftrightarrow \quad y^2 + 2y - 8 = 0 \quad \Leftrightarrow \quad (y+4)(y-2) = 0 \quad \Leftrightarrow \quad y = -4$ or 2. So the points of intersection are $(24, -4)$ and $(0, 2)$.

**54.**

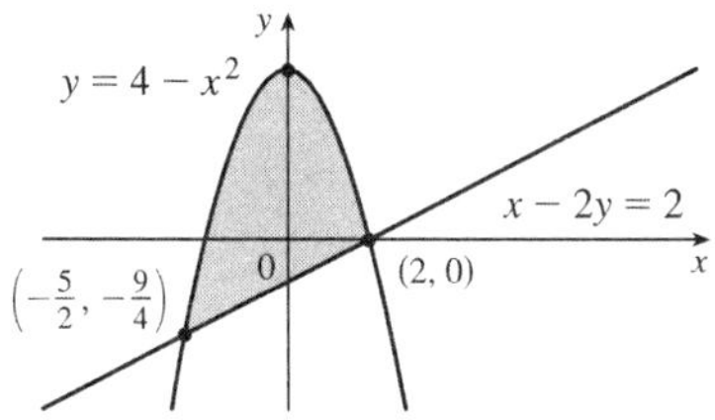


$y = 4 - x^2$, $x - 2y = 2$. Substitute $y$ from the first equation into the second: $x - 2(4 - x^2) = 2 \quad \Leftrightarrow \quad 2x^2 + x - 10 = 0$ $\Leftrightarrow \quad (2x+5)(x-2) = 0 \quad \Leftrightarrow \quad x = -\frac{5}{2}$ or 2. So the points of intersection are $\left(-\frac{5}{2}, -\frac{9}{4}\right)$ and $(2, 0)$.

**55.** Differentiating implicitly, $\dfrac{x^2}{a^2} + \dfrac{y^2}{b^2} = 1 \quad \Rightarrow \quad \dfrac{2x}{a^2} + \dfrac{2yy'}{b^2} = 0 \quad \Rightarrow \quad y' = -\dfrac{b^2 x}{a^2 y}$ $(y \neq 0)$. Thus, the slope of the tangent line at $P$ is $-\dfrac{b^2 x_1}{a^2 y_1}$. The slope of $F_1P$ is $\dfrac{y_1}{x_1 + c}$ and of $F_2P$ is $\dfrac{y_1}{x_1 - c}$. By the formula in Problem 15 on text page 260, we have

$$\tan\alpha = \frac{\dfrac{y_1}{x_1 + c} + \dfrac{b^2 x_1}{a^2 y_1}}{1 - \dfrac{b^2 x_1 y_1}{a^2 y_1 (x_1 + c)}} = \frac{a^2 y_1^2 + b^2 x_1 (x_1 + c)}{a^2 y_1 (x_1 + c) - b^2 x_1 y_1}$$

$$= \frac{a^2 b^2 + b^2 c x_1}{c^2 x_1 y_1 + a^2 c y_1} \qquad \left[\begin{array}{c}\text{using } b^2 x_1^2 + a^2 y_1^2 = a^2 b^2 \\ \text{and } a^2 - b^2 = c^2\end{array}\right]$$

$$= \frac{b^2 (c x_1 + a^2)}{c y_1 (c x_1 + a^2)}$$

$$= \frac{b^2}{c y_1}$$

and

$$\tan\beta = \frac{-\dfrac{b^2 x_1}{a^2 y_1} - \dfrac{y_1}{x_1 - c}}{1 - \dfrac{b^2 x_1 y_1}{a^2 y_1 (x_1 - c)}} = \frac{-a^2 y_1^2 - b^2 x_1 (x_1 - c)}{a^2 y_1 (x_1 - c) - b^2 x_1 y_1}$$

$$= \frac{-a^2 b^2 + b^2 c x_1}{c^2 x_1 y_1 - a^2 c y_1} = \frac{b^2 (c x_1 - a^2)}{c y_1 (c x_1 - a^2)}$$

$$= \frac{b^2}{c y_1}$$

Thus, $\alpha = \beta$.

# C TRIGONOMETRY

**1.** (a) $210° = 210\left(\frac{\pi}{180}\right) = \frac{7\pi}{6}$ rad (b) $9° = 9\left(\frac{\pi}{180}\right) = \frac{\pi}{20}$ rad

**2.** (a) $-315° = -315\left(\frac{\pi}{180}\right) = -\frac{7\pi}{4}$ rad (b) $36° = 36\left(\frac{\pi}{180}\right) = \frac{\pi}{5}$ rad

**3.** (a) $4\pi$ rad $= 4\pi\left(\frac{180}{\pi}\right) = 720°$ (b) $-\frac{3\pi}{8}$ rad $= -\frac{3\pi}{8}\left(\frac{180}{\pi}\right) = -67.5°$

**4.** (a) $-\frac{7\pi}{2}$ rad $= -\frac{7\pi}{2}\left(\frac{180}{\pi}\right) = -630°$ (b) $\frac{8\pi}{3}$ rad $= \frac{8\pi}{3}\left(\frac{180}{\pi}\right) = 480°$

**5.** Using Formula 3, $a = r\theta = 36 \cdot \frac{\pi}{12} = 3\pi$ cm.

**6.** Using Formula 3, $a = r\theta = 10 \cdot 72\left(\frac{\pi}{180}\right) = 4\pi$ cm.

**7.** Using Formula 3, $\theta = a/r = \frac{1}{1.5} = \frac{2}{3}$ rad $= \frac{2}{3}\left(\frac{180}{\pi}\right) = \left(\frac{120}{\pi}\right)° \approx 38.2°$.

**8.** $a = r\theta \;\Rightarrow\; r = \frac{a}{\theta} = \frac{6}{3\pi/4} = \frac{8}{\pi}$ cm

**9.** (a)

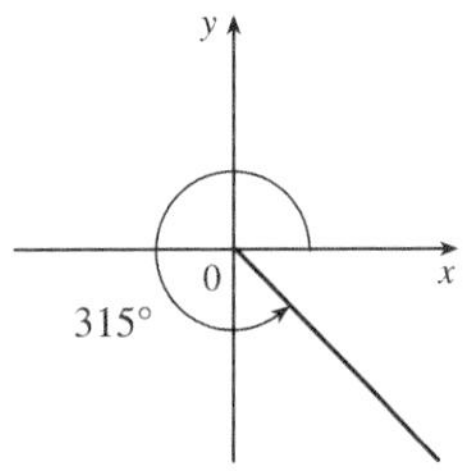


(b)

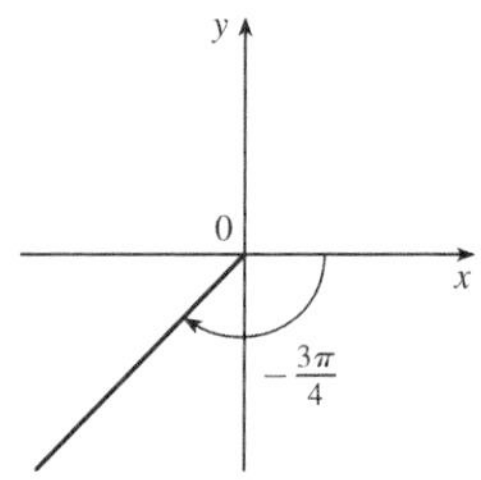


**10.** (a)

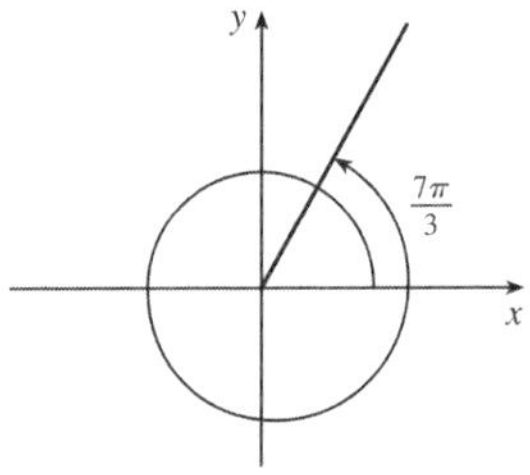


(b)

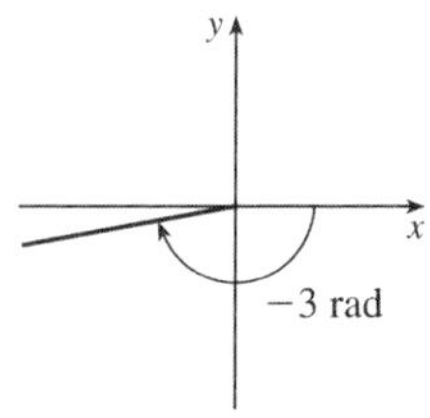


**11.**

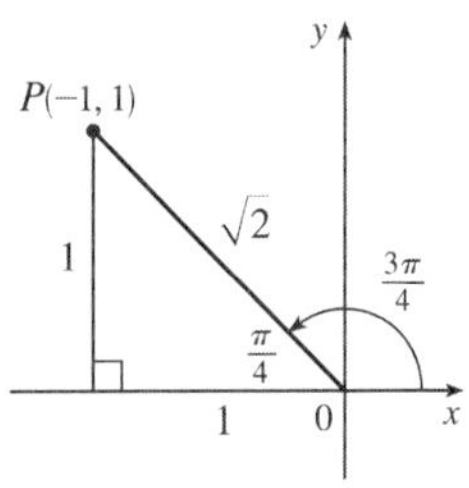


From the diagram we see that a point on the terminal side is $P(-1, 1)$. Therefore, taking $x = -1$, $y = 1$, $r = \sqrt{2}$ in the definitions of the trigonometric ratios, we have $\sin\frac{3\pi}{4} = \frac{1}{\sqrt{2}}$, $\cos\frac{3\pi}{4} = -\frac{1}{\sqrt{2}}$, $\tan\frac{3\pi}{4} = -1$, $\csc\frac{3\pi}{4} = \sqrt{2}$, $\sec\frac{3\pi}{4} = -\sqrt{2}$, and $\cot\frac{3\pi}{4} = -1$.

**12.**

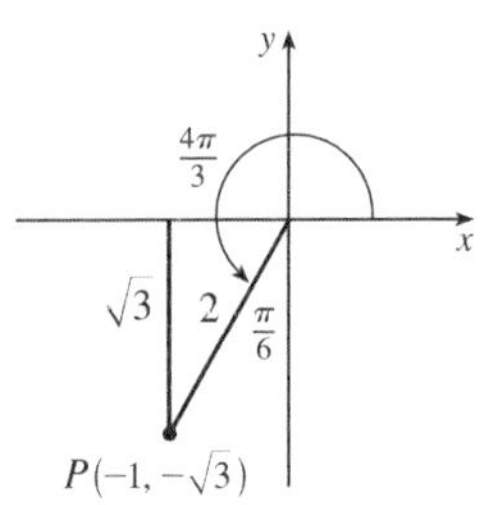


From the diagram and Figure 8, we see that a point on the terminal side is $P\left(-1, -\sqrt{3}\right)$. Therefore, taking $x = -1$, $y = -\sqrt{3}$, $r = 2$ in the definitions of the trigonometric ratios, we have $\sin\frac{4\pi}{3} = -\frac{\sqrt{3}}{2}$, $\cos\frac{4\pi}{3} = -\frac{1}{2}$, $\tan\frac{4\pi}{3} = \sqrt{3}$, $\csc\frac{4\pi}{3} = -\frac{2}{\sqrt{3}}$, $\sec\frac{4\pi}{3} = -2$, and $\cot\frac{4\pi}{3} = \frac{1}{\sqrt{3}}$.

**13.** $\sin\theta = y/r = \frac{3}{5} \;\Rightarrow\; y = 3$, $r = 5$, and $x = \sqrt{r^2 - y^2} = 4$ (since $0 < \theta < \frac{\pi}{2}$). Therefore taking $x = 4$, $y = 3$, $r = 5$ in the definitions of the trigonometric ratios, we have $\cos\theta = \frac{4}{5}$, $\tan\theta = \frac{3}{4}$, $\csc\theta = \frac{5}{3}$, $\sec\theta = \frac{5}{4}$, and $\cot\theta = \frac{4}{3}$.

**14.** Since $0 < \alpha < \frac{\pi}{2}$, $\alpha$ is in the first quadrant where $x$ and $y$ are both positive. Therefore, $\tan\alpha = y/x = \frac{2}{1} \;\Rightarrow\; y = 2$, $x = 1$, and $r = \sqrt{x^2 + y^2} = \sqrt{5}$. Taking $x = 1$, $y = 2$, $r = \sqrt{5}$ in the definitions of the trigonometric ratios, we have $\sin\alpha = \frac{2}{\sqrt{5}}$, $\cos\alpha = \frac{1}{\sqrt{5}}$, $\csc\alpha = \frac{\sqrt{5}}{2}$, $\sec\alpha = \sqrt{5}$, and $\cot\alpha = \frac{1}{2}$.

**15.** $\sin 35° = \dfrac{x}{10} \quad\Rightarrow\quad x = 10\sin 35° \approx 5.73576$ cm

**16.** $\cos 40° = \dfrac{x}{25} \quad\Rightarrow\quad x = 25\cos 40° \approx 19.15111$ cm

**17.** $\tan\dfrac{2\pi}{5} = \dfrac{x}{8} \quad\Rightarrow\quad x = 8\tan\dfrac{2\pi}{5} \approx 24.62147$ cm

**18.** $\cos\dfrac{3\pi}{8} = \dfrac{22}{x} \quad\Rightarrow\quad x = \dfrac{22}{\cos\dfrac{3\pi}{8}} \approx 57.48877$ cm

**19.**

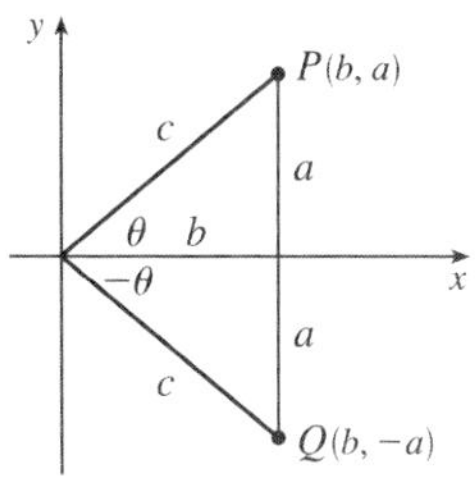


(a) From the diagram we see that $\sin\theta = \dfrac{y}{r} = \dfrac{a}{c}$, and

$$\sin(-\theta) = \frac{-a}{c} = -\frac{a}{c} = -\sin\theta.$$

(b) Again from the diagram we see that $\cos\theta = \dfrac{x}{r} = \dfrac{b}{c} = \cos(-\theta)$.

**20.** (a) Using (12a) and (12b), we have

$$\tan(x+y) = \frac{\sin(x+y)}{\cos(x+y)} = \frac{\sin x\cos y + \cos x\sin y}{\cos x\cos y - \sin x\sin y} = \frac{\dfrac{\sin x\cos y}{\cos x\cos y} + \dfrac{\cos x\sin y}{\cos x\cos y}}{\dfrac{\cos x\cos y}{\cos x\cos y} - \dfrac{\sin x\sin y}{\cos x\cos y}} = \frac{\tan x + \tan y}{1 - \tan x\tan y}$$

(b) From (10a) and (10b), we have $\tan(-\theta) = -\tan\theta$, so (14a) implies that

$$\tan(x-y) = \tan(x+(-y)) = \frac{\tan x + \tan(-y)}{1 - \tan x\tan(-y)} = \frac{\tan x - \tan y}{1 + \tan x\tan y}$$

**21.** Using (12a), we have $\sin\left(\frac{\pi}{2} + x\right) = \sin\frac{\pi}{2}\cos x + \cos\frac{\pi}{2}\sin x = 1\cdot\cos x + 0\cdot\sin x = \cos x$.

**22.** Using (13a), we have $\sin(\pi - x) = \sin\pi\cos x - \cos\pi\,\sin x = 0\cdot\cos x - (-1)\sin x = \sin x$.

**23.** Using (6), we have $\sin\theta\cot\theta = \sin\theta\cdot\dfrac{\cos\theta}{\sin\theta} = \cos\theta$.

**24.** $(\sin x + \cos x)^2 = \sin^2 x + 2\sin x\cos x + \cos^2 x = \left(\sin^2 x + \cos^2 x\right) + \sin 2x$ [by (15a)]

$= 1 + \sin 2x$ [by (7)]

**25.** Using (14a), we have $\tan 2\theta = \tan(\theta+\theta) = \dfrac{\tan\theta + \tan\theta}{1 - \tan\theta\tan\theta} = \dfrac{2\tan\theta}{1 - \tan^2\theta}$.

**26.** We use (12b) with $x = 2\theta$, $y = \theta$ to get

$$\begin{aligned}\cos 3\theta &= \cos(2\theta + \theta) = \cos 2\theta\cos\theta - \sin 2\theta\sin\theta\\ &= \left(2\cos^2\theta - 1\right)\cos\theta - 2\sin^2\theta\cos\theta \quad \text{[by (16a) and (15a)]}\\ &= \left(2\cos^2\theta - 1\right)\cos\theta - 2\left(1 - \cos^2\theta\right)\cos\theta \quad \text{[by (7)]}\\ &= 2\cos^3\theta - \cos\theta - 2\cos\theta + 2\cos^3\theta = 4\cos^3\theta - 3\cos\theta\end{aligned}$$

**27.** Since $\sin x = \frac{1}{3}$ we can label the opposite side as having length 1, the hypotenuse as having length 3, and use the Pythagorean Theorem to get that the adjacent side has length $\sqrt{8}$. Then, from the diagram, $\cos x = \frac{\sqrt{8}}{3}$. Similarly we have that $\sin y = \frac{3}{5}$. Now use (12a):

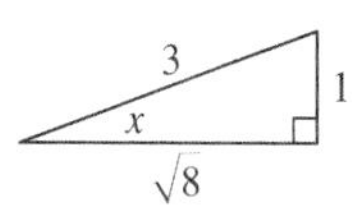


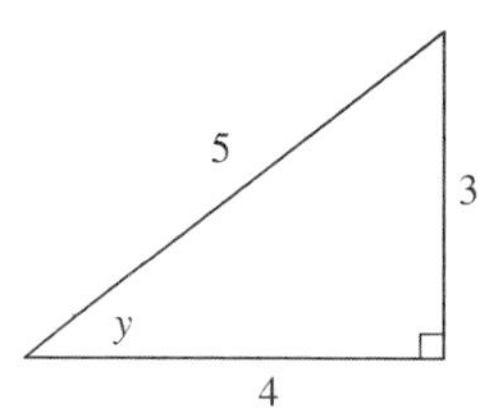


$\sin(x+y) = \sin x\cos y + \cos x\,\sin y = \frac{1}{3}\cdot\frac{4}{5} + \frac{\sqrt{8}}{3}\cdot\frac{3}{5} = \frac{4}{15} + \frac{3\sqrt{8}}{15} = \frac{4+6\sqrt{2}}{15}$.

**28.** Using (16a) with $\cos y = \frac{4}{5}$, we have $\cos 2y = 2\cos^2 y - 1 = 2\left(\frac{4}{5}\right)^2 - 1 = \frac{32}{25} - 1 = \frac{7}{25}$.

**29.** $2\cos x - 1 = 0 \quad \Leftrightarrow \quad \cos x = \frac{1}{2} \quad \Rightarrow \quad x = \frac{\pi}{3}, \frac{5\pi}{3}$ for $x \in [0, 2\pi]$.

**30.** $2\sin^2 x = 1 \quad \Leftrightarrow \quad \sin^2 x = \frac{1}{2} \quad \Leftrightarrow \quad \sin x = \pm\frac{1}{\sqrt{2}} \quad \Rightarrow \quad x = \frac{\pi}{4}, \frac{3\pi}{4}, \frac{5\pi}{4}, \frac{7\pi}{4}$.

**31.** Using (15a), we have $\sin 2x = \cos x \quad \Leftrightarrow \quad 2\sin x\cos x - \cos x = 0 \quad \Leftrightarrow \quad \cos x(2\sin x - 1) = 0 \quad \Leftrightarrow \quad \cos x = 0$ or $2\sin x - 1 = 0 \quad \Rightarrow \quad x = \frac{\pi}{2}, \frac{3\pi}{2}$ or $\sin x = \frac{1}{2} \quad \Rightarrow \quad x = \frac{\pi}{6}$ or $\frac{5\pi}{6}$. Therefore, the solutions are $x = \frac{\pi}{6}, \frac{\pi}{2}, \frac{5\pi}{6}, \frac{3\pi}{2}$.

**32.** $|\tan x| = 1 \quad \Leftrightarrow \quad \tan x = -1$ or $\tan x = 1 \quad \Leftrightarrow \quad x = \frac{3\pi}{4}, \frac{7\pi}{4}$ or $x = \frac{\pi}{4}, \frac{5\pi}{4}$.

**33.** We know that $\sin x = \frac{1}{2}$ when $x = \frac{\pi}{6}$ or $\frac{5\pi}{6}$, and from Figure 13(a), we see that $\sin x \le \frac{1}{2} \quad \Rightarrow \quad 0 \le x \le \frac{\pi}{6}$ or $\frac{5\pi}{6} \le x \le 2\pi$ for $x \in [0, 2\pi]$.

**34.** $2\cos x + 1 > 0 \quad \Rightarrow \quad 2\cos x > -1 \quad \Rightarrow \quad \cos x > -\frac{1}{2}$. $\cos x = -\frac{1}{2}$ when $x = \frac{2\pi}{3}, \frac{4\pi}{3}$ and from Figure 13(b), we see that $\cos x > -\frac{1}{2}$ when $0 \le x < \frac{2\pi}{3}$, $\frac{4\pi}{3} < x \le 2\pi$.

**35.** $\tan x = -1$ when $x = \frac{3\pi}{4}, \frac{7\pi}{4}$, and $\tan x = 1$ when $x = \frac{\pi}{4}$ or $\frac{5\pi}{4}$. From Figure 14 we see that $-1 < \tan x < 1 \quad \Rightarrow$ $0 \le x < \frac{\pi}{4}$, $\frac{3\pi}{4} < x < \frac{5\pi}{4}$, and $\frac{7\pi}{4} < x \le 2\pi$.

**36.** We know that $\sin x = \cos x$ when $x = \frac{\pi}{4}, \frac{5\pi}{4}$, and from the diagram we see that $\sin x > \cos x$ when $\frac{\pi}{4} < x < \frac{5\pi}{4}$.

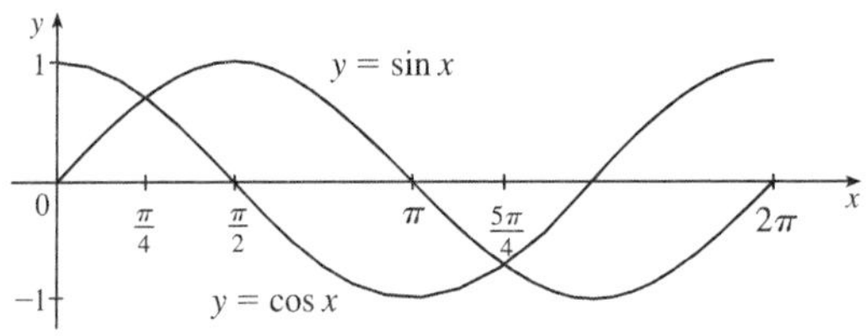


**37.** $y = \cos\left(x - \frac{\pi}{3}\right)$. We start with the graph of $y = \cos x$ and shift it $\frac{\pi}{3}$ units to the right.

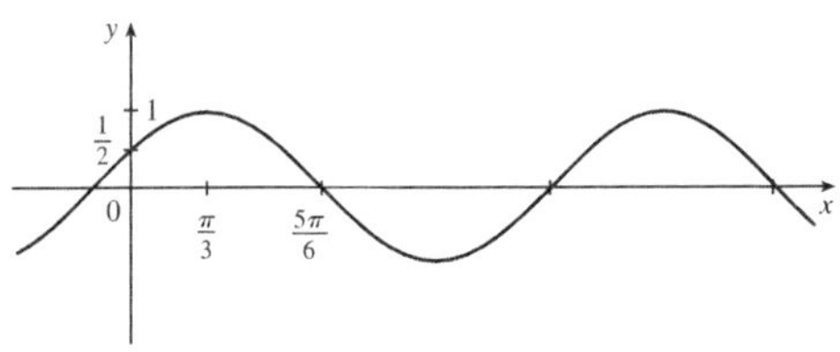


**38.** $y = \tan 2x$. Start with the graph of $y = \tan x$ with period $\pi$ and compress it to a period of $\frac{\pi}{2}$.

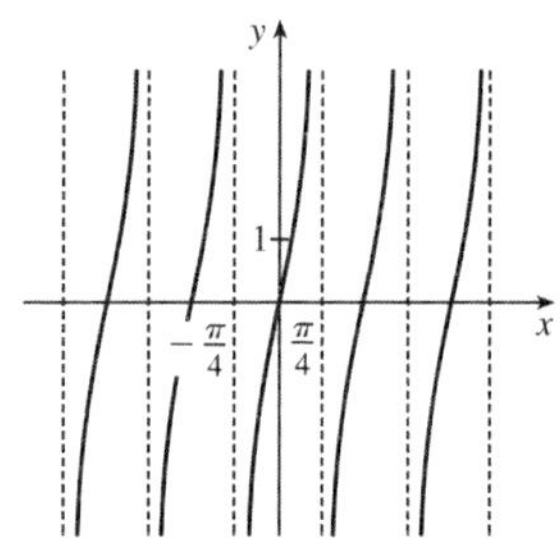


**39.** $y = \frac{1}{3}\tan\left(x - \frac{\pi}{2}\right)$. We start with the graph of $y = \tan x$, shift it $\frac{\pi}{2}$ units to the right and compress it to $\frac{1}{3}$ of its original vertical size.

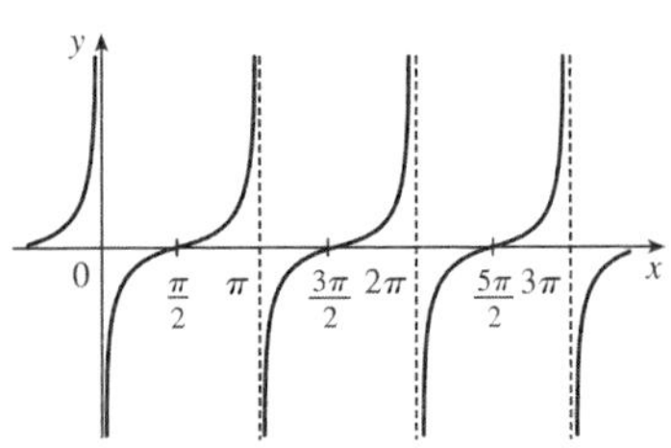


**40.** $y = |\sin x|$. We start with the graph of $y = \sin x$ and reflect the parts below the $x$-axis about the $x$-axis.

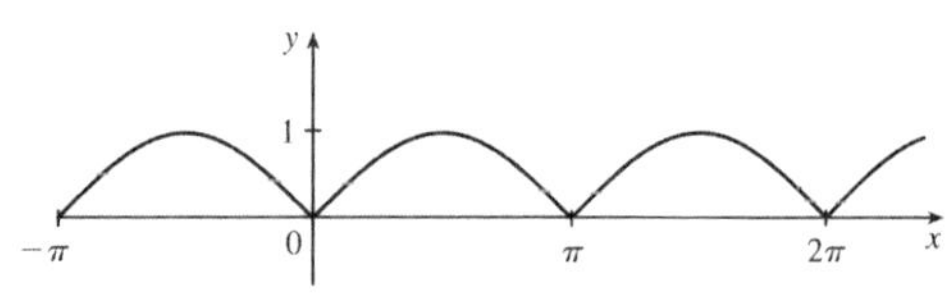


**41.** (a) $\sin^{-1}(0.5) = \frac{\pi}{6}$ since $\sin\frac{\pi}{6} = 0.5$ and $-\frac{\pi}{2} \le \frac{\pi}{6} \le \frac{\pi}{2}$.

(b) $\arctan(-1) = -\frac{\pi}{4}$ since $\tan\left(-\frac{\pi}{4}\right) = -1$ and $-\frac{\pi}{4}$ is in $\left(-\frac{\pi}{2}, \frac{\pi}{2}\right)$.

**42.** (a) $\tan^{-1}\sqrt{3} = \frac{\pi}{3}$ since $\tan\frac{\pi}{3} = \sqrt{3}$ and $\frac{\pi}{3}$ is in $\left(-\frac{\pi}{2}, \frac{\pi}{2}\right)$.

(b) $\arcsin 1 = \frac{\pi}{2}$ since $\sin\frac{\pi}{2} = 1$ and $\frac{\pi}{2}$ is in $\left[-\frac{\pi}{2}, \frac{\pi}{2}\right]$.

**43.** (a) $\sin\left(\sin^{-1}(0.7)\right) = 0.7$ since $0.7$ is in $[-1, 1]$.

(b) $\arcsin\left(\sin\frac{5\pi}{4}\right) = \arcsin\left(-\frac{1}{\sqrt{2}}\right) = -\frac{\pi}{4}$

**44.** (a) Let $\theta = \arctan 2$, so $\tan\theta = 2$. Now $\sec^2\theta = 1 + \tan^2\theta = 1 + 2^2 = 5 \;\Rightarrow\; \sec\theta = \sqrt{5}$ ($\sec\theta > 0$ since $0 < \theta < \pi/2$). Thus, $\sec(\arctan 2) = \sec\theta = \sqrt{5}$.

(b) Let $\theta = \sin^{-1}\frac{3}{5}$. Then $\sin\theta = \frac{3}{5} \;\Rightarrow\; \cos\theta = \sqrt{1 - \left(\frac{3}{5}\right)^2} = \frac{4}{5}$, so $\sin\left(2\sin^{-1}\frac{3}{5}\right) = \sin 2\theta = 2\sin\theta\cos\theta = 2\cdot\frac{3}{5}\cdot\frac{4}{5} = \frac{24}{25}$.

**45.** Let $y = \sin^{-1}x$. Then $-\frac{\pi}{2} \le y \le \frac{\pi}{2} \;\Rightarrow\; \cos y \ge 0$, so $\cos\left(\sin^{-1}x\right) = \cos y = \sqrt{1 - \sin^2 y} = \sqrt{1 - x^2}$.

**46.** (a) For the restricted cosine function, we have domain $= [0, \pi]$ and range $= [-1, 1]$. So for the inverse cosine function, domain $= [-1, 1]$ and range $= [0, \pi]$.

(b)

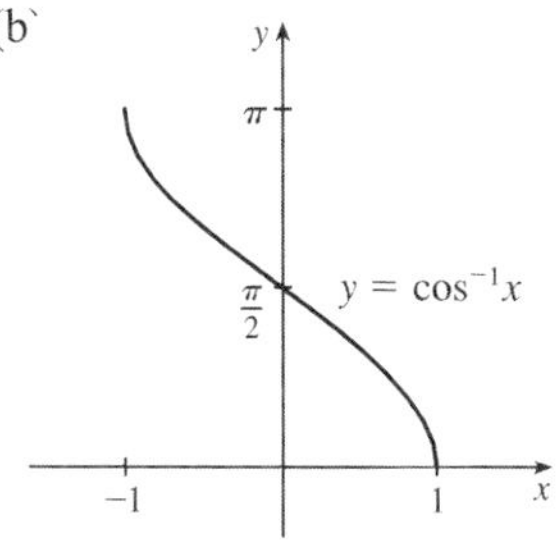


**47.** $g(x) = \sin^{-1}(3x + 1)$.

Domain $(g) = \{x \mid -1 \le 3x + 1 \le 1\} = \{x \mid -2 \le 3x \le 0\} = \left\{x \mid -\frac{2}{3} \le x \le 0\right\} = \left[-\frac{2}{3}, 0\right]$.

Range $(g) = \left\{y \mid -\frac{\pi}{2} \le y \le \frac{\pi}{2}\right\} = \left[-\frac{\pi}{2}, \frac{\pi}{2}\right]$.

**48.** (a) $f(x) = \sin\left(\sin^{-1}x\right)$

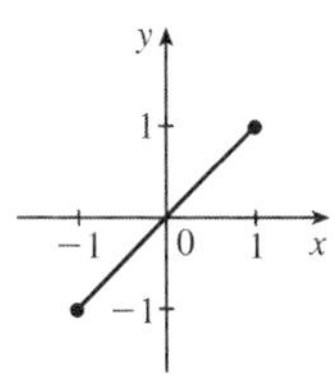


Since one function undoes what the other one does, we get the identity function, $y = x$, on the restricted domain $-1 \le x \le 1$.

(b) $g(x) = \sin^{-1}(\sin x)$

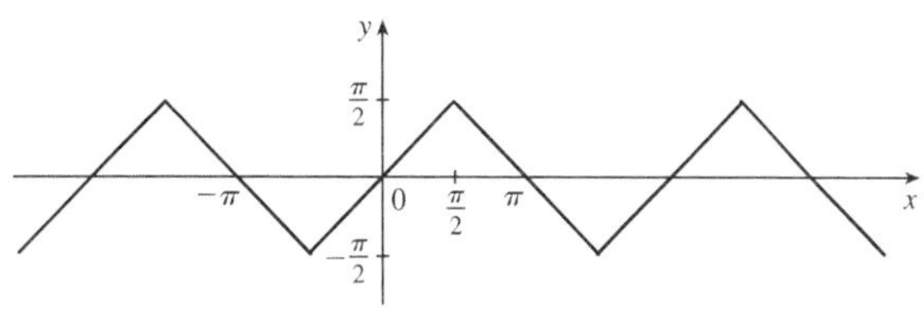


This is similar to part (a), but with domain $\mathbb{R}$. Equations for $g$ on intervals of the form $\left(-\frac{\pi}{2} + \pi n, \frac{\pi}{2} + \pi n\right)$, for any integer $n$, can be found using $g(x) = (-1)^n x + (-1)^{n+1} n\pi$. The sine function is monotonic on each of these intervals, and hence, so is $g$ (but in a linear fashion).

**49.** From the figure in the text, we see that $x = b\cos\theta$, $y = b\sin\theta$, and from the distance formula we have that the distance $c$ from $(x, y)$ to $(a, 0)$ is $c = \sqrt{(x - a)^2 + (y - 0)^2} \;\Rightarrow$

$$\begin{aligned} c^2 &= (b\cos\theta - a)^2 + (b\sin\theta)^2 = b^2\cos^2\theta - 2ab\cos\theta + a^2 + b^2\sin^2\theta \\ &= a^2 + b^2\left(\cos^2\theta + \sin^2\theta\right) - 2ab\cos\theta = a^2 + b^2 - 2ab\cos\theta \quad \text{[by (7)]} \end{aligned}$$

**50.**
$$\begin{aligned} |AB|^2 &= |AC|^2 + |BC|^2 - 2\,|AC|\,|BC|\cos\angle C \\ &= (820)^2 + (910)^2 - 2(820)(910)\cos 103^\circ \\ &\approx 1{,}836{,}217 \;\Rightarrow\; |AB| \approx 1355 \text{ m} \end{aligned}$$

**51.** Using the Law of Cosines, we have $c^2 = 1^2 + 1^2 - 2(1)(1)\cos(\alpha - \beta) = 2[1 - \cos(\alpha - \beta)]$. Now, using the distance formula, $c^2 = |AB|^2 = (\cos\alpha - \cos\beta)^2 + (\sin\alpha - \sin\beta)^2$. Equating these two expressions for $c^2$, we get $2[1 - \cos(\alpha - \beta)] = \cos^2\alpha + \sin^2\alpha + \cos^2\beta + \sin^2\beta - 2\cos\alpha\cos\beta - 2\sin\alpha\sin\beta \Rightarrow$ $1 - \cos(\alpha - \beta) = 1 - \cos\alpha\cos\beta - \sin\alpha\sin\beta \Rightarrow \cos(\alpha - \beta) = \cos\alpha\cos\beta + \sin\alpha\sin\beta$.

**52.** $\cos(x + y) = \cos(x - (-y)) = \cos x\cos(-y) + \sin x\sin(-y)$

$= \cos x\cos y - \sin x\sin y$ [using Equations (10a) and (10b)]

**53.** In Exercise 52 we used the subtraction formula for cosine to prove the addition formula for cosine. Using that formula with $x = \frac{\pi}{2} - \alpha$, $y = \beta$, we get $\cos\left[\left(\frac{\pi}{2} - \alpha\right) + \beta\right] = \cos\left(\frac{\pi}{2} - \alpha\right)\cos\beta - \sin\left(\frac{\pi}{2} - \alpha\right)\sin\beta \Rightarrow$ $\cos\left[\frac{\pi}{2} - (\alpha - \beta)\right] = \cos\left(\frac{\pi}{2} - \alpha\right)\cos\beta - \sin\left(\frac{\pi}{2} - \alpha\right)\sin\beta$. Now we use the identities given in the problem, $\cos\left(\frac{\pi}{2} - \theta\right) = \sin\theta$ and $\sin\left(\frac{\pi}{2} - \theta\right) = \cos\theta$, to get $\sin(\alpha - \beta) = \sin\alpha\cos\beta - \cos\alpha\sin\beta$.

**54.** (a) If $0 < \theta < \frac{\pi}{2}$, we have the case depicted in the first diagram. In this case, we see that the height of the triangle is $h = a\sin\theta$. If $\frac{\pi}{2} \le \theta < \pi$, we have the case depicted in the second diagram. In this case, the height of the triangle is $h = a\sin(\pi - \theta) = a\sin\theta$ (by the identity proved in Exercise 22). So in either case, the area of the triangle is $\frac{1}{2}bh = \frac{1}{2}ab\sin\theta$.

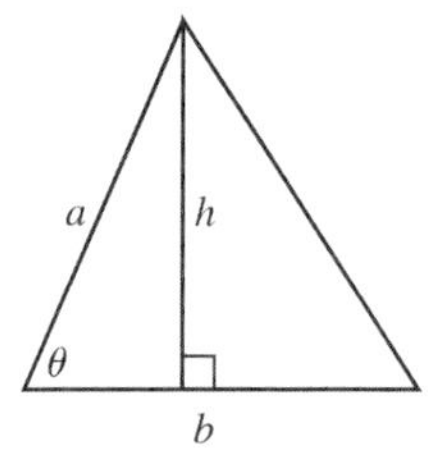


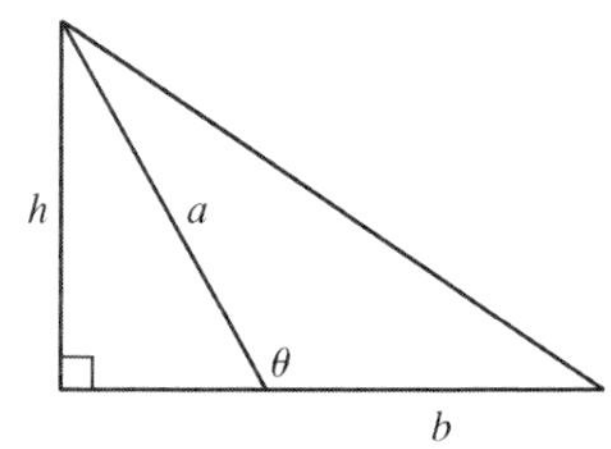


(b) Using the formula from part (a), the area of the triangle is $\frac{1}{2}(10)(3)\sin 107° \approx 14.34457\text{ cm}^2$.

## D PRECISE DEFINITIONS OF LIMITS

**1.** On the left side of $x = 2$, we need $|x - 2| < \left|\frac{10}{7} - 2\right| = \frac{4}{7}$. On the right side, we need $|x - 2| < \left|\frac{10}{3} - 2\right| = \frac{4}{3}$. For both of these conditions to be satisfied at once, we need the more restrictive of the two to hold, that is, $|x - 2| < \frac{4}{7}$. So we can choose $\delta = \frac{4}{7}$, or any smaller positive number.

**2.** The left-hand question mark is the positive solution of $x^2 = \frac{1}{2}$, that is, $x = \frac{1}{\sqrt{2}}$, and the right-hand question mark is the positive solution of $x^2 = \frac{3}{2}$, that is, $x = \sqrt{\frac{3}{2}}$. On the left side, we need $|x - 1| < \left|\frac{1}{\sqrt{2}} - 1\right| \approx 0.292$ (rounding down to be safe). On the right side, we need $|x - 1| < \left|\sqrt{\frac{3}{2}} - 1\right| \approx 0.224$. The more restrictive of these two conditions must apply, so we choose $\delta = 0.224$ (or any smaller positive number).

**3.** $\left|\sqrt{4x + 1} - 3\right| < 0.5 \quad\Leftrightarrow\quad 2.5 < \sqrt{4x + 1} < 3.5$. We plot the three parts of this inequality on the same screen and identify the $x$-coordinates of the points of intersection using the cursor. It appears that the inequality holds for $1.3125 \le x \le 2.8125$. Since $|2 - 1.3125| = 0.6875$ and $|2 - 2.8125| = 0.8125$, we choose $0 < \delta < \min\{0.6875, 0.8125\} = 0.6875$.

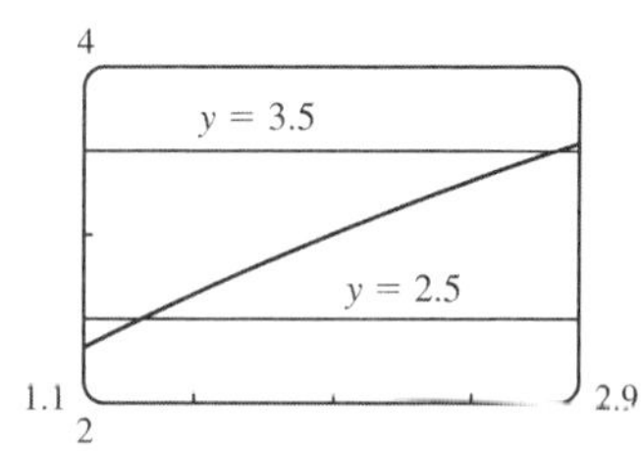


**4.** $\left|\sin x - \frac{1}{2}\right| < 0.1 \quad\Leftrightarrow\quad 0.4 < \sin x < 0.6$. From the graph, we see that for this inequality to hold, we need $A \le x \le B$ with $A \approx 0.412$ and $B \approx 0.644$. So since $\left|\frac{\pi}{6} - A\right| \approx 0.112$ and $\left|\frac{\pi}{6} - B\right| \approx 0.120$, we choose $0 < \delta \le \min\{0.112, 0.120\} = 0.112$.

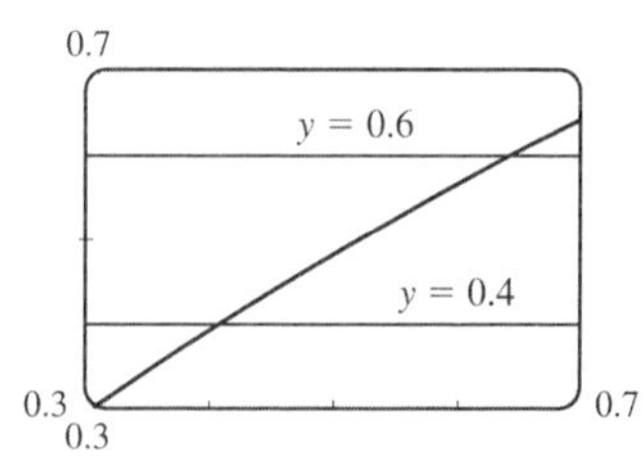

**5.** For $\varepsilon = 1$, the definition of a limit requires that we find $\delta$ such that $\left|(4 + x - 3x^3) - 2\right| < 1 \quad \Leftrightarrow$ $1 < 4 + x - 3x^3 < 3$ whenever $0 < |x - 1| < \delta$. If we plot the graphs of $y = 1$, $y = 4 + x - 3x^3$ and $y = 3$ on the same screen, we see that we need $0.86 \leq x \leq 1.11$. So since $|1 - 0.86| = 0.14$ and $|1 - 1.11| = 0.11$, we choose $\delta = 0.11$ (or any smaller positive number). For $\varepsilon = 0.1$, we must find $\delta$ such that $\left|(4 + x - 3x^3) - 2\right| < 0.1 \quad \Leftrightarrow \quad 1.9 < 4 + x - 3x^3 < 2.1$ whenever $0 < |x - 1| < \delta$. From the graph, we see that we need $0.988 \leq x \leq 1.012$. So since $|1 - 0.988| = 0.012$ and $|1 - 1.012| = 0.012$, we choose $\delta = 0.012$ (or any smaller positive number) for the inequality to hold.

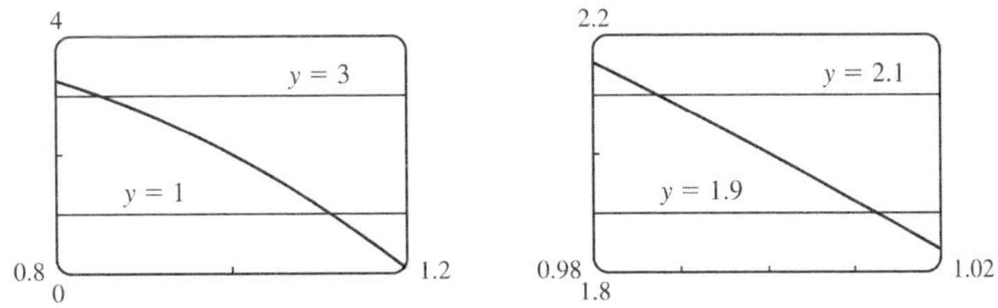


**6.** For $\varepsilon = 0.5$, the definition of a limit requires that we find $\delta$ such that $\left|\dfrac{e^x - 1}{x} - 1\right| < 0.5 \quad \Leftrightarrow$ $0.5 < \dfrac{e^x - 1}{x} < 1.5$ whenever $0 < |x - 0| < \delta$. If we plot the graphs of $y = 0.5$, $y = \dfrac{e^x - 1}{x}$, and $y = 1.5$ on the same screen, we see that we need $-1.59 \leq x \leq 0.76$. So since $|0 - (-1.59)| = 1.59$ and $|0 - 0.76| = 0.76$, we choose $\delta = 0.76$ (or any smaller positive number). For $\varepsilon = 0.1$, we must find $\delta$ such that $\left|\dfrac{e^x - 1}{x} - 1\right| < 0.1 \quad \Leftrightarrow \quad 0.9 < \dfrac{e^x - 1}{x} < 1.1$ whenever $0 < |x - 0| < \delta$. From the graph, we see that we need $-0.21 \leq x \leq 0.18$. So since $|0 - (-0.21)| = 0.21$ and $|0 - 0.18| = 0.18$, we choose $\delta = 0.18$ (or any smaller positive number) for the inequality to hold.

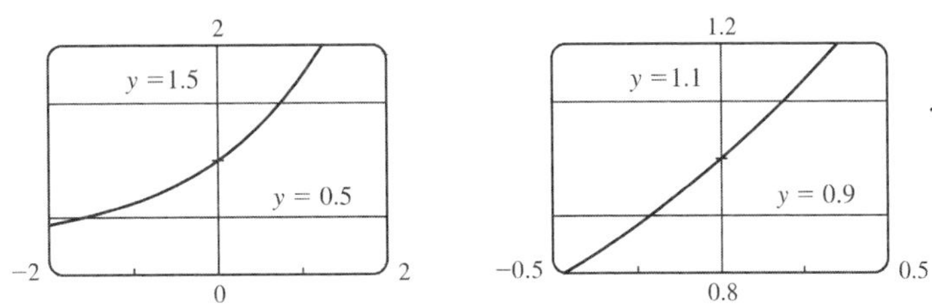


**7.** Given $\varepsilon > 0$, we need $\delta > 0$ such that if $|x| < \delta$ then $\left|x^3 - 0\right| < \varepsilon \quad \Leftrightarrow \quad |x|^3 < \varepsilon \quad \Leftrightarrow \quad |x| < \sqrt[3]{\varepsilon}$. Take $\delta = \sqrt[3]{\varepsilon}$. Then $|x - 0| < \delta \quad \Rightarrow \quad \left|x^3 - 0\right| < \delta^3 = \varepsilon$. Thus, $\lim\limits_{x \to 0} x^3 = 0$ by the definition of a limit.

**8.** (a) We must restrict the open interval $(a - \delta, a + \delta)$ in Definition 1 to the right half of the interval, $(a, a + \delta)$. Hence, $\lim\limits_{x \to a^+} f(x) = L$ if for every number $\varepsilon > 0$ there is a corresponding number $\delta > 0$ such that $|f(x) - L| < \varepsilon$ whenever $a < x < a + \delta$.

(b) 1. *Guessing a value for $\delta$.* Let $\varepsilon$ be a given positive number. Here $a = 0$ and $L = 0$, so we want to find a number $\delta$ such that $\left|\sqrt{x} - 0\right| < \varepsilon$ whenever $0 < x < \delta$, that is, $\sqrt{x} < \varepsilon$ whenever $0 < x < \delta$ or, squaring both sides of the inequality $\sqrt{x} < \varepsilon$, we get $x < \varepsilon^2$ whenever $0 < x < \delta$. This suggests that we should choose $\delta = \varepsilon^2$.

2. *Showing that this $\delta$ works.* Given $\varepsilon > 0$, let $\delta = \varepsilon^2$. If $0 < x < \delta$, then $\sqrt{x} < \sqrt{\delta} = \sqrt{\varepsilon^2} = \varepsilon$, so $\left|\sqrt{x} - 0\right| < \varepsilon$. According to the definition in part (a), this shows that $\lim\limits_{x \to 0^+} \sqrt{x} = 0$.

**9.** Given $\varepsilon > 0$, we need $\delta > 0$ such that if $|x - 2| < \delta$, then

$|(3x - 2) - 4| < \varepsilon \quad \Leftrightarrow \quad |3x - 6| < \varepsilon \Leftrightarrow 3\,|x - 2| < \varepsilon \Leftrightarrow$

$|x - 2| < \varepsilon/3$. So if we choose $\delta = \varepsilon/3$, then

$|x - 2| < \delta \quad \Rightarrow \quad |(3x - 2) - 4| < \varepsilon$. Thus, $\lim\limits_{x \to 2}(3x - 2) = 4$ by the definition of a limit.

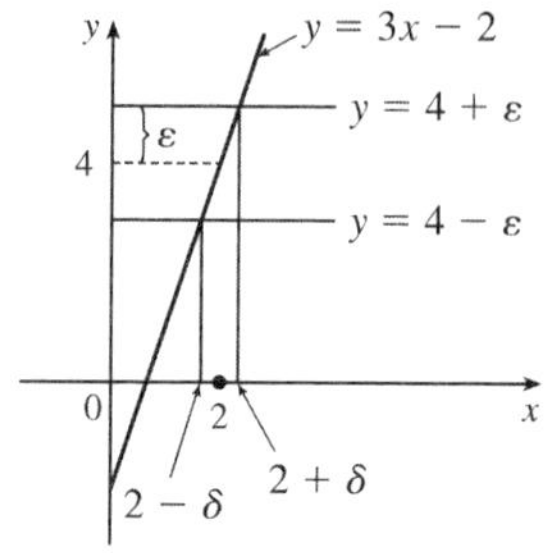


**10.** Given $\varepsilon > 0$, we need $\delta > 0$ such that if $|x - 4| < \delta$, then

$|(5 - 2x) - (-3)| < \varepsilon \quad \Leftrightarrow \quad |-2x + 8| < \varepsilon \quad \Leftrightarrow \quad 2\,|x - 4| < \varepsilon \quad \Leftrightarrow \quad |x - 4| < \varepsilon/2$. So if we choose $\delta = \varepsilon/2$, then

$|x - 4| < \delta \quad \Rightarrow \quad |(5 - 2x) - (-3)| < \varepsilon$. Thus,

$\lim\limits_{x \to 4}(5 - 2x) = -3$ by the definition of a limit.

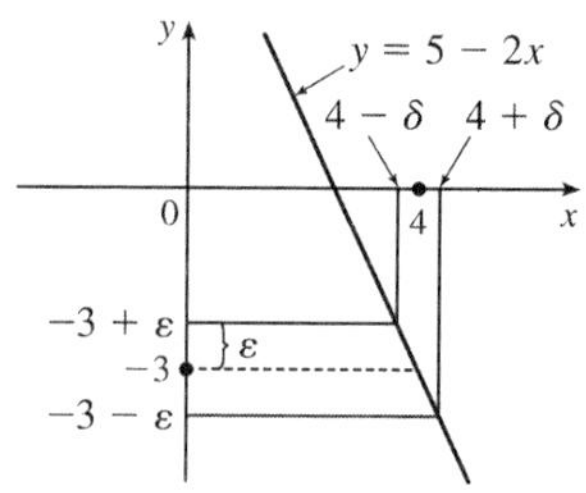


**11.** (a) $A = \pi r^2$ and $A = 1000\text{ cm}^2 \quad \Rightarrow \quad \pi r^2 = 1000 \quad \Rightarrow \quad r^2 = \frac{1000}{\pi} \quad \Rightarrow$

$r = \sqrt{\frac{1000}{\pi}} \quad (r > 0) \quad \approx 17.8412$ cm.

(b) $|A - 1000| \le 5 \quad \Rightarrow \quad -5 \le \pi r^2 - 1000 \le 5 \quad \Rightarrow \quad 1000 - 5 \le \pi r^2 \le 1000 + 5 \quad \Rightarrow$

$\sqrt{\frac{995}{\pi}} \le r \le \sqrt{\frac{1005}{\pi}} \quad \Rightarrow \quad 17.7966 \le r \le 17.8858$. $\sqrt{\frac{1000}{\pi}} - \sqrt{\frac{995}{\pi}} \approx 0.04466$ and $\sqrt{\frac{1005}{\pi}} - \sqrt{\frac{1000}{\pi}} \approx 0.04455$. So if the machinist gets the radius within 0.0445 cm of 17.8412, the area will be within 5 cm$^2$ of 1000.

(c) $x$ is the radius, $f(x)$ is the area, $a$ is the target radius given in part (a), $L$ is the target area (1000), $\varepsilon$ is the tolerance in the area (5), and $\delta$ is the tolerance in the radius given in part (b).

**12.** (a) $T = 0.1w^2 + 2.155w + 20$ and $T = 200 \quad \Rightarrow \quad 0.1w^2 + 2.155w + 20 = 200 \quad \Rightarrow$ (by the quadratic formula or from the graph) $w \approx 33.0$ watts ($w > 0$)

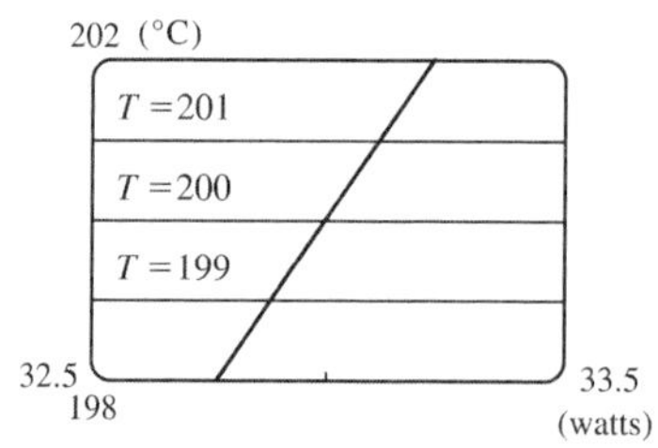


(b) From the graph, $199 \le T \le 201 \quad \Rightarrow \quad 32.89 < w < 33.11$.

(c) $x$ is the input power, $f(x)$ is the temperature, $a$ is the target input power given in part (a), $L$ is the target temperature (200), $\varepsilon$ is the tolerance in the temperature (1), and $\delta$ is the tolerance in the power input in watts indicated in part (b) (0.11 watts).

**13.** $\left|\dfrac{6x^2 + 5x - 3}{2x^2 - 1} - 3\right| < 0.2 \quad \Leftrightarrow \quad 2.8 < \dfrac{6x^2 + 5x - 3}{2x^2 - 1} < 3.2$. So we graph the three parts of this inequality on the same screen, and find that the curve $y = \dfrac{6x^2 + 5x - 3}{2x^2 - 1}$ seems to lie between the lines $y = 2.8$ and $y = 3.2$ whenever $x > 12.5$. So we can choose $N = 13$ (or any larger number), so that the inequality holds whenever $x \ge N$.

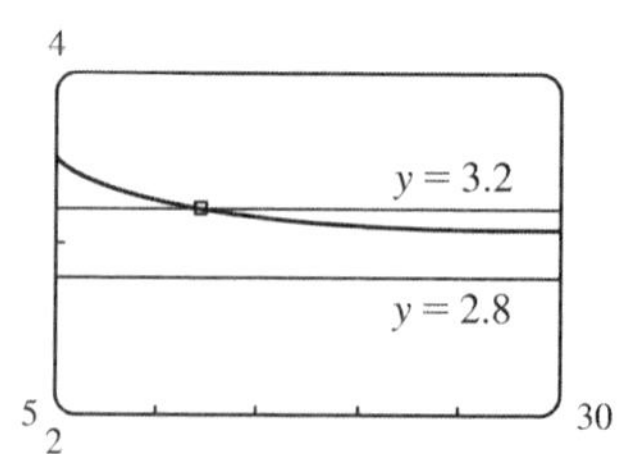

**14.** For $\varepsilon = 0.5$, we must find $N$ such that whenever $x \geq N$, we have $\left| \dfrac{\sqrt{4x^2+1}}{x+1} - 2 \right| < 0.5 \quad \Leftrightarrow \quad 1.5 < \dfrac{\sqrt{4x^2+1}}{x+1} < 2.5$. We graph the three parts of this inequality on the same screen, and find that it holds whenever $x > 2.82$. So we choose $N = 3$ (or any larger number). For $\varepsilon = 0.1$, we must have $1.9 < \dfrac{\sqrt{4x^2+1}}{x+1} < 2.1$, and the graphs show that this holds whenever $x > 18.9$. So we choose $N = 19$ (or any larger number).

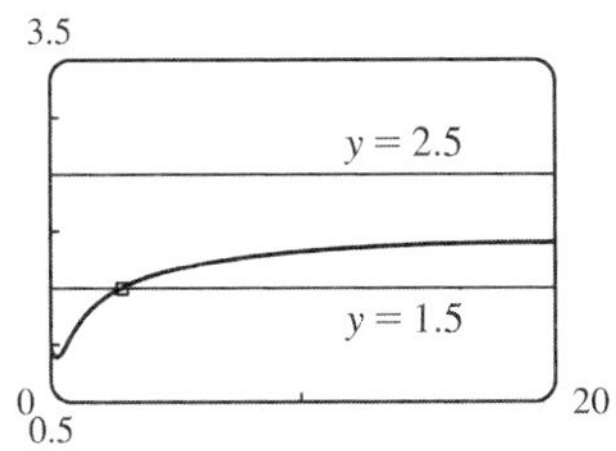


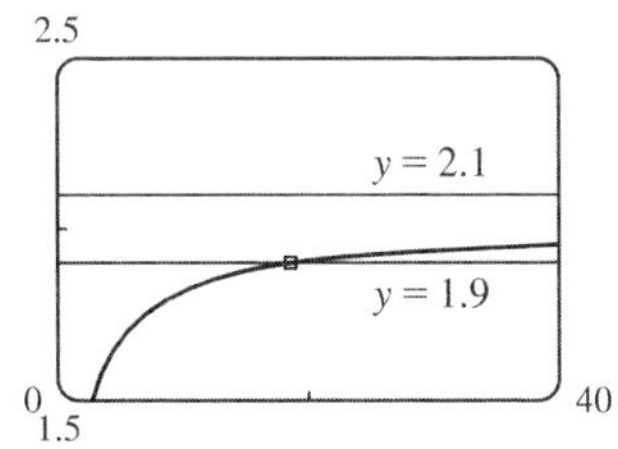


**15.** (a) $1/x^2 < 0.0001 \quad \Leftrightarrow \quad x^2 > 1/0.0001 = 10{,}000 \quad \Leftrightarrow \quad x > 100 \quad (x > 0)$

(b) If $\varepsilon > 0$ is given, then $1/x^2 < \varepsilon \quad \Leftrightarrow \quad x^2 > 1/\varepsilon \quad \Leftrightarrow \quad x > 1/\sqrt{\varepsilon}$. Let $N = 1/\sqrt{\varepsilon}$ . Then $x > N \quad \Rightarrow \quad x > \dfrac{1}{\sqrt{\varepsilon}} \quad \Rightarrow$ $\left| \dfrac{1}{x^2} - 0 \right| = \dfrac{1}{x^2} < \varepsilon$, so $\lim\limits_{x \to \infty} \dfrac{1}{x^2} = 0$.

**16.** (a) $1/x^2 < 0.0001 \quad \Leftrightarrow \quad x^2 > 1/0.0001 = 10{,}000 \quad \Leftrightarrow \quad x > 100 \quad (x > 0)$

(b) If $\varepsilon > 0$ is given, then $1/x^2 < \varepsilon \quad \Leftrightarrow \quad x^2 > 1/\varepsilon \quad \Leftrightarrow \quad x > 1/\sqrt{\varepsilon}$. Let $N = 1/\sqrt{\varepsilon}$ . Then $x > N \quad \Rightarrow \quad x > \dfrac{1}{\sqrt{\varepsilon}} \quad \Rightarrow$ $\left| \dfrac{1}{x^2} - 0 \right| = \dfrac{1}{x^2} < \varepsilon$, so $\lim\limits_{x \to \infty} \dfrac{1}{x^2} = 0$.

**17.** (a)

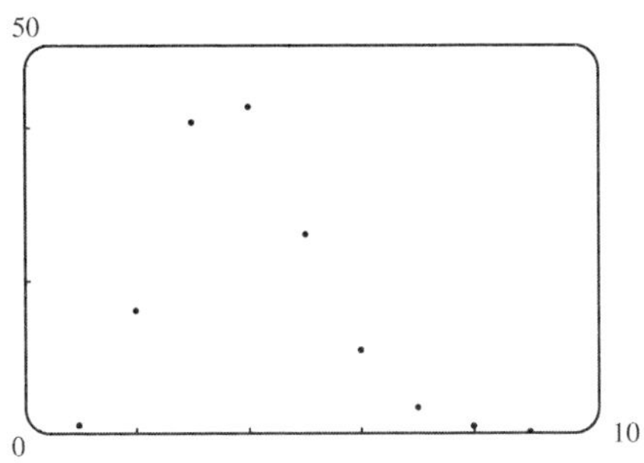


From the graph, it appears that the sequence $\left\{ \dfrac{n^5}{n!} \right\}$ converges to 0, that is, $\lim\limits_{n \to \infty} \dfrac{n^5}{n!} = 0$.

(b)

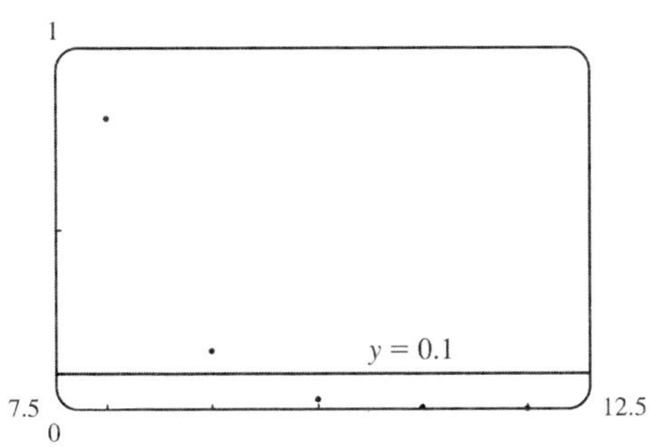


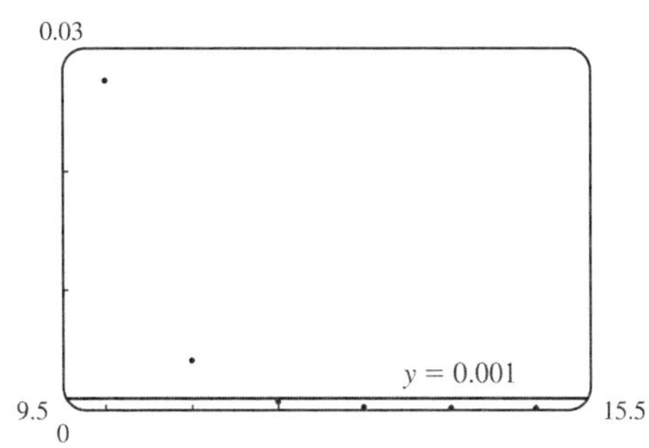


From the first graph, it seems that the smallest possible value of $N$ corresponding to $\varepsilon = 0.1$ is 9, since $n^5/n! < 0.1$ whenever $n \geq 10$, but $9^5/9! > 0.1$. From the second graph, it seems that for $\varepsilon = 0.001$, the smallest possible value for $N$ is 11 since $n^5/n! < 0.001$ whenever $n \geq 12$.

**18.** Let $\varepsilon > 0$ and let $N$ be any positive integer larger than $\ln(\varepsilon)/\ln|r|$. If $n > N$, then $n > \ln(\varepsilon)/\ln|r| \Rightarrow n\ln|r| < \ln\varepsilon$ [since $|r| < 1 \Rightarrow \ln|r| < 0$] $\Rightarrow \ln(|r|^n) < \ln\varepsilon \Rightarrow |r|^n < \varepsilon \Rightarrow |r^n - 0| < \varepsilon$, and so by Definition 3, $\lim\limits_{n\to\infty} r^n = 0$.

**19.** If $\lim\limits_{n\to\infty} |a_n| = 0$, then $\lim\limits_{n\to\infty} (-|a_n|) = 0$, and since $-|a_n| \le a_n \le |a_n|$, we have that $\lim\limits_{n\to\infty} a_n = 0$ by the Squeeze Theorem.

**20.** Let $M$ be any positive number. Then $n^3 > M \Leftrightarrow n > \sqrt[3]{M}$. So if we take $N = \sqrt[3]{M}$, then $n^3 > M$ whenever $n > N$. By Definition 4, $\lim\limits_{n\to\infty} n^3 = \infty$.

## F SIGMA NOTATION

**1.** $\sum\limits_{i=1}^{5} \sqrt{i} = \sqrt{1} + \sqrt{2} + \sqrt{3} + \sqrt{4} + \sqrt{5}$

**2.** $\sum\limits_{i=1}^{6} \dfrac{1}{i+1} = \dfrac{1}{2} + \dfrac{1}{3} + \dfrac{1}{4} + \dfrac{1}{5} + \dfrac{1}{6} + \dfrac{1}{7}$

**3.** $\sum\limits_{i=4}^{6} 3^i = 3^4 + 3^5 + 3^6$

**4.** $\sum\limits_{i=4}^{6} i^3 = 4^3 + 5^3 + 6^3$

**5.** $\sum\limits_{k=0}^{4} \dfrac{2k-1}{2k+1} = -1 + \dfrac{1}{3} + \dfrac{3}{5} + \dfrac{5}{7} + \dfrac{7}{9}$

**6.** $\sum\limits_{k=5}^{8} x^k = x^5 + x^6 + x^7 + x^8$

**7.** $\sum\limits_{i=1}^{n} i^{10} = 1^{10} + 2^{10} + 3^{10} + \cdots + n^{10}$

**8.** $\sum\limits_{j=n}^{n+3} j^2 = n^2 + (n+1)^2 + (n+2)^2 + (n+3)^2$

**9.** $\sum\limits_{j=0}^{n-1} (-1)^j = 1 - 1 + 1 - 1 + \cdots + (-1)^{n-1}$

**10.** $\sum\limits_{i=1}^{n} f(x_i)\,\Delta x_i = f(x_1)\,\Delta x_1 + f(x_2)\,\Delta x_2 + f(x_3)\,\Delta x_3 + \cdots + f(x_n)\,\Delta x_n$

**11.** $1 + 2 + 3 + 4 + \cdots + 10 = \sum\limits_{i=1}^{10} i$

**12.** $\sqrt{3} + \sqrt{4} + \sqrt{5} + \sqrt{6} + \sqrt{7} = \sum\limits_{i=3}^{7} \sqrt{i}$

**13.** $\dfrac{1}{2} + \dfrac{2}{3} + \dfrac{3}{4} + \dfrac{4}{5} + \cdots + \dfrac{19}{20} = \sum\limits_{i=1}^{19} \dfrac{i}{i+1}$

**14.** $\dfrac{3}{7} + \dfrac{4}{8} + \dfrac{5}{9} + \dfrac{6}{10} + \cdots + \dfrac{23}{27} = \sum\limits_{i=3}^{23} \dfrac{i}{i+4}$

**15.** $2 + 4 + 6 + 8 + \cdots + 2n = \sum\limits_{i=1}^{n} 2i$

**16.** $1 + 3 + 5 + 7 + \cdots + (2n-1) = \sum\limits_{i=1}^{n} (2i-1)$

**17.** $1 + 2 + 4 + 8 + 16 + 32 = \sum\limits_{i=0}^{5} 2^i$

**18.** $\dfrac{1}{1} + \dfrac{1}{4} + \dfrac{1}{9} + \dfrac{1}{16} + \dfrac{1}{25} + \dfrac{1}{36} = \sum\limits_{i=1}^{6} \dfrac{1}{i^2}$

**19.** $x + x^2 + x^3 + \cdots + x^n = \sum\limits_{i=1}^{n} x^i$

**20.** $1 - x + x^2 - x^3 + \cdots + (-1)^n x^n = \sum\limits_{i=0}^{n} (-1)^i x^i$

**21.** $\sum\limits_{i=4}^{8} (3i-2) = [3(4)-2] + [3(5)-2] + [3(6)-2] + [3(7)-2] + [3(8)-2] = 10 + 13 + 16 + 19 + 22 = 80$

**22.** $\sum\limits_{i=3}^{6} i(i+2) = 3\cdot 5 + 4\cdot 6 + 5\cdot 7 + 6\cdot 8 = 15 + 24 + 35 + 48 = 122$

**23.** $\sum_{j=1}^{6} 3^{j+1} = 3^2 + 3^3 + 3^4 + 3^5 + 3^6 + 3^7 = 9 + 27 + 81 + 243 + 729 + 2187 = 3276$

(For a more general method, see Exercise 47.)

**24.** $\sum_{k=0}^{8} \cos k\pi = \cos 0 + \cos \pi + \cos 2\pi + \cos 3\pi + \cos 4\pi + \cos 5\pi + \cos 6\pi + \cos 7\pi + \cos 8\pi$

$= 1 - 1 + 1 - 1 + 1 - 1 + 1 - 1 + 1 = 1$

**25.** $\sum_{n=1}^{20} (-1)^n = -1 + 1 - 1 + 1 - 1 + 1 - 1 + 1 - 1 + 1 - 1 + 1 - 1 + 1 - 1 + 1 - 1 + 1 - 1 + 1 = 0$

**26.** $\sum_{i=1}^{100} 4 = \underbrace{4 + 4 + 4 + \cdots + 4}_{\text{(100 summands)}} = 100 \cdot 4 = 400$

**27.** $\sum_{i=0}^{4} (2^i + i^2) = (1 + 0) + (2 + 1) + (4 + 4) + (8 + 9) + (16 + 16) = 61$

**28.** $\sum_{i=-2}^{4} 2^{3-i} = 2^5 + 2^4 + 2^3 + 2^2 + 2^1 + 2^0 + 2^{-1} = 63.5$

**29.** $\sum_{i=1}^{n} 2i = 2 \sum_{i=1}^{n} i = 2 \cdot \frac{n(n+1)}{2}$ [by Theorem 3(c)] $= n(n+1)$

**30.** $\sum_{i=1}^{n} (2 - 5i) = \sum_{i=1}^{n} 2 - \sum_{i=1}^{n} 5i = 2n - 5 \sum_{i=1}^{n} i = 2n - \frac{5n(n+1)}{2} = \frac{4n}{2} - \frac{5n^2 + 5n}{2} = -\frac{n(5n+1)}{2}$

**31.**
$$\begin{aligned}\sum_{i=1}^{n} (i^2 + 3i + 4) &= \sum_{i=1}^{n} i^2 + 3 \sum_{i=1}^{n} i + \sum_{i=1}^{n} 4 = \frac{n(n+1)(2n+1)}{6} + \frac{3n(n+1)}{2} + 4n \\ &= \tfrac{1}{6}\left[(2n^3 + 3n^2 + n) + (9n^2 + 9n) + 24n\right] = \tfrac{1}{6}\left(2n^3 + 12n^2 + 34n\right) \\ &= \tfrac{1}{3}n(n^2 + 6n + 17)\end{aligned}$$

**32.**
$$\begin{aligned}\sum_{i=1}^{n} (3 + 2i)^2 &= \sum_{i=1}^{n} (9 + 12i + 4i^2) = \sum_{i=1}^{n} 9 + 12 \sum_{i=1}^{n} i + 4 \sum_{i=1}^{n} i^2 \\ &= 9n + 6n(n+1) + \frac{2n(n+1)(2n+1)}{3} = \frac{27n + 18n^2 + 18n + 4n^3 + 6n^2 + 2n}{3} \\ &= \tfrac{1}{3}\left(4n^3 + 24n^2 + 47n\right) = \tfrac{1}{3}n(4n^2 + 24n + 47)\end{aligned}$$

**33.**
$$\begin{aligned}\sum_{i=1}^{n} (i+1)(i+2) &= \sum_{i=1}^{n} (i^2 + 3i + 2) = \sum_{i=1}^{n} i^2 + 3 \sum_{i=1}^{n} i + \sum_{i=1}^{n} 2 \\ &= \frac{n(n+1)(2n+1)}{6} + \frac{3n(n+1)}{2} + 2n = \frac{n(n+1)}{6}\left[(2n+1) + 9\right] + 2n \\ &= \frac{n(n+1)}{3}(n+5) + 2n = \frac{n}{3}\left[(n+1)(n+5) + 6\right] = \frac{n}{3}\left(n^2 + 6n + 11\right)\end{aligned}$$

**34.**
$$\begin{aligned}\sum_{i=1}^{n} i(i+1)(i+2) &= \sum_{i=1}^{n} \left(i^3 + 3i^2 + 2i\right) = \sum_{i=1}^{n} i^3 + 3 \sum_{i=1}^{n} i^2 + 2 \sum_{i=1}^{n} i \\ &= \left[\frac{n(n+1)}{2}\right]^2 + \frac{3n(n+1)(2n+1)}{6} + \frac{2n(n+1)}{2} \\ &= n(n+1)\left[\frac{n(n+1)}{4} + \frac{2n+1}{2} + 1\right] = \frac{n(n+1)}{4}\left(n^2 + n + 4n + 2 + 4\right) \\ &= \frac{n(n+1)}{4}\left(n^2 + 5n + 6\right) = \frac{n(n+1)(n+2)(n+3)}{4}\end{aligned}$$

**35.** $\sum_{i=1}^{n}(i^3 - i - 2) = \sum_{i=1}^{n} i^3 - \sum_{i=1}^{n} i - \sum_{i=1}^{n} 2 = \left[\frac{n(n+1)}{2}\right]^2 - \frac{n(n+1)}{2} - 2n$

$$= \tfrac{1}{4}n(n+1)\,[n(n+1) - 2] - 2n = \tfrac{1}{4}n(n+1)(n+2)(n-1) - 2n$$
$$= \tfrac{1}{4}n\,[(n+1)(n-1)(n+2) - 8] = \tfrac{1}{4}n\left[(n^2-1)(n+2) - 8\right] = \tfrac{1}{4}n(n^3 + 2n^2 - n - 10)$$

**36.** By Theorem 3(c) we have that $\sum_{i=1}^{n} i = \frac{n(n+1)}{2} = 78 \Leftrightarrow n(n+1) = 156 \Leftrightarrow n^2 + n - 156 = 0 \Leftrightarrow$ $(n+13)(n-12) = 0 \Leftrightarrow n = 12$ or $-13$. But $n = -13$ produces a negative answer for the sum, so $n = 12$.

**37.** By Theorem 2(a) and Example 3, $\sum_{i=1}^{n} c = c\sum_{i=1}^{n} 1 = cn$.

**38.** Let $S_n$ be the statement that $\sum_{i=1}^{n} i^3 = \left[\frac{n(n+1)}{2}\right]^2$.

1. $S_1$ is true because $1^3 = \left(\frac{1\cdot 2}{2}\right)^2$.

2. Assume $S_k$ is true. Then $\sum_{i=1}^{k} i^3 = \left[\frac{k(k+1)}{2}\right]^2$, so

$$\sum_{i=1}^{k+1} i^3 = \left[\frac{k(k+1)}{2}\right]^2 + (k+1)^3 = \frac{(k+1)^2}{4}\left[k^2 + 4(k+1)\right] = \frac{(k+1)^2}{4}(k+2)^2$$
$$= \left(\frac{(k+1)\,[(k+1)+1]}{2}\right)^2$$

showing that $S_{k+1}$ is true.

Therefore, $S_n$ is true for all $n$ by mathematical induction.

**39.** $\sum_{i=1}^{n}\left[(i+1)^4 - i^4\right] = (2^4 - 1^4) + (3^4 - 2^4) + (4^4 - 3^4) + \cdots + \left[(n+1)^4 - n^4\right]$

$$= (n+1)^4 - 1^4 = n^4 + 4n^3 + 6n^2 + 4n$$

On the other hand,

$$\sum_{i=1}^{n}\left[(i+1)^4 - i^4\right] = \sum_{i=1}^{n}(4i^3 + 6i^2 + 4i + 1) = 4\sum_{i=1}^{n} i^3 + 6\sum_{i=1}^{n} i^2 + 4\sum_{i=1}^{n} i + \sum_{i=1}^{n} 1$$
$$= 4S + n(n+1)(2n+1) + 2n(n+1) + n \quad \left[\text{where } S = \sum_{i=1}^{n} i^3\right]$$
$$= 4S + 2n^3 + 3n^2 + n + 2n^2 + 2n + n = 4S + 2n^3 + 5n^2 + 4n$$

Thus, $n^4 + 4n^3 + 6n^2 + 4n = 4S + 2n^3 + 5n^2 + 4n$, from which it follows that $4S = n^4 + 2n^3 + n^2 = n^2(n^2 + 2n + 1) = n^2(n+1)^2$ and $S = \left[\frac{n(n+1)}{2}\right]^2$.

**40.** The area of $G_i$ is

$$\left(\sum_{k=1}^{i} k\right)^2 - \left(\sum_{k=1}^{i-1} k\right)^2 = \left[\frac{i(i+1)}{2}\right]^2 - \left[\frac{(i-1)i}{2}\right]^2 = \frac{i^2}{4}\left[(i+1)^2 - (i-1)^2\right]$$
$$= \frac{i^2}{4}\left[(i^2 + 2i + 1) - (i^2 - 2i + 1)\right] = \frac{i^2}{4}(4i) = i^3$$

Thus, the area of $ABCD$ is $\sum_{i=1}^{n} i^3 = \left[\frac{n(n+1)}{2}\right]^2$.

**41.** (a) $\sum_{i=1}^{n}\left[i^4-(i-1)^4\right]=\left(1^4-0^4\right)+\left(2^4-1^4\right)+\left(3^4-2^4\right)+\cdots+\left[n^4-(n-1)^4\right]=n^4-0=n^4$

(b) $\sum_{i=1}^{100}\left(5^i-5^{i-1}\right)=\left(5^1-5^0\right)+\left(5^2-5^1\right)+\left(5^3-5^2\right)+\cdots+\left(5^{100}-5^{99}\right)=5^{100}-5^0=5^{100}-1$

(c) $\sum_{i=3}^{99}\left(\frac{1}{i}-\frac{1}{i+1}\right)=\left(\frac{1}{3}-\frac{1}{4}\right)+\left(\frac{1}{4}-\frac{1}{5}\right)+\left(\frac{1}{5}-\frac{1}{6}\right)+\cdots+\left(\frac{1}{99}-\frac{1}{100}\right)=\frac{1}{3}-\frac{1}{100}=\frac{97}{300}$

(d) $\sum_{i=1}^{n}(a_i-a_{i-1})=(a_1-a_0)+(a_2-a_1)+(a_3-a_2)+\cdots+(a_n-a_{n-1})=a_n-a_0$

**42.** Summing the inequalities $-|a_i|\le a_i\le |a_i|$ for $i=1,2,\ldots,n$, we get $-\sum_{i=1}^{n}|a_i|\le\sum_{i=1}^{n}a_i\le\sum_{i=1}^{n}|a_i|$. Since $|x|\le c\ \Leftrightarrow$ $-c\le x\le c$, we have $\left|\sum_{i=1}^{n}a_i\right|\le\sum_{i=1}^{n}|a_i|$. *Another method:* Use mathematical induction.

**43.** $\lim_{n\to\infty}\sum_{i=1}^{n}\frac{1}{n}\left(\frac{i}{n}\right)^2=\lim_{n\to\infty}\frac{1}{n^3}\sum_{i=1}^{n}i^2=\lim_{n\to\infty}\frac{1}{n^3}\frac{n(n+1)(2n+1)}{6}=\lim_{n\to\infty}\frac{1}{6}\left(1+\frac{1}{n}\right)\left(2+\frac{1}{n}\right)$
$=\frac{1}{6}(1)(2)=\frac{1}{3}$

**44.** $\lim_{n\to\infty}\sum_{i=1}^{n}\frac{1}{n}\left[\left(\frac{i}{n}\right)^3+1\right]=\lim_{n\to\infty}\sum_{i=1}^{n}\left[\frac{i^3}{n^4}+\frac{1}{n}\right]=\lim_{n\to\infty}\left[\frac{1}{n^4}\sum_{i=1}^{n}i^3+\frac{1}{n}\sum_{i=1}^{n}1\right]$
$=\lim_{n\to\infty}\left[\frac{1}{n^4}\left(\frac{n(n+1)}{2}\right)^2+\frac{1}{n}(n)\right]=\lim_{n\to\infty}\frac{1}{4}\left(1+\frac{1}{n}\right)^2+1=\frac{1}{4}+1=\frac{5}{4}$

**45.** $\lim_{n\to\infty}\sum_{i=1}^{n}\frac{2}{n}\left[\left(\frac{2i}{n}\right)^3+5\left(\frac{2i}{n}\right)\right]=\lim_{n\to\infty}\sum_{i=1}^{n}\left[\frac{16}{n^4}i^3+\frac{20}{n^2}i\right]=\lim_{n\to\infty}\left[\frac{16}{n^4}\sum_{i=1}^{n}i^3+\frac{20}{n^2}\sum_{i=1}^{n}i\right]$
$=\lim_{n\to\infty}\left[\frac{16}{n^4}\frac{n^2(n+1)^2}{4}+\frac{20}{n^2}\frac{n(n+1)}{2}\right]=\lim_{n\to\infty}\left[\frac{4(n+1)^2}{n^2}+\frac{10n(n+1)}{n^2}\right]$
$=\lim_{n\to\infty}\left[4\left(1+\frac{1}{n}\right)^2+10\left(1+\frac{1}{n}\right)\right]=4\cdot 1+10\cdot 1=14$

**46.** $\lim_{n\to\infty}\sum_{i=1}^{n}\frac{3}{n}\left[\left(1+\frac{3i}{n}\right)^3-2\left(1+\frac{3i}{n}\right)\right]=\lim_{n\to\infty}\sum_{i=1}^{n}\frac{3}{n}\left[1+\frac{9i}{n}+\frac{27i^2}{n^2}+\frac{27i^3}{n^3}-2-\frac{6i}{n}\right]$
$=\lim_{n\to\infty}\sum_{i=1}^{n}\left[\frac{81}{n^4}i^3+\frac{81}{n^3}i^2+\frac{9}{n^2}i-\frac{3}{n}\right]$
$=\lim_{n\to\infty}\left[\frac{81}{n^4}\frac{n^2(n+1)^2}{4}+\frac{81}{n^3}\frac{n(n+1)(2n+1)}{6}+\frac{9}{n^2}\frac{n(n+1)}{2}-\frac{3}{n}n\right]$
$=\lim_{n\to\infty}\left[\frac{81}{4}\left(1+\frac{1}{n}\right)^2+\frac{27}{2}\left(1+\frac{1}{n}\right)\left(2+\frac{1}{n}\right)+\frac{9}{2}\left(1+\frac{1}{n}\right)-3\right]=\frac{81}{4}+\frac{54}{2}+\frac{9}{2}-3=\frac{195}{4}$

**47.** Let $S=\sum_{i=1}^{n}ar^{i-1}=a+ar+ar^2+\cdots+ar^{n-1}$. Multiplying both sides by $r$ gives us $rS=ar+ar^2+\cdots+ar^{n-1}+ar^n$. Subtracting the first equation from the second, we find $(r-1)S=ar^n-a=a(r^n-1)$, so $S=\frac{a(r^n-1)}{r-1}$ (since $r\neq 1$).

**48.** $\sum_{i=1}^{n}\frac{3}{2^{i-1}}=3\sum_{i=1}^{n}\left(\frac{1}{2}\right)^{i-1}=\frac{3\left[\left(\frac{1}{2}\right)^n-1\right]}{\frac{1}{2}-1}$ [using Exercise 47 with $a=3$ and $r=\frac{1}{2}$] $=6\left[1-\left(\frac{1}{2}\right)^n\right]$

**49.** $\sum_{i=1}^{n}(2i+2^i)=2\sum_{i=1}^{n}i+\sum_{i=1}^{n}2\cdot 2^{i-1}=2\frac{n(n+1)}{2}+\frac{2(2^n-1)}{2-1}=2^{n+1}+n^2+n-2.$

For the first sum we have used Theorem 3(c), and for the second, Exercise 47 with $a=r=2$.

**50.** $\sum_{i=1}^{m}\left[\sum_{j=1}^{n}(i+j)\right]=\sum_{i=1}^{m}\left[\sum_{j=1}^{n}i+\sum_{j=1}^{n}j\right]$ [Theorem 2(b)] $=\sum_{i=1}^{m}\left[ni+\frac{n(n+1)}{2}\right]$ [Theorem 3(b) and 3(c)]

$=\sum_{i=1}^{m}ni+\sum_{i=1}^{m}\frac{n(n+1)}{2}=\frac{nm(m+1)}{2}+\frac{nm(n+1)}{2}=\frac{nm}{2}(m+n+2)$

## G INTEGRATION OF RATIONAL FUNCTIONS BY PARTIAL FRACTIONS

**1.** (a) $\frac{2x}{(x+3)(3x+1)}=\frac{A}{x+3}+\frac{B}{3x+1}$

(b) $\frac{1}{x^3+2x^2+x}=\frac{1}{x(x^2+2x+1)}=\frac{1}{x(x+1)^2}=\frac{A}{x}+\frac{B}{x+1}+\frac{C}{(x+1)^2}$

**2.** (a) $\frac{x-1}{x^3+x^2}=\frac{x-1}{x^2(x+1)}=\frac{A}{x}+\frac{B}{x^2}+\frac{C}{x+1}$

(b) $\frac{x-1}{x^3+x}=\frac{x-1}{x(x^2+1)}=\frac{A}{x}+\frac{Bx+C}{x^2+1}$

**3.** (a) $\frac{2}{x^2+3x-4}=\frac{2}{(x+4)(x-1)}=\frac{A}{x+4}+\frac{B}{x-1}$

(b) $x^2+x+1$ is irreducible, so $\frac{x^2}{(x-1)(x^2+x+1)}=\frac{A}{x-1}+\frac{Bx+C}{x^2+x+1}$.

**4.** (a) $\frac{x^3}{x^2+4x+3}=x-4+\frac{13x+12}{x^2+4x+3}=x-4+\frac{13x+12}{(x+1)(x+3)}=x-4+\frac{A}{x+1}+\frac{B}{x+3}$

(b) $\frac{2x+1}{(x+1)^3(x^2+4)^2}=\frac{A}{x+1}+\frac{B}{(x+1)^2}+\frac{C}{(x+1)^3}+\frac{Dx+E}{x^2+4}+\frac{Fx+G}{(x^2+4)^2}$

**5.** (a) $\frac{x^4}{x^4-1}=\frac{(x^4-1)+1}{x^4-1}=1+\frac{1}{x^4-1}$ [or use long division] $=1+\frac{1}{(x^2-1)(x^2+1)}$

$=1+\frac{1}{(x-1)(x+1)(x^2+1)}=1+\frac{A}{x-1}+\frac{B}{x+1}+\frac{Cx+D}{x^2+1}$

(b) $\frac{t^4+t^2+1}{(t^2+1)(t^2+4)^2}=\frac{At+B}{t^2+1}+\frac{Ct+D}{t^2+4}+\frac{Et+F}{(t^2+4)^2}$

**6.** (a) $\frac{x^4}{(x^3+x)(x^2-x+3)}=\frac{x^4}{x(x^2+1)(x^2-x+3)}=\frac{x^3}{(x^2+1)(x^2-x+3)}=\frac{Ax+B}{x^2+1}+\frac{Cx+D}{x^2-x+3}$

(b) $\frac{1}{x^6-x^3}=\frac{1}{x^3(x^3-1)}=\frac{1}{x^3(x-1)(x^2+x+1)}=\frac{A}{x}+\frac{B}{x^2}+\frac{C}{x^3}+\frac{D}{x-1}+\frac{Ex+F}{x^2+x+1}$

**7.** $\int\frac{x}{x-6}\,dx=\int\frac{(x-6)+6}{x-6}\,dx=\int\left(1+\frac{6}{x-6}\right)dx=x+6\ln|x-6|+C$

**8.** $\int\frac{r^2}{r+4}\,dr=\int\left(\frac{r^2-16}{r+4}+\frac{16}{r+4}\right)dr=\int\left(r-4+\frac{16}{r+4}\right)dr$ [or use long division]

$=\frac{1}{2}r^2-4r+16\ln|r+4|+C$

**9.** $\dfrac{x-9}{(x+5)(x-2)} = \dfrac{A}{x+5} + \dfrac{B}{x-2}$. Multiply both sides by $(x+5)(x-2)$ to get $x-9 = A(x-2) + B(x+5)$. Substituting 2 for $x$ gives $-7 = 7B \Leftrightarrow B = -1$. Substituting $-5$ for $x$ gives $-14 = -7A \Leftrightarrow A = 2$. Thus,

$$\int \frac{x-9}{(x+5)(x-2)}\,dx = \int \left(\frac{2}{x+5} + \frac{-1}{x-2}\right) dx = 2\ln|x+5| - \ln|x-2| + C$$

**10.** $\dfrac{1}{(t+4)(t-1)} = \dfrac{A}{t+4} + \dfrac{B}{t-1} \;\Rightarrow\; 1 = A(t-1) + B(t+4)$.
$t = 1 \;\Rightarrow\; 1 = 5B \;\Rightarrow\; B = \frac{1}{5}$. $t = -4 \;\Rightarrow\; 1 = -5A \;\Rightarrow\; A = -\frac{1}{5}$. Thus,

$$\int \frac{1}{(t+4)(t-1)}\,dt = \int \left(\frac{-1/5}{t+4} + \frac{1/5}{t-1}\right) dt = -\tfrac{1}{5}\ln|t+4| + \tfrac{1}{5}\ln|t-1| + C \quad \text{or} \quad \frac{1}{5}\ln\left|\frac{t-1}{t+4}\right| + C$$

**11.** $\dfrac{1}{x^2-1} = \dfrac{1}{(x+1)(x-1)} = \dfrac{A}{x+1} + \dfrac{B}{x-1}$. Multiply both sides by $(x+1)(x-1)$ to get $1 = A(x-1) + B(x+1)$. Substituting 1 for $x$ gives $1 = 2B \Leftrightarrow B = \frac{1}{2}$. Substituting $-1$ for $x$ gives $1 = -2A \Leftrightarrow A = -\frac{1}{2}$. Thus,

$$\int_2^3 \frac{1}{x^2-1}\,dx = \int_2^3 \left(\frac{-1/2}{x+1} + \frac{1/2}{x-1}\right) dx = \left[-\tfrac{1}{2}\ln|x+1| + \tfrac{1}{2}\ln|x-1|\right]_2^3$$
$$= \left(-\tfrac{1}{2}\ln 4 + \tfrac{1}{2}\ln 2\right) - \left(-\tfrac{1}{2}\ln 3 + \tfrac{1}{2}\ln 1\right) = \tfrac{1}{2}(\ln 2 + \ln 3 - \ln 4) \quad \left[\text{or } \tfrac{1}{2}\ln\tfrac{3}{2}\right]$$

**12.** $\dfrac{x-1}{x^2+3x+2} = \dfrac{A}{x+1} + \dfrac{B}{x+2}$. Multiply both sides by $(x+1)(x+2)$ to get $x-1 = A(x+2) + B(x+1)$. Substituting $-2$ for $x$ gives $-3 = -B \Leftrightarrow B = 3$. Substituting $-1$ for $x$ gives $-2 = A$. Thus,

$$\int_0^1 \frac{x-1}{x^2+3x+2}\,dx = \int_0^1 \left(\frac{-2}{x+1} + \frac{3}{x+2}\right) dx = \left[-2\ln|x+1| + 3\ln|x+2|\right]_0^1$$
$$= (-2\ln 2 + 3\ln 3) - (-2\ln 1 + 3\ln 2) = 3\ln 3 - 5\ln 2 \quad \left[\text{or } \ln\tfrac{27}{32}\right]$$

**13.** $\displaystyle\int \frac{ax}{x^2-bx}\,dx = \int \frac{ax}{x(x-b)}\,dx = \int \frac{a}{x-b}\,dx = a\ln|x-b| + C$

**14.** If $a \neq b$, $\dfrac{1}{(x+a)(x+b)} = \dfrac{1}{b-a}\left(\dfrac{1}{x+a} - \dfrac{1}{x+b}\right)$, so if $a \neq b$, then

$$\int \frac{dx}{(x+a)(x+b)} = \frac{1}{b-a}\left(\ln|x+a| - \ln|x+b|\right) + C = \frac{1}{b-a}\ln\left|\frac{x+a}{x+b}\right| + C$$

If $a = b$, then $\displaystyle\int \frac{dx}{(x+a)^2} = -\frac{1}{x+a} + C$.

**15.** $\dfrac{2x+3}{(x+1)^2} = \dfrac{A}{x+1} + \dfrac{B}{(x+1)^2} \;\Rightarrow\; 2x+3 = A(x+1) + B$. Take $x = -1$ to get $B = 1$, and equate coefficients of $x$ to get $A = 2$. Now

$$\int_0^1 \frac{2x+3}{(x+1)^2}\,dx = \int_0^1 \left[\frac{2}{x+1} + \frac{1}{(x+1)^2}\right] dx = \left[2\ln(x+1) - \frac{1}{x+1}\right]_0^1$$
$$= 2\ln 2 - \tfrac{1}{2} - (2\ln 1 - 1) = 2\ln 2 + \tfrac{1}{2}$$

**16.** $\dfrac{x^3-4x-10}{x^2-x-6} = x+1+\dfrac{3x-4}{(x-3)(x+2)}$. Write $\dfrac{3x-4}{(x-3)(x+2)} = \dfrac{A}{x-3}+\dfrac{B}{x+2}$. Then $3x-4 = A(x+2)+B(x-3)$. Taking $x=3$ and $x=-2$, we get $5=5A \Leftrightarrow A=1$ and $-10=-5B \Leftrightarrow B=2$, so

$$\int_0^1 \frac{x^3-4x-10}{x^2-x-6}\,dx = \int_0^1 \left(x+1+\frac{1}{x-3}+\frac{2}{x+2}\right)dx = \left[\frac{1}{2}x^2+x+\ln|x-3|+2\ln(x+2)\right]_0^1$$
$$= \left(\tfrac{1}{2}+1+\ln 2+2\ln 3\right)-(0+0+\ln 3+2\ln 2) = \tfrac{3}{2}+\ln 3-\ln 2 = \tfrac{3}{2}+\ln\tfrac{3}{2}$$

**17.** $\dfrac{4y^2-7y-12}{y(y+2)(y-3)} = \dfrac{A}{y}+\dfrac{B}{y+2}+\dfrac{C}{y-3} \Rightarrow 4y^2-7y-12 = A(y+2)(y-3)+By(y-3)+Cy(y+2)$. Setting $y=0$ gives $-12=-6A$, so $A=2$. Setting $y=-2$ gives $18=10B$, so $B=\frac{9}{5}$. Setting $y=3$ gives $3=15C$, so $C=\frac{1}{5}$. Now

$$\int_1^2 \frac{4y^2-7y-12}{y(y+2)(y-3)}\,dy = \int_1^2\left(\frac{2}{y}+\frac{9/5}{y+2}+\frac{1/5}{y-3}\right)dy = \left[2\ln|y|+\tfrac{9}{5}\ln|y+2|+\tfrac{1}{5}\ln|y-3|\right]_1^2$$
$$= 2\ln 2+\tfrac{9}{5}\ln 4+\tfrac{1}{5}\ln 1-2\ln 1-\tfrac{9}{5}\ln 3-\tfrac{1}{5}\ln 2$$
$$= 2\ln 2+\tfrac{18}{5}\ln 2-\tfrac{1}{5}\ln 2-\tfrac{9}{5}\ln 3 = \tfrac{27}{5}\ln 2-\tfrac{9}{5}\ln 3 = \tfrac{9}{5}(3\ln 2-\ln 3) = \tfrac{9}{5}\ln\tfrac{8}{3}$$

**18.** $\dfrac{x^2+2x-1}{x^3-x} = \dfrac{x^2+2x-1}{x(x+1)(x-1)} = \dfrac{A}{x}+\dfrac{B}{x+1}+\dfrac{C}{x-1}$. Multiply both sides by $x(x+1)(x-1)$ to get $x^2+2x-1 = A(x+1)(x-1)+Bx(x-1)+Cx(x+1)$. Substituting 0 for $x$ gives $-1=-A \Leftrightarrow A=1$. Substituting $-1$ for $x$ gives $-2=2B \Leftrightarrow B=-1$. Substituting 1 for $x$ gives $2=2C \Leftrightarrow C=1$. Thus,

$$\int\frac{x^2+2x-1}{x^3-x}\,dx = \int\left(\frac{1}{x}-\frac{1}{x+1}+\frac{1}{x-1}\right)dx = \ln|x|-\ln|x+1|+\ln|x-1|+C = \ln\left|\frac{x(x-1)}{x+1}\right|+C.$$

**19.** $\dfrac{1}{(x+5)^2(x-1)} = \dfrac{A}{x+5}+\dfrac{B}{(x+5)^2}+\dfrac{C}{x-1} \Rightarrow 1 = A(x+5)(x-1)+B(x-1)+C(x+5)^2$. Setting $x=-5$ gives $1=-6B$, so $B=-\frac{1}{6}$. Setting $x=1$ gives $1=36C$, so $C=\frac{1}{36}$. Setting $x=-2$ gives $1 = A(3)(-3)+B(-3)+C(3^2) = -9A-3B+9C = -9A+\frac{1}{2}+\frac{1}{4} = -9A+\frac{3}{4}$, so $9A=-\frac{1}{4}$ and $A=-\frac{1}{36}$. Now

$$\int\frac{1}{(x+5)^2(x-1)}\,dx = \int\left[\frac{-1/36}{x+5}-\frac{1/6}{(x+5)^2}+\frac{1/36}{x-1}\right]dx$$
$$= -\frac{1}{36}\ln|x+5|+\frac{1}{6(x+5)}+\frac{1}{36}\ln|x-1|+C$$

**20.** $\dfrac{x^2}{(x-3)(x+2)^2} = \dfrac{A}{x-3}+\dfrac{B}{x+2}+\dfrac{C}{(x+2)^2} \Rightarrow x^2 = A(x+2)^2+B(x-3)(x+2)+C(x-3)$. Setting $x=3$ gives $A=\frac{9}{25}$. Take $x=-2$ to get $C=-\frac{4}{5}$, and equate the coefficients of $x^2$ to get $1=A+B \Rightarrow B=\frac{16}{25}$. Then

$$\int\frac{x^2}{(x-3)(x+2)^2}\,dx = \int\left[\frac{9/25}{x-3}+\frac{16/25}{x+2}-\frac{4/5}{(x+2)^2}\right]dx$$
$$= \frac{9}{25}\ln|x-3|+\frac{16}{25}\ln|x+2|+\frac{4}{5(x+2)}+C$$

**21.** $\dfrac{5x^2+3x-2}{x^3+2x^2} = \dfrac{5x^2+3x-2}{x^2(x+2)} = \dfrac{A}{x} + \dfrac{B}{x^2} + \dfrac{C}{x+2}$. Multiply by $x^2(x+2)$ to get $5x^2+3x-2 = Ax(x+2) + B(x+2) + Cx^2$. Set $x=-2$ to get $C=3$, and take $x=0$ to get $B=-1$. Equating the coefficients of $x^2$ gives $5 = A + C \;\Rightarrow\; A = 2$. So

$$\int \frac{5x^2+3x-2}{x^3+2x^2}\,dx = \int\left(\frac{2}{x} - \frac{1}{x^2} + \frac{3}{x+2}\right)dx = 2\ln|x| + \frac{1}{x} + 3\ln|x+2| + C.$$

**22.** $\dfrac{x^2-x+6}{x^3+3x} = \dfrac{x^2-x+6}{x(x^2+3)} = \dfrac{A}{x} + \dfrac{Bx+C}{x^2+3}$. Multiply by $x(x^2+3)$ to get $x^2-x+6 = A(x^2+3) + (Bx+C)x$. Substituting 0 for $x$ gives $6 = 3A \;\Leftrightarrow\; A = 2$. The coefficients of the $x^2$-terms must be equal, so $1 = A + B \;\Rightarrow\; B = 1 - 2 = -1$. The coefficients of the $x$-terms must be equal, so $-1 = C$. Thus,

$$\begin{aligned}\int \frac{x^2-x+6}{x^3+3x}\,dx &= \int\left(\frac{2}{x} + \frac{-x-1}{x^2+3}\right)dx = \int\left(\frac{2}{x} - \frac{x}{x^2+3} - \frac{1}{x^2+3}\right)dx \\ &= 2\ln|x| - \frac{1}{2}\ln(x^2+3) - \frac{1}{\sqrt{3}}\tan^{-1}\frac{x}{\sqrt{3}} + C\end{aligned}$$

**23.** $\dfrac{10}{(x-1)(x^2+9)} = \dfrac{A}{x-1} + \dfrac{Bx+C}{x^2+9}$. Multiply both sides by $(x-1)(x^2+9)$ to get $10 = A(x^2+9) + (Bx+C)(x-1)$ $(*)$. Substituting 1 for $x$ gives $10 = 10A \;\Leftrightarrow\; A = 1$. Substituting 0 for $x$ gives $10 = 9A - C \;\Rightarrow\; C = 9(1) - 10 = -1$. The coefficients of the $x^2$-terms in $(*)$ must be equal, so $0 = A + B \;\Rightarrow\; B = -1$. Thus,

$$\begin{aligned}\int \frac{10}{(x-1)(x^2+9)}\,dx &= \int\left(\frac{1}{x-1} + \frac{-x-1}{x^2+9}\right)dx = \int\left(\frac{1}{x-1} - \frac{x}{x^2+9} - \frac{1}{x^2+9}\right)dx \\ &= \ln|x-1| - \tfrac{1}{2}\ln(x^2+9)\ [\text{let } u = x^2+9] - \tfrac{1}{3}\tan^{-1}\left(\tfrac{x}{3}\right)\ [\text{Formula 10}] + C\end{aligned}$$

**24.** $\dfrac{x^2-2x-1}{(x-1)^2(x^2+1)} = \dfrac{A}{x-1} + \dfrac{B}{(x-1)^2} + \dfrac{Cx+D}{x^2+1} \;\Rightarrow$ $x^2-2x-1 = A(x-1)(x^2+1) + B(x^2+1) + (Cx+D)(x-1)^2$. Setting $x=1$ gives $B=-1$. Equating the coefficients of $x^3$ gives $A = -C$. Equating the constant terms gives $-1 = -A - 1 + D$, so $D = A$, and setting $x = 2$ gives $-1 = 5A - 5 - 2A + A$ or $A = 1$. We have

$$\begin{aligned}\int \frac{x^2-2x-1}{(x-1)^2(x^2+1)}\,dx &= \int\left[\frac{1}{x-1} - \frac{1}{(x-1)^2} - \frac{x-1}{x^2+1}\right]dx \\ &= \ln|x-1| + \frac{1}{x-1} - \tfrac{1}{2}\ln(x^2+1) + \tan^{-1}x + C\end{aligned}$$

**25.** $\dfrac{x^3+x^2+2x+1}{(x^2+1)(x^2+2)} = \dfrac{Ax+B}{x^2+1} + \dfrac{Cx+D}{x^2+2}$. Multiply both sides by $(x^2+1)(x^2+2)$ to get

$x^3+x^2+2x+1 = (Ax+B)(x^2+2) + (Cx+D)(x^2+1) \quad \Leftrightarrow$

$x^3+x^2+2x+1 = (Ax^3+Bx^2+2Ax+2B) + (Cx^3+Dx^2+Cx+D) \quad \Leftrightarrow$

$x^3+x^2+2x+1 = (A+C)x^3 + (B+D)x^2 + (2A+C)x + (2B+D)$. Comparing coefficients gives us the following system of equations:

$$A+C=1 \quad \textbf{(1)} \qquad B+D=1 \quad \textbf{(2)}$$
$$2A+C=2 \quad \textbf{(3)} \qquad 2B+D=1 \quad \textbf{(4)}$$

Subtracting equation **(1)** from equation **(3)** gives us $A=1$, so $C=0$. Subtracting equation **(2)** from equation **(4)** gives us $B=0$, so $D=1$. Thus, $I = \displaystyle\int \frac{x^3+x^2+2x+1}{(x^2+1)(x^2+2)}\,dx = \int \left(\frac{x}{x^2+1} + \frac{1}{x^2+2}\right) dx$. For $\displaystyle\int \frac{x}{x^2+1}\,dx$, let $u = x^2+1$ so $du = 2x\,dx$ and then $\displaystyle\int \frac{x}{x^2+1}\,dx = \frac{1}{2}\int \frac{1}{u}\,du = \frac{1}{2}\ln|u| + C = \frac{1}{2}\ln(x^2+1) + C$. For $\displaystyle\int \frac{1}{x^2+2}\,dx$, use Formula 10 with $a=\sqrt{2}$. So $\displaystyle\int \frac{1}{x^2+2}\,dx = \int \frac{1}{x^2+(\sqrt{2})^2}\,dx = \frac{1}{\sqrt{2}}\tan^{-1}\frac{x}{\sqrt{2}} + C$.

Thus, $I = \dfrac{1}{2}\ln(x^2+1) + \dfrac{1}{\sqrt{2}}\tan^{-1}\dfrac{x}{\sqrt{2}} + C$.

**26.** $\dfrac{x^3-2x^2+x+1}{x^4+5x^2+4} = \dfrac{x^3-2x^2+x+1}{(x^2+1)(x^2+4)} = \dfrac{Ax+B}{x^2+1} + \dfrac{Cx+D}{x^2+4} \quad \Rightarrow$

$x^3-2x^2+x+1 = (Ax+B)(x^2+4) + (Cx+D)(x^2+1)$. Equating coefficients gives $A+C=1$, $B+D=-2$, $4A+C=1$, $4B+D=1 \quad \Rightarrow \quad A=0$, $C=1$, $B=1$, $D=-3$. Now

$$\int \frac{x^3-2x^2+x+1}{x^4+5x^2+4}\,dx = \int \frac{dx}{x^2+1} + \int \frac{x-3}{x^2+4}\,dx = \tan^{-1}x + \tfrac{1}{2}\ln(x^2+4) - \tfrac{3}{2}\tan^{-1}(x/2) + C.$$

**27.**
$$\begin{aligned}\int \frac{x+4}{x^2+2x+5}\,dx &= \int \frac{x+1}{x^2+2x+5}\,dx + \int \frac{3}{x^2+2x+5}\,dx = \frac{1}{2}\int \frac{(2x+2)\,dx}{x^2+2x+5} + \int \frac{3\,dx}{(x+1)^2+4} \\ &= \frac{1}{2}\ln|x^2+2x+5| + 3\int \frac{2\,du}{4(u^2+1)} \qquad \begin{bmatrix}\text{where } x+1=2u, \\ \text{and } dx = 2\,du\end{bmatrix} \\ &= \frac{1}{2}\ln(x^2+2x+5) + \frac{3}{2}\tan^{-1}u + C = \frac{1}{2}\ln(x^2+2x+5) + \frac{3}{2}\tan^{-1}\left(\frac{x+1}{2}\right) + C\end{aligned}$$

**28.**
$$\begin{aligned}\int_0^1 \frac{x}{x^2+4x+13}\,dx &= \int_0^1 \frac{\frac{1}{2}(2x+4)}{x^2+4x+13}\,dx - 2\int_0^1 \frac{dx}{(x+2)^2+9} \\ &= \frac{1}{2}\int_{13}^{18} \frac{dy}{y} - 2\int_{2/3}^{1} \frac{3\,du}{9u^2+9} \qquad \begin{bmatrix}\text{where } y = x^2+4x+13,\ dy = (2x+4)\,dx, \\ x+2 = 3u, \text{ and } dx = 3\,du\end{bmatrix} \\ &= \tfrac{1}{2}\left[\ln y\right]_{13}^{18} - \tfrac{2}{3}\left[\tan^{-1}u\right]_{2/3}^{1} = \tfrac{1}{2}\ln\tfrac{18}{13} - \tfrac{2}{3}\left(\tfrac{\pi}{4} - \tan^{-1}\left(\tfrac{2}{3}\right)\right) \\ &= \tfrac{1}{2}\ln\tfrac{18}{13} - \tfrac{\pi}{6} + \tfrac{2}{3}\tan^{-1}\left(\tfrac{2}{3}\right)\end{aligned}$$

**29.** $\dfrac{1}{x^3-1} = \dfrac{1}{(x-1)(x^2+x+1)} = \dfrac{A}{x-1} + \dfrac{Bx+C}{x^2+x+1} \quad\Rightarrow\quad 1 = A(x^2+x+1) + (Bx+C)(x-1).$

Take $x = 1$ to get $A = \frac{1}{3}$. Equating coefficients of $x^2$ and then comparing the constant terms, we get $0 = \frac{1}{3} + B$, $1 = \frac{1}{3} - C$, so $B = -\frac{1}{3}$, $C = -\frac{2}{3}$ $\Rightarrow$

$$\begin{aligned}\int \frac{1}{x^3-1}\,dx &= \int \frac{\frac{1}{3}}{x-1}\,dx + \int \frac{-\frac{1}{3}x - \frac{2}{3}}{x^2+x+1}\,dx = \tfrac{1}{3}\ln|x-1| - \frac{1}{3}\int \frac{x+2}{x^2+x+1}\,dx \\ &= \tfrac{1}{3}\ln|x-1| - \frac{1}{3}\int \frac{x+1/2}{x^2+x+1}\,dx - \frac{1}{3}\int \frac{(3/2)\,dx}{(x+1/2)^2+3/4} \\ &= \tfrac{1}{3}\ln|x-1| - \tfrac{1}{6}\ln\left(x^2+x+1\right) - \tfrac{1}{2}\left(\tfrac{2}{\sqrt{3}}\right)\tan^{-1}\left(\frac{x+\frac{1}{2}}{\sqrt{3}/2}\right) + K \\ &= \tfrac{1}{3}\ln|x-1| - \tfrac{1}{6}\ln\left(x^2+x+1\right) - \tfrac{1}{\sqrt{3}}\tan^{-1}\left(\tfrac{1}{\sqrt{3}}(2x+1)\right) + K\end{aligned}$$

**30.** $\dfrac{x^3}{x^3+1} = \dfrac{(x^3+1)-1}{x^3+1} = 1 - \dfrac{1}{x^3+1} = 1 - \left(\dfrac{A}{x+1} + \dfrac{Bx+C}{x^2-x+1}\right) \quad\Rightarrow\quad 1 = A(x^2-x+1) + (Bx+C)(x+1).$

Equate the terms of degree 2, 1 and 0 to get $0 = A + B$, $0 = -A + B + C$, $1 = A + C$. Solve the three equations to get $A = \frac{1}{3}$, $B = -\frac{1}{3}$, and $C = \frac{2}{3}$. So

$$\begin{aligned}\int \frac{x^3}{x^3+1}\,dx &= \int \left[1 - \frac{\frac{1}{3}}{x+1} + \frac{\frac{1}{3}x - \frac{2}{3}}{x^2-x+1}\right] dx \\ &= x - \tfrac{1}{3}\ln|x+1| + \frac{1}{6}\int \frac{2x-1}{x^2-x+1}\,dx - \frac{1}{2}\int \frac{dx}{\left(x-\frac{1}{2}\right)^2 + \frac{3}{4}} \\ &= x - \tfrac{1}{3}\ln|x+1| + \tfrac{1}{6}\ln\left(x^2-x+1\right) - \tfrac{1}{\sqrt{3}}\tan^{-1}\left(\tfrac{1}{\sqrt{3}}(2x-1)\right) + K\end{aligned}$$

**31.** $\dfrac{1}{x^4-x^2} = \dfrac{1}{x^2(x-1)(x+1)} = \dfrac{A}{x} + \dfrac{B}{x^2} + \dfrac{C}{x-1} + \dfrac{D}{x+1}$. Multiply by $x^2(x-1)(x+1)$ to get $1 = Ax(x-1)(x+1) + B(x-1)(x+1) + Cx^2(x+1) + Dx^2(x-1)$. Setting $x = 1$ gives $C = \frac{1}{2}$, taking $x = -1$ gives $D = -\frac{1}{2}$. Equating the coefficients of $x^3$ gives $0 = A + C + D = A$. Finally, setting $x = 0$ yields $B = -1$. Now

$$\int \frac{dx}{x^4-x^2} = \int \left[\frac{-1}{x^2} + \frac{1/2}{x-1} - \frac{1/2}{x+1}\right] dx = \frac{1}{x} + \tfrac{1}{2}\ln\left|\frac{x-1}{x+1}\right| + C.$$

**32.** Let $u = x^4 + 5x^2 + 4 \quad\Rightarrow\quad du = (4x^3 + 10x)\,dx = 2(2x^3 + 5x)\,dx$, so

$$\int_0^1 \frac{2x^3+5x}{x^4+5x^2+4}\,dx = \frac{1}{2}\int_4^{10} \frac{du}{u} = \tfrac{1}{2}\Big[\ln|u|\Big]_4^{10} = \tfrac{1}{2}(\ln 10 - \ln 4) = \tfrac{1}{2}\ln\tfrac{5}{2}.$$

**33.**
$$\begin{aligned}\int \frac{x-3}{(x^2+2x+4)^2}\,dx &= \int \frac{x-3}{\left[(x+1)^2+3\right]^2}\,dx = \int \frac{u-4}{(u^2+3)^2}\,du \qquad [\text{with } u = x+1] \\ &= \int \frac{u\,du}{(u^2+3)^2} - 4\int \frac{du}{(u^2+3)^2} = \frac{1}{2}\int \frac{dv}{v^2} - 4\int \frac{\sqrt{3}\sec^2\theta\,d\theta}{9\sec^4\theta} \qquad \left[\begin{array}{l} v = u^2+3 \text{ in the first integral;} \\ u = \sqrt{3}\tan\theta \text{ in the second} \end{array}\right] \\ &= \frac{-1}{2v} - \frac{4\sqrt{3}}{9}\int \cos^2\theta\,d\theta = \frac{-1}{2(u^2+3)} - \frac{2\sqrt{3}}{9}(\theta + \sin\theta\cos\theta) + C \\ &= \frac{-1}{2(x^2+2x+4)} - \frac{2\sqrt{3}}{9}\left[\tan^{-1}\left(\frac{x+1}{\sqrt{3}}\right) + \frac{\sqrt{3}(x+1)}{x^2+2x+4}\right] + C \\ &= \frac{-1}{2(x^2+2x+4)} - \frac{2\sqrt{3}}{9}\tan^{-1}\left(\frac{x+1}{\sqrt{3}}\right) - \frac{2(x+1)}{3(x^2+2x+4)} + C\end{aligned}$$

**34.** $\dfrac{x^4+1}{x(x^2+1)^2} = \dfrac{A}{x} + \dfrac{Bx+C}{x^2+1} + \dfrac{Dx+E}{(x^2+1)^2} \Rightarrow x^4+1 = A(x^2+1)^2 + (Bx+C)x(x^2+1) + (Dx+E)x.$

Setting $x=0$ gives $A=1$, and equating the coefficients of $x^4$ gives $1=A+B$, so $B=0$. Now

$\dfrac{C}{x^2+1} + \dfrac{Dx+E}{(x^2+1)^2} = \dfrac{x^4+1}{x(x^2+1)^2} - \dfrac{1}{x} = \dfrac{1}{x}\left[\dfrac{x^4+1-(x^4+2x^2+1)}{(x^2+1)^2}\right] = \dfrac{-2x}{(x^2+1)^2}$, so we can take $C=0$, $D=-2$,

and $E=0$. Hence, $\displaystyle\int \frac{x^4+1}{x(x^2+1)^2}\,dx = \int\left[\frac{1}{x} - \frac{2x}{(x^2+1)^2}\right]dx = \ln|x| + \frac{1}{x^2+1} + C.$

**35.** Let $u=\sqrt{x}$, so $u^2=x$ and $dx = 2u\,du$. Thus,

$$\int_9^{16}\frac{\sqrt{x}}{x-4}\,dx = \int_3^4 \frac{u}{u^2-4}2u\,du = 2\int_3^4\frac{u^2}{u^2-4}\,du = 2\int_3^4\left(1+\frac{4}{u^2-4}\right)du \quad \text{[by long division]}$$

$$= 2 + 8\int_3^4\frac{du}{(u+2)(u-2)} \quad (*)$$

Multiply $\dfrac{1}{(u+2)(u-2)} = \dfrac{A}{u+2} + \dfrac{B}{u-2}$ by $(u+2)(u-2)$ to get $1 = A(u-2)+B(u+2)$. Equating coefficients we get $A+B=0$ and $-2A+2B=1$. Solving gives us $B=\frac{1}{4}$ and $A=-\frac{1}{4}$, so $\dfrac{1}{(u+2)(u-2)} = \dfrac{-1/4}{u+2} + \dfrac{1/4}{u-2}$ and $(*)$ is

$$\begin{aligned} 2+8\int_3^4\left(\frac{-1/4}{u+2}+\frac{1/4}{u-2}\right)du &= 2+8\Big[-\tfrac{1}{4}\ln|u+2| + \tfrac{1}{4}\ln|u-2|\Big]_3^4 \\ &= 2+\Big[2\ln|u-2| - 2\ln|u+2|\Big]_3^4 = 2+2\left[\ln\left|\frac{u-2}{u+2}\right|\right]_3^4 \\ &= 2+2\left(\ln\tfrac{2}{6} - \ln\tfrac{1}{5}\right) = 2+2\ln\tfrac{2/6}{1/5} \\ &= 2+2\ln\tfrac{5}{3} \text{ or } 2+\ln\left(\tfrac{5}{3}\right)^2 = 2+\ln\tfrac{25}{9} \end{aligned}$$

**36.** Let $u=\sqrt[3]{x}$. Then $x=u^3$, $dx = 3u^2\,du \Rightarrow$

$$\int_0^1\frac{1}{1+\sqrt[3]{x}}\,dx = \int_0^1\frac{3u^2\,du}{1+u} = \int_0^1\left(3u-3+\frac{3}{1+u}\right)du = \left[\tfrac{3}{2}u^2 - 3u + 3\ln(1+u)\right]_0^1 = 3\left(\ln 2 - \tfrac{1}{2}\right).$$

**37.** Let $u=e^x$. Then $x=\ln u$, $dx = \dfrac{du}{u} \Rightarrow$

$$\begin{aligned}\int\frac{e^{2x}\,dx}{e^{2x}+3e^x+2} &= \int\frac{u^2\,(du/u)}{u^2+3u+2} = \int\frac{u\,du}{(u+1)(u+2)} = \int\left[\frac{-1}{u+1}+\frac{2}{u+2}\right]du \\ &= 2\ln|u+2| - \ln|u+1| + C = \ln\left[(e^x+2)^2/(e^x+1)\right] + C\end{aligned}$$

**38.** Let $u=\sin x$. Then $du = \cos x\,dx \Rightarrow$

$$\begin{aligned}\int\frac{\cos x\,dx}{\sin^2 x+\sin x} &= \int\frac{du}{u^2+u} = \int\frac{du}{u(u+1)} = \int\left[\frac{1}{u} - \frac{1}{u+1}\right]du \\ &= \ln\left|\frac{u}{u+1}\right| + C = \ln\left|\frac{\sin x}{1+\sin x}\right| + C\end{aligned}$$

**39.** 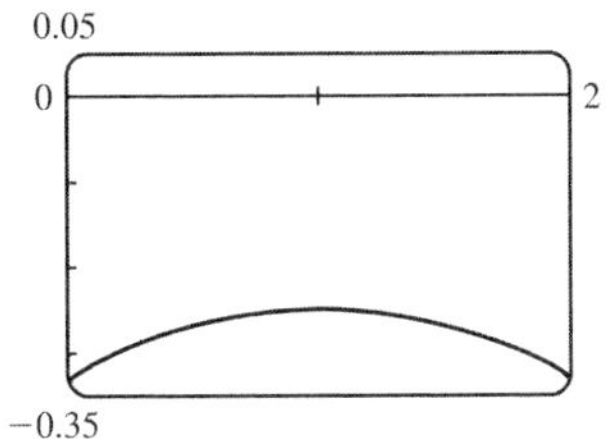


From the graph, we see that the integral will be negative, and we guess that the area is about the same as that of a rectangle with width 2 and height 0.3, so we estimate the integral to be $-(2 \cdot 0.3) = -0.6$. Now $\dfrac{1}{x^2 - 2x - 3} = \dfrac{1}{(x-3)(x+1)} = \dfrac{A}{x-3} + \dfrac{B}{x+1} \quad \Leftrightarrow$ $1 = (A+B)x + A - 3B$, so $A = -B$ and $A - 3B = 1 \quad \Leftrightarrow \quad A = \frac{1}{4}$ and $B = -\frac{1}{4}$, so the integral becomes

$$\int_0^2 \frac{dx}{x^2 - 2x - 3} = \frac{1}{4}\int_0^2 \frac{dx}{x-3} - \frac{1}{4}\int_0^2 \frac{dx}{x+1} = \tfrac{1}{4}\Big[\ln|x-3| - \ln|x+1|\Big]_0^2$$
$$= \frac{1}{4}\left[\ln\left|\frac{x-3}{x+1}\right|\right]_0^2 = \tfrac{1}{4}\left(\ln\tfrac{1}{3} - \ln 3\right) = -\tfrac{1}{2}\ln 3 \approx -0.55$$

**40.** $\dfrac{1}{x^3 - 2x^2} = \dfrac{1}{x^2(x-2)} = \dfrac{A}{x} + \dfrac{B}{x^2} + \dfrac{C}{x-2} \Rightarrow 1 = (A+C)x^2 + (B-2A)x - 2B$, so $A + C = B - 2A = 0$ and $-2B = 1 \quad \Rightarrow \quad B = -\frac{1}{2}$, $A = -\frac{1}{4}$, and $C = \frac{1}{4}$. So the general antiderivative of $\dfrac{1}{x^3 - 2x^2}$ is

$$\int \frac{dx}{x^3 - 2x^2} = -\frac{1}{4}\int \frac{dx}{x} - \frac{1}{2}\int \frac{dx}{x^2} + \frac{1}{4}\int \frac{dx}{x-2}$$
$$= -\tfrac{1}{4}\ln|x| - \tfrac{1}{2}(-1/x) + \tfrac{1}{4}\ln|x-2| + C$$
$$= \frac{1}{4}\ln\left|\frac{x-2}{x}\right| + \frac{1}{2x} + C$$

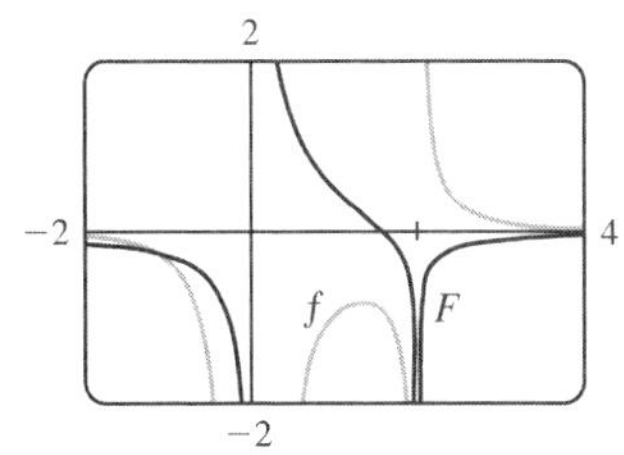


We plot this function with $C = 0$ on the same screen as $y = \dfrac{1}{x^3 - 2x^2}$.

**41.** $\dfrac{P+S}{P\,[(r-1)P - S]} = \dfrac{A}{P} + \dfrac{B}{(r-1)P - S} \quad \Rightarrow \quad P + S = A\,[(r-1)P - S] + BP = [(r-1)A + B]\,P - AS \quad \Rightarrow$ $(r-1)A + B = 1,\ -A = 1 \quad \Rightarrow \quad A = -1,\ B = r$. Now

$$t = \int \frac{P+S}{P\,[(r-1)P - S]}\,dP = \int \left[\frac{-1}{P} + \frac{r}{(r-1)P - S}\right] dP = -\int \frac{dP}{P} + \frac{r}{r-1}\int \frac{r-1}{(r-1)P - S}\,dP$$

so $t = -\ln P + \dfrac{r}{r-1}\ln|(r-1)P - S| + C$. Here $r = 0.10$ and $S = 900$, so

$$t = -\ln P + \tfrac{0.1}{-0.9}\ln|-0.9P - 900| + C = -\ln P - \tfrac{1}{9}\ln(|-1|\,|0.9P + 900|)$$
$$= -\ln P - \tfrac{1}{9}\ln(0.9P + 900) + C$$

When $t = 0$, $P = 10{,}000$, so $0 = -\ln 10{,}000 - \frac{1}{9}\ln(9900) + C$. Thus, $C = \ln 10{,}000 + \frac{1}{9}\ln 9900$ [$\approx 10.2326$], so our equation becomes

$$t = \ln 10{,}000 - \ln P + \tfrac{1}{9}\ln 9900 - \tfrac{1}{9}\ln(0.9P + 900) = \ln\frac{10{,}000}{P} + \frac{1}{9}\ln\frac{9900}{0.9P + 900}$$
$$= \ln\frac{10{,}000}{P} + \frac{1}{9}\ln\frac{1100}{0.1P + 100} = \ln\frac{10{,}000}{P} + \frac{1}{9}\ln\frac{11{,}000}{P + 1000}$$

**42.** The area of a cross-section is a disk with radius $\dfrac{1}{x^2+3x+2}$, so the volume is

$V = \pi \int_0^1 \left[\dfrac{1}{x^2+3x+2}\right]^2 dx = \pi \int_0^1 \dfrac{dx}{(x+1)^2(x+2)^2}$. To evaluate the integral, we

use partial fractions: $\dfrac{1}{(x+1)^2(x+2)^2} = \dfrac{A}{x+1} + \dfrac{B}{(x+1)^2} + \dfrac{C}{x+2} + \dfrac{D}{(x+2)^2} \quad \Rightarrow$

$1 = A(x+1)(x+2)^2 + B(x+2)^2 + C(x+1)^2(x+2) + D(x+1)^2$. We set $x = -1$, giving $B = 1$, then set $x = -2$, giving $D = 1$. Now equating coefficients of $x^3$ gives $A = -C$, and then equating constants gives $1 = 4A + 4 + 2(-A) + 1 \quad \Rightarrow \quad A = -2 \quad \Rightarrow \quad C = 2$. So the expression becomes

$$V = \pi \int_0^1 \left[\frac{-2}{x+1} + \frac{1}{(x+1)^2} + \frac{2}{(x+2)} + \frac{1}{(x+2)^2}\right] dx = \pi \left[2 \ln\left|\frac{x+2}{x+1}\right| - \frac{1}{x+1} - \frac{1}{x+2}\right]_0^1$$

$$= \pi\left[\left(2\ln\tfrac{3}{2} - \tfrac{1}{2} - \tfrac{1}{3}\right) - \left(2\ln 2 - 1 - \tfrac{1}{2}\right)\right] = \pi\left(2\ln\tfrac{3/2}{2} + \tfrac{2}{3}\right) = \pi\left(\tfrac{2}{3} + \ln\tfrac{9}{16}\right).$$

**43.** (a) In Maple, we define $f(x)$, and then use `convert(f,parfrac,x);` to obtain

$$f(x) = \frac{24{,}110/4879}{5x+2} - \frac{668/323}{2x+1} - \frac{9438/80{,}155}{3x-7} + \frac{(22{,}098x + 48{,}935)/260{,}015}{x^2+x+5}.$$

In Mathematica, we use the command `Apart`, and in Derive, we use `Expand`.

(b) $$\int f(x)\,dx = \tfrac{24{,}110}{4879}\cdot\tfrac{1}{5}\ln|5x+2| - \tfrac{668}{323}\cdot\tfrac{1}{2}\ln|2x+1| - \tfrac{9438}{80{,}155}\cdot\tfrac{1}{3}\ln|3x-7|$$

$$+ \frac{1}{260{,}015}\int \frac{22{,}098\left(x+\frac{1}{2}\right) + 37{,}886}{\left(x+\frac{1}{2}\right)^2 + \frac{19}{4}}\,dx + C$$

$$= \tfrac{24{,}110}{4879}\cdot\tfrac{1}{5}\ln|5x+2| - \tfrac{668}{323}\cdot\tfrac{1}{2}\ln|2x+1| - \tfrac{9438}{80{,}155}\cdot\tfrac{1}{3}\ln|3x-7|$$

$$+ \tfrac{1}{260{,}015}\left[22{,}098\cdot\tfrac{1}{2}\ln\left(x^2+x+5\right) + 37{,}886\cdot\sqrt{\tfrac{4}{19}}\tan^{-1}\left(\tfrac{1}{\sqrt{19/4}}\left(x+\tfrac{1}{2}\right)\right)\right] + C$$

$$= \tfrac{4822}{4879}\ln|5x+2| - \tfrac{334}{323}\ln|2x+1| - \tfrac{3146}{80{,}155}\ln|3x-7| + \tfrac{11{,}049}{260{,}015}\ln\left(x^2+x+5\right)$$

$$+ \tfrac{75{,}772}{260{,}015\sqrt{19}}\tan^{-1}\left[\tfrac{1}{\sqrt{19}}(2x+1)\right] + C$$

Using a CAS, we get

$$\frac{4822\ln(5x+2)}{4879} - \frac{334\ln(2x+1)}{323} - \frac{3146\ln(3x-7)}{80{,}155} + \frac{11{,}049\ln\left(x^2+x+5\right)}{260{,}015} + \frac{3988\sqrt{19}}{260{,}015}\tan^{-1}\left[\frac{\sqrt{19}}{19}(2x+1)\right]$$

The main difference in this answer is that the absolute value signs and the constant of integration have been omitted. Also, the fractions have been reduced and the denominators rationalized.

**44.** (a) In Maple, we define $f(x)$, and then use `convert(f,parfrac,x);` to get

$$f(x) = \frac{5828/1815}{(5x-2)^2} - \frac{59{,}096/19{,}965}{5x-2} + \frac{2(2843x+816)/3993}{2x^2+1} + \frac{(313x-251)/363}{(2x^2+1)^2}.$$

In Mathematica, we use the command `Apart`, and in Derive, we use `Expand`.

(b) As we saw in Exercise 43, computer algebra systems omit the absolute value signs in $\int (1/y)\,dy = \ln|y|$. So we use the CAS to integrate the expression in part (a) and add the necessary absolute value signs and constant of integration to get

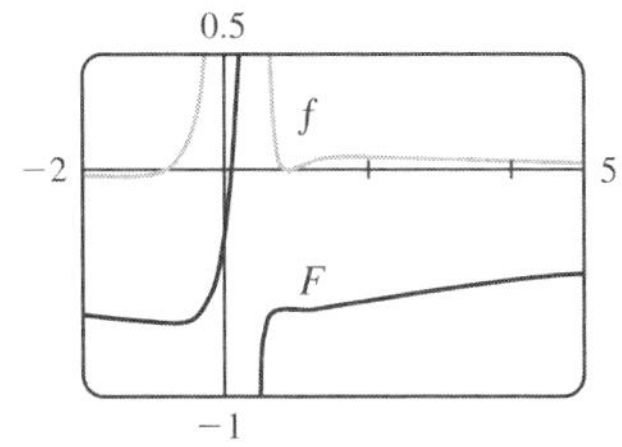


$$\int f(x)\,dx = -\frac{5828}{9075(5x-2)} - \frac{59{,}096\ln|5x-2|}{99{,}825} + \frac{2843\ln(2x^2+1)}{7986}$$
$$+\frac{503}{15{,}972}\sqrt{2}\tan^{-1}\left(\sqrt{2}\,x\right) - \frac{1}{2904}\,\frac{1004x+626}{2x^2+1} + C$$

(c) From the graph, we see that $f$ goes from negative to positive at $x \approx -0.78$, then back to negative at $x \approx 0.8$, and finally back to positive at $x = 1$. Also, $\lim_{x\to 0.4} f(x) = \infty$. So we see (by the First Derivative Test) that $\int f(x)\,dx$ has minima at $x \approx -0.78$ and $x = 1$, and a maximum at $x \approx 0.80$, and that $\int f(x)\,dx$ is unbounded as $x \to 0.4$. Note also that just to the right of $x = 0.4$, $f$ has large values, so $\int f(x)\,dx$ increases rapidly, but slows down as $f$ drops toward 0. $\int f(x)\,dx$ decreases from about 0.8 to 1, then increases slowly since $f$ stays small and positive.

**45.** There are only finitely many values of $x$ where $Q(x) = 0$ (assuming that $Q$ is not the zero polynomial). At all other values of $x$, $F(x)/Q(x) = G(x)/Q(x)$, so $F(x) = G(x)$. In other words, the values of $F$ and $G$ agree at all except perhaps finitely many values of $x$. By continuity of $F$ and $G$, the polynomials $F$ and $G$ must agree at those values of $x$ too.

More explicitly: if $a$ is a value of $x$ such that $Q(a) = 0$, then $Q(x) \neq 0$ for all $x$ sufficiently close to $a$. Thus,

$$F(a) = \lim_{x\to a} F(x) \quad [\text{by continuity of } F] \quad = \lim_{x\to a} G(x) \quad [\text{whenever } Q(x) \neq 0]$$
$$= G(a) \quad [\text{by continuity of } G]$$

**46.** Let $f(x) = ax^2 + bx + c$. We calculate the partial fraction decomposition of $\dfrac{f(x)}{x^2(x+1)^3}$. Since $f(0) = 1$, we must have $c = 1$, so $\dfrac{f(x)}{x^2(x+1)^3} = \dfrac{ax^2+bx+1}{x^2(x+1)^3} = \dfrac{A}{x} + \dfrac{B}{x^2} + \dfrac{C}{x+1} + \dfrac{D}{(x+1)^2} + \dfrac{E}{(x+1)^3}$. Now in order for the integral not to contain any logarithms (that is, in order for it to be a rational function), we must have $A = C = 0$, so $ax^2 + bx + 1 = B(x+1)^3 + Dx^2(x+1) + Ex^2$. Equating constant terms gives $B = 1$, then equating coefficients of $x$ gives $3B = b \Rightarrow b = 3$. This is the quantity we are looking for, since $f'(0) = b$.

# H POLAR COORDINATES

## H.1 Curves in Polar Coordinates

**1.** (a) By adding $2\pi$ to $\frac{\pi}{2}$, we obtain the point $\left(1, \frac{5\pi}{2}\right)$. The direction opposite $\frac{\pi}{2}$ is $\frac{3\pi}{2}$, so $\left(-1, \frac{3\pi}{2}\right)$ is a point that satisfies the $r < 0$ requirement.

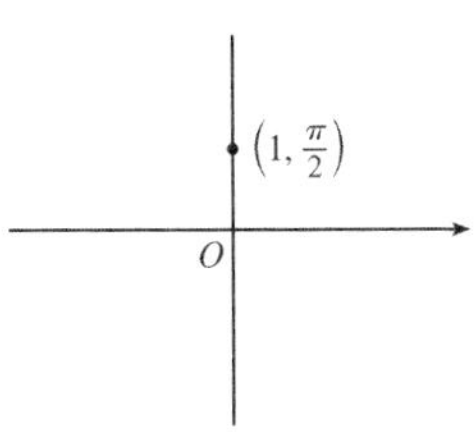


(b) $\left(-2, \frac{\pi}{4}\right)$

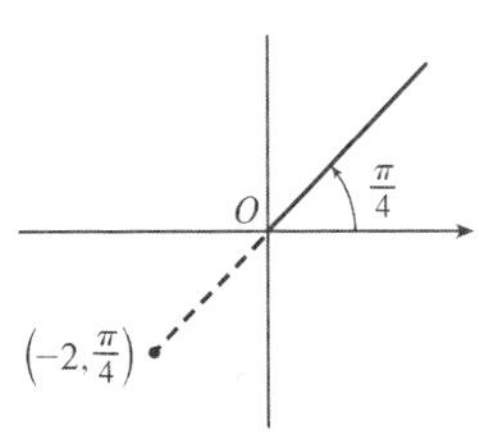


$\left(2, \frac{5\pi}{4}\right), \left(-2, \frac{9\pi}{4}\right)$

(c) $(3, 2)$

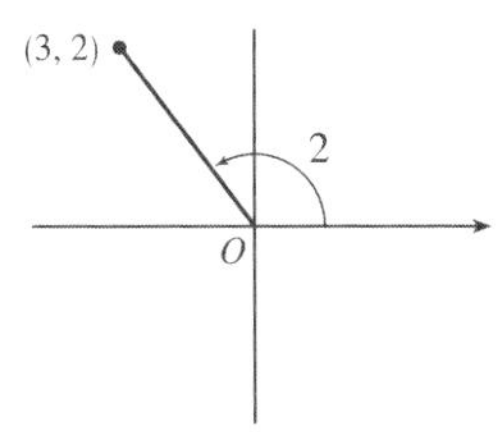


$(3, 2 + 2\pi), (-3, 2 + \pi)$

2. (a) $(3, 0)$

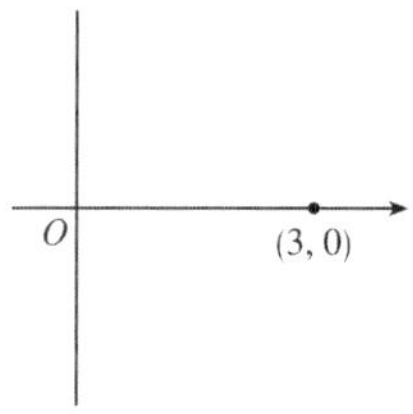


$(3, 2\pi), (-3, \pi)$

(b) $\left(2, -\frac{\pi}{7}\right)$

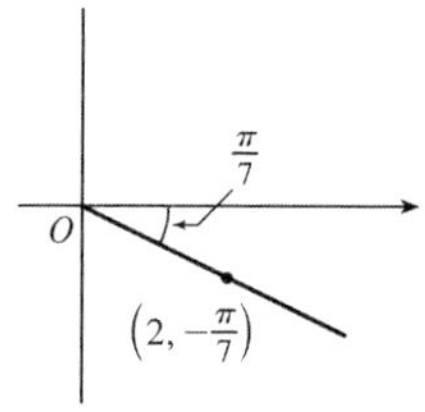


$\left(2, \frac{13\pi}{7}\right), \left(-2, \frac{6\pi}{7}\right)$

(c) $\left(-1, -\frac{\pi}{2}\right)$

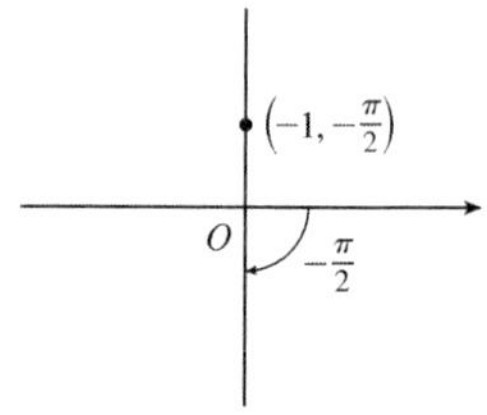


$\left(1, \frac{\pi}{2}\right), \left(-1, \frac{3\pi}{2}\right)$

3. (a)

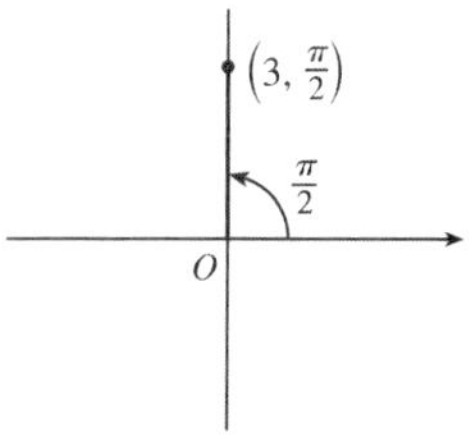


$x = 3\cos\frac{\pi}{2} = 3(0) = 0$ and $y = 3\sin\frac{\pi}{2} = 3(1) = 3$ give us the Cartesian coordinates $(0, 3)$.

(b)

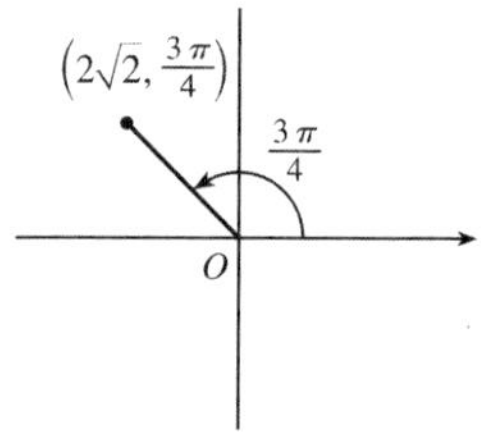


$x = 2\sqrt{2}\cos\frac{3\pi}{4} = 2\sqrt{2}\left(-\frac{1}{\sqrt{2}}\right) = -2$ and $y = 2\sqrt{2}\sin\frac{3\pi}{4} = 2\sqrt{2}\left(\frac{1}{\sqrt{2}}\right) = 2$ give us $(-2, 2)$.

(c)

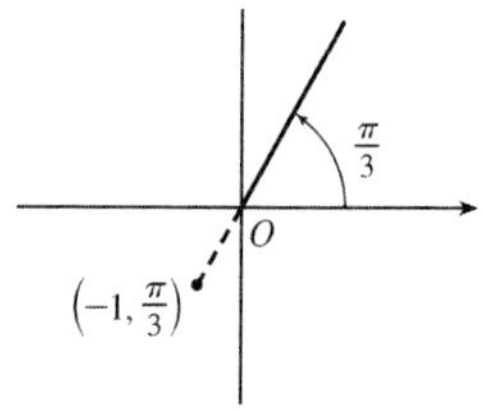


$x = -1\cos\frac{\pi}{3} = -\frac{1}{2}$ and $y = -1\sin\frac{\pi}{3} = -\frac{\sqrt{3}}{2}$ give us $\left(-\frac{1}{2}, -\frac{\sqrt{3}}{2}\right)$.

4. (a)

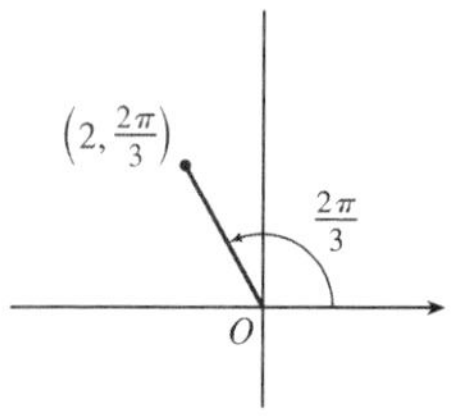


$x = 2\cos\frac{2\pi}{3} = -1,$
$y = 2\sin\frac{2\pi}{3} = \sqrt{3}$

(b)

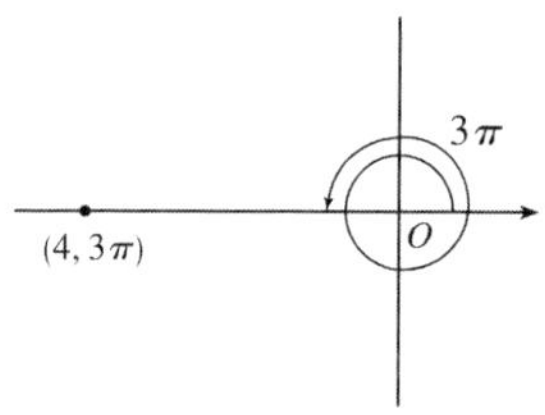


$x = 4\cos 3\pi = -4,$
$y = 4\sin 3\pi = 0$

(c)

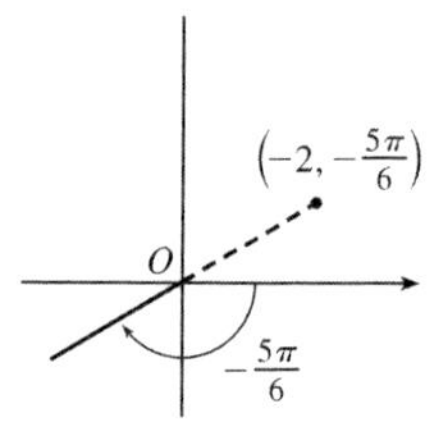


$x = -2\cos\left(-\frac{5\pi}{6}\right) = \sqrt{3},$
$y = -2\sin\left(-\frac{5\pi}{6}\right) = 1$

5. (a) $x = 1$ and $y = 1 \quad\Rightarrow\quad r = \sqrt{1^2 + 1^2} = \sqrt{2}$ and $\theta = \tan^{-1}\left(\frac{1}{1}\right) = \frac{\pi}{4}$. Since $(1, 1)$ is in the first quadrant, the polar coordinates are (i) $\left(\sqrt{2}, \frac{\pi}{4}\right)$ and (ii) $\left(-\sqrt{2}, \frac{5\pi}{4}\right)$.

(b) $x = 2\sqrt{3}$ and $y = -2 \quad\Rightarrow\quad r = \sqrt{\left(2\sqrt{3}\right)^2 + (-2)^2} = \sqrt{12 + 4} = \sqrt{16} = 4$ and $\theta = \tan^{-1}\left(-\frac{2}{2\sqrt{3}}\right) = \tan^{-1}\left(-\frac{1}{\sqrt{3}}\right) = -\frac{\pi}{6}$. Since $\left(2\sqrt{3}, -2\right)$ is in the fourth quadrant and $0 \le \theta \le 2\pi$, the polar coordinates are (i) $\left(4, \frac{11\pi}{6}\right)$ and (ii) $\left(-4, \frac{5\pi}{6}\right)$.

6. (a) $(x, y) = \left(-1, -\sqrt{3}\right)$, $r = \sqrt{1 + 3} = 2$, $\tan\theta = y/x = \sqrt{3}$ and $(x, y)$ is in the third quadrant, so $\theta = \frac{4\pi}{3}$. The polar coordinates are (i) $\left(2, \frac{4\pi}{3}\right)$ and (ii) $\left(-2, \frac{\pi}{3}\right)$.

(b) $(x, y) = (-2, 3)$, $r = \sqrt{4 + 9} = \sqrt{13}$, $\tan\theta = y/x = -\frac{3}{2}$ and $(x, y)$ is in the second quadrant, so $\theta = \tan^{-1}\left(-\frac{3}{2}\right) + \pi$. The polar coordinates are (i) $\left(\sqrt{13}, \theta\right)$ and (ii) $\left(-\sqrt{13}, \theta + \pi\right)$.

**7.** The curves $r = 1$ and $r = 2$ represent circles with center $O$ and radii 1 and 2. The region in the plane satisfying $1 \le r \le 2$ consists of both circles and the shaded region between them in the figure.

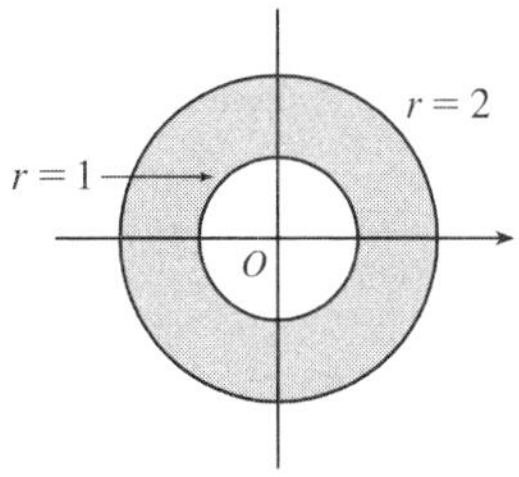


**8.** $r \ge 0, \quad \pi/3 \le \theta \le 2\pi/3$

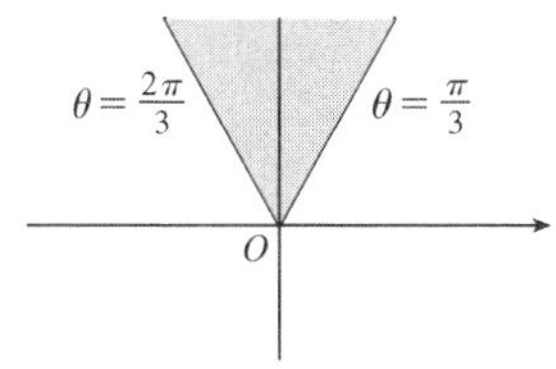


**9.** The region satisfying $0 \le r < 4$ and $-\pi/2 \le \theta < \pi/6$ does not include the circle $r = 4$ nor the line $\theta = \frac{\pi}{6}$.

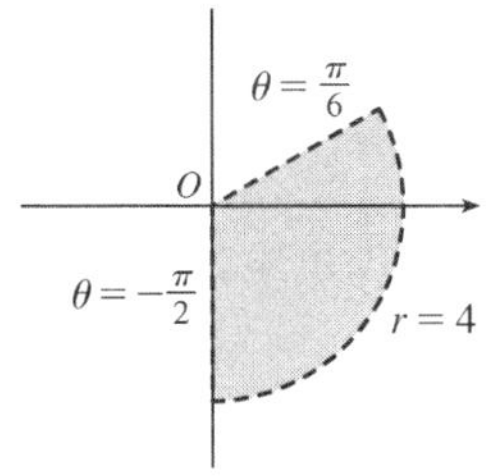


**10.** $2 < r \le 5, \quad 3\pi/4 < \theta < 5\pi/4$

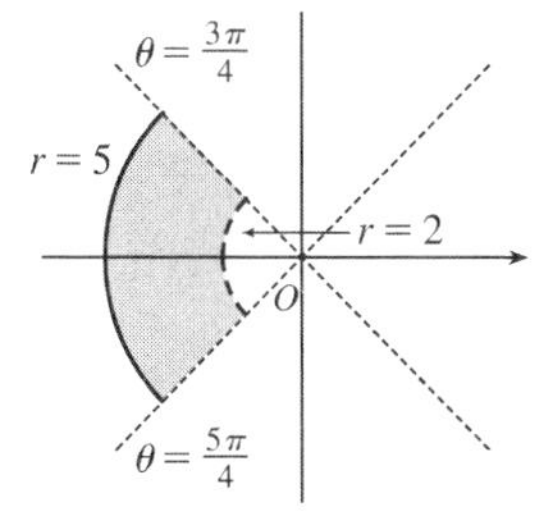


**11.** $2 < r < 3$, $\frac{5\pi}{3} \le \theta \le \frac{7\pi}{3}$

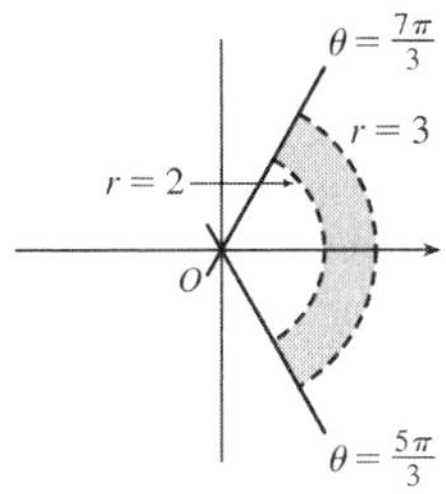


**12.** $-1 \le r \le 1$, $\frac{\pi}{4} \le \theta \le \frac{3\pi}{4}$

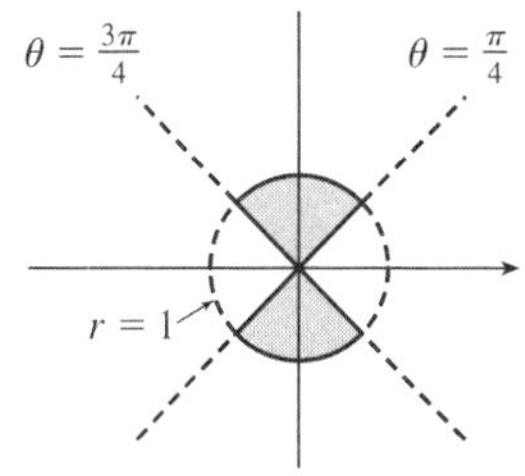


**13.** $r = 3\sin\theta \;\Rightarrow\; r^2 = 3r\sin\theta \;\Leftrightarrow\; x^2 + y^2 = 3y \;\Leftrightarrow\; x^2 + \left(y - \frac{3}{2}\right)^2 = \left(\frac{3}{2}\right)^2$, a circle of radius $\frac{3}{2}$ centered at $\left(0, \frac{3}{2}\right)$. The first two equations are actually equivalent since $r^2 = 3r\sin\theta \;\Rightarrow\; r(r - 3\sin\theta) = 0 \;\Rightarrow\; r = 0$ or $r = 3\sin\theta$. But $r = 3\sin\theta$ gives the point $r = 0$ (the pole) when $\theta = 0$. Thus, the single equation $r = 3\sin\theta$ is equivalent to the compound condition ($r = 0$ or $r = 3\sin\theta$).

**14.** $r = 2\sin\theta + 2\cos\theta \;\Rightarrow\; r^2 = 2r\sin\theta + 2r\cos\theta \;\Leftrightarrow\; x^2 + y^2 = 2y + 2x \;\Leftrightarrow\; (x^2 - 2x + 1) + (y^2 - 2y + 1) = 2$ $\;\Leftrightarrow\; (x-1)^2 + (y-1)^2 = 2$. The first implication is reversible since $r^2 = 2r\sin\theta + 2r\cos\theta \;\Rightarrow\; r = 0$ or $r = 2\sin\theta + 2\cos\theta$, but the curve $r = 2\sin\theta + 2\cos\theta$ passes through the pole ($r = 0$) when $\theta = -\frac{\pi}{4}$, so $r = 2\sin\theta + 2\cos\theta$ includes the single point of $r = 0$. The curve is a circle of radius $\sqrt{2}$, centered at $(1, 1)$.

**15.** $r = \csc\theta \;\Leftrightarrow\; r = \dfrac{1}{\sin\theta} \;\Leftrightarrow\; r\sin\theta = 1 \;\Leftrightarrow\; y = 1$, a horizontal line 1 unit above the $x$-axis.

**16.** $r = \tan\theta\sec\theta = \dfrac{\sin\theta}{\cos^2\theta} \;\Rightarrow\; r\cos^2\theta = \sin\theta \;\Leftrightarrow\; (r\cos\theta)^2 = r\sin\theta \;\Leftrightarrow\; x^2 = y$, a parabola with vertex at the origin opening upward. The first implication is reversible since $\cos\theta = 0$ would imply $\sin\theta = r\cos^2\theta = 0$, contradicting the fact that $\cos^2\theta + \sin^2\theta = 1$.

**17.** $x = -y^2 \;\Leftrightarrow\; r\cos\theta = -r^2\sin^2\theta \;\Leftrightarrow\; \cos\theta = -r\sin^2\theta \;\Leftrightarrow\; r = -\dfrac{\cos\theta}{\sin^2\theta} = -\cot\theta\,\csc\theta$.

**18.** $x + y = 9 \;\Leftrightarrow\; r\cos\theta + r\sin\theta = 9 \;\Leftrightarrow\; r = 9/(\cos\theta + \sin\theta)$.

**19.** $x^2 + y^2 = 2cx \;\Leftrightarrow\; r^2 = 2cr\cos\theta \;\Leftrightarrow\; r^2 - 2cr\cos\theta = 0 \;\Leftrightarrow\; r(r - 2c\cos\theta) = 0 \;\Leftrightarrow\; r = 0$ or $r = 2c\cos\theta$. $r = 0$ is included in $r = 2c\cos\theta$ when $\theta = \frac{\pi}{2} + n\pi$, so the curve is represented by the single equation $r = 2c\cos\theta$.

**20.** $x^2 - y^2 = 1 \;\Leftrightarrow\; (r\cos\theta)^2 - (r\sin\theta)^2 = 1 \;\Leftrightarrow\; r^2(\cos^2\theta - \sin^2\theta) = 1 \;\Leftrightarrow\; r^2\cos 2\theta = 1 \;\Rightarrow\; r^2 = \sec 2\theta$

**21.** (a) The description leads immediately to the polar equation $\theta = \frac{\pi}{6}$, and the Cartesian equation $y = \tan\left(\frac{\pi}{6}\right)x = \frac{1}{\sqrt{3}}x$ is slightly more difficult to derive.

(b) The easier description here is the Cartesian equation $x = 3$.

**22.** (a) Because its center is not at the origin, it is more easily described by its Cartesian equation, $(x-2)^2+(y-3)^2=5^2$.

(b) This circle is more easily given in polar coordinates: $r=4$. The Cartesian equation is also simple: $x^2+y^2=16$.

**23.** $\theta=-\pi/6$

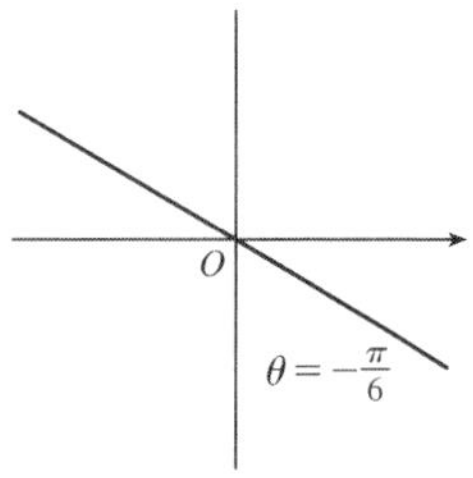


**24.** $r^2-3r+2=0 \quad\Leftrightarrow\quad (r-1)(r-2)=0 \quad\Leftrightarrow$ $r=1$ or $r=2$

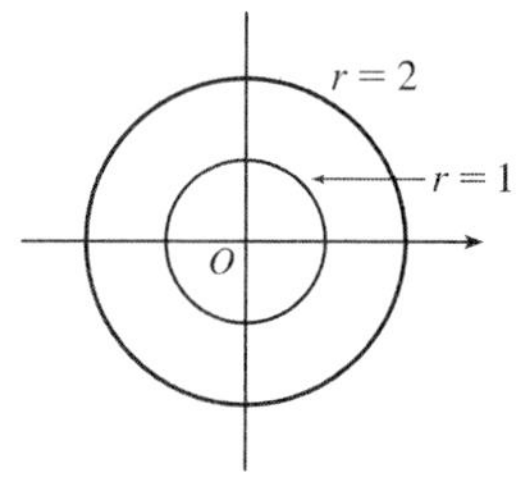


**25.** $r=\sin\theta \quad\Leftrightarrow\quad r^2=r\sin\theta \quad\Leftrightarrow\quad x^2+y^2=y \quad\Leftrightarrow$ $x^2+\left(y-\frac{1}{2}\right)^2=\left(\frac{1}{2}\right)^2$. The reasoning here is the same as in Exercise 13. This is a circle of radius $\frac{1}{2}$ centered at $\left(0,\frac{1}{2}\right)$.

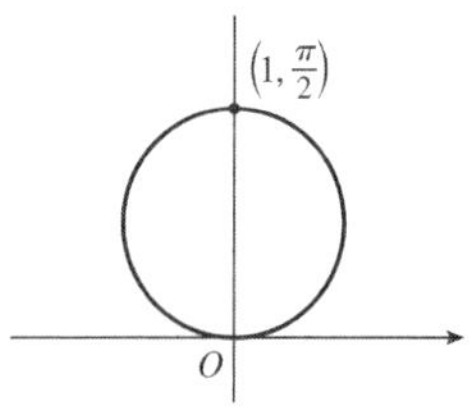


**26.** $r=-3\cos\theta \quad\Leftrightarrow\quad r^2=-3r\cos\theta \quad\Leftrightarrow$ $x^2+y^2=-3x \quad\Leftrightarrow\quad \left(x+\frac{3}{2}\right)^2+y^2=\left(\frac{3}{2}\right)^2$. This curve is a circle of radius $\frac{3}{2}$ centered at $\left(-\frac{3}{2},0\right)$.

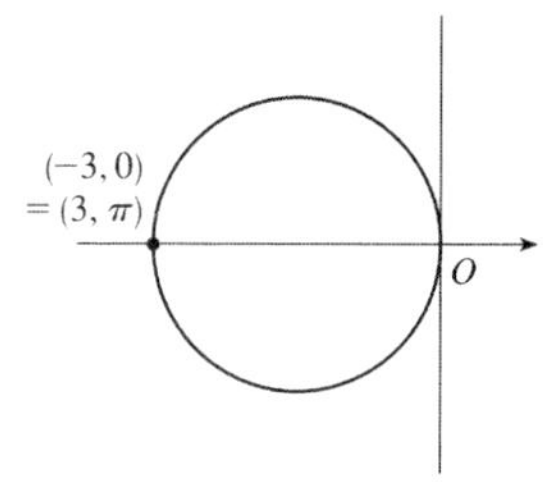


**27.** $r=2(1-\sin\theta)$. This curve is a cardioid.

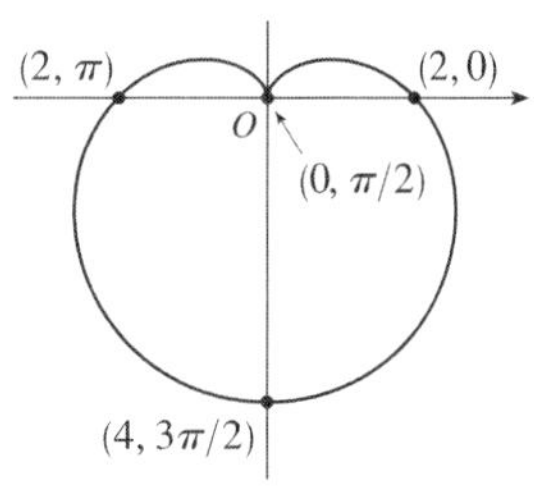


**28.** $r=1-3\cos\theta$. This is a limaçon.

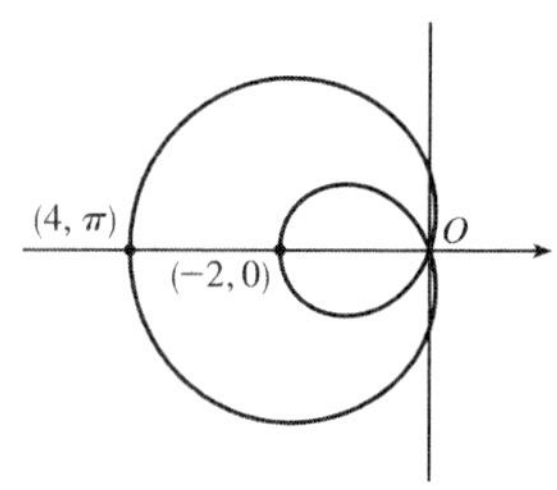


**29.** $r=\theta,\quad \theta\geq 0$

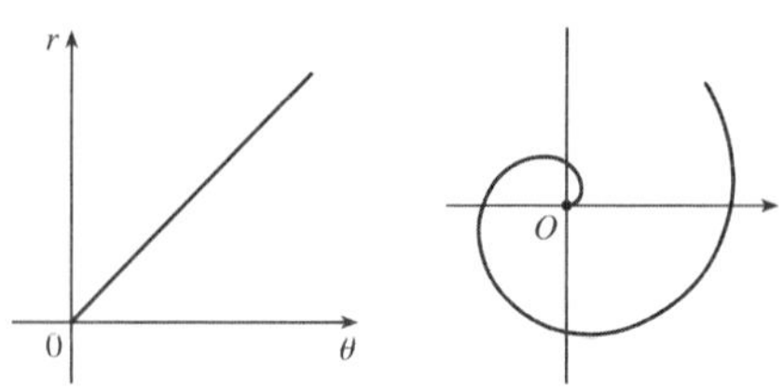


**30.** $r=\ln\theta,\quad \theta\geq 1$

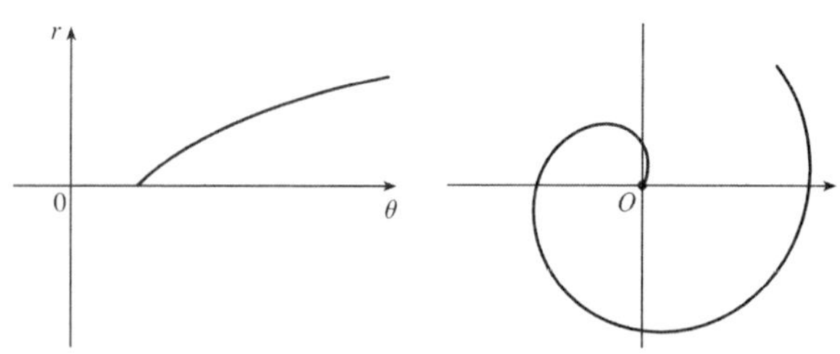


**31.** $r=\sin 2\theta$

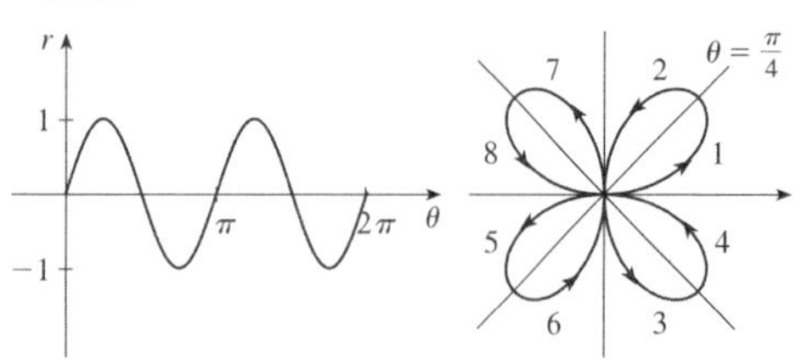


**32.** $r=2\cos 3\theta$

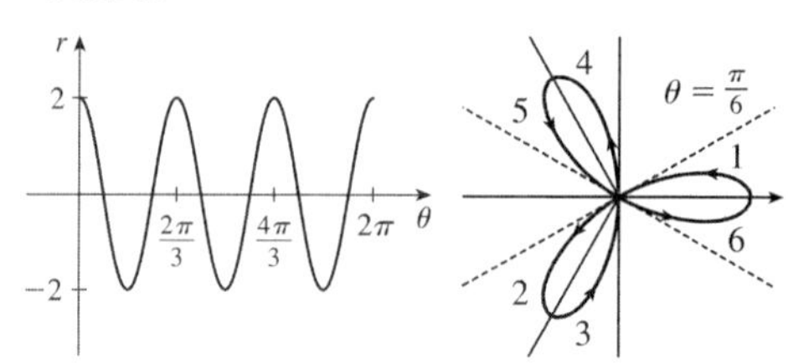

**33.** $r = 2\cos 4\theta$

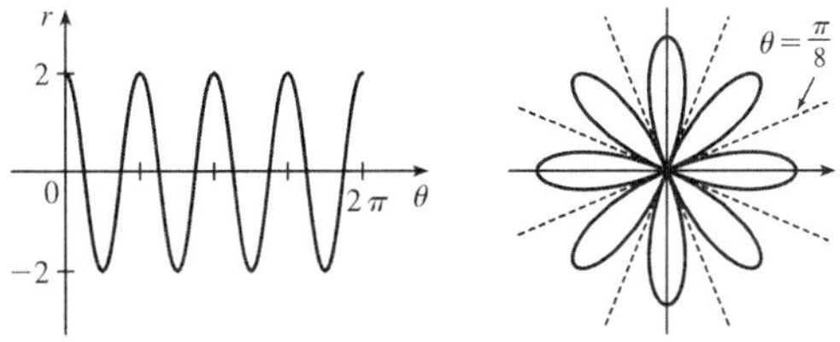


**34.** $r = \sin 5\theta$

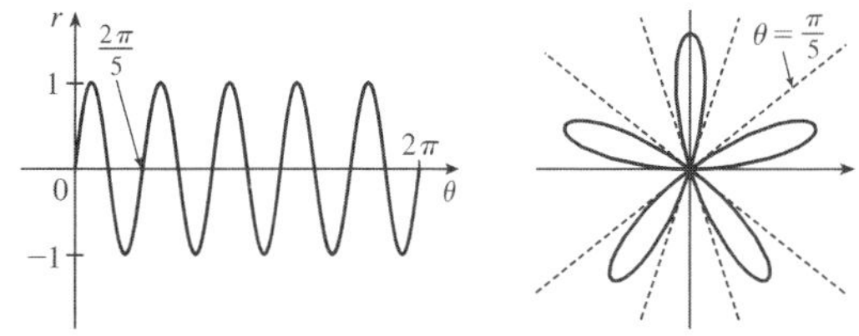


**35.** $r^2 = 4\cos 2\theta$

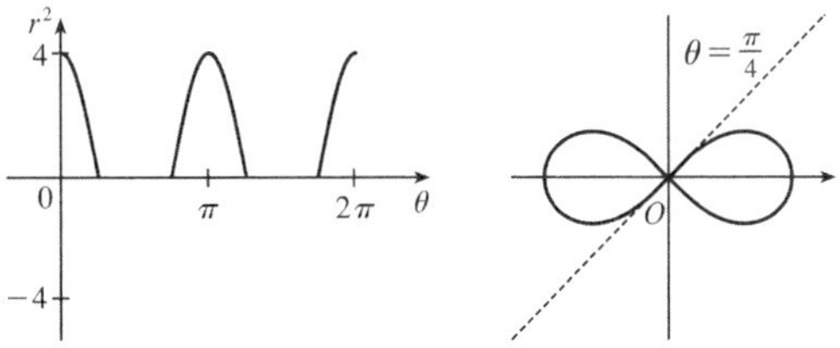


**36.** $r^2 = \sin 2\theta$

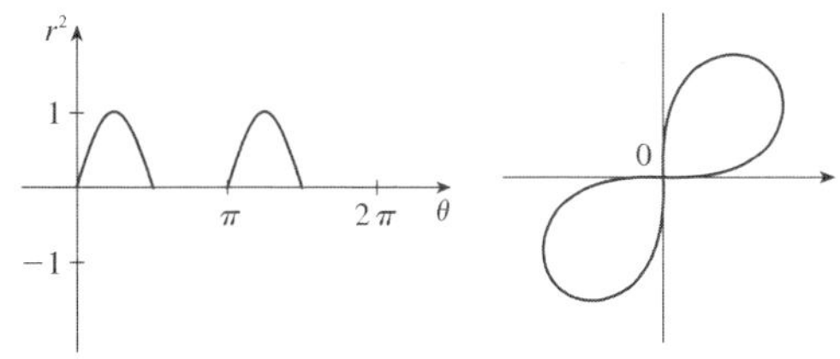


**37.** $r = 2\cos\left(\frac{3}{2}\theta\right)$

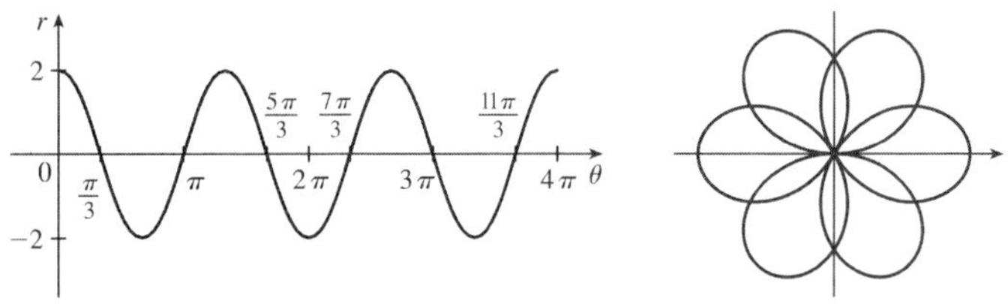


**38.** $r^2\theta = 1 \quad \Leftrightarrow \quad r = \pm 1/\sqrt{\theta}$ for $\theta > 0$

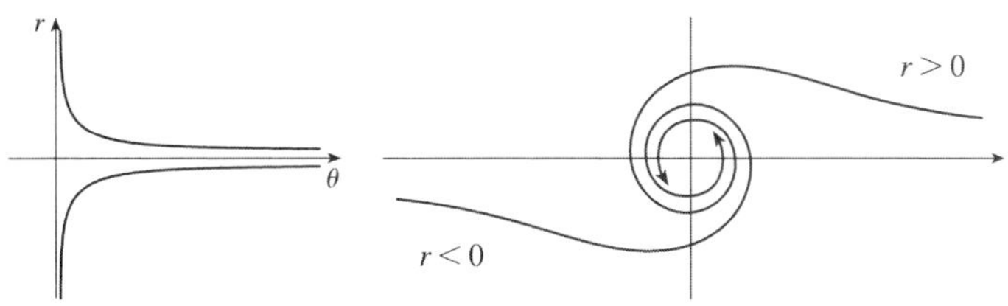


**39.** $r = 1 + 2\cos 2\theta$

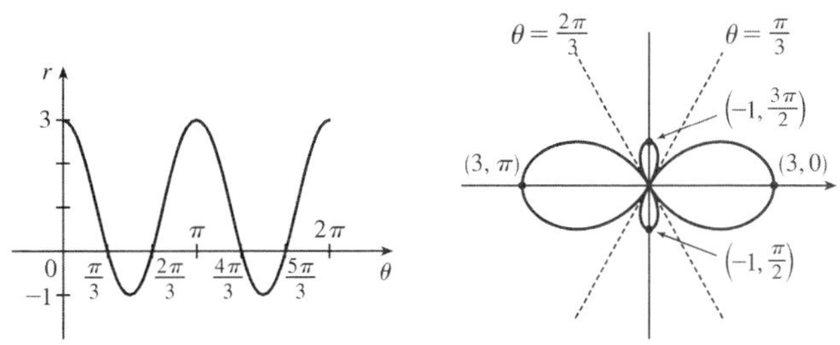


**40.** $r = 1 + 2\cos(\theta/2)$

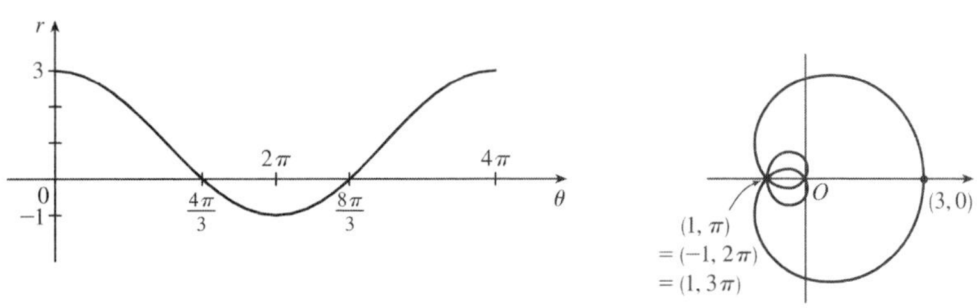

**41.** For $\theta = 0, \pi$, and $2\pi$, $r$ has its minimum value of about 0.5. For $\theta = \frac{\pi}{2}$ and $\frac{3\pi}{2}$, $r$ attains its maximum value of 2. We see that the graph has a similar shape for $0 \le \theta \le \pi$ and $\pi \le \theta \le 2\pi$.

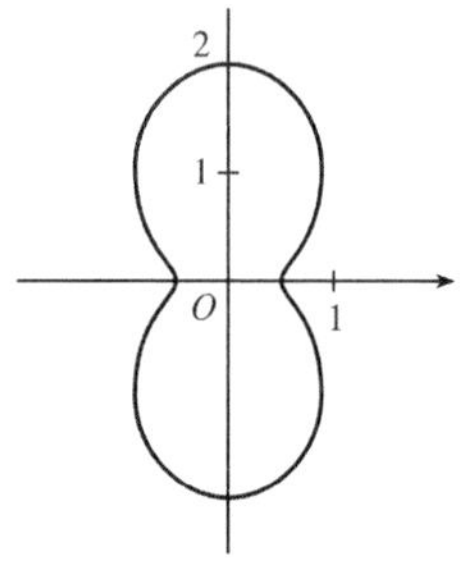


**42.**

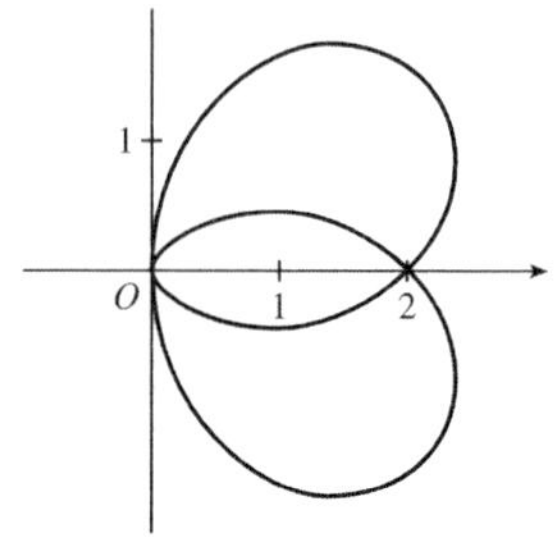


**43.** $x = (r)\cos\theta = (4 + 2\sec\theta)\cos\theta = 4\cos\theta + 2$. Now, $r \to \infty \Rightarrow$ $(4 + 2\sec\theta) \to \infty \Rightarrow \theta \to \left(\frac{\pi}{2}\right)^-$ or $\theta \to \left(\frac{3\pi}{2}\right)^+$ (since we need only consider $0 \le \theta < 2\pi$), so $\lim\limits_{r\to\infty} x = \lim\limits_{\theta\to\pi/2^-} (4\cos\theta + 2) = 2$. Also, $r \to -\infty \Rightarrow$ $(4 + 2\sec\theta) \to -\infty \Rightarrow \theta \to \left(\frac{\pi}{2}\right)^+$ or $\theta \to \left(\frac{3\pi}{2}\right)^-$, so $\lim\limits_{r\to-\infty} x = \lim\limits_{\theta\to\pi/2^+} (4\cos\theta + 2) = 2$. Therefore, $\lim\limits_{r\to\pm\infty} x = 2 \Rightarrow x = 2$ is a vertical asymptote.

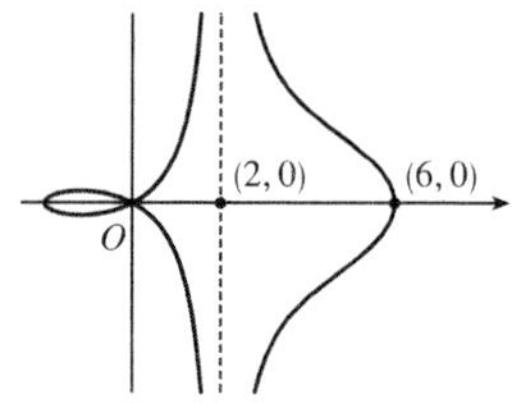


**44.** To show that $x = 1$ is an asymptote we must prove $\lim\limits_{r\to\pm\infty} x = 1$.

$x = (r)\cos\theta = (\sin\theta\,\tan\theta)\cos\theta = \sin^2\theta$. Now, $r \to \infty \Rightarrow$ $\sin\theta\,\tan\theta \to \infty \Rightarrow \theta \to \left(\frac{\pi}{2}\right)^-$, so $\lim\limits_{r\to\infty} x = \lim\limits_{\theta\to\pi/2^-} \sin^2\theta = 1$.

Also, $r \to -\infty \Rightarrow \sin\theta\,\tan\theta \to -\infty \Rightarrow \theta \to \left(\frac{\pi}{2}\right)^+$, so $\lim\limits_{r\to-\infty} x = \lim\limits_{\theta\to\pi/2^+} \sin^2\theta = 1$. Therefore, $\lim\limits_{r\to\pm\infty} x = 1 \Rightarrow x = 1$ is a vertical asymptote. Also notice that $x = \sin^2\theta \ge 0$ for all $\theta$, and $x = \sin^2\theta \le 1$ for all $\theta$. And $x \ne 1$, since the curve is not defined at odd multiples of $\frac{\pi}{2}$. Therefore, the curve lies entirely within the vertical strip $0 \le x < 1$.

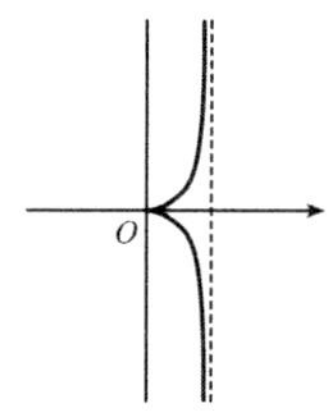


**45.** (a) We see that the curve crosses itself at the origin, where $r = 0$ (in fact the inner loop corresponds to negative $r$-values,) so we solve the equation of the limaçon for $r = 0 \Leftrightarrow c\sin\theta = -1 \Leftrightarrow \sin\theta = -1/c$. Now if $|c| < 1$, then this equation has no solution and hence there is no inner loop. But if $c < -1$, then on the interval $(0, 2\pi)$ the equation has the two solutions $\theta = \sin^{-1}(-1/c)$ and $\theta = \pi - \sin^{-1}(-1/c)$, and if $c > 1$, the solutions are $\theta = \pi + \sin^{-1}(1/c)$ and $\theta = 2\pi - \sin^{-1}(1/c)$. In each case, $r < 0$ for $\theta$ between the two solutions, indicating a loop.

(b) For $0 < c < 1$, the dimple (if it exists) is characterized by the fact that $y$ has a local maximum at $\theta = \frac{3\pi}{2}$. So we determine for what $c$-values $\dfrac{d^2y}{d\theta^2}$ is negative at $\theta = \frac{3\pi}{2}$, since by the Second Derivative Test this indicates a maximum:

$y = r\sin\theta = \sin\theta + c\sin^2\theta \Rightarrow \dfrac{dy}{d\theta} = \cos\theta + 2c\sin\theta\cos\theta = \cos\theta + c\sin 2\theta \Rightarrow \dfrac{d^2y}{d\theta^2} = -\sin\theta + 2c\cos 2\theta$.

At $\theta = \frac{3\pi}{2}$, this is equal to $-(-1) + 2c(-1) = 1 - 2c$, which is negative only for $c > \frac{1}{2}$. A similar argument shows that for $-1 < c < 0$, $y$ only has a local minimum at $\theta = \frac{\pi}{2}$ (indicating a dimple) for $c < -\frac{1}{2}$.

**46.** (a) $r = \sin(\theta/2)$. This equation must correspond to one of II, III or VI, since these are the only graphs which are bounded. In fact it must be VI, since this is the only graph which is completed after a rotation of exactly $4\pi$.

(b) $r = \sin(\theta/4)$. This equation must correspond to III, since this is the only graph which is completed after a rotation of exactly $8\pi$.

(c) $r = \sec(3\theta)$. This must correspond to IV, since the graph is unbounded at $\theta = \frac{\pi}{6}, \frac{\pi}{2}, \frac{2\pi}{3}$, and so on.

(d) $r = \theta \sin\theta$. This must correspond to V. Note that $r = 0$ whenever $\theta$ is a multiple of $\pi$. This graph is unbounded, and each time $\theta$ moves through an interval of $2\pi$, the same basic shape is repeated (because of the periodic $\sin\theta$ factor) but it gets larger each time (since $\theta$ increases each time we go around.)

(e) $r = 1 + 4\cos 5\theta$. This corresponds to II, since it is bounded, has fivefold rotational symmetry, and takes only one takes only one rotation through $2\pi$ to be complete.

(f) $r = 1/\sqrt{\theta}$. This corresponds to I, since it is unbounded at $\theta = 0$, and $r$ decreases as $\theta$ increases; in fact $r \to 0$ as $\theta \to \infty$.

**47.** $r = 2\sin\theta \quad\Rightarrow\quad x = r\cos\theta = 2\sin\theta\cos\theta = \sin 2\theta,\ y = r\sin\theta = 2\sin^2\theta \quad\Rightarrow$

$$\frac{dy}{dx} = \frac{dy/d\theta}{dx/d\theta} = \frac{2 \cdot 2\sin\theta\cos\theta}{\cos 2\theta \cdot 2} = \frac{\sin 2\theta}{\cos 2\theta} = \tan 2\theta$$

When $\theta = \frac{\pi}{6}$, $\frac{dy}{dx} = \tan\left(2 \cdot \frac{\pi}{6}\right) = \tan\frac{\pi}{3} = \sqrt{3}$.

**48.** $r = 2 - \sin\theta \quad\Rightarrow\quad x = r\cos\theta = (2 - \sin\theta)\cos\theta,\ y = r\sin\theta = (2 - \sin\theta)\sin\theta \quad\Rightarrow$

$$\frac{dy}{dx} = \frac{dy/d\theta}{dx/d\theta} = \frac{(2 - \sin\theta)\cos\theta + \sin\theta(-\cos\theta)}{(2 - \sin\theta)(-\sin\theta) + \cos\theta(-\cos\theta)} = \frac{2\cos\theta - 2\sin\theta\cos\theta}{-2\sin\theta + \sin^2\theta - \cos^2\theta} = \frac{2\cos\theta - \sin 2\theta}{-2\sin\theta - \cos 2\theta}$$

When $\theta = \frac{\pi}{3}$, $\frac{dy}{dx} = \frac{2(1/2) - (\sqrt{3}/2)}{-2(\sqrt{3}/2) - (-1/2)} = \frac{1 - \sqrt{3}/2}{-\sqrt{3} + 1/2} \cdot \frac{2}{2} = \frac{2 - \sqrt{3}}{1 - 2\sqrt{3}}$.

**49.** $r = 1/\theta \quad\Rightarrow\quad x = r\cos\theta = (\cos\theta)/\theta,\ y = r\sin\theta = (\sin\theta)/\theta \quad\Rightarrow$

$$\frac{dy}{dx} = \frac{dy/d\theta}{dx/d\theta} = \frac{\sin\theta(-1/\theta^2) + (1/\theta)\cos\theta}{\cos\theta(-1/\theta^2) - (1/\theta)\sin\theta} \cdot \frac{\theta^2}{\theta^2} = \frac{-\sin\theta + \theta\cos\theta}{-\cos\theta - \theta\sin\theta}$$

When $\theta = \pi$, $\frac{dy}{dx} = \frac{-0 + \pi(-1)}{-(-1) - \pi(0)} = \frac{-\pi}{1} = -\pi$.

**50.** $r = \sin 3\theta \quad\Rightarrow\quad x = r\cos\theta = \sin 3\theta\cos\theta,\ y = r\sin\theta = \sin 3\theta\sin\theta \quad\Rightarrow$

$$\frac{dy}{dx} = \frac{dy/d\theta}{dx/d\theta} = \frac{3\cos 3\theta\sin\theta + \sin 3\theta\cos\theta}{3\cos 3\theta\cos\theta - \sin 3\theta\sin\theta}$$

When $\theta = \frac{\pi}{6}$, $\frac{dy}{dx} = \frac{3(0)(1/2) + 1(\sqrt{3}/2)}{3(0)(\sqrt{3}/2) - 1(1/2)} = \frac{\sqrt{3}/2}{-1/2} = -\sqrt{3}$.

**51.** $r = 3\cos\theta \quad\Rightarrow\quad x = r\cos\theta = 3\cos\theta\cos\theta,\ y = r\sin\theta = 3\cos\theta\sin\theta \quad\Rightarrow$
$dy/d\theta = -3\sin^2\theta + 3\cos^2\theta = 3\cos 2\theta = 0 \quad\Rightarrow\quad 2\theta = \frac{\pi}{2}$ or $\frac{3\pi}{2} \quad\Leftrightarrow\quad \theta = \frac{\pi}{4}$ or $\frac{3\pi}{4}$. So the tangent is horizontal at $\left(\frac{3}{\sqrt{2}}, \frac{\pi}{4}\right)$ and $\left(-\frac{3}{\sqrt{2}}, \frac{3\pi}{4}\right)$ $\left[\text{same as } \left(\frac{3}{\sqrt{2}}, -\frac{\pi}{4}\right)\right]$. $dx/d\theta = -6\sin\theta\cos\theta = -3\sin 2\theta = 0 \quad\Rightarrow\quad 2\theta = 0$ or $\pi \quad\Leftrightarrow$
$\theta = 0$ or $\frac{\pi}{2}$. So the tangent is vertical at $(3, 0)$ and $\left(0, \frac{\pi}{2}\right)$.

**52.** $r = e^\theta \quad\Rightarrow\quad x = r\cos\theta = e^\theta\cos\theta,\ y = r\sin\theta = e^\theta\sin\theta \quad\Rightarrow$
$dy/d\theta = e^\theta\sin\theta + e^\theta\cos\theta = e^\theta(\sin\theta + \cos\theta) = 0 \quad\Rightarrow\quad \sin\theta = -\cos\theta \quad\Rightarrow\quad \tan\theta = -1 \quad\Rightarrow$
$\theta = -\frac{1}{4}\pi + n\pi$ ($n$ any integer) $\Rightarrow$ horizontal tangents at $\left(e^{\pi(n-1/4)}, \pi\left(n - \frac{1}{4}\right)\right)$.
$dx/d\theta = e^\theta\cos\theta - e^\theta\sin\theta = e^\theta(\cos\theta - \sin\theta) = 0 \quad\Rightarrow\quad \sin\theta = \cos\theta \quad\Rightarrow\quad \tan\theta = 1 \quad\Rightarrow$
$\theta = \frac{1}{4}\pi + n\pi$ ($n$ any integer) $\Rightarrow$ vertical tangents at $\left(e^{\pi(n+1/4)}, \pi\left(n + \frac{1}{4}\right)\right)$.

**53.** $r = 1 + \cos\theta \;\Rightarrow\; x = r\cos\theta = \cos\theta(1+\cos\theta),\; y = r\sin\theta = \sin\theta(1+\cos\theta) \;\Rightarrow$
$dy/d\theta = (1+\cos\theta)\cos\theta - \sin^2\theta = 2\cos^2\theta + \cos\theta - 1 = (2\cos\theta - 1)(\cos\theta + 1) = 0 \;\Rightarrow$
$\cos\theta = \frac{1}{2}$ or $-1 \;\Rightarrow\; \theta = \frac{\pi}{3}, \pi$, or $\frac{5\pi}{3} \;\Rightarrow$ horizontal tangent at $\left(\frac{3}{2}, \frac{\pi}{3}\right)$, $(0, \pi)$, and $\left(\frac{3}{2}, \frac{5\pi}{3}\right)$.
$dx/d\theta = -(1+\cos\theta)\sin\theta - \cos\theta\sin\theta = -\sin\theta(1+2\cos\theta) = 0 \;\Rightarrow\; \sin\theta = 0$ or $\cos\theta = -\frac{1}{2} \;\Rightarrow$
$\theta = 0, \pi, \frac{2\pi}{3}$, or $\frac{4\pi}{3} \;\Rightarrow$ vertical tangent at $(2, 0)$, $\left(\frac{1}{2}, \frac{2\pi}{3}\right)$, and $\left(\frac{1}{2}, \frac{4\pi}{3}\right)$.
Note that the tangent is horizontal, not vertical when $\theta = \pi$, since $\lim\limits_{\theta\to\pi} \dfrac{dy/d\theta}{dx/d\theta} = 0$.

**54.** By differentiating implicitly, $r^2 = \sin 2\theta \;\Rightarrow\; 2r\,(dr/d\theta) = 2\cos 2\theta \;\Rightarrow$
$dr/d\theta = (1/r)\cos 2\theta$, so $dy/d\theta = (dr/d\theta)\sin\theta + r\cos\theta \;\Rightarrow$

$$\frac{dy}{d\theta} = \frac{1}{r}\cos 2\theta\,\sin\theta + r\cos\theta = \frac{1}{r}\left(\cos 2\theta\,\sin\theta + r^2\cos\theta\right)$$
$$= \frac{1}{r}\left(\cos 2\theta\,\sin\theta + \sin 2\theta\,\cos\theta\right) = \frac{1}{r}\sin 3\theta$$

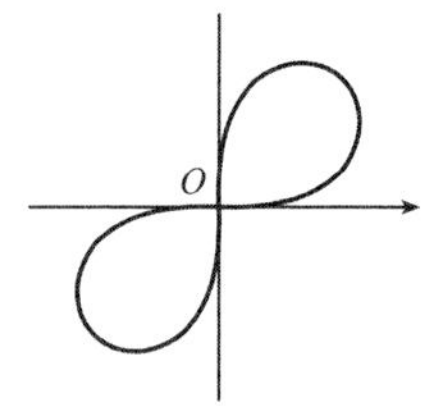


This is 0 when $\sin 3\theta = 0 \;\Rightarrow\; \theta = 0, \frac{\pi}{3}$ or $\frac{4\pi}{3}$ (restricting $\theta$ to the domain of the lemniscate), so there are horizontal tangents at $\left(\sqrt[4]{\frac{3}{4}}, \frac{\pi}{3}\right)$, $\left(\sqrt[4]{\frac{3}{4}}, \frac{4\pi}{3}\right)$ and $(0, 0)$. Similarly, $dx/d\theta = (1/r)\cos 3\theta = 0$ when $\theta = \frac{\pi}{6}$ or $\frac{7\pi}{6}$, so there are vertical tangents at $\left(\sqrt[4]{\frac{3}{4}}, \frac{\pi}{6}\right)$ and $\left(\sqrt[4]{\frac{3}{4}}, \frac{7\pi}{6}\right)$ [and $(0, 0)$].

**55.** $r = a\sin\theta + b\cos\theta \;\Rightarrow\; r^2 = ar\sin\theta + br\cos\theta \;\Rightarrow\; x^2 + y^2 = ay + bx \;\Rightarrow$
$x^2 - bx + \left(\frac{1}{2}b\right)^2 + y^2 - ay + \left(\frac{1}{2}a\right)^2 = \left(\frac{1}{2}b\right)^2 + \left(\frac{1}{2}a\right)^2 \;\Rightarrow\; \left(x - \frac{1}{2}b\right)^2 + \left(y - \frac{1}{2}a\right)^2 = \frac{1}{4}(a^2 + b^2)$, and this is a circle with center $\left(\frac{1}{2}b, \frac{1}{2}a\right)$ and radius $\frac{1}{2}\sqrt{a^2 + b^2}$.

**56.** These curves are circles which intersect at the origin and at $\left(\frac{1}{\sqrt{2}}a, \frac{\pi}{4}\right)$. At the origin, the first circle has a horizontal tangent and the second a vertical one, so the tangents are perpendicular here. For the first circle ($r = a\sin\theta$), $dy/d\theta = a\cos\theta\sin\theta + a\sin\theta\cos\theta = a\sin 2\theta = a$ at $\theta = \frac{\pi}{4}$ and $dx/d\theta = a\cos^2\theta - a\sin^2\theta = a\cos 2\theta = 0$ at $\theta = \frac{\pi}{4}$, so the tangent here is vertical. Similarly, for the second circle ($r = a\cos\theta$), $dy/d\theta = a\cos 2\theta = 0$ and $dx/d\theta = -a\sin 2\theta = -a$ at $\theta = \frac{\pi}{4}$, so the tangent is horizontal, and again the tangents are perpendicular.

Note for Exercises 57–60: Maple is able to plot polar curves using the `polarplot` command, or using the `coords=polar` option in a regular `plot` command. In Mathematica, use `PolarPlot`. In Derive, change to `Polar` under `Options State`. If your graphing device cannot plot polar equations, you must convert to parametric equations. For example, in Exercise 57, $x = r\cos\theta = \left[e^{\sin\theta} - 2\cos(4\theta)\right]\cos\theta$, $y = r\sin\theta = \left[e^{\sin\theta} - 2\cos(4\theta)\right]\sin\theta$.

**57.** $r = e^{\sin\theta} - 2\cos(4\theta)$. The parameter interval is $[0, 2\pi]$.

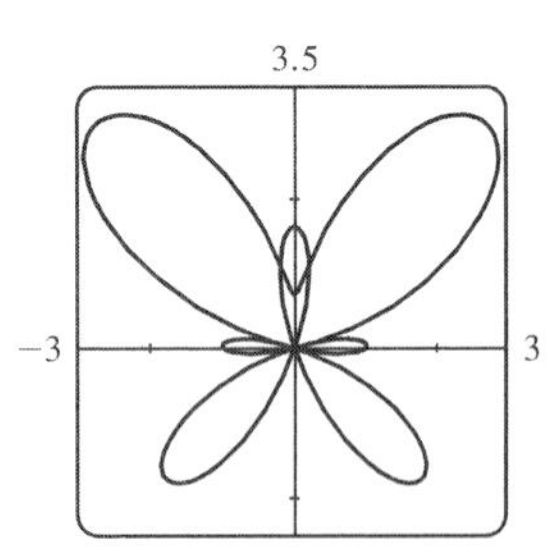


**58.** $r = \sin^2(4\theta) + \cos(4\theta)$. The parameter interval is $[0, 2\pi]$.

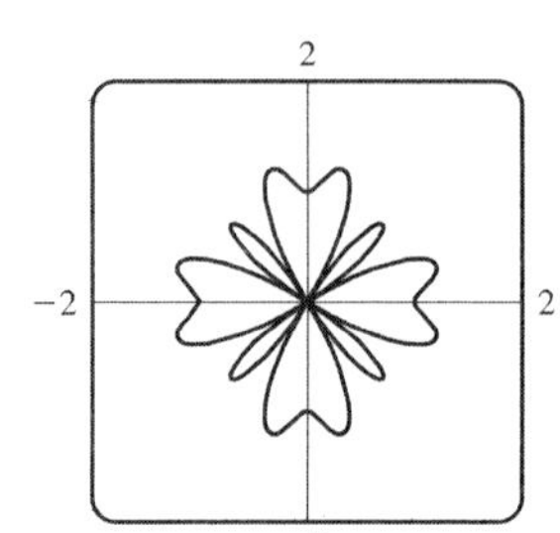

**59.** $r = 2 - 5\sin(\theta/6)$. The parameter interval is $[-6\pi, 6\pi]$.

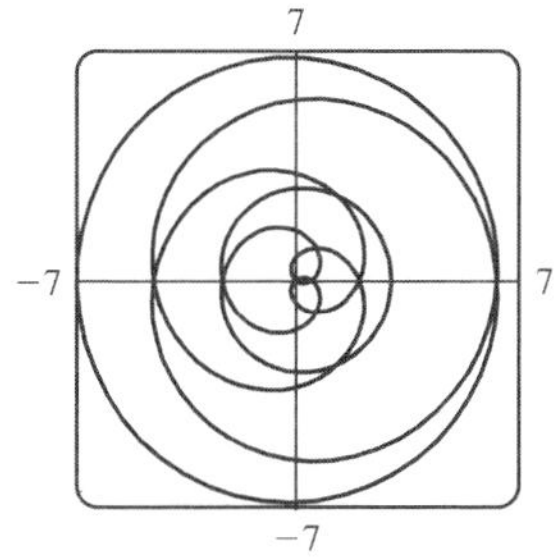


**60.** $r = \cos(\theta/2) + \cos(\theta/3)$. The parameter interval is $[-6\pi, 6\pi]$.

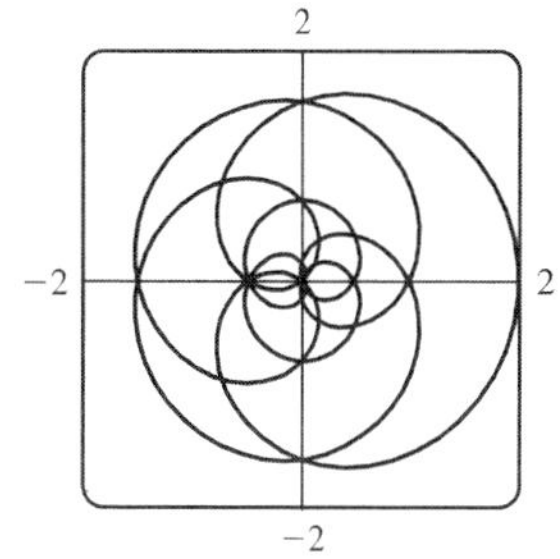


**61.**

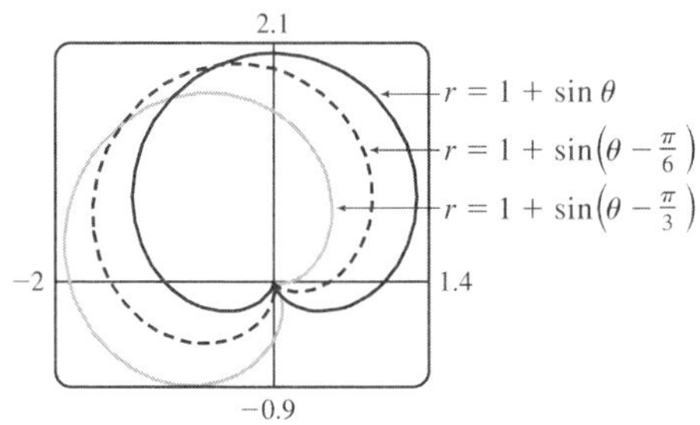


It appears that the graph of $r = 1 + \sin\left(\theta - \frac{\pi}{6}\right)$ is the same shape as the graph of $r = 1 + \sin\theta$, but rotated counterclockwise about the origin by $\frac{\pi}{6}$. Similarly, the graph of $r = 1 + \sin\left(\theta - \frac{\pi}{3}\right)$ is rotated by $\frac{\pi}{3}$. In general, the graph of $r = f(\theta - \alpha)$ is the same shape as that of $r = f(\theta)$, but rotated counterclockwise through $\alpha$ about the origin. That is, for any point $(r_0, \theta_0)$ on the curve $r = f(\theta)$, the point $(r_0, \theta_0 + \alpha)$ is on the curve $r = f(\theta - \alpha)$, since $r_0 = f(\theta_0) = f((\theta_0 + \alpha) - \alpha)$.

**62.**

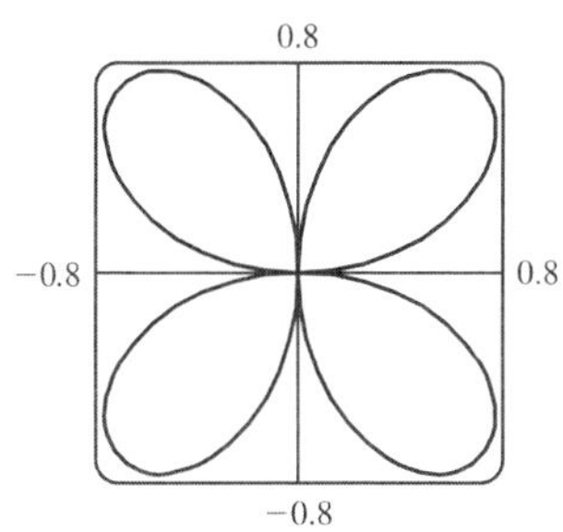


From the graph, the highest points seem to have $y \approx 0.77$. To find the exact value, we solve $dy/d\theta = 0$. $y = r\sin\theta = \sin\theta\sin 2\theta \quad\Rightarrow$

$$\begin{aligned} dy/d\theta &= 2\sin\theta\,\cos 2\theta + \cos\theta\,\sin 2\theta \\ &= 2\sin\theta\left(2\cos^2\theta - 1\right) + \cos\theta\left(2\sin\theta\,\cos\theta\right) \\ &= 2\sin\theta\left(3\cos^2\theta - 1\right) \end{aligned}$$

In the first quadrant, this is 0 when $\cos\theta = \frac{1}{\sqrt{3}} \quad\Leftrightarrow\quad \sin\theta = \sqrt{\frac{2}{3}} \quad\Leftrightarrow\quad y = 2\sin^2\theta\cos\theta = 2\cdot\frac{2}{3}\cdot\frac{1}{\sqrt{3}} = \frac{4\sqrt{3}}{9} \approx 0.77$.

**63.** (a) $r = \sin n\theta$. From the graphs, it seems that when $n$ is even, the number of loops in the curve (called a rose) is $2n$, and when $n$ is odd, the number of loops is simply $n$.

This is because in the case of $n$ odd, every point on the graph is traversed twice, due to the fact that

$$r(\theta + \pi) = \sin\left[n(\theta + \pi)\right] = \sin n\theta\,\cos n\pi + \cos n\theta\,\sin n\pi = \begin{cases} \sin n\theta & \text{if } n \text{ is even} \\ -\sin n\theta & \text{if } n \text{ is odd} \end{cases}$$

$n = 2$ $\qquad n = 3$ $\qquad n = 4$ $\qquad n = 5$

(b) The graph of $r = |\sin n\theta|$ has $2n$ loops whether $n$ is odd or even, since $r(\theta + \pi) = r(\theta)$.

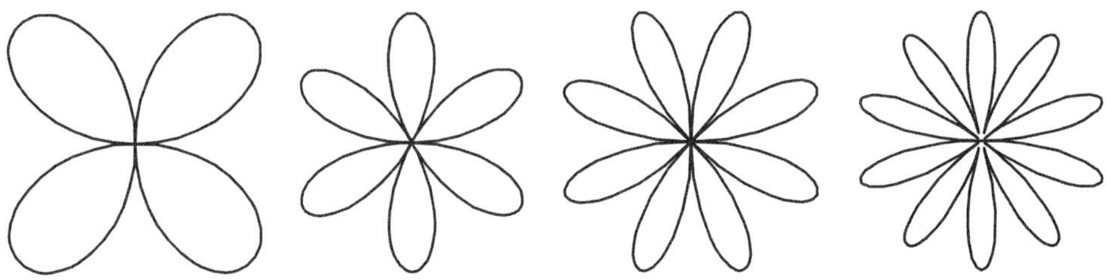

$n = 2$ $n = 3$ $n = 4$ $n = 5$

**64.** $r = 1 + c \sin n\theta$. We vary $n$ while keeping $c$ constant at 2. As $n$ changes, the curves change in the same way as those in Exercise 63: the number of loops increases. Note that if $n$ is even, the smaller loops are outside the larger ones; if $n$ is odd, they are inside.

$c = 2$

$n = 2$ $n = 3$ $n = 4$ $n = 5$

Now we vary $c$ while keeping $n = 3$. As $c$ increases toward 0, the entire graph gets smaller (the graphs below are not to scale) and the smaller loops shrink in relation to the large ones. At $c = -1$, the small loops disappear entirely, and for $-1 < c < 1$, the graph is a simple, closed curve (at $c = 0$ it is a circle). As $c$ continues to increase, the same changes are seen, but in reverse order, since $1 + (-c) \sin n\theta = 1 + c \sin n(\theta + \pi)$, so the graph for $c = c_0$ is the same as that for $c = -c_0$, with a rotation through $\pi$. As $c \to \infty$, the smaller loops get relatively closer in size to the large ones. Note that the distance between the outermost points of corresponding inner and outer loops is always 2.
Maple's `animate` command (or Mathematica's `Animate`) is very useful for seeing the changes that occur as $c$ varies.

$n = 3$

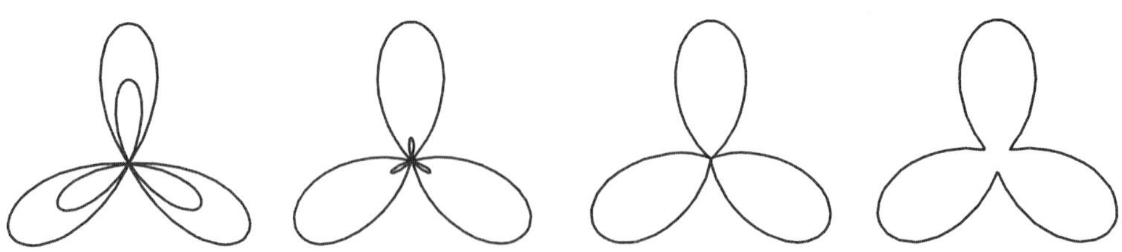

$c = -4$ $c = -1.4$ $c = -1$ $c = -0.8$

$c = -0.2$ $c = 0$ $c = 0.5$ $c = 8$

**65.** $r = \dfrac{1 - a\cos\theta}{1 + a\cos\theta}$. We start with $a = 0$, since in this case the curve is simply the circle $r = 1$.

As $a$ increases, the graph moves to the left, and its right side becomes flattened. As $a$ increases through about $0.4$, the right side seems to grow a dimple, which upon closer investigation (with narrower $\theta$-ranges) seems to appear at $a \approx 0.42$ (the actual value is $\sqrt{2} - 1$). As $a \to 1$, this dimple becomes more pronounced, and the curve begins to stretch out horizontally, until at $a = 1$ the denominator vanishes at $\theta = \pi$, and the dimple becomes an actual cusp. For $a > 1$ we must choose our parameter interval carefully, since $r \to \infty$ as $1 + a\cos\theta \to 0 \quad \Leftrightarrow \quad \theta \to \pm\cos^{-1}(-1/a)$. As $a$ increases from 1, the curve splits into two parts. The left part has a loop, which grows larger as $a$ increases, and the right part grows broader vertically, and its left tip develops a dimple when $a \approx 2.42$ (actually, $\sqrt{2} + 1$). As $a$ increases, the dimple grows more and more pronounced. If $a < 0$, we get the same graph as we do for the corresponding positive $a$-value, but with a rotation through $\pi$ about the pole, as happened when $c$ was replaced with $-c$ in Exercise 64.

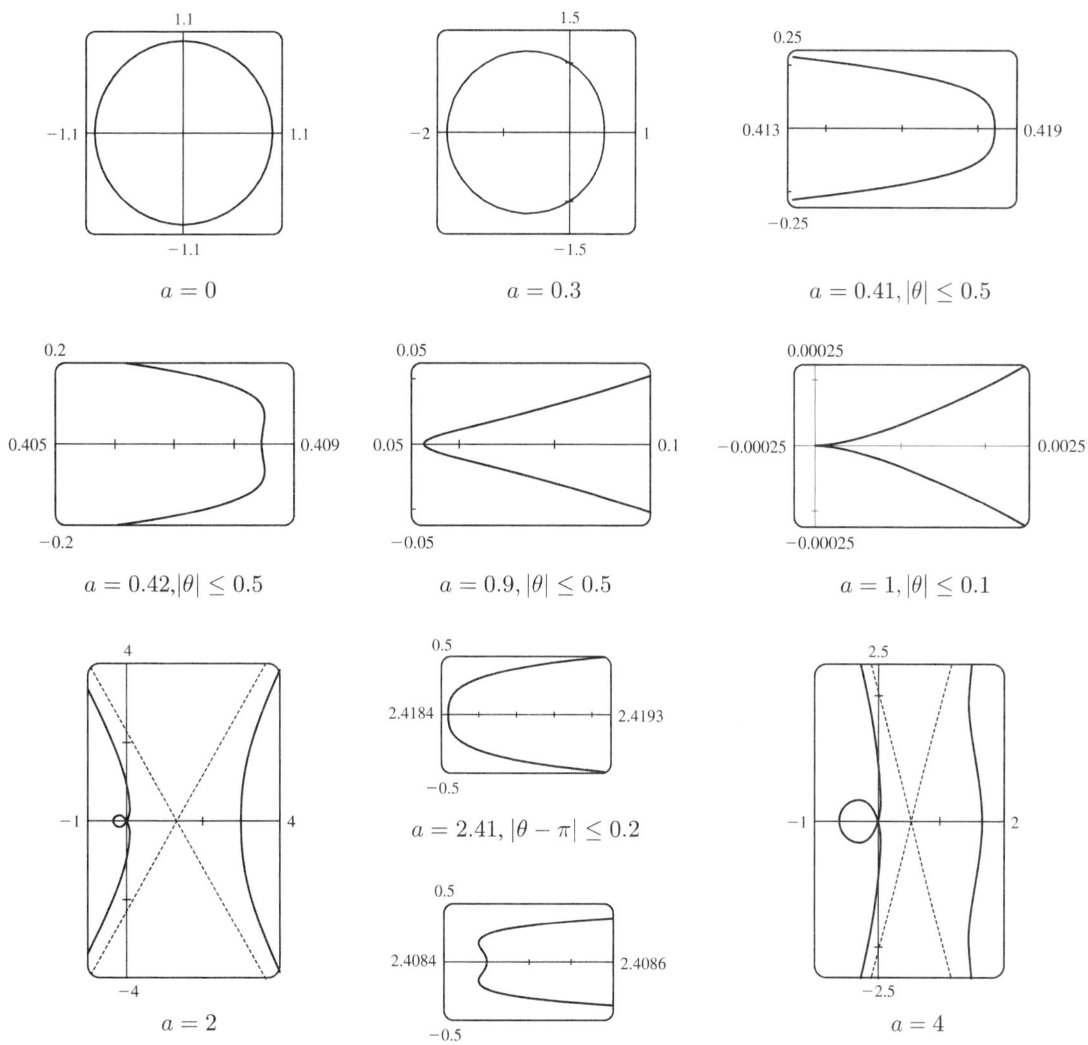


$a = 0$ $\quad$ $a = 0.3$ $\quad$ $a = 0.41, |\theta| \le 0.5$

$a = 0.42, |\theta| \le 0.5$ $\quad$ $a = 0.9, |\theta| \le 0.5$ $\quad$ $a = 1, |\theta| \le 0.1$

$a = 2$ $\quad$ $a = 2.41, |\theta - \pi| \le 0.2$ $\quad$ $a = 2.42, |\theta - \pi| \le 0.2$ $\quad$ $a = 4$

**66.** Most graphing devices cannot plot implicit polar equations, so we must first find an explicit expression (or expressions) for $r$ in terms of $\theta$, $a$, and $c$. We note that the given equation is a quadratic in $r^2$, so we use the quadratic formula and find that

$$\begin{aligned} r^2 &= \frac{2c^2\cos 2\theta \pm \sqrt{4c^4\cos^2 2\theta - 4\left(c^4 - a^4\right)}}{2} \\ &= c^2\cos 2\theta \pm \sqrt{a^4 - c^4\sin^2 2\theta} \end{aligned}$$

so $r = \pm\sqrt{c^2\cos 2\theta \pm \sqrt{a^4 - c^4\sin^2 2\theta}}$. So for each graph, we must plot four curves to be sure of plotting all the points which satisfy the given equation. Note that all four functions have period $\pi$.

We start with the case $a = c = 1$, and the resulting curve resembles the symbol for infinity. If we let $a$ decrease, the curve splits into two symmetric parts, and as $a$ decreases further, the parts become smaller, further apart, and rounder. If instead we let $a$ increase from 1, the two lobes of the curve join together, and as $a$ increases further they continue to merge, until at $a \approx 1.4$, the graph no longer has dimples, and has an oval shape. As $a \to \infty$, the oval becomes larger and rounder, since the $c^2$ and $c^4$ terms lose their significance. Note that the shape of the graph seems to depend only on the ratio $c/a$, while the size of the graph varies as $c$ and $a$ jointly increase.

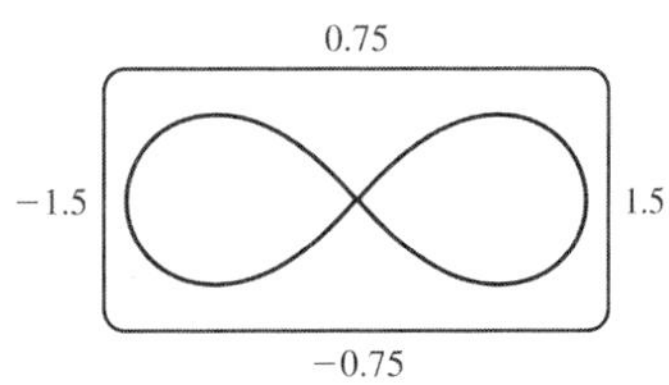

$(a, c) = (1, 1)$

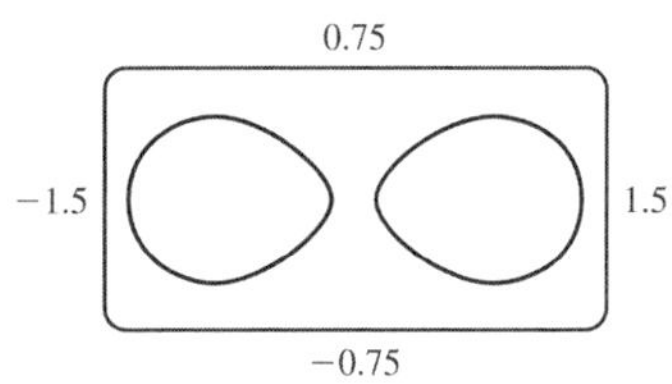

$(a, c) = (0.99, 1)$

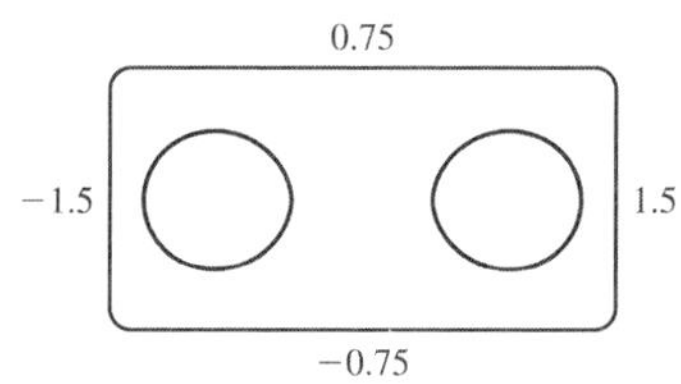

$(a, c) = (0.9, 1)$

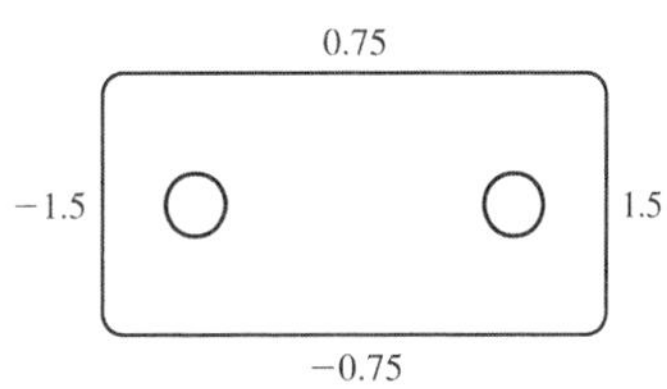

$(a, c) = (0.6, 1)$

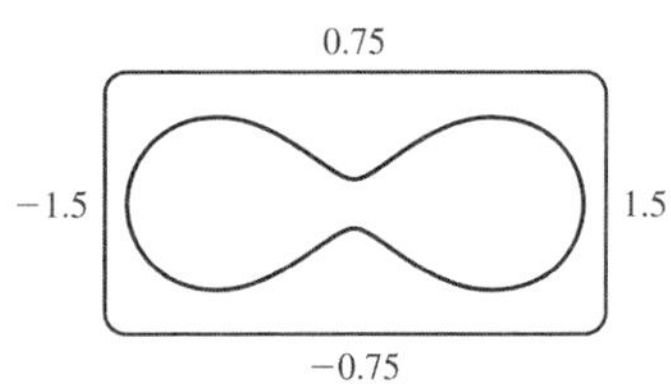

$(a, c) = (1.01, 1)$

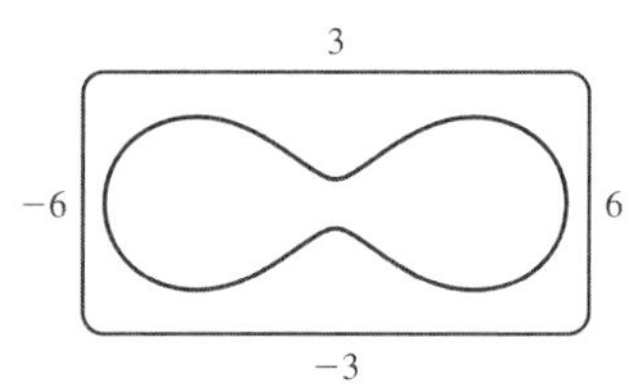

$(a, c) = (4.04, 4)$

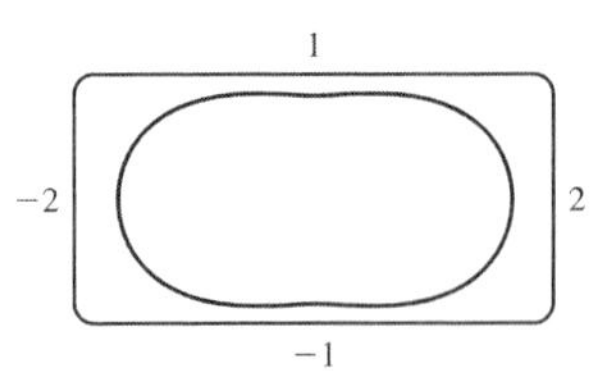

$(a, c) = (1.3, 1)$

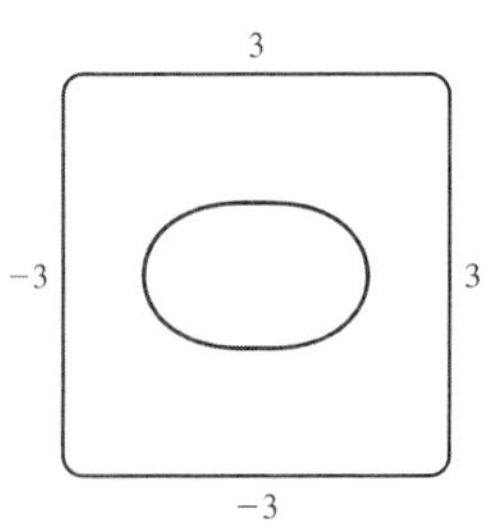

$(a, c) = (1.5, 1)$

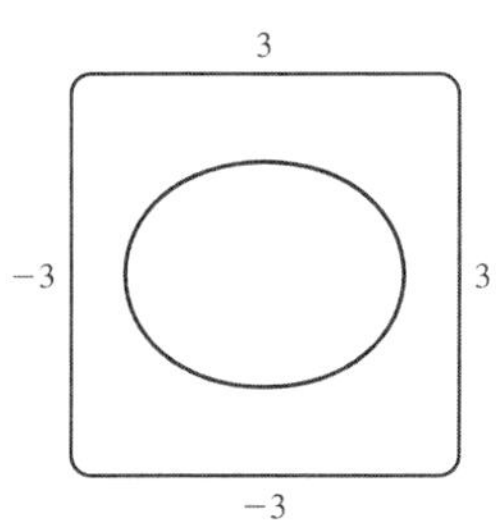

$(a, c) = (2, 1)$

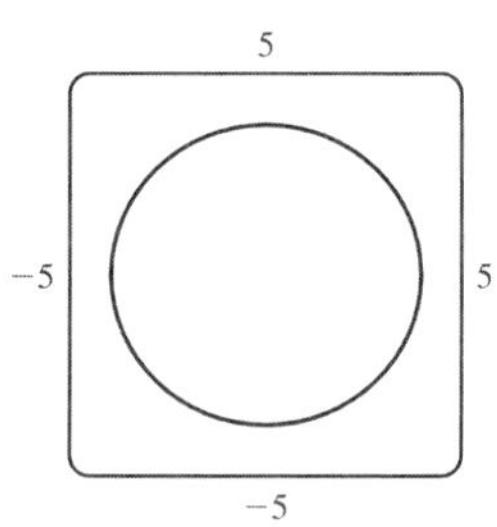

$(a, c) = (4, 1)$

**67.** $$\tan\psi = \tan(\phi - \theta) = \frac{\tan\phi - \tan\theta}{1 + \tan\phi\,\tan\theta} = \frac{\dfrac{dy}{dx} - \tan\theta}{1 + \dfrac{dy}{dx}\tan\theta} = \frac{\dfrac{dy/d\theta}{dx/d\theta} - \tan\theta}{1 + \dfrac{dy/d\theta}{dx/d\theta}\tan\theta}$$

$$= \frac{\dfrac{dy}{d\theta} - \dfrac{dx}{d\theta}\tan\theta}{\dfrac{dx}{d\theta} + \dfrac{dy}{d\theta}\tan\theta} = \frac{\left(\dfrac{dr}{d\theta}\sin\theta + r\cos\theta\right) - \tan\theta\left(\dfrac{dr}{d\theta}\cos\theta - r\sin\theta\right)}{\left(\dfrac{dr}{d\theta}\cos\theta - r\sin\theta\right) + \tan\theta\left(\dfrac{dr}{d\theta}\sin\theta + r\cos\theta\right)}$$

$$= \frac{r\cos\theta + r\cdot\dfrac{\sin^2\theta}{\cos\theta}}{\dfrac{dr}{d\theta}\cos\theta + \dfrac{dr}{d\theta}\cdot\dfrac{\sin^2\theta}{\cos\theta}} = \frac{r\cos^2\theta + r\sin^2\theta}{\dfrac{dr}{d\theta}\cos^2\theta + \dfrac{dr}{d\theta}\sin^2\theta} = \frac{r}{dr/d\theta}$$

**68.** (a) $r = e^\theta \;\Rightarrow\; dr/d\theta = e^\theta$, so by Exercise 67, $\tan\psi = r/e^\theta = 1 \;\Rightarrow\; \psi = \arctan 1 = \frac{\pi}{4}$.

(b) The Cartesian equation of the tangent line at $(1, 0)$ is $y = x - 1$, and that of the tangent line at $\left(0, e^{\pi/2}\right)$ is $y = e^{\pi/2} - x$.

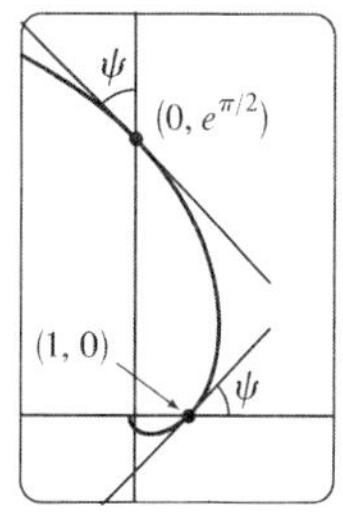


(c) Let $a$ be the tangent of the angle between the tangent and radial lines, that is, $a = \tan\psi$. Then, by Exercise 67, $a = \dfrac{r}{dr/d\theta} \;\Rightarrow\; \dfrac{dr}{d\theta} = \dfrac{1}{a}r \;\Rightarrow$ $r = Ce^{\theta/a}$ (by Theorem 7.4.2).

## H.2 Areas and Lengths in Polar Coordinates

**1.** $r = \sqrt{\theta}, 0 \le \theta \le \frac{\pi}{4}$. $A = \int_0^{\pi/4} \frac{1}{2}r^2\,d\theta = \int_0^{\pi/4} \frac{1}{2}\left(\sqrt{\theta}\right)^2 d\theta = \int_0^{\pi/4} \frac{1}{2}\theta\,d\theta = \left[\frac{1}{4}\theta^2\right]_0^{\pi/4} = \frac{1}{64}\pi^2$

**2.** $r = e^{\theta/2}, \pi \le \theta \le 2\pi$. $A = \int_\pi^{2\pi} \frac{1}{2}\left(e^{\theta/2}\right)^2 d\theta = \int_\pi^{2\pi} \frac{1}{2}e^\theta\,d\theta = \frac{1}{2}\left[e^\theta\right]_\pi^{2\pi} = \frac{1}{2}(e^{2\pi} - e^\pi)$

**3.** $r = \sin\theta, \frac{\pi}{3} \le \theta \le \frac{2\pi}{3}$.

$$A = \int_{\pi/3}^{2\pi/3} \tfrac{1}{2}\sin^2\theta\,d\theta = \tfrac{1}{4}\int_{\pi/3}^{2\pi/3} (1 - \cos 2\theta)\,d\theta = \tfrac{1}{4}\left[\theta - \tfrac{1}{2}\sin 2\theta\right]_{\pi/3}^{2\pi/3}$$

$$= \tfrac{1}{4}\left[\tfrac{2\pi}{3} - \tfrac{1}{2}\sin\tfrac{4\pi}{3} - \tfrac{\pi}{3} + \tfrac{1}{2}\sin\tfrac{2\pi}{3}\right] = \tfrac{1}{4}\left[\tfrac{2\pi}{3} - \tfrac{1}{2}\left(-\tfrac{\sqrt{3}}{2}\right) - \tfrac{\pi}{3} + \tfrac{1}{2}\left(\tfrac{\sqrt{3}}{2}\right)\right] = \tfrac{1}{4}\left(\tfrac{\pi}{3} + \tfrac{\sqrt{3}}{2}\right) = \tfrac{\pi}{12} + \tfrac{\sqrt{3}}{8}$$

**4.** $r = \sqrt{\sin\theta}, 0 \le \theta \le \pi$. $A = \int_0^\pi \frac{1}{2}\left(\sqrt{\sin\theta}\right)^2 d\theta = \int_0^\pi \frac{1}{2}\sin\theta\,d\theta = \left[-\frac{1}{2}\cos\theta\right]_0^\pi = \frac{1}{2} + \frac{1}{2} = 1$

**5.** $r = \theta, 0 \le \theta \le \pi$. $A = \int_0^\pi \frac{1}{2}\theta^2 d\theta = \left[\frac{1}{6}\theta^3\right]_0^\pi = \frac{1}{6}\pi^3$

**6.** $r = 1 + \sin\theta, \frac{\pi}{2} \le \theta \le \pi$.

$$A = \int_{\pi/2}^{\pi} \tfrac{1}{2}(1 + \sin\theta)^2\,d\theta = \tfrac{1}{2}\int_{\pi/2}^{\pi} \left(1 + 2\sin\theta + \sin^2\theta\right) d\theta = \tfrac{1}{2}\int_{\pi/2}^{\pi} \left[1 + 2\sin\theta + \tfrac{1}{2}(1 - \cos 2\theta)\right] d\theta$$

$$= \tfrac{1}{2}\left[\theta - 2\cos\theta + \tfrac{1}{2}\theta - \tfrac{1}{4}\sin 2\theta\right]_{\pi/2}^{\pi} = \tfrac{1}{2}\left[\pi + 2 + \tfrac{\pi}{2} - 0 - \left(\tfrac{\pi}{2} - 0 + \tfrac{\pi}{4} - 0\right)\right] = \tfrac{1}{2}\left(\tfrac{3\pi}{4} + 2\right) = \tfrac{3\pi}{8} + 1$$

**7.** $r = 4 + 3\sin\theta,\ -\frac{\pi}{2} \le \theta \le \frac{\pi}{2}$.

$$A = \int_{-\pi/2}^{\pi/2} \tfrac{1}{2}(4+3\sin\theta)^2 d\theta = \tfrac{1}{2}\int_{-\pi/2}^{\pi/2} \left(16 + 24\sin\theta + 9\sin^2\theta\right) d\theta$$

$$= \tfrac{1}{2}\int_{-\pi/2}^{\pi/2} \left(16 + 9\sin^2\theta\right) d\theta \quad \text{[by Theorem 5.5.6(b)]}$$

$$= \tfrac{1}{2}\cdot 2\int_{0}^{\pi/2} \left[16 + 9\cdot\tfrac{1}{2}(1-\cos 2\theta)\right] d\theta \quad \text{[by Theorem 5.5.6(a)]}$$

$$= \int_0^{\pi/2} \left(\tfrac{41}{2} - \tfrac{9}{2}\cos 2\theta\right) d\theta = \left[\tfrac{41}{2}\theta - \tfrac{9}{4}\sin 2\theta\right]_0^{\pi/2} = \left(\tfrac{41\pi}{4} - 0\right) - (0-0) = \tfrac{41\pi}{4}$$

**8.** $r = \sin 4\theta,\ 0 \le \theta \le \frac{\pi}{4}$. $A = \int_0^{\pi/4} \frac{1}{2}\sin^2 4\theta\, d\theta = \int_0^{\pi/4} \frac{1}{4}(1-\cos 8\theta)\, d\theta = \left[\frac{1}{4}\theta - \frac{1}{32}\sin 8\theta\right]_0^{\pi/4} = \frac{\pi}{16}$

**9.** The curve goes through the pole when $\theta = \pi/4$, so we'll find the area for $0 \le \theta \le \pi/4$ and multiply it by 4.

$A = 4\int_0^{\pi/4} \frac{1}{2}r^2\, d\theta = 2\int_0^{\pi/4} (4\cos 2\theta)\, d\theta$

$= 8\int_0^{\pi/4} \cos 2\theta\, d\theta = 4\left[\sin 2\theta\right]_0^{\pi/4} = 4$

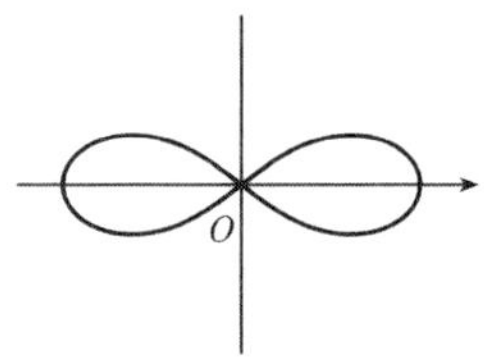


**10.** $A = \int_0^{2\pi} \frac{1}{2}r^2\, d\theta = \int_0^{2\pi} \frac{1}{2}\left[3(1+\cos\theta)\right]^2 d\theta$

$= \frac{9}{2}\int_0^{2\pi} \left(1 + 2\cos\theta + \cos^2\theta\right) d\theta$

$= \frac{9}{2}\int_0^{2\pi} \left[1 + 2\cos\theta + \frac{1}{2}(1+\cos 2\theta)\right] d\theta$

$= \frac{9}{2}\left[\frac{3}{2}\theta + 2\sin\theta + \frac{1}{4}\sin 2\theta\right]_0^{2\pi} = \frac{27}{2}\pi$

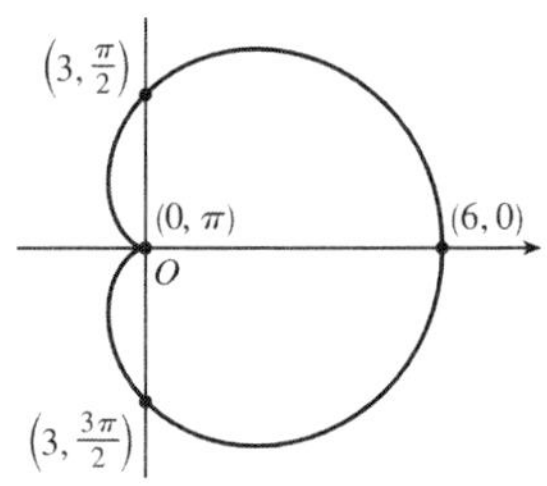


**11.** One-sixth of the area lies above the polar axis and is bounded by the curve $r = 2\cos 3\theta$ for $\theta = 0$ to $\theta = \pi/6$.

$A = 6\int_0^{\pi/6} \frac{1}{2}(2\cos 3\theta)^2\, d\theta = 12\int_0^{\pi/6} \cos^2 3\theta\, d\theta$

$= \frac{12}{2}\int_0^{\pi/6} (1 + \cos 6\theta) d\theta$

$= 6\left[\theta + \frac{1}{6}\sin 6\theta\right]_0^{\pi/6} = 6\left(\frac{\pi}{6}\right) = \pi$

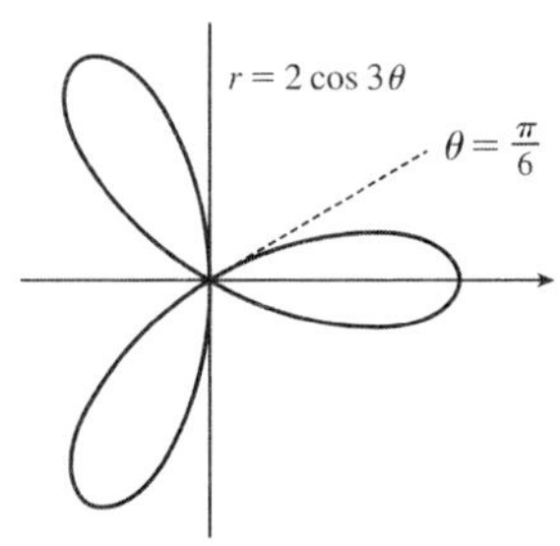


**12.** $A = \int_0^{2\pi} \frac{1}{2}(2+\cos 2\theta)^2\, d\theta = \frac{1}{2}\int_0^{2\pi} (4 + 4\cos 2\theta + \cos^2 2\theta)\, d\theta$

$= \frac{1}{2}\int_0^{2\pi} \left(4 + 4\cos 2\theta + \frac{1}{2} + \frac{1}{2}\cos 4\theta\right) d\theta$

$= \frac{1}{2}\left[\frac{9}{2}\theta + 2\sin 2\theta + \frac{1}{8}\sin 4\theta\right]_0^{2\pi}$

$= \frac{1}{2}(9\pi) = \frac{9\pi}{2}$

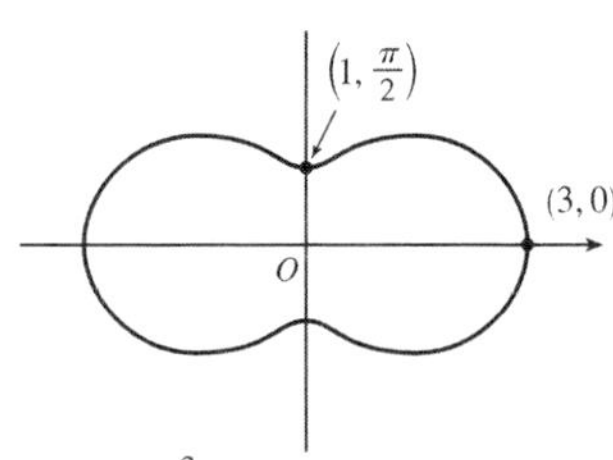


**13.** $A = \int_0^{2\pi} \frac{1}{2}(1 + 2\sin 6\theta)^2\, d\theta = \frac{1}{2}\int_0^{2\pi} (1 + 4\sin 6\theta + 4\sin^2 6\theta) d\theta$

$= \frac{1}{2}\int_0^{2\pi} \left[1 + 4\sin 6\theta + 4\cdot\frac{1}{2}(1-\cos 12\theta)\right] d\theta$

$= \frac{1}{2}\int_0^{2\pi} (3 + 4\sin 6\theta - 2\cos 12\theta)\, d\theta$

$= \frac{1}{2}\left[3\theta - \frac{2}{3}\cos 6\theta - \frac{1}{6}\sin 12\theta\right]_0^{2\pi}$

$= \frac{1}{2}\left[\left(6\pi - \frac{2}{3} - 0\right) - \left(0 - \frac{2}{3} - 0\right)\right] = 3\pi.$

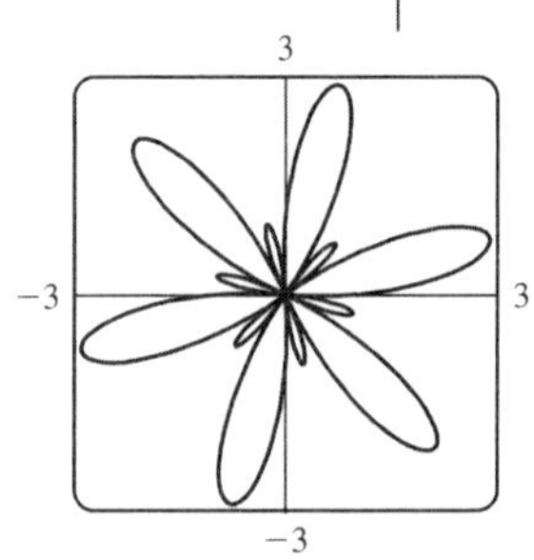

**14.** $A = \int_0^{\pi} \frac{1}{2}(2\sin\theta + 3\sin 9\theta)^2\,d\theta = 2\int_0^{\pi/2} \frac{1}{2}(2\sin\theta + 3\sin 9\theta)^2\,d\theta$

$= \int_0^{\pi/2}(4\sin^2\theta + 12\sin\theta\sin 9\theta + 9\sin^2 9\theta)\,d\theta$

$= \int_0^{\pi/2}\left[2(1-\cos 2\theta) + 12\cdot\frac{1}{2}(\cos(\theta - 9\theta) - \cos(\theta + 9\theta)) + \frac{9}{2}(1 - \cos 18\theta)\right]d\theta$

[integration by parts could be used for $\int \sin\theta\sin 9\theta\,d\theta$]

$= \int_0^{\pi/2}(2 - 2\cos 2\theta + 6\cos 8\theta - 6\cos 10\theta + \frac{9}{2} - \frac{9}{2}\cos 18\theta)\,d\theta$

$= \left[\frac{13}{2}\theta - \sin 2\theta + \frac{3}{4}\sin 8\theta - \frac{3}{5}\sin 10\theta - \frac{1}{4}\sin 18\theta\right]_0^{\pi/2} = \frac{13}{4}\pi$

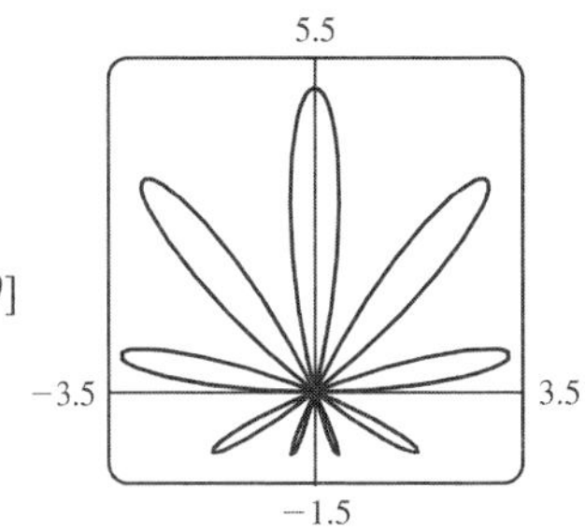


**15.** The shaded loop is traced out from $\theta = 0$ to $\theta = \pi/2$.

$$A = \int_0^{\pi/2} \tfrac{1}{2}r^2\,d\theta = \tfrac{1}{2}\int_0^{\pi/2}\sin^2 2\theta\,d\theta$$

$$= \tfrac{1}{2}\int_0^{\pi/2}\tfrac{1}{2}(1 - \cos 4\theta)\,d\theta = \tfrac{1}{4}\left[\theta - \tfrac{1}{4}\sin 4\theta\right]_0^{\pi/2} = \tfrac{1}{4}\left(\tfrac{\pi}{2}\right) = \tfrac{\pi}{8}$$

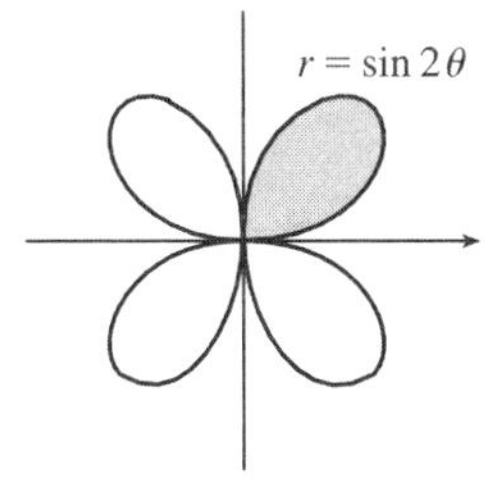


**16.** $A = \displaystyle\int_0^{\pi/3}\tfrac{1}{2}(4\sin 3\theta)^2\,d\theta = 8\int_0^{\pi/3}\sin^2 3\theta\,d\theta$

$= 4\displaystyle\int_0^{\pi/3}(1 - \cos 6\theta)\,d\theta = 4\left[\theta - \tfrac{1}{6}\sin 6\theta\right]_0^{\pi/3} = \tfrac{4\pi}{3}$

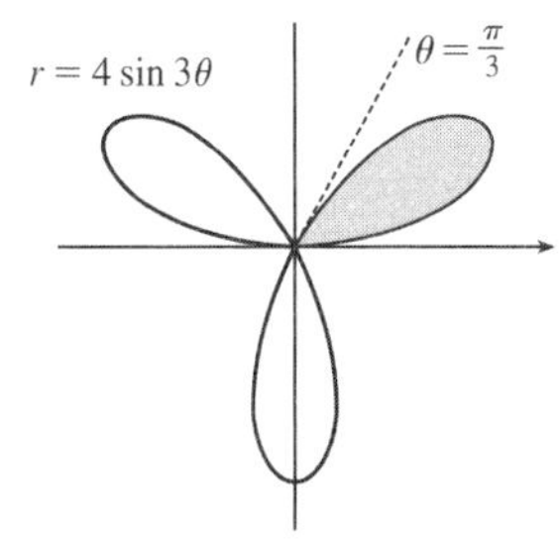


**17.**

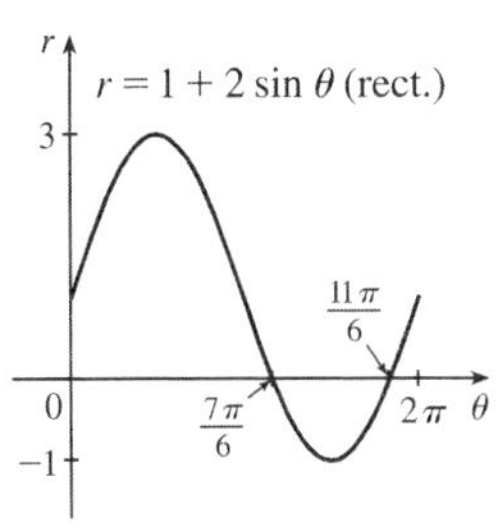


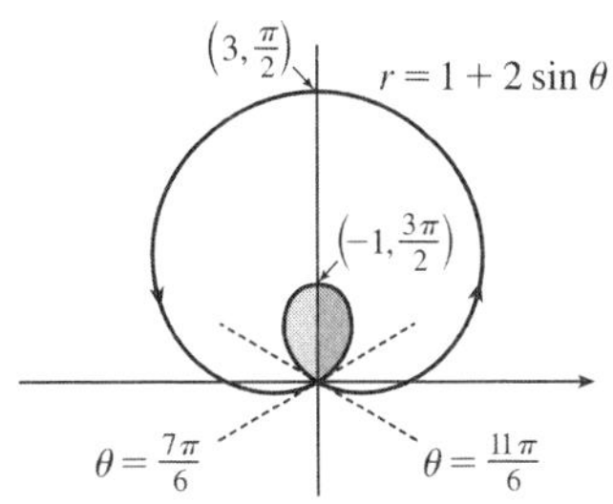


This is a limaçon, with inner loop traced out between $\theta = \frac{7\pi}{6}$ and $\frac{11\pi}{6}$ [found by solving $r = 0$].

$$A = 2\int_{7\pi/6}^{3\pi/2}\tfrac{1}{2}(1 + 2\sin\theta)^2\,d\theta = \int_{7\pi/6}^{3\pi/2}(1 + 4\sin\theta + 4\sin^2\theta)\,d\theta = \int_{7\pi/6}^{3\pi/2}\left[1 + 4\sin\theta + 4\cdot\tfrac{1}{2}(1 - \cos 2\theta)\right]d\theta$$

$$= [\theta - 4\cos\theta + 2\theta - \sin 2\theta]_{7\pi/6}^{3\pi/2} = \left(\tfrac{9\pi}{2}\right) - \left(\tfrac{7\pi}{2} + 2\sqrt{3} - \tfrac{\sqrt{3}}{2}\right) = \pi - \tfrac{3\sqrt{3}}{2}$$

**18.** To determine when the strophoid $r = 2\cos\theta - \sec\theta$ passes through the pole, we solve

$r = 0 \;\Rightarrow\; 2\cos\theta - \dfrac{1}{\cos\theta} = 0 \;\Rightarrow\; 2\cos^2\theta - 1 = 0 \;\Rightarrow\; \cos^2\theta = \dfrac{1}{2} \;\Rightarrow$

$\cos\theta = \pm\dfrac{1}{\sqrt{2}} \;\Rightarrow\; \theta = \frac{\pi}{4}$ or $\theta = \frac{3\pi}{4}$ for $0 \le \theta \le \pi$ with $\theta \neq \frac{\pi}{2}$.

$A = 2\int_0^{\pi/4}\frac{1}{2}(2\cos\theta - \sec\theta)^2\,d\theta = \int_0^{\pi/4}(4\cos^2\theta - 4 + \sec^2\theta)\,d\theta$

$= \int_0^{\pi/4}\left[4\cdot\frac{1}{2}(1 + \cos 2\theta) - 4 + \sec^2\theta\right]d\theta = \int_0^{\pi/4}(-2 + 2\cos 2\theta + \sec^2\theta)\,d\theta$

$= [-2\theta + \sin 2\theta + \tan\theta]_0^{\pi/4} = \left(-\frac{\pi}{2} + 1 + 1\right) - 0 = 2 - \frac{\pi}{2}$

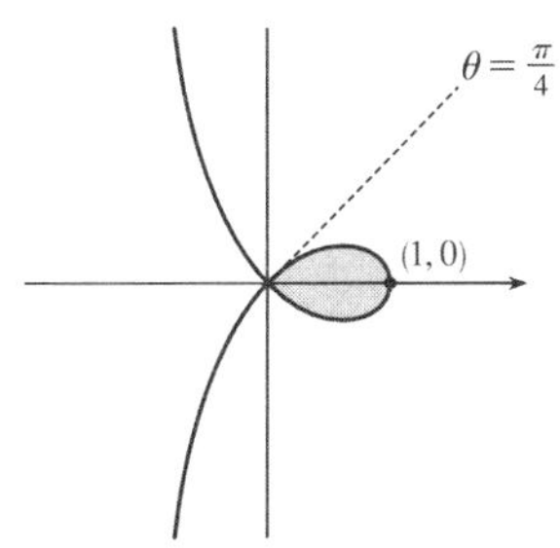

**19.** $4\sin\theta = 2 \Leftrightarrow \sin\theta = \frac{1}{2} \Leftrightarrow \theta = \frac{\pi}{6}$ or $\frac{5\pi}{6}$ (for $0 \le \theta \le 2\pi$). We'll subtract the unshaded area from the shaded area for $\pi/6 \le \theta \le \pi/2$ and double that value.

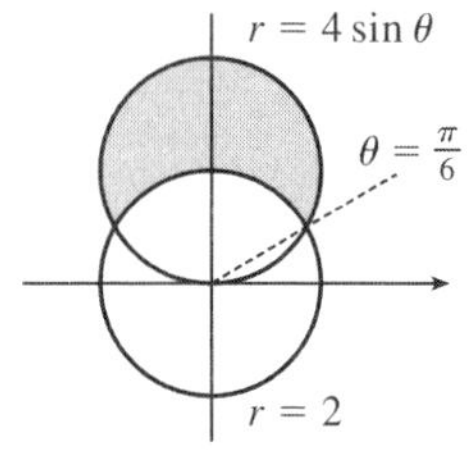


$$A = 2\int_{\pi/6}^{\pi/2} \tfrac{1}{2}(4\sin\theta)^2\,d\theta - 2\int_{\pi/6}^{\pi/2} \tfrac{1}{2}(2)^2\,d\theta = 2\int_{\pi/6}^{\pi/2} \tfrac{1}{2}\left[(4\sin\theta)^2 - 2^2\right]d\theta$$

$$= \int_{\pi/6}^{\pi/2} (16\sin^2\theta - 4)\,d\theta = \int_{\pi/6}^{\pi/2} [8(1-\cos 2\theta) - 4]\,d\theta$$

$$= \int_{\pi/6}^{\pi/2} (4 - 8\cos 2\theta)\,d\theta = [4\theta - 4\sin 2\theta]_{\pi/6}^{\pi/2} = (2\pi - 0) - \left(\tfrac{2\pi}{3} - 4\cdot\tfrac{\sqrt{3}}{2}\right) = \tfrac{4}{3}\pi + 2\sqrt{3}$$

**20.** $1 - \sin\theta = 1 \Rightarrow \sin\theta = 0 \Rightarrow \theta = 0$ or $\pi \Rightarrow$

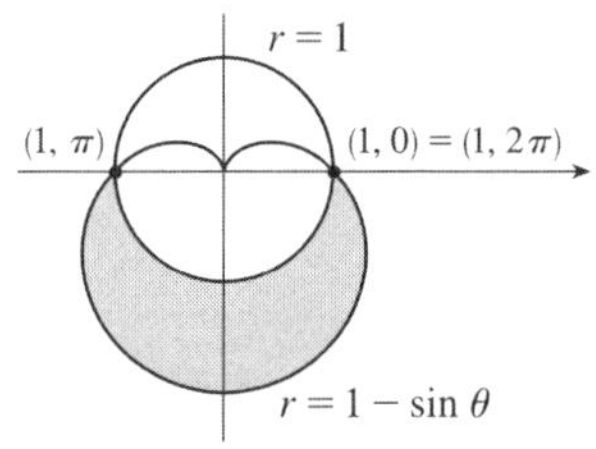


$$A = \int_{\pi}^{2\pi} \tfrac{1}{2}\left[(1-\sin\theta)^2 - 1\right]d\theta = \tfrac{1}{2}\int_{\pi}^{2\pi} (\sin^2\theta - 2\sin\theta)\,d\theta$$

$$= \tfrac{1}{4}\int_{\pi}^{2\pi} (1 - \cos 2\theta - 4\sin\theta)\,d\theta = \tfrac{1}{4}\left[\theta - \tfrac{1}{2}\sin 2\theta + 4\cos\theta\right]_{\pi}^{2\pi}$$

$$= \tfrac{1}{4}\pi + 2$$

**21.** $3\cos\theta = 1 + \cos\theta \Leftrightarrow \cos\theta = \frac{1}{2} \Rightarrow \theta = \frac{\pi}{3}$ or $-\frac{\pi}{3}$.

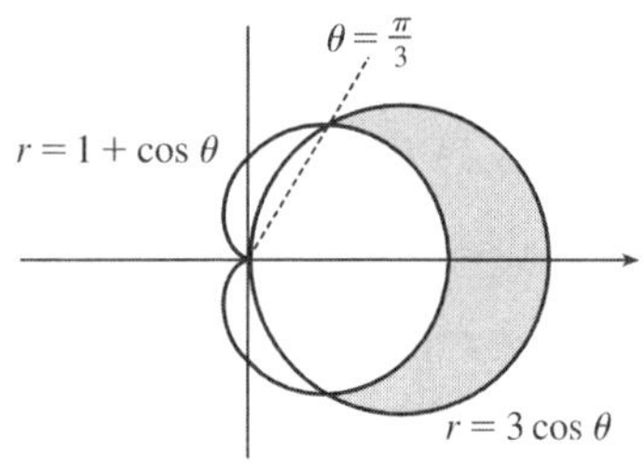


$$A = 2\int_0^{\pi/3} \tfrac{1}{2}\left[(3\cos\theta)^2 - (1+\cos\theta)^2\right]d\theta$$

$$= \int_0^{\pi/3} \left(8\cos^2\theta - 2\cos\theta - 1\right)d\theta = \int_0^{\pi/3} [4(1+\cos 2\theta) - 2\cos\theta - 1]\,d\theta$$

$$= \int_0^{\pi/3} (3 + 4\cos 2\theta - 2\cos\theta)\,d\theta = [3\theta + 2\sin 2\theta - 2\sin\theta]_0^{\pi/3}$$

$$= \pi + \sqrt{3} - \sqrt{3} = \pi$$

**22.** To find the shaded area $A$, we'll find the area $A_1$ inside the curve $r = 2 + \sin\theta$ and subtract $\pi\left(\frac{3}{2}\right)^2$ since $r = 3\sin\theta$ is a circle with radius $\frac{3}{2}$.

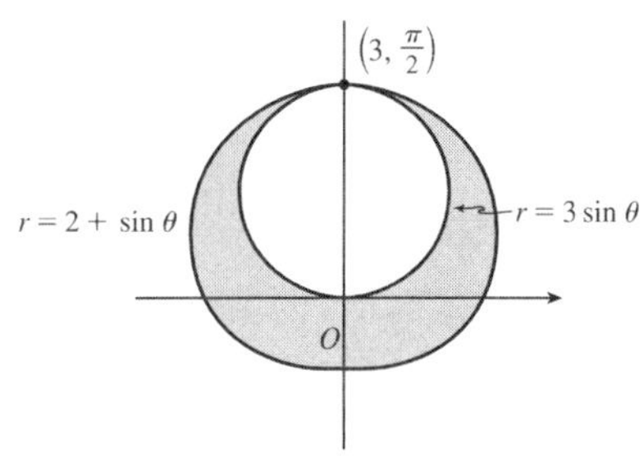


$$A_1 = \int_0^{2\pi} \tfrac{1}{2}(2+\sin\theta)^2\,d\theta = \tfrac{1}{2}\int_0^{2\pi} (4 + 4\sin\theta + \sin^2\theta)\,d\theta$$

$$= \tfrac{1}{2}\int_0^{2\pi} \left[4 + 4\sin\theta + \tfrac{1}{2}\cdot(1-\cos 2\theta)\right]d\theta$$

$$= \tfrac{1}{2}\int_0^{2\pi} \left(\tfrac{9}{2} + 4\sin\theta - \tfrac{1}{2}\cos 2\theta\right)d\theta$$

$$= \tfrac{1}{2}\left[\tfrac{9}{2}\theta - 4\cos\theta - \tfrac{1}{4}\sin 2\theta\right]_0^{2\pi} = \tfrac{1}{2}\left[(9\pi - 4) - (-4)\right] = \tfrac{9\pi}{2}$$

So $A = A_1 - \frac{9\pi}{4} = \frac{9\pi}{2} - \frac{9\pi}{4} = \frac{9\pi}{4}$.

**23.** $A = 2\int_0^{\pi/4} \frac{1}{2}\sin^2\theta\,d\theta = \displaystyle\int_0^{\pi/4} \tfrac{1}{2}(1 - \cos 2\theta)\,d\theta$

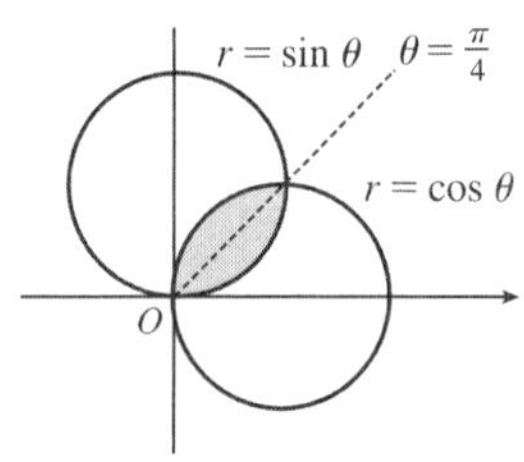


$$= \tfrac{1}{2}\left[\theta - \tfrac{1}{2}\sin 2\theta\right]_0^{\pi/4} = \tfrac{1}{2}\left[\left(\tfrac{\pi}{4} - \tfrac{1}{2}\cdot 1\right) - (0 - 0)\right]$$

$$= \tfrac{1}{8}\pi - \tfrac{1}{4}$$

**24.** $r = \sin 2\theta$ takes on both positive and negative values.
$\sin\theta = \pm\sin 2\theta = \pm 2\sin\theta\,\cos\theta \;\Rightarrow\; \sin\theta\,(1 \pm 2\cos\theta) = 0$. From the figure we can see that the intersections occur where $\cos\theta = \pm\frac{1}{2}$, or $\theta = \frac{\pi}{3}$ and $\frac{2\pi}{3}$.

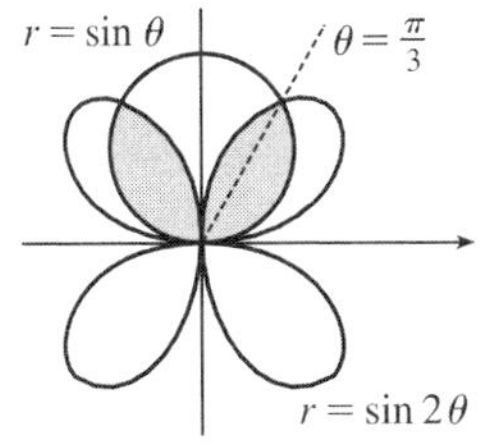


$$A = 2\left[\int_0^{\pi/3} \tfrac{1}{2}\sin^2\theta\,d\theta + \int_{\pi/3}^{\pi/2} \tfrac{1}{2}\sin^2 2\theta\,d\theta\right]$$
$$= \int_0^{\pi/3} \tfrac{1}{2}(1-\cos 2\theta)\,d\theta + \int_{\pi/3}^{\pi/2} \tfrac{1}{2}(1-\cos 4\theta)\,d\theta$$
$$= \tfrac{1}{2}\left[\theta - \tfrac{1}{2}\sin 2\theta\right]_0^{\pi/3} + \tfrac{1}{2}\left[\theta - \tfrac{1}{4}\sin 4\theta\right]_{\pi/3}^{\pi/2} = \tfrac{4\pi - 3\sqrt{3}}{16}$$

**25.** $\sin 2\theta = \cos 2\theta \;\Rightarrow\; \dfrac{\sin 2\theta}{\cos 2\theta} = 1 \;\Rightarrow\; \tan 2\theta = 1 \;\Rightarrow\; 2\theta = \frac{\pi}{4} \;\Rightarrow$
$\theta = \frac{\pi}{8} \;\Rightarrow$

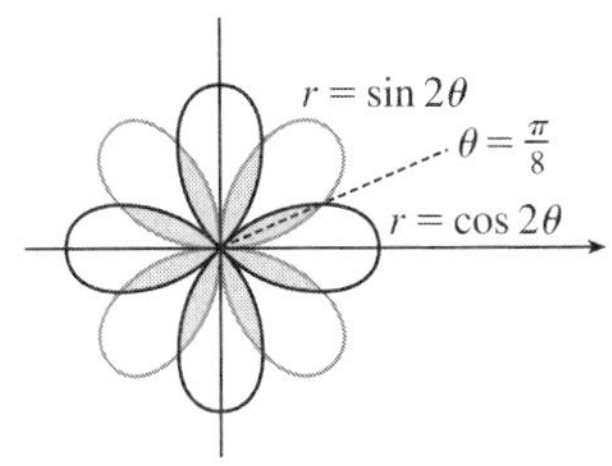


$$A = 8\cdot 2\int_0^{\pi/8} \tfrac{1}{2}\sin^2 2\theta\,d\theta = 8\int_0^{\pi/8} \tfrac{1}{2}(1-\cos 4\theta)\,d\theta$$
$$= 4\left[\theta - \tfrac{1}{4}\sin 4\theta\right]_0^{\pi/8} = 4\left(\tfrac{\pi}{8} - \tfrac{1}{4}\cdot 1\right) = \tfrac{1}{2}\pi - 1$$

**26.** $2\sin 2\theta = 1^2 \;\Rightarrow\; \sin 2\theta = \frac{1}{2} \;\Rightarrow\; 2\theta = \frac{\pi}{6}$ or $\frac{5\pi}{6} \;\Rightarrow\; \theta = \frac{\pi}{12}$ or $\frac{5\pi}{12}$.

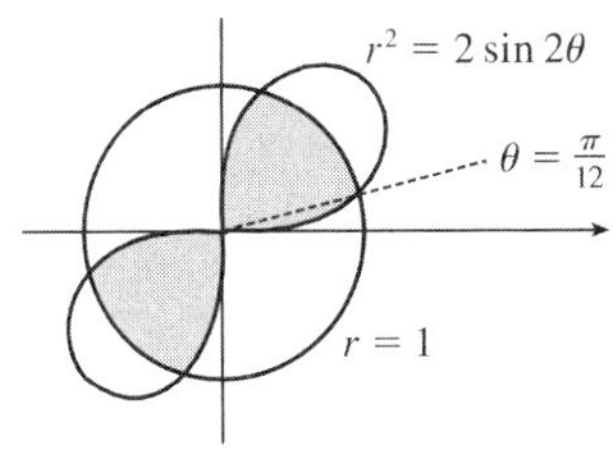


$$A = 4\left[\int_0^{\pi/12} \tfrac{1}{2}\cdot 2\sin 2\theta\,d\theta + \int_{\pi/12}^{\pi/4} \tfrac{1}{2}(1^2)\,d\theta\right]$$
$$= \left[-2\cos 2\theta\right]_0^{\pi/12} + \left[2\theta\right]_{\pi/12}^{\pi/4} = -2\left(\tfrac{\sqrt{3}}{2} - 1\right) + 2\left(\tfrac{1}{4}\pi - \tfrac{1}{12}\pi\right)$$
$$= 2 - \sqrt{3} + \tfrac{\pi}{3}$$

**27.** The darker shaded region (from $\theta = 0$ to $\theta = 2\pi/3$) represents $\frac{1}{2}$ of the desired area plus $\frac{1}{2}$ of the area of the inner loop. From this area, we'll subtract $\frac{1}{2}$ of the area of the inner loop (the lighter shaded region from $\theta = 2\pi/3$ to $\theta = \pi$), and then double that difference to obtain the desired area.

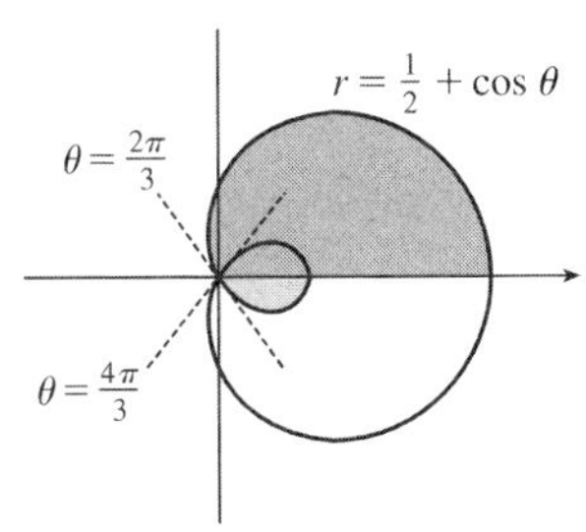


$$A = 2\left[\int_0^{2\pi/3} \tfrac{1}{2}\left(\tfrac{1}{2} + \cos\theta\right)^2 d\theta - \int_{2\pi/3}^{\pi} \tfrac{1}{2}\left(\tfrac{1}{2} + \cos\theta\right)^2 d\theta\right]$$
$$= \int_0^{2\pi/3} \left(\tfrac{1}{4} + \cos\theta + \cos^2\theta\right) d\theta - \int_{2\pi/3}^{\pi} \left(\tfrac{1}{4} + \cos\theta + \cos^2\theta\right) d\theta$$
$$= \int_0^{2\pi/3} \left[\tfrac{1}{4} + \cos\theta + \tfrac{1}{2}(1+\cos 2\theta)\right] d\theta$$
$$- \int_{2\pi/3}^{\pi} \left[\tfrac{1}{4} + \cos\theta + \tfrac{1}{2}(1+\cos 2\theta)\right] d\theta$$
$$= \left[\frac{\theta}{4} + \sin\theta + \frac{\theta}{2} + \frac{\sin 2\theta}{4}\right]_0^{2\pi/3} - \left[\frac{\theta}{4} + \sin\theta + \frac{\theta}{2} + \frac{\sin 2\theta}{4}\right]_{2\pi/3}^{\pi}$$
$$= \left(\tfrac{\pi}{6} + \tfrac{\sqrt{3}}{2} + \tfrac{\pi}{3} - \tfrac{\sqrt{3}}{8}\right) - \left(\tfrac{\pi}{4} + \tfrac{\pi}{2}\right) + \left(\tfrac{\pi}{6} + \tfrac{\sqrt{3}}{2} + \tfrac{\pi}{3} - \tfrac{\sqrt{3}}{8}\right)$$
$$= \tfrac{\pi}{4} + \tfrac{3}{4}\sqrt{3} = \tfrac{1}{4}\left(\pi + 3\sqrt{3}\right)$$

**28.** 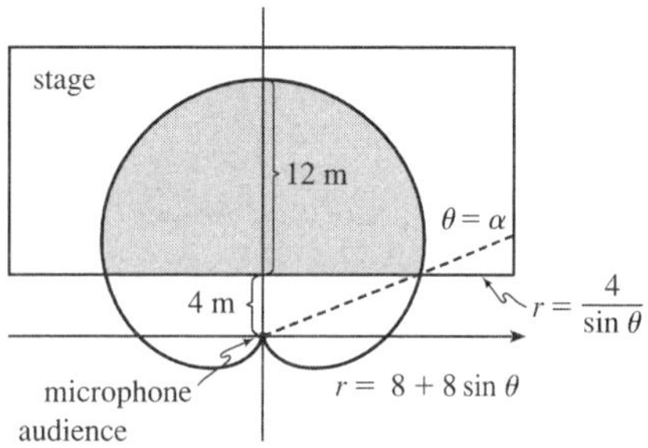


We need to find the shaded area $A$ in the figure. The horizontal line representing the front of the stage has equation $y = 4 \;\Leftrightarrow\; r\sin\theta = 4 \;\Rightarrow\; r = 4/\sin\theta$. This line intersects the curve $r = 8 + 8\sin\theta$ when $8 + 8\sin\theta = \dfrac{4}{\sin\theta} \;\Rightarrow\; 8\sin\theta + 8\sin^2\theta = 4 \;\Rightarrow\; 2\sin^2\theta + 2\sin\theta - 1 = 0 \;\Rightarrow$

$$\sin\theta = \frac{-2 \pm \sqrt{4+8}}{4} = \frac{-2 \pm 2\sqrt{3}}{4} = \frac{-1+\sqrt{3}}{2} \quad \text{[the other value is less than } -1] \quad \Rightarrow \quad \theta = \sin^{-1}\!\left(\frac{\sqrt{3}-1}{2}\right).$$

This angle is about $21.5^\circ$ and is denoted by $\alpha$ in the figure.

$$\begin{aligned}
A &= 2\int_\alpha^{\pi/2} \frac{1}{2}(8+8\sin\theta)^2\,d\theta - 2\int_\alpha^{\pi/2}\frac{1}{2}(4\csc\theta)^2\,d\theta = 64\int_\alpha^{\pi/2}(1 + 2\sin\theta + \sin^2\theta)\,d\theta - 16\int_\alpha^{\pi/2}\csc^2\theta\,d\theta \\
&= 64\int_\alpha^{\pi/2}\left(1 + 2\sin\theta + \frac{1}{2} - \frac{1}{2}\cos 2\theta\right)d\theta + 16\int_\alpha^{\pi/2}(-\csc^2\theta)\,d\theta \\
&= 64\left[\tfrac{3}{2}\theta - 2\cos\theta - \tfrac{1}{4}\sin 2\theta\right]_\alpha^{\pi/2} + 16\left[\cot\theta\right]_\alpha^{\pi/2} = 16\left[6\theta - 8\cos\theta - \sin 2\theta + \cot\theta\right]_\alpha^{\pi/a} \\
&= 16[(3\pi - 0 - 0 + 0) - (6\alpha - 8\cos\alpha - \sin 2\alpha + \cot\alpha)] = 48\pi - 96\alpha + 128\cos\alpha + 16\sin 2\alpha - 16\cot\alpha
\end{aligned}$$

From the figure, $x^2 + \left(\sqrt{3}-1\right)^2 = 2^2 \;\Rightarrow\; x^2 = 4 - \left(3 - 2\sqrt{3} + 1\right) \;\Rightarrow$ $x^2 = 2\sqrt{3} = \sqrt{12}$, so $x = \sqrt{2\sqrt{3}} = \sqrt[4]{12}$. Using the trigonometric relationships for a right triangle and the identity $\sin 2\alpha = 2\sin\alpha\cos\alpha$, we continue:

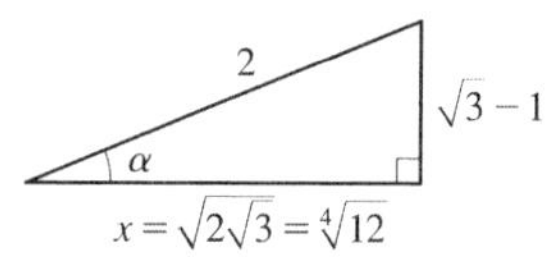


$$\begin{aligned}
A &= 48\pi - 96\alpha + 128\cdot\frac{\sqrt[4]{12}}{2} + 16\cdot 2\cdot\frac{\sqrt{3}-1}{2}\cdot\frac{\sqrt[4]{12}}{2} - 16\cdot\frac{\sqrt[4]{12}}{\sqrt{3}-1}\cdot\frac{\sqrt{3}+1}{\sqrt{3}+1} \\
&= 48\pi - 96\alpha + 64\sqrt[4]{12} + 8\sqrt[4]{12}\left(\sqrt{3}-1\right) - 8\sqrt[4]{12}\left(\sqrt{3}+1\right) \\
&= 48\pi + 48\sqrt[4]{12} - 96\sin^{-1}\!\left(\frac{\sqrt{3}-1}{2}\right) \approx 204.16\text{ m}^2
\end{aligned}$$

**29.** The curves intersect at the pole since $\left(0, \frac{\pi}{2}\right)$ satisfies $r = \cos\theta$ and $(0,0)$ satisfies $r = 1 - \cos\theta$. Now $\cos\theta = 1 - \cos\theta \;\Rightarrow\; 2\cos\theta = 1 \;\Rightarrow\; \cos\theta = \frac{1}{2} \;\Rightarrow\; \theta = \frac{\pi}{3}$ or $\frac{5\pi}{3} \;\Rightarrow$ the other intersection points are $\left(\frac{1}{2}, \frac{\pi}{3}\right)$ and $\left(\frac{1}{2}, \frac{5\pi}{3}\right)$.

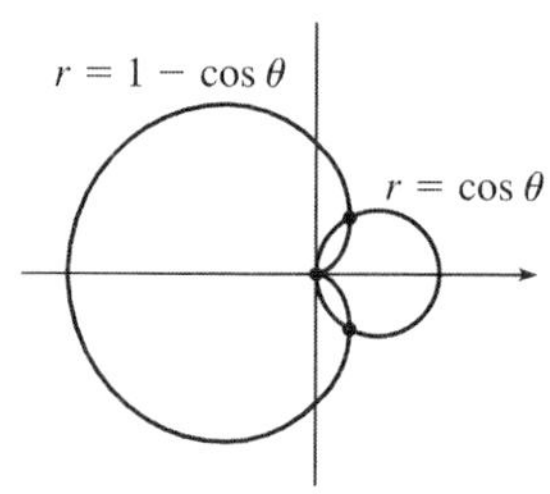


**30.** Clearly the pole lies on both curves. $\sin 3\theta = \cos 3\theta \;\Rightarrow\; \tan 3\theta = 1 \;\Rightarrow\; 3\theta = \frac{\pi}{4} + n\pi$ ($n$ any integer) $\Rightarrow\; \theta = \frac{\pi}{12} + \frac{\pi}{3}n \;\Rightarrow\; \theta = \frac{\pi}{12}, \frac{5\pi}{12}$, or $\frac{3\pi}{4}$, so the three remaining intersection points are $\left(\frac{1}{\sqrt{2}}, \frac{\pi}{12}\right)$, $\left(-\frac{1}{\sqrt{2}}, \frac{5\pi}{12}\right)$, and $\left(\frac{1}{\sqrt{2}}, \frac{3\pi}{4}\right)$.

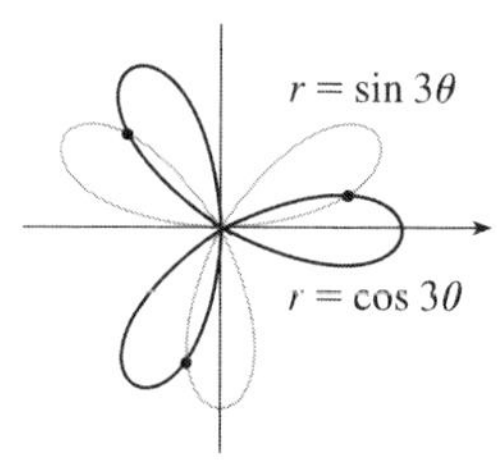


**31.** The pole is a point of intersection. $\sin\theta = \sin 2\theta = 2\sin\theta\cos\theta \;\Leftrightarrow\; \sin\theta\,(1 - 2\cos\theta) = 0 \;\Leftrightarrow\; \sin\theta = 0$ or $\cos\theta = \frac{1}{2} \;\Rightarrow\; \theta = 0, \pi, \frac{\pi}{3}$, or $-\frac{\pi}{3} \;\Rightarrow$ the other intersection points are $\left(\frac{\sqrt{3}}{2}, \frac{\pi}{3}\right)$ and $\left(\frac{\sqrt{3}}{2}, \frac{2\pi}{3}\right)$ (by symmetry).

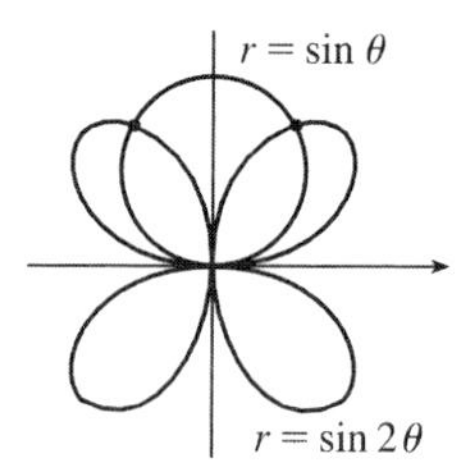

**32.** Clearly the pole is a point of intersection. $\sin 2\theta = \cos 2\theta \Rightarrow \tan 2\theta = 1 \Rightarrow 2\theta = \frac{\pi}{4} + 2n\pi$ (since $\sin 2\theta$ and $\cos 2\theta$ must be positive in the equations) $\Rightarrow \theta = \frac{\pi}{8} + n\pi \Rightarrow \theta = \frac{\pi}{8}$ or $\frac{9\pi}{8}$. So the curves also intersect at $\left(\frac{1}{\sqrt[4]{2}}, \frac{\pi}{8}\right)$ and $\left(\frac{1}{\sqrt[4]{2}}, \frac{9\pi}{8}\right)$.

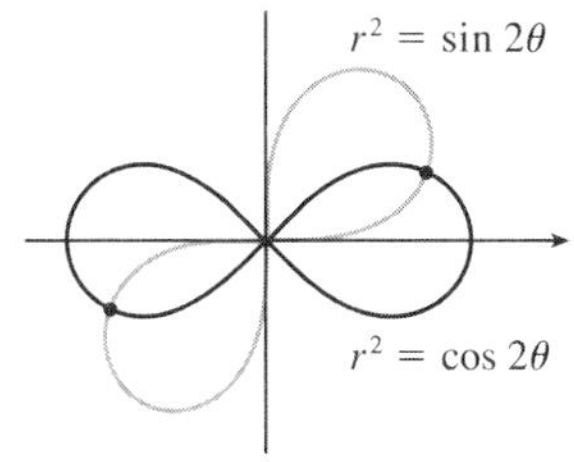


**33.**

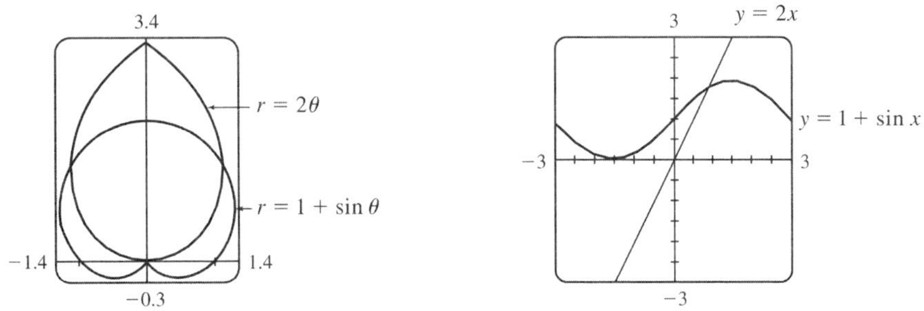


From the first graph, we see that the pole is one point of intersection. By zooming in or using the cursor, we find the $\theta$-values of the intersection points to be $\alpha \approx 0.88786 \approx 0.89$ and $\pi - \alpha \approx 2.25$. (The first of these values may be more easily estimated by plotting $y = 1 + \sin x$ and $y = 2x$ in rectangular coordinates; see the second graph.)
By symmetry, the total area contained is twice the area contained in the first quadrant, that is,

$$A = 2\int_0^{\alpha} \tfrac{1}{2}(2\theta)^2\,d\theta + 2\int_{\alpha}^{\pi/2} \tfrac{1}{2}(1+\sin\theta)^2\,d\theta = \int_0^{\alpha} 4\theta^2\,d\theta + \int_{\alpha}^{\pi/2}\left[1 + 2\sin\theta + \tfrac{1}{2}(1-\cos 2\theta)\right]d\theta$$

$$= \left[\tfrac{4}{3}\theta^3\right]_0^{\alpha} + \left[\theta - 2\cos\theta + \left(\tfrac{1}{2}\theta - \tfrac{1}{4}\sin 2\theta\right)\right]_{\alpha}^{\pi/2}$$

$$= \tfrac{4}{3}\alpha^3 + \left[\left(\tfrac{\pi}{2} + \tfrac{\pi}{4}\right) - \left(\alpha - 2\cos\alpha + \tfrac{1}{2}\alpha - \tfrac{1}{4}\sin 2\alpha\right)\right] \approx 3.4645$$

**34.**

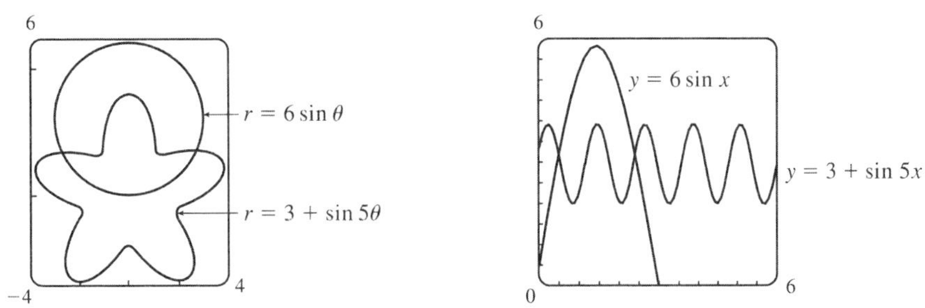


From the first graph, it appears that the $\theta$-values of the points of intersection are $\alpha \approx 0.57504 \approx 0.58$ and $\pi - \alpha \approx 2.57$. (These values may be more easily estimated by plotting $y = 3 + \sin 5x$ and $y = 6\sin x$ in rectangular coordinates; see the second graph.) By symmetry, the total area enclosed in both curves is

$$A = 2\int_0^{\alpha} \tfrac{1}{2}(6\sin\theta)^2\,d\theta + 2\int_{\alpha}^{\pi/2} \tfrac{1}{2}(3+\sin 5\theta)^2\,d\theta = \int_0^{\alpha} 36\sin^2\theta\,d\theta + \int_{\alpha}^{\pi/2}\left(9 + 6\sin 5\theta + \sin^2 5\theta\right)d\theta$$

$$= \int_0^{\alpha} 36\cdot\tfrac{1}{2}(1-\cos 2\theta)\,d\theta + \int_{\alpha}^{\pi/2}\left[9 + 6\sin 5\theta + \tfrac{1}{2}(1-\cos 10\theta)\right]d\theta$$

$$= \left[36\left(\tfrac{1}{2}\theta - \tfrac{1}{4}\sin 2\theta\right)\right]_0^{\alpha} + \left[9\theta - \tfrac{6}{5}\cos 5\theta + \left(\tfrac{1}{2}\theta - \tfrac{1}{20}\sin 10\theta\right)\right]_{\alpha}^{\pi/2} \approx 10.41$$

**35.** $L = \int_a^b \sqrt{r^2 + (dr/d\theta)^2}\, d\theta = \int_0^{\pi/3} \sqrt{(3\sin\theta)^2 + (3\cos\theta)^2}\, d\theta = \int_0^{\pi/3} \sqrt{9(\sin^2\theta + \cos^2\theta)}\, d\theta$

$= 3\int_0^{\pi/3} d\theta = 3\big[\,\theta\,\big]_0^{\pi/3} = 3\left(\frac{\pi}{3}\right) = \pi.$

As a check, note that the circumference of a circle with radius $\frac{3}{2}$ is $2\pi\left(\frac{3}{2}\right) = 3\pi$, and since $\theta = 0$ to $\pi = \frac{\pi}{3}$ traces out $\frac{1}{3}$ of the circle (from $\theta = 0$ to $\theta = \pi$), $\frac{1}{3}(3\pi) = \pi$.

**36.** $L = \int_a^b \sqrt{r^2 + (dr/d\theta)^2}\, d\theta = \int_0^{2\pi} \sqrt{(e^{2\theta})^2 + (2e^{2\theta})^2}\, d\theta = \int_0^{2\pi} \sqrt{e^{4\theta} + 4e^{4\theta}}\, d\theta = \int_0^{2\pi} \sqrt{5e^{4\theta}}\, d\theta$

$= \sqrt{5}\int_0^{2\pi} e^{2\theta}\, d\theta = \frac{\sqrt{5}}{2}\left[e^{2\theta}\right]_0^{2\pi} = \frac{\sqrt{5}}{2}\left(e^{4\pi} - 1\right)$

**37.** $L = \int_a^b \sqrt{r^2 + (dr/d\theta)^2}\, d\theta = \int_0^{2\pi} \sqrt{(\theta^2)^2 + (2\theta)^2}\, d\theta = \int_0^{2\pi} \sqrt{\theta^4 + 4\theta^2}\, d\theta$

$= \int_0^{2\pi} \sqrt{\theta^2(\theta^2 + 4)}\, d\theta = \int_0^{2\pi} \theta\sqrt{\theta^2 + 4}\, d\theta$

Now let $u = \theta^2 + 4$, so that $du = 2\theta\, d\theta$ $\left[\theta\, d\theta = \frac{1}{2}\, du\right]$ and

$$\int_0^{2\pi} \theta\sqrt{\theta^2 + 4}\, d\theta = \int_4^{4\pi^2+4} \tfrac{1}{2}\sqrt{u}\, du = \frac{1}{2}\cdot\frac{2}{3}\left[u^{3/2}\right]_4^{4(\pi^2+1)} = \frac{1}{3}\left[4^{3/2}(\pi^2+1)^{3/2} - 4^{3/2}\right] = \frac{8}{3}\left[(\pi^2+1)^{3/2} - 1\right]$$

**38.** $L = \int_a^b \sqrt{r^2 + (dr/d\theta)^2}\, d\theta = \int_0^{2\pi} \sqrt{\theta^2 + 1}\, d\theta \overset{21}{=} \left[\frac{\theta}{2}\sqrt{\theta^2 + 1} + \frac{1}{2}\ln\left(\theta + \sqrt{\theta^2 + 1}\right)\right]_0^{2\pi}$

$= \pi\sqrt{4\pi^2 + 1} + \frac{1}{2}\ln\left(2\pi + \sqrt{4\pi^2 + 1}\right)$

**39.** The curve $r = 3\sin 2\theta$ is completely traced with $0 \le \theta \le 2\pi$.

$r^2 + \left(\frac{dr}{d\theta}\right)^2 = (3\sin 2\theta)^2 + (6\cos 2\theta)^2 \quad\Rightarrow$

$L = \int_0^{2\pi} \sqrt{9\sin^2 2\theta + 36\cos^2 2\theta}\, d\theta \approx 29.0653$

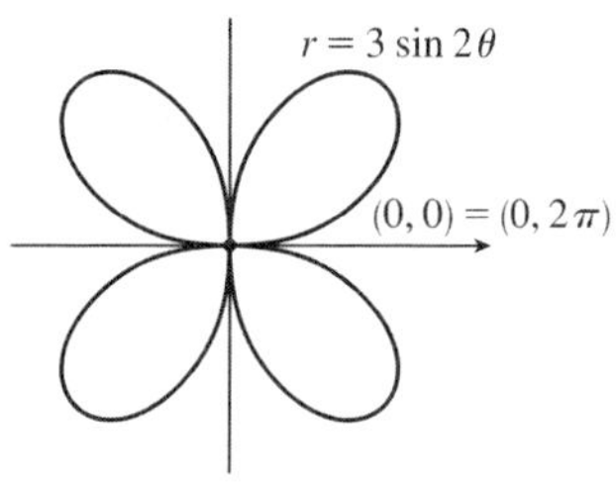


**40.** The curve $r = 4\sin 3\theta$ is completely traced with $0 \le \theta \le \pi$.

$r^2 + \left(\frac{dr}{d\theta}\right)^2 = (4\sin 3\theta)^2 + (12\cos 3\theta)^2 \quad\Rightarrow$

$L = \int_0^{\pi} \sqrt{16\sin^2 3\theta + 144\cos^2 3\theta}\, d\theta \approx 26.7298$

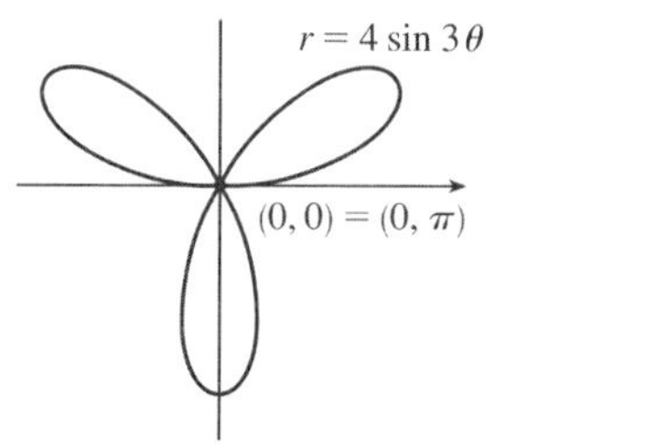


## DISCOVERY PROJECT Conic Sections in Polar Coordinates

**1.** We see from Figure 1 that $|PF| = r$ and $|Pl| = d - r\cos\theta$. Thus, the condition $|PF|\,/|Pl| = e$, or $|PF| = e\,|Pl|$, becomes $r = e(d - r\cos\theta)$.

**2.** If we square both sides of $r = e(d - r\cos\theta)$ and convert to rectangular coordinates, we get $x^2 + y^2 = e^2\,(d - x)^2 = e^2\left(d^2 - 2dx + x^2\right)$ or $\left(1 - e^2\right)x^2 + 2de^2x + y^2 = e^2d^2$. After completing the square, we have $\left(x + \dfrac{e^2d}{1 - e^2}\right)^2 + \dfrac{y^2}{1 - e^2} = \dfrac{e^2d^2}{(1 - e^2)^2}$ (1). If $e < 1$, we recognize Equation 1 as the equation of an ellipse. In fact, it is of the form $\dfrac{(x - h)^2}{a^2} + \dfrac{y^2}{b^2} = 1$, where $h = -\dfrac{e^2d}{1 - e^2}$, $a^2 = \dfrac{e^2d^2}{(1 - e^2)^2}$ $(\star)$, and $b^2 = \dfrac{e^2d^2}{1 - e^2}$.

**3.** If $e > 1$, then $1 - e^2 < 0$ and we see that Equation 1 represents a hyperbola. Just as we did in Problem 2, we could rewrite Equation 1 in the form $\dfrac{(x-h)^2}{a^2} - \dfrac{y^2}{b^2} = 1$ and see that $e = \dfrac{c}{a}$, where $c^2 = a^2 + b^2$.

**4.** By solving $r = e(d - r\cos\theta)$ for $r$, we see that the polar equation of the conic shown in Figure 1 can be written as $r = \dfrac{ed}{1 + e\cos\theta}$.

**5.** (a) $r = \dfrac{4}{1 + 3\cos\theta} \quad \Rightarrow \quad e = 3 > 1 \quad \Rightarrow$ hyperbola;

$ed = 4 \quad \Rightarrow \quad d = \frac{4}{3} \quad \Rightarrow$ directrix $x = \frac{4}{3}$; vertices $(1, 0)$ and $(-2, \pi) = (2, 0)$; center $\left(\frac{3}{2}, 0\right)$; asymptotes parallel to $\theta = \pm\cos^{-1}\left(-\frac{1}{3}\right)$

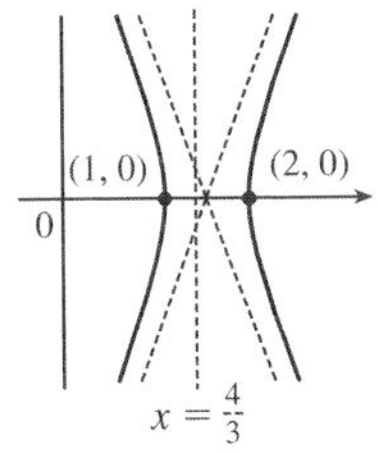


(b) $r = \dfrac{8}{3 + 3\cos\theta} = \dfrac{\frac{8}{3}}{1 + \cos\theta} \quad \Rightarrow$

$e = 1 \quad \Rightarrow$ parabola; $ed = \frac{8}{3} \quad \Rightarrow$

$d = \frac{8}{3} \quad \Rightarrow$ directrix $x = \frac{8}{3}$; vertex $\left(\frac{4}{3}, 0\right)$

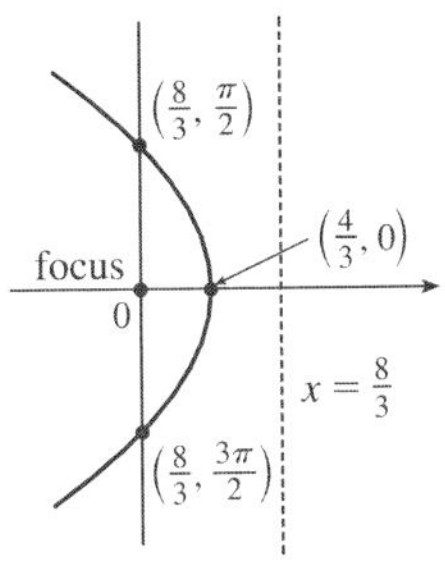


(c) $r = \dfrac{2}{2 + \cos\theta} = \dfrac{1}{1 + \frac{1}{2}\cos\theta} \quad \Rightarrow$

$e = \frac{1}{2} < 1 \quad \Rightarrow$ ellipse;

$ed = 1 \quad \Rightarrow \quad d = 2 \quad \Rightarrow$ directrix $x = 2$;

vertices $\left(\frac{2}{3}, 0\right)$ and $(2, \pi) = (-2, 0)$.

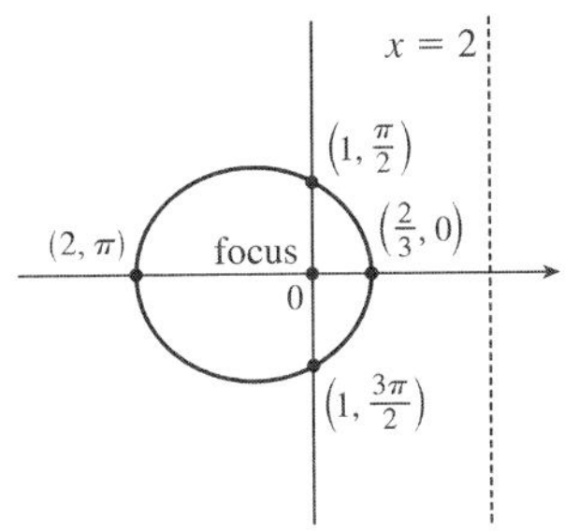


**6.** For $e < 1$ the curve is an ellipse. It is nearly circular when $e$ is close to 0. As $e$ increases, the graph is stretched out to the right, and grows larger (that is, its right-hand focus moves to the right while its left-hand focus remains at the origin.) At $e = 1$, the curve becomes a parabola with focus at the origin.

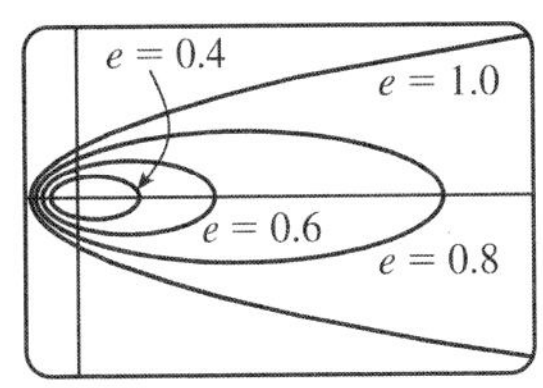


**7.** (a) If the directrix is $x = d$, then $r = \dfrac{ed}{1 + e\cos\theta}$ and, from ($\star$) in Problem 2, $a^2 = \dfrac{e^2d^2}{(1-e^2)^2} \quad \Rightarrow \quad ed = a(1 - e^2)$.

Therefore, $r = \dfrac{a(1 - e^2)}{1 + e\cos\theta}$.

(b) $e = 0.017$ and the length of the major axis $= 2a = 2.99 \times 10^8 \quad \Rightarrow \quad a = 1.495 \times 10^8$. Therefore,

$$r = \frac{1.495 \times 10^8\left[1 - (0.017)^2\right]}{1 + 0.017\cos\theta} \approx \frac{1.495 \times 10^8}{1 + 0.017\cos\theta}.$$

**8.** (a) The Sun is at point $F$ in Figure 1 so that perihelion is in the positive $x$-direction and aphelion is in the negative $x$-direction.

At perihelion, $\theta = 0$, so $r = \dfrac{a(1 - e^2)}{1 + e\cos 0} = \dfrac{a(1-e)(1+e)}{1+e} = a(1 - e)$.

At aphelion, $\theta = \pi$, so $r = \dfrac{a(1 - e^2)}{1 + e\cos\pi} = \dfrac{a(1-e)(1+e)}{1-e} = a(1 + e)$.

(b) At perihelion, $r = a(1-e) \approx (1.495 \times 10^8)(1-0.017) \approx 1.47 \times 10^8$ km.
At aphelion, $r = a(1+e) \approx (1.495 \times 10^8)(1+0.017) \approx 1.52 \times 10^8$ km.

**9.** (a) The minimum distance is at perihelion, where
$4.6 \times 10^7 = r = a(1-e) = a(1-0.206) = a(0.794) \quad \Rightarrow \quad a = 4.6 \times 10^7/0.794$. So the maximum distance, which is at aphelion, is $r = a(1+e) = (4.6 \times 10^7/0.794)(1.206) \approx 7.0 \times 10^7$ km.

(b) From part (a), we have $e = 0.206$ and $a(1-e) = 4.6 \times 10^7$ km. Thus, $a = 4.6 \times 10^7/0.794$. From Problem 7, we can write the equation of Mercury's orbit as $r = a\dfrac{1-e^2}{1+e\cos\theta}$. So since $\dfrac{dr}{d\theta} = \dfrac{a(1-e^2)\,e\sin\theta}{(1-e\cos\theta)^2} \quad \Rightarrow$
$r^2 + \left(\dfrac{dr}{d\theta}\right)^2 = \dfrac{a^2(1-e^2)^2}{(1+e\cos\theta)^2} + \dfrac{a^2(1-e^2)^2 e^2\sin^2\theta}{(1+e\cos\theta)^4} = \dfrac{a^2(1-e^2)^2}{(1+e\cos\theta)^4}(1+2e\cos\theta+e^2)$, the length of the orbit is

$$L = \int_0^{2\pi} \sqrt{r^2 + (dr/d\theta)^2}\, d\theta = a(1-e^2)\int_0^{2\pi} \frac{\sqrt{1+e^2+2e\cos\theta}}{(1+e\cos\theta)^2}\, d\theta \approx 3.6 \times 10^8 \text{ km.}$$

This seems reasonable, since Mercury's orbit is nearly circular, and the circumference of a circle of radius $a$ is $2\pi a \approx 3.6 \times 10^8$ km.

## I COMPLEX NUMBERS

**1.** $(5-6i)+(3+2i) = (5+3)+(-6+2)i = 8+(-4)i = 8-4i$

**2.** $(4-\frac{1}{2}i)-(9+\frac{5}{2}i) = (4-9)+(-\frac{1}{2}-\frac{5}{2})\,i = -5+(-3)i = -5-3i$

**3.** $(2+5i)(4-i) = 2(4)+2(-i)+(5i)(4)+(5i)(-i) = 8-2i+20i-5i^2$
$= 8+18i-5(-1) = 8+18i+5 = 13+18i$

**4.** $(1-2i)(8-3i) = 8-3i-16i+6(-1) = 2-19i$

**5.** $\overline{12+7i} = 12-7i$

**6.** $2i(\frac{1}{2}-i) = i-2(-1) = 2+i \quad \Rightarrow \quad \overline{2i(\frac{1}{2}-i)} = \overline{2+i} = 2-i$

**7.** $\dfrac{1+4i}{3+2i} = \dfrac{1+4i}{3+2i}\cdot\dfrac{3-2i}{3-2i} = \dfrac{3-2i+12i-8(-1)}{3^2+2^2} = \dfrac{11+10i}{13} = \dfrac{11}{13}+\dfrac{10}{13}i$

**8.** $\dfrac{3+2i}{1-4i} = \dfrac{3+2i}{1-4i}\cdot\dfrac{1+4i}{1+4i} = \dfrac{3+12i+2i+8(-1)}{1^2+4^2} = \dfrac{-5+14i}{17} = -\dfrac{5}{17}+\dfrac{14}{17}i$

**9.** $\dfrac{1}{1+i} = \dfrac{1}{1+i}\cdot\dfrac{1-i}{1-i} = \dfrac{1-i}{1-(-1)} = \dfrac{1-i}{2} = \dfrac{1}{2}-\dfrac{1}{2}i$

**10.** $\dfrac{3}{4-3i} = \dfrac{3}{4-3i}\cdot\dfrac{4+3i}{4+3i} = \dfrac{12+9i}{16-9(-1)} = \dfrac{12}{25}+\dfrac{9}{25}i$

**11.** $i^3 = i^2 \cdot i = (-1)i = -i$

**12.** $i^{100} = (i^2)^{50} = (-1)^{50} = 1$

**13.** $\sqrt{-25} = \sqrt{25}\,i = 5i$

**14.** $\sqrt{-3}\,\sqrt{-12} = \sqrt{3}\,i\,\sqrt{12}\,i = \sqrt{3\cdot 12}\,i^2 = \sqrt{36}\,(-1) = -6$

**15.** $\overline{12-5i} = 12+15i$ and $|12-15i| = \sqrt{12^2+(-5)^2} = \sqrt{144+25} = \sqrt{169} = 13$

**16.** $\overline{-1+2\sqrt{2}\,i} = -1-2\sqrt{2}\,i$ and $\left|-1+2\sqrt{2}\,i\right| = \sqrt{(-1)^2+\left(2\sqrt{2}\right)^2} = \sqrt{1+8} = \sqrt{9} = 3$

**17.** $\overline{-4i} = \overline{0-4i} = 0+4i = 4i$ and $|-4i| = \sqrt{0^2+(-4)^2} = \sqrt{16} = 4$

**18.** Let $z = a+bi$, $w = c+di$.

(a) $\overline{z+w} = \overline{(a+bi)+(c+di)} = \overline{(a+c)+(b+d)i}$
$= (a+c)-(b+d)i = (a-bi)+(c-di) = \overline{z}+\overline{w}$

(b) $\overline{zw} = \overline{(a+bi)(c+di)} = \overline{(ac-bd)+(ad+bc)i} = (ac-bd)-(ad+bc)i$.
On the other hand, $\overline{z}\,\overline{w} = (a-bi)(c-di) = (ac-bd)-(ad+bc)i = \overline{zw}$.

(c) Use mathematical induction and part (b): Let $S_n$ be the statement that $\overline{z^n} = \overline{z}^{\,n}$. $S_1$ is true because $\overline{z^1} = \overline{z} = \overline{z}^{\,1}$. Assume $S_k$ is true, that is $\overline{z^k} = \overline{z}^{\,k}$. Then $\overline{z^{k+1}} = \overline{z^{1+k}} = \overline{zz^k} = \overline{z}\,\overline{z^k}$ [part (b) with $w = z^k$] $= \overline{z}^1\overline{z}^k = \overline{z}^{1+k} = \overline{z}^{k+1}$, which shows that $S_{k+1}$ is true. Therefore, by mathematical induction, $\overline{z^n} = \overline{z}^n$ for every positive integer $n$.
*Another proof:* Use part (b) with $w = z$, and mathematical induction.

**19.** $4x^2+9=0 \;\Leftrightarrow\; 4x^2=-9 \;\Leftrightarrow\; x^2=-\frac{9}{4} \;\Leftrightarrow\; x=\pm\sqrt{-\frac{9}{4}}=\pm\sqrt{\frac{9}{4}}\,i=\pm\frac{3}{2}i.$

**20.** $x^4=1 \;\Leftrightarrow\; x^4-1=0 \;\Leftrightarrow\; \left(x^2-1\right)\left(x^2+1\right)=0 \;\Leftrightarrow\; x^2-1=0$ or $x^2+1=0 \;\Leftrightarrow$
$x=\pm1$ or $x=\pm i$.

**21.** By the quadratic formula, $x^2+2x+5=0 \;\Leftrightarrow\; x=\dfrac{-2\pm\sqrt{2^2-4(1)(5)}}{2(1)}=\dfrac{-2\pm\sqrt{-16}}{2}=\dfrac{-2\pm4i}{2}=-1\pm2i.$

**22.** $2x^2-2x+1=0 \;\Leftrightarrow\; x=\dfrac{-(-2)\pm\sqrt{(-2)^2-4(2)(1)}}{2(2)}=\dfrac{2\pm\sqrt{-4}}{4}=\dfrac{2\pm2i}{4}=\dfrac{1}{2}\pm\dfrac{1}{2}i$

**23.** By the quadratic formula, $z^2+z+2=0 \;\Leftrightarrow\; z=\dfrac{-1\pm\sqrt{1^2-4(1)(2)}}{2(1)}=\dfrac{-1\pm\sqrt{-7}}{2}=-\dfrac{1}{2}\pm\dfrac{\sqrt{7}}{2}i.$

**24.** $z^2+\frac{1}{2}z+\frac{1}{4}=0 \;\Leftrightarrow\; 4z^2+2z+1=0 \;\Leftrightarrow$

$$z=\frac{-2\pm\sqrt{2^2-4(4)(1)}}{2(4)}=\frac{-2\pm\sqrt{-12}}{8}=\frac{-2\pm2\sqrt{3}\,i}{8}=-\frac{1}{4}\pm\frac{\sqrt{3}}{4}i$$

**25.** For $z=-3+3i$, $r=\sqrt{(-3)^2+3^2}=3\sqrt{2}$ and $\tan\theta=\frac{3}{-3}=-1 \;\Rightarrow\; \theta=\frac{3\pi}{4}$ (since $z$ lies in the second quadrant).
Therefore, $-3+3i=3\sqrt{2}\left(\cos\frac{3\pi}{4}+i\sin\frac{3\pi}{4}\right)$.

**26.** For $z=1-\sqrt{3}\,i$, $r=\sqrt{1^2+\left(-\sqrt{3}\right)^2}=2$ and $\tan\theta=\frac{-\sqrt{3}}{1}=-\sqrt{3} \;\Rightarrow\; \theta=\frac{5\pi}{3}$ (since $z$ lies in the fourth quadrant).
Therefore, $1-\sqrt{3}\,i=2\left(\cos\frac{5\pi}{3}+i\sin\frac{5\pi}{3}\right)$.

**27.** For $z=3+4i$, $r=\sqrt{3^2+4^2}=5$ and $\tan\theta=\frac{4}{3} \;\Rightarrow\; \theta=\tan^{-1}\left(\frac{4}{3}\right)$ (since $z$ lies in the first quadrant). Therefore,
$3+4i=5\left[\cos\left(\tan^{-1}\frac{4}{3}\right)+i\sin\left(\tan^{-1}\frac{4}{3}\right)\right]$.

**28.** For $z=8i$, $r=\sqrt{0^2+8^2}=8$ and $\tan\theta=\frac{8}{0}$ is undefined, so $\theta=\frac{\pi}{2}$ (since $z$ lies on the positive imaginary axis). Therefore,
$8i=8\left(\cos\frac{\pi}{2}+i\sin\frac{\pi}{2}\right)$.

**29.** For $z = \sqrt{3} + i$, $r = \sqrt{(\sqrt{3})^2 + 1^2} = 2$ and $\tan\theta = \frac{1}{\sqrt{3}} \Rightarrow \theta = \frac{\pi}{6} \Rightarrow z = 2\left(\cos\frac{\pi}{6} + i\sin\frac{\pi}{6}\right)$.

For $w = 1 + \sqrt{3}\,i$, $r = 2$ and $\tan\theta = \sqrt{3} \Rightarrow \theta = \frac{\pi}{3} \Rightarrow w = 2\left(\cos\frac{\pi}{3} + i\sin\frac{\pi}{3}\right)$.

Therefore, $zw = 2 \cdot 2\left[\cos\left(\frac{\pi}{6} + \frac{\pi}{3}\right) + i\sin\left(\frac{\pi}{6} + \frac{\pi}{3}\right)\right] = 4\left(\cos\frac{\pi}{2} + i\sin\frac{\pi}{2}\right)$,

$z/w = \frac{2}{2}\left[\cos\left(\frac{\pi}{6} - \frac{\pi}{3}\right) + i\sin\left(\frac{\pi}{6} - \frac{\pi}{3}\right)\right] = \cos\left(-\frac{\pi}{6}\right) + i\sin\left(-\frac{\pi}{6}\right)$, and $1 = 1 + 0i = 1(\cos 0 + i\sin 0) \Rightarrow$

$1/z = \frac{1}{2}\left[\cos\left(0 - \frac{\pi}{6}\right) + i\sin\left(0 - \frac{\pi}{6}\right)\right] = \frac{1}{2}\left[\cos\left(-\frac{\pi}{6}\right) + i\sin\left(-\frac{\pi}{6}\right)\right]$. For $1/z$, we could also use the formula that precedes Example 5 to obtain $1/z = \frac{1}{2}\left(\cos\frac{\pi}{6} - i\sin\frac{\pi}{6}\right)$.

**30.** For $z = 4\sqrt{3} - 4i$, $r = \sqrt{(4\sqrt{3})^2 + (-4)^2} = \sqrt{64} = 8$ and $\tan\theta = \frac{-4}{4\sqrt{3}} = -\frac{1}{\sqrt{3}} \Rightarrow \theta = \frac{11\pi}{6} \Rightarrow$

$z = 8\left(\cos\frac{11\pi}{6} + i\sin\frac{11\pi}{6}\right)$. For $w = 8i$, $r = \sqrt{0^2 + 8^2} = 8$ and $\tan\theta = \frac{8}{0}$ is undefined, so $\theta = \frac{\pi}{2} \Rightarrow$

$w = 8\left(\cos\frac{\pi}{2} + i\sin\frac{\pi}{2}\right)$. Therefore, $zw = 8 \cdot 8\left[\cos\left(\frac{11\pi}{6} + \frac{\pi}{2}\right) + i\sin\left(\frac{11\pi}{6} + \frac{\pi}{2}\right)\right] = 64\left(\cos\frac{\pi}{3} + i\sin\frac{\pi}{3}\right)$,

$z/w = \frac{8}{8}\left[\cos\left(\frac{11\pi}{6} - \frac{\pi}{2}\right) + i\sin\left(\frac{11\pi}{6} - \frac{\pi}{2}\right)\right] = \cos\frac{4\pi}{3} + i\sin\frac{4\pi}{3}$, and

$1 = 1 + 0i = 1(\cos 0 + i\sin 0) \Rightarrow 1/z = \frac{1}{8}\left[\cos\left(0 - \frac{11\pi}{6}\right) + i\sin\left(0 - \frac{11\pi}{6}\right)\right] = \frac{1}{8}\left[\cos\left(\frac{\pi}{6}\right) + i\sin\left(\frac{\pi}{6}\right)\right]$.

For $1/z$, we could also use the formula that precedes Example 5 to obtain $1/z = \frac{1}{8}\left(\cos\frac{11\pi}{6} - i\sin\frac{11\pi}{6}\right)$.

**31.** For $z = 2\sqrt{3} - 2i$, $r = \sqrt{(2\sqrt{3})^2 + (-2)^2} = 4$ and $\tan\theta = \frac{-2}{2\sqrt{3}} = -\frac{1}{\sqrt{3}} \Rightarrow \theta = -\frac{\pi}{6} \Rightarrow$

$z = 4\left[\cos\left(-\frac{\pi}{6}\right) + i\sin\left(-\frac{\pi}{6}\right)\right]$. For $w = -1 + i$, $r = \sqrt{2}$, $\tan\theta = \frac{1}{-1} = -1 \Rightarrow \theta = \frac{3\pi}{4} \Rightarrow$

$w = \sqrt{2}\left(\cos\frac{3\pi}{4} + i\sin\frac{3\pi}{4}\right)$. Therefore, $zw = 4\sqrt{2}\left[\cos\left(-\frac{\pi}{6} + \frac{3\pi}{4}\right) + i\sin\left(-\frac{\pi}{6} + \frac{3\pi}{4}\right)\right] = 4\sqrt{2}\left(\cos\frac{7\pi}{12} + i\sin\frac{7\pi}{12}\right)$,

$$z/w = \frac{4}{\sqrt{2}}\left[\cos\left(-\frac{\pi}{6} - \frac{3\pi}{4}\right) + i\sin\left(-\frac{\pi}{6} - \frac{3\pi}{4}\right)\right] = \frac{4}{\sqrt{2}}\left[\cos\left(-\frac{11\pi}{12}\right) + i\sin\left(-\frac{11\pi}{12}\right)\right]$$

$$= 2\sqrt{2}\left(\cos\frac{13\pi}{12} + i\sin\frac{13\pi}{12}\right), \text{ and}$$

$1/z = \frac{1}{4}\left[\cos\left(-\frac{\pi}{6}\right) - i\sin\left(-\frac{\pi}{6}\right)\right] = \frac{1}{4}\left(\cos\frac{\pi}{6} + i\sin\frac{\pi}{6}\right)$.

**32.** For $z = 4(\sqrt{3} + i) = 4\sqrt{3} + 4i$, $r = \sqrt{(4\sqrt{3})^2 + 4^2} = \sqrt{64} = 8$ and $\tan\theta = \frac{4}{4\sqrt{3}} = \frac{1}{\sqrt{3}} \Rightarrow \theta = \frac{\pi}{6} \Rightarrow$

$z = 8\left(\cos\frac{\pi}{6} + i\sin\frac{\pi}{6}\right)$. For $w = -3 - 3i$, $r = \sqrt{(-3)^2 + (-3)^2} = \sqrt{18} = 3\sqrt{2}$ and $\tan\theta = \frac{-3}{-3} = 1 \Rightarrow \theta = \frac{5\pi}{4} \Rightarrow$

$w = 3\sqrt{2}\left(\cos\frac{5\pi}{4} + i\sin\frac{5\pi}{4}\right)$. Therefore, $zw = 8 \cdot 3\sqrt{2}\left[\cos\left(\frac{\pi}{6} + \frac{5\pi}{4}\right) + i\sin\left(\frac{\pi}{6} + \frac{5\pi}{4}\right)\right] = 24\sqrt{2}\left(\cos\frac{17\pi}{12} + i\sin\frac{17\pi}{12}\right)$,

$z/w = \frac{8}{3\sqrt{2}}\left[\cos\left(\frac{\pi}{6} - \frac{5\pi}{4}\right) + i\sin\left(\frac{\pi}{6} - \frac{5\pi}{4}\right)\right] = \frac{4\sqrt{2}}{3}\left[\cos\left(-\frac{13\pi}{12}\right) + i\sin\left(-\frac{13\pi}{12}\right)\right]$, and $1/z = \frac{1}{8}\left(\cos\frac{\pi}{6} - i\sin\frac{\pi}{6}\right)$.

**33.** For $z = 1 + i$, $r = \sqrt{2}$ and $\tan\theta = \frac{1}{1} = 1 \Rightarrow \theta = \frac{\pi}{4} \Rightarrow z = \sqrt{2}\left(\cos\frac{\pi}{4} + i\sin\frac{\pi}{4}\right)$. So by De Moivre's Theorem,

$$(1+i)^{20} = \left[\sqrt{2}\left(\cos\frac{\pi}{4} + i\sin\frac{\pi}{4}\right)\right]^{20} = \left(2^{1/2}\right)^{20}\left(\cos\frac{20\cdot\pi}{4} + i\sin\frac{20\cdot\pi}{4}\right)$$

$$= 2^{10}(\cos 5\pi + i\sin 5\pi) = 2^{10}[-1 + i(0)] = -2^{10} = -1024$$

**34.** For $z = 1 - \sqrt{3}\,i$, $r = \sqrt{1^2 + (-\sqrt{3})^2} = 2$ and $\tan\theta = \frac{-\sqrt{3}}{1} = -\sqrt{3} \Rightarrow \theta = \frac{5\pi}{3} \Rightarrow z = 2\left(\cos\frac{5\pi}{3} + i\sin\frac{5\pi}{3}\right)$.

So by De Moivre's Theorem,

$$\left(1 - \sqrt{3}\,i\right)^5 = \left[2\left(\cos\frac{5\pi}{3} + i\sin\frac{5\pi}{3}\right)\right]^5 = 2^5\left(\cos\frac{5\cdot5\pi}{3} + i\sin\frac{5\cdot5\pi}{3}\right)$$

$$= 2^5\left(\cos\frac{\pi}{3} + i\sin\frac{\pi}{3}\right) = 32\left(\frac{1}{2} + \frac{\sqrt{3}}{2}\,i\right) = 16 + 16\sqrt{3}\,i$$

**35.** For $z = 2\sqrt{3} + 2i$, $r = \sqrt{(2\sqrt{3})^2 + 2^2} = \sqrt{16} = 4$ and $\tan\theta = \frac{2}{2\sqrt{3}} = \frac{1}{\sqrt{3}} \Rightarrow \theta = \frac{\pi}{6} \Rightarrow z = 4\left(\cos\frac{\pi}{6} + i\sin\frac{\pi}{6}\right)$.

So by De Moivre's Theorem,

$$\left(2\sqrt{3} + 2i\right)^5 = \left[4\left(\cos\frac{\pi}{6} + i\sin\frac{\pi}{6}\right)\right]^5 = 4^5\left(\cos\frac{5\pi}{6} + i\sin\frac{5\pi}{6}\right) = 1024\left[-\frac{\sqrt{3}}{2} + \frac{1}{2}i\right] = -512\sqrt{3} + 512i$$

**36.** For $z = 1 - i$, $r = \sqrt{2}$ and $\tan\theta = \frac{-1}{1} = -1 \quad\Rightarrow\quad \theta = \frac{7\pi}{4} \quad\Rightarrow\quad z = \sqrt{2}\left(\cos\frac{7\pi}{4} + i\sin\frac{7\pi}{4}\right) \quad\Rightarrow$

$$\begin{aligned}(1-i)^8 &= \left[\sqrt{2}\left(\cos\tfrac{7\pi}{4} + i\sin\tfrac{7\pi}{4}\right)\right]^8 = 2^4\left(\cos\tfrac{8\cdot 7\pi}{4} + i\sin\tfrac{8\cdot 7\pi}{4}\right)\\ &= 16(\cos 14\pi + i\sin 14\pi) = 16(1+0i) = 16\end{aligned}$$

**37.** $1 = 1 + 0i = 1\,(\cos 0 + i\sin 0)$. Using Equation 3 with $r = 1$, $n = 8$, and $\theta = 0$, we have

$w_k = 1^{1/8}\left[\cos\left(\dfrac{0 + 2k\pi}{8}\right) + i\sin\left(\dfrac{0 + 2k\pi}{8}\right)\right] = \cos\dfrac{k\pi}{4} + i\sin\dfrac{k\pi}{4}$, where $k = 0, 1, 2, \ldots, 7$.

$w_0 = 1(\cos 0 + i\sin 0) = 1$, $w_1 = 1\left(\cos\frac{\pi}{4} + i\sin\frac{\pi}{4}\right) = \frac{1}{\sqrt{2}} + \frac{1}{\sqrt{2}}i$,

$w_2 = 1\left(\cos\frac{\pi}{2} + i\sin\frac{\pi}{2}\right) = i$, $w_3 = 1\left(\cos\frac{3\pi}{4} + i\sin\frac{3\pi}{4}\right) = -\frac{1}{\sqrt{2}} + \frac{1}{\sqrt{2}}i$,

$w_4 = 1(\cos\pi + i\sin\pi) = -1$, $w_5 = 1\left(\cos\frac{5\pi}{4} + i\sin\frac{5\pi}{4}\right) = -\frac{1}{\sqrt{2}} - \frac{1}{\sqrt{2}}i$,

$w_6 = 1\left(\cos\frac{3\pi}{2} + i\sin\frac{3\pi}{2}\right) = -i$, $w_7 = 1\left(\cos\frac{7\pi}{4} + i\sin\frac{7\pi}{4}\right) = \frac{1}{\sqrt{2}} - \frac{1}{\sqrt{2}}i$

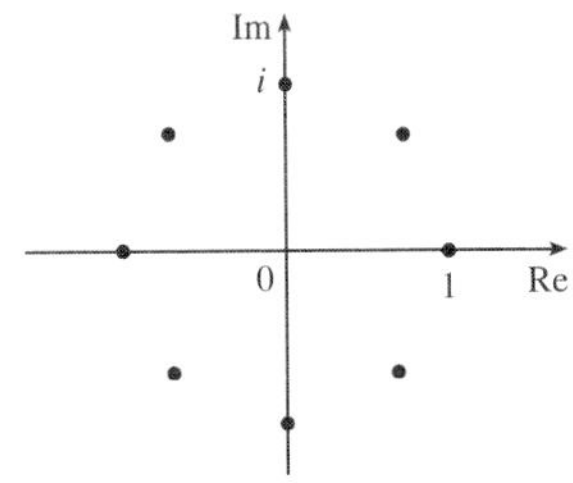


**38.** $32 = 32 + 0i = 32\,(\cos 0 + i\sin 0)$. Using Equation 3 with $r = 32$, $n = 5$, and $\theta = 0$, we have

$w_k = 32^{1/5}\left[\cos\left(\dfrac{0 + 2k\pi}{5}\right) + i\sin\left(\dfrac{0 + 2k\pi}{5}\right)\right] = 2\left(\cos\frac{2}{5}\pi k + i\sin\frac{2}{5}\pi k\right)$, where $k = 0, 1, 2, 3, 4$.

$w_0 = 2(\cos 0 + i\sin 0) = 2$

$w_1 = 2\left(\cos\frac{2\pi}{5} + i\sin\frac{2\pi}{5}\right)$

$w_2 = 2\left(\cos\frac{4\pi}{5} + i\sin\frac{4\pi}{5}\right)$

$w_3 = 2\left(\cos\frac{6\pi}{5} + i\sin\frac{6\pi}{5}\right)$

$w_4 = 2\left(\cos\frac{8\pi}{5} + i\sin\frac{8\pi}{5}\right)$

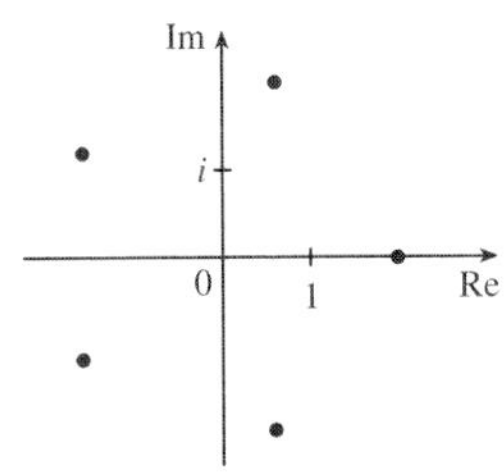


**39.** $i = 0 + i = 1\left(\cos\frac{\pi}{2} + i\sin\frac{\pi}{2}\right)$. Using Equation 3 with $r = 1$, $n = 3$, and $\theta = \frac{\pi}{2}$, we have

$w_k = 1^{1/3}\left[\cos\left(\dfrac{\frac{\pi}{2} + 2k\pi}{3}\right) + i\sin\left(\dfrac{\frac{\pi}{2} + 2k\pi}{3}\right)\right]$, where $k = 0, 1, 2$.

$w_0 = \left(\cos\frac{\pi}{6} + i\sin\frac{\pi}{6}\right) = \frac{\sqrt{3}}{2} + \frac{1}{2}i$

$w_1 = \left(\cos\frac{5\pi}{6} + i\sin\frac{5\pi}{6}\right) = -\frac{\sqrt{3}}{2} + \frac{1}{2}i$

$w_2 = \left(\cos\frac{9\pi}{6} + i\sin\frac{9\pi}{6}\right) = -i$

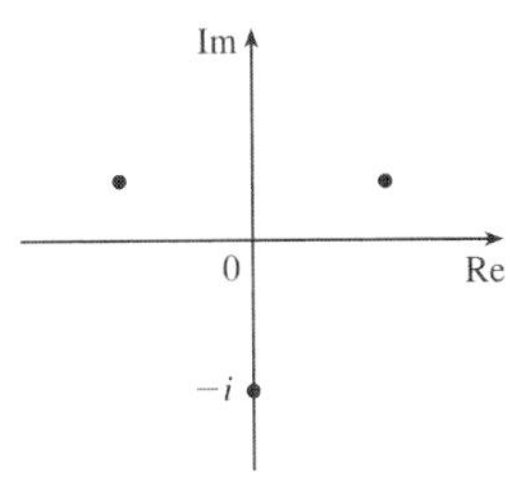


**40.** $1 + i = \sqrt{2}\left(\cos\frac{\pi}{4} + i\sin\frac{\pi}{4}\right)$. Using Equation 3 with $r = \sqrt{2}$, $n = 3$, and $\theta = \frac{\pi}{4}$, we have

$w_k = \left(\sqrt{2}\right)^{1/3}\left[\cos\left(\dfrac{\frac{\pi}{4} + 2k\pi}{3}\right) + i\sin\left(\dfrac{\frac{\pi}{4} + 2k\pi}{3}\right)\right]$, where $k = 0, 1, 2$.

$w_0 = 2^{1/6}\left(\cos\frac{\pi}{12} + i\sin\frac{\pi}{12}\right)$

$w_1 = 2^{1/6}\left(\cos\frac{3\pi}{4} + i\sin\frac{3\pi}{4}\right) = 2^{1/6}\left(-\frac{1}{\sqrt{2}} + \frac{1}{\sqrt{2}}i\right) = -2^{-1/3} + 2^{-1/3}i$

$w_2 = 2^{1/6}\left(\cos\frac{17\pi}{12} + i\sin\frac{17\pi}{12}\right)$

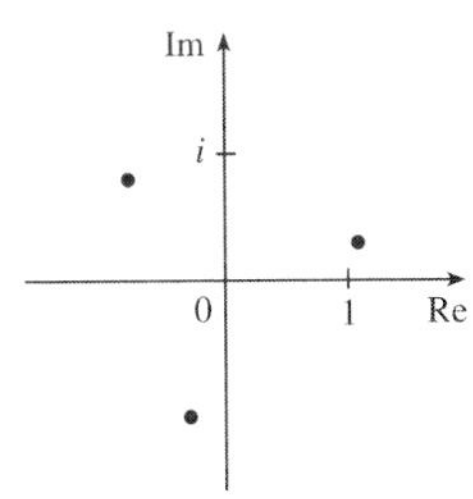

**41.** Using Euler's formula (6) with $y = \frac{\pi}{2}$, we have $e^{i\pi/2} = \cos\frac{\pi}{2} + i\sin\frac{\pi}{2} = 0 + 1i = i$.

**42.** Using Euler's formula (6) with $y = 2\pi$, we have $e^{2\pi i} = \cos 2\pi + i\sin 2\pi = 1$.

**43.** Using Euler's formula (6) with $y = \dfrac{\pi}{3}$, we have $e^{i\pi/3} = \cos\dfrac{\pi}{3} + i\sin\dfrac{\pi}{3} = \dfrac{1}{2} + \dfrac{\sqrt{3}}{2}i$.

**44.** Using Euler's formula (6) with $y = -\pi$, we have $e^{-i\pi} = \cos(-\pi) + i\sin(-\pi) = -1$.

**45.** Using Equation 7 with $x = 2$ and $y = \pi$, we have $e^{2+i\pi} = e^2e^{i\pi} = e^2(\cos\pi + i\sin\pi) = e^2(-1+0) = -e^2$.

**46.** Using Equation 7 with $x = \pi$ and $y = 1$, we have $e^{\pi+i} = e^\pi \cdot e^{1i} = e^\pi(\cos 1 + i\sin 1) = e^\pi\cos 1 + (e^\pi\sin 1)i$.

**47.** Take $r = 1$ and $n = 3$ in De Moivre's Theorem to get

$$\begin{aligned}
[1(\cos\theta + i\sin\theta)]^3 &= 1^3(\cos 3\theta + i\sin 3\theta) \\
(\cos\theta + i\sin\theta)^3 &= \cos 3\theta + i\sin 3\theta \\
\cos^3\theta + 3(\cos^2\theta)(i\sin\theta) + 3(\cos\theta)(i\sin\theta)^2 + (i\sin\theta)^3 &= \cos 3\theta + i\sin 3\theta \\
\cos^3\theta + (3\cos^2\theta\,\sin\theta)i - 3\cos\theta\,\sin^2\theta - (\sin^3\theta)i &= \cos 3\theta + i\sin 3\theta \\
(\cos^3\theta - 3\sin^2\theta\,\cos\theta) + (3\sin\theta\,\cos^2\theta - \sin^3\theta)i &= \cos 3\theta + i\sin 3\theta
\end{aligned}$$

Equating real and imaginary parts gives

$$\cos 3\theta = \cos^3\theta - 3\sin^2\theta\,\cos\theta \quad\text{and}\quad \sin 3\theta = 3\sin\theta\,\cos^2\theta - \sin^3\theta$$

**48.** Using Formula 6,

$$\begin{aligned}
e^{ix} + e^{-ix} &= (\cos x + i\sin x) + [\cos(-x) + i\sin(-x)] \\
&= \cos x + i\sin x + \cos x - i\sin x = 2\cos x
\end{aligned}$$

Thus, $\cos x = \dfrac{e^{ix} + e^{-ix}}{2}$.

Similarly,

$$\begin{aligned}
e^{ix} - e^{-ix} &= (\cos x + i\sin x) - [\cos(-x) + i\sin(-x)] \\
&= \cos x + i\sin x - \cos x - (-i\sin x) = 2i\sin x
\end{aligned}$$

Therefore, $\sin x = \dfrac{e^{ix} - e^{-ix}}{2i}$.

**49.** $F(x) = e^{rx} = e^{(a+bi)x} = e^{ax+bxi} = e^{ax}(\cos bx + i\sin bx) = e^{ax}\cos bx + i(e^{ax}\sin bx) \quad\Rightarrow$

$$\begin{aligned}
F'(x) &= (e^{ax}\cos bx)' + i(e^{ax}\sin bx)' \\
&= (ae^{ax}\cos bx - be^{ax}\sin bx) + i(ae^{ax}\sin bx + be^{ax}\cos bx) \\
&= a\left[e^{ax}(\cos bx + i\sin bx)\right] + b\left[e^{ax}(-\sin bx + i\cos bx)\right] \\
&= ae^{rx} + b\left[e^{ax}(i^2\sin bx + i\cos bx)\right] \\
&= ae^{rx} + bi\left[e^{ax}(\cos bx + i\sin bx)\right] = ae^{rx} + bie^{rx} = (a+bi)e^{rx} = re^{rx}
\end{aligned}$$

**50.** (a) From Exercise 49, $F(x) = e^{(1+i)x} \quad \Rightarrow \quad F'(x) = (1+i)e^{(1+i)x}$. So

$$\int e^{(1+i)x}\,dx = \frac{1}{1+i}\int F'(x)\,dx = \frac{1}{1+i}F(x) + C = \frac{1-i}{2}F(x) + C = \frac{1-i}{2}e^{(1+i)x} + C$$

(b) $\int e^{(1+i)x}\,dx = \int e^x e^{ix}\,dx = \int e^x(\cos x + i\sin x)\,dx = \int e^x \cos x\,dx + i\int e^x \sin x\,dx$ **(1)**.

Also,

$$\begin{aligned} \frac{1-i}{2}e^{(1+i)x} &= \tfrac{1}{2}e^{(1+i)x} - \tfrac{1}{2}ie^{(1+i)x} = \tfrac{1}{2}e^{x+ix} - \tfrac{1}{2}ie^{x+ix} \\ &= \tfrac{1}{2}e^x(\cos x + i\sin x) - \tfrac{1}{2}ie^x(\cos x + i\sin x) \\ &= \tfrac{1}{2}e^x\cos x + \tfrac{1}{2}e^x\sin x + \tfrac{1}{2}ie^x\sin x - \tfrac{1}{2}ie^x\cos x \\ &= \tfrac{1}{2}e^x(\cos x + \sin x) + i\left[\tfrac{1}{2}e^x(\sin x - \cos x)\right] \quad \textbf{(2)} \end{aligned}$$

Equating the real and imaginary parts in **(1)** and **(2)**, we see that $\int e^x \cos x\,dx = \frac{1}{2}e^x(\cos x + \sin x) + C$ and $\int e^x \sin x\,dx = \frac{1}{2}e^x(\sin x - \cos x) + C$.